TRIGONOMETRY

Definition of the Six Trigonometric Functions

Right triangle definitions, where $0 < \theta < \pi/2$.

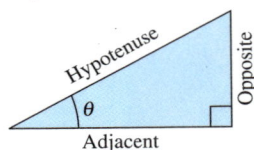

$$\sin\theta = \frac{\text{opp}}{\text{hyp}} \qquad \csc\theta = \frac{\text{hyp}}{\text{opp}}$$

$$\cos\theta = \frac{\text{adj}}{\text{hyp}} \qquad \sec\theta = \frac{\text{hyp}}{\text{adj}}$$

$$\tan\theta = \frac{\text{opp}}{\text{adj}} \qquad \cot\theta = \frac{\text{adj}}{\text{opp}}$$

Circular function definitions, where θ is any angle.

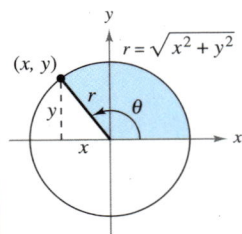

$$r = \sqrt{x^2 + y^2}$$

$$\sin\theta = \frac{y}{r} \qquad \csc\theta = \frac{r}{y}$$

$$\cos\theta = \frac{x}{r} \qquad \sec\theta = \frac{r}{x}$$

$$\tan\theta = \frac{y}{x} \qquad \cot\theta = \frac{x}{y}$$

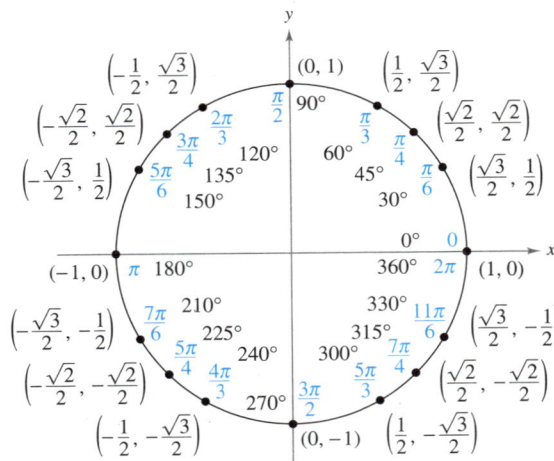

Reciprocal Identities

$$\sin x = \frac{1}{\csc x} \qquad \sec x = \frac{1}{\cos x} \qquad \tan x = \frac{1}{\cot x}$$

$$\csc x = \frac{1}{\sin x} \qquad \cos x = \frac{1}{\sec x} \qquad \cot x = \frac{1}{\tan x}$$

Quotient Identities

$$\tan x = \frac{\sin x}{\cos x} \qquad \cot x = \frac{\cos x}{\sin x}$$

Pythagorean Identities

$$\sin^2 x + \cos^2 x = 1$$
$$1 + \tan^2 x = \sec^2 x \qquad 1 + \cot^2 x = \csc^2 x$$

Cofunction Identities

$$\sin\left(\frac{\pi}{2} - x\right) = \cos x \qquad \cos\left(\frac{\pi}{2} - x\right) = \sin x$$

$$\csc\left(\frac{\pi}{2} - x\right) = \sec x \qquad \tan\left(\frac{\pi}{2} - x\right) = \cot x$$

$$\sec\left(\frac{\pi}{2} - x\right) = \csc x \qquad \cot\left(\frac{\pi}{2} - x\right) = \tan x$$

Even/Odd Identities

$$\sin(-x) = -\sin x \qquad \cos(-x) = \cos x$$
$$\csc(-x) = -\csc x \qquad \tan(-x) = -\tan x$$
$$\sec(-x) = \sec x \qquad \cot(-x) = -\cot x$$

Sum and Difference Formulas

$$\sin(u \pm v) = \sin u \cos v \pm \cos u \sin v$$
$$\cos(u \pm v) = \cos u \cos v \mp \sin u \sin v$$

$$\tan(u \pm v) = \frac{\tan u \pm \tan v}{1 \mp \tan u \tan v}$$

Double-Angle Formulas

$$\sin 2u = 2\sin u \cos u$$
$$\cos 2u = \cos^2 u - \sin^2 u = 2\cos^2 u - 1 = 1 - 2\sin^2 u$$

$$\tan 2u = \frac{2\tan u}{1 - \tan^2 u}$$

Power-Reducing Formulas

$$\sin^2 u = \frac{1 - \cos 2u}{2}$$

$$\cos^2 u = \frac{1 + \cos 2u}{2}$$

$$\tan^2 u = \frac{1 - \cos 2u}{1 + \cos 2u}$$

Sum-to-Product Formulas

$$\sin u + \sin v = 2\sin\left(\frac{u + v}{2}\right)\cos\left(\frac{u - v}{2}\right)$$

$$\sin u - \sin v = 2\cos\left(\frac{u + v}{2}\right)\sin\left(\frac{u - v}{2}\right)$$

$$\cos u + \cos v = 2\cos\left(\frac{u + v}{2}\right)\cos\left(\frac{u - v}{2}\right)$$

$$\cos u - \cos v = -2\sin\left(\frac{u + v}{2}\right)\sin\left(\frac{u - v}{2}\right)$$

Product-to-Sum Formulas

$$\sin u \sin v = \frac{1}{2}[\cos(u - v) - \cos(u + v)]$$

$$\cos u \cos v = \frac{1}{2}[\cos(u - v) + \cos(u + v)]$$

$$\sin u \cos v = \frac{1}{2}[\sin(u + v) + \sin(u - v)]$$

$$\cos u \sin v = \frac{1}{2}[\sin(u + v) - \sin(u - v)]$$

Calculus

10e

Calculus

10e

Ron Larson
The Pennsylvania State University
The Behrend College

Bruce Edwards
University of Florida

BROOKS/COLE
CENGAGE Learning

Australia • Brazil • Japan • Korea • Mexico • Singapore • Spain • United Kingdom • United States

BROOKS/COLE
CENGAGE Learning™

Calculus
Tenth Edition

Ron Larson

Publisher: Liz Covello

Senior Development Editor: Carolyn Lewis

Assistant Editor: Liza Neustaetter

Editorial Assistant: Stephanie Kreuz

Associate Media Editor: Guanglei Zhang

Senior Content Project Manager: Jessica Rasile

Art Director: Linda May

Rights Acquisition Specialist: Shalice Shah-Caldwell

Manufacturing Planner: Doug Bertke

Text/Cover Designer: Larson Texts, Inc.

Compositor: Larson Texts, Inc.

Cover Image: Larson Texts, Inc.

For product information and technology assistance, contact us at
Cengage Learning Customer & Sales Support, 1-800-354-9706.
For permission to use material from this text or product,
submit all requests online at **www.cengage.com/permissions.**
Further permissions questions can be emailed to
permissionrequest@cengage.com.

Library of Congress Control Number: 2012948320

ISBN-13: 978-1-285-05709-5
ISBN-10: 1-285-05709-0

Brooks/Cole
20 Channel Center Street
Boston, MA 02210
USA

Cengage Learning is a leading provider of customized learning solutions with office locations around the globe, including Singapore, the United Kingdom, Australia, Mexico, Brazil, and Japan. Locate your local office at: **international.cengage.com/region**

Cengage Learning products are represented in Canada by Nelson Education, Ltd.

For your course and learning solutions, visit **www.cengage.com.**

Purchase any of our products at your local college store or at our preferred online store **www.cengagebrain.com.**

Instructors: Please visit **login.cengage.com** and log in to access instructor-specific resources.

Printed in the United States of America
1 2 3 4 5 6 7 16 15 14 13 12

Contents

*Available at the text-specific website *www.cengagebrain.com*

Preface

Welcome to *Calculus*, Tenth Edition. We are proud to present this new edition to you. As with all editions, we have been able to incorporate many useful comments from you, our user. For this edition, we have introduced some new features and revised others. You will still find what you expect – a pedagogically sound, mathematically precise, and comprehensive textbook.

We are pleased and excited to offer you something brand new with this edition – a companion website at LarsonCalculus.com. This site offers many resources that will help you as you study calculus. All of these resources are just a click away.

Our goal for every edition of this textbook is to provide you with the tools you need to master calculus. We hope that you find the changes in this edition, together with **LarsonCalculus.com,** will accomplish just that.

In each exercise set, be sure to notice the reference to **CalcChat.com.** At this free site, you can download a step-by-step solution to any odd-numbered exercise. Also, you can talk to a tutor, free of charge, during the hours posted at the site. Over the years, thousands of students have visited the site for help. We use all of this information to help guide each revision of the exercises and solutions.

New To This Edition

NEW LarsonCalculus.com

This companion website offers multiple tools and resources to supplement your learning. Access to these features is free. Watch videos explaining concepts or proofs from the book, explore examples, view three-dimensional graphs, download articles from math journals and much more.

NEW Chapter Opener

Each Chapter Opener highlights real-life applications used in the examples and exercises.

NEW Interactive Examples

Examples throughout the book are accompanied by Interactive Examples at LarsonCalculus.com. These interactive examples use Wolfram's free CDF Player and allow you to explore calculus by manipulating functions or graphs, and observing the results.

NEW Proof Videos

Watch videos of co-author Bruce Edwards as he explains the proofs of theorems in *Calculus*, Tenth Edition at LarsonCalculus.com.

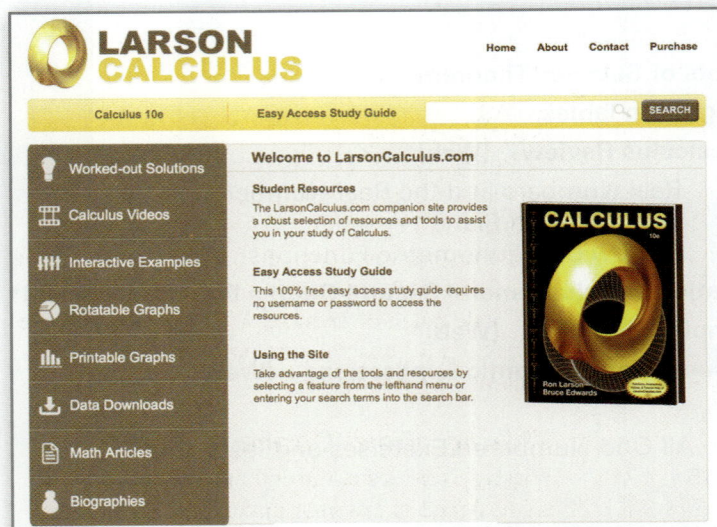

NEW How Do You See It?

The How Do You See It? feature in each section presents a real-life problem that you will solve by visual inspection using the concepts learned in the lesson. This exercise is excellent for classroom discussion or test preparation.

REVISED Remark

These hints and tips reinforce or expand upon concepts, help you learn how to study mathematics, caution you about common errors, address special cases, or show alternative or additional steps to a solution of an example.

REVISED Exercise Sets

The exercise sets have been carefully and extensively examined to ensure they are rigorous and relevant and include all topics our users have suggested. The exercises have been reorganized and titled so you can better see the connections between examples and exercises. Multi-step, real-life exercises reinforce problem-solving skills and mastery of concepts by giving students the opportunity to apply the concepts in real-life situations.

118. HOW DO YOU SEE IT?   The figure shows the graphs of the position, velocity, and acceleration functions of a particle.

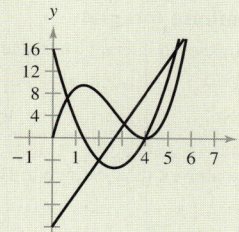

(a) Copy the graphs of the functions shown. Identify each graph. Explain your reasoning. To print an enlarged copy of the graph, go to *MathGraphs.com*.

(b) On your sketch, identify when the particle speeds up and when it slows down. Explain your reasoning.

Table of Content Changes

Appendix A (Proofs of Selected Theorems) now appears in video format at LarsonCalculus.com. The proofs also appear in text form at CengageBrain.com.

Trusted Features

Applications

Carefully chosen applied exercises and examples are included throughout to address the question, "When will I use this?" These applications are pulled from diverse sources, such as current events, world data, industry trends, and more, and relate to a wide range of interests. Understanding where calculus is (or can be) used promotes fuller understanding of the material.

Writing about Concepts

Writing exercises at the end of each section are designed to test your understanding of basic concepts in each section, encouraging you to verbalize and write answers and promote technical communication skills that will be invaluable in your future careers.

Theorems

Theorems provide the conceptual framework for calculus. Theorems
are clearly stated and separated from the rest of the text by boxes for quick
visual reference. Key proofs often follow the theorem and can be found
at LarsonCalculus.com.

Definition of Definite Integral

If f is defined on the closed interval $[a, b]$ and the limit of Riemann sums over
partitions Δ

$$\lim_{\|\Delta\| \to 0} \sum_{i=1}^{n} f(c_i)\, \Delta x_i$$

exists (as described above), then f is said to be **integrable** on $[a, b]$ and the
limit is denoted by

$$\lim_{\|\Delta\| \to 0} \sum_{i=1}^{n} f(c_i)\, \Delta x_i = \int_a^b f(x)\, dx.$$

The limit is called the **definite integral** of f from a to b. The number a is the
lower limit of integration, and the number b is the **upper limit** of integration.

Definitions

As with theorems, definitions are clearly
stated using precise, formal wording and
are separated from the text by boxes for
quick visual reference.

Explorations

Explorations provide unique challenges
to study concepts that have not yet been
formally covered in the text. They allow
you to learn by discovery and introduce
topics related to ones presently being
studied. Exploring topics in this way
encourages you to think outside the box.

Historical Notes and Biographies

Historical Notes provide you with background
information on the foundations of calculus.
The Biographies introduce you to the people
who created and contributed to calculus.

Technology

Throughout the book, technology boxes show
you how to use technology to solve problems
and explore concepts of calculus. These tips
also point out some pitfalls of using technology.

Section Projects

Projects appear in selected sections and encourage
you to explore applications related to the topics
you are studying. They provide an interesting
and engaging way for you and other students
to work and investigate ideas collaboratively.

Putnam Exam Challenges

Putnam Exam questions appear in selected
sections. These actual Putnam Exam questions
will challenge you and push the limits of
your understanding of calculus.

SECTION PROJECT

St. Louis Arch

The Gateway Arch in St. Louis, Missouri, was constructed using
the hyperbolic cosine function. The equation used for construction
was

$$y = 693.8597 - 68.7672 \cosh 0.0100333x,$$
$$-299.2239 \le x \le 299.2239$$

where x and y are measured in feet. Cross sections of the arch are
equilateral triangles, and (x, y) traces the path of the centers of mass
of the cross-sectional triangles. For each value of x, the area of the
cross-sectional triangle is

$$A = 125.1406 \cosh 0.0100333x.$$

*(Source: Owner's Manual for the Gateway Arch, Saint Louis, MO,
by William Thayer)*

(a) How high above
    the ground is the
    center of the highest
    triangle? (At ground
    level, $y = 0$.)

(b) What is the height
    of the arch? (*Hint:*
    For an equilateral
    triangle, $A = \sqrt{3}c^2$,
    where c is one-half the base of the triangle, and the center of
    mass of the triangle is located at two-thirds the height of the
    triangle.)

(c) How wide is the arch at ground level?

Additional Resources

Student Resources

- **Student Solutions Manual for Calculus of a Single Variable**
  (Chapters P–10 of *Calculus*): ISBN 1-285-08571-X

  Student Solutions Manual for Multivariable Calculus
  (Chapters 11–16 of *Calculus*): ISBN 1-285-08575-2

  These manuals contain worked-out solutions for all odd-numbered exercises.

  ENHANCED WebAssign *www.webassign.net*
  Printed Access Card: ISBN 0-538-73807-3
  Online Access Code: ISBN 1-285-18421-1
  Enhanced WebAssign is designed for you to do your homework online. This proven and reliable system uses pedagogy and content found in this text, and then enhances it to help you learn calculus more effectively. Automatically graded homework allows you to focus on your learning and get interactive study assistance outside of class. Enhanced WebAssign for *Calculus*, 10e contains the Cengage YouBook, an interactive eBook that contains video clips, highlighting and note-taking features, and more!

  CourseMate

  CourseMate is a perfect study tool for bringing concepts to life with interactive learning, study, and exam preparation tools that support the printed textbook. CourseMate includes: an interactive eBook, videos, quizzes, flashcards, and more!

- **CengageBrain.com**—To access additional materials including CourseMate, visit www.cengagebrain.com. At the CengageBrain.com home page, search for the ISBN of your title (from the back cover of your book) using the search box at the top of the page. This will take you to the product page where these resources can be found.

Instructor Resources

ENHANCED WebAssign *www.webassign.net*
Exclusively from Cengage Learning, Enhanced WebAssign offers an extensive online program for *Calculus*, 10e to encourage the practice that is so critical for concept mastery. The meticulously crafted pedagogy and exercises in our proven texts become even more effective in Enhanced WebAssign, supplemented by multimedia tutorial support and immediate feedback as students complete their assignments. Key features include:

- Thousands of homework problems that match your textbook's end-of-section exercises
- Opportunities for students to review prerequisite skills and content both at the start of the course and at the beginning of each section
- Read It eBook pages, Watch It Videos, Master It tutorials, and Chat About It links
- A customizable Cengage YouBook with highlighting, note-taking, and search features, as well as links to multimedia resources
- Personal Study Plans (based on diagnostic quizzing) that identify chapter topics that students will need to master
- A WebAssign Answer Evaluator that recognizes and accepts equivalent mathematical responses in the same way you grade assignments
- A *Show My Work* feature that gives you the option of seeing students' detailed solutions
- Lecture videos, and more!

- **Cengage Customizable YouBook**—YouBook is an eBook that is both interactive and customizable! Containing all the content from *Calculus*, 10e, YouBook features a text edit tool that allows you to modify the textbook narrative as needed. With YouBook, you can quickly re-order entire sections and chapters or hide any content you don't teach to create an eBook that perfectly matches your syllabus. You can further customize the text by adding instructor-created or YouTube video links. Additional media assets include: video clips, highlighting and note-taking features, and more! YouBook is available within Enhanced WebAssign.

- **Complete Solutions Manual for Calculus of a Single Variable, Volume 1**
  (Chapters P–6 of *Calculus*): ISBN 1-285-08576-0

  Complete Solutions Manual for Calculus of a Single Variable, Volume 2
  (Chapters 7–10 of *Calculus*): ISBN 1-285-08577-9

  Complete Solutions Manual for Multivariable Calculus
  (Chapters 11–16 of *Calculus*): ISBN 1-285-08580-9

  The *Complete Solutions Manuals* contain worked-out solutions to all exercises in the text.

- **Solution Builder** (www.cengage.com/solutionbuilder)— This online instructor database offers complete worked-out solutions to all exercises in the text, allowing you to create customized, secure solutions printouts (in PDF format) matched exactly to the problems you assign in class.

- **PowerLecture** (ISBN 1-285-08583-3)—This comprehensive instructor DVD includes resources such as an electronic version of the Instructor's Resource Guide, complete pre-built PowerPoint® lectures, all art from the text in both jpeg and PowerPoint formats, ExamView® algorithmic computerized testing software, JoinIn™ content for audience response systems (clickers), and a link to Solution Builder.

- **ExamView Computerized Testing**— Create, deliver, and customize tests in print and online formats with ExamView®, an easy-to-use assessment and tutorial software. ExamView for *Calculus*, 10e contains hundreds of algorithmic multiple-choice and short answer test items. ExamView® is available on the PowerLecture DVD.

- **Instructor's Resource Guide** (ISBN 1-285-09074-8)—This robust manual contains an abundance of resources keyed to the textbook by chapter and section, including chapter summaries and teaching strategies. An electronic version of the Instructor's Resource Guide is available on the PowerLecture DVD.

CourseMate

CourseMate is a perfect study tool for students, and requires no set up from you. CourseMate brings course concepts to life with interactive learning, study, and exam preparation tools that support the printed textbook. CourseMate for *Calculus*, 10e includes: an interactive eBook, videos, quizzes, flashcards, and more! For instructors, CourseMate includes Engagement Tracker, a first-of-its kind tool that monitors student engagement.

- **CengageBrain.com**—To access additional course materials including CourseMate, please visit http://login.cengage.com. At the CengageBrain.com home page, search for the ISBN of your title (from the back cover of your book) using the search box at the top of the page. This will take you to the product page where these resources can be found.

Acknowledgements

We would like to thank the many people who have helped us at various stages of *Calculus* over the last 39 years. Their encouragement, criticisms, and suggestions have been invaluable.

Reviewers of the Tenth Edition

Denis Bell, *University of Northern Florida*; Abraham Biggs, *Broward Community College*; Jesse Blosser, *Eastern Mennonite School*; Mark Brittenham, *University of Nebraska*; Mingxiang Chen, *North Carolina A & T State University*; Marcia Kleinz, *Atlantic Cape Community College*; Maxine Lifshitz, *Friends Academy*; Bill Meisel, *Florida State College at Jacksonville*; Martha Nega, *Georgia Perimeter College*; Laura Ritter, *Southern Polytechnic State University*; Chia-Lin Wu, *Richard Stockton College of New Jersey*

Reviewers of Previous Editions

Stan Adamski, *Owens Community College*; Alexander Arhangelskii, *Ohio University;* Seth G. Armstrong, *Southern Utah University;* Jim Ball, *Indiana State University;* Marcelle Bessman, *Jacksonville University;* Linda A. Bolte, *Eastern Washington University;* James Braselton, *Georgia Southern University;* Harvey Braverman, *Middlesex County College;* Tim Chappell, *Penn Valley Community College;* Oiyin Pauline Chow, *Harrisburg Area Community College;* Julie M. Clark, *Hollins University;* P.S. Crooke, *Vanderbilt University;* Jim Dotzler, *Nassau Community College;* Murray Eisenberg, *University of Massachusetts at Amherst;* Donna Flint, *South Dakota State University;* Michael Frantz, *University of La Verne;* Sudhir Goel, *Valdosta State University;* Arek Goetz, *San Francisco State University;* Donna J. Gorton, *Butler County Community College;* John Gosselin, *University of Georgia;* Shahryar Heydari, *Piedmont College;* Guy Hogan, *Norfolk State University;* Ashok Kumar, *Valdosta State University;* Kevin J. Leith, *Albuquerque Community College;* Douglas B. Meade, *University of South Carolina;* Teri Murphy, *University of Oklahoma;* Darren Narayan, *Rochester Institute of Technology;* Susan A. Natale, *The Ursuline School, NY;* Terence H. Perciante, *Wheaton College;* James Pommersheim, *Reed College;* Leland E. Rogers, *Pepperdine University;* Paul Seeburger, *Monroe Community College;* Edith A. Silver, *Mercer County Community College;* Howard Speier, *Chandler-Gilbert Community College;* Desmond Stephens, *Florida A&M University;* Jianzhong Su, *University of Texas at Arlington;* Patrick Ward, *Illinois Central College;* Diane Zych, *Erie Community College*

Many thanks to Robert Hostetler, The Behrend College, The Pennsylvania State University, and David Heyd, The Behrend College, The Pennsylvania State University, for their significant contributions to previous editions of this text.

We would also like to thank the staff at Larson Texts, Inc., who assisted in preparing the manuscript, rendering the art package, typesetting, and proofreading the pages and supplements.

On a personal level, we are grateful to our wives, Deanna Gilbert Larson and Consuelo Edwards, for their love, patience, and support. Also, a special note of thanks goes out to R. Scott O'Neil.

If you have suggestions for improving this text, please feel free to write to us. Over the years we have received many useful comments from both instructors and students, and we value these very much.

Ron Larson

Bruce Edwards

Your Course. Your Way.

Calculus Textbook Options

The traditional calculus course is available in a variety of textbook configurations to address the different ways instructors teach—and students take—their classes.

The book can be customized to meet your individual needs and is available through CengageBrain.com.

TOPICS COVERED	APPROACH			
	Late Transcendental Functions	Early Transcendental Functions	Accelerated coverage	Integrated coverage
3-semester	Calculus 10e	Calculus Early Transcendental Functions 5e	Essential Calculus	
Single Variable Only	Calculus 10e Single Variable	Calculus: Early Transcendental Functions 5e Single Variable		Calculus I with Precalculus 3e
Multivariable	Calculus 10e Multivariable	Calculus 10e Multivariable		
Custom All of these textbook choices can be customized to fit the individual needs of your course.	Calculus 10e	Calculus: Early Transcendental Functions 5e	Essential Calculus	Calculus I with Precalculus 3e

Calculus

10e

P Preparation for Calculus

Automobile Aerodynamics *(Exercise 96, p. 30)*

Hours of Daylight *(Example 3, p. 33)*

Conveyor Design *(Exercise 23, p. 16)*

Cell Phone Subscribers *(Exercise 68, p. 9)*

Modeling Carbon Dioxide Concentration *(Example 6, p. 7)*

P.1   Graphs and Models

- Sketch the graph of an equation.
- Find the intercepts of a graph.
- Test a graph for symmetry with respect to an axis and the origin.
- Find the points of intersection of two graphs.
- Interpret mathematical models for real-life data.

The Graph of an Equation

In 1637, the French mathematician René Descartes revolutionized the study of mathematics by combining its two major fields—algebra and geometry. With Descartes's coordinate plane, geometric concepts could be formulated analytically and algebraic concepts could be viewed graphically. The power of this approach was such that within a century of its introduction, much of calculus had been developed.

The same approach can be followed in your study of calculus. That is, by viewing calculus from multiple perspectives—*graphically*, *analytically*, and *numerically*—you will increase your understanding of core concepts.

Consider the equation $3x + y = 7$. The point $(2, 1)$ is a **solution point** of the equation because the equation is satisfied (is true) when 2 is substituted for x and 1 is substituted for y. This equation has many other solutions, such as $(1, 4)$ and $(0, 7)$. To find other solutions systematically, solve the original equation for y.

$$y = 7 - 3x \qquad \text{Analytic approach}$$

Then construct a **table of values** by substituting several values of x.

x	0	1	2	3	4
y	7	4	1	-2	-5

Numerical approach

From the table, you can see that $(0, 7)$, $(1, 4)$, $(2, 1)$, $(3, -2)$, and $(4, -5)$ are solutions of the original equation $3x + y = 7$. Like many equations, this equation has an infinite number of solutions. The set of all solution points is the **graph** of the equation, as shown in Figure P.1. Note that the sketch shown in Figure P.1 is referred to as the graph of $3x + y = 7$, even though it really represents only a *portion* of the graph. The entire graph would extend beyond the page.

In this course, you will study many sketching techniques. The simplest is point plotting—that is, you plot points until the basic shape of the graph seems apparent.

Graphical approach: $3x + y = 7$
Figure P.1

EXAMPLE 1   Sketching a Graph by Point Plotting

To sketch the graph of $y = x^2 - 2$, first construct a table of values. Next, plot the points shown in the table. Then connect the points with a smooth curve, as shown in Figure P.2. This graph is a **parabola.** It is one of the conics you will study in Chapter 10.

x	-2	-1	0	1	2	3
y	2	-1	-2	-1	2	7

RENÉ DESCARTES (1596–1650)

Descartes made many contributions to philosophy, science, and mathematics. The idea of representing points in the plane by pairs of real numbers and representing curves in the plane by equations was described by Descartes in his book *La Géométrie*, published in 1637. *See LarsonCalculus.com to read more of this biography.*

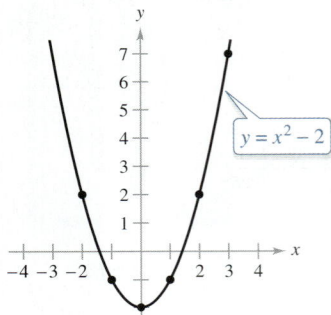

The parabola $y = x^2 - 2$
Figure P.2

One disadvantage of point plotting is that to get a good idea about the shape of a graph, you may need to plot many points. With only a few points, you could badly misrepresent the graph. For instance, to sketch the graph of

$$y = \frac{1}{30}x(39 - 10x^2 + x^4)$$

you plot five points:

$$(-3, -3), \quad (-1, -1), \quad (0, 0), \quad (1, 1), \quad \text{and} \quad (3, 3)$$

as shown in Figure P.3(a). From these five points, you might conclude that the graph is a line. This, however, is not correct. By plotting several more points, you can see that the graph is more complicated, as shown in Figure P.3(b).

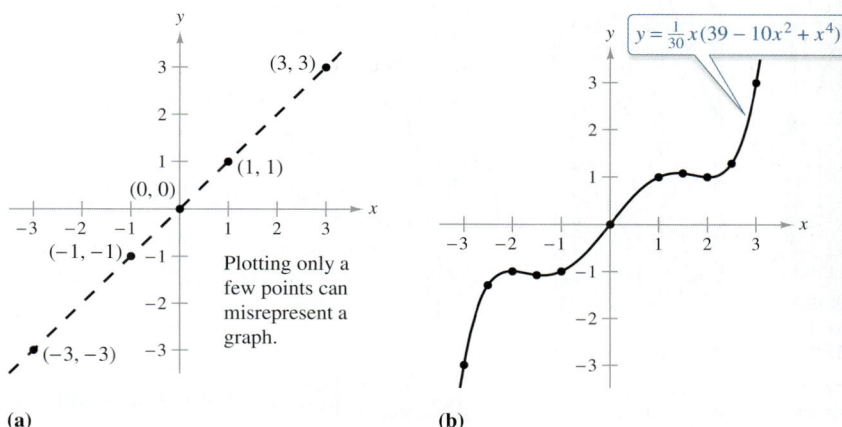

(a)

(b)

Plotting only a few points can misrepresent a graph.

Figure P.3

> **TECHNOLOGY** Graphing an equation has been made easier by technology. Even with technology, however, it is possible to misrepresent a graph badly. For instance, each of the graphing utility* screens in Figure P.4 shows a portion of the graph of

$$y = x^3 - x^2 - 25.$$

From the screen on the left, you might assume that the graph is a line. From the screen on the right, however, you can see that the graph is not a line. So, whether you are sketching a graph by hand or using a graphing utility, you must realize that different "viewing windows" can produce very different views of a graph. In choosing a viewing window, your goal is to show a view of the graph that fits well in the context of the problem.

Graphing utility screens of $y = x^3 - x^2 - 25$

Figure P.4

*In this text, the term *graphing utility* means either a graphing calculator, such as the TI-Nspire, or computer graphing software, such as *Maple* or *Mathematica*.

Exploration

Comparing Graphical and Analytic Approaches Use a graphing utility to graph each equation. In each case, find a viewing window that shows the important characteristics of the graph.

a. $y = x^3 - 3x^2 + 2x + 5$

b. $y = x^3 - 3x^2 + 2x + 25$

c. $y = -x^3 - 3x^2 + 20x + 5$

d. $y = 3x^3 - 40x^2 + 50x - 45$

e. $y = -(x + 12)^3$

f. $y = (x - 2)(x - 4)(x - 6)$

A purely graphical approach to this problem would involve a simple "guess, check, and revise" strategy. What types of things do you think an analytic approach might involve? For instance, does the graph have symmetry? Does the graph have turns? If so, where are they? As you proceed through Chapters 1, 2, and 3 of this text, you will study many new analytic tools that will help you analyze graphs of equations such as these.

Intercepts of a Graph

REMARK Some texts denote the x-intercept as the x-coordinate of the point $(a, 0)$ rather than the point itself. Unless it is necessary to make a distinction, when the term *intercept* is used in this text, it will mean either the point or the coordinate.

Two types of solution points that are especially useful in graphing an equation are those having zero as their x- or y-coordinate. Such points are called **intercepts** because they are the points at which the graph intersects the x- or y-axis. The point $(a, 0)$ is an **x-intercept** of the graph of an equation when it is a solution point of the equation. To find the x-intercepts of a graph, let y be zero and solve the equation for x. The point $(0, b)$ is a **y-intercept** of the graph of an equation when it is a solution point of the equation. To find the y-intercepts of a graph, let x be zero and solve the equation for y.

It is possible for a graph to have no intercepts, or it might have several. For instance, consider the four graphs shown in Figure P.5.

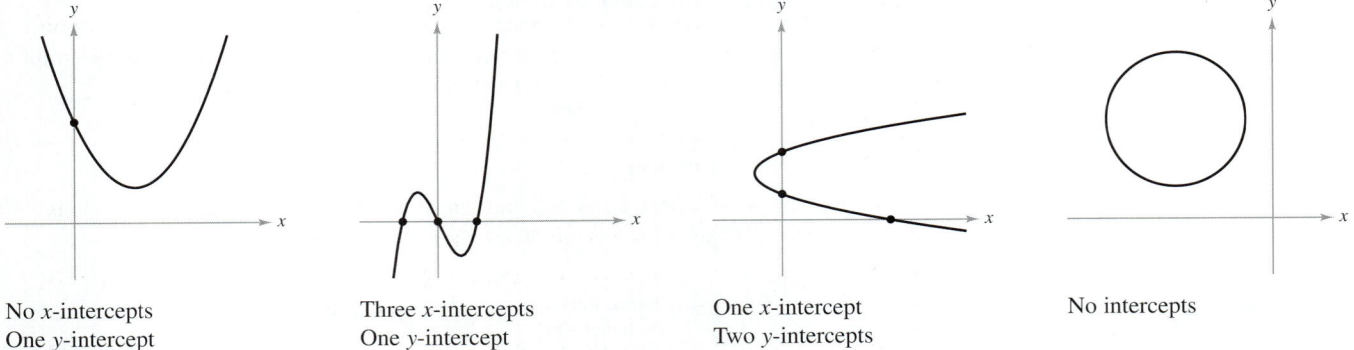

No x-intercepts
One y-intercept

Three x-intercepts
One y-intercept

One x-intercept
Two y-intercepts

No intercepts

Figure P.5

EXAMPLE 2   **Finding x- and y-Intercepts**

Find the x- and y-intercepts of the graph of $y = x^3 - 4x$.

Solution   To find the x-intercepts, let y be zero and solve for x.

$$x^3 - 4x = 0 \qquad \text{Let } y \text{ be zero.}$$
$$x(x - 2)(x + 2) = 0 \qquad \text{Factor.}$$
$$x = 0, 2, \text{ or } -2 \qquad \text{Solve for } x.$$

Because this equation has three solutions, you can conclude that the graph has three x-intercepts:

$$(0, 0), \quad (2, 0), \quad \text{and} \quad (-2, 0). \qquad x\text{-intercepts}$$

To find the y-intercepts, let x be zero. Doing this produces $y = 0$. So, the y-intercept is

$$(0, 0). \qquad y\text{-intercept}$$

(See Figure P.6.)

▷ **TECHNOLOGY** Example 2 uses an analytic approach to finding intercepts. When an analytic approach is not possible, you can use a graphical approach by finding the points at which the graph intersects the axes. Use the *trace* feature of a graphing utility to approximate the intercepts of the graph of the equation in Example 2. Note that your utility may have a built-in program that can find the x-intercepts of a graph. (Your utility may call this the *root* or *zero* feature.) If so, use the program to find the x-intercepts of the graph of the equation in Example 2.

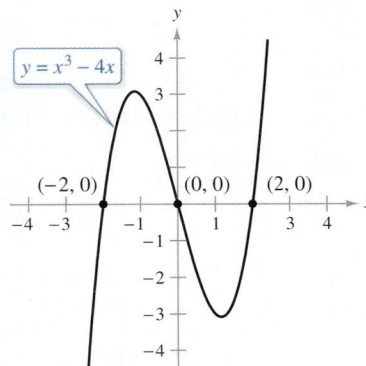

Intercepts of a graph
Figure P.6

Symmetry of a Graph

Knowing the symmetry of a graph before attempting to sketch it is useful because you need only half as many points to sketch the graph. The three types of symmetry listed below can be used to help sketch the graphs of equations (see Figure P.7).

1. A graph is **symmetric with respect to the y-axis** if, whenever (x, y) is a point on the graph, then $(-x, y)$ is also a point on the graph. This means that the portion of the graph to the left of the y-axis is a mirror image of the portion to the right of the y-axis.

2. A graph is **symmetric with respect to the x-axis** if, whenever (x, y) is a point on the graph, then $(x, -y)$ is also a point on the graph. This means that the portion of the graph below the x-axis is a mirror image of the portion above the x-axis.

3. A graph is **symmetric with respect to the origin** if, whenever (x, y) is a point on the graph, then $(-x, -y)$ is also a point on the graph. This means that the graph is unchanged by a rotation of 180° about the origin.

> ### Tests for Symmetry
>
> 1. The graph of an equation in x and y is symmetric with respect to the y-axis when replacing x by $-x$ yields an equivalent equation.
>
> 2. The graph of an equation in x and y is symmetric with respect to the x-axis when replacing y by $-y$ yields an equivalent equation.
>
> 3. The graph of an equation in x and y is symmetric with respect to the origin when replacing x by $-x$ and y by $-y$ yields an equivalent equation.

The graph of a polynomial has symmetry with respect to the y-axis when each term has an even exponent (or is a constant). For instance, the graph of

$$y = 2x^4 - x^2 + 2$$

has symmetry with respect to the y-axis. Similarly, the graph of a polynomial has symmetry with respect to the origin when each term has an odd exponent, as illustrated in Example 3.

EXAMPLE 3   **Testing for Symmetry**

Test the graph of $y = 2x^3 - x$ for symmetry with respect to (a) the y-axis and (b) the origin.

Solution

a.

$y = 2x^3 - x$	Write original equation.
$y = 2(-x)^3 - (-x)$	Replace x by $-x$.
$y = -2x^3 + x$	Simplify. It is not an equivalent equation.

Because replacing x by $-x$ does *not* yield an equivalent equation, you can conclude that the graph of $y = 2x^3 - x$ is *not* symmetric with respect to the y-axis.

b.

$y = 2x^3 - x$	Write original equation.
$-y = 2(-x)^3 - (-x)$	Replace x by $-x$ and y by $-y$.
$-y = -2x^3 + x$	Simplify.
$y = 2x^3 - x$	Equivalent equation

Because replacing x by $-x$ and y by $-y$ yields an equivalent equation, you can conclude that the graph of $y = 2x^3 - x$ is symmetric with respect to the origin, as shown in Figure P.8.

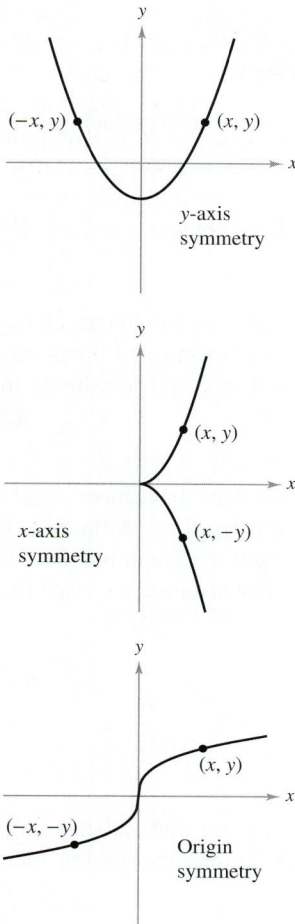

y-axis symmetry

x-axis symmetry

Origin symmetry

Figure P.7

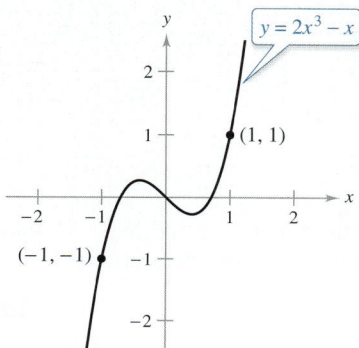

Origin symmetry
Figure P.8

EXAMPLE 4   **Using Intercepts and Symmetry to Sketch a Graph**

⋮ ▷ *See LarsonCalculus.com for an interactive version of this type of example.*

Sketch the graph of $x - y^2 = 1$.

Solution   The graph is symmetric with respect to the x-axis because replacing y by $-y$ yields an equivalent equation.

$$x - y^2 = 1 \qquad \text{Write original equation.}$$
$$x - (-y)^2 = 1 \qquad \text{Replace } y \text{ by } -y.$$
$$x - y^2 = 1 \qquad \text{Equivalent equation}$$

This means that the portion of the graph below the x-axis is a mirror image of the portion above the x-axis. To sketch the graph, first plot the x-intercept and the points above the x-axis. Then reflect in the x-axis to obtain the entire graph, as shown in Figure P.9.

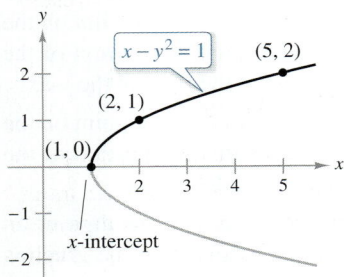

Figure P.9

▷ **TECHNOLOGY**   Graphing utilities are designed so that they most easily graph equations in which y is a function of x (see Section P.3 for a definition of **function**). To graph other types of equations, you need to split the graph into two or more parts *or* you need to use a different graphing mode. For instance, to graph the equation in Example 4, you can split it into two parts.

$$y_1 = \sqrt{x - 1} \qquad \text{Top portion of graph}$$
$$y_2 = -\sqrt{x - 1} \qquad \text{Bottom portion of graph}$$

Points of Intersection

A **point of intersection** of the graphs of two equations is a point that satisfies both equations. You can find the point(s) of intersection of two graphs by solving their equations simultaneously.

EXAMPLE 5   **Finding Points of Intersection**

Find all points of intersection of the graphs of

$$x^2 - y = 3 \quad \text{and} \quad x - y = 1.$$

Solution   Begin by sketching the graphs of both equations in the *same* rectangular coordinate system, as shown in Figure P.10. From the figure, it appears that the graphs have two points of intersection. You can find these two points as follows.

$$y = x^2 - 3 \qquad \text{Solve first equation for } y.$$
$$y = x - 1 \qquad \text{Solve second equation for } y.$$
$$x^2 - 3 = x - 1 \qquad \text{Equate } y\text{-values.}$$
$$x^2 - x - 2 = 0 \qquad \text{Write in general form.}$$
$$(x - 2)(x + 1) = 0 \qquad \text{Factor.}$$
$$x = 2 \text{ or } -1 \qquad \text{Solve for } x.$$

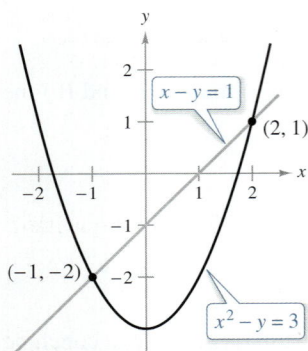

Two points of intersection
Figure P.10

The corresponding values of y are obtained by substituting $x = 2$ and $x = -1$ into either of the original equations. Doing this produces two points of intersection:

$$(2, 1) \quad \text{and} \quad (-1, -2). \qquad \text{Points of intersection}$$

You can check the points of intersection in Example 5 by substituting into *both* of the original equations or by using the *intersect* feature of a graphing utility.

Mathematical Models

Real-life applications of mathematics often use equations as **mathematical models.** In developing a mathematical model to represent actual data, you should strive for two (often conflicting) goals: accuracy and simplicity. That is, you want the model to be simple enough to be workable, yet accurate enough to produce meaningful results. Section P.4 explores these goals more completely.

The Mauna Loa Observatory in Hawaii has been measuring the increasing concentration of carbon dioxide in Earth's atmosphere since 1958.

EXAMPLE 6 **Comparing Two Mathematical Models**

The Mauna Loa Observatory in Hawaii records the carbon dioxide concentration y (in parts per million) in Earth's atmosphere. The January readings for various years are shown in Figure P.11. In the July 1990 issue of *Scientific American*, these data were used to predict the carbon dioxide level in Earth's atmosphere in the year 2035, using the quadratic model

$$y = 0.018t^2 + 0.70t + 316.2 \qquad \text{\color{red}Quadratic model for 1960–1990 data}$$

where $t = 0$ represents 1960, as shown in Figure P.11(a). The data shown in Figure P.11(b) represent the years 1980 through 2010 and can be modeled by

$$y = 1.68t + 303.5 \qquad \text{\color{red}Linear model for 1980–2010 data}$$

where $t = 0$ represents 1960. What was the prediction given in the *Scientific American* article in 1990? Given the new data for 1990 through 2010, does this prediction for the year 2035 seem accurate?

(a)

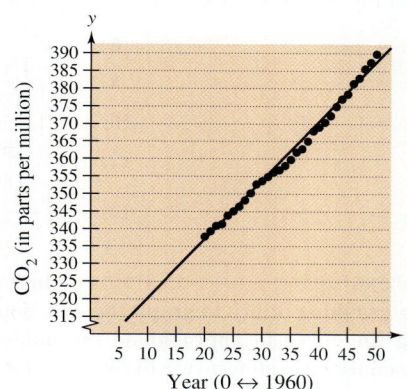

(b)

Figure P.11

Solution To answer the first question, substitute $t = 75$ (for 2035) into the quadratic model.

$$y = 0.018(75)^2 + 0.70(75) + 316.2 = 469.95 \qquad \text{\color{red}Quadratic model}$$

So, the prediction in the *Scientific American* article was that the carbon dioxide concentration in Earth's atmosphere would reach about 470 parts per million in the year 2035. Using the linear model for the 1980–2010 data, the prediction for the year 2035 is

$$y = 1.68(75) + 303.5 = 429.5. \qquad \text{\color{red}Linear model}$$

So, based on the linear model for 1980–2010, it appears that the 1990 prediction was too high. ∎

The models in Example 6 were developed using a procedure called *least squares regression* (see Section 13.9). The quadratic and linear models have correlations given by $r^2 \approx 0.997$ and $r^2 \approx 0.994$, respectively. The closer r^2 is to 1, the "better" the model.

P.1 Exercises

See **CalcChat.com** for tutorial help and worked-out solutions to odd-numbered exercises.

Matching In Exercises 1–4, match the equation with its graph. [The graphs are labeled (a), (b), (c), and (d).]

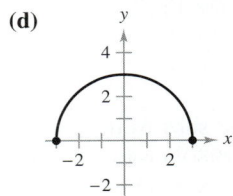

(a)

(b)

(c)

(d)

1. $y = -\frac{3}{2}x + 3$

2. $y = \sqrt{9 - x^2}$

3. $y = 3 - x^2$

4. $y = x^3 - x$

Sketching a Graph by Point Plotting In Exercises 5–14, sketch the graph of the equation by point plotting.

5. $y = \frac{1}{2}x + 2$

6. $y = 5 - 2x$

7. $y = 4 - x^2$

8. $y = (x - 3)^2$

9. $y = |x + 2|$

10. $y = |x| - 1$

11. $y = \sqrt{x} - 6$

12. $y = \sqrt{x + 2}$

13. $y = \frac{3}{x}$

14. $y = \frac{1}{x + 2}$

Approximating Solution Points In Exercises 15 and 16, use a graphing utility to graph the equation. Move the cursor along the curve to approximate the unknown coordinate of each solution point accurate to two decimal places.

15. $y = \sqrt{5 - x}$

(a) $(2, y)$

(b) $(x, 3)$

16. $y = x^5 - 5x$

(a) $(-0.5, y)$

(b) $(x, -4)$

Finding Intercepts In Exercises 17–26, find any intercepts.

17. $y = 2x - 5$

18. $y = 4x^2 + 3$

19. $y = x^2 + x - 2$

20. $y^2 = x^3 - 4x$

21. $y = x\sqrt{16 - x^2}$

22. $y = (x - 1)\sqrt{x^2 + 1}$

23. $y = \frac{2 - \sqrt{x}}{5x + 1}$

24. $y = \frac{x^2 + 3x}{(3x + 1)^2}$

25. $x^2y - x^2 + 4y = 0$

26. $y = 2x - \sqrt{x^2 + 1}$

Testing for Symmetry In Exercises 27–38, test for symmetry with respect to each axis and to the origin.

27. $y = x^2 - 6$

28. $y = x^2 - x$

29. $y^2 = x^3 - 8x$

30. $y = x^3 + x$

31. $xy = 4$

32. $xy^2 = -10$

33. $y = 4 - \sqrt{x + 3}$

34. $xy - \sqrt{4 - x^2} = 0$

35. $y = \frac{x}{x^2 + 1}$

36. $y = \frac{x^2}{x^2 + 1}$

37. $y = |x^3 + x|$

38. $|y| - x = 3$

Using Intercepts and Symmetry to Sketch a Graph In Exercises 39–56, find any intercepts and test for symmetry. Then sketch the graph of the equation.

39. $y = 2 - 3x$

40. $y = \frac{2}{3}x + 1$

41. $y = 9 - x^2$

42. $y = 2x^2 + x$

43. $y = x^3 + 2$

44. $y = x^3 - 4x$

45. $y = x\sqrt{x + 5}$

46. $y = \sqrt{25 - x^2}$

47. $x = y^3$

48. $x = y^2 - 4$

49. $y = \frac{8}{x}$

50. $y = \frac{10}{x^2 + 1}$

51. $y = 6 - |x|$

52. $y = |6 - x|$

53. $y^2 - x = 9$

54. $x^2 + 4y^2 = 4$

55. $x + 3y^2 = 6$

56. $3x - 4y^2 = 8$

Finding Points of Intersection In Exercises 57–62, find the points of intersection of the graphs of the equations.

57. $x + y = 8$

$4x - y = 7$

58. $3x - 2y = -4$

$4x + 2y = -10$

59. $x^2 + y = 6$

$x + y = 4$

60. $x = 3 - y^2$

$y = x - 1$

61. $x^2 + y^2 = 5$

$x - y = 1$

62. $x^2 + y^2 = 25$

$-3x + y = 15$

Finding Points of Intersection In Exercises 63–66, use a graphing utility to find the points of intersection of the graphs. Check your results analytically.

63. $y = x^3 - 2x^2 + x - 1$

$y = -x^2 + 3x - 1$

64. $y = x^4 - 2x^2 + 1$

$y = 1 - x^2$

65. $y = \sqrt{x + 6}$

$y = \sqrt{-x^2 - 4x}$

66. $y = -|2x - 3| + 6$

$y = 6 - x$

The symbol ⊞ indicates an exercise in which you are instructed to use graphing technology or a symbolic computer algebra system. The solutions of other exercises may also be facilitated by the use of appropriate technology.

67. Modeling Data The table shows the Gross Domestic Product, or GDP (in trillions of dollars), for selected years. *(Source: U.S. Bureau of Economic Analysis)*

Year	1980	1985	1990	1995
GDP	2.8	4.2	5.8	7.4

Year	2000	2005	2010
GDP	10.0	12.6	14.5

(a) Use the regression capabilities of a graphing utility to find a mathematical model of the form $y = at^2 + bt + c$ for the data. In the model, y represents the GDP (in trillions of dollars) and t represents the year, with $t = 0$ corresponding to 1980.

(b) Use a graphing utility to plot the data and graph the model. Compare the data with the model.

(c) Use the model to predict the GDP in the year 2020.

68. Modeling Data

The table shows the numbers of cellular phone subscribers (in millions) in the United States for selected years. *(Source: CTIA-The Wireless)*

Year	1995	1998	2001	2004	2007	2010
Number	34	69	128	182	255	303

(a) Use the regression capabilities of a graphing utility to find a mathematical model of the form $y = at^2 + bt + c$ for the data. In the model, y represents the number of subscribers (in millions) and t represents the year, with $t = 5$ corresponding to 1995.

(b) Use a graphing utility to plot the data and graph the model. Compare the data with the model.

(c) Use the model to predict the number of cellular phone subscribers in the United States in the year 2020.

69. Break-Even Point Find the sales necessary to break even ($R = C$) when the cost C of producing x units is $C = 2.04x + 5600$ and the revenue R from selling x units is $R = 3.29x$.

70. Copper Wire The resistance y in ohms of 1000 feet of solid copper wire at 77°F can be approximated by the model

$$y = \frac{10{,}770}{x^2} - 0.37, \quad 5 \le x \le 100$$

where x is the diameter of the wire in mils (0.001 in.). Use a graphing utility to graph the model. By about what factor is the resistance changed when the diameter of the wire is doubled?

71. Using Solution Points For what values of k does the graph of $y = kx^3$ pass through the point?

(a) $(1, 4)$ (b) $(-2, 1)$ (c) $(0, 0)$ (d) $(-1, -1)$

72. Using Solution Points For what values of k does the graph of $y^2 = 4kx$ pass through the point?

(a) $(1, 1)$ (b) $(2, 4)$ (c) $(0, 0)$ (d) $(3, 3)$

WRITING ABOUT CONCEPTS

Writing Equations In Exercises 73 and 74, write an equation whose graph has the indicated property. (There may be more than one correct answer.)

73. The graph has intercepts at $x = -4$, $x = 3$, and $x = 8$.

74. The graph has intercepts at $x = -\frac{3}{2}$, $x = 4$, and $x = \frac{5}{2}$.

75. Proof

(a) Prove that if a graph is symmetric with respect to the x-axis and to the y-axis, then it is symmetric with respect to the origin. Give an example to show that the converse is not true.

(b) Prove that if a graph is symmetric with respect to one axis and to the origin, then it is symmetric with respect to the other axis.

76. HOW DO YOU SEE IT? Use the graphs of the two equations to answer the questions below.

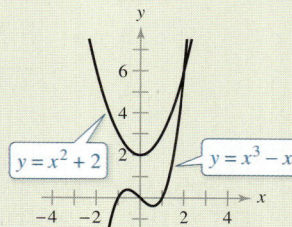

(a) What are the intercepts for each equation?

(b) Determine the symmetry for each equation.

(c) Determine the point of intersection of the two equations.

True or False? In Exercises 77–80, determine whether the statement is true or false. If it is false, explain why or give an example that shows it is false.

77. If $(-4, -5)$ is a point on a graph that is symmetric with respect to the x-axis, then $(4, -5)$ is also a point on the graph.

78. If $(-4, -5)$ is a point on a graph that is symmetric with respect to the y-axis, then $(4, -5)$ is also a point on the graph.

79. If $b^2 - 4ac > 0$ and $a \ne 0$, then the graph of $y = ax^2 + bx + c$ has two x-intercepts.

80. If $b^2 - 4ac = 0$ and $a \ne 0$, then the graph of $y = ax^2 + bx + c$ has only one x-intercept.

P.2 Linear Models and Rates of Change

■ Find the slope of a line passing through two points.
■ Write the equation of a line with a given point and slope.
■ Interpret slope as a ratio or as a rate in a real-life application.
■ Sketch the graph of a linear equation in slope-intercept form.
■ Write equations of lines that are parallel or perpendicular to a given line.

The Slope of a Line

The **slope** of a nonvertical line is a measure of the number of units the line rises (or falls) vertically for each unit of horizontal change from left to right. Consider the two points (x_1, y_1) and (x_2, y_2) on the line in Figure P.12. As you move from left to right along this line, a vertical change of

$$\Delta y = y_2 - y_1 \qquad \text{Change in } y$$

units corresponds to a horizontal change of

$$\Delta x = x_2 - x_1 \qquad \text{Change in } x$$

units. (Δ is the Greek uppercase letter *delta*, and the symbols Δy and Δx are read "delta y" and "delta x.")

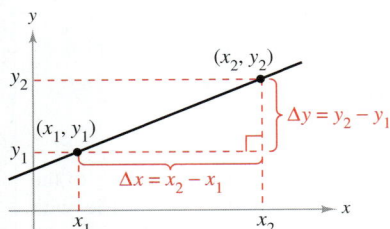

$\Delta y = y_2 - y_1 = $ change in y
$\Delta x = x_2 - x_1 = $ change in x
Figure P.12

Definition of the Slope of a Line

The **slope** m of the nonvertical line passing through (x_1, y_1) and (x_2, y_2) is

$$m = \frac{\Delta y}{\Delta x} = \frac{y_2 - y_1}{x_2 - x_1}, \quad x_1 \neq x_2.$$

Slope is not defined for vertical lines.

When using the formula for slope, note that

$$\frac{y_2 - y_1}{x_2 - x_1} = \frac{-(y_1 - y_2)}{-(x_1 - x_2)} = \frac{y_1 - y_2}{x_1 - x_2}.$$

So, it does not matter in which order you subtract *as long as* you are consistent and both "subtracted coordinates" come from the same point.

Figure P.13 shows four lines: one has a positive slope, one has a slope of zero, one has a negative slope, and one has an "undefined" slope. In general, the greater the absolute value of the slope of a line, the steeper the line. For instance, in Figure P.13, the line with a slope of -5 is steeper than the line with a slope of $\frac{1}{5}$.

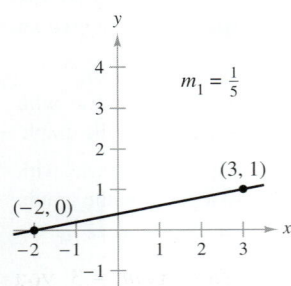

If m is positive, then the line rises from left to right.

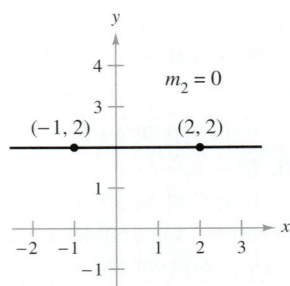

If m is zero, then the line is horizontal.

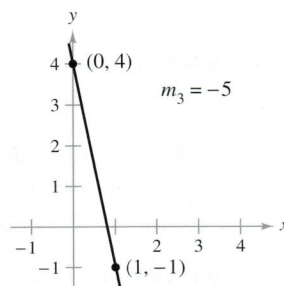

If m is negative, then the line falls from left to right.

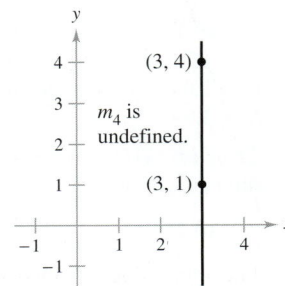

If m is undefined, then the line is vertical.

Figure P.13

Equations of Lines

Any two points on a nonvertical line can be used to calculate its slope. This can be verified from the similar triangles shown in Figure P.14. (Recall that the ratios of corresponding sides of similar triangles are equal.)

$$m = \frac{y_2^* - y_1^*}{x_2^* - x_1^*} = \frac{y_2 - y_1}{x_2 - x_1}$$

Any two points on a nonvertical line can be used to determine its slope.
Figure P.14

If (x_1, y_1) is a point on a nonvertical line that has a slope of m and (x, y) is *any other* point on the line, then

$$\frac{y - y_1}{x - x_1} = m.$$

This equation in the variables x and y can be rewritten in the form

$$y - y_1 = m(x - x_1)$$

which is the **point-slope form** of the equation of a line.

Point-Slope Form of the Equation of a Line

The **point-slope form** of the equation of the line that passes through the point (x_1, y_1) and has a slope of m is

$$y - y_1 = m(x - x_1).$$

REMARK Remember that only nonvertical lines have a slope. Consequently, vertical lines cannot be written in point-slope form. For instance, the equation of the vertical line passing through the point $(1, -2)$ is $x = 1$.

EXAMPLE 1   Finding an Equation of a Line

Find an equation of the line that has a slope of 3 and passes through the point $(1, -2)$. Then sketch the line.

Solution

$$y - y_1 = m(x - x_1) \quad \text{Point-slope form}$$
$$y - (-2) = 3(x - 1) \quad \text{Substitute } -2 \text{ for } y_1, 1 \text{ for } x_1, \text{ and } 3 \text{ for } m.$$
$$y + 2 = 3x - 3 \quad \text{Simplify.}$$
$$y = 3x - 5 \quad \text{Solve for } y.$$

To sketch the line, first plot the point $(1, -2)$. Then, because the slope is $m = 3$, you can locate a second point on the line by moving one unit to the right and three units upward, as shown in Figure P.15.

Exploration

Investigating Equations of Lines Use a graphing utility to graph each of the linear equations. Which point is common to all seven lines? Which value in the equation determines the slope of each line?

a. $y - 4 = -2(x + 1)$
b. $y - 4 = -1(x + 1)$
c. $y - 4 = -\frac{1}{2}(x + 1)$
d. $y - 4 = 0(x + 1)$
e. $y - 4 = \frac{1}{2}(x + 1)$
f. $y - 4 = 1(x + 1)$
g. $y - 4 = 2(x + 1)$

Use your results to write an equation of a line passing through $(-1, 4)$ with a slope of m.

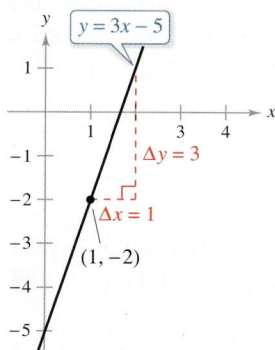

The line with a slope of 3 passing through the point $(1, -2)$
Figure P.15

Ratios and Rates of Change

The slope of a line can be interpreted as either a *ratio* or a *rate*. If the *x*- and *y*-axes have the same unit of measure, then the slope has no units and is a **ratio.** If the *x*- and *y*-axes have different units of measure, then the slope is a rate or **rate of change.** In your study of calculus, you will encounter applications involving both interpretations of slope.

EXAMPLE 2   **Using Slope as a Ratio**

The maximum recommended slope of a wheelchair ramp is $\frac{1}{12}$. A business installs a wheelchair ramp that rises to a height of 22 inches over a length of 24 feet, as shown in Figure P.16. Is the ramp steeper than recommended?   *(Source: ADA Standards for Accessible Design)*

Figure P.16

Solution   The length of the ramp is 24 feet or $12(24) = 288$ inches. The slope of the ramp is the ratio of its height (the rise) to its length (the run).

$$\text{Slope of ramp} = \frac{\text{rise}}{\text{run}}$$

$$= \frac{22 \text{ in.}}{288 \text{ in.}}$$

$$\approx 0.076$$

Because the slope of the ramp is less than $\frac{1}{12} \approx 0.083$, the ramp is not steeper than recommended. Note that the slope is a ratio and has no units.

EXAMPLE 3   **Using Slope as a Rate of Change**

The population of Colorado was about 4,302,000 in 2000 and about 5,029,000 in 2010. Find the average rate of change of the population over this 10-year period. What will the population of Colorado be in 2020?   *(Source: U.S. Census Bureau)*

Solution   Over this 10-year period, the average rate of change of the population of Colorado was

$$\text{Rate of change} = \frac{\text{change in population}}{\text{change in years}}$$

$$= \frac{5,029,000 - 4,302,000}{2010 - 2000}$$

$$= 72,700 \text{ people per year.}$$

Assuming that Colorado's population continues to increase at this same rate for the next 10 years, it will have a 2020 population of about 5,756,000 (see Figure P.17).

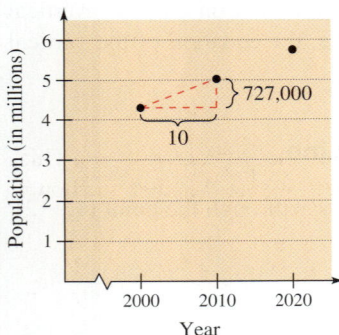

Population of Colorado
Figure P.17

The rate of change found in Example 3 is an **average rate of change.** An average rate of change is always calculated over an interval. In this case, the interval is [2000, 2010]. In Chapter 2, you will study another type of rate of change called an *instantaneous rate of change.*

Graphing Linear Models

Many problems in coordinate geometry can be classified into two basic categories.

1. Given a graph (or parts of it), find its equation.
2. Given an equation, sketch its graph.

For lines, problems in the first category can be solved by using the point-slope form. The point-slope form, however, is not especially useful for solving problems in the second category. The form that is better suited to sketching the graph of a line is the **slope-intercept** form of the equation of a line.

The Slope-Intercept Form of the Equation of a Line

The graph of the linear equation

$$y = mx + b \qquad \text{Slope-intercept form}$$

is a line whose slope is m and whose y-intercept is $(0, b)$.

EXAMPLE 4   **Sketching Lines in the Plane**

Sketch the graph of each equation.

a. $y = 2x + 1$

b. $y = 2$

c. $3y + x - 6 = 0$

Solution

a. Because $b = 1$, the y-intercept is $(0, 1)$. Because the slope is $m = 2$, you know that the line rises two units for each unit it moves to the right, as shown in Figure P.18(a).

b. By writing the equation $y = 2$ in slope-intercept form

$$y = (0)x + 2$$

you can see that the slope is $m = 0$ and the y-intercept is $(0, 2)$. Because the slope is zero, you know that the line is horizontal, as shown in Figure P.18(b).

c. Begin by writing the equation in slope-intercept form.

$$3y + x - 6 = 0 \qquad \text{Write original equation.}$$
$$3y = -x + 6 \qquad \text{Isolate } y\text{-term on the left.}$$
$$y = -\tfrac{1}{3}x + 2 \qquad \text{Slope-intercept form}$$

In this form, you can see that the y-intercept is $(0, 2)$ and the slope is $m = -\tfrac{1}{3}$. This means that the line falls one unit for every three units it moves to the right, as shown in Figure P.18(c).

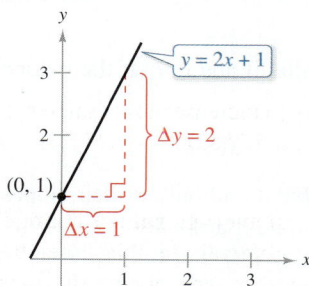

(a) $m = 2$; line rises

(b) $m = 0$; line is horizontal

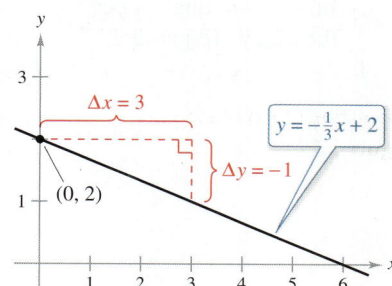

(c) $m = -\tfrac{1}{3}$; line falls

Figure P.18

Because the slope of a vertical line is not defined, its equation cannot be written in slope-intercept form. However, the equation of any line can be written in the **general form**

$$Ax + By + C = 0 \qquad \text{General form of the equation of a line}$$

where A and B are not *both* zero. For instance, the vertical line

$$x = a \qquad \text{Vertical line}$$

can be represented by the general form

$$x - a = 0. \qquad \text{General form}$$

SUMMARY OF EQUATIONS OF LINES

1. General form: $Ax + By + C = 0$
2. Vertical line: $x = a$
3. Horizontal line: $y = b$
4. Slope-intercept form: $y = mx + b$
5. Point-slope form: $y - y_1 = m(x - x_1)$

Parallel and Perpendicular Lines

The slope of a line is a convenient tool for determining whether two lines are parallel or perpendicular, as shown in Figure P.19. Specifically, nonvertical lines with the same slope are parallel, and nonvertical lines whose slopes are negative reciprocals are perpendicular.

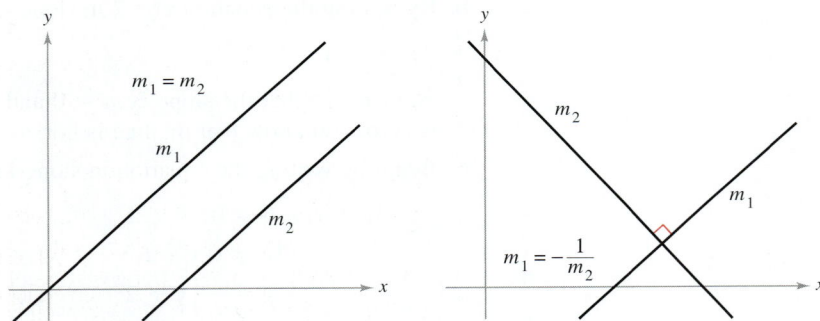

Parallel lines

Perpendicular lines

Figure P.19

REMARK In mathematics, the phrase "if and only if" is a way of stating two implications in one statement. For instance, the first statement at the right could be rewritten as the following two implications.

a. If two distinct nonvertical lines are parallel, then their slopes are equal.

b. If two distinct nonvertical lines have equal slopes, then they are parallel.

Parallel and Perpendicular Lines

1. Two distinct nonvertical lines are **parallel** if and only if their slopes are equal—that is, if and only if

$$m_1 = m_2. \qquad \text{Parallel} \Longleftrightarrow \text{Slopes are equal.}$$

2. Two nonvertical lines are **perpendicular** if and only if their slopes are negative reciprocals of each other—that is, if and only if

$$m_1 = -\frac{1}{m_2}. \qquad \text{Perpendicular} \Longleftrightarrow \text{Slopes are negative reciprocals.}$$

EXAMPLE 5   **Finding Parallel and Perpendicular Lines**

• • • ▷ *See LarsonCalculus.com for an interactive version of this type of example.*

Find the general forms of the equations of the lines that pass through the point $(2, -1)$ and are (a) parallel to and (b) perpendicular to the line $2x - 3y = 5$.

Solution   Begin by writing the linear equation $2x - 3y = 5$ in slope-intercept form.

$$2x - 3y = 5 \qquad \text{Write original equation.}$$
$$y = \tfrac{2}{3}x - \tfrac{5}{3} \qquad \text{Slope-intercept form}$$

So, the given line has a slope of $m = \tfrac{2}{3}$. (See Figure P.20.)

a. The line through $(2, -1)$ that is parallel to the given line also has a slope of $\tfrac{2}{3}$.

$$y - y_1 = m(x - x_1) \qquad \text{Point-slope form}$$
$$y - (-1) = \tfrac{2}{3}(x - 2) \qquad \text{Substitute.}$$
$$3(y + 1) = 2(x - 2) \qquad \text{Simplify.}$$
$$3y + 3 = 2x - 4 \qquad \text{Distributive Property}$$
$$2x - 3y - 7 = 0 \qquad \text{General form}$$

Note the similarity to the equation of the given line, $2x - 3y = 5$.

b. Using the negative reciprocal of the slope of the given line, you can determine that the slope of a line perpendicular to the given line is $-\tfrac{3}{2}$.

$$y - y_1 = m(x - x_1) \qquad \text{Point-slope form}$$
$$y - (-1) = -\tfrac{3}{2}(x - 2) \qquad \text{Substitute.}$$
$$2(y + 1) = -3(x - 2) \qquad \text{Simplify.}$$
$$2y + 2 = -3x + 6 \qquad \text{Distributive Property}$$
$$3x + 2y - 4 = 0 \qquad \text{General form}$$

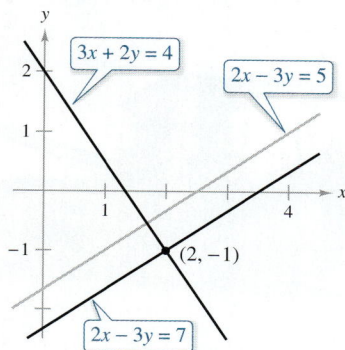

Lines parallel and perpendicular to $2x - 3y = 5$
Figure P.20

▷ **TECHNOLOGY PITFALL**   The slope of a line will appear distorted if you use different tick-mark spacing on the x- and y-axes. For instance, the graphing utility screens in Figures P.21(a) and P.21(b) both show the lines

$$y = 2x \quad \text{and} \quad y = -\tfrac{1}{2}x + 3.$$

Because these lines have slopes that are negative reciprocals, they must be perpendicular. In Figure P.21(a), however, the lines don't appear to be perpendicular because the tick-mark spacing on the x-axis is not the same as that on the y-axis. In Figure P.21(b), the lines appear perpendicular because the tick-mark spacing on the x-axis is the same as on the y-axis. This type of viewing window is said to have a *square setting*.

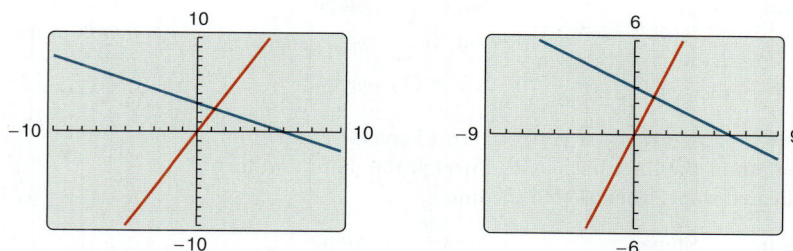

(a) Tick-mark spacing on the x-axis is not the same as tick-mark spacing on the y-axis.

(b) Tick-mark spacing on the x-axis is the same as tick-mark spacing on the y-axis.

Figure P.21

P.2 Exercises

See **CalcChat.com** for tutorial help and worked-out solutions to odd-numbered exercises.

Estimating Slope In Exercises 1–4, estimate the slope of the line from its graph. To print an enlarged copy of the graph, go to *MathGraphs.com*.

1.

2.

3.

4.

Finding the Slope of a Line In Exercises 5–10, plot the pair of points and find the slope of the line passing through them.

5. $(3, -4), (5, 2)$

6. $(1, 1), (-2, 7)$

7. $(4, 6), (4, 1)$

8. $(3, -5), (5, -5)$

9. $\left(-\frac{1}{2}, \frac{2}{3}\right), \left(-\frac{3}{4}, \frac{1}{6}\right)$

10. $\left(\frac{7}{8}, \frac{3}{4}\right), \left(\frac{5}{4}, -\frac{1}{4}\right)$

Sketching Lines In Exercises 11 and 12, sketch the lines through the point with the indicated slopes. Make the sketches on the same set of coordinate axes.

	Point	Slopes			
11.	$(3, 4)$	(a) 1	(b) -2	(c) $-\frac{3}{2}$	(d) Undefined
12.	$(-2, 5)$	(a) 3	(b) -3	(c) $\frac{1}{3}$	(d) 0

Finding Points on a Line In Exercises 13–16, use the point on the line and the slope of the line to find three additional points that the line passes through. (There is more than one correct answer.)

	Point	Slope		Point	Slope
13.	$(6, 2)$	$m = 0$	**14.**	$(-4, 3)$	m is undefined.
15.	$(1, 7)$	$m = -3$	**16.**	$(-2, -2)$	$m = 2$

Finding an Equation of a Line In Exercises 17–22, find an equation of the line that passes through the point and has the indicated slope. Then sketch the line.

	Point	Slope		Point	Slope
17.	$(0, 3)$	$m = \frac{3}{4}$	**18.**	$(-5, -2)$	m is undefined.
19.	$(0, 0)$	$m = \frac{2}{3}$	**20.**	$(0, 4)$	$m = 0$
21.	$(3, -2)$	$m = 3$	**22.**	$(-2, 4)$	$m = -\frac{3}{5}$

23. Conveyor Design

A moving conveyor is built to rise 1 meter for each 3 meters of horizontal change.

(a) Find the slope of the conveyor.

(b) Suppose the conveyor runs between two floors in a factory. Find the length of the conveyor when the vertical distance between floors is 10 feet.

24. Modeling Data The table shows the populations y (in millions) of the United States for 2004 through 2009. The variable t represents the time in years, with $t = 4$ corresponding to 2004. *(Source: U.S. Census Bureau)*

t	4	5	6	7	8	9
y	293.0	295.8	298.6	301.6	304.4	307.0

(a) Plot the data by hand and connect adjacent points with a line segment.

(b) Use the slope of each line segment to determine the year when the population increased least rapidly.

(c) Find the average rate of change of the population of the United States from 2004 through 2009.

(d) Use the average rate of change of the population to predict the population of the United States in 2020.

Finding the Slope and y-Intercept In Exercises 25–30, find the slope and the *y*-intercept (if possible) of the line.

25. $y = 4x - 3$

26. $-x + y = 1$

27. $x + 5y = 20$

28. $6x - 5y = 15$

29. $x = 4$

30. $y = -1$

Sketching a Line in the Plane In Exercises 31–38, sketch a graph of the equation.

31. $y = -3$

32. $x = 4$

33. $y = -2x + 1$

34. $y = \frac{1}{3}x - 1$

35. $y - 2 = \frac{3}{2}(x - 1)$

36. $y - 1 = 3(x + 4)$

37. $2x - y - 3 = 0$

38. $x + 2y + 6 = 0$

Finding an Equation of a Line In Exercises 39–46, find an equation of the line that passes through the points. Then sketch the line.

39. $(0, 0), (4, 8)$

40. $(-2, -2), (1, 7)$

41. $(2, 8), (5, 0)$ **42.** $(-3, 6), (1, 2)$

43. $(6, 3), (6, 8)$ **44.** $(1, -2), (3, -2)$

45. $\left(\frac{1}{2}, \frac{7}{2}\right), \left(0, \frac{3}{4}\right)$ **46.** $\left(\frac{7}{8}, \frac{3}{4}\right), \left(\frac{5}{4}, -\frac{1}{4}\right)$

47. Find an equation of the vertical line with x-intercept at 3.

48. Show that the line with intercepts $(a, 0)$ and $(0, b)$ has the following equation.

$$\frac{x}{a} + \frac{y}{b} = 1, \quad a \neq 0, b \neq 0$$

Writing an Equation in General Form **In Exercises 49–54, use the result of Exercise 48 to write an equation of the line in general form.**

49. x-intercept: $(2, 0)$

    y-intercept: $(0, 3)$

50. x-intercept: $\left(-\frac{2}{3}, 0\right)$

    y-intercept: $(0, -2)$

51. Point on line: $(1, 2)$

    x-intercept: $(a, 0)$

    y-intercept: $(0, a)$

    $(a \neq 0)$

52. Point on line: $(-3, 4)$

    x-intercept: $(a, 0)$

    y-intercept: $(0, a)$

    $(a \neq 0)$

53. Point on line: $(9, -2)$

    x-intercept: $(2a, 0)$

    y-intercept: $(0, a)$

    $(a \neq 0)$

54. Point on line: $\left(-\frac{2}{3}, -2\right)$

    x-intercept: $(a, 0)$

    y-intercept: $(0, -a)$

    $(a \neq 0)$

Finding Parallel and Perpendicular Lines **In Exercises 55–62, write the general forms of the equations of the lines through the point (a) parallel to the given line and (b) perpendicular to the given line.**

	Point	Line		Point	Line
55.	$(-7, -2)$	$x = 1$	**56.**	$(-1, 0)$	$y = -3$
57.	$(2, 5)$	$x - y = -2$	**58.**	$(-3, 2)$	$x + y = 7$
59.	$(2, 1)$	$4x - 2y = 3$	**60.**	$\left(\frac{5}{6}, -\frac{1}{2}\right)$	$7x + 4y = 8$
61.	$\left(\frac{3}{4}, \frac{7}{8}\right)$	$5x - 3y = 0$	**62.**	$(4, -5)$	$3x + 4y = 7$

Rate of Change **In Exercises 63–66, you are given the dollar value of a product in 2012 *and* the rate at which the value of the product is expected to change during the next 5 years. Write a linear equation that gives the dollar value V of the product in terms of the year t. (Let $t = 0$ represent 2010.)**

	2012 Value	Rate
63.	$1850	$250 increase per year
64.	$156	$4.50 increase per year
65.	$17,200	$1600 decrease per year
66.	$245,000	$5600 decrease per year

Collinear Points **In Exercises 67 and 68, determine whether the points are collinear. (Three points are *collinear* if they lie on the same line.)**

67. $(-2, 1), (-1, 0), (2, -2)$

68. $(0, 4), (7, -6), (-5, 11)$

WRITING ABOUT CONCEPTS

Finding Points of Intersection In Exercises 69–71, find the coordinates of the point of intersection of the given segments. Explain your reasoning.

69.

Perpendicular bisectors

70.

Medians

71.

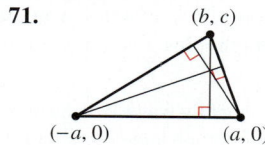

Altitudes

72. Show that the points of intersection in Exercises 69, 70, and 71 are collinear.

73. Analyzing a Line A line is represented by the equation $ax + by = 4$.

(a) When is the line parallel to the x-axis?

(b) When is the line parallel to the y-axis?

(c) Give values for a and b such that the line has a slope of $\frac{5}{8}$.

(d) Give values for a and b such that the line is perpendicular to $y = \frac{2}{5}x + 3$.

(e) Give values for a and b such that the line coincides with the graph of $5x + 6y = 8$.

74. HOW DO YOU SEE IT? Use the graphs of the equations to answer the questions below.

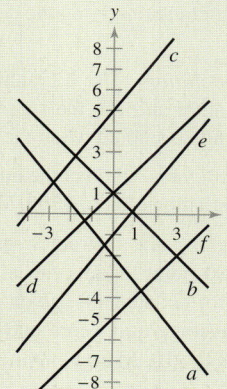

(a) Which lines have a positive slope?

(b) Which lines have a negative slope?

(c) Which lines appear parallel?

(d) Which lines appear perpendicular?

75. Temperature Conversion Find a linear equation that expresses the relationship between the temperature in degrees Celsius C and degrees Fahrenheit F. Use the fact that water freezes at 0°C (32°F) and boils at 100°C (212°F). Use the equation to convert 72°F to degrees Celsius.

76. Reimbursed Expenses A company reimburses its sales representatives $200 per day for lodging and meals plus $0.51 per mile driven. Write a linear equation giving the daily cost C to the company in terms of x, the number of miles driven. How much does it cost the company if a sales representative drives 137 miles on a given day?

77. Choosing a Job As a salesperson, you receive a monthly salary of $2000, plus a commission of 7% of sales. You are offered a new job at $2300 per month, plus a commission of 5% of sales.

  (a) Write linear equations for your monthly wage W in terms of your monthly sales s for your current job and your job offer.

  (b) Use a graphing utility to graph each equation and find the point of intersection. What does it signify?

  (c) You think you can sell $20,000 worth of a product per month. Should you change jobs? Explain.

78. Straight-Line Depreciation A small business purchases a piece of equipment for $875. After 5 years, the equipment will be outdated, having no value.

  (a) Write a linear equation giving the value y of the equipment in terms of the time x (in years), $0 \leq x \leq 5$.

  (b) Find the value of the equipment when $x = 2$.

  (c) Estimate (to two-decimal-place accuracy) the time when the value of the equipment is $200.

79. Apartment Rental A real estate office manages an apartment complex with 50 units. When the rent is $780 per month, all 50 units are occupied. However, when the rent is $825, the average number of occupied units drops to 47. Assume that the relationship between the monthly rent p and the demand x is linear. (*Note:* The term *demand* refers to the number of occupied units.)

  (a) Write a linear equation giving the demand x in terms of the rent p.

  (b) *Linear extrapolation* Use a graphing utility to graph the demand equation and use the *trace* feature to predict the number of units occupied when the rent is raised to $855.

  (c) *Linear interpolation* Predict the number of units occupied when the rent is lowered to $795. Verify graphically.

80. Modeling Data An instructor gives regular 20-point quizzes and 100-point exams in a mathematics course. Average scores for six students, given as ordered pairs (x, y), where x is the average quiz score and y is the average exam score, are $(18, 87)$, $(10, 55)$, $(19, 96)$, $(16, 79)$, $(13, 76)$, and $(15, 82)$.

  (a) Use the regression capabilities of a graphing utility to find the least squares regression line for the data.

  (b) Use a graphing utility to plot the points and graph the regression line in the same viewing window.

  (c) Use the regression line to predict the average exam score for a student with an average quiz score of 17.

  (d) Interpret the meaning of the slope of the regression line.

  (e) The instructor adds 4 points to the average exam score of everyone in the class. Describe the changes in the positions of the plotted points and the change in the equation of the line.

81. Tangent Line Find an equation of the line tangent to the circle $x^2 + y^2 = 169$ at the point $(5, 12)$.

82. Tangent Line Find an equation of the line tangent to the circle $(x - 1)^2 + (y - 1)^2 = 25$ at the point $(4, -3)$.

Distance In Exercises 83–86, find the distance between the point and line, or between the lines, using the formula for the distance between the point (x_1, y_1) and the line $Ax + By + C = 0$.

$$\text{Distance} = \frac{|Ax_1 + By_1 + C|}{\sqrt{A^2 + B^2}}$$

83. Point: $(-2, 1)$
  Line: $x - y - 2 = 0$

84. Point: $(2, 3)$
  Line: $4x + 3y = 10$

85. Line: $x + y = 1$
  Line: $x + y = 5$

86. Line: $3x - 4y = 1$
  Line: $3x - 4y = 10$

87. Distance Show that the distance between the point (x_1, y_1) and the line $Ax + By + C = 0$ is

$$\text{Distance} = \frac{|Ax_1 + By_1 + C|}{\sqrt{A^2 + B^2}}.$$

88. Distance Write the distance d between the point $(3, 1)$ and the line $y = mx + 4$ in terms of m. Use a graphing utility to graph the equation. When is the distance 0? Explain the result geometrically.

89. Proof Prove that the diagonals of a rhombus intersect at right angles. (A rhombus is a quadrilateral with sides of equal lengths.)

90. Proof Prove that the figure formed by connecting consecutive midpoints of the sides of any quadrilateral is a parallelogram.

91. Proof Prove that if the points (x_1, y_1) and (x_2, y_2) lie on the same line as (x_1^*, y_1^*) and (x_2^*, y_2^*), then

$$\frac{y_2^* - y_1^*}{x_2^* - x_1^*} = \frac{y_2 - y_1}{x_2 - x_1}.$$

Assume $x_1 \neq x_2$ and $x_1^* \neq x_2^*$.

92. Proof Prove that if the slopes of two nonvertical lines are negative reciprocals of each other, then the lines are perpendicular.

True or False? In Exercises 93–96, determine whether the statement is true or false. If it is false, explain why or give an example that shows it is false.

93. The lines represented by $ax + by = c_1$ and $bx - ay = c_2$ are perpendicular. Assume $a \neq 0$ and $b \neq 0$.

94. It is possible for two lines with positive slopes to be perpendicular to each other.

95. If a line contains points in both the first and third quadrants, then its slope must be positive.

96. The equation of any line can be written in general form.

P.3   Functions and Their Graphs

- Use function notation to represent and evaluate a function.
- Find the domain and range of a function.
- Sketch the graph of a function.
- Identify different types of transformations of functions.
- Classify functions and recognize combinations of functions.

Functions and Function Notation

A **relation** between two sets X and Y is a set of ordered pairs, each of the form (x, y), where x is a member of X and y is a member of Y. A **function** from X to Y is a relation between X and Y that has the property that any two ordered pairs with the same x-value also have the same y-value. The variable x is the **independent variable,** and the variable y is the **dependent variable.**

Many real-life situations can be modeled by functions. For instance, the area A of a circle is a function of the circle's radius r.

$$A = \pi r^2 \qquad \text{\textcolor{red}{\textit{A is a function of r.}}}$$

In this case, r is the independent variable and A is the dependent variable.

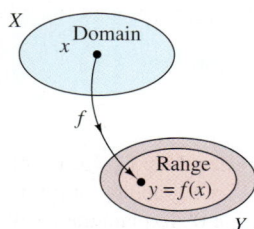

X
Domain
x

f

Range
$y = f(x)$

Y

A real-valued function f of a real variable

Figure P.22

> ### Definition of a Real-Valued Function of a Real Variable
>
> Let X and Y be sets of real numbers. A **real-valued function f of a real variable x** from X to Y is a correspondence that assigns to each number x in X exactly one number y in Y.
>
> The **domain** of f is the set X. The number y is the **image** of x under f and is denoted by $f(x)$, which is called the **value of f at x.** The **range** of f is a subset of Y and consists of all images of numbers in X (see Figure P.22).

Functions can be specified in a variety of ways. In this text, however, you will concentrate primarily on functions that are given by equations involving the dependent and independent variables. For instance, the equation

$$x^2 + 2y = 1 \qquad \text{\textcolor{red}{Equation in implicit form}}$$

defines y, the dependent variable, as a function of x, the independent variable. To **evaluate** this function (that is, to find the y-value that corresponds to a given x-value), it is convenient to isolate y on the left side of the equation.

$$y = \tfrac{1}{2}(1 - x^2) \qquad \text{\textcolor{red}{Equation in explicit form}}$$

Using f as the name of the function, you can write this equation as

$$f(x) = \tfrac{1}{2}(1 - x^2). \qquad \text{\textcolor{red}{Function notation}}$$

The original equation

$$x^2 + 2y = 1$$

implicitly defines y as a function of x. When you solve the equation for y, you are writing the equation in **explicit** form.

Function notation has the advantage of clearly identifying the dependent variable as $f(x)$ while at the same time telling you that x is the independent variable and that the function itself is "f." The symbol $f(x)$ is read "f of x." Function notation allows you to be less wordy. Instead of asking "What is the value of y that corresponds to $x = 3$?" you can ask "What is $f(3)$?"

FUNCTION NOTATION

The word *function* was first used by Gottfried Wilhelm Leibniz in 1694 as a term to denote any quantity connected with a curve, such as the coordinates of a point on a curve or the slope of a curve. Forty years later, Leonhard Euler used the word "function" to describe any expression made up of a variable and some constants. He introduced the notation $y = f(x)$.

In an equation that defines a function of x, the role of the variable x is simply that of a placeholder. For instance, the function

$$f(x) = 2x^2 - 4x + 1$$

can be described by the form

$$f(\boxed{}) = 2(\boxed{})^2 - 4(\boxed{}) + 1$$

where rectangles are used instead of x. To evaluate $f(-2)$, replace each rectangle with -2.

$$\begin{aligned} f(-2) &= 2(-2)^2 - 4(-2) + 1 & &\text{Substitute } -2 \text{ for } x. \\ &= 2(4) + 8 + 1 & &\text{Simplify.} \\ &= 17 & &\text{Simplify.} \end{aligned}$$

Although f is often used as a convenient function name and x as the independent variable, you can use other symbols. For instance, these three equations all define the same function.

$$\begin{aligned} f(x) &= x^2 - 4x + 7 & &\text{Function name is } f, \text{ independent variable is } x. \\ f(t) &= t^2 - 4t + 7 & &\text{Function name is } f, \text{ independent variable is } t. \\ g(s) &= s^2 - 4s + 7 & &\text{Function name is } g, \text{ independent variable is } s. \end{aligned}$$

EXAMPLE 1 Evaluating a Function

For the function f defined by $f(x) = x^2 + 7$, evaluate each expression.

a. $f(3a)$ **b.** $f(b - 1)$ **c.** $\dfrac{f(x + \Delta x) - f(x)}{\Delta x}$

Solution

REMARK The expression in Example 1(c) is called a *difference quotient* and has a special significance in calculus. You will learn more about this in Chapter 2.

a.
$$\begin{aligned} f(3a) &= (3a)^2 + 7 & &\text{Substitute } 3a \text{ for } x. \\ &= 9a^2 + 7 & &\text{Simplify.} \end{aligned}$$

b.
$$\begin{aligned} f(b - 1) &= (b - 1)^2 + 7 & &\text{Substitute } b - 1 \text{ for } x. \\ &= b^2 - 2b + 1 + 7 & &\text{Expand binomial.} \\ &= b^2 - 2b + 8 & &\text{Simplify.} \end{aligned}$$

c.
$$\begin{aligned} \frac{f(x + \Delta x) - f(x)}{\Delta x} &= \frac{[(x + \Delta x)^2 + 7] - (x^2 + 7)}{\Delta x} \\ &= \frac{x^2 + 2x\Delta x + (\Delta x)^2 + 7 - x^2 - 7}{\Delta x} \\ &= \frac{2x\Delta x + (\Delta x)^2}{\Delta x} \\ &= \frac{\cancel{\Delta x}(2x + \Delta x)}{\cancel{\Delta x}} \\ &= 2x + \Delta x, \quad \Delta x \neq 0 \end{aligned}$$

In calculus, it is important to specify the domain of a function or expression clearly. For instance, in Example 1(c), the two expressions

$$\frac{f(x + \Delta x) - f(x)}{\Delta x} \quad \text{and} \quad 2x + \Delta x, \quad \Delta x \neq 0$$

are equivalent because $\Delta x = 0$ is excluded from the domain of each expression. Without a stated domain restriction, the two expressions would not be equivalent.

The Domain and Range of a Function

The domain of a function can be described explicitly, or it may be described *implicitly* by an equation used to define the function. The implied domain is the set of all real numbers for which the equation is defined, whereas an explicitly defined domain is one that is given along with the function. For example, the function

$$f(x) = \frac{1}{x^2 - 4}, \quad 4 \le x \le 5$$

has an explicitly defined domain given by $\{x: 4 \le x \le 5\}$. On the other hand, the function

$$g(x) = \frac{1}{x^2 - 4}$$

has an implied domain that is the set $\{x: x \ne \pm 2\}$.

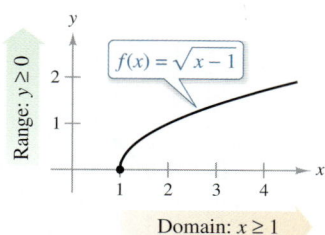

$f(x) = \sqrt{x - 1}$

Range: $y \ge 0$

Domain: $x \ge 1$

(a) The domain of f is $[1, \infty)$, and the range is $[0, \infty)$.

EXAMPLE 2 **Finding the Domain and Range of a Function**

a. The domain of the function

$$f(x) = \sqrt{x - 1}$$

is the set of all x-values for which $x - 1 \ge 0$, which is the interval $[1, \infty)$. To find the range, observe that $f(x) = \sqrt{x - 1}$ is never negative. So, the range is the interval $[0, \infty)$, as shown in Figure P.23(a).

b. The domain of the tangent function

$$f(x) = \tan x$$

is the set of all x-values such that

$$x \ne \frac{\pi}{2} + n\pi, \quad n \text{ is an integer.} \qquad \text{Domain of tangent function}$$

The range of this function is the set of all real numbers, as shown in Figure P.23(b). For a review of the characteristics of this and other trigonometric functions, see Appendix C.

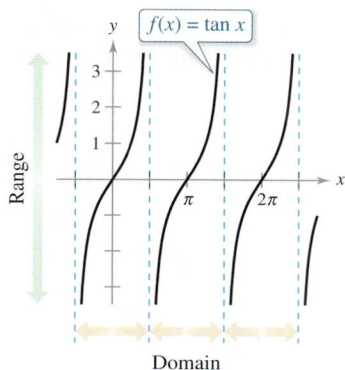

$f(x) = \tan x$

Range

Domain

(b) The domain of f is all x-values such that $x \ne \frac{\pi}{2} + n\pi$, and the range is $(-\infty, \infty)$.

Figure P.23

EXAMPLE 3 **A Function Defined by More than One Equation**

For the piecewise-defined function

$$f(x) = \begin{cases} 1 - x, & x < 1 \\ \sqrt{x - 1}, & x \ge 1 \end{cases}$$

f is defined for $x < 1$ and $x \ge 1$. So, the domain is the set of all real numbers. On the portion of the domain for which $x \ge 1$, the function behaves as in Example 2(a). For $x < 1$, the values of $1 - x$ are positive. So, the range of the function is the interval $[0, \infty)$. (See Figure P.24.)

$f(x) = \begin{cases} 1 - x, & x < 1 \\ \sqrt{x - 1}, & x \ge 1 \end{cases}$

Range: $y \ge 0$

Domain: all real x

The domain of f is $(-\infty, \infty)$, and the range is $[0, \infty)$.

Figure P.24

A function from X to Y is **one-to-one** when to each y-value in the range there corresponds exactly one x-value in the domain. For instance, the function in Example 2(a) is one-to-one, whereas the functions in Examples 2(b) and 3 are not one-to-one. A function from X to Y is **onto** when its range consists of all of Y.

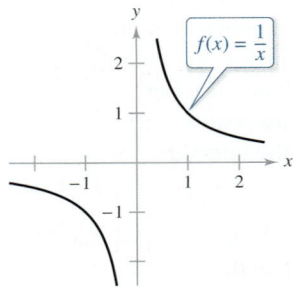

The Graph of a Function

The graph of a function
Figure P.25

The graph of the function $y = f(x)$ consists of all points $(x, f(x))$, where x is in the domain of f. In Figure P.25, note that

$$x = \text{the directed distance from the } y\text{-axis}$$

and

$$f(x) = \text{the directed distance from the } x\text{-axis.}$$

A vertical line can intersect the graph of a function of x at most *once*. This observation provides a convenient visual test, called the **Vertical Line Test,** for functions of x. That is, a graph in the coordinate plane is the graph of a function of x if and only if no vertical line intersects the graph at more than one point. For example, in Figure P.26(a), you can see that the graph does not define y as a function of x because a vertical line intersects the graph twice, whereas in Figures P.26(b) and (c), the graphs do define y as a function of x.

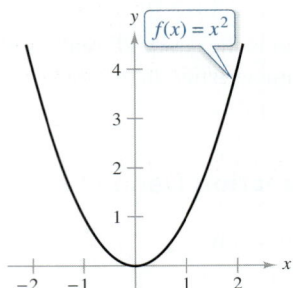

(a) Not a function of x **(b)** A function of x **(c)** A function of x

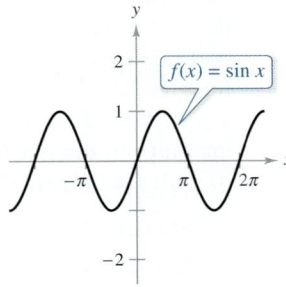

Figure P.26

Figure P.27 shows the graphs of eight basic functions. You should be able to recognize these graphs. (Graphs of the other four basic trigonometric functions are shown in Appendix C.)

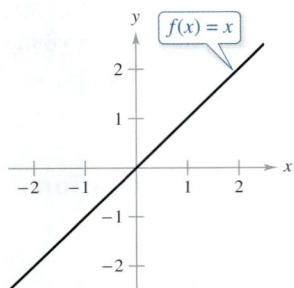

Identity function   Squaring function   Cubing function   Square root function

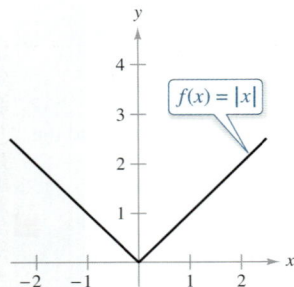

Absolute value function   Rational function   Sine function   Cosine function

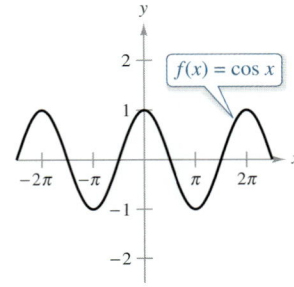

The graphs of eight basic functions
Figure P.27

Transformations of Functions

Some families of graphs have the same basic shape. For example, compare the graph of $y = x^2$ with the graphs of the four other quadratic functions shown in Figure P.28.

(a) Vertical shift upward

(b) Horizontal shift to the left

(c) Reflection

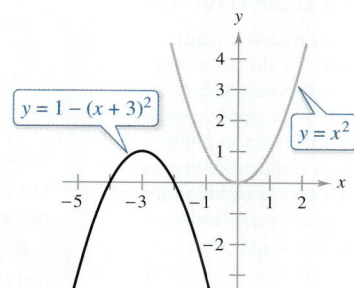

(d) Shift left, reflect, and shift upward

Figure P.28

Each of the graphs in Figure P.28 is a **transformation** of the graph of $y = x^2$. The three basic types of transformations illustrated by these graphs are vertical shifts, horizontal shifts, and reflections. Function notation lends itself well to describing transformations of graphs in the plane. For instance, using

$f(x) = x^2$ Original function

as the original function, the transformations shown in Figure P.28 can be represented by these equations.

a. $y = f(x) + 2$ Vertical shift up two units

b. $y = f(x + 2)$ Horizontal shift to the left two units

c. $y = -f(x)$ Reflection about the x-axis

d. $y = -f(x + 3) + 1$ Shift left three units, reflect about the x-axis, and shift up one unit

---

Basic Types of Transformations ($c > 0$)

Original graph:	$y = f(x)$
Horizontal shift c units to the **right:**	$y = f(x - c)$
Horizontal shift c units to the **left:**	$y = f(x + c)$
Vertical shift c units **downward:**	$y = f(x) - c$
Vertical shift c units **upward:**	$y = f(x) + c$
Reflection (about the x-axis):	$y = -f(x)$
Reflection (about the y-axis):	$y = f(-x)$
Reflection (about the origin):	$y = -f(-x)$

Classifications and Combinations of Functions

The modern notion of a function is derived from the efforts of many seventeenth- and eighteenth-century mathematicians. Of particular note was Leonhard Euler, who introduced the function notation $y = f(x)$. By the end of the eighteenth century, mathematicians and scientists had concluded that many real-world phenomena could be represented by mathematical models taken from a collection of functions called **elementary functions.** Elementary functions fall into three categories.

1. Algebraic functions (polynomial, radical, rational)
2. Trigonometric functions (sine, cosine, tangent, and so on)
3. Exponential and logarithmic functions

You can review the trigonometric functions in Appendix C. The other nonalgebraic functions, such as the inverse trigonometric functions and the exponential and logarithmic functions, are introduced in Chapter 5.

The most common type of algebraic function is a **polynomial function**

$$f(x) = a_n x^n + a_{n-1} x^{n-1} + \cdots + a_2 x^2 + a_1 x + a_0$$

where n is a nonnegative integer. The numbers a_i are **coefficients,** with a_n the **leading coefficient** and a_0 the **constant term** of the polynomial function. If $a_n \neq 0$, then n is the **degree** of the polynomial function. The zero polynomial $f(x) = 0$ is not assigned a degree. It is common practice to use subscript notation for coefficients of general polynomial functions, but for polynomial functions of low degree, these simpler forms are often used. (Note that $a \neq 0$.)

Zeroth degree:	$f(x) = a$	Constant function
First degree:	$f(x) = ax + b$	Linear function
Second degree:	$f(x) = ax^2 + bx + c$	Quadratic function
Third degree:	$f(x) = ax^3 + bx^2 + cx + d$	Cubic function

Although the graph of a nonconstant polynomial function can have several turns, eventually the graph will rise or fall without bound as x moves to the right or left. Whether the graph of

$$f(x) = a_n x^n + a_{n-1} x^{n-1} + \cdots + a_2 x^2 + a_1 x + a_0$$

eventually rises or falls can be determined by the function's degree (odd or even) and by the leading coefficient a_n, as indicated in Figure P.29. Note that the dashed portions of the graphs indicate that the **Leading Coefficient Test** determines *only* the right and left behavior of the graph.

LEONHARD EULER (1707–1783)

In addition to making major contributions to almost every branch of mathematics, Euler was one of the first to apply calculus to real-life problems in physics. His extensive published writings include such topics as shipbuilding, acoustics, optics, astronomy, mechanics, and magnetism. *See LarsonCalculus.com to read more of this biography.*

■ **FOR FURTHER INFORMATION** For more on the history of the concept of a function, see the article "Evolution of the Function Concept: A Brief Survey" by Israel Kleiner in *The College Mathematics Journal.* To view this article, go to *MathArticles.com.*

Graphs of polynomial functions of even degree

Graphs of polynomial functions of odd degree

The Leading Coefficient Test for polynomial functions
Figure P.29

Just as a rational number can be written as the quotient of two integers, a **rational function** can be written as the quotient of two polynomials. Specifically, a function f is rational when it has the form

$$f(x) = \frac{p(x)}{q(x)}, \quad q(x) \neq 0$$

where $p(x)$ and $q(x)$ are polynomials.

Polynomial functions and rational functions are examples of **algebraic functions.** An algebraic function of x is one that can be expressed as a finite number of sums, differences, multiples, quotients, and radicals involving x^n. For example,

$$f(x) = \sqrt{x + 1}$$

is algebraic. Functions that are not algebraic are **transcendental.** For instance, the trigonometric functions are transcendental.

Two functions can be combined in various ways to create new functions. For example, given $f(x) = 2x - 3$ and $g(x) = x^2 + 1$, you can form the functions shown.

$$(f + g)(x) = f(x) + g(x) = (2x - 3) + (x^2 + 1) \qquad \text{Sum}$$
$$(f - g)(x) = f(x) - g(x) = (2x - 3) - (x^2 + 1) \qquad \text{Difference}$$
$$(fg)(x) = f(x)g(x) = (2x - 3)(x^2 + 1) \qquad \text{Product}$$
$$(f/g)(x) = \frac{f(x)}{g(x)} = \frac{2x - 3}{x^2 + 1} \qquad \text{Quotient}$$

You can combine two functions in yet another way, called **composition.** The resulting function is called a **composite function.**

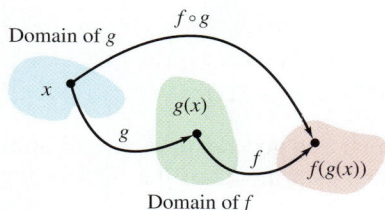

Domain of g

$f \circ g$

x

$g(x)$

g

f

$f(g(x))$

Domain of f

The domain of the composite function $f \circ g$

Figure P.30

Definition of Composite Function

Let f and g be functions. The function $(f \circ g)(x) = f(g(x))$ is the **composite** of f with g. The domain of $f \circ g$ is the set of all x in the domain of g such that $g(x)$ is in the domain of f (see Figure P.30).

The composite of f with g is generally not the same as the composite of g with f. This is shown in the next example.

EXAMPLE 4 Finding Composite Functions

▷ *See LarsonCalculus.com for an interactive version of this type of example.*

For $f(x) = 2x - 3$ and $g(x) = \cos x$, find each composite function.

a. $f \circ g$ **b.** $g \circ f$

Solution

a. $(f \circ g)(x) = f(g(x))$ Definition of $f \circ g$

$\qquad\qquad = f(\cos x)$ Substitute $\cos x$ for $g(x)$.

$\qquad\qquad = 2(\cos x) - 3$ Definition of $f(x)$

$\qquad\qquad = 2 \cos x - 3$ Simplify.

b. $(g \circ f)(x) = g(f(x))$ Definition of $g \circ f$

$\qquad\qquad = g(2x - 3)$ Substitute $2x - 3$ for $f(x)$.

$\qquad\qquad = \cos(2x - 3)$ Definition of $g(x)$

Note that $(f \circ g)(x) \neq (g \circ f)(x)$.

In Section P.1, an *x*-intercept of a graph was defined to be a point $(a, 0)$ at which the graph crosses the *x*-axis. If the graph represents a function f, then the number a is a **zero** of f. In other words, *the zeros of a function f are the solutions of the equation* $f(x) = 0$. For example, the function

$$f(x) = x - 4$$

has a zero at $x = 4$ because $f(4) = 0$.

In Section P.1, you also studied different types of symmetry. In the terminology of functions, a function is **even** when its graph is symmetric with respect to the *y*-axis, and is **odd** when its graph is symmetric with respect to the origin. The symmetry tests in Section P.1 yield the following test for even and odd functions.

Test for Even and Odd Functions

The function $y = f(x)$ is **even** when

$$f(-x) = f(x).$$

The function $y = f(x)$ is **odd** when

$$f(-x) = -f(x).$$

EXAMPLE 5 **Even and Odd Functions and Zeros of Functions**

Determine whether each function is even, odd, or neither. Then find the zeros of the function.

a. $f(x) = x^3 - x$ **b.** $g(x) = 1 + \cos x$

Solution

a. This function is odd because

$$f(-x) = (-x)^3 - (-x) = -x^3 + x = -(x^3 - x) = -f(x).$$

The zeros of f are

$$x^3 - x = 0 \qquad \text{Let } f(x) = 0.$$
$$x(x^2 - 1) = 0 \qquad \text{Factor.}$$
$$x(x - 1)(x + 1) = 0 \qquad \text{Factor.}$$
$$x = 0, 1, -1. \qquad \text{Zeros of } f$$

See Figure P.31(a).

b. This function is even because

$$g(-x) = 1 + \cos(-x) = 1 + \cos x = g(x). \qquad \cos(-x) = \cos(x)$$

The zeros of g are

$$1 + \cos x = 0 \qquad \text{Let } g(x) = 0.$$
$$\cos x = -1 \qquad \text{Subtract 1 from each side.}$$
$$x = (2n + 1)\pi, \; n \text{ is an integer.} \qquad \text{Zeros of } g$$

See Figure P.31(b).

Each function in Example 5 is either even or odd. However, some functions, such as

$$f(x) = x^2 + x + 1$$

are neither even nor odd.

(a) Odd function

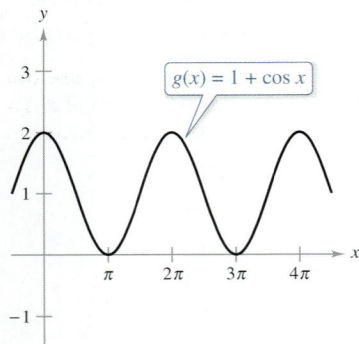

(b) Even function

Figure P.31

P.3 Exercises

See **CalcChat.com** for tutorial help and worked-out solutions to odd-numbered exercises.

Evaluating a Function In Exercises 1–10, evaluate the function at the given value(s) of the independent variable. Simplify the results.

1. $f(x) = 7x - 4$
 (a) $f(0)$ (b) $f(-3)$
 (c) $f(b)$ (d) $f(x - 1)$

2. $f(x) = \sqrt{x + 5}$
 (a) $f(-4)$ (b) $f(11)$
 (c) $f(4)$ (d) $f(x + \Delta x)$

3. $g(x) = 5 - x^2$
 (a) $g(0)$ (b) $g(\sqrt{5})$
 (c) $g(-2)$ (d) $g(t - 1)$

4. $g(x) = x^2(x - 4)$
 (a) $g(4)$ (b) $g(\frac{3}{2})$
 (c) $g(c)$ (d) $g(t + 4)$

5. $f(x) = \cos 2x$
 (a) $f(0)$ (b) $f\left(-\frac{\pi}{4}\right)$
 (c) $f\left(\frac{\pi}{3}\right)$ (d) $f(\pi)$

6. $f(x) = \sin x$
 (a) $f(\pi)$ (b) $f\left(\frac{5\pi}{4}\right)$
 (c) $f\left(\frac{2\pi}{3}\right)$ (d) $f\left(-\frac{\pi}{6}\right)$

7. $f(x) = x^3$
 $\dfrac{f(x + \Delta x) - f(x)}{\Delta x}$

8. $f(x) = 3x - 1$
 $\dfrac{f(x) - f(1)}{x - 1}$

9. $f(x) = \dfrac{1}{\sqrt{x - 1}}$
 $\dfrac{f(x) - f(2)}{x - 2}$

10. $f(x) = x^3 - x$
 $\dfrac{f(x) - f(1)}{x - 1}$

Finding the Domain and Range of a Function In Exercises 11–22, find the domain and range of the function.

11. $f(x) = 4x^2$

12. $g(x) = x^2 - 5$

13. $f(x) = x^3$

14. $h(x) = 4 - x^2$

15. $g(x) = \sqrt{6x}$

16. $h(x) = -\sqrt{x + 3}$

17. $f(x) = \sqrt{16 - x^2}$

18. $f(x) = |x - 3|$

19. $f(t) = \sec \dfrac{\pi t}{4}$

20. $h(t) = \cot t$

21. $f(x) = \dfrac{3}{x}$

22. $f(x) = \dfrac{x - 2}{x + 4}$

Finding the Domain of a Function In Exercises 23–28, find the domain of the function.

23. $f(x) = \sqrt{x} + \sqrt{1 - x}$

24. $f(x) = \sqrt{x^2 - 3x + 2}$

25. $g(x) = \dfrac{2}{1 - \cos x}$

26. $h(x) = \dfrac{1}{\sin x - (1/2)}$

27. $f(x) = \dfrac{1}{|x + 3|}$

28. $g(x) = \dfrac{1}{|x^2 - 4|}$

Finding the Domain and Range of a Piecewise Function In Exercises 29–32, evaluate the function as indicated. Determine its domain and range.

29. $f(x) = \begin{cases} 2x + 1, & x < 0 \\ 2x + 2, & x \geq 0 \end{cases}$
 (a) $f(-1)$ (b) $f(0)$ (c) $f(2)$ (d) $f(t^2 + 1)$

30. $f(x) = \begin{cases} x^2 + 2, & x \leq 1 \\ 2x^2 + 2, & x > 1 \end{cases}$
 (a) $f(-2)$ (b) $f(0)$ (c) $f(1)$ (d) $f(s^2 + 2)$

31. $f(x) = \begin{cases} |x| + 1, & x < 1 \\ -x + 1, & x \geq 1 \end{cases}$
 (a) $f(-3)$ (b) $f(1)$ (c) $f(3)$ (d) $f(b^2 + 1)$

32. $f(x) = \begin{cases} \sqrt{x + 4}, & x \leq 5 \\ (x - 5)^2, & x > 5 \end{cases}$
 (a) $f(-3)$ (b) $f(0)$ (c) $f(5)$ (d) $f(10)$

Sketching a Graph of a Function In Exercises 33–40, sketch a graph of the function and find its domain and range. Use a graphing utility to verify your graph.

33. $f(x) = 4 - x$

34. $g(x) = \dfrac{4}{x}$

35. $h(x) = \sqrt{x - 6}$

36. $f(x) = \frac{1}{4}x^3 + 3$

37. $f(x) = \sqrt{9 - x^2}$

38. $f(x) = x + \sqrt{4 - x^2}$

39. $g(t) = 3 \sin \pi t$

40. $h(\theta) = -5 \cos \dfrac{\theta}{2}$

WRITING ABOUT CONCEPTS

41. Describing a Graph
The graph of the distance that a student drives in a 10-minute trip to school is shown in the figure. Give a verbal description of the characteristics of the student's drive to school.

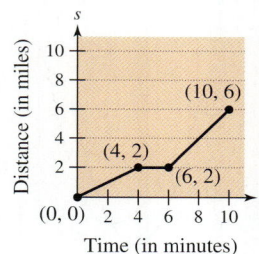

42. Sketching a Graph A student who commutes 27 miles to attend college remembers, after driving a few minutes, that a term paper that is due has been forgotten. Driving faster than usual, the student returns home, picks up the paper, and once again starts toward school. Sketch a possible graph of the student's distance from home as a function of time.

Using the Vertical Line Test In Exercises 43–46, use the Vertical Line Test to determine whether y is a function of x. To print an enlarged copy of the graph, go to *MathGraphs.com*.

43. $x - y^2 = 0$

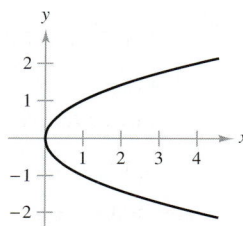

44. $\sqrt{x^2 - 4} - y = 0$

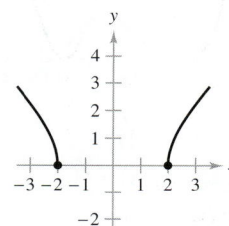

45. $y = \begin{cases} x + 1, & x \le 0 \\ -x + 2, & x > 0 \end{cases}$ **46.** $x^2 + y^2 = 4$

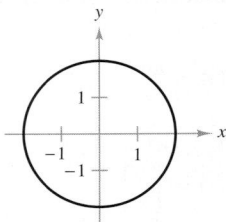

Deciding Whether an Equation Is a Function In Exercises 47–50, determine whether y is a function of x.

47. $x^2 + y^2 = 16$ **48.** $x^2 + y = 16$

49. $y^2 = x^2 - 1$ **50.** $x^2 y - x^2 + 4y = 0$

Transformation of a Function In Exercises 51–54, the graph shows one of the eight basic functions on page 22 and a transformation of the function. Describe the transformation. Then use your description to write an equation for the transformation.

51.

52.

53.

54.

Matching In Exercises 55–60, use the graph of $y = f(x)$ to match the function with its graph.

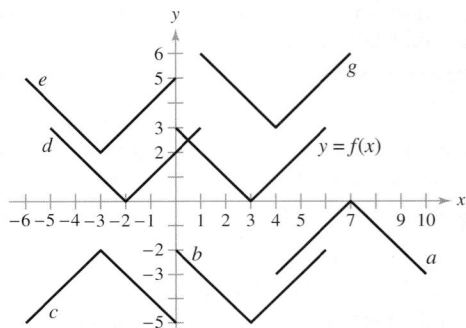

55. $y = f(x + 5)$ **56.** $y = f(x) - 5$

57. $y = -f(-x) - 2$ **58.** $y = -f(x - 4)$

59. $y = f(x + 6) + 2$ **60.** $y = f(x - 1) + 3$

61. Sketching Transformations Use the graph of f shown in the figure to sketch the graph of each function. To print an enlarged copy of the graph, go to *MathGraphs.com*.

(a) $f(x + 3)$ (b) $f(x - 1)$
(c) $f(x) + 2$ (d) $f(x) - 4$
(e) $3f(x)$ (f) $\frac{1}{4}f(x)$
(g) $-f(x)$ (h) $-f(-x)$

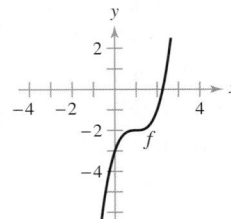

62. Sketching Transformations Use the graph of f shown in the figure to sketch the graph of each function. To print an enlarged copy of the graph, go to *MathGraphs.com*.

(a) $f(x - 4)$ (b) $f(x + 2)$
(c) $f(x) + 4$ (d) $f(x) - 1$
(e) $2f(x)$ (f) $\frac{1}{2}f(x)$
(g) $f(-x)$ (h) $-f(x)$

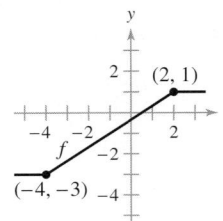

Combinations of Functions In Exercises 63 and 64, find (a) $f(x) + g(x)$, (b) $f(x) - g(x)$, (c) $f(x) \cdot g(x)$, and (d) $f(x)/g(x)$.

63. $f(x) = 3x - 4$ **64.** $f(x) = x^2 + 5x + 4$
    $g(x) = 4$ $g(x) = x + 1$

65. Evaluating Composite Functions Given $f(x) = \sqrt{x}$ and $g(x) = x^2 - 1$, evaluate each expression.

(a) $f(g(1))$ (b) $g(f(1))$ (c) $g(f(0))$
(d) $f(g(-4))$ (e) $f(g(x))$ (f) $g(f(x))$

66. Evaluating Composite Functions Given $f(x) = \sin x$ and $g(x) = \pi x$, evaluate each expression.

(a) $f(g(2))$ (b) $f\left(g\left(\frac{1}{2}\right)\right)$ (c) $g(f(0))$
(d) $g\left(f\left(\frac{\pi}{4}\right)\right)$ (e) $f(g(x))$ (f) $g(f(x))$

Finding Composite Functions In Exercises 67–70, find the composite functions $f \circ g$ and $g \circ f$. Find the domain of each composite function. Are the two composite functions equal?

67. $f(x) = x^2$, $g(x) = \sqrt{x}$ **68.** $f(x) = x^2 - 1$, $g(x) = \cos x$

69. $f(x) = \dfrac{3}{x}$, $g(x) = x^2 - 1$ **70.** $f(x) = \dfrac{1}{x}$, $g(x) = \sqrt{x + 2}$

71. Evaluating Composite Functions Use the graphs of f and g to evaluate each expression. If the result is undefined, explain why.

(a) $(f \circ g)(3)$ (b) $g(f(2))$
(c) $g(f(5))$ (d) $(f \circ g)(-3)$
(e) $(g \circ f)(-1)$ (f) $f(g(-1))$

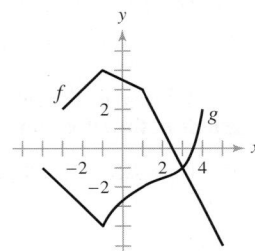

72. Ripples A pebble is dropped into a calm pond, causing ripples in the form of concentric circles. The radius (in feet) of the outer ripple is given by $r(t) = 0.6t$, where t is the time in seconds after the pebble strikes the water. The area of the circle is given by the function $A(r) = \pi r^2$. Find and interpret $(A \circ r)(t)$.

Think About It In Exercises 73 and 74, $F(x) = f \circ g \circ h$. Identify functions for f, g, and h. (There are many correct answers.)

73. $F(x) = \sqrt{2x - 2}$

74. $F(x) = -4\sin(1 - x)$

Think About It In Exercises 75 and 76, find the coordinates of a second point on the graph of a function f when the given point is on the graph and the function is (a) even and (b) odd.

75. $\left(-\frac{3}{2}, 4\right)$

76. $(4, 9)$

77. Even and Odd Functions The graphs of f, g, and h are shown in the figure. Decide whether each function is even, odd, or neither.

Figure for 77

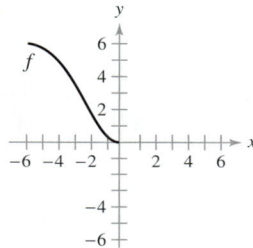

Figure for 78

78. Even and Odd Functions The domain of the function f shown in the figure is $-6 \le x \le 6$.

(a) Complete the graph of f given that f is even.

(b) Complete the graph of f given that f is odd.

Even and Odd Functions and Zeros of Functions In Exercises 79–82, determine whether the function is even, odd, or neither. Then find the zeros of the function. Use a graphing utility to verify your result.

79. $f(x) = x^2(4 - x^2)$

80. $f(x) = \sqrt[3]{x}$

81. $f(x) = x\cos x$

82. $f(x) = \sin^2 x$

Writing Functions In Exercises 83–86, write an equation for a function that has the given graph.

83. Line segment connecting $(-2, 4)$ and $(0, -6)$

84. Line segment connecting $(3, 1)$ and $(5, 8)$

85. The bottom half of the parabola $x + y^2 = 0$

86. The bottom half of the circle $x^2 + y^2 = 36$

Sketching a Graph In Exercises 87–90, sketch a possible graph of the situation.

87. The speed of an airplane as a function of time during a 5-hour flight

88. The height of a baseball as a function of horizontal distance during a home run

89. The amount of a certain brand of sneaker sold by a sporting goods store as a function of the price of the sneaker

90. The value of a new car as a function of time over a period of 8 years

91. Domain Find the value of c such that the domain of $f(x) = \sqrt{c - x^2}$ is $[-5, 5]$.

92. Domain Find all values of c such that the domain of

$$f(x) = \frac{x + 3}{x^2 + 3cx + 6}$$

is the set of all real numbers.

93. Graphical Reasoning An electronically controlled thermostat is programmed to lower the temperature during the night automatically (see figure). The temperature T in degrees Celsius is given in terms of t, the time in hours on a 24-hour clock.

(a) Approximate $T(4)$ and $T(15)$.

(b) The thermostat is reprogrammed to produce a temperature $H(t) = T(t - 1)$. How does this change the temperature? Explain.

(c) The thermostat is reprogrammed to produce a temperature $H(t) = T(t) - 1$. How does this change the temperature? Explain.

94. HOW DO YOU SEE IT? Water runs into a vase of height 30 centimeters at a constant rate. The vase is full after 5 seconds. Use this information and the shape of the vase shown to answer the questions when d is the depth of the water in centimeters and t is the time in seconds (see figure).

(a) Explain why d is a function of t.

(b) Determine the domain and range of the function.

(c) Sketch a possible graph of the function.

(d) Use the graph in part (c) to approximate $d(4)$. What does this represent?

95. Modeling Data The table shows the average numbers of acres per farm in the United States for selected years. (*Source: U.S. Department of Agriculture*)

Year	1960	1970	1980	1990	2000	2010
Acreage	297	374	429	460	436	418

(a) Plot the data, where A is the acreage and t is the time in years, with $t = 0$ corresponding to 1960. Sketch a freehand curve that approximates the data.

(b) Use the curve in part (a) to approximate $A(25)$.

96. Automobile Aerodynamics

The horsepower H required to overcome wind drag on a certain automobile is approximated by

$$H(x) = 0.002x^2 + 0.005x - 0.029, \quad 10 \le x \le 100$$

where x is the speed of the car in miles per hour.

(a) Use a graphing utility to graph H.

(b) Rewrite the power function so that x represents the speed in kilometers per hour. [Find $H(x/1.6)$.]

97. Think About It Write the function $f(x) = |x| + |x - 2|$ without using absolute value signs. (For a review of absolute value, see Appendix C.)

98. Writing Use a graphing utility to graph the polynomial functions $p_1(x) = x^3 - x + 1$ and $p_2(x) = x^3 - x$. How many zeros does each function have? Is there a cubic polynomial that has no zeros? Explain.

99. Proof Prove that the function is odd.

$$f(x) = a_{2n+1}x^{2n+1} + \cdots + a_3x^3 + a_1x$$

100. Proof Prove that the function is even.

$$f(x) = a_{2n}x^{2n} + a_{2n-2}x^{2n-2} + \cdots + a_2x^2 + a_0$$

101. Proof Prove that the product of two even (or two odd) functions is even.

102. Proof Prove that the product of an odd function and an even function is odd.

103. Length A right triangle is formed in the first quadrant by the x- and y-axes and a line through the point $(3, 2)$ (see figure). Write the length L of the hypotenuse as a function of x.

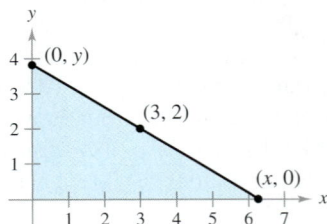

104. Volume An open box of maximum volume is to be made from a square piece of material 24 centimeters on a side by cutting equal squares from the corners and turning up the sides (see figure).

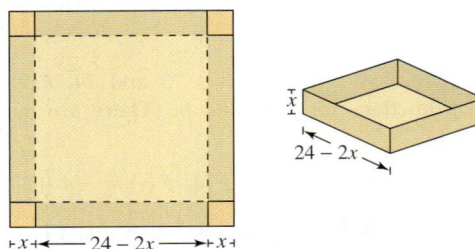

(a) Write the volume V as a function of x, the length of the corner squares. What is the domain of the function?

(b) Use a graphing utility to graph the volume function and approximate the dimensions of the box that yield a maximum volume.

(c) Use the *table* feature of a graphing utility to verify your answer in part (b). (The first two rows of the table are shown.)

Height, x	Length and Width	Volume, V
1	$24 - 2(1)$	$1[24 - 2(1)]^2 = 484$
2	$24 - 2(2)$	$2[24 - 2(2)]^2 = 800$

True or False? In Exercises 105–110, determine whether the statement is true or false. If it is false, explain why or give an example that shows it is false.

105. If $f(a) = f(b)$, then $a = b$.

106. A vertical line can intersect the graph of a function at most once.

107. If $f(x) = f(-x)$ for all x in the domain of f, then the graph of f is symmetric with respect to the y-axis.

108. If f is a function, then

$$f(ax) = af(x).$$

109. The graph of a function of x cannot have symmetry with respect to the x-axis.

110. If the domain of a function consists of a single number, then its range must also consist of only one number.

PUTNAM EXAM CHALLENGE

111. Let R be the region consisting of the points (x, y) of the Cartesian plane satisfying both $|x| - |y| \le 1$ and $|y| \le 1$. Sketch the region R and find its area.

112. Consider a polynomial $f(x)$ with real coefficients having the property $f(g(x)) = g(f(x))$ for every polynomial $g(x)$ with real coefficients. Determine and prove the nature of $f(x)$.

These problems were composed by the Committee on the Putnam Prize Competition. © The Mathematical Association of America. All rights reserved.

P.4   Fitting Models to Data

■ Fit a linear model to a real-life data set.
■ Fit a quadratic model to a real-life data set.
■ Fit a trigonometric model to a real-life data set.

Fitting a Linear Model to Data

A basic premise of science is that much of the physical world can be described mathematically and that many physical phenomena are predictable. This scientific outlook was part of the scientific revolution that took place in Europe during the late 1500s. Two early publications connected with this revolution were *On the Revolutions of the Heavenly Spheres* by the Polish astronomer Nicolaus Copernicus and *On the Fabric of the Human Body* by the Belgian anatomist Andreas Vesalius. Each of these books was published in 1543, and each broke with prior tradition by suggesting the use of a scientific method rather than unquestioned reliance on authority.

One basic technique of modern science is gathering data and then describing the data with a mathematical model. For instance, the data in Example 1 are inspired by Leonardo da Vinci's famous drawing that indicates that a person's height and arm span are equal.

A computer graphics drawing based on the pen and ink drawing of Leonardo da Vinci's famous study of human proportions, called *Vitruvian Man*

EXAMPLE 1   Fitting a Linear Model to Data

• • • ▷ *See LarsonCalculus.com for an interactive version of this type of example.*

A class of 28 people collected the data shown below, which represent their heights x and arm spans y (rounded to the nearest inch).

(60, 61), (65, 65), (68, 67), (72, 73), (61, 62), (63, 63), (70, 71),
(75, 74), (71, 72), (62, 60), (65, 65), (66, 68), (62, 62), (72, 73),
(70, 70), (69, 68), (69, 70), (60, 61), (63, 63), (64, 64), (71, 71),
(68, 67), (69, 70), (70, 72), (65, 65), (64, 63), (71, 70), (67, 67)

Find a linear model to represent these data.

Solution There are different ways to model these data with an equation. The simplest would be to observe that x and y are about the same and list the model as simply $y = x$. A more careful analysis would be to use a procedure from statistics called linear regression. (You will study this procedure in Section 13.9.) The least squares regression line for these data is

$$y = 1.006x - 0.23.$$ Least squares regression line

The graph of the model and the data are shown in Figure P.32. From this model, you can see that a person's arm span tends to be about the same as his or her height. ■

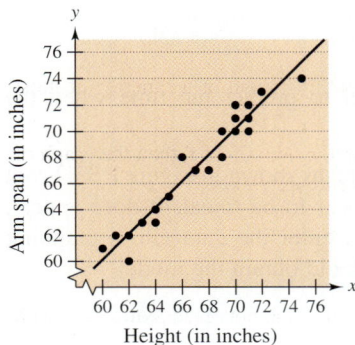

Linear model and data
Figure P.32

▷ **TECHNOLOGY** Many graphing utilities have built-in least squares regression programs. Typically, you enter the data into the calculator and then run the linear regression program. The program usually displays the slope and y-intercept of the best-fitting line and the *correlation coefficient r*. The correlation coefficient gives a measure of how well the data can be modeled by a line. The closer $|r|$ is to 1, the better the data can be modeled by a line. For instance, the correlation coefficient for the model in Example 1 is $r \approx 0.97$, which indicates that the linear model is a good fit for the data. If the r-value is positive, then the variables have a positive correlation, as in Example 1. If the r-value is negative, then the variables have a negative correlation.

Fitting a Quadratic Model to Data

A function that gives the height s of a falling object in terms of the time t is called a *position function*. If air resistance is not considered, then the position of a falling object can be modeled by

$$s(t) = \tfrac{1}{2}gt^2 + v_0t + s_0$$

where g is the acceleration due to gravity, v_0 is the initial velocity, and s_0 is the initial height. The value of g depends on where the object is dropped. On Earth, g is approximately -32 feet per second per second, or -9.8 meters per second per second.

To discover the value of g experimentally, you could record the heights of a falling object at several increments, as shown in Example 2.

EXAMPLE 2   Fitting a Quadratic Model to Data

A basketball is dropped from a height of about $5\tfrac{1}{4}$ feet. The height of the basketball is recorded 23 times at intervals of about 0.02 second. The results are shown in the table.

Time	0.0	0.02	0.04	0.06	0.08	0.099996
Height	5.23594	5.20353	5.16031	5.0991	5.02707	4.95146

Time	0.119996	0.139992	0.159988	0.179988	0.199984	0.219984
Height	4.85062	4.74979	4.63096	4.50132	4.35728	4.19523

Time	0.23998	0.25993	0.27998	0.299976	0.319972	0.339961
Height	4.02958	3.84593	3.65507	3.44981	3.23375	3.01048

Time	0.359961	0.379951	0.399941	0.419941	0.439941
Height	2.76921	2.52074	2.25786	1.98058	1.63488

Find a model to fit these data. Then use the model to predict the time when the basketball will hit the ground.

Solution   Begin by sketching a scatter plot of the data, as shown in Figure P.33. From the scatter plot, you can see that the data do not appear to be linear. It does appear, however, that they might be quadratic. To check this, enter the data into a graphing utility that has a quadratic regression program. You should obtain the model

$$s = -15.45t^2 - 1.302t + 5.2340. \qquad \text{\color{red}Least squares regression quadratic}$$

Using this model, you can predict the time when the basketball hits the ground by substituting 0 for s and solving the resulting equation for t.

$$0 = -15.45t^2 - 1.302t + 5.2340 \qquad \text{\color{red}Let } s = 0.$$

$$t = \frac{-b \pm \sqrt{b^2 - 4ac}}{2a} \qquad \text{\color{red}Quadratic Formula}$$

$$t = \frac{-(-1.302) \pm \sqrt{(-1.302)^2 - 4(-15.45)(5.2340)}}{2(-15.45)} \qquad \text{\color{red}Substitute } a = -15.45, \; b = -1.302, \text{ and } c = 5.2340.$$

$$t \approx 0.54 \qquad \text{\color{red}Choose positive solution.}$$

The solution is about 0.54 second. In other words, the basketball will continue to fall for about 0.1 second more before hitting the ground. (Note that the experimental value of g is $\tfrac{1}{2}g = -15.45$, or $g = -30.90$ feet per second per second.) ∎

Scatter plot of data
Figure P.33

Fitting a Trigonometric Model to Data

What is mathematical modeling? This is one of the questions that is asked in the book *Guide to Mathematical Modelling*. Here is part of the answer.*

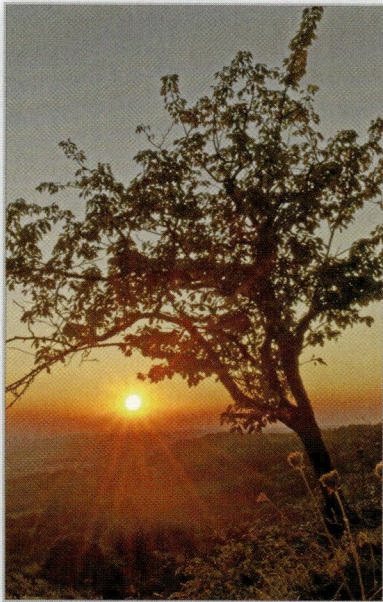

1. Mathematical modeling consists of applying your mathematical skills to obtain useful answers to real problems.

2. Learning to apply mathematical skills is very different from learning mathematics itself.

3. Models are used in a very wide range of applications, some of which do not appear initially to be mathematical in nature.

4. Models often allow quick and cheap evaluation of alternatives, leading to optimal solutions that are not otherwise obvious.

5. There are no precise rules in mathematical modeling and no "correct" answers.

6. Modeling can be learned only by *doing*.

The amount of daylight received by locations on Earth varies with the time of year.

REMARK For a review of trigonometric functions, see Appendix C.

EXAMPLE 3 **Fitting a Trigonometric Model to Data**

The number of hours of daylight on a given day on Earth depends on the latitude and the time of year. Here are the numbers of minutes of daylight at a location of 20°N latitude on the longest and shortest days of the year: June 21, 801 minutes; December 22, 655 minutes. Use these data to write a model for the amount of daylight d (in minutes) on each day of the year at a location of 20°N latitude. How could you check the accuracy of your model?

Solution Here is one way to create a model. You can hypothesize that the model is a sine function whose period is 365 days. Using the given data, you can conclude that the amplitude of the graph is $(801 - 655)/2$, or 73. So, one possible model is

$$d = 728 - 73 \sin\left(\frac{2\pi t}{365} + \frac{\pi}{2}\right).$$

In this model, t represents the number of each day of the year, with December 22 represented by $t = 0$. A graph of this model is shown in Figure P.34. To check the accuracy of this model, a weather almanac was used to find the numbers of minutes of daylight on different days of the year at the location of 20°N latitude.

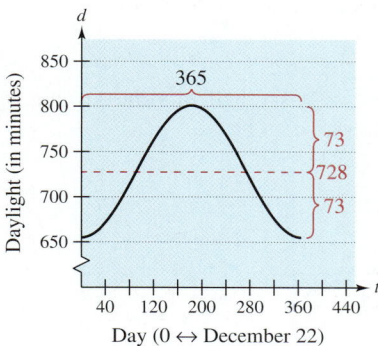

Date	Value of t	Actual Daylight	Daylight Given by Model
Dec 22	0	655 min	655 min
Jan 1	10	657 min	656 min
Feb 1	41	676 min	672 min
Mar 1	69	705 min	701 min
Apr 1	100	740 min	739 min
May 1	130	772 min	773 min
Jun 1	161	796 min	796 min
Jun 21	181	801 min	801 min
Jul 1	191	799 min	800 min
Aug 1	222	782 min	785 min
Sep 1	253	752 min	754 min
Oct 1	283	718 min	716 min
Nov 1	314	685 min	681 min
Dec 1	344	661 min	660 min

Graph of model
Figure P.34

You can see that the model is fairly accurate.

* Text from Dilwyn Edwards and Mike Hamson, *Guide to Mathematical Modelling* (Boca Raton: CRC Press, 1990), p. 4. Used by permission of the authors.

hjschneider/iStockphoto.com

P.4 Exercises

See **CalcChat.com** for tutorial help and worked-out solutions to odd-numbered exercises.

1. Wages Each ordered pair gives the average weekly wage x for federal government workers and the average weekly wage y for state government workers for 2001 through 2009. *(Source: U.S. Bureau of Labor Statistics)*

(941, 727), (1001, 754), (1043, 770), (1111, 791), (1151, 812),

(1198, 844), (1248, 883), (1275, 923), (1303, 937)

(a) Plot the data. From the graph, do the data appear to be approximately linear?

(b) Visually find a linear model for the data. Graph the model.

(c) Use the model to approximate y when $x = 1075$.

2. Quiz Scores The ordered pairs represent the scores on two consecutive 15-point quizzes for a class of 15 students.

(7, 13), (9, 7), (14, 14), (15, 15), (10, 15), (9, 7), (11, 14), (7, 14),

(14, 11), (14, 15), (8, 10), (15, 9), (10, 11), (9, 10), (11, 10)

(a) Plot the data. From the graph, does the relationship between consecutive scores appear to be approximately linear?

(b) If the data appear to be approximately linear, find a linear model for the data. If not, give some possible explanations.

3. Hooke's Law Hooke's Law states that the force F required to compress or stretch a spring (within its elastic limits) is proportional to the distance d that the spring is compressed or stretched from its original length. That is, $F = kd$, where k is a measure of the stiffness of the spring and is called the *spring constant*. The table shows the elongation d in centimeters of a spring when a force of F newtons is applied.

F	20	40	60	80	100
d	1.4	2.5	4.0	5.3	6.6

(a) Use the regression capabilities of a graphing utility to find a linear model for the data.

(b) Use a graphing utility to plot the data and graph the model. How well does the model fit the data? Explain.

(c) Use the model to estimate the elongation of the spring when a force of 55 newtons is applied.

4. Falling Object In an experiment, students measured the speed s (in meters per second) of a falling object t seconds after it was released. The results are shown in the table.

t	0	1	2	3	4
s	0	11.0	19.4	29.2	39.4

(a) Use the regression capabilities of a graphing utility to find a linear model for the data.

(b) Use a graphing utility to plot the data and graph the model. How well does the model fit the data? Explain.

(c) Use the model to estimate the speed of the object after 2.5 seconds.

5. Energy Consumption and Gross National Product The data show the per capita energy consumptions (in millions of Btu) and the per capita gross national incomes (in thousands of U.S. dollars) for several countries in 2008. *(Source: U.S. Energy Information Administration and The World Bank)*

Argentina	(81, 7.19)	India	(17, 1.04)
Australia	(274, 40.24)	Italy	(136, 35.46)
Bangladesh	(6, 0.52)	Japan	(172, 38.13)
Brazil	(54, 7.30)	Mexico	(66, 9.99)
Canada	(422, 43.64)	Poland	(101, 11.73)
Ecuador	(35, 3.69)	Turkey	(57, 9.02)
Hungary	(110, 12.81)	Venezuela	(121, 9.23)

(a) Use the regression capabilities of a graphing utility to find a linear model for the data. What is the correlation coefficient?

(b) Use a graphing utility to plot the data and graph the model.

(c) Interpret the graph in part (b). Use the graph to identify the three countries that differ most from the linear model.

(d) Delete the data for the three countries identified in part (c). Fit a linear model to the remaining data and give the correlation coefficient.

6. HOW DO YOU SEE IT? Determine whether the data can be modeled by a linear function, a quadratic function, or a trigonometric function, or that there appears to be no relationship between x and y.

7. Beam Strength Students in a lab measured the breaking strength S (in pounds) of wood 2 inches thick, x inches high, and 12 inches long. The results are shown in the table.

x	4	6	8	10	12
S	2370	5460	10,310	16,250	23,860

(a) Use the regression capabilities of a graphing utility to find a quadratic model for the data.

(b) Use a graphing utility to plot the data and graph the model.

(c) Use the model to approximate the breaking strength when $x = 2$.

(d) How many times greater is the breaking strength for a 4-inch-high board than for a 2-inch-high board?

(e) How many times greater is the breaking strength for a 12-inch-high board than for a 6-inch-high board? When the height of a board increases by a factor, does the breaking strength increase by the same factor? Explain.

8. Car Performance The time t (in seconds) required to attain a speed of s miles per hour from a standing start for a Volkswagen Passat is shown in the table. *(Source: Car & Driver)*

s	30	40	50	60	70	80	90
t	2.7	3.8	4.9	6.3	8.0	9.9	12.2

(a) Use the regression capabilities of a graphing utility to find a quadratic model for the data.

(b) Use a graphing utility to plot the data and graph the model.

(c) Use the graph in part (b) to state why the model is not appropriate for determining the times required to attain speeds of less than 20 miles per hour.

(d) Because the test began from a standing start, add the point $(0, 0)$ to the data. Fit a quadratic model to the revised data and graph the new model.

(e) Does the quadratic model in part (d) more accurately model the behavior of the car? Explain.

9. Engine Performance A V8 car engine is coupled to a dynamometer, and the horsepower y is measured at different engine speeds x (in thousands of revolutions per minute). The results are shown in the table.

x	1	2	3	4	5	6
y	40	85	140	200	225	245

(a) Use the regression capabilities of a graphing utility to find a cubic model for the data.

(b) Use a graphing utility to plot the data and graph the model.

(c) Use the model to approximate the horsepower when the engine is running at 4500 revolutions per minute.

10. Boiling Temperature The table shows the temperatures T (in degrees Fahrenheit) at which water boils at selected pressures p (in pounds per square inch). *(Source: Standard Handbook for Mechanical Engineers)*

p	5	10	14.696 (1 atmosphere)	20
T	162.24°	193.21°	212.00°	227.96°

p	30	40	60	80	100
T	250.33°	267.25°	292.71°	312.03°	327.81°

(a) Use the regression capabilities of a graphing utility to find a cubic model for the data.

(b) Use a graphing utility to plot the data and graph the model.

(c) Use the graph to estimate the pressure required for the boiling point of water to exceed 300°F.

(d) Explain why the model would not be accurate for pressures exceeding 100 pounds per square inch.

11. Automobile Costs The data in the table show the variable costs of operating an automobile in the United States for 2000 through 2010, where t is the year, with $t = 0$ corresponding to 2000. The functions y_1, y_2, and y_3 represent the costs in cents per mile for gas, maintenance, and tires, respectively. *(Source: Bureau of Transportation Statistics)*

t	y_1	y_2	y_3
0	6.9	3.6	1.7
1	7.9	3.9	1.8
2	5.9	4.1	1.8
3	7.2	4.1	1.8
4	6.5	5.4	0.7
5	9.5	4.9	0.7
6	8.9	4.9	0.7
7	11.7	4.6	0.7
8	10.1	4.6	0.8
9	11.4	4.5	0.8
10	12.3	4.4	1.0

(a) Use the regression capabilities of a graphing utility to find cubic models for y_1 and y_3, and a quadratic model for y_2.

(b) Use a graphing utility to graph y_1, y_2, y_3, and $y_1 + y_2 + y_3$ in the same viewing window. Use the model to estimate the total variable cost per mile in 2014.

12. Health Maintenance Organizations The bar graph shows the numbers of people N (in millions) receiving care in HMOs for the years 1994 through 2008. *(Source: HealthLeaders-InterStudy)*

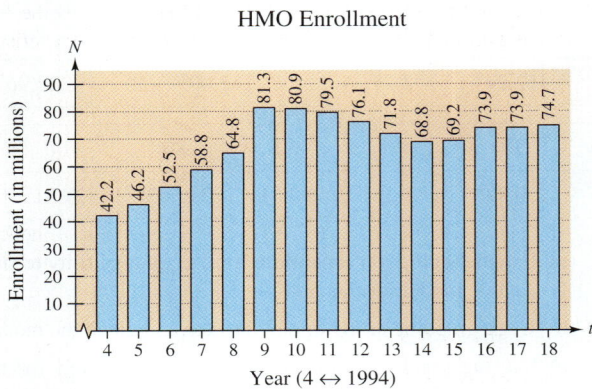

HMO Enrollment

(a) Let t be the time in years, with $t = 4$ corresponding to 1994. Use the regression capabilities of a graphing utility to find linear and cubic models for the data.

(b) Use a graphing utility to plot the data and graph the linear and cubic models.

(c) Use the graphs in part (b) to determine which is the better model.

(d) Use a graphing utility to find and graph a quadratic model for the data. How well does the model fit the data? Explain.

(e) Use the linear and cubic models to estimate the number of people receiving care in HMOs in the year 2014. What do you notice?

(f) Use a graphing utility to find other models for the data. Which models do you think best represent the data? Explain.

13. Harmonic Motion The motion of an oscillating weight suspended by a spring was measured by a motion detector. The data collected and the approximate maximum (positive and negative) displacements from equilibrium are shown in the figure. The displacement y is measured in centimeters, and the time t is measured in seconds.

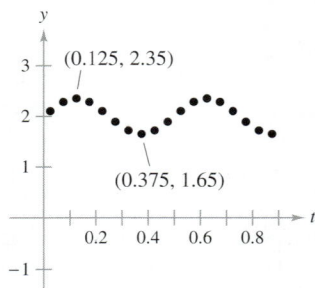

(a) Is y a function of t? Explain.

(b) Approximate the amplitude and period of the oscillations.

(c) Find a model for the data.

(d) Use a graphing utility to graph the model in part (c). Compare the result with the data in the figure.

14. Temperature The table shows the normal daily high temperatures for Miami M and Syracuse S (in degrees Fahrenheit) for month t, with $t = 1$ corresponding to January. *(Source: National Oceanic and Atmospheric Administration)*

t	1	2	3	4	5	6
M	76.5	77.7	80.7	83.8	87.2	89.5
S	31.4	33.5	43.1	55.7	68.5	77.0

t	7	8	9	10	11	12
M	90.9	90.6	89.0	85.4	81.2	77.5
S	81.7	79.6	71.4	59.8	47.4	36.3

(a) A model for Miami is

$$M(t) = 83.70 + 7.46 \sin(0.4912t - 1.95).$$

Find a model for Syracuse.

(b) Use a graphing utility to plot the data and graph the model for Miami. How well does the model fit?

(c) Use a graphing utility to plot the data and graph the model for Syracuse. How well does the model fit?

(d) Use the models to estimate the average annual temperature in each city. Which term of the model did you use? Explain.

(e) What is the period of each model? Is it what you expected? Explain.

(f) Which city has a greater variability in temperature throughout the year? Which factor of the models determines this variability? Explain.

WRITING ABOUT CONCEPTS

Modeling Data In Exercises 15 and 16, describe a possible real-life situation for each data set. Then describe how a model could be used in the real-life setting.

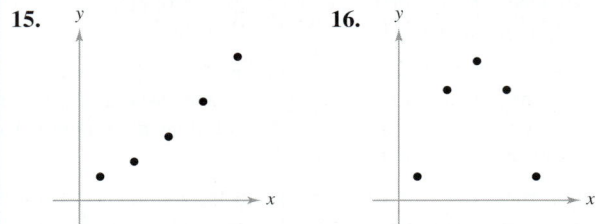

15.

16.

PUTNAM EXAM CHALLENGE

17. For $i = 1, 2$, let T_i be a triangle with side lengths a_i, b_i, c_i, and area A_i. Suppose that $a_1 \leq a_2$, $b_1 \leq b_2$, $c_1 \leq c_2$, and that T_2 is an acute triangle. Does it follow that $A_1 \leq A_2$?

This problem was composed by the Committee on the Putnam Prize Competition.
© The Mathematical Association of America. All rights reserved.

Finding Intercepts   In Exercises 1–4, find any intercepts.

1. $y = 5x - 8$
2. $y = x^2 - 8x + 12$
3. $y = \dfrac{x - 3}{x - 4}$
4. $y = (x - 3)\sqrt{x + 4}$

Testing for Symmetry   In Exercises 5–8, test for symmetry with respect to each axis and to the origin.

5. $y = x^2 + 4x$
6. $y = x^4 - x^2 + 3$
7. $y^2 = x^2 - 5$
8. $xy = -2$

Using Intercepts and Symmetry to Sketch a Graph
In Exercises 9–14, sketch the graph of the equation. Identify any intercepts and test for symmetry.

9. $y = -\frac{1}{2}x + 3$
10. $y = -x^2 + 4$
11. $y = x^3 - 4x$
12. $y^2 = 9 - x$
13. $y = 2\sqrt{4 - x}$
14. $y = |x - 4| - 4$

Finding Points of Intersection   In Exercises 15–18, find the points of intersection of the graphs of the equations.

15. $5x + 3y = -1$
    $x - y = -5$
16. $2x + 4y = 9$
    $6x - 4y = 7$
17. $x - y = -5$
    $x^2 - y = 1$
18. $x^2 + y^2 = 1$
    $-x + y = 1$

Finding the Slope of a Line   In Exercises 19 and 20, plot the points and find the slope of the line passing through them.

19. $\left(\frac{3}{2}, 1\right), \left(5, \frac{5}{2}\right)$
20. $(-7, 8), (-1, 8)$

Finding an Equation of a Line   In Exercises 21–24, find an equation of the line that passes through the point and has the indicated slope. Then sketch the line.

	Point	Slope		Point	Slope
21.	$(3, -5)$	$m = \frac{7}{4}$	22.	$(-8, 1)$	m is undefined.
23.	$(-3, 0)$	$m = -\frac{2}{3}$	24.	$(5, 4)$	$m = 0$

Sketching Lines in the Plane   In Exercises 25–28, use the slope and y-intercept to sketch a graph of the equation.

25. $y = 6$
26. $x = -3$
27. $y = 4x - 2$
28. $3x + 2y = 12$

Finding an Equation of a Line   In Exercises 29 and 30, find an equation of the line that passes through the points. Then sketch the line.

29. $(0, 0), (8, 2)$
30. $(-5, 5), (10, -1)$

31. **Finding Equations of Lines**   Find equations of the lines passing through $(-3, 5)$ and having the following characteristics.
    (a) Slope of $\frac{7}{16}$
    (b) Parallel to the line $5x - 3y = 3$
    (c) Perpendicular to the line $3x + 4y = 8$
    (d) Parallel to the y-axis

32. **Finding Equations of Lines**   Find equations of the lines passing through $(2, 4)$ and having the following characteristics.
    (a) Slope of $-\frac{2}{3}$
    (b) Perpendicular to the line $x + y = 0$
    (c) Passing through the point $(6, 1)$
    (d) Parallel to the x-axis

33. **Rate of Change**   The purchase price of a new machine is $12,500, and its value will decrease by $850 per year. Use this information to write a linear equation that gives the value V of the machine t years after it is purchased. Find its value at the end of 3 years.

34. **Break-Even Analysis**   A contractor purchases a piece of equipment for $36,500 that costs an average of $9.25 per hour for fuel and maintenance. The equipment operator is paid $13.50 per hour, and customers are charged $30 per hour.
    (a) Write an equation for the cost C of operating this equipment for t hours.
    (b) Write an equation for the revenue R derived from t hours of use.
    (c) Find the break-even point for this equipment by finding the time at which $R = C$.

Evaluating a Function   In Exercises 35–38, evaluate the function at the given value(s) of the independent variable. Simplify the result.

35. $f(x) = 5x + 4$
    (a) $f(0)$
    (b) $f(5)$
    (c) $f(-3)$
    (d) $f(t + 1)$
36. $f(x) = x^3 - 2x$
    (a) $f(-3)$
    (b) $f(2)$
    (c) $f(-1)$
    (d) $f(c - 1)$
37. $f(x) = 4x^2$
    $\dfrac{f(x + \Delta x) - f(x)}{\Delta x}$
38. $f(x) = 2x - 6$
    $\dfrac{f(x) - f(-1)}{x - 1}$

Finding the Domain and Range of a Function   In Exercises 39–42, find the domain and range of the function.

39. $f(x) = x^2 + 3$
40. $g(x) = \sqrt{6 - x}$
41. $f(x) = -|x + 1|$
42. $h(x) = \dfrac{2}{x + 1}$

Using the Vertical Line Test In Exercises 43–46, sketch the graph of the equation and use the Vertical Line Test to determine whether y is a function of x.

43. $x - y^2 = 6$

44. $x^2 - y = 0$

45. $y = \dfrac{|x - 2|}{x - 2}$

46. $x = 9 - y^2$

47. Transformations of Functions Use a graphing utility to graph $f(x) = x^3 - 3x^2$. Use the graph to write a formula for the function g shown in the figure. To print an enlarged copy of the graph, go to *MathGraphs.com*.

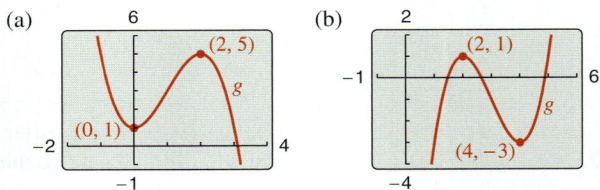

(a)

(b)

48. Conjecture

(a) Use a graphing utility to graph the functions f, g, and h in the same viewing window. Write a description of any similarities and differences you observe among the graphs.

Odd powers: $f(x) = x$, $g(x) = x^3$, $h(x) = x^5$

Even powers: $f(x) = x^2$, $g(x) = x^4$, $h(x) = x^6$

(b) Use the result in part (a) to make a conjecture about the graphs of the functions $y = x^7$ and $y = x^8$. Use a graphing utility to verify your conjecture.

49. Think About It Use the results of Exercise 48 to guess the shapes of the graphs of the functions f, g, and h. Then use a graphing utility to graph each function and compare the result with your guess.

(a) $f(x) = x^2(x - 6)^2$

(b) $g(x) = x^3(x - 6)^2$

(c) $h(x) = x^3(x - 6)^3$

50. Think About It What is the minimum degree of the polynomial function whose graph approximates the given graph? What sign must the leading coefficient have?

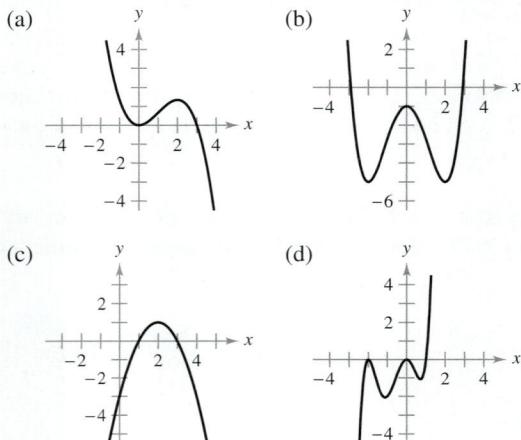

(a)

(b)

(c)

(d)

51. Stress Test A machine part was tested by bending it x centimeters 10 times per minute until the time y (in hours) of failure. The results are recorded in the table.

x	3	6	9	12	15
y	61	56	53	55	48

x	18	21	24	27	30
y	35	36	33	44	23

(a) Use the regression capabilities of a graphing utility to find a linear model for the data.

(b) Use a graphing utility to plot the data and graph the model.

(c) Use the graph to determine whether there may have been an error made in conducting one of the tests or in recording the results. If so, eliminate the erroneous point and find the model for the remaining data.

52. Median Income The data in the table show the median income y (in thousands of dollars) for males of various ages x in the United States in 2009. *(Source: U.S. Census Bureau)*

x	20	30	40	50	60	70
y	10.0	31.9	42.2	44.7	41.3	25.9

(a) Use the regression capabilities of a graphing utility to find a quadratic model for the data.

(b) Use a graphing utility to plot the data and graph the model.

(c) Use the model to approximate the median income for a male who is 26 years old.

(d) Use the model to approximate the median income for a male who is 34 years old.

53. Harmonic Motion The motion of an oscillating weight suspended by a spring was measured by a motion detector. The data collected and the approximate maximum (positive and negative) displacements from equilibrium are shown in the figure. The displacement y is measured in feet, and the time t is measured in seconds.

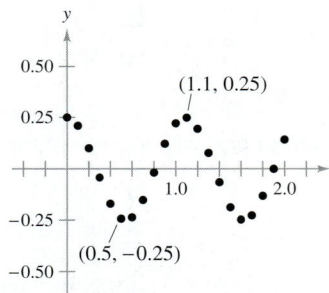

(a) Is y a function of t? Explain.

(b) Approximate the amplitude and period of the oscillations.

(c) Find a model for the data.

(d) Use a graphing utility to graph the model in part (c). Compare the result with the data in the figure.

P.S. Problem Solving

See **CalcChat.com** for tutorial help and worked-out solutions to odd-numbered exercises.

1. **Finding Tangent Lines**   Consider the circle

   $$x^2 + y^2 - 6x - 8y = 0,$$

   as shown in the figure.

   (a) Find the center and radius of the circle.

   (b) Find an equation of the tangent line to the circle at the point $(0, 0)$.

   (c) Find an equation of the tangent line to the circle at the point $(6, 0)$.

   (d) Where do the two tangent lines intersect?

      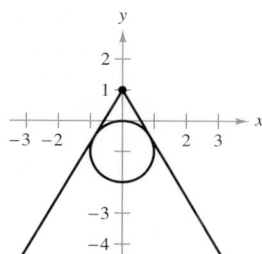

   Figure for 1   Figure for 2

2. **Finding Tangent Lines**   There are two tangent lines from the point $(0, 1)$ to the circle $x^2 + (y + 1)^2 = 1$ (see figure). Find equations of these two lines by using the fact that each tangent line intersects the circle at *exactly* one point.

3. **Heaviside Function**   The Heaviside function $H(x)$ is widely used in engineering applications.

   $$H(x) = \begin{cases} 1, & x \geq 0 \\ 0, & x < 0 \end{cases}$$

   Sketch the graph of the Heaviside function and the graphs of the following functions by hand.

   (a) $H(x) - 2$ (b) $H(x - 2)$ (c) $-H(x)$

   (d) $H(-x)$ (e) $\frac{1}{2}H(x)$ (f) $-H(x - 2) + 2$

   OLIVER HEAVISIDE (1850–1925)

   Heaviside was a British mathematician and physicist who contributed to the field of applied mathematics, especially applications of mathematics to electrical engineering. The *Heaviside function* is a classic type of "on-off" function that has applications to electricity and computer science.

4. **Sketching Transformations**   Consider the graph of the function f shown below. Use this graph to sketch the graphs of the following functions. To print an enlarged copy of the graph, go to *MathGraphs.com*.

   (a) $f(x + 1)$ (b) $f(x) + 1$
   (c) $2f(x)$ (d) $f(-x)$
   (e) $-f(x)$ (f) $|f(x)|$
   (g) $f(|x|)$

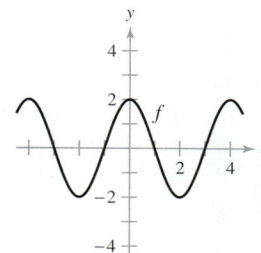

5. **Maximum Area**   A rancher plans to fence a rectangular pasture adjacent to a river. The rancher has 100 meters of fencing, and no fencing is needed along the river (see figure).

   (a) Write the area A of the pasture as a function of x, the length of the side parallel to the river. What is the domain of A?

   (b) Graph the area function and estimate the dimensions that yield the maximum amount of area for the pasture.

   (c) Find the dimensions that yield the maximum amount of area for the pasture by completing the square.

      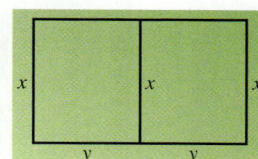

   Figure for 5   Figure for 6

6. **Maximum Area**   A rancher has 300 feet of fencing to enclose two adjacent pastures (see figure).

   (a) Write the total area A of the two pastures as a function of x. What is the domain of A?

   (b) Graph the area function and estimate the dimensions that yield the maximum amount of area for the pastures.

   (c) Find the dimensions that yield the maximum amount of area for the pastures by completing the square.

7. **Writing a Function**   You are in a boat 2 miles from the nearest point on the coast. You are to go to a point Q located 3 miles down the coast and 1 mile inland (see figure). You can row at 2 miles per hour and walk at 4 miles per hour. Write the total time T of the trip as a function of x.

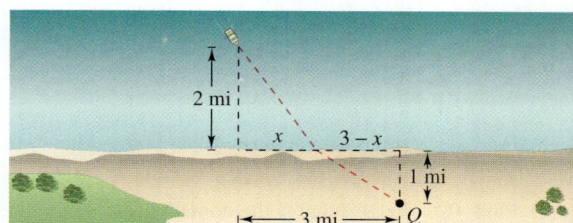

8. **Average Speed** You drive to the beach at a rate of 120 kilometers per hour. On the return trip, you drive at a rate of 60 kilometers per hour. What is your average speed for the entire trip? Explain your reasoning.

9. **Slope of a Tangent Line** One of the fundamental themes of calculus is to find the slope of the tangent line to a curve at a point. To see how this can be done, consider the point $(2, 4)$ on the graph of $f(x) = x^2$ (see figure).

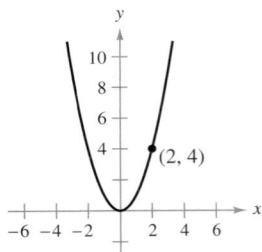

(a) Find the slope of the line joining $(2, 4)$ and $(3, 9)$. Is the slope of the tangent line at $(2, 4)$ greater than or less than this number?

(b) Find the slope of the line joining $(2, 4)$ and $(1, 1)$. Is the slope of the tangent line at $(2, 4)$ greater than or less than this number?

(c) Find the slope of the line joining $(2, 4)$ and $(2.1, 4.41)$. Is the slope of the tangent line at $(2, 4)$ greater than or less than this number?

(d) Find the slope of the line joining $(2, 4)$ and $(2 + h, f(2 + h))$ in terms of the nonzero number h. Verify that $h = 1, -1$, and 0.1 yield the solutions to parts (a)–(c) above.

(e) What is the slope of the tangent line at $(2, 4)$? Explain how you arrived at your answer.

10. **Slope of a Tangent Line** Sketch the graph of the function $f(x) = \sqrt{x}$ and label the point $(4, 2)$ on the graph.

(a) Find the slope of the line joining $(4, 2)$ and $(9, 3)$. Is the slope of the tangent line at $(4, 2)$ greater than or less than this number?

(b) Find the slope of the line joining $(4, 2)$ and $(1, 1)$. Is the slope of the tangent line at $(4, 2)$ greater than or less than this number?

(c) Find the slope of the line joining $(4, 2)$ and $(4.41, 2.1)$. Is the slope of the tangent line at $(4, 2)$ greater than or less than this number?

(d) Find the slope of the line joining $(4, 2)$ and $(4 + h, f(4 + h))$ in terms of the nonzero number h.

(e) What is the slope of the tangent line at $(4, 2)$? Explain how you arrived at your answer.

11. **Composite Functions** Let $f(x) = \dfrac{1}{1 - x}$.

(a) What are the domain and range of f?

(b) Find the composition $f(f(x))$. What is the domain of this function?

(c) Find $f(f(f(x)))$. What is the domain of this function?

(d) Graph $f(f(f(x)))$. Is the graph a line? Why or why not?

12. **Graphing an Equation** Explain how you would graph the equation

$$y + |y| = x + |x|.$$

Then sketch the graph.

13. **Sound Intensity** A large room contains two speakers that are 3 meters apart. The sound intensity I of one speaker is twice that of the other, as shown in the figure. (To print an enlarged copy of the graph, go to *MathGraphs.com.*) Suppose the listener is free to move about the room to find those positions that receive equal amounts of sound from both speakers. Such a location satisfies two conditions: (1) the sound intensity at the listener's position is directly proportional to the sound level of a source, and (2) the sound intensity is inversely proportional to the square of the distance from the source.

(a) Find the points on the x-axis that receive equal amounts of sound from both speakers.

(b) Find and graph the equation of all locations (x, y) where one could stand and receive equal amounts of sound from both speakers.

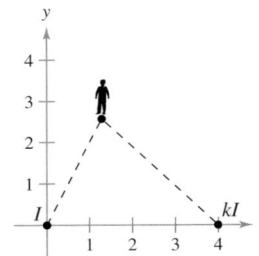

Figure for 13   Figure for 14

14. **Sound Intensity** Suppose the speakers in Exercise 13 are 4 meters apart and the sound intensity of one speaker is k times that of the other, as shown in the figure. To print an enlarged copy of the graph, go to *MathGraphs.com.*

(a) Find the equation of all locations (x, y) where one could stand and receive equal amounts of sound from both speakers.

(b) Graph the equation for the case $k = 3$.

(c) Describe the set of locations of equal sound as k becomes very large.

15. **Lemniscate** Let d_1 and d_2 be the distances from the point (x, y) to the points $(-1, 0)$ and $(1, 0)$, respectively, as shown in the figure. Show that the equation of the graph of all points (x, y) satisfying $d_1 d_2 = 1$ is

$$(x^2 + y^2)^2 = 2(x^2 - y^2).$$

This curve is called a **lemniscate.** Graph the lemniscate and identify three points on the graph.

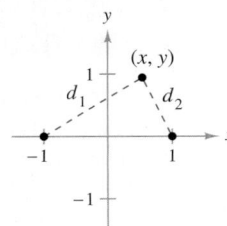

1 Limits and Their Properties

Inventory Management *(Exercise 110, p. 81)*

Average Speed *(Exercise 62, p. 89)*

Free-Falling Object *(Exercises 101 and 102, p. 69)*

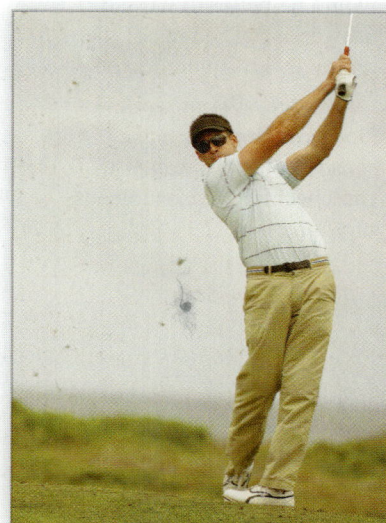

Sports *(Exercise 62, p. 57)*

Bicyclist *(Exercise 3, p. 47)*

1.1 A Preview of Calculus

- Understand what calculus is and how it compares with precalculus.
- Understand that the tangent line problem is basic to calculus.
- Understand that the area problem is also basic to calculus.

What Is Calculus?

Calculus is the mathematics of change. For instance, calculus is the mathematics of velocities, accelerations, tangent lines, slopes, areas, volumes, arc lengths, centroids, curvatures, and a variety of other concepts that have enabled scientists, engineers, and economists to model real-life situations.

Although precalculus mathematics also deals with velocities, accelerations, tangent lines, slopes, and so on, there is a fundamental difference between precalculus mathematics and calculus. Precalculus mathematics is more static, whereas calculus is more dynamic. Here are some examples.

- An object traveling at a constant velocity can be analyzed with precalculus mathematics. To analyze the velocity of an accelerating object, you need calculus.
- The slope of a line can be analyzed with precalculus mathematics. To analyze the slope of a curve, you need calculus.
- The curvature of a circle is constant and can be analyzed with precalculus mathematics. To analyze the variable curvature of a general curve, you need calculus.
- The area of a rectangle can be analyzed with precalculus mathematics. To analyze the area under a general curve, you need calculus.

Each of these situations involves the same general strategy—the reformulation of precalculus mathematics through the use of a limit process. So, one way to answer the question "What is calculus?" is to say that calculus is a "limit machine" that involves three stages. The first stage is precalculus mathematics, such as the slope of a line or the area of a rectangle. The second stage is the limit process, and the third stage is a new calculus formulation, such as a derivative or integral.

| Precalculus mathematics | ⇒ | Limit process | ⇒ | Calculus |

Some students try to learn calculus as if it were simply a collection of new formulas. This is unfortunate. If you reduce calculus to the memorization of differentiation and integration formulas, you will miss a great deal of understanding, self-confidence, and satisfaction.

On the next two pages are listed some familiar precalculus concepts coupled with their calculus counterparts. Throughout the text, your goal should be to learn how precalculus formulas and techniques are used as building blocks to produce the more general calculus formulas and techniques. Don't worry if you are unfamiliar with some of the "old formulas" listed on the next two pages—you will be reviewing all of them.

As you proceed through this text, come back to this discussion repeatedly. Try to keep track of where you are relative to the three stages involved in the study of calculus. For instance, note how these chapters relate to the three stages.

Chapter P: Preparation for Calculus	Precalculus
Chapter 1: Limits and Their Properties	Limit process
Chapter 2: Differentiation	Calculus

This cycle is repeated many times on a smaller scale throughout the text.

Without Calculus	**With Differential Calculus**
Value of $f(x)$ when $x = c$	Limit of $f(x)$ as x approaches c
Slope of a line	Slope of a curve
Secant line to a curve	Tangent line to a curve
Average rate of change between $t = a$ and $t = b$	Instantaneous rate of change at $t = c$
Curvature of a circle	Curvature of a curve
Height of a curve when $x = c$	Maximum height of a curve on an interval
Tangent plane to a sphere	Tangent plane to a surface
Direction of motion along a line	Direction of motion along a curve

Without Calculus		With Integral Calculus	
Area of a rectangle		Area under a curve	
Work done by a constant force		Work done by a variable force	
Center of a rectangle		Centroid of a region	
Length of a line segment		Length of an arc	
Surface area of a cylinder		Surface area of a solid of revolution	
Mass of a solid of constant density		Mass of a solid of variable density	
Volume of a rectangular solid		Volume of a region under a surface	
Sum of a finite number of terms	$a_1 + a_2 + \cdots + a_n = S$	Sum of an infinite number of terms	$a_1 + a_2 + a_3 + \cdots = S$

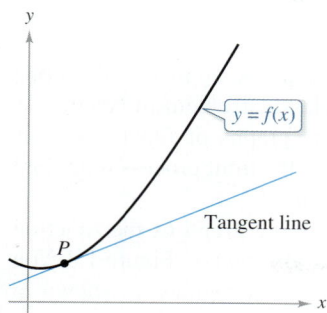

The tangent line to the graph of f at P
Figure 1.1

The Tangent Line Problem

The notion of a limit is fundamental to the study of calculus. The following brief descriptions of two classic problems in calculus—*the tangent line problem* and *the area problem*—should give you some idea of the way limits are used in calculus.

In the tangent line problem, you are given a function f and a point P on its graph and are asked to find an equation of the tangent line to the graph at point P, as shown in Figure 1.1.

Except for cases involving a vertical tangent line, the problem of finding the **tangent line** at a point P is equivalent to finding the *slope* of the tangent line at P. You can approximate this slope by using a line through the point of tangency and a second point on the curve, as shown in Figure 1.2(a). Such a line is called a **secant line.** If $P(c, f(c))$ is the point of tangency and

$$Q(c + \Delta x, f(c + \Delta x))$$

is a second point on the graph of f, then the slope of the secant line through these two points can be found using precalculus and is

$$m_{sec} = \frac{f(c + \Delta x) - f(c)}{c + \Delta x - c} = \frac{f(c + \Delta x) - f(c)}{\Delta x}.$$

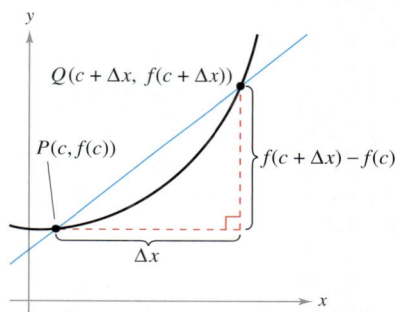

(a) The secant line through $(c, f(c))$ and $(c + \Delta x, f(c + \Delta x))$

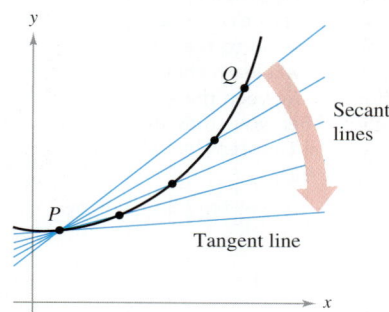

(b) As Q approaches P, the secant lines approach the tangent line.

Figure 1.2

As point Q approaches point P, the slopes of the secant lines approach the slope of the tangent line, as shown in Figure 1.2(b). When such a "limiting position" exists, the slope of the tangent line is said to be the **limit** of the slopes of the secant lines. (Much more will be said about this important calculus concept in Chapter 2.)

**GRACE CHISHOLM YOUNG
(1868–1944)**

Grace Chisholm Young received her degree in mathematics from Girton College in Cambridge, England. Her early work was published under the name of William Young, her husband. Between 1914 and 1916, Grace Young published work on the foundations of calculus that won her the Gamble Prize from Girton College.

Exploration

The following points lie on the graph of $f(x) = x^2$.

$Q_1(1.5, f(1.5))$, $Q_2(1.1, f(1.1))$, $Q_3(1.01, f(1.01))$,
$Q_4(1.001, f(1.001))$, $Q_5(1.0001, f(1.0001))$

Each successive point gets closer to the point $P(1, 1)$. Find the slopes of the secant lines through Q_1 and P, Q_2 and P, and so on. Graph these secant lines on a graphing utility. Then use your results to estimate the slope of the tangent line to the graph of f at the point P.

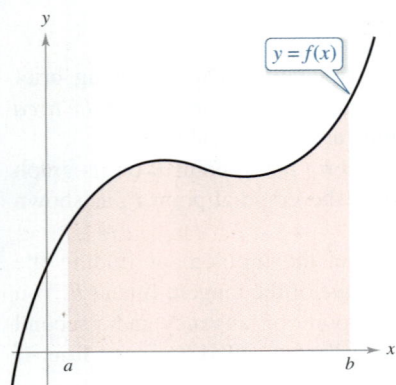

Area under a curve
Figure 1.3

The Area Problem

In the tangent line problem, you saw how the limit process can be applied to the slope of a line to find the slope of a general curve. A second classic problem in calculus is finding the area of a plane region that is bounded by the graphs of functions. This problem can also be solved with a limit process. In this case, the limit process is applied to the area of a rectangle to find the area of a general region.

As a simple example, consider the region bounded by the graph of the function $y = f(x)$, the x-axis, and the vertical lines $x = a$ and $x = b$, as shown in Figure 1.3. You can approximate the area of the region with several rectangular regions, as shown in Figure 1.4. As you increase the number of rectangles, the approximation tends to become better and better because the amount of area missed by the rectangles decreases. Your goal is to determine the limit of the sum of the areas of the rectangles as the number of rectangles increases without bound.

HISTORICAL NOTE

In one of the most astounding events ever to occur in mathematics, it was discovered that the tangent line problem and the area problem are closely related. This discovery led to the birth of calculus. You will learn about the relationship between these two problems when you study the Fundamental Theorem of Calculus in Chapter 4.

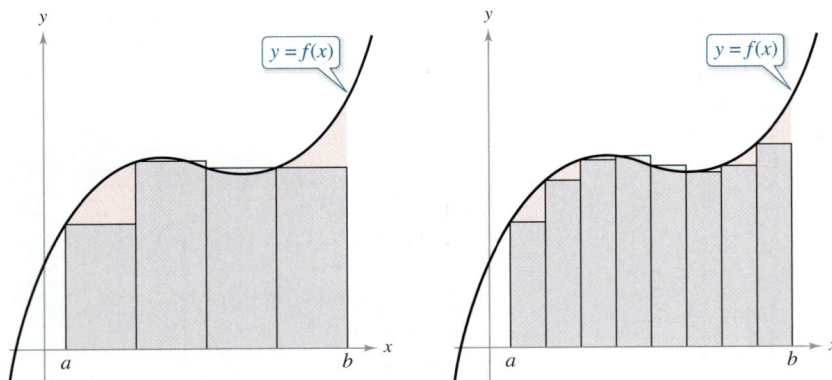

Approximation using four rectangles

Approximation using eight rectangles
Figure 1.4

Exploration

Consider the region bounded by the graphs of

$$f(x) = x^2, \quad y = 0, \quad \text{and} \quad x = 1$$

as shown in part (a) of the figure. The area of the region can be approximated by two sets of rectangles—one set inscribed within the region and the other set circumscribed over the region, as shown in parts (b) and (c). Find the sum of the areas of each set of rectangles. Then use your results to approximate the area of the region.

(a) Bounded region

(b) Inscribed rectangles

(c) Circumscribed rectangles

1.1 Exercises

See CalcChat.com for tutorial help and worked-out solutions to odd-numbered exercises.

Precalculus or Calculus In Exercises 1–5, decide whether the problem can be solved using precalculus or whether calculus is required. If the problem can be solved using precalculus, solve it. If the problem seems to require calculus, explain your reasoning and use a graphical or numerical approach to estimate the solution.

1. Find the distance traveled in 15 seconds by an object traveling at a constant velocity of 20 feet per second.

2. Find the distance traveled in 15 seconds by an object moving with a velocity of $v(t) = 20 + 7\cos t$ feet per second.

3. **Rate of Change**

   A bicyclist is riding on a path modeled by the function $f(x) = 0.04(8x - x^2)$, where x and $f(x)$ are measured in miles (see figure). Find the rate of change of elevation at $x = 2$.

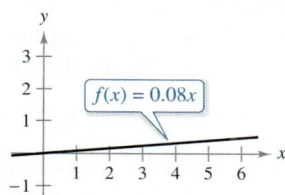

   $f(x) = 0.04(8x - x^2)$

4. A bicyclist is riding on a path modeled by the function $f(x) = 0.08x$, where x and $f(x)$ are measured in miles (see figure). Find the rate of change of elevation at $x = 2$.

   $f(x) = 0.08x$

5. Find the area of the shaded region.

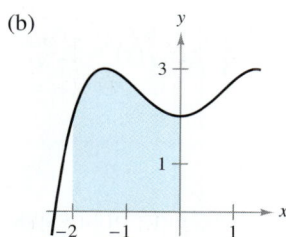

   (a) $(2, 4)$ $(0, 0)$ $(5, 0)$
   (b)

6. **Secant Lines** Consider the function

   $f(x) = \sqrt{x}$

   and the point $P(4, 2)$ on the graph of f.

   (a) Graph f and the secant lines passing through $P(4, 2)$ and $Q(x, f(x))$ for x-values of 1, 3, and 5.

   (b) Find the slope of each secant line.

   (c) Use the results of part (b) to estimate the slope of the tangent line to the graph of f at $P(4, 2)$. Describe how to improve your approximation of the slope.

7. **Secant Lines** Consider the function $f(x) = 6x - x^2$ and the point $P(2, 8)$ on the graph of f.

   (a) Graph f and the secant lines passing through $P(2, 8)$ and $Q(x, f(x))$ for x-values of 3, 2.5, and 1.5.

   (b) Find the slope of each secant line.

   (c) Use the results of part (b) to estimate the slope of the tangent line to the graph of f at $P(2, 8)$. Describe how to improve your approximation of the slope.

8. **HOW DO YOU SEE IT?** How would you describe the instantaneous rate of change of an automobile's position on a highway?

9. **Approximating Area** Use the rectangles in each graph to approximate the area of the region bounded by $y = 5/x$, $y = 0$, $x = 1$, and $x = 5$. Describe how you could continue this process to obtain a more accurate approximation of the area.

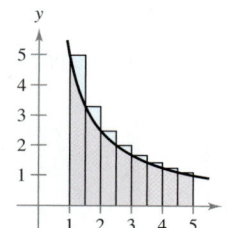

WRITING ABOUT CONCEPTS

10. **Approximating the Length of a Curve** Consider the length of the graph of $f(x) = 5/x$ from $(1, 5)$ to $(5, 1)$.

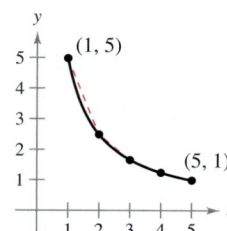

    (a) Approximate the length of the curve by finding the distance between its two endpoints, as shown in the first figure.

    (b) Approximate the length of the curve by finding the sum of the lengths of four line segments, as shown in the second figure.

    (c) Describe how you could continue this process to obtain a more accurate approximation of the length of the curve.

1.2 Finding Limits Graphically and Numerically

■ Estimate a limit using a numerical or graphical approach.
■ Learn different ways that a limit can fail to exist.
■ Study and use a formal definition of limit.

An Introduction to Limits

To sketch the graph of the function

$$f(x) = \frac{x^3 - 1}{x - 1}$$

for values other than $x = 1$, you can use standard curve-sketching techniques. At $x = 1$, however, it is not clear what to expect. To get an idea of the behavior of the graph of f near $x = 1$, you can use two sets of x-values—one set that approaches 1 from the left and one set that approaches 1 from the right, as shown in the table.

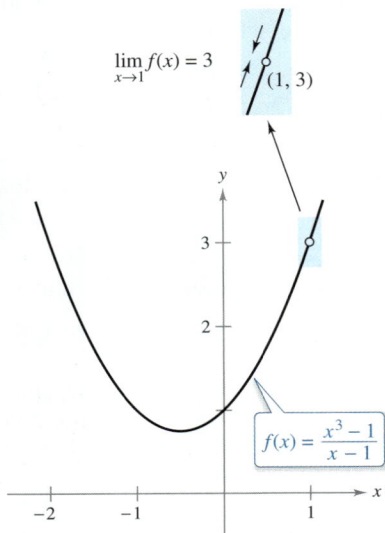

$\lim_{x \to 1} f(x) = 3$

$(1, 3)$

$f(x) = \dfrac{x^3 - 1}{x - 1}$

The limit of $f(x)$ as x approaches 1 is 3.
Figure 1.5

	x approaches 1 from the left.					*x approaches 1 from the right.*			
x	0.75	0.9	0.99	0.999	1	1.001	1.01	1.1	1.25
$f(x)$	2.313	2.710	2.970	2.997	?	3.003	3.030	3.310	3.813

	f(x) approaches 3.	*f(x) approaches 3.*

The graph of f is a parabola that has a gap at the point $(1, 3)$, as shown in Figure 1.5. Although x cannot equal 1, you can move arbitrarily close to 1, and as a result $f(x)$ moves arbitrarily close to 3. Using limit notation, you can write

$$\lim_{x \to 1} f(x) = 3.$$ This is read as "the limit of $f(x)$ as x approaches 1 is 3."

This discussion leads to an informal definition of limit. If $f(x)$ becomes arbitrarily close to a single number L as x approaches c from either side, then the **limit** of $f(x)$, as x approaches c, is L. This limit is written as

$$\lim_{x \to c} f(x) = L.$$

Exploration

The discussion above gives an example of how you can estimate a limit *numerically* by constructing a table and *graphically* by drawing a graph. Estimate the following limit numerically by completing the table.

$$\lim_{x \to 2} \frac{x^2 - 3x + 2}{x - 2}$$

x	1.75	1.9	1.99	1.999	2	2.001	2.01	2.1	2.25
$f(x)$	?	?	?	?	?	?	?	?	?

Then use a graphing utility to estimate the limit graphically.

EXAMPLE 1 **Estimating a Limit Numerically**

Evaluate the function $f(x) = x/(\sqrt{x+1} - 1)$ at several x-values near 0 and use the results to estimate the limit

$$\lim_{x \to 0} \frac{x}{\sqrt{x+1} - 1}.$$

Solution The table lists the values of $f(x)$ for several x-values near 0.

	x approaches 0 from the left.				x approaches 0 from the right.		
x	-0.01	-0.001	-0.0001	0	0.0001	0.001	0.01
$f(x)$	1.99499	1.99950	1.99995	?	2.00005	2.00050	2.00499
	$f(x)$ approaches 2.				$f(x)$ approaches 2.		

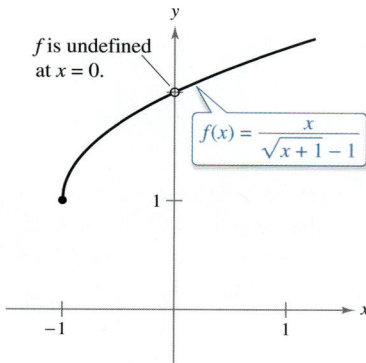

f is undefined at $x = 0$.

$$f(x) = \frac{x}{\sqrt{x+1} - 1}$$

The limit of $f(x)$ as x approaches 0 is 2.
Figure 1.6

From the results shown in the table, you can estimate the limit to be 2. This limit is reinforced by the graph of f (see Figure 1.6).

In Example 1, note that the function is undefined at $x = 0$, and yet $f(x)$ appears to be approaching a limit as x approaches 0. This often happens, and it is important to realize that *the existence or nonexistence of $f(x)$ at $x = c$ has no bearing on the existence of the limit of $f(x)$ as x approaches c.*

EXAMPLE 2 **Finding a Limit**

Find the limit of $f(x)$ as x approaches 2, where

$$f(x) = \begin{cases} 1, & x \neq 2 \\ 0, & x = 2 \end{cases}.$$

Solution Because $f(x) = 1$ for all x other than $x = 2$, you can estimate that the limit is 1, as shown in Figure 1.7. So, you can write

$$\lim_{x \to 2} f(x) = 1.$$

The fact that $f(2) = 0$ has no bearing on the existence or value of the limit as x approaches 2. For instance, as x approaches 2, the function

$$g(x) = \begin{cases} 1, & x \neq 2 \\ 2, & x = 2 \end{cases}$$

has the same limit as f.

$$f(x) = \begin{cases} 1, & x \neq 2 \\ 0, & x = 2 \end{cases}$$

The limit of $f(x)$ as x approaches 2 is 1.
Figure 1.7

So far in this section, you have been estimating limits numerically and graphically. Each of these approaches produces an estimate of the limit. In Section 1.3, you will study analytic techniques for evaluating limits. Throughout the course, try to develop a habit of using this three-pronged approach to problem solving.

1. Numerical approach   Construct a table of values.

2. Graphical approach   Draw a graph by hand or using technology.

3. Analytic approach   Use algebra or calculus.

Limits That Fail to Exist

In the next three examples, you will examine some limits that fail to exist.

EXAMPLE 3   **Different Right and Left Behavior**

Show that the limit $\lim\limits_{x \to 0} \dfrac{|x|}{x}$ does not exist.

Solution   Consider the graph of the function

$$f(x) = \frac{|x|}{x}.$$

In Figure 1.8 and from the definition of absolute value,

$$|x| = \begin{cases} x, & x \ge 0 \\ -x, & x < 0 \end{cases} \qquad \text{Definition of absolute value}$$

you can see that

$$\frac{|x|}{x} = \begin{cases} 1, & x > 0 \\ -1, & x < 0 \end{cases}.$$

So, no matter how close x gets to 0, there will be both positive and negative x-values that yield $f(x) = 1$ or $f(x) = -1$. Specifically, if δ (the lowercase Greek letter *delta*) is a positive number, then for x-values satisfying the inequality $0 < |x| < \delta$, you can classify the values of $|x|/x$ as

$$(-\delta, 0) \qquad \text{or} \qquad (0, \delta).$$

Negative x-values yield $|x|/x = -1$.

Positive x-values yield $|x|/x = 1$.

Because $|x|/x$ approaches a different number from the right side of 0 than it approaches from the left side, the limit $\lim\limits_{x \to 0} (|x|/x)$ does not exist.

EXAMPLE 4   **Unbounded Behavior**

Discuss the existence of the limit $\lim\limits_{x \to 0} \dfrac{1}{x^2}$.

Solution   Consider the graph of the function

$$f(x) = \frac{1}{x^2}.$$

In Figure 1.9, you can see that as x approaches 0 from either the right or the left, $f(x)$ increases without bound. This means that by choosing x close enough to 0, you can force $f(x)$ to be as large as you want. For instance, $f(x)$ will be greater than 100 when you choose x within $\frac{1}{10}$ of 0. That is,

$$0 < |x| < \frac{1}{10} \implies f(x) = \frac{1}{x^2} > 100.$$

Similarly, you can force $f(x)$ to be greater than 1,000,000, as shown.

$$0 < |x| < \frac{1}{1000} \implies f(x) = \frac{1}{x^2} > 1,000,000$$

Because $f(x)$ does not become arbitrarily close to a single number L as x approaches 0, you can conclude that the limit does not exist.

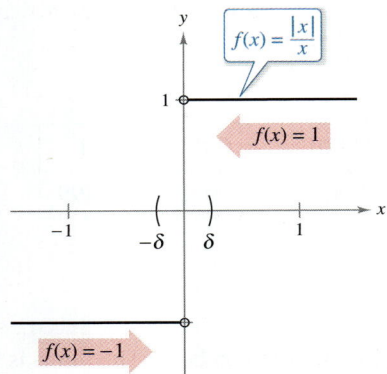

$\lim\limits_{x \to 0} f(x)$ does not exist.
Figure 1.8

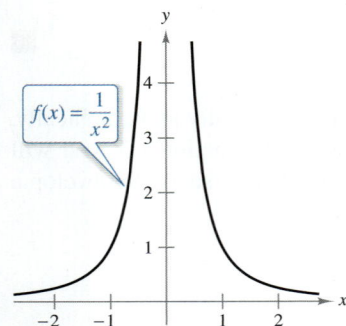

$\lim\limits_{x \to 0} f(x)$ does not exist.
Figure 1.9

EXAMPLE 5   **Oscillating Behavior**

····▷ *See LarsonCalculus.com for an interactive version of this type of example.*

Discuss the existence of the limit $\lim\limits_{x \to 0} \sin \dfrac{1}{x}$.

Solution   Let $f(x) = \sin(1/x)$. In Figure 1.10, you can see that as x approaches 0, $f(x)$ oscillates between -1 and 1. So, the limit does not exist because no matter how small you choose δ, it is possible to choose x_1 and x_2 within δ units of 0 such that $\sin(1/x_1) = 1$ and $\sin(1/x_2) = -1$, as shown in the table.

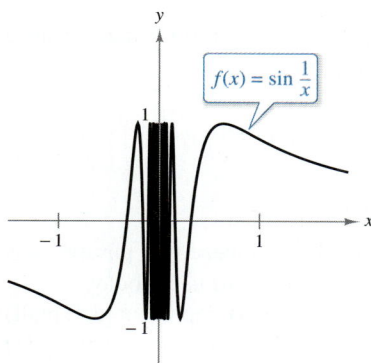

x	$\dfrac{2}{\pi}$	$\dfrac{2}{3\pi}$	$\dfrac{2}{5\pi}$	$\dfrac{2}{7\pi}$	$\dfrac{2}{9\pi}$	$\dfrac{2}{11\pi}$	$x \to 0$
$\sin \dfrac{1}{x}$	1	-1	1	-1	1	-1	Limit does not exist.

$\lim\limits_{x \to 0} f(x)$ does not exist.
Figure 1.10

Common Types of Behavior Associated with Nonexistence of a Limit

1. $f(x)$ approaches a different number from the right side of c than it approaches from the left side.

2. $f(x)$ increases or decreases without bound as x approaches c.

3. $f(x)$ oscillates between two fixed values as x approaches c.

There are many other interesting functions that have unusual limit behavior. An often cited one is the *Dirichlet function*

$$f(x) = \begin{cases} 0, & \text{if } x \text{ is rational} \\ 1, & \text{if } x \text{ is irrational} \end{cases}.$$

Because this function has *no limit* at any real number c, it is *not continuous* at any real number c. You will study continuity more closely in Section 1.4.

▷ **TECHNOLOGY PITFALL**   When you use a graphing utility to investigate the behavior of a function near the x-value at which you are trying to evaluate a limit, remember that you can't always trust the pictures that graphing utilities draw. When you use a graphing utility to graph the function in Example 5 over an interval containing 0, you will most likely obtain an incorrect graph such as that shown in Figure 1.11. The reason that a graphing utility can't show the correct graph is that the graph has infinitely many oscillations over any interval that contains 0.

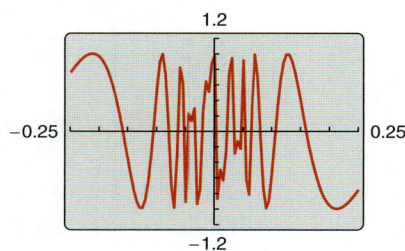

Incorrect graph of $f(x) = \sin(1/x)$
Figure 1.11

**PETER GUSTAV DIRICHLET
(1805–1859)**

In the early development of calculus, the definition of a function was much more restricted than it is today, and "functions" such as the Dirichlet function would not have been considered. The modern definition of function is attributed to the German mathematician Peter Gustav Dirichlet.
See LarsonCalculus.com to read more of this biography.

■ **FOR FURTHER INFORMATION**
For more on the introduction of
rigor to calculus, see "Who Gave
You the Epsilon? Cauchy and the
Origins of Rigorous Calculus"
by Judith V. Grabiner in *The
American Mathematical Monthly.*
To view this article, go to
MathArticles.com.

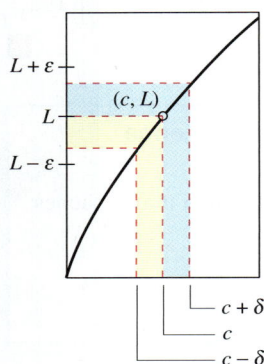

The ε-δ definition of the limit of $f(x)$
as x approaches c
Figure 1.12

A Formal Definition of Limit

Consider again the informal definition of limit. If $f(x)$ becomes arbitrarily close to a single number L as x approaches c from either side, then the limit of $f(x)$ as x approaches c is L, written as

$$\lim_{x \to c} f(x) = L.$$

At first glance, this definition looks fairly technical. Even so, it is informal because exact meanings have not yet been given to the two phrases

"$f(x)$ becomes arbitrarily close to L"

and

"x approaches c."

The first person to assign mathematically rigorous meanings to these two phrases was Augustin-Louis Cauchy. His **ε-δ definition of limit** is the standard used today.

In Figure 1.12, let ε (the lowercase Greek letter *epsilon*) represent a (small) positive number. Then the phrase "$f(x)$ becomes arbitrarily close to L" means that $f(x)$ lies in the interval $(L - \varepsilon, L + \varepsilon)$. Using absolute value, you can write this as

$$|f(x) - L| < \varepsilon.$$

Similarly, the phrase "x approaches c" means that there exists a positive number δ such that x lies in either the interval $(c - \delta, c)$ or the interval $(c, c + \delta)$. This fact can be concisely expressed by the double inequality

$$0 < |x - c| < \delta.$$

The first inequality

$$0 < |x - c| \qquad \text{The distance between } x \text{ and } c \text{ is more than 0.}$$

expresses the fact that $x \neq c$. The second inequality

$$|x - c| < \delta \qquad x \text{ is within } \delta \text{ units of } c.$$

says that x is within a distance δ of c.

> **Definition of Limit**
>
> Let f be a function defined on an open interval containing c (except possibly at c), and let L be a real number. The statement
>
> $$\lim_{x \to c} f(x) = L$$
>
> means that for each $\varepsilon > 0$ there exists a $\delta > 0$ such that if
>
> $$0 < |x - c| < \delta$$
>
> then
>
> $$|f(x) - L| < \varepsilon.$$

REMARK Throughout this text, the expression

$$\lim_{x \to c} f(x) = L$$

implies two statements—the limit exists *and* the limit is L.

Some functions do not have limits as x approaches c, but those that do cannot have two different limits as x approaches c. That is, *if the limit of a function exists, then the limit is unique* (see Exercise 75).

The next three examples should help you develop a better understanding of the ε-δ definition of limit.

EXAMPLE 6   Finding a δ for a Given ε

Given the limit

$$\lim_{x \to 3} (2x - 5) = 1$$

find δ such that

$$|(2x - 5) - 1| < 0.01$$

whenever

$$0 < |x - 3| < \delta.$$

• • REMARK In Example 6, note that 0.005 is the *largest* value of δ that will guarantee

$$|(2x - 5) - 1| < 0.01$$

whenever

$$0 < |x - 3| < \delta.$$

Any *smaller* positive value of δ would also work.

Solution In this problem, you are working with a given value of ε—namely, $\varepsilon = 0.01$. To find an appropriate δ, try to establish a connection between the absolute values

$$|(2x - 5) - 1| \quad \text{and} \quad |x - 3|.$$

Notice that

$$|(2x - 5) - 1| = |2x - 6| = 2|x - 3|.$$

Because the inequality $|(2x - 5) - 1| < 0.01$ is equivalent to $2|x - 3| < 0.01$, you can choose

$$\delta = \tfrac{1}{2}(0.01) = 0.005.$$

This choice works because

$$0 < |x - 3| < 0.005$$

implies that

$$|(2x - 5) - 1| = 2|x - 3| < 2(0.005) = 0.01.$$

As you can see in Figure 1.13, for x-values within 0.005 of 3 ($x \neq 3$), the values of $f(x)$ are within 0.01 of 1.

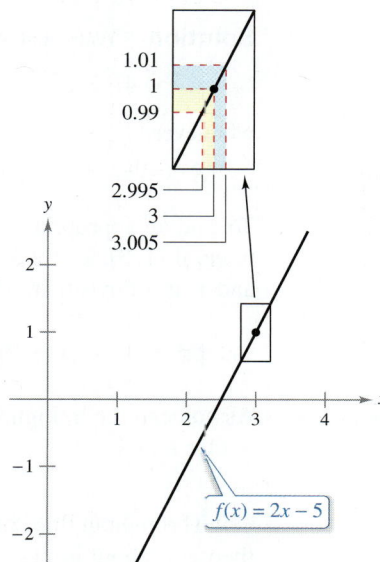

The limit of $f(x)$ as x approaches 3 is 1.
Figure 1.13

In Example 6, you found a δ-value for a *given* ε. This does not prove the existence of the limit. To do that, you must prove that you can find a δ for *any* ε, as shown in the next example.

EXAMPLE 7 **Using the ε-δ Definition of Limit**

Use the ε-δ definition of limit to prove that

$$\lim_{x \to 2} (3x - 2) = 4.$$

Solution You must show that for each ε > 0, there exists a δ > 0 such that

$$|(3x - 2) - 4| < \varepsilon$$

whenever

$$0 < |x - 2| < \delta.$$

Because your choice of δ depends on ε, you need to establish a connection between the absolute values $|(3x - 2) - 4|$ and $|x - 2|$.

$$|(3x - 2) - 4| = |3x - 6| = 3|x - 2|$$

So, for a given ε > 0, you can choose δ = ε/3. This choice works because

$$0 < |x - 2| < \delta = \frac{\varepsilon}{3}$$

implies that

$$|(3x - 2) - 4| = 3|x - 2| < 3\left(\frac{\varepsilon}{3}\right) = \varepsilon.$$

As you can see in Figure 1.14, for x-values within δ of 2 ($x \neq 2$), the values of $f(x)$ are within ε of 4.

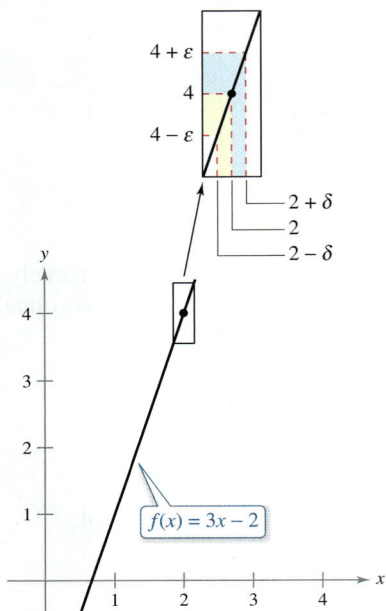

The limit of $f(x)$ as x approaches 2 is 4.
Figure 1.14

EXAMPLE 8 **Using the ε-δ Definition of Limit**

Use the ε-δ definition of limit to prove that

$$\lim_{x \to 2} x^2 = 4.$$

Solution You must show that for each ε > 0, there exists a δ > 0 such that

$$|x^2 - 4| < \varepsilon$$

whenever

$$0 < |x - 2| < \delta.$$

To find an appropriate δ, begin by writing $|x^2 - 4| = |x - 2||x + 2|$. For all x in the interval $(1, 3)$, $x + 2 < 5$ and thus $|x + 2| < 5$. So, letting δ be the minimum of ε/5 and 1, it follows that, whenever $0 < |x - 2| < \delta$, you have

$$|x^2 - 4| = |x - 2||x + 2| < \left(\frac{\varepsilon}{5}\right)(5) = \varepsilon.$$

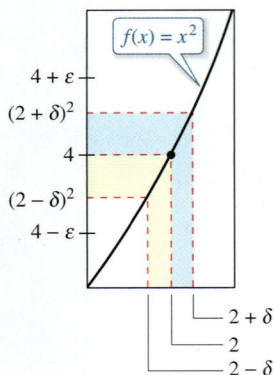

The limit of $f(x)$ as x approaches 2 is 4.
Figure 1.15

As you can see in Figure 1.15, for x-values within δ of 2 ($x \neq 2$), the values of $f(x)$ are within ε of 4.

Throughout this chapter, you will use the ε-δ definition of limit primarily to prove theorems about limits and to establish the existence or nonexistence of particular types of limits. For *finding* limits, you will learn techniques that are easier to use than the ε-δ definition of limit.

1.2 Exercises

See **CalcChat.com** for tutorial help and worked-out solutions to odd-numbered exercises.

Estimating a Limit Numerically In Exercises 1–6, complete the table and use the result to estimate the limit. Use a graphing utility to graph the function to confirm your result.

1. $\lim\limits_{x \to 4} \dfrac{x - 4}{x^2 - 3x - 4}$

x	3.9	3.99	3.999	4	4.001	4.01	4.1
$f(x)$				?			

2. $\lim\limits_{x \to 3} \dfrac{x - 3}{x^2 - 9}$

x	2.9	2.99	2.999	3	3.001	3.01	3.1
$f(x)$				?			

3. $\lim\limits_{x \to 0} \dfrac{\sqrt{x + 1} - 1}{x}$

x	−0.1	−0.01	−0.001	0	0.001	0.01	0.1
$f(x)$				?			

4. $\lim\limits_{x \to 3} \dfrac{[1/(x + 1)] - (1/4)}{x - 3}$

x	2.9	2.99	2.999	3	3.001	3.01	3.1
$f(x)$				?			

5. $\lim\limits_{x \to 0} \dfrac{\sin x}{x}$

x	−0.1	−0.01	−0.001	0	0.001	0.01	0.1
$f(x)$				?			

6. $\lim\limits_{x \to 0} \dfrac{\cos x - 1}{x}$

x	−0.1	−0.01	−0.001	0	0.001	0.01	0.1
$f(x)$				?			

Estimating a Limit Numerically In Exercises 7–14, create a table of values for the function and use the result to estimate the limit. Use a graphing utility to graph the function to confirm your result.

7. $\lim\limits_{x \to 1} \dfrac{x - 2}{x^2 + x - 6}$

8. $\lim\limits_{x \to -4} \dfrac{x + 4}{x^2 + 9x + 20}$

9. $\lim\limits_{x \to 1} \dfrac{x^4 - 1}{x^6 - 1}$

10. $\lim\limits_{x \to -3} \dfrac{x^3 + 27}{x + 3}$

11. $\lim\limits_{x \to -6} \dfrac{\sqrt{10 - x} - 4}{x + 6}$

12. $\lim\limits_{x \to 2} \dfrac{[x/(x + 1)] - (2/3)}{x - 2}$

13. $\lim\limits_{x \to 0} \dfrac{\sin 2x}{x}$

14. $\lim\limits_{x \to 0} \dfrac{\tan x}{\tan 2x}$

Finding a Limit Graphically In Exercises 15–22, use the graph to find the limit (if it exists). If the limit does not exist, explain why.

15. $\lim\limits_{x \to 3} (4 - x)$

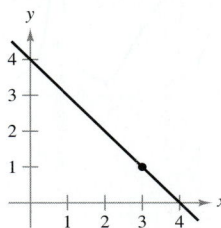

16. $\lim\limits_{x \to 0} \sec x$

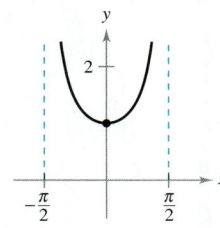

17. $\lim\limits_{x \to 2} f(x)$

$$f(x) = \begin{cases} 4 - x, & x \neq 2 \\ 0, & x = 2 \end{cases}$$

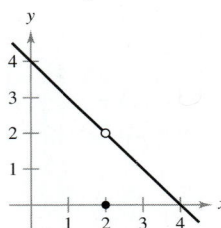

18. $\lim\limits_{x \to 1} f(x)$

$$f(x) = \begin{cases} x^2 + 3, & x \neq 1 \\ 2, & x = 1 \end{cases}$$

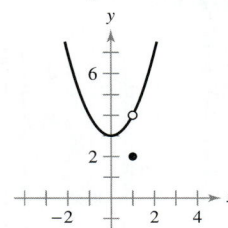

19. $\lim\limits_{x \to 2} \dfrac{|x - 2|}{x - 2}$

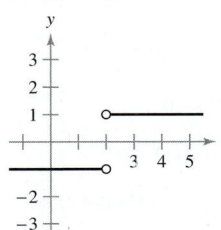

20. $\lim\limits_{x \to 5} \dfrac{2}{x - 5}$

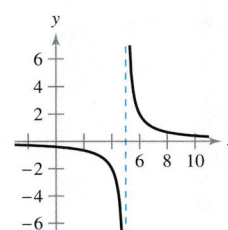

21. $\lim\limits_{x \to 0} \cos \dfrac{1}{x}$

22. $\lim\limits_{x \to \pi/2} \tan x$

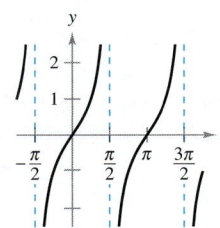

Graphical Reasoning In Exercises 23 and 24, use the graph of the function f to decide whether the value of the given quantity exists. If it does, find it. If not, explain why.

23. (a) $f(1)$

(b) $\lim_{x \to 1} f(x)$

(c) $f(4)$

(d) $\lim_{x \to 4} f(x)$

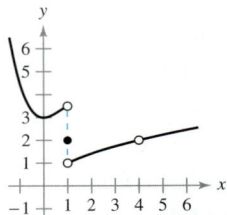

24. (a) $f(-2)$

(b) $\lim_{x \to -2} f(x)$

(c) $f(0)$

(d) $\lim_{x \to 0} f(x)$

(e) $f(2)$

(f) $\lim_{x \to 2} f(x)$

(g) $f(4)$

(h) $\lim_{x \to 4} f(x)$

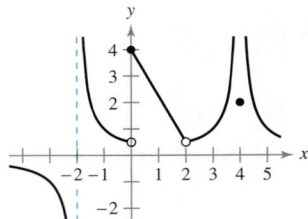

Limits of a Piecewise Function In Exercises 25 and 26, sketch the graph of f. Then identify the values of c for which $\lim_{x \to c} f(x)$ exists.

25. $f(x) = \begin{cases} x^2, & x \le 2 \\ 8 - 2x, & 2 < x < 4 \\ 4, & x \ge 4 \end{cases}$

26. $f(x) = \begin{cases} \sin x, & x < 0 \\ 1 - \cos x, & 0 \le x \le \pi \\ \cos x, & x > \pi \end{cases}$

Sketching a Graph In Exercises 27 and 28, sketch a graph of a function f that satisfies the given values. (There are many correct answers.)

27. $f(0)$ is undefined.

$\lim_{x \to 0} f(x) = 4$

$f(2) = 6$

$\lim_{x \to 2} f(x) = 3$

28. $f(-2) = 0$

$f(2) = 0$

$\lim_{x \to -2} f(x) = 0$

$\lim_{x \to 2} f(x)$ does not exist.

29. Finding a δ for a Given ε The graph of $f(x) = x + 1$ is shown in the figure. Find δ such that if $0 < |x - 2| < \delta$, then $|f(x) - 3| < 0.4$.

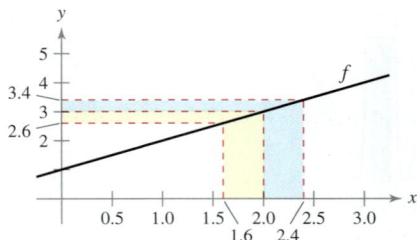

30. Finding a δ for a Given ε The graph of

$$f(x) = \frac{1}{x - 1}$$

is shown in the figure. Find δ such that if $0 < |x - 2| < \delta$, then $|f(x) - 1| < 0.01$.

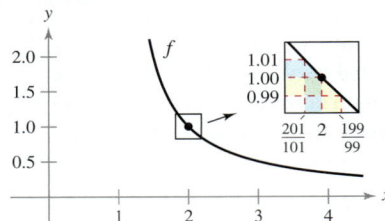

31. Finding a δ for a Given ε The graph of

$$f(x) = 2 - \frac{1}{x}$$

is shown in the figure. Find δ such that if $0 < |x - 1| < \delta$, then $|f(x) - 1| < 0.1$.

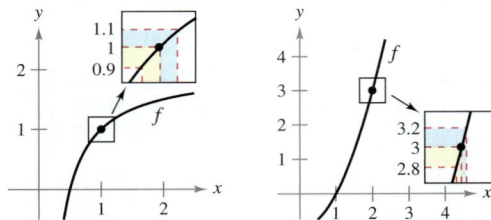

Figure for 31   Figure for 32

32. Finding a δ for a Given ε The graph of

$$f(x) = x^2 - 1$$

is shown in the figure. Find δ such that if $0 < |x - 2| < \delta$, then $|f(x) - 3| < 0.2$.

Finding a δ for a Given ε In Exercises 33–36, find the limit L. Then find $\delta > 0$ such that $|f(x) - L| < 0.01$ whenever $0 < |x - c| < \delta$.

33. $\lim_{x \to 2} (3x + 2)$

34. $\lim_{x \to 6} \left(6 - \frac{x}{3}\right)$

35. $\lim_{x \to 2} (x^2 - 3)$

36. $\lim_{x \to 4} (x^2 + 6)$

Using the ε-δ Definition of Limit In Exercises 37–48, find the limit L. Then use the ε-δ definition to prove that the limit is L.

37. $\lim_{x \to 4} (x + 2)$

38. $\lim_{x \to -2} (4x + 5)$

39. $\lim_{x \to -4} \left(\frac{1}{2}x - 1\right)$

40. $\lim_{x \to 3} \left(\frac{3}{4}x + 1\right)$

41. $\lim_{x \to 6} 3$

42. $\lim_{x \to 2} (-1)$

43. $\lim_{x \to 0} \sqrt[3]{x}$

44. $\lim_{x \to 4} \sqrt{x}$

45. $\lim_{x \to -5} |x - 5|$

46. $\lim_{x \to 3} |x - 3|$

47. $\lim_{x \to 1} (x^2 + 1)$

48. $\lim_{x \to -4} (x^2 + 4x)$

49. Finding a Limit What is the limit of $f(x) = 4$ as x approaches π?

50. Finding a Limit What is the limit of $g(x) = x$ as x approaches π?

Writing In Exercises 51–54, use a graphing utility to graph the function and estimate the limit (if it exists). What is the domain of the function? Can you detect a possible error in determining the domain of a function solely by analyzing the graph generated by a graphing utility? Write a short paragraph about the importance of examining a function analytically as well as graphically.

51. $f(x) = \dfrac{\sqrt{x+5} - 3}{x - 4}$

$\lim\limits_{x \to 4} f(x)$

52. $f(x) = \dfrac{x - 3}{x^2 - 4x + 3}$

$\lim\limits_{x \to 3} f(x)$

53. $f(x) = \dfrac{x - 9}{\sqrt{x} - 3}$

$\lim\limits_{x \to 9} f(x)$

54. $f(x) = \dfrac{x - 3}{x^2 - 9}$

$\lim\limits_{x \to 3} f(x)$

55. Modeling Data For a long distance phone call, a hotel charges $9.99 for the first minute and $0.79 for each additional minute or fraction thereof. A formula for the cost is given by

$$C(t) = 9.99 - 0.79 [\![-(t - 1)]\!]$$

where t is the time in minutes.

(*Note:* $[\![x]\!] = $ greatest integer n such that $n \leq x$. For example, $[\![3.2]\!] = 3$ and $[\![-1.6]\!] = -2$.)

(a) Use a graphing utility to graph the cost function for $0 < t \leq 6$.

(b) Use the graph to complete the table and observe the behavior of the function as t approaches 3.5. Use the graph and the table to find $\lim\limits_{t \to 3.5} C(t)$.

t	3	3.3	3.4	3.5	3.6	3.7	4
C				?			

(c) Use the graph to complete the table and observe the behavior of the function as t approaches 3.

t	2	2.5	2.9	3	3.1	3.5	4
C				?			

Does the limit of $C(t)$ as t approaches 3 exist? Explain.

56. Repeat Exercise 55 for

$$C(t) = 5.79 - 0.99 [\![-(t - 1)]\!].$$

WRITING ABOUT CONCEPTS

57. Describing Notation Write a brief description of the meaning of the notation

$$\lim\limits_{x \to 8} f(x) = 25.$$

58. Using the Definition of Limit The definition of limit on page 52 requires that f is a function defined on an open interval containing c, except possibly at c. Why is this requirement necessary?

59. Limits That Fail to Exist Identify three types of behavior associated with the nonexistence of a limit. Illustrate each type with a graph of a function.

60. Comparing Functions and Limits

(a) If $f(2) = 4$, can you conclude anything about the limit of $f(x)$ as x approaches 2? Explain your reasoning.

(b) If the limit of $f(x)$ as x approaches 2 is 4, can you conclude anything about $f(2)$? Explain your reasoning.

61. Jewelry A jeweler resizes a ring so that its inner circumference is 6 centimeters.

(a) What is the radius of the ring?

(b) The inner circumference of the ring varies between 5.5 centimeters and 6.5 centimeters. How does the radius vary?

(c) Use the ε-δ definition of limit to describe this situation. Identify ε and δ.

62. Sports

A sporting goods manufacturer designs a golf ball having a volume of 2.48 cubic inches.

(a) What is the radius of the golf ball?

(b) The volume of the golf ball varies between 2.45 cubic inches and 2.51 cubic inches. How does the radius vary?

(c) Use the ε-δ definition of limit to describe this situation. Identify ε and δ.

63. Estimating a Limit Consider the function

$$f(x) = (1 + x)^{1/x}.$$

Estimate

$$\lim\limits_{x \to 0} (1 + x)^{1/x}$$

by evaluating f at x-values near 0. Sketch the graph of f.

The symbol 〽 indicates an exercise in which you are instructed to use graphing technology or a symbolic computer algebra system. The solutions of other exercises may also be facilitated by the use of appropriate technology.

64. Estimating a Limit Consider the function

$$f(x) = \frac{|x + 1| - |x - 1|}{x}.$$

Estimate

$$\lim_{x \to 0} \frac{|x + 1| - |x - 1|}{x}$$

by evaluating f at x-values near 0. Sketch the graph of f.

65. Graphical Analysis The statement

$$\lim_{x \to 2} \frac{x^2 - 4}{x - 2} = 4$$

means that for each $\varepsilon > 0$ there corresponds a $\delta > 0$ such that if $0 < |x - 2| < \delta$, then

$$\left| \frac{x^2 - 4}{x - 2} - 4 \right| < \varepsilon.$$

If $\varepsilon = 0.001$, then

$$\left| \frac{x^2 - 4}{x - 2} - 4 \right| < 0.001.$$

Use a graphing utility to graph each side of this inequality. Use the *zoom* feature to find an interval $(2 - \delta, 2 + \delta)$ such that the graph of the left side is below the graph of the right side of the inequality.

66. HOW DO YOU SEE IT? Use the graph of f to identify the values of c for which $\lim_{x \to c} f(x)$ exists.

(a)

(b)

True or False? In Exercises 67–70, determine whether the statement is true or false. If it is false, explain why or give an example that shows it is false.

67. If f is undefined at $x = c$, then the limit of $f(x)$ as x approaches c does not exist.

68. If the limit of $f(x)$ as x approaches c is 0, then there must exist a number k such that $f(k) < 0.001$.

69. If $f(c) = L$, then $\lim_{x \to c} f(x) = L$.

70. If $\lim_{x \to c} f(x) = L$, then $f(c) = L$.

Determining a Limit In Exercises 71 and 72, consider the function $f(x) = \sqrt{x}$.

71. Is $\lim_{x \to 0.25} \sqrt{x} = 0.5$ a true statement? Explain.

72. Is $\lim_{x \to 0} \sqrt{x} = 0$ a true statement? Explain.

73. Evaluating Limits Use a graphing utility to evaluate

$$\lim_{x \to 0} \frac{\sin nx}{x}$$

for several values of n. What do you notice?

74. Evaluating Limits Use a graphing utility to evaluate

$$\lim_{x \to 0} \frac{\tan nx}{x}$$

for several values of n. What do you notice?

75. Proof Prove that if the limit of $f(x)$ as x approaches c exists, then the limit must be unique. [*Hint:* Let $\lim_{x \to c} f(x) = L_1$ and $\lim_{x \to c} f(x) = L_2$ and prove that $L_1 = L_2$.]

76. Proof Consider the line $f(x) = mx + b$, where $m \neq 0$. Use the ε-δ definition of limit to prove that $\lim_{x \to c} f(x) = mc + b$.

77. Proof Prove that

$$\lim_{x \to c} f(x) = L$$

is equivalent to

$$\lim_{x \to c} [f(x) - L] = 0.$$

78. Proof

(a) Given that

$$\lim_{x \to 0} (3x + 1)(3x - 1)x^2 + 0.01 = 0.01$$

prove that there exists an open interval (a, b) containing 0 such that $(3x + 1)(3x - 1)x^2 + 0.01 > 0$ for all $x \neq 0$ in (a, b).

(b) Given that $\lim_{x \to c} g(x) = L$, where $L > 0$, prove that there exists an open interval (a, b) containing c such that $g(x) > 0$ for all $x \neq c$ in (a, b).

PUTNAM EXAM CHALLENGE

79. Inscribe a rectangle of base b and height h in a circle of radius one, and inscribe an isosceles triangle in a region of the circle cut off by one base of the rectangle (with that side as the base of the triangle). For what value of h do the rectangle and triangle have the same area?

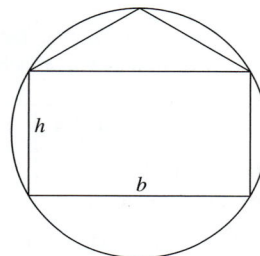

80. A right circular cone has base of radius 1 and height 3. A cube is inscribed in the cone so that one face of the cube is contained in the base of the cone. What is the side-length of the cube?

These problems were composed by the Committee on the Putnam Prize Competition.
© The Mathematical Association of America. All rights reserved.

1.3 Evaluating Limits Analytically

- ■ Evaluate a limit using properties of limits.
- ■ Develop and use a strategy for finding limits.
- ■ Evaluate a limit using the dividing out technique.
- ■ Evaluate a limit using the rationalizing technique.
- ■ Evaluate a limit using the Squeeze Theorem.

Properties of Limits

In Section 1.2, you learned that the limit of $f(x)$ as x approaches c does not depend on the value of f at $x = c$. It may happen, however, that the limit is precisely $f(c)$. In such cases, the limit can be evaluated by **direct substitution.** That is,

$$\lim_{x \to c} f(x) = f(c). \qquad \text{Substitute } c \text{ for } x.$$

Such *well-behaved* functions are **continuous at c.** You will examine this concept more closely in Section 1.4.

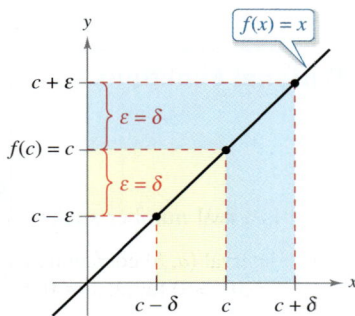

Figure 1.16

THEOREM 1.1 Some Basic Limits

Let b and c be real numbers, and let n be a positive integer.

1. $\lim_{x \to c} b = b$ 2. $\lim_{x \to c} x = c$ 3. $\lim_{x \to c} x^n = c^n$

Proof The proofs of Properties 1 and 3 of Theorem 1.1 are left as exercises (see Exercises 107 and 108). To prove Property 2, you need to show that for each $\varepsilon > 0$ there exists a $\delta > 0$ such that $|x - c| < \varepsilon$ whenever $0 < |x - c| < \delta$. To do this, choose $\delta = \varepsilon$. The second inequality then implies the first, as shown in Figure 1.16. *See LarsonCalculus.com for Bruce Edwards's video of this proof.*

> ● ● **REMARK** When encountering new notations or symbols in mathematics, be sure you know how the notations are read. For instance, the limit in Example 1(c) is read as "the limit of x^2 as x approaches 2 is 4."

EXAMPLE 1 Evaluating Basic Limits

a. $\lim_{x \to 2} 3 = 3$ **b.** $\lim_{x \to -4} x = -4$ **c.** $\lim_{x \to 2} x^2 = 2^2 = 4$

> ● ● **REMARK** The proof of Property 1 is left as an exercise (see Exercise 109).

THEOREM 1.2 Properties of Limits

Let b and c be real numbers, let n be a positive integer, and let f and g be functions with the limits

$$\lim_{x \to c} f(x) = L \quad \text{and} \quad \lim_{x \to c} g(x) = K.$$

1. Scalar multiple: $\lim_{x \to c} [b f(x)] = bL$

2. Sum or difference: $\lim_{x \to c} [f(x) \pm g(x)] = L \pm K$

3. Product: $\lim_{x \to c} [f(x)g(x)] = LK$

4. Quotient: $\lim_{x \to c} \dfrac{f(x)}{g(x)} = \dfrac{L}{K}, \quad K \neq 0$

5. Power: $\lim_{x \to c} [f(x)]^n = L^n$

A proof of this theorem is given in Appendix A. *See LarsonCalculus.com for Bruce Edwards's video of this proof.*

EXAMPLE 2   **The Limit of a Polynomial**

Find the limit: $\lim\limits_{x \to 2} (4x^2 + 3)$.

Solution

$$\lim_{x \to 2} (4x^2 + 3) = \lim_{x \to 2} 4x^2 + \lim_{x \to 2} 3 \qquad \text{Property 2, Theorem 1.2}$$

$$= 4 \left(\lim_{x \to 2} x^2 \right) + \lim_{x \to 2} 3 \qquad \text{Property 1, Theorem 1.2}$$

$$= 4(2^2) + 3 \qquad \text{Properties 1 and 3, Theorem 1.1}$$

$$= 19 \qquad \text{Simplify.}$$

In Example 2, note that the limit (as x approaches 2) of the *polynomial function* $p(x) = 4x^2 + 3$ is simply the value of p at $x = 2$.

$$\lim_{x \to 2} p(x) = p(2) = 4(2^2) + 3 = 19$$

This *direct substitution* property is valid for all polynomial and rational functions with nonzero denominators.

THEOREM 1.3   Limits of Polynomial and Rational Functions

If p is a polynomial function and c is a real number, then

$$\lim_{x \to c} p(x) = p(c).$$

If r is a rational function given by $r(x) = p(x)/q(x)$ and c is a real number such that $q(c) \neq 0$, then

$$\lim_{x \to c} r(x) = r(c) = \frac{p(c)}{q(c)}.$$

EXAMPLE 3   **The Limit of a Rational Function**

Find the limit: $\lim\limits_{x \to 1} \dfrac{x^2 + x + 2}{x + 1}$.

Solution   Because the denominator is not 0 when $x = 1$, you can apply Theorem 1.3 to obtain

$$\lim_{x \to 1} \frac{x^2 + x + 2}{x + 1} = \frac{1^2 + 1 + 2}{1 + 1} = \frac{4}{2} = 2.$$

Polynomial functions and rational functions are two of the three basic types of algebraic functions. The next theorem deals with the limit of the third type of algebraic function—one that involves a radical.

THE SQUARE ROOT SYMBOL

The first use of a symbol to denote the square root can be traced to the sixteenth century. Mathematicians first used the symbol $\sqrt{}$, which had only two strokes. This symbol was chosen because it resembled a lowercase r, to stand for the Latin word *radix*, meaning root.

THEOREM 1.4   The Limit of a Function Involving a Radical

Let n be a positive integer. The limit below is valid for all c when n is odd, and is valid for $c > 0$ when n is even.

$$\lim_{x \to c} \sqrt[n]{x} = \sqrt[n]{c}$$

A proof of this theorem is given in Appendix A.
See LarsonCalculus.com for Bruce Edwards's video of this proof.

The next theorem greatly expands your ability to evaluate limits because it shows how to analyze the limit of a composite function.

> **THEOREM 1.5 The Limit of a Composite Function**
>
> If f and g are functions such that $\lim_{x \to c} g(x) = L$ and $\lim_{x \to L} f(x) = f(L)$, then
>
> $$\lim_{x \to c} f(g(x)) = f\left(\lim_{x \to c} g(x)\right) = f(L).$$
>
> A proof of this theorem is given in Appendix A.
> *See LarsonCalculus.com for Bruce Edwards's video of this proof.*

EXAMPLE 4   The Limit of a Composite Function

⋯▷ *See LarsonCalculus.com for an interactive version of this type of example.*

Find the limit.

a. $\lim_{x \to 0} \sqrt{x^2 + 4}$ **b.** $\lim_{x \to 3} \sqrt[3]{2x^2 - 10}$

Solution

a. Because

$$\lim_{x \to 0} (x^2 + 4) = 0^2 + 4 = 4 \quad \text{and} \quad \lim_{x \to 4} \sqrt{x} = \sqrt{4} = 2$$

you can conclude that

$$\lim_{x \to 0} \sqrt{x^2 + 4} = \sqrt{4} = 2.$$

b. Because

$$\lim_{x \to 3} (2x^2 - 10) = 2(3^2) - 10 = 8 \quad \text{and} \quad \lim_{x \to 8} \sqrt[3]{x} = \sqrt[3]{8} = 2$$

you can conclude that

$$\lim_{x \to 3} \sqrt[3]{2x^2 - 10} = \sqrt[3]{8} = 2.$$

You have seen that the limits of many algebraic functions can be evaluated by direct substitution. The six basic trigonometric functions also exhibit this desirable quality, as shown in the next theorem (presented without proof).

> **THEOREM 1.6 Limits of Trigonometric Functions**
>
> Let c be a real number in the domain of the given trigonometric function.
>
> **1.** $\lim_{x \to c} \sin x = \sin c$ **2.** $\lim_{x \to c} \cos x = \cos c$ **3.** $\lim_{x \to c} \tan x = \tan c$
>
> **4.** $\lim_{x \to c} \cot x = \cot c$ **5.** $\lim_{x \to c} \sec x = \sec c$ **6.** $\lim_{x \to c} \csc x = \csc c$

EXAMPLE 5   Limits of Trigonometric Functions

a. $\lim_{x \to 0} \tan x = \tan(0) = 0$

b. $\lim_{x \to \pi} (x \cos x) = \left(\lim_{x \to \pi} x\right)\left(\lim_{x \to \pi} \cos x\right) = \pi \cos(\pi) = -\pi$

c. $\lim_{x \to 0} \sin^2 x = \lim_{x \to 0} (\sin x)^2 = 0^2 = 0$

A Strategy for Finding Limits

On the previous three pages, you studied several types of functions whose limits can be evaluated by direct substitution. This knowledge, together with the next theorem, can be used to develop a strategy for finding limits.

> **THEOREM 1.7   Functions That Agree at All but One Point**
>
> Let c be a real number, and let $f(x) = g(x)$ for all $x \neq c$ in an open interval containing c. If the limit of $g(x)$ as x approaches c exists, then the limit of $f(x)$ also exists and
>
> $$\lim_{x \to c} f(x) = \lim_{x \to c} g(x).$$
>
> A proof of this theorem is given in Appendix A.
> *See LarsonCalculus.com for Bruce Edwards's video of this proof.*

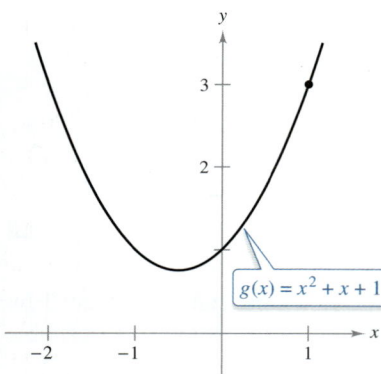

EXAMPLE 6   **Finding the Limit of a Function**

Find the limit.

$$\lim_{x \to 1} \frac{x^3 - 1}{x - 1}$$

Solution   Let $f(x) = (x^3 - 1)/(x - 1)$. By factoring and dividing out like factors, you can rewrite f as

$$f(x) = \frac{(x - 1)(x^2 + x + 1)}{(x - 1)} = x^2 + x + 1 = g(x), \quad x \neq 1.$$

So, for all x-values other than $x = 1$, the functions f and g agree, as shown in Figure 1.17. Because $\lim_{x \to 1} g(x)$ exists, you can apply Theorem 1.7 to conclude that f and g have the same limit at $x = 1$.

$$\lim_{x \to 1} \frac{x^3 - 1}{x - 1} = \lim_{x \to 1} \frac{(x - 1)(x^2 + x + 1)}{x - 1} \qquad \text{Factor.}$$

$$= \lim_{x \to 1} \frac{(x - 1)(x^2 + x + 1)}{x - 1} \qquad \text{Divide out like factors.}$$

$$= \lim_{x \to 1} (x^2 + x + 1) \qquad \text{Apply Theorem 1.7.}$$

$$= 1^2 + 1 + 1 \qquad \text{Use direct substitution.}$$

$$= 3 \qquad \text{Simplify.}$$

$f(x) = \dfrac{x^3 - 1}{x - 1}$

$g(x) = x^2 + x + 1$

f and g agree at all but one point.
Figure 1.17

• • **REMARK** When applying this strategy for finding a limit, remember that some functions do not have a limit (as x approaches c). For instance, the limit below does not exist.

$$\lim_{x \to 1} \frac{x^3 + 1}{x - 1}$$

A Strategy for Finding Limits

1. Learn to recognize which limits can be evaluated by direct substitution. (These limits are listed in Theorems 1.1 through 1.6.)

2. When the limit of $f(x)$ as x approaches c *cannot* be evaluated by direct substitution, try to find a function g that agrees with f for all x other than $x = c$. [Choose g such that the limit of $g(x)$ *can* be evaluated by direct substitution.] Then apply Theorem 1.7 to conclude *analytically* that

$$\lim_{x \to c} f(x) = \lim_{x \to c} g(x) = g(c).$$

3. Use a *graph* or *table* to reinforce your conclusion.

Dividing Out Technique

One procedure for finding a limit analytically is the **dividing out technique.** This technique involves dividing out common factors, as shown in Example 7.

EXAMPLE 7   Dividing Out Technique

∙∙∙▷ *See LarsonCalculus.com for an interactive version of this type of example.*

Find the limit: $\lim\limits_{x\to-3}\dfrac{x^2+x-6}{x+3}$.

Solution   Although you are taking the limit of a rational function, you *cannot* apply Theorem 1.3 because the limit of the denominator is 0.

$$\lim_{x\to-3}\frac{x^2+x-6}{x+3}$$

$$\lim_{x\to-3}(x^2+x-6)=0$$

Direct substitution fails.

$$\lim_{x\to-3}(x+3)=0$$

Because the limit of the numerator is also 0, the numerator and denominator have a *common factor* of $(x+3)$. So, for all $x\neq-3$, you can divide out this factor to obtain

$$f(x)=\frac{x^2+x-6}{x+3}=\frac{(x+3)(x-2)}{x+3}=x-2=g(x),\quad x\neq-3.$$

Using Theorem 1.7, it follows that

$$\lim_{x\to-3}\frac{x^2+x-6}{x+3}=\lim_{x\to-3}(x-2)\qquad\text{Apply Theorem 1.7.}$$

$$=-5.\qquad\text{Use direct substitution.}$$

This result is shown graphically in Figure 1.18. Note that the graph of the function f coincides with the graph of the function $g(x)=x-2$, except that the graph of f has a gap at the point $(-3,-5)$.

In Example 7, direct substitution produced the meaningless fractional form $0/0$. An expression such as $0/0$ is called an **indeterminate form** because you cannot (from the form alone) determine the limit. When you try to evaluate a limit and encounter this form, remember that you must rewrite the fraction so that the new denominator does not have 0 as its limit. One way to do this is to *divide out like factors.* Another way is to use the *rationalizing technique* shown on the next page.

> **REMARK** In the solution to Example 7, be sure you see the usefulness of the Factor Theorem of Algebra. This theorem states that if c is a zero of a polynomial function, then $(x-c)$ is a factor of the polynomial. So, when you apply direct substitution to a rational function and obtain
>
> $$r(c)=\frac{p(c)}{q(c)}=\frac{0}{0}$$
>
> you can conclude that $(x-c)$ must be a common factor of both $p(x)$ and $q(x)$.

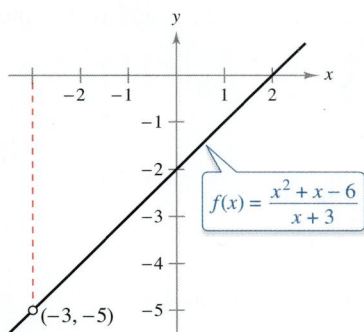

$f(x)=\dfrac{x^2+x-6}{x+3}$

$(-3,-5)$

f is undefined when $x=-3$.
Figure 1.18

▷ **TECHNOLOGY PITFALL** A graphing utility can give misleading information about the graph of a function. For instance, try graphing the function from Example 7

$$f(x)=\frac{x^2+x-6}{x+3}$$

on a standard viewing window (see Figure 1.19). On most graphing utilities, the graph appears to be defined at every real number. However, because f is undefined when $x=-3$, you know that the graph of f has a hole at $x=-3$. You can verify this on a graphing utility using the *trace* or *table* feature.

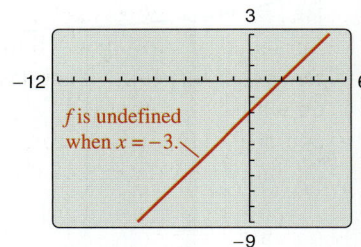

f is undefined when $x=-3$.

Misleading graph of f
Figure 1.19

Rationalizing Technique

Another way to find a limit analytically is the **rationalizing technique,** which involves rationalizing the numerator of a fractional expression. Recall that rationalizing the numerator means multiplying the numerator and denominator by the conjugate of the numerator. For instance, to rationalize the numerator of

$$\frac{\sqrt{x+4}}{x}$$

multiply the numerator and denominator by the conjugate of $\sqrt{x} + 4$, which is

$$\sqrt{x} - 4.$$

EXAMPLE 8 **Rationalizing Technique**

Find the limit: $\displaystyle\lim_{x \to 0} \frac{\sqrt{x+1}-1}{x}$.

Solution By direct substitution, you obtain the indeterminate form 0/0.

$$\lim_{x \to 0} \left(\sqrt{x+1}-1\right) = 0$$

$$\lim_{x \to 0} \frac{\sqrt{x+1}-1}{x}$$

Direct substitution fails.

$$\lim_{x \to 0} x = 0$$

In this case, you can rewrite the fraction by rationalizing the numerator.

$$\frac{\sqrt{x+1}-1}{x} = \left(\frac{\sqrt{x+1}-1}{x}\right)\left(\frac{\sqrt{x+1}+1}{\sqrt{x+1}+1}\right)$$

$$= \frac{(x+1)-1}{x\left(\sqrt{x+1}+1\right)}$$

$$= \frac{\cancel{x}}{\cancel{x}\left(\sqrt{x+1}+1\right)}$$

$$= \frac{1}{\sqrt{x+1}+1}, \quad x \neq 0$$

Now, using Theorem 1.7, you can evaluate the limit as shown.

$$\lim_{x \to 0} \frac{\sqrt{x+1}-1}{x} = \lim_{x \to 0} \frac{1}{\sqrt{x+1}+1}$$

$$= \frac{1}{1+1}$$

$$= \frac{1}{2}$$

A table or a graph can reinforce your conclusion that the limit is $\frac{1}{2}$. (See Figure 1.20.)

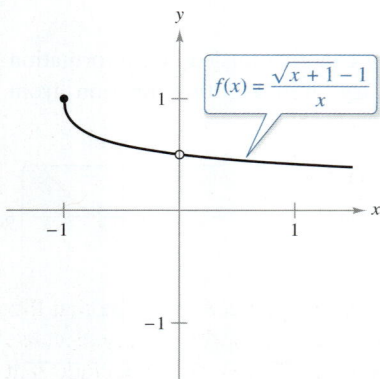

> •• **REMARK** The rationalizing technique for evaluating limits is based on multiplication by a convenient form of 1. In Example 8, the convenient form is
> $$1 = \frac{\sqrt{x+1}+1}{\sqrt{x+1}+1}.$$

$f(x) = \dfrac{\sqrt{x+1}-1}{x}$

The limit of $f(x)$ as x approaches 0 is $\frac{1}{2}$.
Figure 1.20

	x approaches 0 from the left.				x approaches 0 from the right.			
x	-0.25	-0.1	-0.01	-0.001	0	0.001	0.01	0.1	0.25
$f(x)$	0.5359	0.5132	0.5013	0.5001	?	0.4999	0.4988	0.4881	0.4721

$f(x)$ approaches 0.5.	$f(x)$ approaches 0.5.

The Squeeze Theorem

$h(x) \le f(x) \le g(x)$

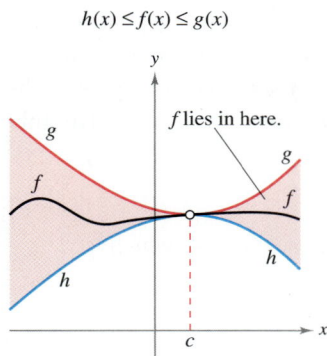

The Squeeze Theorem
Figure 1.21

The next theorem concerns the limit of a function that is squeezed between two other functions, each of which has the same limit at a given x-value, as shown in Figure 1.21.

THEOREM 1.8 The Squeeze Theorem

If $h(x) \le f(x) \le g(x)$ for all x in an open interval containing c, except possibly at c itself, and if

$$\lim_{x \to c} h(x) = L = \lim_{x \to c} g(x)$$

then $\lim_{x \to c} f(x)$ exists and is equal to L.

A proof of this theorem is given in Appendix A.

See LarsonCalculus.com for Bruce Edwards's video of this proof.

You can see the usefulness of the Squeeze Theorem (also called the Sandwich Theorem or the Pinching Theorem) in the proof of Theorem 1.9.

THEOREM 1.9 Two Special Trigonometric Limits

1. $\lim_{x \to 0} \dfrac{\sin x}{x} = 1$ **2.** $\lim_{x \to 0} \dfrac{1 - \cos x}{x} = 0$

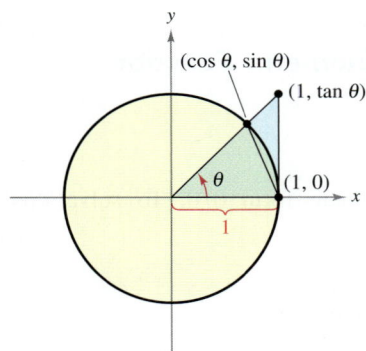

A circular sector is used to prove Theorem 1.9.
Figure 1.22

Proof The proof of the second limit is left as an exercise (see Exercise 121). To avoid the confusion of two different uses of x, the proof of the first limit is presented using the variable θ, where θ is an acute positive angle *measured in radians*. Figure 1.22 shows a circular sector that is squeezed between two triangles.

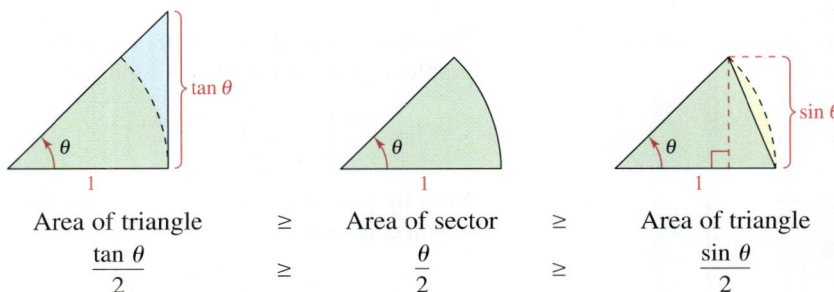

Area of triangle	$\ge$	Area of sector	$\ge$	Area of triangle
$\dfrac{\tan \theta}{2}$	$\ge$	$\dfrac{\theta}{2}$	$\ge$	$\dfrac{\sin \theta}{2}$

Multiplying each expression by $2/\sin \theta$ produces

$$\frac{1}{\cos \theta} \ge \frac{\theta}{\sin \theta} \ge 1$$

and taking reciprocals and reversing the inequalities yields

$$\cos \theta \le \frac{\sin \theta}{\theta} \le 1.$$

Because $\cos \theta = \cos(-\theta)$ and $(\sin \theta)/\theta = [\sin(-\theta)]/(-\theta)$, you can conclude that this inequality is valid for *all* nonzero θ in the open interval $(-\pi/2, \pi/2)$. Finally, because $\lim_{\theta \to 0} \cos \theta = 1$ and $\lim_{\theta \to 0} 1 = 1$, you can apply the Squeeze Theorem to conclude that $\lim_{\theta \to 0} (\sin \theta)/\theta = 1$. *See LarsonCalculus.com for Bruce Edwards's video of this proof.*

EXAMPLE 9 **A Limit Involving a Trigonometric Function**

Find the limit: $\displaystyle\lim_{x \to 0} \frac{\tan x}{x}$.

Solution Direct substitution yields the indeterminate form $0/0$. To solve this problem, you can write $\tan x$ as $(\sin x)/(\cos x)$ and obtain

$$\lim_{x \to 0} \frac{\tan x}{x} = \lim_{x \to 0} \left(\frac{\sin x}{x}\right)\left(\frac{1}{\cos x}\right).$$

Now, because

$$\lim_{x \to 0} \frac{\sin x}{x} = 1$$

and

$$\lim_{x \to 0} \frac{1}{\cos x} = 1$$

you can obtain

$$\lim_{x \to 0} \frac{\tan x}{x} = \left(\lim_{x \to 0} \frac{\sin x}{x}\right)\left(\lim_{x \to 0} \frac{1}{\cos x}\right)$$
$$= (1)(1)$$
$$= 1.$$

(See Figure 1.23.)

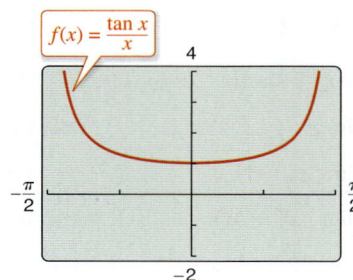

$f(x) = \dfrac{\tan x}{x}$

The limit of $f(x)$ as x approaches 0 is 1.
Figure 1.23

> • **REMARK** Be sure you understand the mathematical conventions regarding parentheses and trigonometric functions. For instance, in Example 10, $\sin 4x$ means $\sin(4x)$.

EXAMPLE 10 **A Limit Involving a Trigonometric Function**

Find the limit: $\displaystyle\lim_{x \to 0} \frac{\sin 4x}{x}$.

Solution Direct substitution yields the indeterminate form $0/0$. To solve this problem, you can rewrite the limit as

$$\lim_{x \to 0} \frac{\sin 4x}{x} = 4\left(\lim_{x \to 0} \frac{\sin 4x}{4x}\right). \qquad \text{Multiply and divide by 4.}$$

Now, by letting $y = 4x$ and observing that x approaches 0 if and only if y approaches 0, you can write

$$\lim_{x \to 0} \frac{\sin 4x}{x} = 4\left(\lim_{x \to 0} \frac{\sin 4x}{4x}\right)$$
$$= 4\left(\lim_{y \to 0} \frac{\sin y}{y}\right) \qquad \text{Let } y = 4x.$$
$$= 4(1) \qquad \text{Apply Theorem 1.9(1).}$$
$$= 4.$$

(See Figure 1.24.)

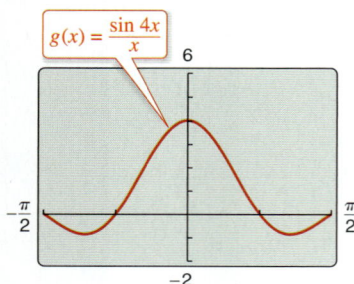

$g(x) = \dfrac{\sin 4x}{x}$

The limit of $g(x)$ as x approaches 0 is 4.
Figure 1.24

> ▷ **TECHNOLOGY** Use a graphing utility to confirm the limits in the examples and in the exercise set. For instance, Figures 1.23 and 1.24 show the graphs of
>
> $$f(x) = \frac{\tan x}{x} \quad \text{and} \quad g(x) = \frac{\sin 4x}{x}.$$
>
> Note that the first graph appears to contain the point $(0, 1)$ and the second graph appears to contain the point $(0, 4)$, which lends support to the conclusions obtained in Examples 9 and 10.

1.3 Exercises

See CalcChat.com for tutorial help and worked-out solutions to odd-numbered exercises.

Estimating Limits In Exercises 1–4, use a graphing utility to graph the function and visually estimate the limits.

1. $h(x) = -x^2 + 4x$

  (a) $\lim_{x \to 4} h(x)$

  (b) $\lim_{x \to -1} h(x)$

2. $g(x) = \dfrac{12(\sqrt{x} - 3)}{x - 9}$

  (a) $\lim_{x \to 4} g(x)$

  (b) $\lim_{x \to 9} g(x)$

3. $f(x) = x \cos x$

  (a) $\lim_{x \to 0} f(x)$

  (b) $\lim_{x \to \pi/3} f(x)$

4. $f(t) = t|t - 4|$

  (a) $\lim_{t \to 4} f(t)$

  (b) $\lim_{t \to -1} f(t)$

Finding a Limit In Exercises 5–22, find the limit.

5. $\lim_{x \to 2} x^3$

6. $\lim_{x \to -3} x^4$

7. $\lim_{x \to 0} (2x - 1)$

8. $\lim_{x \to -4} (2x + 3)$

9. $\lim_{x \to -3} (x^2 + 3x)$

10. $\lim_{x \to 2} (-x^3 + 1)$

11. $\lim_{x \to -3} (2x^2 + 4x + 1)$

12. $\lim_{x \to 1} (2x^3 - 6x + 5)$

13. $\lim_{x \to 3} \sqrt{x + 1}$

14. $\lim_{x \to 2} \sqrt[3]{12x + 3}$

15. $\lim_{x \to -4} (x + 3)^2$

16. $\lim_{x \to 0} (3x - 2)^4$

17. $\lim_{x \to 2} \dfrac{1}{x}$

18. $\lim_{x \to -5} \dfrac{5}{x + 3}$

19. $\lim_{x \to 1} \dfrac{x}{x^2 + 4}$

20. $\lim_{x \to 1} \dfrac{3x + 5}{x + 1}$

21. $\lim_{x \to 7} \dfrac{3x}{\sqrt{x + 2}}$

22. $\lim_{x \to 3} \dfrac{\sqrt{x + 6}}{x + 2}$

Finding Limits In Exercises 23–26, find the limits.

23. $f(x) = 5 - x, \ g(x) = x^3$

  (a) $\lim_{x \to 1} f(x)$ (b) $\lim_{x \to 4} g(x)$ (c) $\lim_{x \to 1} g(f(x))$

24. $f(x) = x + 7, \ g(x) = x^2$

  (a) $\lim_{x \to -3} f(x)$ (b) $\lim_{x \to 4} g(x)$ (c) $\lim_{x \to -3} g(f(x))$

25. $f(x) = 4 - x^2, \ g(x) = \sqrt{x + 1}$

  (a) $\lim_{x \to 1} f(x)$ (b) $\lim_{x \to 3} g(x)$ (c) $\lim_{x \to 1} g(f(x))$

26. $f(x) = 2x^2 - 3x + 1, \ g(x) = \sqrt[3]{x + 6}$

  (a) $\lim_{x \to 4} f(x)$ (b) $\lim_{x \to 21} g(x)$ (c) $\lim_{x \to 4} g(f(x))$

Finding a Limit of a Trigonometric Function In Exercises 27–36, find the limit of the trigonometric function.

27. $\lim_{x \to \pi/2} \sin x$

28. $\lim_{x \to \pi} \tan x$

29. $\lim_{x \to 1} \cos \dfrac{\pi x}{3}$

30. $\lim_{x \to 2} \sin \dfrac{\pi x}{2}$

31. $\lim_{x \to 0} \sec 2x$

32. $\lim_{x \to \pi} \cos 3x$

33. $\lim_{x \to 5\pi/6} \sin x$

34. $\lim_{x \to 5\pi/3} \cos x$

35. $\lim_{x \to 3} \tan\left(\dfrac{\pi x}{4}\right)$

36. $\lim_{x \to 7} \sec\left(\dfrac{\pi x}{6}\right)$

Evaluating Limits In Exercises 37–40, use the information to evaluate the limits.

37. $\lim_{x \to c} f(x) = 3$

  $\lim_{x \to c} g(x) = 2$

  (a) $\lim_{x \to c} [5g(x)]$

  (b) $\lim_{x \to c} [f(x) + g(x)]$

  (c) $\lim_{x \to c} [f(x)g(x)]$

  (d) $\lim_{x \to c} \dfrac{f(x)}{g(x)}$

38. $\lim_{x \to c} f(x) = 2$

  $\lim_{x \to c} g(x) = \frac{3}{4}$

  (a) $\lim_{x \to c} [4f(x)]$

  (b) $\lim_{x \to c} [f(x) + g(x)]$

  (c) $\lim_{x \to c} [f(x)g(x)]$

  (d) $\lim_{x \to c} \dfrac{f(x)}{g(x)}$

39. $\lim_{x \to c} f(x) = 4$

  (a) $\lim_{x \to c} [f(x)]^3$

  (b) $\lim_{x \to c} \sqrt{f(x)}$

  (c) $\lim_{x \to c} [3f(x)]$

  (d) $\lim_{x \to c} [f(x)]^{3/2}$

40. $\lim_{x \to c} f(x) = 27$

  (a) $\lim_{x \to c} \sqrt[3]{f(x)}$

  (b) $\lim_{x \to c} \dfrac{f(x)}{18}$

  (c) $\lim_{x \to c} [f(x)]^2$

  (d) $\lim_{x \to c} [f(x)]^{2/3}$

Finding a Limit In Exercises 41–46, write a simpler function that agrees with the given function at all but one point. Then find the limit of the function. Use a graphing utility to confirm your result.

41. $\lim_{x \to 0} \dfrac{x^2 + 3x}{x}$

42. $\lim_{x \to 0} \dfrac{x^4 - 5x^2}{x^2}$

43. $\lim_{x \to -1} \dfrac{x^2 - 1}{x + 1}$

44. $\lim_{x \to -2} \dfrac{3x^2 + 5x - 2}{x + 2}$

45. $\lim_{x \to 2} \dfrac{x^3 - 8}{x - 2}$

46. $\lim_{x \to -1} \dfrac{x^3 + 1}{x + 1}$

Finding a Limit In Exercises 47–62, find the limit.

47. $\lim_{x \to 0} \dfrac{x}{x^2 - x}$

48. $\lim_{x \to 0} \dfrac{2x}{x^2 + 4x}$

49. $\lim_{x \to 4} \dfrac{x - 4}{x^2 - 16}$

50. $\lim_{x \to 5} \dfrac{5 - x}{x^2 - 25}$

51. $\lim_{x \to -3} \dfrac{x^2 + x - 6}{x^2 - 9}$

52. $\lim_{x \to 2} \dfrac{x^2 + 2x - 8}{x^2 - x - 2}$

53. $\lim_{x \to 4} \dfrac{\sqrt{x + 5} - 3}{x - 4}$

54. $\lim_{x \to 3} \dfrac{\sqrt{x + 1} - 2}{x - 3}$

55. $\lim_{x \to 0} \dfrac{\sqrt{x + 5} - \sqrt{5}}{x}$

56. $\lim_{x \to 0} \dfrac{\sqrt{2 + x} - \sqrt{2}}{x}$

57. $\lim_{x \to 0} \dfrac{[1/(3 + x)] - (1/3)}{x}$

58. $\lim_{x \to 0} \dfrac{[1/(x + 4)] - (1/4)}{x}$

59. $\lim\limits_{\Delta x \to 0} \dfrac{2(x + \Delta x) - 2x}{\Delta x}$

60. $\lim\limits_{\Delta x \to 0} \dfrac{(x + \Delta x)^2 - x^2}{\Delta x}$

61. $\lim\limits_{\Delta x \to 0} \dfrac{(x + \Delta x)^2 - 2(x + \Delta x) + 1 - (x^2 - 2x + 1)}{\Delta x}$

62. $\lim\limits_{\Delta x \to 0} \dfrac{(x + \Delta x)^3 - x^3}{\Delta x}$

Finding a Limit of a Trigonometric Function In Exercises 63–74, find the limit of the trigonometric function.

63. $\lim\limits_{x \to 0} \dfrac{\sin x}{5x}$

64. $\lim\limits_{x \to 0} \dfrac{3(1 - \cos x)}{x}$

65. $\lim\limits_{x \to 0} \dfrac{\sin x(1 - \cos x)}{x^2}$

66. $\lim\limits_{\theta \to 0} \dfrac{\cos \theta \tan \theta}{\theta}$

67. $\lim\limits_{x \to 0} \dfrac{\sin^2 x}{x}$

68. $\lim\limits_{x \to 0} \dfrac{\tan^2 x}{x}$

69. $\lim\limits_{h \to 0} \dfrac{(1 - \cos h)^2}{h}$

70. $\lim\limits_{\phi \to \pi} \phi \sec \phi$

71. $\lim\limits_{x \to \pi/2} \dfrac{\cos x}{\cot x}$

72. $\lim\limits_{x \to \pi/4} \dfrac{1 - \tan x}{\sin x - \cos x}$

73. $\lim\limits_{t \to 0} \dfrac{\sin 3t}{2t}$

74. $\lim\limits_{x \to 0} \dfrac{\sin 2x}{\sin 3x}$ $\left[Hint: \text{ Find } \lim\limits_{x \to 0} \left(\dfrac{2 \sin 2x}{2x} \right) \left(\dfrac{3x}{3 \sin 3x} \right). \right]$

Graphical, Numerical, and Analytic Analysis In Exercises 75–82, use a graphing utility to graph the function and estimate the limit. Use a table to reinforce your conclusion. Then find the limit by analytic methods.

75. $\lim\limits_{x \to 0} \dfrac{\sqrt{x + 2} - \sqrt{2}}{x}$

76. $\lim\limits_{x \to 16} \dfrac{4 - \sqrt{x}}{x - 16}$

77. $\lim\limits_{x \to 0} \dfrac{[1/(2 + x)] - (1/2)}{x}$

78. $\lim\limits_{x \to 2} \dfrac{x^5 - 32}{x - 2}$

79. $\lim\limits_{t \to 0} \dfrac{\sin 3t}{t}$

80. $\lim\limits_{x \to 0} \dfrac{\cos x - 1}{2x^2}$

81. $\lim\limits_{x \to 0} \dfrac{\sin x^2}{x}$

82. $\lim\limits_{x \to 0} \dfrac{\sin x}{\sqrt[3]{x}}$

Finding a Limit In Exercises 83–88, find

$$\lim\limits_{\Delta x \to 0} \dfrac{f(x + \Delta x) - f(x)}{\Delta x}.$$

83. $f(x) = 3x - 2$

84. $f(x) = -6x + 3$

85. $f(x) = x^2 - 4x$

86. $f(x) = \sqrt{x}$

87. $f(x) = \dfrac{1}{x + 3}$

88. $f(x) = \dfrac{1}{x^2}$

Using the Squeeze Theorem In Exercises 89 and 90, use the Squeeze Theorem to find $\lim\limits_{x \to c} f(x)$.

89. $c = 0$

$4 - x^2 \leq f(x) \leq 4 + x^2$

90. $c = a$

$b - |x - a| \leq f(x) \leq b + |x - a|$

Using the Squeeze Theorem In Exercises 91–94, use a graphing utility to graph the given function and the equations $y = |x|$ and $y = -|x|$ in the same viewing window. Using the graphs to observe the Squeeze Theorem visually, find $\lim\limits_{x \to 0} f(x)$.

91. $f(x) = |x| \sin x$

92. $f(x) = |x| \cos x$

93. $f(x) = x \sin \dfrac{1}{x}$

94. $h(x) = x \cos \dfrac{1}{x}$

WRITING ABOUT CONCEPTS

95. Functions That Agree at All but One Point

(a) In the context of finding limits, discuss what is meant by two functions that agree at all but one point.

(b) Give an example of two functions that agree at all but one point.

96. Indeterminate Form What is meant by an indeterminate form?

97. Squeeze Theorem In your own words, explain the Squeeze Theorem.

---

98. HOW DO YOU SEE IT? Would you use the dividing out technique or the rationalizing technique to find the limit of the function? Explain your reasoning.

(a) $\lim\limits_{x \to -2} \dfrac{x^2 + x - 2}{x + 2}$

(b) $\lim\limits_{x \to 0} \dfrac{\sqrt{x + 4} - 2}{x}$

99. Writing Use a graphing utility to graph

$$f(x) = x, \quad g(x) = \sin x, \quad \text{and} \quad h(x) = \dfrac{\sin x}{x}$$

in the same viewing window. Compare the magnitudes of $f(x)$ and $g(x)$ when x is close to 0. Use the comparison to write a short paragraph explaining why

$$\lim\limits_{x \to 0} h(x) = 1.$$

100. Writing Use a graphing utility to graph

$$f(x) = x, \quad g(x) = \sin^2 x, \quad \text{and} \quad h(x) = \dfrac{\sin^2 x}{x}$$

in the same viewing window. Compare the magnitudes of $f(x)$ and $g(x)$ when x is close to 0. Use the comparison to write a short paragraph explaining why

$$\lim\limits_{x \to 0} h(x) = 0.$$

Kevin Fleming/Corbis

Free-Falling Object

In Exercises 101 and 102, use the position function $s(t) = -16t^2 + 500$, which gives the height (in feet) of an object that has fallen for t seconds from a height of 500 feet. The velocity at time $t = a$ seconds is given by

$$\lim_{t \to a} \frac{s(a) - s(t)}{a - t}.$$

101. A construction worker drops a full paint can from a height of 500 feet. How fast will the paint can be falling after 2 seconds?

102. A construction worker drops a full paint can from a height of 500 feet. When will the paint can hit the ground? At what velocity will the paint can impact the ground?

Free-Falling Object In Exercises 103 and 104, use the position function $s(t) = -4.9t^2 + 200$, which gives the height (in meters) of an object that has fallen for t seconds from a height of 200 meters. The velocity at time $t = a$ seconds is given by

$$\lim_{t \to a} \frac{s(a) - s(t)}{a - t}.$$

103. Find the velocity of the object when $t = 3$.

104. At what velocity will the object impact the ground?

105. Finding Functions Find two functions f and g such that $\lim_{x \to 0} f(x)$ and $\lim_{x \to 0} g(x)$ do not exist, but

$$\lim_{x \to 0} [f(x) + g(x)]$$

does exist.

106. Proof Prove that if $\lim_{x \to c} f(x)$ exists and $\lim_{x \to c} [f(x) + g(x)]$ does not exist, then $\lim_{x \to c} g(x)$ does not exist.

107. Proof Prove Property 1 of Theorem 1.1.

108. Proof Prove Property 3 of Theorem 1.1. (You may use Property 3 of Theorem 1.2.)

109. Proof Prove Property 1 of Theorem 1.2.

110. Proof Prove that if $\lim_{x \to c} f(x) = 0$, then $\lim_{x \to c} |f(x)| = 0$.

111. Proof Prove that if $\lim_{x \to c} f(x) = 0$ and $|g(x)| \le M$ for a fixed number M and all $x \ne c$, then $\lim_{x \to c} f(x)g(x) = 0$.

112. Proof

(a) Prove that if $\lim_{x \to c} |f(x)| = 0$, then $\lim_{x \to c} f(x) = 0$.

(*Note:* This is the converse of Exercise 110.)

(b) Prove that if $\lim_{x \to c} f(x) = L$, then $\lim_{x \to c} |f(x)| = |L|$.

[*Hint:* Use the inequality $||f(x)| - |L|| \le |f(x) - L|$.]

113. Think About It Find a function f to show that the converse of Exercise 112(b) is not true. [*Hint:* Find a function f such that $\lim_{x \to c} |f(x)| = |L|$ but $\lim_{x \to c} f(x)$ does not exist.]

114. Think About It When using a graphing utility to generate a table to approximate

$$\lim_{x \to 0} \frac{\sin x}{x}$$

a student concluded that the limit was 0.01745 rather than 1. Determine the probable cause of the error.

True or False? In Exercises 115–120, determine whether the statement is true or false. If it is false, explain why or give an example that shows it is false.

115. $\lim_{x \to 0} \dfrac{|x|}{x} = 1$

116. $\lim_{x \to \pi} \dfrac{\sin x}{x} = 1$

117. If $f(x) = g(x)$ for all real numbers other than $x = 0$, and $\lim_{x \to 0} f(x) = L$, then $\lim_{x \to 0} g(x) = L$.

118. If $\lim_{x \to c} f(x) = L$, then $f(c) = L$.

119. $\lim_{x \to 2} f(x) = 3$, where $f(x) = \begin{cases} 3, & x \le 2 \\ 0, & x > 2 \end{cases}$

120. If $f(x) < g(x)$ for all $x \ne a$, then $\lim_{x \to a} f(x) < \lim_{x \to a} g(x)$.

121. Proof Prove the second part of Theorem 1.9.

$$\lim_{x \to 0} \frac{1 - \cos x}{x} = 0$$

122. Piecewise Functions Let

$$f(x) = \begin{cases} 0, & \text{if } x \text{ is rational} \\ 1, & \text{if } x \text{ is irrational} \end{cases}$$

and

$$g(x) = \begin{cases} 0, & \text{if } x \text{ is rational} \\ x, & \text{if } x \text{ is irrational} \end{cases}.$$

Find (if possible) $\lim_{x \to 0} f(x)$ and $\lim_{x \to 0} g(x)$.

123. Graphical Reasoning Consider $f(x) = \dfrac{\sec x - 1}{x^2}$.

(a) Find the domain of f.

(b) Use a graphing utility to graph f. Is the domain of f obvious from the graph? If not, explain.

(c) Use the graph of f to approximate $\lim_{x \to 0} f(x)$.

(d) Confirm your answer to part (c) analytically.

124. Approximation

(a) Find $\lim_{x \to 0} \dfrac{1 - \cos x}{x^2}$.

(b) Use your answer to part (a) to derive the approximation $\cos x \approx 1 - \frac{1}{2}x^2$ for x near 0.

(c) Use your answer to part (b) to approximate $\cos(0.1)$.

(d) Use a calculator to approximate $\cos(0.1)$ to four decimal places. Compare the result with part (c).

1.4 Continuity and One-Sided Limits

- Determine continuity at a point and continuity on an open interval.
- Determine one-sided limits and continuity on a closed interval.
- Use properties of continuity.
- Understand and use the Intermediate Value Theorem.

Continuity at a Point and on an Open Interval

In mathematics, the term *continuous* has much the same meaning as it has in everyday usage. Informally, to say that a function f is continuous at $x = c$ means that there is no interruption in the graph of f at c. That is, its graph is unbroken at c, and there are no holes, jumps, or gaps. Figure 1.25 identifies three values of x at which the graph of f is *not* continuous. At all other points in the interval (a, b), the graph of f is uninterrupted and **continuous.**

Exploration

Informally, you might say that a function is *continuous* on an open interval when its graph can be drawn with a pencil without lifting the pencil from the paper. Use a graphing utility to graph each function on the given interval. From the graphs, which functions would you say are continuous on the interval? Do you think you can trust the results you obtained graphically? Explain your reasoning.

Function	Interval
a. $y = x^2 + 1$	$(-3, 3)$
b. $y = \dfrac{1}{x - 2}$	$(-3, 3)$
c. $y = \dfrac{\sin x}{x}$	$(-\pi, \pi)$
d. $y = \dfrac{x^2 - 4}{x + 2}$	$(-3, 3)$

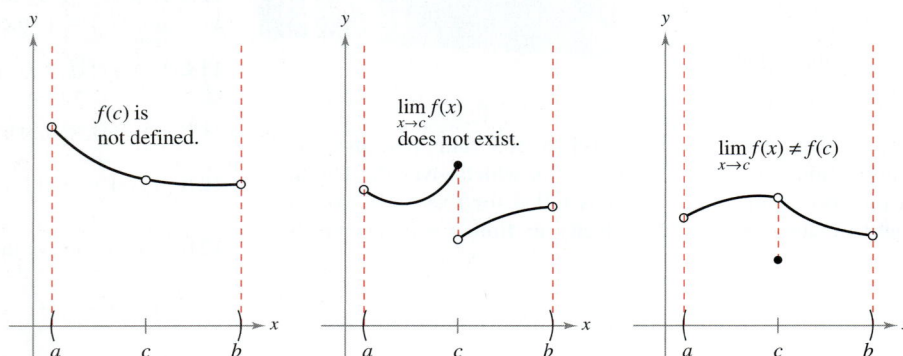

Three conditions exist for which the graph of f is not continuous at $x = c$.
Figure 1.25

In Figure 1.25, it appears that continuity at $x = c$ can be destroyed by any one of three conditions.

1. The function is not defined at $x = c$.
2. The limit of $f(x)$ does not exist at $x = c$.
3. The limit of $f(x)$ exists at $x = c$, but it is not equal to $f(c)$.

If *none* of the three conditions is true, then the function f is called **continuous at c,** as indicated in the important definition below.

■ **FOR FURTHER INFORMATION**
For more information on the concept of continuity, see the article "Leibniz and the Spell of the Continuous" by Hardy Grant in *The College Mathematics Journal*. To view this article, go to *MathArticles.com*.

> ### Definition of Continuity
>
> *Continuity at a Point*
> A function f is **continuous at c** when these three conditions are met.
>
> 1. $f(c)$ is defined.
> 2. $\lim\limits_{x \to c} f(x)$ exists.
> 3. $\lim\limits_{x \to c} f(x) = f(c)$
>
> *Continuity on an Open Interval*
> A function is **continuous on an open interval (a, b)** when the function is continuous at each point in the interval. A function that is continuous on the entire real number line $(-\infty, \infty)$ is **everywhere continuous.**

Consider an open interval I that contains a real number c. If a function f is defined on I (except possibly at c), and f is not continuous at c, then f is said to have a **discontinuity** at c. Discontinuities fall into two categories: **removable** and **nonremovable**. A discontinuity at c is called removable when f can be made continuous by appropriately defining (or redefining) $f(c)$. For instance, the functions shown in Figures 1.26(a) and (c) have removable discontinuities at c and the function shown in Figure 1.26(b) has a nonremovable discontinuity at c.

EXAMPLE 1   **Continuity of a Function**

Discuss the continuity of each function.

a. $f(x) = \dfrac{1}{x}$ **b.** $g(x) = \dfrac{x^2 - 1}{x - 1}$ **c.** $h(x) = \begin{cases} x + 1, & x \le 0 \\ x^2 + 1, & x > 0 \end{cases}$ **d.** $y = \sin x$

Solution

a. The domain of f is all nonzero real numbers. From Theorem 1.3, you can conclude that f is continuous at every x-value in its domain. At $x = 0$, f has a nonremovable discontinuity, as shown in Figure 1.27(a). In other words, there is no way to define $f(0)$ so as to make the function continuous at $x = 0$.

b. The domain of g is all real numbers except $x = 1$. From Theorem 1.3, you can conclude that g is continuous at every x-value in its domain. At $x = 1$, the function has a removable discontinuity, as shown in Figure 1.27(b). By defining $g(1)$ as 2, the "redefined" function is continuous for all real numbers.

c. The domain of h is all real numbers. The function h is continuous on $(-\infty, 0)$ and $(0, \infty)$, and, because

$$\lim_{x \to 0} h(x) = 1$$

h is continuous on the entire real number line, as shown in Figure 1.27(c).

d. The domain of y is all real numbers. From Theorem 1.6, you can conclude that the function is continuous on its entire domain, $(-\infty, \infty)$, as shown in Figure 1.27(d).

(a) Removable discontinuity

(b) Nonremovable discontinuity

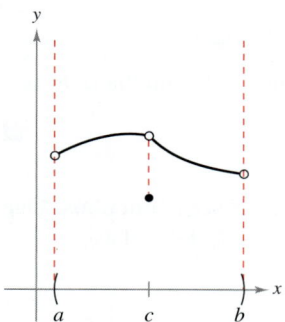

(c) Removable discontinuity

Figure 1.26

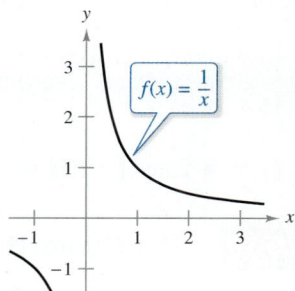

(a) Nonremovable discontinuity at $x = 0$

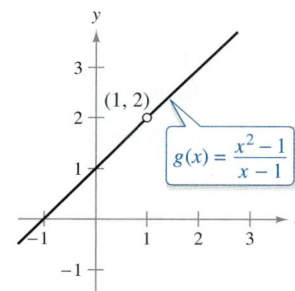

(b) Removable discontinuity at $x = 1$

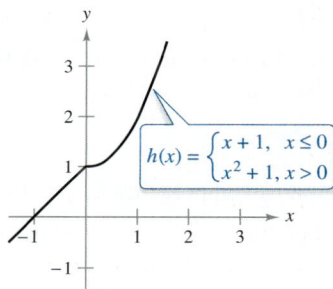

(c) Continuous on entire real number line

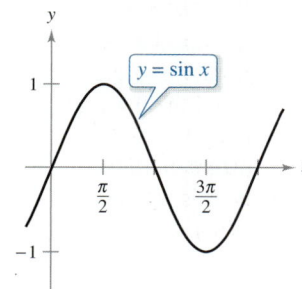

(d) Continuous on entire real number line

Figure 1.27

• REMARK Some people may refer to the function in Example 1(a) as "discontinuous." We have found that this terminology can be confusing. Rather than saying that the function is discontinuous, we prefer to say that it has a discontinuity at $x = 0$.

(a) Limit as x approaches c from the right.

(b) Limit as x approaches c from the left.
Figure 1.28

One-Sided Limits and Continuity on a Closed Interval

To understand continuity on a closed interval, you first need to look at a different type of limit called a **one-sided limit.** For instance, the **limit from the right** (or right-hand limit) means that x approaches c from values greater than c [see Figure 1.28(a)]. This limit is denoted as

$$\lim_{x \to c^+} f(x) = L. \qquad \text{Limit from the right}$$

Similarly, the **limit from the left** (or left-hand limit) means that x approaches c from values less than c [see Figure 1.28(b)]. This limit is denoted as

$$\lim_{x \to c^-} f(x) = L. \qquad \text{Limit from the left}$$

One-sided limits are useful in taking limits of functions involving radicals. For instance, if n is an even integer, then

$$\lim_{x \to 0^+} \sqrt[n]{x} = 0.$$

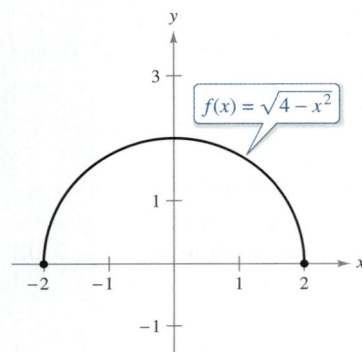

The limit of $f(x)$ as x approaches -2 from the right is 0.
Figure 1.29

EXAMPLE 2   A One-Sided Limit

Find the limit of $f(x) = \sqrt{4 - x^2}$ as x approaches -2 from the right.

Solution   As shown in Figure 1.29, the limit as x approaches -2 from the right is

$$\lim_{x \to -2^+} \sqrt{4 - x^2} = 0.$$

One-sided limits can be used to investigate the behavior of **step functions.** One common type of step function is the **greatest integer function** $[\![x]\!]$, defined as

$$[\![x]\!] = \text{greatest integer } n \text{ such that } n \le x. \qquad \text{Greatest integer function}$$

For instance, $[\![2.5]\!] = 2$ and $[\![-2.5]\!] = -3$.

EXAMPLE 3   The Greatest Integer Function

Find the limit of the greatest integer function $f(x) = [\![x]\!]$ as x approaches 0 from the left and from the right.

Solution   As shown in Figure 1.30, the limit as x approaches 0 *from the left* is

$$\lim_{x \to 0^-} [\![x]\!] = -1$$

and the limit as x approaches 0 *from the right* is

$$\lim_{x \to 0^+} [\![x]\!] = 0.$$

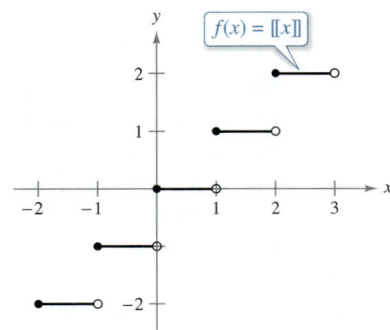

Greatest integer function
Figure 1.30

The greatest integer function has a discontinuity at zero because the left- and right-hand limits at zero are different. By similar reasoning, you can see that the greatest integer function has a discontinuity at any integer n.

When the limit from the left is not equal to the limit from the right, the (two-sided) limit *does not exist*. The next theorem makes this more explicit. The proof of this theorem follows directly from the definition of a one-sided limit.

THEOREM 1.10   The Existence of a Limit

Let f be a function, and let c and L be real numbers. The limit of $f(x)$ as x approaches c is L if and only if

$$\lim_{x \to c^-} f(x) = L \quad \text{and} \quad \lim_{x \to c^+} f(x) = L.$$

The concept of a one-sided limit allows you to extend the definition of continuity to closed intervals. Basically, a function is continuous on a closed interval when it is continuous in the interior of the interval and exhibits one-sided continuity at the endpoints. This is stated formally in the next definition.

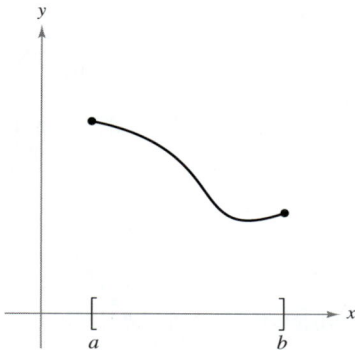

y

a   *b*   x

Continuous function on a closed interval
Figure 1.31

Definition of Continuity on a Closed Interval

A function f is **continuous on the closed interval $[a, b]$** when f is continuous on the open interval (a, b) and

$$\lim_{x \to a^+} f(x) = f(a)$$

and

$$\lim_{x \to b^-} f(x) = f(b).$$

The function f is **continuous from the right** at a and **continuous from the left** at b (see Figure 1.31).

Similar definitions can be made to cover continuity on intervals of the form $(a, b]$ and $[a, b)$ that are neither open nor closed, or on infinite intervals. For example,

$$f(x) = \sqrt{x}$$

is continuous on the infinite interval $[0, \infty)$, and the function

$$g(x) = \sqrt{2 - x}$$

is continuous on the infinite interval $(-\infty, 2]$.

EXAMPLE 4   Continuity on a Closed Interval

Discuss the continuity of

$$f(x) = \sqrt{1 - x^2}.$$

Solution The domain of f is the closed interval $[-1, 1]$. At all points in the open interval $(-1, 1)$, the continuity of f follows from Theorems 1.4 and 1.5. Moreover, because

$$\lim_{x \to -1^+} \sqrt{1 - x^2} = 0 = f(-1) \qquad \text{\color{red}Continuous from the right}$$

and

$$\lim_{x \to 1^-} \sqrt{1 - x^2} = 0 = f(1) \qquad \text{\color{red}Continuous from the left}$$

you can conclude that f is continuous on the closed interval $[-1, 1]$, as shown in Figure 1.32.

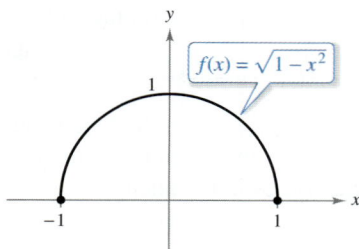

$f(x) = \sqrt{1 - x^2}$

f is continuous on $[-1, 1]$.
Figure 1.32

The next example shows how a one-sided limit can be used to determine the value of absolute zero on the Kelvin scale.

> **EXAMPLE 5**   **Charles's Law and Absolute Zero**

··REMARK   Charles's Law for gases (assuming constant pressure) can be stated as

$$V = kT$$

where V is volume, k is a constant, and T is temperature.

On the Kelvin scale, *absolute zero* is the temperature 0 K. Although temperatures very close to 0 K have been produced in laboratories, absolute zero has never been attained. In fact, evidence suggests that absolute zero *cannot* be attained. How did scientists determine that 0 K is the "lower limit" of the temperature of matter? What is absolute zero on the Celsius scale?

Solution   The determination of absolute zero stems from the work of the French physicist Jacques Charles (1746–1823). Charles discovered that the volume of gas at a constant pressure increases linearly with the temperature of the gas. The table illustrates this relationship between volume and temperature. To generate the values in the table, one mole of hydrogen is held at a constant pressure of one atmosphere. The volume V is approximated and is measured in liters, and the temperature T is measured in degrees Celsius.

T	-40	-20	0	20	40	60	80
V	19.1482	20.7908	22.4334	24.0760	25.7186	27.3612	29.0038

The points represented by the table are shown in Figure 1.33. Moreover, by using the points in the table, you can determine that T and V are related by the linear equation

$$V = 0.08213T + 22.4334.$$

Solving for T, you get an equation for the temperature of the gas.

$$T = \frac{V - 22.4334}{0.08213}$$

By reasoning that the volume of the gas can approach 0 (but can never equal or go below 0), you can determine that the "least possible temperature" is

$$\lim_{V \to 0^+} T = \lim_{V \to 0^+} \frac{V - 22.4334}{0.08213}$$

$$= \frac{0 - 22.4334}{0.08213} \qquad \text{Use direct substitution.}$$

$$\approx -273.15.$$

The volume of hydrogen gas depends on its temperature.

Figure 1.33

So, absolute zero on the Kelvin scale (0 K) is approximately $-273.15°$ on the Celsius scale.

In 2003, researchers at the Massachusetts Institute of Technology used lasers and evaporation to produce a super-cold gas in which atoms overlap. This gas is called a Bose-Einstein condensate. They measured a temperature of about 450 pK (picokelvin), or approximately $-273.14999999955°C$. *(Source: Science magazine, September 12, 2003)*

The table below shows the temperatures in Example 5 converted to the Fahrenheit scale. Try repeating the solution shown in Example 5 using these temperatures and volumes. Use the result to find the value of absolute zero on the Fahrenheit scale.

T	-40	-4	32	68	104	140	176
V	19.1482	20.7908	22.4334	24.0760	25.7186	27.3612	29.0038

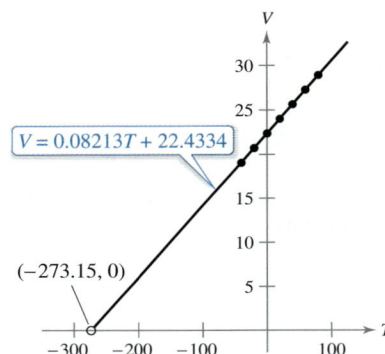

Properties of Continuity

In Section 1.3, you studied several properties of limits. Each of those properties yields a corresponding property pertaining to the continuity of a function. For instance, Theorem 1.11 follows directly from Theorem 1.2.

AUGUSTIN-LOUIS CAUCHY (1789–1857)

The concept of a continuous function was first introduced by Augustin-Louis Cauchy in 1821. The definition given in his text *Cours d'Analyse* stated that indefinite small changes in y were the result of indefinite small changes in x. "... $f(x)$ will be called a *continuous* function if ... the numerical values of the difference $f(x + \alpha) - f(x)$ decrease indefinitely with those of α ...."
See LarsonCalculus.com to read more of this biography.

> **THEOREM 1.11   Properties of Continuity**
>
> If b is a real number and f and g are continuous at $x = c$, then the functions listed below are also continuous at c.
>
> **1.** Scalar multiple: bf **2.** Sum or difference: $f \pm g$
>
> **3.** Product: fg **4.** Quotient: $\dfrac{f}{g}, \quad g(c) \neq 0$
>
> A proof of this theorem is given in Appendix A.
> *See LarsonCalculus.com for Bruce Edwards's video of this proof.*

It is important for you to be able to recognize functions that are continuous at every point in their domains. The list below summarizes the functions you have studied so far that are continuous at every point in their domains.

1. Polynomial: $p(x) = a_n x^n + a_{n-1} x^{n-1} + \cdots + a_1 x + a_0$

2. Rational: $r(x) = \dfrac{p(x)}{q(x)}, \quad q(x) \neq 0$

3. Radical: $f(x) = \sqrt[n]{x}$

4. Trigonometric: $\sin x, \cos x, \tan x, \cot x, \sec x, \csc x$

By combining Theorem 1.11 with this list, you can conclude that a wide variety of elementary functions are continuous at every point in their domains.

EXAMPLE 6   Applying Properties of Continuity

▷ *See LarsonCalculus.com for an interactive version of this type of example.*

By Theorem 1.11, it follows that each of the functions below is continuous at every point in its domain.

$$f(x) = x + \sin x, \quad f(x) = 3 \tan x, \quad f(x) = \frac{x^2 + 1}{\cos x}$$

The next theorem, which is a consequence of Theorem 1.5, allows you to determine the continuity of *composite* functions such as

$$f(x) = \sin 3x, \quad f(x) = \sqrt{x^2 + 1}, \quad \text{and} \quad f(x) = \tan \frac{1}{x}.$$

REMARK   One consequence of Theorem 1.12 is that when f and g satisfy the given conditions, you can determine the limit of $f(g(x))$ as x approaches c to be

$$\lim_{x \to c} f(g(x)) = f(g(c)).$$

> **THEOREM 1.12   Continuity of a Composite Function**
>
> If g is continuous at c and f is continuous at $g(c)$, then the composite function given by $(f \circ g)(x) = f(g(x))$ is continuous at c.

Proof By the definition of continuity, $\lim\limits_{x \to c} g(x) = g(c)$ and $\lim\limits_{x \to g(c)} f(x) = f(g(c))$.

Apply Theorem 1.5 with $L = g(c)$ to obtain $\lim\limits_{x \to c} f(g(x)) = f\left(\lim\limits_{x \to c} g(x)\right) = f(g(c))$. So, $(f \circ g)(x) = f(g(x))$ is continuous at c.

See LarsonCalculus.com for Bruce Edwards's video of this proof.

EXAMPLE 7   **Testing for Continuity**

Describe the interval(s) on which each function is continuous.

a. $f(x) = \tan x$ **b.** $g(x) = \begin{cases} \sin \dfrac{1}{x}, & x \neq 0 \\ 0, & x = 0 \end{cases}$ **c.** $h(x) = \begin{cases} x \sin \dfrac{1}{x}, & x \neq 0 \\ 0, & x = 0 \end{cases}$

Solution

a. The tangent function $f(x) = \tan x$ is undefined at

$$x = \frac{\pi}{2} + n\pi, \quad n \text{ is an integer.}$$

At all other points, f is continuous. So, $f(x) = \tan x$ is continuous on the open intervals

$$\cdots, \left(-\frac{3\pi}{2}, -\frac{\pi}{2}\right), \left(-\frac{\pi}{2}, \frac{\pi}{2}\right), \left(\frac{\pi}{2}, \frac{3\pi}{2}\right), \cdots$$

as shown in Figure 1.34(a).

b. Because $y = 1/x$ is continuous except at $x = 0$ and the sine function is continuous for all real values of x, it follows from Theorem 1.12 that

$$y = \sin \frac{1}{x}$$

is continuous at all real values except $x = 0$. At $x = 0$, the limit of $g(x)$ does not exist (see Example 5, Section 1.2). So, g is continuous on the intervals $(-\infty, 0)$ and $(0, \infty)$, as shown in Figure 1.34(b).

c. This function is similar to the function in part (b) except that the oscillations are damped by the factor x. Using the Squeeze Theorem, you obtain

$$-|x| \leq x \sin \frac{1}{x} \leq |x|, \quad x \neq 0$$

and you can conclude that

$$\lim_{x \to 0} h(x) = 0.$$

So, h is continuous on the entire real number line, as shown in Figure 1.34(c).

(a) f is continuous on each open interval in its domain.

$f(x) = \tan x$

(b) g is continuous on $(-\infty, 0)$ and $(0, \infty)$.

$g(x) = \begin{cases} \sin \dfrac{1}{x}, & x \neq 0 \\ 0, & x = 0 \end{cases}$

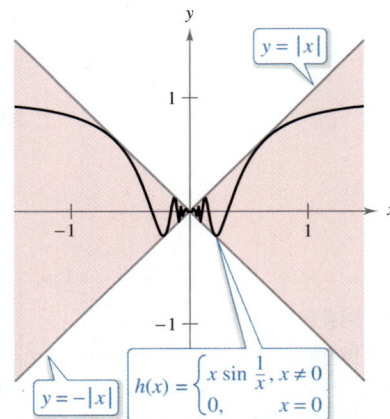

(c) h is continuous on the entire real number line.

$y = |x|$

$y = -|x|$

$h(x) = \begin{cases} x \sin \dfrac{1}{x}, & x \neq 0 \\ 0, & x = 0 \end{cases}$

Figure 1.34

The Intermediate Value Theorem

Theorem 1.13 is an important theorem concerning the behavior of functions that are continuous on a closed interval.

> **THEOREM 1.13** **Intermediate Value Theorem**
>
> If f is continuous on the closed interval $[a, b]$, $f(a) \neq f(b)$, and k is any number between $f(a)$ and $f(b)$, then there is at least one number c in $[a, b]$ such that
>
> $$f(c) = k.$$

REMARK The Intermediate Value Theorem tells you that at least one number c exists, but it does not provide a method for finding c. Such theorems are called **existence theorems.** By referring to a text on advanced calculus, you will find that a proof of this theorem is based on a property of real numbers called *completeness.* The Intermediate Value Theorem states that for a continuous function f, if x takes on all values between a and b, then $f(x)$ must take on all values between $f(a)$ and $f(b)$.

As an example of the application of the Intermediate Value Theorem, consider a person's height. A girl is 5 feet tall on her thirteenth birthday and 5 feet 7 inches tall on her fourteenth birthday. Then, for any height h between 5 feet and 5 feet 7 inches, there must have been a time t when her height was exactly h. This seems reasonable because human growth is continuous and a person's height does not abruptly change from one value to another.

The Intermediate Value Theorem guarantees the existence of *at least one* number c in the closed interval $[a, b]$. There may, of course, be more than one number c such that

$$f(c) = k$$

as shown in Figure 1.35. A function that is not continuous does not necessarily exhibit the intermediate value property. For example, the graph of the function shown in Figure 1.36 jumps over the horizontal line

$$y = k$$

and for this function there is no value of c in $[a, b]$ such that $f(c) = k$.

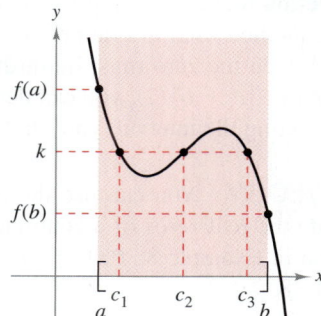

f is continuous on $[a, b]$.
[There exist three c's such that $f(c) = k$.]
Figure 1.35

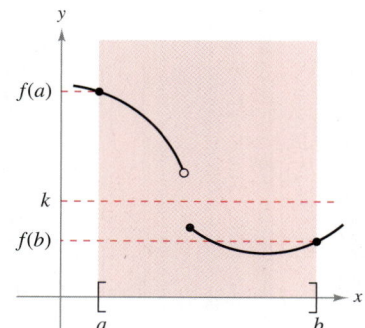

f is not continuous on $[a, b]$.
[There are no c's such that $f(c) = k$.]
Figure 1.36

The Intermediate Value Theorem often can be used to locate the zeros of a function that is continuous on a closed interval. Specifically, if f is continuous on $[a, b]$ and $f(a)$ and $f(b)$ differ in sign, then the Intermediate Value Theorem guarantees the existence of at least one zero of f in the closed interval $[a, b]$.

EXAMPLE 8   **An Application of the Intermediate Value Theorem**

Use the Intermediate Value Theorem to show that the polynomial function

$$f(x) = x^3 + 2x - 1$$

has a zero in the interval $[0, 1]$.

Solution   Note that f is continuous on the closed interval $[0, 1]$. Because

$$f(0) = 0^3 + 2(0) - 1 = -1 \quad \text{and} \quad f(1) = 1^3 + 2(1) - 1 = 2$$

it follows that $f(0) < 0$ and $f(1) > 0$. You can therefore apply the Intermediate Value Theorem to conclude that there must be some c in $[0, 1]$ such that

$$f(c) = 0 \qquad \text{\textit{f} has a zero in the closed interval } [0, 1].$$

as shown in Figure 1.37.

f is continuous on $[0, 1]$ with $f(0) < 0$ and $f(1) > 0$.
Figure 1.37

The **bisection method** for approximating the real zeros of a continuous function is similar to the method used in Example 8. If you know that a zero exists in the closed interval $[a, b]$, then the zero must lie in the interval $[a, (a + b)/2]$ or $[(a + b)/2, b]$. From the sign of $f([a + b]/2)$, you can determine which interval contains the zero. By repeatedly bisecting the interval, you can "close in" on the zero of the function.

▷ **TECHNOLOGY**   You can use the *root* or *zero* feature of a graphing utility to approximate the real zeros of a continuous function. Using this feature, the zero of the function in Example 8, $f(x) = x^3 + 2x - 1$, is approximately 0.453, as shown in Figure 1.38.

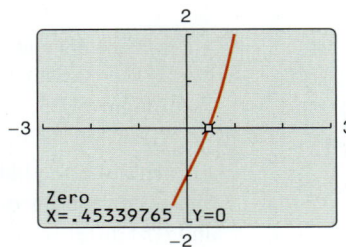

Zero of $f(x) = x^3 + 2x - 1$
Figure 1.38

1.4 Exercises

See **CalcChat.com** for tutorial help and worked-out solutions to odd-numbered exercises.

Limits and Continuity In Exercises 1–6, use the graph to determine the limit, and discuss the continuity of the function.

(a) $\lim\limits_{x \to c^+} f(x)$ (b) $\lim\limits_{x \to c^-} f(x)$ (c) $\lim\limits_{x \to c} f(x)$

1.

2.

3.

4.

5.

6.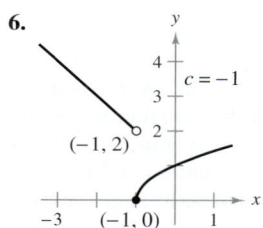

Finding a Limit In Exercises 7–26, find the limit (if it exists). If it does not exist, explain why.

7. $\lim\limits_{x \to 8^+} \dfrac{1}{x + 8}$

8. $\lim\limits_{x \to 2^-} \dfrac{2}{x + 2}$

9. $\lim\limits_{x \to 5^+} \dfrac{x - 5}{x^2 - 25}$

10. $\lim\limits_{x \to 4^+} \dfrac{4 - x}{x^2 - 16}$

11. $\lim\limits_{x \to -3^-} \dfrac{x}{\sqrt{x^2 - 9}}$

12. $\lim\limits_{x \to 4^-} \dfrac{\sqrt{x} - 2}{x - 4}$

13. $\lim\limits_{x \to 0^-} \dfrac{|x|}{x}$

14. $\lim\limits_{x \to 10^+} \dfrac{|x - 10|}{x - 10}$

15. $\lim\limits_{\Delta x \to 0^-} \dfrac{\dfrac{1}{x + \Delta x} - \dfrac{1}{x}}{\Delta x}$

16. $\lim\limits_{\Delta x \to 0^+} \dfrac{(x + \Delta x)^2 + x + \Delta x - (x^2 + x)}{\Delta x}$

17. $\lim\limits_{x \to 3^-} f(x)$, where $f(x) = \begin{cases} \dfrac{x + 2}{2}, & x \le 3 \\ \dfrac{12 - 2x}{3}, & x > 3 \end{cases}$

18. $\lim\limits_{x \to 3^-} f(x)$, where $f(x) = \begin{cases} x^2 - 4x + 6, & x < 3 \\ -x^2 + 4x - 2, & x \ge 3 \end{cases}$

19. $\lim\limits_{x \to 1} f(x)$, where $f(x) = \begin{cases} x^3 + 1, & x < 1 \\ x + 1, & x \ge 1 \end{cases}$

20. $\lim\limits_{x \to 1^+} f(x)$, where $f(x) = \begin{cases} x, & x \le 1 \\ 1 - x, & x > 1 \end{cases}$

21. $\lim\limits_{x \to \pi} \cot x$

22. $\lim\limits_{x \to \pi/2} \sec x$

23. $\lim\limits_{x \to 4^-} (5[\![x]\!] - 7)$

24. $\lim\limits_{x \to 2^+} (2x - [\![x]\!])$

25. $\lim\limits_{x \to 3} (2 - [\![-x]\!])$

26. $\lim\limits_{x \to 1} \left(1 - \left[\!\left[-\dfrac{x}{2}\right]\!\right]\right)$

Continuity of a Function In Exercises 27–30, discuss the continuity of each function.

27. $f(x) = \dfrac{1}{x^2 - 4}$

28. $f(x) = \dfrac{x^2 - 1}{x + 1}$

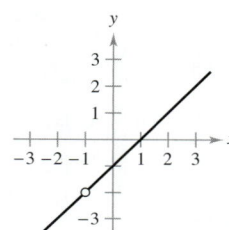

29. $f(x) = \frac{1}{2}[\![x]\!] - x$

30. $f(x) = \begin{cases} x, & x < 1 \\ 2, & x = 1 \\ 2x - 1, & x > 1 \end{cases}$

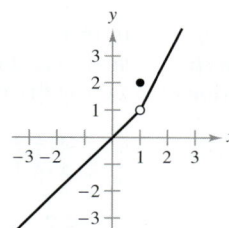

Continuity on a Closed Interval In Exercises 31–34, discuss the continuity of the function on the closed interval.

Function	Interval
31. $g(x) = \sqrt{49 - x^2}$	$[-7, 7]$
32. $f(t) = 3 - \sqrt{9 - t^2}$	$[-3, 3]$
33. $f(x) = \begin{cases} 3 - x, & x \le 0 \\ 3 + \frac{1}{2}x, & x > 0 \end{cases}$	$[-1, 4]$
34. $g(x) = \dfrac{1}{x^2 - 4}$	$[-1, 2]$

Removable and Nonremovable Discontinuities In Exercises 35–60, find the x-values (if any) at which f is not continuous. Which of the discontinuities are removable?

35. $f(x) = \dfrac{6}{x}$

36. $f(x) = \dfrac{4}{x - 6}$

37. $f(x) = x^2 - 9$

38. $f(x) = x^2 - 4x + 4$

39. $f(x) = \dfrac{1}{4 - x^2}$

40. $f(x) = \dfrac{1}{x^2 + 1}$

41. $f(x) = 3x - \cos x$

42. $f(x) = \cos \dfrac{\pi x}{2}$

43. $f(x) = \dfrac{x}{x^2 - x}$

44. $f(x) = \dfrac{x}{x^2 - 4}$

45. $f(x) = \dfrac{x}{x^2 + 1}$

46. $f(x) = \dfrac{x - 5}{x^2 - 25}$

47. $f(x) = \dfrac{x + 2}{x^2 - 3x - 10}$

48. $f(x) = \dfrac{x + 2}{x^2 - x - 6}$

49. $f(x) = \dfrac{|x + 7|}{x + 7}$

50. $f(x) = \dfrac{|x - 5|}{x - 5}$

51. $f(x) = \begin{cases} x, & x \le 1 \\ x^2, & x > 1 \end{cases}$

52. $f(x) = \begin{cases} -2x + 3, & x < 1 \\ x^2, & x \ge 1 \end{cases}$

53. $f(x) = \begin{cases} \frac{1}{2}x + 1, & x \le 2 \\ 3 - x, & x > 2 \end{cases}$

54. $f(x) = \begin{cases} -2x, & x \le 2 \\ x^2 - 4x + 1, & x > 2 \end{cases}$

55. $f(x) = \begin{cases} \tan \frac{\pi x}{4}, & |x| < 1 \\ x, & |x| \ge 1 \end{cases}$

56. $f(x) = \begin{cases} \csc \frac{\pi x}{6}, & |x - 3| \le 2 \\ 2, & |x - 3| > 2 \end{cases}$

57. $f(x) = \csc 2x$

58. $f(x) = \tan \dfrac{\pi x}{2}$

59. $f(x) = [\![x - 8]\!]$

60. $f(x) = 5 - [\![x]\!]$

Making a Function Continuous In Exercises 61–66, find the constant a, or the constants a and b, such that the function is continuous on the entire real number line.

61. $f(x) = \begin{cases} 3x^2, & x \ge 1 \\ ax - 4, & x < 1 \end{cases}$

62. $f(x) = \begin{cases} 3x^3, & x \le 1 \\ ax + 5, & x > 1 \end{cases}$

63. $f(x) = \begin{cases} x^3, & x \le 2 \\ ax^2, & x > 2 \end{cases}$

64. $g(x) = \begin{cases} \dfrac{4 \sin x}{x}, & x < 0 \\ a - 2x, & x \ge 0 \end{cases}$

65. $f(x) = \begin{cases} 2, & x \le -1 \\ ax + b, & -1 < x < 3 \\ -2, & x \ge 3 \end{cases}$

66. $g(x) = \begin{cases} \dfrac{x^2 - a^2}{x - a}, & x \ne a \\ 8, & x = a \end{cases}$

Continuity of a Composite Function In Exercises 67–72, discuss the continuity of the composite function $h(x) = f(g(x))$.

67. $f(x) = x^2$

$g(x) = x - 1$

68. $f(x) = 5x + 1$

$g(x) = x^3$

69. $f(x) = \dfrac{1}{x - 6}$

$g(x) = x^2 + 5$

70. $f(x) = \dfrac{1}{\sqrt{x}}$

$g(x) = x - 1$

71. $f(x) = \tan x$

$g(x) = \dfrac{x}{2}$

72. $f(x) = \sin x$

$g(x) = x^2$

Finding Discontinuities In Exercises 73–76, use a graphing utility to graph the function. Use the graph to determine any x-values at which the function is not continuous.

73. $f(x) = [\![x]\!] - x$

74. $h(x) = \dfrac{1}{x^2 + 2x - 15}$

75. $g(x) = \begin{cases} x^2 - 3x, & x > 4 \\ 2x - 5, & x \le 4 \end{cases}$

76. $f(x) = \begin{cases} \dfrac{\cos x - 1}{x}, & x < 0 \\ 5x, & x \ge 0 \end{cases}$

Testing for Continuity In Exercises 77–84, describe the interval(s) on which the function is continuous.

77. $f(x) = \dfrac{x}{x^2 + x + 2}$

78. $f(x) = \dfrac{x + 1}{\sqrt{x}}$

79. $f(x) = 3 - \sqrt{x}$

80. $f(x) = x\sqrt{x + 3}$

81. $f(x) = \sec \dfrac{\pi x}{4}$

82. $f(x) = \cos \dfrac{1}{x}$

83. $f(x) = \begin{cases} \dfrac{x^2 - 1}{x - 1}, & x \ne 1 \\ 2, & x = 1 \end{cases}$

84. $f(x) = \begin{cases} 2x - 4, & x \ne 3 \\ 1, & x = 3 \end{cases}$

Writing In Exercises 85 and 86, use a graphing utility to graph the function on the interval $[-4, 4]$. Does the graph of the function appear to be continuous on this interval? Is the function continuous on $[-4, 4]$? Write a short paragraph about the importance of examining a function analytically as well as graphically.

85. $f(x) = \dfrac{\sin x}{x}$

86. $f(x) = \dfrac{x^3 - 8}{x - 2}$

Writing In Exercises 87–90, explain why the function has a zero in the given interval.

Function	Interval
87. $f(x) = \frac{1}{12}x^4 - x^3 + 4$	$[1, 2]$
88. $f(x) = x^3 + 5x - 3$	$[0, 1]$
89. $f(x) = x^2 - 2 - \cos x$	$[0, \pi]$
90. $f(x) = -\dfrac{5}{x} + \tan\left(\dfrac{\pi x}{10}\right)$	$[1, 4]$

Using the Intermediate Value Theorem In Exercises 91–94, use the Intermediate Value Theorem and a graphing utility to approximate the zero of the function in the interval $[0, 1]$. Repeatedly "zoom in" on the graph of the function to approximate the zero accurate to two decimal places. Use the *zero* or *root* feature of the graphing utility to approximate the zero accurate to four decimal places.

91. $f(x) = x^3 + x - 1$

92. $f(x) = x^4 - x^2 + 3x - 1$

93. $g(t) = 2 \cos t - 3t$

94. $h(\theta) = \tan \theta + 3\theta - 4$

Using the Intermediate Value Theorem In Exercises 95–98, verify that the Intermediate Value Theorem applies to the indicated interval and find the value of c guaranteed by the theorem.

95. $f(x) = x^2 + x - 1$, $[0, 5]$, $f(c) = 11$

96. $f(x) = x^2 - 6x + 8$, $[0, 3]$, $f(c) = 0$

97. $f(x) = x^3 - x^2 + x - 2$, $[0, 3]$, $f(c) = 4$

98. $f(x) = \dfrac{x^2 + x}{x - 1}$, $\left[\dfrac{5}{2}, 4\right]$, $f(c) = 6$

WRITING ABOUT CONCEPTS

99. Using the Definition of Continuity State how continuity is destroyed at $x = c$ for each of the following graphs.

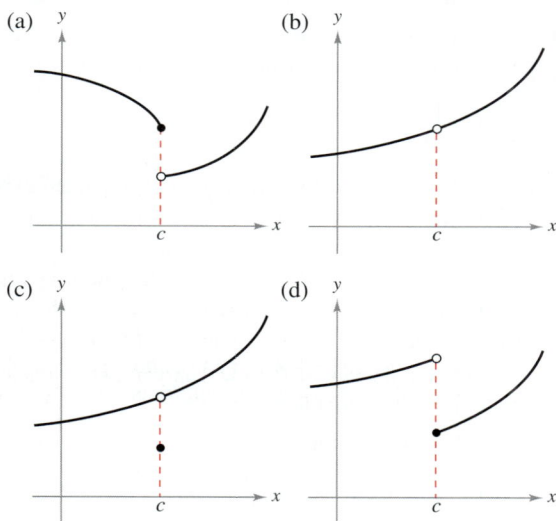

(a)

(b)

(c)

(d)

100. Sketching a Graph Sketch the graph of any function f such that

$$\lim_{x \to 3^+} f(x) = 1 \quad \text{and} \quad \lim_{x \to 3^-} f(x) = 0.$$

Is the function continuous at $x = 3$? Explain.

101. Continuity of Combinations of Functions If the functions f and g are continuous for all real x, is $f + g$ always continuous for all real x? Is f/g always continuous for all real x? If either is not continuous, give an example to verify your conclusion.

102. Removable and Nonremovable Discontinuities Describe the difference between a discontinuity that is removable and one that is nonremovable. In your explanation, give examples of the following descriptions.

(a) A function with a nonremovable discontinuity at $x = 4$

(b) A function with a removable discontinuity at $x = -4$

(c) A function that has both of the characteristics described in parts (a) and (b)

True or False? In Exercises 103–106, determine whether the statement is true or false. If it is false, explain why or give an example that shows it is false.

103. If $\lim\limits_{x \to c} f(x) = L$ and $f(c) = L$, then f is continuous at c.

104. If $f(x) = g(x)$ for $x \neq c$ and $f(c) \neq g(c)$, then either f or g is not continuous at c.

105. A rational function can have infinitely many x-values at which it is not continuous.

106. The function

$$f(x) = \frac{|x - 1|}{x - 1}$$

is continuous on $(-\infty, \infty)$.

107. Think About It Describe how the functions

$$f(x) = 3 + [\![x]\!] \quad \text{and} \quad g(x) = 3 - [\![-x]\!]$$

differ.

108. HOW DO YOU SEE IT? Every day you dissolve 28 ounces of chlorine in a swimming pool. The graph shows the amount of chlorine $f(t)$ in the pool after t days. Estimate and interpret $\lim\limits_{t \to 4^-} f(t)$ and $\lim\limits_{t \to 4^+} f(t)$.

109. Telephone Charges A long distance phone service charges $0.40 for the first 10 minutes and $0.05 for each additional minute or fraction thereof. Use the greatest integer function to write the cost C of a call in terms of time t (in minutes). Sketch the graph of this function and discuss its continuity.

110. Inventory Management

The number of units in inventory in a small company is given by

$$N(t) = 25\left(2\left[\!\left[\frac{t + 2}{2}\right]\!\right] - t\right)$$

where t is the time in months. Sketch the graph of this function and discuss its continuity. How often must this company replenish its inventory?

111. Déjà Vu At 8:00 A.M. on Saturday, a man begins running up the side of a mountain to his weekend campsite (see figure). On Sunday morning at 8:00 A.M., he runs back down the mountain. It takes him 20 minutes to run up, but only 10 minutes to run down. At some point on the way down, he realizes that he passed the same place at exactly the same time on Saturday. Prove that he is correct. [*Hint:* Let $s(t)$ and $r(t)$ be the position functions for the runs up and down, and apply the Intermediate Value Theorem to the function $f(t) = s(t) - r(t)$.]

Saturday 8:00 A.M.   Sunday 8:00 A.M.

Not drawn to scale

112. Volume Use the Intermediate Value Theorem to show that for all spheres with radii in the interval $[5, 8]$, there is one with a volume of 1500 cubic centimeters.

113. Proof Prove that if f is continuous and has no zeros on $[a, b]$, then either

$$f(x) > 0 \text{ for all } x \text{ in } [a, b] \quad \text{or} \quad f(x) < 0 \text{ for all } x \text{ in } [a, b].$$

114. Dirichlet Function Show that the Dirichlet function

$$f(x) = \begin{cases} 0, & \text{if } x \text{ is rational} \\ 1, & \text{if } x \text{ is irrational} \end{cases}$$

is not continuous at any real number.

115. Continuity of a Function Show that the function

$$f(x) = \begin{cases} 0, & \text{if } x \text{ is rational} \\ kx, & \text{if } x \text{ is irrational} \end{cases}$$

is continuous only at $x = 0$. (Assume that k is any nonzero real number.)

116. Signum Function The **signum function** is defined by

$$\text{sgn}(x) = \begin{cases} -1, & x < 0 \\ 0, & x = 0. \\ 1, & x > 0 \end{cases}$$

Sketch a graph of $\text{sgn}(x)$ and find the following (if possible).

(a) $\lim\limits_{x \to 0^-} \text{sgn}(x)$ (b) $\lim\limits_{x \to 0^+} \text{sgn}(x)$ (c) $\lim\limits_{x \to 0} \text{sgn}(x)$

117. Modeling Data The table lists the speeds S (in feet per second) of a falling object at various times t (in seconds).

t	0	5	10	15	20	25	30
S	0	48.2	53.5	55.2	55.9	56.2	56.3

(a) Create a line graph of the data.

(b) Does there appear to be a limiting speed of the object? If there is a limiting speed, identify a possible cause.

118. Creating Models A swimmer crosses a pool of width b by swimming in a straight line from $(0, 0)$ to $(2b, b)$. (See figure.)

(a) Let f be a function defined as the y-coordinate of the point on the long side of the pool that is nearest the swimmer at any given time during the swimmer's crossing of the pool. Determine the function f and sketch its graph. Is f continuous? Explain.

(b) Let g be the minimum distance between the swimmer and the long sides of the pool. Determine the function g and sketch its graph. Is g continuous? Explain.

119. Making a Function Continuous Find all values of c such that f is continuous on $(-\infty, \infty)$.

$$f(x) = \begin{cases} 1 - x^2, & x \le c \\ x, & x > c \end{cases}$$

120. Proof Prove that for any real number y there exists x in $(-\pi/2, \pi/2)$ such that $\tan x = y$.

121. Making a Function Continuous Let

$$f(x) = \frac{\sqrt{x + c^2} - c}{x}, \quad c > 0.$$

What is the domain of f? How can you define f at $x = 0$ in order for f to be continuous there?

122. Proof Prove that if

$$\lim_{\Delta x \to 0} f(c + \Delta x) = f(c)$$

then f is continuous at c.

123. Continuity of a Function Discuss the continuity of the function $h(x) = x \llbracket x \rrbracket$.

124. Proof

(a) Let $f_1(x)$ and $f_2(x)$ be continuous on the closed interval $[a, b]$. If $f_1(a) < f_2(a)$ and $f_1(b) > f_2(b)$, prove that there exists c between a and b such that $f_1(c) = f_2(c)$.

(b) Show that there exists c in $\left[0, \frac{\pi}{2}\right]$ such that $\cos x = x$. Use a graphing utility to approximate c to three decimal places.

PUTNAM EXAM CHALLENGE

125. Prove or disprove: If x and y are real numbers with $y \ge 0$ and $y(y + 1) \le (x + 1)^2$, then $y(y - 1) \le x^2$.

126. Determine all polynomials $P(x)$ such that

$$P(x^2 + 1) = (P(x))^2 + 1 \text{ and } P(0) = 0.$$

These problems were composed by the Committee on the Putnam Prize Competition.
© The Mathematical Association of America. All rights reserved.

1.5 Infinite Limits

■ Determine infinite limits from the left and from the right.
■ Find and sketch the vertical asymptotes of the graph of a function.

Infinite Limits

Consider the function $f(x) = 3/(x - 2)$. From Figure 1.39 and the table, you can see that $f(x)$ *decreases without bound* as x approaches 2 from the left, and $f(x)$ *increases without bound* as x approaches 2 from the right.

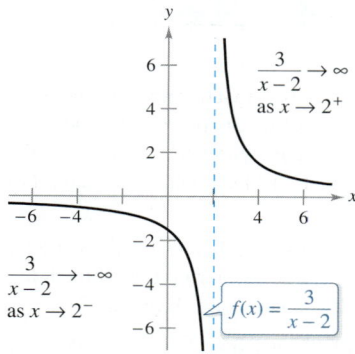

$\dfrac{3}{x-2} \to \infty$ as $x \to 2^+$

$\dfrac{3}{x-2} \to -\infty$ as $x \to 2^-$

$f(x) = \dfrac{3}{x-2}$

$f(x)$ increases and decreases without bound as x approaches 2.

Figure 1.39

	x approaches 2 from the left.					x approaches 2 from the right.			
x	1.5	1.9	1.99	1.999	2	2.001	2.01	2.1	2.5
$f(x)$	-6	-30	-300	-3000	?	3000	300	30	6

$f(x)$ decreases without bound.	$f(x)$ increases without bound.

This behavior is denoted as

$$\lim_{x \to 2^-} \frac{3}{x - 2} = -\infty \qquad \text{\textcolor{red}{$f(x)$ decreases without bound as x approaches 2 from the left.}}$$

and

$$\lim_{x \to 2^+} \frac{3}{x - 2} = \infty. \qquad \text{\textcolor{red}{$f(x)$ increases without bound as x approaches 2 from the right.}}$$

The symbols ∞ and $-\infty$ refer to positive infinity and negative infinity, respectively. These symbols do not represent real numbers. They are convenient symbols used to describe unbounded conditions more concisely. A limit in which $f(x)$ increases or decreases without bound as x approaches c is called an **infinite limit.**

$\lim_{x \to c} f(x) = \infty$

M

$\delta \quad \delta$

c

Infinite limits
Figure 1.40

Definition of Infinite Limits

Let f be a function that is defined at every real number in some open interval containing c (except possibly at c itself). The statement

$$\lim_{x \to c} f(x) = \infty$$

means that for each $M > 0$ there exists a $\delta > 0$ such that $f(x) > M$ whenever $0 < |x - c| < \delta$ (see Figure 1.40). Similarly, the statement

$$\lim_{x \to c} f(x) = -\infty$$

means that for each $N < 0$ there exists a $\delta > 0$ such that $f(x) < N$ whenever

$$0 < |x - c| < \delta.$$

To define the **infinite limit from the left,** replace $0 < |x - c| < \delta$ by $c - \delta < x < c$. To define the **infinite limit from the right,** replace $0 < |x - c| < \delta$ by $c < x < c + \delta$.

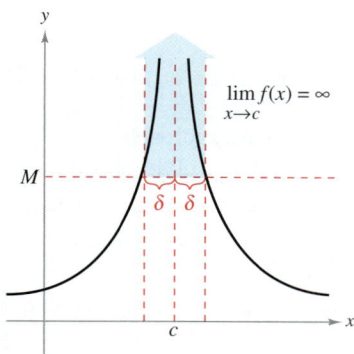

Be sure you see that the equal sign in the statement $\lim f(x) = \infty$ does not mean that the limit exists! On the contrary, it tells you how the limit **fails to exist** by denoting the unbounded behavior of $f(x)$ as x approaches c.

Exploration

Use a graphing utility to graph each function. For each function, analytically find the single real number c that is not in the domain. Then graphically find the limit (if it exists) of $f(x)$ as x approaches c from the left and from the right.

a. $f(x) = \dfrac{3}{x - 4}$

b. $f(x) = \dfrac{1}{2 - x}$

c. $f(x) = \dfrac{2}{(x - 3)^2}$

d. $f(x) = \dfrac{-3}{(x + 2)^2}$

EXAMPLE 1   **Determining Infinite Limits from a Graph**

Determine the limit of each function shown in Figure 1.41 as x approaches 1 from the left and from the right.

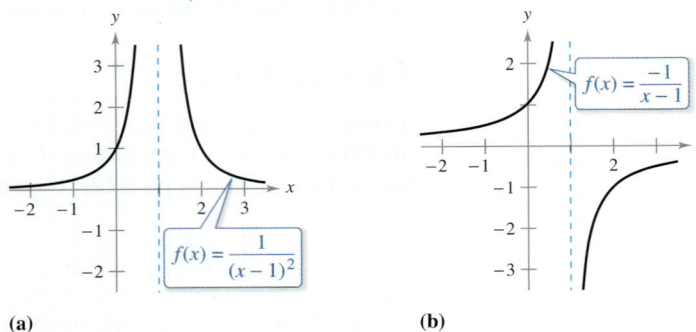

(a)

(b)

Each graph has an asymptote at $x = 1$.

Figure 1.41

Solution

a. When x approaches 1 from the left or the right, $(x - 1)^2$ is a small positive number. Thus, the quotient $1/(x - 1)^2$ is a large positive number, and $f(x)$ approaches infinity from each side of $x = 1$. So, you can conclude that

$$\lim_{x \to 1} \frac{1}{(x - 1)^2} = \infty. \qquad \text{\textcolor{red}{Limit from each side is infinity.}}$$

Figure 1.41(a) confirms this analysis.

b. When x approaches 1 from the left, $x - 1$ is a small negative number. Thus, the quotient $-1/(x - 1)$ is a large positive number, and $f(x)$ approaches infinity from the left of $x = 1$. So, you can conclude that

$$\lim_{x \to 1^-} \frac{-1}{x - 1} = \infty. \qquad \text{\textcolor{red}{Limit from the left side is infinity.}}$$

When x approaches 1 from the right, $x - 1$ is a small positive number. Thus, the quotient $-1/(x - 1)$ is a large negative number, and $f(x)$ approaches negative infinity from the right of $x = 1$. So, you can conclude that

$$\lim_{x \to 1^+} \frac{-1}{x - 1} = -\infty. \qquad \text{\textcolor{red}{Limit from the right side is negative infinity.}}$$

Figure 1.41(b) confirms this analysis.

▷ **TECHNOLOGY**   Remember that you can use a numerical approach to analyze a limit. For instance, you can use a graphing utility to create a table of values to analyze the limit in Example 1(a), as shown in Figure 1.42.

Enter x-values using *ask* mode.

X	Y₁
.9	100
.99	10000
.999	1E6
1	ERROR
1.001	1E6
1.01	10000
1.1	100

X = 1

As x approaches 1 from the left, $f(x)$ increases without bound.

As x approaches 1 from the right, $f(x)$ increases without bound.

Figure 1.42

Use a graphing utility to make a table of values to analyze the limit in Example 1(b).

Vertical Asymptotes

If it were possible to extend the graphs in Figure 1.41 toward positive and negative infinity, you would see that each graph becomes arbitrarily close to the vertical line $x = 1$. This line is a **vertical asymptote** of the graph of f. (You will study other types of asymptotes in Sections 3.5 and 3.6.)

····▷

REMARK If the graph of a function f has a vertical asymptote at $x = c$, then f is *not continuous* at c.

> ### Definition of Vertical Asymptote
>
> If $f(x)$ approaches infinity (or negative infinity) as x approaches c from the right or the left, then the line $x = c$ is a **vertical asymptote** of the graph of f.

In Example 1, note that each of the functions is a *quotient* and that the vertical asymptote occurs at a number at which the denominator is 0 (and the numerator is not 0). The next theorem generalizes this observation.

> ### THEOREM 1.14 Vertical Asymptotes
>
> Let f and g be continuous on an open interval containing c. If $f(c) \neq 0$, $g(c) = 0$, and there exists an open interval containing c such that $g(x) \neq 0$ for all $x \neq c$ in the interval, then the graph of the function
>
> $$h(x) = \frac{f(x)}{g(x)}$$
>
> has a vertical asymptote at $x = c$.
>
> A proof of this theorem is given in Appendix A.
>
> *See LarsonCalculus.com for Bruce Edwards's video of this proof.*

$f(x) = \dfrac{1}{2(x+1)}$

(a)

$f(x) = \dfrac{x^2+1}{x^2-1}$

(b)

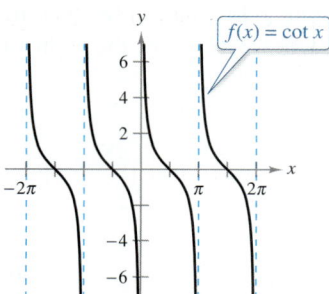

$f(x) = \cot x$

(c)

Functions with vertical asymptotes

Figure 1.43

EXAMPLE 2 Finding Vertical Asymptotes

····▷ *See LarsonCalculus.com for an interactive version of this type of example.*

a. When $x = -1$, the denominator of

$$f(x) = \frac{1}{2(x+1)}$$

is 0 and the numerator is not 0. So, by Theorem 1.14, you can conclude that $x = -1$ is a vertical asymptote, as shown in Figure 1.43(a).

b. By factoring the denominator as

$$f(x) = \frac{x^2+1}{x^2-1} = \frac{x^2+1}{(x-1)(x+1)}$$

you can see that the denominator is 0 at $x = -1$ and $x = 1$. Also, because the numerator is not 0 at these two points, you can apply Theorem 1.14 to conclude that the graph of f has two vertical asymptotes, as shown in Figure 1.43(b).

c. By writing the cotangent function in the form

$$f(x) = \cot x = \frac{\cos x}{\sin x}$$

you can apply Theorem 1.14 to conclude that vertical asymptotes occur at all values of x such that $\sin x = 0$ and $\cos x \neq 0$, as shown in Figure 1.43(c). So, the graph of this function has infinitely many vertical asymptotes. These asymptotes occur at $x = n\pi$, where n is an integer.

Theorem 1.14 requires that the value of the numerator at $x = c$ be nonzero. When both the numerator and the denominator are 0 at $x = c$, you obtain the *indeterminate form* $0/0$, and you cannot determine the limit behavior at $x = c$ without further investigation, as illustrated in Example 3.

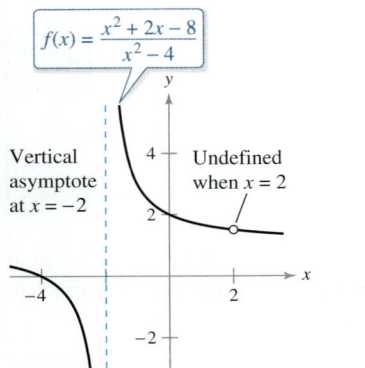

$f(x) = \dfrac{x^2 + 2x - 8}{x^2 - 4}$

Vertical asymptote at $x = -2$

Undefined when $x = 2$

$f(x)$ increases and decreases without bound as x approaches -2.

Figure 1.44

EXAMPLE 3 **A Rational Function with Common Factors**

Determine all vertical asymptotes of the graph of

$$f(x) = \frac{x^2 + 2x - 8}{x^2 - 4}.$$

Solution Begin by simplifying the expression, as shown.

$$f(x) = \frac{x^2 + 2x - 8}{x^2 - 4}$$

$$= \frac{(x + 4)(x - 2)}{(x + 2)(x - 2)}$$

$$= \frac{x + 4}{x + 2}, \quad x \neq 2$$

At all x-values other than $x = 2$, the graph of f coincides with the graph of $g(x) = (x + 4)/(x + 2)$. So, you can apply Theorem 1.14 to g to conclude that there is a vertical asymptote at $x = -2$, as shown in Figure 1.44. From the graph, you can see that

$$\lim_{x \to -2^-} \frac{x^2 + 2x - 8}{x^2 - 4} = -\infty \quad \text{and} \quad \lim_{x \to -2^+} \frac{x^2 + 2x - 8}{x^2 - 4} = \infty.$$

Note that $x = 2$ is *not* a vertical asymptote.

EXAMPLE 4 **Determining Infinite Limits**

Find each limit.

$$\lim_{x \to 1^-} \frac{x^2 - 3x}{x - 1} \quad \text{and} \quad \lim_{x \to 1^+} \frac{x^2 - 3x}{x - 1}$$

Solution Because the denominator is 0 when $x = 1$ (and the numerator is not zero), you know that the graph of

$$f(x) = \frac{x^2 - 3x}{x - 1}$$

has a vertical asymptote at $x = 1$. This means that each of the given limits is either ∞ or $-\infty$. You can determine the result by analyzing f at values of x close to 1, or by using a graphing utility. From the graph of f shown in Figure 1.45, you can see that the graph approaches ∞ from the left of $x = 1$ and approaches $-\infty$ from the right of $x = 1$. So, you can conclude that

$$\lim_{x \to 1^-} \frac{x^2 - 3x}{x - 1} = \infty \qquad \text{The limit from the left is infinity.}$$

and

$$\lim_{x \to 1^+} \frac{x^2 - 3x}{x - 1} = -\infty. \qquad \text{The limit from the right is negative infinity.}$$

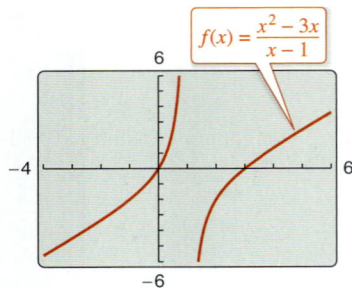

$f(x) = \dfrac{x^2 - 3x}{x - 1}$

f has a vertical asymptote at $x = 1$.

Figure 1.45

▷ **TECHNOLOGY PITFALL** When using a graphing utility, be careful to interpret correctly the graph of a function with a vertical asymptote—some graphing utilities have difficulty drawing this type of graph.

THEOREM 1.15   **Properties of Infinite Limits**

Let c and L be real numbers, and let f and g be functions such that

$$\lim_{x \to c} f(x) = \infty \quad \text{and} \quad \lim_{x \to c} g(x) = L.$$

1. Sum or difference: $\lim\limits_{x \to c} [f(x) \pm g(x)] = \infty$

2. Product: $\lim\limits_{x \to c} [f(x)g(x)] = \infty, \quad L > 0$

   $\lim\limits_{x \to c} [f(x)g(x)] = -\infty, \quad L < 0$

3. Quotient: $\lim\limits_{x \to c} \dfrac{g(x)}{f(x)} = 0$

Similar properties hold for one-sided limits and for functions for which the limit of $f(x)$ as x approaches c is $-\infty$ [see Example 5(d)].

Proof   Here is a proof of the sum property. (The proofs of the remaining properties are left as an exercise [see Exercise 70].) To show that the limit of $f(x) + g(x)$ is infinite, choose $M > 0$. You then need to find $\delta > 0$ such that $[f(x) + g(x)] > M$ whenever $0 < |x - c| < \delta$. For simplicity's sake, you can assume L is positive. Let $M_1 = M + 1$. Because the limit of $f(x)$ is infinite, there exists δ_1 such that $f(x) > M_1$ whenever $0 < |x - c| < \delta_1$. Also, because the limit of $g(x)$ is L, there exists δ_2 such that $|g(x) - L| < 1$ whenever $0 < |x - c| < \delta_2$. By letting δ be the smaller of δ_1 and δ_2, you can conclude that $0 < |x - c| < \delta$ implies $f(x) > M + 1$ and $|g(x) - L| < 1$. The second of these two inequalities implies that $g(x) > L - 1$, and, adding this to the first inequality, you can write

$$f(x) + g(x) > (M + 1) + (L - 1) = M + L > M.$$

So, you can conclude that

$$\lim_{x \to c} [f(x) + g(x)] = \infty.$$

See LarsonCalculus.com for Bruce Edwards's video of this proof.

EXAMPLE 5   **Determining Limits**

a. Because $\lim\limits_{x \to 0} 1 = 1$ and $\lim\limits_{x \to 0} \dfrac{1}{x^2} = \infty$, you can write

$$\lim_{x \to 0} \left(1 + \frac{1}{x^2} \right) = \infty. \qquad \text{\color{red}Property 1, Theorem 1.15}$$

b. Because $\lim\limits_{x \to 1^-} (x^2 + 1) = 2$ and $\lim\limits_{x \to 1^-} (\cot \pi x) = -\infty$, you can write

$$\lim_{x \to 1^-} \frac{x^2 + 1}{\cot \pi x} = 0. \qquad \text{\color{red}Property 3, Theorem 1.15}$$

c. Because $\lim\limits_{x \to 0^+} 3 = 3$ and $\lim\limits_{x \to 0^+} \cot x = \infty$, you can write

$$\lim_{x \to 0^+} 3 \cot x = \infty. \qquad \text{\color{red}Property 2, Theorem 1.15}$$

REMARK   Note that the solution to Example 5(d) uses Property 1 from Theorem 1.15 for which the limit of $f(x)$ as x approaches c is $-\infty$.

d. Because $\lim\limits_{x \to 0^-} x^2 = 0$ and $\lim\limits_{x \to 0^-} \dfrac{1}{x} = -\infty$, you can write

$$\lim_{x \to 0^-} \left(x^2 + \frac{1}{x} \right) = -\infty. \qquad \text{\color{red}Property 1, Theorem 1.15}$$

1.5   Exercises

See CalcChat.com for tutorial help and worked-out solutions to odd-numbered exercises.

Determining Infinite Limits from a Graph   In Exercises 1–4, determine whether $f(x)$ approaches ∞ or $-\infty$ as x approaches -2 from the left and from the right.

1. $f(x) = 2\left|\dfrac{x}{x^2 - 4}\right|$

2. $f(x) = \dfrac{1}{x + 2}$

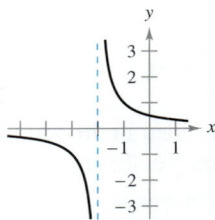

3. $f(x) = \tan\dfrac{\pi x}{4}$

4. $f(x) = \sec\dfrac{\pi x}{4}$

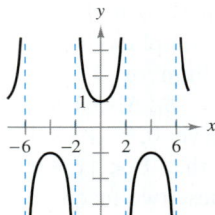

Determining Infinite Limits   In Exercises 5–8, determine whether $f(x)$ approaches ∞ or $-\infty$ as x approaches 4 from the left and from the right.

5. $f(x) = \dfrac{1}{x - 4}$

6. $f(x) = \dfrac{-1}{x - 4}$

7. $f(x) = \dfrac{1}{(x - 4)^2}$

8. $f(x) = \dfrac{-1}{(x - 4)^2}$

Numerical and Graphical Analysis   In Exercises 9–12, determine whether $f(x)$ approaches ∞ or $-\infty$ as x approaches -3 from the left and from the right by completing the table. Use a graphing utility to graph the function to confirm your answer.

x	-3.5	-3.1	-3.01	-3.001	-3
$f(x)$					?

x	-2.999	-2.99	-2.9	-2.5
$f(x)$				

9. $f(x) = \dfrac{1}{x^2 - 9}$

10. $f(x) = \dfrac{x}{x^2 - 9}$

11. $f(x) = \dfrac{x^2}{x^2 - 9}$

12. $f(x) = \cot\dfrac{\pi x}{3}$

Finding Vertical Asymptotes   In Exercises 13–28, find the vertical asymptotes (if any) of the graph of the function.

13. $f(x) = \dfrac{1}{x^2}$

14. $f(x) = \dfrac{2}{(x - 3)^3}$

15. $f(x) = \dfrac{x^2}{x^2 - 4}$

16. $f(x) = \dfrac{3x}{x^2 + 9}$

17. $g(t) = \dfrac{t - 1}{t^2 + 1}$

18. $h(s) = \dfrac{3s + 4}{s^2 - 16}$

19. $f(x) = \dfrac{3}{x^2 + x - 2}$

20. $g(x) = \dfrac{x^3 - 8}{x - 2}$

21. $f(x) = \dfrac{4x^2 + 4x - 24}{x^4 - 2x^3 - 9x^2 + 18x}$

22. $h(x) = \dfrac{x^2 - 9}{x^3 + 3x^2 - x - 3}$

23. $f(x) = \dfrac{x^2 - 2x - 15}{x^3 - 5x^2 + x - 5}$

24. $h(t) = \dfrac{t^2 - 2t}{t^4 - 16}$

25. $f(x) = \csc \pi x$

26. $f(x) = \tan \pi x$

27. $s(t) = \dfrac{t}{\sin t}$

28. $g(\theta) = \dfrac{\tan \theta}{\theta}$

Vertical Asymptote or Removable Discontinuity   In Exercises 29–32, determine whether the graph of the function has a vertical asymptote or a removable discontinuity at $x = -1$. Graph the function using a graphing utility to confirm your answer.

29. $f(x) = \dfrac{x^2 - 1}{x + 1}$

30. $f(x) = \dfrac{x^2 - 2x - 8}{x + 1}$

31. $f(x) = \dfrac{x^2 + 1}{x + 1}$

32. $f(x) = \dfrac{\sin(x + 1)}{x + 1}$

Finding a One-Sided Limit   In Exercises 33–48, find the one-sided limit (if it exists).

33. $\displaystyle\lim_{x \to -1^+} \dfrac{1}{x + 1}$

34. $\displaystyle\lim_{x \to 1^-} \dfrac{-1}{(x - 1)^2}$

35. $\displaystyle\lim_{x \to 2^+} \dfrac{x}{x - 2}$

36. $\displaystyle\lim_{x \to 2^-} \dfrac{x^2}{x^2 + 4}$

37. $\displaystyle\lim_{x \to -3^-} \dfrac{x + 3}{x^2 + x - 6}$

38. $\displaystyle\lim_{x \to (-1/2)^+} \dfrac{6x^2 + x - 1}{4x^2 - 4x - 3}$

39. $\displaystyle\lim_{x \to 0^-} \left(1 + \dfrac{1}{x}\right)$

40. $\displaystyle\lim_{x \to 0^+} \left(6 - \dfrac{1}{x^3}\right)$

41. $\displaystyle\lim_{x \to -4^-} \left(x^2 + \dfrac{2}{x + 4}\right)$

42. $\displaystyle\lim_{x \to 3^+} \left(\dfrac{x}{3} + \cot\dfrac{\pi x}{2}\right)$

43. $\displaystyle\lim_{x \to 0^+} \dfrac{2}{\sin x}$

44. $\displaystyle\lim_{x \to (\pi/2)^+} \dfrac{-2}{\cos x}$

45. $\displaystyle\lim_{x \to \pi^+} \dfrac{\sqrt{x}}{\csc x}$

46. $\displaystyle\lim_{x \to 0^-} \dfrac{x + 2}{\cot x}$

47. $\displaystyle\lim_{x\to(1/2)^-} x \sec \pi x$ **48.** $\displaystyle\lim_{x\to(1/2)^+} x^2 \tan \pi x$

One-Sided Limit In Exercises 49–52, use a graphing utility to graph the function and determine the one-sided limit.

49. $f(x) = \dfrac{x^2 + x + 1}{x^3 - 1}$ **50.** $f(x) = \dfrac{x^3 - 1}{x^2 + x + 1}$

$\displaystyle\lim_{x\to1^+} f(x)$ $\displaystyle\lim_{x\to1^-} f(x)$

51. $f(x) = \dfrac{1}{x^2 - 25}$ **52.** $f(x) = \sec \dfrac{\pi x}{8}$

$\displaystyle\lim_{x\to5^-} f(x)$ $\displaystyle\lim_{x\to4^+} f(x)$

WRITING ABOUT CONCEPTS

53. Infinite Limit In your own words, describe the meaning of an infinite limit. Is ∞ a real number?

54. Asymptote In your own words, describe what is meant by an asymptote of a graph.

55. Writing a Rational Function Write a rational function with vertical asymptotes at $x = 6$ and $x = -2$, and with a zero at $x = 3$.

56. Rational Function Does the graph of every rational function have a vertical asymptote? Explain.

57. Sketching a Graph Use the graph of the function f (see figure) to sketch the graph of $g(x) = 1/f(x)$ on the interval $[-2, 3]$. To print an enlarged copy of the graph, go to *MathGraphs.com*.

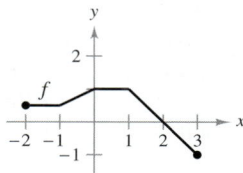

58. Relativity According to the theory of relativity, the mass m of a particle depends on its velocity v. That is,

$$m = \frac{m_0}{\sqrt{1 - (v^2/c^2)}}$$

where m_0 is the mass when the particle is at rest and c is the speed of light. Find the limit of the mass as v approaches c from the left.

59. Numerical and Graphical Analysis Use a graphing utility to complete the table for each function and graph each function to estimate the limit. What is the value of the limit when the power of x in the denominator is greater than 3?

x	1	0.5	0.2	0.1	0.01	0.001	0.0001
$f(x)$							

(a) $\displaystyle\lim_{x\to0^+} \dfrac{x - \sin x}{x}$ (b) $\displaystyle\lim_{x\to0^+} \dfrac{x - \sin x}{x^2}$

(c) $\displaystyle\lim_{x\to0^+} \dfrac{x - \sin x}{x^3}$ (d) $\displaystyle\lim_{x\to0^+} \dfrac{x - \sin x}{x^4}$

60. HOW DO YOU SEE IT? For a quantity of gas at a constant temperature, the pressure P is inversely proportional to the volume V. What is the limit of P as V approaches 0 from the right? Explain what this means in the context of the problem.

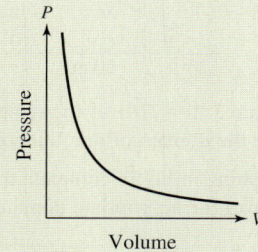

61. Rate of Change A 25-foot ladder is leaning against a house (see figure). If the base of the ladder is pulled away from the house at a rate of 2 feet per second, then the top will move down the wall at a rate of

$$r = \frac{2x}{\sqrt{625 - x^2}} \text{ ft/sec}$$

where x is the distance between the base of the ladder and the house, and r is the rate in feet per second.

(a) Find the rate r when x is 7 feet.

(b) Find the rate r when x is 15 feet.

(c) Find the limit of r as x approaches 25 from the left.

62. Average Speed

On a trip of d miles to another city, a truck driver's average speed was x miles per hour. On the return trip, the average speed was y miles per hour. The average speed for the round trip was 50 miles per hour.

(a) Verify that

$$y = \frac{25x}{x - 25}.$$

What is the domain?

(b) Complete the table.

x	30	40	50	60
y				

Are the values of y different than you expected? Explain.

(c) Find the limit of y as x approaches 25 from the right and interpret its meaning.

63. Numerical and Graphical Analysis   Consider the shaded region outside the sector of a circle of radius 10 meters and inside a right triangle (see figure).

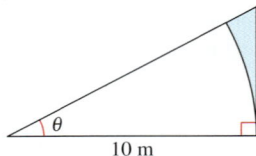

(a) Write the area $A = f(\theta)$ of the region as a function of θ. Determine the domain of the function.

(b) Use a graphing utility to complete the table and graph the function over the appropriate domain.

θ	0.3	0.6	0.9	1.2	1.5
$f(\theta)$					

(c) Find the limit of A as θ approaches $\pi/2$ from the left.

64. Numerical and Graphical Reasoning   A crossed belt connects a 20-centimeter pulley (10-cm radius) on an electric motor with a 40-centimeter pulley (20-cm radius) on a saw arbor (see figure). The electric motor runs at 1700 revolutions per minute.

(a) Determine the number of revolutions per minute of the saw.

(b) How does crossing the belt affect the saw in relation to the motor?

(c) Let L be the total length of the belt. Write L as a function of ϕ, where ϕ is measured in radians. What is the domain of the function? (*Hint:* Add the lengths of the straight sections of the belt and the length of the belt around each pulley.)

(d) Use a graphing utility to complete the table.

ϕ	0.3	0.6	0.9	1.2	1.5
L					

(e) Use a graphing utility to graph the function over the appropriate domain.

(f) Find $\lim\limits_{\phi \to (\pi/2)^-} L$. Use a geometric argument as the basis of a second method of finding this limit.

(g) Find $\lim\limits_{\phi \to 0^+} L$.

True or False?   **In Exercises 65–68, determine whether the statement is true or false. If it is false, explain why or give an example that shows it is false.**

65. The graph of a rational function has at least one vertical asymptote.

66. The graphs of polynomial functions have no vertical asymptotes.

67. The graphs of trigonometric functions have no vertical asymptotes.

68. If f has a vertical asymptote at $x = 0$, then f is undefined at $x = 0$.

69. Finding Functions   Find functions f and g such that $\lim\limits_{x \to c} f(x) = \infty$ and $\lim\limits_{x \to c} g(x) = \infty$, but $\lim\limits_{x \to c} [f(x) - g(x)] \neq 0$.

70. Proof   Prove the difference, product, and quotient properties in Theorem 1.15.

71. Proof   Prove that if $\lim\limits_{x \to c} f(x) = \infty$, then $\lim\limits_{x \to c} \dfrac{1}{f(x)} = 0$.

72. Proof   Prove that if

$$\lim_{x \to c} \frac{1}{f(x)} = 0$$

then $\lim\limits_{x \to c} f(x)$ does not exist.

Infinite Limits   **In Exercises 73 and 74, use the ε–δ definition of infinite limits to prove the statement.**

73. $\lim\limits_{x \to 3^+} \dfrac{1}{x - 3} = \infty$

74. $\lim\limits_{x \to 5^-} \dfrac{1}{x - 5} = -\infty$

SECTION PROJECT

Graphs and Limits of Trigonometric Functions

Recall from Theorem 1.9 that the limit of $f(x) = (\sin x)/x$ as x approaches 0 is 1.

(a) Use a graphing utility to graph the function f on the interval $-\pi \le x \le \pi$. Explain how the graph helps confirm this theorem.

(b) Explain how you could use a table of values to confirm the value of this limit numerically.

(c) Graph $g(x) = \sin x$ by hand. Sketch a tangent line at the point $(0, 0)$ and visually estimate the slope of this tangent line.

(d) Let $(x, \sin x)$ be a point on the graph of g near $(0, 0)$, and write a formula for the slope of the secant line joining $(x, \sin x)$ and $(0, 0)$. Evaluate this formula at $x = 0.1$ and $x = 0.01$. Then find the exact slope of the tangent line to g at the point $(0, 0)$.

(e) Sketch the graph of the cosine function $h(x) = \cos x$. What is the slope of the tangent line at the point $(0, 1)$? Use limits to find this slope analytically.

(f) Find the slope of the tangent line to $k(x) = \tan x$ at $(0, 0)$.

Review Exercises   See CalcChat.com for tutorial help and worked-out solutions to odd-numbered exercises.

Precalculus or Calculus In Exercises 1 and 2, determine whether the problem can be solved using precalculus or whether calculus is required. If the problem can be solved using precalculus, solve it. If the problem seems to require calculus, explain your reasoning and use a graphical or numerical approach to estimate the solution.

1. Find the distance between the points $(1, 1)$ and $(3, 9)$ along the curve $y = x^2$.

2. Find the distance between the points $(1, 1)$ and $(3, 9)$ along the line $y = 4x - 3$.

Estimating a Limit Numerically In Exercises 3 and 4, complete the table and use the result to estimate the limit. Use a graphing utility to graph the function to confirm your result.

3. $\lim\limits_{x \to 3} \dfrac{x - 3}{x^2 - 7x + 12}$

x	2.9	2.99	2.999	3	3.001	3.01	3.1
$f(x)$				?			

4. $\lim\limits_{x \to 0} \dfrac{\sqrt{x + 4} - 2}{x}$

x	-0.1	-0.01	-0.001	0	0.001	0.01	0.1
$f(x)$				?			

Finding a Limit Graphically In Exercises 5 and 6, use the graph to find the limit (if it exists). If the limit does not exist, explain why.

5. $h(x) = \dfrac{4x - x^2}{x}$

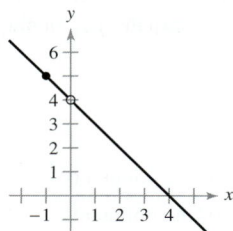

6. $g(x) = \dfrac{-2x}{x - 3}$

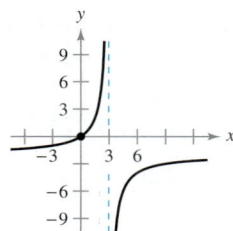

(a) $\lim\limits_{x \to 0} h(x)$ (b) $\lim\limits_{x \to -1} h(x)$ (a) $\lim\limits_{x \to 3} g(x)$ (b) $\lim\limits_{x \to 0} g(x)$

Using the ϵ–δ Definition of a Limit In Exercises 7–10, find the limit L. Then use the ϵ–δ definition to prove that the limit is L.

7. $\lim\limits_{x \to 1} (x + 4)$

8. $\lim\limits_{x \to 9} \sqrt{x}$

9. $\lim\limits_{x \to 2} (1 - x^2)$

10. $\lim\limits_{x \to 5} 9$

Finding a Limit In Exercises 11–28, find the limit.

11. $\lim\limits_{x \to -6} x^2$

12. $\lim\limits_{x \to 0} (5x - 3)$

13. $\lim\limits_{t \to 4} \sqrt{t + 2}$

14. $\lim\limits_{x \to -5} \sqrt[3]{x - 3}$

15. $\lim\limits_{x \to 6} (x - 2)^2$

16. $\lim\limits_{x \to 7} (x - 4)^3$

17. $\lim\limits_{x \to 4} \dfrac{4}{x - 1}$

18. $\lim\limits_{x \to 2} \dfrac{x}{x^2 + 1}$

19. $\lim\limits_{x \to -2} \dfrac{t + 2}{t^2 - 4}$

20. $\lim\limits_{x \to 4} \dfrac{t^2 - 16}{t - 4}$

21. $\lim\limits_{x \to 4} \dfrac{\sqrt{x - 3} - 1}{x - 4}$

22. $\lim\limits_{x \to 0} \dfrac{\sqrt{4 + x} - 2}{x}$

23. $\lim\limits_{x \to 0} \dfrac{[1/(x + 1)] - 1}{x}$

24. $\lim\limits_{s \to 0} \dfrac{(1/\sqrt{1 + s}) - 1}{s}$

25. $\lim\limits_{x \to 0} \dfrac{1 - \cos x}{\sin x}$

26. $\lim\limits_{x \to \pi/4} \dfrac{4x}{\tan x}$

27. $\lim\limits_{\Delta x \to 0} \dfrac{\sin[(\pi/6) + \Delta x] - (1/2)}{\Delta x}$

   [*Hint:* $\sin(\theta + \phi) = \sin\theta\cos\phi + \cos\theta\sin\phi$]

28. $\lim\limits_{\Delta x \to 0} \dfrac{\cos(\pi + \Delta x) + 1}{\Delta x}$

   [*Hint:* $\cos(\theta + \phi) = \cos\theta\cos\phi - \sin\theta\sin\phi$]

Evaluating a Limit In Exercises 29–32, evaluate the limit given $\lim\limits_{x \to c} f(x) = -6$ and $\lim\limits_{x \to c} g(x) = \frac{1}{2}$.

29. $\lim\limits_{x \to c} [f(x)g(x)]$

30. $\lim\limits_{x \to c} \dfrac{f(x)}{g(x)}$

31. $\lim\limits_{x \to c} [f(x) + 2g(x)]$

32. $\lim\limits_{x \to c} [f(x)]^2$

Graphical, Numerical, and Analytic Analysis In Exercises 33–36, use a graphing utility to graph the function and estimate the limit. Use a table to reinforce your conclusion. Then find the limit by analytic methods.

33. $\lim\limits_{x \to 0} \dfrac{\sqrt{2x + 9} - 3}{x}$

34. $\lim\limits_{x \to 0} \dfrac{[1/(x + 4)] - (1/4)}{x}$

35. $\lim\limits_{x \to -5} \dfrac{x^3 + 125}{x + 5}$

36. $\lim\limits_{x \to 0} \dfrac{\cos x - 1}{x}$

Free-Falling Object In Exercises 37 and 38, use the position function $s(t) = -4.9t^2 + 250$, which gives the height (in meters) of an object that has fallen for t seconds from a height of 250 meters. The velocity at time $t = a$ seconds is given by

$$\lim\limits_{t \to a} \dfrac{s(a) - s(t)}{a - t}.$$

37. Find the velocity of the object when $t = 4$.

38. At what velocity will the object impact the ground?

Finding a Limit In Exercises 39–48, find the limit (if it exists). If it does not exist, explain why.

39. $\lim\limits_{x \to 3^+} \dfrac{1}{x + 3}$

40. $\lim\limits_{x \to 6^-} \dfrac{x - 6}{x^2 - 36}$

41. $\lim\limits_{x \to 4^-} \dfrac{\sqrt{x} - 2}{x - 4}$

42. $\lim\limits_{x \to 3^-} \dfrac{|x - 3|}{x - 3}$

43. $\lim\limits_{x \to 2^-} f(x)$, where $f(x) = \begin{cases} (x-2)^2, & x \le 2 \\ 2 - x, & x > 2 \end{cases}$

44. $\lim\limits_{x \to 1^+} g(x)$, where $g(x) = \begin{cases} \sqrt{1-x}, & x \le 1 \\ x + 1, & x > 1 \end{cases}$

45. $\lim\limits_{t \to 1} h(t)$, where $h(t) = \begin{cases} t^3 + 1, & t < 1 \\ \frac{1}{2}(t+1), & t \ge 1 \end{cases}$

46. $\lim\limits_{s \to -2} f(s)$, where $f(s) = \begin{cases} -s^2 - 4s - 2, & s \le -2 \\ s^2 + 4s + 6, & s > -2 \end{cases}$

47. $\lim\limits_{x \to 2^-} (2[\![x]\!] + 1)$

48. $\lim\limits_{x \to 4} [\![x - 1]\!]$

Removable and Nonremovable Discontinuities In Exercises 49–54, find the x-values (if any) at which f is not continuous. Which of the discontinuities are removable?

49. $f(x) = x^2 - 4$

50. $f(x) = x^2 - x + 20$

51. $f(x) = \dfrac{4}{x-5}$

52. $f(x) = \dfrac{1}{x^2 - 9}$

53. $f(x) = \dfrac{x}{x^3 - x}$

54. $f(x) = \dfrac{x+3}{x^2 - 3x - 18}$

55. Making a Function Continuous Determine the value of c such that the function is continuous on the entire real number line.

$$f(x) = \begin{cases} x + 3, & x \le 2 \\ cx + 6, & x > 2 \end{cases}$$

56. Making a Function Continuous Determine the values of b and c such that the function is continuous on the entire real number line.

$$f(x) = \begin{cases} x + 1, & 1 < x < 3 \\ x^2 + bx + c, & |x - 2| \ge 1 \end{cases}$$

Testing for Continuity In Exercises 57–62, describe the intervals on which the function is continuous.

57. $f(x) = -3x^2 + 7$

58. $f(x) = \dfrac{4x^2 + 7x - 2}{x + 2}$

59. $f(x) = \sqrt{x - 4}$

60. $f(x) = [\![x + 3]\!]$

61. $f(x) = \begin{cases} \dfrac{3x^2 - x - 2}{x - 1}, & x \ne 1 \\ 0, & x = 1 \end{cases}$

62. $f(x) = \begin{cases} 5 - x, & x \le 2 \\ 2x - 3, & x > 2 \end{cases}$

63. Using the Intermediate Value Theorem Use the Intermediate Value Theorem to show that $f(x) = 2x^3 - 3$ has a zero in the interval $[1, 2]$.

64. Delivery Charges The cost of sending an overnight package from New York to Atlanta is $12.80 for the first pound and $2.50 for each additional pound or fraction thereof. Use the greatest integer function to create a model for the cost C of overnight delivery of a package weighing x pounds. Sketch the graph of this function and discuss its continuity.

65. Finding Limits Let

$$f(x) = \frac{x^2 - 4}{|x - 2|}.$$

Find each limit (if it exists).

(a) $\lim\limits_{x \to 2^-} f(x)$ (b) $\lim\limits_{x \to 2^+} f(x)$ (c) $\lim\limits_{x \to 2} f(x)$

66. Finding Limits Let $f(x) = \sqrt{x(x-1)}$.

(a) Find the domain of f.

(b) Find $\lim\limits_{x \to 0^-} f(x)$.

(c) Find $\lim\limits_{x \to 1^+} f(x)$.

Finding Vertical Asymptotes In Exercises 67–72, find the vertical asymptotes (if any) of the graph of the function.

67. $f(x) = \dfrac{3}{x}$

68. $f(x) = \dfrac{5}{(x-2)^4}$

69. $f(x) = \dfrac{x^3}{x^2 - 9}$

70. $h(x) = \dfrac{6x}{36 - x^2}$

71. $g(x) = \dfrac{2x + 1}{x^2 - 64}$

72. $f(x) = \csc \pi x$

Finding a One-Sided Limit In Exercises 73–82, find the one-sided limit (if it exists).

73. $\lim\limits_{x \to 1^-} \dfrac{x^2 + 2x + 1}{x - 1}$

74. $\lim\limits_{x \to (1/2)^+} \dfrac{x}{2x - 1}$

75. $\lim\limits_{x \to -1^+} \dfrac{x + 1}{x^3 + 1}$

76. $\lim\limits_{x \to -1^-} \dfrac{x + 1}{x^4 - 1}$

77. $\lim\limits_{x \to 0^+} \left(x - \dfrac{1}{x^3} \right)$

78. $\lim\limits_{x \to 2^-} \dfrac{1}{\sqrt[3]{x^2 - 4}}$

79. $\lim\limits_{x \to 0^+} \dfrac{\sin 4x}{5x}$

80. $\lim\limits_{x \to 0^+} \dfrac{\sec x}{x}$

81. $\lim\limits_{x \to 0^+} \dfrac{\csc 2x}{x}$

82. $\lim\limits_{x \to 0^-} \dfrac{\cos^2 x}{x}$

83. Environment A utility company burns coal to generate electricity. The cost C in dollars of removing $p\%$ of the air pollutants in the stack emissions is

$$C = \frac{80{,}000p}{100 - p}, \quad 0 \le p < 100.$$

(a) Find the cost of removing 15% of the pollutants.

(b) Find the cost of removing 50% of the pollutants.

(c) Find the cost of removing 90% of the pollutants.

(d) Find the limit of C as p approaches 100 from the left and interpret its meaning.

84. Limits and Continuity The function f is defined as shown.

$$f(x) = \frac{\tan 2x}{x}, \quad x \ne 0$$

(a) Find $\lim\limits_{x \to 0} \dfrac{\tan 2x}{x}$ (if it exists).

(b) Can the function f be defined at $x = 0$ such that it is continuous at $x = 0$?

P.S. Problem Solving

See **CalcChat.com** for tutorial help and worked-out solutions to odd-numbered exercises.

1. **Perimeter** Let $P(x, y)$ be a point on the parabola $y = x^2$ in the first quadrant. Consider the triangle $\triangle PAO$ formed by P, $A(0, 1)$, and the origin $O(0, 0)$, and the triangle $\triangle PBO$ formed by P, $B(1, 0)$, and the origin.

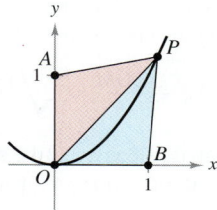

(a) Write the perimeter of each triangle in terms of x.

(b) Let $r(x)$ be the ratio of the perimeters of the two triangles,

$$r(x) = \frac{\text{Perimeter } \triangle PAO}{\text{Perimeter } \triangle PBO}.$$

Complete the table. Calculate $\lim\limits_{x \to 0^+} r(x)$.

x	4	2	1	0.1	0.01
Perimeter $\triangle PAO$					
Perimeter $\triangle PBO$					
$r(x)$					

2. **Area** Let $P(x, y)$ be a point on the parabola $y = x^2$ in the first quadrant. Consider the triangle $\triangle PAO$ formed by P, $A(0, 1)$, and the origin $O(0, 0)$, and the triangle $\triangle PBO$ formed by P, $B(1, 0)$, and the origin.

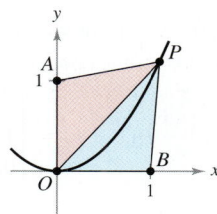

(a) Write the area of each triangle in terms of x.

(b) Let $a(x)$ be the ratio of the areas of the two triangles,

$$a(x) = \frac{\text{Area } \triangle PBO}{\text{Area } \triangle PAO}.$$

Complete the table. Calculate $\lim\limits_{x \to 0^+} a(x)$.

x	4	2	1	0.1	0.01
Area $\triangle PAO$					
Area $\triangle PBO$					
$a(x)$					

3. **Area of a Circle**

(a) Find the area of a regular hexagon inscribed in a circle of radius 1. How close is this area to that of the circle?

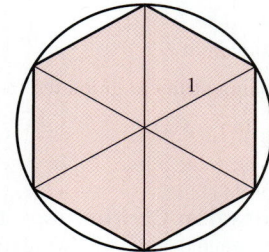

(b) Find the area A_n of an n-sided regular polygon inscribed in a circle of radius 1. Write your answer as a function of n.

(c) Complete the table. What number does A_n approach as n gets larger and larger?

n	6	12	24	48	96
A_n					

4. **Tangent Line** Let $P(3, 4)$ be a point on the circle $x^2 + y^2 = 25$.

(a) What is the slope of the line joining P and $O(0, 0)$?

(b) Find an equation of the tangent line to the circle at P.

(c) Let $Q(x, y)$ be another point on the circle in the first quadrant. Find the slope m_x of the line joining P and Q in terms of x.

(d) Calculate $\lim\limits_{x \to 3} m_x$. How does this number relate to your answer in part (b)?

Figure for 4

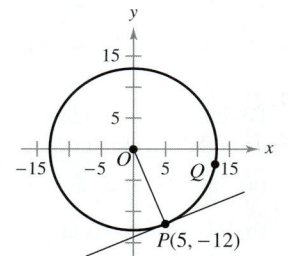

Figure for 5

5. **Tangent Line** Let $P(5, -12)$ be a point on the circle $x^2 + y^2 = 169$.

(a) What is the slope of the line joining P and $O(0, 0)$?

(b) Find an equation of the tangent line to the circle at P.

(c) Let $Q(x, y)$ be another point on the circle in the fourth quadrant. Find the slope m_x of the line joining P and Q in terms of x.

(d) Calculate $\lim\limits_{x \to 5} m_x$. How does this number relate to your answer in part (b)?

6. Finding Values Find the values of the constants a and b such that

$$\lim_{x \to 0} \frac{\sqrt{a + bx} - \sqrt{3}}{x} = \sqrt{3}.$$

7. Finding Limits Consider the function

$$f(x) = \frac{\sqrt{3 + x^{1/3}} - 2}{x - 1}.$$

(a) Find the domain of f.

(b) Use a graphing utility to graph the function.

(c) Calculate $\displaystyle\lim_{x \to -27^+} f(x)$.

(d) Calculate $\displaystyle\lim_{x \to 1} f(x)$.

8. Making a Function Continuous Determine all values of the constant a such that the following function is continuous for all real numbers.

$$f(x) = \begin{cases} \dfrac{ax}{\tan x}, & x \ge 0 \\ a^2 - 2, & x < 0 \end{cases}$$

9. Choosing Graphs Consider the graphs of the four functions g_1, g_2, g_3, and g_4.

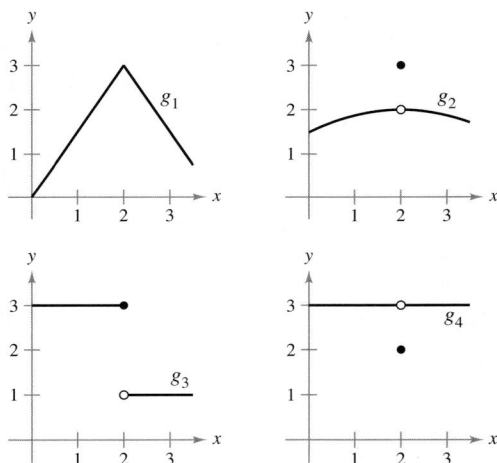

For each given condition of the function f, which of the graphs could be the graph of f?

(a) $\displaystyle\lim_{x \to 2} f(x) = 3$

(b) f is continuous at 2.

(c) $\displaystyle\lim_{x \to 2^-} f(x) = 3$

10. Limits and Continuity Sketch the graph of the function

$$f(x) = \left[\!\!\left[\frac{1}{x} \right]\!\!\right].$$

(a) Evaluate $f\left(\frac{1}{4}\right)$, $f(3)$, and $f(1)$.

(b) Evaluate the limits $\displaystyle\lim_{x \to 1^-} f(x)$, $\displaystyle\lim_{x \to 1^+} f(x)$, $\displaystyle\lim_{x \to 0^-} f(x)$, and $\displaystyle\lim_{x \to 0^+} f(x)$.

(c) Discuss the continuity of the function.

11. Limits and Continuity Sketch the graph of the function $f(x) = [\![x]\!] + [\![-x]\!]$.

(a) Evaluate $f(1)$, $f(0)$, $f\left(\frac{1}{2}\right)$, and $f(-2.7)$.

(b) Evaluate the limits $\displaystyle\lim_{x \to 1^-} f(x)$, $\displaystyle\lim_{x \to 1^+} f(x)$, and $\displaystyle\lim_{x \to 1/2} f(x)$.

(c) Discuss the continuity of the function.

12. Escape Velocity To escape Earth's gravitational field, a rocket must be launched with an initial velocity called the **escape velocity.** A rocket launched from the surface of Earth has velocity v (in miles per second) given by

$$v = \sqrt{\frac{2GM}{r} + v_0^2 - \frac{2GM}{R}} \approx \sqrt{\frac{192{,}000}{r} + v_0^2 - 48}$$

where v_0 is the initial velocity, r is the distance from the rocket to the center of Earth, G is the gravitational constant, M is the mass of Earth, and R is the radius of Earth (approximately 4000 miles).

(a) Find the value of v_0 for which you obtain an infinite limit for r as v approaches zero. This value of v_0 is the escape velocity for Earth.

(b) A rocket launched from the surface of the moon has velocity v (in miles per second) given by

$$v = \sqrt{\frac{1920}{r} + v_0^2 - 2.17}.$$

Find the escape velocity for the moon.

(c) A rocket launched from the surface of a planet has velocity v (in miles per second) given by

$$v = \sqrt{\frac{10{,}600}{r} + v_0^2 - 6.99}.$$

Find the escape velocity for this planet. Is the mass of this planet larger or smaller than that of Earth? (Assume that the mean density of this planet is the same as that of Earth.)

13. Pulse Function For positive numbers $a < b$, the **pulse function** is defined as

$$P_{a,b}(x) = H(x - a) - H(x - b) = \begin{cases} 0, & x < a \\ 1, & a \le x < b \\ 0, & x \ge b \end{cases}$$

where $H(x) = \begin{cases} 1, & x \ge 0 \\ 0, & x < 0 \end{cases}$ is the Heaviside function.

(a) Sketch the graph of the pulse function.

(b) Find the following limits:

(i) $\displaystyle\lim_{x \to a^+} P_{a,b}(x)$ (ii) $\displaystyle\lim_{x \to a^-} P_{a,b}(x)$

(iii) $\displaystyle\lim_{x \to b^+} P_{a,b}(x)$ (iv) $\displaystyle\lim_{x \to b^-} P_{a,b}(x)$

(c) Discuss the continuity of the pulse function.

(d) Why is $U(x) = \dfrac{1}{b - a} P_{a,b}(x)$ called the **unit pulse function?**

14. Proof Let a be a nonzero constant. Prove that if $\displaystyle\lim_{x \to 0} f(x) = L$, then $\displaystyle\lim_{x \to 0} f(ax) = L$. Show by means of an example that a must be nonzero.

2 Differentiation

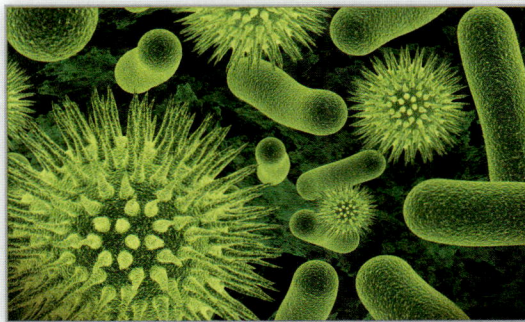

Bacteria *(Exercise 111, p. 139)*

Rate of Change
(Example 2, p. 149)

Acceleration Due to Gravity *(Example 10, p. 124)*

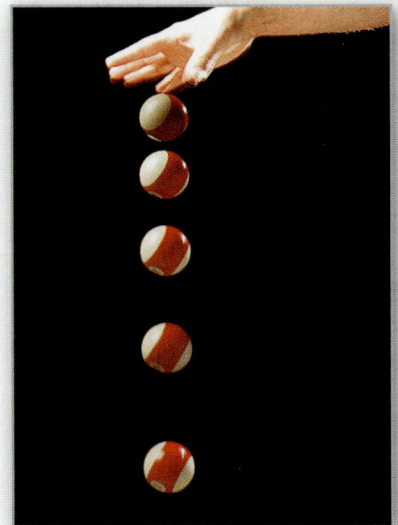

Velocity of a Falling Object
(Example 9, p. 112)

Stopping Distance *(Exercise 107, p. 117)*

95

2.1 The Derivative and the Tangent Line Problem

■ Find the slope of the tangent line to a curve at a point.
■ Use the limit definition to find the derivative of a function.
■ Understand the relationship between differentiability and continuity.

The Tangent Line Problem

Calculus grew out of four major problems that European mathematicians were working on during the seventeenth century.

1. The tangent line problem (Section 1.1 and this section)

2. The velocity and acceleration problem (Sections 2.2 and 2.3)

3. The minimum and maximum problem (Section 3.1)

4. The area problem (Sections 1.1 and 4.2)

Each problem involves the notion of a limit, and calculus can be introduced with any of the four problems.

A brief introduction to the tangent line problem is given in Section 1.1. Although partial solutions to this problem were given by Pierre de Fermat (1601–1665), René Descartes (1596–1650), Christian Huygens (1629–1695), and Isaac Barrow (1630–1677), credit for the first general solution is usually given to Isaac Newton (1642–1727) and Gottfried Leibniz (1646–1716). Newton's work on this problem stemmed from his interest in optics and light refraction.

What does it mean to say that a line is tangent to a curve at a point? For a circle, the tangent line at a point P is the line that is perpendicular to the radial line at point P, as shown in Figure 2.1.

For a general curve, however, the problem is more difficult. For instance, how would you define the tangent lines shown in Figure 2.2? You might say that a line is tangent to a curve at a point P when it touches, but does not cross, the curve at point P. This definition would work for the first curve shown in Figure 2.2, but not for the second. *Or* you might say that a line is tangent to a curve when the line touches or intersects the curve at exactly one point. This definition would work for a circle, but not for more general curves, as the third curve in Figure 2.2 shows.

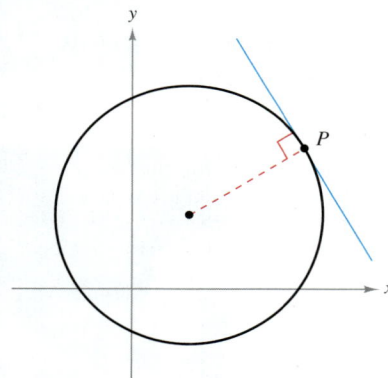

ISAAC NEWTON (1642–1727)

In addition to his work in calculus, Newton made revolutionary contributions to physics, including the Law of Universal Gravitation and his three laws of motion.
See LarsonCalculus.com to read more of this biography.

Tangent line to a circle
Figure 2.1

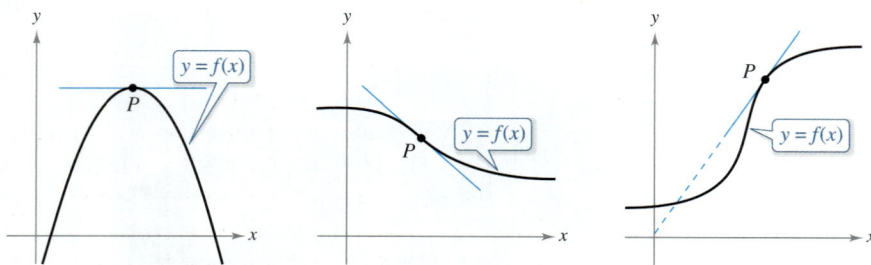

Exploration

Use a graphing utility to graph $f(x) = 2x^3 - 4x^2 + 3x - 5$. On the same screen, graph $y = x - 5$, $y = 2x - 5$, and $y = 3x - 5$. Which of these lines, if any, appears to be tangent to the graph of f at the point $(0, -5)$? Explain your reasoning.

Tangent line to a curve at a point
Figure 2.2

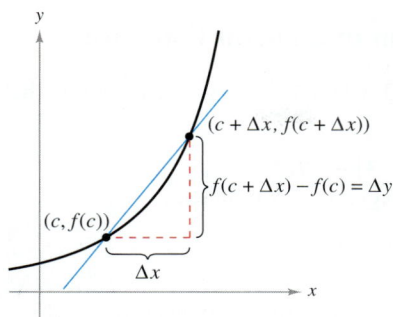

The secant line through $(c, f(c))$ and $(c + \Delta x, f(c + \Delta x))$

Figure 2.3

Essentially, the problem of finding the tangent line at a point P boils down to the problem of finding the *slope* of the tangent line at point P. You can approximate this slope using a **secant line*** through the point of tangency and a second point on the curve, as shown in Figure 2.3. If $(c, f(c))$ is the point of tangency and

$$(c + \Delta x, f(c + \Delta x))$$

is a second point on the graph of f, then the slope of the secant line through the two points is given by substitution into the slope formula

$$m = \frac{y_2 - y_1}{x_2 - x_1}$$

$$m_{\text{sec}} = \frac{f(c + \Delta x) - f(c)}{(c + \Delta x) - c} \qquad \begin{array}{l}\text{Change in } y \\ \text{Change in } x\end{array}$$

$$m_{\text{sec}} = \frac{f(c + \Delta x) - f(c)}{\Delta x}. \qquad \text{Slope of secant line}$$

The right-hand side of this equation is a **difference quotient.** The denominator Δx is the **change in x,** and the numerator

$$\Delta y = f(c + \Delta x) - f(c)$$

is the **change in y.**

The beauty of this procedure is that you can obtain more and more accurate approximations of the slope of the tangent line by choosing points closer and closer to the point of tangency, as shown in Figure 2.4.

THE TANGENT LINE PROBLEM

In 1637, mathematician René Descartes stated this about the tangent line problem:

"And I dare say that this is not only the most useful and general problem in geometry that I know, but even that I ever desire to know."

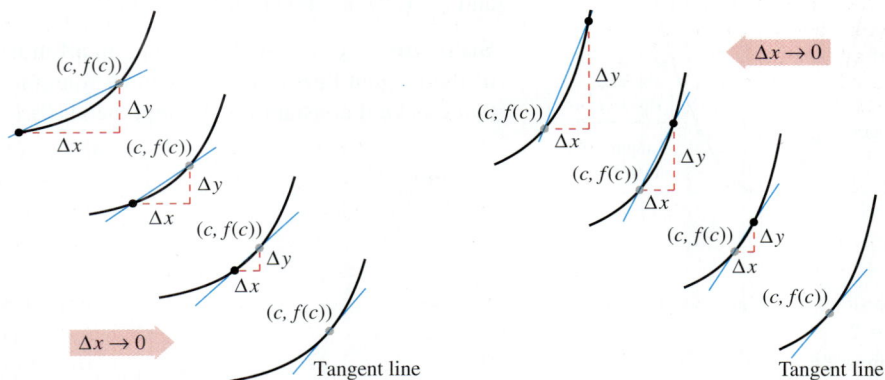

Tangent line approximations
Figure 2.4

Definition of Tangent Line with Slope m

If f is defined on an open interval containing c, and if the limit

$$\lim_{\Delta x \to 0} \frac{\Delta y}{\Delta x} = \lim_{\Delta x \to 0} \frac{f(c + \Delta x) - f(c)}{\Delta x} = m$$

exists, then the line passing through $(c, f(c))$ with slope m is the **tangent line** to the graph of f at the point $(c, f(c))$.

The slope of the tangent line to the graph of f at the point $(c, f(c))$ is also called the **slope of the graph of f at $x = c$.**

---

* This use of the word *secant* comes from the Latin *secare*, meaning to cut, and is not a reference to the trigonometric function of the same name.

EXAMPLE 1   The Slope of the Graph of a Linear Function

To find the slope of the graph of $f(x) = 2x - 3$ when $c = 2$, you can apply the definition of the slope of a tangent line, as shown.

$$\lim_{\Delta x \to 0} \frac{f(2 + \Delta x) - f(2)}{\Delta x} = \lim_{\Delta x \to 0} \frac{[2(2 + \Delta x) - 3] - [2(2) - 3]}{\Delta x}$$

$$= \lim_{\Delta x \to 0} \frac{4 + 2\Delta x - 3 - 4 + 3}{\Delta x}$$

$$= \lim_{\Delta x \to 0} \frac{2\Delta x}{\Delta x}$$

$$= \lim_{\Delta x \to 0} 2$$

$$= 2$$

The slope of f at $(c, f(c)) = (2, 1)$ is $m = 2$, as shown in Figure 2.5. Notice that the limit definition of the slope of f agrees with the definition of the slope of a line as discussed in Section P.2.

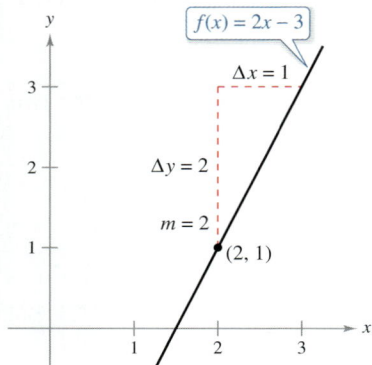

The slope of f at $(2, 1)$ is $m = 2$.
Figure 2.5

The graph of a linear function has the same slope at any point. This is not true of nonlinear functions, as shown in the next example.

EXAMPLE 2   Tangent Lines to the Graph of a Nonlinear Function

Find the slopes of the tangent lines to the graph of $f(x) = x^2 + 1$ at the points $(0, 1)$ and $(-1, 2)$, as shown in Figure 2.6.

Solution   Let $(c, f(c))$ represent an arbitrary point on the graph of f. Then the slope of the tangent line at $(c, f(c))$ can be found as shown below. [Note in the limit process that c is held constant (as Δx approaches 0).]

$$\lim_{\Delta x \to 0} \frac{f(c + \Delta x) - f(c)}{\Delta x} = \lim_{\Delta x \to 0} \frac{[(c + \Delta x)^2 + 1] - (c^2 + 1)}{\Delta x}$$

$$= \lim_{\Delta x \to 0} \frac{c^2 + 2c(\Delta x) + (\Delta x)^2 + 1 - c^2 - 1}{\Delta x}$$

$$= \lim_{\Delta x \to 0} \frac{2c(\Delta x) + (\Delta x)^2}{\Delta x}$$

$$= \lim_{\Delta x \to 0} (2c + \Delta x)$$

$$= 2c$$

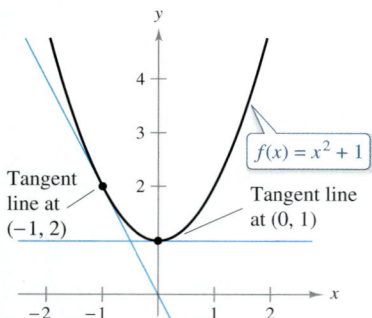

The slope of f at any point $(c, f(c))$ is $m = 2c$.
Figure 2.6

So, the slope at *any* point $(c, f(c))$ on the graph of f is $m = 2c$. At the point $(0, 1)$, the slope is $m = 2(0) = 0$, and at $(-1, 2)$, the slope is $m = 2(-1) = -2$.

The definition of a tangent line to a curve does not cover the possibility of a vertical tangent line. For vertical tangent lines, you can use the following definition. If f is continuous at c and

$$\lim_{\Delta x \to 0} \frac{f(c + \Delta x) - f(c)}{\Delta x} = \infty \quad \text{or} \quad \lim_{\Delta x \to 0} \frac{f(c + \Delta x) - f(c)}{\Delta x} = -\infty$$

then the vertical line $x = c$ passing through $(c, f(c))$ is a **vertical tangent line** to the graph of f. For example, the function shown in Figure 2.7 has a vertical tangent line at $(c, f(c))$. When the domain of f is the closed interval $[a, b]$, you can extend the definition of a vertical tangent line to include the endpoints by considering continuity and limits from the right (for $x = a$) and from the left (for $x = b$).

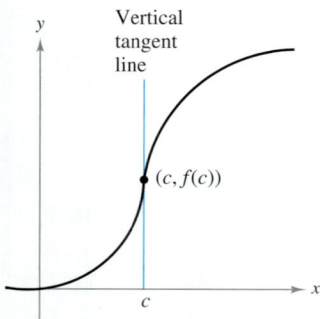

The graph of f has a vertical tangent line at $(c, f(c))$.
Figure 2.7

The Derivative of a Function

You have now arrived at a crucial point in the study of calculus. The limit used to define the slope of a tangent line is also used to define one of the two fundamental operations of calculus—**differentiation.**

Definition of the Derivative of a Function

The **derivative** of f at x is

$$f'(x) = \lim_{\Delta x \to 0} \frac{f(x + \Delta x) - f(x)}{\Delta x}$$

provided the limit exists. For all x for which this limit exists, f' is a function of x.

· · **REMARK** The notation $f'(x)$ is read as "f prime of x."

Be sure you see that the derivative of a function of x is also a function of x. This "new" function gives the slope of the tangent line to the graph of f at the point $(x, f(x))$, provided that the graph has a tangent line at this point. The derivative can also be used to determine the **instantaneous rate of change** (or simply the **rate of change**) of one variable with respect to another.

The process of finding the derivative of a function is called **differentiation.** A function is **differentiable** at x when its derivative exists at x and is **differentiable on an open interval** (a, b) when it is differentiable at every point in the interval.

In addition to $f'(x)$, other notations are used to denote the derivative of $y = f(x)$. The most common are

$$f'(x), \quad \frac{dy}{dx}, \quad y', \quad \frac{d}{dx}[f(x)], \quad D_x[y].$$

Notation for derivatives

■ FOR FURTHER INFORMATION For more information on the crediting of mathematical discoveries to the first "discoverers," see the article "Mathematical Firsts— Who Done It?" by Richard H. Williams and Roy D. Mazzagatti in *Mathematics Teacher*. To view this article, go to *MathArticles.com*.

The notation dy/dx is read as "the derivative of y *with respect to* x" or simply "dy, dx." Using limit notation, you can write

$$\frac{dy}{dx} = \lim_{\Delta x \to 0} \frac{\Delta y}{\Delta x} = \lim_{\Delta x \to 0} \frac{f(x + \Delta x) - f(x)}{\Delta x} = f'(x).$$

EXAMPLE 3   **Finding the Derivative by the Limit Process**

· · · ▷ *See LarsonCalculus.com for an interactive version of this type of example.*

To find the derivative of $f(x) = x^3 + 2x$, use the definition of the derivative as shown.

$$f'(x) = \lim_{\Delta x \to 0} \frac{f(x + \Delta x) - f(x)}{\Delta x} \qquad \text{Definition of derivative}$$

$$= \lim_{\Delta x \to 0} \frac{(x + \Delta x)^3 + 2(x + \Delta x) - (x^3 + 2x)}{\Delta x}$$

$$= \lim_{\Delta x \to 0} \frac{x^3 + 3x^2 \Delta x + 3x(\Delta x)^2 + (\Delta x)^3 + 2x + 2\Delta x - x^3 - 2x}{\Delta x}$$

$$= \lim_{\Delta x \to 0} \frac{3x^2 \Delta x + 3x(\Delta x)^2 + (\Delta x)^3 + 2\Delta x}{\Delta x}$$

$$= \lim_{\Delta x \to 0} \frac{\Delta x[3x^2 + 3x\Delta x + (\Delta x)^2 + 2]}{\Delta x}$$

$$= \lim_{\Delta x \to 0} [3x^2 + 3x\Delta x + (\Delta x)^2 + 2]$$

$$= 3x^2 + 2$$

· · **REMARK** When using the definition to find a derivative of a function, the key is to rewrite the difference quotient so that Δx does not occur as a factor of the denominator.

· · · · · · · · · · · · ·▷
·· REMARK Remember that the derivative of a function f is itself a function, which can be used to find the slope of the tangent line at the point $(x, f(x))$ on the graph of f.

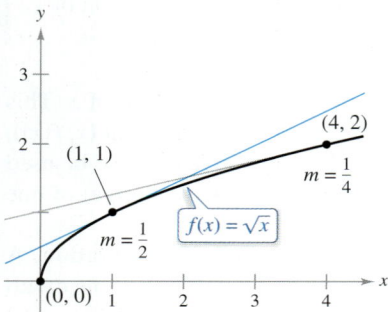

EXAMPLE 4   Using the Derivative to Find the Slope at a Point

Find $f'(x)$ for $f(x) = \sqrt{x}$. Then find the slopes of the graph of f at the points $(1, 1)$ and $(4, 2)$. Discuss the behavior of f at $(0, 0)$.

Solution   Use the procedure for rationalizing numerators, as discussed in Section 1.3.

$$f'(x) = \lim_{\Delta x \to 0} \frac{f(x + \Delta x) - f(x)}{\Delta x} \qquad \text{Definition of derivative}$$

$$= \lim_{\Delta x \to 0} \frac{\sqrt{x + \Delta x} - \sqrt{x}}{\Delta x}$$

$$= \lim_{\Delta x \to 0} \left(\frac{\sqrt{x + \Delta x} - \sqrt{x}}{\Delta x} \right)\left(\frac{\sqrt{x + \Delta x} + \sqrt{x}}{\sqrt{x + \Delta x} + \sqrt{x}} \right)$$

$$= \lim_{\Delta x \to 0} \frac{(x + \Delta x) - x}{\Delta x(\sqrt{x + \Delta x} + \sqrt{x})}$$

$$= \lim_{\Delta x \to 0} \frac{\Delta x}{\Delta x(\sqrt{x + \Delta x} + \sqrt{x})}$$

$$= \lim_{\Delta x \to 0} \frac{1}{\sqrt{x + \Delta x} + \sqrt{x}}$$

$$= \frac{1}{2\sqrt{x}}, \quad x > 0$$

The slope of f at $(x, f(x))$, $x > 0$, is $m = 1/(2\sqrt{x})$.
Figure 2.8

At the point $(1, 1)$, the slope is $f'(1) = \frac{1}{2}$. At the point $(4, 2)$, the slope is $f'(4) = \frac{1}{4}$. See Figure 2.8. At the point $(0, 0)$, the slope is undefined. Moreover, the graph of f has a vertical tangent line at $(0, 0)$.

· · · · · · · · · · · · ·▷
·· REMARK   In many applications, it is convenient to use a variable other than x as the independent variable, as shown in Example 5.

EXAMPLE 5   Finding the Derivative of a Function

··▷ *See LarsonCalculus.com for an interactive version of this type of example.*

Find the derivative with respect to t for the function $y = 2/t$.

Solution   Considering $y = f(t)$, you obtain

$$\frac{dy}{dt} = \lim_{\Delta t \to 0} \frac{f(t + \Delta t) - f(t)}{\Delta t} \qquad \text{Definition of derivative}$$

$$= \lim_{\Delta t \to 0} \frac{\dfrac{2}{t + \Delta t} - \dfrac{2}{t}}{\Delta t} \qquad f(t + \Delta t) = \frac{2}{t + \Delta t} \text{ and } f(t) = \frac{2}{t}$$

$$= \lim_{\Delta t \to 0} \frac{\dfrac{2t - 2(t + \Delta t)}{t(t + \Delta t)}}{\Delta t} \qquad \text{Combine fractions in numerator.}$$

$$= \lim_{\Delta t \to 0} \frac{-2\Delta t}{\Delta t(t)(t + \Delta t)} \qquad \text{Divide out common factor of } \Delta t.$$

$$= \lim_{\Delta t \to 0} \frac{-2}{t(t + \Delta t)} \qquad \text{Simplify.}$$

$$= -\frac{2}{t^2}. \qquad \text{Evaluate limit as } \Delta t \to 0.$$

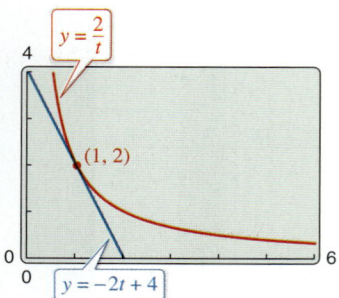

At the point $(1, 2)$, the line $y = -2t + 4$ is tangent to the graph of $y = 2/t$.
Figure 2.9

▷ **TECHNOLOGY**   A graphing utility can be used to reinforce the result given in Example 5. For instance, using the formula $dy/dt = -2/t^2$, you know that the slope of the graph of $y = 2/t$ at the point $(1, 2)$ is $m = -2$. Using the point-slope form, you can find that the equation of the tangent line to the graph at $(1, 2)$ is

$$y - 2 = -2(t - 1) \quad \text{or} \quad y = -2t + 4$$

as shown in Figure 2.9.

Differentiability and Continuity

The alternative limit form of the derivative shown below is useful in investigating the relationship between differentiability and continuity. The derivative of f at c is

$$f'(c) = \lim_{x \to c} \frac{f(x) - f(c)}{x - c}$$

Alternative form of derivative

provided this limit exists (see Figure 2.10).

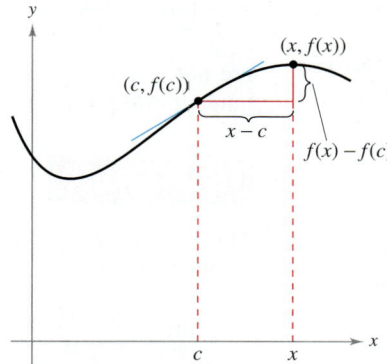

As x approaches c, the secant line approaches the tangent line.
Figure 2.10

Note that the existence of the limit in this alternative form requires that the one-sided limits

$$\lim_{x \to c^-} \frac{f(x) - f(c)}{x - c}$$

and

$$\lim_{x \to c^+} \frac{f(x) - f(c)}{x - c}$$

exist and are equal. These one-sided limits are called the **derivatives from the left and from the right,** respectively. It follows that f is **differentiable on the closed interval** $[a, b]$ when it is differentiable on (a, b) and when the derivative from the right at a and the derivative from the left at b both exist.

When a function is not continuous at $x = c$, it is also not differentiable at $x = c$. For instance, the greatest integer function

$$f(x) = [\![x]\!]$$

is not continuous at $x = 0$, and so it is not differentiable at $x = 0$ (see Figure 2.11). You can verify this by observing that

$$\lim_{x \to 0^-} \frac{f(x) - f(0)}{x - 0} = \lim_{x \to 0^-} \frac{[\![x]\!] - 0}{x} = \infty$$

Derivative from the left

and

$$\lim_{x \to 0^+} \frac{f(x) - f(0)}{x - 0} = \lim_{x \to 0^+} \frac{[\![x]\!] - 0}{x} = 0.$$

Derivative from the right

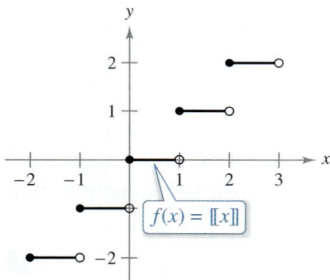

$$f(x) = [\![x]\!]$$

The greatest integer function is not differentiable at $x = 0$ because it is not continuous at $x = 0$.
Figure 2.11

Although it is true that differentiability implies continuity (as shown in Theorem 2.1 on the next page), the converse is not true. That is, it is possible for a function to be continuous at $x = c$ and *not* differentiable at $x = c$. Examples 6 and 7 illustrate this possibility.

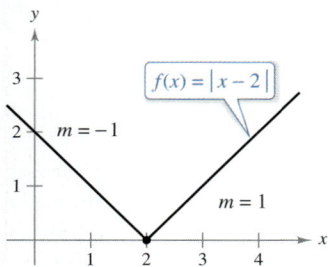

f is not differentiable at $x = 2$ because the derivatives from the left and from the right are not equal.

Figure 2.12

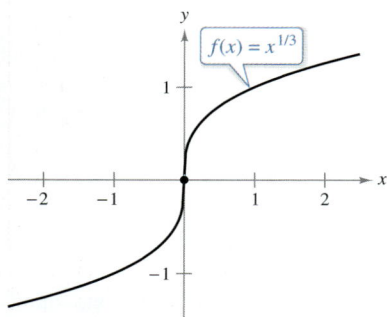

f is not differentiable at $x = 0$ because f has a vertical tangent line at $x = 0$.

Figure 2.13

EXAMPLE 6 **A Graph with a Sharp Turn**

•••▷ See LarsonCalculus.com for an interactive version of this type of example.

The function $f(x) = |x - 2|$, shown in Figure 2.12, is continuous at $x = 2$. The one-sided limits, however,

$$\lim_{x \to 2^-} \frac{f(x) - f(2)}{x - 2} = \lim_{x \to 2^-} \frac{|x - 2| - 0}{x - 2} = -1 \qquad \text{Derivative from the left}$$

and

$$\lim_{x \to 2^+} \frac{f(x) - f(2)}{x - 2} = \lim_{x \to 2^+} \frac{|x - 2| - 0}{x - 2} = 1 \qquad \text{Derivative from the right}$$

are not equal. So, f is not differentiable at $x = 2$ and the graph of f does not have a tangent line at the point $(2, 0)$.

EXAMPLE 7 **A Graph with a Vertical Tangent Line**

The function $f(x) = x^{1/3}$ is continuous at $x = 0$, as shown in Figure 2.13. However, because the limit

$$\lim_{x \to 0} \frac{f(x) - f(0)}{x - 0} = \lim_{x \to 0} \frac{x^{1/3} - 0}{x} = \lim_{x \to 0} \frac{1}{x^{2/3}} = \infty$$

is infinite, you can conclude that the tangent line is vertical at $x = 0$. So, f is not differentiable at $x = 0$.

From Examples 6 and 7, you can see that a function is not differentiable at a point at which its graph has a sharp turn *or* a vertical tangent line.

THEOREM 2.1 **Differentiability Implies Continuity**

If f is differentiable at $x = c$, then f is continuous at $x = c$.

Proof You can prove that f is continuous at $x = c$ by showing that $f(x)$ approaches $f(c)$ as $x \to c$. To do this, use the differentiability of f at $x = c$ and consider the following limit.

$$\lim_{x \to c} [f(x) - f(c)] = \lim_{x \to c} \left[(x - c) \left(\frac{f(x) - f(c)}{x - c} \right) \right]$$

$$= \left[\lim_{x \to c} (x - c) \right] \left[\lim_{x \to c} \frac{f(x) - f(c)}{x - c} \right]$$

$$= (0)[f'(c)]$$

$$= 0$$

Because the difference $f(x) - f(c)$ approaches zero as $x \to c$, you can conclude that $\lim_{x \to c} f(x) = f(c)$. So, f is continuous at $x = c$.

See LarsonCalculus.com for Bruce Edwards's video of this proof.

▷ **TECHNOLOGY** Some graphing utilities, such as *Maple*, *Mathematica*, and the *TI-nspire*, perform symbolic differentiation. Others perform *numerical differentiation* by finding values of derivatives using the formula

$$f'(x) \approx \frac{f(x + \Delta x) - f(x - \Delta x)}{2\Delta x}$$

where Δx is a small number such as 0.001. Can you see any problems with this definition? For instance, using this definition, what is the value of the derivative of $f(x) = |x|$ when $x = 0$?

The relationship between continuity and differentiability is summarized below.

1. If a function is differentiable at $x = c$, then it is continuous at $x = c$. So, differentiability implies continuity.

2. It is possible for a function to be continuous at $x = c$ and not be differentiable at $x = c$. So, continuity does not imply differentiability (see Example 6).

2.1   Exercises

See **CalcChat.com** for tutorial help and worked-out solutions to odd-numbered exercises.

Estimating Slope   In Exercises 1 and 2, estimate the slope of the graph at the points (x_1, y_1) and (x_2, y_2).

1.

2.

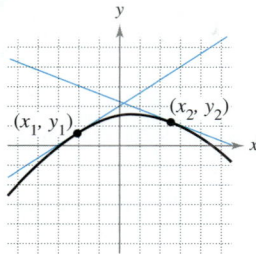

Slopes of Secant Lines   In Exercises 3 and 4, use the graph shown in the figure. To print an enlarged copy of the graph, go to *MathGraphs.com*.

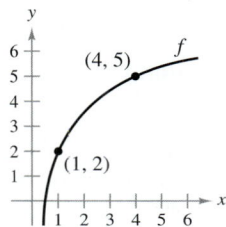

3. Identify or sketch each of the quantities on the figure.

(a) $f(1)$ and $f(4)$ (b) $f(4) - f(1)$

(c) $y = \dfrac{f(4) - f(1)}{4 - 1}(x - 1) + f(1)$

4. Insert the proper inequality symbol ($<$ or $>$) between the given quantities.

(a) $\dfrac{f(4) - f(1)}{4 - 1}$ ▨ $\dfrac{f(4) - f(3)}{4 - 3}$

(b) $\dfrac{f(4) - f(1)}{4 - 1}$ ▨ $f'(1)$

Finding the Slope of a Tangent Line   In Exercises 5–10, find the slope of the tangent line to the graph of the function at the given point.

5. $f(x) = 3 - 5x$, $(-1, 8)$ **6.** $g(x) = \frac{3}{2}x + 1$, $(-2, -2)$

7. $g(x) = x^2 - 9$, $(2, -5)$ **8.** $f(x) = 5 - x^2$, $(3, -4)$

9. $f(t) = 3t - t^2$, $(0, 0)$ **10.** $h(t) = t^2 + 4t$, $(1, 5)$

Finding the Derivative by the Limit Process   In Exercises 11–24, find the derivative of the function by the limit process.

11. $f(x) = 7$ **12.** $g(x) = -3$

13. $f(x) = -10x$ **14.** $f(x) = 7x - 3$

15. $h(s) = 3 + \frac{2}{3}s$ **16.** $f(x) = 5 - \frac{2}{3}x$

17. $f(x) = x^2 + x - 3$ **18.** $f(x) = x^2 - 5$

19. $f(x) = x^3 - 12x$ **20.** $f(x) = x^3 + x^2$

21. $f(x) = \dfrac{1}{x - 1}$ **22.** $f(x) = \dfrac{1}{x^2}$

23. $f(x) = \sqrt{x + 4}$ **24.** $f(x) = \dfrac{4}{\sqrt{x}}$

Finding an Equation of a Tangent Line   In Exercises 25–32, (a) find an equation of the tangent line to the graph of f at the given point, (b) use a graphing utility to graph the function and its tangent line at the point, and (c) use the *derivative* feature of a graphing utility to confirm your results.

25. $f(x) = x^2 + 3$, $(-1, 4)$ **26.** $f(x) = x^2 + 2x - 1$, $(1, 2)$

27. $f(x) = x^3$, $(2, 8)$ **28.** $f(x) = x^3 + 1$, $(-1, 0)$

29. $f(x) = \sqrt{x}$, $(1, 1)$ **30.** $f(x) = \sqrt{x - 1}$, $(5, 2)$

31. $f(x) = x + \dfrac{4}{x}$, $(-4, -5)$ **32.** $f(x) = \dfrac{6}{x + 2}$, $(0, 3)$

Finding an Equation of a Tangent Line   In Exercises 33–38, find an equation of the line that is tangent to the graph of f **and** parallel to the given line.

Function	Line
33. $f(x) = x^2$	$2x - y + 1 = 0$
34. $f(x) = 2x^2$	$4x + y + 3 = 0$
35. $f(x) = x^3$	$3x - y + 1 = 0$
36. $f(x) = x^3 + 2$	$3x - y - 4 = 0$
37. $f(x) = \dfrac{1}{\sqrt{x}}$	$x + 2y - 6 = 0$
38. $f(x) = \dfrac{1}{\sqrt{x - 1}}$	$x + 2y + 7 = 0$

WRITING ABOUT CONCEPTS

Sketching a Derivative   In Exercises 39–44, sketch the graph of f'. Explain how you found your answer.

39.

40.

41.

42.

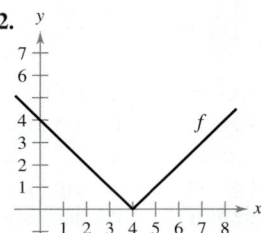

WRITING ABOUT CONCEPTS (continued)

43.

44.

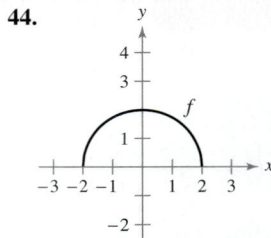

45. Sketching a Graph Sketch a graph of a function whose derivative is always negative. Explain how you found the answer.

46. Sketching a Graph Sketch a graph of a function whose derivative is always positive. Explain how you found the answer.

47. Using a Tangent Line The tangent line to the graph of $y = g(x)$ at the point $(4, 5)$ passes through the point $(7, 0)$. Find $g(4)$ and $g'(4)$.

48. Using a Tangent Line The tangent line to the graph of $y = h(x)$ at the point $(-1, 4)$ passes through the point $(3, 6)$. Find $h(-1)$ and $h'(-1)$.

Working Backwards In Exercises 49–52, the limit represents $f'(c)$ for a function f and a number c. Find f and c.

49. $\displaystyle\lim_{\Delta x \to 0} \frac{[5 - 3(1 + \Delta x)] - 2}{\Delta x}$

50. $\displaystyle\lim_{\Delta x \to 0} \frac{(-2 + \Delta x)^3 + 8}{\Delta x}$

51. $\displaystyle\lim_{x \to 6} \frac{-x^2 + 36}{x - 6}$

52. $\displaystyle\lim_{x \to 9} \frac{2\sqrt{x} - 6}{x - 9}$

Writing a Function Using Derivatives In Exercises 53 and 54, identify a function f that has the given characteristics. Then sketch the function.

53. $f(0) = 2; f'(x) = -3$ for $-\infty < x < \infty$

54. $f(0) = 4; f'(0) = 0; f'(x) < 0$ for $x < 0; f'(x) > 0$ for $x > 0$

Finding an Equation of a Tangent Line In Exercises 55 and 56, find equations of the two tangent lines to the graph of f that pass through the indicated point.

55. $f(x) = 4x - x^2$

56. $f(x) = x^2$

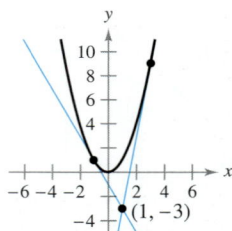

57. Graphical Reasoning Use a graphing utility to graph each function and its tangent lines at $x = -1, x = 0$, and $x = 1$. Based on the results, determine whether the slopes of tangent lines to the graph of a function at different values of x are always distinct.

(a) $f(x) = x^2$ (b) $g(x) = x^3$

58. HOW DO YOU SEE IT? The figure shows the graph of g'.

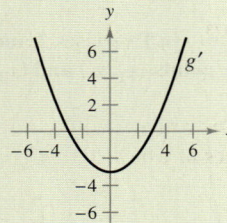

(a) $g'(0) = $ ▢   (b) $g'(3) = $ ▢

(c) What can you conclude about the graph of g knowing that $g'(1) = -\frac{8}{3}$?

(d) What can you conclude about the graph of g knowing that $g'(-4) = \frac{7}{3}$?

(e) Is $g(6) - g(4)$ positive or negative? Explain.

(f) Is it possible to find $g(2)$ from the graph? Explain.

59. Graphical Reasoning Consider the function $f(x) = \frac{1}{2}x^2$.

(a) Use a graphing utility to graph the function and estimate the values of $f'(0), f'(\frac{1}{2}), f'(1)$, and $f'(2)$.

(b) Use your results from part (a) to determine the values of $f'(-\frac{1}{2}), f'(-1)$, and $f'(-2)$.

(c) Sketch a possible graph of f'.

(d) Use the definition of derivative to find $f'(x)$.

60. Graphical Reasoning Consider the function $f(x) = \frac{1}{3}x^3$.

(a) Use a graphing utility to graph the function and estimate the values of $f'(0), f'(\frac{1}{2}), f'(1), f'(2)$, and $f'(3)$.

(b) Use your results from part (a) to determine the values of $f'(-\frac{1}{2}), f'(-1), f'(-2)$, and $f'(-3)$.

(c) Sketch a possible graph of f'.

(d) Use the definition of derivative to find $f'(x)$.

Graphical Reasoning In Exercises 61 and 62, use a graphing utility to graph the functions f and g in the same viewing window, where

$$g(x) = \frac{f(x + 0.01) - f(x)}{0.01}.$$

Label the graphs and describe the relationship between them.

61. $f(x) = 2x - x^2$

62. $f(x) = 3\sqrt{x}$

Approximating a Derivative In Exercises 63 and 64, evaluate $f(2)$ and $f(2.1)$ and use the results to approximate $f'(2)$.

63. $f(x) = x(4 - x)$

64. $f(x) = \frac{1}{4}x^3$

Using the Alternative Form of the Derivative In Exercises 65–74, use the alternative form of the derivative to find the derivative at $x = c$ (if it exists).

65. $f(x) = x^2 - 5, \quad c = 3$

66. $g(x) = x^2 - x, \quad c = 1$

67. $f(x) = x^3 + 2x^2 + 1, \quad c = -2$

68. $f(x) = x^3 + 6x$, $c = 2$

69. $g(x) = \sqrt{|x|}$, $c = 0$ **70.** $f(x) = 3/x$, $c = 4$

71. $f(x) = (x - 6)^{2/3}$, $c = 6$

72. $g(x) = (x + 3)^{1/3}$, $c = -3$

73. $h(x) = |x + 7|$, $c = -7$ **74.** $f(x) = |x - 6|$, $c = 6$

Determining Differentiability In Exercises 75–80, describe the *x*-values at which *f* is differentiable.

75. $f(x) = \dfrac{2}{x - 3}$ **76.** $f(x) = |x^2 - 9|$

    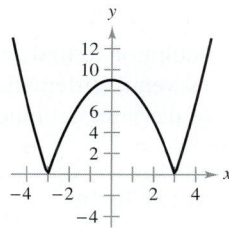

77. $f(x) = (x + 4)^{2/3}$ **78.** $f(x) = \dfrac{x^2}{x^2 - 4}$

    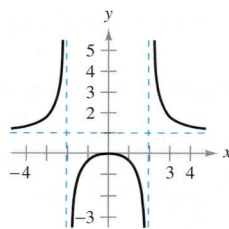

79. $f(x) = \sqrt{x - 1}$ **80.** $f(x) = \begin{cases} x^2 - 4, & x \le 0 \\ 4 - x^2, & x > 0 \end{cases}$

    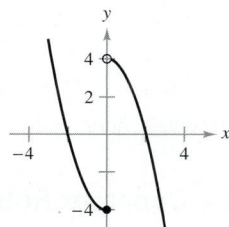

Graphical Reasoning In Exercises 81–84, use a graphing utility to graph the function and find the *x*-values at which *f* is differentiable.

81. $f(x) = |x - 5|$ **82.** $f(x) = \dfrac{4x}{x - 3}$

83. $f(x) = x^{2/5}$

84. $f(x) = \begin{cases} x^3 - 3x^2 + 3x, & x \le 1 \\ x^2 - 2x, & x > 1 \end{cases}$

Determining Differentiability In Exercises 85–88, find the derivatives from the left and from the right at $x = 1$ (if they exist). Is the function differentiable at $x = 1$?

85. $f(x) = |x - 1|$ **86.** $f(x) = \sqrt{1 - x^2}$

87. $f(x) = \begin{cases} (x - 1)^3, & x \le 1 \\ (x - 1)^2, & x > 1 \end{cases}$ **88.** $f(x) = \begin{cases} x, & x \le 1 \\ x^2, & x > 1 \end{cases}$

Determining Differentiability In Exercises 89 and 90, determine whether the function is differentiable at $x = 2$.

89. $f(x) = \begin{cases} x^2 + 1, & x \le 2 \\ 4x - 3, & x > 2 \end{cases}$ **90.** $f(x) = \begin{cases} \frac{1}{2}x + 1, & x < 2 \\ \sqrt{2x}, & x \ge 2 \end{cases}$

91. Graphical Reasoning A line with slope m passes through the point $(0, 4)$ and has the equation $y = mx + 4$.

(a) Write the distance d between the line and the point $(3, 1)$ as a function of m.

(b) Use a graphing utility to graph the function d in part (a). Based on the graph, is the function differentiable at every value of m? If not, where is it not differentiable?

92. Conjecture Consider the functions $f(x) = x^2$ and $g(x) = x^3$.

(a) Graph f and f' on the same set of axes.

(b) Graph g and g' on the same set of axes.

(c) Identify a pattern between f and g and their respective derivatives. Use the pattern to make a conjecture about $h'(x)$ if $h(x) = x^n$, where n is an integer and $n \ge 2$.

(d) Find $f'(x)$ if $f(x) = x^4$. Compare the result with the conjecture in part (c). Is this a proof of your conjecture? Explain.

True or False? In Exercises 93–96, determine whether the statement is true or false. If it is false, explain why or give an example that shows it is false.

93. The slope of the tangent line to the differentiable function f at the point $(2, f(2))$ is

$$\frac{f(2 + \Delta x) - f(2)}{\Delta x}.$$

94. If a function is continuous at a point, then it is differentiable at that point.

95. If a function has derivatives from both the right and the left at a point, then it is differentiable at that point.

96. If a function is differentiable at a point, then it is continuous at that point.

97. Differentiability and Continuity Let

$$f(x) = \begin{cases} x \sin \dfrac{1}{x}, & x \ne 0 \\ 0, & x = 0 \end{cases}$$

and

$$g(x) = \begin{cases} x^2 \sin \dfrac{1}{x}, & x \ne 0 \\ 0, & x = 0 \end{cases}.$$

Show that f is continuous, but not differentiable, at $x = 0$. Show that g is differentiable at 0, and find $g'(0)$.

98. Writing Use a graphing utility to graph the two functions $f(x) = x^2 + 1$ and $g(x) = |x| + 1$ in the same viewing window. Use the *zoom* and *trace* features to analyze the graphs near the point $(0, 1)$. What do you observe? Which function is differentiable at this point? Write a short paragraph describing the geometric significance of differentiability at a point.

2.2 Basic Differentiation Rules and Rates of Change

- Find the derivative of a function using the Constant Rule.
- Find the derivative of a function using the Power Rule.
- Find the derivative of a function using the Constant Multiple Rule.
- Find the derivative of a function using the Sum and Difference Rules.
- Find the derivatives of the sine function and of the cosine function.
- Use derivatives to find rates of change.

The Constant Rule

In Section 2.1, you used the limit definition to find derivatives. In this and the next two sections, you will be introduced to several "differentiation rules" that allow you to find derivatives without the *direct* use of the limit definition.

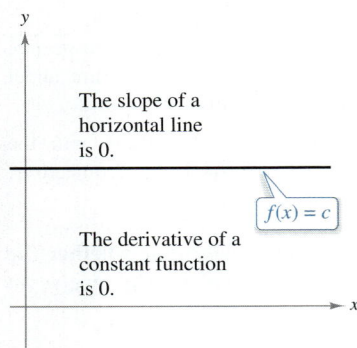

> **THEOREM 2.2   The Constant Rule**
>
> The derivative of a constant function is 0. That is, if c is a real number, then
>
> $$\frac{d}{dx}[c] = 0. \qquad \text{(See Figure 2.14.)}$$

The slope of a horizontal line is 0.

$f(x) = c$

The derivative of a constant function is 0.

Notice that the Constant Rule is equivalent to saying that the slope of a horizontal line is 0. This demonstrates the relationship between slope and derivative.

Figure 2.14

Proof   Let $f(x) = c$. Then, by the limit definition of the derivative,

$$\frac{d}{dx}[c] = f'(x)$$
$$= \lim_{\Delta x \to 0} \frac{f(x + \Delta x) - f(x)}{\Delta x}$$
$$= \lim_{\Delta x \to 0} \frac{c - c}{\Delta x}$$
$$= \lim_{\Delta x \to 0} 0$$
$$= 0.$$

See LarsonCalculus.com for Bruce Edwards's video of this proof.

EXAMPLE 1   **Using the Constant Rule**

Function	Derivative
a. $y = 7$	$dy/dx = 0$
b. $f(x) = 0$	$f'(x) = 0$
c. $s(t) = -3$	$s'(t) = 0$
d. $y = k\pi^2$, k is constant	$y' = 0$

Exploration

Writing a Conjecture   Use the definition of the derivative given in Section 2.1 to find the derivative of each function. What patterns do you see? Use your results to write a conjecture about the derivative of $f(x) = x^n$.

a. $f(x) = x^1$ **b.** $f(x) = x^2$ **c.** $f(x) = x^3$

d. $f(x) = x^4$ **e.** $f(x) = x^{1/2}$ **f.** $f(x) = x^{-1}$

The Power Rule

Before proving the next rule, it is important to review the procedure for expanding a binomial.

$$(x + \Delta x)^2 = x^2 + 2x\Delta x + (\Delta x)^2$$
$$(x + \Delta x)^3 = x^3 + 3x^2\Delta x + 3x(\Delta x)^2 + (\Delta x)^3$$
$$(x + \Delta x)^4 = x^4 + 4x^3\Delta x + 6x^2(\Delta x)^2 + 4x(\Delta x)^3 + (\Delta x)^4$$
$$(x + \Delta x)^5 = x^5 + 5x^4\Delta x + 10x^3(\Delta x)^2 + 10x^2(\Delta x)^3 + 5x(\Delta x)^4 + (\Delta x)^5$$

The general binomial expansion for a positive integer n is

$$(x + \Delta x)^n = x^n + nx^{n-1}(\Delta x) + \underbrace{\frac{n(n-1)x^{n-2}}{2}(\Delta x)^2 + \cdots + (\Delta x)^n}.$$

$(\Delta x)^2$ is a factor of these terms.

This binomial expansion is used in proving a special case of the Power Rule.

THEOREM 2.3   The Power Rule

If n is a rational number, then the function $f(x) = x^n$ is differentiable and

$$\frac{d}{dx}[x^n] = nx^{n-1}.$$

For f to be differentiable at $x = 0$, n must be a number such that x^{n-1} is defined on an interval containing 0.

REMARK From Example 7 in Section 2.1, you know that the function $f(x) = x^{1/3}$ is defined at $x = 0$, but is not differentiable at $x = 0$. This is because $x^{-2/3}$ is not defined on an interval containing 0.

Proof   If n is a positive integer greater than 1, then the binomial expansion produces

$$\frac{d}{dx}[x^n] = \lim_{\Delta x \to 0} \frac{(x + \Delta x)^n - x^n}{\Delta x}$$

$$= \lim_{\Delta x \to 0} \frac{x^n + nx^{n-1}(\Delta x) + \frac{n(n-1)x^{n-2}}{2}(\Delta x)^2 + \cdots + (\Delta x)^n - x^n}{\Delta x}$$

$$= \lim_{\Delta x \to 0} \left[nx^{n-1} + \frac{n(n-1)x^{n-2}}{2}(\Delta x) + \cdots + (\Delta x)^{n-1} \right]$$

$$= nx^{n-1} + 0 + \cdots + 0$$

$$= nx^{n-1}.$$

This proves the case for which n is a positive integer greater than 1. It is left to you to prove the case for $n = 1$. Example 7 in Section 2.3 proves the case for which n is a negative integer. In Exercise 71 in Section 2.5, you are asked to prove the case for which n is rational. (In Section 5.5, the Power Rule will be extended to cover irrational values of n.)
See LarsonCalculus.com for Bruce Edwards's video of this proof.

When using the Power Rule, the case for which $n = 1$ is best thought of as a separate differentiation rule. That is,

$$\frac{d}{dx}[x] = 1.$$ Power Rule when $n = 1$

This rule is consistent with the fact that the slope of the line $y = x$ is 1, as shown in Figure 2.15.

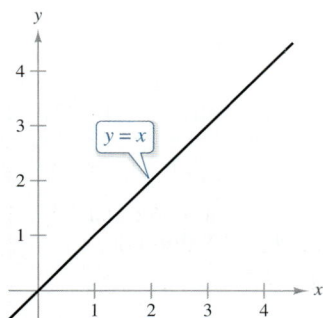

The slope of the line $y = x$ is 1.
Figure 2.15

| EXAMPLE 2 | **Using the Power Rule** |

| **Function** | **Derivative** |

a. $f(x) = x^3$ $f'(x) = 3x^2$

b. $g(x) = \sqrt[3]{x}$ $g'(x) = \dfrac{d}{dx}[x^{1/3}] = \dfrac{1}{3}x^{-2/3} = \dfrac{1}{3x^{2/3}}$

c. $y = \dfrac{1}{x^2}$ $\dfrac{dy}{dx} = \dfrac{d}{dx}[x^{-2}] = (-2)x^{-3} = -\dfrac{2}{x^3}$ ■

In Example 2(c), note that *before* differentiating, $1/x^2$ was rewritten as x^{-2}. Rewriting is the first step in *many* differentiation problems.

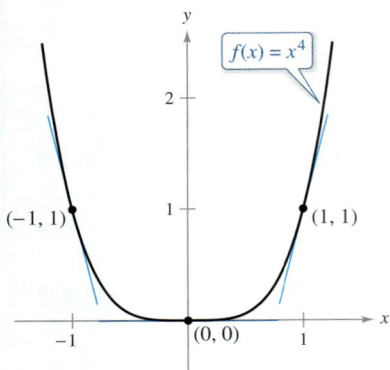

Given:	Rewrite:	Differentiate:	Simplify:
$y = \dfrac{1}{x^2}$	$y = x^{-2}$	$\dfrac{dy}{dx} = (-2)x^{-3}$	$\dfrac{dy}{dx} = -\dfrac{2}{x^3}$

| EXAMPLE 3 | **Finding the Slope of a Graph** |

⋯▷ *See LarsonCalculus.com for an interactive version of this type of example.*

Find the slope of the graph of

$$f(x) = x^4$$

for each value of x.

a. $x = -1$ **b.** $x = 0$ **c.** $x = 1$

Solution   The slope of a graph at a point is the value of the derivative at that point. The derivative of f is $f'(x) = 4x^3$.

a. When $x = -1$, the slope is $f'(-1) = 4(-1)^3 = -4$.   Slope is negative.

b. When $x = 0$, the slope is $f'(0) = 4(0)^3 = 0$.   Slope is zero.

c. When $x = 1$, the slope is $f'(1) = 4(1)^3 = 4$.   Slope is positive.

See Figure 2.16.

Note that the slope of the graph is negative at the point $(-1, 1)$, the slope is zero at the point $(0, 0)$, and the slope is positive at the point $(1, 1)$.
Figure 2.16

| EXAMPLE 4 | **Finding an Equation of a Tangent Line** |

⋯▷ *See LarsonCalculus.com for an interactive version of this type of example.*

Find an equation of the tangent line to the graph of $f(x) = x^2$ when $x = -2$.

Solution   To find the *point* on the graph of f, evaluate the original function at $x = -2$.

$$(-2, f(-2)) = (-2, 4)$$ Point on graph

To find the *slope* of the graph when $x = -2$, evaluate the derivative, $f'(x) = 2x$, at $x = -2$.

$$m = f'(-2) = -4$$ Slope of graph at $(-2, 4)$

Now, using the point-slope form of the equation of a line, you can write

$$y - y_1 = m(x - x_1)$$ Point-slope form

$$y - 4 = -4[x - (-2)]$$ Substitute for y_1, m, and x_1.

$$y = -4x - 4.$$ Simplify.

See Figure 2.17.

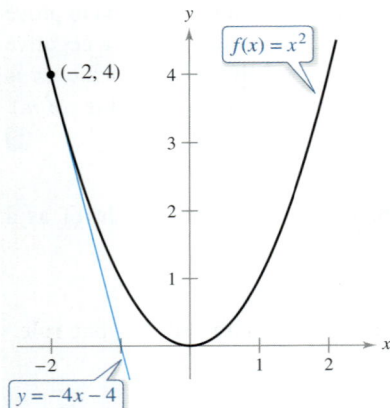

The line $y = -4x - 4$ is tangent to the graph of $f(x) = x^2$ at the point $(-2, 4)$.
Figure 2.17

■

The Constant Multiple Rule

THEOREM 2.4   The Constant Multiple Rule

If f is a differentiable function and c is a real number, then cf is also differentiable and $\dfrac{d}{dx}[cf(x)] = cf'(x)$.

Proof

$$\frac{d}{dx}[cf(x)] = \lim_{\Delta x \to 0} \frac{cf(x + \Delta x) - cf(x)}{\Delta x} \qquad \text{Definition of derivative}$$

$$= \lim_{\Delta x \to 0} c\left[\frac{f(x + \Delta x) - f(x)}{\Delta x}\right]$$

$$= c\left[\lim_{\Delta x \to 0} \frac{f(x + \Delta x) - f(x)}{\Delta x}\right] \qquad \text{Apply Theorem 1.2.}$$

$$= cf'(x)$$

See LarsonCalculus.com for Bruce Edwards's video of this proof.

Informally, the Constant Multiple Rule states that constants can be factored out of the differentiation process, even when the constants appear in the denominator.

$$\frac{d}{dx}[cf(x)] = c\frac{d}{dx}[(\;)f(x)] = cf'(x)$$

$$\frac{d}{dx}\left[\frac{f(x)}{c}\right] = \frac{d}{dx}\left[\left(\frac{1}{c}\right)f(x)\right] = \left(\frac{1}{c}\right)\frac{d}{dx}[(\;)f(x)] = \left(\frac{1}{c}\right)f'(x)$$

EXAMPLE 5   **Using the Constant Multiple Rule**

Function	Derivative

a. $y = 5x^3$ $\qquad \dfrac{dy}{dx} = \dfrac{d}{dx}[5x^3] = 5\dfrac{d}{dx}[x^3] = 5(3)x^2 = 15x^2$

b. $y = \dfrac{2}{x}$ $\qquad \dfrac{dy}{dx} = \dfrac{d}{dx}[2x^{-1}] = 2\dfrac{d}{dx}[x^{-1}] = 2(-1)x^{-2} = -\dfrac{2}{x^2}$

c. $f(t) = \dfrac{4t^2}{5}$ $\qquad f'(t) = \dfrac{d}{dt}\left[\dfrac{4}{5}t^2\right] = \dfrac{4}{5}\dfrac{d}{dt}[t^2] = \dfrac{4}{5}(2t) = \dfrac{8}{5}t$

d. $y = 2\sqrt{x}$ $\qquad \dfrac{dy}{dx} = \dfrac{d}{dx}[2x^{1/2}] = 2\left(\dfrac{1}{2}x^{-1/2}\right) = x^{-1/2} = \dfrac{1}{\sqrt{x}}$

e. $y = \dfrac{1}{2\sqrt[3]{x^2}}$ $\qquad \dfrac{dy}{dx} = \dfrac{d}{dx}\left[\dfrac{1}{2}x^{-2/3}\right] = \dfrac{1}{2}\left(-\dfrac{2}{3}\right)x^{-5/3} = -\dfrac{1}{3x^{5/3}}$

f. $y = -\dfrac{3x}{2}$ $\qquad y' = \dfrac{d}{dx}\left[-\dfrac{3}{2}x\right] = -\dfrac{3}{2}(1) = -\dfrac{3}{2}$

REMARK Before differentiating functions involving radicals, rewrite the function with rational exponents.

The Constant Multiple Rule and the Power Rule can be combined into one rule. The combination rule is

$$\frac{d}{dx}[cx^n] = cnx^{n-1}.$$

EXAMPLE 6 **Using Parentheses When Differentiating**

Original Function	Rewrite	Differentiate	Simplify
a. $y = \dfrac{5}{2x^3}$	$y = \dfrac{5}{2}(x^{-3})$	$y' = \dfrac{5}{2}(-3x^{-4})$	$y' = -\dfrac{15}{2x^4}$
b. $y = \dfrac{5}{(2x)^3}$	$y = \dfrac{5}{8}(x^{-3})$	$y' = \dfrac{5}{8}(-3x^{-4})$	$y' = -\dfrac{15}{8x^4}$
c. $y = \dfrac{7}{3x^{-2}}$	$y = \dfrac{7}{3}(x^2)$	$y' = \dfrac{7}{3}(2x)$	$y' = \dfrac{14x}{3}$
d. $y = \dfrac{7}{(3x)^{-2}}$	$y = 63(x^2)$	$y' = 63(2x)$	$y' = 126x$

The Sum and Difference Rules

> **THEOREM 2.5** **The Sum and Difference Rules**
>
> The sum (or difference) of two differentiable functions f and g is itself differentiable. Moreover, the derivative of $f + g$ (or $f - g$) is the sum (or difference) of the derivatives of f and g.
>
> $$\frac{d}{dx}[f(x) + g(x)] = f'(x) + g'(x) \qquad \text{Sum Rule}$$
>
> $$\frac{d}{dx}[f(x) - g(x)] = f'(x) - g'(x) \qquad \text{Difference Rule}$$

Proof A proof of the Sum Rule follows from Theorem 1.2. (The Difference Rule can be proved in a similar way.)

$$\frac{d}{dx}[f(x) + g(x)] = \lim_{\Delta x \to 0} \frac{[f(x + \Delta x) + g(x + \Delta x)] - [f(x) + g(x)]}{\Delta x}$$

$$= \lim_{\Delta x \to 0} \frac{f(x + \Delta x) + g(x + \Delta x) - f(x) - g(x)}{\Delta x}$$

$$= \lim_{\Delta x \to 0} \left[\frac{f(x + \Delta x) - f(x)}{\Delta x} + \frac{g(x + \Delta x) - g(x)}{\Delta x} \right]$$

$$= \lim_{\Delta x \to 0} \frac{f(x + \Delta x) - f(x)}{\Delta x} + \lim_{\Delta x \to 0} \frac{g(x + \Delta x) - g(x)}{\Delta x}$$

$$= f'(x) + g'(x)$$

See LarsonCalculus.com for Bruce Edwards's video of this proof.

The Sum and Difference Rules can be extended to any finite number of functions. For instance, if $F(x) = f(x) + g(x) - h(x)$, then $F'(x) = f'(x) + g'(x) - h'(x)$.

•• **REMARK** In Example 7(c), note that before differentiating,

$$\frac{3x^2 - x + 1}{x}$$

was rewritten as

$$3x - 1 + \frac{1}{x}.$$

EXAMPLE 7 **Using the Sum and Difference Rules**

Function	Derivative
a. $f(x) = x^3 - 4x + 5$	$f'(x) = 3x^2 - 4$
b. $g(x) = -\dfrac{x^4}{2} + 3x^3 - 2x$	$g'(x) = -2x^3 + 9x^2 - 2$
c. $y = \dfrac{3x^2 - x + 1}{x} = 3x - 1 + \dfrac{1}{x}$	$y' = 3 - \dfrac{1}{x^2} = \dfrac{3x^2 - 1}{x^2}$

FOR FURTHER INFORMATION
For the outline of a geometric proof of the derivatives of the sine and cosine functions, see the article "The Spider's Spacewalk Derivation of sin′ and cos′" by Tim Hesterberg in *The College Mathematics Journal*. To view this article, go to *MathArticles.com*.

Derivatives of the Sine and Cosine Functions

In Section 1.3, you studied the limits

$$\lim_{\Delta x \to 0} \frac{\sin \Delta x}{\Delta x} = 1 \quad \text{and} \quad \lim_{\Delta x \to 0} \frac{1 - \cos \Delta x}{\Delta x} = 0.$$

These two limits can be used to prove differentiation rules for the sine and cosine functions. (The derivatives of the other four trigonometric functions are discussed in Section 2.3.)

THEOREM 2.6   Derivatives of Sine and Cosine Functions

$$\frac{d}{dx}[\sin x] = \cos x \qquad \frac{d}{dx}[\cos x] = -\sin x$$

Proof   Here is a proof of the first rule. (The proof of the second rule is left as an exercise [see Exercise 118].)

$$\frac{d}{dx}[\sin x] = \lim_{\Delta x \to 0} \frac{\sin(x + \Delta x) - \sin x}{\Delta x} \qquad \text{Definition of derivative}$$

$$= \lim_{\Delta x \to 0} \frac{\sin x \cos \Delta x + \cos x \sin \Delta x - \sin x}{\Delta x}$$

$$= \lim_{\Delta x \to 0} \frac{\cos x \sin \Delta x - (\sin x)(1 - \cos \Delta x)}{\Delta x}$$

$$= \lim_{\Delta x \to 0} \left[(\cos x)\left(\frac{\sin \Delta x}{\Delta x}\right) - (\sin x)\left(\frac{1 - \cos \Delta x}{\Delta x}\right) \right]$$

$$= \cos x \left(\lim_{\Delta x \to 0} \frac{\sin \Delta x}{\Delta x} \right) - \sin x \left(\lim_{\Delta x \to 0} \frac{1 - \cos \Delta x}{\Delta x} \right)$$

$$= (\cos x)(1) - (\sin x)(0)$$

$$= \cos x$$

This differentiation rule is shown graphically in Figure 2.18. Note that for each x, the *slope* of the sine curve is equal to the value of the cosine.
See LarsonCalculus.com for Bruce Edwards's video of this proof.

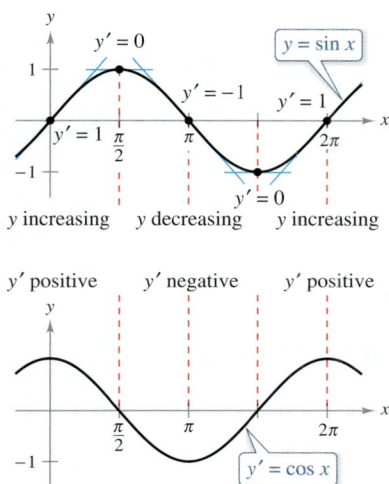

The derivative of the sine function is the cosine function.
Figure 2.18

EXAMPLE 8   Derivatives Involving Sines and Cosines

⋯▷ *See LarsonCalculus.com for an interactive version of this type of example.*

Function	Derivative
a. $y = 2 \sin x$	$y' = 2 \cos x$
b. $y = \dfrac{\sin x}{2} = \dfrac{1}{2} \sin x$	$y' = \dfrac{1}{2} \cos x = \dfrac{\cos x}{2}$
c. $y = x + \cos x$	$y' = 1 - \sin x$
d. $\cos x - \dfrac{\pi}{3} \sin x$	$-\sin x - \dfrac{\pi}{3} \cos x$

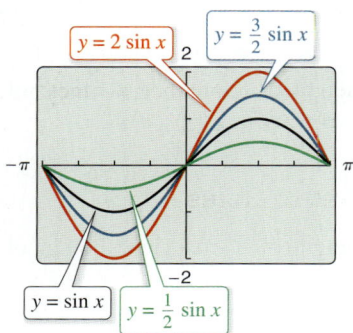

$$\frac{d}{dx}[a \sin x] = a \cos x$$

Figure 2.19

▷ **TECHNOLOGY**   A graphing utility can provide insight into the interpretation of a derivative. For instance, Figure 2.19 shows the graphs of

$$y = a \sin x$$

for $a = \frac{1}{2}, 1, \frac{3}{2},$ and 2. Estimate the slope of each graph at the point $(0, 0)$. Then verify your estimates analytically by evaluating the derivative of each function when $x = 0$.

Rates of Change

You have seen how the derivative is used to determine slope. The derivative can also be used to determine the rate of change of one variable with respect to another. Applications involving rates of change, sometimes referred to as instantaneous rates of change, occur in a wide variety of fields. A few examples are population growth rates, production rates, water flow rates, velocity, and acceleration.

A common use for rate of change is to describe the motion of an object moving in a straight line. In such problems, it is customary to use either a horizontal or a vertical line with a designated origin to represent the line of motion. On such lines, movement to the right (or upward) is considered to be in the positive direction, and movement to the left (or downward) is considered to be in the negative direction.

The function s that gives the position (relative to the origin) of an object as a function of time t is called a **position function.** If, over a period of time Δt, the object changes its position by the amount

$$\Delta s = s(t + \Delta t) - s(t)$$

then, by the familiar formula

$$\text{Rate} = \frac{\text{distance}}{\text{time}}$$

the **average velocity** is

$$\frac{\text{Change in distance}}{\text{Change in time}} = \frac{\Delta s}{\Delta t}. \qquad \text{Average velocity}$$

EXAMPLE 9 Finding Average Velocity of a Falling Object

A billiard ball is dropped from a height of 100 feet. The ball's height s at time t is the position function

$$s = -16t^2 + 100 \qquad \text{Position function}$$

where s is measured in feet and t is measured in seconds. Find the average velocity over each of the following time intervals.

a. $[1, 2]$ **b.** $[1, 1.5]$ **c.** $[1, 1.1]$

Solution

a. For the interval $[1, 2]$, the object falls from a height of $s(1) = -16(1)^2 + 100 = 84$ feet to a height of $s(2) = -16(2)^2 + 100 = 36$ feet. The average velocity is

$$\frac{\Delta s}{\Delta t} = \frac{36 - 84}{2 - 1} = \frac{-48}{1} = -48 \text{ feet per second.}$$

b. For the interval $[1, 1.5]$, the object falls from a height of 84 feet to a height of $s(1.5) = -16(1.5)^2 + 100 = 64$ feet. The average velocity is

$$\frac{\Delta s}{\Delta t} = \frac{64 - 84}{1.5 - 1} = \frac{-20}{0.5} = -40 \text{ feet per second.}$$

c. For the interval $[1, 1.1]$, the object falls from a height of 84 feet to a height of $s(1.1) = -16(1.1)^2 + 100 = 80.64$ feet. The average velocity is

$$\frac{\Delta s}{\Delta t} = \frac{80.64 - 84}{1.1 - 1} = \frac{-3.36}{0.1} = -33.6 \text{ feet per second.}$$

Note that the average velocities are *negative*, indicating that the object is moving downward.

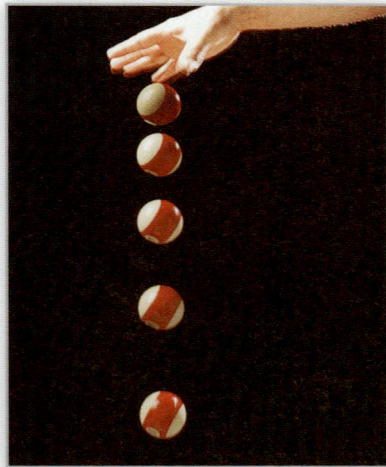

Time-lapse photograph of a free-falling billiard ball

Richard Megna/Fundamental Photographs

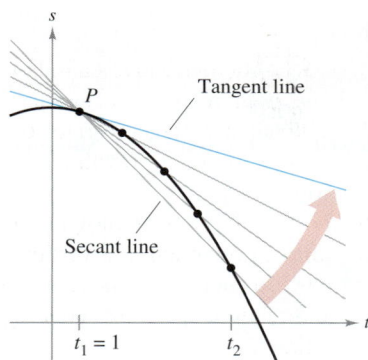

The average velocity between t_1 and t_2 is the slope of the secant line, and the instantaneous velocity at t_1 is the slope of the tangent line.
Figure 2.20

Suppose that in Example 9, you wanted to find the *instantaneous* velocity (or simply the velocity) of the object when $t = 1$. Just as you can approximate the slope of the tangent line by calculating the slope of the secant line, you can approximate the velocity at $t = 1$ by calculating the average velocity over a small interval $[1, 1 + \Delta t]$ (see Figure 2.20). By taking the limit as Δt approaches zero, you obtain the velocity when $t = 1$. Try doing this—you will find that the velocity when $t = 1$ is -32 feet per second.

In general, if $s = s(t)$ is the position function for an object moving along a straight line, then the **velocity** of the object at time t is

$$v(t) = \lim_{\Delta t \to 0} \frac{s(t + \Delta t) - s(t)}{\Delta t} = s'(t). \qquad \text{Velocity function}$$

In other words, the velocity function is the derivative of the position function. Velocity can be negative, zero, or positive. The **speed** of an object is the absolute value of its velocity. Speed cannot be negative.

The position of a free-falling object (neglecting air resistance) under the influence of gravity can be represented by the equation

$$s(t) = \frac{1}{2}gt^2 + v_0 t + s_0 \qquad \text{Position function}$$

where s_0 is the initial height of the object, v_0 is the initial velocity of the object, and g is the acceleration due to gravity. On Earth, the value of g is approximately -32 feet per second per second or -9.8 meters per second per second.

EXAMPLE 10   Using the Derivative to Find Velocity

At time $t = 0$, a diver jumps from a platform diving board that is 32 feet above the water (see Figure 2.21). Because the initial velocity of the diver is 16 feet per second, the position of the diver is

$$s(t) = -16t^2 + 16t + 32 \qquad \text{Position function}$$

where s is measured in feet and t is measured in seconds.

a. When does the diver hit the water?

b. What is the diver's velocity at impact?

Solution

a. To find the time t when the diver hits the water, let $s = 0$ and solve for t.

$$-16t^2 + 16t + 32 = 0 \qquad \text{Set position function equal to 0.}$$
$$-16(t + 1)(t - 2) = 0 \qquad \text{Factor.}$$
$$t = -1 \text{ or } 2 \qquad \text{Solve for } t.$$

Velocity is positive when an object is rising, and is negative when an object is falling. Notice that the diver moves upward for the first half-second because the velocity is positive for $0 < t < \frac{1}{2}$. When the velocity is 0, the diver has reached the maximum height of the dive.
Figure 2.21

Because $t \geq 0$, choose the positive value to conclude that the diver hits the water at $t = 2$ seconds.

b. The velocity at time t is given by the derivative

$$s'(t) = -32t + 16. \qquad \text{Velocity function}$$

So, the velocity at time $t = 2$ is

$$s'(2) = -32(2) + 16 = -48 \text{ feet per second.} \qquad \blacksquare$$

2.2 Exercises

See **CalcChat.com** for tutorial help and worked-out solutions to odd-numbered exercises.

Estimating Slope In Exercises 1 and 2, use the graph to estimate the slope of the tangent line to $y = x^n$ at the point $(1, 1)$. Verify your answer analytically. To print an enlarged copy of the graph, go to *MathGraphs.com*.

1. (a) $y = x^{1/2}$ (b) $y = x^3$

 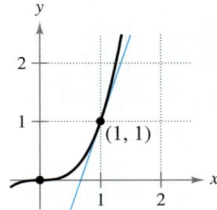

2. (a) $y = x^{-1/2}$ (b) $y = x^{-1}$

 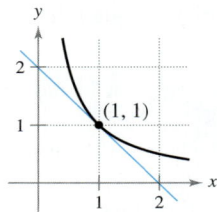

Finding a Derivative In Exercises 3–24, use the rules of differentiation to find the derivative of the function.

3. $y = 12$

4. $f(x) = -9$

5. $y = x^7$

6. $y = x^{12}$

7. $y = \dfrac{1}{x^5}$

8. $y = \dfrac{3}{x^7}$

9. $f(x) = \sqrt[5]{x}$

10. $g(x) = \sqrt[4]{x}$

11. $f(x) = x + 11$

12. $g(x) = 6x + 3$

13. $f(t) = -2t^2 + 3t - 6$

14. $y = t^2 - 3t + 1$

15. $g(x) = x^2 + 4x^3$

16. $y = 4x - 3x^3$

17. $s(t) = t^3 + 5t^2 - 3t + 8$

18. $y = 2x^3 + 6x^2 - 1$

19. $y = \dfrac{\pi}{2} \sin \theta - \cos \theta$

20. $g(t) = \pi \cos t$

21. $y = x^2 - \frac{1}{2} \cos x$

22. $y = 7 + \sin x$

23. $y = \dfrac{1}{x} - 3 \sin x$

24. $y = \dfrac{5}{(2x)^3} + 2 \cos x$

Rewriting a Function Before Differentiating In Exercises 25–30, complete the table to find the derivative of the function.

Original Function	Rewrite	Differentiate	Simplify
25. $y = \dfrac{5}{2x^2}$			
26. $y = \dfrac{3}{2x^4}$			
27. $y = \dfrac{6}{(5x)^3}$			

Original Function	Rewrite	Differentiate	Simplify
28. $y = \dfrac{\pi}{(3x)^2}$			
29. $y = \dfrac{\sqrt{x}}{x}$			
30. $y = \dfrac{4}{x^{-3}}$			

Finding the Slope of a Graph In Exercises 31–38, find the slope of the graph of the function at the given point. Use the *derivative* feature of a graphing utility to confirm your results.

Function	Point
31. $f(x) = \dfrac{8}{x^2}$	$(2, 2)$
32. $f(t) = 2 - \dfrac{4}{t}$	$(4, 1)$
33. $f(x) = -\frac{1}{2} + \frac{7}{5}x^3$	$\left(0, -\frac{1}{2}\right)$
34. $y = 2x^4 - 3$	$(1, -1)$
35. $y = (4x + 1)^2$	$(0, 1)$
36. $f(x) = 2(x - 4)^2$	$(2, 8)$
37. $f(\theta) = 4 \sin \theta - \theta$	$(0, 0)$
38. $g(t) = -2 \cos t + 5$	$(\pi, 7)$

Finding a Derivative In Exercises 39–52, find the derivative of the function.

39. $f(x) = x^2 + 5 - 3x^{-2}$

40. $f(x) = x^3 - 2x + 3x^{-3}$

41. $g(t) = t^2 - \dfrac{4}{t^3}$

42. $f(x) = 8x + \dfrac{3}{x^2}$

43. $f(x) = \dfrac{4x^3 + 3x^2}{x}$

44. $f(x) = \dfrac{2x^4 - x}{x^3}$

45. $f(x) = \dfrac{x^3 - 3x^2 + 4}{x^2}$

46. $h(x) = \dfrac{4x^3 + 2x + 5}{x}$

47. $y = x(x^2 + 1)$

48. $y = x^2(2x^2 - 3x)$

49. $f(x) = \sqrt{x} - 6\sqrt[3]{x}$

50. $f(t) = t^{2/3} - t^{1/3} + 4$

51. $f(x) = 6\sqrt{x} + 5 \cos x$

52. $f(x) = \dfrac{2}{\sqrt[3]{x}} + 3 \cos x$

Finding an Equation of a Tangent Line In Exercises 53–56, (a) find an equation of the tangent line to the graph of f at the given point, (b) use a graphing utility to graph the function and its tangent line at the point, and (c) use the *derivative* feature of a graphing utility to confirm your results.

Function	Point
53. $y = x^4 - 3x^2 + 2$	$(1, 0)$
54. $y = x^3 - 3x$	$(2, 2)$
55. $f(x) = \dfrac{2}{\sqrt[4]{x^3}}$	$(1, 2)$
56. $y = (x - 2)(x^2 + 3x)$	$(1, -4)$

Horizontal Tangent Line In Exercises 57–62, determine the point(s) (if any) at which the graph of the function has a horizontal tangent line.

57. $y = x^4 - 2x^2 + 3$ **58.** $y = x^3 + x$

59. $y = \dfrac{1}{x^2}$ **60.** $y = x^2 + 9$

61. $y = x + \sin x, \quad 0 \le x < 2\pi$

62. $y = \sqrt{3}x + 2\cos x, \quad 0 \le x < 2\pi$

Finding a Value In Exercises 63–68, find k such that the line is tangent to the graph of the function.

Function	Line
63. $f(x) = k - x^2$	$y = -6x + 1$
64. $f(x) = kx^2$	$y = -2x + 3$
65. $f(x) = \dfrac{k}{x}$	$y = -\dfrac{3}{4}x + 3$
66. $f(x) = k\sqrt{x}$	$y = x + 4$
67. $f(x) = kx^3$	$y = x + 1$
68. $f(x) = kx^4$	$y = 4x - 1$

69. Sketching a Graph Sketch the graph of a function f such that $f' > 0$ for all x and the rate of change of the function is decreasing.

70. HOW DO YOU SEE IT? Use the graph of f to answer each question. To print an enlarged copy of the graph, go to *MathGraphs.com*.

(a) Between which two consecutive points is the average rate of change of the function greatest?

(b) Is the average rate of change of the function between A and B greater than or less than the instantaneous rate of change at B?

(c) Sketch a tangent line to the graph between C and D such that the slope of the tangent line is the same as the average rate of change of the function between C and D.

WRITING ABOUT CONCEPTS

Exploring a Relationship In Exercises 71–74, the relationship between f and g is given. Explain the relationship between f' and g'.

71. $g(x) = f(x) + 6$ **72.** $g(x) = 2f(x)$

73. $g(x) = -5f(x)$ **74.** $g(x) = 3f(x) - 1$

WRITING ABOUT CONCEPTS (continued)

A Function and Its Derivative In Exercises 75 and 76, the graphs of a function f and its derivative f' are shown in the same set of coordinate axes. Label the graphs as f or f' and write a short paragraph stating the criteria you used in making your selection. To print an enlarged copy of the graph, go to *MathGraphs.com*.

75.   **76.**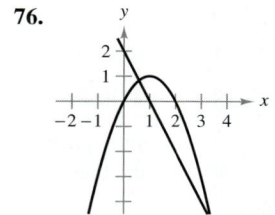

77. Finding Equations of Tangent Lines Sketch the graphs of $y = x^2$ and $y = -x^2 + 6x - 5$, and sketch the two lines that are tangent to both graphs. Find equations of these lines.

78. Tangent Lines Show that the graphs of the two equations

$$y = x \quad \text{and} \quad y = \dfrac{1}{x}$$

have tangent lines that are perpendicular to each other at their point of intersection.

79. Tangent Line Show that the graph of the function

$$f(x) = 3x + \sin x + 2$$

does not have a horizontal tangent line.

80. Tangent Line Show that the graph of the function

$$f(x) = x^5 + 3x^3 + 5x$$

does not have a tangent line with a slope of 3.

Finding an Equation of a Tangent Line In Exercises 81 and 82, find an equation of the tangent line to the graph of the function f through the point (x_0, y_0) not on the graph. To find the point of tangency (x, y) on the graph of f, solve the equation

$$f'(x) = \dfrac{y_0 - y}{x_0 - x}.$$

81. $f(x) = \sqrt{x}$ **82.** $f(x) = \dfrac{2}{x}$

$(x_0, y_0) = (-4, 0)$ $(x_0, y_0) = (5, 0)$

83. Linear Approximation Use a graphing utility with a square window setting to zoom in on the graph of

$$f(x) = 4 - \tfrac{1}{2}x^2$$

to approximate $f'(1)$. Use the derivative to find $f'(1)$.

84. Linear Approximation Use a graphing utility with a square window setting to zoom in on the graph of

$$f(x) = 4\sqrt{x} + 1$$

to approximate $f'(4)$. Use the derivative to find $f'(4)$.

85. Linear Approximation Consider the function $f(x) = x^{3/2}$ with the solution point $(4, 8)$.

(a) Use a graphing utility to graph f. Use the *zoom* feature to obtain successive magnifications of the graph in the neighborhood of the point $(4, 8)$. After zooming in a few times, the graph should appear nearly linear. Use the *trace* feature to determine the coordinates of a point near $(4, 8)$. Find an equation of the secant line $S(x)$ through the two points.

(b) Find the equation of the line

$$T(x) = f'(4)(x - 4) + f(4)$$

tangent to the graph of f passing through the given point. Why are the linear functions S and T nearly the same?

(c) Use a graphing utility to graph f and T in the same set of coordinate axes. Note that T is a good approximation of f when x is close to 4. What happens to the accuracy of the approximation as you move farther away from the point of tangency?

(d) Demonstrate the conclusion in part (c) by completing the table.

Δx	-3	-2	-1	-0.5	-0.1	0
$f(4 + \Delta x)$						
$T(4 + \Delta x)$						

Δx	0.1	0.5	1	2	3
$f(4 + \Delta x)$					
$T(4 + \Delta x)$					

86. Linear Approximation Repeat Exercise 85 for the function $f(x) = x^3$, where $T(x)$ is the line tangent to the graph at the point $(1, 1)$. Explain why the accuracy of the linear approximation decreases more rapidly than in Exercise 85.

True or False? In Exercises 87–92, determine whether the statement is true or false. If it is false, explain why or give an example that shows it is false.

87. If $f'(x) = g'(x)$, then $f(x) = g(x)$.

88. If $f(x) = g(x) + c$, then $f'(x) = g'(x)$.

89. If $y = \pi^2$, then $dy/dx = 2\pi$.

90. If $y = x/\pi$, then $dy/dx = 1/\pi$.

91. If $g(x) = 3f(x)$, then $g'(x) = 3f'(x)$.

92. If $f(x) = \dfrac{1}{x^n}$, then $f'(x) = \dfrac{1}{nx^{n-1}}$.

Finding Rates of Change In Exercises 93–96, find the average rate of change of the function over the given interval. Compare this average rate of change with the instantaneous rates of change at the endpoints of the interval.

93. $f(t) = 4t + 5$, $[1, 2]$

94. $f(t) = t^2 - 7$, $[3, 3.1]$

95. $f(x) = \dfrac{-1}{x}$, $[1, 2]$

96. $f(x) = \sin x$, $\left[0, \dfrac{\pi}{6}\right]$

Vertical Motion In Exercises 97 and 98, use the position function $s(t) = -16t^2 + v_0 t + s_0$ for free-falling objects.

97. A silver dollar is dropped from the top of a building that is 1362 feet tall.

(a) Determine the position and velocity functions for the coin.

(b) Determine the average velocity on the interval $[1, 2]$.

(c) Find the instantaneous velocities when $t = 1$ and $t = 2$.

(d) Find the time required for the coin to reach ground level.

(e) Find the velocity of the coin at impact.

98. A ball is thrown straight down from the top of a 220-foot building with an initial velocity of -22 feet per second. What is its velocity after 3 seconds? What is its velocity after falling 108 feet?

Vertical Motion In Exercises 99 and 100, use the position function $s(t) = -4.9t^2 + v_0 t + s_0$ for free-falling objects.

99. A projectile is shot upward from the surface of Earth with an initial velocity of 120 meters per second. What is its velocity after 5 seconds? After 10 seconds?

100. To estimate the height of a building, a stone is dropped from the top of the building into a pool of water at ground level. The splash is seen 5.6 seconds after the stone is dropped. What is the height of the building?

Think About It In Exercises 101 and 102, the graph of a position function is shown. It represents the distance in miles that a person drives during a 10-minute trip to work. Make a sketch of the corresponding velocity function.

101.

102.

Think About It In Exercises 103 and 104, the graph of a velocity function is shown. It represents the velocity in miles per hour during a 10-minute trip to work. Make a sketch of the corresponding position function.

103.

104.

105. Volume The volume of a cube with sides of length s is given by $V = s^3$. Find the rate of change of the volume with respect to s when $s = 6$ centimeters.

106. Area The area of a square with sides of length s is given by $A = s^2$. Find the rate of change of the area with respect to s when $s = 6$ meters.

107. Modeling Data

The stopping distance of an automobile, on dry, level pavement, traveling at a speed v (in kilometers per hour) is the distance R (in meters) the car travels during the reaction time of the driver plus the distance B (in meters) the car travels after the brakes are applied (see figure). The table shows the results of an experiment.

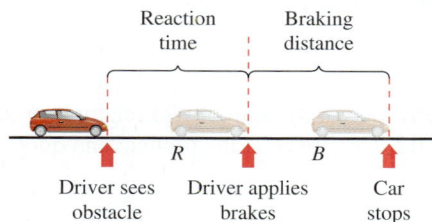

Reaction time | Braking distance

R | B

Driver sees obstacle | Driver applies brakes | Car stops

Speed, v	20	40	60	80	100
Reaction Time Distance, R	8.3	16.7	25.0	33.3	41.7
Braking Time Distance, B	2.3	9.0	20.2	35.8	55.9

(a) Use the regression capabilities of a graphing utility to find a linear model for reaction time distance R.

(b) Use the regression capabilities of a graphing utility to find a quadratic model for braking time distance B.

(c) Determine the polynomial giving the total stopping distance T.

(d) Use a graphing utility to graph the functions R, B, and T in the same viewing window.

(e) Find the derivative of T and the rates of change of the total stopping distance for $v = 40$, $v = 80$, and $v = 100$.

(f) Use the results of this exercise to draw conclusions about the total stopping distance as speed increases.

108. Fuel Cost   A car is driven 15,000 miles a year and gets x miles per gallon. Assume that the average fuel cost is $3.48 per gallon. Find the annual cost of fuel C as a function of x and use this function to complete the table.

x	10	15	20	25	30	35	40
C							
dC/dx							

Who would benefit more from a one-mile-per-gallon increase in fuel efficiency—the driver of a car that gets 15 miles per gallon, or the driver of a car that gets 35 miles per gallon? Explain.

109. Velocity   Verify that the average velocity over the time interval $[t_0 - \Delta t, t_0 + \Delta t]$ is the same as the instantaneous velocity at $t = t_0$ for the position function

$$s(t) = -\frac{1}{2}at^2 + c.$$

110. Inventory Management   The annual inventory cost C for a manufacturer is

$$C = \frac{1,008,000}{Q} + 6.3Q$$

where Q is the order size when the inventory is replenished. Find the change in annual cost when Q is increased from 350 to 351, and compare this with the instantaneous rate of change when $Q = 350$.

111. Finding an Equation of a Parabola   Find an equation of the parabola $y = ax^2 + bx + c$ that passes through $(0, 1)$ and is tangent to the line $y = x - 1$ at $(1, 0)$.

112. Proof   Let (a, b) be an arbitrary point on the graph of $y = 1/x$, $x > 0$. Prove that the area of the triangle formed by the tangent line through (a, b) and the coordinate axes is 2.

113. Finding Equation(s) of Tangent Line(s)   Find the equation(s) of the tangent line(s) to the graph of the curve $y = x^3 - 9x$ through the point $(1, -9)$ not on the graph.

114. Finding Equation(s) of Tangent Line(s)   Find the equation(s) of the tangent line(s) to the graph of the parabola $y = x^2$ through the given point not on the graph.

(a) $(0, a)$ (b) $(a, 0)$

Are there any restrictions on the constant a?

Making a Function Differentiable   **In Exercises 115 and 116, find a and b such that f is differentiable everywhere.**

115. $f(x) = \begin{cases} ax^3, & x \le 2 \\ x^2 + b, & x > 2 \end{cases}$

116. $f(x) = \begin{cases} \cos x, & x < 0 \\ ax + b, & x \ge 0 \end{cases}$

117. Determining Differentiability   Where are the functions $f_1(x) = |\sin x|$ and $f_2(x) = \sin |x|$ differentiable?

118. Proof   Prove that $\dfrac{d}{dx}[\cos x] = -\sin x$.

■ **FOR FURTHER INFORMATION**   For a geometric interpretation of the derivatives of trigonometric functions, see the article "Sines and Cosines of the Times" by Victor J. Katz in *Math Horizons*. To view this article, go to *MathArticles.com*.

PUTNAM EXAM CHALLENGE

119. Find all differentiable functions $f : \mathbb{R} \to \mathbb{R}$ such that

$$f'(x) = \frac{f(x + n) - f(x)}{n}$$

for all real numbers x and all positive integers n.

2.3   Product and Quotient Rules and Higher-Order Derivatives

- **Find the derivative of a function using the Product Rule.**
- **Find the derivative of a function using the Quotient Rule.**
- **Find the derivative of a trigonometric function.**
- **Find a higher-order derivative of a function.**

The Product Rule

In Section 2.2, you learned that the derivative of the sum of two functions is simply the sum of their derivatives. The rules for the derivatives of the product and quotient of two functions are not as simple.

> **REMARK**   A version of the Product Rule that some people prefer is
>
> $$\frac{d}{dx}[f(x)g(x)] = f'(x)g(x) + f(x)g'(x).$$
>
> The advantage of this form is that it generalizes easily to products of three or more factors.

THEOREM 2.7   The Product Rule

The product of two differentiable functions f and g is itself differentiable. Moreover, the derivative of fg is the first function times the derivative of the second, plus the second function times the derivative of the first.

$$\frac{d}{dx}[f(x)g(x)] = f(x)g'(x) + g(x)f'(x)$$

Proof   Some mathematical proofs, such as the proof of the Sum Rule, are straightforward. Others involve clever steps that may appear unmotivated to a reader. This proof involves such a step—subtracting and adding the same quantity—which is shown in color.

$$\frac{d}{dx}[f(x)g(x)] = \lim_{\Delta x \to 0} \frac{f(x + \Delta x)g(x + \Delta x) - f(x)g(x)}{\Delta x}$$

$$= \lim_{\Delta x \to 0} \frac{f(x + \Delta x)g(x + \Delta x) - f(x + \Delta x)g(x) + f(x + \Delta x)g(x) - f(x)g(x)}{\Delta x}$$

$$= \lim_{\Delta x \to 0} \left[f(x + \Delta x)\frac{g(x + \Delta x) - g(x)}{\Delta x} + g(x)\frac{f(x + \Delta x) - f(x)}{\Delta x} \right]$$

$$= \lim_{\Delta x \to 0} \left[f(x + \Delta x)\frac{g(x + \Delta x) - g(x)}{\Delta x} \right] + \lim_{\Delta x \to 0} \left[g(x)\frac{f(x + \Delta x) - f(x)}{\Delta x} \right]$$

$$= \lim_{\Delta x \to 0} f(x + \Delta x) \cdot \lim_{\Delta x \to 0} \frac{g(x + \Delta x) - g(x)}{\Delta x} + \lim_{\Delta x \to 0} g(x) \cdot \lim_{\Delta x \to 0} \frac{f(x + \Delta x) - f(x)}{\Delta x}$$

$$= f(x)g'(x) + g(x)f'(x)$$

Note that $\lim_{\Delta x \to 0} f(x + \Delta x) = f(x)$ because f is given to be differentiable and therefore is continuous.

See LarsonCalculus.com for Bruce Edwards's video of this proof.

> **REMARK**   The proof of the Product Rule for products of more than two factors is left as an exercise (see Exercise 137).

The Product Rule can be extended to cover products involving more than two factors. For example, if f, g, and h are differentiable functions of x, then

$$\frac{d}{dx}[f(x)g(x)h(x)] = f'(x)g(x)h(x) + f(x)g'(x)h(x) + f(x)g(x)h'(x).$$

So, the derivative of $y = x^2 \sin x \cos x$ is

$$\frac{dy}{dx} = 2x \sin x \cos x + x^2 \cos x \cos x + x^2 \sin x(-\sin x)$$

$$= 2x \sin x \cos x + x^2(\cos^2 x - \sin^2 x).$$

The derivative of a product of two functions is not (in general) given by the product of the derivatives of the two functions. To see this, try comparing the product of the derivatives of

$$f(x) = 3x - 2x^2$$

and

$$g(x) = 5 + 4x$$

with the derivative in Example 1.

EXAMPLE 1   **Using the Product Rule**

Find the derivative of $h(x) = (3x - 2x^2)(5 + 4x)$.

Solution

$$h'(x) = \overbrace{(3x - 2x^2)}^{\text{First}} \overbrace{\frac{d}{dx}[5 + 4x]}^{\substack{\text{Derivative of}\\ \text{second}}} + \overbrace{(5 + 4x)}^{\text{Second}} \overbrace{\frac{d}{dx}[3x - 2x^2]}^{\substack{\text{Derivative}\\ \text{of first}}} \quad\quad \text{Apply Product Rule.}$$

$$= (3x - 2x^2)(4) + (5 + 4x)(3 - 4x)$$
$$= (12x - 8x^2) + (15 - 8x - 16x^2)$$
$$= -24x^2 + 4x + 15$$

In Example 1, you have the option of finding the derivative with or without the Product Rule. To find the derivative without the Product Rule, you can write

$$D_x[(3x - 2x^2)(5 + 4x)] = D_x[-8x^3 + 2x^2 + 15x]$$
$$= -24x^2 + 4x + 15.$$

In the next example, you must use the Product Rule.

EXAMPLE 2   **Using the Product Rule**

Find the derivative of $y = 3x^2 \sin x$.

Solution

$$\frac{d}{dx}[3x^2 \sin x] = 3x^2 \frac{d}{dx}[\sin x] + \sin x \frac{d}{dx}[3x^2] \quad\quad \text{Apply Product Rule.}$$

$$= 3x^2 \cos x + (\sin x)(6x)$$
$$= 3x^2 \cos x + 6x \sin x$$
$$= 3x(x \cos x + 2 \sin x)$$

• • REMARK In Example 3, notice that you use the Product Rule when both factors of the product are variable, and you use the Constant Multiple Rule when one of the factors is a constant.

EXAMPLE 3   **Using the Product Rule**

Find the derivative of $y = 2x \cos x - 2 \sin x$.

Solution

$$\frac{dy}{dx} = \overbrace{(2x)\left(\frac{d}{dx}[\cos x]\right) + (\cos x)\left(\frac{d}{dx}[2x]\right)}^{\text{Product Rule}} - \overbrace{2\frac{d}{dx}[\sin x]}^{\text{Constant Multiple Rule}}$$

$$= (2x)(-\sin x) + (\cos x)(2) - 2(\cos x)$$
$$= -2x \sin x$$

The Quotient Rule

> **THEOREM 2.8   The Quotient Rule**
>
> The quotient f/g of two differentiable functions f and g is itself differentiable at all values of x for which $g(x) \neq 0$. Moreover, the derivative of f/g is given by the denominator times the derivative of the numerator minus the numerator times the derivative of the denominator, all divided by the square of the denominator.
>
> $$\frac{d}{dx}\left[\frac{f(x)}{g(x)}\right] = \frac{g(x)f'(x) - f(x)g'(x)}{[g(x)]^2}, \quad g(x) \neq 0$$

REMARK From the Quotient Rule, you can see that the derivative of a quotient is not (in general) the quotient of the derivatives.

Proof   As with the proof of Theorem 2.7, the key to this proof is subtracting and adding the same quantity.

$$\frac{d}{dx}\left[\frac{f(x)}{g(x)}\right] = \lim_{\Delta x \to 0} \frac{\dfrac{f(x + \Delta x)}{g(x + \Delta x)} - \dfrac{f(x)}{g(x)}}{\Delta x} \qquad \text{Definition of derivative}$$

$$= \lim_{\Delta x \to 0} \frac{g(x)f(x + \Delta x) - f(x)g(x + \Delta x)}{\Delta x g(x)g(x + \Delta x)}$$

$$= \lim_{\Delta x \to 0} \frac{g(x)f(x + \Delta x) - f(x)g(x) + f(x)g(x) - f(x)g(x + \Delta x)}{\Delta x g(x)g(x + \Delta x)}$$

$$= \frac{\displaystyle\lim_{\Delta x \to 0} \frac{g(x)[f(x + \Delta x) - f(x)]}{\Delta x} - \lim_{\Delta x \to 0} \frac{f(x)[g(x + \Delta x) - g(x)]}{\Delta x}}{\displaystyle\lim_{\Delta x \to 0}\, [g(x)g(x + \Delta x)]}$$

$$= \frac{g(x)\left[\displaystyle\lim_{\Delta x \to 0} \frac{f(x + \Delta x) - f(x)}{\Delta x}\right] - f(x)\left[\displaystyle\lim_{\Delta x \to 0} \frac{g(x + \Delta x) - g(x)}{\Delta x}\right]}{\displaystyle\lim_{\Delta x \to 0}\, [g(x)g(x + \Delta x)]}$$

$$= \frac{g(x)f'(x) - f(x)g'(x)}{[g(x)]^2}$$

Note that $\displaystyle\lim_{\Delta x \to 0} g(x + \Delta x) = g(x)$ because g is given to be differentiable and therefore is continuous.

See LarsonCalculus.com for Bruce Edwards's video of this proof.

TECHNOLOGY A graphing utility can be used to compare the graph of a function with the graph of its derivative. For instance, in Figure 2.22, the graph of the function in Example 4 appears to have two points that have horizontal tangent lines. What are the values of y' at these two points?

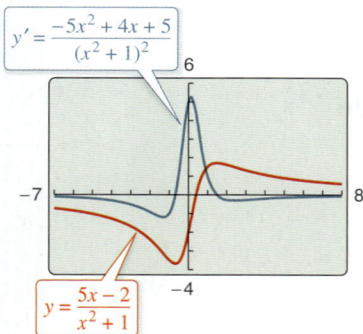

$$y' = \frac{-5x^2 + 4x + 5}{(x^2 + 1)^2}$$

$$y = \frac{5x - 2}{x^2 + 1}$$

Graphical comparison of a function and its derivative

Figure 2.22

EXAMPLE 4   Using the Quotient Rule

Find the derivative of $y = \dfrac{5x - 2}{x^2 + 1}$.

Solution

$$\frac{d}{dx}\left[\frac{5x - 2}{x^2 + 1}\right] = \frac{(x^2 + 1)\dfrac{d}{dx}[5x - 2] - (5x - 2)\dfrac{d}{dx}[x^2 + 1]}{(x^2 + 1)^2} \qquad \text{Apply Quotient Rule.}$$

$$= \frac{(x^2 + 1)(5) - (5x - 2)(2x)}{(x^2 + 1)^2}$$

$$= \frac{(5x^2 + 5) - (10x^2 - 4x)}{(x^2 + 1)^2}$$

$$= \frac{-5x^2 + 4x + 5}{(x^2 + 1)^2}$$

Note the use of parentheses in Example 4. A liberal use of parentheses is recommended for *all* types of differentiation problems. For instance, with the Quotient Rule, it is a good idea to enclose all factors and derivatives in parentheses, and to pay special attention to the subtraction required in the numerator.

When differentiation rules were introduced in the preceding section, the need for rewriting *before* differentiating was emphasized. The next example illustrates this point with the Quotient Rule.

EXAMPLE 5   **Rewriting Before Differentiating**

Find an equation of the tangent line to the graph of $f(x) = \dfrac{3 - (1/x)}{x + 5}$ at $(-1, 1)$.

Solution   Begin by rewriting the function.

$$f(x) = \frac{3 - (1/x)}{x + 5} \qquad \text{Write original function.}$$

$$= \frac{x\left(3 - \dfrac{1}{x}\right)}{x(x + 5)} \qquad \text{Multiply numerator and denominator by } x.$$

$$= \frac{3x - 1}{x^2 + 5x} \qquad \text{Rewrite.}$$

Next, apply the Quotient Rule.

$$f'(x) = \frac{(x^2 + 5x)(3) - (3x - 1)(2x + 5)}{(x^2 + 5x)^2} \qquad \text{Quotient Rule}$$

$$= \frac{(3x^2 + 15x) - (6x^2 + 13x - 5)}{(x^2 + 5x)^2}$$

$$= \frac{-3x^2 + 2x + 5}{(x^2 + 5x)^2} \qquad \text{Simplify.}$$

To find the slope at $(-1, 1)$, evaluate $f'(-1)$.

$$f'(-1) = 0 \qquad \text{Slope of graph at } (-1, 1)$$

Then, using the point-slope form of the equation of a line, you can determine that the equation of the tangent line at $(-1, 1)$ is $y = 1$. See Figure 2.23.

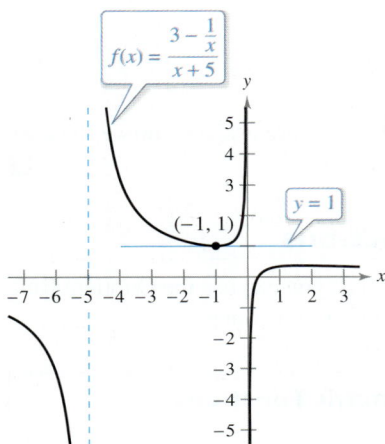

The line $y = 1$ is tangent to the graph of $f(x)$ at the point $(-1, 1)$.

Figure 2.23

Not every quotient needs to be differentiated by the Quotient Rule. For instance, each quotient in the next example can be considered as the product of a constant times a function of x. In such cases, it is more convenient to use the Constant Multiple Rule.

EXAMPLE 6   **Using the Constant Multiple Rule**

REMARK   To see the benefit of using the Constant Multiple Rule for some quotients, try using the Quotient Rule to differentiate the functions in Example 6—you should obtain the same results, but with more work.

Original Function	Rewrite	Differentiate	Simplify
a. $y = \dfrac{x^2 + 3x}{6}$	$y = \dfrac{1}{6}(x^2 + 3x)$	$y' = \dfrac{1}{6}(2x + 3)$	$y' = \dfrac{2x + 3}{6}$
b. $y = \dfrac{5x^4}{8}$	$y = \dfrac{5}{8}x^4$	$y' = \dfrac{5}{8}(4x^3)$	$y' = \dfrac{5}{2}x^3$
c. $y = \dfrac{-3(3x - 2x^2)}{7x}$	$y = -\dfrac{3}{7}(3 - 2x)$	$y' = -\dfrac{3}{7}(-2)$	$y' = \dfrac{6}{7}$
d. $y = \dfrac{9}{5x^2}$	$y = \dfrac{9}{5}(x^{-2})$	$y' = \dfrac{9}{5}(-2x^{-3})$	$y' = -\dfrac{18}{5x^3}$

In Section 2.2, the Power Rule was proved only for the case in which the exponent n is a positive integer greater than 1. The next example extends the proof to include negative integer exponents.

EXAMPLE 7   **Power Rule: Negative Integer Exponents**

If n is a negative integer, then there exists a positive integer k such that $n = -k$. So, by the Quotient Rule, you can write

$$\frac{d}{dx}[x^n] = \frac{d}{dx}\left[\frac{1}{x^k}\right]$$

$$= \frac{x^k(0) - (1)(kx^{k-1})}{(x^k)^2} \qquad \text{Quotient Rule and Power Rule}$$

$$= \frac{0 - kx^{k-1}}{x^{2k}}$$

$$= -kx^{-k-1}$$

$$= nx^{n-1}. \qquad n = -k$$

So, the Power Rule

$$\frac{d}{dx}[x^n] = nx^{n-1} \qquad \text{Power Rule}$$

is valid for any integer. In Exercise 71 in Section 2.5, you are asked to prove the case for which n is any rational number.

Derivatives of Trigonometric Functions

Knowing the derivatives of the sine and cosine functions, you can use the Quotient Rule to find the derivatives of the four remaining trigonometric functions.

THEOREM 2.9   Derivatives of Trigonometric Functions

$$\frac{d}{dx}[\tan x] = \sec^2 x \qquad\qquad \frac{d}{dx}[\cot x] = -\csc^2 x$$

$$\frac{d}{dx}[\sec x] = \sec x \tan x \qquad\qquad \frac{d}{dx}[\csc x] = -\csc x \cot x$$

REMARK In the proof of Theorem 2.9, note the use of the trigonometric identities

$$\sin^2 x + \cos^2 x = 1$$

and

$$\sec x = \frac{1}{\cos x}.$$

These trigonometric identities and others are listed in Appendix C and on the formula cards for this text.

Proof Considering $\tan x = (\sin x)/(\cos x)$ and applying the Quotient Rule, you obtain

$$\frac{d}{dx}[\tan x] = \frac{d}{dx}\left[\frac{\sin x}{\cos x}\right]$$

$$= \frac{(\cos x)(\cos x) - (\sin x)(-\sin x)}{\cos^2 x} \qquad \text{Apply Quotient Rule.}$$

$$= \frac{\cos^2 x + \sin^2 x}{\cos^2 x}$$

$$= \frac{1}{\cos^2 x}$$

$$= \sec^2 x.$$

See LarsonCalculus.com for Bruce Edwards's video of this proof.

The proofs of the other three parts of the theorem are left as an exercise (see Exercise 87).

EXAMPLE 8 **Differentiating Trigonometric Functions**

· · · · ▷ *See LarsonCalculus.com for an interactive version of this type of example.*

Function	Derivative
a. $y = x - \tan x$	$\dfrac{dy}{dx} = 1 - \sec^2 x$
b. $y = x \sec x$	$y' = x(\sec x \tan x) + (\sec x)(1)$
	$= (\sec x)(1 + x \tan x)$

EXAMPLE 9 **Different Forms of a Derivative**

Differentiate both forms of

$$y = \frac{1 - \cos x}{\sin x} = \csc x - \cot x.$$

Solution

First form: $\quad y = \dfrac{1 - \cos x}{\sin x}$

$$y' = \frac{(\sin x)(\sin x) - (1 - \cos x)(\cos x)}{\sin^2 x}$$

$$= \frac{\sin^2 x - \cos x + \cos^2 x}{\sin^2 x}$$

$$= \frac{1 - \cos x}{\sin^2 x} \qquad \qquad \textcolor{red}{\sin^2 x + \cos^2 x = 1}$$

Second form: $\quad y = \csc x - \cot x$

$$y' = -\csc x \cot x + \csc^2 x$$

To show that the two derivatives are equal, you can write

$$\frac{1 - \cos x}{\sin^2 x} = \frac{1}{\sin^2 x} - \frac{\cos x}{\sin^2 x}$$

$$= \frac{1}{\sin^2 x} - \left(\frac{1}{\sin x}\right)\left(\frac{\cos x}{\sin x}\right)$$

$$= \csc^2 x - \csc x \cot x.$$

· · · · · · · · · · · · · · · ▷

·· **REMARK** Because of trigonometric identities, the derivative of a trigonometric function can take many forms. This presents a challenge when you are trying to match your answers to those given in the back of the text.

The summary below shows that much of the work in obtaining a simplified form of a derivative occurs *after* differentiating. Note that two characteristics of a simplified form are the absence of negative exponents and the combining of like terms.

	$f'(x)$ After Differentiating	$f'(x)$ After Simplifying
Example 1	$(3x - 2x^2)(4) + (5 + 4x)(3 - 4x)$	$-24x^2 + 4x + 15$
Example 3	$(2x)(-\sin x) + (\cos x)(2) - 2(\cos x)$	$-2x \sin x$
Example 4	$\dfrac{(x^2 + 1)(5) - (5x - 2)(2x)}{(x^2 + 1)^2}$	$\dfrac{-5x^2 + 4x + 5}{(x^2 + 1)^2}$
Example 5	$\dfrac{(x^2 + 5x)(3) - (3x - 1)(2x + 5)}{(x^2 + 5x)^2}$	$\dfrac{-3x^2 + 2x + 5}{(x^2 + 5x)^2}$
Example 6	$\dfrac{(\sin x)(\sin x) - (1 - \cos x)(\cos x)}{\sin^2 x}$	$\dfrac{1 - \cos x}{\sin^2 x}$

Higher-Order Derivatives

Just as you can obtain a velocity function by differentiating a position function, you can obtain an **acceleration** function by differentiating a velocity function. Another way of looking at this is that you can obtain an acceleration function by differentiating a position function *twice*.

$$s(t) \qquad \text{Position function}$$
$$v(t) = s'(t) \qquad \text{Velocity function}$$
$$a(t) = v'(t) = s''(t) \qquad \text{Acceleration function}$$

The function $a(t)$ is the **second derivative** of $s(t)$ and is denoted by $s''(t)$.

The second derivative is an example of a **higher-order derivative.** You can define derivatives of any positive integer order. For instance, the **third derivative** is the derivative of the second derivative. Higher-order derivatives are denoted as shown below.

First derivative: $\quad y', \quad f'(x), \quad \dfrac{dy}{dx}, \quad \dfrac{d}{dx}[f(x)], \quad D_x[y]$

Second derivative: $\ y'', \quad f''(x), \quad \dfrac{d^2y}{dx^2}, \quad \dfrac{d^2}{dx^2}[f(x)], \quad D_x^2[y]$

Third derivative: $\quad y''', \quad f'''(x), \quad \dfrac{d^3y}{dx^3}, \quad \dfrac{d^3}{dx^3}[f(x)], \quad D_x^3[y]$

Fourth derivative: $\ y^{(4)}, \quad f^{(4)}(x), \quad \dfrac{d^4y}{dx^4}, \quad \dfrac{d^4}{dx^4}[f(x)], \quad D_x^4[y]$

$$\vdots$$

nth derivative: $\quad y^{(n)}, \quad f^{(n)}(x), \quad \dfrac{d^ny}{dx^n}, \quad \dfrac{d^n}{dx^n}[f(x)], \quad D_x^n[y]$

REMARK The second derivative of a function is the derivative of the first derivative of the function.

EXAMPLE 10 Finding the Acceleration Due to Gravity

Because the moon has no atmosphere, a falling object on the moon encounters no air resistance. In 1971, astronaut David Scott demonstrated that a feather and a hammer fall at the same rate on the moon. The position function for each of these falling objects is

$$s(t) = -0.81t^2 + 2$$

where $s(t)$ is the height in meters and t is the time in seconds, as shown in the figure at the right. What is the ratio of Earth's gravitational force to the moon's?

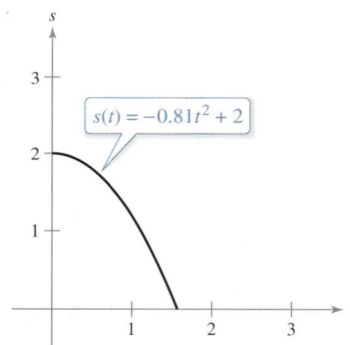

$s(t) = -0.81t^2 + 2$

The moon's mass is 7.349×10^{22} kilograms, and Earth's mass is 5.976×10^{24} kilograms. The moon's radius is 1737 kilometers, and Earth's radius is 6378 kilometers. Because the gravitational force on the surface of a planet is directly proportional to its mass and inversely proportional to the square of its radius, the ratio of the gravitational force on Earth to the gravitational force on the moon is

$$\frac{(5.976 \times 10^{24})/6378^2}{(7.349 \times 10^{22})/1737^2} \approx 6.0.$$

Solution To find the acceleration, differentiate the position function twice.

$$s(t) = -0.81t^2 + 2 \qquad \text{Position function}$$
$$s'(t) = -1.62t \qquad \text{Velocity function}$$
$$s''(t) = -1.62 \qquad \text{Acceleration function}$$

So, the acceleration due to gravity on the moon is -1.62 meters per second per second. Because the acceleration due to gravity on Earth is -9.8 meters per second per second, the ratio of Earth's gravitational force to the moon's is

$$\frac{\text{Earth's gravitational force}}{\text{Moon's gravitational force}} = \frac{-9.8}{-1.62}$$
$$\approx 6.0.$$

NASA

2.3 Exercises

See **CalcChat.com** for tutorial help and worked-out solutions to odd-numbered exercises.

Using the Product Rule In Exercises 1–6, use the Product Rule to find the derivative of the function.

1. $g(x) = (x^2 + 3)(x^2 - 4x)$
2. $y = (3x - 4)(x^3 + 5)$
3. $h(t) = \sqrt{t}(1 - t^2)$
4. $g(s) = \sqrt{s}(s^2 + 8)$
5. $f(x) = x^3 \cos x$
6. $g(x) = \sqrt{x} \sin x$

Using the Quotient Rule In Exercises 7–12, use the Quotient Rule to find the derivative of the function.

7. $f(x) = \dfrac{x}{x^2 + 1}$
8. $g(t) = \dfrac{3t^2 - 1}{2t + 5}$
9. $h(x) = \dfrac{\sqrt{x}}{x^3 + 1}$
10. $f(x) = \dfrac{x^2}{2\sqrt{x} + 1}$
11. $g(x) = \dfrac{\sin x}{x^2}$
12. $f(t) = \dfrac{\cos t}{t^3}$

Finding and Evaluating a Derivative In Exercises 13–18, find $f'(x)$ and $f'(c)$.

Function	Value of c
13. $f(x) = (x^3 + 4x)(3x^2 + 2x - 5)$	$c = 0$
14. $y = (x^2 - 3x + 2)(x^3 + 1)$	$c = 2$
15. $f(x) = \dfrac{x^2 - 4}{x - 3}$	$c = 1$
16. $f(x) = \dfrac{x - 4}{x + 4}$	$c = 3$
17. $f(x) = x \cos x$	$c = \dfrac{\pi}{4}$
18. $f(x) = \dfrac{\sin x}{x}$	$c = \dfrac{\pi}{6}$

Using the Constant Multiple Rule In Exercises 19–24, complete the table to find the derivative of the function without using the Quotient Rule.

Function	Rewrite	Differentiate	Simplify
19. $y = \dfrac{x^2 + 3x}{7}$			
20. $y = \dfrac{5x^2 - 3}{4}$			
21. $y = \dfrac{6}{7x^2}$			
22. $y = \dfrac{10}{3x^3}$			
23. $y = \dfrac{4x^{3/2}}{x}$			
24. $y = \dfrac{2x}{x^{1/3}}$			

Finding a Derivative In Exercises 25–38, find the derivative of the algebraic function.

25. $f(x) = \dfrac{4 - 3x - x^2}{x^2 - 1}$
26. $f(x) = \dfrac{x^2 + 5x + 6}{x^2 - 4}$
27. $f(x) = x\left(1 - \dfrac{4}{x + 3}\right)$
28. $f(x) = x^4\left(1 - \dfrac{2}{x + 1}\right)$
29. $f(x) = \dfrac{3x - 1}{\sqrt{x}}$
30. $f(x) = \sqrt[3]{x}(\sqrt{x} + 3)$
31. $h(s) = (s^3 - 2)^2$
32. $h(x) = (x^2 + 3)^3$
33. $f(x) = \dfrac{2 - \dfrac{1}{x}}{x - 3}$
34. $g(x) = x^2\left(\dfrac{2}{x} - \dfrac{1}{x + 1}\right)$
35. $f(x) = (2x^3 + 5x)(x - 3)(x + 2)$
36. $f(x) = (x^3 - x)(x^2 + 2)(x^2 + x - 1)$
37. $f(x) = \dfrac{x^2 + c^2}{x^2 - c^2}$, c is a constant
38. $f(x) = \dfrac{c^2 - x^2}{c^2 + x^2}$, c is a constant

Finding a Derivative of a Trigonometric Function In Exercises 39–54, find the derivative of the trigonometric function.

39. $f(t) = t^2 \sin t$
40. $f(\theta) = (\theta + 1) \cos \theta$
41. $f(t) = \dfrac{\cos t}{t}$
42. $f(x) = \dfrac{\sin x}{x^3}$
43. $f(x) = -x + \tan x$
44. $y = x + \cot x$
45. $g(t) = \sqrt[4]{t} + 6 \csc t$
46. $h(x) = \dfrac{1}{x} - 12 \sec x$
47. $y = \dfrac{3(1 - \sin x)}{2 \cos x}$
48. $y = \dfrac{\sec x}{x}$
49. $y = -\csc x - \sin x$
50. $y = x \sin x + \cos x$
51. $f(x) = x^2 \tan x$
52. $f(x) = \sin x \cos x$
53. $y = 2x \sin x + x^2 \cos x$
54. $h(\theta) = 5\theta \sec \theta + \theta \tan \theta$

Finding a Derivative Using Technology In Exercises 55–58, use a computer algebra system to find the derivative of the function.

55. $g(x) = \left(\dfrac{x + 1}{x + 2}\right)(2x - 5)$
56. $f(x) = \left(\dfrac{x^2 - x - 3}{x^2 + 1}\right)(x^2 + x + 1)$
57. $g(\theta) = \dfrac{\theta}{1 - \sin \theta}$
58. $f(\theta) = \dfrac{\sin \theta}{1 - \cos \theta}$

Evaluating a Derivative In Exercises 59–62, evaluate the derivative of the function at the given point. Use a graphing utility to verify your result.

Function	Point
59. $y = \dfrac{1 + \csc x}{1 - \csc x}$	$\left(\dfrac{\pi}{6}, -3\right)$
60. $f(x) = \tan x \cot x$	$(1, 1)$
61. $h(t) = \dfrac{\sec t}{t}$	$\left(\pi, -\dfrac{1}{\pi}\right)$
62. $f(x) = \sin x(\sin x + \cos x)$	$\left(\dfrac{\pi}{4}, 1\right)$

Finding an Equation of a Tangent Line In Exercises 63–68, (a) find an equation of the tangent line to the graph of f at the given point, (b) use a graphing utility to graph the function and its tangent line at the point, and (c) use the *derivative* feature of a graphing utility to confirm your results.

63. $f(x) = (x^3 + 4x - 1)(x - 2)$, $(1, -4)$

64. $f(x) = (x - 2)(x^2 + 4)$, $(1, -5)$

65. $f(x) = \dfrac{x}{x + 4}$, $(-5, 5)$ **66.** $f(x) = \dfrac{x + 3}{x - 3}$, $(4, 7)$

67. $f(x) = \tan x$, $\left(\dfrac{\pi}{4}, 1\right)$ **68.** $f(x) = \sec x$, $\left(\dfrac{\pi}{3}, 2\right)$

Famous Curves In Exercises 69–72, find an equation of the tangent line to the graph at the given point. (The graphs in Exercises 69 and 70 are called *Witches of Agnesi*. The graphs in Exercises 71 and 72 are called *serpentines*.)

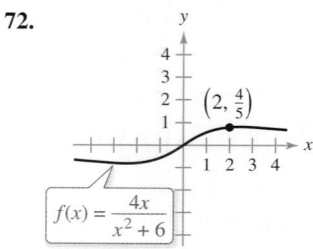

69.

$f(x) = \dfrac{8}{x^2 + 4}$

$(2, 1)$

70.

$f(x) = \dfrac{27}{x^2 + 9}$

$\left(-3, \dfrac{3}{2}\right)$

71.

$f(x) = \dfrac{16x}{x^2 + 16}$

$\left(-2, -\dfrac{8}{5}\right)$

72.

$\left(2, \dfrac{4}{5}\right)$

$f(x) = \dfrac{4x}{x^2 + 6}$

Horizontal Tangent Line In Exercises 73–76, determine the point(s) at which the graph of the function has a horizontal tangent line.

73. $f(x) = \dfrac{2x - 1}{x^2}$ **74.** $f(x) = \dfrac{x^2}{x^2 + 1}$

75. $f(x) = \dfrac{x^2}{x - 1}$ **76.** $f(x) = \dfrac{x - 4}{x^2 - 7}$

77. Tangent Lines Find equations of the tangent lines to the graph of $f(x) = (x + 1)/(x - 1)$ that are parallel to the line $2y + x = 6$. Then graph the function and the tangent lines.

78. Tangent Lines Find equations of the tangent lines to the graph of $f(x) = x/(x - 1)$ that pass through the point $(-1, 5)$. Then graph the function and the tangent lines.

Exploring a Relationship In Exercises 79 and 80, verify that $f'(x) = g'(x)$, and explain the relationship between f and g.

79. $f(x) = \dfrac{3x}{x + 2}$, $g(x) = \dfrac{5x + 4}{x + 2}$

80. $f(x) = \dfrac{\sin x - 3x}{x}$, $g(x) = \dfrac{\sin x + 2x}{x}$

Evaluating Derivatives In Exercises 81 and 82, use the graphs of f and g. Let $p(x) = f(x)g(x)$ and $q(x) = f(x)/g(x)$.

81. (a) Find $p'(1)$.
(b) Find $q'(4)$.

82. (a) Find $p'(4)$.
(b) Find $q'(7)$.

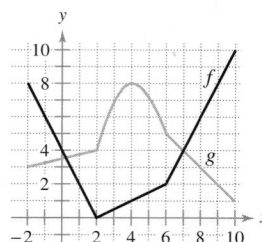

83. Area The length of a rectangle is given by $6t + 5$ and its height is $\sqrt{t}$, where t is time in seconds and the dimensions are in centimeters. Find the rate of change of the area with respect to time.

84. Volume The radius of a right circular cylinder is given by $\sqrt{t + 2}$ and its height is $\frac{1}{2}\sqrt{t}$, where t is time in seconds and the dimensions are in inches. Find the rate of change of the volume with respect to time.

85. Inventory Replenishment The ordering and transportation cost C for the components used in manufacturing a product is

$$C = 100\left(\dfrac{200}{x^2} + \dfrac{x}{x + 30}\right), \quad x \geq 1$$

where C is measured in thousands of dollars and x is the order size in hundreds. Find the rate of change of C with respect to x when (a) $x = 10$, (b) $x = 15$, and (c) $x = 20$. What do these rates of change imply about increasing order size?

86. Population Growth A population of 500 bacteria is introduced into a culture and grows in number according to the equation

$$P(t) = 500\left(1 + \dfrac{4t}{50 + t^2}\right)$$

where t is measured in hours. Find the rate at which the population is growing when $t = 2$.

87. Proof Prove the following differentiation rules.

(a) $\dfrac{d}{dx}[\sec x] = \sec x \tan x$

(b) $\dfrac{d}{dx}[\csc x] = -\csc x \cot x$

(c) $\dfrac{d}{dx}[\cot x] = -\csc^2 x$

88. Rate of Change Determine whether there exist any values of x in the interval $[0, 2\pi)$ such that the rate of change of $f(x) = \sec x$ and the rate of change of $g(x) = \csc x$ are equal.

89. Modeling Data The table shows the health care expenditures h (in billions of dollars) in the United States and the population p (in millions) of the United States for the years 2004 through 2009. The year is represented by t, with $t = 4$ corresponding to 2004. *(Source: U.S. Centers for Medicare & Medicaid Services and U.S. Census Bureau)*

Year, t	4	5	6	7	8	9
h	1773	1890	2017	2135	2234	2330
p	293	296	299	302	305	307

(a) Use a graphing utility to find linear models for the health care expenditures $h(t)$ and the population $p(t)$.

(b) Use a graphing utility to graph each model found in part (a).

(c) Find $A = h(t)/p(t)$, then graph A using a graphing utility. What does this function represent?

(d) Find and interpret $A'(t)$ in the context of these data.

90. Satellites When satellites observe Earth, they can scan only part of Earth's surface. Some satellites have sensors that can measure the angle θ shown in the figure. Let h represent the satellite's distance from Earth's surface, and let r represent Earth's radius.

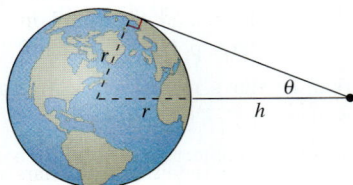

(a) Show that $h = r(\csc \theta - 1)$.

(b) Find the rate at which h is changing with respect to θ when $\theta = 30°$. (Assume $r = 3960$ miles.)

Finding a Second Derivative In Exercises 91–98, find the second derivative of the function.

91. $f(x) = x^4 + 2x^3 - 3x^2 - x$

92. $f(x) = 4x^5 - 2x^3 + 5x^2$

93. $f(x) = 4x^{3/2}$

94. $f(x) = x^2 + 3x^{-3}$

95. $f(x) = \dfrac{x}{x - 1}$

96. $f(x) = \dfrac{x^2 + 3x}{x - 4}$

97. $f(x) = x \sin x$

98. $f(x) = \sec x$

Finding a Higher-Order Derivative In Exercises 99–102, find the given higher-order derivative.

99. $f'(x) = x^2$, $\quad f''(x)$

100. $f''(x) = 2 - \dfrac{2}{x}$, $\quad f'''(x)$

101. $f'''(x) = 2\sqrt{x}$, $\quad f^{(4)}(x)$

102. $f^{(4)}(x) = 2x + 1$, $\quad f^{(6)}(x)$

Using Relationships In Exercises 103–106, use the given information to find $f'(2)$.

$g(2) = 3 \quad$ and $\quad g'(2) = -2$

$h(2) = -1 \quad$ and $\quad h'(2) = 4$

103. $f(x) = 2g(x) + h(x)$

104. $f(x) = 4 - h(x)$

105. $f(x) = \dfrac{g(x)}{h(x)}$

106. $f(x) = g(x)h(x)$

WRITING ABOUT CONCEPTS

107. Sketching a Graph Sketch the graph of a differentiable function f such that $f(2) = 0$, $f' < 0$ for $-\infty < x < 2$, and $f' > 0$ for $2 < x < \infty$. Explain how you found your answer.

108. Sketching a Graph Sketch the graph of a differentiable function f such that $f > 0$ and $f' < 0$ for all real numbers x. Explain how you found your answer.

Identifying Graphs In Exercises 109 and 110, the graphs of f, f', and f'' are shown on the same set of coordinate axes. Identify each graph. Explain your reasoning. To print an enlarged copy of the graph, go to *MathGraphs.com*.

109.

110.

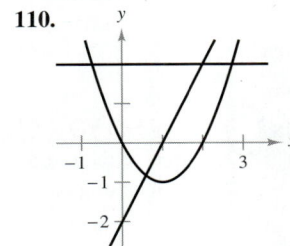

Sketching Graphs In Exercises 111–114, the graph of f is shown. Sketch the graphs of f' and f''. To print an enlarged copy of the graph, go to *MathGraphs.com*.

111.

112.

113.

114.

115. Acceleration The velocity of an object in meters per second is

$$v(t) = 36 - t^2$$

for $0 \le t \le 6$. Find the velocity and acceleration of the object when $t = 3$. What can be said about the speed of the object when the velocity and acceleration have opposite signs?

116. Acceleration The velocity of an automobile starting from rest is

$$v(t) = \frac{100t}{2t + 15}$$

where v is measured in feet per second. Find the acceleration at (a) 5 seconds, (b) 10 seconds, and (c) 20 seconds.

117. Stopping Distance A car is traveling at a rate of 66 feet per second (45 miles per hour) when the brakes are applied. The position function for the car is $s(t) = -8.25t^2 + 66t$, where s is measured in feet and t is measured in seconds. Use this function to complete the table, and find the average velocity during each time interval.

t	0	1	2	3	4
$s(t)$					
$v(t)$					
$a(t)$					

118. HOW DO YOU SEE IT? The figure shows the graphs of the position, velocity, and acceleration functions of a particle.

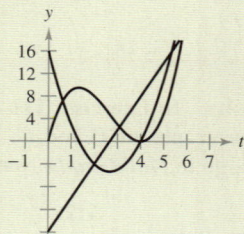

(a) Copy the graphs of the functions shown. Identify each graph. Explain your reasoning. To print an enlarged copy of the graph, go to *MathGraphs.com*.

(b) On your sketch, identify when the particle speeds up and when it slows down. Explain your reasoning.

Finding a Pattern In Exercises 119 and 120, develop a general rule for $f^{(n)}(x)$ given $f(x)$.

119. $f(x) = x^n$

120. $f(x) = \dfrac{1}{x}$

121. Finding a Pattern Consider the function $f(x) = g(x)h(x)$.

(a) Use the Product Rule to generate rules for finding $f''(x)$, $f'''(x)$, and $f^{(4)}(x)$.

(b) Use the results of part (a) to write a general rule for $f^{(n)}(x)$.

122. Finding a Pattern Develop a general rule for $[xf(x)]^{(n)}$, where f is a differentiable function of x.

Finding a Pattern In Exercises 123 and 124, find the derivatives of the function f for $n = 1, 2, 3,$ and 4. Use the results to write a general rule for $f'(x)$ in terms of n.

123. $f(x) = x^n \sin x$

124. $f(x) = \dfrac{\cos x}{x^n}$

Differential Equations In Exercises 125–128, verify that the function satisfies the differential equation.

Function	Differential Equation
125. $y = \dfrac{1}{x}, x > 0$	$x^3 y'' + 2x^2 y' = 0$
126. $y = 2x^3 - 6x + 10$	$-y''' - xy'' - 2y' = -24x^2$
127. $y = 2 \sin x + 3$	$y'' + y = 3$
128. $y = 3 \cos x + \sin x$	$y'' + y = 0$

True or False? In Exercises 129–134, determine whether the statement is true or false. If it is false, explain why or give an example that shows it is false.

129. If $y = f(x)g(x)$, then $\dfrac{dy}{dx} = f'(x)g'(x)$.

130. If $y = (x + 1)(x + 2)(x + 3)(x + 4)$, then $\dfrac{d^5 y}{dx^5} = 0$.

131. If $f'(c)$ and $g'(c)$ are zero and $h(x) = f(x)g(x)$, then $h'(c) = 0$.

132. If $f(x)$ is an nth-degree polynomial, then $f^{(n+1)}(x) = 0$.

133. The second derivative represents the rate of change of the first derivative.

134. If the velocity of an object is constant, then its acceleration is zero.

135. Absolute Value Find the derivative of $f(x) = x|x|$. Does $f''(0)$ exist? (*Hint:* Rewrite the function as a piecewise function and then differentiate each part.)

136. Think About It Let f and g be functions whose first and second derivatives exist on an interval I. Which of the following formulas is (are) true?

(a) $fg'' - f''g = (fg' - f'g)'$ (b) $fg'' + f''g = (fg)''$

137. Proof Use the Product Rule twice to prove that if f, g, and h are differentiable functions of x, then

$$\frac{d}{dx}[f(x)g(x)h(x)] = f'(x)g(x)h(x) + f(x)g'(x)h(x) + f(x)g(x)h'(x).$$

2.4   The Chain Rule

■ Find the derivative of a composite function using the Chain Rule.
■ Find the derivative of a function using the General Power Rule.
■ Simplify the derivative of a function using algebra.
■ Find the derivative of a trigonometric function using the Chain Rule.

The Chain Rule

This text has yet to discuss one of the most powerful differentiation rules—the **Chain Rule.** This rule deals with composite functions and adds a surprising versatility to the rules discussed in the two previous sections. For example, compare the functions shown below. Those on the left can be differentiated without the Chain Rule, and those on the right are best differentiated with the Chain Rule.

Without the Chain Rule	With the Chain Rule
$y = x^2 + 1$	$y = \sqrt{x^2 + 1}$
$y = \sin x$	$y = \sin 6x$
$y = 3x + 2$	$y = (3x + 2)^5$
$y = x + \tan x$	$y = x + \tan x^2$

Basically, the Chain Rule states that if y changes dy/du times as fast as u, and u changes du/dx times as fast as x, then y changes $(dy/du)(du/dx)$ times as fast as x.

EXAMPLE 1   The Derivative of a Composite Function

A set of gears is constructed, as shown in Figure 2.24, such that the second and third gears are on the same axle. As the first axle revolves, it drives the second axle, which in turn drives the third axle. Let y, u, and x represent the numbers of revolutions per minute of the first, second, and third axles, respectively. Find dy/du, du/dx, and dy/dx, and show that

$$\frac{dy}{dx} = \frac{dy}{du} \cdot \frac{du}{dx}.$$

Solution   Because the circumference of the second gear is three times that of the first, the first axle must make three revolutions to turn the second axle once. Similarly, the second axle must make two revolutions to turn the third axle once, and you can write

$$\frac{dy}{du} = 3 \quad \text{and} \quad \frac{du}{dx} = 2.$$

Combining these two results, you know that the first axle must make six revolutions to turn the third axle once. So, you can write

$$\frac{dy}{dx} = \boxed{\begin{array}{c}\text{Rate of change of first axle}\\\text{with respect to second axle}\end{array}} \cdot \boxed{\begin{array}{c}\text{Rate of change of second axle}\\\text{with respect to third axle}\end{array}}$$

$$= \frac{dy}{du} \cdot \frac{du}{dx}$$

$$= 3 \cdot 2$$

$$= 6$$

$$= \boxed{\begin{array}{c}\text{Rate of change of first axle}\\\text{with respect to third axle}\end{array}}.$$

In other words, the rate of change of y with respect to x is the product of the rate of change of y with respect to u and the rate of change of u with respect to x.

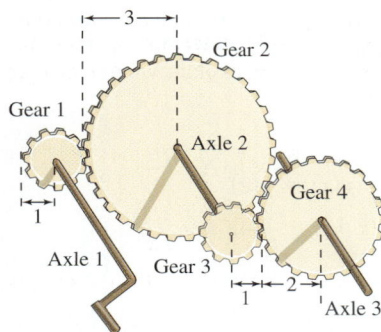

Axle 1: y revolutions per minute
Axle 2: u revolutions per minute
Axle 3: x revolutions per minute
Figure 2.24

Exploration

Using the Chain Rule Each of the following functions can be differentiated using rules that you studied in Sections 2.2 and 2.3. For each function, find the derivative using those rules. Then find the derivative using the Chain Rule. Compare your results. Which method is simpler?

a. $\dfrac{2}{3x + 1}$

b. $(x + 2)^3$

c. $\sin 2x$

Example 1 illustrates a simple case of the Chain Rule. The general rule is stated in the next theorem.

> **THEOREM 2.10   The Chain Rule**
>
> If $y = f(u)$ is a differentiable function of u and $u = g(x)$ is a differentiable function of x, then $y = f(g(x))$ is a differentiable function of x and
>
> $$\frac{dy}{dx} = \frac{dy}{du} \cdot \frac{du}{dx}$$
>
> or, equivalently,
>
> $$\frac{d}{dx}[f(g(x))] = f'(g(x))g'(x).$$

Proof   Let $h(x) = f(g(x))$. Then, using the alternative form of the derivative, you need to show that, for $x = c$,

$$h'(c) = f'(g(c))g'(c).$$

An important consideration in this proof is the behavior of g as x approaches c. A problem occurs when there are values of x, other than c, such that

$$g(x) = g(c).$$

Appendix A shows how to use the differentiability of f and g to overcome this problem. For now, assume that $g(x) \neq g(c)$ for values of x other than c. In the proofs of the Product Rule and the Quotient Rule, the same quantity was added and subtracted to obtain the desired form. This proof uses a similar technique—multiplying and dividing by the same (nonzero) quantity. Note that because g is differentiable, it is also continuous, and it follows that $g(x)$ approaches $g(c)$ as x approaches c.

$$h'(c) = \lim_{x \to c} \frac{f(g(x)) - f(g(c))}{x - c} \qquad \color{red}{\text{Alternative form of derivative}}$$

$$= \lim_{x \to c} \left[\frac{f(g(x)) - f(g(c))}{x - c} \cdot \frac{g(x) - g(c)}{g(x) - g(c)} \right], \quad g(x) \neq g(c)$$

$$= \lim_{x \to c} \left[\frac{f(g(x)) - f(g(c))}{g(x) - g(c)} \cdot \frac{g(x) - g(c)}{x - c} \right]$$

$$= \left[\lim_{x \to c} \frac{f(g(x)) - f(g(c))}{g(x) - g(c)} \right]\left[\lim_{x \to c} \frac{g(x) - g(c)}{x - c} \right]$$

$$= f'(g(c))g'(c)$$

See LarsonCalculus.com for Bruce Edwards's video of this proof.

• • REMARK The alternative limit form of the derivative was given at the end of Section 2.1.

When applying the Chain Rule, it is helpful to think of the composite function $f \circ g$ as having two parts—an inner part and an outer part.

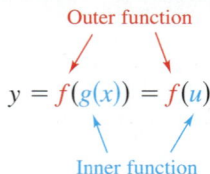

$$y = f(g(x)) = f(u)$$

The derivative of $y = f(u)$ is the derivative of the outer function (at the inner function u) *times* the derivative of the inner function.

$$y' = f'(u) \cdot u'$$

EXAMPLE 2 Decomposition of a Composite Function

$y = f(g(x))$	$u = g(x)$	$y = f(u)$
a. $y = \dfrac{1}{x+1}$	$u = x + 1$	$y = \dfrac{1}{u}$
b. $y = \sin 2x$	$u = 2x$	$y = \sin u$
c. $y = \sqrt{3x^2 - x + 1}$	$u = 3x^2 - x + 1$	$y = \sqrt{u}$
d. $y = \tan^2 x$	$u = \tan x$	$y = u^2$

EXAMPLE 3 Using the Chain Rule

Find dy/dx for

$$y = (x^2 + 1)^3.$$

Solution For this function, you can consider the inside function to be $u = x^2 + 1$ and the outer function to be $y = u^3$. By the Chain Rule, you obtain

$$\frac{dy}{dx} = \underbrace{3(x^2 + 1)^2}_{\frac{dy}{du}}\underbrace{(2x)}_{\frac{du}{dx}} = 6x(x^2 + 1)^2.$$

> **REMARK** You could also solve the problem in Example 3 without using the Chain Rule by observing that
>
> $y = x^6 + 3x^4 + 3x^2 + 1$
>
> and
>
> $y' = 6x^5 + 12x^3 + 6x.$
>
> Verify that this is the same as the derivative in Example 3. Which method would you use to find
>
> $\dfrac{d}{dx}(x^2 + 1)^{50}$?

The General Power Rule

The function in Example 3 is an example of one of the most common types of composite functions, $y = [u(x)]^n$. The rule for differentiating such functions is called the **General Power Rule,** and it is a special case of the Chain Rule.

THEOREM 2.11 The General Power Rule

If $y = [u(x)]^n$, where u is a differentiable function of x and n is a rational number, then

$$\frac{dy}{dx} = n[u(x)]^{n-1}\frac{du}{dx}$$

or, equivalently,

$$\frac{d}{dx}[u^n] = nu^{n-1}u'.$$

Proof Because $y = [u(x)]^n = u^n$, you apply the Chain Rule to obtain

$$\frac{dy}{dx} = \left(\frac{dy}{du}\right)\left(\frac{du}{dx}\right)$$

$$= \frac{d}{du}[u^n]\frac{du}{dx}.$$

By the (Simple) Power Rule in Section 2.2, you have $D_u[u^n] = nu^{n-1}$, and it follows that

$$\frac{dy}{dx} = nu^{n-1}\frac{du}{dx}.$$

See LarsonCalculus.com for Bruce Edwards's video of this proof.

EXAMPLE 4   **Applying the General Power Rule**

Find the derivative of $f(x) = (3x - 2x^2)^3$.

Solution   Let $u = 3x - 2x^2$. Then

$$f(x) = (3x - 2x^2)^3 = u^3$$

and, by the General Power Rule, the derivative is

$$\overbrace{f'(x) = 3}^{n} \overbrace{(3x - 2x^2)^2}^{u^{n-1}} \overbrace{\frac{d}{dx}[3x - 2x^2]}^{u'} \qquad \text{Apply General Power Rule.}$$

$$= 3(3x - 2x^2)^2(3 - 4x). \qquad \text{Differentiate } 3x - 2x^2.$$

EXAMPLE 5   **Differentiating Functions Involving Radicals**

Find all points on the graph of

$$f(x) = \sqrt[3]{(x^2 - 1)^2}$$

for which $f'(x) = 0$ and those for which $f'(x)$ does not exist.

Solution   Begin by rewriting the function as

$$f(x) = (x^2 - 1)^{2/3}.$$

Then, applying the General Power Rule (with $u = x^2 - 1$) produces

$$\overset{n}{|} \overbrace{f'(x) = \frac{2}{3}}\ \overbrace{(x^2 - 1)^{-1/3}}^{u^{n-1}} \overbrace{(2x)}^{u'} \qquad \text{Apply General Power Rule.}$$

$$= \frac{4x}{3\sqrt[3]{x^2 - 1}}. \qquad \text{Write in radical form.}$$

So, $f'(x) = 0$ when $x = 0$, and $f'(x)$ does not exist when $x = \pm 1$, as shown in Figure 2.25.

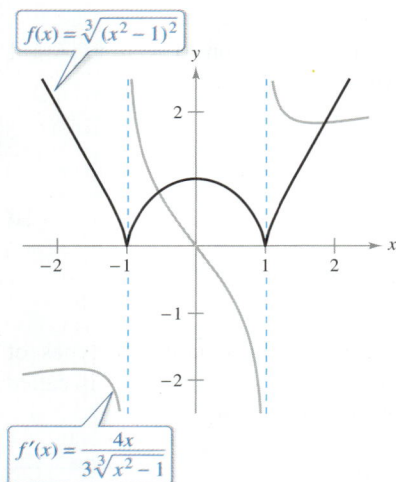

$f(x) = \sqrt[3]{(x^2 - 1)^2}$

$f'(x) = \dfrac{4x}{3\sqrt[3]{x^2 - 1}}$

The derivative of f is 0 at $x = 0$ and is undefined at $x = \pm 1$.
Figure 2.25

EXAMPLE 6   **Differentiating Quotients: Constant Numerators**

Differentiate the function

$$g(t) = \frac{-7}{(2t - 3)^2}.$$

Solution   Begin by rewriting the function as

$$g(t) = -7(2t - 3)^{-2}.$$

Then, applying the General Power Rule (with $u = 2t - 3$) produces

$$\overbrace{g'(t) = (-7)(-2)}^{n}\overbrace{(2t - 3)^{-3}}^{u^{n-1}}\overbrace{(2)}^{u'} \qquad \text{Apply General Power Rule.}$$

Constant
Multiple Rule

$$= 28(2t - 3)^{-3} \qquad \text{Simplify.}$$

$$= \frac{28}{(2t - 3)^3}. \qquad \text{Write with positive exponent.}$$

• • REMARK Try differentiating the function in Example 6 using the Quotient Rule. You should obtain the same result, but using the Quotient Rule is less efficient than using the General Power Rule.

Simplifying Derivatives

The next three examples demonstrate techniques for simplifying the "raw derivatives" of functions involving products, quotients, and composites.

EXAMPLE 7   Simplifying by Factoring Out the Least Powers

Find the derivative of $f(x) = x^2\sqrt{1 - x^2}$.

Solution

$$f(x) = x^2\sqrt{1 - x^2} \qquad \text{Write original function.}$$

$$= x^2(1 - x^2)^{1/2} \qquad \text{Rewrite.}$$

$$f'(x) = x^2 \frac{d}{dx}\left[(1 - x^2)^{1/2}\right] + (1 - x^2)^{1/2} \frac{d}{dx}\left[x^2\right] \qquad \text{Product Rule}$$

$$= x^2\left[\frac{1}{2}(1 - x^2)^{-1/2}(-2x)\right] + (1 - x^2)^{1/2}(2x) \qquad \text{General Power Rule}$$

$$= -x^3(1 - x^2)^{-1/2} + 2x(1 - x^2)^{1/2} \qquad \text{Simplify.}$$

$$= x(1 - x^2)^{-1/2}\left[-x^2(1) + 2(1 - x^2)\right] \qquad \text{Factor.}$$

$$= \frac{x(2 - 3x^2)}{\sqrt{1 - x^2}} \qquad \text{Simplify.}$$

EXAMPLE 8   Simplifying the Derivative of a Quotient

▷ **TECHNOLOGY** Symbolic differentiation utilities are capable of differentiating very complicated functions. Often, however, the result is given in unsimplified form. If you have access to such a utility, use it to find the derivatives of the functions given in Examples 7, 8, and 9. Then compare the results with those given in these examples.

$$f(x) = \frac{x}{\sqrt[3]{x^2 + 4}} \qquad \text{Original function}$$

$$= \frac{x}{(x^2 + 4)^{1/3}} \qquad \text{Rewrite.}$$

$$f'(x) = \frac{(x^2 + 4)^{1/3}(1) - x(1/3)(x^2 + 4)^{-2/3}(2x)}{(x^2 + 4)^{2/3}} \qquad \text{Quotient Rule}$$

$$= \frac{1}{3}(x^2 + 4)^{-2/3}\left[\frac{3(x^2 + 4) - (2x^2)(1)}{(x^2 + 4)^{2/3}}\right] \qquad \text{Factor.}$$

$$= \frac{x^2 + 12}{3(x^2 + 4)^{4/3}} \qquad \text{Simplify.}$$

EXAMPLE 9   Simplifying the Derivative of a Power

• • • • ▷ *See LarsonCalculus.com for an interactive version of this type of example.*

$$y = \left(\frac{3x - 1}{x^2 + 3}\right)^2 \qquad \text{Original function}$$

$$\overset{n}{} \overset{u^{n-1}}{\left(\frac{3x-1}{x^2+3}\right)} \overset{u'}{\frac{d}{dx}\left[\frac{3x-1}{x^2+3}\right]}$$

$$y' = 2\left(\frac{3x - 1}{x^2 + 3}\right)\frac{d}{dx}\left[\frac{3x - 1}{x^2 + 3}\right] \qquad \text{General Power Rule}$$

$$= \left[\frac{2(3x - 1)}{x^2 + 3}\right]\left[\frac{(x^2 + 3)(3) - (3x - 1)(2x)}{(x^2 + 3)^2}\right] \qquad \text{Quotient Rule}$$

$$= \frac{2(3x - 1)(3x^2 + 9 - 6x^2 + 2x)}{(x^2 + 3)^3} \qquad \text{Multiply.}$$

$$= \frac{2(3x - 1)(-3x^2 + 2x + 9)}{(x^2 + 3)^3} \qquad \text{Simplify.}$$

Trigonometric Functions and the Chain Rule

The "Chain Rule versions" of the derivatives of the six trigonometric functions are shown below.

$$\frac{d}{dx}[\sin u] = (\cos u)u' \qquad\qquad \frac{d}{dx}[\cos u] = -(\sin u)u'$$

$$\frac{d}{dx}[\tan u] = (\sec^2 u)u' \qquad\qquad \frac{d}{dx}[\cot u] = -(\csc^2 u)u'$$

$$\frac{d}{dx}[\sec u] = (\sec u \tan u)u' \qquad \frac{d}{dx}[\csc u] = -(\csc u \cot u)u'$$

EXAMPLE 10 The Chain Rule and Trigonometric Functions

a. $y = \sin 2x$ $\qquad y' = \overset{\cos u}{\cos 2x}\, \overset{u'}{\frac{d}{dx}[2x]} = (\cos 2x)(2) = 2\cos 2x$

b. $y = \cos(x-1)$ $\qquad y' = \overset{-(\sin u)}{-\sin(x-1)}\, \overset{u'}{\frac{d}{dx}[x-1]} = -\sin(x-1)$

c. $y = \tan 3x$ $\qquad y' = \overset{(\sec^2 u)}{\sec^2 3x}\, \overset{u'}{\frac{d}{dx}[3x]} = (\sec^2 3x)(3) = 3\sec^2(3x)$

Be sure you understand the mathematical conventions regarding parentheses and trigonometric functions. For instance, in Example 10(a), $\sin 2x$ is written to mean $\sin(2x)$.

EXAMPLE 11 Parentheses and Trigonometric Functions

a. $y = \cos 3x^2 = \cos(3x^2)$ $\qquad y' = (-\sin 3x^2)(6x) = -6x\sin 3x^2$

b. $y = (\cos 3)x^2$ $\qquad\qquad\qquad y' = (\cos 3)(2x) = 2x\cos 3$

c. $y = \cos(3x)^2 = \cos(9x^2)$ $\qquad y' = (-\sin 9x^2)(18x) = -18x\sin 9x^2$

d. $y = \cos^2 x = (\cos x)^2$ $\qquad\quad y' = 2(\cos x)(-\sin x) = -2\cos x\sin x$

e. $y = \sqrt{\cos x} = (\cos x)^{1/2}$ $\qquad y' = \frac{1}{2}(\cos x)^{-1/2}(-\sin x) = -\frac{\sin x}{2\sqrt{\cos x}}$

To find the derivative of a function of the form $k(x) = f(g(h(x)))$, you need to apply the Chain Rule twice, as shown in Example 12.

EXAMPLE 12 Repeated Application of the Chain Rule

$$f(t) = \sin^3 4t \qquad\qquad\qquad \text{Original function}$$

$$ = (\sin 4t)^3 \qquad\qquad\qquad \text{Rewrite.}$$

$$f'(t) = 3(\sin 4t)^2 \frac{d}{dt}[\sin 4t] \qquad \text{Apply Chain Rule once.}$$

$$ = 3(\sin 4t)^2(\cos 4t)\frac{d}{dt}[4t] \qquad \text{Apply Chain Rule a second time.}$$

$$ = 3(\sin 4t)^2(\cos 4t)(4)$$

$$ = 12\sin^2 4t \cos 4t \qquad\qquad \text{Simplify.}$$

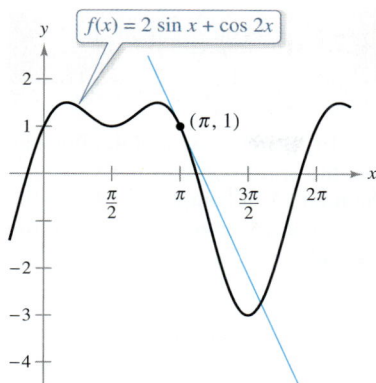

$f(x) = 2 \sin x + \cos 2x$

$(\pi, 1)$

Figure 2.26

EXAMPLE 13   Tangent Line of a Trigonometric Function

Find an equation of the tangent line to the graph of $f(x) = 2 \sin x + \cos 2x$ at the point $(\pi, 1)$, as shown in Figure 2.26. Then determine all values of x in the interval $(0, 2\pi)$ at which the graph of f has a horizontal tangent.

Solution   Begin by finding $f'(x)$.

$f(x) = 2 \sin x + \cos 2x$	Write original function.
$f'(x) = 2 \cos x + (-\sin 2x)(2)$	Apply Chain Rule to $\cos 2x$.
$= 2 \cos x - 2 \sin 2x$	Simplify.

To find the equation of the tangent line at $(\pi, 1)$, evaluate $f'(\pi)$.

$f'(\pi) = 2 \cos \pi - 2 \sin 2\pi$	Substitute.
$= -2$	Slope of graph at $(\pi, 1)$

Now, using the point-slope form of the equation of a line, you can write

$y - y_1 = m(x - x_1)$	Point-slope form
$y - 1 = -2(x - \pi)$	Substitute for y_1, m, and x_1.
$y = 1 - 2x + 2\pi.$	Equation of tangent line at $(\pi, 1)$

You can then determine that $f'(x) = 0$ when $x = \dfrac{\pi}{6}, \dfrac{\pi}{2}, \dfrac{5\pi}{6}$, and $\dfrac{3\pi}{2}$. So, f has horizontal tangents at $x = \dfrac{\pi}{6}, \dfrac{\pi}{2}, \dfrac{5\pi}{6}$, and $\dfrac{3\pi}{2}$. ■

This section concludes with a summary of the differentiation rules studied so far. To become skilled at differentiation, you should memorize each rule in words, not symbols. As an aid to memorization, note that the cofunctions (cosine, cotangent, and cosecant) require a negative sign as part of their derivatives.

SUMMARY OF DIFFERENTIATION RULES

General Differentiation Rules   Let f, g, and u be differentiable functions of x.

Constant Multiple Rule:

$$\frac{d}{dx}[cf] = cf'$$

Sum or Difference Rule:

$$\frac{d}{dx}[f \pm g] = f' \pm g'$$

Product Rule:

$$\frac{d}{dx}[fg] = fg' + gf'$$

Quotient Rule:

$$\frac{d}{dx}\left[\frac{f}{g}\right] = \frac{gf' - fg'}{g^2}$$

Derivatives of Algebraic Functions

Constant Rule:

$$\frac{d}{dx}[c] = 0$$

(Simple) Power Rule:

$$\frac{d}{dx}[x^n] = nx^{n-1}, \quad \frac{d}{dx}[x] = 1$$

Derivatives of Trigonometric Functions

$$\frac{d}{dx}[\sin x] = \cos x \qquad \frac{d}{dx}[\tan x] = \sec^2 x \qquad \frac{d}{dx}[\sec x] = \sec x \tan x$$

$$\frac{d}{dx}[\cos x] = -\sin x \qquad \frac{d}{dx}[\cot x] = -\csc^2 x \qquad \frac{d}{dx}[\csc x] = -\csc x \cot x$$

Chain Rule

Chain Rule:

$$\frac{d}{dx}[f(u)] = f'(u)\, u'$$

General Power Rule:

$$\frac{d}{dx}[u^n] = nu^{n-1} u'$$

2.4 Exercises

See CalcChat.com for tutorial help and worked-out solutions to odd-numbered exercises.

Decomposition of a Composite Function In Exercises 1–6, complete the table.

$y = f(g(x))$	$u = g(x)$	$y = f(u)$
1. $y = (5x - 8)^4$		
2. $y = \dfrac{1}{\sqrt{x + 1}}$		
3. $y = \sqrt{x^3 - 7}$		
4. $y = 3 \tan(\pi x^2)$		
5. $y = \csc^3 x$		
6. $y = \sin \dfrac{5x}{2}$		

Finding a Derivative In Exercises 7–34, find the derivative of the function.

7. $y = (4x - 1)^3$
8. $y = 5(2 - x^3)^4$
9. $g(x) = 3(4 - 9x)^4$
10. $f(t) = (9t + 2)^{2/3}$
11. $f(t) = \sqrt{5 - t}$
12. $g(x) = \sqrt{4 - 3x^2}$
13. $y = \sqrt[3]{6x^2 + 1}$
14. $f(x) = \sqrt{x^2 - 4x + 2}$
15. $y = 2\sqrt[4]{9 - x^2}$
16. $f(x) = \sqrt[3]{12x - 5}$
17. $y = \dfrac{1}{x - 2}$
18. $s(t) = \dfrac{1}{4 - 5t - t^2}$
19. $f(t) = \left(\dfrac{1}{t - 3}\right)^2$
20. $y = -\dfrac{3}{(t - 2)^4}$
21. $y = \dfrac{1}{\sqrt{3x + 5}}$
22. $g(t) = \dfrac{1}{\sqrt{t^2 - 2}}$
23. $f(x) = x^2(x - 2)^4$
24. $f(x) = x(2x - 5)^3$
25. $y = x\sqrt{1 - x^2}$
26. $y = \frac{1}{2}x^2\sqrt{16 - x^2}$
27. $y = \dfrac{x}{\sqrt{x^2 + 1}}$
28. $y = \dfrac{x}{\sqrt{x^4 + 4}}$
29. $g(x) = \left(\dfrac{x + 5}{x^2 + 2}\right)^2$
30. $h(t) = \left(\dfrac{t^2}{t^3 + 2}\right)^2$
31. $f(v) = \left(\dfrac{1 - 2v}{1 + v}\right)^3$
32. $g(x) = \left(\dfrac{3x^2 - 2}{2x + 3}\right)^3$
33. $f(x) = ((x^2 + 3)^5 + x)^2$
34. $g(x) = (2 + (x^2 + 1)^4)^3$

Finding a Derivative Using Technology In Exercises 35–40, use a computer algebra system to find the derivative of the function. Then use the utility to graph the function and its derivative on the same set of coordinate axes. Describe the behavior of the function that corresponds to any zeros of the graph of the derivative.

35. $y = \dfrac{\sqrt{x + 1}}{x^2 + 1}$
36. $y = \sqrt{\dfrac{2x}{x + 1}}$
37. $y = \sqrt{\dfrac{x + 1}{x}}$
38. $g(x) = \sqrt{x - 1} + \sqrt{x + 1}$
39. $y = \dfrac{\cos \pi x + 1}{x}$
40. $y = x^2 \tan \dfrac{1}{x}$

Slope of a Tangent Line In Exercises 41 and 42, find the slope of the tangent line to the sine function at the origin. Compare this value with the number of complete cycles in the interval $[0, 2\pi]$. What can you conclude about the slope of the sine function $\sin ax$ at the origin?

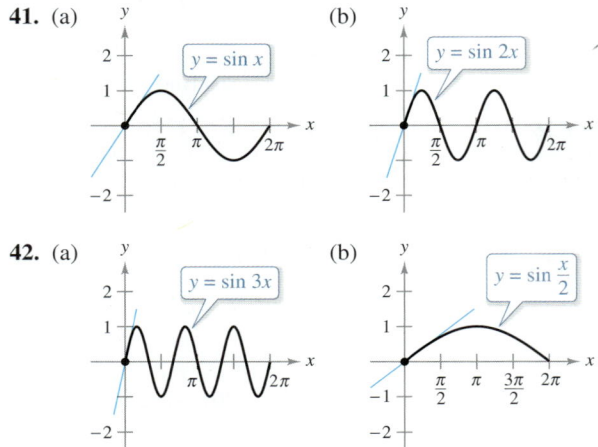

41. (a)

$y = \sin x$

(b)

$y = \sin 2x$

42. (a)

$y = \sin 3x$

(b)

$y = \sin \dfrac{x}{2}$

Finding a Derivative In Exercises 43–64, find the derivative of the function.

43. $y = \cos 4x$
44. $y = \sin \pi x$
45. $g(x) = 5 \tan 3x$
46. $h(x) = \sec x^2$
47. $y = \sin(\pi x)^2$
48. $y = \cos(1 - 2x)^2$
49. $h(x) = \sin 2x \cos 2x$
50. $g(\theta) = \sec(\frac{1}{2}\theta) \tan(\frac{1}{2}\theta)$
51. $f(x) = \dfrac{\cot x}{\sin x}$
52. $g(v) = \dfrac{\cos v}{\csc v}$
53. $y = 4 \sec^2 x$
54. $g(t) = 5 \cos^2 \pi t$
55. $f(\theta) = \tan^2 5\theta$
56. $g(\theta) = \cos^2 8\theta$
57. $f(\theta) = \frac{1}{4} \sin^2 2\theta$
58. $h(t) = 2 \cot^2(\pi t + 2)$
59. $f(t) = 3 \sec^2(\pi t - 1)$
60. $y = 3x - 5 \cos(\pi x)^2$
61. $y = \sqrt{x} + \frac{1}{4} \sin(2x)^2$
62. $y = \sin \sqrt[3]{x} + \sqrt[3]{\sin x}$
63. $y = \sin(\tan 2x)$
64. $y = \cos\sqrt{\sin(\tan \pi x)}$

Evaluating a Derivative In Exercises 65–72, find and evaluate the derivative of the function at the given point. Use a graphing utility to verify your result.

65. $y = \sqrt{x^2 + 8x}$, $(1, 3)$
66. $y = \sqrt[5]{3x^3 + 4x}$, $(2, 2)$
67. $f(x) = \dfrac{5}{x^3 - 2}$, $\left(-2, -\dfrac{1}{2}\right)$
68. $f(x) = \dfrac{1}{(x^2 - 3x)^2}$, $\left(4, \dfrac{1}{16}\right)$
69. $f(t) = \dfrac{3t + 2}{t - 1}$, $(0, -2)$
70. $f(x) = \dfrac{x + 4}{2x - 5}$, $(9, 1)$
71. $y = 26 - \sec^3 4x$, $(0, 25)$
72. $y = \dfrac{1}{x} + \sqrt{\cos x}$, $\left(\dfrac{\pi}{2}, \dfrac{2}{\pi}\right)$

Finding an Equation of a Tangent Line In Exercises 73–80, (a) find an equation of the tangent line to the graph of f at the given point, (b) use a graphing utility to graph the function and its tangent line at the point, and (c) use the *derivative* feature of the graphing utility to confirm your results.

73. $f(x) = \sqrt{2x^2 - 7},\ (4, 5)$ **74.** $f(x) = \frac{1}{3}x\sqrt{x^2 + 5},\ (2, 2)$

75. $y = (4x^3 + 3)^2,\ (-1, 1)$ **76.** $f(x) = (9 - x^2)^{2/3},\ (1, 4)$

77. $f(x) = \sin 2x,\ (\pi, 0)$ **78.** $y = \cos 3x,\ \left(\frac{\pi}{4}, -\frac{\sqrt{2}}{2}\right)$

79. $f(x) = \tan^2 x,\ \left(\frac{\pi}{4}, 1\right)$ **80.** $y = 2\tan^3 x,\ \left(\frac{\pi}{4}, 2\right)$

Famous Curves In Exercises 81 and 82, find an equation of the tangent line to the graph at the given point. Then use a graphing utility to graph the function and its tangent line in the same viewing window.

81. Top half of circle   **82.** Bullet-nose curve

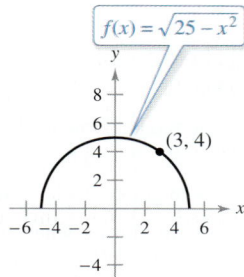

$f(x) = \sqrt{25 - x^2}$; $(3, 4)$

$f(x) = \dfrac{|x|}{\sqrt{2 - x^2}}$; $(1, 1)$

83. Horizontal Tangent Line Determine the point(s) in the interval $(0, 2\pi)$ at which the graph of

$$f(x) = 2\cos x + \sin 2x$$

has a horizontal tangent.

84. Horizontal Tangent Line Determine the point(s) at which the graph of

$$f(x) = \frac{x}{\sqrt{2x - 1}}$$

has a horizontal tangent.

Finding a Second Derivative In Exercises 85–90, find the second derivative of the function.

85. $f(x) = 5(2 - 7x)^4$ **86.** $f(x) = 6(x^3 + 4)^3$

87. $f(x) = \dfrac{1}{x - 6}$ **88.** $f(x) = \dfrac{8}{(x - 2)^2}$

89. $f(x) = \sin x^2$ **90.** $f(x) = \sec^2 \pi x$

Evaluating a Second Derivative In Exercises 91–94, evaluate the second derivative of the function at the given point. Use a computer algebra system to verify your result.

91. $h(x) = \frac{1}{9}(3x + 1)^3,\ \left(1, \frac{64}{9}\right)$ **92.** $f(x) = \dfrac{1}{\sqrt{x + 4}},\ \left(0, \frac{1}{2}\right)$

93. $f(x) = \cos x^2,\ (0, 1)$ **94.** $g(t) = \tan 2t,\ \left(\frac{\pi}{6}, \sqrt{3}\right)$

WRITING ABOUT CONCEPTS

Identifying Graphs In Exercises 95–98, the graphs of a function f and its derivative f' are shown. Label the graphs as f or f' and write a short paragraph stating the criteria you used in making your selection. To print an enlarged copy of the graph, go to *MathGraphs.com*.

95.   **96.**

97.   **98.**

Describing a Relationship In Exercises 99 and 100, the relationship between f and g is given. Explain the relationship between f' and g'.

99. $g(x) = f(3x)$ **100.** $g(x) = f(x^2)$

101. Think About It The table shows some values of the derivative of an unknown function f. Complete the table by finding the derivative of each transformation of f, if possible.

(a) $g(x) = f(x) - 2$

(b) $h(x) = 2f(x)$

(c) $r(x) = f(-3x)$

(d) $s(x) = f(x + 2)$

x	-2	-1	0	1	2	3
$f'(x)$	4	$\frac{2}{3}$	$-\frac{1}{3}$	-1	-2	-4
$g'(x)$						
$h'(x)$						
$r'(x)$						
$s'(x)$						

102. Using Relationships Given that $g(5) = -3$, $g'(5) = 6$, $h(5) = 3$, and $h'(5) = -2$, find $f'(5)$ for each of the following, if possible. If it is not possible, state what additional information is required.

(a) $f(x) = g(x)h(x)$ (b) $f(x) = g(h(x))$

(c) $f(x) = \dfrac{g(x)}{h(x)}$ (d) $f(x) = [g(x)]^3$

Finding Derivatives   In Exercises 103 and 104, the graphs of f and g are shown. Let $h(x) = f(g(x))$ and $s(x) = g(f(x))$. Find each derivative, if it exists. If the derivative does not exist, explain why.

103. (a) Find $h'(1)$.

(b) Find $s'(5)$.

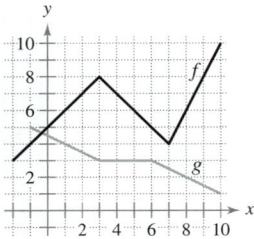

104. (a) Find $h'(3)$.

(b) Find $s'(9)$.

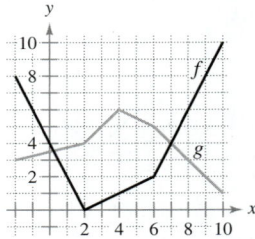

105. Doppler Effect   The frequency F of a fire truck siren heard by a stationary observer is

$$F = \frac{132{,}400}{331 \pm v}$$

where $\pm v$ represents the velocity of the accelerating fire truck in meters per second (see figure). Find the rate of change of F with respect to v when

(a) the fire truck is approaching at a velocity of 30 meters per second (use $-v$).

(b) the fire truck is moving away at a velocity of 30 meters per second (use $+v$).

$$F = \frac{132{,}400}{331 + v} \qquad\qquad F = \frac{132{,}400}{331 - v}$$

106. Harmonic Motion   The displacement from equilibrium of an object in harmonic motion on the end of a spring is

$$y = \frac{1}{3}\cos 12t - \frac{1}{4}\sin 12t$$

where y is measured in feet and t is the time in seconds. Determine the position and velocity of the object when $t = \pi/8$.

107. Pendulum   A 15-centimeter pendulum moves according to the equation $\theta = 0.2\cos 8t$, where θ is the angular displacement from the vertical in radians and t is the time in seconds. Determine the maximum angular displacement and the rate of change of θ when $t = 3$ seconds.

108. Wave Motion   A buoy oscillates in simple harmonic motion $y = A\cos \omega t$ as waves move past it. The buoy moves a total of 3.5 feet (vertically) from its low point to its high point. It returns to its high point every 10 seconds.

(a) Write an equation describing the motion of the buoy if it is at its high point at $t = 0$.

(b) Determine the velocity of the buoy as a function of t.

109. Modeling Data   The normal daily maximum temperatures T (in degrees Fahrenheit) for Chicago, Illinois, are shown in the table. *(Source: National Oceanic and Atmospheric Administration)*

Month	Jan	Feb	Mar	Apr
Temperature	29.6	34.7	46.1	58.0

Month	May	Jun	Jul	Aug
Temperature	69.9	79.2	83.5	81.2

Month	Sep	Oct	Nov	Dec
Temperature	73.9	62.1	47.1	34.4

(a) Use a graphing utility to plot the data and find a model for the data of the form

$$T(t) = a + b\sin(ct - d)$$

where T is the temperature and t is the time in months, with $t = 1$ corresponding to January.

(b) Use a graphing utility to graph the model. How well does the model fit the data?

(c) Find T' and use a graphing utility to graph the derivative.

(d) Based on the graph of the derivative, during what times does the temperature change most rapidly? Most slowly? Do your answers agree with your observations of the temperature changes? Explain.

110. HOW DO YOU SEE IT?   The cost C (in dollars) of producing x units of a product is $C = 60x + 1350$. For one week, management determined that the number of units produced x at the end of t hours can be modeled by $x = -1.6t^3 + 19t^2 - 0.5t - 1$. The graph shows the cost C in terms of the time t.

Cost of Producing a Product

(a) Using the graph, which is greater, the rate of change of the cost after 1 hour or the rate of change of the cost after 4 hours?

(b) Explain why the cost function is not increasing at a constant rate during the eight-hour shift.

111. Biology

The number N of bacteria in a culture after t days is modeled by

$$N = 400\left[1 - \frac{3}{(t^2 + 2)^2}\right].$$

Find the rate of change of N with respect to t when (a) $t = 0$, (b) $t = 1$, (c) $t = 2$, (d) $t = 3$, and (e) $t = 4$. (f) What can you conclude?

112. Depreciation The value V of a machine t years after it is purchased is inversely proportional to the square root of $t + 1$. The initial value of the machine is $10,000.

(a) Write V as a function of t.

(b) Find the rate of depreciation when $t = 1$.

(c) Find the rate of depreciation when $t = 3$.

113. Finding a Pattern Consider the function $f(x) = \sin \beta x$, where β is a constant.

(a) Find the first-, second-, third-, and fourth-order derivatives of the function.

(b) Verify that the function and its second derivative satisfy the equation $f''(x) + \beta^2 f(x) = 0$.

(c) Use the results of part (a) to write general rules for the even- and odd-order derivatives $f^{(2k)}(x)$ and $f^{(2k-1)}(x)$. [*Hint:* $(-1)^k$ is positive if k is even and negative if k is odd.]

114. Conjecture Let f be a differentiable function of period p.

(a) Is the function f' periodic? Verify your answer.

(b) Consider the function $g(x) = f(2x)$. Is the function $g'(x)$ periodic? Verify your answer.

115. Think About It Let $r(x) = f(g(x))$ and $s(x) = g(f(x))$, where f and g are shown in the figure. Find (a) $r'(1)$ and (b) $s'(4)$.

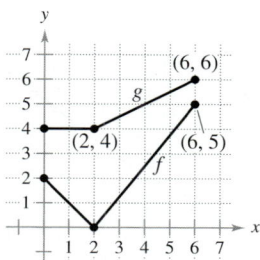

116. Using Trigonometric Functions

(a) Find the derivative of the function $g(x) = \sin^2 x + \cos^2 x$ in two ways.

(b) For $f(x) = \sec^2 x$ and $g(x) = \tan^2 x$, show that

$$f'(x) = g'(x).$$

117. Even and Odd Functions

(a) Show that the derivative of an odd function is even. That is, if $f(-x) = -f(x)$, then $f'(-x) = f'(x)$.

(b) Show that the derivative of an even function is odd. That is, if $f(-x) = f(x)$, then $f'(-x) = -f'(x)$.

118. Proof Let u be a differentiable function of x. Use the fact that $|u| = \sqrt{u^2}$ to prove that

$$\frac{d}{dx}[|u|] = u'\frac{u}{|u|}, \quad u \neq 0.$$

Using Absolute Value In Exercises 119–122, use the result of Exercise 118 to find the derivative of the function.

119. $g(x) = |3x - 5|$

120. $f(x) = |x^2 - 9|$

121. $h(x) = |x| \cos x$

122. $f(x) = |\sin x|$

Linear and Quadratic Approximations The linear and quadratic approximations of a function f at $x = a$ are

$$P_1(x) = f'(a)(x - a) + f(a) \quad \text{and}$$
$$P_2(x) = \tfrac{1}{2}f''(a)(x - a)^2 + f'(a)(x - a) + f(a).$$

In Exercises 123 and 124, (a) find the specified linear and quadratic approximations of f, (b) use a graphing utility to graph f and the approximations, (c) determine whether P_1 or P_2 is the better approximation, and (d) state how the accuracy changes as you move farther from $x = a$.

123. $f(x) = \tan x; \quad a = \dfrac{\pi}{4}$

124. $f(x) = \sec x; \quad a = \dfrac{\pi}{6}$

True or False? In Exercises 125–128, determine whether the statement is true or false. If it is false, explain why or give an example that shows it is false.

125. If $y = (1 - x)^{1/2}$, then $y' = \tfrac{1}{2}(1 - x)^{-1/2}$.

126. If $f(x) = \sin^2(2x)$, then $f'(x) = 2(\sin 2x)(\cos 2x)$.

127. If y is a differentiable function of u, and u is a differentiable function of x, then y is a differentiable function of x.

128. If y is a differentiable function of u, u is a differentiable function of v, and v is a differentiable function of x, then

$$\frac{dy}{dx} = \frac{dy}{du}\frac{du}{dv}\frac{dv}{dx}.$$

PUTNAM EXAM CHALLENGE

129. Let $f(x) = a_1 \sin x + a_2 \sin 2x + \cdots + a_n \sin nx$, where $a_1, a_2, \ldots, a_n$ are real numbers and where n is a positive integer. Given that $|f(x)| \leq |\sin x|$ for all real x, prove that $|a_1 + 2a_2 + \cdots + na_n| \leq 1$.

130. Let k be a fixed positive integer. The nth derivative of $\dfrac{1}{x^k - 1}$ has the form $\dfrac{P_n(x)}{(x^k - 1)^{n+1}}$ where $P_n(x)$ is a polynomial. Find $P_n(1)$.

These problems were composed by the Committee on the Putnam Prize Competition.
© The Mathematical Association of America. All rights reserved.

2.5 Implicit Differentiation

■ Distinguish between functions written in implicit form and explicit form.
■ Use implicit differentiation to find the derivative of a function.

Implicit and Explicit Functions

Up to this point in the text, most functions have been expressed in **explicit form.** For example, in the equation $y = 3x^2 - 5$, the variable y is explicitly written as a function of x. Some functions, however, are only implied by an equation. For instance, the function $y = 1/x$ is defined **implicitly** by the equation

$$xy = 1. \qquad \qquad \text{Implicit form}$$

To find dy/dx for this equation, you can write y explicitly as a function of x and then differentiate.

Implicit Form	Explicit Form	Derivative
$xy = 1$	$y = \dfrac{1}{x} = x^{-1}$	$\dfrac{dy}{dx} = -x^{-2} = -\dfrac{1}{x^2}$

This strategy works whenever you can solve for the function explicitly. You cannot, however, use this procedure when you are unable to solve for y as a function of x. For instance, how would you find dy/dx for the equation

$$x^2 - 2y^3 + 4y = 2?$$

For this equation, it is difficult to express y as a function of x explicitly. To find dy/dx, you can use **implicit differentiation.**

To understand how to find dy/dx implicitly, you must realize that the differentiation is taking place *with respect to x.* This means that when you differentiate terms involving x alone, you can differentiate as usual. However, when you differentiate terms involving y, you must apply the Chain Rule, because you are assuming that y is defined implicitly as a differentiable function of x.

EXAMPLE 1   **Differentiating with Respect to x**

a. $\dfrac{d}{dx}[x^3] = 3x^2$ Variables agree: use Simple Power Rule.

Variables agree

b. $\dfrac{d}{dx}[y^3] = 3y^2 \dfrac{dy}{dx}$ Variables disagree: use Chain Rule.

u^n $\qquad nu^{n-1}$ u'

Variables disagree

c. $\dfrac{d}{dx}[x + 3y] = 1 + 3\dfrac{dy}{dx}$ Chain Rule: $\dfrac{d}{dx}[3y] = 3y'$

d. $\dfrac{d}{dx}[xy^2] = x\dfrac{d}{dx}[y^2] + y^2\dfrac{d}{dx}[x]$ Product Rule

$\qquad = x\left(2y\dfrac{dy}{dx}\right) + y^2(1)$ Chain Rule

$\qquad = 2xy\dfrac{dy}{dx} + y^2$ Simplify.

Implicit Differentiation

> ## GUIDELINES FOR IMPLICIT DIFFERENTIATION
>
> 1. Differentiate both sides of the equation *with respect to x.*
> 2. Collect all terms involving dy/dx on the left side of the equation and move all other terms to the right side of the equation.
> 3. Factor dy/dx out of the left side of the equation.
> 4. Solve for dy/dx.

In Example 2, note that implicit differentiation can produce an expression for dy/dx that contains both x and y.

EXAMPLE 2   Implicit Differentiation

Find dy/dx given that $y^3 + y^2 - 5y - x^2 = -4$.

Solution

1. Differentiate both sides of the equation with respect to x.

$$\frac{d}{dx}[y^3 + y^2 - 5y - x^2] = \frac{d}{dx}[-4]$$

$$\frac{d}{dx}[y^3] + \frac{d}{dx}[y^2] - \frac{d}{dx}[5y] - \frac{d}{dx}[x^2] = \frac{d}{dx}[-4]$$

$$3y^2\frac{dy}{dx} + 2y\frac{dy}{dx} - 5\frac{dy}{dx} - 2x = 0$$

2. Collect the dy/dx terms on the left side of the equation and move all other terms to the right side of the equation.

$$3y^2\frac{dy}{dx} + 2y\frac{dy}{dx} - 5\frac{dy}{dx} = 2x$$

3. Factor dy/dx out of the left side of the equation.

$$\frac{dy}{dx}(3y^2 + 2y - 5) = 2x$$

4. Solve for dy/dx by dividing by $(3y^2 + 2y - 5)$.

$$\frac{dy}{dx} = \frac{2x}{3y^2 + 2y - 5}$$

To see how you can use an *implicit derivative*, consider the graph shown in Figure 2.27. From the graph, you can see that y is not a function of x. Even so, the derivative found in Example 2 gives a formula for the slope of the tangent line at a point on this graph. The slopes at several points on the graph are shown below the graph.

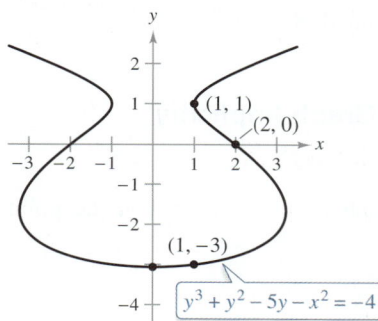

Point on Graph	Slope of Graph
$(2, 0)$	$-\frac{4}{5}$
$(1, -3)$	$\frac{1}{8}$
$x = 0$	0
$(1, 1)$	Undefined

The implicit equation

$$y^3 + y^2 - 5y - x^2 = -4$$

has the derivative

$$\frac{dy}{dx} = \frac{2x}{3y^2 + 2y - 5}.$$

Figure 2.27

▷ **TECHNOLOGY**   With most graphing utilities, it is easy to graph an equation that explicitly represents y as a function of x. Graphing other equations, however, can require some ingenuity. For instance, to graph the equation given in Example 2, use a graphing utility, set in *parametric* mode, to graph the parametric representations $x = \sqrt{t^3 + t^2 - 5t + 4}$, $y = t$, and $x = -\sqrt{t^3 + t^2 - 5t + 4}$, $y = t$, for $-5 \le t \le 5$. How does the result compare with the graph shown in Figure 2.27?

(a)

(b)

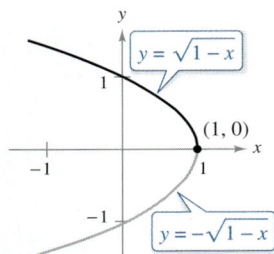

(c)
Some graph segments can be represented by differentiable functions.
Figure 2.28

It is meaningless to solve for dy/dx in an equation that has no solution points. (For example, $x^2 + y^2 = -4$ has no solution points.) If, however, a segment of a graph can be represented by a differentiable function, then dy/dx will have meaning as the slope at each point on the segment. Recall that a function is not differentiable at (a) points with vertical tangents and (b) points at which the function is not continuous.

EXAMPLE 3   **Graphs and Differentiable Functions**

If possible, represent y as a differentiable function of x.

a. $x^2 + y^2 = 0$ **b.** $x^2 + y^2 = 1$ **c.** $x + y^2 = 1$

Solution

a. The graph of this equation is a single point. So, it does not define y as a differentiable function of x. See Figure 2.28(a).

b. The graph of this equation is the unit circle centered at $(0, 0)$. The upper semicircle is given by the differentiable function

$$y = \sqrt{1 - x^2}, \quad -1 < x < 1$$

and the lower semicircle is given by the differentiable function

$$y = -\sqrt{1 - x^2}, \quad -1 < x < 1.$$

At the points $(-1, 0)$ and $(1, 0)$, the slope of the graph is undefined. See Figure 2.28(b).

c. The upper half of this parabola is given by the differentiable function

$$y = \sqrt{1 - x}, \quad x < 1$$

and the lower half of this parabola is given by the differentiable function

$$y = -\sqrt{1 - x}, \quad x < 1.$$

At the point $(1, 0)$, the slope of the graph is undefined. See Figure 2.28(c).

EXAMPLE 4   **Finding the Slope of a Graph Implicitly**

⋅ ⋅ ⋅▷ *See LarsonCalculus.com for an interactive version of this type of example.*

Determine the slope of the tangent line to the graph of $x^2 + 4y^2 = 4$ at the point $\left(\sqrt{2}, -1/\sqrt{2}\right)$. See Figure 2.29.

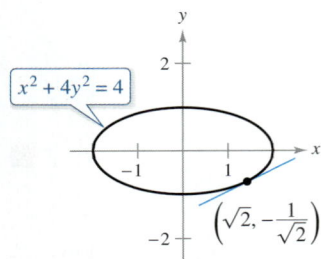

Figure 2.29

Solution

$x^2 + 4y^2 = 4$	Write original equation.
$2x + 8y\dfrac{dy}{dx} = 0$	Differentiate with respect to x.
$\dfrac{dy}{dx} = \dfrac{-2x}{8y}$	Solve for $\dfrac{dy}{dx}$.
$= \dfrac{-x}{4y}$	Simplify.

So, at $\left(\sqrt{2}, -1/\sqrt{2}\right)$, the slope is

$$\frac{dy}{dx} = \frac{-\sqrt{2}}{-4/\sqrt{2}} = \frac{1}{2}. \qquad \text{Evaluate } \frac{dy}{dx} \text{ when } x = \sqrt{2} \text{ and } y = -\frac{1}{\sqrt{2}}.$$

REMARK To see the benefit of implicit differentiation, try doing Example 4 using the explicit function $y = -\frac{1}{2}\sqrt{4 - x^2}$.

EXAMPLE 5 **Finding the Slope of a Graph Implicitly**

Determine the slope of the graph of

$$3(x^2 + y^2)^2 = 100xy$$

at the point $(3, 1)$.

Solution

$$\frac{d}{dx}[3(x^2 + y^2)^2] = \frac{d}{dx}[100xy]$$

$$3(2)(x^2 + y^2)\left(2x + 2y\frac{dy}{dx}\right) = 100\left[x\frac{dy}{dx} + y(1)\right]$$

$$12y(x^2 + y^2)\frac{dy}{dx} - 100x\frac{dy}{dx} = 100y - 12x(x^2 + y^2)$$

$$[12y(x^2 + y^2) - 100x]\frac{dy}{dx} = 100y - 12x(x^2 + y^2)$$

$$\frac{dy}{dx} = \frac{100y - 12x(x^2 + y^2)}{-100x + 12y(x^2 + y^2)}$$

$$= \frac{25y - 3x(x^2 + y^2)}{-25x + 3y(x^2 + y^2)}$$

At the point $(3, 1)$, the slope of the graph is

$$\frac{dy}{dx} = \frac{25(1) - 3(3)(3^2 + 1^2)}{-25(3) + 3(1)(3^2 + 1^2)} = \frac{25 - 90}{-75 + 30} = \frac{-65}{-45} = \frac{13}{9}$$

as shown in Figure 2.30. This graph is called a **lemniscate**.

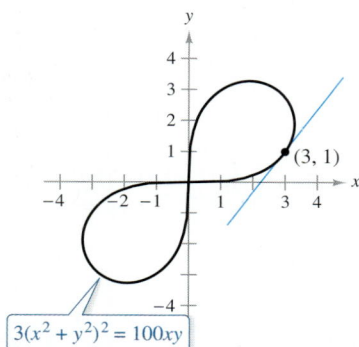

$3(x^2 + y^2)^2 = 100xy$

Lemniscate
Figure 2.30

EXAMPLE 6 **Determining a Differentiable Function**

Find dy/dx implicitly for the equation $\sin y = x$. Then find the largest interval of the form $-a < y < a$ on which y is a differentiable function of x (see Figure 2.31).

Solution

$$\frac{d}{dx}[\sin y] = \frac{d}{dx}[x]$$

$$\cos y\frac{dy}{dx} = 1$$

$$\frac{dy}{dx} = \frac{1}{\cos y}$$

The largest interval about the origin for which y is a differentiable function of x is $-\pi/2 < y < \pi/2$. To see this, note that $\cos y$ is positive for all y in this interval and is 0 at the endpoints. When you restrict y to the interval $-\pi/2 < y < \pi/2$, you should be able to write dy/dx explicitly as a function of x. To do this, you can use

$$\cos y = \sqrt{1 - \sin^2 y}$$

$$= \sqrt{1 - x^2}, \quad -\frac{\pi}{2} < y < \frac{\pi}{2}$$

and conclude that

$$\frac{dy}{dx} = \frac{1}{\sqrt{1 - x^2}}.$$

You will study this example further when inverse trigonometric functions are defined in Section 5.6.

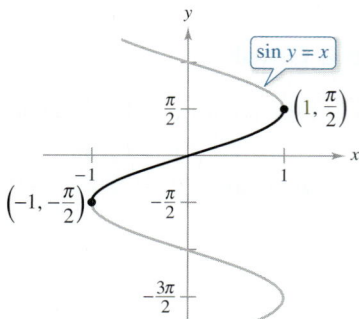

$\sin y = x$

The derivative is $\dfrac{dy}{dx} = \dfrac{1}{\sqrt{1 - x^2}}$.

Figure 2.31

With implicit differentiation, the form of the derivative often can be simplified (as in Example 6) by an appropriate use of the *original* equation. A similar technique can be used to find and simplify higher-order derivatives obtained implicitly.

EXAMPLE 7 Finding the Second Derivative Implicitly

Given $x^2 + y^2 = 25$, find $\dfrac{d^2y}{dx^2}$.

Solution Differentiating each term with respect to x produces

$$2x + 2y\frac{dy}{dx} = 0$$

$$2y\frac{dy}{dx} = -2x$$

$$\frac{dy}{dx} = \frac{-2x}{2y}$$

$$= -\frac{x}{y}.$$

Differentiating a second time with respect to x yields

$$\frac{d^2y}{dx^2} = -\frac{(y)(1) - (x)(dy/dx)}{y^2} \qquad \text{Quotient Rule}$$

$$= -\frac{y - (x)(-x/y)}{y^2} \qquad \text{Substitute } -\frac{x}{y} \text{ for } \frac{dy}{dx}.$$

$$= -\frac{y^2 + x^2}{y^2} \qquad \text{Simplify.}$$

$$= -\frac{25}{y^3}. \qquad \text{Substitute 25 for } x^2 + y^2.$$

EXAMPLE 8 Finding a Tangent Line to a Graph

Find the tangent line to the graph of $x^2(x^2 + y^2) = y^2$ at the point $\left(\sqrt{2}/2,\ \sqrt{2}/2\right)$, as shown in Figure 2.32.

Solution By rewriting and differentiating implicitly, you obtain

$$x^4 + x^2y^2 - y^2 = 0$$

$$4x^3 + x^2\left(2y\frac{dy}{dx}\right) + 2xy^2 - 2y\frac{dy}{dx} = 0$$

$$2y(x^2 - 1)\frac{dy}{dx} = -2x(2x^2 + y^2)$$

$$\frac{dy}{dx} = \frac{x(2x^2 + y^2)}{y(1 - x^2)}.$$

At the point $\left(\sqrt{2}/2,\ \sqrt{2}/2\right)$, the slope is

$$\frac{dy}{dx} = \frac{\left(\sqrt{2}/2\right)[2(1/2) + (1/2)]}{\left(\sqrt{2}/2\right)[1 - (1/2)]} = \frac{3/2}{1/2} = 3$$

and the equation of the tangent line at this point is

$$y - \frac{\sqrt{2}}{2} = 3\left(x - \frac{\sqrt{2}}{2}\right)$$

$$y = 3x - \sqrt{2}.$$

ISAAC BARROW (1630–1677)

The graph in Figure 2.32 is called the **kappa curve** because it resembles the Greek letter kappa, κ. The general solution for the tangent line to this curve was discovered by the English mathematician Isaac Barrow. Newton was Barrow's student, and they corresponded frequently regarding their work in the early development of calculus. *See LarsonCalculus.com to read more of this biography.*

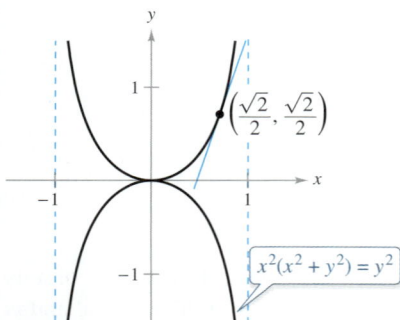

The kappa curve
Figure 2.32

2.5 Exercises

See **CalcChat.com** for tutorial help and worked-out solutions to odd-numbered exercises.

Finding a Derivative In Exercises 1–16, find dy/dx by implicit differentiation.

1. $x^2 + y^2 = 9$

2. $x^2 - y^2 = 25$

3. $x^{1/2} + y^{1/2} = 16$

4. $2x^3 + 3y^3 = 64$

5. $x^3 - xy + y^2 = 7$

6. $x^2y + y^2x = -2$

7. $x^3y^3 - y = x$

8. $\sqrt{xy} = x^2y + 1$

9. $x^3 - 3x^2y + 2xy^2 = 12$

10. $4 \cos x \sin y = 1$

11. $\sin x + 2 \cos 2y = 1$

12. $(\sin \pi x + \cos \pi y)^2 = 2$

13. $\sin x = x(1 + \tan y)$

14. $\cot y = x - y$

15. $y = \sin xy$

16. $x = \sec \dfrac{1}{y}$

Finding Derivatives Implicity and Explicitly In Exercises 17–20, (a) find two explicit functions by solving the equation for y in terms of x, (b) sketch the graph of the equation and label the parts given by the corresponding explicit functions, (c) differentiate the explicit functions, and (d) find dy/dx implicitly and show that the result is equivalent to that of part (c).

17. $x^2 + y^2 = 64$

18. $25x^2 + 36y^2 = 300$

19. $16y^2 - x^2 = 16$

20. $x^2 + y^2 - 4x + 6y + 9 = 0$

Finding and Evaluating a Derivative In Exercises 21–28, find dy/dx by implicit differentiation and evaluate the derivative at the given point.

21. $xy = 6$, $(-6, -1)$

22. $y^3 - x^2 = 4$, $(2, 2)$

23. $y^2 = \dfrac{x^2 - 49}{x^2 + 49}$, $(7, 0)$

24. $x^{2/3} + y^{2/3} = 5$, $(8, 1)$

25. $(x + y)^3 = x^3 + y^3$, $(-1, 1)$

26. $x^3 + y^3 = 6xy - 1$, $(2, 3)$

27. $\tan(x + y) = x$, $(0, 0)$

28. $x \cos y = 1$, $\left(2, \dfrac{\pi}{3}\right)$

Famous Curves In Exercises 29–32, find the slope of the tangent line to the graph at the given point.

29. Witch of Agnesi:

$(x^2 + 4)y = 8$

Point: $(2, 1)$

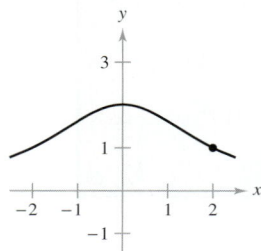

30. Cissoid:

$(4 - x)y^2 = x^3$

Point: $(2, 2)$

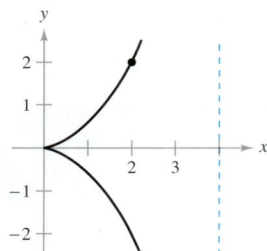

31. Bifolium:

$(x^2 + y^2)^2 = 4x^2y$

Point: $(1, 1)$

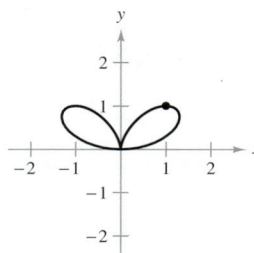

32. Folium of Descartes:

$x^3 + y^3 - 6xy = 0$

Point: $\left(\dfrac{4}{3}, \dfrac{8}{3}\right)$

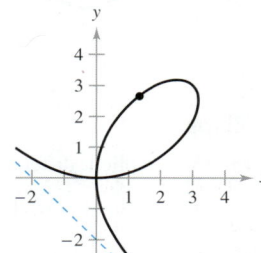

Famous Curves In Exercises 33–40, find an equation of the tangent line to the graph at the given point. To print an enlarged copy of the graph, go to *MathGraphs.com*.

33. Parabola

$(y - 3)^2 = 4(x - 5)$

$(6, 1)$

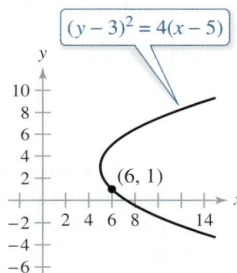

34. Circle

$(x + 2)^2 + (y - 3)^2 = 37$

$(4, 4)$

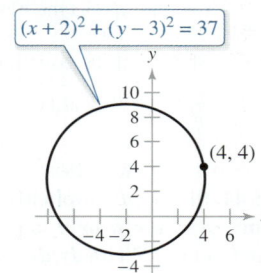

35. Rotated hyperbola

$xy = 1$

$(1, 1)$

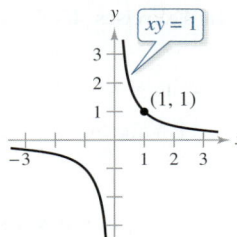

36. Rotated ellipse

$7x^2 - 6\sqrt{3}xy + 13y^2 - 16 = 0$

$(\sqrt{3}, 1)$

37. Cruciform

$x^2y^2 - 9x^2 - 4y^2 = 0$

$(-4, 2\sqrt{3})$

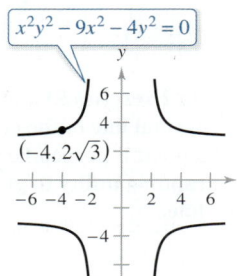

38. Astroid

$x^{2/3} + y^{2/3} = 5$

$(8, 1)$

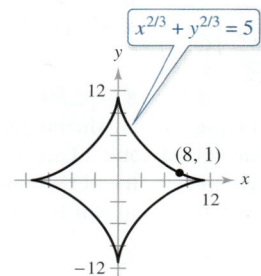

39. Lemniscate

$$3(x^2 + y^2)^2 = 100(x^2 - y^2)$$

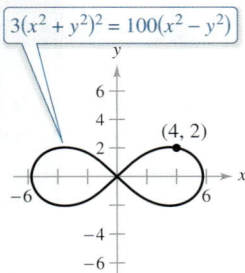

(4, 2)

40. Kappa curve

$$y^2(x^2 + y^2) = 2x^2$$

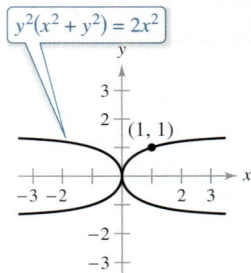

(1, 1)

41. Ellipse

(a) Use implicit differentiation to find an equation of the tangent line to the ellipse $\dfrac{x^2}{2} + \dfrac{y^2}{8} = 1$ at $(1, 2)$.

(b) Show that the equation of the tangent line to the ellipse $\dfrac{x^2}{a^2} + \dfrac{y^2}{b^2} = 1$ at (x_0, y_0) is $\dfrac{x_0 x}{a^2} + \dfrac{y_0 y}{b^2} = 1$.

42. Hyperbola

(a) Use implicit differentiation to find an equation of the tangent line to the hyperbola $\dfrac{x^2}{6} - \dfrac{y^2}{8} = 1$ at $(3, -2)$.

(b) Show that the equation of the tangent line to the hyperbola $\dfrac{x^2}{a^2} - \dfrac{y^2}{b^2} = 1$ at (x_0, y_0) is $\dfrac{x_0 x}{a^2} - \dfrac{y_0 y}{b^2} = 1$.

Determining a Differentiable Function In Exercises 43 and 44, find dy/dx implicitly and find the largest interval of the form $-a < y < a$ or $0 < y < a$ such that y is a differentiable function of x. Write dy/dx as a function of x.

43. $\tan y = x$

44. $\cos y = x$

Finding a Second Derivative In Exercises 45–50, find d^2y/dx^2 implicitly in terms of x and y.

45. $x^2 + y^2 = 4$

46. $x^2 y - 4x = 5$

47. $x^2 - y^2 = 36$

48. $xy - 1 = 2x + y^2$

49. $y^2 = x^3$

50. $y^3 = 4x$

Finding an Equation of a Tangent Line In Exercises 51 and 52, use a graphing utility to graph the equation. Find an equation of the tangent line to the graph at the given point and graph the tangent line in the same viewing window.

51. $\sqrt{x} + \sqrt{y} = 5$, $(9, 4)$

52. $y^2 = \dfrac{x - 1}{x^2 + 1}$, $\left(2, \dfrac{\sqrt{5}}{5}\right)$

Tangent Lines and Normal Lines In Exercises 53 and 54, find equations for the tangent line and normal line to the circle at each given point. (The *normal line* at a point is perpendicular to the tangent line at the point.) Use a graphing utility to graph the equation, tangent line, and normal line.

53. $x^2 + y^2 = 25$

$(4, 3), (-3, 4)$

54. $x^2 + y^2 = 36$

$(6, 0), \left(5, \sqrt{11}\right)$

55. Normal Lines Show that the normal line at any point on the circle $x^2 + y^2 = r^2$ passes through the origin.

56. Circles Two circles of radius 4 are tangent to the graph of $y^2 = 4x$ at the point $(1, 2)$. Find equations of these two circles.

Vertical and Horizontal Tangent Lines In Exercises 57 and 58, find the points at which the graph of the equation has a vertical or horizontal tangent line.

57. $25x^2 + 16y^2 + 200x - 160y + 400 = 0$

58. $4x^2 + y^2 - 8x + 4y + 4 = 0$

Orthogonal Trajectories In Exercises 59–62, use a graphing utility to sketch the intersecting graphs of the equations and show that they are orthogonal. [Two graphs are *orthogonal* if at their point(s) of intersection, their tangent lines are perpendicular to each other.]

59. $2x^2 + y^2 = 6$

$y^2 = 4x$

60. $y^2 = x^3$

$2x^2 + 3y^2 = 5$

61. $x + y = 0$

$x = \sin y$

62. $x^3 = 3(y - 1)$

$x(3y - 29) = 3$

Orthogonal Trajectories In Exercises 63 and 64, verify that the two families of curves are orthogonal, where C and K are real numbers. Use a graphing utility to graph the two families for two values of C and two values of K.

63. $xy = C$, $x^2 - y^2 = K$

64. $x^2 + y^2 = C^2$, $y = Kx$

WRITING ABOUT CONCEPTS

65. Explicit and Implicit Functions Describe the difference between the explicit form of a function and an implicit equation. Give an example of each.

66. Implicit Differentiation In your own words, state the guidelines for implicit differentiation.

67. Orthogonal Trajectories The figure below shows the topographic map carried by a group of hikers. The hikers are in a wooded area on top of the hill shown on the map, and they decide to follow the path of steepest descent (orthogonal trajectories to the contours on the map). Draw their routes if they start from point A and if they start from point B. Their goal is to reach the road along the top of the map. Which starting point should they use? To print an enlarged copy of the map, go to *MathGraphs.com*.

68. HOW DO YOU SEE IT? Use the graph to answer the questions.

$$y^3 - 9y^2 + 27y + 5x^2 = 47$$

(a) Which is greater, the slope of the tangent line at $x = -3$ or the slope of the tangent line at $x = -1$?

(b) Estimate the point(s) where the graph has a vertical tangent line.

(c) Estimate the point(s) where the graph has a horizontal tangent line.

69. Finding Equations of Tangent Lines Consider the equation $x^4 = 4(4x^2 - y^2)$.

(a) Use a graphing utility to graph the equation.

(b) Find and graph the four tangent lines to the curve for $y = 3$.

(c) Find the exact coordinates of the point of intersection of the two tangent lines in the first quadrant.

70. Tangent Lines and Intercepts Let L be any tangent line to the curve

$$\sqrt{x} + \sqrt{y} = \sqrt{c}.$$

Show that the sum of the x- and y-intercepts of L is c.

71. Proof Prove (Theorem 2.3) that

$$\frac{d}{dx}[x^n] = nx^{n-1}$$

for the case in which n is a rational number. (*Hint:* Write $y = x^{p/q}$ in the form $y^q = x^p$ and differentiate implicitly. Assume that p and q are integers, where $q > 0$.)

72. Slope Find all points on the circle $x^2 + y^2 = 100$ where the slope is $\frac{3}{4}$.

73. Tangent Lines Find equations of both tangent lines to the graph of the ellipse $\frac{x^2}{4} + \frac{y^2}{9} = 1$ that pass through the point $(4, 0)$ not on the graph.

74. Normals to a Parabola The graph shows the normal lines from the point $(2, 0)$ to the graph of the parabola $x = y^2$. How many normal lines are there from the point $(x_0, 0)$ to the graph of the parabola if (a) $x_0 = \frac{1}{4}$, (b) $x_0 = \frac{1}{2}$, and (c) $x_0 = 1$? For what value of x_0 are two of the normal lines perpendicular to each other?

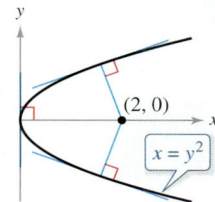

75. Normal Lines (a) Find an equation of the normal line to the ellipse $\frac{x^2}{32} + \frac{y^2}{8} = 1$ at the point $(4, 2)$. (b) Use a graphing utility to graph the ellipse and the normal line. (c) At what other point does the normal line intersect the ellipse?

SECTION PROJECT

Optical Illusions

In each graph below, an optical illusion is created by having lines intersect a family of curves. In each case, the lines appear to be curved. Find the value of dy/dx for the given values of x and y.

(a) Circles: $x^2 + y^2 = C^2$
$x = 3, y = 4, C = 5$

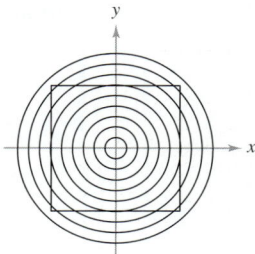

(b) Hyperbolas: $xy = C$
$x = 1, y = 4, C = 4$

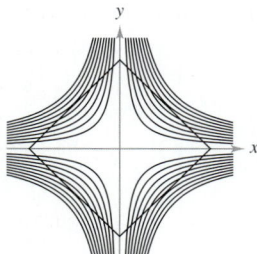

(c) Lines: $ax = by$
$x = \sqrt{3}, y = 3,$
$a = \sqrt{3}, b = 1$

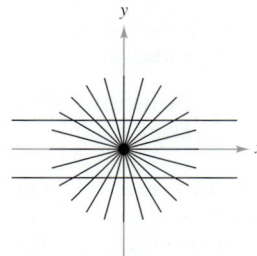

(d) Cosine curves: $y = C \cos x$
$x = \frac{\pi}{3}, y = \frac{1}{3}, C = \frac{2}{3}$

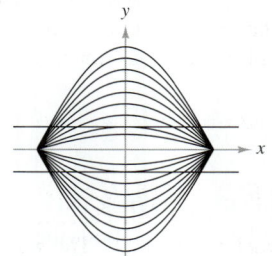

FOR FURTHER INFORMATION For more information on the mathematics of optical illusions, see the article "Descriptive Models for Perception of Optical Illusions" by David A. Smith in *The UMAP Journal.*

2.6 Related Rates

- Find a related rate.
- Use related rates to solve real-life problems.

Finding Related Rates

You have seen how the Chain Rule can be used to find dy/dx implicitly. Another important use of the Chain Rule is to find the rates of change of two or more related variables that are changing with respect to *time*.

For example, when water is drained out of a conical tank (see Figure 2.33), the volume V, the radius r, and the height h of the water level are all functions of time t. Knowing that these variables are related by the equation

$$V = \frac{\pi}{3} r^2 h \qquad\qquad \text{Original equation}$$

you can differentiate implicitly with respect to t to obtain the **related-rate** equation

$$\frac{d}{dt}[V] = \frac{d}{dt}\left[\frac{\pi}{3} r^2 h\right]$$

$$\frac{dV}{dt} = \frac{\pi}{3}\left[r^2 \frac{dh}{dt} + h\left(2r \frac{dr}{dt}\right)\right] \qquad \text{Differentiate with respect to } t.$$

$$= \frac{\pi}{3}\left(r^2 \frac{dh}{dt} + 2rh \frac{dr}{dt}\right).$$

From this equation, you can see that the rate of change of V is related to the rates of change of both h and r.

Volume is related to radius and height.
Figure 2.33

> ### Exploration
>
> ***Finding a Related Rate***   In the conical tank shown in Figure 2.33, the height of the water level is changing at a rate of -0.2 foot per minute and the radius is changing at a rate of -0.1 foot per minute. What is the rate of change in the volume when the radius is $r = 1$ foot and the height is $h = 2$ feet? Does the rate of change in the volume depend on the values of r and h? Explain.

EXAMPLE 1   Two Rates That Are Related

The variables x and y are both differentiable functions of t and are related by the equation $y = x^2 + 3$. Find dy/dt when $x = 1$, given that $dx/dt = 2$ when $x = 1$.

Solution   Using the Chain Rule, you can differentiate both sides of the equation *with respect to t.*

$$y = x^2 + 3 \qquad\qquad \text{Write original equation.}$$

$$\frac{d}{dt}[y] = \frac{d}{dt}[x^2 + 3] \qquad\qquad \text{Differentiate with respect to } t.$$

$$\frac{dy}{dt} = 2x \frac{dx}{dt} \qquad\qquad \text{Chain Rule}$$

When $x = 1$ and $dx/dt = 2$, you have

$$\frac{dy}{dt} = 2(1)(2) = 4.$$

■ FOR FURTHER INFORMATION To learn more about the history of related-rate problems, see the article "The Lengthening Shadow: The Story of Related Rates" by Bill Austin, Don Barry, and David Berman in *Mathematics Magazine*. To view this article, go to *MathArticles.com*.

Problem Solving with Related Rates

In Example 1, you were *given* an equation that related the variables x and y and were asked to find the rate of change of y when $x = 1$.

Equation: $y = x^2 + 3$

Given rate: $\dfrac{dx}{dt} = 2$ when $x = 1$

Find: $\dfrac{dy}{dt}$ when $x = 1$

In each of the remaining examples in this section, you must *create* a mathematical model from a verbal description.

EXAMPLE 2   **Ripples in a Pond**

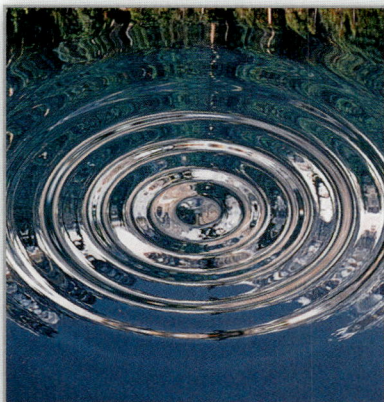

A pebble is dropped into a calm pond, causing ripples in the form of concentric circles, as shown in Figure 2.34. The radius r of the outer ripple is increasing at a constant rate of 1 foot per second. When the radius is 4 feet, at what rate is the total area A of the disturbed water changing?

Solution   The variables r and A are related by $A = \pi r^2$. The rate of change of the radius r is $dr/dt = 1$.

Equation: $A = \pi r^2$

Given rate: $\dfrac{dr}{dt} = 1$

Find: $\dfrac{dA}{dt}$ when $r = 4$

With this information, you can proceed as in Example 1.

$$\frac{d}{dt}[A] = \frac{d}{dt}[\pi r^2] \qquad \text{Differentiate with respect to } t.$$

$$\frac{dA}{dt} = 2\pi r \frac{dr}{dt} \qquad \text{Chain Rule}$$

$$= 2\pi(4)(1) \qquad \text{Substitute 4 for } r \text{ and 1 for } \frac{dr}{dt}.$$

$$= 8\pi \text{ square feet per second} \qquad \text{Simplify.}$$

When the radius is 4 feet, the area is changing at a rate of 8π square feet per second.

Total area increases as the outer radius increases.
Figure 2.34

••**REMARK**   When using these guidelines, be sure you perform Step 3 before Step 4. Substituting the known values of the variables before differentiating will produce an inappropriate derivative.

GUIDELINES FOR SOLVING RELATED-RATE PROBLEMS

1. Identify all *given* quantities and quantities *to be determined*. Make a sketch and label the quantities.
2. Write an equation involving the variables whose rates of change either are given or are to be determined.
3. Using the Chain Rule, implicitly differentiate both sides of the equation *with respect to time t*.
4. *After* completing Step 3, substitute into the resulting equation all known values for the variables and their rates of change. Then solve for the required rate of change.

The table below lists examples of mathematical models involving rates of change. For instance, the rate of change in the first example is the velocity of a car.

Verbal Statement	Mathematical Model
The velocity of a car after traveling for 1 hour is 50 miles per hour.	x = distance traveled $\dfrac{dx}{dt} = 50$ mi/h when $t = 1$
Water is being pumped into a swimming pool at a rate of 10 cubic meters per hour.	V = volume of water in pool $\dfrac{dV}{dt} = 10$ m³/h
A gear is revolving at a rate of 25 revolutions per minute (1 revolution = 2π rad).	θ = angle of revolution $\dfrac{d\theta}{dt} = 25(2\pi)$ rad/min
A population of bacteria is increasing at a rate of 2000 per hour.	x = number in population $\dfrac{dx}{dt} = 2000$ bacteria per hour

EXAMPLE 3   **An Inflating Balloon**

Air is being pumped into a spherical balloon (see Figure 2.35) at a rate of 4.5 cubic feet per minute. Find the rate of change of the radius when the radius is 2 feet.

Solution   Let V be the volume of the balloon, and let r be its radius. Because the volume is increasing at a rate of 4.5 cubic feet per minute, you know that at time t the rate of change of the volume is $dV/dt = \frac{9}{2}$. So, the problem can be stated as shown.

Given rate: $\dfrac{dV}{dt} = \dfrac{9}{2}$ (constant rate)

Find: $\dfrac{dr}{dt}$ when $r = 2$

To find the rate of change of the radius, you must find an equation that relates the radius r to the volume V.

Equation: $V = \dfrac{4}{3}\pi r^3$ Volume of a sphere

Differentiating both sides of the equation with respect to t produces

$$\frac{dV}{dt} = 4\pi r^2 \frac{dr}{dt} \qquad \text{\color{red}Differentiate with respect to } t.$$

$$\frac{dr}{dt} = \frac{1}{4\pi r^2}\left(\frac{dV}{dt}\right). \qquad \text{\color{red}Solve for } \frac{dr}{dt}.$$

Finally, when $r = 2$, the rate of change of the radius is

$$\frac{dr}{dt} = \frac{1}{4\pi(2)^2}\left(\frac{9}{2}\right) \approx 0.09 \text{ foot per minute.}$$

Inflating a balloon
Figure 2.35

In Example 3, note that the volume is increasing at a *constant* rate, but the radius is increasing at a *variable* rate. Just because two rates are related does not mean that they are proportional. In this particular case, the radius is growing more and more slowly as t increases. Do you see why?

EXAMPLE 4 The Speed of an Airplane Tracked by Radar

▷ See LarsonCalculus.com for an interactive version of this type of example.

An airplane is flying on a flight path that will take it directly over a radar tracking station, as shown in Figure 2.36. The distance s is decreasing at a rate of 400 miles per hour when $s = 10$ miles. What is the speed of the plane?

Solution Let x be the horizontal distance from the station, as shown in Figure 2.36. Notice that when $s = 10$, $x = \sqrt{10^2 - 36} = 8$.

Given rate: $ds/dt = -400$ when $s = 10$

Find: dx/dt when $s = 10$ and $x = 8$

You can find the velocity of the plane as shown.

Equation: $x^2 + 6^2 = s^2$ Pythagorean Theorem

$$2x\frac{dx}{dt} = 2s\frac{ds}{dt}$$ Differentiate with respect to t.

$$\frac{dx}{dt} = \frac{s}{x}\left(\frac{ds}{dt}\right)$$ Solve for $\frac{dx}{dt}$.

$$= \frac{10}{8}(-400)$$ Substitute for s, x, and $\frac{ds}{dt}$.

$$= -500 \text{ miles per hour}$$ Simplify.

▷ Because the velocity is -500 miles per hour, the *speed* is 500 miles per hour. ■

REMARK The velocity in Example 4 is negative because x represents a distance that is decreasing.

An airplane is flying at an altitude of 6 miles, s miles from the station.
Figure 2.36

EXAMPLE 5 A Changing Angle of Elevation

Find the rate of change in the angle of elevation of the camera shown in Figure 2.37 at 10 seconds after lift-off.

Solution Let θ be the angle of elevation, as shown in Figure 2.37. When $t = 10$, the height s of the rocket is $s = 50t^2 = 50(10)^2 = 5000$ feet.

Given rate: $ds/dt = 100t = $ velocity of rocket

Find: $d\theta/dt$ when $t = 10$ and $s = 5000$

Using Figure 2.37, you can relate s and θ by the equation $\tan\theta = s/2000$.

Equation: $\tan\theta = \dfrac{s}{2000}$ See Figure 2.37.

$$(\sec^2\theta)\frac{d\theta}{dt} = \frac{1}{2000}\left(\frac{ds}{dt}\right)$$ Differentiate with respect to t.

$$\frac{d\theta}{dt} = \cos^2\theta\,\frac{100t}{2000}$$ Substitute $100t$ for $\frac{ds}{dt}$.

$$= \left(\frac{2000}{\sqrt{s^2 + 2000^2}}\right)^2 \frac{100t}{2000}$$ $\cos\theta = \dfrac{2000}{\sqrt{s^2 + 2000^2}}$

When $t = 10$ and $s = 5000$, you have

$$\frac{d\theta}{dt} = \frac{2000(100)(10)}{5000^2 + 2000^2} = \frac{2}{29} \text{ radian per second.}$$

So, when $t = 10$, θ is changing at a rate of $\frac{2}{29}$ radian per second. ■

$\tan\theta = \dfrac{s}{2000}$

A television camera at ground level is filming the lift-off of a rocket that is rising vertically according to the position equation $s = 50t^2$, where s is measured in feet and t is measured in seconds. The camera is 2000 feet from the launch pad.
Figure 2.37

EXAMPLE 6 **The Velocity of a Piston**

In the engine shown in Figure 2.38, a 7-inch connecting rod is fastened to a crank of radius 3 inches. The crankshaft rotates counterclockwise at a constant rate of 200 revolutions per minute. Find the velocity of the piston when $\theta = \pi/3$.

The velocity of a piston is related to the angle of the crankshaft.
Figure 2.38

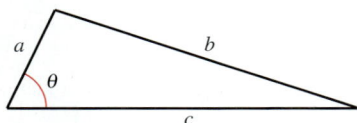

Law of Cosines:
$b^2 = a^2 + c^2 - 2ac \cos \theta$
Figure 2.39

Solution Label the distances as shown in Figure 2.38. Because a complete revolution corresponds to 2π radians, it follows that $d\theta/dt = 200(2\pi) = 400\pi$ radians per minute.

Given rate: $\dfrac{d\theta}{dt} = 400\pi$ (constant rate)

Find: $\dfrac{dx}{dt}$ when $\theta = \dfrac{\pi}{3}$

You can use the Law of Cosines (see Figure 2.39) to find an equation that relates x and θ.

Equation:

$$7^2 = 3^2 + x^2 - 2(3)(x) \cos \theta$$

$$0 = 2x\frac{dx}{dt} - 6\left(-x \sin \theta \frac{d\theta}{dt} + \cos \theta \frac{dx}{dt}\right)$$

$$(6 \cos \theta - 2x)\frac{dx}{dt} = 6x \sin \theta \frac{d\theta}{dt}$$

$$\frac{dx}{dt} = \frac{6x \sin \theta}{6 \cos \theta - 2x}\left(\frac{d\theta}{dt}\right)$$

When $\theta = \pi/3$, you can solve for x as shown.

$$7^2 = 3^2 + x^2 - 2(3)(x) \cos \frac{\pi}{3}$$

$$49 = 9 + x^2 - 6x\left(\frac{1}{2}\right)$$

$$0 = x^2 - 3x - 40$$

$$0 = (x - 8)(x + 5)$$

$$x = 8 \qquad\qquad \text{Choose positive solution.}$$

So, when $x = 8$ and $\theta = \pi/3$, the velocity of the piston is

$$\frac{dx}{dt} = \frac{6(8)\left(\sqrt{3}/2\right)}{6(1/2) - 16}(400\pi)$$

$$= \frac{9600\pi\sqrt{3}}{-13}$$

$$\approx -4018 \text{ inches per minute.}$$

REMARK The velocity in Example 6 is negative because x represents a distance that is decreasing.

2.6 Exercises

See CalcChat.com for tutorial help and worked-out solutions to odd-numbered exercises.

Using Related Rates In Exercises 1–4, assume that x and y are both differentiable functions of t and find the required values of dy/dt and dx/dt.

Equation	Find	Given
1. $y = \sqrt{x}$	(a) $\dfrac{dy}{dt}$ when $x = 4$	$\dfrac{dx}{dt} = 3$
	(b) $\dfrac{dx}{dt}$ when $x = 25$	$\dfrac{dy}{dt} = 2$
2. $y = 3x^2 - 5x$	(a) $\dfrac{dy}{dt}$ when $x = 3$	$\dfrac{dx}{dt} = 2$
	(b) $\dfrac{dx}{dt}$ when $x = 2$	$\dfrac{dy}{dt} = 4$
3. $xy = 4$	(a) $\dfrac{dy}{dt}$ when $x = 8$	$\dfrac{dx}{dt} = 10$
	(b) $\dfrac{dx}{dt}$ when $x = 1$	$\dfrac{dy}{dt} = -6$
4. $x^2 + y^2 = 25$	(a) $\dfrac{dy}{dt}$ when $x = 3, y = 4$	$\dfrac{dx}{dt} = 8$
	(b) $\dfrac{dx}{dt}$ when $x = 4, y = 3$	$\dfrac{dy}{dt} = -2$

Moving Point In Exercises 5–8, a point is moving along the graph of the given function at the rate dx/dt. Find dy/dt for the given values of x.

5. $y = 2x^2 + 1$; $\dfrac{dx}{dt} = 2$ centimeters per second

(a) $x = -1$ (b) $x = 0$ (c) $x = 1$

6. $y = \dfrac{1}{1 + x^2}$; $\dfrac{dx}{dt} = 6$ inches per second

(a) $x = -2$ (b) $x = 0$ (c) $x = 2$

7. $y = \tan x$; $\dfrac{dx}{dt} = 3$ feet per second

(a) $x = -\dfrac{\pi}{3}$ (b) $x = -\dfrac{\pi}{4}$ (c) $x = 0$

8. $y = \cos x$; $\dfrac{dx}{dt} = 4$ centimeters per second

(a) $x = \dfrac{\pi}{6}$ (b) $x = \dfrac{\pi}{4}$ (c) $x = \dfrac{\pi}{3}$

WRITING ABOUT CONCEPTS

9. Related Rates Consider the linear function

$y = ax + b$.

If x changes at a constant rate, does y change at a constant rate? If so, does it change at the same rate as x? Explain.

10. Related Rates In your own words, state the guidelines for solving related-rate problems.

11. Area The radius r of a circle is increasing at a rate of 4 centimeters per minute. Find the rates of change of the area when (a) $r = 8$ centimeters and (b) $r = 32$ centimeters.

12. Area The included angle of the two sides of constant equal length s of an isosceles triangle is θ.

(a) Show that the area of the triangle is given by $A = \frac{1}{2}s^2 \sin \theta$.

(b) The angle θ is increasing at the rate of $\frac{1}{2}$ radian per minute. Find the rates of change of the area when $\theta = \pi/6$ and $\theta = \pi/3$.

(c) Explain why the rate of change of the area of the triangle is not constant even though $d\theta/dt$ is constant.

13. Volume The radius r of a sphere is increasing at a rate of 3 inches per minute.

(a) Find the rates of change of the volume when $r = 9$ inches and $r = 36$ inches.

(b) Explain why the rate of change of the volume of the sphere is not constant even though dr/dt is constant.

14. Volume A spherical balloon is inflated with gas at the rate of 800 cubic centimeters per minute. How fast is the radius of the balloon increasing at the instant the radius is (a) 30 centimeters and (b) 60 centimeters?

15. Volume All edges of a cube are expanding at a rate of 6 centimeters per second. How fast is the volume changing when each edge is (a) 2 centimeters and (b) 10 centimeters?

16. Surface Area All edges of a cube are expanding at a rate of 6 centimeters per second. How fast is the surface area changing when each edge is (a) 2 centimeters and (b) 10 centimeters?

17. Volume At a sand and gravel plant, sand is falling off a conveyor and onto a conical pile at a rate of 10 cubic feet per minute. The diameter of the base of the cone is approximately three times the altitude. At what rate is the height of the pile changing when the pile is 15 feet high? (*Hint:* The formula for the volume of a cone is $V = \frac{1}{3}\pi r^2 h$.)

18. Depth A conical tank (with vertex down) is 10 feet across the top and 12 feet deep. Water is flowing into the tank at a rate of 10 cubic feet per minute. Find the rate of change of the depth of the water when the water is 8 feet deep.

19. Depth A swimming pool is 12 meters long, 6 meters wide, 1 meter deep at the shallow end, and 3 meters deep at the deep end (see figure). Water is being pumped into the pool at $\frac{1}{4}$ cubic meter per minute, and there is 1 meter of water at the deep end.

(a) What percent of the pool is filled?

(b) At what rate is the water level rising?

20. Depth A trough is 12 feet long and 3 feet across the top (see figure). Its ends are isosceles triangles with altitudes of 3 feet.

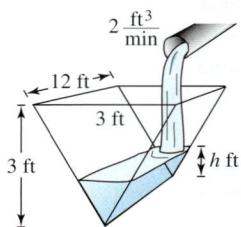

(a) Water is being pumped into the trough at 2 cubic feet per minute. How fast is the water level rising when the depth h is 1 foot?

(b) The water is rising at a rate of $\frac{3}{8}$ inch per minute when $h = 2$. Determine the rate at which water is being pumped into the trough.

21. Moving Ladder A ladder 25 feet long is leaning against the wall of a house (see figure). The base of the ladder is pulled away from the wall at a rate of 2 feet per second.

(a) How fast is the top of the ladder moving down the wall when its base is 7 feet, 15 feet, and 24 feet from the wall?

(b) Consider the triangle formed by the side of the house, the ladder, and the ground. Find the rate at which the area of the triangle is changing when the base of the ladder is 7 feet from the wall.

(c) Find the rate at which the angle between the ladder and the wall of the house is changing when the base of the ladder is 7 feet from the wall.

Figure for 21

Figure for 22

■ **FOR FURTHER INFORMATION** For more information on the mathematics of moving ladders, see the article "The Falling Ladder Paradox" by Paul Scholten and Andrew Simoson in *The College Mathematics Journal.* To view this article, go to *MathArticles.com.*

22. Construction A construction worker pulls a five-meter plank up the side of a building under construction by means of a rope tied to one end of the plank (see figure). Assume the opposite end of the plank follows a path perpendicular to the wall of the building and the worker pulls the rope at a rate of 0.15 meter per second. How fast is the end of the plank sliding along the ground when it is 2.5 meters from the wall of the building?

23. Construction A winch at the top of a 12-meter building pulls a pipe of the same length to a vertical position, as shown in the figure. The winch pulls in rope at a rate of -0.2 meter per second. Find the rate of vertical change and the rate of horizontal change at the end of the pipe when $y = 6$.

Figure for 23

Figure for 24

24. Boating A boat is pulled into a dock by means of a winch 12 feet above the deck of the boat (see figure).

(a) The winch pulls in rope at a rate of 4 feet per second. Determine the speed of the boat when there is 13 feet of rope out. What happens to the speed of the boat as it gets closer to the dock?

(b) Suppose the boat is moving at a constant rate of 4 feet per second. Determine the speed at which the winch pulls in rope when there is a total of 13 feet of rope out. What happens to the speed at which the winch pulls in rope as the boat gets closer to the dock?

25. Air Traffic Control An air traffic controller spots two planes at the same altitude converging on a point as they fly at right angles to each other (see figure). One plane is 225 miles from the point moving at 450 miles per hour. The other plane is 300 miles from the point moving at 600 miles per hour.

(a) At what rate is the distance between the planes decreasing?

(b) How much time does the air traffic controller have to get one of the planes on a different flight path?

Figure for 25

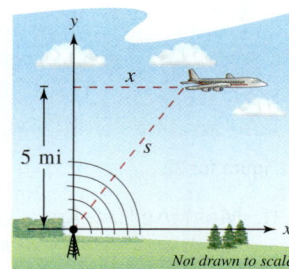

Figure for 26

26. Air Traffic Control An airplane is flying at an altitude of 5 miles and passes directly over a radar antenna (see figure). When the plane is 10 miles away ($s = 10$), the radar detects that the distance s is changing at a rate of 240 miles per hour. What is the speed of the plane?

27. Sports A baseball diamond has the shape of a square with sides 90 feet long (see figure). A player running from second base to third base at a speed of 25 feet per second is 20 feet from third base. At what rate is the player's distance s from home plate changing?

Figure for 27 and 28   Figure for 29

28. Sports For the baseball diamond in Exercise 27, suppose the player is running from first base to second base at a speed of 25 feet per second. Find the rate at which the distance from home plate is changing when the player is 20 feet from second base.

29. Shadow Length A man 6 feet tall walks at a rate of 5 feet per second away from a light that is 15 feet above the ground (see figure).

(a) When he is 10 feet from the base of the light, at what rate is the tip of his shadow moving?

(b) When he is 10 feet from the base of the light, at what rate is the length of his shadow changing?

30. Shadow Length Repeat Exercise 29 for a man 6 feet tall walking at a rate of 5 feet per second *toward* a light that is 20 feet above the ground (see figure).

Figure for 30   Figure for 31

31. Machine Design The endpoints of a movable rod of length 1 meter have coordinates $(x, 0)$ and $(0, y)$ (see figure). The position of the end on the x-axis is

$$x(t) = \frac{1}{2} \sin \frac{\pi t}{6}$$

where t is the time in seconds.

(a) Find the time of one complete cycle of the rod.

(b) What is the lowest point reached by the end of the rod on the y-axis?

(c) Find the speed of the y-axis endpoint when the x-axis endpoint is $\left(\frac{1}{4}, 0\right)$.

32. Machine Design Repeat Exercise 31 for a position function of $x(t) = \frac{3}{5} \sin \pi t$. Use the point $\left(\frac{3}{10}, 0\right)$ for part (c).

33. Evaporation As a spherical raindrop falls, it reaches a layer of dry air and begins to evaporate at a rate that is proportional to its surface area $(S = 4\pi r^2)$. Show that the radius of the raindrop decreases at a constant rate.

34. **HOW DO YOU SEE IT?** Using the graph of f, (a) determine whether dy/dt is positive or negative given that dx/dt is negative, and (b) determine whether dx/dt is positive or negative given that dy/dt is positive.

(i)   (ii)

35. Electricity The combined electrical resistance R of two resistors R_1 and R_2, connected in parallel, is given by

$$\frac{1}{R} = \frac{1}{R_1} + \frac{1}{R_2}$$

where R, R_1, and R_2 are measured in ohms. R_1 and R_2 are increasing at rates of 1 and 1.5 ohms per second, respectively. At what rate is R changing when $R_1 = 50$ ohms and $R_2 = 75$ ohms?

36. Adiabatic Expansion When a certain polyatomic gas undergoes adiabatic expansion, its pressure p and volume V satisfy the equation $pV^{1.3} = k$, where k is a constant. Find the relationship between the related rates dp/dt and dV/dt.

37. Roadway Design Cars on a certain roadway travel on a circular arc of radius r. In order not to rely on friction alone to overcome the centrifugal force, the road is banked at an angle of magnitude θ from the horizontal (see figure). The banking angle must satisfy the equation $rg \tan \theta = v^2$, where v is the velocity of the cars and $g = 32$ feet per second per second is the acceleration due to gravity. Find the relationship between the related rates dv/dt and $d\theta/dt$.

38. Angle of Elevation A balloon rises at a rate of 4 meters per second from a point on the ground 50 meters from an observer. Find the rate of change of the angle of elevation of the balloon from the observer when the balloon is 50 meters above the ground.

39. Angle of Elevation A fish is reeled in at a rate of 1 foot per second from a point 10 feet above the water (see figure). At what rate is the angle θ between the line and the water changing when there is a total of 25 feet of line from the end of the rod to the water?

Figure for 39

Figure for 40

40. Angle of Elevation An airplane flies at an altitude of 5 miles toward a point directly over an observer (see figure). The speed of the plane is 600 miles per hour. Find the rates at which the angle of elevation θ is changing when the angle is (a) $\theta = 30°$, (b) $\theta = 60°$, and (c) $\theta = 75°$.

41. Linear vs. Angular Speed A patrol car is parked 50 feet from a long warehouse (see figure). The revolving light on top of the car turns at a rate of 30 revolutions per minute. How fast is the light beam moving along the wall when the beam makes angles of (a) $\theta = 30°$, (b) $\theta = 60°$, and (c) $\theta = 70°$ with the perpendicular line from the light to the wall?

Figure for 41

Figure for 42

42. Linear vs. Angular Speed A wheel of radius 30 centimeters revolves at a rate of 10 revolutions per second. A dot is painted at a point P on the rim of the wheel (see figure).

(a) Find dx/dt as a function of θ.

(b) Use a graphing utility to graph the function in part (a).

(c) When is the absolute value of the rate of change of x greatest? When is it least?

(d) Find dx/dt when $\theta = 30°$ and $\theta = 60°$.

43. Flight Control An airplane is flying in still air with an airspeed of 275 miles per hour. The plane is climbing at an angle of 18°. Find the rate at which it is gaining altitude.

44. Security Camera A security camera is centered 50 feet above a 100-foot hallway (see figure). It is easiest to design the camera with a constant angular rate of rotation, but this results in recording the images of the surveillance area at a variable rate. So, it is desirable to design a system with a variable rate of rotation and a constant rate of movement of the scanning beam along the hallway. Find a model for the variable rate of rotation when $|dx/dt| = 2$ feet per second.

Figure for 44

45. Think About It Describe the relationship between the rate of change of y and the rate of change of x in each expression. Assume all variables and derivatives are positive.

(a) $\dfrac{dy}{dt} = 3\dfrac{dx}{dt}$ (b) $\dfrac{dy}{dt} = x(L - x)\dfrac{dx}{dt}, \quad 0 \le x \le L$

Acceleration In Exercises 46 and 47, find the acceleration of the specified object. (*Hint:* Recall that if a variable is changing at a constant rate, its acceleration is zero.)

46. Find the acceleration of the top of the ladder described in Exercise 21 when the base of the ladder is 7 feet from the wall.

47. Find the acceleration of the boat in Exercise 24(a) when there is a total of 13 feet of rope out.

48. Modeling Data The table shows the numbers (in millions) of single women (never married) s and married women m in the civilian work force in the United States for the years 2003 through 2010. (*Source: U.S. Bureau of Labor Statistics*)

Year	2003	2004	2005	2006
s	18.4	18.6	19.2	19.5
m	36.0	35.8	35.9	36.3

Year	2007	2008	2009	2010
s	19.7	20.2	20.2	20.6
m	36.9	37.2	37.3	36.7

(a) Use the regression capabilities of a graphing utility to find a model of the form $m(s) = as^3 + bs^2 + cs + d$ for the data, where t is the time in years, with $t = 3$ corresponding to 2003.

(b) Find dm/dt. Then use the model to estimate dm/dt for $t = 7$ when it is predicted that the number of single women in the work force will increase at the rate of 0.75 million per year.

49. Moving Shadow A ball is dropped from a height of 20 meters, 12 meters away from the top of a 20-meter lamppost (see figure). The ball's shadow, caused by the light at the top of the lamppost, is moving along the level ground. How fast is the shadow moving 1 second after the ball is released? (*Submitted by Dennis Gittinger, St. Philips College, San Antonio, TX*)

Review Exercises   See CalcChat.com for tutorial help and worked-out solutions to odd-numbered exercises.

Finding the Derivative by the Limit Process In Exercises 1–4, find the derivative of the function by the limit process.

1. $f(x) = 12$

2. $f(x) = 5x - 4$

3. $f(x) = x^2 - 4x + 5$

4. $f(x) = \dfrac{6}{x}$

Using the Alternative Form of the Derivative In Exercises 5 and 6, use the alternative form of the derivative to find the derivative at $x = c$ (if it exists).

5. $g(x) = 2x^2 - 3x$, $c = 2$

6. $f(x) = \dfrac{1}{x + 4}$, $c = 3$

Determining Differentiability In Exercises 7 and 8, describe the x-values at which f is differentiable.

7. $f(x) = (x - 3)^{2/5}$

8. $f(x) = \dfrac{3x}{x + 1}$

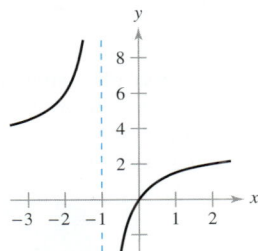

Finding a Derivative In Exercises 9–20, use the rules of differentiation to find the derivative of the function.

9. $y = 25$

10. $f(t) = 4t^4$

11. $f(x) = x^3 - 11x^2$

12. $g(s) = 3s^5 - 2s^4$

13. $h(x) = 6\sqrt{x} + 3\sqrt[3]{x}$

14. $f(x) = x^{1/2} - x^{-1/2}$

15. $g(t) = \dfrac{2}{3t^2}$

16. $h(x) = \dfrac{8}{5x^4}$

17. $f(\theta) = 4\theta - 5\sin\theta$

18. $g(\alpha) = 4\cos\alpha + 6$

19. $f(\theta) = 3\cos\theta - \dfrac{\sin\theta}{4}$

20. $g(\alpha) = \dfrac{5\sin\alpha}{3} - 2\alpha$

Finding the Slope of a Graph In Exercises 21–24, find the slope of the graph of the functions at the given point.

21. $f(x) = \dfrac{27}{x^3}$, $(3, 1)$

22. $f(x) = 3x^2 - 4x$, $(1, -1)$

23. $f(x) = 2x^4 - 8$, $(0, -8)$

24. $f(\theta) = 3\cos\theta - 2\theta$, $(0, 3)$

25. **Vibrating String** When a guitar string is plucked, it vibrates with a frequency of $F = 200\sqrt{T}$, where F is measured in vibrations per second and the tension T is measured in pounds. Find the rates of change of F when (a) $T = 4$ and (b) $T = 9$.

26. **Volume** The surface area of a cube with sides of length ℓ is given by $S = 6\ell^2$. Find the rates of change of the surface area with respect to ℓ when (a) $\ell = 3$ inches and (b) $\ell = 5$ inches.

Vertical Motion In Exercises 27 and 28, use the position function $s(t) = -16t^2 + v_0 t + s_0$ for free-falling objects.

27. A ball is thrown straight down from the top of a 600-foot building with an initial velocity of -30 feet per second.

(a) Determine the position and velocity functions for the ball.

(b) Determine the average velocity on the interval $[1, 3]$.

(c) Find the instantaneous velocities when $t = 1$ and $t = 3$.

(d) Find the time required for the ball to reach ground level.

(e) Find the velocity of the ball at impact.

28. To estimate the height of a building, a weight is dropped from the top of the building into a pool at ground level. The splash is seen 9.2 seconds after the weight is dropped. What is the height (in feet) of the building?

Finding a Derivative In Exercises 29–40, use the Product Rule or the Quotient Rule to find the derivative of the function.

29. $f(x) = (5x^2 + 8)(x^2 - 4x - 6)$

30. $g(x) = (2x^3 + 5x)(3x - 4)$

31. $h(x) = \sqrt{x}\sin x$

32. $f(t) = 2t^5 \cos t$

33. $f(x) = \dfrac{x^2 + x - 1}{x^2 - 1}$

34. $f(x) = \dfrac{2x + 7}{x^2 + 4}$

35. $y = \dfrac{x^4}{\cos x}$

36. $y = \dfrac{\sin x}{x^4}$

37. $y = 3x^2 \sec x$

38. $y = 2x - x^2 \tan x$

39. $y = x\cos x - \sin x$

40. $g(x) = 3x\sin x + x^2 \cos x$

Finding an Equation of a Tangent Line In Exercises 41–44, find an equation of the tangent line to the graph of f at the given point.

41. $f(x) = (x + 2)(x^2 + 5)$, $(-1, 6)$

42. $f(x) = (x - 4)(x^2 + 6x - 1)$, $(0, 4)$

43. $f(x) = \dfrac{x + 1}{x - 1}$, $\left(\dfrac{1}{2}, -3\right)$

44. $f(x) = \dfrac{1 + \cos x}{1 - \cos x}$, $\left(\dfrac{\pi}{2}, 1\right)$

Finding a Second Derivative In Exercises 45–50, find the second derivative of the function.

45. $g(t) = -8t^3 - 5t + 12$

46. $h(x) = 6x^{-2} + 7x^2$

47. $f(x) = 15x^{5/2}$

48. $f(x) = 20\sqrt[5]{x}$

49. $f(\theta) = 3\tan\theta$

50. $h(t) = 10\cos t - 15\sin t$

51. Acceleration The velocity of an object in meters per second is $v(t) = 20 - t^2$, $0 \le t \le 6$. Find the velocity and acceleration of the object when $t = 3$.

52. Acceleration The velocity of an automobile starting from rest is

$$v(t) = \frac{90t}{4t + 10}$$

where v is measured in feet per second. Find the acceleration at (a) 1 second, (b) 5 seconds, and (c) 10 seconds.

Finding a Derivative In Exercises 53–64, find the derivative of the function.

53. $y = (7x + 3)^4$

54. $y = (x^2 - 6)^3$

55. $y = \dfrac{1}{x^2 + 4}$

56. $f(x) = \dfrac{1}{(5x + 1)^2}$

57. $y = 5\cos(9x + 1)$

58. $y = 1 - \cos 2x + 2\cos^2 x$

59. $y = \dfrac{x}{2} - \dfrac{\sin 2x}{4}$

60. $y = \dfrac{\sec^7 x}{7} - \dfrac{\sec^5 x}{5}$

61. $y = x(6x + 1)^5$

62. $f(s) = (s^2 - 1)^{5/2}(s^3 + 5)$

63. $f(x) = \dfrac{3x}{\sqrt{x^2 + 1}}$

64. $h(x) = \left(\dfrac{x + 5}{x^2 + 3}\right)^2$

Evaluating a Derivative In Exercises 65–70, find and evaluate the derivative of the function at the given point.

65. $f(x) = \sqrt{1 - x^3}$, $(-2, 3)$

66. $f(x) = \sqrt[3]{x^2 - 1}$, $(3, 2)$

67. $f(x) = \dfrac{4}{x^2 + 1}$, $(-1, 2)$

68. $f(x) = \dfrac{3x + 1}{4x - 3}$, $(4, 1)$

69. $y = \dfrac{1}{2}\csc 2x$, $\left(\dfrac{\pi}{4}, \dfrac{1}{2}\right)$

70. $y = \csc 3x + \cot 3x$, $\left(\dfrac{\pi}{6}, 1\right)$

Finding a Second Derivative In Exercises 71–74, find the second derivative of the function.

71. $y = (8x + 5)^3$

72. $y = \dfrac{1}{5x + 1}$

73. $f(x) = \cot x$

74. $y = \sin^2 x$

75. Refrigeration The temperature T (in degrees Fahrenheit) of food in a freezer is

$$T = \frac{700}{t^2 + 4t + 10}$$

where t is the time in hours. Find the rate of change of T with respect to t at each of the following times.

(a) $t = 1$ (b) $t = 3$ (c) $t = 5$ (d) $t = 10$

76. Harmonic Motion The displacement from equilibrium of an object in harmonic motion on the end of a spring is

$$y = \frac{1}{4}\cos 8t - \frac{1}{4}\sin 8t$$

where y is measured in feet and t is the time in seconds. Determine the position and velocity of the object when $t = \pi/4$.

Finding a Derivative In Exercises 77–82, find dy/dx by implicit differentiation.

77. $x^2 + y^2 = 64$

78. $x^2 + 4xy - y^3 = 6$

79. $x^3y - xy^3 = 4$

80. $\sqrt{xy} = x - 4y$

81. $x \sin y = y \cos x$

82. $\cos(x + y) = x$

Tangent Lines and Normal Lines In Exercises 83 and 84, find equations for the tangent line and the normal line to the graph of the equation at the given point. (The *normal line* at a point is perpendicular to the tangent line at the point.) Use a graphing utility to graph the equation, the tangent line, and the normal line.

83. $x^2 + y^2 = 10$, $(3, 1)$

84. $x^2 - y^2 = 20$, $(6, 4)$

85. Rate of Change A point moves along the curve $y = \sqrt{x}$ in such a way that the y-value is increasing at a rate of 2 units per second. At what rate is x changing for each of the following values?

(a) $x = \frac{1}{2}$ (b) $x = 1$ (c) $x = 4$

86. Surface Area All edges of a cube are expanding at a rate of 8 centimeters per second. How fast is the surface area changing when each edge is 6.5 centimeters?

87. Linear vs. Angular Speed A rotating beacon is located 1 kilometer off a straight shoreline (see figure). The beacon rotates at a rate of 3 revolutions per minute. How fast (in kilometers per hour) does the beam of light appear to be moving to a viewer who is $\frac{1}{2}$ kilometer down the shoreline?

Not drawn to scale

88. Moving Shadow A sandbag is dropped from a balloon at a height of 60 meters when the angle of elevation to the sun is 30° (see figure). The position of the sandbag is

$$s(t) = 60 - 4.9t^2.$$

Find the rate at which the shadow of the sandbag is traveling along the ground when the sandbag is at a height of 35 meters.

Position: $s(t) = 60 - 4.9t^2$

60 m

Rays

30°

Shadow's path

P.S. Problem Solving

See **CalcChat.com** for tutorial help and worked-out solutions to odd-numbered exercises.

1. Finding Equations of Circles   Consider the graph of the parabola $y = x^2$.

(a) Find the radius r of the largest possible circle centered on the y-axis that is tangent to the parabola at the origin, as shown in the figure. This circle is called the **circle of curvature** (see Section 12.5). Find the equation of this circle. Use a graphing utility to graph the circle and parabola in the same viewing window to verify your answer.

(b) Find the center $(0, b)$ of the circle of radius 1 centered on the y-axis that is tangent to the parabola at two points, as shown in the figure. Find the equation of this circle. Use a graphing utility to graph the circle and parabola in the same viewing window to verify your answer.

Figure for 1(a)

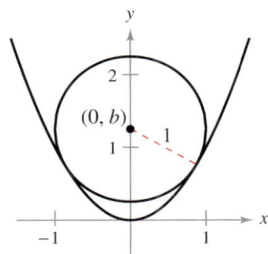

Figure for 1(b)

2. Finding Equations of Tangent Lines   Graph the two parabolas

$$y = x^2 \quad \text{and} \quad y = -x^2 + 2x - 5$$

in the same coordinate plane. Find equations of the two lines that are simultaneously tangent to both parabolas.

3. Finding a Polynomial   Find a third-degree polynomial $p(x)$ that is tangent to the line $y = 14x - 13$ at the point $(1, 1)$, and tangent to the line $y = -2x - 5$ at the point $(-1, -3)$.

4. Finding a Function   Find a function of the form $f(x) = a + b \cos cx$ that is tangent to the line $y = 1$ at the point $(0, 1)$, and tangent to the line

$$y = x + \frac{3}{2} - \frac{\pi}{4}$$

at the point $\left(\dfrac{\pi}{4}, \dfrac{3}{2}\right)$.

5. Tangent Lines and Normal Lines

(a) Find an equation of the tangent line to the parabola $y = x^2$ at the point $(2, 4)$.

(b) Find an equation of the normal line to $y = x^2$ at the point $(2, 4)$. (The *normal line* at a point is perpendicular to the tangent line at the point.) Where does this line intersect the parabola a second time?

(c) Find equations of the tangent line and normal line to $y = x^2$ at the point $(0, 0)$.

(d) Prove that for any point $(a, b) \neq (0, 0)$ on the parabola $y = x^2$, the normal line intersects the graph a second time.

6. Finding Polynomials

(a) Find the polynomial $P_1(x) = a_0 + a_1 x$ whose value and slope agree with the value and slope of $f(x) = \cos x$ at the point $x = 0$.

(b) Find the polynomial $P_2(x) = a_0 + a_1 x + a_2 x^2$ whose value and first two derivatives agree with the value and first two derivatives of $f(x) = \cos x$ at the point $x = 0$. This polynomial is called the second-degree Taylor polynomial of $f(x) = \cos x$ at $x = 0$.

(c) Complete the table comparing the values of $f(x) = \cos x$ and $P_2(x)$. What do you observe?

x	-1.0	-0.1	-0.001	0	0.001	0.1	1.0
$\cos x$							
$P_2(x)$							

(d) Find the third-degree Taylor polynomial of $f(x) = \sin x$ at $x = 0$.

7. Famous Curve   The graph of the **eight curve**

$$x^4 = a^2(x^2 - y^2), \quad a \neq 0$$

is shown below.

(a) Explain how you could use a graphing utility to graph this curve.

(b) Use a graphing utility to graph the curve for various values of the constant a. Describe how a affects the shape of the curve.

(c) Determine the points on the curve at which the tangent line is horizontal.

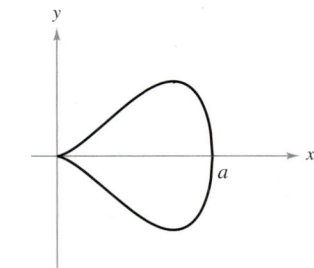

Figure for 7

Figure for 8

8. Famous Curve   The graph of the **pear-shaped quartic**

$$b^2 y^2 = x^3(a - x), \quad a, b > 0$$

is shown above.

(a) Explain how you could use a graphing utility to graph this curve.

(b) Use a graphing utility to graph the curve for various values of the constants a and b. Describe how a and b affect the shape of the curve.

(c) Determine the points on the curve at which the tangent line is horizontal.

9. Shadow Length   A man 6 feet tall walks at a rate of 5 feet per second toward a streetlight that is 30 feet high (see figure). The man's 3-foot-tall child follows at the same speed, but 10 feet behind the man. At times, the shadow behind the child is caused by the man, and at other times, by the child.

(a) Suppose the man is 90 feet from the streetlight. Show that the man's shadow extends beyond the child's shadow.

(b) Suppose the man is 60 feet from the streetlight. Show that the child's shadow extends beyond the man's shadow.

(c) Determine the distance d from the man to the streetlight at which the tips of the two shadows are exactly the same distance from the streetlight.

(d) Determine how fast the tip of the man's shadow is moving as a function of x, the distance between the man and the streetlight. Discuss the continuity of this shadow speed function.

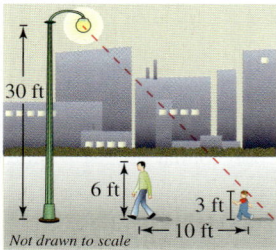
30 ft
6 ft
3 ft
10 ft
Not drawn to scale

Figure for 9

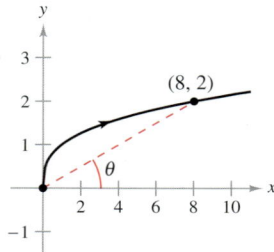
(8, 2)
θ

Figure for 10

10. Moving Point   A particle is moving along the graph of $y = \sqrt[3]{x}$ (see figure). When $x = 8$, the y-component of the position of the particle is increasing at the rate of 1 centimeter per second.

(a) How fast is the x-component changing at this moment?

(b) How fast is the distance from the origin changing at this moment?

(c) How fast is the angle of inclination θ changing at this moment?

11. Projectile Motion   An astronaut standing on the moon throws a rock upward. The height of the rock is

$$s = -\frac{27}{10}t^2 + 27t + 6$$

where s is measured in feet and t is measured in seconds.

(a) Find expressions for the velocity and acceleration of the rock.

(b) Find the time when the rock is at its highest point by finding the time when the velocity is zero. What is the height of the rock at this time?

(c) How does the acceleration of the rock compare with the acceleration due to gravity on Earth?

12. Proof   Let E be a function satisfying $E(0) = E'(0) = 1$. Prove that if $E(a + b) = E(a)E(b)$ for all a and b, then E is differentiable and $E'(x) = E(x)$ for all x. Find an example of a function satisfying $E(a + b) = E(a)E(b)$.

13. Proof   Let L be a differentiable function for all x. Prove that if $L(a + b) = L(a) + L(b)$ for all a and b, then $L'(x) = L'(0)$ for all x. What does the graph of L look like?

14. Radians and Degrees   The fundamental limit

$$\lim_{x \to 0} \frac{\sin x}{x} = 1$$

assumes that x is measured in radians. Suppose you assume that x is measured in degrees instead of radians.

(a) Set your calculator to *degree* mode and complete the table.

z (in degrees)	0.1	0.01	0.0001
$\dfrac{\sin z}{z}$			

(b) Use the table to estimate

$$\lim_{z \to 0} \frac{\sin z}{z}$$

for z in degrees. What is the exact value of this limit? (*Hint:* $180° = \pi$ radians)

(c) Use the limit definition of the derivative to find

$$\frac{d}{dz} \sin z$$

for z in degrees.

(d) Define the new functions $S(z) = \sin(cz)$ and $C(z) = \cos(cz)$, where $c = \pi/180$. Find $S(90)$ and $C(180)$. Use the Chain Rule to calculate

$$\frac{d}{dz} S(z).$$

(e) Explain why calculus is made easier by using radians instead of degrees.

15. Acceleration and Jerk   If a is the acceleration of an object, then the *jerk* j is defined by $j = a'(t)$.

(a) Use this definition to give a physical interpretation of j.

(b) Find j for the slowing vehicle in Exercise 117 in Section 2.3 and interpret the result.

(c) The figure shows the graphs of the position, velocity, acceleration, and jerk functions of a vehicle. Identify each graph and explain your reasoning.

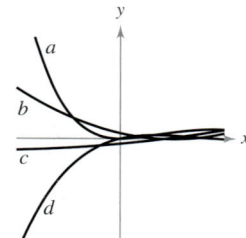
a
b
c
d

3 Applications of Differentiation

Offshore Oil Well (Exercise 39, p. 222)

Estimation of Error
(Example 3, p. 233)

Engine Efficiency (Exercise 85, p. 204)

Path of a Projectile
(Example 5, p. 182)

Speed (Exercise 57, p. 175)

161

3.1 Extrema on an Interval

■ Understand the definition of extrema of a function on an interval.
■ Understand the definition of relative extrema of a function on an open interval.
■ Find extrema on a closed interval.

Extrema of a Function

In calculus, much effort is devoted to determining the behavior of a function f on an interval I. Does f have a maximum value on I? Does it have a minimum value? Where is the function increasing? Where is it decreasing? In this chapter, you will learn how derivatives can be used to answer these questions. You will also see why these questions are important in real-life applications.

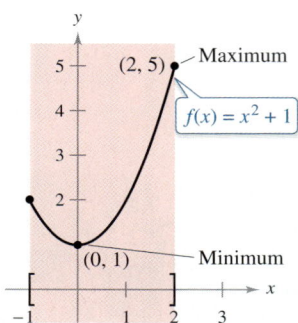

(a) f is continuous, $[-1, 2]$ is closed.

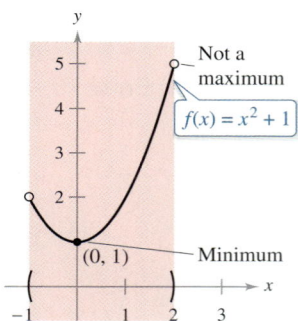

(b) f is continuous, $(-1, 2)$ is open.

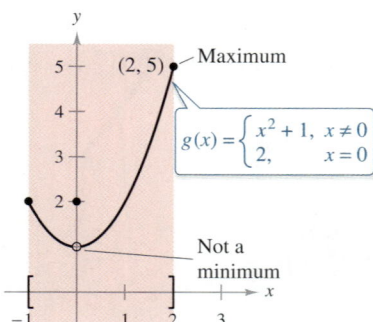

(c) g is not continuous, $[-1, 2]$ is closed.

Figure 3.1

Definition of Extrema

Let f be defined on an interval I containing c.

1. $f(c)$ is the **minimum of f on I** when $f(c) \le f(x)$ for all x in I.
2. $f(c)$ is the **maximum of f on I** when $f(c) \ge f(x)$ for all x in I.

The minimum and maximum of a function on an interval are the **extreme values,** or **extrema** (the singular form of extrema is extremum), of the function on the interval. The minimum and maximum of a function on an interval are also called the **absolute minimum** and **absolute maximum,** or the **global minimum** and **global maximum,** on the interval. Extrema can occur at interior points or endpoints of an interval (see Figure 3.1). Extrema that occur at the endpoints are called **endpoint extrema.**

A function need not have a minimum or a maximum on an interval. For instance, in Figure 3.1(a) and (b), you can see that the function $f(x) = x^2 + 1$ has both a minimum and a maximum on the closed interval $[-1, 2]$, but does not have a maximum on the open interval $(-1, 2)$. Moreover, in Figure 3.1(c), you can see that continuity (or the lack of it) can affect the existence of an extremum on the interval. This suggests the theorem below. (Although the Extreme Value Theorem is intuitively plausible, a proof of this theorem is not within the scope of this text.)

THEOREM 3.1 The Extreme Value Theorem

If f is continuous on a closed interval $[a, b]$, then f has both a minimum and a maximum on the interval.

Exploration

Finding Minimum and Maximum Values The Extreme Value Theorem (like the Intermediate Value Theorem) is an *existence theorem* because it tells of the existence of minimum and maximum values but does not show how to find these values. Use the *minimum* and *maximum* features of a graphing utility to find the extrema of each function. In each case, do you think the x-values are exact or approximate? Explain your reasoning.

a. $f(x) = x^2 - 4x + 5$ on the closed interval $[-1, 3]$
b. $f(x) = x^3 - 2x^2 - 3x - 2$ on the closed interval $[-1, 3]$

Relative Extrema and Critical Numbers

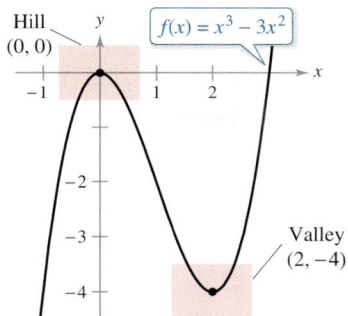

Hill
(0, 0)

$f(x) = x^3 - 3x^2$

Valley
(2, −4)

f has a relative maximum at (0, 0) and a relative minimum at (2, −4).

Figure 3.2

In Figure 3.2, the graph of $f(x) = x^3 - 3x^2$ has a **relative maximum** at the point (0, 0) and a **relative minimum** at the point (2, −4). Informally, for a continuous function, you can think of a relative maximum as occurring on a "hill" on the graph, and a relative minimum as occurring in a "valley" on the graph. Such a hill and valley can occur in two ways. When the hill (or valley) is smooth and rounded, the graph has a horizontal tangent line at the high point (or low point). When the hill (or valley) is sharp and peaked, the graph represents a function that is not differentiable at the high point (or low point).

Definition of Relative Extrema

1. If there is an open interval containing *c* on which $f(c)$ is a maximum, then $f(c)$ is called a **relative maximum** of *f*, or you can say that *f* has a **relative maximum at $(c, f(c))$.**

2. If there is an open interval containing *c* on which $f(c)$ is a minimum, then $f(c)$ is called a **relative minimum** of *f*, or you can say that *f* has a **relative minimum at $(c, f(c))$.**

The plural of relative maximum is relative maxima, and the plural of relative minimum is relative minima. Relative maximum and relative minimum are sometimes called **local maximum** and **local minimum,** respectively.

Example 1 examines the derivatives of functions at *given* relative extrema. (Much more is said about *finding* the relative extrema of a function in Section 3.3.)

EXAMPLE 1   **The Value of the Derivative at Relative Extrema**

Find the value of the derivative at each relative extremum shown in Figure 3.3.

Solution

a. The derivative of $f(x) = \dfrac{9(x^2 - 3)}{x^3}$ is

$$f'(x) = \frac{x^3(18x) - (9)(x^2 - 3)(3x^2)}{(x^3)^2} \qquad \text{Differentiate using Quotient Rule.}$$

$$= \frac{9(9 - x^2)}{x^4}. \qquad \text{Simplify.}$$

At the point (3, 2), the value of the derivative is $f'(3) = 0$ [see Figure 3.3(a)].

b. At $x = 0$, the derivative of $f(x) = |x|$ *does not exist* because the following one-sided limits differ [see Figure 3.3(b)].

$$\lim_{x \to 0^-} \frac{f(x) - f(0)}{x - 0} = \lim_{x \to 0^-} \frac{|x|}{x} = -1 \qquad \text{Limit from the left}$$

$$\lim_{x \to 0^+} \frac{f(x) - f(0)}{x - 0} = \lim_{x \to 0^+} \frac{|x|}{x} = 1 \qquad \text{Limit from the right}$$

c. The derivative of $f(x) = \sin x$ is

$$f'(x) = \cos x.$$

At the point $(\pi/2, 1)$, the value of the derivative is $f'(\pi/2) = \cos(\pi/2) = 0$. At the point $(3\pi/2, -1)$, the value of the derivative is $f'(3\pi/2) = \cos(3\pi/2) = 0$ [see Figure 3.3(c)].

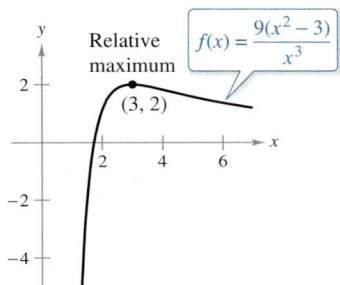

Relative maximum

$f(x) = \dfrac{9(x^2 - 3)}{x^3}$

(3, 2)

(a) $f'(3) = 0$

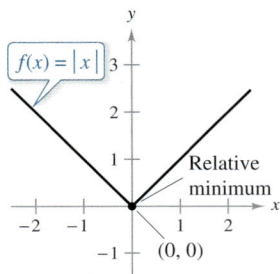

$f(x) = |x|$

Relative minimum

(0, 0)

(b) $f'(0)$ does not exist.

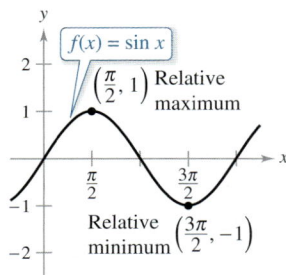

$f(x) = \sin x$

$\left(\dfrac{\pi}{2}, 1\right)$ Relative maximum

Relative minimum $\left(\dfrac{3\pi}{2}, -1\right)$

(c) $f'\left(\dfrac{\pi}{2}\right) = 0; f'\left(\dfrac{3\pi}{2}\right) = 0$

Figure 3.3

Note in Example 1 that at each relative extremum, the derivative either is zero or does not exist. The *x*-values at these special points are called **critical numbers.** Figure 3.4 illustrates the two types of critical numbers. Notice in the definition that the critical number *c* has to be in the domain of *f*, but *c* does not have to be in the domain of *f'*.

Definition of a Critical Number

Let *f* be defined at *c*. If $f'(c) = 0$ or if *f* is not differentiable at *c*, then *c* is a **critical number** of *f*.

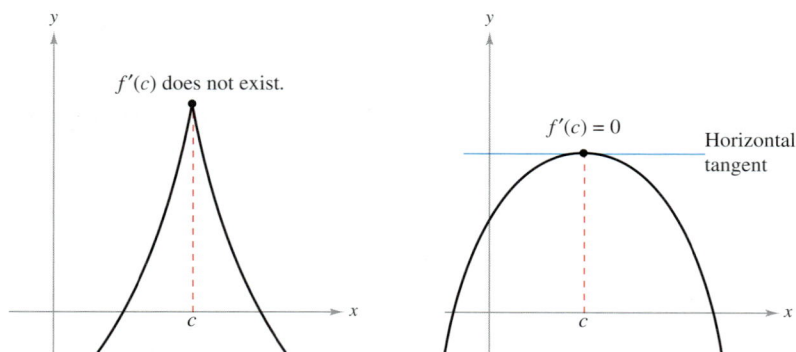

$f'(c)$ does not exist.

$f'(c) = 0$

Horizontal tangent

c is a critical number of *f*.
Figure 3.4

THEOREM 3.2   Relative Extrema Occur Only at Critical Numbers

If *f* has a relative minimum or relative maximum at $x = c$, then *c* is a critical number of *f*.

PIERRE DE FERMAT (1601–1665)

For Fermat, who was trained as a lawyer, mathematics was more of a hobby than a profession. Nevertheless, Fermat made many contributions to analytic geometry, number theory, calculus, and probability. In letters to friends, he wrote of many of the fundamental ideas of calculus, long before Newton or Leibniz. For instance, Theorem 3.2 is sometimes attributed to Fermat. *See LarsonCalculus.com to read more of this biography.*

Proof

Case 1: If *f* is *not* differentiable at $x = c$, then, by definition, *c* is a critical number of *f* and the theorem is valid.

Case 2: If *f* is differentiable at $x = c$, then $f'(c)$ must be positive, negative, or 0. Suppose $f'(c)$ is positive. Then

$$f'(c) = \lim_{x \to c} \frac{f(x) - f(c)}{x - c} > 0$$

which implies that there exists an interval (a, b) containing *c* such that

$$\frac{f(x) - f(c)}{x - c} > 0, \text{ for all } x \neq c \text{ in } (a, b).$$ [See Exercise 78(b), Section 1.2.]

Because this quotient is positive, the signs of the denominator and numerator must agree. This produces the following inequalities for *x*-values in the interval (a, b).

Left of c: $x < c$ and $f(x) < f(c)$ $\Longrightarrow$ $f(c)$ is not a relative minimum.

Right of c: $x > c$ and $f(x) > f(c)$ $\Longrightarrow$ $f(c)$ is not a relative maximum.

So, the assumption that $f'(c) > 0$ contradicts the hypothesis that $f(c)$ is a relative extremum. Assuming that $f'(c) < 0$ produces a similar contradiction, you are left with only one possibility—namely, $f'(c) = 0$. So, by definition, *c* is a critical number of *f* and the theorem is valid.

See LarsonCalculus.com for Bruce Edwards's video of this proof.

Finding Extrema on a Closed Interval

Theorem 3.2 states that the relative extrema of a function can occur *only* at the critical numbers of the function. Knowing this, you can use the following guidelines to find extrema on a closed interval.

GUIDELINES FOR FINDING EXTREMA ON A CLOSED INTERVAL

To find the extrema of a continuous function f on a closed interval $[a, b]$, use these steps.

1. Find the critical numbers of f in (a, b).
2. Evaluate f at each critical number in (a, b).
3. Evaluate f at each endpoint of $[a, b]$.
4. The least of these values is the minimum. The greatest is the maximum.

The next three examples show how to apply these guidelines. Be sure you see that finding the critical numbers of the function is only part of the procedure. Evaluating the function at the critical numbers *and* the endpoints is the other part.

EXAMPLE 2   **Finding Extrema on a Closed Interval**

Find the extrema of

$$f(x) = 3x^4 - 4x^3$$

on the interval $[-1, 2]$.

Solution   Begin by differentiating the function.

$$f(x) = 3x^4 - 4x^3 \qquad \text{Write original function.}$$
$$f'(x) = 12x^3 - 12x^2 \qquad \text{Differentiate.}$$

To find the critical numbers of f in the interval $(-1, 2)$, you must find all x-values for which $f'(x) = 0$ and all x-values for which $f'(x)$ does not exist.

$$12x^3 - 12x^2 = 0 \qquad \text{Set } f'(x) \text{ equal to 0.}$$
$$12x^2(x - 1) = 0 \qquad \text{Factor.}$$
$$x = 0, 1 \qquad \text{Critical numbers}$$

Because f' is defined for all x, you can conclude that these are the only critical numbers of f. By evaluating f at these two critical numbers and at the endpoints of $[-1, 2]$, you can determine that the maximum is $f(2) = 16$ and the minimum is $f(1) = -1$, as shown in the table. The graph of f is shown in Figure 3.5.

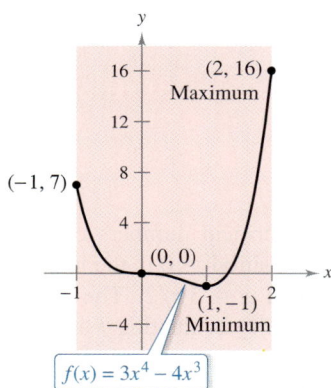

On the closed interval $[-1, 2]$, f has a minimum at $(1, -1)$ and a maximum at $(2, 16)$.
Figure 3.5

Left Endpoint	Critical Number	Critical Number	Right Endpoint
$f(-1) = 7$	$f(0) = 0$	$f(1) = -1$ Minimum	$f(2) = 16$ Maximum

In Figure 3.5, note that the critical number $x = 0$ does not yield a relative minimum or a relative maximum. This tells you that the converse of Theorem 3.2 is not true. In other words, *the critical numbers of a function need not produce relative extrema.*

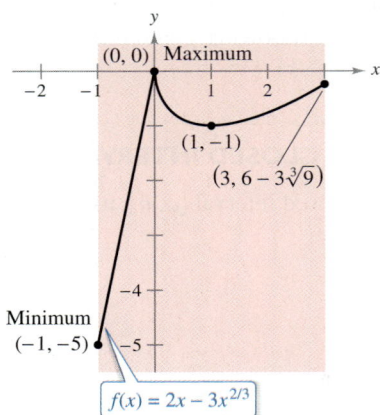

On the closed interval $[-1, 3]$, f has a minimum at $(-1, -5)$ and a maximum at $(0, 0)$.

Figure 3.6

EXAMPLE 3 **Finding Extrema on a Closed Interval**

Find the extrema of $f(x) = 2x - 3x^{2/3}$ on the interval $[-1, 3]$.

Solution Begin by differentiating the function.

$$f(x) = 2x - 3x^{2/3} \qquad \text{Write original function.}$$

$$f'(x) = 2 - \frac{2}{x^{1/3}} \qquad \text{Differentiate.}$$

$$= 2\left(\frac{x^{1/3} - 1}{x^{1/3}}\right) \qquad \text{Simplify.}$$

From this derivative, you can see that the function has two critical numbers in the interval $(-1, 3)$. The number 1 is a critical number because $f'(1) = 0$, and the number 0 is a critical number because $f'(0)$ does not exist. By evaluating f at these two numbers and at the endpoints of the interval, you can conclude that the minimum is $f(-1) = -5$ and the maximum is $f(0) = 0$, as shown in the table. The graph of f is shown in Figure 3.6.

Left Endpoint	Critical Number	Critical Number	Right Endpoint
$f(-1) = -5$ Minimum	$f(0) = 0$ Maximum	$f(1) = -1$	$f(3) = 6 - 3\sqrt[3]{9} \approx -0.24$

EXAMPLE 4 **Finding Extrema on a Closed Interval**

• • • • ▷ *See LarsonCalculus.com for an interactive version of this type of example.*

Find the extrema of

$$f(x) = 2 \sin x - \cos 2x$$

on the interval $[0, 2\pi]$.

Solution Begin by differentiating the function.

$$f(x) = 2 \sin x - \cos 2x \qquad \text{Write original function.}$$

$$f'(x) = 2 \cos x + 2 \sin 2x \qquad \text{Differentiate.}$$

$$= 2 \cos x + 4 \cos x \sin x \qquad \sin 2x = 2 \cos x \sin x$$

$$= 2(\cos x)(1 + 2 \sin x) \qquad \text{Factor.}$$

Because f is differentiable for all real x, you can find all critical numbers of f by finding the zeros of its derivative. Considering $2(\cos x)(1 + 2 \sin x) = 0$ in the interval $(0, 2\pi)$, the factor $\cos x$ is zero when $x = \pi/2$ and when $x = 3\pi/2$. The factor $(1 + 2 \sin x)$ is zero when $x = 7\pi/6$ and when $x = 11\pi/6$. By evaluating f at these four critical numbers and at the endpoints of the interval, you can conclude that the maximum is $f(\pi/2) = 3$ and the minimum occurs at *two* points, $f(7\pi/6) = -3/2$ and $f(11\pi/6) = -3/2$, as shown in the table. The graph is shown in Figure 3.7.

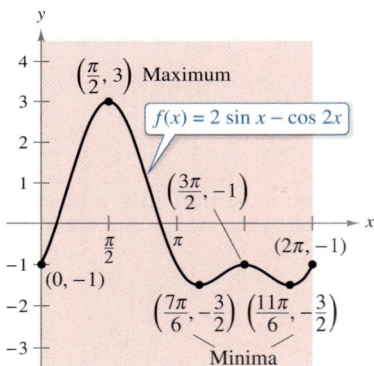

On the closed interval $[0, 2\pi]$, f has two minima at $(7\pi/6, -3/2)$ and $(11\pi/6, -3/2)$ and a maximum at $(\pi/2, 3)$.

Figure 3.7

Left Endpoint	Critical Number	Critical Number	Critical Number	Critical Number	Right Endpoint
$f(0) = -1$	$f\left(\dfrac{\pi}{2}\right) = 3$ Maximum	$f\left(\dfrac{7\pi}{6}\right) = -\dfrac{3}{2}$ Minimum	$f\left(\dfrac{3\pi}{2}\right) = -1$	$f\left(\dfrac{11\pi}{6}\right) = -\dfrac{3}{2}$ Minimum	$f(2\pi) = -1$

3.1 Exercises

See **CalcChat.com** for tutorial help and worked-out solutions to odd-numbered exercises.

Finding the Value of the Derivative at Relative Extrema
In Exercises 1–6, find the value of the derivative (if it exists) at each indicated extremum.

1. $f(x) = \dfrac{x^2}{x^2 + 4}$

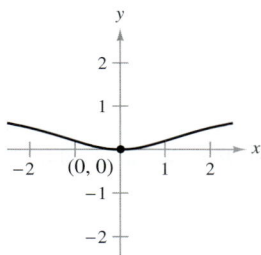

2. $f(x) = \cos \dfrac{\pi x}{2}$

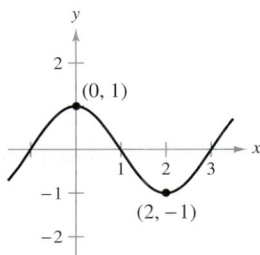

3. $g(x) = x + \dfrac{4}{x^2}$

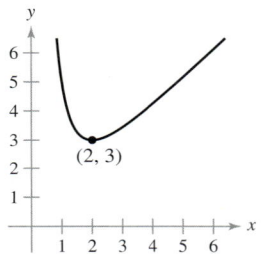

4. $f(x) = -3x\sqrt{x + 1}$

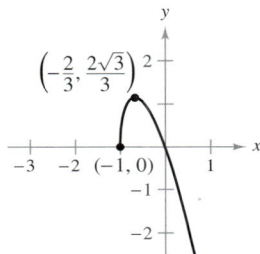

5. $f(x) = (x + 2)^{2/3}$

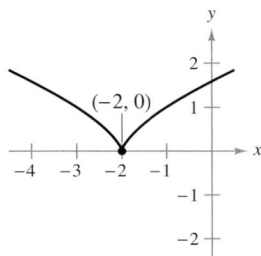

6. $f(x) = 4 - |x|$

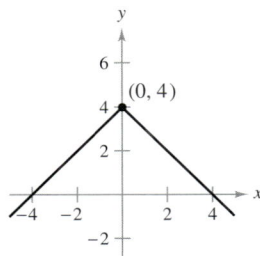

Approximating Critical Numbers In Exercises 7–10, approximate the critical numbers of the function shown in the graph. Determine whether the function has a relative maximum, a relative minimum, an absolute maximum, an absolute minimum, or none of these at each critical number on the interval shown.

7.

8.

9.

10.

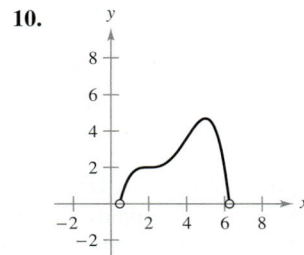

Finding Critical Numbers In Exercises 11–16, find the critical numbers of the function.

11. $f(x) = x^3 - 3x^2$

12. $g(x) = x^4 - 8x^2$

13. $g(t) = t\sqrt{4 - t},\ t < 3$

14. $f(x) = \dfrac{4x}{x^2 + 1}$

15. $h(x) = \sin^2 x + \cos x$
$0 < x < 2\pi$

16. $f(\theta) = 2 \sec \theta + \tan \theta$
$0 < \theta < 2\pi$

Finding Extrema on a Closed Interval In Exercises 17–36, find the absolute extrema of the function on the closed interval.

17. $f(x) = 3 - x,\ [-1, 2]$

18. $f(x) = \dfrac{3}{4}x + 2,\ [0, 4]$

19. $g(x) = 2x^2 - 8x,\ [0, 6]$

20. $h(x) = 5 - x^2,\ [-3, 1]$

21. $f(x) = x^3 - \dfrac{3}{2}x^2,\ [-1, 2]$

22. $f(x) = 2x^3 - 6x,\ [0, 3]$

23. $y = 3x^{2/3} - 2x,\ [-1, 1]$

24. $g(x) = \sqrt[3]{x},\ [-8, 8]$

25. $g(t) = \dfrac{t^2}{t^2 + 3},\ [-1, 1]$

26. $f(x) = \dfrac{2x}{x^2 + 1},\ [-2, 2]$

27. $h(s) = \dfrac{1}{s - 2},\ [0, 1]$

28. $h(t) = \dfrac{t}{t + 3},\ [-1, 6]$

29. $y = 3 - |t - 3|,\ [-1, 5]$

30. $g(x) = |x + 4|,\ [-7, 1]$

31. $f(x) = [\![x]\!],\ [-2, 2]$

32. $h(x) = [\![2 - x]\!],\ [-2, 2]$

33. $f(x) = \sin x,\ \left[\dfrac{5\pi}{6}, \dfrac{11\pi}{6}\right]$

34. $g(x) = \sec x,\ \left[-\dfrac{\pi}{6}, \dfrac{\pi}{3}\right]$

35. $y = 3 \cos x,\ [0, 2\pi]$

36. $y = \tan\left(\dfrac{\pi x}{8}\right),\ [0, 2]$

Finding Extrema on an Interval In Exercises 37–40, find the absolute extrema of the function (if any exist) on each interval.

37. $f(x) = 2x - 3$
(a) $[0, 2]$ (b) $[0, 2)$
(c) $(0, 2]$ (d) $(0, 2)$

38. $f(x) = 5 - x$
(a) $[1, 4]$ (b) $[1, 4)$
(c) $(1, 4]$ (d) $(1, 4)$

39. $f(x) = x^2 - 2x$
(a) $[-1, 2]$ (b) $(1, 3]$
(c) $(0, 2)$ (d) $[1, 4)$

40. $f(x) = \sqrt{4 - x^2}$
(a) $[-2, 2]$ (b) $[-2, 0)$
(c) $(-2, 2)$ (d) $[1, 2)$

Finding Absolute Extrema In Exercises 41–44, use a graphing utility to graph the function and find the absolute extrema of the function on the given interval.

41. $f(x) = \dfrac{3}{x-1}, \quad (1, 4]$ **42.** $f(x) = \dfrac{2}{2-x}, \quad [0, 2)$

43. $f(x) = x^4 - 2x^3 + x + 1, \quad [-1, 3]$

44. $f(x) = \sqrt{x} + \cos\dfrac{x}{2}, \quad [0, 2\pi]$

Finding Extrema Using Technology In Exercises 45 and 46, (a) use a computer algebra system to graph the function and approximate any absolute extrema on the given interval. (b) Use the utility to find any critical numbers, and use them to find any absolute extrema not located at the endpoints. Compare the results with those in part (a).

45. $f(x) = 3.2x^5 + 5x^3 - 3.5x, \quad [0, 1]$

46. $f(x) = \dfrac{4}{3}x\sqrt{3 - x}, \quad [0, 3]$

Finding Maximum Values Using Technology In Exercises 47 and 48, use a computer algebra system to find the maximum value of $|f''(x)|$ on the closed interval. (This value is used in the error estimate for the Trapezoidal Rule, as discussed in Section 4.6.)

47. $f(x) = \sqrt{1 + x^3}, \quad [0, 2]$ **48.** $f(x) = \dfrac{1}{x^2 + 1}, \quad \left[\dfrac{1}{2}, 3\right]$

Finding Maximum Values Using Technology In Exercises 49 and 50, use a computer algebra system to find the maximum value of $|f^{(4)}(x)|$ on the closed interval. (This value is used in the error estimate for Simpson's Rule, as discussed in Section 4.6.)

49. $f(x) = (x + 1)^{2/3}, \quad [0, 2]$

50. $f(x) = \dfrac{1}{x^2 + 1}, \quad [-1, 1]$

51. Writing Write a short paragraph explaining why a continuous function on an open interval may not have a maximum or minimum. Illustrate your explanation with a sketch of the graph of such a function.

52. **HOW DO YOU SEE IT?** Determine whether each labeled point is an absolute maximum or minimum, a relative maximum or minimum, or none of these.

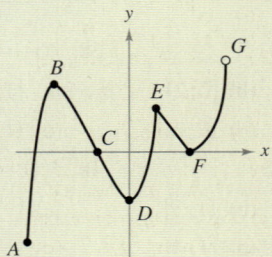

WRITING ABOUT CONCEPTS

Creating the Graph of a Function In Exercises 53 and 54, graph a function on the interval $[-2, 5]$ having the given characteristics.

53. Absolute maximum at $x = -2$
Absolute minimum at $x = 1$
Relative maximum at $x = 3$

54. Relative minimum at $x = -1$
Critical number (but no extremum) at $x = 0$
Absolute maximum at $x = 2$
Absolute minimum at $x = 5$

Using Graphs In Exercises 55–58, determine from the graph whether f has a minimum in the open interval (a, b).

55. (a)   (b)

56. (a)   (b)

57. (a)   (b)

58. (a)   (b)

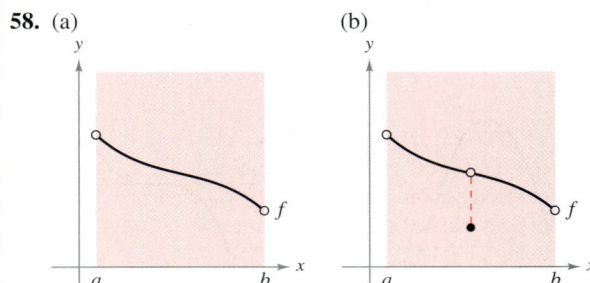

59. Power The formula for the power output P of a battery is

$$P = VI - RI^2$$

where V is the electromotive force in volts, R is the resistance in ohms, and I is the current in amperes. Find the current that corresponds to a maximum value of P in a battery for which $V = 12$ volts and $R = 0.5$ ohm. Assume that a 15-ampere fuse bounds the output in the interval $0 \leq I \leq 15$. Could the power output be increased by replacing the 15-ampere fuse with a 20-ampere fuse? Explain.

60. Lawn Sprinkler A lawn sprinkler is constructed in such a way that $d\theta/dt$ is constant, where θ ranges between 45° and 135° (see figure). The distance the water travels horizontally is

$$x = \frac{v^2 \sin 2\theta}{32}, \quad 45° \leq \theta \leq 135°$$

where v is the speed of the water. Find dx/dt and explain why this lawn sprinkler does not water evenly. What part of the lawn receives the most water?

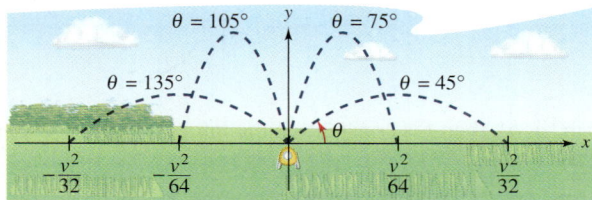

Water sprinkler: $45° \leq \theta \leq 135°$

■ **FOR FURTHER INFORMATION** For more information on the "calculus of lawn sprinklers," see the article "Design of an Oscillating Sprinkler" by Bart Braden in *Mathematics Magazine*. To view this article, go to *MathArticles.com*.

61. Honeycomb The surface area of a cell in a honeycomb is

$$S = 6hs + \frac{3s^2}{2}\left(\frac{\sqrt{3} - \cos \theta}{\sin \theta}\right)$$

where h and s are positive constants and θ is the angle at which the upper faces meet the altitude of the cell (see figure). Find the angle θ ($\pi/6 \leq \theta \leq \pi/2$) that minimizes the surface area S.

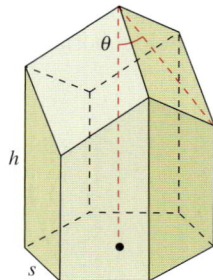

■ **FOR FURTHER INFORMATION** For more information on the geometric structure of a honeycomb cell, see the article "The Design of Honeycombs" by Anthony L. Peressini in UMAP Module 502, published by COMAP, Inc., Suite 210, 57 Bedford Street, Lexington, MA.

62. Highway Design In order to build a highway, it is necessary to fill a section of a valley where the grades (slopes) of the sides are 9% and 6% (see figure). The top of the filled region will have the shape of a parabolic arc that is tangent to the two slopes at the points A and B. The horizontal distances from A to the y-axis and from B to the y-axis are both 500 feet.

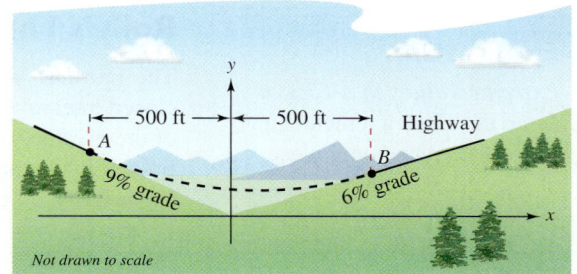

(a) Find the coordinates of A and B.

(b) Find a quadratic function $y = ax^2 + bx + c$ for $-500 \leq x \leq 500$ that describes the top of the filled region.

(c) Construct a table giving the depths d of the fill for $x = -500, -400, -300, -200, -100, 0, 100, 200, 300, 400,$ and 500.

(d) What will be the lowest point on the completed highway? Will it be directly over the point where the two hillsides come together?

True or False? In Exercises 63–66, determine whether the statement is true or false. If it is false, explain why or give an example that shows it is false.

63. The maximum of a function that is continuous on a closed interval can occur at two different values in the interval.

64. If a function is continuous on a closed interval, then it must have a minimum on the interval.

65. If $x = c$ is a critical number of the function f, then it is also a critical number of the function $g(x) = f(x) + k$, where k is a constant.

66. If $x = c$ is a critical number of the function f, then it is also a critical number of the function $g(x) = f(x - k)$, where k is a constant.

67. Functions Let the function f be differentiable on an interval I containing c. If f has a maximum value at $x = c$, show that $-f$ has a minimum value at $x = c$.

68. Critical Numbers Consider the cubic function $f(x) = ax^3 + bx^2 + cx + d$, where $a \neq 0$. Show that f can have zero, one, or two critical numbers and give an example of each case.

PUTNAM EXAM CHALLENGE

69. Determine all real numbers $a > 0$ for which there exists a nonnegative continuous function $f(x)$ defined on $[0, a]$ with the property that the region $R = \{(x, y); 0 \leq x \leq a, 0 \leq y \leq f(x)\}$ has perimeter k units and area k square units for some real number k.

This problem was composed by the Committee on the Putnam Prize Competition.
© The Mathematical Association of America. All rights reserved.

3.2 Rolle's Theorem and the Mean Value Theorem

- Understand and use Rolle's Theorem.
- Understand and use the Mean Value Theorem.

Rolle's Theorem

The Extreme Value Theorem (see Section 3.1) states that a continuous function on a closed interval $[a, b]$ must have both a minimum and a maximum on the interval. Both of these values, however, can occur at the endpoints. **Rolle's Theorem,** named after the French mathematician Michel Rolle (1652–1719), gives conditions that guarantee the existence of an extreme value in the *interior* of a closed interval.

> **THEOREM 3.3   Rolle's Theorem**
>
> Let f be continuous on the closed interval $[a, b]$ and differentiable on the open interval (a, b). If $f(a) = f(b)$, then there is at least one number c in (a, b) such that $f'(c) = 0$.

Proof   Let $f(a) = d = f(b)$.

Case 1: If $f(x) = d$ for all x in $[a, b]$, then f is constant on the interval and, by Theorem 2.2, $f'(x) = 0$ for all x in (a, b).

Case 2: Consider $f(x) > d$ for some x in (a, b). By the Extreme Value Theorem, you know that f has a maximum at some c in the interval. Moreover, because $f(c) > d$, this maximum does not occur at either endpoint. So, f has a maximum in the *open* interval (a, b). This implies that $f(c)$ is a *relative* maximum and, by Theorem 3.2, c is a critical number of f. Finally, because f is differentiable at c, you can conclude that $f'(c) = 0$.

Case 3: When $f(x) < d$ for some x in (a, b), you can use an argument similar to that in Case 2, but involving the minimum instead of the maximum.

See LarsonCalculus.com for Bruce Edwards's video of this proof.

From Rolle's Theorem, you can see that if a function f is continuous on $[a, b]$ and differentiable on (a, b), and if $f(a) = f(b)$, then there must be at least one x-value between a and b at which the graph of f has a horizontal tangent [see Figure 3.8(a)]. When the differentiability requirement is dropped from Rolle's Theorem, f will still have a critical number in (a, b), but it may not yield a horizontal tangent. Such a case is shown in Figure 3.8(b).

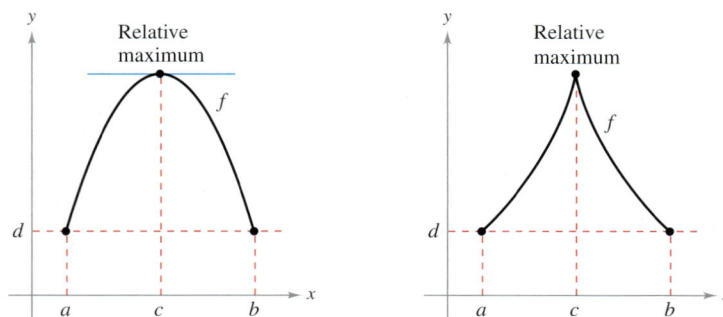

(a) f is continuous on $[a, b]$ and differentiable on (a, b).

(b) f is continuous on $[a, b]$.

Figure 3.8

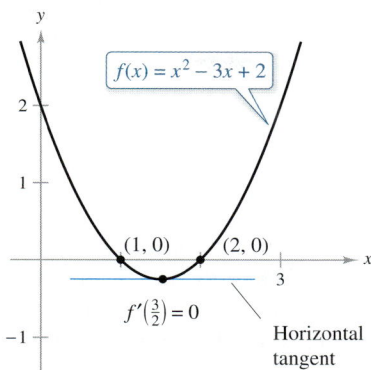

The x-value for which $f'(x) = 0$ is between the two x-intercepts.
Figure 3.9

Illustrating Rolle's Theorem

Find the two x-intercepts of

$$f(x) = x^2 - 3x + 2$$

and show that $f'(x) = 0$ at some point between the two x-intercepts.

Solution   Note that f is differentiable on the entire real number line. Setting $f(x)$ equal to 0 produces

$$x^2 - 3x + 2 = 0 \qquad \text{Set } f(x) \text{ equal to 0.}$$
$$(x - 1)(x - 2) = 0 \qquad \text{Factor.}$$
$$x = 1, 2. \qquad x\text{-values for which } f'(x) = 0$$

So, $f(1) = f(2) = 0$, and from Rolle's Theorem you know that there *exists* at least one c in the interval $(1, 2)$ such that $f'(c) = 0$. To *find* such a c, differentiate f to obtain

$$f'(x) = 2x - 3 \qquad \text{Differentiate.}$$

and then determine that $f'(x) = 0$ when $x = \frac{3}{2}$. Note that this x-value lies in the open interval $(1, 2)$, as shown in Figure 3.9.

Rolle's Theorem states that when f satisfies the conditions of the theorem, there must be *at least* one point between a and b at which the derivative is 0. There may, of course, be more than one such point, as shown in the next example.

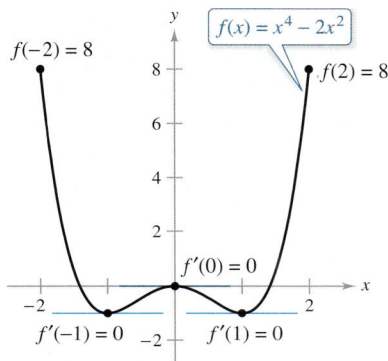

$f'(x) = 0$ for more than one x-value in the interval $(-2, 2)$.
Figure 3.10

EXAMPLE 2   **Illustrating Rolle's Theorem**

Let $f(x) = x^4 - 2x^2$. Find all values of c in the interval $(-2, 2)$ such that $f'(c) = 0$.

Solution   To begin, note that the function satisfies the conditions of Rolle's Theorem. That is, f is continuous on the interval $[-2, 2]$ and differentiable on the interval $(-2, 2)$. Moreover, because $f(-2) = f(2) = 8$, you can conclude that there exists at least one c in $(-2, 2)$ such that $f'(c) = 0$. Because

$$f'(x) = 4x^3 - 4x \qquad \text{Differentiate.}$$

setting the derivative equal to 0 produces

$$4x^3 - 4x = 0 \qquad \text{Set } f'(x) \text{ equal to 0.}$$
$$4x(x - 1)(x + 1) = 0 \qquad \text{Factor.}$$
$$x = 0, 1, -1. \qquad x\text{-values for which } f'(x) = 0$$

So, in the interval $(-2, 2)$, the derivative is zero at three different values of x, as shown in Figure 3.10.

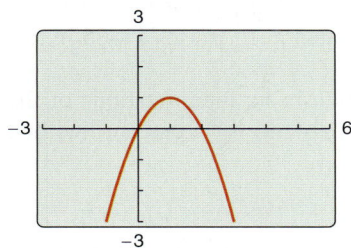

Figure 3.11

▷ **TECHNOLOGY PITFALL**   A graphing utility can be used to indicate whether the points on the graphs in Examples 1 and 2 are relative minima or relative maxima of the functions. When using a graphing utility, however, you should keep in mind that it can give misleading pictures of graphs. For example, use a graphing utility to graph

$$f(x) = 1 - (x - 1)^2 - \frac{1}{1000(x - 1)^{1/7} + 1}.$$

With most viewing windows, it appears that the function has a maximum of 1 when $x = 1$ (see Figure 3.11). By evaluating the function at $x = 1$, however, you can see that $f(1) = 0$. To determine the behavior of this function near $x = 1$, you need to examine the graph analytically to get the complete picture.

The Mean Value Theorem

Rolle's Theorem can be used to prove another theorem—the **Mean Value Theorem.**

> **THEOREM 3.4** **The Mean Value Theorem**
>
> If f is continuous on the closed interval $[a, b]$ and differentiable on the open
> interval (a, b), then there exists a number c in (a, b) such that
>
> $$f'(c) = \frac{f(b) - f(a)}{b - a}.$$

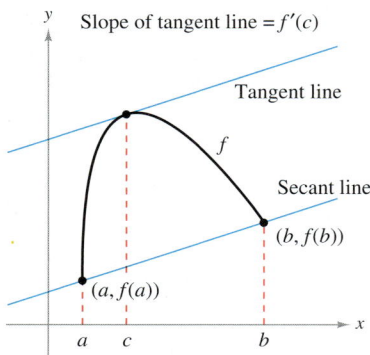

Figure 3.12

Proof Refer to Figure 3.12. The equation of the secant line that passes through the
points $(a, f(a))$ and $(b, f(b))$ is

$$y = \left[\frac{f(b) - f(a)}{b - a}\right](x - a) + f(a).$$

Let $g(x)$ be the difference between $f(x)$ and y. Then

$$g(x) = f(x) - y$$
$$= f(x) - \left[\frac{f(b) - f(a)}{b - a}\right](x - a) - f(a).$$

By evaluating g at a and b, you can see that

$$g(a) = 0 = g(b).$$

Because f is continuous on $[a, b]$, it follows that g is also continuous on $[a, b]$.
Furthermore, because f is differentiable, g is also differentiable, and you can apply
Rolle's Theorem to the function g. So, there exists a number c in (a, b) such that
$g'(c) = 0$, which implies that

$$g'(c) = 0$$
$$f'(c) - \frac{f(b) - f(a)}{b - a} = 0.$$

So, there exists a number c in (a, b) such that

$$f'(c) = \frac{f(b) - f(a)}{b - a}.$$

See LarsonCalculus.com for Bruce Edwards's video of this proof. ■

**JOSEPH-LOUIS LAGRANGE
(1736–1813)**

The Mean Value Theorem was
first proved by the famous mathematician Joseph-Louis Lagrange.
Born in Italy, Lagrange held a
position in the court of Frederick
the Great in Berlin for 20 years.
*See LarsonCalculus.com to read
more of this biography.*

Although the Mean Value Theorem can be used directly in problem solving, it is
used more often to prove other theorems. In fact, some people consider this to be the
most important theorem in calculus—it is closely related to the Fundamental Theorem
of Calculus discussed in Section 4.4. For now, you can get an idea of the versatility
of the Mean Value Theorem by looking at the results stated in Exercises 77–85 in this
section.

The Mean Value Theorem has implications for both basic interpretations of the
derivative. Geometrically, the theorem guarantees the existence of a tangent line that
is parallel to the secant line through the points

$$(a, f(a)) \quad \text{and} \quad (b, f(b)),$$

as shown in Figure 3.12. Example 3 illustrates this geometric interpretation of the Mean
Value Theorem. In terms of rates of change, the Mean Value Theorem implies that there
must be a point in the open interval (a, b) at which the instantaneous rate of change is
equal to the average rate of change over the interval $[a, b]$. This is illustrated in Example 4.

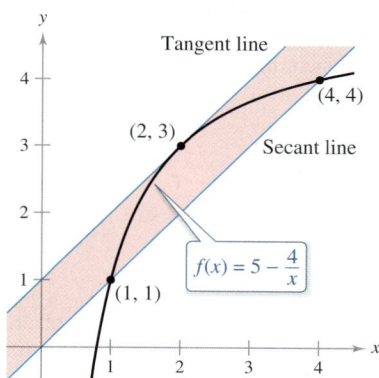

The tangent line at (2, 3) is parallel to the secant line through (1, 1) and (4, 4).

Figure 3.13

EXAMPLE 3   **Finding a Tangent Line**

:···▷ *See LarsonCalculus.com for an interactive version of this type of example.*

For $f(x) = 5 - (4/x)$, find all values of c in the open interval $(1, 4)$ such that

$$f'(c) = \frac{f(4) - f(1)}{4 - 1}.$$

Solution   The slope of the secant line through $(1, f(1))$ and $(4, f(4))$ is

$$\frac{f(4) - f(1)}{4 - 1} = \frac{4 - 1}{4 - 1} = 1.$$ Slope of secant line

Note that the function satisfies the conditions of the Mean Value Theorem. That is, f is continuous on the interval $[1, 4]$ and differentiable on the interval $(1, 4)$. So, there exists at least one number c in $(1, 4)$ such that $f'(c) = 1$. Solving the equation $f'(x) = 1$ yields

$$\frac{4}{x^2} = 1$$ Set $f'(x)$ equal to 1.

which implies that

$$x = \pm 2.$$

So, in the interval $(1, 4)$, you can conclude that $c = 2$, as shown in Figure 3.13.

EXAMPLE 4   **Finding an Instantaneous Rate of Change**

At some time t, the instantaneous velocity is equal to the average velocity over 4 minutes.

Figure 3.14

Two stationary patrol cars equipped with radar are 5 miles apart on a highway, as shown in Figure 3.14. As a truck passes the first patrol car, its speed is clocked at 55 miles per hour. Four minutes later, when the truck passes the second patrol car, its speed is clocked at 50 miles per hour. Prove that the truck must have exceeded the speed limit (of 55 miles per hour) at some time during the 4 minutes.

Solution   Let $t = 0$ be the time (in hours) when the truck passes the first patrol car. The time when the truck passes the second patrol car is

$$t = \frac{4}{60} = \frac{1}{15} \text{ hour.}$$

By letting $s(t)$ represent the distance (in miles) traveled by the truck, you have $s(0) = 0$ and $s\left(\frac{1}{15}\right) = 5$. So, the average velocity of the truck over the five-mile stretch of highway is

$$\text{Average velocity} = \frac{s(1/15) - s(0)}{(1/15) - 0} = \frac{5}{1/15} = 75 \text{ miles per hour.}$$

Assuming that the position function is differentiable, you can apply the Mean Value Theorem to conclude that the truck must have been traveling at a rate of 75 miles per hour sometime during the 4 minutes.

A useful alternative form of the Mean Value Theorem is: If f is continuous on $[a, b]$ and differentiable on (a, b), then there exists a number c in (a, b) such that

$$f(b) = f(a) + (b - a)f'(c).$$ Alternative form of Mean Value Theorem

When doing the exercises for this section, keep in mind that polynomial functions, rational functions, and trigonometric functions are differentiable at all points in their domains.

3.2 Exercises

Writing In Exercises 1–4, explain why Rolle's Theorem does not apply to the function even though there exist a and b such that $f(a) = f(b)$.

1. $f(x) = \left|\dfrac{1}{x}\right|$, $[-1, 1]$ **2.** $f(x) = \cot\dfrac{x}{2}$, $[\pi, 3\pi]$

3. $f(x) = 1 - |x - 1|$, $[0, 2]$ **4.** $f(x) = \sqrt{(2 - x^{2/3})^3}$, $[-1, 1]$

Intercepts and Derivatives In Exercises 5–8, find the two x-intercepts of the function f and show that $f'(x) = 0$ at some point between the two x-intercepts.

5. $f(x) = x^2 - x - 2$ **6.** $f(x) = x^2 + 6x$

7. $f(x) = x\sqrt{x + 4}$ **8.** $f(x) = -3x\sqrt{x + 1}$

Using Rolle's Theorem In Exercises 9–22, determine whether Rolle's Theorem can be applied to f on the closed interval $[a, b]$. If Rolle's Theorem can be applied, find all values of c in the open interval (a, b) such that $f'(c) = 0$. If Rolle's Theorem cannot be applied, explain why not.

9. $f(x) = -x^2 + 3x$, $[0, 3]$

10. $f(x) = x^2 - 8x + 5$, $[2, 6]$

11. $f(x) = (x - 1)(x - 2)(x - 3)$, $[1, 3]$

12. $f(x) = (x - 4)(x + 2)^2$, $[-2, 4]$

13. $f(x) = x^{2/3} - 1$, $[-8, 8]$ **14.** $f(x) = 3 - |x - 3|$, $[0, 6]$

15. $f(x) = \dfrac{x^2 - 2x - 3}{x + 2}$, $[-1, 3]$

16. $f(x) = \dfrac{x^2 - 1}{x}$, $[-1, 1]$

17. $f(x) = \sin x$, $[0, 2\pi]$ **18.** $f(x) = \cos x$, $[0, 2\pi]$

19. $f(x) = \sin 3x$, $\left[0, \dfrac{\pi}{3}\right]$ **20.** $f(x) = \cos 2x$, $[-\pi, \pi]$

21. $f(x) = \tan x$, $[0, \pi]$ **22.** $f(x) = \sec x$, $[\pi, 2\pi]$

Using Rolle's Theorem In Exercises 23–26, use a graphing utility to graph the function on the closed interval $[a, b]$. Determine whether Rolle's Theorem can be applied to f on the interval and, if so, find all values of c in the open interval (a, b) such that $f'(c) = 0$.

23. $f(x) = |x| - 1$, $[-1, 1]$ **24.** $f(x) = x - x^{1/3}$, $[0, 1]$

25. $f(x) = x - \tan \pi x$, $\left[-\dfrac{1}{4}, \dfrac{1}{4}\right]$

26. $f(x) = \dfrac{x}{2} - \sin\dfrac{\pi x}{6}$, $[-1, 0]$

27. Vertical Motion The height of a ball t seconds after it is thrown upward from a height of 6 feet and with an initial velocity of 48 feet per second is $f(t) = -16t^2 + 48t + 6$.

(a) Verify that $f(1) = f(2)$.

(b) According to Rolle's Theorem, what must the velocity be at some time in the interval $(1, 2)$? Find that time.

28. Reorder Costs The ordering and transportation cost C for components used in a manufacturing process is approximated by

$$C(x) = 10\left(\dfrac{1}{x} + \dfrac{x}{x + 3}\right)$$

where C is measured in thousands of dollars and x is the order size in hundreds.

(a) Verify that $C(3) = C(6)$.

(b) According to Rolle's Theorem, the rate of change of the cost must be 0 for some order size in the interval $(3, 6)$. Find that order size.

Mean Value Theorem In Exercises 29 and 30, copy the graph and sketch the secant line to the graph through the points $(a, f(a))$ and $(b, f(b))$. Then sketch any tangent lines to the graph for each value of c guaranteed by the Mean Value Theorem. To print an enlarged copy of the graph, go to *MathGraphs.com*.

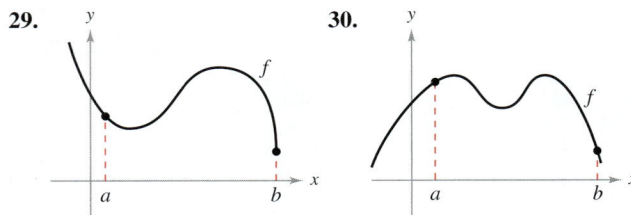

29.

30.

Writing In Exercises 31–34, explain why the Mean Value Theorem does not apply to the function f on the interval $[0, 6]$.

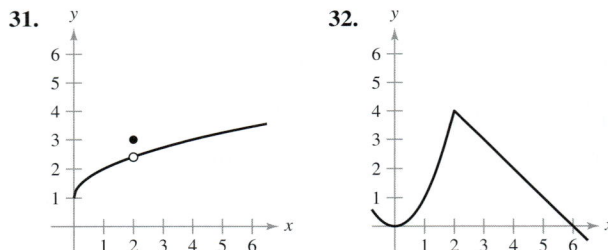

31.

32.

33. $f(x) = \dfrac{1}{x - 3}$ **34.** $f(x) = |x - 3|$

35. Mean Value Theorem Consider the graph of the function $f(x) = -x^2 + 5$ (see figure on next page).

(a) Find the equation of the secant line joining the points $(-1, 4)$ and $(2, 1)$.

(b) Use the Mean Value Theorem to determine a point c in the interval $(-1, 2)$ such that the tangent line at c is parallel to the secant line.

(c) Find the equation of the tangent line through c.

(d) Then use a graphing utility to graph f, the secant line, and the tangent line.

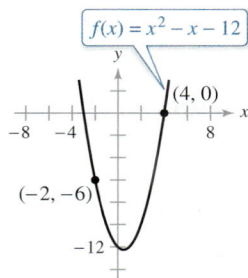

Figure for 35   Figure for 36

36. Mean Value Theorem   Consider the graph of the function $f(x) = x^2 - x - 12$ (see figure).

(a) Find the equation of the secant line joining the points $(-2, -6)$ and $(4, 0)$.

(b) Use the Mean Value Theorem to determine a point c in the interval $(-2, 4)$ such that the tangent line at c is parallel to the secant line.

(c) Find the equation of the tangent line through c.

(d) Then use a graphing utility to graph f, the secant line, and the tangent line.

Using the Mean Value Theorem   In Exercises 37–46, determine whether the Mean Value Theorem can be applied to f on the closed interval $[a, b]$. If the Mean Value Theorem can be applied, find all values of c in the open interval (a, b) such that

$$f'(c) = \frac{f(b) - f(a)}{b - a}.$$

If the Mean Value Theorem cannot be applied, explain why not.

37. $f(x) = x^2$, $[-2, 1]$ **38.** $f(x) = 2x^3$, $[0, 6]$

39. $f(x) = x^3 + 2x$, $[-1, 1]$

40. $f(x) = x^4 - 8x$, $[0, 2]$

41. $f(x) = x^{2/3}$, $[0, 1]$ **42.** $f(x) = \dfrac{x + 1}{x}$, $[-1, 2]$

43. $f(x) = |2x + 1|$, $[-1, 3]$

44. $f(x) = \sqrt{2 - x}$, $[-7, 2]$

45. $f(x) = \sin x$, $[0, \pi]$

46. $f(x) = \cos x + \tan x$, $[0, \pi]$

Using the Mean Value Theorem   In Exercises 47–50, use a graphing utility to (a) graph the function f on the given interval, (b) find and graph the secant line through points on the graph of f at the endpoints of the given interval, and (c) find and graph any tangent lines to the graph of f that are parallel to the secant line.

47. $f(x) = \dfrac{x}{x + 1}$, $\left[-\dfrac{1}{2}, 2\right]$

48. $f(x) = x - 2 \sin x$, $[-\pi, \pi]$

49. $f(x) = \sqrt{x}$, $[1, 9]$

50. $f(x) = x^4 - 2x^3 + x^2$, $[0, 6]$

Andrew Barker/Shutterstock.com

51. Vertical Motion   The height of an object t seconds after it is dropped from a height of 300 meters is

$$s(t) = -4.9t^2 + 300.$$

(a) Find the average velocity of the object during the first 3 seconds.

(b) Use the Mean Value Theorem to verify that at some time during the first 3 seconds of fall, the instantaneous velocity equals the average velocity. Find that time.

52. Sales   A company introduces a new product for which the number of units sold S is

$$S(t) = 200\left(5 - \frac{9}{2 + t}\right)$$

where t is the time in months.

(a) Find the average rate of change of $S(t)$ during the first year.

(b) During what month of the first year does $S'(t)$ equal the average rate of change?

WRITING ABOUT CONCEPTS

53. Converse of Rolle's Theorem   Let f be continuous on $[a, b]$ and differentiable on (a, b). If there exists c in (a, b) such that $f'(c) = 0$, does it follow that $f(a) = f(b)$? Explain.

54. Rolle's Theorem   Let f be continuous on $[a, b]$ and differentiable on (a, b). Also, suppose that $f(a) = f(b)$ and that c is a real number in the interval such that $f'(c) = 0$. Find an interval for the function g over which Rolle's Theorem can be applied, and find the corresponding critical number of g (k is a constant).

(a) $g(x) = f(x) + k$ (b) $g(x) = f(x - k)$

(c) $g(x) = f(kx)$

55. Rolle's Theorem   The function

$$f(x) = \begin{cases} 0, & x = 0 \\ 1 - x, & 0 < x \leq 1 \end{cases}$$

is differentiable on $(0, 1)$ and satisfies $f(0) = f(1)$. However, its derivative is never zero on $(0, 1)$. Does this contradict Rolle's Theorem? Explain.

56. Mean Value Theorem   Can you find a function f such that $f(-2) = -2$, $f(2) = 6$, and $f'(x) < 1$ for all x? Why or why not?

57. Speed

A plane begins its take-off at 2:00 P.M. on a 2500-mile flight. After 5.5 hours, the plane arrives at its destination. Explain why there are at least two times during the flight when the speed of the plane is 400 miles per hour.

58. Temperature When an object is removed from a furnace and placed in an environment with a constant temperature of 90°F, its core temperature is 1500°F. Five hours later, the core temperature is 390°F. Explain why there must exist a time in the interval when the temperature is decreasing at a rate of 222°F per hour.

59. Velocity Two bicyclists begin a race at 8:00 A.M. They both finish the race 2 hours and 15 minutes later. Prove that at some time during the race, the bicyclists are traveling at the same velocity.

60. Acceleration At 9:13 A.M., a sports car is traveling 35 miles per hour. Two minutes later, the car is traveling 85 miles per hour. Prove that at some time during this two-minute interval, the car's acceleration is exactly 1500 miles per hour squared.

61. Using a Function Consider the function

$$f(x) = 3 \cos^2\left(\frac{\pi x}{2}\right).$$

(a) Use a graphing utility to graph f and f'.

(b) Is f a continuous function? Is f' a continuous function?

(c) Does Rolle's Theorem apply on the interval $[-1, 1]$? Does it apply on the interval $[1, 2]$? Explain.

(d) Evaluate, if possible, $\lim\limits_{x \to 3^-} f'(x)$ and $\lim\limits_{x \to 3^+} f'(x)$.

62. HOW DO YOU SEE IT? The figure shows two parts of the graph of a continuous differentiable function f on $[-10, 4]$. The derivative f' is also continuous. To print an enlarged copy of the graph, go to *MathGraphs.com*.

(a) Explain why f must have at least one zero in $[-10, 4]$.

(b) Explain why f' must also have at least one zero in the interval $[-10, 4]$. What are these zeros called?

(c) Make a possible sketch of the function with one zero of f' on the interval $[-10, 4]$.

Think About It In Exercises 63 and 64, sketch the graph of an arbitrary function f that satisfies the given condition but does not satisfy the conditions of the Mean Value Theorem on the interval $[-5, 5]$.

63. f is continuous on $[-5, 5]$.

64. f is not continuous on $[-5, 5]$.

Finding a Solution In Exercises 65–68, use the Intermediate Value Theorem and Rolle's Theorem to prove that the equation has exactly one real solution.

65. $x^5 + x^3 + x + 1 = 0$ **66.** $2x^5 + 7x - 1 = 0$

67. $3x + 1 - \sin x = 0$ **68.** $2x - 2 - \cos x = 0$

Differential Equation In Exercises 69–72, find a function f that has the derivative $f'(x)$ and whose graph passes through the given point. Explain your reasoning.

69. $f'(x) = 0$, $(2, 5)$ **70.** $f'(x) = 4$, $(0, 1)$

71. $f'(x) = 2x$, $(1, 0)$ **72.** $f'(x) = 6x - 1$, $(2, 7)$

True or False? In Exercises 73–76, determine whether the statement is true or false. If it is false, explain why or give an example that shows it is false.

73. The Mean Value Theorem can be applied to

$$f(x) = \frac{1}{x}$$

on the interval $[-1, 1]$.

74. If the graph of a function has three x-intercepts, then it must have at least two points at which its tangent line is horizontal.

75. If the graph of a polynomial function has three x-intercepts, then it must have at least two points at which its tangent line is horizontal.

76. If $f'(x) = 0$ for all x in the domain of f, then f is a constant function.

77. Proof Prove that if $a > 0$ and n is any positive integer, then the polynomial function $p(x) = x^{2n+1} + ax + b$ cannot have two real roots.

78. Proof Prove that if $f'(x) = 0$ for all x in an interval (a, b), then f is constant on (a, b).

79. Proof Let $p(x) = Ax^2 + Bx + C$. Prove that for any interval $[a, b]$, the value c guaranteed by the Mean Value Theorem is the midpoint of the interval.

80. Using Rolle's Theorem

(a) Let $f(x) = x^2$ and $g(x) = -x^3 + x^2 + 3x + 2$. Then $f(-1) = g(-1)$ and $f(2) = g(2)$. Show that there is at least one value c in the interval $(-1, 2)$ where the tangent line to f at $(c, f(c))$ is parallel to the tangent line to g at $(c, g(c))$. Identify c.

(b) Let f and g be differentiable functions on $[a, b]$ where $f(a) = g(a)$ and $f(b) = g(b)$. Show that there is at least one value c in the interval (a, b) where the tangent line to f at $(c, f(c))$ is parallel to the tangent line to g at $(c, g(c))$.

81. Proof Prove that if f is differentiable on $(-\infty, \infty)$ and $f'(x) < 1$ for all real numbers, then f has at most one fixed point. A fixed point of a function f is a real number c such that $f(c) = c$.

82. Fixed Point Use the result of Exercise 81 to show that $f(x) = \frac{1}{2}\cos x$ has at most one fixed point.

83. Proof Prove that $|\cos a - \cos b| \le |a - b|$ for all a and b.

84. Proof Prove that $|\sin a - \sin b| \le |a - b|$ for all a and b.

85. Using the Mean Value Theorem Let $0 < a < b$. Use the Mean Value Theorem to show that

$$\sqrt{b} - \sqrt{a} < \frac{b - a}{2\sqrt{a}}.$$

3.3 Increasing and Decreasing Functions and the First Derivative Test

- Determine intervals on which a function is increasing or decreasing.
- Apply the First Derivative Test to find relative extrema of a function.

Increasing and Decreasing Functions

In this section, you will learn how derivatives can be used to *classify* relative extrema as either relative minima or relative maxima. First, it is important to define increasing and decreasing functions.

> ### Definitions of Increasing and Decreasing Functions
>
> A function f is **increasing** on an interval when, for any two numbers x_1 and x_2 in the interval, $x_1 < x_2$ implies $f(x_1) < f(x_2)$.
>
> A function f is **decreasing** on an interval when, for any two numbers x_1 and x_2 in the interval, $x_1 < x_2$ implies $f(x_1) > f(x_2)$.

A function is increasing when, *as x moves to the right*, its graph moves up, and is decreasing when its graph moves down. For example, the function in Figure 3.15 is decreasing on the interval $(-\infty, a)$, is constant on the interval (a, b), and is increasing on the interval (b, ∞). As shown in Theorem 3.5 below, a positive derivative implies that the function is increasing, a negative derivative implies that the function is decreasing, and a zero derivative on an entire interval implies that the function is constant on that interval.

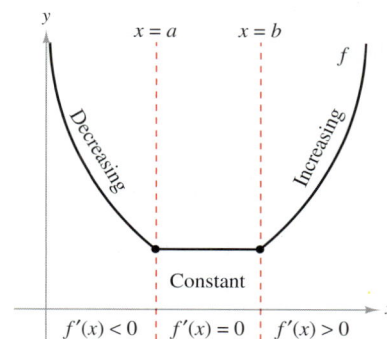

The derivative is related to the slope of a function.

Figure 3.15

> ### THEOREM 3.5 Test for Increasing and Decreasing Functions
>
> Let f be a function that is continuous on the closed interval $[a, b]$ and differentiable on the open interval (a, b).
>
> 1. If $f'(x) > 0$ for all x in (a, b), then f is increasing on $[a, b]$.
> 2. If $f'(x) < 0$ for all x in (a, b), then f is decreasing on $[a, b]$.
> 3. If $f'(x) = 0$ for all x in (a, b), then f is constant on $[a, b]$.

• • REMARK The conclusions in the first two cases of Theorem 3.5 are valid even when $f'(x) = 0$ at a finite number of x-values in (a, b).

Proof To prove the first case, assume that $f'(x) > 0$ for all x in the interval (a, b) and let $x_1 < x_2$ be any two points in the interval. By the Mean Value Theorem, you know that there exists a number c such that $x_1 < c < x_2$, and

$$f'(c) = \frac{f(x_2) - f(x_1)}{x_2 - x_1}.$$

Because $f'(c) > 0$ and $x_2 - x_1 > 0$, you know that $f(x_2) - f(x_1) > 0$, which implies that $f(x_1) < f(x_2)$. So, f is increasing on the interval. The second case has a similar proof (see Exercise 97), and the third case is a consequence of Exercise 78 in Section 3.2.

See LarsonCalculus.com for Bruce Edwards's video of this proof.

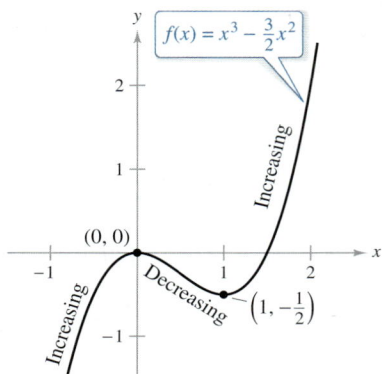

EXAMPLE 1 **Intervals on Which f Is Increasing or Decreasing**

Find the open intervals on which $f(x) = x^3 - \frac{3}{2}x^2$ is increasing or decreasing.

Solution Note that f is differentiable on the entire real number line and the derivative of f is

$$f(x) = x^3 - \tfrac{3}{2}x^2 \qquad \text{Write original function.}$$
$$f'(x) = 3x^2 - 3x. \qquad \text{Differentiate.}$$

To determine the critical numbers of f, set $f'(x)$ equal to zero.

$$3x^2 - 3x = 0 \qquad \text{Set } f'(x) \text{ equal to 0.}$$
$$3(x)(x - 1) = 0 \qquad \text{Factor.}$$
$$x = 0, 1 \qquad \text{Critical numbers}$$

Because there are no points for which f' does not exist, you can conclude that $x = 0$ and $x = 1$ are the only critical numbers. The table summarizes the testing of the three intervals determined by these two critical numbers.

Interval	$-\infty < x < 0$	$0 < x < 1$	$1 < x < \infty$
Test Value	$x = -1$	$x = \frac{1}{2}$	$x = 2$
Sign of $f'(x)$	$f'(-1) = 6 > 0$	$f'\left(\frac{1}{2}\right) = -\frac{3}{4} < 0$	$f'(2) = 6 > 0$
Conclusion	Increasing	Decreasing	Increasing

By Theorem 3.5, f is increasing on the intervals $(-\infty, 0)$ and $(1, \infty)$ and decreasing on the interval $(0, 1)$, as shown in Figure 3.16. ■

Figure 3.16

Example 1 gives you one instance of how to find intervals on which a function is increasing or decreasing. The guidelines below summarize the steps followed in that example.

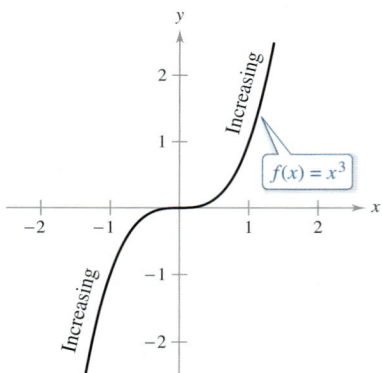

(a) Strictly monotonic function

GUIDELINES FOR FINDING INTERVALS ON WHICH A FUNCTION IS INCREASING OR DECREASING

Let f be continuous on the interval (a, b). To find the open intervals on which f is increasing or decreasing, use the following steps.

1. Locate the critical numbers of f in (a, b), and use these numbers to determine test intervals.
2. Determine the sign of $f'(x)$ at one test value in each of the intervals.
3. Use Theorem 3.5 to determine whether f is increasing or decreasing on each interval.

These guidelines are also valid when the interval (a, b) is replaced by an interval of the form $(-\infty, b)$, (a, ∞), or $(-\infty, \infty)$.

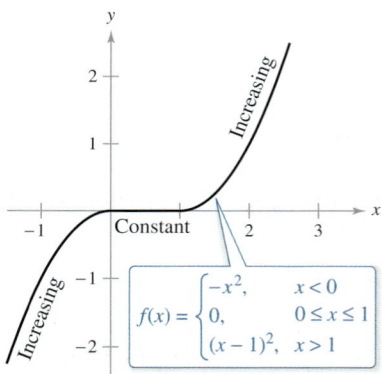

(b) Not strictly monotonic

Figure 3.17

A function is **strictly monotonic** on an interval when it is either increasing on the entire interval or decreasing on the entire interval. For instance, the function $f(x) = x^3$ is strictly monotonic on the entire real number line because it is increasing on the entire real number line, as shown in Figure 3.17(a). The function shown in Figure 3.17(b) is not strictly monotonic on the entire real number line because it is constant on the interval $[0, 1]$.

The First Derivative Test

After you have determined the intervals on
which a function is increasing or decreasing,
it is not difficult to locate the relative extrema
of the function. For instance, in Figure 3.18
(from Example 1), the function

$$f(x) = x^3 - \frac{3}{2}x^2$$

has a relative maximum at the point $(0, 0)$
because f is increasing immediately to the left
of $x = 0$ and decreasing immediately to the
right of $x = 0$. Similarly, f has a relative
minimum at the point $\left(1, -\frac{1}{2}\right)$ because f is
decreasing immediately to the left of $x = 1$
and increasing immediately to the right of
$x = 1$. The next theorem makes this more explicit.

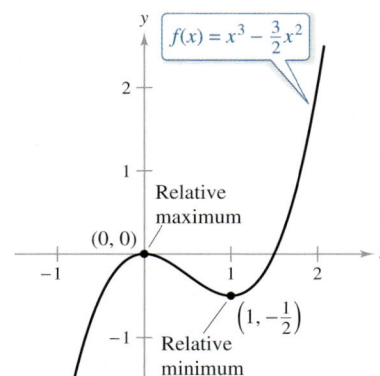

$f(x) = x^3 - \frac{3}{2}x^2$

Relative
maximum

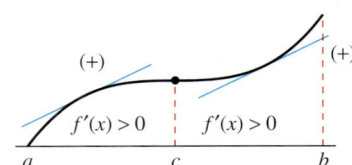

$(0, 0)$

$\left(1, -\frac{1}{2}\right)$

Relative
minimum

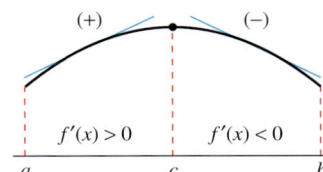

Relative extrema of f
Figure 3.18

THEOREM 3.6   The First Derivative Test

Let c be a critical number of a function f that is continuous on an open
interval I containing c. If f is differentiable on the interval, except possibly
at c, then $f(c)$ can be classified as follows.

1. If $f'(x)$ changes from negative to positive at c, then f has a *relative minimum*
   at $(c, f(c))$.
2. If $f'(x)$ changes from positive to negative at c, then f has a *relative maximum*
   at $(c, f(c))$.
3. If $f'(x)$ is positive on both sides of c or negative on both sides of c, then $f(c)$
   is neither a relative minimum nor a relative maximum.

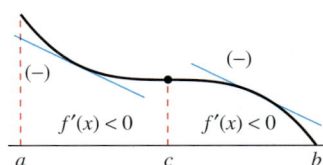

$f'(x) < 0$ $f'(x) > 0$
Relative minimum

$f'(x) > 0$ $f'(x) < 0$
Relative maximum

$f'(x) > 0$ $f'(x) > 0$

$f'(x) < 0$ $f'(x) < 0$

Neither relative minimum nor relative maximum

Proof   Assume that $f'(x)$ changes from negative to positive at c. Then there exist a and
b in I such that

$$f'(x) < 0 \text{ for all } x \text{ in } (a, c) \quad \text{and} \quad f'(x) > 0 \text{ for all } x \text{ in } (c, b).$$

By Theorem 3.5, f is decreasing on $[a, c]$ and increasing on $[c, b]$. So, $f(c)$ is a
minimum of f on the open interval (a, b) and, consequently, a relative minimum of f.
This proves the first case of the theorem. The second case can be proved in a similar
way (see Exercise 98).

See LarsonCalculus.com for Bruce Edwards's video of this proof.

EXAMPLE 2   Applying the First Derivative Test

Find the relative extrema of $f(x) = \frac{1}{2}x - \sin x$ in the interval $(0, 2\pi)$.

Solution Note that f is continuous on the interval $(0, 2\pi)$. The derivative of f is $f'(x) = \frac{1}{2} - \cos x$. To determine the critical numbers of f in this interval, set $f'(x)$ equal to 0.

$$\frac{1}{2} - \cos x = 0 \qquad \text{Set } f'(x) \text{ equal to 0.}$$

$$\cos x = \frac{1}{2}$$

$$x = \frac{\pi}{3}, \frac{5\pi}{3} \qquad \text{Critical numbers}$$

Because there are no points for which f' does not exist, you can conclude that $x = \pi/3$ and $x = 5\pi/3$ are the only critical numbers. The table summarizes the testing of the three intervals determined by these two critical numbers. By applying the First Derivative Test, you can conclude that f has a relative minimum at the point where $x = \pi/3$ and a relative maximum at the point where $x = 5\pi/3$, as shown in Figure 3.19.

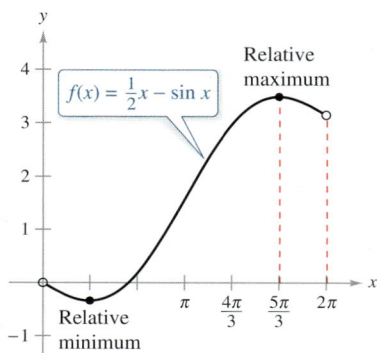

$f(x) = \frac{1}{2}x - \sin x$

A relative minimum occurs where f changes from decreasing to increasing, and a relative maximum occurs where f changes from increasing to decreasing.
Figure 3.19

Interval	$0 < x < \frac{\pi}{3}$	$\frac{\pi}{3} < x < \frac{5\pi}{3}$	$\frac{5\pi}{3} < x < 2\pi$
Test Value	$x = \frac{\pi}{4}$	$x = \pi$	$x = \frac{7\pi}{4}$
Sign of $f'(x)$	$f'\left(\frac{\pi}{4}\right) < 0$	$f'(\pi) > 0$	$f'\left(\frac{7\pi}{4}\right) < 0$
Conclusion	Decreasing	Increasing	Decreasing

EXAMPLE 3   Applying the First Derivative Test

Find the relative extrema of $f(x) = (x^2 - 4)^{2/3}$.

Solution Begin by noting that f is continuous on the entire real number line. The derivative of f

$$f'(x) = \frac{2}{3}(x^2 - 4)^{-1/3}(2x) \qquad \text{General Power Rule}$$

$$= \frac{4x}{3(x^2 - 4)^{1/3}} \qquad \text{Simplify.}$$

is 0 when $x = 0$ and does not exist when $x = \pm 2$. So, the critical numbers are $x = -2$, $x = 0$, and $x = 2$. The table summarizes the testing of the four intervals determined by these three critical numbers. By applying the First Derivative Test, you can conclude that f has a relative minimum at the point $(-2, 0)$, a relative maximum at the point $(0, \sqrt[3]{16})$, and another relative minimum at the point $(2, 0)$, as shown in Figure 3.20.

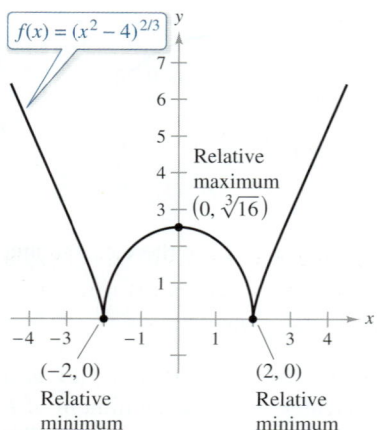

$f(x) = (x^2 - 4)^{2/3}$

Figure 3.20

Interval	$-\infty < x < -2$	$-2 < x < 0$	$0 < x < 2$	$2 < x < \infty$
Test Value	$x = -3$	$x = -1$	$x = 1$	$x = 3$
Sign of $f'(x)$	$f'(-3) < 0$	$f'(-1) > 0$	$f'(1) < 0$	$f'(3) > 0$
Conclusion	Decreasing	Increasing	Decreasing	Increasing

Note that in Examples 1 and 2, the given functions are differentiable on the entire real number line. For such functions, the only critical numbers are those for which $f'(x) = 0$. Example 3 concerns a function that has two types of critical numbers—those for which $f'(x) = 0$ and those for which f is not differentiable.

When using the First Derivative Test, be sure to consider the domain of the function. For instance, in the next example, the function

$$f(x) = \frac{x^4 + 1}{x^2}$$

is not defined when $x = 0$. This x-value must be used with the critical numbers to determine the test intervals.

EXAMPLE 4   Applying the First Derivative Test

· · · · ▷ *See LarsonCalculus.com for an interactive version of this type of example.*

Find the relative extrema of $f(x) = \dfrac{x^4 + 1}{x^2}$.

Solution   Note that f is not defined when $x = 0$.

$f(x) = x^2 + x^{-2}$	Rewrite original function.
$f'(x) = 2x - 2x^{-3}$	Differentiate.
$\quad = 2x - \dfrac{2}{x^3}$	Rewrite with positive exponent.
$\quad = \dfrac{2(x^4 - 1)}{x^3}$	Simplify.
$\quad = \dfrac{2(x^2 + 1)(x - 1)(x + 1)}{x^3}$	Factor.

So, $f'(x)$ is zero at $x = \pm 1$. Moreover, because $x = 0$ is not in the domain of f, you should use this x-value along with the critical numbers to determine the test intervals.

$x = \pm 1$	Critical numbers, $f'(\pm 1) = 0$
$x = 0$	0 is not in the domain of f.

The table summarizes the testing of the four intervals determined by these three x-values. By applying the First Derivative Test, you can conclude that f has one relative minimum at the point $(-1, 2)$ and another at the point $(1, 2)$, as shown in Figure 3.21.

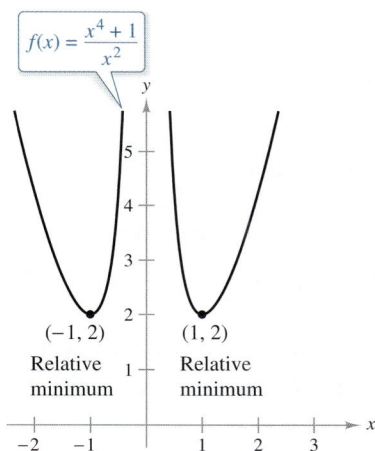

$f(x) = \dfrac{x^4 + 1}{x^2}$

$(-1, 2)$ Relative minimum $(1, 2)$ Relative minimum

x-values that are not in the domain of f, as well as critical numbers, determine test intervals for f'.

Figure 3.21

Interval	$-\infty < x < -1$	$-1 < x < 0$	$0 < x < 1$	$1 < x < \infty$
Test Value	$x = -2$	$x = -\frac{1}{2}$	$x = \frac{1}{2}$	$x = 2$
Sign of $f'(x)$	$f'(-2) < 0$	$f'\left(-\frac{1}{2}\right) > 0$	$f'\left(\frac{1}{2}\right) < 0$	$f'(2) > 0$
Conclusion	Decreasing	Increasing	Decreasing	Increasing

▷ **TECHNOLOGY**   The most difficult step in applying the First Derivative Test is finding the values for which the derivative is equal to 0. For instance, the values of x for which the derivative of

$$f(x) = \frac{x^4 + 1}{x^2 + 1}$$

is equal to zero are $x = 0$ and $x = \pm\sqrt{\sqrt{2} - 1}$. If you have access to technology that can perform symbolic differentiation and solve equations, use it to apply the First Derivative Test to this function.

EXAMPLE 5 **The Path of a Projectile**

Neglecting air resistance, the path of a projectile that is propelled at an angle θ is

$$y = \frac{g \sec^2 \theta}{2v_0^2}x^2 + (\tan \theta)x + h, \quad 0 \le \theta \le \frac{\pi}{2}$$

where y is the height, x is the horizontal distance, g is the acceleration due to gravity, v_0 is the initial velocity, and h is the initial height. (This equation is derived in Section 12.3.) Let $g = -32$ feet per second per second, $v_0 = 24$ feet per second, and $h = 9$ feet. What value of θ will produce a maximum horizontal distance?

Solution To find the distance the projectile travels, let $y = 0$, $g = -32$, $v_0 = 24$, and $h = 9$. Then substitute these values in the given equation as shown.

$$\frac{g \sec^2 \theta}{2v_0^2}x^2 + (\tan \theta)x + h = y$$

$$\frac{-32 \sec^2 \theta}{2(24^2)}x^2 + (\tan \theta)x + 9 = 0$$

$$-\frac{\sec^2 \theta}{36}x^2 + (\tan \theta)x + 9 = 0$$

Next, solve for x using the Quadratic Formula with $a = -\sec^2 \theta/36$, $b = \tan \theta$, and $c = 9$.

$$x = \frac{-b \pm \sqrt{b^2 - 4ac}}{2a}$$

$$x = \frac{-\tan \theta \pm \sqrt{(\tan \theta)^2 - 4(-\sec^2 \theta/36)(9)}}{2(-\sec^2 \theta/36)}$$

$$x = \frac{-\tan \theta \pm \sqrt{\tan^2 \theta + \sec^2 \theta}}{-\sec^2 \theta/18}$$

$$x = 18 \cos \theta \left(\sin \theta + \sqrt{\sin^2 \theta + 1} \right), \quad x \ge 0$$

At this point, you need to find the value of θ that produces a maximum value of x. Applying the First Derivative Test by hand would be very tedious. Using technology to solve the equation $dx/d\theta = 0$, however, eliminates most of the messy computations. The result is that the maximum value of x occurs when

$$\theta \approx 0.61548 \text{ radian}, \quad \text{or} \quad 35.3°.$$

This conclusion is reinforced by sketching the path of the projectile for different values of θ, as shown in Figure 3.22. Of the three paths shown, note that the distance traveled is greatest for $\theta = 35°$.

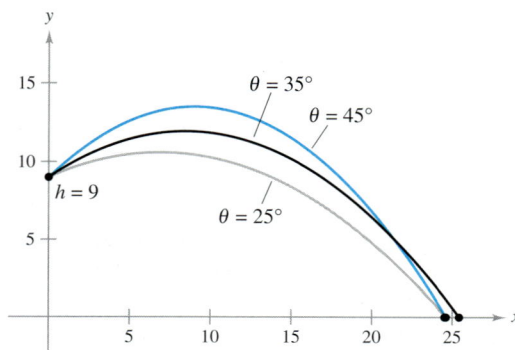

The path of a projectile with initial angle θ
Figure 3.22

3.3 Exercises

See **CalcChat.com** for tutorial help and worked-out solutions to odd-numbered exercises.

Using a Graph In Exercises 1 and 2, use the graph of f to find (a) the largest open interval on which f is increasing, and (b) the largest open interval on which f is decreasing.

1.

2.

Using a Graph In Exercises 3–8, use the graph to estimate the open intervals on which the function is increasing or decreasing. Then find the open intervals analytically.

3. $f(x) = x^2 - 6x + 8$

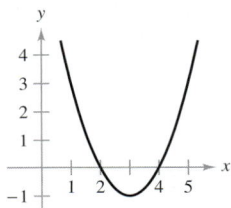

4. $y = -(x + 1)^2$

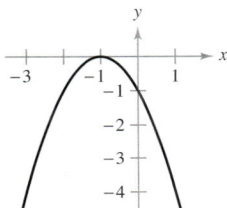

5. $y = \dfrac{x^3}{4} - 3x$

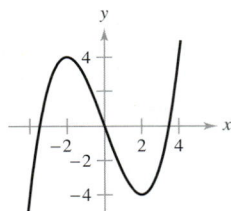

6. $f(x) = x^4 - 2x^2$

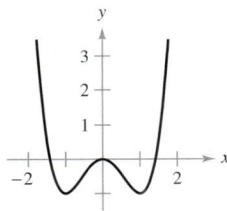

7. $f(x) = \dfrac{1}{(x + 1)^2}$

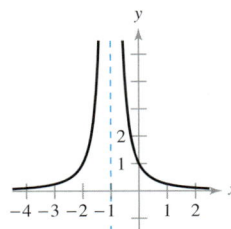

8. $y = \dfrac{x^2}{2x - 1}$

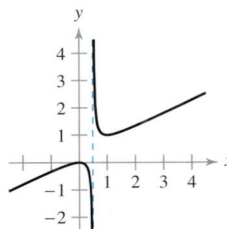

Intervals on Which f Is Increasing or Decreasing In Exercises 9–16, identify the open intervals on which the function is increasing or decreasing.

9. $g(x) = x^2 - 2x - 8$

10. $h(x) = 12x - x^3$

11. $y = x\sqrt{16 - x^2}$

12. $y = x + \dfrac{9}{x}$

13. $f(x) = \sin x - 1, \quad 0 < x < 2\pi$

14. $h(x) = \cos \dfrac{x}{2}, \quad 0 < x < 2\pi$

15. $y = x - 2\cos x, \quad 0 < x < 2\pi$

16. $f(x) = \sin^2 x + \sin x, \quad 0 < x < 2\pi$

Applying the First Derivative Test In Exercises 17–40, (a) find the critical numbers of f (if any), (b) find the open interval(s) on which the function is increasing or decreasing, (c) apply the First Derivative Test to identify all relative extrema, and (d) use a graphing utility to confirm your results.

17. $f(x) = x^2 - 4x$

18. $f(x) = x^2 + 6x + 10$

19. $f(x) = -2x^2 + 4x + 3$

20. $f(x) = -3x^2 - 4x - 2$

21. $f(x) = 2x^3 + 3x^2 - 12x$

22. $f(x) = x^3 - 6x^2 + 15$

23. $f(x) = (x - 1)^2(x + 3)$

24. $f(x) = (x + 2)^2(x - 1)$

25. $f(x) = \dfrac{x^5 - 5x}{5}$

26. $f(x) = x^4 - 32x + 4$

27. $f(x) = x^{1/3} + 1$

28. $f(x) = x^{2/3} - 4$

29. $f(x) = (x + 2)^{2/3}$

30. $f(x) = (x - 3)^{1/3}$

31. $f(x) = 5 - |x - 5|$

32. $f(x) = |x + 3| - 1$

33. $f(x) = 2x + \dfrac{1}{x}$

34. $f(x) = \dfrac{x}{x - 5}$

35. $f(x) = \dfrac{x^2}{x^2 - 9}$

36. $f(x) = \dfrac{x^2 - 2x + 1}{x + 1}$

37. $f(x) = \begin{cases} 4 - x^2, & x \le 0 \\ -2x, & x > 0 \end{cases}$

38. $f(x) = \begin{cases} 2x + 1, & x \le -1 \\ x^2 - 2, & x > -1 \end{cases}$

39. $f(x) = \begin{cases} 3x + 1, & x \le 1 \\ 5 - x^2, & x > 1 \end{cases}$

40. $f(x) = \begin{cases} -x^3 + 1, & x \le 0 \\ -x^2 + 2x, & x > 0 \end{cases}$

Applying the First Derivative Test In Exercises 41–48, consider the function on the interval $(0, 2\pi)$. For each function, (a) find the open interval(s) on which the function is increasing or decreasing, (b) apply the First Derivative Test to identify all relative extrema, and (c) use a graphing utility to confirm your results.

41. $f(x) = \dfrac{x}{2} + \cos x$

42. $f(x) = \sin x \cos x + 5$

43. $f(x) = \sin x + \cos x$

44. $f(x) = x + 2\sin x$

45. $f(x) = \cos^2(2x)$

46. $f(x) = \sin x - \sqrt{3}\cos x$

47. $f(x) = \sin^2 x + \sin x$

48. $f(x) = \dfrac{\sin x}{1 + \cos^2 x}$

Finding and Analyzing Derivatives Using Technology
In Exercises 49–54, (a) use a computer algebra system to differentiate the function, (b) sketch the graphs of f and f' on the same set of coordinate axes over the given interval, (c) find the critical numbers of f in the open interval, and (d) find the interval(s) on which f' is positive and the interval(s) on which it is negative. Compare the behavior of f and the sign of f'.

49. $f(x) = 2x\sqrt{9 - x^2}, \quad [-3, 3]$

50. $f(x) = 10\left(5 - \sqrt{x^2 - 3x + 16}\right), \quad [0, 5]$

51. $f(t) = t^2 \sin t, \quad [0, 2\pi]$

52. $f(x) = \dfrac{x}{2} + \cos\dfrac{x}{2}, \quad [0, 4\pi]$

53. $f(x) = -3 \sin\dfrac{x}{3}, \quad [0, 6\pi]$

54. $f(x) = 2 \sin 3x + 4 \cos 3x, \quad [0, \pi]$

Comparing Functions In Exercises 55 and 56, use symmetry, extrema, and zeros to sketch the graph of f. How do the functions f and g differ?

55. $f(x) = \dfrac{x^5 - 4x^3 + 3x}{x^2 - 1}$

$g(x) = x(x^2 - 3)$

56. $f(t) = \cos^2 t - \sin^2 t$

$g(t) = 1 - 2 \sin^2 t$

Think About It In Exercises 57–62, the graph of f is shown in the figure. Sketch a graph of the derivative of f. To print an enlarged copy of the graph, go to *MathGraphs.com*.

57.

58.

59.

60.

61.

62.

WRITING ABOUT CONCEPTS

Transformations of Functions In Exercises 63–68, assume that f is differentiable for all x. The signs of f' are as follows.

$f'(x) > 0$ on $(-\infty, -4)$

$f'(x) < 0$ on $(-4, 6)$

$f'(x) > 0$ on $(6, \infty)$

Supply the appropriate inequality sign for the indicated value of c.

Function	Sign of $g'(c)$
63. $g(x) = f(x) + 5$	$g'(0) \;\rule{1cm}{0.4pt}\; 0$
64. $g(x) = 3f(x) - 3$	$g'(-5) \;\rule{1cm}{0.4pt}\; 0$
65. $g(x) = -f(x)$	$g'(-6) \;\rule{1cm}{0.4pt}\; 0$
66. $g(x) = -f(x)$	$g'(0) \;\rule{1cm}{0.4pt}\; 0$
67. $g(x) = f(x - 10)$	$g'(0) \;\rule{1cm}{0.4pt}\; 0$
68. $g(x) = f(x - 10)$	$g'(8) \;\rule{1cm}{0.4pt}\; 0$

69. Sketching a Graph Sketch the graph of the arbitrary function f such that

$$f'(x) \begin{cases} > 0, & x < 4 \\ \text{undefined}, & x = 4 \\ < 0, & x > 4 \end{cases}.$$

70. HOW DO YOU SEE IT? Use the graph of f' to (a) identify the critical numbers of f, (b) identify the open interval(s) on which f is increasing or decreasing, and (c) determine whether f has a relative maximum, a relative minimum, or neither at each critical number.

(i)

(ii)

(iii)

(iv)

71. Analyzing a Critical Number   A differentiable function f has one critical number at $x = 5$. Identify the relative extrema of f at the critical number when $f'(4) = -2.5$ and $f'(6) = 3$.

72. Analyzing a Critical Number   A differentiable function f has one critical number at $x = 2$. Identify the relative extrema of f at the critical number when $f'(1) = 2$ and $f'(3) = 6$.

Think About It   In Exercises 73 and 74, the function f is differentiable on the indicated interval. The table shows $f'(x)$ for selected values of x. (a) Sketch the graph of f, (b) approximate the critical numbers, and (c) identify the relative extrema.

73. f is differentiable on $[-1, 1]$.

x	-1	-0.75	-0.50	-0.25	0
$f'(x)$	-10	-3.2	-0.5	0.8	5.6

x	0.25	0.50	0.75	1
$f'(x)$	3.6	-0.2	-6.7	-20.1

74. f is differentiable on $[0, \pi]$.

x	0	$\pi/6$	$\pi/4$	$\pi/3$	$\pi/2$
$f'(x)$	3.14	-0.23	-2.45	-3.11	0.69

x	$2\pi/3$	$3\pi/4$	$5\pi/6$	π
$f'(x)$	3.00	1.37	-1.14	-2.84

75. Rolling a Ball Bearing   A ball bearing is placed on an inclined plane and begins to roll. The angle of elevation of the plane is θ. The distance (in meters) the ball bearing rolls in t seconds is $s(t) = 4.9(\sin \theta)t^2$.

(a) Determine the speed of the ball bearing after t seconds.

(b) Complete the table and use it to determine the value of θ that produces the maximum speed at a particular time.

θ	0	$\pi/4$	$\pi/3$	$\pi/2$	$2\pi/3$	$3\pi/4$	π
$s'(t)$							

76. Modeling Data   The end-of-year assets of the Medicare Hospital Insurance Trust Fund (in billions of dollars) for the years 1999 through 2010 are shown.

1999: 141.4; 2000: 177.5; 2001: 208.7; 2002: 234.8;
2003: 256.0; 2004: 269.3; 2005: 285.8; 2006: 305.4
2007: 326.0; 2008: 321.3; 2009: 304.2; 2010: 271.9

(Source: U.S. Centers for Medicare and Medicaid Services)

(a) Use the regression capabilities of a graphing utility to find a model of the form $M = at^4 + bt^3 + ct^2 + dt + e$ for the data. (Let $t = 9$ represent 1999.)

(b) Use a graphing utility to plot the data and graph the model.

(c) Find the maximum value of the model and compare the result with the actual data.

77. Numerical, Graphical, and Analytic Analysis   The concentration C of a chemical in the bloodstream t hours after injection into muscle tissue is

$$C(t) = \frac{3t}{27 + t^3}, \quad t \geq 0.$$

(a) Complete the table and use it to approximate the time when the concentration is greatest.

t	0	0.5	1	1.5	2	2.5	3
$C(t)$							

(b) Use a graphing utility to graph the concentration function and use the graph to approximate the time when the concentration is greatest.

(c) Use calculus to determine analytically the time when the concentration is greatest.

78. Numerical, Graphical, and Analytic Analysis   Consider the functions $f(x) = x$ and $g(x) = \sin x$ on the interval $(0, \pi)$.

(a) Complete the table and make a conjecture about which is the greater function on the interval $(0, \pi)$.

x	0.5	1	1.5	2	2.5	3
$f(x)$						
$g(x)$						

(b) Use a graphing utility to graph the functions and use the graphs to make a conjecture about which is the greater function on the interval $(0, \pi)$.

(c) Prove that $f(x) > g(x)$ on the interval $(0, \pi)$. [*Hint*: Show that $h'(x) > 0$, where $h = f - g$.]

79. Trachea Contraction   Coughing forces the trachea (windpipe) to contract, which affects the velocity v of the air passing through the trachea. The velocity of the air during coughing is

$$v = k(R - r)r^2, \quad 0 \leq r < R$$

where k is a constant, R is the normal radius of the trachea, and r is the radius during coughing. What radius will produce the maximum air velocity?

80. Electrical Resistance   The resistance R of a certain type of resistor is

$$R = \sqrt{0.001T^4 - 4T + 100}$$

where R is measured in ohms and the temperature T is measured in degrees Celsius.

(a) Use a computer algebra system to find dR/dT and the critical number of the function. Determine the minimum resistance for this type of resistor.

(b) Use a graphing utility to graph the function R and use the graph to approximate the minimum resistance for this type of resistor.

Motion Along a Line In Exercises 81–84, the function $s(t)$ describes the motion of a particle along a line. For each function, (a) find the velocity function of the particle at any time $t \geq 0$, (b) identify the time interval(s) in which the particle is moving in a positive direction, (c) identify the time interval(s) in which the particle is moving in a negative direction, and (d) identify the time(s) at which the particle changes direction.

81. $s(t) = 6t - t^2$

82. $s(t) = t^2 - 7t + 10$

83. $s(t) = t^3 - 5t^2 + 4t$

84. $s(t) = t^3 - 20t^2 + 128t - 280$

Motion Along a Line In Exercises 85 and 86, the graph shows the position of a particle moving along a line. Describe how the particle's position changes with respect to time.

85.

86.

Creating Polynomial Functions In Exercises 87–90, find a polynomial function

$$f(x) = a_n x^n + a_{n-1} x^{n-1} + \cdots + a_2 x^2 + a_1 x + a_0$$

that has only the specified extrema. (a) Determine the minimum degree of the function and give the criteria you used in determining the degree. (b) Using the fact that the coordinates of the extrema are solution points of the function, and that the x-coordinates are critical numbers, determine a system of linear equations whose solution yields the coefficients of the required function. (c) Use a graphing utility to solve the system of equations and determine the function. (d) Use a graphing utility to confirm your result graphically.

87. Relative minimum: $(0, 0)$; Relative maximum: $(2, 2)$

88. Relative minimum: $(0, 0)$; Relative maximum: $(4, 1000)$

89. Relative minima: $(0, 0)$, $(4, 0)$; Relative maximum: $(2, 4)$

90. Relative minimum: $(1, 2)$; Relative maxima: $(-1, 4)$, $(3, 4)$

True or False? In Exercises 91–96, determine whether the statement is true or false. If it is false, explain why or give an example that shows it is false.

91. The sum of two increasing functions is increasing.

92. The product of two increasing functions is increasing.

93. Every nth-degree polynomial has $(n - 1)$ critical numbers.

94. An nth-degree polynomial has at most $(n - 1)$ critical numbers.

95. There is a relative maximum or minimum at each critical number.

96. The relative maxima of the function f are $f(1) = 4$ and $f(3) = 10$. Therefore, f has at least one minimum for some x in the interval $(1, 3)$.

97. Proof Prove the second case of Theorem 3.5.

98. Proof Prove the second case of Theorem 3.6.

99. Proof Use the definitions of increasing and decreasing functions to prove that $f(x) = x^3$ is increasing on $(-\infty, \infty)$.

100. Proof Use the definitions of increasing and decreasing functions to prove that

$$f(x) = \frac{1}{x}$$

is decreasing on $(0, \infty)$.

PUTNAM EXAM CHALLENGE

101. Find the minimum value of

$$\left| \sin x + \cos x + \tan x + \cot x + \sec x + \csc x \right|$$

for real numbers x.

This problem was composed by the Committee on the Putnam Prize Competition.
© The Mathematical Association of America. All rights reserved.

SECTION PROJECT

Rainbows

Rainbows are formed when light strikes raindrops and is reflected and refracted, as shown in the figure. (This figure shows a cross section of a spherical raindrop.) The Law of Refraction states that

$$\frac{\sin \alpha}{\sin \beta} = k$$

where $k \approx 1.33$ (for water). The angle of deflection is given by $D = \pi + 2\alpha - 4\beta$.

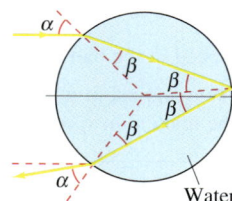

(a) Use a graphing utility to graph

$$D = \pi + 2\alpha - 4 \sin^{-1}\left(\frac{\sin \alpha}{k}\right), \quad 0 \leq \alpha \leq \frac{\pi}{2}.$$

(b) Prove that the minimum angle of deflection occurs when

$$\cos \alpha = \sqrt{\frac{k^2 - 1}{3}}.$$

For water, what is the minimum angle of deflection $D_{\min}$? (The angle $\pi - D_{\min}$ is called the *rainbow angle*.) What value of α produces this minimum angle? (A ray of sunlight that strikes a raindrop at this angle, α, is called a *rainbow ray*.)

FOR FURTHER INFORMATION For more information about the mathematics of rainbows, see the article "Somewhere Within the Rainbow" by Steven Janke in *The UMAP Journal*.

3.4 Concavity and the Second Derivative Test

■ Determine intervals on which a function is concave upward or concave downward.
■ Find any points of inflection of the graph of a function.
■ Apply the Second Derivative Test to find relative extrema of a function.

Concavity

You have already seen that locating the intervals in which a function f increases or decreases helps to describe its graph. In this section, you will see how locating the intervals in which f' increases or decreases can be used to determine where the graph of f is *curving upward* or *curving downward*.

Definition of Concavity

Let f be differentiable on an open interval I. The graph of f is **concave upward** on I when f' is increasing on the interval and **concave downward** on I when f' is decreasing on the interval.

The following graphical interpretation of concavity is useful. (See Appendix A for a proof of these results.) *See LarsonCalculus.com for Bruce Edwards's video of this proof.*

1. Let f be differentiable on an open interval I. If the graph of f is concave *upward* on I, then the graph of f lies *above* all of its tangent lines on I. [See Figure 3.23(a).]

2. Let f be differentiable on an open interval I. If the graph of f is concave *downward* on I, then the graph of f lies *below* all of its tangent lines on I. [See Figure 3.23(b).]

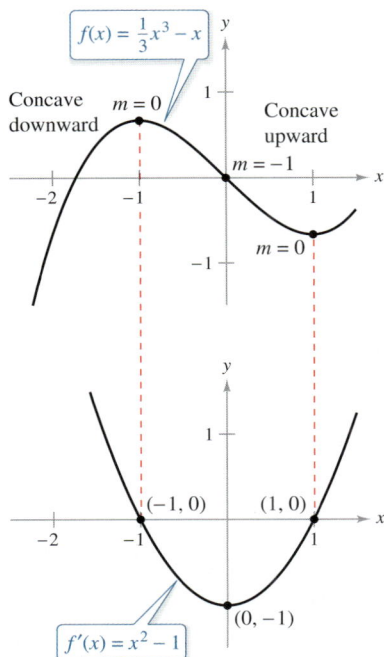

$f(x) = \frac{1}{3}x^3 - x$

Concave downward $m = 0$ Concave upward $m = -1$ $m = 0$

$f'(x) = x^2 - 1$

$(-1, 0)$ $(1, 0)$ $(0, -1)$

f' is decreasing. f' is increasing.

The concavity of f is related to the slope of the derivative.

Figure 3.24

Concave upward, f' is increasing.

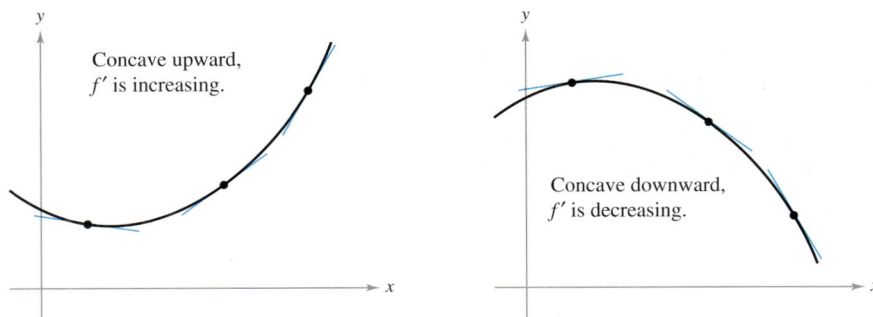

(a) The graph of f lies above its tangent lines.

Concave downward, f' is decreasing.

(b) The graph of f lies below its tangent lines.

Figure 3.23

To find the open intervals on which the graph of a function f is concave upward or concave downward, you need to find the intervals on which f' is increasing or decreasing. For instance, the graph of

$$f(x) = \frac{1}{3}x^3 - x$$

is concave downward on the open interval $(-\infty, 0)$ because

$$f'(x) = x^2 - 1$$

is decreasing there. (See Figure 3.24.) Similarly, the graph of f is concave upward on the interval $(0, \infty)$ because f' is increasing on $(0, \infty)$.

The next theorem shows how to use the *second* derivative of a function f to determine intervals on which the graph of f is concave upward or concave downward. A proof of this theorem follows directly from Theorem 3.5 and the definition of concavity.

● ● ● ● ● ● ● ● ● ● ● ● ● ● ● ● ▷

● ● **REMARK** A third case of Theorem 3.7 could be that if $f''(x) = 0$ for all x in I, then f is linear. Note, however, that concavity is not defined for a line. In other words, a straight line is neither concave upward nor concave downward.

> **THEOREM 3.7 Test for Concavity**
>
> Let f be a function whose second derivative exists on an open interval I.
>
> 1. If $f''(x) > 0$ for all x in I, then the graph of f is concave upward on I.
> 2. If $f''(x) < 0$ for all x in I, then the graph of f is concave downward on I.
>
> A proof of this theorem is given in Appendix A.
> *See LarsonCalculus.com for Bruce Edwards's video of this proof.*

To apply Theorem 3.7, locate the x-values at which $f''(x) = 0$ or f'' does not exist. Use these x-values to determine test intervals. Finally, test the sign of $f''(x)$ in each of the test intervals.

EXAMPLE 1 **Determining Concavity**

Determine the open intervals on which the graph of

$$f(x) = \frac{6}{x^2 + 3}$$

is concave upward or downward.

Solution Begin by observing that f is continuous on the entire real number line. Next, find the second derivative of f.

$$f(x) = 6(x^2 + 3)^{-1} \qquad \text{Rewrite original function.}$$
$$f'(x) = (-6)(x^2 + 3)^{-2}(2x) \qquad \text{Differentiate.}$$
$$= \frac{-12x}{(x^2 + 3)^2} \qquad \text{First derivative}$$
$$f''(x) = \frac{(x^2 + 3)^2(-12) - (-12x)(2)(x^2 + 3)(2x)}{(x^2 + 3)^4} \qquad \text{Differentiate.}$$
$$= \frac{36(x^2 - 1)}{(x^2 + 3)^3} \qquad \text{Second derivative}$$

Because $f''(x) = 0$ when $x = \pm 1$ and f'' is defined on the entire real number line, you should test f'' in the intervals $(-\infty, -1)$, $(-1, 1)$, and $(1, \infty)$. The results are shown in the table and in Figure 3.25.

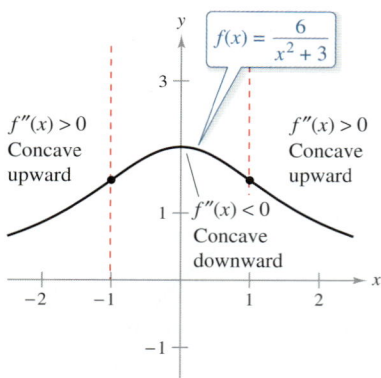

$f(x) = \dfrac{6}{x^2 + 3}$

$f''(x) > 0$
Concave upward

$f''(x) > 0$
Concave upward

$f''(x) < 0$
Concave downward

From the sign of f'', you can determine the concavity of the graph of f.
Figure 3.25

Interval	$-\infty < x < -1$	$-1 < x < 1$	$1 < x < \infty$
Test Value	$x = -2$	$x = 0$	$x = 2$
Sign of $f''(x)$	$f''(-2) > 0$	$f''(0) < 0$	$f''(2) > 0$
Conclusion	Concave upward	Concave downward	Concave upward

The function given in Example 1 is continuous on the entire real number line. When there are x-values at which the function is not continuous, these values should be used, along with the points at which $f''(x) = 0$ or $f''(x)$ does not exist, to form the test intervals.

EXAMPLE 2   **Determining Concavity**

Determine the open intervals on which the graph of

$$f(x) = \frac{x^2 + 1}{x^2 - 4}$$

is concave upward or concave downward.

Solution   Differentiating twice produces the following.

$$f(x) = \frac{x^2 + 1}{x^2 - 4} \qquad \text{Write original function.}$$

$$f'(x) = \frac{(x^2 - 4)(2x) - (x^2 + 1)(2x)}{(x^2 - 4)^2} \qquad \text{Differentiate.}$$

$$= \frac{-10x}{(x^2 - 4)^2} \qquad \text{First derivative}$$

$$f''(x) = \frac{(x^2 - 4)^2(-10) - (-10x)(2)(x^2 - 4)(2x)}{(x^2 - 4)^4} \qquad \text{Differentiate.}$$

$$= \frac{10(3x^2 + 4)}{(x^2 - 4)^3} \qquad \text{Second derivative}$$

There are no points at which $f''(x) = 0$, but at $x = \pm 2$, the function f is not continuous. So, test for concavity in the intervals $(-\infty, -2)$, $(-2, 2)$, and $(2, \infty)$, as shown in the table. The graph of f is shown in Figure 3.26.

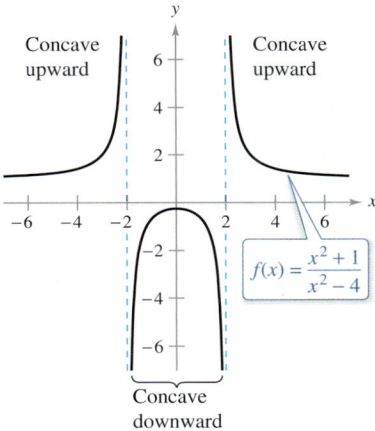

Interval	$-\infty < x < -2$	$-2 < x < 2$	$2 < x < \infty$
Test Value	$x = -3$	$x = 0$	$x = 3$
Sign of $f''(x)$	$f''(-3) > 0$	$f''(0) < 0$	$f''(3) > 0$
Conclusion	Concave upward	Concave downward	Concave upward

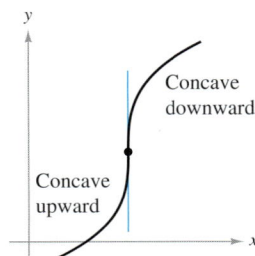

Figure 3.26

Points of Inflection

The graph in Figure 3.25 has two points at which the concavity changes. If the tangent line to the graph exists at such a point, then that point is a **point of inflection.** Three types of points of inflection are shown in Figure 3.27.

Definition of Point of Inflection

Let f be a function that is continuous on an open interval, and let c be a point in the interval. If the graph of f has a tangent line at this point $(c, f(c))$, then this point is a **point of inflection** of the graph of f when the concavity of f changes from upward to downward (or downward to upward) at the point.

REMARK   The definition of *point of inflection* requires that the tangent line exists at the point of inflection. Some books do not require this. For instance, we do not consider the function

$$f(x) = \begin{cases} x^3, & x < 0 \\ x^2 + 2x, & x \geq 0 \end{cases}$$

to have a point of inflection at the origin, even though the concavity of the graph changes from concave downward to concave upward.

The concavity of f changes at a point of inflection. Note that the graph crosses its tangent line at a point of inflection.

Figure 3.27

To locate *possible* points of inflection, you can determine the values of x for which $f''(x) = 0$ or $f''(x)$ does not exist. This is similar to the procedure for locating relative extrema of f.

THEOREM 3.8   Points of Inflection

If $(c, f(c))$ is a point of inflection of the graph of f, then either $f''(c) = 0$ or f'' does not exist at $x = c$.

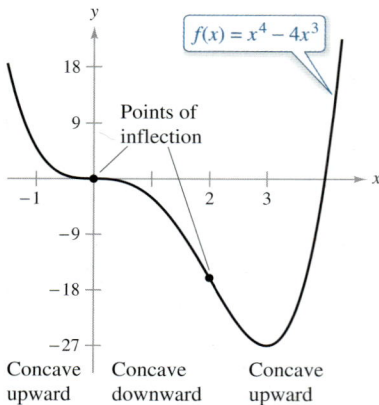

$f(x) = x^4 - 4x^3$

Points of inflection

Concave upward   Concave downward   Concave upward

Points of inflection can occur where $f''(x) = 0$ or f'' does not exist.

Figure 3.28

EXAMPLE 3   Finding Points of Inflection

Determine the points of inflection and discuss the concavity of the graph of

$$f(x) = x^4 - 4x^3.$$

Solution   Differentiating twice produces the following.

$f(x) = x^4 - 4x^3$	Write original function.
$f'(x) = 4x^3 - 12x^2$	Find first derivative.
$f''(x) = 12x^2 - 24x = 12x(x - 2)$	Find second derivative.

Setting $f''(x) = 0$, you can determine that the possible points of inflection occur at $x = 0$ and $x = 2$. By testing the intervals determined by these x-values, you can conclude that they both yield points of inflection. A summary of this testing is shown in the table, and the graph of f is shown in Figure 3.28.

Interval	$-\infty < x < 0$	$0 < x < 2$	$2 < x < \infty$
Test Value	$x = -1$	$x = 1$	$x = 3$
Sign of $f''(x)$	$f''(-1) > 0$	$f''(1) < 0$	$f''(3) > 0$
Conclusion	Concave upward	Concave downward	Concave upward

The converse of Theorem 3.8 is not generally true. That is, it is possible for the second derivative to be 0 at a point that is *not* a point of inflection. For instance, the graph of $f(x) = x^4$ is shown in Figure 3.29. The second derivative is 0 when $x = 0$, but the point $(0, 0)$ is not a point of inflection because the graph of f is concave upward in both intervals $-\infty < x < 0$ and $0 < x < \infty$.

Exploration

Consider a general cubic function of the form

$$f(x) = ax^3 + bx^2 + cx + d.$$

You know that the value of d has a bearing on the location of the graph but has no bearing on the value of the first derivative at given values of x. Graphically, this is true because changes in the value of d shift the graph up or down but do not change its basic shape. Use a graphing utility to graph several cubics with different values of c. Then give a graphical explanation of why changes in c do not affect the values of the second derivative.

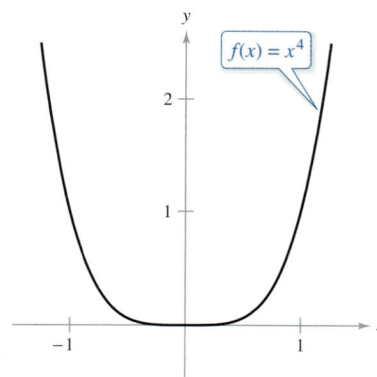

$f(x) = x^4$

$f''(x) = 0$, but $(0, 0)$ is not a point of inflection.
Figure 3.29

The Second Derivative Test

In addition to testing for concavity, the second derivative can be used to perform a simple test for relative maxima and minima. The test is based on the fact that if the graph of a function f is concave upward on an open interval containing c, and $f'(c) = 0$, then $f(c)$ must be a relative minimum of f. Similarly, if the graph of a function f is concave downward on an open interval containing c, and $f'(c) = 0$, then $f(c)$ must be a relative maximum of f (see Figure 3.30).

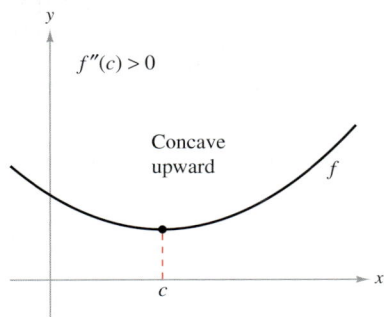

If $f'(c) = 0$ and $f''(c) > 0$, then $f(c)$ is a relative minimum.

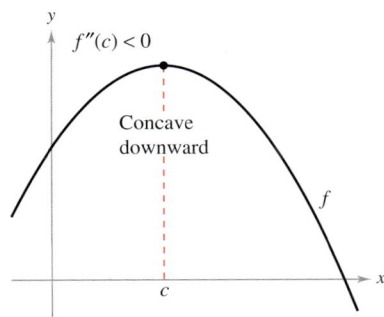

If $f'(c) = 0$ and $f''(c) < 0$, then $f(c)$ is a relative maximum.
Figure 3.30

THEOREM 3.9   Second Derivative Test

Let f be a function such that $f'(c) = 0$ and the second derivative of f exists on an open interval containing c.

1. If $f''(c) > 0$, then f has a relative minimum at $(c, f(c))$.
2. If $f''(c) < 0$, then f has a relative maximum at $(c, f(c))$.

If $f''(c) = 0$, then the test fails. That is, f may have a relative maximum, a relative minimum, or neither. In such cases, you can use the First Derivative Test.

Proof If $f'(c) = 0$ and $f''(c) > 0$, then there exists an open interval I containing c for which

$$\frac{f'(x) - f'(c)}{x - c} = \frac{f'(x)}{x - c} > 0$$

for all $x \neq c$ in I. If $x < c$, then $x - c < 0$ and $f'(x) < 0$. Also, if $x > c$, then $x - c > 0$ and $f'(x) > 0$. So, $f'(x)$ changes from negative to positive at c, and the First Derivative Test implies that $f(c)$ is a relative minimum. A proof of the second case is left to you. *See LarsonCalculus.com for Bruce Edwards's video of this proof.*

EXAMPLE 4   Using the Second Derivative Test

• • • • ▷ *See LarsonCalculus.com for an interactive version of this type of example.*

Find the relative extrema of

$$f(x) = -3x^5 + 5x^3.$$

Solution Begin by finding the first derivative of f.

$$f'(x) = -15x^4 + 15x^2 = 15x^2(1 - x^2)$$

From this derivative, you can see that $x = -1, 0$, and 1 are the only critical numbers of f. By finding the second derivative

$$f''(x) = -60x^3 + 30x = 30x(1 - 2x^2)$$

you can apply the Second Derivative Test as shown below.

$(0, 0)$ is neither a relative minimum nor a relative maximum.
Figure 3.31

Point	$(-1, -2)$	$(0, 0)$	$(1, 2)$
Sign of $f''(x)$	$f''(-1) > 0$	$f''(0) = 0$	$f''(1) < 0$
Conclusion	Relative minimum	Test fails	Relative maximum

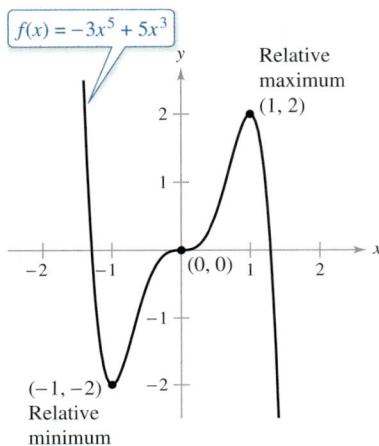

Because the Second Derivative Test fails at $(0, 0)$, you can use the First Derivative Test and observe that f increases to the left and right of $x = 0$. So, $(0, 0)$ is neither a relative minimum nor a relative maximum (even though the graph has a horizontal tangent line at this point). The graph of f is shown in Figure 3.31.

3.4 Exercises

See **CalcChat.com** for tutorial help and worked-out solutions to odd-numbered exercises.

Using a Graph In Exercises 1 and 2, the graph of f is shown. State the signs of f' and f'' on the interval $(0, 2)$.

1.

2.

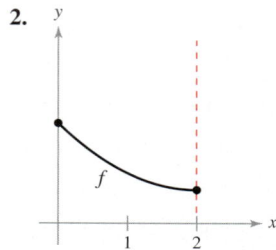

Determining Concavity In Exercises 3–14, determine the open intervals on which the graph is concave upward or concave downward.

3. $y = x^2 - x - 2$

4. $g(x) = 3x^2 - x^3$

5. $f(x) = -x^3 + 6x^2 - 9x - 1$

6. $h(x) = x^5 - 5x + 2$

7. $f(x) = \dfrac{24}{x^2 + 12}$

8. $f(x) = \dfrac{2x^2}{3x^2 + 1}$

9. $f(x) = \dfrac{x^2 + 1}{x^2 - 1}$

10. $y = \dfrac{-3x^5 + 40x^3 + 135x}{270}$

11. $g(x) = \dfrac{x^2 + 4}{4 - x^2}$

12. $h(x) = \dfrac{x^2 - 1}{2x - 1}$

13. $y = 2x - \tan x, \quad \left(-\dfrac{\pi}{2}, \dfrac{\pi}{2}\right)$

14. $y = x + \dfrac{2}{\sin x}, \quad (-\pi, \pi)$

Finding Points of Inflection In Exercises 15–30, find the points of inflection and discuss the concavity of the graph of the function.

15. $f(x) = x^3 - 6x^2 + 12x$

16. $f(x) = -x^3 + 6x^2 - 5$

17. $f(x) = \frac{1}{2}x^4 + 2x^3$

18. $f(x) = 4 - x - 3x^4$

19. $f(x) = x(x - 4)^3$

20. $f(x) = (x - 2)^3(x - 1)$

21. $f(x) = x\sqrt{x + 3}$

22. $f(x) = x\sqrt{9 - x}$

23. $f(x) = \dfrac{4}{x^2 + 1}$

24. $f(x) = \dfrac{x + 3}{\sqrt{x}}$

25. $f(x) = \sin \dfrac{x}{2}, \quad [0, 4\pi]$

26. $f(x) = 2 \csc \dfrac{3x}{2}, \quad (0, 2\pi)$

27. $f(x) = \sec\left(x - \dfrac{\pi}{2}\right), \quad (0, 4\pi)$

28. $f(x) = \sin x + \cos x, \quad [0, 2\pi]$

29. $f(x) = 2 \sin x + \sin 2x, \quad [0, 2\pi]$

30. $f(x) = x + 2 \cos x, \quad [0, 2\pi]$

Using the Second Derivative Test In Exercises 31–42, find all relative extrema. Use the Second Derivative Test where applicable.

31. $f(x) = 6x - x^2$

32. $f(x) = x^2 + 3x - 8$

33. $f(x) = x^3 - 3x^2 + 3$

34. $f(x) = -x^3 + 7x^2 - 15x$

35. $f(x) = x^4 - 4x^3 + 2$

36. $f(x) = -x^4 + 4x^3 + 8x^2$

37. $f(x) = x^{2/3} - 3$

38. $f(x) = \sqrt{x^2 + 1}$

39. $f(x) = x + \dfrac{4}{x}$

40. $f(x) = \dfrac{x}{x - 1}$

41. $f(x) = \cos x - x, \quad [0, 4\pi]$

42. $f(x) = 2 \sin x + \cos 2x, \quad [0, 2\pi]$

Finding Extrema and Points of Inflection Using Technology In Exercises 43–46, use a computer algebra system to analyze the function over the given interval. (a) Find the first and second derivatives of the function. (b) Find any relative extrema and points of inflection. (c) Graph f, f', and f'' on the same set of coordinate axes and state the relationship between the behavior of f and the signs of f' and f''.

43. $f(x) = 0.2x^2(x - 3)^3, \quad [-1, 4]$

44. $f(x) = x^2\sqrt{6 - x^2}, \quad [-\sqrt{6}, \sqrt{6}]$

45. $f(x) = \sin x - \frac{1}{3} \sin 3x + \frac{1}{5} \sin 5x, \quad [0, \pi]$

46. $f(x) = \sqrt{2x} \sin x, \quad [0, 2\pi]$

WRITING ABOUT CONCEPTS

47. Sketching a Graph Consider a function f such that f' is increasing. Sketch graphs of f for (a) $f' < 0$ and (b) $f' > 0$.

48. Sketching a Graph Consider a function f such that f' is decreasing. Sketch graphs of f for (a) $f' < 0$ and (b) $f' > 0$.

49. Sketching a Graph Sketch the graph of a function f that does *not* have a point of inflection at $(c, f(c))$ even though $f''(c) = 0$.

50. Think About It S represents weekly sales of a product. What can be said of S' and S'' for each of the following statements?

(a) The rate of change of sales is increasing.

(b) Sales are increasing at a slower rate.

(c) The rate of change of sales is constant.

(d) Sales are steady.

(e) Sales are declining, but at a slower rate.

(f) Sales have bottomed out and have started to rise.

Sketching Graphs In Exercises 51 and 52, the graph of f is shown. Graph f, f', and f'' on the same set of coordinate axes. To print an enlarged copy of the graph, go to *MathGraphs.com*.

51.

52.

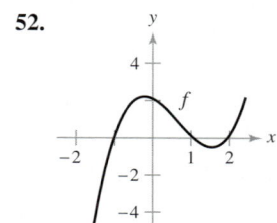

Think About It In Exercises 53–56, sketch the graph of a function f having the given characteristics.

53. $f(2) = f(4) = 0$
$f'(x) < 0$ for $x < 3$
$f'(3)$ does not exist.
$f'(x) > 0$ for $x > 3$
$f''(x) < 0, x \neq 3$

54. $f(0) = f(2) = 0$
$f'(x) > 0$ for $x < 1$
$f'(1) = 0$
$f'(x) < 0$ for $x > 1$
$f''(x) < 0$

55. $f(2) = f(4) = 0$
$f'(x) > 0$ for $x < 3$
$f'(3)$ does not exist.
$f'(x) < 0$ for $x > 3$
$f''(x) > 0, x \neq 3$

56. $f(0) = f(2) = 0$
$f'(x) < 0$ for $x < 1$
$f'(1) = 0$
$f'(x) > 0$ for $x > 1$
$f''(x) > 0$

57. Think About It The figure shows the graph of f''. Sketch a graph of f. (The answer is not unique.) To print an enlarged copy of the graph, go to *MathGraphs.com*.

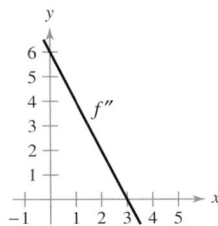

58. HOW DO YOU SEE IT? Water is running into the vase shown in the figure at a constant rate.

(a) Graph the depth d of water in the vase as a function of time.

(b) Does the function have any extrema? Explain.

(c) Interpret the inflection points of the graph of d.

59. Conjecture Consider the function
$$f(x) = (x - 2)^n.$$

(a) Use a graphing utility to graph f for $n = 1, 2, 3,$ and 4. Use the graphs to make a conjecture about the relationship between n and any inflection points of the graph of f.

(b) Verify your conjecture in part (a).

60. Inflection Point Consider the function $f(x) = \sqrt[3]{x}$.

(a) Graph the function and identify the inflection point.

(b) Does $f''(x)$ exist at the inflection point? Explain.

Finding a Cubic Function In Exercises 61 and 62, find a, b, c, and d such that the cubic
$$f(x) = ax^3 + bx^2 + cx + d$$
satisfies the given conditions.

61. Relative maximum: $(3, 3)$
Relative minimum: $(5, 1)$
Inflection point: $(4, 2)$

62. Relative maximum: $(2, 4)$
Relative minimum: $(4, 2)$
Inflection point: $(3, 3)$

63. Aircraft Glide Path A small aircraft starts its descent from an altitude of 1 mile, 4 miles west of the runway (see figure).

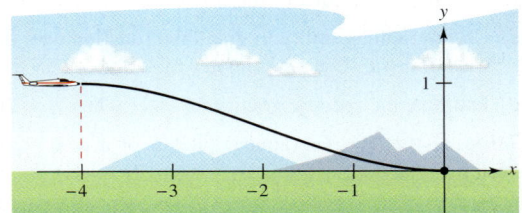

(a) Find the cubic $f(x) = ax^3 + bx^2 + cx + d$ on the interval $[-4, 0]$ that describes a smooth glide path for the landing.

(b) The function in part (a) models the glide path of the plane. When would the plane be descending at the greatest rate?

■ **FOR FURTHER INFORMATION** For more information on this type of modeling, see the article "How Not to Land at Lake Tahoe!" by Richard Barshinger in *The American Mathematical Monthly*. To view this article, go to *MathArticles.com*.

64. Highway Design A section of highway connecting two hillsides with grades of 6% and 4% is to be built between two points that are separated by a horizontal distance of 2000 feet (see figure). At the point where the two hillsides come together, there is a 50-foot difference in elevation.

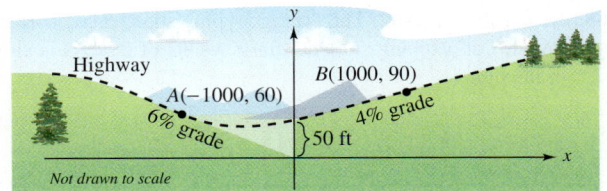

(a) Design a section of highway connecting the hillsides modeled by the function
$$f(x) = ax^3 + bx^2 + cx + d, \quad -1000 \leq x \leq 1000.$$
At points A and B, the slope of the model must match the grade of the hillside.

(b) Use a graphing utility to graph the model.

(c) Use a graphing utility to graph the derivative of the model.

(d) Determine the grade at the steepest part of the transitional section of the highway.

65. Average Cost A manufacturer has determined that the total cost C of operating a factory is

$$C = 0.5x^2 + 15x + 5000$$

where x is the number of units produced. At what level of production will the average cost per unit be minimized? (The average cost per unit is C/x.)

66. Specific Gravity A model for the specific gravity of water S is

$$S = \frac{5.755}{10^8}T^3 - \frac{8.521}{10^6}T^2 + \frac{6.540}{10^5}T + 0.99987, \; 0 < T < 25$$

where T is the water temperature in degrees Celsius.

(a) Use the second derivative to determine the concavity of S.

(b) Use a computer algebra system to find the coordinates of the maximum value of the function.

(c) Use a graphing utility to graph the function over the specified domain. (Use a setting in which $0.996 \le S \le 1.001$.)

(d) Estimate the specific gravity of water when $T = 20°$.

67. Sales Growth The annual sales S of a new product are given by

$$S = \frac{5000t^2}{8 + t^2}, \quad 0 \le t \le 3$$

where t is time in years.

(a) Complete the table. Then use it to estimate when the annual sales are increasing at the greatest rate.

t	0.5	1	1.5	2	2.5	3
S						

(b) Use a graphing utility to graph the function S. Then use the graph to estimate when the annual sales are increasing at the greatest rate.

(c) Find the exact time when the annual sales are increasing at the greatest rate.

68. Modeling Data The average typing speed S (in words per minute) of a typing student after t weeks of lessons is shown in the table.

t	5	10	15	20	25	30
S	38	56	79	90	93	94

A model for the data is

$$S = \frac{100t^2}{65 + t^2}, \quad t > 0.$$

(a) Use a graphing utility to plot the data and graph the model.

(b) Use the second derivative to determine the concavity of S. Compare the result with the graph in part (a).

(c) What is the sign of the first derivative for $t > 0$? By combining this information with the concavity of the model, what inferences can be made about the typing speed as t increases?

Linear and Quadratic Approximations In Exercises 69–72, use a graphing utility to graph the function. Then graph the linear and quadratic approximations

$$P_1(x) = f(a) + f'(a)(x - a)$$

and

$$P_2(x) = f(a) + f'(a)(x - a) + \tfrac{1}{2}f''(a)(x - a)^2$$

in the same viewing window. Compare the values of f, P_1, and P_2 and their first derivatives at $x = a$. How do the approximations change as you move farther away from $x = a$?

Function		Value of a
69. $f(x) = 2(\sin x + \cos x)$		$a = \dfrac{\pi}{4}$
70. $f(x) = 2(\sin x + \cos x)$		$a = 0$
71. $f(x) = \sqrt{1 - x}$		$a = 0$
72. $f(x) = \dfrac{\sqrt{x}}{x - 1}$		$a = 2$

73. Determining Concavity Use a graphing utility to graph

$$y = x \sin \frac{1}{x}.$$

Show that the graph is concave downward to the right of

$$x = \frac{1}{\pi}.$$

74. Point of Inflection and Extrema Show that the point of inflection of

$$f(x) = x(x - 6)^2$$

lies midway between the relative extrema of f.

True or False? In Exercises 75–78, determine whether the statement is true or false. If it is false, explain why or give an example that shows it is false.

75. The graph of every cubic polynomial has precisely one point of inflection.

76. The graph of

$$f(x) = \frac{1}{x}$$

is concave downward for $x < 0$ and concave upward for $x > 0$, and thus it has a point of inflection at $x = 0$.

77. If $f'(c) > 0$, then f is concave upward at $x = c$.

78. If $f''(2) = 0$, then the graph of f must have a point of inflection at $x = 2$.

Proof In Exercises 79 and 80, let f and g represent differentiable functions such that $f'' \ne 0$ and $g'' \ne 0$.

79. Show that if f and g are concave upward on the interval (a, b), then $f + g$ is also concave upward on (a, b).

80. Prove that if f and g are positive, increasing, and concave upward on the interval (a, b), then fg is also concave upward on (a, b).

3.5 Limits at Infinity

- Determine (finite) limits at infinity.
- Determine the horizontal asymptotes, if any, of the graph of a function.
- Determine infinite limits at infinity.

Limits at Infinity

This section discusses the "end behavior" of a function on an *infinite* interval. Consider the graph of

$$f(x) = \frac{3x^2}{x^2 + 1}$$

$f(x) = \dfrac{3x^2}{x^2 + 1}$

$f(x) \to 3$ as $x \to -\infty$

$f(x) \to 3$ as $x \to \infty$

The limit of $f(x)$ as x approaches $-\infty$ or ∞ is 3.

Figure 3.32

as shown in Figure 3.32. Graphically, you can see that the values of $f(x)$ appear to approach 3 as x increases without bound or decreases without bound. You can come to the same conclusions numerically, as shown in the table.

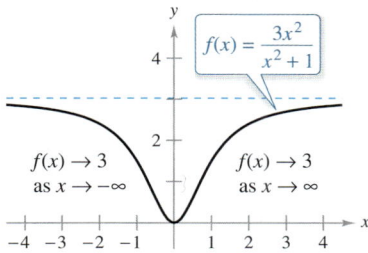

	x decreases without bound.				x increases without bound.		
x	$-\infty \leftarrow$	-100	-10	-1	0	1	10	100	$\to \infty$
$f(x)$	$3 \leftarrow$	2.9997	2.9703	1.5	0	1.5	2.9703	2.9997	$\to 3$

$f(x)$ approaches 3. *$f(x)$ approaches 3.*

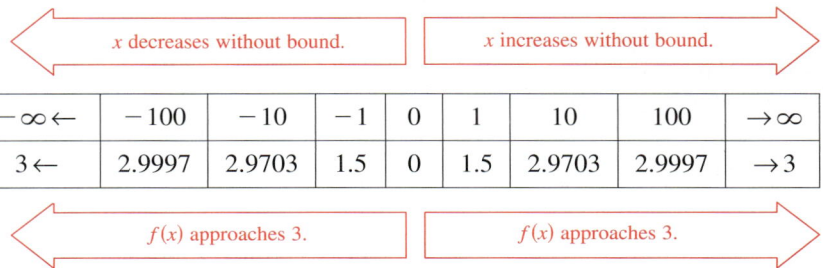

The table suggests that the value of $f(x)$ approaches 3 as x increases without bound ($x \to \infty$). Similarly, $f(x)$ approaches 3 as x decreases without bound ($x \to -\infty$). These **limits at infinity** are denoted by

$$\lim_{x \to -\infty} f(x) = 3 \qquad \text{Limit at negative infinity}$$

and

$$\lim_{x \to \infty} f(x) = 3. \qquad \text{Limit at positive infinity}$$

To say that a statement is true as x increases *without bound* means that for some (large) real number M, the statement is true for *all* x in the interval $\{x: x > M\}$. The next definition uses this concept.

• • REMARK The statement $\lim\limits_{x \to -\infty} f(x) = L$ or $\lim\limits_{x \to \infty} f(x) = L$ means that the limit exists *and* the limit is equal to L.

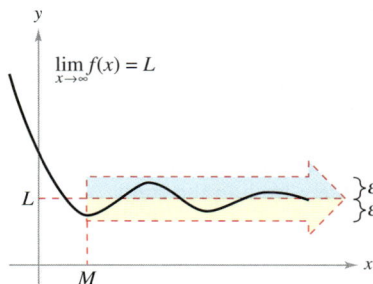

Definition of Limits at Infinity

Let L be a real number.

1. The statement $\lim\limits_{x \to \infty} f(x) = L$ means that for each $\varepsilon > 0$ there exists an $M > 0$ such that $|f(x) - L| < \varepsilon$ whenever $x > M$.
2. The statement $\lim\limits_{x \to -\infty} f(x) = L$ means that for each $\varepsilon > 0$ there exists an $N < 0$ such that $|f(x) - L| < \varepsilon$ whenever $x < N$.

$\lim\limits_{x \to \infty} f(x) = L$

L

M

$f(x)$ is within ε units of L as $x \to \infty$.

Figure 3.33

The definition of a limit at infinity is shown in Figure 3.33. In this figure, note that for a given positive number ε, there exists a positive number M such that, for $x > M$, the graph of f will lie between the horizontal lines

$$y = L + \varepsilon \quad \text{and} \quad y = L - \varepsilon.$$

Exploration

Use a graphing utility to graph

$$f(x) = \frac{2x^2 + 4x - 6}{3x^2 + 2x - 16}.$$

Describe all the important features of the graph. Can you find a single viewing window that shows all of these features clearly? Explain your reasoning.

What are the horizontal asymptotes of the graph? How far to the right do you have to move on the graph so that the graph is within 0.001 unit of its horizontal asymptote? Explain your reasoning.

Horizontal Asymptotes

In Figure 3.33, the graph of f approaches the line $y = L$ as x increases without bound. The line $y = L$ is called a **horizontal asymptote** of the graph of f.

Definition of a Horizontal Asymptote

The line $y = L$ is a **horizontal asymptote** of the graph of f when

$$\lim_{x \to -\infty} f(x) = L \quad \text{or} \quad \lim_{x \to \infty} f(x) = L.$$

Note that from this definition, it follows that the graph of a *function* of x can have at most two horizontal asymptotes—one to the right and one to the left.

Limits at infinity have many of the same properties of limits discussed in Section 1.3. For example, if $\lim\limits_{x \to \infty} f(x)$ and $\lim\limits_{x \to \infty} g(x)$ both exist, then

$$\lim_{x \to \infty} [f(x) + g(x)] = \lim_{x \to \infty} f(x) + \lim_{x \to \infty} g(x)$$

and

$$\lim_{x \to \infty} [f(x)g(x)] = \left[\lim_{x \to \infty} f(x)\right]\left[\lim_{x \to \infty} g(x)\right].$$

Similar properties hold for limits at $-\infty$.

When evaluating limits at infinity, the next theorem is helpful.

THEOREM 3.10   Limits at Infinity

If r is a positive rational number and c is any real number, then

$$\lim_{x \to \infty} \frac{c}{x^r} = 0.$$

Furthermore, if x^r is defined when $x < 0$, then

$$\lim_{x \to -\infty} \frac{c}{x^r} = 0.$$

A proof of this theorem is given in Appendix A.
See LarsonCalculus.com for Bruce Edwards's video of this proof.

EXAMPLE 1   **Finding a Limit at Infinity**

Find the limit: $\lim\limits_{x \to \infty} \left(5 - \dfrac{2}{x^2}\right)$.

Solution   Using Theorem 3.10, you can write

$$\lim_{x \to \infty} \left(5 - \frac{2}{x^2}\right) = \lim_{x \to \infty} 5 - \lim_{x \to \infty} \frac{2}{x^2} \qquad \text{Property of limits}$$

$$= 5 - 0$$

$$= 5.$$

So, the line $y = 5$ is a horizontal asymptote to the right. By finding the limit

$$\lim_{x \to -\infty} \left(5 - \frac{2}{x^2}\right) \qquad \text{Limit as } x \to -\infty.$$

you can see that $y = 5$ is also a horizontal asymptote to the left. The graph of the function is shown in Figure 3.34.

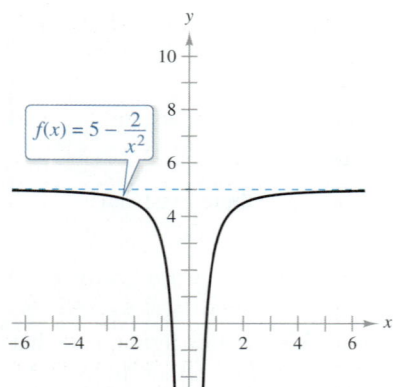

$f(x) = 5 - \dfrac{2}{x^2}$

$y = 5$ is a horizontal asymptote.
Figure 3.34

EXAMPLE 2   **Finding a Limit at Infinity**

Find the limit: $\displaystyle\lim_{x\to\infty}\frac{2x-1}{x+1}$.

Solution   Note that both the numerator and the denominator approach infinity as x approaches infinity.

$$\lim_{x\to\infty}\frac{2x-1}{x+1}$$

$$\lim_{x\to\infty}(2x-1)\to\infty$$

$$\lim_{x\to\infty}(x+1)\to\infty$$

REMARK   When you encounter an indeterminate form such as the one in Example 2, you should divide the numerator and denominator by the highest power of x in the *denominator*.

This results in $\dfrac{\infty}{\infty}$, an **indeterminate form.** To resolve this problem, you can divide both the numerator and the denominator by x. After dividing, the limit may be evaluated as shown.

$$\lim_{x\to\infty}\frac{2x-1}{x+1}=\lim_{x\to\infty}\frac{\dfrac{2x-1}{x}}{\dfrac{x+1}{x}} \qquad \text{Divide numerator and denominator by } x.$$

$$=\lim_{x\to\infty}\frac{2-\dfrac{1}{x}}{1+\dfrac{1}{x}} \qquad \text{Simplify.}$$

$$=\frac{\displaystyle\lim_{x\to\infty}2-\lim_{x\to\infty}\dfrac{1}{x}}{\displaystyle\lim_{x\to\infty}1+\lim_{x\to\infty}\dfrac{1}{x}} \qquad \text{Take limits of numerator and denominator.}$$

$$=\frac{2-0}{1+0} \qquad \text{Apply Theorem 3.10.}$$

$$=2$$

So, the line $y=2$ is a horizontal asymptote to the right. By taking the limit as $x\to-\infty$, you can see that $y=2$ is also a horizontal asymptote to the left. The graph of the function is shown in Figure 3.35.

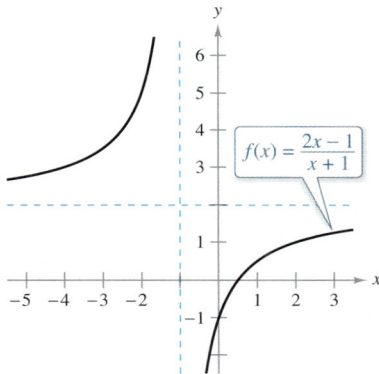

$y=2$ is a horizontal asymptote.

Figure 3.35

▷ TECHNOLOGY   You can test the reasonableness of the limit found in Example 2 by evaluating $f(x)$ for a few large positive values of x. For instance,

$$f(100)\approx1.9703, \quad f(1000)\approx1.9970,$$
$$\text{and}\quad f(10{,}000)\approx1.9997.$$

Another way to test the reasonableness of the limit is to use a graphing utility. For instance, in Figure 3.36, the graph of

$$f(x)=\frac{2x-1}{x+1}$$

is shown with the horizontal line $y=2$. Note that as x increases, the graph of f moves closer and closer to its horizontal asymptote.

As x increases, the graph of f moves closer and closer to the line $y=2$.

Figure 3.36

| EXAMPLE 3 | **A Comparison of Three Rational Functions** |

⋅ ⋅ ⋅▷ See LarsonCalculus.com for an interactive version of this type of example.

Find each limit.

a. $\displaystyle\lim_{x\to\infty}\frac{2x+5}{3x^2+1}$ **b.** $\displaystyle\lim_{x\to\infty}\frac{2x^2+5}{3x^2+1}$ **c.** $\displaystyle\lim_{x\to\infty}\frac{2x^3+5}{3x^2+1}$

Solution   In each case, attempting to evaluate the limit produces the indeterminate form ∞/∞.

a. Divide both the numerator and the denominator by x^2.

$$\lim_{x\to\infty}\frac{2x+5}{3x^2+1}=\lim_{x\to\infty}\frac{(2/x)+(5/x^2)}{3+(1/x^2)}=\frac{0+0}{3+0}=\frac{0}{3}=0$$

b. Divide both the numerator and the denominator by x^2.

$$\lim_{x\to\infty}\frac{2x^2+5}{3x^2+1}=\lim_{x\to\infty}\frac{2+(5/x^2)}{3+(1/x^2)}=\frac{2+0}{3+0}=\frac{2}{3}$$

c. Divide both the numerator and the denominator by x^2.

$$\lim_{x\to\infty}\frac{2x^3+5}{3x^2+1}=\lim_{x\to\infty}\frac{2x+(5/x^2)}{3+(1/x^2)}=\frac{\infty}{3}$$

You can conclude that the limit *does not exist* because the numerator increases without bound while the denominator approaches 3. ■

Example 3 suggests the guidelines below for finding limits at infinity of rational functions. Use these guidelines to check the results in Example 3.

GUIDELINES FOR FINDING LIMITS AT $\pm\infty$ OF RATIONAL FUNCTIONS

1. If the degree of the numerator is *less than* the degree of the denominator, then the limit of the rational function is 0.

2. If the degree of the numerator is *equal to* the degree of the denominator, then the limit of the rational function is the ratio of the leading coefficients.

3. If the degree of the numerator is *greater than* the degree of the denominator, then the limit of the rational function does not exist.

The guidelines for finding limits at infinity of rational functions seem reasonable when you consider that for large values of x, the highest-power term of the rational function is the most "influential" in determining the limit. For instance,

$$\lim_{x\to\infty}\frac{1}{x^2+1}$$

is 0 because the denominator overpowers the numerator as x increases or decreases without bound, as shown in Figure 3.37.

The function shown in Figure 3.37 is a special case of a type of curve studied by the Italian mathematician Maria Gaetana Agnesi. The general form of this function is

$$f(x)=\frac{8a^3}{x^2+4a^2} \qquad \text{Witch of Agnesi}$$

and, through a mistranslation of the Italian word *vertéré*, the curve has come to be known as the Witch of Agnesi. Agnesi's work with this curve first appeared in a comprehensive text on calculus that was published in 1748.

**MARIA GAETANA AGNESI
(1718–1799)**

Agnesi was one of a handful of women to receive credit for significant contributions to mathematics before the twentieth century. In her early twenties, she wrote the first text that included both differential and integral calculus. By age 30, she was an honorary member of the faculty at the University of Bologna.
See LarsonCalculus.com to read more of this biography.
For more information on the contributions of women to mathematics, see the article "Why Women Succeed in Mathematics" by Mona Fabricant, Sylvia Svitak, and Patricia Clark Kenschaft in *Mathematics Teacher*. To view this article, go to *MathArticles.com*.

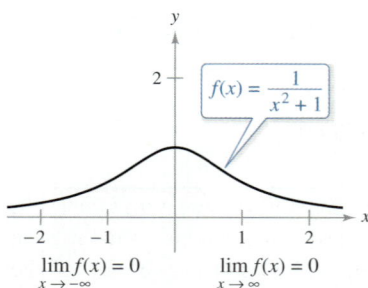

$$\lim_{x\to-\infty}f(x)=0 \qquad \lim_{x\to\infty}f(x)=0$$

f has a horizontal asymptote at $y=0$.
Figure 3.37

In Figure 3.37, you can see that the function

$$f(x) = \frac{1}{x^2 + 1}$$

approaches the same horizontal asymptote to the right and to the left. This is always true of rational functions. Functions that are not rational, however, may approach different horizontal asymptotes to the right and to the left. This is demonstrated in Example 4.

EXAMPLE 4   **A Function with Two Horizontal Asymptotes**

Find each limit.

a. $\displaystyle \lim_{x \to \infty} \frac{3x - 2}{\sqrt{2x^2 + 1}}$ **b.** $\displaystyle \lim_{x \to -\infty} \frac{3x - 2}{\sqrt{2x^2 + 1}}$

Solution

a. For $x > 0$, you can write $x = \sqrt{x^2}$. So, dividing both the numerator and the denominator by x produces

$$\frac{3x - 2}{\sqrt{2x^2 + 1}} = \frac{\dfrac{3x - 2}{x}}{\dfrac{\sqrt{2x^2 + 1}}{\sqrt{x^2}}} = \frac{3 - \dfrac{2}{x}}{\sqrt{\dfrac{2x^2 + 1}{x^2}}} = \frac{3 - \dfrac{2}{x}}{\sqrt{2 + \dfrac{1}{x^2}}}$$

and you can take the limit as follows.

$$\lim_{x \to \infty} \frac{3x - 2}{\sqrt{2x^2 + 1}} = \lim_{x \to \infty} \frac{3 - \dfrac{2}{x}}{\sqrt{2 + \dfrac{1}{x^2}}} = \frac{3 - 0}{\sqrt{2 + 0}} = \frac{3}{\sqrt{2}}$$

b. For $x < 0$, you can write $x = -\sqrt{x^2}$. So, dividing both the numerator and the denominator by x produces

$$\frac{3x - 2}{\sqrt{2x^2 + 1}} = \frac{\dfrac{3x - 2}{x}}{\dfrac{\sqrt{2x^2 + 1}}{-\sqrt{x^2}}} = \frac{3 - \dfrac{2}{x}}{-\sqrt{\dfrac{2x^2 + 1}{x^2}}} = \frac{3 - \dfrac{2}{x}}{-\sqrt{2 + \dfrac{1}{x^2}}}$$

and you can take the limit as follows.

$$\lim_{x \to -\infty} \frac{3x - 2}{\sqrt{2x^2 + 1}} = \lim_{x \to -\infty} \frac{3 - \dfrac{2}{x}}{-\sqrt{2 + \dfrac{1}{x^2}}} = \frac{3 - 0}{-\sqrt{2 + 0}} = -\frac{3}{\sqrt{2}}$$

The graph of $f(x) = (3x - 2)/\sqrt{2x^2 + 1}$ is shown in Figure 3.38.

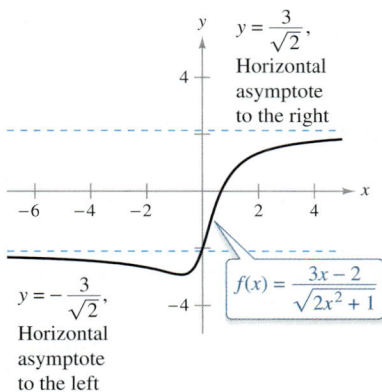

$y = \dfrac{3}{\sqrt{2}}$, Horizontal asymptote to the right

$y = -\dfrac{3}{\sqrt{2}}$, Horizontal asymptote to the left

$f(x) = \dfrac{3x - 2}{\sqrt{2x^2 + 1}}$

Functions that are not rational may have different right and left horizontal asymptotes.
Figure 3.38

▷ **TECHNOLOGY PITFALL**   If you use a graphing utility to estimate a limit, be sure that you also confirm the estimate analytically—the pictures shown by a graphing utility can be misleading. For instance, Figure 3.39 shows one view of the graph of

$$y = \frac{2x^3 + 1000x^2 + x}{x^3 + 1000x^2 + x + 1000}.$$

From this view, one could be convinced that the graph has $y = 1$ as a horizontal asymptote. An analytical approach shows that the horizontal asymptote is actually $y = 2$. Confirm this by enlarging the viewing window on the graphing utility.

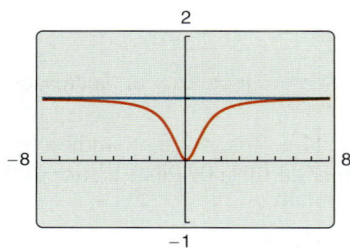

The horizontal asymptote appears to be the line $y = 1$, but it is actually the line $y = 2$.
Figure 3.39

In Section 1.3 (Example 9), you saw how the Squeeze Theorem can be used to evaluate limits involving trigonometric functions. This theorem is also valid for limits at infinity.

EXAMPLE 5 Limits Involving Trigonometric Functions

Find each limit.

a. $\displaystyle\lim_{x\to\infty} \sin x$ **b.** $\displaystyle\lim_{x\to\infty} \frac{\sin x}{x}$

Solution

a. As x approaches infinity, the sine function oscillates between 1 and -1. So, this limit does not exist.

b. Because $-1 \le \sin x \le 1$, it follows that for $x > 0$,

$$-\frac{1}{x} \le \frac{\sin x}{x} \le \frac{1}{x}$$

where

$$\lim_{x\to\infty}\left(-\frac{1}{x}\right) = 0 \quad\text{and}\quad \lim_{x\to\infty}\frac{1}{x} = 0.$$

So, by the Squeeze Theorem, you can obtain

$$\lim_{x\to\infty} \frac{\sin x}{x} = 0$$

as shown in Figure 3.40.

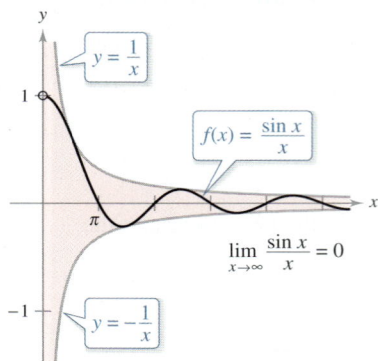

As x increases without bound, $f(x)$ approaches 0.

Figure 3.40

EXAMPLE 6 Oxygen Level in a Pond

Let $f(t)$ measure the level of oxygen in a pond, where $f(t) = 1$ is the normal (unpolluted) level and the time t is measured in weeks. When $t = 0$, organic waste is dumped into the pond, and as the waste material oxidizes, the level of oxygen in the pond is

$$f(t) = \frac{t^2 - t + 1}{t^2 + 1}.$$

What percent of the normal level of oxygen exists in the pond after 1 week? After 2 weeks? After 10 weeks? What is the limit as t approaches infinity?

Solution When $t = 1, 2,$ and 10, the levels of oxygen are as shown.

$$f(1) = \frac{1^2 - 1 + 1}{1^2 + 1} = \frac{1}{2} = 50\% \qquad \text{1 week}$$

$$f(2) = \frac{2^2 - 2 + 1}{2^2 + 1} = \frac{3}{5} = 60\% \qquad \text{2 weeks}$$

$$f(10) = \frac{10^2 - 10 + 1}{10^2 + 1} = \frac{91}{101} \approx 90.1\% \qquad \text{10 weeks}$$

To find the limit as t approaches infinity, you can use the guidelines on page 198, or you can divide the numerator and the denominator by t^2 to obtain

$$\lim_{t\to\infty} \frac{t^2 - t + 1}{t^2 + 1} = \lim_{t\to\infty} \frac{1 - (1/t) + (1/t^2)}{1 + (1/t^2)} = \frac{1 - 0 + 0}{1 + 0} = 1 = 100\%.$$

The level of oxygen in a pond approaches the normal level of 1 as t approaches ∞.

Figure 3.41

See Figure 3.41.

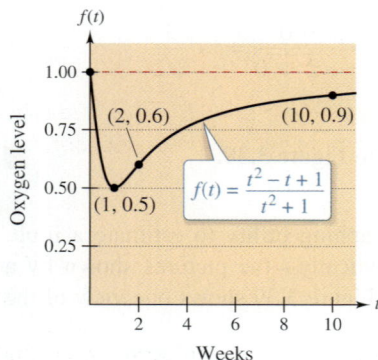

Infinite Limits at Infinity

Many functions do not approach a finite limit as x increases (or decreases) without bound. For instance, no polynomial function has a finite limit at infinity. The next definition is used to describe the behavior of polynomial and other functions at infinity.

• • • • • • • • • • • • • ▷

• • REMARK Determining whether a function has an infinite limit at infinity is useful in analyzing the "end behavior" of its graph. You will see examples of this in Section 3.6 on curve sketching.

> ### Definition of Infinite Limits at Infinity
>
> Let f be a function defined on the interval (a, ∞).
>
> 1. The statement $\displaystyle\lim_{x\to\infty} f(x) = \infty$ means that for each positive number M, there is a corresponding number $N > 0$ such that $f(x) > M$ whenever $x > N$.
> 2. The statement $\displaystyle\lim_{x\to\infty} f(x) = -\infty$ means that for each negative number M, there is a corresponding number $N > 0$ such that $f(x) < M$ whenever $x > N$.
>
> Similar definitions can be given for the statements
> $$\lim_{x\to-\infty} f(x) = \infty \quad \text{and} \quad \lim_{x\to-\infty} f(x) = -\infty.$$

EXAMPLE 7   Finding Infinite Limits at Infinity

Find each limit.

a. $\displaystyle\lim_{x\to\infty} x^3$ **b.** $\displaystyle\lim_{x\to-\infty} x^3$

Solution

a. As x increases without bound, x^3 also increases without bound. So, you can write
$$\lim_{x\to\infty} x^3 = \infty.$$

b. As x decreases without bound, x^3 also decreases without bound. So, you can write
$$\lim_{x\to-\infty} x^3 = -\infty.$$

The graph of $f(x) = x^3$ in Figure 3.42 illustrates these two results. These results agree with the Leading Coefficient Test for polynomial functions as described in Section P.3.

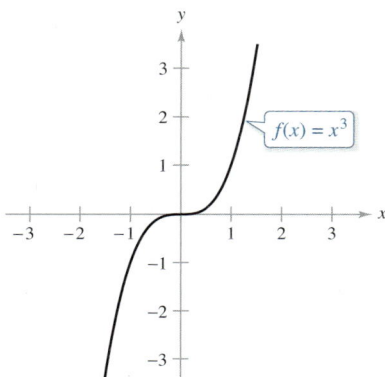

Figure 3.42

$f(x) = x^3$

EXAMPLE 8   Finding Infinite Limits at Infinity

Find each limit.

a. $\displaystyle\lim_{x\to\infty} \frac{2x^2 - 4x}{x + 1}$ **b.** $\displaystyle\lim_{x\to-\infty} \frac{2x^2 - 4x}{x + 1}$

Solution   One way to evaluate each of these limits is to use long division to rewrite the improper rational function as the sum of a polynomial and a rational function.

a. $\displaystyle\lim_{x\to\infty} \frac{2x^2 - 4x}{x + 1} = \lim_{x\to\infty} \left(2x - 6 + \frac{6}{x + 1} \right) = \infty$

b. $\displaystyle\lim_{x\to-\infty} \frac{2x^2 - 4x}{x + 1} = \lim_{x\to-\infty} \left(2x - 6 + \frac{6}{x + 1} \right) = -\infty$

The statements above can be interpreted as saying that as x approaches $\pm\infty$, the function $f(x) = (2x^2 - 4x)/(x + 1)$ behaves like the function $g(x) = 2x - 6$. In Section 3.6, you will see that this is graphically described by saying that the line $y = 2x - 6$ is a slant asymptote of the graph of f, as shown in Figure 3.43.

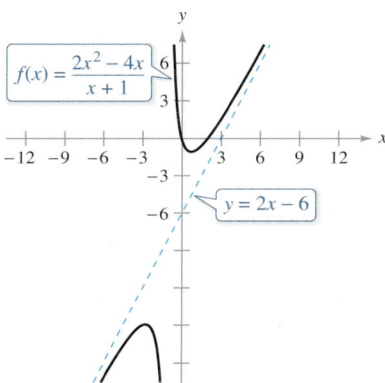

$f(x) = \dfrac{2x^2 - 4x}{x + 1}$

$y = 2x - 6$

Figure 3.43

Matching In Exercises 1–6, match the function with one of the graphs [(a), (b), (c), (d), (e), or (f)] using horizontal asymptotes as an aid.

(a)

(b)

(c)

(d)

(e)

(f)

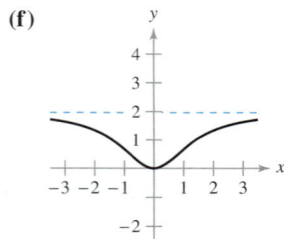

1. $f(x) = \dfrac{2x^2}{x^2 + 2}$

2. $f(x) = \dfrac{2x}{\sqrt{x^2 + 2}}$

3. $f(x) = \dfrac{x}{x^2 + 2}$

4. $f(x) = 2 + \dfrac{x^2}{x^4 + 1}$

5. $f(x) = \dfrac{4 \sin x}{x^2 + 1}$

6. $f(x) = \dfrac{2x^2 - 3x + 5}{x^2 + 1}$

Numerical and Graphical Analysis In Exercises 7–12, use a graphing utility to complete the table and estimate the limit as x approaches infinity. Then use a graphing utility to graph the function and estimate the limit graphically.

x	10^0	10^1	10^2	10^3	10^4	10^5	10^6
$f(x)$							

7. $f(x) = \dfrac{4x + 3}{2x - 1}$

8. $f(x) = \dfrac{2x^2}{x + 1}$

9. $f(x) = \dfrac{-6x}{\sqrt{4x^2 + 5}}$

10. $f(x) = \dfrac{10}{\sqrt{2x^2 - 1}}$

11. $f(x) = 5 - \dfrac{1}{x^2 + 1}$

12. $f(x) = 4 + \dfrac{3}{x^2 + 2}$

Finding Limits at Infinity In Exercises 13 and 14, find $\lim\limits_{x \to \infty} h(x)$, if possible.

13. $f(x) = 5x^3 - 3x^2 + 10x$

  (a) $h(x) = \dfrac{f(x)}{x^2}$

  (b) $h(x) = \dfrac{f(x)}{x^3}$

  (c) $h(x) = \dfrac{f(x)}{x^4}$

14. $f(x) = -4x^2 + 2x - 5$

  (a) $h(x) = \dfrac{f(x)}{x}$

  (b) $h(x) = \dfrac{f(x)}{x^2}$

  (c) $h(x) = \dfrac{f(x)}{x^3}$

Finding Limits at Infinity In Exercises 15–18, find each limit, if possible.

15. (a) $\lim\limits_{x \to \infty} \dfrac{x^2 + 2}{x^3 - 1}$

   (b) $\lim\limits_{x \to \infty} \dfrac{x^2 + 2}{x^2 - 1}$

   (c) $\lim\limits_{x \to \infty} \dfrac{x^2 + 2}{x - 1}$

16. (a) $\lim\limits_{x \to \infty} \dfrac{3 - 2x}{3x^3 - 1}$

   (b) $\lim\limits_{x \to \infty} \dfrac{3 - 2x}{3x - 1}$

   (c) $\lim\limits_{x \to \infty} \dfrac{3 - 2x^2}{3x - 1}$

17. (a) $\lim\limits_{x \to \infty} \dfrac{5 - 2x^{3/2}}{3x^2 - 4}$

   (b) $\lim\limits_{x \to \infty} \dfrac{5 - 2x^{3/2}}{3x^{3/2} - 4}$

   (c) $\lim\limits_{x \to \infty} \dfrac{5 - 2x^{3/2}}{3x - 4}$

18. (a) $\lim\limits_{x \to \infty} \dfrac{5x^{3/2}}{4x^2 + 1}$

   (b) $\lim\limits_{x \to \infty} \dfrac{5x^{3/2}}{4x^{3/2} + 1}$

   (c) $\lim\limits_{x \to \infty} \dfrac{5x^{3/2}}{4\sqrt{x} + 1}$

Finding a Limit In Exercises 19–38, find the limit.

19. $\lim\limits_{x \to \infty} \left(4 + \dfrac{3}{x}\right)$

20. $\lim\limits_{x \to -\infty} \left(\dfrac{5}{x} - \dfrac{x}{3}\right)$

21. $\lim\limits_{x \to \infty} \dfrac{2x - 1}{3x + 2}$

22. $\lim\limits_{x \to -\infty} \dfrac{4x^2 + 5}{x^2 + 3}$

23. $\lim\limits_{x \to \infty} \dfrac{x}{x^2 - 1}$

24. $\lim\limits_{x \to \infty} \dfrac{5x^3 + 1}{10x^3 - 3x^2 + 7}$

25. $\lim\limits_{x \to -\infty} \dfrac{5x^2}{x + 3}$

26. $\lim\limits_{x \to -\infty} \dfrac{x^3 - 4}{x^2 + 1}$

27. $\lim\limits_{x \to -\infty} \dfrac{x}{\sqrt{x^2 - x}}$

28. $\lim\limits_{x \to -\infty} \dfrac{x}{\sqrt{x^2 + 1}}$

29. $\lim\limits_{x \to -\infty} \dfrac{2x + 1}{\sqrt{x^2 - x}}$

30. $\lim\limits_{x \to \infty} \dfrac{5x^2 + 2}{\sqrt{x^2 + 3}}$

31. $\lim\limits_{x \to \infty} \dfrac{\sqrt{x^2 - 1}}{2x - 1}$

32. $\lim\limits_{x \to -\infty} \dfrac{\sqrt{x^4 - 1}}{x^3 - 1}$

33. $\lim\limits_{x \to \infty} \dfrac{x + 1}{(x^2 + 1)^{1/3}}$

34. $\lim\limits_{x \to -\infty} \dfrac{2x}{(x^6 - 1)^{1/3}}$

35. $\lim\limits_{x \to \infty} \dfrac{1}{2x + \sin x}$

36. $\lim\limits_{x \to \infty} \cos \dfrac{1}{x}$

37. $\lim\limits_{x \to \infty} \dfrac{\sin 2x}{x}$

38. $\lim\limits_{x \to \infty} \dfrac{x - \cos x}{x}$

Horizontal Asymptotes In Exercises 39–42, use a graphing utility to graph the function and identify any horizontal asymptotes.

39. $f(x) = \dfrac{|x|}{x + 1}$

40. $f(x) = \dfrac{|3x + 2|}{x - 2}$

41. $f(x) = \dfrac{3x}{\sqrt{x^2 + 2}}$

42. $f(x) = \dfrac{\sqrt{9x^2 - 2}}{2x + 1}$

Finding a Limit In Exercises 43 and 44, find the limit. (*Hint:* Let $x = 1/t$ and find the limit as $t \to 0^+$.)

43. $\displaystyle\lim_{x \to \infty} x \sin \dfrac{1}{x}$

44. $\displaystyle\lim_{x \to \infty} x \tan \dfrac{1}{x}$

Finding a Limit In Exercises 45–48, find the limit. (*Hint:* Treat the expression as a fraction whose denominator is 1, and rationalize the numerator.) Use a graphing utility to verify your result.

45. $\displaystyle\lim_{x \to -\infty} \left(x + \sqrt{x^2 + 3} \right)$

46. $\displaystyle\lim_{x \to \infty} \left(x - \sqrt{x^2 + x} \right)$

47. $\displaystyle\lim_{x \to -\infty} \left(3x + \sqrt{9x^2 - x} \right)$

48. $\displaystyle\lim_{x \to \infty} \left(4x - \sqrt{16x^2 - x} \right)$

Numerical, Graphical, and Analytic Analysis In Exercises 49–52, use a graphing utility to complete the table and estimate the limit as x approaches infinity. Then use a graphing utility to graph the function and estimate the limit. Finally, find the limit analytically and compare your results with the estimates.

x	10^0	10^1	10^2	10^3	10^4	10^5	10^6
$f(x)$							

49. $f(x) = x - \sqrt{x(x - 1)}$

50. $f(x) = x^2 - x\sqrt{x(x - 1)}$

51. $f(x) = x \sin \dfrac{1}{2x}$

52. $f(x) = \dfrac{x + 1}{x\sqrt{x}}$

WRITING ABOUT CONCEPTS

Writing In Exercises 53 and 54, describe in your own words what the statement means.

53. $\displaystyle\lim_{x \to \infty} f(x) = 4$

54. $\displaystyle\lim_{x \to -\infty} f(x) = 2$

55. Sketching a Graph Sketch a graph of a differentiable function f that satisfies the following conditions and has $x = 2$ as its only critical number.

$f'(x) < 0$ for $x < 2$

$f'(x) > 0$ for $x > 2$

$\displaystyle\lim_{x \to -\infty} f(x) = 6$

$\displaystyle\lim_{x \to \infty} f(x) = 6$

56. Points of Inflection Is it possible to sketch a graph of a function that satisfies the conditions of Exercise 55 and has *no* points of inflection? Explain.

WRITING ABOUT CONCEPTS (continued)

57. Using Symmetry to Find Limits If f is a continuous function such that $\displaystyle\lim_{x \to \infty} f(x) = 5$, find, if possible, $\displaystyle\lim_{x \to -\infty} f(x)$ for each specified condition.

(a) The graph of f is symmetric with respect to the y-axis.

(b) The graph of f is symmetric with respect to the origin.

58. A Function and Its Derivative The graph of a function f is shown below. To print an enlarged copy of the graph, go to *MathGraphs.com*.

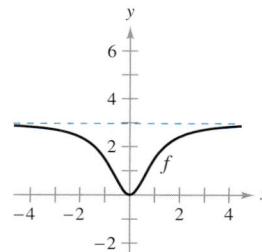

(a) Sketch f'.

(b) Use the graphs to estimate $\displaystyle\lim_{x \to \infty} f(x)$ and $\displaystyle\lim_{x \to \infty} f'(x)$.

(c) Explain the answers you gave in part (b).

Sketching a Graph In Exercises 59–74, sketch the graph of the equation using extrema, intercepts, symmetry, and asymptotes. Then use a graphing utility to verify your result.

59. $y = \dfrac{x}{1 - x}$

60. $y = \dfrac{x - 4}{x - 3}$

61. $y = \dfrac{x + 1}{x^2 - 4}$

62. $y = \dfrac{2x}{9 - x^2}$

63. $y = \dfrac{x^2}{x^2 + 16}$

64. $y = \dfrac{2x^2}{x^2 - 4}$

65. $xy^2 = 9$

66. $x^2 y = 9$

67. $y = \dfrac{3x}{x - 1}$

68. $y = \dfrac{3x}{1 - x^2}$

69. $y = 2 - \dfrac{3}{x^2}$

70. $y = 1 - \dfrac{1}{x}$

71. $y = 3 + \dfrac{2}{x}$

72. $y = \dfrac{4}{x^2} + 1$

73. $y = \dfrac{x^3}{\sqrt{x^2 - 4}}$

74. $y = \dfrac{x}{\sqrt{x^2 - 4}}$

Analyzing a Graph Using Technology In Exercises 75–82, use a computer algebra system to analyze the graph of the function. Label any extrema and/or asymptotes that exist.

75. $f(x) = 9 - \dfrac{5}{x^2}$

76. $f(x) = \dfrac{1}{x^2 - x - 2}$

77. $f(x) = \dfrac{x - 2}{x^2 - 4x + 3}$

78. $f(x) = \dfrac{x + 1}{x^2 + x + 1}$

79. $f(x) = \dfrac{3x}{\sqrt{4x^2 + 1}}$

80. $g(x) = \dfrac{2x}{\sqrt{3x^2 + 1}}$

81. $g(x) = \sin\left(\dfrac{x}{x-2}\right), \quad x > 3$

82. $f(x) = \dfrac{2 \sin 2x}{x}$

Comparing Functions In Exercises 83 and 84, (a) use a graphing utility to graph f and g in the same viewing window, (b) verify algebraically that f and g represent the same function, and (c) zoom out sufficiently far so that the graph appears as a line. What equation does this line appear to have? (Note that the points at which the function is not continuous are not readily seen when you zoom out.)

83. $f(x) = \dfrac{x^3 - 3x^2 + 2}{x(x-3)}$

$g(x) = x + \dfrac{2}{x(x-3)}$

84. $f(x) = -\dfrac{x^3 - 2x^2 + 2}{2x^2}$

$g(x) = -\dfrac{1}{2}x + 1 - \dfrac{1}{x^2}$

85. Engine Efficiency

The efficiency of an internal combustion engine is

$$\text{Efficiency (\%)} = 100\left[1 - \dfrac{1}{(v_1/v_2)^c}\right]$$

where v_1/v_2 is the ratio of the uncompressed gas to the compressed gas and c is a positive constant dependent on the engine design. Find the limit of the efficiency as the compression ratio approaches infinity.

86. Average Cost A business has a cost of $C = 0.5x + 500$ for producing x units. The average cost per unit is

$$\overline{C} = \dfrac{C}{x}.$$

Find the limit of $\overline{C}$ as x approaches infinity.

87. Physics Newton's First Law of Motion and Einstein's Special Theory of Relativity differ concerning a particle's behavior as its velocity approaches the speed of light c. In the graph, functions N and E represent the velocity v, with respect to time t, of a particle accelerated by a constant force as predicted by Newton and Einstein, respectively. Write limit statements that describe these two theories.

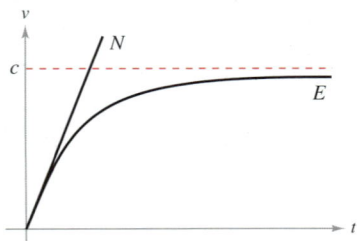

88. **HOW DO YOU SEE IT?** The graph shows the temperature T, in degrees Fahrenheit, of molten glass t seconds after it is removed from a kiln.

(a) Find $\lim\limits_{t \to 0^+} T$. What does this limit represent?

(b) Find $\lim\limits_{t \to \infty} T$. What does this limit represent?

(c) Will the temperature of the glass ever actually reach room temperature? Why?

89. Modeling Data The average typing speeds S (in words per minute) of a typing student after t weeks of lessons are shown in the table.

t	5	10	15	20	25	30
S	28	56	79	90	93	94

A model for the data is $S = \dfrac{100t^2}{65 + t^2}, \; t > 0$.

(a) Use a graphing utility to plot the data and graph the model.

(b) Does there appear to be a limiting typing speed? Explain.

90. Modeling Data A heat probe is attached to the heat exchanger of a heating system. The temperature T (in degrees Celsius) is recorded t seconds after the furnace is started. The results for the first 2 minutes are recorded in the table.

t	0	15	30	45	60
T	25.2°	36.9°	45.5°	51.4°	56.0°

t	75	90	105	120
T	59.6°	62.0°	64.0°	65.2°

(a) Use the regression capabilities of a graphing utility to find a model of the form $T_1 = at^2 + bt + c$ for the data.

(b) Use a graphing utility to graph T_1.

(c) A rational model for the data is

$$T_2 = \dfrac{1451 + 86t}{58 + t}.$$

Use a graphing utility to graph T_2.

(d) Find $T_1(0)$ and $T_2(0)$.

(e) Find $\lim\limits_{t \to \infty} T_2$.

(f) Interpret the result in part (e) in the context of the problem. Is it possible to do this type of analysis using T_1? Explain.

Straight 8 Photography/Shutterstock.com

91. Using the Definition of Limits at Infinity The graph of

$$f(x) = \frac{2x^2}{x^2 + 2}$$

is shown.

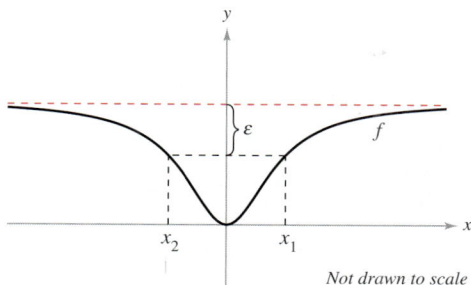

Not drawn to scale

(a) Find $L = \lim\limits_{x \to \infty} f(x)$.

(b) Determine x_1 and x_2 in terms of ε.

(c) Determine M, where $M > 0$, such that $|f(x) - L| < \varepsilon$ for $x > M$.

(d) Determine N, where $N < 0$, such that $|f(x) - L| < \varepsilon$ for $x < N$.

92. Using the Definition of Limits at Infinity The graph of

$$f(x) = \frac{6x}{\sqrt{x^2 + 2}}$$

is shown.

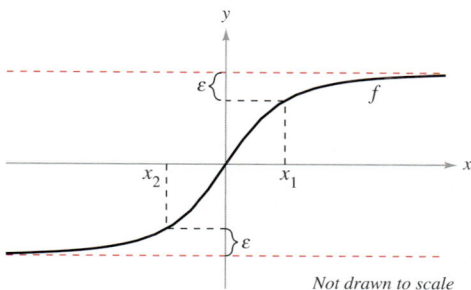

Not drawn to scale

(a) Find $L = \lim\limits_{x \to \infty} f(x)$ and $K = \lim\limits_{x \to -\infty} f(x)$.

(b) Determine x_1 and x_2 in terms of ε.

(c) Determine M, where $M > 0$, such that $|f(x) - L| < \varepsilon$ for $x > M$.

(d) Determine N, where $N < 0$, such that $|f(x) - K| < \varepsilon$ for $x < N$.

93. Using the Definition of Limits at Infinity Consider

$$\lim_{x \to \infty} \frac{3x}{\sqrt{x^2 + 3}}.$$

(a) Use the definition of limits at infinity to find values of M that correspond to $\varepsilon = 0.5$.

(b) Use the definition of limits at infinity to find values of M that correspond to $\varepsilon = 0.1$.

94. Using the Definition of Limits at Infinity Consider

$$\lim_{x \to -\infty} \frac{3x}{\sqrt{x^2 + 3}}.$$

(a) Use the definition of limits at infinity to find values of N that correspond to $\varepsilon = 0.5$.

(b) Use the definition of limits at infinity to find values of N that correspond to $\varepsilon = 0.1$.

Proof In Exercises 95–98, use the definition of limits at infinity to prove the limit.

95. $\lim\limits_{x \to \infty} \dfrac{1}{x^2} = 0$ **96.** $\lim\limits_{x \to \infty} \dfrac{2}{\sqrt{x}} = 0$

97. $\lim\limits_{x \to -\infty} \dfrac{1}{x^3} = 0$ **98.** $\lim\limits_{x \to -\infty} \dfrac{1}{x - 2} = 0$

99. Distance A line with slope m passes through the point $(0, 4)$.

(a) Write the shortest distance d between the line and the point $(3, 1)$ as a function of m.

(b) Use a graphing utility to graph the equation in part (a).

(c) Find $\lim\limits_{m \to \infty} d(m)$ and $\lim\limits_{m \to -\infty} d(m)$. Interpret the results geometrically.

100. Distance A line with slope m passes through the point $(0, -2)$.

(a) Write the shortest distance d between the line and the point $(4, 2)$ as a function of m.

(b) Use a graphing utility to graph the equation in part (a).

(c) Find $\lim\limits_{m \to \infty} d(m)$ and $\lim\limits_{m \to -\infty} d(m)$. Interpret the results geometrically.

101. Proof Prove that if

$$p(x) = a_n x^n + \cdots + a_1 x + a_0$$

and

$$q(x) = b_m x^m + \cdots + b_1 x + b_0$$

where $a_n \neq 0$ and $b_m \neq 0$, then

$$\lim_{x \to \infty} \frac{p(x)}{q(x)} = \begin{cases} 0, & n < m \\ \dfrac{a_n}{b_m}, & n = m. \\ \pm\infty, & n > m \end{cases}$$

102. Proof Use the definition of infinite limits at infinity to prove that $\lim\limits_{x \to \infty} x^3 = \infty$.

True or False? In Exercises 103 and 104, determine whether the statement is true or false. If it is false, explain why or give an example that shows it is false.

103. If $f'(x) > 0$ for all real numbers x, then f increases without bound.

104. If $f''(x) < 0$ for all real numbers x, then f decreases without bound.

3.6 A Summary of Curve Sketching

■ Analyze and sketch the graph of a function.

Analyzing the Graph of a Function

It would be difficult to overstate the importance of using graphs in mathematics. Descartes's introduction of analytic geometry contributed significantly to the rapid advances in calculus that began during the mid-seventeenth century. In the words of Lagrange, "As long as algebra and geometry traveled separate paths their advance was slow and their applications limited. But when these two sciences joined company, they drew from each other fresh vitality and thenceforth marched on at a rapid pace toward perfection."

So far, you have studied several concepts that are useful in analyzing the graph of a function.

- x-intercepts and y-intercepts   (Section P.1)
- Symmetry   (Section P.1)
- Domain and range   (Section P.3)
- Continuity   (Section 1.4)
- Vertical asymptotes   (Section 1.5)
- Differentiability   (Section 2.1)
- Relative extrema   (Section 3.1)
- Concavity   (Section 3.4)
- Points of inflection   (Section 3.4)
- Horizontal asymptotes   (Section 3.5)
- Infinite limits at infinity   (Section 3.5)

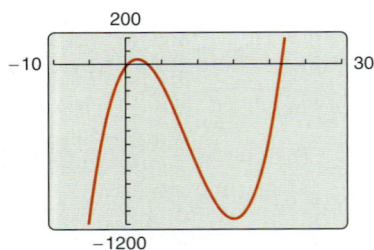

When you are sketching the graph of a function, either by hand or with a graphing utility, remember that normally you cannot show the *entire* graph. The decision as to which part of the graph you choose to show is often crucial. For instance, which of the viewing windows in Figure 3.44 better represents the graph of

$$f(x) = x^3 - 25x^2 + 74x - 20?$$

By seeing both views, it is clear that the second viewing window gives a more complete representation of the graph. But would a third viewing window reveal other interesting portions of the graph? To answer this, you need to use calculus to interpret the first and second derivatives. Here are some guidelines for determining a good viewing window for the graph of a function.

Different viewing windows for the graph of $f(x) = x^3 - 25x^2 + 74x - 20$
Figure 3.44

▷ **GUIDELINES FOR ANALYZING THE GRAPH OF A FUNCTION**

1. Determine the domain and range of the function.
2. Determine the intercepts, asymptotes, and symmetry of the graph.
3. Locate the x-values for which $f'(x)$ and $f''(x)$ either are zero or do not exist. Use the results to determine relative extrema and points of inflection.

•••••• **REMARK** In these guidelines, note the importance of *algebra* (as well as calculus) for solving the equations

$$f(x) = 0, \quad f'(x) = 0, \quad \text{and} \quad f''(x) = 0.$$

EXAMPLE 1   **Sketching the Graph of a Rational Function**

Analyze and sketch the graph of

$$f(x) = \frac{2(x^2 - 9)}{x^2 - 4}.$$

Solution

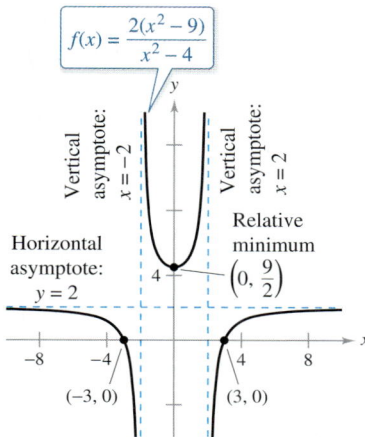

$$f(x) = \frac{2(x^2 - 9)}{x^2 - 4}$$

Vertical asymptote: $x = -2$
Vertical asymptote: $x = 2$
Horizontal asymptote: $y = 2$
Relative minimum $\left(0, \frac{9}{2}\right)$
$(-3, 0)$ $(3, 0)$

Using calculus, you can be certain that you have determined all characteristics of the graph of f.
Figure 3.45

First derivative:	$f'(x) = \dfrac{20x}{(x^2 - 4)^2}$
Second derivative:	$f''(x) = \dfrac{-20(3x^2 + 4)}{(x^2 - 4)^3}$
x-intercepts:	$(-3, 0), (3, 0)$
y-intercept:	$\left(0, \frac{9}{2}\right)$
Vertical asymptotes:	$x = -2, x = 2$
Horizontal asymptote:	$y = 2$
Critical number:	$x = 0$
Possible points of inflection:	None
Domain:	All real numbers except $x = \pm 2$
Symmetry:	With respect to y-axis
Test intervals:	$(-\infty, -2), (-2, 0), (0, 2), (2, \infty)$

The table shows how the test intervals are used to determine several characteristics of the graph. The graph of f is shown in Figure 3.45.

	$f(x)$	$f'(x)$	$f''(x)$	Characteristic of Graph
$-\infty < x < -2$		$-$	$-$	Decreasing, concave downward
$x = -2$	Undef.	Undef.	Undef.	Vertical asymptote
$-2 < x < 0$		$-$	$+$	Decreasing, concave upward
$x = 0$	$\frac{9}{2}$	0	$+$	Relative minimum
$0 < x < 2$		$+$	$+$	Increasing, concave upward
$x = 2$	Undef.	Undef.	Undef.	Vertical asymptote
$2 < x < \infty$		$+$	$-$	Increasing, concave downward

■ FOR FURTHER INFORMATION
For more information on the use of technology to graph rational functions, see the article "Graphs of Rational Functions for Computer Assisted Calculus" by Stan Byrd and Terry Walters in *The College Mathematics Journal*. To view this article, go to *MathArticles.com*.

Be sure you understand all of the implications of creating a table such as that shown in Example 1. By using calculus, you can be *sure* that the graph has no relative extrema or points of inflection other than those shown in Figure 3.45.

▷ **TECHNOLOGY PITFALL**   Without using the type of analysis outlined in Example 1, it is easy to obtain an incomplete view of a graph's basic characteristics. For instance, Figure 3.46 shows a view of the graph of

$$g(x) = \frac{2(x^2 - 9)(x - 20)}{(x^2 - 4)(x - 21)}.$$

From this view, it appears that the graph of g is about the same as the graph of f shown in Figure 3.45. The graphs of these two functions, however, differ significantly. Try enlarging the viewing window to see the differences.

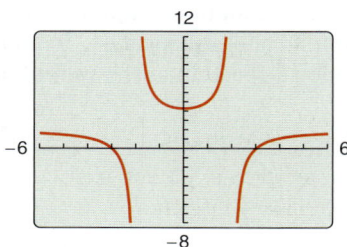

By not using calculus, you may overlook important characteristics of the graph of g.
Figure 3.46

EXAMPLE 2   **Sketching the Graph of a Rational Function**

Analyze and sketch the graph of $f(x) = \dfrac{x^2 - 2x + 4}{x - 2}$.

Solution

First derivative: $f'(x) = \dfrac{x(x - 4)}{(x - 2)^2}$

Second derivative: $f''(x) = \dfrac{8}{(x - 2)^3}$

x-intercepts: None

y-intercept: $(0, -2)$

Vertical asymptote: $x = 2$

Horizontal asymptotes: None

End behavior: $\displaystyle\lim_{x \to -\infty} f(x) = -\infty, \ \lim_{x \to \infty} f(x) = \infty$

Critical numbers: $x = 0, \ x = 4$

Possible points of inflection: None

Domain: All real numbers except $x = 2$

Test intervals: $(-\infty, 0), \ (0, 2), \ (2, 4), \ (4, \infty)$

The analysis of the graph of f is shown in the table, and the graph is shown in Figure 3.47.

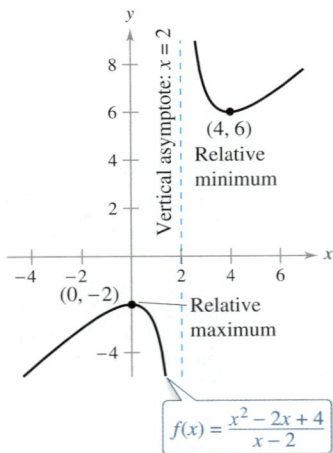

$f(x) = \dfrac{x^2 - 2x + 4}{x - 2}$

Figure 3.47

	$f(x)$	$f'(x)$	$f''(x)$	Characteristic of Graph
$-\infty < x < 0$		$+$	$-$	Increasing, concave downward
$x = 0$	-2	0	$-$	Relative maximum
$0 < x < 2$		$-$	$-$	Decreasing, concave downward
$x = 2$	Undef.	Undef.	Undef.	Vertical asymptote
$2 < x < 4$		$-$	$+$	Decreasing, concave upward
$x = 4$	6	0	$+$	Relative minimum
$4 < x < \infty$		$+$	$+$	Increasing, concave upward

Although the graph of the function in Example 2 has no horizontal asymptote, it does have a slant asymptote. The graph of a rational function (having no common factors and whose denominator is of degree 1 or greater) has a **slant asymptote** when the degree of the numerator exceeds the degree of the denominator by exactly 1. To find the slant asymptote, use long division to rewrite the rational function as the sum of a first-degree polynomial and another rational function.

$f(x) = \dfrac{x^2 - 2x + 4}{x - 2}$ Write original equation.

$= x + \dfrac{4}{x - 2}$ Rewrite using long division.

In Figure 3.48, note that the graph of f approaches the slant asymptote $y = x$ as x approaches $-\infty$ or ∞.

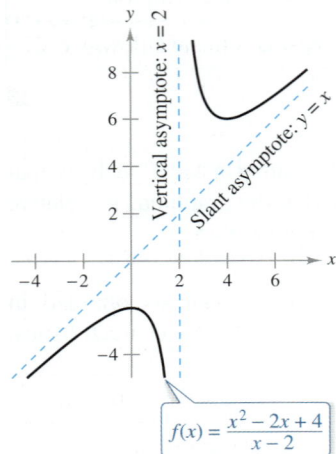

$f(x) = \dfrac{x^2 - 2x + 4}{x - 2}$

A slant asymptote

Figure 3.48

EXAMPLE 3 Sketching the Graph of a Radical Function

Analyze and sketch the graph of $f(x) = \dfrac{x}{\sqrt{x^2 + 2}}$.

Solution

$$f'(x) = \frac{2}{(x^2 + 2)^{3/2}} \qquad \text{Find first derivative.}$$

$$f''(x) = -\frac{6x}{(x^2 + 2)^{5/2}} \qquad \text{Find second derivative.}$$

The graph has only one intercept, $(0, 0)$. It has no vertical asymptotes, but it has two horizontal asymptotes: $y = 1$ (to the right) and $y = -1$ (to the left). The function has no critical numbers and one possible point of inflection (at $x = 0$). The domain of the function is all real numbers, and the graph is symmetric with respect to the origin. The analysis of the graph of f is shown in the table, and the graph is shown in Figure 3.49.

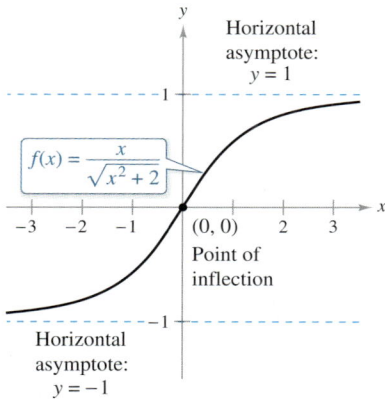

Horizontal asymptote: $y = 1$

$f(x) = \dfrac{x}{\sqrt{x^2 + 2}}$

$(0, 0)$ Point of inflection

Horizontal asymptote: $y = -1$

Figure 3.49

	$f(x)$	$f'(x)$	$f''(x)$	Characteristic of Graph
$-\infty < x < 0$		$+$	$+$	Increasing, concave upward
$x = 0$	0	$\dfrac{1}{\sqrt{2}}$	0	Point of inflection
$0 < x < \infty$		$+$	$-$	Increasing, concave downward

EXAMPLE 4 Sketching the Graph of a Radical Function

Analyze and sketch the graph of $f(x) = 2x^{5/3} - 5x^{4/3}$.

Solution

$$f'(x) = \frac{10}{3}x^{1/3}(x^{1/3} - 2) \qquad \text{Find first derivative.}$$

$$f''(x) = \frac{20(x^{1/3} - 1)}{9x^{2/3}} \qquad \text{Find second derivative.}$$

The function has two intercepts: $(0, 0)$ and $\left(\frac{125}{8}, 0\right)$. There are no horizontal or vertical asymptotes. The function has two critical numbers ($x = 0$ and $x = 8$) and two possible points of inflection ($x = 0$ and $x = 1$). The domain is all real numbers. The analysis of the graph of f is shown in the table, and the graph is shown in Figure 3.50.

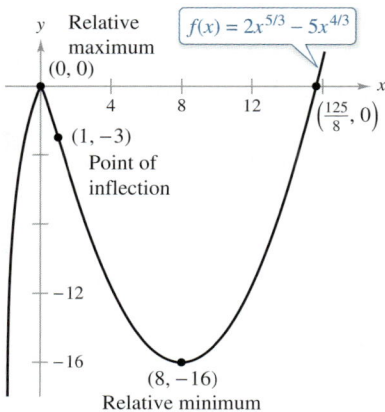

Relative maximum $(0, 0)$

$f(x) = 2x^{5/3} - 5x^{4/3}$

$\left(\frac{125}{8}, 0\right)$

$(1, -3)$ Point of inflection

$(8, -16)$ Relative minimum

Figure 3.50

	$f(x)$	$f'(x)$	$f''(x)$	Characteristic of Graph
$-\infty < x < 0$		$+$	$-$	Increasing, concave downward
$x = 0$	0	0	Undef.	Relative maximum
$0 < x < 1$		$-$	$-$	Decreasing, concave downward
$x = 1$	-3	$-$	0	Point of inflection
$1 < x < 8$		$-$	$+$	Decreasing, concave upward
$x = 8$	-16	0	$+$	Relative minimum
$8 < x < \infty$		$+$	$+$	Increasing, concave upward

> **EXAMPLE 5** **Sketching the Graph of a Polynomial Function**

⋮ ⊳ *See LarsonCalculus.com for an interactive version of this type of example.*

Analyze and sketch the graph of

$$f(x) = x^4 - 12x^3 + 48x^2 - 64x.$$

Solution Begin by factoring to obtain

$$
\begin{aligned}
f(x) &= x^4 - 12x^3 + 48x^2 - 64x \\
&= x(x - 4)^3.
\end{aligned}
$$

Then, using the factored form of $f(x)$, you can perform the following analysis.

First derivative:	$f'(x) = 4(x - 1)(x - 4)^2$
Second derivative:	$f''(x) = 12(x - 4)(x - 2)$
x-intercepts:	$(0, 0),\ (4, 0)$
y-intercept:	$(0, 0)$
Vertical asymptotes:	None
Horizontal asymptotes:	None
End behavior:	$\lim\limits_{x \to -\infty} f(x) = \infty,\ \lim\limits_{x \to \infty} f(x) = \infty$
Critical numbers:	$x = 1,\ x = 4$
Possible points of inflection:	$x = 2,\ x = 4$
Domain:	All real numbers
Test intervals:	$(-\infty, 1),\ (1, 2),\ (2, 4),\ (4, \infty)$

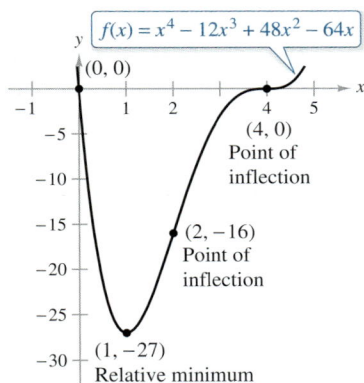

The analysis of the graph of f is shown in the table, and the graph is shown in Figure 3.51(a). Using a computer algebra system such as *Maple* [see Figure 3.51(b)] can help you verify your analysis.

(a)

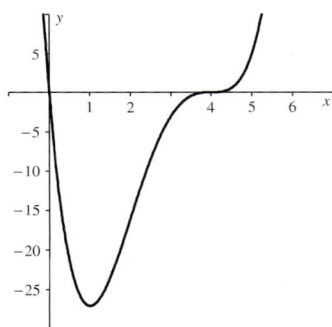

Generated by Maple

(b)
A polynomial function of even degree must have at least one relative extremum.
Figure 3.51

	$f(x)$	$f'(x)$	$f''(x)$	Characteristic of Graph
$-\infty < x < 1$		$-$	$+$	Decreasing, concave upward
$x = 1$	-27	0	$+$	Relative minimum
$1 < x < 2$		$+$	$+$	Increasing, concave upward
$x = 2$	-16	$+$	0	Point of inflection
$2 < x < 4$		$+$	$-$	Increasing, concave downward
$x = 4$	0	0	0	Point of inflection
$4 < x < \infty$		$+$	$+$	Increasing, concave upward

The fourth-degree polynomial function in Example 5 has one relative minimum and no relative maxima. In general, a polynomial function of degree n can have *at most* $n - 1$ relative extrema, and *at most* $n - 2$ points of inflection. Moreover, polynomial functions of even degree must have *at least* one relative extremum.

Remember from the Leading Coefficient Test described in Section P.3 that the "end behavior" of the graph of a polynomial function is determined by its leading coefficient and its degree. For instance, because the polynomial in Example 5 has a positive leading coefficient, the graph rises to the right. Moreover, because the degree is even, the graph also rises to the left.

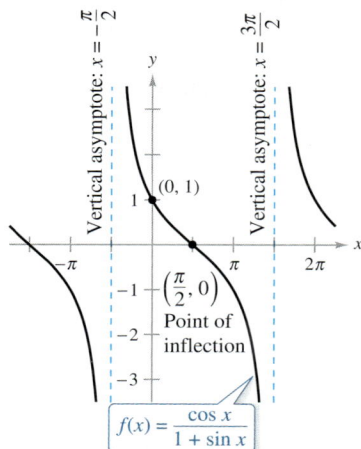

$f(x) = \dfrac{\cos x}{1 + \sin x}$

(a)

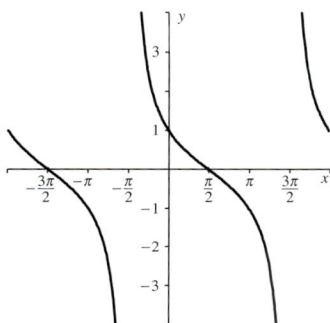

Generated by Maple

(b)

Figure 3.52

| EXAMPLE 6 | **Sketching the Graph of a Trigonometric Function** |

Analyze and sketch the graph of $f(x) = (\cos x)/(1 + \sin x)$.

Solution Because the function has a period of 2π, you can restrict the analysis of the graph to any interval of length 2π. For convenience, choose $(-\pi/2, 3\pi/2)$.

$$\textit{First derivative:} \quad f'(x) = -\frac{1}{1 + \sin x}$$

$$\textit{Second derivative:} \quad f''(x) = \frac{\cos x}{(1 + \sin x)^2}$$

$$\textit{Period:} \quad 2\pi$$

$$\textit{x-intercept:} \quad \left(\frac{\pi}{2}, 0\right)$$

$$\textit{y-intercept:} \quad (0, 1)$$

$$\textit{Vertical asymptotes:} \quad x = -\frac{\pi}{2}, x = \frac{3\pi}{2} \qquad \text{See Remark below.}$$

$$\textit{Horizontal asymptotes:} \quad \text{None}$$

$$\textit{Critical numbers:} \quad \text{None}$$

$$\textit{Possible points of inflection:} \quad x = \frac{\pi}{2}$$

$$\textit{Domain:} \quad \text{All real numbers except } x = \frac{3 + 4n}{2}\pi$$

$$\textit{Test intervals:} \quad \left(-\frac{\pi}{2}, \frac{\pi}{2}\right), \left(\frac{\pi}{2}, \frac{3\pi}{2}\right)$$

The analysis of the graph of f on the interval $(-\pi/2, 3\pi/2)$ is shown in the table, and the graph is shown in Figure 3.52(a). Compare this with the graph generated by the computer algebra system *Maple* in Figure 3.52(b).

	$f(x)$	$f'(x)$	$f''(x)$	Characteristic of Graph
$x = -\dfrac{\pi}{2}$	Undef.	Undef.	Undef.	Vertical asymptote
$-\dfrac{\pi}{2} < x < \dfrac{\pi}{2}$		$-$	$+$	Decreasing, concave upward
$x = \dfrac{\pi}{2}$	0	$-\frac{1}{2}$	0	Point of inflection
$\dfrac{\pi}{2} < x < \dfrac{3\pi}{2}$		$-$	$-$	Decreasing, concave downward
$x = \dfrac{3\pi}{2}$	Undef.	Undef.	Undef.	Vertical asymptote

REMARK By substituting $-\pi/2$ or $3\pi/2$ into the function, you obtain the form $0/0$. This is called an indeterminate form, which you will study in Section 8.7. To determine that the function has vertical asymptotes at these two values, rewrite f as

$$f(x) = \frac{\cos x}{1 + \sin x} = \frac{(\cos x)(1 - \sin x)}{(1 + \sin x)(1 - \sin x)} = \frac{(\cos x)(1 - \sin x)}{\cos^2 x} = \frac{1 - \sin x}{\cos x}.$$

In this form, it is clear that the graph of f has vertical asymptotes at $x = -\pi/2$ and $3\pi/2$.

3.6 Exercises

Matching In Exercises 1–4, match the graph of f in the left column with that of its derivative in the right column.

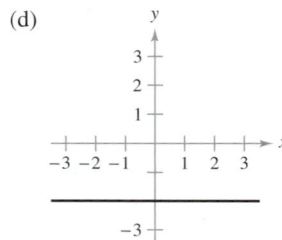

Graph of f	Graph of f'
1.	**(a)**

2. **(b)**

3. **(c)**

4. **(d)**

Analyzing the Graph of a Function In Exercises 5–24, analyze and sketch a graph of the function. Label any intercepts, relative extrema, points of inflection, and asymptotes. Use a graphing utility to verify your results.

5. $y = \dfrac{1}{x-2} - 3$

6. $y = \dfrac{x}{x^2+1}$

7. $y = \dfrac{x^2}{x^2+3}$

8. $y = \dfrac{x^2+1}{x^2-4}$

9. $y = \dfrac{3x}{x^2-1}$

10. $f(x) = \dfrac{x-3}{x}$

11. $f(x) = x + \dfrac{32}{x^2}$

12. $f(x) = \dfrac{x^3}{x^2-9}$

13. $y = \dfrac{x^2 - 6x + 12}{x - 4}$

14. $y = \dfrac{-x^2 - 4x - 7}{x + 3}$

15. $y = x\sqrt{4 - x}$

16. $g(x) = x\sqrt{9 - x^2}$

17. $y = 3x^{2/3} - 2x$

18. $y = (x + 1)^2 - 3(x + 1)^{2/3}$

19. $y = 2 - x - x^3$

20. $y = -\tfrac{1}{3}(x^3 - 3x + 2)$

21. $y = 3x^4 + 4x^3$

22. $y = -2x^4 + 3x^2$

23. $y = x^5 - 5x$

24. $y = (x - 1)^5$

Analyzing the Graph of a Function Using Technology
In Exercises 25–34, use a computer algebra system to analyze and graph the function. Identify any relative extrema, points of inflection, and asymptotes.

25. $f(x) = \dfrac{20x}{x^2+1} - \dfrac{1}{x}$

26. $f(x) = x + \dfrac{4}{x^2+1}$

27. $f(x) = \dfrac{-2x}{\sqrt{x^2+7}}$

28. $f(x) = \dfrac{4x}{\sqrt{x^2+15}}$

29. $f(x) = 2x - 4\sin x, \quad 0 \le x \le 2\pi$

30. $f(x) = -x + 2\cos x, \quad 0 \le x \le 2\pi$

31. $y = \cos x - \tfrac{1}{4}\cos 2x, \quad 0 \le x \le 2\pi$

32. $y = 2x - \tan x, \quad -\dfrac{\pi}{2} < x < \dfrac{\pi}{2}$

33. $y = 2(\csc x + \sec x), \quad 0 < x < \dfrac{\pi}{2}$

34. $g(x) = x \cot x, \quad -2\pi < x < 2\pi$

WRITING ABOUT CONCEPTS

35. Using a Derivative Let $f'(t) < 0$ for all t in the interval $(2, 8)$. Explain why $f(3) > f(5)$.

36. Using a Derivative Let $f(0) = 3$ and $2 \le f'(x) \le 4$ for all x in the interval $[-5, 5]$. Determine the greatest and least possible values of $f(2)$.

Identifying Graphs In Exercises 37 and 38, the graphs of f, f', and f'' are shown on the same set of coordinate axes. Which is which? Explain your reasoning. To print an enlarged copy of the graph, go to *MathGraphs.com*.

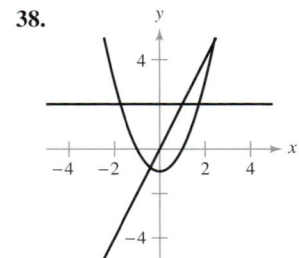

37. **38.**

Horizontal and Vertical Asymptotes   **In Exercises 39–42, use a graphing utility to graph the function. Use the graph to determine whether it is possible for the graph of a function to cross its horizontal asymptote. Do you think it is possible for the graph of a function to cross its vertical asymptote? Why or why not?**

39. $f(x) = \dfrac{4(x-1)^2}{x^2 - 4x + 5}$ **40.** $g(x) = \dfrac{3x^4 - 5x + 3}{x^4 + 1}$

41. $h(x) = \dfrac{\sin 2x}{x}$ **42.** $f(x) = \dfrac{\cos 3x}{4x}$

Examining a Function   **In Exercises 43 and 44, use a graphing utility to graph the function. Explain why there is no vertical asymptote when a superficial examination of the function may indicate that there should be one.**

43. $h(x) = \dfrac{6 - 2x}{3 - x}$ **44.** $g(x) = \dfrac{x^2 + x - 2}{x - 1}$

Slant Asymptote   **In Exercises 45–48, use a graphing utility to graph the function and determine the slant asymptote of the graph. Zoom out repeatedly and describe how the graph on the display appears to change. Why does this occur?**

45. $f(x) = -\dfrac{x^2 - 3x - 1}{x - 2}$ **46.** $g(x) = \dfrac{2x^2 - 8x - 15}{x - 5}$

47. $f(x) = \dfrac{2x^3}{x^2 + 1}$ **48.** $h(x) = \dfrac{-x^3 + x^2 + 4}{x^2}$

Graphical Reasoning   **In Exercises 49–52, use the graph of f' to sketch a graph of f and the graph of f''. To print an enlarged copy of the graph, go to *MathGraphs.com*.**

49.

50.

51.

52.

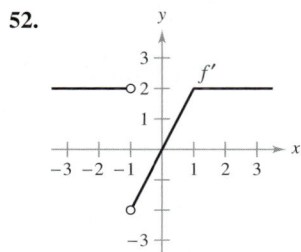

(Submitted by Bill Fox, Moberly Area Community College, Moberly, MO)

53. Graphical Reasoning   Consider the function

$$f(x) = \dfrac{\cos^2 \pi x}{\sqrt{x^2 + 1}}, \quad 0 < x < 4.$$

(a) Use a computer algebra system to graph the function and use the graph to approximate the critical numbers visually.

(b) Use a computer algebra system to find f' and approximate the critical numbers. Are the results the same as the visual approximation in part (a)? Explain.

54. Graphical Reasoning   Consider the function

$$f(x) = \tan(\sin \pi x).$$

(a) Use a graphing utility to graph the function.

(b) Identify any symmetry of the graph.

(c) Is the function periodic? If so, what is the period?

(d) Identify any extrema on $(-1, 1)$.

(e) Use a graphing utility to determine the concavity of the graph on $(0, 1)$.

Think About It   **In Exercises 55–58, create a function whose graph has the given characteristics. (There is more than one correct answer.)**

55. Vertical asymptote: $x = 3$

Horizontal asymptote: $y = 0$

56. Vertical asymptote: $x = -5$

Horizontal asymptote: None

57. Vertical asymptote: $x = 3$

Slant asymptote: $y = 3x + 2$

58. Vertical asymptote: $x = 2$

Slant asymptote: $y = -x$

59. Graphical Reasoning   Identify the real numbers x_0, x_1, x_2, x_3, and x_4 in the figure such that each of the following is true.

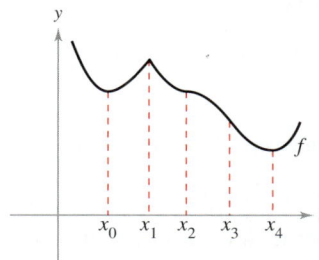

(a) $f'(x) = 0$

(b) $f''(x) = 0$

(c) $f'(x)$ does not exist.

(d) f has a relative maximum.

(e) f has a point of inflection.

60. **HOW DO YOU SEE IT?** The graph of f is shown in the figure.

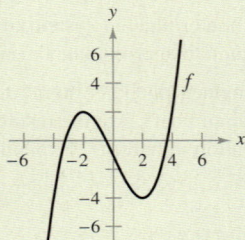

(a) For which values of x is $f'(x)$ zero? Positive? Negative? What do these values mean?

(b) For which values of x is $f''(x)$ zero? Positive? Negative? What do these values mean?

(c) On what open interval is f' an increasing function?

(d) For which value of x is $f'(x)$ minimum? For this value of x, how does the rate of change of f compare with the rates of change of f for other values of x? Explain.

61. Investigation Let $P(x_0, y_0)$ be an arbitrary point on the graph of f such that $f'(x_0) \neq 0$, as shown in the figure. Verify each statement.

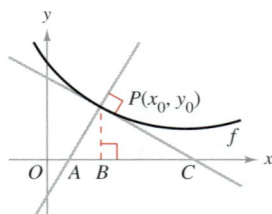

(a) The x-intercept of the tangent line is

$$\left(x_0 - \frac{f(x_0)}{f'(x_0)}, 0 \right).$$

(b) The y-intercept of the tangent line is

$$(0, f(x_0) - x_0 f'(x_0)).$$

(c) The x-intercept of the normal line is

$$(x_0 + f(x_0) f'(x_0), 0).$$

(d) The y-intercept of the normal line is

$$\left(0, y_0 + \frac{x_0}{f'(x_0)} \right).$$

(e) $|BC| = \left| \dfrac{f(x_0)}{f'(x_0)} \right|$

(f) $|PC| = \left| \dfrac{f(x_0) \sqrt{1 + [f'(x_0)]^2}}{f'(x_0)} \right|$

(g) $|AB| = |f(x_0) f'(x_0)|$

(h) $|AP| = |f(x_0)| \sqrt{1 + [f'(x_0)]^2}$

62. Investigation Consider the function

$$f(x) = \frac{2x^n}{x^4 + 1}$$

for nonnegative integer values of n.

(a) Discuss the relationship between the value of n and the symmetry of the graph.

(b) For which values of n will the x-axis be the horizontal asymptote?

(c) For which value of n will $y = 2$ be the horizontal asymptote?

(d) What is the asymptote of the graph when $n = 5$?

(e) Use a graphing utility to graph f for the indicated values of n in the table. Use the graph to determine the number of extrema M and the number of inflection points N of the graph.

n	0	1	2	3	4	5
M						
N						

63. Graphical Reasoning Consider the function

$$f(x) = \frac{ax}{(x - b)^2}.$$

Determine the effect on the graph of f as a and b are changed. Consider cases where a and b are both positive or both negative, and cases where a and b have opposite signs.

64. Graphical Reasoning Consider the function

$$f(x) = \frac{1}{2}(ax)^2 - ax, \quad a \neq 0.$$

(a) Determine the changes (if any) in the intercepts, extrema, and concavity of the graph of f when a is varied.

(b) In the same viewing window, use a graphing utility to graph the function for four different values of a.

Slant Asymptotes In Exercises 65 and 66, the graph of the function has two slant asymptotes. Identify each slant asymptote. Then graph the function and its asymptotes.

65. $y = \sqrt{4 + 16x^2}$ **66.** $y = \sqrt{x^2 + 6x}$

PUTNAM EXAM CHALLENGE

67. Let $f(x)$ be defined for $a \leq x \leq b$. Assuming appropriate properties of continuity and derivability, prove for $a < x < b$ that

$$\frac{\dfrac{f(x) - f(a)}{x - a} - \dfrac{f(b) - f(a)}{b - a}}{x - b} = \frac{1}{2} f''(\varepsilon),$$

where ε is some number between a and b.

3.7 Optimization Problems

■ Solve applied minimum and maximum problems.

Applied Minimum and Maximum Problems

One of the most common applications of calculus involves the determination of minimum and maximum values. Consider how frequently you hear or read terms such as greatest profit, least cost, least time, greatest voltage, optimum size, least size, greatest strength, and greatest distance. Before outlining a general problem-solving strategy for such problems, consider the next example.

EXAMPLE 1 Finding Maximum Volume

A manufacturer wants to design an open box having a square base and a surface area of 108 square inches, as shown in Figure 3.53. What dimensions will produce a box with maximum volume?

Solution Because the box has a square base, its volume is

$$V = x^2 h. \qquad \text{Primary equation}$$

This equation is called the **primary equation** because it gives a formula for the quantity to be optimized. The surface area of the box is

$$S = (\text{area of base}) + (\text{area of four sides})$$
$$108 = x^2 + 4xh. \qquad \text{Secondary equation}$$

Because V is to be maximized, you want to write V as a function of just one variable. To do this, you can solve the equation $x^2 + 4xh = 108$ for h in terms of x to obtain $h = (108 - x^2)/(4x)$. Substituting into the primary equation produces

$$V = x^2 h \qquad \text{Function of two variables}$$
$$= x^2 \left(\frac{108 - x^2}{4x} \right) \qquad \text{Substitute for } h.$$
$$= 27x - \frac{x^3}{4}. \qquad \text{Function of one variable}$$

Before finding which x-value will yield a maximum value of V, you should determine the *feasible domain*. That is, what values of x make sense in this problem? You know that $V \geq 0$. You also know that x must be nonnegative and that the area of the base $(A = x^2)$ is at most 108. So, the feasible domain is

$$0 \leq x \leq \sqrt{108}. \qquad \text{Feasible domain}$$

To maximize V, find the critical numbers of the volume function on the interval $\left(0, \sqrt{108} \right)$.

$$\frac{dV}{dx} = 27 - \frac{3x^2}{4} \qquad \text{Differentiate with respect to } x.$$
$$27 - \frac{3x^2}{4} = 0 \qquad \text{Set derivative equal to 0.}$$
$$3x^2 = 108 \qquad \text{Simplify.}$$
$$x = \pm 6 \qquad \text{Critical numbers}$$

So, the critical numbers are $x = \pm 6$. You do not need to consider $x = -6$ because it is outside the domain. Evaluating V at the critical number $x = 6$ and at the endpoints of the domain produces $V(0) = 0$, $V(6) = 108$, and $V\left(\sqrt{108} \right) = 0$. So, V is maximum when $x = 6$, and the dimensions of the box are 6 inches by 6 inches by 3 inches. ■

Open box with square base:
$S = x^2 + 4xh = 108$
Figure 3.53

▷ TECHNOLOGY You can verify your answer in Example 1 by using a graphing utility to graph the volume function

$$V = 27x - \frac{x^3}{4}.$$

Use a viewing window in which $0 \leq x \leq \sqrt{108} \approx 10.4$ and $0 \leq y \leq 120$, and use the *maximum* or *trace* feature to determine the maximum value of V.

In Example 1, you should realize that there are infinitely many open boxes having 108 square inches of surface area. To begin solving the problem, you might ask yourself which basic shape would seem to yield a maximum volume. Should the box be tall, squat, or nearly cubical?

You might even try calculating a few volumes, as shown in Figure 3.54, to see if you can get a better feeling for what the optimum dimensions should be. Remember that you are not ready to begin solving a problem until you have clearly identified what the problem is.

Volume = $74\frac{1}{4}$

Volume = 92

Volume = $103\frac{3}{4}$

$5 \times 5 \times 4\frac{3}{20}$

$4 \times 4 \times 5\frac{3}{4}$

$3 \times 3 \times 8\frac{1}{4}$

Volume = 108

Volume = 88

$6 \times 6 \times 3$

$8 \times 8 \times 1\frac{3}{8}$

Which box has the greatest volume?

Figure 3.54

Example 1 illustrates the following guidelines for solving applied minimum and maximum problems.

GUIDELINES FOR SOLVING APPLIED MINIMUM AND MAXIMUM PROBLEMS

1. Identify all *given* quantities and all quantities *to be determined*. If possible, make a sketch.

2. Write a **primary equation** for the quantity that is to be maximized or minimized. (A review of several useful formulas from geometry is presented inside the back cover.)

3. Reduce the primary equation to one having a *single independent variable*. This may involve the use of **secondary equations** relating the independent variables of the primary equation.

4. Determine the feasible domain of the primary equation. That is, determine the values for which the stated problem makes sense.

5. Determine the desired maximum or minimum value by the calculus techniques discussed in Sections 3.1 through 3.4.

• • • • • ▷ REMARK For Step 5, recall that to determine the maximum or minimum value of a continuous function f on a closed interval, you should compare the values of f at its critical numbers with the values of f at the endpoints of the interval.

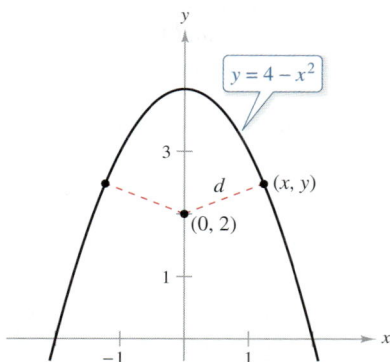

The quantity to be minimized is distance: $d = \sqrt{(x-0)^2 + (y-2)^2}$.

Figure 3.55

EXAMPLE 2 **Finding Minimum Distance**

···▷ *See LarsonCalculus.com for an interactive version of this type of example.*

Which points on the graph of $y = 4 - x^2$ are closest to the point $(0, 2)$?

Solution Figure 3.55 shows that there are two points at a minimum distance from the point $(0, 2)$. The distance between the point $(0, 2)$ and a point (x, y) on the graph of $y = 4 - x^2$ is

$$d = \sqrt{(x-0)^2 + (y-2)^2}. \qquad \text{Primary equation}$$

Using the secondary equation $y = 4 - x^2$, you can rewrite the primary equation as

$$d = \sqrt{x^2 + (4 - x^2 - 2)^2}$$
$$= \sqrt{x^4 - 3x^2 + 4}.$$

Because d is smallest when the expression inside the radical is smallest, you need only find the critical numbers of $f(x) = x^4 - 3x^2 + 4$. Note that the domain of f is the entire real number line. So, there are no endpoints of the domain to consider. Moreover, the derivative of f

$$f'(x) = 4x^3 - 6x$$
$$= 2x(2x^2 - 3)$$

is zero when

$$x = 0, \ \sqrt{\frac{3}{2}}, \ -\sqrt{\frac{3}{2}}.$$

Testing these critical numbers using the First Derivative Test verifies that $x = 0$ yields a relative maximum, whereas both $x = \sqrt{3/2}$ and $x = -\sqrt{3/2}$ yield a minimum distance. So, the closest points are $\left(\sqrt{3/2}, 5/2\right)$ and $\left(-\sqrt{3/2}, 5/2\right)$.

EXAMPLE 3 **Finding Minimum Area**

The quantity to be minimized is area: $A = (x+3)(y+2)$.

Figure 3.56

A rectangular page is to contain 24 square inches of print. The margins at the top and bottom of the page are to be $1\frac{1}{2}$ inches, and the margins on the left and right are to be 1 inch (see Figure 3.56). What should the dimensions of the page be so that the least amount of paper is used?

Solution Let A be the area to be minimized.

$$A = (x+3)(y+2) \qquad \text{Primary equation}$$

The printed area inside the margins is

$$24 = xy. \qquad \text{Secondary equation}$$

Solving this equation for y produces $y = 24/x$. Substitution into the primary equation produces

$$A = (x+3)\left(\frac{24}{x} + 2\right) = 30 + 2x + \frac{72}{x}. \qquad \text{Function of one variable}$$

Because x must be positive, you are interested only in values of A for $x > 0$. To find the critical numbers, differentiate with respect to x

$$\frac{dA}{dx} = 2 - \frac{72}{x^2}$$

and note that the derivative is zero when $x^2 = 36$, or $x = \pm 6$. So, the critical numbers are $x = \pm 6$. You do not have to consider $x = -6$ because it is outside the domain. The First Derivative Test confirms that A is a minimum when $x = 6$. So, $y = \frac{24}{6} = 4$ and the dimensions of the page should be $x + 3 = 9$ inches by $y + 2 = 6$ inches.

Finding Minimum Length

Two posts, one 12 feet high and the other 28 feet high, stand 30 feet apart. They are to be stayed by two wires, attached to a single stake, running from ground level to the top of each post. Where should the stake be placed to use the least amount of wire?

Solution Let W be the wire length to be minimized. Using Figure 3.57, you can write

$$W = y + z. \qquad \text{Primary equation}$$

In this problem, rather than solving for y in terms of z (or vice versa), you can solve for both y and z in terms of a third variable x, as shown in Figure 3.57. From the Pythagorean Theorem, you obtain

$$x^2 + 12^2 = y^2$$
$$(30 - x)^2 + 28^2 = z^2$$

which implies that

$$y = \sqrt{x^2 + 144}$$
$$z = \sqrt{x^2 - 60x + 1684}.$$

The quantity to be minimized is length. From the diagram, you can see that x varies between 0 and 30.

Figure 3.57

So, you can rewrite the primary equation as

$$W = y + z$$
$$= \sqrt{x^2 + 144} + \sqrt{x^2 - 60x + 1684}, \quad 0 \le x \le 30.$$

Differentiating W with respect to x yields

$$\frac{dW}{dx} = \frac{x}{\sqrt{x^2 + 144}} + \frac{x - 30}{\sqrt{x^2 - 60x + 1684}}.$$

By letting $dW/dx = 0$, you obtain

$$\frac{x}{\sqrt{x^2 + 144}} + \frac{x - 30}{\sqrt{x^2 - 60x + 1684}} = 0$$
$$x\sqrt{x^2 - 60x + 1684} = (30 - x)\sqrt{x^2 + 144}$$
$$x^2(x^2 - 60x + 1684) = (30 - x)^2(x^2 + 144)$$
$$x^4 - 60x^3 + 1684x^2 = x^4 - 60x^3 + 1044x^2 - 8640x + 129{,}600$$
$$640x^2 + 8640x - 129{,}600 = 0$$
$$320(x - 9)(2x + 45) = 0$$
$$x = 9, \ -22.5.$$

Because $x = -22.5$ is not in the domain and

$$W(0) \approx 53.04, \quad W(9) = 50, \quad \text{and} \quad W(30) \approx 60.31$$

you can conclude that the wire should be staked at 9 feet from the 12-foot pole. ■

▷ **TECHNOLOGY** From Example 4, you can see that applied optimization problems can involve a lot of algebra. If you have access to a graphing utility, you can confirm that $x = 9$ yields a minimum value of W by graphing

$$W = \sqrt{x^2 + 144} + \sqrt{x^2 - 60x + 1684}$$

as shown in Figure 3.58.

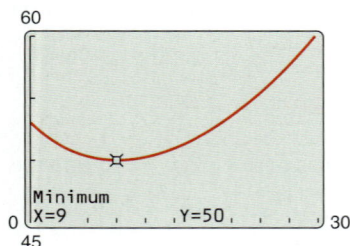

You can confirm the minimum value of W with a graphing utility.

Figure 3.58

In each of the first four examples, the extreme value occurred at a critical number. Although this happens often, remember that an extreme value can also occur at an endpoint of an interval, as shown in Example 5.

| EXAMPLE 5 | **An Endpoint Maximum** |

Four feet of wire is to be used to form a square and a circle. How much of the wire should be used for the square and how much should be used for the circle to enclose the maximum total area?

Solution The total area (see Figure 3.59) is

$$A = \text{(area of square)} + \text{(area of circle)}$$

$$A = x^2 + \pi r^2. \qquad \text{Primary equation}$$

Because the total length of wire is 4 feet, you obtain

$$4 = \text{(perimeter of square)} + \text{(circumference of circle)}$$

$$4 = 4x + 2\pi r.$$

So, $r = 2(1 - x)/\pi$, and by substituting into the primary equation you have

$$A = x^2 + \pi \left[\frac{2(1 - x)}{\pi} \right]^2$$

$$= x^2 + \frac{4(1 - x)^2}{\pi}$$

$$= \frac{1}{\pi}[(\pi + 4)x^2 - 8x + 4].$$

The feasible domain is $0 \le x \le 1$, restricted by the square's perimeter. Because

$$\frac{dA}{dx} = \frac{2(\pi + 4)x - 8}{\pi}$$

the only critical number in $(0, 1)$ is $x = 4/(\pi + 4) \approx 0.56$. So, using

$$A(0) \approx 1.273, \quad A(0.56) \approx 0.56, \quad \text{and} \quad A(1) = 1$$

you can conclude that the maximum area occurs when $x = 0$. That is, *all* the wire is used for the circle. ∎

The quantity to be maximized is area:
$A = x^2 + \pi r^2$.

Figure 3.59

Exploration

What would the answer be if Example 5 asked for the dimensions needed to enclose the *minimum* total area?

Before doing the section exercises, review the primary equations developed in the first five examples. As applications go, these five examples are fairly simple, and yet the resulting primary equations are quite complicated.

$$V = 27x - \frac{x^3}{4} \qquad \text{Example 1}$$

$$d = \sqrt{x^4 - 3x^2 + 4} \qquad \text{Example 2}$$

$$A = 30 + 2x + \frac{72}{x} \qquad \text{Example 3}$$

$$W = \sqrt{x^2 + 144} + \sqrt{x^2 - 60x + 1684} \qquad \text{Example 4}$$

$$A = \frac{1}{\pi}[(\pi + 4)x^2 - 8x + 4] \qquad \text{Example 5}$$

You must expect that real-life applications often involve equations that are *at least as complicated* as these five. Remember that one of the main goals of this course is to learn to use calculus to analyze equations that initially seem formidable.

3.7 Exercises

See **CalcChat.com** for tutorial help and worked-out solutions to odd-numbered exercises.

1. **Numerical, Graphical, and Analytic Analysis** Find two positive numbers whose sum is 110 and whose product is a maximum.

   (a) Analytically complete six rows of a table such as the one below. (The first two rows are shown.)

First Number, x	Second Number	Product, P
10	$110 - 10$	$10(110 - 10) = 1000$
20	$110 - 20$	$20(110 - 20) = 1800$

   (b) Use a graphing utility to generate additional rows of the table. Use the table to estimate the solution. (*Hint:* Use the table feature of the graphing utility.)

   (c) Write the product P as a function of x.

   (d) Use a graphing utility to graph the function in part (c) and estimate the solution from the graph.

   (e) Use calculus to find the critical number of the function in part (c). Then find the two numbers.

2. **Numerical, Graphical, and Analytic Analysis** An open box of maximum volume is to be made from a square piece of material, 24 inches on a side, by cutting equal squares from the corners and turning up the sides (see figure).

   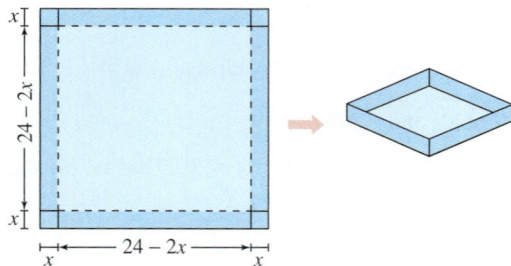

   (a) Analytically complete six rows of a table such as the one below. (The first two rows are shown.) Use the table to guess the maximum volume.

Height, x	Length and Width	Volume, V
1	$24 - 2(1)$	$1[24 - 2(1)]^2 = 484$
2	$24 - 2(2)$	$2[24 - 2(2)]^2 = 800$

   (b) Write the volume V as a function of x.

   (c) Use calculus to find the critical number of the function in part (b) and find the maximum value.

   (d) Use a graphing utility to graph the function in part (b) and verify the maximum volume from the graph.

Finding Numbers In Exercises 3–8, find two positive numbers that satisfy the given requirements.

3. The sum is S and the product is a maximum.

4. The product is 185 and the sum is a minimum.

5. The product is 147 and the sum of the first number plus three times the second number is a minimum.

6. The second number is the reciprocal of the first number and the sum is a minimum.

7. The sum of the first number and twice the second number is 108 and the product is a maximum.

8. The sum of the first number squared and the second number is 54 and the product is a maximum.

Maximum Area In Exercises 9 and 10, find the length and width of a rectangle that has the given perimeter and a maximum area.

9. Perimeter: 80 meters

10. Perimeter: P units

Minimum Perimeter In Exercises 11 and 12, find the length and width of a rectangle that has the given area and a minimum perimeter.

11. Area: 32 square feet

12. Area: A square centimeters

Minimum Distance In Exercises 13–16, find the point on the graph of the function that is closest to the given point.

13. $f(x) = x^2$, $\left(2, \frac{1}{2}\right)$

14. $f(x) = (x - 1)^2$, $(-5, 3)$

15. $f(x) = \sqrt{x}$, $(4, 0)$

16. $f(x) = \sqrt{x - 8}$, $(12, 0)$

17. **Minimum Area** A rectangular page is to contain 30 square inches of print. The margins on each side are 1 inch. Find the dimensions of the page such that the least amount of paper is used.

18. **Minimum Area** A rectangular page is to contain 36 square inches of print. The margins on each side are $1\frac{1}{2}$ inches. Find the dimensions of the page such that the least amount of paper is used.

19. **Minimum Length** A farmer plans to fence a rectangular pasture adjacent to a river (see figure). The pasture must contain 245,000 square meters in order to provide enough grass for the herd. No fencing is needed along the river. What dimensions will require the least amount of fencing?

20. Maximum Volume A rectangular solid (with a square base) has a surface area of 337.5 square centimeters. Find the dimensions that will result in a solid with maximum volume.

21. Maximum Area A Norman window is constructed by adjoining a semicircle to the top of an ordinary rectangular window (see figure). Find the dimensions of a Norman window of maximum area when the total perimeter is 16 feet.

22. Maximum Area A rectangle is bounded by the x- and y-axes and the graph of $y = (6 - x)/2$ (see figure). What length and width should the rectangle have so that its area is a maximum?

Figure for 22

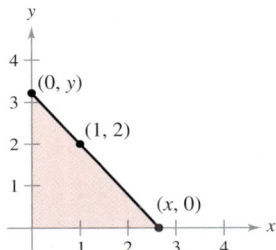

Figure for 23

23. Minimum Length and Minimum Area A right triangle is formed in the first quadrant by the x- and y-axes and a line through the point $(1, 2)$ (see figure).

(a) Write the length L of the hypotenuse as a function of x.

(b) Use a graphing utility to approximate x graphically such that the length of the hypotenuse is a minimum.

(c) Find the vertices of the triangle such that its area is a minimum.

24. Maximum Area Find the area of the largest isosceles triangle that can be inscribed in a circle of radius 6 (see figure).

(a) Solve by writing the area as a function of h.

(b) Solve by writing the area as a function of α.

(c) Identify the type of triangle of maximum area.

Figure for 24

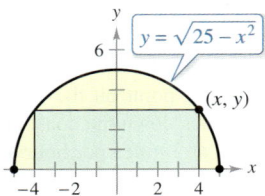

Figure for 25

25. Maximum Area A rectangle is bounded by the x-axis and the semicircle

$$y = \sqrt{25 - x^2}$$

(see figure). What length and width should the rectangle have so that its area is a maximum?

26. Maximum Area Find the dimensions of the largest rectangle that can be inscribed in a semicircle of radius r (see Exercise 25).

27. Numerical, Graphical, and Analytic Analysis An exercise room consists of a rectangle with a semicircle on each end. A 200-meter running track runs around the outside of the room.

(a) Draw a figure to represent the problem. Let x and y represent the length and width of the rectangle.

(b) Analytically complete six rows of a table such as the one below. (The first two rows are shown.) Use the table to guess the maximum area of the rectangular region.

Length, x	Width, y	Area, xy
10	$\frac{2}{\pi}(100 - 10)$	$(10)\frac{2}{\pi}(100 - 10) \approx 573$
20	$\frac{2}{\pi}(100 - 20)$	$(20)\frac{2}{\pi}(100 - 20) \approx 1019$

(c) Write the area A as a function of x.

(d) Use calculus to find the critical number of the function in part (c) and find the maximum value.

(e) Use a graphing utility to graph the function in part (c) and verify the maximum area from the graph.

28. Numerical, Graphical, and Analytic Analysis A right circular cylinder is designed to hold 22 cubic inches of a soft drink (approximately 12 fluid ounces).

(a) Analytically complete six rows of a table such as the one below. (The first two rows are shown.)

Radius, r	Height	Surface Area, S
0.2	$\dfrac{22}{\pi(0.2)^2}$	$2\pi(0.2)\left[0.2 + \dfrac{22}{\pi(0.2)^2}\right] \approx 220.3$
0.4	$\dfrac{22}{\pi(0.4)^2}$	$2\pi(0.4)\left[0.4 + \dfrac{22}{\pi(0.4)^2}\right] \approx 111.0$

(b) Use a graphing utility to generate additional rows of the table. Use the table to estimate the minimum surface area. (*Hint:* Use the *table* feature of the graphing utility.)

(c) Write the surface area S as a function of r.

(d) Use a graphing utility to graph the function in part (c) and estimate the minimum surface area from the graph.

(e) Use calculus to find the critical number of the function in part (c) and find dimensions that will yield the minimum surface area.

29. Maximum Volume A rectangular package to be sent by a postal service can have a maximum combined length and girth (perimeter of a cross section) of 108 inches (see figure). Find the dimensions of the package of maximum volume that can be sent. (Assume the cross section is square.)

30. Maximum Volume Rework Exercise 29 for a cylindrical package. (The cross section is circular.)

WRITING ABOUT CONCEPTS

31. Surface Area and Volume A shampoo bottle is a right circular cylinder. Because the surface area of the bottle does not change when it is squeezed, is it true that the volume remains the same? Explain.

32. Area and Perimeter The perimeter of a rectangle is 20 feet. Of all possible dimensions, the maximum area is 25 square feet when its length and width are both 5 feet. Are there dimensions that yield a minimum area? Explain.

---

33. Minimum Surface Area A solid is formed by adjoining two hemispheres to the ends of a right circular cylinder. The total volume of the solid is 14 cubic centimeters. Find the radius of the cylinder that produces the minimum surface area.

34. Minimum Cost An industrial tank of the shape described in Exercise 33 must have a volume of 4000 cubic feet. The hemispherical ends cost twice as much per square foot of surface area as the sides. Find the dimensions that will minimize cost.

35. Minimum Area The sum of the perimeters of an equilateral triangle and a square is 10. Find the dimensions of the triangle and the square that produce a minimum total area.

36. Maximum Area Twenty feet of wire is to be used to form two figures. In each of the following cases, how much wire should be used for each figure so that the total enclosed area is maximum?

(a) Equilateral triangle and square

(b) Square and regular pentagon

(c) Regular pentagon and regular hexagon

(d) Regular hexagon and circle

What can you conclude from this pattern? {*Hint:* The area of a regular polygon with n sides of length x is $A = (n/4)[\cot(\pi/n)]x^2$.}

37. Beam Strength A wooden beam has a rectangular cross section of height h and width w (see figure). The strength S of the beam is directly proportional to the width and the square of the height. What are the dimensions of the strongest beam that can be cut from a round log of diameter 20 inches? (*Hint:* $S = kh^2w$, where k is the proportionality constant.)

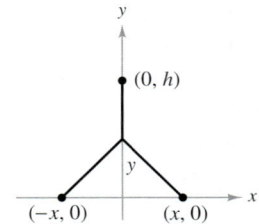

Figure for 37   Figure for 38

38. Minimum Length Two factories are located at the coordinates $(-x, 0)$ and $(x, 0)$, and their power supply is at $(0, h)$ (see figure). Find y such that the total length of power line from the power supply to the factories is a minimum.

39. Minimum Cost

An offshore oil well is 2 kilometers off the coast. The refinery is 4 kilometers down the coast. Laying pipe in the ocean is twice as expensive as laying it on land. What path should the pipe follow in order to minimize the cost?

40. Illumination A light source is located over the center of a circular table of diameter 4 feet (see figure). Find the height h of the light source such that the illumination I at the perimeter of the table is maximum when

$$I = \frac{k \sin \alpha}{s^2}$$

where s is the slant height, α is the angle at which the light strikes the table, and k is a constant.

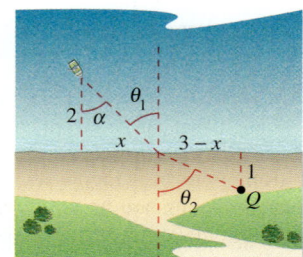

Figure for 40   Figure for 41

41. Minimum Time A man is in a boat 2 miles from the nearest point on the coast. He is to go to a point Q, located 3 miles down the coast and 1 mile inland (see figure). He can row at 2 miles per hour and walk at 4 miles per hour. Toward what point on the coast should he row in order to reach point Q in the least time?

42. Minimum Time The conditions are the same as in Exercise 41 except that the man can row at v_1 miles per hour and walk at v_2 miles per hour. If θ_1 and θ_2 are the magnitudes of the angles, show that the man will reach point Q in the least time when

$$\frac{\sin \theta_1}{v_1} = \frac{\sin \theta_2}{v_2}.$$

43. Minimum Distance Sketch the graph of $f(x) = 2 - 2 \sin x$ on the interval $[0, \pi/2]$.

(a) Find the distance from the origin to the y-intercept and the distance from the origin to the x-intercept.

(b) Write the distance d from the origin to a point on the graph of f as a function of x. Use your graphing utility to graph d and find the minimum distance.

(c) Use calculus and the *zero* or *root* feature of a graphing utility to find the value of x that minimizes the function d on the interval $[0, \pi/2]$. What is the minimum distance?

(Submitted by Tim Chapell, Penn Valley Community College, Kansas City, MO)

44. Minimum Time When light waves traveling in a transparent medium strike the surface of a second transparent medium, they change direction. This change of direction is called *refraction* and is defined by **Snell's Law of Refraction,**

$$\frac{\sin \theta_1}{v_1} = \frac{\sin \theta_2}{v_2}$$

where θ_1 and θ_2 are the magnitudes of the angles shown in the figure and v_1 and v_2 are the velocities of light in the two media. Show that this problem is equivalent to that in Exercise 42, and that light waves traveling from P to Q follow the path of minimum time.

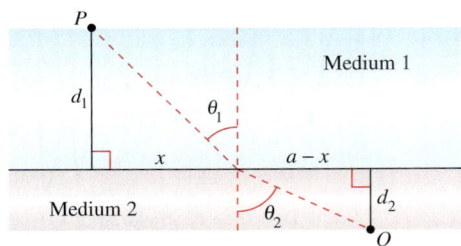

45. Maximum Volume A sector with central angle θ is cut from a circle of radius 12 inches (see figure), and the edges of the sector are brought together to form a cone. Find the magnitude of θ such that the volume of the cone is a maximum.

Figure for 45

Figure for 46

46. Numerical, Graphical, and Analytic Analysis The cross sections of an irrigation canal are isosceles trapezoids of which three sides are 8 feet long (see figure). Determine the angle of elevation θ of the sides such that the area of the cross sections is a maximum by completing the following.

(a) Analytically complete six rows of a table such as the one below. (The first two rows are shown.)

Base 1	Base 2	Altitude	Area
8	$8 + 16 \cos 10°$	$8 \sin 10°$	≈ 22.1
8	$8 + 16 \cos 20°$	$8 \sin 20°$	≈ 42.5

(b) Use a graphing utility to generate additional rows of the table and estimate the maximum cross-sectional area. (*Hint:* Use the *table* feature of the graphing utility.)

(c) Write the cross-sectional area A as a function of θ.

(d) Use calculus to find the critical number of the function in part (c) and find the angle that will yield the maximum cross-sectional area.

(e) Use a graphing utility to graph the function in part (c) and verify the maximum cross-sectional area.

47. Maximum Profit Assume that the amount of money deposited in a bank is proportional to the square of the interest rate the bank pays on this money. Furthermore, the bank can reinvest this money at 12%. Find the interest rate the bank should pay to maximize profit. (Use the simple interest formula.)

48. HOW DO YOU SEE IT? The graph shows the profit P (in thousands of dollars) of a company in terms of its advertising cost x (in thousands of dollars).

(a) Estimate the interval on which the profit is increasing.

(b) Estimate the interval on which the profit is decreasing.

(c) Estimate the amount of money the company should spend on advertising in order to yield a maximum profit.

(d) The *point of diminishing returns* is the point at which the rate of growth of the profit function begins to decline. Estimate the point of diminishing returns.

Minimum Distance In Exercises 49–51, consider a fuel distribution center located at the origin of the rectangular coordinate system (units in miles; see figures). The center supplies three factories with coordinates $(4, 1)$, $(5, 6)$, and $(10, 3)$. A trunk line will run from the distribution center along the line $y = mx$, and feeder lines will run to the three factories. The objective is to find m such that the lengths of the feeder lines are minimized.

49. Minimize the sum of the squares of the lengths of the vertical feeder lines (see figure) given by

$$S_1 = (4m - 1)^2 + (5m - 6)^2 + (10m - 3)^2.$$

Find the equation of the trunk line by this method and then determine the sum of the lengths of the feeder lines.

50. Minimize the sum of the absolute values of the lengths of the vertical feeder lines (see figure) given by

$$S_2 = |4m - 1| + |5m - 6| + |10m - 3|.$$

Find the equation of the trunk line by this method and then determine the sum of the lengths of the feeder lines. (*Hint*: Use a graphing utility to graph the function S_2 and approximate the required critical number.)

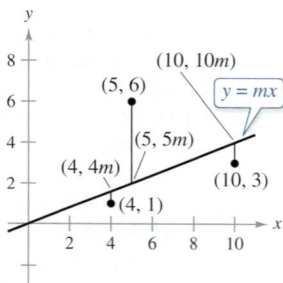

Figure for 49 and 50   Figure for 51

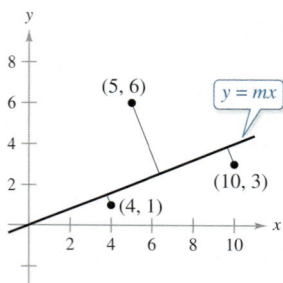

51. Minimize the sum of the perpendicular distances (see figure and Exercises 83–86 in Section P.2) from the trunk line to the factories given by

$$S_3 = \frac{|4m - 1|}{\sqrt{m^2 + 1}} + \frac{|5m - 6|}{\sqrt{m^2 + 1}} + \frac{|10m - 3|}{\sqrt{m^2 + 1}}.$$

Find the equation of the trunk line by this method and then determine the sum of the lengths of the feeder lines. (*Hint*: Use a graphing utility to graph the function S_3 and approximate the required critical number.)

52. **Maximum Area** Consider a symmetric cross inscribed in a circle of radius r (see figure).

(a) Write the area A of the cross as a function of x and find the value of x that maximizes the area.

(b) Write the area A of the cross as a function of θ and find the value of θ that maximizes the area.

(c) Show that the critical numbers of parts (a) and (b) yield the same maximum area. What is that area?

PUTNAM EXAM CHALLENGE

53. Find, with explanation, the maximum value of $f(x) = x^3 - 3x$ on the set of all real numbers x satisfying $x^4 + 36 \leq 13x^2$.

54. Find the minimum value of

$$\frac{(x + 1/x)^6 - (x^6 + 1/x^6) - 2}{(x + 1/x)^3 + (x^3 + 1/x^3)} \quad \text{for} \quad x > 0.$$

These problems were composed by the Committee on the Putnam Prize Competition. © The Mathematical Association of America. All rights reserved.

SECTION PROJECT

Connecticut River

Whenever the Connecticut River reaches a level of 105 feet above sea level, two Northampton, Massachusetts, flood control station operators begin a round-the-clock river watch. Every 2 hours, they check the height of the river, using a scale marked off in tenths of a foot, and record the data in a log book. In the spring of 1996, the flood watch lasted from April 4, when the river reached 105 feet and was rising at 0.2 foot per hour, until April 25, when the level subsided again to 105 feet. Between those dates, their log shows that the river rose and fell several times, at one point coming close to the 115-foot mark. If the river had reached 115 feet, the city would have closed down Mount Tom Road (Route 5, south of Northampton).

The graph below shows the rate of change of the level of the river during one portion of the flood watch. Use the graph to answer each question.

Day (0 ↔ 12:01 A.M. April 14)

(a) On what date was the river rising most rapidly? How do you know?

(b) On what date was the river falling most rapidly? How do you know?

(c) There were two dates in a row on which the river rose, then fell, then rose again during the course of the day. On which days did this occur, and how do you know?

(d) At 1 minute past midnight, April 14, the river level was 111.0 feet. Estimate its height 24 hours later and 48 hours later. Explain how you made your estimates.

(e) The river crested at 114.4 feet. On what date do you think this occurred?

(Submitted by Mary Murphy, Smith College, Northampton, MA)

3.8 Newton's Method

■ Approximate a zero of a function using Newton's Method.

Newton's Method

In this section, you will study a technique for approximating the real zeros of a function. The technique is called **Newton's Method**, and it uses tangent lines to approximate the graph of the function near its x-intercepts.

To see how Newton's Method works, consider a function f that is continuous on the interval $[a, b]$ and differentiable on the interval (a, b). If $f(a)$ and $f(b)$ differ in sign, then, by the Intermediate Value Theorem, f must have at least one zero in the interval (a, b). To estimate this zero, you choose

$$x = x_1 \qquad \text{First estimate}$$

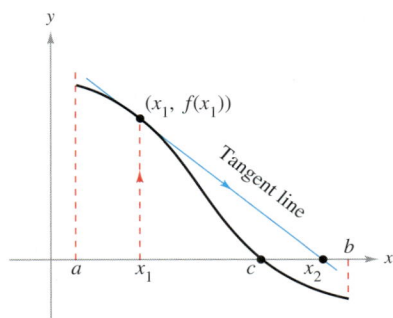

(a)

as shown in Figure 3.60(a). Newton's Method is based on the assumption that the graph of f and the tangent line at $(x_1, f(x_1))$ both cross the x-axis at *about* the same point. Because you can easily calculate the x-intercept for this tangent line, you can use it as a second (and, usually, better) estimate of the zero of f. The tangent line passes through the point $(x_1, f(x_1))$ with a slope of $f'(x_1)$. In point-slope form, the equation of the tangent line is

$$y - f(x_1) = f'(x_1)(x - x_1)$$
$$y = f'(x_1)(x - x_1) + f(x_1).$$

Letting $y = 0$ and solving for x produces

$$x = x_1 - \frac{f(x_1)}{f'(x_1)}.$$

So, from the initial estimate x_1, you obtain a new estimate

$$x_2 = x_1 - \frac{f(x_1)}{f'(x_1)}. \qquad \text{Second estimate [See Figure 3.60(b).]}$$

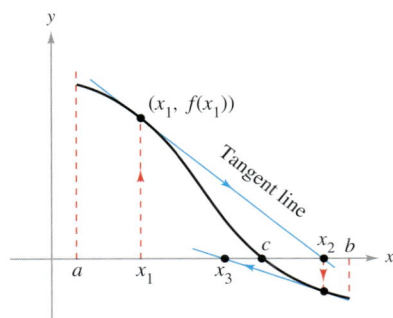

(b)
The x-intercept of the tangent line approximates the zero of f.
Figure 3.60

You can improve on x_2 and calculate yet a third estimate

$$x_3 = x_2 - \frac{f(x_2)}{f'(x_2)}. \qquad \text{Third estimate}$$

Repeated application of this process is called Newton's Method.

NEWTON'S METHOD

Isaac Newton first described the method for approximating the real zeros of a function in his text *Method of Fluxions.* Although the book was written in 1671, it was not published until 1736. Meanwhile, in 1690, Joseph Raphson (1648–1715) published a paper describing a method for approximating the real zeros of a function that was very similar to Newton's. For this reason, the method is often referred to as the Newton-Raphson method.

Newton's Method for Approximating the Zeros of a Function

Let $f(c) = 0$, where f is differentiable on an open interval containing c. Then, to approximate c, use these steps.

1. Make an initial estimate x_1 that is close to c. (A graph is helpful.)
2. Determine a new approximation

$$x_{n+1} = x_n - \frac{f(x_n)}{f'(x_n)}.$$

3. When $|x_n - x_{n+1}|$ is within the desired accuracy, let x_{n+1} serve as the final approximation. Otherwise, return to Step 2 and calculate a new approximation.

Each successive application of this procedure is called an **iteration.**

· · · · · · · · · · · · · · · ▷

·· REMARK For many functions, just a few iterations of Newton's Method will produce approximations having very small errors, as shown in Example 1.

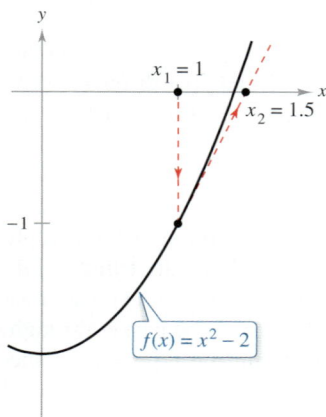

The first iteration of Newton's Method
Figure 3.61

EXAMPLE 1 Using Newton's Method

Calculate three iterations of Newton's Method to approximate a zero of $f(x) = x^2 - 2$. Use $x_1 = 1$ as the initial guess.

Solution Because $f(x) = x^2 - 2$, you have $f'(x) = 2x$, and the iterative formula is

$$x_{n+1} = x_n - \frac{f(x_n)}{f'(x_n)} = x_n - \frac{x_n^2 - 2}{2x_n}.$$

The calculations for three iterations are shown in the table.

n	x_n	$f(x_n)$	$f'(x_n)$	$\dfrac{f(x_n)}{f'(x_n)}$	$x_n - \dfrac{f(x_n)}{f'(x_n)}$
1	1.000000	-1.000000	2.000000	-0.500000	1.500000
2	1.500000	0.250000	3.000000	0.083333	1.416667
3	1.416667	0.006945	2.833334	0.002451	1.414216
4	1.414216				

Of course, in this case you know that the two zeros of the function are $\pm\sqrt{2}$. To six decimal places, $\sqrt{2} = 1.414214$. So, after only three iterations of Newton's Method, you have obtained an approximation that is within 0.000002 of an actual root. The first iteration of this process is shown in Figure 3.61.

EXAMPLE 2 Using Newton's Method

· · · ·▷ *See LarsonCalculus.com for an interactive version of this type of example.*

Use Newton's Method to approximate the zeros of

$$f(x) = 2x^3 + x^2 - x + 1.$$

Continue the iterations until two successive approximations differ by less than 0.0001.

Solution Begin by sketching a graph of f, as shown in Figure 3.62. From the graph, you can observe that the function has only one zero, which occurs near $x = -1.2$. Next, differentiate f and form the iterative formula

$$x_{n+1} = x_n - \frac{f(x_n)}{f'(x_n)} = x_n - \frac{2x_n^3 + x_n^2 - x_n + 1}{6x_n^2 + 2x_n - 1}.$$

The calculations are shown in the table.

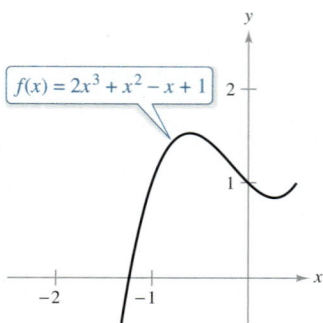

After three iterations of Newton's Method, the zero of f is approximated to the desired accuracy.
Figure 3.62

n	x_n	$f(x_n)$	$f'(x_n)$	$\dfrac{f(x_n)}{f'(x_n)}$	$x_n - \dfrac{f(x_n)}{f'(x_n)}$
1	-1.20000	0.18400	5.24000	0.03511	-1.23511
2	-1.23511	-0.00771	5.68276	-0.00136	-1.23375
3	-1.23375	0.00001	5.66533	0.00000	-1.23375
4	-1.23375				

Because two successive approximations differ by less than the required 0.0001, you can estimate the zero of f to be -1.23375.

FOR FURTHER INFORMATION
For more on when Newton's
Method fails, see the article
"No Fooling! Newton's Method
Can Be Fooled" by Peter Horton
in *Mathematics Magazine*. To
view this article, go to
MathArticles.com.

When, as in Examples 1 and 2, the approximations approach a limit, the sequence $x_1, x_2, x_3, \ldots, x_n, \ldots$ is said to **converge.** Moreover, when the limit is c, it can be shown that c must be a zero of f.

Newton's Method does not always yield a convergent sequence. One way it can fail to do so is shown in Figure 3.63. Because Newton's Method involves division by $f'(x_n)$, it is clear that the method will fail when the derivative is zero for any x_n in the sequence. When you encounter this problem, you can usually overcome it by choosing a different value for x_1. Another way Newton's Method can fail is shown in the next example.

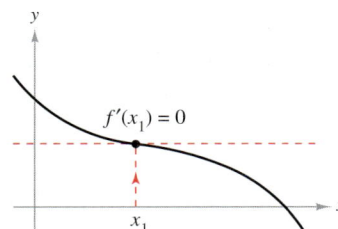

Newton's Method fails to converge when $f'(x_n) = 0$.
Figure 3.63

EXAMPLE 3 **An Example in Which Newton's Method Fails**

The function $f(x) = x^{1/3}$ is not differentiable at $x = 0$. Show that Newton's Method fails to converge using $x_1 = 0.1$.

Solution Because $f'(x) = \frac{1}{3}x^{-2/3}$, the iterative formula is

$$x_{n+1} = x_n - \frac{f(x_n)}{f'(x_n)} = x_n - \frac{x_n^{1/3}}{\frac{1}{3}x_n^{-2/3}} = x_n - 3x_n = -2x_n.$$

The calculations are shown in the table. This table and Figure 3.64 indicate that x_n continues to increase in magnitude as $n \to \infty$, and so the limit of the sequence does not exist.

n	x_n	$f(x_n)$	$f'(x_n)$	$\dfrac{f(x_n)}{f'(x_n)}$	$x_n - \dfrac{f(x_n)}{f'(x_n)}$
1	0.10000	0.46416	1.54720	0.30000	−0.20000
2	−0.20000	−0.58480	0.97467	−0.60000	0.40000
3	0.40000	0.73681	0.61401	1.20000	−0.80000
4	−0.80000	−0.92832	0.3680	−2.40000	1.60000

REMARK In Example 3, the initial estimate $x_1 = 0.1$ fails to produce a convergent sequence. Try showing that Newton's Method also fails for every other choice of x_1 (other than the actual zero).

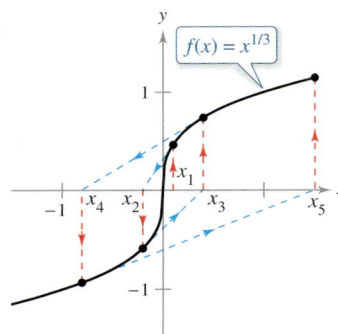

Newton's Method fails to converge for every x-value other than the actual zero of f.
Figure 3.64

It can be shown that a condition sufficient to produce convergence of Newton's Method to a zero of f is that

$$\left|\frac{f(x)\,f''(x)}{[f'(x)]^2}\right| < 1 \qquad \text{Condition for convergence}$$

on an open interval containing the zero. For instance, in Example 1, this test would yield

$$f(x) = x^2 - 2, \quad f'(x) = 2x, \quad f''(x) = 2,$$

and

$$\left|\frac{f(x)\,f''(x)}{[f'(x)]^2}\right| = \left|\frac{(x^2 - 2)(2)}{4x^2}\right| = \left|\frac{1}{2} - \frac{1}{x^2}\right|. \qquad \text{Example 1}$$

On the interval $(1, 3)$, this quantity is less than 1 and therefore the convergence of Newton's Method is guaranteed. On the other hand, in Example 3, you have

$$f(x) = x^{1/3}, \quad f'(x) = \frac{1}{3}x^{-2/3}, \quad f''(x) = -\frac{2}{9}x^{-5/3}$$

and

$$\left|\frac{f(x)\,f''(x)}{[f'(x)]^2}\right| = \left|\frac{x^{1/3}(-2/9)(x^{-5/3})}{(1/9)(x^{-4/3})}\right| = 2 \qquad \text{Example 3}$$

which is not less than 1 for any value of x, so you cannot conclude that Newton's Method will converge.

You have learned several techniques for finding the zeros of functions. The zeros of some functions, such as

$$f(x) = x^3 - 2x^2 - x + 2$$

can be found by simple algebraic techniques, such as factoring. The zeros of other functions, such as

$$f(x) = x^3 - x + 1$$

cannot be found by *elementary* algebraic methods. This particular function has only one real zero, and by using more advanced algebraic techniques, you can determine the zero to be

$$x = -\sqrt[3]{\frac{3 - \sqrt{23/3}}{6}} - \sqrt[3]{\frac{3 + \sqrt{23/3}}{6}}.$$

Because the *exact* solution is written in terms of square roots and cube roots, it is called a **solution by radicals.**

The determination of radical solutions of a polynomial equation is one of the fundamental problems of algebra. The earliest such result is the Quadratic Formula, which dates back at least to Babylonian times. The general formula for the zeros of a cubic function was developed much later. In the sixteenth century, an Italian mathematician, Jerome Cardan, published a method for finding radical solutions to cubic and quartic equations. Then, for 300 years, the problem of finding a general quintic formula remained open. Finally, in the nineteenth century, the problem was answered independently by two young mathematicians. Niels Henrik Abel, a Norwegian mathematician, and Evariste Galois, a French mathematician, proved that it is not possible to solve a *general* fifth- (or higher-) degree polynomial equation by radicals. Of course, you can solve particular fifth-degree equations, such as

$$x^5 - 1 = 0$$

but Abel and Galois were able to show that no general *radical* solution exists.

NIELS HENRIK ABEL (1802–1829)

EVARISTE GALOIS (1811–1832)

Although the lives of both Abel and Galois were brief, their work in the fields of analysis and abstract algebra was far-reaching. *See LarsonCalculus.com to read a biography about each of these mathematicians.*

3.8 Exercises

See **CalcChat.com** for tutorial help and worked-out solutions to odd-numbered exercises.

Using Newton's Method In Exercises 1–4, complete two iterations of Newton's Method to approximate a zero of the function using the given initial guess.

1. $f(x) = x^2 - 5$, $x_1 = 2.2$

2. $f(x) = x^3 - 3$, $x_1 = 1.4$

3. $f(x) = \cos x$, $x_1 = 1.6$

4. $f(x) = \tan x$, $x_1 = 0.1$

Using Newton's Method In Exercises 5–14, approximate the zero(s) of the function. Use Newton's Method and continue the process until two successive approximations differ by less than 0.001. Then find the zero(s) using a graphing utility and compare the results.

5. $f(x) = x^3 + 4$

6. $f(x) = 2 - x^3$

7. $f(x) = x^3 + x - 1$

8. $f(x) = x^5 + x - 1$

9. $f(x) = 5\sqrt{x - 1} - 2x$

10. $f(x) = x - 2\sqrt{x + 1}$

11. $f(x) = x^3 - 3.9x^2 + 4.79x - 1.881$

12. $f(x) = x^4 + x^3 - 1$

13. $f(x) = 1 - x + \sin x$

14. $f(x) = x^3 - \cos x$

Finding Point(s) of Intersection In Exercises 15–18, apply Newton's Method to approximate the x-value(s) of the indicated point(s) of intersection of the two graphs. Continue the process until two successive approximations differ by less than 0.001. [*Hint:* Let $h(x) = f(x) - g(x)$.]

15. $f(x) = 2x + 1$

$g(x) = \sqrt{x + 4}$

16. $f(x) = 3 - x$

$g(x) = \dfrac{1}{x^2 + 1}$

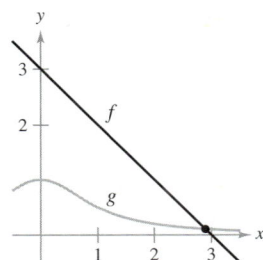

17. $f(x) = x$

$g(x) = \tan x$

18. $f(x) = x^2$

$g(x) = \cos x$

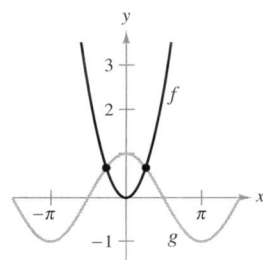

19. **Mechanic's Rule** The Mechanic's Rule for approximating $\sqrt{a}$, $a > 0$, is

$$x_{n+1} = \frac{1}{2}\left(x_n + \frac{a}{x_n}\right), \quad n = 1, 2, 3 \ldots$$

where x_1 is an approximation of $\sqrt{a}$.

(a) Use Newton's Method and the function $f(x) = x^2 - a$ to derive the Mechanic's Rule.

(b) Use the Mechanic's Rule to approximate $\sqrt{5}$ and $\sqrt{7}$ to three decimal places.

20. **Approximating Radicals**

(a) Use Newton's Method and the function $f(x) = x^n - a$ to obtain a general rule for approximating $x = \sqrt[n]{a}$.

(b) Use the general rule found in part (a) to approximate $\sqrt[4]{6}$ and $\sqrt[3]{15}$ to three decimal places.

Failure of Newton's Method In Exercises 21 and 22, apply Newton's Method using the given initial guess, and explain why the method fails.

21. $y = 2x^3 - 6x^2 + 6x - 1$, $x_1 = 1$

Figure for 21

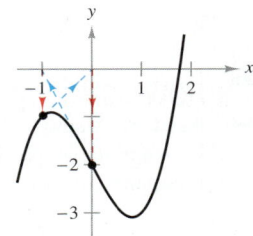

Figure for 22

22. $y = x^3 - 2x - 2$, $x_1 = 0$

Fixed Point In Exercises 23 and 24, approximate the fixed point of the function to two decimal places. [A *fixed point* x_0 of a function f is a value of x such that $f(x_0) = x_0$.]

23. $f(x) = \cos x$

24. $f(x) = \cot x$, $0 < x < \pi$

25. **Approximating Reciprocals** Use Newton's Method to show that the equation

$$x_{n+1} = x_n(2 - ax_n)$$

can be used to approximate $1/a$ when x_1 is an initial guess of the reciprocal of a. Note that this method of approximating reciprocals uses only the operations of multiplication and subtraction. (*Hint:* Consider

$$f(x) = \frac{1}{x} - a.)$$

26. **Approximating Reciprocals** Use the result of Exercise 25 to approximate (a) $\frac{1}{3}$ and (b) $\frac{1}{11}$ to three decimal places.

WRITING ABOUT CONCEPTS

27. Using Newton's Method Consider the function $f(x) = x^3 - 3x^2 + 3$.

(a) Use a graphing utility to graph f.

(b) Use Newton's Method to approximate a zero with $x_1 = 1$ as an initial guess.

(c) Repeat part (b) using $x_1 = \frac{1}{4}$ as an initial guess and observe that the result is different.

(d) To understand why the results in parts (b) and (c) are different, sketch the tangent lines to the graph of f at the points $(1, f(1))$ and $\left(\frac{1}{4}, f\left(\frac{1}{4}\right)\right)$. Find the x-intercept of each tangent line and compare the intercepts with the first iteration of Newton's Method using the respective initial guesses.

(e) Write a short paragraph summarizing how Newton's Method works. Use the results of this exercise to describe why it is important to select the initial guess carefully.

28. Using Newton's Method Repeat the steps in Exercise 27 for the function $f(x) = \sin x$ with initial guesses of $x_1 = 1.8$ and $x_1 = 3$.

29. Newton's Method In your own words and using a sketch, describe Newton's Method for approximating the zeros of a function.

30. HOW DO YOU SEE IT? For what value(s) will Newton's Method fail to converge for the function shown in the graph? Explain your reasoning.

Using Newton's Method Exercises 31–33 present problems similar to exercises from the previous sections of this chapter. In each case, use Newton's Method to approximate the solution.

31. Minimum Distance Find the point on the graph of $f(x) = 4 - x^2$ that is closest to the point $(1, 0)$.

32. Medicine The concentration C of a chemical in the bloodstream t hours after injection into muscle tissue is given by

$$C = \frac{3t^2 + t}{50 + t^3}.$$

When is the concentration the greatest?

33. Minimum Time You are in a boat 2 miles from the nearest point on the coast (see figure). You are to go to a point Q that is 3 miles down the coast and 1 mile inland. You can row at 3 miles per hour and walk at 4 miles per hour. Toward what point on the coast should you row in order to reach Q in the least time?

34. Crime The total number of arrests T (in thousands) for all males ages 15 to 24 in 2010 is approximated by the model

$$T = 0.2988x^4 - 22.625x^3 + 628.49x^2 - 7565.9x + 33,478$$

for $15 \le x \le 24$, where x is the age in years (see figure). Approximate the two ages that had total arrests of 300 thousand. (*Source: U.S. Department of Justice*)

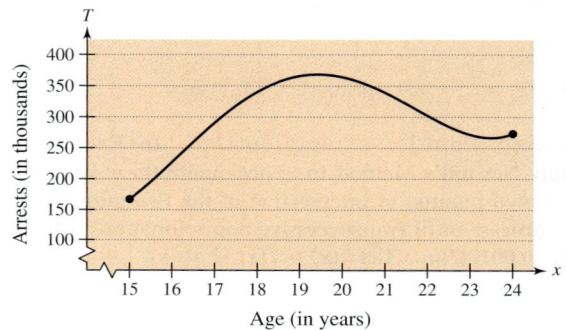

True or False? In Exercises 35–38, determine whether the statement is true or false. If it is false, explain why or give an example that shows it is false.

35. The zeros of $f(x) = \dfrac{p(x)}{q(x)}$ coincide with the zeros of $p(x)$.

36. If the coefficients of a polynomial function are all positive, then the polynomial has no positive zeros.

37. If $f(x)$ is a cubic polynomial such that $f'(x)$ is never zero, then any initial guess will force Newton's Method to converge to the zero of f.

38. The roots of $\sqrt{f(x)} = 0$ coincide with the roots of $f(x) = 0$.

39. Tangent Lines The graph of $f(x) = -\sin x$ has infinitely many tangent lines that pass through the origin. Use Newton's Method to approximate to three decimal places the slope of the tangent line having the greatest slope.

40. Point of Tangency The graph of $f(x) = \cos x$ and a tangent line to f through the origin are shown. Find the coordinates of the point of tangency to three decimal places.

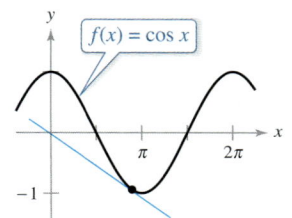

3.9 Differentials

- Understand the concept of a tangent line approximation.
- Compare the value of the differential, dy, with the actual change in y, Δy.
- Estimate a propagated error using a differential.
- Find the differential of a function using differentiation formulas.

Tangent Line Approximations

Newton's Method (Section 3.8) is an example of the use of a tangent line to approximate the graph of a function. In this section, you will study other situations in which the graph of a function can be approximated by a straight line.

To begin, consider a function f that is differentiable at c. The equation for the tangent line at the point $(c, f(c))$ is

$$y - f(c) = f'(c)(x - c)$$

$$\boxed{y = f(c) + f'(c)(x - c)}$$

and is called the **tangent line approximation** (or **linear approximation**) of f at c. Because c is a constant, y is a linear function of x. Moreover, by restricting the values of x to those sufficiently close to c, the values of y can be used as approximations (to any desired degree of accuracy) of the values of the function f. In other words, as x approaches c, the limit of y is $f(c)$.

EXAMPLE 1 Using a Tangent Line Approximation

• • • ▷ *See LarsonCalculus.com for an interactive version of this type of example.*

Find the tangent line approximation of $f(x) = 1 + \sin x$ at the point $(0, 1)$. Then use a table to compare the y-values of the linear function with those of $f(x)$ on an open interval containing $x = 0$.

Solution The derivative of f is

$$f'(x) = \cos x. \qquad \text{First derivative}$$

So, the equation of the tangent line to the graph of f at the point $(0, 1)$ is

$$y = f(0) + f'(0)(x - 0)$$
$$y = 1 + (1)(x - 0)$$
$$y = 1 + x. \qquad \text{Tangent line approximation}$$

The table compares the values of y given by this linear approximation with the values of $f(x)$ near $x = 0$. Notice that the closer x is to 0, the better the approximation. This conclusion is reinforced by the graph shown in Figure 3.65.

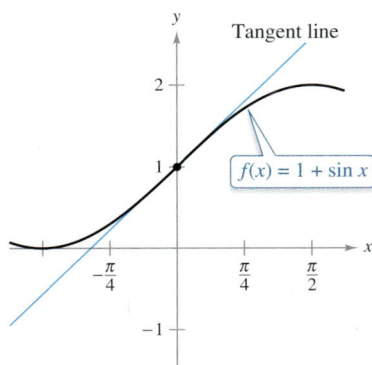

Tangent line

$f(x) = 1 + \sin x$

The tangent line approximation of f at the point $(0, 1)$

Figure 3.65

x	-0.5	-0.1	-0.01	0	0.01	0.1	0.5
$f(x) = 1 + \sin x$	0.521	0.9002	0.9900002	1	1.0099998	1.0998	1.479
$y = 1 + x$	0.5	0.9	0.99	1	1.01	1.1	1.5

REMARK Be sure you see that this linear approximation of $f(x) = 1 + \sin x$ depends on the point of tangency. At a different point on the graph of f, you would obtain a different tangent line approximation.

Differentials

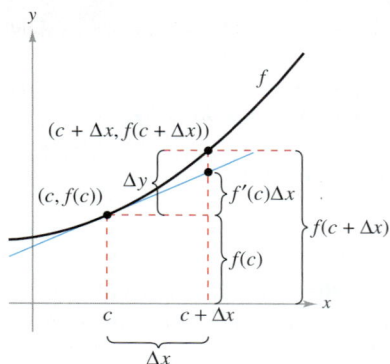

When Δx is small,
$\Delta y = f(c + \Delta x) - f(c)$ is
approximated by $f'(c)\Delta x$.

Figure 3.66

When the tangent line to the graph of f at the point $(c, f(c))$

$$y = f(c) + f'(c)(x - c) \qquad \text{Tangent line at } (c, f(c))$$

is used as an approximation of the graph of f, the quantity $x - c$ is called the change in x, and is denoted by Δx, as shown in Figure 3.66. When Δx is small, the change in y (denoted by Δy) can be approximated as shown.

$$\Delta y = f(c + \Delta x) - f(c) \qquad \text{Actual change in } y$$
$$\approx f'(c)\Delta x \qquad \text{Approximate change in } y$$

For such an approximation, the quantity Δx is traditionally denoted by dx, and is called the **differential of x.** The expression $f'(x)\,dx$ is denoted by dy, and is called the **differential of y.**

> **Definition of Differentials**
>
> Let $y = f(x)$ represent a function that is differentiable on an open interval containing x. The **differential of x** (denoted by dx) is any nonzero real number. The **differential of y** (denoted by dy) is
>
> $$dy = f'(x)\,dx.$$

In many types of applications, the differential of y can be used as an approximation of the change in y. That is,

$$\Delta y \approx dy \qquad \text{or} \qquad \Delta y \approx f'(x)\,dx.$$

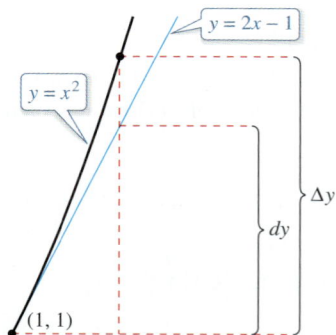

The change in y, Δy, is approximated by the differential of y, dy.

Figure 3.67

EXAMPLE 2   Comparing Δy and dy

Let $y = x^2$. Find dy when $x = 1$ and $dx = 0.01$. Compare this value with Δy for $x = 1$ and $\Delta x = 0.01$.

Solution   Because $y = f(x) = x^2$, you have $f'(x) = 2x$, and the differential dy is

$$dy = f'(x)\,dx = f'(1)(0.01) = 2(0.01) = 0.02. \qquad \text{Differential of } y$$

Now, using $\Delta x = 0.01$, the change in y is

$$\Delta y = f(x + \Delta x) - f(x) = f(1.01) - f(1) = (1.01)^2 - 1^2 = 0.0201.$$

Figure 3.67 shows the geometric comparison of dy and Δy. Try comparing other values of dy and Δy. You will see that the values become closer to each other as dx (or Δx) approaches 0.

In Example 2, the tangent line to the graph of $f(x) = x^2$ at $x = 1$ is

$$y = 2x - 1. \qquad \text{Tangent line to the graph of } f \text{ at } x = 1.$$

For x-values near 1, this line is close to the graph of f, as shown in Figure 3.67 and in the table.

x	0.5	0.9	0.99	1	1.01	1.1	1.5
$f(x) = x^2$	0.25	0.81	0.9801	1	1.0201	1.21	2.25
$y = 2x - 1$	0	0.8	0.98	1	1.02	1.2	2

Error Propagation

Physicists and engineers tend to make liberal use of the approximation of Δy by dy. One way this occurs in practice is in the estimation of errors propagated by physical measuring devices. For example, if you let x represent the measured value of a variable and let $x + \Delta x$ represent the exact value, then Δx is the *error in measurement*. Finally, if the measured value x is used to compute another value $f(x)$, then the difference between $f(x + \Delta x)$ and $f(x)$ is the **propagated error.**

$$\underbrace{f(\overbrace{x + \Delta x}^{\text{Measurement error}})}_{\substack{\text{Exact} \\ \text{value}}} - \underbrace{f(x)}_{\substack{\text{Measured} \\ \text{value}}} = \overbrace{\Delta y}^{\substack{\text{Propagated} \\ \text{error}}}$$

EXAMPLE 3   **Estimation of Error**

The measured radius of a ball bearing is 0.7 inch, as shown in the figure. The measurement is correct to within 0.01 inch. Estimate the propagated error in the volume V of the ball bearing.

Solution   The formula for the volume of a sphere is

$$V = \frac{4}{3}\pi r^3$$

0.7

Ball bearing with measured radius that is correct to within 0.01 inch.

where r is the radius of the sphere. So, you can write

$$r = 0.7 \qquad \textcolor{red}{\text{Measured radius}}$$

and

$$-0.01 \le \Delta r \le 0.01. \qquad \textcolor{red}{\text{Possible error}}$$

To approximate the propagated error in the volume, differentiate V to obtain $dV/dr = 4\pi r^2$ and write

$$\begin{aligned}
\Delta V &\approx dV & \textcolor{red}{\text{Approximate } \Delta V \text{ by } dV.}\\
&= 4\pi r^2 \, dr \\
&= 4\pi(0.7)^2(\pm 0.01) & \textcolor{red}{\text{Substitute for } r \text{ and } dr.}\\
&\approx \pm 0.06158 \text{ cubic inch.}
\end{aligned}$$

So, the volume has a propagated error of about 0.06 cubic inch. ∎

Would you say that the propagated error in Example 3 is large or small? The answer is best given in *relative* terms by comparing dV with V. The ratio

$$\begin{aligned}
\frac{dV}{V} &= \frac{4\pi r^2 \, dr}{\frac{4}{3}\pi r^3} & \textcolor{red}{\text{Ratio of } dV \text{ to } V}\\
&= \frac{3 \, dr}{r} & \textcolor{red}{\text{Simplify.}}\\
&\approx \frac{3}{0.7}(\pm 0.01) & \textcolor{red}{\text{Substitute for } dr \text{ and } r.}\\
&\approx \pm 0.0429
\end{aligned}$$

is called the **relative error.** The corresponding **percent error** is approximately 4.29%.

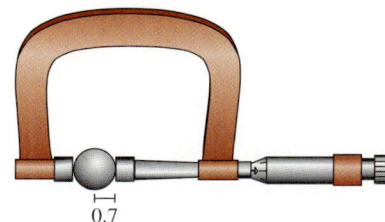

Calculating Differentials

Each of the differentiation rules that you studied in Chapter 2 can be written in **differential form.** For example, let u and v be differentiable functions of x. By the definition of differentials, you have

$$du = u' \, dx$$

and

$$dv = v' \, dx.$$

So, you can write the differential form of the Product Rule as shown below.

$$d[uv] = \frac{d}{dx}[uv] \, dx \qquad \text{Differential of } uv.$$
$$= [uv' + vu'] \, dx \qquad \text{Product Rule}$$
$$= uv' \, dx + vu' \, dx$$
$$= u \, dv + v \, du$$

Differential Formulas

Let u and v be differentiable functions of x.

Constant multiple:	$d[cu] = c \, du$
Sum or difference:	$d[u \pm v] = du \pm dv$
Product:	$d[uv] = u \, dv + v \, du$
Quotient:	$d\left[\dfrac{u}{v}\right] = \dfrac{v \, du - u \, dv}{v^2}$

EXAMPLE 4 **Finding Differentials**

Function	Derivative	Differential
a. $y = x^2$	$\dfrac{dy}{dx} = 2x$	$dy = 2x \, dx$
b. $y = \sqrt{x}$	$\dfrac{dy}{dx} = \dfrac{1}{2\sqrt{x}}$	$dy = \dfrac{dx}{2\sqrt{x}}$
c. $y = 2 \sin x$	$\dfrac{dy}{dx} = 2 \cos x$	$dy = 2 \cos x \, dx$
d. $y = x \cos x$	$\dfrac{dy}{dx} = -x \sin x + \cos x$	$dy = (-x \sin x + \cos x) \, dx$
e. $y = \dfrac{1}{x}$	$\dfrac{dy}{dx} = -\dfrac{1}{x^2}$	$dy = -\dfrac{dx}{x^2}$

GOTTFRIED WILHELM LEIBNIZ (1646–1716)

Both Leibniz and Newton are credited with creating calculus. It was Leibniz, however, who tried to broaden calculus by developing rules and formal notation. He often spent days choosing an appropriate notation for a new concept.
See LarsonCalculus.com to read more of this biography.

The notation in Example 4 is called the **Leibniz notation** for derivatives and differentials, named after the German mathematician Gottfried Wilhelm Leibniz. The beauty of this notation is that it provides an easy way to remember several important calculus formulas by making it seem as though the formulas were derived from algebraic manipulations of differentials. For instance, in Leibniz notation, the *Chain Rule*

$$\frac{dy}{dx} = \frac{dy}{du} \frac{du}{dx}$$

would appear to be true because the du's divide out. Even though this reasoning is *incorrect*, the notation does help one remember the Chain Rule.

EXAMPLE 5   Finding the Differential of a Composite Function

$$y = f(x) = \sin 3x \qquad \text{Original function}$$
$$f'(x) = 3 \cos 3x \qquad \text{Apply Chain Rule.}$$
$$dy = f'(x)\, dx = 3 \cos 3x\, dx \qquad \text{Differential form}$$

EXAMPLE 6   Finding the Differential of a Composite Function

$$y = f(x) = (x^2 + 1)^{1/2} \qquad \text{Original function}$$
$$f'(x) = \frac{1}{2}(x^2 + 1)^{-1/2}(2x) = \frac{x}{\sqrt{x^2 + 1}} \qquad \text{Apply Chain Rule.}$$
$$dy = f'(x)\, dx = \frac{x}{\sqrt{x^2 + 1}}\, dx \qquad \text{Differential form}$$

Differentials can be used to approximate function values. To do this for the function given by $y = f(x)$, use the formula

$$f(x + \Delta x) \approx f(x) + dy = f(x) + f'(x)\, dx$$

which is derived from the approximation

$$\Delta y = f(x + \Delta x) - f(x) \approx dy.$$

The key to using this formula is to choose a value for x that makes the calculations easier, as shown in Example 7.

• REMARK   This formula is equivalent to the tangent line approximation given earlier in this section.

EXAMPLE 7   Approximating Function Values

Use differentials to approximate $\sqrt{16.5}$.

Solution   Using $f(x) = \sqrt{x}$, you can write

$$f(x + \Delta x) \approx f(x) + f'(x)\, dx = \sqrt{x} + \frac{1}{2\sqrt{x}}\, dx.$$

Now, choosing $x = 16$ and $dx = 0.5$, you obtain the following approximation.

$$f(x + \Delta x) = \sqrt{16.5} \approx \sqrt{16} + \frac{1}{2\sqrt{16}}(0.5) = 4 + \left(\frac{1}{8}\right)\left(\frac{1}{2}\right) = 4.0625$$

The tangent line approximation to $f(x) = \sqrt{x}$ at $x = 16$ is the line $g(x) = \frac{1}{8}x + 2$. For x-values near 16, the graphs of f and g are close together, as shown in Figure 3.68. For instance,

$$f(16.5) = \sqrt{16.5} \approx 4.0620$$

and

$$g(16.5) = \frac{1}{8}(16.5) + 2 = 4.0625.$$

In fact, if you use a graphing utility to zoom in near the point of tangency $(16, 4)$, you will see that the two graphs appear to coincide. Notice also that as you move farther away from the point of tangency, the linear approximation becomes less accurate.

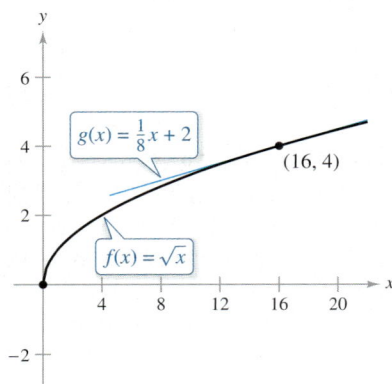

Figure 3.68

3.9 Exercises

See **CalcChat.com** for tutorial help and worked-out solutions to odd-numbered exercises.

Using a Tangent Line Approximation In Exercises 1–6, find the tangent line approximation T to the graph of f at the given point. Use this linear approximation to complete the table.

x	1.9	1.99	2	2.01	2.1
$f(x)$					
$T(x)$					

1. $f(x) = x^2$, $(2, 4)$ **2.** $f(x) = \dfrac{6}{x^2}$, $\left(2, \dfrac{3}{2}\right)$

3. $f(x) = x^5$, $(2, 32)$ **4.** $f(x) = \sqrt{x}$, $\left(2, \sqrt{2}\right)$

5. $f(x) = \sin x$, $(2, \sin 2)$ **6.** $f(x) = \csc x$, $(2, \csc 2)$

Comparing Δy and dy In Exercises 7–10, use the information to evaluate and compare Δy and dy.

Function	x-Value	Differential of x
7. $y = x^3$	$x = 1$	$\Delta x = dx = 0.1$
8. $y = 6 - 2x^2$	$x = -2$	$\Delta x = dx = 0.1$
9. $y = x^4 + 1$	$x = -1$	$\Delta x = dx = 0.01$
10. $y = 2 - x^4$	$x = 2$	$\Delta x = dx = 0.01$

Finding a Differential In Exercises 11–20, find the differential dy of the given function.

11. $y = 3x^2 - 4$ **12.** $y = 3x^{2/3}$

13. $y = x \tan x$ **14.** $y = \csc 2x$

15. $y = \dfrac{x + 1}{2x - 1}$ **16.** $y = \sqrt{x} + \dfrac{1}{\sqrt{x}}$

17. $y = \sqrt{9 - x^2}$ **18.** $y = x\sqrt{1 - x^2}$

19. $y = 3x - \sin^2 x$ **20.** $y = \dfrac{\sec^2 x}{x^2 + 1}$

Using Differentials In Exercises 21 and 22, use differentials and the graph of f to approximate (a) $f(1.9)$ and (b) $f(2.04)$. To print an enlarged copy of the graph, go to *MathGraphs.com*.

21.

22.
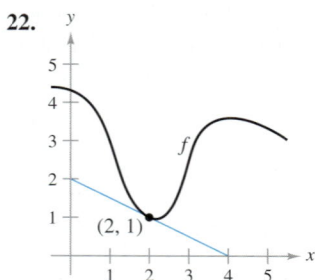

Using Differentials In Exercises 23 and 24, use differentials and the graph of g' to approximate (a) $g(2.93)$ and (b) $g(3.1)$ given that $g(3) = 8$.

23.

24.
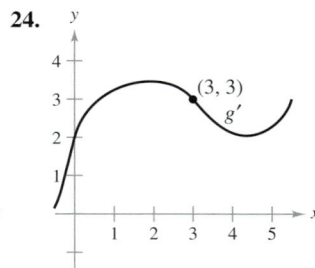

25. Area The measurement of the side of a square floor tile is 10 inches, with a possible error of $\frac{1}{32}$ inch.

  (a) Use differentials to approximate the possible propagated error in computing the area of the square.

  (b) Approximate the percent error in computing the area of the square.

26. Area The measurement of the radius of a circle is 16 inches, with a possible error of $\frac{1}{4}$ inch.

  (a) Use differentials to approximate the possible propagated error in computing the area of the circle.

  (b) Approximate the percent error in computing the area of the circle.

27. Area The measurements of the base and altitude of a triangle are found to be 36 and 50 centimeters, respectively. The possible error in each measurement is 0.25 centimeter.

  (a) Use differentials to approximate the possible propagated error in computing the area of the triangle.

  (b) Approximate the percent error in computing the area of the triangle.

28. Circumference The measurement of the circumference of a circle is found to be 64 centimeters, with a possible error of 0.9 centimeter.

  (a) Approximate the percent error in computing the area of the circle.

  (b) Estimate the maximum allowable percent error in measuring the circumference if the error in computing the area cannot exceed 3%.

29. Volume and Surface Area The measurement of the edge of a cube is found to be 15 inches, with a possible error of 0.03 inch.

  (a) Use differentials to approximate the possible propagated error in computing the volume of the cube.

  (b) Use differentials to approximate the possible propagated error in computing the surface area of the cube.

  (c) Approximate the percent errors in parts (a) and (b).

30. Volume and Surface Area   The radius of a spherical balloon is measured as 8 inches, with a possible error of 0.02 inch.

(a) Use differentials to approximate the possible propagated error in computing the volume of the sphere.

(b) Use differentials to approximate the possible propagated error in computing the surface area of the sphere.

(c) Approximate the percent errors in parts (a) and (b).

31. Stopping Distance   The total stopping distance T of a vehicle is

$$T = 2.5x + 0.5x^2$$

where T is in feet and x is the speed in miles per hour. Approximate the change and percent change in total stopping distance as speed changes from $x = 25$ to $x = 26$ miles per hour.

32. **HOW DO YOU SEE IT?**   The graph shows the profit P (in dollars) from selling x units of an item. Use the graph to determine which is greater, the change in profit when the production level changes from 400 to 401 units or the change in profit when the production level changes from 900 to 901 units. Explain your reasoning

Number of units

33. Pendulum   The period of a pendulum is given by

$$T = 2\pi\sqrt{\frac{L}{g}}$$

where L is the length of the pendulum in feet, g is the acceleration due to gravity, and T is the time in seconds. The pendulum has been subjected to an increase in temperature such that the length has increased by $\frac{1}{2}\%$.

(a) Find the approximate percent change in the period.

(b) Using the result in part (a), find the approximate error in this pendulum clock in 1 day.

34. Ohm's Law   A current of I amperes passes through a resistor of R ohms. **Ohm's Law** states that the voltage E applied to the resistor is

$$E = IR.$$

The voltage is constant. Show that the magnitude of the relative error in R caused by a change in I is equal in magnitude to the relative error in I.

35. Projectile Motion   The range R of a projectile is

$$R = \frac{v_0^2}{32}(\sin 2\theta)$$

where v_0 is the initial velocity in feet per second and θ is the angle of elevation. Use differentials to approximate the change in the range when $v_0 = 2500$ feet per second and θ is changed from $10°$ to $11°$.

36. Surveying   A surveyor standing 50 feet from the base of a large tree measures the angle of elevation to the top of the tree as $71.5°$. How accurately must the angle be measured if the percent error in estimating the height of the tree is to be less than 6%?

Approximating Function Values   In Exercises 37–40, use differentials to approximate the value of the expression. Compare your answer with that of a calculator.

37. $\sqrt{99.4}$ **38.** $\sqrt[3]{26}$

39. $\sqrt[4]{624}$ **40.** $(2.99)^3$

Verifying a Tangent Line Approximation   In Exercises 41 and 42, verify the tangent line approximation of the function at the given point. Then use a graphing utility to graph the function and its approximation in the same viewing window.

Function	Approximation	Point
41. $f(x) = \sqrt{x + 4}$	$y = 2 + \dfrac{x}{4}$	$(0, 2)$
42. $f(x) = \tan x$	$y = x$	$(0, 0)$

WRITING ABOUT CONCEPTS

43. Comparing Δy and dy   Describe the change in accuracy of dy as an approximation for Δy when Δx is decreased.

44. Describing Terms   When using differentials, what is meant by the terms *propagated error*, *relative error*, and *percent error*?

Using Differentials   In Exercises 45 and 46, give a short explanation of why the approximation is valid.

45. $\sqrt{4.02} \approx 2 + \frac{1}{4}(0.02)$ **46.** $\tan 0.05 \approx 0 + 1(0.05)$

True or False?   In Exercises 47–50, determine whether the statement is true or false. If it is false, explain why or give an example that shows it is false.

47. If $y = x + c$, then $dy = dx$.

48. If $y = ax + b$, then $\dfrac{\Delta y}{\Delta x} = \dfrac{dy}{dx}$.

49. If y is differentiable, then $\lim\limits_{\Delta x \to 0} (\Delta y - dy) = 0$.

50. If $y = f(x)$, f is increasing and differentiable, and $\Delta x > 0$, then $\Delta y \geq dy$.

Review Exercises

See **CalcChat.com** for tutorial help and worked-out solutions to odd-numbered exercises.

Finding Extrema on a Closed Interval In Exercises 1–8, find the absolute extrema of the function on the closed interval.

1. $f(x) = x^2 + 5x$, $[-4, 0]$
2. $f(x) = x^3 + 6x^2$, $[-6, 1]$
3. $f(x) = \sqrt{x} - 2$, $[0, 4]$
4. $h(x) = 3\sqrt{x} - x$, $[0, 9]$
5. $f(x) = \dfrac{4x}{x^2 + 9}$, $[-4, 4]$
6. $f(x) = \dfrac{x}{\sqrt{x^2 + 1}}$, $[0, 2]$
7. $g(x) = 2x + 5\cos x$, $[0, 2\pi]$
8. $f(x) = \sin 2x$, $[0, 2\pi]$

Using Rolle's Theorem In Exercises 9–12, determine whether Rolle's Theorem can be applied to f on the closed interval $[a, b]$. If Rolle's Theorem can be applied, find all values of c in the open interval (a, b) such that $f'(c) = 0$. If Rolle's Theorem cannot be applied, explain why not.

9. $f(x) = 2x^2 - 7$, $[0, 4]$
10. $f(x) = (x - 2)(x + 3)^2$, $[-3, 2]$
11. $f(x) = \dfrac{x^2}{1 - x^2}$, $[-2, 2]$
12. $f(x) = \sin 2x$, $[-\pi, \pi]$

Using the Mean Value Theorem In Exercises 13–18, determine whether the Mean Value Theorem can be applied to f on the closed interval $[a, b]$. If the Mean Value Theorem can be applied, find all values of c in the open interval (a, b) such that

$$f'(c) = \frac{f(b) - f(a)}{b - a}.$$

If the Mean Value Theorem cannot be applied, explain why not.

13. $f(x) = x^{2/3}$, $[1, 8]$
14. $f(x) = \dfrac{1}{x}$, $[1, 4]$
15. $f(x) = |5 - x|$, $[2, 6]$
16. $f(x) = 2x - 3\sqrt{x}$, $[-1, 1]$
17. $f(x) = x - \cos x$, $\left[-\dfrac{\pi}{2}, \dfrac{\pi}{2}\right]$
18. $f(x) = \sqrt{x} - 2x$, $[0, 4]$

19. **Mean Value Theorem** Can the Mean Value Theorem be applied to the function

$$f(x) = \frac{1}{x^2}$$

on the interval $[-2, 1]$? Explain.

20. **Using the Mean Value Theorem**

    (a) For the function $f(x) = Ax^2 + Bx + C$, determine the value of c guaranteed by the Mean Value Theorem on the interval $[x_1, x_2]$.

    (b) Demonstrate the result of part (a) for $f(x) = 2x^2 - 3x + 1$ on the interval $[0, 4]$.

Intervals on Which f Is Increasing or Decreasing In Exercises 21–26, identify the open intervals on which the function is increasing or decreasing.

21. $f(x) = x^2 + 3x - 12$
22. $h(x) = (x + 2)^{1/3} + 8$
23. $f(x) = (x - 1)^2(x - 3)$
24. $g(x) = (x + 1)^3$
25. $h(x) = \sqrt{x}(x - 3)$, $x > 0$
26. $f(x) = \sin x + \cos x$, $[0, 2\pi]$

Applying the First Derivative Test In Exercises 27–34, (a) find the critical numbers of f (if any), (b) find the open interval(s) on which the function is increasing or decreasing, (c) apply the First Derivative Test to identify all relative extrema, and (d) use a graphing utility to confirm your results.

27. $f(x) = x^2 - 6x + 5$
28. $f(x) = 4x^3 - 5x$
29. $h(t) = \dfrac{1}{4}t^4 - 8t$
30. $g(x) = \dfrac{x^3 - 8x}{4}$
31. $f(x) = \dfrac{x + 4}{x^2}$
32. $f(x) = \dfrac{x^2 - 3x - 4}{x - 2}$
33. $f(x) = \cos x - \sin x$, $(0, 2\pi)$
34. $g(x) = \dfrac{3}{2}\sin\left(\dfrac{\pi x}{2} - 1\right)$, $[0, 4]$

Finding Points of Inflection In Exercises 35–40, find the points of inflection and discuss the concavity of the graph of the function.

35. $f(x) = x^3 - 9x^2$
36. $f(x) = 6x^4 - x^2$
37. $g(x) = x\sqrt{x + 5}$
38. $f(x) = 3x - 5x^3$
39. $f(x) = x + \cos x$, $[0, 2\pi]$
40. $f(x) = \tan\dfrac{x}{4}$, $(0, 2\pi)$

Using the Second Derivative Test In Exercises 41–46, find all relative extrema. Use the Second Derivative Test where applicable.

41. $f(x) = (x + 9)^2$
42. $f(x) = 2x^3 + 11x^2 - 8x - 12$
43. $g(x) = 2x^2(1 - x^2)$
44. $h(t) = t - 4\sqrt{t + 1}$

45. $f(x) = 2x + \dfrac{18}{x}$

46. $h(x) = x - 2\cos x, \quad [0, 4\pi]$

Think About It   In Exercises 47 and 48, sketch the graph of a function f having the given characteristics.

47. $f(0) = f(6) = 0$

$f'(3) = f'(5) = 0$

$f'(x) > 0$ for $x < 3$

$f'(x) > 0$ for $3 < x < 5$

$f'(x) < 0$ for $x > 5$

$f''(x) < 0$ for $x < 3$ or $x > 4$

$f''(x) > 0$ for $3 < x < 4$

48. $f(0) = 4, \ f(6) = 0$

$f'(x) < 0$ for $x < 2$ or $x > 4$

$f'(2)$ does not exist.

$f'(4) = 0$

$f'(x) > 0$ for $2 < x < 4$

$f''(x) < 0$ for $x \neq 2$

49. Writing   A newspaper headline states that "The rate of growth of the national deficit is decreasing." What does this mean? What does it imply about the graph of the deficit as a function of time?

50. Inventory Cost   The cost of inventory C depends on the ordering and storage costs according to the inventory model

$$C = \left(\frac{Q}{x}\right)s + \left(\frac{x}{2}\right)r.$$

Determine the order size that will minimize the cost, assuming that sales occur at a constant rate, Q is the number of units sold per year, r is the cost of storing one unit for one year, s is the cost of placing an order, and x is the number of units per order.

51. Modeling Data   Outlays for national defense D (in billions of dollars) for selected years from 1970 through 2010 are shown in the table, where t is time in years, with $t = 0$ corresponding to 1970. (*Source: U.S. Office of Management and Budget*)

t	0	5	10	15	20
D	81.7	86.5	134.0	252.7	299.3

t	25	30	35	40
D	272.1	294.4	495.3	693.6

(a) Use the regression capabilities of a graphing utility to find a model of the form

$$D = at^4 + bt^3 + ct^2 + dt + e$$

for the data.

(b) Use a graphing utility to plot the data and graph the model.

(c) For the years shown in the table, when does the model indicate that the outlay for national defense was at a maximum? When was it at a minimum?

(d) For the years shown in the table, when does the model indicate that the outlay for national defense was increasing at the greatest rate?

52. Modeling Data   The manager of a store recorded the annual sales S (in thousands of dollars) of a product over a period of 7 years, as shown in the table, where t is the time in years, with $t = 6$ corresponding to 2006.

t	6	7	8	9	10	11	12
S	5.4	6.9	11.5	15.5	19.0	22.0	23.6

(a) Use the regression capabilities of a graphing utility to find a model of the form

$$S = at^3 + bt^2 + ct + d$$

for the data.

(b) Use a graphing utility to plot the data and graph the model.

(c) Use calculus and the model to find the time t when sales were increasing at the greatest rate.

(d) Do you think the model would be accurate for predicting future sales? Explain.

Finding a Limit   In Exercises 53–62, find the limit.

53. $\displaystyle\lim_{x \to \infty} \left(8 + \frac{1}{x}\right)$

54. $\displaystyle\lim_{x \to -\infty} \frac{1 - 4x}{x + 1}$

55. $\displaystyle\lim_{x \to \infty} \frac{2x^2}{3x^2 + 5}$

56. $\displaystyle\lim_{x \to \infty} \frac{4x^3}{x^4 + 3}$

57. $\displaystyle\lim_{x \to -\infty} \frac{3x^2}{x + 5}$

58. $\displaystyle\lim_{x \to -\infty} \frac{\sqrt{x^2 + x}}{-2x}$

59. $\displaystyle\lim_{x \to \infty} \frac{5\cos x}{x}$

60. $\displaystyle\lim_{x \to \infty} \frac{x^3}{\sqrt{x^2 + 2}}$

61. $\displaystyle\lim_{x \to -\infty} \frac{6x}{x + \cos x}$

62. $\displaystyle\lim_{x \to -\infty} \frac{x}{2\sin x}$

Horizontal Asymptotes   In Exercises 63–66, use a graphing utility to graph the function and identify any horizontal asymptotes.

63. $f(x) = \dfrac{3}{x} - 2$

64. $g(x) = \dfrac{5x^2}{x^2 + 2}$

65. $h(x) = \dfrac{2x + 3}{x - 4}$

66. $f(x) = \dfrac{3x}{\sqrt{x^2 + 2}}$

Analyzing the Graph of a Function   In Exercises 67–76, analyze and sketch a graph of the function. Label any intercepts, relative extrema, points of inflection, and asymptotes. Use a graphing utility to verify your results.

67. $f(x) = 4x - x^2$

68. $f(x) = 4x^3 - x^4$

69. $f(x) = x\sqrt{16 - x^2}$

70. $f(x) = (x^2 - 4)^2$

71. $f(x) = x^{1/3}(x + 3)^{2/3}$

72. $f(x) = (x - 3)(x + 2)^3$

73. $f(x) = \dfrac{5 - 3x}{x - 2}$

74. $f(x) = \dfrac{2x}{1 + x^2}$

75. $f(x) = x^3 + x + \dfrac{4}{x}$

76. $f(x) = x^2 + \dfrac{1}{x}$

77. Maximum Area A rancher has 400 feet of fencing with which to enclose two adjacent rectangular corrals (see figure). What dimensions should be used so that the enclosed area will be a maximum?

78. Maximum Area Find the dimensions of the rectangle of maximum area, with sides parallel to the coordinate axes, that can be inscribed in the ellipse given by

$$\frac{x^2}{144} + \frac{y^2}{16} = 1.$$

79. Minimum Length A right triangle in the first quadrant has the coordinate axes as sides, and the hypotenuse passes through the point $(1, 8)$. Find the vertices of the triangle such that the length of the hypotenuse is minimum.

80. Minimum Length The wall of a building is to be braced by a beam that must pass over a parallel fence 5 feet high and 4 feet from the building. Find the length of the shortest beam that can be used.

81. Maximum Length Find the length of the longest pipe that can be carried level around a right-angle corner at the intersection of two corridors of widths 4 feet and 6 feet.

82. Maximum Length A hallway of width 6 feet meets a hallway of width 9 feet at right angles. Find the length of the longest pipe that can be carried level around this corner. [*Hint:* If L is the length of the pipe, show that

$$L = 6 \csc \theta + 9 \csc\left(\frac{\pi}{2} - \theta\right)$$

where θ is the angle between the pipe and the wall of the narrower hallway.]

83. Maximum Volume Find the volume of the largest right circular cone that can be inscribed in a sphere of radius r.

84. Maximum Volume Find the volume of the largest right circular cylinder that can be inscribed in a sphere of radius r.

Using Newton's Method In Exercises 85–88, approximate the zero(s) of the function. Use Newton's Method and continue the process until two successive approximations differ by less than 0.001. Then find the zero(s) using a graphing utility and compare the results.

85. $f(x) = x^3 - 3x - 1$

86. $f(x) = x^3 + 2x + 1$

87. $f(x) = x^4 + x^3 - 3x^2 + 2$

88. $f(x) = 3\sqrt{x - 1} - x$

Finding Point(s) of Intersection In Exercises 89 and 90, apply Newton's Method to approximate the x-value(s) of the indicated point(s) of intersection of the two graphs. Continue the process until two successive approximations differ by less than 0.001. [*Hint:* Let $h(x) = f(x) - g(x)$.]

89. $f(x) = 1 - x$
$\quad\;\; g(x) = x^5 + 2$

90. $f(x) = \sin x$
$\quad\;\; g(x) = x^2 - 2x + 1$

   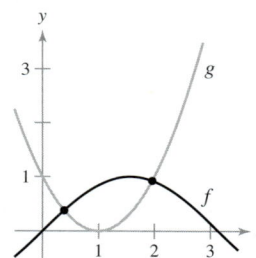

Comparing Δy and dy In Exercises 91 and 92, use the information to evaluate and compare Δy and dy.

Function	x-Value	Differential of x
91. $y = 0.5x^2$	$x = 3$	$\Delta x = dx = 0.01$
92. $y = x^3 - 6x$	$x = 2$	$\Delta x = dx = 0.1$

Finding a Differential In Exercises 93 and 94, find the differential dy of the given function.

93. $y = x(1 - \cos x)$ **94.** $y = \sqrt{36 - x^2}$

95. Volume and Surface Area The radius of a sphere is measured as 9 centimeters, with a possible error of 0.025 centimeter.

   (a) Use differentials to approximate the possible propagated error in computing the volume of the sphere.

   (b) Use differentials to approximate the possible propagated error in computing the surface area of the sphere.

   (c) Approximate the percent errors in parts (a) and (b).

96. Demand Function A company finds that the demand for its commodity is

$$p = 75 - \frac{1}{4}x$$

where p is the price in dollars and x is the number of units. Find and compare the values of Δp and dp as x changes from 7 to 8.

P.S. Problem Solving

See **CalcChat.com** for tutorial help and worked-out solutions to odd-numbered exercises.

1. **Relative Extrema** Graph the fourth-degree polynomial

$$p(x) = x^4 + ax^2 + 1$$

for various values of the constant a.

(a) Determine the values of a for which p has exactly one relative minimum.

(b) Determine the values of a for which p has exactly one relative maximum.

(c) Determine the values of a for which p has exactly two relative minima.

(d) Show that the graph of p cannot have exactly two relative extrema.

2. **Relative Extrema**

(a) Graph the fourth-degree polynomial $p(x) = ax^4 - 6x^2$ for $a = -3, -2, -1, 0, 1, 2$, and 3. For what values of the constant a does p have a relative minimum or relative maximum?

(b) Show that p has a relative maximum for all values of the constant a.

(c) Determine analytically the values of a for which p has a relative minimum.

(d) Let $(x, y) = (x, p(x))$ be a relative extremum of p. Show that (x, y) lies on the graph of $y = -3x^2$. Verify this result graphically by graphing $y = -3x^2$ together with the seven curves from part (a).

3. **Relative Minimum** Let

$$f(x) = \frac{c}{x} + x^2.$$

Determine all values of the constant c such that f has a relative minimum, but no relative maximum.

4. **Points of Inflection**

(a) Let $f(x) = ax^2 + bx + c$, $a \neq 0$, be a quadratic polynomial. How many points of inflection does the graph of f have?

(b) Let $f(x) = ax^3 + bx^2 + cx + d$, $a \neq 0$, be a cubic polynomial. How many points of inflection does the graph of f have?

(c) Suppose the function $y = f(x)$ satisfies the equation

$$\frac{dy}{dx} = ky\left(1 - \frac{y}{L}\right)$$

where k and L are positive constants. Show that the graph of f has a point of inflection at the point where $y = L/2$. (This equation is called the **logistic differential equation.**)

5. **Extended Mean Value Theorem** Prove the following **Extended Mean Value Theorem.** If f and f' are continuous on the closed interval $[a, b]$, and if f'' exists in the open interval (a, b), then there exists a number c in (a, b) such that

$$f(b) = f(a) + f'(a)(b - a) + \frac{1}{2}f''(c)(b - a)^2.$$

6. **Illumination** The amount of illumination of a surface is proportional to the intensity of the light source, inversely proportional to the square of the distance from the light source, and proportional to sin θ, where θ is the angle at which the light strikes the surface. A rectangular room measures 10 feet by 24 feet, with a 10-foot ceiling (see figure). Determine the height at which the light should be placed to allow the corners of the floor to receive as much light as possible.

7. **Minimum Distance** Consider a room in the shape of a cube, 4 meters on each side. A bug at point P wants to walk to point Q at the opposite corner, as shown in the figure. Use calculus to determine the shortest path. Explain how you can solve this problem without calculus. (*Hint:* Consider the two walls as one wall.)

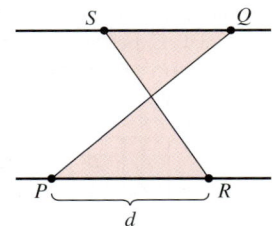

Figure for 7   Figure for 8

8. **Areas of Triangles** The line joining P and Q crosses the two parallel lines, as shown in the figure. The point R is d units from P. How far from Q should the point S be positioned so that the sum of the areas of the two shaded triangles is a minimum? So that the sum is a maximum?

9. **Mean Value Theorem** Determine the values a, b, and c such that the function f satisfies the hypotheses of the Mean Value Theorem on the interval $[0, 3]$.

$$f(x) = \begin{cases} 1, & x = 0 \\ ax + b, & 0 < x \leq 1 \\ x^2 + 4x + c, & 1 < x \leq 3 \end{cases}$$

10. **Mean Value Theorem** Determine the values a, b, c, and d such that the function f satisfies the hypotheses of the Mean Value Theorem on the interval $[-1, 2]$.

$$f(x) = \begin{cases} a, & x = -1 \\ 2, & -1 < x \leq 0 \\ bx^2 + c, & 0 < x \leq 1 \\ dx + 4, & 1 < x \leq 2 \end{cases}$$

11. Proof Let f and g be functions that are continuous on $[a, b]$ and differentiable on (a, b). Prove that if $f(a) = g(a)$ and $g'(x) > f'(x)$ for all x in (a, b), then $g(b) > f(b)$.

12. Proof

(a) Prove that $\lim\limits_{x \to \infty} x^2 = \infty$.

(b) Prove that $\lim\limits_{x \to \infty}\left(\dfrac{1}{x^2}\right) = 0$.

(c) Let L be a real number. Prove that if $\lim\limits_{x \to \infty} f(x) = L$, then

$$\lim_{y \to 0^+} f\left(\frac{1}{y}\right) = L.$$

13. Tangent Lines Find the point on the graph of

$$y = \frac{1}{1 + x^2}$$

(see figure) where the tangent line has the greatest slope, and the point where the tangent line has the least slope.

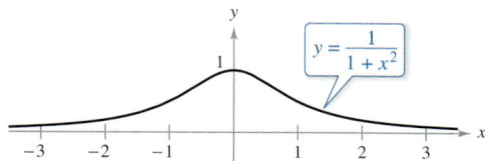

14. Stopping Distance The police department must determine the speed limit on a bridge such that the flow rate of cars is maximum per unit time. The greater the speed limit, the farther apart the cars must be in order to keep a safe stopping distance. Experimental data on the stopping distances d (in meters) for various speeds v (in kilometers per hour) are shown in the table.

v	20	40	60	80	100
d	5.1	13.7	27.2	44.2	66.4

(a) Convert the speeds v in the table to speeds s in meters per second. Use the regression capabilities of a graphing utility to find a model of the form $d(s) = as^2 + bs + c$ for the data.

(b) Consider two consecutive vehicles of average length 5.5 meters, traveling at a safe speed on the bridge. Let T be the difference between the times (in seconds) when the front bumpers of the vehicles pass a given point on the bridge. Verify that this difference in times is given by

$$T = \frac{d(s)}{s} + \frac{5.5}{s}.$$

(c) Use a graphing utility to graph the function T and estimate the speed s that minimizes the time between vehicles.

(d) Use calculus to determine the speed that minimizes T. What is the minimum value of T? Convert the required speed to kilometers per hour.

(e) Find the optimal distance between vehicles for the posted speed limit determined in part (d).

15. Darboux's Theorem Prove Darboux's Theorem: Let f be differentiable on the closed interval $[a, b]$ such that $f'(a) = y_1$ and $f'(b) = y_2$. If d lies between y_1 and y_2, then there exists c in (a, b) such that $f'(c) = d$.

16. Maximum Area The figures show a rectangle, a circle, and a semicircle inscribed in a triangle bounded by the coordinate axes and the first-quadrant portion of the line with intercepts $(3, 0)$ and $(0, 4)$. Find the dimensions of each inscribed figure such that its area is maximum. State whether calculus was helpful in finding the required dimensions. Explain your reasoning.

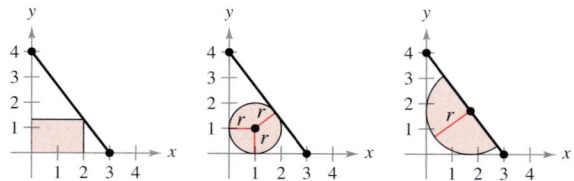

17. Point of Inflection Show that the cubic polynomial $p(x) = ax^3 + bx^2 + cx + d$ has exactly one point of inflection (x_0, y_0), where

$$x_0 = \frac{-b}{3a} \quad \text{and} \quad y_0 = \frac{2b^3}{27a^2} - \frac{bc}{3a} + d.$$

Use this formula to find the point of inflection of $p(x) = x^3 - 3x^2 + 2$.

18. Minimum Length A legal-sized sheet of paper (8.5 inches by 14 inches) is folded so that corner P touches the opposite 14-inch edge at R (see figure). $\left(\textit{Note: } PQ = \sqrt{C^2 - x^2}.\right)$

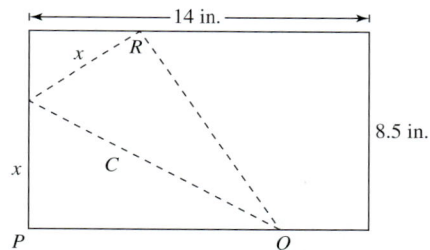

(a) Show that $C^2 = \dfrac{2x^3}{2x - 8.5}$.

(b) What is the domain of C?

(c) Determine the x-value that minimizes C.

(d) Determine the minimum length C.

19. Quadratic Approximation The polynomial

$$P(x) = c_0 + c_1(x - a) + c_2(x - a)^2$$

is the quadratic approximation of the function f at $(a, f(a))$ when $P(a) = f(a)$, $P'(a) = f'(a)$, and $P''(a) = f''(a)$.

(a) Find the quadratic approximation of

$$f(x) = \frac{x}{x + 1}$$

at $(0, 0)$.

(b) Use a graphing utility to graph $P(x)$ and $f(x)$ in the same viewing window.

4 Integration

Electricity *(Exercise 84, p. 303)*

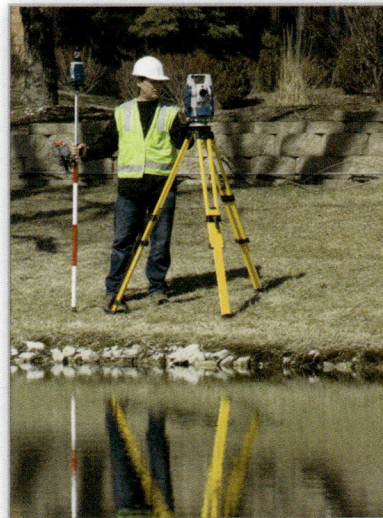

Surveying
(Exercise 39, p. 311)

The Speed of Sound *(Example 5, p. 282)*

Amount of Chemical
Flowing into a Tank
(Example 9, p. 286)

Grand Canyon *(Exercise 58, p. 252)*

4.1 Antiderivatives and Indefinite Integration

- Write the general solution of a differential equation and use indefinite integral notation for antiderivatives.
- Use basic integration rules to find antiderivatives.
- Find a particular solution of a differential equation.

Antiderivatives

To find a function F whose derivative is $f(x) = 3x^2$, you might use your knowledge of derivatives to conclude that

$$F(x) = x^3 \quad \text{because} \quad \frac{d}{dx}[x^3] = 3x^2.$$

The function F is an *antiderivative* of f.

> **Definition of Antiderivative**
>
> A function F is an **antiderivative** of f on an interval I when $F'(x) = f(x)$ for all x in I.

Note that F is called *an* antiderivative of f, rather than *the* antiderivative of f. To see why, observe that

$$F_1(x) = x^3, \quad F_2(x) = x^3 - 5, \quad \text{and} \quad F_3(x) = x^3 + 97$$

are all antiderivatives of $f(x) = 3x^2$. In fact, for any constant C, the function $F(x) = x^3 + C$ is an antiderivative of f.

> **THEOREM 4.1 Representation of Antiderivatives**
>
> If F is an antiderivative of f on an interval I, then G is an antiderivative of f on the interval I if and only if G is of the form $G(x) = F(x) + C$, for all x in I where C is a constant.

Proof The proof of Theorem 4.1 in one direction is straightforward. That is, if $G(x) = F(x) + C$, $F'(x) = f(x)$, and C is a constant, then

$$G'(x) = \frac{d}{dx}[F(x) + C] = F'(x) + 0 = f(x).$$

To prove this theorem in the other direction, assume that G is an antiderivative of f. Define a function H such that

$$H(x) = G(x) - F(x).$$

For any two points a and b $(a < b)$ in the interval, H is continuous on $[a, b]$ and differentiable on (a, b). By the Mean Value Theorem,

$$H'(c) = \frac{H(b) - H(a)}{b - a}$$

for some c in (a, b). However, $H'(c) = 0$, so $H(a) = H(b)$. Because a and b are arbitrary points in the interval, you know that H is a constant function C. So, $G(x) - F(x) = C$ and it follows that $G(x) = F(x) + C$.

See LarsonCalculus.com for Bruce Edwards's video of this proof.

Using Theorem 4.1, you can represent the entire family of antiderivatives of a function by adding a constant to a *known* antiderivative. For example, knowing that

$$D_x[x^2] = 2x$$

you can represent the family of *all* antiderivatives of $f(x) = 2x$ by

$$G(x) = x^2 + C \qquad \text{Family of all antiderivatives of } f(x) = 2x$$

where C is a constant. The constant C is called the **constant of integration.** The family of functions represented by G is the **general antiderivative** of f, and $G(x) = x^2 + C$ is the **general solution** of the *differential equation*

$$G'(x) = 2x. \qquad \text{Differential equation}$$

A **differential equation** in x and y is an equation that involves x, y, and derivatives of y. For instance,

$$y' = 3x \quad \text{and} \quad y' = x^2 + 1$$

are examples of differential equations.

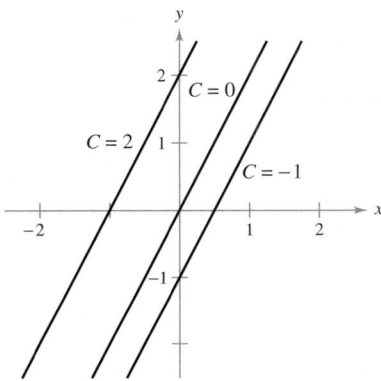

Functions of the form $y = 2x + C$

Figure 4.1

EXAMPLE 1   Solving a Differential Equation

Find the general solution of the differential equation $y' = 2$.

Solution   To begin, you need to find a function whose derivative is 2. One such function is

$$y = 2x. \qquad \text{2x is } an \text{ antiderivative of 2.}$$

Now, you can use Theorem 4.1 to conclude that the general solution of the differential equation is

$$y = 2x + C. \qquad \text{General solution}$$

The graphs of several functions of the form $y = 2x + C$ are shown in Figure 4.1. ■

When solving a differential equation of the form

$$\frac{dy}{dx} = f(x)$$

it is convenient to write it in the equivalent differential form

$$dy = f(x)\,dx.$$

The operation of finding all solutions of this equation is called **antidifferentiation** (or **indefinite integration**) and is denoted by an integral sign $\int$. The general solution is denoted by

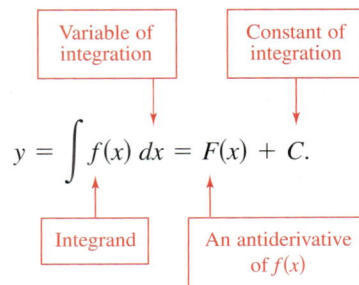

Variable of integration   Constant of integration

$$y = \int f(x)\,dx = F(x) + C.$$

Integrand   An antiderivative of $f(x)$

▷

REMARK   In this text, the notation $\int f(x)\,dx = F(x) + C$ means that F is an antiderivative of f on an interval.

The expression $\int f(x)\,dx$ is read as the *antiderivative of f with respect to x*. So, the differential dx serves to identify x as the variable of integration. The term **indefinite integral** is a synonym for antiderivative.

Basic Integration Rules

The inverse nature of integration and differentiation can be verified by substituting $F'(x)$ for $f(x)$ in the indefinite integration definition to obtain

$$\int F'(x)\, dx = F(x) + C.$$

Integration is the "inverse" of differentiation.

Moreover, if $\int f(x)\, dx = F(x) + C$, then

$$\frac{d}{dx}\left[\int f(x)\, dx\right] = f(x).$$

Differentiation is the "inverse" of integration.

These two equations allow you to obtain integration formulas directly from differentiation formulas, as shown in the following summary.

Basic Integration Rules

Differentiation Formula	Integration Formula
$\frac{d}{dx}[C] = 0$	$\int 0\, dx = C$
$\frac{d}{dx}[kx] = k$	$\int k\, dx = kx + C$
$\frac{d}{dx}[kf(x)] = kf'(x)$	$\int kf(x)\, dx = k\int f(x)\, dx$
$\frac{d}{dx}[f(x) \pm g(x)] = f'(x) \pm g'(x)$	$\int [f(x) \pm g(x)]\, dx = \int f(x)\, dx \pm \int g(x)\, dx$
$\frac{d}{dx}[x^n] = nx^{n-1}$	$\int x^n\, dx = \frac{x^{n+1}}{n+1} + C, \quad n \neq -1$ Power Rule
$\frac{d}{dx}[\sin x] = \cos x$	$\int \cos x\, dx = \sin x + C$
$\frac{d}{dx}[\cos x] = -\sin x$	$\int \sin x\, dx = -\cos x + C$
$\frac{d}{dx}[\tan x] = \sec^2 x$	$\int \sec^2 x\, dx = \tan x + C$
$\frac{d}{dx}[\sec x] = \sec x \tan x$	$\int \sec x \tan x\, dx = \sec x + C$
$\frac{d}{dx}[\cot x] = -\csc^2 x$	$\int \csc^2 x\, dx = -\cot x + C$
$\frac{d}{dx}[\csc x] = -\csc x \cot x$	$\int \csc x \cot x\, dx = -\csc x + C$

Note that the Power Rule for Integration has the restriction that $n \neq -1$. The evaluation of

$$\int \frac{1}{x}\, dx$$

must wait until the introduction of the natural logarithmic function in Chapter 5.

REMARK In Example 2, note that the general pattern of integration is similar to that of differentiation.

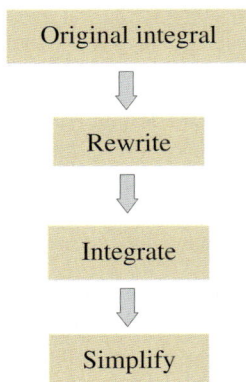

Original integral

↓

Rewrite

↓

Integrate

↓

Simplify

EXAMPLE 2 Describing Antiderivatives

$$\int 3x \, dx = 3\int x \, dx \qquad \text{Constant Multiple Rule}$$
$$= 3\int x^1 \, dx \qquad \text{Rewrite } x \text{ as } x^1.$$
$$= 3\left(\frac{x^2}{2}\right) + C \qquad \text{Power Rule } (n = 1)$$
$$= \frac{3}{2}x^2 + C \qquad \text{Simplify.}$$

The antiderivatives of $3x$ are of the form $\frac{3}{2}x^2 + C$, where C is any constant.

When indefinite integrals are evaluated, a strict application of the basic integration rules tends to produce complicated constants of integration. For instance, in Example 2, the solution could have been written as

$$\int 3x \, dx = 3\int x \, dx = 3\left(\frac{x^2}{2} + C\right) = \frac{3}{2}x^2 + 3C.$$

Because C represents *any* constant, it is both cumbersome and unnecessary to write $3C$ as the constant of integration. So, $\frac{3}{2}x^2 + 3C$ is written in the simpler form $\frac{3}{2}x^2 + C$.

TECHNOLOGY Some software programs, such as *Maple* and *Mathematica*, are capable of performing integration symbolically. If you have access to such a symbolic integration utility, try using it to evaluate the indefinite integrals in Example 3.

EXAMPLE 3 Rewriting Before Integrating

See LarsonCalculus.com for an interactive version of this type of example.

	Original Integral	Rewrite	Integrate	Simplify
a.	$\int \frac{1}{x^3} \, dx$	$\int x^{-3} \, dx$	$\frac{x^{-2}}{-2} + C$	$-\frac{1}{2x^2} + C$
b.	$\int \sqrt{x} \, dx$	$\int x^{1/2} \, dx$	$\frac{x^{3/2}}{3/2} + C$	$\frac{2}{3}x^{3/2} + C$
c.	$\int 2\sin x \, dx$	$2\int \sin x \, dx$	$2(-\cos x) + C$	$-2\cos x + C$

REMARK The basic integration rules allow you to integrate any polynomial function.

EXAMPLE 4 Integrating Polynomial Functions

a. $\int dx = \int 1 \, dx$ \qquad Integrand is understood to be 1.
$$= x + C \qquad \text{Integrate.}$$

b. $\int (x + 2) \, dx = \int x \, dx + \int 2 \, dx$
$$= \frac{x^2}{2} + C_1 + 2x + C_2 \qquad \text{Integrate.}$$
$$= \frac{x^2}{2} + 2x + C \qquad C = C_1 + C_2$$

The second line in the solution is usually omitted.

c. $\int (3x^4 - 5x^2 + x) \, dx = 3\left(\frac{x^5}{5}\right) - 5\left(\frac{x^3}{3}\right) + \frac{x^2}{2} + C$
$$= \frac{3}{5}x^5 - \frac{5}{3}x^3 + \frac{1}{2}x^2 + C$$

· · · · · · · · · · · · · · · ▷

· · **REMARK** Before you begin
the exercise set, be sure you
realize that one of the most
important steps in integration
is *rewriting the integrand* in
a form that fits one of the
basic integration rules.

EXAMPLE 5 **Rewriting Before Integrating**

$$\int \frac{x+1}{\sqrt{x}}\, dx = \int \left(\frac{x}{\sqrt{x}} + \frac{1}{\sqrt{x}} \right) dx \qquad \text{Rewrite as two fractions.}$$

$$= \int (x^{1/2} + x^{-1/2})\, dx \qquad \text{Rewrite with fractional exponents.}$$

$$= \frac{x^{3/2}}{3/2} + \frac{x^{1/2}}{1/2} + C \qquad \text{Integrate.}$$

$$= \frac{2}{3}x^{3/2} + 2x^{1/2} + C \qquad \text{Simplify.}$$

$$= \frac{2}{3}\sqrt{x}(x + 3) + C$$

When integrating quotients, do not integrate the numerator and denominator
separately. This is no more valid in integration than it is in differentiation. For instance,
in Example 5, be sure you understand that

$$\int \frac{x+1}{\sqrt{x}}\, dx = \frac{2}{3}\sqrt{x}(x + 3) + C$$

is not the same as

$$\frac{\int (x+1)\, dx}{\int \sqrt{x}\, dx} = \frac{\frac{1}{2}x^2 + x + C_1}{\frac{2}{3}x\sqrt{x} + C_2}.$$

EXAMPLE 6 **Rewriting Before Integrating**

$$\int \frac{\sin x}{\cos^2 x}\, dx = \int \left(\frac{1}{\cos x} \right) \left(\frac{\sin x}{\cos x} \right) dx \qquad \text{Rewrite as a product.}$$

$$= \int \sec x \tan x\, dx \qquad \text{Rewrite using trigonometric identities.}$$

$$= \sec x + C \qquad \text{Integrate.}$$

EXAMPLE 7 **Rewriting Before Integrating**

	Original Integral	Rewrite	Integrate	Simplify
a.	$\int \frac{2}{\sqrt{x}}\, dx$	$2\int x^{-1/2}\, dx$	$2\left(\frac{x^{1/2}}{1/2} \right) + C$	$4x^{1/2} + C$
b.	$\int (t^2 + 1)^2\, dt$	$\int (t^4 + 2t^2 + 1)\, dt$	$\frac{t^5}{5} + 2\left(\frac{t^3}{3} \right) + t + C$	$\frac{1}{5}t^5 + \frac{2}{3}t^3 + t + C$
c.	$\int \frac{x^3 + 3}{x^2}\, dx$	$\int (x + 3x^{-2})\, dx$	$\frac{x^2}{2} + 3\left(\frac{x^{-1}}{-1} \right) + C$	$\frac{1}{2}x^2 - \frac{3}{x} + C$
d.	$\int \sqrt[3]{x}(x - 4)\, dx$	$\int (x^{4/3} - 4x^{1/3})\, dx$	$\frac{x^{7/3}}{7/3} - 4\left(\frac{x^{4/3}}{4/3} \right) + C$	$\frac{3}{7}x^{7/3} - 3x^{4/3} + C$

As you do the exercises, note that you can check your answer to an
antidifferentiation problem by differentiating. For instance, in Example 7(a), you can
check that $4x^{1/2} + C$ is the correct antiderivative by differentiating the answer to obtain

$$D_x[4x^{1/2} + C] = 4\left(\frac{1}{2} \right)x^{-1/2} = \frac{2}{\sqrt{x}}. \qquad \text{Use differentiation to check antiderivative.}$$

Initial Conditions and Particular Solutions

You have already seen that the equation $y = \int f(x)\, dx$ has many solutions (each differing from the others by a constant). This means that the graphs of any two antiderivatives of f are vertical translations of each other. For example, Figure 4.2 shows the graphs of several antiderivatives of the form

$$y = \int (3x^2 - 1)\, dx = x^3 - x + C \qquad \text{General solution}$$

for various integer values of C. Each of these antiderivatives is a solution of the differential equation

$$\frac{dy}{dx} = 3x^2 - 1.$$

In many applications of integration, you are given enough information to determine a **particular solution.** To do this, you need only know the value of $y = F(x)$ for one value of x. This information is called an **initial condition.** For example, in Figure 4.2, only one curve passes through the point $(2, 4)$. To find this curve, you can use the general solution

$$F(x) = x^3 - x + C \qquad \text{General solution}$$

and the initial condition

$$F(2) = 4. \qquad \text{Initial condition}$$

By using the initial condition in the general solution, you can determine that

$$F(2) = 8 - 2 + C = 4$$

which implies that $C = -2$. So, you obtain

$$F(x) = x^3 - x - 2. \qquad \text{Particular solution}$$

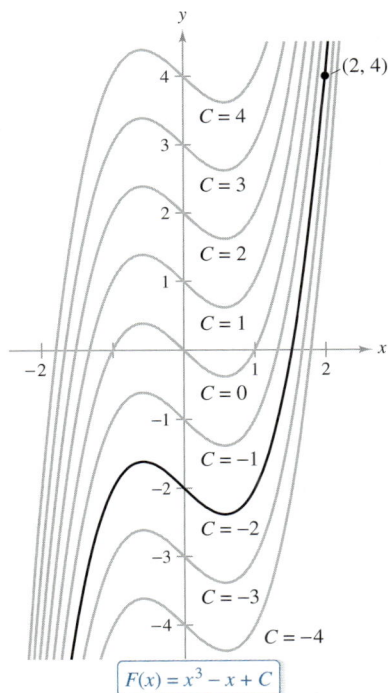

The particular solution that satisfies the initial condition $F(2) = 4$ is $F(x) = x^3 - x - 2$.
Figure 4.2

$$\boxed{F(x) = x^3 - x + C}$$

EXAMPLE 8 **Finding a Particular Solution**

Find the general solution of

$$F'(x) = \frac{1}{x^2}, \quad x > 0$$

and find the particular solution that satisfies the initial condition $F(1) = 0$.

Solution To find the general solution, integrate to obtain

$$F(x) = \int \frac{1}{x^2}\, dx \qquad\qquad F(x) = \int F'(x)\, dx$$

$$= \int x^{-2}\, dx \qquad\qquad \text{Rewrite as a power.}$$

$$= \frac{x^{-1}}{-1} + C \qquad\qquad \text{Integrate.}$$

$$= -\frac{1}{x} + C, \quad x > 0. \qquad \text{General solution}$$

Using the initial condition $F(1) = 0$, you can solve for C as follows.

$$F(1) = -\frac{1}{1} + C = 0 \quad \Longrightarrow \quad C = 1$$

So, the particular solution, as shown in Figure 4.3, is

$$F(x) = -\frac{1}{x} + 1, \quad x > 0. \qquad \text{Particular solution}$$

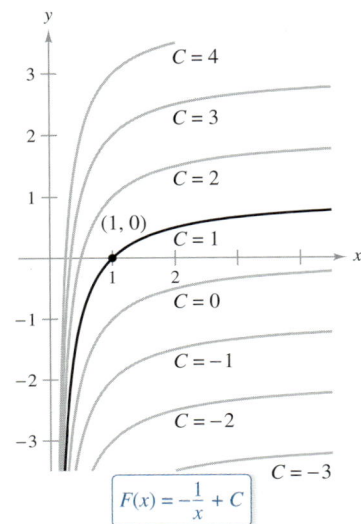

The particular solution that satisfies the initial condition $F(1) = 0$ is $F(x) = -(1/x) + 1, x > 0$.
Figure 4.3

$$\boxed{F(x) = -\frac{1}{x} + C}$$

So far in this section, you have been using x as the variable of integration. In applications, it is often convenient to use a different variable. For instance, in the next example, involving *time*, the variable of integration is t.

EXAMPLE 9 **Solving a Vertical Motion Problem**

A ball is thrown upward with an initial velocity of 64 feet per second from an initial height of 80 feet.

a. Find the position function giving the height s as a function of the time t.

b. When does the ball hit the ground?

Solution

a. Let $t = 0$ represent the initial time. The two given initial conditions can be written as follows.

$$s(0) = 80 \qquad \text{Initial height is 80 feet.}$$
$$s'(0) = 64 \qquad \text{Initial velocity is 64 feet per second.}$$

Using -32 feet per second per second as the acceleration due to gravity, you can write

$$s''(t) = -32$$
$$s'(t) = \int s''(t)\, dt = \int -32\, dt = -32t + C_1.$$

Using the initial velocity, you obtain $s'(0) = 64 = -32(0) + C_1$, which implies that $C_1 = 64$. Next, by integrating $s'(t)$, you obtain

$$s(t) = \int s'(t)\, dt = \int (-32t + 64)\, dt = -16t^2 + 64t + C_2.$$

Using the initial height, you obtain

$$s(0) = 80 = -16(0^2) + 64(0) + C_2$$

which implies that $C_2 = 80$. So, the position function is

$$s(t) = -16t^2 + 64t + 80. \qquad \text{See Figure 4.4.}$$

b. Using the position function found in part (a), you can find the time at which the ball hits the ground by solving the equation $s(t) = 0$.

$$-16t^2 + 64t + 80 = 0$$
$$-16(t + 1)(t - 5) = 0$$
$$t = -1, 5$$

Because t must be positive, you can conclude that the ball hits the ground 5 seconds after it was thrown. ■

In Example 9, note that the position function has the form

$$s(t) = \frac{1}{2}gt^2 + v_0 t + s_0$$

where $g = -32$, v_0 is the initial velocity, and s_0 is the initial height, as presented in Section 2.2.

Example 9 shows how to use calculus to analyze vertical motion problems in which the acceleration is determined by a gravitational force. You can use a similar strategy to analyze other linear motion problems (vertical or horizontal) in which the acceleration (or deceleration) is the result of some other force, as you will see in Exercises 61–68.

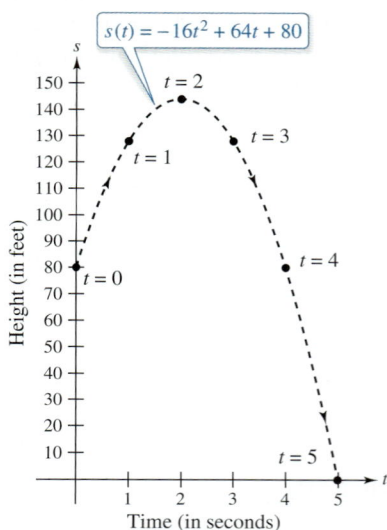

$s(t) = -16t^2 + 64t + 80$

Height of a ball at time t

Figure 4.4

4.1 Exercises

See **CalcChat.com** for tutorial help and worked-out solutions to odd-numbered exercises.

Integration and Differentiation In Exercises 1 and 2, verify the statement by showing that the derivative of the right side equals the integrand of the left side.

1. $\int \left(-\dfrac{6}{x^4} \right) dx = \dfrac{2}{x^3} + C$

2. $\int \left(8x^3 + \dfrac{1}{2x^2} \right) dx = 2x^4 - \dfrac{1}{2x} + C$

Solving a Differential Equation In Exercises 3–6, find the general solution of the differential equation and check the result by differentiation.

3. $\dfrac{dy}{dt} = 9t^2$

4. $\dfrac{dy}{dt} = 5$

5. $\dfrac{dy}{dx} = x^{3/2}$

6. $\dfrac{dy}{dx} = 2x^{-3}$

Rewriting Before Integrating In Exercises 7–10, complete the table to find the indefinite integral.

Original Integral	Rewrite	Integrate	Simplify
7. $\int \sqrt[3]{x}\, dx$			
8. $\int \dfrac{1}{4x^2}\, dx$			
9. $\int \dfrac{1}{x\sqrt{x}}\, dx$			
10. $\int \dfrac{1}{(3x)^2}\, dx$			

Finding an Indefinite Integral In Exercises 11–32, find the indefinite integral and check the result by differentiation.

11. $\int (x + 7)\, dx$

12. $\int (13 - x)\, dx$

13. $\int (x^5 + 1)\, dx$

14. $\int (8x^3 - 9x^2 + 4)\, dx$

15. $\int (x^{3/2} + 2x + 1)\, dx$

16. $\int \left(\sqrt{x} + \dfrac{1}{2\sqrt{x}} \right) dx$

17. $\int \sqrt[3]{x^2}\, dx$

18. $\int \left(\sqrt[4]{x^3} + 1 \right) dx$

19. $\int \dfrac{1}{x^5}\, dx$

20. $\int \dfrac{3}{x^7}\, dx$

21. $\int \dfrac{x + 6}{\sqrt{x}}\, dx$

22. $\int \dfrac{x^4 - 3x^2 + 5}{x^4}\, dx$

23. $\int (x + 1)(3x - 2)\, dx$

24. $\int (4t^2 + 3)^2\, dt$

25. $\int (5\cos x + 4\sin x)\, dx$

26. $\int (t^2 - \cos t)\, dt$

27. $\int (1 - \csc t \cot t)\, dt$

28. $\int (\theta^2 + \sec^2 \theta)\, d\theta$

29. $\int (\sec^2 \theta - \sin \theta)\, d\theta$

30. $\int \sec y\,(\tan y - \sec y)\, dy$

31. $\int (\tan^2 y + 1)\, dy$

32. $\int (4x - \csc^2 x)\, dx$

Sketching a Graph In Exercises 33 and 34, the graph of the derivative of a function is given. Sketch the graphs of *two* functions that have the given derivative. (There is more than one correct answer.) To print an enlarged copy of the graph, go to *MathGraphs.com*.

33.

34.

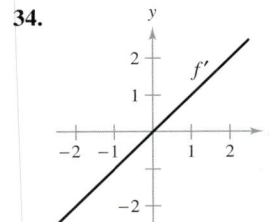

Finding a Particular Solution In Exercises 35–42, find the particular solution that satisfies the differential equation and the initial condition.

35. $f'(x) = 6x, \ f(0) = 8$

36. $g'(x) = 4x^2, \ g(-1) = 3$

37. $h'(t) = 8t^3 + 5, \ h(1) = -4$

38. $f'(s) = 10s - 12s^3, \ f(3) = 2$

39. $f''(x) = 2, \ f'(2) = 5, \ f(2) = 10$

40. $f''(x) = x^2, \ f'(0) = 8, \ f(0) = 4$

41. $f''(x) = x^{-3/2}, \ f'(4) = 2, \ f(0) = 0$

42. $f''(x) = \sin x, \ f'(0) = 1, \ f(0) = 6$

Slope Field In Exercises 43 and 44, a differential equation, a point, and a slope field are given. A *slope field* (or *direction field*) consists of line segments with slopes given by the differential equation. These line segments give a visual perspective of the slopes of the solutions of the differential equation. (a) Sketch two approximate solutions of the differential equation on the slope field, one of which passes through the indicated point. (To print an enlarged copy of the graph, go to *MathGraphs.com*.) (b) Use integration to find the particular solution of the differential equation and use a graphing utility to graph the solution. Compare the result with the sketches in part (a).

43. $\dfrac{dy}{dx} = x^2 - 1, \ (-1, 3)$

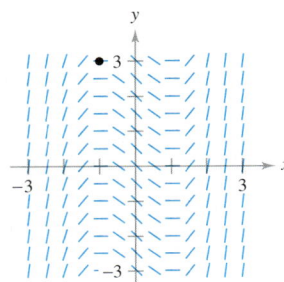

44. $\dfrac{dy}{dx} = -\dfrac{1}{x^2}, \ x > 0, \ (1, 3)$

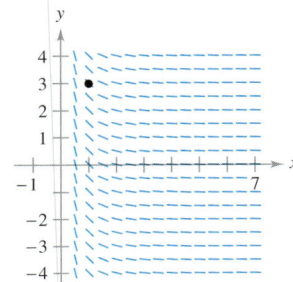

Slope Field In Exercises 45 and 46, (a) use a graphing utility to graph a slope field for the differential equation, (b) use integration and the given point to find the particular solution of the differential equation, and (c) graph the solution and the slope field in the same viewing window.

45. $\dfrac{dy}{dx} = 2x, \ (-2, -2)$ **46.** $\dfrac{dy}{dx} = 2\sqrt{x}, \ (4, 12)$

WRITING ABOUT CONCEPTS

47. Antiderivatives and Indefinite Integrals What is the difference, if any, between finding the antiderivative of $f(x)$ and evaluating the integral $\int f(x)\,dx$?

48. Comparing Functions Consider $f(x) = \tan^2 x$ and $g(x) = \sec^2 x$. What do you notice about the derivatives of $f(x)$ and $g(x)$? What can you conclude about the relationship between $f(x)$ and $g(x)$?

49. Sketching Graphs The graphs of f and f' each pass through the origin. Use the graph of f'' shown in the figure to sketch the graphs of f and f'. To print an enlarged copy of the graph, go to *MathGraphs.com*.

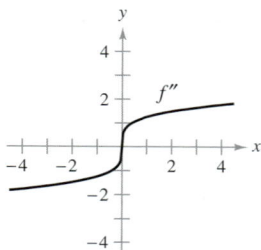

---

50. **HOW DO YOU SEE IT?** Use the graph of f' shown in the figure to answer the following.

(a) Approximate the slope of f at $x = 4$. Explain.

(b) Is it possible that $f(2) = -1$? Explain.

(c) Is $f(5) - f(4) > 0$? Explain.

(d) Approximate the value of x where f is maximum. Explain.

(e) Approximate any open intervals in which the graph of f is concave upward and any open intervals in which it is concave downward. Approximate the x-coordinates of any points of inflection.

51. Tree Growth An evergreen nursery usually sells a certain type of shrub after 6 years of growth and shaping. The growth rate during those 6 years is approximated by $dh/dt = 1.5t + 5$, where t is the time in years and h is the height in centimeters. The seedlings are 12 centimeters tall when planted ($t = 0$).

(a) Find the height after t years.

(b) How tall are the shrubs when they are sold?

52. Population Growth The rate of growth dP/dt of a population of bacteria is proportional to the square root of t, where P is the population size and t is the time in days ($0 \le t \le 10$). That is,

$$\frac{dP}{dt} = k\sqrt{t}.$$

The initial size of the population is 500. After 1 day the population has grown to 600. Estimate the population after 7 days.

Vertical Motion In Exercises 53–55, use $a(t) = -32$ feet per second per second as the acceleration due to gravity. (Neglect air resistance.)

53. A ball is thrown vertically upward from a height of 6 feet with an initial velocity of 60 feet per second. How high will the ball go?

54. With what initial velocity must an object be thrown upward (from ground level) to reach the top of the Washington Monument (approximately 550 feet)?

55. A balloon, rising vertically with a velocity of 16 feet per second, releases a sandbag at the instant it is 64 feet above the ground.

(a) How many seconds after its release will the bag strike the ground?

(b) At what velocity will it hit the ground?

Vertical Motion In Exercises 56–58, use $a(t) = -9.8$ meters per second per second as the acceleration due to gravity. (Neglect air resistance.)

56. A baseball is thrown upward from a height of 2 meters with an initial velocity of 10 meters per second. Determine its maximum height.

57. With what initial velocity must an object be thrown upward (from a height of 2 meters) to reach a maximum height of 200 meters?

58. Grand Canyon

The Grand Canyon is 1800 meters deep at its deepest point. A rock is dropped from the rim above this point. Write the height of the rock as a function of the time t in seconds. How long will it take the rock to hit the canyon floor?

59. Lunar Gravity   On the moon, the acceleration due to gravity is -1.6 meters per second per second. A stone is dropped from a cliff on the moon and hits the surface of the moon 20 seconds later. How far did it fall? What was its velocity at impact?

60. Escape Velocity   The minimum velocity required for an object to escape Earth's gravitational pull is obtained from the solution of the equation

$$\int v \, dv = -GM \int \frac{1}{y^2} \, dy$$

where v is the velocity of the object projected from Earth, y is the distance from the center of Earth, G is the gravitational constant, and M is the mass of Earth. Show that v and y are related by the equation

$$v^2 = v_0^2 + 2GM\left(\frac{1}{y} - \frac{1}{R}\right)$$

where v_0 is the initial velocity of the object and R is the radius of Earth.

Rectilinear Motion   In Exercises 61–64, consider a particle moving along the x-axis where $x(t)$ is the position of the particle at time t, $x'(t)$ is its velocity, and $x''(t)$ is its acceleration.

61. $x(t) = t^3 - 6t^2 + 9t - 2, \quad 0 \le t \le 5$

 (a) Find the velocity and acceleration of the particle.

 (b) Find the open t-intervals on which the particle is moving to the right.

 (c) Find the velocity of the particle when the acceleration is 0.

62. Repeat Exercise 61 for the position function

$$x(t) = (t - 1)(t - 3)^2, \quad 0 \le t \le 5.$$

63. A particle moves along the x-axis at a velocity of $v(t) = 1/\sqrt{t}$, $t > 0$. At time $t = 1$, its position is $x = 4$. Find the acceleration and position functions for the particle.

64. A particle, initially at rest, moves along the x-axis such that its acceleration at time $t > 0$ is given by $a(t) = \cos t$. At the time $t = 0$, its position is $x = 3$.

 (a) Find the velocity and position functions for the particle.

 (b) Find the values of t for which the particle is at rest.

65. Acceleration   The maker of an automobile advertises that it takes 13 seconds to accelerate from 25 kilometers per hour to 80 kilometers per hour. Assume the acceleration is constant.

 (a) Find the acceleration in meters per second per second.

 (b) Find the distance the car travels during the 13 seconds.

66. Deceleration   A car traveling at 45 miles per hour is brought to a stop, at constant deceleration, 132 feet from where the brakes are applied.

 (a) How far has the car moved when its speed has been reduced to 30 miles per hour?

 (b) How far has the car moved when its speed has been reduced to 15 miles per hour?

 (c) Draw the real number line from 0 to 132. Plot the points found in parts (a) and (b). What can you conclude?

67. Acceleration   At the instant the traffic light turns green, a car that has been waiting at an intersection starts with a constant acceleration of 6 feet per second per second. At the same instant, a truck traveling with a constant velocity of 30 feet per second passes the car.

 (a) How far beyond its starting point will the car pass the truck?

 (b) How fast will the car be traveling when it passes the truck?

68. Acceleration   Assume that a fully loaded plane starting from rest has a constant acceleration while moving down a runway. The plane requires 0.7 mile of runway and a speed of 160 miles per hour in order to lift off. What is the plane's acceleration?

True or False?   In Exercises 69–74, determine whether the statement is true or false. If it is false, explain why or give an example that shows it is false.

69. The antiderivative of $f(x)$ is unique.

70. Each antiderivative of an nth-degree polynomial function is an $(n + 1)$th-degree polynomial function.

71. If $p(x)$ is a polynomial function, then p has exactly one antiderivative whose graph contains the origin.

72. If $F(x)$ and $G(x)$ are antiderivatives of $f(x)$, then

$$F(x) = G(x) + C.$$

73. If $f'(x) = g(x)$, then $\int g(x) \, dx = f(x) + C.$

74. $\int f(x)g(x) \, dx = \int f(x) \, dx \int g(x) \, dx$

75. Horizontal Tangent   Find a function f such that the graph of f has a horizontal tangent at $(2, 0)$ and $f''(x) = 2x$.

76. Finding a Function   The graph of f' is shown. Find and sketch the graph of f given that f is continuous and $f(0) = 1$.

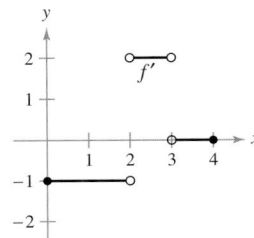

77. Proof   Let $s(x)$ and $c(x)$ be two functions satisfying $s'(x) = c(x)$ and $c'(x) = -s(x)$ for all x. If $s(0) = 0$ and $c(0) = 1$, prove that $[s(x)]^2 + [c(x)]^2 = 1$.

PUTNAM EXAM CHALLENGE

78. Suppose f and g are non-constant, differentiable, real-valued functions defined on $(-\infty, \infty)$. Furthermore, suppose that for each pair of real numbers x and y,

$$f(x + y) = f(x)f(y) - g(x)g(y) \quad \text{and}$$
$$g(x + y) = f(x)g(y) + g(x)f(y).$$

If $f'(0) = 0$, prove that $(f(x))^2 + (g(x))^2 = 1$ for all x.

4.2 Area

- Use sigma notation to write and evaluate a sum.
- Understand the concept of area.
- Approximate the area of a plane region.
- Find the area of a plane region using limits.

Sigma Notation

In the preceding section, you studied antidifferentiation. In this section, you will look further into a problem introduced in Section 1.1—that of finding the area of a region in the plane. At first glance, these two ideas may seem unrelated, but you will discover in Section 4.4 that they are closely related by an extremely important theorem called the Fundamental Theorem of Calculus.

This section begins by introducing a concise notation for sums. This notation is called **sigma notation** because it uses the uppercase Greek letter sigma, written as Σ.

Sigma Notation

The sum of n terms $a_1, a_2, a_3, \ldots, a_n$ is written as

$$\sum_{i=1}^{n} a_i = a_1 + a_2 + a_3 + \cdots + a_n$$

where i is the **index of summation**, a_i is the **ith term** of the sum, and the **upper and lower bounds of summation** are n and 1.

REMARK The upper and lower bounds must be constant with respect to the index of summation. However, the lower bound doesn't have to be 1. Any integer less than or equal to the upper bound is legitimate.

EXAMPLE 1 **Examples of Sigma Notation**

a. $\displaystyle\sum_{i=1}^{6} i = 1 + 2 + 3 + 4 + 5 + 6$

b. $\displaystyle\sum_{i=0}^{5} (i + 1) = 1 + 2 + 3 + 4 + 5 + 6$

c. $\displaystyle\sum_{j=3}^{7} j^2 = 3^2 + 4^2 + 5^2 + 6^2 + 7^2$

d. $\displaystyle\sum_{j=1}^{5} \frac{1}{\sqrt{j}} = \frac{1}{\sqrt{1}} + \frac{1}{\sqrt{2}} + \frac{1}{\sqrt{3}} + \frac{1}{\sqrt{4}} + \frac{1}{\sqrt{5}}$

e. $\displaystyle\sum_{k=1}^{n} \frac{1}{n}(k^2 + 1) = \frac{1}{n}(1^2 + 1) + \frac{1}{n}(2^2 + 1) + \cdots + \frac{1}{n}(n^2 + 1)$

f. $\displaystyle\sum_{i=1}^{n} f(x_i) \Delta x = f(x_1) \Delta x + f(x_2) \Delta x + \cdots + f(x_n) \Delta x$

From parts (a) and (b), notice that the same sum can be represented in different ways using sigma notation. ∎

Although any variable can be used as the index of summation, i, j, and k are often used. Notice in Example 1 that the index of summation does not appear in the terms of the expanded sum.

FOR FURTHER INFORMATION
For a geometric interpretation of summation formulas, see the article "Looking at $\displaystyle\sum_{k=1}^{n} k$ and $\displaystyle\sum_{k=1}^{n} k^2$ Geometrically" by Eric Hegblom in *Mathematics Teacher*. To view this article, go to *MathArticles.com*.

The properties of summation shown below can be derived using the Associative and Commutative Properties of Addition and the Distributive Property of Addition over Multiplication. (In the first property, k is a constant.)

1. $\displaystyle\sum_{i=1}^{n} ka_i = k\sum_{i=1}^{n} a_i$ **2.** $\displaystyle\sum_{i=1}^{n}(a_i \pm b_i) = \sum_{i=1}^{n}a_i \pm \sum_{i=1}^{n}b_i$

The next theorem lists some useful formulas for sums of powers.

THEOREM 4.2   Summation Formulas

1. $\displaystyle\sum_{i=1}^{n} c = cn, c$ is a constant   **2.** $\displaystyle\sum_{i=1}^{n} i = \frac{n(n+1)}{2}$

3. $\displaystyle\sum_{i=1}^{n} i^2 = \frac{n(n+1)(2n+1)}{6}$ **4.** $\displaystyle\sum_{i=1}^{n} i^3 = \frac{n^2(n+1)^2}{4}$

A proof of this theorem is given in Appendix A.
See LarsonCalculus.com for Bruce Edwards's video of this proof.

EXAMPLE 2   Evaluating a Sum

Evaluate $\displaystyle\sum_{i=1}^{n} \frac{i+1}{n^2}$ for $n = 10, 100, 1000,$ and $10,000$.

Solution

$$\sum_{i=1}^{n}\frac{i+1}{n^2} = \frac{1}{n^2}\sum_{i=1}^{n}(i+1) \qquad \text{Factor the constant } 1/n^2 \text{ out of sum.}$$

$$= \frac{1}{n^2}\left(\sum_{i=1}^{n}i + \sum_{i=1}^{n}1\right) \qquad \text{Write as two sums.}$$

$$= \frac{1}{n^2}\left[\frac{n(n+1)}{2} + n\right] \qquad \text{Apply Theorem 4.2.}$$

$$= \frac{1}{n^2}\left[\frac{n^2+3n}{2}\right] \qquad \text{Simplify.}$$

$$= \frac{n+3}{2n} \qquad \text{Simplify.}$$

Now you can evaluate the sum by substituting the appropriate values of n, as shown in the table below.

n	10	100	1000	10,000
$\displaystyle\sum_{i=1}^{n}\frac{i+1}{n^2} = \frac{n+3}{2n}$	0.65000	0.51500	0.50150	0.50015

In the table, note that the sum appears to approach a limit as n increases. Although the discussion of limits at infinity in Section 3.5 applies to a variable x, where x can be any real number, many of the same results hold true for limits involving the variable n, where n is restricted to positive integer values. So, to find the limit of $(n+3)/2n$ as n approaches infinity, you can write

$$\lim_{n\to\infty}\frac{n+3}{2n} = \lim_{n\to\infty}\left(\frac{n}{2n}+\frac{3}{2n}\right) = \lim_{n\to\infty}\left(\frac{1}{2}+\frac{3}{2n}\right) = \frac{1}{2}+0 = \frac{1}{2}.$$

THE SUM OF THE FIRST 100 INTEGERS

A teacher of Carl Friedrich Gauss (1777–1855) asked him to add all the integers from 1 to 100. When Gauss returned with the correct answer after only a few moments, the teacher could only look at him in astounded silence. This is what Gauss did:

$$\begin{array}{r} 1 + 2 + 3 + \cdots + 100 \\ 100 + 99 + 98 + \cdots + 1 \\ \hline 101 + 101 + 101 + \cdots + 101 \end{array}$$

$$\frac{100 \times 101}{2} = 5050$$

This is generalized by Theorem 4.2, Property 2, where

$$\sum_{i=1}^{100} i = \frac{100(101)}{2} = 5050.$$

Area

In Euclidean geometry, the simplest type of plane region is a rectangle. Although people often say that the *formula* for the area of a rectangle is

$$A = bh$$

it is actually more proper to say that this is the *definition* of the **area of a rectangle.**

From this definition, you can develop formulas for the areas of many other plane regions. For example, to determine the area of a triangle, you can form a rectangle whose area is twice that of the triangle, as shown in Figure 4.5. Once you know how to find the area of a triangle, you can determine the area of any polygon by subdividing the polygon into triangular regions, as shown in Figure 4.6.

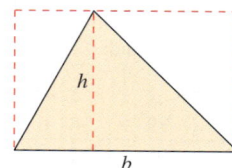

Triangle: $A = \frac{1}{2}bh$
Figure 4.5

Parallelogram
Figure 4.6

Hexagon

Polygon

ARCHIMEDES (287–212 B.C.)

Archimedes used the method of exhaustion to derive formulas for the areas of ellipses, parabolic segments, and sectors of a spiral. He is considered to have been the greatest applied mathematician of antiquity.

See LarsonCalculus.com to read more of this biography.

Finding the areas of regions other than polygons is more difficult. The ancient Greeks were able to determine formulas for the areas of some general regions (principally those bounded by conics) by the *exhaustion* method. The clearest description of this method was given by Archimedes. Essentially, the method is a limiting process in which the area is squeezed between two polygons—one inscribed in the region and one circumscribed about the region.

For instance, in Figure 4.7, the area of a circular region is approximated by an n-sided inscribed polygon and an n-sided circumscribed polygon. For each value of n, the area of the inscribed polygon is less than the area of the circle, and the area of the circumscribed polygon is greater than the area of the circle. Moreover, as n increases, the areas of both polygons become better and better approximations of the area of the circle.

■ **FOR FURTHER INFORMATION**
For an alternative development of the formula for the area of a circle, see the article "Proof Without Words: Area of a Disk is πR^2" by Russell Jay Hendel in *Mathematics Magazine.* To view this article, go to *MathArticles.com.*

$n = 6$

$n = 12$

The exhaustion method for finding the area of a circular region
Figure 4.7

A process that is similar to that used by Archimedes to determine the area of a plane region is used in the remaining examples in this section.

The Area of a Plane Region

Recall from Section 1.1 that the origins of calculus are connected to two classic problems: the tangent line problem and the area problem. Example 3 begins the investigation of the area problem.

EXAMPLE 3   **Approximating the Area of a Plane Region**

Use the five rectangles in Figure 4.8(a) and (b) to find *two* approximations of the area of the region lying between the graph of

$$f(x) = -x^2 + 5$$

and the *x*-axis between $x = 0$ and $x = 2$.

Solution

a. The right endpoints of the five intervals are

$$\frac{2}{5}i \qquad \text{Right endpoints}$$

where $i = 1, 2, 3, 4, 5$. The width of each rectangle is $\frac{2}{5}$, and the height of each rectangle can be obtained by evaluating f at the right endpoint of each interval.

$$\left[0, \frac{2}{5}\right], \left[\frac{2}{5}, \frac{4}{5}\right], \left[\frac{4}{5}, \frac{6}{5}\right], \left[\frac{6}{5}, \frac{8}{5}\right], \left[\frac{8}{5}, \frac{10}{5}\right]$$

Evaluate f at the right endpoints of these intervals.

The sum of the areas of the five rectangles is

$$\sum_{i=1}^{5} f\left(\frac{2i}{5}\right)\left(\frac{2}{5}\right) = \sum_{i=1}^{5}\left[-\left(\frac{2i}{5}\right)^2 + 5\right]\left(\frac{2}{5}\right) = \frac{162}{25} = 6.48.$$

Because each of the five rectangles lies inside the parabolic region, you can conclude that the area of the parabolic region is greater than 6.48.

b. The left endpoints of the five intervals are

$$\frac{2}{5}(i - 1) \qquad \text{Left endpoints}$$

where $i = 1, 2, 3, 4, 5$. The width of each rectangle is $\frac{2}{5}$, and the height of each rectangle can be obtained by evaluating f at the left endpoint of each interval. So, the sum is

$$\sum_{i=1}^{5} f\left(\frac{2i - 2}{5}\right)\left(\frac{2}{5}\right) = \sum_{i=1}^{5}\left[-\left(\frac{2i - 2}{5}\right)^2 + 5\right]\left(\frac{2}{5}\right) = \frac{202}{25} = 8.08.$$

Because the parabolic region lies within the union of the five rectangular regions, you can conclude that the area of the parabolic region is less than 8.08.

By combining the results in parts (a) and (b), you can conclude that

$$6.48 < (\text{Area of region}) < 8.08.$$

By increasing the number of rectangles used in Example 3, you can obtain closer and closer approximations of the area of the region. For instance, using 25 rectangles of width $\frac{2}{25}$ each, you can conclude that

$$7.1712 < (\text{Area of region}) < 7.4912.$$

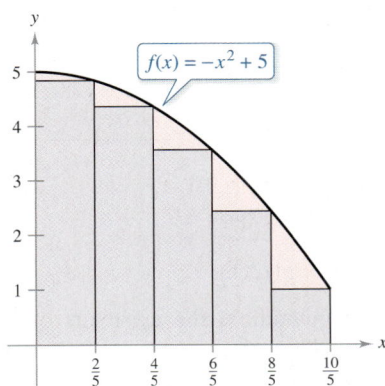

(a) The area of the parabolic region is greater than the area of the rectangles.

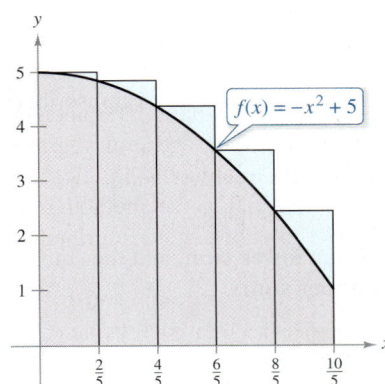

(b) The area of the parabolic region is less than the area of the rectangles.

Figure 4.8

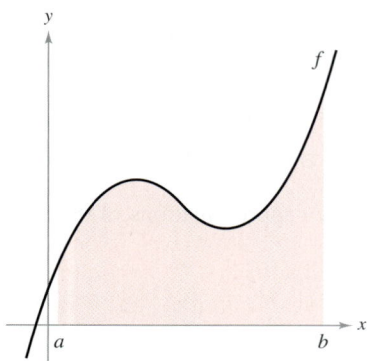

The region under a curve
Figure 4.9

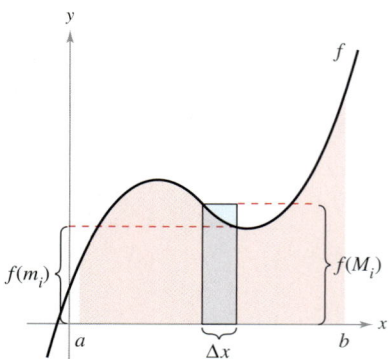

The interval $[a, b]$ is divided into n subintervals of width $\Delta x = \dfrac{b - a}{n}$.

Figure 4.10

Upper and Lower Sums

The procedure used in Example 3 can be generalized as follows. Consider a plane region bounded above by the graph of a nonnegative, continuous function

$$y = f(x)$$

as shown in Figure 4.9. The region is bounded below by the x-axis, and the left and right boundaries of the region are the vertical lines $x = a$ and $x = b$.

To approximate the area of the region, begin by subdividing the interval $[a, b]$ into n subintervals, each of width

$$\Delta x = \frac{b - a}{n}$$

as shown in Figure 4.10. The endpoints of the intervals are

$$\overbrace{a = x_0}^{} \qquad \overbrace{x_1}^{} \qquad \overbrace{x_2}^{} \qquad \overbrace{x_n = b}^{}$$
$$a + 0(\Delta x) < a + 1(\Delta x) < a + 2(\Delta x) < \cdots < a + n(\Delta x).$$

Because f is continuous, the Extreme Value Theorem guarantees the existence of a minimum and a maximum value of $f(x)$ in *each* subinterval.

$$f(m_i) = \text{Minimum value of } f(x) \text{ in } i\text{th subinterval}$$
$$f(M_i) = \text{Maximum value of } f(x) \text{ in } i\text{th subinterval}$$

Next, define an **inscribed rectangle** lying *inside* the ith subregion and a **circumscribed rectangle** extending *outside* the ith subregion. The height of the ith inscribed rectangle is $f(m_i)$ and the height of the ith circumscribed rectangle is $f(M_i)$. For *each* i, the area of the inscribed rectangle is less than or equal to the area of the circumscribed rectangle.

$$\begin{pmatrix} \text{Area of inscribed} \\ \text{rectangle} \end{pmatrix} = f(m_i)\,\Delta x \le f(M_i)\,\Delta x = \begin{pmatrix} \text{Area of circumscribed} \\ \text{rectangle} \end{pmatrix}$$

The sum of the areas of the inscribed rectangles is called a **lower sum,** and the sum of the areas of the circumscribed rectangles is called an **upper sum.**

$$\text{Lower sum} = s(n) = \sum_{i=1}^{n} f(m_i)\,\Delta x \qquad \text{Area of inscribed rectangles}$$

$$\text{Upper sum} = S(n) = \sum_{i=1}^{n} f(M_i)\,\Delta x \qquad \text{Area of circumscribed rectangles}$$

From Figure 4.11, you can see that the lower sum $s(n)$ is less than or equal to the upper sum $S(n)$. Moreover, the actual area of the region lies between these two sums.

$$s(n) \le (\text{Area of region}) \le S(n)$$

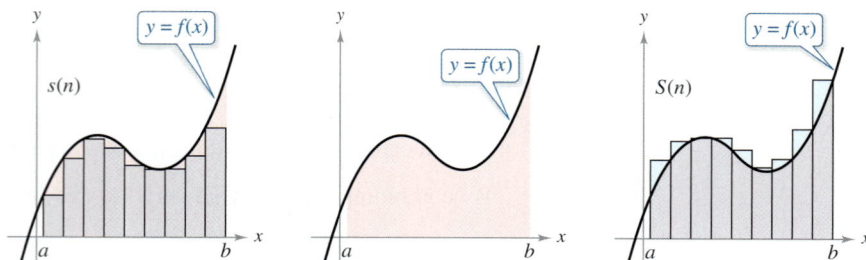

Area of inscribed rectangles is less than area of region.

Area of region

Area of circumscribed rectangles is greater than area of region.

Figure 4.11

EXAMPLE 4   Finding Upper and Lower Sums for a Region

Find the upper and lower sums for the region bounded by the graph of $f(x) = x^2$ and the x-axis between $x = 0$ and $x = 2$.

Solution   To begin, partition the interval $[0, 2]$ into n subintervals, each of width

$$\Delta x = \frac{b - a}{n} = \frac{2 - 0}{n} = \frac{2}{n}.$$

Figure 4.12 shows the endpoints of the subintervals and several inscribed and circumscribed rectangles. Because f is increasing on the interval $[0, 2]$, the minimum value on each subinterval occurs at the left endpoint, and the maximum value occurs at the right endpoint.

Left Endpoints

$$m_i = 0 + (i - 1)\left(\frac{2}{n}\right) = \frac{2(i - 1)}{n}$$

Right Endpoints

$$M_i = 0 + i\left(\frac{2}{n}\right) = \frac{2i}{n}$$

Inscribed rectangles

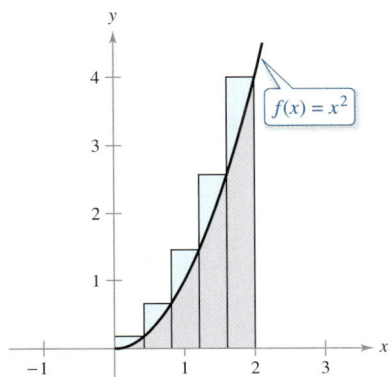

Circumscribed rectangles
Figure 4.12

Using the left endpoints, the lower sum is

$$\begin{aligned}
s(n) &= \sum_{i=1}^{n} f(m_i)\, \Delta x \\
&= \sum_{i=1}^{n} f\left[\frac{2(i-1)}{n}\right]\left(\frac{2}{n}\right) \\
&= \sum_{i=1}^{n} \left[\frac{2(i-1)}{n}\right]^2\left(\frac{2}{n}\right) \\
&= \sum_{i=1}^{n} \left(\frac{8}{n^3}\right)(i^2 - 2i + 1) \\
&= \frac{8}{n^3}\left(\sum_{i=1}^{n} i^2 - 2\sum_{i=1}^{n} i + \sum_{i=1}^{n} 1\right) \\
&= \frac{8}{n^3}\left\{\frac{n(n+1)(2n+1)}{6} - 2\left[\frac{n(n+1)}{2}\right] + n\right\} \\
&= \frac{4}{3n^3}(2n^3 - 3n^2 + n) \\
&= \frac{8}{3} - \frac{4}{n} + \frac{4}{3n^2}. \qquad \text{Lower sum}
\end{aligned}$$

Using the right endpoints, the upper sum is

$$\begin{aligned}
S(n) &= \sum_{i=1}^{n} f(M_i)\, \Delta x \\
&= \sum_{i=1}^{n} f\left(\frac{2i}{n}\right)\left(\frac{2}{n}\right) \\
&= \sum_{i=1}^{n} \left(\frac{2i}{n}\right)^2\left(\frac{2}{n}\right) \\
&= \sum_{i=1}^{n} \left(\frac{8}{n^3}\right)i^2 \\
&= \frac{8}{n^3}\left[\frac{n(n+1)(2n+1)}{6}\right] \\
&= \frac{4}{3n^3}(2n^3 + 3n^2 + n) \\
&= \frac{8}{3} + \frac{4}{n} + \frac{4}{3n^2}. \qquad \text{Upper sum}
\end{aligned}$$

Exploration

For the region given in Example 4, evaluate the lower sum

$$s(n) = \frac{8}{3} - \frac{4}{n} + \frac{4}{3n^2}$$

and the upper sum

$$S(n) = \frac{8}{3} + \frac{4}{n} + \frac{4}{3n^2}$$

for $n = 10$, 100, and 1000. Use your results to determine the area of the region.

Example 4 illustrates some important things about lower and upper sums. First, notice that for any value of n, the lower sum is less than (or equal to) the upper sum.

$$s(n) = \frac{8}{3} - \frac{4}{n} + \frac{4}{3n^2} < \frac{8}{3} + \frac{4}{n} + \frac{4}{3n^2} = S(n)$$

Second, the difference between these two sums lessens as n increases. In fact, when you take the limits as $n \to \infty$, both the lower sum and the upper sum approach $\frac{8}{3}$.

$$\lim_{n \to \infty} s(n) = \lim_{n \to \infty} \left(\frac{8}{3} - \frac{4}{n} + \frac{4}{3n^2} \right) = \frac{8}{3} \qquad \text{Lower sum limit}$$

and

$$\lim_{n \to \infty} S(n) = \lim_{n \to \infty} \left(\frac{8}{3} + \frac{4}{n} + \frac{4}{3n^2} \right) = \frac{8}{3} \qquad \text{Upper sum limit}$$

The next theorem shows that the equivalence of the limits (as $n \to \infty$) of the upper and lower sums is not mere coincidence. It is true for all functions that are continuous and nonnegative on the closed interval $[a, b]$. The proof of this theorem is best left to a course in advanced calculus.

THEOREM 4.3   Limits of the Lower and Upper Sums

Let f be continuous and nonnegative on the interval $[a, b]$. The limits as $n \to \infty$ of both the lower and upper sums exist and are equal to each other. That is,

$$\lim_{n \to \infty} s(n) = \lim_{n \to \infty} \sum_{i=1}^{n} f(m_i) \Delta x$$

$$= \lim_{n \to \infty} \sum_{i=1}^{n} f(M_i) \Delta x$$

$$= \lim_{n \to \infty} S(n)$$

where $\Delta x = (b - a)/n$ and $f(m_i)$ and $f(M_i)$ are the minimum and maximum values of f on the subinterval.

In Theorem 4.3, the same limit is attained for both the minimum value $f(m_i)$ and the maximum value $f(M_i)$. So, it follows from the Squeeze Theorem (Theorem 1.8) that the choice of x in the ith subinterval does not affect the limit. This means that you are free to choose an *arbitrary* x-value in the ith subinterval, as shown in the *definition of the area of a region in the plane*.

Definition of the Area of a Region in the Plane

Let f be continuous and nonnegative on the interval $[a, b]$. (See Figure 4.13.) The area of the region bounded by the graph of f, the x-axis, and the vertical lines $x = a$ and $x = b$ is

$$\text{Area} = \lim_{n \to \infty} \sum_{i=1}^{n} f(c_i) \Delta x$$

where $x_{i-1} \le c_i \le x_i$ and

$$\Delta x = \frac{b - a}{n}.$$

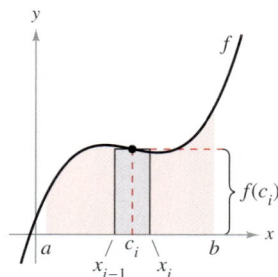

The width of the ith subinterval is $\Delta x = x_i - x_{i-1}$.

Figure 4.13

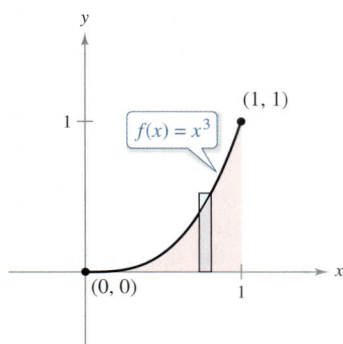

The area of the region bounded by the graph of f, the x-axis, $x = 0$, and $x = 1$ is $\frac{1}{4}$.

Figure 4.14

EXAMPLE 5   **Finding Area by the Limit Definition**

Find the area of the region bounded by the graph $f(x) = x^3$, the x-axis, and the vertical lines $x = 0$ and $x = 1$, as shown in Figure 4.14.

Solution   Begin by noting that f is continuous and nonnegative on the interval $[0, 1]$. Next, partition the interval $[0, 1]$ into n subintervals, each of width $\Delta x = 1/n$. According to the definition of area, you can choose any x-value in the ith subinterval. For this example, the right endpoints $c_i = i/n$ are convenient.

$$\text{Area} = \lim_{n \to \infty} \sum_{i=1}^{n} f(c_i)\, \Delta x$$

$$= \lim_{n \to \infty} \sum_{i=1}^{n} \left(\frac{i}{n}\right)^3 \left(\frac{1}{n}\right) \qquad \text{Right endpoints: } c_i = \frac{i}{n}$$

$$= \lim_{n \to \infty} \frac{1}{n^4} \sum_{i=1}^{n} i^3$$

$$= \lim_{n \to \infty} \frac{1}{n^4} \left[\frac{n^2(n+1)^2}{4}\right]$$

$$= \lim_{n \to \infty} \left(\frac{1}{4} + \frac{1}{2n} + \frac{1}{4n^2}\right)$$

$$= \frac{1}{4}$$

The area of the region is $\frac{1}{4}$.

EXAMPLE 6   **Finding Area by the Limit Definition**

∙ ∙ ∙ ▷ *See LarsonCalculus.com for an interactive version of this type of example.*

Find the area of the region bounded by the graph of $f(x) = 4 - x^2$, the x-axis, and the vertical lines $x = 1$ and $x = 2$, as shown in Figure 4.15.

Solution   Note that the function f is continuous and nonnegative on the interval $[1, 2]$. So, begin by partitioning the interval into n subintervals, each of width $\Delta x = 1/n$. Choosing the right endpoint

$$c_i = a + i\Delta x = 1 + \frac{i}{n} \qquad \text{Right endpoints}$$

of each subinterval, you obtain

$$\text{Area} = \lim_{n \to \infty} \sum_{i=1}^{n} f(c_i)\, \Delta x$$

$$= \lim_{n \to \infty} \sum_{i=1}^{n} \left[4 - \left(1 + \frac{i}{n}\right)^2\right]\left(\frac{1}{n}\right)$$

$$= \lim_{n \to \infty} \sum_{i=1}^{n} \left(3 - \frac{2i}{n} - \frac{i^2}{n^2}\right)\left(\frac{1}{n}\right)$$

$$= \lim_{n \to \infty} \left(\frac{1}{n}\sum_{i=1}^{n} 3 - \frac{2}{n^2}\sum_{i=1}^{n} i - \frac{1}{n^3}\sum_{i=1}^{n} i^2\right)$$

$$= \lim_{n \to \infty} \left[3 - \left(1 + \frac{1}{n}\right) - \left(\frac{1}{3} + \frac{1}{2n} + \frac{1}{6n^2}\right)\right]$$

$$= 3 - 1 - \frac{1}{3}$$

$$= \frac{5}{3}.$$

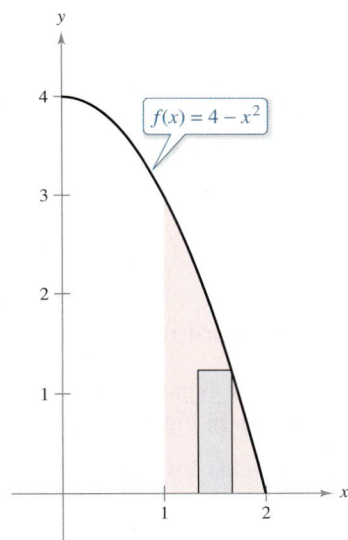

The area of the region bounded by the graph of f, the x-axis, $x = 1$, and $x = 2$ is $\frac{5}{3}$.

Figure 4.15

The area of the region is $\frac{5}{3}$.

The next example looks at a region that is bounded by the y-axis (rather than by the x-axis).

EXAMPLE 7 **A Region Bounded by the y-axis**

Find the area of the region bounded by the graph of $f(y) = y^2$ and the y-axis for $0 \le y \le 1$, as shown in Figure 4.16.

Solution When f is a continuous, nonnegative function of y, you can still use the same basic procedure shown in Examples 5 and 6. Begin by partitioning the interval $[0, 1]$ into n subintervals, each of width $\Delta y = 1/n$. Then, using the upper endpoints $c_i = i/n$, you obtain

$$\text{Area} = \lim_{n \to \infty} \sum_{i=1}^{n} f(c_i)\,\Delta y$$

$$= \lim_{n \to \infty} \sum_{i=1}^{n} \left(\frac{i}{n}\right)^2 \left(\frac{1}{n}\right) \qquad \text{Upper endpoints: } c_i = \frac{i}{n}$$

$$= \lim_{n \to \infty} \frac{1}{n^3} \sum_{i=1}^{n} i^2$$

$$= \lim_{n \to \infty} \frac{1}{n^3} \left[\frac{n(n+1)(2n+1)}{6}\right]$$

$$= \lim_{n \to \infty} \left(\frac{1}{3} + \frac{1}{2n} + \frac{1}{6n^2}\right)$$

$$= \frac{1}{3}.$$

The area of the region is $\frac{1}{3}$.

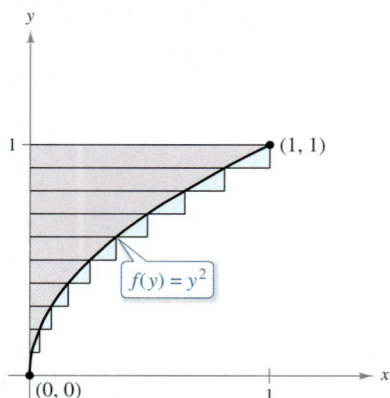

The area of the region bounded by the graph of f and the y-axis for $0 \le y \le 1$ is $\frac{1}{3}$.

Figure 4.16

In Examples 5, 6, and 7, c_i is chosen to be a value that is convenient for calculating the limit. Because each limit gives the exact area for *any* c_i, there is no need to find values that give good approximations when n is small. For an *approximation*, however, you should try to find a value of c_i that gives a good approximation of the area of the ith subregion. In general, a good value to choose is the midpoint of the interval, $c_i = (x_i + x_{i-1})/2$, and apply the **Midpoint Rule.**

$$\text{Area} \approx \sum_{i=1}^{n} f\left(\frac{x_i + x_{i-1}}{2}\right)\Delta x. \qquad \text{Midpoint Rule}$$

• REMARK You will learn about other approximation methods in Section 4.6. One of the methods, the Trapezoidal Rule, is similar to the Midpoint Rule.

EXAMPLE 8 **Approximating Area with the Midpoint Rule**

Use the Midpoint Rule with $n = 4$ to approximate the area of the region bounded by the graph of $f(x) = \sin x$ and the x-axis for $0 \le x \le \pi$, as shown in Figure 4.17.

Solution For $n = 4$, $\Delta x = \pi/4$. The midpoints of the subregions are shown below.

$$c_1 = \frac{0 + (\pi/4)}{2} = \frac{\pi}{8} \qquad\qquad c_2 = \frac{(\pi/4) + (\pi/2)}{2} = \frac{3\pi}{8}$$

$$c_3 = \frac{(\pi/2) + (3\pi/4)}{2} = \frac{5\pi}{8} \qquad\qquad c_4 = \frac{(3\pi/4) + \pi}{2} = \frac{7\pi}{8}$$

So, the area is approximated by

$$\text{Area} \approx \sum_{i=1}^{n} f(c_i)\,\Delta x = \sum_{i=1}^{4} (\sin c_i)\left(\frac{\pi}{4}\right) = \frac{\pi}{4}\left(\sin\frac{\pi}{8} + \sin\frac{3\pi}{8} + \sin\frac{5\pi}{8} + \sin\frac{7\pi}{8}\right)$$

which is about 2.052.

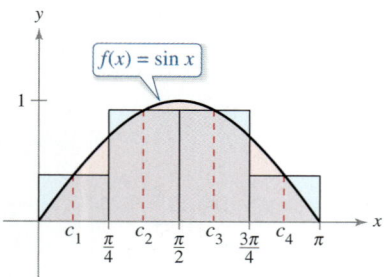

The area of the region bounded by the graph of $f(x) = \sin x$ and the x-axis for $0 \le x \le \pi$ is about 2.052.

Figure 4.17

4.2 Exercises

See CalcChat.com for tutorial help and worked-out solutions to odd-numbered exercises.

Finding a Sum In Exercises 1–6, find the sum. Use the summation capabilities of a graphing utility to verify your result.

1. $\displaystyle\sum_{i=1}^{6} (3i + 2)$

2. $\displaystyle\sum_{k=3}^{9} (k^2 + 1)$

3. $\displaystyle\sum_{k=0}^{4} \frac{1}{k^2 + 1}$

4. $\displaystyle\sum_{j=4}^{6} \frac{3}{j}$

5. $\displaystyle\sum_{k=1}^{4} c$

6. $\displaystyle\sum_{i=1}^{4} [(i - 1)^2 + (i + 1)^3]$

Using Sigma Notation In Exercises 7–12, use sigma notation to write the sum.

7. $\dfrac{1}{5(1)} + \dfrac{1}{5(2)} + \dfrac{1}{5(3)} + \cdots + \dfrac{1}{5(11)}$

8. $\dfrac{9}{1 + 1} + \dfrac{9}{1 + 2} + \dfrac{9}{1 + 3} + \cdots + \dfrac{9}{1 + 14}$

9. $\left[7\left(\dfrac{1}{6}\right) + 5\right] + \left[7\left(\dfrac{2}{6}\right) + 5\right] + \cdots + \left[7\left(\dfrac{6}{6}\right) + 5\right]$

10. $\left[1 - \left(\dfrac{1}{4}\right)^2\right] + \left[1 - \left(\dfrac{2}{4}\right)^2\right] + \cdots + \left[1 - \left(\dfrac{4}{4}\right)^2\right]$

11. $\left[\left(\dfrac{2}{n}\right)^3 - \dfrac{2}{n}\right]\left(\dfrac{2}{n}\right) + \cdots + \left[\left(\dfrac{2n}{n}\right)^3 - \dfrac{2n}{n}\right]\left(\dfrac{2}{n}\right)$

12. $\left[2\left(1 + \dfrac{3}{n}\right)^2\right]\left(\dfrac{3}{n}\right) + \cdots + \left[2\left(1 + \dfrac{3n}{n}\right)^2\right]\left(\dfrac{3}{n}\right)$

Evaluating a Sum In Exercises 13–20, use the properties of summation and Theorem 4.2 to evaluate the sum. Use the summation capabilities of a graphing utility to verify your result.

13. $\displaystyle\sum_{i=1}^{12} 7$

14. $\displaystyle\sum_{i=1}^{30} -18$

15. $\displaystyle\sum_{i=1}^{24} 4i$

16. $\displaystyle\sum_{i=1}^{16} (5i - 4)$

17. $\displaystyle\sum_{i=1}^{20} (i - 1)^2$

18. $\displaystyle\sum_{i=1}^{10} (i^2 - 1)$

19. $\displaystyle\sum_{i=1}^{15} i(i - 1)^2$

20. $\displaystyle\sum_{i=1}^{25} (i^3 - 2i)$

Evaluating a Sum In Exercises 21–24, use the summation formulas to rewrite the expression without the summation notation. Use the result to find the sums for $n = 10, 100, 1000$, and 10,000.

21. $\displaystyle\sum_{i=1}^{n} \frac{2i + 1}{n^2}$

22. $\displaystyle\sum_{j=1}^{n} \frac{7j + 4}{n^2}$

23. $\displaystyle\sum_{k=1}^{n} \frac{6k(k - 1)}{n^3}$

24. $\displaystyle\sum_{i=1}^{n} \frac{2i^3 - 3i}{n^4}$

Approximating the Area of a Plane Region In Exercises 25–30, use left and right endpoints and the given number of rectangles to find two approximations of the area of the region between the graph of the function and the x-axis over the given interval.

25. $f(x) = 2x + 5$, $[0, 2]$, 4 rectangles

26. $f(x) = 9 - x$, $[2, 4]$, 6 rectangles

27. $g(x) = 2x^2 - x - 1$, $[2, 5]$, 6 rectangles

28. $g(x) = x^2 + 1$, $[1, 3]$, 8 rectangles

29. $f(x) = \cos x$, $\left[0, \dfrac{\pi}{2}\right]$, 4 rectangles

30. $g(x) = \sin x$, $[0, \pi]$, 6 rectangles

Using Upper and Lower Sums In Exercises 31 and 32, bound the area of the shaded region by approximating the upper and lower sums. Use rectangles of width 1.

31.

32.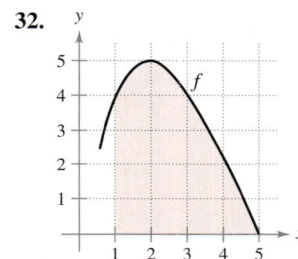

Finding Upper and Lower Sums for a Region In Exercises 33–36, use upper and lower sums to approximate the area of the region using the given number of subintervals (of equal width).

33. $y = \sqrt{x}$

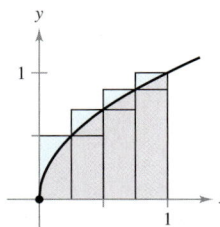

34. $y = \sqrt{x} + 2$

35. $y = \dfrac{1}{x}$

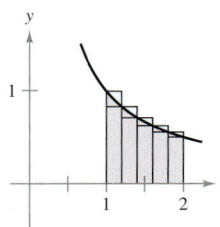

36. $y = \sqrt{1 - x^2}$

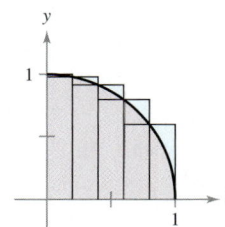

Finding a Limit In Exercises 37–42, find a formula for the sum of n terms. Use the formula to find the limit as $n \to \infty$.

37. $\lim\limits_{n\to\infty} \sum\limits_{i=1}^{n} \dfrac{24i}{n^2}$

38. $\lim\limits_{n\to\infty} \sum\limits_{i=1}^{n} \left(\dfrac{3i}{n}\right)\left(\dfrac{3}{n}\right)$

39. $\lim\limits_{n\to\infty} \sum\limits_{i=1}^{n} \dfrac{1}{n^3}(i-1)^2$

40. $\lim\limits_{n\to\infty} \sum\limits_{i=1}^{n} \left(1+\dfrac{2i}{n}\right)^2\left(\dfrac{2}{n}\right)$

41. $\lim\limits_{n\to\infty} \sum\limits_{i=1}^{n} \left(1+\dfrac{i}{n}\right)\left(\dfrac{2}{n}\right)$

42. $\lim\limits_{n\to\infty} \sum\limits_{i=1}^{n} \left(2+\dfrac{3i}{n}\right)^3\left(\dfrac{3}{n}\right)$

43. **Numerical Reasoning** Consider a triangle of area 2 bounded by the graphs of $y = x$, $y = 0$, and $x = 2$.

(a) Sketch the region.

(b) Divide the interval $[0, 2]$ into n subintervals of equal width and show that the endpoints are

$$0 < 1\left(\dfrac{2}{n}\right) < \cdots < (n-1)\left(\dfrac{2}{n}\right) < n\left(\dfrac{2}{n}\right).$$

(c) Show that $s(n) = \sum\limits_{i=1}^{n} \left[(i-1)\left(\dfrac{2}{n}\right)\right]\left(\dfrac{2}{n}\right)$.

(d) Show that $S(n) = \sum\limits_{i=1}^{n} \left[i\left(\dfrac{2}{n}\right)\right]\left(\dfrac{2}{n}\right)$.

(e) Complete the table.

n	5	10	50	100
$s(n)$				
$S(n)$				

(f) Show that $\lim\limits_{n\to\infty} s(n) = \lim\limits_{n\to\infty} S(n) = 2$.

44. **Numerical Reasoning** Consider a trapezoid of area 4 bounded by the graphs of $y = x$, $y = 0$, $x = 1$, and $x = 3$.

(a) Sketch the region.

(b) Divide the interval $[1, 3]$ into n subintervals of equal width and show that the endpoints are

$$1 < 1 + 1\left(\dfrac{2}{n}\right) < \cdots < 1 + (n-1)\left(\dfrac{2}{n}\right) < 1 + n\left(\dfrac{2}{n}\right).$$

(c) Show that $s(n) = \sum\limits_{i=1}^{n} \left[1+(i-1)\left(\dfrac{2}{n}\right)\right]\left(\dfrac{2}{n}\right)$.

(d) Show that $S(n) = \sum\limits_{i=1}^{n} \left[1+i\left(\dfrac{2}{n}\right)\right]\left(\dfrac{2}{n}\right)$.

(e) Complete the table.

n	5	10	50	100
$s(n)$				
$S(n)$				

(f) Show that $\lim\limits_{n\to\infty} s(n) = \lim\limits_{n\to\infty} S(n) = 4$.

Finding Area by the Limit Definition In Exercises 45–54, use the limit process to find the area of the region bounded by the graph of the function and the x-axis over the given interval. Sketch the region.

45. $y = -4x + 5$, $[0, 1]$

46. $y = 3x - 2$, $[2, 5]$

47. $y = x^2 + 2$, $[0, 1]$

48. $y = 3x^2 + 1$, $[0, 2]$

49. $y = 25 - x^2$, $[1, 4]$

50. $y = 4 - x^2$, $[-2, 2]$

51. $y = 27 - x^3$, $[1, 3]$

52. $y = 2x - x^3$, $[0, 1]$

53. $y = x^2 - x^3$, $[-1, 1]$

54. $y = 2x^3 - x^2$, $[1, 2]$

Finding Area by the Limit Definition In Exercises 55–60, use the limit process to find the area of the region bounded by the graph of the function and the y-axis over the given y-interval. Sketch the region.

55. $f(y) = 4y$, $0 \le y \le 2$

56. $g(y) = \frac{1}{2}y$, $2 \le y \le 4$

57. $f(y) = y^2$, $0 \le y \le 5$

58. $f(y) = 4y - y^2$, $1 \le y \le 2$

59. $g(y) = 4y^2 - y^3$, $1 \le y \le 3$

60. $h(y) = y^3 + 1$, $1 \le y \le 2$

Approximating Area with the Midpoint Rule In Exercises 61–64, use the Midpoint Rule with $n = 4$ to approximate the area of the region bounded by the graph of the function and the x-axis over the given interval.

61. $f(x) = x^2 + 3$, $[0, 2]$

62. $f(x) = x^2 + 4x$, $[0, 4]$

63. $f(x) = \tan x$, $\left[0, \dfrac{\pi}{4}\right]$

64. $f(x) = \cos x$, $\left[0, \dfrac{\pi}{2}\right]$

WRITING ABOUT CONCEPTS

Approximation In Exercises 65 and 66, determine which value best approximates the area of the region between the x-axis and the graph of the function over the given interval. (Make your selection on the basis of a sketch of the region, not by performing calculations.)

65. $f(x) = 4 - x^2$, $[0, 2]$

(a) -2 (b) 6 (c) 10 (d) 3 (e) 8

66. $f(x) = \sin \dfrac{\pi x}{4}$, $[0, 4]$

(a) 3 (b) 1 (c) -2 (d) 8 (e) 6

67. **Upper and Lower Sums** In your own words and using appropriate figures, describe the methods of upper sums and lower sums in approximating the area of a region.

68. **Area of a Region in the Plane** Give the definition of the area of a region in the plane.

69. Graphical Reasoning Consider the region bounded by the graphs of $f(x) = 8x/(x + 1)$, $x = 0$, $x = 4$, and $y = 0$, as shown in the figure. To print an enlarged copy of the graph, go to *MathGraphs.com*.

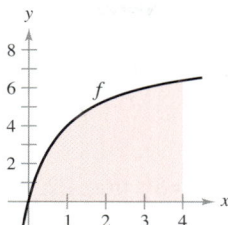

(a) Redraw the figure, and complete and shade the rectangles representing the lower sum when $n = 4$. Find this lower sum.

(b) Redraw the figure, and complete and shade the rectangles representing the upper sum when $n = 4$. Find this upper sum.

(c) Redraw the figure, and complete and shade the rectangles whose heights are determined by the functional values at the midpoint of each subinterval when $n = 4$. Find this sum using the Midpoint Rule.

(d) Verify the following formulas for approximating the area of the region using n subintervals of equal width.

Lower sum: $s(n) = \sum_{i=1}^{n} f\left[(i - 1)\dfrac{4}{n}\right]\left(\dfrac{4}{n}\right)$

Upper sum: $S(n) = \sum_{i=1}^{n} f\left[(i)\dfrac{4}{n}\right]\left(\dfrac{4}{n}\right)$

Midpoint Rule: $M(n) = \sum_{i=1}^{n} f\left[\left(i - \dfrac{1}{2}\right)\dfrac{4}{n}\right]\left(\dfrac{4}{n}\right)$

(e) Use a graphing utility to create a table of values of $s(n)$, $S(n)$, and $M(n)$ for $n = 4, 8, 20, 100$, and 200.

(f) Explain why $s(n)$ increases and $S(n)$ decreases for increasing values of n, as shown in the table in part (e).

70. HOW DO YOU SEE IT? The function shown in the graph below is increasing on the interval $[1, 4]$. The interval will be divided into 12 subintervals.

(a) What are the left endpoints of the first and last subintervals?

(b) What are the right endpoints of the first two subintervals?

(c) When using the right endpoints, do the rectangles lie above or below the graph of the function?

(d) What can you conclude about the heights of the rectangles when the function is constant on the given interval?

True or False? In Exercises 71 and 72, determine whether the statement is true or false. If it is false, explain why or give an example that shows it is false.

71. The sum of the first n positive integers is $n(n + 1)/2$.

72. If f is continuous and nonnegative on $[a, b]$, then the limits as $n \to \infty$ of its lower sum $s(n)$ and upper sum $S(n)$ both exist and are equal.

73. Writing Use the figure to write a short paragraph explaining why the formula $1 + 2 + \cdots + n = \frac{1}{2}n(n + 1)$ is valid for all positive integers n.

Figure for 73

Figure for 74

74. Graphical Reasoning Consider an n-sided regular polygon inscribed in a circle of radius r. Join the vertices of the polygon to the center of the circle, forming n congruent triangles (see figure).

(a) Determine the central angle θ in terms of n.

(b) Show that the area of each triangle is $\frac{1}{2}r^2 \sin \theta$.

(c) Let A_n be the sum of the areas of the n triangles. Find $\lim_{n \to \infty} A_n$.

75. Building Blocks A child places n cubic building blocks in a row to form the base of a triangular design (see figure). Each successive row contains two fewer blocks than the preceding row. Find a formula for the number of blocks used in the design. (*Hint:* The number of building blocks in the design depends on whether n is odd or even.)

n is even.

76. Proof Prove each formula by mathematical induction. (You may need to review the method of proof by induction from a precalculus text.)

(a) $\sum_{i=1}^{n} 2i = n(n + 1)$ (b) $\sum_{i=1}^{n} i^3 = \dfrac{n^2(n + 1)^2}{4}$

PUTNAM EXAM CHALLENGE

77. A dart, thrown at random, hits a square target. Assuming that any two parts of the target of equal area are equally likely to be hit, find the probability that the point hit is nearer to the center than to any edge. Write your answer in the form $\left(a\sqrt{b} + c\right)/d$, where a, b, c, and d are integers.

This problem was composed by the Committee on the Putnam Prize Competition.
© The Mathematical Association of America. All rights reserved.

4.3 Riemann Sums and Definite Integrals

■ Understand the definition of a Riemann sum.
■ Evaluate a definite integral using limits.
■ Evaluate a definite integral using properties of definite integrals.

Riemann Sums

In the definition of area given in Section 4.2, the partitions have subintervals of *equal width*. This was done only for computational convenience. The next example shows that it is not necessary to have subintervals of equal width.

EXAMPLE 1   **A Partition with Subintervals of Unequal Widths**

Consider the region bounded by the graph of

$$f(x) = \sqrt{x}$$

and the x-axis for $0 \le x \le 1$, as shown in Figure 4.18. Evaluate the limit

$$\lim_{n \to \infty} \sum_{i=1}^{n} f(c_i)\, \Delta x_i$$

where c_i is the right endpoint of the partition given by $c_i = i^2/n^2$ and Δx_i is the width of the ith interval.

Solution   The width of the ith interval is

$$\Delta x_i = \frac{i^2}{n^2} - \frac{(i-1)^2}{n^2}$$

$$= \frac{i^2 - i^2 + 2i - 1}{n^2}$$

$$= \frac{2i - 1}{n^2}.$$

So, the limit is

$$\lim_{n \to \infty} \sum_{i=1}^{n} f(c_i)\, \Delta x_i = \lim_{n \to \infty} \sum_{i=1}^{n} \sqrt{\frac{i^2}{n^2}} \left(\frac{2i-1}{n^2}\right)$$

$$= \lim_{n \to \infty} \frac{1}{n^3} \sum_{i=1}^{n} (2i^2 - i)$$

$$= \lim_{n \to \infty} \frac{1}{n^3}\left[2\left(\frac{n(n+1)(2n+1)}{6}\right) - \frac{n(n+1)}{2}\right]$$

$$= \lim_{n \to \infty} \frac{4n^3 + 3n^2 - n}{6n^3}$$

$$= \frac{2}{3}.$$

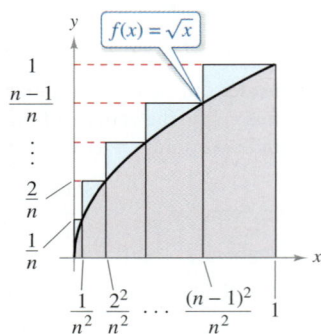

The subintervals do not have equal widths.
Figure 4.18

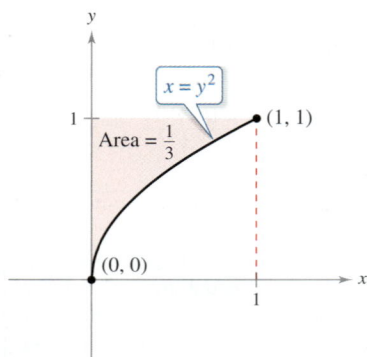

The area of the region bounded by the graph of $x = y^2$ and the y-axis for $0 \le y \le 1$ is $\frac{1}{3}$.
Figure 4.19

From Example 7 in Section 4.2, you know that the region shown in Figure 4.19 has an area of $\frac{1}{3}$. Because the square bounded by $0 \le x \le 1$ and $0 \le y \le 1$ has an area of 1, you can conclude that the area of the region shown in Figure 4.18 has an area of $\frac{2}{3}$. This agrees with the limit found in Example 1, even though that example used a partition having subintervals of unequal widths. The reason this particular partition gave the proper area is that as n increases, the *width of the largest subinterval approaches zero*. This is a key feature of the development of definite integrals.

In Section 4.2, the limit of a sum was used to define the area of a region in the plane. Finding area by this means is only one of *many* applications involving the limit of a sum. A similar approach can be used to determine quantities as diverse as arc lengths, average values, centroids, volumes, work, and surface areas. The next definition is named after Georg Friedrich Bernhard Riemann. Although the definite integral had been defined and used long before Riemann's time, he generalized the concept to cover a broader category of functions.

In the definition of a Riemann sum below, note that the function f has no restrictions other than being defined on the interval $[a, b]$. (In Section 4.2, the function f was assumed to be continuous and nonnegative because you were finding the area under a curve.)

Definition of Riemann Sum

Let f be defined on the closed interval $[a, b]$, and let Δ be a partition of $[a, b]$ given by

$$a = x_0 < x_1 < x_2 < \ldots < x_{n-1} < x_n = b$$

where Δx_i is the width of the ith subinterval

$$[x_{i-1}, x_i].$$ *ith subinterval*

If c_i is *any* point in the ith subinterval, then the sum

$$\sum_{i=1}^{n} f(c_i)\, \Delta x_i, \quad x_{i-1} \le c_i \le x_i$$

is called a **Riemann sum** of f for the partition Δ. (The sums in Section 4.2 are examples of Riemann sums, but there are more general Riemann sums than those covered there.)

The width of the largest subinterval of a partition Δ is the **norm** of the partition and is denoted by $\|\Delta\|$. If every subinterval is of equal width, then the partition is **regular** and the norm is denoted by

$$\|\Delta\| = \Delta x = \frac{b-a}{n}.$$ *Regular partition*

For a general partition, the norm is related to the number of subintervals of $[a, b]$ in the following way.

$$\frac{b-a}{\|\Delta\|} \le n$$ *General partition*

So, the number of subintervals in a partition approaches infinity as the norm of the partition approaches 0. That is, $\|\Delta\| \to 0$ implies that $n \to \infty$.

The converse of this statement is not true. For example, let Δ_n be the partition of the interval $[0, 1]$ given by

$$0 < \frac{1}{2^n} < \frac{1}{2^{n-1}} < \cdots < \frac{1}{8} < \frac{1}{4} < \frac{1}{2} < 1.$$

As shown in Figure 4.20, for any positive value of n, the norm of the partition Δ_n is $\frac{1}{2}$. So, letting n approach infinity does not force $\|\Delta\|$ to approach 0. In a regular partition, however, the statements

$$\|\Delta\| \to 0 \quad \text{and} \quad n \to \infty$$

are equivalent.

$\|\Delta\| = \frac{1}{2}$

$n \to \infty$ does not imply that $\|\Delta\| \to 0$.

Figure 4.20

Definite Integrals

■ **FOR FURTHER INFORMATION**
For insight into the history of the
definite integral, see the article
"The Evolution of Integration" by
A. Shenitzer and J. Steprāns in *The
American Mathematical Monthly.*
To view this article, go to
MathArticles.com.

To define the definite integral, consider the limit

$$\lim_{\|\Delta\|\to 0} \sum_{i=1}^{n} f(c_i)\,\Delta x_i = L.$$

To say that this limit exists means there exists a real number L such that for each $\varepsilon > 0$,
there exists a $\delta > 0$ such that for every partition with $\|\Delta\| < \delta$, it follows that

$$\left| L - \sum_{i=1}^{n} f(c_i)\,\Delta x_i \right| < \varepsilon$$

regardless of the choice of c_i in the ith subinterval of each partition Δ.

Definition of Definite Integral

If f is defined on the closed interval $[a, b]$ and the limit of Riemann sums over
partitions Δ

$$\lim_{\|\Delta\|\to 0} \sum_{i=1}^{n} f(c_i)\,\Delta x_i$$

exists (as described above), then f is said to be **integrable** on $[a, b]$ and the
limit is denoted by

$$\lim_{\|\Delta\|\to 0} \sum_{i=1}^{n} f(c_i)\,\Delta x_i = \int_{a}^{b} f(x)\,dx.$$

The limit is called the **definite integral** of f from a to b. The number a is the
lower limit of integration, and the number b is the **upper limit** of integration.

REMARK Later in this
chapter, you will learn
convenient methods for
calculating $\int_{a}^{b} f(x)\,dx$ for
continuous functions. For now,
you must use the limit definition.

It is not a coincidence that the notation for definite integrals is similar to that used
for indefinite integrals. You will see why in the next section when the Fundamental
Theorem of Calculus is introduced. For now, it is important to see that definite integrals
and indefinite integrals are different concepts. A definite integral is a *number*, whereas
an indefinite integral is a *family of functions.*

Though Riemann sums were defined for functions with very few restrictions, a
sufficient condition for a function f to be integrable on $[a, b]$ is that it is continuous on
$[a, b]$. A proof of this theorem is beyond the scope of this text.

THEOREM 4.4   Continuity Implies Integrability

If a function f is continuous on the closed interval $[a, b]$, then f is integrable
on $[a, b]$. That is, $\int_{a}^{b} f(x)\,dx$ exists.

Exploration

The Converse of Theorem 4.4   Is the converse of Theorem 4.4 true? That is,
when a function is integrable, does it have to be continuous? Explain your
reasoning and give examples.

Describe the relationships among continuity, differentiability, and
integrability. Which is the strongest condition? Which is the weakest? Which
conditions imply other conditions?

EXAMPLE 2   **Evaluating a Definite Integral as a Limit**

Evaluate the definite integral $\displaystyle\int_{-2}^{1} 2x \, dx$.

Solution   The function $f(x) = 2x$ is integrable on the interval $[-2, 1]$ because it is continuous on $[-2, 1]$. Moreover, the definition of integrability implies that any partition whose norm approaches 0 can be used to determine the limit. For computational convenience, define Δ by subdividing $[-2, 1]$ into n subintervals of equal width

$$\Delta x_i = \Delta x = \frac{b - a}{n} = \frac{3}{n}.$$

Choosing c_i as the right endpoint of each subinterval produces

$$c_i = a + i(\Delta x) = -2 + \frac{3i}{n}.$$

So, the definite integral is

$$\int_{-2}^{1} 2x \, dx = \lim_{\|\Delta\| \to 0} \sum_{i=1}^{n} f(c_i) \, \Delta x_i$$

$$= \lim_{n \to \infty} \sum_{i=1}^{n} f(c_i) \, \Delta x$$

$$= \lim_{n \to \infty} \sum_{i=1}^{n} 2\left(-2 + \frac{3i}{n}\right)\left(\frac{3}{n}\right)$$

$$= \lim_{n \to \infty} \frac{6}{n} \sum_{i=1}^{n} \left(-2 + \frac{3i}{n}\right)$$

$$= \lim_{n \to \infty} \frac{6}{n}\left(-2\sum_{i=1}^{n} 1 + \frac{3}{n}\sum_{i=1}^{n} i\right)$$

$$= \lim_{n \to \infty} \frac{6}{n}\left\{-2n + \frac{3}{n}\left[\frac{n(n + 1)}{2}\right]\right\}$$

$$= \lim_{n \to \infty} \left(-12 + 9 + \frac{9}{n}\right)$$

$$= -3.$$

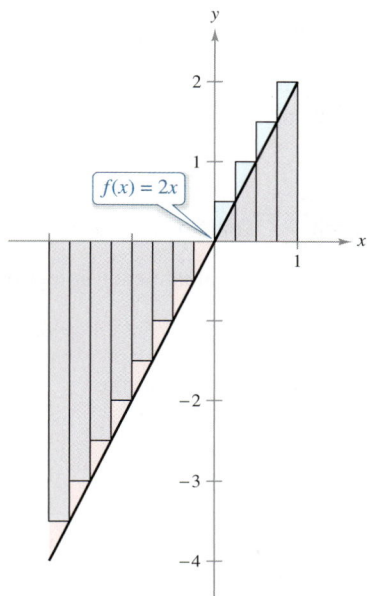

Because the definite integral is negative, it does not represent the area of the region.

Figure 4.21

Because the definite integral in Example 2 is negative, it *does not* represent the area of the region shown in Figure 4.21. Definite integrals can be positive, negative, or zero. For a definite integral to be interpreted as an area (as defined in Section 4.2), the function f must be continuous and nonnegative on $[a, b]$, as stated in the next theorem. The proof of this theorem is straightforward—you simply use the definition of area given in Section 4.2, because it is a Riemann sum.

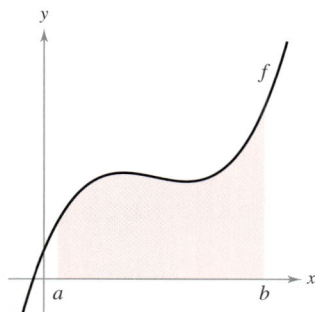

You can use a definite integral to find the area of the region bounded by the graph of f, the x-axis, $x = a$, and $x = b$.

Figure 4.22

THEOREM 4.5   The Definite Integral as the Area of a Region

If f is continuous and nonnegative on the closed interval $[a, b]$, then the area of the region bounded by the graph of f, the x-axis, and the vertical lines $x = a$ and $x = b$ is

$$\text{Area} = \int_{a}^{b} f(x) \, dx.$$

(See Figure 4.22.)

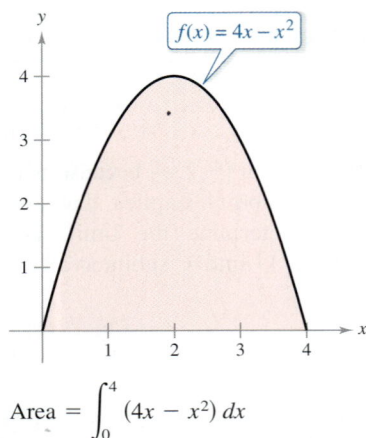

$$\text{Area} = \int_0^4 (4x - x^2)\, dx$$

Figure 4.23

As an example of Theorem 4.5, consider the region bounded by the graph of

$$f(x) = 4x - x^2$$

and the x-axis, as shown in Figure 4.23. Because f is continuous and nonnegative on the closed interval $[0, 4]$, the area of the region is

$$\text{Area} = \int_0^4 (4x - x^2)\, dx.$$

A straightforward technique for evaluating a definite integral such as this will be discussed in Section 4.4. For now, however, you can evaluate a definite integral in two ways—you can use the limit definition *or* you can check to see whether the definite integral represents the area of a common geometric region, such as a rectangle, triangle, or semicircle.

EXAMPLE 3 **Areas of Common Geometric Figures**

Sketch the region corresponding to each definite integral. Then evaluate each integral using a geometric formula.

a. $\displaystyle\int_1^3 4\, dx$ **b.** $\displaystyle\int_0^3 (x + 2)\, dx$ **c.** $\displaystyle\int_{-2}^2 \sqrt{4 - x^2}\, dx$

Solution A sketch of each region is shown in Figure 4.24.

a. This region is a rectangle of height 4 and width 2.

$$\int_1^3 4\, dx = (\text{Area of rectangle}) = 4(2) = 8$$

b. This region is a trapezoid with an altitude of 3 and parallel bases of lengths 2 and 5. The formula for the area of a trapezoid is $\frac{1}{2}h(b_1 + b_2)$.

$$\int_0^3 (x + 2)\, dx = (\text{Area of trapezoid}) = \frac{1}{2}(3)(2 + 5) = \frac{21}{2}$$

c. This region is a semicircle of radius 2. The formula for the area of a semicircle is $\frac{1}{2}\pi r^2$.

$$\int_{-2}^2 \sqrt{4 - x^2}\, dx = (\text{Area of semicircle}) = \frac{1}{2}\pi(2^2) = 2\pi$$

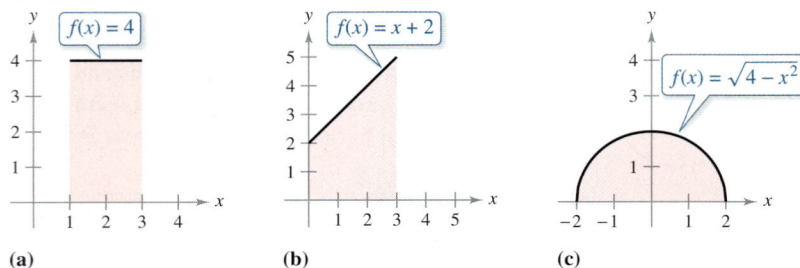

(a)   (b)   (c)

Figure 4.24

The variable of integration in a definite integral is sometimes called a *dummy variable* because it can be replaced by any other variable without changing the value of the integral. For instance, the definite integrals

$$\int_0^3 (x + 2)\, dx \quad \text{and} \quad \int_0^3 (t + 2)\, dt$$

have the same value.

Properties of Definite Integrals

The definition of the definite integral of f on the interval $[a, b]$ specifies that $a < b$. Now, however, it is convenient to extend the definition to cover cases in which $a = b$ or $a > b$. Geometrically, the next two definitions seem reasonable. For instance, it makes sense to define the area of a region of zero width and finite height to be 0.

Definitions of Two Special Definite Integrals

1. If f is defined at $x = a$, then $\int_a^a f(x)\, dx = 0$.

2. If f is integrable on $[a, b]$, then $\int_b^a f(x)\, dx = -\int_a^b f(x)\, dx$.

EXAMPLE 4 **Evaluating Definite Integrals**

$\cdots\triangleright$ *See LarsonCalculus.com for an interactive version of this type of example.*

Evaluate each definite integral.

a. $\int_\pi^\pi \sin x\, dx$ **b.** $\int_3^0 (x + 2)\, dx$

Solution

a. Because the sine function is defined at $x = \pi$, and the upper and lower limits of integration are equal, you can write

$$\int_\pi^\pi \sin x\, dx = 0.$$

b. The integral $\int_3^0 (x + 2)\, dx$ is the same as that given in Example 3(b) except that the upper and lower limits are interchanged. Because the integral in Example 3(b) has a value of $\frac{21}{2}$, you can write

$$\int_3^0 (x + 2)\, dx = -\int_0^3 (x + 2)\, dx = -\frac{21}{2}.$$

In Figure 4.25, the larger region can be divided at $x = c$ into two subregions whose intersection is a line segment. Because the line segment has zero area, it follows that the area of the larger region is equal to the sum of the areas of the two smaller regions.

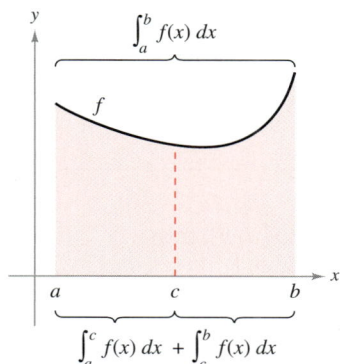

THEOREM 4.6 **Additive Interval Property**

If f is integrable on the three closed intervals determined by a, b, and c, then

$$\int_a^b f(x)\, dx = \int_a^c f(x)\, dx + \int_c^b f(x)\, dx.$$

$\int_a^b f(x)\, dx$

f

$\int_a^c f(x)\, dx + \int_c^b f(x)\, dx$

Figure 4.25

EXAMPLE 5 **Using the Additive Interval Property**

$$\int_{-1}^1 |x|\, dx = \int_{-1}^0 -x\, dx + \int_0^1 x\, dx \qquad \text{Theorem 4.6}$$

$$= \frac{1}{2} + \frac{1}{2} \qquad \text{Area of a triangle}$$

$$= 1$$

Because the definite integral is defined as the limit of a sum, it inherits the properties of summation given at the top of page 255.

THEOREM 4.7   Properties of Definite Integrals

If f and g are integrable on $[a, b]$ and k is a constant, then the functions kf and $f \pm g$ are integrable on $[a, b]$, and

1. $\displaystyle \int_a^b kf(x)\, dx = k \int_a^b f(x)\, dx$

2. $\displaystyle \int_a^b \left[f(x) \pm g(x) \right] dx = \int_a^b f(x)\, dx \pm \int_a^b g(x)\, dx.$

• • **REMARK**   Property 2 of Theorem 4.7 can be extended to cover any finite number of functions (see Example 6).

EXAMPLE 6   **Evaluation of a Definite Integral**

Evaluate $\displaystyle \int_1^3 (-x^2 + 4x - 3)\, dx$ using each of the following values.

$$\int_1^3 x^2\, dx = \frac{26}{3}, \qquad \int_1^3 x\, dx = 4, \qquad \int_1^3 dx = 2$$

Solution

$$\int_1^3 (-x^2 + 4x - 3)\, dx = \int_1^3 (-x^2)\, dx + \int_1^3 4x\, dx + \int_1^3 (-3)\, dx$$
$$= -\int_1^3 x^2\, dx + 4\int_1^3 x\, dx - 3\int_1^3 dx$$
$$= -\left(\frac{26}{3}\right) + 4(4) - 3(2)$$
$$= \frac{4}{3}$$

If f and g are continuous on the closed interval $[a, b]$ and $0 \le f(x) \le g(x)$ for $a \le x \le b$, then the following properties are true. First, the area of the region bounded by the graph of f and the x-axis (between a and b) must be nonnegative. Second, this area must be less than or equal to the area of the region bounded by the graph of g and the x-axis (between a and b), as shown in Figure 4.26. These two properties are generalized in Theorem 4.8.

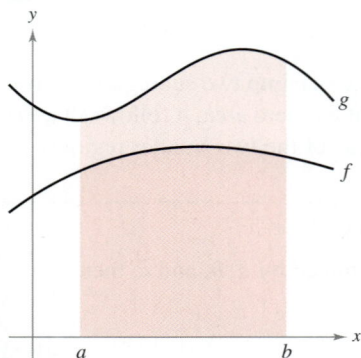

$\displaystyle \int_a^b f(x)\, dx \le \int_a^b g(x)\, dx$

Figure 4.26

THEOREM 4.8   Preservation of Inequality

1. If f is integrable and nonnegative on the closed interval $[a, b]$, then

$$0 \le \int_a^b f(x)\, dx.$$

2. If f and g are integrable on the closed interval $[a, b]$ and $f(x) \le g(x)$ for every x in $[a, b]$, then

$$\int_a^b f(x)\, dx \le \int_a^b g(x)\, dx.$$

A proof of this theorem is given in Appendix A.
See LarsonCalculus.com for Bruce Edwards's video of this proof.

4.3 Exercises

See **CalcChat.com** for tutorial help and worked-out solutions to odd-numbered exercises.

Evaluating a Limit In Exercises 1 and 2, use Example 1 as a model to evaluate the limit

$$\lim_{n \to \infty} \sum_{i=1}^{n} f(c_i)\,\Delta x_i$$

over the region bounded by the graphs of the equations.

1. $f(x) = \sqrt{x}, \quad y = 0, \quad x = 0, \quad x = 3$

$\left(\text{Hint: Let } c_i = \dfrac{3i^2}{n^2}.\right)$

2. $f(x) = \sqrt[3]{x}, \quad y = 0, \quad x = 0, \quad x = 1$

$\left(\text{Hint: Let } c_i = \dfrac{i^3}{n^3}.\right)$

Evaluating a Definite Integral as a Limit In Exercises 3–8, evaluate the definite integral by the limit definition.

3. $\displaystyle\int_{2}^{6} 8\,dx$

4. $\displaystyle\int_{-2}^{3} x\,dx$

5. $\displaystyle\int_{-1}^{1} x^3\,dx$

6. $\displaystyle\int_{1}^{4} 4x^2\,dx$

7. $\displaystyle\int_{1}^{2} (x^2 + 1)\,dx$

8. $\displaystyle\int_{-2}^{1} (2x^2 + 3)\,dx$

Writing a Limit as a Definite Integral In Exercises 9–12, write the limit as a definite integral on the interval $[a, b]$, where c_i is any point in the ith subinterval.

Limit	Interval
9. $\displaystyle\lim_{\|\Delta\| \to 0} \sum_{i=1}^{n} (3c_i + 10)\,\Delta x_i$	$[-1, 5]$
10. $\displaystyle\lim_{\|\Delta\| \to 0} \sum_{i=1}^{n} 6c_i(4 - c_i)^2\,\Delta x_i$	$[0, 4]$
11. $\displaystyle\lim_{\|\Delta\| \to 0} \sum_{i=1}^{n} \sqrt{c_i^2 + 4}\,\Delta x_i$	$[0, 3]$
12. $\displaystyle\lim_{\|\Delta\| \to 0} \sum_{i=1}^{n} \left(\dfrac{3}{c_i^2}\right)\Delta x_i$	$[1, 3]$

Writing a Definite Integral In Exercises 13–22, set up a definite integral that yields the area of the region. (Do not evaluate the integral.)

13. $f(x) = 5$

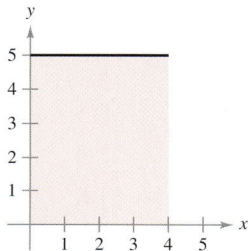

14. $f(x) = 6 - 3x$

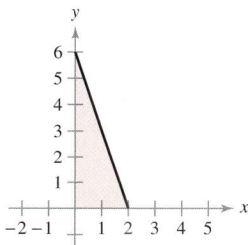

15. $f(x) = 4 - |x|$

16. $f(x) = x^2$

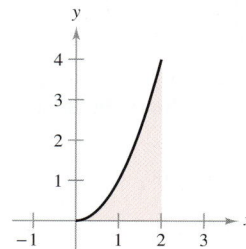

17. $f(x) = 25 - x^2$

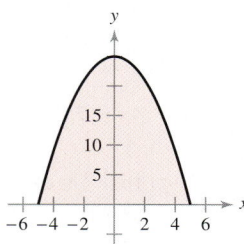

18. $f(x) = \dfrac{4}{x^2 + 2}$

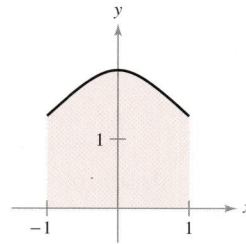

19. $f(x) = \cos x$

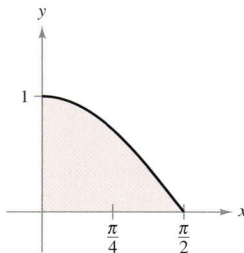

20. $f(x) = \tan x$

21. $g(y) = y^3$

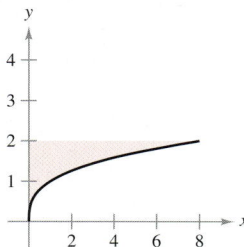

22. $f(y) = (y - 2)^2$

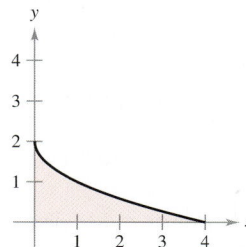

Evaluating a Definite Integral Using a Geometric Formula In Exercises 23–32, sketch the region whose area is given by the definite integral. Then use a geometric formula to evaluate the integral ($a > 0, r > 0$).

23. $\displaystyle\int_{0}^{3} 4\,dx$

24. $\displaystyle\int_{-4}^{6} 6\,dx$

25. $\displaystyle\int_{0}^{4} x\,dx$

26. $\displaystyle\int_{0}^{8} \dfrac{x}{4}\,dx$

27. $\displaystyle\int_0^2 (3x+4)\,dx$ **28.** $\displaystyle\int_0^3 (8-2x)\,dx$

29. $\displaystyle\int_{-1}^1 (1-|x|)\,dx$ **30.** $\displaystyle\int_{-a}^a (a-|x|)\,dx$

31. $\displaystyle\int_{-7}^7 \sqrt{49-x^2}\,dx$ **32.** $\displaystyle\int_{-r}^r \sqrt{r^2-x^2}\,dx$

Using Properties of Definite Integrals In Exercises 33–40, evaluate the integral using the following values.

$$\int_2^4 x^3\,dx = 60, \qquad \int_2^4 x\,dx = 6, \qquad \int_2^4 dx = 2$$

33. $\displaystyle\int_4^2 x\,dx$ **34.** $\displaystyle\int_2^2 x^3\,dx$

35. $\displaystyle\int_2^4 8x\,dx$ **36.** $\displaystyle\int_2^4 25\,dx$

37. $\displaystyle\int_2^4 (x-9)\,dx$ **38.** $\displaystyle\int_2^4 (x^3+4)\,dx$

39. $\displaystyle\int_2^4 \left(\tfrac{1}{2}x^3 - 3x + 2\right)\,dx$ **40.** $\displaystyle\int_2^4 (10+4x-3x^3)\,dx$

41. Using Properties of Definite Integrals Given

$$\int_0^5 f(x)\,dx = 10 \quad \text{and} \quad \int_5^7 f(x)\,dx = 3$$

evaluate

(a) $\displaystyle\int_0^7 f(x)\,dx.$ (b) $\displaystyle\int_5^0 f(x)\,dx.$

(c) $\displaystyle\int_5^5 f(x)\,dx.$ (d) $\displaystyle\int_0^5 3f(x)\,dx.$

42. Using Properties of Definite Integrals Given

$$\int_0^3 f(x)\,dx = 4 \quad \text{and} \quad \int_3^6 f(x)\,dx = -1$$

evaluate

(a) $\displaystyle\int_0^6 f(x)\,dx.$ (b) $\displaystyle\int_6^3 f(x)\,dx.$

(c) $\displaystyle\int_3^3 f(x)\,dx.$ (d) $\displaystyle\int_3^6 -5f(x)\,dx.$

43. Using Properties of Definite Integrals Given

$$\int_2^6 f(x)\,dx = 10 \quad \text{and} \quad \int_2^6 g(x)\,dx = -2$$

evaluate

(a) $\displaystyle\int_2^6 [f(x)+g(x)]\,dx.$ (b) $\displaystyle\int_2^6 [g(x)-f(x)]\,dx.$

(c) $\displaystyle\int_2^6 2g(x)\,dx.$ (d) $\displaystyle\int_2^6 3f(x)\,dx.$

44. Using Properties of Definite Integrals Given

$$\int_{-1}^1 f(x)\,dx = 0 \quad \text{and} \quad \int_0^1 f(x)\,dx = 5$$

evaluate

(a) $\displaystyle\int_{-1}^0 f(x)\,dx.$ (b) $\displaystyle\int_0^1 f(x)\,dx - \int_{-1}^0 f(x)\,dx.$

(c) $\displaystyle\int_{-1}^1 3f(x)\,dx.$ (d) $\displaystyle\int_0^1 3f(x)\,dx.$

45. Estimating a Definite Integral Use the table of values to find lower and upper estimates of

$$\int_0^{10} f(x)\,dx.$$

Assume that f is a decreasing function.

x	0	2	4	6	8	10
$f(x)$	32	24	12	-4	-20	-36

46. Estimating a Definite Integral Use the table of values to estimate

$$\int_0^6 f(x)\,dx.$$

Use three equal subintervals and the (a) left endpoints, (b) right endpoints, and (c) midpoints. When f is an increasing function, how does each estimate compare with the actual value? Explain your reasoning.

x	0	1	2	3	4	5	6
$f(x)$	-6	0	8	18	30	50	80

47. Think About It The graph of f consists of line segments and a semicircle, as shown in the figure. Evaluate each definite integral by using geometric formulas.

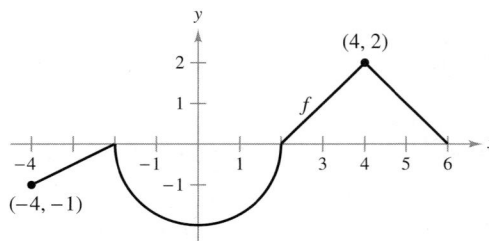

(a) $\displaystyle\int_0^2 f(x)\,dx$ (b) $\displaystyle\int_2^6 f(x)\,dx$

(c) $\displaystyle\int_{-4}^2 f(x)\,dx$ (d) $\displaystyle\int_{-4}^6 f(x)\,dx$

(e) $\displaystyle\int_{-4}^6 |f(x)|\,dx$ (f) $\displaystyle\int_{-4}^6 [f(x)+2]\,dx$

48. Think About It The graph of f consists of line segments, as shown in the figure. Evaluate each definite integral by using geometric formulas.

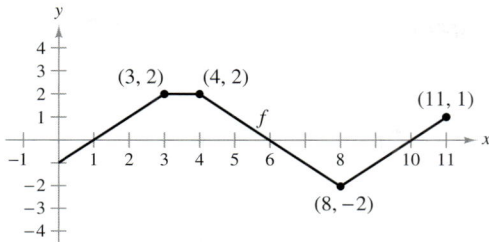

(a) $\displaystyle\int_0^1 -f(x)\,dx$ (b) $\displaystyle\int_3^4 3f(x)\,dx$

(c) $\displaystyle\int_0^7 f(x)\,dx$ (d) $\displaystyle\int_5^{11} f(x)\,dx$

(e) $\displaystyle\int_0^{11} f(x)\,dx$ (f) $\displaystyle\int_4^{10} f(x)\,dx$

49. Think About It Consider the function f that is continuous on the interval $[-5, 5]$ and for which

$$\int_0^5 f(x)\,dx = 4.$$

Evaluate each integral.

(a) $\displaystyle\int_0^5 [f(x) + 2]\,dx$ (b) $\displaystyle\int_{-2}^3 f(x + 2)\,dx$

(c) $\displaystyle\int_{-5}^5 f(x)\,dx$ (f is even.) (d) $\displaystyle\int_{-5}^5 f(x)\,dx$ (f is odd.)

50. HOW DO YOU SEE IT? Use the figure to fill in the blank with the symbol $<$, $>$, or $=$. Explain your reasoning.

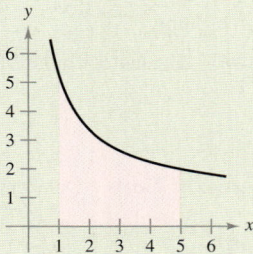

(a) The interval $[1, 5]$ is partitioned into n subintervals of equal width Δx, and x_i is the left endpoint of the ith subinterval.

$$\sum_{i=1}^n f(x_i)\,\Delta x \quad\rule{1cm}{0.4pt}\quad \int_1^5 f(x)\,dx$$

(b) The interval $[1, 5]$ is partitioned into n subintervals of equal width Δx, and x_i is the right endpoint of the ith subinterval.

$$\sum_{i=1}^n f(x_i)\,\Delta x \quad\rule{1cm}{0.4pt}\quad \int_1^5 f(x)\,dx$$

51. Think About It A function f is defined below. Use geometric formulas to find $\int_0^8 f(x)\,dx$.

$$f(x) = \begin{cases} 4, & x < 4 \\ x, & x \geq 4 \end{cases}$$

52. Think About It A function f is defined below. Use geometric formulas to find $\int_0^{12} f(x)\,dx$.

$$f(x) = \begin{cases} 6, & x > 6 \\ -\frac{1}{2}x + 9, & x \leq 6 \end{cases}$$

WRITING ABOUT CONCEPTS

Approximation In Exercises 53–56, determine which value best approximates the definite integral. Make your selection on the basis of a sketch.

53. $\displaystyle\int_0^4 \sqrt{x}\,dx$

(a) 5 (b) -3 (c) 10 (d) 2 (e) 8

54. $\displaystyle\int_0^{1/2} 4\cos \pi x\,dx$

(a) 4 (b) $\frac{4}{3}$ (c) 16 (d) 2π (e) -6

55. $\displaystyle\int_0^1 2\sin \pi x\,dx$

(a) 6 (b) $\frac{1}{2}$ (c) 4 (d) $\frac{5}{4}$

56. $\displaystyle\int_0^9 \left(1 + \sqrt{x}\right)dx$

(a) -3 (b) 9 (c) 27 (d) 3

57. Determining Integrability Determine whether the function

$$f(x) = \frac{1}{x - 4}$$

is integrable on the interval $[3, 5]$. Explain.

58. Finding a Function Give an example of a function that is integrable on the interval $[-1, 1]$, but not continuous on $[-1, 1]$.

Finding Values In Exercises 59–62, find possible values of a and b that make the statement true. If possible, use a graph to support your answer. (There may be more than one correct answer.)

59. $\displaystyle\int_{-2}^1 f(x)\,dx + \int_1^5 f(x)\,dx = \int_a^b f(x)\,dx$

60. $\displaystyle\int_{-3}^3 f(x)\,dx + \int_3^6 f(x)\,dx - \int_a^b f(x)\,dx = \int_{-1}^6 f(x)\,dx$

61. $\displaystyle\int_a^b \sin x\,dx < 0$

62. $\displaystyle\int_a^b \cos x\,dx = 0$

True or False? In Exercises 63–68, determine whether the statement is true or false. If it is false, explain why or give an example that shows it is false.

63. $\int_a^b [f(x) + g(x)] \, dx = \int_a^b f(x) \, dx + \int_a^b g(x) \, dx$

64. $\int_a^b f(x)g(x) \, dx = \left[\int_a^b f(x) \, dx \right]\left[\int_a^b g(x) \, dx \right]$

65. If the norm of a partition approaches zero, then the number of subintervals approaches infinity.

66. If f is increasing on $[a, b]$, then the minimum value of $f(x)$ on $[a, b]$ is $f(a)$.

67. The value of

$$\int_a^b f(x) \, dx$$

must be positive.

68. The value of

$$\int_2^2 \sin (x^2) \, dx$$

is 0.

69. Finding a Riemann Sum Find the Riemann sum for $f(x) = x^2 + 3x$ over the interval $[0, 8]$, where

$$x_0 = 0, \quad x_1 = 1, \quad x_2 = 3, \quad x_3 = 7, \quad \text{and} \quad x_4 = 8$$

and where

$$c_1 = 1, \quad c_2 = 2, \quad c_3 = 5, \quad \text{and} \quad c_4 = 8.$$

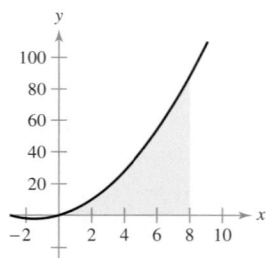

70. Finding a Riemann Sum Find the Riemann sum for $f(x) = \sin x$ over the interval $[0, 2\pi]$, where

$$x_0 = 0, \quad x_1 = \frac{\pi}{4}, \quad x_2 = \frac{\pi}{3}, \quad x_3 = \pi, \quad \text{and} \quad x_4 = 2\pi,$$

and where

$$c_1 = \frac{\pi}{6}, \quad c_2 = \frac{\pi}{3}, \quad c_3 = \frac{2\pi}{3}, \quad \text{and} \quad c_4 = \frac{3\pi}{2}.$$

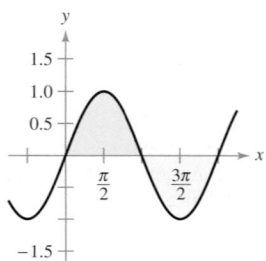

71. Proof Prove that $\int_a^b x \, dx = \dfrac{b^2 - a^2}{2}$.

72. Proof Prove that $\int_a^b x^2 \, dx = \dfrac{b^3 - a^3}{3}$.

73. Think About It Determine whether the Dirichlet function

$$f(x) = \begin{cases} 1, & x \text{ is rational} \\ 0, & x \text{ is irrational} \end{cases}$$

is integrable on the interval $[0, 1]$. Explain.

74. Finding a Definite Integral The function

$$f(x) = \begin{cases} 0, & x = 0 \\ \dfrac{1}{x}, & 0 < x \leq 1 \end{cases}$$

is defined on $[0, 1]$, as shown in the figure. Show that

$$\int_0^1 f(x) \, dx$$

does not exist. Why doesn't this contradict Theorem 4.4?

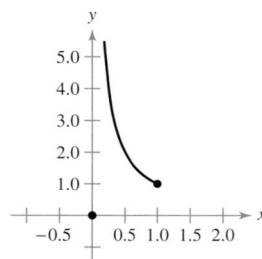

75. Finding Values Find the constants a and b that maximize the value of

$$\int_a^b (1 - x^2) \, dx.$$

Explain your reasoning.

76. Step Function Evaluate, if possible, the integral

$$\int_0^2 [\![x]\!] \, dx.$$

77. Using a Riemann Sum Determine

$$\lim_{n \to \infty} \frac{1}{n^3}[1^2 + 2^2 + 3^2 + \cdots + n^2]$$

by using an appropriate Riemann sum.

PUTNAM EXAM CHALLENGE

78. For each continuous function $f: [0, 1] \to \mathbb{R}$, let

$$I(f) = \int_0^1 x^2 f(x) \, dx \quad \text{and} \quad J(x) = \int_0^1 x (f(x))^2 \, dx.$$

Find the maximum value of $I(f) - J(f)$ over all such functions f.

This problem was composed by the Committee on the Putnam Prize Competition.
© The Mathematical Association of America. All rights reserved.

4.4 The Fundamental Theorem of Calculus

■ Evaluate a definite integral using the Fundamental Theorem of Calculus.
■ Understand and use the Mean Value Theorem for Integrals.
■ Find the average value of a function over a closed interval.
■ Understand and use the Second Fundamental Theorem of Calculus.
■ Understand and use the Net Change Theorem.

The Fundamental Theorem of Calculus

You have now been introduced to the two major branches of calculus: differential calculus (introduced with the tangent line problem) and integral calculus (introduced with the area problem). So far, these two problems might seem unrelated—but there is a very close connection. The connection was discovered independently by Isaac Newton and Gottfried Leibniz and is stated in the **Fundamental Theorem of Calculus.**

Informally, the theorem states that differentiation and (definite) integration are inverse operations, in the same sense that division and multiplication are inverse operations. To see how Newton and Leibniz might have anticipated this relationship, consider the approximations shown in Figure 4.27. The slope of the tangent line was defined using the *quotient* $\Delta y/\Delta x$ (the slope of the secant line). Similarly, the area of a region under a curve was defined using the *product* $\Delta y \Delta x$ (the area of a rectangle). So, at least in the primitive approximation stage, the operations of differentiation and definite integration appear to have an inverse relationship in the same sense that division and multiplication are inverse operations. The Fundamental Theorem of Calculus states that the limit processes (used to define the derivative and definite integral) preserve this inverse relationship.

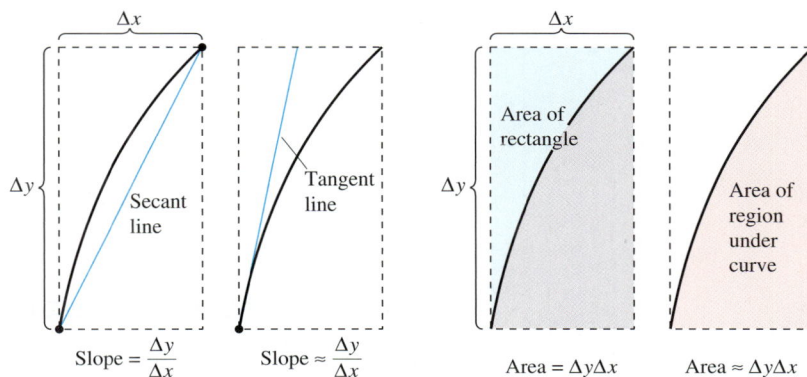

(a) Differentiation   (b) Definite integration
Differentiation and definite integration have an "inverse" relationship.
Figure 4.27

ANTIDIFFERENTIATION AND DEFINITE INTEGRATION

Throughout this chapter, you have been using the integral sign to denote an antiderivative (a family of functions) and a definite integral (a number).

Antidifferentiation: $\displaystyle\int f(x)\,dx$ Definite integration: $\displaystyle\int_a^b f(x)\,dx$

The use of the same symbol for both operations makes it appear that they are related. In the early work with calculus, however, it was not known that the two operations were related. The symbol $\int$ was first applied to the definite integral by Leibniz and was derived from the letter S. (Leibniz calculated area as an infinite sum, thus, the letter S.)

> **THEOREM 4.9   The Fundamental Theorem of Calculus**
>
> If a function f is continuous on the closed interval $[a, b]$ and F is an antiderivative of f on the interval $[a, b]$, then
>
> $$\int_a^b f(x)\, dx = F(b) - F(a).$$

Proof   The key to the proof is writing the difference $F(b) - F(a)$ in a convenient form. Let Δ be any partition of $[a, b]$.

$$a = x_0 < x_1 < x_2 < \cdots < x_{n-1} < x_n = b$$

By pairwise subtraction and addition of like terms, you can write

$$F(b) - F(a) = F(x_n) - F(x_{n-1}) + F(x_{n-1}) - \cdots - F(x_1) + F(x_1) - F(x_0)$$

$$= \sum_{i=1}^{n} [F(x_i) - F(x_{i-1})].$$

By the Mean Value Theorem, you know that there exists a number c_i in the ith subinterval such that

$$F'(c_i) = \frac{F(x_i) - F(x_{i-1})}{x_i - x_{i-1}}.$$

Because $F'(c_i) = f(c_i)$, you can let $\Delta x_i = x_i - x_{i-1}$ and obtain

$$F(b) - F(a) = \sum_{i=1}^{n} f(c_i)\, \Delta x_i.$$

This important equation tells you that by repeatedly applying the Mean Value Theorem, you can always find a collection of c_i's such that the *constant* $F(b) - F(a)$ is a Riemann sum of f on $[a, b]$ for any partition. Theorem 4.4 guarantees that the limit of Riemann sums over the partition with $\|\Delta\| \to 0$ exists. So, taking the limit (as $\|\Delta\| \to 0$) produces

$$F(b) - F(a) = \int_a^b f(x)\, dx.$$

See LarsonCalculus.com for Bruce Edwards's video of this proof.

> **GUIDELINES FOR USING THE FUNDAMENTAL THEOREM OF CALCULUS**
>
> 1. *Provided you can find* an antiderivative of f, you now have a way to evaluate a definite integral without having to use the limit of a sum.
> 2. When applying the Fundamental Theorem of Calculus, the notation shown below is convenient.
>
> $$\int_a^b f(x)\, dx = F(x) \Big]_a^b = F(b) - F(a)$$
>
> For instance, to evaluate $\int_1^3 x^3\, dx$, you can write
>
> $$\int_1^3 x^3\, dx = \frac{x^4}{4} \Big]_1^3 = \frac{3^4}{4} - \frac{1^4}{4} = \frac{81}{4} - \frac{1}{4} = 20.$$
>
> 3. It is not necessary to include a constant of integration C in the antiderivative.
>
> $$\int_a^b f(x)\, dx = \left[F(x) + C \right]_a^b = [F(b) + C] - [F(a) + C] = F(b) - F(a)$$

EXAMPLE 1   **Evaluating a Definite Integral**

⋮⋯▷ *See LarsonCalculus.com for an interactive version of this type of example.*

Evaluate each definite integral.

a. $\displaystyle\int_{1}^{2} (x^2 - 3)\, dx$ **b.** $\displaystyle\int_{1}^{4} 3\sqrt{x}\, dx$ **c.** $\displaystyle\int_{0}^{\pi/4} \sec^2 x\, dx$

Solution

a. $\displaystyle\int_{1}^{2} (x^2 - 3)\, dx = \left[\frac{x^3}{3} - 3x\right]_{1}^{2} = \left(\frac{8}{3} - 6\right) - \left(\frac{1}{3} - 3\right) = -\frac{2}{3}$

b. $\displaystyle\int_{1}^{4} 3\sqrt{x}\, dx = 3\int_{1}^{4} x^{1/2}\, dx = 3\left[\frac{x^{3/2}}{3/2}\right]_{1}^{4} = 2(4)^{3/2} - 2(1)^{3/2} = 14$

c. $\displaystyle\int_{0}^{\pi/4} \sec^2 x\, dx = \tan x\,\Big]_{0}^{\pi/4} = 1 - 0 = 1$

EXAMPLE 2   **A Definite Integral Involving Absolute Value**

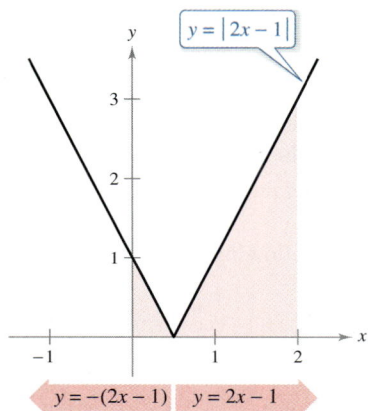

y = |2x − 1|

$y = -(2x - 1)$ $y = 2x - 1$

The definite integral of y on $[0, 2]$ is $\frac{5}{2}$.
Figure 4.28

Evaluate $\displaystyle\int_{0}^{2} |2x - 1|\, dx$.

Solution   Using Figure 4.28 and the definition of absolute value, you can rewrite the integrand as shown.

$$|2x - 1| = \begin{cases} -(2x - 1), & x < \frac{1}{2} \\ 2x - 1, & x \ge \frac{1}{2} \end{cases}$$

From this, you can rewrite the integral in two parts.

$$\int_{0}^{2} |2x - 1|\, dx = \int_{0}^{1/2} -(2x - 1)\, dx + \int_{1/2}^{2} (2x - 1)\, dx$$

$$= \left[-x^2 + x\right]_{0}^{1/2} + \left[x^2 - x\right]_{1/2}^{2}$$

$$= \left(-\frac{1}{4} + \frac{1}{2}\right) - (0 + 0) + (4 - 2) - \left(\frac{1}{4} - \frac{1}{2}\right)$$

$$= \frac{5}{2}$$

EXAMPLE 3   **Using the Fundamental Theorem to Find Area**

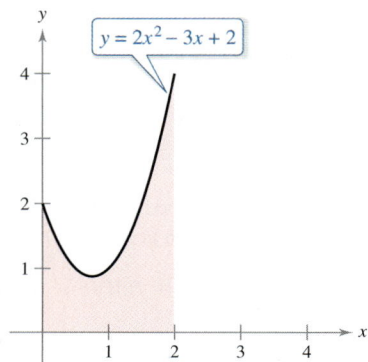

$y = 2x^2 - 3x + 2$

The area of the region bounded by the graph of y, the x-axis, $x = 0$, and $x = 2$ is $\frac{10}{3}$.
Figure 4.29

Find the area of the region bounded by the graph of

$$y = 2x^2 - 3x + 2$$

the x-axis, and the vertical lines $x = 0$ and $x = 2$, as shown in Figure 4.29.

Solution   Note that $y > 0$ on the interval $[0, 2]$.

$$\text{Area} = \int_{0}^{2} (2x^2 - 3x + 2)\, dx \qquad \text{Integrate between } x = 0 \text{ and } x = 2.$$

$$= \left[\frac{2x^3}{3} - \frac{3x^2}{2} + 2x\right]_{0}^{2} \qquad \text{Find antiderivative.}$$

$$= \left(\frac{16}{3} - 6 + 4\right) - (0 - 0 + 0) \qquad \text{Apply Fundamental Theorem.}$$

$$= \frac{10}{3} \qquad \text{Simplify.}$$

The Mean Value Theorem for Integrals

In Section 4.2, you saw that the area of a region under a curve is greater than the area of an inscribed rectangle and less than the area of a circumscribed rectangle. The Mean Value Theorem for Integrals states that somewhere "between" the inscribed and circumscribed rectangles, there is a rectangle whose area is precisely equal to the area of the region under the curve, as shown in Figure 4.30.

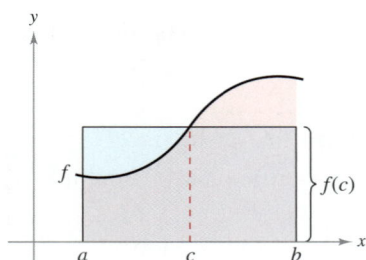

Mean value rectangle:

$$f(c)(b - a) = \int_a^b f(x)\, dx$$

Figure 4.30

THEOREM 4.10   Mean Value Theorem for Integrals

If f is continuous on the closed interval $[a, b]$, then there exists a number c in the closed interval $[a, b]$ such that

$$\int_a^b f(x)\, dx = f(c)(b - a).$$

Proof

Case 1: If f is constant on the interval $[a, b]$, then the theorem is clearly valid because c can be any point in $[a, b]$.

Case 2: If f is not constant on $[a, b]$, then, by the Extreme Value Theorem, you can choose $f(m)$ and $f(M)$ to be the minimum and maximum values of f on $[a, b]$. Because

$$f(m) \leq f(x) \leq f(M)$$

for all x in $[a, b]$, you can apply Theorem 4.8 to write the following.

$$\int_a^b f(m)\, dx \leq \int_a^b f(x)\, dx \leq \int_a^b f(M)\, dx \qquad \text{See Figure 4.31.}$$

$$f(m)(b - a) \leq \int_a^b f(x)\, dx \leq f(M)(b - a) \qquad \text{Apply Fundamental Theorem.}$$

$$f(m) \leq \frac{1}{b - a}\int_a^b f(x)\, dx \leq f(M) \qquad \text{Divide by } b - a.$$

From the third inequality, you can apply the Intermediate Value Theorem to conclude that there exists some c in $[a, b]$ such that

$$f(c) = \frac{1}{b - a}\int_a^b f(x)\, dx \quad \text{or} \quad f(c)(b - a) = \int_a^b f(x)\, dx.$$

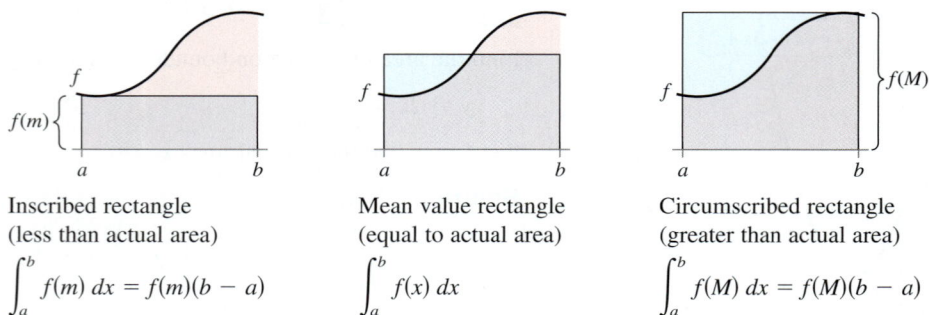

Inscribed rectangle
(less than actual area)

$$\int_a^b f(m)\, dx = f(m)(b - a)$$

Mean value rectangle
(equal to actual area)

$$\int_a^b f(x)\, dx$$

Circumscribed rectangle
(greater than actual area)

$$\int_a^b f(M)\, dx = f(M)(b - a)$$

Figure 4.31

See LarsonCalculus.com for Bruce Edwards's video of this proof.

Notice that Theorem 4.10 does not specify how to determine c. It merely guarantees the existence of at least one number c in the interval.

Average Value of a Function

The value of $f(c)$ given in the Mean Value Theorem for Integrals is called the **average value** of f on the interval $[a, b]$.

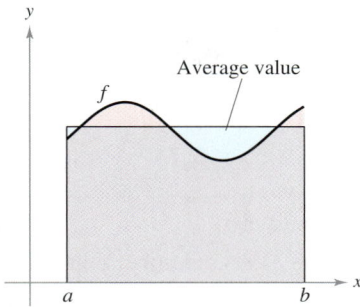

Average value $= \dfrac{1}{b-a}\displaystyle\int_a^b f(x)\,dx$

Figure 4.32

Definition of the Average Value of a Function on an Interval

If f is integrable on the closed interval $[a, b]$, then the **average value** of f on the interval is

$$\frac{1}{b-a}\int_a^b f(x)\,dx. \qquad \text{See Figure 4.32.}$$

To see why the average value of f is defined in this way, partition $[a, b]$ into n subintervals of equal width

$$\Delta x = \frac{b-a}{n}.$$

If c_i is any point in the ith subinterval, then the arithmetic average (or mean) of the function values at the c_i's is

$$a_n = \frac{1}{n}\big[f(c_1) + f(c_2) + \cdots + f(c_n)\big]. \qquad \text{Average of } f(c_1), \ldots, f(c_n)$$

By multiplying and dividing by $(b - a)$, you can write the average as

$$a_n = \frac{1}{n}\sum_{i=1}^{n} f(c_i)\left(\frac{b-a}{b-a}\right)$$

$$= \frac{1}{b-a}\sum_{i=1}^{n} f(c_i)\left(\frac{b-a}{n}\right)$$

$$= \frac{1}{b-a}\sum_{i=1}^{n} f(c_i)\,\Delta x.$$

Finally, taking the limit as $n \to \infty$ produces the average value of f on the interval $[a, b]$, as given in the definition above. In Figure 4.32, notice that the area of the region under the graph of f is equal to the area of the rectangle whose height is the average value.

This development of the average value of a function on an interval is only one of many practical uses of definite integrals to represent summation processes. In Chapter 7, you will study other applications, such as volume, arc length, centers of mass, and work.

EXAMPLE 4   **Finding the Average Value of a Function**

Find the average value of $f(x) = 3x^2 - 2x$ on the interval $[1, 4]$.

Solution   The average value is

$$\frac{1}{b-a}\int_a^b f(x)\,dx = \frac{1}{4-1}\int_1^4 (3x^2 - 2x)\,dx$$

$$= \frac{1}{3}\Big[x^3 - x^2\Big]_1^4$$

$$= \frac{1}{3}\big[64 - 16 - (1 - 1)\big]$$

$$= \frac{48}{3}$$

$$= 16. \qquad \text{See Figure 4.33.}$$

Figure 4.33

The first person to fly at a speed greater than the speed of sound was Charles Yeager. On October 14, 1947, Yeager was clocked at 295.9 meters per second at an altitude of 12.2 kilometers. If Yeager had been flying at an altitude below 11.275 kilometers, this speed would not have "broken the sound barrier." The photo shows an F/A-18F Super Hornet, a supersonic twin-engine strike fighter. A "green Hornet" using a 50/50 mixture of biofuel made from camelina oil became the first U.S. naval tactical aircraft to exceed 1 mach.

EXAMPLE 5 **The Speed of Sound**

At different altitudes in Earth's atmosphere, sound travels at different speeds. The speed of sound $s(x)$ (in meters per second) can be modeled by

$$s(x) = \begin{cases} -4x + 341, & 0 \le x < 11.5 \\ 295, & 11.5 \le x < 22 \\ \frac{3}{4}x + 278.5, & 22 \le x < 32 \\ \frac{3}{2}x + 254.5, & 32 \le x < 50 \\ -\frac{3}{2}x + 404.5, & 50 \le x \le 80 \end{cases}$$

where x is the altitude in kilometers (see Figure 4.34). What is the average speed of sound over the interval $[0, 80]$?

Speed of sound depends on altitude.
Figure 4.34

Solution Begin by integrating $s(x)$ over the interval $[0, 80]$. To do this, you can break the integral into five parts.

$$\int_0^{11.5} s(x)\, dx = \int_0^{11.5} (-4x + 341)\, dx = \left[-2x^2 + 341x \right]_0^{11.5} = 3657$$

$$\int_{11.5}^{22} s(x)\, dx = \int_{11.5}^{22} 295\, dx = \left[295x \right]_{11.5}^{22} = 3097.5$$

$$\int_{22}^{32} s(x)\, dx = \int_{22}^{32} \left(\tfrac{3}{4}x + 278.5\right) dx = \left[\tfrac{3}{8}x^2 + 278.5x \right]_{22}^{32} = 2987.5$$

$$\int_{32}^{50} s(x)\, dx = \int_{32}^{50} \left(\tfrac{3}{2}x + 254.5\right) dx = \left[\tfrac{3}{4}x^2 + 254.5x \right]_{32}^{50} = 5688$$

$$\int_{50}^{80} s(x)\, dx = \int_{50}^{80} \left(-\tfrac{3}{2}x + 404.5\right) dx = \left[-\tfrac{3}{4}x^2 + 404.5x \right]_{50}^{80} = 9210$$

By adding the values of the five integrals, you have

$$\int_0^{80} s(x)\, dx = 24,640.$$

So, the average speed of sound from an altitude of 0 kilometers to an altitude of 80 kilometers is

$$\text{Average speed} = \frac{1}{80} \int_0^{80} s(x)\, dx = \frac{24,640}{80} = 308 \text{ meters per second.} \qquad \blacksquare$$

The Second Fundamental Theorem of Calculus

Earlier you saw that the definite integral of f on the interval $[a, b]$ was defined using the constant b as the upper limit of integration and x as the variable of integration. However, a slightly different situation may arise in which the variable x is used in the upper limit of integration. To avoid the confusion of using x in two different ways, t is temporarily used as the variable of integration. (Remember that the definite integral is *not* a function of its variable of integration.)

The Definite Integral as a Number

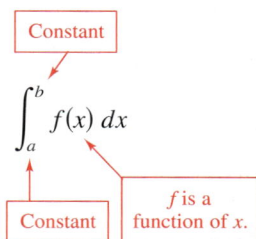

The Definite Integral as a Function of x

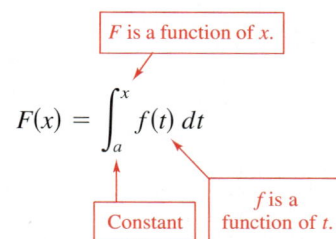

Exploration

Use a graphing utility to graph the function

$$F(x) = \int_0^x \cos t \, dt$$

for $0 \le x \le \pi$. Do you recognize this graph? Explain.

EXAMPLE 6   **The Definite Integral as a Function**

Evaluate the function

$$F(x) = \int_0^x \cos t \, dt$$

at $x = 0, \dfrac{\pi}{6}, \dfrac{\pi}{4}, \dfrac{\pi}{3}$, and $\dfrac{\pi}{2}$.

Solution   You could evaluate five different definite integrals, one for each of the given upper limits. However, it is much simpler to fix x (as a constant) temporarily to obtain

$$
\begin{aligned}
\int_0^x \cos t \, dt &= \sin t \Big]_0^x \\
&= \sin x - \sin 0 \\
&= \sin x.
\end{aligned}
$$

Now, using $F(x) = \sin x$, you can obtain the results shown in Figure 4.35.

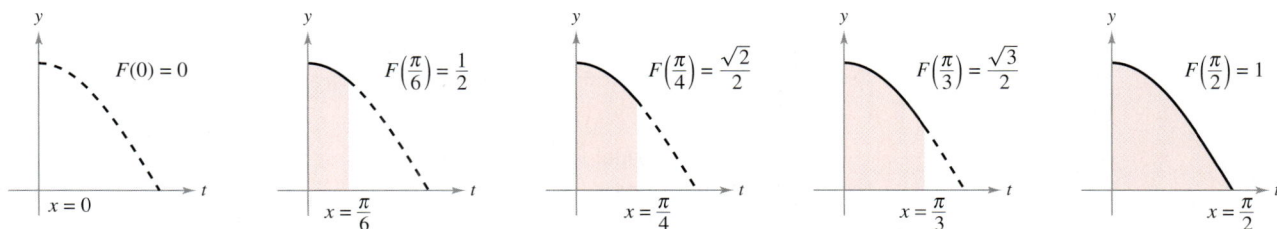

$F(x) = \displaystyle\int_0^x \cos t \, dt$ is the area under the curve $f(t) = \cos t$ from 0 to x.

Figure 4.35

You can think of the function $F(x)$ as *accumulating* the area under the curve $f(t) = \cos t$ from $t = 0$ to $t = x$. For $x = 0$, the area is 0 and $F(0) = 0$. For $x = \pi/2$, $F(\pi/2) = 1$ gives the accumulated area under the cosine curve on the entire interval $[0, \pi/2]$. This interpretation of an integral as an **accumulation function** is used often in applications of integration.

In Example 6, note that the derivative of F is the original integrand (with only the variable changed). That is,

$$\frac{d}{dx}[F(x)] = \frac{d}{dx}[\sin x] = \frac{d}{dx}\left[\int_0^x \cos t \, dt\right] = \cos x.$$

This result is generalized in the next theorem, called the **Second Fundamental Theorem of Calculus.**

THEOREM 4.11 **The Second Fundamental Theorem of Calculus**

If f is continuous on an open interval I containing a, then, for every x in the interval,

$$\frac{d}{dx}\left[\int_a^x f(t) \, dt\right] = f(x).$$

Proof Begin by defining F as

$$F(x) = \int_a^x f(t) \, dt.$$

Then, by the definition of the derivative, you can write

$$F'(x) = \lim_{\Delta x \to 0} \frac{F(x + \Delta x) - F(x)}{\Delta x}$$

$$= \lim_{\Delta x \to 0} \frac{1}{\Delta x}\left[\int_a^{x+\Delta x} f(t) \, dt - \int_a^x f(t) \, dt\right]$$

$$= \lim_{\Delta x \to 0} \frac{1}{\Delta x}\left[\int_a^{x+\Delta x} f(t) \, dt + \int_x^a f(t) \, dt\right]$$

$$= \lim_{\Delta x \to 0} \frac{1}{\Delta x}\left[\int_x^{x+\Delta x} f(t) \, dt\right].$$

From the Mean Value Theorem for Integrals (assuming $\Delta x > 0$), you know there exists a number c in the interval $[x, x + \Delta x]$ such that the integral in the expression above is equal to $f(c) \Delta x$. Moreover, because $x \le c \le x + \Delta x$, it follows that $c \to x$ as $\Delta x \to 0$. So, you obtain

$$F'(x) = \lim_{\Delta x \to 0}\left[\frac{1}{\Delta x} f(c) \Delta x\right] = \lim_{\Delta x \to 0} f(c) = f(x).$$

A similar argument can be made for $\Delta x < 0$.

See LarsonCalculus.com for Bruce Edwards's video of this proof.

Using the area model for definite integrals, the approximation

$$f(x) \Delta x \approx \int_x^{x+\Delta x} f(t) \, dt$$

can be viewed as saying that the area of the rectangle of height $f(x)$ and width Δx is approximately equal to the area of the region lying between the graph of f and the x-axis on the interval

$$[x, x + \Delta x]$$

as shown in the figure at the right.

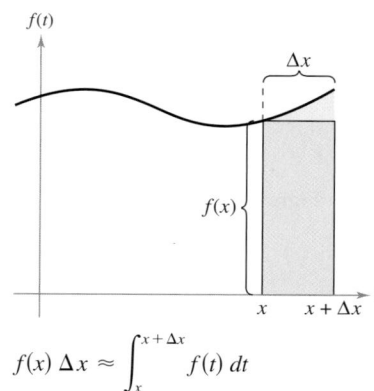

$$f(x) \Delta x \approx \int_x^{x+\Delta x} f(t) \, dt$$

Note that the Second Fundamental Theorem of Calculus tells you that when a function is continuous, you can be sure that it has an antiderivative. This antiderivative need not, however, be an elementary function. (Recall the discussion of elementary functions in Section P.3.)

EXAMPLE 7   **The Second Fundamental Theorem of Calculus**

Evaluate $\dfrac{d}{dx}\left[\displaystyle\int_0^x \sqrt{t^2+1}\,dt\right]$.

Solution   Note that $f(t)=\sqrt{t^2+1}$ is continuous on the entire real number line. So, using the Second Fundamental Theorem of Calculus, you can write

$$\frac{d}{dx}\left[\int_0^x \sqrt{t^2+1}\,dt\right]=\sqrt{x^2+1}.$$

The differentiation shown in Example 7 is a straightforward application of the Second Fundamental Theorem of Calculus. The next example shows how this theorem can be combined with the Chain Rule to find the derivative of a function.

EXAMPLE 8   **The Second Fundamental Theorem of Calculus**

Find the derivative of $F(x)=\displaystyle\int_{\pi/2}^{x^3}\cos t\,dt$.

Solution   Using $u=x^3$, you can apply the Second Fundamental Theorem of Calculus with the Chain Rule as shown.

$$\begin{aligned}
F'(x)&=\frac{dF}{du}\frac{du}{dx} &&\text{Chain Rule}\\
&=\frac{d}{du}[F(x)]\frac{du}{dx} &&\text{Definition of }\frac{dF}{du}\\
&=\frac{d}{du}\left[\int_{\pi/2}^{x^3}\cos t\,dt\right]\frac{du}{dx} &&\text{Substitute }\int_{\pi/2}^{x^3}\cos t\,dt\text{ for }F(x).\\
&=\frac{d}{du}\left[\int_{\pi/2}^{u}\cos t\,dt\right]\frac{du}{dx} &&\text{Substitute }u\text{ for }x^3.\\
&=(\cos u)(3x^2) &&\text{Apply Second Fundamental Theorem of Calculus.}\\
&=(\cos x^3)(3x^2) &&\text{Rewrite as function of }x.
\end{aligned}$$

Because the integrand in Example 8 is easily integrated, you can verify the derivative as follows.

$$\begin{aligned}
F(x)&=\int_{\pi/2}^{x^3}\cos t\,dt\\
&=\sin t\,\Big]_{\pi/2}^{x^3}\\
&=\sin x^3-\sin\frac{\pi}{2}\\
&=\sin x^3-1
\end{aligned}$$

In this form, you can apply the Power Rule to verify that the derivative of F is the same as that obtained in Example 8.

$$\frac{d}{dx}[\sin x^3-1]=(\cos x^3)(3x^2) \qquad \text{Derivative of }F$$

Net Change Theorem

The Fundamental Theorem of Calculus (Theorem 4.9) states that if f is continuous on the closed interval $[a, b]$ and F is an antiderivative of f on $[a, b]$, then

$$\int_a^b f(x)\, dx = F(b) - F(a).$$

But because $F'(x) = f(x)$, this statement can be rewritten as

$$\int_a^b F'(x)\, dx = F(b) - F(a)$$

where the quantity $F(b) - F(a)$ represents the *net change of F* on the interval $[a, b]$.

THEOREM 4.12   The Net Change Theorem

The definite integral of the rate of change of quantity $F'(x)$ gives the total change, or **net change,** in that quantity on the interval $[a, b]$.

$$\int_a^b F'(x)\, dx = F(b) - F(a) \qquad \text{Net change of } F$$

EXAMPLE 9   Using the Net Change Theorem

A chemical flows into a storage tank at a rate of $(180 + 3t)$ liters per minute, where t is the time in minutes and $0 \le t \le 60$. Find the amount of the chemical that flows into the tank during the first 20 minutes.

Solution   Let $c(t)$ be the amount of the chemical in the tank at time t. Then $c'(t)$ represents the rate at which the chemical flows into the tank at time t. During the first 20 minutes, the amount that flows into the tank is

$$\int_0^{20} c'(t)\, dt = \int_0^{20} (180 + 3t)\, dt$$

$$= \left[180t + \frac{3}{2}t^2 \right]_0^{20}$$

$$= 3600 + 600$$

$$= 4200.$$

So, the amount that flows into the tank during the first 20 minutes is 4200 liters.

Another way to illustrate the Net Change Theorem is to examine the velocity of a particle moving along a straight line, where $s(t)$ is the position at time t. Then its velocity is $v(t) = s'(t)$ and

$$\int_a^b v(t)\, dt = s(b) - s(a).$$

This definite integral represents the net change in position, or **displacement,** of the particle.

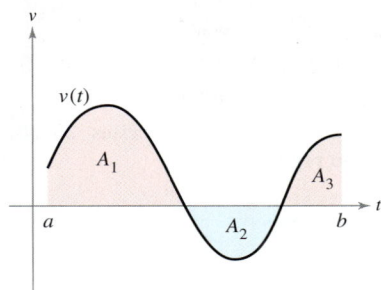

$v(t)$

A_1

A_3

a A_2 b

$A_1, A_2,$ and A_3 are the areas of the shaded regions.

Figure 4.36

When calculating the *total* distance traveled by the particle, you must consider the intervals where $v(t) \leq 0$ and the intervals where $v(t) \geq 0$. When $v(t) \leq 0$, the particle moves to the left, and when $v(t) \geq 0$, the particle moves to the right. To calculate the total distance traveled, integrate the absolute value of velocity $|v(t)|$. So, the **displacement** of the particle on the interval $[a, b]$ is

$$\text{Displacement on } [a, b] = \int_a^b v(t)\, dt = A_1 - A_2 + A_3$$

and the **total distance traveled** by the particle on $[a, b]$ is

$$\text{Total distance traveled on } [a, b] = \int_a^b |v(t)|\, dt = A_1 + A_2 + A_3.$$

(See Figure 4.36.)

EXAMPLE 10   **Solving a Particle Motion Problem**

The velocity (in feet per second) of a particle moving along a line is

$$v(t) = t^3 - 10t^2 + 29t - 20$$

where t is the time in seconds.

a. What is the displacement of the particle on the time interval $1 \leq t \leq 5$?

b. What is the total distance traveled by the particle on the time interval $1 \leq t \leq 5$?

Solution

a. By definition, you know that the displacement is

$$\int_1^5 v(t)\, dt = \int_1^5 (t^3 - 10t^2 + 29t - 20)\, dt$$
$$= \left[\frac{t^4}{4} - \frac{10}{3}t^3 + \frac{29}{2}t^2 - 20t\right]_1^5$$
$$= \frac{25}{12} - \left(-\frac{103}{12}\right)$$
$$= \frac{128}{12}$$
$$= \frac{32}{3}.$$

So, the particle moves $\frac{32}{3}$ feet to the right.

b. To find the total distance traveled, calculate $\int_1^5 |v(t)|\, dt$. Using Figure 4.37 and the fact that $v(t)$ can be factored as $(t - 1)(t - 4)(t - 5)$, you can determine that $v(t) \geq 0$ on $[1, 4]$ and $v(t) \leq 0$ on $[4, 5]$. So, the total distance traveled is

$$\int_1^5 |v(t)|\, dt = \int_1^4 v(t)\, dt - \int_4^5 v(t)\, dt$$
$$= \int_1^4 (t^3 - 10t^2 + 29t - 20)\, dt - \int_4^5 (t^3 - 10t^2 + 29t - 20)\, dt$$
$$= \left[\frac{t^4}{4} - \frac{10}{3}t^3 + \frac{29}{2}t^2 - 20t\right]_1^4 - \left[\frac{t^4}{4} - \frac{10}{3}t^3 + \frac{29}{2}t^2 - 20t\right]_4^5$$
$$= \frac{45}{4} - \left(-\frac{7}{12}\right)$$
$$= \frac{71}{6} \text{ feet.}$$

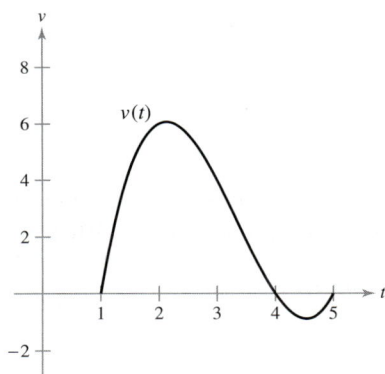

$v(t)$

Figure 4.37

4.4 Exercises

See CalcChat.com for tutorial help and worked-out solutions to odd-numbered exercises.

Graphical Reasoning In Exercises 1–4, use a graphing utility to graph the integrand. Use the graph to determine whether the definite integral is positive, negative, or zero.

1. $\int_0^\pi \dfrac{4}{x^2 + 1}\, dx$

2. $\int_0^\pi \cos x \, dx$

3. $\int_{-2}^{2} x\sqrt{x^2 + 1}\, dx$

4. $\int_{-2}^{2} x\sqrt{2 - x}\, dx$

Evaluating a Definite Integral In Exercises 5–34, evaluate the definite integral. Use a graphing utility to verify your result.

5. $\int_0^2 6x \, dx$

6. $\int_{-3}^{1} 8\, dt$

7. $\int_{-1}^{0} (2x - 1)\, dx$

8. $\int_{-1}^{2} (7 - 3t)\, dt$

9. $\int_{-1}^{1} (t^2 - 2)\, dt$

10. $\int_1^2 (6x^2 - 3x)\, dx$

11. $\int_0^1 (2t - 1)^2\, dt$

12. $\int_1^3 (4x^3 - 3x^2)\, dx$

13. $\int_1^2 \left(\dfrac{3}{x^2} - 1\right) dx$

14. $\int_{-2}^{-1} \left(u - \dfrac{1}{u^2}\right) du$

15. $\int_1^4 \dfrac{u - 2}{\sqrt{u}}\, du$

16. $\int_{-8}^{8} x^{1/3}\, dx$

17. $\int_{-1}^{1} (\sqrt[3]{t} - 2)\, dt$

18. $\int_1^8 \sqrt{\dfrac{2}{x}}\, dx$

19. $\int_0^1 \dfrac{x - \sqrt{x}}{3}\, dx$

20. $\int_0^2 (2 - t)\sqrt{t}\, dt$

21. $\int_{-1}^{0} (t^{1/3} - t^{2/3})\, dt$

22. $\int_{-8}^{-1} \dfrac{x - x^2}{2\sqrt[3]{x}}\, dx$

23. $\int_0^5 |2x - 5|\, dx$

24. $\int_1^4 (3 - |x - 3|)\, dx$

25. $\int_0^4 |x^2 - 9|\, dx$

26. $\int_0^4 |x^2 - 4x + 3|\, dx$

27. $\int_0^\pi (1 + \sin x)\, dx$

28. $\int_0^\pi (2 + \cos x)\, dx$

29. $\int_0^{\pi/4} \dfrac{1 - \sin^2\theta}{\cos^2\theta}\, d\theta$

30. $\int_0^{\pi/4} \dfrac{\sec^2\theta}{\tan^2\theta + 1}\, d\theta$

31. $\int_{-\pi/6}^{\pi/6} \sec^2 x \, dx$

32. $\int_{\pi/4}^{\pi/2} (2 - \csc^2 x)\, dx$

33. $\int_{-\pi/3}^{\pi/3} 4 \sec \theta \tan \theta \, d\theta$

34. $\int_{-\pi/2}^{\pi/2} (2t + \cos t)\, dt$

Finding the Area of a Region In Exercises 35–38, determine the area of the given region.

35. $y = x - x^2$

36. $y = \dfrac{1}{x^2}$

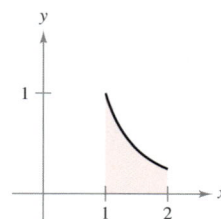

37. $y = \cos x$

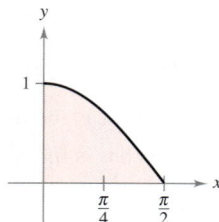

38. $y = x + \sin x$

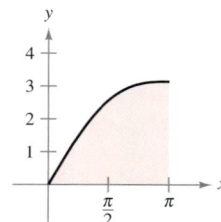

Finding the Area of a Region In Exercises 39–44, find the area of the region bounded by the graphs of the equations.

39. $y = 5x^2 + 2$, $\quad x = 0$, $\quad x = 2$, $\quad y = 0$

40. $y = x^3 + x$, $\quad x = 2$, $\quad y = 0$

41. $y = 1 + \sqrt[3]{x}$, $\quad x = 0$, $\quad x = 8$, $\quad y = 0$

42. $y = 2\sqrt{x} - x$, $\quad y = 0$

43. $y = -x^2 + 4x$, $\quad y = 0$

44. $y = 1 - x^4$, $\quad y = 0$

Using the Mean Value Theorem for Integrals In Exercises 45–50, find the value(s) of c guaranteed by the Mean Value Theorem for Integrals for the function over the given interval.

45. $f(x) = x^3$, $\quad [0, 3]$

46. $f(x) = \sqrt{x}$, $\quad [4, 9]$

47. $y = \dfrac{x^2}{4}$, $\quad [0, 6]$

48. $f(x) = \dfrac{9}{x^3}$, $\quad [1, 3]$

49. $f(x) = 2 \sec^2 x$, $\quad \left[-\dfrac{\pi}{4}, \dfrac{\pi}{4}\right]$

50. $f(x) = \cos x$, $\quad \left[-\dfrac{\pi}{3}, \dfrac{\pi}{3}\right]$

Finding the Average Value of a Function In Exercises 51–56, find the average value of the function over the given interval and all values of x in the interval for which the function equals its average value.

51. $f(x) = 9 - x^2$, $\quad [-3, 3]$

52. $f(x) = \dfrac{4(x^2 + 1)}{x^2}$, $\quad [1, 3]$

53. $f(x) = x^3$, $\quad [0, 1]$

54. $f(x) = 4x^3 - 3x^2$, $\quad [0, 1]$

55. $f(x) = \sin x$, $\quad [0, \pi]$

56. $f(x) = \cos x$, $\quad \left[0, \dfrac{\pi}{2}\right]$

57. Velocity   The graph shows the velocity, in feet per second, of a car accelerating from rest. Use the graph to estimate the distance the car travels in 8 seconds.

Figure for 57

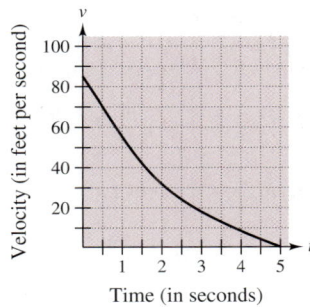

Figure for 58

58. Velocity   The graph shows the velocity, in feet per second, of a decelerating car after the driver applies the brakes. Use the graph to estimate how far the car travels before it comes to a stop.

WRITING ABOUT CONCEPTS

59. Using a Graph   The graph of f is shown in the figure.

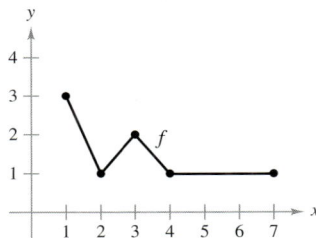

(a) Evaluate $\int_1^7 f(x)\, dx$.

(b) Determine the average value of f on the interval $[1, 7]$.

(c) Determine the answers to parts (a) and (b) when the graph is translated two units upward.

60. Rate of Growth   Let $r'(t)$ represent the rate of growth of a dog, in pounds per year. What does $r(t)$ represent? What does $\int_2^6 r'(t)\, dt$ represent about the dog?

---

61. Force   The force F (in newtons) of a hydraulic cylinder in a press is proportional to the square of sec x, where x is the distance (in meters) that the cylinder is extended in its cycle. The domain of F is $[0, \pi/3]$, and $F(0) = 500$.

(a) Find F as a function of x.

(b) Find the average force exerted by the press over the interval $[0, \pi/3]$.

62. Blood Flow   The velocity v of the flow of blood at a distance r from the central axis of an artery of radius R is

$$v = k(R^2 - r^2)$$

where k is the constant of proportionality. Find the average rate of flow of blood along a radius of the artery. (Use 0 and R as the limits of integration.)

---

63. Respiratory Cycle   The volume V, in liters, of air in the lungs during a five-second respiratory cycle is approximated by the model $V = 0.1729t + 0.1522t^2 - 0.0374t^3$, where t is the time in seconds. Approximate the average volume of air in the lungs during one cycle.

64. Average Sales   A company fits a model to the monthly sales data for a seasonal product. The model is

$$S(t) = \frac{t}{4} + 1.8 + 0.5 \sin\left(\frac{\pi t}{6}\right), \quad 0 \le t \le 24$$

where S is sales (in thousands) and t is time in months.

(a) Use a graphing utility to graph $f(t) = 0.5 \sin(\pi t/6)$ for $0 \le t \le 24$. Use the graph to explain why the average value of $f(t)$ is 0 over the interval.

(b) Use a graphing utility to graph $S(t)$ and the line $g(t) = t/4 + 1.8$ in the same viewing window. Use the graph and the result of part (a) to explain why g is called the *trend line*.

65. Modeling Data   An experimental vehicle is tested on a straight track. It starts from rest, and its velocity v (in meters per second) is recorded every 10 seconds for 1 minute (see table).

t	0	10	20	30	40	50	60
v	0	5	21	40	62	78	83

(a) Use a graphing utility to find a model of the form $v = at^3 + bt^2 + ct + d$ for the data.

(b) Use a graphing utility to plot the data and graph the model.

(c) Use the Fundamental Theorem of Calculus to approximate the distance traveled by the vehicle during the test.

---

66. HOW DO YOU SEE IT?   The graph of f is shown in the figure. The shaded region A has an area of 1.5, and $\int_0^6 f(x)\, dx = 3.5$. Use this information to fill in the blanks.

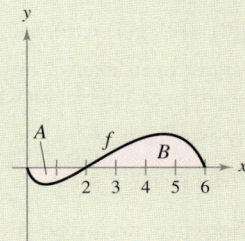

(a) $\displaystyle\int_0^2 f(x)\, dx = \underline{\quad}$ (b) $\displaystyle\int_2^6 f(x)\, dx = \underline{\quad}$

(c) $\displaystyle\int_0^6 |f(x)|\, dx = \underline{\quad}$ (d) $\displaystyle\int_0^2 -2f(x)\, dx = \underline{\quad}$

(e) $\displaystyle\int_0^6 [2 + f(x)]\, dx = \underline{\quad}$

(f) The average value of f over the interval $[0, 6]$ is $\underline{\quad}$.

Evaluating a Definite Integral In Exercises 67–72, find F as a function of x and evaluate it at $x = 2$, $x = 5$, and $x = 8$.

67. $F(x) = \displaystyle\int_0^x (4t - 7) \, dt$

68. $F(x) = \displaystyle\int_2^x (t^3 + 2t - 2) \, dt$

69. $F(x) = \displaystyle\int_1^x \frac{20}{v^2} \, dv$ **70.** $F(x) = \displaystyle\int_2^x -\frac{2}{t^3} \, dt$

71. $F(x) = \displaystyle\int_1^x \cos \theta \, d\theta$ **72.** $F(x) = \displaystyle\int_0^x \sin \theta \, d\theta$

73. Analyzing a Function Let

$$g(x) = \int_0^x f(t) \, dt$$

where f is the function whose graph is shown in the figure.

(a) Estimate $g(0)$, $g(2)$, $g(4)$, $g(6)$, and $g(8)$.

(b) Find the largest open interval on which g is increasing. Find the largest open interval on which g is decreasing.

(c) Identify any extrema of g.

(d) Sketch a rough graph of g.

Figure for 73

Figure for 74

74. Analyzing a Function Let

$$g(x) = \int_0^x f(t) \, dt$$

where f is the function whose graph is shown in the figure.

(a) Estimate $g(0)$, $g(2)$, $g(4)$, $g(6)$, and $g(8)$.

(b) Find the largest open interval on which g is increasing. Find the largest open interval on which g is decreasing.

(c) Identify any extrema of g.

(d) Sketch a rough graph of g.

Finding and Checking an Integral In Exercises 75–80, (a) integrate to find F as a function of x, and (b) demonstrate the Second Fundamental Theorem of Calculus by differentiating the result in part (a).

75. $F(x) = \displaystyle\int_0^x (t + 2) \, dt$ **76.** $F(x) = \displaystyle\int_0^x t(t^2 + 1) \, dt$

77. $F(x) = \displaystyle\int_8^x \sqrt[3]{t} \, dt$ **78.** $F(x) = \displaystyle\int_4^x \sqrt{t} \, dt$

79. $F(x) = \displaystyle\int_{\pi/4}^x \sec^2 t \, dt$ **80.** $F(x) = \displaystyle\int_{\pi/3}^x \sec t \tan t \, dt$

Using the Second Fundamental Theorem of Calculus In Exercises 81–86, use the Second Fundamental Theorem of Calculus to find $F'(x)$.

81. $F(x) = \displaystyle\int_{-2}^x (t^2 - 2t) \, dt$ **82.** $F(x) = \displaystyle\int_1^x \frac{t^2}{t^2 + 1} \, dt$

83. $F(x) = \displaystyle\int_{-1}^x \sqrt{t^4 + 1} \, dt$ **84.** $F(x) = \displaystyle\int_1^x \sqrt[4]{t} \, dt$

85. $F(x) = \displaystyle\int_0^x t \cos t \, dt$ **86.** $F(x) = \displaystyle\int_0^x \sec^3 t \, dt$

Finding a Derivative In Exercises 87–92, find $F'(x)$.

87. $F(x) = \displaystyle\int_x^{x+2} (4t + 1) \, dt$ **88.** $F(x) = \displaystyle\int_{-x}^x t^3 \, dt$

89. $F(x) = \displaystyle\int_0^{\sin x} \sqrt{t} \, dt$ **90.** $F(x) = \displaystyle\int_2^{x^2} \frac{1}{t^3} \, dt$

91. $F(x) = \displaystyle\int_0^{x^3} \sin t^2 \, dt$ **92.** $F(x) = \displaystyle\int_0^{x^2} \sin \theta^2 \, d\theta$

93. Graphical Analysis Sketch an approximate graph of g on the interval $0 \le x \le 4$, where

$$g(x) = \int_0^x f(t) \, dt.$$

Identify the x-coordinate of an extremum of g. To print an enlarged copy of the graph, go to *MathGraphs.com*

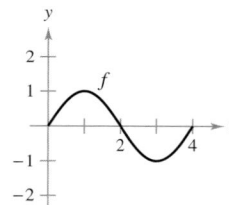

94. Area The area A between the graph of the function

$$g(t) = 4 - \frac{4}{t^2}$$

and the t-axis over the interval $[1, x]$ is

$$A(x) = \int_1^x \left(4 - \frac{4}{t^2}\right) dt.$$

(a) Find the horizontal asymptote of the graph of g.

(b) Integrate to find A as a function of x. Does the graph of A have a horizontal asymptote? Explain.

Particle Motion In Exercises 95–100, the velocity function, in feet per second, is given for a particle moving along a straight line. Find (a) the displacement and (b) the total distance that the particle travels over the given interval.

95. $v(t) = 5t - 7$, $0 \le t \le 3$

96. $v(t) = t^2 - t - 12$, $1 \le t \le 5$

97. $v(t) = t^3 - 10t^2 + 27t - 18$, $1 \le t \le 7$

98. $v(t) = t^3 - 8t^2 + 15t$, $0 \le t \le 5$

99. $v(t) = \dfrac{1}{\sqrt{t}}, \quad 1 \le t \le 4$

100. $v(t) = \cos t, \quad 0 \le t \le 3\pi$

101. Particle Motion A particle is moving along the x-axis. The position of the particle at time t is given by

$$x(t) = t^3 - 6t^2 + 9t - 2, \quad 0 \le t \le 5.$$

Find the total distance the particle travels in 5 units of time.

102. Particle Motion Repeat Exercise 101 for the position function given by

$$x(t) = (t - 1)(t - 3)^2, \quad 0 \le t \le 5.$$

103. Water Flow Water flows from a storage tank at a rate of $(500 - 5t)$ liters per minute. Find the amount of water that flows out of the tank during the first 18 minutes.

104. Oil Leak At 1:00 P.M., oil begins leaking from a tank at a rate of $(4 + 0.75t)$ gallons per hour.

(a) How much oil is lost from 1:00 P.M. to 4:00 P.M.?

(b) How much oil is lost from 4:00 P.M. to 7:00 P.M.?

(c) Compare your answers to parts (a) and (b). What do you notice?

Error Analysis In Exercises 105–108, describe why the statement is incorrect.

105. $\displaystyle\int_{-1}^{1} x^{-2}\, dx = \left[-x^{-1}\right]_{-1}^{1} = (-1) - 1 = -2$

106. $\displaystyle\int_{-2}^{1} \frac{2}{x^3}\, dx = \left[\frac{1}{x^2}\right]_{-2}^{1} = -\frac{3}{4}$

107. $\displaystyle\int_{\pi/4}^{3\pi/4} \sec^2 x\, dx = \left[\tan x\right]_{\pi/4}^{3\pi/4} = -2$

108. $\displaystyle\int_{\pi/2}^{3\pi/2} \csc x \cot x\, dx = \left[-\csc x\right]_{\pi/2}^{3\pi/2} = 2$

109. Buffon's Needle Experiment A horizontal plane is ruled with parallel lines 2 inches apart. A two-inch needle is tossed randomly onto the plane. The probability that the needle will touch a line is

$$P = \frac{2}{\pi} \int_{0}^{\pi/2} \sin \theta\, d\theta$$

where θ is the acute angle between the needle and any one of the parallel lines. Find this probability.

110. Proof Prove that

$$\frac{d}{dx}\left[\int_{u(x)}^{v(x)} f(t)\, dt\right] = f(v(x))v'(x) - f(u(x))u'(x).$$

True or False? In Exercises 111 and 112, determine whether the statement is true or false. If it is false, explain why or give an example that shows it is false.

111. If $F'(x) = G'(x)$ on the interval $[a, b]$, then

$$F(b) - F(a) = G(b) - G(a).$$

112. If f is continuous on $[a, b]$, then f is integrable on $[a, b]$.

113. Analyzing a Function Show that the function

$$f(x) = \int_{0}^{1/x} \frac{1}{t^2 + 1}\, dt + \int_{0}^{x} \frac{1}{t^2 + 1}\, dt$$

is constant for $x > 0$.

114. Finding a Function Find the function $f(x)$ and all values of c such that

$$\int_{c}^{x} f(t)\, dt = x^2 + x - 2.$$

115. Finding Values Let

$$G(x) = \int_{0}^{x} \left[s \int_{0}^{s} f(t)\, dt \right] ds$$

where f is continuous for all real t. Find (a) $G(0)$, (b) $G'(0)$, (c) $G''(x)$, and (d) $G''(0)$.

SECTION PROJECT

Demonstrating the Fundamental Theorem

Use a graphing utility to graph the function

$$y_1 = \sin^2 t$$

on the interval $0 \le t \le \pi$. Let $F(x)$ be the following function of x.

$$F(x) = \int_{0}^{x} \sin^2 t\, dt$$

(a) Complete the table. Explain why the values of F are increasing.

x	0	$\dfrac{\pi}{6}$	$\dfrac{\pi}{3}$	$\dfrac{\pi}{2}$	$\dfrac{2\pi}{3}$	$\dfrac{5\pi}{6}$	π
$F(x)$							

(b) Use the integration capabilities of a graphing utility to graph F.

(c) Use the differentiation capabilities of a graphing utility to graph $F'(x)$. How is this graph related to the graph in part (b)?

(d) Verify that the derivative of

$$y = \frac{1}{2} t - \frac{1}{4} \sin 2t$$

is $\sin^2 t$. Graph y and write a short paragraph about how this graph is related to those in parts (b) and (c).

4.5 Integration by Substitution

- Use pattern recognition to find an indefinite integral.
- Use a change of variables to find an indefinite integral.
- Use the General Power Rule for Integration to find an indefinite integral.
- Use a change of variables to evaluate a definite integral.
- Evaluate a definite integral involving an even or odd function.

Pattern Recognition

In this section, you will study techniques for integrating composite functions. The discussion is split into two parts—*pattern recognition* and *change of variables*. Both techniques involve a **u-substitution.** With pattern recognition, you perform the substitution mentally, and with change of variables, you write the substitution steps.

The role of substitution in integration is comparable to the role of the Chain Rule in differentiation. Recall that for the differentiable functions

$$y = F(u) \quad \text{and} \quad u = g(x)$$

the Chain Rule states that

$$\frac{d}{dx}[F(g(x))] = F'(g(x))g'(x).$$

From the definition of an antiderivative, it follows that

$$\int F'(g(x))g'(x)\,dx = F(g(x)) + C.$$

These results are summarized in the next theorem.

REMARK The statement of Theorem 4.13 doesn't tell how to distinguish between $f(g(x))$ and $g'(x)$ in the integrand. As you become more experienced at integration, your skill in doing this will increase. Of course, part of the key is familiarity with derivatives.

> **THEOREM 4.13 Antidifferentiation of a Composite Function**
>
> Let g be a function whose range is an interval I, and let f be a function that is continuous on I. If g is differentiable on its domain and F is an antiderivative of f on I, then
>
> $$\int f(g(x))g'(x)\,dx = F(g(x)) + C.$$
>
> Letting $u = g(x)$ gives $du = g'(x)\,dx$ and
>
> $$\int f(u)\,du = F(u) + C.$$

Examples 1 and 2 show how to apply Theorem 4.13 *directly*, by recognizing the presence of $f(g(x))$ and $g'(x)$. Note that the composite function in the integrand has an *outside function* f and an *inside function* g. Moreover, the derivative $g'(x)$ is present as a factor of the integrand.

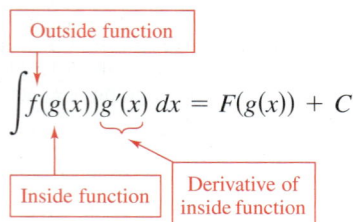

$$\int f(g(x))g'(x)\,dx = F(g(x)) + C$$

Outside function / Inside function / Derivative of inside function

EXAMPLE 1   Recognizing the $f(g(x))g'(x)$ Pattern

Find $\displaystyle\int (x^2 + 1)^2(2x)\, dx$.

Solution   Letting $g(x) = x^2 + 1$, you obtain

$$g'(x) = 2x$$

and

$$f(g(x)) = f(x^2 + 1) = (x^2 + 1)^2.$$

From this, you can recognize that the integrand follows the $f(g(x))g'(x)$ pattern. Using the Power Rule for Integration and Theorem 4.13, you can write

$$\int \overbrace{(x^2 + 1)^2}^{f(g(x))}\overbrace{(2x)}^{g'(x)}\, dx = \frac{1}{3}(x^2 + 1)^3 + C.$$

Try using the Chain Rule to check that the derivative of $\frac{1}{3}(x^2 + 1)^3 + C$ is the integrand of the original integral.

EXAMPLE 2   Recognizing the $f(g(x))g'(x)$ Pattern

Find $\displaystyle\int 5 \cos 5x\, dx$.

Solution   Letting $g(x) = 5x$, you obtain

$$g'(x) = 5$$

and

$$f(g(x)) = f(5x) = \cos 5x.$$

▷ **TECHNOLOGY** Try using a computer algebra system, such as *Maple, Mathematica,* or the *TI-Nspire*, to solve the integrals given in Examples 1 and 2. Do you obtain the same antiderivatives that are listed in the examples?

From this, you can recognize that the integrand follows the $f(g(x))g'(x)$ pattern. Using the Cosine Rule for Integration and Theorem 4.13, you can write

$$\int \overbrace{(\cos 5x)}^{f(g(x))}\overbrace{(5)}^{g'(x)}\, dx = \sin 5x + C.$$

You can check this by differentiating $\sin 5x + C$ to obtain the original integrand. ◼

Exploration

Recognizing Patterns   The integrand in each of the integrals labeled (a)–(c) fits the pattern $f(g(x))g'(x)$. Identify the pattern and use the result to evaluate the integral.

a. $\displaystyle\int 2x(x^2 + 1)^4\, dx$ **b.** $\displaystyle\int 3x^2\sqrt{x^3 + 1}\, dx$ **c.** $\displaystyle\int \sec^2 x(\tan x + 3)\, dx$

The integrals labeled (d)–(f) are similar to (a)–(c). Show how you can multiply and divide by a constant to evaluate these integrals.

d. $\displaystyle\int x(x^2 + 1)^4\, dx$ **e.** $\displaystyle\int x^2\sqrt{x^3 + 1}\, dx$ **f.** $\displaystyle\int 2 \sec^2 x(\tan x + 3)\, dx$

The integrands in Examples 1 and 2 fit the $f(g(x))g'(x)$ pattern exactly—you only had to recognize the pattern. You can extend this technique considerably with the Constant Multiple Rule

$$\int kf(x)\, dx = k \int f(x)\, dx.$$

Many integrands contain the essential part (the variable part) of $g'(x)$ but are missing a constant multiple. In such cases, you can multiply and divide by the necessary constant multiple, as shown in Example 3.

EXAMPLE 3   **Multiplying and Dividing by a Constant**

Find the indefinite integral.

$$\int x(x^2 + 1)^2\, dx$$

Solution   This is similar to the integral given in Example 1, except that the integrand is missing a factor of 2. Recognizing that $2x$ is the derivative of $x^2 + 1$, you can let

$$g(x) = x^2 + 1$$

and supply the $2x$ as shown.

$$\int x(x^2 + 1)^2\, dx = \int (x^2 + 1)^2 \left(\frac{1}{2}\right)(2x)\, dx \qquad \text{Multiply and divide by 2.}$$

$$= \frac{1}{2}\int \overbrace{(x^2 + 1)^2}^{f(g(x))}\ \overbrace{(2x)}^{g'(x)}\, dx \qquad \text{Constant Multiple Rule}$$

$$= \frac{1}{2}\left[\frac{(x^2 + 1)^3}{3}\right] + C \qquad \text{Integrate.}$$

$$= \frac{1}{6}(x^2 + 1)^3 + C \qquad \text{Simplify.}$$

In practice, most people would not write as many steps as are shown in Example 3. For instance, you could evaluate the integral by simply writing

$$\int x(x^2 + 1)^2\, dx = \frac{1}{2}\int (x^2 + 1)^2\,(2x)\, dx$$

$$= \frac{1}{2}\left[\frac{(x^2 + 1)^3}{3}\right] + C$$

$$= \frac{1}{6}(x^2 + 1)^3 + C.$$

Be sure you see that the *Constant* Multiple Rule applies only to *constants*. You cannot multiply and divide by a variable and then move the variable outside the integral sign. For instance,

$$\int (x^2 + 1)^2\, dx \neq \frac{1}{2x}\int (x^2 + 1)^2\,(2x)\, dx.$$

After all, if it were legitimate to move variable quantities outside the integral sign, you could move the entire integrand out and simplify the whole process. But the result would be incorrect.

Change of Variables

With a formal **change of variables,** you completely rewrite the integral in terms of u and du (or any other convenient variable). Although this procedure can involve more written steps than the pattern recognition illustrated in Examples 1 to 3, it is useful for complicated integrands. The change of variables technique uses the Leibniz notation for the differential. That is, if $u = g(x)$, then $du = g'(x)\,dx$, and the integral in Theorem 4.13 takes the form

$$\int f(g(x))g'(x)\,dx = \int f(u)\,du = F(u) + C.$$

EXAMPLE 4   Change of Variables

Find $\int \sqrt{2x - 1}\,dx$.

Solution   First, let u be the inner function, $u = 2x - 1$. Then calculate the differential du to be $du = 2\,dx$. Now, using $\sqrt{2x - 1} = \sqrt{u}$ and $dx = du/2$, substitute to obtain

$$\int \sqrt{2x - 1}\,dx = \int \sqrt{u}\left(\frac{du}{2}\right) \qquad \text{Integral in terms of } u$$

$$= \frac{1}{2}\int u^{1/2}\,du \qquad \text{Constant Multiple Rule}$$

$$= \frac{1}{2}\left(\frac{u^{3/2}}{3/2}\right) + C \qquad \text{Antiderivative in terms of } u$$

$$= \frac{1}{3}u^{3/2} + C \qquad \text{Simplify.}$$

$$= \frac{1}{3}(2x - 1)^{3/2} + C. \qquad \text{Antiderivative in terms of } x$$

• • REMARK   Because integration is usually more difficult than differentiation, you should always check your answer to an integration problem by differentiating. For instance, in Example 4, you should differentiate $\frac{1}{3}(2x - 1)^{3/2} + C$ to verify that you obtain the original integrand.

EXAMPLE 5   Change of Variables

• • • ▷   *See LarsonCalculus.com for an interactive version of this type of example.*

Find $\int x\sqrt{2x - 1}\,dx$.

Solution   As in the previous example, let $u = 2x - 1$ and obtain $dx = du/2$. Because the integrand contains a factor of x, you must also solve for x in terms of u, as shown.

$$u = 2x - 1 \quad \Longrightarrow \quad x = \frac{u + 1}{2} \qquad \text{Solve for } x \text{ in terms of } u.$$

Now, using substitution, you obtain

$$\int x\sqrt{2x - 1}\,dx = \int \left(\frac{u + 1}{2}\right)u^{1/2}\left(\frac{du}{2}\right)$$

$$= \frac{1}{4}\int (u^{3/2} + u^{1/2})\,du$$

$$= \frac{1}{4}\left(\frac{u^{5/2}}{5/2} + \frac{u^{3/2}}{3/2}\right) + C$$

$$= \frac{1}{10}(2x - 1)^{5/2} + \frac{1}{6}(2x - 1)^{3/2} + C.$$

To complete the change of variables in Example 5, you solved for x in terms of u. Sometimes this is very difficult. Fortunately, it is not always necessary, as shown in the next example.

EXAMPLE 6 **Change of Variables**

Find $\displaystyle\int \sin^2 3x \cos 3x\, dx$.

Solution Because $\sin^2 3x = (\sin 3x)^2$, you can let $u = \sin 3x$. Then

$$du = (\cos 3x)(3)\, dx.$$

Now, because $\cos 3x\, dx$ is part of the original integral, you can write

$$\frac{du}{3} = \cos 3x\, dx.$$

Substituting u and $du/3$ in the original integral yields

$$\int \sin^2 3x \cos 3x\, dx = \int u^2 \frac{du}{3}$$

$$= \frac{1}{3}\int u^2\, du$$

$$= \frac{1}{3}\left(\frac{u^3}{3}\right) + C$$

$$= \frac{1}{9} \sin^3 3x + C.$$

You can check this by differentiating.

$$\frac{d}{dx}\left[\frac{1}{9}\sin^3 3x + C\right] = \left(\frac{1}{9}\right)(3)(\sin 3x)^2(\cos 3x)(3)$$

$$= \sin^2 3x \cos 3x$$

Because differentiation produces the original integrand, you know that you have obtained the correct antiderivative.

- **REMARK** When making a change of variables, be sure that your answer is written using the same variables as in the original integrand. For instance, in Example 6, you should not leave your answer as

$$\frac{1}{9}u^3 + C$$

but rather, you should replace u by $\sin 3x$.

The steps used for integration by substitution are summarized in the following guidelines.

GUIDELINES FOR MAKING A CHANGE OF VARIABLES

1. Choose a substitution $u = g(x)$. Usually, it is best to choose the *inner* part of a composite function, such as a quantity raised to a power.
2. Compute $du = g'(x)\, dx$.
3. Rewrite the integral in terms of the variable u.
4. Find the resulting integral in terms of u.
5. Replace u by $g(x)$ to obtain an antiderivative in terms of x.
6. Check your answers by differentiating.

So far, you have seen two techniques for applying substitution, and you will see more techniques in the remainder of this section. Each technique differs slightly from the others. You should remember, however, that the goal is the same with each technique—*you are trying to find an antiderivative of the integrand.*

The General Power Rule for Integration

One of the most common u-substitutions involves quantities in the integrand that are raised to a power. Because of the importance of this type of substitution, it is given a special name—the **General Power Rule for Integration.** A proof of this rule follows directly from the (simple) Power Rule for Integration, together with Theorem 4.13.

> **THEOREM 4.14   The General Power Rule for Integration**
>
> If g is a differentiable function of x, then
>
> $$\int [g(x)]^n g'(x)\, dx = \frac{[g(x)]^{n+1}}{n+1} + C, \quad n \neq -1.$$
>
> Equivalently, if $u = g(x)$, then
>
> $$\int u^n\, du = \frac{u^{n+1}}{n+1} + C, \quad n \neq -1.$$

EXAMPLE 7   Substitution and the General Power Rule

a. $\displaystyle\int 3(3x-1)^4\, dx = \int \overbrace{(3x-1)^4}^{u^4}\overbrace{(3)}^{du}\, dx = \overbrace{\frac{(3x-1)^5}{5}}^{u^5/5} + C$

b. $\displaystyle\int (2x+1)(x^2+x)\, dx = \int \overbrace{(x^2+x)^1}^{u^1}\overbrace{(2x+1)}^{du}\, dx = \overbrace{\frac{(x^2+x)^2}{2}}^{u^2/2} + C$

c. $\displaystyle\int 3x^2\sqrt{x^3-2}\, dx = \int \overbrace{(x^3-2)^{1/2}}^{u^{1/2}}\overbrace{(3x^2)}^{du}\, dx = \overbrace{\frac{(x^3-2)^{3/2}}{3/2}}^{u^{3/2}/(3/2)} + C = \frac{2}{3}(x^3-2)^{3/2} + C$

d. $\displaystyle\int \frac{-4x}{(1-2x^2)^2}\, dx = \int \overbrace{(1-2x^2)^{-2}}^{u^{-2}}\overbrace{(-4x)}^{du}\, dx = \overbrace{\frac{(1-2x^2)^{-1}}{-1}}^{u^{-1}/(-1)} + C = -\frac{1}{1-2x^2} + C$

e. $\displaystyle\int \cos^2 x \sin x\, dx = -\int \overbrace{(\cos x)^2}^{u^2}\overbrace{(-\sin x)}^{du}\, dx = -\overbrace{\frac{(\cos x)^3}{3}}^{u^3/3} + C$ ∎

Some integrals whose integrands involve quantities raised to powers cannot be found by the General Power Rule. Consider the two integrals

$$\int x(x^2+1)^2\, dx \quad \text{and} \quad \int (x^2+1)^2\, dx.$$

The substitution

$$u = x^2 + 1$$

works in the first integral, but not in the second. In the second, the substitution fails because the integrand lacks the factor x needed for du. Fortunately, *for this particular integral,* you can expand the integrand as

$$(x^2+1)^2 = x^4 + 2x^2 + 1$$

and use the (simple) Power Rule to integrate each term.

Change of Variables for Definite Integrals

When using u-substitution with a definite integral, it is often convenient to determine the limits of integration for the variable u rather than to convert the antiderivative back to the variable x and evaluate at the original limits. This change of variables is stated explicitly in the next theorem. The proof follows from Theorem 4.13 combined with the Fundamental Theorem of Calculus.

> **THEOREM 4.15   Change of Variables for Definite Integrals**
>
> If the function $u = g(x)$ has a continuous derivative on the closed interval $[a, b]$ and f is continuous on the range of g, then
> $$\int_a^b f(g(x))g'(x)\, dx = \int_{g(a)}^{g(b)} f(u)\, du.$$

EXAMPLE 8   Change of Variables

Evaluate $\displaystyle\int_0^1 x(x^2 + 1)^3\, dx.$

Solution   To evaluate this integral, let $u = x^2 + 1$. Then, you obtain

$$u = x^2 + 1 \quad\Longrightarrow\quad du = 2x\, dx.$$

Before substituting, determine the new upper and lower limits of integration.

Lower Limit   **Upper Limit**

When $x = 0,\ u = 0^2 + 1 = 1$.   When $x = 1,\ u = 1^2 + 1 = 2$.

Now, you can substitute to obtain

$$\int_0^1 x(x^2+1)^3\, dx = \frac{1}{2}\int_0^1 (x^2+1)^3 (2x)\, dx \qquad \text{Integration limits for } x$$

$$= \frac{1}{2}\int_1^2 u^3\, du \qquad \text{Integration limits for } u$$

$$= \frac{1}{2}\left[\frac{u^4}{4}\right]_1^2$$

$$= \frac{1}{2}\left(4 - \frac{1}{4}\right)$$

$$= \frac{15}{8}.$$

Notice that you obtain the same result when you rewrite the antiderivative $\frac{1}{2}(u^4/4)$ in terms of the variable x and evaluate the definite integral at the original limits of integration, as shown below.

$$\frac{1}{2}\left[\frac{u^4}{4}\right]_1^2 = \frac{1}{2}\left[\frac{(x^2+1)^4}{4}\right]_0^1$$

$$= \frac{1}{2}\left(4 - \frac{1}{4}\right)$$

$$= \frac{15}{8}$$

EXAMPLE 9   **Change of Variables**

Evaluate the definite integral.

$$\int_1^5 \frac{x}{\sqrt{2x-1}}\, dx$$

Solution   To evaluate this integral, let $u = \sqrt{2x-1}$. Then, you obtain

$$u^2 = 2x - 1$$
$$u^2 + 1 = 2x$$
$$\frac{u^2 + 1}{2} = x$$
$$u\, du = dx. \qquad \text{Differentiate each side.}$$

Before substituting, determine the new upper and lower limits of integration.

Lower Limit | **Upper Limit**
When $x = 1$, $u = \sqrt{2-1} = 1$. | When $x = 5$, $u = \sqrt{10-1} = 3$.

Now, substitute to obtain

$$\int_1^5 \frac{x}{\sqrt{2x-1}}\, dx = \int_1^3 \frac{1}{u}\left(\frac{u^2+1}{2}\right) u\, du$$

$$= \frac{1}{2}\int_1^3 (u^2 + 1)\, du$$

$$= \frac{1}{2}\left[\frac{u^3}{3} + u\right]_1^3$$

$$= \frac{1}{2}\left(9 + 3 - \frac{1}{3} - 1\right)$$

$$= \frac{16}{3}.$$

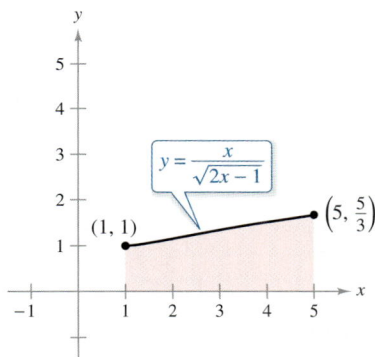

$y = \dfrac{x}{\sqrt{2x-1}}$

$(1, 1)$ $\left(5, \frac{5}{3}\right)$

The region before substitution has an area of $\frac{16}{3}$.

Figure 4.38

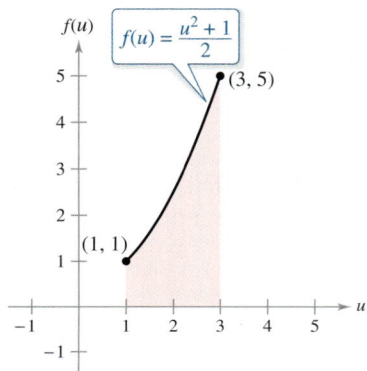

$f(u) = \dfrac{u^2 + 1}{2}$

$(3, 5)$

$(1, 1)$

The region after substitution has an area of $\frac{16}{3}$.

Figure 4.39

Geometrically, you can interpret the equation

$$\int_1^5 \frac{x}{\sqrt{2x-1}}\, dx = \int_1^3 \frac{u^2 + 1}{2}\, du$$

to mean that the two *different* regions shown in Figures 4.38 and 4.39 have the *same* area.

When evaluating definite integrals by substitution, it is possible for the upper limit of integration of the u-variable form to be smaller than the lower limit. When this happens, don't rearrange the limits. Simply evaluate as usual. For example, after substituting $u = \sqrt{1-x}$ in the integral

$$\int_0^1 x^2(1-x)^{1/2}\, dx$$

you obtain $u = \sqrt{1-1} = 0$ when $x = 1$, and $u = \sqrt{1-0} = 1$ when $x = 0$. So, the correct u-variable form of this integral is

$$-2\int_1^0 (1-u^2)^2 u^2\, du.$$

Expanding the integrand, you can evaluate this integral as shown.

$$-2\int_1^0 (u^2 - 2u^4 + u^6)\, du = -2\left[\frac{u^3}{3} - \frac{2u^5}{5} + \frac{u^7}{7}\right]_1^0 = -2\left(-\frac{1}{3} + \frac{2}{5} - \frac{1}{7}\right) = \frac{16}{105}$$

Integration of Even and Odd Functions

Even with a change of variables, integration can be difficult. Occasionally, you can simplify the evaluation of a definite integral over an interval that is symmetric about the y-axis or about the origin by recognizing the integrand to be an even or odd function (see Figure 4.40).

Even function

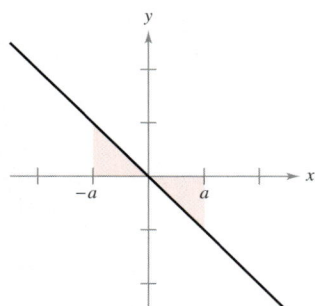

Odd function
Figure 4.40

> **THEOREM 4.16   Integration of Even and Odd Functions**
>
> Let f be integrable on the closed interval $[a, -a]$.
>
> 1. If f is an *even* function, then $\displaystyle\int_{-a}^{a} f(x)\,dx = 2\int_{0}^{a} f(x)\,dx.$
>
> 2. If f is an *odd* function, then $\displaystyle\int_{-a}^{a} f(x)\,dx = 0.$

Proof   Here is the proof of the first property. (The proof of the second property is left to you [see Exercise 99].) Because f is even, you know that $f(x) = f(-x)$. Using Theorem 4.13 with the substitution $u = -x$ produces

$$\int_{-a}^{0} f(x)\,dx = \int_{a}^{0} f(-u)(du) = -\int_{a}^{0} f(u)\,du = \int_{0}^{a} f(u)\,du = \int_{0}^{a} f(x)\,dx.$$

Finally, using Theorem 4.6, you obtain

$$\int_{-a}^{a} f(x)\,dx = \int_{-a}^{0} f(x)\,dx + \int_{0}^{a} f(x)\,dx$$

$$= \int_{0}^{a} f(x)\,dx + \int_{0}^{a} f(x)\,dx$$

$$= 2\int_{0}^{a} f(x)\,dx.$$

See LarsonCalculus.com for Bruce Edwards's video of this proof.

EXAMPLE 10   Integration of an Odd Function

Evaluate the definite integral.

$$\int_{-\pi/2}^{\pi/2} (\sin^3 x \cos x + \sin x \cos x)\,dx$$

Solution   Letting $f(x) = \sin^3 x \cos x + \sin x \cos x$ produces

$$f(-x) = \sin^3(-x)\cos(-x) + \sin(-x)\cos(-x)$$

$$= -\sin^3 x \cos x - \sin x \cos x$$

$$= -f(x).$$

So, f is an odd function, and because f is symmetric about the origin over $[-\pi/2, \pi/2]$, you can apply Theorem 4.16 to conclude that

$$\int_{-\pi/2}^{\pi/2} (\sin^3 x \cos x + \sin x \cos x)\,dx = 0.$$

$f(x) = \sin^3 x \cos x + \sin x \cos x$

Because f is an odd function,
$$\int_{-\pi/2}^{\pi/2} f(x)\,dx = 0.$$
Figure 4.41

From Figure 4.41, you can see that the two regions on either side of the y-axis have the same area. However, because one lies below the x-axis and one lies above it, integration produces a cancellation effect. (More will be said about areas below the x-axis in Section 7.1.)

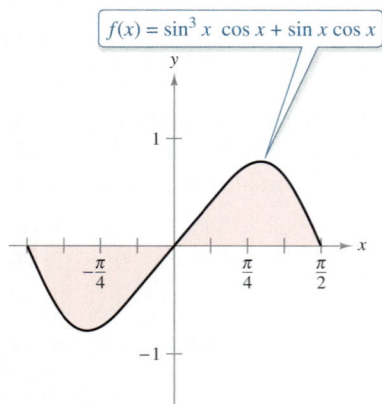

4.5 Exercises

Finding u and du In Exercises 1–4, complete the table by identifying u and du for the integral.

$\int f(g(x))g'(x)\,dx$	$u = g(x)$	$du = g'(x)\,dx$

1. $\int (8x^2 + 1)^2(16x)\,dx$

2. $\int x^2\sqrt{x^3 + 1}\,dx$

3. $\int \tan^2 x\sec^2 x\,dx$

4. $\int \dfrac{\cos x}{\sin^2 x}\,dx$

Finding an Indefinite Integral In Exercises 5–26, find the indefinite integral and check the result by differentiation.

5. $\int (1 + 6x)^4(6)\,dx$

6. $\int (x^2 - 9)^3(2x)\,dx$

7. $\int \sqrt{25 - x^2}\,(-2x)\,dx$

8. $\int \sqrt[3]{3 - 4x^2}(-8x)\,dx$

9. $\int x^3(x^4 + 3)^2\,dx$

10. $\int x^2(6 - x^3)^5\,dx$

11. $\int x^2(x^3 - 1)^4\,dx$

12. $\int x(5x^2 + 4)^3\,dx$

13. $\int t\sqrt{t^2 + 2}\,dt$

14. $\int t^3\sqrt{2t^4 + 3}\,dt$

15. $\int 5x\sqrt[3]{1 - x^2}\,dx$

16. $\int u^2\sqrt{u^3 + 2}\,du$

17. $\int \dfrac{x}{(1 - x^2)^3}\,dx$

18. $\int \dfrac{x^3}{(1 + x^4)^2}\,dx$

19. $\int \dfrac{x^2}{(1 + x^3)^2}\,dx$

20. $\int \dfrac{6x^2}{(4x^3 - 9)^3}\,dx$

21. $\int \dfrac{x}{\sqrt{1 - x^2}}\,dx$

22. $\int \dfrac{x^3}{\sqrt{1 + x^4}}\,dx$

23. $\int \left(1 + \dfrac{1}{t}\right)^3\left(\dfrac{1}{t^2}\right)dt$

24. $\int \left[x^2 + \dfrac{1}{(3x)^2}\right]dx$

25. $\int \dfrac{1}{\sqrt{2x}}\,dx$

26. $\int \dfrac{x}{\sqrt[3]{5x^2}}\,dx$

Differential Equation In Exercises 27–30, solve the differential equation.

27. $\dfrac{dy}{dx} = 4x + \dfrac{4x}{\sqrt{16 - x^2}}$

28. $\dfrac{dy}{dx} = \dfrac{10x^2}{\sqrt{1 + x^3}}$

29. $\dfrac{dy}{dx} = \dfrac{x + 1}{(x^2 + 2x - 3)^2}$

30. $\dfrac{dy}{dx} = \dfrac{x - 4}{\sqrt{x^2 - 8x + 1}}$

Slope Field In Exercises 31 and 32, a differential equation, a point, and a slope field are given. A *slope field* consists of line segments with slopes given by the differential equation. These line segments give a visual perspective of the directions of the solutions of the differential equation. (a) Sketch two approximate solutions of the differential equation on the slope field, one of which passes through the given point. (To print an enlarged copy of the graph, go to *MathGraphs.com*.) (b) Use integration to find the particular solution of the differential equation and use a graphing utility to graph the solution. Compare the result with the sketches in part (a).

31. $\dfrac{dy}{dx} = x\sqrt{4 - x^2}$

(2, 2)

32. $\dfrac{dy}{dx} = x^2(x^3 - 1)^2$

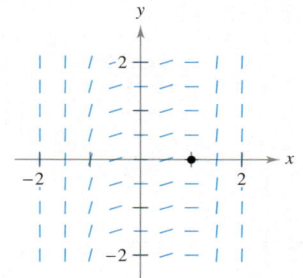

(1, 0)

Finding an Indefinite Integral In Exercises 33–42, find the indefinite integral.

33. $\int \pi \sin \pi x\,dx$

34. $\int \sin 4x\,dx$

35. $\int \cos 8x\,dx$

36. $\int \csc^2\left(\dfrac{x}{2}\right)dx$

37. $\int \dfrac{1}{\theta^2}\cos\dfrac{1}{\theta}\,d\theta$

38. $\int x\sin x^2\,dx$

39. $\int \sin 2x\cos 2x\,dx$

40. $\int \sqrt{\tan x}\sec^2 x\,dx$

41. $\int \dfrac{\csc^2 x}{\cot^3 x}\,dx$

42. $\int \dfrac{\sin x}{\cos^3 x}\,dx$

Finding an Equation In Exercises 43–46, find an equation for the function f that has the given derivative and whose graph passes through the given point.

Derivative	Point
43. $f'(x) = -\sin\dfrac{x}{2}$	(0, 6)
44. $f'(x) = \sec^2(2x)$	$\left(\dfrac{\pi}{2}, 2\right)$
45. $f'(x) = 2x(4x^2 - 10)^2$	(2, 10)
46. $f'(x) = -2x\sqrt{8 - x^2}$	(2, 7)

Change of Variables In Exercises 47–54, find the indefinite integral by the method shown in Example 5.

47. $\displaystyle\int x\sqrt{x+6}\,dx$

48. $\displaystyle\int x\sqrt{3x-4}\,dx$

49. $\displaystyle\int x^2\sqrt{1-x}\,dx$

50. $\displaystyle\int (x+1)\sqrt{2-x}\,dx$

51. $\displaystyle\int \frac{x^2-1}{\sqrt{2x-1}}\,dx$

52. $\displaystyle\int \frac{2x+1}{\sqrt{x+4}}\,dx$

53. $\displaystyle\int \frac{-x}{(x+1)-\sqrt{x+1}}\,dx$

54. $\displaystyle\int t\sqrt[3]{t+10}\,dt$

Evaluating a Definite Integral In Exercises 55–62, evaluate the definite integral. Use a graphing utility to verify your result.

55. $\displaystyle\int_{-1}^{1} x(x^2+1)^3\,dx$

56. $\displaystyle\int_{0}^{1} x^3(2x^4+1)^2\,dx$

57. $\displaystyle\int_{1}^{2} 2x^2\sqrt{x^3+1}\,dx$

58. $\displaystyle\int_{0}^{1} x\sqrt{1-x^2}\,dx$

59. $\displaystyle\int_{0}^{4} \frac{1}{\sqrt{2x+1}}\,dx$

60. $\displaystyle\int_{0}^{2} \frac{x}{\sqrt{1+2x^2}}\,dx$

61. $\displaystyle\int_{1}^{9} \frac{1}{\sqrt{x}\,(1+\sqrt{x})^2}\,dx$

62. $\displaystyle\int_{1}^{5} \frac{x}{\sqrt{2x-1}}\,dx$

Differential Equation In Exercises 63 and 64, the graph of a function f is shown. Use the differential equation and the given point to find an equation of the function.

63. $\dfrac{dy}{dx}=18x^2(2x^3+1)^2$

64. $\dfrac{dy}{dx}=\dfrac{-48}{(3x+5)^3}$

Finding the Area of a Region In Exercises 65–68, find the area of the region. Use a graphing utility to verify your result.

65. $\displaystyle\int_{0}^{7} x\sqrt[3]{x+1}\,dx$

66. $\displaystyle\int_{-2}^{6} x^2\sqrt[3]{x+2}\,dx$

67. $\displaystyle\int_{\pi/2}^{2\pi/3} \sec^2\left(\frac{x}{2}\right)dx$

68. $\displaystyle\int_{\pi/12}^{\pi/4} \csc 2x \cot 2x\,dx$

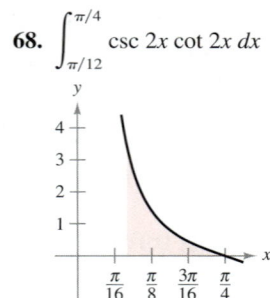

Even and Odd Functions In Exercises 69–72, evaluate the integral using the properties of even and odd functions as an aid.

69. $\displaystyle\int_{-2}^{2} x^2(x^2+1)\,dx$

70. $\displaystyle\int_{-2}^{2} x(x^2+1)^3\,dx$

71. $\displaystyle\int_{-\pi/2}^{\pi/2} \sin^2 x\cos x\,dx$

72. $\displaystyle\int_{-\pi/2}^{\pi/2} \sin x\cos x\,dx$

73. Using an Even Function Use $\int_0^4 x^2\,dx=\frac{64}{3}$ to evaluate each definite integral without using the Fundamental Theorem of Calculus.

(a) $\displaystyle\int_{-4}^{0} x^2\,dx$
(b) $\displaystyle\int_{-4}^{4} x^2\,dx$
(c) $\displaystyle\int_{0}^{4} -x^2\,dx$
(d) $\displaystyle\int_{-4}^{0} 3x^2\,dx$

74. Using Symmetry Use the symmetry of the graphs of the sine and cosine functions as an aid in evaluating each definite integral.

(a) $\displaystyle\int_{-\pi/4}^{\pi/4} \sin x\,dx$
(b) $\displaystyle\int_{-\pi/4}^{\pi/4} \cos x\,dx$
(c) $\displaystyle\int_{-\pi/2}^{\pi/2} \cos x\,dx$
(d) $\displaystyle\int_{-\pi/2}^{\pi/2} \sin x\cos x\,dx$

Even and Odd Functions In Exercises 75 and 76, write the integral as the sum of the integral of an odd function and the integral of an even function. Use this simplification to evaluate the integral.

75. $\displaystyle\int_{-3}^{3} (x^3+4x^2-3x-6)\,dx$

76. $\displaystyle\int_{-\pi/2}^{\pi/2} (\sin 4x+\cos 4x)\,dx$

WRITING ABOUT CONCEPTS

77. Using Substitution Describe why

$$\int x(5-x^2)^3\,dx \neq \int u^3\,du$$

where $u=5-x^2$.

78. Analyzing the Integrand Without integrating, explain why

$$\int_{-2}^{2} x(x^2+1)^2\,dx = 0.$$

WRITING ABOUT CONCEPTS (continued)

79. Choosing an Integral   You are asked to find one of the integrals. Which one would you choose? Explain.

(a) $\displaystyle\int \sqrt{x^3 + 1}\, dx$ or $\displaystyle\int x^2 \sqrt{x^3 + 1}\, dx$

(b) $\displaystyle\int \tan(3x) \sec^2(3x)\, dx$ or $\displaystyle\int \tan(3x)\, dx$

80. Comparing Methods   Find the indefinite integral in two ways. Explain any difference in the forms of the answers.

(a) $\displaystyle\int (2x - 1)^2\, dx$ (b) $\displaystyle\int \tan x \sec^2 x\, dx$

81. Depreciation   The rate of depreciation dV/dt of a machine is inversely proportional to the square of $(t + 1)$, where V is the value of the machine t years after it was purchased. The initial value of the machine was \$500,000, and its value decreased \$100,000 in the first year. Estimate its value after 4 years.

82. **HOW DO YOU SEE IT?**   The graph shows the flow rate of water at a pumping station for one day.

Hour (0 ↔ midnight)

(a) Approximate the maximum flow rate at the pumping station. At what time does this occur?

(b) Explain how you can find the amount of water used during the day.

(c) Approximate the two-hour period when the least amount of water is used. Explain your reasoning.

83. Sales   The sales S (in thousands of units) of a seasonal product are given by the model

$$S = 74.50 + 43.75 \sin \frac{\pi t}{6}$$

where t is the time in months, with $t = 1$ corresponding to January. Find the average sales for each time period.

(a) The first quarter ($0 \le t \le 3$)

(b) The second quarter ($3 \le t \le 6$)

(c) The entire year ($0 \le t \le 12$)

Molodec/Shutterstock.com

84. Electricity

The oscillating current in an electrical circuit is

$$I = 2 \sin(60\pi t) + \cos(120\pi t)$$

where I is measured in amperes and t is measured in seconds. Find the average current for each time interval.

(a) $0 \le t \le \dfrac{1}{60}$

(b) $0 \le t \le \dfrac{1}{240}$

(c) $0 \le t \le \dfrac{1}{30}$

Probability   **In Exercises 85 and 86, the function**

$$f(x) = kx^n(1 - x)^m, \quad 0 \le x \le 1$$

where $n > 0$, $m > 0$, and k is a constant, can be used to represent various probability distributions. If k is chosen such that

$$\int_0^1 f(x)\, dx = 1$$

then the probability that x will fall between a and b ($0 \le a \le b \le 1$) is

$$P_{a,b} = \int_a^b f(x)\, dx.$$

85. The probability that a person will remember between $100a\%$ and $100b\%$ of material learned in an experiment is

$$P_{a,b} = \int_a^b \frac{15}{4} x\sqrt{1 - x}\, dx$$

where x represents the proportion remembered. (See figure.)

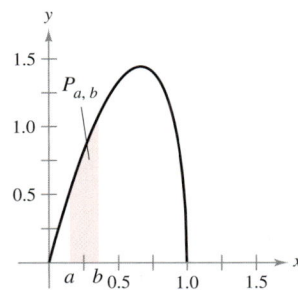

(a) For a randomly chosen individual, what is the probability that he or she will recall between 50% and 75% of the material?

(b) What is the median percent recall? That is, for what value of b is it true that the probability of recalling 0 to b is 0.5?

86. The probability that ore samples taken from a region contain between $100a\%$ and $100b\%$ iron is

$$P_{a,\,b} = \int_a^b \frac{1155}{32} x^3 (1 - x)^{3/2}\,dx$$

where x represents the proportion of iron. (See figure.) What is the probability that a sample will contain between

(a) 0% and 25% iron?   (b) 50% and 100% iron?

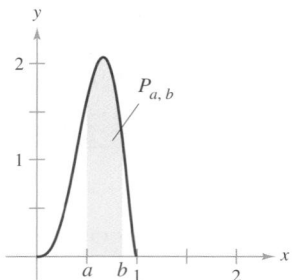

87. Graphical Analysis Consider the functions f and g, where

$$f(x) = 6 \sin x \cos^2 x \quad \text{and} \quad g(t) = \int_0^t f(x)\,dx.$$

(a) Use a graphing utility to graph f and g in the same viewing window.

(b) Explain why g is nonnegative.

(c) Identify the points on the graph of g that correspond to the extrema of f.

(d) Does each of the zeros of f correspond to an extremum of g? Explain.

(e) Consider the function

$$h(t) = \int_{\pi/2}^t f(x)\,dx.$$

Use a graphing utility to graph h. What is the relationship between g and h? Verify your conjecture.

88. Finding a Limit Using a Definite Integral Find

$$\lim_{n \to +\infty} \sum_{i=1}^n \frac{\sin(i\pi/n)}{n}$$

by evaluating an appropriate definite integral over the interval $[0, 1]$.

89. Rewriting Integrals

(a) Show that $\displaystyle\int_0^1 x^2(1 - x)^5\,dx = \int_0^1 x^5(1 - x)^2\,dx.$

(b) Show that $\displaystyle\int_0^1 x^a(1 - x)^b\,dx = \int_0^1 x^b(1 - x)^a\,dx.$

90. Rewriting Integrals

(a) Show that $\displaystyle\int_0^{\pi/2} \sin^2 x\,dx = \int_0^{\pi/2} \cos^2 x\,dx.$

(b) Show that $\displaystyle\int_0^{\pi/2} \sin^n x\,dx = \int_0^{\pi/2} \cos^n x\,dx$, where n is a positive integer.

True or False? In Exercises 91–96, determine whether the statement is true or false. If it is false, explain why or give an example that shows it is false.

91. $\displaystyle\int (2x + 1)^2\,dx = \frac{1}{3}(2x + 1)^3 + C$

92. $\displaystyle\int x(x^2 + 1)\,dx = \frac{1}{2}x^2\left(\frac{1}{3}x^3 + x\right) + C$

93. $\displaystyle\int_{-10}^{10} (ax^3 + bx^2 + cx + d)\,dx = 2\int_0^{10} (bx^2 + d)\,dx$

94. $\displaystyle\int_a^b \sin x\,dx = \int_a^{b+2\pi} \sin x\,dx$

95. $\displaystyle 4\int \sin x \cos x\,dx = -\cos 2x + C$

96. $\displaystyle\int \sin^2 2x \cos 2x\,dx = \frac{1}{3}\sin^3 2x + C$

97. Rewriting Integrals Assume that f is continuous everywhere and that c is a constant. Show that

$$\int_{ca}^{cb} f(x)\,dx = c\int_a^b f(cx)\,dx.$$

98. Integration and Differentiation

(a) Verify that $\displaystyle \sin u - u \cos u + C = \int u \sin u\,du.$

(b) Use part (a) to show that $\displaystyle\int_0^{\pi^2} \sin \sqrt{x}\,dx = 2\pi.$

99. Proof Complete the proof of Theorem 4.16.

100. Rewriting Integrals Show that if f is continuous on the entire real number line, then

$$\int_a^b f(x + h)\,dx = \int_{a+h}^{b+h} f(x)\,dx.$$

PUTNAM EXAM CHALLENGE

101. If $a_0, a_1, \ldots, a_n$ are real numbers satisfying

$$\frac{a_0}{1} + \frac{a_1}{2} + \cdots + \frac{a_n}{n + 1} = 0,$$

show that the equation

$$a_0 + a_1 x + a_2 x^2 + \cdots + a_n x^n = 0$$

has at least one real root.

102. Find all the continuous positive functions $f(x)$, for $0 \le x \le 1$, such that

$$\int_0^1 f(x)\,dx = 1$$

$$\int_0^1 f(x)x\,dx = \alpha$$

$$\int_0^1 f(x)x^2\,dx = \alpha^2$$

where α is a given real number.

These problems were composed by the Committee on the Putnam Prize Competition.
© The Mathematical Association of America. All rights reserved.

4.6 Numerical Integration

- Approximate a definite integral using the Trapezoidal Rule.
- Approximate a definite integral using Simpson's Rule.
- Analyze the approximate errors in the Trapezoidal Rule and Simpson's Rule.

The Trapezoidal Rule

Some elementary functions simply do not have antiderivatives that are elementary functions. For example, there is no elementary function that has any of the following functions as its derivative.

$$\sqrt[3]{x}\sqrt{1-x}, \qquad \sqrt{x}\cos x, \qquad \frac{\cos x}{x}, \qquad \sqrt{1-x^3}, \qquad \sin x^2$$

If you need to evaluate a definite integral involving a function whose antiderivative cannot be found, then while the Fundamental Theorem of Calculus is still true, it cannot be easily applied. In this case, it is easier to resort to an approximation technique. Two such techniques are described in this section.

One way to approximate a definite integral is to use n trapezoids, as shown in Figure 4.42. In the development of this method, assume that f is continuous and positive on the interval $[a, b]$. So, the definite integral

$$\int_a^b f(x)\, dx$$

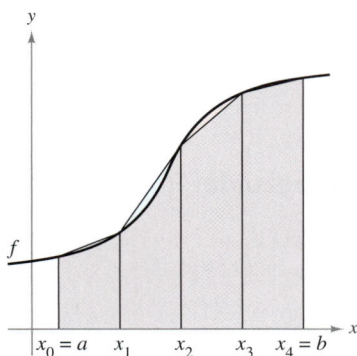

The area of the region can be approximated using four trapezoids.
Figure 4.42

represents the area of the region bounded by the graph of f and the x-axis, from $x = a$ to $x = b$. First, partition the interval $[a, b]$ into n subintervals, each of width $\Delta x = (b - a)/n$, such that

$$a = x_0 < x_1 < x_2 < \cdots < x_n = b.$$

Then form a trapezoid for each subinterval (see Figure 4.43). The area of the ith trapezoid is

$$\text{Area of } i\text{th trapezoid} = \left[\frac{f(x_{i-1}) + f(x_i)}{2}\right]\left(\frac{b-a}{n}\right).$$

This implies that the sum of the areas of the n trapezoids is

$$\text{Area} = \left(\frac{b-a}{n}\right)\left[\frac{f(x_0) + f(x_1)}{2} + \cdots + \frac{f(x_{n-1}) + f(x_n)}{2}\right]$$

$$= \left(\frac{b-a}{2n}\right)\left[f(x_0) + f(x_1) + f(x_1) + f(x_2) + \cdots + f(x_{n-1}) + f(x_n)\right]$$

$$= \left(\frac{b-a}{2n}\right)\left[f(x_0) + 2f(x_1) + 2f(x_2) + \cdots + 2f(x_{n-1}) + f(x_n)\right].$$

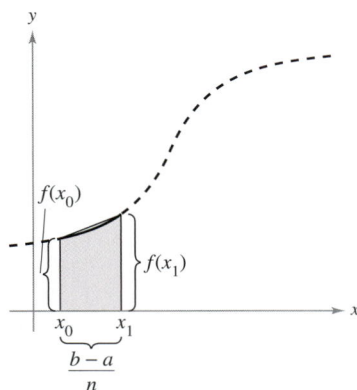

The area of the first trapezoid is
$$\left[\frac{f(x_0) + f(x_1)}{2}\right]\left(\frac{b-a}{n}\right).$$
Figure 4.43

Letting $\Delta x = (b - a)/n$, you can take the limit as $n \to \infty$ to obtain

$$\lim_{n\to\infty}\left(\frac{b-a}{2n}\right)\left[f(x_0) + 2f(x_1) + \cdots + 2f(x_{n-1}) + f(x_n)\right]$$

$$= \lim_{n\to\infty}\left[\frac{[f(a) - f(b)]\,\Delta x}{2} + \sum_{i=1}^{n} f(x_i)\,\Delta x\right]$$

$$= \lim_{n\to\infty}\frac{[f(a) - f(b)](b - a)}{2n} + \lim_{n\to\infty}\sum_{i=1}^{n} f(x_i)\,\Delta x$$

$$= 0 + \int_a^b f(x)\, dx.$$

The result is summarized in the next theorem.

THEOREM 4.17 The Trapezoidal Rule

Let f be continuous on $[a, b]$. The Trapezoidal Rule for approximating $\int_a^b f(x)\, dx$ is

$$\int_a^b f(x)\, dx \approx \frac{b-a}{2n}\left[f(x_0) + 2f(x_1) + 2f(x_2) + \cdots + 2f(x_{n-1}) + f(x_n)\right].$$

Moreover, as $n \to \infty$, the right-hand side approaches $\int_a^b f(x)\, dx$.

REMARK Observe that the coefficients in the Trapezoidal Rule have the following pattern.

$$1 \quad 2 \quad 2 \quad 2 \quad \cdots \quad 2 \quad 2 \quad 1$$

EXAMPLE 1 **Approximation with the Trapezoidal Rule**

Use the Trapezoidal Rule to approximate

$$\int_0^\pi \sin x\, dx.$$

Compare the results for $n = 4$ and $n = 8$, as shown in Figure 4.44.

Solution When $n = 4$, $\Delta x = \pi/4$, and you obtain

$$\int_0^\pi \sin x\, dx \approx \frac{\pi}{8}\left(\sin 0 + 2\sin\frac{\pi}{4} + 2\sin\frac{\pi}{2} + 2\sin\frac{3\pi}{4} + \sin \pi\right)$$

$$= \frac{\pi}{8}\left(0 + \sqrt{2} + 2 + \sqrt{2} + 0\right)$$

$$= \frac{\pi(1 + \sqrt{2})}{4}$$

$$\approx 1.896.$$

When $n = 8$, $\Delta x = \pi/8$, and you obtain

$$\int_0^\pi \sin x\, dx \approx \frac{\pi}{16}\left(\sin 0 + 2\sin\frac{\pi}{8} + 2\sin\frac{\pi}{4} + 2\sin\frac{3\pi}{8} + 2\sin\frac{\pi}{2}\right.$$

$$\left. + 2\sin\frac{5\pi}{8} + 2\sin\frac{3\pi}{4} + 2\sin\frac{7\pi}{8} + \sin \pi\right)$$

$$= \frac{\pi}{16}\left(2 + 2\sqrt{2} + 4\sin\frac{\pi}{8} + 4\sin\frac{3\pi}{8}\right)$$

$$\approx 1.974.$$

For this particular integral, you could have found an antiderivative and determined that the exact area of the region is 2.

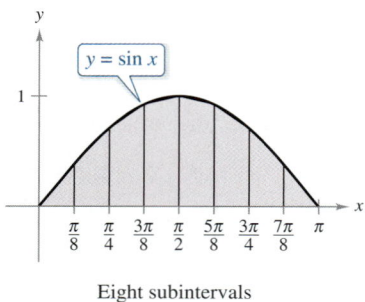

Four subintervals

Eight subintervals
Trapezoidal approximations
Figure 4.44

▷ **TECHNOLOGY** Most graphing utilities and computer algebra systems have built-in programs that can be used to approximate the value of a definite integral. Try using such a program to approximate the integral in Example 1. How close is your approximation? When you use such a program, you need to be aware of its limitations. Often, you are given no indication of the degree of accuracy of the approximation. Other times, you may be given an approximation that is completely wrong. For instance, try using a built-in numerical integration program to evaluate

$$\int_{-1}^2 \frac{1}{x}\, dx.$$

Your calculator should give an error message. Does yours?

It is interesting to compare the Trapezoidal Rule with the Midpoint Rule given in Section 4.2. For the Trapezoidal Rule, you average the function values at the endpoints of the subintervals, but for the Midpoint Rule, you take the function values of the subinterval midpoints.

$$\int_a^b f(x)\,dx \approx \sum_{i=1}^n f\!\left(\frac{x_i + x_{i-1}}{2}\right)\Delta x \qquad \text{Midpoint Rule}$$

$$\int_a^b f(x)\,dx \approx \sum_{i=1}^n \left(\frac{f(x_i) + f(x_{i-1})}{2}\right)\Delta x \qquad \text{Trapezoidal Rule}$$

There are two important points that should be made concerning the Trapezoidal Rule (or the Midpoint Rule). First, the approximation tends to become more accurate as n increases. For instance, in Example 1, when $n = 16$, the Trapezoidal Rule yields an approximation of 1.994. Second, although you could have used the Fundamental Theorem to evaluate the integral in Example 1, this theorem cannot be used to evaluate an integral as simple as $\int_0^\pi \sin x^2\,dx$ because $\sin x^2$ has no elementary antiderivative. Yet, the Trapezoidal Rule can be applied to estimate this integral.

Simpson's Rule

One way to view the trapezoidal approximation of a definite integral is to say that on each subinterval, you approximate f by a *first*-degree polynomial. In Simpson's Rule, named after the English mathematician Thomas Simpson (1710–1761), you take this procedure one step further and approximate f by *second*-degree polynomials.

Before presenting Simpson's Rule, consider the next theorem for evaluating integrals of polynomials of degree 2 (or less).

THEOREM 4.18 Integral of $p(x) = Ax^2 + Bx + C$

If $p(x) = Ax^2 + Bx + C$, then

$$\int_a^b p(x)\,dx = \left(\frac{b-a}{6}\right)\left[p(a) + 4p\!\left(\frac{a+b}{2}\right) + p(b)\right].$$

Proof

$$\int_a^b p(x)\,dx = \int_a^b (Ax^2 + Bx + C)\,dx$$
$$= \left[\frac{Ax^3}{3} + \frac{Bx^2}{2} + Cx\right]_a^b$$
$$= \frac{A(b^3 - a^3)}{3} + \frac{B(b^2 - a^2)}{2} + C(b-a)$$
$$= \left(\frac{b-a}{6}\right)[2A(a^2 + ab + b^2) + 3B(b+a) + 6C]$$

By expansion and collection of terms, the expression inside the brackets becomes

$$\underbrace{(Aa^2 + Ba + C)}_{p(a)} + \underbrace{4\left[A\left(\frac{b+a}{2}\right)^2 + B\left(\frac{b+a}{2}\right) + C\right]}_{4p\left(\frac{a+b}{2}\right)} + \underbrace{(Ab^2 + Bb + C)}_{p(b)}$$

and you can write

$$\int_a^b p(x)\,dx = \left(\frac{b-a}{6}\right)\left[p(a) + 4p\!\left(\frac{a+b}{2}\right) + p(b)\right].$$

See LarsonCalculus.com for Bruce Edwards's video of this proof.

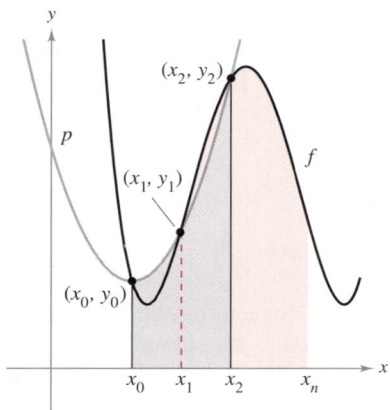

$$\int_{x_0}^{x_2} p(x)\, dx \approx \int_{x_0}^{x_2} f(x)\, dx$$

Figure 4.45

To develop Simpson's Rule for approximating a definite integral, you again partition the interval $[a, b]$ into n subintervals, each of width $\Delta x = (b - a)/n$. This time, however, n is required to be even, and the subintervals are grouped in pairs such that

$$a = x_0 < x_1 < x_2 < x_3 < x_4 < \cdots < x_{n-2} < x_{n-1} < x_n = b.$$

$$\underbrace{\qquad}_{[x_0, x_2]} \quad \underbrace{\qquad}_{[x_2, x_4]} \qquad \underbrace{\qquad}_{[x_{n-2}, x_n]}$$

On each (double) subinterval $[x_{i-2}, x_i]$, you can approximate f by a polynomial p of degree less than or equal to 2. (See Exercise 47.) For example, on the subinterval $[x_0, x_2]$, choose the polynomial of least degree passing through the points (x_0, y_0), (x_1, y_1), and (x_2, y_2), as shown in Figure 4.45. Now, using p as an approximation of f on this subinterval, you have, by Theorem 4.18,

$$\int_{x_0}^{x_2} f(x)\, dx \approx \int_{x_0}^{x_2} p(x)\, dx$$

$$= \frac{x_2 - x_0}{6}\left[p(x_0) + 4p\left(\frac{x_0 + x_2}{2}\right) + p(x_2)\right]$$

$$= \frac{2[(b-a)/n]}{6}\left[p(x_0) + 4p(x_1) + p(x_2)\right]$$

$$= \frac{b-a}{3n}\left[f(x_0) + 4f(x_1) + f(x_2)\right].$$

Repeating this procedure on the entire interval $[a, b]$ produces the next theorem.

. . . . ▷

REMARK Observe that the coefficients in Simpson's Rule have the following pattern.

1 4 2 4 2 4 . . . 4 2 4 1

THEOREM 4.19 **Simpson's Rule**

Let f be continuous on $[a, b]$ and let n be an even integer. Simpson's Rule for approximating $\int_a^b f(x)\, dx$ is

$$\int_a^b f(x)\, dx \approx \frac{b-a}{3n}\left[f(x_0) + 4f(x_1) + 2f(x_2) + 4f(x_3) + \cdots \right.$$

$$\left. + 4f(x_{n-1}) + f(x_n)\right].$$

Moreover, as $n \to \infty$, the right-hand side approaches $\int_a^b f(x)\, dx$.

REMARK In Section 4.2, Example 8, the Midpoint Rule with $n = 4$ approximates $\int_0^\pi \sin x\, dx$ as 2.052. In Example 1, the Trapezoidal Rule with $n = 4$ gives an approximation of 1.896. In Example 2, Simpson's Rule with $n = 4$ gives an approximation of 2.005. The antiderivative would produce the true value of 2.

. . . . ▷

In Example 1, the Trapezoidal Rule was used to estimate $\int_0^\pi \sin x\, dx$. In the next example, Simpson's Rule is applied to the same integral.

EXAMPLE 2 **Approximation with Simpson's Rule**

. . . . ▷ *See LarsonCalculus.com for an interactive version of this type of example.*

Use Simpson's Rule to approximate

$$\int_0^\pi \sin x\, dx.$$

Compare the results for $n = 4$ and $n = 8$.

Solution When $n = 4$, you have

$$\int_0^\pi \sin x\, dx \approx \frac{\pi}{12}\left(\sin 0 + 4\sin\frac{\pi}{4} + 2\sin\frac{\pi}{2} + 4\sin\frac{3\pi}{4} + \sin\pi\right) \approx 2.005.$$

When $n = 8$, you have $\int_0^\pi \sin x\, dx \approx 2.0003.$ ∎

■ FOR FURTHER INFORMATION
For proofs of the formulas used for estimating the errors involved in the use of the Midpoint Rule and Simpson's Rule, see the article "Elementary Proofs of Error Estimates for the Midpoint and Simpson's Rules" by Edward C. Fazekas, Jr. and Peter R. Mercer in *Mathematics Magazine*. To view this article, go to *MathArticles.com*.

Error Analysis

When you use an approximation technique, it is important to know how accurate you can expect the approximation to be. The next theorem, which is listed without proof, gives the formulas for estimating the errors involved in the use of Simpson's Rule and the Trapezoidal Rule. In general, when using an approximation, you can think of the error E as the difference between $\int_a^b f(x)\,dx$ and the approximation.

THEOREM 4.20   Errors in the Trapezoidal Rule and Simpson's Rule

If f has a continuous second derivative on $[a, b]$, then the error E in approximating $\int_a^b f(x)\,dx$ by the Trapezoidal Rule is

$$|E| \le \frac{(b-a)^3}{12n^2}\left[\max|f''(x)|\right], \quad a \le x \le b. \qquad \text{Trapezoidal Rule}$$

Moreover, if f has a continuous fourth derivative on $[a, b]$, then the error E in approximating $\int_a^b f(x)\,dx$ by Simpson's Rule is

$$|E| \le \frac{(b-a)^5}{180n^4}\left[\max|f^{(4)}(x)|\right], \quad a \le x \le b. \qquad \text{Simpson's Rule}$$

Theorem 4.20 states that the errors generated by the Trapezoidal Rule and Simpson's Rule have upper bounds dependent on the extreme values of $f''(x)$ and $f^{(4)}(x)$ in the interval $[a, b]$. Furthermore, these errors can be made arbitrarily small by *increasing n*, provided that f'' and $f^{(4)}$ are continuous and therefore bounded in $[a, b]$.

▷ TECHNOLOGY   If you have access to a computer algebra system, use it to evaluate the definite integral in Example 3. You should obtain a value of

$$\int_0^1 \sqrt{1+x^2}\,dx$$
$$= \frac{1}{2}\left[\sqrt{2} + \ln\left(1+\sqrt{2}\right)\right]$$
$$\approx 1.14779.$$

(The symbol "ln" represents the natural logarithmic function, which you will study in Section 5.1.)

EXAMPLE 3   The Approximate Error in the Trapezoidal Rule

Determine a value of n such that the Trapezoidal Rule will approximate the value of

$$\int_0^1 \sqrt{1+x^2}\,dx$$

with an error that is less than or equal to 0.01.

Solution   Begin by letting $f(x) = \sqrt{1+x^2}$ and finding the second derivative of f.

$$f'(x) = x(1+x^2)^{-1/2} \quad \text{and} \quad f''(x) = (1+x^2)^{-3/2}$$

The maximum value of $|f''(x)|$ on the interval $[0, 1]$ is $|f''(0)| = 1$. So, by Theorem 4.20, you can write

$$|E| \le \frac{(b-a)^3}{12n^2}|f''(0)| = \frac{1}{12n^2}(1) = \frac{1}{12n^2}.$$

To obtain an error E that is less than 0.01, you must choose n such that $1/(12n^2) \le 1/100$.

$$100 \le 12n^2 \implies n \ge \sqrt{\tfrac{100}{12}} \approx 2.89$$

So, you can choose $n = 3$ (because n must be greater than or equal to 2.89) and apply the Trapezoidal Rule, as shown in Figure 4.46, to obtain

$$\int_0^1 \sqrt{1+x^2}\,dx \approx \frac{1}{6}\left[\sqrt{1+0^2} + 2\sqrt{1+\left(\tfrac{1}{3}\right)^2} + 2\sqrt{1+\left(\tfrac{2}{3}\right)^2} + \sqrt{1+1^2}\right]$$
$$\approx 1.154.$$

So, by adding and subtracting the error from this estimate, you know that

$$1.144 \le \int_0^1 \sqrt{1+x^2}\,dx \le 1.164.$$

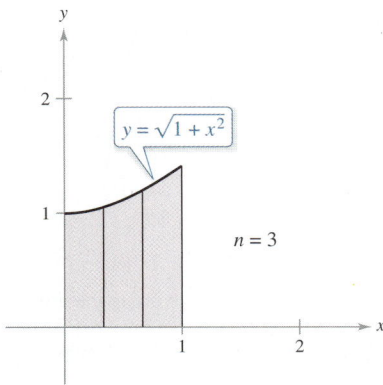

$$1.144 \le \int_0^1 \sqrt{1+x^2}\,dx \le 1.164$$
Figure 4.46

4.6 Exercises

See **CalcChat.com** for tutorial help and worked-out solutions to odd-numbered exercises.

Using the Trapezoidal Rule and Simpson's Rule In Exercises 1–10, use the Trapezoidal Rule and Simpson's Rule to approximate the value of the definite integral for the given value of *n*. Round your answer to four decimal places and compare the results with the exact value of the definite integral.

1. $\int_0^2 x^2 \, dx, \quad n = 4$

2. $\int_1^2 \left(\frac{x^2}{4} + 1\right) dx, \quad n = 4$

3. $\int_0^2 x^3 \, dx, \quad n = 4$

4. $\int_2^3 \frac{2}{x^2} \, dx, \quad n = 4$

5. $\int_1^3 x^3 \, dx, \quad n = 6$

6. $\int_0^8 \sqrt[3]{x} \, dx, \quad n = 8$

7. $\int_4^9 \sqrt{x} \, dx, \quad n = 8$

8. $\int_1^4 (4 - x^2) \, dx, \quad n = 6$

9. $\int_0^1 \frac{2}{(x + 2)^2} \, dx, \quad n = 4$

10. $\int_0^2 x\sqrt{x^2 + 1} \, dx, \quad n = 4$

Using the Trapezoidal Rule and Simpson's Rule In Exercises 11–20, approximate the definite integral using the Trapezoidal Rule and Simpson's Rule with *n* = 4. Compare these results with the approximation of the integral using a graphing utility.

11. $\int_0^2 \sqrt{1 + x^3} \, dx$

12. $\int_0^2 \frac{1}{\sqrt{1 + x^3}} \, dx$

13. $\int_0^1 \sqrt{x} \sqrt{1 - x} \, dx$

14. $\int_{\pi/2}^{\pi} \sqrt{x} \sin x \, dx$

15. $\int_0^{\sqrt{\pi/2}} \sin x^2 \, dx$

16. $\int_0^{\sqrt{\pi/4}} \tan x^2 \, dx$

17. $\int_3^{3.1} \cos x^2 \, dx$

18. $\int_0^{\pi/2} \sqrt{1 + \sin^2 x} \, dx$

19. $\int_0^{\pi/4} x \tan x \, dx$

20. $\int_0^{\pi} f(x) \, dx, \quad f(x) = \begin{cases} \frac{\sin x}{x}, & x > 0 \\ 1, & x = 0 \end{cases}$

WRITING ABOUT CONCEPTS

21. Polynomial Approximations The Trapezoidal Rule and Simpson's Rule yield approximations of a definite integral $\int_a^b f(x) \, dx$ based on polynomial approximations of *f*. What is the degree of the polynomials used for each?

22. Describing an Error Describe the size of the error when the Trapezoidal Rule is used to approximate $\int_a^b f(x) \, dx$ when $f(x)$ is a linear function. Use a graph to explain your answer.

Estimating Errors In Exercises 23–26, use the error formulas in Theorem 4.20 to estimate the errors in approximating the integral, with *n* = 4, using (a) the Trapezoidal Rule and (b) Simpson's Rule.

23. $\int_1^3 2x^3 \, dx$

24. $\int_3^5 (5x + 2) \, dx$

25. $\int_2^4 \frac{1}{(x - 1)^2} \, dx$

26. $\int_0^{\pi} \cos x \, dx$

Estimating Errors In Exercises 27–30, use the error formulas in Theorem 4.20 to find *n* such that the error in the approximation of the definite integral is less than or equal to 0.00001 using (a) the Trapezoidal Rule and (b) Simpson's Rule.

27. $\int_1^3 \frac{1}{x} \, dx$

28. $\int_0^1 \frac{1}{1 + x} \, dx$

29. $\int_0^2 \sqrt{x + 2} \, dx$

30. $\int_0^{\pi/2} \sin x \, dx$

Estimating Errors Using Technology In Exercises 31–34, use a computer algebra system and the error formulas to find *n* such that the error in the approximation of the definite integral is less than or equal to 0.00001 using (a) the Trapezoidal Rule and (b) Simpson's Rule.

31. $\int_0^2 \sqrt{1 + x} \, dx$

32. $\int_0^2 (x + 1)^{2/3} \, dx$

33. $\int_0^1 \tan x^2 \, dx$

34. $\int_0^1 \sin x^2 \, dx$

35. Finding the Area of a Region Approximate the area of the shaded region using

(a) the Trapezoidal Rule with *n* = 4.

(b) Simpson's Rule with *n* = 4.

Figure for 35

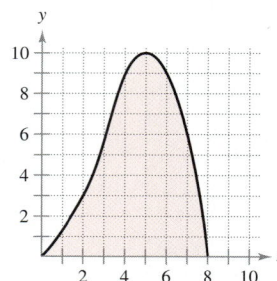
Figure for 36

36. Finding the Area of a Region Approximate the area of the shaded region using

(a) the Trapezoidal Rule with *n* = 8.

(b) Simpson's Rule with *n* = 8.

37. Area Use Simpson's Rule with *n* = 14 to approximate the area of the region bounded by the graphs of $y = \sqrt{x} \cos x$, $y = 0$, $x = 0$, and $x = \pi/2$.

38. Circumference   The **elliptic integral**

$$8\sqrt{3}\int_{0}^{\pi/2}\sqrt{1-\tfrac{2}{3}\sin^2\theta}\,d\theta$$

gives the circumference of an ellipse. Use Simpson's Rule with $n = 8$ to approximate the circumference.

39. Surveying

Use the Trapezoidal Rule to estimate the number of square meters of land, where x and y are measured in meters, as shown in the figure. The land is bounded by a stream and two straight roads that meet at right angles.

x	0	100	200	300	400	500
y	125	125	120	112	90	90

x	600	700	800	900	1000
y	95	88	75	35	0

40. **HOW DO YOU SEE IT?**   The function $f(x)$ is concave upward on the interval $[0, 2]$ and the function $g(x)$ is concave downward on the interval $[0, 2]$.

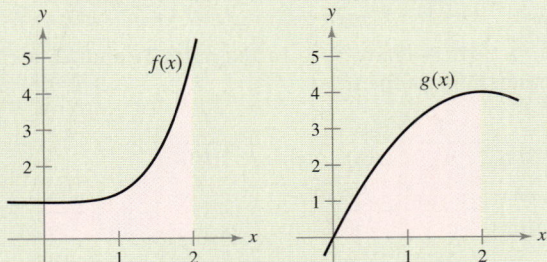

(a) Using the Trapezoidal Rule with $n = 4$, which integral would be overestimated? Which integral would be underestimated? Explain your reasoning.

(b) Which rule would you use for more accurate approximations of $\int_0^2 f(x)\,dx$ and $\int_0^2 g(x)\,dx$, the Trapezoidal Rule or Simpson's Rule? Explain your reasoning.

41. Work   To determine the size of the motor required to operate a press, a company must know the amount of work done when the press moves an object linearly 5 feet. The variable force to move the object is

$$F(x) = 100x\sqrt{125 - x^3}$$

where F is given in pounds and x gives the position of the unit in feet. Use Simpson's Rule with $n = 12$ to approximate the work W (in foot-pounds) done through one cycle when

$$W = \int_{0}^{5} F(x)\,dx.$$

42. Approximating a Function   The table lists several measurements gathered in an experiment to approximate an unknown continuous function $y = f(x)$.

x	0.00	0.25	0.50	0.75	1.00
y	4.32	4.36	4.58	5.79	6.14

x	1.25	1.50	1.75	2.00
y	7.25	7.64	8.08	8.14

(a) Approximate the integral

$$\int_{0}^{2} f(x)\,dx$$

using the Trapezoidal Rule and Simpson's Rule.

(b) Use a graphing utility to find a model of the form $y = ax^3 + bx^2 + cx + d$ for the data. Integrate the resulting polynomial over $[0, 2]$ and compare the result with the integral from part (a).

Approximation of Pi   In Exercises 43 and 44, use Simpson's Rule with $n = 6$ to approximate π using the given equation. (In Section 5.7, you will be able to evaluate the integral using inverse trigonometric functions.)

43. $\pi = \int_{0}^{1/2}\dfrac{6}{\sqrt{1-x^2}}\,dx$ **44.** $\pi = \int_{0}^{1}\dfrac{4}{1+x^2}\,dx$

45. Using Simpson's Rule   Use Simpson's Rule with $n = 10$ and a computer algebra system to approximate t in the integral equation

$$\int_{0}^{t}\sin\sqrt{x}\,dx = 2.$$

46. Proof   Prove that Simpson's Rule is exact when approximating the integral of a cubic polynomial function, and demonstrate the result with $n = 4$ for

$$\int_{0}^{1} x^3\,dx.$$

47. Proof   Prove that you can find a polynomial

$$p(x) = Ax^2 + Bx + C$$

that passes through any three points (x_1, y_1), (x_2, y_2), and (x_3, y_3), where the x_i's are distinct.

Review Exercises
See CalcChat.com for tutorial help and worked-out solutions to odd-numbered exercises.

Finding an Indefinite Integral In Exercises 1–8, find the indefinite integral.

1. $\int (x - 6)\, dx$

2. $\int (x^4 + 3)\, dx$

3. $\int (4x^2 + x + 3)\, dx$

4. $\int \frac{6}{\sqrt[3]{x}}\, dx$

5. $\int \frac{x^4 + 8}{x^3}\, dx$

6. $\int \frac{x^2 + 2x - 6}{x^4}$

7. $\int (2x - 9 \sin x)\, dx$

8. $\int (5 \cos x - 2 \sec^2 x)\, dx$

Finding a Particular Solution In Exercises 9–12, find the particular solution that satisfies the differential equation and the initial condition.

9. $f'(x) = -6x$, $f(1) = -2$

10. $f'(x) = 9x^2 + 1$, $f(0) = 7$

11. $f''(x) = 24x$, $f'(-1) = 7$, $f(1) = -4$

12. $f''(x) = 2 \cos x$, $f'(0) = 4$, $f(0) = -5$

Slope Field In Exercises 13 and 14, a differential equation, a point, and a slope field are given. (a) Sketch two approximate solutions of the differential equation on the slope field, one of which passes through the indicated point. (To print an enlarged copy of the graph, go to *MathGraphs.com*.) (b) Use integration to find the particular solution of the differential equation and use a graphing utility to graph the solution.

13. $\frac{dy}{dx} = 2x - 4$, $(4, -2)$

14. $\frac{dy}{dx} = \frac{1}{2}x^2 - 2x$, $(6, 2)$

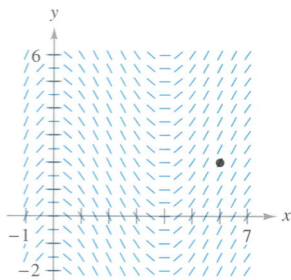

15. **Velocity and Acceleration** A ball is thrown vertically upward from ground level with an initial velocity of 96 feet per second. Use $a(t) = -32$ feet per second per second as the acceleration due to gravity. (Neglect air resistance.)

(a) How long will it take the ball to rise to its maximum height? What is the maximum height?

(b) After how many seconds is the velocity of the ball one-half the initial velocity?

(c) What is the height of the ball when its velocity is one-half the initial velocity?

16. **Velocity and Acceleration** The speed of a car traveling in a straight line is reduced from 45 to 30 miles per hour in a distance of 264 feet. Find the distance in which the car can be brought to rest from 30 miles per hour, assuming the same constant deceleration.

17. **Velocity and Acceleration** An airplane taking off from a runway travels 3600 feet before lifting off. The airplane starts from rest, moves with constant acceleration, and makes the run in 30 seconds. With what speed does it lift off?

18. **Modeling Data** The table shows the velocities (in miles per hour) of two cars on an entrance ramp to an interstate highway. The time t is in seconds.

t	v_1	v_2
0	0	0
5	2.5	21
10	7	38
15	16	51
20	29	60
25	45	64
30	65	65

(a) Rewrite the velocities in feet per second.

(b) Use the regression capabilities of a graphing utility to find quadratic models for the data in part (a).

(c) Approximate the distance traveled by each car during the 30 seconds. Explain the difference in the distances.

Finding a Sum In Exercises 19 and 20, find the sum. Use the summation capabilities of a graphing utility to verify your result.

19. $\sum_{i=1}^{5} (5i - 3)$

20. $\sum_{k=0}^{3} (k^2 + 1)$

Using Sigma Notation In Exercises 21 and 22, use sigma notation to write the sum.

21. $\frac{1}{3(1)} + \frac{1}{3(2)} + \frac{1}{3(3)} + \cdots + \frac{1}{3(10)}$

22. $\left(\frac{3}{n}\right)\left(\frac{1+1}{n}\right)^2 + \left(\frac{3}{n}\right)\left(\frac{2+1}{n}\right)^2 + \cdots + \left(\frac{3}{n}\right)\left(\frac{n+1}{n}\right)^2$

Evaluating a Sum In Exercises 23–28, use the properties of summation and Theorem 4.2 to evaluate the sum.

23. $\sum_{i=1}^{24} 8$

24. $\sum_{i=1}^{75} 5i$

25. $\sum_{i=1}^{20} 2i$

26. $\sum_{i=1}^{30} (3i - 4)$

27. $\sum_{i=1}^{20} (i + 1)^2$

28. $\sum_{i=1}^{12} i(i^2 - 1)$

Finding Upper and Lower Sums for a Region In Exercises 29 and 30, use upper and lower sums to approximate the area of the region using the given number of subintervals (of equal width.)

29. $y = \dfrac{10}{x^2 + 1}$

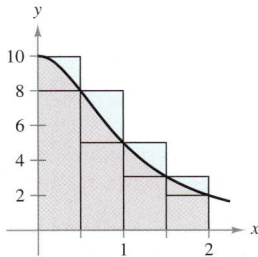

30. $y = 9 - \dfrac{1}{4}x^2$

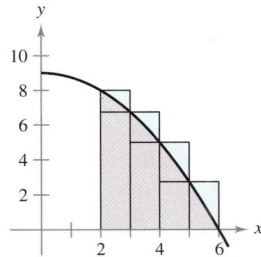

Finding Area by the Limit Definition In Exercises 31–34, use the limit process to find the area of the region bounded by the graph of the function and the *x*-axis over the given interval. Sketch the region.

31. $y = 8 - 2x,\quad [0, 3]$

32. $y = x^2 + 3,\quad [0, 2]$

33. $y = 5 - x^2,\quad [-2, 1]$

34. $y = \frac{1}{4}x^3,\quad [2, 4]$

35. Finding Area by the Limit Definition Use the limit process to find the area of the region bounded by $x = 5y - y^2$, $x = 0$, $y = 2$, and $y = 5$.

36. Upper and Lower Sums Consider the region bounded by $y = mx$, $y = 0$, $x = 0$, and $x = b$.

(a) Find the upper and lower sums to approximate the area of the region when $\Delta x = b/4$.

(b) Find the upper and lower sums to approximate the area of the region when $\Delta x = b/n$.

(c) Find the area of the region by letting n approach infinity in both sums in part (b). Show that, in each case, you obtain the formula for the area of a triangle.

Writing a Definite Integral In Exercises 37 and 38, set up a definite integral that yields the area of the region. (Do not evaluate the integral.)

37. $f(x) = 2x + 8$

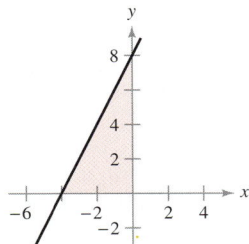

38. $f(x) = 100 - x^2$

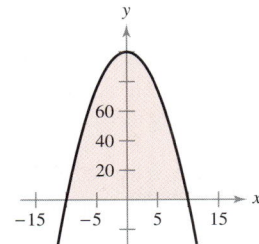

Evaluating a Definite Integral Using a Geometric Formula In Exercises 39 and 40, sketch the region whose area is given by the definite integral. Then use a geometric formula to evaluate the integral.

39. $\displaystyle \int_0^5 (5 - |x - 5|)\, dx$

40. $\displaystyle \int_{-6}^6 \sqrt{36 - x^2}\, dx$

41. Using Properties of Definite Integrals Given

$$\int_4^8 f(x)\, dx = 12 \quad \text{and} \quad \int_4^8 g(x)\, dx = 5$$

evaluate

(a) $\displaystyle \int_4^8 [f(x) + g(x)]\, dx.$

(b) $\displaystyle \int_4^8 [f(x) - g(x)]\, dx.$

(c) $\displaystyle \int_4^8 [2f(x) - 3g(x)]\, dx.$

(d) $\displaystyle \int_4^8 7f(x)\, dx.$

42. Using Properties of Definite Integrals Given

$$\int_0^3 f(x)\, dx = 4 \quad \text{and} \quad \int_3^6 f(x)\, dx = -1$$

evaluate

(a) $\displaystyle \int_0^6 f(x)\, dx.$

(b) $\displaystyle \int_6^3 f(x)\, dx.$

(c) $\displaystyle \int_4^4 f(x)\, dx.$

(d) $\displaystyle \int_3^6 -10 f(x)\, dx.$

Evaluating a Definite Integral In Exercises 43–50, use the Fundamental Theorem of Calculus to evaluate the definite integral.

43. $\displaystyle \int_0^8 (3 + x)\, dx$

44. $\displaystyle \int_2^3 (t^2 - 1)\, dt$

45. $\displaystyle \int_{-1}^1 (4t^3 - 2t)\, dt$

46. $\displaystyle \int_2^3 (x^4 + 4x - 6)\, dx$

47. $\displaystyle \int_4^9 x\sqrt{x}\, dx$

48. $\displaystyle \int_1^4 \left(\frac{1}{x^3} + x\right) dx$

49. $\displaystyle \int_0^{3\pi/4} \sin\theta\, d\theta$

50. $\displaystyle \int_{-\pi/4}^{\pi/4} \sec^2 t\, dt$

Finding the Area of a Region In Exercises 51 and 52, determine the area of the given region.

51. $y = \sin x$

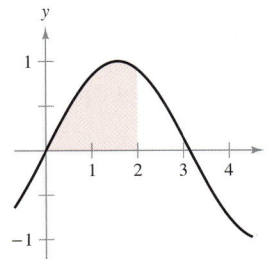

52. $y = x + \cos x$

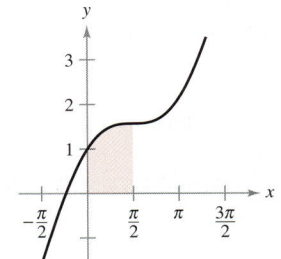

Finding the Area of a Region In Exercises 53–56, find the area of the region bounded by the graphs of the equations.

53. $y = 8 - x,\ x = 0,\ x = 6,\ y = 0$

54. $y = -x^2 + x + 6,\ y = 0$

55. $y = x - x^3,\ x = 0,\ x = 1,\ y = 0$

56. $y = \sqrt{x}(1 - x),\ y = 0$

314   Chapter 4   Integration

Finding the Average Value of a Function   **In Exercises 57 and 58, find the average value of the function over the given interval and all values of x in the interval for which the function equals its average value.**

57. $f(x) = \dfrac{1}{\sqrt{x}}, \quad [4, 9]$ **58.** $f(x) = x^3, \quad [0, 2]$

Using the Second Fundamental Theorem of Calculus **In Exercises 59–62, use the Second Fundamental Theorem of Calculus to find $F'(x)$.**

59. $F(x) = \displaystyle\int_0^x t^2 \sqrt{1 + t^3}\, dt$ **60.** $F(x) = \displaystyle\int_1^x \dfrac{1}{t^2}\, dt$

61. $F(x) = \displaystyle\int_{-3}^x (t^2 + 3t + 2)\, dt$

62. $F(x) = \displaystyle\int_0^x \csc^2 t\, dt$

Finding an Indefinite Integral   **In Exercises 63–72, find the indefinite integral.**

63. $\displaystyle\int \dfrac{x^2}{\sqrt{x^3 + 3}}\, dx$ **64.** $\displaystyle\int 6x^3 \sqrt{3x^4 + 2}\, dx$

65. $\displaystyle\int x(1 - 3x^2)^4\, dx$ **66.** $\displaystyle\int \dfrac{x + 4}{(x^2 + 8x - 7)^2}\, dx$

67. $\displaystyle\int \sin^3 x \cos x\, dx$ **68.** $\displaystyle\int x \sin 3x^2\, dx$

69. $\displaystyle\int \dfrac{\cos \theta}{\sqrt{1 - \sin \theta}}\, d\theta$

70. $\displaystyle\int \dfrac{\sin x}{\sqrt{\cos x}}\, dx$

71. $\displaystyle\int (1 + \sec \pi x)^2 \sec \pi x \tan \pi x\, dx$

72. $\displaystyle\int \sec 2x \tan 2x\, dx$

Slope Field   **In Exercises 73 and 74, a differential equation, a point, and a slope field are given. (a) Sketch two approximate solutions of the differential equation on the slope field, one of which passes through the given point. (To print an enlarged copy of the graph, go to *MathGraphs.com*.) (b) Use integration to find the particular solution of the differential equation and use a graphing utility to graph the solution.**

73. $\dfrac{dy}{dx} = x\sqrt{9 - x^2}, \quad (0, -4)$ **74.** $\dfrac{dy}{dx} = -\dfrac{1}{2}x \sin(x^2), \quad (0, 0)$

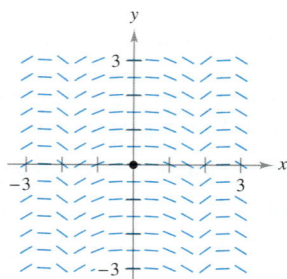

Evaluating a Definite Integral   **In Exercises 75–82, evaluate the definite integral. Use a graphing utility to verify your result.**

75. $\displaystyle\int_0^1 (3x + 1)^5\, dx$ **76.** $\displaystyle\int_0^1 x^2(x^3 - 2)^3\, dx$

77. $\displaystyle\int_0^3 \dfrac{1}{\sqrt{1 + x}}\, dx$ **78.** $\displaystyle\int_3^6 \dfrac{x}{3\sqrt{x^2 - 8}}\, dx$

79. $2\pi \displaystyle\int_0^1 (y + 1)\sqrt{1 - y}\, dy$ **80.** $2\pi \displaystyle\int_{-1}^0 x^2 \sqrt{x + 1}\, dx$

81. $\displaystyle\int_0^\pi \cos \dfrac{x}{2}\, dx$ **82.** $\displaystyle\int_{-\pi/4}^{\pi/4} \sin 2x\, dx$

Finding the Area of a Region   **In Exercises 83 and 84, find the area of the region. Use a graphing utility to verify your result.**

83. $\displaystyle\int_1^9 x \sqrt[3]{x - 1}\, dx$ **84.** $\displaystyle\int_0^{\pi/2} [\cos x + \sin(2x)]\, dx$

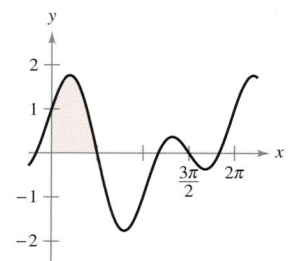

85. Using an Even Function   Use $\displaystyle\int_0^2 x^4\, dx = \dfrac{32}{5}$ to evaluate each definite integral without using the Fundamental Theorem of Calculus.

(a) $\displaystyle\int_{-2}^2 x^4\, dx$ (b) $\displaystyle\int_{-2}^0 x^4\, dx$

(c) $\displaystyle\int_0^2 3x^4\, dx$ (d) $\displaystyle\int_{-2}^0 -5x^4\, dx$

86. Respiratory Cycle   After exercising for a few minutes, a person has a respiratory cycle for which the rate of air intake is

$$v = 1.75 \sin \dfrac{\pi t}{2}.$$

Find the volume, in liters, of air inhaled during one cycle by integrating the function over the interval $[0, 2]$.

Using the Trapezoidal Rule and Simpson's Rule **In Exercises 87–90, approximate the definite integral using the Trapezoidal Rule and Simpson's Rule with $n = 4$. Compare these results with the approximation of the integral using a graphing utility.**

87. $\displaystyle\int_2^3 \dfrac{2}{1 + x^2}\, dx$ **88.** $\displaystyle\int_0^1 \dfrac{x^{3/2}}{3 - x^2}\, dx$

89. $\displaystyle\int_0^{\pi/2} \sqrt{x} \cos x\, dx$ **90.** $\displaystyle\int_0^\pi \sqrt{1 + \sin^2 x}\, dx$

P.S. Problem Solving

See **CalcChat.com** for tutorial help and worked-out solutions to odd-numbered exercises.

1. **Using a Function**   Let $L(x) = \int_1^x \frac{1}{t}\, dt,\ x > 0$.

   (a) Find $L(1)$.

   (b) Find $L'(x)$ and $L'(1)$.

   (c) Use a graphing utility to approximate the value of x (to three decimal places) for which $L(x) = 1$.

   (d) Prove that $L(x_1 x_2) = L(x_1) + L(x_2)$ for all positive values of x_1 and x_2.

2. **Parabolic Arch**   Archimedes showed that the area of a parabolic arch is equal to $\frac{2}{3}$ the product of the base and the height (see figure).

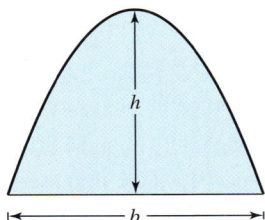

   (a) Graph the parabolic arch bounded by $y = 9 - x^2$ and the x-axis. Use an appropriate integral to find the area A.

   (b) Find the base and height of the arch and verify Archimedes' formula.

   (c) Prove Archimedes' formula for a general parabola.

Evaluating a Sum and a Limit   In Exercises 3 and 4, (a) write the area under the graph of the given function defined on the given interval as a limit. Then (b) evaluate the sum in part (a), and (c) evaluate the limit using the result of part (b).

3. $y = x^4 - 4x^3 + 4x^2,\quad [0, 2]$

$$\left(Hint:\ \sum_{i=1}^{n} i^4 = \frac{n(n+1)(2n+1)(3n^2+3n-1)}{30} \right)$$

4. $y = \frac{1}{2}x^5 + 2x^3,\quad [0, 2]$

$$\left(Hint:\ \sum_{i=1}^{n} i^5 = \frac{n^2(n+1)^2(2n^2+2n-1)}{12} \right)$$

5. **Fresnel Function**   The **Fresnel function** S is defined by the integral

$$S(x) = \int_0^x \sin\left(\frac{\pi t^2}{2}\right) dt.$$

   (a) Graph the function $y = \sin\left(\frac{\pi x^2}{2}\right)$ on the interval $[0, 3]$.

   (b) Use the graph in part (a) to sketch the graph of S on the interval $[0, 3]$.

   (c) Locate all relative extrema of S on the interval $(0, 3)$.

   (d) Locate all points of inflection of S on the interval $(0, 3)$.

6. **Approximation**   The **Two-Point Gaussian Quadrature Approximation** for f is

$$\int_{-1}^{1} f(x)\, dx \approx f\left(-\frac{1}{\sqrt{3}}\right) + f\left(\frac{1}{\sqrt{3}}\right).$$

   (a) Use this formula to approximate

$$\int_{-1}^{1} \cos x\, dx.$$

   Find the error of the approximation.

   (b) Use this formula to approximate

$$\int_{-1}^{1} \frac{1}{1 + x^2}\, dx.$$

   (c) Prove that the Two-Point Gaussian Quadrature Approximation is exact for all polynomials of degree 3 or less.

7. **Extrema and Points of Inflection**   The graph of the function f consists of the three line segments joining the points $(0, 0)$, $(2, -2)$, $(6, 2)$, and $(8, 3)$. The function F is defined by the integral

$$F(x) = \int_0^x f(t)\, dt.$$

   (a) Sketch the graph of f.

   (b) Complete the table.

x	0	1	2	3	4	5	6	7	8
$F(x)$									

   (c) Find the extrema of F on the interval $[0, 8]$.

   (d) Determine all points of inflection of F on the interval $(0, 8)$.

8. **Falling Objects**   Galileo Galilei (1564–1642) stated the following proposition concerning falling objects:

   The time in which any space is traversed by a uniformly accelerating body is equal to the time in which that same space would be traversed by the same body moving at a uniform speed whose value is the mean of the highest speed of the accelerating body and the speed just before acceleration began.

   Use the techniques of this chapter to verify this proposition.

9. **Proof**   Prove $\int_0^x f(t)(x - t)\, dt = \int_0^x \left(\int_0^t f(v)\, dv \right) dt$.

10. **Proof**   Prove $\int_a^b f(x)f'(x)\, dx = \frac{1}{2}([f(b)]^2 - [f(a)]^2)$.

11. **Riemann Sum**   Use an appropriate Riemann sum to evaluate the limit

$$\lim_{n \to \infty} \frac{\sqrt{1} + \sqrt{2} + \sqrt{3} + \cdots + \sqrt{n}}{n^{3/2}}.$$

12. Riemann Sum Use an appropriate Riemann sum to evaluate the limit

$$\lim_{n \to \infty} \frac{1^5 + 2^5 + 3^5 + \cdots + n^5}{n^6}.$$

13. Proof Suppose that f is integrable on $[a, b]$ and $0 < m \le f(x) \le M$ for all x in the interval $[a, b]$. Prove that

$$m(a - b) \le \int_a^b f(x)\, dx \le M(b - a).$$

Use this result to estimate $\int_0^1 \sqrt{1 + x^4}\, dx$.

14. Using a Continuous Function Let f be continuous on the interval $[0, b]$, where $f(x) + f(b - x) \ne 0$ on $[0, b]$.

(a) Show that

$$\int_0^b \frac{f(x)}{f(x) + f(b - x)}\, dx = \frac{b}{2}.$$

(b) Use the result in part (a) to evaluate

$$\int_0^1 \frac{\sin x}{\sin(1 - x) + \sin x}\, dx.$$

(c) Use the result in part (a) to evaluate

$$\int_0^3 \frac{\sqrt{x}}{\sqrt{x} + \sqrt{3 - x}}\, dx.$$

15. Velocity and Acceleration A car travels in a straight line for 1 hour. Its velocity v in miles per hour at six-minute intervals is shown in the table.

t (hours)	0	0.1	0.2	0.3	0.4	0.5
v (mi/h)	0	10	20	40	60	50

t (hours)	0.6	0.7	0.8	0.9	1.0
v (mi/h)	40	35	40	50	65

(a) Produce a reasonable graph of the velocity function v by graphing these points and connecting them with a smooth curve.

(b) Find the open intervals over which the acceleration a is positive.

(c) Find the average acceleration of the car (in miles per hour squared) over the interval $[0, 0.4]$.

(d) What does the integral

$$\int_0^1 v(t)\, dt$$

signify? Approximate this integral using the Trapezoidal Rule with five subintervals.

(e) Approximate the acceleration at $t = 0.8$.

16. Proof Prove that if f is a continuous function on a closed interval $[a, b]$, then

$$\left| \int_a^b f(x)\, dx \right| \le \int_a^b |f(x)|\, dx.$$

17. Verifying a Sum Verify that

$$\sum_{i=1}^{n} i^2 = \frac{n(n + 1)(2n + 1)}{6}$$

by showing the following.

(a) $(1 + i)^3 - i^3 = 3i^2 + 3i + 1$

(b) $(n + 1)^3 = \sum_{i=1}^{n} (3i^2 + 3i + 1) + 1$

(c) $\sum_{i=1}^{n} i^2 = \frac{n(n + 1)(2n + 1)}{6}$

18. Sine Integral Function The **sine integral function**

$$\mathrm{Si}(x) = \int_0^x \frac{\sin t}{t}\, dt$$

is often used in engineering. The function

$$f(t) = \frac{\sin t}{t}$$

is not defined at $t = 0$, but its limit is 1 as $t \to 0$. So, define $f(0) = 1$. Then f is continuous everywhere.

(a) Use a graphing utility to graph $\mathrm{Si}(x)$.

(b) At what values of x does $\mathrm{Si}(x)$ have relative maxima?

(c) Find the coordinates of the first inflection point where $x > 0$.

(d) Decide whether $\mathrm{Si}(x)$ has any horizontal asymptotes. If so, identify each.

19. Comparing Methods Let

$$I = \int_0^4 f(x)\, dx$$

where f is shown in the figure. Let $L(n)$ and $R(n)$ represent the Riemann sums using the left-hand endpoints and right-hand endpoints of n subintervals of equal width. (Assume n is even.) Let $T(n)$ and $S(n)$ be the corresponding values of the Trapezoidal Rule and Simpson's Rule.

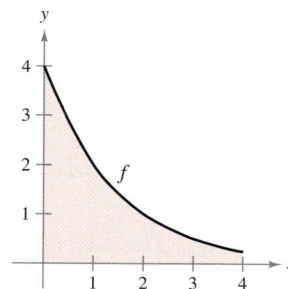

(a) For any n, list $L(n)$, $R(n)$, $T(n)$, and I in increasing order.

(b) Approximate $S(4)$.

20. Minimizing an Integral Determine the limits of integration where $a \le b$ such that

$$\int_a^b (x^2 - 16)\, dx$$

has minimal value.

5 Logarithmic, Exponential, and Other Transcendental Functions

Radioactive Half-Life Model *(Example 1, p. 356)*

St. Louis Arch
(Section Project, p. 392)

Atmospheric Pressure *(Exercise 85, p. 353)*

Heat Transfer
(Exercise 99, p. 336)

Sound Intensity *(Exercise 104, p. 327)*

317

5.1 The Natural Logarithmic Function: Differentiation

■ Develop and use properties of the natural logarithmic function.
■ Understand the definition of the number e.
■ Find derivatives of functions involving the natural logarithmic function.

The Natural Logarithmic Function

Recall that the General Power Rule

$$\int x^n \, dx = \frac{x^{n+1}}{n+1} + C, \quad n \neq -1 \qquad \text{General Power Rule}$$

has an important disclaimer—it doesn't apply when $n = -1$. Consequently, you have not yet found an antiderivative for the function $f(x) = 1/x$. In this section, you will use the Second Fundamental Theorem of Calculus to *define* such a function. This antiderivative is a function that you have not encountered previously in the text. It is neither algebraic nor trigonometric, but falls into a new class of functions called *logarithmic functions*. This particular function is the **natural logarithmic function.**

JOHN NAPIER (1550–1617)

Logarithms were invented by the Scottish mathematician John Napier. Napier coined the term *logarithm*, from the two Greek words *logos* (or ratio) and *arithmos* (or number), to describe the theory that he spent 20 years developing and that first appeared in the book *Mirifici Logarithmorum canonis descriptio* (A Description of the Marvelous Rule of Logarithms). Although he did not introduce the *natural* logarithmic function, it is sometimes called the *Napierian* logarithm.

See LarsonCalculus.com to read more of this biography.

Definition of the Natural Logarithmic Function

The **natural logarithmic function** is defined by

$$\ln x = \int_1^x \frac{1}{t} \, dt, \quad x > 0.$$

The domain of the natural logarithmic function is the set of all positive real numbers.

From this definition, you can see that $\ln x$ is positive for $x > 1$ and negative for $0 < x < 1$, as shown in Figure 5.1. Moreover, $\ln(1) = 0$, because the upper and lower limits of integration are equal when $x = 1$.

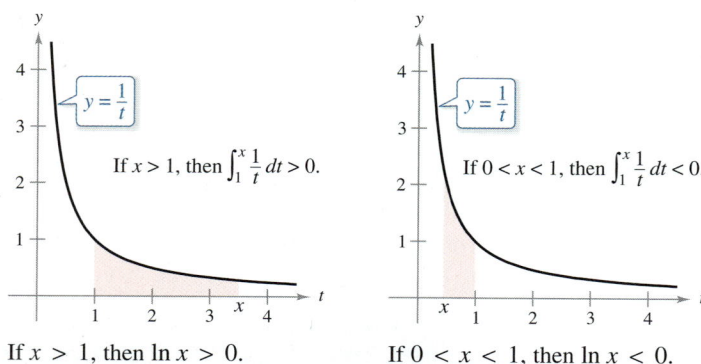

If $x > 1$, then $\int_1^x \frac{1}{t} \, dt > 0.$

If $x > 1$, then $\ln x > 0.$

If $0 < x < 1$, then $\int_1^x \frac{1}{t} \, dt < 0.$

If $0 < x < 1$, then $\ln x < 0.$

Figure 5.1

Exploration

Graphing the Natural Logarithmic Function Using *only* the definition of the natural logarithmic function, sketch a graph of the function. Explain your reasoning.

Each small line segment has a slope of $\frac{1}{x}$.

Figure 5.2

To sketch the graph of $y = \ln x$, you can think of the natural logarithmic function as an *antiderivative* given by the differential equation

$$\frac{dy}{dx} = \frac{1}{x}.$$

Figure 5.2 is a computer-generated graph, called a *slope field (or direction field)*, showing small line segments of slope $1/x$. The graph of $y = \ln x$ is the solution that passes through the point $(1, 0)$. (You will study slope fields in Section 6.1.)

The next theorem lists some basic properties of the natural logarithmic function.

THEOREM 5.1 Properties of the Natural Logarithmic Function

The natural logarithmic function has the following properties.

1. The domain is $(0, \infty)$ and the range is $(-\infty, \infty)$.
2. The function is continuous, increasing, and one-to-one.
3. The graph is concave downward.

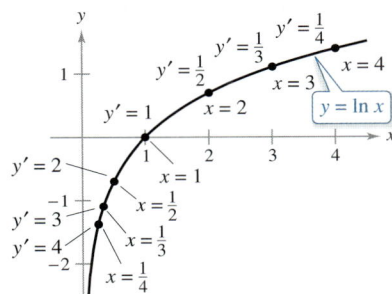

The natural logarithmic function is increasing, and its graph is concave downward.

Figure 5.3

Proof The domain of $f(x) = \ln x$ is $(0, \infty)$ by definition. Moreover, the function is continuous because it is differentiable. It is increasing because its derivative

$$f'(x) = \frac{1}{x} \qquad \text{First derivative}$$

is positive for $x > 0$, as shown in Figure 5.3. It is concave downward because

$$f''(x) = -\frac{1}{x^2} \qquad \text{Second derivative}$$

is negative for $x > 0$. The proof that f is one-to-one is given in Appendix A. The following limits imply that its range is the entire real number line.

$$\lim_{x \to 0^+} \ln x = -\infty$$

and

$$\lim_{x \to \infty} \ln x = \infty$$

Verification of these two limits is given in Appendix A.

See LarsonCalculus.com for Bruce Edwards's video of this proof.

Using the definition of the natural logarithmic function, you can prove several important properties involving operations with natural logarithms. If you are already familiar with logarithms, you will recognize that these properties are characteristic of all logarithms.

THEOREM 5.2 Logarithmic Properties

If a and b are positive numbers and n is rational, then the following properties are true.

1. $\ln(1) = 0$
2. $\ln(ab) = \ln a + \ln b$
3. $\ln(a^n) = n \ln a$
4. $\ln\left(\frac{a}{b}\right) = \ln a - \ln b$

Proof The first property has already been discussed. The proof of the second property follows from the fact that two antiderivatives of the same function differ at most by a constant. From the Second Fundamental Theorem of Calculus and the definition of the natural logarithmic function, you know that

$$\frac{d}{dx}[\ln x] = \frac{d}{dx}\left[\int_1^x \frac{1}{t}\,dt\right] = \frac{1}{x}.$$

So, consider the two derivatives

$$\frac{d}{dx}[\ln(ax)] = \frac{a}{ax} = \frac{1}{x}$$

and

$$\frac{d}{dx}[\ln a + \ln x] = 0 + \frac{1}{x} = \frac{1}{x}.$$

Because $\ln(ax)$ and $(\ln a + \ln x)$ are both antiderivatives of $1/x$, they must differ at most by a constant.

$$\ln(ax) = \ln a + \ln x + C$$

By letting $x = 1$, you can see that $C = 0$. The third property can be proved similarly by comparing the derivatives of $\ln(x^n)$ and $n \ln x$. Finally, using the second and third properties, you can prove the fourth property.

$$\ln\left(\frac{a}{b}\right) = \ln[a(b^{-1})] = \ln a + \ln(b^{-1}) = \ln a - \ln b$$

See LarsonCalculus.com for Bruce Edwards's video of this proof.

Example 1 shows how logarithmic properties can be used to expand logarithmic expressions.

EXAMPLE 1 **Expanding Logarithmic Expressions**

a. $\ln\dfrac{10}{9} = \ln 10 - \ln 9$ Property 4

b. $\ln\sqrt{3x + 2} = \ln(3x + 2)^{1/2}$ Rewrite with rational exponent.

$$= \frac{1}{2}\ln(3x + 2) \qquad \text{Property 3}$$

c. $\ln\dfrac{6x}{5} = \ln(6x) - \ln 5$ Property 4

$$= \ln 6 + \ln x - \ln 5 \qquad \text{Property 2}$$

d. $\ln\dfrac{(x^2 + 3)^2}{x\sqrt[3]{x^2 + 1}} = \ln(x^2 + 3)^2 - \ln\left(x\sqrt[3]{x^2 + 1}\right)$

$$= 2\ln(x^2 + 3) - [\ln x + \ln(x^2 + 1)^{1/3}]$$

$$= 2\ln(x^2 + 3) - \ln x - \ln(x^2 + 1)^{1/3}$$

$$= 2\ln(x^2 + 3) - \ln x - \frac{1}{3}\ln(x^2 + 1)$$

When using the properties of logarithms to rewrite logarithmic functions, you must check to see whether the domain of the rewritten function is the same as the domain of the original. For instance, the domain of $f(x) = \ln x^2$ is all real numbers except $x = 0$, and the domain of $g(x) = 2 \ln x$ is all positive real numbers. (See Figure 5.4.)

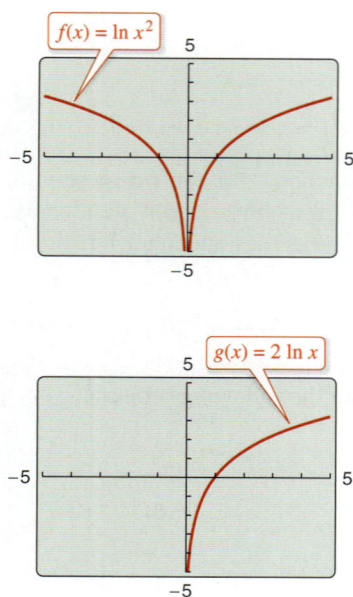

$f(x) = \ln x^2$

$g(x) = 2 \ln x$

Figure 5.4

The Number e

It is likely that you have studied logarithms in an algebra course. There, without the benefit of calculus, logarithms would have been defined in terms of a **base** number. For example, common logarithms have a base of 10 and therefore $\log_{10} 10 = 1$. (You will learn more about this in Section 5.5.)

The **base for the natural logarithm** is defined using the fact that the natural logarithmic function is continuous, is one-to-one, and has a range of $(-\infty, \infty)$. So, there must be a unique real number x such that $\ln x = 1$, as shown in Figure 5.5. This number is denoted by the letter e. It can be shown that e is irrational and has the following decimal approximation.

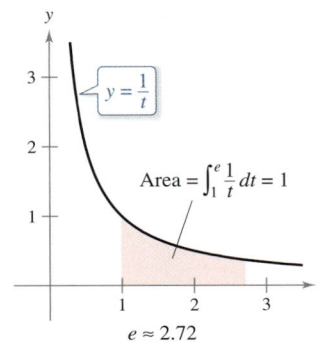

Area $= \int_1^e \frac{1}{t} dt = 1$

$e \approx 2.72$

e is the base for the natural logarithm because $\ln e = 1$.
Figure 5.5

$$e \approx 2.71828182846$$

Definition of e

The letter e denotes the positive real number such that

$$\ln e = \int_1^e \frac{1}{t} dt = 1.$$

FOR FURTHER INFORMATION   To learn more about the number e, see the article "Unexpected Occurrences of the Number e" by Harris S. Shultz and Bill Leonard in *Mathematics Magazine*. To view this article, go to *MathArticles.com*.

Once you know that $\ln e = 1$, you can use logarithmic properties to evaluate the natural logarithms of several other numbers. For example, by using the property

$$\ln(e^n) = n \ln e$$
$$= n(1)$$
$$= n$$

you can evaluate $\ln(e^n)$ for various values of n, as shown in the table and in Figure 5.6.

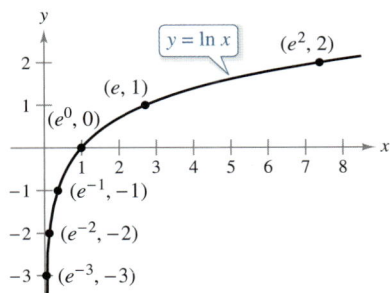

$y = \ln x$ $(e^2, 2)$

$(e, 1)$

$(e^0, 0)$

$(e^{-1}, -1)$

$(e^{-2}, -2)$

$(e^{-3}, -3)$

If $x = e^n$, then $\ln x = n$.
Figure 5.6

x	$\frac{1}{e^3} \approx 0.050$	$\frac{1}{e^2} \approx 0.135$	$\frac{1}{e} \approx 0.368$	$e^0 = 1$	$e \approx 2.718$	$e^2 \approx 7.389$
$\ln x$	-3	-2	-1	0	1	2

The logarithms shown in the table above are convenient because the x-values are integer powers of e. Most logarithmic expressions are, however, best evaluated with a calculator.

EXAMPLE 2   **Evaluating Natural Logarithmic Expressions**

a. $\ln 2 \approx 0.693$

b. $\ln 32 \approx 3.466$

c. $\ln 0.1 \approx -2.303$

The Derivative of the Natural Logarithmic Function

The derivative of the natural logarithmic function is given in Theorem 5.3. The first part of the theorem follows from the definition of the natural logarithmic function as an antiderivative. The second part of the theorem is simply the Chain Rule version of the first part.

THEOREM 5.3 Derivative of the Natural Logarithmic Function

Let u be a differentiable function of x.

1. $\dfrac{d}{dx}[\ln x] = \dfrac{1}{x}, \quad x > 0$

2. $\dfrac{d}{dx}[\ln u] = \dfrac{1}{u}\dfrac{du}{dx} = \dfrac{u'}{u}, \quad u > 0$

EXAMPLE 3 Differentiation of Logarithmic Functions

$\vdots$ $\cdots\triangleright$ *See LarsonCalculus.com for an interactive version of this type of example.*

a. $\dfrac{d}{dx}[\ln(2x)] = \dfrac{u'}{u} = \dfrac{2}{2x} = \dfrac{1}{x}$ $\qquad\qquad u = 2x$

b. $\dfrac{d}{dx}[\ln(x^2 + 1)] = \dfrac{u'}{u} = \dfrac{2x}{x^2 + 1}$ $\qquad\qquad u = x^2 + 1$

c. $\dfrac{d}{dx}[x \ln x] = x\left(\dfrac{d}{dx}[\ln x]\right) + (\ln x)\left(\dfrac{d}{dx}[x]\right)$ $\qquad$ Product Rule

$\qquad\qquad = x\left(\dfrac{1}{x}\right) + (\ln x)(1)$

$\qquad\qquad = 1 + \ln x$

d. $\dfrac{d}{dx}[(\ln x)^3] = 3(\ln x)^2 \dfrac{d}{dx}[\ln x]$ $\qquad\qquad$ Chain Rule

$\qquad\qquad = 3(\ln x)^2 \dfrac{1}{x}$

Napier used logarithmic properties to simplify *calculations* involving products, quotients, and powers. Of course, given the availability of calculators, there is now little need for this particular application of logarithms. However, there is great value in using logarithmic properties to simplify *differentiation* involving products, quotients, and powers.

EXAMPLE 4 Logarithmic Properties as Aids to Differentiation

Differentiate

$$f(x) = \ln\sqrt{x + 1}.$$

Solution Because

$$f(x) = \ln\sqrt{x + 1} = \ln(x + 1)^{1/2} = \frac{1}{2}\ln(x + 1) \qquad \text{Rewrite before differentiating.}$$

you can write

$$f'(x) = \frac{1}{2}\left(\frac{1}{x + 1}\right) = \frac{1}{2(x + 1)}. \qquad \text{Differentiate.}$$

EXAMPLE 5 **Logarithmic Properties as Aids to Differentiation**

Differentiate

$$f(x) = \ln \frac{x(x^2 + 1)^2}{\sqrt{2x^3 - 1}}.$$

Solution Because

$$f(x) = \ln \frac{x(x^2 + 1)^2}{\sqrt{2x^3 - 1}} \qquad \text{Write original function.}$$

$$= \ln x + 2\ln(x^2 + 1) - \frac{1}{2}\ln(2x^3 - 1) \qquad \text{Rewrite before differentiating.}$$

you can write

$$f'(x) = \frac{1}{x} + 2\left(\frac{2x}{x^2 + 1}\right) - \frac{1}{2}\left(\frac{6x^2}{2x^3 - 1}\right) \qquad \text{Differentiate.}$$

$$= \frac{1}{x} + \frac{4x}{x^2 + 1} - \frac{3x^2}{2x^3 - 1}. \qquad \text{Simplify.}$$

In Examples 4 and 5, be sure you see the benefit of applying logarithmic properties *before* differentiating. Consider, for instance, the difficulty of direct differentiation of the function given in Example 5.

On occasion, it is convenient to use logarithms as aids in differentiating *nonlogarithmic* functions. This procedure is called **logarithmic differentiation.**

EXAMPLE 6 **Logarithmic Differentiation**

Find the derivative of

$$y = \frac{(x - 2)^2}{\sqrt{x^2 + 1}}, \quad x \neq 2.$$

Solution Note that $y > 0$ for all $x \neq 2$. So, $\ln y$ is defined. Begin by taking the natural logarithm of each side of the equation. Then apply logarithmic properties and differentiate implicitly. Finally, solve for y'.

$$y = \frac{(x - 2)^2}{\sqrt{x^2 + 1}}, \quad x \neq 2 \qquad \text{Write original equation.}$$

$$\ln y = \ln \frac{(x - 2)^2}{\sqrt{x^2 + 1}} \qquad \text{Take natural log of each side.}$$

$$\ln y = 2\ln(x - 2) - \frac{1}{2}\ln(x^2 + 1) \qquad \text{Logarithmic properties}$$

$$\frac{y'}{y} = 2\left(\frac{1}{x - 2}\right) - \frac{1}{2}\left(\frac{2x}{x^2 + 1}\right) \qquad \text{Differentiate.}$$

$$\frac{y'}{y} = \frac{x^2 + 2x + 2}{(x - 2)(x^2 + 1)} \qquad \text{Simplify.}$$

$$y' = y\left[\frac{x^2 + 2x + 2}{(x - 2)(x^2 + 1)}\right] \qquad \text{Solve for } y'.$$

$$y' = \frac{(x - 2)^2}{\sqrt{x^2 + 1}}\left[\frac{x^2 + 2x + 2}{(x - 2)(x^2 + 1)}\right] \qquad \text{Substitute for } y.$$

$$y' = \frac{(x - 2)(x^2 + 2x + 2)}{(x^2 + 1)^{3/2}} \qquad \text{Simplify.}$$

Because the natural logarithm is undefined for negative numbers, you will often encounter expressions of the form $\ln|u|$. The next theorem states that you can differentiate functions of the form $y = \ln|u|$ as though the absolute value notation was not present.

THEOREM 5.4   Derivative Involving Absolute Value

If u is a differentiable function of x such that $u \neq 0$, then

$$\frac{d}{dx}[\ln|u|] = \frac{u'}{u}.$$

Proof If $u > 0$, then $|u| = u$, and the result follows from Theorem 5.3. If $u < 0$, then $|u| = -u$, and you have

$$\frac{d}{dx}[\ln|u|] = \frac{d}{dx}[\ln(-u)]$$

$$= \frac{-u'}{-u}$$

$$= \frac{u'}{u}.$$

See LarsonCalculus.com for Bruce Edwards's video of this proof.

EXAMPLE 7   **Derivative Involving Absolute Value**

Find the derivative of

$$f(x) = \ln|\cos x|.$$

Solution Using Theorem 5.4, let $u = \cos x$ and write

$$\frac{d}{dx}[\ln|\cos x|] = \frac{u'}{u} \qquad \frac{d}{dx}[\ln|u|] = \frac{u'}{u}$$

$$= \frac{-\sin x}{\cos x} \qquad u = \cos x$$

$$= -\tan x. \qquad \text{Simplify.}$$

EXAMPLE 8   **Finding Relative Extrema**

Locate the relative extrema of

$$y = \ln(x^2 + 2x + 3).$$

Solution Differentiating y, you obtain

$$\frac{dy}{dx} = \frac{2x + 2}{x^2 + 2x + 3}.$$

Because $dy/dx = 0$ when $x = -1$, you can apply the First Derivative Test and conclude that the point $(-1, \ln 2)$ is a relative minimum. Because there are no other critical points, it follows that this is the only relative extremum. (See Figure 5.7.)

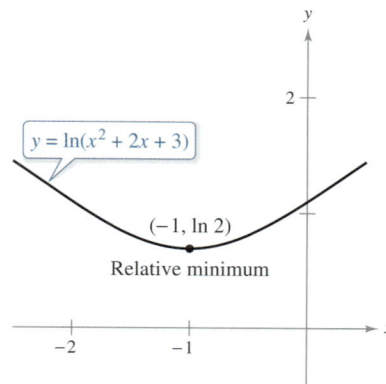

The derivative of y changes from negative to positive at $x = -1$.

Figure 5.7

Evaluating a Logarithm In Exercises 1–4, use a graphing utility to evaluate the logarithm by (a) using the natural logarithm key and (b) using the integration capabilities to evaluate the integral $\int_1^x (1/t)\, dt$.

1. $\ln 45$ 2. $\ln 8.3$
3. $\ln 0.8$ 4. $\ln 0.6$

Matching In Exercises 5–8, match the function with its graph. [The graphs are labeled (a), (b), (c), and (d).]

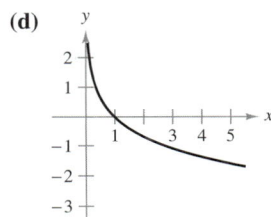

(a)

(b)

(c)

(d)

5. $f(x) = \ln x + 1$ 6. $f(x) = -\ln x$
7. $f(x) = \ln(x - 1)$ 8. $f(x) = -\ln(-x)$

Sketching a Graph In Exercises 9–16, sketch the graph of the function and state its domain.

9. $f(x) = 3 \ln x$ 10. $f(x) = -2 \ln x$
11. $f(x) = \ln 2x$ 12. $f(x) = \ln|x|$
13. $f(x) = \ln (x - 3)$ 14. $f(x) = \ln x - 4$
15. $h(x) = \ln(x + 2)$ 16. $f(x) = \ln(x - 2) + 1$

Using Properties of Logarithms In Exercises 17 and 18, use the properties of logarithms to approximate the indicated logarithms, given that $\ln 2 \approx 0.6931$ and $\ln 3 \approx 1.0986$.

17. (a) $\ln 6$ (b) $\ln \frac{2}{3}$ (c) $\ln 81$ (d) $\ln \sqrt{3}$
18. (a) $\ln 0.25$ (b) $\ln 24$ (c) $\ln \sqrt[3]{12}$ (d) $\ln \frac{1}{72}$

Expanding a Logarithmic Expression In Exercises 19–28, use the properties of logarithms to expand the logarithmic expression.

19. $\ln \dfrac{x}{4}$ 20. $\ln \sqrt{x^5}$

21. $\ln \dfrac{xy}{z}$ 22. $\ln(xyz)$

23. $\ln\left(x\sqrt{x^2 + 5}\right)$ 24. $\ln \sqrt{a - 1}$

25. $\ln \sqrt{\dfrac{x - 1}{x}}$ 26. $\ln(3e^2)$

27. $\ln z(z - 1)^2$ 28. $\ln \dfrac{1}{e}$

Condensing a Logarithmic Expression In Exercises 29–34, write the expression as a logarithm of a single quantity.

29. $\ln(x - 2) - \ln(x + 2)$
30. $3 \ln x + 2 \ln y - 4 \ln z$
31. $\frac{1}{3}[2 \ln(x + 3) + \ln x - \ln(x^2 - 1)]$
32. $2[\ln x - \ln(x + 1) - \ln(x - 1)]$
33. $2 \ln 3 - \frac{1}{2} \ln(x^2 + 1)$
34. $\frac{3}{2}[\ln(x^2 + 1) - \ln(x + 1) - \ln(x - 1)]$

Verifying Properties of Logarithms In Exercises 35 and 36, (a) verify that $f = g$ by using a graphing utility to graph f and g in the same viewing window and (b) verify that $f = g$ algebraically.

35. $f(x) = \ln \dfrac{x^2}{4}, \quad x > 0, \quad g(x) = 2 \ln x - \ln 4$

36. $f(x) = \ln \sqrt{x(x^2 + 1)}, \quad g(x) = \frac{1}{2}[\ln x + \ln(x^2 + 1)]$

Finding a Limit In Exercises 37–40, find the limit.

37. $\displaystyle \lim_{x \to 3^+} \ln(x - 3)$ 38. $\displaystyle \lim_{x \to 6^-} \ln(6 - x)$

39. $\displaystyle \lim_{x \to 2^-} \ln[x^2(3 - x)]$ 40. $\displaystyle \lim_{x \to 5^+} \ln \dfrac{x}{\sqrt{x - 4}}$

Finding a Derivative In Exercises 41–64, find the derivative of the function.

41. $f(x) = \ln(3x)$ 42. $f(x) = \ln(x - 1)$
43. $g(x) = \ln x^2$ 44. $h(x) = \ln(2x^2 + 1)$
45. $y = (\ln x)^4$ 46. $y = x^2 \ln x$
47. $y = \ln(t + 1)^2$ 48. $y = \ln \sqrt{x^2 - 4}$
49. $y = \ln\left(x\sqrt{x^2 - 1}\right)$ 50. $y = \ln[t(t^2 + 3)^3]$
51. $f(x) = \ln\left(\dfrac{x}{x^2 + 1}\right)$ 52. $f(x) = \ln\left(\dfrac{2x}{x + 3}\right)$
53. $g(t) = \dfrac{\ln t}{t^2}$ 54. $h(t) = \dfrac{\ln t}{t}$
55. $y = \ln(\ln x^2)$ 56. $y = \ln(\ln x)$
57. $y = \ln \sqrt{\dfrac{x + 1}{x - 1}}$ 58. $y = \ln \sqrt[3]{\dfrac{x - 1}{x + 1}}$
59. $f(x) = \ln\left(\dfrac{\sqrt{4 + x^2}}{x}\right)$ 60. $f(x) = \ln\left(x + \sqrt{4 + x^2}\right)$
61. $y = \ln|\sin x|$ 62. $y = \ln|\csc x|$
63. $y = \ln\left|\dfrac{\cos x}{\cos x - 1}\right|$ 64. $y = \ln|\sec x + \tan x|$

Finding an Equation of a Tangent Line In Exercises 65–72, (a) find an equation of the tangent line to the graph of f at the given point, (b) use a graphing utility to graph the function and its tangent line at the point, and (c) use the *derivative* feature of a graphing utility to confirm your results.

65. $y = \ln x^4$, $(1, 0)$

66. $y = \ln x^{3/2}$, $(1, 0)$

67. $f(x) = 3x^2 - \ln x$, $(1, 3)$

68. $f(x) = 4 - x^2 - \ln(\frac{1}{2}x + 1)$, $(0, 4)$

69. $f(x) = \ln\sqrt{1 + \sin^2 x}$, $\left(\frac{\pi}{4}, \ln\sqrt{\frac{3}{2}}\right)$

70. $f(x) = \sin 2x \ln x^2$, $(1, 0)$

71. $f(x) = x^3 \ln x$, $(1, 0)$

72. $f(x) = \frac{1}{2}x \ln x^2$, $(-1, 0)$

Finding a Derivative Implicitly In Exercises 73–76, use implicit differentiation to find dy/dx.

73. $x^2 - 3 \ln y + y^2 = 10$ **74.** $\ln xy + 5x = 30$

75. $4x^3 + \ln y^2 + 2y = 2x$ **76.** $4xy + \ln x^2 y = 7$

Differential Equation In Exercises 77 and 78, show that the function is a solution of the differential equation.

Function	Differential Equation
77. $y = 2 \ln x + 3$	$xy'' + y' = 0$
78. $y = x \ln x - 4x$	$x + y - xy' = 0$

Relative Extrema and Points of Inflection In Exercises 79–84, locate any relative extrema and points of inflection. Use a graphing utility to confirm your results.

79. $y = \dfrac{x^2}{2} - \ln x$ **80.** $y = 2x - \ln(2x)$

81. $y = x \ln x$ **82.** $y = \dfrac{\ln x}{x}$

83. $y = \dfrac{x}{\ln x}$ **84.** $y = x^2 \ln \dfrac{x}{4}$

Linear and Quadratic Approximation In Exercises 85 and 86, use a graphing utility to graph the function. Then graph

$$P_1(x) = f(1) + f'(1)(x - 1)$$

and

$$P_2(x) = f(1) + f'(1)(x - 1) + \tfrac{1}{2}f''(1)(x - 1)^2$$

in the same viewing window. Compare the values of f, P_1, P_2, and their first derivatives at $x = 1$.

85. $f(x) = \ln x$ **86.** $f(x) = x \ln x$

Using Newton's Method In Exercises 87 and 88, use Newton's Method to approximate, to three decimal places, the x-coordinate of the point of intersection of the graphs of the two equations. Use a graphing utility to verify your result.

87. $y = \ln x$, $y = -x$ **88.** $y = \ln x$, $y = 3 - x$

Logarithmic Differentiation In Exercises 89–94, use logarithmic differentiation to find dy/dx.

89. $y = x\sqrt{x^2 + 1}$, $x > 0$

90. $y = \sqrt{x^2(x + 1)(x + 2)}$, $x > 0$

91. $y = \dfrac{x^2\sqrt{3x - 2}}{(x + 1)^2}$, $x > \dfrac{2}{3}$

92. $y = \sqrt{\dfrac{x^2 - 1}{x^2 + 1}}$, $x > 1$

93. $y = \dfrac{x(x - 1)^{3/2}}{\sqrt{x + 1}}$, $x > 1$

94. $y = \dfrac{(x + 1)(x - 2)}{(x - 1)(x + 2)}$, $x > 2$

WRITING ABOUT CONCEPTS

95. Properties In your own words, state the properties of the natural logarithmic function.

96. Base Define the base for the natural logarithmic function.

97. Comparing Functions Let f be a function that is positive and differentiable on the entire real number line. Let $g(x) = \ln f(x)$.

(a) When g is increasing, must f be increasing? Explain.

(b) When the graph of f is concave upward, must the graph of g be concave upward? Explain.

---

98. HOW DO YOU SEE IT? The graph shows the temperature T (in °C) of an object h hours after it is removed from a furnace.

(a) Find $\lim\limits_{h \to \infty} T$. What does this limit represent?

(b) When is the temperature changing most rapidly?

True or False? In Exercises 99–102, determine whether the statement is true or false. If it is false, explain why or give an example that shows it is false.

99. $\ln(x + 25) = \ln x + \ln 25$

100. $\ln xy = \ln x \ln y$

101. If $y = \ln \pi$, then $y' = 1/\pi$.

102. If $y = \ln e$, then $y' = 1$.

103. Home Mortgage The term t (in years) of a $200,000 home mortgage at 7.5% interest can be approximated by

$$t = 13.375 \ln\left(\frac{x}{x - 1250}\right), \quad x > 1250$$

where x is the monthly payment in dollars.

(a) Use a graphing utility to graph the model.

(b) Use the model to approximate the term of a home mortgage for which the monthly payment is $1398.43. What is the total amount paid?

(c) Use the model to approximate the term of a home mortgage for which the monthly payment is $1611.19. What is the total amount paid?

(d) Find the instantaneous rates of change of t with respect to x when $x = \$1398.43$ and $x = \$1611.19$.

(e) Write a short paragraph describing the benefit of the higher monthly payment.

104. Sound Intensity

The relationship between the number of decibels β and the intensity of a sound I in watts per centimeter squared is

$$\beta = \frac{10}{\ln 10} \ln\left(\frac{I}{10^{-16}}\right).$$

(a) Use the properties of logarithms to write the formula in simpler form.

(b) Determine the number of decibels of a sound with an intensity of 10^{-5} watt per square centimeter.

105. Modeling Data The table shows the temperatures T (in °F) at which water boils at selected pressures p (in pounds per square inch). (*Source: Standard Handbook of Mechanical Engineers*)

p	5	10	14.696 (1 atm)	20
T	162.24°	193.21°	212.00°	227.96°

p	30	40	60	80	100
T	250.33°	267.25°	292.71°	312.03°	327.81°

A model that approximates the data is

$$T = 87.97 + 34.96 \ln p + 7.91 \sqrt{p}.$$

(a) Use a graphing utility to plot the data and graph the model.

(b) Find the rates of change of T with respect to p when $p = 10$ and $p = 70$.

(c) Use a graphing utility to graph T'. Find $\lim_{p \to \infty} T'(p)$ and interpret the result in the context of the problem.

106. Modeling Data The atmospheric pressure decreases with increasing altitude. At sea level, the average air pressure is one atmosphere (1.033227 kilograms per square centimeter). The table shows the pressures p (in atmospheres) at selected altitudes h (in kilometers).

h	0	5	10	15	20	25
p	1	0.55	0.25	0.12	0.06	0.02

(a) Use a graphing utility to find a model of the form $p = a + b \ln h$ for the data. Explain why the result is an error message.

(b) Use a graphing utility to find the logarithmic model $h = a + b \ln p$ for the data.

(c) Use a graphing utility to plot the data and graph the model.

(d) Use the model to estimate the altitude when $p = 0.75$.

(e) Use the model to estimate the pressure when $h = 13$.

(f) Use the model to find the rates of change of pressure when $h = 5$ and $h = 20$. Interpret the results.

107. Tractrix A person walking along a dock drags a boat by a 10-meter rope. The boat travels along a path known as a *tractrix* (see figure). The equation of this path is

$$y = 10 \ln\left(\frac{10 + \sqrt{100 - x^2}}{x}\right) - \sqrt{100 - x^2}.$$

(a) Use a graphing utility to graph the function.

(b) What are the slopes of this path when $x = 5$ and $x = 9$?

(c) What does the slope of the path approach as $x \to 10$?

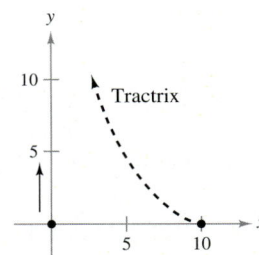

108. Prime Number Theorem There are 25 prime numbers less than 100. The **Prime Number Theorem** states that the number of primes less than x approaches

$$p(x) \approx \frac{x}{\ln x}.$$

Use this approximation to estimate the rate (in primes per 100 integers) at which the prime numbers occur when

(a) $x = 1000$.

(b) $x = 1,000,000$.

(c) $x = 1,000,000,000$.

109. Conjecture Use a graphing utility to graph f and g in the same viewing window and determine which is increasing at the greater rate for large values of x. What can you conclude about the rate of growth of the natural logarithmic function?

(a) $f(x) = \ln x$, $g(x) = \sqrt{x}$

(b) $f(x) = \ln x$, $g(x) = \sqrt[4]{x}$

Christopher Dodge/Shutterstock.com

5.2 The Natural Logarithmic Function: Integration

■ Use the Log Rule for Integration to integrate a rational function.
■ Integrate trigonometric functions.

Log Rule for Integration

The differentiation rules

$$\frac{d}{dx}[\ln|x|] = \frac{1}{x} \quad \text{and} \quad \frac{d}{dx}[\ln|u|] = \frac{u'}{u}$$

that you studied in the preceding section produce the following integration rule.

THEOREM 5.5   Log Rule for Integration

Let u be a differentiable function of x.

1. $\displaystyle\int \frac{1}{x}\, dx = \ln|x| + C$ **2.** $\displaystyle\int \frac{1}{u}\, du = \ln|u| + C$

Because $du = u'\, dx$, the second formula can also be written as

$$\int \frac{u'}{u}\, dx = \ln|u| + C. \qquad \text{Alternative form of Log Rule}$$

EXAMPLE 1   Using the Log Rule for Integration

$$\int \frac{2}{x}\, dx = 2\int \frac{1}{x}\, dx \qquad \text{Constant Multiple Rule}$$
$$= 2\ln|x| + C \qquad \text{Log Rule for Integration}$$
$$= \ln(x^2) + C \qquad \text{Property of logarithms}$$

Because x^2 cannot be negative, the absolute value notation is unnecessary in the final form of the antiderivative.

EXAMPLE 2   Using the Log Rule with a Change of Variables

Find $\displaystyle\int \frac{1}{4x - 1}\, dx$.

Solution   If you let $u = 4x - 1$, then $du = 4\, dx$.

$$\int \frac{1}{4x - 1}\, dx = \frac{1}{4}\int \left(\frac{1}{4x - 1}\right) 4\, dx \qquad \text{Multiply and divide by 4.}$$
$$= \frac{1}{4}\int \frac{1}{u}\, du \qquad \text{Substitute: } u = 4x - 1.$$
$$= \frac{1}{4}\ln|u| + C \qquad \text{Apply Log Rule.}$$
$$= \frac{1}{4}\ln|4x - 1| + C \qquad \text{Back-substitute.}$$

Exploration

Integrating Rational Functions

Early in Chapter 4, you learned rules that allowed you to integrate *any* polynomial function. The Log Rule presented in this section goes a long way toward enabling you to integrate rational functions. For instance, each of the following functions can be integrated with the Log Rule.

$\dfrac{2}{x}$	Example 1
$\dfrac{1}{4x - 1}$	Example 2
$\dfrac{x}{x^2 + 1}$	Example 3
$\dfrac{3x^2 + 1}{x^3 + x}$	Example 4(a)
$\dfrac{x + 1}{x^2 + 2x}$	Example 4(c)
$\dfrac{1}{3x + 2}$	Example 4(d)
$\dfrac{x^2 + x + 1}{x^2 + 1}$	Example 5
$\dfrac{2x}{(x + 1)^2}$	Example 6

There are still some rational functions that cannot be integrated using the Log Rule. Give examples of these functions, and explain your reasoning.

Example 3 uses the alternative form of the Log Rule. To apply this rule, look for quotients in which the numerator is the derivative of the denominator.

EXAMPLE 3   Finding Area with the Log Rule

Find the area of the region bounded by the graph of

$$y = \frac{x}{x^2 + 1}$$

the x-axis, and the line $x = 3$.

Solution   In Figure 5.8, you can see that the area of the region is given by the definite integral

$$\int_0^3 \frac{x}{x^2 + 1} \, dx.$$

If you let $u = x^2 + 1$, then $u' = 2x$. To apply the Log Rule, multiply and divide by 2 as shown.

$$\int_0^3 \frac{x}{x^2 + 1} \, dx = \frac{1}{2} \int_0^3 \frac{2x}{x^2 + 1} \, dx \qquad \text{Multiply and divide by 2.}$$

$$= \frac{1}{2} \left[\ln(x^2 + 1) \right]_0^3 \qquad \int \frac{u'}{u} \, dx = \ln|u| + C$$

$$= \frac{1}{2} (\ln 10 - \ln 1)$$

$$= \frac{1}{2} \ln 10 \qquad \ln 1 = 0$$

$$\approx 1.151$$

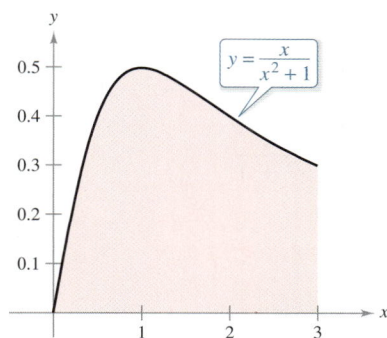

Area $= \displaystyle\int_0^3 \frac{x}{x^2 + 1} \, dx$

The area of the region bounded by the graph of y, the x-axis, and $x = 3$ is $\frac{1}{2} \ln 10$.

Figure 5.8

EXAMPLE 4   Recognizing Quotient Forms of the Log Rule

a. $\displaystyle\int \frac{3x^2 + 1}{x^3 + x} \, dx = \ln|x^3 + x| + C \qquad u = x^3 + x$

b. $\displaystyle\int \frac{\sec^2 x}{\tan x} \, dx = \ln|\tan x| + C \qquad u = \tan x$

c. $\displaystyle\int \frac{x + 1}{x^2 + 2x} \, dx = \frac{1}{2} \int \frac{2x + 2}{x^2 + 2x} \, dx \qquad u = x^2 + 2x$

$$= \frac{1}{2} \ln|x^2 + 2x| + C$$

d. $\displaystyle\int \frac{1}{3x + 2} \, dx = \frac{1}{3} \int \frac{3}{3x + 2} \, dx \qquad u = 3x + 2$

$$= \frac{1}{3} \ln|3x + 2| + C$$

With antiderivatives involving logarithms, it is easy to obtain forms that look quite different but are still equivalent. For instance, both

$$\ln\left|(3x + 2)^{1/3}\right| + C$$

and

$$\ln|3x + 2|^{1/3} + C$$

are equivalent to the antiderivative listed in Example 4(d).

Integrals to which the Log Rule can be applied often appear in disguised form. For instance, when a rational function has a *numerator of degree greater than or equal to that of the denominator*, division may reveal a form to which you can apply the Log Rule. This is shown in Example 5.

EXAMPLE 5 **Using Long Division Before Integrating**

⋯▷ *See LarsonCalculus.com for an interactive version of this type of example.*

Find the indefinite integral.

$$\int \frac{x^2 + x + 1}{x^2 + 1}\, dx$$

Solution Begin by using long division to rewrite the integrand.

$$\frac{x^2 + x + 1}{x^2 + 1} \implies x^2 + 1 \overline{\smash{)}\, x^2 + x + 1} \implies 1 + \frac{x}{x^2 + 1}$$
$$\underline{x^2 \qquad + 1}$$
$$x$$

Now, you can integrate to obtain

$$\int \frac{x^2 + x + 1}{x^2 + 1}\, dx = \int \left(1 + \frac{x}{x^2 + 1}\right) dx \qquad \text{Rewrite using long division.}$$

$$= \int dx + \frac{1}{2}\int \frac{2x}{x^2 + 1}\, dx \qquad \text{Rewrite as two integrals.}$$

$$= x + \frac{1}{2}\ln(x^2 + 1) + C. \qquad \text{Integrate.}$$

Check this result by differentiating to obtain the original integrand. ■

The next example presents another instance in which the use of the Log Rule is disguised. In this case, a change of variables helps you recognize the Log Rule.

EXAMPLE 6 **Change of Variables with the Log Rule**

Find the indefinite integral.

$$\int \frac{2x}{(x + 1)^2}\, dx$$

Solution If you let $u = x + 1$, then $du = dx$ and $x = u - 1$.

$$\int \frac{2x}{(x + 1)^2}\, dx = \int \frac{2(u - 1)}{u^2}\, du \qquad \text{Substitute.}$$

$$= 2\int \left(\frac{u}{u^2} - \frac{1}{u^2}\right) du \qquad \text{Rewrite as two fractions.}$$

$$= 2\int \frac{du}{u} - 2\int u^{-2}\, du \qquad \text{Rewrite as two integrals.}$$

$$= 2 \ln|u| - 2\left(\frac{u^{-1}}{-1}\right) + C \qquad \text{Integrate.}$$

$$= 2 \ln|u| + \frac{2}{u} + C \qquad \text{Simplify.}$$

$$= 2 \ln|x + 1| + \frac{2}{x + 1} + C \qquad \text{Back-substitute.}$$

▷ **TECHNOLOGY** If you have access to a computer algebra system, use it to find the indefinite integrals in Examples 5 and 6. How does the form of the antiderivative that it gives you compare with that given in Examples 5 and 6?

Check this result by differentiating to obtain the original integrand. ■

As you study the methods shown in Examples 5 and 6, be aware that both methods involve rewriting a disguised integrand so that it fits one or more of the basic integration formulas. Throughout the remaining sections of Chapter 5 and in Chapter 8, much time will be devoted to integration techniques. To master these techniques, you must recognize the "form-fitting" nature of integration. In this sense, integration is not nearly as straightforward as differentiation. Differentiation takes the form

"Here is the question; what is the answer?"

Integration is more like

"Here is the answer; what is the question?"

Here are some guidelines you can use for integration.

GUIDELINES FOR INTEGRATION

1. Learn a basic list of integration formulas. (Including those given in this section, you now have 12 formulas: the Power Rule, the Log Rule, and 10 trigonometric rules. By the end of Section 5.7, this list will have expanded to 20 basic rules.)

2. Find an integration formula that resembles all or part of the integrand, and, by trial and error, find a choice of u that will make the integrand conform to the formula.

3. When you cannot find a u-substitution that works, try altering the integrand. You might try a trigonometric identity, multiplication and division by the same quantity, addition and subtraction of the same quantity, or long division. Be creative.

4. If you have access to computer software that will find antiderivatives symbolically, use it.

EXAMPLE 7   ***u*-Substitution and the Log Rule**

Solve the differential equation $\dfrac{dy}{dx} = \dfrac{1}{x \ln x}$.

Solution   The solution can be written as an indefinite integral.

$$y = \int \frac{1}{x \ln x}\, dx$$

Because the integrand is a quotient whose denominator is raised to the first power, you should try the Log Rule. There are three basic choices for u. The choices

$$u = x \quad \text{and} \quad u = x \ln x$$

fail to fit the u'/u form of the Log Rule. However, the third choice does fit. Letting $u = \ln x$ produces $u' = 1/x$, and you obtain the following.

$$\int \frac{1}{x \ln x}\, dx = \int \frac{1/x}{\ln x}\, dx \qquad \text{Divide numerator and denominator by } x.$$

$$= \int \frac{u'}{u}\, dx \qquad \text{Substitute: } u = \ln x.$$

$$= \ln|u| + C \qquad \text{Apply Log Rule.}$$

$$= \ln|\ln x| + C \qquad \text{Back-substitute.}$$

So, the solution is $y = \ln|\ln x| + C$.

• • REMARK   Keep in mind that you can check your answer to an integration problem by differentiating the answer. For instance, in Example 7, the derivative of $y = \ln|\ln x| + C$ is $y' = 1/(x \ln x)$.

Integrals of Trigonometric Functions

In Section 4.1, you looked at six trigonometric integration rules—the six that correspond directly to differentiation rules. With the Log Rule, you can now complete the set of basic trigonometric integration formulas.

EXAMPLE 8 Using a Trigonometric Identity

Find $\int \tan x \, dx$.

Solution This integral does not seem to fit any formulas on our basic list. However, by using a trigonometric identity, you obtain

$$\int \tan x \, dx = \int \frac{\sin x}{\cos x} \, dx.$$

Knowing that $D_x[\cos x] = -\sin x$, you can let $u = \cos x$ and write

$$\int \tan x \, dx = -\int \frac{-\sin x}{\cos x} \, dx \qquad \text{Apply trigonometric identity and multiply and divide by } -1.$$

$$= -\int \frac{u'}{u} \, dx \qquad \text{Substitute: } u = \cos x.$$

$$= -\ln|u| + C \qquad \text{Apply Log Rule.}$$

$$= -\ln|\cos x| + C. \qquad \text{Back-substitute.}$$

Example 8 uses a trigonometric identity to derive an integration rule for the tangent function. The next example takes a rather unusual step (multiplying and dividing by the same quantity) to derive an integration rule for the secant function.

EXAMPLE 9 Derivation of the Secant Formula

Find $\int \sec x \, dx$.

Solution Consider the following procedure.

$$\int \sec x \, dx = \int \sec x \left(\frac{\sec x + \tan x}{\sec x + \tan x} \right) dx$$

$$= \int \frac{\sec^2 x + \sec x \tan x}{\sec x + \tan x} \, dx$$

Letting u be the denominator of this quotient produces

$$u = \sec x + \tan x$$

and

$$u' = \sec x \tan x + \sec^2 x.$$

So, you can conclude that

$$\int \sec x \, dx = \int \frac{\sec^2 x + \sec x \tan x}{\sec x + \tan x} \, dx \qquad \text{Rewrite integrand.}$$

$$= \int \frac{u'}{u} \, dx \qquad \text{Substitute: } u = \sec x + \tan x.$$

$$= \ln|u| + C \qquad \text{Apply Log Rule.}$$

$$= \ln|\sec x + \tan x| + C. \qquad \text{Back-substitute.}$$

With the results of Examples 8 and 9, you now have integration formulas for $\sin x$, $\cos x$, $\tan x$, and $\sec x$. The integrals of the six basic trigonometric functions are summarized below. (For proofs of $\cot u$ and $\csc u$, see Exercises 87 and 88.)

> **INTEGRALS OF THE SIX BASIC TRIGONOMETRIC FUNCTIONS**
>
> $$\int \sin u \, du = -\cos u + C \qquad \int \cos u \, du = \sin u + C$$
>
> $$\int \tan u \, du = -\ln|\cos u| + C \qquad \int \cot u \, du = \ln|\sin u| + C$$
>
> $$\int \sec u \, du = \ln|\sec u + \tan u| + C \quad \int \csc u \, du = -\ln|\csc u + \cot u| + C$$

• • REMARK Using trigonometric identities and properties of logarithms, you could rewrite these six integration rules in other forms. For instance, you could write

$$\int \csc u \, du$$
$$= \ln|\csc u - \cot u| + C.$$

(See Exercises 89–92.)

EXAMPLE 10 Integrating Trigonometric Functions

Evaluate $\displaystyle\int_0^{\pi/4} \sqrt{1 + \tan^2 x} \, dx$.

Solution Using $1 + \tan^2 x = \sec^2 x$, you can write

$$\int_0^{\pi/4} \sqrt{1 + \tan^2 x} \, dx = \int_0^{\pi/4} \sqrt{\sec^2 x} \, dx$$

$$= \int_0^{\pi/4} \sec x \, dx \qquad \sec x \geq 0 \text{ for } 0 \leq x \leq \frac{\pi}{4}.$$

$$= \ln|\sec x + \tan x| \Big]_0^{\pi/4}$$

$$= \ln(\sqrt{2} + 1) - \ln 1$$

$$\approx 0.881.$$

EXAMPLE 11 Finding an Average Value

Find the average value of

$$f(x) = \tan x$$

on the interval $[0, \pi/4]$.

Solution

$$\text{Average value} = \frac{1}{(\pi/4) - 0} \int_0^{\pi/4} \tan x \, dx \qquad {\color{red}\text{Average value} = \frac{1}{b-a}\int_a^b f(x)\,dx}$$

$$= \frac{4}{\pi} \int_0^{\pi/4} \tan x \, dx \qquad {\color{red}\text{Simplify.}}$$

$$= \frac{4}{\pi} \Big[-\ln|\cos x| \Big]_0^{\pi/4} \qquad {\color{red}\text{Integrate.}}$$

$$= -\frac{4}{\pi} \left[\ln\left(\frac{\sqrt{2}}{2}\right) - \ln(1) \right]$$

$$= -\frac{4}{\pi} \ln\left(\frac{\sqrt{2}}{2}\right)$$

$$\approx 0.441$$

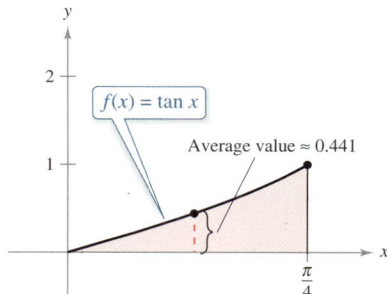

Figure 5.9

The average value is about 0.441, as shown in Figure 5.9.

Finding an Indefinite Integral In Exercises 1–26, find the indefinite integral.

1. $\displaystyle\int \frac{5}{x}\,dx$

2. $\displaystyle\int \frac{10}{x}\,dx$

3. $\displaystyle\int \frac{1}{x+1}\,dx$

4. $\displaystyle\int \frac{1}{x-5}\,dx$

5. $\displaystyle\int \frac{1}{2x+5}\,dx$

6. $\displaystyle\int \frac{9}{5-4x}\,dx$

7. $\displaystyle\int \frac{x}{x^2-3}\,dx$

8. $\displaystyle\int \frac{x^2}{5-x^3}\,dx$

9. $\displaystyle\int \frac{4x^3+3}{x^4+3x}\,dx$

10. $\displaystyle\int \frac{x^2-2x}{x^3-3x^2}\,dx$

11. $\displaystyle\int \frac{x^2-4}{x}\,dx$

12. $\displaystyle\int \frac{x^3-8x}{x^2}\,dx$

13. $\displaystyle\int \frac{x^2+2x+3}{x^3+3x^2+9x}\,dx$

14. $\displaystyle\int \frac{x^2+4x}{x^3+6x^2+5}\,dx$

15. $\displaystyle\int \frac{x^2-3x+2}{x+1}\,dx$

16. $\displaystyle\int \frac{2x^2+7x-3}{x-2}\,dx$

17. $\displaystyle\int \frac{x^3-3x^2+5}{x-3}\,dx$

18. $\displaystyle\int \frac{x^3-6x-20}{x+5}\,dx$

19. $\displaystyle\int \frac{x^4+x-4}{x^2+2}\,dx$

20. $\displaystyle\int \frac{x^3-4x^2-4x+20}{x^2-5}\,dx$

21. $\displaystyle\int \frac{(\ln x)^2}{x}\,dx$

22. $\displaystyle\int \frac{1}{x\ln x^3}\,dx$

23. $\displaystyle\int \frac{1}{\sqrt{x}(1-3\sqrt{x})}\,dx$

24. $\displaystyle\int \frac{1}{x^{2/3}(1+x^{1/3})}\,dx$

25. $\displaystyle\int \frac{2x}{(x-1)^2}\,dx$

26. $\displaystyle\int \frac{x(x-2)}{(x-1)^3}\,dx$

Finding an Indefinite Integral by u-Substitution In Exercises 27–30, find the indefinite integral by u-substitution. (*Hint:* Let u be the denominator of the integrand.)

27. $\displaystyle\int \frac{1}{1+\sqrt{2x}}\,dx$

28. $\displaystyle\int \frac{1}{1+\sqrt{3x}}\,dx$

29. $\displaystyle\int \frac{\sqrt{x}}{\sqrt{x}-3}\,dx$

30. $\displaystyle\int \frac{\sqrt[3]{x}}{\sqrt[3]{x}-1}\,dx$

Finding an Indefinite Integral of a Trigonometric Function In Exercises 31–40, find the indefinite integral.

31. $\displaystyle\int \cot\frac{\theta}{3}\,d\theta$

32. $\displaystyle\int \tan 5\theta\,d\theta$

33. $\displaystyle\int \csc 2x\,dx$

34. $\displaystyle\int \sec\frac{x}{2}\,dx$

35. $\displaystyle\int (\cos 3\theta - 1)\,d\theta$

36. $\displaystyle\int \left(2 - \tan\frac{\theta}{4}\right)d\theta$

37. $\displaystyle\int \frac{\cos t}{1+\sin t}\,dt$

38. $\displaystyle\int \frac{\csc^2 t}{\cot t}\,dt$

39. $\displaystyle\int \frac{\sec x\tan x}{\sec x - 1}\,dx$

40. $\displaystyle\int (\sec 2x + \tan 2x)\,dx$

Differential Equation In Exercises 41–44, solve the differential equation. Use a graphing utility to graph three solutions, one of which passes through the given point.

41. $\dfrac{dy}{dx} = \dfrac{3}{2-x}$, $\quad(1,0)$

42. $\dfrac{dy}{dx} = \dfrac{x-2}{x}$, $\quad(-1,0)$

43. $\dfrac{dy}{dx} = \dfrac{2x}{x^2-9}$, $\quad(0,4)$

44. $\dfrac{dr}{dt} = \dfrac{\sec^2 t}{\tan t + 1}$, $\quad(\pi,4)$

Finding a Particular Solution In Exercises 45 and 46, find the particular solution that satisfies the differential equation and the initial equations.

45. $f''(x) = \dfrac{2}{x^2}, f'(1) = 1, f(1) = 1, x > 0$

46. $f''(x) = -\dfrac{4}{(x-1)^2} - 2, f'(2) = 0, f(2) = 3, x > 1$

Slope Field In Exercises 47 and 48, a differential equation, a point, and a slope field are given. (a) Sketch two approximate solutions of the differential equation on the slope field, one of which passes through the given point. (b) Use integration to find the particular solution of the differential equation and use a graphing utility to graph the solution. Compare the result with the sketches in part (a). To print an enlarged copy of the graph, go to *MathGraphs.com*.

47. $\dfrac{dy}{dx} = \dfrac{1}{x+2}$, $\quad(0,1)$

48. $\dfrac{dy}{dx} = \dfrac{\ln x}{x}$, $\quad(1,-2)$

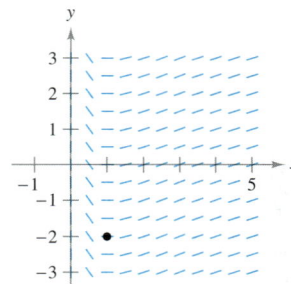

Evaluating a Definite Integral In Exercises 49–56, evaluate the definite integral. Use a graphing utility to verify your result.

49. $\displaystyle\int_0^4 \frac{5}{3x+1}\,dx$

50. $\displaystyle\int_{-1}^1 \frac{1}{2x+3}\,dx$

51. $\displaystyle\int_1^e \frac{(1 + \ln x)^2}{x}\,dx$

52. $\displaystyle\int_e^{e^2} \frac{1}{x \ln x}\,dx$

53. $\displaystyle\int_0^2 \frac{x^2 - 2}{x + 1}\,dx$

54. $\displaystyle\int_0^1 \frac{x - 1}{x + 1}\,dx$

55. $\displaystyle\int_1^2 \frac{1 - \cos\theta}{\theta - \sin\theta}\,d\theta$

56. $\displaystyle\int_{\pi/8}^{\pi/4} (\csc 2\theta - \cot 2\theta)\,d\theta$

Using Technology to Find an Integral In Exercises 57–62, use a computer algebra system to find or evaluate the integral.

57. $\displaystyle\int \frac{1}{1 + \sqrt{x}}\,dx$

58. $\displaystyle\int \frac{1 - \sqrt{x}}{1 + \sqrt{x}}\,dx$

59. $\displaystyle\int \frac{\sqrt{x}}{x - 1}\,dx$

60. $\displaystyle\int \frac{x^2}{x - 1}\,dx$

61. $\displaystyle\int_{\pi/4}^{\pi/2} (\csc x - \sin x)\,dx$

62. $\displaystyle\int_{-\pi/4}^{\pi/4} \frac{\sin^2 x - \cos^2 x}{\cos x}\,dx$

Finding a Derivative In Exercises 63–66, find $F'(x)$.

63. $\displaystyle F(x) = \int_1^x \frac{1}{t}\,dt$

64. $\displaystyle F(x) = \int_0^x \tan t\,dt$

65. $\displaystyle F(x) = \int_1^{3x} \frac{1}{t}\,dt$

66. $\displaystyle F(x) = \int_1^{x^2} \frac{1}{t}\,dt$

Area In Exercises 67–70, find the area of the given region. Use a graphing utility to verify your result.

67. $y = \dfrac{6}{x}$

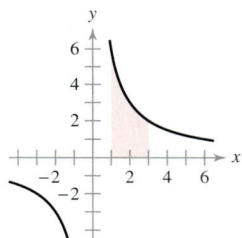

68. $y = \dfrac{2}{x \ln x}$

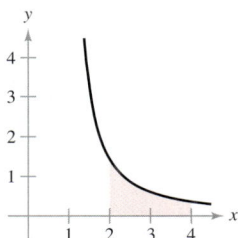

69. $y = \tan x$

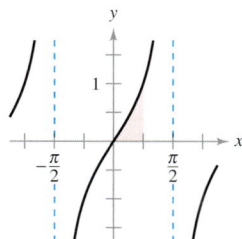

70. $y = \dfrac{\sin x}{1 + \cos x}$

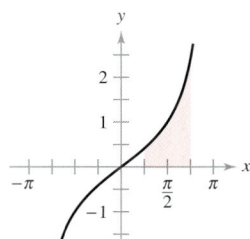

Area In Exercises 71–74, find the area of the region bounded by the graphs of the equations. Use a graphing utility to verify your result.

71. $y = \dfrac{x^2 + 4}{x}$, $\quad x = 1$, $\quad x = 4$, $\quad y = 0$

72. $y = \dfrac{5x}{x^2 + 2}$, $\quad x = 1$, $\quad x = 5$, $\quad y = 0$

73. $y = 2\sec\dfrac{\pi x}{6}$, $\quad x = 0$, $\quad x = 2$, $\quad y = 0$

74. $y = 2x - \tan 0.3x$, $\quad x = 1$, $\quad x = 4$, $\quad y = 0$

Numerical Integration In Exercises 75–78, use the Trapezoidal Rule and Simpson's Rule to approximate the value of the definite integral. Let $n = 4$ and round your answer to four decimal places. Use a graphing utility to verify your result.

75. $\displaystyle\int_1^5 \frac{12}{x}\,dx$

76. $\displaystyle\int_0^4 \frac{8x}{x^2 + 4}\,dx$

77. $\displaystyle\int_2^6 \ln x\,dx$

78. $\displaystyle\int_{-\pi/3}^{\pi/3} \sec x\,dx$

WRITING ABOUT CONCEPTS

Choosing a Formula In Exercises 79–82, state the integration formula you would use to perform the integration. Do not integrate.

79. $\displaystyle\int \sqrt[3]{x}\,dx$

80. $\displaystyle\int \frac{x}{(x^2 + 4)^3}\,dx$

81. $\displaystyle\int \frac{x}{x^2 + 4}\,dx$

82. $\displaystyle\int \frac{\sec^2 x}{\tan x}\,dx$

Approximation In Exercises 83 and 84, determine which value best approximates the area of the region between the x-axis and the graph of the function over the given interval. (Make your selection on the basis of a sketch of the region, not by performing any calculations.)

83. $f(x) = \sec x$, $\quad [0, 1]$

(a) 6   (b) -6 (c) $\frac{1}{2}$ (d) 1.25   (e) 3

84. $f(x) = \dfrac{2x}{x^2 + 1}$, $\quad [0, 4]$

(a) 3   (b) 7   (c) -2 (d) 5   (e) 1

---

85. Finding a Value Find a value of x such that

$$\int_1^x \frac{3}{t}\,dt = \int_{1/4}^x \frac{1}{t}\,dt.$$

86. Finding a Value Find a value of x such that

$$\int_1^x \frac{1}{t}\,dt$$

is equal to (a) ln 5 and (b) 1.

87. Proof Prove that

$$\int \cot u\,du = \ln|\sin u| + C.$$

88. Proof Prove that

$$\int \csc u\,du = -\ln|\csc u + \cot u| + C.$$

Using Properties of Logarithms and Trigonometric Identities In Exercises 89–92, show that the two formulas are equivalent.

89. $\int \tan x \, dx = -\ln|\cos x| + C$

$\int \tan x \, dx = \ln|\sec x| + C$

90. $\int \cot x \, dx = \ln|\sin x| + C$

$\int \cot x \, dx = -\ln|\csc x| + C$

91. $\int \sec x \, dx = \ln|\sec x + \tan x| + C$

$\int \sec x \, dx = -\ln|\sec x - \tan x| + C$

92. $\int \csc x \, dx = -\ln|\csc x + \cot x| + C$

$\int \csc x \, dx = \ln|\csc x - \cot x| + C$

Finding the Average Value of a Function In Exercises 93–96, find the average value of the function over the given interval.

93. $f(x) = \dfrac{8}{x^2}, \quad [2, 4]$ **94.** $f(x) = \dfrac{4(x+1)}{x^2}, \quad [2, 4]$

95. $f(x) = \dfrac{2 \ln x}{x}, \quad [1, e]$

96. $f(x) = \sec \dfrac{\pi x}{6}, \quad [0, 2]$

97. Population Growth A population of bacteria P is changing at a rate of

$$\frac{dP}{dt} = \frac{3000}{1 + 0.25t}$$

where t is the time in days. The initial population (when $t = 0$) is 1000. Write an equation that gives the population at any time t. Then find the population when $t = 3$ days.

98. Sales The rate of change in sales S is inversely proportional to time t ($t > 1$), measured in weeks. Find S as a function of t when the sales after 2 and 4 weeks are 200 units and 300 units, respectively.

99. Heat Transfer

Find the time required for an object to cool from 300°F to 250°F by evaluating

$$t = \frac{10}{\ln 2} \int_{250}^{300} \frac{1}{T - 100} \, dT$$

where t is time in minutes.

100. Average Price The demand equation for a product is

$$p = \frac{90{,}000}{400 + 3x}$$

where p is the price (in dollars) and x is the number of units (in thousands). Find the average price p on the interval $40 \le x \le 50$.

101. Area and Slope Graph the function

$$f(x) = \frac{x}{1 + x^2}$$

on the interval $[0, \infty)$.

(a) Find the area bounded by the graph of f and the line $y = \frac{1}{2}x$.

(b) Determine the values of the slope m such that the line $y = mx$ and the graph of f enclose a finite region.

(c) Calculate the area of this region as a function of m.

102. **HOW DO YOU SEE IT?** Use the graph of f' shown in the figure to answer the following.

(a) Approximate the slope of f at $x = -1$. Explain.

(b) Approximate any open intervals in which the graph of f is increasing and any open intervals in which it is decreasing. Explain.

True or False? In Exercises 103–106, determine whether the statement is true or false. If it is false, explain why or give an example that shows it is false.

103. $(\ln x)^{1/2} = \frac{1}{2} \ln x$

104. $\int \ln x \, dx = (1/x) + C$

105. $\displaystyle\int \frac{1}{x} \, dx = \ln|cx|, \quad c \ne 0$

106. $\displaystyle\int_{-1}^{2} \frac{1}{x} \, dx = \left[\ln|x| \right]_{-1}^{2} = \ln 2 - \ln 1 = \ln 2$

107. Napier's Inequality For $0 < x < y$, show that

$$\frac{1}{y} < \frac{\ln y - \ln x}{y - x} < \frac{1}{x}.$$

108. Proof Prove that the function

$$F(x) = \int_{x}^{2x} \frac{1}{t} \, dt$$

is constant on the interval $(0, \infty)$.

5.3   Inverse Functions

■ Verify that one function is the inverse function of another function.
■ Determine whether a function has an inverse function.
■ Find the derivative of an inverse function.

Inverse Functions

Recall from Section P.3 that a function can be represented by a set of ordered pairs. For instance, the function $f(x) = x + 3$ from $A = \{1, 2, 3, 4\}$ to $B = \{4, 5, 6, 7\}$ can be written as

$$f: \{(1, 4), (2, 5), (3, 6), (4, 7)\}.$$

By interchanging the first and second coordinates of each ordered pair, you can form the **inverse function** of f. This function is denoted by f^{-1}. It is a function from B to A, and can be written as

$$f^{-1}: \{(4, 1), (5, 2), (6, 3), (7, 4)\}.$$

Note that the domain of f is equal to the range of f^{-1}, and vice versa, as shown in Figure 5.10. The functions f and f^{-1} have the effect of "undoing" each other. That is, when you form the composition of f with f^{-1} or the composition of f^{-1} with f, you obtain the identity function.

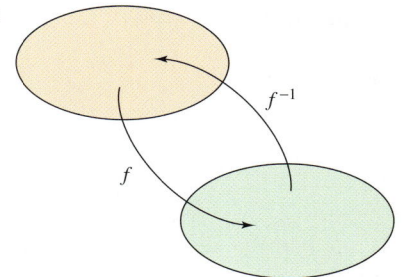

$$f(f^{-1}(x)) = x \quad \text{and} \quad f^{-1}(f(x)) = x$$

Domain of f = range of f^{-1}
Domain of f^{-1} = range of f
Figure 5.10

> **• •REMARK**   Although the notation used to denote an inverse function resembles *exponential notation*, it is a different use of -1 as a superscript. That is, in general,
>
> $$f^{-1}(x) \neq \frac{1}{f(x)}.$$

Exploration

Finding Inverse Functions
Explain how to "undo" each of the functions below. Then use your explanation to write the inverse function of f.

a. $f(x) = x - 5$
b. $f(x) = 6x$
c. $f(x) = \dfrac{x}{2}$
d. $f(x) = 3x + 2$
e. $f(x) = x^3$
f. $f(x) = 4(x - 2)$

Use a graphing utility to graph each function and its inverse function in the same "square" viewing window. What observation can you make about each pair of graphs?

Definition of Inverse Function

A function g is the **inverse function** of the function f when

$$f(g(x)) = x \text{ for each } x \text{ in the domain of } g$$

and

$$g(f(x)) = x \text{ for each } x \text{ in the domain of } f.$$

The function g is denoted by f^{-1} (read "f inverse").

Here are some important observations about inverse functions.

1. If g is the inverse function of f, then f is the inverse function of g.
2. The domain of f^{-1} is equal to the range of f, and the range of f^{-1} is equal to the domain of f.
3. A function need not have an inverse function, but when it does, the inverse function is unique (see Exercise 96).

You can think of f^{-1} as undoing what has been done by f. For example, subtraction can be used to undo addition, and division can be used to undo multiplication. So,

$$f(x) = x + c \quad \text{and} \quad f^{-1}(x) = x - c \qquad \text{Subtraction can be used to undo addition.}$$

are inverse functions of each other and

$$f(x) = cx \quad \text{and} \quad f^{-1}(x) = \frac{x}{c}, \quad c \neq 0 \qquad \text{Division can be used to undo multiplication.}$$

are inverse functions of each other.

EXAMPLE 1 **Verifying Inverse Functions**

Show that the functions are inverse functions of each other.

$$f(x) = 2x^3 - 1 \quad \text{and} \quad g(x) = \sqrt[3]{\frac{x+1}{2}}$$

••**REMARK** In Example 1, try comparing the functions f and g verbally.

  For f: First cube x, then multiply by 2, then subtract 1.

  For g: First add 1, then divide by 2, then take the cube root.

Do you see the "undoing pattern"?

Solution Because the domains and ranges of both f and g consist of all real numbers, you can conclude that both composite functions exist for all x. The composition of f with g is

$$f(g(x)) = 2\left(\sqrt[3]{\frac{x+1}{2}}\right)^3 - 1$$

$$= 2\left(\frac{x+1}{2}\right) - 1$$

$$= x + 1 - 1$$

$$= x.$$

The composition of g with f is

$$g(f(x)) = \sqrt[3]{\frac{(2x^3 - 1) + 1}{2}}$$

$$= \sqrt[3]{\frac{2x^3}{2}}$$

$$= \sqrt[3]{x^3}$$

$$= x.$$

Because $f(g(x)) = x$ and $g(f(x)) = x$, you can conclude that f and g are inverse functions of each other (see Figure 5.11).

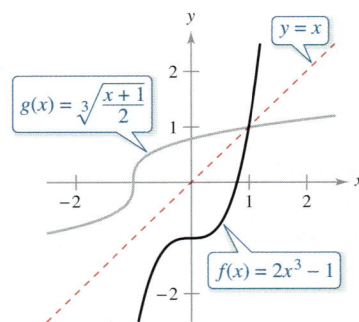

f and g are inverse functions of each other.

Figure 5.11

In Figure 5.11, the graphs of f and $g = f^{-1}$ appear to be mirror images of each other with respect to the line $y = x$. The graph of f^{-1} is a **reflection** of the graph of f in the line $y = x$. This idea is generalized in the next theorem.

THEOREM 5.6 **Reflective Property of Inverse Functions**

The graph of f contains the point (a, b) if and only if the graph of f^{-1} contains the point (b, a).

Proof If (a, b) is on the graph of f, then $f(a) = b$, and you can write

$$f^{-1}(b) = f^{-1}(f(a)) = a.$$

So, (b, a) is on the graph of f^{-1}, as shown in Figure 5.12. A similar argument will prove the theorem in the other direction.

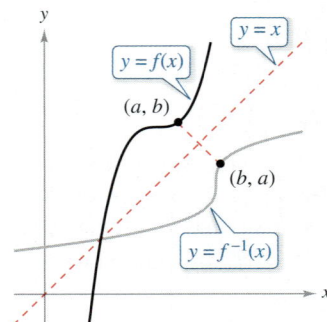

The graph of f^{-1} is a reflection of the graph of f in the line $y = x$.

Figure 5.12

See LarsonCalculus.com for Bruce Edwards's video of this proof.

Existence of an Inverse Function

Not every function has an inverse function, and Theorem 5.6 suggests a graphical test for those that do—the **Horizontal Line Test** for an inverse function. This test states that a function f has an inverse function if and only if every horizontal line intersects the graph of f at most once (see Figure 5.13). The next theorem formally states why the Horizontal Line Test is valid. (Recall from Section 3.3 that a function is *strictly monotonic* when it is either increasing on its entire domain or decreasing on its entire domain.)

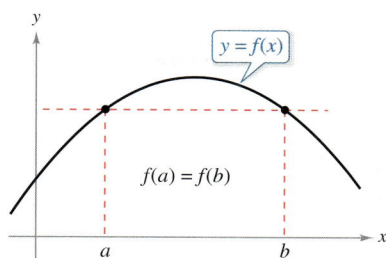

If a horizontal line intersects the graph of f twice, then f is not one-to-one.
Figure 5.13

> **THEOREM 5.7   The Existence of an Inverse Function**
>
> 1. A function has an inverse function if and only if it is one-to-one.
> 2. If f is strictly monotonic on its entire domain, then it is one-to-one and therefore has an inverse function.

Proof   The proof of the first part of the theorem is left as an exercise (See Exercise 97). To prove the second part of the theorem, recall from Section P.3 that f is one-to-one when for x_1 and x_2 in its domain

$$x_1 \neq x_2 \quad \Longrightarrow \quad f(x_1) \neq f(x_2).$$

Now, choose x_1 and x_2 in the domain of f. If $x_1 \neq x_2$, then, because f is strictly monotonic, it follows that either $f(x_1) < f(x_2)$ or $f(x_1) > f(x_2)$. In either case, $f(x_1) \neq f(x_2)$. So, f is one-to-one on the interval.

See LarsonCalculus.com for Bruce Edwards's video of this proof.

> **EXAMPLE 2**   **The Existence of an Inverse Function**
>
> **a.** From the graph of $f(x) = x^3 + x - 1$ shown in Figure 5.14(a), it appears that f is increasing over its entire domain. To verify this, note that the derivative, $f'(x) = 3x^2 + 1$, is positive for all real values of x. So, f is strictly monotonic, and it must have an inverse function.
>
> **b.** From the graph of $f(x) = x^3 - x + 1$ shown in Figure 5.14(b), you can see that the function does not pass the Horizontal Line Test. In other words, it is not one-to-one. For instance, f has the same value when $x = -1$, 0, and 1.
>
> $$f(-1) = f(1) = f(0) = 1 \qquad \text{Not one-to-one}$$
>
> So, by Theorem 5.7, f does not have an inverse function.

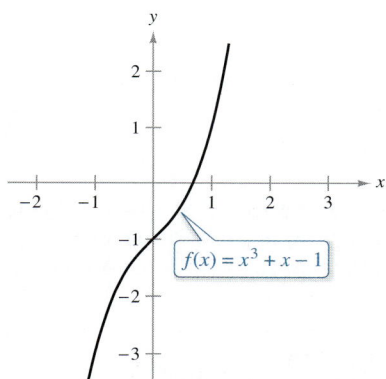

(a) Because f is increasing over its entire domain, it has an inverse function.

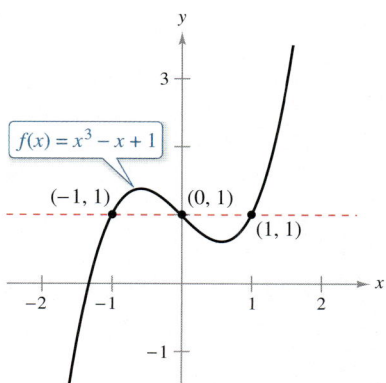

(b) Because f is not one-to-one, it does not have an inverse function.
Figure 5.14

Often, it is easier to prove that a function *has* an inverse function than to find the inverse function. For instance, it would be difficult algebraically to find the inverse function of the function in Example 2(a).

> **GUIDELINES FOR FINDING AN INVERSE FUNCTION**
>
> 1. Use Theorem 5.7 to determine whether the function $y = f(x)$ has an inverse function.
> 2. Solve for x as a function of y: $x = g(y) = f^{-1}(y)$.
> 3. Interchange x and y. The resulting equation is $y = f^{-1}(x)$.
> 4. Define the domain of f^{-1} as the range of f.
> 5. Verify that $f(f^{-1}(x)) = x$ and $f^{-1}(f(x)) = x$.

EXAMPLE 3 Finding an Inverse Function

Find the inverse function of $f(x) = \sqrt{2x - 3}$.

Solution From the graph of f in Figure 5.15, it appears that f is increasing over its entire domain, $[3/2, \infty)$. To verify this, note that

$$f'(x) = \frac{1}{\sqrt{2x - 3}}$$

is positive on the domain of f. So, f is strictly monotonic, and it must have an inverse function. To find an equation for the inverse function, let $y = f(x)$, and solve for x in terms of y.

$$\sqrt{2x - 3} = y \qquad \text{Let } y = f(x).$$
$$2x - 3 = y^2 \qquad \text{Square each side.}$$
$$x = \frac{y^2 + 3}{2} \qquad \text{Solve for } x.$$
$$y = \frac{x^2 + 3}{2} \qquad \text{Interchange } x \text{ and } y.$$
$$f^{-1}(x) = \frac{x^2 + 3}{2} \qquad \text{Replace } y \text{ by } f^{-1}(x).$$

The domain of f^{-1} is the range of f, which is $[0, \infty)$. You can verify this result as shown.

$$f(f^{-1}(x)) = \sqrt{2\left(\frac{x^2 + 3}{2}\right) - 3} = \sqrt{x^2} = x, \quad x \geq 0$$

$$f^{-1}(f(x)) = \frac{\left(\sqrt{2x - 3}\right)^2 + 3}{2} = \frac{2x - 3 + 3}{2} = x, \quad x \geq \frac{3}{2}$$

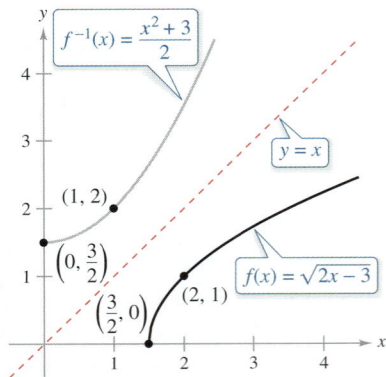

The domain of f^{-1}, $[0, \infty)$, is the range of f.
Figure 5.15

Theorem 5.7 is useful in the next type of problem. You are given a function that is *not* one-to-one on its domain. By restricting the domain to an interval on which the function is strictly monotonic, you can conclude that the new function *is* one-to-one on the restricted domain.

EXAMPLE 4 Testing Whether a Function Is One-to-One

····▷ *See LarsonCalculus.com for an interactive version of this type of example.*

Show that the sine function

$$f(x) = \sin x$$

is not one-to-one on the entire real number line. Then show that $[-\pi/2, \pi/2]$ is the largest interval, centered at the origin, on which f is strictly monotonic.

Solution It is clear that f is not one-to-one, because many different x-values yield the same y-value. For instance,

$$\sin(0) = 0 = \sin(\pi).$$

Moreover, f is increasing on the open interval $(-\pi/2, \pi/2)$, because its derivative

$$f'(x) = \cos x$$

is positive there. Finally, because the left and right endpoints correspond to relative extrema of the sine function, you can conclude that f is increasing on the closed interval $[-\pi/2, \pi/2]$ *and* that on any larger interval the function is not strictly monotonic (see Figure 5.16).

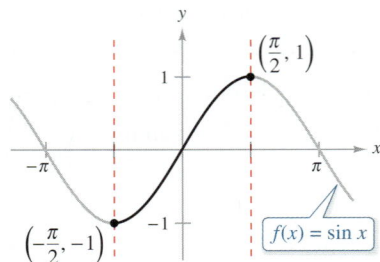

f is one-to-one on the interval $[-\pi/2, \pi/2]$.
Figure 5.16

Derivative of an Inverse Function

The next two theorems discuss the derivative of an inverse function. The reasonableness of Theorem 5.8 follows from the reflective property of inverse functions, as shown in Figure 5.12.

THEOREM 5.8 Continuity and Differentiability of Inverse Functions

Let f be a function whose domain is an interval I. If f has an inverse function, then the following statements are true.

1. If f is continuous on its domain, then f^{-1} is continuous on its domain.
2. If f is increasing on its domain, then f^{-1} is increasing on its domain.
3. If f is decreasing on its domain, then f^{-1} is decreasing on its domain.
4. If f is differentiable on an interval containing c and $f'(c) \neq 0$, then f^{-1} is differentiable at $f(c)$.

A proof of this theorem is given in Appendix A.
See LarsonCalculus.com for Bruce Edwards's video of this proof.

Exploration

Graph the inverse functions $f(x) = x^3$ and $g(x) = x^{1/3}$. Calculate the slopes of f at $(1, 1)$, $(2, 8)$, and $(3, 27)$, and the slopes of g at $(1, 1)$, $(8, 2)$, and $(27, 3)$. What do you observe? What happens at $(0, 0)$?

THEOREM 5.9 The Derivative of an Inverse Function

Let f be a function that is differentiable on an interval I. If f has an inverse function g, then g is differentiable at any x for which $f'(g(x)) \neq 0$. Moreover,

$$g'(x) = \frac{1}{f'(g(x))}, \quad f'(g(x)) \neq 0.$$

A proof of this theorem is given in Appendix A.
See LarsonCalculus.com for Bruce Edwards's video of this proof.

EXAMPLE 5   **Evaluating the Derivative of an Inverse Function**

Let $f(x) = \frac{1}{4}x^3 + x - 1$. (a) What is the value of $f^{-1}(x)$ when $x = 3$? (b) What is the value of $(f^{-1})'(x)$ when $x = 3$?

Solution   Notice that f is one-to-one and therefore has an inverse function.

a. Because $f(x) = 3$ when $x = 2$, you know that $f^{-1}(3) = 2$.

b. Because the function f is differentiable and has an inverse function, you can apply Theorem 5.9 (with $g = f^{-1}$) to write

$$(f^{-1})'(3) = \frac{1}{f'(f^{-1}(3))} = \frac{1}{f'(2)}.$$

Moreover, using $f'(x) = \frac{3}{4}x^2 + 1$, you can conclude that

$$(f^{-1})'(3) = \frac{1}{f'(2)} = \frac{1}{\frac{3}{4}(2^2) + 1} = \frac{1}{4}.$$

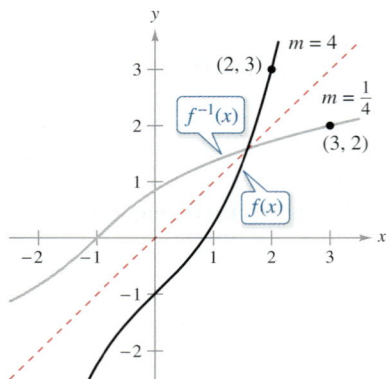

The graphs of the inverse functions f and f^{-1} have reciprocal slopes at points (a, b) and (b, a).

Figure 5.17

In Example 5, note that at the point $(2, 3)$, the slope of the graph of f is 4, and at the point $(3, 2)$, the slope of the graph of f^{-1} is

$$m = \frac{1}{4}$$

as shown in Figure 5.17. In general, if $y = g(x) = f^{-1}(x)$, then $f(y) = x$ and $f'(y) = \dfrac{dx}{dy}$. It follows from Theorem 5.9 that

$$g'(x) = \frac{dy}{dx} = \frac{1}{f'(g(x))} = \frac{1}{f'(y)} = \frac{1}{(dx/dy)}.$$

This reciprocal relationship is sometimes written as

$$\frac{dy}{dx} = \frac{1}{dx/dy}.$$

EXAMPLE 6   **Graphs of Inverse Functions Have Reciprocal Slopes**

Let $f(x) = x^2$ (for $x \geq 0$), and let $f^{-1}(x) = \sqrt{x}$. Show that the slopes of the graphs of f and f^{-1} are reciprocals at each of the following points.

a. $(2, 4)$ and $(4, 2)$ **b.** $(3, 9)$ and $(9, 3)$

Solution   The derivatives of f and f^{-1} are

$$f'(x) = 2x \quad \text{and} \quad (f^{-1})'(x) = \frac{1}{2\sqrt{x}}.$$

a. At $(2, 4)$, the slope of the graph of f is $f'(2) = 2(2) = 4$. At $(4, 2)$, the slope of the graph of f^{-1} is

$$(f^{-1})'(4) = \frac{1}{2\sqrt{4}} = \frac{1}{2(2)} = \frac{1}{4}.$$

b. At $(3, 9)$, the slope of the graph of f is $f'(3) = 2(3) = 6$. At $(9, 3)$, the slope of the graph of f^{-1} is

$$(f^{-1})'(9) = \frac{1}{2\sqrt{9}} = \frac{1}{2(3)} = \frac{1}{6}.$$

So, in both cases, the slopes are reciprocals, as shown in Figure 5.18.

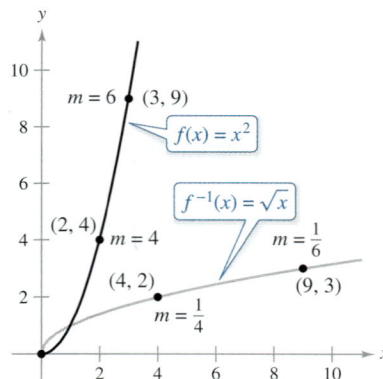

At $(0, 0)$, the derivative of f is 0, and the derivative of f^{-1} does not exist.

Figure 5.18

5.3 Exercises

See **CalcChat.com** for tutorial help and worked-out solutions to odd-numbered exercises.

Verifying Inverse Functions   In Exercises 1–8, show that f and g are inverse functions (a) analytically and (b) graphically.

1. $f(x) = 5x + 1,$ $\qquad$ $g(x) = \dfrac{x - 1}{5}$

2. $f(x) = 3 - 4x,$ $\qquad$ $g(x) = \dfrac{3 - x}{4}$

3. $f(x) = x^3,$ $\qquad$ $g(x) = \sqrt[3]{x}$

4. $f(x) = 1 - x^3,$ $\qquad$ $g(x) = \sqrt[3]{1 - x}$

5. $f(x) = \sqrt{x - 4},$ $\qquad$ $g(x) = x^2 + 4, \quad x \ge 0$

6. $f(x) = 16 - x^2, \quad x \ge 0,$ $\quad$ $g(x) = \sqrt{16 - x}$

7. $f(x) = \dfrac{1}{x},$ $\qquad$ $g(x) = \dfrac{1}{x}$

8. $f(x) = \dfrac{1}{1 + x}, \quad x \ge 0,$ $\quad$ $g(x) = \dfrac{1 - x}{x}, \quad 0 < x \le 1$

Matching   In Exercises 9–12, match the graph of the function with the graph of its inverse function. [The graphs of the inverse functions are labeled (a), (b), (c), and (d).]

(a)

(b)

(c)

(d)

9.

10.

11.

12.

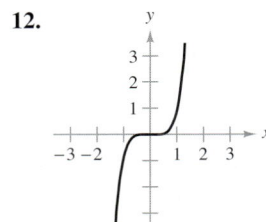

Using the Horizontal Line Test   In Exercises 13–22, use a graphing utility to graph the function. Then use the Horizontal Line Test to determine whether the function is one-to-one on its entire domain and therefore has an inverse function.

13. $f(x) = \frac{3}{4}x + 6$ $\qquad$ 14. $f(x) = 5x - 3$

15. $f(\theta) = \sin \theta$ $\qquad$ 16. $f(x) = \dfrac{6x}{x^2 + 4}$

17. $h(s) = \dfrac{1}{s - 2} - 3$ $\qquad$ 18. $g(t) = \dfrac{1}{\sqrt{t^2 + 1}}$

19. $f(x) = \ln x$ $\qquad$ 20. $f(x) = 5x\sqrt{x - 1}$

21. $g(x) = (x + 5)^3$ $\qquad$ 22. $h(x) = |x + 4| - |x - 4|$

Determining Whether a Function Has an Inverse Function   In Exercises 23–28, use the derivative to determine whether the function is strictly monotonic on its entire domain and therefore has an inverse function.

23. $f(x) = 2 - x - x^3$ $\qquad$ 24. $f(x) = x^3 - 6x^2 + 12x$

25. $f(x) = \dfrac{x^4}{4} - 2x^2$ $\qquad$ 26. $f(x) = x^5 + 2x^3$

27. $f(x) = \ln(x - 3)$ $\qquad$ 28. $f(x) = \cos \dfrac{3x}{2}$

Verifying a Function Has an Inverse Function   In Exercises 29–34, show that f is strictly monotonic on the given interval and therefore has an inverse function on that interval.

29. $f(x) = (x - 4)^2, \quad [4, \infty)$

30. $f(x) = |x + 2|, \quad [-2, \infty)$

31. $f(x) = \dfrac{4}{x^2}, \quad (0, \infty)$

32. $f(x) = \cot x, \quad (0, \pi)$

33. $f(x) = \cos x, \quad [0, \pi]$

34. $f(x) = \sec x, \quad \left[0, \dfrac{\pi}{2}\right)$

Finding an Inverse Function   In Exercises 35–46, (a) find the inverse function of f, (b) graph f and f^{-1} on the same set of coordinate axes, (c) describe the relationship between the graphs, and (d) state the domain and range of f and f^{-1}.

35. $f(x) = 2x - 3$ $\qquad$ 36. $f(x) = 7 - 4x$

37. $f(x) = x^5$ $\qquad$ 38. $f(x) = x^3 - 1$

39. $f(x) = \sqrt{x}$ $\qquad$ 40. $f(x) = x^2, \quad x \ge 0$

41. $f(x) = \sqrt{4 - x^2}, \quad 0 \le x \le 2$

42. $f(x) = \sqrt{x^2 - 4}, \quad x \ge 2$

43. $f(x) = \sqrt[3]{x - 1}$ $\qquad$ 44. $f(x) = x^{2/3}, \quad x \ge 0$

45. $f(x) = \dfrac{x}{\sqrt{x^2 + 7}}$ $\qquad$ 46. $f(x) = \dfrac{x + 2}{x}$

Finding an Inverse Function In Exercises 47 and 48, use the graph of the function f to make a table of values for the given points. Then make a second table that can be used to find f^{-1}, and sketch the graph of f^{-1}. To print an enlarged copy of the graph, go to *MathGraphs.com*.

47.

48.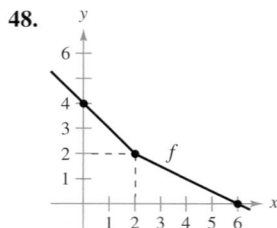

49. Cost You need 50 pounds of two commodities costing $1.25 and $1.60 per pound.

(a) Verify that the total cost is $y = 1.25x + 1.60(50 - x)$, where x is the number of pounds of the less expensive commodity.

(b) Find the inverse function of the cost function. What does each variable represent in the inverse function?

(c) What is the domain of the inverse function? Validate or explain your answer using the context of the problem.

(d) Determine the number of pounds of the less expensive commodity purchased when the total cost is $73.

50. Temperature The formula $C = \frac{5}{9}(F - 32)$, where $F \geq -459.6$, represents Celsius temperature C as a function of Fahrenheit temperature F.

(a) Find the inverse function of C.

(b) What does the inverse function represent?

(c) What is the domain of the inverse function? Validate or explain your answer using the context of the problem.

(d) The temperature is 22°C. What is the corresponding temperature in degrees Fahrenheit?

Testing Whether a Function Is One-to-One In Exercises 51–54, determine whether the function is one-to-one. If it is, find its inverse function.

51. $f(x) = \sqrt{x - 2}$ **52.** $f(x) = -3$

53. $f(x) = |x - 2|, \quad x \leq 2$ **54.** $f(x) = ax + b, \quad a \neq 0$

Making a Function One-to-One In Exercises 55–58, delete part of the domain so that the function that remains is one-to-one. Find the inverse function of the remaining function and give the domain of the inverse function. (*Note:* There is more than one correct answer.)

55. $f(x) = (x - 3)^2$ **56.** $f(x) = 16 - x^4$

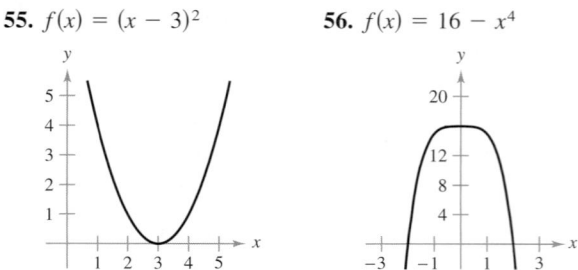

57. $f(x) = |x + 3|$ **58.** $f(x) = |x - 3|$

    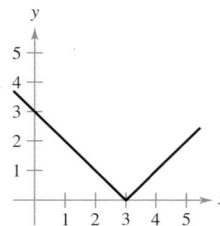

Think About It In Exercises 59–62, decide whether the function has an inverse function. If so, what is the inverse function?

59. $g(t)$ is the volume of water that has passed through a water line t minutes after a control valve is opened.

60. $h(t)$ is the height of the tide t hours after midnight, where $0 \leq t < 24$.

61. $C(t)$ is the cost of a long distance call lasting t minutes.

62. $A(r)$ is the area of a circle of radius r.

Evaluating the Derivative of an Inverse Function In Exercises 63–70, verify that f has an inverse. Then use the function f and the given real number a to find $(f^{-1})'(a)$. (*Hint:* See Example 5.)

63. $f(x) = 5 - 2x^3, \quad a = 7$

64. $f(x) = x^3 + 2x - 1, \quad a = 2$

65. $f(x) = \frac{1}{27}(x^5 + 2x^3), \quad a = -11$

66. $f(x) = \sqrt{x - 4}, \quad a = 2$

67. $f(x) = \sin x, \quad -\frac{\pi}{2} \leq x \leq \frac{\pi}{2}, \quad a = \frac{1}{2}$

68. $f(x) = \cos 2x, \quad 0 \leq x \leq \frac{\pi}{2}, \quad a = 1$

69. $f(x) = \frac{x + 6}{x - 2}, \quad x > 2, \quad a = 3$

70. $f(x) = \frac{x + 3}{x + 1}, \quad x > -1, \quad a = 2$

Using Inverse Functions In Exercises 71–74, (a) find the domains of f and f^{-1}, (b) find the ranges of f and f^{-1}, (c) graph f and f^{-1}, and (d) show that the slopes of the graphs of f and f^{-1} are reciprocals at the given points.

Functions	Point
71. $f(x) = x^3$	$\left(\frac{1}{2}, \frac{1}{8}\right)$
$f^{-1}(x) = \sqrt[3]{x}$	$\left(\frac{1}{8}, \frac{1}{2}\right)$
72. $f(x) = 3 - 4x$	$(1, -1)$
$f^{-1}(x) = \dfrac{3 - x}{4}$	$(-1, 1)$
73. $f(x) = \sqrt{x - 4}$	$(5, 1)$
$f^{-1}(x) = x^2 + 4, \quad x \geq 0$	$(1, 5)$
74. $f(x) = \dfrac{4}{1 + x^2}, \quad x \geq 0$	$(1, 2)$
$f^{-1}(x) = \sqrt{\dfrac{4 - x}{x}}$	$(2, 1)$

Using Composite and Inverse Functions In Exercises 75–78, use the functions $f(x) = \frac{1}{8}x - 3$ and $g(x) = x^3$ to find the given value.

75. $(f^{-1} \circ g^{-1})(1)$

76. $(g^{-1} \circ f^{-1})(-3)$

77. $(f^{-1} \circ f^{-1})(6)$

78. $(g^{-1} \circ g^{-1})(-4)$

Using Composite and Inverse Functions In Exercises 79–82, use the functions $f(x) = x + 4$ and $g(x) = 2x - 5$ to find the given function.

79. $g^{-1} \circ f^{-1}$

80. $f^{-1} \circ g^{-1}$

81. $(f \circ g)^{-1}$

82. $(g \circ f)^{-1}$

WRITING ABOUT CONCEPTS

83. In Your Own Words Describe how to find the inverse function of a one-to-one function given by an equation in x and y. Give an example.

84. A Function and Its Inverse Describe the relationship between the graph of a function and the graph of its inverse function.

Explaining Why a Function Is Not One-to-One In Exercises 85 and 86, the derivative of the function has the same sign for all x in its domain, but the function is not one-to-one. Explain.

85. $f(x) = \tan x$

86. $f(x) = \dfrac{x}{x^2 - 4}$

87. Think About It The function $f(x) = k(2 - x - x^3)$ is one-to-one and $f^{-1}(3) = -2$. Find k.

88. HOW DO YOU SEE IT? Use the information in the graph of f below.

(a) What is the slope of the tangent line to the graph of f^{-1} at the point $\left(-\frac{1}{2}, -1\right)$? Explain.

(b) What is the slope of the tangent line to the graph of f^{-1} at the point $(1, 2)$? Explain.

True or False? In Exercises 89–92, determine whether the statement is true or false. If it is false, explain why or give an example that shows it is false.

89. If f is an even function, then f^{-1} exists.

90. If the inverse function of f exists, then the y-intercept of f is an x-intercept of f^{-1}.

91. If $f(x) = x^n$, where n is odd, then f^{-1} exists.

92. There exists no function f such that $f = f^{-1}$.

93. Making a Function One-to-One

(a) Show that $f(x) = 2x^3 + 3x^2 - 36x$ is not one-to-one on $(-\infty, \infty)$.

(b) Determine the greatest value c such that f is one-to-one on $(-c, c)$.

94. Proof Let f and g be one-to-one functions. Prove that

(a) $f \circ g$ is one-to-one.

(b) $(f \circ g)^{-1}(x) = (g^{-1} \circ f^{-1})(x)$.

95. Proof Prove that if f has an inverse function, then $(f^{-1})^{-1} = f$.

96. Proof Prove that if a function has an inverse function, then the inverse function is unique.

97. Proof Prove that a function has an inverse function if and only if it is one-to-one.

98. Using Theorem 5.7 Is the converse of the second part of Theorem 5.7 true? That is, if a function is one-to-one (and therefore has an inverse function), then must the function be strictly monotonic? If so, prove it. If not, give a counterexample.

99. Concavity Let f be twice-differentiable and one-to-one on an open interval I. Show that its inverse function g satisfies

$$g''(x) = -\frac{f''(g(x))}{[f'(g(x))]^3}.$$

When f is increasing and concave downward, what is the concavity of $f^{-1} = g$?

100. Derivative of an Inverse Function Let

$$f(x) = \int_2^x \frac{dt}{\sqrt{1 + t^4}}.$$

Find $(f^{-1})'(0)$.

101. Derivative of an Inverse Function Show that

$$f(x) = \int_2^x \sqrt{1 + t^2}\, dt$$

is one-to-one and find

$$(f^{-1})'(0).$$

102. Inverse Function Let

$$y = \frac{x - 2}{x - 1}.$$

Show that y is its own inverse function. What can you conclude about the graph of f? Explain.

103. Using a Function Let $f(x) = \dfrac{ax + b}{cx + d}$.

(a) Show that f is one-to-one if and only if $bc - ad \neq 0$.

(b) Given $bc - ad \neq 0$, find f^{-1}.

(c) Determine the values of a, b, c, and d such that $f = f^{-1}$.

5.4 Exponential Functions: Differentiation and Integration

■ Develop properties of the natural exponential function.
■ Differentiate natural exponential functions.
■ Integrate natural exponential functions.

The Natural Exponential Function

The function $f(x) = \ln x$ is increasing on its entire domain, and therefore it has an inverse function f^{-1}. The domain of f^{-1} is the set of all real numbers, and the range is the set of positive real numbers, as shown in Figure 5.19. So, for any real number x,

$$f(f^{-1}(x)) = \ln[f^{-1}(x)] = x. \qquad \text{\textcolor{red}{x is any real number.}}$$

If x is rational, then

$$\ln(e^x) = x \ln e = x(1) = x. \qquad \text{\textcolor{red}{x is a rational number.}}$$

Because the natural logarithmic function is one-to-one, you can conclude that $f^{-1}(x)$ and e^x agree for *rational* values of x. The next definition extends the meaning of e^x to include *all* real values of x.

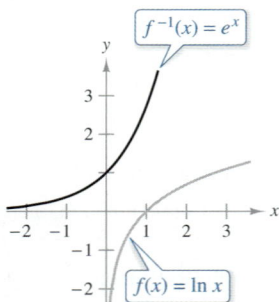

The inverse function of the natural logarithmic function is the natural exponential function.

Figure 5.19

Definition of the Natural Exponential Function

The inverse function of the natural logarithmic function $f(x) = \ln x$ is called the **natural exponential function** and is denoted by

$$f^{-1}(x) = e^x.$$

That is,

$$y = e^x \quad \text{if and only if} \quad x = \ln y.$$

THE NUMBER e

The symbol e was first used by mathematician Leonhard Euler to represent the base of natural logarithms in a letter to another mathematician, Christian Goldbach, in 1731.

The inverse relationship between the natural logarithmic function and the natural exponential function can be summarized as shown.

$$\ln(e^x) = x \quad \text{and} \quad e^{\ln x} = x \qquad \text{\textcolor{red}{Inverse relationship}}$$

EXAMPLE 1   Solving an Exponential Equation

Solve $7 = e^{x+1}$.

Solution You can convert from exponential form to logarithmic form by *taking the natural logarithm of each side* of the equation.

$$7 = e^{x+1} \qquad \text{\textcolor{red}{Write original equation.}}$$
$$\ln 7 = \ln(e^{x+1}) \qquad \text{\textcolor{red}{Take natural logarithm of each side.}}$$
$$\ln 7 = x + 1 \qquad \text{\textcolor{red}{Apply inverse property.}}$$
$$-1 + \ln 7 = x \qquad \text{\textcolor{red}{Solve for x.}}$$

So, the solution is $-1 + \ln 7 \approx -0.946$. You can check this solution as shown.

$$7 = e^{x+1} \qquad \text{\textcolor{red}{Write original equation.}}$$
$$7 \overset{?}{=} e^{(-1+\ln 7)+1} \qquad \text{\textcolor{red}{Substitute $-1 + \ln 7$ for x in original equation.}}$$
$$7 \overset{?}{=} e^{\ln 7} \qquad \text{\textcolor{red}{Simplify.}}$$
$$7 = 7 \checkmark \qquad \text{\textcolor{red}{Solution checks.}}$$

EXAMPLE 2 **Solving a Logarithmic Equation**

Solve $\ln(2x - 3) = 5$.

Solution To convert from logarithmic form to exponential form, you can *exponentiate each side* of the logarithmic equation.

$$\ln(2x - 3) = 5 \qquad \text{Write original equation.}$$
$$e^{\ln(2x-3)} = e^5 \qquad \text{Exponentiate each side.}$$
$$2x - 3 = e^5 \qquad \text{Apply inverse property.}$$
$$x = \tfrac{1}{2}(e^5 + 3) \qquad \text{Solve for } x.$$
$$x \approx 75.707 \qquad \text{Use a calculator.}$$

The familiar rules for operating with rational exponents can be extended to the natural exponential function, as shown in the next theorem.

THEOREM 5.10   Operations with Exponential Functions

Let a and b be any real numbers.

1. $e^a e^b = e^{a+b}$ **2.** $\dfrac{e^a}{e^b} = e^{a-b}$

Proof To prove Property 1, you can write

$$\ln(e^a e^b) = \ln(e^a) + \ln(e^b) = a + b = \ln(e^{a+b}).$$

Because the natural logarithmic function is one-to-one, you can conclude that

$$e^a e^b = e^{a+b}.$$

The proof of the other property is given in Appendix A.

See LarsonCalculus.com for Bruce Edwards's video of this proof.

In Section 5.3, you learned that an inverse function f^{-1} shares many properties with f. So, the natural exponential function inherits the properties listed below from the natural logarithmic function.

Properties of the Natural Exponential Function

1. The domain of $f(x) = e^x$ is

$$(-\infty, \infty)$$

and the range is

$$(0, \infty).$$

2. The function $f(x) = e^x$ is continuous, increasing, and one-to-one on its entire domain.

3. The graph of $f(x) = e^x$ is concave upward on its entire domain.

4. $\lim\limits_{x \to -\infty} e^x = 0$

5. $\lim\limits_{x \to \infty} e^x = \infty$

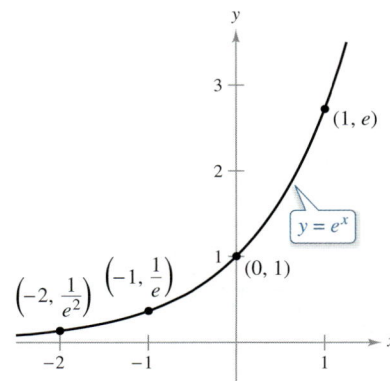

The natural exponential function is increasing, and its graph is concave upward.

Derivatives of Exponential Functions

One of the most intriguing (and useful) characteristics of the natural exponential function is that *it is its own derivative*. In other words, it is a solution of the differential equation $y' = y$. This result is stated in the next theorem.

> **THEOREM 5.11 Derivatives of the Natural Exponential Function**
>
> Let u be a differentiable function of x.
>
> 1. $\dfrac{d}{dx}[e^x] = e^x$
>
> 2. $\dfrac{d}{dx}[e^u] = e^u \dfrac{du}{dx}$

REMARK You can interpret this theorem geometrically by saying that the slope of the graph of $f(x) = e^x$ at any point (x, e^x) is equal to the y-coordinate of the point.

Proof To prove Property 1, use the fact that $\ln e^x = x$, and differentiate each side of the equation.

$$\ln e^x = x \qquad \text{Definition of exponential function}$$
$$\frac{d}{dx}[\ln e^x] = \frac{d}{dx}[x] \qquad \text{Differentiate each side with respect to } x.$$
$$\frac{1}{e^x}\frac{d}{dx}[e^x] = 1$$
$$\frac{d}{dx}[e^x] = e^x$$

The derivative of e^u follows from the Chain Rule.

See LarsonCalculus.com for Bruce Edwards's video of this proof.

FOR FURTHER INFORMATION To find out about derivatives of exponential functions of order 1/2, see the article "A Child's Garden of Fractional Derivatives" by Marcia Kleinz and Thomas J. Osler in *The College Mathematics Journal*. To view this article, go to *MathArticles.com*.

EXAMPLE 3 Differentiating Exponential Functions

Find the derivative of each function.

a. $y = e^{2x-1}$ **b.** $y = e^{-3/x}$

Solution

a. $\dfrac{d}{dx}[e^{2x-1}] = e^u \dfrac{du}{dx} = 2e^{2x-1}$ $u = 2x - 1$

b. $\dfrac{d}{dx}[e^{-3/x}] = e^u \dfrac{du}{dx} = \left(\dfrac{3}{x^2}\right)e^{-3/x} = \dfrac{3e^{-3/x}}{x^2}$ $u = -\dfrac{3}{x}$

EXAMPLE 4 Locating Relative Extrema

Find the relative extrema of

$$f(x) = xe^x.$$

Solution The derivative of f is

$$f'(x) = x(e^x) + e^x(1) \qquad \text{Product Rule}$$
$$= e^x(x + 1).$$

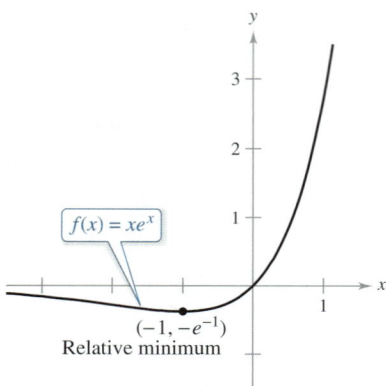

$f(x) = xe^x$

$(-1, -e^{-1})$
Relative minimum

The derivative of f changes from negative to positive at $x = -1$.

Figure 5.20

Because e^x is never 0, the derivative is 0 only when $x = -1$. Moreover, by the First Derivative Test, you can determine that this corresponds to a relative minimum, as shown in Figure 5.20. Because the derivative $f'(x) = e^x(x + 1)$ is defined for all x, there are no other critical points.

| EXAMPLE 5 | **The Standard Normal Probability Density Function** |

⋮ ⋅ ⋅ ⋅ ▷ *See LarsonCalculus.com for an interactive version of this type of example.*

Show that the *standard normal probability density function*

$$f(x) = \frac{1}{\sqrt{2\pi}}e^{-x^2/2}$$

has points of inflection when $x = \pm 1$.

Solution To locate possible points of inflection, find the x-values for which the second derivative is 0.

$$f(x) = \frac{1}{\sqrt{2\pi}}e^{-x^2/2} \qquad \text{Write original function.}$$

$$f'(x) = \frac{1}{\sqrt{2\pi}}(-x)e^{-x^2/2} \qquad \text{First derivative}$$

$$f''(x) = \frac{1}{\sqrt{2\pi}}[(-x)(-x)e^{-x^2/2} + (-1)e^{-x^2/2}] \qquad \text{Product Rule}$$

$$= \frac{1}{\sqrt{2\pi}}(e^{-x^2/2})(x^2 - 1) \qquad \text{Second derivative}$$

So, $f''(x) = 0$ when $x = \pm 1$, and you can apply the techniques of Chapter 3 to conclude that these values yield the two points of inflection shown in Figure 5.21.

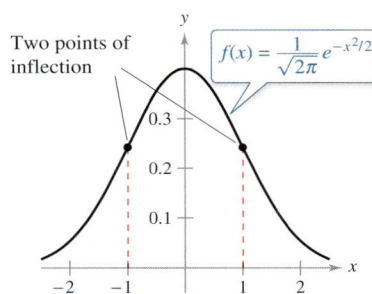

The bell-shaped curve given by a standard normal probability density function
Figure 5.21

> **REMARK** The general form of a normal probability density function (whose mean is 0) is
>
> $$f(x) = \frac{1}{\sigma\sqrt{2\pi}}e^{-x^2/(2\sigma^2)}$$
>
> where σ is the standard deviation (σ is the lowercase Greek letter sigma). This "bell-shaped curve" has points of inflection when $x = \pm\sigma$.

| EXAMPLE 6 | **Population of California** |

The projected populations y (in thousands) of California from 2015 through 2030 can be modeled by

$$y = 34{,}696e^{0.0097t}$$

where t represents the year, with $t = 15$ corresponding to 2015. At what rate will the population be changing in 2020? (*Source: U.S. Census Bureau*)

Solution The derivative of the model is

$$y' = (0.0097)(34{,}696)e^{0.0097t}$$
$$\approx 336.55e^{0.0097t}.$$

By evaluating the derivative when $t = 20$, you can estimate that the rate of change in 2020 will be about

408.6 thousand people per year.

Integrals of Exponential Functions

Each differentiation formula in Theorem 5.11 has a corresponding integration formula.

THEOREM 5.12 Integration Rules for Exponential Functions

Let u be a differentiable function of x.

1. $\displaystyle\int e^x \, dx = e^x + C$ **2.** $\displaystyle\int e^u \, du = e^u + C$

EXAMPLE 7   **Integrating Exponential Functions**

Find the indefinite integral.

$$\int e^{3x+1} \, dx$$

Solution   If you let $u = 3x + 1$, then $du = 3 \, dx$.

$$\int e^{3x+1} \, dx = \frac{1}{3} \int e^{3x+1}(3) \, dx \qquad \text{Multiply and divide by 3.}$$

$$= \frac{1}{3} \int e^u \, du \qquad \text{Substitute: } u = 3x + 1.$$

$$= \frac{1}{3} e^u + C \qquad \text{Apply Exponential Rule.}$$

$$= \frac{e^{3x+1}}{3} + C \qquad \text{Back-substitute.}$$

REMARK   In Example 7, the missing *constant* factor 3 was introduced to create $du = 3 \, dx$. However, remember that you cannot introduce a missing *variable* factor in the integrand. For instance,

$$\int e^{-x^2} \, dx \neq \frac{1}{x} \int e^{-x^2}(x \, dx).$$

EXAMPLE 8   **Integrating Exponential Functions**

Find the indefinite integral.

$$\int 5xe^{-x^2} \, dx$$

Solution   If you let $u = -x^2$, then $du = -2x \, dx$ or $x \, dx = -du/2$.

$$\int 5xe^{-x^2} \, dx = \int 5e^{-x^2}(x \, dx) \qquad \text{Regroup integrand.}$$

$$= \int 5e^u \left(-\frac{du}{2}\right) \qquad \text{Substitute: } u = -x^2.$$

$$= -\frac{5}{2} \int e^u \, du \qquad \text{Constant Multiple Rule}$$

$$= -\frac{5}{2} e^u + C \qquad \text{Apply Exponential Rule.}$$

$$= -\frac{5}{2} e^{-x^2} + C \qquad \text{Back-substitute.}$$

EXAMPLE 9 **Integrating Exponential Functions**

Find each indefinite integral.

a. $\displaystyle\int \frac{e^{1/x}}{x^2}\, dx$ **b.** $\displaystyle\int \sin x\, e^{\cos x}\, dx$

Solution

a. $\displaystyle\int \frac{e^{1/x}}{x^2}\, dx = -\int \overbrace{e^{1/x}}^{e^u} \overbrace{\left(-\frac{1}{x^2}\right) dx}^{du}$ $u = \dfrac{1}{x}$

$$= -e^{1/x} + C$$

b. $\displaystyle\int \sin x\, e^{\cos x}\, dx = -\int \overbrace{e^{\cos x}}^{e^u}\, \overbrace{(-\sin x\, dx)}^{du}$ $u = \cos x$

$$= -e^{\cos x} + C$$

EXAMPLE 10 **Finding Areas Bounded by Exponential Functions**

Evaluate each definite integral.

a. $\displaystyle\int_0^1 e^{-x}\, dx$ **b.** $\displaystyle\int_0^1 \frac{e^x}{1 + e^x}\, dx$ **c.** $\displaystyle\int_{-1}^0 \left[e^x \cos(e^x)\right] dx$

Solution

a. $\displaystyle\int_0^1 e^{-x}\, dx = -e^{-x}\Big]_0^1$ See Figure 5.22(a).

$$= -e^{-1} - (-1)$$

$$= 1 - \frac{1}{e}$$

$$\approx 0.632$$

b. $\displaystyle\int_0^1 \frac{e^x}{1 + e^x}\, dx = \ln(1 + e^x)\Big]_0^1$ See Figure 5.22(b).

$$= \ln(1 + e) - \ln 2$$

$$\approx 0.620$$

c. $\displaystyle\int_{-1}^0 \left[e^x \cos(e^x)\right] dx = \sin(e^x)\Big]_{-1}^0$ See Figure 5.22(c).

$$= \sin 1 - \sin(e^{-1})$$

$$\approx 0.482$$

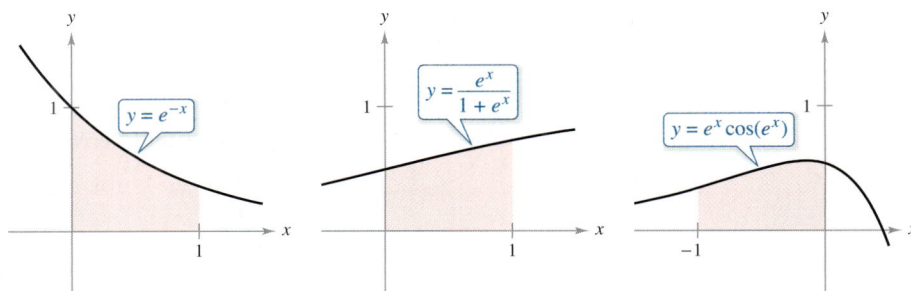

(a)   (b)   (c)

Figure 5.22

5.4 Exercises

See CalcChat.com for tutorial help and worked-out solutions to odd-numbered exercises.

Solving an Exponential or Logarithmic Equation In Exercises 1–16, solve for x accurate to three decimal places.

1. $e^{\ln x} = 4$

2. $e^{\ln 3x} = 24$

3. $e^x = 12$

4. $5e^x = 36$

5. $9 - 2e^x = 7$

6. $8e^x - 12 = 7$

7. $50e^{-x} = 30$

8. $100e^{-2x} = 35$

9. $\dfrac{800}{100 - e^{x/2}} = 50$

10. $\dfrac{5000}{1 + e^{2x}} = 2$

11. $\ln x = 2$

12. $\ln x^2 = 10$

13. $\ln(x - 3) = 2$

14. $\ln 4x = 1$

15. $\ln\sqrt{x + 2} = 1$

16. $\ln(x - 2)^2 = 12$

Sketching a Graph In Exercises 17–22, sketch the graph of the function.

17. $y = e^{-x}$

18. $y = \frac{1}{2}e^x$

19. $y = e^x + 2$

20. $y = e^{x-1}$

21. $y = e^{-x^2}$

22. $y = e^{-x/2}$

23. Comparing Graphs Use a graphing utility to graph $f(x) = e^x$ and the given function in the same viewing window. How are the two graphs related?

(a) $g(x) = e^{x-2}$ (b) $h(x) = -\frac{1}{2}e^x$ (c) $q(x) = e^{-x} + 3$

24. Asymptotes Use a graphing utility to graph the function. Use the graph to determine any asymptotes of the function.

(a) $f(x) = \dfrac{8}{1 + e^{-0.5x}}$ (b) $g(x) = \dfrac{8}{1 + e^{-0.5/x}}$

Matching In Exercises 25–28, match the equation with the correct graph. Assume that a and C are positive real numbers. [The graphs are labeled (a), (b), (c), and (d).]

(a)

(b)

(c)

(d)
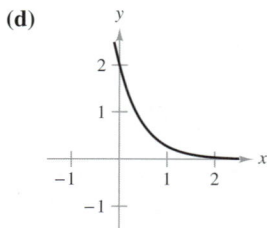

25. $y = Ce^{ax}$

26. $y = Ce^{-ax}$

27. $y = C(1 - e^{-ax})$

28. $y = \dfrac{C}{1 + e^{-ax}}$

Inverse Functions In Exercises 29–32, illustrate that the functions are inverses of each other by graphing both functions on the same set of coordinate axes.

29. $f(x) = e^{2x}$
  $g(x) = \ln\sqrt{x}$

30. $f(x) = e^{x/3}$
  $g(x) = \ln x^3$

31. $f(x) = e^x - 1$
  $g(x) = \ln(x + 1)$

32. $f(x) = e^{x-1}$
  $g(x) = 1 + \ln x$

Finding a Derivative In Exercises 33–54, find the derivative.

33. $f(x) = e^{2x}$

34. $y = e^{-8x}$

35. $y = e^{\sqrt{x}}$

36. $y = e^{-2x^3}$

37. $y = e^{x-4}$

38. $y = 5e^{x^2+5}$

39. $y = e^x \ln x$

40. $y = xe^{4x}$

41. $y = x^3 e^x$

42. $y = x^2 e^{-x}$

43. $g(t) = (e^{-t} + e^t)^3$

44. $g(t) = e^{-3/t^2}$

45. $y = \ln(1 + e^{2x})$

46. $y = \ln\left(\dfrac{1 + e^x}{1 - e^x}\right)$

47. $y = \dfrac{2}{e^x + e^{-x}}$

48. $y = \dfrac{e^x - e^{-x}}{2}$

49. $y = \dfrac{e^x + 1}{e^x - 1}$

50. $y = \dfrac{e^{2x}}{e^{2x} + 1}$

51. $y = e^x(\sin x + \cos x)$

52. $y = e^{2x}\tan 2x$

53. $F(x) = \displaystyle\int_{\pi}^{\ln x} \cos e^t \, dt$

54. $F(x) = \displaystyle\int_{0}^{e^{2x}} \ln(t + 1) \, dt$

Finding an Equation of a Tangent Line In Exercises 55–62, find an equation of the tangent line to the graph of the function at the given point.

55. $f(x) = e^{3x}$, $(0, 1)$

56. $f(x) = e^{-2x}$, $(0, 1)$

57. $f(x) = e^{1-x}$, $(1, 1)$

58. $y = e^{-2x+x^2}$, $(2, 1)$

59. $f(x) = e^{-x}\ln x$, $(1, 0)$

60. $y = \ln\dfrac{e^x + e^{-x}}{2}$, $(0, 0)$

61. $y = x^2 e^x - 2xe^x + 2e^x$, $(1, e)$

62. $y = xe^x - e^x$, $(1, 0)$

Implicit Differentiation In Exercises 63 and 64, use implicit differentiation to find dy/dx.

63. $xe^y - 10x + 3y = 0$

64. $e^{xy} + x^2 - y^2 = 10$

Finding the Equation of a Tangent Line In Exercises 65 and 66, find an equation of the tangent line to the graph of the function at the given point.

65. $xe^y + ye^x = 1$, $(0, 1)$

66. $1 + \ln xy = e^{x-y}$, $(1, 1)$

Finding a Second Derivative In Exercises 67 and 68, find the second derivative of the function.

67. $f(x) = (3 + 2x)e^{-3x}$ **68.** $g(x) = \sqrt{x} + e^x \ln x$

Differential Equation In Exercises 69 and 70, show that the function $y = f(x)$ is a solution of the differential equation.

69. $y = 4e^{-x}$

$y'' - y = 0$

70. $y = e^{3x} + e^{-3x}$

$y'' - 9y = 0$

Finding Extrema and Points of Inflection In Exercises 71–78, find the extrema and the points of inflection (if any) exist) of the function. Use a graphing utility to graph the function and confirm your results.

71. $f(x) = \dfrac{e^x + e^{-x}}{2}$ **72.** $f(x) = \dfrac{e^x - e^{-x}}{2}$

73. $g(x) = \dfrac{1}{\sqrt{2\pi}} e^{-(x-2)^2/2}$ **74.** $g(x) = \dfrac{1}{\sqrt{2\pi}} e^{-(x-3)^2/2}$

75. $f(x) = x^2 e^{-x}$ **76.** $f(x) = xe^{-x}$

77. $g(t) = 1 + (2 + t)e^{-t}$ **78.** $f(x) = -2 + e^{3x}(4 - 2x)$

79. Area Find the area of the largest rectangle that can be inscribed under the curve $y = e^{-x^2}$ in the first and second quadrants.

80. Area Perform the following steps to find the maximum area of the rectangle shown in the figure.

$f(x) = 10xe^{-x}$

(a) Solve for c in the equation $f(c) = f(c + x)$.

(b) Use the result in part (a) to write the area A as a function of x. [*Hint:* $A = xf(c)$]

(c) Use a graphing utility to graph the area function. Use the graph to approximate the dimensions of the rectangle of maximum area. Determine the maximum area.

(d) Use a graphing utility to graph the expression for c found in part (a). Use the graph to approximate

$$\lim_{x \to 0^+} c \quad \text{and} \quad \lim_{x \to \infty} c.$$

Use this result to describe the changes in dimensions and position of the rectangle for $0 < x < \infty$.

81. Finding an Equation of a Tangent Line Find a point on the graph of the function $f(x) = e^{2x}$ such that the tangent line to the graph at that point passes through the origin. Use a graphing utility to graph f and the tangent line in the same viewing window.

82. **HOW DO YOU SEE IT?** The figure shows the graphs of f and g, where a is a positive real number. Identify the open interval(s) on which the graphs of f and g are (a) increasing or decreasing, and (b) concave upward or concave downward.

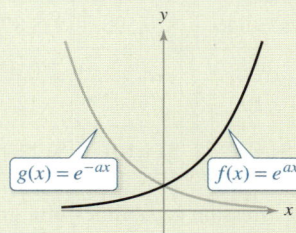

$g(x) = e^{-ax}$ $f(x) = e^{ax}$

83. Depreciation The value V of an item t years after it is purchased is $V = 15{,}000e^{-0.6286t}$, $0 \le t \le 10$.

(a) Use a graphing utility to graph the function.

(b) Find the rates of change of V with respect to t when $t = 1$ and $t = 5$.

(c) Use a graphing utility to graph the tangent lines to the function when $t = 1$ and $t = 5$.

84. Harmonic Motion The displacement from equilibrium of a mass oscillating on the end of a spring suspended from a ceiling is $y = 1.56e^{-0.22t} \cos 4.9t$, where y is the displacement (in feet) and t is the time (in seconds). Use a graphing utility to graph the displacement function on the interval $[0, 10]$. Find a value of t past which the displacement is less than 3 inches from equilibrium.

85. Atmospheric Pressure

A meteorologist measures the atmospheric pressure P (in kilograms per square meter) at altitude h (in kilometers). The data are shown below.

h	0	5	10	15	20
P	10,332	5583	2376	1240	517

(a) Use a graphing utility to plot the points $(h, \ln P)$. Use the regression capabilities of the graphing utility to find a linear model for the revised data points.

(b) The line in part (a) has the form $\ln P = ah + b$. Write the equation in exponential form.

(c) Use a graphing utility to plot the original data and graph the exponential model in part (b).

(d) Find the rate of change of the pressure when $h = 5$ and $h = 18$.

86. Modeling Data The table lists the approximate values V of a mid-sized sedan for the years 2006 through 2012. The variable t represents the time (in years), with $t = 6$ corresponding to 2006.

t	6	7	8	9
V	\$23,046	\$20,596	\$18,851	\$17,001

t	10	11	12
V	\$15,226	\$14,101	\$12,841

(a) Use the regression capabilities of a graphing utility to fit linear and quadratic models to the data. Plot the data and graph the models.

(b) What does the slope represent in the linear model in part (a)?

(c) Use the regression capabilities of a graphing utility to fit an exponential model to the data.

(d) Determine the horizontal asymptote of the exponential model found in part (c). Interpret its meaning in the context of the problem.

(e) Use the exponential model to find the rate of decrease in the value of the sedan when $t = 7$ and $t = 11$.

Linear and Quadratic Approximation In Exercises 87 and 88, use a graphing utility to graph the function. Then graph

$$P_1(x) = f(0) + f'(0)(x - 0) \quad \text{and}$$

$$P_2(x) = f(0) + f'(0)(x - 0) + \tfrac{1}{2}f''(0)(x - 0)^2$$

in the same viewing window. Compare the values of f, P_1, P_2, and their first derivatives at $x = 0$.

87. $f(x) = e^x$

88. $f(x) = e^{x/2}$

Stirling's Formula For large values of n,

$$n! = 1 \cdot 2 \cdot 3 \cdot 4 \cdots (n-1) \cdot n$$

can be approximated by Stirling's Formula,

$$n! \approx \left(\frac{n}{e}\right)^n \sqrt{2\pi n}.$$

In Exercises 89 and 90, find the exact value of $n!$, and then approximate $n!$ using Stirling's Formula.

89. $n = 12$

90. $n = 15$

Finding an Indefinite Integral In Exercises 91–108, find the indefinite integral.

91. $\displaystyle\int e^{5x}(5)\,dx$

92. $\displaystyle\int e^{-x^4}(-4x^3)\,dx$

93. $\displaystyle\int e^{2x-1}\,dx$

94. $\displaystyle\int e^{1-3x}\,dx$

95. $\displaystyle\int x^2 e^{x^3}\,dx$

96. $\displaystyle\int e^x(e^x + 1)^2\,dx$

97. $\displaystyle\int \frac{e^{\sqrt{x}}}{\sqrt{x}}\,dx$

98. $\displaystyle\int \frac{e^{1/x^2}}{x^3}\,dx$

99. $\displaystyle\int \frac{e^{-x}}{1 + e^{-x}}\,dx$

100. $\displaystyle\int \frac{e^{2x}}{1 + e^{2x}}\,dx$

101. $\displaystyle\int e^x \sqrt{1 - e^x}\,dx$

102. $\displaystyle\int \frac{e^x - e^{-x}}{e^x + e^{-x}}\,dx$

103. $\displaystyle\int \frac{e^x + e^{-x}}{e^x - e^{-x}}\,dx$

104. $\displaystyle\int \frac{2e^x - 2e^{-x}}{(e^x + e^{-x})^2}\,dx$

105. $\displaystyle\int \frac{5 - e^x}{e^{2x}}\,dx$

106. $\displaystyle\int \frac{e^{2x} + 2e^x + 1}{e^x}\,dx$

107. $\displaystyle\int e^{-x} \tan(e^{-x})\,dx$

108. $\displaystyle\int e^{2x} \csc(e^{2x})\,dx$

Evaluating a Definite Integral In Exercises 109–118, evaluate the definite integral. Use a graphing utility to verify your result.

109. $\displaystyle\int_0^1 e^{-2x}\,dx$

110. $\displaystyle\int_1^2 e^{5x-3}\,dx$

111. $\displaystyle\int_0^1 xe^{-x^2}\,dx$

112. $\displaystyle\int_{-2}^0 x^2 e^{x^3/2}\,dx$

113. $\displaystyle\int_1^3 \frac{e^{3/x}}{x^2}\,dx$

114. $\displaystyle\int_0^{\sqrt{2}} xe^{-(x^2/2)}\,dx$

115. $\displaystyle\int_0^3 \frac{2e^{2x}}{1 + e^{2x}}\,dx$

116. $\displaystyle\int_0^1 \frac{e^x}{5 - e^x}\,dx$

117. $\displaystyle\int_0^{\pi/2} e^{\sin \pi x} \cos \pi x\,dx$

118. $\displaystyle\int_{\pi/3}^{\pi/2} e^{\sec 2x} \sec 2x \tan 2x\,dx$

Slope Field In Exercises 119 and 120, a differential equation, a point, and a slope field are given. (a) Sketch two approximate solutions of the differential equation on the slope field, one of which passes through the given point. (b) Use integration to find the particular solution of the differential equation and use a graphing utility to graph the solution. Compare the result with the sketches in part (a). To print an enlarged copy of the graph, go to *MathGraphs.com*.

119. $\dfrac{dy}{dx} = 2e^{-x/2}, \quad (0, 1)$

120. $\dfrac{dy}{dx} = xe^{-0.2x^2}, \quad \left(0, -\dfrac{3}{2}\right)$

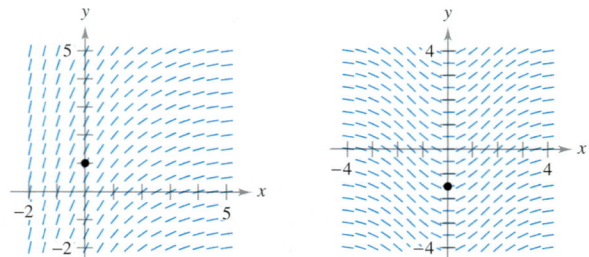

Differential Equation In Exercises 121 and 122, solve the differential equation.

121. $\dfrac{dy}{dx} = xe^{ax^2}$

122. $\dfrac{dy}{dx} = (e^x - e^{-x})^2$

Differential Equation In Exercises 123 and 124, find the particular solution that satisfies the initial conditions.

123. $f''(x) = \frac{1}{2}(e^x + e^{-x})$,
$f(0) = 1, f'(0) = 0$

124. $f''(x) = \sin x + e^{2x}$,
$f(0) = \frac{1}{4}, f'(0) = \frac{1}{2}$

Area In Exercises 125–128, find the area of the region bounded by the graphs of the equations. Use a graphing utility to graph the region and verify your result.

125. $y = e^x, y = 0, x = 0, x = 5$

126. $y = e^{-2x}, y = 0, x = -1, x = 3$

127. $y = xe^{-x^2/4}, y = 0, x = 0, x = \sqrt{6}$

128. $y = e^{-2x} + 2, y = 0, x = 0, x = 2$

Numerical Integration In Exercises 129 and 130, approximate the integral using the Midpoint Rule, the Trapezoidal Rule, and Simpson's Rule with $n = 12$. Use a graphing utility to verify your results.

129. $\displaystyle\int_0^4 \sqrt{x}\, e^x \, dx$

130. $\displaystyle\int_0^2 2xe^{-x} \, dx$

131. **Probability** A car battery has an average lifetime of 48 months with a standard deviation of 6 months. The battery lives are normally distributed. The probability that a given battery will last between 48 months and 60 months is

$$0.0065 \int_{48}^{60} e^{-0.0139(t-48)^2} \, dt.$$

Use the integration capabilities of a graphing utility to approximate the integral. Interpret the resulting probability.

132. **Probability** The median waiting time (in minutes) for people waiting for service in a convenience store is given by the solution of the equation

$$\int_0^x 0.3e^{-0.3t} \, dt = \frac{1}{2}.$$

What is the median waiting time?

133. **Using the Area of a Region** Find the value of a such that the area bounded by $y = e^{-x}$, the x-axis, $x = -a$, and $x = a$ is $\frac{8}{3}$.

134. **Modeling Data** A valve on a storage tank is opened for 4 hours to release a chemical in a manufacturing process. The flow rate R (in liters per hour) at time t (in hours) is given in the table.

t	0	1	2	3	4
R	425	240	118	71	36

(a) Use the regression capabilities of a graphing utility to find a linear model for the points $(t, \ln R)$. Write the resulting equation of the form $\ln R = at + b$ in exponential form.

(b) Use a graphing utility to plot the data and graph the exponential model.

(c) Use the definite integral to approximate the number of liters of chemical released during the 4 hours.

WRITING ABOUT CONCEPTS

135. **Properties of the Natural Exponential Function** In your own words, state the properties of the natural exponential function.

136. **A Function and Its Derivative** Is there a function f such that $f(x) = f'(x)$? If so, identify it.

137. **Choosing a Function** Without integrating, state the integration formula you can use to integrate each of the following.

(a) $\displaystyle\int \frac{e^x}{e^x + 1} \, dx$

(b) $\displaystyle\int xe^{x^2} \, dx$

138. **Analyzing a Graph** Consider the function

$$f(x) = \frac{2}{1 + e^{1/x}}.$$

(a) Use a graphing utility to graph f.

(b) Write a short paragraph explaining why the graph has a horizontal asymptote at $y = 1$ and why the function has a nonremovable discontinuity at $x = 0$.

139. **Deriving an Inequality** Given $e^x \geq 1$ for $x \geq 0$, it follows that

$$\int_0^x e^t \, dt \geq \int_0^x 1 \, dt.$$

Perform this integration to derive the inequality

$$e^x \geq 1 + x$$

for $x \geq 0$.

140. **Solving an Equation** Find, to three decimal places, the value of x such that $e^{-x} = x$. (Use Newton's Method or the *zero* or *root* feature of a graphing utility.)

141. **Horizontal Motion** The position function of a particle moving along the x-axis is $x(t) = Ae^{kt} + Be^{-kt}$, where A, B, and k are positive constants.

(a) During what times t is the particle closest to the origin?

(b) Show that the acceleration of the particle is proportional to the position of the particle. What is the constant of proportionality?

142. **Analyzing a Function** Let $f(x) = \dfrac{\ln x}{x}$.

(a) Graph f on $(0, \infty)$ and show that f is strictly decreasing on (e, ∞).

(b) Show that if $e \leq A < B$, then $A^B > B^A$.

(c) Use part (b) to show that $e^\pi > \pi^e$.

143. **Finding the Maximum Rate of Change** Verify that the function

$$y = \frac{L}{1 + ae^{-x/b}}, \quad a > 0, \quad b > 0, \quad L > 0$$

increases at a maximum rate when $y = L/2$.

5.5   Bases Other than *e* and Applications

■ Define exponential functions that have bases other than e.
■ Differentiate and integrate exponential functions that have bases other than e.
■ Use exponential functions to model compound interest and exponential growth.

Bases Other than e

The **base** of the natural exponential function is *e*. This "natural" base can be used to assign a meaning to a general base *a*.

Definition of Exponential Function to Base *a*

If *a* is a positive real number ($a \neq 1$) and *x* is any real number, then the **exponential function to the base *a*** is denoted by a^x and is defined by

$$a^x = e^{(\ln a)x}.$$

If $a = 1$, then $y = 1^x = 1$ is a constant function.

These functions obey the usual laws of exponents. For instance, here are some familiar properties.

1. $a^0 = 1$ **2.** $a^x a^y = a^{x+y}$ **3.** $\dfrac{a^x}{a^y} = a^{x-y}$ **4.** $(a^x)^y = a^{xy}$

When modeling the half-life of a radioactive sample, it is convenient to use $\frac{1}{2}$ as the base of the exponential model. (*Half-life* is the number of years required for half of the atoms in a sample of radioactive material to decay.)

EXAMPLE 1   **Radioactive Half-Life Model**

The half-life of carbon-14 is about 5715 years. A sample contains 1 gram of carbon-14. How much will be present in 10,000 years?

Solution   Let $t = 0$ represent the present time and let *y* represent the amount (in grams) of carbon-14 in the sample. Using a base of $\frac{1}{2}$, you can model *y* by the equation

$$y = \left(\frac{1}{2}\right)^{t/5715}.$$

Notice that when $t = 5715$, the amount is reduced to half of the original amount.

$$y = \left(\frac{1}{2}\right)^{5715/5715} = \frac{1}{2} \text{ gram}$$

When $t = 11{,}430$, the amount is reduced to a quarter of the original amount, and so on. To find the amount of carbon-14 after 10,000 years, substitute 10,000 for *t*.

$$y = \left(\frac{1}{2}\right)^{10{,}000/5715}$$

$$\approx 0.30 \text{ gram}$$

The graph of *y* is shown in Figure 5.23.

The half-life of carbon-14 is about 5715 years.
Figure 5.23

Carbon dating uses the radioisotope carbon-14 to estimate the age of dead organic materials. The method is based on the decay rate of carbon-14 (see Example 1), a compound organisms take in when they are alive.

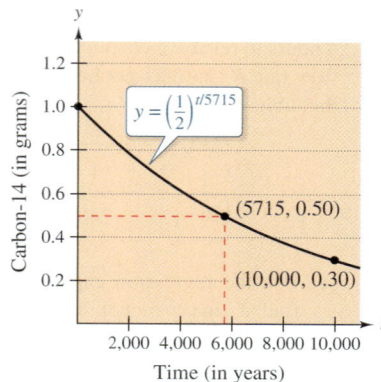

Logarithmic functions to bases other than *e* can be defined in much the same way as exponential functions to other bases are defined.

• • • • • • • • • • • • • • • • • • ▷

• • REMARK In precalculus, you learned that $\log_a x$ is the value to which *a* must be raised to produce *x*. This agrees with the definition at the right because

$$a^{\log_a x} = a^{(1/\ln a)\ln x}$$
$$= (e^{\ln a})^{(1/\ln a)\ln x}$$
$$= e^{(\ln a/\ln a)\ln x}$$
$$= e^{\ln x}$$
$$= x.$$

Definition of Logarithmic Function to Base *a*

If *a* is a positive real number $(a \neq 1)$ and *x* is any positive real number, then the **logarithmic function to the base *a*** is denoted by $\log_a x$ and is defined as

$$\log_a x = \frac{1}{\ln a} \ln x.$$

Logarithmic functions to the base *a* have properties similar to those of the natural logarithmic function given in Theorem 5.2. (Assume *x* and *y* are positive numbers and *n* is rational.)

1. $\log_a 1 = 0$ Log of 1

2. $\log_a xy = \log_a x + \log_a y$ Log of a product

3. $\log_a x^n = n \log_a x$ Log of a power

4. $\log_a \dfrac{x}{y} = \log_a x - \log_a y$ Log of a quotient

From the definitions of the exponential and logarithmic functions to the base *a*, it follows that $f(x) = a^x$ and $g(x) = \log_a x$ are inverse functions of each other.

Properties of Inverse Functions

1. $y = a^x$ if and only if $x = \log_a y$

2. $a^{\log_a x} = x$, for $x > 0$

3. $\log_a a^x = x$, for all *x*

The logarithmic function to the base 10 is called the **common logarithmic function.** So, for common logarithms,

$$y = 10^x \quad \text{if and only if} \quad x = \log_{10} y. \qquad \text{Property of Inverse Functions}$$

EXAMPLE 2   **Bases Other than *e***

Solve for *x* in each equation.

a. $3^x = \dfrac{1}{81}$ **b.** $\log_2 x = -4$

Solution

a. To solve this equation, you can apply the logarithmic function to the base 3 to each side of the equation.

$$3^x = \frac{1}{81}$$
$$\log_3 3^x = \log_3 \frac{1}{81}$$
$$x = \log_3 3^{-4}$$
$$x = -4$$

b. To solve this equation, you can apply the exponential function to the base 2 to each side of the equation.

$$\log_2 x = -4$$
$$2^{\log_2 x} = 2^{-4}$$
$$x = \frac{1}{2^4}$$
$$x = \frac{1}{16}$$

Differentiation and Integration

To differentiate exponential and logarithmic functions to other bases, you have three options: (1) use the definitions of a^x and $\log_a x$ and differentiate using the rules for the natural exponential and logarithmic functions, (2) use logarithmic differentiation, or (3) use the differentiation rules for bases other than e given in the next theorem.

▷

REMARK These differentiation rules are similar to those for the natural exponential function and the natural logarithmic function. In fact, they differ only by the constant factors $\ln a$ and $1/\ln a$. This points out one reason why, for calculus, e is the most convenient base.

THEOREM 5.13   Derivatives for Bases Other than e

Let a be a positive real number $(a \neq 1)$, and let u be a differentiable function of x.

1. $\dfrac{d}{dx}[a^x] = (\ln a)a^x$ 2. $\dfrac{d}{dx}[a^u] = (\ln a)a^u \dfrac{du}{dx}$

3. $\dfrac{d}{dx}[\log_a x] = \dfrac{1}{(\ln a)x}$ 4. $\dfrac{d}{dx}[\log_a u] = \dfrac{1}{(\ln a)u}\dfrac{du}{dx}$

Proof   By definition, $a^x = e^{(\ln a)x}$. So, you can prove the first rule by letting $u = (\ln a)x$ and differentiating with base e to obtain

$$\frac{d}{dx}[a^x] = \frac{d}{dx}[e^{(\ln a)x}] = e^u \frac{du}{dx} = e^{(\ln a)x}(\ln a) = (\ln a)a^x.$$

To prove the third rule, you can write

$$\frac{d}{dx}[\log_a x] = \frac{d}{dx}\left[\frac{1}{\ln a}\ln x\right] = \frac{1}{\ln a}\left(\frac{1}{x}\right) = \frac{1}{(\ln a)x}.$$

The second and fourth rules are simply the Chain Rule versions of the first and third rules.
See LarsonCalculus.com for Bruce Edwards's video of this proof.

EXAMPLE 3   **Differentiating Functions to Other Bases**

Find the derivative of each function.

a. $y = 2^x$ **b.** $y = 2^{3x}$ **c.** $y = \log_{10} \cos x$ **d.** $y = \log_3 \dfrac{\sqrt{x}}{x + 5}$

Solution

a. $y' = \dfrac{d}{dx}[2^x] = (\ln 2)2^x$

▷

b. $y' = \dfrac{d}{dx}[2^{3x}] = (\ln 2)2^{3x}(3) = (3 \ln 2)2^{3x}$

REMARK Try writing 2^{3x} as 8^x and differentiating to see that you obtain the same result.

c. $y' = \dfrac{d}{dx}[\log_{10} \cos x] = \dfrac{-\sin x}{(\ln 10)\cos x} = -\dfrac{1}{\ln 10}\tan x$

d. Before differentiating, rewrite the function using logarithmic properties.

$$y = \log_3 \frac{\sqrt{x}}{x + 5} = \frac{1}{2}\log_3 x - \log_3(x + 5)$$

Next, apply Theorem 5.13 to differentiate the function.

$$y' = \frac{d}{dx}\left[\frac{1}{2}\log_3 x - \log_3(x + 5)\right]$$

$$= \frac{1}{2(\ln 3)x} - \frac{1}{(\ln 3)(x + 5)}$$

$$= \frac{5 - x}{2(\ln 3)x(x + 5)}$$

Occasionally, an integrand involves an exponential function to a base other than *e*. When this occurs, there are two options: (1) convert to base *e* using the formula $a^x = e^{(\ln a)x}$ and then integrate, or (2) integrate directly, using the integration formula

$$\int a^x\, dx = \left(\frac{1}{\ln a}\right)a^x + C$$

which follows from Theorem 5.13.

EXAMPLE 4   **Integrating an Exponential Function to Another Base**

Find $\int 2^x\, dx$.

Solution

$$\int 2^x\, dx = \frac{1}{\ln 2} 2^x + C$$

When the Power Rule, $D_x[x^n] = nx^{n-1}$, was introduced in Chapter 2, the exponent *n* was required to be a rational number. Now the rule is extended to cover any real value of *n*. Try to prove this theorem using logarithmic differentiation.

THEOREM 5.14   **The Power Rule for Real Exponents**

Let *n* be any number, and let *u* be a differentiable function of *x*.

1. $\dfrac{d}{dx}[x^n] = nx^{n-1}$ **2.** $\dfrac{d}{dx}[u^n] = nu^{n-1}\dfrac{du}{dx}$

The next example compares the derivatives of four types of functions. Each function uses a different differentiation formula, depending on whether the base and the exponent are constants or variables.

EXAMPLE 5   **Comparing Variables and Constants**

a. $\dfrac{d}{dx}[e^e] = 0$ Constant Rule

b. $\dfrac{d}{dx}[e^x] = e^x$ Exponential Rule

c. $\dfrac{d}{dx}[x^e] = ex^{e-1}$ Power Rule

d. $y = x^x$ Logarithmic differentiation

$\ln y = \ln x^x$

$\ln y = x \ln x$

$\dfrac{y'}{y} = x\left(\dfrac{1}{x}\right) + (\ln x)(1)$

$\dfrac{y'}{y} = 1 + \ln x$

$y' = y(1 + \ln x)$

$y' = x^x(1 + \ln x)$

REMARK   Be sure you see that there is no simple differentiation rule for calculating the derivative of $y = x^x$. In general, when $y = u(x)^{v(x)}$, you need to use logarithmic differentiation.

Applications of Exponential Functions

An amount of P dollars is deposited in an account at an annual interest rate r (in decimal form). What is the balance in the account at the end of 1 year? The answer depends on the number of times n the interest is compounded according to the formula

$$A = P\left(1 + \frac{r}{n}\right)^n.$$

n	A
1	$1080.00
2	$1081.60
4	$1082.43
12	$1083.00
365	$1083.28

For instance, the result for a deposit of $1000 at 8% interest compounded n times a year is shown in the table at the right.

As n increases, the balance A approaches a limit. To develop this limit, use the next theorem. To test the reasonableness of this theorem, try evaluating

$$\left(\frac{x+1}{x}\right)^x$$

for several values of x, as shown in the table at the left.

x	$\left(\dfrac{x+1}{x}\right)^x$
10	2.59374
100	2.70481
1000	2.71692
10,000	2.71815
100,000	2.71827
1,000,000	2.71828

THEOREM 5.15 A Limit Involving e

$$\lim_{x\to\infty}\left(1 + \frac{1}{x}\right)^x = \lim_{x\to\infty}\left(\frac{x+1}{x}\right)^x = e$$

A proof of this theorem is given in Appendix A.
See LarsonCalculus.com for Bruce Edwards's video of this proof.

Given Theorem 5.15, take another look at the formula for the balance A in an account in which the interest is compounded n times per year. By taking the limit as n approaches infinity, you obtain

$$A = \lim_{n\to\infty} P\left(1 + \frac{r}{n}\right)^n \qquad \text{Take limit as } n \to \infty.$$

$$= P\lim_{n\to\infty}\left[\left(1 + \frac{1}{n/r}\right)^{n/r}\right]^r \qquad \text{Rewrite.}$$

$$= P\left[\lim_{x\to\infty}\left(1 + \frac{1}{x}\right)^x\right]^r \qquad \text{Let } x = n/r. \text{ Then } x \to \infty \text{ as } n \to \infty.$$

$$= Pe^r. \qquad \text{Apply Theorem 5.15.}$$

This limit produces the balance after 1 year of **continuous compounding.** So, for a deposit of $1000 at 8% interest compounded continuously, the balance at the end of 1 year would be

$$A = 1000e^{0.08} \approx \$1083.29.$$

SUMMARY OF COMPOUND INTEREST FORMULAS

Let P = amount of deposit, t = number of years, A = balance after t years, r = annual interest rate (decimal form), and n = number of compoundings per year.

1. Compounded n times per year: $A = P\left(1 + \dfrac{r}{n}\right)^{nt}$

2. Compounded continuously: $A = Pe^{rt}$

EXAMPLE 6   **Continuous, Quarterly, and Monthly Compounding**

• • • • ▷ *See LarsonCalculus.com for an interactive version of this type of example.*

A deposit of $2500 is made in an account that pays an annual interest rate of 5%. Find the balance in the account at the end of 5 years when the interest is compounded (a) quarterly, (b) monthly, and (c) continuously.

Solution

a. $A = P\left(1 + \dfrac{r}{n}\right)^{nt}$ ⟶ Compounded quarterly

$\quad = 2500\left(1 + \dfrac{0.05}{4}\right)^{4(5)}$

$\quad = 2500(1.0125)^{20}$

$\quad \approx \$3205.09$

b. $A = P\left(1 + \dfrac{r}{n}\right)^{nt}$ ⟶ Compounded monthly

$\quad = 2500\left(1 + \dfrac{0.05}{12}\right)^{12(5)}$

$\quad \approx 2500(1.0041667)^{60}$

$\quad \approx \$3208.40$

c. $A = Pe^{rt}$ ⟶ Compounded continuously

$\quad = 2500\left[e^{0.05(5)}\right]$

$\quad = 2500e^{0.25}$

$\quad \approx \$3210.06$

EXAMPLE 7   **Bacterial Culture Growth**

A bacterial culture is growing according to the *logistic growth function*

$$y = \frac{1.25}{1 + 0.25e^{-0.4t}}, \quad t \geq 0$$

where *y* is the weight of the culture in grams and *t* is the time in hours. Find the weight of the culture after (a) 0 hours, (b) 1 hour, and (c) 10 hours. (d) What is the limit as *t* approaches infinity?

Solution

a. When $t = 0$, $y = \dfrac{1.25}{1 + 0.25e^{-0.4(0)}}$

$\qquad\qquad\qquad = 1 \text{ gram.}$

b. When $t = 1$, $y = \dfrac{1.25}{1 + 0.25e^{-0.4(1)}}$

$\qquad\qquad\qquad \approx 1.071 \text{ grams.}$

c. When $t = 10$, $y = \dfrac{1.25}{1 + 0.25e^{-0.4(10)}}$

$\qquad\qquad\qquad \approx 1.244 \text{ grams.}$

d. Taking the limit as *t* approaches infinity, you obtain

$$\lim_{t \to \infty} \frac{1.25}{1 + 0.25e^{-0.4t}} = \frac{1.25}{1 + 0} = 1.25 \text{ grams.}$$

The graph of the function is shown in Figure 5.24.

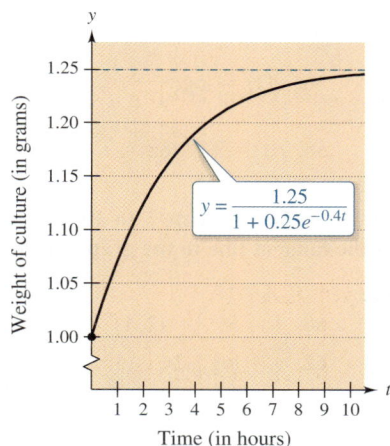

The limit of the weight of the culture as $t \to \infty$ is 1.25 grams.

Figure 5.24

Evaluating a Logarithmic Expression   In Exercises 1–4, evaluate the expression without using a calculator.

1. $\log_2 \frac{1}{8}$

2. $\log_{27} 9$

3. $\log_7 1$

4. $\log_a \frac{1}{a}$

Exponential and Logarithmic Forms of Equations   In Exercises 5–8, write the exponential equation as a logarithmic equation or vice versa.

5. (a) $2^3 = 8$

   (b) $3^{-1} = \frac{1}{3}$

6. (a) $27^{2/3} = 9$

   (b) $16^{3/4} = 8$

7. (a) $\log_{10} 0.01 = -2$

   (b) $\log_{0.5} 8 = -3$

8. (a) $\log_3 \frac{1}{9} = -2$

   (b) $49^{1/2} = 7$

Sketching a Graph   In Exercises 9–14, sketch the graph of the function by hand.

9. $y = 2^x$

10. $y = 4^{x-1}$

11. $y = \left(\frac{1}{3}\right)^x$

12. $y = 2^{x^2}$

13. $h(x) = 5^{x-2}$

14. $y = 3^{-|x|}$

Matching   In Exercises 15–18, match the function with its graph. [The graphs are labeled (a), (b), (c), and (d).]

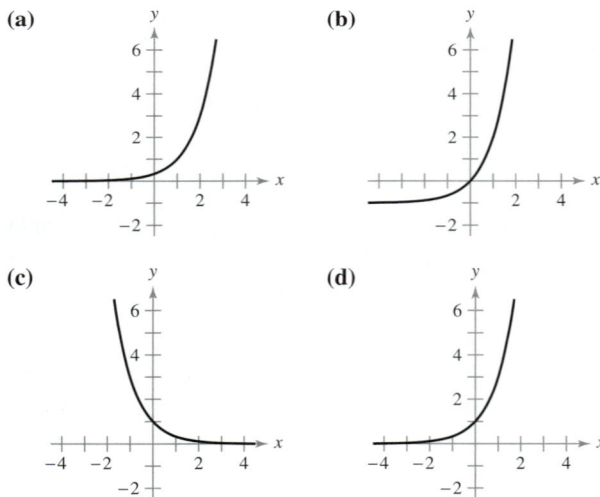

(a)

(b)

(c)

(d)

15. $f(x) = 3^x$

16. $f(x) = 3^{-x}$

17. $f(x) = 3^x - 1$

18. $f(x) = 3^{x-1}$

Solving an Equation   In Exercises 19–24, solve for x or b.

19. (a) $\log_{10} 1000 = x$

   (b) $\log_{10} 0.1 = x$

20. (a) $\log_3 \frac{1}{81} = x$

   (b) $\log_6 36 = x$

21. (a) $\log_3 x = -1$

   (b) $\log_2 x = -4$

22. (a) $\log_b 27 = 3$

   (b) $\log_b 125 = 3$

23. (a) $x^2 - x = \log_5 25$

   (b) $3x + 5 = \log_2 64$

24. (a) $\log_3 x + \log_3(x - 2) = 1$

   (b) $\log_{10}(x + 3) - \log_{10} x = 1$

Solving an Equation   In Exercises 25–34, solve the equation accurate to three decimal places.

25. $3^{2x} = 75$

26. $5^{6x} = 8320$

27. $2^{3-z} = 625$

28. $3(5^{x-1}) = 86$

29. $\left(1 + \frac{0.09}{12}\right)^{12t} = 3$

30. $\left(1 + \frac{0.10}{365}\right)^{365t} = 2$

31. $\log_2(x - 1) = 5$

32. $\log_{10}(t - 3) = 2.6$

33. $\log_3 x^2 = 4.5$

34. $\log_5 \sqrt{x - 4} = 3.2$

Verifying Inverse Functions   In Exercises 35 and 36, illustrate that the functions are inverse functions of each other by sketching their graphs on the same set of coordinate axes.

35. $f(x) = 4^x$

   $g(x) = \log_4 x$

36. $f(x) = 3^x$

   $g(x) = \log_3 x$

Finding a Derivative   In Exercises 37–58, find the derivative of the function. (*Hint:* In some exercises, you may find it helpful to apply logarithmic properties *before* differentiating.)

37. $f(x) = 4^x$

38. $f(x) = 3^{4x}$

39. $y = 5^{-4x}$

40. $y = 6^{3x-4}$

41. $f(x) = x\, 9^x$

42. $y = x(6^{-2x})$

43. $g(t) = t^2 2^t$

44. $f(t) = \frac{3^{2t}}{t}$

45. $h(\theta) = 2^{-\theta} \cos \pi\theta$

46. $g(\alpha) = 5^{-\alpha/2} \sin 2\alpha$

47. $y = \log_4(5x + 1)$

48. $y = \log_3(x^2 - 3x)$

49. $h(t) = \log_5(4 - t)^2$

50. $g(t) = \log_2(t^2 + 7)^3$

51. $y = \log_5 \sqrt{x^2 - 1}$

52. $f(x) = \log_2 \sqrt[3]{2x + 1}$

53. $f(x) = \log_2 \frac{x^2}{x - 1}$

54. $y = \log_{10} \frac{x^2 - 1}{x}$

55. $h(x) = \log_3 \frac{x\sqrt{x - 1}}{2}$

56. $g(x) = \log_5 \frac{4}{x^2 \sqrt{1 - x}}$

57. $g(t) = \frac{10 \log_4 t}{t}$

58. $f(t) = t^{3/2} \log_2 \sqrt{t + 1}$

Finding an Equation of a Tangent Line   In Exercises 59–62, find an equation of the tangent line to the graph of the function at the given point.

59. $y = 2^{-x}$, $(-1, 2)$

60. $y = 5^{x-2}$, $(2, 1)$

61. $y = \log_3 x$, $(27, 3)$

62. $y = \log_{10} 2x$, $(5, 1)$

Logarithmic Differentiation   In Exercises 63–66, use logarithmic differentiation to find dy/dx.

63. $y = x^{2/x}$

64. $y = x^{x-1}$

65. $y = (x - 2)^{x+1}$

66. $y = (1 + x)^{1/x}$

Finding an Equation of a Tangent Line   In Exercises 67–70, find an equation of the tangent line to the graph of the function at the given point.

67. $y = x^{\sin x}$, $\left(\dfrac{\pi}{2}, \dfrac{\pi}{2}\right)$ **68.** $y = (\sin x)^{2x}$, $\left(\dfrac{\pi}{2}, 1\right)$

69. $y = (\ln x)^{\cos x}$, $(e, 1)$ **70.** $y = x^{1/x}$, $(1, 1)$

Finding an Indefinite Integral   In Exercises 71–78, find the indefinite integral.

71. $\displaystyle\int 3^x \, dx$ **72.** $\displaystyle\int 8^{-x} \, dx$

73. $\displaystyle\int (x^2 + 2^{-x}) \, dx$ **74.** $\displaystyle\int (x^4 + 5^x) \, dx$

75. $\displaystyle\int x(5^{-x^2}) \, dx$ **76.** $\displaystyle\int (x + 4)6^{(x+4)^2} \, dx$

77. $\displaystyle\int \dfrac{3^{2x}}{1 + 3^{2x}} \, dx$ **78.** $\displaystyle\int 2^{\sin x} \cos x \, dx$

Evaluating a Definite Integral   In Exercises 79–82, evaluate the definite integral.

79. $\displaystyle\int_{-1}^{2} 2^x \, dx$ **80.** $\displaystyle\int_{-4}^{4} 3^{x/4} \, dx$

81. $\displaystyle\int_{0}^{1} (5^x - 3^x) \, dx$ **82.** $\displaystyle\int_{1}^{3} (7^x - 4^x) \, dx$

Area   In Exercises 83 and 84, find the area of the region bounded by the graphs of the equations.

83. $y = 3^x$, $y = 0$, $x = 0$, $x = 3$

84. $y = 3^{\cos x} \sin x$, $y = 0$, $x = 0$, $x = \pi$

WRITING ABOUT CONCEPTS

85. Analyzing a Logarithmic Equation   Consider the function $f(x) = \log_{10} x$.

  (a) What is the domain of f?

  (b) Find f^{-1}.

  (c) Let x be a real number between 1000 and 10,000. Determine the interval in which $f(x)$ will be found.

  (d) Determine the interval in which x will be found if $f(x)$ is negative.

  (e) When $f(x)$ is increased by one unit, x must have been increased by what factor?

  (f) Find the ratio of x_1 to x_2 given that $f(x_1) = 3n$ and $f(x_2) = n$.

86. Comparing Rates of Growth   Order the functions

$$f(x) = \log_2 x, \ g(x) = x^x, \ h(x) = x^2, \text{ and } k(x) = 2^x$$

from the one with the greatest rate of growth to the one with the least rate of growth for large values of x.

---

87. Inflation   When the annual rate of inflation averages 5% over the next 10 years, the approximate cost C of goods or services during any year in that decade is

$$C(t) = P(1.05)^t$$

where t is the time in years and P is the present cost.

  (a) The price of an oil change for your car is presently $24.95. Estimate the price 10 years from now.

  (b) Find the rates of change of C with respect to t when $t = 1$ and $t = 8$.

  (c) Verify that the rate of change of C is proportional to C. What is the constant of proportionality?

88. Depreciation   After t years, the value of a car purchased for $25,000 is

$$V(t) = 25{,}000\left(\tfrac{3}{4}\right)^t.$$

  (a) Use a graphing utility to graph the function and determine the value of the car 2 years after it was purchased.

  (b) Find the rates of change of V with respect to t when $t = 1$ and $t = 4$.

  (c) Use a graphing utility to graph $V'(t)$ and determine the horizontal asymptote of $V'(t)$. Interpret its meaning in the context of the problem.

Compound Interest   In Exercises 89–92, complete the table by determining the balance A for P dollars invested at rate r for t years and compounded n times per year.

n	1	2	4	12	365	Continuous Compounding
A						

89. $P = \$1000$ **90.** $P = \$2500$

    $r = 3\tfrac{1}{2}\%$ $r = 6\%$

    $t = 10$ years $t = 20$ years

91. $P = \$1000$ **92.** $P = \$4000$

    $r = 5\%$ $r = 4\%$

    $t = 30$ years $t = 15$ years

Compound Interest   In Exercises 93–96, complete the table by determining the amount of money P (present value) that should be invested at rate r to produce a balance of $100,000 in t years.

t	1	10	20	30	40	50
P						

93. $r = 5\%$ **94.** $r = 3\%$

    Compounded continuously   Compounded continuously

95. $r = 5\%$ **96.** $r = 2\%$

    Compounded monthly   Compounded daily

97. Compound Interest Assume that you can earn 6% on an investment, compounded daily. Which of the following options would yield the greatest balance after 8 years?

(a) $20,000 now (b) $30,000 after 8 years

(c) $8000 now and $20,000 after 4 years

(d) $9000 now, $9000 after 4 years, and $9000 after 8 years

98. Compound Interest Consider a deposit of $100 placed in an account for 20 years at r% compounded continuously. Use a graphing utility to graph the exponential functions describing the growth of the investment over the 20 years for the following interest rates. Compare the ending balances for the three rates.

(a) $r = 3\%$ (b) $r = 5\%$ (c) $r = 6\%$

99. Timber Yield The yield V (in millions of cubic feet per acre) for a stand of timber at age t is $V = 6.7e^{(-48.1)/t}$, where t is measured in years.

(a) Find the limiting volume of wood per acre as t approaches infinity.

(b) Find the rates at which the yield is changing when $t = 20$ years and $t = 60$ years.

100. HOW DO YOU SEE IT? The graph shows the proportion P of correct responses after n trials in a group project in learning theory.

$$P = \frac{0.86}{1 + e^{-0.25n}}$$

(a) What is the limiting proportion of correct responses as n approaches infinity?

(b) What happens to the rate of change of the proportion in the long run?

101. Population Growth A lake is stocked with 500 fish, and the population increases according to the logistic curve

$$p(t) = \frac{10,000}{1 + 19e^{-t/5}}$$

where t is measured in months.

(a) Use a graphing utility to graph the function.

(b) What is the limiting size of the fish population?

(c) At what rates is the fish population changing at the end of 1 month and at the end of 10 months?

(d) After how many months is the population increasing most rapidly?

102. Modeling Data The breaking strengths B (in tons) of steel cables of various diameters d (in inches) are shown in the table.

d	0.50	0.75	1.00	1.25	1.50	1.75
B	9.85	21.8	38.3	59.2	84.4	114.0

(a) Use the regression capabilities of a graphing utility to fit an exponential model to the data.

(b) Use a graphing utility to plot the data and graph the model.

(c) Find the rates of growth of the model when $d = 0.8$ and $d = 1.5$.

103. Comparing Models The numbers y of pancreas transplants in the United States for the years 2004 through 2010 are shown in the table, with $x = 4$ corresponding to 2004. (*Source: Organ Procurement and Transplantation Network*)

x	4	5	6	7	8	9	10
y	603	542	466	468	436	376	350

(a) Use the regression capabilities of a graphing utility to find the following models for the data.

$y_1 = ax + b$ $y_2 = a + b \ln x$

$y_3 = ab^x$ $y_4 = ax^b$

(b) Use a graphing utility to plot the data and graph each of the models. Which model do you think best fits the data?

(c) Interpret the slope of the linear model in the context of the problem.

(d) Find the rate of change of each of the models for the year 2008. Which model is decreasing at the greatest rate in 2008?

104. An Approximation of e Complete the table to demonstrate that e can also be defined as

$$\lim_{x \to 0^+} (1 + x)^{1/x}.$$

x	1	10^{-1}	10^{-2}	10^{-4}	10^{-6}
$(1 + x)^{1/x}$					

Modeling Data In Exercises 105 and 106, find an exponential function that fits the experimental data collected over time t.

105.

t	0	1	2	3	4
y	1200.00	720.00	432.00	259.20	155.52

106.

t	0	1	2	3	4
y	600.00	630.00	661.50	694.58	729.30

Using Properties of Exponents In Exercises 107–110, find the exact value of the expression.

107. $5^{1/\ln 5}$

108. $6^{\ln 10/\ln 6}$

109. $9^{1/\ln 3}$

110. $32^{1/\ln 2}$

True or False? In Exercises 111–116, determine whether the statement is true or false. If it is false, explain why or give an example that shows it is false.

111. $e = \dfrac{271{,}801}{99{,}900}$

112. If $f(x) = \ln x$, then $f(e^{n+1}) - f(e^n) = 1$ for any value of n.

113. The functions $f(x) = 2 + e^x$ and $g(x) = \ln(x - 2)$ are inverse functions of each other.

114. The exponential function $y = Ce^x$ is a solution of the differential equation

$$\frac{d^n y}{dx^n} = y, \quad n = 1, 2, 3, \ldots.$$

115. The graphs of $f(x) = e^x$ and $g(x) = e^{-x}$ meet at right angles.

116. If $f(x) = g(x)e^x$, then the only zeros of f are the zeros of g.

117. Comparing Functions

(a) Show that $(2^3)^2 \neq 2^{(3^2)}$.

(b) Are $f(x) = (x^x)^x$ and $g(x) = x^{(x^x)}$ the same function? Why or why not?

(c) Find $f'(x)$ and $g'(x)$.

118. Finding an Inverse Function Let

$$f(x) = \frac{a^x - 1}{a^x + 1}$$

for $a > 0$, $a \neq 1$. Show that f has an inverse function. Then find f^{-1}.

119. Logistic Differential Equation Show that solving the logistic differential equation

$$\frac{dy}{dt} = \frac{8}{25} y \left(\frac{5}{4} - y \right), \quad y(0) = 1$$

results in the logistic growth function in Example 7.

$$\left[Hint: \ \frac{1}{y\left(\frac{5}{4} - y\right)} = \frac{4}{5}\left(\frac{1}{y} + \frac{1}{\frac{5}{4} - y} \right) \right]$$

120. Using Properties of Exponents Given the exponential function $f(x) = a^x$, show that

(a) $f(u + v) = f(u) \cdot f(v)$.

(b) $f(2x) = [f(x)]^2$.

121. Tangent Lines

(a) Determine y' given $y^x = x^y$.

(b) Find the slope of the tangent line to the graph of $y^x = x^y$ at each of the following points.

    (i) (c, c) (ii) $(2, 4)$ (iii) $(4, 2)$

(c) At what points on the graph of $y^x = x^y$ does the tangent line not exist?

SECTION PROJECT

Using Graphing Utilities to Estimate Slope

Let $f(x) = \begin{cases} |x|^x, & x \neq 0 \\ 1, & x = 0. \end{cases}$

(a) Use a graphing utility to graph f in the viewing window $-3 \leq x \leq 3$, $-2 \leq y \leq 2$. What is the domain of f?

(b) Use the *zoom* and *trace* features of a graphing utility to estimate

$$\lim_{x \to 0} f(x).$$

(c) Write a short paragraph explaining why the function f is continuous for all real numbers.

(d) Visually estimate the slope of f at the point $(0, 1)$.

(e) Explain why the derivative of a function can be approximated by the formula

$$\frac{f(x + \Delta x) - f(x - \Delta x)}{2\Delta x}$$

for small values of Δx. Use this formula to approximate the slope of f at the point $(0, 1)$.

$$f'(0) \approx \frac{f(0 + \Delta x) - f(0 - \Delta x)}{2\Delta x}$$

$$= \frac{f(\Delta x) - f(-\Delta x)}{2\Delta x}$$

What do you think the slope of the graph of f is at $(0, 1)$?

(f) Find a formula for the derivative of f and determine $f'(0)$. Write a short paragraph explaining how a graphing utility might lead you to approximate the slope of a graph incorrectly.

(g) Use your formula for the derivative of f to find the relative extrema of f. Verify your answer using a graphing utility.

■ **FOR FURTHER INFORMATION** For more information on using graphing utilities to estimate slope, see the article "Computer-Aided Delusions" by Richard L. Hall in *The College Mathematics Journal*. To view this article, go to *MathArticles.com*.

5.6 Inverse Trigonometric Functions: Differentiation

■ Develop properties of the six inverse trigonometric functions.
■ Differentiate an inverse trigonometric function.
■ Review the basic differentiation rules for elementary functions.

$y = \sin x$
Domain: $[-\pi/2, \pi/2]$
Range: $[-1, 1]$

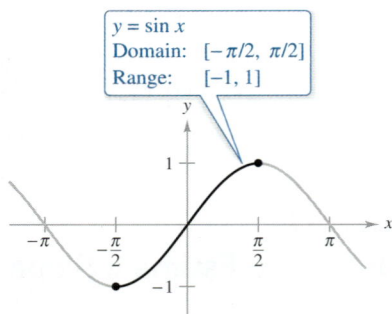

The sine function is one-to-one on $[-\pi/2, \pi/2]$.
Figure 5.25

Inverse Trigonometric Functions

This section begins with a rather surprising statement: *None of the six basic trigonometric functions has an inverse function.* This statement is true because all six trigonometric functions are periodic and therefore are not one-to-one. In this section, you will examine these six functions to see whether their domains can be redefined in such a way that they will have inverse functions on the *restricted domains*.

In Example 4 of Section 5.3, you saw that the sine function is increasing (and therefore is one-to-one) on the interval

$$\left[-\frac{\pi}{2}, \frac{\pi}{2} \right]$$

as shown in Figure 5.25. On this interval, you can define the inverse of the *restricted* sine function as

$y = \arcsin x$ if and only if $\sin y = x$

where $-1 \le x \le 1$ and $-\pi/2 \le \arcsin x \le \pi/2$.

Under suitable restrictions, each of the six trigonometric functions is one-to-one and so has an inverse function, as shown in the next definition. (Note that the term "iff" is used to represent the phrase "if and only if.")

Definitions of Inverse Trigonometric Functions

Function	Domain	Range
$y = \arcsin x$ iff $\sin y = x$	$-1 \le x \le 1$	$-\dfrac{\pi}{2} \le y \le \dfrac{\pi}{2}$
$y = \arccos x$ iff $\cos y = x$	$-1 \le x \le 1$	$0 \le y \le \pi$
$y = \arctan x$ iff $\tan y = x$	$-\infty < x < \infty$	$-\dfrac{\pi}{2} < y < \dfrac{\pi}{2}$
$y = \text{arccot } x$ iff $\cot y = x$	$-\infty < x < \infty$	$0 < y < \pi$
$y = \text{arcsec } x$ iff $\sec y = x$	$\lvert x \rvert \ge 1$	$0 \le y \le \pi, \quad y \ne \dfrac{\pi}{2}$
$y = \text{arccsc } x$ iff $\csc y = x$	$\lvert x \rvert \ge 1$	$-\dfrac{\pi}{2} \le y \le \dfrac{\pi}{2}, \quad y \ne 0$

• • • • • • • • • • • • • • • • ▷

•• **REMARK**   The term "arcsin x" is read as "the arcsine of x" or sometimes "the angle whose sine is x." An alternative notation for the inverse sine function is "$\sin^{-1} x$."

Exploration

The Inverse Secant Function   In the definitions of the inverse trigonometric functions, the inverse secant function is defined by restricting the domain of the secant function to the intervals $[0, \pi/2) \cup (\pi/2, \pi]$. Most other texts and reference books agree with this, but some disagree. What other domains might make sense? Explain your reasoning graphically. Most calculators do not have a key for the inverse secant function. How can you use a calculator to evaluate the inverse secant function?

The graphs of the six inverse trigonometric functions are shown in Figure 5.26.

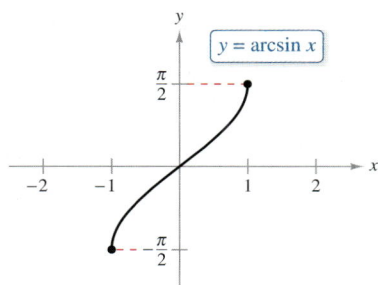

Domain: $[-1, 1]$
Range: $[-\pi/2, \pi/2]$

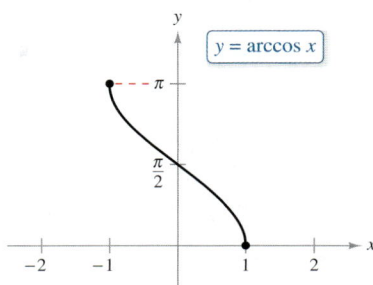

Domain: $[-1, 1]$
Range: $[0, \pi]$

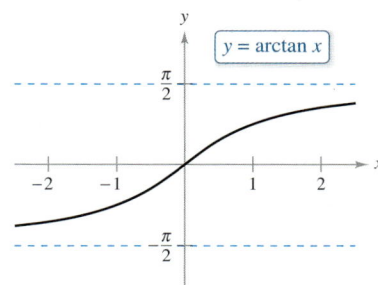

Domain: $(-\infty, \infty)$
Range: $(-\pi/2, \pi/2)$

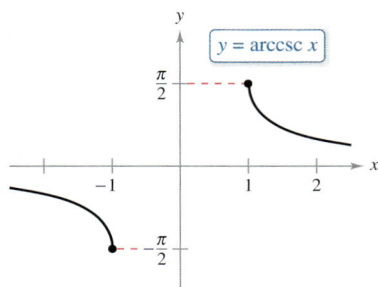

Domain: $(-\infty, -1] \cup [1, \infty)$
Range: $[-\pi/2, 0) \cup (0, \pi/2]$

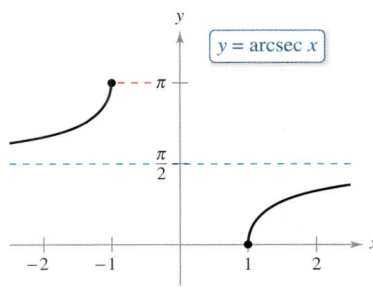

Domain: $(-\infty, -1] \cup [1, \infty)$
Range: $[0, \pi/2) \cup (\pi/2, \pi]$

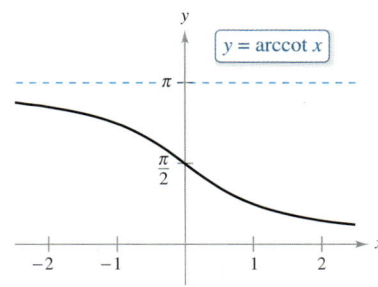

Domain: $(-\infty, \infty)$
Range: $(0, \pi)$

Figure 5.26

When evaluating inverse trigonometric functions, remember that they denote angles in *radian measure*.

EXAMPLE 1   **Evaluating Inverse Trigonometric Functions**

Evaluate each function.

a. $\arcsin\left(-\dfrac{1}{2}\right)$ **b.** $\arccos 0$ **c.** $\arctan \sqrt{3}$ **d.** $\arcsin(0.3)$

Solution

a. By definition, $y = \arcsin\left(-\frac{1}{2}\right)$ implies that $\sin y = -\frac{1}{2}$. In the interval $[-\pi/2, \pi/2]$, the correct value of y is $-\pi/6$.

$$\arcsin\left(-\frac{1}{2}\right) = -\frac{\pi}{6}$$

b. By definition, $y = \arccos 0$ implies that $\cos y = 0$. In the interval $[0, \pi]$, you have $y = \pi/2$.

$$\arccos 0 = \frac{\pi}{2}$$

c. By definition, $y = \arctan \sqrt{3}$ implies that $\tan y = \sqrt{3}$. In the interval $(-\pi/2, \pi/2)$, you have $y = \pi/3$.

$$\arctan \sqrt{3} = \frac{\pi}{3}$$

d. Using a calculator set in *radian* mode produces

$$\arcsin(0.3) \approx 0.305.$$

Inverse functions have the properties $f(f^{-1}(x)) = x$ and $f^{-1}(f(x)) = x$. When applying these properties to inverse trigonometric functions, remember that the trigonometric functions have inverse functions only in restricted domains. For x-values outside these domains, these two properties do not hold. For example, $\arcsin(\sin \pi)$ is equal to 0, not π.

Properties of Inverse Trigonometric Functions

If $-1 \le x \le 1$ and $-\pi/2 \le y \le \pi/2$, then

$$\sin(\arcsin x) = x \quad \text{and} \quad \arcsin(\sin y) = y.$$

If $-\pi/2 < y < \pi/2$, then

$$\tan(\arctan x) = x \quad \text{and} \quad \arctan(\tan y) = y.$$

If $|x| \ge 1$ and $0 \le y < \pi/2$ or $\pi/2 < y \le \pi$, then

$$\sec(\arcsec x) = x \quad \text{and} \quad \arcsec(\sec y) = y.$$

Similar properties hold for the other inverse trigonometric functions.

EXAMPLE 2 **Solving an Equation**

$$\arctan(2x - 3) = \frac{\pi}{4} \qquad \text{Original equation}$$

$$\tan[\arctan(2x - 3)] = \tan\frac{\pi}{4} \qquad \text{Take tangent of each side.}$$

$$2x - 3 = 1 \qquad \tan(\arctan x) = x$$

$$x = 2 \qquad \text{Solve for } x.$$

Some problems in calculus require that you evaluate expressions such as $\cos(\arcsin x)$, as shown in Example 3.

EXAMPLE 3 **Using Right Triangles**

a. Given $y = \arcsin x$, where $0 < y < \pi/2$, find $\cos y$.
b. Given $y = \arcsec(\sqrt{5}/2)$, find $\tan y$.

Solution

a. Because $y = \arcsin x$, you know that $\sin y = x$. This relationship between x and y can be represented by a right triangle, as shown in the figure at the right.

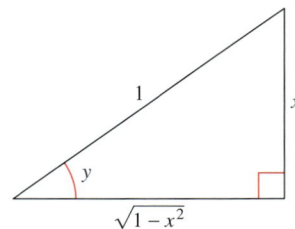

$$\cos y = \cos(\arcsin x) = \frac{\text{adj.}}{\text{hyp.}} = \sqrt{1 - x^2}$$

(This result is also valid for $-\pi/2 < y < 0$.)

b. Use the right triangle shown in the figure at the left.

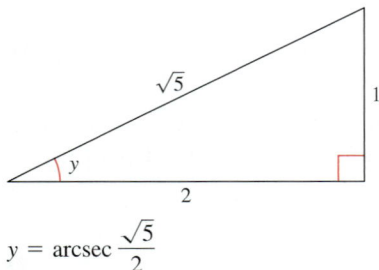

$$\tan y = \tan\left[\arcsec\left(\frac{\sqrt{5}}{2}\right)\right]$$

$$= \frac{\text{opp.}}{\text{adj.}}$$

$$= \frac{1}{2}$$

$y = \arcsin x$

$y = \arcsec\dfrac{\sqrt{5}}{2}$

Derivatives of Inverse Trigonometric Functions

In Section 5.1, you saw that the derivative of the *transcendental* function $f(x) = \ln x$ is the *algebraic* function $f'(x) = 1/x$. You will now see that the derivatives of the inverse trigonometric functions also are algebraic (even though the inverse trigonometric functions are themselves transcendental).

The next theorem lists the derivatives of the six inverse trigonometric functions. Note that the derivatives of arccos u, arccot u, and arccsc u are the *negatives* of the derivatives of arcsin u, arctan u, and arcsec u, respectively.

THEOREM 5.16 Derivatives of Inverse Trigonometric Functions

Let u be a differentiable function of x.

$$\frac{d}{dx}[\arcsin u] = \frac{u'}{\sqrt{1 - u^2}} \qquad \frac{d}{dx}[\arccos u] = \frac{-u'}{\sqrt{1 - u^2}}$$

$$\frac{d}{dx}[\arctan u] = \frac{u'}{1 + u^2} \qquad \frac{d}{dx}[\operatorname{arccot} u] = \frac{-u'}{1 + u^2}$$

$$\frac{d}{dx}[\operatorname{arcsec} u] = \frac{u'}{|u|\sqrt{u^2 - 1}} \qquad \frac{d}{dx}[\operatorname{arccsc} u] = \frac{-u'}{|u|\sqrt{u^2 - 1}}$$

Proofs for arcsin u and arccos u are given in Appendix A. [The proofs for the other rules are left as an exercise (see Exercise 98).]

See LarsonCalculus.com for Bruce Edwards's video of this proof.

▷ **TECHNOLOGY** If your graphing utility does not have the arcsecant function, you can obtain its graph using

$$f(x) = \operatorname{arcsec} x = \arccos \frac{1}{x}.$$

EXAMPLE 4 Differentiating Inverse Trigonometric Functions

a. $\dfrac{d}{dx}[\arcsin(2x)] = \dfrac{2}{\sqrt{1 - (2x)^2}} = \dfrac{2}{\sqrt{1 - 4x^2}}$

b. $\dfrac{d}{dx}[\arctan(3x)] = \dfrac{3}{1 + (3x)^2} = \dfrac{3}{1 + 9x^2}$

c. $\dfrac{d}{dx}[\arcsin \sqrt{x}] = \dfrac{(1/2)\, x^{-1/2}}{\sqrt{1 - x}} = \dfrac{1}{2\sqrt{x}\sqrt{1 - x}} = \dfrac{1}{2\sqrt{x - x^2}}$

d. $\dfrac{d}{dx}[\operatorname{arcsec} e^{2x}] = \dfrac{2e^{2x}}{e^{2x}\sqrt{(e^{2x})^2 - 1}} = \dfrac{2e^{2x}}{e^{2x}\sqrt{e^{4x} - 1}} = \dfrac{2}{\sqrt{e^{4x} - 1}}$

The absolute value sign is not necessary because $e^{2x} > 0$.

EXAMPLE 5 A Derivative That Can Be Simplified

$$y = \arcsin x + x\sqrt{1 - x^2}$$

$$y' = \frac{1}{\sqrt{1 - x^2}} + x\left(\frac{1}{2}\right)(-2x)(1 - x^2)^{-1/2} + \sqrt{1 - x^2}$$

$$= \frac{1}{\sqrt{1 - x^2}} - \frac{x^2}{\sqrt{1 - x^2}} + \sqrt{1 - x^2}$$

$$= \sqrt{1 - x^2} + \sqrt{1 - x^2}$$

$$= 2\sqrt{1 - x^2}$$

From Example 5, you can see one of the benefits of inverse trigonometric functions—they can be used to integrate common algebraic functions. For instance, from the result shown in the example, it follows that

$$\int \sqrt{1 - x^2}\, dx = \frac{1}{2}\left(\arcsin x + x\sqrt{1 - x^2}\right).$$

EXAMPLE 6   **Analyzing an Inverse Trigonometric Graph**

Analyze the graph of $y = (\arctan x)^2$.

Solution   From the derivative

$$y' = 2(\arctan x)\left(\frac{1}{1 + x^2}\right)$$

$$= \frac{2 \arctan x}{1 + x^2}$$

you can see that the only critical number is $x = 0$. By the First Derivative Test, this value corresponds to a relative minimum. From the second derivative

$$y'' = \frac{(1 + x^2)\left(\dfrac{2}{1 + x^2}\right) - (2 \arctan x)(2x)}{(1 + x^2)^2}$$

$$= \frac{2(1 - 2x \arctan x)}{(1 + x^2)^2}$$

it follows that points of inflection occur when $2x \arctan x = 1$. Using Newton's Method, these points occur when $x \approx \pm 0.765$. Finally, because

$$\lim_{x \to \pm\infty} (\arctan x)^2 = \frac{\pi^2}{4}$$

it follows that the graph has a horizontal asymptote at $y = \pi^2/4$. The graph is shown in Figure 5.27.

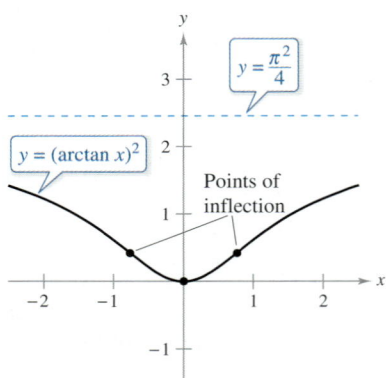

The graph of $y = (\arctan x)^2$ has a horizontal asymptote at $y = \pi^2/4$.
Figure 5.27

EXAMPLE 7   **Maximizing an Angle**

••••▷ *See LarsonCalculus.com for an interactive version of this type of example.*

A photographer is taking a picture of a painting hung in an art gallery. The height of the painting is 4 feet. The camera lens is 1 foot below the lower edge of the painting, as shown in the figure at the right. How far should the camera be from the painting to maximize the angle subtended by the camera lens?

Solution   In the figure, let β be the angle to be maximized.

$$\beta = \theta - \alpha$$

$$= \text{arccot}\,\frac{x}{5} - \text{arccot}\,x$$

Differentiating produces

$$\frac{d\beta}{dx} = \frac{-1/5}{1 + (x^2/25)} - \frac{-1}{1 + x^2}$$

$$= \frac{-5}{25 + x^2} + \frac{1}{1 + x^2}$$

$$= \frac{4(5 - x^2)}{(25 + x^2)(1 + x^2)}.$$

Because $d\beta/dx = 0$ when $x = \sqrt{5}$, you can conclude from the First Derivative Test that this distance yields a maximum value of β. So, the distance is $x \approx 2.236$ feet and the angle is $\beta \approx 0.7297$ radian $\approx 41.81°$.

Not drawn to scale

The camera should be 2.236 feet from the painting to maximize the angle β.

Review of Basic Differentiation Rules

In the 1600s, Europe was ushered into the scientific age by such great thinkers as Descartes, Galileo, Huygens, Newton, and Kepler. These men believed that nature is governed by basic laws—laws that can, for the most part, be written in terms of mathematical equations. One of the most influential publications of this period—*Dialogue on the Great World Systems*, by Galileo Galilei—has become a classic description of modern scientific thought.

As mathematics has developed during the past few hundred years, a small number of elementary functions have proven sufficient for modeling most* phenomena in physics, chemistry, biology, engineering, economics, and a variety of other fields. An **elementary function** is a function from the following list or one that can be formed as the sum, product, quotient, or composition of functions in the list.

Algebraic Functions	**Transcendental Functions**
Polynomial functions	Logarithmic functions
Rational functions	Exponential functions
Functions involving radicals	Trigonometric functions
	Inverse trigonometric functions

With the differentiation rules introduced so far in the text, you can differentiate *any* elementary function. For convenience, these differentiation rules are summarized below.

GALILEO GALILEI (1564–1642)

Galileo's approach to science departed from the accepted Aristotelian view that nature had describable *qualities*, such as "fluidity" and "potentiality." He chose to describe the physical world in terms of measurable *quantities*, such as time, distance, force, and mass.
See LarsonCalculus.com to read more of this biography.

BASIC DIFFERENTIATION RULES FOR ELEMENTARY FUNCTIONS

1. $\dfrac{d}{dx}[cu] = cu'$

2. $\dfrac{d}{dx}[u \pm v] = u' \pm v'$

3. $\dfrac{d}{dx}[uv] = uv' + vu'$

4. $\dfrac{d}{dx}\left[\dfrac{u}{v}\right] = \dfrac{vu' - uv'}{v^2}$

5. $\dfrac{d}{dx}[c] = 0$

6. $\dfrac{d}{dx}[u^n] = nu^{n-1}u'$

7. $\dfrac{d}{dx}[x] = 1$

8. $\dfrac{d}{dx}[|u|] = \dfrac{u}{|u|}(u'), \quad u \neq 0$

9. $\dfrac{d}{dx}[\ln u] = \dfrac{u'}{u}$

10. $\dfrac{d}{dx}[e^u] = e^u u'$

11. $\dfrac{d}{dx}[\log_a u] = \dfrac{u'}{(\ln a)u}$

12. $\dfrac{d}{dx}[a^u] = (\ln a)a^u u'$

13. $\dfrac{d}{dx}[\sin u] = (\cos u)u'$

14. $\dfrac{d}{dx}[\cos u] = -(\sin u)u'$

15. $\dfrac{d}{dx}[\tan u] = (\sec^2 u)u'$

16. $\dfrac{d}{dx}[\cot u] = -(\csc^2 u)u'$

17. $\dfrac{d}{dx}[\sec u] = (\sec u \tan u)u'$

18. $\dfrac{d}{dx}[\csc u] = -(\csc u \cot u)u'$

19. $\dfrac{d}{dx}[\arcsin u] = \dfrac{u'}{\sqrt{1 - u^2}}$

20. $\dfrac{d}{dx}[\arccos u] = \dfrac{-u'}{\sqrt{1 - u^2}}$

21. $\dfrac{d}{dx}[\arctan u] = \dfrac{u'}{1 + u^2}$

22. $\dfrac{d}{dx}[\operatorname{arccot} u] = \dfrac{-u'}{1 + u^2}$

23. $\dfrac{d}{dx}[\operatorname{arcsec} u] = \dfrac{u'}{|u|\sqrt{u^2 - 1}}$

24. $\dfrac{d}{dx}[\operatorname{arccsc} u] = \dfrac{-u'}{|u|\sqrt{u^2 - 1}}$

---

* Some important functions used in engineering and science (such as Bessel functions and gamma functions) are not elementary functions.

5.6 Exercises

See CalcChat.com for tutorial help and worked-out solutions to odd-numbered exercises.

Finding Coordinates In Exercises 1 and 2, determine the missing coordinates of the points on the graph of the function.

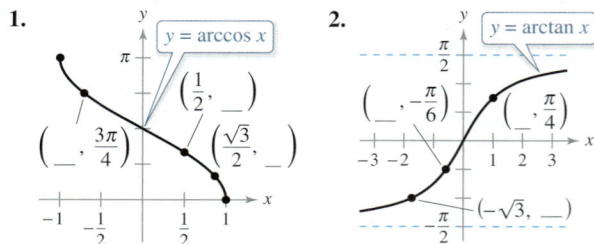

1.

2.

Evaluating Inverse Trigonometric Functions In Exercises 3–10, evaluate the expression without using a calculator.

3. $\arcsin \frac{1}{2}$

4. $\arcsin 0$

5. $\arccos \frac{1}{2}$

6. $\arccos 1$

7. $\arctan \dfrac{\sqrt{3}}{3}$

8. $\operatorname{arccot}\left(-\sqrt{3}\right)$

9. $\operatorname{arccsc}\left(-\sqrt{2}\right)$

10. $\operatorname{arcsec}\left(-\sqrt{2}\right)$

Approximating Inverse Trigonometric Functions In Exercises 11–14, use a calculator to approximate the value. Round your answer to two decimal places.

11. $\arccos(-0.8)$

12. $\arcsin(-0.39)$

13. $\operatorname{arcsec} 1.269$

14. $\arctan(-5)$

Using a Right Triangle In Exercises 15–20, use the figure to write the expression in algebraic form given $y = \arccos x$, where $0 < y < \pi/2$.

15. $\cos y$

16. $\sin y$

17. $\tan y$

18. $\cot y$

19. $\sec y$

20. $\csc y$

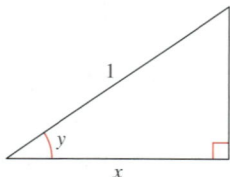

Evaluating an Expression In Exercises 21–24, evaluate each expression without using a calculator. (*Hint:* See Example 3.)

21. (a) $\sin\left(\arctan \dfrac{3}{4}\right)$

   (b) $\sec\left(\arcsin \dfrac{4}{5}\right)$

22. (a) $\tan\left(\arccos \dfrac{\sqrt{2}}{2}\right)$

   (b) $\cos\left(\arcsin \dfrac{5}{13}\right)$

23. (a) $\cot\left[\arcsin\left(-\dfrac{1}{2}\right)\right]$

   (b) $\csc\left[\arctan\left(-\dfrac{5}{12}\right)\right]$

24. (a) $\sec\left[\arctan\left(-\dfrac{3}{5}\right)\right]$

   (b) $\tan\left[\arcsin\left(-\dfrac{5}{6}\right)\right]$

Simplifying an Expression Using a Right Triangle In Exercises 25–32, write the expression in algebraic form. (*Hint:* Sketch a right triangle, as demonstrated in Example 3.)

25. $\cos(\arcsin 2x)$

26. $\sec(\arctan 4x)$

27. $\sin(\operatorname{arcsec} x)$

28. $\cos(\operatorname{arccot} x)$

29. $\tan\left(\operatorname{arcsec} \dfrac{x}{3}\right)$

30. $\sec[\arcsin(x-1)]$

31. $\csc\left(\arctan \dfrac{x}{\sqrt{2}}\right)$

32. $\cos\left(\arcsin \dfrac{x-h}{r}\right)$

Solving an Equation In Exercises 33–36, solve the equation for x.

33. $\arcsin(3x - \pi) = \frac{1}{2}$

34. $\arctan(2x - 5) = -1$

35. $\arcsin\sqrt{2x} = \arccos\sqrt{x}$

36. $\arccos x = \operatorname{arcsec} x$

Verifying Identities In Exercises 37 and 38, verify each identity.

37. (a) $\operatorname{arccsc} x = \arcsin \dfrac{1}{x}, \quad x \geq 1$

   (b) $\arctan x + \arctan \dfrac{1}{x} = \dfrac{\pi}{2}, \quad x > 0$

38. (a) $\arcsin(-x) = -\arcsin x, \quad |x| \leq 1$

   (b) $\arccos(-x) = \pi - \arccos x, \quad |x| \leq 1$

Finding a Derivative In Exercises 39–58, find the derivative of the function.

39. $f(x) = 2\arcsin(x - 1)$

40. $f(t) = \arcsin t^2$

41. $g(x) = 3\arccos \dfrac{x}{2}$

42. $f(x) = \operatorname{arcsec} 2x$

43. $f(x) = \arctan e^x$

44. $f(x) = \arctan\sqrt{x}$

45. $g(x) = \dfrac{\arcsin 3x}{x}$

46. $h(x) = x^2 \arctan 5x$

47. $h(t) = \sin(\arccos t)$

48. $f(x) = \arcsin x + \arccos x$

49. $y = 2x\arccos x - 2\sqrt{1 - x^2}$

50. $y = \ln(t^2 + 4) - \dfrac{1}{2}\arctan \dfrac{t}{2}$

51. $y = \dfrac{1}{2}\left(\dfrac{1}{2}\ln \dfrac{x + 1}{x - 1} + \arctan x\right)$

52. $y = \dfrac{1}{2}\left[x\sqrt{4 - x^2} + 4\arcsin\left(\dfrac{x}{2}\right)\right]$

53. $y = x\arcsin x + \sqrt{1 - x^2}$

54. $y = x\arctan 2x - \dfrac{1}{4}\ln(1 + 4x^2)$

55. $y = 8\arcsin \dfrac{x}{4} - \dfrac{x\sqrt{16 - x^2}}{2}$

56. $y = 25\arcsin \dfrac{x}{5} - x\sqrt{25 - x^2}$

57. $y = \arctan x + \dfrac{x}{1 + x^2}$

58. $y = \arctan \dfrac{x}{2} - \dfrac{1}{2(x^2 + 4)}$

Finding an Equation of a Tangent Line In Exercises 59–64, find an equation of the tangent line to the graph of the function at the given point.

59. $y = 2 \arcsin x$, $\left(\dfrac{1}{2}, \dfrac{\pi}{3}\right)$

60. $y = \dfrac{1}{2} \arccos x$, $\left(-\dfrac{\sqrt{2}}{2}, \dfrac{3\pi}{8}\right)$

61. $y = \arctan \dfrac{x}{2}$, $\left(2, \dfrac{\pi}{4}\right)$

62. $y = \text{arcsec } 4x$, $\left(\dfrac{\sqrt{2}}{4}, \dfrac{\pi}{4}\right)$

63. $y = 4x \arccos(x - 1)$, $(1, 2\pi)$

64. $y = 3x \arcsin x$, $\left(\dfrac{1}{2}, \dfrac{\pi}{4}\right)$

Linear and Quadratic Approximations In Exercises 65–68, use a computer algebra system to find the linear approximation

$$P_1(x) = f(a) + f'(a)(x - a)$$

and the quadratic approximation

$$P_2(x) = f(a) + f'(a)(x - a) + \tfrac{1}{2}f''(a)(x - a)^2$$

of the function f at $x = a$. Sketch the graph of the function and its linear and quadratic approximations.

65. $f(x) = \arctan x$, $a = 0$ **66.** $f(x) = \arccos x$, $a = 0$

67. $f(x) = \arcsin x$, $a = \dfrac{1}{2}$ **68.** $f(x) = \arctan x$, $a = 1$

Finding Relative Extrema In Exercises 69–72, find any relative extrema of the function.

69. $f(x) = \text{arcsec } x - x$ **70.** $f(x) = \arcsin x - 2x$

71. $f(x) = \arctan x - \arctan(x - 4)$

72. $h(x) = \arcsin x - 2 \arctan x$

Analyzing an Inverse Trigonometric Graph In Exercises 73–76, analyze and sketch a graph of the function. Identify any relative extrema, points of inflection, and asymptotes. Use a graphing utility to verify your results.

73. $f(x) = \arcsin(x - 1)$ **74.** $f(x) = \arctan x + \dfrac{\pi}{2}$

75. $f(x) = \text{arcsec } 2x$ **76.** $f(x) = \arccos \dfrac{x}{4}$

Implicit Differentiation In Exercises 77–80, use implicit differentiation to find an equation of the tangent line to the graph of the equation at the given point.

77. $x^2 + x \arctan y = y - 1$, $\left(-\dfrac{\pi}{4}, 1\right)$

78. $\arctan(xy) = \arcsin(x + y)$, $(0, 0)$

79. $\arcsin x + \arcsin y = \dfrac{\pi}{2}$, $\left(\dfrac{\sqrt{2}}{2}, \dfrac{\sqrt{2}}{2}\right)$

80. $\arctan(x + y) = y^2 + \dfrac{\pi}{4}$, $(1, 0)$

83. Finding Values

(a) Use a graphing utility to evaluate $\arcsin(\arcsin 0.5)$ and $\arcsin(\arcsin 1)$.

(b) Let

$$f(x) = \arcsin(\arcsin x).$$

Find the values of x in the interval $-1 \le x \le 1$ such that $f(x)$ is a real number.

84. HOW DO YOU SEE IT? The graphs of $f(x) = \sin x$ and $g(x) = \cos x$ are shown below.

(a) Explain whether the points

$$\left(-\dfrac{\sqrt{2}}{2}, -\dfrac{\pi}{4}\right), \quad (0, 0) \quad \text{and} \quad \left(\dfrac{\sqrt{3}}{2}, \dfrac{2\pi}{3}\right)$$

lie on the graph of $y = \arcsin x$.

(b) Explain whether the points

$$\left(-\dfrac{1}{2}, \dfrac{2\pi}{3}\right), \quad \left(0, \dfrac{\pi}{2}\right), \quad \text{and} \quad \left(\dfrac{1}{2}, -\dfrac{\pi}{3}\right)$$

lie on the graph of $y = \arccos x$.

True or False? In Exercises 85–90, determine whether the statement is true or false. If it is false, explain why or give an example that shows it is false.

85. Because $\cos\left(-\dfrac{\pi}{3}\right) = \dfrac{1}{2}$, it follows that $\arccos \dfrac{1}{2} = -\dfrac{\pi}{3}$.

86. $\arcsin \dfrac{\pi}{4} = \dfrac{\sqrt{2}}{2}$

87. The slope of the graph of the inverse tangent function is positive for all x.

88. The range of $y = \arcsin x$ is $[0, \pi]$.

89. $\dfrac{d}{dx}[\arctan(\tan x)] = 1$ for all x in the domain.

90. $\arcsin^2 x + \arccos^2 x = 1$

91. Angular Rate of Change An airplane flies at an altitude of 5 miles toward a point directly over an observer. Consider θ and x as shown in the figure.

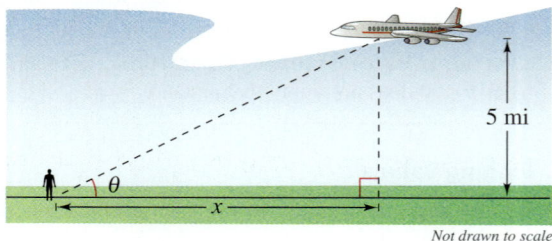

Not drawn to scale

(a) Write θ as a function of x.

(b) The speed of the plane is 400 miles per hour. Find $d\theta/dt$ when $x = 10$ miles and $x = 3$ miles.

92. Writing Repeat Exercise 91 for an altitude of 3 miles and describe how the altitude affects the rate of change of θ.

93. Angular Rate of Change In a free-fall experiment, an object is dropped from a height of 256 feet. A camera on the ground 500 feet from the point of impact records the fall of the object (see figure).

(a) Find the position function that yields the height of the object at time t, assuming the object is released at time $t = 0$. At what time will the object reach ground level?

(b) Find the rates of change of the angle of elevation of the camera when $t = 1$ and $t = 2$.

Figure for 93

Figure for 94

94. Angular Rate of Change A television camera at ground level is filming the lift-off of a rocket at a point 800 meters from the launch pad. Let θ be the angle of elevation of the rocket and let s be the distance between the camera and the rocket (see figure). Write θ as a function of s for the period of time when the rocket is moving vertically. Differentiate the result to find $d\theta/dt$ in terms of s and ds/dt.

95. Maximizing an Angle A billboard 85 feet wide is perpendicular to a straight road and is 40 feet from the road (see figure). Find the point on the road at which the angle θ subtended by the billboard is a maximum.

Figure for 95

Figure for 96

96. Angular Speed A patrol car is parked 50 feet from a long warehouse (see figure). The revolving light on top of the car turns at a rate of 30 revolutions per minute. Write θ as a function of x. How fast is the light beam moving along the wall when the beam makes an angle of $\theta = 45°$ with the line perpendicular from the light to the wall?

97. Proof

(a) Prove that $\arctan x + \arctan y = \arctan \dfrac{x + y}{1 - xy}$, $xy \neq 1$.

(b) Use the formula in part (a) to show that

$$\arctan \frac{1}{2} + \arctan \frac{1}{3} = \frac{\pi}{4}.$$

98. Proof Prove each differentiation formula.

(a) $\dfrac{d}{dx}[\arctan u] = \dfrac{u'}{1 + u^2}$

(b) $\dfrac{d}{dx}[\text{arccot } u] = \dfrac{-u'}{1 + u^2}$

(c) $\dfrac{d}{dx}[\text{arcsec } u] = \dfrac{u'}{|u|\sqrt{u^2 - 1}}$

(d) $\dfrac{d}{dx}[\text{arccsc } u] = \dfrac{-u'}{|u|\sqrt{u^2 - 1}}$

99. Describing a Graph

(a) Graph the function $f(x) = \arccos x + \arcsin x$ on the interval $[-1, 1]$.

(b) Describe the graph of f.

(c) Verify the result of part (b) analytically.

100. Think About It Use a graphing utility to graph $f(x) = \sin x$ and $g(x) = \arcsin(\sin x)$.

(a) Why isn't the graph of g the line $y = x$?

(b) Determine the extrema of g.

101. Maximizing an Angle In the figure, find the value of c in the interval $[0, 4]$ on the x-axis that maximizes angle θ.

Figure for 101

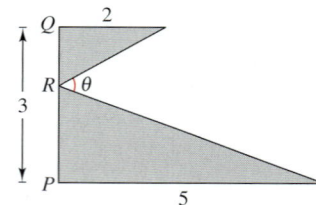

Figure for 102

102. Finding a Distance In the figure, find PR such that $0 \leq PR \leq 3$ and $m \angle \theta$ is a maximum.

103. Proof Prove that $\arcsin x = \arctan\left(\dfrac{x}{\sqrt{1 - x^2}}\right)$, $|x| < 1$.

104. Inverse Secant Function Some calculus textbooks define the inverse secant function using the range $[0, \pi/2) \cup [\pi, 3\pi/2)$.

(a) Sketch the graph $y = \text{arcsec } x$ using this range.

(b) Show that $y' = \dfrac{1}{x\sqrt{x^2 - 1}}$.

5.7 Inverse Trigonometric Functions: Integration

- Integrate functions whose antiderivatives involve inverse trigonometric functions.
- Use the method of completing the square to integrate a function.
- Review the basic integration rules involving elementary functions.

Integrals Involving Inverse Trigonometric Functions

The derivatives of the six inverse trigonometric functions fall into three pairs. In each pair, the derivative of one function is the negative of the other. For example,

$$\frac{d}{dx}[\arcsin x] = \frac{1}{\sqrt{1-x^2}}$$

and

$$\frac{d}{dx}[\arccos x] = -\frac{1}{\sqrt{1-x^2}}.$$

When listing the *antiderivative* that corresponds to each of the inverse trigonometric functions, you need to use only one member from each pair. It is conventional to use $\arcsin x$ as the antiderivative of $1/\sqrt{1-x^2}$, rather than $-\arccos x$. The next theorem gives one antiderivative formula for each of the three pairs. The proofs of these integration rules are left to you (see Exercises 75–77).

■ **FOR FURTHER INFORMATION**
For a detailed proof of rule 2 of Theorem 5.17, see the article "A Direct Proof of the Integral Formula for Arctangent" by Arnold J. Insel in *The College Mathematics Journal*. To view this article, go to *MathArticles.com*.

THEOREM 5.17 Integrals Involving Inverse Trigonometric Functions

Let u be a differentiable function of x, and let $a > 0$.

1. $\displaystyle\int \frac{du}{\sqrt{a^2 - u^2}} = \arcsin \frac{u}{a} + C$

2. $\displaystyle\int \frac{du}{a^2 + u^2} = \frac{1}{a}\arctan \frac{u}{a} + C$

3. $\displaystyle\int \frac{du}{u\sqrt{u^2 - a^2}} = \frac{1}{a}\operatorname{arcsec} \frac{|u|}{a} + C$

EXAMPLE 1 Integration with Inverse Trigonometric Functions

a. $\displaystyle\int \frac{dx}{\sqrt{4 - x^2}} = \arcsin \frac{x}{2} + C$

b. $\displaystyle\int \frac{dx}{2 + 9x^2} = \frac{1}{3}\int \frac{3\,dx}{\left(\sqrt{2}\right)^2 + (3x)^2}$ $u = 3x,\ a = \sqrt{2}$

$\qquad = \frac{1}{3\sqrt{2}}\arctan \frac{3x}{\sqrt{2}} + C$

c. $\displaystyle\int \frac{dx}{x\sqrt{4x^2 - 9}} = \int \frac{2\,dx}{2x\sqrt{(2x)^2 - 3^2}}$ $u = 2x,\ a = 3$

$\qquad = \frac{1}{3}\operatorname{arcsec} \frac{|2x|}{3} + C$

The integrals in Example 1 are fairly straightforward applications of integration formulas. Unfortunately, this is not typical. The integration formulas for inverse trigonometric functions can be disguised in many ways.

EXAMPLE 2 **Integration by Substitution**

Find $\displaystyle\int \frac{dx}{\sqrt{e^{2x} - 1}}$.

Solution As it stands, this integral doesn't fit any of the three inverse trigonometric formulas. Using the substitution $u = e^x$, however, produces

$$u = e^x \implies du = e^x\,dx \implies dx = \frac{du}{e^x} = \frac{du}{u}.$$

With this substitution, you can integrate as shown.

$$\int \frac{dx}{\sqrt{e^{2x} - 1}} = \int \frac{dx}{\sqrt{(e^x)^2 - 1}} \qquad \text{Write } e^{2x} \text{ as } (e^x)^2.$$

$$= \int \frac{du/u}{\sqrt{u^2 - 1}} \qquad \text{Substitute.}$$

$$= \int \frac{du}{u\sqrt{u^2 - 1}} \qquad \text{Rewrite to fit Arcsecant Rule.}$$

$$= \operatorname{arcsec} \frac{|u|}{1} + C \qquad \text{Apply Arcsecant Rule.}$$

$$= \operatorname{arcsec} e^x + C \qquad \text{Back-substitute.}$$

▷ **TECHNOLOGY PITFALL** A symbolic integration utility can be useful for integrating functions such as the one in Example 2. In some cases, however, the utility may fail to find an antiderivative for two reasons. First, some elementary functions do not have antiderivatives that are elementary functions. Second, every utility has limitations—you might have entered a function that the utility was not programmed to handle. You should also remember that antiderivatives involving trigonometric functions or logarithmic functions can be written in many different forms. For instance, one utility found the integral in Example 2 to be

$$\int \frac{dx}{\sqrt{e^{2x} - 1}} = \arctan \sqrt{e^{2x} - 1} + C.$$

Try showing that this antiderivative is equivalent to the one found in Example 2.

EXAMPLE 3 **Rewriting as the Sum of Two Quotients**

Find $\displaystyle\int \frac{x + 2}{\sqrt{4 - x^2}}\,dx$.

Solution This integral does not appear to fit any of the basic integration formulas. By splitting the integrand into two parts, however, you can see that the first part can be found with the Power Rule and the second part yields an inverse sine function.

$$\int \frac{x + 2}{\sqrt{4 - x^2}}\,dx = \int \frac{x}{\sqrt{4 - x^2}}\,dx + \int \frac{2}{\sqrt{4 - x^2}}\,dx$$

$$= -\frac{1}{2}\int (4 - x^2)^{-1/2}(-2x)\,dx + 2\int \frac{1}{\sqrt{4 - x^2}}\,dx$$

$$= -\frac{1}{2}\left[\frac{(4 - x^2)^{1/2}}{1/2}\right] + 2\arcsin\frac{x}{2} + C$$

$$= -\sqrt{4 - x^2} + 2\arcsin\frac{x}{2} + C$$

Completing the Square

Completing the square helps when quadratic functions are involved in the integrand. For example, the quadratic $x^2 + bx + c$ can be written as the difference of two squares by adding and subtracting $(b/2)^2$.

$$x^2 + bx + c = x^2 + bx + \left(\frac{b}{2}\right)^2 - \left(\frac{b}{2}\right)^2 + c = \left(x + \frac{b}{2}\right)^2 - \left(\frac{b}{2}\right)^2 + c$$

EXAMPLE 4 Completing the Square

> •••▷ *See LarsonCalculus.com for an interactive version of this type of example.*

Find $\displaystyle\int \frac{dx}{x^2 - 4x + 7}$.

Solution You can write the denominator as the sum of two squares, as shown.

$$x^2 - 4x + 7 = (x^2 - 4x + 4) - 4 + 7 = (x - 2)^2 + 3 = u^2 + a^2$$

Now, in this completed square form, let $u = x - 2$ and $a = \sqrt{3}$.

$$\int \frac{dx}{x^2 - 4x + 7} = \int \frac{dx}{(x - 2)^2 + 3} = \frac{1}{\sqrt{3}} \arctan \frac{x - 2}{\sqrt{3}} + C$$

When the leading coefficient is not 1, it helps to factor before completing the square. For instance, you can complete the square of $2x^2 - 8x + 10$ by factoring first.

$$2x^2 - 8x + 10 = 2(x^2 - 4x + 5)$$
$$= 2(x^2 - 4x + 4 - 4 + 5)$$
$$= 2[(x - 2)^2 + 1]$$

To complete the square when the coefficient of x^2 is negative, use the same factoring process shown above. For instance, you can complete the square for $3x - x^2$ as shown.

$$3x - x^2 = -(x^2 - 3x) = -\left[x^2 - 3x + \left(\tfrac{3}{2}\right)^2 - \left(\tfrac{3}{2}\right)^2\right] = \left(\tfrac{3}{2}\right)^2 - \left(x - \tfrac{3}{2}\right)^2$$

EXAMPLE 5 Completing the Square

Find the area of the region bounded by the graph of

$$f(x) = \frac{1}{\sqrt{3x - x^2}}$$

the x-axis, and the lines $x = \frac{3}{2}$ and $x = \frac{9}{4}$.

Solution In Figure 5.28, you can see that the area is

$$\text{Area} = \int_{3/2}^{9/4} \frac{1}{\sqrt{3x - x^2}}\, dx$$

$$= \int_{3/2}^{9/4} \frac{dx}{\sqrt{(3/2)^2 - [x - (3/2)]^2}} \qquad \text{\color{red}{Use completed square form derived above.}}$$

$$= \arcsin \frac{x - (3/2)}{3/2}\Bigg]_{3/2}^{9/4}$$

$$= \arcsin \frac{1}{2} - \arcsin 0$$

$$= \frac{\pi}{6}$$

$$\approx 0.524.$$

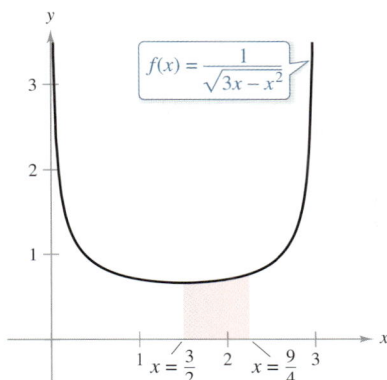

The area of the region bounded by the graph of f, the x-axis, $x = \frac{3}{2}$, and $x = \frac{9}{4}$ is $\pi/6$.

Figure 5.28

▷ **TECHNOLOGY** With definite integrals such as the one given in Example 5, remember that you can resort to a numerical solution. For instance, applying Simpson's Rule (with $n = 12$) to the integral in the example, you obtain

$$\int_{3/2}^{9/4} \frac{1}{\sqrt{3x - x^2}}\, dx \approx 0.523599.$$

This differs from the exact value of the integral ($\pi/6 \approx 0.5235988$) by less than one-millionth.

Review of Basic Integration Rules

You have now completed the introduction of the **basic integration rules.** To be efficient at applying these rules, you should have practiced enough so that each rule is committed to memory.

BASIC INTEGRATION RULES ($a > 0$)

1. $\int kf(u)\,du = k\int f(u)\,du$

2. $\int [f(u) \pm g(u)]\,du = \int f(u)\,du \pm \int g(u)\,du$

3. $\int du = u + C$

4. $\int u^n\,du = \dfrac{u^{n+1}}{n+1} + C, \quad n \neq -1$

5. $\int \dfrac{du}{u} = \ln|u| + C$

6. $\int e^u\,du = e^u + C$

7. $\int a^u\,du = \left(\dfrac{1}{\ln a}\right)a^u + C$

8. $\int \sin u\,du = -\cos u + C$

9. $\int \cos u\,du = \sin u + C$

10. $\int \tan u\,du = -\ln|\cos u| + C$

11. $\int \cot u\,du = \ln|\sin u| + C$

12. $\int \sec u\,du = \ln|\sec u + \tan u| + C$

13. $\int \csc u\,du = -\ln|\csc u + \cot u| + C$

14. $\int \sec^2 u\,du = \tan u + C$

15. $\int \csc^2 u\,du = -\cot u + C$

16. $\int \sec u \tan u\,du = \sec u + C$

17. $\int \csc u \cot u\,du = -\csc u + C$

18. $\int \dfrac{du}{\sqrt{a^2 - u^2}} = \arcsin \dfrac{u}{a} + C$

19. $\int \dfrac{du}{a^2 + u^2} = \dfrac{1}{a}\arctan \dfrac{u}{a} + C$

20. $\int \dfrac{du}{u\sqrt{u^2 - a^2}} = \dfrac{1}{a}\operatorname{arcsec} \dfrac{|u|}{a} + C$

You can learn a lot about the nature of integration by comparing this list with the summary of differentiation rules given in the preceding section. For differentiation, you now have rules that allow you to differentiate *any* elementary function. For integration, this is far from true.

The integration rules listed above are primarily those that were happened on during the development of differentiation rules. So far, you have not learned any rules or techniques for finding the antiderivative of a general product or quotient, the natural logarithmic function, or the inverse trigonometric functions. More important, you cannot apply any of the rules in this list unless you can create the proper *du* corresponding to the *u* in the formula. The point is that you need to work more on integration techniques, which you will do in Chapter 8. The next two examples should give you a better feeling for the integration problems that you *can* and *cannot* solve with the techniques and rules you now know.

EXAMPLE 6 **Comparing Integration Problems**

Find as many of the following integrals as you can using the formulas and techniques you have studied so far in the text.

a. $\displaystyle\int \frac{dx}{x\sqrt{x^2-1}}$

b. $\displaystyle\int \frac{x\,dx}{\sqrt{x^2-1}}$

c. $\displaystyle\int \frac{dx}{\sqrt{x^2-1}}$

Solution

a. You *can* find this integral (it fits the Arcsecant Rule).

$$\int \frac{dx}{x\sqrt{x^2-1}} = \operatorname{arcsec}|x| + C$$

b. You *can* find this integral (it fits the Power Rule).

$$\int \frac{x\,dx}{\sqrt{x^2-1}} = \frac{1}{2}\int (x^2-1)^{-1/2}(2x)\,dx$$

$$= \frac{1}{2}\left[\frac{(x^2-1)^{1/2}}{1/2}\right] + C$$

$$= \sqrt{x^2-1} + C$$

c. You *cannot* find this integral using the techniques you have studied so far. (You should scan the list of basic integration rules to verify this conclusion.)

EXAMPLE 7 **Comparing Integration Problems**

Find as many of the following integrals as you can using the formulas and techniques you have studied so far in the text.

a. $\displaystyle\int \frac{dx}{x\ln x}$

b. $\displaystyle\int \frac{\ln x\,dx}{x}$

c. $\displaystyle\int \ln x\,dx$

Solution

a. You *can* find this integral (it fits the Log Rule).

$$\int \frac{dx}{x\ln x} = \int \frac{1/x}{\ln x}\,dx$$

$$= \ln|\ln x| + C$$

b. You *can* find this integral (it fits the Power Rule).

$$\int \frac{\ln x\,dx}{x} = \int \left(\frac{1}{x}\right)(\ln x)^1\,dx$$

$$= \frac{(\ln x)^2}{2} + C$$

•• **REMARK** Note in Examples 6 and 7 that the *simplest* functions are the ones that you cannot yet integrate.

c. You *cannot* find this integral using the techniques you have studied so far. ∎

5.7 Exercises

See **CalcChat.com** for tutorial help and worked-out solutions to odd-numbered exercises.

Finding an Indefinite Integral In Exercises 1–20, find the indefinite integral.

1. $\int \dfrac{dx}{\sqrt{9 - x^2}}$

2. $\int \dfrac{dx}{\sqrt{1 - 4x^2}}$

3. $\int \dfrac{1}{x\sqrt{4x^2 - 1}} \, dx$

4. $\int \dfrac{12}{1 + 9x^2} \, dx$

5. $\int \dfrac{1}{\sqrt{1 - (x + 1)^2}} \, dx$

6. $\int \dfrac{1}{4 + (x - 3)^2} \, dx$

7. $\int \dfrac{t}{\sqrt{1 - t^4}} \, dt$

8. $\int \dfrac{1}{x\sqrt{x^4 - 4}} \, dx$

9. $\int \dfrac{t}{t^4 + 25} \, dt$

10. $\int \dfrac{1}{x\sqrt{1 - (\ln x)^2}} \, dx$

11. $\int \dfrac{e^{2x}}{4 + e^{4x}} \, dx$

12. $\int \dfrac{2}{x\sqrt{9x^2 - 25}} \, dx$

13. $\int \dfrac{\sec^2 x}{\sqrt{25 - \tan^2 x}} \, dx$

14. $\int \dfrac{\sin x}{7 + \cos^2 x} \, dx$

15. $\int \dfrac{1}{\sqrt{x}\sqrt{1 - x}} \, dx$

16. $\int \dfrac{3}{2\sqrt{x}(1 + x)} \, dx$

17. $\int \dfrac{x - 3}{x^2 + 1} \, dx$

18. $\int \dfrac{x^2 + 3}{x\sqrt{x^2 - 4}} \, dx$

19. $\int \dfrac{x + 5}{\sqrt{9 - (x - 3)^2}} \, dx$

20. $\int \dfrac{x - 2}{(x + 1)^2 + 4} \, dx$

Evaluating a Definite Integral In Exercises 21–32, evaluate the definite integral.

21. $\int_0^{1/6} \dfrac{3}{\sqrt{1 - 9x^2}} \, dx$

22. $\int_0^{\sqrt{2}} \dfrac{1}{\sqrt{4 - x^2}} \, dx$

23. $\int_0^{\sqrt{3}/2} \dfrac{1}{1 + 4x^2} \, dx$

24. $\int_{\sqrt{3}}^{3} \dfrac{1}{x\sqrt{4x^2 - 9}} \, dx$

25. $\int_3^6 \dfrac{1}{25 + (x - 3)^2} \, dx$

26. $\int_1^4 \dfrac{1}{x\sqrt{16x^2 - 5}} \, dx$

27. $\int_0^{\ln 5} \dfrac{e^x}{1 + e^{2x}} \, dx$

28. $\int_{\ln 2}^{\ln 4} \dfrac{e^{-x}}{\sqrt{1 - e^{-2x}}} \, dx$

29. $\int_{\pi/2}^{\pi} \dfrac{\sin x}{1 + \cos^2 x} \, dx$

30. $\int_0^{\pi/2} \dfrac{\cos x}{1 + \sin^2 x} \, dx$

31. $\int_0^{1/\sqrt{2}} \dfrac{\arcsin x}{\sqrt{1 - x^2}} \, dx$

32. $\int_0^{1/\sqrt{2}} \dfrac{\arccos x}{\sqrt{1 - x^2}} \, dx$

Completing the Square In Exercises 33–42, find or evaluate the integral by completing the square.

33. $\int_0^2 \dfrac{dx}{x^2 - 2x + 2}$

34. $\int_{-2}^2 \dfrac{dx}{x^2 + 4x + 13}$

35. $\int \dfrac{2x}{x^2 + 6x + 13} \, dx$

36. $\int \dfrac{2x - 5}{x^2 + 2x + 2} \, dx$

37. $\int \dfrac{1}{\sqrt{-x^2 - 4x}} \, dx$

38. $\int \dfrac{2}{\sqrt{-x^2 + 4x}} \, dx$

39. $\int_2^3 \dfrac{2x - 3}{\sqrt{4x - x^2}} \, dx$

40. $\int \dfrac{1}{(x - 1)\sqrt{x^2 - 2x}} \, dx$

41. $\int \dfrac{x}{x^4 + 2x^2 + 2} \, dx$

42. $\int \dfrac{x}{\sqrt{9 + 8x^2 - x^4}} \, dx$

Integration by Substitution In Exercises 43–46, use the specified substitution to find or evaluate the integral.

43. $\int \sqrt{e^t - 3} \, dt$

$u = \sqrt{e^t - 3}$

44. $\int \dfrac{\sqrt{x - 2}}{x + 1} \, dx$

$u = \sqrt{x - 2}$

45. $\int_1^3 \dfrac{dx}{\sqrt{x}(1 + x)}$

$u = \sqrt{x}$

46. $\int_0^1 \dfrac{dx}{2\sqrt{3 - x}\sqrt{x + 1}}$

$u = \sqrt{x + 1}$

WRITING ABOUT CONCEPTS

Comparing Integration Problems In Exercises 47–50, determine which of the integrals can be found using the basic integration formulas you have studied so far in the text.

47. (a) $\int \dfrac{1}{\sqrt{1 - x^2}} \, dx$

(b) $\int \dfrac{x}{\sqrt{1 - x^2}} \, dx$

(c) $\int \dfrac{1}{x\sqrt{1 - x^2}} \, dx$

48. (a) $\int e^{x^2} \, dx$

(b) $\int x e^{x^2} \, dx$

(c) $\int \dfrac{1}{x^2} e^{1/x} \, dx$

49. (a) $\int \sqrt{x - 1} \, dx$

(b) $\int x\sqrt{x - 1} \, dx$

(c) $\int \dfrac{x}{\sqrt{x - 1}} \, dx$

50. (a) $\int \dfrac{1}{1 + x^4} \, dx$

(b) $\int \dfrac{x}{1 + x^4} \, dx$

(c) $\int \dfrac{x^3}{1 + x^4} \, dx$

51. Finding an Integral Decide whether you can find the integral

$$\int \dfrac{2 \, dx}{\sqrt{x^2 + 4}}$$

using the formulas and techniques you have studied so far. Explain your reasoning.

52. HOW DO YOU SEE IT? Using the graph, which value best approximates the area of the region between the x-axis and the function over the interval $\left[-\frac{1}{2}, \frac{1}{2}\right]$? Explain.

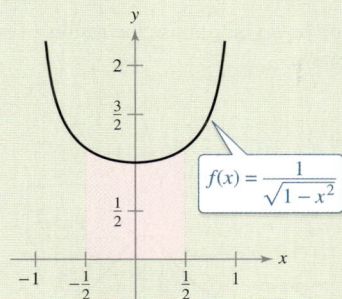

$$f(x) = \frac{1}{\sqrt{1-x^2}}$$

(a) -3 (b) $\frac{1}{2}$ (c) 1 (d) 2 (e) 4

Differential Equation In Exercises 53 and 54, use the differential equation and the specified initial condition to find y.

53. $\dfrac{dy}{dx} = \dfrac{1}{\sqrt{4-x^2}}$

$y(0) = \pi$

54. $\dfrac{dy}{dx} = \dfrac{1}{4+x^2}$

$y(2) = \pi$

Slope Field In Exercises 55 and 56, a differential equation, a point, and a slope field are given. (a) Sketch two approximate solutions of the differential equation on the slope field, one of which passes through the given point. (b) Use integration to find the particular solution of the differential equation and use a graphing utility to graph the solution. Compare the result with the sketches in part (a). To print an enlarged copy of the graph, go to *MathGraphs.com.*

55. $\dfrac{dy}{dx} = \dfrac{2}{9+x^2}$, $(0, 2)$

56. $\dfrac{dy}{dx} = \dfrac{2}{\sqrt{25-x^2}}$, $(5, \pi)$

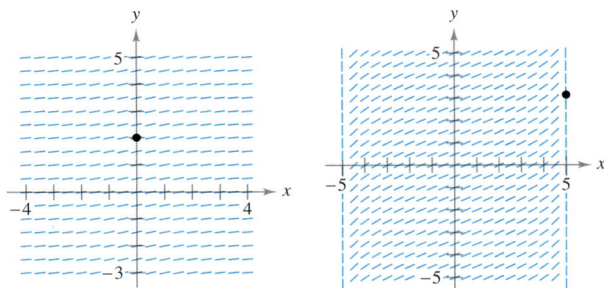

Slope Field In Exercises 57–60, use a computer algebra system to graph the slope field for the differential equation and graph the solution satisfying the specified initial condition.

57. $\dfrac{dy}{dx} = \dfrac{10}{x\sqrt{x^2-1}}$

$y(3) = 0$

58. $\dfrac{dy}{dx} = \dfrac{1}{12+x^2}$

$y(4) = 2$

59. $\dfrac{dy}{dx} = \dfrac{2y}{\sqrt{16-x^2}}$

$y(0) = 2$

60. $\dfrac{dy}{dx} = \dfrac{\sqrt{y}}{1+x^2}$

$y(0) = 4$

Area In Exercises 61–66, find the area of the region.

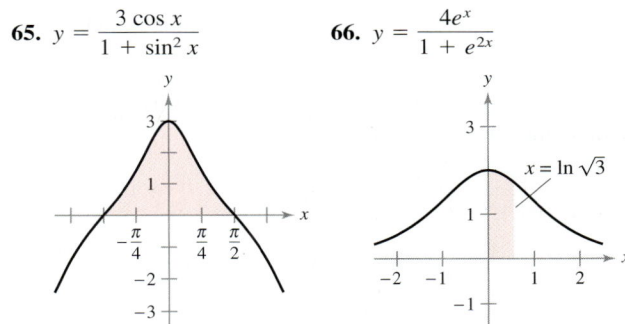

61. $y = \dfrac{2}{\sqrt{4-x^2}}$

62. $y = \dfrac{1}{x\sqrt{x^2-1}}$

$x = \dfrac{2}{\sqrt{2}}$

63. $y = \dfrac{1}{x^2-2x+5}$

64. $y = \dfrac{2}{x^2+4x+8}$

65. $y = \dfrac{3\cos x}{1+\sin^2 x}$

66. $y = \dfrac{4e^x}{1+e^{2x}}$

$x = \ln\sqrt{3}$

67. Area

(a) Sketch the region whose area is represented by

$$\int_0^1 \arcsin x \, dx.$$

(b) Use the integration capabilities of a graphing utility to approximate the area.

(c) Find the exact area analytically.

68. Approximating Pi

(a) Show that

$$\int_0^1 \frac{4}{1+x^2} \, dx = \pi.$$

(b) Approximate the number π using Simpson's Rule (with $n = 6$) and the integral in part (a).

(c) Approximate the number π by using the integration capabilities of a graphing utility.

69. Investigation   Consider the function

$$F(x) = \frac{1}{2}\int_x^{x+2} \frac{2}{t^2 + 1}\, dt.$$

(a) Write a short paragraph giving a geometric interpretation of the function $F(x)$ relative to the function

$$f(x) = \frac{2}{x^2 + 1}.$$

Use what you have written to guess the value of x that will make F maximum.

(b) Perform the specified integration to find an alternative form of $F(x)$. Use calculus to locate the value of x that will make F maximum and compare the result with your guess in part (a).

70. Comparing Integrals   Consider the integral

$$\int \frac{1}{\sqrt{6x - x^2}}\, dx.$$

(a) Find the integral by completing the square of the radicand.

(b) Find the integral by making the substitution $u = \sqrt{x}$.

(c) The antiderivatives in parts (a) and (b) appear to be significantly different. Use a graphing utility to graph each antiderivative in the same viewing window and determine the relationship between them. Find the domain of each.

True or False?   In Exercises 71–74, determine whether the statement is true or false. If it is false, explain why or give an example that shows it is false.

71. $\displaystyle\int \frac{dx}{3x\sqrt{9x^2 - 16}} = \frac{1}{4}\operatorname{arcsec}\frac{3x}{4} + C$

72. $\displaystyle\int \frac{dx}{25 + x^2} = \frac{1}{25}\arctan\frac{x}{25} + C$

73. $\displaystyle\int \frac{dx}{\sqrt{4 - x^2}} = -\arccos\frac{x}{2} + C$

74. One way to find $\displaystyle\int \frac{2e^{2x}}{\sqrt{9 - e^{2x}}}\, dx$ is to use the Arcsine Rule.

Verifying an Integration Rule   In Exercises 75–77, verify the rule by differentiating. Let $a > 0$.

75. $\displaystyle\int \frac{du}{\sqrt{a^2 - u^2}} = \arcsin\frac{u}{a} + C$

76. $\displaystyle\int \frac{du}{a^2 + u^2} = \frac{1}{a}\arctan\frac{u}{a} + C$

77. $\displaystyle\int \frac{du}{u\sqrt{u^2 - a^2}} = \frac{1}{a}\operatorname{arcsec}\frac{|u|}{a} + C$

78. Proof   Graph

$$y_1 = \frac{x}{1 + x^2}, \quad y_2 = \arctan x, \quad \text{and} \quad y_3 = x$$

on $[0, 10]$. Prove that

$$\frac{x}{1 + x^2} < \arctan x < x \quad \text{for} \quad x > 0.$$

79. Numerical Integration

(a) Write an integral that represents the area of the region in the figure.

(b) Use the Trapezoidal Rule with $n = 8$ to estimate the area of the region.

(c) Explain how you can use the results of parts (a) and (b) to estimate π.

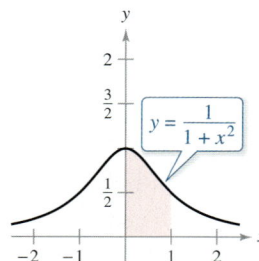

80. Vertical Motion   An object is projected upward from ground level with an initial velocity of 500 feet per second. In this exercise, the goal is to analyze the motion of the object during its upward flight.

(a) If air resistance is neglected, find the velocity of the object as a function of time. Use a graphing utility to graph this function.

(b) Use the result of part (a) to find the position function and determine the maximum height attained by the object.

(c) If the air resistance is proportional to the square of the velocity, you obtain the equation

$$\frac{dv}{dt} = -(32 + kv^2)$$

where -32 feet per second per second is the acceleration due to gravity and k is a constant. Find the velocity as a function of time by solving the equation

$$\int \frac{dv}{32 + kv^2} = -\int dt.$$

(d) Use a graphing utility to graph the velocity function $v(t)$ in part (c) for $k = 0.001$. Use the graph to approximate the time t_0 at which the object reaches its maximum height.

(e) Use the integration capabilities of a graphing utility to approximate the integral

$$\int_0^{t_0} v(t)\, dt$$

where $v(t)$ and t_0 are those found in part (d). This is the approximation of the maximum height of the object.

(f) Explain the difference between the results in parts (b) and (e).

■ **FOR FURTHER INFORMATION**   For more information on this topic, see the article "What Goes Up Must Come Down; Will Air Resistance Make It Return Sooner, or Later?" by John Lekner in *Mathematics Magazine*. To view this article, go to *MathArticles.com*.

5.8 Hyperbolic Functions

- Develop properties of hyperbolic functions.
- Differentiate and integrate hyperbolic functions.
- Develop properties of inverse hyperbolic functions.
- Differentiate and integrate functions involving inverse hyperbolic functions.

Hyperbolic Functions

In this section, you will look briefly at a special class of exponential functions called **hyperbolic functions.** The name *hyperbolic function* arose from comparison of the area of a semicircular region, as shown in Figure 5.29, with the area of a region under a hyperbola, as shown in Figure 5.30.

Circle: $x^2 + y^2 = 1$
Figure 5.29

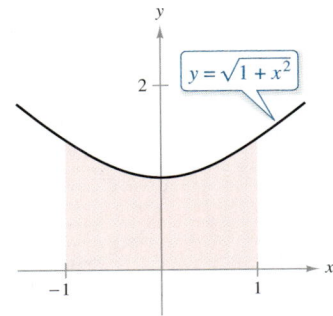

Hyperbola: $-x^2 + y^2 = 1$
Figure 5.30

The integral for the semicircular region involves an inverse trigonometric (circular) function:

$$\int_{-1}^{1} \sqrt{1 - x^2}\, dx = \frac{1}{2}\left[x\sqrt{1 - x^2} + \arcsin x \right]_{-1}^{1} = \frac{\pi}{2} \approx 1.571.$$

The integral for the hyperbolic region involves an inverse hyperbolic function:

$$\int_{-1}^{1} \sqrt{1 + x^2}\, dx = \frac{1}{2}\left[x\sqrt{1 + x^2} + \sinh^{-1}x \right]_{-1}^{1} \approx 2.296.$$

This is only one of many ways in which the hyperbolic functions are similar to the trigonometric functions.

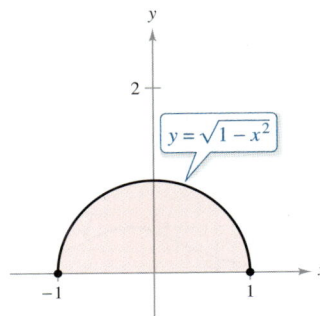

REMARK The notation $\sinh x$ is read as "the hyperbolic sine of x," $\cosh x$ as "the hyperbolic cosine of x," and so on.

Definitions of the Hyperbolic Functions

$$\sinh x = \frac{e^x - e^{-x}}{2} \qquad \operatorname{csch} x = \frac{1}{\sinh x},\quad x \neq 0$$

$$\cosh x = \frac{e^x + e^{-x}}{2} \qquad \operatorname{sech} x = \frac{1}{\cosh x}$$

$$\tanh x = \frac{\sinh x}{\cosh x} \qquad \coth x = \frac{1}{\tanh x},\quad x \neq 0$$

FOR FURTHER INFORMATION For more information on the development of hyperbolic functions, see the article "An Introduction to Hyperbolic Functions in Elementary Calculus" by Jerome Rosenthal in *Mathematics Teacher.* To view this article, go to *MathArticles.com.*

The graphs of the six hyperbolic functions and their domains and ranges are shown in Figure 5.31. Note that the graph of sinh x can be obtained by adding the corresponding y-coordinates of the exponential functions $f(x) = \frac{1}{2}e^x$ and $g(x) = -\frac{1}{2}e^{-x}$. Likewise, the graph of cosh x can be obtained by adding the corresponding y-coordinates of the exponential functions $f(x) = \frac{1}{2}e^x$ and $h(x) = \frac{1}{2}e^{-x}$.

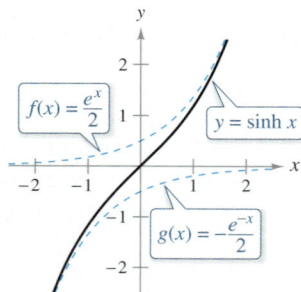

Domain: $(-\infty, \infty)$
Range: $(-\infty, \infty)$

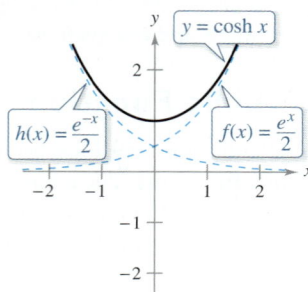

Domain: $(-\infty, \infty)$
Range: $[1, \infty)$

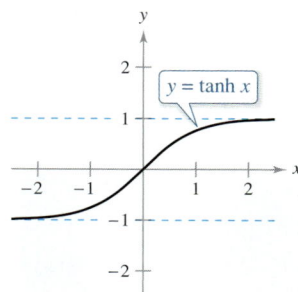

Domain: $(-\infty, \infty)$
Range: $(-1, 1)$

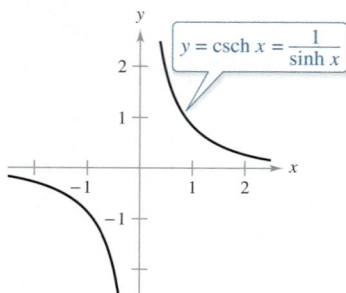

Domain: $(-\infty, 0) \cup (0, \infty)$
Range: $(-\infty, 0) \cup (0, \infty)$

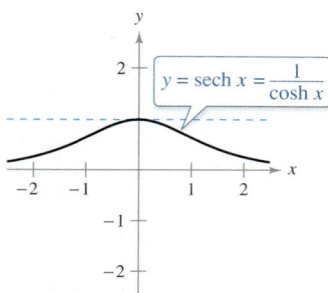

Domain: $(-\infty, \infty)$
Range: $(0, 1]$

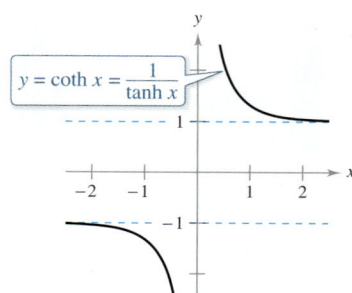

Domain: $(-\infty, 0) \cup (0, \infty)$
Range: $(-\infty, -1) \cup (1, \infty)$

Figure 5.31

Many of the trigonometric identities have corresponding *hyperbolic identities*. For instance,

$$\cosh^2 x - \sinh^2 x = \left(\frac{e^x + e^{-x}}{2}\right)^2 - \left(\frac{e^x - e^{-x}}{2}\right)^2$$

$$= \frac{e^{2x} + 2 + e^{-2x}}{4} - \frac{e^{2x} - 2 + e^{-2x}}{4}$$

$$= \frac{4}{4}$$

$$= 1.$$

FOR FURTHER INFORMATION
To understand geometrically the relationship between the hyperbolic and exponential functions, see the article "A Short Proof Linking the Hyperbolic and Exponential Functions" by Michael J. Seery in *The AMATYC Review*.

HYPERBOLIC IDENTITIES

$\cosh^2 x - \sinh^2 x = 1$

$\tanh^2 x + \text{sech}^2 x = 1$

$\coth^2 x - \text{csch}^2 x = 1$

$\sinh^2 x = \dfrac{-1 + \cosh 2x}{2}$

$\sinh 2x = 2 \sinh x \cosh x$

$\sinh(x + y) = \sinh x \cosh y + \cosh x \sinh y$

$\sinh(x - y) = \sinh x \cosh y - \cosh x \sinh y$

$\cosh(x + y) = \cosh x \cosh y + \sinh x \sinh y$

$\cosh(x - y) = \cosh x \cosh y - \sinh x \sinh y$

$\cosh^2 x = \dfrac{1 + \cosh 2x}{2}$

$\cosh 2x = \cosh^2 x + \sinh^2 x$

Differentiation and Integration of Hyperbolic Functions

Because the hyperbolic functions are written in terms of e^x and e^{-x}, you can easily derive rules for their derivatives. The next theorem lists these derivatives with the corresponding integration rules.

THEOREM 5.18   Derivatives and Integrals of Hyperbolic Functions

Let u be a differentiable function of x.

$$\frac{d}{dx}[\sinh u] = (\cosh u)u' \qquad \int \cosh u\, du = \sinh u + C$$

$$\frac{d}{dx}[\cosh u] = (\sinh u)u' \qquad \int \sinh u\, du = \cosh u + C$$

$$\frac{d}{dx}[\tanh u] = (\operatorname{sech}^2 u)u' \qquad \int \operatorname{sech}^2 u\, du = \tanh u + C$$

$$\frac{d}{dx}[\coth u] = -(\operatorname{csch}^2 u)u' \qquad \int \operatorname{csch}^2 u\, du = -\coth u + C$$

$$\frac{d}{dx}[\operatorname{sech} u] = -(\operatorname{sech} u \tanh u)u' \qquad \int \operatorname{sech} u \tanh u\, du = -\operatorname{sech} u + C$$

$$\frac{d}{dx}[\operatorname{csch} u] = -(\operatorname{csch} u \coth u)u' \qquad \int \operatorname{csch} u \coth u\, du = -\operatorname{csch} u + C$$

Proof   Here is a proof of two of the differentiation rules. (You are asked to prove some of the other differentiation rules in Exercises 103–105.)

$$\frac{d}{dx}[\sinh x] = \frac{d}{dx}\left[\frac{e^x - e^{-x}}{2}\right]$$
$$= \frac{e^x + e^{-x}}{2}$$
$$= \cosh x$$

$$\frac{d}{dx}[\tanh x] = \frac{d}{dx}\left[\frac{\sinh x}{\cosh x}\right]$$
$$= \frac{\cosh x(\cosh x) - \sinh x(\sinh x)}{\cosh^2 x}$$
$$= \frac{1}{\cosh^2 x}$$
$$= \operatorname{sech}^2 x$$

See LarsonCalculus.com for Bruce Edwards's video of this proof.

EXAMPLE 1   Differentiation of Hyperbolic Functions

a. $\dfrac{d}{dx}[\sinh(x^2 - 3)] = 2x \cosh(x^2 - 3)$

b. $\dfrac{d}{dx}[\ln(\cosh x)] = \dfrac{\sinh x}{\cosh x} = \tanh x$

c. $\dfrac{d}{dx}[x \sinh x - \cosh x] = x \cosh x + \sinh x - \sinh x = x \cosh x$

d. $\dfrac{d}{dx}[(x - 1)\cosh x - \sin x] = (x - 1)\sinh x + \cosh x - \cosh x = (x - 1)\sinh x$

$f(x) = (x - 1)\cosh x - \sinh x$

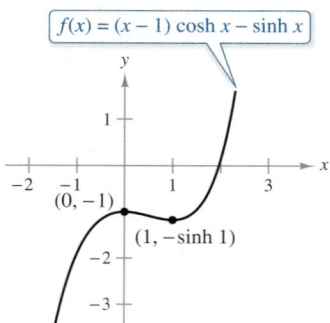

$f''(0) < 0$, so $(0, -1)$ is a relative maximum. $f''(1) > 0$, so $(1, -\sinh 1)$ is a relative minimum.

Figure 5.32

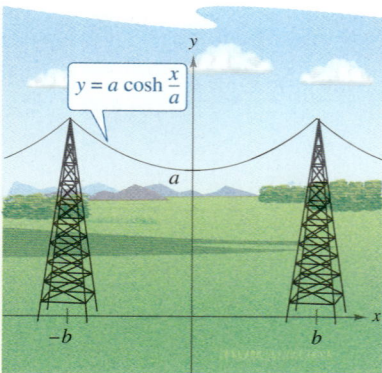

Catenary
Figure 5.33

◼ FOR FURTHER INFORMATION
In Example 3, the cable is a catenary between two supports at the same height. To learn about the shape of a cable hanging between supports of different heights, see the article "Reexamining the Catenary" by Paul Cella in *The College Mathematics Journal*. To view this article, go to *MathArticles.com*.

EXAMPLE 2 **Finding Relative Extrema**

Find the relative extrema of

$$f(x) = (x - 1)\cosh x - \sinh x.$$

Solution Using the result of Example 1(d), set the first derivative of f equal to 0.

$$(x - 1)\sinh x = 0$$

So, the critical numbers are $x = 1$ and $x = 0$. Using the Second Derivative Test, you can verify that the point $(0, -1)$ yields a relative maximum and the point $(1, -\sinh 1)$ yields a relative minimum, as shown in Figure 5.32. Try using a graphing utility to confirm this result. If your graphing utility does not have hyperbolic functions, you can use exponential functions, as shown.

$$f(x) = (x - 1)\left(\frac{1}{2}\right)(e^x + e^{-x}) - \frac{1}{2}(e^x - e^{-x})$$

$$= \frac{1}{2}(xe^x + xe^{-x} - e^x - e^{-x} - e^x + e^{-x})$$

$$= \frac{1}{2}(xe^x + xe^{-x} - 2e^x)$$

When a uniform flexible cable, such as a telephone wire, is suspended from two points, it takes the shape of a *catenary*, as discussed in Example 3.

EXAMPLE 3 **Hanging Power Cables**

• • • ▷ *See LarsonCalculus.com for an interactive version of this type of example.*

Power cables are suspended between two towers, forming the catenary shown in Figure 5.33. The equation for this catenary is

$$y = a\cosh\frac{x}{a}.$$

The distance between the two towers is $2b$. Find the slope of the catenary at the point where the cable meets the right-hand tower.

Solution Differentiating produces

$$y' = a\left(\frac{1}{a}\right)\sinh\frac{x}{a} = \sinh\frac{x}{a}.$$

At the point $(b, a\cosh(b/a))$, the slope (from the left) is $m = \sinh\frac{b}{a}$.

EXAMPLE 4 **Integrating a Hyperbolic Function**

Find $\int \cosh 2x \sinh^2 2x \, dx$.

Solution

$$\int \cosh 2x \sinh^2 2x \, dx = \frac{1}{2}\int (\sinh 2x)^2(2\cosh 2x)\, dx \qquad u = \sinh 2x$$

$$= \frac{1}{2}\left[\frac{(\sinh 2x)^3}{3}\right] + C$$

$$= \frac{\sinh^3 2x}{6} + C$$

Inverse Hyperbolic Functions

Unlike trigonometric functions, hyperbolic functions are not periodic. In fact, by looking back at Figure 5.31, you can see that four of the six hyperbolic functions are actually one-to-one (the hyperbolic sine, tangent, cosecant, and cotangent). So, you can apply Theorem 5.7 to conclude that these four functions have inverse functions. The other two (the hyperbolic cosine and secant) are one-to-one when their domains are restricted to the positive real numbers, and for this restricted domain they also have inverse functions. Because the hyperbolic functions are defined in terms of exponential functions, it is not surprising to find that the inverse hyperbolic functions can be written in terms of logarithmic functions, as shown in Theorem 5.19.

THEOREM 5.19   Inverse Hyperbolic Functions

Function	**Domain**
$\sinh^{-1} x = \ln\left(x + \sqrt{x^2 + 1}\right)$	$(-\infty, \infty)$
$\cosh^{-1} x = \ln\left(x + \sqrt{x^2 - 1}\right)$	$[1, \infty)$
$\tanh^{-1} x = \dfrac{1}{2} \ln \dfrac{1 + x}{1 - x}$	$(-1, 1)$
$\coth^{-1} x = \dfrac{1}{2} \ln \dfrac{x + 1}{x - 1}$	$(-\infty, -1) \cup (1, \infty)$
$\operatorname{sech}^{-1} x = \ln \dfrac{1 + \sqrt{1 - x^2}}{x}$	$(0, 1]$
$\operatorname{csch}^{-1} x = \ln\left(\dfrac{1}{x} + \dfrac{\sqrt{1 + x^2}}{|x|}\right)$	$(-\infty, 0) \cup (0, \infty)$

Proof   The proof of this theorem is a straightforward application of the properties of the exponential and logarithmic functions. For example, for

$$f(x) = \sinh x = \frac{e^x - e^{-x}}{2}$$

and

$$g(x) = \ln(x + \sqrt{x^2 + 1})$$

you can show that

$$f(g(x)) = x \quad \text{and} \quad g(f(x)) = x$$

which implies that g is the inverse function of f.

See LarsonCalculus.com for Bruce Edwards's video of this proof.

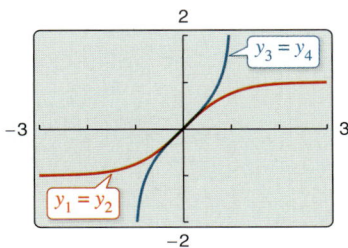

Graphs of the hyperbolic tangent function and the inverse hyperbolic tangent function

Figure 5.34

▷ **TECHNOLOGY**   You can use a graphing utility to confirm graphically the results of Theorem 5.19. For instance, graph the following functions.

$y_1 = \tanh x$ Hyperbolic tangent

$y_2 = \dfrac{e^x - e^{-x}}{e^x + e^{-x}}$ Definition of hyperbolic tangent

$y_3 = \tanh^{-1} x$ Inverse hyperbolic tangent

$y_4 = \dfrac{1}{2} \ln \dfrac{1 + x}{1 - x}$ Definition of inverse hyperbolic tangent

The resulting display is shown in Figure 5.34. As you watch the graphs being traced out, notice that $y_1 = y_2$ and $y_3 = y_4$. Also notice that the graph of y_1 is the reflection of the graph of y_3 in the line $y = x$.

The graphs of the inverse hyperbolic functions are shown in Figure 5.35.

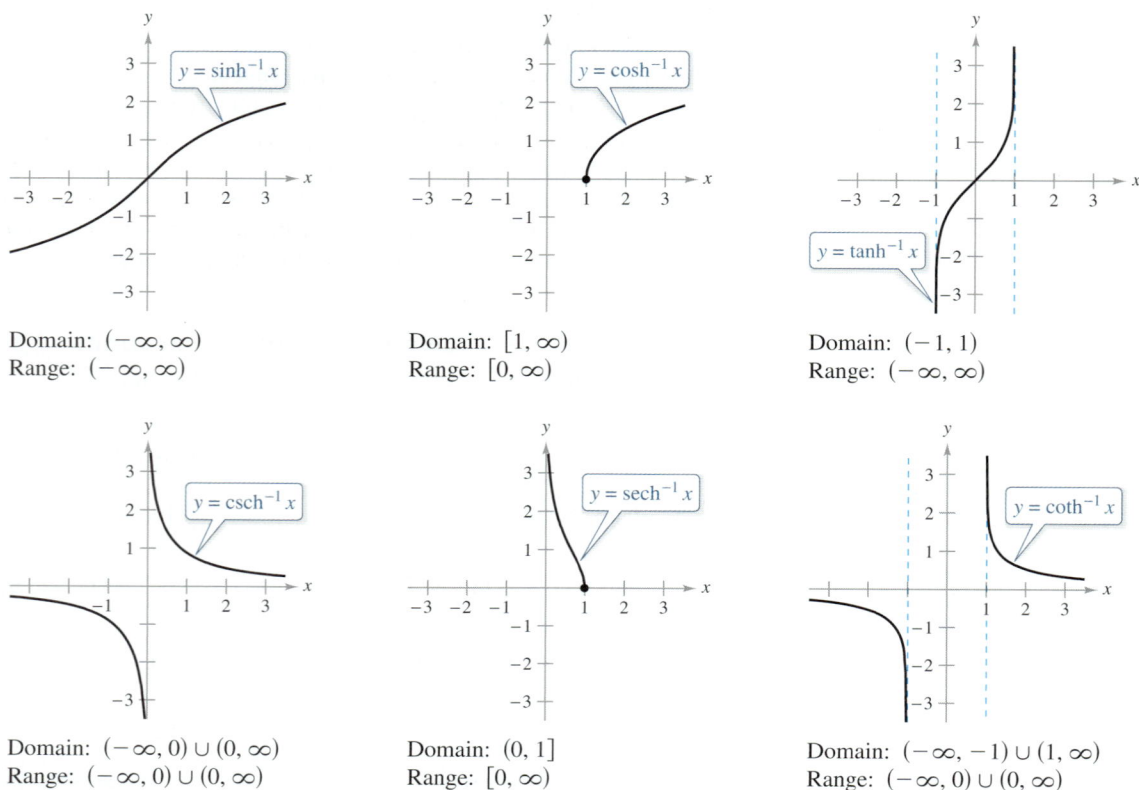

$y = \sinh^{-1} x$

Domain: $(-\infty, \infty)$
Range: $(-\infty, \infty)$

$y = \cosh^{-1} x$

Domain: $[1, \infty)$
Range: $[0, \infty)$

$y = \tanh^{-1} x$

Domain: $(-1, 1)$
Range: $(-\infty, \infty)$

$y = \operatorname{csch}^{-1} x$

Domain: $(-\infty, 0) \cup (0, \infty)$
Range: $(-\infty, 0) \cup (0, \infty)$

$y = \operatorname{sech}^{-1} x$

Domain: $(0, 1]$
Range: $[0, \infty)$

$y = \coth^{-1} x$

Domain: $(-\infty, -1) \cup (1, \infty)$
Range: $(-\infty, 0) \cup (0, \infty)$

Figure 5.35

The inverse hyperbolic secant can be used to define a curve called a *tractrix* or *pursuit curve*, as discussed in Example 5.

EXAMPLE 5   A Tractrix

A person is holding a rope that is tied to a boat, as shown in Figure 5.36. As the person walks along the dock, the boat travels along a **tractrix,** given by the equation

$$y = a \operatorname{sech}^{-1} \frac{x}{a} - \sqrt{a^2 - x^2}$$

where a is the length of the rope. For $a = 20$ feet, find the distance the person must walk to bring the boat to a position 5 feet from the dock.

Solution   In Figure 5.36, notice that the distance the person has walked is

$$y_1 = y + \sqrt{20^2 - x^2}$$

$$= \left(20 \operatorname{sech}^{-1} \frac{x}{20} - \sqrt{20^2 - x^2}\right) + \sqrt{20^2 - x^2}$$

$$= 20 \operatorname{sech}^{-1} \frac{x}{20}.$$

When $x = 5$, this distance is

$$y_1 = 20 \operatorname{sech}^{-1} \frac{5}{20} = 20 \ln \frac{1 + \sqrt{1 - (1/4)^2}}{1/4} = 20 \ln(4 + \sqrt{15}) \approx 41.27 \text{ feet.}$$

So, the person must walk about 41.27 feet to bring the boat to a position 5 feet from the dock.

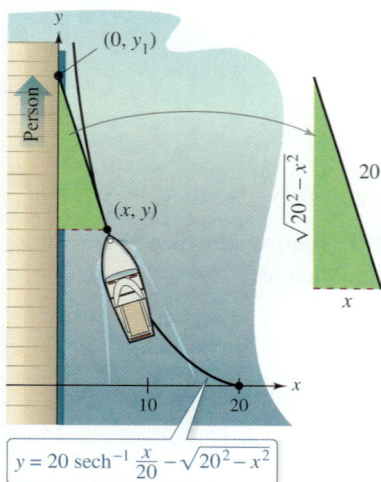

$y = 20 \operatorname{sech}^{-1} \frac{x}{20} - \sqrt{20^2 - x^2}$

A person must walk about 41.27 feet to bring the boat to a position 5 feet from the dock.

Figure 5.36

Inverse Hyperbolic Functions: Differentiation and Integration

The derivatives of the inverse hyperbolic functions, which resemble the derivatives of the inverse trigonometric functions, are listed in Theorem 5.20 with the corresponding integration formulas (in logarithmic form). You can verify each of these formulas by applying the logarithmic definitions of the inverse hyperbolic functions. (See Exercises 106–108.)

THEOREM 5.20   Differentiation and Integration Involving Inverse Hyperbolic Functions

Let u be a differentiable function of x.

$$\frac{d}{dx}[\sinh^{-1} u] = \frac{u'}{\sqrt{u^2 + 1}} \qquad\qquad \frac{d}{dx}[\cosh^{-1} u] = \frac{u'}{\sqrt{u^2 - 1}}$$

$$\frac{d}{dx}[\tanh^{-1} u] = \frac{u'}{1 - u^2} \qquad\qquad \frac{d}{dx}[\coth^{-1} u] = \frac{u'}{1 - u^2}$$

$$\frac{d}{dx}[\operatorname{sech}^{-1} u] = \frac{-u'}{u\sqrt{1 - u^2}} \qquad\qquad \frac{d}{dx}[\operatorname{csch}^{-1} u] = \frac{-u'}{|u|\sqrt{1 + u^2}}$$

$$\int \frac{du}{\sqrt{u^2 \pm a^2}} = \ln\left(u + \sqrt{u^2 \pm a^2}\right) + C$$

$$\int \frac{du}{a^2 - u^2} = \frac{1}{2a} \ln\left|\frac{a + u}{a - u}\right| + C$$

$$\int \frac{du}{u\sqrt{a^2 \pm u^2}} = -\frac{1}{a} \ln \frac{a + \sqrt{a^2 \pm u^2}}{|u|} + C$$

EXAMPLE 6   Differentiation of Inverse Hyperbolic Functions

a. $\dfrac{d}{dx}\left[\sinh^{-1}(2x)\right] = \dfrac{2}{\sqrt{(2x)^2 + 1}}$

$$= \frac{2}{\sqrt{4x^2 + 1}}$$

b. $\dfrac{d}{dx}\left[\tanh^{-1}(x^3)\right] = \dfrac{3x^2}{1 - (x^3)^2}$

$$= \frac{3x^2}{1 - x^6}$$

EXAMPLE 7   Integration Using Inverse Hyperbolic Functions

• • REMARK Let $a = 2$ and $u = 3x$.

a. $\displaystyle\int \frac{dx}{x\sqrt{4 - 9x^2}} = \int \frac{3\,dx}{(3x)\sqrt{4 - 9x^2}}$ $\qquad$ $\displaystyle\int \frac{du}{u\sqrt{a^2 - u^2}}$

$$= -\frac{1}{2} \ln \frac{2 + \sqrt{4 - 9x^2}}{|3x|} + C \qquad -\frac{1}{a} \ln \frac{a + \sqrt{a^2 - u^2}}{|u|} + C$$

• • REMARK Let $a = \sqrt{5}$ and $u = 2x$.

b. $\displaystyle\int \frac{dx}{5 - 4x^2} = \frac{1}{2}\int \frac{2\,dx}{(\sqrt{5})^2 - (2x)^2}$ $\qquad$ $\displaystyle\int \frac{du}{a^2 - u^2}$

$$= \frac{1}{2}\left(\frac{1}{2\sqrt{5}} \ln\left|\frac{\sqrt{5} + 2x}{\sqrt{5} - 2x}\right|\right) + C \qquad \frac{1}{2a} \ln\left|\frac{a + u}{a - u}\right| + C$$

$$= \frac{1}{4\sqrt{5}} \ln\left|\frac{\sqrt{5} + 2x}{\sqrt{5} - 2x}\right| + C$$

5.8 Exercises

See **CalcChat.com** for tutorial help and worked-out solutions to odd-numbered exercises.

Evaluating a Function In Exercises 1–6, evaluate the function. If the value is not a rational number, round your answer to three decimal places.

1. (a) $\sinh 3$

(b) $\tanh(-2)$

2. (a) $\cosh 0$

(b) $\operatorname{sech} 1$

3. (a) $\operatorname{csch}(\ln 2)$

(b) $\coth(\ln 5)$

4. (a) $\sinh^{-1} 0$

(b) $\tanh^{-1} 0$

5. (a) $\cosh^{-1} 2$

(b) $\operatorname{sech}^{-1} \frac{2}{3}$

6. (a) $\operatorname{csch}^{-1} 2$

(b) $\coth^{-1} 3$

Verifying an Identity In Exercises 7–14, verify the identity.

7. $\tanh^2 x + \operatorname{sech}^2 x = 1$

8. $\coth^2 x - \operatorname{csch}^2 x = 1$

9. $\cosh^2 x = \dfrac{1 + \cosh 2x}{2}$

10. $\sinh^2 x = \dfrac{-1 + \cosh 2x}{2}$

11. $\sinh 2x = 2 \sinh x \cosh x$

12. $e^{2x} = \sinh 2x + \cosh 2x$

13. $\sinh(x + y) = \sinh x \cosh y + \cosh x \sinh y$

14. $\cosh x + \cosh y = 2 \cosh \dfrac{x + y}{2} \cosh \dfrac{x - y}{2}$

Finding Values of Hyperbolic Functions In Exercises 15 and 16, use the value of the given hyperbolic function to find the values of the other hyperbolic functions at x.

15. $\sinh x = \dfrac{3}{2}$

16. $\tanh x = \dfrac{1}{2}$

Finding a Limit In Exercises 17–22, find the limit.

17. $\lim\limits_{x \to \infty} \sinh x$

18. $\lim\limits_{x \to -\infty} \tanh x$

19. $\lim\limits_{x \to \infty} \operatorname{sech} x$

20. $\lim\limits_{x \to -\infty} \operatorname{csch} x$

21. $\lim\limits_{x \to 0} \dfrac{\sinh x}{x}$

22. $\lim\limits_{x \to 0^-} \coth x$

Finding a Derivative In Exercises 23–32, find the derivative of the function.

23. $f(x) = \sinh 3x$

24. $f(x) = \cosh(8x + 1)$

25. $y = \operatorname{sech}(5x^2)$

26. $f(x) = \tanh(4x^2 + 3x)$

27. $f(x) = \ln(\sinh x)$

28. $y = \ln\left(\tanh \dfrac{x}{2}\right)$

29. $h(x) = \dfrac{1}{4} \sinh 2x - \dfrac{x}{2}$

30. $y = x \cosh x - \sinh x$

31. $f(t) = \arctan(\sinh t)$

32. $g(x) = \operatorname{sech}^2 3x$

Finding an Equation of a Tangent Line In Exercises 33–36, find an equation of the tangent line to the graph of the function at the given point.

33. $y = \sinh(1 - x^2)$, $(1, 0)$

34. $y = x^{\cosh x}$, $(1, 1)$

35. $y = (\cosh x - \sinh x)^2$, $(0, 1)$

36. $y = e^{\sinh x}$, $(0, 1)$

Finding Relative Extrema In Exercises 37–40, find any relative extrema of the function. Use a graphing utility to confirm your result.

37. $f(x) = \sin x \sinh x - \cos x \cosh x$, $-4 \le x \le 4$

38. $f(x) = x \sinh(x - 1) - \cosh(x - 1)$

39. $g(x) = x \operatorname{sech} x$

40. $h(x) = 2 \tanh x - x$

Catenary In Exercises 41 and 42, a model for a power cable suspended between two towers is given. (a) Graph the model, (b) find the heights of the cable at the towers and at the midpoint between the towers, and (c) find the slope of the model at the point where the cable meets the right-hand tower.

41. $y = 10 + 15 \cosh \dfrac{x}{15}$, $-15 \le x \le 15$

42. $y = 18 + 25 \cosh \dfrac{x}{25}$, $-25 \le x \le 25$

Finding an Indefinite Integral In Exercises 43–54, find the indefinite integral.

43. $\displaystyle\int \cosh 2x \, dx$

44. $\displaystyle\int \operatorname{sech}^2(3x) \, dx$

45. $\displaystyle\int \sinh(1 - 2x) \, dx$

46. $\displaystyle\int \dfrac{\cosh \sqrt{x}}{\sqrt{x}} \, dx$

47. $\displaystyle\int \cosh^2(x - 1) \sinh(x - 1) \, dx$

48. $\displaystyle\int \dfrac{\sinh x}{1 + \sinh^2 x} \, dx$

49. $\displaystyle\int \dfrac{\cosh x}{\sinh x} \, dx$

50. $\displaystyle\int \operatorname{sech}^2(2x - 1) \, dx$

51. $\displaystyle\int x \operatorname{csch}^2 \dfrac{x^2}{2} \, dx$

52. $\displaystyle\int \operatorname{sech}^3 x \tanh x \, dx$

53. $\displaystyle\int \dfrac{\operatorname{csch}(1/x) \coth(1/x)}{x^2} \, dx$

54. $\displaystyle\int \dfrac{\cosh x}{\sqrt{9 - \sinh^2 x}} \, dx$

Evaluating a Definite Integral In Exercises 55–60, evaluate the integral.

55. $\displaystyle\int_0^{\ln 2} \tanh x \, dx$

56. $\displaystyle\int_0^1 \cosh^2 x \, dx$

57. $\displaystyle\int_0^4 \dfrac{1}{25 - x^2} \, dx$

58. $\displaystyle\int_0^4 \dfrac{1}{\sqrt{25 - x^2}} \, dx$

59. $\displaystyle\int_0^{\sqrt{2}/4} \dfrac{2}{\sqrt{1 - 4x^2}} \, dx$

60. $\displaystyle\int_0^{\ln 2} 2e^{-x} \cosh x \, dx$

61. Comparing Functions Discuss several ways in which the hyperbolic functions are similar to the trigonometric functions.

62. Hyperbolic Functions Which hyperbolic functions take on only positive values? Which hyperbolic functions are increasing on their domains?

63. Comparing Derivative Formulas Which hyperbolic derivative formulas differ from their trigonometric counterparts by a minus sign?

64. **HOW DO YOU SEE IT?** Use the graphs of f and g shown in the figures to answer the following.

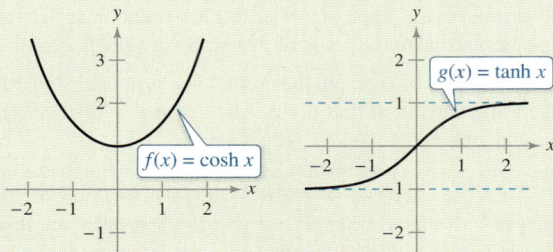

(a) Identify the open interval(s) on which the graphs of f and g are increasing or decreasing.

(b) Identify the open interval(s) on which the graphs of f and g are concave upward or concave downward.

Finding a Derivative In Exercises 65–74, find the derivative of the function.

65. $y = \cosh^{-1}(3x)$

66. $y = \tanh^{-1}\dfrac{x}{2}$

67. $y = \tanh^{-1}\sqrt{x}$ **68.** $f(x) = \coth^{-1}(x^2)$

69. $y = \sinh^{-1}(\tan x)$ **70.** $y = \tanh^{-1}(\sin 2x)$

71. $y = (\operatorname{csch}^{-1} x)^2$

72. $y = \operatorname{sech}^{-1}(\cos 2x), \quad 0 < x < \pi/4$

73. $y = 2x\sinh^{-1}(2x) - \sqrt{1 + 4x^2}$

74. $y = x\tanh^{-1} x + \ln\sqrt{1 - x^2}$

Finding an Indefinite Integral In Exercises 75–82, find the indefinite integral using the formulas from Theorem 5.20.

75. $\displaystyle\int \frac{1}{3 - 9x^2}\,dx$ **76.** $\displaystyle\int \frac{1}{2x\sqrt{1 - 4x^2}}\,dx$

77. $\displaystyle\int \frac{1}{\sqrt{1 + e^{2x}}}\,dx$ **78.** $\displaystyle\int \frac{x}{9 - x^4}\,dx$

79. $\displaystyle\int \frac{1}{\sqrt{x}\sqrt{1 + x}}\,dx$ **80.** $\displaystyle\int \frac{\sqrt{x}}{\sqrt{1 + x^3}}\,dx$

81. $\displaystyle\int \frac{-1}{4x - x^2}\,dx$ **82.** $\displaystyle\int \frac{dx}{(x + 2)\sqrt{x^2 + 4x + 8}}$

Evaluating a Definite Integral In Exercises 83–86, evaluate the definite integral using the formulas from Theorem 5.20.

83. $\displaystyle\int_3^7 \frac{1}{\sqrt{x^2 - 4}}\,dx$ **84.** $\displaystyle\int_1^3 \frac{1}{x\sqrt{4 + x^2}}\,dx$

85. $\displaystyle\int_{-1}^1 \frac{1}{16 - 9x^2}\,dx$ **86.** $\displaystyle\int_0^1 \frac{1}{\sqrt{25x^2 + 1}}\,dx$

Differential Equation In Exercises 87–90, solve the differential equation.

87. $\dfrac{dy}{dx} = \dfrac{1}{\sqrt{80 + 8x - 16x^2}}$

88. $\dfrac{dy}{dx} = \dfrac{1}{(x - 1)\sqrt{-4x^2 + 8x - 1}}$

89. $\dfrac{dy}{dx} = \dfrac{x^3 - 21x}{5 + 4x - x^2}$

90. $\dfrac{dy}{dx} = \dfrac{1 - 2x}{4x - x^2}$

Area In Exercises 91–94, find the area of the region.

91. $y = \operatorname{sech}\dfrac{x}{2}$ **92.** $y = \tanh 2x$

    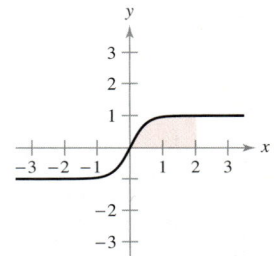

93. $y = \dfrac{5x}{\sqrt{x^4 + 1}}$ **94.** $y = \dfrac{6}{\sqrt{x^2 - 4}}$

    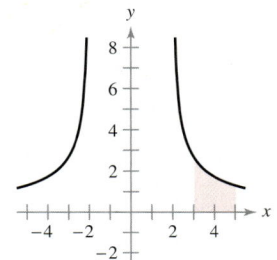

95. Chemical Reactions Chemicals A and B combine in a 3-to-1 ratio to form a compound. The amount of compound x being produced at any time t is proportional to the unchanged amounts of A and B remaining in the solution. So, when 3 kilograms of A is mixed with 2 kilograms of B, you have

$$\frac{dx}{dt} = k\left(3 - \frac{3x}{4}\right)\left(2 - \frac{x}{4}\right) = \frac{3k}{16}(x^2 - 12x + 32).$$

One kilogram of the compound is formed after 10 minutes. Find the amount formed after 20 minutes by solving the equation

$$\int \frac{3k}{16}\,dt = \int \frac{dx}{x^2 - 12x + 32}.$$

96. Vertical Motion An object is dropped from a height of 400 feet.

(a) Find the velocity of the object as a function of time (neglect air resistance on the object).

(b) Use the result in part (a) to find the position function.

(c) If the air resistance is proportional to the square of the velocity, then $dv/dt = -32 + kv^2$, where -32 feet per second per second is the acceleration due to gravity and k is a constant. Show that the velocity v as a function of time is $v(t) = -\sqrt{32/k}\tanh\left(\sqrt{32k}\,t\right)$ by performing $\int dv/(32 - kv^2) = -\int dt$ and simplifying the result.

(d) Use the result of part (c) to find $\lim\limits_{t\to\infty} v(t)$ and give its interpretation.

(e) Integrate the velocity function in part (c) and find the position s of the object as a function of t. Use a graphing utility to graph the position function when $k = 0.01$ and the position function in part (b) in the same viewing window. Estimate the additional time required for the object to reach ground level when air resistance is not neglected.

(f) Give a written description of what you believe would happen if k were increased. Then test your assertion with a particular value of k.

97. Tractrix Consider the equation of the tractrix

$$y = a\,\text{sech}^{-1}(x/a) - \sqrt{a^2 - x^2}, \quad a > 0.$$

(a) Find dy/dx.

(b) Let L be the tangent line to the tractrix at the point P. When L intersects the y-axis at the point Q, show that the distance between P and Q is a.

98. Tractrix Show that the boat in Example 5 is always pointing toward the person.

99. Proof Prove that

$$\tanh^{-1} x = \frac{1}{2}\ln\!\left(\frac{1+x}{1-x}\right), \quad -1 < x < 1.$$

100. Proof Prove that

$$\sinh^{-1} t = \ln\!\left(t + \sqrt{t^2 + 1}\right).$$

101. Using a Right Triangle Show that

$$\arctan(\sinh x) = \arcsin(\tanh x).$$

102. Integration Let $x > 0$ and $b > 0$. Show that

$$\int_{-b}^{b} e^{xt}\,dt = \frac{2\sinh bx}{x}.$$

Proof In Exercises 103–105, prove the differentiation formula.

103. $\dfrac{d}{dx}[\cosh x] = \sinh x$

104. $\dfrac{d}{dx}[\coth x] = -\text{csch}^2 x$

105. $\dfrac{d}{dx}[\text{sech } x] = -\text{sech } x \tanh x$

Verifying a Differentiation Rule In Exercises 106–108, verify the differentiation formula.

106. $\dfrac{d}{dx}[\cosh^{-1} x] = \dfrac{1}{\sqrt{x^2 - 1}}$

107. $\dfrac{d}{dx}[\sinh^{-1} x] = \dfrac{1}{\sqrt{x^2 + 1}}$

108. $\dfrac{d}{dx}[\text{sech}^{-1} x] = \dfrac{-1}{x\sqrt{1 - x^2}}$

PUTNAM EXAM CHALLENGE

109. From the vertex $(0, c)$ of the catenary $y = c\cosh(x/c)$ a line L is drawn perpendicular to the tangent to the catenary at point P. Prove that the length of L intercepted by the axes is equal to the ordinate y of the point P.

110. Prove or disprove: there is at least one straight line normal to the graph of $y = \cosh x$ at a point $(a, \cosh a)$ and also normal to the graph of $y = \sinh x$ at a point $(c, \sinh c)$.

[At a point on a graph, the normal line is the perpendicular to the tangent at that point. Also, $\cosh x = (e^x + e^{-x})/2$ and $\sinh x = (e^x - e^{-x})/2$.]

These problems were composed by the Committee on the Putnam Prize Competition.
© The Mathematical Association of America. All rights reserved.

SECTION PROJECT

St. Louis Arch

The Gateway Arch in St. Louis, Missouri, was constructed using the hyperbolic cosine function. The equation used for construction was

$$y = 693.8597 - 68.7672\cosh 0.0100333x,$$
$$-299.2239 \le x \le 299.2239$$

where x and y are measured in feet. Cross sections of the arch are equilateral triangles, and (x, y) traces the path of the centers of mass of the cross-sectional triangles. For each value of x, the area of the cross-sectional triangle is

$$A = 125.1406\cosh 0.0100333x.$$

(Source: Owner's Manual for the Gateway Arch, Saint Louis, MO, by William Thayer)

(a) How high above the ground is the center of the highest triangle? (At ground level, $y = 0$.)

(b) What is the height of the arch? (*Hint:* For an equilateral triangle, $A = \sqrt{3}c^2$, where c is one-half the base of the triangle, and the center of mass of the triangle is located at two-thirds the height of the triangle.)

(c) How wide is the arch at ground level?

Review Exercises

See **CalcChat.com** for tutorial help and worked-out solutions to odd-numbered exercises.

Sketching a Graph In Exercises 1 and 2, sketch the graph of the function and state its domain.

1. $f(x) = \ln x - 3$

2. $f(x) = \ln(x + 3)$

Expanding a Logarithmic Expression In Exercises 3 and 4, use the properties of logarithms to expand the logarithmic expression.

3. $\ln \sqrt[5]{\dfrac{4x^2 - 1}{4x^2 + 1}}$

4. $\ln[(x^2 + 1)(x - 1)]$

Condensing a Logarithmic Expression In Exercises 5 and 6, write the expression as the logarithm of a single quantity.

5. $\ln 3 + \frac{1}{3}\ln(4 - x^2) - \ln x$

6. $3[\ln x - 2\ln(x^2 + 1)] + 2\ln 5$

Finding a Derivative In Exercises 7–12, find the derivative of the function.

7. $g(x) = \ln \sqrt{2x}$

8. $f(x) = \ln(3x^2 + 2x)$

9. $f(x) = x\sqrt{\ln x}$

10. $f(x) = [\ln(2x)]^3$

11. $y = \ln \sqrt{\dfrac{x^2 + 4}{x^2 - 4}}$

12. $y = \ln\left(\dfrac{4x}{x - 6}\right)$

Finding an Equation of a Tangent Line In Exercises 13 and 14, find an equation of the tangent line to the graph of the function at the given point.

13. $y = \ln(2 + x) + \dfrac{2}{2 + x}, \quad (-1, 2)$

14. $y = 2x^2 + \ln x^2, \quad (1, 2)$

Finding an Indefinite Integral In Exercises 15–18, find the indefinite integral.

15. $\displaystyle\int \dfrac{1}{7x - 2}\,dx$

16. $\displaystyle\int \dfrac{x^2}{x^3 + 1}\,dx$

17. $\displaystyle\int \dfrac{\sin x}{1 + \cos x}\,dx$

18. $\displaystyle\int \dfrac{\ln \sqrt{x}}{x}\,dx$

Evaluating a Definite Integral In Exercises 19–22, evaluate the definite integral.

19. $\displaystyle\int_1^4 \dfrac{2x + 1}{2x}\,dx$

20. $\displaystyle\int_1^e \dfrac{\ln x}{x}\,dx$

21. $\displaystyle\int_0^{\pi/3} \sec \theta\,d\theta$

22. $\displaystyle\int_0^{\pi} \tan \dfrac{\theta}{3}\,d\theta$

Finding an Inverse Function In Exercises 23–28, (a) find the inverse function of f, (b) graph f and f^{-1} on the same set of coordinate axes, (c) verify that $f^{-1}(f(x)) = x$ and $f(f^{-1}(x)) = x$, and (d) state the domains and ranges of f and f^{-1}.

23. $f(x) = \frac{1}{2}x - 3$

24. $f(x) = 5x - 7$

25. $f(x) = \sqrt{x + 1}$

26. $f(x) = x^3 + 2$

27. $f(x) = \sqrt[3]{x + 1}$

28. $f(x) = x^2 - 5, \quad x \geq 0$

Evaluating the Derivative of an Inverse Function In Exercises 29–32, verify that f has an inverse. Then use the function f and the given real number a to find $(f^{-1})'(a)$. (*Hint:* Use Theorem 5.9.)

29. $f(x) = x^3 + 2, \quad a = -1$

30. $f(x) = x\sqrt{x - 3}, \quad a = 4$

31. $f(x) = \tan x, \quad -\dfrac{\pi}{4} \leq x \leq \dfrac{\pi}{4}, \quad a = \dfrac{\sqrt{3}}{3}$

32. $f(x) = \cos x, \quad 0 \leq x \leq \pi, \quad a = 0$

Solving an Exponential or Logarithmic Equation In Exercises 33–36, solve for x accurate to three decimal places.

33. $e^{3x} = 30$

34. $-4 + 3e^{-2x} = 6$

35. $\ln \sqrt{x + 1} = 2$

36. $\ln x + \ln(x - 3) = 0$

Finding a Derivative In Exercises 37–42, find the derivative of the function.

37. $g(t) = t^2 e^t$

38. $g(x) = \ln \dfrac{e^x}{1 + e^x}$

39. $y = \sqrt{e^{2x} + e^{-2x}}$

40. $h(z) = e^{-z^2/2}$

41. $g(x) = \dfrac{x^2}{e^x}$

42. $y = 3e^{-3/t}$

Finding an Equation of a Tangent Line In Exercises 43 and 44, find an equation of the tangent line to the graph of the function at the given point.

43. $f(x) = e^{6x}, \quad (0, 1)$

44. $f(x) = e^{x-4}, \quad (4, 1)$

Implicit Differentiation In Exercises 45 and 46, use implicit differentiation to find dy/dx.

45. $y \ln x + y^2 = 0$

46. $\cos x^2 = xe^y$

Finding an Indefinite Integral In Exercises 47–50, find the indefinite integral.

47. $\displaystyle\int xe^{1-x^2}\,dx$

48. $\displaystyle\int x^2 e^{x^3+1}\,dx$

49. $\displaystyle\int \dfrac{e^{4x} - e^{2x} + 1}{e^x}\,dx$

50. $\displaystyle\int \dfrac{e^{2x} - e^{-2x}}{e^{2x} + e^{-2x}}\,dx$

Evaluating a Definite Integral In Exercises 51–54, evaluate the definite integral.

51. $\int_0^1 xe^{-3x^2}\, dx$

52. $\int_{1/2}^2 \frac{e^{1/x}}{x^2}\, dx$

53. $\int_1^3 \frac{e^x}{e^x - 1}\, dx$

54. $\int_0^2 \frac{e^{2x}}{e^{2x} + 1}\, dx$

55. Area Find the area of the region bounded by the graphs of

$y = 2e^{-x}, \quad y = 0, \quad x = 0, \quad$ and $\quad x = 2.$

56. Depreciation The value V of an item t years after it is purchased is $V = 9000e^{-0.6t}$ for $0 \le t \le 5.$

(a) Use a graphing utility to graph the function.

(b) Find the rates of change of V with respect to t when $t = 1$ and $t = 4.$

(c) Use a graphing utility to graph the tangent lines to the function when $t = 1$ and $t = 4.$

Sketching a Graph In Exercises 57 and 58, sketch the graph of the function by hand.

57. $y = 3^{x/2}$

58. $y = \left(\frac{1}{4}\right)^x$

Finding a Derivative In Exercises 59–64, find the derivative of the function.

59. $f(x) = 3^{x-1}$

60. $f(x) = 5^{3x}$

61. $y = x^{2x+1}$

62. $f(x) = x(4^{-3x})$

63. $g(x) = \log_3 \sqrt{1 - x}$

64. $h(x) = \log_5 \frac{x}{x - 1}$

Finding an Indefinite Integral In Exercises 65 and 66, find the indefinite integral.

65. $\int (x + 1)5^{(x+1)^2}\, dx$

66. $\int \frac{2^{-1/t}}{t^2}\, dt$

67. Climb Rate The time t (in minutes) for a small plane to climb to an altitude of h feet is

$$t = 50 \log_{10} \frac{18{,}000}{18{,}000 - h}$$

where 18,000 feet is the plane's absolute ceiling.

(a) Determine the domain of the function appropriate for the context of the problem.

(b) Use a graphing utility to graph the time function and identify any asymptotes.

(c) Find the time when the altitude is increasing at the greatest rate.

68. Compound Interest

(a) How large a deposit, at 5% interest compounded continuously, must be made to obtain a balance of $10,000 in 15 years?

(b) A deposit earns interest at a rate of r percent compounded continuously and doubles in value in 10 years. Find $r.$

Evaluating an Expression In Exercises 69 and 70, evaluate each expression without using a calculator. (*Hint:* Make a sketch of a right triangle.)

69. (a) $\sin\left(\arcsin \frac{1}{2}\right)$

(b) $\cos\left(\arcsin \frac{1}{2}\right)$

70. (a) $\tan(\text{arccot } 2)$

(b) $\cos\left(\text{arcsec } \sqrt{5}\right)$

Finding a Derivative In Exercises 71–76, find the derivative of the function.

71. $y = \tan(\arcsin x)$

72. $y = \arctan(2x^2 - 3)$

73. $y = x \text{ arcsec } x$

74. $y = \frac{1}{2} \arctan e^{2x}$

75. $y = x(\arcsin x)^2 - 2x + 2\sqrt{1 - x^2} \arcsin x$

76. $y = \sqrt{x^2 - 4} - 2 \text{ arcsec } \frac{x}{2}, \quad 2 < x < 4$

Finding an Indefinite Integral In Exercises 77–82, find the indefinite integral.

77. $\int \frac{1}{e^{2x} + e^{-2x}}\, dx$

78. $\int \frac{1}{3 + 25x^2}\, dx$

79. $\int \frac{x}{\sqrt{1 - x^4}}\, dx$

80. $\int \frac{1}{x\sqrt{9x^2 - 49}}\, dx$

81. $\int \frac{\arctan(x/2)}{4 + x^2}\, dx$

82. $\int \frac{\arcsin 2x}{\sqrt{1 - 4x^2}}\, dx$

Area In Exercises 83 and 84, find the area of the region.

83. $y = \frac{4 - x}{\sqrt{4 - x^2}}$

84. $y = \frac{6}{16 + x^2}$

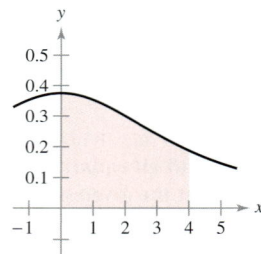

Finding a Derivative In Exercises 85–90, find the derivative of the function.

85. $y = \text{sech}(4x - 1)$

86. $y = 2x - \cosh \sqrt{x}$

87. $y = \coth(8x^2)$

88. $y = \ln(\cosh x)$

89. $y = \sinh^{-1}(4x)$

90. $y = x \tanh^{-1} 2x$

Finding an Indefinite Integral In Exercises 91–96, find the indefinite integral.

91. $\int x^2 \text{ sech}^2 x^3\, dx$

92. $\int \sinh 6x\, dx$

93. $\int \frac{\text{sech}^2 x}{\tanh x}\, dx$

94. $\int \text{csch}^4(3x)\coth(3x)\, dx$

95. $\int \frac{1}{9 - 4x^2}\, dx$

96. $\int \frac{x}{\sqrt{x^4 - 1}}\, dx$

P.S. Problem Solving

See **CalcChat.com** for tutorial help and worked-out solutions to odd-numbered exercises.

1. Approximation To approximate e^x, you can use a function of the form

$$f(x) = \frac{a + bx}{1 + cx}.$$

(This function is known as a **Padé approximation.**) The values of $f(0), f'(0)$, and $f''(0)$ are equal to the corresponding values of e^x. Show that these values are equal to 1 and find the values of a, b, and c such that $f(0) = f'(0) = f''(0) = 1$. Then use a graphing utility to compare the graphs of f and e^x.

2. Symmetry Recall that the graph of a function $y = f(x)$ is symmetric with respect to the origin if, whenever (x, y) is a point on the graph, $(-x, -y)$ is also a point on the graph. The graph of the function $y = f(x)$ is **symmetric with respect to the point (a, b)** if, whenever $(a - x, b - y)$ is a point on the graph, $(a + x, b + y)$ is also a point on the graph, as shown in the figure.

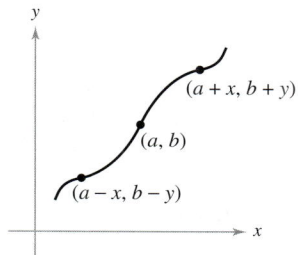

(a) Sketch the graph of $y = \sin x$ on the interval $[0, 2\pi]$. Write a short paragraph explaining how the symmetry of the graph with respect to the point $(\pi, 0)$ allows you to conclude that

$$\int_0^{2\pi} \sin x \, dx = 0.$$

(b) Sketch the graph of $y = \sin x + 2$ on the interval $[0, 2\pi]$. Use the symmetry of the graph with respect to the point $(\pi, 2)$ to evaluate the integral

$$\int_0^{2\pi} (\sin x + 2) \, dx.$$

(c) Sketch the graph of $y = \arccos x$ on the interval $[-1, 1]$. Use the symmetry of the graph to evaluate the integral

$$\int_{-1}^{1} \arccos x \, dx.$$

(d) Evaluate the integral $\displaystyle\int_0^{\pi/2} \frac{1}{1 + (\tan x)^{\sqrt{2}}} \, dx$.

3. Proof

(a) Use a graphing utility to graph $f(x) = \dfrac{\ln(x + 1)}{x}$ on the interval $[-1, 1]$.

(b) Use the graph to estimate $\displaystyle\lim_{x \to 0} f(x)$.

(c) Use the definition of derivative to prove your answer to part (b).

4. Using a Function Let $f(x) = \sin(\ln x)$.

(a) Determine the domain of the function f.

(b) Find two values of x satisfying $f(x) = 1$.

(c) Find two values of x satisfying $f(x) = -1$.

(d) What is the range of the function f?

(e) Calculate $f'(x)$ and use calculus to find the maximum value of f on the interval $[1, 10]$.

(f) Use a graphing utility to graph f in the viewing window $[0, 5] \times [-2, 2]$ and estimate $\displaystyle\lim_{x \to 0^+} f(x)$, if it exists.

(g) Determine $\displaystyle\lim_{x \to 0^+} f(x)$ analytically, if it exists.

5. Intersection Graph the exponential function $y = a^x$ for $a = 0.5, 1.2$, and 2.0. Which of these curves intersects the line $y = x$? Determine all positive numbers a for which the curve $y = a^x$ intersects the line $y = x$.

6. Areas and Angles

(a) Let $P(\cos t, \sin t)$ be a point on the unit circle $x^2 + y^2 = 1$ in the first quadrant (see figure). Show that t is equal to twice the area of the shaded circular sector AOP.

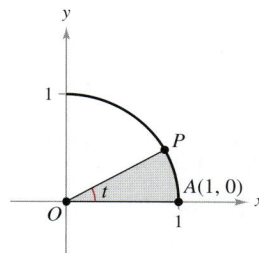

(b) Let $P(\cosh t, \sinh t)$ be a point on the unit hyperbola $x^2 - y^2 = 1$ in the first quadrant (see figure). Show that t is equal to twice the area of the shaded region AOP. Begin by showing that the area of the shaded region AOP is given by the formula

$$A(t) = \frac{1}{2} \cosh t \sinh t - \int_1^{\cosh t} \sqrt{x^2 - 1} \, dx.$$

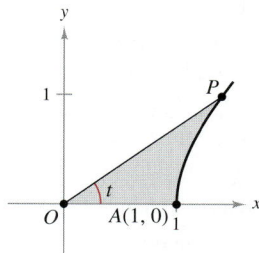

7. Mean Value Theorem Apply the Mean Value Theorem to the function $f(x) = \ln x$ on the closed interval $[1, e]$. Find the value of c in the open interval $(1, e)$ such that

$$f'(c) = \frac{f(e) - f(1)}{e - 1}.$$

8. Decreasing Function Show that $f(x) = \dfrac{\ln x^n}{x}$ is a decreasing function for $x > e$ and $n > 0$.

9. Area Consider the three regions A, B, and C determined by the graph of $f(x) = \arcsin x$, as shown in the figure.

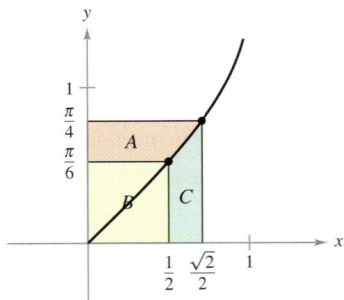

(a) Calculate the areas of regions A and B.

(b) Use your answers in part (a) to evaluate the integral

$$\int_{1/2}^{\sqrt{2}/2} \arcsin x \, dx.$$

(c) Use the methods in part (a) to evaluate the integral

$$\int_{1}^{3} \ln x \, dx.$$

(d) Use the methods in part (a) to evaluate the integral

$$\int_{1}^{\sqrt{3}} \arctan x \, dx.$$

10. Distance Let L be the tangent line to the graph of the function $y = \ln x$ at the point (a, b). Show that the distance between b and c is always equal to 1.

Figure for 10

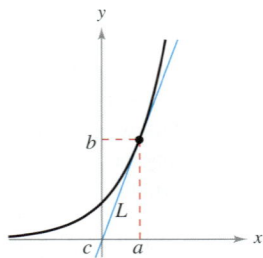

Figure for 11

11. Distance Let L be the tangent line to the graph of the function $y = e^x$ at the point (a, b). Show that the distance between a and c is always equal to 1.

12. Gudermannian Function The **Gudermannian function** of x is $\text{gd}(x) = \arctan(\sinh x)$.

(a) Graph gd using a graphing utility.

(b) Show that gd is an odd function.

(c) Show that gd is monotonic and therefore has an inverse.

(d) Find the inflection point of gd.

(e) Verify that $\text{gd}(x) = \arcsin(\tanh x)$.

(f) Verify that $\text{gd}(x) = \displaystyle\int_{0}^{x} \dfrac{dt}{\cosh t}$.

13. Area Use integration by substitution to find the area under the curve

$$y = \dfrac{1}{\sqrt{x} + x}$$

between $x = 1$ and $x = 4$.

14. Area Use integration by substitution to find the area under the curve

$$y = \dfrac{1}{\sin^2 x + 4\cos^2 x}$$

between $x = 0$ and $x = \dfrac{\pi}{4}$.

15. Approximating a Function

(a) Use a graphing utility to compare the graph of the function $y = e^x$ with the graph of each given function.

(i) $y_1 = 1 + \dfrac{x}{1!}$

(ii) $y_2 = 1 + \dfrac{x}{1!} + \dfrac{x^2}{2!}$

(iii) $y_3 = 1 + \dfrac{x}{1!} + \dfrac{x^2}{2!} + \dfrac{x^3}{3!}$

(b) Identify the pattern of successive polynomials in part (a), extend the pattern one more term, and compare the graph of the resulting polynomial function with the graph of $y = e^x$.

(c) What do you think this pattern implies?

16. Mortgage A \$120,000 home mortgage for 35 years at $9\frac{1}{2}\%$ has a monthly payment of \$985.93. Part of the monthly payment goes for the interest charge on the unpaid balance, and the remainder of the payment is used to reduce the principal. The amount that goes for interest is

$$u = M - \left(M - \dfrac{Pr}{12}\right)\left(1 + \dfrac{r}{12}\right)^{12t}$$

and the amount that goes toward reduction of the principal is

$$v = \left(M - \dfrac{Pr}{12}\right)\left(1 + \dfrac{r}{12}\right)^{12t}.$$

In these formulas, P is the amount of the mortgage, r is the interest rate (in decimal form), M is the monthly payment, and t is the time in years.

(a) Use a graphing utility to graph each function in the same viewing window. (The viewing window should show all 35 years of mortgage payments.)

(b) In the early years of the mortgage, the larger part of the monthly payment goes for what purpose? Approximate the time when the monthly payment is evenly divided between interest and principal reduction.

(c) Use the graphs in part (a) to make a conjecture about the relationship between the slopes of the tangent lines to the two curves for a specified value of t. Give an analytical argument to verify your conjecture. Find $u'(15)$ and $v'(15)$.

(d) Repeat parts (a) and (b) for a repayment period of 20 years ($M = \$1118.56$). What can you conclude?

6 Differential Equations

6.1 Slope Fields and Euler's Method
6.2 Differential Equations: Growth and Decay
6.3 Separation of Variables and the Logistic Equation
6.4 First-Order Linear Differential Equations

Sailing *(Exercise 65, p. 423)*

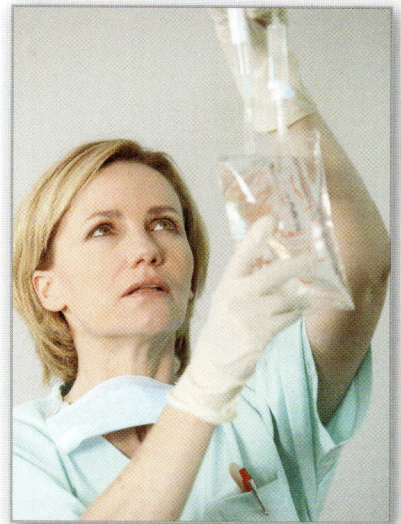

Intravenous Feeding
(Exercise 30, p. 429)

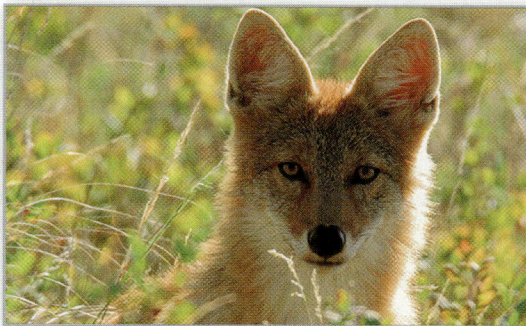

Wildlife Population *(Example 4, p. 417)*

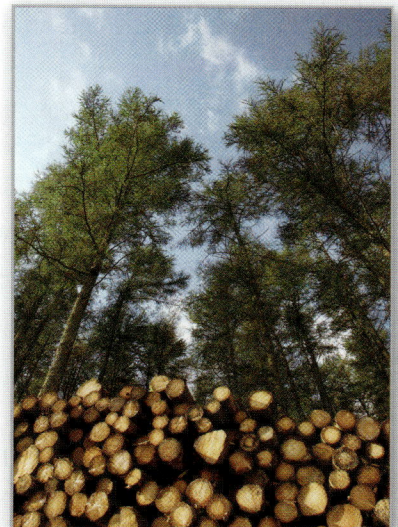

Forestry
(Exercise 62, p. 414)

Radioactive Decay *(Example 3, p. 409)*

6.1 Slope Fields and Euler's Method

- Use initial conditions to find particular solutions of differential equations.
- Use slope fields to approximate solutions of differential equations.
- Use Euler's Method to approximate solutions of differential equations.

General and Particular Solutions

In this text, you will learn that physical phenomena can be described by differential equations. Recall that a **differential equation** in x and y is an equation that involves x, y, and derivatives of y. For example,

$$2xy' - 3y = 0 \qquad \text{\color{red}{Differential equation}}$$

is a differential equation. In Section 6.2, you will see that problems involving radioactive decay, population growth, and Newton's Law of Cooling can be formulated in terms of differential equations.

A function $y = f(x)$ is called a **solution** of a differential equation if the equation is satisfied when y and its derivatives are replaced by $f(x)$ and its derivatives. For example, differentiation and substitution would show that $y = e^{-2x}$ is a solution of the differential equation $y' + 2y = 0$. It can be shown that every solution of this differential equation is of the form

$$y = Ce^{-2x} \qquad \text{\color{red}{General solution of } } y' + 2y = 0$$

where C is any real number. This solution is called the **general solution.** Some differential equations have **singular solutions** that cannot be written as special cases of the general solution. Such solutions, however, are not considered in this text. The **order** of a differential equation is determined by the highest-order derivative in the equation. For instance, $y' = 4y$ is a first-order differential equation. First-order linear differential equations are discussed in Section 6.4.

In Section 4.1, Example 9, you saw that the second-order differential equation $s''(t) = -32$ has the general solution

$$s(t) = -16t^2 + C_1 t + C_2 \qquad \text{\color{red}{General solution of } } s''(t) = -32$$

which contains two arbitrary constants. It can be shown that a differential equation of order n has a general solution with n arbitrary constants.

EXAMPLE 1   Verifying Solutions

Determine whether the function is a solution of the differential equation $y'' - y = 0$.

a. $y = \sin x$ **b.** $y = 4e^{-x}$ **c.** $y = Ce^x$

Solution

a. Because $y = \sin x$, $y' = \cos x$, and $y'' = -\sin x$, it follows that

$$y'' - y = -\sin x - \sin x = -2 \sin x \neq 0.$$

So, $y = \sin x$ is *not* a solution.

b. Because $y = 4e^{-x}$, $y' = -4e^{-x}$, and $y'' = 4e^{-x}$, it follows that

$$y'' - y = 4e^{-x} - 4e^{-x} = 0.$$

So, $y = 4e^{-x}$ is a solution.

c. Because $y = Ce^x$, $y' = Ce^x$, and $y'' = Ce^x$, it follows that

$$y'' - y = Ce^x - Ce^x = 0.$$

So, $y = Ce^x$ is a solution for any value of C.

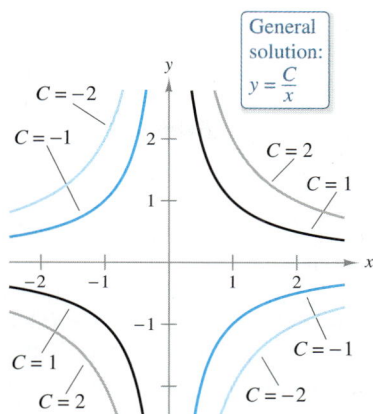

Solution curves for $xy' + y = 0$
Figure 6.1

Geometrically, the general solution of a first-order differential equation represents a family of curves known as **solution curves,** one for each value assigned to the arbitrary constant. For instance, you can verify that every function of the form

$$y = \frac{C}{x} \qquad \text{\color{red}General solution of } xy' + y = 0$$

is a solution of the differential equation

$$xy' + y = 0.$$

Figure 6.1 shows four of the solution curves corresponding to different values of C.

As discussed in Section 4.1, **particular solutions** of a differential equation are obtained from **initial conditions** that give the values of the dependent variable or one of its derivatives for particular values of the independent variable. The term "initial condition" stems from the fact that, often in problems involving time, the value of the dependent variable or one of its derivatives is known at the *initial* time $t = 0$. For instance, the second-order differential equation

$$s''(t) = -32$$

having the general solution

$$s(t) = -16t^2 + C_1 t + C_2 \qquad \text{\color{red}General solution of } s''(t) = -32$$

might have the following initial conditions.

$$s(0) = 80, \quad s'(0) = 64 \qquad \text{\color{red}Initial conditions}$$

In this case, the initial conditions yield the particular solution

$$s(t) = -16t^2 + 64t + 80. \qquad \text{\color{red}Particular solution}$$

EXAMPLE 2   **Finding a Particular Solution**

⋯▷ *See LarsonCalculus.com for an interactive version of this type of example.*

For the differential equation

$$xy' - 3y = 0$$

verify that $y = Cx^3$ is a solution. Then find the particular solution determined by the initial condition $y = 2$ when $x = -3$.

Solution   You know that $y = Cx^3$ is a solution because $y' = 3Cx^2$ and

$$xy' - 3y = x(3Cx^2) - 3(Cx^3) = 0.$$

Furthermore, the initial condition $y = 2$ when $x = -3$ yields

$$y = Cx^3 \qquad \text{\color{red}General solution}$$
$$2 = C(-3)^3 \qquad \text{\color{red}Substitute initial condition.}$$
$$-\frac{2}{27} = C \qquad \text{\color{red}Solve for } C.$$

and you can conclude that the particular solution is

$$y = -\frac{2x^3}{27}. \qquad \text{\color{red}Particular solution}$$

Try checking this solution by substituting for y and y' in the original differential equation. ◼

Note that to determine a particular solution, the number of initial conditions must match the number of constants in the general solution.

Slope Fields

Solving a differential equation analytically can be difficult or even impossible. However, there is a graphical approach you can use to learn a lot about the solution of a differential equation. Consider a differential equation of the form

$$y' = F(x, y) \qquad \text{Differential equation}$$

where $F(x, y)$ is some expression in x and y. At each point (x, y) in the xy-plane where F is defined, the differential equation determines the slope $y' = F(x, y)$ of the solution at that point. If you draw short line segments with slope $F(x, y)$ at selected points (x, y) in the domain of F, then these line segments form a **slope field,** or a *direction field,* for the differential equation $y' = F(x, y)$. Each line segment has the same slope as the solution curve through that point. A slope field shows the general shape of all the solutions and can be helpful in getting a visual perspective of the directions of the solutions of a differential equation.

EXAMPLE 3 **Sketching a Slope Field**

Sketch a slope field for the differential equation $y' = x - y$ for the points $(-1, 1)$, $(0, 1)$, and $(1, 1)$.

Solution The slope of the solution curve at any point (x, y) is

$$F(x, y) = x - y. \qquad \text{Slope at } (x, y).$$

So, the slope at each point can be found as shown.

Slope at $(-1, 1)$: $y' = -1 - 1 = -2$

Slope at $(0, 1)$: $\quad y' = 0 - 1 = -1$

Slope at $(1, 1)$: $\quad y' = 1 - 1 = 0$

Draw short line segments at the three points with their respective slopes, as shown in Figure 6.2.

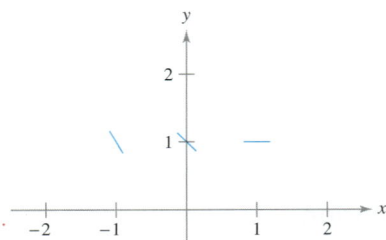

Figure 6.2

EXAMPLE 4 **Identifying Slope Fields for Differential Equations**

Match each slope field with its differential equation.

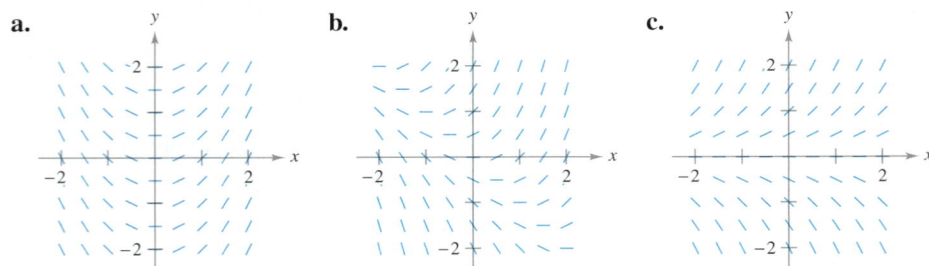

a.

b.

c.

i. $y' = x + y$ **ii.** $y' = x$ **iii.** $y' = y$

Solution

a. You can see that the slope at any point along the y-axis is 0. The only equation that satisfies this condition is $y' = x$. So, the graph matches equation (ii).

b. You can see that the slope at the point $(1, -1)$ is 0. The only equation that satisfies this condition is $y' = x + y$. So, the graph matches equation (i).

c. You can see that the slope at any point along the x-axis is 0. The only equation that satisfies this condition is $y' = y$. So, the graph matches equation (iii).

A solution curve of a differential equation $y' = F(x, y)$ is simply a curve in the xy-plane whose tangent line at each point (x, y) has slope equal to $F(x, y)$. This is illustrated in Example 5.

EXAMPLE 5 Sketching a Solution Using a Slope Field

Sketch a slope field for the differential equation

$$y' = 2x + y.$$

Use the slope field to sketch the solution that passes through the point $(1, 1)$.

Solution Make a table showing the slopes at several points. The table shown is a small sample. The slopes at many other points should be calculated to get a representative slope field.

x	-2	-2	-1	-1	0	0	1	1	2	2
y	-1	1	-1	1	-1	1	-1	1	-1	1
$y' = 2x + y$	-5	-3	-3	-1	-1	1	1	3	3	5

Next, draw line segments at the points with their respective slopes, as shown in Figure 6.3.

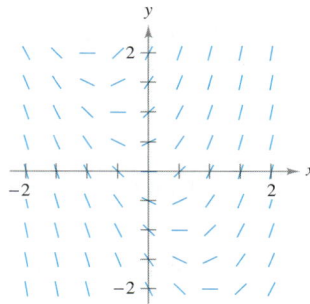

Slope field for $y' = 2x + y$
Figure 6.3

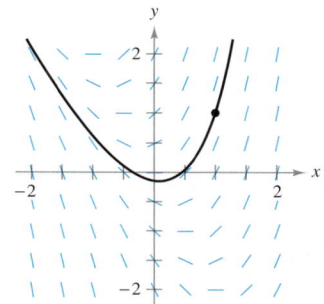

Particular solution for $y' = 2x + y$
passing through $(1, 1)$
Figure 6.4

After the slope field is drawn, start at the initial point $(1, 1)$ and move to the right in the direction of the line segment. Continue to draw the solution curve so that it moves parallel to the nearby line segments. Do the same to the left of $(1, 1)$. The resulting solution is shown in Figure 6.4.

In Example 5, note that the slope field shows that y' increases to infinity as x increases.

▷ **TECHNOLOGY** Drawing a slope field by hand is tedious. In practice, slope fields are usually drawn using a graphing utility. If you have access to a graphing utility that can graph slope fields, try graphing the slope field for the differential equation in Example 5. One example of a slope field drawn by a graphing utility is shown at the right.

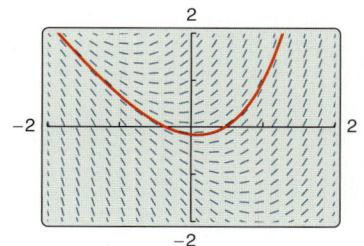

Generated by Maple.

Euler's Method

Euler's Method is a numerical approach to approximating the particular solution of the differential equation

$$y' = F(x, y)$$

that passes through the point (x_0, y_0). From the given information, you know that the graph of the solution passes through the point (x_0, y_0) and has a slope of $F(x_0, y_0)$ at this point. This gives you a "starting point" for approximating the solution.

From this starting point, you can proceed in the direction indicated by the slope. Using a small step h, move along the tangent line until you arrive at the point (x_1, y_1), where

$$x_1 = x_0 + h \quad \text{and} \quad y_1 = y_0 + hF(x_0, y_0)$$

as shown in Figure 6.5. Then, using (x_1, y_1) as a new starting point, you can repeat the process to obtain a second point (x_2, y_2). The values of x_i and y_i are shown below.

$$x_1 = x_0 + h \qquad\qquad y_1 = y_0 + hF(x_0, y_0)$$
$$x_2 = x_1 + h \qquad\qquad y_2 = y_1 + hF(x_1, y_1)$$
$$\vdots \qquad\qquad\qquad\qquad \vdots$$
$$x_n = x_{n-1} + h \qquad\quad y_n = y_{n-1} + hF(x_{n-1}, y_{n-1})$$

When using this method, note that you can obtain better approximations of the exact solution by choosing smaller and smaller step sizes.

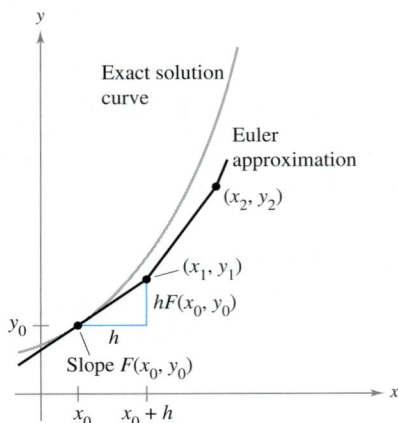

Figure 6.5

EXAMPLE 6   **Approximating a Solution Using Euler's Method**

Use Euler's Method to approximate the particular solution of the differential equation

$$y' = x - y$$

passing through the point $(0, 1)$. Use a step of $h = 0.1$.

Solution   Using $h = 0.1$, $x_0 = 0$, $y_0 = 1$, and $F(x, y) = x - y$, you have

$$x_0 = 0, \quad x_1 = 0.1, \quad x_2 = 0.2, \quad x_3 = 0.3,$$

and the first three approximations are

$$y_1 = y_0 + hF(x_0, y_0) = 1 + (0.1)(0 - 1) = 0.9$$
$$y_2 = y_1 + hF(x_1, y_1) = 0.9 + (0.1)(0.1 - 0.9) = 0.82$$
$$y_3 = y_2 + hF(x_2, y_2) = 0.82 + (0.1)(0.2 - 0.82) = 0.758.$$

The first ten approximations are shown in the table. You can plot these values to see a graph of the approximate solution, as shown in Figure 6.6.

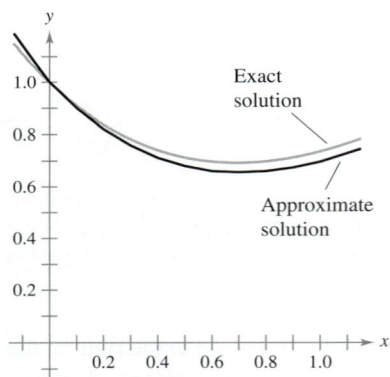

Figure 6.6

n	0	1	2	3	4	5	6	7	8	9	10
x_n	0	0.1	0.2	0.3	0.4	0.5	0.6	0.7	0.8	0.9	1.0
y_n	1	0.900	0.820	0.758	0.712	0.681	0.663	0.657	0.661	0.675	0.697

For the differential equation in Example 6, you can verify the exact solution to be the equation

$$y = x - 1 + 2e^{-x}.$$

Figure 6.6 compares this exact solution with the approximate solution obtained in Example 6.

6.1 Exercises

See **CalcChat.com** for tutorial help and worked-out solutions to odd-numbered exercises.

Verifying a Solution In Exercises 1–8, verify the solution of the differential equation.

Solution	Differential Equation
1. $y = Ce^{4x}$	$y' = 4y$
2. $y = e^{-2x}$	$3y' + 5y = -e^{-2x}$
3. $x^2 + y^2 = Cy$	$y' = \dfrac{2xy}{x^2 - y^2}$
4. $y^2 - 2\ln y = x^2$	$\dfrac{dy}{dx} = \dfrac{xy}{y^2 - 1}$
5. $y = C_1 \sin x - C_2 \cos x$	$y'' + y = 0$
6. $y = C_1 e^{-x} \cos x + C_2 e^{-x} \sin x$	$y'' + 2y' + 2y = 0$
7. $y = -\cos x \ln|\sec x + \tan x|$	$y'' + y = \tan x$
8. $y = \frac{2}{5}(e^{-4x} + e^x)$	$y'' + 4y' = 2e^x$

Verifying a Particular Solution In Exercises 9–12, verify the particular solution of the differential equation.

Solution	Differential Equation and Initial Condition
9. $y = \sin x \cos x - \cos^2 x$	$2y + y' = 2\sin(2x) - 1$ $y\left(\dfrac{\pi}{4}\right) = 0$
10. $y = 6x - 4\sin x + 1$	$y' = 6 - 4\cos x$ $y(0) = 1$
11. $y = 4e^{-6x^2}$	$y' = -12xy$ $y(0) = 4$
12. $y = e^{-\cos x}$	$y' = y \sin x$ $y\left(\dfrac{\pi}{2}\right) = 1$

Determining a Solution In Exercises 13–20, determine whether the function is a solution of the differential equation $y^{(4)} - 16y = 0$.

13. $y = 3\cos x$ 14. $y = 2\sin x$

15. $y = 3\cos 2x$ 16. $y = 3\sin 2x$

17. $y = e^{-2x}$ 18. $y = 5\ln x$

19. $y = C_1 e^{2x} + C_2 e^{-2x} + C_3 \sin 2x + C_4 \cos 2x$

20. $y = 3e^{2x} - 4\sin 2x$

Determining a Solution In Exercises 21–28, determine whether the function is a solution of the differential equation $xy' - 2y = x^3 e^x$.

21. $y = x^2$ 22. $y = x^3$

23. $y = x^2 e^x$ 24. $y = x^2(2 + e^x)$

25. $y = \sin x$ 26. $y = \cos x$

27. $y = \ln x$ 28. $y = x^2 e^x - 5x^2$

Finding a Particular Solution In Exercises 29–32, some of the curves corresponding to different values of C in the general solution of the differential equation are shown in the graph. Find the particular solution that passes through the point shown on the graph.

29. $y^2 = Ce^{-x/2}$
    $2y' + y = 0$

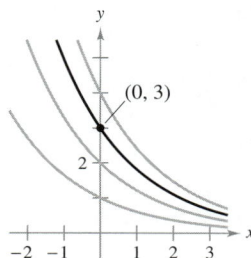

30. $y(x^2 + y) = C$
    $2xy + (x^2 + 2y)y' = 0$

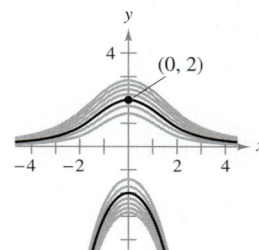

31. $y^2 = Cx^3$
    $2xy' - 3y = 0$

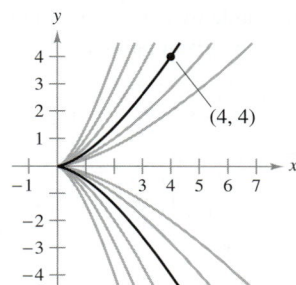

32. $2x^2 - y^2 = C$
    $yy' - 2x = 0$

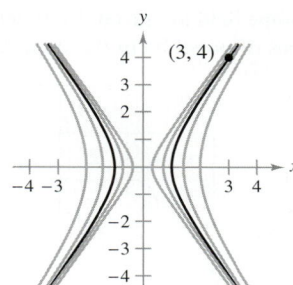

Graphs of Particular Solutions In Exercises 33 and 34, the general solution of the differential equation is given. Use a graphing utility to graph the particular solutions for the given values of C.

33. $4yy' - x = 0$
    $4y^2 - x^2 = C$
    $C = 0, C = \pm 1, C = \pm 4$

34. $yy' + x = 0$
    $x^2 + y^2 = C$
    $C = 0, C = 1, C = 4$

Finding a Particular Solution In Exercises 35–40, verify that the general solution satisfies the differential equation. Then find the particular solution that satisfies the initial condition(s).

35. $y = Ce^{-2x}$
    $y' + 2y = 0$
    $y = 3$ when $x = 0$

36. $3x^2 + 2y^2 = C$
    $3x + 2yy' = 0$
    $y = 3$ when $x = 1$

37. $y = C_1 \sin 3x + C_2 \cos 3x$
    $y'' + 9y = 0$
    $y = 2$ when $x = \dfrac{\pi}{6}$
    $y' = 1$ when $x = \dfrac{\pi}{6}$

38. $y = C_1 + C_2 \ln x$
    $xy'' + y' = 0$
    $y = 0$ when $x = 2$
    $y' = \dfrac{1}{2}$ when $x = 2$

39. $y = C_1x + C_2x^3$

$x^2y'' - 3xy' + 3y = 0$

$y = 0$ when $x = 2$

$y' = 4$ when $x = 2$

40. $y = e^{2x/3}(C_1 + C_2x)$

$9y'' - 12y' + 4y = 0$

$y = 4$ when $x = 0$

$y = 0$ when $x = 3$

Finding a General Solution In Exercises 41–52, use integration to find a general solution of the differential equation.

41. $\dfrac{dy}{dx} = 6x^2$

42. $\dfrac{dy}{dx} = 10x^4 - 2x^3$

43. $\dfrac{dy}{dx} = \dfrac{x}{1 + x^2}$

44. $\dfrac{dy}{dx} = \dfrac{e^x}{4 + e^x}$

45. $\dfrac{dy}{dx} = \dfrac{x - 2}{x}$

46. $\dfrac{dy}{dx} = x\cos x^2$

47. $\dfrac{dy}{dx} = \sin 2x$

48. $\dfrac{dy}{dx} = \tan^2 x$

49. $\dfrac{dy}{dx} = x\sqrt{x - 6}$

50. $\dfrac{dy}{dx} = 2x\sqrt{4x^2 + 1}$

51. $\dfrac{dy}{dx} = xe^{x^2}$

52. $\dfrac{dy}{dx} = 5e^{-x/2}$

Slope Field In Exercises 53–56, a differential equation and its slope field are given. Complete the table by determining the slopes (if possible) in the slope field at the given points.

x	-4	-2	0	2	4	8
y	2	0	4	4	6	8
dy/dx						

53. $\dfrac{dy}{dx} = \dfrac{2x}{y}$

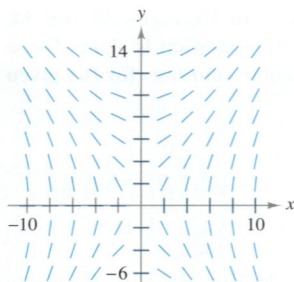

54. $\dfrac{dy}{dx} = y - x$

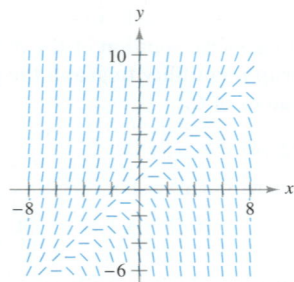

55. $\dfrac{dy}{dx} = x\cos\dfrac{\pi y}{8}$

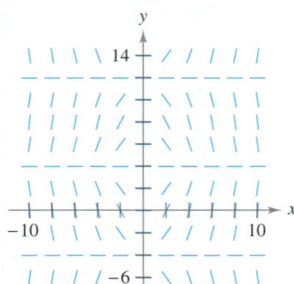

56. $\dfrac{dy}{dx} = \tan\left(\dfrac{\pi y}{6}\right)$

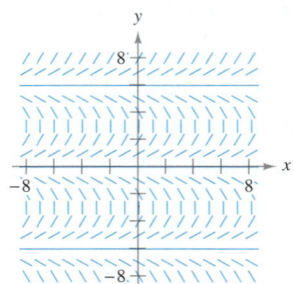

Matching In Exercises 57–60, match the differential equation with its slope field. [The slope fields are labeled (a), (b), (c), and (d).]

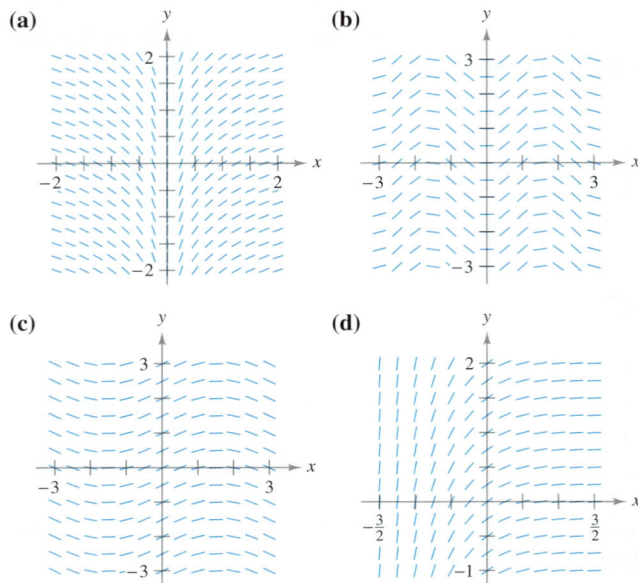

(a)

(b)

(c)

(d)

57. $\dfrac{dy}{dx} = \sin(2x)$

58. $\dfrac{dy}{dx} = \dfrac{1}{2}\cos x$

59. $\dfrac{dy}{dx} = e^{-2x}$

60. $\dfrac{dy}{dx} = \dfrac{1}{x}$

Slope Field In Exercises 61–64, (a) sketch the slope field for the differential equation, (b) use the slope field to sketch the solution that passes through the given point, and (c) discuss the graph of the solution as $x \to \infty$ and $x \to -\infty$. Use a graphing utility to verify your results. To print a blank graph, go to *MathGraphs.com.*

61. $y' = 3 - x$, $(4, 2)$

62. $y' = \frac{1}{3}x^2 - \frac{1}{2}x$, $(1, 1)$

63. $y' = y - 4x$, $(2, 2)$

64. $y' = y + xy$, $(0, -4)$

65. Slope Field Use the slope field for the differential equation $y' = 1/x$, where $x > 0$, to sketch the graph of the solution that satisfies each given initial condition. Then make a conjecture about the behavior of a particular solution of $y' = 1/x$ as $x \to \infty$. To print an enlarged copy of the graph, go to *MathGraphs.com.*

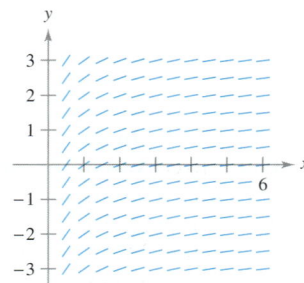

(a) $(1, 0)$ (b) $(2, -1)$

66. Slope Field Use the slope field for the differential equation $y' = 1/y$, where $y > 0$, to sketch the graph of the solution that satisfies each given initial condition. Then make a conjecture about the behavior of a particular solution of $y' = 1/y$ as $x \to \infty$. To print an enlarged copy of the graph, go to *MathGraphs.com*.

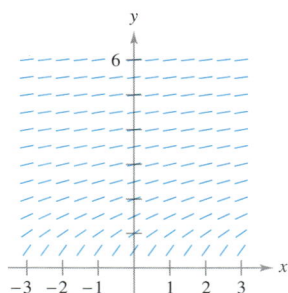

(a) $(0, 1)$ (b) $(1, 1)$

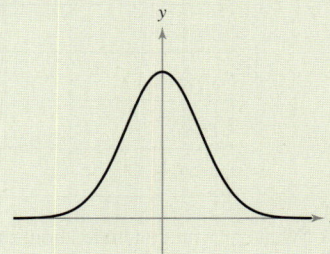 **Slope Field** In Exercises 67–72, use a computer algebra system to (a) graph the slope field for the differential equation and (b) graph the solution satisfying the specified initial condition.

67. $\dfrac{dy}{dx} = 0.25y, \quad y(0) = 4$

68. $\dfrac{dy}{dx} = 4 - y, \quad y(0) = 6$

69. $\dfrac{dy}{dx} = 0.02y(10 - y), \quad y(0) = 2$

70. $\dfrac{dy}{dx} = 0.2x(2 - y), \quad y(0) = 9$

71. $\dfrac{dy}{dx} = 0.4y(3 - x), \quad y(0) = 1$

72. $\dfrac{dy}{dx} = \dfrac{1}{2}e^{-x/8} \sin \dfrac{\pi y}{4}, \quad y(0) = 2$

Euler's Method In Exercises 73–78, use Euler's Method to make a table of values for the approximate solution of the differential equation with the specified initial value. Use n steps of size h.

73. $y' = x + y, \quad y(0) = 2, \quad n = 10, \quad h = 0.1$

74. $y' = x + y, \quad y(0) = 2, \quad n = 20, \quad h = 0.05$

75. $y' = 3x - 2y, \quad y(0) = 3, \quad n = 10, \quad h = 0.05$

76. $y' = 0.5x(3 - y), \quad y(0) = 1, \quad n = 5, \quad h = 0.4$

77. $y' = e^{xy}, \quad y(0) = 1, \quad n = 10, \quad h = 0.1$

78. $y' = \cos x + \sin y, \quad y(0) = 5, \quad n = 10, \quad h = 0.1$

Euler's Method In Exercises 79–81, complete the table using the exact solution of the differential equation and two approximations obtained using Euler's Method to approximate the particular solution of the differential equation. Use $h = 0.2$ and $h = 0.1$, and compute each approximation to four decimal places.

x	0	0.2	0.4	0.6	0.8	1
$y(x)$ (exact)						
$y(x)$ ($h = 0.2$)						
$y(x)$ ($h = 0.1$)						

Table for 79–81

Differential Equation	Initial Condition	Exact Solution
79. $\dfrac{dy}{dx} = y$	$(0, 3)$	$y = 3e^x$
80. $\dfrac{dy}{dx} = \dfrac{2x}{y}$	$(0, 2)$	$y = \sqrt{2x^2 + 4}$
81. $\dfrac{dy}{dx} = y + \cos(x)$	$(0, 0)$	$y = \dfrac{1}{2}(\sin x - \cos x + e^x)$

82. Euler's Method Compare the values of the approximations in Exercises 79–81 with the values given by the exact solution. How does the error change as h increases?

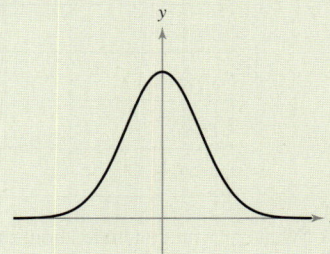 **83. Temperature** At time $t = 0$ minutes, the temperature of an object is $140°F$. The temperature of the object is changing at the rate given by the differential equation

$$\frac{dy}{dt} = -\frac{1}{2}(y - 72).$$

(a) Use a graphing utility and Euler's Method to approximate the particular solutions of this differential equation at $t = 1, 2,$ and 3. Use a step size of $h = 0.1$. (A graphing utility program for Euler's Method is available at the website *college.hmco.com*.)

(b) Compare your results with the exact solution

$$y = 72 + 68e^{-t/2}.$$

(c) Repeat parts (a) and (b) using a step size of $h = 0.05$. Compare the results.

84. HOW DO YOU SEE IT? The graph shows a solution of one of the following differential equations. Determine the correct equation. Explain your reasoning.

(a) $y' = xy$

(b) $y' = \dfrac{4x}{y}$

(c) $y' = -4xy$

(d) $y' = 4 - xy$

True or False? In Exercises 89–92, determine whether the statement is true or false. If it is false, explain why or give an example that shows it is false.

89. If $y = f(x)$ is a solution of a first-order differential equation, then $y = f(x) + C$ is also a solution.

90. The general solution of a differential equation is $y = -4.9x^2 + C_1x + C_2$. To find a particular solution, you must be given two initial conditions.

91. Slope fields represent the general solutions of differential equations.

92. A slope field shows that the slope at the point $(1, 1)$ is 6. This slope field represents the family of solutions for the differential equation $y' = 4x + 2y$.

93. Errors and Euler's Method The exact solution of the differential equation

$$\frac{dy}{dx} = -2y$$

where $y(0) = 4$, is $y = 4e^{-2x}$.

(a) Use a graphing utility to complete the table, where y is the exact value of the solution, y_1 is the approximate solution using Euler's Method with $h = 0.1$, y_2 is the approximate solution using Euler's Method with $h = 0.2$, e_1 is the absolute error $|y - y_1|$, e_2 is the absolute error $|y - y_2|$, and r is the ratio e_1/e_2.

x	0	0.2	0.4	0.6	0.8	1
y						
y_1						
y_2						
e_1						
e_2						
r						

(b) What can you conclude about the ratio r as h changes?

(c) Predict the absolute error when $h = 0.05$.

94. Errors and Euler's Method Repeat Exercise 93 for which the exact solution of the differential equation

$$\frac{dy}{dx} = x - y$$

where $y(0) = 1$, is $y = x - 1 + 2e^{-x}$.

95. Electric Circuit The diagram shows a simple electric circuit consisting of a power source, a resistor, and an inductor.

A model of the current I, in amperes (A), at time t is given by the first-order differential equation

$$L\frac{dI}{dt} + RI = E(t)$$

where $E(t)$ is the voltage (V) produced by the power source, R is the resistance, in ohms (Ω), and L is the inductance, in henrys (H). Suppose the electric circuit consists of a 24-V power source, a 12-Ω resistor, and a 4-H inductor.

(a) Sketch a slope field for the differential equation.

(b) What is the limiting value of the current? Explain.

96. Think About It It is known that $y = e^{kt}$ is a solution of the differential equation $y'' - 16y = 0$. Find the values of k.

97. Think About It It is known that $y = A \sin \omega t$ is a solution of the differential equation $y'' + 16y = 0$. Find the values of ω.

PUTNAM EXAM CHALLENGE

98. Let f be a twice-differentiable real-valued function satisfying

$$f(x) + f''(x) = -xg(x)f'(x)$$

where $g(x) \geq 0$ for all real x. Prove that $|f(x)|$ is bounded.

99. Prove that if the family of integral curves of the differential equation

$$\frac{dy}{dx} + p(x)y = q(x), \quad p(x) \cdot q(x) \neq 0$$

is cut by the line $x = k$, the tangents at the points of intersection are concurrent.

6.2 Differential Equations: Growth and Decay

■ Use separation of variables to solve a simple differential equation.
■ Use exponential functions to model growth and decay in applied problems.

Differential Equations

In Section 6.1, you learned to analyze the solutions visually of differential equations using slope fields and to approximate solutions numerically using Euler's Method. Analytically, you have learned to solve only two types of differential equations—those of the forms $y' = f(x)$ and $y'' = f(x)$. In this section, you will learn how to solve a more general type of differential equation. The strategy is to rewrite the equation so that each variable occurs on only one side of the equation. This strategy is called *separation of variables*. (You will study this strategy in detail in Section 6.3.)

EXAMPLE 1   Solving a Differential Equation

$$y' = \frac{2x}{y} \qquad \text{Original equation}$$

$$yy' = 2x \qquad \text{Multiply both sides by } y.$$

$$\int yy' \, dx = \int 2x \, dx \qquad \text{Integrate with respect to } x.$$

$$\int y \, dy = \int 2x \, dx \qquad dy = y' \, dx$$

$$\frac{1}{2} y^2 = x^2 + C_1 \qquad \text{Apply Power Rule.}$$

$$y^2 - 2x^2 = C \qquad \text{Rewrite, letting } C = 2C_1.$$

REMARK You can use implicit differentiation to check the solution in Example 1.

▷ So, the general solution is $y^2 - 2x^2 = C$. ■

When you integrate both sides of the equation in Example 1, you don't need to add a constant of integration to both sides. When you do, you still obtain the same result.

$$\int y \, dy = \int 2x \, dx$$

$$\frac{1}{2} y^2 + C_2 = x^2 + C_3$$

$$\frac{1}{2} y^2 = x^2 + (C_3 - C_2)$$

$$\frac{1}{2} y^2 = x^2 + C_1$$

Some people prefer to use Leibniz notation and differentials when applying separation of variables. The solution to Example 1 is shown below using this notation.

$$\frac{dy}{dx} = \frac{2x}{y}$$

$$y \, dy = 2x \, dx$$

$$\int y \, dy = \int 2x \, dx$$

$$\frac{1}{2} y^2 = x^2 + C_1$$

$$y^2 - 2x^2 = C$$

Exploration

In Example 1, the general solution of the differential equation is

$$y^2 - 2x^2 = C.$$

Use a graphing utility to sketch the particular solutions for $C = \pm 2$, $C = \pm 1$, and $C = 0$. Describe the solutions graphically. Is the following statement true of each solution?

The slope of the graph at the point (x, y) is equal to twice the ratio of x and y.

Explain your reasoning. Are all curves for which this statement is true represented by the general solution?

Growth and Decay Models

In many applications, the rate of change of a variable y is proportional to the value of y. When y is a function of time t, the proportion can be written as shown.

Rate of change of y is   proportional to y.

$$\frac{dy}{dt} = ky$$

The general solution of this differential equation is given in the next theorem.

THEOREM 6.1   **Exponential Growth and Decay Model**

If y is a differentiable function of t such that $y > 0$ and $y' = ky$ for some constant k, then

$$y = Ce^{kt}$$

where C is the **initial value** of y, and k is the **proportionality constant.** **Exponential growth** occurs when $k > 0$, and **exponential decay** occurs when $k < 0$.

Proof

$$y' = ky \qquad \text{Write original equation.}$$

$$\frac{y'}{y} = k \qquad \text{Separate variables.}$$

$$\int \frac{y'}{y}\, dt = \int k\, dt \qquad \text{Integrate with respect to } t.$$

$$\int \frac{1}{y}\, dy = \int k\, dt \qquad dy = y'\, dt$$

$$\ln y = kt + C_1 \qquad \text{Find antiderivative of each side.}$$

$$y = e^{kt}e^{C_1} \qquad \text{Solve for } y.$$

$$y = Ce^{kt} \qquad \text{Let } C = e^{C_1}.$$

So, all solutions of $y' = ky$ are of the form $y = Ce^{kt}$. Remember that you can differentiate the function $y = Ce^{kt}$ with respect to t to verify that $y' = ky$.

See LarsonCalculus.com for Bruce Edwards's video of this proof.

EXAMPLE 2   **Using an Exponential Growth Model**

The rate of change of y is proportional to y. When $t = 0$, $y = 2$, and when $t = 2$, $y = 4$. What is the value of y when $t = 3$?

Solution Because $y' = ky$, you know that y and t are related by the equation $y = Ce^{kt}$. You can find the values of the constants C and k by applying the initial conditions.

$$2 = Ce^0 \implies C = 2 \qquad \text{When } t = 0, y = 2.$$

$$4 = 2e^{2k} \implies k = \frac{1}{2}\ln 2 \approx 0.3466 \qquad \text{When } t = 2, y = 4.$$

So, the model is $y = 2e^{0.3466t}$. When $t = 3$, the value of y is $2e^{0.3466(3)} \approx 5.657$ (see Figure 6.7).

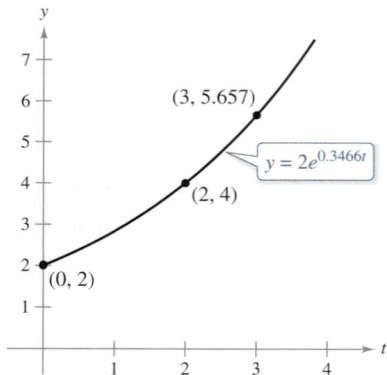

(3, 5.657)

$y = 2e^{0.3466t}$

(2, 4)

(0, 2)

If the rate of change of y is proportional to y, then y follows an exponential model.

Figure 6.7

Using logarithmic properties, the value of k in Example 2 can also be written as $\ln\sqrt{2}$. So, the model becomes $y = 2e^{(\ln\sqrt{2})t}$, which can be rewritten as $y = 2\left(\sqrt{2}\right)^t$.

▷ **TECHNOLOGY**   Most graphing utilities have curve-fitting capabilities that can be used to find models that represent data. Use the *exponential regression* feature of a graphing utility and the information in Example 2 to find a model for the data. How does your model compare with the given model?

Radioactive decay is measured in terms of *half-life*—the number of years required for half of the atoms in a sample of radioactive material to decay. The rate of decay is proportional to the amount present. The half-lives of some common radioactive isotopes are listed below.

Uranium (^{238}U)	4,470,000,000 years
Plutonium (^{239}Pu)	24,100 years
Carbon (^{14}C)	5715 years
Radium (^{226}Ra)	1599 years
Einsteinium (^{254}Es)	276 days
Radon (^{222}Rn)	3.82 days
Nobelium (^{257}No)	25 seconds

EXAMPLE 3   **Radioactive Decay**

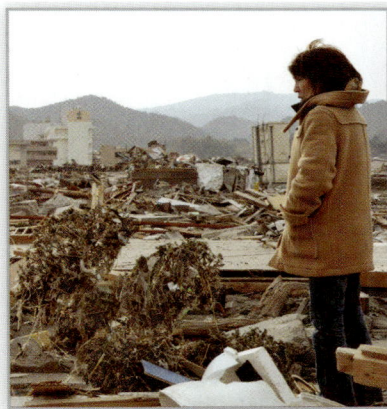

Ten grams of the plutonium isotope ^{239}Pu were released in a nuclear accident. How long will it take for the 10 grams to decay to 1 gram?

Solution   Let y represent the mass (in grams) of the plutonium. Because the rate of decay is proportional to y, you know that

$$y = Ce^{kt}$$

where t is the time in years. To find the values of the constants C and k, apply the initial conditions. Using the fact that $y = 10$ when $t = 0$, you can write

$$10 = Ce^{k(0)} \implies 10 = Ce^0$$

which implies that $C = 10$. Next, using the fact that the half-life of ^{239}Pu is 24,100 years, you have $y = 10/2 = 5$ when $t = 24,100$, so you can write

$$5 = 10e^{k(24,100)}$$

$$\frac{1}{2} = e^{24,100k}$$

$$\frac{1}{24,100} \ln \frac{1}{2} = k$$

$$-0.000028761 \approx k.$$

So, the model is

$$y = 10e^{-0.000028761t}. \qquad \text{\textcolor{red}{Half-life model}}$$

To find the time it would take for 10 grams to decay to 1 gram, you can solve for t in the equation

$$1 = 10e^{-0.000028761t}.$$

The solution is approximately 80,059 years.   ∎

The Fukushima Daiichi nuclear disaster occurred after an earthquake and tsunami. Several of the reactors at the plant experienced full meltdowns.

▷ **REMARK**   The exponential decay model in Example 3 could also be written as $y = 10\left(\frac{1}{2}\right)^{t/24,100}$. This model is much easier to derive, but for some applications it is not as convenient to use.

From Example 3, notice that in an exponential growth or decay problem, it is easy to solve for C when you are given the value of y at $t = 0$. The next example demonstrates a procedure for solving for C and k when you do not know the value of y at $t = 0$.

⋯▷ *See LarsonCalculus.com for an interactive version of this type of example.*

An experimental population of fruit flies increases according to the law of exponential growth. There were 100 flies after the second day of the experiment and 300 flies after the fourth day. Approximately how many flies were in the original population?

Solution   Let $y = Ce^{kt}$ be the number of flies at time t, where t is measured in days. Note that y is continuous, whereas the number of flies is discrete. Because $y = 100$ when $t = 2$ and $y = 300$ when $t = 4$, you can write

$$100 = Ce^{2k} \quad \text{and} \quad 300 = Ce^{4k}.$$

From the first equation, you know that

$$C = 100e^{-2k}.$$

Substituting this value into the second equation produces the following.

$$300 = 100e^{-2k}e^{4k}$$
$$300 = 100e^{2k}$$
$$3 = e^{2k}$$
$$\ln 3 = 2k$$
$$\frac{1}{2}\ln 3 = k$$
$$0.5493 \approx k$$

So, the exponential growth model is

$$y = Ce^{0.5493t}.$$

To solve for C, reapply the condition $y = 100$ when $t = 2$ and obtain

$$100 = Ce^{0.5493(2)}$$
$$C = 100e^{-1.0986}$$
$$C \approx 33.$$

So, the original population (when $t = 0$) consisted of approximately $y = C = 33$ flies, as shown in Figure 6.8.

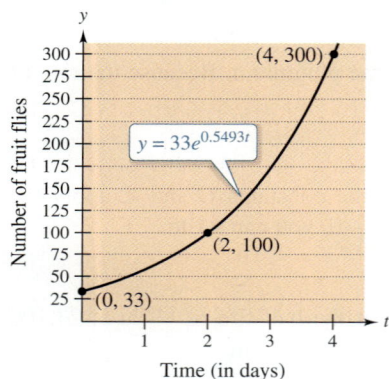

Figure 6.8

Four months after it stops advertising, a manufacturing company notices that its sales have dropped from 100,000 units per month to 80,000 units per month. The sales follow an exponential pattern of decline. What will the sales be after another 2 months?

Solution   Use the exponential decay model $y = Ce^{kt}$, where t is measured in months. From the initial condition ($t = 0$), you know that $C = 100,000$. Moreover, because $y = 80,000$ when $t = 4$, you have

$$80,000 = 100,000e^{4k}$$
$$0.8 = e^{4k}$$
$$\ln(0.8) = 4k$$
$$-0.0558 \approx k.$$

So, after 2 more months ($t = 6$), you can expect the monthly sales rate to be

$$y = 100,000e^{-0.0558(6)}$$
$$\approx 71,500 \text{ units.}$$

See Figure 6.9.

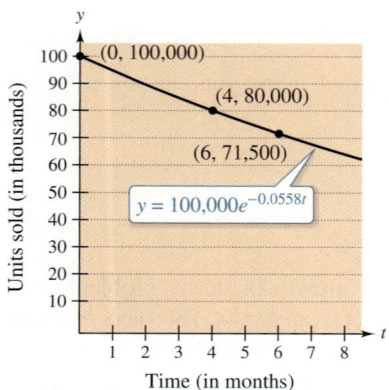

Figure 6.9

In Examples 2 through 5, you did not actually have to solve the differential equation $y' = ky$. (This was done once in the proof of Theorem 6.1.) The next example demonstrates a problem whose solution involves the separation of variables technique. The example concerns **Newton's Law of Cooling,** which states that the rate of change in the temperature of an object is proportional to the difference between the object's temperature and the temperature of the surrounding medium.

EXAMPLE 6   **Newton's Law of Cooling**

Let y represent the temperature (in °F) of an object in a room whose temperature is kept at a constant 60°. The object cools from 100° to 90° in 10 minutes. How much longer will it take for the temperature of the object to decrease to 80°?

Solution   From Newton's Law of Cooling, you know that the rate of change in y is proportional to the difference between y and 60. This can be written as

$$y' = k(y - 60), \quad 80 \le y \le 100.$$

To solve this differential equation, use separation of variables, as shown.

$$\frac{dy}{dt} = k(y - 60) \qquad \text{\color{red}Differential equation}$$

$$\left(\frac{1}{y - 60}\right) dy = k\, dt \qquad \text{\color{red}Separate variables.}$$

$$\int \frac{1}{y - 60}\, dy = \int k\, dt \qquad \text{\color{red}Integrate each side.}$$

$$\ln|y - 60| = kt + C_1 \qquad \text{\color{red}Find antiderivative of each side.}$$

Because $y > 60$, $|y - 60| = y - 60$, and you can omit the absolute value signs. Using exponential notation, you have

$$y - 60 = e^{kt + C_1}$$

$$y = 60 + Ce^{kt}. \qquad \text{\color{red}}C = e^{C_1}$$

Using $y = 100$ when $t = 0$, you obtain

$$100 = 60 + Ce^{k(0)} = 60 + C$$

which implies that $C = 40$. Because $y = 90$ when $t = 10$,

$$90 = 60 + 40e^{k(10)}$$

$$30 = 40e^{10k}$$

$$k = \frac{1}{10} \ln \frac{3}{4}.$$

So, $k \approx -0.02877$ and the model is

$$y = 60 + 40e^{-0.02877t}. \qquad \text{\color{red}Cooling model}$$

When $y = 80$, you obtain

$$80 = 60 + 40e^{-0.02877t}$$

$$20 = 40e^{-0.02877t}$$

$$\frac{1}{2} = e^{-0.02877t}$$

$$\ln \frac{1}{2} = -0.02877t$$

$$t \approx 24.09 \text{ minutes.}$$

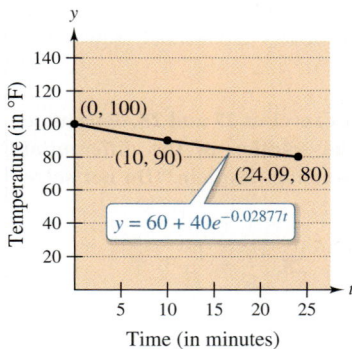

Figure 6.10

So, it will require about 14.09 *more* minutes for the object to cool to a temperature of 80° (see Figure 6.10).

6.2 Exercises

See **CalcChat.com** for tutorial help and worked-out solutions to odd-numbered exercises.

Solving a Differential Equation In Exercises 1–10, solve the differential equation.

1. $\dfrac{dy}{dx} = x + 3$

2. $\dfrac{dy}{dx} = 5 - 8x$

3. $\dfrac{dy}{dx} = y + 3$

4. $\dfrac{dy}{dx} = 6 - y$

5. $y' = \dfrac{5x}{y}$

6. $y' = -\dfrac{\sqrt{x}}{4y}$

7. $y' = \sqrt{x}\, y$

8. $y' = x(1 + y)$

9. $(1 + x^2)y' - 2xy = 0$

10. $xy + y' = 100x$

Writing and Solving a Differential Equation In Exercises 11 and 12, write and solve the differential equation that models the verbal statement.

11. The rate of change of Q with respect to t is inversely proportional to the square of t.

12. The rate of change of P with respect to t is proportional to $25 - t$.

Slope Field In Exercises 13 and 14, a differential equation, a point, and a slope field are given. (a) Sketch two approximate solutions of the differential equation on the slope field, one of which passes through the given point. (b) Use integration to find the particular solution of the differential equation and use a graphing utility to graph the solution. Compare the result with the sketch in part (a). To print an enlarged copy of the graph, go to *MathGraphs.com.*

13. $\dfrac{dy}{dx} = x(6 - y), \quad (0, 0)$

14. $\dfrac{dy}{dx} = xy, \quad \left(0, \dfrac{1}{2}\right)$

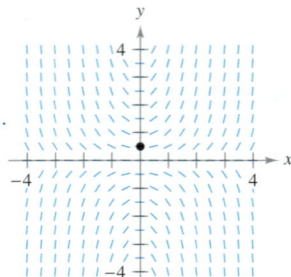

Finding a Particular Solution In Exercises 15–18, find the function $y = f(t)$ passing through the point $(0, 10)$ with the given first derivative. Use a graphing utility to graph the solution.

15. $\dfrac{dy}{dt} = \dfrac{1}{2}t$

16. $\dfrac{dy}{dt} = -9\sqrt{t}$

17. $\dfrac{dy}{dt} = -\dfrac{1}{2}y$

18. $\dfrac{dy}{dt} = \dfrac{3}{4}y$

Writing and Solving a Differential Equation In Exercises 19 and 20, write and solve the differential equation that models the verbal statement. Evaluate the solution at the specified value of the independent variable.

19. The rate of change of N is proportional to N. When $t = 0$, $N = 250$, and when $t = 1$, $N = 400$. What is the value of N when $t = 4$?

20. The rate of change of P is proportional to P. When $t = 0$, $P = 5000$, and when $t = 1$, $P = 4750$. What is the value of P when $t = 5$?

Finding an Exponential Function In Exercises 21–24, find the exponential function $y = Ce^{kt}$ that passes through the two given points.

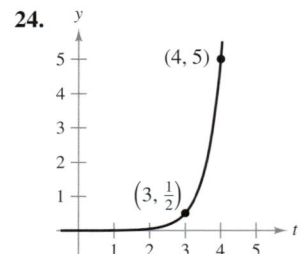

21.

22.

23.

24.

WRITING ABOUT CONCEPTS

25. Describing Values Describe what the values of C and k represent in the exponential growth and decay model, $y = Ce^{kt}$.

26. Exponential Growth and Decay Give the differential equation that models exponential growth and decay.

Increasing Function In Exercises 27 and 28, determine the quadrants in which the solution of the differential equation is an increasing function. Explain. (Do not solve the differential equation.)

27. $\dfrac{dy}{dx} = \dfrac{1}{2}xy$

28. $\dfrac{dy}{dx} = \dfrac{1}{2}x^2y$

Radioactive Decay In Exercises 29–36, complete the table for the radioactive isotope.

Isotope	Half-life (in years)	Initial Quantity	Amount After 1000 Years	Amount After 10,000 Years
29. ^{226}Ra	1599	20 g		
30. ^{226}Ra	1599		1.5 g	
31. ^{226}Ra	1599			0.1 g
32. ^{14}C	5715			3 g
33. ^{14}C	5715	5 g		
34. ^{14}C	5715		1.6 g	
35. ^{239}Pu	24,100		2.1 g	
36. ^{239}Pu	24,100			0.4 g

37. Radioactive Decay Radioactive radium has a half-life of approximately 1599 years. What percent of a given amount remains after 100 years?

38. Carbon Dating Carbon-14 dating assumes that the carbon dioxide on Earth today has the same radioactive content as it did centuries ago. If this is true, the amount of ^{14}C absorbed by a tree that grew several centuries ago should be the same as the amount of ^{14}C absorbed by a tree growing today. A piece of ancient charcoal contains only 15% as much of the radioactive carbon as a piece of modern charcoal. How long ago was the tree burned to make the ancient charcoal? (The half-life of ^{14}C is 5715 years.)

Compound Interest In Exercises 39–44, complete the table for a savings account in which interest is compounded continuously.

Initial Investment	Annual Rate	Time to Double	Amount After 10 Years
39. $4000	6%		
40. $18,000	$5\frac{1}{2}$%		
41. $750		$7\frac{3}{4}$ yr	
42. $12,500		20 yr	
43. $500			$1292.85
44. $6000			$8950.95

Compound Interest In Exercises 45–48, find the principal P that must be invested at rate r, compounded monthly, so that $1,000,000 will be available for retirement in t years.

45. $r = 7\frac{1}{2}$%, $t = 20$

46. $r = 6$%, $t = 40$

47. $r = 8$%, $t = 35$

48. $r = 9$%, $t = 25$

Compound Interest In Exercises 49 and 50, find the time necessary for $1000 to double when it is invested at a rate of r compounded (a) annually, (b) monthly, (c) daily, and (d) continuously.

49. $r = 7$%   **50.** $r = 5.5$%

Population In Exercises 51–54, the population (in millions) of a country in 2011 and the expected continuous annual rate of change k of the population are given. (*Source: U.S. Census Bureau, International Data Base*)

(a) Find the exponential growth model

$$P = Ce^{kt}$$

for the population by letting $t = 0$ correspond to 2010.

(b) Use the model to predict the population of the country in 2020.

(c) Discuss the relationship between the sign of k and the change in population for the country.

	Country	2011 Population	k
51.	Latvia	2.2	-0.006
52.	Egypt	82.1	0.020
53.	Uganda	34.6	0.036
54.	Hungary	10.0	-0.002

55. Modeling Data One hundred bacteria are started in a culture and the number N of bacteria is counted each hour for 5 hours. The results are shown in the table, where t is the time in hours.

t	0	1	2	3	4	5
N	100	126	151	198	243	297

(a) Use the regression capabilities of a graphing utility to find an exponential model for the data.

(b) Use the model to estimate the time required for the population to quadruple in size.

56. Bacteria Growth The number of bacteria in a culture is increasing according to the law of exponential growth. There are 125 bacteria in the culture after 2 hours and 350 bacteria after 4 hours.

(a) Find the initial population.

(b) Write an exponential growth model for the bacteria population. Let t represent time in hours.

(c) Use the model to determine the number of bacteria after 8 hours.

(d) After how many hours will the bacteria count be 25,000?

57. Learning Curve The management at a certain factory has found that a worker can produce at most 30 units in a day. The learning curve for the number of units N produced per day after a new employee has worked t days is

$$N = 30(1 - e^{kt}).$$

After 20 days on the job, a particular worker produces 19 units.

(a) Find the learning curve for this worker.

(b) How many days should pass before this worker is producing 25 units per day?

58. Learning Curve Suppose the management in Exercise 57 requires a new employee to produce at least 20 units per day after 30 days on the job.

(a) Find the learning curve that describes this minimum requirement.

(b) Find the number of days before a minimal achiever is producing 25 units per day.

59. Insect Population

(a) Suppose an insect population increases by a constant number each month. Explain why the number of insects can be represented by a linear function.

(b) Suppose an insect population increases by a constant percentage each month. Explain why the number of insects can be represented by an exponential function.

60. HOW DO YOU SEE IT? The functions f and g are both of the form $y = Ce^{kt}$.

(a) Do the functions f and g represent exponential growth or exponential decay? Explain.

(b) Assume both functions have the same value of C. Which function has a greater value of k? Explain.

61. Modeling Data The table shows the resident populations P (in millions) of the United States from 1920 to 2010. (*Source: U.S. Census Bureau*)

Year	1920	1930	1940	1950	1960
Population, P	106	123	132	151	179

Year	1970	1980	1990	2000	2010
Population, P	203	227	249	281	309

(a) Use the 1920 and 1930 data to find an exponential model P_1 for the data. Let $t = 0$ represent 1920.

(b) Use a graphing utility to find an exponential model P_2 for all the data. Let $t = 0$ represent 1920.

(c) Use a graphing utility to plot the data and graph models P_1 and P_2 in the same viewing window. Compare the actual data with the predictions. Which model better fits the data?

(d) Use the model chosen in part (c) to estimate when the resident population will be 400 million.

Stephen Aaron Rees/Shutterstock.com

62. Forestry

The value of a tract of timber is

$$V(t) = 100,000e^{0.8\sqrt{t}}$$

where t is the time in years, with $t = 0$ corresponding to 2010. If money earns interest continuously at 10%, then the present value of the timber at any time t is

$$A(t) = V(t)e^{-0.10t}.$$

Find the year in which the timber should be harvested to maximize the present value function.

63. Sound Intensity The level of sound β (in decibels) with an intensity of I is

$$\beta(I) = 10 \log_{10}\left(\frac{I}{I_0}\right)$$

where I_0 is an intensity of 10^{-16} watt per square centimeter, corresponding roughly to the faintest sound that can be heard. Determine $\beta(I)$ for the following.

(a) $I = 10^{-14}$ watt per square centimeter (whisper)

(b) $I = 10^{-9}$ watt per square centimeter (busy street corner)

(c) $I = 10^{-6.5}$ watt per square centimeter (air hammer)

(d) $I = 10^{-4}$ watt per square centimeter (threshold of pain)

64. Noise Level With the installation of noise suppression materials, the noise level in an auditorium was reduced from 93 to 80 decibels. Use the function in Exercise 63 to find the percent decrease in the intensity level of the noise as a result of the installation of these materials.

65. Newton's Law of Cooling When an object is removed from a furnace and placed in an environment with a constant temperature of 80°F, its core temperature is 1500°F. One hour after it is removed, the core temperature is 1120°F. Find the core temperature 5 hours after the object is removed from the furnace.

66. Newton's Law of Cooling A container of hot liquid is placed in a freezer that is kept at a constant temperature of 20°F. The initial temperature of the liquid is 160°F. After 5 minutes, the liquid's temperature is 60°F. How much longer will it take for its temperature to decrease to 30°F?

True or False? In Exercises 67–70, determine whether the statement is true or false. If it is false, explain why or give an example that shows it is false.

67. In exponential growth, the rate of growth is constant.

68. In linear growth, the rate of growth is constant.

69. If prices are rising at a rate of 0.5% per month, then they are rising at a rate of 6% per year.

70. The differential equation modeling exponential growth is $dy/dx = ky$, where k is a constant.

6.3 Separation of Variables and the Logistic Equation

■ Recognize and solve differential equations that can be solved by separation of variables.
■ Use differential equations to model and solve applied problems.
■ Solve and analyze logistic differential equations.

Separation of Variables

Consider a differential equation that can be written in the form

$$M(x) + N(y)\frac{dy}{dx} = 0$$

where M is a continuous function of x alone and N is a continuous function of y alone. As you saw in Section 6.2, for this type of equation, all x terms can be collected with dx and all y terms with dy, and a solution can be obtained by integration. Such equations are said to be **separable,** and the solution procedure is called **separation of variables.** Below are some examples of differential equations that are separable.

Original Differential Equation	Rewritten with Variables Separated
$x^2 + 3y\dfrac{dy}{dx} = 0$	$3y\,dy = -x^2\,dx$
$(\sin x)y' = \cos x$	$dy = \cot x\,dx$
$\dfrac{xy'}{e^y + 1} = 2$	$\dfrac{1}{e^y + 1}\,dy = \dfrac{2}{x}\,dx$

EXAMPLE 1   Separation of Variables

▷ *See LarsonCalculus.com for an interactive version of this type of example.*

Find the general solution of

$$(x^2 + 4)\frac{dy}{dx} = xy.$$

Solution To begin, note that $y = 0$ is a solution. To find other solutions, assume that $y \neq 0$ and separate variables as shown.

$$(x^2 + 4)\,dy = xy\,dx \qquad \text{Differential form}$$
$$\frac{dy}{y} = \frac{x}{x^2 + 4}\,dx \qquad \text{Separate variables.}$$

Now, integrate to obtain

$$\int \frac{dy}{y} = \int \frac{x}{x^2 + 4}\,dx \qquad \text{Integrate.}$$
$$\ln|y| = \frac{1}{2}\ln(x^2 + 4) + C_1$$
$$\ln|y| = \ln\sqrt{x^2 + 4} + C_1$$
$$|y| = e^{C_1}\sqrt{x^2 + 4}$$
$$y = \pm e^{C_1}\sqrt{x^2 + 4}.$$

Because $y = 0$ is also a solution, you can write the general solution as

$$y = C\sqrt{x^2 + 4}. \qquad \text{General solution}$$

• REMARK Be sure to check your solutions throughout this chapter. In Example 1, you can check the solution

$$y = C\sqrt{x^2 + 4}$$

by differentiating and substituting into the original equation.

$$(x^2 + 4)\frac{dy}{dx} = xy$$
$$(x^2 + 4)\frac{Cx}{\sqrt{x^2 + 4}} \stackrel{?}{=} x\left(C\sqrt{x^2 + 4}\right)$$
$$Cx\sqrt{x^2 + 4} = Cx\sqrt{x^2 + 4}$$

So, the solution checks.

In some cases, it is not feasible to write the general solution in the explicit form $y = f(x)$. The next example illustrates such a solution. Implicit differentiation can be used to verify this solution.

■ **FOR FURTHER INFORMATION**
For an example (from engineering) of a differential equation that is separable, see the article "Designing a Rose Cutter" by J. S. Hartzler in *The College Mathematics Journal*. To view this article, go to *MathArticles.com*.

EXAMPLE 2   **Finding a Particular Solution**

Given the initial condition $y(0) = 1$, find the particular solution of the equation

$$xy\,dx + e^{-x^2}(y^2 - 1)\,dy = 0.$$

Solution   Note that $y = 0$ is a solution of the differential equation—but this solution does not satisfy the initial condition. So, you can assume that $y \neq 0$. To separate variables, you must rid the first term of y and the second term of e^{-x^2}. So, you should multiply by e^{x^2}/y and obtain the following.

$$xy\,dx + e^{-x^2}(y^2 - 1)\,dy = 0$$
$$e^{-x^2}(y^2 - 1)\,dy = -xy\,dx$$
$$\int \left(y - \frac{1}{y}\right) dy = \int -xe^{x^2}\,dx$$
$$\frac{y^2}{2} - \ln|y| = -\frac{1}{2}e^{x^2} + C$$

From the initial condition $y(0) = 1$, you have

$$\frac{1}{2} - 0 = -\frac{1}{2} + C$$

which implies that $C = 1$. So, the particular solution has the implicit form

$$\frac{y^2}{2} - \ln|y| = -\frac{1}{2}e^{x^2} + 1$$
$$y^2 - \ln y^2 + e^{x^2} = 2.$$

You can check this by differentiating and rewriting to get the original equation.

EXAMPLE 3   **Finding a Particular Solution Curve**

Find the equation of the curve that passes through the point $(1, 3)$ and has a slope of y/x^2 at any point (x, y).

Solution   Because the slope of the curve is y/x^2, you have

$$\frac{dy}{dx} = \frac{y}{x^2}$$

with the initial condition $y(1) = 3$. Separating variables and integrating produces

$$\int \frac{dy}{y} = \int \frac{dx}{x^2}, \quad y \neq 0$$
$$\ln|y| = -\frac{1}{x} + C_1$$
$$y = e^{-(1/x)+C_1}$$
$$y = Ce^{-1/x}.$$

Because $y = 3$ when $x = 1$, it follows that $3 = Ce^{-1}$ and $C = 3e$. So, the equation of the specified curve is

$$y = (3e)e^{-1/x} \quad \Longrightarrow \quad y = 3e^{(x-1)/x}, \quad x > 0.$$

Because the solution is not defined at $x = 0$ and the initial condition is given at $x = 1$, x is restricted to positive values. See Figure 6.11.

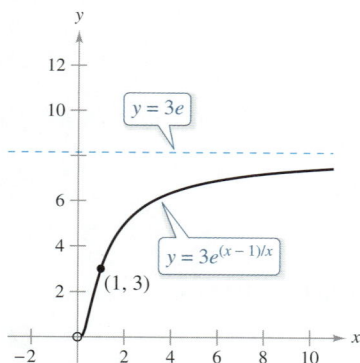

Figure 6.11

Applications

EXAMPLE 4   **Wildlife Population**

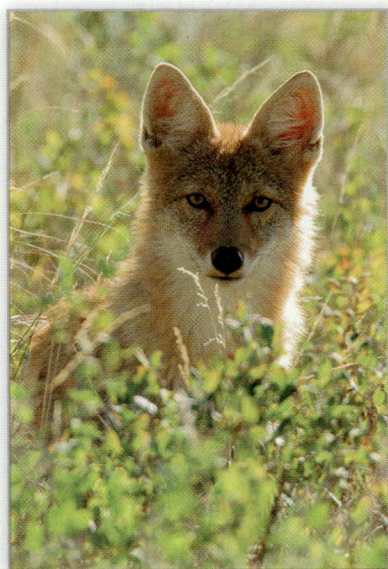

The rate of change of the number of coyotes $N(t)$ in a population is directly proportional to $650 - N(t)$, where t is the time in years. When $t = 0$, the population is 300, and when $t = 2$, the population has increased to 500. Find the population when $t = 3$.

Solution   Because the rate of change of the population is proportional to $650 - N(t)$, or $650 - N$, you can write the differential equation

$$\frac{dN}{dt} = k(650 - N).$$

You can solve this equation using separation of variables.

$$dN = k(650 - N)\, dt \qquad \text{\color{red}Differential form}$$

$$\frac{dN}{650 - N} = k\, dt \qquad \text{\color{red}Separate variables.}$$

$$-\ln|650 - N| = kt + C_1 \qquad \text{\color{red}Integrate.}$$

$$\ln|650 - N| = -kt - C_1$$

$$650 - N = e^{-kt - C_1} \qquad \text{\color{red}Assume } N < 650.$$

$$N = 650 - Ce^{-kt} \qquad \text{\color{red}General solution}$$

Using $N = 300$ when $t = 0$, you can conclude that $C = 350$, which produces

$$N = 650 - 350e^{-kt}.$$

Then, using $N = 500$ when $t = 2$, it follows that

$$500 = 650 - 350e^{-2k} \quad \Longrightarrow \quad e^{-2k} = \frac{3}{7} \quad \Longrightarrow \quad k \approx 0.4236.$$

So, the model for the coyote population is

$$N = 650 - 350e^{-0.4236t}. \qquad \text{\color{red}Model for population}$$

When $t = 3$, you can approximate the population to be

$$N = 650 - 350e^{-0.4236(3)}$$

$$\approx 552 \text{ coyotes.}$$

The model for the population is shown in Figure 6.12. Note that $N = 650$ is the horizontal asymptote of the graph and is the *carrying capacity* of the model. You will learn more about carrying capacity later in this section.

Figure 6.12

A common problem in electrostatics, thermodynamics, and hydrodynamics involves finding a family of curves, each of which is orthogonal to all members of a given family of curves. For example, Figure 6.13 shows a family of circles

$$x^2 + y^2 = C \qquad \text{Family of circles}$$

each of which intersects the lines in the family

$$y = Kx \qquad \text{Family of lines}$$

at right angles. Two such families of curves are said to be **mutually orthogonal,** and each curve in one of the families is called an **orthogonal trajectory** of the other family. In electrostatics, lines of force are orthogonal to the *equipotential curves.* In thermodynamics, the flow of heat across a plane surface is orthogonal to the *isothermal curves.* In hydrodynamics, the flow (stream) lines are orthogonal trajectories of the *velocity potential curves.*

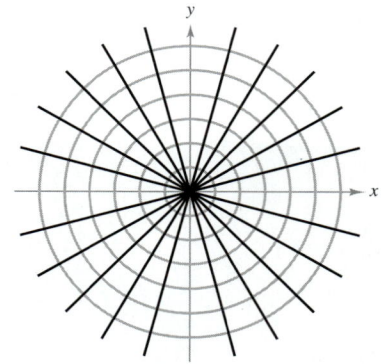

Each line $y = Kx$ is an orthogonal trajectory of the family of circles.
Figure 6.13

EXAMPLE 5   **Finding Orthogonal Trajectories**

Describe the orthogonal trajectories for the family of curves given by

$$y = \frac{C}{x}$$

for $C \neq 0$. Sketch several members of each family.

Solution   First, solve the given equation for C and write $xy = C$. Then, by differentiating implicitly with respect to x, you obtain the differential equation

$$x\frac{dy}{dx} + y = 0 \qquad \text{Differential equation}$$

$$x\frac{dy}{dx} = -y$$

$$\frac{dy}{dx} = -\frac{y}{x}. \qquad \text{Slope of given family}$$

Because dy/dx represents the slope of the given family of curves at (x, y), it follows that the orthogonal family has the negative reciprocal slope x/y. So,

$$\frac{dy}{dx} = \frac{x}{y}. \qquad \text{Slope of orthogonal family}$$

Now you can find the orthogonal family by separating variables and integrating.

$$\int y\,dy = \int x\,dx$$

$$\frac{y^2}{2} = \frac{x^2}{2} + C_1$$

$$y^2 - x^2 = K$$

The centers are at the origin, and the transverse axes are vertical for $K > 0$ and horizontal for $K < 0$. When $K = 0$, the orthogonal trajectories are the lines $y = \pm x$. When $K \neq 0$, the orthogonal trajectories are hyperbolas. Several trajectories are shown in Figure 6.14.

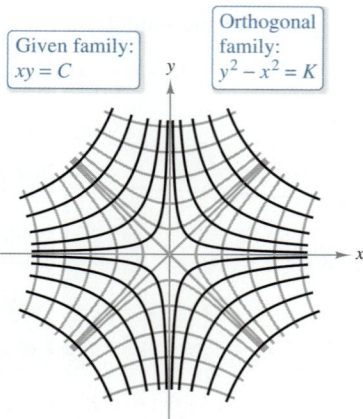

Given family: $xy = C$

Orthogonal family: $y^2 - x^2 = K$

Orthogonal trajectories
Figure 6.14

Logistic Differential Equation

In Section 6.2, the exponential growth model was derived from the fact that the rate of change of a variable y is proportional to the value of y. You observed that the differential equation $dy/dt = ky$ has the general solution $y = Ce^{kt}$. Exponential growth is unlimited, but when describing a population, there often exists some upper limit L past which growth cannot occur. This upper limit L is called the **carrying capacity,** which is the maximum population $y(t)$ that can be sustained or supported as time t increases. A model that is often used to describe this type of growth is the **logistic differential equation**

$$\frac{dy}{dt} = ky\left(1 - \frac{y}{L}\right) \qquad \text{Logistic differential equation}$$

where k and L are positive constants. A population that satisfies this equation does not grow without bound, but approaches the carrying capacity L as t increases.

From the equation, you can see that if y is between 0 and the carrying capacity L, then $dy/dt > 0$, and the population increases. If y is greater than L, then $dy/dt < 0$, and the population decreases. The graph of the function y is called the *logistic curve,* as shown in Figure 6.15.

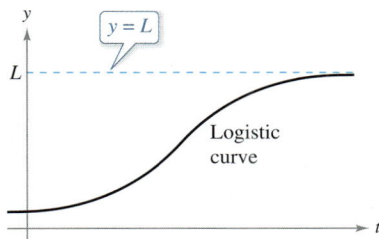

Note that as $t \to \infty$, $y \to L$.
Figure 6.15

EXAMPLE 6   Deriving the General Solution

Solve the logistic differential equation

$$\frac{dy}{dt} = ky\left(1 - \frac{y}{L}\right).$$

Solution   Begin by separating variables.

$\dfrac{dy}{dt} = ky\left(1 - \dfrac{y}{L}\right)$	Write differential equation.
$\dfrac{1}{y(1 - y/L)}dy = k\,dt$	Separate variables.
$\displaystyle\int \dfrac{1}{y(1 - y/L)}dy = \int k\,dt$	Integrate each side.
$\displaystyle\int \left(\dfrac{1}{y} + \dfrac{1}{L - y}\right)dy = \int k\,dt$	Rewrite left side using partial fractions.
$\ln\lvert y \rvert - \ln\lvert L - y \rvert = kt + C$	Find antiderivative of each side.
$\ln\left\lvert \dfrac{L - y}{y} \right\rvert = -kt - C$	Multiply each side by -1 and simplify.
$\left\lvert \dfrac{L - y}{y} \right\rvert = e^{-kt - C}$	Exponentiate each side.
$\left\lvert \dfrac{L - y}{y} \right\rvert = e^{-C}e^{-kt}$	Property of exponents
$\dfrac{L - y}{y} = be^{-kt}$	Let $\pm e^{-C} = b$.

REMARK A review of the method of partial fractions is given in Section 8.5.

Solving this equation for y produces $y = \dfrac{L}{1 + be^{-kt}}$.

From Example 6, you can conclude that all solutions of the logistic differential equation are of the general form

$$y = \frac{L}{1 + be^{-kt}}.$$

Exploration

Use a graphing utility to investigate the effects of the values of L, b, and k on the graph of

$$y = \frac{L}{1 + be^{-kt}}.$$

Include some examples to support your results.

EXAMPLE 7   **Solving a Logistic Differential Equation**

A state game commission releases 40 elk into a game refuge. After 5 years, the elk population is 104. The commission believes that the environment can support no more than 4000 elk. The growth rate of the elk population p is

$$\frac{dp}{dt} = kp\left(1 - \frac{p}{4000}\right), \quad 40 \le p \le 4000$$

where t is the number of years.

a. Write a model for the elk population in terms of t.

b. Graph the slope field for the differential equation and the solution that passes through the point $(0, 40)$.

c. Use the model to estimate the elk population after 15 years.

d. Find the limit of the model as $t \to \infty$.

Solution

a. You know that $L = 4000$. So, the solution of the equation is of the form

$$p = \frac{4000}{1 + be^{-kt}}.$$

Because $p(0) = 40$, you can solve for b as follows.

$$40 = \frac{4000}{1 + be^{-k(0)}} \implies 40 = \frac{4000}{1 + b} \implies b = 99$$

Then, because $p = 104$ when $t = 5$, you can solve for k.

$$104 = \frac{4000}{1 + 99e^{-k(5)}} \implies k \approx 0.194$$

So, a model for the elk population is

$$p = \frac{4000}{1 + 99e^{-0.194t}}.$$

b. Using a graphing utility, you can graph the slope field for

$$\frac{dp}{dt} = 0.194p\left(1 - \frac{p}{4000}\right)$$

and the solution that passes through $(0, 40)$, as shown in Figure 6.16.

c. To estimate the elk population after 15 years, substitute 15 for t in the model.

$$p = \frac{4000}{1 + 99e^{-0.194(15)}} \qquad \text{Substitute 15 for } t.$$

$$= \frac{4000}{1 + 99e^{-2.91}} \qquad \text{Simplify.}$$

$$\approx 626$$

d. As t increases without bound, the denominator of

$$\frac{4000}{1 + 99e^{-0.194t}}$$

gets closer and closer to 1. So,

$$\lim_{t \to \infty} \frac{4000}{1 + 99e^{-0.194t}} = 4000.$$

Slope field for

$$\frac{dp}{dt} = 0.194p\left(1 - \frac{p}{4000}\right)$$

and the solution passing through $(0, 40)$

Figure 6.16

6.3 Exercises

See **CalcChat.com** for tutorial help and worked-out solutions to odd-numbered exercises.

Finding a General Solution Using Separation of Variables In Exercises 1–14, find the general solution of the differential equation.

1. $\dfrac{dy}{dx} = \dfrac{x}{y}$

2. $\dfrac{dy}{dx} = \dfrac{3x^2}{y^2}$

3. $x^2 + 5y\dfrac{dy}{dx} = 0$

4. $\dfrac{dy}{dx} = \dfrac{6 - x^2}{2y^3}$

5. $\dfrac{dr}{ds} = 0.75r$

6. $\dfrac{dr}{ds} = 0.75s$

7. $(2 + x)y' = 3y$

8. $xy' = y$

9. $yy' = 4\sin x$

10. $yy' = -8\cos(\pi x)$

11. $\sqrt{1 - 4x^2}\,y' = x$

12. $\sqrt{x^2 - 16}\,y' = 11x$

13. $y \ln x - xy' = 0$

14. $12yy' - 7e^x = 0$

Finding a Particular Solution Using Separation of Variables In Exercises 15–24, find the particular solution that satisfies the initial condition.

Differential Equation	Initial Condition
15. $yy' - 2e^x = 0$	$y(0) = 3$
16. $\sqrt{x} + \sqrt{y}\,y' = 0$	$y(1) = 9$
17. $y(x + 1) + y' = 0$	$y(-2) = 1$
18. $2xy' - \ln x^2 = 0$	$y(1) = 2$
19. $y(1 + x^2)y' - x(1 + y^2) = 0$	$y(0) = \sqrt{3}$
20. $y\sqrt{1 - x^2}\,y' - x\sqrt{1 - y^2} = 0$	$y(0) = 1$
21. $\dfrac{du}{dv} = uv \sin v^2$	$u(0) = 1$
22. $\dfrac{dr}{ds} = e^{r - 2s}$	$r(0) = 0$
23. $dP - kP\,dt = 0$	$P(0) = P_0$
24. $dT + k(T - 70)\,dt = 0$	$T(0) = 140$

Finding a Particular Solution In Exercises 25–28, find an equation of the graph that passes through the point and has the given slope.

25. $(0, 2)$, $y' = \dfrac{x}{4y}$

26. $(1, 1)$, $y' = -\dfrac{9x}{16y}$

27. $(9, 1)$, $y' = \dfrac{y}{2x}$

28. $(8, 2)$, $y' = \dfrac{2y}{3x}$

Using Slope In Exercises 29 and 30, find all functions f having the indicated property.

29. The tangent to the graph of f at the point (x, y) intersects the x-axis at $(x + 2, 0)$.

30. All tangents to the graph of f pass through the origin.

Slope Field In Exercises 31 and 32, sketch a few solutions of the differential equation on the slope field and then find the general solution analytically. To print an enlarged copy of the graph, go to *MathGraphs.com*.

31. $\dfrac{dy}{dx} = x$

32. $\dfrac{dy}{dx} = -\dfrac{x}{y}$

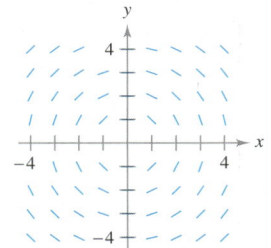

Slope Field In Exercises 33–36, (a) write a differential equation for the statement, (b) match the differential equation with a possible slope field, and (c) verify your result by using a graphing utility to graph a slope field for the differential equation. [The slope fields are labeled (a), (b), (c), and (d).] To print an enlarged copy of the graph, go to *MathGraphs.com*.

(a)

(b)

(c)

(d)

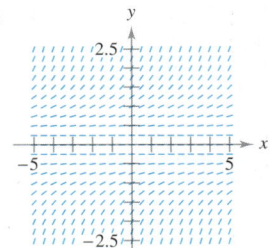

33. The rate of change of y with respect to x is proportional to the difference between y and 4.

34. The rate of change of y with respect to x is proportional to the difference between x and 4.

35. The rate of change of y with respect to x is proportional to the product of y and the difference between y and 4.

36. The rate of change of y with respect to x is proportional to y^2.

37. Radioactive Decay The rate of decomposition of radioactive radium is proportional to the amount present at any time. The half-life of radioactive radium is 1599 years. What percent of a present amount will remain after 50 years?

38. Chemical Reaction In a chemical reaction, a certain compound changes into another compound at a rate proportional to the unchanged amount. There is 40 grams of the original compound initially and 35 grams after 1 hour. When will 75 percent of the compound be changed?

39. Weight Gain A calf that weighs 60 pounds at birth gains weight at the rate

$$\frac{dw}{dt} = k(1200 - w)$$

where w is weight in pounds and t is time in years.

(a) Solve the differential equation.

(b) Use a graphing utility to graph the particular solutions for $k = 0.8$, 0.9, and 1.

(c) The animal is sold when its weight reaches 800 pounds. Find the time of sale for each of the models in part (b).

(d) What is the maximum weight of the animal for each of the models in part (b)?

40. Weight Gain A calf that weighs w_0 pounds at birth gains weight at the rate $dw/dt = 1200 - w$, where w is weight in pounds and t is time in years. Solve the differential equation.

Finding Orthogonal Trajectories In Exercises 41–46, find the orthogonal trajectories of the family. Use a graphing utility to graph several members of each family.

41. $x^2 + y^2 = C$

42. $x^2 - 2y^2 = C$

43. $x^2 = Cy$

44. $y^2 = 2Cx$

45. $y^2 = Cx^3$

46. $y = Ce^x$

Matching In Exercises 47–50, match the logistic equation with its graph. [The graphs are labeled (a), (b), (c), and (d).]

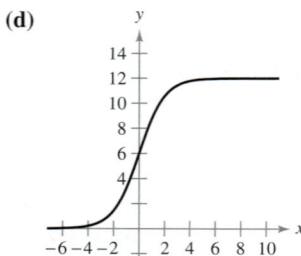

(a)

(b)

(c)

(d)

47. $y = \dfrac{12}{1 + e^{-x}}$

48. $y = \dfrac{12}{1 + 3e^{-x}}$

49. $y = \dfrac{12}{1 + \frac{1}{2}e^{-x}}$

50. $y = \dfrac{12}{1 + e^{-2x}}$

Using a Logistic Equation In Exercises 51 and 52, the logistic equation models the growth of a population. Use the equation to (a) find the value of k, (b) find the carrying capacity, (c) find the initial population, (d) determine when the population will reach 50% of its carrying capacity, and (e) write a logistic differential equation that has the solution $P(t)$.

51. $P(t) = \dfrac{2100}{1 + 29e^{-0.75t}}$

52. $P(t) = \dfrac{5000}{1 + 39e^{-0.2t}}$

Using a Logistic Differential Equation In Exercises 53 and 54, the logistic differential equation models the growth rate of a population. Use the equation to (a) find the value of k, (b) find the carrying capacity, (c) graph a slope field using a computer algebra system, and (d) determine the value of P at which the population growth rate is the greatest.

53. $\dfrac{dP}{dt} = 3P\left(1 - \dfrac{P}{100}\right)$

54. $\dfrac{dP}{dt} = 0.1P - 0.0004P^2$

Solving a Logistic Differential Equation In Exercises 55–58, find the logistic equation that passes through the given point.

55. $\dfrac{dy}{dt} = y\left(1 - \dfrac{y}{36}\right)$, $(0, 4)$

56. $\dfrac{dy}{dt} = 2.8y\left(1 - \dfrac{y}{10}\right)$, $(0, 7)$

57. $\dfrac{dy}{dt} = \dfrac{4y}{5} - \dfrac{y^2}{150}$, $(0, 8)$

58. $\dfrac{dy}{dt} = \dfrac{3y}{20} - \dfrac{y^2}{1600}$, $(0, 15)$

59. Endangered Species A conservation organization releases 25 Florida panthers into a game preserve. After 2 years, there are 39 panthers in the preserve. The Florida preserve has a carrying capacity of 200 panthers.

(a) Write a logistic equation that models the population of panthers in the preserve.

(b) Find the population after 5 years.

(c) When will the population reach 100?

(d) Write a logistic differential equation that models the growth rate of the panther population. Then repeat part (b) using Euler's Method with a step size of $h = 1$. Compare the approximation with the exact answer.

(e) At what time is the panther population growing most rapidly? Explain.

60. Bacteria Growth At time $t = 0$, a bacterial culture weighs 1 gram. Two hours later, the culture weighs 4 grams. The maximum weight of the culture is 20 grams.

(a) Write a logistic equation that models the weight of the bacterial culture.

(b) Find the culture's weight after 5 hours.

(c) When will the culture's weight reach 18 grams?

(d) Write a logistic differential equation that models the growth rate of the culture's weight. Then repeat part (b) using Euler's Method with a step size of $h = 1$. Compare the approximation with the exact answer.

(e) At what time is the culture's weight increasing most rapidly? Explain.

61. Separation of Variables In your own words, describe how to recognize and solve differential equations that can be solved by separation of variables.

62. Mutually Orthogonal In your own words, describe the relationship between two families of curves that are mutually orthogonal.

63. Finding a Derivative Show that if

$$y = \frac{1}{1 + be^{-kt}}$$

then

$$\frac{dy}{dt} = ky(1 - y).$$

64. Point of Inflection For any logistic growth curve, show that the point of inflection occurs at $y = L/2$ when the solution starts below the carrying capacity L.

• 65. Sailing

Ignoring resistance, a sailboat starting from rest accelerates (dv/dt) at a rate proportional to the difference between the velocities of the wind and the boat.

(a) The wind is blowing at 20 knots, and after 1 half-hour, the boat is moving at 10 knots. Write the velocity v as a function of time t.

(b) Use the result of part (a) to write the distance traveled by the boat as a function of time.

66. HOW DO YOU SEE IT? The growth of a population is modeled by a logistic equation as shown in the graph below. What happens to the rate of growth as the population increases? What do you think causes this to occur in real-life situations, such as animal or human populations?

Determining if a Function Is Homogeneous In Exercises 67–74, determine whether the function is homogeneous, and if it is, determine its degree. A function $f(x, y)$ is *homogeneous of degree n* if $f(tx, ty) = t^n f(x, y)$.

67. $f(x, y) = x^3 - 4xy^2 + y^3$

68. $f(x, y) = x^3 + 3x^2y^2 - 2y^2$

69. $f(x, y) = \dfrac{x^2 y^2}{\sqrt{x^2 + y^2}}$

70. $f(x, y) = \dfrac{xy}{\sqrt{x^2 + y^2}}$

71. $f(x, y) = 2 \ln xy$

72. $f(x, y) = \tan(x + y)$

73. $f(x, y) = 2 \ln \dfrac{x}{y}$

74. $f(x, y) = \tan \dfrac{y}{x}$

Solving a Homogeneous Differential Equation In Exercises 75–80, solve the homogeneous differential equation in terms of x and y. A *homogeneous differential equation* is an equation of the form $M(x, y) \, dx + N(x, y) \, dy = 0$, where M and N are homogeneous functions of the same degree. To solve an equation of this form by the method of separation of variables, use the substitutions $y = vx$ and $dy = x \, dv + v \, dx$.

75. $(x + y) \, dx - 2x \, dy = 0$

76. $(x^3 + y^3) \, dx - xy^2 \, dy = 0$

77. $(x - y) \, dx - (x + y) \, dy = 0$

78. $(x^2 + y^2) \, dx - 2xy \, dy = 0$

79. $xy \, dx + (y^2 - x^2) \, dy = 0$

80. $(2x + 3y) \, dx - x \, dy = 0$

True or False? In Exercises 81–83, determine whether the statement is true or false. If it is false, explain why or give an example that shows it is false.

81. The function $y = 0$ is always a solution of a differential equation that can be solved by separation of variables.

82. The differential equation $y' = xy - 2y + x - 2$ can be written in separated variables form.

83. The families $x^2 + y^2 = 2Cy$ and $x^2 + y^2 = 2Kx$ are mutually orthogonal.

PUTNAM EXAM CHALLENGE

84. A not uncommon calculus mistake is to believe that the product rule for derivatives says that $(fg)' = f'g'$. If $f(x) = e^{x^2}$, determine, with proof, whether there exists an open interval (a, b) and a nonzero function g defined on (a, b) such that this wrong product rule is true for x in (a, b).

This problem was composed by the Committee on the Putnam Prize Competition.
© The Mathematical Association of America. All rights reserved.

6.4 First-Order Linear Differential Equations

■ Solve a first-order linear differential equation, and use linear differential equations to solve applied problems.

First-Order Linear Differential Equations

In this section, you will see how to solve a very important class of first-order differential equations—first-order linear differential equations.

> ### Definition of First-Order Linear Differential Equation
>
> A **first-order linear differential equation** is an equation of the form
>
> $$\frac{dy}{dx} + P(x)y = Q(x)$$
>
> where P and Q are continuous functions of x. This first-order linear differential equation is said to be in **standard form.**

To solve a linear differential equation, write it in standard form to identify the functions $P(x)$ and $Q(x)$. Then integrate $P(x)$ and form the expression

$$u(x) = e^{\int P(x)\, dx} \qquad \text{Integrating factor}$$

which is called an **integrating factor.** The general solution of the equation is

$$y = \frac{1}{u(x)} \int Q(x)u(x)\, dx. \qquad \text{General solution}$$

It is instructive to see why the integrating factor helps solve a linear differential equation of the form $y' + P(x)y = Q(x)$. When both sides of the equation are multiplied by the integrating factor $u(x) = e^{\int P(x)\,dx}$, the left-hand side becomes the derivative of a product.

$$y'e^{\int P(x)\,dx} + P(x)ye^{\int P(x)\,dx} = Q(x)e^{\int P(x)\,dx}$$
$$\left[ye^{\int P(x)\,dx}\right]' = Q(x)e^{\int P(x)\,dx}$$

Integrating both sides of this second equation and dividing by $u(x)$ produce the general solution.

EXAMPLE 1 **Solving a Linear Differential Equation**

Find the general solution of

$$y' + y = e^x.$$

Solution For this equation, $P(x) = 1$ and $Q(x) = e^x$. So, the integrating factor is

$$u(x) = e^{\int P(x)\,dx} = e^{\int dx} = e^x.$$

This implies that the general solution is

$$y = \frac{1}{u(x)} \int Q(x)u(x)\, dx$$
$$= \frac{1}{e^x} \int e^x(e^x)\, dx$$
$$= e^{-x}\left(\frac{1}{2}e^{2x} + C\right)$$
$$= \frac{1}{2}e^x + Ce^{-x}.$$

Courtesy of the Visual Collections, Canaday Library, Bryn Mawr College.

THEOREM 6.2 Solution of a First-Order Linear Differential Equation

An integrating factor for the first-order linear differential equation

$$y' + P(x)y = Q(x)$$

is $u(x) = e^{\int P(x)\,dx}$. The solution of the differential equation is

$$ye^{\int P(x)\,dx} = \int Q(x)e^{\int P(x)\,dx}\,dx + C.$$

EXAMPLE 2   **Solving a First-Order Linear Differential Equation**

▷ *See LarsonCalculus.com for an interactive version of this type of example.*

Find the general solution of $xy' - 2y = x^2$.

Solution   The standard form of the equation is

$$y' + \left(-\frac{2}{x}\right)y = x. \qquad \text{Standard form}$$

So, $P(x) = -2/x$, and you have

$$\int P(x)\,dx = -\int \frac{2}{x}\,dx = -\ln x^2$$

which implies that the integrating factor is

$$e^{\int P(x)\,dx} = e^{-\ln x^2} = \frac{1}{e^{\ln x^2}} = \frac{1}{x^2}. \qquad \text{Integrating factor}$$

So, multiplying each side of the standard form by $1/x^2$ yields

$$\frac{y'}{x^2} - \frac{2y}{x^3} = \frac{1}{x}$$

$$\frac{d}{dx}\left[\frac{y}{x^2}\right] = \frac{1}{x}$$

$$\frac{y}{x^2} = \int \frac{1}{x}\,dx$$

$$\frac{y}{x^2} = \ln|x| + C$$

$$y = x^2(\ln|x| + C). \qquad \text{General solution}$$

Several solution curves (for $C = -2, -1, 0, 1, 2, 3,$ and 4) are shown in Figure 6.17.

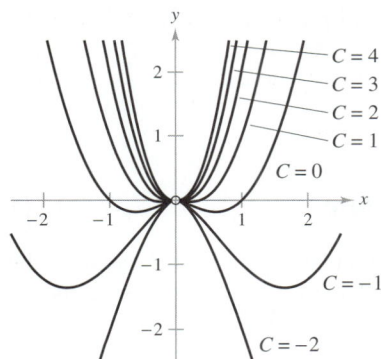

Figure 6.17

In most falling-body problems discussed so far in the text, air resistance has been neglected. The next example includes this factor. In the example, the air resistance on the falling object is assumed to be proportional to its velocity v. If g is the gravitational constant, the downward force F on a falling object of mass m is given by the difference $mg - kv$. If a is the acceleration of the object, then by Newton's Second Law of Motion,

$$F = ma = m\frac{dv}{dt}$$

which yields the following differential equation.

$$m\frac{dv}{dt} = mg - kv \quad \Longrightarrow \quad \frac{dv}{dt} + \frac{kv}{m} = g$$

A Falling Object with Air Resistance

An object of mass m is dropped from a hovering helicopter. The air resistance is proportional to the velocity of the object. Find the velocity of the object as a function of time t.

Solution   The velocity v satisfies the equation

$$\frac{dv}{dt} + \frac{kv}{m} = g.$$ g = gravitational constant, k = constant of proportionality

Letting $b = k/m$, you can *separate variables* to obtain

$$dv = (g - bv)\, dt$$

$$\int \frac{dv}{g - bv} = \int dt$$

$$-\frac{1}{b} \ln|g - bv| = t + C_1$$

$$\ln|g - bv| = -bt - bC_1$$

$$g - bv = Ce^{-bt}.$$ $C = e^{-bC_1}$

Because the object was dropped, $v = 0$ when $t = 0$; so $g = C$, and it follows that

$$-bv = -g + ge^{-bt} \quad \Longrightarrow \quad v = \frac{g - ge^{-bt}}{b} = \frac{mg}{k}(1 - e^{-kt/m}).$$

> **·· REMARK** Notice in Example 3 that the velocity approaches a limit of mg/k as a result of the air resistance. For falling-body problems in which air resistance is neglected, the velocity increases without bound.

Figure 6.18

A simple electric circuit consists of an electric current I (in amperes), a resistance R (in ohms), an inductance L (in henrys), and a constant electromotive force E (in volts), as shown in Figure 6.18. According to Kirchhoff's Second Law, if the switch S is closed when $t = 0$, then the applied electromotive force (voltage) is equal to the sum of the voltage drops in the rest of the circuit. This, in turn, means that the current I satisfies the differential equation

$$L\frac{dI}{dt} + RI = E.$$

An Electric Circuit Problem

Find the current I as a function of time t (in seconds), given that I satisfies the differential equation $L(dI/dt) + RI = \sin 2t$, where R and L are nonzero constants.

Solution   In standard form, the given linear equation is

$$\frac{dI}{dt} + \frac{R}{L}I = \frac{1}{L}\sin 2t.$$

Let $P(t) = R/L$, so that $e^{\int P(t)\, dt} = e^{(R/L)t}$, and, by Theorem 6.2,

$$Ie^{(R/L)t} = \frac{1}{L}\int e^{(R/L)t} \sin 2t\, dt$$

$$= \frac{1}{4L^2 + R^2} e^{(R/L)t}(R \sin 2t - 2L \cos 2t) + C.$$

So, the general solution is

$$I = e^{-(R/L)t}\left[\frac{1}{4L^2 + R^2} e^{(R/L)t}(R \sin 2t - 2L \cos 2t) + C \right]$$

$$= \frac{1}{4L^2 + R^2}(R \sin 2t - 2L \cos 2t) + Ce^{-(R/L)t}.$$

> ▷ **TECHNOLOGY** The integral in Example 4 was found using a computer algebra system. If you have access to *Maple*, *Mathematica*, or the *TI-Nspire*, try using it to integrate
>
> $$\frac{1}{L}\int e^{(R/L)t} \sin 2t\, dt.$$
>
> In Chapter 8, you will learn how to integrate functions of this type using integration by parts.

One type of problem that can be described in terms of a differential equation involves chemical mixtures, as illustrated in the next example.

EXAMPLE 5 **A Mixture Problem**

4 gal/min

5 gal/min

Figure 6.19

A tank contains 50 gallons of a solution composed of 90% water and 10% alcohol. A second solution containing 50% water and 50% alcohol is added to the tank at the rate of 4 gallons per minute. As the second solution is being added, the tank is being drained at a rate of 5 gallons per minute, as shown in Figure 6.19. The solution in the tank is stirred constantly. How much alcohol is in the tank after 10 minutes?

Solution Let y be the number of gallons of alcohol in the tank at any time t. You know that $y = 5$ when $t = 0$. Because the number of gallons of solution in the tank at any time is $50 - t$, and the tank loses 5 gallons of solution per minute, it must lose

$$\left(\frac{5}{50 - t}\right)y$$

gallons of alcohol per minute. Furthermore, because the tank is gaining 2 gallons of alcohol per minute, the rate of change of alcohol in the tank is

$$\frac{dy}{dt} = 2 - \left(\frac{5}{50 - t}\right)y \quad \Longrightarrow \quad \frac{dy}{dt} + \left(\frac{5}{50 - t}\right)y = 2.$$

To solve this linear differential equation, let

$$P(t) = \frac{5}{50 - t}$$

and obtain

$$\int P(t)\, dt = \int \frac{5}{50 - t}\, dt = -5\ln|50 - t|.$$

Because $t < 50$, you can drop the absolute value signs and conclude that

$$e^{\int P(t)\, dt} = e^{-5\ln(50 - t)} = \frac{1}{(50 - t)^5}.$$

So, the general solution is

$$\frac{y}{(50 - t)^5} = \int \frac{2}{(50 - t)^5}\, dt$$

$$\frac{y}{(50 - t)^5} = \frac{1}{2(50 - t)^4} + C$$

$$y = \frac{50 - t}{2} + C(50 - t)^5.$$

Because $y = 5$ when $t = 0$, you have

$$5 = \frac{50}{2} + C(50)^5 \quad \Longrightarrow \quad -\frac{20}{50^5} = C$$

which means that the particular solution is

$$y = \frac{50 - t}{2} - 20\left(\frac{50 - t}{50}\right)^5.$$

Finally, when $t = 10$, the amount of alcohol in the tank is

$$y = \frac{50 - 10}{2} - 20\left(\frac{50 - 10}{50}\right)^5 \approx 13.45\text{ gal}$$

which represents a solution containing 33.6% alcohol.

6.4 Exercises

See **CalcChat.com** for tutorial help and worked-out solutions to odd-numbered exercises.

Determining Whether a Differential Equation Is Linear
In Exercises 1–4, determine whether the differential equation is linear. Explain your reasoning.

1. $x^3 y' + xy = e^x + 1$

2. $2xy - y' \ln x = y$

3. $y' - y \sin x = xy^2$

4. $\dfrac{2 - y'}{y} = 5x$

Solving a First-Order Linear Differential Equation In Exercises 5–14, solve the first-order linear differential equation.

5. $\dfrac{dy}{dx} + \left(\dfrac{1}{x}\right)y = 6x + 2$

6. $\dfrac{dy}{dx} + \left(\dfrac{2}{x}\right)y = 3x - 5$

7. $y' - y = 16$

8. $y' + 2xy = 10x$

9. $(y + 1) \cos x \, dx - dy = 0$

10. $(y - 1) \sin x \, dx - dy = 0$

11. $(x - 1)y' + y = x^2 - 1$

12. $y' + 3y = e^{3x}$

13. $y' - 3x^2 y = e^{x^3}$

14. $y' + y \tan x = \sec x$

Slope Field In Exercises 15 and 16, (a) sketch an approximate solution of the differential equation satisfying the given initial condition by hand on the slope field, (b) find the particular solution that satisfies the given initial condition, and (c) use a graphing utility to graph the particular solution. Compare the graph with the hand-drawn graph in part (a). To print an enlarged copy of the graph, go to *MathGraphs.com*.

15. $\dfrac{dy}{dx} = e^x - y$,

    $(0, 1)$

16. $y' + \left(\dfrac{1}{x}\right)y = \sin x^2$,

    $(\sqrt{\pi}, 0)$

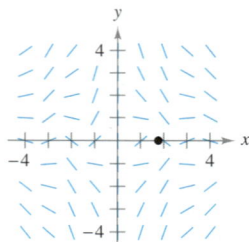

Finding a Particular Solution In Exercises 17–24, find the particular solution of the differential equation that satisfies the initial condition.

Differential Equation	Initial Condition
17. $y' \cos^2 x + y - 1 = 0$	$y(0) = 5$
18. $x^3 y' + 2y = e^{1/x^2}$	$y(1) = e$
19. $y' + y \tan x = \sec x + \cos x$	$y(0) = 1$
20. $y' + y \sec x = \sec x$	$y(0) = 4$

Differential Equation	Initial Condition
21. $y' + \left(\dfrac{1}{x}\right)y = 0$	$y(2) = 2$
22. $y' + (2x - 1)y = 0$	$y(1) = 2$
23. $x \, dy = (x + y + 2) \, dx$	$y(1) = 10$
24. $2xy' - y = x^3 - x$	$y(4) = 2$

25. **Population Growth** When predicting population growth, demographers must consider birth and death rates as well as the net change caused by the difference between the rates of immigration and emigration. Let P be the population at time t and let N be the net increase per unit time resulting from the difference between immigration and emigration. So, the rate of growth of the population is given by

$$\frac{dP}{dt} = kP + N$$

where N is constant. Solve this differential equation to find P as a function of time, when at time $t = 0$ the size of the population is P_0.

26. **Investment Growth** A large corporation starts at time $t = 0$ to invest part of its receipts continuously at a rate of P dollars per year in a fund for future corporate expansion. Assume that the fund earns r percent interest per year compounded continuously. So, the rate of growth of the amount A in the fund is given by

$$\frac{dA}{dt} = rA + P$$

where $A = 0$ when $t = 0$. Solve this differential equation for A as a function of t.

Investment Growth In Exercises 27 and 28, use the result of Exercise 26.

27. Find A for the following.

    (a) $P = \$275{,}000$, $r = 8\%$, $t = 10$ years

    (b) $P = \$550{,}000$, $r = 5.9\%$, $t = 25$ years

28. Find t if the corporation needs $\$1{,}000{,}000$ and it can invest $\$125{,}000$ per year in a fund earning 8% interest compounded continuously.

29. **Learning Curve** The management at a certain factory has found that the maximum number of units a worker can produce in a day is 75. The rate of increase in the number of units N produced with respect to time t in days by a new employee is proportional to $75 - N$.

    (a) Determine the differential equation describing the rate of change of performance with respect to time.

    (b) Solve the differential equation from part (a).

    (c) Find the particular solution for a new employee who produced 20 units on the first day at the factory and 35 units on the twentieth day.

30. Intravenous Feeding

Glucose is added intravenously to the bloodstream at the rate of q units per minute, and the body removes glucose from the bloodstream at a rate proportional to the amount present. Assume that $Q(t)$ is the amount of glucose in the bloodstream at time t.

(a) Determine the differential equation describing the rate of change of glucose in the bloodstream with respect to time.

(b) Solve the differential equation from part (a), letting $Q = Q_0$ when $t = 0$.

(c) Find the limit of $Q(t)$ as $t \to \infty$.

Falling Object In Exercises 31 and 32, consider an eight-pound object dropped from a height of 5000 feet, where the air resistance is proportional to the velocity.

31. Write the velocity of the object as a function of time when the velocity after 5 seconds is approximately -101 feet per second. What is the limiting value of the velocity function?

32. Use the result of Exercise 31 to write the position of the object as a function of time. Approximate the velocity of the object when it reaches ground level.

Electric Circuits In Exercises 33 and 34, use the differential equation for electric circuits given by

$$L\frac{dI}{dt} + RI = E.$$

In this equation, I is the current, R is the resistance, L is the inductance, and E is the electromotive force (voltage).

33. Solve the differential equation for the current given a constant voltage E_0.

34. Use the result of Exercise 33 to find the equation for the current when $I(0) = 0$, $E_0 = 120$ volts, $R = 600$ ohms, and $L = 4$ henrys. When does the current reach 90% of its limiting value?

Mixture In Exercises 35–38, consider a tank that at time $t = 0$ contains v_0 gallons of a solution of which, by weight, q_0 pounds is soluble concentrate. Another solution containing q_1 pounds of the concentrate per gallon is running into the tank at the rate of r_1 gallons per minute. The solution in the tank is kept well stirred and is withdrawn at the rate of r_2 gallons per minute.

35. Let Q be the amount of concentrate in the solution at any time t. Show that

$$\frac{dQ}{dt} + \frac{r_2 Q}{v_0 + (r_1 - r_2)t} = q_1 r_1.$$

36. Let Q be the amount of concentrate in the solution at any time t. Write the differential equation for the rate of change of Q with respect to t when $r_1 = r_2 = r$.

37. A 200-gallon tank is full of a solution containing 25 pounds of concentrate. Starting at time $t = 0$, distilled water is admitted to the tank at a rate of 10 gallons per minute, and the well-stirred solution is withdrawn at the same rate.

(a) Find the amount of concentrate Q in the solution as a function of t.

(b) Find the time at which the amount of concentrate in the tank reaches 15 pounds.

(c) Find the quantity of the concentrate in the solution as $t \to \infty$.

38. A 200-gallon tank is half full of distilled water. At time $t = 0$, a solution containing 0.5 pound of concentrate per gallon enters the tank at the rate of 5 gallons per minute, and the well-stirred mixture is withdrawn at the rate of 3 gallons per minute.

(a) At what time will the tank be full?

(b) At the time the tank is full, how many pounds of concentrate will it contain?

(c) Repeat parts (a) and (b), assuming that the solution entering the tank contains 1 pound of concentrate per gallon.

39. Using an Integrating Factor The expression $u(x)$ is an integrating factor for $y' + P(x)y = Q(x)$. Which of the following is equal to $u'(x)$? Verify your answer.

(a) $P(x)u(x)$ (b) $P'(x)u(x)$

(c) $Q(x)u(x)$ (d) $Q'(x)u(x)$

40. HOW DO YOU SEE IT? The graph shows the amount of concentrate Q (in pounds) in a solution in a tank at time t (in minutes) as a solution with concentrate enters the tank, is well stirred, and is withdrawn from the tank.

(a) How much concentrate is in the tank at time $t = 0$?

(b) Which is greater, the rate of solution into the tank, or the rate of solution withdrawn from the tank? Explain.

(c) At what time is there no concentrate in the tank? What does this mean?

WRITING ABOUT CONCEPTS

41. Standard Form Give the standard form of a first-order linear differential equation. What is its integrating factor?

42. First-Order What does the term "first-order" refer to in a first-order linear differential equation?

Matching In Exercises 43–46, match the differential equation with its solution.

Differential Equation	Solution
43. $y' - 2x = 0$	(a) $y = Ce^{x^2}$
44. $y' - 2y = 0$	(b) $y = -\frac{1}{2} + Ce^{x^2}$
45. $y' - 2xy = 0$	(c) $y = x^2 + C$
46. $y' - 2xy = x$	(d) $y = Ce^{2x}$

Slope Field   In Exercises 47–50, (a) use a graphing utility to graph the slope field for the differential equation, (b) find the particular solutions of the differential equation passing through the given points, and (c) use a graphing utility to graph the particular solutions on the slope field.

Differential Equation	Points
47. $\dfrac{dy}{dx} - \dfrac{1}{x}y = x^2$	$(-2, 4),\ (2, 8)$
48. $\dfrac{dy}{dx} + 4x^3 y = x^3$	$\left(0, \dfrac{7}{2}\right),\ \left(0, -\dfrac{1}{2}\right)$
49. $\dfrac{dy}{dx} + (\cot x)y = 2$	$(1, 1),\ (3, -1)$
50. $\dfrac{dy}{dx} + 2xy = xy^2$	$(0, 3),\ (0, 1)$

Solving a First-Order Linear Differential Equation   In Exercises 51–58, solve the first-order differential equation by any appropriate method.

51. $\dfrac{dy}{dx} = \dfrac{e^{2x+y}}{e^{x-y}}$

52. $\dfrac{dy}{dx} = \dfrac{x - 3}{y(y + 4)}$

53. $y \cos x - \cos x + \dfrac{dy}{dx} = 0$

54. $y' = 2x\sqrt{1 - y^2}$

55. $(2y - e^x)\,dx + x\,dy = 0$

56. $(x + y)\,dx - x\,dy = 0$

57. $3(y - 4x^2)\,dx + x\,dy = 0$

58. $x\,dx + (y + e^y)(x^2 + 1)\,dy = 0$

Solving a Bernoulli Differential Equation   In Exercises 59–66, solve the Bernoulli differential equation. The *Bernoulli equation* is a well-known nonlinear equation of the form

$$y' + P(x)y = Q(x)y^n$$

that can be reduced to a linear form by a substitution. The general solution of a Bernoulli equation is

$$y^{1-n}e^{\int (1-n)P(x)\,dx} = \int (1 - n)Q(x)e^{\int (1-n)P(x)\,dx}\,dx + C.$$

59. $y' + 3x^2 y = x^2 y^3$

60. $y' + xy = xy^{-1}$

61. $y' + \left(\dfrac{1}{x}\right)y = xy^2$

62. $y' + \left(\dfrac{1}{x}\right)y = x\sqrt{y}$

63. $xy' + y = xy^3$

64. $y' - y = y^3$

65. $y' - y = e^x \sqrt[3]{y}$

66. $yy' - 2y^2 = e^x$

True or False?   In Exercises 67 and 68, determine whether the statement is true or false. If it is false, explain why or give an example that shows it is false.

67. $y' + x\sqrt{y} = x^2$ is a first-order linear differential equation.

68. $y' + xy = e^x y$ is a first-order linear differential equation.

SECTION PROJECT

Weight Loss

A person's weight depends on both the number of calories consumed and the energy used. Moreover, the amount of energy used depends on a person's weight—the average amount of energy used by a person is 17.5 calories per pound per day. So, the more weight a person loses, the less energy a person uses (assuming that the person maintains a constant level of activity). An equation that can be used to model weight loss is

$$\frac{dw}{dt} = \frac{C}{3500} - \frac{17.5}{3500}w$$

where w is the person's weight (in pounds), t is the time in days, and C is the constant daily calorie consumption.

(a) Find the general solution of the differential equation.

(b) Consider a person who weighs 180 pounds and begins a diet of 2500 calories per day. How long will it take the person to lose 10 pounds? How long will it take the person to lose 35 pounds?

(c) Use a graphing utility to graph the solution. What is the "limiting" weight of the person?

(d) Repeat parts (b) and (c) for a person who weighs 200 pounds when the diet is started.

■ **FOR FURTHER INFORMATION**   For more information on modeling weight loss, see the article "A Linear Diet Model" by Arthur C. Segal in *The College Mathematics Journal*.

1. Determining a Solution Determine whether the function $y = x^3$ is a solution of the differential equation $2xy' + 4y = 10x^3$.

2. Determining a Solution Determine whether the function $y = 2 \sin 2x$ is a solution of the differential equation $y''' - 8y = 0$.

Finding a General Solution In Exercises 3–8, use integration to find a general solution of the differential equation.

3. $\dfrac{dy}{dx} = 4x^2 + 7$

4. $\dfrac{dy}{dx} = 3x^3 - 8x$

5. $\dfrac{dy}{dx} = \cos 2x$

6. $\dfrac{dy}{dx} = 2 \sin x$

7. $\dfrac{dy}{dx} = e^{2-x}$

8. $\dfrac{dy}{dx} = 2e^{3x}$

Slope Field In Exercises 9 and 10, a differential equation and its slope field are given. Complete the table by determining the slopes (if possible) in the slope field at the given points.

x	-4	-2	0	2	4	8
y	2	0	4	4	6	8
dy/dx						

9. $\dfrac{dy}{dx} = 2x - y$

10. $\dfrac{dy}{dx} = x \sin\left(\dfrac{\pi y}{4}\right)$

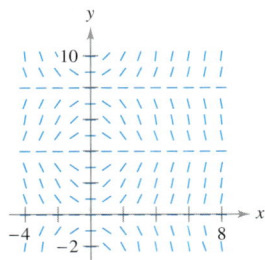

Slope Field In Exercises 11 and 12, (a) sketch the slope field for the differential equation, and (b) use the slope field to sketch the solution that passes through the given point. Use a graphing utility to verify your results. To print a blank graph, go to *MathGraphs.com*.

11. $y' = 2x^2 - x$, $(0, 2)$

12. $y' = y + 4x$, $(-1, 1)$

Euler's Method In Exercises 13 and 14, use Euler's Method to make a table of values for the approximate solution of the differential equation with the specified initial value. Use n steps of size h.

13. $y' = x - y$, $y(0) = 4$, $n = 10$, $h = 0.05$

14. $y' = 5x - 2y$, $y(0) = 2$, $n = 10$, $h = 0.1$

Solving a Differential Equation In Exercises 15–20, solve the differential equation.

15. $\dfrac{dy}{dx} = 2x - 5x^2$

16. $\dfrac{dy}{dx} = y + 8$

17. $\dfrac{dy}{dx} = (3 + y)^2$

18. $\dfrac{dy}{dx} = 10\sqrt{y}$

19. $(2 + x)y' - xy = 0$

20. $xy' - (x + 1)y = 0$

Writing and Solving a Differential Equation In Exercises 21 and 22, write and solve the differential equation that models the verbal statement.

21. The rate of change of y with respect to t is inversely proportional to the cube of t.

22. The rate of change of y with respect to t is proportional to $50 - t$.

Finding an Exponential Function In Exercises 23–26, find the exponential function $y = Ce^{kt}$ that passes through the two points.

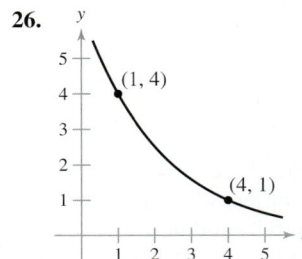

23.

24.

25.

26.

27. Air Pressure Under ideal conditions, air pressure decreases continuously with the height above sea level at a rate proportional to the pressure at that height. The barometer reads 30 inches at sea level and 15 inches at 18,000 feet. Find the barometric pressure at 35,000 feet.

28. Radioactive Decay Radioactive radium has a half-life of approximately 1599 years. The initial quantity is 15 grams. How much remains after 750 years?

29. Population Growth A population grows continuously at the rate of 1.85%. How long will it take the population to double?

30. Compound Interest Find the balance in an account when $1000 is deposited for 8 years at an interest rate of 4% compounded continuously.

31. Sales The sales S (in thousands of units) of a new product after it has been on the market for t years is given by

$$S = Ce^{k/t}.$$

(a) Find S as a function of t when 5000 units have been sold after 1 year and the saturation point for the market is 30,000 units (that is, $\lim\limits_{t \to \infty} S = 30$).

(b) How many units will have been sold after 5 years?

32. Sales The sales S (in thousands of units) of a new product after it has been on the market for t years is given by

$$S = 25(1 - e^{kt}).$$

(a) Find S as a function of t when 4000 units have been sold after 1 year.

(b) How many units will saturate this market?

(c) How many units will have been sold after 5 years?

Finding a General Solution Using Separation of Variables In Exercises 33–36, find the general solution of the differential equation.

33. $\dfrac{dy}{dx} = \dfrac{5x}{y}$ **34.** $\dfrac{dy}{dx} = \dfrac{x^3}{2y^2}$

35. $y' - 16xy = 0$ **36.** $y' - e^y \sin x = 0$

Finding a Particular Solution Using Separation of Variables In Exercises 37–40, find the particular solution that satisfies the initial condition.

Differential Equation	Initial Condition
37. $y^3 y' - 3x = 0$	$y(2) = 2$
38. $yy' - 5e^{2x} = 0$	$y(0) = -3$
39. $y^3(x^4 + 1)y' - x^3(y^4 + 1) = 0$	$y(0) = 1$
40. $yy' - x \cos x^2 = 0$	$y(0) = -2$

Slope Field In Exercises 41 and 42, sketch a few solutions of the differential equation on the slope field and then find the general solution analytically. To print an enlarged copy of the graph, go to *MathGraphs.com*.

41. $\dfrac{dy}{dx} = -\dfrac{4x}{y}$ **42.** $\dfrac{dy}{dx} = 3 - 2y$

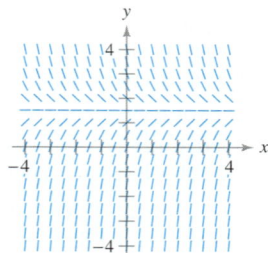

Using a Logistic Equation In Exercises 43 and 44, the logistic equation models the growth of a population. Use the equation to (a) find the value of k, (b) find the carrying capacity, (c) find the initial population, (d) determine when the population will reach 50% of its carrying capacity, and (e) write a logistic differential equation that has the solution $P(t)$.

43. $P(t) = \dfrac{5250}{1 + 34e^{-0.55t}}$

44. $P(t) = \dfrac{4800}{1 + 14e^{-0.15t}}$

Solving a Logistic Differential Equation In Exercises 45 and 46, find the logistic equation that passes through the given point.

45. $\dfrac{dy}{dt} = y\left(1 - \dfrac{y}{80}\right),$ $(0, 8)$

46. $\dfrac{dy}{dt} = 1.76y\left(1 - \dfrac{y}{8}\right),$ $(0, 3)$

47. Environment A conservation department releases 1200 brook trout into a lake. It is estimated that the carrying capacity of the lake for the species is 20,400. After the first year, there are 2000 brook trout in the lake.

(a) Write a logistic equation that models the number of brook trout in the lake.

(b) Find the number of brook trout in the lake after 8 years.

(c) When will the number of brook trout reach 10,000?

48. Environment Write a logistic differential equation that models the growth rate of the brook trout population in Exercise 47. Then repeat part (b) using Euler's Method with a step size of $h = 1$. Compare the approximation with the exact answer.

Solving a First-Order Linear Differential Equation In Exercises 49–54, solve the first-order linear differential equation.

49. $y' - y = 10$

50. $e^x y' + 4e^x y = 1$

51. $4y' = e^{x/4} + y$

52. $\dfrac{dy}{dx} - \dfrac{5y}{x^2} = \dfrac{1}{x^2}$

53. $(x - 2)y' + y = 1$

54. $(x + 3)y' + 2y = 2(x + 3)^2$

Finding a Particular Solution In Exercises 55 and 56, find the particular solution of the differential equation that satisfies the initial condition.

Differential Equation	Initial Condition
55. $y' + 5y = e^{5x}$	$y(0) = 3$
56. $y' - \left(\dfrac{3}{x}\right)y = 2x^3$	$y(1) = 1$

P.S. Problem Solving

See **CalcChat.com** for tutorial help and
worked-out solutions to odd-numbered exercises.

1. Doomsday Equation   The differential equation

$$\frac{dy}{dt} = ky^{1+\varepsilon}$$

where k and ε are positive constants, is called the **doomsday equation.**

(a) Solve the doomsday equation

$$\frac{dy}{dt} = y^{1.01}$$

given that $y(0) = 1$. Find the time T at which

$$\lim_{t \to T^-} y(t) = \infty.$$

(b) Solve the doomsday equation

$$\frac{dy}{dt} = ky^{1+\varepsilon}$$

given that $y(0) = y_0$. Explain why this equation is called the doomsday equation.

2. Sales   Let S represent sales of a new product (in thousands of units), let L represent the maximum level of sales (in thousands of units), and let t represent time (in months). The rate of change of S with respect to t varies jointly as the product of S and $L - S$.

(a) Write the differential equation for the sales model when $L = 100$, $S = 10$ when $t = 0$, and $S = 20$ when $t = 1$. Verify that

$$S = \frac{L}{1 + Ce^{-kt}}.$$

(b) At what time is the growth in sales increasing most rapidly?

(c) Use a graphing utility to graph the sales function.

(d) Sketch the solution from part (a) on the slope field shown in the figure below. To print an enlarged copy of the graph, go to *MathGraphs.com*.

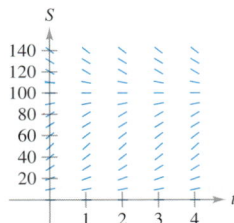

(e) Assume the estimated maximum level of sales is correct. Use the slope field to describe the shape of the solution curves for sales when, at some period of time, sales exceed L.

3. Gompertz Equation   Another model that can be used to represent population growth is the **Gompertz equation,** which is the solution of the differential equation

$$\frac{dy}{dt} = k \ln\left(\frac{L}{y}\right) y$$

where k is a constant and L is the carrying capacity.

(a) Solve the differential equation.

(b) Use a graphing utility to graph the slope field for the differential equation when $k = 0.05$ and $L = 1000$.

(c) Describe the behavior of the graph as $t \to \infty$.

(d) Graph the equation you found in part (a) for $L = 5000$, $y_0 = 500$, and $k = 0.02$. Determine the concavity of the graph and how it compares with the general solution of the logistic differential equation.

4. Error Using Product Rule   Although it is true for some functions f and g, a common mistake in calculus is to believe that the Product Rule for derivatives is $(fg)' = f'g'$.

(a) Given $g(x) = x$, find f such that $(fg)' = f'g'$.

(b) Given an arbitrary function g, find a function f such that $(fg)' = f'g'$.

(c) Describe what happens if $g(x) = e^x$.

5. Torricelli's Law   **Torricelli's Law** states that water will flow from an opening at the bottom of a tank with the same speed that it would attain falling from the surface of the water to the opening. One of the forms of Torricelli's Law is

$$A(h)\frac{dh}{dt} = -k\sqrt{2gh}$$

where h is the height of the water in the tank, k is the area of the opening at the bottom of the tank, $A(h)$ is the horizontal cross-sectional area at height h, and g is the acceleration due to gravity ($g \approx 32$ feet per second per second). A hemispherical water tank has a radius of 6 feet. When the tank is full, a circular valve with a radius of 1 inch is opened at the bottom, as shown in the figure. How long will it take for the tank to drain completely?

6. Torricelli's Law The cylindrical water tank shown in the figure has a height of 18 feet. When the tank is full, a circular valve is opened at the bottom of the tank. After 30 minutes, the depth of the water is 12 feet.

(a) Using Torricelli's Law, how long will it take for the tank to drain completely?

(b) What is the depth of the water in the tank after 1 hour?

7. Torricelli's Law Suppose the tank in Exercise 6 has a height of 20 feet and a radius of 8 feet, and the valve is circular with a radius of 2 inches. The tank is full when the valve is opened. How long will it take for the tank to drain completely?

8. Rewriting the Logistic Equation Show that the logistic equation

$$y = \frac{L}{1 + be^{-kt}}$$

can be written as

$$y = \frac{1}{2}L\left[1 + \tanh\left(\frac{1}{2}k\left(t - \frac{\ln b}{k}\right)\right)\right].$$

What can you conclude about the graph of the logistic equation?

9. Biomass Biomass is a measure of the amount of living matter in an ecosystem. Suppose the biomass $s(t)$ in a given ecosystem increases at a rate of about 3.5 tons per year, and decreases by about 1.9% per year. This situation can be modeled by the differential equation

$$\frac{ds}{dt} = 3.5 - 0.019s.$$

(a) Solve the differential equation.

(b) Use a graphing utility to graph the slope field for the differential equation. What do you notice?

(c) Explain what happens as $t \to \infty$.

Medical Science In Exercises 10–12, a medical researcher wants to determine the concentration C (in moles per liter) of a tracer drug injected into a moving fluid. Solve this problem by considering a single-compartment dilution model (see figure). Assume that the fluid is continuously mixed and that the volume of the fluid in the compartment is constant.

Figure for 10–12

10. If the tracer is injected instantaneously at time $t = 0$, then the concentration of the fluid in the compartment begins diluting according to the differential equation

$$\frac{dC}{dt} = \left(-\frac{R}{V}\right)C$$

where $C = C_0$ when $t = 0$.

(a) Solve this differential equation to find the concentration C as a function of time t.

(b) Find the limit of C as $t \to \infty$.

11. Use the solution of the differential equation in Exercise 10 to find the concentration C as a function of time t, and use a graphing utility to graph the function.

(a) $V = 2$ liters, $R = 0.5$ liter per minute, and $C_0 = 0.6$ mole per liter

(b) $V = 2$ liters, $R = 1.5$ liters per minute, and $C_0 = 0.6$ mole per liter

12. In Exercises 10 and 11, it was assumed that there was a single initial injection of the tracer drug into the compartment. Now consider the case in which the tracer is continuously injected (beginning at $t = 0$) at the rate of Q moles per minute. Considering Q to be negligible compared with R, use the differential equation

$$\frac{dC}{dt} = \frac{Q}{V} - \left(\frac{R}{V}\right)C$$

where $C = 0$ when $t = 0$.

(a) Solve this differential equation to find the concentration C as a function of time t.

(b) Find the limit of C as $t \to \infty$.

7 Applications of Integration

Moving a Space Module into Orbit *(Example 3, p. 480)*

Tidal Energy
(Section Project, p. 485)

Saturn *(Section Project, p. 465)*

Water Tower
(Exercise 66, p. 455)

Building Design *(Exercise 79, p. 445)*

435

7.1 Area of a Region Between Two Curves

■ Find the area of a region between two curves using integration.
■ Find the area of a region between intersecting curves using integration.
■ Describe integration as an accumulation process.

Area of a Region Between Two Curves

With a few modifications, you can extend the application of definite integrals from the area of a region *under* a curve to the area of a region *between* two curves. Consider two functions f and g that are continuous on the interval $[a, b]$. Also, the graphs of both f and g lie above the x-axis, and the graph of g lies below the graph of f, as shown in Figure 7.1. You can geometrically interpret the area of the region between the graphs as the area of the region under the graph of g subtracted from the area of the region under the graph of f, as shown in Figure 7.2.

Figure 7.1

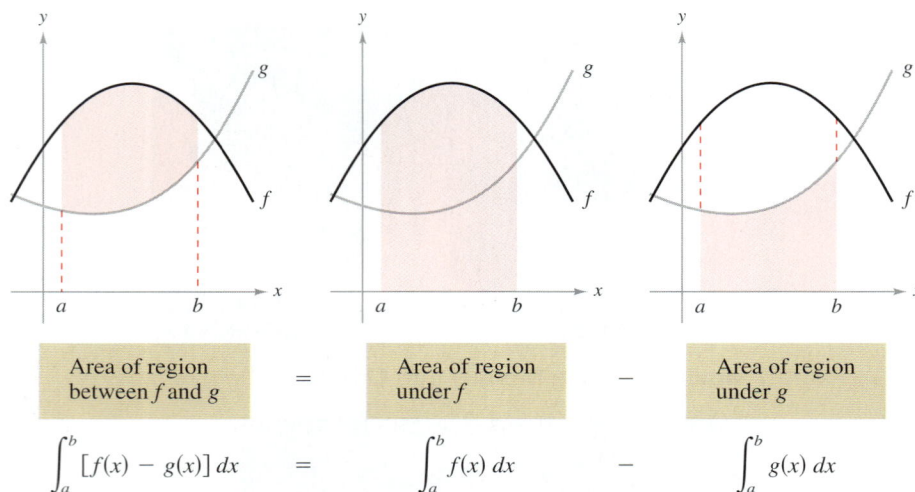

| Area of region between f and g | = | Area of region under f | − | Area of region under g |

$$\int_a^b [f(x) - g(x)]\, dx = \int_a^b f(x)\, dx - \int_a^b g(x)\, dx$$

Figure 7.2

To verify the reasonableness of the result shown in Figure 7.2, you can partition the interval $[a, b]$ into n subintervals, each of width Δx. Then, as shown in Figure 7.3, sketch a **representative rectangle** of width Δx and height $f(x_i) - g(x_i)$, where x_i is in the ith subinterval. The area of this representative rectangle is

$$\Delta A_i = (\text{height})(\text{width}) = [f(x_i) - g(x_i)]\Delta x.$$

By adding the areas of the n rectangles and taking the limit as $\|\Delta\| \to 0$ $(n \to \infty)$, you obtain

$$\lim_{n \to \infty} \sum_{i=1}^n [f(x_i) - g(x_i)]\Delta x.$$

• • REMARK Recall from Section 4.3 that $\|\Delta\|$ is the norm of the partition. In a regular partition, the statements $\|\Delta\| \to 0$ and $n \to \infty$ are equivalent.

Because f and g are continuous on $[a, b]$, $f - g$ is also continuous on $[a, b]$ and the limit exists. So, the area of the region is

$$\text{Area} = \lim_{n \to \infty} \sum_{i=1}^n [f(x_i) - g(x_i)]\Delta x$$

$$= \int_a^b [f(x) - g(x)]\, dx.$$

Representative rectangle
Height: $f(x_i) - g(x_i)$
Width: Δx

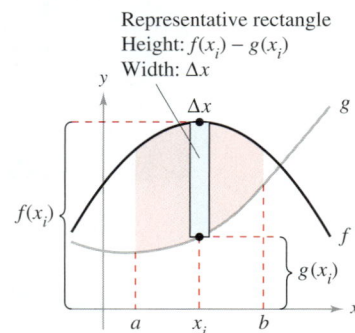

Figure 7.3

Area of a Region Between Two Curves

If f and g are continuous on $[a, b]$ and $g(x) \leq f(x)$ for all x in $[a, b]$, then the area of the region bounded by the graphs of f and g and the vertical lines $x = a$ and $x = b$ is

$$A = \int_a^b [f(x) - g(x)]\,dx.$$

In Figure 7.1, the graphs of f and g are shown above the x-axis. This, however, is not necessary. The same integrand $[f(x) - g(x)]$ can be used as long as f and g are continuous and $g(x) \leq f(x)$ for all x in the interval $[a, b]$. This is summarized graphically in Figure 7.4. Notice in Figure 7.4 that the height of a representative rectangle is $f(x) - g(x)$ regardless of the relative position of the x-axis.

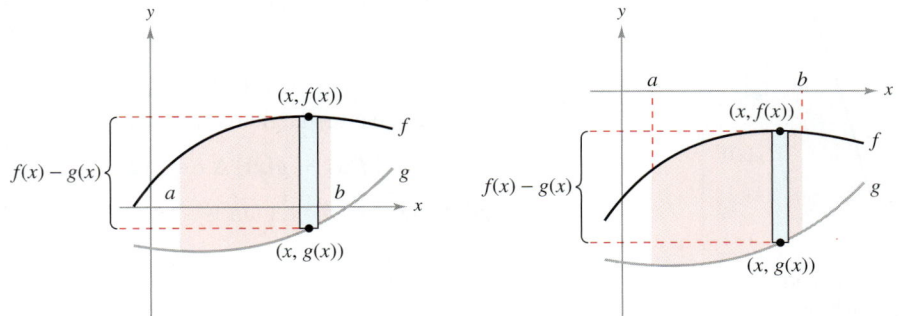

Figure 7.4

Representative rectangles are used throughout this chapter in various applications of integration. A vertical rectangle (of width Δx) implies integration with respect to x, whereas a horizontal rectangle (of width Δy) implies integration with respect to y.

EXAMPLE 1   Finding the Area of a Region Between Two Curves

Find the area of the region bounded by the graphs of $y = x^2 + 2$, $y = -x$, $x = 0$, and $x = 1$.

Solution   Let $g(x) = -x$ and $f(x) = x^2 + 2$. Then $g(x) \leq f(x)$ for all x in $[0, 1]$, as shown in Figure 7.5. So, the area of the representative rectangle is

$$\Delta A = [f(x) - g(x)]\,\Delta x$$
$$= [(x^2 + 2) - (-x)]\,\Delta x$$

and the area of the region is

$$A = \int_a^b [f(x) - g(x)]\,dx$$
$$= \int_0^1 [(x^2 + 2) - (-x)]\,dx$$
$$= \left[\frac{x^3}{3} + \frac{x^2}{2} + 2x\right]_0^1$$
$$= \frac{1}{3} + \frac{1}{2} + 2$$
$$= \frac{17}{6}.$$

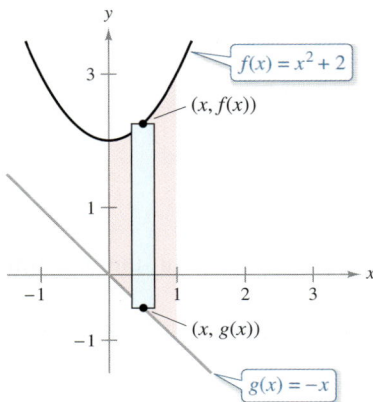

Region bounded by the graph of f, the graph of g, $x = 0$, and $x = 1$

Figure 7.5

Area of a Region Between Intersecting Curves

In Example 1, the graphs of $f(x) = x^2 + 2$ and $g(x) = -x$ do not intersect, and the values of a and b are given explicitly. A more common problem involves the area of a region bounded by two *intersecting* graphs, where the values of a and b must be calculated.

EXAMPLE 2   **A Region Lying Between Two Intersecting Graphs**

Find the area of the region bounded by the graphs of $f(x) = 2 - x^2$ and $g(x) = x$.

Solution   In Figure 7.6, notice that the graphs of f and g have two points of intersection. To find the x-coordinates of these points, set $f(x)$ and $g(x)$ equal to each other and solve for x.

$$2 - x^2 = x \qquad \text{Set } f(x) \text{ equal to } g(x).$$
$$-x^2 - x + 2 = 0 \qquad \text{Write in general form.}$$
$$-(x + 2)(x - 1) = 0 \qquad \text{Factor.}$$
$$x = -2 \text{ or } 1 \qquad \text{Solve for } x.$$

So, $a = -2$ and $b = 1$. Because $g(x) \le f(x)$ for all x in the interval $[-2, 1]$, the representative rectangle has an area of

$$\Delta A = [f(x) - g(x)]\,\Delta x = [(2 - x^2) - x]\,\Delta x$$

and the area of the region is

$$A = \int_{-2}^{1} [(2 - x^2) - x]\,dx$$
$$= \left[-\frac{x^3}{3} - \frac{x^2}{2} + 2x \right]_{-2}^{1}$$
$$= \frac{9}{2}.$$

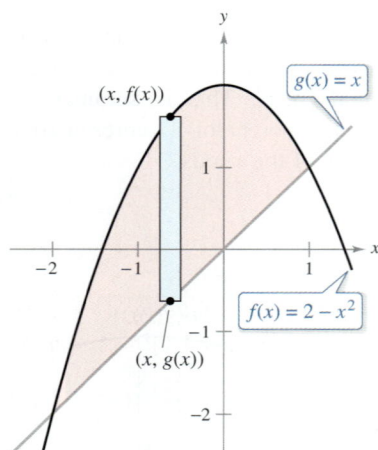

Region bounded by the graph of f and the graph of g

Figure 7.6

EXAMPLE 3   **A Region Lying Between Two Intersecting Graphs**

The sine and cosine curves intersect infinitely many times, bounding regions of equal areas, as shown in Figure 7.7. Find the area of one of these regions.

Solution   Let $g(x) = \cos x$ and $f(x) = \sin x$. Then $g(x) \le f(x)$ for all x in the interval corresponding to the shaded region in Figure 7.7. To find the two points of intersection on this interval, set $f(x)$ and $g(x)$ equal to each other and solve for x.

$$\sin x = \cos x \qquad \text{Set } f(x) \text{ equal to } g(x).$$
$$\frac{\sin x}{\cos x} = 1 \qquad \text{Divide each side by } \cos x.$$
$$\tan x = 1 \qquad \text{Trigonometric identity}$$
$$x = \frac{\pi}{4} \text{ or } \frac{5\pi}{4}, \quad 0 \le x \le 2\pi \qquad \text{Solve for } x.$$

So, $a = \pi/4$ and $b = 5\pi/4$. Because $\sin x \ge \cos x$ for all x in the interval $[\pi/4, 5\pi/4]$, the area of the region is

$$A = \int_{\pi/4}^{5\pi/4} [\sin x - \cos x]\,dx$$
$$= \left[-\cos x - \sin x \right]_{\pi/4}^{5\pi/4}$$
$$= 2\sqrt{2}.$$

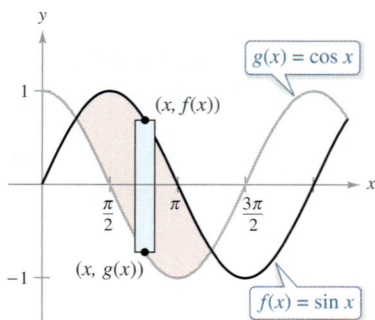

One of the regions bounded by the graphs of the sine and cosine functions

Figure 7.7

To find the area of the region between two curves that intersect at *more* than two points, first determine all points of intersection. Then check to see which curve is above the other in each interval determined by these points, as shown in Example 4.

EXAMPLE 4   Curves That Intersect at More than Two Points

$\cdots\triangleright$ *See LarsonCalculus.com for an interactive version of this type of example.*

Find the area of the region between the graphs of

$$f(x) = 3x^3 - x^2 - 10x \quad \text{and} \quad g(x) = -x^2 + 2x.$$

Solution   Begin by setting $f(x)$ and $g(x)$ equal to each other and solving for x. This yields the x-values at all points of intersection of the two graphs.

$$3x^3 - x^2 - 10x = -x^2 + 2x \qquad \text{Set } f(x) \text{ equal to } g(x).$$
$$3x^3 - 12x = 0 \qquad \text{Write in general form.}$$
$$3x(x - 2)(x + 2) = 0 \qquad \text{Factor.}$$
$$x = -2, 0, 2 \qquad \text{Solve for } x.$$

So, the two graphs intersect when $x = -2, 0$, and 2. In Figure 7.8, notice that $g(x) \le f(x)$ on the interval $[-2, 0]$. The two graphs switch at the origin, however, and $f(x) \le g(x)$ on the interval $[0, 2]$. So, you need two integrals—one for the interval $[-2, 0]$ and one for the interval $[0, 2]$.

$$A = \int_{-2}^{0} [f(x) - g(x)]\, dx + \int_{0}^{2} [g(x) - f(x)]\, dx$$
$$= \int_{-2}^{0} (3x^3 - 12x)\, dx + \int_{0}^{2} (-3x^3 + 12x)\, dx$$
$$= \left[\frac{3x^4}{4} - 6x^2 \right]_{-2}^{0} + \left[\frac{-3x^4}{4} + 6x^2 \right]_{0}^{2}$$
$$= -(12 - 24) + (-12 + 24)$$
$$= 24$$

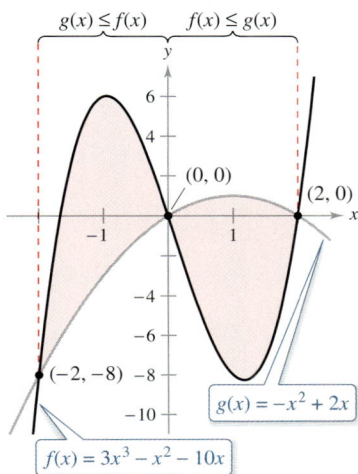

On $[-2, 0]$, $g(x) \le f(x)$, and on $[0, 2]$, $f(x) \le g(x)$.
Figure 7.8

$\cdots\cdots$**REMARK**   In Example 4, notice that you obtain an incorrect result when you integrate from -2 to 2. Such integration produces

$$\int_{-2}^{2} [f(x) - g(x)]\, dx = \int_{-2}^{2} (3x^3 - 12x)\, dx$$
$$= 0.$$

When the graph of a function of y is a boundary of a region, it is often convenient to use representative rectangles that are *horizontal* and find the area by integrating with respect to y. In general, to determine the area between two curves, you can use

$$A = \int_{x_1}^{x_2} [(\text{top curve}) - (\text{bottom curve})]\, dx \qquad \text{Vertical rectangles}$$

in variable x

or

$$A = \int_{y_1}^{y_2} [(\text{right curve}) - (\text{left curve})]\, dy \qquad \text{Horizontal rectangles}$$

in variable y

where (x_1, y_1) and (x_2, y_2) are either adjacent points of intersection of the two curves involved or points on the specified boundary lines.

EXAMPLE 5 **Horizontal Representative Rectangles**

Find the area of the region bounded by the graphs of $x = 3 - y^2$ and $x = y + 1$.

Solution Consider

$$g(y) = 3 - y^2 \quad \text{and} \quad f(y) = y + 1.$$

These two curves intersect when $y = -2$ and $y = 1$, as shown in Figure 7.9. Because $f(y) \leq g(y)$ on this interval, you have

$$\Delta A = [g(y) - f(y)] \Delta y = [(3 - y^2) - (y + 1)] \Delta y.$$

So, the area is

$$
\begin{aligned}
A &= \int_{-2}^{1} [(3 - y^2) - (y + 1)] \, dy \\
&= \int_{-2}^{1} (-y^2 - y + 2) \, dy \\
&= \left[\frac{-y^3}{3} - \frac{y^2}{2} + 2y \right]_{-2}^{1} \\
&= \left(-\frac{1}{3} - \frac{1}{2} + 2 \right) - \left(\frac{8}{3} - 2 - 4 \right) \\
&= \frac{9}{2}.
\end{aligned}
$$

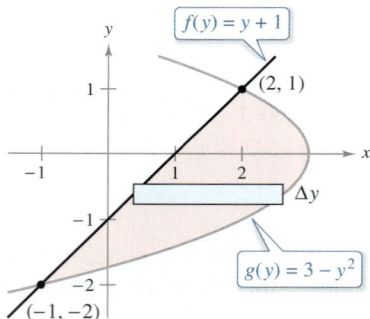

Horizontal rectangles (integration with respect to y)

Figure 7.9

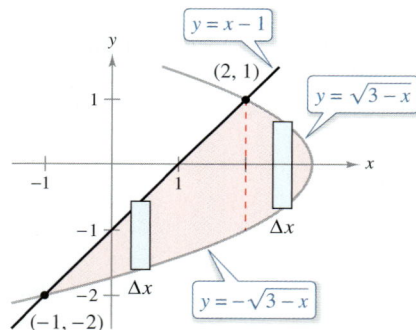

Vertical rectangles (integration with respect to x)

Figure 7.10

In Example 5, notice that by integrating with respect to y, you need only one integral. To integrate with respect to x, you would need two integrals because the upper boundary changes at $x = 2$, as shown in Figure 7.10.

$$
\begin{aligned}
A &= \int_{-1}^{2} \left[(x - 1) + \sqrt{3 - x} \right] dx + \int_{2}^{3} \left(\sqrt{3 - x} + \sqrt{3 - x} \right) dx \\
&= \int_{-1}^{2} \left[x - 1 + (3 - x)^{1/2} \right] dx + 2 \int_{2}^{3} (3 - x)^{1/2} \, dx \\
&= \left[\frac{x^2}{2} - x - \frac{(3 - x)^{3/2}}{3/2} \right]_{-1}^{2} - 2 \left[\frac{(3 - x)^{3/2}}{3/2} \right]_{2}^{3} \\
&= \left(2 - 2 - \frac{2}{3} \right) - \left(\frac{1}{2} + 1 - \frac{16}{3} \right) - 2(0) + 2 \left(\frac{2}{3} \right) \\
&= \frac{9}{2}
\end{aligned}
$$

Integration as an Accumulation Process

In this section, the integration formula for the area between two curves was developed by using a rectangle as the *representative element*. For each new application in the remaining sections of this chapter, an appropriate representative element will be constructed using precalculus formulas you already know. Each integration formula will then be obtained by summing or accumulating these representative elements.

| Known precalculus formula | ⇨ | Representative element | ⇨ | New integration formula |

For example, the area formula in this section was developed as follows.

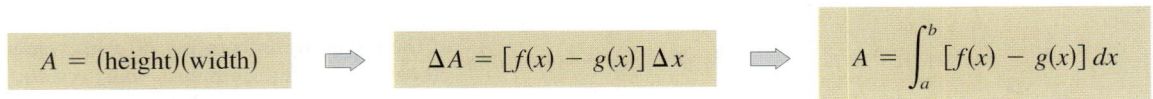

| $A = (\text{height})(\text{width})$ | ⇨ | $\Delta A = [f(x) - g(x)] \Delta x$ | ⇨ | $A = \displaystyle\int_a^b [f(x) - g(x)]\, dx$ |

EXAMPLE 6   **Integration as an Accumulation Process**

Find the area of the region bounded by the graph of $y = 4 - x^2$ and the x-axis. Describe the integration as an accumulation process.

Solution   The area of the region is

$$A = \int_{-2}^{2} (4 - x^2)\, dx.$$

You can think of the integration as an accumulation of the areas of the rectangles formed as the representative rectangle slides from $x = -2$ to $x = 2$, as shown in Figure 7.11.

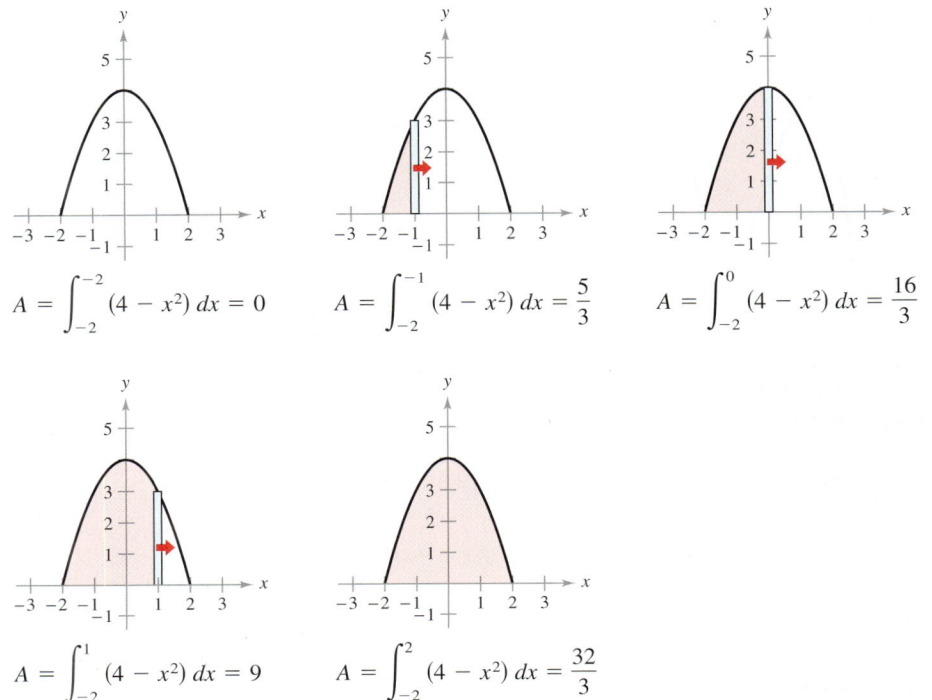

$$A = \int_{-2}^{-2} (4 - x^2)\, dx = 0 \qquad A = \int_{-2}^{-1} (4 - x^2)\, dx = \frac{5}{3} \qquad A = \int_{-2}^{0} (4 - x^2)\, dx = \frac{16}{3}$$

$$A = \int_{-2}^{1} (4 - x^2)\, dx = 9 \qquad A = \int_{-2}^{2} (4 - x^2)\, dx = \frac{32}{3}$$

Figure 7.11

7.1 Exercises

Writing a Definite Integral In Exercises 1–6, set up the definite integral that gives the area of the region.

1. $y_1 = x^2 - 6x$
$y_2 = 0$

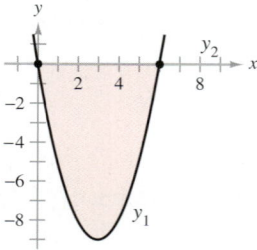

2. $y_1 = x^2 + 2x + 1$
$y_2 = 2x + 5$

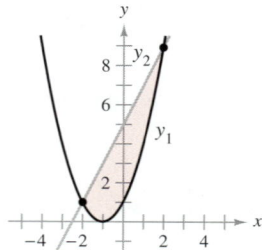

3. $y_1 = x^2 - 4x + 3$
$y_2 = -x^2 + 2x + 3$

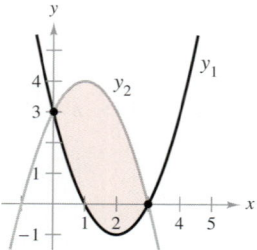

4. $y_1 = x^2$
$y_2 = x^3$

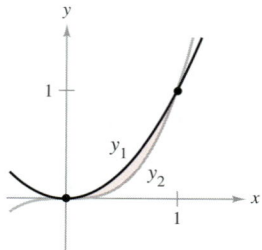

5. $y_1 = 3(x^3 - x)$
$y_2 = 0$

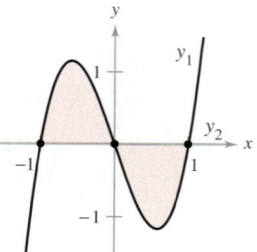

6. $y_1 = (x - 1)^3$
$y_2 = x - 1$

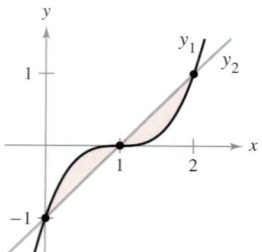

Finding a Region In Exercises 7–12, the integrand of the definite integral is a difference of two functions. Sketch the graph of each function and shade the region whose area is represented by the integral.

7. $\int_0^4 \left[(x + 1) - \frac{x}{2} \right] dx$

8. $\int_{-1}^1 [(2 - x^2) - x^2] \, dx$

9. $\int_2^3 \left[\left(\frac{x^3}{3} - x \right) - \frac{x}{3} \right] dx$

10. $\int_{-\pi/4}^{\pi/4} (\sec^2 x - \cos x) \, dx$

11. $\int_{-2}^1 [(2 - y) - y^2] \, dy$

12. $\int_0^4 \left(2\sqrt{y} - y \right) dy$

Think About It In Exercises 13 and 14, determine which value best approximates the area of the region bounded by the graphs of f and g. (Make your selection on the basis of a sketch of the region and not by performing any calculations.)

13. $f(x) = x + 1$, $g(x) = (x - 1)^2$
   (a) -2 (b) 2 (c) 10 (d) 4 (e) 8

14. $f(x) = 2 - \frac{1}{2}x$, $g(x) = 2 - \sqrt{x}$
   (a) 1 (b) 6 (c) -3 (d) 3 (e) 4

Comparing Methods In Exercises 15 and 16, find the area of the region by integrating (a) with respect to x and (b) with respect to y. (c) Compare your results. Which method is simpler? In general, will this method always be simpler than the other one? Why or why not?

15. $x = 4 - y^2$
$x = y - 2$

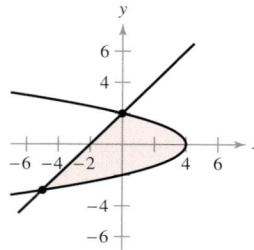

16. $y = x^2$
$y = 6 - x$

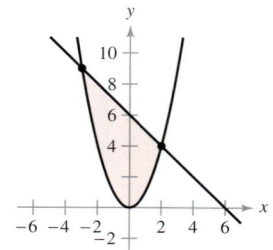

Finding the Area of a Region In Exercises 17–30, sketch the region bounded by the graphs of the equations and find the area of the region.

17. $y = x^2 - 1$, $y = -x + 2$, $x = 0$, $x = 1$

18. $y = -x^3 + 2$, $y = x - 3$, $x = -1$, $x = 1$

19. $f(x) = x^2 + 2x$, $g(x) = x + 2$

20. $y = -x^2 + 3x + 1$, $y = -x + 1$

21. $y = x$, $y = 2 - x$, $y = 0$

22. $y = \frac{4}{x^3}$, $y = 0$, $x = 1$, $x = 4$

23. $f(x) = \sqrt{x} + 3$, $g(x) = \frac{1}{2}x + 3$

24. $f(x) = \sqrt[3]{x - 1}$, $g(x) = x - 1$

25. $f(y) = y^2$, $g(y) = y + 2$

26. $f(y) = y(2 - y)$, $g(y) = -y$

27. $f(y) = y^2 + 1$, $g(y) = 0$, $y = -1$, $y = 2$

28. $f(y) = \dfrac{y}{\sqrt{16 - y^2}}$, $g(y) = 0$, $y = 3$

29. $f(x) = \dfrac{10}{x}$, $x = 0$, $y = 2$, $y = 10$

30. $g(x) = \dfrac{4}{2 - x}$, $y = 4$, $x = 0$

Finding the Area of a Region In Exercises 31–36, (a) use a graphing utility to graph the region bounded by the graphs of the equations, (b) find the area of the region analytically, and (c) use the integration capabilities of the graphing utility to verify your results.

31. $f(x) = x(x^2 - 3x + 3)$, $g(x) = x^2$

32. $y = x^4 - 2x^2$, $y = 2x^2$

33. $f(x) = x^4 - 4x^2$, $g(x) = x^2 - 4$

34. $f(x) = x^4 - 9x^2$, $g(x) = x^3 - 9x$

35. $f(x) = \dfrac{1}{1 + x^2}$, $g(x) = \dfrac{1}{2}x^2$

36. $f(x) = \dfrac{6x}{x^2 + 1}$, $y = 0$, $0 \le x \le 3$

Finding the Area of a Region In Exercises 37–42, sketch the region bounded by the graphs of the functions and find the area of the region.

37. $f(x) = \cos x$, $g(x) = 2 - \cos x$, $0 \le x \le 2\pi$

38. $f(x) = \sin x$, $g(x) = \cos 2x$, $-\dfrac{\pi}{2} \le x \le \dfrac{\pi}{6}$

39. $f(x) = 2 \sin x$, $g(x) = \tan x$, $-\dfrac{\pi}{3} \le x \le \dfrac{\pi}{3}$

40. $f(x) = \sec \dfrac{\pi x}{4} \tan \dfrac{\pi x}{4}$, $g(x) = \left(\sqrt{2} - 4\right)x + 4$, $x = 0$

41. $f(x) = xe^{-x^2}$, $y = 0$, $0 \le x \le 1$

42. $f(x) = 2^x$, $g(x) = \dfrac{3}{2}x + 1$

Finding the Area of a Region In Exercises 43–46, (a) use a graphing utility to graph the region bounded by the graphs of the equations, (b) find the area of the region, and (c) use the integration capabilities of the graphing utility to verify your results.

43. $f(x) = 2 \sin x + \sin 2x$, $y = 0$, $0 \le x \le \pi$

44. $f(x) = 2 \sin x + \cos 2x$, $y = 0$, $0 < x \le \pi$

45. $f(x) = \dfrac{1}{x^2}e^{1/x}$, $y = 0$, $1 \le x \le 3$

46. $g(x) = \dfrac{4 \ln x}{x}$, $y = 0$, $x = 5$

Finding the Area of a Region In Exercises 47–50, (a) use a graphing utility to graph the region bounded by the graphs of the equations, (b) explain why the area of the region is difficult to find by hand, and (c) use the integration capabilities of the graphing utility to approximate the area to four decimal places.

47. $y = \sqrt{\dfrac{x^3}{4 - x}}$, $y = 0$, $x = 3$

48. $y = \sqrt{x}\, e^x$, $y = 0$, $x = 0$, $x = 1$

49. $y = x^2$, $y = 4 \cos x$

50. $y = x^2$, $y = \sqrt{3 + x}$

Integration as an Accumulation Process In Exercises 51–54, find the accumulation function F. Then evaluate F at each value of the independent variable and graphically show the area given by each value of F.

51. $F(x) = \displaystyle\int_0^x \left(\dfrac{1}{2}t + 1\right) dt$ (a) $F(0)$ (b) $F(2)$ (c) $F(6)$

52. $F(x) = \displaystyle\int_0^x \left(\dfrac{1}{2}t^2 + 2\right) dt$ (a) $F(0)$ (b) $F(4)$ (c) $F(6)$

53. $F(\alpha) = \displaystyle\int_{-1}^{\alpha} \cos \dfrac{\pi\theta}{2}\, d\theta$ (a) $F(-1)$ (b) $F(0)$ (c) $F\left(\dfrac{1}{2}\right)$

54. $F(y) = \displaystyle\int_{-1}^{y} 4e^{x/2}\, dx$ (a) $F(-1)$ (b) $F(0)$ (c) $F(4)$

Finding the Area of a Figure In Exercises 55–58, use integration to find the area of the figure having the given vertices.

55. $(2, -3)$, $(4, 6)$, $(6, 1)$

56. $(0, 0)$, $(6, 0)$, $(4, 3)$

57. $(0, 2)$, $(4, 2)$, $(0, -2)$, $(-4, -2)$

58. $(0, 0)$, $(1, 2)$, $(3, -2)$, $(1, -3)$

59. **Numerical Integration** Estimate the surface area of the golf green using (a) the Trapezoidal Rule and (b) Simpson's Rule.

60. **Numerical Integration** Estimate the surface area of the oil spill using (a) the Trapezoidal Rule and (b) Simpson's Rule.

Using a Tangent Line In Exercises 61–64, set up and evaluate the definite integral that gives the area of the region bounded by the graph of the function and the tangent line to the graph at the given point.

61. $f(x) = x^3$, $(1, 1)$

62. $y = x^3 - 2x$, $(-1, 1)$

63. $f(x) = \dfrac{1}{x^2 + 1}$, $\left(1, \dfrac{1}{2}\right)$

64. $y = \dfrac{2}{1 + 4x^2}$, $\left(\dfrac{1}{2}, 1\right)$

65. Area Between Curves The graphs of $y = 1 - x^2$ and $y = x^4 - 2x^2 + 1$ intersect at three points. However, the area between the curves *can* be found by a single integral. Explain why this is so, and write an integral for this area.

66. Using Symmetry The area of the region bounded by the graphs of $y = x^3$ and $y = x$ *cannot* be found by the single integral $\int_{-1}^{1} (x^3 - x)\, dx$. Explain why this is so. Use symmetry to write a single integral that does represent the area.

67. Interpreting Integrals Two cars with velocities v_1 and v_2 are tested on a straight track (in meters per second). Consider the following.

$$\int_0^5 [v_1(t) - v_2(t)]\, dt = 10 \qquad \int_0^{10} [v_1(t) - v_2(t)]\, dt = 30$$

$$\int_{20}^{30} [v_1(t) - v_2(t)]\, dt = -5$$

(a) Write a verbal interpretation of each integral.

(b) Is it possible to determine the distance between the two cars when $t = 5$ seconds? Why or why not?

(c) Assume both cars start at the same time and place. Which car is ahead when $t = 10$ seconds? How far ahead is the car?

(d) Suppose Car 1 has velocity v_1 and is ahead of Car 2 by 13 meters when $t = 20$ seconds. How far ahead or behind is Car 1 when $t = 30$ seconds?

68. HOW DO YOU SEE IT? A state legislature is debating two proposals for eliminating the annual budget deficits after 10 years. The rate of decrease of the deficits for each proposal is shown in the figure.

(a) What does the area between the two curves represent?

(b) From the viewpoint of minimizing the cumulative state deficit, which is the better proposal? Explain.

Dividing a Region In Exercises 69 and 70, find b such that the line $y = b$ divides the region bounded by the graphs of the two equations into two regions of equal area.

69. $y = 9 - x^2$, $y = 0$ **70.** $y = 9 - |x|$, $y = 0$

Dividing a Region In Exercises 71 and 72, find a such that the line $x = a$ divides the region bounded by the graphs of the equations into two regions of equal area.

71. $y = x$, $y = 4$, $x = 0$ **72.** $y^2 = 4 - x$, $x = 0$

Limits and Integrals In Exercises 73 and 74, evaluate the limit and sketch the graph of the region whose area is represented by the limit.

73. $\lim\limits_{\|\Delta\| \to 0} \sum\limits_{i=1}^{n} (x_i - x_i^2)\, \Delta x$, where $x_i = \dfrac{i}{n}$ and $\Delta x = \dfrac{1}{n}$

74. $\lim\limits_{\|\Delta\| \to 0} \sum\limits_{i=1}^{n} (4 - x_i^2)\, \Delta x$, where $x_i = -2 + \dfrac{4i}{n}$ and $\Delta x = \dfrac{4}{n}$

Revenue In Exercises 75 and 76, two models R_1 and R_2 are given for revenue (in billions of dollars) for a large corporation. Both models are estimates of revenues from 2015 through 2020, with $t = 15$ corresponding to 2015. Which model projects the greater revenue? How much more total revenue does that model project over the six-year period?

75. $R_1 = 7.21 + 0.58t$

$R_2 = 7.21 + 0.45t$

76. $R_1 = 7.21 + 0.26t + 0.02t^2$

$R_2 = 7.21 + 0.1t + 0.01t^2$

77. Lorenz Curve Economists use *Lorenz curves* to illustrate the distribution of income in a country. A Lorenz curve, $y = f(x)$, represents the actual income distribution in the country. In this model, x represents percents of families in the country and y represents percents of total income. The model $y = x$ represents a country in which each family has the same income. The area between these two models, where $0 \le x \le 100$, indicates a country's "income inequality." The table lists percents of income y for selected percents of families x in a country.

x	10	20	30	40	50
y	3.35	6.07	9.17	13.39	19.45

x	60	70	80	90
y	28.03	39.77	55.28	75.12

(a) Use a graphing utility to find a quadratic model for the Lorenz curve.

(b) Plot the data and graph the model.

(c) Graph the model $y = x$. How does this model compare with the model in part (a)?

(d) Use the integration capabilities of a graphing utility to approximate the "income inequality."

78. Profit The chief financial officer of a company reports that profits for the past fiscal year were $15.9 million. The officer predicts that profits for the next 5 years will grow at a continuous annual rate somewhere between $3\frac{1}{2}\%$ and 5%. Estimate the cumulative difference in total profit over the 5 years based on the predicted range of growth rates.

79. Building Design

Concrete sections for a new building have the dimensions (in meters) and shape shown in the figure.

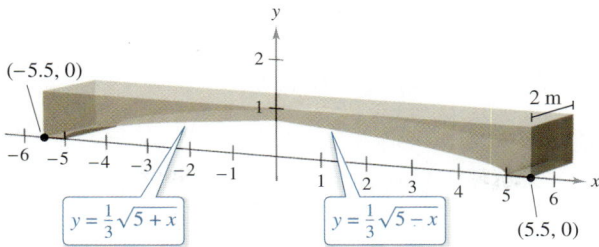

(−5.5, 0)

2 m

$y = \frac{1}{3}\sqrt{5 + x}$ $y = \frac{1}{3}\sqrt{5 - x}$

(5.5, 0)

(a) Find the area of the face of the section superimposed on the rectangular coordinate system.

(b) Find the volume of concrete in one of the sections by multiplying the area in part (a) by 2 meters.

(c) One cubic meter of concrete weighs 5000 pounds. Find the weight of the section.

80. Mechanical Design   The surface of a machine part is the region between the graphs of $y_1 = |x|$ and $y_2 = 0.08x^2 + k$ (see figure).

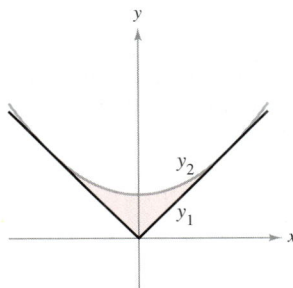

(a) Find k where the parabola is tangent to the graph of y_1.

(b) Find the area of the surface of the machine part.

81. Area   Find the area between the graph of $y = \sin x$ and the line segment joining the points $(0, 0)$ and $\left(\frac{7\pi}{6}, -\frac{1}{2}\right)$, as shown in the figure.

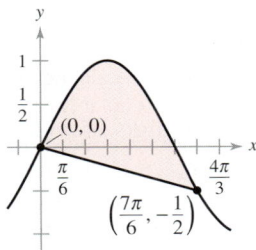

82. Area   Let $a > 0$ and $b > 0$. Show that the area of the ellipse $\frac{x^2}{a^2} + \frac{y^2}{b^2} = 1$ is πab (see figure).

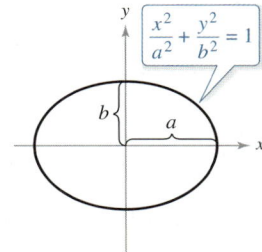

True or False?   **In Exercises 83–86, determine whether the statement is true or false. If it is false, explain why or give an example that shows it is false.**

83. If the area of the region bounded by the graphs of f and g is 1, then the area of the region bounded by the graphs of $h(x) = f(x) + C$ and $k(x) = g(x) + C$ is also 1.

84. If

$$\int_a^b [f(x) - g(x)]\, dx = A$$

then

$$\int_a^b [g(x) - f(x)]\, dx = -A.$$

85. If the graphs of f and g intersect midway between $x = a$ and $x = b$, then

$$\int_a^b [f(x) - g(x)]\, dx = 0.$$

86. The line

$$y = \left(1 - \sqrt[3]{0.5}\right)x$$

divides the region under the curve

$$f(x) = x(1 - x)$$

on $[0, 1]$ into two regions of equal area.

PUTNAM EXAM CHALLENGE

87. The horizontal line $y = c$ intersects the curve $y = 2x - 3x^3$ in the first quadrant as shown in the figure. Find c so that the areas of the two shaded regions are equal.

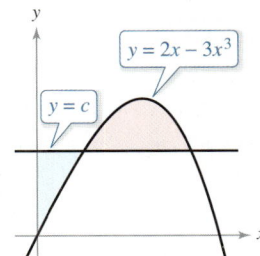

This problem was composed by the Committee on the Putnam Prize Competition.
© The Mathematical Association of America. All rights reserved.

7.2 Volume: The Disk Method

■ Find the volume of a solid of revolution using the disk method.
■ Find the volume of a solid of revolution using the washer method.
■ Find the volume of a solid with known cross sections.

The Disk Method

You have already learned that area is only one of the *many* applications of the definite integral. Another important application is its use in finding the volume of a three-dimensional solid. In this section, you will study a particular type of three-dimensional solid—one whose cross sections are similar. Solids of revolution are used commonly in engineering and manufacturing. Some examples are axles, funnels, pills, bottles, and pistons, as shown in Figure 7.12.

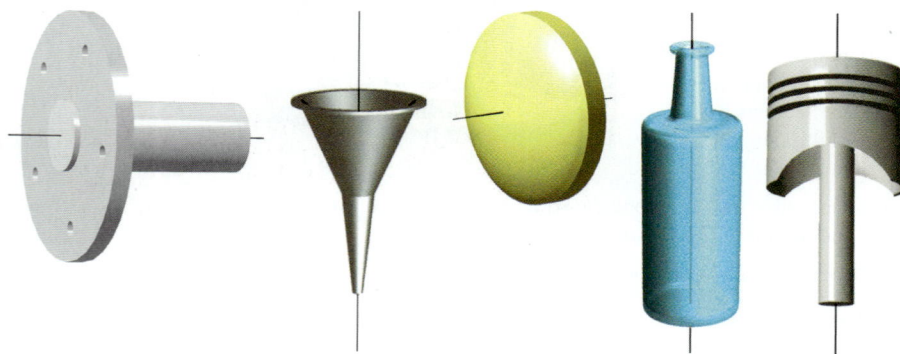

Solids of revolution
Figure 7.12

When a region in the plane is revolved about a line, the resulting solid is a **solid of revolution,** and the line is called the **axis of revolution.** The simplest such solid is a right circular cylinder or **disk,** which is formed by revolving a rectangle about an axis adjacent to one side of the rectangle, as shown in Figure 7.13. The volume of such a disk is

Volume of disk = (area of disk)(width of disk)
$$= \pi R^2 w$$

where R is the radius of the disk and w is the width.

To see how to use the volume of a disk to find the volume of a general solid of revolution, consider a solid of revolution formed by revolving the plane region in Figure 7.14 about the indicated axis. To determine the volume of this solid, consider a representative rectangle in the plane region. When this rectangle is revolved about the axis of revolution, it generates a representative disk whose volume is

$$\Delta V = \pi R^2 \Delta x.$$

Approximating the volume of the solid by n such disks of width Δx and radius $R(x_i)$ produces

$$\text{Volume of solid} \approx \sum_{i=1}^{n} \pi [R(x_i)]^2 \Delta x$$
$$= \pi \sum_{i=1}^{n} [R(x_i)]^2 \Delta x.$$

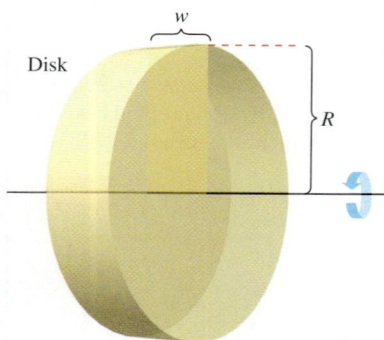

Volume of a disk: $\pi R^2 w$
Figure 7.13

Disk method
Figure 7.14

This approximation appears to become better and better as $\|\Delta\| \to 0$ ($n \to \infty$). So, you can define the volume of the solid as

$$\text{Volume of solid} = \lim_{\|\Delta\| \to 0} \pi \sum_{i=1}^{n} [R(x_i)]^2 \, \Delta x = \pi \int_{a}^{b} [R(x)]^2 \, dx.$$

Schematically, the disk method looks like this.

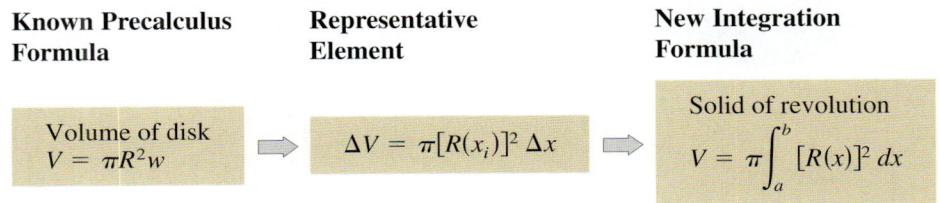

Known Precalculus Formula	**Representative Element**	**New Integration Formula**
Volume of disk $V = \pi R^2 w$	$\Delta V = \pi [R(x_i)]^2 \, \Delta x$	Solid of revolution $V = \pi \int_{a}^{b} [R(x)]^2 \, dx$

A similar formula can be derived when the axis of revolution is vertical.

THE DISK METHOD

To find the volume of a solid of revolution with the **disk method,** use one of the formulas below. (See Figure 7.15.)

Horizontal Axis of Revolution	**Vertical Axis of Revolution**
$\text{Volume} = V = \pi \displaystyle\int_{a}^{b} [R(x)]^2 \, dx$	$\text{Volume} = V = \pi \displaystyle\int_{c}^{d} [R(y)]^2 \, dy$

•• **REMARK** In Figure 7.15, note that you can determine the variable of integration by placing a representative rectangle in the plane region "perpendicular" to the axis of revolution. When the width of the rectangle is Δx, integrate with respect to x, and when the width of the rectangle is Δy, integrate with respect to y.

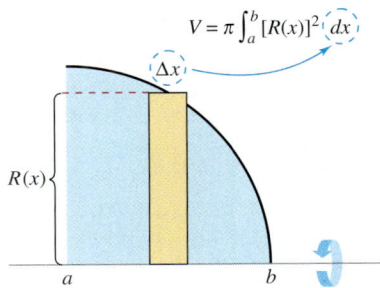

Horizontal axis of revolution
Vertical axis of revolution
Figure 7.15

The simplest application of the disk method involves a plane region bounded by the graph of f and the x-axis. When the axis of revolution is the x-axis, the radius $R(x)$ is simply $f(x)$.

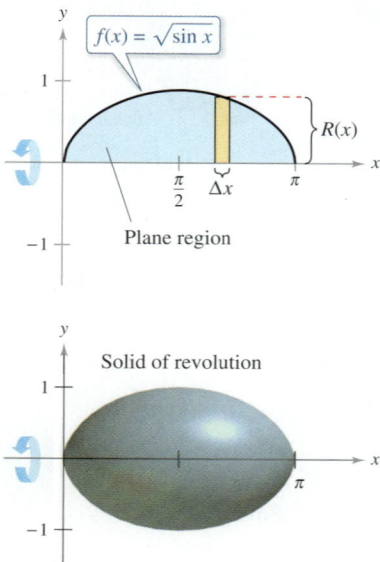

$f(x) = \sqrt{\sin x}$

$R(x)$

$\dfrac{\pi}{2}$ Δx π

Plane region

Solid of revolution

π

Figure 7.16

EXAMPLE 1 Using the Disk Method

Find the volume of the solid formed by revolving the region bounded by the graph of

$$f(x) = \sqrt{\sin x}$$

and the x-axis $(0 \le x \le \pi)$ about the x-axis.

Solution From the representative rectangle in the upper graph in Figure 7.16, you can see that the radius of this solid is

$$R(x) = f(x)$$
$$= \sqrt{\sin x}.$$

So, the volume of the solid of revolution is

$$V = \pi \int_a^b [R(x)]^2\, dx \qquad \text{Apply disk method.}$$

$$= \pi \int_0^\pi \left(\sqrt{\sin x}\right)^2 dx \qquad \text{Substitute } \sqrt{\sin x} \text{ for } R(x).$$

$$= \pi \int_0^\pi \sin x\, dx \qquad \text{Simplify.}$$

$$= \pi \Big[-\cos x \Big]_0^\pi \qquad \text{Integrate.}$$

$$= \pi(1 + 1)$$

$$= 2\pi.$$

EXAMPLE 2 Using a Line That Is Not a Coordinate Axis

Find the volume of the solid formed by revolving the region bounded by the graphs of

$$f(x) = 2 - x^2$$

and $g(x) = 1$ about the line $y = 1$, as shown in Figure 7.17.

Solution By equating $f(x)$ and $g(x)$, you can determine that the two graphs intersect when $x = \pm 1$. To find the radius, subtract $g(x)$ from $f(x)$.

$$R(x) = f(x) - g(x)$$
$$= (2 - x^2) - 1$$
$$= 1 - x^2$$

To find the volume, integrate between -1 and 1.

$$V = \pi \int_a^b [R(x)]^2\, dx \qquad \text{Apply disk method.}$$

$$= \pi \int_{-1}^1 (1 - x^2)^2\, dx \qquad \text{Substitute } 1 - x^2 \text{ for } R(x).$$

$$= \pi \int_{-1}^1 (1 - 2x^2 + x^4)\, dx \qquad \text{Simplify.}$$

$$= \pi \left[x - \frac{2x^3}{3} + \frac{x^5}{5} \right]_{-1}^1 \qquad \text{Integrate.}$$

$$= \frac{16\pi}{15}$$

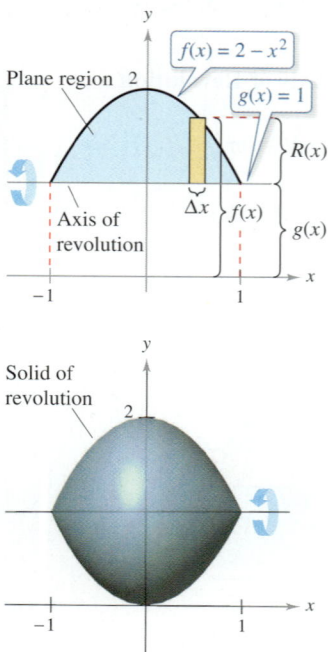

$f(x) = 2 - x^2$

Plane region 2

$g(x) = 1$

$R(x)$

Axis of revolution

Δx $f(x)$

$g(x)$

-1 1

Solid of revolution 2

-1 1

Figure 7.17

The Washer Method

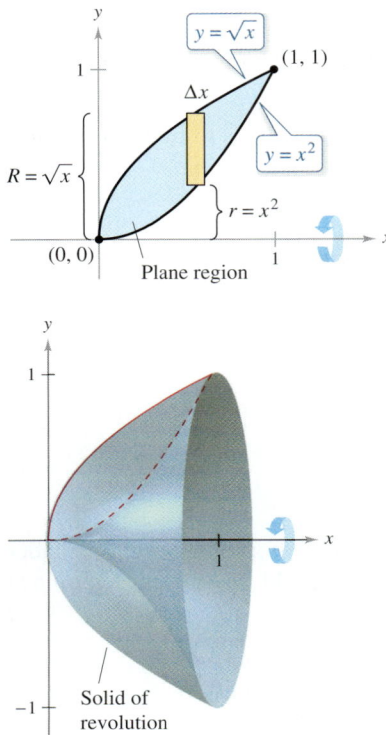

The disk method can be extended to cover solids of revolution with holes by replacing the representative disk with a representative **washer.** The washer is formed by revolving a rectangle about an axis, as shown in Figure 7.18. If r and R are the inner and outer radii of the washer and w is the width of the washer, then the volume is

$$\text{Volume of washer} = \pi(R^2 - r^2)w.$$

To see how this concept can be used to find the volume of a solid of revolution, consider a region bounded by an **outer radius** $R(x)$ and an **inner radius** $r(x)$, as shown in Figure 7.19. If the region is revolved about its axis of revolution, then the volume of the resulting solid is

$$V = \pi \int_a^b \left([R(x)]^2 - [r(x)]^2\right) dx. \qquad \text{Washer method}$$

Note that the integral involving the inner radius represents the volume of the hole and is *subtracted* from the integral involving the outer radius.

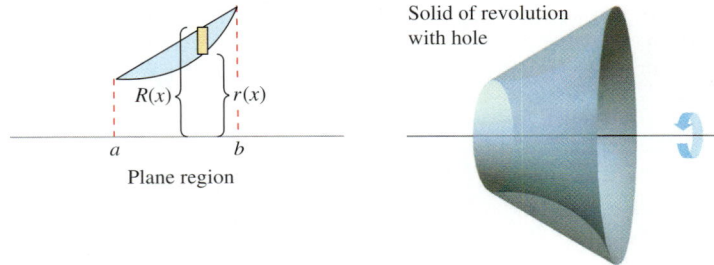

Figure 7.18

Figure 7.19

EXAMPLE 3   **Using the Washer Method**

Find the volume of the solid formed by revolving the region bounded by the graphs of

$$y = \sqrt{x} \quad \text{and} \quad y = x^2$$

about the x-axis, as shown in Figure 7.20.

Solution   In Figure 7.20, you can see that the outer and inner radii are as follows.

$$R(x) = \sqrt{x} \qquad \qquad \text{Outer radius}$$
$$r(x) = x^2 \qquad \qquad \text{Inner radius}$$

Integrating between 0 and 1 produces

$$V = \pi \int_a^b \left([R(x)]^2 - [r(x)]^2\right) dx \qquad \text{Apply washer method.}$$

$$= \pi \int_0^1 \left[\left(\sqrt{x}\right)^2 - (x^2)^2\right] dx \qquad \text{Substitute } \sqrt{x} \text{ for } R(x) \text{ and } x^2 \text{ for } r(x).$$

$$= \pi \int_0^1 (x - x^4) \, dx \qquad \text{Simplify.}$$

$$= \pi \left[\frac{x^2}{2} - \frac{x^5}{5}\right]_0^1 \qquad \text{Integrate.}$$

$$= \frac{3\pi}{10}.$$

Figure 7.20

In each example so far, the axis of revolution has been *horizontal* and you have integrated with respect to *x*. In the next example, the axis of revolution is *vertical* and you integrate with respect to *y*. In this example, you need two separate integrals to compute the volume.

EXAMPLE 4 **Integrating with Respect to y, Two-Integral Case**

Find the volume of the solid formed by revolving the region bounded by the graphs of

$$y = x^2 + 1, \quad y = 0, \quad x = 0, \quad \text{and} \quad x = 1$$

about the *y*-axis, as shown in Figure 7.21.

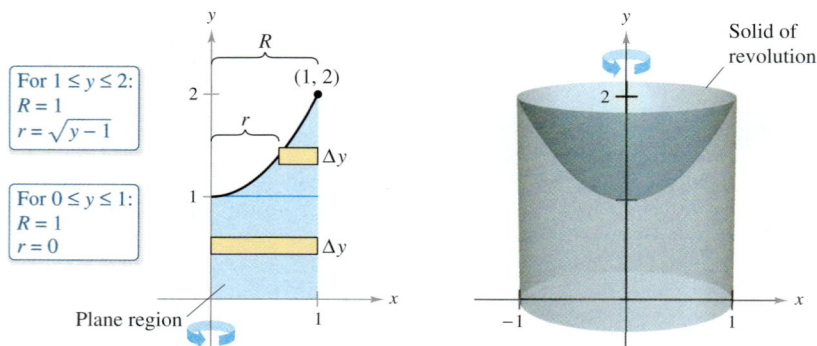

Figure 7.21

Solution For the region shown in Figure 7.21, the outer radius is simply $R = 1$. There is, however, no convenient formula that represents the inner radius. When $0 \le y \le 1$, $r = 0$, but when $1 \le y \le 2$, r is determined by the equation $y = x^2 + 1$, which implies that $r = \sqrt{y - 1}$.

$$r(y) = \begin{cases} 0, & 0 \le y \le 1 \\ \sqrt{y - 1}, & 1 \le y \le 2 \end{cases}$$

Using this definition of the inner radius, you can use two integrals to find the volume.

$$V = \pi \int_0^1 (1^2 - 0^2)\, dy + \pi \int_1^2 \left[1^2 - \left(\sqrt{y - 1}\right)^2\right] dy \qquad \text{Apply washer method.}$$

$$= \pi \int_0^1 1\, dy + \pi \int_1^2 (2 - y)\, dy \qquad \text{Simplify.}$$

$$= \pi \Big[y\Big]_0^1 + \pi \left[2y - \frac{y^2}{2}\right]_1^2 \qquad \text{Integrate.}$$

$$= \pi + \pi\left(4 - 2 - 2 + \frac{1}{2}\right)$$

$$= \frac{3\pi}{2}$$

Note that the first integral $\pi \int_0^1 1\, dy$ represents the volume of a right circular cylinder of radius 1 and height 1. This portion of the volume could have been determined without using calculus. ∎

▷ **TECHNOLOGY** Some graphing utilities have the capability of generating (or have built-in software capable of generating) a solid of revolution. If you have access to such a utility, use it to graph some of the solids of revolution described in this section. For instance, the solid in Example 4 might appear like that shown in Figure 7.22.

Generated by Mathematica

Figure 7.22

Solid of revolution

(a)

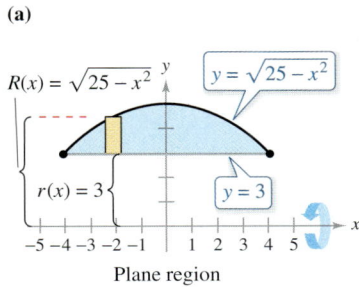

Plane region

(b)

Figure 7.23

Manufacturing

: • • • ▷ *See LarsonCalculus.com for an interactive version of this type of example.*

A manufacturer drills a hole through the center of a metal sphere of radius 5 inches, as shown in Figure 7.23(a). The hole has a radius of 3 inches. What is the volume of the resulting metal ring?

Solution   You can imagine the ring to be generated by a segment of the circle whose equation is $x^2 + y^2 = 25$, as shown in Figure 7.23(b). Because the radius of the hole is 3 inches, you can let $y = 3$ and solve the equation $x^2 + y^2 = 25$ to determine that the limits of integration are $x = \pm 4$. So, the inner and outer radii are $r(x) = 3$ and $R(x) = \sqrt{25 - x^2}$, and the volume is

$$
\begin{aligned}
V &= \pi \int_a^b ([R(x)]^2 - [r(x)]^2)\, dx \\
&= \pi \int_{-4}^{4} \left[\left(\sqrt{25 - x^2} \right)^2 - (3)^2 \right] dx \\
&= \pi \int_{-4}^{4} (16 - x^2)\, dx \\
&= \pi \left[16x - \frac{x^3}{3} \right]_{-4}^{4} \\
&= \frac{256\pi}{3} \text{ cubic inches.}
\end{aligned}
$$

Solids with Known Cross Sections

With the disk method, you can find the volume of a solid having a circular cross section whose area is $A = \pi R^2$. This method can be generalized to solids of any shape, as long as you know a formula for the area of an arbitrary cross section. Some common cross sections are squares, rectangles, triangles, semicircles, and trapezoids.

VOLUMES OF SOLIDS WITH KNOWN CROSS SECTIONS

1. For cross sections of area $A(x)$ taken perpendicular to the x-axis,

$$ \text{Volume} = \int_a^b A(x)\, dx. \qquad \text{See Figure 7.24(a).} $$

2. For cross sections of area $A(y)$ taken perpendicular to the y-axis,

$$ \text{Volume} = \int_c^d A(y)\, dy. \qquad \text{See Figure 7.24(b).} $$

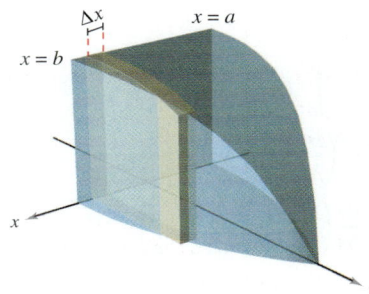

(a) Cross sections perpendicular to x-axis

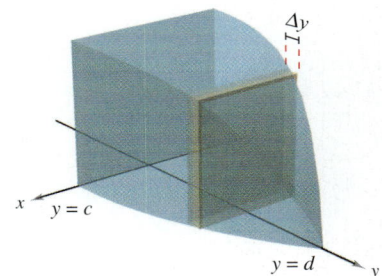

(b) Cross sections perpendicular to y-axis

Figure 7.24

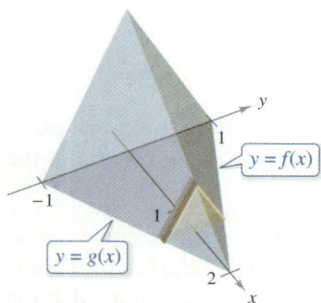

Cross sections are equilateral triangles.

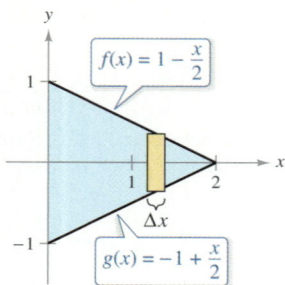

Triangular base in *xy*-plane
Figure 7.25

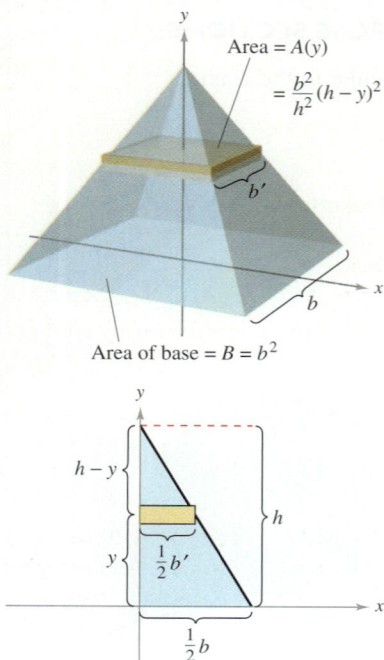

Area of base = $B = b^2$

Figure 7.26

EXAMPLE 6 **Triangular Cross Sections**

Find the volume of the solid shown in Figure 7.25. The base of the solid is the region bounded by the lines

$$f(x) = 1 - \frac{x}{2}, \quad g(x) = -1 + \frac{x}{2}, \quad \text{and} \quad x = 0.$$

The cross sections perpendicular to the *x*-axis are equilateral triangles.

Solution The base and area of each triangular cross section are as follows.

$$\text{Base} = \left(1 - \frac{x}{2}\right) - \left(-1 + \frac{x}{2}\right) = 2 - x \qquad \textcolor{red}{\text{Length of base}}$$

$$\text{Area} = \frac{\sqrt{3}}{4}(\text{base})^2 \qquad \textcolor{red}{\text{Area of equilateral triangle}}$$

$$A(x) = \frac{\sqrt{3}}{4}(2 - x)^2 \qquad \textcolor{red}{\text{Area of cross section}}$$

Because *x* ranges from 0 to 2, the volume of the solid is

$$V = \int_a^b A(x)\, dx = \int_0^2 \frac{\sqrt{3}}{4}(2 - x)^2\, dx = -\frac{\sqrt{3}}{4}\left[\frac{(2 - x)^3}{3}\right]_0^2 = \frac{2\sqrt{3}}{3}.$$

EXAMPLE 7 **An Application to Geometry**

Prove that the volume of a pyramid with a square base is

$$V = \frac{1}{3}hB$$

where *h* is the height of the pyramid and *B* is the area of the base.

Solution As shown in Figure 7.26, you can intersect the pyramid with a plane parallel to the base at height *y* to form a square cross section whose sides are of length *b′*. Using similar triangles, you can show that

$$\frac{b'}{b} = \frac{h - y}{h} \quad \text{or} \quad b' = \frac{b}{h}(h - y)$$

where *b* is the length of the sides of the base of the pyramid. So,

$$A(y) = (b')^2 = \frac{b^2}{h^2}(h - y)^2.$$

Integrating between 0 and *h* produces

$$V = \int_0^h A(y)\, dy$$

$$= \int_0^h \frac{b^2}{h^2}(h - y)^2\, dy$$

$$= \frac{b^2}{h^2}\int_0^h (h - y)^2\, dy$$

$$= -\left(\frac{b^2}{h^2}\right)\left[\frac{(h - y)^3}{3}\right]_0^h$$

$$= \frac{b^2}{h^2}\left(\frac{h^3}{3}\right)$$

$$= \frac{1}{3}hB. \qquad \textcolor{red}{B = b^2}$$

7.2 Exercises

See **CalcChat.com** for tutorial help and worked-out solutions to odd-numbered exercises.

Finding the Volume of a Solid In Exercises 1–6, set up and evaluate the integral that gives the volume of the solid formed by revolving the region about the *x*-axis.

1. $y = -x + 1$

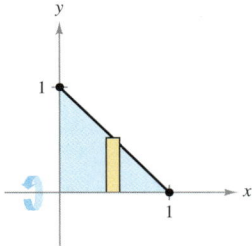

2. $y = 4 - x^2$

3. $y = \sqrt{x}$

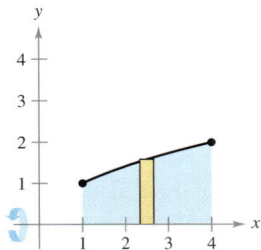

4. $y = \sqrt{9 - x^2}$

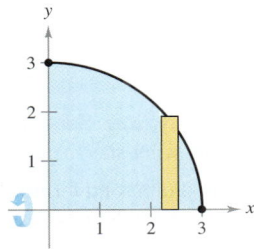

5. $y = x^2, \ y = x^5$

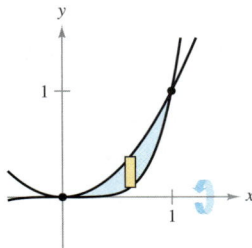

6. $y = 2, \ y = 4 - \dfrac{x^2}{4}$

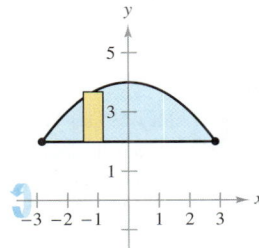

Finding the Volume of a Solid In Exercises 7–10, set up and evaluate the integral that gives the volume of the solid formed by revolving the region about the *y*-axis.

7. $y = x^2$

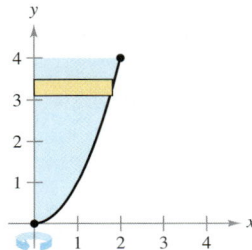

8. $y = \sqrt{16 - x^2}$

9. $y = x^{2/3}$

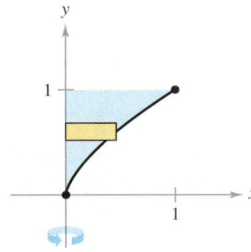

10. $x = -y^2 + 4y$

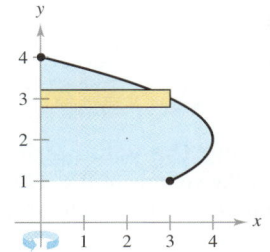

Finding the Volume of a Solid In Exercises 11–14, find the volumes of the solids generated by revolving the region bounded by the graphs of the equations about the given lines.

11. $y = \sqrt{x}, \ y = 0, \ x = 3$

  (a) the *x*-axis   (b) the *y*-axis

  (c) the line $x = 3$ (d) the line $x = 6$

12. $y = 2x^2, \ y = 0, \ x = 2$

  (a) the *y*-axis   (b) the *x*-axis

  (c) the line $y = 8$ (d) the line $x = 2$

13. $y = x^2, \ y = 4x - x^2$

  (a) the *x*-axis   (b) the line $y = 6$

14. $y = 4 + 2x - x^2, \ y = 4 - x$

  (a) the *x*-axis   (b) the line $y = 1$

Finding the Volume of a Solid In Exercises 15–18, find the volume of the solid generated by revolving the region bounded by the graphs of the equations about the line $y = 4$.

15. $y = x, \ y = 3, \ x = 0$ **16.** $y = \frac{1}{2}x^3, \ y = 4, \ x = 0$

17. $y = \dfrac{3}{1 + x}, \ y = 0, \ x = 0, \ x = 3$

18. $y = \sec x, \ y = 0, \ 0 \le x \le \dfrac{\pi}{3}$

Finding the Volume of a Solid In Exercises 19–22, find the volume of the solid generated by revolving the region bounded by the graphs of the equations about the line $x = 5$.

19. $y = x, \ y = 0, \ y = 4, \ x = 5$

20. $y = 3 - x, \ y = 0, \ y = 2, \ x = 0$

21. $x = y^2, \ x = 4$

22. $xy = 3, \ y = 1, \ y = 4, \ x = 5$

Finding the Volume of a Solid In Exercises 23–30, find the volume of the solid generated by revolving the region bounded by the graphs of the equations about the *x*-axis.

23. $y = \dfrac{1}{\sqrt{x + 1}}, \ y = 0, \ x = 0, \ x = 4$

24. $y = x\sqrt{4 - x^2}, \ y = 0$

25. $y = \dfrac{1}{x}$, $y = 0$, $x = 1$, $x = 3$

26. $y = \dfrac{2}{x+1}$, $y = 0$, $x = 0$, $x = 6$

27. $y = e^{-x}$, $y = 0$, $x = 0$, $y = 1$

28. $y = e^{x/4}$, $y = 0$, $x = 0$, $x = 6$

29. $y = x^2 + 1$, $y = -x^2 + 2x + 5$, $x = 0$, $x = 3$

30. $y = \sqrt{x}$, $y = -\frac{1}{2}x + 4$, $x = 0$, $x = 8$

Finding the Volume of a Solid In Exercises 31 and 32, find the volume of the solid generated by revolving the region bounded by the graphs of the equations about the y-axis.

31. $y = 3(2 - x)$, $y = 0$, $x = 0$

32. $y = 9 - x^2$, $y = 0$, $x = 2$, $x = 3$

Finding the Volume of a Solid In Exercises 33–36, find the volume of the solid generated by revolving the region bounded by the graphs of the equations about the x-axis. Verify your results using the integration capabilities of a graphing utility.

33. $y = \sin x$, $y = 0$, $x = 0$, $x = \pi$

34. $y = \cos 2x$, $y = 0$, $x = 0$, $x = \dfrac{\pi}{4}$

35. $y = e^{x-1}$, $y = 0$, $x = 1$, $x = 2$

36. $y = e^{x/2} + e^{-x/2}$, $y = 0$, $x = -1$, $x = 2$

Finding the Volume of a Solid In Exercises 37–40, use the integration capabilities of a graphing utility to approximate the volume of the solid generated by revolving the region bounded by the graphs of the equations about the x-axis.

37. $y = e^{-x^2}$, $y = 0$, $x = 0$, $x = 2$

38. $y = \ln x$, $y = 0$, $x = 1$, $x = 3$

39. $y = 2\arctan(0.2x)$, $y = 0$, $x = 0$, $x = 5$

40. $y = \sqrt{2x}$, $y = x^2$

Finding the Volume of a Solid In Exercises 41–48, find the volume generated by rotating the given region about the specified line.

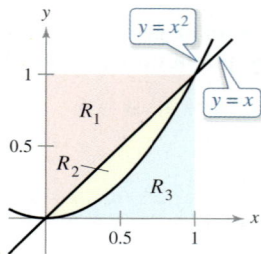

41. R_1 about $x = 0$
42. R_1 about $x = 1$
43. R_2 about $y = 0$
44. R_2 about $y = 1$
45. R_3 about $x = 0$
46. R_3 about $x = 1$
47. R_2 about $x = 0$
48. R_2 about $x = 1$

WRITING ABOUT CONCEPTS

Describing a Solid In Exercises 49 and 50, the integral represents the volume of a solid. Describe the solid.

49. $\pi \displaystyle\int_0^{\pi/2} \sin^2 x\, dx$

50. $\pi \displaystyle\int_2^4 y^4\, dy$

51. **Comparing Volumes** A region bounded by the parabola $y = 4x - x^2$ and the x-axis is revolved about the x-axis. A second region bounded by the parabola $y = 4 - x^2$ and the x-axis is revolved about the x-axis. Without integrating, how do the volumes of the two solids compare? Explain.

52. **Comparing Volumes** The region in the figure is revolved about the indicated axes and line. Order the volumes of the resulting solids from least to greatest. Explain your reasoning.
(a) x-axis
(b) y-axis
(c) $x = 3$

53. **Analyzing Statements** Discuss the validity of the following statements.
(a) For a solid formed by rotating the region under a graph about the x-axis, the cross sections perpendicular to the x-axis are circular disks.
(b) For a solid formed by rotating the region between two graphs about the x-axis, the cross sections perpendicular to the x-axis are circular disks.

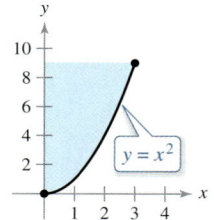

54. **HOW DO YOU SEE IT?** Use the graph to match the integral for the volume with the axis of rotation.

(a) $V = \pi \displaystyle\int_0^b (a^2 - [f(y)]^2)\, dy$ (i) x-axis

(b) $V = \pi \displaystyle\int_0^a (b^2 - [b - f(x)]^2)\, dx$ (ii) y-axis

(c) $V = \pi \displaystyle\int_0^a [f(x)]^2\, dx$ (iii) $x = a$

(d) $V = \pi \displaystyle\int_0^b [a - f(y)]^2\, dy$ (iv) $y = b$

Dividing a Solid   In Exercises 55 and 56, consider the solid formed by revolving the region bounded by $y = \sqrt{x}$, $y = 0$, and $x = 4$ about the x-axis.

55. Find the value of x in the interval $[0, 4]$ that divides the solid into two parts of equal volume.

56. Find the values of x in the interval $[0, 4]$ that divide the solid into three parts of equal volume.

57. Manufacturing   A manufacturer drills a hole through the center of a metal sphere of radius R. The hole has a radius r. Find the volume of the resulting ring.

58. Manufacturing   For the metal sphere in Exercise 57, let $R = 6$. What value of r will produce a ring whose volume is exactly half the volume of the sphere?

59. Volume of a Cone   Use the disk method to verify that the volume of a right circular cone is $\frac{1}{3}\pi r^2 h$, where r is the radius of the base and h is the height.

60. Volume of a Sphere   Use the disk method to verify that the volume of a sphere is $\frac{4}{3}\pi r^3$, where r is the radius.

61. Using a Cone   A cone of height H with a base of radius r is cut by a plane parallel to and h units above the base, where $h < H$. Find the volume of the solid (frustum of a cone) below the plane.

62. Using a Sphere   A sphere of radius r is cut by a plane h units above the equator, where $h < r$. Find the volume of the solid (spherical segment) above the plane.

63. Volume of a Fuel Tank   A tank on the wing of a jet aircraft is formed by revolving the region bounded by the graph of $y = \frac{1}{8}x^2\sqrt{2 - x}$ and the x-axis ($0 \leq x \leq 2$) about the x-axis, where x and y are measured in meters. Use a graphing utility to graph the function and find the volume of the tank.

64. Volume of a Lab Glass   A glass container can be modeled by revolving the graph of

$$y = \begin{cases} \sqrt{0.1x^3 - 2.2x^2 + 10.9x + 22.2}, & 0 \leq x \leq 11.5 \\ 2.95, & 11.5 < x \leq 15 \end{cases}$$

about the x-axis, where x and y are measured in centimeters. Use a graphing utility to graph the function and find the volume of the container.

65. Finding Volumes of a Solid   Find the volumes of the solids (see figures) generated if the upper half of the ellipse $9x^2 + 25y^2 = 225$ is revolved about (a) the x-axis to form a prolate spheroid (shaped like a football), and (b) the y-axis to form an oblate spheroid (shaped like half of a candy).

Figure for 65(a)

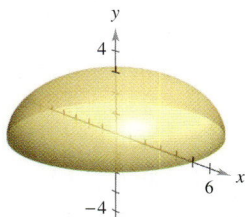
Figure for 65(b)

66. Water Tower

A tank on a water tower is a sphere of radius 50 feet. Determine the depths of the water when the tank is filled to one-fourth and three-fourths of its total capacity. (*Note:* Use the *zero* or *root* feature of a graphing utility after evaluating the definite integral.)

67. Minimum Volume   The arc of $y = 4 - (x^2/4)$ on the interval $[0, 4]$ is revolved about the line $y = b$ (see figure).

(a) Find the volume of the resulting solid as a function of b.

(b) Use a graphing utility to graph the function in part (a), and use the graph to approximate the value of b that minimizes the volume of the solid.

(c) Use calculus to find the value of b that minimizes the volume of the solid, and compare the result with the answer to part (b).

Figure for 67

Figure for 68

68. Modeling Data   A draftsman is asked to determine the amount of material required to produce a machine part (see figure). The diameters d of the part at equally spaced points x are listed in the table. The measurements are listed in centimeters.

x	0	1	2	3	4	5
d	4.2	3.8	4.2	4.7	5.2	5.7

x	6	7	8	9	10
d	5.8	5.4	4.9	4.4	4.6

(a) Use these data with Simpson's Rule to approximate the volume of the part.

(b) Use the regression capabilities of a graphing utility to find a fourth-degree polynomial through the points representing the radius of the solid. Plot the data and graph the model.

(c) Use a graphing utility to approximate the definite integral yielding the volume of the part. Compare the result with the answer to part (a).

69. Think About It Match each integral with the solid whose volume it represents, and give the dimensions of each solid.

(a) Right circular cylinder   (b) Ellipsoid

(c) Sphere   (d) Right circular cone   (e) Torus

(i) $\pi \int_0^h \left(\dfrac{rx}{h}\right)^2 dx$ (ii) $\pi \int_0^h r^2 \, dx$

(iii) $\pi \int_{-r}^r \left(\sqrt{r^2 - x^2}\right)^2 dx$

(iv) $\pi \int_{-b}^b \left(a\sqrt{1 - \dfrac{x^2}{b^2}}\right)^2 dx$

(v) $\pi \int_{-r}^r \left[\left(R + \sqrt{r^2 - x^2}\right)^2 - \left(R - \sqrt{r^2 - x^2}\right)^2\right] dx$

70. Cavalieri's Theorem Prove that if two solids have equal altitudes and all plane sections parallel to their bases and at equal distances from their bases have equal areas, then the solids have the same volume (see figure).

Area of R_1 = area of R_2

71. Using Cross Sections Find the volumes of the solids whose bases are bounded by the graphs of $y = x + 1$ and $y = x^2 - 1$, with the indicated cross sections taken perpendicular to the x-axis.

(a) Squares   (b) Rectangles of height 1

72. Using Cross Sections Find the volumes of the solids whose bases are bounded by the circle $x^2 + y^2 = 4$, with the indicated cross sections taken perpendicular to the x-axis.

(a) Squares   (b) Equilateral triangles

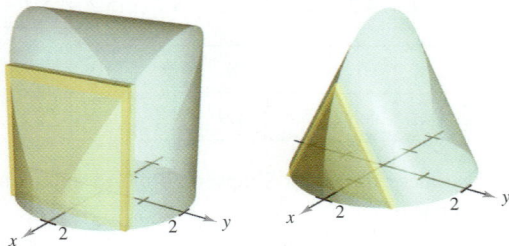

(c) Semicircles   (d) Isosceles right triangles

73. Using Cross Sections Find the volume of the solid of intersection (the solid common to both) of the two right circular cylinders of radius r whose axes meet at right angles (see figure).

Two intersecting cylinders   Solid of intersection

■ **FOR FURTHER INFORMATION** For more information on this problem, see the article "Estimating the Volumes of Solid Figures with Curved Surfaces" by Donald Cohen in *Mathematics Teacher*. To view this article, go to *MathArticles.com*.

74. Using Cross Sections The solid shown in the figure has cross sections bounded by the graph of $|x|^a + |y|^a = 1$, where $1 \le a \le 2$.

(a) Describe the cross section when $a = 1$ and $a = 2$.

(b) Describe a procedure for approximating the volume of the solid.

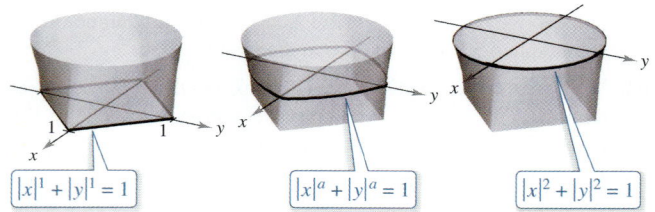

$\boxed{|x|^1 + |y|^1 = 1}$ $\boxed{|x|^a + |y|^a = 1}$ $\boxed{|x|^2 + |y|^2 = 1}$

75. Volume of a Wedge Two planes cut a right circular cylinder to form a wedge. One plane is perpendicular to the axis of the cylinder and the second makes an angle of θ degrees with the first (see figure).

(a) Find the volume of the wedge if $\theta = 45°$.

(b) Find the volume of the wedge for an arbitrary angle θ. Assuming that the cylinder has sufficient length, how does the volume of the wedge change as θ increases from $0°$ to $90°$?

    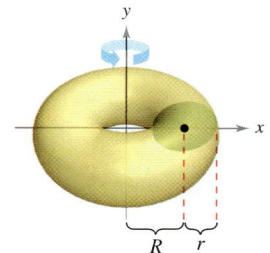

Figure for 75   Figure for 76

76. Volume of a Torus

(a) Show that the volume of the torus shown in the figure is given by the integral $8\pi R \int_0^r \sqrt{r^2 - y^2} \, dy$, where $R > r > 0$.

(b) Find the volume of the torus.

7.3   Volume: The Shell Method

■ Find the volume of a solid of revolution using the shell method.
■ Compare the uses of the disk method and the shell method.

The Shell Method

In this section, you will study an alternative method for finding the volume of a solid of revolution. This method is called the **shell method** because it uses cylindrical shells. A comparison of the advantages of the disk and shell methods is given later in this section.

To begin, consider a representative rectangle as shown in Figure 7.27, where w is the width of the rectangle, h is the height of the rectangle, and p is the distance between the axis of revolution and the *center* of the rectangle. When this rectangle is revolved about its axis of revolution, it forms a cylindrical shell (or tube) of thickness w. To find the volume of this shell, consider two cylinders. The radius of the larger cylinder corresponds to the outer radius of the shell, and the radius of the smaller cylinder corresponds to the inner radius of the shell. Because p is the average radius of the shell, you know the outer radius is

$$p + \frac{w}{2} \qquad \text{Outer radius}$$

and the inner radius is

$$p - \frac{w}{2}. \qquad \text{Inner radius}$$

So, the volume of the shell is

$$\text{Volume of shell} = (\text{volume of cylinder}) - (\text{volume of hole})$$
$$= \pi\left(p + \frac{w}{2}\right)^2 h - \pi\left(p - \frac{w}{2}\right)^2 h$$
$$= 2\pi p h w$$
$$= 2\pi(\text{average radius})(\text{height})(\text{thickness}).$$

Figure 7.27

You can use this formula to find the volume of a solid of revolution. For instance, the plane region in Figure 7.28 is revolved about a line to form the indicated solid. Consider a horizontal rectangle of width Δy. As the plane region is revolved about a line parallel to the x-axis, the rectangle generates a representative shell whose volume is

$$\Delta V = 2\pi[p(y)h(y)]\,\Delta y.$$

You can approximate the volume of the solid by n such shells of thickness Δy, height $h(y_i)$, and average radius $p(y_i)$.

$$\text{Volume of solid} \approx \sum_{i=1}^{n} 2\pi[p(y_i)h(y_i)]\Delta y = 2\pi \sum_{i=1}^{n}[p(y_i)h(y_i)]\Delta y$$

This approximation appears to become better and better as $\|\Delta\| \to 0$ $(n \to \infty)$. So, the volume of the solid is

$$\text{Volume of solid} = \lim_{\|\Delta\|\to 0} 2\pi \sum_{i=1}^{n}[p(y_i)h(y_i)]\Delta y$$
$$= 2\pi \int_{c}^{d}[p(y)h(y)]\,dy.$$

Figure 7.28

THE SHELL METHOD

To find the volume of a solid of revolution with the **shell method,** use one of the formulas below. (See Figure 7.29.)

Horizontal Axis of Revolution   **Vertical Axis of Revolution**

$$\text{Volume} = V = 2\pi \int_{c}^{d} p(y)h(y)\, dy \qquad \text{Volume} = V = 2\pi \int_{a}^{b} p(x)h(x)\, dx$$

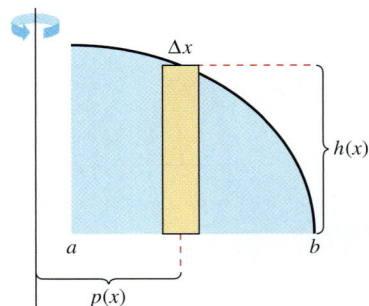

Horizontal axis of revolution   Vertical axis of revolution

Figure 7.29

EXAMPLE 1   **Using the Shell Method to Find Volume**

Find the volume of the solid of revolution formed by revolving the region bounded by

$$y = x - x^3$$

and the x-axis $(0 \le x \le 1)$ about the y-axis.

Solution   Because the axis of revolution is vertical, use a vertical representative rectangle, as shown in Figure 7.30. The width Δx indicates that x is the variable of integration. The distance from the center of the rectangle to the axis of revolution is $p(x) = x$, and the height of the rectangle is

$$h(x) = x - x^3.$$

Because x ranges from 0 to 1, apply the shell method to find the volume of the solid.

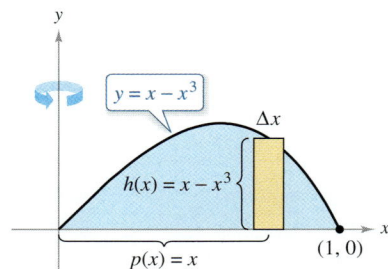

Figure 7.30

$$V = 2\pi \int_{a}^{b} p(x)h(x)\, dx$$

$$= 2\pi \int_{0}^{1} x(x - x^3)\, dx$$

$$= 2\pi \int_{0}^{1} (-x^4 + x^2)\, dx \qquad \text{Simplify.}$$

$$= 2\pi \left[-\frac{x^5}{5} + \frac{x^3}{3} \right]_{0}^{1} \qquad \text{Integrate.}$$

$$= 2\pi \left(-\frac{1}{5} + \frac{1}{3} \right)$$

$$= \frac{4\pi}{15}$$

EXAMPLE 2	**Using the Shell Method to Find Volume**

Find the volume of the solid of revolution formed by revolving the region bounded by the graph of

$$x = e^{-y^2}$$

and the y-axis $(0 \leq y \leq 1)$ about the x-axis.

Solution   Because the axis of revolution is horizontal, use a horizontal representative rectangle, as shown in Figure 7.31. The width Δy indicates that y is the variable of integration. The distance from the center of the rectangle to the axis of revolution is $p(y) = y$, and the height of the rectangle is $h(y) = e^{-y^2}$. Because y ranges from 0 to 1, the volume of the solid is

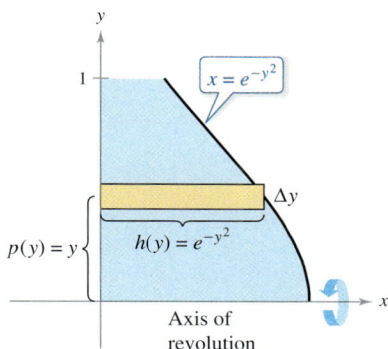

$$V = 2\pi \int_c^d p(y)h(y)\, dy \qquad \text{Apply shell method.}$$

$$= 2\pi \int_0^1 y e^{-y^2}\, dy$$

$$= -\pi \left[e^{-y^2} \right]_0^1 \qquad \text{Integrate.}$$

$$= \pi \left(1 - \frac{1}{e} \right)$$

$$\approx 1.986.$$

Figure 7.31

Exploration

To see the advantage of using the shell method in Example 2, solve the equation $x = e^{-y^2}$ for y.

$$y = \begin{cases} 1, & 0 \leq x \leq 1/e \\ \sqrt{-\ln x}, & 1/e < x \leq 1 \end{cases}$$

Then use this equation to find the volume using the disk method.

Comparison of Disk and Shell Methods

The disk and shell methods can be distinguished as follows. For the disk method, the representative rectangle is always *perpendicular* to the axis of revolution, whereas for the shell method, the representative rectangle is always *parallel* to the axis of revolution, as shown in Figure 7.32.

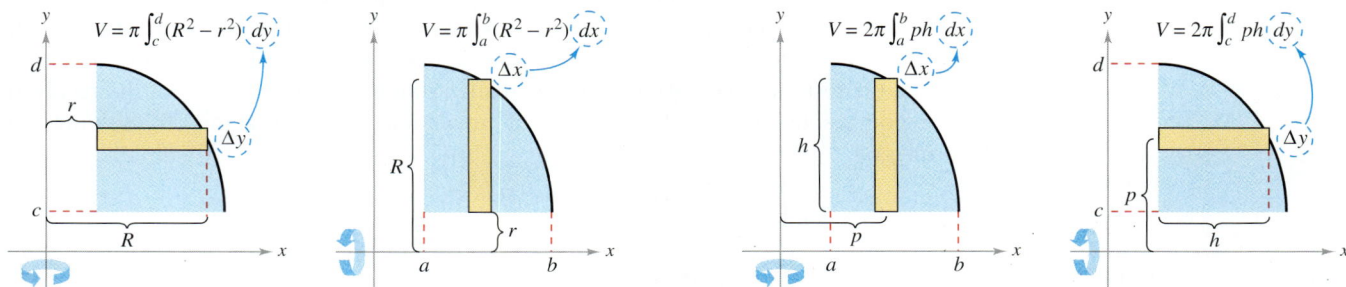

Vertical axis of revolution	Horizontal axis of revolution	Vertical axis of revolution	Horizontal axis of revolution

Disk method: Representative rectangle is perpendicular to the axis of revolution.

Shell method: Representative rectangle is parallel to the axis of revolution.

Figure 7.32

Often, one method is more convenient to use than the other. The next example illustrates a case in which the shell method is preferable.

EXAMPLE 3 **Shell Method Preferable**

$\cdots\cdots\triangleright$ *See LarsonCalculus.com for an interactive version of this type of example.*

Find the volume of the solid formed by revolving the region bounded by the graphs of

$$y = x^2 + 1, \quad y = 0, \quad x = 0, \quad \text{and} \quad x = 1$$

about the y-axis.

Solution In Example 4 in Section 7.2, you saw that the washer method requires two integrals to determine the volume of this solid. See Figure 7.33(a).

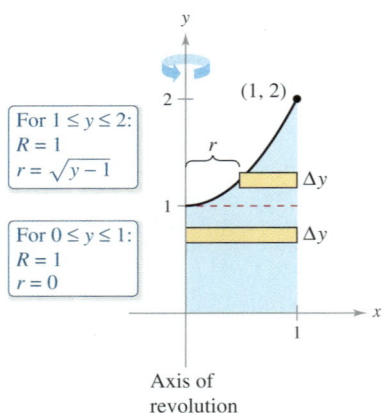

$$V = \pi \int_0^1 (1^2 - 0^2)\, dy + \pi \int_1^2 \left[1^2 - \left(\sqrt{y-1}\right)^2\right] dy \qquad \text{Apply washer method.}$$

$$= \pi \int_0^1 1\, dy + \pi \int_1^2 (2 - y)\, dy \qquad\qquad \text{Simplify.}$$

$$= \pi\left[y\right]_0^1 + \pi\left[2y - \frac{y^2}{2}\right]_1^2 \qquad\qquad \text{Integrate.}$$

$$= \pi + \pi\left(4 - 2 - 2 + \frac{1}{2}\right)$$

$$= \frac{3\pi}{2}$$

In Figure 7.33(b), you can see that the shell method requires only one integral to find the volume.

$$V = 2\pi \int_a^b p(x)h(x)\, dx \qquad\qquad \text{Apply shell method.}$$

$$= 2\pi \int_0^1 x(x^2 + 1)\, dx$$

$$= 2\pi\left[\frac{x^4}{4} + \frac{x^2}{2}\right]_0^1 \qquad\qquad \text{Integrate.}$$

$$= 2\pi\left(\frac{3}{4}\right)$$

$$= \frac{3\pi}{2}$$

Consider the solid formed by revolving the region in Example 3 about the vertical line $x = 1$. Would the resulting solid of revolution have a greater volume or a smaller volume than the solid in Example 3? Without integrating, you should be able to reason that the resulting solid would have a smaller volume because "more" of the revolved region would be closer to the axis of revolution. To confirm this, try solving the integral

$$V = 2\pi \int_0^1 (1 - x)(x^2 + 1)\, dx \qquad\qquad p(x) = 1 - x$$

which gives the volume of the solid.

■ **FOR FURTHER INFORMATION** To learn more about the disk and shell methods, see the article "The Disk and Shell Method" by Charles A. Cable in *The American Mathematical Monthly.* To view this article, go to *MathArticles.com.*

For $1 \le y \le 2$:
$R = 1$
$r = \sqrt{y - 1}$

For $0 \le y \le 1$:
$R = 1$
$r = 0$

(1, 2)

r

Δy

Δy

Axis of revolution

(a) Disk method

(1, 2)

$p(x) = x$

$h(x) = x^2 + 1$

Δx

Axis of revolution

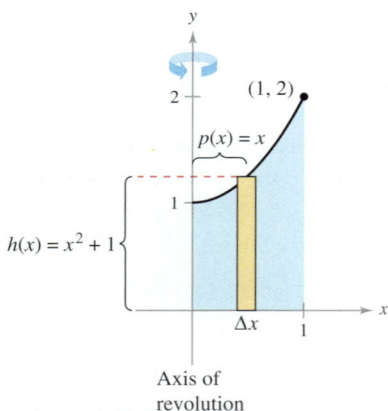

(b) Shell method

Figure 7.33

Figure 7.34

Disk method
Figure 7.35

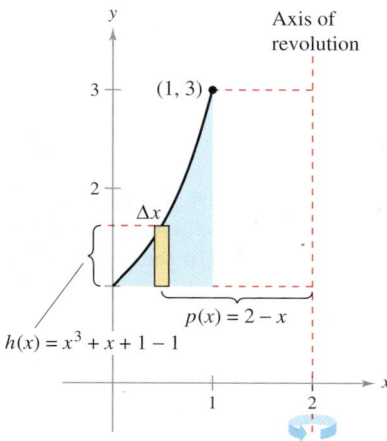

Figure 7.36

| EXAMPLE 4 | **Volume of a Pontoon** |

A pontoon is to be made in the shape shown in Figure 7.34. The pontoon is designed by rotating the graph of

$$y = 1 - \frac{x^2}{16}, \quad -4 \leq x \leq 4$$

about the x-axis, where x and y are measured in feet. Find the volume of the pontoon.

Solution Refer to Figure 7.35 and use the disk method as shown.

$$V = \pi \int_{-4}^{4} \left(1 - \frac{x^2}{16}\right)^2 dx \qquad \text{Apply disk method.}$$

$$= \pi \int_{-4}^{4} \left(1 - \frac{x^2}{8} + \frac{x^4}{256}\right) dx \qquad \text{Simplify.}$$

$$= \pi \left[x - \frac{x^3}{24} + \frac{x^5}{1280}\right]_{-4}^{4} \qquad \text{Integrate.}$$

$$= \frac{64\pi}{15}$$

$$\approx 13.4 \text{ cubic feet}$$

To use the shell method in Example 4, you would have to solve for x in terms of y in the equation

$$y = 1 - \frac{x^2}{16}$$

and then evaluate an integral that requires a u-substitution.

Sometimes, solving for x is very difficult (or even impossible). In such cases, you must use a vertical rectangle (of width Δx), thus making x the variable of integration. The position (horizontal or vertical) of the axis of revolution then determines the method to be used. This is shown in Example 5.

| EXAMPLE 5 | **Shell Method Necessary** |

Find the volume of the solid formed by revolving the region bounded by the graphs of $y = x^3 + x + 1$, $y = 1$, and $x = 1$ about the line $x = 2$, as shown in Figure 7.36.

Solution In the equation $y = x^3 + x + 1$, you cannot easily solve for x in terms of y. (See the discussion at the end of Section 3.8.) Therefore, the variable of integration must be x, and you should choose a vertical representative rectangle. Because the rectangle is parallel to the axis of revolution, use the shell method.

$$V = 2\pi \int_a^b p(x)h(x)\, dx \qquad \text{Apply shell method.}$$

$$= 2\pi \int_0^1 (2-x)(x^3 + x + 1 - 1)\, dx$$

$$= 2\pi \int_0^1 (-x^4 + 2x^3 - x^2 + 2x)\, dx \qquad \text{Simplify.}$$

$$= 2\pi \left[-\frac{x^5}{5} + \frac{x^4}{2} - \frac{x^3}{3} + x^2\right]_0^1 \qquad \text{Integrate.}$$

$$= 2\pi\left(-\frac{1}{5} + \frac{1}{2} - \frac{1}{3} + 1\right)$$

$$= \frac{29\pi}{15}$$

7.3 Exercises

See **CalcChat.com** for tutorial help and worked-out solutions to odd-numbered exercises.

Finding the Volume of a Solid In Exercises 1–14, use the shell method to set up and evaluate the integral that gives the volume of the solid generated by revolving the plane region about the *y*-axis.

1. $y = x$

2. $y = 1 - x$

3. $y = \sqrt{x}$

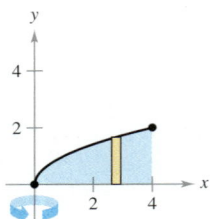

4. $y = \frac{1}{2}x^2 + 1$

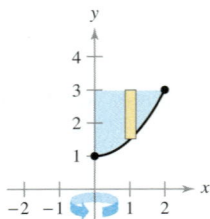

5. $y = \frac{1}{4}x^2$, $y = 0$, $x = 4$ **6.** $y = \frac{1}{2}x^3$, $y = 0$, $x = 3$

7. $y = x^2$, $y = 4x - x^2$ **8.** $y = 9 - x^2$, $y = 0$

9. $y = 4x - x^2$, $x = 0$, $y = 4$

10. $y = x^{3/2}$, $y = 8$, $x = 0$

11. $y = \sqrt{x - 2}$, $y = 0$, $x = 4$

12. $y = -x^2 + 1$, $y = 0$

13. $y = \frac{1}{\sqrt{2\pi}} e^{-x^2/2}$, $y = 0$, $x = 0$, $x = 1$

14. $y = \begin{cases} \dfrac{\sin x}{x}, & x > 0 \\ 1, & x = 0 \end{cases}$, $y = 0$, $x = 0$, $x = \pi$

Finding the Volume of a Solid In Exercises 15–22, use the shell method to set up and evaluate the integral that gives the volume of the solid generated by revolving the plane region about the *x*-axis.

15. $y = x$

16. $y = 1 - x$

17. $y = \dfrac{1}{x}$

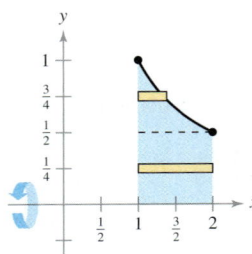

18. $x + y^2 = 16$

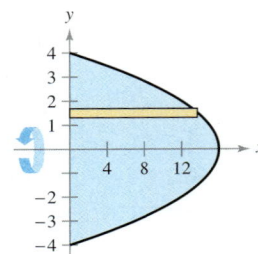

19. $y = x^3$, $x = 0$, $y = 8$ **20.** $y = 4x^2$, $x = 0$, $y = 4$

21. $x + y = 4$, $y = x$, $y = 0$

22. $y = \sqrt{x + 2}$, $y = x$, $y = 0$

Finding the Volume of a Solid In Exercises 23–26, use the shell method to find the volume of the solid generated by revolving the plane region about the given line.

23. $y = 2x - x^2$, $y = 0$, about the line $x = 4$

24. $y = \sqrt{x}$, $y = 0$, $x = 4$, about the line $x = 6$

25. $y = x^2$, $y = 4x - x^2$, about the line $x = 4$

26. $y = \frac{1}{3}x^3$, $y = 6x - x^2$, about the line $x = 3$

Choosing a Method In Exercises 27 and 28, decide whether it is more convenient to use the disk method or the shell method to find the volume of the solid of revolution. Explain your reasoning. (Do not find the volume.)

27. $(y - 2)^2 = 4 - x$

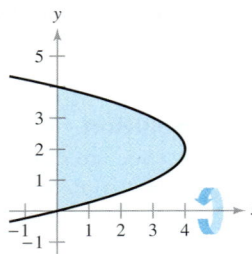

28. $y = 4 - e^x$

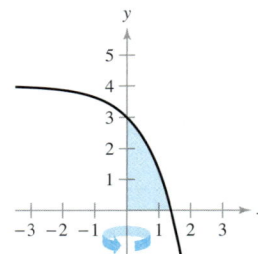

Choosing a Method In Exercises 29–32, use the disk method *or* the shell method to find the volumes of the solids generated by revolving the region bounded by the graphs of the equations about the given lines.

29. $y = x^3$, $y = 0$, $x = 2$

(a) the *x*-axis   (b) the *y*-axis   (c) the line $x = 4$

30. $y = \dfrac{10}{x^2}$, $y = 0$, $x = 1$, $x = 5$

(a) the *x*-axis   (b) the *y*-axis   (c) the line $y = 10$

31. $x^{1/2} + y^{1/2} = a^{1/2}$, $x = 0$, $y = 0$

    (a) the x-axis   (b) the y-axis   (c) the line $x = a$

32. $x^{2/3} + y^{2/3} = a^{2/3}$, $a > 0$ (hypocycloid)

    (a) the x-axis   (b) the y-axis

Finding the Volume of a Solid In Exercises 33–36, (a) use a graphing utility to graph the plane region bounded by the graphs of the equations, and (b) use the integration capabilities of the graphing utility to approximate the volume of the solid generated by revolving the region about the y-axis.

33. $x^{4/3} + y^{4/3} = 1$, $x = 0$, $y = 0$,   first quadrant

34. $y = \sqrt{1 - x^3}$, $y = 0$, $x = 0$

35. $y = \sqrt[3]{(x - 2)^2(x - 6)^2}$, $y = 0$, $x = 2$, $x = 6$

36. $y = \dfrac{2}{1 + e^{1/x}}$, $y = 0$, $x = 1$, $x = 3$

WRITING ABOUT CONCEPTS

37. Representative Rectangles Consider a solid that is generated by revolving a plane region about the y-axis. Describe the position of a representative rectangle when using (a) the shell method and (b) the disk method to find the volume of the solid.

38. Describing Cylindrical Shells Consider the plane region bounded by the graphs of

$$y = k, \quad y = 0, \quad x = 0, \quad \text{and} \quad x = b$$

where $k > 0$ and $b > 0$. What are the heights and radii of the cylinders generated when this region is revolved about (a) the x-axis and (b) the y-axis?

Comparing Integrals In Exercises 39 and 40, give a geometric argument that explains why the integrals have equal values.

39. $\pi \displaystyle\int_1^5 (x - 1)\, dx = 2\pi \displaystyle\int_0^2 y[5 - (y^2 + 1)]\, dy$

40. $\pi \displaystyle\int_0^2 [16 - (2y)^2]\, dy = 2\pi \displaystyle\int_0^4 x\left(\dfrac{x}{2}\right) dx$

41. Comparing Volumes The region in the figure is revolved about the indicated axes and line. Order the volumes of the resulting solids from least to greatest. Explain your reasoning.

    (a) x-axis   (b) y-axis   (c) $x = 4$

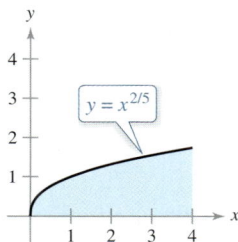

$y = x^{2/5}$

42. HOW DO YOU SEE IT? Use the graph to answer the following.

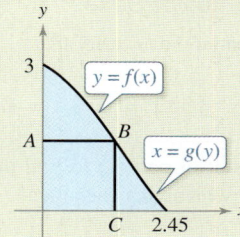

    (a) Describe the figure generated by revolving segment AB about the y-axis.

    (b) Describe the figure generated by revolving segment BC about the y-axis.

    (c) Assume the curve in the figure can be described as $y = f(x)$ or $x = g(y)$. A solid is generated by revolving the region bounded by the curve, $y = 0$, and $x = 0$ about the y-axis. Set up integrals to find the volume of this solid using the disk method and the shell method. (Do not integrate.)

Analyzing an Integral In Exercises 43–46, the integral represents the volume of a solid of revolution. Identify (a) the plane region that is revolved and (b) the axis of revolution.

43. $2\pi \displaystyle\int_0^2 x^3\, dx$

44. $2\pi \displaystyle\int_0^1 y - y^{3/2}\, dy$

45. $2\pi \displaystyle\int_0^6 (y + 2)\sqrt{6 - y}\, dy$

46. $2\pi \displaystyle\int_0^1 (4 - x)e^x\, dx$

47. Machine Part A solid is generated by revolving the region bounded by $y = \frac{1}{2}x^2$ and $y = 2$ about the y-axis. A hole, centered along the axis of revolution, is drilled through this solid so that one-fourth of the volume is removed. Find the diameter of the hole.

48. Machine Part A solid is generated by revolving the region bounded by $y = \sqrt{9 - x^2}$ and $y = 0$ about the y-axis. A hole, centered along the axis of revolution, is drilled through this solid so that one-third of the volume is removed. Find the diameter of the hole.

49. Volume of a Torus A torus is formed by revolving the region bounded by the circle $x^2 + y^2 = 1$ about the line $x = 2$ (see figure). Find the volume of this "doughnut-shaped" solid. (*Hint:* The integral $\int_{-1}^1 \sqrt{1 - x^2}\, dx$ represents the area of a semicircle.)

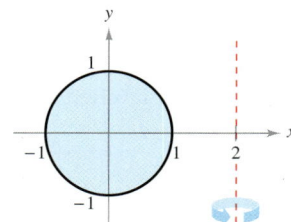

50. Volume of a Torus Repeat Exercise 49 for a torus formed by revolving the region bounded by the circle $x^2 + y^2 = r^2$ about the line $x = R$, where $r < R$.

51. Finding Volumes of Solids

(a) Use differentiation to verify that

$$\int x \sin x \, dx = \sin x - x \cos x + C.$$

(b) Use the result of part (a) to find the volume of the solid generated by revolving each plane region about the y-axis.

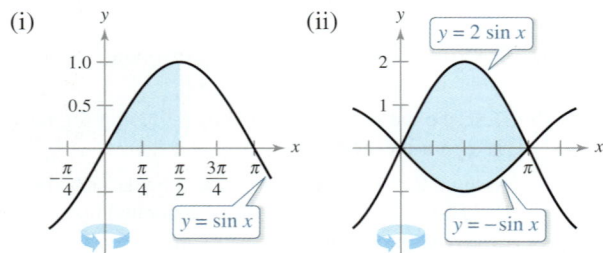

(i)

(ii)

52. Finding Volumes of Solids

(a) Use differentiation to verify that

$$\int x \cos x \, dx = \cos x + x \sin x + C.$$

(b) Use the result of part (a) to find the volume of the solid generated by revolving each plane region about the y-axis. (*Hint:* Begin by approximating the points of intersection.)

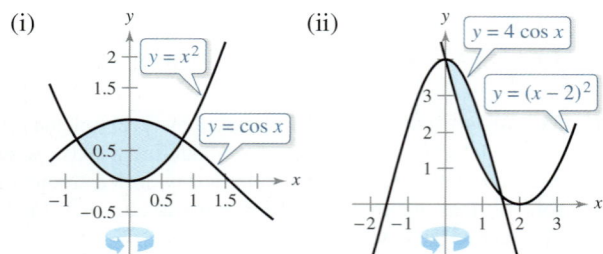

(i)

(ii)

53. Volume of a Segment of a Sphere Let a sphere of radius r be cut by a plane, thereby forming a segment of height h. Show that the volume of this segment is

$$\frac{1}{3}\pi h^2(3r - h).$$

54. Volume of an Ellipsoid Consider the plane region bounded by the graph of

$$\left(\frac{x}{a}\right)^2 + \left(\frac{y}{b}\right)^2 = 1$$

where $a > 0$ and $b > 0$. Show that the volume of the ellipsoid formed when this region is revolved about the y-axis is

$$\frac{4}{3}\pi a^2 b.$$

What is the volume when the region is revolved about the x-axis?

55. Exploration Consider the region bounded by the graphs of $y = ax^n$, $y = ab^n$, and $x = 0$ (see figure).

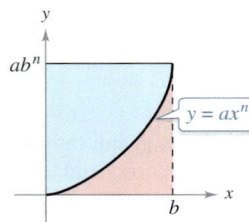

(a) Find the ratio $R_1(n)$ of the area of the region to the area of the circumscribed rectangle.

(b) Find $\lim_{n \to \infty} R_1(n)$ and compare the result with the area of the circumscribed rectangle.

(c) Find the volume of the solid of revolution formed by revolving the region about the y-axis. Find the ratio $R_2(n)$ of this volume to the volume of the circumscribed right circular cylinder.

(d) Find $\lim_{n \to \infty} R_2(n)$ and compare the result with the volume of the circumscribed cylinder.

(e) Use the results of parts (b) and (d) to make a conjecture about the shape of the graph of $y = ax^n$ $(0 \le x \le b)$ as $n \to \infty$.

56. Think About It Match each integral with the solid whose volume it represents, and give the dimensions of each solid.

(a) Right circular cone   (b) Torus   (c) Sphere

(d) Right circular cylinder   (e) Ellipsoid

(i) $2\pi \displaystyle\int_0^r hx \, dx$ (ii) $2\pi \displaystyle\int_0^r hx\left(1 - \frac{x}{r}\right) dx$

(iii) $2\pi \displaystyle\int_0^r 2x\sqrt{r^2 - x^2} \, dx$ (iv) $2\pi \displaystyle\int_0^b 2ax\sqrt{1 - \frac{x^2}{b^2}} \, dx$

(v) $2\pi \displaystyle\int_{-r}^r (R - x)\left(2\sqrt{r^2 - x^2}\right) dx$

57. Volume of a Storage Shed A storage shed has a circular base of diameter 80 feet. Starting at the center, the interior height is measured every 10 feet and recorded in the table (see figure).

x	Height
0	50
10	45
20	40
30	20
40	0

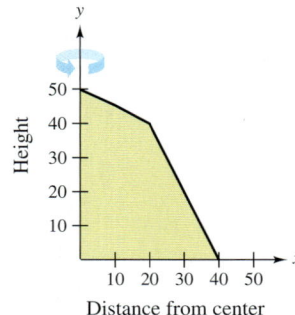

(a) Use Simpson's Rule to approximate the volume of the shed.

(b) Note that the roof line consists of two line segments. Find the equations of the line segments and use integration to find the volume of the shed.

58. Modeling Data A pond is approximately circular, with a diameter of 400 feet. Starting at the center, the depth of the water is measured every 25 feet and recorded in the table (see figure).

x	0	25	50
Depth	20	19	19

x	75	100	125
Depth	17	15	14

x	150	175	200
Depth	10	6	0

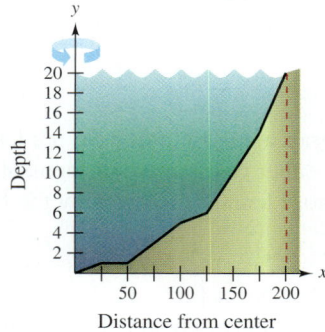

Distance from center

(a) Use Simpson's Rule to approximate the volume of water in the pond.

(b) Use the regression capabilities of a graphing utility to find a quadratic model for the depths recorded in the table. Use the graphing utility to plot the depths and graph the model.

(c) Use the integration capabilities of a graphing utility and the model in part (b) to approximate the volume of water in the pond.

(d) Use the result of part (c) to approximate the number of gallons of water in the pond. (*Hint:* 1 cubic foot of water is approximately 7.48 gallons.)

59. Equal Volumes Let V_1 and V_2 be the volumes of the solids that result when the plane region bounded by $y = 1/x$, $y = 0$, $x = \frac{1}{4}$, and $x = c$ (where $c > \frac{1}{4}$) is revolved about the x-axis and the y-axis, respectively. Find the value of c for which $V_1 = V_2$.

60. Volume of a Segment of a Paraboloid The region bounded by $y = r^2 - x^2$, $y = 0$, and $x = 0$ is revolved about the y-axis to form a paraboloid. A hole, centered along the axis of revolution, is drilled through this solid. The hole has a radius k, $0 < k < r$. Find the volume of the resulting ring (a) by integrating with respect to x and (b) by integrating with respect to y.

61. Finding Volumes of Solids Consider the graph of $y^2 = x(4 - x)^2$ (see figure). Find the volumes of the solids that are generated when the loop of this graph is revolved about (a) the x-axis, (b) the y-axis, and (c) the line $x = 4$.

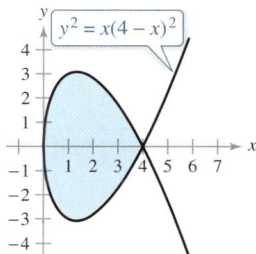

$y^2 = x(4 - x)^2$

Saturn

The Oblateness of Saturn Saturn is the most oblate of the planets in our solar system. Its equatorial radius is 60,268 kilometers and its polar radius is 54,364 kilometers. The color-enhanced photograph of Saturn was taken by Voyager 1. In the photograph, the oblateness of Saturn is clearly visible.

(a) Find the ratio of the volumes of the sphere and the oblate ellipsoid shown below.

(b) If a planet were spherical and had the same volume as Saturn, what would its radius be?

Computer model of "spherical Saturn," whose equatorial radius is equal to its polar radius. The equation of the cross section passing through the pole is

$$x^2 + y^2 = 60{,}268^2.$$

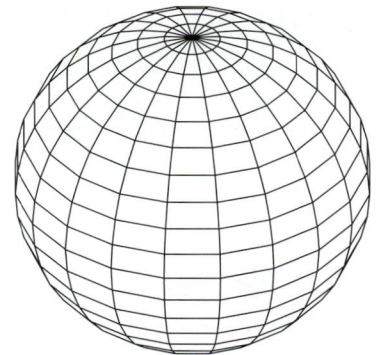

Computer model of "oblate Saturn," whose equatorial radius is greater than its polar radius. The equation of the cross section passing through the pole is

$$\frac{x^2}{60{,}268^2} + \frac{y^2}{54{,}364^2} = 1.$$

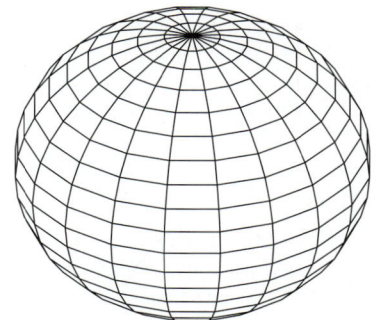

NASA

7.4 Arc Length and Surfaces of Revolution

■ Find the arc length of a smooth curve.
■ Find the area of a surface of revolution.

Arc Length

In this section, definite integrals are used to find the arc lengths of curves and the areas of surfaces of revolution. In either case, an arc (a segment of a curve) is approximated by straight line segments whose lengths are given by the familiar Distance Formula

$$d = \sqrt{(x_2 - x_1)^2 + (y_2 - y_1)^2}.$$

A **rectifiable** curve is one that has a finite arc length. You will see that a sufficient condition for the graph of a function f to be rectifiable between $(a, f(a))$ and $(b, f(b))$ is that f' be continuous on $[a, b]$. Such a function is **continuously differentiable** on $[a, b]$, and its graph on the interval $[a, b]$ is a **smooth curve.**

Consider a function $y = f(x)$ that is continuously differentiable on the interval $[a, b]$. You can approximate the graph of f by n line segments whose endpoints are determined by the partition

$$a = x_0 < x_1 < x_2 < \cdots < x_n = b$$

as shown in Figure 7.37. By letting $\Delta x_i = x_i - x_{i-1}$ and $\Delta y_i = y_i - y_{i-1}$, you can approximate the length of the graph by

$$
\begin{aligned}
s &\approx \sum_{i=1}^{n} \sqrt{(x_i - x_{i-1})^2 + (y_i - y_{i-1})^2} \\
&= \sum_{i=1}^{n} \sqrt{(\Delta x_i)^2 + (\Delta y_i)^2} \\
&= \sum_{i=1}^{n} \sqrt{(\Delta x_i)^2 + \left(\frac{\Delta y_i}{\Delta x_i}\right)^2 (\Delta x_i)^2} \\
&= \sum_{i=1}^{n} \sqrt{1 + \left(\frac{\Delta y_i}{\Delta x_i}\right)^2} \, (\Delta x_i).
\end{aligned}
$$

This approximation appears to become better and better as $\|\Delta\| \to 0$ ($n \to \infty$). So, the length of the graph is

$$s = \lim_{\|\Delta\| \to 0} \sum_{i=1}^{n} \sqrt{1 + \left(\frac{\Delta y_i}{\Delta x_i}\right)^2} \, (\Delta x_i).$$

Because $f'(x)$ exists for each x in (x_{i-1}, x_i), the Mean Value Theorem guarantees the existence of c_i in (x_{i-1}, x_i) such that

$$f(x_i) - f(x_{i-1}) = f'(c_i)(x_i - x_{i-1})$$
$$\frac{f(x_i) - f(x_{i-1})}{x_i - x_{i-1}} = f'(c_i)$$
$$\frac{\Delta y_i}{\Delta x_i} = f'(c_i).$$

Because f' is continuous on $[a, b]$, it follows that $\sqrt{1 + [f'(x)]^2}$ is also continuous (and therefore integrable) on $[a, b]$, which implies that

$$
\begin{aligned}
s &= \lim_{\|\Delta\| \to 0} \sum_{i=1}^{n} \sqrt{1 + [f'(c_i)]^2} \, (\Delta x_i) \\
&= \int_a^b \sqrt{1 + [f'(x)]^2} \, dx
\end{aligned}
$$

where s is called the **arc length** of f between a and b.

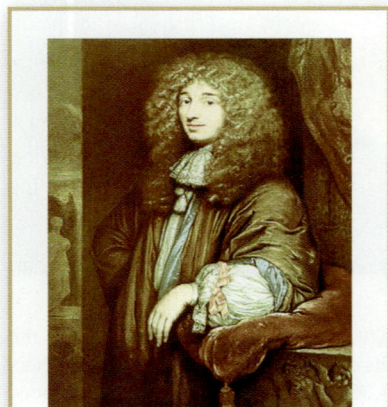

CHRISTIAN HUYGENS (1629–1695)

The Dutch mathematician Christian Huygens, who invented the pendulum clock, and James Gregory (1638–1675), a Scottish mathematician, both made early contributions to the problem of finding the length of a rectifiable curve.

See LarsonCalculus.com to read more of this biography.

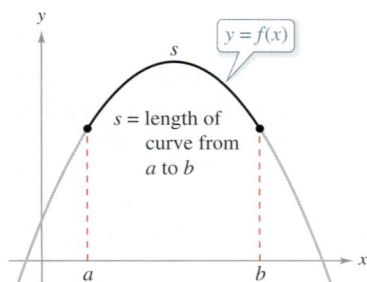

Figure 7.37

Bettmann/Corbis

Definition of Arc Length

Let the function $y = f(x)$ represent a smooth curve on the interval $[a, b]$. The **arc length** of f between a and b is

$$s = \int_a^b \sqrt{1 + [f'(x)]^2}\, dx.$$

Similarly, for a smooth curve $x = g(y)$, the **arc length** of g between c and d is

$$s = \int_c^d \sqrt{1 + [g'(y)]^2}\, dy.$$

■ **FOR FURTHER INFORMATION**   To see how arc length can be used to define trigonometric functions, see the article "Trigonometry Requires Calculus, Not Vice Versa" by Yves Nievergelt in *UMAP Modules*.

Because the definition of arc length can be applied to a linear function, you can check to see that this new definition agrees with the standard Distance Formula for the length of a line segment. This is shown in Example 1.

EXAMPLE 1   **The Length of a Line Segment**

Find the arc length from (x_1, y_1) to (x_2, y_2) on the graph of

$$f(x) = mx + b$$

as shown in Figure 7.38.

Solution   Because

$$m = f'(x) = \frac{y_2 - y_1}{x_2 - x_1}$$

it follows that

$$
\begin{aligned}
s &= \int_{x_1}^{x_2} \sqrt{1 + [f'(x)]^2}\, dx && \text{\color{red}Formula for arc length}\\[2mm]
&= \int_{x_1}^{x_2} \sqrt{1 + \left(\frac{y_2 - y_1}{x_2 - x_1}\right)^2}\, dx \\[2mm]
&= \sqrt{\frac{(x_2 - x_1)^2 + (y_2 - y_1)^2}{(x_2 - x_1)^2}} \, (x)\Bigg]_{x_1}^{x_2} && \text{\color{red}Integrate and simplify.}\\[2mm]
&= \sqrt{\frac{(x_2 - x_1)^2 + (y_2 - y_1)^2}{(x_2 - x_1)^2}} \, (x_2 - x_1) \\[2mm]
&= \sqrt{(x_2 - x_1)^2 + (y_2 - y_1)^2}
\end{aligned}
$$

which is the formula for the distance between two points in the plane.   ■

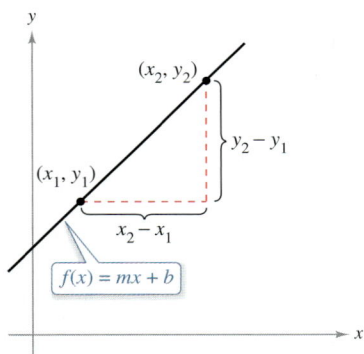

The formula for the arc length of the graph of f from (x_1, y_1) to (x_2, y_2) is the same as the standard Distance Formula.
Figure 7.38

▷ **TECHNOLOGY**   Definite integrals representing arc length often are very difficult to evaluate. In this section, a few examples are presented. In the next chapter, with more advanced integration techniques, you will be able to tackle more difficult arc length problems. In the meantime, remember that you can always use a numerical integration program to approximate an arc length. For instance, use the numerical integration feature of a graphing utility to approximate arc lengths in Examples 2 and 3.

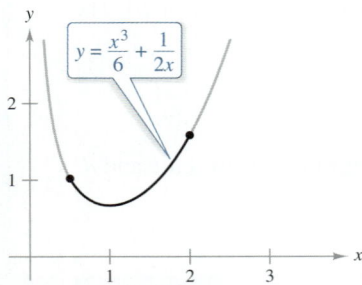

$y = \dfrac{x^3}{6} + \dfrac{1}{2x}$

The arc length of the graph of y on $\left[\frac{1}{2}, 2\right]$

Figure 7.39

EXAMPLE 2 **Finding Arc Length**

Find the arc length of the graph of

$$y = \frac{x^3}{6} + \frac{1}{2x}$$

on the interval $\left[\frac{1}{2}, 2\right]$, as shown in Figure 7.39.

Solution Using

$$\frac{dy}{dx} = \frac{3x^2}{6} - \frac{1}{2x^2} = \frac{1}{2}\left(x^2 - \frac{1}{x^2}\right)$$

yields an arc length of

$$
\begin{aligned}
s &= \int_a^b \sqrt{1 + \left(\frac{dy}{dx}\right)^2}\, dx && \text{Formula for arc length}\\
&= \int_{1/2}^{2} \sqrt{1 + \left[\frac{1}{2}\left(x^2 - \frac{1}{x^2}\right)\right]^2}\, dx \\
&= \int_{1/2}^{2} \sqrt{\frac{1}{4}\left(x^4 + 2 + \frac{1}{x^4}\right)}\, dx \\
&= \int_{1/2}^{2} \frac{1}{2}\left(x^2 + \frac{1}{x^2}\right) dx && \text{Simplify.}\\
&= \frac{1}{2}\left[\frac{x^3}{3} - \frac{1}{x}\right]_{1/2}^{2} && \text{Integrate.}\\
&= \frac{1}{2}\left(\frac{13}{6} + \frac{47}{24}\right) \\
&= \frac{33}{16}.
\end{aligned}
$$

EXAMPLE 3 **Finding Arc Length**

Find the arc length of the graph of $(y - 1)^3 = x^2$ on the interval $[0, 8]$, as shown in Figure 7.40.

Solution Begin by solving for x in terms of y: $x = \pm(y - 1)^{3/2}$. Choosing the positive value of x produces

$$\frac{dx}{dy} = \frac{3}{2}(y - 1)^{1/2}.$$

The x-interval $[0, 8]$ corresponds to the y-interval $[1, 5]$, and the arc length is

$$
\begin{aligned}
s &= \int_c^d \sqrt{1 + \left(\frac{dx}{dy}\right)^2}\, dy && \text{Formula for arc length}\\
&= \int_{1}^{5} \sqrt{1 + \left[\frac{3}{2}(y - 1)^{1/2}\right]^2}\, dy \\
&= \int_{1}^{5} \sqrt{\frac{9}{4}y - \frac{5}{4}}\, dy \\
&= \frac{1}{2}\int_{1}^{5} \sqrt{9y - 5}\, dy && \text{Simplify.}\\
&= \frac{1}{18}\left[\frac{(9y - 5)^{3/2}}{3/2}\right]_{1}^{5} && \text{Integrate.}\\
&= \frac{1}{27}(40^{3/2} - 4^{3/2}) \\
&\approx 9.073.
\end{aligned}
$$

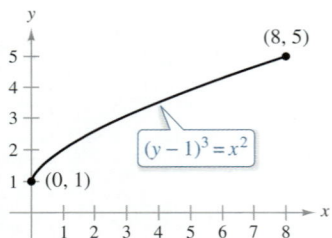

$(y - 1)^3 = x^2$

The arc length of the graph of y on $[0, 8]$

Figure 7.40

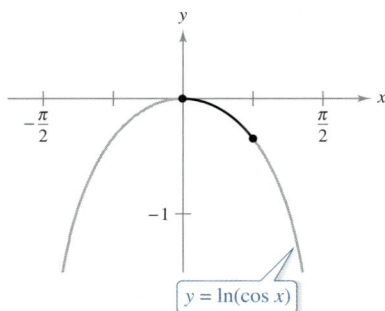

The arc length of the graph of y on $\left[0, \dfrac{\pi}{4}\right]$

Figure 7.41

EXAMPLE 4   **Finding Arc Length**

• • • ▷ *See LarsonCalculus.com for an interactive version of this type of example.*

Find the arc length of the graph of

$$y = \ln(\cos x)$$

from $x = 0$ to $x = \pi/4$, as shown in Figure 7.41.

Solution   Using

$$\frac{dy}{dx} = -\frac{\sin x}{\cos x} = -\tan x$$

yields an arc length of

$$
\begin{aligned}
s &= \int_a^b \sqrt{1 + \left(\frac{dy}{dx}\right)^2}\, dx && \text{Formula for arc length} \\
&= \int_0^{\pi/4} \sqrt{1 + \tan^2 x}\, dx \\
&= \int_0^{\pi/4} \sqrt{\sec^2 x}\, dx && \text{Trigonometric identity} \\
&= \int_0^{\pi/4} \sec x\, dx && \text{Simplify.} \\
&= \Big[\ln|\sec x + \tan x|\Big]_0^{\pi/4} && \text{Integrate.} \\
&= \ln\left(\sqrt{2} + 1\right) - \ln 1 \\
&\approx 0.881.
\end{aligned}
$$

EXAMPLE 5   **Length of a Cable**

An electric cable is hung between two towers that are 200 feet apart, as shown in Figure 7.42. The cable takes the shape of a catenary whose equation is

$$y = 75(e^{x/150} + e^{-x/150}) = 150 \cosh \frac{x}{150}.$$

Find the arc length of the cable between the two towers.

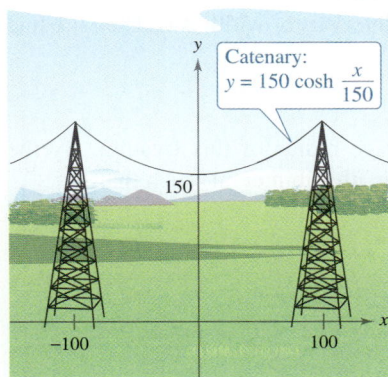

Figure 7.42

Solution   Because $y' = \frac{1}{2}(e^{x/150} - e^{-x/150})$, you can write

$$(y')^2 = \frac{1}{4}(e^{x/75} - 2 + e^{-x/75})$$

and

$$1 + (y')^2 = \frac{1}{4}(e^{x/75} + 2 + e^{-x/75}) = \left[\frac{1}{2}(e^{x/150} + e^{-x/150})\right]^2.$$

Therefore, the arc length of the cable is

$$
\begin{aligned}
s &= \int_a^b \sqrt{1 + (y')^2}\, dx && \text{Formula for arc length} \\
&= \frac{1}{2}\int_{-100}^{100} (e^{x/150} + e^{-x/150})\, dx \\
&= 75\left[e^{x/150} - e^{-x/150}\right]_{-100}^{100} && \text{Integrate.} \\
&= 150(e^{2/3} - e^{-2/3}) \\
&\approx 215 \text{ feet.}
\end{aligned}
$$

Area of a Surface of Revolution

In Sections 7.2 and 7.3, integration was used to calculate the volume of a solid of revolution. You will now look at a procedure for finding the area of a surface of revolution.

Definition of Surface of Revolution

When the graph of a continuous function is revolved about a line, the resulting surface is a **surface of revolution.**

The area of a surface of revolution is derived from the formula for the lateral surface area of the frustum of a right circular cone. Consider the line segment in the figure at the right, where L is the length of the line segment, r_1 is the radius at the left end of the line segment, and r_2 is the radius at the right end of the line segment. When the line segment is revolved about its axis of revolution, it forms a frustum of a right circular cone, with

$$S = 2\pi rL \qquad \text{Lateral surface area of frustum}$$

where

$$r = \frac{1}{2}(r_1 + r_2). \qquad \text{Average radius of frustum}$$

(In Exercise 54, you are asked to verify the formula for S.)

Consider a function f that has a continuous derivative on the interval $[a, b]$. The graph of f is revolved about the x-axis to form a surface of revolution, as shown in Figure 7.43. Let Δ be a partition of $[a, b]$, with subintervals of width Δx_i. Then the line segment of length

$$\Delta L_i = \sqrt{\Delta x_i^2 + \Delta y_i^2}$$

generates a frustum of a cone. Let r_i be the average radius of this frustum. By the Intermediate Value Theorem, a point d_i exists (in the ith subinterval) such that

$$r_i = f(d_i).$$

The lateral surface area ΔS_i of the frustum is

$$
\begin{aligned}
\Delta S_i &= 2\pi r_i \, \Delta L_i \\
&= 2\pi f(d_i)\sqrt{\Delta x_i^2 + \Delta y_i^2} \\
&= 2\pi f(d_i)\sqrt{1 + \left(\frac{\Delta y_i}{\Delta x_i}\right)^2}\, \Delta x_i.
\end{aligned}
$$

Figure 7.43

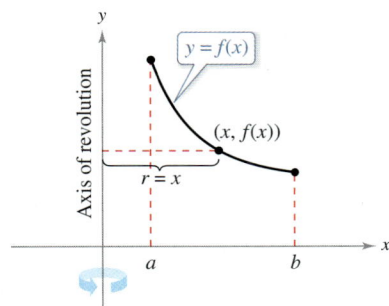

Figure 7.44

By the Mean Value Theorem, a point c_i exists in (x_{i-1}, x_i) such that

$$f'(c_i) = \frac{f(x_i) - f(x_{i-1})}{x_i - x_{i-1}}$$

$$= \frac{\Delta y_i}{\Delta x_i}.$$

So, $\Delta S_i = 2\pi f(d_i)\sqrt{1 + [f'(c_i)]^2}\,\Delta x_i$, and the total surface area can be approximated by

$$S \approx 2\pi \sum_{i=1}^{n} f(d_i)\sqrt{1 + [f'(c_i)]^2}\,\Delta x_i.$$

It can be shown that the limit of the right side as $\|\Delta\| \to 0$ $(n \to \infty)$ is

$$S = 2\pi \int_a^b f(x)\sqrt{1 + [f'(x)]^2}\,dx.$$

In a similar manner, if the graph of f is revolved about the y-axis, then S is

$$S = 2\pi \int_a^b x\sqrt{1 + [f'(x)]^2}\,dx.$$

In these two formulas for S, you can regard the products $2\pi f(x)$ and $2\pi x$ as the circumferences of the circles traced by a point (x, y) on the graph of f as it is revolved about the x-axis and the y-axis (Figure 7.44). In one case, the radius is $r = f(x)$, and in the other case, the radius is $r = x$. Moreover, by appropriately adjusting r, you can generalize the formula for surface area to cover *any* horizontal or vertical axis of revolution, as indicated in the next definition.

Definition of the Area of a Surface of Revolution

Let $y = f(x)$ have a continuous derivative on the interval $[a, b]$. The area S of the surface of revolution formed by revolving the graph of f about a horizontal or vertical axis is

$$S = 2\pi \int_a^b r(x)\sqrt{1 + [f'(x)]^2}\,dx \qquad \text{\textit{y} is a function of \textit{x}.}$$

where $r(x)$ is the distance between the graph of f and the axis of revolution. If $x = g(y)$ on the interval $[c, d]$, then the surface area is

$$S = 2\pi \int_c^d r(y)\sqrt{1 + [g'(y)]^2}\,dy \qquad \text{\textit{x} is a function of \textit{y}.}$$

where $r(y)$ is the distance between the graph of g and the axis of revolution.

The formulas in this definition are sometimes written as

$$S = 2\pi \int_a^b r(x)\,ds \qquad \text{\textit{y} is a function of \textit{x}.}$$

and

$$S = 2\pi \int_c^d r(y)\,ds \qquad \text{\textit{x} is a function of \textit{y}.}$$

where

$$ds = \sqrt{1 + [f'(x)]^2}\,dx \quad \text{and} \quad ds = \sqrt{1 + [g'(y)]^2}\,dy,$$

respectively.

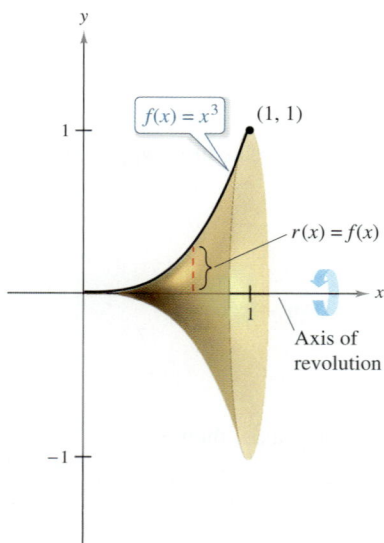

$f(x) = x^3$ (1, 1)

$r(x) = f(x)$

Axis of revolution

Figure 7.45

EXAMPLE 6 **The Area of a Surface of Revolution**

Find the area of the surface formed by revolving the graph of $f(x) = x^3$ on the interval $[0, 1]$ about the x-axis, as shown in Figure 7.45.

Solution The distance between the x-axis and the graph of f is $r(x) = f(x)$, and because $f'(x) = 3x^2$, the surface area is

$$S = 2\pi \int_a^b r(x)\sqrt{1 + [f'(x)]^2}\, dx \qquad \text{Formula for surface area}$$

$$= 2\pi \int_0^1 x^3 \sqrt{1 + (3x^2)^2}\, dx$$

$$= \frac{2\pi}{36} \int_0^1 (36x^3)(1 + 9x^4)^{1/2}\, dx \qquad \text{Simplify.}$$

$$= \frac{\pi}{18}\left[\frac{(1 + 9x^4)^{3/2}}{3/2} \right]_0^1 \qquad \text{Integrate.}$$

$$= \frac{\pi}{27}(10^{3/2} - 1)$$

$$\approx 3.563.$$

EXAMPLE 7 **The Area of a Surface of Revolution**

Find the area of the surface formed by revolving the graph of $f(x) = x^2$ on the interval $\left[0, \sqrt{2}\right]$ about the y-axis, as shown in the figure below.

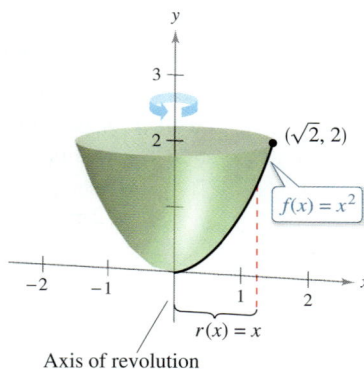

$(\sqrt{2}, 2)$

$f(x) = x^2$

$r(x) = x$

Axis of revolution

Solution In this case, the distance between the graph of f and the y-axis is $r(x) = x$. Using $f'(x) = 2x$ and the formula for surface area, you can determine that

$$S = 2\pi \int_a^b r(x)\sqrt{1 + [f'(x)]^2}\, dx$$

$$= 2\pi \int_0^{\sqrt{2}} x\sqrt{1 + (2x)^2}\, dx$$

$$= \frac{2\pi}{8} \int_0^{\sqrt{2}} (1 + 4x^2)^{1/2}(8x)\, dx \qquad \text{Simplify.}$$

$$= \frac{\pi}{4}\left[\frac{(1 + 4x^2)^{3/2}}{3/2} \right]_0^{\sqrt{2}} \qquad \text{Integrate.}$$

$$= \frac{\pi}{6}\left[(1 + 8)^{3/2} - 1 \right]$$

$$= \frac{13\pi}{3}$$

$$\approx 13.614.$$

Finding Distance Using Two Methods   In Exercises 1 and 2, find the distance between the points using (a) the Distance Formula and (b) integration.

1. $(0, 0)$, $(8, 15)$ **2.** $(1, 2)$, $(7, 10)$

Finding Arc Length   In Exercises 3–16, find the arc length of the graph of the function over the indicated interval.

3. $y = \frac{2}{3}(x^2 + 1)^{3/2}$ **4.** $y = \frac{x^3}{6} + \frac{1}{2x}$

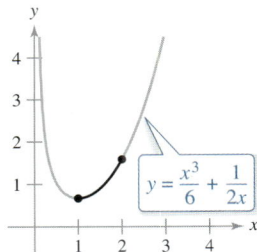

$y = \frac{2}{3}(x^2 + 1)^{3/2}$

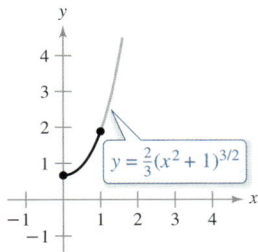

$y = \frac{x^3}{6} + \frac{1}{2x}$

5. $y = \frac{2}{3}x^{3/2} + 1$ **6.** $y = 2x^{3/2} + 3$

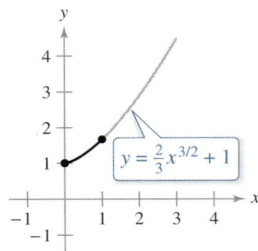

$y = \frac{2}{3}x^{3/2} + 1$

$y = 2x^{3/2} + 3$

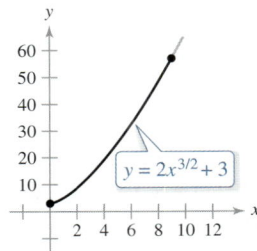

7. $y = \frac{3}{2}x^{2/3}$, $[1, 8]$ **8.** $y = \frac{x^4}{8} + \frac{1}{4x^2}$, $[1, 3]$

9. $y = \frac{x^5}{10} + \frac{1}{6x^3}$, $[2, 5]$ **10.** $y = \frac{3}{2}x^{2/3} + 4$, $[1, 27]$

11. $y = \ln(\sin x)$, $\left[\frac{\pi}{4}, \frac{3\pi}{4}\right]$ **12.** $y = \ln(\cos x)$, $\left[0, \frac{\pi}{3}\right]$

13. $y = \frac{1}{2}(e^x + e^{-x})$, $[0, 2]$

14. $y = \ln\left(\frac{e^x + 1}{e^x - 1}\right)$, $[\ln 2, \ln 3]$

15. $x = \frac{1}{3}(y^2 + 2)^{3/2}$, $0 \le y \le 4$

16. $x = \frac{1}{3}\sqrt{y}\,(y - 3)$, $1 \le y \le 4$

Finding Arc Length   In Exercises 17–26, (a) sketch the graph of the function, highlighting the part indicated by the given interval, (b) find a definite integral that represents the arc length of the curve over the indicated interval and observe that the integral cannot be evaluated with the techniques studied so far, and (c) use the integration capabilities of a graphing utility to approximate the arc length.

17. $y = 4 - x^2$, $0 \le x \le 2$

18. $y = x^2 + x - 2$, $-2 \le x \le 1$

19. $y = \frac{1}{x}$, $1 \le x \le 3$

20. $y = \frac{1}{x + 1}$, $0 \le x \le 1$

21. $y = \sin x$, $0 \le x \le \pi$

22. $y = \cos x$, $-\frac{\pi}{2} \le x \le \frac{\pi}{2}$

23. $x = e^{-y}$, $0 \le y \le 2$

24. $y = \ln x$, $1 \le x \le 5$

25. $y = 2 \arctan x$, $0 \le x \le 1$

26. $x = \sqrt{36 - y^2}$, $0 \le y \le 3$

Approximation   In Exercises 27 and 28, determine which value best approximates the length of the arc represented by the integral. (Make your selection on the basis of a sketch of the arc, *not* by performing any calculations.)

27. $\displaystyle\int_0^2 \sqrt{1 + \left[\frac{d}{dx}\left(\frac{5}{x^2 + 1}\right)\right]^2}\, dx$

   (a) 25   (b) 5   (c) 2   (d) -4 (e) 3

28. $\displaystyle\int_0^{\pi/4} \sqrt{1 + \left[\frac{d}{dx}(\tan x)\right]^2}\, dx$

   (a) 3   (b) -2 (c) 4   (d) $\frac{4\pi}{3}$ (e) 1

Approximation   In Exercises 29 and 30, approximate the arc length of the graph of the function over the interval $[0, 4]$ in four ways. (a) Use the Distance Formula to find the distance between the endpoints of the arc. (b) Use the Distance Formula to find the lengths of the four line segments connecting the points on the arc when $x = 0$, $x = 1$, $x = 2$, $x = 3$, and $x = 4$. Find the sum of the four lengths. (c) Use Simpson's Rule with $n = 10$ to approximate the integral yielding the indicated arc length. (d) Use the integration capabilities of a graphing utility to approximate the integral yielding the indicated arc length.

29. $f(x) = x^3$ **30.** $f(x) = (x^2 - 4)^2$

31. Length of a Catenary   Electrical wires suspended between two towers form a catenary (see figure) modeled by the equation

$$y = 20 \cosh \frac{x}{20},   -20 \le x \le 20$$

where x and y are measured in meters. The towers are 40 meters apart. Find the length of the suspended cable.

32. Roof Area A barn is 100 feet long and 40 feet wide (see figure). A cross section of the roof is the inverted catenary $y = 31 - 10(e^{x/20} + e^{-x/20})$. Find the number of square feet of roofing on the barn.

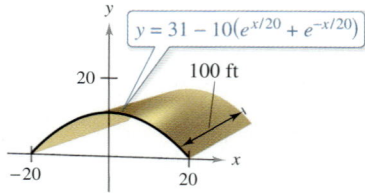

33. Length of Gateway Arch The Gateway Arch in St. Louis, Missouri, is modeled by

$$y = 693.8597 - 68.7672 \cosh 0.0100333x,$$
$$-299.2239 \le x \le 299.2239.$$

(See Section 5.8, Section Project: St. Louis Arch.) Use the integration capabilities of a graphing utility to approximate the length of this curve (see figure).

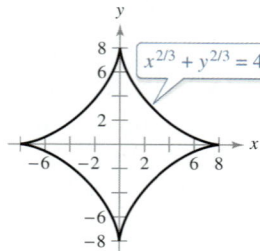

Figure for 33   **Figure for 34**

34. Astroid Find the total length of the graph of the astroid $x^{2/3} + y^{2/3} = 4$.

35. Arc Length of a Sector of a Circle Find the arc length from $(0, 3)$ clockwise to $(2, \sqrt{5})$ along the circle $x^2 + y^2 = 9$.

36. Arc Length of a Sector of a Circle Find the arc length from $(-3, 4)$ clockwise to $(4, 3)$ along the circle $x^2 + y^2 = 25$. Show that the result is one-fourth the circumference of the circle.

Finding the Area of a Surface of Revolution In Exercises 37–42, set up and evaluate the definite integral for the area of the surface generated by revolving the curve about the x-axis.

37. $y = \dfrac{1}{3}x^3$ **38.** $y = 2\sqrt{x}$

39. $y = \dfrac{x^3}{6} + \dfrac{1}{2x}, \quad 1 \le x \le 2$

40. $y = 3x, \quad 0 \le x \le 3$

41. $y = \sqrt{4 - x^2}, \quad -1 \le x \le 1$

42. $y = \sqrt{9 - x^2}, \quad -2 \le x \le 2$

Finding the Area of a Surface of Revolution In Exercises 43–46, set up and evaluate the definite integral for the area of the surface generated by revolving the curve about the y-axis.

43. $y = \sqrt[3]{x} + 2$ **44.** $y = 9 - x^2$

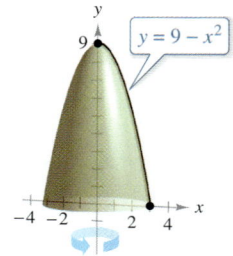

45. $y = 1 - \dfrac{x^2}{4}, \quad 0 \le x \le 2$ **46.** $y = \dfrac{x}{2} + 3, \quad 1 \le x \le 5$

Finding the Area of a Surface of Revolution In Exercises 47 and 48, use the integration capabilities of a graphing utility to approximate the surface area of the solid of revolution.

Function	Interval	Axis of Revolution
47. $y = \sin x$	$[0, \pi]$	x-axis
48. $y = \ln x$	$[1, e]$	y-axis

WRITING ABOUT CONCEPTS

49. Rectifiable Curve Define a rectifiable curve.

50. Precalculus and Calculus What precalculus formula and representative element are used to develop the integration formula for arc length?

51. Precalculus and Calculus What precalculus formula and representative element are used to develop the integration formula for the area of a surface of revolution?

52. HOW DO YOU SEE IT? The graphs of the functions f_1 and f_2 on the interval $[a, b]$ are shown in the figure. The graph of each function is revolved about the x-axis. Which surface of revolution has the greater surface area? Explain.

53. Think About It The figure shows the graphs of the functions $y_1 = x$, $y_2 = \frac{1}{2}x^{3/2}$, $y_3 = \frac{1}{4}x^2$, and $y_4 = \frac{1}{8}x^{5/2}$ on the interval $[0, 4]$. To print an enlarged copy of the graph, go to *MathGraphs.com*.

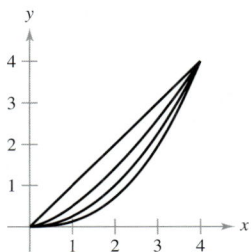

(a) Label the functions.

(b) List the functions in order of increasing arc length.

(c) Verify your answer in part (b) by using the integration capabilities of a graphing utility to approximate each arc length accurate to three decimal places.

54. Verifying a Formula

(a) Given a circular sector with radius L and central angle θ (see figure), show that the area of the sector is given by

$$S = \frac{1}{2}L^2\theta.$$

(b) By joining the straight-line edges of the sector in part (a), a right circular cone is formed (see figure) and the lateral surface area of the cone is the same as the area of the sector. Show that the area is $S = \pi r L$, where r is the radius of the base of the cone. (*Hint:* The arc length of the sector equals the circumference of the base of the cone.)

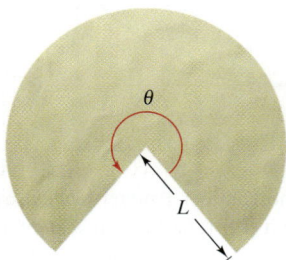

Figure for 54(a)   **Figure for 54(b)**

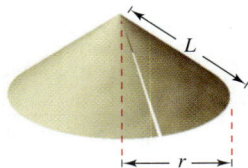

(c) Use the result of part (b) to verify that the formula for the lateral surface area of the frustum of a cone with slant height L and radii r_1 and r_2 (see figure) is $S = \pi(r_1 + r_2)L$. (*Note:* This formula was used to develop the integral for finding the surface area of a surface of revolution.)

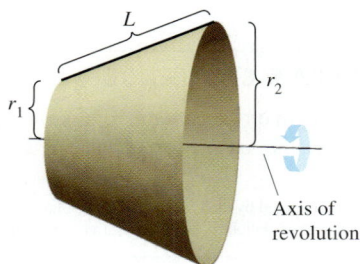

55. Lateral Surface Area of a Cone A right circular cone is generated by revolving the region bounded by $y = 3x/4$, $y = 3$, and $x = 0$ about the y-axis. Find the lateral surface area of the cone.

56. Lateral Surface Area of a Cone A right circular cone is generated by revolving the region bounded by $y = hx/r$, $y = h$, and $x = 0$ about the y-axis. Verify that the lateral surface area of the cone is $S = \pi r\sqrt{r^2 + h^2}$.

57. Using a Sphere Find the area of the zone of a sphere formed by revolving the graph of $y = \sqrt{9 - x^2}$, $0 \le x \le 2$, about the y-axis.

58. Using a Sphere Find the area of the zone of a sphere formed by revolving the graph of $y = \sqrt{r^2 - x^2}$, $0 \le x \le a$, about the y-axis. Assume that $a < r$.

59. Modeling Data The circumference C (in inches) of a vase is measured at three-inch intervals starting at its base. The measurements are shown in the table, where y is the vertical distance in inches from the base.

y	0	3	6	9	12	15	18
C	50	65.5	70	66	58	51	48

(a) Use the data to approximate the volume of the vase by summing the volumes of approximating disks.

(b) Use the data to approximate the outside surface area (excluding the base) of the vase by summing the outside surface areas of approximating frustums of right circular cones.

(c) Use the regression capabilities of a graphing utility to find a cubic model for the points (y, r), where $r = C/(2\pi)$. Use the graphing utility to plot the points and graph the model.

(d) Use the model in part (c) and the integration capabilities of a graphing utility to approximate the volume and outside surface area of the vase. Compare the results with your answers in parts (a) and (b).

60. Modeling Data Property bounded by two perpendicular roads and a stream is shown in the figure. All distances are measured in feet.

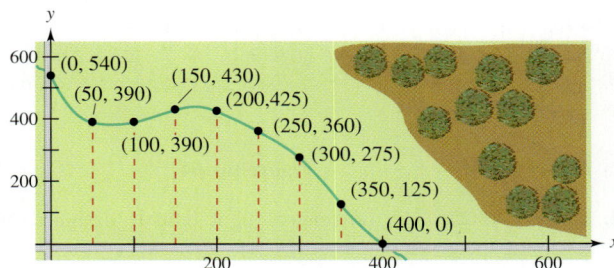

(a) Use the regression capabilities of a graphing utility to fit a fourth-degree polynomial to the path of the stream.

(b) Use the model in part (a) to approximate the area of the property in acres.

(c) Use the integration capabilities of a graphing utility to find the length of the stream that bounds the property.

61. Volume and Surface Area   Let R be the region bounded by $y = 1/x$, the x-axis, $x = 1$, and $x = b$, where $b > 1$. Let D be the solid formed when R is revolved about the x-axis.

(a) Find the volume V of D.

(b) Write the surface area S as an integral.

(c) Show that V approaches a finite limit as $b \to \infty$.

(d) Show that $S \to \infty$ as $b \to \infty$.

62. Think About It   Consider the equation $\dfrac{x^2}{9} + \dfrac{y^2}{4} = 1$.

(a) Use a graphing utility to graph the equation.

(b) Set up the definite integral for finding the first-quadrant arc length of the graph in part (a).

(c) Compare the interval of integration in part (b) and the domain of the integrand. Is it possible to evaluate the definite integral? Is it possible to use Simpson's Rule to evaluate the definite integral? Explain. (You will learn how to evaluate this type of integral in Section 8.8.)

Approximating Arc Length or Surface Area   In Exercises 63–66, set up the definite integral for finding the indicated arc length or surface area. Then use the integration capabilities of a graphing utility to approximate the arc length or surface area. (You will learn how to evaluate this type of integral in Section 8.8.)

63. Length of Pursuit   A fleeing object leaves the origin and moves up the y-axis (see figure). At the same time, a pursuer leaves the point $(1, 0)$ and always moves toward the fleeing object. The pursuer's speed is twice that of the fleeing object. The equation of the path is modeled by

$$y = \frac{1}{3}(x^{3/2} - 3x^{1/2} + 2).$$

How far has the fleeing object traveled when it is caught? Show that the pursuer has traveled twice as far.

Figure for 63   **Figure for 64**

64. Bulb Design   An ornamental light bulb is designed by revolving the graph of

$$y = \frac{1}{3}x^{1/2} - x^{3/2}, \quad 0 \le x \le \frac{1}{3},$$

about the x-axis, where x and y are measured in feet (see figure). Find the surface area of the bulb and use the result to approximate the amount of glass needed to make the bulb. (Assume that the glass is 0.015 inch thick.)

65. Astroid   Find the area of the surface formed by revolving the portion in the first quadrant of the graph of $x^{2/3} + y^{2/3} = 4$, $0 \le y \le 8$, about the y-axis.

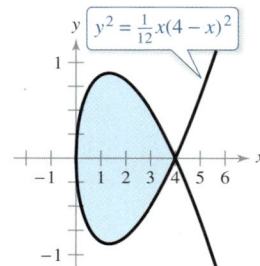

Figure for 65   **Figure for 66**

66. Using a Loop   Consider the graph of

$$y^2 = \frac{1}{12}x(4 - x)^2$$

shown in the figure. Find the area of the surface formed when the loop of this graph is revolved about the x-axis.

67. Suspension Bridge   A cable for a suspension bridge has the shape of a parabola with equation $y = kx^2$. Let h represent the height of the cable from its lowest point to its highest point and let $2w$ represent the total span of the bridge (see figure). Show that the length C of the cable is given by

$$C = 2\int_0^w \sqrt{1 + (4h^2/w^4)x^2}\, dx.$$

68. Suspension Bridge   The Humber Bridge, located in the United Kingdom and opened in 1981, has a main span of about 1400 meters. Each of its towers has a height of about 155 meters. Use these dimensions, the integral in Exercise 67, and the integration capabilities of a graphing utility to approximate the length of a parabolic cable along the main span.

69. Arc Length and Area   Let C be the curve given by $f(x) = \cosh x$ for $0 \le x \le t$, where $t > 0$. Show that the arc length of C is equal to the area bounded by C and the x-axis. Identify another curve on the interval $0 \le x \le t$ with this property.

PUTNAM EXAM CHALLENGE

70. Find the length of the curve $y^2 = x^3$ from the origin to the point where the tangent makes an angle of $45°$ with the x-axis.

7.5 Work

■ Find the work done by a constant force.
■ Find the work done by a variable force.

Work Done by a Constant Force

The concept of work is important to scientists and engineers for determining the energy needed to perform various jobs. For instance, it is useful to know the amount of work done when a crane lifts a steel girder, when a spring is compressed, when a rocket is propelled into the air, or when a truck pulls a load along a highway.

In general, **work** is done by a force when it moves an object. If the force applied to the object is *constant,* then the definition of work is as follows.

Definition of Work Done by a Constant Force

If an object is moved a distance D in the direction of an applied constant force F, then the **work** W done by the force is defined as $W = FD$.

There are four fundamental types of forces—gravitational, electromagnetic, strong nuclear, and weak nuclear. A **force** can be thought of as a *push* or a *pull*; a force changes the state of rest or state of motion of a body. For gravitational forces on Earth, it is common to use units of measure corresponding to the weight of an object.

EXAMPLE 1   Lifting an Object

Determine the work done in lifting a 50-pound object 4 feet.

Solution   The magnitude of the required force F is the weight of the object, as shown in Figure 7.46. So, the work done in lifting the object 4 feet is

$W = FD$ Work = (force)(distance)

$\quad = 50(4)$ Force = 50 pounds, distance = 4 feet

$\quad = 200$ foot-pounds.

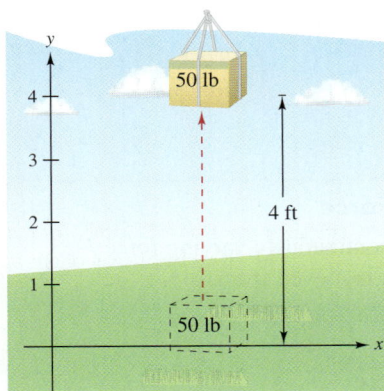

The work done in lifting a 50-pound object 4 feet is 200 foot-pounds.

Figure 7.46

In the U.S. measurement system, work is typically expressed in foot-pounds (ft-lb), inch-pounds, or foot-tons. In the International System of Units (SI), the basic unit of force is the **newton**—the force required to produce an acceleration of 1 meter per second per second on a mass of 1 kilogram. In this system, work is typically expressed in newton-meters, also called joules. In another system, the centimeter-gram-second (C-G-S) system, the basic unit of force is the **dyne**—the force required to produce an acceleration of 1 centimeter per second per second on a mass of 1 gram. In this system, work is typically expressed in dyne-centimeters (ergs) or newton-meters (joules).

Exploration

How Much Work?   In Example 1, 200 foot-pounds of work was needed to lift the 50-pound object 4 feet vertically off the ground. After lifting the object, you carry it a horizontal distance of 4 feet. Would this require an additional 200 foot-pounds of work? Explain your reasoning.

Work Done by a Variable Force

In Example 1, the force involved was *constant*. When a *variable* force is applied to an object, calculus is needed to determine the work done, because the amount of force changes as the object changes position. For instance, the force required to compress a spring increases as the spring is compressed.

Consider an object that is moved along a straight line from $x = a$ to $x = b$ by a continuously varying force $F(x)$. Let Δ be a partition that divides the interval $[a, b]$ into n subintervals determined by

$$a = x_0 < x_1 < x_2 < \cdots < x_n = b$$

and let $\Delta x_i = x_i - x_{i-1}$. For each i, choose c_i such that

$$x_{i-1} \leq c_i \leq x_i.$$

Then at c_i, the force is $F(c_i)$. Because F is continuous, you can approximate the work done in moving the object through the ith subinterval by the increment

$$\Delta W_i = F(c_i)\,\Delta x_i$$

as shown in Figure 7.47. So, the total work done as the object moves from a to b is approximated by

$$W \approx \sum_{i=1}^{n} \Delta W_i$$

$$= \sum_{i=1}^{n} F(c_i)\,\Delta x_i.$$

This approximation appears to become better and better as $\|\Delta\| \to 0$ $(n \to \infty)$. So, the work done is

$$W = \lim_{\|\Delta\| \to 0} \sum_{i=1}^{n} F(c_i)\,\Delta x_i$$

$$= \int_a^b F(x)\,dx.$$

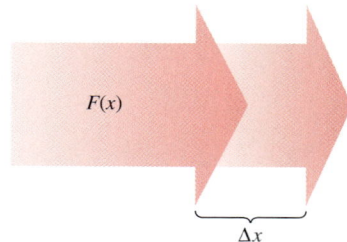

$F(x)$

Δx

The amount of force changes as an object changes position (Δx).
Figure 7.47

Definition of Work Done by a Variable Force

If an object is moved along a straight line by a continuously varying force $F(x)$, then the **work** W done by the force as the object is moved from

$$x = a \quad \text{to} \quad x = b$$

is given by

$$W = \lim_{\|\Delta\| \to 0} \sum_{i=1}^{n} \Delta W_i$$

$$= \int_a^b F(x)\,dx.$$

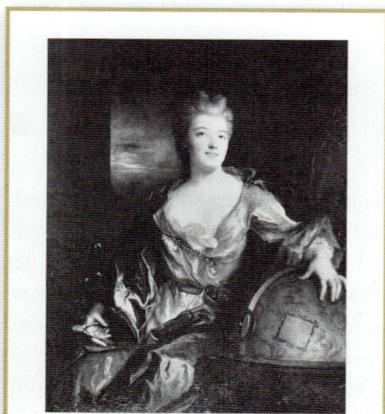

EMILIE DE BRETEUIL (1706–1749)
A major work by Breteuil was the translation of Newton's "Philosophiae Naturalis Principia Mathematica" into French. Her translation and commentary greatly contributed to the acceptance of Newtonian science in Europe.
See LarsonCalculus.com to read more of this biography.

The remaining examples in this section use some well-known physical laws. The discoveries of many of these laws occurred during the same period in which calculus was being developed. In fact, during the seventeenth and eighteenth centuries, there was little difference between physicists and mathematicians. One such physicist-mathematician was Emilie de Breteuil. Breteuil was instrumental in synthesizing the work of many other scientists, including Newton, Leibniz, Huygens, Kepler, and Descartes. Her physics text *Institutions* was widely used for many years.

Bettmann/Corbis

The three laws of physics listed below were developed by Robert Hooke (1635–1703), Isaac Newton (1642–1727), and Charles Coulomb (1736–1806).

1. **Hooke's Law:** The force F required to compress or stretch a spring (within its elastic limits) is proportional to the distance d that the spring is compressed or stretched from its original length. That is,

$$F = kd$$

where the constant of proportionality k (the spring constant) depends on the specific nature of the spring.

2. **Newton's Law of Universal Gravitation:** The force F of attraction between two particles of masses m_1 and m_2 is proportional to the product of the masses and inversely proportional to the square of the distance d between the two particles. That is,

$$F = G \frac{m_1 m_2}{d^2}.$$

When m_1 and m_2 are in kilograms and d in meters, F will be in newtons for a value of $G = 6.67 \times 10^{-11}$ cubic meter per kilogram-second squared, where G is the **gravitational constant.**

3. **Coulomb's Law:** The force F between two charges q_1 and q_2 in a vacuum is proportional to the product of the charges and inversely proportional to the square of the distance d between the two charges. That is,

$$F = k \frac{q_1 q_2}{d^2}.$$

When q_1 and q_2 are given in electrostatic units and d in centimeters, F will be in dynes for a value of $k = 1$.

EXAMPLE 2 Compressing a Spring

:··· ▷ *See LarsonCalculus.com for an interactive version of this type of example.*

A force of 750 pounds compresses a spring 3 inches from its natural length of 15 inches. Find the work done in compressing the spring an additional 3 inches.

Solution By Hooke's Law, the force $F(x)$ required to compress the spring x units (from its natural length) is $F(x) = kx$. Because $F(3) = 750$, it follows that

$$F(3) = (k)(3) \implies 750 = 3k \implies 250 = k.$$

So, $F(x) = 250x$, as shown in Figure 7.48. To find the increment of work, assume that the force required to compress the spring over a small increment Δx is nearly constant. So, the increment of work is

$$\Delta W = (\text{force})(\text{distance increment}) = (250x)\, \Delta x.$$

Because the spring is compressed from $x = 3$ to $x = 6$ inches less than its natural length, the work required is

$$W = \int_a^b F(x)\, dx = \int_3^6 250x\, dx = 125x^2 \Big]_3^6 = 4500 - 1125 = 3375 \text{ inch-pounds.}$$

Note that you do *not* integrate from $x = 0$ to $x = 6$ because you were asked to determine the work done in compressing the spring an *additional* 3 inches (not including the first 3 inches).

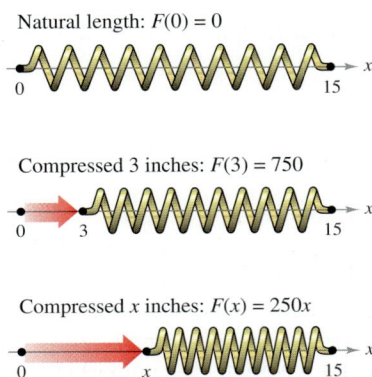

Natural length: $F(0) = 0$

Compressed 3 inches: $F(3) = 750$

Compressed x inches: $F(x) = 250x$

Figure 7.48

In 2011, China launched an 8.5-ton space module. The module will be used to conduct tests as China prepares to build a space station between 2020 and 2022.

EXAMPLE 3 **Moving a Space Module into Orbit**

A space module weighs 15 metric tons on the surface of Earth. How much work is done in propelling the module to a height of 800 miles above Earth, as shown in Figure 7.49? (Use 4000 miles as the radius of Earth. Do not consider the effect of air resistance or the weight of the propellant.)

Solution Because the weight of a body varies inversely as the square of its distance from the center of Earth, the force $F(x)$ exerted by gravity is

$$F(x) = \frac{C}{x^2}$$

Not drawn to scale

Figure 7.49

where C is the constant of proportionality. Because the module weighs 15 metric tons on the surface of Earth and the radius of Earth is approximately 4000 miles, you have

$$15 = \frac{C}{(4000)^2} \quad \Longrightarrow \quad 240{,}000{,}000 = C.$$

So, the increment of work is

$$\Delta W = (\text{force})(\text{distance increment}) = \frac{240{,}000{,}000}{x^2} \Delta x.$$

Finally, because the module is propelled from $x = 4000$ to $x = 4800$ miles, the total work done is

$$W = \int_a^b F(x)\, dx \qquad\qquad \text{Formula for work}$$

$$= \int_{4000}^{4800} \frac{240{,}000{,}000}{x^2}\, dx$$

$$= \frac{-240{,}000{,}000}{x} \Big]_{4000}^{4800} \qquad\qquad \text{Integrate.}$$

$$= -50{,}000 + 60{,}000$$

$$= 10{,}000 \ \text{mile-tons}$$

$$\approx 1.164 \times 10^{11} \ \text{foot-pounds}.$$

In SI units, using a conversion factor of 1 foot-pound ≈ 1.35582 joules, the work done is

$$W \approx 1.578 \times 10^{11} \ \text{joules}.$$

The solutions to Examples 2 and 3 conform to our development of work as the summation of increments in the form

$$\Delta W = (\text{force})(\text{distance increment}) = (F)(\Delta x).$$

Another way to formulate the increment of work is

$$\Delta W = (\text{force increment})(\text{distance}) = (\Delta F)(x).$$

This second interpretation of ΔW is useful in problems involving the movement of nonrigid substances such as fluids and chains.

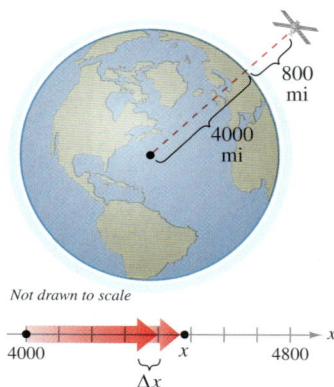

EXAMPLE 4   **Emptying a Tank of Oil**

A spherical tank of radius 8 feet is half full of oil that weighs 50 pounds per cubic foot. Find the work required to pump oil out through a hole in the top of the tank.

Solution   Consider the oil to be subdivided into disks of thickness Δy and radius x, as shown in Figure 7.50. Because the increment of force for each disk is given by its weight, you have

$$\Delta F = \text{weight}$$
$$= \left(\frac{50 \text{ pounds}}{\text{cubic foot}}\right)(\text{volume})$$
$$= 50(\pi x^2 \Delta y) \text{ pounds.}$$

For a circle of radius 8 and center at $(0, 8)$, you have

$$x^2 + (y - 8)^2 = 8^2$$
$$x^2 = 16y - y^2$$

and you can write the force increment as

$$\Delta F = 50(\pi x^2 \Delta y)$$
$$= 50\pi(16y - y^2) \Delta y.$$

In Figure 7.50, note that a disk y feet from the bottom of the tank must be moved a distance of $(16 - y)$ feet. So, the increment of work is

$$\Delta W = \Delta F(16 - y)$$
$$= 50\pi(16y - y^2) \Delta y (16 - y)$$
$$= 50\pi(256y - 32y^2 + y^3) \Delta y.$$

Because the tank is half full, y ranges from 0 to 8, and the work required to empty the tank is

$$W = \int_0^8 50\pi(256y - 32y^2 + y^3) \, dy$$
$$= 50\pi\left[128y^2 - \frac{32}{3}y^3 + \frac{y^4}{4}\right]_0^8$$
$$= 50\pi\left(\frac{11{,}264}{3}\right)$$
$$\approx 589{,}782 \text{ foot-pounds.}$$

Figure 7.50

To estimate the reasonableness of the result in Example 4, consider that the weight of the oil in the tank is

$$\left(\frac{1}{2}\right)(\text{volume})(\text{density}) = \frac{1}{2}\left(\frac{4}{3}\pi 8^3\right)(50) \approx 53{,}616.5 \text{ pounds}$$

Lifting the entire half-tank of oil 8 feet would involve work of

$$W = FD \qquad\qquad \text{\color{red}{Formula for work done by a constant force}}$$
$$\approx (53{,}616.5)(8)$$
$$= 428{,}932 \text{ foot-pounds.}$$

Because the oil is actually lifted between 8 and 16 feet, it seems reasonable that the work done is about 589,782 foot-pounds.

Work required to raise one end of the chain
Figure 7.51

Work done by expanding gas
Figure 7.52

EXAMPLE 5 **Lifting a Chain**

A 20-foot chain weighing 5 pounds per foot is lying coiled on the ground. How much work is required to raise one end of the chain to a height of 20 feet so that it is fully extended, as shown in Figure 7.51?

Solution Imagine that the chain is divided into small sections, each of length Δy. Then the weight of each section is the increment of force

$$\Delta F = (\text{weight}) = \left(\frac{5 \text{ pounds}}{\text{foot}}\right)(\text{length}) = 5 \, \Delta y.$$

Because a typical section (initially on the ground) is raised to a height of y, the increment of work is

$$\Delta W = (\text{force increment})(\text{distance}) = (5 \, \Delta y)y = 5y \, \Delta y.$$

Because y ranges from 0 to 20, the total work is

$$W = \int_0^{20} 5y \, dy = \frac{5y^2}{2}\bigg]_0^{20} = \frac{5(400)}{2} = 1000 \text{ foot-pounds.} \quad \blacksquare$$

In the next example, you will consider a piston of radius r in a cylindrical casing, as shown in Figure 7.52. As the gas in the cylinder expands, the piston moves, and work is done. If p represents the pressure of the gas (in pounds per square foot) against the piston head and V represents the volume of the gas (in cubic feet), then the work increment involved in moving the piston Δx feet is

$$\Delta W = (\text{force})(\text{distance increment}) = F(\Delta x) = p(\pi r^2) \, \Delta x = p \, \Delta V.$$

So, as the volume of the gas expands from V_0 to V_1, the work done in moving the piston is

$$W = \int_{V_0}^{V_1} p \, dV.$$

Assuming the pressure of the gas to be inversely proportional to its volume, you have $p = k/V$ and the integral for work becomes

$$W = \int_{V_0}^{V_1} \frac{k}{V} \, dV.$$

EXAMPLE 6 **Work Done by an Expanding Gas**

A quantity of gas with an initial volume of 1 cubic foot and a pressure of 500 pounds per square foot expands to a volume of 2 cubic feet. Find the work done by the gas. (Assume that the pressure is inversely proportional to the volume.)

Solution Because $p = k/V$ and $p = 500$ when $V = 1$, you have $k = 500$. So, the work is

$$W = \int_{V_0}^{V_1} \frac{k}{V} \, dV$$

$$= \int_1^2 \frac{500}{V} \, dV$$

$$= 500 \ln|V| \bigg]_1^2$$

$$\approx 346.6 \text{ foot-pounds.} \quad \blacksquare$$

7.5 Exercises

See **CalcChat.com** for tutorial help and worked-out solutions to odd-numbered exercises.

Constant Force In Exercises 1–4, determine the work done by the constant force.

1. A 1200-pound steel beam is lifted 40 feet.

2. An electric hoist lifts a 2500-pound car 6 feet.

3. A force of 112 newtons is required to slide a cement block 8 meters in a construction project.

4. The locomotive of a freight train pulls its cars with a constant force of 9 tons a distance of one-half mile.

Hooke's Law In Exercises 5–10, use Hooke's Law to determine the variable force in the spring problem.

5. A force of 5 pounds compresses a 15-inch spring a total of 3 inches. How much work is done in compressing the spring 7 inches?

6. A force of 250 newtons stretches a spring 30 centimeters. How much work is done in stretching the spring from 20 centimeters to 50 centimeters?

7. A force of 20 pounds stretches a spring 9 inches in an exercise machine. Find the work done in stretching the spring 1 foot from its natural position.

8. An overhead garage door has two springs, one on each side of the door. A force of 15 pounds is required to stretch each spring 1 foot. Because of the pulley system, the springs stretch only one-half the distance the door travels. The door moves a total of 8 feet, and the springs are at their natural length when the door is open. Find the work done by the pair of springs.

9. Eighteen foot-pounds of work is required to stretch a spring 4 inches from its natural length. Find the work required to stretch the spring an additional 3 inches.

10. Seven and one-half foot-pounds of work is required to compress a spring 2 inches from its natural length. Find the work required to compress the spring an additional one-half inch.

11. **Propulsion** Neglecting air resistance and the weight of the propellant, determine the work done in propelling a five-ton satellite to a height of (a) 100 miles above Earth and (b) 300 miles above Earth.

12. **Propulsion** Use the information in Exercise 11 to write the work W of the propulsion system as a function of the height h of the satellite above Earth. Find the limit (if it exists) of W as h approaches infinity.

13. **Propulsion** Neglecting air resistance and the weight of the propellant, determine the work done in propelling a 10-ton satellite to a height of (a) 11,000 miles above Earth and (b) 22,000 miles above Earth.

14. **Propulsion** A lunar module weighs 12 tons on the surface of Earth. How much work is done in propelling the module from the surface of the moon to a height of 50 miles? Consider the radius of the moon to be 1100 miles and its force of gravity to be one-sixth that of Earth.

15. **Pumping Water** A rectangular tank with a base 4 feet by 5 feet and a height of 4 feet is full of water (see figure). The water weighs 62.4 pounds per cubic foot. How much work is done in pumping water out over the top edge in order to empty (a) half of the tank and (b) all of the tank?

16. **Think About It** Explain why the answer in part (b) of Exercise 15 is not twice the answer in part (a).

17. **Pumping Water** A cylindrical water tank 4 meters high with a radius of 2 meters is buried so that the top of the tank is 1 meter below ground level (see figure). How much work is done in pumping a full tank of water up to ground level? (The water weighs 9800 newtons per cubic meter.)

Figure for 17

Figure for 18

18. **Pumping Water** Suppose the tank in Exercise 17 is located on a tower so that the bottom of the tank is 10 meters above the level of a stream (see figure). How much work is done in filling the tank half full of water through a hole in the bottom, using water from the stream?

19. **Pumping Water** An open tank has the shape of a right circular cone (see figure). The tank is 8 feet across the top and 6 feet high. How much work is done in emptying the tank by pumping the water over the top edge?

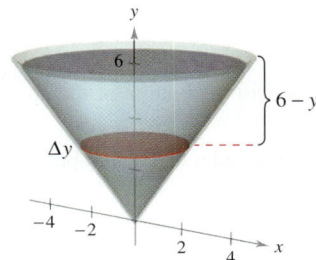

20. Pumping Water Water is pumped in through the bottom of the tank in Exercise 19. How much work is done to fill the tank

(a) to a depth of 2 feet?

(b) from a depth of 4 feet to a depth of 6 feet?

21. Pumping Water A hemispherical tank of radius 6 feet is positioned so that its base is circular. How much work is required to fill the tank with water through a hole in the base when the water source is at the base?

22. Pumping Diesel Fuel The fuel tank on a truck has trapezoidal cross sections with the dimensions (in feet) shown in the figure. Assume that the engine is approximately 3 feet above the top of the fuel tank and that diesel fuel weighs approximately 53.1 pounds per cubic foot. Find the work done by the fuel pump in raising a full tank of fuel to the level of the engine.

Pumping Gasoline In Exercises 23 and 24, find the work done in pumping gasoline that weighs 42 pounds per cubic foot. (*Hint:* Evaluate one integral by a geometric formula and the other by observing that the integrand is an odd function.)

23. A cylindrical gasoline tank 3 feet in diameter and 4 feet long is carried on the back of a truck and is used to fuel tractors. The axis of the tank is horizontal. The opening on the tractor tank is 5 feet above the top of the tank in the truck. Find the work done in pumping the entire contents of the fuel tank into the tractor.

24. The top of a cylindrical gasoline storage tank at a service station is 4 feet below ground level. The axis of the tank is horizontal and its diameter and length are 5 feet and 12 feet, respectively. Find the work done in pumping the entire contents of the full tank to a height of 3 feet above ground level.

Lifting a Chain In Exercises 25–28, consider a 20-foot chain that weighs 3 pounds per foot hanging from a winch 20 feet above ground level. Find the work done by the winch in winding up the specified amount of chain.

25. Wind up the entire chain.

26. Wind up one-third of the chain.

27. Run the winch until the bottom of the chain is at the 10-foot level.

28. Wind up the entire chain with a 500-pound load attached to it.

Lifting a Chain In Exercises 29 and 30, consider a 15-foot hanging chain that weighs 3 pounds per foot. Find the work done in lifting the chain vertically to the indicated position.

29. Take the bottom of the chain and raise it to the 15-foot level, leaving the chain doubled and still hanging vertically (see figure).

30. Repeat Exercise 29 raising the bottom of the chain to the 12-foot level.

WRITING ABOUT CONCEPTS

31. Work by a Constant Force State the definition of work done by a constant force.

32. Work by a Variable Force State the definition of work done by a variable force.

33. Work Which of the following requires more work? Explain your reasoning.

(a) A 60-pound box of books is lifted 3 feet.

(b) A 60-pound box of books is held 3 feet in the air for 2 minutes.

---

34. HOW DO YOU SEE IT? The graphs show the force F_i (in pounds) required to move an object 9 feet along the x-axis. Order the force functions from the one that yields the least work to the one that yields the most work without doing any calculations. Explain your reasoning.

35. Ordering Forces Verify your answer to Exercise 34 by calculating the work for each force function.

36. Electric Force Two electrons repel each other with a force that varies inversely as the square of the distance between them. One electron is fixed at the point $(2, 4)$. Find the work done in moving the second electron from $(-2, 4)$ to $(1, 4)$.

Boyle's Law **In Exercises 37 and 38, find the work done by the gas for the given volume and pressure. Assume that the pressure is inversely proportional to the volume. (See Example 6.)**

37. A quantity of gas with an initial volume of 2 cubic feet and a pressure of 1000 pounds per square foot expands to a volume of 3 cubic feet.

38. A quantity of gas with an initial volume of 1 cubic foot and a pressure of 2500 pounds per square foot expands to a volume of 3 cubic feet.

Hydraulic Press **In Exercises 39–42, use the integration capabilities of a graphing utility to approximate the work done by a press in a manufacturing process. A model for the variable force F (in pounds) and the distance x (in feet) the press moves is given.**

Force	Interval
39. $F(x) = 1000[1.8 - \ln(x + 1)]$	$0 \le x \le 5$
40. $F(x) = \dfrac{e^{x^2} - 1}{100}$	$0 \le x \le 4$
41. $F(x) = 100x\sqrt{125 - x^3}$	$0 \le x \le 5$
42. $F(x) = 1000 \sinh x$	$0 \le x \le 2$

43. Modeling Data The hydraulic cylinder on a woodsplitter has a 4-inch bore (diameter) and a stroke of 2 feet. The hydraulic pump creates a maximum pressure of 2000 pounds per square inch. Therefore, the maximum force created by the cylinder is $2000(\pi 2^2) = 8000\pi$ pounds.

(a) Find the work done through one extension of the cylinder, given that the maximum force is required.

(b) The force exerted in splitting a piece of wood is variable. Measurements of the force obtained in splitting a piece of wood are shown in the table. The variable x measures the extension of the cylinder in feet, and F is the force in pounds. Use Simpson's Rule to approximate the work done in splitting the piece of wood.

x	0	$\frac{1}{3}$	$\frac{2}{3}$	1	$\frac{4}{3}$	$\frac{5}{3}$	2
$F(x)$	0	20,000	22,000	15,000	10,000	5000	0

(c) Use the regression capabilities of a graphing utility to find a fourth-degree polynomial model for the data. Plot the data and graph the model.

(d) Use the model in part (c) to approximate the extension of the cylinder when the force is maximum.

(e) Use the model in part (c) to approximate the work done in splitting the piece of wood.

Andrew J. Martinez/Photo Researchers, Inc

SECTION PROJECT

Tidal Energy

Tidal power plants use "tidal energy" to produce electrical energy. To construct a tidal power plant, a dam is built to separate a basin from the sea. Electrical energy is produced as the water flows back and forth between the basin and the sea. The amount of "natural energy" produced depends on the volume of the basin and the tidal range—the vertical distance between high and low tides. (Several natural basins have tidal ranges in excess of 15 feet; the Bay of Fundy in Nova Scotia has a tidal range of 53 feet.)

$y = \frac{1}{40,000} x^2$

(a) Consider a basin with a rectangular base, as shown in the figure. The basin has a tidal range of 25 feet, with low tide corresponding to $y = 0$. How much water does the basin hold at high tide?

(b) The amount of energy produced during the filling (or the emptying) of the basin is proportional to the amount of work required to fill (or empty) the basin. How much work is required to fill the basin with seawater? (Use a seawater density of 64 pounds per cubic foot.)

The Bay of Fundy in Nova Scotia has an extreme tidal range, as displayed in the greatly contrasting photos above.

■ **FOR FURTHER INFORMATION** For more information on tidal power, see the article "LaRance: Six Years of Operating a Tidal Power Plant in France" by J. Cotillon in *Water Power Magazine*.

7.6 Moments, Centers of Mass, and Centroids

- ■ Understand the definition of mass.
- ■ Find the center of mass in a one-dimensional system.
- ■ Find the center of mass in a two-dimensional system.
- ■ Find the center of mass of a planar lamina.
- ■ Use the Theorem of Pappus to find the volume of a solid of revolution.

Mass

In this section, you will study several important applications of integration that are related to mass. Mass is a measure of a body's resistance to changes in motion, and is independent of the particular gravitational system in which the body is located. However, because so many applications involving mass occur on Earth's surface, an object's mass is sometimes equated with its weight. This is not technically correct. Weight is a type of force and as such is dependent on gravity. Force and mass are related by the equation

$$\text{Force} = (\text{mass})(\text{acceleration}).$$

The table below lists some commonly used measures of mass and force, together with their conversion factors.

System of Measurement	Measure of Mass	Measure of Force
U.S.	Slug	Pound $= (\text{slug})(\text{ft/sec}^2)$
International	Kilogram	Newton $= (\text{kilogram})(\text{m/sec}^2)$
C-G-S	Gram	Dyne $= (\text{gram})(\text{cm/sec}^2)$

Conversions:

1 pound $= 4.448$ newtons	1 slug $= 14.59$ kilograms
1 newton $= 0.2248$ pound	1 kilogram $= 0.06852$ slug
1 dyne $= 0.000002248$ pound	1 gram $= 0.00006852$ slug
1 dyne $= 0.00001$ newton	1 foot $= 0.3048$ meter

EXAMPLE 1   **Mass on the Surface of Earth**

Find the mass (in slugs) of an object whose weight at sea level is 1 pound.

Solution   Use 32 feet per second per second as the acceleration due to gravity.

$$\text{Mass} = \frac{\text{force}}{\text{acceleration}} \qquad \textcolor{red}{\text{Force} = (\text{mass})(\text{acceleration})}$$

$$= \frac{1 \text{ pound}}{32 \text{ feet per second per second}}$$

$$= 0.03125 \frac{\text{pound}}{\text{foot per second per second}}$$

$$= 0.03125 \text{ slug}$$

Because many applications involving mass occur on Earth's surface, this amount of mass is called a **pound mass.**

Center of Mass in a One-Dimensional System

You will now consider two types of moments of a mass—the **moment about a point** and the **moment about a line.** To define these two moments, consider an idealized situation in which a mass m is concentrated at a point. If x is the distance between this point mass and another point P, then the **moment of m about the point P is**

$$\boxed{\text{Moment} = mx}$$

and x is the **length of the moment arm.**

The concept of moment can be demonstrated simply by a seesaw, as shown in Figure 7.53. A child of mass 20 kilograms sits 2 meters to the left of fulcrum P, and an older child of mass 30 kilograms sits 2 meters to the right of P. From experience, you know that the seesaw will begin to rotate clockwise, moving the larger child down. This rotation occurs because the moment produced by the child on the left is less than the moment produced by the child on the right.

Left moment $= (20)(2) = 40$ kilogram-meters

Right moment $= (30)(2) = 60$ kilogram-meters

The seesaw will balance when the left and the right moments are equal.
Figure 7.53

To balance the seesaw, the two moments must be equal. For example, if the larger child moved to a position $\frac{4}{3}$ meters from the fulcrum, then the seesaw would balance, because each child would produce a moment of 40 kilogram-meters.

To generalize this, you can introduce a coordinate line on which the origin corresponds to the fulcrum, as shown in Figure 7.54. Several point masses are located on the x-axis. The measure of the tendency of this system to rotate about the origin is the **moment about the origin,** and it is defined as the sum of the n products $m_i x_i$. The moment about the origin is denoted by M_0 and can be written as

$$M_0 = m_1 x_1 + m_2 x_2 + \cdots + m_n x_n.$$

If M_0 is 0, then the system is said to be in **equilibrium.**

If $m_1 x_1 + m_2 x_2 + \cdots + m_n x_n = 0$, then the system is in equilibrium.
Figure 7.54

For a system that is not in equilibrium, the **center of mass** is defined as the point $\bar{x}$ at which the fulcrum could be relocated to attain equilibrium. If the system were translated $\bar{x}$ units, then each coordinate x_i would become

$$(x_i - \bar{x})$$

and because the moment of the translated system is 0, you have

$$\sum_{i=1}^{n} m_i(x_i - \bar{x}) = \sum_{i=1}^{n} m_i x_i - \sum_{i=1}^{n} m_i \bar{x} = 0.$$

Solving for $\bar{x}$ produces

$$\bar{x} = \frac{\displaystyle\sum_{i=1}^{n} m_i x_i}{\displaystyle\sum_{i=1}^{n} m_i} = \frac{\text{moment of system about origin}}{\text{total mass of system}}.$$

When $m_1 x_1 + m_2 x_2 + \cdots + m_n x_n = 0$, the system is in equilibrium.

> **Moments and Center of Mass: One-Dimensional System**
>
> Let the point masses $m_1, m_2, \ldots, m_n$ be located at $x_1, x_2, \ldots, x_n$.
>
> **1.** The **moment about the origin** is
>
> $$M_0 = m_1 x_1 + m_2 x_2 + \cdots + m_n x_n.$$
>
> **2.** The **center of mass** is
>
> $$\bar{x} = \frac{M_0}{m}$$
>
> where $m = m_1 + m_2 + \cdots + m_n$ is the **total mass** of the system.

EXAMPLE 2   **The Center of Mass of a Linear System**

Find the center of mass of the linear system shown in Figure 7.55.

Figure 7.55

Solution   The moment about the origin is

$$
\begin{aligned}
M_0 &= m_1 x_1 + m_2 x_2 + m_3 x_3 + m_4 x_4 \\
&= 10(-5) + 15(0) + 5(4) + 10(7) \\
&= -50 + 0 + 20 + 70 \\
&= 40.
\end{aligned}
$$

Because the total mass of the system is

$$m = 10 + 15 + 5 + 10 = 40$$

the center of mass is

$$\bar{x} = \frac{M_0}{m} = \frac{40}{40} = 1.$$

Note that the point masses will be in equilibrium when the fulcrum is located at $x = 1$.

Rather than define the moment of a mass, you could define the moment of a *force*. In this context, the center of mass is called the **center of gravity**. Consider a system of point masses $m_1, m_2, \ldots, m_n$ that is located at $x_1, x_2, \ldots, x_n$. Then, because

$$\text{force} = (\text{mass})(\text{acceleration})$$

the total force of the system is

$$F = m_1 a + m_2 a + \cdots + m_n a = ma.$$

The **torque** (moment) about the origin is

$$T_0 = (m_1 a) x_1 + (m_2 a) x_2 + \cdots + (m_n a) x_n = M_0 a$$

and the **center of gravity** is

$$\frac{T_0}{F} = \frac{M_0 a}{ma} = \frac{M_0}{m} = \bar{x}.$$

So, the center of gravity and the center of mass have the same location.

Center of Mass in a Two-Dimensional System

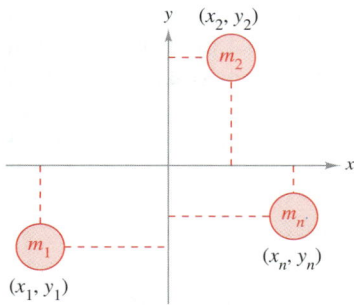

In a two-dimensional system, there is a moment about the y-axis M_y and a moment about the x-axis M_x.

Figure 7.56

You can extend the concept of moment to two dimensions by considering a system of masses located in the xy-plane at the points $(x_1, y_1), (x_2, y_2), \ldots, (x_n, y_n)$, as shown in Figure 7.56. Rather than defining a single moment (with respect to the origin), two moments are defined—one with respect to the x-axis and one with respect to the y-axis.

Moment and Center of Mass: Two-Dimensional System

Let the point masses $m_1, m_2, \ldots, m_n$ be located at $(x_1, y_1), (x_2, y_2), \ldots, (x_n, y_n)$.

1. The **moment about the y-axis** is
$$M_y = m_1 x_1 + m_2 x_2 + \ldots m_n x_n.$$

2. The **moment about the x-axis** is
$$M_x = m_1 y_1 + m_2 y_2 + \ldots m_n y_n.$$

3. The **center of mass** $(\bar{x}, \bar{y})$ (or **center of gravity**) is
$$\bar{x} = \frac{M_y}{m} \quad \text{and} \quad \bar{y} = \frac{M_x}{m}.$$

where
$$m = m_1 + m_2 + \ldots + m_n$$

is the **total mass** of the system.

The moment of a system of masses in the plane can be taken about any horizontal or vertical line. In general, the moment about a line is the sum of the product of the masses and the *directed distances* from the points to the line.

$$\text{Moment} = m_1(y_1 - b) + m_2(y_2 - b) + \cdots + m_n(y_n - b) \quad \text{Horizontal line } y = b$$
$$\text{Moment} = m_1(x_1 - a) + m_2(x_2 - a) + \cdots + m_n(x_n - a) \quad \text{Vertical line } x = a$$

EXAMPLE 3   The Center of Mass of a Two-Dimensional System

Find the center of mass of a system of point masses $m_1 = 6, m_2 = 3, m_3 = 2$, and $m_4 = 9$, located at
$$(3, -2), \quad (0, 0), \quad (-5, 3), \quad \text{and} \quad (4, 2)$$
as shown in Figure 7.57.

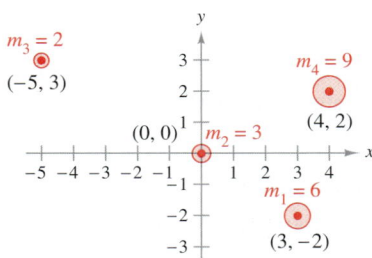

Figure 7.57

Solution

$$m = 6 \quad + 3 \quad + 2 \quad + 9 \quad = 20 \quad \text{Mass}$$
$$M_y = 6(3) \quad + 3(0) + 2(-5) + 9(4) = 44 \quad \text{Moment about y-axis}$$
$$M_x = 6(-2) + 3(0) + 2(3) \quad + 9(2) = 12 \quad \text{Moment about x-axis}$$

So,
$$\bar{x} = \frac{M_y}{m} = \frac{44}{20} = \frac{11}{5}$$

and
$$\bar{y} = \frac{M_x}{m} = \frac{12}{20} = \frac{3}{5}.$$

The center of mass is $\left(\frac{11}{5}, \frac{3}{5}\right)$.

Center of Mass of a Planar Lamina

You can think of the center of mass $(\bar{x}, \bar{y})$ of a lamina as its balancing point. For a circular lamina, the center of mass is the center of the circle. For a rectangular lamina, the center of mass is the center of the rectangle.

Figure 7.58

So far in this section, you have assumed the total mass of a system to be distributed at discrete points in a plane or on a line. Now consider a thin, flat plate of material of constant density called a **planar lamina** (see Figure 7.58). **Density** is a measure of mass per unit of volume, such as grams per cubic centimeter. For planar laminas, however, density is considered to be a measure of mass per unit of area. Density is denoted by ρ, the lowercase Greek letter rho.

Consider an irregularly shaped planar lamina of uniform density ρ, bounded by the graphs of $y = f(x)$, $y = g(x)$, and $a \le x \le b$, as shown in Figure 7.59. The mass of this region is

$$
\begin{aligned}
m &= (\text{density})(\text{area}) \\
&= \rho \int_a^b [f(x) - g(x)]\, dx \\
&= \rho A
\end{aligned}
$$

Planar lamina of uniform density ρ
Figure 7.59

where A is the area of the region. To find the center of mass of this lamina, partition the interval $[a, b]$ into n subintervals of equal width Δx. Let x_i be the center of the ith subinterval. You can approximate the portion of the lamina lying in the ith subinterval by a rectangle whose height is $h = f(x_i) - g(x_i)$. Because the density of the rectangle is ρ, its mass is

$$
m_i = \underset{\text{Density}}{\rho} \underset{\text{Height}}{[f(x_i) - g(x_i)]} \underset{\text{Width}}{\Delta x}.
$$

Now, considering this mass to be located at the center (x_i, y_i) of the rectangle, the directed distance from the x-axis to (x_i, y_i) is $y_i = [f(x_i) + g(x_i)]/2$. So, the moment of m_i about the x-axis is

$$
\begin{aligned}
\text{Moment} &= (\text{mass})(\text{distance}) \\
&= m_i y_i \\
&= \rho [f(x_i) - g(x_i)] \Delta x \left[\frac{f(x_i) + g(x_i)}{2} \right].
\end{aligned}
$$

Summing the moments and taking the limit as $n \to \infty$ suggest the definitions below.

Moments and Center of Mass of a Planar Lamina

Let f and g be continuous functions such that $f(x) \ge g(x)$ on $[a, b]$, and consider the planar lamina of uniform density ρ bounded by the graphs of $y = f(x)$, $y = g(x)$, and $a \le x \le b$.

1. The **moments about the x- and y-axes** are

$$
M_x = \rho \int_a^b \left[\frac{f(x) + g(x)}{2} \right] [f(x) - g(x)]\, dx
$$

$$
M_y = \rho \int_a^b x[f(x) - g(x)]\, dx.
$$

2. The **center of mass** $(\bar{x}, \bar{y})$ is given by $\bar{x} = \dfrac{M_y}{m}$ and $\bar{y} = \dfrac{M_x}{m}$, where $m = \rho \int_a^b [f(x) - g(x)]\, dx$ is the mass of the lamina.

EXAMPLE 4   **The Center of Mass of a Planar Lamina**

··· ▷ *See LarsonCalculus.com for an interactive version of this type of example.*

Find the center of mass of the lamina of uniform density ρ bounded by the graph of $f(x) = 4 - x^2$ and the x-axis.

Solution   Because the center of mass lies on the axis of symmetry, you know that $\bar{x} = 0$. Moreover, the mass of the lamina is

$$m = \rho \int_{-2}^{2} (4 - x^2)\, dx$$

$$= \rho \left[4x - \frac{x^3}{3} \right]_{-2}^{2}$$

$$= \frac{32\rho}{3}.$$

To find the moment about the x-axis, place a representative rectangle in the region, as shown in the figure at the right. The distance from the x-axis to the center of this rectangle is

$$y_i = \frac{f(x)}{2} = \frac{4 - x^2}{2}.$$

Because the mass of the representative rectangle is

$$\rho f(x)\, \Delta x = \rho(4 - x^2)\, \Delta x$$

you have

$$M_x = \rho \int_{-2}^{2} \frac{4 - x^2}{2} (4 - x^2)\, dx$$

$$= \frac{\rho}{2} \int_{-2}^{2} (16 - 8x^2 + x^4)\, dx$$

$$= \frac{\rho}{2} \left[16x - \frac{8x^3}{3} + \frac{x^5}{5} \right]_{-2}^{2}$$

$$= \frac{256\rho}{15}$$

and $\bar{y}$ is

$$\bar{y} = \frac{M_x}{m} = \frac{256\rho/15}{32\rho/3} = \frac{8}{5}.$$

So, the center of mass (the balancing point) of the lamina is $\left(0, \frac{8}{5}\right)$, as shown in Figure 7.60.

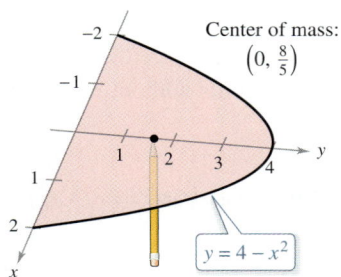

The center of mass is the balancing point.

Figure 7.60

The density ρ in Example 4 is a common factor of both the moments and the mass, and as such divides out of the quotients representing the coordinates of the center of mass. So, the center of mass of a lamina of *uniform* density depends only on the shape of the lamina and not on its density. For this reason, the point

$(\bar{x}, \bar{y})$ Center of mass or centroid

is sometimes called the center of mass of a *region* in the plane, or the **centroid** of the region. In other words, to find the centroid of a region in the plane, you simply assume that the region has a constant density of $\rho = 1$ and compute the corresponding center of mass.

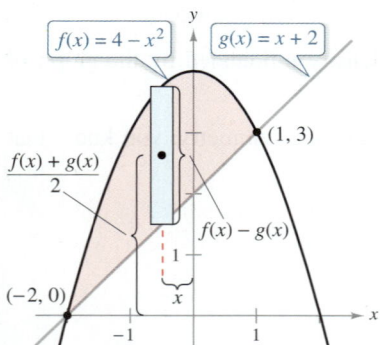

$f(x) = 4 - x^2$ $g(x) = x + 2$

$\dfrac{f(x) + g(x)}{2}$

$(1, 3)$

$f(x) - g(x)$

$(-2, 0)$

x

Figure 7.61

| EXAMPLE 5 | **The Centroid of a Plane Region** |

Find the centroid of the region bounded by the graphs of $f(x) = 4 - x^2$ and $g(x) = x + 2$.

Solution   The two graphs intersect at the points $(-2, 0)$ and $(1, 3)$, as shown in Figure 7.61. So, the area of the region is

$$A = \int_{-2}^{1} [f(x) - g(x)] \, dx = \int_{-2}^{1} (2 - x - x^2) \, dx = \frac{9}{2}.$$

The centroid $(\bar{x}, \bar{y})$ of the region has the following coordinates.

$$\bar{x} = \frac{1}{A} \int_{-2}^{1} x[(4 - x^2) - (x + 2)] \, dx$$

$$= \frac{2}{9} \int_{-2}^{1} (-x^3 - x^2 + 2x) \, dx$$

$$= \frac{2}{9} \left[-\frac{x^4}{4} - \frac{x^3}{3} + x^2 \right]_{-2}^{1}$$

$$= -\frac{1}{2}$$

$$\bar{y} = \frac{1}{A} \int_{-2}^{1} \left[\frac{(4 - x^2) + (x + 2)}{2} \right] [(4 - x^2) - (x + 2)] \, dx$$

$$= \frac{2}{9} \left(\frac{1}{2} \right) \int_{-2}^{1} (-x^2 + x + 6)(-x^2 - x + 2) \, dx$$

$$= \frac{1}{9} \int_{-2}^{1} (x^4 - 9x^2 - 4x + 12) \, dx$$

$$= \frac{1}{9} \left[\frac{x^5}{5} - 3x^3 - 2x^2 + 12x \right]_{-2}^{1}$$

$$= \frac{12}{5}$$

So, the centroid of the region is $(\bar{x}, \bar{y}) = \left(-\frac{1}{2}, \frac{12}{5} \right)$. ◼

For simple plane regions, you may be able to find the centroids without resorting to integration.

| EXAMPLE 6 | **The Centroid of a Simple Plane Region** |

Find the centroid of the region shown in Figure 7.62(a).

Solution   By superimposing a coordinate system on the region, as shown in Figure 7.62(b), you can locate the centroids of the three rectangles at

$$\left(\frac{1}{2}, \frac{3}{2} \right), \quad \left(\frac{5}{2}, \frac{1}{2} \right), \quad \text{and} \quad (5, 1).$$

Using these three points, you can find the centroid of the region.

$$A = \text{area of region} = 3 + 3 + 4 = 10$$

$$\bar{x} = \frac{(1/2)(3) + (5/2)(3) + (5)(4)}{10} = \frac{29}{10} = 2.9$$

$$\bar{y} = \frac{(3/2)(3) + (1/2)(3) + (1)(4)}{10} = \frac{10}{10} = 1$$

(a) Original region

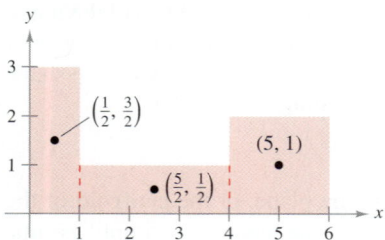

(b) The centroids of the three rectangles

Figure 7.62

So, the centroid of the region is $(2.9, 1)$. Notice that $(2.9, 1)$ is not the "average" of $\left(\frac{1}{2}, \frac{3}{2} \right)$, $\left(\frac{5}{2}, \frac{1}{2} \right)$, and $(5, 1)$. ◼

Theorem of Pappus

The final topic in this section is a useful theorem credited to Pappus of Alexandria (ca. 300 A.D.), a Greek mathematician whose eight-volume *Mathematical Collection* is a record of much of classical Greek mathematics. You are asked to prove this theorem in Section 14.4.

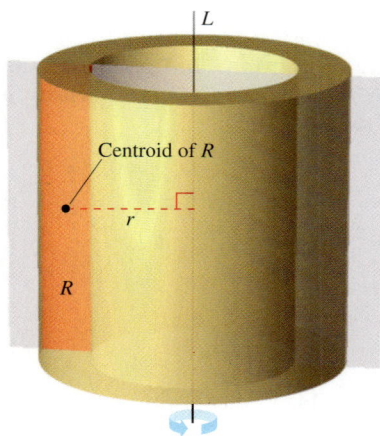

The volume V is $2\pi rA$, where A is the area of region R.

Figure 7.63

THEOREM 7.1 The Theorem of Pappus

Let R be a region in a plane and let L be a line in the same plane such that L does not intersect the interior of R, as shown in Figure 7.63. If r is the distance between the centroid of R and the line, then the volume V of the solid of revolution formed by revolving R about the line is

$$V = 2\pi rA$$

where A is the area of R. (Note that $2\pi r$ is the distance traveled by the centroid as the region is revolved about the line.)

The Theorem of Pappus can be used to find the volume of a torus, as shown in the next example. Recall that a torus is a doughnut-shaped solid formed by revolving a circular region about a line that lies in the same plane as the circle (but does not intersect the circle).

EXAMPLE 7 **Finding Volume by the Theorem of Pappus**

Find the volume of the torus shown in Figure 7.64(a), which was formed by revolving the circular region bounded by

$$(x - 2)^2 + y^2 = 1$$

about the y-axis, as shown in Figure 7.64(b).

Torus

(a)

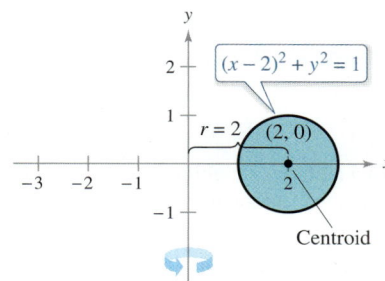

(b)

Figure 7.64

Solution In Figure 7.67(b), you can see that the centroid of the circular region is $(2, 0)$. So, the distance between the centroid and the axis of revolution is

$$r = 2.$$

Because the area of the circular region is $A = \pi$, the volume of the torus is

$$
\begin{aligned}
V &= 2\pi rA \\
&= 2\pi(2)(\pi) \\
&= 4\pi^2 \\
&\approx 39.5.
\end{aligned}
$$

Exploration

Use the shell method to show that the volume of the torus in Example 7 is

$$V = \int_1^3 4\pi x \sqrt{1 - (x - 2)^2}\, dx.$$

Evaluate this integral using a graphing utility. Does your answer agree with the one in Example 7?

7.6 Exercises

Center of Mass of a Linear System In Exercises 1–4, find the center of mass of the point masses lying on the x-axis.

1. $m_1 = 7, m_2 = 3, m_3 = 5$

$x_1 = -5, x_2 = 0, x_3 = 3$

2. $m_1 = 7, m_2 = 4, m_3 = 3, m_4 = 8$

$x_1 = -3, x_2 = -2, x_3 = 5, x_4 = 4$

3. $m_1 = 1, m_2 = 3, m_3 = 2, m_4 = 9, m_5 = 5$

$x_1 = 6, x_2 = 10, x_3 = 3, x_4 = 2, x_5 = 4$

4. $m_1 = 8, m_2 = 5, m_3 = 5, m_4 = 12, m_5 = 2$

$x_1 = -2, x_2 = 6, x_3 = 0, x_4 = 3, x_5 = -5$

5. Graphical Reasoning

(a) Translate each point mass in Exercise 3 to the right four units and determine the resulting center of mass.

(b) Translate each point mass in Exercise 4 to the left two units and determine the resulting center of mass.

6. Conjecture Use the result of Exercise 5 to make a conjecture about the change in the center of mass that results when each point mass is translated k units horizontally.

Statics Problems In Exercises 7 and 8, consider a beam of length L with a fulcrum x feet from one end (see figure). There are objects with weights W_1 and W_2 placed on opposite ends of the beam. Find x such that the system is in equilibrium.

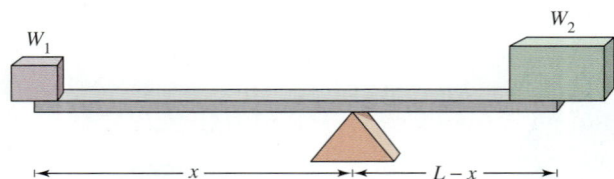

7. Two children weighing 48 pounds and 72 pounds are going to play on a seesaw that is 10 feet long.

8. In order to move a 600-pound rock, a person weighing 200 pounds wants to balance it on a beam that is 5 feet long.

Center of Mass of a Two-Dimensional System In Exercises 9–12, find the center of mass of the given system of point masses.

9.

m_i	5	1	3
(x_i, y_i)	$(2, 2)$	$(-3, 1)$	$(1, -4)$

10.

m_i	10	2	5
(x_i, y_i)	$(1, -1)$	$(5, 5)$	$(-4, 0)$

11.

m_i	12	6	4.5	15
(x_i, y_i)	$(2, 3)$	$(-1, 5)$	$(6, 8)$	$(2, -2)$

12.

m_i	3	4	2	1	6
(x_i, y_i)	$(-2, -3)$	$(5, 5)$	$(7, 1)$	$(0, 0)$	$(-3, 0)$

Center of Mass of a Planar Lamina In Exercises 13–26, find M_x, M_y, and $(\bar{x}, \bar{y})$ for the laminas of uniform density ρ bounded by the graphs of the equations.

13. $y = \frac{1}{2}x, y = 0, x = 2$ **14.** $y = 6 - x, y = 0, x = 0$

15. $y = \sqrt{x}, y = 0, x = 4$ **16.** $y = \frac{1}{2}x^2, y = 0, x = 2$

17. $y = x^2, y = x^3$ **18.** $y = \sqrt{x}, y = \frac{1}{2}x$

19. $y = -x^2 + 4x + 2, y = x + 2$

20. $y = \sqrt{x} + 1, y = \frac{1}{3}x + 1$

21. $y = x^{2/3}, y = 0, x = 8$ **22.** $y = x^{2/3}, y = 4$

23. $x = 4 - y^2, x = 0$ **24.** $x = 3y - y^2, x = 0$

25. $x = -y, x = 2y - y^2$ **26.** $x = y + 2, x = y^2$

Approximating a Centroid In Exercises 27–30, use a graphing utility to graph the region bounded by the graphs of the equations. Use the integration capabilities of the graphing utility to approximate the centroid of the region.

27. $y = 10x\sqrt{125 - x^3}, y = 0$

28. $y = xe^{-x/2}, y = 0, x = 0, \ x = 4$

29. Prefabricated End Section of a Building

$y = 5\sqrt[3]{400 - x^2}, y = 0$

30. Witch of Agnesi

$y = \dfrac{8}{x^2 + 4}, y = 0, x = -2, x = 2$

Finding the Center of Mass In Exercises 31–34, introduce an appropriate coordinate system and find the coordinates of the center of mass of the planar lamina. (The answer depends on the position of the coordinate system.)

31.

32.

33.

34.

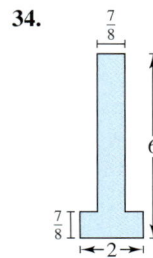

35. Finding the Center of Mass Find the center of mass of the lamina in Exercise 31 when the circular portion of the lamina has twice the density of the square portion of the lamina.

36. Finding the Center of Mass Find the center of mass of the lamina in Exercise 31 when the square portion of the lamina has twice the density of the circular portion of the lamina.

Finding Volume by the Theorem of Pappus In Exercises 37–40, use the Theorem of Pappus to find the volume of the solid of revolution.

37. The torus formed by revolving the circle

$$(x - 5)^2 + y^2 = 16$$

about the y-axis

38. The torus formed by revolving the circle

$$x^2 + (y - 3)^2 = 4$$

about the x-axis

39. The solid formed by revolving the region bounded by the graphs of $y = x$, $y = 4$, and $x = 0$ about the x-axis

40. The solid formed by revolving the region bounded by the graphs of $y = 2\sqrt{x - 2}$, $y = 0$, and $x = 6$ about the y-axis

---

WRITING ABOUT CONCEPTS

41. Center of Mass Let the point masses m_1, $m_2, \ldots, m_n$ be located at (x_1, y_1), $(x_2, y_2), \ldots, (x_n, y_n)$. Define the center of mass $(\bar{x}, \bar{y})$.

42. Planar Lamina What is a planar lamina? Describe what is meant by the center of mass $(\bar{x}, \bar{y})$ of a planar lamina.

43. Theorem of Pappus State the Theorem of Pappus.

---

44. HOW DO YOU SEE IT? The centroid of the plane region bounded by the graphs of $y = f(x)$, $y = 0$, $x = 0$, and $x = 3$ is $(1.2, 1.4)$. Is it possible to find the centroid of each of the regions bounded by the graphs of the following sets of equations? If so, identify the centroid and explain your answer.

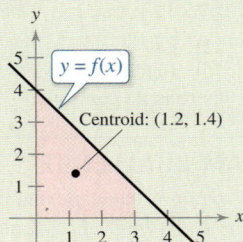

(a) $y = f(x) + 2$, $y = 2$, $x = 0$, and $x = 3$

(b) $y = f(x - 2)$, $y = 0$, $x = 2$, and $x = 5$

(c) $y = -f(x)$, $y = 0$, $x = 0$, and $x = 3$

(d) $y = f(x)$, $y = 0$, $x = 2$, and $x = 4$

Centroid of a Common Region In Exercises 45–50, find and/or verify the centroid of the common region used in engineering.

45. Triangle Show that the centroid of the triangle with vertices $(-a, 0)$, $(a, 0)$, and (b, c) is the point of intersection of the medians (see figure).

Figure for 45

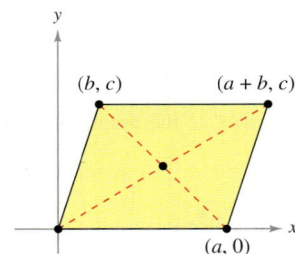

Figure for 46

46. Parallelogram Show that the centroid of the parallelogram with vertices $(0, 0)$, $(a, 0)$, (b, c), and $(a + b, c)$ is the point of intersection of the diagonals (see figure).

47. Trapezoid Find the centroid of the trapezoid with vertices $(0, 0)$, $(0, a)$, (c, b), and $(c, 0)$. Show that it is the intersection of the line connecting the midpoints of the parallel sides and the line connecting the extended parallel sides, as shown in the figure.

Figure for 47

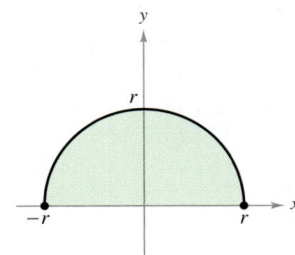

Figure for 48

48. Semicircle Find the centroid of the region bounded by the graphs of $y = \sqrt{r^2 - x^2}$ and $y = 0$ (see figure).

49. Semiellipse Find the centroid of the region bounded by the graphs of $y = \dfrac{b}{a}\sqrt{a^2 - x^2}$ and $y = 0$ (see figure).

Figure for 49

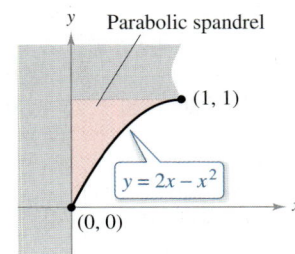

Figure for 50

50. Parabolic Spandrel Find the centroid of the **parabolic spandrel** shown in the figure.

51. Graphical Reasoning   Consider the region bounded by the graphs of $y = x^2$ and $y = b$, where $b > 0$.

(a) Sketch a graph of the region.

(b) Use the graph in part (a) to determine $\bar{x}$. Explain.

(c) Set up the integral for finding M_y. Because of the form of the integrand, the value of the integral can be obtained without integrating. What is the form of the integrand? What is the value of the integral? Compare with the result in part (b).

(d) Use the graph in part (a) to determine whether $\bar{y} > \dfrac{b}{2}$ or $\bar{y} < \dfrac{b}{2}$. Explain.

(e) Use integration to verify your answer in part (d).

52. Graphical and Numerical Reasoning   Consider the region bounded by the graphs of $y = x^{2n}$ and $y = b$, where $b > 0$ and n is a positive integer.

(a) Sketch a graph of the region.

(b) Set up the integral for finding M_y. Because of the form of the integrand, the value of the integral can be obtained without integrating. What is the form of the integrand? What is the value of the integral and what is the value of $\bar{x}$?

(c) Use the graph in part (a) to determine whether $\bar{y} > \dfrac{b}{2}$ or $\bar{y} < \dfrac{b}{2}$. Explain.

(d) Use integration to find $\bar{y}$ as a function of n.

(e) Use the result of part (d) to complete the table.

n	1	2	3	4
$\bar{y}$				

(f) Find $\lim\limits_{n \to \infty} \bar{y}$.

(g) Give a geometric explanation of the result in part (f).

53. Modeling Data   The manufacturer of glass for a window in a conversion van needs to approximate its center of mass. A coordinate system is superimposed on a prototype of the glass (see figure). The measurements (in centimeters) for the right half of the symmetric piece of glass are listed in the table.

x	0	10	20	30	40
y	30	29	26	20	0

(a) Use Simpson's Rule to approximate the center of mass of the glass.

(b) Use the regression capabilities of a graphing utility to find a fourth-degree polynomial model for the data.

(c) Use the integration capabilities of a graphing utility and the model to approximate the center of mass of the glass. Compare with the result in part (a).

54. Modeling Data   The manufacturer of a boat needs to approximate the center of mass of a section of the hull. A coordinate system is superimposed on a prototype (see figure). The measurements (in feet) for the right half of the symmetric prototype are listed in the table.

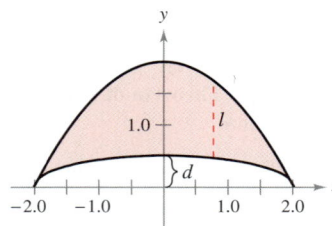

x	0	0.5	1.0	1.5	2
l	1.50	1.45	1.30	0.99	0
d	0.50	0.48	0.43	0.33	0

(a) Use Simpson's Rule to approximate the center of mass of the hull section.

(b) Use the regression capabilities of a graphing utility to find fourth-degree polynomial models for both curves shown in the figure. Plot the data and graph the models.

(c) Use the integration capabilities of a graphing utility and the models to approximate the center of mass of the hull section. Compare with the result in part (a).

Second Theorem of Pappus   **In Exercises 55 and 56, use the *Second Theorem of Pappus*, which is stated as follows. If a segment of a plane curve C is revolved about an axis that does not intersect the curve (except possibly at its endpoints), the area S of the resulting surface of revolution is equal to the product of the length of C times the distance d traveled by the centroid of C.**

55. A sphere is formed by revolving the graph of $y = \sqrt{r^2 - x^2}$ about the x-axis. Use the formula for surface area, $S = 4\pi r^2$, to find the centroid of the semicircle $y = \sqrt{r^2 - x^2}$.

56. A torus is formed by revolving the graph of $(x - 1)^2 + y^2 = 1$ about the y-axis. Find the surface area of the torus.

57. Finding a Centroid   Let $n \geq 1$ be constant, and consider the region bounded by $f(x) = x^n$, the x-axis, and $x = 1$. Find the centroid of this region. As $n \to \infty$, what does the region look like, and where is its centroid?

PUTNAM EXAM CHALLENGE

58. Let V be the region in the cartesian plane consisting of all points (x, y) satisfying the simultaneous conditions $|x| \leq y \leq |x| + 3$ and $y \leq 4$. Find the centroid $(\bar{x}, \bar{y})$ of V.

7.7 Fluid Pressure and Fluid Force

■ Find fluid pressure and fluid force.

Fluid Pressure and Fluid Force

Swimmers know that the deeper an object is submerged in a fluid, the greater the pressure on the object. **Pressure** is defined as the force per unit of area over the surface of a body. For example, because a column of water that is 10 feet in height and 1 inch square weighs 4.3 pounds, the *fluid pressure* at a depth of 10 feet of water is 4.3 pounds per square inch.* At 20 feet, this would increase to 8.6 pounds per square inch, and in general the pressure is proportional to the depth of the object in the fluid.

BLAISE PASCAL (1623–1662)

Pascal is well known for his work in many areas of mathematics and physics, and also for his influence on Leibniz. Although much of Pascal's work in calculus was intuitive and lacked the rigor of modern mathematics, he nevertheless anticipated many important results.
See LarsonCalculus.com to read more of this biography.

Definition of Fluid Pressure

The **pressure** on an object at depth h in a liquid is

Pressure $= P = wh$

where w is the weight-density of the liquid per unit of volume.

Below are some common weight-densities of fluids in pounds per cubic foot.

Ethyl alcohol	49.4
Gasoline	41.0–43.0
Glycerin	78.6
Kerosene	51.2
Mercury	849.0
Seawater	64.0
Water	62.4

When calculating fluid pressure, you can use an important (and rather surprising) physical law called **Pascal's Principle,** named after the French mathematician Blaise Pascal. Pascal's Principle states that the pressure exerted by a fluid at a depth h is transmitted equally *in all directions*. For example, in Figure 7.65, the pressure at the indicated depth is the same for all three objects. Because fluid pressure is given in terms of force per unit area ($P = F/A$), the fluid force on a *submerged horizontal* surface of area A is

Fluid force $= F = PA =$ (pressure)(area).

The pressure at h is the same for all three objects.
Figure 7.65

---

* The total pressure on an object in 10 feet of water would also include the pressure due to Earth's atmosphere. At sea level, atmospheric pressure is approximately 14.7 pounds per square inch.

The fluid force on a horizontal metal sheet is equal to the fluid pressure times the area.

Figure 7.66

EXAMPLE 1 **Fluid Force on a Submerged Sheet**

Find the fluid force on a rectangular metal sheet measuring 3 feet by 4 feet that is submerged in 6 feet of water, as shown in Figure 7.66.

Solution Because the weight-density of water is 62.4 pounds per cubic foot and the sheet is submerged in 6 feet of water, the fluid pressure is

$$P = (62.4)(6) \qquad\qquad P = wh$$

$$= 374.4 \text{ pounds per square foot.}$$

Because the total area of the sheet is $A = (3)(4) = 12$ square feet, the fluid force is

$$F = PA$$

$$= \left(374.4 \ \frac{\text{pounds}}{\text{square foot}}\right)(12 \text{ square feet})$$

$$= 4492.8 \text{ pounds.}$$

This result is independent of the size of the body of water. The fluid force would be the same in a swimming pool or lake.

In Example 1, the fact that the sheet is rectangular and horizontal means that you do not need the methods of calculus to solve the problem. Consider a surface that is submerged vertically in a fluid. This problem is more difficult because the pressure is not constant over the surface.

Consider a vertical plate that is submerged in a fluid of weight-density w (per unit of volume), as shown in Figure 7.67. To determine the total force against *one side* of the region from depth c to depth d, you can subdivide the interval $[c, d]$ into n subintervals, each of width Δy. Next, consider the representative rectangle of width Δy and length $L(y_i)$, where y_i is in the ith subinterval. The force against this representative rectangle is

$$\Delta F_i = w(\text{depth})(\text{area})$$

$$= wh(y_i)L(y_i) \, \Delta y.$$

The force against n such rectangles is

$$\sum_{i=1}^{n} \Delta F_i = w \sum_{i=1}^{n} h(y_i)L(y_i) \, \Delta y.$$

Calculus methods must be used to find the fluid force on a vertical metal plate.

Figure 7.67

Note that w is considered to be constant and is factored out of the summation. Therefore, taking the limit as $\|\Delta\| \to 0$ ($n \to \infty$) suggests the next definition.

Definition of Force Exerted by a Fluid

The **force F exerted by a fluid** of constant weight-density w (per unit of volume) against a submerged vertical plane region from $y = c$ to $y = d$ is

$$F = w \lim_{\|\Delta\| \to 0} \sum_{i=1}^{n} h(y_i)L(y_i) \, \Delta y$$

$$= w \int_{c}^{d} h(y)L(y) \, dy$$

where $h(y)$ is the depth of the fluid at y and $L(y)$ is the horizontal length of the region at y.

No

EXAMPLE 2 **Fluid Force on a Vertical Surface**

•••▷ *See LarsonCalculus.com for an interactive version of this type of example.*

A vertical gate in a dam has the shape of an isosceles trapezoid 8 feet across the top and 6 feet across the bottom, with a height of 5 feet, as shown in Figure 7.68(a). What is the fluid force on the gate when the top of the gate is 4 feet below the surface of the water?

Solution In setting up a mathematical model for this problem, you are at liberty to locate the *x*- and *y*-axes in several different ways. A convenient approach is to let the *y*-axis bisect the gate and place the *x*-axis at the surface of the water, as shown in Figure 7.68(b). So, the depth of the water at *y* in feet is

$$\text{Depth} = h(y) = -y.$$

To find the length $L(y)$ of the region at *y*, find the equation of the line forming the right side of the gate. Because this line passes through the points $(3, -9)$ and $(4, -4)$, its equation is

$$y - (-9) = \frac{-4 - (-9)}{4 - 3}(x - 3)$$
$$y + 9 = 5(x - 3)$$
$$y = 5x - 24$$
$$x = \frac{y + 24}{5}.$$

(a) Water gate in a dam

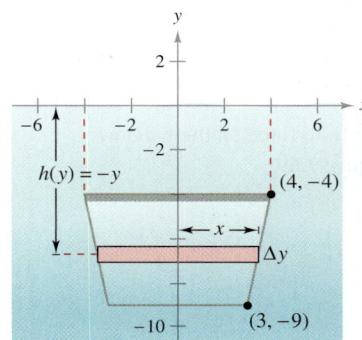

(b) The fluid force against the gate
Figure 7.68

In Figure 7.68(b) you can see that the length of the region at *y* is

$$\text{Length} = 2x = \frac{2}{5}(y + 24) = L(y).$$

Finally, by integrating from $y = -9$ to $y = -4$, you can calculate the fluid force to be

$$F = w \int_c^d h(y)L(y)\, dy$$
$$= 62.4 \int_{-9}^{-4} (-y)\left(\frac{2}{5}\right)(y + 24)\, dy$$
$$= -62.4 \left(\frac{2}{5}\right)\int_{-9}^{-4} (y^2 + 24y)\, dy$$
$$= -62.4 \left(\frac{2}{5}\right)\left[\frac{y^3}{3} + 12y^2\right]_{-9}^{-4}$$
$$= -62.4 \left(\frac{2}{5}\right)\left(\frac{-1675}{3}\right)$$
$$= 13{,}936 \text{ pounds.}$$

In Example 2, the *x*-axis coincided with the surface of the water. This was convenient, but arbitrary. In choosing a coordinate system to represent a physical situation, you should consider various possibilities. Often you can simplify the calculations in a problem by locating the coordinate system to take advantage of special characteristics of the problem, such as symmetry.

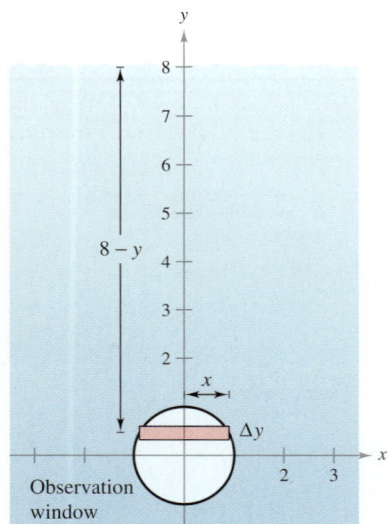

The fluid force on the window

Figure 7.69

EXAMPLE 3 **Fluid Force on a Vertical Surface**

A circular observation window on a marine science ship has a radius of 1 foot, and the center of the window is 8 feet below water level, as shown in Figure 7.69. What is the fluid force on the window?

Solution   To take advantage of symmetry, locate a coordinate system such that the origin coincides with the center of the window, as shown in Figure 7.69. The depth at y is then

$$\text{Depth} = h(y) = 8 - y.$$

The horizontal length of the window is $2x$, and you can use the equation for the circle, $x^2 + y^2 = 1$, to solve for x as shown.

$$\text{Length} = 2x$$
$$= 2\sqrt{1 - y^2} = L(y)$$

Finally, because y ranges from -1 to 1, and using 64 pounds per cubic foot as the weight-density of seawater, you have

$$F = w \int_c^d h(y)L(y)\,dy$$
$$= 64 \int_{-1}^1 (8 - y)(2)\sqrt{1 - y^2}\,dy.$$

Initially it looks as though this integral would be difficult to solve. However, when you break the integral into two parts and apply symmetry, the solution is simpler.

$$F = 64(16)\int_{-1}^1 \sqrt{1 - y^2}\,dy - 64(2)\int_{-1}^1 y\sqrt{1 - y^2}\,dy$$

The second integral is 0 (because the integrand is odd and the limits of integration are symmetric with respect to the origin). Moreover, by recognizing that the first integral represents the area of a semicircle of radius 1, you obtain

$$F = 64(16)\left(\frac{\pi}{2}\right) - 64(2)(0)$$
$$= 512\pi$$
$$\approx 1608.5 \text{ pounds.}$$

So, the fluid force on the window is about 1608.5 pounds. ◼

▷ **TECHNOLOGY**   To confirm the result obtained in Example 3, you might have considered using Simpson's Rule to approximate the value of

$$128 \int_{-1}^1 (8 - x)\sqrt{1 - x^2}\,dx.$$

From the graph of

$$f(x) = (8 - x)\sqrt{1 - x^2}$$

however, you can see that f is not differentiable when $x = \pm 1$ (see figure at the right). This means that you cannot apply Theorem 4.20 from Section 4.6 to determine the potential error in Simpson's Rule. Without knowing the potential error, the approximation is of little value. Use a graphing utility to approximate the integral.

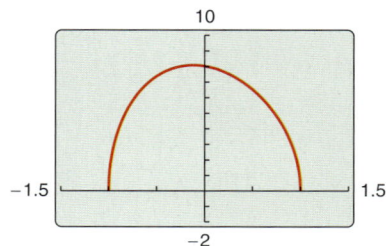

f is not differentiable at $x = \pm 1$.

7.7 Exercises

See **CalcChat.com** for tutorial help and worked-out solutions to odd-numbered exercises.

Force on a Submerged Sheet In Exercises 1–4, the area of the top side of a piece of sheet metal is given. The sheet metal is submerged horizontally in 8 feet of water. Find the fluid force on the top side.

1. 3 square feet
2. 8 square feet
3. 10 square feet
4. 25 square feet

Buoyant Force In Exercises 5 and 6, find the buoyant force of a rectangular solid of the given dimensions submerged in water so that the top side is parallel to the surface of the water. The buoyant force is the difference between the fluid forces on the top and bottom sides of the solid.

5.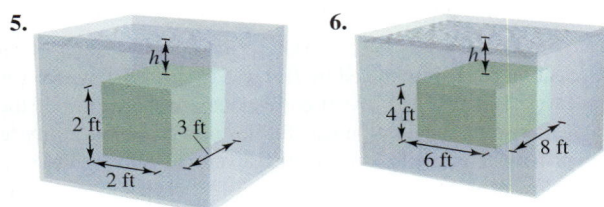

6.

Fluid Force on a Tank Wall In Exercises 7–12, find the fluid force on the vertical side of the tank, where the dimensions are given in feet. Assume that the tank is full of water.

7. Rectangle

8. Triangle

9. Trapezoid

10. Semicircle

11. Parabola, $y = x^2$

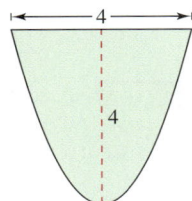

12. Semiellipse,
$$y = -\frac{1}{2}\sqrt{36 - 9x^2}$$

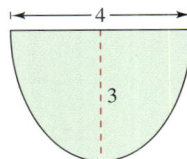

Fluid Force of Water In Exercises 13–16, find the fluid force on the vertical plate submerged in water, where the dimensions are given in meters and the weight-density of water is 9800 newtons per cubic meter.

13. Square

14. Square

15. Triangle

16. Rectangle

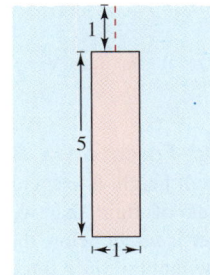

Force on a Concrete Form In Exercises 17–20, the figure is the vertical side of a form for poured concrete that weighs 140.7 pounds per cubic foot. Determine the force on this part of the concrete form.

17. Rectangle

18. Semiellipse,
$$y = -\frac{3}{4}\sqrt{16 - x^2}$$

19. Rectangle

20. Triangle

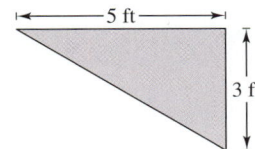

21. **Fluid Force of Gasoline** A cylindrical gasoline tank is placed so that the axis of the cylinder is horizontal. Find the fluid force on a circular end of the tank when the tank is half full, where the diameter is 3 feet and the gasoline weighs 42 pounds per cubic foot.

22. Fluid Force of Gasoline Repeat Exercise 21 for a tank that is full. (Evaluate one integral by a geometric formula and the other by observing that the integrand is an odd function.)

23. Fluid Force on a Circular Plate A circular plate of radius r feet is submerged vertically in a tank of fluid that weighs w pounds per cubic foot. The center of the circle is k feet below the surface of the fluid, where $k > r$. Show that the fluid force on the surface of the plate is

$$F = wk(\pi r^2).$$

(Evaluate one integral by a geometric formula and the other by observing that the integrand is an odd function.)

24. Fluid Force on a Circular Plate Use the result of Exercise 23 to find the fluid force on the circular plate shown in each figure. Assume the plates are in the wall of a tank filled with water and the measurements are given in feet.

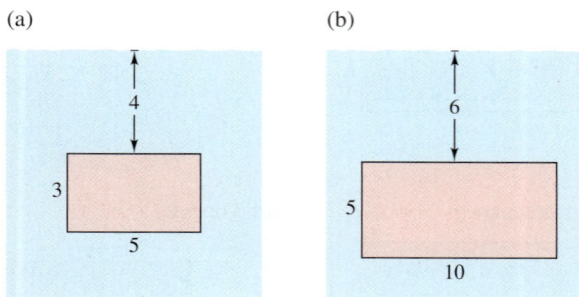

(a)

(b)

25. Fluid Force on a Rectangular Plate A rectangular plate of height h feet and base b feet is submerged vertically in a tank of fluid that weighs w pounds per cubic foot. The center is k feet below the surface of the fluid, where $k > h/2$. Show that the fluid force on the surface of the plate is

$$F = wkhb.$$

26. Fluid Force on a Rectangular Plate Use the result of Exercise 25 to find the fluid force on the rectangular plate shown in each figure. Assume the plates are in the wall of a tank filled with water and the measurements are given in feet.

(a)

(b)

27. Submarine Porthole A square porthole on a vertical side of a submarine (submerged in seawater) has an area of 1 square foot. Find the fluid force on the porthole, assuming that the center of the square is 15 feet below the surface.

28. Submarine Porthole Repeat Exercise 27 for a circular porthole that has a diameter of 1 foot. The center is 15 feet below the surface.

29. Modeling Data The vertical stern of a boat with a superimposed coordinate system is shown in the figure. The table shows the widths w of the stern (in feet) at indicated values of y. Find the fluid force against the stern.

y	0	$\frac{1}{2}$	1	$\frac{3}{2}$	2	$\frac{5}{2}$	3	$\frac{7}{2}$	4
w	0	3	5	8	9	10	10.25	10.5	10.5

30. Irrigation Canal Gate The vertical cross section of an irrigation canal is modeled by $f(x) = 5x^2/(x^2 + 4)$, where x is measured in feet and $x = 0$ corresponds to the center of the canal. Use the integration capabilities of a graphing utility to approximate the fluid force against a vertical gate used to stop the flow of water when the water is 3 feet deep.

WRITING ABOUT CONCEPTS

31. Think About It Approximate the depth of the water in the tank in Exercise 7 if the fluid force is one-half as great as when the tank is full. Explain why the answer is not $\frac{3}{2}$.

32. Fluid Pressure and Fluid Force

(a) Define fluid pressure.

(b) Define fluid force against a submerged vertical plane region.

33. Fluid Pressure Explain why fluid pressure on a surface is calculated using horizontal representative rectangles instead of vertical representative rectangles.

34. HOW DO YOU SEE IT? Two identical semicircular windows are placed at the same depth in the vertical wall of an aquarium (see figure). Which is subjected to the greater fluid force? Explain.

Review Exercises

See **CalcChat.com** for tutorial help and worked-out solutions to odd-numbered exercises.

Finding the Area of a Region In Exercises 1–10, sketch the region bounded by the graphs of the equations and find the area of the region.

1. $y = 6 - \frac{1}{2}x^2$, $y = \frac{3}{4}x$, $x = -2$, $x = 2$

2. $y = \frac{1}{x^2}$, $y = 4$, $x = 5$

3. $y = \frac{1}{x^2 + 1}$, $y = 0$, $x = -1$, $x = 1$

4. $x = y^2 - 2y$, $x = -1$, $y = 0$

5. $y = x$, $y = x^3$

6. $x = y^2 + 1$, $x = y + 3$

7. $y = e^x$, $y = e^2$, $x = 0$

8. $y = \csc x$, $y = 2$, $\frac{\pi}{6} \le x \le \frac{5\pi}{6}$

9. $y = \sin x$, $y = \cos x$, $\frac{\pi}{4} \le x \le \frac{5\pi}{4}$

10. $x = \cos y$, $x = \frac{1}{2}$, $\frac{\pi}{3} \le y \le \frac{7\pi}{3}$

Finding the Area of a Region In Exercises 11–14, use a graphing utility to graph the region bounded by the graphs of the equations, and use the integration capabilities of the graphing utility to find the area of the region.

11. $y = x^2 - 8x + 3$, $y = 3 + 8x - x^2$

12. $y = x^2 - 4x + 3$, $y = x^3$, $x = 0$

13. $\sqrt{x} + \sqrt{y} = 1$, $y = 0$, $x = 0$

14. $y = x^4 - 2x^2$, $y = 2x^2$

15. **Numerical Integration** Estimate the surface area of the pond using (a) the Trapezoidal Rule and (b) Simpson's Rule.

16. **Revenue** The models $R_1 = 6.4 + 0.2t + 0.01t^2$ and $R_2 = 8.4 + 0.35t$ give the revenue (in billions of dollars) for a large corporation. Both models are estimates of the revenues from 2015 through 2020, with $t = 15$ corresponding to 2015. Which model projects the greater revenue? How much more total revenue does that model project over the six-year period?

Finding the Volume of a Solid In Exercises 17–22, use the disk method *or* the shell method to find the volumes of the solids generated by revolving the region bounded by the graphs of the equations about the given line(s).

17. $y = x$, $y = 0$, $x = 3$
    (a) the x-axis   (b) the y-axis
    (c) the line $x = 3$ (d) the line $x = 6$

18. $y = \sqrt{x}$, $y = 2$, $x = 0$
    (a) the x-axis   (b) the line $y = 2$
    (c) the y-axis   (d) the line $x = -1$

19. $y = \frac{1}{x^4 + 1}$, $y = 0$, $x = 0$, $x = 1$

    revolved about the y-axis

20. $y = \frac{1}{\sqrt{1 + x^2}}$, $y = 0$, $x = -1$, $x = 1$

    revolved about the x-axis

21. $y = \frac{1}{x^2}$, $y = 0$, $x = 2$, $x = 5$

    revolved about the y-axis

22. $y = e^{-x}$, $y = 0$, $x = 0$, $x = 1$

    revolved about the x-axis

23. **Depth of Gasoline in a Tank** A gasoline tank is an oblate spheroid generated by revolving the region bounded by the graph of

    $$\frac{x^2}{16} + \frac{y^2}{9} = 1$$

    about the y-axis, where x and y are measured in feet. Find the depth of the gasoline in the tank when it is filled to one-fourth its capacity.

24. **Using Cross Sections** Find the volume of the solid whose base is bounded by the circle $x^2 + y^2 = 9$ and the cross sections perpendicular to the x-axis are equilateral triangles.

Finding Arc Length In Exercises 25 and 26, find the arc length of the graph of the function over the indicated interval.

25. $f(x) = \frac{4}{5}x^{5/4}$, $[0, 4]$ 26. $y = \frac{1}{6}x^3 + \frac{1}{2x}$, $[1, 3]$

27. **Length of a Catenary** A cable of a suspension bridge forms a catenary modeled by the equation

    $$y = 300 \cosh\left(\frac{x}{2000}\right) - 280, \quad -2000 \le x \le 2000$$

    where x and y are measured in feet. Use the integration capabilities of a graphing utility to approximate the length of the cable.

28. Approximation Determine which value best approximates the length of the arc represented by the integral

$$\int_0^1 \sqrt{1 + \left[\frac{d}{dx}\left(\frac{4}{x+1}\right)\right]^2}\, dx.$$

(Make your selection on the basis of a sketch of the arc and *not* by performing any calculations.)

(a) 10   (b) -5 (c) 2   (d) 4   (e) 1

29. Surface Area Use integration to find the lateral surface area of a right circular cone of height 4 and radius 3.

30. Surface Area The region bounded by the graphs of $y = 2\sqrt{x}$, $y = 0$, $x = 3$, and $x = 8$ is revolved about the x-axis. Find the surface area of the solid generated.

31. Work A force of 5 pounds is needed to stretch a spring 1 inch from its natural position. Find the work done in stretching the spring from its natural length of 10 inches to a length of 15 inches.

32. Work A force of 50 pounds is needed to stretch a spring 1 inch from its natural position. Find the work done in stretching the spring from its natural length of 10 inches to double that length.

33. Work A water well has an 8-inch casing (diameter) and is 190 feet deep. The water is 25 feet from the top of the well. Determine the amount of work done in pumping the well dry, assuming that no water enters it while it is being pumped.

34. Boyle's Law A quantity of gas with an initial volume of 2 cubic feet and a pressure of 800 pounds per square foot expands to a volume of 3 cubic feet. Find the work done by the gas. Assume that the pressure is inversely proportional to the volume.

35. Work A chain 10 feet long weighs 4 pounds per foot and is hung from a platform 20 feet above the ground. How much work is required to raise the entire chain to the 20-foot level?

36. Work A windlass, 200 feet above ground level on the top of a building, uses a cable weighing 5 pounds per foot. Find the work done in winding up the cable when

(a) one end is at ground level.

(b) there is a 300-pound load attached to the end of the cable.

37. Work The work done by a variable force in a press is 80 foot-pounds. The press moves a distance of 4 feet, and the force is a quadratic of the form $F = ax^2$. Find a.

38. Work Find the work done by the force F shown in the figure.

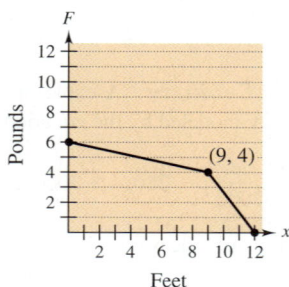

39. Center of Mass of a Linear System Find the center of mass of the point masses lying on the x-axis.

$m_1 = 8$, $m_2 = 12$, $m_3 = 6$, $m_4 = 14$

$x_1 = -1$, $x_2 = 2$, $x_3 = 5$, $x_4 = 7$

40. Center of Mass of a Two-Dimensional System Find the center of mass of the given system of point masses.

m_i	3	2	6	9
(x_i, y_i)	$(2, 1)$	$(-3, 2)$	$(4, -1)$	$(6, 5)$

Finding a Centroid In Exercises 41 and 42, find the centroid of the region bounded by the graphs of the equations.

41. $y = x^2$, $y = 2x + 3$ **42.** $y = x^{2/3}$, $y = \frac{1}{2}x$

43. Centroid A blade on an industrial fan has the configuration of a semicircle attached to a trapezoid (see figure). Find the centroid of the blade.

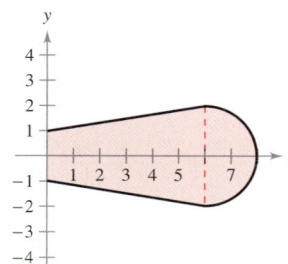

44. Finding Volume Use the Theorem of Pappus to find the volume of the torus formed by revolving the circle $(x - 4)^2 + y^2 = 4$ about the y-axis.

45. Fluid Force of Seawater Find the fluid force on the vertical plate submerged in seawater (see figure).

Figure for 45   **Figure for 46**

46. Force on a Concrete Form The figure is the vertical side of a form for poured concrete that weights 140.7 pounds per cubic foot. Determine the force on this part of the concrete form.

47. Fluid Force A swimming pool is 5 feet deep at one end and 10 feet deep at the other, and the bottom is an inclined plane. The length and width of the pool are 40 feet and 20 feet. If the pool is full of water, what is the fluid force on each of the vertical walls?

P.S. Problem Solving

See **CalcChat.com** for tutorial help and worked-out solutions to odd-numbered exercises.

1. Finding a Limit Let R be the area of the region in the first quadrant bounded by the parabola $y = x^2$ and the line $y = cx$, $c > 0$. Let T be the area of the triangle AOB. Calculate the limit

$$\lim_{c \to 0^+} \frac{T}{R}.$$

Figure for 1

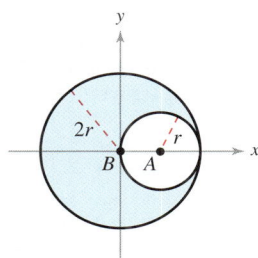

Figure for 2

2. Center of Mass of a Lamina Let L be the lamina of uniform density $\rho = 1$ obtained by removing circle A of radius r from circle B of radius $2r$ (see figure).

(a) Show that $M_x = 0$ for L.

(b) Show that M_y for L is equal to $(M_y$ for $B) - (M_y$ for $A)$.

(c) Find M_y for B and M_y for A. Then use part (b) to compute M_y for L.

(d) What is the center of mass of L?

3. Dividing a Region Let R be the region bounded by the parabola $y = x - x^2$ and the x-axis. Find the equation of the line $y = mx$ that divides this region into two regions of equal area.

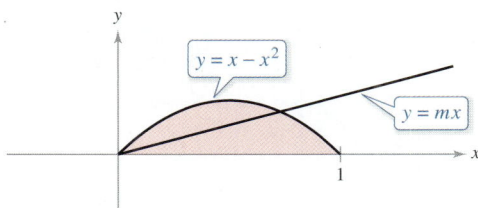

4. Volume A hole is cut through the center of a sphere of radius r (see figure). The height of the remaining spherical ring is h. Find the volume of the ring and show that it is independent of the radius of the sphere.

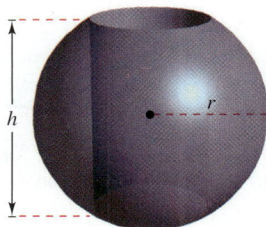

5. Surface Area Graph the curve

$$8y^2 = x^2(1 - x^2).$$

Use a computer algebra system to find the surface area of the solid of revolution obtained by revolving the curve about the y-axis.

6. Torus

(a) A torus is formed by revolving the region bounded by the circle

$$(x - 2)^2 + y^2 = 1$$

about the y-axis (see figure). Use the disk method to calculate the volume of the torus.

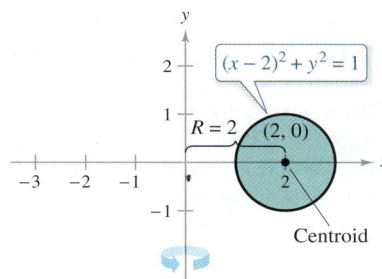

(b) Use the disk method to find the volume of the general torus when the circle has radius r and its center is R units from the axis of rotation.

7. Volume A rectangle R of length ℓ and width w is revolved about the line L (see figure). Find the volume of the resulting solid of revolution.

Figure for 7

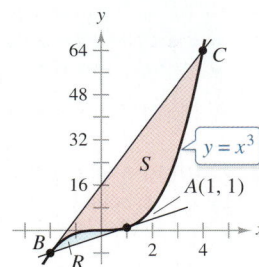

Figure for 8

8. Comparing Areas of Regions

(a) The tangent line to the curve $y = x^3$ at the point $A(1, 1)$ intersects the curve at another point B. Let R be the area of the region bounded by the curve and the tangent line. The tangent line at B intersects the curve at another point C (see figure). Let S be the area of the region bounded by the curve and this second tangent line. How are the areas R and S related?

(b) Repeat the construction in part (a) by selecting an arbitrary point A on the curve $y = x^3$. Show that the two areas R and S are always related in the same way.

9. Using Arc Length The graph of $y = f(x)$ passes through the origin. The arc length of the curve from $(0, 0)$ to $(x, f(x))$ is given by

$$s(x) = \int_0^x \sqrt{1 + e^t}\, dt.$$

Identify the function f.

10. Using a Function Let f be rectifiable on the interval $[a, b]$, and let

$$s(x) = \int_a^x \sqrt{1 + [f'(t)]^2}\, dt.$$

(a) Find $\dfrac{ds}{dx}$.

(b) Find ds and $(ds)^2$.

(c) Find $s(x)$ on $[1, 3]$ when $f(t) = t^{3/2}$.

(d) Use the function and interval in part (c) to calculate $s(2)$ and describe what it signifies.

11. Archimedes' Principle **Archimedes' Principle** states that the upward or buoyant force on an object within a fluid is equal to the weight of the fluid that the object displaces. For a partially submerged object, you can obtain information about the relative densities of the floating object and the fluid by observing how much of the object is above and below the surface. You can also determine the size of a floating object if you know the amount that is above the surface and the relative densities. You can see the top of a floating iceberg (see figure). The density of ocean water is 1.03×10^3 kilograms per cubic meter, and that of ice is 0.92×10^3 kilograms per cubic meter. What percent of the total iceberg is below the surface?

12. Finding a Centroid Sketch the region bounded on the left by $x = 1$, bounded above by $y = 1/x^3$, and bounded below by $y = -1/x^3$.

(a) Find the centroid of the region for $1 \le x \le 6$.

(b) Find the centroid of the region for $1 \le x \le b$.

(c) Where is the centroid as $b \to \infty$?

13. Finding a Centroid Sketch the region to the right of the y-axis, bounded above by $y = 1/x^4$, and bounded below by $y = -1/x^4$.

(a) Find the centroid of the region for $1 \le x \le 6$.

(b) Find the centroid of the region for $1 \le x \le b$.

(c) Where is the centroid as $b \to \infty$?

14. Work Find the work done by each force F.

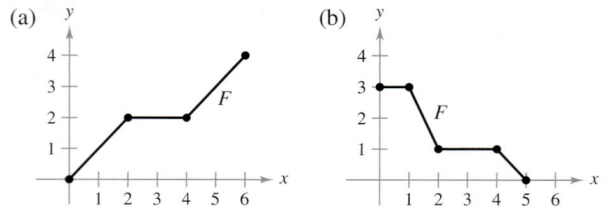

(a)

(b)

Consumer and Producer Surplus In Exercises 15 and 16, find the consumer surplus and producer surplus for the given demand $[p_1(x)]$ and supply $[p_2(x)]$ curves. The consumer surplus and producer surplus are represented by the areas shown in the figure.

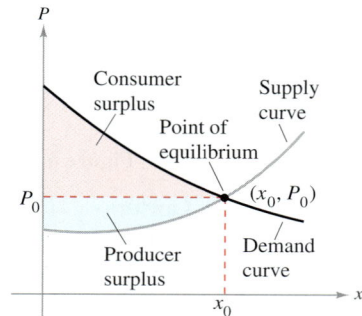

15. $p_1(x) = 50 - 0.5x$, $\quad p_2(x) = 0.125x$

16. $p_1(x) = 1000 - 0.4x^2$, $\quad p_2(x) = 42x$

17. Fluid Force A swimming pool is 20 feet wide, 40 feet long, 4 feet deep at one end, and 8 feet deep at the other end (see figures). The bottom is an inclined plane. Find the fluid force on each vertical wall.

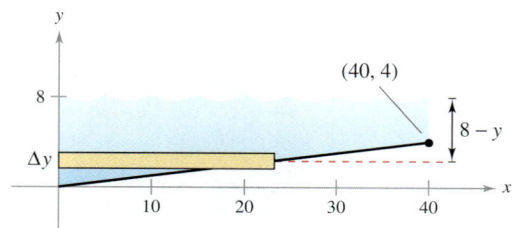

8 Integration Techniques, L'Hôpital's Rule, and Improper Integrals

Chemical Reaction *(Exercise 50, p. 550)*

Sending a Space Module into Orbit *(Example 5, p. 571)*

Fluid Force *(Exercise 69, p. 541)*

Power Lines *(Section Project, p. 532)*

Memory Model *(Exercise 88, p. 523)*

507

8.1 Basic Integration Rules

■ Review procedures for fitting an integrand to one of the basic integration rules.

REVIEW OF BASIC INTEGRATION RULES
($a > 0$)

1. $\int kf(u)\, du = k \int f(u)\, du$

2. $\int [f(u) \pm g(u)]\, du =$
   $\int f(u)\, du \pm \int g(u)\, du$

3. $\int du = u + C$

4. $\int u^n\, du = \dfrac{u^{n+1}}{n+1} + C,$
   $n \neq -1$

5. $\int \dfrac{du}{u} = \ln|u| + C$

6. $\int e^u\, du = e^u + C$

7. $\int a^u\, du = \left(\dfrac{1}{\ln a}\right) a^u + C$

8. $\int \sin u\, du = -\cos u + C$

9. $\int \cos u\, du = \sin u + C$

10. $\int \tan u\, du = -\ln|\cos u| + C$

11. $\int \cot u\, du = \ln|\sin u| + C$

12. $\int \sec u\, du =$
    $\ln|\sec u + \tan u| + C$

13. $\int \csc u\, du =$
    $-\ln|\csc u + \cot u| + C$

14. $\int \sec^2 u\, du = \tan u + C$

15. $\int \csc^2 u\, du = -\cot u + C$

16. $\int \sec u \tan u\, du = \sec u + C$

17. $\int \csc u \cot u\, du = -\csc u + C$

18. $\int \dfrac{du}{\sqrt{a^2 - u^2}} = \arcsin \dfrac{u}{a} + C$

19. $\int \dfrac{du}{a^2 + u^2} = \dfrac{1}{a} \arctan \dfrac{u}{a} + C$

20. $\int \dfrac{du}{u\sqrt{u^2 - a^2}} = \dfrac{1}{a} \text{arcsec} \dfrac{|u|}{a} + C$

Fitting Integrands to Basic Integration Rules

In this chapter, you will study several integration techniques that greatly expand the set of integrals to which the basic integration rules can be applied. These rules are reviewed at the left. A major step in solving any integration problem is recognizing which basic integration rule to use.

EXAMPLE 1   A Comparison of Three Similar Integrals

⋯▷ *See LarsonCalculus.com for an interactive version of this type of example.*

Find each integral.

a. $\displaystyle\int \frac{4}{x^2 + 9}\, dx$ **b.** $\displaystyle\int \frac{4x}{x^2 + 9}\, dx$ **c.** $\displaystyle\int \frac{4x^2}{x^2 + 9}\, dx$

Solution

a. Use the Arctangent Rule and let $u = x$ and $a = 3$.

$$\int \frac{4}{x^2 + 9}\, dx = 4 \int \frac{1}{x^2 + 3^2}\, dx \qquad \text{Constant Multiple Rule}$$

$$= 4\left(\frac{1}{3} \arctan \frac{x}{3}\right) + C \qquad \text{Arctangent Rule}$$

$$= \frac{4}{3} \arctan \frac{x}{3} + C \qquad \text{Simplify.}$$

b. The Arctangent Rule does not apply because the numerator contains a factor of x. Consider the Log Rule and let $u = x^2 + 9$. Then $du = 2x\, dx$, and you have

$$\int \frac{4x}{x^2 + 9}\, dx = 2 \int \frac{2x\, dx}{x^2 + 9} \qquad \text{Constant Multiple Rule}$$

$$= 2 \int \frac{du}{u} \qquad \text{Substitution: } u = x^2 + 9$$

$$= 2 \ln|u| + C \qquad \text{Log Rule}$$

$$= 2 \ln(x^2 + 9) + C. \qquad \text{Rewrite as a function of } x.$$

c. Because the degree of the numerator is equal to the degree of the denominator, you should first use division to rewrite the improper rational function as the sum of a polynomial and a proper rational function.

$$\int \frac{4x^2}{x^2 + 9}\, dx = \int \left(4 + \frac{-36}{x^2 + 9}\right) dx \qquad \text{Rewrite using long division.}$$

$$= \int 4\, dx - 36 \int \frac{1}{x^2 + 9}\, dx \qquad \text{Write as two integrals.}$$

$$= 4x - 36\left(\frac{1}{3} \arctan \frac{x}{3}\right) + C \qquad \text{Integrate.}$$

$$= 4x - 12 \arctan \frac{x}{3} + C \qquad \text{Simplify.}$$

Note in Example 1(c) that some algebra is required before applying any integration rules, and more than one rule is needed to evaluate the resulting integral.

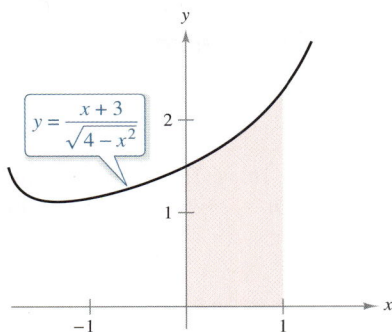

$$y = \frac{x + 3}{\sqrt{4 - x^2}}$$

The area of the region is approximately 1.839.

Figure 8.1

EXAMPLE 2 **Using Two Basic Rules to Solve a Single Integral**

Evaluate $\displaystyle\int_0^1 \frac{x + 3}{\sqrt{4 - x^2}}\, dx$.

Solution Begin by writing the integral as the sum of two integrals. Then apply the Power Rule and the Arcsine Rule.

$$\int_0^1 \frac{x + 3}{\sqrt{4 - x^2}}\, dx = \int_0^1 \frac{x}{\sqrt{4 - x^2}}\, dx + \int_0^1 \frac{3}{\sqrt{4 - x^2}}\, dx$$

$$= -\frac{1}{2}\int_0^1 (4 - x^2)^{-1/2}(-2x)\, dx + 3\int_0^1 \frac{1}{\sqrt{2^2 - x^2}}\, dx$$

$$= \left[-(4 - x^2)^{1/2} + 3\arcsin\frac{x}{2} \right]_0^1$$

$$= \left(-\sqrt{3} + \frac{\pi}{2} \right) - (-2 + 0)$$

$$\approx 1.839 \qquad \text{See Figure 8.1.}$$

▷ **TECHNOLOGY** Simpson's Rule can be used to give a good approximation of the value of the integral in Example 2 (for $n = 10$, the approximation is 1.839). When using numerical integration, however, you should be aware that Simpson's Rule does not always give good approximations when one or both of the limits of integration are near a vertical asymptote. For instance, using the Fundamental Theorem of Calculus, you can obtain

$$\int_0^{1.99} \frac{x + 3}{\sqrt{4 - x^2}}\, dx \approx 6.213.$$

For $n = 10$, Simpson's Rule gives an approximation of 6.889.

Rules 18, 19, and 20 of the basic integration rules on the preceding page all have expressions involving the sum or difference of two squares:

$$a^2 - u^2, \quad a^2 + u^2, \quad \text{and} \quad u^2 - a^2.$$

These expressions are often apparent after a u-substitution, as shown in Example 3.

Exploration

A Comparison of Three Similar Integrals Which, if any, of the integrals listed below can be evaluated using the 20 basic integration rules? For any that can be evaluated, do so. For any that cannot, explain why not.

a. $\displaystyle\int \frac{3}{\sqrt{1 - x^2}}\, dx$

b. $\displaystyle\int \frac{3x}{\sqrt{1 - x^2}}\, dx$

c. $\displaystyle\int \frac{3x^2}{\sqrt{1 - x^2}}\, dx$

EXAMPLE 3 **A Substitution Involving $a^2 - u^2$**

Find $\displaystyle\int \frac{x^2}{\sqrt{16 - x^6}}\, dx$.

Solution Because the radical in the denominator can be written in the form

$$\sqrt{a^2 - u^2} = \sqrt{4^2 - (x^3)^2}$$

you can try the substitution $u = x^3$. Then $du = 3x^2\, dx$, and you have

$$\int \frac{x^2}{\sqrt{16 - x^6}}\, dx = \frac{1}{3}\int \frac{3x^2\, dx}{\sqrt{16 - (x^3)^2}} \qquad \text{Rewrite integral.}$$

$$= \frac{1}{3}\int \frac{du}{\sqrt{4^2 - u^2}} \qquad \text{Substitution: } u = x^3$$

$$= \frac{1}{3}\arcsin\frac{u}{4} + C \qquad \text{Arcsine Rule}$$

$$= \frac{1}{3}\arcsin\frac{x^3}{4} + C. \qquad \text{Rewrite as a function of } x.$$

Two of the most commonly overlooked integration rules are the Log Rule and the Power Rule. Notice in the next two examples how these two integration rules can be disguised.

EXAMPLE 4 A Disguised Form of the Log Rule

Find $\displaystyle\int \frac{1}{1 + e^x}\, dx$.

Solution The integral does not appear to fit any of the basic rules. The quotient form, however, suggests the Log Rule. If you let $u = 1 + e^x$, then $du = e^x\, dx$. You can obtain the required du by adding and subtracting e^x in the numerator.

$$\int \frac{1}{1 + e^x}\, dx = \int \frac{1 + e^x - e^x}{1 + e^x}\, dx \qquad \text{Add and subtract } e^x \text{ in numerator.}$$

$$= \int \left(\frac{1 + e^x}{1 + e^x} - \frac{e^x}{1 + e^x} \right) dx \qquad \text{Rewrite as two fractions.}$$

$$= \int dx - \int \frac{e^x\, dx}{1 + e^x} \qquad \text{Rewrite as two integrals.}$$

$$= x - \ln(1 + e^x) + C \qquad \text{Integrate.} \quad\blacksquare$$

> **REMARK** Remember that you can separate numerators but not denominators. Watch out for this common error when fitting integrands to basic rules. For instance, you cannot separate denominators in Example 4.
>
> $$\frac{1}{1 + e^x} \neq \frac{1}{1} + \frac{1}{e^x}$$

There is usually more than one way to solve an integration problem. For instance, in Example 4, try integrating by multiplying the numerator and denominator by e^{-x} to obtain an integral of the form $-\int du/u$. See if you can get the same answer by this procedure. (Be careful: the answer will appear in a different form.)

EXAMPLE 5 A Disguised Form of the Power Rule

Find $\displaystyle\int (\cot x)[\ln(\sin x)]\, dx$.

Solution Again, the integral does not appear to fit any of the basic rules. However, considering the two primary choices for u

$$u = \cot x \quad \text{or} \quad u = \ln(\sin x)$$

you can see that the second choice is the appropriate one because

$$u = \ln(\sin x) \quad \text{and} \quad du = \frac{\cos x}{\sin x}\, dx = \cot x\, dx.$$

So,

$$\int (\cot x)[\ln(\sin x)]\, dx = \int u\, du \qquad \text{Substitution: } u = \ln(\sin x)$$

$$= \frac{u^2}{2} + C \qquad \text{Integrate.}$$

$$= \frac{1}{2}[\ln(\sin x)]^2 + C. \qquad \text{Rewrite as a function of } x. \quad\blacksquare$$

In Example 5, try *checking* that the derivative of

$$\frac{1}{2}[\ln(\sin x)]^2 + C$$

is the integrand of the original integral.

Trigonometric identities can often be used to fit integrals to one of the basic integration rules.

Using Trigonometric Identities

Find $\int \tan^2 2x \, dx$.

▷ **TECHNOLOGY** If you have access to a computer algebra system, try using it to evaluate the integrals in this section. Compare the *forms* of the antiderivatives given by the software with the forms obtained by hand. Sometimes the forms will be the same, but often they will differ. For instance, why is the antiderivative $\ln 2x + C$ equivalent to the antiderivative $\ln x + C$?

Solution   Note that $\tan^2 u$ is not in the list of basic integration rules. However, $\sec^2 u$ is in the list. This suggests the trigonometric identity $\tan^2 u = \sec^2 u - 1$. If you let $u = 2x$, then $du = 2\,dx$ and

$$\int \tan^2 2x \, dx = \frac{1}{2}\int \tan^2 u \, du \qquad \text{Substitution: } u = 2x$$

$$= \frac{1}{2}\int (\sec^2 u - 1)\, du \qquad \text{Trigonometric identity}$$

$$= \frac{1}{2}\int \sec^2 u \, du - \frac{1}{2}\int du \qquad \text{Rewrite as two integrals.}$$

$$= \frac{1}{2}\tan u - \frac{u}{2} + C \qquad \text{Integrate.}$$

$$= \frac{1}{2}\tan 2x - x + C. \qquad \text{Rewrite as a function of } x.$$

This section concludes with a summary of the common procedures for fitting integrands to the basic integration rules.

PROCEDURES FOR FITTING INTEGRANDS TO BASIC INTEGRATION RULES

Technique	Example
Expand (numerator).	$(1 + e^x)^2 = 1 + 2e^x + e^{2x}$
Separate numerator.	$\dfrac{1+x}{x^2+1} = \dfrac{1}{x^2+1} + \dfrac{x}{x^2+1}$
Complete the square.	$\dfrac{1}{\sqrt{2x-x^2}} = \dfrac{1}{\sqrt{1-(x-1)^2}}$
Divide improper rational function.	$\dfrac{x^2}{x^2+1} = 1 - \dfrac{1}{x^2+1}$
Add and subtract terms in numerator.	$\dfrac{2x}{x^2+2x+1} = \dfrac{2x+2-2}{x^2+2x+1}$
	$= \dfrac{2x+2}{x^2+2x+1} - \dfrac{2}{(x+1)^2}$
Use trigonometric identities.	$\cot^2 x = \csc^2 x - 1$
Multiply and divide by Pythagorean conjugate.	$\dfrac{1}{1+\sin x} = \left(\dfrac{1}{1+\sin x}\right)\left(\dfrac{1-\sin x}{1-\sin x}\right)$
	$= \dfrac{1-\sin x}{1-\sin^2 x}$
	$= \dfrac{1-\sin x}{\cos^2 x}$
	$= \sec^2 x - \dfrac{\sin x}{\cos^2 x}$

8.1 Exercises

See **CalcChat.com** for tutorial help and worked-out solutions to odd-numbered exercises.

Choosing an Antiderivative In Exercises 1–4, select the correct antiderivative.

1. $\dfrac{dy}{dx} = \dfrac{x}{\sqrt{x^2 + 1}}$

   (a) $2\sqrt{x^2 + 1} + C$ (b) $\sqrt{x^2 + 1} + C$

   (c) $\frac{1}{2}\sqrt{x^2 + 1} + C$ (d) $\ln(x^2 + 1) + C$

2. $\dfrac{dy}{dx} = \dfrac{x}{x^2 + 1}$

   (a) $\ln\sqrt{x^2 + 1} + C$ (b) $\dfrac{2x}{(x^2 + 1)^2} + C$

   (c) $\arctan x + C$ (d) $\ln(x^2 + 1) + C$

3. $\dfrac{dy}{dx} = \dfrac{1}{x^2 + 1}$

   (a) $\ln\sqrt{x^2 + 1} + C$ (b) $\dfrac{2x}{(x^2 + 1)^2} + C$

   (c) $\arctan x + C$ (d) $\ln(x^2 + 1) + C$

4. $\dfrac{dy}{dx} = x\cos(x^2 + 1)$

   (a) $2x\sin(x^2 + 1) + C$ (b) $-\frac{1}{2}\sin(x^2 + 1) + C$

   (c) $\frac{1}{2}\sin(x^2 + 1) + C$ (d) $-2x\sin(x^2 + 1) + C$

Choosing a Formula In Exercises 5–14, select the basic integration formula you can use to find the integral, and identify u and a when appropriate.

5. $\displaystyle\int (5x - 3)^4 \, dx$

6. $\displaystyle\int \dfrac{2t + 1}{t^2 + t - 4} \, dt$

7. $\displaystyle\int \dfrac{1}{\sqrt{x}(1 - 2\sqrt{x})} \, dx$

8. $\displaystyle\int \dfrac{2}{(2t - 1)^2 + 4} \, dt$

9. $\displaystyle\int \dfrac{3}{\sqrt{1 - t^2}} \, dt$

10. $\displaystyle\int \dfrac{-2x}{\sqrt{x^2 - 4}} \, dx$

11. $\displaystyle\int t\sin t^2 \, dt$

12. $\displaystyle\int \sec 5x \tan 5x \, dx$

13. $\displaystyle\int (\cos x)e^{\sin x} \, dx$

14. $\displaystyle\int \dfrac{1}{x\sqrt{x^2 - 4}} \, dx$

Finding an Indefinite Integral In Exercises 15–46, find the indefinite integral.

15. $\displaystyle\int 14(x - 5)^6 \, dx$

16. $\displaystyle\int \dfrac{5}{(t + 6)^3} \, dt$

17. $\displaystyle\int \dfrac{7}{(z - 10)^7} \, dz$

18. $\displaystyle\int t^3\sqrt{t^4 + 1} \, dt$

19. $\displaystyle\int \left[v + \dfrac{1}{(3v - 1)^3} \right] dv$

20. $\displaystyle\int \left[4x - \dfrac{2}{(2x + 3)^2} \right] dx$

21. $\displaystyle\int \dfrac{t^2 - 3}{-t^3 + 9t + 1} \, dt$

22. $\displaystyle\int \dfrac{x + 1}{\sqrt{3x^2 + 6x}} \, dx$

23. $\displaystyle\int \dfrac{x^2}{x - 1} \, dx$

24. $\displaystyle\int \dfrac{3x}{x + 4} \, dx$

25. $\displaystyle\int \dfrac{e^x}{1 + e^x} \, dx$

26. $\displaystyle\int \left(\dfrac{1}{2x + 5} - \dfrac{1}{2x - 5} \right) dx$

27. $\displaystyle\int (5 + 4x^2)^2 \, dx$

28. $\displaystyle\int x\left(3 + \dfrac{2}{x} \right)^2 dx$

29. $\displaystyle\int x\cos 2\pi x^2 \, dx$

30. $\displaystyle\int \csc \pi x \cot \pi x \, dx$

31. $\displaystyle\int \dfrac{\sin x}{\sqrt{\cos x}} \, dx$

32. $\displaystyle\int \csc^2 x e^{\cot x} \, dx$

33. $\displaystyle\int \dfrac{2}{e^{-x} + 1} \, dx$

34. $\displaystyle\int \dfrac{2}{7e^x + 4} \, dx$

35. $\displaystyle\int \dfrac{\ln x^2}{x} \, dx$

36. $\displaystyle\int (\tan x)[\ln(\cos x)] \, dx$

37. $\displaystyle\int \dfrac{1 + \cos \alpha}{\sin \alpha} \, d\alpha$

38. $\displaystyle\int \dfrac{1}{\cos \theta - 1} \, d\theta$

39. $\displaystyle\int \dfrac{-1}{\sqrt{1 - (4t + 1)^2}} \, dt$

40. $\displaystyle\int \dfrac{1}{25 + 4x^2} \, dx$

41. $\displaystyle\int \dfrac{\tan(2/t)}{t^2} \, dt$

42. $\displaystyle\int \dfrac{e^{1/t}}{t^2} \, dt$

43. $\displaystyle\int \dfrac{6}{\sqrt{10x - x^2}} \, dx$

44. $\displaystyle\int \dfrac{1}{(x - 1)\sqrt{4x^2 - 8x + 3}} \, dx$

45. $\displaystyle\int \dfrac{4}{4x^2 + 4x + 65} \, dx$

46. $\displaystyle\int \dfrac{1}{x^2 - 4x + 9} \, dx$

Slope Field In Exercises 47 and 48, a differential equation, a point, and a slope field are given. (a) Sketch two approximate solutions of the differential equation on the slope field, one of which passes through the given point. (b) Use integration to find the particular solution of the differential equation and use a graphing utility to graph the solution. Compare the result with the sketches in part (a). To print an enlarged copy of the graph, go to *MathGraphs.com*.

47. $\dfrac{ds}{dt} = \dfrac{t}{\sqrt{1 - t^4}}$

$\left(0, -\dfrac{1}{2} \right)$

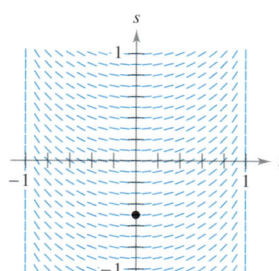

48. $\dfrac{dy}{dx} = \dfrac{1}{\sqrt{4x - x^2}}$

$\left(2, \dfrac{1}{2} \right)$

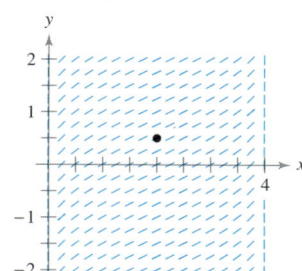

Slope Field In Exercises 49 and 50, use a computer algebra system to graph the slope field for the differential equation and graph the solution through the specified initial condition.

49. $\dfrac{dy}{dx} = 0.8y,\ y(0) = 4$

50. $\dfrac{dy}{dx} = 5 - y,\ y(0) = 1$

Differential Equation In Exercises 51–56, solve the differential equation.

51. $\dfrac{dy}{dx} = (e^x + 5)^2$

52. $\dfrac{dy}{dx} = (4 - e^{2x})^2$

53. $\dfrac{dr}{dt} = \dfrac{10e^t}{\sqrt{1 - e^{2t}}}$

54. $\dfrac{dr}{dt} = \dfrac{(1 + e^t)^2}{e^{3t}}$

55. $(4 + \tan^2 x)y' = \sec^2 x$

56. $y' = \dfrac{1}{x\sqrt{4x^2 - 9}}$

Evaluating a Definite Integral In Exercises 57–64, evaluate the definite integral. Use the integration capabilities of a graphing utility to verify your result.

57. $\displaystyle\int_0^{\pi/4} \cos 2x\, dx$

58. $\displaystyle\int_0^{\pi} \sin^2 t \cos t\, dt$

59. $\displaystyle\int_0^1 xe^{-x^2}\, dx$

60. $\displaystyle\int_1^e \dfrac{1 - \ln x}{x}\, dx$

61. $\displaystyle\int_0^8 \dfrac{2x}{\sqrt{x^2 + 36}}\, dx$

62. $\displaystyle\int_1^3 \dfrac{2x^2 + 3x - 2}{x}\, dx$

63. $\displaystyle\int_0^{2/\sqrt{3}} \dfrac{1}{4 + 9x^2}\, dx$

64. $\displaystyle\int_0^7 \dfrac{1}{\sqrt{100 - x^2}}\, dx$

Area In Exercises 65–68, find the area of the region.

65. $y = (-4x + 6)^{3/2}$

66. $y = \dfrac{3x + 2}{x^2 + 9}$

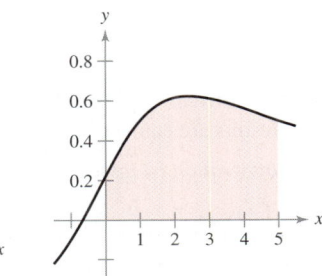

67. $y^2 = x^2(1 - x^2)$

68. $y = \sin 2x$

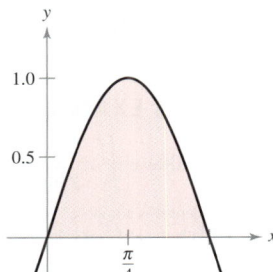

Finding an Integral Using Technology In Exercises 69–72, use a computer algebra system to find the integral. Use the computer algebra system to graph two antiderivatives. Describe the relationship between the graphs of the two antiderivatives.

69. $\displaystyle\int \dfrac{1}{x^2 + 4x + 13}\, dx$

70. $\displaystyle\int \dfrac{x - 2}{x^2 + 4x + 13}\, dx$

71. $\displaystyle\int \dfrac{1}{1 + \sin\theta}\, d\theta$

72. $\displaystyle\int \left(\dfrac{e^x + e^{-x}}{2}\right)^3 dx$

WRITING ABOUT CONCEPTS

Choosing a Formula In Exercises 73–76, state the integration formula you would use to perform the integration. Explain why you chose that formula. Do not integrate.

73. $\displaystyle\int x(x^2 + 1)^3\, dx$

74. $\displaystyle\int x \sec(x^2 + 1) \tan(x^2 + 1)\, dx$

75. $\displaystyle\int \dfrac{x}{x^2 + 1}\, dx$

76. $\displaystyle\int \dfrac{1}{x^2 + 1}\, dx$

77. Finding Constants Determine the constants a and b such that

$$\sin x + \cos x = a \sin(x + b).$$

Use this result to integrate

$$\int \dfrac{dx}{\sin x + \cos x}.$$

78. Deriving a Rule Show that

$$\sec x = \dfrac{\sin x}{\cos x} + \dfrac{\cos x}{1 + \sin x}.$$

Then use this identity to derive the basic integration rule

$$\int \sec x\, dx = \ln|\sec x + \tan x| + C.$$

79. Area The graphs of $f(x) = x$ and $g(x) = ax^2$ intersect at the points $(0, 0)$ and $(1/a, 1/a)$. Find $a\ (a > 0)$ such that the area of the region bounded by the graphs of these two functions is $\frac{2}{3}$.

80. Think About It When evaluating

$$\int_{-1}^1 x^2\, dx$$

is it appropriate to substitute

$$u = x^2,\quad x = \sqrt{u},\quad \text{and}\quad dx = \dfrac{du}{2\sqrt{u}}$$

to obtain

$$\dfrac{1}{2}\int_1^1 \sqrt{u}\, du = 0?$$

Explain.

81. Comparing Antiderivatives

(a) Explain why the antiderivative $y_1 = e^{x+C_1}$ is equivalent to the antiderivative $y_2 = Ce^x$.

(b) Explain why the antiderivative $y_1 = \sec^2 x + C_1$ is equivalent to the antiderivative $y_2 = \tan^2 x + C$.

82. HOW DO YOU SEE IT? Using the graph, is

$$\int_0^5 f(x)\, dx$$ positive or negative? Explain.

$f(x) = \frac{1}{5}(x^3 - 7x^2 + 10x)$

Approximation In Exercises 83 and 84, determine which value best approximates the area of the region between the *x*-axis and the function over the given interval. (Make your selection on the basis of a sketch of the region and *not* by integrating.)

83. $f(x) = \dfrac{4x}{x^2 + 1}$, $[0, 2]$

(a) 3   (b) 1   (c) -8 (d) 8   (e) 10

84. $f(x) = \dfrac{4}{x^2 + 1}$, $[0, 2]$

(a) 3   (b) 1   (c) -4 (d) 4   (e) 10

Interpreting Integrals In Exercises 85 and 86, (a) sketch the region whose area is given by the integral, (b) sketch the solid whose volume is given by the integral when the disk method is used, and (c) sketch the solid whose volume is given by the integral when the shell method is used. (There is more than one correct answer for each part.)

85. $\displaystyle\int_0^2 2\pi x^2 \, dx$ **86.** $\displaystyle\int_0^4 \pi y \, dy$

87. Volume The region bounded by $y = e^{-x^2}$, $y = 0$, $x = 0$, and $x = b$ ($b > 0$) is revolved about the *y*-axis.

(a) Find the volume of the solid generated when $b = 1$.

(b) Find b such that the volume of the generated solid is $\frac{4}{3}$ cubic units.

88. Volume Consider the region bounded by the graphs of $x = 0$, $y = \cos x^2$, $y = \sin x^2$, and $x = \sqrt{\pi}/2$. Find the volume of the solid generated by revolving the region about the *y*-axis.

89. Arc Length Find the arc length of the graph of $y = \ln(\sin x)$ from $x = \pi/4$ to $x = \pi/2$.

90. Arc Length Find the arc length of the graph of $y = \ln(\cos x)$ from $x = 0$ to $x = \pi/3$.

91. Surface Area Find the area of the surface formed by revolving the graph of $y = 2\sqrt{x}$ on the interval $[0, 9]$ about the *x*-axis.

92. Centroid Find the *x*-coordinate of the centroid of the region bounded by the graphs of

$$y = \frac{5}{\sqrt{25 - x^2}}, \quad y = 0, \quad x = 0, \quad \text{and} \quad x = 4.$$

Average Value of a Function In Exercises 93 and 94, find the average value of the function over the given interval.

93. $f(x) = \dfrac{1}{1 + x^2}$, $-3 \le x \le 3$

94. $f(x) = \sin nx$, $0 \le x \le \pi/n$, n is a positive integer.

Arc Length In Exercises 95 and 96, use the integration capabilities of a graphing utility to approximate the arc length of the curve over the given interval.

95. $y = \tan \pi x$, $\left[0, \frac{1}{4}\right]$ **96.** $y = x^{2/3}$, $[1, 8]$

97. Finding a Pattern

(a) Find $\displaystyle\int \cos^3 x \, dx$.

(b) Find $\displaystyle\int \cos^5 x \, dx$.

(c) Find $\displaystyle\int \cos^7 x \, dx$.

(d) Explain how to find $\int \cos^{15} x \, dx$ without actually integrating.

98. Finding a Pattern

(a) Write $\int \tan^3 x \, dx$ in terms of $\int \tan x \, dx$. Then find $\int \tan^3 x \, dx$.

(b) Write $\int \tan^5 x \, dx$ in terms of $\int \tan^3 x \, dx$.

(c) Write $\int \tan^{2k+1} x \, dx$, where k is a positive integer, in terms of $\int \tan^{2k-1} x \, dx$.

(d) Explain how to find $\int \tan^{15} x \, dx$ without actually integrating.

99. Methods of Integration Show that the following results are equivalent.

Integration by tables:

$$\int \sqrt{x^2 + 1} \, dx = \frac{1}{2}\left(x\sqrt{x^2 + 1} + \ln\left|x + \sqrt{x^2 + 1}\right|\right) + C$$

Integration by computer algebra system:

$$\int \sqrt{x^2 + 1} \, dx = \frac{1}{2}\left[x\sqrt{x^2 + 1} + \text{arcsinh}(x)\right] + C$$

PUTNAM EXAM CHALLENGE

100. Evaluate $\displaystyle\int_2^4 \frac{\sqrt{\ln(9 - x)}\, dx}{\sqrt{\ln(9 - x)} + \sqrt{\ln(x + 3)}}$.

8.2 Integration by Parts

■ Find an antiderivative using integration by parts.

Integration by Parts

In this section, you will study an important integration technique called **integration by parts**. This technique can be applied to a wide variety of functions and is particularly useful for integrands involving *products* of algebraic and transcendental functions. For instance, integration by parts works well with integrals such as

$$\int x \ln x \, dx, \quad \int x^2 e^x \, dx, \quad \text{and} \quad \int e^x \sin x \, dx.$$

Integration by parts is based on the formula for the derivative of a product

$$\frac{d}{dx}[uv] = u \frac{dv}{dx} + v \frac{du}{dx}$$

$$= uv' + vu'$$

where both u and v are differentiable functions of x. When u' and v' are continuous, you can integrate both sides of this equation to obtain

$$uv = \int uv' \, dx + \int vu' \, dx$$

$$= \int u \, dv + \int v \, du.$$

By rewriting this equation, you obtain the next theorem.

> **THEOREM 8.1   Integration by Parts**
>
> If u and v are functions of x and have continuous derivatives, then
>
> $$\int u \, dv = uv - \int v \, du.$$

This formula expresses the original integral in terms of another integral. Depending on the choices of u and dv, it may be easier to evaluate the second integral than the original one. Because the choices of u and dv are critical in the integration by parts process, the guidelines below are provided.

> **GUIDELINES FOR INTEGRATION BY PARTS**
>
> **1.** Try letting dv be the most complicated portion of the integrand that fits a basic integration rule. Then u will be the remaining factor(s) of the integrand.
>
> **2.** Try letting u be the portion of the integrand whose derivative is a function simpler than u. Then dv will be the remaining factor(s) of the integrand.
>
> Note that dv always includes the dx of the original integrand.

When using integration by parts, note that you can first choose dv or first choose u. After you choose, however, the choice of the other factor is determined—it must be the remaining portion of the integrand. Also note that dv must contain the differential dx of the original integral.

Exploration

Proof Without Words   Here is a different approach to proving the formula for integration by parts. This approach is from "Proof Without Words: Integration by Parts" by Roger B. Nelsen, *Mathematics Magazine,* 64, No. 2, April 1991, p. 130, by permission of the author.

Area ▭ + Area ▭ = $qs - pr$

$$\int_r^s u \, dv + \int_q^p v \, du = \Big[uv \Big]_{(p,\,r)}^{(q,\,s)}$$

$$\int_r^s u \, dv = \Big[uv \Big]_{(p,\,r)}^{(q,\,s)} - \int_q^p v \, du$$

Explain how this graph proves the theorem. Which notation in this proof is unfamiliar? What do you think it means?

EXAMPLE 1 **Integration by Parts**

Find $\displaystyle\int xe^x\,dx$.

Solution To apply integration by parts, you need to write the integral in the form $\int u\,dv$. There are several ways to do this.

$$\int \underbrace{(x)}_{u}\,\underbrace{(e^x\,dx)}_{dv}, \quad \int \underbrace{(e^x)}_{u}\,\underbrace{(x\,dx)}_{dv}, \quad \int \underbrace{(1)}_{u}\,\underbrace{(xe^x\,dx)}_{dv}, \quad \int \underbrace{(xe^x)}_{u}\,\underbrace{(dx)}_{dv}$$

The guidelines on the preceding page suggest the first option because the derivative of $u = x$ is simpler than x, and $dv = e^x\,dx$ is the most complicated portion of the integrand that fits a basic integration formula.

$$dv = e^x\,dx \implies v = \int dv = \int e^x\,dx = e^x$$
$$u = x \implies du = dx$$

Now, integration by parts produces

$$\int u\,dv = uv - \int v\,du \qquad \text{Integration by parts formula}$$
$$\int xe^x\,dx = xe^x - \int e^x\,dx \qquad \text{Substitute.}$$
$$= xe^x - e^x + C. \qquad \text{Integrate.}$$

To check this, differentiate $xe^x - e^x + C$ to see that you obtain the original integrand.

> • **REMARK** In Example 1, note that it is not necessary to include a constant of integration when solving
>
> $$v = \int e^x\,dx = e^x + C_1.$$
>
> To illustrate this, replace $v = e^x$ by $v = e^x + C_1$ and apply integration by parts to see that you obtain the same result.

EXAMPLE 2 **Integration by Parts**

Find $\displaystyle\int x^2 \ln x\,dx$.

Solution In this case, x^2 is more easily integrated than $\ln x$. Furthermore, the derivative of $\ln x$ is simpler than $\ln x$. So, you should let $dv = x^2\,dx$.

$$dv = x^2\,dx \implies v = \int x^2\,dx = \frac{x^3}{3}$$
$$u = \ln x \implies du = \frac{1}{x}\,dx$$

Integration by parts produces

$$\int u\,dv = uv - \int v\,du \qquad \text{Integration by parts formula}$$
$$\int x^2 \ln x\,dx = \frac{x^3}{3}\ln x - \int \left(\frac{x^3}{3}\right)\left(\frac{1}{x}\right)dx \qquad \text{Substitute.}$$
$$= \frac{x^3}{3}\ln x - \frac{1}{3}\int x^2\,dx \qquad \text{Simplify.}$$
$$= \frac{x^3}{3}\ln x - \frac{x^3}{9} + C. \qquad \text{Integrate.}$$

You can check this result by differentiating.

$$\frac{d}{dx}\left[\frac{x^3}{3}\ln x - \frac{x^3}{9} + C\right] = \frac{x^3}{3}\left(\frac{1}{x}\right) + (\ln x)(x^2) - \frac{x^2}{3} = x^2 \ln x$$

> ▷ **TECHNOLOGY** Try graphing
>
> $$\int x^2 \ln x\,dx \text{ and } \frac{x^3}{3}\ln x - \frac{x^3}{9}$$
>
> on your graphing utility. Do you get the same graph? (This may take a while, so be patient.)

One surprising application of integration by parts involves integrands consisting of single terms, such as

$$\int \ln x \, dx \quad \text{or} \quad \int \arcsin x \, dx.$$

In these cases, try letting $dv = dx$, as shown in the next example.

EXAMPLE 3   An Integrand with a Single Term

Evaluate $\displaystyle\int_0^1 \arcsin x \, dx$.

Solution   Let $dv = dx$.

$$dv = dx \quad \implies \quad v = \int dx = x$$

$$u = \arcsin x \quad \implies \quad du = \frac{1}{\sqrt{1 - x^2}} \, dx$$

Integration by parts produces

$$\int u \, dv = uv - \int v \, du \qquad \text{Integration by parts formula}$$

$$\int \arcsin x \, dx = x \arcsin x - \int \frac{x}{\sqrt{1 - x^2}} \, dx \qquad \text{Substitute.}$$

$$= x \arcsin x + \frac{1}{2} \int (1 - x^2)^{-1/2} (-2x) \, dx \qquad \text{Rewrite.}$$

$$= x \arcsin x + \sqrt{1 - x^2} + C. \qquad \text{Integrate.}$$

Using this antiderivative, you can evaluate the definite integral as shown.

$$\int_0^1 \arcsin x \, dx = \left[x \arcsin x + \sqrt{1 - x^2} \right]_0^1$$

$$= \frac{\pi}{2} - 1$$

$$\approx 0.571$$

The area represented by this definite integral is shown in Figure 8.2.

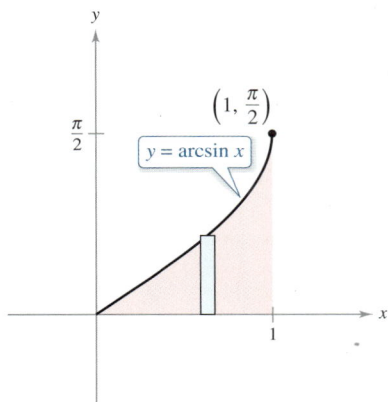

The area of the region is approximately 0.571.

Figure 8.2

▷ **TECHNOLOGY**   Remember that there are two ways to use technology to evaluate a definite integral: (1) you can use a numerical approximation such as the Trapezoidal Rule or Simpson's Rule, or (2) you can use a computer algebra system to find the antiderivative and then apply the Fundamental Theorem of Calculus. Both methods have shortcomings. To find the possible error when using a numerical method, the integrand must have a second derivative (Trapezoidal Rule) or a fourth derivative (Simpson's Rule) in the interval of integration; the integrand in Example 3 fails to meet either of these requirements. To apply the Fundamental Theorem of Calculus, the symbolic integration utility must be able to find the antiderivative.

■ **FOR FURTHER INFORMATION**   To see how integration by parts is used to prove Stirling's approximation

$$\ln(n!) = n \ln n - n$$

see the article "The Validity of Stirling's Approximation: A Physical Chemistry Project" by A. S. Wallner and K. A. Brandt in *Journal of Chemical Education*.

Some integrals require repeated use of the integration by parts formula.

EXAMPLE 4   **Repeated Use of Integration by Parts**

Find $\int x^2 \sin x \, dx$.

Solution The factors x^2 and $\sin x$ are equally easy to integrate. However, the derivative of x^2 becomes simpler, whereas the derivative of $\sin x$ does not. So, you should let $u = x^2$.

$$dv = \sin x \, dx \implies v = \int \sin x \, dx = -\cos x$$

$$u = x^2 \implies du = 2x \, dx$$

Now, integration by parts produces

$$\int x^2 \sin x \, dx = -x^2 \cos x + \int 2x \cos x \, dx. \qquad \text{First use of integration by parts}$$

This first use of integration by parts has succeeded in simplifying the original integral, but the integral on the right still doesn't fit a basic integration rule. To evaluate that integral, you can apply integration by parts again. This time, let $u = 2x$.

$$dv = \cos x \, dx \implies v = \int \cos x \, dx = \sin x$$

$$u = 2x \implies du = 2 \, dx$$

Now, integration by parts produces

$$\int 2x \cos x \, dx = 2x \sin x - \int 2 \sin x \, dx \qquad \text{Second use of integration by parts}$$

$$= 2x \sin x + 2 \cos x + C.$$

Combining these two results, you can write

$$\int x^2 \sin x \, dx = -x^2 \cos x + 2x \sin x + 2 \cos x + C. \qquad \blacksquare$$

When making repeated applications of integration by parts, you need to be careful not to interchange the substitutions in successive applications. For instance, in Example 4, the first substitution was $u = x^2$ and $dv = \sin x \, dx$. If, in the second application, you had switched the substitution to $u = \cos x$ and $dv = 2x$, you would have obtained

$$\int x^2 \sin x \, dx = -x^2 \cos x + \int 2x \cos x \, dx$$

$$= -x^2 \cos x + x^2 \cos x + \int x^2 \sin x \, dx$$

$$= \int x^2 \sin x \, dx$$

thereby undoing the previous integration and returning to the *original* integral. When making repeated applications of integration by parts, you should also watch for the appearance of a *constant multiple* of the original integral. For instance, this occurs when you use integration by parts to evaluate $\int e^x \cos 2x \, dx$, and it also occurs in Example 5 on the next page.

The integral in Example 5 is an important one. In Section 8.4 (Example 5), you will see that it is used to find the arc length of a parabolic segment.

EXAMPLE 5   Integration by Parts

Find $\displaystyle\int \sec^3 x\, dx$.

Solution   The most complicated portion of the integrand that can be easily integrated is $\sec^2 x$, so you should let $dv = \sec^2 x\, dx$ and $u = \sec x$.

$$dv = \sec^2 x\, dx \quad \Longrightarrow \quad v = \int \sec^2 x\, dx = \tan x$$

$$u = \sec x \quad \Longrightarrow \quad du = \sec x \tan x\, dx$$

Integration by parts produces

$$\int u\, dv = uv - \int v\, du$$ Integration by parts formula

$$\int \sec^3 x\, dx = \sec x \tan x - \int \sec x \tan^2 x\, dx$$ Substitute.

$$\int \sec^3 x\, dx = \sec x \tan x - \int \sec x(\sec^2 x - 1)\, dx$$ Trigonometric identity

$$\int \sec^3 x\, dx = \sec x \tan x - \int \sec^3 x\, dx + \int \sec x\, dx$$ Rewrite.

$$2\int \sec^3 x\, dx = \sec x \tan x + \int \sec x\, dx$$ Collect like integrals.

$$2\int \sec^3 x\, dx = \sec x \tan x + \ln|\sec x + \tan x| + C$$ Integrate.

$$\int \sec^3 x\, dx = \frac{1}{2}\sec x \tan x + \frac{1}{2}\ln|\sec x + \tan x| + C.$$ Divide by 2.

EXAMPLE 6   Finding a Centroid

A machine part is modeled by the region bounded by the graph of $y = \sin x$ and the x-axis, $0 \le x \le \pi/2$, as shown in Figure 8.3. Find the centroid of this region.

Solution   Begin by finding the area of the region.

$$A = \int_0^{\pi/2} \sin x\, dx = \left[-\cos x\right]_0^{\pi/2} = 1$$

Now, you can find the coordinates of the centroid. To evaluate the integral for $\bar{y}$, first rewrite the integrand using the trigonometric identity $\sin^2 x = (1 - \cos 2x)/2$.

$$\bar{y} = \frac{1}{A}\int_0^{\pi/2} \frac{\sin x}{2}(\sin x)\, dx = \frac{1}{4}\int_0^{\pi/2}(1 - \cos 2x)\, dx = \frac{1}{4}\left[x - \frac{\sin 2x}{2}\right]_0^{\pi/2} = \frac{\pi}{8}$$

You can evaluate the integral for $\bar{x}$, $(1/A)\int_0^{\pi/2} x \sin x\, dx$, with integration by parts. To do this, let $dv = \sin x\, dx$ and $u = x$. This produces $v = -\cos x$ and $du = dx$, and you can write

$$\int x \sin x\, dx = -x \cos x + \int \cos x\, dx = -x \cos x + \sin x + C.$$

Finally, you can determine $\bar{x}$ to be

$$\bar{x} = \frac{1}{A}\int_0^{\pi/2} x \sin x\, dx = \left[-x \cos x + \sin x\right]_0^{\pi/2} = 1.$$

So, the centroid of the region is $(1, \pi/8)$.

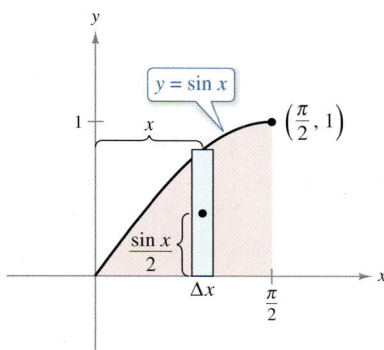

Figure 8.3

As you gain experience in using integration by parts, your skill in determining u and dv will increase. The next summary lists several common integrals with suggestions for the choices of u and dv.

• • • • • • • • • • • • • • • • • ▷
•
• • **REMARK** You can use the acronym LIATE as a guideline for choosing u in integration by parts. In order, check the integrand for the following.

Is there a **L**ogarithmic part?

Is there an **I**nverse trigonometric part?

Is there an **A**lgebraic part?

Is there a **T**rigonometric part?

Is there an **E**xponential part?

> **SUMMARY: COMMON INTEGRALS USING INTEGRATION BY PARTS**
>
> **1.** For integrals of the form
> $$\int x^n e^{ax}\, dx, \quad \int x^n \sin ax\, dx, \quad \text{or} \quad \int x^n \cos ax\, dx$$
> let $u = x^n$ and let $dv = e^{ax}\, dx$, $\sin ax\, dx$, or $\cos ax\, dx$.
>
> **2.** For integrals of the form
> $$\int x^n \ln x\, dx, \quad \int x^n \arcsin ax\, dx, \quad \text{or} \quad \int x^n \arctan ax\, dx$$
> let $u = \ln x$, $\arcsin ax$, or $\arctan ax$ and let $dv = x^n\, dx$.
>
> **3.** For integrals of the form
> $$\int e^{ax} \sin bx\, dx \quad \text{or} \quad \int e^{ax} \cos bx\, dx$$
> let $u = \sin bx$ or $\cos bx$ and let $dv = e^{ax}\, dx$.

In problems involving repeated applications of integration by parts, a tabular method, illustrated in Example 7, can help to organize the work. This method works well for integrals of the form

$$\int x^n \sin ax\, dx, \quad \int x^n \cos ax\, dx, \quad \text{and} \quad \int x^n e^{ax}\, dx.$$

EXAMPLE 7 **Using the Tabular Method**

• • • • ▷ *See LarsonCalculus.com for an interactive version of this type of example.*

Find $\displaystyle\int x^2 \sin 4x\, dx$.

Solution Begin as usual by letting $u = x^2$ and $dv = v'\, dx = \sin 4x\, dx$. Next, create a table consisting of three columns, as shown.

Alternate Signs	u and Its Derivatives	v' and Its Antiderivatives
+	x^2	$\sin 4x$
−	$2x$	$-\frac{1}{4}\cos 4x$
+	2	$-\frac{1}{16}\sin 4x$
−	0	$\frac{1}{64}\cos 4x$

↑
Differentiate until you obtain 0 as a derivative.

■ **FOR FURTHER INFORMATION**
For more information on the tabular method, see the article "Tabular Integration by Parts" by David Horowitz in *The College Mathematics Journal*, and the article "More on Tabular Integration by Parts" by Leonard Gillman in *The College Mathematics Journal*. To view these articles, go to *MathArticles.com*.

The solution is obtained by adding the signed products of the diagonal entries:

$$\int x^2 \sin 4x\, dx = -\frac{1}{4}x^2 \cos 4x + \frac{1}{8}x \sin 4x + \frac{1}{32}\cos 4x + C.$$

8.2 Exercises

See **CalcChat.com** for tutorial help and worked-out solutions to odd-numbered exercises.

Setting Up Integration by Parts In Exercises 1–6, identify u and dv for finding the integral using integration by parts. (Do not evaluate the integral.)

1. $\int xe^{2x}\, dx$

2. $\int x^2 e^{2x}\, dx$

3. $\int (\ln x)^2\, dx$

4. $\int \ln 5x\, dx$

5. $\int x \sec^2 x\, dx$

6. $\int x^2 \cos x\, dx$

Using Integration by Parts In Exercises 7–10, evaluate the integral using integration by parts with the given choices of u and dv.

7. $\int x^3 \ln x\, dx;\ u = \ln x,\ dv = x^3\, dx$

8. $\int (4x + 7)e^x\, dx;\ u = 4x + 7,\ dv = e^x\, dx$

9. $\int x \sin 3x\, dx;\ u = x,\ dv = \sin 3x\, dx$

10. $\int x \cos 4x\, dx;\ u = x,\ dv = \cos 4x\, dx$

Finding an Indefinite Integral In Exercises 11–30, find the indefinite integral. (*Note:* Solve by the simplest method— not all require integration by parts.)

11. $\int xe^{-4x}\, dx$

12. $\int \dfrac{5x}{e^{2x}}\, dx$

13. $\int x^3 e^x\, dx$

14. $\int \dfrac{e^{1/t}}{t^2}\, dt$

15. $\int t \ln(t + 1)\, dt$

16. $\int x^5 \ln 3x\, dx$

17. $\int \dfrac{(\ln x)^2}{x}\, dx$

18. $\int \dfrac{\ln x}{x^3}\, dx$

19. $\int \dfrac{xe^{2x}}{(2x + 1)^2}\, dx$

20. $\int \dfrac{x^3 e^{x^2}}{(x^2 + 1)^2}\, dx$

21. $\int x\sqrt{x - 5}\, dx$

22. $\int \dfrac{x}{\sqrt{6x + 1}}\, dx$

23. $\int x \cos x\, dx$

24. $\int t \csc t \cot t\, dt$

25. $\int x^3 \sin x\, dx$

26. $\int x^2 \cos x\, dx$

27. $\int \arctan x\, dx$

28. $\int 4 \arccos x\, dx$

29. $\int e^{-3x} \sin 5x\, dx$

30. $\int e^{4x} \cos 2x\, dx$

Differential Equation In Exercises 31–34, solve the differential equation.

31. $y' = \ln x$

32. $y' = \arctan \dfrac{x}{2}$

33. $\dfrac{dy}{dt} = \dfrac{t^2}{\sqrt{3 + 5t}}$

34. $\dfrac{dy}{dx} = x^2\sqrt{x - 3}$

Slope Field In Exercises 35 and 36, a differential equation, a point, and a slope field are given. (a) Sketch two approximate solutions of the differential equation on the slope field, one of which passes through the given point. (b) Use integration to find the particular solution of the differential equation and use a graphing utility to graph the solution. Compare the result with the sketches in part (a). To print an enlarged copy of the graph, go to *MathGraphs.com*.

35. $\dfrac{dy}{dx} = x\sqrt{y} \cos x,\ (0, 4)$

36. $\dfrac{dy}{dx} = e^{-x/3} \sin 2x,\ \left(0, -\frac{18}{37}\right)$

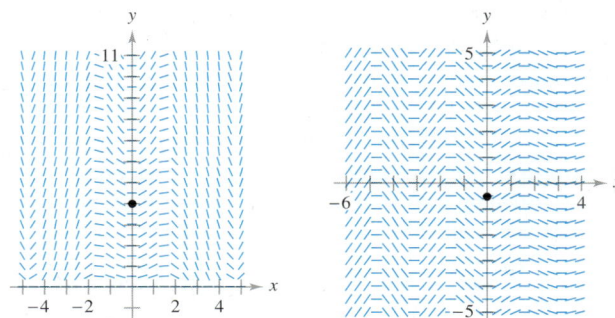

Slope Field In Exercises 37 and 38, use a computer algebra system to graph the slope field for the differential equation and graph the solution through the specified initial condition.

37. $\dfrac{dy}{dx} = \dfrac{x}{y}e^{x/8},\ y(0) = 2$

38. $\dfrac{dy}{dx} = \dfrac{x}{y}\sin x,\ y(0) = 4$

Evaluating a Definite Integral In Exercises 39–48, evaluate the definite integral. Use a graphing utility to confirm your result.

39. $\int_0^3 xe^{x/2}\, dx$

40. $\int_0^2 x^2 e^{-2x}\, dx$

41. $\int_0^{\pi/4} x \cos 2x\, dx$

42. $\int_0^{\pi} x \sin 2x\, dx$

43. $\int_0^{1/2} \arccos x\, dx$

44. $\int_0^1 x \arcsin x^2\, dx$

45. $\int_0^1 e^x \sin x\, dx$

46. $\int_0^1 \ln(4 + x^2)\, dx$

47. $\int_2^4 x \arcsec x\, dx$

48. $\int_0^{\pi/8} x \sec^2 2x\, dx$

Using the Tabular Method In Exercises 49–54, use the tabular method to find the integral.

49. $\int x^2 e^{2x}\, dx$

50. $\int x^3 e^{-2x}\, dx$

51. $\int x^3 \sin x\, dx$

52. $\int x^3 \cos 2x\, dx$

53. $\displaystyle\int x \sec^2 x \, dx$ **54.** $\displaystyle\int x^2(x-2)^{3/2} \, dx$

Using Two Methods Together   In Exercises 55–58, find the indefinite integral by using substitution followed by integration by parts.

55. $\displaystyle\int \sin \sqrt{x} \, dx$ **56.** $\displaystyle\int 2x^3 \cos x^2 \, dx$

57. $\displaystyle\int x^5 e^{x^2} \, dx$ **58.** $\displaystyle\int e^{\sqrt{2x}} \, dx$

WRITING ABOUT CONCEPTS

59. Integration by Parts

(a) Integration by parts is based on what differentiation rule? Explain.

(b) In your own words, state how you determine which parts of the integrand should be u and dv.

60. Integration by Parts   When evaluating $\int x \sin x \, dx$, explain how letting $u = \sin x$ and $dv = x \, dx$ makes the solution more difficult to find.

61. Integration by Parts   State whether you would use integration by parts to evaluate each integral. If so, identify what you would use for u and dv. Explain your reasoning.

(a) $\displaystyle\int \frac{\ln x}{x} \, dx$ (b) $\displaystyle\int x \ln x \, dx$ (c) $\displaystyle\int x^2 e^{-3x} \, dx$

(d) $\displaystyle\int 2xe^{x^2} \, dx$ (e) $\displaystyle\int \frac{x}{\sqrt{x+1}} \, dx$ (f) $\displaystyle\int \frac{x}{\sqrt{x^2+1}} \, dx$

---

62. HOW DO YOU SEE IT?   Use the graph of f' shown in the figure to answer the following.

$f'(x) = x \ln x$

(a) Approximate the slope of f at $x = 2$. Explain.

(b) Approximate any open intervals in which the graph of f is increasing and any open intervals in which it is decreasing. Explain.

---

63. Using Two Methods   Integrate $\displaystyle\int \frac{x^3}{\sqrt{4+x^2}} \, dx$

(a) by parts, letting $dv = \dfrac{x}{\sqrt{4+x^2}} \, dx$.

(b) by substitution, letting $u = 4 + x^2$.

64. Using Two Methods   Integrate $\displaystyle\int x\sqrt{4-x} \, dx$

(a) by parts, letting $dv = \sqrt{4-x} \, dx$.

(b) by substitution, letting $u = 4 - x$.

Finding a General Rule   In Exercises 65 and 66, use a computer algebra system to find the integrals for $n = 0, 1, 2,$ and 3. Use the result to obtain a general rule for the integrals for any positive integer n and test your results for $n = 4$.

65. $\displaystyle\int x^n \ln x \, dx$ **66.** $\displaystyle\int x^n e^x \, dx$

Proof   In Exercises 67–72, use integration by parts to prove the formula. (For Exercises 67–70, assume that n is a positive integer.)

67. $\displaystyle\int x^n \sin x \, dx = -x^n \cos x + n \int x^{n-1} \cos x \, dx$

68. $\displaystyle\int x^n \cos x \, dx = x^n \sin x - n \int x^{n-1} \sin x \, dx$

69. $\displaystyle\int x^n \ln x \, dx = \frac{x^{n+1}}{(n+1)^2}\left[-1 + (n+1)\ln x\right] + C$

70. $\displaystyle\int x^n e^{ax} \, dx = \frac{x^n e^{ax}}{a} - \frac{n}{a}\int x^{n-1} e^{ax} \, dx$

71. $\displaystyle\int e^{ax} \sin bx \, dx = \frac{e^{ax}(a \sin bx - b \cos bx)}{a^2 + b^2} + C$

72. $\displaystyle\int e^{ax} \cos bx \, dx = \frac{e^{ax}(a \cos bx + b \sin bx)}{a^2 + b^2} + C$

Using Formulas   In Exercises 73–78, find the integral by using the appropriate formula from Exercises 67–72.

73. $\displaystyle\int x^2 \sin x \, dx$ **74.** $\displaystyle\int x^2 \cos x \, dx$

75. $\displaystyle\int x^5 \ln x \, dx$ **76.** $\displaystyle\int x^3 e^{2x} \, dx$

77. $\displaystyle\int e^{-3x} \sin 4x \, dx$ **78.** $\displaystyle\int e^{2x} \cos 3x \, dx$

Area   In Exercises 79–82, use a graphing utility to graph the region bounded by the graphs of the equations. Then find the area of the region analytically.

79. $y = 2xe^{-x}, \quad y = 0, \quad x = 3$

80. $y = \dfrac{1}{10}xe^{3x}, \quad y = 0, \quad x = 0, \quad x = 2$

81. $y = e^{-x} \sin \pi x, \quad y = 0, \quad x = 1$

82. $y = x^3 \ln x, \quad y = 0, \quad x = 1, \quad x = 3$

83. Area, Volume, and Centroid   Given the region bounded by the graphs of $y = \ln x$, $y = 0$, and $x = e$, find

(a) the area of the region.

(b) the volume of the solid generated by revolving the region about the x-axis.

(c) the volume of the solid generated by revolving the region about the y-axis.

(d) the centroid of the region.

84. Area, Volume, and Centroid   Given the region bounded by the graphs of $y = x \sin x$, $y = 0$, $x = 0$, and $x = \pi$, find

(a) the area of the region.

(b) the volume of the solid generated by revolving the region about the x-axis.

(c) the volume of the solid generated by revolving the region about the y-axis.

(d) the centroid of the region.

85. Centroid   Find the centroid of the region bounded by the graphs of $y = \arcsin x$, $x = 0$, and $y = \pi/2$. How is this problem related to Example 6 in this section?

86. Centroid   Find the centroid of the region bounded by the graphs of $f(x) = x^2$, $g(x) = 2^x$, $x = 2$, and $x = 4$.

87. Average Displacement   A damping force affects the vibration of a spring so that the displacement of the spring is given by

$$y = e^{-4t}(\cos 2t + 5 \sin 2t).$$

Find the average value of y on the interval from $t = 0$ to $t = \pi$.

• • 88. Memory Model • • • • • • • • • • • • • • • •

A model for the ability M of a child to memorize, measured on a scale from 0 to 10, is given by

$$M = 1 + 1.6t \ln t, \quad 0 < t \le 4$$

where t is the child's age in years. Find the average value of this model

(a) between the child's first and second birthdays.

(b) between the child's third and fourth birthdays.

Present Value   In Exercises 89 and 90, find the present value P of a continuous income flow of $c(t)$ dollars per year for

$$P = \int_0^{t_1} c(t)e^{-rt}\, dt$$

where t_1 is the time in years and r is the annual interest rate compounded continuously.

89. $c(t) = 100{,}000 + 4000t$, $r = 5\%$, $t_1 = 10$

90. $c(t) = 30{,}000 + 500t$, $r = 7\%$, $t_1 = 5$

Integrals Used to Find Fourier Coefficients   In Exercises 91 and 92, verify the value of the definite integral, where n is a positive integer.

91. $\displaystyle\int_{-\pi}^{\pi} x \sin nx\, dx = \begin{cases} \dfrac{2\pi}{n}, & n \text{ is odd} \\[2mm] -\dfrac{2\pi}{n}, & n \text{ is even} \end{cases}$

92. $\displaystyle\int_{-\pi}^{\pi} x^2 \cos nx\, dx = \dfrac{(-1)^n\, 4\pi}{n^2}$

93. Vibrating String   A string stretched between the two points $(0, 0)$ and $(2, 0)$ is plucked by displacing the string h units at its midpoint. The motion of the string is modeled by a **Fourier Sine Series** whose coefficients are given by

$$b_n = h \int_0^1 x \sin \frac{n\pi x}{2}\, dx + h \int_1^2 (-x + 2) \sin \frac{n\pi x}{2}\, dx.$$

Find b_n.

94. Euler's Method   Consider the differential equation $f'(x) = xe^{-x}$ with the initial condition $f(0) = 0$.

(a) Use integration to solve the differential equation.

(b) Use a graphing utility to graph the solution of the differential equation.

(c) Use Euler's Method with $h = 0.05$, and the recursive capabilities of a graphing utility, to generate the first 80 points of the graph of the approximate solution. Use the graphing utility to plot the points. Compare the result with the graph in part (b).

(d) Repeat part (c) using $h = 0.1$ and generate the first 40 points.

(e) Why is the result in part (c) a better approximation of the solution than the result in part (d)?

Euler's Method   In Exercises 95 and 96, consider the differential equation and repeat parts (a)–(d) of Exercise 94.

95. $f'(x) = 3x \sin(2x)$ **96.** $f'(x) = \cos \sqrt{x}$

   $f(0) = 0$ $f(0) = 1$

97. Think About It   Give a geometric explanation of why

$$\int_0^{\pi/2} x \sin x\, dx \le \int_0^{\pi/2} x\, dx.$$

Verify the inequality by evaluating the integrals.

98. Finding a Pattern   Find the area bounded by the graphs of $y = x \sin x$ and $y = 0$ over each interval.

(a) $[0, \pi]$ (b) $[\pi, 2\pi]$ (c) $[2\pi, 3\pi]$

Describe any patterns that you notice. What is the area between the graphs of $y = x \sin x$ and $y = 0$ over the interval $[n\pi, (n + 1)\pi]$, where n is any nonnegative integer? Explain.

99. Finding an Error   Find the fallacy in the following argument that $0 = 1$.

$$dv = dx \implies v = \int dx = x$$

$$u = \frac{1}{x} \implies du = -\frac{1}{x^2}\, dx$$

$$0 + \int \frac{dx}{x} = \left(\frac{1}{x}\right)(x) - \int \left(-\frac{1}{x^2}\right)(x)\, dx$$

$$= 1 + \int \frac{dx}{x}$$

So, $0 = 1$.

8.3 Trigonometric Integrals

- Solve trigonometric integrals involving powers of sine and cosine.
- Solve trigonometric integrals involving powers of secant and tangent.
- Solve trigonometric integrals involving sine-cosine products with different angles.

Integrals Involving Powers of Sine and Cosine

> **SHEILA SCOTT MACINTYRE (1910–1960)**
>
> Sheila Scott Macintyre published her first paper on the asymptotic periods of integral functions in 1935. She completed her doctorate work at Aberdeen University, where she taught. In 1958 she accepted a visiting research fellowship at the University of Cincinnati.

In this section, you will study techniques for evaluating integrals of the form

$$\int \sin^m x \cos^n x \, dx \quad \text{and} \quad \int \sec^m x \tan^n x \, dx$$

where either m or n is a positive integer. To find antiderivatives for these forms, try to break them into combinations of trigonometric integrals to which you can apply the Power Rule.

For instance, you can evaluate

$$\int \sin^5 x \cos x \, dx$$

with the Power Rule by letting $u = \sin x$. Then, $du = \cos x \, dx$ and you have

$$\int \sin^5 x \cos x \, dx = \int u^5 \, du = \frac{u^6}{6} + C = \frac{\sin^6 x}{6} + C.$$

To break up $\int \sin^m x \cos^n x \, dx$ into forms to which you can apply the Power Rule, use the following identities.

$\sin^2 x + \cos^2 x = 1$	Pythagorean identity
$\sin^2 x = \dfrac{1 - \cos 2x}{2}$	Half-angle identity for $\sin^2 x$
$\cos^2 x = \dfrac{1 + \cos 2x}{2}$	Half-angle identity for $\cos^2 x$

GUIDELINES FOR EVALUATING INTEGRALS INVOLVING POWERS OF SINE AND COSINE

1. When the power of the sine is odd and positive, save one sine factor and convert the remaining factors to cosines. Then, expand and integrate.

$$\int \sin^{2k+1} x \cos^n x \, dx = \int \overbrace{(\sin^2 x)^k}^{\text{Convert to cosines}} \cos^n x \underbrace{\sin x \, dx}_{\text{Save for } du} = \int (1 - \cos^2 x)^k \cos^n x \sin x \, dx$$

2. When the power of the cosine is odd and positive, save one cosine factor and convert the remaining factors to sines. Then, expand and integrate.

$$\int \sin^m x \cos^{2k+1} x \, dx = \int \sin^m x \overbrace{(\cos^2 x)^k}^{\text{Convert to sines}} \underbrace{\cos x \, dx}_{\text{Save for } du} = \int \sin^m x \, (1 - \sin^2 x)^k \cos x \, dx$$

3. When the powers of both the sine and cosine are even and nonnegative, make repeated use of the identities

$$\sin^2 x = \frac{1 - \cos 2x}{2} \quad \text{and} \quad \cos^2 x = \frac{1 + \cos 2x}{2}$$

to convert the integrand to odd powers of the cosine. Then proceed as in the second guideline.

EXAMPLE 1 **Power of Sine Is Odd and Positive**

Find $\int \sin^3 x \cos^4 x \, dx$.

Solution Because you expect to use the Power Rule with $u = \cos x$, *save one sine factor* to form du and convert the remaining sine factors to cosines.

$$\int \sin^3 x \cos^4 x \, dx = \int \sin^2 x \cos^4 x (\sin x) \, dx \qquad \text{Rewrite.}$$

$$= \int (1 - \cos^2 x) \cos^4 x \sin x \, dx \qquad \text{Trigonometric identity}$$

$$= \int (\cos^4 x - \cos^6 x) \sin x \, dx \qquad \text{Multiply.}$$

$$= \int \cos^4 x \sin x \, dx - \int \cos^6 x \sin x \, dx \qquad \text{Rewrite.}$$

$$= -\int \cos^4 x (-\sin x) \, dx + \int \cos^6 x (-\sin x) \, dx$$

$$= -\frac{\cos^5 x}{5} + \frac{\cos^7 x}{7} + C \qquad \text{Integrate.} \quad \blacksquare$$

▷ **TECHNOLOGY** A computer algebra system used to find the integral in Example 1 yielded the following.

$$\int \sin^3 x \cos^4 x \, dx = -\cos^5 x \left(\frac{1}{7} \sin^2 x + \frac{2}{35} \right) + C$$

Is this equivalent to the result obtained in Example 1?

In Example 1, *both* of the powers m and n happened to be positive integers. This strategy will work as long as either m or n is odd and positive. For instance, in the next example, the power of the cosine is 3, but the power of the sine is $-\frac{1}{2}$.

EXAMPLE 2 **Power of Cosine Is Odd and Positive**

••••▷ *See LarsonCalculus.com for an interactive version of this type of example.*

Evaluate $\displaystyle\int_{\pi/6}^{\pi/3} \frac{\cos^3 x}{\sqrt{\sin x}} \, dx$.

Solution Because you expect to use the Power Rule with $u = \sin x$, *save one cosine factor* to form du and convert the remaining cosine factors to sines.

$$\int_{\pi/6}^{\pi/3} \frac{\cos^3 x}{\sqrt{\sin x}} \, dx = \int_{\pi/6}^{\pi/3} \frac{\cos^2 x \cos x}{\sqrt{\sin x}} \, dx$$

$$= \int_{\pi/6}^{\pi/3} \frac{(1 - \sin^2 x)(\cos x)}{\sqrt{\sin x}} \, dx$$

$$= \int_{\pi/6}^{\pi/3} \left[(\sin x)^{-1/2} - (\sin x)^{3/2} \right] \cos x \, dx$$

$$= \left[\frac{(\sin x)^{1/2}}{1/2} - \frac{(\sin x)^{5/2}}{5/2} \right]_{\pi/6}^{\pi/3}$$

$$= 2 \left(\frac{\sqrt{3}}{2} \right)^{1/2} - \frac{2}{5} \left(\frac{\sqrt{3}}{2} \right)^{5/2} - \sqrt{2} + \frac{\sqrt{32}}{80}$$

$$\approx 0.239$$

Figure 8.4 shows the region whose area is represented by this integral. $\blacksquare$

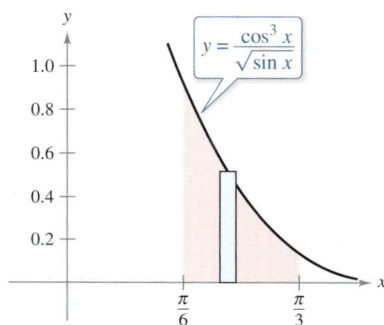

$$y = \frac{\cos^3 x}{\sqrt{\sin x}}$$

The area of the region is approximately 0.239.

Figure 8.4

EXAMPLE 3 **Power of Cosine Is Even and Nonnegative**

Find $\displaystyle\int \cos^4 x \, dx$.

Solution Because m and n are both even and nonnegative ($m = 0$), you can replace $\cos^4 x$ by

$$\left(\frac{1 + \cos 2x}{2}\right)^2.$$

So, you can integrate as shown.

$$\int \cos^4 x \, dx = \int \left(\frac{1 + \cos 2x}{2}\right)^2 dx \qquad \text{Half-angle identity}$$

$$= \int \left(\frac{1}{4} + \frac{\cos 2x}{2} + \frac{\cos^2 2x}{4}\right) dx \qquad \text{Expand.}$$

$$= \int \left[\frac{1}{4} + \frac{\cos 2x}{2} + \frac{1}{4}\left(\frac{1 + \cos 4x}{2}\right)\right] dx \qquad \text{Half-angle identity}$$

$$= \frac{3}{8}\int dx + \frac{1}{4}\int 2\cos 2x \, dx + \frac{1}{32}\int 4\cos 4x \, dx \qquad \text{Rewrite.}$$

$$= \frac{3x}{8} + \frac{\sin 2x}{4} + \frac{\sin 4x}{32} + C \qquad \text{Integrate.}$$

Use a symbolic differentiation utility to verify this. Can you simplify the derivative to obtain the original integrand?

In Example 3, when you evaluate the definite integral from 0 to $\pi/2$, you obtain

$$\int_0^{\pi/2} \cos^4 x \, dx = \left[\frac{3x}{8} + \frac{\sin 2x}{4} + \frac{\sin 4x}{32}\right]_0^{\pi/2}$$

$$= \left(\frac{3\pi}{16} + 0 + 0\right) - (0 + 0 + 0)$$

$$= \frac{3\pi}{16}.$$

Note that the only term that contributes to the solution is

$$\frac{3x}{8}.$$

This observation is generalized in the following formulas developed by John Wallis (1616–1703).

JOHN WALLIS (1616–1703)

Wallis did much of his work in calculus prior to Newton and Leibniz, and he influenced the thinking of both of these men. Wallis is also credited with introducing the present symbol (∞) for infinity.
See LarsonCalculus.com to read more of this biography.

Wallis's Formulas

1. If n is odd ($n \geq 3$), then

$$\int_0^{\pi/2} \cos^n x \, dx = \left(\frac{2}{3}\right)\left(\frac{4}{5}\right)\left(\frac{6}{7}\right) \cdots \left(\frac{n-1}{n}\right).$$

2. If n is even ($n \geq 2$), then

$$\int_0^{\pi/2} \cos^n x \, dx = \left(\frac{1}{2}\right)\left(\frac{3}{4}\right)\left(\frac{5}{6}\right) \cdots \left(\frac{n-1}{n}\right)\left(\frac{\pi}{2}\right).$$

These formulas are also valid when $\cos^n x$ is replaced by $\sin^n x$. (You are asked to prove both formulas in Exercise 88.)

Integrals Involving Powers of Secant and Tangent

The guidelines below can help you evaluate integrals of the form

$$\int \sec^m x \tan^n x \, dx.$$

GUIDELINES FOR EVALUATING INTEGRALS INVOLVING POWERS OF SECANT AND TANGENT

1. When the power of the secant is even and positive, save a secant-squared factor and convert the remaining factors to tangents. Then, expand and integrate.

$$\int \overbrace{\sec^{2k}}^{\text{Even}} x \tan^n x \, dx = \int \overbrace{(\sec^2 x)^{k-1}}^{\text{Convert to tangents}} \tan^n x \overbrace{\sec^2 x \, dx}^{\text{Save for } du} = \int (1 + \tan^2 x)^{k-1} \tan^n x \sec^2 x \, dx$$

2. When the power of the tangent is odd and positive, save a secant-tangent factor and convert the remaining factors to secants. Then, expand and integrate.

$$\int \sec^m x \overbrace{\tan^{2k+1}}^{\text{Odd}} x \, dx = \int \sec^{m-1} x \overbrace{(\tan^2 x)^k}^{\text{Convert to secants}} \overbrace{\sec x \tan x \, dx}^{\text{Save for } du} = \int \sec^{m-1} x (\sec^2 x - 1)^k \sec x \tan x \, dx$$

3. When there are no secant factors and the power of the tangent is even and positive, convert a tangent-squared factor to a secant-squared factor, then expand and repeat if necessary.

$$\int \tan^n x \, dx = \int \tan^{n-2} x \overbrace{(\tan^2 x)}^{\text{Convert to secants}} \, dx = \int \tan^{n-2} x (\sec^2 x - 1) \, dx$$

4. When the integral is of the form

$$\int \sec^m x \, dx$$

where m is odd and positive, use integration by parts, as illustrated in Example 5 in Section 8.2.

5. When none of the first four guidelines applies, try converting to sines and cosines.

EXAMPLE 4   **Power of Tangent Is Odd and Positive**

Find $\displaystyle\int \frac{\tan^3 x}{\sqrt{\sec x}} \, dx.$

Solution   Because you expect to use the Power Rule with $u = \sec x$, *save a factor of* $(\sec x \tan x)$ *to form* du and convert the remaining tangent factors to secants.

$$\int \frac{\tan^3 x}{\sqrt{\sec x}} \, dx = \int (\sec x)^{-1/2} \tan^3 x \, dx$$

$$= \int (\sec x)^{-3/2} (\tan^2 x)(\sec x \tan x) \, dx$$

$$= \int (\sec x)^{-3/2} (\sec^2 x - 1)(\sec x \tan x) \, dx$$

$$= \int [(\sec x)^{1/2} - (\sec x)^{-3/2}](\sec x \tan x) \, dx$$

$$= \frac{2}{3}(\sec x)^{3/2} + 2(\sec x)^{-1/2} + C$$

EXAMPLE 5   **Power of Secant Is Even and Positive**

Find $\displaystyle\int \sec^4 3x \tan^3 3x \, dx$.

Solution   Let $u = \tan 3x$, then $du = 3 \sec^2 3x \, dx$ and you can write

$$\int \sec^4 3x \tan^3 3x \, dx = \int \sec^2 3x \tan^3 3x (\sec^2 3x) \, dx$$

$$= \int (1 + \tan^2 3x) \tan^3 3x (\sec^2 3x) \, dx$$

$$= \frac{1}{3}\int (\tan^3 3x + \tan^5 3x)(3 \sec^2 3x) \, dx$$

$$= \frac{1}{3}\left(\frac{\tan^4 3x}{4} + \frac{\tan^6 3x}{6}\right) + C$$

$$= \frac{\tan^4 3x}{12} + \frac{\tan^6 3x}{18} + C.$$

In Example 5, the power of the tangent is odd and positive. So, you could also find the integral using the procedure described in the second guideline on page 527. In Exercises 69 and 70, you are asked to show that the results obtained by these two procedures differ only by a constant.

EXAMPLE 6   **Power of Tangent Is Even**

Evaluate $\displaystyle\int_0^{\pi/4} \tan^4 x \, dx$.

Solution   Because there are no secant factors, you can begin by converting a tangent-squared factor to a secant-squared factor.

$$\int \tan^4 x \, dx = \int \tan^2 x (\tan^2 x) \, dx$$

$$= \int \tan^2 x (\sec^2 x - 1) \, dx$$

$$= \int \tan^2 x \sec^2 x \, dx - \int \tan^2 x \, dx$$

$$= \int \tan^2 x \sec^2 x \, dx - \int (\sec^2 x - 1) \, dx$$

$$= \frac{\tan^3 x}{3} - \tan x + x + C$$

Next, evaluate the definite integral.

$$\int_0^{\pi/4} \tan^4 x \, dx = \left[\frac{\tan^3 x}{3} - \tan x + x\right]_0^{\pi/4}$$

$$= \frac{1}{3} - 1 + \frac{\pi}{4}$$

$$\approx 0.119$$

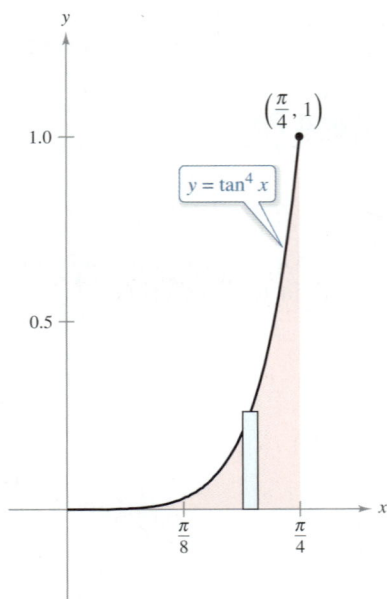

$y = \tan^4 x$

$\left(\frac{\pi}{4}, 1\right)$

The area of the region is approximately 0.119.

Figure 8.5

The area represented by the definite integral is shown in Figure 8.5. Try using Simpson's Rule to approximate this integral. With $n = 18$, you should obtain an approximation that is within 0.00001 of the actual value.

For integrals involving powers of cotangents and cosecants, you can follow a strategy similar to that used for powers of tangents and secants. Also, when integrating trigonometric functions, remember that it sometimes helps to convert the entire integrand to powers of sines and cosines.

EXAMPLE 7   **Converting to Sines and Cosines**

Find $\displaystyle\int \frac{\sec x}{\tan^2 x}\, dx$.

Solution   Because the first four guidelines on page 527 do not apply, try converting the integrand to sines and cosines. In this case, you are able to integrate the resulting powers of sine and cosine as shown.

$$\int \frac{\sec x}{\tan^2 x}\, dx = \int \left(\frac{1}{\cos x}\right)\left(\frac{\cos x}{\sin x}\right)^2 dx$$

$$= \int (\sin x)^{-2}(\cos x)\, dx$$

$$= -(\sin x)^{-1} + C$$

$$= -\csc x + C$$

Integrals Involving Sine-Cosine Products with Different Angles

Integrals involving the products of sines and cosines of two *different* angles occur in many applications. In such instances, you can use the following product-to-sum identities.

$$\sin mx \sin nx = \frac{1}{2}\big(\cos\big[(m-n)x\big] - \cos\big[(m+n)x\big]\big)$$

$$\sin mx \cos nx = \frac{1}{2}\big(\sin\big[(m-n)x\big] + \sin\big[(m+n)x\big]\big)$$

$$\cos mx \cos nx = \frac{1}{2}\big(\cos\big[(m-n)x\big] + \cos\big[(m+n)x\big]\big)$$

EXAMPLE 8   **Using Product-to-Sum Identities**

Find $\displaystyle\int \sin 5x \cos 4x\, dx$.

Solution   Considering the second product-to-sum identity above, you can write

$$\int \sin 5x \cos 4x\, dx = \frac{1}{2}\int (\sin x + \sin 9x)\, dx$$

$$= \frac{1}{2}\left(-\cos x - \frac{\cos 9x}{9}\right) + C$$

$$= -\frac{\cos x}{2} - \frac{\cos 9x}{18} + C.$$

◼ FOR FURTHER INFORMATION

To learn more about integrals involving sine-cosine products with different angles, see the article "Integrals of Products of Sine and Cosine with Different Arguments" by Sherrie J. Nicol in *The College Mathematics Journal.* To view this article, go to *MathArticles.com.*

8.3 Exercises

See **CalcChat.com** for tutorial help and worked-out solutions to odd-numbered exercises.

Finding an Indefinite Integral Involving Sine and Cosine In Exercises 1–12, find the indefinite integral.

1. $\displaystyle\int \cos^5 x \sin x \, dx$

2. $\displaystyle\int \cos^3 x \sin^4 x \, dx$

3. $\displaystyle\int \sin^7 2x \cos 2x \, dx$

4. $\displaystyle\int \sin^3 3x \, dx$

5. $\displaystyle\int \sin^3 x \cos^2 x \, dx$

6. $\displaystyle\int \cos^3 \frac{x}{3} \, dx$

7. $\displaystyle\int \sin^3 2\theta \sqrt{\cos 2\theta} \, d\theta$

8. $\displaystyle\int \frac{\cos^5 t}{\sqrt{\sin t}} \, dt$

9. $\displaystyle\int \cos^2 3x \, dx$

10. $\displaystyle\int \sin^4 6\theta \, d\theta$

11. $\displaystyle\int x \sin^2 x \, dx$

12. $\displaystyle\int x^2 \sin^2 x \, dx$

Using Wallis's Formulas In Exercises 13–18, use Wallis's Formulas to evaluate the integral.

13. $\displaystyle\int_0^{\pi/2} \cos^7 x \, dx$

14. $\displaystyle\int_0^{\pi/2} \cos^9 x \, dx$

15. $\displaystyle\int_0^{\pi/2} \cos^{10} x \, dx$

16. $\displaystyle\int_0^{\pi/2} \sin^5 x \, dx$

17. $\displaystyle\int_0^{\pi/2} \sin^6 x \, dx$

18. $\displaystyle\int_0^{\pi/2} \sin^8 x \, dx$

Finding an Indefinite Integral Involving Secant and Tangent In Exercises 19–32, find the indefinite integral.

19. $\displaystyle\int \sec 4x \, dx$

20. $\displaystyle\int \sec^4 2x \, dx$

21. $\displaystyle\int \sec^3 \pi x \, dx$

22. $\displaystyle\int \tan^6 3x \, dx$

23. $\displaystyle\int \tan^5 \frac{x}{2} \, dx$

24. $\displaystyle\int \tan^3 \frac{\pi x}{2} \sec^2 \frac{\pi x}{2} \, dx$

25. $\displaystyle\int \tan^3 2t \sec^3 2t \, dt$

26. $\displaystyle\int \tan^5 2x \sec^4 2x \, dx$

27. $\displaystyle\int \sec^6 4x \tan 4x \, dx$

28. $\displaystyle\int \sec^2 \frac{x}{2} \tan \frac{x}{2} \, dx$

29. $\displaystyle\int \sec^5 x \tan^3 x \, dx$

30. $\displaystyle\int \tan^3 3x \, dx$

31. $\displaystyle\int \frac{\tan^2 x}{\sec x} \, dx$

32. $\displaystyle\int \frac{\tan^2 x}{\sec^5 x} \, dx$

Differential Equation In Exercises 33–36, solve the differential equation.

33. $\dfrac{dr}{d\theta} = \sin^4 \pi\theta$

34. $\dfrac{ds}{d\alpha} = \sin^2 \dfrac{\alpha}{2} \cos^2 \dfrac{\alpha}{2}$

35. $y' = \tan^3 3x \sec 3x$

36. $y' = \sqrt{\tan x} \sec^4 x$

Slope Field In Exercises 37 and 38, a differential equation, a point, and a slope field are given. (a) Sketch two approximate solutions of the differential equation on the slope field, one of which passes through the given point. (b) Use integration to find the particular solution of the differential equation and use a graphing utility to graph the solution. Compare the result with the sketches in part (a). To print an enlarged copy of the graph, go to *MathGraphs.com*.

37. $\dfrac{dy}{dx} = \sin^2 x$, $(0, 0)$

38. $\dfrac{dy}{dx} = \sec^2 x \tan^2 x$, $\left(0, -\dfrac{1}{4}\right)$

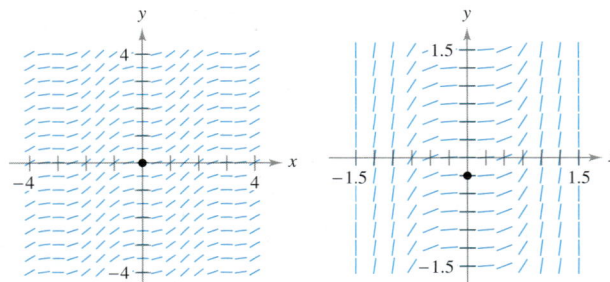

Slope Field In Exercises 39 and 40, use a computer algebra system to graph the slope field for the differential equation, and graph the solution through the specified initial condition.

39. $\dfrac{dy}{dx} = \dfrac{3 \sin x}{y}$, $y(0) = 2$

40. $\dfrac{dy}{dx} = 3\sqrt{y} \tan^2 x$, $y(0) = 3$

Using Product-to-Sum Identities In Exercises 41–46, find the indefinite integral.

41. $\displaystyle\int \cos 2x \cos 6x \, dx$

42. $\displaystyle\int \cos 5\theta \cos 3\theta \, d\theta$

43. $\displaystyle\int \sin 2x \cos 4x \, dx$

44. $\displaystyle\int \sin(-7x) \cos 6x \, dx$

45. $\displaystyle\int \sin \theta \sin 3\theta \, d\theta$

46. $\displaystyle\int \sin 5x \sin 4x \, dx$

Finding an Indefinite Integral In Exercises 47–56, find the indefinite integral. Use a computer algebra system to confirm your result.

47. $\displaystyle\int \cot^3 2x \, dx$

48. $\displaystyle\int \tan^5 \frac{x}{4} \sec^4 \frac{x}{4} \, dx$

49. $\displaystyle\int \csc^4 3x \, dx$

50. $\displaystyle\int \cot^3 \frac{x}{2} \csc^4 \frac{x}{2} \, dx$

51. $\displaystyle\int \frac{\cot^2 t}{\csc t} \, dt$

52. $\displaystyle\int \frac{\cot^3 t}{\csc t} \, dt$

53. $\displaystyle\int \frac{1}{\sec x \tan x} \, dx$

54. $\displaystyle\int \frac{\sin^2 x - \cos^2 x}{\cos x} \, dx$

55. $\displaystyle\int (\tan^4 t - \sec^4 t) \, dt$

56. $\displaystyle\int \frac{1 - \sec t}{\cos t - 1} \, dt$

Evaluating a Definite Integral In Exercises 57–64, evaluate the definite integral.

57. $\displaystyle\int_{-\pi}^{\pi} \sin^2 x \, dx$

58. $\displaystyle\int_{0}^{\pi/3} \tan^2 x \, dx$

59. $\displaystyle\int_{0}^{\pi/4} 6 \tan^3 x \, dx$

60. $\displaystyle\int_{0}^{\pi/3} \sec^{3/2} x \tan x \, dx$

61. $\displaystyle\int_{0}^{\pi/2} \frac{\cos t}{1 + \sin t} \, dt$

62. $\displaystyle\int_{\pi/6}^{\pi/3} \sin 6x \cos 4x \, dx$

63. $\displaystyle\int_{-\pi/2}^{\pi/2} 3 \cos^3 x \, dx$

64. $\displaystyle\int_{-\pi/2}^{\pi/2} (\sin^2 x + 1) \, dx$

WRITING ABOUT CONCEPTS

65. Describing How to Find an Integral In your own words, describe how you would integrate $\int \sin^m x \cos^n x \, dx$ for each condition.

(a) m is positive and odd. (b) n is positive and odd.

(c) m and n are both positive and even.

66. Describing How to Find an Integral In your own words, describe how you would integrate $\int \sec^m x \tan^n x \, dx$ for each condition.

(a) m is positive and even. (b) n is positive and odd.

(c) n is positive and even, and there are no secant factors.

(d) m is positive and odd, and there are no tangent factors.

---

67. Comparing Methods Evaluate $\int \sin x \cos x \, dx$ using the given method. Explain how your answers differ for each method.

(a) Substitution where $u = \sin x$

(b) Substitution where $u = \cos x$

(c) Integration by parts

(d) Using the identity $\sin 2x = 2 \sin x \cos x$

68. HOW DO YOU SEE IT? Use the graph of f' shown in the figure to answer the following.

(a) Using the interval shown in the graph, approximate the value(s) of x where f is maximum. Explain.

(b) Using the interval shown in the graph, approximate the value(s) of x where f is minimum. Explain.

Comparing Methods In Exercises 69 and 70, (a) find the indefinite integral in two different ways. (b) Use a graphing utility to graph the antiderivative (without the constant of integration) obtained by each method to show that the results differ only by a constant. (c) Verify analytically that the results differ only by a constant.

69. $\displaystyle\int \sec^4 3x \tan^3 3x \, dx$

70. $\displaystyle\int \sec^2 x \tan x \, dx$

Area In Exercises 71–74, find the area of the region bounded by the graphs of the equations.

71. $y = \sin x, \quad y = \sin^3 x, \quad x = 0, \quad x = \dfrac{\pi}{2}$

72. $y = \sin^2 \pi x, \quad y = 0, \quad x = 0, \quad x = 1$

73. $y = \cos^2 x, \quad y = \sin^2 x, \quad x = -\dfrac{\pi}{4}, \quad x = \dfrac{\pi}{4}$

74. $y = \cos^2 x, \quad y = \sin x \cos x, \quad x = -\dfrac{\pi}{2}, \quad x = \dfrac{\pi}{4}$

Volume In Exercises 75 and 76, find the volume of the solid generated by revolving the region bounded by the graphs of the equations about the x-axis.

75. $y = \tan x, \quad y = 0, \quad x = -\dfrac{\pi}{4}, \quad x = \dfrac{\pi}{4}$

76. $y = \cos \dfrac{x}{2}, \quad y = \sin \dfrac{x}{2}, \quad x = 0, \quad x = \dfrac{\pi}{2}$

Volume and Centroid In Exercises 77 and 78, for the region bounded by the graphs of the equations, find (a) the volume of the solid formed by revolving the region about the x-axis and (b) the centroid of the region.

77. $y = \sin x, \quad y = 0, \quad x = 0, \quad x = \pi$

78. $y = \cos x, \quad y = 0, \quad x = 0, \quad x = \dfrac{\pi}{2}$

Verifying a Reduction Formula In Exercises 79–82, use integration by parts to verify the reduction formula.

79. $\displaystyle\int \sin^n x \, dx = -\frac{\sin^{n-1} x \cos x}{n} + \frac{n-1}{n} \int \sin^{n-2} x \, dx$

80. $\displaystyle\int \cos^n x \, dx = \frac{\cos^{n-1} x \sin x}{n} + \frac{n-1}{n} \int \cos^{n-2} x \, dx$

81. $\displaystyle\int \cos^m x \sin^n x \, dx = -\frac{\cos^{m+1} x \sin^{n-1} x}{m+n} + $
$\dfrac{n-1}{m+n} \displaystyle\int \cos^m x \sin^{n-2} x \, dx$

82. $\displaystyle\int \sec^n x \, dx = \frac{1}{n-1} \sec^{n-2} x \tan x + \frac{n-2}{n-1} \int \sec^{n-2} x \, dx$

Using Formulas In Exercises 83–86, use the results of Exercises 79–82 to find the integral.

83. $\displaystyle\int \sin^5 x \, dx$

84. $\displaystyle\int \cos^4 x \, dx$

85. $\displaystyle\int \sec^4 (2\pi x/5)\, dx$ **86.** $\displaystyle\int \sin^4 x \cos^2 x\, dx$

87. Modeling Data The table shows the normal maximum (high) and minimum (low) temperatures (in degrees Fahrenheit) in Erie, Pennsylvania, for each month of the year. *(Source: NOAA)*

Month	Jan	Feb	Mar	Apr	May	Jun
Max	33.5	35.4	44.7	55.6	67.4	76.2
Min	20.3	20.9	28.2	37.9	48.7	58.5

Month	Jul	Aug	Sep	Oct	Nov	Dec
Max	80.4	79.0	72.0	61.0	49.3	38.6
Min	63.7	62.7	55.9	45.5	36.4	26.8

The maximum and minimum temperatures can be modeled by

$$f(t) = a_0 + a_1 \cos\frac{\pi t}{6} + b_1 \sin\frac{\pi t}{6}$$

where $t = 0$ corresponds to January 1 and a_0, a_1, and b_1 are as follows.

$$a_0 = \frac{1}{12}\int_0^{12} f(t)\, dt \qquad a_1 = \frac{1}{6}\int_0^{12} f(t) \cos\frac{\pi t}{6}\, dt$$

$$b_1 = \frac{1}{6}\int_0^{12} f(t) \sin\frac{\pi t}{6}\, dt$$

(a) Approximate the model $H(t)$ for the maximum temperatures. (*Hint:* Use Simpson's Rule to approximate the integrals and use the January data twice.)

(b) Repeat part (a) for a model $L(t)$ for the minimum temperature data.

(c) Use a graphing utility to graph each model. During what part of the year is the difference between the maximum and minimum temperatures greatest?

88. Wallis's Formulas Use the result of Exercise 80 to prove the following versions of Wallis's Formulas.

(a) If n is odd ($n \geq 3$), then

$$\int_0^{\pi/2} \cos^n x\, dx = \left(\frac{2}{3}\right)\left(\frac{4}{5}\right)\left(\frac{6}{7}\right) \cdots \left(\frac{n-1}{n}\right).$$

(b) If n is even ($n \geq 2$), then

$$\int_0^{\pi/2} \cos^n x\, dx = \left(\frac{1}{2}\right)\left(\frac{3}{4}\right)\left(\frac{5}{6}\right) \cdots \left(\frac{n-1}{n}\right)\left(\frac{\pi}{2}\right).$$

89. Orthogonal Functions The **inner product** of two functions f and g on $[a, b]$ is given by

$$\langle f, g\rangle = \int_a^b f(x)g(x)\, dx.$$

Two distinct functions f and g are said to be **orthogonal** if $\langle f, g\rangle = 0$. Show that the following set of functions is orthogonal on $[-\pi, \pi]$.

$$\{\sin x, \sin 2x, \sin 3x, \ldots, \cos x, \cos 2x, \cos 3x, \ldots\}$$

Victor Soares/Shutterstock.com

90. Fourier Series The following sum is a *finite Fourier series*.

$$f(x) = \sum_{i=1}^{N} a_i \sin ix$$

$$= a_1 \sin x + a_2 \sin 2x + a_3 \sin 3x + \cdots + a_N \sin Nx$$

(a) Use Exercise 89 to show that the nth coefficient a_n is given by $a_n = (1/\pi)\displaystyle\int_{-\pi}^{\pi} f(x) \sin nx\, dx.$

(b) Let $f(x) = x$. Find a_1, a_2, and a_3.

SECTION PROJECT

Power Lines

Power lines are constructed by stringing wire between supports and adjusting the tension on each span. The wire hangs between supports in the shape of a catenary, as shown in the figure.

Let T be the tension (in pounds) on a span of wire, let u be the density (in pounds per foot), let $g \approx 32.2$ be the acceleration due to gravity (in feet per second per second), and let L be the distance (in feet) between the supports. Then the equation of the catenary is

$$y = \frac{T}{ug}\left(\cosh\frac{ugx}{T} - 1\right),$$ where x and y are measured in feet.

(a) Find the length of the wire between two spans.

(b) To measure the tension in a span, power line workers use the *return wave method*. The wire is struck at one support, creating a wave in the line, and the time t (in seconds) it takes for the wave to make a round trip is measured. The velocity v (in feet per second) is given by $v = \sqrt{T/u}$. How long does it take the wave to make a round trip between supports?

(c) The sag s (in inches) can be obtained by evaluating y when $x = L/2$ in the equation for the catenary (and multiplying by 12). In practice, however, power line workers use the "lineman's equation" given by $s \approx 12.075t^2$. Use the fact that

$$\cosh\frac{ugL}{2T} + 1 \approx 2$$

to derive this equation.

■ **FOR FURTHER INFORMATION** To learn more about the mathematics of power lines, see the article "Constructing Power Lines" by Thomas O'Neil in *The UMAP Journal.*

8.4 Trigonometric Substitution

■ Use trigonometric substitution to solve an integral.
■ Use integrals to model and solve real-life applications.

Trigonometric Substitution

Now that you can evaluate integrals involving powers of trigonometric functions, you can use **trigonometric substitution** to evaluate integrals involving the radicals

$$\sqrt{a^2 - u^2}, \quad \sqrt{a^2 + u^2}, \quad \text{and} \quad \sqrt{u^2 - a^2}.$$

The objective with trigonometric substitution is to eliminate the radical in the integrand. You do this by using the Pythagorean identities.

$$\cos^2 \theta = 1 - \sin^2 \theta$$
$$\sec^2 \theta = 1 + \tan^2 \theta$$
$$\tan^2 \theta = \sec^2 \theta - 1$$

For example, for $a > 0$, let $u = a \sin \theta$, where $-\pi/2 \le \theta \le \pi/2$. Then

$$\begin{aligned}
\sqrt{a^2 - u^2} &= \sqrt{a^2 - a^2 \sin^2 \theta} \\
&= \sqrt{a^2(1 - \sin^2 \theta)} \\
&= \sqrt{a^2 \cos^2 \theta} \\
&= a \cos \theta.
\end{aligned}$$

Note that $\cos \theta \ge 0$, because $-\pi/2 \le \theta \le \pi/2$.

Trigonometric Substitution ($a > 0$)

1. For integrals involving $\sqrt{a^2 - u^2}$, let

   $u = a \sin \theta.$

   Then $\sqrt{a^2 - u^2} = a \cos \theta$, where

   $-\pi/2 \le \theta \le \pi/2.$

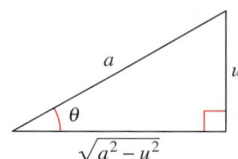

2. For integrals involving $\sqrt{a^2 + u^2}$, let

   $u = a \tan \theta.$

   Then $\sqrt{a^2 + u^2} = a \sec \theta$, where

   $-\pi/2 < \theta < \pi/2.$

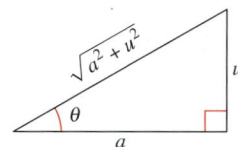

3. For integrals involving $\sqrt{u^2 - a^2}$, let

   $u = a \sec \theta.$

   Then

   $$\sqrt{u^2 - a^2} = \begin{cases} a \tan \theta \text{ for } u > a, \text{ where } 0 \le \theta \le \pi/2 \\ -a \tan \theta \text{ for } u < -a, \text{ where } \pi/2 < \theta \le \pi. \end{cases}$$

Exploration

Integrating a Radical Function   Up to this point in the text, you have not evaluated the integral

$$\int_{-1}^{1} \sqrt{1 - x^2}\, dx.$$

From geometry, you should be able to find the exact value of this integral—what is it? Using numerical integration, with Simpson's Rule or the Trapezoidal Rule, you can't be sure of the accuracy of the approximation. Why?

Try finding the exact value using the substitution

$x = \sin \theta$

and

$dx = \cos \theta\, d\theta.$

Does your answer agree with the value you obtained using geometry?

The restrictions on θ ensure that the function that defines the substitution is one-to-one. In fact, these are the same intervals over which the arcsine, arctangent, and arcsecant are defined.

EXAMPLE 1 **Trigonometric Substitution:** $u = a \sin \theta$

Find $\displaystyle\int \frac{dx}{x^2\sqrt{9-x^2}}$.

Solution First, note that none of the basic integration rules applies. To use trigonometric substitution, you should observe that

$$\sqrt{9-x^2}$$

is of the form $\sqrt{a^2-u^2}$. So, you can use the substitution

$$x = a \sin \theta = 3 \sin \theta.$$

Using differentiation and the triangle shown in Figure 8.6, you obtain

$$dx = 3 \cos \theta \, d\theta, \quad \sqrt{9-x^2} = 3 \cos \theta, \quad \text{and} \quad x^2 = 9 \sin^2 \theta.$$

So, trigonometric substitution yields

$$\int \frac{dx}{x^2\sqrt{9-x^2}} = \int \frac{3\cos\theta \, d\theta}{(9\sin^2\theta)(3\cos\theta)} \qquad \text{Substitute.}$$

$$= \frac{1}{9}\int \frac{d\theta}{\sin^2\theta} \qquad \text{Simplify.}$$

$$= \frac{1}{9}\int \csc^2\theta \, d\theta \qquad \text{Trigonometric identity}$$

$$= -\frac{1}{9}\cot\theta + C \qquad \text{Apply Cosecant Rule.}$$

$$= -\frac{1}{9}\left(\frac{\sqrt{9-x^2}}{x}\right) + C \qquad \text{Substitute for } \cot\theta.$$

$$= -\frac{\sqrt{9-x^2}}{9x} + C.$$

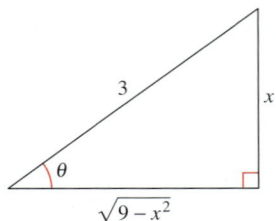

Note that the triangle in Figure 8.6 can be used to convert the θ's back to x's, as shown.

$$\cot\theta = \frac{\text{adj.}}{\text{opp.}}$$

$$= \frac{\sqrt{9-x^2}}{x}$$

$\sin\theta = \dfrac{x}{3}$, $\cot\theta = \dfrac{\sqrt{9-x^2}}{x}$

Figure 8.6

▷ **TECHNOLOGY** Use a computer algebra system to find each indefinite integral.

$$\int \frac{dx}{\sqrt{9-x^2}} \qquad \int \frac{dx}{x\sqrt{9-x^2}}$$

$$\int \frac{dx}{x^2\sqrt{9-x^2}} \qquad \int \frac{dx}{x^3\sqrt{9-x^2}}$$

Then use trigonometric substitution to duplicate the results obtained with the computer algebra system.

In Chapter 5, you saw how the inverse hyperbolic functions can be used to evaluate the integrals

$$\int \frac{du}{\sqrt{u^2\pm a^2}}, \quad \int \frac{du}{a^2-u^2}, \quad \text{and} \quad \int \frac{du}{u\sqrt{a^2\pm u^2}}.$$

You can also evaluate these integrals using trigonometric substitution. This is shown in the next example.

EXAMPLE 2   Trigonometric Substitution: $u = a \tan \theta$

Find $\displaystyle\int \frac{dx}{\sqrt{4x^2 + 1}}$.

Solution   Let $u = 2x$, $a = 1$, and $2x = \tan\theta$, as shown in Figure 8.7. Then,

$$dx = \frac{1}{2}\sec^2\theta \, d\theta \quad \text{and} \quad \sqrt{4x^2 + 1} = \sec\theta.$$

Trigonometric substitution produces

$$\int \frac{1}{\sqrt{4x^2 + 1}} \, dx = \frac{1}{2}\int \frac{\sec^2\theta \, d\theta}{\sec\theta} \qquad \text{Substitute.}$$

$$= \frac{1}{2}\int \sec\theta \, d\theta \qquad \text{Simplify.}$$

$$= \frac{1}{2}\ln|\sec\theta + \tan\theta| + C \qquad \text{Apply Secant Rule.}$$

$$= \frac{1}{2}\ln\left|\sqrt{4x^2 + 1} + 2x\right| + C. \qquad \text{Back-substitute.}$$

Try checking this result with a computer algebra system. Is the result given in this form or in the form of an inverse hyperbolic function?

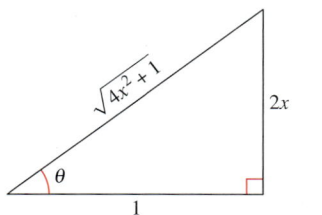

$\tan\theta = 2x$, $\sec\theta = \sqrt{4x^2 + 1}$
Figure 8.7

You can extend the use of trigonometric substitution to cover integrals involving expressions such as $(a^2 - u^2)^{n/2}$ by writing the expression as

$$(a^2 - u^2)^{n/2} = \left(\sqrt{a^2 - u^2}\right)^n.$$

EXAMPLE 3   Trigonometric Substitution: Rational Powers

$\cdots\cdots\triangleright$ *See LarsonCalculus.com for an interactive version of this type of example.*

Find $\displaystyle\int \frac{dx}{(x^2 + 1)^{3/2}}$.

Solution   Begin by writing $(x^2 + 1)^{3/2}$ as

$$\left(\sqrt{x^2 + 1}\right)^3.$$

Then, let $a = 1$ and $u = x = \tan\theta$, as shown in Figure 8.8. Using

$$dx = \sec^2\theta \, d\theta \quad \text{and} \quad \sqrt{x^2 + 1} = \sec\theta$$

you can apply trigonometric substitution, as shown.

$$\int \frac{dx}{(x^2 + 1)^{3/2}} = \int \frac{dx}{\left(\sqrt{x^2 + 1}\right)^3} \qquad \text{Rewrite denominator.}$$

$$= \int \frac{\sec^2\theta \, d\theta}{\sec^3\theta} \qquad \text{Substitute.}$$

$$= \int \frac{d\theta}{\sec\theta} \qquad \text{Simplify.}$$

$$= \int \cos\theta \, d\theta \qquad \text{Trigonometric identity}$$

$$= \sin\theta + C \qquad \text{Apply Cosine Rule.}$$

$$= \frac{x}{\sqrt{x^2 + 1}} + C \qquad \text{Back-substitute.}$$

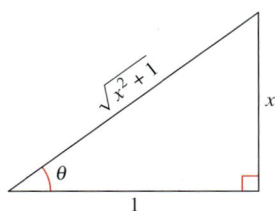

$\tan\theta = x$, $\sin\theta = \dfrac{x}{\sqrt{x^2 + 1}}$

Figure 8.8

For definite integrals, it is often convenient to determine integration limits for θ that avoid converting back to x. You might want to review this procedure in Section 4.5, Examples 8 and 9.

EXAMPLE 4   **Converting the Limits of Integration**

Evaluate $\displaystyle\int_{\sqrt{3}}^{2} \frac{\sqrt{x^2 - 3}}{x} \, dx$.

Solution   Because $\sqrt{x^2 - 3}$ has the form $\sqrt{u^2 - a^2}$, you can consider

$$u = x, \quad a = \sqrt{3}, \quad \text{and} \quad x = \sqrt{3} \sec \theta$$

as shown in Figure 8.9. Then,

$$dx = \sqrt{3} \sec \theta \tan \theta \, d\theta \quad \text{and} \quad \sqrt{x^2 - 3} = \sqrt{3} \tan \theta.$$

To determine the upper and lower limits of integration, use the substitution $x = \sqrt{3} \sec \theta$, as shown.

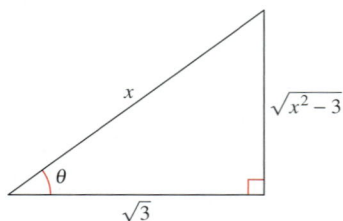

$\sec \theta = \dfrac{x}{\sqrt{3}}$, $\tan \theta = \dfrac{\sqrt{x^2 - 3}}{\sqrt{3}}$

Figure 8.9

Lower Limit	**Upper Limit**
When $x = \sqrt{3}$, $\sec \theta = 1$	When $x = 2$, $\sec \theta = \dfrac{2}{\sqrt{3}}$
and $\theta = 0$.	and $\theta = \dfrac{\pi}{6}$.

So, you have

Integration limits for x Integration limits for θ

$$\int_{\sqrt{3}}^{2} \frac{\sqrt{x^2 - 3}}{x} \, dx = \int_{0}^{\pi/6} \frac{\left(\sqrt{3} \tan \theta\right)\left(\sqrt{3} \sec \theta \tan \theta\right) d\theta}{\sqrt{3} \sec \theta}$$

$$= \int_{0}^{\pi/6} \sqrt{3} \tan^2 \theta \, d\theta$$

$$= \sqrt{3} \int_{0}^{\pi/6} (\sec^2 \theta - 1) \, d\theta$$

$$= \sqrt{3} \Big[\tan \theta - \theta \Big]_{0}^{\pi/6}$$

$$= \sqrt{3} \left(\frac{1}{\sqrt{3}} - \frac{\pi}{6} \right)$$

$$= 1 - \frac{\sqrt{3}\,\pi}{6}$$

$$\approx 0.0931.$$

In Example 4, try converting back to the variable x and evaluating the antiderivative at the original limits of integration. You should obtain

$$\int_{\sqrt{3}}^{2} \frac{\sqrt{x^2 - 3}}{x} \, dx = \sqrt{3} \left[\frac{\sqrt{x^2 - 3}}{\sqrt{3}} - \operatorname{arcsec} \frac{x}{\sqrt{3}} \right]_{\sqrt{3}}^{2}$$

$$= \sqrt{3} \left(\frac{1}{\sqrt{3}} - \frac{\pi}{6} \right)$$

$$\approx 0.0931.$$

When using trigonometric substitution to evaluate definite integrals, you must be careful to check that the values of θ lie in the intervals discussed at the beginning of this section. For instance, if in Example 4 you had been asked to evaluate the definite integral

$$\int_{-2}^{-\sqrt{3}} \frac{\sqrt{x^2 - 3}}{x}\, dx$$

then using $u = x$ and $a = \sqrt{3}$ in the interval $\left[-2, -\sqrt{3}\right]$ would imply that $u < -a$. So, when determining the upper and lower limits of integration, you would have to choose θ such that $\pi/2 < \theta \leq \pi$. In this case, the integral would be evaluated as shown.

$$\begin{aligned}
\int_{-2}^{-\sqrt{3}} \frac{\sqrt{x^2 - 3}}{x}\, dx &= \int_{5\pi/6}^{\pi} \frac{\left(-\sqrt{3}\tan\theta\right)\left(\sqrt{3}\sec\theta\tan\theta\right) d\theta}{\sqrt{3}\sec\theta} \\
&= \int_{5\pi/6}^{\pi} -\sqrt{3}\tan^2\theta\, d\theta \\
&= -\sqrt{3}\int_{5\pi/6}^{\pi} (\sec^2\theta - 1)\, d\theta \\
&= -\sqrt{3}\Big[\tan\theta - \theta\Big]_{5\pi/6}^{\pi} \\
&= -\sqrt{3}\left[(0 - \pi) - \left(-\frac{1}{\sqrt{3}} - \frac{5\pi}{6}\right)\right] \\
&= -1 + \frac{\sqrt{3}\pi}{6} \\
&\approx -0.0931
\end{aligned}$$

Trigonometric substitution can be used with completing the square. For instance, try finding the integral

$$\int \sqrt{x^2 - 2x}\, dx.$$

To begin, you could complete the square and write the integral as

$$\int \sqrt{(x - 1)^2 - 1^2}\, dx.$$

Because the integrand has the form

$$\sqrt{u^2 - a^2}$$

with $u = x - 1$ and $a = 1$, you can now use trigonometric substitution to find the integral.

Trigonometric substitution can be used to evaluate the three integrals listed in the next theorem. These integrals will be encountered several times in the remainder of the text. When this happens, we will simply refer to this theorem. (In Exercise 71, you are asked to verify the formulas given in the theorem.)

THEOREM 8.2 **Special Integration Formulas ($a > 0$)**

1. $\displaystyle\int \sqrt{a^2 - u^2}\, du = \frac{1}{2}\left(a^2 \arcsin\frac{u}{a} + u\sqrt{a^2 - u^2}\right) + C$

2. $\displaystyle\int \sqrt{u^2 - a^2}\, du = \frac{1}{2}\left(u\sqrt{u^2 - a^2} - a^2 \ln\left|u + \sqrt{u^2 - a^2}\right|\right) + C, \quad u > a$

3. $\displaystyle\int \sqrt{u^2 + a^2}\, du = \frac{1}{2}\left(u\sqrt{u^2 + a^2} + a^2 \ln\left|u + \sqrt{u^2 + a^2}\right|\right) + C$

Applications

EXAMPLE 5 **Finding Arc Length**

Find the arc length of the graph of $f(x) = \frac{1}{2}x^2$ from $x = 0$ to $x = 1$ (see Figure 8.10).

Solution Refer to the arc length formula in Section 7.4.

$$s = \int_0^1 \sqrt{1 + [f'(x)]^2}\, dx \qquad \text{Formula for arc length}$$

$$= \int_0^1 \sqrt{1 + x^2}\, dx \qquad f'(x) = x$$

$$= \int_0^{\pi/4} \sec^3 \theta \, d\theta \qquad \text{Let } a = 1 \text{ and } x = \tan \theta.$$

$$= \frac{1}{2}\left[\sec \theta \tan \theta + \ln|\sec \theta + \tan \theta| \right]_0^{\pi/4} \qquad \text{Example 5, Section 8.2}$$

$$= \frac{1}{2}\left[\sqrt{2} + \ln(\sqrt{2} + 1) \right]$$

$$\approx 1.148$$

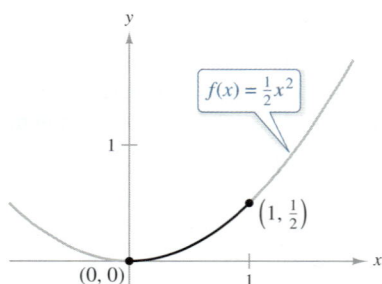

The arc length of the curve from $(0, 0)$ to $\left(1, \frac{1}{2}\right)$

Figure 8.10

EXAMPLE 6 **Comparing Two Fluid Forces**

A sealed barrel of oil (weighing 48 pounds per cubic foot) is floating in seawater (weighing 64 pounds per cubic foot), as shown in Figures 8.11 and 8.12. (The barrel is not completely full of oil. With the barrel lying on its side, the top 0.2 foot of the barrel is empty.) Compare the fluid forces against one end of the barrel from the inside and from the outside.

Solution In Figure 8.12, locate the coordinate system with the origin at the center of the circle

$$x^2 + y^2 = 1.$$

To find the fluid force against an end of the barrel *from the inside*, integrate between -1 and 0.8 (using a weight of $w = 48$).

$$F = w \int_c^d h(y)L(y)\, dy \qquad \text{General equation (See Section 7.7.)}$$

$$F_{\text{inside}} = 48 \int_{-1}^{0.8} (0.8 - y)(2)\sqrt{1 - y^2}\, dy$$

$$= 76.8 \int_{-1}^{0.8} \sqrt{1 - y^2}\, dy - 96 \int_{-1}^{0.8} y\sqrt{1 - y^2}\, dy$$

To find the fluid force *from the outside*, integrate between -1 and 0.4 (using a weight of $w = 64$).

$$F_{\text{outside}} = 64 \int_{-1}^{0.4} (0.4 - y)(2)\sqrt{1 - y^2}\, dy$$

$$= 51.2 \int_{-1}^{0.4} \sqrt{1 - y^2}\, dy - 128 \int_{-1}^{0.4} y\sqrt{1 - y^2}\, dy$$

The details of integration are left for you to complete in Exercise 70. Intuitively, would you say that the force from the oil (the inside) or the force from the seawater (the outside) is greater? By evaluating these two integrals, you can determine that

$$F_{\text{inside}} \approx 121.3 \text{ pounds} \quad \text{and} \quad F_{\text{outside}} \approx 93.0 \text{ pounds.}$$

The barrel is not quite full of oil—the top 0.2 foot of the barrel is empty.

Figure 8.11

Figure 8.12

8.4 Exercises

See **CalcChat.com** for tutorial help and worked-out solutions to odd-numbered exercises.

Trigonometric Substitution In Exercises 1–4, state the trigonometric substitution you would use to find the indefinite integral. Do not integrate.

1. $\displaystyle\int (9 + x^2)^{-2}\, dx$

2. $\displaystyle\int \sqrt{4 - x^2}\, dx$

3. $\displaystyle\int \frac{x^2}{\sqrt{25 - x^2}}\, dx$

4. $\displaystyle\int x^2(x^2 - 25)^{3/2}\, dx$

Using Trigonometric Substitution In Exercises 5–8, find the indefinite integral using the substitution $x = 4 \sin \theta$.

5. $\displaystyle\int \frac{1}{(16 - x^2)^{3/2}}\, dx$

6. $\displaystyle\int \frac{4}{x^2\sqrt{16 - x^2}}\, dx$

7. $\displaystyle\int \frac{\sqrt{16 - x^2}}{x}\, dx$

8. $\displaystyle\int \frac{x^3}{\sqrt{16 - x^2}}\, dx$

Using Trigonometric Substitution In Exercises 9–12, find the indefinite integral using the substitution $x = 5 \sec \theta$.

9. $\displaystyle\int \frac{1}{\sqrt{x^2 - 25}}\, dx$

10. $\displaystyle\int \frac{\sqrt{x^2 - 25}}{x}\, dx$

11. $\displaystyle\int x^3\sqrt{x^2 - 25}\, dx$

12. $\displaystyle\int \frac{x^3}{\sqrt{x^2 - 25}}\, dx$

Using Trigonometric Substitution In Exercises 13–16, find the indefinite integral using the substitution $x = \tan \theta$.

13. $\displaystyle\int x\sqrt{1 + x^2}\, dx$

14. $\displaystyle\int \frac{9x^3}{\sqrt{1 + x^2}}\, dx$

15. $\displaystyle\int \frac{1}{(1 + x^2)^2}\, dx$

16. $\displaystyle\int \frac{x^2}{(1 + x^2)^2}\, dx$

Using Formulas In Exercises 17–20, use the Special Integration Formulas (Theorem 8.2) to find the indefinite integral.

17. $\displaystyle\int \sqrt{9 + 16x^2}\, dx$

18. $\displaystyle\int \sqrt{4 + x^2}\, dx$

19. $\displaystyle\int \sqrt{25 - 4x^2}\, dx$

20. $\displaystyle\int \sqrt{5x^2 - 1}\, dx$

Finding an Indefinite Integral In Exercises 21–36, find the indefinite integral.

21. $\displaystyle\int \frac{1}{\sqrt{16 - x^2}}\, dx$

22. $\displaystyle\int \frac{x^2}{\sqrt{36 - x^2}}\, dx$

23. $\displaystyle\int \sqrt{16 - 4x^2}\, dx$

24. $\displaystyle\int \frac{1}{\sqrt{x^2 - 4}}\, dx$

25. $\displaystyle\int \frac{\sqrt{1 - x^2}}{x^4}\, dx$

26. $\displaystyle\int \frac{\sqrt{25x^2 + 4}}{x^4}\, dx$

27. $\displaystyle\int \frac{1}{x\sqrt{4x^2 + 9}}\, dx$

28. $\displaystyle\int \frac{1}{x\sqrt{9x^2 + 1}}\, dx$

29. $\displaystyle\int \frac{-3x}{(x^2 + 3)^{3/2}}\, dx$

30. $\displaystyle\int \frac{1}{(x^2 + 5)^{3/2}}\, dx$

31. $\displaystyle\int e^x\sqrt{1 - e^{2x}}\, dx$

32. $\displaystyle\int \frac{\sqrt{1 - x}}{\sqrt{x}}\, dx$

33. $\displaystyle\int \frac{1}{4 + 4x^2 + x^4}\, dx$

34. $\displaystyle\int \frac{x^3 + x + 1}{x^4 + 2x^2 + 1}\, dx$

35. $\displaystyle\int \text{arcsec } 2x\, dx, \quad x > \frac{1}{2}$

36. $\displaystyle\int x \arcsin x\, dx$

Completing the Square In Exercises 37–40, complete the square and find the indefinite integral.

37. $\displaystyle\int \frac{1}{\sqrt{4x - x^2}}\, dx$

38. $\displaystyle\int \frac{x^2}{\sqrt{2x - x^2}}\, dx$

39. $\displaystyle\int \frac{x}{\sqrt{x^2 + 6x + 12}}\, dx$

40. $\displaystyle\int \frac{x}{\sqrt{x^2 - 6x + 5}}\, dx$

Converting Limits of Integration In Exercises 41–46, evaluate the definite integral using (a) the given integration limits and (b) the limits obtained by trigonometric substitution.

41. $\displaystyle\int_0^{\sqrt{3}/2} \frac{t^2}{(1 - t^2)^{3/2}}\, dt$

42. $\displaystyle\int_0^{\sqrt{3}/2} \frac{1}{(1 - t^2)^{5/2}}\, dt$

43. $\displaystyle\int_0^3 \frac{x^3}{\sqrt{x^2 + 9}}\, dx$

44. $\displaystyle\int_0^{3/5} \sqrt{9 - 25x^2}\, dx$

45. $\displaystyle\int_4^6 \frac{x^2}{\sqrt{x^2 - 9}}\, dx$

46. $\displaystyle\int_4^8 \frac{\sqrt{x^2 - 16}}{x}\, dx$

WRITING ABOUT CONCEPTS

47. Trigonometric Substitution State the substitution you would make if you used trigonometric substitution for an integral involving the given radical, where $a > 0$. Explain your reasoning.

(a) $\sqrt{a^2 - u^2}$

(b) $\sqrt{a^2 + u^2}$

(c) $\sqrt{u^2 - a^2}$

48. Choosing a Method State the method of integration you would use to perform each integration. Explain why you chose that method. Do not integrate.

(a) $\displaystyle\int x\sqrt{x^2 + 1}\, dx$

(b) $\displaystyle\int x^2\sqrt{x^2 - 1}\, dx$

49. Comparing Methods

(a) Find the integral $\displaystyle\int \frac{x}{x^2 + 9}\, dx$ using u-substitution. Then find the integral using trigonometric substitution. Discuss the results.

(b) Find the integral $\displaystyle\int \frac{x^2}{x^2 + 9}\, dx$ algebraically using $x^2 = (x^2 + 9) - 9$. Then find the integral using trigonometric substitution. Discuss the results.

50. **HOW DO YOU SEE IT?**   Use the graph of f' shown in the figure to answer the following.

$$f'(x) = \frac{\sqrt{x^2 + 4}}{2x}$$

(a) Identify the open interval(s) on which the graph of f is increasing or decreasing. Explain.

(b) Identify the open interval(s) on which the graph of f is concave upward or concave downward. Explain.

True or False?   **In Exercises 51–54, determine whether the statement is true or false. If it is false, explain why or give an example that shows it is false.**

51. If $x = \sin \theta$, then

$$\int \frac{dx}{\sqrt{1 - x^2}} = \int d\theta.$$

52. If $x = \sec \theta$, then

$$\int \frac{\sqrt{x^2 - 1}}{x} \, dx = \int \sec \theta \tan \theta \, d\theta.$$

53. If $x = \tan \theta$, then

$$\int_0^{\sqrt{3}} \frac{dx}{(1 + x^2)^{3/2}} = \int_0^{4\pi/3} \cos \theta \, d\theta.$$

54. If $x = \sin \theta$, then

$$\int_{-1}^{1} x^2 \sqrt{1 - x^2} \, dx = 2 \int_0^{\pi/2} \sin^2 \theta \cos^2 \theta \, d\theta.$$

55. Area   Find the area enclosed by the ellipse $\dfrac{x^2}{a^2} + \dfrac{y^2}{b^2} = 1$ shown in the figure.

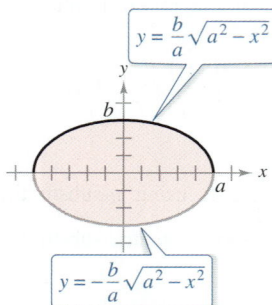

$$y = \frac{b}{a}\sqrt{a^2 - x^2}$$

$$y = -\frac{b}{a}\sqrt{a^2 - x^2}$$

Figure for 55

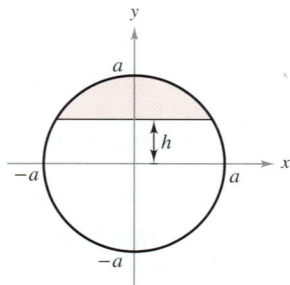

Figure for 56

56. Area   Find the area of the shaded region of the circle of radius a when the chord is h units ($0 < h < a$) from the center of the circle (see figure).

57. Mechanical Design   The surface of a machine part is the region between the graphs of $y = |x|$ and $x^2 + (y - k)^2 = 25$ (see figure).

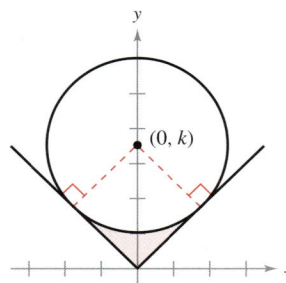

(a) Find k when the circle is tangent to the graph of $y = |x|$.

(b) Find the area of the surface of the machine part.

(c) Find the area of the surface of the machine part as a function of the radius r of the circle.

58. Volume   The axis of a storage tank in the form of a right circular cylinder is horizontal (see figure). The radius and length of the tank are 1 meter and 3 meters, respectively.

(a) Determine the volume of fluid in the tank as a function of its depth d.

(b) Use a graphing utility to graph the function in part (a).

(c) Design a dip stick for the tank with markings of $\frac{1}{4}$, $\frac{1}{2}$, and $\frac{3}{4}$.

(d) Fluid is entering the tank at a rate of $\frac{1}{4}$ cubic meter per minute. Determine the rate of change of the depth of the fluid as a function of its depth d.

(e) Use a graphing utility to graph the function in part (d). When will the rate of change of the depth be minimum? Does this agree with your intuition? Explain.

Volume of a Torus   **In Exercises 59 and 60, find the volume of the torus generated by revolving the region bounded by the graph of the circle about the y-axis.**

59. $(x - 3)^2 + y^2 = 1$

60. $(x - h)^2 + y^2 = r^2$, $h > r$

Arc Length   **In Exercises 61 and 62, find the arc length of the curve over the given interval.**

61. $y = \ln x$, $[1, 5]$

62. $y = \frac{1}{2}x^2$, $[0, 4]$

63. Arc Length   Show that the length of one arch of the sine curve is equal to the length of one arch of the cosine curve.

64. Conjecture

(a) Find formulas for the distances between $(0, 0)$ and (a, a^2) along the line between these points and along the parabola $y = x^2$.

(b) Use the formulas from part (a) to find the distances for $a = 1$ and $a = 10$.

(c) Make a conjecture about the difference between the two distances as a increases.

Centroid **In Exercises 65 and 66, find the centroid of the region determined by the graphs of the inequalities.**

65. $y \le 3/\sqrt{x^2 + 9}, y \ge 0, x \ge -4, x \le 4$

66. $y \le \frac{1}{4}x^2, (x - 4)^2 + y^2 \le 16, y \ge 0$

67. Surface Area Find the surface area of the solid generated by revolving the region bounded by the graphs of $y = x^2$, $y = 0$, $x = 0$, and $x = \sqrt{2}$ about the x-axis.

68. Field Strength The field strength H of a magnet of length $2L$ on a particle r units from the center of the magnet is

$$H = \frac{2mL}{(r^2 + L^2)^{3/2}}$$

where $\pm m$ are the poles of the magnet (see figure). Find the average field strength as the particle moves from 0 to R units from the center by evaluating the integral

$$\frac{1}{R}\int_0^R \frac{2mL}{(r^2 + L^2)^{3/2}}\, dr.$$

69. Fluid Force

Find the fluid force on a circular observation window of radius 1 foot in a vertical wall of a large water-filled tank at a fish hatchery when the center of the window is (a) 3 feet and (b) d feet $(d > 1)$ below the water's surface (see figure). Use trigonometric substitution to evaluate the one integral. Water weighs 62.4 pounds per cubic foot. (Recall that in Section 7.7 in a similar problem, you evaluated one integral by a geometric formula and the other by observing that the integrand was odd.)

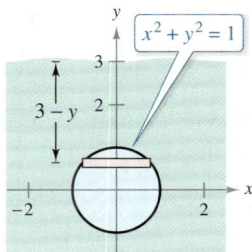

70. Fluid Force Evaluate the following two integrals, which yield the fluid forces given in Example 6.

(a) $F_{inside} = 48 \int_{-1}^{0.8} (0.8 - y)(2)\sqrt{1 - y^2}\, dy$

(b) $F_{outside} = 64 \int_{-1}^{0.4} (0.4 - y)(2)\sqrt{1 - y^2}\, dy$

71. Verifying Formulas Use trigonometric substitution to verify the integration formulas given in Theorem 8.2.

72. Arc Length Show that the arc length of the graph of $y = \sin x$ on the interval $[0, 2\pi]$ is equal to the circumference of the ellipse $x^2 + 2y^2 = 2$ (see figure).

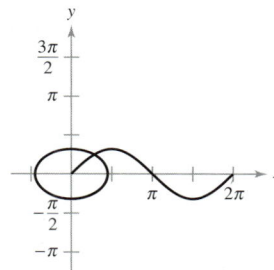

73. Area of a Lune The crescent-shaped region bounded by two circles forms a *lune* (see figure). Find the area of the lune given that the radius of the smaller circle is 3 and the radius of the larger circle is 5.

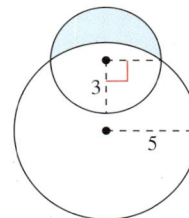

74. Area Two circles of radius 3, with centers at $(-2, 0)$ and $(2, 0)$, intersect as shown in the figure. Find the area of the shaded region.

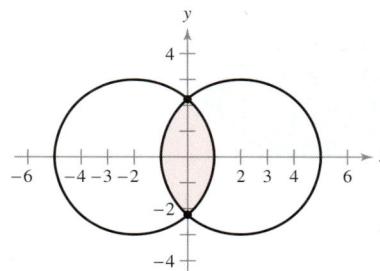

PUTNAM EXAM CHALLENGE

75. Evaluate

$$\int_0^1 \frac{\ln(x + 1)}{x^2 + 1}\, dx.$$

8.5 Partial Fractions

■ Understand the concept of partial fraction decomposition.
■ Use partial fraction decomposition with linear factors to integrate rational functions.
■ Use partial fraction decomposition with quadratic factors to integrate rational functions.

Partial Fractions

This section examines a procedure for decomposing a rational function into simpler rational functions to which you can apply the basic integration formulas. This procedure is called the **method of partial fractions.** To see the benefit of the method of partial fractions, consider the integral

$$\int \frac{1}{x^2 - 5x + 6}\, dx.$$

To evaluate this integral *without* partial fractions, you can complete the square and use trigonometric substitution (see Figure 8.13) to obtain

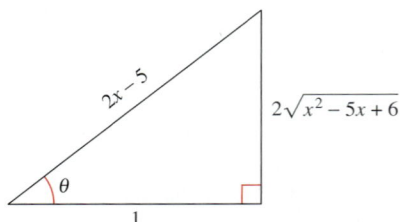

$$\int \frac{1}{x^2 - 5x + 6}\, dx = \int \frac{dx}{(x - 5/2)^2 - (1/2)^2} \qquad a = \tfrac{1}{2},\, x - \tfrac{5}{2} = \tfrac{1}{2}\sec\theta$$

$$= \int \frac{(1/2)\sec\theta\tan\theta\, d\theta}{(1/4)\tan^2\theta} \qquad dx = \tfrac{1}{2}\sec\theta\tan\theta\, d\theta$$

$$= 2\int \csc\theta\, d\theta$$

$$= 2\ln|\csc\theta - \cot\theta| + C$$

$$= 2\ln\left|\frac{2x - 5}{2\sqrt{x^2 - 5x + 6}} - \frac{1}{2\sqrt{x^2 - 5x + 6}}\right| + C$$

$$= 2\ln\left|\frac{x - 3}{\sqrt{x^2 - 5x + 6}}\right| + C$$

$$= 2\ln\left|\frac{\sqrt{x - 3}}{\sqrt{x - 2}}\right| + C$$

$$= \ln\left|\frac{x - 3}{x - 2}\right| + C$$

$$= \ln|x - 3| - \ln|x - 2| + C.$$

$\sec\theta = 2x - 5$
Figure 8.13

Now, suppose you had observed that

$$\frac{1}{x^2 - 5x + 6} = \frac{1}{x - 3} - \frac{1}{x - 2}. \qquad \text{Partial fraction decomposition}$$

Then you could evaluate the integral, as shown.

$$\int \frac{1}{x^2 - 5x + 6}\, dx = \int \left(\frac{1}{x - 3} - \frac{1}{x - 2}\right) dx$$

$$= \ln|x - 3| - \ln|x - 2| + C$$

This method is clearly preferable to trigonometric substitution. Its use, however, depends on the ability to factor the denominator, $x^2 - 5x + 6$, and to find the **partial fractions**

$$\frac{1}{x - 3} \quad \text{and} \quad -\frac{1}{x - 2}.$$

In this section, you will study techniques for finding partial fraction decompositions.

JOHN BERNOULLI (1667–1748)

The method of partial fractions was introduced by John Bernoulli, a Swiss mathematician who was instrumental in the early development of calculus. John Bernoulli was a professor at the University of Basel and taught many outstanding students, the most famous of whom was Leonhard Euler.

See LarsonCalculus.com to read more of this biography.

Recall from algebra that every polynomial with real coefficients can be factored into linear and irreducible quadratic factors.* For instance, the polynomial

$$x^5 + x^4 - x - 1$$

can be written as

$$
\begin{aligned}
x^5 + x^4 - x - 1 &= x^4(x + 1) - (x + 1) \\
&= (x^4 - 1)(x + 1) \\
&= (x^2 + 1)(x^2 - 1)(x + 1) \\
&= (x^2 + 1)(x + 1)(x - 1)(x + 1) \\
&= (x - 1)(x + 1)^2(x^2 + 1)
\end{aligned}
$$

where $(x - 1)$ is a linear factor, $(x + 1)^2$ is a repeated linear factor, and $(x^2 + 1)$ is an irreducible quadratic factor. Using this factorization, you can write the partial fraction decomposition of the rational expression

$$\frac{N(x)}{x^5 + x^4 - x - 1}$$

where $N(x)$ is a polynomial of degree less than 5, as shown.

$$\frac{N(x)}{(x - 1)(x + 1)^2(x^2 + 1)} = \frac{A}{x - 1} + \frac{B}{x + 1} + \frac{C}{(x + 1)^2} + \frac{Dx + E}{x^2 + 1}$$

REMARK In precalculus, you learned how to combine functions such as

$$\frac{1}{x - 2} + \frac{-1}{x + 3} = \frac{5}{(x - 2)(x + 3)}.$$

The method of partial fractions shows you how to reverse this process.

$$\frac{5}{(x - 2)(x + 3)} = \frac{?}{x - 2} + \frac{?}{x + 3}$$

Decomposition of $N(x)/D(x)$ into Partial Fractions

1. **Divide when improper:** When $N(x)/D(x)$ is an improper fraction (that is, when the degree of the numerator is greater than or equal to the degree of the denominator), divide the denominator into the numerator to obtain

$$\frac{N(x)}{D(x)} = (\text{a polynomial}) + \frac{N_1(x)}{D(x)}$$

   where the degree of $N_1(x)$ is less than the degree of $D(x)$. Then apply Steps 2, 3, and 4 to the proper rational expression $N_1(x)/D(x)$.

2. **Factor denominator:** Completely factor the denominator into factors of the form

$$(px + q)^m \quad \text{and} \quad (ax^2 + bx + c)^n$$

   where $ax^2 + bx + c$ is irreducible.

3. **Linear factors:** For each factor of the form $(px + q)^m$, the partial fraction decomposition must include the following sum of m fractions.

$$\frac{A_1}{(px + q)} + \frac{A_2}{(px + q)^2} + \cdots + \frac{A_m}{(px + q)^m}$$

4. **Quadratic factors:** For each factor of the form $(ax^2 + bx + c)^n$, the partial fraction decomposition must include the following sum of n fractions.

$$\frac{B_1x + C_1}{ax^2 + bx + c} + \frac{B_2x + C_2}{(ax^2 + bx + c)^2} + \cdots + \frac{B_nx + C_n}{(ax^2 + bx + c)^n}$$

---

* For a review of factorization techniques, see *Precalculus*, 9th edition, or *Precalculus: Real Mathematics, Real People*, 6th edition, both by Ron Larson (Boston, Massachusetts: Brooks/Cole, Cengage Learning, 2014 and 2012, respectively).

Linear Factors

Algebraic techniques for determining the constants in the numerators of a partial fraction decomposition with linear or repeated linear factors are shown in Examples 1 and 2.

EXAMPLE 1   **Distinct Linear Factors**

Write the partial fraction decomposition for

$$\frac{1}{x^2 - 5x + 6}.$$

Solution   Because $x^2 - 5x + 6 = (x - 3)(x - 2)$, you should include one partial fraction for each factor and write

$$\frac{1}{x^2 - 5x + 6} = \frac{A}{x - 3} + \frac{B}{x - 2}$$

where A and B are to be determined. Multiplying this equation by the least common denominator $(x - 3)(x - 2)$ yields the **basic equation**

$$1 = A(x - 2) + B(x - 3). \qquad \text{\color{red}Basic equation}$$

Because this equation is to be true for all x, you can substitute any *convenient* values for x to obtain equations in A and B. The most convenient values are the ones that make particular factors equal to 0.

To solve for A, let $x = 3$.

$$1 = A(3 - 2) + B(3 - 3) \qquad \text{\color{red}Let } x = 3 \text{ in basic equation.}$$
$$1 = A(1) + B(0)$$
$$1 = A$$

To solve for B, let $x = 2$.

$$1 = A(2 - 2) + B(2 - 3) \qquad \text{\color{red}Let } x = 2 \text{ in basic equation.}$$
$$1 = A(0) + B(-1)$$
$$-1 = B$$

So, the decomposition is

$$\frac{1}{x^2 - 5x + 6} = \frac{1}{x - 3} - \frac{1}{x - 2}$$

as shown at the beginning of this section.

> **REMARK**   Note that the substitutions for x in Example 1 are chosen for their convenience in determining values for A and B; $x = 3$ is chosen to eliminate the term $B(x - 3)$, and $x = 2$ is chosen to eliminate the term $A(x - 2)$. The goal is to make *convenient* substitutions whenever possible.

■ **FOR FURTHER INFORMATION** To learn a different method for finding partial fraction decompositions, called the Heaviside Method, see the article "Calculus to Algebra Connections in Partial Fraction Decomposition" by Joseph Wiener and Will Watkins in *The AMATYC Review.*

Be sure you see that the method of partial fractions is practical only for integrals of rational functions whose denominators factor "nicely." For instance, when the denominator in Example 1 is changed to

$$x^2 - 5x + 5$$

its factorization as

$$x^2 - 5x + 5 = \left[x - \frac{5 + \sqrt{5}}{2} \right]\left[x - \frac{5 - \sqrt{5}}{2} \right]$$

would be too cumbersome to use with partial fractions. In such cases, you should use completing the square or a computer algebra system to perform the integration. When you do this, you should obtain

$$\int \frac{1}{x^2 - 5x + 5}\, dx = \frac{\sqrt{5}}{5}\ln\left| 2x - \sqrt{5} - 5 \right| - \frac{\sqrt{5}}{5}\ln\left| 2x + \sqrt{5} - 5 \right| + C.$$

EXAMPLE 2   **Repeated Linear Factors**

Find $\int \dfrac{5x^2 + 20x + 6}{x^3 + 2x^2 + x}\, dx.$

FOR FURTHER INFORMATION
For an alternative approach to using partial fractions, see the article "A Shortcut in Partial Fractions" by Xun-Cheng Huang in *The College Mathematics Journal.*

Solution   Because

$$x^3 + 2x^2 + x = x(x^2 + 2x + 1) = x(x + 1)^2$$

you should include one fraction for *each power* of x and $(x + 1)$ and write

$$\frac{5x^2 + 20x + 6}{x(x + 1)^2} = \frac{A}{x} + \frac{B}{x + 1} + \frac{C}{(x + 1)^2}.$$

Multiplying by the least common denominator $x(x + 1)^2$ yields the *basic equation*

$$5x^2 + 20x + 6 = A(x + 1)^2 + Bx(x + 1) + Cx. \qquad \text{Basic equation}$$

To solve for A, let $x = 0$. This eliminates the B and C terms and yields

$$6 = A(1) + 0 + 0$$
$$6 = A.$$

To solve for C, let $x = -1$. This eliminates the A and B terms and yields

$$5 - 20 + 6 = 0 + 0 - C$$
$$9 = C.$$

The most convenient choices for x have been used, so to find the value of B, you can use *any other value* of x along with the calculated values of A and C. Using $x = 1$, $A = 6$, and $C = 9$ produces

$$5 + 20 + 6 = A(4) + B(2) + C$$
$$31 = 6(4) + 2B + 9$$
$$-2 = 2B$$
$$-1 = B.$$

So, it follows that

$$\int \frac{5x^2 + 20x + 6}{x(x + 1)^2}\, dx = \int \left(\frac{6}{x} - \frac{1}{x + 1} + \frac{9}{(x + 1)^2} \right) dx$$
$$= 6\ln|x| - \ln|x + 1| + 9\frac{(x + 1)^{-1}}{-1} + C$$
$$= \ln\left| \frac{x^6}{x + 1} \right| - \frac{9}{x + 1} + C.$$

Try checking this result by differentiating. Include algebra in your check, simplifying the derivative until you have obtained the original integrand.

It is necessary to make as many substitutions for x as there are unknowns $(A, B, C, \ldots)$ to be determined. For instance, in Example 2, three substitutions $(x = 0, x = -1,$ and $x = 1)$ were made to solve for A, B, and C.

▷ **TECHNOLOGY**   Most computer algebra systems, such as *Maple, Mathematica,* and the *TI-nSpire,* can be used to convert a rational function to its partial fraction decomposition. For instance, using *Mathematica,* you obtain the following.

$$\text{Apart}\,[(5 * x \wedge 2 + 20 * x + 6)/(x * (x + 1) \wedge 2), x]$$
$$\frac{6}{x} + \frac{9}{(1 + x)^2} - \frac{1}{1 + x}$$

Quadratic Factors

When using the method of partial fractions with *linear* factors, a convenient choice of x immediately yields a value for one of the coefficients. With *quadratic* factors, a system of linear equations usually has to be solved, regardless of the choice of x.

EXAMPLE 3   **Distinct Linear and Quadratic Factors**

$\cdots \triangleright$ *See LarsonCalculus.com for an interactive version of this type of example.*

Find $\displaystyle\int \frac{2x^3 - 4x - 8}{(x^2 - x)(x^2 + 4)}\, dx.$

Solution   Because

$$(x^2 - x)(x^2 + 4) = x(x - 1)(x^2 + 4)$$

you should include one partial fraction for each factor and write

$$\frac{2x^3 - 4x - 8}{x(x - 1)(x^2 + 4)} = \frac{A}{x} + \frac{B}{x - 1} + \frac{Cx + D}{x^2 + 4}.$$

Multiplying by the least common denominator

$$x(x - 1)(x^2 + 4)$$

yields the *basic equation*

$$2x^3 - 4x - 8 = A(x - 1)(x^2 + 4) + Bx(x^2 + 4) + (Cx + D)(x)(x - 1).$$

To solve for A, let $x = 0$ and obtain

$$-8 = A(-1)(4) + 0 + 0$$
$$2 = A.$$

To solve for B, let $x = 1$ and obtain

$$-10 = 0 + B(5) + 0$$
$$-2 = B.$$

At this point, C and D are yet to be determined. You can find these remaining constants by choosing two other values for x and solving the resulting system of linear equations. Using $x = -1$, $A = 2$, and $B = -2$, you can write

$$-6 = (2)(-2)(5) + (-2)(-1)(5) + (-C + D)(-1)(-2)$$
$$2 = -C + D.$$

For $x = 2$, you have

$$0 = (2)(1)(8) + (-2)(2)(8) + (2C + D)(2)(1)$$
$$8 = 2C + D.$$

Solving the linear system by subtracting the first equation from the second

$$-C + D = 2$$
$$2C + D = 8$$

yields $C = 2$. Consequently, $D = 4$, and it follows that

$$\int \frac{2x^3 - 4x - 8}{x(x - 1)(x^2 + 4)}\, dx = \int \left(\frac{2}{x} - \frac{2}{x - 1} + \frac{2x}{x^2 + 4} + \frac{4}{x^2 + 4}\right) dx$$

$$= 2\ln|x| - 2\ln|x - 1| + \ln(x^2 + 4) + 2\arctan\frac{x}{2} + C.$$

In Examples 1, 2, and 3, the solution of the basic equation began with substituting values of x that made the linear factors equal to 0. This method works well when the partial fraction decomposition involves linear factors. When the decomposition involves only quadratic factors, however, an alternative procedure is often more convenient. For instance, try writing the right side of the basic equation in polynomial form and *equating the coefficients* of like terms. This method is shown in Example 4.

EXAMPLE 4   Repeated Quadratic Factors

Find $\displaystyle\int \frac{8x^3 + 13x}{(x^2 + 2)^2}\, dx$.

Solution   Include one partial fraction for each power of $(x^2 + 2)$ and write

$$\frac{8x^3 + 13x}{(x^2 + 2)^2} = \frac{Ax + B}{x^2 + 2} + \frac{Cx + D}{(x^2 + 2)^2}.$$

Multiplying by the least common denominator $(x^2 + 2)^2$ yields the *basic equation*

$$8x^3 + 13x = (Ax + B)(x^2 + 2) + Cx + D.$$

Expanding the basic equation and collecting like terms produces

$$8x^3 + 13x = Ax^3 + 2Ax + Bx^2 + 2B + Cx + D$$
$$8x^3 + 13x = Ax^3 + Bx^2 + (2A + C)x + (2B + D).$$

Now, you can equate the coefficients of like terms on opposite sides of the equation.

$$8 = A \qquad\qquad 0 = 2B + D$$

$$8x^3 + 0x^2 + 13x + 0 = Ax^3 + Bx^2 + (2A + C)x + (2B + D)$$

$$0 = B$$

$$13 = 2A + C$$

Using the known values $A = 8$ and $B = 0$, you can write

$$13 = 2A + C \implies 13 = 2(8) + C \implies -3 = C$$
$$0 = 2B + D \implies 0 = 2(0) + D \implies 0 = D.$$

Finally, you can conclude that

$$\int \frac{8x^3 + 13x}{(x^2 + 2)^2}\, dx = \int \left(\frac{8x}{x^2 + 2} + \frac{-3x}{(x^2 + 2)^2} \right) dx$$

$$= 4 \ln(x^2 + 2) + \frac{3}{2(x^2 + 2)} + C.$$

▷ **TECHNOLOGY**   You can use a graphing utility to confirm the decomposition found in Example 4. To do this, graph

$$y_1 = \frac{8x^3 + 13x}{(x^2 + 2)^2}$$

and

$$y_2 = \frac{8x}{x^2 + 2} + \frac{-3x}{(x^2 + 2)^2}$$

in the same viewing window. The graphs should be identical, as shown at the right.

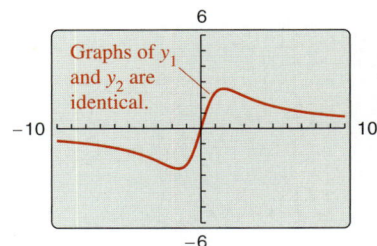

Graphs of y_1 and y_2 are identical.

When integrating rational expressions, keep in mind that for *improper* rational expressions such as

$$\frac{N(x)}{D(x)} = \frac{2x^3 + x^2 - 7x + 7}{x^2 + x - 2}$$

you must first divide to obtain

$$\frac{N(x)}{D(x)} = 2x - 1 + \frac{-2x + 5}{x^2 + x - 2}.$$

The proper rational expression is then decomposed into its partial fractions by the usual methods.

Here are some guidelines for solving the basic equation that is obtained in a partial fraction decomposition.

GUIDELINES FOR SOLVING THE BASIC EQUATION

Linear Factors

1. Substitute the roots of the distinct linear factors in the basic equation.

2. For repeated linear factors, use the coefficients determined in the first guideline to rewrite the basic equation. Then substitute other convenient values of x and solve for the remaining coefficients.

Quadratic Factors

1. Expand the basic equation.

2. Collect terms according to powers of x.

3. Equate the coefficients of like powers to obtain a system of linear equations involving A, B, C, and so on.

4. Solve the system of linear equations.

■ FOR FURTHER INFORMATION
To read about another method of evaluating integrals of rational functions, see the article "Alternate Approach to Partial Fractions to Evaluate Integrals of Rational Functions" by N. R. Nandakumar and Michael J. Bossé in *The Pi Mu Epsilon Journal*. To view this article, go to *MathArticles.com*.

Before concluding this section, here are a few things you should remember. First, it is not necessary to use the partial fractions technique on all rational functions. For instance, the following integral is evaluated more easily by the Log Rule.

$$\int \frac{x^2 + 1}{x^3 + 3x - 4} \, dx = \frac{1}{3} \int \frac{3x^2 + 3}{x^3 + 3x - 4} \, dx$$

$$= \frac{1}{3} \ln|x^3 + 3x - 4| + C$$

Second, when the integrand is not in reduced form, reducing it may eliminate the need for partial fractions, as shown in the following integral.

$$\int \frac{x^2 - x - 2}{x^3 - 2x - 4} \, dx = \int \frac{(x + 1)(x - 2)}{(x - 2)(x^2 + 2x + 2)} \, dx$$

$$= \int \frac{x + 1}{x^2 + 2x + 2} \, dx$$

$$= \frac{1}{2} \ln|x^2 + 2x + 2| + C$$

Finally, partial fractions can be used with some quotients involving transcendental functions. For instance, the substitution $u = \sin x$ allows you to write

$$\int \frac{\cos x}{\sin x(\sin x - 1)} \, dx = \int \frac{du}{u(u - 1)}. \qquad u = \sin x, \, du = \cos x \, dx$$

8.5 Exercises

Partial Fraction Decomposition   In Exercises 1–4, write the form of the partial fraction decomposition of the rational expression. Do not solve for the constants.

1. $\dfrac{4}{x^2 - 8x}$

2. $\dfrac{2x^2 + 1}{(x - 3)^3}$

3. $\dfrac{2x - 3}{x^3 + 10x}$

4. $\dfrac{2x - 1}{x(x^2 + 1)^2}$

Using Partial Fractions   In Exercises 5–22, use partial fractions to find the indefinite integral.

5. $\displaystyle\int \dfrac{1}{x^2 - 9}\,dx$

6. $\displaystyle\int \dfrac{2}{9x^2 - 1}\,dx$

7. $\displaystyle\int \dfrac{5}{x^2 + 3x - 4}\,dx$

8. $\displaystyle\int \dfrac{3 - x}{3x^2 - 2x - 1}\,dx$

9. $\displaystyle\int \dfrac{x^2 + 12x + 12}{x^3 - 4x}\,dx$

10. $\displaystyle\int \dfrac{x^3 - x + 3}{x^2 + x - 2}\,dx$

11. $\displaystyle\int \dfrac{2x^3 - 4x^2 - 15x + 5}{x^2 - 2x - 8}\,dx$

12. $\displaystyle\int \dfrac{x + 2}{x^2 + 5x}\,dx$

13. $\displaystyle\int \dfrac{4x^2 + 2x - 1}{x^3 + x^2}\,dx$

14. $\displaystyle\int \dfrac{5x - 2}{(x - 2)^2}\,dx$

15. $\displaystyle\int \dfrac{x^2 + 3x - 4}{x^3 - 4x^2 + 4x}\,dx$

16. $\displaystyle\int \dfrac{8x}{x^3 + x^2 - x - 1}\,dx$

17. $\displaystyle\int \dfrac{x^2 - 1}{x^3 + x}\,dx$

18. $\displaystyle\int \dfrac{6x}{x^3 - 8}\,dx$

19. $\displaystyle\int \dfrac{x^2}{x^4 - 2x^2 - 8}\,dx$

20. $\displaystyle\int \dfrac{x}{16x^4 - 1}\,dx$

21. $\displaystyle\int \dfrac{x^2 + 5}{x^3 - x^2 + x + 3}\,dx$

22. $\displaystyle\int \dfrac{x^2 + 6x + 4}{x^4 + 8x^2 + 16}\,dx$

Evaluating a Definite Integral   In Exercises 23–26, evaluate the definite integral. Use a graphing utility to verify your result.

23. $\displaystyle\int_0^2 \dfrac{3}{4x^2 + 5x + 1}\,dx$

24. $\displaystyle\int_1^5 \dfrac{x - 1}{x^2(x + 1)}\,dx$

25. $\displaystyle\int_1^2 \dfrac{x + 1}{x(x^2 + 1)}\,dx$

26. $\displaystyle\int_0^1 \dfrac{x^2 - x}{x^2 + x + 1}\,dx$

Finding an Indefinite Integral   In Exercises 27–34, use substitution and partial fractions to find the indefinite integral.

27. $\displaystyle\int \dfrac{\sin x}{\cos x + \cos^2 x}\,dx$

28. $\displaystyle\int \dfrac{5 \cos x}{\sin^2 x + 3 \sin x - 4}\,dx$

29. $\displaystyle\int \dfrac{\sec^2 x}{\tan^2 x + 5 \tan x + 6}\,dx$

30. $\displaystyle\int \dfrac{\sec^2 x}{\tan x(\tan x + 1)}\,dx$

31. $\displaystyle\int \dfrac{e^x}{(e^x - 1)(e^x + 4)}\,dx$

32. $\displaystyle\int \dfrac{e^x}{(e^{2x} + 1)(e^x - 1)}\,dx$

33. $\displaystyle\int \dfrac{\sqrt{x}}{x - 4}\,dx$

34. $\displaystyle\int \dfrac{1}{\sqrt{x} - \sqrt[3]{x}}\,dx$

Verifying a Formula   In Exercises 35–38, use the method of partial fractions to verify the integration formula.

35. $\displaystyle\int \dfrac{1}{x(a + bx)}\,dx = \dfrac{1}{a}\ln\left|\dfrac{x}{a + bx}\right| + C$

36. $\displaystyle\int \dfrac{1}{a^2 - x^2}\,dx = \dfrac{1}{2a}\ln\left|\dfrac{a + x}{a - x}\right| + C$

37. $\displaystyle\int \dfrac{x}{(a + bx)^2}\,dx = \dfrac{1}{b^2}\left(\dfrac{a}{a + bx} + \ln|a + bx|\right) + C$

38. $\displaystyle\int \dfrac{1}{x^2(a + bx)}\,dx = -\dfrac{1}{ax} - \dfrac{b}{a^2}\ln\left|\dfrac{x}{a + bx}\right| + C$

WRITING ABOUT CONCEPTS

39. Using Partial Fractions   What is the first step when integrating $\displaystyle\int \dfrac{x^3}{x - 5}\,dx$? Explain.

40. Decomposition   Describe the decomposition of the proper rational function $N(x)/D(x)$ (a) for $D(x) = (px + q)^m$ and (b) for $D(x) = (ax^2 + bx + c)^n$ where $ax^2 + bx + c$ is irreducible. Explain why you chose that method.

41. Choosing a Method   State the method you would use to evaluate each integral. Explain why you chose that method. Do not integrate.

(a) $\displaystyle\int \dfrac{x + 1}{x^2 + 2x - 8}\,dx$ (b) $\displaystyle\int \dfrac{7x + 4}{x^2 + 2x - 8}\,dx$

(c) $\displaystyle\int \dfrac{4}{x^2 + 2x + 5}\,dx$

---

42. **HOW DO YOU SEE IT?**   Use the graph of f' shown in the figure to answer the following.

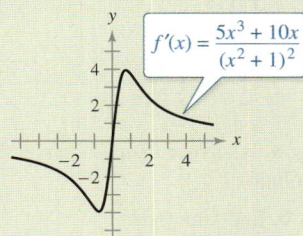

$$f'(x) = \dfrac{5x^3 + 10x}{(x^2 + 1)^2}$$

(a) Is $f(3) - f(2) > 0$? Explain.

(b) Which is greater, the area under the graph of f' from 1 to 2, or the area under the graph of f' from 3 to 4?

---

43. Area   Find the area of the region bounded by the graphs of $y = 12/(x^2 + 5x + 6)$, $y = 0$, $x = 0$, and $x = 1$.

44. Area   Find the area of the region bounded by the graphs of $y = 7/(16 - x^2)$ and $y = 1$.

45. Modeling Data The predicted cost C (in hundreds of thousands of dollars) for a company to remove $p\%$ of a chemical from its waste water is shown in the table.

P	0	10	20	30	40
C	0	0.7	1.0	1.3	1.7

P	50	60	70	80	90
C	2.0	2.7	3.6	5.5	11.2

A model for the data is given by $C = \dfrac{124p}{(10 + p)(100 - p)}$ for $0 \le p < 100$. Use the model to find the average cost of removing between 75% and 80% of the chemical.

46. Logistic Growth In Chapter 6, the exponential growth equation was derived from the assumption that the rate of growth was proportional to the existing quantity. In practice, there often exists some upper limit L past which growth cannot occur. In such cases, you assume the rate of growth to be proportional not only to the existing quantity, but also to the difference between the existing quantity y and the upper limit L. That is, $dy/dt = ky(L - y)$. In integral form, you can write this relationship as

$$\int \frac{dy}{y(L - y)} = \int k \, dt.$$

(a) A slope field for the differential equation $dy/dt = y(3 - y)$ is shown. Draw a possible solution to the differential equation when $y(0) = 5$, and another when $y(0) = \frac{1}{2}$. To print an enlarged copy of the graph, go to *MathGraphs.com*.

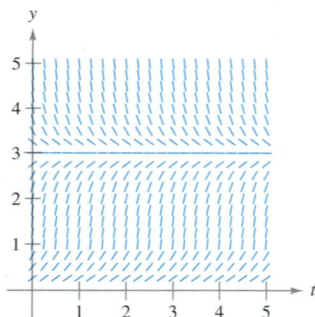

(b) Where $y(0)$ is greater than 3, what is the sign of the slope of the solution?

(c) For $y > 0$, find $\lim\limits_{t \to \infty} y(t)$.

(d) Evaluate the two given integrals and solve for y as a function of t, where y_0 is the initial quantity.

(e) Use the result of part (d) to find and graph the solutions in part (a). Use a graphing utility to graph the solutions and compare the results with the solutions in part (a).

(f) The graph of the function y is a **logistic curve.** Show that the rate of growth is maximum at the point of inflection, and that this occurs when $y = L/2$.

dextroza/Shutterstock.com

47. Volume and Centroid Consider the region bounded by the graphs of $y = 2x/(x^2 + 1)$, $y = 0$, $x = 0$, and $x = 3$. Find the volume of the solid generated by revolving the region about the x-axis. Find the centroid of the region.

48. Volume Consider the region bounded by the graph of

$$y^2 = \frac{(2 - x)^2}{(1 + x)^2}$$

on the interval $[0, 1]$. Find the volume of the solid generated by revolving this region about the x-axis.

49. Epidemic Model A single infected individual enters a community of n susceptible individuals. Let x be the number of newly infected individuals at time t. The common epidemic model assumes that the disease spreads at a rate proportional to the product of the total number infected and the number not yet infected. So, $dx/dt = k(x + 1)(n - x)$ and you obtain

$$\int \frac{1}{(x + 1)(n - x)} \, dx = \int k \, dt.$$

Solve for x as a function of t.

50. Chemical Reaction

In a chemical reaction, one unit of compound Y and one unit of compound Z are converted into a single unit of compound X. Let x be the amount of compound X formed. The rate of formation of X is proportional to the product of the amounts of unconverted compounds Y and Z. So, $dx/dt = k(y_0 - x)(z_0 - x)$, where y_0 and z_0 are the initial amounts of compounds Y and Z. From this equation, you obtain

$$\int \frac{1}{(y_0 - x)(z_0 - x)} \, dx = \int k \, dt.$$

(a) Perform the two integrations and solve for x in terms of t.

(b) Use the result of part (a) to find x as $t \to \infty$ for (1) $y_0 < z_0$, (2) $y_0 > z_0$, and (3) $y_0 = z_0$.

51. Using Two Methods Evaluate

$$\int_0^1 \frac{x}{1 + x^4} \, dx$$

in two different ways, one of which is partial fractions.

PUTNAM EXAM CHALLENGE

52. Prove $\dfrac{22}{7} - \pi = \displaystyle\int_0^1 \frac{x^4(1 - x)^4}{1 + x^2} \, dx$.

This problem was composed by the Committee on the Putnam Prize Competition.
© The Mathematical Association of America. All rights reserved.

8.6 Integration by Tables and Other Integration Techniques

- ■ Evaluate an indefinite integral using a table of integrals.
- ■ Evaluate an indefinite integral using reduction formulas.
- ■ Evaluate an indefinite integral involving rational functions of sine and cosine.

Integration by Tables

So far in this chapter, you have studied several integration techniques that can be used with the basic integration rules. But merely knowing *how* to use the various techniques is not enough. You also need to know *when* to use them. Integration is first and foremost a problem of recognition. That is, you must recognize which rule or technique to apply to obtain an antiderivative. Frequently, a slight alteration of an integrand will require a different integration technique (or produce a function whose antiderivative is not an elementary function), as shown below.

$$\int x \ln x \, dx = \frac{x^2}{2} \ln x - \frac{x^2}{4} + C \qquad \text{Integration by parts}$$

$$\int \frac{\ln x}{x} \, dx = \frac{(\ln x)^2}{2} + C \qquad \text{Power Rule}$$

$$\int \frac{1}{x \ln x} \, dx = \ln|\ln x| + C \qquad \text{Log Rule}$$

$$\int \frac{x}{\ln x} \, dx = ? \qquad \text{Not an elementary function}$$

▷ **TECHNOLOGY** A computer algebra system consists, in part, of a database of integration formulas. The primary difference between using a computer algebra system and using tables of integrals is that with a computer algebra system, the computer searches through the database to find a fit. With integration tables, *you* must do the searching.

Many people find tables of integrals to be a valuable supplement to the integration techniques discussed in this chapter. Tables of common integrals can be found in Appendix B. **Integration by tables** is not a "cure-all" for all of the difficulties that can accompany integration—using tables of integrals requires considerable thought and insight and often involves substitution.

Each integration formula in Appendix B can be developed using one or more of the techniques in this chapter. You should try to verify several of the formulas. For instance, Formula 4

$$\int \frac{u}{(a + bu)^2} \, du = \frac{1}{b^2} \left(\frac{a}{a + bu} + \ln|a + bu| \right) + C \qquad \text{Formula 4}$$

can be verified using the method of partial fractions, Formula 19

$$\int \frac{\sqrt{a + bu}}{u} \, du = 2\sqrt{a + bu} + a \int \frac{du}{u\sqrt{a + bu}} \qquad \text{Formula 19}$$

can be verified using integration by parts, and Formula 84

$$\int \frac{1}{1 + e^u} \, du = u - \ln(1 + e^u) + C \qquad \text{Formula 84}$$

can be verified using substitution. Note that the integrals in Appendix B are classified according to the form of the integrand. Several of the forms are listed below.

u^n	$(a + bu)$
$(a + bu + cu^2)$	$\sqrt{a + bu}$
$(a^2 \pm u^2)$	$\sqrt{u^2 \pm a^2}$
$\sqrt{a^2 - u^2}$	Trigonometric functions
Inverse trigonometric functions	Exponential functions
Logarithmic functions	

Exploration

Use the tables of integrals in Appendix B and the substitution

$$u = \sqrt{x - 1}$$

to evaluate the integral in Example 1. When you do this, you should obtain

$$\int \frac{dx}{x\sqrt{x - 1}} = \int \frac{2\,du}{u^2 + 1}.$$

Does this produce the same result as that obtained in Example 1?

EXAMPLE 1 Integration by Tables

Find $\displaystyle\int \frac{dx}{x\sqrt{x - 1}}$.

Solution Because the expression inside the radical is linear, you should consider forms involving $\sqrt{a + bu}$.

$$\int \frac{du}{u\sqrt{a + bu}} = \frac{2}{\sqrt{-a}} \arctan \sqrt{\frac{a + bu}{-a}} + C \qquad \text{Formula 17 } (a < 0)$$

Let $a = -1$, $b = 1$, and $u = x$. Then $du = dx$, and you can write

$$\int \frac{dx}{x\sqrt{x - 1}} = 2 \arctan \sqrt{x - 1} + C.$$

EXAMPLE 2 Integration by Tables

$\cdots\blacktriangleright$ *See LarsonCalculus.com for an interactive version of this type of example.*

Find $\displaystyle\int x\sqrt{x^4 - 9}\,dx$.

Solution Because the radical has the form $\sqrt{u^2 - a^2}$, you should consider Formula 26.

$$\int \sqrt{u^2 - a^2}\,du = \frac{1}{2}\left(u\sqrt{u^2 - a^2} - a^2 \ln\left|u + \sqrt{u^2 - a^2}\right|\right) + C$$

Let $u = x^2$ and $a = 3$. Then $du = 2x\,dx$, and you have

$$\int x\sqrt{x^4 - 9}\,dx = \frac{1}{2}\int \sqrt{(x^2)^2 - 3^2}\,(2x)\,dx$$

$$= \frac{1}{4}\left(x^2\sqrt{x^4 - 9} - 9 \ln\left|x^2 + \sqrt{x^4 - 9}\right|\right) + C.$$

EXAMPLE 3 Integration by Tables

Evaluate $\displaystyle\int_0^2 \frac{x}{1 + e^{-x^2}}\,dx$.

Solution Of the forms involving e^u, consider the formula

$$\int \frac{du}{1 + e^u} = u - \ln(1 + e^u) + C. \qquad \text{Formula 84}$$

Let $u = -x^2$. Then $du = -2x\,dx$, and you have

$$\int \frac{x}{1 + e^{-x^2}}\,dx = -\frac{1}{2}\int \frac{-2x\,dx}{1 + e^{-x^2}}$$

$$= -\frac{1}{2}\left[-x^2 - \ln\left(1 + e^{-x^2}\right)\right] + C$$

$$= \frac{1}{2}\left[x^2 + \ln\left(1 + e^{-x^2}\right)\right] + C.$$

So, the value of the definite integral is

$$\int_0^2 \frac{x}{1 + e^{-x^2}}\,dx = \frac{1}{2}\left[x^2 + \ln\left(1 + e^{-x^2}\right)\right]_0^2 = \frac{1}{2}[4 + \ln\left(1 + e^{-4}\right) - \ln 2] \approx 1.66.$$

Figure 8.14 shows the region whose area is represented by this integral.

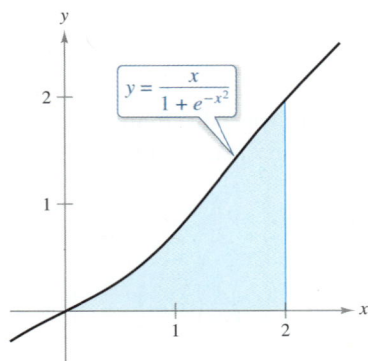

$y = \dfrac{x}{1 + e^{-x^2}}$

Figure 8.14

Reduction Formulas

Several of the integrals in the integration tables have the form

$$\int f(x)\, dx = g(x) + \int h(x)\, dx.$$

Such integration formulas are called **reduction formulas** because they reduce a given integral to the sum of a function and a simpler integral.

EXAMPLE 4   **Using a Reduction Formula**

Find $\displaystyle\int x^3 \sin x\, dx$.

Solution   Consider the three formulas listed below.

$$\int u \sin u\, du = \sin u - u \cos u + C \qquad\qquad \text{Formula 52}$$

$$\int u^n \sin u\, du = -u^n \cos u + n \int u^{n-1} \cos u\, du \qquad \text{Formula 54}$$

$$\int u^n \cos u\, du = u^n \sin u - n \int u^{n-1} \sin u\, du \qquad \text{Formula 55}$$

Using Formula 54, Formula 55, and then Formula 52 produces

$$\int x^3 \sin x\, dx = -x^3 \cos x + 3 \int x^2 \cos x\, dx$$

$$= -x^3 \cos x + 3 \left(x^2 \sin x - 2 \int x \sin x\, dx \right)$$

$$= -x^3 \cos x + 3x^2 \sin x + 6x \cos x - 6 \sin x + C.$$

EXAMPLE 5   **Using a Reduction Formula**

Find $\displaystyle\int \frac{\sqrt{3 - 5x}}{2x}\, dx$.

Solution   Consider the two formulas listed below.

$$\int \frac{du}{u\sqrt{a + bu}} = \frac{1}{\sqrt{a}} \ln \left| \frac{\sqrt{a + bu} - \sqrt{a}}{\sqrt{a + bu} + \sqrt{a}} \right| + C \qquad \text{Formula 17 } (a > 0)$$

$$\int \frac{\sqrt{a + bu}}{u}\, du = 2\sqrt{a + bu} + a \int \frac{du}{u\sqrt{a + bu}} \qquad \text{Formula 19}$$

Using Formula 19, with $a = 3$, $b = -5$, and $u = x$, produces

$$\frac{1}{2} \int \frac{\sqrt{3 - 5x}}{x}\, dx = \frac{1}{2} \left(2\sqrt{3 - 5x} + 3 \int \frac{dx}{x\sqrt{3 - 5x}} \right)$$

$$= \sqrt{3 - 5x} + \frac{3}{2} \int \frac{dx}{x\sqrt{3 - 5x}}.$$

Using Formula 17, with $a = 3$, $b = -5$, and $u = x$, produces

$$\int \frac{\sqrt{3 - 5x}}{2x}\, dx = \sqrt{3 - 5x} + \frac{3}{2} \left(\frac{1}{\sqrt{3}} \ln \left| \frac{\sqrt{3 - 5x} - \sqrt{3}}{\sqrt{3 - 5x} + \sqrt{3}} \right| \right) + C$$

$$= \sqrt{3 - 5x} + \frac{\sqrt{3}}{2} \ln \left| \frac{\sqrt{3 - 5x} - \sqrt{3}}{\sqrt{3 - 5x} + \sqrt{3}} \right| + C.$$

▷ **TECHNOLOGY**   Sometimes when you use computer algebra systems, you obtain results that look very different, but are actually equivalent. Here is how two different systems evaluated the integral in Example 5.

Maple

$$\sqrt{3 - 5x} - $$
$$\sqrt{3}\, \text{arctanh}\!\left(\tfrac{1}{3}\sqrt{3 - 5x}\sqrt{3}\right)$$

Mathematica

$$\sqrt{3 - 5x} - $$
$$\sqrt{3}\, \text{ArcTanh}\!\left[\sqrt{1 - \frac{5x}{3}} \right]$$

Notice that computer algebra systems do not include a constant of integration.

Rational Functions of Sine and Cosine

EXAMPLE 6 **Integration by Tables**

Find $\displaystyle\int \frac{\sin 2x}{2 + \cos x}\, dx$.

Solution Substituting $2 \sin x \cos x$ for $\sin 2x$ produces

$$\int \frac{\sin 2x}{2 + \cos x}\, dx = 2 \int \frac{\sin x \cos x}{2 + \cos x}\, dx.$$

A check of the forms involving $\sin u$ or $\cos u$ in Appendix B shows that none of those listed applies. So, you can consider forms involving $a + bu$. For example,

$$\int \frac{u\, du}{a + bu} = \frac{1}{b^2}(bu - a \ln|a + bu|) + C. \qquad \text{Formula 3}$$

Let $a = 2$, $b = 1$, and $u = \cos x$. Then $du = -\sin x\, dx$, and you have

$$\begin{aligned}
2 \int \frac{\sin x \cos x}{2 + \cos x}\, dx &= -2 \int \frac{\cos x(-\sin x\, dx)}{2 + \cos x} \\
&= -2(\cos x - 2 \ln|2 + \cos x|) + C \\
&= -2 \cos x + 4 \ln|2 + \cos x| + C.
\end{aligned}$$

Example 6 involves a rational expression of $\sin x$ and $\cos x$. When you are unable to find an integral of this form in the integration tables, try using the following special substitution to convert the trigonometric expression to a standard rational expression.

Substitution for Rational Functions of Sine and Cosine

For integrals involving rational functions of sine and cosine, the substitution

$$u = \frac{\sin x}{1 + \cos x} = \tan \frac{x}{2}$$

yields

$$\cos x = \frac{1 - u^2}{1 + u^2}, \quad \sin x = \frac{2u}{1 + u^2}, \quad \text{and} \quad dx = \frac{2\, du}{1 + u^2}.$$

Proof From the substitution for u, it follows that

$$u^2 = \frac{\sin^2 x}{(1 + \cos x)^2} = \frac{1 - \cos^2 x}{(1 + \cos x)^2} = \frac{1 - \cos x}{1 + \cos x}.$$

Solving for $\cos x$ produces $\cos x = (1 - u^2)/(1 + u^2)$. To find $\sin x$, write $u = \sin x/(1 + \cos x)$ as

$$\sin x = u(1 + \cos x) = u\left(1 + \frac{1 - u^2}{1 + u^2}\right) = \frac{2u}{1 + u^2}.$$

Finally, to find dx, consider $u = \tan(x/2)$. Then you have $\arctan u = x/2$ and

$$dx = \frac{2\, du}{1 + u^2}.$$

See LarsonCalculus.com for Bruce Edwards's video of this proof.

8.6 Exercises

Integration by Tables In Exercises 1 and 2, use a table of integrals with forms involving $a + bu$ to find the indefinite integral.

1. $\displaystyle\int \frac{x^2}{5 + x}\, dx$

2. $\displaystyle\int \frac{2}{x^2(4 + 3x)^2}\, dx$

Integration by Tables In Exercises 3 and 4, use a table of integrals with forms involving $\sqrt{a^2 - u^2}$ to find the indefinite integral.

3. $\displaystyle\int \frac{1}{x^2\sqrt{1 - x^2}}\, dx$

4. $\displaystyle\int \frac{\sqrt{64 - x^4}}{x}\, dx$

Integration by Tables In Exercises 5–8, use a table of integrals with forms involving the trigonometric functions to find the indefinite integral.

5. $\displaystyle\int \cos^4 3x\, dx$

6. $\displaystyle\int \frac{\sin^4 \sqrt{x}}{\sqrt{x}}\, dx$

7. $\displaystyle\int \frac{1}{\sqrt{x}\left(1 - \cos \sqrt{x}\right)}\, dx$

8. $\displaystyle\int \frac{1}{1 + \cot 4x}\, dx$

Integration by Tables In Exercises 9 and 10, use a table of integrals with forms involving e^u to find the indefinite integral.

9. $\displaystyle\int \frac{1}{1 + e^{2x}}\, dx$

10. $\displaystyle\int e^{-4x} \sin 3x\, dx$

Integration by Tables In Exercises 11 and 12, use a table of integrals with forms involving $\ln u$ to find the indefinite integral.

11. $\displaystyle\int x^7 \ln x\, dx$

12. $\displaystyle\int (\ln x)^3\, dx$

Using Two Methods In Exercises 13–16, find the indefinite integral (a) using integration tables and (b) using the given method.

	Integral	Method
13.	$\displaystyle\int x^2 e^{3x}\, dx$	Integration by parts
14.	$\displaystyle\int x^5 \ln x\, dx$	Integration by parts
15.	$\displaystyle\int \frac{1}{x^2(x + 1)}\, dx$	Partial fractions
16.	$\displaystyle\int \frac{1}{x^2 - 36}\, dx$	Partial fractions

Finding an Indefinite Integral In Exercises 17–38, use integration tables to find the indefinite integral.

17. $\displaystyle\int x \, \text{arccsc}(x^2 + 1)\, dx$

18. $\displaystyle\int \arcsin 4x\, dx$

19. $\displaystyle\int \frac{1}{x^2\sqrt{x^2 - 4}}\, dx$

20. $\displaystyle\int \frac{1}{x^2 + 4x + 8}\, dx$

21. $\displaystyle\int \frac{4x}{(2 - 5x)^2}\, dx$

22. $\displaystyle\int \frac{\theta^3}{1 + \sin \theta^4}\, d\theta$

23. $\displaystyle\int e^x \arccos e^x\, dx$

24. $\displaystyle\int \frac{e^x}{1 - \tan e^x}\, dx$

25. $\displaystyle\int \frac{x}{1 - \sec x^2}\, dx$

26. $\displaystyle\int \frac{1}{t[1 + (\ln t)^2]}\, dt$

27. $\displaystyle\int \frac{\cos \theta}{3 + 2 \sin \theta + \sin^2 \theta}\, d\theta$

28. $\displaystyle\int x^2 \sqrt{2 + 9x^2}\, dx$

29. $\displaystyle\int \frac{1}{x^2\sqrt{2 + 9x^2}}\, dx$

30. $\displaystyle\int \sqrt{x} \arctan x^{3/2}\, dx$

31. $\displaystyle\int \frac{\ln x}{x(3 + 2 \ln x)}\, dx$

32. $\displaystyle\int \frac{e^x}{(1 - e^{2x})^{3/2}}\, dx$

33. $\displaystyle\int \frac{x}{(x^2 - 6x + 10)^2}\, dx$

34. $\displaystyle\int \sqrt{\frac{5 - x}{5 + x}}\, dx$

35. $\displaystyle\int \frac{x}{\sqrt{x^4 - 6x^2 + 5}}\, dx$

36. $\displaystyle\int \frac{\cos x}{\sqrt{\sin^2 x + 1}}\, dx$

37. $\displaystyle\int \frac{e^{3x}}{(1 + e^x)^3}\, dx$

38. $\displaystyle\int \cot^4 \theta\, d\theta$

Evaluating a Definite Integral In Exercises 39–46, use integration tables to evaluate the definite integral.

39. $\displaystyle\int_0^1 x e^{x^2}\, dx$

40. $\displaystyle\int_0^4 \frac{x}{\sqrt{3 + 2x}}\, dx$

41. $\displaystyle\int_1^2 x^4 \ln x\, dx$

42. $\displaystyle\int_0^{\pi/2} x \sin 2x\, dx$

43. $\displaystyle\int_{-\pi/2}^{\pi/2} \frac{\cos x}{1 + \sin^2 x}\, dx$

44. $\displaystyle\int_0^5 \frac{x^2}{(5 + 2x)^2}\, dx$

45. $\displaystyle\int_0^{\pi/2} t^3 \cos t\, dt$

46. $\displaystyle\int_0^3 \sqrt{x^2 + 16}\, dx$

Verifying a Formula In Exercises 47–52, verify the integration formula.

47. $\displaystyle\int \frac{u^2}{(a + bu)^2}\, du = \frac{1}{b^3}\left(bu - \frac{a^2}{a + bu} - 2a \ln|a + bu|\right) + C$

48. $\displaystyle\int \frac{u^n}{\sqrt{a + bu}}\, du = \frac{2}{(2n + 1)b}\left(u^n \sqrt{a + bu} - na \int \frac{u^{n-1}}{\sqrt{a + bu}}\, du\right)$

49. $\displaystyle\int \frac{1}{(u^2 \pm a^2)^{3/2}}\, du = \frac{\pm u}{a^2\sqrt{u^2 \pm a^2}} + C$

50. $\displaystyle\int u^n \cos u\, du = u^n \sin u - n \int u^{n-1} \sin u\, du$

51. $\displaystyle\int \arctan u\, du = u \arctan u - \ln \sqrt{1 + u^2} + C$

52. $\displaystyle\int (\ln u)^n\, du = u(\ln u)^n - n \int (\ln u)^{n-1}\, du$

Finding or Evaluating an Integral In Exercises 53–60, find or evaluate the integral.

53. $\int \dfrac{1}{2 - 3\sin\theta}\, d\theta$

54. $\int \dfrac{\sin\theta}{1 + \cos^2\theta}\, d\theta$

55. $\int_0^{\pi/2} \dfrac{1}{1 + \sin\theta + \cos\theta}\, d\theta$

56. $\int_0^{\pi/2} \dfrac{1}{3 - 2\cos\theta}\, d\theta$

57. $\int \dfrac{\sin\theta}{3 - 2\cos\theta}\, d\theta$

58. $\int \dfrac{\cos\theta}{1 + \cos\theta}\, d\theta$

59. $\int \dfrac{\sin\sqrt{\theta}}{\sqrt{\theta}}\, d\theta$

60. $\int \dfrac{4}{\csc\theta - \cot\theta}\, d\theta$

Area In Exercises 61 and 62, find the area of the region bounded by the graphs of the equations.

61. $y = \dfrac{x}{\sqrt{x+3}},\ y = 0,\ x = 6$

62. $y = \dfrac{x}{1 + e^{x^2}},\ y = 0,\ x = 2$

WRITING ABOUT CONCEPTS

63. Finding a Pattern

(a) Evaluate $\int x^n \ln x\, dx$ for $n = 1, 2,$ and 3. Describe any patterns you notice.

(b) Write a general rule for evaluating the integral in part (a), for an integer $n \geq 1$.

64. Reduction Formula Describe what is meant by a reduction formula. Give an example.

65. Choosing a Method State (if possible) the method or integration formula you would use to find the antiderivative. Explain why you chose that method or formula. Do not integrate.

(a) $\int \dfrac{e^x}{e^{2x}+1}\, dx$ (b) $\int \dfrac{e^x}{e^x+1}\, dx$ (c) $\int xe^{x^2}\, dx$

(d) $\int xe^x\, dx$ (e) $\int e^{x^2}\, dx$ (f) $\int e^{2x}\sqrt{e^{2x}+1}\, dx$

66. **HOW DO YOU SEE IT?** Use the graph of f' shown in the figure to answer the following.

$f'(x) = -0.15x\sqrt{x^4+9}$

(a) Approximate the slope of f at $x = -1$. Explain.

(b) Approximate any open intervals in which the graph of f is increasing and any open intervals in which it is decreasing. Explain.

True or False? In Exercises 67 and 68, determine whether the statement is true or false. If it is false, explain why or give an example that shows it is false.

67. To use a table of integrals, the integral you are evaluating must appear in the table.

68. When using a table of integrals, you may have to make substitutions to rewrite your integral in the form in which it appears in the table.

69. Work A hydraulic cylinder on an industrial machine pushes a steel block a distance of x feet ($0 \leq x \leq 5$), where the variable force required is $F(x) = 2000xe^{-x}$ pounds. Find the work done in pushing the block the full 5 feet through the machine.

70. Work Repeat Exercise 69, using $F(x) = \dfrac{500x}{\sqrt{26 - x^2}}$ pounds.

71. Volume Consider the region bounded by the graphs of

$y = x\sqrt{16 - x^2},\quad y = 0,\quad x = 0,\quad \text{and}\quad x = 4.$

Find the volume of the solid generated by revolving the region about the y-axis.

72. Building Design The cross section of a precast concrete beam for a building is bounded by the graphs of the equations

$x = \dfrac{2}{\sqrt{1 + y^2}},\quad x = \dfrac{-2}{\sqrt{1 + y^2}},\quad y = 0,\quad \text{and}\quad y = 3$

where x and y are measured in feet. The length of the beam is 20 feet (see figure).

(a) Find the volume V and the weight W of the beam. Assume the concrete weighs 148 pounds per cubic foot.

(b) Find the centroid of a cross section of the beam.

73. Population A population is growing according to the logistic model

$N = \dfrac{5000}{1 + e^{4.8 - 1.9t}}$

where t is the time in days. Find the average population over the interval $[0, 2]$.

PUTNAM EXAM CHALLENGE

74. Evaluate $\displaystyle\int_0^{\pi/2} \dfrac{dx}{1 + (\tan x)^{\sqrt{2}}}.$

8.7 Indeterminate Forms and L'Hôpital's Rule

- Recognize limits that produce indeterminate forms.
- Apply L'Hôpital's Rule to evaluate a limit.

Indeterminate Forms

Recall that the forms $0/0$ and ∞/∞ are called *indeterminate* because they do not guarantee that a limit exists, nor do they indicate what the limit is, if one does exist. When you encountered one of these indeterminate forms earlier in the text, you attempted to rewrite the expression by using various algebraic techniques.

Indeterminate Form	Limit	Algebraic Technique
$\dfrac{0}{0}$	$\displaystyle\lim_{x\to-1}\frac{2x^2-2}{x+1}=\lim_{x\to-1}2(x-1)$ $\qquad\qquad = -4$	Divide numerator and denominator by $(x+1)$.
$\dfrac{\infty}{\infty}$	$\displaystyle\lim_{x\to\infty}\frac{3x^2-1}{2x^2+1}=\lim_{x\to\infty}\frac{3-(1/x^2)}{2+(1/x^2)}$ $\qquad\qquad = \dfrac{3}{2}$	Divide numerator and denominator by x^2.

Occasionally, you can extend these algebraic techniques to find limits of transcendental functions. For instance, the limit

$$\lim_{x\to0}\frac{e^{2x}-1}{e^x-1}$$

produces the indeterminate form $0/0$. Factoring and then dividing produces

$$\lim_{x\to0}\frac{e^{2x}-1}{e^x-1}=\lim_{x\to0}\frac{(e^x+1)(e^x-1)}{e^x-1}$$
$$=\lim_{x\to0}(e^x+1)$$
$$=2.$$

Not all indeterminate forms, however, can be evaluated by algebraic manipulation. This is often true when *both* algebraic and transcendental functions are involved. For instance, the limit

$$\lim_{x\to0}\frac{e^{2x}-1}{x}$$

produces the indeterminate form $0/0$. Rewriting the expression to obtain

$$\lim_{x\to0}\left(\frac{e^{2x}}{x}-\frac{1}{x}\right)$$

merely produces another indeterminate form, $\infty-\infty$. Of course, you could use technology to estimate the limit, as shown in the table and in Figure 8.15. From the table and the graph, the limit appears to be 2. (This limit will be verified in Example 1.)

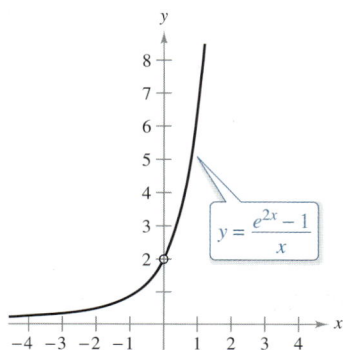

$$y=\frac{e^{2x}-1}{x}$$

The limit as x approaches 0 appears to be 2.

Figure 8.15

x	-1	-0.1	-0.01	-0.001	0	0.001	0.01	0.1	1
$\dfrac{e^{2x}-1}{x}$	0.865	1.813	1.980	1.998	?	2.002	2.020	2.214	6.389

L'Hôpital's Rule

To find the limit illustrated in Figure 8.15, you can use a theorem called **L'Hôpital's Rule.** This theorem states that under certain conditions, the limit of the quotient $f(x)/g(x)$ is determined by the limit of the quotient of the derivatives

$$\frac{f'(x)}{g'(x)}.$$

To prove this theorem, you can use a more general result called the **Extended Mean Value Theorem.**

THEOREM 8.3   The Extended Mean Value Theorem

If f and g are differentiable on an open interval (a, b) and continuous on $[a, b]$ such that $g'(x) \neq 0$ for any x in (a, b), then there exists a point c in (a, b) such that

$$\frac{f'(c)}{g'(c)} = \frac{f(b) - f(a)}{g(b) - g(a)}.$$

A proof of this theorem is given in Appendix A.
See LarsonCalculus.com for Bruce Edwards's video of this proof.

To see why Theorem 8.3 is called the Extended Mean Value Theorem, consider the special case in which $g(x) = x$. For this case, you obtain the "standard" Mean Value Theorem as presented in Section 3.2.

THEOREM 8.4   L'Hôpital's Rule

Let f and g be functions that are differentiable on an open interval (a, b) containing c, except possibly at c itself. Assume that $g'(x) \neq 0$ for all x in (a, b), except possibly at c itself. If the limit of $f(x)/g(x)$ as x approaches c produces the indeterminate form $0/0$, then

$$\lim_{x \to c} \frac{f(x)}{g(x)} = \lim_{x \to c} \frac{f'(x)}{g'(x)}$$

provided the limit on the right exists (or is infinite). This result also applies when the limit of $f(x)/g(x)$ as x approaches c produces any one of the indeterminate forms ∞/∞, $(-\infty)/\infty$, $\infty/(-\infty)$, or $(-\infty)/(-\infty)$.

A proof of this theorem is given in Appendix A.
See LarsonCalculus.com for Bruce Edwards's video of this proof.

GUILLAUME L'HÔPITAL (1661–1704)

L'Hôpital's Rule is named after the French mathematician Guillaume François Antoine de L'Hôpital. L'Hôpital is credited with writing the first text on differential calculus (in 1696) in which the rule publicly appeared. It was recently discovered that the rule and its proof were written in a letter from John Bernoulli to L'Hôpital. "… I acknowledge that I owe very much to the bright minds of the Bernoulli brothers. … I have made free use of their discoveries …," said L'Hôpital.
See LarsonCalculus.com to read more of this biography.

■ FOR FURTHER INFORMATION
To enhance your understanding of the necessity of the restriction that $g'(x)$ be nonzero for all x in (a, b), except possibly at c, see the article "Counterexamples to L'Hôpital's Rule" by R. P. Boas in *The American Mathematical Monthly.* To view this article, go to *MathArticles.com.*

People occasionally use L'Hôpital's Rule incorrectly by applying the Quotient Rule to $f(x)/g(x)$. Be sure you see that the rule involves

$$\frac{f'(x)}{g'(x)}$$

not the derivative of $f(x)/g(x)$.

L'Hôpital's Rule can also be applied to one-sided limits. For instance, if the limit of $f(x)/g(x)$ as x approaches c *from the right* produces the indeterminate form $0/0$, then

$$\lim_{x \to c^+} \frac{f(x)}{g(x)} = \lim_{x \to c^+} \frac{f'(x)}{g'(x)}$$

provided the limit exists (or is infinite).

Exploration

Numerical and Graphical Approaches Use a numerical or a graphical approach to approximate each limit.

a. $\lim\limits_{x\to 0} \dfrac{2^{2x}-1}{x}$

b. $\lim\limits_{x\to 0} \dfrac{3^{2x}-1}{x}$

c. $\lim\limits_{x\to 0} \dfrac{4^{2x}-1}{x}$

d. $\lim\limits_{x\to 0} \dfrac{5^{2x}-1}{x}$

What pattern do you observe? Does an analytic approach have an advantage for determining these limits? If so, explain your reasoning.

EXAMPLE 1 **Indeterminate Form 0/0**

Evaluate $\lim\limits_{x\to 0} \dfrac{e^{2x}-1}{x}$.

Solution Because direct substitution results in the indeterminate form 0/0

$$\lim_{x\to 0}\frac{e^{2x}-1}{x} \qquad \lim_{x\to 0}(e^{2x}-1)=0 \qquad \lim_{x\to 0}x=0$$

you can apply L'Hôpital's Rule, as shown below.

$$\lim_{x\to 0}\frac{e^{2x}-1}{x} = \lim_{x\to 0}\frac{\frac{d}{dx}[e^{2x}-1]}{\frac{d}{dx}[x]} \qquad \text{Apply L'Hôpital's Rule.}$$

$$= \lim_{x\to 0}\frac{2e^{2x}}{1} \qquad \text{Differentiate numerator and denominator.}$$

$$= 2 \qquad \text{Evaluate the limit.}$$

In the solution to Example 1, note that you actually do not know that the first limit is equal to the second limit until you have shown that the second limit exists. In other words, if the second limit had not existed, then it would not have been permissible to apply L'Hôpital's Rule.

Another form of L'Hôpital's Rule states that if the limit of $f(x)/g(x)$ as x approaches ∞ (or $-\infty$) produces the indeterminate form 0/0 or ∞/∞, then

$$\lim_{x\to\infty}\frac{f(x)}{g(x)} = \lim_{x\to\infty}\frac{f'(x)}{g'(x)}$$

provided the limit on the right exists.

EXAMPLE 2 **Indeterminate Form ∞/∞**

Evaluate $\lim\limits_{x\to\infty} \dfrac{\ln x}{x}$.

Solution Because direct substitution results in the indeterminate form ∞/∞, you can apply L'Hôpital's Rule to obtain

$$\lim_{x\to\infty}\frac{\ln x}{x} = \lim_{x\to\infty}\frac{\frac{d}{dx}[\ln x]}{\frac{d}{dx}[x]} \qquad \text{Apply L'Hôpital's Rule.}$$

$$= \lim_{x\to\infty}\frac{1}{x} \qquad \text{Differentiate numerator and denominator.}$$

$$= 0. \qquad \text{Evaluate the limit.}$$

▷ **TECHNOLOGY** Use a graphing utility to graph $y_1 = \ln x$ and $y_2 = x$ in the same viewing window. Which function grows faster as x approaches ∞? How is this observation related to Example 2?

Occasionally it is necessary to apply L'Hôpital's Rule more than once to remove an indeterminate form, as shown in Example 3.

■ **FOR FURTHER INFORMATION**
To read about the connection between Leonhard Euler and Guillaume L'Hôpital, see the article "When Euler Met l'Hôpital" by William Dunham in *Mathematics Magazine*. To view this article, go to *MathArticles.com*.

EXAMPLE 3   **Applying L'Hôpital's Rule More than Once**

Evaluate $\displaystyle\lim_{x \to -\infty} \frac{x^2}{e^{-x}}$.

Solution   Because direct substitution results in the indeterminate form ∞/∞, you can apply L'Hôpital's Rule.

$$\lim_{x \to -\infty} \frac{x^2}{e^{-x}} = \lim_{x \to -\infty} \frac{\dfrac{d}{dx}[x^2]}{\dfrac{d}{dx}[e^{-x}]} = \lim_{x \to -\infty} \frac{2x}{-e^{-x}}$$

This limit yields the indeterminate form $(-\infty)/(-\infty)$, so you can apply L'Hôpital's Rule again to obtain

$$\lim_{x \to -\infty} \frac{2x}{-e^{-x}} = \lim_{x \to -\infty} \frac{\dfrac{d}{dx}[2x]}{\dfrac{d}{dx}[-e^{-x}]} = \lim_{x \to -\infty} \frac{2}{e^{-x}} = 0.$$

In addition to the forms $0/0$ and ∞/∞, there are other indeterminate forms such as $0 \cdot \infty$, 1^∞, ∞^0, 0^0, and $\infty - \infty$. For example, consider the following four limits that lead to the indeterminate form $0 \cdot \infty$.

$$\lim_{x \to 0} \left(\frac{1}{x}\right)(x), \qquad \lim_{x \to 0} \left(\frac{2}{x}\right)(x), \qquad \lim_{x \to \infty} \left(\frac{1}{e^x}\right)(x), \qquad \lim_{x \to \infty} \left(\frac{1}{x}\right)(e^x)$$

$$\underbrace{\qquad\qquad}_{\text{Limit is 1.}} \qquad \underbrace{\qquad\qquad}_{\text{Limit is 2.}} \qquad \underbrace{\qquad\qquad}_{\text{Limit is 0.}} \qquad \underbrace{\qquad\qquad}_{\text{Limit is } \infty.}$$

Because each limit is different, it is clear that the form $0 \cdot \infty$ is indeterminate in the sense that it does not determine the value (or even the existence) of the limit. The remaining examples in this section show methods for evaluating these forms. Basically, you attempt to convert each of these forms to $0/0$ or ∞/∞ so that L'Hôpital's Rule can be applied.

EXAMPLE 4   **Indeterminate Form $0 \cdot \infty$**

Evaluate $\displaystyle\lim_{x \to \infty} e^{-x} \sqrt{x}$.

Solution   Because direct substitution produces the indeterminate form $0 \cdot \infty$, you should try to rewrite the limit to fit the form $0/0$ or ∞/∞. In this case, you can rewrite the limit to fit the second form.

$$\lim_{x \to \infty} e^{-x} \sqrt{x} = \lim_{x \to \infty} \frac{\sqrt{x}}{e^x}$$

Now, by L'Hôpital's Rule, you have

$$\lim_{x \to \infty} \frac{\sqrt{x}}{e^x} = \lim_{x \to \infty} \frac{1/(2\sqrt{x})}{e^x} \qquad \text{Differentiate numerator and denominator.}$$

$$= \lim_{x \to \infty} \frac{1}{2\sqrt{x}\,e^x} \qquad \text{Simplify.}$$

$$= 0. \qquad \text{Evaluate the limit.}$$

When rewriting a limit in one of the forms $0/0$ or ∞/∞ does not seem to work, try the other form. For instance, in Example 4, you can write the limit as

$$\lim_{x \to \infty} e^{-x}\sqrt{x} = \lim_{x \to \infty} \frac{e^{-x}}{x^{-1/2}}$$

which yields the indeterminate form $0/0$. As it happens, applying L'Hôpital's Rule to this limit produces

$$\lim_{x \to \infty} \frac{e^{-x}}{x^{-1/2}} = \lim_{x \to \infty} \frac{-e^{-x}}{-1/(2x^{3/2})}$$

which also yields the indeterminate form $0/0$.

The indeterminate forms 1^∞, ∞^0, and 0^0 arise from limits of functions that have variable bases and variable exponents. When you previously encountered this type of function, you used logarithmic differentiation to find the derivative. You can use a similar procedure when taking limits, as shown in the next example.

EXAMPLE 5   Indeterminate Form 1^∞

Evaluate $\displaystyle\lim_{x \to \infty} \left(1 + \frac{1}{x}\right)^x$.

Solution   Because direct substitution yields the indeterminate form 1^∞, you can proceed as follows. To begin, assume that the limit exists and is equal to y.

$$y = \lim_{x \to \infty} \left(1 + \frac{1}{x}\right)^x$$

Taking the natural logarithm of each side produces

$$\ln y = \ln\left[\lim_{x \to \infty} \left(1 + \frac{1}{x}\right)^x\right].$$

Because the natural logarithmic function is continuous, you can write

$$\ln y = \lim_{x \to \infty} \left[x \ln\left(1 + \frac{1}{x}\right)\right] \qquad \text{Indeterminate form } \infty \cdot 0$$

$$= \lim_{x \to \infty} \left(\frac{\ln[1 + (1/x)]}{1/x}\right) \qquad \text{Indeterminate form } 0/0$$

$$= \lim_{x \to \infty} \left(\frac{(-1/x^2)\{1/[1 + (1/x)]\}}{-1/x^2}\right) \qquad \text{L'Hôpital's Rule}$$

$$= \lim_{x \to \infty} \frac{1}{1 + (1/x)}$$

$$= 1.$$

Now, because you have shown that

$$\ln y = 1$$

you can conclude that

$$y = e$$

and obtain

$$\lim_{x \to \infty} \left(1 + \frac{1}{x}\right)^x = e.$$

You can use a graphing utility to confirm this result, as shown in Figure 8.16.

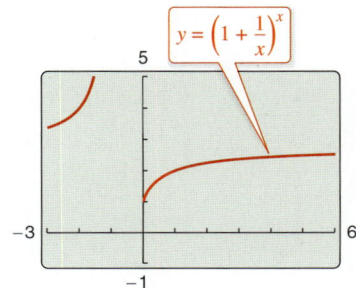

The limit of $[1 + (1/x)]^x$ as x approaches infinity is e.
Figure 8.16

L'Hôpital's Rule can also be applied to one-sided limits, as demonstrated in Examples 6 and 7.

EXAMPLE 6 Indeterminate Form 0^0

·····▷ *See LarsonCalculus.com for an interactive version of this type of example.*

Evaluate $\lim\limits_{x \to 0^+} (\sin x)^x$.

Solution Because direct substitution produces the indeterminate form 0^0, you can proceed as shown below. To begin, assume that the limit exists and is equal to y.

$$y = \lim_{x \to 0^+} (\sin x)^x \qquad \text{Indeterminate form } 0^0$$

$$\ln y = \ln\left[\lim_{x \to 0^+} (\sin x)^x \right] \qquad \text{Take natural log of each side.}$$

$$= \lim_{x \to 0^+} \left[\ln(\sin x)^x \right] \qquad \text{Continuity}$$

$$= \lim_{x \to 0^+} \left[x \ln(\sin x) \right] \qquad \text{Indeterminate form } 0 \cdot (-\infty)$$

$$= \lim_{x \to 0^+} \frac{\ln(\sin x)}{1/x} \qquad \text{Indeterminate form } -\infty/\infty$$

$$= \lim_{x \to 0^+} \frac{\cot x}{-1/x^2} \qquad \text{L'Hôpital's Rule}$$

$$= \lim_{x \to 0^+} \frac{-x^2}{\tan x} \qquad \text{Indeterminate form } 0/0$$

$$= \lim_{x \to 0^+} \frac{-2x}{\sec^2 x} \qquad \text{L'Hôpital's Rule}$$

$$= 0$$

Now, because $\ln y = 0$, you can conclude that $y = e^0 = 1$, and it follows that

$$\lim_{x \to 0^+} (\sin x)^x = 1.$$

▷ **TECHNOLOGY** When evaluating complicated limits such as the one in Example 6, it is helpful to check the reasonableness of the solution with a graphing utility. For instance, the calculations in the table and the graph in the figure (see below) are consistent with the conclusion that $(\sin x)^x$ approaches 1 as x approaches 0 from the right.

x	1.0	0.1	0.01	0.001	0.0001	0.00001
$(\sin x)^x$	0.8415	0.7942	0.9550	0.9931	0.9991	0.9999

Use a graphing utility to estimate the limits $\lim\limits_{x \to 0} (1 - \cos x)^x$ and $\lim\limits_{x \to 0^+} (\tan x)^x$. Then try to verify your estimates analytically.

$y = (\sin x)^x$

The limit of $(\sin x)^x$ is 1 as x approaches 0 from the right.

EXAMPLE 7 Indeterminate Form $\infty - \infty$

Evaluate $\lim\limits_{x \to 1^+} \left(\dfrac{1}{\ln x} - \dfrac{1}{x-1} \right)$.

Solution Because direct substitution yields the indeterminate form $\infty - \infty$, you should try to rewrite the expression to produce a form to which you can apply L'Hôpital's Rule. In this case, you can combine the two fractions to obtain

$$\lim_{x \to 1^+} \left(\frac{1}{\ln x} - \frac{1}{x-1} \right) = \lim_{x \to 1^+} \left[\frac{x-1-\ln x}{(x-1)\ln x} \right].$$

Now, because direct substitution produces the indeterminate form $0/0$, you can apply L'Hôpital's Rule to obtain

$$\lim_{x \to 1^+} \left(\frac{1}{\ln x} - \frac{1}{x-1} \right) = \lim_{x \to 1^+} \frac{\dfrac{d}{dx}[x-1-\ln x]}{\dfrac{d}{dx}[(x-1)\ln x]}$$

$$= \lim_{x \to 1^+} \left[\frac{1-(1/x)}{(x-1)(1/x)+\ln x} \right]$$

$$= \lim_{x \to 1^+} \left(\frac{x-1}{x-1+x\ln x} \right).$$

This limit also yields the indeterminate form $0/0$, so you can apply L'Hôpital's Rule again to obtain

$$\lim_{x \to 1^+} \left(\frac{1}{\ln x} - \frac{1}{x-1} \right) = \lim_{x \to 1^+} \left[\frac{1}{1+x(1/x)+\ln x} \right] = \frac{1}{2}.$$ ∎

The forms $0/0$, ∞/∞, $\infty - \infty$, $0 \cdot \infty$, 0^0, 1^∞, and ∞^0 have been identified as *indeterminate*. There are similar forms that you should recognize as "determinate."

$\infty + \infty \to \infty$ Limit is positive infinity.
$-\infty - \infty \to -\infty$ Limit is negative infinity.
$0^\infty \to 0$ Limit is zero.
$0^{-\infty} \to \infty$ Limit is positive infinity.

(You are asked to verify two of these in Exercises 108 and 109.)

As a final comment, remember that L'Hôpital's Rule can be applied only to quotients leading to the indeterminate forms $0/0$ and ∞/∞. For instance, the application of L'Hôpital's Rule shown below is *incorrect*.

$$\lim_{x \to 0} \frac{e^x}{x} = \lim_{x \to 0} \frac{e^x}{1} = 1 \qquad \text{Incorrect use of L'Hôpital's Rule}$$

The reason this application is incorrect is that, even though the limit of the denominator is 0, the limit of the numerator is 1, which means that the hypotheses of L'Hôpital's Rule have not been satisfied.

Exploration

In each of the examples presented in this section, L'Hôpital's Rule is used to find a limit that exists. It can also be used to conclude that a limit is infinite. For instance, try using L'Hôpital's Rule to show that $\lim\limits_{x \to \infty} e^x/x = \infty$.

8.7 Exercises

See **CalcChat.com** for tutorial help and worked-out solutions to odd-numbered exercises.

Numerical and Graphical Analysis In Exercises 1–4, complete the table and use the result to estimate the limit. Use a graphing utility to graph the function to support your result.

1. $\lim\limits_{x\to 0} \dfrac{\sin 4x}{\sin 3x}$

x	-0.1	-0.01	-0.001	0.001	0.01	0.1
$f(x)$						

2. $\lim\limits_{x\to 0} \dfrac{1 - e^x}{x}$

x	-0.1	-0.01	-0.001	0.001	0.01	0.1
$f(x)$						

3. $\lim\limits_{x\to \infty} x^5 e^{-x/100}$

x	1	10	10^2	10^3	10^4	10^5
$f(x)$						

4. $\lim\limits_{x\to \infty} \dfrac{6x}{\sqrt{3x^2 - 2x}}$

x	1	10	10^2	10^3	10^4	10^5
$f(x)$						

Using Two Methods In Exercises 5–10, evaluate the limit (a) using techniques from Chapters 1 and 3 and (b) using L'Hôpital's Rule.

5. $\lim\limits_{x\to 4} \dfrac{3(x - 4)}{x^2 - 16}$

6. $\lim\limits_{x\to -4} \dfrac{2x^2 + 13x + 20}{x + 4}$

7. $\lim\limits_{x\to 6} \dfrac{\sqrt{x + 10} - 4}{x - 6}$

8. $\lim\limits_{x\to 0} \dfrac{\sin 6x}{4x}$

9. $\lim\limits_{x\to \infty} \dfrac{5x^2 - 3x + 1}{3x^2 - 5}$

10. $\lim\limits_{x\to \infty} \dfrac{4x - 3}{5x^2 + 1}$

Evaluating a Limit In Exercises 11–42, evaluate the limit, using L'Hôpital's Rule if necessary.

11. $\lim\limits_{x\to 3} \dfrac{x^2 - 2x - 3}{x - 3}$

12. $\lim\limits_{x\to -2} \dfrac{x^2 - 3x - 10}{x + 2}$

13. $\lim\limits_{x\to 0} \dfrac{\sqrt{25 - x^2} - 5}{x}$

14. $\lim\limits_{x\to 5^-} \dfrac{\sqrt{25 - x^2}}{x - 5}$

15. $\lim\limits_{x\to 0^+} \dfrac{e^x - (1 + x)}{x^3}$

16. $\lim\limits_{x\to 1} \dfrac{\ln x^3}{x^2 - 1}$

17. $\lim\limits_{x\to 1} \dfrac{x^{11} - 1}{x^4 - 1}$

18. $\lim\limits_{x\to 1} \dfrac{x^a - 1}{x^b - 1}$, where $a, b \neq 0$

19. $\lim\limits_{x\to 0} \dfrac{\sin 3x}{\sin 5x}$

20. $\lim\limits_{x\to 0} \dfrac{\sin ax}{\sin bx}$, where $a, b \neq 0$

21. $\lim\limits_{x\to 0} \dfrac{\arcsin x}{x}$

22. $\lim\limits_{x\to 1} \dfrac{\arctan x - (\pi/4)}{x - 1}$

23. $\lim\limits_{x\to \infty} \dfrac{5x^2 + 3x - 1}{4x^2 + 5}$

24. $\lim\limits_{x\to \infty} \dfrac{5x + 3}{x^3 - 6x + 2}$

25. $\lim\limits_{x\to \infty} \dfrac{x^2 + 4x + 7}{x - 6}$

26. $\lim\limits_{x\to \infty} \dfrac{x^3}{x + 2}$

27. $\lim\limits_{x\to \infty} \dfrac{x^3}{e^{x/2}}$

28. $\lim\limits_{x\to \infty} \dfrac{x^3}{e^{x^2}}$

29. $\lim\limits_{x\to \infty} \dfrac{x}{\sqrt{x^2 + 1}}$

30. $\lim\limits_{x\to \infty} \dfrac{x^2}{\sqrt{x^2 + 1}}$

31. $\lim\limits_{x\to \infty} \dfrac{\cos x}{x}$

32. $\lim\limits_{x\to \infty} \dfrac{\sin x}{x - \pi}$

33. $\lim\limits_{x\to \infty} \dfrac{\ln x}{x^2}$

34. $\lim\limits_{x\to \infty} \dfrac{\ln x^4}{x^3}$

35. $\lim\limits_{x\to \infty} \dfrac{e^x}{x^4}$

36. $\lim\limits_{x\to \infty} \dfrac{e^{x/2}}{x}$

37. $\lim\limits_{x\to 0} \dfrac{\sin 5x}{\tan 9x}$

38. $\lim\limits_{x\to 1} \dfrac{\ln x}{\sin \pi x}$

39. $\lim\limits_{x\to 0} \dfrac{\arctan x}{\sin x}$

40. $\lim\limits_{x\to 0} \dfrac{x}{\arctan 2x}$

41. $\lim\limits_{x\to \infty} \dfrac{\int_1^x \ln(e^{4t-1})\, dt}{x}$

42. $\lim\limits_{x\to 1^+} \dfrac{\int_1^x \cos \theta\, d\theta}{x - 1}$

Evaluating a Limit In Exercises 43–60, (a) describe the type of indeterminate form (if any) that is obtained by direct substitution. (b) Evaluate the limit, using L'Hôpital's Rule if necessary. (c) Use a graphing utility to graph the function and verify the result in part (b).

43. $\lim\limits_{x\to \infty} x \ln x$

44. $\lim\limits_{x\to 0^+} x^3 \cot x$

45. $\lim\limits_{x\to \infty} \left(x \sin \dfrac{1}{x}\right)$

46. $\lim\limits_{x\to \infty} x \tan \dfrac{1}{x}$

47. $\lim\limits_{x\to 0^+} x^{1/x}$

48. $\lim\limits_{x\to 0^+} (e^x + x)^{2/x}$

49. $\lim\limits_{x\to \infty} x^{1/x}$

50. $\lim\limits_{x\to \infty} \left(1 + \dfrac{1}{x}\right)^x$

51. $\lim\limits_{x\to 0^+} (1 + x)^{1/x}$

52. $\lim\limits_{x\to \infty} (1 + x)^{1/x}$

53. $\lim\limits_{x\to 0^+} [3(x)^{x/2}]$

54. $\lim\limits_{x\to 4^+} [3(x - 4)]^{x-4}$

55. $\lim\limits_{x\to 1^+} (\ln x)^{x-1}$

56. $\lim\limits_{x\to 0^+} \left[\cos\left(\dfrac{\pi}{2} - x\right)\right]^x$

57. $\lim\limits_{x\to 2^+} \left(\dfrac{8}{x^2 - 4} - \dfrac{x}{x - 2}\right)$

58. $\lim\limits_{x\to 2^+} \left(\dfrac{1}{x^2 - 4} - \dfrac{\sqrt{x - 1}}{x^2 - 4}\right)$

59. $\lim\limits_{x\to 1^+} \left(\dfrac{3}{\ln x} - \dfrac{2}{x - 1}\right)$

60. $\lim\limits_{x\to 0^+} \left(\dfrac{10}{x} - \dfrac{3}{x^2}\right)$

WRITING ABOUT CONCEPTS

61. Indeterminate Forms List six different indeterminate forms.

62. L'Hôpital's Rule State L'Hôpital's Rule.

63. Finding Functions Find differentiable functions f and g that satisfy the specified condition such that

$$\lim_{x \to 5} f(x) = 0 \quad \text{and} \quad \lim_{x \to 5} g(x) = 0.$$

Explain how you obtained your answers. (*Note:* There are many correct answers.)

(a) $\displaystyle \lim_{x \to 5} \frac{f(x)}{g(x)} = 10$ (b) $\displaystyle \lim_{x \to 5} \frac{f(x)}{g(x)} = 0$

(c) $\displaystyle \lim_{x \to 5} \frac{f(x)}{g(x)} = \infty$

64. Finding Functions Find differentiable functions f and g such that

$$\lim_{x \to \infty} f(x) = \lim_{x \to \infty} g(x) = \infty \quad \text{and} \quad \lim_{x \to \infty} [f(x) - g(x)] = 25.$$

Explain how you obtained your answers. (*Note:* There are many correct answers.)

65. L'Hôpital's Rule Determine which of the following limits can be evaluated using L'Hôpital's Rule. Explain your reasoning. Do not evaluate the limit.

(a) $\displaystyle \lim_{x \to 2} \frac{x - 2}{x^3 - x - 6}$ (b) $\displaystyle \lim_{x \to 0} \frac{x^2 - 4x}{2x - 1}$

(c) $\displaystyle \lim_{x \to \infty} \frac{x^3}{e^x}$ (d) $\displaystyle \lim_{x \to 3} \frac{e^{x^2} - e^9}{x - 3}$

(e) $\displaystyle \lim_{x \to 1} \frac{\cos \pi x}{\ln x}$ (f) $\displaystyle \lim_{x \to 1} \frac{1 + x(\ln x - 1)}{(x - 1) \ln x}$

66. HOW DO YOU SEE IT? Use the graph of f to find the limit.

$$f(x) = \frac{3}{\ln x} - \frac{4}{x - 1}$$

(a) $\displaystyle \lim_{x \to 1^-} f(x)$ (b) $\displaystyle \lim_{x \to 1^+} f(x)$ (c) $\displaystyle \lim_{x \to 1} f(x)$

67. Numerical Approach Complete the table to show that x eventually "overpowers" $(\ln x)^4$.

x	10	10^2	10^4	10^6	10^8	10^{10}
$\dfrac{(\ln x)^4}{x}$						

68. Numerical Approach Complete the table to show that e^x eventually "overpowers" x^5.

x	1	5	10	20	30	40	50	100
$\dfrac{e^x}{x^5}$								

Comparing Functions In Exercises 69–74, use L'Hôpital's Rule to determine the comparative rates of increase of the functions $f(x) = x^m$, $g(x) = e^{nx}$, and $h(x) = (\ln x)^n$, where $n > 0$, $m > 0$, and $x \to \infty$.

69. $\displaystyle \lim_{x \to \infty} \frac{x^2}{e^{5x}}$ **70.** $\displaystyle \lim_{x \to \infty} \frac{x^3}{e^{2x}}$

71. $\displaystyle \lim_{x \to \infty} \frac{(\ln x)^3}{x}$ **72.** $\displaystyle \lim_{x \to \infty} \frac{(\ln x)^2}{x^3}$

73. $\displaystyle \lim_{x \to \infty} \frac{(\ln x)^n}{x^m}$ **74.** $\displaystyle \lim_{x \to \infty} \frac{x^m}{e^{nx}}$

Asymptotes and Relative Extrema In Exercises 75–78, find any asymptotes and relative extrema that may exist and use a graphing utility to graph the function. (*Hint:* Some of the limits required in finding asymptotes have been found in previous exercises.)

75. $y = x^{1/x}$, $x > 0$ **76.** $y = x^x$, $x > 0$

77. $y = 2xe^{-x}$ **78.** $y = \dfrac{\ln x}{x}$

Think About It In Exercises 79–82, L'Hôpital's Rule is used incorrectly. Describe the error.

79. $\displaystyle \lim_{x \to 2} \frac{3x^2 + 4x + 1}{x^2 - x - 2} = \lim_{x \to 2} \frac{6x + 4}{2x - 1} = \lim_{x \to 2} \frac{6}{2} = 3$

80. $\displaystyle \lim_{x \to 0} \frac{e^{2x} - 1}{e^x} = \lim_{x \to 0} \frac{2e^{2x}}{e^x}$

$$= \lim_{x \to 0} 2e^x$$
$$= 2$$

81. $\displaystyle \lim_{x \to \infty} \frac{e^{-x}}{1 + e^{-x}} = \lim_{x \to \infty} \frac{-e^{-x}}{-e^{-x}}$

$$= \lim_{x \to \infty} 1$$
$$= 1$$

82. $\displaystyle \lim_{x \to \infty} x \cos \frac{1}{x} = \lim_{x \to \infty} \frac{\cos(1/x)}{1/x}$

$$= \lim_{x \to \infty} \frac{[-\sin(1/x)](1/x^2)}{-1/x^2}$$
$$= 0$$

Analytical Approach In Exercises 83 and 84, (a) explain why L'Hôpital's Rule cannot be used to find the limit, (b) find the limit analytically, and (c) use a graphing utility to graph the function and approximate the limit from the graph. Compare the result with that in part (b).

83. $\displaystyle \lim_{x \to \infty} \frac{x}{\sqrt{x^2 + 1}}$ **84.** $\displaystyle \lim_{x \to \pi/2^-} \frac{\tan x}{\sec x}$

Graphical Analysis In Exercises 85 and 86, graph $f(x)/g(x)$ and $f'(x)/g'(x)$ near $x = 0$. What do you notice about these ratios as $x \to 0$? How does this illustrate L'Hôpital's Rule?

85. $f(x) = \sin 3x$, $g(x) = \sin 4x$

86. $f(x) = e^{3x} - 1$, $g(x) = x$

87. Velocity in a Resisting Medium The velocity v of an object falling through a resisting medium such as air or water is given by

$$v = \frac{32}{k}\left(1 - e^{-kt} + \frac{v_0 k e^{-kt}}{32}\right)$$

where v_0 is the initial velocity, t is the time in seconds, and k is the resistance constant of the medium. Use L'Hôpital's Rule to find the formula for the velocity of a falling body in a vacuum by fixing v_0 and t and letting k approach zero. (Assume that the downward direction is positive.)

88. Compound Interest The formula for the amount A in a savings account compounded n times per year for t years at an interest rate r and an initial deposit of P is given by

$$A = P\left(1 + \frac{r}{n}\right)^{nt}.$$

Use L'Hôpital's Rule to show that the limiting formula as the number of compoundings per year approaches infinity is given by $A = Pe^{rt}$.

89. The Gamma Function The Gamma Function $\Gamma(n)$ is defined in terms of the integral of the function given by $f(x) = x^{n-1}e^{-x}$, $n > 0$. Show that for any fixed value of n, the limit of $f(x)$ as x approaches infinity is zero.

90. Tractrix A person moves from the origin along the positive y-axis pulling a weight at the end of a 12-meter rope (see figure). Initially, the weight is located at the point $(12, 0)$.

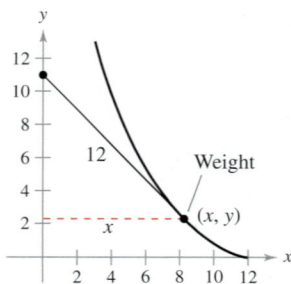

(a) Show that the slope of the tangent line of the path of the weight is

$$\frac{dy}{dx} = -\frac{\sqrt{144 - x^2}}{x}.$$

(b) Use the result of part (a) to find the equation of the path of the weight. Use a graphing utility to graph the path and compare it with the figure.

(c) Find any vertical asymptotes of the graph in part (b).

(d) When the person has reached the point $(0, 12)$, how far has the weight moved?

Extended Mean Value Theorem In Exercises 91–94, apply the Extended Mean Value Theorem to the functions f and g on the given interval. Find all values c in the interval (a, b) such that

$$\frac{f'(c)}{g'(c)} = \frac{f(b) - f(a)}{g(b) - g(a)}.$$

Functions	Interval
91. $f(x) = x^3$, $g(x) = x^2 + 1$	$[0, 1]$
92. $f(x) = \dfrac{1}{x}$, $g(x) = x^2 - 4$	$[1, 2]$
93. $f(x) = \sin x$, $g(x) = \cos x$	$\left[0, \dfrac{\pi}{2}\right]$
94. $f(x) = \ln x$, $g(x) = x^3$	$[1, 4]$

True or False? In Exercises 95–98, determine whether the statement is true or false. If it is false, explain why or give an example that shows it is false.

95. $\displaystyle\lim_{x \to 0}\left[\frac{x^2 + x + 1}{x}\right] = \lim_{x \to 0}\left[\frac{2x + 1}{1}\right] = 1$

96. If $y = \dfrac{e^x}{x^2}$, then $y' = \dfrac{e^x}{2x}$.

97. If $p(x)$ is a polynomial, then $\displaystyle\lim_{x \to \infty}\frac{p(x)}{e^x} = 0$.

98. If $\displaystyle\lim_{x \to \infty}\frac{f(x)}{g(x)} = 1$, then $\displaystyle\lim_{x \to \infty}[f(x) - g(x)] = 0$.

99. Area Find the limit, as x approaches 0, of the ratio of the area of the triangle to the total shaded area in the figure.

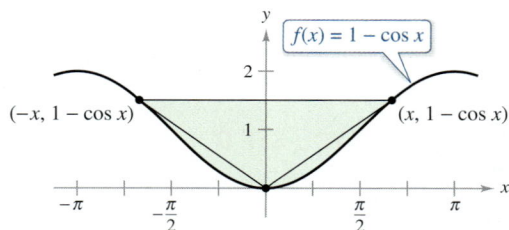

100. Finding a Limit In Section 1.3, a geometric argument (see figure) was used to prove that

$$\lim_{\theta \to 0}\frac{\sin \theta}{\theta} = 1.$$

(a) Write the area of $\triangle ABD$ in terms of θ.

(b) Write the area of the shaded region in terms of θ.

(c) Write the ratio R of the area of $\triangle ABD$ to that of the shaded region.

(d) Find $\displaystyle\lim_{\theta \to 0} R$.

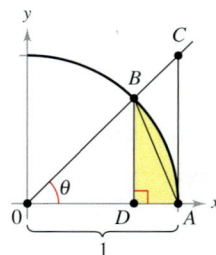

Continuous Function In Exercises 101 and 102, find the value of c that makes the function continuous at $x = 0$.

101. $f(x) = \begin{cases} \dfrac{4x - 2\sin 2x}{2x^3}, & x \neq 0 \\ c, & x = 0 \end{cases}$

102. $f(x) = \begin{cases} (e^x + x)^{1/x}, & x \neq 0 \\ c, & x = 0 \end{cases}$

103. Finding Values Find the values of a and b such that

$$\lim_{x \to 0} \frac{a - \cos bx}{x^2} = 2.$$

104. Evaluating a Limit Use a graphing utility to graph

$$f(x) = \frac{x^k - 1}{k}$$

for $k = 1, 0.1,$ and 0.01. Then evaluate the limit

$$\lim_{k \to 0^+} \frac{x^k - 1}{k}.$$

105. Finding a Derivative

(a) Let $f'(x)$ be continuous. Show that

$$\lim_{h \to 0} \frac{f(x + h) - f(x - h)}{2h} = f'(x).$$

(b) Explain the result of part (a) graphically.

106. Finding a Second Derivative Let $f''(x)$ be continuous. Show that

$$\lim_{h \to 0} \frac{f(x + h) - 2f(x) + f(x - h)}{h^2} = f''(x).$$

107. Evaluating a Limit Consider the limit $\lim_{x \to 0^+} (-x \ln x)$.

(a) Describe the type of indeterminate form that is obtained by direct substitution.

(b) Evaluate the limit. Use a graphing utility to verify the result.

■ **FOR FURTHER INFORMATION** For a geometric approach to this exercise, see the article "A Geometric Proof of $\lim_{d \to 0^+} (-d \ln d) = 0$" by John H. Mathews in *The College Mathematics Journal*. To view this article, go to *MathArticles.com*.

108. Proof Prove that if $f(x) \geq 0$, $\lim_{x \to a} f(x) = 0$, and $\lim_{x \to a} g(x) = \infty$, then $\lim_{x \to a} f(x)^{g(x)} = 0$.

109. Proof Prove that if $f(x) \geq 0$, $\lim_{x \to a} f(x) = 0$, and $\lim_{x \to a} g(x) = -\infty$, then $\lim_{x \to a} f(x)^{g(x)} = \infty$.

110. Proof Prove the following generalization of the Mean Value Theorem. If f is twice differentiable on the closed interval $[a, b]$, then

$$f(b) - f(a) = f'(a)(b - a) - \int_a^b f''(t)(t - b)\, dt.$$

111. Indeterminate Forms Show that the indeterminate forms 0^0, ∞^0, and 1^∞ do not always have a value of 1 by evaluating each limit.

(a) $\lim_{x \to 0^+} x^{\ln 2/(1 + \ln x)}$

(b) $\lim_{x \to \infty} x^{\ln 2/(1 + \ln x)}$

(c) $\lim_{x \to 0} (x + 1)^{(\ln 2)/x}$

112. Calculus History In L'Hôpital's 1696 calculus textbook, he illustrated his rule using the limit of the function

$$f(x) = \frac{\sqrt{2a^3 x - x^4} - a\sqrt[3]{a^2 x}}{a - \sqrt[4]{ax^3}}$$

as x approaches a, $a > 0$. Find this limit.

113. Finding a Limit Consider the function

$$h(x) = \frac{x + \sin x}{x}.$$

(a) Use a graphing utility to graph the function. Then use the *zoom* and *trace* features to investigate $\lim_{x \to \infty} h(x)$.

(b) Find $\lim_{x \to \infty} h(x)$ analytically by writing

$$h(x) = \frac{x}{x} + \frac{\sin x}{x}.$$

(c) Can you use L'Hôpital's Rule to find $\lim_{x \to \infty} h(x)$? Explain your reasoning.

114. Evaluating a Limit Let $f(x) = x + x\sin x$ and $g(x) = x^2 - 4$.

(a) Show that $\lim_{x \to \infty} \dfrac{f(x)}{g(x)} = 0$.

(b) Show that $\lim_{x \to \infty} f(x) = \infty$ and $\lim_{x \to \infty} g(x) = \infty$.

(c) Evaluate the limit

$$\lim_{x \to \infty} \frac{f'(x)}{g'(x)}.$$

What do you notice?

(d) Do your answers to parts (a) through (c) contradict L'Hôpital's Rule? Explain your reasoning.

PUTNAM EXAM CHALLENGE

115. Evaluate $\lim_{x \to \infty} \left[\dfrac{1}{x} \cdot \dfrac{a^x - 1}{a - 1} \right]^{1/x}$ where $a > 0$, $a \neq 1$.

8.8 Improper Integrals

■ Evaluate an improper integral that has an infinite limit of integration.
■ Evaluate an improper integral that has an infinite discontinuity.

Improper Integrals with Infinite Limits of Integration

The definition of a definite integral

$$\int_a^b f(x)\, dx$$

requires that the interval $[a, b]$ be finite. Furthermore, the Fundamental Theorem of Calculus, by which you have been evaluating definite integrals, requires that f be continuous on $[a, b]$. In this section, you will study a procedure for evaluating integrals that do not satisfy these requirements—usually because either one or both of the limits of integration are infinite, or because f has a finite number of infinite discontinuities in the interval $[a, b]$. Integrals that possess either property are **improper integrals.** Note that a function f is said to have an **infinite discontinuity** at c when, *from the right or left,*

$$\lim_{x \to c} f(x) = \infty \quad \text{or} \quad \lim_{x \to c} f(x) = -\infty.$$

To get an idea of how to evaluate an improper integral, consider the integral

$$\int_1^b \frac{dx}{x^2} = -\frac{1}{x}\Big]_1^b = -\frac{1}{b} + 1 = 1 - \frac{1}{b}$$

which can be interpreted as the area of the shaded region shown in Figure 8.17. Taking the limit as $b \to \infty$ produces

$$\int_1^\infty \frac{dx}{x^2} = \lim_{b \to \infty}\left(\int_1^b \frac{dx}{x^2}\right) = \lim_{b \to \infty}\left(1 - \frac{1}{b}\right) = 1.$$

This improper integral can be interpreted as the area of the *unbounded* region between the graph of $f(x) = 1/x^2$ and the x-axis (to the right of $x = 1$).

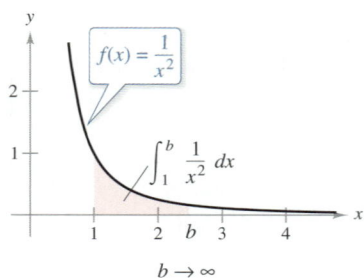

$f(x) = \dfrac{1}{x^2}$

$\displaystyle\int_1^b \frac{1}{x^2}\, dx$

$b \to \infty$

The unbounded region has an area of 1.
Figure 8.17

Definition of Improper Integrals with Infinite Integration Limits

1. If f is continuous on the interval $[a, \infty)$, then

$$\int_a^\infty f(x)\, dx = \lim_{b \to \infty} \int_a^b f(x)\, dx.$$

2. If f is continuous on the interval $(-\infty, b]$, then

$$\int_{-\infty}^b f(x)\, dx = \lim_{a \to -\infty} \int_a^b f(x)\, dx.$$

3. If f is continuous on the interval $(-\infty, \infty)$, then

$$\int_{-\infty}^\infty f(x)\, dx = \int_{-\infty}^c f(x)\, dx + \int_c^\infty f(x)\, dx$$

where c is any real number (see Exercise 111).

In the first two cases, the improper integral **converges** when the limit exists—otherwise, the improper integral **diverges.** In the third case, the improper integral on the left diverges when either of the improper integrals on the right diverges.

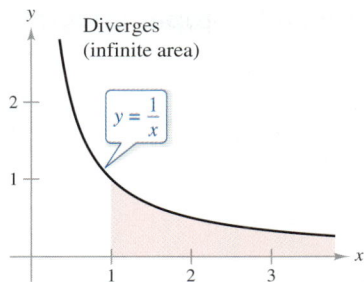

Diverges (infinite area)

$y = \dfrac{1}{x}$

This unbounded region has an infinite area.
Figure 8.18

EXAMPLE 1   **An Improper Integral That Diverges**

Evaluate $\displaystyle\int_1^{\infty} \frac{dx}{x}$.

Solution

$$\int_1^{\infty} \frac{dx}{x} = \lim_{b \to \infty} \int_1^{b} \frac{dx}{x} \qquad \text{Take limit as } b \to \infty.$$

$$= \lim_{b \to \infty} \left[\ln x \right]_1^{b} \qquad \text{Apply Log Rule.}$$

$$= \lim_{b \to \infty} (\ln b - 0) \qquad \text{Apply Fundamental Theorem of Calculus.}$$

$$= \infty \qquad \text{Evaluate limit.}$$

The limit does not exist. So, you can conclude that the improper integral diverges. See Figure 8.18.

Try comparing the regions shown in Figures 8.17 and 8.18. They look similar, yet the region in Figure 8.17 has a finite area of 1 and the region in Figure 8.18 has an infinite area.

EXAMPLE 2   **Improper Integrals That Converge**

Evaluate each improper integral.

a. $\displaystyle\int_0^{\infty} e^{-x}\, dx$

b. $\displaystyle\int_0^{\infty} \frac{1}{x^2 + 1}\, dx$

Solution

a.
$$\int_0^{\infty} e^{-x}\, dx = \lim_{b \to \infty} \int_0^{b} e^{-x}\, dx$$
$$= \lim_{b \to \infty} \left[-e^{-x} \right]_0^{b}$$
$$= \lim_{b \to \infty} (-e^{-b} + 1)$$
$$= 1$$

See Figure 8.19.

b.
$$\int_0^{\infty} \frac{1}{x^2 + 1}\, dx = \lim_{b \to \infty} \int_0^{\infty} \frac{1}{x^2 + 1}\, dx$$
$$= \lim_{b \to \infty} \left[\arctan x \right]_0^{b}$$
$$= \lim_{b \to \infty} \arctan b$$
$$= \frac{\pi}{2}$$

See Figure 8.20.

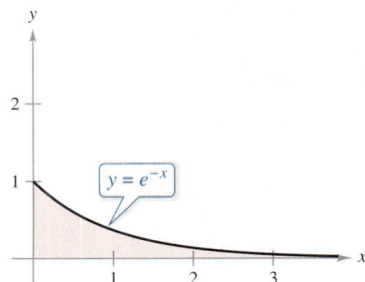

$y = e^{-x}$

The area of the unbounded region is 1.
Figure 8.19

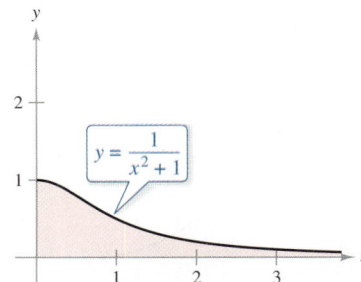

$y = \dfrac{1}{x^2 + 1}$

The area of the unbounded region is $\pi/2$.
Figure 8.20

In the next example, note how L'Hôpital's Rule can be used to evaluate an improper integral.

EXAMPLE 3 **Using L'Hôpital's Rule with an Improper Integral**

Evaluate $\int_1^\infty (1 - x)e^{-x}\, dx$.

Solution Use integration by parts, with $dv = e^{-x}\, dx$ and $u = (1 - x)$.

$$\int (1 - x)e^{-x}\, dx = -e^{-x}(1 - x) - \int e^{-x}\, dx$$
$$= -e^{-x} + xe^{-x} + e^{-x} + C$$
$$= xe^{-x} + C$$

Now, apply the definition of an improper integral.

$$\int_1^\infty (1 - x)e^{-x}\, dx = \lim_{b\to\infty}\left[xe^{-x}\right]_1^b$$
$$= \lim_{b\to\infty}\left(\frac{b}{e^b} - \frac{1}{e}\right)$$
$$= \lim_{b\to\infty}\frac{b}{e^b} - \lim_{b\to\infty}\frac{1}{e}$$

For the first limit, use L'Hôpital's Rule.

$$\lim_{b\to\infty}\frac{b}{e^b} = \lim_{b\to\infty}\frac{1}{e^b} = 0$$

So, you can conclude that

$$\int_1^\infty (1 - x)e^{-x}\, dx = \lim_{b\to\infty}\frac{b}{e^b} - \lim_{b\to\infty}\frac{1}{e}$$
$$= 0 - \frac{1}{e}$$
$$= -\frac{1}{e}. \qquad \text{See Figure 8.21.}$$

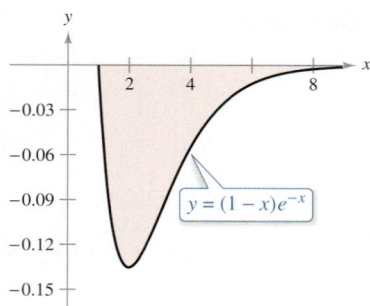

The area of the unbounded region is $|-1/e|$.

Figure 8.21

EXAMPLE 4 **Infinite Upper and Lower Limits of Integration**

Evaluate $\int_{-\infty}^\infty \frac{e^x}{1 + e^{2x}}\, dx$.

Solution Note that the integrand is continuous on $(-\infty, \infty)$. To evaluate the integral, you can break it into two parts, choosing $c = 0$ as a convenient value.

$$\int_{-\infty}^\infty \frac{e^x}{1 + e^{2x}}\, dx = \int_{-\infty}^0 \frac{e^x}{1 + e^{2x}}\, dx + \int_0^\infty \frac{e^x}{1 + e^{2x}}\, dx$$
$$= \lim_{b\to-\infty}\left[\arctan e^x\right]_b^0 + \lim_{b\to\infty}\left[\arctan e^x\right]_0^b$$
$$= \lim_{b\to-\infty}\left(\frac{\pi}{4} - \arctan e^b\right) + \lim_{b\to\infty}\left(\arctan e^b - \frac{\pi}{4}\right)$$
$$= \frac{\pi}{4} - 0 + \frac{\pi}{2} - \frac{\pi}{4}$$
$$= \frac{\pi}{2} \qquad \text{See Figure 8.22.}$$

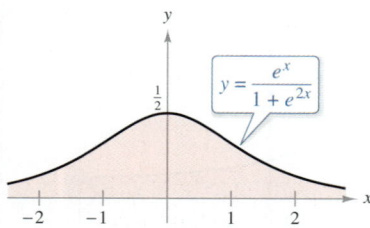

The area of the unbounded region is $\pi/2$.

Figure 8.22

EXAMPLE 5	Sending a Space Module into Orbit

In Example 3 in Section 7.5, you found that it would require 10,000 mile-tons of work to propel a 15-metric-ton space module to a height of 800 miles above Earth. How much work is required to propel the module an unlimited distance away from Earth's surface?

Solution   At first you might think that an infinite amount of work would be required. But if this were the case, it would be impossible to send rockets into outer space. Because this has been done, the work required must be finite. You can determine the work in the following manner. Using the integral in Example 3, Section 7.5, replace the upper bound of 4800 miles by ∞ and write

$$
\begin{aligned}
W &= \int_{4000}^{\infty} \frac{240{,}000{,}000}{x^2}\, dx \\
&= \lim_{b \to \infty} \left[-\frac{240{,}000{,}000}{x} \right]_{4000}^{b} \\
&= \lim_{b \to \infty} \left(-\frac{240{,}000{,}000}{b} + \frac{240{,}000{,}000}{4000} \right) \\
&= 60{,}000 \text{ mile-tons} \\
&\approx 6.984 \times 10^{11} \text{ foot-pounds.}
\end{aligned}
$$

In SI units, using a conversion factor of

$$1 \text{ foot-pound} \approx 1.35582 \text{ joules}$$

the work done is $W \approx 9.469 \times 10^{11}$ joules.

The work required to move a 15-metric-ton space module an unlimited distance away from Earth is about 6.984×10^{11} foot-pounds.

Improper Integrals with Infinite Discontinuities

The second basic type of improper integral is one that has an infinite discontinuity *at or between* the limits of integration.

Definition of Improper Integrals with Infinite Discontinuities

1. If f is continuous on the interval $[a, b)$ and has an infinite discontinuity at b, then

$$\int_{a}^{b} f(x)\, dx = \lim_{c \to b^{-}} \int_{a}^{c} f(x)\, dx.$$

2. If f is continuous on the interval $(a, b]$ and has an infinite discontinuity at a, then

$$\int_{a}^{b} f(x)\, dx = \lim_{c \to a^{+}} \int_{c}^{b} f(x)\, dx.$$

3. If f is continuous on the interval $[a, b]$, except for some c in (a, b) at which f has an infinite discontinuity, then

$$\int_{a}^{b} f(x)\, dx = \int_{a}^{c} f(x)\, dx + \int_{c}^{b} f(x)\, dx.$$

In the first two cases, the improper integral **converges** when the limit exists—otherwise, the improper integral **diverges**. In the third case, the improper integral on the left diverges when either of the improper integrals on the right diverges.

| EXAMPLE 6 | **An Improper Integral with an Infinite Discontinuity** |

Evaluate $\displaystyle\int_0^1 \frac{dx}{\sqrt[3]{x}}$.

Solution The integrand has an infinite discontinuity at $x = 0$, as shown in Figure 8.23. You can evaluate this integral as shown below.

$$\int_0^1 x^{-1/3}\, dx = \lim_{b \to 0^+} \left[\frac{x^{2/3}}{2/3} \right]_b^1$$

$$= \lim_{b \to 0^+} \frac{3}{2}(1 - b^{2/3})$$

$$= \frac{3}{2}$$

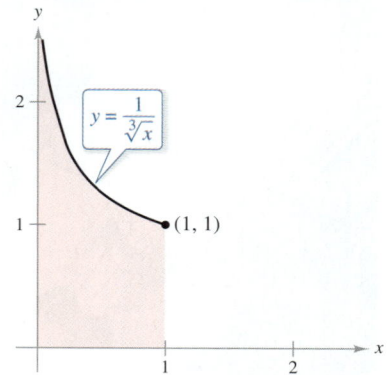

Infinite discontinuity at $x = 0$
Figure 8.23

| EXAMPLE 7 | **An Improper Integral That Diverges** |

Evaluate $\displaystyle\int_0^2 \frac{dx}{x^3}$.

Solution Because the integrand has an infinite discontinuity at $x = 0$, you can write

$$\int_0^2 \frac{dx}{x^3} = \lim_{b \to 0^+} \left[-\frac{1}{2x^2} \right]_b^2$$

$$= \lim_{b \to 0^+} \left(-\frac{1}{8} + \frac{1}{2b^2} \right)$$

$$= \infty.$$

So, you can conclude that the improper integral diverges.

| EXAMPLE 8 | **An Improper Integral with an Interior Discontinuity** |

Evaluate $\displaystyle\int_{-1}^2 \frac{dx}{x^3}$.

Solution This integral is improper because the integrand has an infinite discontinuity at the interior point $x = 0$, as shown in Figure 8.24. So, you can write

$$\int_{-1}^2 \frac{dx}{x^3} = \int_{-1}^0 \frac{dx}{x^3} + \int_0^2 \frac{dx}{x^3}.$$

From Example 7, you know that the second integral diverges. So, the original improper integral also diverges.

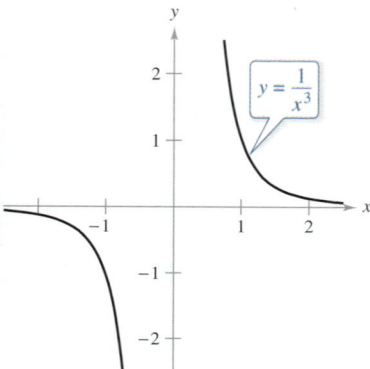

Remember to check for infinite discontinuities at interior points as well as at endpoints when determining whether an integral is improper. For instance, if you had not recognized that the integral in Example 8 was improper, you would have obtained the *incorrect* result

$$\int_{-1}^2 \frac{dx}{x^3} \overset{?}{=} \left[\frac{-1}{2x^2} \right]_{-1}^2 = -\frac{1}{8} + \frac{1}{2} = \frac{3}{8}. \qquad \text{Incorrect evaluation}$$

The improper integral $\displaystyle\int_{-1}^2 \frac{dx}{x^3}$ diverges.
Figure 8.24

The integral in the next example is improper for *two* reasons. One limit of integration is infinite, and the integrand has an infinite discontinuity at the outer limit of integration.

EXAMPLE 9   A Doubly Improper Integral

> See LarsonCalculus.com for an interactive version of this type of example.

Evaluate $\displaystyle \int_0^\infty \frac{dx}{\sqrt{x}(x+1)}$.

Solution   To evaluate this integral, split it at a convenient point (say, $x = 1$) and write

$$\int_0^\infty \frac{dx}{\sqrt{x}(x+1)} = \int_0^1 \frac{dx}{\sqrt{x}(x+1)} + \int_1^\infty \frac{dx}{\sqrt{x}(x+1)}$$

$$= \lim_{b \to 0^+} \left[2 \arctan \sqrt{x} \right]_b^1 + \lim_{c \to \infty} \left[2 \arctan \sqrt{x} \right]_1^c$$

$$= \lim_{b \to 0^+} (2 \arctan 1 - 2 \arctan \sqrt{b}) + \lim_{c \to \infty} (2 \arctan \sqrt{c} - 2 \arctan 1)$$

$$= 2\left(\frac{\pi}{4}\right) - 0 + 2\left(\frac{\pi}{2}\right) - 2\left(\frac{\pi}{4}\right)$$

$$= \pi.$$

See Figure 8.25.

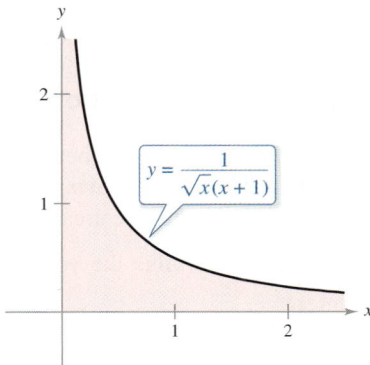

The area of the unbounded region is π.
Figure 8.25

$y = \dfrac{1}{\sqrt{x}(x+1)}$

EXAMPLE 10   An Application Involving Arc Length

Use the formula for arc length to show that the circumference of the circle $x^2 + y^2 = 1$ is 2π.

Solution   To simplify the work, consider the quarter circle given by $y = \sqrt{1 - x^2}$, where $0 \le x \le 1$. The function y is differentiable for any x in this interval except $x = 1$. Therefore, the arc length of the quarter circle is given by the improper integral

$$s = \int_0^1 \sqrt{1 + (y')^2}\, dx$$

$$= \int_0^1 \sqrt{1 + \left(\frac{-x}{\sqrt{1 - x^2}}\right)^2}\, dx$$

$$= \int_0^1 \frac{dx}{\sqrt{1 - x^2}}.$$

This integral is improper because it has an infinite discontinuity at $x = 1$. So, you can write

$$s = \int_0^1 \frac{dx}{\sqrt{1 - x^2}}$$

$$= \lim_{b \to 1^-} \left[\arcsin x \right]_0^b$$

$$= \lim_{b \to 1^-} (\arcsin b - \arcsin 0)$$

$$= \frac{\pi}{2} - 0$$

$$= \frac{\pi}{2}.$$

Finally, multiplying by 4, you can conclude that the circumference of the circle is $4s = 2\pi$, as shown in Figure 8.26.

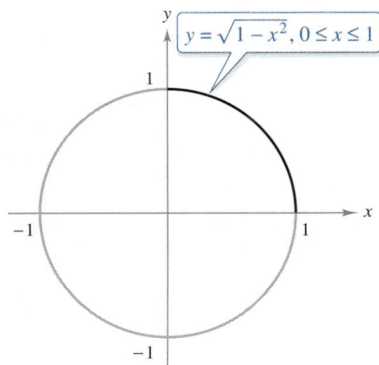

The circumference of the circle is 2π.
Figure 8.26

$y = \sqrt{1 - x^2},\, 0 \le x \le 1$

This section concludes with a useful theorem describing the convergence or divergence of a common type of improper integral. The proof of this theorem is left as an exercise (see Exercise 49).

THEOREM 8.5 **A Special Type of Improper Integral**

$$\int_1^\infty \frac{dx}{x^p} = \begin{cases} \dfrac{1}{p-1}, & p > 1 \\ \text{diverges}, & p \le 1 \end{cases}$$

EXAMPLE 11 **An Application Involving a Solid of Revolution**

■ **FOR FURTHER INFORMATION**
For further investigation of solids that have finite volumes and infinite surface areas, see the article "Supersolids: Solids Having Finite Volume and Infinite Surfaces" by William P. Love in *Mathematics Teacher.* To view this article, go to *MathArticles.com.*

The solid formed by revolving (about the x-axis) the *unbounded* region lying between the graph of $f(x) = 1/x$ and the x-axis $(x \ge 1)$ is called **Gabriel's Horn.** (See Figure 8.27.) Show that this solid has a finite volume and an infinite surface area.

Solution Using the disk method and Theorem 8.5, you can determine the volume to be

$$V = \pi \int_1^\infty \left(\frac{1}{x}\right)^2 dx \qquad \textcolor{red}{\text{Theorem 8.5, } p = 2 > 1}$$

$$= \pi\left(\frac{1}{2-1}\right)$$

$$= \pi.$$

The surface area is given by

$$S = 2\pi \int_1^\infty f(x)\sqrt{1 + [f'(x)]^2}\, dx = 2\pi \int_1^\infty \frac{1}{x}\sqrt{1 + \frac{1}{x^4}}\, dx.$$

Because

$$\sqrt{1 + \frac{1}{x^4}} > 1$$

on the interval $[1, \infty)$, and the improper integral

$$\int_1^\infty \frac{1}{x}\, dx$$

diverges, you can conclude that the improper integral

$$\int_1^\infty \frac{1}{x}\sqrt{1 + \frac{1}{x^4}}\, dx$$

also diverges. (See Exercise 52.) So, the surface area is infinite.

■ **FOR FURTHER INFORMATION**
To learn about another function that has a finite volume and an infinite surface area, see the article "Gabriel's Wedding Cake" by Julian F. Fleron in *The College Mathematics Journal.* To view this article, go to *MathArticles.com.*

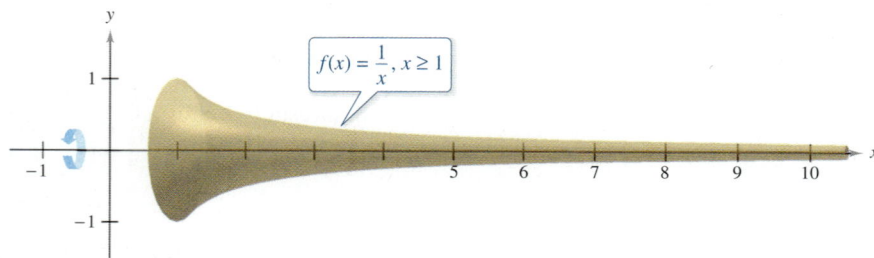

$f(x) = \dfrac{1}{x},\ x \ge 1$

Gabriel's Horn has a finite volume and an infinite surface area.
Figure 8.27

8.8 Exercises

See CalcChat.com for tutorial help and worked-out solutions to odd-numbered exercises.

Determining Whether an Integral Is Improper In Exercises 1–8, decide whether the integral is improper. Explain your reasoning.

1. $\displaystyle\int_0^1 \frac{dx}{5x-3}$

2. $\displaystyle\int_1^2 \frac{dx}{x^3}$

3. $\displaystyle\int_0^1 \frac{2x-5}{x^2-5x+6}\,dx$

4. $\displaystyle\int_1^\infty \ln(x^2)\,dx$

5. $\displaystyle\int_0^2 e^{-x}\,dx$

6. $\displaystyle\int_0^\infty \cos x\,dx$

7. $\displaystyle\int_{-\infty}^\infty \frac{\sin x}{4+x^2}\,dx$

8. $\displaystyle\int_0^{\pi/4} \csc x\,dx$

Evaluating an Improper Integral In Exercises 9–12, explain why the integral is improper and determine whether it diverges or converges. Evaluate the integral if it converges.

9. $\displaystyle\int_0^4 \frac{1}{\sqrt{x}}\,dx$

10. $\displaystyle\int_3^4 \frac{1}{(x-3)^{3/2}}\,dx$

11. $\displaystyle\int_0^2 \frac{1}{(x-1)^2}\,dx$

12. $\displaystyle\int_{-\infty}^0 e^{3x}\,dx$

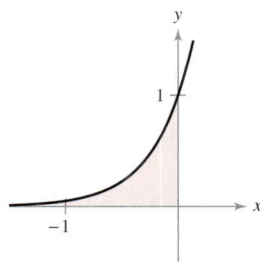

Writing In Exercises 13–16, explain why the evaluation of the integral is *incorrect*. Use the integration capabilities of a graphing utility to attempt to evaluate the integral. Determine whether the utility gives the correct answer.

13. $\displaystyle\int_{-1}^1 \frac{1}{x^2}\,dx = -2$

14. $\displaystyle\int_{-2}^2 \frac{-2}{(x-1)^3}\,dx = \frac{8}{9}$

15. $\displaystyle\int_0^\infty e^{-x}\,dx = 0$

16. $\displaystyle\int_0^\pi \sec x\,dx = 0$

Evaluating an Improper Integral In Exercises 17–32, determine whether the improper integral diverges or converges. Evaluate the integral if it converges.

17. $\displaystyle\int_1^\infty \frac{1}{x^3}\,dx$

18. $\displaystyle\int_1^\infty \frac{6}{x^4}\,dx$

19. $\displaystyle\int_1^\infty \frac{3}{\sqrt[3]{x}}\,dx$

20. $\displaystyle\int_1^\infty \frac{4}{\sqrt[4]{x}}\,dx$

21. $\displaystyle\int_{-\infty}^0 xe^{-4x}\,dx$

22. $\displaystyle\int_0^\infty xe^{-x/3}\,dx$

23. $\displaystyle\int_0^\infty x^2 e^{-x}\,dx$

24. $\displaystyle\int_0^\infty e^{-x}\cos x\,dx$

25. $\displaystyle\int_4^\infty \frac{1}{x(\ln x)^3}\,dx$

26. $\displaystyle\int_1^\infty \frac{\ln x}{x}\,dx$

27. $\displaystyle\int_{-\infty}^\infty \frac{4}{16+x^2}\,dx$

28. $\displaystyle\int_0^\infty \frac{x^3}{(x^2+1)^2}\,dx$

29. $\displaystyle\int_0^\infty \frac{1}{e^x+e^{-x}}\,dx$

30. $\displaystyle\int_0^\infty \frac{e^x}{1+e^x}\,dx$

31. $\displaystyle\int_0^\infty \cos \pi x\,dx$

32. $\displaystyle\int_0^\infty \sin \frac{x}{2}\,dx$

Evaluating an Improper Integral In Exercises 33–48, determine whether the improper integral diverges or converges. Evaluate the integral if it converges, and check your results with the results obtained by using the integration capabilities of a graphing utility.

33. $\displaystyle\int_0^1 \frac{1}{x^2}\,dx$

34. $\displaystyle\int_0^5 \frac{10}{x}\,dx$

35. $\displaystyle\int_0^2 \frac{1}{\sqrt[3]{x-1}}\,dx$

36. $\displaystyle\int_0^8 \frac{3}{\sqrt{8-x}}\,dx$

37. $\displaystyle\int_0^1 x\ln x\,dx$

38. $\displaystyle\int_0^e \ln x^2\,dx$

39. $\displaystyle\int_0^{\pi/2} \tan\theta\,d\theta$

40. $\displaystyle\int_0^{\pi/2} \sec\theta\,d\theta$

41. $\displaystyle\int_2^4 \frac{2}{x\sqrt{x^2-4}}\,dx$

42. $\displaystyle\int_3^6 \frac{1}{\sqrt{36-x^2}}\,dx$

43. $\displaystyle\int_3^5 \frac{1}{\sqrt{x^2-9}}\,dx$

44. $\displaystyle\int_0^5 \frac{1}{25-x^2}\,dx$

45. $\displaystyle\int_3^\infty \frac{1}{x\sqrt{x^2-9}}\,dx$

46. $\displaystyle\int_4^\infty \frac{\sqrt{x^2-16}}{x^2}\,dx$

47. $\displaystyle\int_0^\infty \frac{4}{\sqrt{x}(x+6)}\,dx$

48. $\displaystyle\int_1^\infty \frac{1}{x\ln x}\,dx$

Finding Values In Exercises 49 and 50, determine all values of *p* for which the improper integral converges.

49. $\displaystyle\int_1^\infty \frac{1}{x^p}\,dx$

50. $\displaystyle\int_0^1 \frac{1}{x^p}\,dx$

51. Mathematical Induction Use mathematical induction to verify that the following integral converges for any positive integer n.

$$\int_0^\infty x^n e^{-x}\, dx$$

52. Comparison Test for Improper Integrals In some cases, it is impossible to find the exact value of an improper integral, but it is important to determine whether the integral converges or diverges. Suppose the functions f and g are continuous and $0 \le g(x) \le f(x)$ on the interval $[a, \infty)$. It can be shown that if $\int_a^\infty f(x)\, dx$ converges, then $\int_a^\infty g(x)\, dx$ also converges, and if $\int_a^\infty g(x)\, dx$ diverges, then $\int_a^\infty f(x)\, dx$ also diverges. This is known as the Comparison Test for improper integrals.

(a) Use the Comparison Test to determine whether $\int_1^\infty e^{-x^2}\, dx$ converges or diverges. (*Hint:* Use the fact that $e^{-x^2} \le e^{-x}$ for $x \ge 1$.)

(b) Use the Comparison Test to determine whether $\int_1^\infty \dfrac{1}{x^5 + 1}\, dx$ converges or diverges. (*Hint:* Use the fact that $\dfrac{1}{x^5 + 1} \le \dfrac{1}{x^5}$ for $x \ge 1$.)

Convergence or Divergence In Exercises 53–62, use the results of Exercises 49–52 to determine whether the improper integral converges or diverges.

53. $\displaystyle\int_0^1 \frac{1}{x^5}\, dx$

54. $\displaystyle\int_0^1 \frac{1}{\sqrt[5]{x}}\, dx$

55. $\displaystyle\int_1^\infty \frac{1}{x^5}\, dx$

56. $\displaystyle\int_0^\infty x^4 e^{-x}\, dx$

57. $\displaystyle\int_1^\infty \frac{1}{x^2 + 5}\, dx$

58. $\displaystyle\int_2^\infty \frac{1}{\sqrt{x - 1}}\, dx$

59. $\displaystyle\int_2^\infty \frac{1}{\sqrt[3]{x(x - 1)}}\, dx$

60. $\displaystyle\int_1^\infty \frac{1}{\sqrt{x}(x + 1)}\, dx$

61. $\displaystyle\int_1^\infty \frac{1 - \sin x}{x^2}\, dx$

62. $\displaystyle\int_0^\infty \frac{1}{e^x + x}\, dx$

WRITING ABOUT CONCEPTS

63. Improper Integrals Describe the different types of improper integrals.

64. Improper Integrals Define the terms *converges* and *diverges* when working with improper integrals.

65. Improper Integral Explain why $\displaystyle\int_{-1}^1 \frac{1}{x^3}\, dx \ne 0$.

66. Improper Integral Consider the integral

$$\int_0^3 \frac{10}{x^2 - 2x}\, dx.$$

To determine the convergence or divergence of the integral, how many improper integrals must be analyzed? What must be true of each of these integrals if the given integral converges?

Area In Exercises 67–70, find the area of the unbounded shaded region.

67. $y = e^x, \quad -\infty < x \le 1$

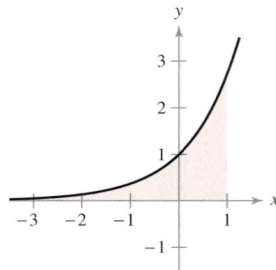

68. $y = -\ln x$

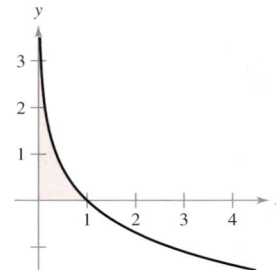

69. Witch of Agnesi:

$$y = \frac{1}{x^2 + 1}$$

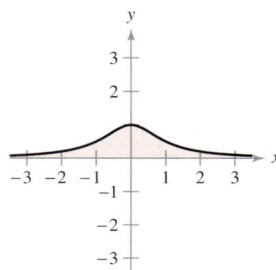

70. Witch of Agnesi:

$$y = \frac{8}{x^2 + 4}$$

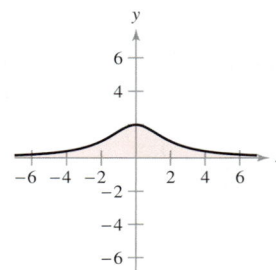

Area and Volume In Exercises 71 and 72, consider the region satisfying the inequalities. (a) Find the area of the region. (b) Find the volume of the solid generated by revolving the region about the x-axis. (c) Find the volume of the solid generated by revolving the region about the y-axis.

71. $y \le e^{-x}, \ y \ge 0, \ x \ge 0$

72. $y \le \dfrac{1}{x^2}, \ y \ge 0, \ x \ge 1$

73. Arc Length Sketch the graph of the hypocycloid of four cusps $x^{2/3} + y^{2/3} = 4$ and find its perimeter.

74. Arc Length Find the arc length of the graph of $y = \sqrt{16 - x^2}$ over the interval $[0, 4]$.

75. Surface Area The region bounded by $(x - 2)^2 + y^2 = 1$ is revolved about the y-axis to form a torus. Find the surface area of the torus.

76. Surface Area Find the area of the surface formed by revolving the graph of $y = 2e^{-x}$ on the interval $[0, \infty)$ about the x-axis.

Propulsion In Exercises 77 and 78, use the weight of the rocket to answer each question. (Use 4000 miles as the radius of Earth and do not consider the effect of air resistance.)

(a) How much work is required to propel the rocket an unlimited distance away from Earth's surface?

(b) How far has the rocket traveled when half the total work has occurred?

77. 5-ton rocket

78. 10-ton rocket

Probability   A nonnegative function f is called a *probability density function* if

$$\int_{-\infty}^{\infty} f(t)\, dt = 1.$$

The probability that x lies between a and b is given by

$$P(a \le x \le b) = \int_a^b f(t)\, dt.$$

The expected value of x is given by

$$E(x) = \int_{-\infty}^{\infty} t f(t)\, dt.$$

In Exercises 79 and 80, (a) show that the nonnegative function is a probability density function, (b) find $P(0 \le x \le 4)$, and (c) find $E(x)$.

79. $f(t) = \begin{cases} \frac{1}{7}e^{-t/7}, & t \ge 0 \\ 0, & t < 0 \end{cases}$ **80.** $f(t) = \begin{cases} \frac{2}{5}e^{-2t/5}, & t \ge 0 \\ 0, & t < 0 \end{cases}$

Capitalized Cost   In Exercises 81 and 82, find the capitalized cost C of an asset (a) for $n = 5$ years, (b) for $n = 10$ years, and (c) forever. The capitalized cost is given by

$$C = C_0 + \int_0^n c(t)e^{-rt}\, dt$$

where C_0 is the original investment, t is the time in years, r is the annual interest rate compounded continuously, and $c(t)$ is the annual cost of maintenance.

81. $C_0 = \$650,000$ **82.** $C_0 = \$650,000$

   $c(t) = \$25,000$ $c(t) = \$25,000(1 + 0.08t)$

   $r = 0.06$ $r = 0.06$

83. Electromagnetic Theory   The magnetic potential P at a point on the axis of a circular coil is given by

$$P = \frac{2\pi NIr}{k} \int_c^{\infty} \frac{1}{(r^2 + x^2)^{3/2}}\, dx$$

where N, I, r, k, and c are constants. Find P.

84. Gravitational Force   A "semi-infinite" uniform rod occupies the nonnegative x-axis. The rod has a linear density δ, which means that a segment of length dx has a mass of $\delta\, dx$. A particle of mass M is located at the point $(-a, 0)$. The gravitational force F that the rod exerts on the mass is given by

$$F = \int_0^{\infty} \frac{GM\delta}{(a + x)^2}\, dx$$

where G is the gravitational constant. Find F.

True or False?   In Exercises 85–88, determine whether the statement is true or false. If it is false, explain why or give an example that shows it is false.

85. If f is continuous on $[0, \infty)$ and $\lim\limits_{x \to \infty} f(x) = 0$, then $\int_0^{\infty} f(x)\, dx$ converges.

86. If f is continuous on $[0, \infty)$ and $\int_0^{\infty} f(x)\, dx$ diverges, then $\lim\limits_{x \to \infty} f(x) \ne 0$.

87. If f' is continuous on $[0, \infty)$ and $\lim\limits_{x \to \infty} f(x) = 0$, then

$$\int_0^{\infty} f'(x)\, dx = -f(0).$$

88. If the graph of f is symmetric with respect to the origin or the y-axis, then $\int_0^{\infty} f(x)\, dx$ converges if and only if $\int_{-\infty}^{\infty} f(x)\, dx$ converges.

89. Comparing Integrals

(a) Show that $\int_{-\infty}^{\infty} \sin x\, dx$ diverges.

(b) Show that $\lim\limits_{a \to \infty} \int_{-a}^{a} \sin x\, dx = 0$.

(c) What do parts (a) and (b) show about the definition of improper integrals?

90. Making an Integral Improper   For each integral, find a nonnegative real number b that makes the integral improper. Explain your reasoning.

(a) $\int_0^b \frac{1}{x^2 - 9}\, dx$ (b) $\int_0^b \frac{1}{\sqrt{4 - x}}\, dx$

(c) $\int_0^b \frac{x}{x^2 - 7x + 12}\, dx$ (d) $\int_b^{10} \ln x\, dx$

(e) $\int_0^b \tan 2x\, dx$ (f) $\int_0^b \frac{\cos x}{1 - \sin x}\, dx$

91. Writing

(a) The improper integrals

$$\int_1^{\infty} \frac{1}{x}\, dx \quad \text{and} \quad \int_1^{\infty} \frac{1}{x^2}\, dx$$

diverge and converge, respectively. Describe the essential differences between the integrands that cause one integral to converge and the other to diverge.

(b) Sketch a graph of the function $y = (\sin x)/x$ over the interval $(1, \infty)$. Use your knowledge of the definite integral to make an inference as to whether the integral

$$\int_1^{\infty} \frac{\sin x}{x}\, dx$$

converges. Give reasons for your answer.

(c) Use one iteration of integration by parts on the integral in part (b) to determine its divergence or convergence.

92. Exploration   Consider the integral

$$\int_0^{\pi/2} \frac{4}{1 + (\tan x)^n}\, dx$$

where n is a positive integer.

(a) Is the integral improper? Explain.

(b) Use a graphing utility to graph the integrand for $n = 2, 4, 8$, and 12.

(c) Use the graphs to approximate the integral as $n \to \infty$.

(d) Use a computer algebra system to evaluate the integral for the values of n in part (b). Make a conjecture about the value of the integral for any positive integer n. Compare your results with your answer in part (c).

93. Normal Probability The mean height of American men between 20 and 29 years old is 70 inches, and the standard deviation is 2.85 inches. A 20- to 29-year-old man is chosen at random from the population. The probability that he is 6 feet tall or taller is

$$P(72 \leq x < \infty) = \int_{72}^{\infty} \frac{1}{2.85\sqrt{2\pi}} e^{-(x-70)^2/6.245} \, dx.$$

(*Source: National Center for Health Statistics*)

(a) Use a graphing utility to graph the integrand. Use the graphing utility to convince yourself that the area between the x-axis and the integrand is 1.

(b) Use a graphing utility to approximate $P(72 \leq x < \infty)$.

(c) Approximate $0.5 - P(70 \leq x \leq 72)$ using a graphing utility. Use the graph in part (a) to explain why this result is the same as the answer in part (b).

94. HOW DO YOU SEE IT? The graph shows the probability density function for a car brand that has a mean fuel efficiency of 26 miles per gallon and a standard deviation of 2.4 miles per gallon.

(a) Which is greater, the probability of choosing a car at random that gets between 26 and 28 miles per gallon or the probability of choosing a car at random that gets between 22 and 24 miles per gallon?

(b) Which is greater, the probability of choosing a car at random that gets between 20 and 22 miles per gallon or the probability of choosing a car at random that gets at least 30 miles per gallon?

Laplace Transforms Let $f(t)$ be a function defined for all positive values of t. The Laplace Transform of $f(t)$ is defined by

$$F(s) = \int_0^{\infty} e^{-st} f(t) \, dt$$

when the improper integral exists. Laplace Transforms are used to solve differential equations. In Exercises 95–102, find the Laplace Transform of the function.

95. $f(t) = 1$

96. $f(t) = t$

97. $f(t) = t^2$

98. $f(t) = e^{at}$

99. $f(t) = \cos at$

100. $f(t) = \sin at$

101. $f(t) = \cosh at$

102. $f(t) = \sinh at$

103. The Gamma Function The Gamma Function $\Gamma(n)$ is defined by

$$\Gamma(n) = \int_0^{\infty} x^{n-1} e^{-x} \, dx, \quad n > 0.$$

(a) Find $\Gamma(1)$, $\Gamma(2)$, and $\Gamma(3)$.

(b) Use integration by parts to show that $\Gamma(n+1) = n\Gamma(n)$.

(c) Write $\Gamma(n)$ using factorial notation where n is a positive integer.

104. Proof Prove that $I_n = \left(\dfrac{n-1}{n+2}\right) I_{n-1}$, where

$$I_n = \int_0^{\infty} \frac{x^{2n-1}}{(x^2+1)^{n+3}} \, dx, \quad n \geq 1.$$

Then evaluate each integral.

(a) $\displaystyle \int_0^{\infty} \frac{x}{(x^2+1)^4} \, dx$

(b) $\displaystyle \int_0^{\infty} \frac{x^3}{(x^2+1)^5} \, dx$

(c) $\displaystyle \int_0^{\infty} \frac{x^5}{(x^2+1)^6} \, dx$

105. Finding a Value For what value of c does the integral

$$\int_0^{\infty} \left(\frac{1}{\sqrt{x^2+1}} - \frac{c}{x+1} \right) dx$$

converge? Evaluate the integral for this value of c.

106. Finding a Value For what value of c does the integral

$$\int_1^{\infty} \left(\frac{cx}{x^2+2} - \frac{1}{3x} \right) dx$$

converge? Evaluate the integral for this value of c.

107. Volume Find the volume of the solid generated by revolving the region bounded by the graph of f about the x-axis.

$$f(x) = \begin{cases} x \ln x, & 0 < x \leq 2 \\ 0, & x = 0 \end{cases}$$

108. Volume Find the volume of the solid generated by revolving the unbounded region lying between $y = -\ln x$ and the y-axis $(y \geq 0)$ about the x-axis.

u-Substitution In Exercises 109 and 110, rewrite the improper integral as a proper integral using the given u-substitution. Then use the Trapezoidal Rule with $n = 5$ to approximate the integral.

109. $\displaystyle \int_0^1 \frac{\sin x}{\sqrt{x}} \, dx, \quad u = \sqrt{x}$

110. $\displaystyle \int_0^1 \frac{\cos x}{\sqrt{1-x}} \, dx, \quad u = \sqrt{1-x}$

111. Rewriting an Integral Let $\displaystyle \int_{-\infty}^{\infty} f(x) \, dx$ be convergent and let a and b be real numbers where $a \neq b$. Show that

$$\int_{-\infty}^{a} f(x) \, dx + \int_a^{\infty} f(x) \, dx = \int_{-\infty}^{b} f(x) \, dx + \int_b^{\infty} f(x) \, dx.$$

Finding or Evaluating an Integral In Exercises 1–8, use the basic integration rules to find or evaluate the integral.

1. $\displaystyle\int x\sqrt{x^2 - 36}\, dx$

2. $\displaystyle\int xe^{x^2 - 1}\, dx$

3. $\displaystyle\int \frac{x}{x^2 - 49}\, dx$

4. $\displaystyle\int \frac{x}{\sqrt[3]{4 - x^2}}\, dx$

5. $\displaystyle\int_1^e \frac{\ln(2x)}{x}\, dx$

6. $\displaystyle\int_{3/2}^2 2x\sqrt{2x - 3}\, dx$

7. $\displaystyle\int \frac{100}{\sqrt{100 - x^2}}\, dx$

8. $\displaystyle\int \frac{2x}{x - 3}\, dx$

Using Integration by Parts In Exercises 9–16, use integration by parts to find the indefinite integral.

9. $\displaystyle\int xe^{3x}\, dx$

10. $\displaystyle\int x^3 e^x\, dx$

11. $\displaystyle\int e^{2x} \sin 3x\, dx$

12. $\displaystyle\int x\sqrt{x - 1}\, dx$

13. $\displaystyle\int x^2 \sin 2x\, dx$

14. $\displaystyle\int \ln\sqrt{x^2 - 4}\, dx$

15. $\displaystyle\int x \arcsin 2x\, dx$

16. $\displaystyle\int \arctan 2x\, dx$

Finding a Trigonometric Integral In Exercises 17–22, find the trigonometric integral.

17. $\displaystyle\int \cos^3(\pi x - 1)\, dx$

18. $\displaystyle\int \sin^2 \frac{\pi x}{2}\, dx$

19. $\displaystyle\int \sec^4 \frac{x}{2}\, dx$

20. $\displaystyle\int \tan\theta \sec^4\theta\, d\theta$

21. $\displaystyle\int \frac{1}{1 - \sin\theta}\, d\theta$

22. $\displaystyle\int \cos 2\theta(\sin\theta + \cos\theta)^2\, d\theta$

Area In Exercises 23 and 24, find the area of the region.

23. $y = \sin^4 x$

24. $y = \sin 3x \cos 2x$

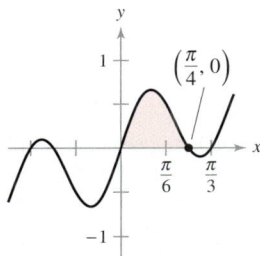

Using Trigonometric Substitution In Exercises 25–30, use trigonometric substitution to find or evaluate the integral.

25. $\displaystyle\int \frac{-12}{x^2\sqrt{4 - x^2}}\, dx$

26. $\displaystyle\int \frac{\sqrt{x^2 - 9}}{x}\, dx, \quad x > 3$

27. $\displaystyle\int \frac{x^3}{\sqrt{4 + x^2}}\, dx$

28. $\displaystyle\int \sqrt{25 - 9x^2}\, dx$

29. $\displaystyle\int_0^1 \frac{6x^3}{\sqrt{16 + x^2}}\, dx$

30. $\displaystyle\int_3^4 x^3\sqrt{x^2 - 9}\, dx$

Using Different Methods In Exercises 31 and 32, find the indefinite integral using each method.

31. $\displaystyle\int \frac{x^3}{\sqrt{4 + x^2}}\, dx$

(a) Trigonometric substitution

(b) Substitution: $u^2 = 4 + x^2$

(c) Integration by parts: $dv = \dfrac{x}{\sqrt{4 + x^2}}\, dx$

32. $\displaystyle\int x\sqrt{4 + x}\, dx$

(a) Trigonometric substitution

(b) Substitution: $u^2 = 4 + x$

(c) Substitution: $u = 4 + x$

(d) Integration by parts: $dv = \sqrt{4 + x}\, dx$

Using Partial Fractions In Exercises 33–38, use partial fractions to find the indefinite integral.

33. $\displaystyle\int \frac{x - 39}{x^2 - x - 12}\, dx$

34. $\displaystyle\int \frac{5x - 2}{x^2 - x}\, dx$

35. $\displaystyle\int \frac{x^2 + 2x}{x^3 - x^2 + x - 1}\, dx$

36. $\displaystyle\int \frac{4x - 2}{3(x - 1)^2}\, dx$

37. $\displaystyle\int \frac{x^2}{x^2 + 5x - 24}\, dx$

38. $\displaystyle\int \frac{\sec^2\theta}{\tan\theta(\tan\theta - 1)}\, d\theta$

Integration by Tables In Exercises 39–46, use integration tables to find or evaluate the integral.

39. $\displaystyle\int \frac{x}{(4 + 5x)^2}\, dx$

40. $\displaystyle\int \frac{x}{\sqrt{4 + 5x}}\, dx$

41. $\displaystyle\int_0^{\sqrt{\pi}/2} \frac{x}{1 + \sin x^2}\, dx$

42. $\displaystyle\int_0^1 \frac{x}{1 + e^{x^2}}\, dx$

43. $\displaystyle\int \frac{x}{x^2 + 4x + 8}\, dx$

44. $\displaystyle\int \frac{3}{2x\sqrt{9x^2 - 1}}\, dx, \quad x > \frac{1}{3}$

45. $\displaystyle\int \frac{1}{\sin \pi x \cos \pi x}\, dx$

46. $\displaystyle\int \frac{1}{1 + \tan \pi x}\, dx$

47. Verifying a Formula Verify the reduction formula

$$\int (\ln x)^n\, dx = x(\ln x)^n - n\int (\ln x)^{n-1}\, dx.$$

48. Verifying a Formula Verify the reduction formula

$$\int \tan^n x\, dx = \frac{1}{n - 1}\tan^{n-1} x - \int \tan^{n-2} x\, dx.$$

Finding an Indefinite Integral   In Exercises 49–56, find the indefinite integral using any method.

49. $\int \theta \sin \theta \cos \theta \, d\theta$

50. $\int \dfrac{\csc \sqrt{2x}}{\sqrt{x}} \, dx$

51. $\int \dfrac{x^{1/4}}{1 + x^{1/2}} \, dx$

52. $\int \sqrt{1 + \sqrt{x}} \, dx$

53. $\int \sqrt{1 + \cos x} \, dx$

54. $\int \dfrac{3x^3 + 4x}{(x^2 + 1)^2} \, dx$

55. $\int \cos x \ln(\sin x) \, dx$

56. $\int (\sin \theta + \cos \theta)^2 \, d\theta$

Differential Equation   In Exercises 57–60, solve the differential equation using any method.

57. $\dfrac{dy}{dx} = \dfrac{25}{x^2 - 25}$

58. $\dfrac{dy}{dx} = \dfrac{\sqrt{4 - x^2}}{2x}$

59. $y' = \ln(x^2 + x)$

60. $y' = \sqrt{1 - \cos \theta}$

Evaluating a Definite Integral   In Exercises 61–66, evaluate the definite integral using any method. Use a graphing utility to verify your result.

61. $\int_2^{\sqrt{5}} x(x^2 - 4)^{3/2} \, dx$

62. $\int_0^1 \dfrac{x}{(x - 2)(x - 4)} \, dx$

63. $\int_1^4 \dfrac{\ln x}{x} \, dx$

64. $\int_0^2 xe^{3x} \, dx$

65. $\int_0^\pi x \sin x \, dx$

66. $\int_0^5 \dfrac{x}{\sqrt{4 + x}} \, dx$

Area   In Exercises 67 and 68, find the area of the region.

67. $y = x\sqrt{4 - x}$

68. $y = \dfrac{1}{25 - x^2}$

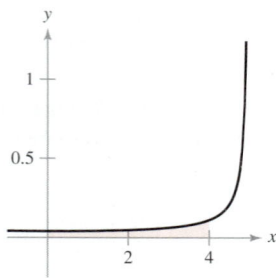

Centroid   In Exercises 69 and 70, find the centroid of the region bounded by the graphs of the equations.

69. $y = \sqrt{1 - x^2}, \quad y = 0$

70. $(x - 1)^2 + y^2 = 1, \quad (x - 4)^2 + y^2 = 4$

Arc Length   In Exercises 71 and 72, approximate to two decimal places the arc length of the curve over the given interval.

Function	Interval
71. $y = \sin x$	$[0, \pi]$
72. $y = \sin^2 x$	$[0, \pi]$

Evaluating a Limit   In Exercises 73–80, use L'Hôpital's Rule to evaluate the limit.

73. $\displaystyle\lim_{x \to 1} \dfrac{(\ln x)^2}{x - 1}$

74. $\displaystyle\lim_{x \to 0} \dfrac{\sin \pi x}{\sin 5 \pi x}$

75. $\displaystyle\lim_{x \to \infty} \dfrac{e^{2x}}{x^2}$

76. $\displaystyle\lim_{x \to \infty} xe^{-x^2}$

77. $\displaystyle\lim_{x \to \infty} (\ln x)^{2/x}$

78. $\displaystyle\lim_{x \to 1^+} (x - 1)^{\ln x}$

79. $\displaystyle\lim_{n \to \infty} 1000\left(1 + \dfrac{0.09}{n}\right)^n$

80. $\displaystyle\lim_{x \to 1^+} \left(\dfrac{2}{\ln x} - \dfrac{2}{x - 1}\right)$

Evaluating an Improper Integral   In Exercises 81–88, determine whether the improper integral diverges or converges. Evaluate the integral if it converges.

81. $\displaystyle\int_0^{16} \dfrac{1}{\sqrt[4]{x}} \, dx$

82. $\displaystyle\int_0^2 \dfrac{7}{x - 2} \, dx$

83. $\displaystyle\int_1^\infty x^2 \ln x \, dx$

84. $\displaystyle\int_0^\infty \dfrac{e^{-1/x}}{x^2} \, dx$

85. $\displaystyle\int_1^\infty \dfrac{\ln x}{x^2} \, dx$

86. $\displaystyle\int_1^\infty \dfrac{1}{\sqrt[4]{x}} \, dx$

87. $\displaystyle\int_2^\infty \dfrac{1}{x\sqrt{x^2 - 4}} \, dx$

88. $\displaystyle\int_0^\infty \dfrac{2}{\sqrt{x}(x + 4)} \, dx$

89. Present Value   The board of directors of a corporation is calculating the price to pay for a business that is forecast to yield a continuous flow of profit of $500,000 per year. The money will earn a nominal rate of 5% per year compounded continuously. What is the present value of the business

(a) for 20 years?

(b) forever (in perpetuity)?

(*Note:* The present value for t_0 years is $\int_0^{t_0} 500{,}000 e^{-0.05t} \, dt$.)

90. Volume   Find the volume of the solid generated by revolving the region bounded by the graphs of $y = xe^{-x}$, $y = 0$, and $x = 0$ about the x-axis.

91. Probability   The average lengths (from beak to tail) of different species of warblers in the eastern United States are approximately normally distributed with a mean of 12.9 centimeters and a standard deviation of 0.95 centimeter (see figure). The probability that a randomly selected warbler has a length between a and b centimeters is

$$P(a \le x \le b) = \dfrac{1}{0.95 \sqrt{2\pi}} \int_a^b e^{-(x - 12.9)^2/1.805} \, dx.$$

Use a graphing utility to approximate the probability that a randomly selected warbler has a length of (a) 13 centimeters or greater and (b) 15 centimeters or greater.   (*Source: Peterson's Field Guide: Eastern Birds*)

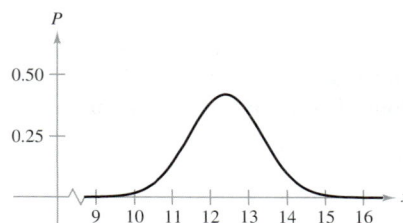

P.S. Problem Solving

See **CalcChat.com** for tutorial help and worked-out solutions to odd-numbered exercises.

1. Wallis's Formulas

(a) Evaluate the integrals

$$\int_{-1}^{1} (1 - x^2)\, dx \quad \text{and} \quad \int_{-1}^{1} (1 - x^2)^2\, dx.$$

(b) Use Wallis's Formulas to prove that

$$\int_{-1}^{1} (1 - x^2)^n\, dx = \frac{2^{2n+1}(n!)^2}{(2n+1)!}$$

for all positive integers n.

2. Proof

(a) Evaluate the integrals

$$\int_{0}^{1} \ln x\, dx \quad \text{and} \quad \int_{0}^{1} (\ln x)^2\, dx.$$

(b) Prove that

$$\int_{0}^{1} (\ln x)^n\, dx = (-1)^n\, n!$$

for all positive integers n.

3. Finding a Value Find the value of the positive constant c such that

$$\lim_{x \to \infty} \left(\frac{x + c}{x - c} \right)^x = 9.$$

4. Finding a Value Find the value of the positive constant c such that

$$\lim_{x \to \infty} \left(\frac{x - c}{x + c} \right)^x = \frac{1}{4}.$$

5. Length The line $x = 1$ is tangent to the unit circle at A. The length of segment QA equals the length of the circular arc $\overset{\frown}{PA}$ (see figure). Show that the length of segment OR approaches 2 as P approaches A.

Figure for 5

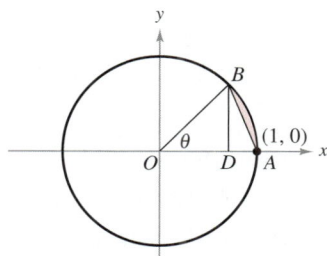

Figure for 6

6. Finding a Limit The segment BD is the height of $\triangle OAB$. Let R be the ratio of the area of $\triangle DAB$ to that of the shaded region formed by deleting $\triangle OAB$ from the circular sector subtended by angle θ (see figure). Find $\lim_{\theta \to 0^+} R$.

7. Area Consider the problem of finding the area of the region bounded by the x-axis, the line $x = 4$, and the curve

$$y = \frac{x^2}{(x^2 + 9)^{3/2}}.$$

(a) Use a graphing utility to graph the region and approximate its area.

(b) Use an appropriate trigonometric substitution to find the exact area.

(c) Use the substitution $x = 3 \sinh u$ to find the exact area and verify that you obtain the same answer as in part (b).

8. Area Use the substitution $u = \tan(x/2)$ to find the area of the shaded region under the graph of $y = \dfrac{1}{2 + \cos x}$ for $0 \le x \le \pi/2$ (see figure).

Figure for 8

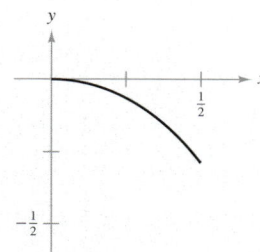

Figure for 9

9. Arc Length Find the arc length of the graph of the function $y = \ln(1 - x^2)$ on the interval $0 \le x \le \frac{1}{2}$ (see figure).

10. Centroid Find the centroid of the region above the x-axis and bounded above by the curve $y = e^{-c^2 x^2}$, where c is a positive constant (see figure).

$$\left(\textit{Hint:} \text{ Show that } \int_{0}^{\infty} e^{-c^2 x^2}\, dx = \frac{1}{c} \int_{0}^{\infty} e^{-x^2}\, dx. \right)$$

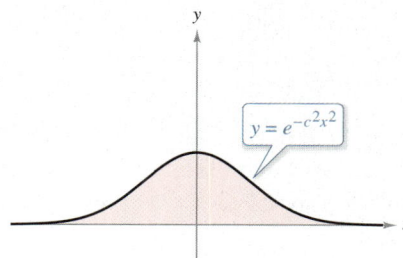

$y = e^{-c^2 x^2}$

11. Finding Limits Use a graphing utility to estimate each limit. Then calculate each limit using L'Hôpital's Rule. What can you conclude about the form $0 \cdot \infty$?

(a) $\displaystyle \lim_{x \to 0^+} \left(\cot x + \frac{1}{x} \right)$ (b) $\displaystyle \lim_{x \to 0^+} \left(\cot x - \frac{1}{x} \right)$

(c) $\displaystyle \lim_{x \to 0^+} \left[\left(\cot x + \frac{1}{x} \right) \left(\cot x - \frac{1}{x} \right) \right]$

12. Inverse Function and Area

(a) Let $y = f^{-1}(x)$ be the inverse function of f. Use integration by parts to derive the formula

$$\int f^{-1}(x)\, dx = xf^{-1}(x) - \int f(y)\, dy.$$

(b) Use the formula in part (a) to find the integral

$$\int \arcsin x\, dx.$$

(c) Use the formula in part (a) to find the area under the graph of $y = \ln x$, $1 \le x \le e$ (see figure).

 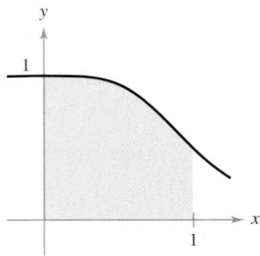

Figure for 12   Figure for 13

13. Area Factor the polynomial $p(x) = x^4 + 1$ and then find the area under the graph of

$$y = \frac{1}{x^4 + 1}, \quad 0 \le x \le 1 \quad \text{(see figure)}.$$

14. Partial Fraction Decomposition Suppose the denominator of a rational function can be factored into distinct linear factors

$$D(x) = (x - c_1)(x - c_2) \cdots (x - c_n)$$

for a positive integer n and distinct real numbers $c_1, c_2, \ldots, c_n$. If N is a polynomial of degree less than n, show that

$$\frac{N(x)}{D(x)} = \frac{P_1}{x - c_1} + \frac{P_2}{x - c_2} + \cdots + \frac{P_n}{x - c_n}$$

where $P_k = N(c_k)/D'(c_k)$ for $k = 1, 2, \ldots, n$. Note that this is the partial fraction decomposition of $N(x)/D(x)$.

15. Partial Fraction Decomposition Use the result of Exercise 14 to find the partial fraction decomposition of

$$\frac{x^3 - 3x^2 + 1}{x^4 - 13x^2 + 12x}.$$

16. Evaluating an Integral

(a) Use the substitution $u = \dfrac{\pi}{2} - x$ to evaluate the integral

$$\int_0^{\pi/2} \frac{\sin x}{\cos x + \sin x}\, dx.$$

(b) Let n be a positive integer. Evaluate the integral

$$\int_0^{\pi/2} \frac{\sin^n x}{\cos^n x + \sin^n x}\, dx.$$

17. Elementary Functions Some elementary functions, such as $f(x) = \sin(x^2)$, do not have antiderivatives that are elementary functions. Joseph Liouville proved that

$$\int \frac{e^x}{x}\, dx$$

does not have an elementary antiderivative. Use this fact to prove that

$$\int \frac{1}{\ln x}\, dx$$

is not elementary.

18. Rocket The velocity v (in feet per second) of a rocket whose initial mass (including fuel) is m is given by

$$v = gt + u \ln \frac{m}{m - rt}, \quad t < \frac{m}{r}$$

where u is the expulsion speed of the fuel, r is the rate at which the fuel is consumed, and $g = -32$ feet per second per second is the acceleration due to gravity. Find the position equation for a rocket for which $m = 50{,}000$ pounds, $u = 12{,}000$ feet per second, and $r = 400$ pounds per second. What is the height of the rocket when $t = 100$ seconds? (Assume that the rocket was fired from ground level and is moving straight upward.)

19. Proof Suppose that $f(a) = f(b) = g(a) = g(b) = 0$ and the second derivatives of f and g are continuous on the closed interval $[a, b]$. Prove that

$$\int_a^b f(x)g''(x)\, dx = \int_a^b f''(x)g(x)\, dx.$$

20. Proof Suppose that $f(a) = f(b) = 0$ and the second derivatives of f exist on the closed interval $[a, b]$. Prove that

$$\int_a^b (x - a)(x - b)f''(x)\, dx = 2\int_a^b f(x)\, dx.$$

21. Approximating an Integral Using the inequality

$$\frac{1}{x^5} + \frac{1}{x^{10}} + \frac{1}{x^{15}} < \frac{1}{x^5 - 1} < \frac{1}{x^5} + \frac{1}{x^{10}} + \frac{2}{x^{15}}$$

for $x \ge 2$, approximate $\displaystyle\int_2^\infty \frac{1}{x^5 - 1}\, dx.$

22. Volume Consider the shaded region between the graph of $y = \sin x$, where $0 \le x \le \pi$, and the line $y = c$, where $0 \le c \le 1$ (see figure). A solid is formed by revolving the region about the line $y = c$.

(a) For what value of c does the solid have minimum volume?

(b) For what value of c does the solid have maximum volume?

9 Infinite Series

Solera Method *(Section Project, p. 618)*

Projectile Motion
(Exercise 84, p. 675)

Sphereflake *(Exercise 86, p. 603)*

Multiplier Effect
(Exercise 73, p. 602)

Compound Interest *(Exercise 67, p. 593)*

9.1 Sequences

■ List the terms of a sequence.
■ Determine whether a sequence converges or diverges.
■ Write a formula for the *n*th term of a sequence.
■ Use properties of monotonic sequences and bounded sequences.

Sequences

In mathematics, the word "sequence" is used in much the same way as it is in ordinary English. Saying that a collection of objects or events is *in sequence* usually means that the collection is ordered in such a way that it has an identified first member, second member, third member, and so on.

Mathematically, a **sequence** is defined as a function whose domain is the set of positive integers. Although a sequence is a function, it is common to represent sequences by subscript notation rather than by the standard function notation. For instance, in the sequence

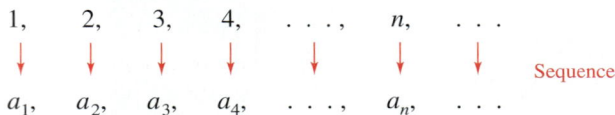

$$1, \quad 2, \quad 3, \quad 4, \quad \ldots, \quad n, \quad \ldots$$

$$\downarrow \quad \downarrow \quad \downarrow \quad \downarrow \quad \quad \downarrow \quad \quad \downarrow \qquad \text{Sequence}$$

$$a_1, \quad a_2, \quad a_3, \quad a_4, \quad \ldots, \quad a_n, \quad \ldots$$

1 is mapped onto a_1, 2 is mapped onto a_2, and so on. The numbers

$$a_1, a_2, a_3, \ldots, a_n, \ldots$$

are the **terms** of the sequence. The number a_n is the **nth term** of the sequence, and the entire sequence is denoted by $\{a_n\}$. Occasionally, it is convenient to begin a sequence with a_0, so that the terms of the sequence become $a_0, a_1, a_2, a_3, \ldots, a_n, \ldots$ and the domain is the set of nonnegative integers.

EXAMPLE 1   Listing the Terms of a Sequence

a. The terms of the sequence $\{a_n\} = \{3 + (-1)^n\}$ are

$$3 + (-1)^1, \ 3 + (-1)^2, \ 3 + (-1)^3, \ 3 + (-1)^4, \ldots$$
$$\quad 2, \qquad\qquad 4, \qquad\qquad 2, \qquad\qquad 4, \qquad \ldots.$$

b. The terms of the sequence $\{b_n\} = \left\{ \dfrac{n}{1 - 2n} \right\}$ are

$$\frac{1}{1 - 2 \cdot 1}, \ \frac{2}{1 - 2 \cdot 2}, \ \frac{3}{1 - 2 \cdot 3}, \ \frac{4}{1 - 2 \cdot 4}, \ \cdots$$

$$-1, \qquad -\frac{2}{3}, \qquad -\frac{3}{5}, \qquad -\frac{4}{7}, \quad \cdots.$$

c. The terms of the sequence $\{c_n\} = \left\{ \dfrac{n^2}{2^n - 1} \right\}$ are

$$\frac{1^2}{2^1 - 1}, \ \frac{2^2}{2^2 - 1}, \ \frac{3^2}{2^3 - 1}, \ \frac{4^2}{2^4 - 1}, \ \cdots$$

$$\frac{1}{1}, \qquad \frac{4}{3}, \qquad \frac{9}{7}, \qquad \frac{16}{15}, \quad \cdots.$$

d. The terms of the **recursively defined** sequence $\{d_n\}$, where $d_1 = 25$ and $d_{n+1} = d_n - 5$, are

$$25, \quad 25 - 5 = 20, \quad 20 - 5 = 15, \quad 15 - 5 = 10, \ldots.$$

Exploration

Finding Patterns Describe a pattern for each of the sequences listed below. Then use your description to write a formula for the *n*th term of each sequence. As *n* increases, do the terms appear to be approaching a limit? Explain your reasoning.

a. $1, \frac{1}{2}, \frac{1}{4}, \frac{1}{8}, \frac{1}{16}, \cdots$

b. $1, \frac{1}{2}, \frac{1}{6}, \frac{1}{24}, \frac{1}{120}, \cdots$

c. $10, \frac{10}{3}, \frac{10}{6}, \frac{10}{10}, \frac{10}{15}, \cdots$

d. $\frac{1}{4}, \frac{4}{9}, \frac{9}{16}, \frac{16}{25}, \frac{25}{36}, \cdots$

e. $\frac{3}{7}, \frac{5}{10}, \frac{7}{13}, \frac{9}{16}, \frac{11}{19}, \cdots$

• • REMARK Some sequences are defined recursively. To define a sequence recursively, you need to be given one or more of the first few terms. All other terms of the sequence are then defined using previous terms, as shown in Example 1(d).

Limit of a Sequence

The primary focus of this chapter concerns sequences whose terms approach limiting values. Such sequences are said to **converge**. For instance, the sequence $\{1/2^n\}$

$$\frac{1}{2}, \frac{1}{4}, \frac{1}{8}, \frac{1}{16}, \frac{1}{32}, \cdots$$

converges to 0, as indicated in the next definition.

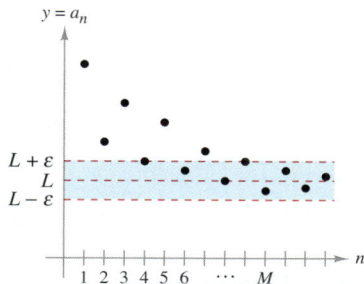

$y = a_n$

$L + \varepsilon$
L
$L - \varepsilon$

1 2 3 4 5 6 ⋯ M

For $n > M$, the terms of the sequence all lie within ε units of L.
Figure 9.1

Definition of the Limit of a Sequence

Let L be a real number. The **limit** of a sequence $\{a_n\}$ is L, written as

$$\lim_{n \to \infty} a_n = L$$

if for each $\varepsilon > 0$, there exists $M > 0$ such that $|a_n - L| < \varepsilon$ whenever $n > M$. If the limit L of a sequence exists, then the sequence **converges** to L. If the limit of a sequence does not exist, then the sequence **diverges.**

Graphically, this definition says that eventually (for $n > M$ and $\varepsilon > 0$), the terms of a sequence that converges to L will lie within the band between the lines $y = L + \varepsilon$ and $y = L - \varepsilon$, as shown in Figure 9.1.

If a sequence $\{a_n\}$ agrees with a function f at every positive integer, and if $f(x)$ approaches a limit L as $x \to \infty$, then the sequence must converge to the same limit L.

REMARK The converse of Theorem 9.1 is not true (see Exercise 84).

THEOREM 9.1   Limit of a Sequence

Let L be a real number. Let f be a function of a real variable such that

$$\lim_{x \to \infty} f(x) = L.$$

If $\{a_n\}$ is a sequence such that $f(n) = a_n$ for every positive integer n, then

$$\lim_{n \to \infty} a_n = L.$$

EXAMPLE 2   Finding the Limit of a Sequence

Find the limit of the sequence whose nth term is $a_n = \left(1 + \frac{1}{n}\right)^n$.

Solution In Theorem 5.15, you learned that

$$\lim_{x \to \infty} \left(1 + \frac{1}{x}\right)^x = e.$$

So, you can apply Theorem 9.1 to conclude that

$$\lim_{n \to \infty} a_n = \lim_{n \to \infty} \left(1 + \frac{1}{n}\right)^n = e.$$

There are different ways in which a sequence can fail to have a limit. One way is that the terms of the sequence increase without bound or decrease without bound. These cases are written symbolically, as shown below.

Terms increase without bound: $\lim_{n \to \infty} a_n = \infty$

Terms decrease without bound: $\lim_{n \to \infty} a_n = -\infty$

The properties of limits of sequences listed in the next theorem parallel those given for limits of functions of a real variable in Section 1.3.

THEOREM 9.2 **Properties of Limits of Sequences**

Let $\lim\limits_{n\to\infty} a_n = L$ and $\lim\limits_{n\to\infty} b_n = K$.

1. $\lim\limits_{n\to\infty} (a_n \pm b_n) = L \pm K$

2. $\lim\limits_{n\to\infty} ca_n = cL$, c is any real number.

3. $\lim\limits_{n\to\infty} (a_n b_n) = LK$

4. $\lim\limits_{n\to\infty} \dfrac{a_n}{b_n} = \dfrac{L}{K}$, $b_n \neq 0$ and $K \neq 0$

EXAMPLE 3 **Determining Convergence or Divergence**

$\vdots$ ▷ *See LarsonCalculus.com for an interactive version of this type of example.*

a. Because the sequence $\{a_n\} = \{3 + (-1)^n\}$ has terms

$$2, 4, 2, 4, \ldots$$ See Example 1(a), page 584.

that alternate between 2 and 4, the limit

$$\lim_{n\to\infty} a_n$$

does not exist. So, the sequence diverges.

b. For $\{b_n\} = \left\{\dfrac{n}{1-2n}\right\}$, divide the numerator and denominator by n to obtain

$$\lim_{n\to\infty} \frac{n}{1-2n} = \lim_{n\to\infty} \frac{1}{(1/n)-2} = -\frac{1}{2}$$ See Example 1(b), page 584.

which implies that the sequence converges to $-\frac{1}{2}$.

EXAMPLE 4 **Using L'Hôpital's Rule to Determine Convergence**

Show that the sequence whose nth term is $a_n = \dfrac{n^2}{2^n - 1}$ converges.

Solution Consider the function of a real variable

$$f(x) = \frac{x^2}{2^x - 1}.$$

Applying L'Hôpital's Rule twice produces

$$\lim_{x\to\infty} \frac{x^2}{2^x - 1} = \lim_{x\to\infty} \frac{2x}{(\ln 2)2^x} = \lim_{x\to\infty} \frac{2}{(\ln 2)^2 2^x} = 0.$$

Because $f(n) = a_n$ for every positive integer, you can apply Theorem 9.1 to conclude that

$$\lim_{n\to\infty} \frac{n^2}{2^n - 1} = 0.$$ See Example 1(c), page 584.

So, the sequence converges to 0.

▷ **TECHNOLOGY** Use a graphing utility to graph the function in Example 4. Notice that as x approaches infinity, the value of the function gets closer and closer to 0. If you have access to a graphing utility that can generate terms of a sequence, try using it to calculate the first 20 terms of the sequence in Example 4. Then view the terms to observe numerically that the sequence converges to 0.

The symbol $n!$ (read "n factorial") is used to simplify some of the formulas developed in this chapter. Let n be a positive integer; then **n factorial** is defined as

$$n! = 1 \cdot 2 \cdot 3 \cdot 4 \cdots (n - 1) \cdot n.$$

As a special case, **zero factorial** is defined as $0! = 1$. From this definition, you can see that $1! = 1, 2! = 1 \cdot 2 = 2, 3! = 1 \cdot 2 \cdot 3 = 6$, and so on. Factorials follow the same conventions for order of operations as exponents. That is, just as $2x^3$ and $(2x)^3$ imply different orders of operations, $2n!$ and $(2n)!$ imply the orders

$$2n! = 2(n!) = 2(1 \cdot 2 \cdot 3 \cdot 4 \cdots n)$$

and

$$(2n)! = 1 \cdot 2 \cdot 3 \cdot 4 \cdots n \cdot (n + 1) \cdots 2n$$

respectively.

Another useful limit theorem that can be rewritten for sequences is the Squeeze Theorem from Section 1.3.

THEOREM 9.3   Squeeze Theorem for Sequences

If $\lim\limits_{n \to \infty} a_n = L = \lim\limits_{n \to \infty} b_n$ and there exists an integer N such that $a_n \leq c_n \leq b_n$ for all $n > N$, then $\lim\limits_{n \to \infty} c_n = L$.

EXAMPLE 5   Using the Squeeze Theorem

Show that the sequence $\{c_n\} = \left\{(-1)^n \dfrac{1}{n!}\right\}$ converges, and find its limit.

Solution   To apply the Squeeze Theorem, you must find two convergent sequences that can be related to $\{c_n\}$. Two possibilities are $a_n = -1/2^n$ and $b_n = 1/2^n$, both of which converge to 0. By comparing the term $n!$ with 2^n, you can see that

$$n! = 1 \cdot 2 \cdot 3 \cdot 4 \cdot 5 \cdot 6 \cdots n = 24 \cdot \underbrace{5 \cdot 6 \cdots n}_{n - 4 \text{ factors}} \qquad (n \geq 4)$$

and

$$2^n = 2 \cdot 2 \cdot 2 \cdot 2 \cdot 2 \cdot 2 \cdots 2 = 16 \cdot \underbrace{2 \cdot 2 \cdots 2}_{n - 4 \text{ factors}}. \qquad (n \geq 4)$$

This implies that for $n \geq 4$, $2^n < n!$, and you have

$$\frac{-1}{2^n} \leq (-1)^n \frac{1}{n!} \leq \frac{1}{2^n}, \quad n \geq 4$$

as shown in Figure 9.2. So, by the Squeeze Theorem, it follows that

$$\lim_{n \to \infty} (-1)^n \frac{1}{n!} = 0.$$

For $n \geq 4$, $(-1)^n/n!$ is squeezed between $-1/2^n$ and $1/2^n$.
Figure 9.2

Example 5 suggests something about the rate at which $n!$ increases as $n \to \infty$. As Figure 9.2 suggests, both $1/2^n$ and $1/n!$ approach 0 as $n \to \infty$. Yet $1/n!$ approaches 0 so much faster than $1/2^n$ does that

$$\lim_{n \to \infty} \frac{1/n!}{1/2^n} = \lim_{n \to \infty} \frac{2^n}{n!} = 0.$$

In fact, it can be shown that for any fixed number k, $\lim\limits_{n \to \infty} (k^n/n!) = 0$. This means that *the factorial function grows faster than any exponential function.*

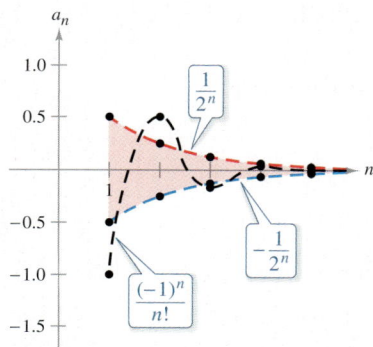

In Example 5, the sequence $\{c_n\}$ has both positive and negative terms. For this sequence, it happens that the sequence of absolute values, $\{|c_n|\}$, also converges to 0. You can show this by the Squeeze Theorem using the inequality

$$0 \le \frac{1}{n!} \le \frac{1}{2^n}, \quad n \ge 4.$$

In such cases, it is often convenient to consider the sequence of absolute values—and then apply Theorem 9.4, which states that if the absolute value sequence converges to 0, then the original signed sequence also converges to 0.

THEOREM 9.4   Absolute Value Theorem

For the sequence $\{a_n\}$, if

$$\lim_{n \to \infty} |a_n| = 0 \quad \text{then} \quad \lim_{n \to \infty} a_n = 0.$$

Proof   Consider the two sequences $\{|a_n|\}$ and $\{-|a_n|\}$. Because both of these sequences converge to 0 and

$$-|a_n| \le a_n \le |a_n|$$

you can use the Squeeze Theorem to conclude that $\{a_n\}$ converges to 0.

See LarsonCalculus.com for Bruce Edwards's video of this proof.

Pattern Recognition for Sequences

Sometimes the terms of a sequence are generated by some rule that does not explicitly identify the nth term of the sequence. In such cases, you may be required to discover a *pattern* in the sequence and to describe the nth term. Once the nth term has been specified, you can investigate the convergence or divergence of the sequence.

EXAMPLE 6   **Finding the nth Term of a Sequence**

Find a sequence $\{a_n\}$ whose first five terms are

$$\frac{2}{1}, \frac{4}{3}, \frac{8}{5}, \frac{16}{7}, \frac{32}{9}, \dots$$

and then determine whether the sequence you have chosen converges or diverges.

Solution   First, note that the numerators are successive powers of 2, and the denominators form the sequence of positive odd integers. By comparing a_n with n, you have the following pattern.

$$\frac{2^1}{1}, \frac{2^2}{3}, \frac{2^3}{5}, \frac{2^4}{7}, \frac{2^5}{9}, \dots, \frac{2^n}{2n-1}, \dots$$

Consider the function of a real variable $f(x) = 2^x/(2x - 1)$. Applying L'Hôpital's Rule produces

$$\lim_{x \to \infty} \frac{2^x}{2x - 1} = \lim_{x \to \infty} \frac{2^x(\ln 2)}{2} = \infty.$$

Next, apply Theorem 9.1 to conclude that

$$\lim_{n \to \infty} \frac{2^n}{2n - 1} = \infty.$$

So, the sequence diverges.

Without a specific rule for generating the terms of a sequence or some knowledge of the context in which the terms of the sequence are obtained, it is not possible to determine the convergence or divergence of the sequence merely from its first several terms. For instance, although the first three terms of the following four sequences are identical, the first two sequences converge to 0, the third sequence converges to $\frac{1}{9}$, and the fourth sequence diverges.

$$\{a_n\}: \frac{1}{2}, \frac{1}{4}, \frac{1}{8}, \frac{1}{16}, \ldots, \frac{1}{2^n}, \ldots$$

$$\{b_n\}: \frac{1}{2}, \frac{1}{4}, \frac{1}{8}, \frac{1}{15}, \ldots, \frac{6}{(n+1)(n^2-n+6)}, \ldots$$

$$\{c_n\}: \frac{1}{2}, \frac{1}{4}, \frac{1}{8}, \frac{7}{62}, \ldots, \frac{n^2-3n+3}{9n^2-25n+18}, \ldots$$

$$\{d_n\}: \frac{1}{2}, \frac{1}{4}, \frac{1}{8}, 0, \ldots, \frac{-n(n+1)(n-4)}{6(n^2+3n-2)}, \ldots$$

The process of determining an nth term from the pattern observed in the first several terms of a sequence is an example of *inductive reasoning*.

EXAMPLE 7 **Finding the nth Term of a Sequence**

Determine the nth term for a sequence whose first five terms are

$$-\frac{2}{1}, \frac{8}{2}, -\frac{26}{6}, \frac{80}{24}, -\frac{242}{120}, \ldots$$

and then decide whether the sequence converges or diverges.

Solution Note that the numerators are 1 less than 3^n.

$$3^1 - 1 = 2 \quad 3^2 - 1 = 8 \quad 3^3 - 1 = 26 \quad 3^4 - 1 = 80 \quad 3^5 - 1 = 242$$

So, you can reason that the numerators are given by the rule

$$3^n - 1.$$

Factoring the denominators produces

$$1 = 1$$
$$2 = 1 \cdot 2$$
$$6 = 1 \cdot 2 \cdot 3$$
$$24 = 1 \cdot 2 \cdot 3 \cdot 4$$

and

$$120 = 1 \cdot 2 \cdot 3 \cdot 4 \cdot 5.$$

This suggests that the denominators are represented by $n!$. Finally, because the signs alternate, you can write the nth term as

$$a_n = (-1)^n \left(\frac{3^n - 1}{n!} \right).$$

From the discussion about the growth of $n!$, it follows that

$$\lim_{n \to \infty} |a_n| = \lim_{n \to \infty} \frac{3^n - 1}{n!} = 0.$$

Applying Theorem 9.4, you can conclude that

$$\lim_{n \to \infty} a_n = 0.$$

So, the sequence $\{a_n\}$ converges to 0.

Monotonic Sequences and Bounded Sequences

So far, you have determined the convergence of a sequence by finding its limit. Even when you cannot determine the limit of a particular sequence, it still may be useful to know whether the sequence converges. Theorem 9.5 (on the next page) provides a test for convergence of sequences without determining the limit. First, some preliminary definitions are given.

> **Definition of Monotonic Sequence**
>
> A sequence $\{a_n\}$ is **monotonic** when its terms are nondecreasing
>
> $$a_1 \leq a_2 \leq a_3 \leq \cdots \leq a_n \leq \cdots$$
>
> or when its terms are nonincreasing
>
> $$a_1 \geq a_2 \geq a_3 \geq \cdots \geq a_n \geq \cdots .$$

(a) Not monotonic

(b) Monotonic

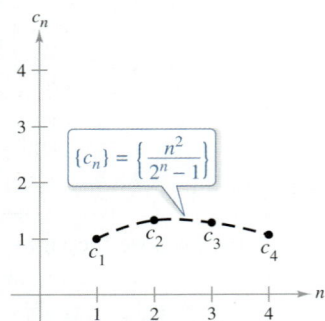

(c) Not monotonic

Figure 9.3

EXAMPLE 8 **Determining Whether a Sequence Is Monotonic**

Determine whether each sequence having the given nth term is monotonic.

a. $a_n = 3 + (-1)^n$

b. $b_n = \dfrac{2n}{1+n}$

c. $c_n = \dfrac{n^2}{2^n - 1}$

Solution

a. This sequence alternates between 2 and 4. So, it is not monotonic.

b. This sequence is monotonic because each successive term is greater than its predecessor. To see this, compare the terms b_n and b_{n+1}. [Note that, because n is positive, you can multiply each side of the inequality by $(1 + n)$ and $(2 + n)$ without reversing the inequality sign.]

$$b_n = \frac{2n}{1+n} \overset{?}{<} \frac{2(n+1)}{1+(n+1)} = b_{n+1}$$

$$2n(2+n) \overset{?}{<} (1+n)(2n+2)$$

$$4n + 2n^2 \overset{?}{<} 2 + 4n + 2n^2$$

$$0 < 2$$

Starting with the final inequality, which is valid, you can reverse the steps to conclude that the original inequality is also valid.

c. This sequence is not monotonic, because the second term is greater than the first term, and greater than the third. (Note that when you drop the first term, the remaining sequence $c_2, c_3, c_4, \ldots$ is monotonic.)

Figure 9.3 graphically illustrates these three sequences. ■

In Example 8(b), another way to see that the sequence is monotonic is to argue that the derivative of the corresponding differentiable function

$$f(x) = \frac{2x}{1+x}$$

is positive for all x. This implies that f is increasing, which in turn implies that $\{b_n\}$ is increasing.

Definition of Bounded Sequence

1. A sequence $\{a_n\}$ is **bounded above** when there is a real number M such that $a_n \leq M$ for all n. The number M is called an **upper bound** of the sequence.
2. A sequence $\{a_n\}$ is **bounded below** when there is a real number N such that $N \leq a_n$ for all n. The number N is called a **lower bound** of the sequence.
3. A sequence $\{a_n\}$ is **bounded** when it is bounded above and bounded below.

Note that all three sequences in Example 3 (and shown in Figure 9.3) are bounded. To see this, note that

$$2 \leq a_n \leq 4, \quad 1 \leq b_n \leq 2, \quad \text{and} \quad 0 \leq c_n \leq \frac{4}{3}.$$

One important property of the real numbers is that they are **complete.** Informally, this means that there are no holes or gaps on the real number line. (The set of rational numbers does not have the completeness property.) The completeness axiom for real numbers can be used to conclude that if a sequence has an upper bound, then it must have a **least upper bound** (an upper bound that is less than all other upper bounds for the sequence). For example, the least upper bound of the sequence $\{a_n\} = \{n/(n+1)\}$,

$$\frac{1}{2}, \frac{2}{3}, \frac{3}{4}, \frac{4}{5}, \cdots, \frac{n}{n+1}, \cdots$$

is 1. The completeness axiom is used in the proof of Theorem 9.5.

THEOREM 9.5   Bounded Monotonic Sequences

If a sequence $\{a_n\}$ is bounded and monotonic, then it converges.

Every bounded, nondecreasing sequence converges.

Figure 9.4

Proof   Assume that the sequence is nondecreasing, as shown in Figure 9.4. For the sake of simplicity, also assume that each term in the sequence is positive. Because the sequence is bounded, there must exist an upper bound M such that

$$a_1 \leq a_2 \leq a_3 \leq \cdots \leq a_n \leq \cdots \leq M.$$

From the completeness axiom, it follows that there is a least upper bound L such that

$$a_1 \leq a_2 \leq a_3 \leq \cdots \leq a_n \leq \cdots \leq L.$$

For $\varepsilon > 0$, it follows that $L - \varepsilon < L$, and therefore $L - \varepsilon$ cannot be an upper bound for the sequence. Consequently, at least one term of $\{a_n\}$ is greater than $L - \varepsilon$. That is, $L - \varepsilon < a_N$ for some positive integer N. Because the terms of $\{a_n\}$ are nondecreasing, it follows that $a_N \leq a_n$ for $n > N$. You now know that $L - \varepsilon < a_N \leq a_n \leq L < L + \varepsilon$, for every $n > N$. It follows that $|a_n - L| < \varepsilon$ for $n > N$, which by definition means that $\{a_n\}$ converges to L. The proof for a nonincreasing sequence is similar (see Exercise 91).

See LarsonCalculus.com for Bruce Edwards's video of this proof.

EXAMPLE 9   Bounded and Monotonic Sequences

a. The sequence $\{a_n\} = \{1/n\}$ is both bounded and monotonic, and so, by Theorem 9.5, it must converge.

b. The divergent sequence $\{b_n\} = \{n^2/(n+1)\}$ is monotonic, but not bounded. (It is bounded below.)

c. The divergent sequence $\{c_n\} = \{(-1)^n\}$ is bounded, but not monotonic.

9.1 Exercises

See **CalcChat.com** for tutorial help and worked-out solutions to odd-numbered exercises.

Listing the Terms of a Sequence In Exercises 1–6, write the first five terms of the sequence.

1. $a_n = 3^n$

2. $a_n = \left(-\dfrac{2}{5}\right)^n$

3. $a_n = \sin \dfrac{n\pi}{2}$

4. $a_n = \dfrac{3n}{n+4}$

5. $a_n = (-1)^{n+1}\left(\dfrac{2}{n}\right)$

6. $a_n = 2 + \dfrac{2}{n} - \dfrac{1}{n^2}$

Listing the Terms of a Sequence In Exercises 7 and 8, write the first five terms of the recursively defined sequence.

7. $a_1 = 3,\ a_{k+1} = 2(a_k - 1)$

8. $a_1 = 6,\ a_{k+1} = \frac{1}{3}a_k^2$

Matching In Exercises 9–12, match the sequence with its graph. [The graphs are labeled (a), (b), (c), and (d).]

(a)

(b)

(c)

(d)

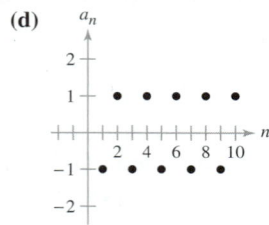

9. $a_n = \dfrac{10}{n+1}$

10. $a_n = \dfrac{10n}{n+1}$

11. $a_n = (-1)^n$

12. $a_n = \dfrac{(-1)^n}{n}$

Writing Terms In Exercises 13–16, write the next two apparent terms of the sequence. Describe the pattern you used to find these terms.

13. 2, 5, 8, 11, . . .

14. 8, 13, 18, 23, 28, . . .

15. 5, 10, 20, 40, . . .

16. $6, -2, \dfrac{2}{3}, -\dfrac{2}{9}, \ldots$

Simplifying Factorials In Exercises 17–20, simplify the ratio of factorials.

17. $\dfrac{(n+1)!}{n!}$

18. $\dfrac{n!}{(n+2)!}$

19. $\dfrac{(2n-1)!}{(2n+1)!}$

20. $\dfrac{(2n+2)!}{(2n)!}$

Finding the Limit of a Sequence In Exercises 21–24, find the limit (if possible) of the sequence.

21. $a_n = \dfrac{5n^2}{n^2 + 2}$

22. $a_n = 6 + \dfrac{2}{n^2}$

23. $a_n = \dfrac{2n}{\sqrt{n^2 + 1}}$

24. $a_n = \cos \dfrac{2}{n}$

Finding the Limit of a Sequence In Exercises 25–28, use a graphing utility to graph the first 10 terms of the sequence. Use the graph to make an inference about the convergence or divergence of the sequence. Verify your inference analytically and, if the sequence converges, find its limit.

25. $a_n = \dfrac{4n+1}{n}$

26. $a_n = \dfrac{1}{n^{3/2}}$

27. $a_n = \sin \dfrac{n\pi}{2}$

28. $a_n = 2 - \dfrac{1}{4^n}$

Determining Convergence or Divergence In Exercises 29–44, determine the convergence or divergence of the sequence with the given nth term. If the sequence converges, find its limit.

29. $a_n = \dfrac{5}{n+2}$

30. $a_n = 8 + \dfrac{5}{n}$

31. $a_n = (-1)^n\left(\dfrac{n}{n+1}\right)$

32. $a_n = \dfrac{1 + (-1)^n}{n^2}$

33. $a_n = \dfrac{10n^2 + 3n + 7}{2n^2 - 6}$

34. $a_n = \dfrac{\sqrt[3]{n}}{\sqrt[3]{n} + 1}$

35. $a_n = \dfrac{\ln(n^3)}{2n}$

36. $a_n = \dfrac{5^n}{3^n}$

37. $a_n = \dfrac{(n+1)!}{n!}$

38. $a_n = \dfrac{(n-2)!}{n!}$

39. $a_n = \dfrac{n^p}{e^n},\ p > 0$

40. $a_n = n \sin \dfrac{1}{n}$

41. $a_n = 2^{1/n}$

42. $a_n = -3^{-n}$

43. $a_n = \dfrac{\sin n}{n}$

44. $a_n = \dfrac{\cos \pi n}{n^2}$

Finding the nth Term of a Sequence In Exercises 45–52, write an expression for the nth term of the sequence. (There is more than one correct answer.)

45. 2, 8, 14, 20, . . .

46. $1, \frac{1}{2}, \frac{1}{6}, \frac{1}{24}, \frac{1}{120}, \ldots$

47. $-2, 1, 6, 13, 22, \ldots$

48. $1, -\frac{1}{4}, \frac{1}{9}, -\frac{1}{16}, \ldots$

49. $\frac{2}{3}, \frac{3}{4}, \frac{4}{5}, \frac{5}{6}, \ldots$

50. 2, 24, 720, 40,320, 3,628,800, . . .

51. $2, 1 + \frac{1}{2}, 1 + \frac{1}{3}, 1 + \frac{1}{4}, 1 + \frac{1}{5}, \ldots$

52. $\dfrac{1}{2 \cdot 3}, \dfrac{2}{3 \cdot 4}, \dfrac{3}{4 \cdot 5}, \dfrac{4}{5 \cdot 6}, \ldots$

Finding Monotonic and Bounded Sequences In Exercises 53–60, determine whether the sequence with the given nth term is monotonic and whether it is bounded. Use a graphing utility to confirm your results.

53. $a_n = 4 - \dfrac{1}{n}$

54. $a_n = \dfrac{3n}{n+2}$

55. $a_n = ne^{-n/2}$

56. $a_n = \left(-\dfrac{2}{3}\right)^n$

57. $a_n = \left(\dfrac{2}{3}\right)^n$

58. $a_n = \left(\dfrac{3}{2}\right)^n$

59. $a_n = \sin\dfrac{n\pi}{6}$

60. $a_n = \dfrac{\cos n}{n}$

Using a Theorem In Exercises 61–64, (a) use Theorem 9.5 to show that the sequence with the given nth term converges, and (b) use a graphing utility to graph the first 10 terms of the sequence and find its limit.

61. $a_n = 7 + \dfrac{1}{n}$

62. $a_n = 5 - \dfrac{2}{n}$

63. $a_n = \dfrac{1}{3}\left(1 - \dfrac{1}{3^n}\right)$

64. $a_n = 2 + \dfrac{1}{5^n}$

65. Increasing Sequence Let $\{a_n\}$ be an increasing sequence such that $2 \le a_n \le 4$. Explain why $\{a_n\}$ has a limit. What can you conclude about the limit?

66. Monotonic Sequence Let $\{a_n\}$ be a monotonic sequence such that $a_n \le 1$. Discuss the convergence of $\{a_n\}$. When $\{a_n\}$ converges, what can you conclude about its limit?

67. Compound Interest

Consider the sequence $\{A_n\}$ whose nth term is given by

$$A_n = P\left(1 + \dfrac{r}{12}\right)^n$$

where P is the principal, A_n is the account balance after n months, and r is the interest rate compounded annually.

(a) Is $\{A_n\}$ a convergent sequence? Explain.

(b) Find the first 10 terms of the sequence when $P = \$10,000$ and $r = 0.055$.

68. Compound Interest A deposit of $100 is made in an account at the beginning of each month at an annual interest rate of 3% compounded monthly. The balance in the account after n months is $A_n = 100(401)(1.0025^n - 1)$.

(a) Compute the first six terms of the sequence $\{A_n\}$.

(b) Find the balance in the account after 5 years by computing the 60th term of the sequence.

(c) Find the balance in the account after 20 years by computing the 240th term of the sequence.

WRITING ABOUT CONCEPTS

69. Sequence Is it possible for a sequence to converge to two different numbers? If so, give an example. If not, explain why not.

70. Defining Terms In your own words, define each of the following.

(a) Sequence   (b) Convergence of a sequence

(c) Monotonic sequence   (d) Bounded sequence

71. Writing a Sequence Give an example of a sequence satisfying the condition or explain why no such sequence exists. (Examples are not unique.)

(a) A monotonically increasing sequence that converges to 10

(b) A monotonically increasing bounded sequence that does not converge

(c) A sequence that converges to $\frac{3}{4}$

(d) An unbounded sequence that converges to 100

72. HOW DO YOU SEE IT? The graphs of two sequences are shown in the figures. Which graph represents the sequence with alternating signs? Explain.

73. Government Expenditures A government program that currently costs taxpayers $4.5 billion per year is cut back by 20 percent per year.

(a) Write an expression for the amount budgeted for this program after n years.

(b) Compute the budgets for the first 4 years.

(c) Determine the convergence or divergence of the sequence of reduced budgets. If the sequence converges, find its limit.

74. Inflation When the rate of inflation is $4\frac{1}{2}\%$ per year and the average price of a car is currently $25,000, the average price after n years is $P_n = \$25,000(1.045)^n$. Compute the average prices for the next 5 years.

75. Using a Sequence Compute the first six terms of the sequence $\{a_n\} = \{\sqrt[n]{n}\}$. If the sequence converges, find its limit.

76. Using a Sequence Compute the first six terms of the sequence

$$\{a_n\} = \left\{\left(1 + \dfrac{1}{n}\right)^n\right\}.$$

If the sequence converges, find its limit.

77. Proof Prove that if $\{s_n\}$ converges to L and $L > 0$, then there exists a number N such that $s_n > 0$ for $n > N$.

78. Modeling Data The amounts of the federal debt a_n (in trillions of dollars) of the United States from 2000 through 2011 are given below as ordered pairs of the form (n, a_n), where n represents the year, with $n = 0$ corresponding to 2000. *(Source: U.S. Office of Management and Budget)*

(0, 5.6), (1, 5.8), (2, 6.2), (3, 6.8), (4, 7.4), (5, 7.9), (6, 8.5), (7, 9.0), (8, 10.0), (9, 11.9), (10, 13.5), (11, 14.8)

(a) Use the regression capabilities of a graphing utility to find a model of the form

$$a_n = bn^2 + cn + d, \quad n = 0, 1, \ldots, 11$$

for the data. Use the graphing utility to plot the points and graph the model.

(b) Use the model to predict the amount of the federal debt in the year 2020.

True or False? In Exercises 79–82, determine whether the statement is true or false. If it is false, explain why or give an example that shows it is false.

79. If $\{a_n\}$ converges to 3 and $\{b_n\}$ converges to 2, then $\{a_n + b_n\}$ converges to 5.

80. If $\{a_n\}$ converges, then $\lim_{n \to \infty} (a_n - a_{n+1}) = 0$.

81. If $\{a_n\}$ converges, then $\{a_n/n\}$ converges to 0.

82. If $\{a_n\}$ diverges and $\{b_n\}$ diverges, then $\{a_n + b_n\}$ diverges.

83. Fibonacci Sequence In a study of the progeny of rabbits, Fibonacci (ca. 1170–ca. 1240) encountered the sequence now bearing his name. The sequence is defined recursively as $a_{n+2} = a_n + a_{n+1}$, where $a_1 = 1$ and $a_2 = 1$.

(a) Write the first 12 terms of the sequence.

(b) Write the first 10 terms of the sequence defined by

$$b_n = \frac{a_{n+1}}{a_n}, \quad n \geq 1.$$

(c) Using the definition in part (b), show that

$$b_n = 1 + \frac{1}{b_{n-1}}.$$

(d) The **golden ratio** ρ can be defined by $\lim_{n \to \infty} b_n = \rho$. Show that

$$\rho = 1 + \frac{1}{\rho}$$

and solve this equation for ρ.

84. Using a Theorem Show that the converse of Theorem 9.1 is not true. [*Hint:* Find a function $f(x)$ such that $f(n) = a_n$ converges, but $\lim_{x \to \infty} f(x)$ does not exist.]

85. Using a Sequence Consider the sequence

$$\sqrt{2}, \sqrt{2 + \sqrt{2}}, \sqrt{2 + \sqrt{2 + \sqrt{2}}}, \ldots.$$

(a) Compute the first five terms of this sequence.

(b) Write a recursion formula for a_n, for $n \geq 2$.

(c) Find $\lim_{n \to \infty} a_n$.

86. Using a Sequence Consider the sequence $\{a_n\}$ where $a_1 = \sqrt{k}$, $a_{n+1} = \sqrt{k + a_n}$, and $k > 0$.

(a) Show that $\{a_n\}$ is increasing and bounded.

(b) Prove that $\lim_{n \to \infty} a_n$ exists.

(c) Find $\lim_{n \to \infty} a_n$.

87. Squeeze Theorem

(a) Show that $\int_1^n \ln x \, dx < \ln(n!)$ for $n \geq 2$.

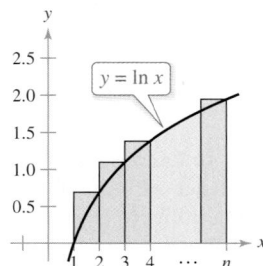

(b) Draw a graph similar to the one above that shows

$$\ln(n!) < \int_1^{n+1} \ln x \, dx.$$

(c) Use the results of parts (a) and (b) to show that

$$\frac{n^n}{e^{n-1}} < n! < \frac{(n+1)^{n+1}}{e^n}, \text{ for } n > 1.$$

(d) Use the Squeeze Theorem for Sequences and the result of part (c) to show that $\lim_{n \to \infty} \left(\sqrt[n]{n!}/n \right) = 1/e$.

(e) Test the result of part (d) for $n = 20, 50$, and 100.

88. Proof Prove, using the definition of the limit of a sequence, that

$$\lim_{n \to \infty} \frac{1}{n^3} = 0.$$

89. Proof Prove, using the definition of the limit of a sequence, that $\lim_{n \to \infty} r^n = 0$ for $-1 < r < 1$.

90. Using a Sequence Find a divergent sequence $\{a_n\}$ such that $\{a_{2n}\}$ converges.

91. Proof Prove Theorem 9.5 for a nonincreasing sequence.

PUTNAM EXAM CHALLENGE

92. Let $\{x_n\}$, $n \geq 0$, be a sequence of nonzero real numbers such that $x_n^2 - x_{n-1} x_{n+1} = 1$ for $n = 1, 2, 3, \ldots$. Prove there exists a real number a such that $x_{n+1} = ax_n - x_{n-1}$ for all $n \geq 1$.

93. Let $T_0 = 2$, $T_1 = 3$, $T_2 = 6$, and for $n \geq 3$,

$$T_n = (n + 4)T_{n-1} - 4nT_{n-2} + (4n - 8)T_{n-3}.$$

The first few terms are

$$2, 3, 6, 14, 40, 152, 784, 5168, 40{,}576$$

Find, with proof, a formula for T_n of the form $T_n = A_n + B_n$, where $\{A_n\}$ and $\{B_n\}$ are well-known sequences.

These problems were composed by the Committee on the Putnam Prize Competition.
© The Mathematical Association of America. All rights reserved..

9.2 Series and Convergence

- Understand the definition of a convergent infinite series.
- Use properties of infinite geometric series.
- Use the *n*th-Term Test for Divergence of an infinite series.

Infinite Series

One important application of infinite sequences is in representing "infinite summations." Informally, if $\{a_n\}$ is an infinite sequence, then

$$\sum_{n=1}^{\infty} a_n = a_1 + a_2 + a_3 + \cdots + a_n + \cdots \qquad \text{Infinite Series}$$

REMARK As you study this chapter, it is important to distinguish between an infinite series and a sequence. A sequence is an ordered collection of numbers

$$a_1, a_2, a_3, \ldots, a_n, \ldots$$

whereas a series is an infinite sum of terms from a sequence

$$a_1 + a_2 + a_3 + \cdots + a_n + \cdots.$$

is an **infinite series** (or simply a **series**). The numbers a_1, a_2, a_3, and so on are the **terms** of the series. For some series, it is convenient to begin the index at $n = 0$ (or some other integer). As a typesetting convention, it is common to represent an infinite series as $\Sigma \, a_n$. In such cases, the starting value for the index must be taken from the context of the statement.

To find the sum of an infinite series, consider the **sequence of partial sums** listed below.

$$S_1 = a_1$$
$$S_2 = a_1 + a_2$$
$$S_3 = a_1 + a_2 + a_3$$
$$S_4 = a_1 + a_2 + a_3 + a_4$$
$$S_5 = a_1 + a_2 + a_3 + a_4 + a_5$$
$$\vdots$$
$$S_n = a_1 + a_2 + a_3 + \cdots + a_n$$

If this sequence of partial sums converges, then the series is said to converge and has the sum indicated in the next definition.

INFINITE SERIES

The study of infinite series was considered a novelty in the fourteenth century. Logician Richard Suiseth, whose nickname was Calculator, solved this problem.

If throughout the first half of a given time interval a variation continues at a certain intensity, throughout the next quarter of the interval at double the intensity, throughout the following eighth at triple the intensity and so ad infinitum; then the average intensity for the whole interval will be the intensity of the variation during the second subinterval (or double the intensity). This is the same as saying that the sum of the infinite series

$$\frac{1}{2} + \frac{2}{4} + \frac{3}{8} + \cdots + \frac{n}{2^n} + \cdots$$

is 2.

Definitions of Convergent and Divergent Series

For the infinite series $\displaystyle\sum_{n=1}^{\infty} a_n$, the ***n*th partial sum** is

$$S_n = a_1 + a_2 + \cdots + a_n.$$

If the sequence of partial sums $\{S_n\}$ converges to S, then the series $\displaystyle\sum_{n=1}^{\infty} a_n$ **converges.** The limit S is called the **sum of the series.**

$$S = a_1 + a_2 + \cdots + a_n + \cdots \qquad S = \sum_{n=1}^{\infty} a_n$$

If $\{S_n\}$ diverges, then the series **diverges.**

As you study this chapter, you will see that there are two basic questions involving infinite series.

- Does a series converge or does it diverge?
- When a series converges, what is its sum?

These questions are not always easy to answer, especially the second one.

▷ **TECHNOLOGY** Figure 9.5 shows the first 15 partial sums of the infinite series in Example 1(a). Notice how the values appear to approach the line $y = 1$.

Figure 9.5

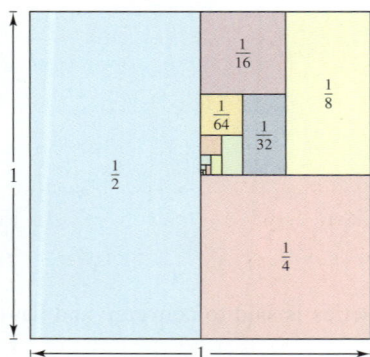

You can determine the partial sums of the series in Example 1(a) geometrically using this figure.

Figure 9.6

FOR FURTHER INFORMATION
To learn more about the partial sums of infinite series, see the article "Six Ways to Sum a Series" by Dan Kalman in *The College Mathematics Journal*. To view this article, go to *MathArticles.com*.

EXAMPLE 1 **Convergent and Divergent Series**

a. The series

$$\sum_{n=1}^{\infty} \frac{1}{2^n} = \frac{1}{2} + \frac{1}{4} + \frac{1}{8} + \frac{1}{16} + \cdots$$

has the partial sums listed below. (You can also determine the partial sums of the series geometrically, as shown in Figure 9.6.)

$$S_1 = \frac{1}{2}$$

$$S_2 = \frac{1}{2} + \frac{1}{4} = \frac{3}{4}$$

$$S_3 = \frac{1}{2} + \frac{1}{4} + \frac{1}{8} = \frac{7}{8}$$

$$\vdots$$

$$S_n = \frac{1}{2} + \frac{1}{4} + \frac{1}{8} + \cdots + \frac{1}{2^n} = \frac{2^n - 1}{2^n}$$

Because

$$\lim_{n \to \infty} \frac{2^n - 1}{2^n} = 1$$

it follows that the series converges and its sum is 1.

b. The nth partial sum of the series

$$\sum_{n=1}^{\infty} \left(\frac{1}{n} - \frac{1}{n+1} \right) = \left(1 - \frac{1}{2} \right) + \left(\frac{1}{2} - \frac{1}{3} \right) + \left(\frac{1}{3} - \frac{1}{4} \right) + \cdots$$

is

$$S_n = 1 - \frac{1}{n+1}.$$

Because the limit of S_n is 1, the series converges and its sum is 1.

c. The series

$$\sum_{n=1}^{\infty} 1 = 1 + 1 + 1 + 1 + \cdots$$

diverges because $S_n = n$ and the sequence of partial sums diverges.

The series in Example 1(b) is a **telescoping series** of the form

$$(b_1 - b_2) + (b_2 - b_3) + (b_3 - b_4) + (b_4 - b_5) + \cdots. \qquad \text{Telescoping series}$$

Note that b_2 is canceled by the second term, b_3 is canceled by the third term, and so on. Because the nth partial sum of this series is

$$S_n = b_1 - b_{n+1}$$

it follows that a telescoping series will converge if and only if b_n approaches a finite number as $n \to \infty$. Moreover, if the series converges, then its sum is

$$S = b_1 - \lim_{n \to \infty} b_{n+1}.$$

EXAMPLE 2   **Writing a Series in Telescoping Form**

Find the sum of the series $\displaystyle\sum_{n=1}^{\infty} \frac{2}{4n^2 - 1}$.

Solution

Using partial fractions, you can write

$$a_n = \frac{2}{4n^2 - 1} = \frac{2}{(2n - 1)(2n + 1)} = \frac{1}{2n - 1} - \frac{1}{2n + 1}.$$

From this telescoping form, you can see that the nth partial sum is

$$S_n = \left(\frac{1}{1} - \frac{1}{3}\right) + \left(\frac{1}{3} - \frac{1}{5}\right) + \cdots + \left(\frac{1}{2n - 1} - \frac{1}{2n + 1}\right) = 1 - \frac{1}{2n + 1}.$$

So, the series converges and its sum is 1. That is,

$$\sum_{n=1}^{\infty} \frac{2}{4n^2 - 1} = \lim_{n \to \infty} S_n = \lim_{n \to \infty} \left(1 - \frac{1}{2n + 1}\right) = 1.$$

Geometric Series

The series in Example 1(a) is a **geometric series.** In general, the series

$$\sum_{n=0}^{\infty} ar^n = a + ar + ar^2 + \cdots + ar^n + \cdots, \quad a \neq 0 \qquad \text{Geometric series}$$

is a **geometric series** with ratio r, $r \neq 0$.

> **THEOREM 9.6   Convergence of a Geometric Series**
>
> A geometric series with ratio r diverges when $|r| \geq 1$. If $0 < |r| < 1$, then the series converges to the sum
>
> $$\sum_{n=0}^{\infty} ar^n = \frac{a}{1 - r}, \quad 0 < |r| < 1.$$

Proof   It is easy to see that the series diverges when $r = \pm 1$. If $r \neq \pm 1$, then

$$S_n = a + ar + ar^2 + \cdots + ar^{n-1}.$$

Multiplication by r yields

$$rS_n = ar + ar^2 + ar^3 + \cdots + ar^n.$$

Subtracting the second equation from the first produces $S_n - rS_n = a - ar^n$. Therefore, $S_n(1 - r) = a(1 - r^n)$, and the nth partial sum is

$$S_n = \frac{a}{1 - r}(1 - r^n).$$

When $0 < |r| < 1$, it follows that $r^n \to 0$ as $n \to \infty$, and you obtain

$$\lim_{n \to \infty} S_n = \lim_{n \to \infty} \left[\frac{a}{1 - r}(1 - r^n)\right] = \frac{a}{1 - r}\left[\lim_{n \to \infty}(1 - r^n)\right] = \frac{a}{1 - r}$$

which means that the series *converges* and its sum is $a/(1 - r)$. It is left to you to show that the series diverges when $|r| > 1$.

See LarsonCalculus.com for Bruce Edwards's video of this proof.

Exploration

In "Proof Without Words," by Benjamin G. Klein and Irl C. Bivens, the authors present the diagram below. Explain why the second statement after the diagram is valid. How is this result related to Theorem 9.6?

$$\Delta PQR \sim \Delta TSP$$

$$1 + r + r^2 + r^3 + \cdots = \frac{1}{1 - r}$$

Exercise taken from "Proof Without Words" by Benjamin G. Klein and Irl C. Bivens, *Mathematics Magazine,* 61, No. 4, October 1988, p. 219, by permission of the authors.

▷ TECHNOLOGY Try using a graphing utility to compute the sum of the first 20 terms of the sequence in Example 3(a). You should obtain a sum of about 5.999994.

EXAMPLE 3   **Convergent and Divergent Geometric Series**

a. The geometric series

$$\sum_{n=0}^{\infty} \frac{3}{2^n} = \sum_{n=0}^{\infty} 3\left(\frac{1}{2}\right)^n = 3(1) + 3\left(\frac{1}{2}\right) + 3\left(\frac{1}{2}\right)^2 + \cdots$$

has a ratio of $r = \frac{1}{2}$ with $a = 3$. Because $0 < |r| < 1$, the series converges and its sum is

$$S = \frac{a}{1 - r} = \frac{3}{1 - (1/2)} = 6.$$

b. The geometric series

$$\sum_{n=0}^{\infty} \left(\frac{3}{2}\right)^n = 1 + \frac{3}{2} + \frac{9}{4} + \frac{27}{8} + \cdots$$

has a ratio of $r = \frac{3}{2}$. Because $|r| \geq 1$, the series diverges.   ■

The formula for the sum of a geometric series can be used to write a repeating decimal as the ratio of two integers, as demonstrated in the next example.

EXAMPLE 4   **A Geometric Series for a Repeating Decimal**

⋯▷ *See LarsonCalculus.com for an interactive version of this type of example.*

Use a geometric series to write $0.\overline{08}$ as the ratio of two integers.

Solution   For the repeating decimal $0.\overline{08}$, you can write

$$0.080808\ldots = \frac{8}{10^2} + \frac{8}{10^4} + \frac{8}{10^6} + \frac{8}{10^8} + \cdots$$

$$= \sum_{n=0}^{\infty} \left(\frac{8}{10^2}\right)\left(\frac{1}{10^2}\right)^n.$$

For this series, you have $a = 8/10^2$ and $r = 1/10^2$. So,

$$0.080808\ldots = \frac{a}{1 - r} = \frac{8/10^2}{1 - (1/10^2)} = \frac{8}{99}.$$

Try dividing 8 by 99 on a calculator to see that it produces $0.\overline{08}$.   ■

The convergence of a series is not affected by the removal of a finite number of terms from the beginning of the series. For instance, the geometric series

$$\sum_{n=4}^{\infty} \left(\frac{1}{2}\right)^n \quad \text{and} \quad \sum_{n=0}^{\infty} \left(\frac{1}{2}\right)^n$$

both converge. Furthermore, because the sum of the second series is

$$\frac{a}{1 - r} = \frac{1}{1 - (1/2)} = 2$$

you can conclude that the sum of the first series is

$$S = 2 - \left[\left(\frac{1}{2}\right)^0 + \left(\frac{1}{2}\right)^1 + \left(\frac{1}{2}\right)^2 + \left(\frac{1}{2}\right)^3\right]$$

$$= 2 - \frac{15}{8}$$

$$= \frac{1}{8}.$$

The properties in the next theorem are direct consequences of the corresponding properties of limits of sequences.

THEOREM 9.7   Properties of Infinite Series

Let $\Sigma\, a_n$ and $\Sigma\, b_n$ be convergent series, and let A, B, and c be real numbers. If $\Sigma\, a_n = A$ and $\Sigma\, b_n = B$, then the following series converge to the indicated sums.

1. $\displaystyle\sum_{n=1}^{\infty} ca_n = cA$

2. $\displaystyle\sum_{n=1}^{\infty} (a_n + b_n) = A + B$

3. $\displaystyle\sum_{n=1}^{\infty} (a_n - b_n) = A - B$

---

*n*th-Term Test for Divergence

The next theorem states that when a series converges, the limit of its nth term must be 0.

· · · · · · · · · · · · · · · ▷

··REMARK   Be sure you see that the converse of Theorem 9.8 is generally not true. That is, if the sequence $\{a_n\}$ converges to 0, then the series $\Sigma\, a_n$ may either converge or diverge.

THEOREM 9.8   Limit of the *n*th Term of a Convergent Series

If $\displaystyle\sum_{n=1}^{\infty} a_n$ converges, then $\displaystyle\lim_{n\to\infty} a_n = 0$.

Proof   Assume that

$$\sum_{n=1}^{\infty} a_n = \lim_{n\to\infty} S_n = L.$$

Then, because $S_n = S_{n-1} + a_n$ and

$$\lim_{n\to\infty} S_n = \lim_{n\to\infty} S_{n-1} = L$$

it follows that

$$\begin{aligned}
L &= \lim_{n\to\infty} S_n \\
&= \lim_{n\to\infty} (S_{n-1} + a_n) \\
&= \lim_{n\to\infty} S_{n-1} + \lim_{n\to\infty} a_n \\
&= L + \lim_{n\to\infty} a_n
\end{aligned}$$

which implies that $\{a_n\}$ converges to 0.

See LarsonCalculus.com for Bruce Edwards's video of this proof.

The contrapositive of Theorem 9.8 provides a useful test for *divergence*. This **nth-Term Test for Divergence** states that if the limit of the nth term of a series does *not* converge to 0, then the series must diverge.

THEOREM 9.9   *n*th-Term Test for Divergence

If $\displaystyle\lim_{n\to\infty} a_n \neq 0$, then $\displaystyle\sum_{n=1}^{\infty} a_n$ diverges.

Using the *n*th-Term Test for Divergence

a. For the series $\sum\limits_{n=0}^{\infty} 2^n$, you have

$$\lim_{n\to\infty} 2^n = \infty.$$

So, the limit of the *n*th term is not 0, and the series diverges.

b. For the series $\sum\limits_{n=1}^{\infty} \dfrac{n!}{2n! + 1}$, you have

$$\lim_{n\to\infty} \frac{n!}{2n! + 1} = \frac{1}{2}.$$

So, the limit of the *n*th term is not 0, and the series diverges.

c. For the series $\sum\limits_{n=1}^{\infty} \dfrac{1}{n}$, you have

$$\lim_{n\to\infty} \frac{1}{n} = 0.$$

Because the limit of the *n*th term is 0, the *n*th-Term Test for Divergence does *not* apply and you can draw no conclusions about convergence or divergence. (In the next section, you will see that this particular series diverges.)

REMARK The series in Example 5(c) will play an important role in this chapter.

$$\sum_{n=1}^{\infty} \frac{1}{n} =$$

$$1 + \frac{1}{2} + \frac{1}{3} + \frac{1}{4} + \cdots$$

You will see that this series diverges even though the *n*th term approaches 0 as *n* approaches ∞.

Bouncing Ball Problem

A ball is dropped from a height of 6 feet and begins bouncing, as shown in Figure 9.7. The height of each bounce is three-fourths the height of the previous bounce. Find the total vertical distance traveled by the ball.

Solution When the ball hits the ground for the first time, it has traveled a distance of $D_1 = 6$ feet. For subsequent bounces, let D_i be the distance traveled up and down. For example, D_2 and D_3 are

$$D_2 = \underbrace{6\left(\frac{3}{4}\right)}_{\text{Up}} + \underbrace{6\left(\frac{3}{4}\right)}_{\text{Down}} = 12\left(\frac{3}{4}\right)$$

and

$$D_3 = \underbrace{6\left(\frac{3}{4}\right)\left(\frac{3}{4}\right)}_{\text{Up}} + \underbrace{6\left(\frac{3}{4}\right)\left(\frac{3}{4}\right)}_{\text{Down}} = 12\left(\frac{3}{4}\right)^2.$$

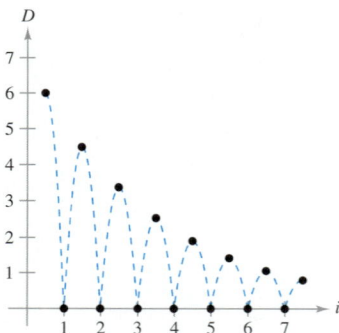

The height of each bounce is three-fourths the height of the preceding bounce.

Figure 9.7

By continuing this process, it can be determined that the total vertical distance is

$$D = 6 + 12\left(\frac{3}{4}\right) + 12\left(\frac{3}{4}\right)^2 + 12\left(\frac{3}{4}\right)^3 + \cdots$$

$$= 6 + 12 \sum_{n=0}^{\infty} \left(\frac{3}{4}\right)^{n+1}$$

$$= 6 + 12\left(\frac{3}{4}\right) \sum_{n=0}^{\infty} \left(\frac{3}{4}\right)^{n}$$

$$= 6 + 9\left[\frac{1}{1 - (3/4)}\right]$$

$$= 6 + 9(4)$$

$$= 42 \text{ feet.}$$

9.2 Exercises   See CalcChat.com for tutorial help and worked-out solutions to odd-numbered exercises.

Finding Partial Sums   In Exercises 1–6, find the sequence of partial sums $S_1, S_2, S_3, S_4,$ and S_5.

1. $1 + \frac{1}{4} + \frac{1}{9} + \frac{1}{16} + \frac{1}{25} + \cdots$

2. $\frac{1}{2 \cdot 3} + \frac{2}{3 \cdot 4} + \frac{3}{4 \cdot 5} + \frac{4}{5 \cdot 6} + \frac{5}{6 \cdot 7} + \cdots$

3. $3 - \frac{9}{2} + \frac{27}{4} - \frac{81}{8} + \frac{243}{16} - \cdots$

4. $1 + \frac{1}{2} + \frac{1}{4} + \frac{1}{6} + \frac{1}{8} + \frac{1}{10} + \cdots$

5. $\sum_{n=1}^{\infty} \frac{3}{2^{n-1}}$ 6. $\sum_{n=1}^{\infty} \frac{(-1)^{n+1}}{n!}$

Verifying Divergence   In Exercises 7–14, verify that the infinite series diverges.

7. $\sum_{n=0}^{\infty} \left(\frac{7}{6}\right)^n$ 8. $\sum_{n=0}^{\infty} 4(-1.05)^n$

9. $\sum_{n=1}^{\infty} \frac{n}{n+1}$ 10. $\sum_{n=1}^{\infty} \frac{n}{2n+3}$

11. $\sum_{n=1}^{\infty} \frac{n^2}{n^2+1}$ 12. $\sum_{n=1}^{\infty} \frac{n}{\sqrt{n^2+1}}$

13. $\sum_{n=1}^{\infty} \frac{2^n+1}{2^{n+1}}$ 14. $\sum_{n=1}^{\infty} \frac{n!}{2^n}$

Verifying Convergence   In Exercises 15–20, verify that the infinite series converges.

15. $\sum_{n=0}^{\infty} \left(\frac{5}{6}\right)^n$ 16. $\sum_{n=1}^{\infty} 2\left(-\frac{1}{2}\right)^n$

17. $\sum_{n=0}^{\infty} (0.9)^n = 1 + 0.9 + 0.81 + 0.729 + \cdots$

18. $\sum_{n=0}^{\infty} (-0.6)^n = 1 - 0.6 + 0.36 - 0.216 + \cdots$

19. $\sum_{n=1}^{\infty} \frac{1}{n(n+1)}$ (Hint: Use partial fractions.)

20. $\sum_{n=1}^{\infty} \frac{1}{n(n+2)}$ (Hint: Use partial fractions.)

✎ **Numerical, Graphical, and Analytic Analysis**   In Exercises 21–24, (a) find the sum of the series, (b) use a graphing utility to find the indicated partial sum S_n and complete the table, (c) use a graphing utility to graph the first 10 terms of the sequence of partial sums and a horizontal line representing the sum, and (d) explain the relationship between the magnitudes of the terms of the series and the rate at which the sequence of partial sums approaches the sum of the series.

n	5	10	20	50	100
S_n					

21. $\sum_{n=1}^{\infty} \frac{6}{n(n+3)}$ 22. $\sum_{n=1}^{\infty} \frac{4}{n(n+4)}$

23. $\sum_{n=1}^{\infty} 2(0.9)^{n-1}$ 24. $\sum_{n=1}^{\infty} 10\left(-\frac{1}{4}\right)^{n-1}$

Finding the Sum of a Convergent Series   In Exercises 25–34, find the sum of the convergent series.

25. $\sum_{n=0}^{\infty} 5\left(\frac{2}{3}\right)^n$ 26. $\sum_{n=0}^{\infty} \left(-\frac{1}{5}\right)^n$

27. $\sum_{n=1}^{\infty} \frac{4}{n(n+2)}$ 28. $\sum_{n=1}^{\infty} \frac{1}{(2n+1)(2n+3)}$

29. $8 + 6 + \frac{9}{2} + \frac{27}{8} + \cdots$ 30. $9 - 3 + 1 - \frac{1}{3} + \cdots$

31. $\sum_{n=0}^{\infty} \left(\frac{1}{2^n} - \frac{1}{3^n}\right)$ 32. $\sum_{n=0}^{\infty} [(0.3)^n + (0.8)^n]$

33. $\sum_{n=1}^{\infty} (\sin 1)^n$ 34. $\sum_{n=1}^{\infty} \frac{1}{9n^2+3n-2}$

Using a Geometric Series   In Exercises 35–40, (a) write the repeating decimal as a geometric series, and (b) write its sum as the ratio of two integers.

35. $0.\overline{4}$ 36. $0.\overline{36}$

37. $0.8\overline{1}$ 38. $0.0\overline{1}$

39. $0.0\overline{75}$ 40. $0.2\overline{15}$

Determining Convergence or Divergence   In Exercises 41–54, determine the convergence or divergence of the series.

41. $\sum_{n=0}^{\infty} (1.075)^n$ 42. $\sum_{n=0}^{\infty} \frac{3^n}{1000}$

43. $\sum_{n=1}^{\infty} \frac{n+10}{10n+1}$ 44. $\sum_{n=1}^{\infty} \frac{4n+1}{3n-1}$

45. $\sum_{n=1}^{\infty} \left(\frac{1}{n} - \frac{1}{n+2}\right)$ 46. $\sum_{n=1}^{\infty} \left(\frac{1}{n+1} - \frac{1}{n+2}\right)$

47. $\sum_{n=1}^{\infty} \frac{3^n}{n^3}$ 48. $\sum_{n=0}^{\infty} \frac{3}{5^n}$

49. $\sum_{n=2}^{\infty} \frac{n}{\ln n}$ 50. $\sum_{n=1}^{\infty} \ln \frac{1}{n}$

51. $\sum_{n=1}^{\infty} \left(1 + \frac{k}{n}\right)^n$ 52. $\sum_{n=1}^{\infty} e^{-n}$

53. $\sum_{n=1}^{\infty} \arctan n$ 54. $\sum_{n=1}^{\infty} \ln\left(\frac{n+1}{n}\right)$

WRITING ABOUT CONCEPTS

55. **Series**   State the definitions of convergent and divergent series.

56. **Sequence and Series**   Describe the difference between $\lim_{n \to \infty} a_n = 5$ and $\sum_{n=1}^{\infty} a_n = 5$.

WRITING ABOUT CONCEPTS (continued)

57. Geometric Series Define a geometric series, state when it converges, and give the formula for the sum of a convergent geometric series.

58. nth-Term Test for Divergence State the nth-Term Test for Divergence.

59. Comparing Series Explain any differences among the following series.

(a) $\sum_{n=1}^{\infty} a_n$ (b) $\sum_{k=1}^{\infty} a_k$ (c) $\sum_{n=1}^{\infty} a_k$

60. Using a Series

(a) You delete a finite number of terms from a divergent series. Will the new series still diverge? Explain your reasoning.

(b) You add a finite number of terms to a convergent series. Will the new series still converge? Explain your reasoning.

Making a Series Converge In Exercises 61–66, find all values of x for which the series converges. For these values of x, write the sum of the series as a function of x.

61. $\sum_{n=1}^{\infty} (3x)^n$

62. $\sum_{n=0}^{\infty} \left(\frac{2}{x}\right)^n$

63. $\sum_{n=1}^{\infty} (x-1)^n$

64. $\sum_{n=0}^{\infty} 5\left(\frac{x-2}{3}\right)^n$

65. $\sum_{n=0}^{\infty} (-1)^n x^n$

66. $\sum_{n=0}^{\infty} (-1)^n x^{2n}$

Using a Geometric Series In Exercises 67 and 68, (a) find the common ratio of the geometric series, (b) write the function that gives the sum of the series, and (c) use a graphing utility to graph the function and the partial sums S_3 and S_5. What do you notice?

67. $1 + x + x^2 + x^3 + \cdots$

68. $1 - \frac{x}{2} + \frac{x^2}{4} - \frac{x^3}{8} + \cdots$

Writing In Exercises 69 and 70, use a graphing utility to determine the first term that is less than 0.0001 in each of the convergent series. Note that the answers are very different. Explain how this will affect the rate at which the series converges.

69. $\sum_{n=1}^{\infty} \frac{1}{n(n+1)}$, $\sum_{n=1}^{\infty} \left(\frac{1}{8}\right)^n$

70. $\sum_{n=1}^{\infty} \frac{1}{2^n}$, $\sum_{n=1}^{\infty} (0.01)^n$

71. Marketing An electronic games manufacturer producing a new product estimates the annual sales to be 8000 units. Each year, 5% of the units that have been sold will become inoperative. So, 8000 units will be in use after 1 year, $[8000 + 0.95(8000)]$ units will be in use after 2 years, and so on. How many units will be in use after n years?

72. Depreciation A company buys a machine for $475,000 that depreciates at a rate of 30% per year. Find a formula for the value of the machine after n years. What is its value after 5 years?

73. Multiplier Effect The total annual spending by tourists in a resort city is $200 million. Approximately 75% of that revenue is again spent in the resort city, and of that amount approximately 75% is again spent in the same city, and so on. Write the geometric series that gives the total amount of spending generated by the $200 million and find the sum of the series.

74. Multiplier Effect Repeat Exercise 73 when the percent of the revenue that is spent again in the city decreases to 60%.

75. Distance A ball is dropped from a height of 16 feet. Each time it drops h feet, it rebounds $0.81h$ feet. Find the total distance traveled by the ball.

76. Time The ball in Exercise 75 takes the following times for each fall.

$s_1 = -16t^2 + 16$, $s_1 = 0$ when $t = 1$
$s_2 = -16t^2 + 16(0.81)$, $s_2 = 0$ when $t = 0.9$
$s_3 = -16t^2 + 16(0.81)^2$, $s_3 = 0$ when $t = (0.9)^2$
$s_4 = -16t^2 + 16(0.81)^3$, $s_4 = 0$ when $t = (0.9)^3$
$\vdots$
$s_n = -16t^2 + 16(0.81)^{n-1}$, $s_n = 0$ when $t = (0.9)^{n-1}$

Beginning with s_2, the ball takes the same amount of time to bounce up as it does to fall, and so the total time elapsed before it comes to rest is given by

$$t = 1 + 2\sum_{n=1}^{\infty} (0.9)^n.$$

Find this total time.

Probability In Exercises 77 and 78, the random variable n represents the number of units of a product sold per day in a store. The probability distribution of n is given by $P(n)$. Find the probability that two units are sold in a given day $[P(2)]$ and show that $P(0) + P(1) + P(2) + P(3) + \cdots = 1$.

77. $P(n) = \frac{1}{2}\left(\frac{1}{2}\right)^n$

78. $P(n) = \frac{1}{3}\left(\frac{2}{3}\right)^n$

79. Probability A fair coin is tossed repeatedly. The probability that the first head occurs on the nth toss is given by $P(n) = \left(\frac{1}{2}\right)^n$, where $n \geq 1$.

(a) Show that $\sum_{n=1}^{\infty} \left(\frac{1}{2}\right)^n = 1$.

(b) The expected number of tosses required until the first head occurs in the experiment is given by

$$\sum_{n=1}^{\infty} n\left(\frac{1}{2}\right)^n.$$

Is this series geometric?

(c) Use a computer algebra system to find the sum in part (b).

80. Probability In an experiment, three people toss a fair coin one at a time until one of them tosses a head. Determine, for each person, the probability that he or she tosses the first head. Verify that the sum of the three probabilities is 1.

81. Area The sides of a square are 16 inches in length. A new square is formed by connecting the midpoints of the sides of the original square, and two of the triangles outside the second square are shaded (see figure). Determine the area of the shaded regions (a) when this process is continued five more times, and (b) when this pattern of shading is continued infinitely.

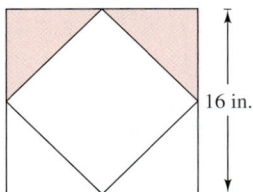

16 in.

Figure for 81

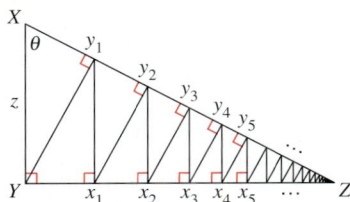

Figure for 82

82. Length A right triangle XYZ is shown above where $|XY| = z$ and $\angle X = \theta$. Line segments are continually drawn to be perpendicular to the triangle, as shown in the figure.

(a) Find the total length of the perpendicular line segments $|Yy_1| + |x_1y_1| + |x_1y_2| + \cdots$ in terms of z and θ.

(b) Find the total length of the perpendicular line segments when $z = 1$ and $\theta = \pi/6$.

Using a Geometric Series In Exercises 83–86, use the formula for the nth partial sum of a geometric series

$$\sum_{i=0}^{n-1} ar^i = \frac{a(1 - r^n)}{1 - r}.$$

83. Present Value The winner of a $2,000,000 sweepstakes will be paid $100,000 per year for 20 years. The money earns 6% interest per year. The present value of the winnings is $\sum_{n=1}^{20} 100,000\left(\frac{1}{1.06}\right)^n$. Compute the present value and interpret its meaning.

84. Annuities When an employee receives a paycheck at the end of each month, P dollars is invested in a retirement account. These deposits are made each month for t years and the account earns interest at the annual percentage rate r. When the interest is compounded monthly, the amount A in the account at the end of t years is

$$A = P + P\left(1 + \frac{r}{12}\right) + \cdots + P\left(1 + \frac{r}{12}\right)^{12t-1}$$

$$= P\left(\frac{12}{r}\right)\left[\left(1 + \frac{r}{12}\right)^{12t} - 1\right].$$

When the interest is compounded continuously, the amount A in the account after t years is

$$A = P + Pe^{r/12} + Pe^{2r/12} + Pe^{(12t-1)r/12}$$

$$= \frac{P(e^{rt} - 1)}{e^{r/12} - 1}.$$

Verify the formulas for the sums given above.

85. Salary You go to work at a company that pays $0.01 for the first day, $0.02 for the second day, $0.04 for the third day, and so on. If the daily wage keeps doubling, what would your total income be for working (a) 29 days, (b) 30 days, and (c) 31 days?

86. Sphereflake

The sphereflake shown below is a computer-generated fractal that was created by Eric Haines. The radius of the large sphere is 1. To the large sphere, nine spheres of radius $\frac{1}{3}$ are attached. To each of these, nine spheres of radius $\frac{1}{9}$ are attached. This process is continued infinitely. Prove that the sphereflake has an infinite surface area.

Annuities In Exercises 87–90, consider making monthly deposits of P dollars in a savings account at an annual interest rate r. Use the results of Exercise 84 to find the balance A after t years when the interest is compounded (a) monthly and (b) continuously.

87. $P = \$45$, $r = 3\%$, $t = 20$ years

88. $P = \$75$, $r = 5.5\%$, $t = 25$ years

89. $P = \$100$, $r = 4\%$, $t = 35$ years

90. $P = \$30$, $r = 6\%$, $t = 50$ years

True or False? In Exercises 91–96, determine whether the statement is true or false. If it is false, explain why or give an example that shows it is false.

91. If $\lim\limits_{n \to \infty} a_n = 0$, then $\sum\limits_{n=1}^{\infty} a_n$ converges.

92. If $\sum\limits_{n=1}^{\infty} a_n = L$, then $\sum\limits_{n=0}^{\infty} a_n = L + a_0$.

93. If $|r| < 1$, then $\sum\limits_{n=1}^{\infty} ar^n = \frac{a}{1 - r}$.

94. The series $\sum\limits_{n=1}^{\infty} \frac{n}{1000(n + 1)}$ diverges.

95. $0.75 = 0.749999\ldots$.

96. Every decimal with a repeating pattern of digits is a rational number.

97. Using Divergent Series Find two divergent series $\Sigma\, a_n$ and $\Sigma\, b_n$ such that $\Sigma(a_n + b_n)$ converges.

98. Proof Given two infinite series $\Sigma\, a_n$ and $\Sigma\, b_n$ such that $\Sigma\, a_n$ converges and $\Sigma\, b_n$ diverges, prove that $\Sigma(a_n + b_n)$ diverges.

99. Fibonacci Sequence The Fibonacci sequence is defined recursively by $a_{n+2} = a_n + a_{n+1}$, where $a_1 = 1$ and $a_2 = 1$.

(a) Show that $\dfrac{1}{a_{n+1}\, a_{n+3}} = \dfrac{1}{a_{n+1}\, a_{n+2}} - \dfrac{1}{a_{n+2}\, a_{n+3}}$.

(b) Show that $\displaystyle\sum_{n=0}^{\infty} \dfrac{1}{a_{n+1}\, a_{n+3}} = 1$.

100. Remainder Let $\Sigma\, a_n$ be a convergent series, and let

$$R_N = a_{N+1} + a_{N+2} + \cdots$$

be the remainder of the series after the first N terms. Prove that $\lim\limits_{N \to \infty} R_N = 0$.

101. Proof Prove that $\dfrac{1}{r} + \dfrac{1}{r^2} + \dfrac{1}{r^3} + \cdots = \dfrac{1}{r-1}$, for $|r| > 1$.

102. HOW DO YOU SEE IT? The figure below represents an informal way of showing that $\displaystyle\sum_{n=1}^{\infty} \dfrac{1}{n^2} < 2$. Explain how the figure implies this conclusion.

■ FOR FURTHER INFORMATION For more on this exercise, see the article "Convergence with Pictures" by P. J. Rippon in *American Mathematical Monthly*.

PUTNAM EXAM CHALLENGE

103. Express $\displaystyle\sum_{k=1}^{\infty} \dfrac{6^k}{(3^{k+1} - 2^{k+1})(3^k - 2^k)}$ as a rational number.

104. Let $f(n)$ be the sum of the first n terms of the sequence 0, $1, 1, 2, 2, 3, 3, 4, \ldots$, where the nth term is given by

$$a_n = \begin{cases} n/2, & \text{if } n \text{ is even} \\ (n-1)/2, & \text{if } n \text{ is odd.} \end{cases}$$

Show that if x and y are positive integers and $x > y$ then $xy = f(x+y) - f(x-y)$.

These problems were composed by the Committee on the Putnam Prize Competition.
© The Mathematical Association of America. All rights reserved.

SECTION PROJECT

Cantor's Disappearing Table

The following procedure shows how to make a table disappear by removing only half of the table!

(a) Original table has a length of L.

(b) Remove $\frac{1}{4}$ of the table centered at the midpoint. Each remaining piece has a length that is less than $\frac{1}{2}L$.

(c) Remove $\frac{1}{8}$ of the table by taking sections of length $\frac{1}{16}L$ from the centers of each of the two remaining pieces. Now, you have removed $\frac{1}{4} + \frac{1}{8}$ of the table. Each remaining piece has a length that is less than $\frac{1}{4}L$.

(d) Remove $\frac{1}{16}$ of the table by taking sections of length $\frac{1}{64}L$ from the centers of each of the four remaining pieces. Now, you have removed $\frac{1}{4} + \frac{1}{8} + \frac{1}{16}$ of the table. Each remaining piece has a length that is less than $\frac{1}{8}L$.

Will continuing this process cause the table to disappear, even though you have only removed half of the table? Why?

■ FOR FURTHER INFORMATION Read the article "Cantor's Disappearing Table" by Larry E. Knop in *The College Mathematics Journal*. To view this article, go to *MathArticles.com*.

9.3 The Integral Test and *p*-Series

■ Use the Integral Test to determine whether an infinite series converges or diverges.
■ Use properties of *p*-series and harmonic series.

The Integral Test

In this and the next section, you will study several convergence tests that apply to series with *positive* terms.

> **THEOREM 9.10 The Integral Test**
>
> If f is positive, continuous, and decreasing for $x \geq 1$ and $a_n = f(n)$, then
>
> $$\sum_{n=1}^{\infty} a_n \quad \text{and} \quad \int_1^{\infty} f(x)\, dx$$
>
> either both converge or both diverge.

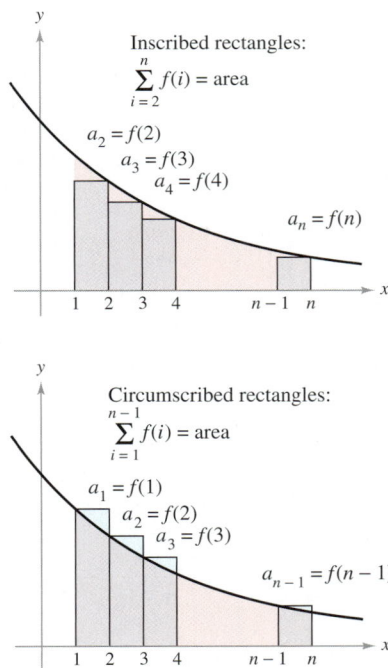

Proof Begin by partitioning the interval $[1, n]$ into $(n - 1)$ unit intervals, as shown in Figure 9.8. The total areas of the inscribed rectangles and the circumscribed rectangles are

$$\sum_{i=2}^{n} f(i) = f(2) + f(3) + \cdots + f(n) \qquad \text{Inscribed area}$$

and

$$\sum_{i=1}^{n-1} f(i) = f(1) + f(2) + \cdots + f(n-1). \qquad \text{Circumscribed area}$$

The exact area under the graph of f from $x = 1$ to $x = n$ lies between the inscribed and circumscribed areas.

$$\sum_{i=2}^{n} f(i) \leq \int_1^n f(x)\, dx \leq \sum_{i=1}^{n-1} f(i)$$

Using the *n*th partial sum, $S_n = f(1) + f(2) + \cdots + f(n)$, you can write this inequality as

$$S_n - f(1) \leq \int_1^n f(x)\, dx \leq S_{n-1}.$$

Now, assuming that $\int_1^{\infty} f(x)\, dx$ converges to L, it follows that for $n \geq 1$

$$S_n - f(1) \leq L \implies S_n \leq L + f(1).$$

Consequently, $\{S_n\}$ is bounded and monotonic, and by Theorem 9.5 it converges. So, $\Sigma\, a_n$ converges. For the other direction of the proof, assume that the improper integral diverges. Then $\int_1^n f(x)\, dx$ approaches infinity as $n \to \infty$, and the inequality $S_{n-1} \geq \int_1^n f(x)\, dx$ implies that $\{S_n\}$ diverges. So, $\Sigma\, a_n$ diverges.
See LarsonCalculus.com for Bruce Edwards's video of this proof.

Remember that the convergence or divergence of $\Sigma\, a_n$ is not affected by deleting the first N terms. Similarly, when the conditions for the Integral Test are satisfied for all $x \geq N > 1$, you can simply use the integral $\int_N^{\infty} f(x)\, dx$ to test for convergence or divergence. (This is illustrated in Example 4.)

Inscribed rectangles: $\sum_{i=2}^{n} f(i) = $ area

Circumscribed rectangles: $\sum_{i=1}^{n-1} f(i) = $ area

Figure 9.8

EXAMPLE 1 **Using the Integral Test**

Apply the Integral Test to the series $\displaystyle\sum_{n=1}^{\infty} \frac{n}{n^2 + 1}$.

Solution The function $f(x) = x/(x^2 + 1)$ is positive and continuous for $x \geq 1$. To determine whether f is decreasing, find the derivative.

$$f'(x) = \frac{(x^2 + 1)(1) - x(2x)}{(x^2 + 1)^2} = \frac{-x^2 + 1}{(x^2 + 1)^2}$$

So, $f'(x) < 0$ for $x > 1$ and it follows that f satisfies the conditions for the Integral Test. You can integrate to obtain

$$
\begin{aligned}
\int_1^{\infty} \frac{x}{x^2 + 1}\, dx &= \frac{1}{2} \int_1^{\infty} \frac{2x}{x^2 + 1}\, dx \\
&= \frac{1}{2} \lim_{b \to \infty} \int_1^b \frac{2x}{x^2 + 1}\, dx \\
&= \frac{1}{2} \lim_{b \to \infty} \left[\ln(x^2 + 1) \right]_1^b \\
&= \frac{1}{2} \lim_{b \to \infty} \left[\ln(b^2 + 1) - \ln 2 \right] \\
&= \infty.
\end{aligned}
$$

So, the series *diverges*.

EXAMPLE 2 **Using the Integral Test**

•••▷ *See LarsonCalculus.com for an interactive version of this type of example.*

Apply the Integral Test to the series $\displaystyle\sum_{n=1}^{\infty} \frac{1}{n^2 + 1}$.

Solution Because $f(x) = 1/(x^2 + 1)$ satisfies the conditions for the Integral Test (check this), you can integrate to obtain

$$
\begin{aligned}
\int_1^{\infty} \frac{1}{x^2 + 1}\, dx &= \lim_{b \to \infty} \int_1^b \frac{1}{x^2 + 1}\, dx \\
&= \lim_{b \to \infty} \left[\arctan x \right]_1^b \\
&= \lim_{b \to \infty} (\arctan b - \arctan 1) \\
&= \frac{\pi}{2} - \frac{\pi}{4} \\
&= \frac{\pi}{4}.
\end{aligned}
$$

So, the series *converges* (see Figure 9.9).

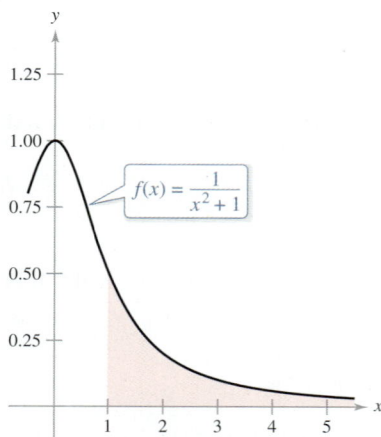

$f(x) = \dfrac{1}{x^2 + 1}$

Because the improper integral converges, the infinite series also converges.
Figure 9.9

In Example 2, the fact that the improper integral converges to $\pi/4$ does not imply that the infinite series converges to $\pi/4$. To approximate the sum of the series, you can use the inequality

$$\sum_{n=1}^{N} \frac{1}{n^2 + 1} \leq \sum_{n=1}^{\infty} \frac{1}{n^2 + 1} \leq \sum_{n=1}^{N} \frac{1}{n^2 + 1} + \int_N^{\infty} \frac{1}{x^2 + 1}\, dx.$$

(See Exercise 54.) The larger the value of N, the better the approximation. For instance, using $N = 200$ produces $1.072 \leq \sum 1/(n^2 + 1) \leq 1.077$.

p-Series and Harmonic Series

In the remainder of this section, you will investigate a second type of series that has a simple arithmetic test for convergence or divergence. A series of the form

$$\sum_{n=1}^{\infty} \frac{1}{n^p} = \frac{1}{1^p} + \frac{1}{2^p} + \frac{1}{3^p} + \cdots \qquad \text{\textcolor{red}{\textit{p}-series}}$$

is a ***p*-series,** where *p* is a positive constant. For $p = 1$, the series

$$\sum_{n=1}^{\infty} \frac{1}{n} = 1 + \frac{1}{2} + \frac{1}{3} + \cdots \qquad \text{\textcolor{red}{Harmonic series}}$$

is the **harmonic** series. A **general harmonic series** is of the form $\Sigma 1/(an + b)$. In music, strings of the same material, diameter, and tension, and whose lengths form a harmonic series, produce harmonic tones.

The Integral Test is convenient for establishing the convergence or divergence of *p*-series. This is shown in the proof of Theorem 9.11.

THEOREM 9.11   Convergence of *p*-Series

The *p*-series

$$\sum_{n=1}^{\infty} \frac{1}{n^p} = \frac{1}{1^p} + \frac{1}{2^p} + \frac{1}{3^p} + \frac{1}{4^p} + \cdots$$

converges for $p > 1$, and diverges for $0 < p \le 1$.

Proof   The proof follows from the Integral Test and from Theorem 8.5, which states that

$$\int_{1}^{\infty} \frac{1}{x^p} \, dx$$

converges for $p > 1$ and diverges for $0 < p \le 1$.

See LarsonCalculus.com for Bruce Edwards's video of this proof. ■

EXAMPLE 3   **Convergent and Divergent *p*-Series**

Discuss the convergence or divergence of (a) the harmonic series and (b) the *p*-series with $p = 2$.

Solution

a. From Theorem 9.11, it follows that the harmonic series

$$\sum_{n=1}^{\infty} \frac{1}{n} = \frac{1}{1} + \frac{1}{2} + \frac{1}{3} + \cdots \qquad \textcolor{red}{p = 1}$$

diverges.

b. From Theorem 9.11, it follows that the *p*-series

$$\sum_{n=1}^{\infty} \frac{1}{n^2} = \frac{1}{1^2} + \frac{1}{2^2} + \frac{1}{3^2} + \cdots \qquad \textcolor{red}{p = 2}$$

converges. ■

The sum of the series in Example 3(b) can be shown to be $\pi^2/6$. (This was proved by Leonhard Euler, but the proof is too difficult to present here.) Be sure you see that the Integral Test does not tell you that the sum of the series is equal to the value of the integral. For instance, the sum of the series in Example 3(b) is

$$\sum_{n=1}^{\infty} \frac{1}{n^2} = \frac{\pi^2}{6} \approx 1.645$$

whereas the value of the corresponding improper integral is

$$\int_1^{\infty} \frac{1}{x^2}\, dx = 1.$$

EXAMPLE 4   **Testing a Series for Convergence**

Determine whether the series

$$\sum_{n=2}^{\infty} \frac{1}{n \ln n}$$

converges or diverges.

Solution   This series is similar to the divergent harmonic series. If its terms were greater than those of the harmonic series, you would expect it to diverge. However, because its terms are less than those of the harmonic series, you are not sure what to expect. The function

$$f(x) = \frac{1}{x \ln x}$$

is positive and continuous for $x \geq 2$. To determine whether f is decreasing, first rewrite f as

$$f(x) = (x \ln x)^{-1}$$

and then find its derivative.

$$f'(x) = (-1)(x \ln x)^{-2}(1 + \ln x) = -\frac{1 + \ln x}{x^2(\ln x)^2}$$

So, $f'(x) < 0$ for $x > 2$ and it follows that f satisfies the conditions for the Integral Test.

$$\int_2^{\infty} \frac{1}{x \ln x}\, dx = \int_2^{\infty} \frac{1/x}{\ln x}\, dx$$
$$= \lim_{b \to \infty} \left[\ln(\ln x)\right]_2^b$$
$$= \lim_{b \to \infty} \left[\ln(\ln b) - \ln(\ln 2)\right]$$
$$= \infty$$

The series diverges. ◼

Note that the infinite series in Example 4 diverges very slowly. For instance, as shown in the table, the sum of the first 10 terms is approximately 1.6878196, whereas the sum of the first 100 terms is just slightly greater: 2.3250871. In fact, the sum of the first 10,000 terms is approximately 3.0150217. You can see that although the infinite series "adds up to infinity," it does so very slowly.

n	11	101	1001	10,001	100,001
S_n	1.6878	2.3251	2.7275	3.0150	3.2382

9.3 Exercises

See CalcChat.com for tutorial help and worked-out solutions to odd-numbered exercises.

Using the Integral Test In Exercises 1–22, confirm that the Integral Test can be applied to the series. Then use the Integral Test to determine the convergence or divergence of the series.

1. $\sum_{n=1}^{\infty} \frac{1}{n+3}$

2. $\sum_{n=1}^{\infty} \frac{2}{3n+5}$

3. $\sum_{n=1}^{\infty} \frac{1}{2^n}$

4. $\sum_{n=1}^{\infty} 3^{-n}$

5. $\sum_{n=1}^{\infty} e^{-n}$

6. $\sum_{n=1}^{\infty} ne^{-n/2}$

7. $\frac{1}{2} + \frac{1}{5} + \frac{1}{10} + \frac{1}{17} + \frac{1}{26} + \cdots$

8. $\frac{1}{3} + \frac{1}{5} + \frac{1}{7} + \frac{1}{9} + \frac{1}{11} + \cdots$

9. $\frac{\ln 2}{2} + \frac{\ln 3}{3} + \frac{\ln 4}{4} + \frac{\ln 5}{5} + \frac{\ln 6}{6} + \cdots$

10. $\frac{\ln 2}{\sqrt{2}} + \frac{\ln 3}{\sqrt{3}} + \frac{\ln 4}{\sqrt{4}} + \frac{\ln 5}{\sqrt{5}} + \frac{\ln 6}{\sqrt{6}} + \cdots$

11. $\frac{1}{\sqrt{1}(\sqrt{1}+1)} + \frac{1}{\sqrt{2}(\sqrt{2}+1)} + \frac{1}{\sqrt{3}(\sqrt{3}+1)}$
$+ \cdots + \frac{1}{\sqrt{n}(\sqrt{n}+1)} + \cdots$

12. $\frac{1}{4} + \frac{2}{7} + \frac{3}{12} + \cdots + \frac{n}{n^2+3} + \cdots$

13. $\sum_{n=1}^{\infty} \frac{\arctan n}{n^2+1}$

14. $\sum_{n=2}^{\infty} \frac{\ln n}{n^3}$

15. $\sum_{n=1}^{\infty} \frac{\ln n}{n^2}$

16. $\sum_{n=2}^{\infty} \frac{1}{n\sqrt{\ln n}}$

17. $\sum_{n=1}^{\infty} \frac{1}{(2n+3)^3}$

18. $\sum_{n=1}^{\infty} \frac{n+2}{n+1}$

19. $\sum_{n=1}^{\infty} \frac{4n}{2n^2+1}$

20. $\sum_{n=1}^{\infty} \frac{1}{\sqrt{n+2}}$

21. $\sum_{n=1}^{\infty} \frac{n}{n^4+1}$

22. $\sum_{n=1}^{\infty} \frac{n}{n^4+2n^2+1}$

Using the Integral Test In Exercises 23 and 24, use the Integral Test to determine the convergence or divergence of the series, where *k* is a positive integer.

23. $\sum_{n=1}^{\infty} \frac{n^{k-1}}{n^k+c}$

24. $\sum_{n=1}^{\infty} n^k e^{-n}$

Requirements of the Integral Test In Exercises 25–28, explain why the Integral Test does not apply to the series.

25. $\sum_{n=1}^{\infty} \frac{(-1)^n}{n}$

26. $\sum_{n=1}^{\infty} e^{-n}\cos n$

27. $\sum_{n=1}^{\infty} \frac{2+\sin n}{n}$

28. $\sum_{n=1}^{\infty} \left(\frac{\sin n}{n}\right)^2$

Using the Integral Test In Exercises 29–32, use the Integral Test to determine the convergence or divergence of the *p*-series.

29. $\sum_{n=1}^{\infty} \frac{1}{n^3}$

30. $\sum_{n=1}^{\infty} \frac{1}{n^{1/2}}$

31. $\sum_{n=1}^{\infty} \frac{1}{n^{1/4}}$

32. $\sum_{n=1}^{\infty} \frac{1}{n^5}$

Using a *p*-Series In Exercises 33–38, use Theorem 9.11 to determine the convergence or divergence of the *p*-series.

33. $\sum_{n=1}^{\infty} \frac{1}{\sqrt[5]{n}}$

34. $\sum_{n=1}^{\infty} \frac{3}{n^{5/3}}$

35. $1 + \frac{1}{2\sqrt{2}} + \frac{1}{3\sqrt{3}} + \frac{1}{4\sqrt{4}} + \frac{1}{5\sqrt{5}} + \cdots$

36. $1 + \frac{1}{\sqrt[3]{4}} + \frac{1}{\sqrt[3]{9}} + \frac{1}{\sqrt[3]{16}} + \frac{1}{\sqrt[3]{25}} + \cdots$

37. $\sum_{n=1}^{\infty} \frac{1}{n^{1.04}}$

38. $\sum_{n=1}^{\infty} \frac{1}{n^{\pi}}$

39. Numerical and Graphical Analysis Use a graphing utility to find the indicated partial sum S_n and complete the table. Then use a graphing utility to graph the first 10 terms of the sequence of partial sums. For each series, compare the rate at which the sequence of partial sums approaches the sum of the series.

n	5	10	20	50	100
S_n					

(a) $\sum_{n=1}^{\infty} 3\left(\frac{1}{5}\right)^{n-1} = \frac{15}{4}$

(b) $\sum_{n=1}^{\infty} \frac{1}{n^2} = \frac{\pi^2}{6}$

40. Numerical Reasoning Because the harmonic series diverges, it follows that for any positive real number *M*, there exists a positive integer *N* such that the partial sum

$$\sum_{n=1}^{N} \frac{1}{n} > M.$$

(a) Use a graphing utility to complete the table.

M	2	4	6	8
N				

(b) As the real number *M* increases in equal increments, does the number *N* increase in equal increments? Explain.

WRITING ABOUT CONCEPTS

41. Integral Test State the Integral Test and give an example of its use.

42. p-Series Define a p-series and state the requirements for its convergence.

43. Using a Series A friend in your calculus class tells you that the following series converges because the terms are very small and approach 0 rapidly. Is your friend correct? Explain.

$$\frac{1}{10,000} + \frac{1}{10,001} + \frac{1}{10,002} + \cdots$$

44. Using a Function Let f be a positive, continuous, and decreasing function for $x \geq 1$, such that $a_n = f(n)$. Use a graph to rank the following quantities in decreasing order. Explain your reasoning.

(a) $\sum_{n=2}^{7} a_n$ (b) $\int_{1}^{7} f(x)\,dx$ (c) $\sum_{n=1}^{6} a_n$

45. Using a Series Use a graph to show that the inequality is true. What can you conclude about the convergence or divergence of the series? Explain.

(a) $\sum_{n=1}^{\infty} \frac{1}{\sqrt{n}} > \int_{1}^{\infty} \frac{1}{\sqrt{x}}\,dx$ (b) $\sum_{n=2}^{\infty} \frac{1}{n^2} < \int_{1}^{\infty} \frac{1}{x^2}\,dx$

46. HOW DO YOU SEE IT? The graphs show the sequences of partial sums of the p-series

$$\sum_{n=1}^{\infty} \frac{1}{n^{0.4}} \quad \text{and} \quad \sum_{n=1}^{\infty} \frac{1}{n^{1.5}}.$$

Using Theorem 9.11, the first series diverges and the second series converges. Explain how the graphs show this.

Finding Values In Exercises 47–52, find the positive values of p for which the series converges.

47. $\sum_{n=2}^{\infty} \frac{1}{n(\ln n)^p}$ **48.** $\sum_{n=2}^{\infty} \frac{\ln n}{n^p}$

49. $\sum_{n=1}^{\infty} \frac{n}{(1+n^2)^p}$ **50.** $\sum_{n=1}^{\infty} n(1+n^2)^p$

51. $\sum_{n=1}^{\infty} \left(\frac{3}{p}\right)^n$ **52.** $\sum_{n=3}^{\infty} \frac{1}{n \ln n[\ln(\ln n)]^p}$

53. Proof Let f be a positive, continuous, and decreasing function for $x \geq 1$, such that $a_n = f(n)$. Prove that if the series

$$\sum_{n=1}^{\infty} a_n$$

converges to S, then the remainder $R_N = S - S_N$ is bounded by

$$0 \leq R_N \leq \int_{N}^{\infty} f(x)\,dx.$$

54. Using a Remainder Show that the result of Exercise 53 can be written as

$$\sum_{n=1}^{N} a_n \leq \sum_{n=1}^{\infty} a_n \leq \sum_{n=1}^{N} a_n + \int_{N}^{\infty} f(x)\,dx.$$

Approximating a Sum In Exercises 55–60, use the result of Exercise 53 to approximate the sum of the convergent series using the indicated number of terms. Include an estimate of the maximum error for your approximation.

55. $\sum_{n=1}^{\infty} \frac{1}{n^2}$, five terms   **56.** $\sum_{n=1}^{\infty} \frac{1}{n^5}$, six terms

57. $\sum_{n=1}^{\infty} \frac{1}{n^2 + 1}$, ten terms

58. $\sum_{n=1}^{\infty} \frac{1}{(n+1)[\ln(n+1)]^3}$, ten terms

59. $\sum_{n=1}^{\infty} ne^{-n^2}$, four terms

60. $\sum_{n=1}^{\infty} e^{-n}$, four terms

Finding a Value In Exercises 61–64, use the result of Exercise 53 to find N such that $R_N \leq 0.001$ for the convergent series.

61. $\sum_{n=1}^{\infty} \frac{1}{n^4}$ **62.** $\sum_{n=1}^{\infty} \frac{1}{n^{3/2}}$

63. $\sum_{n=1}^{\infty} e^{-n/2}$ **64.** $\sum_{n=1}^{\infty} \frac{1}{n^2 + 1}$

65. Comparing Series

(a) Show that $\sum_{n=2}^{\infty} \frac{1}{n^{1.1}}$ converges and $\sum_{n=2}^{\infty} \frac{1}{n \ln n}$ diverges.

(b) Compare the first five terms of each series in part (a).

(c) Find $n > 3$ such that $\frac{1}{n^{1.1}} < \frac{1}{n \ln n}$.

66. Using a p-Series Ten terms are used to approximate a convergent p-series. Therefore, the remainder is a function of p and is

$$0 \leq R_{10}(p) \leq \int_{10}^{\infty} \frac{1}{x^p}\,dx, \quad p > 1.$$

(a) Perform the integration in the inequality.

(b) Use a graphing utility to represent the inequality graphically.

(c) Identify any asymptotes of the error function and interpret their meaning.

67. Euler's Constant Let

$$S_n = \sum_{k=1}^{n} \frac{1}{k} = 1 + \frac{1}{2} + \cdots + \frac{1}{n}.$$

(a) Show that $\ln(n + 1) \le S_n \le 1 + \ln n$.

(b) Show that the sequence $\{a_n\} = \{S_n - \ln n\}$ is bounded.

(c) Show that the sequence $\{a_n\}$ is decreasing.

(d) Show that a_n converges to a limit γ (called Euler's constant).

(e) Approximate γ using a_{100}.

68. Finding a Sum Find the sum of the series

$$\sum_{n=2}^{\infty} \ln\!\left(1 - \frac{1}{n^2}\right).$$

69. Using a Series Consider the series $\displaystyle\sum_{n=2}^{\infty} x^{\ln n}$.

(a) Determine the convergence or divergence of the series for $x = 1$.

(b) Determine the convergence or divergence of the series for $x = 1/e$.

(c) Find the positive values of x for which the series converges.

70. Riemann Zeta Function The **Riemann zeta function** for real numbers is defined for all x for which the series

$$\zeta(x) = \sum_{n=1}^{\infty} n^{-x}$$

converges. Find the domain of the function.

Review In Exercises 71–82, determine the convergence or divergence of the series.

71. $\displaystyle\sum_{n=1}^{\infty} \frac{1}{3n - 2}$

72. $\displaystyle\sum_{n=2}^{\infty} \frac{1}{n\sqrt{n^2 - 1}}$

73. $\displaystyle\sum_{n=1}^{\infty} \frac{1}{n\sqrt[4]{n}}$

74. $\displaystyle 3\sum_{n=1}^{\infty} \frac{1}{n^{0.95}}$

75. $\displaystyle\sum_{n=0}^{\infty} \left(\frac{2}{3}\right)^n$

76. $\displaystyle\sum_{n=0}^{\infty} (1.042)^n$

77. $\displaystyle\sum_{n=1}^{\infty} \frac{n}{\sqrt{n^2 + 1}}$

78. $\displaystyle\sum_{n=1}^{\infty} \left(\frac{1}{n^2} - \frac{1}{n^3}\right)$

79. $\displaystyle\sum_{n=1}^{\infty} \left(1 + \frac{1}{n}\right)^n$

80. $\displaystyle\sum_{n=2}^{\infty} \ln n$

81. $\displaystyle\sum_{n=2}^{\infty} \frac{1}{n(\ln n)^3}$

82. $\displaystyle\sum_{n=2}^{\infty} \frac{\ln n}{n^3}$

SECTION PROJECT

The Harmonic Series

The harmonic series

$$\sum_{n=1}^{\infty} \frac{1}{n} = 1 + \frac{1}{2} + \frac{1}{3} + \frac{1}{4} + \cdots + \frac{1}{n} + \cdots$$

is one of the most important series in this chapter. Even though its terms tend to zero as n increases,

$$\lim_{n \to \infty} \frac{1}{n} = 0$$

the harmonic series diverges. In other words, even though the terms are getting smaller and smaller, the sum "adds up to infinity."

(a) One way to show that the harmonic series diverges is attributed to James Bernoulli. He grouped the terms of the harmonic series as follows:

$$1 + \frac{1}{2} + \underbrace{\frac{1}{3} + \frac{1}{4}}_{> \frac{1}{2}} + \underbrace{\frac{1}{5} + \cdots + \frac{1}{8}}_{> \frac{1}{2}} + \underbrace{\frac{1}{9} + \cdots + \frac{1}{16}}_{> \frac{1}{2}} +$$

$$\underbrace{\frac{1}{17} + \cdots + \frac{1}{32}}_{> \frac{1}{2}} + \cdots$$

Write a short paragraph explaining how you can use this grouping to show that the harmonic series diverges.

(b) Use the proof of the Integral Test, Theorem 9.10, to show that

$$\ln(n + 1) \le 1 + \frac{1}{2} + \frac{1}{3} + \frac{1}{4} + \cdots + \frac{1}{n} \le 1 + \ln n.$$

(c) Use part (b) to determine how many terms M you would need so that

$$\sum_{n=1}^{M} \frac{1}{n} > 50.$$

(d) Show that the sum of the first million terms of the harmonic series is less than 15.

(e) Show that the following inequalities are valid.

$$\ln\frac{21}{10} \le \frac{1}{10} + \frac{1}{11} + \cdots + \frac{1}{20} \le \ln\frac{20}{9}$$

$$\ln\frac{201}{100} \le \frac{1}{100} + \frac{1}{101} + \cdots + \frac{1}{200} \le \ln\frac{200}{99}$$

(f) Use the inequalities in part (e) to find the limit

$$\lim_{m \to \infty} \sum_{n=m}^{2m} \frac{1}{n}.$$

9.4 Comparisons of Series

■ Use the Direct Comparison Test to determine whether a series converges or diverges.
■ Use the Limit Comparison Test to determine whether a series converges or diverges.

Direct Comparison Test

For the convergence tests developed so far, the terms of the series have to be fairly simple and the series must have special characteristics in order for the convergence tests to be applied. A slight deviation from these special characteristics can make a test nonapplicable. For example, in the pairs listed below, the second series cannot be tested by the same convergence test as the first series, even though it is similar to the first.

1. $\sum_{n=0}^{\infty} \frac{1}{2^n}$ is geometric, but $\sum_{n=0}^{\infty} \frac{n}{2^n}$ is not.

2. $\sum_{n=1}^{\infty} \frac{1}{n^3}$ is a p-series, but $\sum_{n=1}^{\infty} \frac{1}{n^3 + 1}$ is not.

3. $a_n = \frac{n}{(n^2 + 3)^2}$ is easily integrated, but $b_n = \frac{n^2}{(n^2 + 3)^2}$ is not.

In this section, you will study two additional tests for positive-term series. These two tests greatly expand the variety of series you are able to test for convergence or divergence. They allow you to *compare* a series having complicated terms with a simpler series whose convergence or divergence is known.

REMARK As stated, the Direct Comparison Test requires that $0 < a_n \le b_n$ for all n. Because the convergence of a series is not dependent on its first several terms, you could modify the test to require only that $0 < a_n \le b_n$ for all n greater than some integer N.

> **THEOREM 9.12 Direct Comparison Test**
>
> Let $0 < a_n \le b_n$ for all n.
>
> 1. If $\sum_{n=1}^{\infty} b_n$ converges, then $\sum_{n=1}^{\infty} a_n$ converges.
>
> 2. If $\sum_{n=1}^{\infty} a_n$ diverges, then $\sum_{n=1}^{\infty} b_n$ diverges.

Proof To prove the first property, let $L = \sum_{n=1}^{\infty} b_n$ and let

$$S_n = a_1 + a_2 + \cdots + a_n.$$

Because $0 < a_n \le b_n$, the sequence $S_1, S_2, S_3, \ldots$ is nondecreasing and bounded above by L; so, it must converge. Because

$$\lim_{n \to \infty} S_n = \sum_{n=1}^{\infty} a_n$$

it follows that $\sum_{n=1}^{\infty} a_n$ converges. The second property is logically equivalent to the first.

See LarsonCalculus.com for Bruce Edwards's video of this proof.

■ **FOR FURTHER INFORMATION** Is the Direct Comparison Test just for nonnegative series? To read about the generalization of this test to real series, see the article "The Comparison Test—Not Just for Nonnegative Series" by Michele Longo and Vincenzo Valori in *Mathematics Magazine*. To view this article, go to *MathArticles.com*.

EXAMPLE 1   **Using the Direct Comparison Test**

Determine the convergence or divergence of

$$\sum_{n=1}^{\infty} \frac{1}{2 + 3^n}.$$

Solution   This series resembles

$$\sum_{n=1}^{\infty} \frac{1}{3^n}.$$ Convergent geometric series

Term-by-term comparison yields

$$a_n = \frac{1}{2 + 3^n} < \frac{1}{3^n} = b_n, \quad n \geq 1.$$

So, by the Direct Comparison Test, the series converges.

EXAMPLE 2   **Using the Direct Comparison Test**

⋮ ▷ *See LarsonCalculus.com for an interactive version of this type of example.*

Determine the convergence or divergence of

$$\sum_{n=1}^{\infty} \frac{1}{2 + \sqrt{n}}.$$

Solution   This series resembles

$$\sum_{n=1}^{\infty} \frac{1}{n^{1/2}}.$$ Divergent *p*-series

Term-by-term comparison yields

$$\frac{1}{2 + \sqrt{n}} \leq \frac{1}{\sqrt{n}}, \quad n \geq 1$$

which *does not* meet the requirements for divergence. (Remember that when term-by-term comparison reveals a series that is *less* than a divergent series, the Direct Comparison Test tells you nothing.) Still expecting the series to diverge, you can compare the series with

$$\sum_{n=1}^{\infty} \frac{1}{n}.$$ Divergent harmonic series

In this case, term-by-term comparison yields

$$a_n = \frac{1}{n} \leq \frac{1}{2 + \sqrt{n}} = b_n, \quad n \geq 4$$

and, by the Direct Comparison Test, the given series diverges. To verify the last inequality, try showing that

$$2 + \sqrt{n} \leq n$$

whenever $n \geq 4$.   ■

Remember that both parts of the Direct Comparison Test require that $0 < a_n \leq b_n$. Informally, the test says the following about the two series with nonnegative terms.

1. If the "larger" series converges, then the "smaller" series must also converge.

2. If the "smaller" series diverges, then the "larger" series must also diverge.

Limit Comparison Test

Sometimes a series closely resembles a *p*-series or a geometric series, yet you cannot establish the term-by-term comparison necessary to apply the Direct Comparison Test. Under these circumstances, you may be able to apply a second comparison test, called the **Limit Comparison Test.**

> **THEOREM 9.13 Limit Comparison Test**
>
> If $a_n > 0$, $b_n > 0$, and
>
> $$\lim_{n\to\infty} \frac{a_n}{b_n} = L$$
>
> where L is *finite and positive*, then
>
> $$\sum_{n=1}^{\infty} a_n \quad \text{and} \quad \sum_{n=1}^{\infty} b_n$$
>
> either both converge or both diverge.

REMARK As with the Direct Comparison Test, the Limit Comparison Test could be modified to require only that a_n and b_n be positive for all n greater than some integer N.

Proof Because $a_n > 0$, $b_n > 0$, and

$$\lim_{n\to\infty} \frac{a_n}{b_n} = L$$

there exists $N > 0$ such that

$$0 < \frac{a_n}{b_n} < L + 1, \quad \text{for } n \geq N.$$

This implies that

$$0 < a_n < (L + 1)b_n.$$

So, by the Direct Comparison Test, the convergence of $\Sigma\, b_n$ implies the convergence of $\Sigma\, a_n$. Similarly, the fact that

$$\lim_{n\to\infty} \frac{b_n}{a_n} = \frac{1}{L}$$

can be used to show that the convergence of $\Sigma\, a_n$ implies the convergence of $\Sigma\, b_n$. *See LarsonCalculus.com for Bruce Edwards's video of this proof.*

EXAMPLE 3   Using the Limit Comparison Test

Show that the general harmonic series below diverges.

$$\sum_{n=1}^{\infty} \frac{1}{an + b}, \quad a > 0, \quad b > 0$$

Solution By comparison with

$$\sum_{n=1}^{\infty} \frac{1}{n} \qquad \text{Divergent harmonic series}$$

you have

$$\lim_{n\to\infty} \frac{1/(an + b)}{1/n} = \lim_{n\to\infty} \frac{n}{an + b} = \frac{1}{a}.$$

Because this limit is greater than 0, you can conclude from the Limit Comparison Test that the series diverges.

The Limit Comparison Test works well for comparing a "messy" algebraic series with a p-series. In choosing an appropriate p-series, you must choose one with an nth term of the same magnitude as the nth term of the given series.

Given Series	Comparison Series	Conclusion
$\displaystyle\sum_{n=1}^{\infty} \frac{1}{3n^2 - 4n + 5}$	$\displaystyle\sum_{n=1}^{\infty} \frac{1}{n^2}$	Both series converge.
$\displaystyle\sum_{n=1}^{\infty} \frac{1}{\sqrt{3n - 2}}$	$\displaystyle\sum_{n=1}^{\infty} \frac{1}{\sqrt{n}}$	Both series diverge.
$\displaystyle\sum_{n=1}^{\infty} \frac{n^2 - 10}{4n^5 + n^3}$	$\displaystyle\sum_{n=1}^{\infty} \frac{n^2}{n^5} = \sum_{n=1}^{\infty} \frac{1}{n^3}$	Both series converge.

In other words, when choosing a series for comparison, you can disregard all but the *highest powers of n* in both the numerator and the denominator.

EXAMPLE 4   Using the Limit Comparison Test

Determine the convergence or divergence of

$$\sum_{n=1}^{\infty} \frac{\sqrt{n}}{n^2 + 1}.$$

Solution   Disregarding all but the highest powers of n in the numerator and the denominator, you can compare the series with

$$\sum_{n=1}^{\infty} \frac{\sqrt{n}}{n^2} = \sum_{n=1}^{\infty} \frac{1}{n^{3/2}}. \qquad \text{\textcolor{red}{Convergent } p\text{\textcolor{red}{-series}}}$$

Because

$$\lim_{n \to \infty} \frac{a_n}{b_n} = \lim_{n \to \infty} \left(\frac{\sqrt{n}}{n^2 + 1} \right) \left(\frac{n^{3/2}}{1} \right)$$

$$= \lim_{n \to \infty} \frac{n^2}{n^2 + 1}$$

$$= 1$$

you can conclude by the Limit Comparison Test that the series converges.

EXAMPLE 5   Using the Limit Comparison Test

Determine the convergence or divergence of

$$\sum_{n=1}^{\infty} \frac{n 2^n}{4n^3 + 1}.$$

Solution   A reasonable comparison would be with the series

$$\sum_{n=1}^{\infty} \frac{2^n}{n^2}. \qquad \text{\textcolor{red}{Divergent series}}$$

Note that this series diverges by the nth-Term Test. From the limit

$$\lim_{n \to \infty} \frac{a_n}{b_n} = \lim_{n \to \infty} \left(\frac{n 2^n}{4n^3 + 1} \right) \left(\frac{n^2}{2^n} \right)$$

$$= \lim_{n \to \infty} \frac{1}{4 + (1/n^3)}$$

$$= \frac{1}{4}$$

you can conclude that the series diverges.

9.4 Exercises

See CalcChat.com for tutorial help and worked-out solutions to odd-numbered exercises.

1. Graphical Analysis The figures show the graphs of the first 10 terms, and the graphs of the first 10 terms of the sequence of partial sums, of each series.

$$\sum_{n=1}^{\infty} \frac{6}{n^{3/2}}, \quad \sum_{n=1}^{\infty} \frac{6}{n^{3/2}+3}, \quad \text{and} \quad \sum_{n=1}^{\infty} \frac{6}{n\sqrt{n^2+0.5}}$$

(a) Identify the series in each figure.

(b) Which series is a *p*-series? Does it converge or diverge?

(c) For the series that are not *p*-series, how do the magnitudes of the terms compare with the magnitudes of the terms of the *p*-series? What conclusion can you draw about the convergence or divergence of the series?

(d) Explain the relationship between the magnitudes of the terms of the series and the magnitudes of the terms of the partial sums.

Graphs of terms   Graphs of partial sums

2. Graphical Analysis The figures show the graphs of the first 10 terms, and the graphs of the first 10 terms of the sequence of partial sums, of each series.

$$\sum_{n=1}^{\infty} \frac{2}{\sqrt{n}}, \quad \sum_{n=1}^{\infty} \frac{2}{\sqrt{n}-0.5}, \quad \text{and} \quad \sum_{n=1}^{\infty} \frac{4}{\sqrt{n}+0.5}$$

(a) Identify the series in each figure.

(b) Which series is a *p*-series? Does it converge or diverge?

(c) For the series that are not *p*-series, how do the magnitudes of the terms compare with the magnitudes of the terms of the *p*-series? What conclusion can you draw about the convergence or divergence of the series?

(d) Explain the relationship between the magnitudes of the terms of the series and the magnitudes of the terms of the partial sums.

Graphs of terms   Graphs of partial sums

Using the Direct Comparison Test In Exercises 3–12, use the Direct Comparison Test to determine the convergence or divergence of the series.

3. $\sum_{n=1}^{\infty} \frac{1}{2n-1}$

4. $\sum_{n=1}^{\infty} \frac{1}{3n^2+2}$

5. $\sum_{n=2}^{\infty} \frac{1}{\sqrt{n}-1}$

6. $\sum_{n=0}^{\infty} \frac{4^n}{5^n+3}$

7. $\sum_{n=2}^{\infty} \frac{\ln n}{n+1}$

8. $\sum_{n=1}^{\infty} \frac{1}{\sqrt{n^3+1}}$

9. $\sum_{n=0}^{\infty} \frac{1}{n!}$

10. $\sum_{n=1}^{\infty} \frac{1}{4\sqrt[3]{n}-1}$

11. $\sum_{n=0}^{\infty} e^{-n^2}$

12. $\sum_{n=1}^{\infty} \frac{3^n}{2^n-1}$

Using the Limit Comparison Test In Exercises 13–22, use the Limit Comparison Test to determine the convergence or divergence of the series.

13. $\sum_{n=1}^{\infty} \frac{n}{n^2+1}$

14. $\sum_{n=1}^{\infty} \frac{5}{4^n+1}$

15. $\sum_{n=0}^{\infty} \frac{1}{\sqrt{n^2+1}}$

16. $\sum_{n=1}^{\infty} \frac{2^n+1}{5^n+1}$

17. $\sum_{n=1}^{\infty} \frac{2n^2-1}{3n^5+2n+1}$

18. $\sum_{n=1}^{\infty} \frac{1}{n^2(n+3)}$

19. $\sum_{n=1}^{\infty} \frac{1}{n\sqrt{n^2+1}}$

20. $\sum_{n=1}^{\infty} \frac{n}{(n+1)2^{n-1}}$

21. $\sum_{n=1}^{\infty} \frac{n^{k-1}}{n^k+1}, \quad k>2$

22. $\sum_{n=1}^{\infty} \sin\frac{1}{n}$

Determining Convergence or Divergence In Exercises 23–30, test for convergence or divergence, using each test at least once. Identify which test was used.

(a) *n*th-Term Test

(b) Geometric Series Test

(c) *p*-Series Test

(d) Telescoping Series Test

(e) Integral Test

(f) Direct Comparison Test

(g) Limit Comparison Test

23. $\sum_{n=1}^{\infty} \frac{\sqrt[3]{n}}{n}$

24. $\sum_{n=0}^{\infty} 5\left(-\frac{4}{3}\right)^n$

25. $\sum_{n=1}^{\infty} \frac{1}{5^n+1}$

26. $\sum_{n=2}^{\infty} \frac{1}{n^3-8}$

27. $\sum_{n=1}^{\infty} \frac{2n}{3n-2}$

28. $\sum_{n=1}^{\infty} \left(\frac{1}{n+1}-\frac{1}{n+2}\right)$

29. $\sum_{n=1}^{\infty} \frac{n}{(n^2+1)^2}$

30. $\sum_{n=1}^{\infty} \frac{3}{n(n+3)}$

31. Using the Limit Comparison Test Use the Limit Comparison Test with the harmonic series to show that the series $\sum a_n$ (where $0 < a_n < a_{n-1}$) diverges when $\lim_{n\to\infty} na_n$ is finite and nonzero.

32. Proof Prove that, if $P(n)$ and $Q(n)$ are polynomials of degree j and k, respectively, then the series

$$\sum_{n=1}^{\infty} \frac{P(n)}{Q(n)}$$

converges if $j < k - 1$ and diverges if $j \geq k - 1$.

Determining Convergence or Divergence In Exercises 33–36, use the polynomial test given in Exercise 32 to determine whether the series converges or diverges.

33. $\frac{1}{2} + \frac{2}{5} + \frac{3}{10} + \frac{4}{17} + \frac{5}{26} + \cdots$

34. $\frac{1}{3} + \frac{1}{8} + \frac{1}{15} + \frac{1}{24} + \frac{1}{35} + \cdots$

35. $\sum_{n=1}^{\infty} \frac{1}{n^3 + 1}$ **36.** $\sum_{n=1}^{\infty} \frac{n^2}{n^3 + 1}$

Verifying Divergence In Exercises 37 and 38, use the divergence test given in Exercise 31 to show that the series diverges.

37. $\sum_{n=1}^{\infty} \frac{n^3}{5n^4 + 3}$ **38.** $\sum_{n=1}^{\infty} \frac{3n^2 + 1}{4n^3 + 2}$

Determining Convergence or Divergence In Exercises 39–42, determine the convergence or divergence of the series.

39. $\frac{1}{200} + \frac{1}{400} + \frac{1}{600} + \frac{1}{800} + \cdots$

40. $\frac{1}{200} + \frac{1}{210} + \frac{1}{220} + \frac{1}{230} + \cdots$

41. $\frac{1}{201} + \frac{1}{204} + \frac{1}{209} + \frac{1}{216} + \cdots$

42. $\frac{1}{201} + \frac{1}{208} + \frac{1}{227} + \frac{1}{264} + \cdots$

WRITING ABOUT CONCEPTS

43. Using Series Review the results of Exercises 39–42. Explain why careful analysis is required to determine the convergence or divergence of a series and why only considering the magnitudes of the terms of a series could be misleading.

44. Direct Comparison Test State the Direct Comparison Test and give an example of its use.

45. Limit Comparison Test State the Limit Comparison Test and give an example of its use.

46. Comparing Series It appears that the terms of the series

$$\frac{1}{1000} + \frac{1}{1001} + \frac{1}{1002} + \frac{1}{1003} + \cdots$$

are less than the corresponding terms of the convergent series

$$1 + \frac{1}{4} + \frac{1}{9} + \frac{1}{16} + \cdots.$$

If the statement above is correct, then the first series converges. Is this correct? Why or why not? Make a statement about how the divergence or convergence of a series is affected by the inclusion or exclusion of the first finite number of terms.

47. Using a Series Consider the series $\sum_{n=1}^{\infty} \frac{1}{(2n-1)^2}$.

(a) Verify that the series converges.

(b) Use a graphing utility to complete the table.

n	5	10	20	50	100
S_n					

(c) The sum of the series is $\pi^2/8$. Find the sum of the series

$$\sum_{n=3}^{\infty} \frac{1}{(2n-1)^2}.$$

(d) Use a graphing utility to find the sum of the series

$$\sum_{n=10}^{\infty} \frac{1}{(2n-1)^2}.$$

48. HOW DO YOU SEE IT? The figure shows the first 20 terms of the convergent series $\sum_{n=1}^{\infty} a_n$ and the first 20 terms of the series $\sum_{n=1}^{\infty} a_n^2$. Identify the two series and explain your reasoning in making the selection.

True or False? In Exercises 49–54, determine whether the statement is true or false. If it is false, explain why or give an example that shows it is false.

49. If $0 < a_n \leq b_n$ and $\sum_{n=1}^{\infty} a_n$ converges, then $\sum_{n=1}^{\infty} b_n$ diverges.

50. If $0 < a_{n+10} \leq b_n$ and $\sum_{n=1}^{\infty} b_n$ converges, then $\sum_{n=1}^{\infty} a_n$ converges.

51. If $a_n + b_n \leq c_n$ and $\sum_{n=1}^{\infty} c_n$ converges, then the series $\sum_{n=1}^{\infty} a_n$ and $\sum_{n=1}^{\infty} b_n$ both converge. (Assume that the terms of all three series are positive.)

52. If $a_n \leq b_n + c_n$ and $\sum_{n=1}^{\infty} a_n$ diverges, then the series $\sum_{n=1}^{\infty} b_n$ and $\sum_{n=1}^{\infty} c_n$ both diverge. (Assume that the terms of all three series are positive.)

53. If $0 < a_n \leq b_n$ and $\sum_{n=1}^{\infty} a_n$ diverges, then $\sum_{n=1}^{\infty} b_n$ diverges.

54. If $0 < a_n \leq b_n$ and $\sum_{n=1}^{\infty} b_n$ diverges, then $\sum_{n=1}^{\infty} a_n$ diverges.

55. Proof Prove that if the nonnegative series

$$\sum_{n=1}^{\infty} a_n \quad \text{and} \quad \sum_{n=1}^{\infty} b_n$$

converge, then so does the series $\sum_{n=1}^{\infty} a_n b_n$.

56. Proof Use the result of Exercise 55 to prove that if the nonnegative series $\sum_{n=1}^{\infty} a_n$ converges, then so does the series

$$\sum_{n=1}^{\infty} a_n^2.$$

57. Finding Series Find two series that demonstrate the result of Exercise 55.

58. Finding Series Find two series that demonstrate the result of Exercise 56.

59. Proof Suppose that $\Sigma\, a_n$ and $\Sigma\, b_n$ are series with positive terms. Prove that if $\lim_{n \to \infty} \dfrac{a_n}{b_n} = 0$ and $\Sigma\, b_n$ converges, $\Sigma\, a_n$ also converges.

60. Proof Suppose that $\Sigma\, a_n$ and $\Sigma\, b_n$ are series with positive terms. Prove that if $\lim_{n \to \infty} \dfrac{a_n}{b_n} = \infty$ and $\Sigma\, b_n$ diverges, $\Sigma\, a_n$ also diverges.

61. Verifying Convergence Use the result of Exercise 59 to show that each series converges.

(a) $\displaystyle\sum_{n=1}^{\infty} \frac{1}{(n+1)^3}$ (b) $\displaystyle\sum_{n=1}^{\infty} \frac{1}{\sqrt{n}\,\pi^n}$

62. Verifying Divergence Use the result of Exercise 60 to show that each series diverges.

(a) $\displaystyle\sum_{n=1}^{\infty} \frac{\ln n}{n}$ (b) $\displaystyle\sum_{n=2}^{\infty} \frac{1}{\ln n}$

63. Proof Suppose that $\Sigma\, a_n$ is a series with positive terms. Prove that if $\Sigma\, a_n$ converges, then $\Sigma \sin a_n$ also converges.

64. Proof Prove that the series

$$\sum_{n=1}^{\infty} \frac{1}{1 + 2 + 3 + \cdots + n}$$

converges.

65. Comparing Series Show that $\displaystyle\sum_{n=1}^{\infty} \frac{\ln n}{n\sqrt{n}}$ converges by comparison with $\displaystyle\sum_{n=1}^{\infty} \frac{1}{n^{5/4}}$.

PUTNAM EXAM CHALLENGE

66. Is the infinite series $\displaystyle\sum_{n=1}^{\infty} \frac{1}{n^{(n+1)/n}}$ convergent? Prove your statement.

67. Prove that if $\displaystyle\sum_{n=1}^{\infty} a_n$ is a convergent series of positive real numbers, then so is $\displaystyle\sum_{n=1}^{\infty} (a_n)^{n/(n+1)}$.

These problems were composed by the Committee on the Putnam Prize Competition.
© The Mathematical Association of America. All rights reserved.

SECTION PROJECT

Solera Method

Most wines are produced entirely from grapes grown in a single year. Sherry, however, is a complex mixture of older wines with new wines. This is done with a sequence of barrels (called a solera) stacked on top of each other, as shown in the photo.

The oldest wine is in the bottom tier of barrels, and the newest is in the top tier. Each year, half of each barrel in the bottom tier is bottled as sherry. The bottom barrels are then refilled with the wine from the barrels above. This process is repeated throughout the solera, with new wine being added to the top barrels.

A mathematical model for the amount of n-year-old wine that is removed from a solera (with k tiers) each year is

$$f(n, k) = \binom{n-1}{k-1}\left(\frac{1}{2}\right)^{n+1}, \quad k \le n.$$

(a) Consider a solera that has five tiers, numbered $k = 1, 2, 3, 4,$ and 5. In 1995 ($n = 0$), half of each barrel in the top tier (tier 1) was refilled with new wine. How much of this wine was removed from the solera in 1996? In 1997? In 1998? . . . In 2010? During which year(s) was the greatest amount of the 1995 wine removed from the solera?

(b) In part (a), let a_n be the amount of 1995 wine that is removed from the solera in year n. Evaluate

$$\sum_{n=0}^{\infty} a_n.$$

■ **FOR FURTHER INFORMATION** See the article "Finding Vintage Concentrations in a Sherry Solera" by Rhodes Peele and John T. MacQueen in the *UMAP Modules.*

Squareplum/Shutterstock.com

9.5 Alternating Series

- Use the Alternating Series Test to determine whether an infinite series converges.
- Use the Alternating Series Remainder to approximate the sum of an alternating series.
- Classify a convergent series as absolutely or conditionally convergent.
- Rearrange an infinite series to obtain a different sum.

Alternating Series

So far, most series you have dealt with have had positive terms. In this section and the next section, you will study series that contain both positive and negative terms. The simplest such series is an **alternating series,** whose terms alternate in sign. For example, the geometric series

$$\sum_{n=0}^{\infty} \left(-\frac{1}{2} \right)^n = \sum_{n=0}^{\infty} (-1)^n \frac{1}{2^n}$$

$$= 1 - \frac{1}{2} + \frac{1}{4} - \frac{1}{8} + \frac{1}{16} - \cdots$$

is an *alternating geometric series* with $r = -\frac{1}{2}$. Alternating series occur in two ways: either the odd terms are negative or the even terms are negative.

THEOREM 9.14   Alternating Series Test

Let $a_n > 0$. The alternating series

$$\sum_{n=1}^{\infty} (-1)^n a_n \quad \text{and} \quad \sum_{n=1}^{\infty} (-1)^{n+1} a_n$$

converge when the two conditions listed below are met.

1. $\displaystyle \lim_{n \to \infty} a_n = 0$

2. $a_{n+1} \le a_n$, for all n

REMARK   The second condition in the Alternating Series Test can be modified to require only that $0 < a_{n+1} \le a_n$ for all n greater than some integer N.

Proof   Consider the alternating series $\sum (-1)^{n+1} a_n$. For this series, the partial sum (where $2n$ is even)

$$S_{2n} = (a_1 - a_2) + (a_3 - a_4) + (a_5 - a_6) + \cdots + (a_{2n-1} - a_{2n})$$

has all nonnegative terms, and therefore $\{S_{2n}\}$ is a nondecreasing sequence. But you can also write

$$S_{2n} = a_1 - (a_2 - a_3) - (a_4 - a_5) - \cdots - (a_{2n-2} - a_{2n-1}) - a_{2n}$$

which implies that $S_{2n} \le a_1$ for every integer n. So, $\{S_{2n}\}$ is a bounded, nondecreasing sequence that converges to some value L. Because $S_{2n-1} - a_{2n} = S_{2n}$ and $a_{2n} \to 0$, you have

$$\lim_{n \to \infty} S_{2n-1} = \lim_{n \to \infty} S_{2n} + \lim_{n \to \infty} a_{2n}$$

$$= L + \lim_{n \to \infty} a_{2n}$$

$$= L.$$

Because both S_{2n} and S_{2n-1} converge to the same limit L, it follows that $\{S_n\}$ also converges to L. Consequently, the given alternating series converges.

See LarsonCalculus.com for Bruce Edwards's video of this proof.

REMARK The series in Example 1 is called the *alternating harmonic series.* More is said about this series in Example 8.

EXAMPLE 1 **Using the Alternating Series Test**

Determine the convergence or divergence of

$$\sum_{n=1}^{\infty} (-1)^{n+1} \frac{1}{n}.$$

Solution Note that $\lim_{n \to \infty} a_n = \lim_{n \to \infty} \frac{1}{n} = 0$. So, the first condition of Theorem 9.14 is satisfied. Also note that the second condition of Theorem 9.14 is satisfied because

$$a_{n+1} = \frac{1}{n+1} \leq \frac{1}{n} = a_n$$

for all n. So, applying the Alternating Series Test, you can conclude that the series converges.

EXAMPLE 2 **Using the Alternating Series Test**

Determine the convergence or divergence of

$$\sum_{n=1}^{\infty} \frac{n}{(-2)^{n-1}}.$$

Solution To apply the Alternating Series Test, note that, for $n \geq 1$,

$$\frac{1}{2} \leq \frac{n}{n+1}$$

$$\frac{2^{n-1}}{2^n} \leq \frac{n}{n+1}$$

$$(n+1)2^{n-1} \leq n2^n$$

$$\frac{n+1}{2^n} \leq \frac{n}{2^{n-1}}.$$

So, $a_{n+1} = (n+1)/2^n \leq n/2^{n-1} = a_n$ for all n. Furthermore, by L'Hôpital's Rule,

$$\lim_{x \to \infty} \frac{x}{2^{x-1}} = \lim_{x \to \infty} \frac{1}{2^{x-1}(\ln 2)} = 0 \implies \lim_{n \to \infty} \frac{n}{2^{n-1}} = 0.$$

Therefore, by the Alternating Series Test, the series converges.

EXAMPLE 3 **When the Alternating Series Test Does Not Apply**

REMARK In Example 3(a), remember that whenever a series does not pass the first condition of the Alternating Series Test, you can use the nth-Term Test for Divergence to conclude that the series diverges.

a. The alternating series

$$\sum_{n=1}^{\infty} \frac{(-1)^{n+1}(n+1)}{n} = \frac{2}{1} - \frac{3}{2} + \frac{4}{3} - \frac{5}{4} + \frac{6}{5} - \cdots$$

passes the second condition of the Alternating Series Test because $a_{n+1} \leq a_n$ for all n. You cannot apply the Alternating Series Test, however, because the series does not pass the first condition. In fact, the series diverges.

b. The alternating series

$$\frac{2}{1} - \frac{1}{1} + \frac{2}{2} - \frac{1}{2} + \frac{2}{3} - \frac{1}{3} + \frac{2}{4} - \frac{1}{4} + \cdots$$

passes the first condition because a_n approaches 0 as $n \to \infty$. You cannot apply the Alternating Series Test, however, because the series does not pass the second condition. To conclude that the series diverges, you can argue that S_{2N} equals the Nth partial sum of the divergent harmonic series. This implies that the sequence of partial sums diverges. So, the series diverges.

Alternating Series Remainder

For a convergent alternating series, the partial sum S_N can be a useful approximation for the sum S of the series. The error involved in using $S \approx S_N$ is the remainder $R_N = S - S_N$.

> ### THEOREM 9.15   Alternating Series Remainder
>
> If a convergent alternating series satisfies the condition $a_{n+1} \leq a_n$, then the absolute value of the remainder R_N involved in approximating the sum S by S_N is less than (or equal to) the first neglected term. That is,
>
> $$|S - S_N| = |R_N| \leq a_{N+1}.$$
>
> A proof of this theorem is given in Appendix A.
>
> *See LarsonCalculus.com for Bruce Edwards's video of this proof.*

EXAMPLE 4   **Approximating the Sum of an Alternating Series**

⋯▷ *See LarsonCalculus.com for an interactive version of this type of example.*

Approximate the sum of the series by its first six terms.

$$\sum_{n=1}^{\infty} (-1)^{n+1} \left(\frac{1}{n!}\right) = \frac{1}{1!} - \frac{1}{2!} + \frac{1}{3!} - \frac{1}{4!} + \frac{1}{5!} - \frac{1}{6!} + \cdots$$

Solution   The series converges by the Alternating Series Test because

$$\frac{1}{(n+1)!} \leq \frac{1}{n!} \quad \text{and} \quad \lim_{n \to \infty} \frac{1}{n!} = 0.$$

The sum of the first six terms is

$$S_6 = 1 - \frac{1}{2} + \frac{1}{6} - \frac{1}{24} + \frac{1}{120} - \frac{1}{720} = \frac{91}{144} \approx 0.63194$$

and, by the Alternating Series Remainder, you have

$$|S - S_6| = |R_6| \leq a_7 = \frac{1}{5040} \approx 0.0002.$$

So, the sum S lies between $0.63194 - 0.0002$ and $0.63194 + 0.0002$, and you have $0.63174 \leq S \leq 0.63214$.

▷ **TECHNOLOGY**   Later, using the techniques in Section 9.10, you will be able to show that the series in Example 4 converges to

$$\frac{e - 1}{e} \approx 0.63212.$$

(See Section 9.10, Exercise 58.) For now, try using a graphing utility to obtain an approximation of the sum of the series. How many terms do you need to obtain an approximation that is within 0.00001 unit of the actual sum?

EXAMPLE 5   **Finding the Number of Terms**

Determine the number of terms required to approximate the sum of the series with an error of less than 0.001.

$$\sum_{n=1}^{\infty} \frac{(-1)^{n+1}}{n^4}$$

Solution   By Theorem 9.15, you know that

$$|R_N| \leq a_{N+1} = \frac{1}{(N+1)^4}.$$

For an error of less than 0.001, N must satisfy the inequality $1/(N+1)^4 < 0.001$.

$$\frac{1}{(N+1)^4} < 0.001 \implies (N+1)^4 > 1000 \implies N > \sqrt[4]{1000} - 1 \approx 4.6$$

So, you will need at least 5 terms. Using 5 terms, the sum is $S \approx S_5 \approx 0.94754$, which has an error of less than 0.001.

Absolute and Conditional Convergence

Occasionally, a series may have both positive and negative terms and not be an alternating series. For instance, the series

$$\sum_{n=1}^{\infty} \frac{\sin n}{n^2} = \frac{\sin 1}{1} + \frac{\sin 2}{4} + \frac{\sin 3}{9} + \cdots$$

has both positive and negative terms, yet it is not an alternating series. One way to obtain some information about the convergence of this series is to investigate the convergence of the series

$$\sum_{n=1}^{\infty} \left| \frac{\sin n}{n^2} \right|.$$

By direct comparison, you have $|\sin n| \leq 1$ for all n, so

$$\left| \frac{\sin n}{n^2} \right| \leq \frac{1}{n^2}, \quad n \geq 1.$$

Therefore, by the Direct Comparison Test, the series $\sum \left| \frac{\sin n}{n^2} \right|$ converges. The next theorem tells you that the original series also converges.

THEOREM 9.16   Absolute Convergence

If the series $\Sigma |a_n|$ converges, then the series $\Sigma\, a_n$ also converges.

Proof Because $0 \leq a_n + |a_n| \leq 2|a_n|$ for all n, the series

$$\sum_{n=1}^{\infty} (a_n + |a_n|)$$

converges by comparison with the convergent series

$$\sum_{n=1}^{\infty} 2|a_n|.$$

Furthermore, because $a_n = (a_n + |a_n|) - |a_n|$, you can write

$$\sum_{n=1}^{\infty} a_n = \sum_{n=1}^{\infty} (a_n + |a_n|) - \sum_{n=1}^{\infty} |a_n|$$

where both series on the right converge. So, it follows that $\Sigma\, a_n$ converges.

See LarsonCalculus.com for Bruce Edwards's video of this proof.

The converse of Theorem 9.16 is not true. For instance, the **alternating harmonic series**

$$\sum_{n=1}^{\infty} \frac{(-1)^{n+1}}{n} = \frac{1}{1} - \frac{1}{2} + \frac{1}{3} - \frac{1}{4} + \cdots$$

converges by the Alternating Series Test. Yet the harmonic series diverges. This type of convergence is called **conditional.**

Definitions of Absolute and Conditional Convergence

1. The series $\Sigma\, a_n$ is **absolutely convergent** when $\Sigma |a_n|$ converges.
2. The series $\Sigma\, a_n$ is **conditionally convergent** when $\Sigma\, a_n$ converges but $\Sigma |a_n|$ diverges.

EXAMPLE 6 **Absolute and Conditional Convergence**

Determine whether each of the series is convergent or divergent. Classify any convergent series as absolutely or conditionally convergent.

a. $\displaystyle\sum_{n=0}^{\infty} \frac{(-1)^n\, n!}{2^n} = \frac{0!}{2^0} - \frac{1!}{2^1} + \frac{2!}{2^2} - \frac{3!}{2^3} + \cdots$

b. $\displaystyle\sum_{n=1}^{\infty} \frac{(-1)^n}{\sqrt{n}} = -\frac{1}{\sqrt{1}} + \frac{1}{\sqrt{2}} - \frac{1}{\sqrt{3}} + \frac{1}{\sqrt{4}} - \cdots$

Solution

a. This is an alternating series, but the Alternating Series Test does not apply because the limit of the nth term is not zero. By the nth-Term Test for Divergence, however, you can conclude that this series diverges.

b. This series can be shown to be convergent by the Alternating Series Test. Moreover, because the p-series

$$\sum_{n=1}^{\infty} \left| \frac{(-1)^n}{\sqrt{n}} \right| = \frac{1}{\sqrt{1}} + \frac{1}{\sqrt{2}} + \frac{1}{\sqrt{3}} + \frac{1}{\sqrt{4}} + \cdots$$

diverges, the given series is *conditionally* convergent.

EXAMPLE 7 **Absolute and Conditional Convergence**

Determine whether each of the series is convergent or divergent. Classify any convergent series as absolutely or conditionally convergent.

a. $\displaystyle\sum_{n=1}^{\infty} \frac{(-1)^{n(n+1)/2}}{3^n} = -\frac{1}{3} - \frac{1}{9} + \frac{1}{27} + \frac{1}{81} - \cdots$

b. $\displaystyle\sum_{n=1}^{\infty} \frac{(-1)^n}{\ln(n+1)} = -\frac{1}{\ln 2} + \frac{1}{\ln 3} - \frac{1}{\ln 4} + \frac{1}{\ln 5} - \cdots$

Solution

a. This is *not* an alternating series (the signs change in pairs). However, note that

$$\sum_{n=1}^{\infty} \left| \frac{(-1)^{n(n+1)/2}}{3^n} \right| = \sum_{n=1}^{\infty} \frac{1}{3^n}$$

is a convergent geometric series, with

$$r = \frac{1}{3}.$$

Consequently, by Theorem 9.16, you can conclude that the given series is *absolutely* convergent (and therefore convergent).

b. In this case, the Alternating Series Test indicates that the series converges. However, the series

$$\sum_{n=1}^{\infty} \left| \frac{(-1)^n}{\ln(n+1)} \right| = \frac{1}{\ln 2} + \frac{1}{\ln 3} + \frac{1}{\ln 4} + \cdots$$

diverges by direct comparison with the terms of the harmonic series. Therefore, the given series is *conditionally* convergent.

■ FOR FURTHER INFORMATION To read more about the convergence of alternating harmonic series, see the article "Almost Alternating Harmonic Series" by Curtis Feist and Ramin Naimi in *The College Mathematics Journal*. To view this article, go to *MathArticles.com*.

Rearrangement of Series

A finite sum such as

$$1 + 3 - 2 + 5 - 4$$

can be rearranged without changing the value of the sum. This is not necessarily true of an infinite series—it depends on whether the series is absolutely convergent or conditionally convergent.

1. If a series is *absolutely convergent,* then its terms can be rearranged in any order without changing the sum of the series.

2. If a series is *conditionally convergent,* then its terms can be rearranged to give a different sum.

The second case is illustrated in Example 8.

EXAMPLE 8 Rearrangement of a Series

■ FOR FURTHER INFORMATION
Georg Friedrich Bernhard Riemann (1826–1866) proved that if $\Sigma \, a_n$ is conditionally convergent and S is any real number, then the terms of the series can be rearranged to converge to S. For more on this topic, see the article "Riemann's Rearrangement Theorem" by Stewart Galanor in *Mathematics Teacher.* To view this article, go to *MathArticles.com.*

The alternating harmonic series converges to ln 2. That is,

$$\sum_{n=1}^{\infty} (-1)^{n+1} \frac{1}{n} = \frac{1}{1} - \frac{1}{2} + \frac{1}{3} - \frac{1}{4} + \cdots = \ln 2. \qquad \text{(See Exercise 55, Section 9.10.)}$$

Rearrange the series to produce a different sum.

Solution Consider the rearrangement below.

$$1 - \frac{1}{2} - \frac{1}{4} + \frac{1}{3} - \frac{1}{6} - \frac{1}{8} + \frac{1}{5} - \frac{1}{10} - \frac{1}{12} + \frac{1}{7} - \frac{1}{14} - \cdots$$

$$= \left(1 - \frac{1}{2}\right) - \frac{1}{4} + \left(\frac{1}{3} - \frac{1}{6}\right) - \frac{1}{8} + \left(\frac{1}{5} - \frac{1}{10}\right) - \frac{1}{12} + \left(\frac{1}{7} - \frac{1}{14}\right) - \cdots$$

$$= \frac{1}{2} - \frac{1}{4} + \frac{1}{6} - \frac{1}{8} + \frac{1}{10} - \frac{1}{12} + \frac{1}{14} - \cdots$$

$$= \frac{1}{2}\left(1 - \frac{1}{2} + \frac{1}{3} - \frac{1}{4} + \frac{1}{5} - \frac{1}{6} + \frac{1}{7} - \cdots\right)$$

$$= \frac{1}{2}(\ln 2)$$

By rearranging the terms, you obtain a sum that is half the original sum. ■

Exploration

In Example 8, you learned that the alternating harmonic series

$$\sum_{n=1}^{\infty} (-1)^{n+1} \frac{1}{n} = 1 - \frac{1}{2} + \frac{1}{3} - \frac{1}{4} + \frac{1}{5} - \frac{1}{6} + \cdots$$

converges to ln 2 ≈ 0.693. Rearrangement of the terms of the series produces a different sum, $\frac{1}{2}$ ln 2 ≈ 0.347.

In this exploration, you will rearrange the terms of the alternating harmonic series in such a way that two positive terms follow each negative term. That is,

$$1 - \frac{1}{2} + \frac{1}{3} + \frac{1}{5} - \frac{1}{4} + \frac{1}{7} + \frac{1}{9} - \frac{1}{6} + \frac{1}{11} + \cdots.$$

Now calculate the partial sums S_4, S_7, S_{10}, S_{13}, S_{16}, and S_{19}. Then estimate the sum of this series to three decimal places.

9.5 Exercises

See **CalcChat.com** for tutorial help and worked-out solutions to odd-numbered exercises.

Numerical and Graphical Analysis In Exercises 1–4, explore the Alternating Series Remainder.

(a) Use a graphing utility to find the indicated partial sum S_n and complete the table.

n	1	2	3	4	5	6	7	8	9	10
S_n										

(b) Use a graphing utility to graph the first 10 terms of the sequence of partial sums and a horizontal line representing the sum.

(c) What pattern exists between the plot of the successive points in part (b) relative to the horizontal line representing the sum of the series? Do the distances between the successive points and the horizontal line increase or decrease?

(d) Discuss the relationship between the answers in part (c) and the Alternating Series Remainder as given in Theorem 9.15.

1. $\sum_{n=1}^{\infty} \frac{(-1)^{n-1}}{2n-1} = \frac{\pi}{4}$

2. $\sum_{n=1}^{\infty} \frac{(-1)^{n-1}}{(n-1)!} = \frac{1}{e}$

3. $\sum_{n=1}^{\infty} \frac{(-1)^{n-1}}{n^2} = \frac{\pi^2}{12}$

4. $\sum_{n=1}^{\infty} \frac{(-1)^{n-1}}{(2n-1)!} = \sin 1$

Determining Convergence or Divergence In Exercises 5–26, determine the convergence or divergence of the series.

5. $\sum_{n=1}^{\infty} \frac{(-1)^{n+1}}{n+1}$

6. $\sum_{n=1}^{\infty} \frac{(-1)^{n+1} n}{3n+2}$

7. $\sum_{n=1}^{\infty} \frac{(-1)^n}{3^n}$

8. $\sum_{n=1}^{\infty} \frac{(-1)^n}{e^n}$

9. $\sum_{n=1}^{\infty} \frac{(-1)^n(5n-1)}{4n+1}$

10. $\sum_{n=1}^{\infty} \frac{(-1)^{n+1} n}{n^2+5}$

11. $\sum_{n=1}^{\infty} \frac{(-1)^n n}{\ln(n+1)}$

12. $\sum_{n=1}^{\infty} \frac{(-1)^n}{\ln(n+1)}$

13. $\sum_{n=1}^{\infty} \frac{(-1)^n}{\sqrt{n}}$

14. $\sum_{n=1}^{\infty} \frac{(-1)^{n+1} n^2}{n^2+4}$

15. $\sum_{n=1}^{\infty} \frac{(-1)^{n+1}(n+1)}{\ln(n+1)}$

16. $\sum_{n=1}^{\infty} \frac{(-1)^{n+1} \ln(n+1)}{n+1}$

17. $\sum_{n=1}^{\infty} \sin \frac{(2n-1)\pi}{2}$

18. $\sum_{n=1}^{\infty} \frac{1}{n} \cos n\pi$

19. $\sum_{n=0}^{\infty} \frac{(-1)^n}{n!}$

20. $\sum_{n=0}^{\infty} \frac{(-1)^n}{(2n+1)!}$

21. $\sum_{n=1}^{\infty} \frac{(-1)^{n+1} \sqrt{n}}{n+2}$

22. $\sum_{n=1}^{\infty} \frac{(-1)^{n+1} \sqrt{n}}{\sqrt[3]{n}}$

23. $\sum_{n=1}^{\infty} \frac{(-1)^{n+1} n!}{1 \cdot 3 \cdot 5 \cdots (2n-1)}$

24. $\sum_{n=1}^{\infty} (-1)^{n+1} \frac{1 \cdot 3 \cdot 5 \cdots (2n-1)}{1 \cdot 4 \cdot 7 \cdots (3n-2)}$

25. $\sum_{n=1}^{\infty} \frac{2(-1)^{n+1}}{e^n - e^{-n}} = \sum_{n=1}^{\infty} (-1)^{n+1} \operatorname{csch} n$

26. $\sum_{n=1}^{\infty} \frac{2(-1)^{n+1}}{e^n + e^{-n}} = \sum_{n=1}^{\infty} (-1)^{n+1} \operatorname{sech} n$

Approximating the Sum of an Alternating Series In Exercises 27–30, approximate the sum of the series by using the first six terms. (See Example 4.)

27. $\sum_{n=0}^{\infty} \frac{(-1)^n 5}{n!}$

28. $\sum_{n=1}^{\infty} \frac{(-1)^{n+1} 4}{\ln(n+1)}$

29. $\sum_{n=1}^{\infty} \frac{(-1)^{n+1} 2}{n^3}$

30. $\sum_{n=1}^{\infty} \frac{(-1)^{n+1} n}{3^n}$

Finding the Number of Terms In Exercises 31–36, use Theorem 9.15 to determine the number of terms required to approximate the sum of the series with an error of less than 0.001.

31. $\sum_{n=1}^{\infty} \frac{(-1)^{n+1}}{n^3}$

32. $\sum_{n=1}^{\infty} \frac{(-1)^{n+1}}{n^2}$

33. $\sum_{n=1}^{\infty} \frac{(-1)^{n+1}}{2n^3-1}$

34. $\sum_{n=1}^{\infty} \frac{(-1)^{n+1}}{n^5}$

35. $\sum_{n=0}^{\infty} \frac{(-1)^n}{n!}$

36. $\sum_{n=0}^{\infty} \frac{(-1)^n}{(2n)!}$

Determining Absolute and Conditional Convergence In Exercises 37–54, determine whether the series converges absolutely or conditionally, or diverges.

37. $\sum_{n=1}^{\infty} \frac{(-1)^n}{2^n}$

38. $\sum_{n=1}^{\infty} \frac{(-1)^{n+1}}{n^2}$

39. $\sum_{n=1}^{\infty} \frac{(-1)^n}{n!}$

40. $\sum_{n=1}^{\infty} \frac{(-1)^{n+1}}{n+3}$

41. $\sum_{n=1}^{\infty} \frac{(-1)^{n+1}}{\sqrt{n}}$

42. $\sum_{n=1}^{\infty} \frac{(-1)^{n+1}}{n\sqrt{n}}$

43. $\sum_{n=1}^{\infty} \frac{(-1)^{n+1} n^2}{(n+1)^2}$

44. $\sum_{n=1}^{\infty} \frac{(-1)^{n+1}(2n+3)}{n+10}$

45. $\sum_{n=2}^{\infty} \frac{(-1)^n}{n \ln n}$

46. $\sum_{n=0}^{\infty} (-1)^n e^{-n^2}$

47. $\sum_{n=2}^{\infty} \frac{(-1)^n n}{n^3-5}$

48. $\sum_{n=1}^{\infty} \frac{(-1)^{n+1}}{n^{4/3}}$

49. $\sum_{n=0}^{\infty} \frac{(-1)^n}{(2n+1)!}$

50. $\sum_{n=0}^{\infty} \frac{(-1)^n}{\sqrt{n+4}}$

51. $\sum_{n=0}^{\infty} \frac{\cos n\pi}{n+1}$

52. $\sum_{n=1}^{\infty} (-1)^{n+1} \arctan n$

53. $\sum_{n=1}^{\infty} \frac{\cos n\pi}{n^2}$

54. $\sum_{n=1}^{\infty} \frac{\sin[(2n-1)\pi/2]}{n}$

55. Alternating Series Define an alternating series.

56. Alternating Series Test State the Alternating Series Test.

57. Alternating Series Remainder Give the remainder after N terms of a convergent alternating series.

58. Absolute and Conditional Convergence In your own words, state the difference between absolute and conditional convergence of an alternating series.

59. Think About It Do you agree with the following statements? Why or why not?

(a) If both $\Sigma\, a_n$ and $\Sigma\,(-a_n)$ converge, then $\Sigma\,|a_n|$ converges.

(b) If $\Sigma\, a_n$ diverges, then $\Sigma\,|a_n|$ diverges.

60. HOW DO YOU SEE IT? The graphs of the sequences of partial sums of two series are shown in the figures. Which graph represents the partial sums of an alternating series? Explain.

(a)

(b)

True or False? In Exercises 61 and 62, determine whether the statement is true or false. If it is false, explain why or give an example that shows it is false.

61. For the alternating series

$$\sum_{n=1}^{\infty} \frac{(-1)^n}{n}$$

the partial sum S_{100} is an overestimate of the sum of the series.

62. If $\Sigma\, a_n$ and $\Sigma\, b_n$ both converge, then $\Sigma\, a_n b_n$ converges.

Finding Values In Exercises 63 and 64, find the values of p for which the series converges.

63. $\displaystyle\sum_{n=1}^{\infty} (-1)^n \left(\frac{1}{n^p}\right)$

64. $\displaystyle\sum_{n=1}^{\infty} (-1)^n \left(\frac{1}{n+p}\right)$

65. Proof Prove that if $\Sigma\,|a_n|$ converges, then $\Sigma\, a_n^2$ converges. Is the converse true? If not, give an example that shows it is false.

66. Finding a Series Use the result of Exercise 63 to give an example of an alternating p-series that converges, but whose corresponding p-series diverges.

67. Finding a Series Give an example of a series that demonstrates the statement you proved in Exercise 65.

68. Finding Values Find all values of x for which the series $\Sigma\,(x^n/n)$ (a) converges absolutely and (b) converges conditionally.

Using a Series In Exercises 69 and 70, use the given series.

(a) Does the series meet the conditions of Theorem 9.14? Explain why or why not.

(b) Does the series converge? If so, what is the sum?

69. $\dfrac{1}{2} - \dfrac{1}{3} + \dfrac{1}{4} - \dfrac{1}{9} + \dfrac{1}{8} - \dfrac{1}{27} + \cdots + \dfrac{1}{2^n} - \dfrac{1}{3^n} + \cdots$

70. $\displaystyle\sum_{n=1}^{\infty} (-1)^{n+1} a_n,\ a_n = \begin{cases} \dfrac{1}{\sqrt{n}}, & \text{if } n \text{ is odd} \\[2mm] \dfrac{1}{n^3}, & \text{if } n \text{ is even} \end{cases}$

Review In Exercises 71–80, test for convergence or divergence and identify the test used.

71. $\displaystyle\sum_{n=1}^{\infty} \frac{10}{n^{3/2}}$

72. $\displaystyle\sum_{n=1}^{\infty} \frac{3}{n^2 + 5}$

73. $\displaystyle\sum_{n=1}^{\infty} \frac{3^n}{n^2}$

74. $\displaystyle\sum_{n=1}^{\infty} \frac{1}{2^n + 1}$

75. $\displaystyle\sum_{n=0}^{\infty} 5\left(\frac{7}{8}\right)^n$

76. $\displaystyle\sum_{n=1}^{\infty} \frac{3n^2}{2n^2 + 1}$

77. $\displaystyle\sum_{n=1}^{\infty} 100 e^{-n/2}$

78. $\displaystyle\sum_{n=0}^{\infty} \frac{(-1)^n}{n + 4}$

79. $\displaystyle\sum_{n=1}^{\infty} \frac{(-1)^{n+1}\, 4}{3n^2 - 1}$

80. $\displaystyle\sum_{n=2}^{\infty} \frac{\ln n}{n}$

81. Describing an Error The following argument, that $0 = 1$, is *incorrect*. Describe the error.

$0 = 0 + 0 + 0 + \cdots$

$= (1 - 1) + (1 - 1) + (1 - 1) + \cdots$

$= 1 + (-1 + 1) + (-1 + 1) + \cdots$

$= 1 + 0 + 0 + \cdots$

$= 1$

PUTNAM EXAM CHALLENGE

82. Assume as known the (true) fact that the alternating harmonic series

(1) $\ 1 - \frac{1}{2} + \frac{1}{3} - \frac{1}{4} + \frac{1}{5} - \frac{1}{6} + \frac{1}{7} - \frac{1}{8} + \cdots$

is convergent, and denote its sum by s. Rearrange the series (1) as follows:

(2) $\ 1 + \frac{1}{3} - \frac{1}{2} + \frac{1}{5} + \frac{1}{7} - \frac{1}{4} + \frac{1}{9} + \frac{1}{11} - \frac{1}{6} + \cdots.$

Assume as known the (true) fact that the series (2) is also convergent, and denote its sum by S. Denote by s_k, S_k the kth partial sum of the series (1) and (2), respectively. Prove the following statements.

(i) $S_{3n} = s_{4n} + \frac{1}{2}s_{2n},$ (ii) $S \neq s$

This problem was composed by the Committee on the Putnam Prize Competition.
© The Mathematical Association of America. All rights reserved.

9.6 The Ratio and Root Tests

- Use the Ratio Test to determine whether a series converges or diverges.
- Use the Root Test to determine whether a series converges or diverges.
- Review the tests for convergence and divergence of an infinite series.

The Ratio Test

This section begins with a test for absolute convergence—the **Ratio Test.**

> **THEOREM 9.17   Ratio Test**
>
> Let $\Sigma\, a_n$ be a series with nonzero terms.
>
> 1. The series $\Sigma\, a_n$ converges absolutely when $\displaystyle\lim_{n\to\infty}\left|\frac{a_{n+1}}{a_n}\right| < 1$.
>
> 2. The series $\Sigma\, a_n$ diverges when $\displaystyle\lim_{n\to\infty}\left|\frac{a_{n+1}}{a_n}\right| > 1$ or $\displaystyle\lim_{n\to\infty}\left|\frac{a_{n+1}}{a_n}\right| = \infty$.
>
> 3. The Ratio Test is inconclusive when $\displaystyle\lim_{n\to\infty}\left|\frac{a_{n+1}}{a_n}\right| = 1$.

Proof   To prove Property 1, assume that

$$\lim_{n\to\infty}\left|\frac{a_{n+1}}{a_n}\right| = r < 1$$

and choose R such that $0 \le r < R < 1$. By the definition of the limit of a sequence, there exists some $N > 0$ such that $|a_{n+1}/a_n| < R$ for all $n > N$. Therefore, you can write the following inequalities.

$$|a_{N+1}| < |a_N|R$$
$$|a_{N+2}| < |a_{N+1}|R < |a_N|R^2$$
$$|a_{N+3}| < |a_{N+2}|R < |a_{N+1}|R^2 < |a_N|R^3$$
$$\vdots$$

The geometric series $\displaystyle\sum_{n=1}^{\infty} |a_N|R^n = |a_N|R + |a_N|R^2 + \cdots + |a_N|R^n + \cdots$ converges, and so, by the Direct Comparison Test, the series

$$\sum_{n=1}^{\infty} |a_{N+n}| = |a_{N+1}| + |a_{N+2}| + \cdots + |a_{N+n}| + \cdots$$

also converges. This in turn implies that the series $\Sigma\, |a_n|$ converges, because discarding a finite number of terms $(n = N - 1)$ does not affect convergence. Consequently, by Theorem 9.16, the series $\Sigma\, a_n$ converges absolutely. The proof of Property 2 is similar and is left as an exercise (see Exercise 99).

See LarsonCalculus.com for Bruce Edwards's video of this proof.

The fact that the Ratio Test is inconclusive when $|a_{n+1}/a_n| \to 1$ can be seen by comparing the two series $\Sigma\, (1/n)$ and $\Sigma\, (1/n^2)$. The first series diverges and the second one converges, but in both cases

$$\lim_{n\to\infty}\left|\frac{a_{n+1}}{a_n}\right| = 1.$$

Although the Ratio Test is not a cure for all ills related to testing for convergence, it is particularly useful for series that *converge rapidly.* Series involving factorials or exponentials are frequently of this type.

EXAMPLE 1 **Using the Ratio Test**

Determine the convergence or divergence of

$$\sum_{n=0}^{\infty} \frac{2^n}{n!}.$$

Solution Because

$$a_n = \frac{2^n}{n!}$$

you can write the following.

$$\lim_{n\to\infty} \left| \frac{a_{n+1}}{a_n} \right| = \lim_{n\to\infty} \left[\frac{2^{n+1}}{(n+1)!} \div \frac{2^n}{n!} \right]$$

$$= \lim_{n\to\infty} \left[\frac{2^{n+1}}{(n+1)!} \cdot \frac{n!}{2^n} \right]$$

$$= \lim_{n\to\infty} \frac{2}{n+1}$$

$$= 0 < 1$$

This series converges because the limit of $|a_{n+1}/a_n|$ is less than 1.

> **REMARK** A step frequently used in applications of the Ratio Test involves simplifying quotients of factorials. In Example 1, for instance, notice that
>
> $$\frac{n!}{(n+1)!} = \frac{n!}{(n+1)n!} = \frac{1}{n+1}.$$

EXAMPLE 2 **Using the Ratio Test**

Determine whether each series converges or diverges.

a. $\displaystyle\sum_{n=0}^{\infty} \frac{n^2 2^{n+1}}{3^n}$ **b.** $\displaystyle\sum_{n=1}^{\infty} \frac{n^n}{n!}$

Solution

a. This series converges because the limit of $|a_{n+1}/a_n|$ is less than 1.

$$\lim_{n\to\infty} \left| \frac{a_{n+1}}{a_n} \right| = \lim_{n\to\infty} \left[(n+1)^2 \left(\frac{2^{n+2}}{3^{n+1}} \right) \left(\frac{3^n}{n^2 2^{n+1}} \right) \right]$$

$$= \lim_{n\to\infty} \frac{2(n+1)^2}{3n^2}$$

$$= \frac{2}{3} < 1$$

b. This series diverges because the limit of $|a_{n+1}/a_n|$ is greater than 1.

$$\lim_{n\to\infty} \left| \frac{a_{n+1}}{a_n} \right| = \lim_{n\to\infty} \left[\frac{(n+1)^{n+1}}{(n+1)!} \left(\frac{n!}{n^n} \right) \right]$$

$$= \lim_{n\to\infty} \left[\frac{(n+1)^{n+1}}{(n+1)} \left(\frac{1}{n^n} \right) \right]$$

$$= \lim_{n\to\infty} \frac{(n+1)^n}{n^n}$$

$$= \lim_{n\to\infty} \left(1 + \frac{1}{n} \right)^n$$

$$= e > 1$$

EXAMPLE 3   **A Failure of the Ratio Test**

· · · · ▷ *See LarsonCalculus.com for an interactive version of this type of example.*

Determine the convergence or divergence of

$$\sum_{n=1}^{\infty} (-1)^n \frac{\sqrt{n}}{n+1}.$$

Solution   The limit of $|a_{n+1}/a_n|$ is equal to 1.

$$\lim_{n\to\infty} \left| \frac{a_{n+1}}{a_n} \right| = \lim_{n\to\infty} \left[\left(\frac{\sqrt{n+1}}{n+2} \right) \left(\frac{n+1}{\sqrt{n}} \right) \right]$$

$$= \lim_{n\to\infty} \left[\sqrt{\frac{n+1}{n}} \left(\frac{n+1}{n+2} \right) \right]$$

$$= \sqrt{1}\,(1)$$

$$= 1$$

So, the Ratio Test is inconclusive. To determine whether the series converges, you need to try a different test. In this case, you can apply the Alternating Series Test. To show that $a_{n+1} \le a_n$, let

$$f(x) = \frac{\sqrt{x}}{x+1}.$$

Then the derivative is

$$f'(x) = \frac{-x+1}{2\sqrt{x}(x+1)^2}.$$

Because the derivative is negative for $x > 1$, you know that f is a decreasing function. Also, by L'Hôpital's Rule,

$$\lim_{x\to\infty} \frac{\sqrt{x}}{x+1} = \lim_{x\to\infty} \frac{1/(2\sqrt{x})}{1}$$

$$= \lim_{x\to\infty} \frac{1}{2\sqrt{x}}$$

$$= 0.$$

Therefore, by the Alternating Series Test, the series converges.   ■

> **REMARK**   The Ratio Test is also inconclusive for any *p*-series.

The series in Example 3 is *conditionally convergent*. This follows from the fact that the series

$$\sum_{n=1}^{\infty} |a_n|$$

diverges $\left(\text{by the Limit Comparison Test with } \Sigma\, 1/\sqrt{n}\right)$, but the series

$$\sum_{n=1}^{\infty} a_n$$

converges.

▷ **TECHNOLOGY**   A graphing utility can reinforce the conclusion that the series in Example 3 converges *conditionally*. By adding the first 100 terms of the series, you obtain a sum of about -0.2. (The sum of the first 100 terms of the series $\Sigma\, |a_n|$ is about 17.)

The Root Test

The next test for convergence or divergence of series works especially well for series involving nth powers. The proof of this theorem is similar to the proof given for the Ratio Test, and is left as an exercise (see Exercise 100).

· · · · · · · · · · · · · · · ▷

•• **REMARK** The Root Test is always inconclusive for any p-series.

THEOREM 9.18   Root Test

1. The series $\Sigma\, a_n$ converges absolutely when $\lim\limits_{n\to\infty} \sqrt[n]{|a_n|} < 1$.

2. The series $\Sigma\, a_n$ diverges when $\lim\limits_{n\to\infty} \sqrt[n]{|a_n|} > 1$ or $\lim\limits_{n\to\infty} \sqrt[n]{|a_n|} = \infty$.

3. The Root Test is inconclusive when $\lim\limits_{n\to\infty} \sqrt[n]{|a_n|} = 1$.

EXAMPLE 4   **Using the Root Test**

Determine the convergence or divergence of

$$\sum_{n=1}^{\infty} \frac{e^{2n}}{n^n}.$$

Solution   You can apply the Root Test as follows.

$$\lim_{n\to\infty} \sqrt[n]{|a_n|} = \lim_{n\to\infty} \sqrt[n]{\frac{e^{2n}}{n^n}}$$

$$= \lim_{n\to\infty} \frac{e^{2n/n}}{n^{n/n}}$$

$$= \lim_{n\to\infty} \frac{e^2}{n}$$

$$= 0 < 1$$

Because this limit is less than 1, you can conclude that the series converges absolutely (and therefore converges). ∎

To see the usefulness of the Root Test for the series in Example 4, try applying the Ratio Test to that series. When you do this, you obtain the following.

$$\lim_{n\to\infty} \left| \frac{a_{n+1}}{a_n} \right| = \lim_{n\to\infty} \left[\frac{e^{2(n+1)}}{(n+1)^{n+1}} \div \frac{e^{2n}}{n^n} \right]$$

$$= \lim_{n\to\infty} \left[\frac{e^{2(n+1)}}{(n+1)^{n+1}} \cdot \frac{n^n}{e^{2n}} \right]$$

$$= \lim_{n\to\infty} e^2 \frac{n^n}{(n+1)^{n+1}}$$

$$= \lim_{n\to\infty} e^2 \left(\frac{n}{n+1} \right)^n \left(\frac{1}{n+1} \right)$$

$$= 0$$

Note that this limit is not as easily evaluated as the limit obtained by the Root Test in Example 4.

■ **FOR FURTHER INFORMATION**   For more information on the usefulness of the Root Test, see the article "*N*! and the Root Test" by Charles C. Mumma II in *The American Mathematical Monthly*. To view this article, go to *MathArticles.com*.

Strategies for Testing Series

You have now studied 10 tests for determining the convergence or divergence of an infinite series. (See the summary in the table on the next page.) Skill in choosing and applying the various tests will come only with practice. Below is a set of guidelines for choosing an appropriate test.

GUIDELINES FOR TESTING A SERIES FOR CONVERGENCE OR DIVERGENCE

1. Does the nth term approach 0? If not, the series diverges.
2. Is the series one of the special types—geometric, p-series, telescoping, or alternating?
3. Can the Integral Test, the Root Test, or the Ratio Test be applied?
4. Can the series be compared favorably to one of the special types?

In some instances, more than one test is applicable. However, your objective should be to learn to choose the most efficient test.

EXAMPLE 5   Applying the Strategies for Testing Series

Determine the convergence or divergence of each series.

a. $\displaystyle\sum_{n=1}^{\infty} \frac{n+1}{3n+1}$ b. $\displaystyle\sum_{n=1}^{\infty} \left(\frac{\pi}{6}\right)^n$ c. $\displaystyle\sum_{n=1}^{\infty} ne^{-n^2}$

d. $\displaystyle\sum_{n=1}^{\infty} \frac{1}{3n+1}$ e. $\displaystyle\sum_{n=1}^{\infty} (-1)^n \frac{3}{4n+1}$ f. $\displaystyle\sum_{n=1}^{\infty} \frac{n!}{10^n}$

g. $\displaystyle\sum_{n=1}^{\infty} \left(\frac{n+1}{2n+1}\right)^n$

Solution

a. For this series, the limit of the nth term is not 0 $\left(a_n \to \frac{1}{3} \text{ as } n \to \infty\right)$. So, by the nth-Term Test, the series diverges.

b. This series is geometric. Moreover, because the ratio of the terms

$$r = \frac{\pi}{6}$$

is less than 1 in absolute value, you can conclude that the series converges.

c. Because the function

$$f(x) = xe^{-x^2}$$

is easily integrated, you can use the Integral Test to conclude that the series converges.

d. The nth term of this series can be compared to the nth term of the harmonic series. After using the Limit Comparison Test, you can conclude that the series diverges.

e. This is an alternating series whose nth term approaches 0. Because $a_{n+1} \le a_n$, you can use the Alternating Series Test to conclude that the series converges.

f. The nth term of this series involves a factorial, which indicates that the Ratio Test may work well. After applying the Ratio Test, you can conclude that the series diverges.

g. The nth term of this series involves a variable that is raised to the nth power, which indicates that the Root Test may work well. After applying the Root Test, you can conclude that the series converges.

SUMMARY OF TESTS FOR SERIES

Test	Series	Condition(s) of Convergence	Condition(s) of Divergence	Comment
nth-Term	$\displaystyle\sum_{n=1}^{\infty} a_n$		$\displaystyle\lim_{n\to\infty} a_n \neq 0$	This test cannot be used to show convergence.
Geometric Series	$\displaystyle\sum_{n=0}^{\infty} ar^n$	$0 < |r| < 1$	$|r| \geq 1$	Sum: $S = \dfrac{a}{1-r}$
Telescoping Series	$\displaystyle\sum_{n=1}^{\infty} (b_n - b_{n+1})$	$\displaystyle\lim_{n\to\infty} b_n = L$		Sum: $S = b_1 - L$
p-Series	$\displaystyle\sum_{n=1}^{\infty} \dfrac{1}{n^p}$	$p > 1$	$0 < p \leq 1$	
Alternating Series	$\displaystyle\sum_{n=1}^{\infty} (-1)^{n-1} a_n$	$0 < a_{n+1} \leq a_n$ and $\displaystyle\lim_{n\to\infty} a_n = 0$		Remainder: $|R_N| \leq a_{N+1}$
Integral (f is continuous, positive, and decreasing)	$\displaystyle\sum_{n=1}^{\infty} a_n,$ $a_n = f(n) \geq 0$	$\displaystyle\int_1^\infty f(x)\,dx$ converges	$\displaystyle\int_1^\infty f(x)\,dx$ diverges	Remainder: $0 < R_N < \displaystyle\int_N^\infty f(x)\,dx$
Root	$\displaystyle\sum_{n=1}^{\infty} a_n$	$\displaystyle\lim_{n\to\infty} \sqrt[n]{|a_n|} < 1$	$\displaystyle\lim_{n\to\infty} \sqrt[n]{|a_n|} > 1$ or $= \infty$	Test is inconclusive when $\displaystyle\lim_{n\to\infty} \sqrt[n]{|a_n|} = 1$.
Ratio	$\displaystyle\sum_{n=1}^{\infty} a_n$	$\displaystyle\lim_{n\to\infty} \left|\dfrac{a_{n+1}}{a_n}\right| < 1$	$\displaystyle\lim_{n\to\infty} \left|\dfrac{a_{n+1}}{a_n}\right| > 1$ or $= \infty$	Test is inconclusive when $\displaystyle\lim_{n\to\infty} \left|\dfrac{a_{n+1}}{a_n}\right| = 1$.
Direct Comparison $(a_n, b_n > 0)$	$\displaystyle\sum_{n=1}^{\infty} a_n$	$0 < a_n \leq b_n$ and $\displaystyle\sum_{n=1}^{\infty} b_n$ converges	$0 < b_n \leq a_n$ and $\displaystyle\sum_{n=1}^{\infty} b_n$ diverges	
Limit Comparison $(a_n, b_n > 0)$	$\displaystyle\sum_{n=1}^{\infty} a_n$	$\displaystyle\lim_{n\to\infty} \dfrac{a_n}{b_n} = L > 0$ and $\displaystyle\sum_{n=1}^{\infty} b_n$ converges	$\displaystyle\lim_{n\to\infty} \dfrac{a_n}{b_n} = L > 0$ and $\displaystyle\sum_{n=1}^{\infty} b_n$ diverges	

9.6   Exercises

See **CalcChat.com** for tutorial help and worked-out solutions to odd-numbered exercises.

Verifying a Formula   In Exercises 1–4, verify the formula.

1. $\dfrac{(n + 1)!}{(n - 2)!} = (n + 1)(n)(n - 1)$

2. $\dfrac{(2k - 2)!}{(2k)!} = \dfrac{1}{(2k)(2k - 1)}$

3. $1 \cdot 3 \cdot 5 \cdots (2k - 1) = \dfrac{(2k)!}{2^k k!}$

4. $\dfrac{1}{1 \cdot 3 \cdot 5 \cdots (2k - 5)} = \dfrac{2^k k!(2k - 3)(2k - 1)}{(2k)!}, \quad k \geq 3$

Matching   In Exercises 5–10, match the series with the graph of its sequence of partial sums. [The graphs are labeled (a), (b), (c), (d), (e), and (f).]

(a)

(b)

(c)

(d)

(e)

(f)
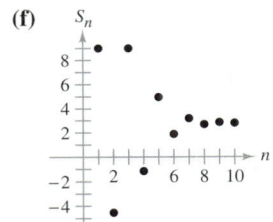

5. $\displaystyle\sum_{n=1}^{\infty} n\left(\dfrac{3}{4}\right)^n$

6. $\displaystyle\sum_{n=1}^{\infty} \left(\dfrac{3}{4}\right)^n\left(\dfrac{1}{n!}\right)$

7. $\displaystyle\sum_{n=1}^{\infty} \dfrac{(-3)^{n+1}}{n!}$

8. $\displaystyle\sum_{n=1}^{\infty} \dfrac{(-1)^{n-1} 4}{(2n)!}$

9. $\displaystyle\sum_{n=1}^{\infty} \left(\dfrac{4n}{5n - 3}\right)^n$

10. $\displaystyle\sum_{n=0}^{\infty} 4e^{-n}$

Numerical, Graphical, and Analytic Analysis   In Exercises 11 and 12, (a) verify that the series converges, (b) use a graphing utility to find the indicated partial sum S_n and complete the table, (c) use a graphing utility to graph the first 10 terms of the sequence of partial sums, (d) use the table to estimate the sum of the series, and (e) explain the relationship between the magnitudes of the terms of the series and the rate at which the sequence of partial sums approaches the sum of the series.

n	5	10	15	20	25
S_n					

11. $\displaystyle\sum_{n=1}^{\infty} n^3\left(\dfrac{1}{2}\right)^n$

12. $\displaystyle\sum_{n=1}^{\infty} \dfrac{n^2 + 1}{n!}$

Using the Ratio Test   In Exercises 13–34, use the Ratio Test to determine the convergence or divergence of the series.

13. $\displaystyle\sum_{n=1}^{\infty} \dfrac{1}{5^n}$

14. $\displaystyle\sum_{n=1}^{\infty} \dfrac{1}{n!}$

15. $\displaystyle\sum_{n=0}^{\infty} \dfrac{n!}{3^n}$

16. $\displaystyle\sum_{n=0}^{\infty} \dfrac{2^n}{n!}$

17. $\displaystyle\sum_{n=1}^{\infty} n\left(\dfrac{6}{5}\right)^n$

18. $\displaystyle\sum_{n=1}^{\infty} n\left(\dfrac{7}{8}\right)^n$

19. $\displaystyle\sum_{n=1}^{\infty} \dfrac{n}{4^n}$

20. $\displaystyle\sum_{n=1}^{\infty} \dfrac{5^n}{n^4}$

21. $\displaystyle\sum_{n=1}^{\infty} \dfrac{n^3}{3^n}$

22. $\displaystyle\sum_{n=1}^{\infty} \dfrac{(-1)^{n+1}(n + 2)}{n(n + 1)}$

23. $\displaystyle\sum_{n=0}^{\infty} \dfrac{(-1)^n 2^n}{n!}$

24. $\displaystyle\sum_{n=1}^{\infty} \dfrac{(-1)^{n-1}(3/2)^n}{n^2}$

25. $\displaystyle\sum_{n=1}^{\infty} \dfrac{n!}{n3^n}$

26. $\displaystyle\sum_{n=1}^{\infty} \dfrac{(2n)!}{n^5}$

27. $\displaystyle\sum_{n=0}^{\infty} \dfrac{e^n}{n!}$

28. $\displaystyle\sum_{n=1}^{\infty} \dfrac{n!}{n^n}$

29. $\displaystyle\sum_{n=0}^{\infty} \dfrac{6^n}{(n + 1)^n}$

30. $\displaystyle\sum_{n=0}^{\infty} \dfrac{(n!)^2}{(3n)!}$

31. $\displaystyle\sum_{n=0}^{\infty} \dfrac{5^n}{2^n + 1}$

32. $\displaystyle\sum_{n=0}^{\infty} \dfrac{(-1)^n 2^{4n}}{(2n + 1)!}$

33. $\displaystyle\sum_{n=0}^{\infty} \dfrac{(-1)^{n+1} n!}{1 \cdot 3 \cdot 5 \cdots (2n + 1)}$

34. $\displaystyle\sum_{n=1}^{\infty} \dfrac{(-1)^n [2 \cdot 4 \cdot 6 \cdots (2n)]}{2 \cdot 5 \cdot 8 \cdots (3n - 1)}$

Using the Root Test In Exercises 35–50, use the Root Test to determine the convergence or divergence of the series.

35. $\displaystyle\sum_{n=1}^{\infty} \frac{1}{5^n}$

36. $\displaystyle\sum_{n=1}^{\infty} \frac{1}{n^n}$

37. $\displaystyle\sum_{n=1}^{\infty} \left(\frac{n}{2n+1}\right)^n$

38. $\displaystyle\sum_{n=1}^{\infty} \left(\frac{2n}{n+1}\right)^n$

39. $\displaystyle\sum_{n=1}^{\infty} \left(\frac{3n+2}{n+3}\right)^n$

40. $\displaystyle\sum_{n=1}^{\infty} \left(\frac{n-2}{5n+1}\right)^n$

41. $\displaystyle\sum_{n=2}^{\infty} \frac{(-1)^n}{(\ln n)^n}$

42. $\displaystyle\sum_{n=1}^{\infty} \left(\frac{-3n}{2n+1}\right)^{3n}$

43. $\displaystyle\sum_{n=1}^{\infty} \left(2\sqrt[n]{n}+1\right)^n$

44. $\displaystyle\sum_{n=0}^{\infty} e^{-3n}$

45. $\displaystyle\sum_{n=1}^{\infty} \frac{n}{3^n}$

46. $\displaystyle\sum_{n=1}^{\infty} \left(\frac{n}{500}\right)^n$

47. $\displaystyle\sum_{n=1}^{\infty} \left(\frac{1}{n}-\frac{1}{n^2}\right)^n$

48. $\displaystyle\sum_{n=1}^{\infty} \left(\frac{\ln n}{n}\right)^n$

49. $\displaystyle\sum_{n=2}^{\infty} \frac{n}{(\ln n)^n}$

50. $\displaystyle\sum_{n=1}^{\infty} \frac{(n!)^n}{(n^n)^2}$

Determining Convergence or Divergence In Exercises 51–68, determine the convergence or divergence of the series using any appropriate test from this chapter. Identify the test used.

51. $\displaystyle\sum_{n=1}^{\infty} \frac{(-1)^{n+1}5}{n}$

52. $\displaystyle\sum_{n=1}^{\infty} \frac{100}{n}$

53. $\displaystyle\sum_{n=1}^{\infty} \frac{3}{n\sqrt{n}}$

54. $\displaystyle\sum_{n=1}^{\infty} \left(\frac{2\pi}{3}\right)^n$

55. $\displaystyle\sum_{n=1}^{\infty} \frac{5n}{2n-1}$

56. $\displaystyle\sum_{n=1}^{\infty} \frac{n}{2n^2+1}$

57. $\displaystyle\sum_{n=1}^{\infty} \frac{(-1)^n 3^{n-2}}{2^n}$

58. $\displaystyle\sum_{n=1}^{\infty} \frac{10}{3\sqrt{n^3}}$

59. $\displaystyle\sum_{n=1}^{\infty} \frac{10n+3}{n2^n}$

60. $\displaystyle\sum_{n=1}^{\infty} \frac{2^n}{4n^2-1}$

61. $\displaystyle\sum_{n=1}^{\infty} \frac{\cos n}{3^n}$

62. $\displaystyle\sum_{n=2}^{\infty} \frac{(-1)^n}{n\ln n}$

63. $\displaystyle\sum_{n=1}^{\infty} \frac{n!}{n7^n}$

64. $\displaystyle\sum_{n=1}^{\infty} \frac{\ln n}{n^2}$

65. $\displaystyle\sum_{n=1}^{\infty} \frac{(-1)^n 3^{n-1}}{n!}$

66. $\displaystyle\sum_{n=1}^{\infty} \frac{(-1)^n 3^n}{n2^n}$

67. $\displaystyle\sum_{n=1}^{\infty} \frac{(-3)^n}{3\cdot5\cdot7\cdots(2n+1)}$

68. $\displaystyle\sum_{n=1}^{\infty} \frac{3\cdot5\cdot7\cdots(2n+1)}{18^n(2n-1)n!}$

Identifying Series In Exercises 69–72, identify the two series that are the same.

69. (a) $\displaystyle\sum_{n=1}^{\infty} \frac{n5^n}{n!}$

(b) $\displaystyle\sum_{n=0}^{\infty} \frac{n5^n}{(n+1)!}$

(c) $\displaystyle\sum_{n=0}^{\infty} \frac{(n+1)5^{n+1}}{(n+1)!}$

70. (a) $\displaystyle\sum_{n=4}^{\infty} n\left(\frac{3}{4}\right)^n$

(b) $\displaystyle\sum_{n=0}^{\infty} (n+1)\left(\frac{3}{4}\right)^n$

(c) $\displaystyle\sum_{n=1}^{\infty} n\left(\frac{3}{4}\right)^{n-1}$

71. (a) $\displaystyle\sum_{n=0}^{\infty} \frac{(-1)^n}{(2n+1)!}$

(b) $\displaystyle\sum_{n=1}^{\infty} \frac{(-1)^{n-1}}{(2n-1)!}$

(c) $\displaystyle\sum_{n=1}^{\infty} \frac{(-1)^{n-1}}{(2n+1)!}$

72. (a) $\displaystyle\sum_{n=2}^{\infty} \frac{(-1)^n}{(n-1)2^{n-1}}$

(b) $\displaystyle\sum_{n=1}^{\infty} \frac{(-1)^{n+1}}{n2^n}$

(c) $\displaystyle\sum_{n=0}^{\infty} \frac{(-1)^{n+1}}{(n+1)2^n}$

Writing an Equivalent Series In Exercises 73 and 74, write an equivalent series with the index of summation beginning at $n=0$.

73. $\displaystyle\sum_{n=1}^{\infty} \frac{n}{7^n}$

74. $\displaystyle\sum_{n=2}^{\infty} \frac{9^n}{(n-2)!}$

Finding the Number of Terms In Exercises 75 and 76, (a) determine the number of terms required to approximate the sum of the series with an error less than 0.0001, and (b) use a graphing utility to approximate the sum of the series with an error less than 0.0001.

75. $\displaystyle\sum_{k=1}^{\infty} \frac{(-3)^k}{2^k k!}$

76. $\displaystyle\sum_{k=0}^{\infty} \frac{(-3)^k}{1\cdot3\cdot5\cdots(2k+1)}$

Using a Recursively Defined Series In Exercises 77–82, the terms of a series $\displaystyle\sum_{n=1}^{\infty} a_n$ are defined recursively. Determine the convergence or divergence of the series. Explain your reasoning.

77. $a_1 = \dfrac{1}{2}, a_{n+1} = \dfrac{4n-1}{3n+2}a_n$

78. $a_1 = 2, a_{n+1} = \dfrac{2n+1}{5n-4}a_n$

79. $a_1 = 1, a_{n+1} = \dfrac{\sin n+1}{\sqrt{n}}a_n$

80. $a_1 = \dfrac{1}{5}, a_{n+1} = \dfrac{\cos n+1}{n}a_n$

81. $a_1 = \dfrac{1}{3}, a_{n+1} = \left(1+\dfrac{1}{n}\right)a_n$

82. $a_1 = \dfrac{1}{4}, a_{n+1} = \sqrt[n]{a_n}$

Using the Ratio Test or Root Test In Exercises 83–86, use the Ratio Test or the Root Test to determine the convergence or divergence of the series.

83. $1 + \dfrac{1\cdot2}{1\cdot3} + \dfrac{1\cdot2\cdot3}{1\cdot3\cdot5} + \dfrac{1\cdot2\cdot3\cdot4}{1\cdot3\cdot5\cdot7} + \cdots$

84. $1 + \dfrac{2}{3} + \dfrac{3}{3^2} + \dfrac{4}{3^3} + \dfrac{5}{3^4} + \dfrac{6}{3^5} + \cdots$

85. $\dfrac{1}{(\ln 3)^3} + \dfrac{1}{(\ln 4)^4} + \dfrac{1}{(\ln 5)^5} + \dfrac{1}{(\ln 6)^6} + \cdots$

86. $1 + \dfrac{1\cdot3}{1\cdot2\cdot3} + \dfrac{1\cdot3\cdot5}{1\cdot2\cdot3\cdot4\cdot5}$
$\qquad + \dfrac{1\cdot3\cdot5\cdot7}{1\cdot2\cdot3\cdot4\cdot5\cdot6\cdot7} + \cdots$

Finding Values   In Exercises 87–92, find the values of x for which the series converges.

87. $\displaystyle\sum_{n=0}^{\infty} 2\left(\frac{x}{3}\right)^n$

88. $\displaystyle\sum_{n=0}^{\infty} \left(\frac{x-3}{5}\right)^n$

89. $\displaystyle\sum_{n=1}^{\infty} \frac{(-1)^n(x+1)^n}{n}$

90. $\displaystyle\sum_{n=0}^{\infty} 3(x-4)^n$

91. $\displaystyle\sum_{n=0}^{\infty} n!\left(\frac{x}{2}\right)^n$

92. $\displaystyle\sum_{n=0}^{\infty} \frac{(x+1)^n}{n!}$

WRITING ABOUT CONCEPTS

93. Ratio Test   State the Ratio Test.

94. Root Test   State the Root Test.

95. Think About It   You are told that the terms of a positive series appear to approach zero rapidly as n approaches infinity. In fact, $a_7 \le 0.0001$. Given no other information, does this imply that the series converges? Support your conclusion with examples.

96. Think About It   What can you conclude about the convergence or divergence of $\Sigma\, a_n$ for each of the following conditions? Explain your reasoning.

(a) $\displaystyle\lim_{n\to\infty}\left|\frac{a_{n+1}}{a_n}\right| = 0$ (b) $\displaystyle\lim_{n\to\infty}\left|\frac{a_{n+1}}{a_n}\right| = 1$

(c) $\displaystyle\lim_{n\to\infty}\left|\frac{a_{n+1}}{a_n}\right| = \frac{3}{2}$ (d) $\displaystyle\lim_{n\to\infty}\sqrt[n]{|a_n|} = 2$

(e) $\displaystyle\lim_{n\to\infty}\sqrt[n]{|a_n|} = 1$ (f) $\displaystyle\lim_{n\to\infty}\sqrt[n]{|a_n|} = e$

97. Using an Alternating Series   Using the Ratio Test, it is determined that an alternating series converges. Does the series converge conditionally or absolutely? Explain.

---

98. HOW DO YOU SEE IT?   The figure shows the first 10 terms of the convergent series $\displaystyle\sum_{n=1}^{\infty} a_n$ and the first 10 terms of the convergent series $\displaystyle\sum_{n=1}^{\infty} \sqrt{a_n}$. Identify the two series and explain your reasoning in making the selection.

---

99. Proof   Prove Property 2 of Theorem 9.17.

100. Proof   Prove Theorem 9.18. (*Hint for Property 1:* If the limit equals $r < 1$, choose a real number R such that $r < R < 1$. By the definitions of the limit, there exists some $N > 0$ such that $\sqrt[n]{|a_n|} < R$ for $n > N$.)

Verifying an Inconclusive Test   In Exercises 101–104, verify that the Ratio Test is inconclusive for the p-series.

101. $\displaystyle\sum_{n=1}^{\infty} \frac{1}{n^{3/2}}$

102. $\displaystyle\sum_{n=1}^{\infty} \frac{1}{n^{1/2}}$

103. $\displaystyle\sum_{n=1}^{\infty} \frac{1}{n^4}$

104. $\displaystyle\sum_{n=1}^{\infty} \frac{1}{n^p}$

105. Verifying an Inconclusive Test   Show that the Root Test is inconclusive for the p-series

$$\sum_{n=1}^{\infty} \frac{1}{n^p}.$$

106. Verifying Inconclusive Tests   Show that the Ratio Test and the Root Test are both inconclusive for the logarithmic p-series

$$\sum_{n=2}^{\infty} \frac{1}{n(\ln n)^p}.$$

107. Using Values   Determine the convergence or divergence of the series

$$\sum_{n=1}^{\infty} \frac{(n!)^2}{(xn)!}$$

when (a) $x = 1$, (b) $x = 2$, (c) $x = 3$, and (d) x is a positive integer.

108. Using a Series   Show that if

$$\sum_{n=1}^{\infty} a_n$$

is absolutely convergent, then

$$\left|\sum_{n=1}^{\infty} a_n\right| \le \sum_{n=1}^{\infty} |a_n|.$$

PUTNAM EXAM CHALLENGE

109. Show that if the series

$$a_1 + a_2 + a_3 + \cdots + a_n + \cdots$$

converges, then the series

$$a_1 + \frac{a_2}{2} + \frac{a_3}{3} + \cdots + \frac{a_n}{n} + \cdots$$

converges also.

110. Is the following series convergent or divergent?

$$1 + \frac{1}{2}\cdot\frac{19}{7} + \frac{2!}{3^2}\left(\frac{19}{7}\right)^2 + \frac{3!}{4^3}\left(\frac{19}{7}\right)^3 + \frac{4!}{5^4}\left(\frac{19}{7}\right)^4 + \cdots$$

These problems were composed by the Committee on the Putnam Prize Competition.
© The Mathematical Association of America. All rights reserved.

9.7 Taylor Polynomials and Approximations

■ Find polynomial approximations of elementary functions and compare them with the elementary functions.
■ Find Taylor and Maclaurin polynomial approximations of elementary functions.
■ Use the remainder of a Taylor polynomial.

Polynomial Approximations of Elementary Functions

The goal of this section is to show how polynomial functions can be used as approximations for other elementary functions. To find a polynomial function P that approximates another function f, begin by choosing a number c in the domain of f at which f and P have the same value. That is,

$$P(c) = f(c). \qquad \text{Graphs of } f \text{ and } P \text{ pass through } (c, f(c)).$$

The approximating polynomial is said to be **expanded about** c or **centered at** c. Geometrically, the requirement that $P(c) = f(c)$ means that the graph of P passes through the point $(c, f(c))$. Of course, there are many polynomials whose graphs pass through the point $(c, f(c))$. Your task is to find a polynomial whose graph resembles the graph of f near this point. One way to do this is to impose the additional requirement that the slope of the polynomial function be the same as the slope of the graph of f at the point $(c, f(c))$.

$$P'(c) = f'(c) \qquad \text{Graphs of } f \text{ and } P \text{ have the same slope at } (c, f(c)).$$

With these two requirements, you can obtain a simple linear approximation of f, as shown in Figure 9.10.

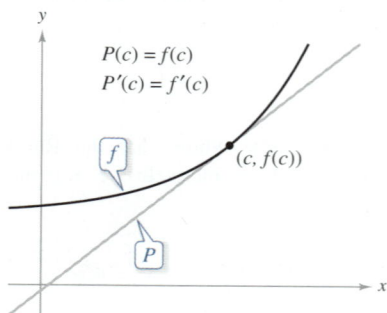

Near $(c, f(c))$, the graph of P can be used to approximate the graph of f.
Figure 9.10

▷

•• REMARK Example 1 is not the first time you have used a linear function to approximate another function. The same procedure was used as the basis for Newton's Method.

EXAMPLE 1 **First-Degree Polynomial Approximation of $f(x) = e^x$**

For the function $f(x) = e^x$, find a first-degree polynomial function $P_1(x) = a_0 + a_1 x$ whose value and slope agree with the value and slope of f at $x = 0$.

Solution Because $f(x) = e^x$ and $f'(x) = e^x$, the value and the slope of f at $x = 0$ are

$$f(0) = e^0 = 1 \qquad \text{Value of } f \text{ at } x = 0$$

and

$$f'(0) = e^0 = 1. \qquad \text{Slope of } f \text{ at } x = 0$$

Because $P_1(x) = a_0 + a_1 x$, you can use the condition that $P_1(0) = f(0)$ to conclude that $a_0 = 1$. Moreover, because $P_1'(x) = a_1$, you can use the condition that $P_1'(0) = f'(0)$ to conclude that $a_1 = 1$. Therefore, $P_1(x) = 1 + x$. Figure 9.11 shows the graphs of $P_1(x) = 1 + x$ and $f(x) = e^x$.

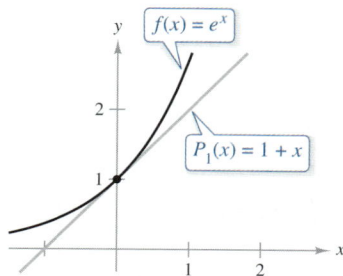

P_1 is the first-degree polynomial approximation of $f(x) = e^x$.
Figure 9.11

In Figure 9.12, you can see that, at points near $(0, 1)$, the graph of the first-degree polynomial function

$$P_1(x) = 1 + x \qquad \text{1st-degree approximation}$$

is reasonably close to the graph of $f(x) = e^x$. As you move away from $(0, 1)$, however, the graphs move farther and farther from each other and the accuracy of the approximation decreases. To improve the approximation, you can impose yet another requirement—that the values of the second derivatives of P and f agree when $x = 0$. The polynomial, P_2, of least degree that satisfies all three requirements $P_2(0) = f(0)$, $P_2'(0) = f'(0)$, and $P_2''(0) = f''(0)$ can be shown to be

$$P_2(x) = 1 + x + \frac{1}{2}x^2. \qquad \text{2nd-degree approximation}$$

Moreover, in Figure 9.12, you can see that P_2 is a better approximation of f than P_1. By requiring that the values of $P_n(x)$ and its first n derivatives match those of $f(x) = e^x$ at $x = 0$, you obtain the nth-degree approximation shown below.

$$P_n(x) = 1 + x + \frac{1}{2}x^2 + \frac{1}{3!}x^3 + \cdots + \frac{1}{n!}x^n \qquad \text{nth-degree approximation}$$

$$\approx e^x$$

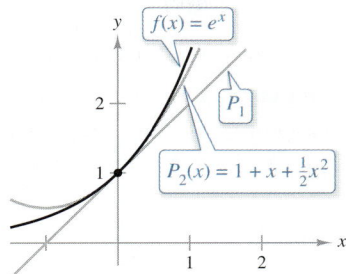

P_2 is the second-degree polynomial approximation of $f(x) = e^x$.
Figure 9.12

EXAMPLE 2　Third-Degree Polynomial Approximation of $f(x) = e^x$

Construct a table comparing the values of the polynomial

$$P_3(x) = 1 + x + \frac{1}{2}x^2 + \frac{1}{3!}x^3 \qquad \text{3rd-degree approximation}$$

with $f(x) = e^x$ for several values of x near 0.

Solution　Using a calculator or a computer, you can obtain the results shown in the table. Note that for $x = 0$, the two functions have the same value, but that as x moves farther away from 0, the accuracy of the approximating polynomial $P_3(x)$ decreases.

x	-1.0	-0.2	-0.1	0	0.1	0.2	1.0
e^x	0.3679	0.81873	0.904837	1	1.105171	1.22140	2.7183
$P_3(x)$	0.3333	0.81867	0.904833	1	1.105167	1.22133	2.6667

▷ **TECHNOLOGY**　A graphing utility can be used to compare the graph of the approximating polynomial with the graph of the function f. For instance, in Figure 9.13, the graph of

$$P_3(x) = 1 + x + \tfrac{1}{2}x^2 + \tfrac{1}{6}x^3 \qquad \text{3rd-degree approximation}$$

is compared with the graph of $f(x) = e^x$. If you have access to a graphing utility, try comparing the graphs of

$$P_4(x) = 1 + x + \tfrac{1}{2}x^2 + \tfrac{1}{6}x^3 + \tfrac{1}{24}x^4 \qquad \text{4th-degree approximation}$$

$$P_5(x) = 1 + x + \tfrac{1}{2}x^2 + \tfrac{1}{6}x^3 + \tfrac{1}{24}x^4 + \tfrac{1}{120}x^5 \qquad \text{5th-degree approximation}$$

and

$$P_6(x) = 1 + x + \tfrac{1}{2}x^2 + \tfrac{1}{6}x^3 + \tfrac{1}{24}x^4 + \tfrac{1}{120}x^5 + \tfrac{1}{720}x^6 \qquad \text{6th-degree approximation}$$

with the graph of f. What do you notice?

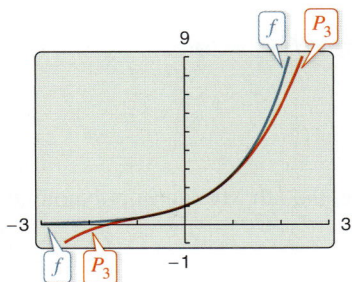

P_3 is the third-degree polynomial approximation of $f(x) = e^x$.
Figure 9.13

Taylor and Maclaurin Polynomials

The polynomial approximation of

$$f(x) = e^x$$

in Example 2 is expanded about $c = 0$. For expansions about an arbitrary value of c, it is convenient to write the polynomial in the form

$$P_n(x) = a_0 + a_1(x - c) + a_2(x - c)^2 + a_3(x - c)^3 + \cdots + a_n(x - c)^n.$$

In this form, repeated differentiation produces

$$P_n'(x) = a_1 + 2a_2(x - c) + 3a_3(x - c)^2 + \cdots + na_n(x - c)^{n-1}$$
$$P_n''(x) = 2a_2 + 2(3a_3)(x - c) + \cdots + n(n - 1)a_n(x - c)^{n-2}$$
$$P_n'''(x) = 2(3a_3) + \cdots + n(n - 1)(n - 2)a_n(x - c)^{n-3}$$
$$\vdots$$
$$P_n^{(n)}(x) = n(n - 1)(n - 2) \cdots (2)(1)a_n.$$

Letting $x = c$, you then obtain

$$P_n(c) = a_0, \qquad P_n'(c) = a_1, \qquad P_n''(c) = 2a_2, \ldots, \qquad P_n^{(n)}(c) = n!a_n$$

and because the values of f and its first n derivatives must agree with the values of P_n and its first n derivatives at $x = c$, it follows that

$$f(c) = a_0, \qquad f'(c) = a_1, \qquad \frac{f''(c)}{2!} = a_2, \quad \ldots, \qquad \frac{f^{(n)}(c)}{n!} = a_n.$$

With these coefficients, you can obtain the following definition of **Taylor polynomials,** named after the English mathematician Brook Taylor, and **Maclaurin polynomials,** named after the English mathematician Colin Maclaurin (1698–1746).

BROOK TAYLOR (1685–1731)

Although Taylor was not the first to seek polynomial approximations of transcendental functions, his account published in 1715 was one of the first comprehensive works on the subject.
See LarsonCalculus.com to read more of this biography.

> **Definitions of nth Taylor Polynomial and nth Maclaurin Polynomial**
>
> If f has n derivatives at c, then the polynomial
>
> $$P_n(x) = f(c) + f'(c)(x - c) + \frac{f''(c)}{2!}(x - c)^2 + \cdots + \frac{f^{(n)}(c)}{n!}(x - c)^n$$
>
> is called the **nth Taylor polynomial for f at c.** If $c = 0$, then
>
> $$P_n(x) = f(0) + f'(0)x + \frac{f''(0)}{2!}x^2 + \frac{f'''(0)}{3!}x^3 + \cdots + \frac{f^{(n)}(0)}{n!}x^n$$
>
> is also called the **nth Maclaurin polynomial for f.**

• • REMARK Maclaurin polynomials are special types of Taylor polynomials for which $c = 0$.

EXAMPLE 3 **A Maclaurin Polynomial for $f(x) = e^x$**

Find the nth Maclaurin polynomial for

$$f(x) = e^x.$$

Solution From the discussion on the preceding page, the nth Maclaurin polynomial for

$$f(x) = e^x$$

is

$$P_n(x) = 1 + x + \frac{1}{2!}x^2 + \frac{1}{3!}x^3 + \cdots + \frac{1}{n!}x^n.$$

■ FOR FURTHER INFORMATION
To see how to use series to obtain other approximations to e, see the article "Novel Series-based Approximations to e" by John Knox and Harlan J. Brothers in *The College Mathematics Journal.* To view this article, go to *MathArticles.com.*

Finding Taylor Polynomials Using Technology In Exercises 31 and 32, use a computer algebra system to find the indicated Taylor polynomials for the function f. Graph the function and the Taylor polynomials.

31. $f(x) = \tan \pi x$

(a) $n = 3$, $c = 0$

(b) $n = 3$, $c = 1/4$

32. $f(x) = \dfrac{1}{x^2 + 1}$

(a) $n = 4$, $c = 0$

(b) $n = 4$, $c = 1$

33. Numerical and Graphical Approximations

(a) Use the Maclaurin polynomials $P_1(x)$, $P_3(x)$, and $P_5(x)$ for $f(x) = \sin x$ to complete the table.

x	0	0.25	0.50	0.75	1.00
$\sin x$	0	0.2474	0.4794	0.6816	0.8415
$P_1(x)$					
$P_3(x)$					
$P_5(x)$					

(b) Use a graphing utility to graph $f(x) = \sin x$ and the Maclaurin polynomials in part (a).

(c) Describe the change in accuracy of a polynomial approximation as the distance from the point where the polynomial is centered increases.

34. Numerical and Graphical Approximations

(a) Use the Taylor polynomials $P_1(x)$, $P_2(x)$, and $P_4(x)$ for $f(x) = e^x$ centered at $c = 1$ to complete the table.

x	1.00	1.25	1.50	1.75	2.00
e^x	e	3.4903	4.4817	5.7546	7.3891
$P_1(x)$					
$P_2(x)$					
$P_4(x)$					

(b) Use a graphing utility to graph $f(x) = e^x$ and the Taylor polynomials in part (a).

(c) Describe the change in accuracy of polynomial approximations as the degree increases.

Numerical and Graphical Approximations In Exercises 35 and 36, (a) find the Maclaurin polynomial $P_3(x)$ for $f(x)$, (b) complete the table for $f(x)$ and $P_3(x)$, and (c) sketch the graphs of $f(x)$ and $P_3(x)$ on the same set of coordinate axes.

x	-0.75	-0.50	-0.25	0	0.25	0.50	0.75
$f(x)$							
$P_3(x)$							

35. $f(x) = \arcsin x$

36. $f(x) = \arctan x$

Identifying Maclaurin Polynomials In Exercises 37–40, the graph of $y = f(x)$ is shown with four of its Maclaurin polynomials. Identify the Maclaurin polynomials and use a graphing utility to confirm your results.

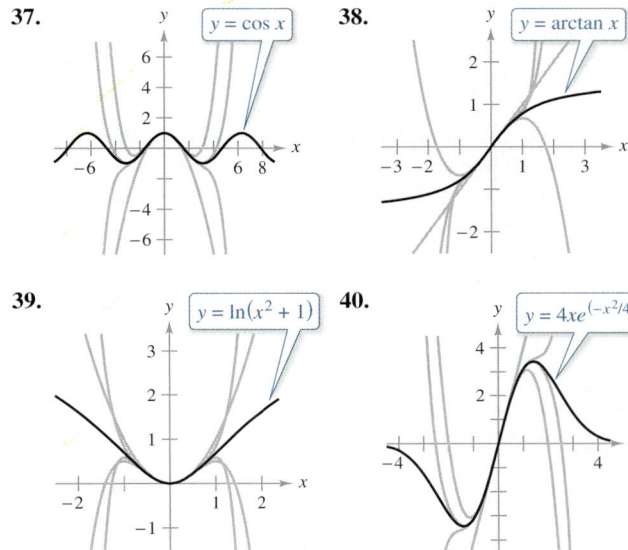

37.

$y = \cos x$

38.

$y = \arctan x$

39.

$y = \ln(x^2 + 1)$

40.

$y = 4xe^{(-x^2/4)}$

Approximating a Function Value In Exercises 41–44, approximate the function at the given value of x, using the polynomial found in the indicated exercise.

41. $f(x) = e^{4x}$, $f\left(\dfrac{1}{4}\right)$, Exercise 13

42. $f(x) = x^2 e^{-x}$, $f\left(\dfrac{1}{5}\right)$, Exercise 20

43. $f(x) = \ln x$, $f(2.1)$, Exercise 29

44. $f(x) = x^2 \cos x$, $f\left(\dfrac{7\pi}{8}\right)$, Exercise 30

Using Taylor's Theorem In Exercises 45–48, use Taylor's Theorem to obtain an upper bound for the error of the approximation. Then calculate the exact value of the error.

45. $\cos(0.3) \approx 1 - \dfrac{(0.3)^2}{2!} + \dfrac{(0.3)^4}{4!}$

46. $e \approx 1 + 1 + \dfrac{1^2}{2!} + \dfrac{1^3}{3!} + \dfrac{1^4}{4!} + \dfrac{1^5}{5!}$

47. $\arcsin(0.4) \approx 0.4 + \dfrac{(0.4)^3}{2 \cdot 3}$

48. $\arctan(0.4) \approx 0.4 - \dfrac{(0.4)^3}{3}$

Finding a Degree In Exercises 49–52, determine the degree of the Maclaurin polynomial required for the error in the approximation of the function at the indicated value of x to be less than 0.001.

49. $\sin(0.3)$

50. $\cos(0.1)$

51. $e^{0.6}$

52. $\ln(1.25)$

Finding a Degree Using Technology In Exercises 53 and 54, determine the degree of the Maclaurin polynomial required for the error in the approximation of the function at the indicated value of x to be less than 0.0001. Use a computer algebra system to obtain and evaluate the required derivative.

53. $f(x) = \ln(x + 1)$, approximate $f(0.5)$.

54. $f(x) = e^{-\pi x}$, approximate $f(1.3)$.

Finding Values In Exercises 55–58, determine the values of x for which the function can be replaced by the Taylor polynomial if the error cannot exceed 0.001.

55. $f(x) = e^x \approx 1 + x + \dfrac{x^2}{2!} + \dfrac{x^3}{3!}, \quad x < 0$

56. $f(x) = \sin x \approx x - \dfrac{x^3}{3!}$

57. $f(x) = \cos x \approx 1 - \dfrac{x^2}{2!} + \dfrac{x^4}{4!}$

58. $f(x) = e^{-2x} \approx 1 - 2x + 2x^2 - \dfrac{4}{3}x^3$

WRITING ABOUT CONCEPTS

59. Polynomial Approximation An elementary function is approximated by a polynomial. In your own words, describe what is meant by saying that the polynomial is *expanded about c* or *centered at c*.

60. Polynomial Approximation When an elementary function f is approximated by a second-degree polynomial P_2 centered at c, what is known about f and P_2 at c? Explain your reasoning.

61. Taylor Polynomial State the definition of an nth-degree Taylor polynomial of f centered at c.

62. Accuracy of a Taylor Polynomial Describe the accuracy of the nth-degree Taylor polynomial of f centered at c as the distance between c and x increases.

63. Accuracy of a Taylor Polynomial In general, how does the accuracy of a Taylor polynomial change as the degree of the polynomial increases? Explain your reasoning.

---

64. HOW DO YOU SEE IT? The graphs show first-, second-, and third-degree polynomial approximations $P_1, P_2,$ and P_3 of a function f. Label the graphs of $P_1, P_2,$ and P_3. To print an enlarged copy of the graph, go to *MathGraphs.com*.

---

65. Comparing Maclaurin Polynomials

(a) Compare the Maclaurin polynomials of degree 4 and degree 5, respectively, for the functions $f(x) = e^x$ and $g(x) = xe^x$. What is the relationship between them?

(b) Use the result in part (a) and the Maclaurin polynomial of degree 5 for $f(x) = \sin x$ to find a Maclaurin polynomial of degree 6 for the function $g(x) = x \sin x$.

(c) Use the result in part (a) and the Maclaurin polynomial of degree 5 for $f(x) = \sin x$ to find a Maclaurin polynomial of degree 4 for the function $g(x) = (\sin x)/x$.

66. Differentiating Maclaurin Polynomials

(a) Differentiate the Maclaurin polynomial of degree 5 for $f(x) = \sin x$ and compare the result with the Maclaurin polynomial of degree 4 for $g(x) = \cos x$.

(b) Differentiate the Maclaurin polynomial of degree 6 for $f(x) = \cos x$ and compare the result with the Maclaurin polynomial of degree 5 for $g(x) = \sin x$.

(c) Differentiate the Maclaurin polynomial of degree 4 for $f(x) = e^x$. Describe the relationship between the two series.

67. Graphical Reasoning The figure shows the graphs of the function $f(x) = \sin(\pi x/4)$ and the second-degree Taylor polynomial $P_2(x) = 1 - (\pi^2/32)(x - 2)^2$ centered at $x = 2$.

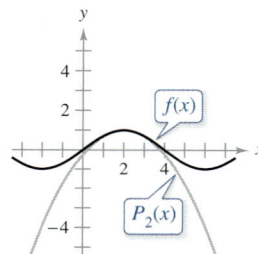

(a) Use the symmetry of the graph of f to write the second-degree Taylor polynomial $Q_2(x)$ for f centered at $x = -2$.

(b) Use a horizontal translation of the result in part (a) to find the second-degree Taylor polynomial $R_2(x)$ for f centered at $x = 6$.

(c) Is it possible to use a horizontal translation of the result in part (a) to write a second-degree Taylor polynomial for f centered at $x = 4$? Explain.

68. Proof Prove that if f is an odd function, then its nth Maclaurin polynomial contains only terms with odd powers of x.

69. Proof Prove that if f is an even function, then its nth Maclaurin polynomial contains only terms with even powers of x.

70. Proof Let $P_n(x)$ be the nth Taylor polynomial for f at c. Prove that $P_n(c) = f(c)$ and $P^{(k)}(c) = f^{(k)}(c)$ for $1 \le k \le n$. (See Exercises 9 and 10.)

71. Writing The proof in Exercise 70 guarantees that the Taylor polynomial and its derivatives agree with the function and its derivatives at $x = c$. Use the graphs and tables in Exercises 33–36 to discuss what happens to the accuracy of the Taylor polynomial as you move away from $x = c$.

9.8 Power Series

■ Understand the definition of a power series.
■ Find the radius and interval of convergence of a power series.
■ Determine the endpoint convergence of a power series.
■ Differentiate and integrate a power series.

Power Series

Exploration

Graphical Reasoning
Use a graphing utility to approximate the graph of each power series near $x = 0$. (Use the first several terms of each series.) Each series represents a well-known function. What is the function?

a. $\displaystyle\sum_{n=0}^{\infty} \frac{(-1)^n x^n}{n!}$

b. $\displaystyle\sum_{n=0}^{\infty} \frac{(-1)^n x^{2n}}{(2n)!}$

c. $\displaystyle\sum_{n=0}^{\infty} \frac{(-1)^n x^{2n+1}}{(2n+1)!}$

d. $\displaystyle\sum_{n=0}^{\infty} \frac{(-1)^n x^{2n+1}}{2n+1}$

e. $\displaystyle\sum_{n=0}^{\infty} \frac{2^n x^n}{n!}$

In Section 9.7, you were introduced to the concept of approximating functions by Taylor polynomials. For instance, the function $f(x) = e^x$ can be *approximated* by its third-degree Maclaurin polynomial

$$e^x \approx 1 + x + \frac{x^2}{2!} + \frac{x^3}{3!}.$$

In that section, you saw that the higher the degree of the approximating polynomial, the better the approximation becomes.

In this and the next two sections, you will see that several important types of functions, including $f(x) = e^x$, can be represented *exactly* by an infinite series called a **power series.** For example, the power series representation for e^x is

$$e^x = 1 + x + \frac{x^2}{2!} + \frac{x^3}{3!} + \cdots + \frac{x^n}{n!} + \cdots.$$

For each real number x, it can be shown that the infinite series on the right converges to the number e^x. Before doing this, however, some preliminary results dealing with power series will be discussed—beginning with the next definition.

REMARK To simplify the notation for power series, assume that $(x - c)^0 = 1$, even when $x = c$.

Definition of Power Series

If x is a variable, then an infinite series of the form

$$\sum_{n=0}^{\infty} a_n x^n = a_0 + a_1 x + a_2 x^2 + a_3 x^3 + \cdots + a_n x^n + \cdots$$

is called a **power series.** More generally, an infinite series of the form

$$\sum_{n=0}^{\infty} a_n (x - c)^n = a_0 + a_1(x - c) + a_2(x - c)^2 + \cdots + a_n(x - c)^n + \cdots$$

is called a **power series centered at c,** where c is a constant.

EXAMPLE 1 Power Series

a. The following power series is centered at 0.

$$\sum_{n=0}^{\infty} \frac{x^n}{n!} = 1 + x + \frac{x^2}{2} + \frac{x^3}{3!} + \cdots$$

b. The following power series is centered at -1.

$$\sum_{n=0}^{\infty} (-1)^n (x + 1)^n = 1 - (x + 1) + (x + 1)^2 - (x + 1)^3 + \cdots$$

c. The following power series is centered at 1.

$$\sum_{n=1}^{\infty} \frac{1}{n}(x - 1)^n = (x - 1) + \frac{1}{2}(x - 1)^2 + \frac{1}{3}(x - 1)^3 + \cdots$$

■

Radius and Interval of Convergence

A power series in x can be viewed as a function of x

$$f(x) = \sum_{n=0}^{\infty} a_n(x - c)^n$$

where the *domain of f* is the set of all x for which the power series converges. Determination of the domain of a power series is the primary concern in this section. Of course, every power series converges at its center c because

$$f(c) = \sum_{n=0}^{\infty} a_n(c - c)^n$$
$$= a_0(1) + 0 + 0 + \cdots + 0 + \cdots$$
$$= a_0.$$

So, c always lies in the domain of f. Theorem 9.20 (see below) states that the domain of a power series can take three basic forms: a single point, an interval centered at c, or the entire real number line, as shown in Figure 9.17.

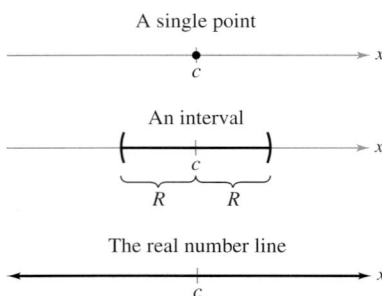

A single point

An interval

The real number line

The domain of a power series has only three basic forms: a single point, an interval centered at c, or the entire real number line.

Figure 9.17

THEOREM 9.20 Convergence of a Power Series

For a power series centered at c, precisely one of the following is true.

1. The series converges only at c.

2. There exists a real number $R > 0$ such that the series converges absolutely for

$$|x - c| < R$$

and diverges for

$$|x - c| > R.$$

3. The series converges absolutely for all x.

The number R is the **radius of convergence** of the power series. If the series converges only at c, then the radius of convergence is $R = 0$. If the series converges for all x, then the radius of convergence is $R = \infty$. The set of all values of x for which the power series converges is the **interval of convergence** of the power series.

A proof of this theorem is given in Appendix A.
See LarsonCalculus.com for Bruce Edwards's video of this proof.

To determine the radius of convergence of a power series, use the Ratio Test, as demonstrated in Examples 2, 3, and 4.

EXAMPLE 2   **Finding the Radius of Convergence**

Find the radius of convergence of $\displaystyle\sum_{n=0}^{\infty} n!x^n$.

Solution   For $x = 0$, you obtain

$$f(0) = \sum_{n=0}^{\infty} n!0^n = 1 + 0 + 0 + \cdots = 1.$$

For any fixed value of x such that $|x| > 0$, let $u_n = n!x^n$. Then

$$\lim_{n\to\infty}\left|\frac{u_{n+1}}{u_n}\right| = \lim_{n\to\infty}\left|\frac{(n+1)!x^{n+1}}{n!x^n}\right|$$
$$= |x| \lim_{n\to\infty} (n+1)$$
$$= \infty.$$

Therefore, by the Ratio Test, the series diverges for $|x| > 0$ and converges only at its center, 0. So, the radius of convergence is $R = 0$.

EXAMPLE 3   **Finding the Radius of Convergence**

Find the radius of convergence of

$$\sum_{n=0}^{\infty} 3(x-2)^n.$$

Solution   For $x \ne 2$, let $u_n = 3(x-2)^n$. Then

$$\lim_{n\to\infty}\left|\frac{u_{n+1}}{u_n}\right| = \lim_{n\to\infty}\left|\frac{3(x-2)^{n+1}}{3(x-2)^n}\right|$$
$$= \lim_{n\to\infty} |x-2|$$
$$= |x-2|.$$

By the Ratio Test, the series converges for $|x-2| < 1$ and diverges for $|x-2| > 1$. Therefore, the radius of convergence of the series is $R = 1$.

EXAMPLE 4   **Finding the Radius of Convergence**

Find the radius of convergence of

$$\sum_{n=0}^{\infty} \frac{(-1)^n x^{2n+1}}{(2n+1)!}.$$

Solution   Let $u_n = (-1)^n x^{2n+1}/(2n+1)!$. Then

$$\lim_{n\to\infty}\left|\frac{u_{n+1}}{u_n}\right| = \lim_{n\to\infty}\left|\frac{\dfrac{(-1)^{n+1} x^{2n+3}}{(2n+3)!}}{\dfrac{(-1)^n x^{2n+1}}{(2n+1)!}}\right|$$
$$= \lim_{n\to\infty} \frac{x^2}{(2n+3)(2n+2)}.$$

For any *fixed* value of x, this limit is 0. So, by the Ratio Test, the series converges for all x. Therefore, the radius of convergence is $R = \infty$.

Endpoint Convergence

Note that for a power series whose radius of convergence is a finite number R, Theorem 9.20 says nothing about the convergence at the *endpoints* of the interval of convergence. Each endpoint must be tested separately for convergence or divergence. As a result, the interval of convergence of a power series can take any one of the six forms shown in Figure 9.18.

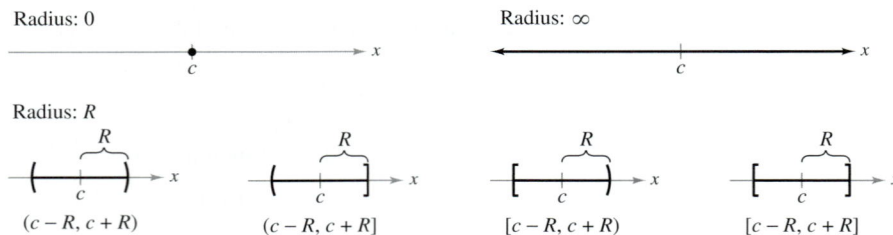

Radius: 0

Radius: ∞

Radius: R

$(c - R, c + R)$ $(c - R, c + R]$ $[c - R, c + R)$ $[c - R, c + R]$

Intervals of convergence
Figure 9.18

EXAMPLE 5   **Finding the Interval of Convergence**

⋯▷ *See LarsonCalculus.com for an interactive version of this type of example.*

Find the interval of convergence of

$$\sum_{n=1}^{\infty} \frac{x^n}{n}.$$

Solution Letting $u_n = x^n/n$ produces

$$\lim_{n\to\infty} \left| \frac{u_{n+1}}{u_n} \right| = \lim_{n\to\infty} \left| \frac{\dfrac{x^{n+1}}{(n+1)}}{\dfrac{x^n}{n}} \right|$$

$$= \lim_{n\to\infty} \left| \frac{nx}{n+1} \right|$$

$$= |x|.$$

So, by the Ratio Test, the radius of convergence is $R = 1$. Moreover, because the series is centered at 0, it converges in the interval $(-1, 1)$. This interval, however, is not necessarily the *interval of convergence*. To determine this, you must test for convergence at each endpoint. When $x = 1$, you obtain the *divergent* harmonic series

$$\sum_{n=1}^{\infty} \frac{1}{n} = \frac{1}{1} + \frac{1}{2} + \frac{1}{3} + \cdots. \qquad \text{Diverges when } x = 1.$$

When $x = -1$, you obtain the *convergent* alternating harmonic series

$$\sum_{n=1}^{\infty} \frac{(-1)^n}{n} = -1 + \frac{1}{2} - \frac{1}{3} + \frac{1}{4} - \cdots. \qquad \text{Converges when } x = -1.$$

So, the interval of convergence for the series is $[-1, 1)$, as shown in Figure 9.19.

Interval: $[-1, 1)$
Radius: $R = 1$

Figure 9.19

EXAMPLE 6   **Finding the Interval of Convergence**

Find the interval of convergence of $\displaystyle\sum_{n=0}^{\infty} \frac{(-1)^n(x+1)^n}{2^n}$.

Solution   Letting $u_n = (-1)^n(x+1)^n/2^n$ produces

$$\lim_{n\to\infty}\left|\frac{u_{n+1}}{u_n}\right| = \lim_{n\to\infty}\left|\frac{\dfrac{(-1)^{n+1}(x+1)^{n+1}}{2^{n+1}}}{\dfrac{(-1)^n(x+1)^n}{2^n}}\right|$$

$$= \lim_{n\to\infty}\left|\frac{2^n(x+1)}{2^{n+1}}\right|$$

$$= \left|\frac{x+1}{2}\right|.$$

By the Ratio Test, the series converges for

$$\left|\frac{x+1}{2}\right| < 1$$

or $|x+1| < 2$. So, the radius of convergence is $R = 2$. Because the series is centered at $x = -1$, it will converge in the interval $(-3, 1)$. Furthermore, at the endpoints, you have

$$\sum_{n=0}^{\infty}\frac{(-1)^n(-2)^n}{2^n} = \sum_{n=0}^{\infty}\frac{2^n}{2^n} = \sum_{n=0}^{\infty}1 \qquad \text{Diverges when } x = -3.$$

and

$$\sum_{n=0}^{\infty}\frac{(-1)^n(2)^n}{2^n} = \sum_{n=0}^{\infty}(-1)^n \qquad \text{Diverges when } x = 1.$$

both of which diverge. So, the interval of convergence is $(-3, 1)$, as shown in Figure 9.20.

Interval: $(-3, 1)$
Radius: $R = 2$

Figure 9.20

EXAMPLE 7   **Finding the Interval of Convergence**

Find the interval of convergence of

$$\sum_{n=1}^{\infty}\frac{x^n}{n^2}.$$

Solution   Letting $u_n = x^n/n^2$ produces

$$\lim_{n\to\infty}\left|\frac{u_{n+1}}{u_n}\right| = \lim_{n\to\infty}\left|\frac{x^{n+1}/(n+1)^2}{x^n/n^2}\right|$$

$$= \lim_{n\to\infty}\left|\frac{n^2 x}{(n+1)^2}\right|$$

$$= |x|.$$

So, the radius of convergence is $R = 1$. Because the series is centered at $x = 0$, it converges in the interval $(-1, 1)$. When $x = 1$, you obtain the convergent p-series

$$\sum_{n=1}^{\infty}\frac{1}{n^2} = \frac{1}{1^2} + \frac{1}{2^2} + \frac{1}{3^2} + \frac{1}{4^2} + \cdots. \qquad \text{Converges when } x = 1.$$

When $x = -1$, you obtain the convergent alternating series

$$\sum_{n=1}^{\infty}\frac{(-1)^n}{n^2} = -\frac{1}{1^2} + \frac{1}{2^2} - \frac{1}{3^2} + \frac{1}{4^2} - \cdots. \qquad \text{Converges when } x = -1.$$

Therefore, the interval of convergence is $[-1, 1]$.

Differentiation and Integration of Power Series

Power series representation of functions has played an important role in the development of calculus. In fact, much of Newton's work with differentiation and integration was done in the context of power series—especially his work with complicated algebraic functions and transcendental functions. Euler, Lagrange, Leibniz, and the Bernoullis all used power series extensively in calculus.

Once you have defined a function with a power series, it is natural to wonder how you can determine the characteristics of the function. Is it continuous? Differentiable? Theorem 9.21, which is stated without proof, answers these questions.

JAMES GREGORY (1638–1675)

One of the earliest mathematicians to work with power series was a Scotsman, James Gregory. He developed a power series method for interpolating table values—a method that was later used by Brook Taylor in the development of Taylor polynomials and Taylor series.

THEOREM 9.21 Properties of Functions Defined by Power Series

If the function

$$f(x) = \sum_{n=0}^{\infty} a_n(x - c)^n$$
$$= a_0 + a_1(x - c) + a_2(x - c)^2 + a_3(x - c)^3 + \cdots$$

has a radius of convergence of $R > 0$, then, on the interval

$$(c - R, c + R)$$

f is differentiable (and therefore continuous). Moreover, the derivative and antiderivative of f are as follows.

1. $f'(x) = \sum_{n=1}^{\infty} na_n(x - c)^{n-1}$
$$= a_1 + 2a_2(x - c) + 3a_3(x - c)^2 + \cdots$$

2. $\int f(x)\, dx = C + \sum_{n=0}^{\infty} a_n \frac{(x - c)^{n+1}}{n + 1}$
$$= C + a_0(x - c) + a_1 \frac{(x - c)^2}{2} + a_2 \frac{(x - c)^3}{3} + \cdots$$

The *radius of convergence* of the series obtained by differentiating or integrating a power series is the same as that of the original power series. The *interval of convergence*, however, may differ as a result of the behavior at the endpoints.

Theorem 9.21 states that, in many ways, a function defined by a power series behaves like a polynomial. It is continuous in its interval of convergence, and both its derivative and its antiderivative can be determined by differentiating and integrating each term of the power series. For instance, the derivative of the power series

$$f(x) = \sum_{n=0}^{\infty} \frac{x^n}{n!}$$
$$= 1 + x + \frac{x^2}{2} + \frac{x^3}{3!} + \frac{x^4}{4!} + \cdots$$

is

$$f'(x) = 1 + (2)\frac{x}{2} + (3)\frac{x^2}{3!} + (4)\frac{x^3}{4!} + \cdots$$
$$= 1 + x + \frac{x^2}{2} + \frac{x^3}{3!} + \frac{x^4}{4!} + \cdots$$
$$= f(x).$$

Notice that $f'(x) = f(x)$. Do you recognize this function?

The Granger Collection

EXAMPLE 8 **Intervals of Convergence for $f(x)$, $f'(x)$, and $\int f(x)\,dx$**

Consider the function

$$f(x) = \sum_{n=1}^{\infty} \frac{x^n}{n} = x + \frac{x^2}{2} + \frac{x^3}{3} + \cdots.$$

Find the interval of convergence for each of the following.

a. $\int f(x)\,dx$ **b.** $f(x)$ **c.** $f'(x)$

Solution By Theorem 9.21, you have

$$f'(x) = \sum_{n=1}^{\infty} x^{n-1}$$
$$= 1 + x + x^2 + x^3 + \cdots$$

and

$$\int f(x)\,dx = C + \sum_{n=1}^{\infty} \frac{x^{n+1}}{n(n+1)}$$
$$= C + \frac{x^2}{1\cdot 2} + \frac{x^3}{2\cdot 3} + \frac{x^4}{3\cdot 4} + \cdots.$$

By the Ratio Test, you can show that each series has a radius of convergence of $R = 1$. Considering the interval $(-1, 1)$, you have the following.

a. For $\int f(x)\,dx$, the series

$$\sum_{n=1}^{\infty} \frac{x^{n+1}}{n(n+1)} \qquad \text{Interval of convergence: } [-1, 1]$$

converges for $x = \pm 1$, and its interval of convergence is $[-1, 1]$. See Figure 9.21(a).

b. For $f(x)$, the series

$$\sum_{n=1}^{\infty} \frac{x^n}{n} \qquad \text{Interval of convergence: } [-1, 1)$$

converges for $x = -1$ and diverges for $x = 1$. So, its interval of convergence is $[-1, 1)$. See Figure 9.21(b).

c. For $f'(x)$, the series

$$\sum_{n=1}^{\infty} x^{n-1} \qquad \text{Interval of convergence: } (-1, 1)$$

diverges for $x = \pm 1$, and its interval of convergence is $(-1, 1)$. See Figure 9.21(c).

Interval: $[-1, 1]$
Radius: $R = 1$

Interval: $[-1, 1)$
Radius: $R = 1$

Interval: $(-1, 1)$
Radius: $R = 1$

(a)   (b)   (c)

Figure 9.21

From Example 8, it appears that of the three series, the one for the derivative, $f'(x)$, is the least likely to converge at the endpoints. In fact, it can be shown that if the series for $f'(x)$ converges at the endpoints

$$x = c \pm R$$

then the series for $f(x)$ will also converge there.

9.8 Exercises

See **CalcChat.com** for tutorial help and worked-out solutions to odd-numbered exercises.

Finding the Center of a Power Series In Exercises 1–4, state where the power series is centered.

1. $\displaystyle\sum_{n=0}^{\infty} n x^n$

2. $\displaystyle\sum_{n=1}^{\infty} \frac{(-1)^n 1 \cdot 3 \cdots (2n-1)}{2^n n!} x^n$

3. $\displaystyle\sum_{n=1}^{\infty} \frac{(x-2)^n}{n^3}$

4. $\displaystyle\sum_{n=0}^{\infty} \frac{(-1)^n (x-\pi)^{2n}}{(2n)!}$

Finding the Radius of Convergence In Exercises 5–10, find the radius of convergence of the power series.

5. $\displaystyle\sum_{n=0}^{\infty} (-1)^n \frac{x^n}{n+1}$

6. $\displaystyle\sum_{n=0}^{\infty} (3x)^n$

7. $\displaystyle\sum_{n=1}^{\infty} \frac{(4x)^n}{n^2}$

8. $\displaystyle\sum_{n=0}^{\infty} \frac{(-1)^n x^n}{5^n}$

9. $\displaystyle\sum_{n=0}^{\infty} \frac{x^{2n}}{(2n)!}$

10. $\displaystyle\sum_{n=0}^{\infty} \frac{(2n)! x^{2n}}{n!}$

Finding the Interval of Convergence In Exercises 11–34, find the interval of convergence of the power series. (Be sure to include a check for convergence at the endpoints of the interval.)

11. $\displaystyle\sum_{n=0}^{\infty} \left(\frac{x}{4}\right)^n$

12. $\displaystyle\sum_{n=0}^{\infty} (2x)^n$

13. $\displaystyle\sum_{n=1}^{\infty} \frac{(-1)^n x^n}{n}$

14. $\displaystyle\sum_{n=0}^{\infty} (-1)^{n+1} (n+1) x^n$

15. $\displaystyle\sum_{n=0}^{\infty} \frac{x^{5n}}{n!}$

16. $\displaystyle\sum_{n=0}^{\infty} \frac{(3x)^n}{(2n)!}$

17. $\displaystyle\sum_{n=0}^{\infty} (2n)! \left(\frac{x}{3}\right)^n$

18. $\displaystyle\sum_{n=0}^{\infty} \frac{(-1)^n x^n}{(n+1)(n+2)}$

19. $\displaystyle\sum_{n=1}^{\infty} \frac{(-1)^{n+1} x^n}{6^n}$

20. $\displaystyle\sum_{n=0}^{\infty} \frac{(-1)^n n! (x-5)^n}{3^n}$

21. $\displaystyle\sum_{n=1}^{\infty} \frac{(-1)^{n+1} (x-4)^n}{n 9^n}$

22. $\displaystyle\sum_{n=0}^{\infty} \frac{(x-3)^{n+1}}{(n+1) 4^{n+1}}$

23. $\displaystyle\sum_{n=0}^{\infty} \frac{(-1)^{n+1} (x-1)^{n+1}}{n+1}$

24. $\displaystyle\sum_{n=1}^{\infty} \frac{(-1)^{n+1} (x-2)^n}{n 2^n}$

25. $\displaystyle\sum_{n=1}^{\infty} \frac{(x-3)^{n-1}}{3^{n-1}}$

26. $\displaystyle\sum_{n=0}^{\infty} \frac{(-1)^n x^{2n+1}}{2n+1}$

27. $\displaystyle\sum_{n=1}^{\infty} \frac{n}{n+1} (-2x)^{n-1}$

28. $\displaystyle\sum_{n=0}^{\infty} \frac{(-1)^n x^{2n}}{n!}$

29. $\displaystyle\sum_{n=0}^{\infty} \frac{x^{3n+1}}{(3n+1)!}$

30. $\displaystyle\sum_{n=1}^{\infty} \frac{n! x^n}{(2n)!}$

31. $\displaystyle\sum_{n=1}^{\infty} \frac{2 \cdot 3 \cdot 4 \cdots (n+1) x^n}{n!}$

32. $\displaystyle\sum_{n=1}^{\infty} \left[\frac{2 \cdot 4 \cdot 6 \cdots 2n}{3 \cdot 5 \cdot 7 \cdots (2n+1)}\right] x^{2n+1}$

33. $\displaystyle\sum_{n=1}^{\infty} \frac{(-1)^{n+1} 3 \cdot 7 \cdot 11 \cdots (4n-1)(x-3)^n}{4^n}$

34. $\displaystyle\sum_{n=1}^{\infty} \frac{n!(x+1)^n}{1 \cdot 3 \cdot 5 \cdots (2n-1)}$

Finding the Radius of Convergence In Exercises 35 and 36, find the radius of convergence of the power series, where $c > 0$ and k is a positive integer.

35. $\displaystyle\sum_{n=1}^{\infty} \frac{(x-c)^{n-1}}{c^{n-1}}$

36. $\displaystyle\sum_{n=0}^{\infty} \frac{(n!)^k x^n}{(kn)!}$

Finding the Interval of Convergence In Exercises 37–40, find the interval of convergence of the power series. (Be sure to include a check for convergence at the endpoints of the interval.)

37. $\displaystyle\sum_{n=0}^{\infty} \left(\frac{x}{k}\right)^n, \quad k > 0$

38. $\displaystyle\sum_{n=1}^{\infty} \frac{(-1)^{n+1} (x-c)^n}{n c^n}$

39. $\displaystyle\sum_{n=1}^{\infty} \frac{k(k+1)(k+2) \cdots (k+n-1) x^n}{n!}, \quad k \geq 1$

40. $\displaystyle\sum_{n=1}^{\infty} \frac{n!(x-c)^n}{1 \cdot 3 \cdot 5 \cdots (2n-1)}$

Writing an Equivalent Series In Exercises 41–44, write an equivalent series with the index of summation beginning at $n = 1$.

41. $\displaystyle\sum_{n=0}^{\infty} \frac{x^n}{n!}$

42. $\displaystyle\sum_{n=0}^{\infty} (-1)^{n+1} (n+1) x^n$

43. $\displaystyle\sum_{n=0}^{\infty} \frac{x^{2n+1}}{(2n+1)!}$

44. $\displaystyle\sum_{n=0}^{\infty} \frac{(-1)^n x^{2n+1}}{2n+1}$

Finding Intervals of Convergence In Exercises 45–48, find the intervals of convergence of (a) $f(x)$, (b) $f'(x)$, (c) $f''(x)$, and (d) $\int f(x)\, dx$. Include a check for convergence at the endpoints of the interval.

45. $f(x) = \displaystyle\sum_{n=0}^{\infty} \left(\frac{x}{3}\right)^n$

46. $f(x) = \displaystyle\sum_{n=1}^{\infty} \frac{(-1)^{n+1} (x-5)^n}{n 5^n}$

47. $f(x) = \displaystyle\sum_{n=0}^{\infty} \frac{(-1)^{n+1} (x-1)^{n+1}}{n+1}$

48. $f(x) = \displaystyle\sum_{n=1}^{\infty} \frac{(-1)^{n+1} (x-2)^n}{n}$

WRITING ABOUT CONCEPTS

49. Power Series   Define a power series centered at c.

50. Radius of Convergence   Describe the radius of convergence of a power series.

51. Interval of Convergence   Describe the interval of convergence of a power series.

52. Domain of a Power Series   Describe the three basic forms of the domain of a power series.

53. Using a Power Series   Describe how to differentiate and integrate a power series with a radius of convergence R. Will the series resulting from the operations of differentiation and integration have a different radius of convergence? Explain.

54. Conditional or Absolute Convergence   Give examples that show that the convergence of a power series at an endpoint of its interval of convergence may be either conditional or absolute. Explain your reasoning.

55. Writing a Power Series   Write a power series that has the indicated interval of convergence. Explain your reasoning.

(a) $(-2, 2)$ (b) $(-1, 1]$
(c) $(-1, 0)$ (d) $[-2, 6)$

---

56.   HOW DO YOU SEE IT?   Match the graph of the first 10 terms of the sequence of partial sums of the series

$$g(x) = \sum_{n=0}^{\infty} \left(\frac{x}{3}\right)^n$$

with the indicated value of the function. [The graphs are labeled (i), (ii), (iii), and (iv).] Explain how you made your choice.

(i) S_n

(ii) S_n

(iii) S_n

(iv) S_n

(a) $g(1)$ (b) $g(2)$
(c) $g(3)$ (d) $g(-2)$

---

57. Using Power Series   Let $f(x) = \displaystyle\sum_{n=0}^{\infty} \frac{(-1)^n x^{2n+1}}{(2n+1)!}$ and $g(x) = \displaystyle\sum_{n=0}^{\infty} \frac{(-1)^n x^{2n}}{(2n)!}$.

(a) Find the intervals of convergence of f and g.

(b) Show that $f'(x) = g(x)$.

(c) Show that $g'(x) = -f(x)$.

(d) Identify the functions f and g.

58. Using a Power Series   Let $f(x) = \displaystyle\sum_{n=0}^{\infty} \frac{x^n}{n!}$.

(a) Find the interval of convergence of f.

(b) Show that $f'(x) = f(x)$.

(c) Show that $f(0) = 1$.

(d) Identify the function f.

Differential Equation   In Exercises 59–64, show that the function represented by the power series is a solution of the differential equation.

59. $y = \displaystyle\sum_{n=0}^{\infty} \frac{(-1)^n x^{2n+1}}{(2n+1)!}$, $y'' + y = 0$

60. $y = \displaystyle\sum_{n=0}^{\infty} \frac{(-1)^n x^{2n}}{(2n)!}$, $y'' + y = 0$

61. $y = \displaystyle\sum_{n=0}^{\infty} \frac{x^{2n+1}}{(2n+1)!}$, $y'' - y = 0$

62. $y = \displaystyle\sum_{n=0}^{\infty} \frac{x^{2n}}{(2n)!}$, $y'' - y = 0$

63. $y = \displaystyle\sum_{n=0}^{\infty} \frac{x^{2n}}{2^n n!}$, $y'' - xy' - y = 0$

64. $y = 1 + \displaystyle\sum_{n=1}^{\infty} \frac{(-1)^n x^{4n}}{2^{2n} n! \cdot 3 \cdot 7 \cdot 11 \cdots (4n-1)}$, $y'' + x^2 y = 0$

65. Bessel Function   The Bessel function of order 0 is

$$J_0(x) = \sum_{k=0}^{\infty} \frac{(-1)^k x^{2k}}{2^{2k}(k!)^2}.$$

(a) Show that the series converges for all x.

(b) Show that the series is a solution of the differential equation $x^2 J_0'' + x J_0' + x^2 J_0 = 0$.

(c) Use a graphing utility to graph the polynomial composed of the first four terms of J_0.

(d) Approximate $\int_0^1 J_0\, dx$ accurate to two decimal places.

66. Bessel Function   The Bessel function of order 1 is

$$J_1(x) = x \sum_{k=0}^{\infty} \frac{(-1)^k x^{2k}}{2^{2k+1} k!(k+1)!}.$$

(a) Show that the series converges for all x.

(b) Show that the series is a solution of the differential equation $x^2 J_1'' + x J_1' + (x^2 - 1) J_1 = 0$.

(c) Use a graphing utility to graph the polynomial composed of the first four terms of J_1.

(d) Show that $J_0'(x) = -J_1(x)$.

67. Investigation The interval of convergence of the geometric series $\sum_{n=0}^{\infty} \left(\frac{x}{4}\right)^n$ is $(-4, 4)$.

(a) Find the sum of the series when $x = \frac{5}{2}$. Use a graphing utility to graph the first six terms of the sequence of partial sums and the horizontal line representing the sum of the series.

(b) Repeat part (a) for $x = -\frac{5}{2}$.

(c) Write a short paragraph comparing the rates of convergence of the partial sums with the sums of the series in parts (a) and (b). How do the plots of the partial sums differ as they converge toward the sum of the series?

(d) Given any positive real number M, there exists a positive integer N such that the partial sum

$$\sum_{n=0}^{N} \left(\frac{5}{4}\right)^n > M.$$

Use a graphing utility to complete the table.

M	10	100	1000	10,000
N				

68. Investigation The interval of convergence of the series $\sum_{n=0}^{\infty} (3x)^n$ is $\left(-\frac{1}{3}, \frac{1}{3}\right)$.

(a) Find the sum of the series when $x = \frac{1}{6}$. Use a graphing utility to graph the first six terms of the sequence of partial sums and the horizontal line representing the sum of the series.

(b) Repeat part (a) for $x = -\frac{1}{6}$.

(c) Write a short paragraph comparing the rates of convergence of the partial sums with the sums of the series in parts (a) and (b). How do the plots of the partial sums differ as they converge toward the sum of the series?

(d) Given any positive real number M, there exists a positive integer N such that the partial sum

$$\sum_{n=0}^{N} \left(3 \cdot \frac{2}{3}\right)^n > M.$$

Use a graphing utility to complete the table.

M	10	100	1000	10,000
N				

Identifying a Function In Exercises 69–72, the series represents a well-known function. Use a computer algebra system to graph the partial sum S_{10} and identify the function from the graph.

69. $f(x) = \sum_{n=0}^{\infty} (-1)^n \frac{x^{2n}}{(2n)!}$

70. $f(x) = \sum_{n=0}^{\infty} (-1)^n \frac{x^{2n+1}}{(2n+1)!}$

71. $f(x) = \sum_{n=0}^{\infty} (-1)^n x^n, \quad -1 < x < 1$

72. $f(x) = \sum_{n=0}^{\infty} (-1)^n \frac{x^{2n+1}}{2n+1}, \quad -1 \le x \le 1$

True or False? In Exercises 73–76, determine whether the statement is true or false. If it is false, explain why or give an example that shows it is false.

73. If the power series $\sum_{n=1}^{\infty} a_n x^n$ converges for $x = 2$, then it also converges for $x = -2$.

74. It is possible to find a power series whose interval of convergence is $[0, \infty)$.

75. If the interval of convergence for $\sum_{n=0}^{\infty} a_n x^n$ is $(-1, 1)$, then the interval of convergence for $\sum_{n=0}^{\infty} a_n (x - 1)^n$ is $(0, 2)$.

76. If $f(x) = \sum_{n=0}^{\infty} a_n x^n$ converges for $|x| < 2$, then

$$\int_0^1 f(x)\, dx = \sum_{n=0}^{\infty} \frac{a_n}{n+1}.$$

77. Proof Prove that the power series

$$\sum_{n=0}^{\infty} \frac{(n+p)!}{n!(n+q)!} x^n$$

has a radius of convergence of $R = \infty$ when p and q are positive integers.

78. Using a Power Series Let

$$g(x) = 1 + 2x + x^2 + 2x^3 + x^4 + \cdots$$

where the coefficients are $c_{2n} = 1$ and $c_{2n+1} = 2$ for $n \ge 0$.

(a) Find the interval of convergence of the series.

(b) Find an explicit formula for $g(x)$.

79. Using a Power Series Let $f(x) = \sum_{n=0}^{\infty} c_n x^n$, where $c_{n+3} = c_n$ for $n \ge 0$.

(a) Find the interval of convergence of the series.

(b) Find an explicit formula for $f(x)$.

80. Proof Prove that if the power series $\sum_{n=0}^{\infty} c_n x^n$ has a radius of convergence of R, then $\sum_{n=0}^{\infty} c_n x^{2n}$ has a radius of convergence of $\sqrt{R}$.

81. Proof For $n > 0$, let $R > 0$ and $c_n > 0$. Prove that if the interval of convergence of the series

$$\sum_{n=0}^{\infty} c_n (x - x_0)^n$$

is $[x_0 - R, x_0 + R]$, then the series converges conditionally at $x_0 - R$.

9.9 Representation of Functions by Power Series

■ Find a geometric power series that represents a function.
■ Construct a power series using series operations.

Geometric Power Series

In this section and the next, you will study several techniques for finding a power series that represents a function. Consider the function

$$f(x) = \frac{1}{1-x}.$$

The form of f closely resembles the sum of a geometric series

$$\sum_{n=0}^{\infty} ar^n = \frac{a}{1-r}, \quad 0 < |r| < 1.$$

In other words, when $a = 1$ and $r = x$, a power series representation for $1/(1-x)$, centered at 0, is

$$\frac{1}{1-x} = \sum_{n=0}^{\infty} ar^n$$
$$= \sum_{n=0}^{\infty} x^n$$
$$= 1 + x + x^2 + x^3 + \cdots, \quad |x| < 1.$$

Of course, this series represents $f(x) = 1/(1-x)$ only on the interval $(-1, 1)$, whereas f is defined for all $x \neq 1$, as shown in Figure 9.22. To represent f in another interval, you must develop a different series. For instance, to obtain the power series centered at -1, you could write

$$\frac{1}{1-x} = \frac{1}{2-(x+1)} = \frac{1/2}{1-[(x+1)/2]} = \frac{a}{1-r}$$

which implies that $a = \frac{1}{2}$ and $r = (x+1)/2$. So, for $|x+1| < 2$, you have

$$\frac{1}{1-x} = \sum_{n=0}^{\infty} \left(\frac{1}{2}\right)\left(\frac{x+1}{2}\right)^n$$
$$= \frac{1}{2}\left[1 + \frac{(x+1)}{2} + \frac{(x+1)^2}{4} + \frac{(x+1)^3}{8} + \cdots\right], \quad |x+1| < 2$$

which converges on the interval $(-3, 1)$.

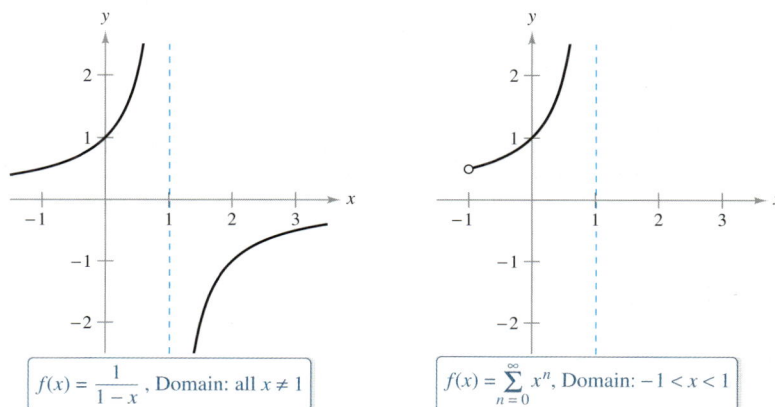

$$f(x) = \frac{1}{1-x}, \text{ Domain: all } x \neq 1$$

$$f(x) = \sum_{n=0}^{\infty} x^n, \text{ Domain: } -1 < x < 1$$

Figure 9.22

EXAMPLE 1   **Finding a Geometric Power Series Centered at 0**

EXAMPLE 1	Finding a Geometric Power Series Centered at 0

Find a power series for $f(x) = \dfrac{4}{x + 2}$, centered at 0.

Solution   Writing $f(x)$ in the form $a/(1 - r)$ produces

$$\frac{4}{2 + x} = \frac{2}{1 - (-x/2)} = \frac{a}{1 - r}$$

which implies that $a = 2$ and

$$r = -\frac{x}{2}.$$

So, the power series for $f(x)$ is

$$\frac{4}{x + 2} = \sum_{n=0}^{\infty} ar^n$$

$$= \sum_{n=0}^{\infty} 2\left(-\frac{x}{2}\right)^n$$

$$= 2\left(1 - \frac{x}{2} + \frac{x^2}{4} - \frac{x^3}{8} + \cdots\right).$$

Long Division

$$2 + x \overline{)\, 4 } \quad \begin{array}{l} 2 - x + \frac{1}{2}x^2 - \frac{1}{4}x^3 + \cdots \end{array}$$

$$\underline{4 + 2x}$$
$$-2x$$
$$\underline{-2x - x^2}$$
$$x^2$$
$$\underline{x^2 + \frac{1}{2}x^3}$$
$$-\frac{1}{2}x^3$$
$$\underline{-\frac{1}{2}x^3 - \frac{1}{4}x^4}$$

This power series converges when

$$\left|-\frac{x}{2}\right| < 1$$

which implies that the interval of convergence is $(-2, 2)$.

Another way to determine a power series for a rational function such as the one in Example 1 is to use long division. For instance, by dividing $2 + x$ into 4, you obtain the result shown at the left.

EXAMPLE 2	Finding a Geometric Power Series Centered at 1

Find a power series for $f(x) = \dfrac{1}{x}$, centered at 1.

Solution   Writing $f(x)$ in the form $a/(1 - r)$ produces

$$\frac{1}{x} = \frac{1}{1 - (-x + 1)} = \frac{a}{1 - r}$$

which implies that $a = 1$ and $r = 1 - x = -(x - 1)$. So, the power series for $f(x)$ is

$$\frac{1}{x} = \sum_{n=0}^{\infty} ar^n$$

$$= \sum_{n=0}^{\infty} [-(x - 1)]^n$$

$$= \sum_{n=0}^{\infty} (-1)^n(x - 1)^n$$

$$= 1 - (x - 1) + (x - 1)^2 - (x - 1)^3 + \cdots.$$

This power series converges when

$$|x - 1| < 1$$

which implies that the interval of convergence is $(0, 2)$.

Operations with Power Series

The versatility of geometric power series will be shown later in this section, following a discussion of power series operations. These operations, used with differentiation and integration, provide a means of developing power series for a variety of elementary functions. (For simplicity, the operations are stated for a series centered at 0.)

Operations with Power Series

Let $f(x) = \displaystyle\sum_{n=0}^{\infty} a_n x^n$ and $g(x) = \displaystyle\sum_{n=0}^{\infty} b_n x^n$.

1. $f(kx) = \displaystyle\sum_{n=0}^{\infty} a_n k^n x^n$

2. $f(x^N) = \displaystyle\sum_{n=0}^{\infty} a_n x^{nN}$

3. $f(x) \pm g(x) = \displaystyle\sum_{n=0}^{\infty} (a_n \pm b_n) x^n$

The operations described above can change the interval of convergence for the resulting series. For example, in the addition shown below, the interval of convergence for the sum is the *intersection* of the intervals of convergence of the two original series.

$$\underbrace{\sum_{n=0}^{\infty} x^n}_{(-1, 1)} + \underbrace{\sum_{n=0}^{\infty} \left(\frac{x}{2}\right)^n}_{(-2, 2)} \underset{=}{} \underbrace{\sum_{n=0}^{\infty} \left(1 + \frac{1}{2^n}\right) x^n}_{(-1, 1)}$$

$$(-1, 1) \cap (-2, 2) = (-1, 1)$$

EXAMPLE 3   **Adding Two Power Series**

Find a power series for

$$f(x) = \frac{3x - 1}{x^2 - 1}$$

centered at 0.

Solution   Using partial fractions, you can write $f(x)$ as

$$\frac{3x - 1}{x^2 - 1} = \frac{2}{x + 1} + \frac{1}{x - 1}.$$

By adding the two geometric power series

$$\frac{2}{x + 1} = \frac{2}{1 - (-x)} = \sum_{n=0}^{\infty} 2(-1)^n x^n, \quad |x| < 1$$

and

$$\frac{1}{x - 1} = \frac{-1}{1 - x} = -\sum_{n=0}^{\infty} x^n, \quad |x| < 1$$

you obtain the power series shown below.

$$\frac{3x - 1}{x^2 - 1} = \sum_{n=0}^{\infty} [2(-1)^n - 1] x^n$$

$$= 1 - 3x + x^2 - 3x^3 + x^4 - \cdots$$

The interval of convergence for this power series is $(-1, 1)$.

EXAMPLE 4   **Finding a Power Series by Integration**

Find a power series for

$$f(x) = \ln x$$

centered at 1.

Solution From Example 2, you know that

$$\frac{1}{x} = \sum_{n=0}^{\infty} (-1)^n (x-1)^n. \qquad \text{Interval of convergence: } (0, 2)$$

Integrating this series produces

$$\ln x = \int \frac{1}{x}\, dx + C$$

$$= C + \sum_{n=0}^{\infty} (-1)^n \frac{(x-1)^{n+1}}{n+1}.$$

By letting $x = 1$, you can conclude that $C = 0$. Therefore,

$$\ln x = \sum_{n=0}^{\infty} (-1)^n \frac{(x-1)^{n+1}}{n+1}$$

$$= \frac{(x-1)}{1} - \frac{(x-1)^2}{2} + \frac{(x-1)^3}{3} - \frac{(x-1)^4}{4} + \cdots. \qquad \begin{array}{l}\text{Interval of}\\ \text{convergence: } (0, 2]\end{array}$$

Note that the series converges at $x = 2$. This is consistent with the observation in the preceding section that integration of a power series may alter the convergence at the endpoints of the interval of convergence. ∎

■ **FOR FURTHER INFORMATION** To read about finding a power series using integration by parts, see the article "Integration by Parts and Infinite Series" by Shelby J. Kilmer in *Mathematics Magazine*. To view this article, go to *MathArticles.com*.

In Section 9.7, Example 4, the fourth-degree Taylor polynomial for the natural logarithmic function

$$\ln x \approx (x-1) - \frac{(x-1)^2}{2} + \frac{(x-1)^3}{3} - \frac{(x-1)^4}{4}$$

was used to approximate $\ln(1.1)$.

$$\ln(1.1) \approx (0.1) - \frac{1}{2}(0.1)^2 + \frac{1}{3}(0.1)^3 - \frac{1}{4}(0.1)^4$$

$$\approx 0.0953083$$

You now know from Example 4 in this section that this polynomial represents the first four terms of the power series for $\ln x$. Moreover, using the Alternating Series Remainder, you can determine that the error in this approximation is less than

$$|R_4| \le |a_5|$$

$$= \frac{1}{5}(0.1)^5$$

$$= 0.000002.$$

During the seventeenth and eighteenth centuries, mathematical tables for logarithms and values of other transcendental functions were computed in this manner. Such numerical techniques are far from outdated, because it is precisely by such means that many modern calculating devices are programmed to evaluate transcendental functions.

EXAMPLE 5 **Finding a Power Series by Integration**

⋯▷ *See LarsonCalculus.com for an interactive version of this type of example.*

Find a power series for

$$g(x) = \arctan x$$

centered at 0.

Solution Because $D_x[\arctan x] = 1/(1 + x^2)$, you can use the series

$$f(x) = \frac{1}{1 + x} = \sum_{n=0}^{\infty} (-1)^n x^n. \qquad \text{Interval of convergence: } (-1, 1)$$

Substituting x^2 for x produces

$$f(x^2) = \frac{1}{1 + x^2} = \sum_{n=0}^{\infty} (-1)^n x^{2n}.$$

Finally, by integrating, you obtain

$$\arctan x = \int \frac{1}{1 + x^2} \, dx + C$$

$$= C + \sum_{n=0}^{\infty} (-1)^n \frac{x^{2n+1}}{2n + 1}$$

$$= \sum_{n=0}^{\infty} (-1)^n \frac{x^{2n+1}}{2n + 1} \qquad \text{Let } x = 0, \text{ then } C = 0.$$

$$= x - \frac{x^3}{3} + \frac{x^5}{5} - \frac{x^7}{7} + \cdots. \qquad \text{Interval of convergence: } (-1, 1)$$

It can be shown that the power series developed for $\arctan x$ in Example 5 also converges (to $\arctan x$) for $x = \pm 1$. For instance, when $x = 1$, you can write

$$\arctan 1 = 1 - \frac{1}{3} + \frac{1}{5} - \frac{1}{7} + \cdots$$

$$= \frac{\pi}{4}.$$

However, this series (developed by James Gregory in 1671) does not give us a practical way of approximating π because it converges so slowly that hundreds of terms would have to be used to obtain reasonable accuracy. Example 6 shows how to use *two* different arctangent series to obtain a very good approximation of π using only a few terms. This approximation was developed by John Machin in 1706.

EXAMPLE 6 **Approximating π with a Series**

Use the trigonometric identity

$$4 \arctan \frac{1}{5} - \arctan \frac{1}{239} = \frac{\pi}{4}$$

to approximate the number π [see Exercise 46(b)].

Solution By using only five terms from each of the series for $\arctan(1/5)$ and $\arctan(1/239)$, you obtain

$$4 \left(4 \arctan \frac{1}{5} - \arctan \frac{1}{239} \right) \approx 3.1415926$$

which agrees with the exact value of π with an error of less than 0.0000001.

SRINIVASA RAMANUJAN (1887–1920)

Series that can be used to approximate π have interested mathematicians for the past 300 years. An amazing series for approximating $1/\pi$ was discovered by the Indian mathematician Srinivasa Ramanujan in 1914 (see Exercise 61). Each successive term of Ramanujan's series adds roughly eight more correct digits to the value of $1/\pi$. For more information about Ramanujan's work, see the article "Ramanujan and Pi" by Jonathan M. Borwein and Peter B. Borwein in *Scientific American*.
See LarsonCalculus.com to read more of this biography.

■ **FOR FURTHER INFORMATION** To read about other methods for approximating π, see the article "Two Methods for Approximating π" by Chien-Lih Hwang in *Mathematics Magazine*. To view this article, go to *MathArticles.com*.

9.9 Exercises

See **CalcChat.com** for tutorial help and worked-out solutions to odd-numbered exercises.

Finding a Geometric Power Series In Exercises 1–4, find a geometric power series for the function, centered at 0, (a) by the technique shown in Examples 1 and 2 and (b) by long division.

1. $f(x) = \dfrac{1}{4 - x}$

2. $f(x) = \dfrac{1}{2 + x}$

3. $f(x) = \dfrac{4}{3 + x}$

4. $f(x) = \dfrac{2}{5 - x}$

Finding a Power Series In Exercises 5–16, find a power series for the function, centered at c, and determine the interval of convergence.

5. $f(x) = \dfrac{1}{3 - x}$, $c = 1$

6. $f(x) = \dfrac{2}{6 - x}$, $c = -2$

7. $f(x) = \dfrac{1}{1 - 3x}$, $c = 0$

8. $h(x) = \dfrac{1}{1 - 5x}$, $c = 0$

9. $g(x) = \dfrac{5}{2x - 3}$, $c = -3$

10. $f(x) = \dfrac{3}{2x - 1}$, $c = 2$

11. $f(x) = \dfrac{3}{3x + 4}$, $c = 0$

12. $f(x) = \dfrac{4}{3x + 2}$, $c = 3$

13. $g(x) = \dfrac{4x}{x^2 + 2x - 3}$, $c = 0$

14. $g(x) = \dfrac{3x - 8}{3x^2 + 5x - 2}$, $c = 0$

15. $f(x) = \dfrac{2}{1 - x^2}$, $c = 0$

16. $f(x) = \dfrac{5}{5 + x^2}$, $c = 0$

Using a Power Series In Exercises 17–26, use the power series

$$\frac{1}{1 + x} = \sum_{n=0}^{\infty} (-1)^n x^n$$

to determine a power series, centered at 0, for the function. Identify the interval of convergence.

17. $h(x) = \dfrac{-2}{x^2 - 1} = \dfrac{1}{1 + x} + \dfrac{1}{1 - x}$

18. $h(x) = \dfrac{x}{x^2 - 1} = \dfrac{1}{2(1 + x)} - \dfrac{1}{2(1 - x)}$

19. $f(x) = -\dfrac{1}{(x + 1)^2} = \dfrac{d}{dx}\left[\dfrac{1}{x + 1}\right]$

20. $f(x) = \dfrac{2}{(x + 1)^3} = \dfrac{d^2}{dx^2}\left[\dfrac{1}{x + 1}\right]$

21. $f(x) = \ln(x + 1) = \displaystyle\int \dfrac{1}{x + 1}\, dx$

22. $f(x) = \ln(1 - x^2) = \displaystyle\int \dfrac{1}{1 + x}\, dx - \int \dfrac{1}{1 - x}\, dx$

23. $g(x) = \dfrac{1}{x^2 + 1}$

24. $f(x) = \ln(x^2 + 1)$

25. $h(x) = \dfrac{1}{4x^2 + 1}$

26. $f(x) = \arctan 2x$

Graphical and Numerical Analysis In Exercises 27 and 28, let

$$S_n = x - \frac{x^2}{2} + \frac{x^3}{3} - \frac{x^4}{4} + \cdots \pm \frac{x^n}{n}.$$

Use a graphing utility to confirm the inequality graphically. Then complete the table to confirm the inequality numerically.

x	0.0	0.2	0.4	0.6	0.8	1.0
S_n						
$\ln(x + 1)$						
S_{n+1}						

27. $S_2 \le \ln(x + 1) \le S_3$

28. $S_4 \le \ln(x + 1) \le S_5$

Approximating a Sum In Exercises 29 and 30, (a) graph several partial sums of the series, (b) find the sum of the series and its radius of convergence, (c) use 50 terms of the series to approximate the sum when $x = 0.5$, and (d) determine what the approximation represents and how good the approximation is.

29. $\displaystyle\sum_{n=1}^{\infty} \dfrac{(-1)^{n+1}(x - 1)^n}{n}$

30. $\displaystyle\sum_{n=0}^{\infty} \dfrac{(-1)^n x^{2n+1}}{(2n + 1)!}$

Approximating a Value In Exercises 31–34, use the series for $f(x) = \arctan x$ to approximate the value, using $R_N \le 0.001$.

31. $\arctan \dfrac{1}{4}$

32. $\displaystyle\int_0^{3/4} \arctan x^2\, dx$

33. $\displaystyle\int_0^{1/2} \dfrac{\arctan x^2}{x}\, dx$

34. $\displaystyle\int_0^{1/2} x^2 \arctan x\, dx$

Using a Power Series In Exercises 35–38, use the power series

$$\frac{1}{1 - x} = \sum_{n=0}^{\infty} x^n, \quad |x| < 1.$$

Find the series representation of the function and determine its interval of convergence.

35. $f(x) = \dfrac{1}{(1 - x)^2}$

36. $f(x) = \dfrac{x}{(1 - x)^2}$

37. $f(x) = \dfrac{1 + x}{(1 - x)^2}$

38. $f(x) = \dfrac{x(1 + x)}{(1 - x)^2}$

39. Probability   A fair coin is tossed repeatedly. The probability that the first head occurs on the nth toss is $P(n) = \left(\frac{1}{2}\right)^n$. When this game is repeated many times, the average number of tosses required until the first head occurs is

$$E(n) = \sum_{n=1}^{\infty} nP(n).$$

(This value is called the *expected value of n*.) Use the results of Exercises 35–38 to find $E(n)$. Is the answer what you expected? Why or why not?

40. Finding the Sum of a Series   Use the results of Exercises 35–38 to find the sum of each series.

(a) $\dfrac{1}{3}\displaystyle\sum_{n=1}^{\infty} n\left(\dfrac{2}{3}\right)^n$

(b) $\dfrac{1}{10}\displaystyle\sum_{n=1}^{\infty} n\left(\dfrac{9}{10}\right)^n$

Writing   In Exercises 41–44, explain how to use the geometric series

$$g(x) = \frac{1}{1-x} = \sum_{n=0}^{\infty} x^n, \quad |x| < 1$$

to find the series for the function. Do not find the series.

41. $f(x) = \dfrac{1}{1+x}$

42. $f(x) = \dfrac{1}{1-x^2}$

43. $f(x) = \dfrac{5}{1+x}$

44. $f(x) = \ln(1-x)$

45. Proof   Prove that

$$\arctan x + \arctan y = \arctan \frac{x+y}{1-xy}$$

for $xy \neq 1$ provided the value of the left side of the equation is between $-\pi/2$ and $\pi/2$.

46. Verifying an Identity   Use the result of Exercise 45 to verify each identity.

(a) $\arctan \dfrac{120}{119} - \arctan \dfrac{1}{239} = \dfrac{\pi}{4}$

(b) $4 \arctan \dfrac{1}{5} - \arctan \dfrac{1}{239} = \dfrac{\pi}{4}$

[*Hint:* Use Exercise 45 twice to find $4 \arctan \frac{1}{5}$. Then use part (a).]

Approximating Pi   In Exercises 47 and 48, (a) verify the given equation, and (b) use the equation and the series for the arctangent to approximate π to two-decimal-place accuracy.

47. $2 \arctan \dfrac{1}{2} - \arctan \dfrac{1}{7} = \dfrac{\pi}{4}$

48. $\arctan \dfrac{1}{2} + \arctan \dfrac{1}{3} = \dfrac{\pi}{4}$

Finding the Sum of a Series   In Exercises 49–54, find the sum of the convergent series by using a well-known function. Identify the function and explain how you obtained the sum.

49. $\displaystyle\sum_{n=1}^{\infty} (-1)^{n+1} \dfrac{1}{2^n n}$

50. $\displaystyle\sum_{n=1}^{\infty} (-1)^{n+1} \dfrac{1}{3^n n}$

51. $\displaystyle\sum_{n=1}^{\infty} (-1)^{n+1} \dfrac{2^n}{5^n n}$

52. $\displaystyle\sum_{n=0}^{\infty} (-1)^n \dfrac{1}{2n+1}$

53. $\displaystyle\sum_{n=0}^{\infty} (-1)^n \dfrac{1}{2^{2n+1}(2n+1)}$

54. $\displaystyle\sum_{n=1}^{\infty} (-1)^{n+1} \dfrac{1}{3^{2n-1}(2n-1)}$

WRITING ABOUT CONCEPTS

55. Using Series   One of the series in Exercises 49–54 converges to its sum at a much lower rate than the other five series. Which is it? Explain why this series converges so slowly. Use a graphing utility to illustrate the rate of convergence.

56. Radius of Convergence   The radius of convergence of the power series $\displaystyle\sum_{n=0}^{\infty} a_n x^n$ is 3. What is the radius of convergence of the series $\displaystyle\sum_{n=1}^{\infty} na_n x^{n-1}$? Explain.

57. Convergence of a Power Series   The power series $\displaystyle\sum_{n=0}^{\infty} a_n x^n$ converges for $|x+1| < 4$. What can you conclude about the series $\displaystyle\sum_{n=0}^{\infty} a_n \dfrac{x^{n+1}}{n+1}$? Explain.

---

58.   HOW DO YOU SEE IT?   The graphs show first-, second-, and third-degree polynomial approximations P_1, P_2, and P_3 of a function f. Label the graphs of P_1, P_2, and P_3. To print an enlarged copy of the graph, go to *MathGraphs.com*.

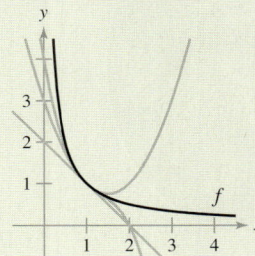

Finding the Sum of a Series   In Exercises 59 and 60, find the sum of the series.

59. $\displaystyle\sum_{n=0}^{\infty} \dfrac{(-1)^n}{3^n(2n+1)}$

60. $\displaystyle\sum_{n=0}^{\infty} \dfrac{(-1)^n \, \pi^{2n+1}}{3^{2n+1}(2n+1)!}$

61. Ramanujan and Pi   Use a graphing utility to show that

$$\frac{\sqrt{8}}{9801} \sum_{n=0}^{\infty} \frac{(4n)!(1103 + 26{,}390n)}{(n!)^4 396^{4n}} = \frac{1}{\pi}.$$

62. Find the Error   Describe why the statement is incorrect.

$$\sum_{n=0}^{\infty} x^n + \sum_{n=0}^{\infty} \left(\frac{x}{5}\right)^n = \sum_{n=0}^{\infty} \left(1 + \frac{1}{5}\right) x^n$$

9.10 Taylor and Maclaurin Series

- Find a Taylor or Maclaurin series for a function.
- Find a binomial series.
- Use a basic list of Taylor series to find other Taylor series.

Taylor Series and Maclaurin Series

In Section 9.9, you derived power series for several functions using geometric series with term-by-term differentiation or integration. In this section, you will study a *general* procedure for deriving the power series for a function that has derivatives of all orders. The next theorem gives the form that *every* convergent power series must take.

> **THEOREM 9.22 The Form of a Convergent Power Series**
>
> If f is represented by a power series $f(x) = \sum a_n(x - c)^n$ for all x in an open interval I containing c, then
>
> $$a_n = \frac{f^{(n)}(c)}{n!}$$
>
> and
>
> $$f(x) = f(c) + f'(c)(x - c) + \frac{f''(c)}{2!}(x - c)^2 + \cdots + \frac{f^{(n)}(c)}{n!}(x - c)^n + \cdots.$$

Proof Consider a power series $\sum a_n(x - c)^n$ that has a radius of convergence R. Then, by Theorem 9.21, you know that the nth derivative of f exists for $|x - c| < R$, and by successive differentiation you obtain the following.

$$f^{(0)}(x) = a_0 + a_1(x - c) + a_2(x - c)^2 + a_3(x - c)^3 + a_4(x - c)^4 + \cdots$$
$$f^{(1)}(x) = a_1 + 2a_2(x - c) + 3a_3(x - c)^2 + 4a_4(x - c)^3 + \cdots$$
$$f^{(2)}(x) = 2a_2 + 3!a_3(x - c) + 4 \cdot 3a_4(x - c)^2 + \cdots$$
$$f^{(3)}(x) = 3!a_3 + 4!a_4(x - c) + \cdots$$
$$\vdots$$
$$f^{(n)}(x) = n!a_n + (n + 1)!a_{n+1}(x - c) + \cdots$$

Evaluating each of these derivatives at $x = c$ yields

$$f^{(0)}(c) = 0!a_0$$
$$f^{(1)}(c) = 1!a_1$$
$$f^{(2)}(c) = 2!a_2$$
$$f^{(3)}(c) = 3!a_3$$

and, in general, $f^{(n)}(c) = n!a_n$. By solving for a_n, you find that the coefficients of the power series representation of $f(x)$ are

$$a_n = \frac{f^{(n)}(c)}{n!}.$$

See LarsonCalculus.com for Bruce Edwards's video of this proof.

Notice that the coefficients of the power series in Theorem 9.22 are precisely the coefficients of the Taylor polynomials for $f(x)$ at c as defined in Section 9.7. For this reason, the series is called the **Taylor series** for $f(x)$ at c.

> **Definition of Taylor and Maclaurin Series**
>
> If a function f has derivatives of all orders at $x = c$, then the series
>
> $$\sum_{n=0}^{\infty} \frac{f^{(n)}(c)}{n!}(x - c)^n = f(c) + f'(c)(x - c) + \cdots + \frac{f^{(n)}(c)}{n!}(x - c)^n + \cdots$$
>
> is called the **Taylor series for $f(x)$ at c.** Moreover, if $c = 0$, then the series is the **Maclaurin series for f.**

When you know the pattern for the coefficients of the Taylor polynomials for a function, you can extend the pattern easily to form the corresponding Taylor series. For instance, in Example 4 in Section 9.7, you found the fourth Taylor polynomial for $\ln x$, centered at 1, to be

$$P_4(x) = (x - 1) - \frac{1}{2}(x - 1)^2 + \frac{1}{3}(x - 1)^3 - \frac{1}{4}(x - 1)^4.$$

From this pattern, you can obtain the Taylor series for $\ln x$ centered at $c = 1$,

$$(x - 1) - \frac{1}{2}(x - 1)^2 + \cdots + \frac{(-1)^{n+1}}{n}(x - 1)^n + \cdots.$$

EXAMPLE 1   **Forming a Power Series**

Use the function

$$f(x) = \sin x$$

to form the Maclaurin series

$$\sum_{n=0}^{\infty} \frac{f^{(n)}(0)}{n!}x^n = f(0) + f'(0)x + \frac{f''(0)}{2!}x^2 + \frac{f^{(3)}(0)}{3!}x^3 + \frac{f^{(4)}(0)}{4!}x^4 + \cdots$$

and determine the interval of convergence.

Solution   Successive differentiation of $f(x)$ yields

$$f(x) = \sin x \qquad\qquad f(0) = \sin 0 = 0$$
$$f'(x) = \cos x \qquad\qquad f'(0) = \cos 0 = 1$$
$$f''(x) = -\sin x \qquad\qquad f''(0) = -\sin 0 = 0$$
$$f^{(3)}(x) = -\cos x \qquad\qquad f^{(3)}(0) = -\cos 0 = -1$$
$$f^{(4)}(x) = \sin x \qquad\qquad f^{(4)}(0) = \sin 0 = 0$$
$$f^{(5)}(x) = \cos x \qquad\qquad f^{(5)}(0) = \cos 0 = 1$$

and so on. The pattern repeats after the third derivative. So, the power series is as follows.

$$\sum_{n=0}^{\infty} \frac{f^{(n)}(0)}{n!}x^n = f(0) + f'(0)x + \frac{f''(0)}{2!}x^2 + \frac{f^{(3)}(0)}{3!}x^3 + \frac{f^{(4)}(0)}{4!}x^4 + \cdots$$

$$\sum_{n=0}^{\infty} \frac{(-1)^n x^{2n+1}}{(2n + 1)!} = 0 + (1)x + \frac{0}{2!}x^2 + \frac{(-1)}{3!}x^3 + \frac{0}{4!}x^4 + \frac{1}{5!}x^5 + \frac{0}{6!}x^6$$

$$+ \frac{(-1)}{7!}x^7 + \cdots$$

$$= x - \frac{x^3}{3!} + \frac{x^5}{5!} - \frac{x^7}{7!} + \cdots$$

By the Ratio Test, you can conclude that this series converges for all x.

$$f(x) = \begin{cases} -1, & x < -\dfrac{\pi}{2} \\ \sin x, & |x| \le \dfrac{\pi}{2} \\ 1, & x > \dfrac{\pi}{2} \end{cases}$$

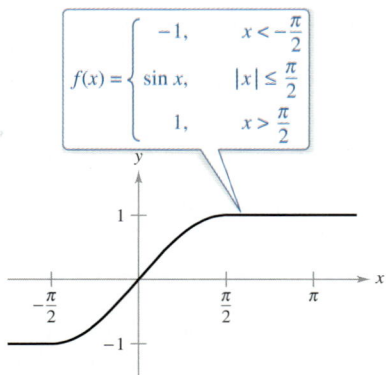

Figure 9.23

Notice that in Example 1, you cannot conclude that the power series converges to $\sin x$ for all x. You can simply conclude that the power series converges to some function, but you are not sure what function it is. This is a subtle, but important, point in dealing with Taylor or Maclaurin series. To persuade yourself that the series

$$f(c) + f'(c)(x - c) + \frac{f''(c)}{2!}(x - c)^2 + \cdots + \frac{f^{(n)}(c)}{n!}(x - c)^n + \cdots$$

might converge to a function other than f, remember that the derivatives are being evaluated at a single point. It can easily happen that another function will agree with the values of $f^{(n)}(x)$ when $x = c$ and disagree at other x-values. For instance, the power series (centered at 0) for the function f shown in Figure 9.23 is the same series as in Example 1. You know that the series converges for all x, and yet it obviously cannot converge to both $f(x)$ and $\sin x$ for all x.

Let f have derivatives of all orders in an open interval I centered at c. The Taylor series for f may fail to converge for some x in I. Or, even when it is convergent, it may fail to have $f(x)$ as its sum. Nevertheless, Theorem 9.19 tells us that for each n,

$$f(x) = f(c) + f'(c)(x - c) + \frac{f''(c)}{2!}(x - c)^2 + \cdots + \frac{f^{(n)}(c)}{n!}(x - c)^n + R_n(x)$$

where

$$R_n(x) = \frac{f^{(n+1)}(z)}{(n + 1)!}(x - c)^{n+1}.$$

Note that in this remainder formula, the particular value of z that makes the remainder formula true depends on the values of x and n. If $R_n \to 0$, then the next theorem tells us that the Taylor series for f actually converges to $f(x)$ for all x in I.

THEOREM 9.23   Convergence of Taylor Series

If $\lim\limits_{n \to \infty} R_n = 0$ for all x in the interval I, then the Taylor series for f converges and equals $f(x)$,

$$f(x) = \sum_{n=0}^{\infty} \frac{f^{(n)}(c)}{n!}(x - c)^n.$$

Proof For a Taylor series, the nth partial sum coincides with the nth Taylor polynomial. That is, $S_n(x) = P_n(x)$. Moreover, because

$$P_n(x) = f(x) - R_n(x)$$

it follows that

$$\lim_{n \to \infty} S_n(x) = \lim_{n \to \infty} P_n(x)$$
$$= \lim_{n \to \infty} [f(x) - R_n(x)]$$
$$= f(x) - \lim_{n \to \infty} R_n(x).$$

So, for a given x, the Taylor series (the sequence of partial sums) converges to $f(x)$ if and only if $R_n(x) \to 0$ as $n \to \infty$.

See LarsonCalculus.com for Bruce Edwards's video of this proof.

Stated another way, Theorem 9.23 says that a power series formed with Taylor coefficients $a_n = f^{(n)}(c)/n!$ converges to the function from which it was derived at precisely those values for which the remainder approaches 0 as $n \to \infty$.

In Example 1, you derived the power series from the sine function and you also concluded that the series converges to some function on the entire real number line. In Example 2, you will see that the series actually converges to $\sin x$. The key observation is that although the value of z is not known, it is possible to obtain an upper bound for

$$\left| f^{(n+1)}(z) \right|.$$

EXAMPLE 2   **A Convergent Maclaurin Series**

Show that the Maclaurin series for

$$f(x) = \sin x$$

converges to $\sin x$ for all x.

Solution   Using the result in Example 1, you need to show that

$$\sin x = x - \frac{x^3}{3!} + \frac{x^5}{5!} - \frac{x^7}{7!} + \cdots + \frac{(-1)^n x^{2n+1}}{(2n+1)!} + \cdots$$

is true for all x. Because

$$f^{(n+1)}(x) = \pm \sin x$$

or

$$f^{(n+1)}(x) = \pm \cos x$$

you know that $\left| f^{(n+1)}(z) \right| \le 1$ for every real number z. Therefore, for any fixed x, you can apply Taylor's Theorem (Theorem 9.19) to conclude that

$$0 \le |R_n(x)| = \left| \frac{f^{(n+1)}(z)}{(n+1)!} x^{n+1} \right| \le \frac{|x|^{n+1}}{(n+1)!}.$$

From the discussion in Section 9.1 regarding the relative rates of convergence of exponential and factorial sequences, it follows that for a fixed x

$$\lim_{n \to \infty} \frac{|x|^{n+1}}{(n+1)!} = 0.$$

Finally, by the Squeeze Theorem, it follows that for all x, $R_n(x) \to 0$ as $n \to \infty$. So, by Theorem 9.23, the Maclaurin series for $\sin x$ converges to $\sin x$ for all x. ∎

Figure 9.24 visually illustrates the convergence of the Maclaurin series for $\sin x$ by comparing the graphs of the Maclaurin polynomials $P_1(x)$, $P_3(x)$, $P_5(x)$, and $P_7(x)$ with the graph of the sine function. Notice that as the degree of the polynomial increases, its graph more closely resembles that of the sine function.

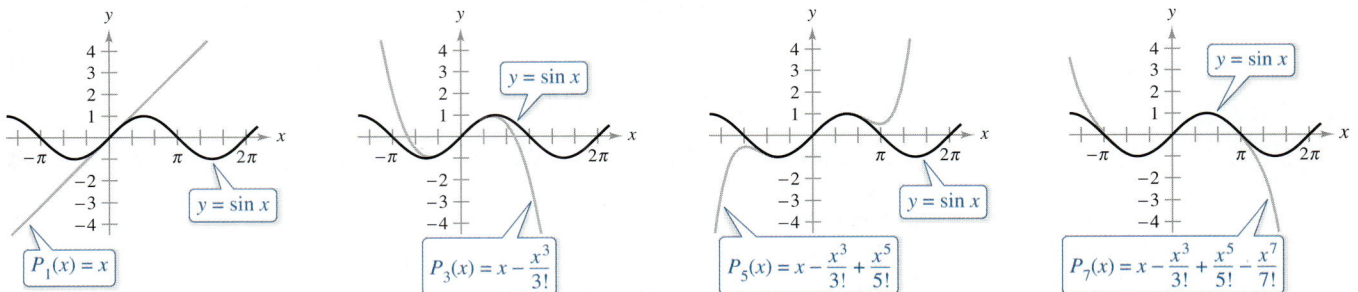

As n increases, the graph of P_n more closely resembles the sine function.
Figure 9.24

The guidelines for finding a Taylor series for $f(x)$ at c are summarized below.

GUIDELINES FOR FINDING A TAYLOR SERIES

1. Differentiate $f(x)$ several times and evaluate each derivative at c.

$$f(c), f'(c), f''(c), f'''(c), \cdots, f^{(n)}(c), \cdots$$

Try to recognize a pattern in these numbers.

2. Use the sequence developed in the first step to form the Taylor coefficients $a_n = f^{(n)}(c)/n!$, and determine the interval of convergence for the resulting power series

$$f(c) + f'(c)(x - c) + \frac{f''(c)}{2!}(x - c)^2 + \cdots + \frac{f^{(n)}(c)}{n!}(x - c)^n + \cdots.$$

3. Within this interval of convergence, determine whether the series converges to $f(x)$.

> • • **REMARK** When you have difficulty recognizing a pattern, remember that you can use Theorem 9.22 to find the Taylor series. Also, you can try using the coefficients of a known Taylor or Maclaurin series, as shown in Example 3.

The direct determination of Taylor or Maclaurin coefficients using successive differentiation can be difficult, and the next example illustrates a shortcut for finding the coefficients indirectly—using the coefficients of a known Taylor or Maclaurin series.

EXAMPLE 3 **Maclaurin Series for a Composite Function**

Find the Maclaurin series for

$$f(x) = \sin x^2.$$

Solution To find the coefficients for this Maclaurin series directly, you must calculate successive derivatives of $f(x) = \sin x^2$. By calculating just the first two,

$$f'(x) = 2x \cos x^2$$

and

$$f''(x) = -4x^2 \sin x^2 + 2 \cos x^2$$

you can see that this task would be quite cumbersome. Fortunately, there is an alternative. First, consider the Maclaurin series for $\sin x$ found in Example 1.

$$g(x) = \sin x$$

$$= x - \frac{x^3}{3!} + \frac{x^5}{5!} - \frac{x^7}{7!} + \cdots$$

Now, because $\sin x^2 = g(x^2)$, you can substitute x^2 for x in the series for $\sin x$ to obtain

$$\sin x^2 = g(x^2)$$

$$= x^2 - \frac{x^6}{3!} + \frac{x^{10}}{5!} - \frac{x^{14}}{7!} + \cdots.$$

Be sure to understand the point illustrated in Example 3. Because direct computation of Taylor or Maclaurin coefficients can be tedious, the most practical way to find a Taylor or Maclaurin series is to develop power series for a *basic list* of elementary functions. From this list, you can determine power series for other functions by the operations of addition, subtraction, multiplication, division, differentiation, integration, and composition with known power series.

Binomial Series

Before presenting the basic list for elementary functions, you will develop one more series—for a function of the form $f(x) = (1 + x)^k$. This produces the **binomial series.**

EXAMPLE 4   Binomial Series

Find the Maclaurin series for $f(x) = (1 + x)^k$ and determine its radius of convergence. Assume that k is not a positive integer and $k \neq 0$.

Solution   By successive differentiation, you have

$$f(x) = (1 + x)^k \qquad\qquad f(0) = 1$$
$$f'(x) = k(1 + x)^{k-1} \qquad\qquad f'(0) = k$$
$$f''(x) = k(k - 1)(1 + x)^{k-2} \qquad\qquad f''(0) = k(k - 1)$$
$$f'''(x) = k(k - 1)(k - 2)(1 + x)^{k-3} \qquad f'''(0) = k(k - 1)(k - 2)$$
$$\vdots \qquad\qquad\qquad\qquad \vdots$$
$$f^{(n)}(x) = k \cdots (k - n + 1)(1 + x)^{k-n} \qquad f^{(n)}(0) = k(k - 1) \ldots (k - n + 1)$$

which produces the series

$$1 + kx + \frac{k(k - 1)x^2}{2} + \cdots + \frac{k(k - 1) \cdots (k - n + 1)x^n}{n!} + \cdots.$$

Because $a_{n+1}/a_n \to 1$, you can apply the Ratio Test to conclude that the radius of convergence is $R = 1$. So, the series converges to some function in the interval $(-1, 1)$. ∎

Note that Example 4 shows that the Taylor series for $(1 + x)^k$ converges to some function in the interval $(-1, 1)$. However, the example does not show that the series actually converges to $(1 + x)^k$. To do this, you could show that the remainder $R_n(x)$ converges to 0, as illustrated in Example 2. You now have enough information to find a binomial series for a function, as shown in the next example.

EXAMPLE 5   Finding a Binomial Series

Find the power series for $f(x) = \sqrt[3]{1 + x}$.

Solution   Using the binomial series

$$(1 + x)^k = 1 + kx + \frac{k(k - 1)x^2}{2!} + \frac{k(k - 1)(k - 2)x^3}{3!} + \cdots$$

let $k = \frac{1}{3}$ and write

$$(1 + x)^{1/3} = 1 + \frac{x}{3} - \frac{2x^2}{3^2 2!} + \frac{2 \cdot 5x^3}{3^3 3!} - \frac{2 \cdot 5 \cdot 8x^4}{3^4 4!} + \cdots$$

which converges for $-1 \leq x \leq 1$. ∎

▷ TECHNOLOGY   Use a graphing utility to confirm the result in Example 5. When you graph the functions

$$f(x) = (1 + x)^{1/3}$$

and

$$P_4(x) = 1 + \frac{x}{3} - \frac{x^2}{9} + \frac{5x^3}{81} - \frac{10x^4}{243}$$

in the same viewing window, you should obtain the result shown in Figure 9.25.

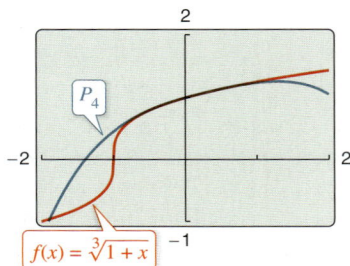

Figure 9.25

Deriving Taylor Series from a Basic List

The list below provides the power series for several elementary functions with the corresponding intervals of convergence.

POWER SERIES FOR ELEMENTARY FUNCTIONS

Function	Interval of Convergence
$\dfrac{1}{x} = 1 - (x-1) + (x-1)^2 - (x-1)^3 + (x-1)^4 - \cdots + (-1)^n(x-1)^n + \cdots$	$0 < x < 2$
$\dfrac{1}{1+x} = 1 - x + x^2 - x^3 + x^4 - x^5 + \cdots + (-1)^n x^n + \cdots$	$-1 < x < 1$
$\ln x = (x-1) - \dfrac{(x-1)^2}{2} + \dfrac{(x-1)^3}{3} - \dfrac{(x-1)^4}{4} + \cdots + \dfrac{(-1)^{n-1}(x-1)^n}{n} + \cdots$	$0 < x \le 2$
$e^x = 1 + x + \dfrac{x^2}{2!} + \dfrac{x^3}{3!} + \dfrac{x^4}{4!} + \dfrac{x^5}{5!} + \cdots + \dfrac{x^n}{n!} + \cdots$	$-\infty < x < \infty$
$\sin x = x - \dfrac{x^3}{3!} + \dfrac{x^5}{5!} - \dfrac{x^7}{7!} + \dfrac{x^9}{9!} - \cdots + \dfrac{(-1)^n x^{2n+1}}{(2n+1)!} + \cdots$	$-\infty < x < \infty$
$\cos x = 1 - \dfrac{x^2}{2!} + \dfrac{x^4}{4!} - \dfrac{x^6}{6!} + \dfrac{x^8}{8!} - \cdots + \dfrac{(-1)^n x^{2n}}{(2n)!} + \cdots$	$-\infty < x < \infty$
$\arctan x = x - \dfrac{x^3}{3} + \dfrac{x^5}{5} - \dfrac{x^7}{7} + \dfrac{x^9}{9} - \cdots + \dfrac{(-1)^n x^{2n+1}}{2n+1} + \cdots$	$-1 \le x \le 1$
$\arcsin x = x + \dfrac{x^3}{2 \cdot 3} + \dfrac{1 \cdot 3x^5}{2 \cdot 4 \cdot 5} + \dfrac{1 \cdot 3 \cdot 5x^7}{2 \cdot 4 \cdot 6 \cdot 7} + \cdots + \dfrac{(2n)!x^{2n+1}}{(2^n n!)^2(2n+1)} + \cdots$	$-1 \le x \le 1$
$(1+x)^k = 1 + kx + \dfrac{k(k-1)x^2}{2!} + \dfrac{k(k-1)(k-2)x^3}{3!} + \dfrac{k(k-1)(k-2)(k-3)x^4}{4!} + \cdots$	$-1 < x < 1*$

* The convergence at $x = \pm 1$ depends on the value of k.

Note that the binomial series is valid for noninteger values of k. Also, when k is a positive integer, the binomial series reduces to a simple binomial expansion.

EXAMPLE 6 **Deriving a Power Series from a Basic List**

Find the power series for

$$f(x) = \cos\sqrt{x}.$$

Solution Using the power series

$$\cos x = 1 - \frac{x^2}{2!} + \frac{x^4}{4!} - \frac{x^6}{6!} + \frac{x^8}{8!} - \cdots$$

you can replace x by

$$\sqrt{x}$$

to obtain the series

$$\cos\sqrt{x} = 1 - \frac{x}{2!} + \frac{x^2}{4!} - \frac{x^3}{6!} + \frac{x^4}{8!} - \cdots.$$

This series converges for all x in the domain of $\cos\sqrt{x}$—that is, for $x \ge 0$.

Power series can be multiplied and divided like polynomials. After finding the first few terms of the product (or quotient), you may be able to recognize a pattern.

EXAMPLE 7   **Multiplication of Power Series**

Find the first three nonzero terms in the Maclaurin series $e^x \arctan x$.

Solution   Using the Maclaurin series for e^x and $\arctan x$ in the table, you have

$$e^x \arctan x = \left(1 + \frac{x}{1!} + \frac{x^2}{2!} + \frac{x^3}{3!} + \frac{x^4}{4!} + \cdots \right)\left(x - \frac{x^3}{3} + \frac{x^5}{5} - \cdots \right).$$

Multiply these expressions and collect like terms as you would in multiplying polynomials.

$$
\begin{array}{l}
1 + x + \dfrac{1}{2}x^2 + \dfrac{1}{6}x^3 + \dfrac{1}{24}x^4 + \cdots \\[2mm]
\quad x \qquad\qquad - \dfrac{1}{3}x^3 \qquad\quad + \dfrac{1}{5}x^5 - \cdots \\[1mm]
\hline
\quad x + \quad x^2 + \dfrac{1}{2}x^3 + \quad\dfrac{1}{6}x^4 + \dfrac{1}{24}x^5 + \cdots \\[2mm]
\qquad\qquad\qquad - \dfrac{1}{3}x^3 - \dfrac{1}{3}x^4 - \dfrac{1}{6}x^5 - \cdots \\[2mm]
\qquad\qquad\qquad\qquad\qquad\qquad + \dfrac{1}{5}x^5 + \cdots \\[1mm]
\hline
\quad x + \quad x^2 + \dfrac{1}{6}x^3 - \dfrac{1}{6}x^4 + \dfrac{3}{40}x^5 + \cdots
\end{array}
$$

So, $e^x \arctan x = x + x^2 + \frac{1}{6}x^3 + \cdots$.

EXAMPLE 8   **Division of Power Series**

Find the first three nonzero terms in the Maclaurin series $\tan x$.

Solution   Using the Maclaurin series for $\sin x$ and $\cos x$ in the table, you have

$$\tan x = \frac{\sin x}{\cos x} = \frac{x - \dfrac{x^3}{3!} + \dfrac{x^5}{5!} - \cdots}{1 - \dfrac{x^2}{2!} + \dfrac{x^4}{4!} - \cdots}.$$

Divide using long division.

$$
\begin{array}{r}
x + \dfrac{1}{3}x^3 + \dfrac{2}{15}x^5 + \cdots \\[1mm]
1 - \dfrac{1}{2}x^2 + \dfrac{1}{24}x^4 - \cdots \overline{\smash{\big)}\ x - \dfrac{1}{6}x^3 + \dfrac{1}{120}x^5 - \cdots} \\[2mm]
\underline{x - \dfrac{1}{2}x^3 + \dfrac{1}{24}x^5 - \cdots} \\[2mm]
\dfrac{1}{3}x^3 - \dfrac{1}{30}x^5 + \cdots \\[2mm]
\underline{\dfrac{1}{3}x^3 - \dfrac{1}{6}x^5 + \cdots} \\[2mm]
\dfrac{2}{15}x^5 + \cdots
\end{array}
$$

So, $\tan x = x + \frac{1}{3}x^3 + \frac{2}{15}x^5 + \cdots$.

EXAMPLE 9 **A Power Series for sin² x**

Find the power series for

$$f(x) = \sin^2 x.$$

Solution Consider rewriting $\sin^2 x$ as

$$\sin^2 x = \frac{1 - \cos 2x}{2} = \frac{1}{2} - \frac{1}{2}\cos 2x.$$

Now, use the series for $\cos x$.

$$\cos x = 1 - \frac{x^2}{2!} + \frac{x^4}{4!} - \frac{x^6}{6!} + \frac{x^8}{8!} - \cdots$$

$$\cos 2x = 1 - \frac{2^2}{2!}x^2 + \frac{2^4}{4!}x^4 - \frac{2^6}{6!}x^6 + \frac{2^8}{8!}x^8 - \cdots$$

$$-\frac{1}{2}\cos 2x = -\frac{1}{2} + \frac{2}{2!}x^2 - \frac{2^3}{4!}x^4 + \frac{2^5}{6!}x^6 - \frac{2^7}{8!}x^8 + \cdots$$

$$\frac{1}{2} - \frac{1}{2}\cos 2x = \frac{1}{2} - \frac{1}{2} + \frac{2}{2!}x^2 - \frac{2^3}{4!}x^4 + \frac{2^5}{6!}x^6 - \frac{2^7}{8!}x^8 + \cdots$$

So, the series for $f(x) = \sin^2 x$ is

$$\sin^2 x = \frac{2}{2!}x^2 - \frac{2^3}{4!}x^4 + \frac{2^5}{6!}x^6 - \frac{2^7}{8!}x^8 + \cdots.$$

This series converges for $-\infty < x < \infty$. ■

As mentioned in the preceding section, power series can be used to obtain tables of values of transcendental functions. They are also useful for estimating the values of definite integrals for which antiderivatives cannot be found. The next example demonstrates this use.

EXAMPLE 10 **Power Series Approximation of a Definite Integral**

••••▷ *See LarsonCalculus.com for an interactive version of this type of example.*

Use a power series to approximate

$$\int_0^1 e^{-x^2}\, dx$$

with an error of less than 0.01.

Solution Replacing x with $-x^2$ in the series for e^x produces the following.

$$e^{-x^2} = 1 - x^2 + \frac{x^4}{2!} - \frac{x^6}{3!} + \frac{x^8}{4!} - \cdots$$

$$\int_0^1 e^{-x^2}\, dx = \left[x - \frac{x^3}{3} + \frac{x^5}{5 \cdot 2!} - \frac{x^7}{7 \cdot 3!} + \frac{x^9}{9 \cdot 4!} - \cdots \right]_0^1$$

$$= 1 - \frac{1}{3} + \frac{1}{10} - \frac{1}{42} + \frac{1}{216} - \cdots$$

Summing the first four terms, you have

$$\int_0^1 e^{-x^2}\, dx \approx 0.74$$

which, by the Alternating Series Test, has an error of less than $\frac{1}{216} \approx 0.005$. ■

9.10 Exercises

See **CalcChat.com** for tutorial help and worked-out solutions to odd-numbered exercises.

Finding a Taylor Series In Exercises 1–12, use the definition of Taylor series to find the Taylor series, centered at c, for the function.

1. $f(x) = e^{2x}, \quad c = 0$

2. $f(x) = e^{-4x}, \quad c = 0$

3. $f(x) = \cos x, \quad c = \dfrac{\pi}{4}$

4. $f(x) = \sin x, \quad c = \dfrac{\pi}{4}$

5. $f(x) = \dfrac{1}{x}, \quad c = 1$

6. $f(x) = \dfrac{1}{1-x}, \quad c = 2$

7. $f(x) = \ln x, \quad c = 1$

8. $f(x) = e^x, \quad c = 1$

9. $f(x) = \sin 3x, \quad c = 0$

10. $f(x) = \ln(x^2 + 1), \quad c = 0$

11. $f(x) = \sec x, \quad c = 0$ (first three nonzero terms)

12. $f(x) = \tan x, \quad c = 0$ (first three nonzero terms)

Proof In Exercises 13–16, prove that the Maclaurin series for the function converges to the function for all x.

13. $f(x) = \cos x$

14. $f(x) = e^{-2x}$

15. $f(x) = \sinh x$

16. $f(x) = \cosh x$

Using a Binomial Series In Exercises 17–26, use the binomial series to find the Maclaurin series for the function.

17. $f(x) = \dfrac{1}{(1+x)^2}$

18. $f(x) = \dfrac{1}{(1+x)^4}$

19. $f(x) = \dfrac{1}{\sqrt{1-x}}$

20. $f(x) = \dfrac{1}{\sqrt{1-x^2}}$

21. $f(x) = \dfrac{1}{\sqrt{4+x^2}}$

22. $f(x) = \dfrac{1}{(2+x)^3}$

23. $f(x) = \sqrt{1+x}$

24. $f(x) = \sqrt[4]{1+x}$

25. $f(x) = \sqrt{1+x^2}$

26. $f(x) = \sqrt{1+x^3}$

Finding a Maclaurin Series In Exercises 27–40, find the Maclaurin series for the function. Use the table of power series for elementary functions on page 670.

27. $f(x) = e^{x^2/2}$

28. $g(x) = e^{-3x}$

29. $f(x) = \ln(1 + x)$

30. $f(x) = \ln(1 + x^2)$

31. $g(x) = \sin 3x$

32. $f(x) = \sin \pi x$

33. $f(x) = \cos 4x$

34. $f(x) = \cos \pi x$

35. $f(x) = \cos x^{3/2}$

36. $g(x) = 2 \sin x^3$

37. $f(x) = \frac{1}{2}(e^x - e^{-x}) = \sinh x$

38. $f(x) = e^x + e^{-x} = 2 \cosh x$

39. $f(x) = \cos^2 x$

40. $f(x) = \sinh^{-1} x = \ln\!\left(x + \sqrt{x^2 + 1}\right)$

$\left(\textit{Hint:} \text{ Integrate the series for } \dfrac{1}{\sqrt{x^2+1}}.\right)$

Finding a Maclaurin Series In Exercises 41–44, find the Maclaurin series for the function. (See Examples 7 and 8.)

41. $f(x) = x \sin x$

42. $h(x) = x \cos x$

43. $g(x) = \begin{cases} \dfrac{\sin x}{x}, & x \neq 0 \\ 1, & x = 0 \end{cases}$

44. $f(x) = \begin{cases} \dfrac{\arcsin x}{x}, & x \neq 0 \\ 1, & x = 0 \end{cases}$

Verifying a Formula In Exercises 45 and 46, use a power series and the fact that $i^2 = -1$ to verify the formula.

45. $g(x) = \dfrac{1}{2i}(e^{ix} - e^{-ix}) = \sin x$

46. $g(x) = \dfrac{1}{2}(e^{ix} + e^{-ix}) = \cos x$

Finding Terms of a Maclaurin Series In Exercises 47–52, find the first four nonzero terms of the Maclaurin series for the function by multiplying or dividing the appropriate power series. Use the table of power series for elementary functions on page 670. Use a graphing utility to graph the function and its corresponding polynomial approximation.

47. $f(x) = e^x \sin x$

48. $g(x) = e^x \cos x$

49. $h(x) = \cos x \ln(1 + x)$

50. $f(x) = e^x \ln(1 + x)$

51. $g(x) = \dfrac{\sin x}{1 + x}$

52. $f(x) = \dfrac{e^x}{1 + x}$

Finding a Maclaurin Series In Exercises 53 and 54, find a Maclaurin series for $f(x)$.

53. $f(x) = \displaystyle\int_0^x (e^{-t^2} - 1)\, dt$

54. $f(x) = \displaystyle\int_0^x \sqrt{1 + t^3}\, dt$

Verifying a Sum In Exercises 55–58, verify the sum. Then use a graphing utility to approximate the sum with an error of less than 0.0001.

55. $\displaystyle\sum_{n=1}^{\infty} (-1)^{n+1} \dfrac{1}{n} = \ln 2$

56. $\displaystyle\sum_{n=0}^{\infty} (-1)^n \left[\dfrac{1}{(2n+1)!}\right] = \sin 1$

57. $\displaystyle\sum_{n=0}^{\infty} \dfrac{2^n}{n!} = e^2$

58. $\displaystyle\sum_{n=1}^{\infty} (-1)^{n-1} \left(\dfrac{1}{n!}\right) = \dfrac{e-1}{e}$

Finding a Limit In Exercises 59–62, use the series representation of the function f to find $\displaystyle\lim_{x \to 0} f(x)$ (if it exists).

59. $f(x) = \dfrac{1 - \cos x}{x}$

60. $f(x) = \dfrac{\sin x}{x}$

61. $f(x) = \dfrac{e^x - 1}{x}$

62. $f(x) = \dfrac{\ln(x + 1)}{x}$

Approximating an Integral In Exercises 63–70, use a power series to approximate the value of the integral with an error of less than 0.0001. (In Exercises 65 and 67, assume that the integrand is defined as 1 when $x = 0$.)

63. $\displaystyle\int_0^1 e^{-x^3}\,dx$

64. $\displaystyle\int_0^{1/4} x\ln(x+1)\,dx$

65. $\displaystyle\int_0^1 \frac{\sin x}{x}\,dx$

66. $\displaystyle\int_0^1 \cos x^2\,dx$

67. $\displaystyle\int_0^{1/2} \frac{\arctan x}{x}\,dx$

68. $\displaystyle\int_0^{1/2} \arctan x^2\,dx$

69. $\displaystyle\int_{0.1}^{0.3} \sqrt{1+x^3}\,dx$

70. $\displaystyle\int_0^{0.2} \sqrt{1+x^2}\,dx$

Area In Exercises 71 and 72, use a power series to approximate the area of the region. Use a graphing utility to verify the result.

71. $\displaystyle\int_0^{\pi/2} \sqrt{x}\cos x\,dx$

72. $\displaystyle\int_{0.5}^1 \cos\sqrt{x}\,dx$

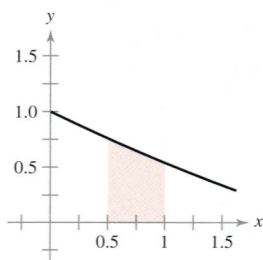

Probability In Exercises 73 and 74, approximate the normal probability with an error of less than 0.0001, where the probability is given by

$$P(a < x < b) = \frac{1}{\sqrt{2\pi}}\int_a^b e^{-x^2/2}\,dx.$$

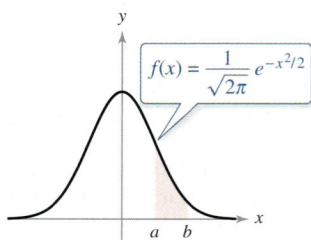

73. $P(0 < x < 1)$ **74.** $P(1 < x < 2)$

Finding a Taylor Polynomial Using Technology In Exercises 75–78, use a computer algebra system to find the fifth-degree Taylor polynomial, centered at c, for the function. Graph the function and the polynomial. Use the graph to determine the largest interval on which the polynomial is a reasonable approximation of the function.

75. $f(x) = x\cos 2x, \quad c = 0$

76. $f(x) = \sin\dfrac{x}{2}\ln(1+x), \quad c = 0$

77. $g(x) = \sqrt{x}\ln x, \quad c = 1$

78. $h(x) = \sqrt[3]{x}\arctan x, \quad c = 1$

WRITING ABOUT CONCEPTS

79. Taylor Series State the guidelines for finding a Taylor series.

80. Binomial Series Define the binomial series. What is its radius of convergence?

81. Finding a Series Explain how to use the series

$$g(x) = e^x = \sum_{n=0}^{\infty} \frac{x^n}{n!}$$

to find the series for each function. Do not find the series.

(a) $f(x) = e^{-x}$ (b) $f(x) = e^{3x}$ (c) $f(x) = xe^x$

---

82. HOW DO YOU SEE IT? Match the polynomial with its graph. [The graphs are labeled (i), (ii), (iii), and (iv).] Factor a common factor from each polynomial and identify the function approximated by the remaining Taylor polynomial.

(a) $y = x^2 - \dfrac{x^4}{3!}$

(b) $y = x - \dfrac{x^3}{2!} + \dfrac{x^5}{4!}$

(c) $y = x + x^2 + \dfrac{x^3}{2!}$

(d) $y = x^2 - x^3 + x^4$

83. Projectile Motion A projectile fired from the ground follows the trajectory given by

$$y = \left(\tan\theta - \frac{g}{kv_0\cos\theta}\right)x - \frac{g}{k^2}\ln\left(1 - \frac{kx}{v_0\cos\theta}\right)$$

where v_0 is the initial speed, θ is the angle of projection, g is the acceleration due to gravity, and k is the drag factor caused by air resistance. Using the power series representation

$$\ln(1 + x) = x - \frac{x^2}{2} + \frac{x^3}{3} - \frac{x^4}{4} + \cdots, \quad -1 < x < 1$$

verify that the trajectory can be rewritten as

$$y = (\tan\theta)x + \frac{gx^2}{2v_0^2\cos^2\theta} + \frac{kgx^3}{3v_0^3\cos^3\theta} + \frac{k^2\,gx^4}{4v_0^4\cos^4\theta} + \cdots.$$

84. Projectile Motion

Use the result of Exercise 83 to determine the series for the path of a projectile launched from ground level at an angle of $\theta = 60°$, with an initial speed of $v_0 = 64$ feet per second and a drag factor of $k = \frac{1}{16}$.

85. Investigation Consider the function f defined by

$$f(x) = \begin{cases} e^{-1/x^2}, & x \neq 0 \\ 0, & x = 0. \end{cases}$$

(a) Sketch a graph of the function.

(b) Use the alternative form of the definition of the derivative (Section 2.1) and L'Hôpital's Rule to show that $f'(0) = 0$. [By continuing this process, it can be shown that $f^{(n)}(0) = 0$ for $n > 1$.]

(c) Using the result in part (b), find the Maclaurin series for f. Does the series converge to f?

86. Investigation

(a) Find the power series centered at 0 for the function

$$f(x) = \frac{\ln(x^2 + 1)}{x^2}.$$

(b) Use a graphing utility to graph f and the eighth-degree Taylor polynomial $P_8(x)$ for f.

(c) Complete the table, where

$$F(x) = \int_0^x \frac{\ln(t^2 + 1)}{t^2}\,dt \quad \text{and} \quad G(x) = \int_0^x P_8(t)\,dt.$$

x	0.25	0.50	0.75	1.00	1.50	2.00
$F(x)$						
$G(x)$						

(d) Describe the relationship between the graphs of f and P_8 and the results given in the table in part (c).

87. Proof Prove that $\displaystyle\lim_{n\to\infty}\frac{x^n}{n!} = 0$ for any real x.

88. Finding a Maclaurin Series Find the Maclaurin series for

$$f(x) = \ln\frac{1 + x}{1 - x}$$

and determine its radius of convergence. Use the first four terms of the series to approximate $\ln 3$.

Evaluating a Binomial Coefficient In Exercises 89–92, evaluate the binomial coefficient using the formula

$$\binom{k}{n} = \frac{k(k - 1)(k - 2)(k - 3)\cdots(k - n + 1)}{n!}$$

where k is a real number, n is a positive integer, and

$$\binom{k}{0} = 1.$$

89. $\dbinom{5}{3}$ **90.** $\dbinom{-2}{2}$

91. $\dbinom{0.5}{4}$ **92.** $\dbinom{-1/3}{5}$

93. Writing a Power Series Write the power series for $(1 + x)^k$ in terms of binomial coefficients.

94. Proof Prove that e is irrational. [*Hint:* Assume that $e = p/q$ is rational (p and q are integers) and consider

$$e = 1 + 1 + \frac{1}{2!} + \cdots + \frac{1}{n!} + \cdots.\Big]$$

95. Using Fibonacci Numbers Show that the Maclaurin series for the function

$$g(x) = \frac{x}{1 - x - x^2}$$

is

$$\sum_{n=1}^{\infty} F_n x^n$$

where F_n is the nth Fibonacci number with $F_1 = F_2 = 1$ and $F_n = F_{n-2} + F_{n-1}$, for $n \geq 3$.

(*Hint:* Write

$$\frac{x}{1 - x - x^2} = a_0 + a_1 x + a_2 x^2 + \cdots$$

and multiply each side of this equation by $1 - x - x^2$.)

PUTNAM EXAM CHALLENGE

96. Assume that $|f(x)| \leq 1$ and $|f''(x)| \leq 1$ for all x on an interval of length at least 2. Show that $|f'(x)| \leq 2$ on the interval.

This problem was composed by the Committee on the Putnam Prize Competition.
© The Mathematical Association of America. All rights reserved.

Review Exercises   See CalcChat.com for tutorial help and worked-out solutions to odd-numbered exercises.

Listing the Terms of a Sequence In Exercises 1–4, write the first five terms of the sequence.

1. $a_n = 5^n$

2. $a_n = \dfrac{3^n}{n!}$

3. $a_n = \left(-\dfrac{1}{4}\right)^n$

4. $a_n = \dfrac{2n}{n+5}$

Matching In Exercises 5–8, match the sequence with its graph. [The graphs are labeled (a), (b), (c), and (d).]

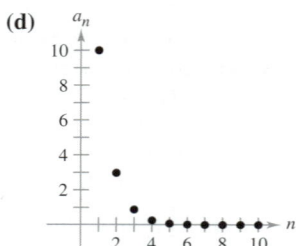

(a)

(b)

(c)

(d)

5. $a_n = 4 + \dfrac{2}{n}$

6. $a_n = 4 - \dfrac{1}{2}n$

7. $a_n = 10(0.3)^{n-1}$

8. $a_n = 6\left(-\dfrac{2}{3}\right)^{n-1}$

Finding the Limit of a Sequence In Exercises 9 and 10, use a graphing utility to graph the first 10 terms of the sequence. Use the graph to make an inference about the convergence or divergence of the sequence. Verify your inference analytically and, if the sequence converges, find its limit.

9. $a_n = \dfrac{5n+2}{n}$

10. $a_n = \sin\dfrac{n\pi}{2}$

Determining Convergence or Divergence In Exercises 11–18, determine the convergence or divergence of the sequence with the given nth term. If the sequence converges, find its limit.

11. $a_n = \left(\dfrac{2}{5}\right)^n + 5$

12. $a_n = 3 - \dfrac{2}{n^2 - 1}$

13. $a_n = \dfrac{n^3 + 1}{n^2}$

14. $a_n = \dfrac{1}{\sqrt{n}}$

15. $a_n = \dfrac{n}{n^2 + 1}$

16. $a_n = \dfrac{n}{\ln n}$

17. $a_n = \sqrt{n+1} - \sqrt{n}$

18. $a_n = \dfrac{\sin\sqrt{n}}{\sqrt{n}}$

Finding the nth Term of a Sequence In Exercises 19–22, write an expression for the nth term of the sequence. (There is more than one correct answer.)

19. 3, 8, 13, 18, 23, . . .

20. $-5, -2, 3, 10, 19, \ldots$

21. $\dfrac{1}{2}, \dfrac{1}{3}, \dfrac{1}{7}, \dfrac{1}{25}, \dfrac{1}{121}, \ldots$

22. $\dfrac{1}{2}, \dfrac{2}{5}, \dfrac{3}{10}, \dfrac{4}{17}, \ldots$

23. Compound Interest A deposit of $8000 is made in an account that earns 5% interest compounded quarterly. The balance in the account after n quarters is

$$A_n = 8000\left(1 + \dfrac{0.05}{4}\right)^n, \quad n = 1, 2, 3, \ldots.$$

(a) Compute the first eight terms of the sequence $\{A_n\}$.

(b) Find the balance in the account after 10 years by computing the 40th term of the sequence.

24. Depreciation A company buys a machine for $175,000. During the next 5 years, the machine will depreciate at a rate of 30% per year. (That is, at the end of each year, the depreciated value will be 70% of what it was at the beginning of the year.)

(a) Find a formula for the nth term of the sequence that gives the value V of the machine t full years after it was purchased.

(b) Find the depreciated value of the machine at the end of 5 full years.

Finding Partial Sums In Exercises 25 and 26, find the sequence of partial sums $S_1, S_2, S_3, S_4,$ and S_5.

25. $3 + \dfrac{3}{2} + 1 + \dfrac{3}{4} + \dfrac{3}{5} + \cdots$

26. $-\dfrac{1}{2} + \dfrac{1}{4} - \dfrac{1}{8} + \dfrac{1}{16} - \dfrac{1}{32} + \cdots$

Numerical, Graphical, and Analytic Analysis In Exercises 27–30, (a) use a graphing utility to find the indicated partial sum S_n and complete the table, and (b) use a graphing utility to graph the first 10 terms of the sequence of partial sums.

n	5	10	15	20	25
S_n					

27. $\displaystyle\sum_{n=1}^{\infty} \left(\dfrac{3}{2}\right)^{n-1}$

28. $\displaystyle\sum_{n=1}^{\infty} \dfrac{(-1)^{n+1}}{2n}$

29. $\displaystyle\sum_{n=1}^{\infty} \dfrac{(-1)^{n+1}}{(2n)!}$

30. $\displaystyle\sum_{n=1}^{\infty} \dfrac{1}{n(n+1)}$

Finding the Sum of a Convergent Series In Exercises 31–34, find the sum of the convergent series.

31. $\displaystyle\sum_{n=0}^{\infty} \left(\frac{2}{5}\right)^n$

32. $\displaystyle\sum_{n=0}^{\infty} \frac{3^{n+2}}{7^n}$

33. $\displaystyle\sum_{n=1}^{\infty} [(0.6)^n + (0.8)^n]$

34. $\displaystyle\sum_{n=0}^{\infty} \left[\left(\frac{2}{3}\right)^n - \frac{1}{(n+1)(n+2)}\right]$

Using a Geometric Series In Exercises 35 and 36, (a) write the repeating decimal as a geometric series, and (b) write its sum as the ratio of two integers.

35. $0.\overline{09}$

36. $0.\overline{64}$

Using Geometric Series or the nth-Term Test In Exercises 37–40, use geometric series or the nth-Term Test to determine the convergence or divergence of the series.

37. $\displaystyle\sum_{n=0}^{\infty} (1.67)^n$

38. $\displaystyle\sum_{n=0}^{\infty} (0.36)^n$

39. $\displaystyle\sum_{n=2}^{\infty} \frac{(-1)^n n}{\ln n}$

40. $\displaystyle\sum_{n=0}^{\infty} \frac{2n+1}{3n+2}$

41. **Distance** A ball is dropped from a height of 8 meters. Each time it drops h meters, it rebounds $0.7h$ meters. Find the total distance traveled by the ball.

42. **Compound Interest** A deposit of \$125 is made at the end of each month for 10 years in an account that pays 3.5% interest, compounded monthly. Determine the balance in the account at the end of 10 years. (*Hint:* Use the result of Section 9.2, Exercise 84.)

Using the Integral Test or a p-Series In Exercises 43–48, use the Integral Test or a p-series to determine the convergence or divergence of the series.

43. $\displaystyle\sum_{n=1}^{\infty} \frac{2}{6n+1}$

44. $\displaystyle\sum_{n=1}^{\infty} \frac{1}{\sqrt[4]{n^3}}$

45. $\displaystyle\sum_{n=1}^{\infty} \frac{1}{n^{5/2}}$

46. $\displaystyle\sum_{n=1}^{\infty} \frac{1}{5^n}$

47. $\displaystyle\sum_{n=1}^{\infty} \left(\frac{1}{n^2} - \frac{1}{n}\right)$

48. $\displaystyle\sum_{n=1}^{\infty} \frac{\ln n}{n^4}$

Using the Direct Comparison Test or the Limit Comparison Test In Exercises 49–54, use the Direct Comparison Test or the Limit Comparison Test to determine the convergence or divergence of the series.

49. $\displaystyle\sum_{n=2}^{\infty} \frac{1}{\sqrt[3]{n}-1}$

50. $\displaystyle\sum_{n=1}^{\infty} \frac{n}{\sqrt{n^3+3n}}$

51. $\displaystyle\sum_{n=1}^{\infty} \frac{1}{\sqrt{n^3+2n}}$

52. $\displaystyle\sum_{n=1}^{\infty} \frac{n+1}{n(n+2)}$

53. $\displaystyle\sum_{n=1}^{\infty} \frac{1\cdot 3\cdot 5\cdots(2n-1)}{2\cdot 4\cdot 6\cdots(2n)}$

54. $\displaystyle\sum_{n=1}^{\infty} \frac{1}{3^n-5}$

Using the Alternating Series Test In Exercises 55–60, use the Alternating Series Test, if applicable, to determine the convergence or divergence of the series.

55. $\displaystyle\sum_{n=1}^{\infty} \frac{(-1)^n}{n^5}$

56. $\displaystyle\sum_{n=1}^{\infty} \frac{(-1)^n(n+1)}{n^2+1}$

57. $\displaystyle\sum_{n=2}^{\infty} \frac{(-1)^n n}{n^2-3}$

58. $\displaystyle\sum_{n=1}^{\infty} \frac{(-1)^n \sqrt{n}}{n+1}$

59. $\displaystyle\sum_{n=4}^{\infty} \frac{(-1)^n n}{n-3}$

60. $\displaystyle\sum_{n=2}^{\infty} \frac{(-1)^n \ln n^3}{n}$

Using the Ratio Test or the Root Test In Exercises 61–66, use the Ratio Test or the Root Test to determine the convergence or divergence of the series.

61. $\displaystyle\sum_{n=1}^{\infty} \left(\frac{3n-1}{2n+5}\right)^n$

62. $\displaystyle\sum_{n=1}^{\infty} \left(\frac{4n}{7n-1}\right)^n$

63. $\displaystyle\sum_{n=1}^{\infty} \frac{n}{e^{n^2}}$

64. $\displaystyle\sum_{n=1}^{\infty} \frac{n!}{e^n}$

65. $\displaystyle\sum_{n=1}^{\infty} \frac{2^n}{n^3}$

66. $\displaystyle\sum_{n=1}^{\infty} \frac{1\cdot 3\cdot 5\cdots(2n-1)}{2\cdot 5\cdot 8\cdots(3n-1)}$

Numerical, Graphical, and Analytic Analysis In Exercises 67 and 68, (a) verify that the series converges, (b) use a graphing utility to find the indicated partial sum S_n and complete the table, (c) use a graphing utility to graph the first 10 terms of the sequence of partial sums, and (d) use the table to estimate the sum of the series.

n	5	10	15	20	25
S_n					

67. $\displaystyle\sum_{n=1}^{\infty} n\left(\frac{3}{5}\right)^n$

68. $\displaystyle\sum_{n=1}^{\infty} \frac{(-1)^{n-1}n}{n^3+5}$

Finding a Maclaurin Polynomial In Exercises 69 and 70, find the nth Maclaurin polynomial for the function.

69. $f(x) = e^{-2x}$, $n = 3$

70. $f(x) = \cos \pi x$, $n = 4$

Finding a Taylor Polynomial In Exercises 71 and 72, find the third-degree Taylor polynomial centered at c.

71. $f(x) = e^{-3x}$, $c = 0$

72. $f(x) = \tan x$, $c = -\dfrac{\pi}{4}$

Finding a Degree In Exercises 73 and 74, determine the degree of the Maclaurin polynomial required for the error in the approximation of the function at the indicated value of x to be less than 0.001.

73. $\cos(0.75)$

74. $e^{-0.25}$

Finding the Interval of Convergence In Exercises 75–80, find the interval of convergence of the power series. (Be sure to include a check for convergence at the endpoints of the interval.)

75. $\displaystyle\sum_{n=0}^{\infty} \left(\frac{x}{10}\right)^n$

76. $\displaystyle\sum_{n=0}^{\infty} (5x)^n$

77. $\displaystyle\sum_{n=0}^{\infty} \frac{(-1)^n(x-2)^n}{(n+1)^2}$

78. $\displaystyle\sum_{n=1}^{\infty} \frac{3^n(x-2)^n}{n}$

79. $\displaystyle\sum_{n=0}^{\infty} n!(x-2)^n$

80. $\displaystyle\sum_{n=0}^{\infty} \frac{(x-2)^n}{2^n}$

Finding Intervals of Convergence In Exercises 81 and 82, find the intervals of convergence of (a) $f(x)$, (b) $f'(x)$, (c) $f''(x)$, and (d) $\int f(x)\,dx$. Include a check for convergence at the endpoints of the interval.

81. $\displaystyle f(x) = \sum_{n=0}^{\infty} \left(\frac{x}{5}\right)^n$

82. $\displaystyle f(x) = \sum_{n=1}^{\infty} \frac{(-1)^{n+1}(x-4)^n}{n}$

Differential Equation In Exercises 83 and 84, show that the function represented by the power series is a solution of the differential equation.

83. $\displaystyle y = \sum_{n=0}^{\infty} (-1)^n \frac{x^{2n}}{4^n(n!)^2}$

$x^2 y'' + xy' + x^2 y = 0$

84. $\displaystyle y = \sum_{n=0}^{\infty} \frac{(-3)^n x^{2n}}{2^n n!}$

$y'' + 3xy' + 3y = 0$

Finding a Geometric Power Series In Exercises 85 and 86, find a geometric power series, centered at 0, for the function.

85. $\displaystyle g(x) = \frac{2}{3-x}$

86. $\displaystyle h(x) = \frac{3}{2+x}$

Finding a Power Series In Exercises 87 and 88, find a power series for the function, centered at c, and determine the interval of convergence.

87. $\displaystyle f(x) = \frac{6}{4-x}, \quad c = 1$

88. $\displaystyle f(x) = \frac{1}{3-2x}, \quad c = 0$

Finding the Sum of a Series In Exercises 89–94, find the sum of the convergent series by using a well-known function. Identify the function and explain how you obtained the sum.

89. $\displaystyle\sum_{n=1}^{\infty} (-1)^{n+1} \frac{1}{4^n n}$

90. $\displaystyle\sum_{n=1}^{\infty} (-1)^{n+1} \frac{1}{5^n n}$

91. $\displaystyle\sum_{n=0}^{\infty} \frac{1}{2^n n!}$

92. $\displaystyle\sum_{n=0}^{\infty} \frac{2^n}{3^n n!}$

93. $\displaystyle\sum_{n=0}^{\infty} (-1)^n \frac{2^{2n}}{3^{2n}(2n)!}$

94. $\displaystyle\sum_{n=0}^{\infty} (-1)^n \frac{1}{3^{2n+1}(2n+1)!}$

Finding a Taylor Series In Exercises 95–102, use the definition of Taylor series to find the Taylor series, centered at c, for the function.

95. $f(x) = \sin x, \quad c = \dfrac{3\pi}{4}$

96. $f(x) = \cos x, \quad c = -\dfrac{\pi}{4}$

97. $f(x) = 3^x, \quad c = 0$

98. $f(x) = \csc x, \quad c = \dfrac{\pi}{2}$ (first three terms)

99. $f(x) = \dfrac{1}{x}, \quad c = -1$

100. $f(x) = \sqrt{x}, \quad c = 4$

101. $g(x) = \sqrt[5]{1+x}, \quad c = 0$

102. $h(x) = \dfrac{1}{(1+x)^3}, \quad c = 0$

103. Forming Maclaurin Series Determine the first four terms of the Maclaurin series for e^{2x}

(a) by using the definition of the Maclaurin series and the formula for the coefficient of the nth term, $a_n = f^{(n)}(0)/n!$.

(b) by replacing x by $2x$ in the series for e^x.

(c) by multiplying the series for e^x by itself, because $e^{2x} = e^x \cdot e^x$.

104. Forming Maclaurin Series Determine the first four terms of the Maclaurin series for $\sin 2x$

(a) by using the definition of the Maclaurin series and the formula for the coefficient of the nth term, $a_n = f^{(n)}(0)/n!$.

(b) by replacing x by $2x$ in the series for $\sin 2x$.

(c) by multiplying 2 by the series for $\sin x$ by the series for $\cos x$, because $\sin 2x = 2 \sin x \cos x$.

Finding a Maclaurin Series In Exercises 105–108, find the Maclaurin series for the function. Use the table of power series for elementary functions on page 670.

105. $f(x) = e^{6x}$

106. $f(x) = \ln(x-1)$

107. $f(x) = \sin 2x$

108. $f(x) = \cos 3x$

Finding a Limit In Exercises 109 and 110, use the series representation of the function f to find $\displaystyle\lim_{x \to 0} f(x)$ (if it exists).

109. $\displaystyle f(x) = \frac{\arctan x}{\sqrt{x}}$

110. $\displaystyle f(x) = \frac{\arcsin x}{x}$

P.S. Problem Solving

See **CalcChat.com** for tutorial help and worked-out solutions to odd-numbered exercises.

1. Cantor Set   The **Cantor set** (Georg Cantor, 1845–1918) is a subset of the unit interval $[0, 1]$. To construct the Cantor set, first remove the middle third $\left(\frac{1}{3}, \frac{2}{3}\right)$ of the interval, leaving two line segments. For the second step, remove the middle third of each of the two remaining segments, leaving four line segments. Continue this procedure indefinitely, as shown in the figure. The Cantor set consists of all numbers in the unit interval $[0, 1]$ that still remain.

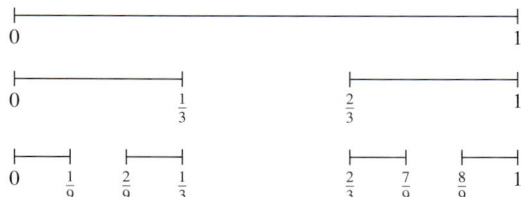

(a) Find the total length of all the line segments that are removed.

(b) Write down three numbers that are in the Cantor set.

(c) Let C_n denote the total length of the remaining line segments after n steps. Find $\lim\limits_{n \to \infty} C_n$.

2. Using Sequences

(a) Given that $\lim\limits_{x \to \infty} a_{2n} = L$ and $\lim\limits_{x \to \infty} a_{2n+1} = L$, show that $\{a_n\}$ is convergent and $\lim\limits_{x \to \infty} a_n = L$.

(b) Let $a_1 = 1$ and $a_{n+1} = 1 + \dfrac{1}{1 + a_n}$. Write out the first eight terms of $\{a_n\}$. Use part (a) to show that $\lim\limits_{x \to \infty} a_n = \sqrt{2}$.

This gives the **continued fraction expansion**

$$\sqrt{2} = 1 + \cfrac{1}{2 + \cfrac{1}{2 + \cdots}}.$$

3. Using a Series   It can be shown that

$$\sum_{n=1}^{\infty} \frac{1}{n^2} = \frac{\pi^2}{6} \text{ [see Section 9.3, page 608]}.$$

Use this fact to show that $\displaystyle\sum_{n=1}^{\infty} \frac{1}{(2n-1)^2} = \frac{\pi^2}{8}$.

4. Finding a Limit   Let T be an equilateral triangle with sides of length 1. Let a_n be the number of circles that can be packed tightly in n rows inside the triangle. For example, $a_1 = 1$, $a_2 = 3$, and $a_3 = 6$, as shown in the figure. Let A_n be the combined area of the a_n circles. Find $\lim\limits_{n \to \infty} A_n$.

5. Using Center of Gravity   Identical blocks of unit length are stacked on top of each other at the edge of a table. The center of gravity of the top block must lie over the block below it, the center of gravity of the top two blocks must lie over the block below them, and so on (see figure).

(a) When there are three blocks, show that it is possible to stack them so that the left edge of the top block extends $\frac{11}{12}$ unit beyond the edge of the table.

(b) Is it possible to stack the blocks so that the right edge of the top block extends beyond the edge of the table?

(c) How far beyond the table can the blocks be stacked?

6. Using Power Series

(a) Consider the power series

$$\sum_{n=0}^{\infty} a_n x^n = 1 + 2x + 3x^2 + x^3 + 2x^4 + 3x^5 + x^6 + \cdots$$

in which the coefficients $a_n = 1, 2, 3, 1, 2, 3, 1, \ldots$ are periodic of period $p = 3$. Find the radius of convergence and the sum of this power series.

(b) Consider a power series

$$\sum_{n=0}^{\infty} a_n x^n$$

in which the coefficients are periodic, $(a_{n+p} = a_p)$, and $a_n > 0$. Find the radius of convergence and the sum of this power series.

7. Finding Sums of Series

(a) Find a power series for the function

$$f(x) = xe^x$$

centered at 0. Use this representation to find the sum of the infinite series

$$\sum_{n=1}^{\infty} \frac{1}{n!(n+2)}.$$

(b) Differentiate the power series for $f(x) = xe^x$. Use the result to find the sum of the infinite series

$$\sum_{n=0}^{\infty} \frac{n+1}{n!}.$$

8. Using the Alternating Series Test The graph of the function

$$f(x) = \begin{cases} 1, & x = 0 \\ \dfrac{\sin x}{x}, & x > 0 \end{cases}$$

is shown below. Use the Alternating Series Test to show that the improper integral $\displaystyle\int_{1}^{\infty} f(x)\, dx$ converges.

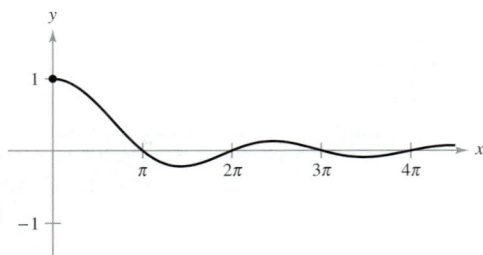

9. Conditional and Absolute Convergence For what values of the positive constants a and b does the following series converge absolutely? For what values does it converge conditionally?

$$a - \frac{b}{2} + \frac{a}{3} - \frac{b}{4} + \frac{a}{5} - \frac{b}{6} + \frac{a}{7} - \frac{b}{8} + \cdots$$

10. Proof

(a) Consider the following sequence of numbers defined recursively.

$$a_1 = 3$$
$$a_2 = \sqrt{3}$$
$$a_3 = \sqrt{3 + \sqrt{3}}$$
$$\vdots$$
$$a_{n+1} = \sqrt{3 + a_n}$$

Write the decimal approximations for the first six terms of this sequence. Prove that the sequence converges, and find its limit.

(b) Consider the following sequence defined recursively by $a_1 = \sqrt{a}$ and $a_{n+1} = \sqrt{a + a_n}$, where $a > 2$.

$$\sqrt{a},\quad \sqrt{a + \sqrt{a}},\quad \sqrt{a + \sqrt{a + \sqrt{a}}},\ \ldots$$

Prove that this sequence converges, and find its limit.

11. Proof Let $\{a_n\}$ be a sequence of positive numbers satisfying $\displaystyle\lim_{n\to\infty} (a_n)^{1/n} = L < \frac{1}{r}$, $r > 0$. Prove that the series $\displaystyle\sum_{n=1}^{\infty} a_n r^n$ converges.

12. Using a Series Consider the infinite series $\displaystyle\sum_{n=1}^{\infty} \frac{1}{2^{n+(-1)^n}}$.

(a) Find the first five terms of the sequence of partial sums.

(b) Show that the Ratio Test is inconclusive for this series.

(c) Use the Root Test to test for the convergence or divergence of this series.

13. Deriving Identities Derive each identity using the appropriate geometric series.

(a) $\dfrac{1}{0.99} = 1.01010101 \ldots$

(b) $\dfrac{1}{0.98} = 1.0204081632 \ldots$

14. Population Consider an idealized population with the characteristic that each member of the population produces one offspring at the end of every time period. Each member has a life span of three time periods and the population begins with 10 newborn members. The following table shows the population during the first five time periods.

	Time Period				
Age Bracket	1	2	3	4	5
0–1	10	10	20	40	70
1–2		10	10	20	40
2–3			10	10	20
Total	10	20	40	70	130

The sequence for the total population has the property that $S_n = S_{n-1} + S_{n-2} + S_{n-3}$, $n > 3$. Find the total population during each of the next five time periods.

15. Spheres Imagine you are stacking an infinite number of spheres of decreasing radii on top of each other, as shown in the figure. The radii of the spheres are 1 meter, $1/\sqrt{2}$ meter, $1/\sqrt{3}$ meter, and so on. The spheres are made of a material that weighs 1 newton per cubic meter.

(a) How high is this infinite stack of spheres?

(b) What is the total surface area of all the spheres in the stack?

(c) Show that the weight of the stack is finite.

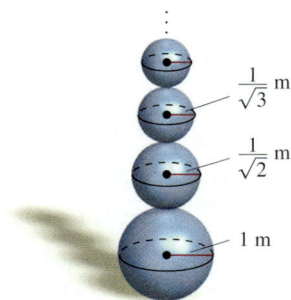

16. Determining Convergence or Divergence

(a) Determine the convergence or divergence of the series

$$\sum_{n=1}^{\infty} \frac{1}{2n}.$$

(b) Determine the convergence or divergence of the series

$$\sum_{n=1}^{\infty} \left(\sin \frac{1}{2n} - \sin \frac{1}{2n+1} \right).$$

10 Conics, Parametric Equations, and Polar Coordinates

Antenna Radiation *(Exercise 47, p. 732)*

Planetary Motion *(Exercise 67, p. 741)*

Anamorphic Art *(Section Project, p. 724)*

Halley's Comet *(Exercise 77, p. 694)*

Architecture *(Exercise 71, p. 694)*

10.1 Conics and Calculus

- Understand the definition of a conic section.
- Analyze and write equations of parabolas using properties of parabolas.
- Analyze and write equations of ellipses using properties of ellipses.
- Analyze and write equations of hyperbolas using properties of hyperbolas.

Conic Sections

Each **conic section** (or simply **conic**) can be described as the intersection of a plane and a double-napped cone. Notice in Figure 10.1 that for the four basic conics, the intersecting plane does not pass through the vertex of the cone. When the plane passes through the vertex, the resulting figure is a **degenerate conic,** as shown in Figure 10.2.

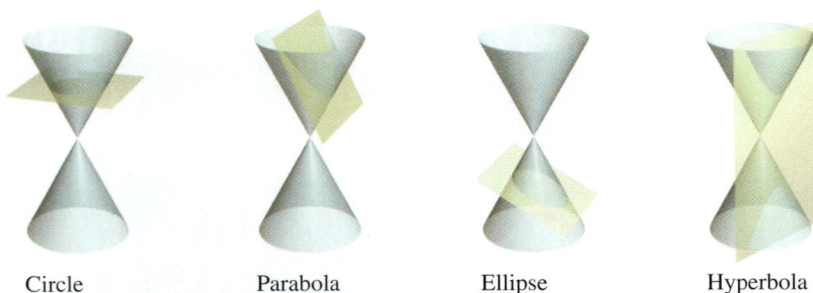

Circle
Conic sections
Figure 10.1

Parabola

Ellipse

Hyperbola

Point
Degenerate conics
Figure 10.2

Line

Two intersecting lines

HYPATIA (370–415 A.D.)

The Greeks discovered conic sections sometime between 600 and 300 B.C. By the beginning of the Alexandrian period, enough was known about conics for Apollonius (262–190 B.C.) to produce an eight-volume work on the subject. Later, toward the end of the Alexandrian period, Hypatia wrote a textbook entitled *On the Conics of Apollonius.* Her death marked the end of major mathematical discoveries in Europe for several hundred years.

   The early Greeks were largely concerned with the geometric properties of conics. It was not until 1900 years later, in the early seventeenth century, that the broader applicability of conics became apparent. Conics then played a prominent role in the development of calculus.

See LarsonCalculus.com to read more of this biography.

There are several ways to study conics. You could begin as the Greeks did, by defining the conics in terms of the intersections of planes and cones, or you could define them algebraically in terms of the general second-degree equation

$$Ax^2 + Bxy + Cy^2 + Dx + Ey + F = 0.$$ General second-degree equation

However, a third approach, in which each of the conics is defined as a **locus** (collection) of points satisfying a certain geometric property, works best. For example, a circle can be defined as the collection of all points (x, y) that are equidistant from a fixed point (h, k). This locus definition easily produces the standard equation of a circle

$$(x - h)^2 + (y - k)^2 = r^2.$$ Standard equation of a circle

For information about rotating second-degree equations in two variables, see Appendix D.

■ **FOR FURTHER INFORMATION** To learn more about the mathematical activities of Hypatia, see the article "Hypatia and Her Mathematics" by Michael A. B. Deakin in *The American Mathematical Monthly.* To view this article, go to *MathArticles.com.*

Parabolas

A **parabola** is the set of all points (x, y) that are equidistant from a fixed line, the **directrix,** and a fixed point, the **focus,** not on the line. The midpoint between the focus and the directrix is the **vertex,** and the line passing through the focus and the vertex is the **axis** of the parabola. Note in Figure 10.3 that a parabola is symmetric with respect to its axis.

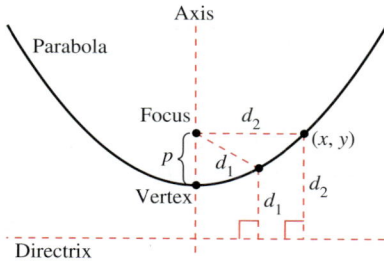

Figure 10.3

THEOREM 10.1 Standard Equation of a Parabola

The **standard form** of the equation of a parabola with vertex (h, k) and directrix $y = k - p$ is

$$(x - h)^2 = 4p(y - k). \qquad \text{Vertical axis}$$

For directrix $x = h - p$, the equation is

$$(y - k)^2 = 4p(x - h). \qquad \text{Horizontal axis}$$

The focus lies on the axis p units (*directed distance*) from the vertex. The coordinates of the focus are as follows.

$$(h, k + p) \qquad \text{Vertical axis}$$
$$(h + p, k) \qquad \text{Horizontal axis}$$

EXAMPLE 1 **Finding the Focus of a Parabola**

Find the focus of the parabola

$$y = \frac{1}{2} - x - \frac{1}{2}x^2.$$

Solution To find the focus, convert to standard form by completing the square.

$$y = \frac{1}{2} - x - \frac{1}{2}x^2 \qquad \text{Write original equation.}$$
$$2y = 1 - 2x - x^2 \qquad \text{Multiply each side by 2.}$$
$$2y = 1 - (x^2 + 2x) \qquad \text{Group terms.}$$
$$2y = 2 - (x^2 + 2x + 1) \qquad \text{Add and subtract 1 on right side.}$$
$$x^2 + 2x + 1 = -2y + 2$$
$$(x + 1)^2 = -2(y - 1) \qquad \text{Write in standard form.}$$

Comparing this equation with

$$(x - h)^2 = 4p(y - k)$$

you can conclude that

$$h = -1, \quad k = 1, \quad \text{and} \quad p = -\frac{1}{2}.$$

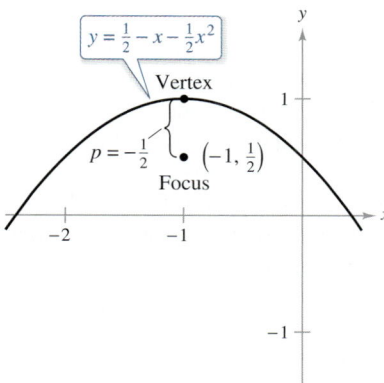

Parabola with a vertical axis, $p < 0$
Figure 10.4

Because p is negative, the parabola opens downward, as shown in Figure 10.4. So, the focus of the parabola is p units from the vertex, or

$$(h, k + p) = \left(-1, \frac{1}{2}\right). \qquad \text{Focus}$$

A line segment that passes through the focus of a parabola and has endpoints on the parabola is called a **focal chord.** The specific focal chord perpendicular to the axis of the parabola is the **latus rectum.** The next example shows how to determine the length of the latus rectum and the length of the corresponding intercepted arc.

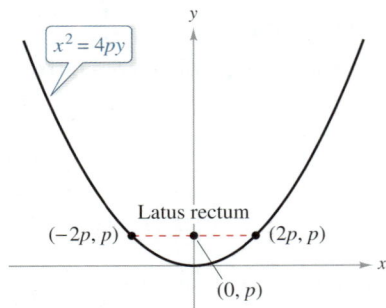

Length of latus rectum: $4p$
Figure 10.5

EXAMPLE 2 **Focal Chord Length and Arc Length**

$\vdots$ ▷ *See LarsonCalculus.com for an interactive version of this type of example.*

Find the length of the latus rectum of the parabola

$$x^2 = 4py.$$

Then find the length of the parabolic arc intercepted by the latus rectum.

Solution Because the latus rectum passes through the focus $(0, p)$ and is perpendicular to the y-axis, the coordinates of its endpoints are

$$(-x, p) \quad \text{and} \quad (x, p).$$

Substituting p for y in the equation of the parabola produces

$$x^2 = 4p(p) \implies x = \pm 2p.$$

So, the endpoints of the latus rectum are $(-2p, p)$ and $(2p, p)$, and you can conclude that its length is $4p$, as shown in Figure 10.5. In contrast, the length of the intercepted arc is

$$s = \int_{-2p}^{2p} \sqrt{1 + (y')^2}\, dx \qquad \text{Use arc length formula.}$$

$$= 2\int_{0}^{2p} \sqrt{1 + \left(\frac{x}{2p}\right)^2}\, dx \qquad y = \frac{x^2}{4p} \implies y' = \frac{x}{2p}$$

$$= \frac{1}{p}\int_{0}^{2p} \sqrt{4p^2 + x^2}\, dx \qquad \text{Simplify.}$$

$$= \frac{1}{2p}\left[x\sqrt{4p^2 + x^2} + 4p^2 \ln\left|x + \sqrt{4p^2 + x^2}\right| \right]_0^{2p} \qquad \text{Theorem 8.2}$$

$$= \frac{1}{2p}\left[2p\sqrt{8p^2} + 4p^2 \ln\left(2p + \sqrt{8p^2}\right) - 4p^2 \ln(2p) \right]$$

$$= 2p\left[\sqrt{2} + \ln\left(1 + \sqrt{2}\right) \right]$$

$$\approx 4.59p.$$

One widely used property of a parabola is its reflective property. In physics, a surface is called **reflective** when the tangent line at any point on the surface makes equal angles with an incoming ray and the resulting outgoing ray. The angle corresponding to the incoming ray is the **angle of incidence,** and the angle corresponding to the outgoing ray is the **angle of reflection.** One example of a reflective surface is a flat mirror.

Another type of reflective surface is that formed by revolving a parabola about its axis. The resulting surface has the property that all incoming rays parallel to the axis are directed through the focus of the parabola. This is the principle behind the design of the parabolic mirrors used in reflecting telescopes. Conversely, all light rays emanating from the focus of a parabolic reflector used in a flashlight are parallel, as shown in Figure 10.6.

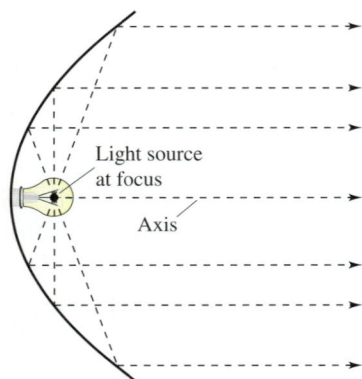

Parabolic reflector: light is reflected in parallel rays.
Figure 10.6

THEOREM 10.2 **Reflective Property of a Parabola**

Let P be a point on a parabola. The tangent line to the parabola at point P makes equal angles with the following two lines.

1. The line passing through P and the focus

2. The line passing through P parallel to the axis of the parabola

Ellipses

NICOLAUS COPERNICUS (1473–1543)

Copernicus began to study planetary motion when he was asked to revise the calendar. At that time, the exact length of the year could not be accurately predicted using the theory that Earth was the center of the universe.
See LarsonCalculus.com to read more of this biography.

More than a thousand years after the close of the Alexandrian period of Greek mathematics, Western civilization finally began a Renaissance of mathematical and scientific discovery. One of the principal figures in this rebirth was the Polish astronomer Nicolaus Copernicus. In his work *On the Revolutions of the Heavenly Spheres,* Copernicus claimed that all of the planets, including Earth, revolved about the sun in circular orbits. Although some of Copernicus's claims were invalid, the controversy set off by his heliocentric theory motivated astronomers to search for a mathematical model to explain the observed movements of the sun and planets. The first to find an accurate model was the German astronomer Johannes Kepler (1571–1630). Kepler discovered that the planets move about the sun in elliptical orbits, with the sun not as the center but as a focal point of the orbit.

The use of ellipses to explain the movements of the planets is only one of many practical and aesthetic uses. As with parabolas, you will begin your study of this second type of conic by defining it as a locus of points. Now, however, *two* focal points are used rather than one.

An **ellipse** is the set of all points (x, y) the sum of whose distances from two distinct fixed points called **foci** is constant. (See Figure 10.7.) The line through the foci intersects the ellipse at two points, called the **vertices.** The chord joining the vertices is the **major axis,** and its midpoint is the **center** of the ellipse. The chord perpendicular to the major axis at the center is the **minor axis** of the ellipse. (See Figure 10.8.)

Figure 10.7

Figure 10.8

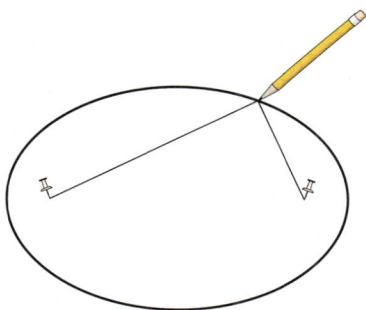

If the ends of a fixed length of string are fastened to the thumbtacks and the string is drawn taut with a pencil, then the path traced by the pencil will be an ellipse.
Figure 10.9

THEOREM 10.3 Standard Equation of an Ellipse

The standard form of the equation of an ellipse with center (h, k) and major and minor axes of lengths $2a$ and $2b$, where $a > b$, is

$$\frac{(x-h)^2}{a^2} + \frac{(y-k)^2}{b^2} = 1 \qquad \text{Major axis is horizontal.}$$

or

$$\frac{(x-h)^2}{b^2} + \frac{(y-k)^2}{a^2} = 1. \qquad \text{Major axis is vertical.}$$

The foci lie on the major axis, c units from the center, with

$$c^2 = a^2 - b^2.$$

You can visualize the definition of an ellipse by imagining two thumbtacks placed at the foci, as shown in Figure 10.9.

■ **FOR FURTHER INFORMATION** To learn about how an ellipse may be "exploded" into a parabola, see the article "Exploding the Ellipse" by Arnold Good in *Mathematics Teacher.* To view this article, go to *MathArticles.com.*

EXAMPLE 3 **Analyzing an Ellipse**

Find the center, vertices, and foci of the ellipse

$$4x^2 + y^2 - 8x + 4y - 8 = 0.$$ General second-degree equation

Solution By completing the square, you can write the original equation in standard form.

$$4x^2 + y^2 - 8x + 4y - 8 = 0$$ Write original equation.

$$4x^2 - 8x + y^2 + 4y = 8$$

$$4(x^2 - 2x + 1) + (y^2 + 4y + 4) = 8 + 4 + 4$$

$$4(x - 1)^2 + (y + 2)^2 = 16$$

$$\frac{(x - 1)^2}{4} + \frac{(y + 2)^2}{16} = 1$$ Write in standard form.

So, the major axis is parallel to the y-axis, where $h = 1$, $k = -2$, $a = 4$, $b = 2$, and $c = \sqrt{16 - 4} = 2\sqrt{3}$. So, you obtain the following.

Center: $(1, -2)$ (h, k)

Vertices: $(1, -6)$ and $(1, 2)$ $(h, k \pm a)$

Foci: $\left(1, -2 - 2\sqrt{3}\right)$ and $\left(1, -2 + 2\sqrt{3}\right)$ $(h, k \pm c)$

The graph of the ellipse is shown in Figure 10.10.

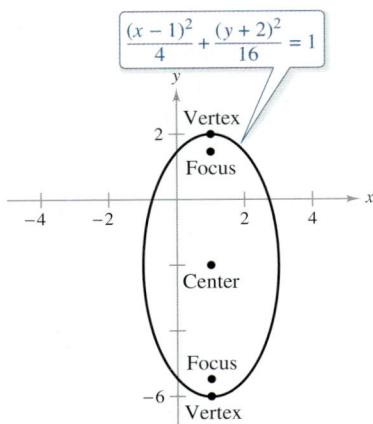

$$\frac{(x - 1)^2}{4} + \frac{(y + 2)^2}{16} = 1$$

Ellipse with a vertical major axis.
Figure 10.10

In Example 3, the constant term in the general second-degree equation is $F = -8$. For a constant term greater than or equal to 8, you would have obtained one of the degenerate cases shown below.

1. $F = 8$, single point, $(1, -2)$: $\dfrac{(x - 1)^2}{4} + \dfrac{(y + 2)^2}{16} = 0$

2. $F > 8$, no solution points: $\dfrac{(x - 1)^2}{4} + \dfrac{(y + 2)^2}{16} < 0$

EXAMPLE 4 **The Orbit of the Moon**

The moon orbits Earth in an elliptical path with the center of Earth at one focus, as shown in Figure 10.11. The major and minor axes of the orbit have lengths of 768,800 kilometers and 767,640 kilometers, respectively. Find the greatest and least distances (the apogee and perigee) from Earth's center to the moon's center.

Solution Begin by solving for a and b.

$$2a = 768,800$$ Length of major axis

$$a = 384,400$$ Solve for a.

$$2b = 767,640$$ Length of minor axis

$$b = 383,820$$ Solve for b.

Now, using these values, you can solve for c as follows.

$$c = \sqrt{a^2 - b^2} \approx 21,108$$

The greatest distance between the center of Earth and the center of the moon is

$$a + c \approx 405,508 \text{ kilometers}$$

and the least distance is

$$a - c \approx 363,292 \text{ kilometers}.$$

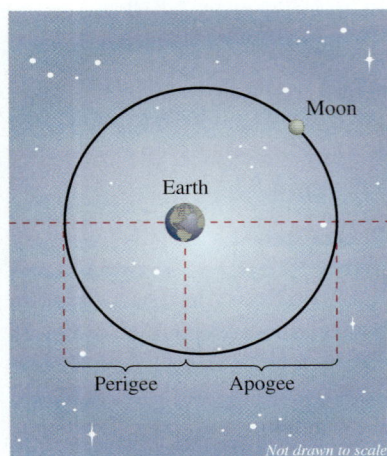

Figure 10.11

■ **FOR FURTHER INFORMATION**
For more information on some uses of the reflective properties of conics, see the article "Parabolic Mirrors, Elliptic and Hyperbolic Lenses" by Mohsen Maesumi in *The American Mathematical Monthly.* Also see the article "The Geometry of Microwave Antennas" by William R. Parzynski in *Mathematics Teacher.*

Theorem 10.2 presented a reflective property of parabolas. Ellipses have a similar reflective property. You are asked to prove the next theorem in Exercise 84.

> **THEOREM 10.4   Reflective Property of an Ellipse**
>
> Let P be a point on an ellipse. The tangent line to the ellipse at point P makes equal angles with the lines through P and the foci.

One of the reasons that astronomers had difficulty detecting that the orbits of the planets are ellipses is that the foci of the planetary orbits are relatively close to the center of the sun, making the orbits nearly circular. To measure the ovalness of an ellipse, you can use the concept of **eccentricity.**

> **Definition of Eccentricity of an Ellipse**
>
> The **eccentricity** e of an ellipse is given by the ratio
>
> $$e = \frac{c}{a}.$$

To see how this ratio is used to describe the shape of an ellipse, note that because the foci of an ellipse are located along the major axis between the vertices and the center, it follows that

$$0 < c < a.$$

For an ellipse that is nearly circular, the foci are close to the center and the ratio c/a is close to 0, and for an elongated ellipse, the foci are close to the vertices and the ratio c/a is close to 1, as shown in Figure 10.12. Note that

$$0 < e < 1$$

for every ellipse.

The orbit of the moon has an eccentricity of $e \approx 0.0549$, and the eccentricities of the eight planetary orbits are listed below.

Mercury:	$e \approx 0.2056$	Jupiter:	$e \approx 0.0484$
Venus:	$e \approx 0.0068$	Saturn:	$e \approx 0.0542$
Earth:	$e \approx 0.0167$	Uranus:	$e \approx 0.0472$
Mars:	$e \approx 0.0934$	Neptune:	$e \approx 0.0086$

You can use integration to show that the area of an ellipse is $A = \pi ab$. For instance, the area of the ellipse

$$\frac{x^2}{a^2} + \frac{y^2}{b^2} = 1$$

is

$$A = 4 \int_0^a \frac{b}{a} \sqrt{a^2 - x^2} \, dx$$

$$= \frac{4b}{a} \int_0^{\pi/2} a^2 \cos^2 \theta \, d\theta. \qquad \text{Trigonometric substitution } x = a \sin \theta$$

However, it is not so simple to find the *circumference* of an ellipse. The next example shows how to use eccentricity to set up an "elliptic integral" for the circumference of an ellipse.

(a) $\dfrac{c}{a}$ is small.

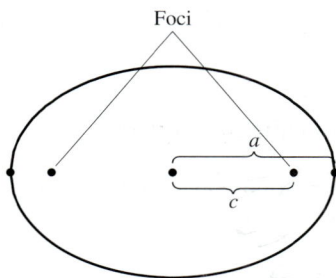

(b) $\dfrac{c}{a}$ is close to 1.

Eccentricity is the ratio $\dfrac{c}{a}$.

Figure 10.12

EXAMPLE 5 **Finding the Circumference of an Ellipse**

‣ *See LarsonCalculus.com for an interactive version of this type of example.*

Show that the circumference of the ellipse $(x^2/a^2) + (y^2/b^2) = 1$ is

$$4a\int_0^{\pi/2} \sqrt{1 - e^2 \sin^2 \theta}\, d\theta. \qquad e = \frac{c}{a}$$

Solution Because the ellipse is symmetric with respect to both the x-axis and the y-axis, you know that its circumference C is four times the arc length of

$$y = \frac{b}{a}\sqrt{a^2 - x^2}$$

in the first quadrant. The function y is differentiable for all x in the interval $[0, a]$ except at $x = a$. So, the circumference is given by the improper integral

$$C = \lim_{d \to a^-} 4\int_0^d \sqrt{1 + (y')^2}\, dx = 4\int_0^a \sqrt{1 + (y')^2}\, dx = 4\int_0^a \sqrt{1 + \frac{b^2 x^2}{a^2(a^2 - x^2)}}\, dx.$$

Using the trigonometric substitution $x = a \sin \theta$, you obtain

$$C = 4\int_0^{\pi/2} \sqrt{1 + \frac{b^2 \sin^2 \theta}{a^2 \cos^2 \theta}}\, (a \cos \theta)\, d\theta$$

$$= 4\int_0^{\pi/2} \sqrt{a^2 \cos^2 \theta + b^2 \sin^2 \theta}\, d\theta$$

$$= 4\int_0^{\pi/2} \sqrt{a^2(1 - \sin^2 \theta) + b^2 \sin^2 \theta}\, d\theta$$

$$= 4\int_0^{\pi/2} \sqrt{a^2 - (a^2 - b^2) \sin^2 \theta}\, d\theta.$$

Because $e^2 = c^2/a^2 = (a^2 - b^2)/a^2$, you can rewrite this integral as

$$C = 4a\int_0^{\pi/2} \sqrt{1 - e^2 \sin^2 \theta}\, d\theta.$$

> **AREA AND CIRCUMFERENCE OF AN ELLIPSE**
>
> In his work with elliptic orbits in the early 1600's, Johannes Kepler successfully developed a formula for the area of an ellipse, $A = \pi ab$. He was less successful, however, in developing a formula for the circumference of an ellipse; the best he could do was to give the approximate formula $C = \pi(a + b)$.

A great deal of time has been devoted to the study of elliptic integrals. Such integrals generally do not have elementary antiderivatives. To find the circumference of an ellipse, you must usually resort to an approximation technique.

EXAMPLE 6 **Approximating the Value of an Elliptic Integral**

Use the elliptic integral in Example 5 to approximate the circumference of the ellipse

$$\frac{x^2}{25} + \frac{y^2}{16} = 1.$$

Solution Because $e^2 = c^2/a^2 = (a^2 - b^2)/a^2 = 9/25$, you have

$$C = (4)(5)\int_0^{\pi/2} \sqrt{1 - \frac{9 \sin^2 \theta}{25}}\, d\theta.$$

Applying Simpson's Rule with $n = 4$ produces

$$C \approx 20\left(\frac{\pi}{6}\right)\left(\frac{1}{4}\right)[1 + 4(0.9733) + 2(0.9055) + 4(0.8323) + 0.8]$$

$$\approx 28.36.$$

So, the ellipse has a circumference of about 28.36 units, as shown in Figure 10.13.

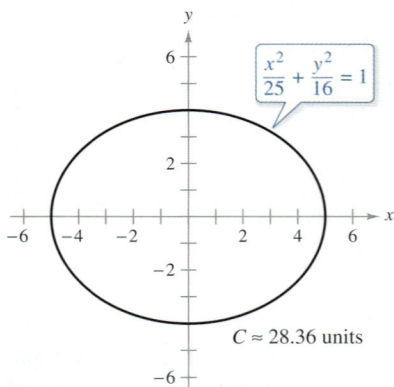

$\frac{x^2}{25} + \frac{y^2}{16} = 1$

$C \approx 28.36$ units

Figure 10.13

Hyperbolas

The definition of a hyperbola is similar to that of an ellipse. For an ellipse, the *sum* of the distances between the foci and a point on the ellipse is fixed, whereas for a hyperbola, the absolute value of the *difference* between these distances is fixed.

A **hyperbola** is the set of all points (x, y) for which the absolute value of the difference between the distances from two distinct fixed points called **foci** is constant. (See Figure 10.14.) The line through the two foci intersects a hyperbola at two points called the **vertices.** The line segment connecting the vertices is the **transverse axis,** and the midpoint of the transverse axis is the **center** of the hyperbola. One distinguishing feature of a hyperbola is that its graph has two separate *branches.*

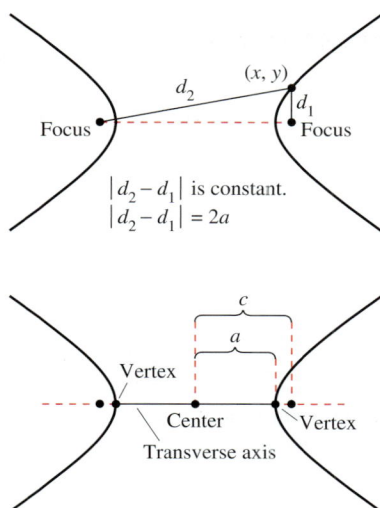

$|d_2 - d_1|$ is constant.
$|d_2 - d_1| = 2a$

Figure 10.14

THEOREM 10.5 Standard Equation of a Hyperbola

The standard form of the equation of a hyperbola with center at (h, k) is

$$\frac{(x - h)^2}{a^2} - \frac{(y - k)^2}{b^2} = 1 \qquad \text{Transverse axis is horizontal.}$$

or

$$\frac{(y - k)^2}{a^2} - \frac{(x - h)^2}{b^2} = 1. \qquad \text{Transverse axis is vertical.}$$

The vertices are a units from the center, and the foci are c units from the center, where $c^2 = a^2 + b^2$.

Note that the constants a, b, and c do not have the same relationship for hyperbolas as they do for ellipses. For hyperbolas, $c^2 = a^2 + b^2$, but for ellipses, $c^2 = a^2 - b^2$.

An important aid in sketching the graph of a hyperbola is the determination of its **asymptotes,** as shown in Figure 10.15. Each hyperbola has two asymptotes that intersect at the center of the hyperbola. The asymptotes pass through the vertices of a rectangle of dimensions $2a$ by $2b$, with its center at (h, k). The line segment of length $2b$ joining

$$(h, k + b)$$

and

$$(h, k - b)$$

is referred to as the **conjugate axis** of the hyperbola.

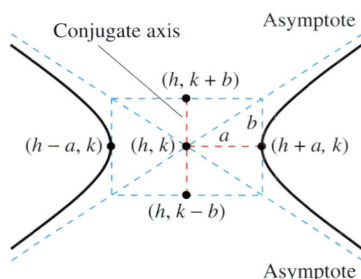

Figure 10.15

THEOREM 10.6 Asymptotes of a Hyperbola

For a *horizontal* transverse axis, the equations of the asymptotes are

$$y = k + \frac{b}{a}(x - h) \quad \text{and} \quad y = k - \frac{b}{a}(x - h).$$

For a *vertical* transverse axis, the equations of the asymptotes are

$$y = k + \frac{a}{b}(x - h) \quad \text{and} \quad y = k - \frac{a}{b}(x - h).$$

In Figure 10.15, you can see that the asymptotes coincide with the diagonals of the rectangle with dimensions $2a$ and $2b$, centered at (h, k). This provides you with a quick means of sketching the asymptotes, which in turn aids in sketching the hyperbola.

$\cdots\cdots\triangleright$ *See LarsonCalculus.com for an interactive version of this type of example.*

EXAMPLE 7 **Using Asymptotes to Sketch a Hyperbola**

Sketch the graph of the hyperbola

$$4x^2 - y^2 = 16.$$

Solution Begin by rewriting the equation in standard form.

$$\frac{x^2}{4} - \frac{y^2}{16} = 1$$

The transverse axis is horizontal and the vertices occur at $(-2, 0)$ and $(2, 0)$. The ends of the conjugate axis occur at $(0, -4)$ and $(0, 4)$. Using these four points, you can sketch the rectangle shown in Figure 10.16(a). By drawing the asymptotes through the corners of this rectangle, you can complete the sketch as shown in Figure 10.16(b).

▷ **TECHNOLOGY** You can use a graphing utility to verify the graph obtained in Example 7 by solving the original equation for y and graphing the following equations.

$$y_1 = \sqrt{4x^2 - 16}$$
$$y_2 = -\sqrt{4x^2 - 16}$$

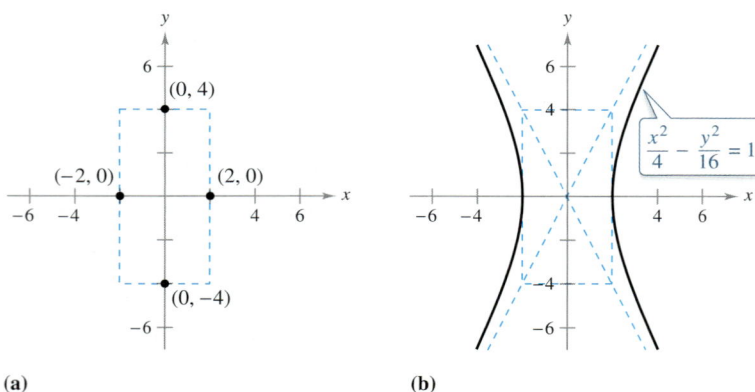

(a)

(b)

Figure 10.16

Definition of Eccentricity of a Hyperbola

The **eccentricity** e of a hyperbola is given by the ratio

$$e = \frac{c}{a}.$$

■ **FOR FURTHER INFORMATION** To read about using a string that traces both elliptic and hyperbolic arcs having the same foci, see the article "Ellipse to Hyperbola: 'With This String I Thee Wed'" by Tom M. Apostol and Mamikon A. Mnatsakanian in *Mathematics Magazine.* To view this article, go to *MathArticles.com.*

As with an ellipse, the **eccentricity** of a hyperbola is $e = c/a$. Because $c > a$ for hyperbolas, it follows that $e > 1$ for hyperbolas. If the eccentricity is large, then the branches of the hyperbola are nearly flat. If the eccentricity is close to 1, then the branches of the hyperbola are more pointed, as shown in Figure 10.17.

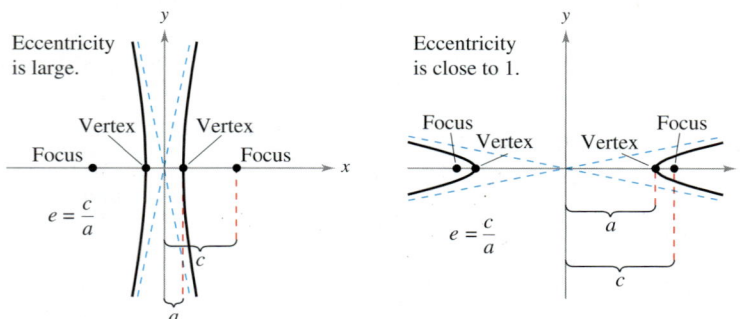

Figure 10.17

The application in Example 8 was developed during World War II. It shows how the properties of hyperbolas can be used in radar and other detection systems.

EXAMPLE 8   **A Hyperbolic Detection System**

Two microphones, 1 mile apart, record an explosion. Microphone A receives the sound 2 seconds before microphone B. Where was the explosion?

Solution   Assuming that sound travels at 1100 feet per second, you know that the explosion took place 2200 feet farther from B than from A, as shown in Figure 10.18. The locus of all points that are 2200 feet closer to A than to B is one branch of the hyperbola

$$\frac{x^2}{a^2} - \frac{y^2}{b^2} = 1$$

where

$$c = \frac{1 \text{ mile}}{2} = \frac{5280 \text{ ft}}{2} = 2640 \text{ feet}$$

and

$$a = \frac{2200 \text{ ft}}{2} = 1100 \text{ feet.}$$

Because $c^2 = a^2 + b^2$, it follows that

$$
\begin{aligned}
b^2 &= c^2 - a^2 \\
&= (2640)^2 - (1100)^2 \\
&= 5{,}759{,}600
\end{aligned}
$$

and you can conclude that the explosion occurred somewhere on the right branch of the hyperbola

$$\frac{x^2}{1{,}210{,}000} - \frac{y^2}{5{,}759{,}600} = 1.$$

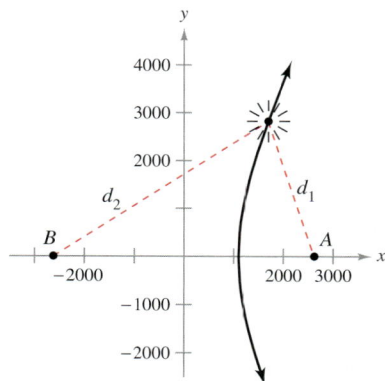

$2c = 5280$
$d_2 - d_1 = 2a = 2200$
Figure 10.18

In Example 8, you were able to determine only the hyperbola on which the explosion occurred, but not the exact location of the explosion. If, however, you had received the sound at a third position C, then two other hyperbolas would be determined. The exact location of the explosion would be the point at which these three hyperbolas intersect.

Another interesting application of conics involves the orbits of comets in our solar system. Of the 610 comets identified prior to 1970, 245 have elliptical orbits, 295 have parabolic orbits, and 70 have hyperbolic orbits. The center of the sun is a focus of each orbit, and each orbit has a vertex at the point at which the comet is closest to the sun. Undoubtedly, many comets with parabolic or hyperbolic orbits have not been identified—such comets pass through our solar system only once. Only comets with elliptical orbits, such as Halley's comet, remain in our solar system.

The type of orbit for a comet can be determined as follows.

1. Ellipse: $v < \sqrt{2GM/p}$
2. Parabola: $v = \sqrt{2GM/p}$
3. Hyperbola: $v > \sqrt{2GM/p}$

In each of the above, p is the distance between one vertex and one focus of the comet's orbit (in meters), v is the velocity of the comet at the vertex (in meters per second), $M \approx 1.989 \times 10^{30}$ kilograms is the mass of the sun, and $G \approx 6.67 \times 10^{-8}$ cubic meters per kilogram-second squared is the gravitational constant.

CAROLINE HERSCHEL (1750–1848)

The first woman to be credited with detecting a new comet was the English astronomer Caroline Herschel. During her life, Caroline Herschel discovered a total of eight new comets.
See LarsonCalculus.com to read more of this biography.

10.1 Exercises

See CalcChat.com for tutorial help and worked-out solutions to odd-numbered exercises.

Matching In Exercises 1–6, match the equation with its graph. [The graphs are labeled (a), (b), (c), (d), (e), and (f).]

(a)

(b)

(c)

(d)

(e)

(f)

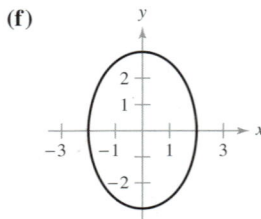

1. $y^2 = 4x$

2. $(x + 4)^2 = -2(y - 2)$

3. $\dfrac{y^2}{16} - \dfrac{x^2}{1} = 1$

4. $\dfrac{(x - 2)^2}{16} + \dfrac{(y + 1)^2}{4} = 1$

5. $\dfrac{x^2}{4} + \dfrac{y^2}{9} = 1$

6. $\dfrac{(x - 2)^2}{9} - \dfrac{y^2}{4} = 1$

Sketching a Parabola In Exercises 7–14, find the vertex, focus, and directrix of the parabola, and sketch its graph.

7. $y^2 = -8x$

8. $x^2 + 6y = 0$

9. $(x + 5) + (y - 3)^2 = 0$

10. $(x - 6)^2 + 8(y + 7) = 0$

11. $y^2 - 4y - 4x = 0$

12. $y^2 + 6y + 8x + 25 = 0$

13. $x^2 + 4x + 4y - 4 = 0$

14. $y^2 + 4y + 8x - 12 = 0$

Finding an Equation of a Parabola In Exercises 15–22, find an equation of the parabola.

15. Vertex: $(5, 4)$

Focus: $(3, 4)$

16. Vertex: $(-2, 1)$

Focus: $(-2, -1)$

17. Vertex: $(0, 5)$

Directrix: $y = -3$

18. Focus: $(2, 2)$

Directrix: $x = -2$

19. Vertex: $(0, 4)$

Points on the parabola:

$(-2, 0), (2, 0)$

20. Vertex: $(2, 4)$

Points on the parabola:

$(0, 0), (4, 0)$

21. Axis is parallel to y-axis; graph passes through $(0, 3)$, $(3, 4)$, and $(4, 11)$.

22. Directrix: $y = -2$; endpoints of latus rectum are $(0, 2)$ and $(8, 2)$.

Sketching an Ellipse In Exercises 23–28, find the center, foci, vertices, and eccentricity of the ellipse, and sketch its graph.

23. $16x^2 + y^2 = 16$

24. $3x^2 + 7y^2 = 63$

25. $\dfrac{(x - 3)^2}{16} + \dfrac{(y - 1)^2}{25} = 1$

26. $(x + 4)^2 + \dfrac{(y + 6)^2}{1/4} = 1$

27. $9x^2 + 4y^2 + 36x - 24y + 36 = 0$

28. $16x^2 + 25y^2 - 64x + 150y + 279 = 0$

Finding an Equation of an Ellipse In Exercises 29–34, find an equation of the ellipse.

29. Center: $(0, 0)$

Focus: $(5, 0)$

Vertex: $(6, 0)$

30. Vertices: $(0, 3), (8, 3)$

Eccentricity: $\frac{3}{4}$

31. Vertices: $(3, 1), (3, 9)$

Minor axis length: 6

32. Foci: $(0, \pm 9)$

Major axis length: 22

33. Center: $(0, 0)$

Major axis: horizontal

Points on the ellipse:

$(3, 1), (4, 0)$

34. Center: $(1, 2)$

Major axis: vertical

Points on the ellipse:

$(1, 6), (3, 2)$

Sketching a Hyperbola In Exercises 35–40, find the center, foci, and vertices of the hyperbola, and sketch its graph using asymptotes as an aid.

35. $\dfrac{x^2}{25} - \dfrac{y^2}{16} = 1$

36. $\dfrac{(y + 3)^2}{225} - \dfrac{(x - 5)^2}{64} = 1$

37. $9x^2 - y^2 - 36x - 6y + 18 = 0$

38. $y^2 - 16x^2 + 64x - 208 = 0$

39. $x^2 - 9y^2 + 2x - 54y - 80 = 0$

40. $9x^2 - 4y^2 + 54x + 8y + 78 = 0$

Finding an Equation of a Hyperbola In Exercises 41–48, find an equation of the hyperbola.

41. Vertices: $(\pm 1, 0)$

Asymptotes: $y = \pm 5x$

42. Vertices: $(0, \pm 4)$

Asymptotes: $y = \pm 2x$

43. Vertices: $(2, \pm 3)$

Point on graph: $(0, 5)$

44. Vertices: $(2, \pm 3)$

Foci: $(2, \pm 5)$

45. Center: $(0, 0)$

Vertex: $(0, 2)$

Focus: $(0, 4)$

46. Center: $(0, 0)$

Vertex: $(6, 0)$

Focus: $(10, 0)$

47. Vertices: $(0, 2), (6, 2)$

Asymptotes: $y = \frac{2}{3}x$

$y = 4 - \frac{2}{3}x$

48. Focus: $(20, 0)$

Asymptotes: $y = \pm\frac{3}{4}x$

Finding Equations of Tangent Lines and Normal Lines
In Exercises 49 and 50, find equations for (a) the tangent lines and (b) the normal lines to the hyperbola for the given value of x.

49. $\dfrac{x^2}{9} - y^2 = 1$, $x = 6$ **50.** $\dfrac{y^2}{4} - \dfrac{x^2}{2} = 1$, $x = 4$

Classifying the Graph of an Equation In Exercises 51–58, classify the graph of the equation as a circle, a parabola, an ellipse, or a hyperbola.

51. $x^2 + 4y^2 - 6x + 16y + 21 = 0$

52. $4x^2 - y^2 - 4x - 3 = 0$

53. $25x^2 - 10x - 200y - 119 = 0$

54. $y^2 - 4y = x + 5$

55. $9x^2 + 9y^2 - 36x + 6y + 34 = 0$

56. $2x(x - y) = y(3 - y - 2x)$

57. $3(x - 1)^2 = 6 + 2(y + 1)^2$

58. $9(x + 3)^2 = 36 - 4(y - 2)^2$

WRITING ABOUT CONCEPTS

59. Parabola

(a) Give the definition of a parabola.

(b) Give the standard forms of a parabola with vertex at (h, k).

(c) In your own words, state the reflective property of a parabola.

60. Ellipse

(a) Give the definition of an ellipse.

(b) Give the standard form of an ellipse with center at (h, k).

61. Hyperbola

(a) Give the definition of a hyperbola.

(b) Give the standard forms of a hyperbola with center at (h, k).

(c) Write equations for the asymptotes of a hyperbola.

62. Eccentricity Define the eccentricity of an ellipse. In your own words, describe how changes in the eccentricity affect the ellipse.

63. Using an Equation Consider the equation

$$9x^2 + 4y^2 - 36x - 24y - 36 = 0.$$

(a) Classify the graph of the equation as a circle, a parabola, an ellipse, or a hyperbola.

(b) Change the $4y^2$-term in the equation to $-4y^2$. Classify the graph of the new equation.

(c) Change the $9x^2$-term in the original equation to $4x^2$. Classify the graph of the new equation.

(d) Describe one way you could change the original equation so that its graph is a parabola.

---

64. HOW DO YOU SEE IT? In parts (a)–(d), describe in words how a plane could intersect with the double-napped cone to form the conic section (see figure).

(a) Circle   (b) Ellipse

(c) Parabola   (d) Hyperbola

65. Solar Collector A solar collector for heating water is constructed with a sheet of stainless steel that is formed into the shape of a parabola (see figure). The water will flow through a pipe that is located at the focus of the parabola. At what distance from the vertex is the pipe?

Not drawn to scale

Figure for 65   Figure for 66

66. Beam Deflection A simply supported beam that is 16 meters long has a load concentrated at the center (see figure). The deflection of the beam at its center is 3 centimeters. Assume that the shape of the deflected beam is parabolic.

(a) Find an equation of the parabola. (Assume that the origin is at the center of the beam.)

(b) How far from the center of the beam is the deflection 1 centimeter?

67. Proof

(a) Prove that any two distinct tangent lines to a parabola intersect.

(b) Demonstrate the result of part (a) by finding the point of intersection of the tangent lines to the parabola $x^2 - 4x - 4y = 0$ at the points $(0, 0)$ and $(6, 3)$.

68. Proof

(a) Prove that if any two tangent lines to a parabola intersect at right angles, their point of intersection must lie on the directrix.

(b) Demonstrate the result of part (a) by showing that the tangent lines to the parabola $x^2 - 4x - 4y + 8 = 0$ at the points $(-2, 5)$ and $\left(3, \frac{5}{4}\right)$ intersect at right angles, and that the point of intersection lies on the directrix.

69. Investigation Sketch the graphs of $x^2 = 4py$ for $p = \frac{1}{4}, \frac{1}{2},$ $1, \frac{3}{2},$ and 2 on the same coordinate axes. Discuss the change in the graphs as p increases.

70. Bridge Design A cable of a suspension bridge is suspended (in the shape of a parabola) between two towers that are 120 meters apart and 20 meters above the roadway (see figure). The cable touches the roadway midway between the towers.

(a) Find an equation for the parabolic shape of the cable.

(b) Find the length of the parabolic cable.

71. Architecture

A church window is bounded above by a parabola and below by the arc of a circle (see figure). Find the surface area of the window.

72. Surface Area A satellite signal receiving dish is formed by revolving the parabola given by $x^2 = 20y$ about the y-axis. The radius of the dish is r feet. Verify that the surface area of the dish is given by

$$2\pi \int_0^r x \sqrt{1 + \left(\frac{x}{10}\right)^2}\, dx = \frac{\pi}{15}[(100 + r^2)^{3/2} - 1000].$$

73. Orbit of Earth Earth moves in an elliptical orbit with the sun at one of the foci. The length of half of the major axis is 149,598,000 kilometers, and the eccentricity is 0.0167. Find the minimum distance (*perihelion*) and the maximum distance (*aphelion*) of Earth from the sun.

74. Satellite Orbit The *apogee* (the point in orbit farthest from Earth) and the *perigee* (the point in orbit closest to Earth) of an elliptical orbit of an Earth satellite are given by A and P. Show that the eccentricity of the orbit is

$$e = \frac{A - P}{A + P}.$$

75. Explorer 18 On November 27, 1963, the United States launched the research satellite Explorer 18. Its low and high points above the surface of Earth were 119 miles and 123,000 miles. Find the eccentricity of its elliptical orbit.

76. Explorer 55 On November 20, 1975, the United States launched the research satellite Explorer 55. Its low and high points above the surface of Earth were 96 miles and 1865 miles. Find the eccentricity of its elliptical orbit.

77. Halley's Comet

Probably the most famous of all comets, Halley's comet, has an elliptical orbit with the sun at one focus. Its maximum distance from the sun is approximately 35.29 AU (1 astronomical unit is approximately 92.956×10^6 miles), and its minimum distance is approximately 0.59 AU. Find the eccentricity of the orbit.

78. Particle Motion Consider a particle traveling clockwise on the elliptical path

$$\frac{x^2}{100} + \frac{y^2}{25} = 1.$$

The particle leaves the orbit at the point $(-8, 3)$ and travels in a straight line tangent to the ellipse. At what point will the particle cross the y-axis?

Area, Volume, and Surface Area In Exercises 79 and 80, find (a) the area of the region bounded by the ellipse, (b) the volume and surface area of the solid generated by revolving the region about its major axis (prolate spheroid), and (c) the volume and surface area of the solid generated by revolving the region about its minor axis (oblate spheroid).

79. $\dfrac{x^2}{4} + \dfrac{y^2}{1} = 1$ **80.** $\dfrac{x^2}{16} + \dfrac{y^2}{9} = 1$

81. Arc Length Use the integration capabilities of a graphing utility to approximate to two-decimal-place accuracy the elliptical integral representing the circumference of the ellipse

$$\frac{x^2}{25} + \frac{y^2}{49} = 1.$$

82. Conjecture

(a) Show that the equation of an ellipse can be written as

$$\frac{(x - h)^2}{a^2} + \frac{(y - k)^2}{a^2(1 - e^2)} = 1.$$

(b) Use a graphing utility to graph the ellipse

$$\frac{(x - 2)^2}{4} + \frac{(y - 3)^2}{4(1 - e^2)} = 1$$

for $e = 0.95$, $e = 0.75$, $e = 0.5$, $e = 0.25$, and $e = 0$.

(c) Use the results of part (b) to make a conjecture about the change in the shape of the ellipse as e approaches 0.

83. Geometry The area of the ellipse in the figure is twice the area of the circle. What is the length of the major axis?

Figure for 83

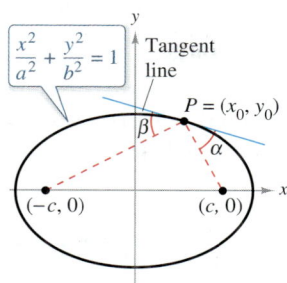

Figure for 84

84. Proof Prove Theorem 10.4 by showing that the tangent line to an ellipse at a point P makes equal angles with lines through P and the foci (see figure). [*Hint:* (1) Find the slope of the tangent line at P, (2) find the slopes of the lines through P and each focus, and (3) use the formula for the tangent of the angle between two lines.]

85. Finding an Equation of a Hyperbola Find an equation of the hyperbola such that for any point on the hyperbola, the difference between its distances from the points $(2, 2)$ and $(10, 2)$ is 6.

86. Hyperbola Consider a hyperbola centered at the origin with a horizontal transverse axis. Use the definition of a hyperbola to derive its standard form:

$$\frac{x^2}{a^2} - \frac{y^2}{b^2} = 1.$$

87. Navigation LORAN (long distance radio navigation) for aircraft and ships uses synchronized pulses transmitted by widely separated transmitting stations. These pulses travel at the speed of light (186,000 miles per second). The difference in the times of arrival of these pulses at an aircraft or ship is constant on a hyperbola having the transmitting stations as foci. Assume that two stations, 300 miles apart, are positioned on a rectangular coordinate system at $(-150, 0)$ and $(150, 0)$ and that a ship is traveling on a path with coordinates $(x, 75)$ (see figure). Find the x-coordinate of the position of the ship if the time difference between the pulses from the transmitting stations is 1000 microseconds (0.001 second).

Figure for 87

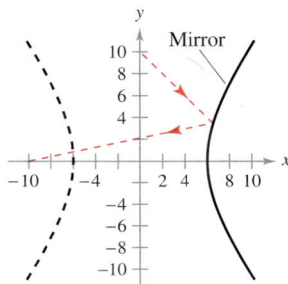

Figure for 88

88. Hyperbolic Mirror A hyperbolic mirror (used in some telescopes) has the property that a light ray directed at the focus will be reflected to the other focus. The mirror in the figure has the equation $(x^2/36) - (y^2/64) = 1$. At which point on the mirror will light from the point $(0, 10)$ be reflected to the other focus?

89. Tangent Line Show that the equation of the tangent line to $\dfrac{x^2}{a^2} - \dfrac{y^2}{b^2} = 1$ at the point (x_0, y_0) is $\left(\dfrac{x_0}{a^2}\right)x - \left(\dfrac{y_0}{b^2}\right)y = 1$.

90. Proof Prove that the graph of the equation

$$Ax^2 + Cy^2 + Dx + Ey + F = 0$$

is one of the following (except in degenerate cases).

Conic	Condition
(a) Circle	$A = C$
(b) Parabola	$A = 0$ or $C = 0$ (but not both)
(c) Ellipse	$AC > 0$
(d) Hyperbola	$AC < 0$

True or False? **In Exercises 91–96, determine whether the statement is true or false. If it is false, explain why or give an example that shows it is false.**

91. It is possible for a parabola to intersect its directrix.

92. The point on a parabola closest to its focus is its vertex.

93. If C is the circumference of the ellipse

$$\frac{x^2}{a^2} + \frac{y^2}{b^2} = 1, \quad b < a$$

then $2\pi b \leq C \leq 2\pi a$.

94. If $D \neq 0$ or $E \neq 0$, then the graph of $y^2 - x^2 + Dx + Ey = 0$ is a hyperbola.

95. If the asymptotes of the hyperbola $(x^2/a^2) - (y^2/b^2) = 1$ intersect at right angles, then $a = b$.

96. Every tangent line to a hyperbola intersects the hyperbola only at the point of tangency.

PUTNAM EXAM CHALLENGE

97. For a point P on an ellipse, let d be the distance from the center of the ellipse to the line tangent to the ellipse at P. Prove that $(PF_1)(PF_2)d^2$ is constant as P varies on the ellipse, where PF_1 and PF_2 are the distances from P to the foci F_1 and F_2 of the ellipse.

98. Find the minimum value of

$$(u - v)^2 + \left(\sqrt{2 - u^2} - \frac{9}{v}\right)^2$$

for $0 < u < \sqrt{2}$ and $v > 0$.

These problems were composed by the Committee on the Putnam Prize Competition.
© The Mathematical Association of America. All rights reserved.

10.2 Plane Curves and Parametric Equations

■ Sketch the graph of a curve given by a set of parametric equations.
■ Eliminate the parameter in a set of parametric equations.
■ Find a set of parametric equations to represent a curve.
■ Understand two classic calculus problems, the tautochrone and brachistochrone problems.

Plane Curves and Parametric Equations

Until now, you have been representing a graph by a single equation involving *two* variables. In this section, you will study situations in which *three* variables are used to represent a curve in the plane.

Consider the path followed by an object that is propelled into the air at an angle of 45°. For an initial velocity of 48 feet per second, the object travels the parabolic path given by

$$y = -\frac{x^2}{72} + x \qquad \text{Rectangular equation}$$

as shown in Figure 10.19. This equation, however, does not tell the whole story. Although it does tell you *where* the object has been, it doesn't tell you *when* the object was at a given point (x, y). To determine this time, you can introduce a third variable t, called a **parameter.** By writing both x and y as functions of t, you obtain the **parametric equations**

$$x = 24\sqrt{2}t \qquad \text{Parametric equation for } x$$

and

$$y = -16t^2 + 24\sqrt{2}t. \qquad \text{Parametric equation for } y$$

From this set of equations, you can determine that at time $t = 0$, the object is at the point $(0, 0)$. Similarly, at time $t = 1$, the object is at the point

$$\left(24\sqrt{2}, 24\sqrt{2} - 16\right)$$

and so on. (You will learn a method for determining this particular set of parametric equations—the equations of motion—later, in Section 12.3.)

For this particular motion problem, x and y are continuous functions of t, and the resulting path is called a **plane curve.**

Rectangular equation:
$$y = -\frac{x^2}{72} + x$$

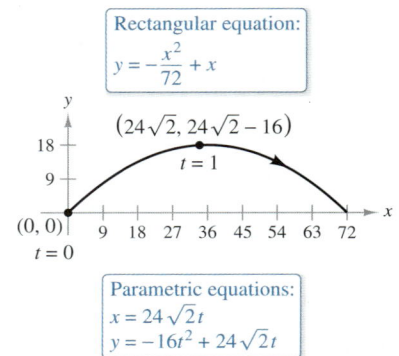

Parametric equations:
$$x = 24\sqrt{2}t$$
$$y = -16t^2 + 24\sqrt{2}t$$

Curvilinear motion: two variables for position, one variable for time
Figure 10.19

•• **REMARK** At times, it is important to distinguish between a graph (the set of points) and a curve (the points together with their defining parametric equations). When it is important, the distinction will be explicit. When it is not important, C will be used to represent either the graph or the curve.

Definition of a Plane Curve

If f and g are continuous functions of t on an interval I, then the equations

$$x = f(t) \quad \text{and} \quad y = g(t)$$

are **parametric equations** and t is the **parameter.** The set of points (x, y) obtained as t varies over the interval I is the **graph** of the parametric equations. Taken together, the parametric equations and the graph are a **plane curve,** denoted by C.

When sketching a curve represented by a set of parametric equations, you can plot points in the *xy*-plane. Each set of coordinates (x, y) is determined from a value chosen for the parameter t. By plotting the resulting points in order of increasing values of t, the curve is traced out in a specific direction. This is called the **orientation** of the curve.

EXAMPLE 1 **Sketching a Curve**

Sketch the curve described by the parametric equations

$$x = f(t) = t^2 - 4$$

and

$$y = g(t) = \frac{t}{2}$$

where $-2 \le t \le 3$.

Solution For values of t on the given interval, the parametric equations yield the points (x, y) shown in the table.

t	-2	-1	0	1	2	3
x	0	-3	-4	-3	0	5
y	-1	$-\frac{1}{2}$	0	$\frac{1}{2}$	1	$\frac{3}{2}$

By plotting these points in order of increasing t and using the continuity of f and g, you obtain the curve C shown in Figure 10.20. Note that the arrows on the curve indicate its orientation as t increases from -2 to 3.

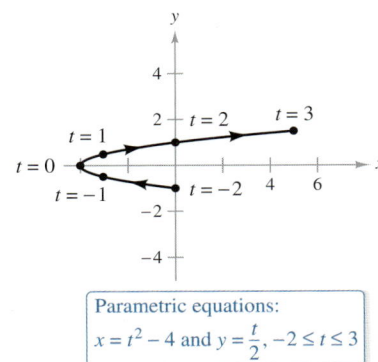

Parametric equations:
$x = t^2 - 4$ and $y = \frac{t}{2}, -2 \le t \le 3$

Figure 10.20

According to the Vertical Line Test, the graph shown in Figure 10.20 does not define y as a function of x. This points out one benefit of parametric equations—they can be used to represent graphs that are more general than graphs of functions.

It often happens that two different sets of parametric equations have the same graph. For instance, the set of parametric equations

$$x = 4t^2 - 4 \quad \text{and} \quad y = t, \quad -1 \le t \le \frac{3}{2}$$

has the same graph as the set given in Example 1. (See Figure 10.21.) However, comparing the values of t in Figures 10.20 and 10.21, you can see that the second graph is traced out more *rapidly* (considering t as time) than the first graph. So, in applications, different parametric representations can be used to represent various *speeds* at which objects travel along a given path.

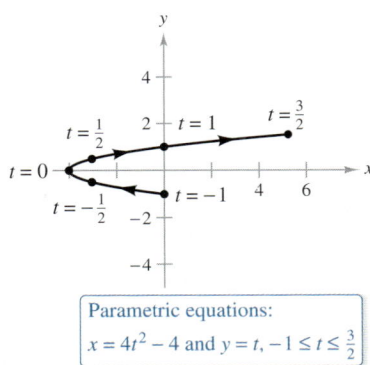

Parametric equations:
$x = 4t^2 - 4$ and $y = t, -1 \le t \le \frac{3}{2}$

Figure 10.21

▷ **TECHNOLOGY** Most graphing utilities have a *parametric* graphing mode. If you have access to such a utility, use it to confirm the graphs shown in Figures 10.20 and 10.21. Does the curve given by the parametric equations

$$x = 4t^2 - 8t \quad \text{and} \quad y = 1 - t, \quad -\frac{1}{2} \le t \le 2$$

represent the same graph as that shown in Figures 10.20 and 10.21? What do you notice about the *orientation* of this curve?

Eliminating the Parameter

Finding a rectangular equation that represents the graph of a set of parametric equations is called **eliminating the parameter.** For instance, you can eliminate the parameter from the set of parametric equations in Example 1 as follows.

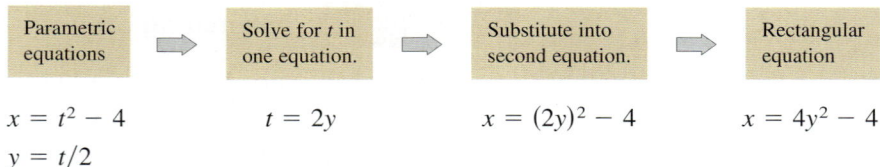

Parametric equations	Solve for t in one equation.	Substitute into second equation.	Rectangular equation
$x = t^2 - 4$	$t = 2y$	$x = (2y)^2 - 4$	$x = 4y^2 - 4$
$y = t/2$			

Once you have eliminated the parameter, you can recognize that the equation $x = 4y^2 - 4$ represents a parabola with a horizontal axis and vertex at $(-4, 0)$, as shown in Figure 10.20.

The range of x and y implied by the parametric equations may be altered by the change to rectangular form. In such instances, the domain of the rectangular equation must be adjusted so that its graph matches the graph of the parametric equations. Such a situation is demonstrated in the next example.

EXAMPLE 2 **Adjusting the Domain**

Sketch the curve represented by the equations

$$x = \frac{1}{\sqrt{t + 1}} \quad \text{and} \quad y = \frac{t}{t + 1}, \quad t > -1$$

by eliminating the parameter and adjusting the domain of the resulting rectangular equation.

Solution Begin by solving one of the parametric equations for t. For instance, you can solve the first equation for t as follows.

$$x = \frac{1}{\sqrt{t + 1}} \qquad \text{Parametric equation for } x$$

$$x^2 = \frac{1}{t + 1} \qquad \text{Square each side.}$$

$$t + 1 = \frac{1}{x^2}$$

$$t = \frac{1}{x^2} - 1$$

$$t = \frac{1 - x^2}{x^2} \qquad \text{Solve for } t.$$

Now, substituting into the parametric equation for y produces

$$y = \frac{t}{t + 1} \qquad \text{Parametric equation for } y$$

$$y = \frac{(1 - x^2)/x^2}{[(1 - x^2)/x^2] + 1} \qquad \text{Substitute } (1 - x^2)/x^2 \text{ for } t.$$

$$y = 1 - x^2. \qquad \text{Simplify.}$$

The rectangular equation, $y = 1 - x^2$, is defined for all values of x, but from the parametric equation for x, you can see that the curve is defined only when $t > -1$. This implies that you should restrict the domain of x to positive values, as shown in Figure 10.22.

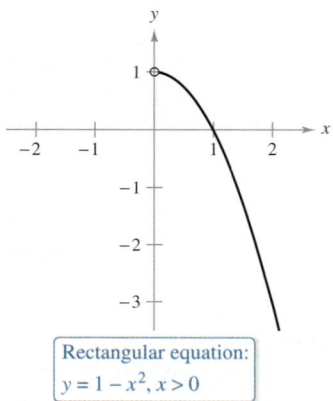

Parametric equations:
$x = \dfrac{1}{\sqrt{t + 1}}, y = \dfrac{t}{t + 1}, t > -1$

Rectangular equation:
$y = 1 - x^2, x > 0$

Figure 10.22

It is not necessary for the parameter in a set of parametric equations to represent time. The next example uses an *angle* as the parameter.

EXAMPLE 3 Using Trigonometry to Eliminate a Parameter

•••▷ *See LarsonCalculus.com for an interactive version of this type of example.*

Sketch the curve represented by

$$x = 3 \cos \theta \quad \text{and} \quad y = 4 \sin \theta, \quad 0 \le \theta \le 2\pi$$

by eliminating the parameter and finding the corresponding rectangular equation.

Solution Begin by solving for $\cos \theta$ and $\sin \theta$ in the given equations.

$$\cos \theta = \frac{x}{3} \qquad\qquad \text{Solve for } \cos \theta.$$

and

$$\sin \theta = \frac{y}{4} \qquad\qquad \text{Solve for } \sin \theta.$$

Next, make use of the identity

$$\sin^2 \theta + \cos^2 \theta = 1$$

to form an equation involving only x and y.

$$\cos^2 \theta + \sin^2 \theta = 1 \qquad\qquad \text{Trigonometric identity}$$

$$\left(\frac{x}{3}\right)^2 + \left(\frac{y}{4}\right)^2 = 1 \qquad\qquad \text{Substitute.}$$

$$\frac{x^2}{9} + \frac{y^2}{16} = 1 \qquad\qquad \text{Rectangular equation}$$

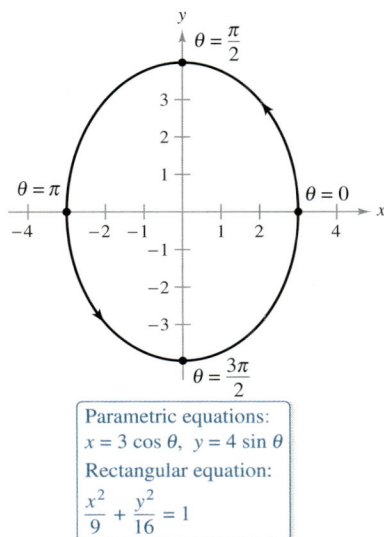

Parametric equations:
$x = 3 \cos \theta, \ y = 4 \sin \theta$
Rectangular equation:
$\dfrac{x^2}{9} + \dfrac{y^2}{16} = 1$

Figure 10.23

From this rectangular equation, you can see that the graph is an ellipse centered at $(0, 0)$, with vertices at $(0, 4)$ and $(0, -4)$ and minor axis of length $2b = 6$, as shown in Figure 10.23. Note that the ellipse is traced out *counterclockwise* as θ varies from 0 to 2π.

Using the technique shown in Example 3, you can conclude that the graph of the parametric equations

$$x = h + a \cos \theta \quad \text{and} \quad y = k + b \sin \theta, \quad 0 \le \theta \le 2\pi$$

is the ellipse (traced counterclockwise) given by

$$\frac{(x - h)^2}{a^2} + \frac{(y - k)^2}{b^2} = 1.$$

The graph of the parametric equations

$$x = h + a \sin \theta \quad \text{and} \quad y = k + b \cos \theta, \quad 0 \le \theta \le 2\pi$$

is also the ellipse (traced clockwise) given by

$$\frac{(x - h)^2}{a^2} + \frac{(y - k)^2}{b^2} = 1.$$

In Examples 2 and 3, it is important to realize that eliminating the parameter is primarily an *aid to curve sketching*. When the parametric equations represent the path of a moving object, the graph alone is not sufficient to describe the object's motion. You still need the parametric equations to tell you the *position*, *direction*, and *speed* at a given time.

▷ **TECHNOLOGY** Use a graphing utility in *parametric* mode to graph several ellipses.

Finding Parametric Equations

The first three examples in this section illustrate techniques for sketching the graph represented by a set of parametric equations. You will now investigate the reverse problem. How can you determine a set of parametric equations for a given graph or a given physical description? From the discussion following Example 1, you know that such a representation is not unique. This is demonstrated further in the next example, which finds two different parametric representations for a given graph.

EXAMPLE 4 **Finding Parametric Equations for a Given Graph**

Find a set of parametric equations that represents the graph of $y = 1 - x^2$, using each of the following parameters.

a. $t = x$ **b.** The slope $m = \dfrac{dy}{dx}$ at the point (x, y)

Solution

a. Letting $x = t$ produces the parametric equations

$$x = t \quad \text{and} \quad y = 1 - x^2 = 1 - t^2.$$

b. To write x and y in terms of the parameter m, you can proceed as follows.

$$m = \frac{dy}{dx}$$

$$m = -2x \qquad \text{Differentiate } y = 1 - x^2.$$

$$x = -\frac{m}{2} \qquad \text{Solve for } x.$$

This produces a parametric equation for x. To obtain a parametric equation for y, substitute $-m/2$ for x in the original equation.

$$y = 1 - x^2 \qquad \text{Write original rectangular equation.}$$

$$y = 1 - \left(-\frac{m}{2}\right)^2 \qquad \text{Substitute } -m/2 \text{ for } x.$$

$$y = 1 - \frac{m^2}{4} \qquad \text{Simplify.}$$

So, the parametric equations are

$$x = -\frac{m}{2} \quad \text{and} \quad y = 1 - \frac{m^2}{4}.$$

In Figure 10.24, note that the resulting curve has a right-to-left orientation as determined by the direction of increasing values of slope m. For part (a), the curve would have the opposite orientation.

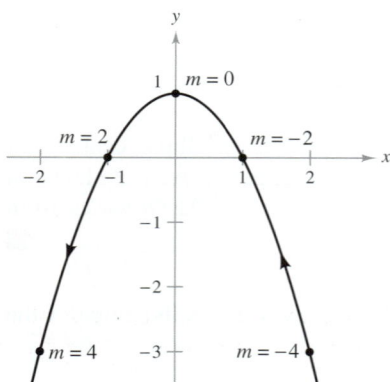

Rectangular equation: $y = 1 - x^2$
Parametric equations:
$x = -\dfrac{m}{2}, y = 1 - \dfrac{m^2}{4}$

Figure 10.24

■ **FOR FURTHER INFORMATION**
To read about other methods for finding parametric equations, see the article "Finding Rational Parametric Curves of Relative Degree One or Two" by Dave Boyles in *The College Mathematics Journal.* To view this article, go to *MathArticles.com.*

▷ **TECHNOLOGY**   To be efficient at using a graphing utility, it is important that you develop skill in representing a graph by a set of parametric equations. The reason for this is that many graphing utilities have only three graphing modes— (1) functions, (2) parametric equations, and (3) polar equations. Most graphing utilities are not programmed to graph a general equation. For instance, suppose you want to graph the hyperbola $x^2 - y^2 = 1$. To graph the hyperbola in *function* mode, you need two equations

$$y = \sqrt{x^2 - 1} \quad \text{and} \quad y = -\sqrt{x^2 - 1}.$$

In *parametric* mode, you can represent the graph by $x = \sec t$ and $y = \tan t$.

■ **FOR FURTHER INFORMATION**
For more information on cycloids, see the article "The Geometry of Rolling Curves" by John Bloom and Lee Whitt in *The American Mathematical Monthly.* To view this article, go to *MathArticles.com.*

EXAMPLE 5　**Parametric Equations for a Cycloid**

Determine the curve traced by a point P on the circumference of a circle of radius a rolling along a straight line in a plane. Such a curve is called a **cycloid.**

Solution　Let the parameter θ be the measure of the circle's rotation, and let the point $P = (x, y)$ begin at the origin. When $\theta = 0$, P is at the origin. When $\theta = \pi$, P is at a maximum point $(\pi a, 2a)$. When $\theta = 2\pi$, P is back on the x-axis at $(2\pi a, 0)$. From Figure 10.25, you can see that $\angle APC = 180° - \theta$. So,

$$\sin \theta = \sin(180° - \theta) = \sin(\angle APC) = \frac{AC}{a} = \frac{BD}{a}$$

$$\cos \theta = -\cos(180° - \theta) = -\cos(\angle APC) = \frac{AP}{-a}$$

which implies that $AP = -a \cos \theta$ and $BD = a \sin \theta$.

Because the circle rolls along the x-axis, you know that $OD = \overset{\frown}{PD} = a\theta$. Furthermore, because $BA = DC = a$, you have

$$x = OD - BD = a\theta - a \sin \theta$$
$$y = BA + AP = a - a \cos \theta.$$

So, the parametric equations are

$$x = a(\theta - \sin \theta) \quad \text{and} \quad y = a(1 - \cos \theta).$$

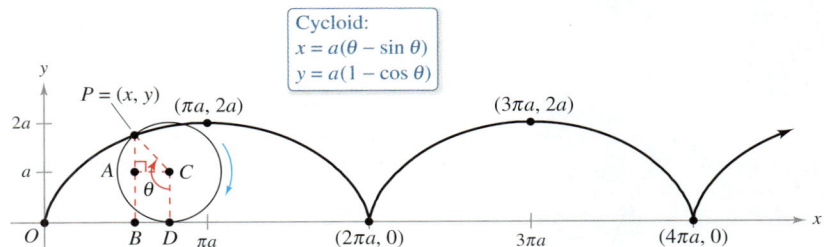

Cycloid:
$x = a(\theta - \sin \theta)$
$y = a(1 - \cos \theta)$

Figure 10.25

▷ **TECHNOLOGY**　Some graphing utilities allow you to simulate the motion of an object that is moving in the plane or in space. If you have access to such a utility, use it to trace out the path of the cycloid shown in Figure 10.25.

The cycloid in Figure 10.25 has sharp corners at the values $x = 2n\pi a$. Notice that the derivatives $x'(\theta)$ and $y'(\theta)$ are both zero at the points for which $\theta = 2n\pi$.

$$x(\theta) = a(\theta - \sin \theta) \qquad y(\theta) = a(1 - \cos \theta)$$
$$x'(\theta) = a - a \cos \theta \qquad y'(\theta) = a \sin \theta$$
$$x'(2n\pi) = 0 \qquad y'(2n\pi) = 0$$

Between these points, the cycloid is called **smooth.**

Definition of a Smooth Curve

A curve C represented by $x = f(t)$ and $y = g(t)$ on an interval I is called **smooth** when f' and g' are continuous on I and not simultaneously 0, except possibly at the endpoints of I. The curve C is called **piecewise smooth** when it is smooth on each subinterval of some partition of I.

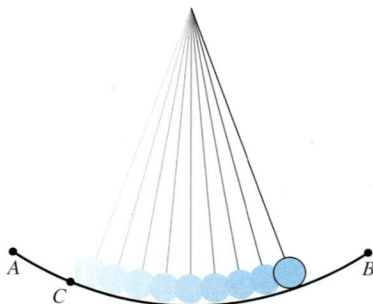

The time required to complete a full swing of the pendulum when starting from point C is only approximately the same as the time required when starting from point A.
Figure 10.26

JAMES BERNOULLI (1654–1705)
James Bernoulli, also called Jacques, was the older brother of John. He was one of several accomplished mathematicians of the Swiss Bernoulli family. James's mathematical accomplishments have given him a prominent place in the early development of calculus.
See LarsonCalculus.com to read more of this biography.

The Tautochrone and Brachistochrone Problems

The curve described in Example 5 is related to one of the most famous pairs of problems in the history of calculus. The first problem (called the **tautochrone problem**) began with Galileo's discovery that the time required to complete a full swing of a pendulum is *approximately* the same whether it makes a large movement at high speed or a small movement at lower speed (see Figure 10.26). Late in his life, Galileo realized that he could use this principle to construct a clock. However, he was not able to conquer the mechanics of actual construction. Christian Huygens (1629–1695) was the first to design and construct a working model. In his work with pendulums, Huygens realized that a pendulum does not take exactly the same time to complete swings of varying lengths. (This doesn't affect a pendulum clock, because the length of the circular arc is kept constant by giving the pendulum a slight boost each time it passes its lowest point.) But, in studying the problem, Huygens discovered that a ball rolling back and forth on an inverted cycloid does complete each cycle in exactly the same time.

The second problem, which was posed by John Bernoulli in 1696, is called the **brachistochrone problem**—in Greek, *brachys* means short and *chronos* means time. The problem was to determine the path down which a particle (such as a ball) will slide from point A to point B in the *shortest time*. Several mathematicians took up the challenge, and the following year the problem was solved by Newton, Leibniz, L'Hôpital, John Bernoulli, and James Bernoulli. As it turns out, the solution is not a straight line from A to B, but an inverted cycloid passing through the points A and B, as shown in Figure 10.27.

An inverted cycloid is the path down which a ball will roll in the shortest time.
Figure 10.27

The amazing part of the solution to the brachistochrone problem is that a particle starting at rest at *any* point C of the cycloid between A and B will take exactly the same time to reach B, as shown in Figure 10.28.

A ball starting at point C takes the same time to reach point B as one that starts at point A.
Figure 10.28

■ **FOR FURTHER INFORMATION** To see a proof of the famous brachistochrone problem, see the article "A New Minimization Proof for the Brachistochrone" by Gary Lawlor in *The American Mathematical Monthly*. To view this article, go to *MathArticles.com*.

10.2 Exercises

See **CalcChat.com** for tutorial help and worked-out solutions to odd-numbered exercises.

Using Parametric Equations In Exercises 1–18, sketch the curve represented by the parametric equations (indicate the orientation of the curve), and write the corresponding rectangular equation by eliminating the parameter.

1. $x = 2t - 3$, $y = 3t + 1$
2. $x = 5 - 4t$, $y = 2 + 5t$

3. $x = t + 1$, $y = t^2$
4. $x = 2t^2$, $y = t^4 + 1$

5. $x = t^3$, $y = \dfrac{t^2}{2}$
6. $x = t^2 + t$, $y = t^2 - t$

7. $x = \sqrt{t}$, $y = t - 5$
8. $x = \sqrt[4]{t}$, $y = 8 - t$

9. $x = t - 3$, $y = \dfrac{t}{t - 3}$

10. $x = 1 + \dfrac{1}{t}$, $y = t - 1$

11. $x = 2t$, $y = |t - 2|$

12. $x = |t - 1|$, $y = t + 2$

13. $x = e^t$, $y = e^{3t} + 1$

14. $x = e^{-t}$, $y = e^{2t} - 1$

15. $x = \sec \theta$, $y = \cos \theta$, $0 \le \theta < \pi/2$, $\pi/2 < \theta \le \pi$

16. $x = \tan^2 \theta$, $y = \sec^2 \theta$

17. $x = 8 \cos \theta$, $y = 8 \sin \theta$

18. $x = 3 \cos \theta$, $y = 7 \sin \theta$

Using Parametric Equations In Exercises 19–30, use a graphing utility to graph the curve represented by the parametric equations (indicate the orientation of the curve). Eliminate the parameter and write the corresponding rectangular equation.

19. $x = 6 \sin 2\theta$
    $y = 4 \cos 2\theta$

20. $x = \cos \theta$
    $y = 2 \sin 2\theta$

21. $x = 4 + 2 \cos \theta$
    $y = -1 + \sin \theta$

22. $x = -2 + 3 \cos \theta$
    $y = -5 + 3 \sin \theta$

23. $x = -3 + 4 \cos \theta$
    $y = 2 + 5 \sin \theta$

24. $x = \sec \theta$
    $y = \tan \theta$

25. $x = 4 \sec \theta$
    $y = 3 \tan \theta$

26. $x = \cos^3 \theta$
    $y = \sin^3 \theta$

27. $x = t^3$, $y = 3 \ln t$
28. $x = \ln 2t$, $y = t^2$

29. $x = e^{-t}$, $y = e^{3t}$
30. $x = e^{2t}$, $y = e^t$

Comparing Plane Curves In Exercises 31–34, determine any differences between the curves of the parametric equations. Are the graphs the same? Are the orientations the same? Are the curves smooth? Explain.

31. (a) $x = t$
       $y = 2t + 1$
    (b) $x = \cos \theta$
       $y = 2 \cos \theta + 1$
    (c) $x = e^{-t}$
       $y = 2e^{-t} + 1$
    (d) $x = e^t$
       $y = 2e^t + 1$

32. (a) $x = 2 \cos \theta$
       $y = 2 \sin \theta$
    (b) $x = \sqrt{4t^2 - 1}/|t|$
       $y = 1/t$
    (c) $x = \sqrt{t}$
       $y = \sqrt{4 - t}$
    (d) $x = -\sqrt{4 - e^{2t}}$
       $y = e^t$

33. (a) $x = \cos \theta$
       $y = 2 \sin^2 \theta$
       $0 < \theta < \pi$
    (b) $x = \cos(-\theta)$
       $y = 2 \sin^2(-\theta)$
       $0 < \theta < \pi$

34. (a) $x = t + 1$, $y = t^3$
    (b) $x = -t + 1$, $y = (-t)^3$

35. **Conjecture**
    (a) Use a graphing utility to graph the curves represented by the two sets of parametric equations.
    $$x = 4 \cos t \qquad x = 4 \cos(-t)$$
    $$y = 3 \sin t \qquad y = 3 \sin(-t)$$
    (b) Describe the change in the graph when the sign of the parameter is changed.
    (c) Make a conjecture about the change in the graph of parametric equations when the sign of the parameter is changed.
    (d) Test your conjecture with another set of parametric equations.

36. **Writing** Review Exercises 31–34 and write a short paragraph describing how the graphs of curves represented by different sets of parametric equations can differ even though eliminating the parameter from each yields the same rectangular equation.

Eliminating a Parameter In Exercises 37–40, eliminate the parameter and obtain the standard form of the rectangular equation.

37. Line through (x_1, y_1) and (x_2, y_2):
    $$x = x_1 + t(x_2 - x_1), \quad y = y_1 + t(y_2 - y_1)$$

38. Circle: $x = h + r \cos \theta$, $y = k + r \sin \theta$

39. Ellipse: $x = h + a \cos \theta$, $y = k + b \sin \theta$

40. Hyperbola: $x = h + a \sec \theta$, $y = k + b \tan \theta$

Writing a Set of Parametric Equations In Exercises 41–48, use the results of Exercises 37–40 to find a set of parametric equations for the line or conic.

41. Line: passes through $(0, 0)$ and $(4, -7)$

42. Line: passes through $(1, 4)$ and $(5, -2)$

43. Circle: center: $(3, 1)$; radius: 2

44. Circle: center: $(-6, 2)$; radius: 4

45. Ellipse: vertices: $(\pm 10, 0)$; foci: $(\pm 8, 0)$

46. Ellipse: vertices: $(4, 7)$, $(4, -3)$; foci: $(4, 5)$, $(4, -1)$

47. Hyperbola: vertices: $(\pm 4, 0)$; foci: $(\pm 5, 0)$

48. Hyperbola: vertices: $(0, \pm 1)$; foci: $(0, \pm 2)$

Finding Parametric Equations In Exercises 49–52, find two different sets of parametric equations for the rectangular equation.

49. $y = 6x - 5$

50. $y = 4/(x - 1)$

51. $y = x^3$

52. $y = x^2$

Finding Parametric Equations In Exercises 53–56, find a set of parametric equations for the rectangular equation that satisfies the given condition.

53. $y = 2x - 5$, $t = 0$ at the point $(3, 1)$

54. $y = 4x + 1$, $t = -1$ at the point $(-2, -7)$

55. $y = x^2$, $t = 4$ at the point $(4, 16)$

56. $y = 4 - x^2$, $t = 1$ at the point $(1, 3)$

Graphing a Plane Curve In Exercises 57–64, use a graphing utility to graph the curve represented by the parametric equations. Indicate the direction of the curve. Identify any points at which the curve is not smooth.

57. Cycloid: $x = 2(\theta - \sin \theta)$, $y = 2(1 - \cos \theta)$

58. Cycloid: $x = \theta + \sin \theta$, $y = 1 - \cos \theta$

59. Prolate cycloid: $x = \theta - \frac{3}{2}\sin \theta$, $y = 1 - \frac{3}{2}\cos \theta$

60. Prolate cycloid: $x = 2\theta - 4\sin \theta$, $y = 2 - 4\cos \theta$

61. Hypocycloid: $x = 3\cos^3 \theta$, $y = 3\sin^3 \theta$

62. Curtate cycloid: $x = 2\theta - \sin \theta$, $y = 2 - \cos \theta$

63. Witch of Agnesi: $x = 2\cot \theta$, $y = 2\sin^2 \theta$

64. Folium of Descartes: $x = 3t/(1 + t^3)$, $y = 3t^2/(1 + t^3)$

> ### WRITING ABOUT CONCEPTS
>
> **65. Plane Curve** State the definition of a plane curve given by parametric equations.
>
> **66. Plane Curve** Explain the process of sketching a plane curve given by parametric equations. What is meant by the orientation of the curve?
>
> **67. Smooth Curve** State the definition of a smooth curve.

68. **HOW DO YOU SEE IT?** Which set of parametric equations is shown in the graph below? Explain your reasoning.

(a) $x = t$ (b) $x = t^2$
   $y = t^2$ $y = t$

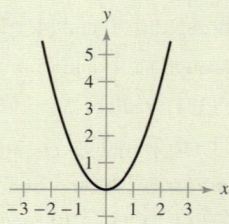

Matching In Exercises 69–72, match each set of parametric equations with the correct graph. [The graphs are labeled (a), (b), (c), and (d).] Explain your reasoning.

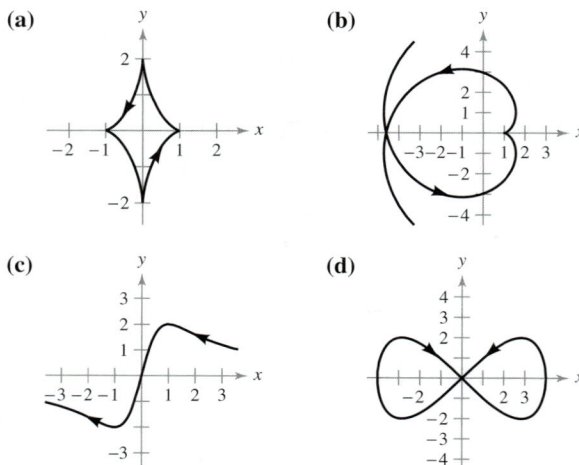

(a) (b) (c) (d)

69. Lissajous curve: $x = 4\cos \theta$, $y = 2\sin 2\theta$

70. Evolute of ellipse: $x = \cos^3 \theta$, $y = 2\sin^3 \theta$

71. Involute of circle: $x = \cos \theta + \theta \sin \theta$, $y = \sin \theta - \theta \cos \theta$

72. Serpentine curve: $x = \cot \theta$, $y = 4\sin \theta \cos \theta$

73. Curtate Cycloid A wheel of radius a rolls along a line without slipping. The curve traced by a point P that is b units from the center $(b < a)$ is called a **curtate cycloid** (see figure). Use the angle θ to find a set of parametric equations for this curve.

Figure for 73

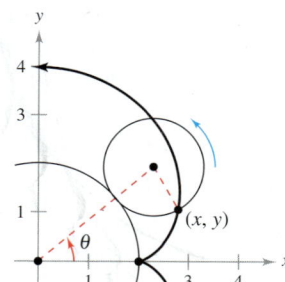

Figure for 74

74. Epicycloid A circle of radius 1 rolls around the outside of a circle of radius 2 without slipping. The curve traced by a point on the circumference of the smaller circle is called an epicycloid (see figure). Use the angle θ to find a set of parametric equations for this curve.

True or False? In Exercises 75–77, determine whether the statement is true or false. If it is false, explain why or give an example that shows it is false.

75. The graph of the parametric equations $x = t^2$ and $y = t^2$ is the line $y = x$.

76. If y is a function of t and x is a function of t, then y is a function of x.

77. The curve represented by the parametric equations $x = t$ and $y = \cos t$ can be written as an equation of the form $y = f(x)$.

78. Translation of a Plane Curve   Consider the parametric equations $x = 8 \cos t$ and $y = 8 \sin t$.

(a) Describe the curve represented by the parametric equations.

(b) How does the curve represented by the parametric equations $x = 8 \cos t + 3$ and $y = 8 \sin t + 6$ compare to the curve described in part (a)?

(c) How does the original curve change when cosine and sine are interchanged?

Projectile Motion   In Exercises 79 and 80, consider a projectile launched at a height h feet above the ground and at an angle θ with the horizontal. When the initial velocity is v_0 feet per second, the path of the projectile is modeled by the parametric equations $x = (v_0 \cos \theta)t$ and $y = h + (v_0 \sin \theta)t - 16t^2$.

79. The center field fence in a ballpark is 10 feet high and 400 feet from home plate. The ball is hit 3 feet above the ground. It leaves the bat at an angle of θ degrees with the horizontal at a speed of 100 miles per hour (see figure).

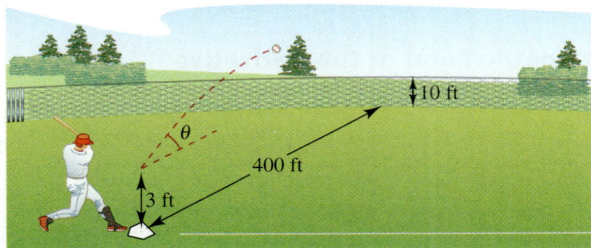

(a) Write a set of parametric equations for the path of the ball.

(b) Use a graphing utility to graph the path of the ball when $\theta = 15°$. Is the hit a home run?

(c) Use a graphing utility to graph the path of the ball when $\theta = 23°$. Is the hit a home run?

(d) Find the minimum angle at which the ball must leave the bat in order for the hit to be a home run.

80. A rectangular equation for the path of a projectile is $y = 5 + x - 0.005x^2$.

(a) Eliminate the parameter t from the position function for the motion of a projectile to show that the rectangular equation is

$$y = -\frac{16 \sec^2 \theta}{v_0^2}x^2 + (\tan \theta)x + h.$$

(b) Use the result of part (a) to find h, v_0, and θ. Find the parametric equations of the path.

(c) Use a graphing utility to graph the rectangular equation for the path of the projectile. Confirm your answer in part (b) by sketching the curve represented by the parametric equations.

(d) Use a graphing utility to approximate the maximum height of the projectile and its range.

SECTION PROJECT

Cycloids

In Greek, the word *cycloid* means *wheel*, the word *hypocycloid* means *under the wheel*, and the word *epicycloid* means *upon the wheel*. Match the hypocycloid or epicycloid with its graph. [The graphs are labeled (a), (b), (c), (d), (e), and (f).]

Hypocycloid, $H(A, B)$

The path traced by a fixed point on a circle of radius B as it rolls around the *inside* of a circle of radius A

$$x = (A - B) \cos t + B \cos\left(\frac{A - B}{B}\right)t$$

$$y = (A - B) \sin t - B \sin\left(\frac{A - B}{B}\right)t$$

Epicycloid, $E(A, B)$

The path traced by a fixed point on a circle of radius B as it rolls around the *outside* of a circle of radius A

$$x = (A + B) \cos t - B \cos\left(\frac{A + B}{B}\right)t$$

$$y = (A + B) \sin t - B \sin\left(\frac{A + B}{B}\right)t$$

I. $H(8, 3)$	II. $E(8, 3)$	III. $H(8, 7)$
IV. $E(24, 3)$	V. $H(24, 7)$	VI. $E(24, 7)$

(a)   (b)

(c)   (d)

(e)   (f)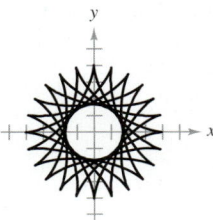

Exercises based on "Mathematical Discovery via Computer Graphics: Hypocycloids and Epicycloids" by Florence S. Gordon and Sheldon P. Gordon, *College Mathematics Journal*, November 1984, p. 441. Used by permission of the authors.

10.3 Parametric Equations and Calculus

■ Find the slope of a tangent line to a curve given by a set of parametric equations.
■ Find the arc length of a curve given by a set of parametric equations.
■ Find the area of a surface of revolution (parametric form).

Slope and Tangent Lines

Now that you can represent a graph in the plane by a set of parametric equations, it is natural to ask how to use calculus to study plane curves. Consider the projectile represented by the parametric equations

$$x = 24\sqrt{2}\,t \quad \text{and} \quad y = -16t^2 + 24\sqrt{2}\,t$$

as shown in Figure 10.29. From the discussion at the beginning of Section 10.2, you know that these equations enable you to locate the position of the projectile at a given time. You also know that the object is initially projected at an angle of 45°, or a slope of $m = \tan 45° = 1$. But how can you find the slope at some other time t? The next theorem answers this question by giving a formula for the slope of the tangent line as a function of t.

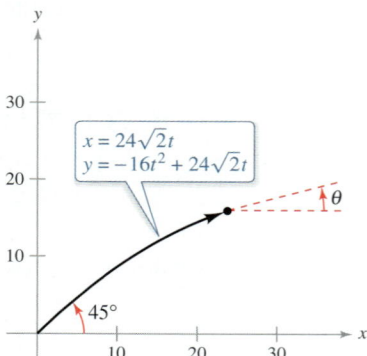

$x = 24\sqrt{2}\,t$
$y = -16t^2 + 24\sqrt{2}\,t$

At time t, the angle of elevation of the projectile is θ.
Figure 10.29

THEOREM 10.7   Parametric Form of the Derivative

If a smooth curve C is given by the equations

$$x = f(t) \quad \text{and} \quad y = g(t)$$

then the slope of C at (x, y) is

$$\frac{dy}{dx} = \frac{dy/dt}{dx/dt}, \quad \frac{dx}{dt} \neq 0.$$

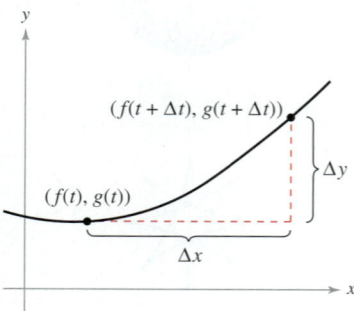

Proof   In Figure 10.30, consider $\Delta t > 0$ and let

$$\Delta y = g(t + \Delta t) - g(t) \quad \text{and} \quad \Delta x = f(t + \Delta t) - f(t).$$

Because $\Delta x \to 0$ as $\Delta t \to 0$, you can write

$$\frac{dy}{dx} = \lim_{\Delta x \to 0} \frac{\Delta y}{\Delta x}$$

$$= \lim_{\Delta t \to 0} \frac{g(t + \Delta t) - g(t)}{f(t + \Delta t) - f(t)}.$$

Dividing both the numerator and denominator by Δt, you can use the differentiability of f and g to conclude that

$$\frac{dy}{dx} = \lim_{\Delta t \to 0} \frac{[g(t + \Delta t) - g(t)]/\Delta t}{[f(t + \Delta t) - f(t)]/\Delta t}$$

$$= \frac{\displaystyle\lim_{\Delta t \to 0} \frac{g(t + \Delta t) - g(t)}{\Delta t}}{\displaystyle\lim_{\Delta t \to 0} \frac{f(t + \Delta t) - f(t)}{\Delta t}}$$

$$= \frac{g'(t)}{f'(t)}$$

$$= \frac{dy/dt}{dx/dt}.$$

$(f(t + \Delta t), g(t + \Delta t))$
$(f(t), g(t))$
Δy
Δx

The slope of the secant line through the points $(f(t), g(t))$ and $(f(t + \Delta t),$ $g(t + \Delta t))$ is $\Delta y / \Delta x$.
Figure 10.30

See LarsonCalculus.com for Bruce Edwards's video of this proof.

EXAMPLE 1 Differentiation and Parametric Form

Find dy/dx for the curve given by $x = \sin t$ and $y = \cos t$.

Solution

$$\frac{dy}{dx} = \frac{dy/dt}{dx/dt}$$

$$= \frac{-\sin t}{\cos t}$$

$$= -\tan t$$

Because dy/dx is a function of t, you can use Theorem 10.7 repeatedly to find *higher-order* derivatives. For instance,

$$\frac{d^2y}{dx^2} = \frac{d}{dx}\left[\frac{dy}{dx}\right] = \frac{\dfrac{d}{dt}\left[\dfrac{dy}{dx}\right]}{dx/dt} \qquad \text{Second derivative}$$

$$\frac{d^3y}{dx^3} = \frac{d}{dx}\left[\frac{d^2y}{dx^2}\right] = \frac{\dfrac{d}{dt}\left[\dfrac{d^2y}{dx^2}\right]}{dx/dt}. \qquad \text{Third derivative}$$

EXAMPLE 2 Finding Slope and Concavity

For the curve given by

$$x = \sqrt{t} \quad \text{and} \quad y = \frac{1}{4}(t^2 - 4), \quad t \geq 0$$

find the slope and concavity at the point $(2, 3)$.

Solution   Because

$$\frac{dy}{dx} = \frac{dy/dt}{dx/dt} = \frac{(1/2)t}{(1/2)t^{-1/2}} = t^{3/2} \qquad \text{Parametric form of first derivative}$$

you can find the second derivative to be

$$\frac{d^2y}{dx^2} = \frac{\dfrac{d}{dt}[dy/dx]}{dx/dt} = \frac{\dfrac{d}{dt}[t^{3/2}]}{dx/dt} = \frac{(3/2)t^{1/2}}{(1/2)t^{-1/2}} = 3t. \qquad \begin{array}{l}\text{Parametric form of second}\\\text{derivative}\end{array}$$

At $(x, y) = (2, 3)$, it follows that $t = 4$, and the slope is

$$\frac{dy}{dx} = (4)^{3/2} = 8.$$

Moreover, when $t = 4$, the second derivative is

$$\frac{d^2y}{dx^2} = 3(4) = 12 > 0$$

and you can conclude that the graph is concave upward at $(2, 3)$, as shown in Figure 10.31.

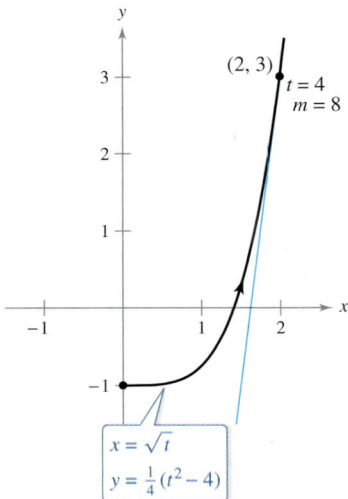

The graph is concave upward at $(2, 3)$ when $t = 4$.
Figure 10.31

Because the parametric equations $x = f(t)$ and $y = g(t)$ need not define y as a function of x, it is possible for a plane curve to loop around and cross itself. At such points, the curve may have more than one tangent line, as shown in the next example.

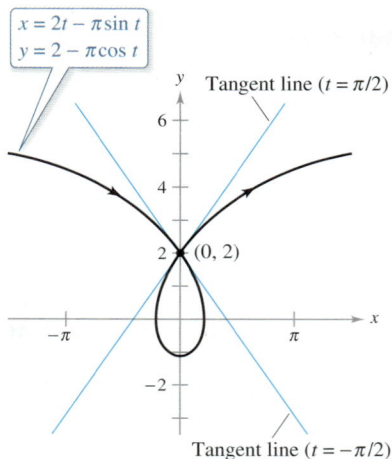

$x = 2t - \pi \sin t$
$y = 2 - \pi \cos t$

This prolate cycloid has two tangent lines at the point $(0, 2)$.
Figure 10.32

EXAMPLE 3   **A Curve with Two Tangent Lines at a Point**

•••▷ *See LarsonCalculus.com for an interactive version of this type of example.*

The **prolate cycloid** given by

$$x = 2t - \pi \sin t \quad \text{and} \quad y = 2 - \pi \cos t$$

crosses itself at the point $(0, 2)$, as shown in Figure 10.32. Find the equations of both tangent lines at this point.

Solution   Because $x = 0$ and $y = 2$ when $t = \pm \pi/2$, and

$$\frac{dy}{dx} = \frac{dy/dt}{dx/dt} = \frac{\pi \sin t}{2 - \pi \cos t}$$

you have $dy/dx = -\pi/2$ when $t = -\pi/2$ and $dy/dx = \pi/2$ when $t = \pi/2$. So, the two tangent lines at $(0, 2)$ are

$$y - 2 = -\left(\frac{\pi}{2}\right)x \qquad \text{Tangent line when } t = -\frac{\pi}{2}$$

and

$$y - 2 = \left(\frac{\pi}{2}\right)x. \qquad \text{Tangent line when } t = \frac{\pi}{2}$$

∎

If $dy/dt = 0$ and $dx/dt \neq 0$ when $t = t_0$, then the curve represented by $x = f(t)$ and $y = g(t)$ has a horizontal tangent at $(f(t_0), g(t_0))$. For instance, in Example 3, the given curve has a horizontal tangent at the point $(0, 2 - \pi)$ (when $t = 0$). Similarly, if $dx/dt = 0$ and $dy/dt \neq 0$ when $t = t_0$, then the curve represented by $x = f(t)$ and $y = g(t)$ has a vertical tangent at $(f(t_0), g(t_0))$.

Arc Length

You have seen how parametric equations can be used to describe the path of a particle moving in the plane. You will now develop a formula for determining the *distance* traveled by the particle along its path.

Recall from Section 7.4 that the formula for the arc length of a curve C given by $y = h(x)$ over the interval $[x_0, x_1]$ is

$$s = \int_{x_0}^{x_1} \sqrt{1 + [h'(x)]^2} \, dx$$

$$= \int_{x_0}^{x_1} \sqrt{1 + \left(\frac{dy}{dx}\right)^2} \, dx.$$

If C is represented by the parametric equations $x = f(t)$ and $y = g(t)$, $a \leq t \leq b$, and if $dx/dt = f'(t) > 0$, then

$$s = \int_{x_0}^{x_1} \sqrt{1 + \left(\frac{dy}{dx}\right)^2} \, dx$$

$$= \int_{x_0}^{x_1} \sqrt{1 + \left(\frac{dy/dt}{dx/dt}\right)^2} \, dx$$

$$= \int_{a}^{b} \sqrt{\frac{(dx/dt)^2 + (dy/dt)^2}{(dx/dt)^2}} \frac{dx}{dt} \, dt$$

$$= \int_{a}^{b} \sqrt{\left(\frac{dx}{dt}\right)^2 + \left(\frac{dy}{dt}\right)^2} \, dt$$

$$= \int_{a}^{b} \sqrt{[f'(t)]^2 + [g'(t)]^2} \, dt.$$

> **THEOREM 10.8 Arc Length in Parametric Form**
>
> If a smooth curve C is given by $x = f(t)$ and $y = g(t)$ such that C does not intersect itself on the interval $a \le t \le b$ (except possibly at the endpoints), then the arc length of C over the interval is given by
>
> $$s = \int_a^b \sqrt{\left(\frac{dx}{dt}\right)^2 + \left(\frac{dy}{dt}\right)^2}\, dt = \int_a^b \sqrt{[f'(t)]^2 + [g'(t)]^2}\, dt.$$

REMARK When applying the arc length formula to a curve, be sure that the curve is traced out only once on the interval of integration. For instance, the circle given by $x = \cos t$ and $y = \sin t$ is traced out once on the interval $0 \le t \le 2\pi$, but is traced out twice on the interval $0 \le t \le 4\pi$.

In the preceding section, you saw that if a circle rolls along a line, then a point on its circumference will trace a path called a cycloid. If the circle rolls around the circumference of another circle, then the path of the point is an **epicycloid**. The next example shows how to find the arc length of an epicycloid.

ARCH OF A CYCLOID

The arc length of an arch of a cycloid was first calculated in 1658 by British architect and mathematician Christopher Wren, famous for rebuilding many buildings and churches in London, including St. Paul's Cathedral.

EXAMPLE 4 Finding Arc Length

A circle of radius 1 rolls around the circumference of a larger circle of radius 4, as shown in Figure 10.33. The epicycloid traced by a point on the circumference of the smaller circle is given by

$$x = 5\cos t - \cos 5t \quad \text{and} \quad y = 5\sin t - \sin 5t.$$

Find the distance traveled by the point in one complete trip about the larger circle.

Solution Before applying Theorem 10.8, note in Figure 10.33 that the curve has sharp points when $t = 0$ and $t = \pi/2$. Between these two points, dx/dt and dy/dt are not simultaneously 0. So, the portion of the curve generated from $t = 0$ to $t = \pi/2$ is smooth. To find the total distance traveled by the point, you can find the arc length of that portion lying in the first quadrant and multiply by 4.

$$s = 4\int_0^{\pi/2} \sqrt{\left(\frac{dx}{dt}\right)^2 + \left(\frac{dy}{dt}\right)^2}\, dt \qquad \text{\color{red}Parametric form for arc length}$$

$$= 4\int_0^{\pi/2} \sqrt{(-5\sin t + 5\sin 5t)^2 + (5\cos t - 5\cos 5t)^2}\, dt$$

$$= 20\int_0^{\pi/2} \sqrt{2 - 2\sin t \sin 5t - 2\cos t \cos 5t}\, dt$$

$$= 20\int_0^{\pi/2} \sqrt{2 - 2\cos 4t}\, dt \qquad \text{\color{red}Difference formula for cosine}$$

$$= 20\int_0^{\pi/2} \sqrt{4\sin^2 2t}\, dt \qquad \text{\color{red}Double-angle formula}$$

$$= 40\int_0^{\pi/2} \sin 2t\, dt$$

$$= -20\Big[\cos 2t\Big]_0^{\pi/2}$$

$$= 40$$

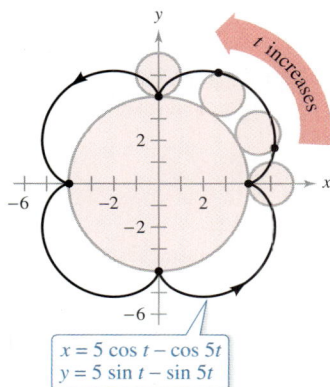

An epicycloid is traced by a point on the smaller circle as it rolls around the larger circle.

$x = 5\cos t - \cos 5t$
$y = 5\sin t - \sin 5t$

Figure 10.33

For the epicycloid shown in Figure 10.33, an arc length of 40 seems about right because the circumference of a circle of radius 6 is

$$2\pi r = 12\pi \approx 37.7.$$

Area of a Surface of Revolution

You can use the formula for the area of a surface of revolution in rectangular form to develop a formula for surface area in parametric form.

THEOREM 10.9 **Area of a Surface of Revolution**

If a smooth curve C given by $x = f(t)$ and $y = g(t)$ does not cross itself on an interval $a \le t \le b$, then the area S of the surface of revolution formed by revolving C about the coordinate axes is given by the following.

1. $S = 2\pi \int_a^b g(t) \sqrt{\left(\dfrac{dx}{dt}\right)^2 + \left(\dfrac{dy}{dt}\right)^2} \, dt$ Revolution about the x-axis: $g(t) \ge 0$

2. $S = 2\pi \int_a^b f(t) \sqrt{\left(\dfrac{dx}{dt}\right)^2 + \left(\dfrac{dy}{dt}\right)^2} \, dt$ Revolution about the y-axis: $f(t) \ge 0$

These formulas may be easier to remember if you think of the differential of arc length as

$$ds = \sqrt{\left(\frac{dx}{dt}\right)^2 + \left(\frac{dy}{dt}\right)^2} \, dt.$$

Then the formulas are written as follows.

1. $S = 2\pi \displaystyle\int_a^b g(t) \, ds$ 2. $S = 2\pi \displaystyle\int_a^b f(t) \, ds$

EXAMPLE 5   **Finding the Area of a Surface of Revolution**

Let C be the arc of the circle $x^2 + y^2 = 9$ from $(3, 0)$ to

$$\left(\frac{3}{2}, \frac{3\sqrt{3}}{2}\right)$$

as shown in Figure 10.34. Find the area of the surface formed by revolving C about the x-axis.

Solution   You can represent C parametrically by the equations

$$x = 3 \cos t \quad \text{and} \quad y = 3 \sin t, \quad 0 \le t \le \pi/3.$$

(Note that you can determine the interval for t by observing that $t = 0$ when $x = 3$ and $t = \pi/3$ when $x = 3/2$.) On this interval, C is smooth and y is nonnegative, and you can apply Theorem 10.9 to obtain a surface area of

$$S = 2\pi \int_0^{\pi/3} (3 \sin t)\sqrt{(-3 \sin t)^2 + (3 \cos t)^2} \, dt \qquad \text{Formula for area of a surface of revolution}$$

$$= 6\pi \int_0^{\pi/3} \sin t \sqrt{9(\sin^2 t + \cos^2 t)} \, dt$$

$$= 6\pi \int_0^{\pi/3} 3 \sin t \, dt \qquad \text{Trigonometric identity}$$

$$= -18\pi \left[\cos t\right]_0^{\pi/3}$$

$$= -18\pi \left(\frac{1}{2} - 1\right)$$

$$= 9\pi.$$

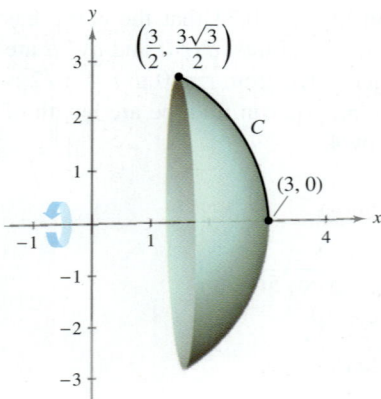

The surface of revolution has a surface area of 9π.

Figure 10.34

10.3 Exercises

See **CalcChat.com** for tutorial help and worked-out solutions to odd-numbered exercises.

Finding a Derivative In Exercises 1–4, find dy/dx.

1. $x = t^2, \ y = 7 - 6t$

2. $x = \sqrt[3]{t}, \ y = 4 - t$

3. $x = \sin^2 \theta, \ y = \cos^2 \theta$

4. $x = 2e^\theta, \ y = e^{-\theta/2}$

Finding Slope and Concavity In Exercises 5–14, find dy/dx and d^2y/dx^2, and find the slope and concavity (if possible) at the given value of the parameter.

Parametric Equations	Parameter
5. $x = 4t, \ y = 3t - 2$	$t = 3$
6. $x = \sqrt{t}, \ y = 3t - 1$	$t = 1$
7. $x = t + 1, \ y = t^2 + 3t$	$t = -1$
8. $x = t^2 + 5t + 4, \ y = 4t$	$t = 0$
9. $x = 4 \cos \theta, \ y = 4 \sin \theta$	$\theta = \dfrac{\pi}{4}$
10. $x = \cos \theta, \ y = 3 \sin \theta$	$\theta = 0$
11. $x = 2 + \sec \theta, \ y = 1 + 2 \tan \theta$	$\theta = \dfrac{\pi}{6}$
12. $x = \sqrt{t}, \ y = \sqrt{t - 1}$	$t = 2$
13. $x = \cos^3 \theta, \ y = \sin^3 \theta$	$\theta = \dfrac{\pi}{4}$
14. $x = \theta - \sin \theta, \ y = 1 - \cos \theta$	$\theta = \pi$

Finding Equations of Tangent Lines In Exercises 15–18, find an equation of the tangent line at each given point on the curve.

15. $x = 2 \cot \theta, \ y = 2 \sin^2 \theta,$

$$\left(-\frac{2}{\sqrt{3}}, \frac{3}{2}\right), (0, 2), \left(2\sqrt{3}, \frac{1}{2}\right)$$

16. $x = 2 - 3 \cos \theta, \ y = 3 + 2 \sin \theta,$

$$(-1, 3), (2, 5), \left(\frac{4 + 3\sqrt{3}}{2}, 2\right)$$

17. $x = t^2 - 4, \ y = t^2 - 2t, \ (0, 0), (-3, -1), (-3, 3)$

18. $x = t^4 + 2, \ y = t^3 + t, \ (2, 0), (3, -2), (18, 10)$

Finding an Equation of a Tangent Line In Exercises 19–22, (a) use a graphing utility to graph the curve represented by the parametric equations, (b) use a graphing utility to find dx/dt, dy/dt, and dy/dx at the given value of the parameter, (c) find an equation of the tangent line to the curve at the given value of the parameter, and (d) use a graphing utility to graph the curve and the tangent line from part (c).

Parametric Equations	Parameter
19. $x = 6t, \ y = t^2 + 4$	$t = 1$
20. $x = t - 2, \ y = \dfrac{1}{t} + 3$	$t = 1$
21. $x = t^2 - t + 2, \ y = t^3 - 3t$	$t = -1$
22. $x = 3t - t^2, \ y = 2t^{3/2}$	$t = \dfrac{1}{4}$

Finding Equations of Tangent Lines In Exercises 23–26, find the equations of the tangent lines at the point where the curve crosses itself.

23. $x = 2 \sin 2t, \ y = 3 \sin t$

24. $x = 2 - \pi \cos t, \ y = 2t - \pi \sin t$

25. $x = t^2 - t, \ y = t^3 - 3t - 1$

26. $x = t^3 - 6t, \ y = t^2$

Horizontal and Vertical Tangency In Exercises 27 and 28, find all points (if any) of horizontal and vertical tangency to the portion of the curve shown.

27. Involute of a circle:

$x = \cos \theta + \theta \sin \theta$

$y = \sin \theta - \theta \cos \theta$

28. $x = 2\theta$

$y = 2(1 - \cos \theta)$

Horizontal and Vertical Tangency In Exercises 29–38, find all points (if any) of horizontal and vertical tangency to the curve. Use a graphing utility to confirm your results.

29. $x = 4 - t, \ y = t^2$

30. $x = t + 1, \ y = t^2 + 3t$

31. $x = t + 4, \ y = t^3 - 3t$

32. $x = t^2 - t + 2, \ y = t^3 - 3t$

33. $x = 3 \cos \theta, \ y = 3 \sin \theta$

34. $x = \cos \theta, \ y = 2 \sin 2\theta$

35. $x = 5 + 3 \cos \theta, \ y = -2 + \sin \theta$

36. $x = 4 \cos^2 \theta, \ y = 2 \sin \theta$

37. $x = \sec \theta, \ y = \tan \theta$

38. $x = \cos^2 \theta, \ y = \cos \theta$

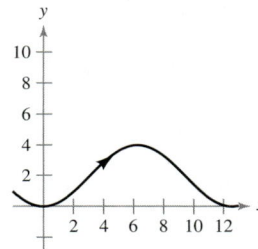

Determining Concavity In Exercises 39–44, determine the open t-intervals on which the curve is concave downward or concave upward.

39. $x = 3t^2, \ y = t^3 - t$

40. $x = 2 + t^2, \ y = t^2 + t^3$

41. $x = 2t + \ln t, \ y = 2t - \ln t$

42. $x = t^2, \ y = \ln t$

43. $x = \sin t, \ y = \cos t, \ 0 < t < \pi$

44. $x = 4 \cos t, \ y = 2 \sin t, \ 0 < t < 2\pi$

Arc Length In Exercises 45–50, find the arc length of the curve on the given interval.

Parametric Equations	Interval
45. $x = 3t + 5$, $y = 7 - 2t$	$-1 \le t \le 3$
46. $x = 6t^2$, $y = 2t^3$	$1 \le t \le 4$
47. $x = e^{-t} \cos t$, $y = e^{-t} \sin t$	$0 \le t \le \dfrac{\pi}{2}$
48. $x = \arcsin t$, $y = \ln\sqrt{1 - t^2}$	$0 \le t \le \tfrac{1}{2}$
49. $x = \sqrt{t}$, $y = 3t - 1$	$0 \le t \le 1$
50. $x = t$, $y = \dfrac{t^5}{10} + \dfrac{1}{6t^3}$	$1 \le t \le 2$

Arc Length In Exercises 51–54, find the arc length of the curve on the interval $[0, 2\pi]$.

51. Hypocycloid perimeter: $x = a \cos^3 \theta$, $y = a \sin^3 \theta$

52. Circle circumference: $x = a \cos \theta$, $y = a \sin \theta$

53. Cycloid arch: $x = a(\theta - \sin \theta)$, $y = a(1 - \cos \theta)$

54. Involute of a circle: $x = \cos \theta + \theta \sin \theta$, $y = \sin \theta - \theta \cos \theta$

55. Path of a Projectile The path of a projectile is modeled by the parametric equations

$$x = (90 \cos 30°)t \quad \text{and} \quad y = (90 \sin 30°)t - 16t^2$$

where x and y are measured in feet.

(a) Use a graphing utility to graph the path of the projectile.

(b) Use a graphing utility to approximate the range of the projectile.

(c) Use the integration capabilities of a graphing utility to approximate the arc length of the path. Compare this result with the range of the projectile.

56. Path of a Projectile When the projectile in Exercise 55 is launched at an angle θ with the horizontal, its parametric equations are

$$x = (90 \cos \theta)t \quad \text{and} \quad y = (90 \sin \theta)t - 16t^2.$$

Use a graphing utility to find the angle that maximizes the range of the projectile. What angle maximizes the arc length of the trajectory?

57. Folium of Descartes Consider the parametric equations

$$x = \frac{4t}{1 + t^3} \quad \text{and} \quad y = \frac{4t^2}{1 + t^3}.$$

(a) Use a graphing utility to graph the curve represented by the parametric equations.

(b) Use a graphing utility to find the points of horizontal tangency to the curve.

(c) Use the integration capabilities of a graphing utility to approximate the arc length of the closed loop. (*Hint:* Use symmetry and integrate over the interval $0 \le t \le 1$.)

58. Witch of Agnesi Consider the parametric equations

$$x = 4 \cot \theta \quad \text{and} \quad y = 4 \sin^2 \theta, \quad -\frac{\pi}{2} \le \theta \le \frac{\pi}{2}.$$

(a) Use a graphing utility to graph the curve represented by the parametric equations.

(b) Use a graphing utility to find the points of horizontal tangency to the curve.

(c) Use the integration capabilities of a graphing utility to approximate the arc length over the interval $\pi/4 \le \theta \le \pi/2$.

59. Writing

(a) Use a graphing utility to graph each set of parametric equations.

$$x = t - \sin t, \quad y = 1 - \cos t, \quad 0 \le t \le 2\pi$$
$$x = 2t - \sin(2t), \quad y = 1 - \cos(2t), \quad 0 \le t \le \pi$$

(b) Compare the graphs of the two sets of parametric equations in part (a). When the curve represents the motion of a particle and t is time, what can you infer about the average speeds of the particle on the paths represented by the two sets of parametric equations?

(c) Without graphing the curve, determine the time required for a particle to traverse the same path as in parts (a) and (b) when the path is modeled by

$$x = \tfrac{1}{2}t - \sin\left(\tfrac{1}{2}t\right) \quad \text{and} \quad y = 1 - \cos\left(\tfrac{1}{2}t\right).$$

60. Writing

(a) Each set of parametric equations represents the motion of a particle. Use a graphing utility to graph each set.

First Particle: $x = 3 \cos t$, $y = 4 \sin t$, $0 \le t \le 2\pi$

Second Particle: $x = 4 \sin t$, $y = 3 \cos t$, $0 \le t \le 2\pi$

(b) Determine the number of points of intersection.

(c) Will the particles ever be at the same place at the same time? If so, identify the point(s).

(d) Explain what happens when the motion of the second particle is represented by

$$x = 2 + 3 \sin t, \quad y = 2 - 4 \cos t, \quad 0 \le t \le 2\pi.$$

Surface Area In Exercises 61–64, write an integral that represents the area of the surface generated by revolving the curve about the x-axis. Use a graphing utility to approximate the integral.

Parametric Equations	Interval
61. $x = 3t$, $y = t + 2$	$0 \le t \le 4$
62. $x = \dfrac{1}{4}t^2$, $y = t + 3$	$0 \le t \le 3$
63. $x = \cos^2 \theta$, $y = \cos \theta$	$0 \le \theta \le \dfrac{\pi}{2}$
64. $x = \theta + \sin \theta$, $y = \theta + \cos \theta$	$0 \le \theta \le \dfrac{\pi}{2}$

Surface Area In Exercises 65–70, find the area of the surface generated by revolving the curve about each given axis.

65. $x = 2t$, $y = 3t$, $0 \le t \le 3$

(a) x-axis (b) y-axis

66. $x = t$, $y = 4 - 2t$, $0 \le t \le 2$

(a) x-axis (b) y-axis

67. $x = 5 \cos \theta$, $y = 5 \sin \theta$, $0 \le \theta \le \dfrac{\pi}{2}$, y-axis

68. $x = \frac{1}{3}t^3$, $y = t + 1$, $1 \le t \le 2$, y-axis

69. $x = a \cos^3 \theta$, $y = a \sin^3 \theta$, $0 \le \theta \le \pi$, x-axis

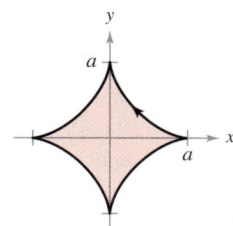

70. $x = a \cos \theta$, $y = b \sin \theta$, $0 \le \theta \le 2\pi$

(a) x-axis (b) y-axis

WRITING ABOUT CONCEPTS

71. Parametric Form of the Derivative Give the parametric form of the derivative.

Mental Math In Exercises 72 and 73, mentally determine dy/dx.

72. $x = t$, $y = 3$

73. $x = t$, $y = 6t - 5$

74. Arc Length Give the integral formula for arc length in parametric form.

75. Surface Area Give the integral formulas for the areas of the surfaces of revolution formed when a smooth curve C is revolved about (a) the x-axis and (b) the y-axis.

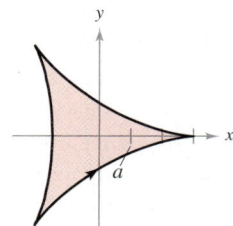

76. HOW DO YOU SEE IT? Using the graph of f, (a) determine whether dy/dt is positive or negative given that dx/dt is negative, and (b) determine whether dx/dt is positive or negative given that dy/dt is positive. Explain your reasoning.

(i)

(ii)

---

77. Integration by Substitution Use integration by substitution to show that if y is a continuous function of x on the interval $a \le x \le b$, where $x = f(t)$ and $y = g(t)$, then

$$\int_a^b y \, dx = \int_{t_1}^{t_2} g(t) f'(t) \, dt$$

where $f(t_1) = a$, $f(t_2) = b$, and both g and f' are continuous on $[t_1, t_2]$.

78. Surface Area A portion of a sphere of radius r is removed by cutting out a circular cone with its vertex at the center of the sphere. The vertex of the cone forms an angle of 2θ. Find the surface area removed from the sphere.

Area In Exercises 79 and 80, find the area of the region. (Use the result of Exercise 77.)

79. $x = 2 \sin^2 \theta$

$y = 2 \sin^2 \theta \tan \theta$

$0 \le \theta < \dfrac{\pi}{2}$

80. $x = 2 \cot \theta$

$y = 2 \sin^2 \theta$

$0 < \theta < \pi$

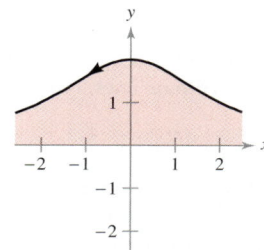

Areas of Simple Closed Curves In Exercises 81–86, use a computer algebra system and the result of Exercise 77 to match the closed curve with its area. (These exercises were based on "The Surveyor's Area Formula" by Bart Braden, *College Mathematics Journal*, September 1986, pp. 335–337, by permission of the author.)

(a) $\frac{8}{3}ab$ (b) $\frac{3}{8}\pi a^2$ (c) $2\pi a^2$

(d) πab (e) $2\pi ab$ (f) $6\pi a^2$

81. Ellipse: $(0 \le t \le 2\pi)$

$x = b \cos t$

$y = a \sin t$

82. Astroid: $(0 \le t \le 2\pi)$

$x = a \cos^3 t$

$y = a \sin^3 t$

83. Cardioid: $(0 \le t \le 2\pi)$

$x = 2a \cos t - a \cos 2t$

$y = 2a \sin t - a \sin 2t$

84. Deltoid: $(0 \le t \le 2\pi)$

$x = 2a \cos t + a \cos 2t$

$y = 2a \sin t - a \sin 2t$

85. Hourglass: $(0 \le t \le 2\pi)$

$x = a \sin 2t$

$y = b \sin t$

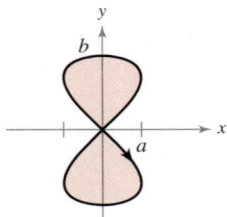

86. Teardrop: $(0 \le t \le 2\pi)$

$x = 2a \cos t - a \sin 2t$

$y = b \sin t$

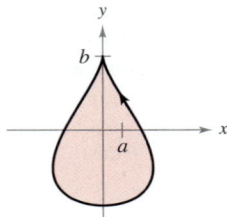

Centroid **In Exercises 87 and 88, find the centroid of the region bounded by the graph of the parametric equations and the coordinate axes. (Use the result of Exercise 77.)**

87. $x = \sqrt{t}, \quad y = 4 - t$

88. $x = \sqrt{4 - t}, \quad y = \sqrt{t}$

Volume **In Exercises 89 and 90, find the volume of the solid formed by revolving the region bounded by the graphs of the given equations about the x-axis. (Use the result of Exercise 77.)**

89. $x = 6 \cos \theta, \quad y = 6 \sin \theta$

90. $x = \cos \theta, \quad y = 3 \sin \theta, \quad a > 0$

91. Cycloid Use the parametric equations

$$x = a(\theta - \sin \theta) \quad \text{and} \quad y = a(1 - \cos \theta), a > 0$$

to answer the following.

(a) Find dy/dx and d^2y/dx^2.

(b) Find the equation of the tangent line at the point where $\theta = \pi/6$.

(c) Find all points (if any) of horizontal tangency.

(d) Determine where the curve is concave upward or concave downward.

(e) Find the length of one arc of the curve.

92. Using Parametric Equations Use the parametric equations

$$x = t^2\sqrt{3} \quad \text{and} \quad y = 3t - \frac{1}{3}t^3$$

to answer the following.

(a) Use a graphing utility to graph the curve on the interval $-3 \le t \le 3$.

(b) Find dy/dx and d^2y/dx^2.

(c) Find the equation of the tangent line at the point $\left(\sqrt{3}, \frac{8}{3}\right)$.

(d) Find the length of the curve.

(e) Find the surface area generated by revolving the curve about the x-axis.

93. Involute of a Circle The involute of a circle is described by the endpoint P of a string that is held taut as it is unwound from a spool that does not turn (see figure). Show that a parametric representation of the involute is

$$x = r(\cos \theta + \theta \sin \theta) \quad \text{and} \quad y = r(\sin \theta - \theta \cos \theta).$$

Figure for 93

Figure for 94

94. Involute of a Circle The figure shows a piece of string tied to a circle with a radius of one unit. The string is just long enough to reach the opposite side of the circle. Find the area that is covered when the string is unwound counterclockwise.

95. Using Parametric Equations

(a) Use a graphing utility to graph the curve given by

$$x = \frac{1 - t^2}{1 + t^2} \quad \text{and} \quad y = \frac{2t}{1 + t^2}$$

where $-20 \le t \le 20$.

(b) Describe the graph and confirm your result analytically.

(c) Discuss the speed at which the curve is traced as t increases from -20 to 20.

96. Tractrix A person moves from the origin along the positive y-axis pulling a weight at the end of a 12-meter rope. Initially, the weight is located at the point $(12, 0)$.

(a) In Exercise 90 of Section 8.7, it was shown that the path of the weight is modeled by the rectangular equation

$$y = -12 \ln\left(\frac{12 - \sqrt{144 - x^2}}{x}\right) - \sqrt{144 - x^2}$$

where $0 < x \le 12$. Use a graphing utility to graph the rectangular equation.

(b) Use a graphing utility to graph the parametric equations

$$x = 12 \operatorname{sech}\frac{t}{12} \quad \text{and} \quad y = t - 12 \tanh\frac{t}{12}$$

where $t \ge 0$. How does this graph compare with the graph in part (a)? Which graph (if either) do you think is a better representation of the path?

(c) Use the parametric equations for the tractrix to verify that the distance from the y-intercept of the tangent line to the point of tangency is independent of the location of the point of tangency.

True or False? **In Exercises 97 and 98, determine whether the statement is true or false. If it is false, explain why or give an example that shows it is false.**

97. If $x = f(t)$ and $y = g(t)$, then $\dfrac{d^2y}{dx^2} = \dfrac{g''(t)}{f''(t)}$.

98. The curve given by $x = t^3$, $y = t^2$ has a horizontal tangent at the origin because $dy/dt = 0$ when $t = 0$.

10.4 Polar Coordinates and Polar Graphs

- Understand the polar coordinate system.
- Rewrite rectangular coordinates and equations in polar form and vice versa.
- Sketch the graph of an equation given in polar form.
- Find the slope of a tangent line to a polar graph.
- Identify several types of special polar graphs.

Polar Coordinates

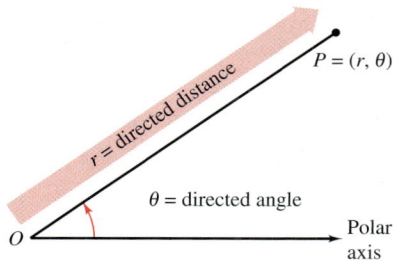

So far, you have been representing graphs as collections of points (x, y) on the rectangular coordinate system. The corresponding equations for these graphs have been in either rectangular or parametric form. In this section, you will study a coordinate system called the **polar coordinate system.**

To form the polar coordinate system in the plane, fix a point O, called the **pole** (or **origin**), and construct from O an initial ray called the **polar axis,** as shown in Figure 10.35. Then each point P in the plane can be assigned **polar coordinates** (r, θ), as follows.

$r =$ *directed distance* from O to P

$\theta =$ *directed angle*, counterclockwise from polar axis to segment $\overline{OP}$

Polar coordinates

Figure 10.35

Figure 10.36 shows three points on the polar coordinate system. Notice that in this system, it is convenient to locate points with respect to a grid of concentric circles intersected by **radial lines** through the pole.

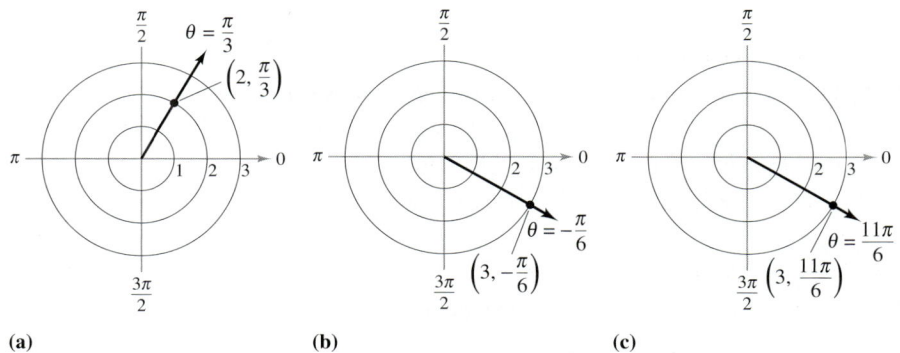

(a)   **(b)**   **(c)**

Figure 10.36

With rectangular coordinates, each point (x, y) has a unique representation. This is not true with polar coordinates. For instance, the coordinates

$$(r, \theta) \quad \text{and} \quad (r, 2\pi + \theta)$$

represent the same point [see parts (b) and (c) in Figure 10.36]. Also, because r is a *directed distance,* the coordinates

$$(r, \theta) \quad \text{and} \quad (-r, \theta + \pi)$$

represent the same point. In general, the point (r, θ) can be written as

$$(r, \theta) = (r, \theta + 2n\pi)$$

or

$$(r, \theta) = (-r, \theta + (2n + 1)\pi)$$

where n is any integer. Moreover, the pole is represented by $(0, \theta)$, where θ is any angle.

POLAR COORDINATES

The mathematician credited with first using polar coordinates was James Bernoulli, who introduced them in 1691. However, there is some evidence that it may have been Isaac Newton who first used them.

Coordinate Conversion

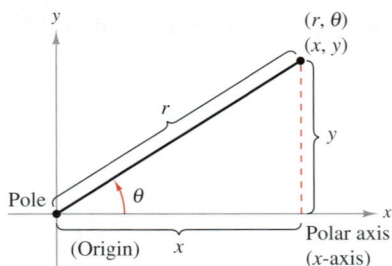

Relating polar and rectangular coordinates
Figure 10.37

To establish the relationship between polar and rectangular coordinates, let the polar axis coincide with the positive x-axis and the pole with the origin, as shown in Figure 10.37. Because (x, y) lies on a circle of radius r, it follows that

$$r^2 = x^2 + y^2.$$

Moreover, for $r > 0$, the definitions of the trigonometric functions imply that

$$\tan \theta = \frac{y}{x}, \quad \cos \theta = \frac{x}{r}, \quad \text{and} \quad \sin \theta = \frac{y}{r}.$$

You can show that the same relationships hold for $r < 0$.

THEOREM 10.10 **Coordinate Conversion**

The polar coordinates (r, θ) of a point are related to the rectangular coordinates (x, y) of the point as follows.

Polar-to-Rectangular	**Rectangular-to-Polar**
$x = r \cos \theta$	$\tan \theta = \dfrac{y}{x}$
$y = r \sin \theta$	$r^2 = x^2 + y^2$

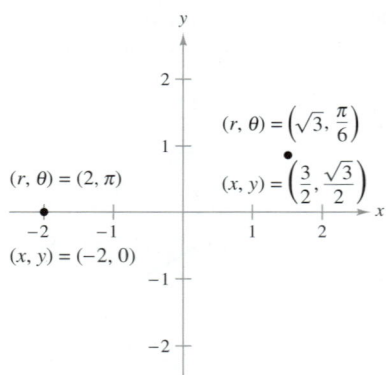

To convert from polar to rectangular coordinates, let $x = r \cos \theta$ and $y = r \sin \theta$.
Figure 10.38

EXAMPLE 1 **Polar-to-Rectangular Conversion**

a. For the point $(r, \theta) = (2, \pi)$,

$$x = r \cos \theta = 2 \cos \pi = -2 \quad \text{and} \quad y = r \sin \theta = 2 \sin \pi = 0.$$

So, the rectangular coordinates are $(x, y) = (-2, 0)$.

b. For the point $(r, \theta) = (\sqrt{3}, \pi/6)$,

$$x = \sqrt{3} \cos \frac{\pi}{6} = \frac{3}{2} \quad \text{and} \quad y = \sqrt{3} \sin \frac{\pi}{6} = \frac{\sqrt{3}}{2}.$$

So, the rectangular coordinates are $(x, y) = (3/2, \sqrt{3}/2)$.

See Figure 10.38.

EXAMPLE 2 **Rectangular-to-Polar Conversion**

a. For the second-quadrant point $(x, y) = (-1, 1)$,

$$\tan \theta = \frac{y}{x} = -1 \quad \Longrightarrow \quad \theta = \frac{3\pi}{4}.$$

Because θ was chosen to be in the same quadrant as (x, y), you should use a positive value of r.

$$r = \sqrt{x^2 + y^2}$$
$$= \sqrt{(-1)^2 + (1)^2}$$
$$= \sqrt{2}$$

This implies that *one* set of polar coordinates is $(r, \theta) = (\sqrt{2}, 3\pi/4)$.

b. Because the point $(x, y) = (0, 2)$ lies on the positive y-axis, choose $\theta = \pi/2$ and $r = 2$, and one set of polar coordinates is $(r, \theta) = (2, \pi/2)$.

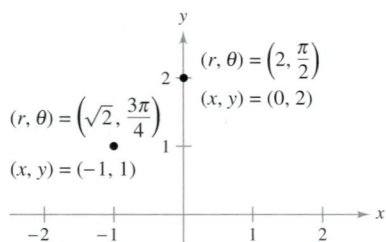

To convert from rectangular to polar coordinates, let $\tan \theta = y/x$ and $r = \sqrt{x^2 + y^2}$.
Figure 10.39

See Figure 10.39.

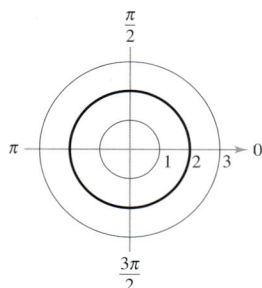

Polar Graphs

One way to sketch the graph of a polar equation is to convert to rectangular coordinates and then sketch the graph of the rectangular equation.

(a) Circle: $r = 2$

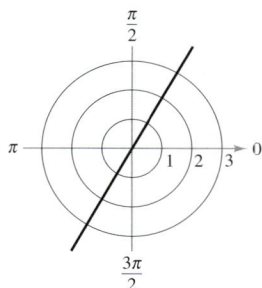

(b) Radial line: $\theta = \dfrac{\pi}{3}$

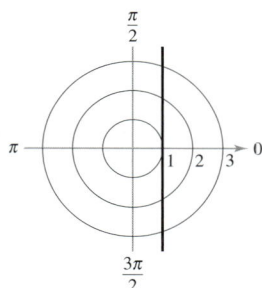

(c) Vertical line: $r = \sec \theta$

Figure 10.40

EXAMPLE 3 **Graphing Polar Equations**

Describe the graph of each polar equation. Confirm each description by converting to a rectangular equation.

a. $r = 2$ **b.** $\theta = \dfrac{\pi}{3}$ **c.** $r = \sec \theta$

Solution

a. The graph of the polar equation $r = 2$ consists of all points that are two units from the pole. So, this graph is a circle centered at the origin with a radius of 2. [See Figure 10.40(a).] You can confirm this by using the relationship $r^2 = x^2 + y^2$ to obtain the rectangular equation

$$x^2 + y^2 = 2^2. \qquad \text{Rectangular equation}$$

b. The graph of the polar equation $\theta = \pi/3$ consists of all points on the line that makes an angle of $\pi/3$ with the positive x-axis. [See Figure 10.40(b).] You can confirm this by using the relationship $\tan \theta = y/x$ to obtain the rectangular equation

$$y = \sqrt{3}x. \qquad \text{Rectangular equation}$$

c. The graph of the polar equation $r = \sec \theta$ is not evident by simple inspection, so you can begin by converting to rectangular form using the relationship $r \cos \theta = x$.

$$r = \sec \theta \qquad \text{Polar equation}$$
$$r \cos \theta = 1$$
$$x = 1 \qquad \text{Rectangular equation}$$

From the rectangular equation, you can see that the graph is a vertical line. [See Figure 10.40(c).]

▷ **TECHNOLOGY**   Sketching the graphs of complicated polar equations *by hand* can be tedious. With technology, however, the task is not difficult. If your graphing utility has a *polar* mode, use it to graph the equations in the exercise set. If your graphing utility doesn't have a *polar* mode, but does have a *parametric* mode, you can graph $r = f(\theta)$ by writing the equation as

$$x = f(\theta) \cos \theta$$
$$y = f(\theta) \sin \theta.$$

For instance, the graph of $r = \frac{1}{2}\theta$ shown in Figure 10.41 was produced with a graphing calculator in parametric mode. This equation was graphed using the parametric equations

$$x = \frac{1}{2}\theta \cos \theta$$

$$y = \frac{1}{2}\theta \sin \theta$$

with the values of θ varying from -4π to 4π. This curve is of the form $r = a\theta$ and is called a **spiral of Archimedes.**

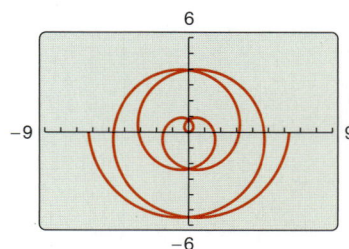

Spiral of Archimedes
Figure 10.41

| EXAMPLE 4 | **Sketching a Polar Graph** |

⋮⋯▷ *See LarsonCalculus.com for an interactive version of this type of example.*

Sketch the graph of $r = 2 \cos 3\theta$.

Solution Begin by writing the polar equation in parametric form.

$$x = 2 \cos 3\theta \cos \theta \quad \text{and} \quad y = 2 \cos 3\theta \sin \theta$$

After some experimentation, you will find that the entire curve, which is called a **rose curve,** can be sketched by letting θ vary from 0 to π, as shown in Figure 10.42. If you try duplicating this graph with a graphing utility, you will find that by letting θ vary from 0 to 2π, you will actually trace the entire curve *twice*.

·· REMARK One way to sketch the graph of $r = 2 \cos 3\theta$ by hand is to make a table of values.

θ	0	$\dfrac{\pi}{6}$	$\dfrac{\pi}{3}$	$\dfrac{\pi}{2}$	$\dfrac{2\pi}{3}$
r	2	0	-2	0	2

By extending the table and plotting the points, you will obtain the curve shown in Example 4.

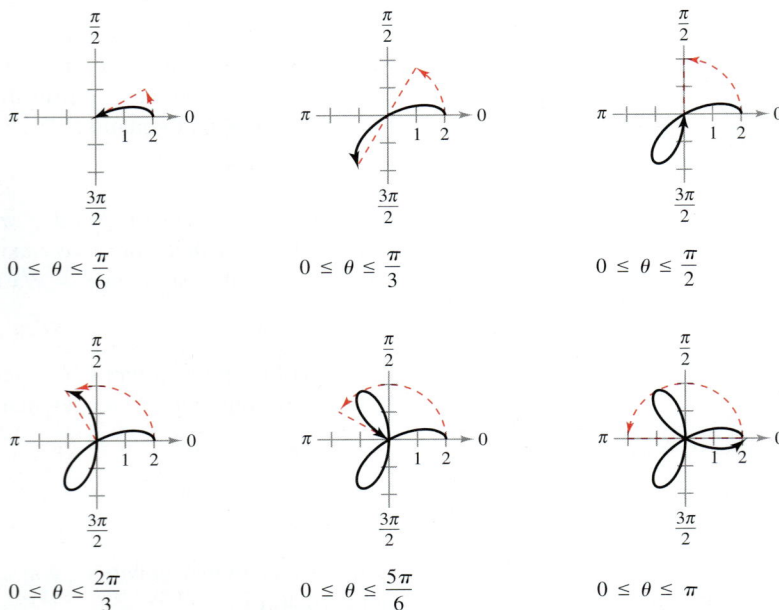

$$0 \le \theta \le \frac{\pi}{6} \qquad 0 \le \theta \le \frac{\pi}{3} \qquad 0 \le \theta \le \frac{\pi}{2}$$

$$0 \le \theta \le \frac{2\pi}{3} \qquad 0 \le \theta \le \frac{5\pi}{6} \qquad 0 \le \theta \le \pi$$

Figure 10.42

Use a graphing utility to experiment with other rose curves. Note that rose curves are of the form

$$r = a \cos n\theta \quad \text{or} \quad r = a \sin n\theta.$$

For instance, Figure 10.43 shows the graphs of two other rose curves.

$r = 2 \sin 5\theta$

$r = 0.5 \cos 2\theta$

Generated by Mathematica

Rose curves
Figure 10.43

Slope and Tangent Lines

To find the slope of a tangent line to a polar graph, consider a differentiable function given by $r = f(\theta)$. To find the slope in polar form, use the parametric equations

$$x = r \cos \theta = f(\theta) \cos \theta \quad \text{and} \quad y = r \sin \theta = f(\theta) \sin \theta.$$

Using the parametric form of dy/dx given in Theorem 10.7, you have

$$\frac{dy}{dx} = \frac{dy/d\theta}{dx/d\theta} = \frac{f(\theta) \cos \theta + f'(\theta) \sin \theta}{-f(\theta) \sin \theta + f'(\theta) \cos \theta}$$

which establishes the next theorem.

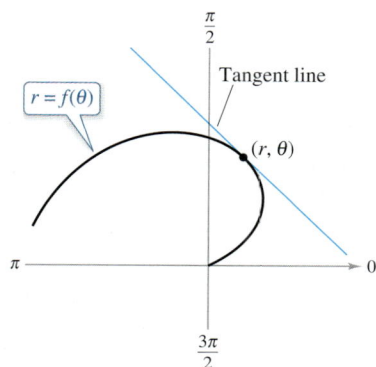

$r = f(\theta)$

Tangent line

(r, θ)

Tangent line to polar curve
Figure 10.44

THEOREM 10.11 Slope in Polar Form

If f is a differentiable function of θ, then the *slope* of the tangent line to the graph of $r = f(\theta)$ at the point (r, θ) is

$$\frac{dy}{dx} = \frac{dy/d\theta}{dx/d\theta} = \frac{f(\theta) \cos \theta + f'(\theta) \sin \theta}{-f(\theta) \sin \theta + f'(\theta) \cos \theta}$$

provided that $dx/d\theta \neq 0$ at (r, θ). (See Figure 10.44.)

From Theorem 10.11, you can make the following observations.

1. Solutions of $\dfrac{dy}{d\theta} = 0$ yield horizontal tangents, provided that $\dfrac{dx}{d\theta} \neq 0$.

2. Solutions of $\dfrac{dx}{d\theta} = 0$ yield vertical tangents, provided that $\dfrac{dy}{d\theta} \neq 0$.

If $dy/d\theta$ and $dx/d\theta$ are *simultaneously* 0, then no conclusion can be drawn about tangent lines.

EXAMPLE 5 **Finding Horizontal and Vertical Tangent Lines**

Find the horizontal and vertical tangent lines of $r = \sin \theta$, $0 \le \theta \le \pi$.

Solution Begin by writing the equation in parametric form.

$$x = r \cos \theta = \sin \theta \cos \theta$$

and

$$y = r \sin \theta = \sin \theta \sin \theta = \sin^2 \theta$$

Next, differentiate x and y with respect to θ and set each derivative equal to 0.

$$\frac{dx}{d\theta} = \cos^2 \theta - \sin^2 \theta = \cos 2\theta = 0 \implies \theta = \frac{\pi}{4}, \frac{3\pi}{4}$$

$$\frac{dy}{d\theta} = 2 \sin \theta \cos \theta = \sin 2\theta = 0 \implies \theta = 0, \frac{\pi}{2}$$

So, the graph has vertical tangent lines at

$$\left(\frac{\sqrt{2}}{2}, \frac{\pi}{4} \right) \quad \text{and} \quad \left(\frac{\sqrt{2}}{2}, \frac{3\pi}{4} \right)$$

and it has horizontal tangent lines at

$$(0, 0) \quad \text{and} \quad \left(1, \frac{\pi}{2} \right)$$

as shown in Figure 10.45.

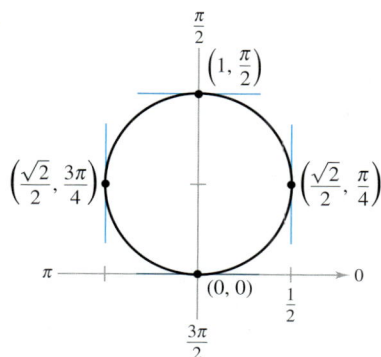

Horizontal and vertical tangent lines of $r = \sin \theta$
Figure 10.45

EXAMPLE 6   **Finding Horizontal and Vertical Tangent Lines**

Find the horizontal and vertical tangents to the graph of $r = 2(1 - \cos \theta)$.

Solution   Let $y = r \sin \theta$ and then differentiate with respect to θ.

$$y = r \sin \theta$$
$$= 2(1 - \cos \theta) \sin \theta$$
$$\frac{dy}{d\theta} = 2[(1 - \cos \theta)(\cos \theta) + \sin \theta(\sin \theta)]$$
$$= 2(\cos \theta - \cos^2 \theta + \sin^2 \theta)$$
$$= 2(\cos \theta - \cos^2 \theta + 1 - \cos^2 \theta)$$
$$= -2(2 \cos^2 \theta - \cos \theta - 1)$$
$$= -2(2 \cos \theta + 1)(\cos \theta - 1)$$

Setting $dy/d\theta$ equal to 0, you can see that $\cos \theta = -\frac{1}{2}$ and $\cos \theta = 1$. So, $dy/d\theta = 0$ when $\theta = 2\pi/3, 4\pi/3$, and 0. Similarly, using $x = r \cos \theta$, you have

$$x = r \cos \theta$$
$$= 2(1 - \cos \theta) \cos \theta$$
$$= 2 \cos \theta - 2 \cos^2 \theta$$
$$\frac{dx}{d\theta} = -2 \sin \theta + 4 \cos \theta \sin \theta$$
$$= 2 \sin \theta(2 \cos \theta - 1).$$

Setting $dx/d\theta$ equal to 0, you can see that $\sin \theta = 0$ and $\cos \theta = \frac{1}{2}$. So, you can conclude that $dx/d\theta = 0$ when $\theta = 0, \pi, \pi/3$, and $5\pi/3$. From these results, and from the graph shown in Figure 10.46, you can conclude that the graph has horizontal tangents at $(3, 2\pi/3)$ and $(3, 4\pi/3)$, and has vertical tangents at $(1, \pi/3), (1, 5\pi/3)$, and $(4, \pi)$. This graph is called a **cardioid.** Note that both derivatives ($dy/d\theta$ and $dx/d\theta$) are 0 when $\theta = 0$. Using this information alone, you don't know whether the graph has a horizontal or vertical tangent line at the pole. From Figure 10.46, however, you can see that the graph has a cusp at the pole.

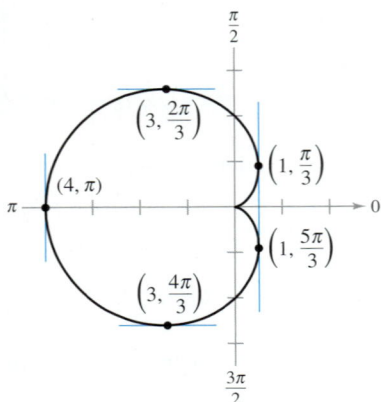

Horizontal and vertical tangent lines of
$r = 2(1 - \cos \theta)$
Figure 10.46

Theorem 10.11 has an important consequence. If the graph of $r = f(\theta)$ passes through the pole when $\theta = \alpha$ and $f'(\alpha) \neq 0$, then the formula for dy/dx simplifies as follows.

$$\frac{dy}{dx} = \frac{f'(\alpha) \sin \alpha + f(\alpha) \cos \alpha}{f'(\alpha) \cos \alpha - f(\alpha) \sin \alpha} = \frac{f'(\alpha) \sin \alpha + 0}{f'(\alpha) \cos \alpha - 0} = \frac{\sin \alpha}{\cos \alpha} = \tan \alpha$$

So, the line $\theta = \alpha$ is tangent to the graph at the pole, $(0, \alpha)$.

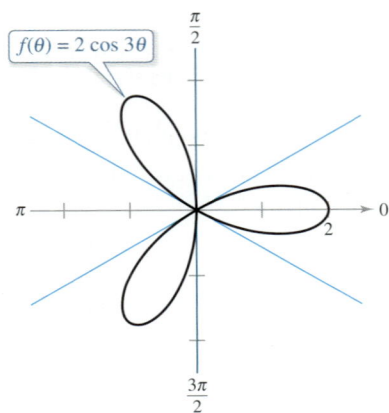

$f(\theta) = 2 \cos 3\theta$

> **THEOREM 10.12   Tangent Lines at the Pole**
>
> If $f(\alpha) = 0$ and $f'(\alpha) \neq 0$, then the line $\theta = \alpha$ is tangent at the pole to the graph of $r = f(\theta)$.

This rose curve has three tangent lines
($\theta = \pi/6, \theta = \pi/2$, and $\theta = 5\pi/6$) at the pole.
Figure 10.47

Theorem 10.12 is useful because it states that the zeros of $r = f(\theta)$ can be used to find the tangent lines at the pole. Note that because a polar curve can cross the pole more than once, it can have more than one tangent line at the pole. For example, the rose curve $f(\theta) = 2 \cos 3\theta$ has three tangent lines at the pole, as shown in Figure 10.47. For this curve, $f(\theta) = 2 \cos 3\theta$ is 0 when θ is $\pi/6, \pi/2$, and $5\pi/6$. Moreover, the derivative $f'(\theta) = -6 \sin 3\theta$ is not 0 for these values of θ.

Special Polar Graphs

Several important types of graphs have equations that are simpler in polar form than in rectangular form. For example, the polar equation of a circle having a radius of a and centered at the origin is simply $r = a$. Later in the text, you will come to appreciate this benefit. For now, several other types of graphs that have simpler equations in polar form are shown below. (Conics are considered in Section 10.6.)

Limaçons

$r = a \pm b \cos \theta$

$r = a \pm b \sin \theta$

$(a > 0, b > 0)$

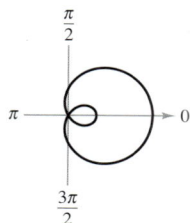

$\dfrac{a}{b} < 1$

Limaçon with inner loop

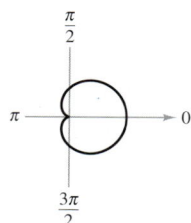

$\dfrac{a}{b} = 1$

Cardioid (heart-shaped)

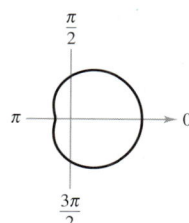

$1 < \dfrac{a}{b} < 2$

Dimpled limaçon

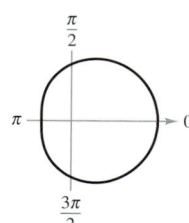

$\dfrac{a}{b} \geq 2$

Convex limaçon

Rose Curves

n petals when n is odd

$2n$ petals when n is even $(n \geq 2)$

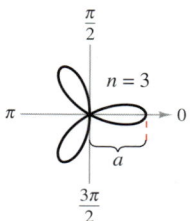

$n = 3$

$r = a \cos n\theta$

Rose curve

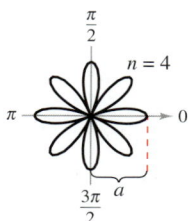

$n = 4$

$r = a \cos n\theta$

Rose curve

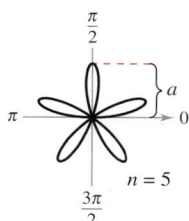

$n = 5$

$r = a \sin n\theta$

Rose curve

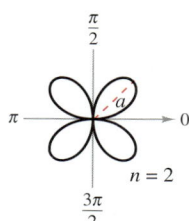

$n = 2$

$r = a \sin n\theta$

Rose curve

Circles and Lemniscates

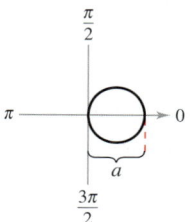

$r = a \cos \theta$

Circle

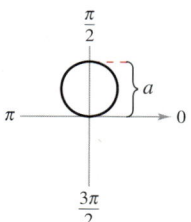

$r = a \sin \theta$

Circle

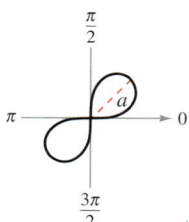

$r^2 = a^2 \sin 2\theta$

Lemniscate

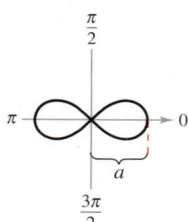

$r^2 = a^2 \cos 2\theta$

Lemniscate

▷ **TECHNOLOGY** The rose curves described above are of the form $r = a \cos n\theta$ or $r = a \sin n\theta$, where n is a positive integer that is greater than or equal to 2. Use a graphing utility to graph

$$r = a \cos n\theta \quad \text{or} \quad r = a \sin n\theta$$

for some noninteger values of n. Are these graphs also rose curves? For example, try sketching the graph of

$$r = \cos \frac{2}{3}\theta, \quad 0 \leq \theta \leq 6\pi.$$

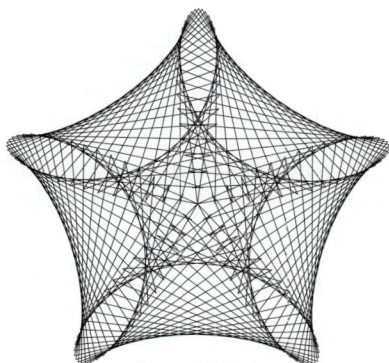

Generated by Maple

■ **FOR FURTHER INFORMATION** For more information on rose curves and related curves, see the article "A Rose is a Rose . . ." by Peter M. Maurer in *The American Mathematical Monthly.* The computer-generated graph at the left is the result of an algorithm that Maurer calls "The Rose." To view this article, go to *MathArticles.com.*

10.4 Exercises

See **CalcChat.com** for tutorial help and worked-out solutions to odd-numbered exercises.

Polar-to-Rectangular Conversion In Exercises 1–10, plot the point in polar coordinates and find the corresponding rectangular coordinates for the point.

1. $\left(8, \dfrac{\pi}{2}\right)$

2. $\left(-2, \dfrac{5\pi}{3}\right)$

3. $\left(-4, -\dfrac{3\pi}{4}\right)$

4. $\left(0, -\dfrac{7\pi}{6}\right)$

5. $\left(7, \dfrac{5\pi}{4}\right)$

6. $\left(-2, \dfrac{11\pi}{6}\right)$

7. $(\sqrt{2}, 2.36)$

8. $(-3, -1.57)$

9. $(-4.5, 3.5)$

10. $(9.25, 1.2)$

Rectangular-to-Polar Conversion In Exercises 11–20, the rectangular coordinates of a point are given. Plot the point and find *two* sets of polar coordinates for the point for $0 \le \theta < 2\pi$.

11. $(2, 2)$

12. $(0, -6)$

13. $(-3, 4)$

14. $(4, -2)$

15. $(-1, -\sqrt{3})$

16. $(3, -\sqrt{3})$

17. $(3, -2)$

18. $(3\sqrt{2}, 3\sqrt{2})$

19. $\left(\dfrac{7}{4}, \dfrac{5}{2}\right)$

20. $(0, -5)$

21. Plotting a Point Plot the point $(4, 3.5)$ when the point is given in

(a) rectangular coordinates.

(b) polar coordinates.

22. Graphical Reasoning

(a) Set the window format of a graphing utility to rectangular coordinates and locate the cursor at any position off the axes. Move the cursor horizontally and vertically. Describe any changes in the displayed coordinates of the points.

(b) Set the window format of a graphing utility to polar coordinates and locate the cursor at any position off the axes. Move the cursor horizontally and vertically. Describe any changes in the displayed coordinates of the points.

(c) Why are the results in parts (a) and (b) different?

Rectangular-to-Polar Conversion In Exercises 23–32, convert the rectangular equation to polar form and sketch its graph.

23. $x^2 + y^2 = 9$

24. $x^2 - y^2 = 9$

25. $x^2 + y^2 = a^2$

26. $x^2 + y^2 - 2ax = 0$

27. $y = 8$

28. $x = 12$

29. $3x - y + 2 = 0$

30. $xy = 4$

31. $y^2 = 9x$

32. $(x^2 + y^2)^2 - 9(x^2 - y^2) = 0$

Polar-to-Rectangular Conversion In Exercises 33–42, convert the polar equation to rectangular form and sketch its graph.

33. $r = 4$

34. $r = -5$

35. $r = 3 \sin \theta$

36. $r = 5 \cos \theta$

37. $r = \theta$

38. $\theta = \dfrac{5\pi}{6}$

39. $r = 3 \sec \theta$

40. $r = 2 \csc \theta$

41. $r = \sec \theta \tan \theta$

42. $r = \cot \theta \csc \theta$

Graphing a Polar Equation In Exercises 43–52, use a graphing utility to graph the polar equation. Find an interval for θ over which the graph is traced *only once*.

43. $r = 2 - 5 \cos \theta$

44. $r = 3(1 - 4 \cos \theta)$

45. $r = 2 + \sin \theta$

46. $r = 4 + 3 \cos \theta$

47. $r = \dfrac{2}{1 + \cos \theta}$

48. $r = \dfrac{2}{4 - 3 \sin \theta}$

49. $r = 2 \cos\left(\dfrac{3\theta}{2}\right)$

50. $r = 3 \sin\left(\dfrac{5\theta}{2}\right)$

51. $r^2 = 4 \sin 2\theta$

52. $r^2 = \dfrac{1}{\theta}$

53. Verifying a Polar Equation Convert the equation

$$r = 2(h \cos \theta + k \sin \theta)$$

to rectangular form and verify that it is the equation of a circle. Find the radius and the rectangular coordinates of the center of the circle.

54. Distance Formula

(a) Verify that the Distance Formula for the distance between the two points (r_1, θ_1) and (r_2, θ_2) in polar coordinates is

$$d = \sqrt{r_1^2 + r_2^2 - 2r_1 r_2 \cos(\theta_1 - \theta_2)}.$$

(b) Describe the positions of the points relative to each other for $\theta_1 = \theta_2$. Simplify the Distance Formula for this case. Is the simplification what you expected? Explain.

(c) Simplify the Distance Formula for $\theta_1 - \theta_2 = 90°$. Is the simplification what you expected? Explain.

(d) Choose two points on the polar coordinate system and find the distance between them. Then choose different polar representations of the same two points and apply the Distance Formula again. Discuss the result.

Distance Formula In Exercises 55–58, use the result of Exercise 54 to approximate the distance between the two points in polar coordinates.

55. $\left(1, \dfrac{5\pi}{6}\right), \quad \left(4, \dfrac{\pi}{3}\right)$

56. $\left(8, \dfrac{7\pi}{4}\right), \quad (5, \pi)$

57. $(2, 0.5), \quad (7, 1.2)$

58. $(4, 2.5), \quad (12, 1)$

Finding Slopes of Tangent Lines In Exercises 59 and 60, find dy/dx and the slopes of the tangent lines shown on the graph of the polar equation.

59. $r = 2 + 3 \sin \theta$

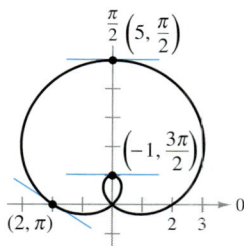

60. $r = 2(1 - \sin \theta)$

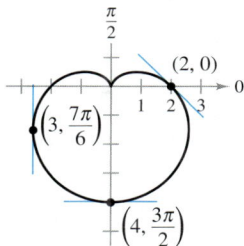

Finding Slopes of Tangent Lines In Exercises 61–64, use a graphing utility to (a) graph the polar equation, (b) draw the tangent line at the given value of θ, and (c) find dy/dx at the given value of θ. (*Hint:* Let the increment between the values of θ equal $\pi/24$.)

61. $r = 3(1 - \cos \theta), \ \theta = \dfrac{\pi}{2}$

62. $r = 3 - 2 \cos \theta, \ \theta = 0$

63. $r = 3 \sin \theta, \ \theta = \dfrac{\pi}{3}$

64. $r = 4, \ \theta = \dfrac{\pi}{4}$

Horizontal and Vertical Tangency In Exercises 65 and 66, find the points of horizontal and vertical tangency (if any) to the polar curve.

65. $r = 1 - \sin \theta$

66. $r = a \sin \theta$

Horizontal Tangency In Exercises 67 and 68, find the points of horizontal tangency (if any) to the polar curve.

67. $r = 2 \csc \theta + 3$

68. $r = a \sin \theta \cos^2 \theta$

Tangent Lines at the Pole In Exercises 69–76, sketch a graph of the polar equation and find the tangents at the pole.

69. $r = 5 \sin \theta$

70. $r = 5 \cos \theta$

71. $r = 2(1 - \sin \theta)$

72. $r = 3(1 - \cos \theta)$

73. $r = 4 \cos 3\theta$

74. $r = -\sin 5\theta$

75. $r = 3 \sin 2\theta$

76. $r = 3 \cos 2\theta$

Sketching a Polar Graph In Exercises 77–88, sketch a graph of the polar equation.

77. $r = 8$

78. $r = 1$

79. $r = 4(1 + \cos \theta)$

80. $r = 1 + \sin \theta$

81. $r = 3 - 2 \cos \theta$

82. $r = 5 - 4 \sin \theta$

83. $r = 3 \csc \theta$

84. $r = \dfrac{6}{2 \sin \theta - 3 \cos \theta}$

85. $r = 2\theta$

86. $r = \dfrac{1}{\theta}$

87. $r^2 = 4 \cos 2\theta$

88. $r^2 = 4 \sin \theta$

Asymptote In Exercises 89–92, use a graphing utility to graph the equation and show that the given line is an asymptote of the graph.

	Name of Graph	Polar Equation	Asymptote
89.	Conchoid	$r = 2 - \sec \theta$	$x = -1$
90.	Conchoid	$r = 2 + \csc \theta$	$y = 1$
91.	Hyperbolic spiral	$r = 2/\theta$	$y = 2$
92.	Strophoid	$r = 2 \cos 2\theta \sec \theta$	$x = -2$

WRITING ABOUT CONCEPTS

93. Comparing Coordinate Systems Describe the differences between the rectangular coordinate system and the polar coordinate system.

94. Coordinate Conversion Give the equations for the coordinate conversion from rectangular to polar coordinates and vice versa.

95. Tangent Lines How are the slopes of tangent lines determined in polar coordinates? What are tangent lines at the pole and how are they determined?

96. HOW DO YOU SEE IT? Identify each special polar graph and write its equation.

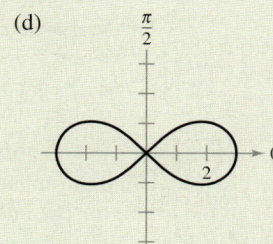

(a)

(b)

(c)

(d)

97. Sketching a Graph Sketch the graph of $r = 4 \sin \theta$ over each interval.

(a) $0 \le \theta \le \dfrac{\pi}{2}$ (b) $\dfrac{\pi}{2} \le \theta \le \pi$ (c) $-\dfrac{\pi}{2} \le \theta \le \dfrac{\pi}{2}$

98. Think About It Use a graphing utility to graph the polar equation $r = 6[1 + \cos(\theta - \phi)]$ for (a) $\phi = 0$, (b) $\phi = \pi/4$, and (c) $\phi = \pi/2$. Use the graphs to describe the effect of the angle ϕ. Write the equation as a function of $\sin \theta$ for part (c).

99. Rotated Curve Verify that if the curve whose polar equation is $r = f(\theta)$ is rotated about the pole through an angle ϕ, then an equation for the rotated curve is $r = f(\theta - \phi)$.

100. Rotated Curve The polar form of an equation of a curve is $r = f(\sin \theta)$. Show that the form becomes

(a) $r = f(-\cos \theta)$ if the curve is rotated counterclockwise $\pi/2$ radians about the pole.

(b) $r = f(-\sin \theta)$ if the curve is rotated counterclockwise π radians about the pole.

(c) $r = f(\cos \theta)$ if the curve is rotated counterclockwise $3\pi/2$ radians about the pole.

Rotated Curve **In Exercises 101–104, use the results of Exercises 99 and 100.**

101. Write an equation for the limaçon $r = 2 - \sin \theta$ after it has been rotated by the given amount. Use a graphing utility to graph the rotated limaçon for (a) $\theta = \pi/4$, (b) $\theta = \pi/2$, (c) $\theta = \pi$, and (d) $\theta = 3\pi/2$.

102. Write an equation for the rose curve $r = 2 \sin 2\theta$ after it has been rotated by the given amount. Verify the results by using a graphing utility to graph the rotated rose curve for (a) $\theta = \pi/6$, (b) $\theta = \pi/2$, (c) $\theta = 2\pi/3$, and (d) $\theta = \pi$.

103. Sketch the graph of each equation.

(a) $r = 1 - \sin \theta$ (b) $r = 1 - \sin\left(\theta - \dfrac{\pi}{4}\right)$

104. Prove that the tangent of the angle $\psi \ (0 \le \psi \le \pi/2)$ between the radial line and the tangent line at the point (r, θ) on the graph of $r = f(\theta)$ (see figure) is given by

$$\tan \psi = \left| \frac{r}{dr/d\theta} \right|.$$

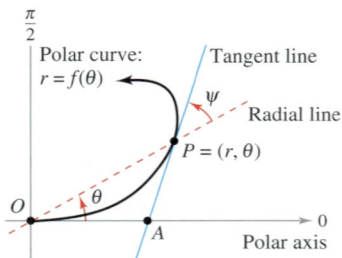

Finding an Angle **In Exercises 105–110, use the result of Exercise 104 to find the angle ψ between the radial and tangent lines to the graph for the indicated value of θ. Use a graphing utility to graph the polar equation, the radial line, and the tangent line for the indicated value of θ. Identify the angle ψ.**

Polar Equation	Value of θ
105. $r = 2(1 - \cos \theta)$	$\theta = \pi$
106. $r = 3(1 - \cos \theta)$	$\theta = \dfrac{3\pi}{4}$
107. $r = 2 \cos 3\theta$	$\theta = \dfrac{\pi}{4}$
108. $r = 4 \sin 2\theta$	$\theta = \dfrac{\pi}{6}$
109. $r = \dfrac{6}{1 - \cos \theta}$	$\theta = \dfrac{2\pi}{3}$
110. $r = 5$	$\theta = \dfrac{\pi}{6}$

True or False? **In Exercises 111–114, determine whether the statement is true or false. If it is false, explain why or give an example that shows it is false.**

111. If (r_1, θ_1) and (r_2, θ_2) represent the same point on the polar coordinate system, then $|r_1| = |r_2|$.

112. If (r, θ_1) and (r, θ_2) represent the same point on the polar coordinate system, then $\theta_1 = \theta_2 + 2\pi n$ for some integer n.

113. If $x > 0$, then the point (x, y) on the rectangular coordinate system can be represented by (r, θ) on the polar coordinate system, where $r = \sqrt{x^2 + y^2}$ and $\theta = \arctan(y/x)$.

114. The polar equations $r = \sin 2\theta$, $r = -\sin 2\theta$, and $r = \sin(-2\theta)$ all have the same graph.

SECTION PROJECT

Anamorphic Art

Anamorphic art appears distorted, but when the art is viewed from a particular point or is viewed with a device such as a mirror, it appears to be normal. Use the anamorphic transformations

$$r = y + 16 \quad \text{and} \quad \theta = -\frac{\pi}{8}x, \quad -\frac{3\pi}{4} \le \theta \le \frac{3\pi}{4}$$

to sketch the transformed polar image of the rectangular graph. When the reflection (in a cylindrical mirror centered at the pole) of each polar image is viewed from the polar axis, the viewer will see the original rectangular image.

(a) $y = 3$ (b) $x = 2$ (c) $y = x + 5$ (d) $x^2 + (y - 5)^2 = 5^2$

This example of anamorphic art is from the Millington-Barnard Collection at the University of Mississippi. When the reflection of the transformed "polar painting" is viewed in the mirror, the viewer sees the distorted art in its proper proportions.

■ **FOR FURTHER INFORMATION** For more information on anamorphic art, see the article "Anamorphisms" by Philip Hickin in the *Mathematical Gazette*.

10.5   Area and Arc Length in Polar Coordinates

- ■ Find the area of a region bounded by a polar graph.
- ■ Find the points of intersection of two polar graphs.
- ■ Find the arc length of a polar graph.
- ■ Find the area of a surface of revolution (polar form).

Area of a Polar Region

The development of a formula for the area of a polar region parallels that for the area of a region on the rectangular coordinate system, but uses sectors of a circle instead of rectangles as the basic elements of area. In Figure 10.48, note that the area of a circular sector of radius r is $\frac{1}{2}\theta r^2$, provided θ is measured in radians.

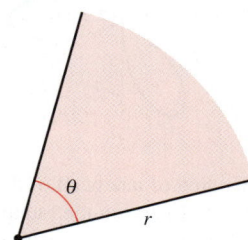

The area of a sector of a circle is $A = \frac{1}{2}\theta r^2$.

Figure 10.48

Consider the function $r = f(\theta)$, where f is continuous and nonnegative on the interval $\alpha \le \theta \le \beta$. The region bounded by the graph of f and the radial lines $\theta = \alpha$ and $\theta = \beta$ is shown in Figure 10.49(a). To find the area of this region, partition the interval $[\alpha, \beta]$ into n equal subintervals

$$\alpha = \theta_0 < \theta_1 < \theta_2 < \cdots < \theta_{n-1} < \theta_n = \beta.$$

Then approximate the area of the region by the sum of the areas of the n sectors, as shown in Figure 10.49(b).

$$\text{Radius of } i\text{th sector} = f(\theta_i)$$

$$\text{Central angle of } i\text{th sector} = \frac{\beta - \alpha}{n} = \Delta\theta$$

$$A \approx \sum_{i=1}^{n} \left(\frac{1}{2}\right) \Delta\theta [f(\theta_i)]^2$$

Taking the limit as $n \to \infty$ produces

$$A = \lim_{n \to \infty} \frac{1}{2} \sum_{i=1}^{n} [f(\theta_i)]^2 \, \Delta\theta$$

$$= \frac{1}{2} \int_{\alpha}^{\beta} [f(\theta)]^2 \, d\theta$$

which leads to the next theorem.

(a)

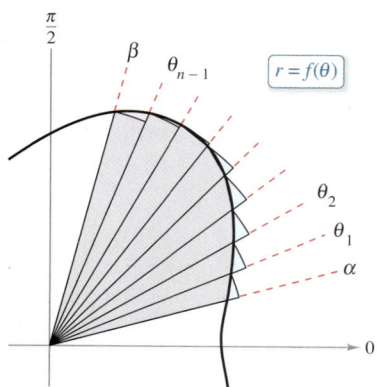

(b)

Figure 10.49

THEOREM 10.13   Area in Polar Coordinates

If f is continuous and nonnegative on the interval $[\alpha, \beta]$, $0 < \beta - \alpha \le 2\pi$, then the area of the region bounded by the graph of $r = f(\theta)$ between the radial lines $\theta = \alpha$ and $\theta = \beta$ is

$$A = \frac{1}{2} \int_{\alpha}^{\beta} [f(\theta)]^2 \, d\theta$$

$$= \frac{1}{2} \int_{\alpha}^{\beta} r^2 \, d\theta. \qquad {\color{red} 0 < \beta - \alpha \le 2\pi}$$

You can use the formula in Theorem 10.13 to find the area of a region bounded by the graph of a continuous *nonpositive* function. The formula is not necessarily valid, however, when f takes on both positive *and* negative values in the interval $[\alpha, \beta]$.

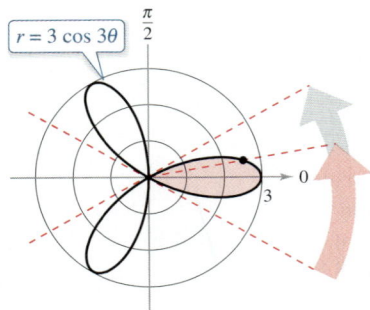

$r = 3 \cos 3\theta$

The area of one petal of the rose curve that lies between the radial lines $\theta = -\pi/6$ and $\theta = \pi/6$ is $3\pi/4$.
Figure 10.50

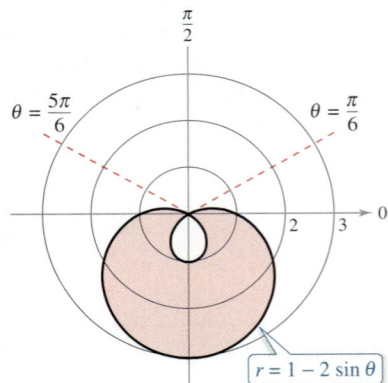

$\theta = \frac{5\pi}{6}$ $\theta = \frac{\pi}{6}$

$r = 1 - 2 \sin \theta$

The area between the inner and outer loops is approximately 8.34.
Figure 10.51

EXAMPLE 1 Finding the Area of a Polar Region

▷ *See LarsonCalculus.com for an interactive version of this type of example.*

Find the area of one petal of the rose curve $r = 3 \cos 3\theta$.

Solution In Figure 10.50, you can see that the petal on the right is traced as θ increases from $-\pi/6$ to $\pi/6$. So, the area is

$$A = \frac{1}{2} \int_\alpha^\beta r^2 \, d\theta = \frac{1}{2} \int_{-\pi/6}^{\pi/6} (3 \cos 3\theta)^2 \, d\theta \quad \text{Use formula for area in polar coordinates.}$$

$$= \frac{9}{2} \int_{-\pi/6}^{\pi/6} \frac{1 + \cos 6\theta}{2} \, d\theta \quad \text{Power-reducing formula}$$

$$= \frac{9}{4} \left[\theta + \frac{\sin 6\theta}{6} \right]_{-\pi/6}^{\pi/6}$$

$$= \frac{9}{4} \left(\frac{\pi}{6} + \frac{\pi}{6} \right)$$

$$= \frac{3\pi}{4}.$$

To find the area of the region lying inside all three petals of the rose curve in Example 1, you could *not* simply integrate between 0 and 2π. By doing this, you would obtain $9\pi/2$, which is twice the area of the three petals. The duplication occurs because the rose curve is traced twice as θ increases from 0 to 2π.

EXAMPLE 2 Finding the Area Bounded by a Single Curve

Find the area of the region lying between the inner and outer loops of the limaçon $r = 1 - 2 \sin \theta$.

Solution In Figure 10.51, note that the inner loop is traced as θ increases from $\pi/6$ to $5\pi/6$. So, the area inside the *inner loop* is

$$A_1 = \frac{1}{2} \int_{\pi/6}^{5\pi/6} (1 - 2 \sin \theta)^2 \, d\theta \quad \text{Use formula for area in polar coordinates.}$$

$$= \frac{1}{2} \int_{\pi/6}^{5\pi/6} (1 - 4 \sin \theta + 4 \sin^2 \theta) \, d\theta$$

$$= \frac{1}{2} \int_{\pi/6}^{5\pi/6} \left[1 - 4 \sin \theta + 4 \left(\frac{1 - \cos 2\theta}{2} \right) \right] d\theta \quad \text{Power-reducing formula}$$

$$= \frac{1}{2} \int_{\pi/6}^{5\pi/6} (3 - 4 \sin \theta - 2 \cos \theta) \, d\theta \quad \text{Simplify.}$$

$$= \frac{1}{2} \left[3\theta + 4 \cos \theta - \sin 2\theta \right]_{\pi/6}^{5\pi/6}$$

$$= \frac{1}{2} (2\pi - 3\sqrt{3})$$

$$= \pi - \frac{3\sqrt{3}}{2}.$$

In a similar way, you can integrate from $5\pi/6$ to $13\pi/6$ to find that the area of the region lying inside the *outer loop* is $A_2 = 2\pi + (3\sqrt{3}/2)$. The area of the region lying between the two loops is the difference of A_2 and A_1.

$$A = A_2 - A_1 = \left(2\pi + \frac{3\sqrt{3}}{2} \right) - \left(\pi - \frac{3\sqrt{3}}{2} \right) = \pi + 3\sqrt{3} \approx 8.34$$

Points of Intersection of Polar Graphs

Because a point may be represented in different ways in polar coordinates, care must be taken in determining the points of intersection of two polar graphs. For example, consider the points of intersection of the graphs of

$$r = 1 - 2 \cos \theta \quad \text{and} \quad r = 1$$

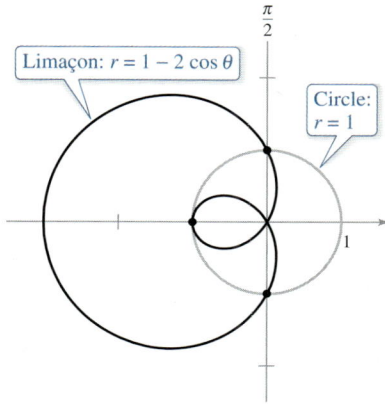

Limaçon: $r = 1 - 2\cos\theta$

Circle: $r = 1$

Three points of intersection: $(1, \pi/2)$, $(-1, 0)$, and $(1, 3\pi/2)$

Figure 10.52

as shown in Figure 10.52. As with rectangular equations, you can attempt to find the points of intersection by solving the two equations simultaneously, as shown.

$r = 1 - 2 \cos \theta$	First equation
$1 = 1 - 2 \cos \theta$	Substitute $r = 1$ from 2nd equation into 1st equation.
$\cos \theta = 0$	Simplify.
$\theta = \dfrac{\pi}{2}, \ \dfrac{3\pi}{2}$	Solve for θ.

The corresponding points of intersection are $(1, \pi/2)$ and $(1, 3\pi/2)$. From Figure 10.52, however, you can see that there is a *third* point of intersection that did not show up when the two polar equations were solved simultaneously. (This is one reason why you should sketch a graph when finding the area of a polar region.) The reason the third point was not found is that it does not occur with the same coordinates in the two graphs. On the graph of $r = 1$, the point occurs with coordinates $(1, \pi)$, but on the graph of

$$r = 1 - 2 \cos \theta$$

the point occurs with coordinates $(-1, 0)$.

In addition to solving equations simultaneously and sketching a graph, note that because the pole can be represented by $(0, \theta)$, where θ is *any* angle, you should check separately for the pole when finding points of intersection.

You can compare the problem of finding points of intersection of two polar graphs with that of finding collision points of two satellites in intersecting orbits about Earth, as shown in Figure 10.53. The satellites will not collide as long as they reach the points of intersection at different times (θ-values). Collisions will occur only at the points of intersection that are "simultaneous points"—those that are reached at the same time (θ-value).

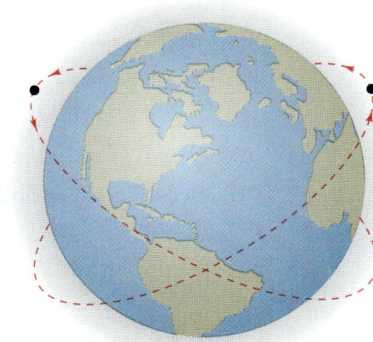

The paths of satellites can cross without causing a collision.

Figure 10.53

■ **FOR FURTHER INFORMATION** For more information on using technology to find points of intersection, see the article "Finding Points of Intersection of Polar-Coordinate Graphs" by Warren W. Esty in *Mathematics Teacher*. To view this article, go to *MathArticles.com*.

> **EXAMPLE 3**   **Finding the Area of a Region Between Two Curves**

Find the area of the region common to the two regions bounded by the curves.

$$r = -6 \cos \theta \qquad \text{Circle}$$

and

$$r = 2 - 2 \cos \theta. \qquad \text{Cardioid}$$

Solution   Because both curves are symmetric with respect to the x-axis, you can work with the upper half-plane, as shown in Figure 10.54. The blue shaded region lies between the circle and the radial line

$$\theta = \frac{2\pi}{3}.$$

Because the circle has coordinates $(0, \pi/2)$ at the pole, you can integrate between $\pi/2$ and $2\pi/3$ to obtain the area of this region. The region that is shaded red is bounded by the radial lines $\theta = 2\pi/3$ and $\theta = \pi$ and the cardioid. So, you can find the area of this second region by integrating between $2\pi/3$ and π. The sum of these two integrals gives the area of the common region lying *above* the radial line $\theta = \pi$.

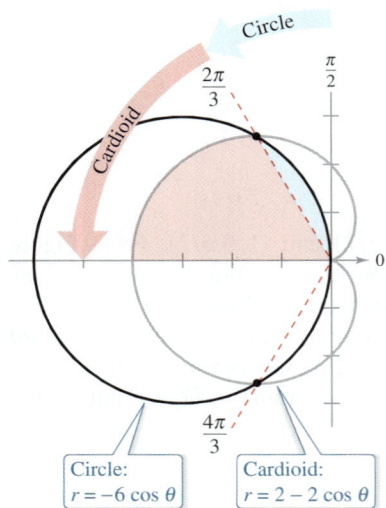

Circle:
$r = -6 \cos \theta$

Cardioid:
$r = 2 - 2 \cos \theta$

Figure 10.54

$$
\overbrace{\phantom{\frac{1}{2}\int_{\pi/2}^{2\pi/3}}}^{\substack{\text{Region between circle} \\ \text{and radial line } \theta = 2\pi/3}}
\qquad
\overbrace{\phantom{\frac{1}{2}\int_{2\pi/3}^{\pi}}}^{\substack{\text{Region between cardioid and} \\ \text{radial lines } \theta = 2\pi/3 \text{ and } \theta = \pi}}
$$

$$\frac{A}{2} = \frac{1}{2}\int_{\pi/2}^{2\pi/3} (-6 \cos \theta)^2 \, d\theta + \frac{1}{2}\int_{2\pi/3}^{\pi} (2 - 2 \cos \theta)^2 \, d\theta$$

$$= 18 \int_{\pi/2}^{2\pi/3} \cos^2 \theta \, d\theta + \frac{1}{2}\int_{2\pi/3}^{\pi} (4 - 8 \cos \theta + 4 \cos^2 \theta) \, d\theta$$

$$= 9 \int_{\pi/2}^{2\pi/3} (1 + \cos 2\theta) \, d\theta + \int_{2\pi/3}^{\pi} (3 - 4 \cos \theta + \cos 2\theta) \, d\theta$$

$$= 9 \left[\theta + \frac{\sin 2\theta}{2} \right]_{\pi/2}^{2\pi/3} + \left[3\theta - 4 \sin \theta + \frac{\sin 2\theta}{2} \right]_{2\pi/3}^{\pi}$$

$$= 9 \left(\frac{2\pi}{3} - \frac{\sqrt{3}}{4} - \frac{\pi}{2} \right) + \left(3\pi - 2\pi + 2\sqrt{3} + \frac{\sqrt{3}}{4} \right)$$

$$= \frac{5\pi}{2}$$

Finally, multiplying by 2, you can conclude that the total area is

$$5\pi \approx 15.7. \qquad \text{Area of region inside circle and cardioid}$$

To check the reasonableness of this result, note that the area of the circular region is

$$\pi r^2 = 9\pi. \qquad \text{Area of circle}$$

So, it seems reasonable that the area of the region lying inside the circle and the cardioid is 5π. ∎

To see the benefit of polar coordinates for finding the area in Example 3, consider the integral below, which gives the comparable area in rectangular coordinates.

$$\frac{A}{2} = \int_{-4}^{-3/2} \sqrt{2\sqrt{1 - 2x} - x^2 - 2x + 2} \, dx + \int_{-3/2}^{0} \sqrt{-x^2 - 6x} \, dx$$

Use the integration capabilities of a graphing utility to show that you obtain the same area as that found in Example 3.

Arc Length in Polar Form

The formula for the length of a polar arc can be obtained from the arc length formula for a curve described by parametric equations. (See Exercise 85.)

(See Exercise 85.)

> **THEOREM 10.14   Arc Length of a Polar Curve**
>
> Let f be a function whose derivative is continuous on an interval $\alpha \leq \theta \leq \beta$. The length of the graph of $r = f(\theta)$ from $\theta = \alpha$ to $\theta = \beta$ is
>
> $$s = \int_\alpha^\beta \sqrt{[f(\theta)]^2 + [f'(\theta)]^2}\, d\theta = \int_\alpha^\beta \sqrt{r^2 + \left(\frac{dr}{d\theta}\right)^2}\, d\theta.$$

• • REMARK   When applying the arc length formula to a polar curve, be sure that the curve is traced out only once on the interval of integration. For instance, the rose curve $r = \cos 3\theta$ is traced out once on the interval $0 \leq \theta \leq \pi$, but is traced out twice on the interval $0 \leq \theta \leq 2\pi$.

EXAMPLE 4   Finding the Length of a Polar Curve

Find the length of the arc from $\theta = 0$ to $\theta = 2\pi$ for the cardioid

$$r = f(\theta) = 2 - 2 \cos \theta$$

as shown in Figure 10.55.

Solution   Because $f'(\theta) = 2 \sin \theta$, you can find the arc length as follows.

$$
\begin{aligned}
s &= \int_\alpha^\beta \sqrt{[f(\theta)]^2 + [f'(\theta)]^2}\, d\theta && \text{\color{red}Formula for arc length of a polar curve}\\[2mm]
&= \int_0^{2\pi} \sqrt{(2 - 2\cos\theta)^2 + (2\sin\theta)^2}\, d\theta \\[2mm]
&= 2\sqrt{2} \int_0^{2\pi} \sqrt{1 - \cos\theta}\, d\theta && \text{\color{red}Simplify.}\\[2mm]
&= 2\sqrt{2} \int_0^{2\pi} \sqrt{2\sin^2 \frac{\theta}{2}}\, d\theta && \text{\color{red}Trigonometric identity}\\[2mm]
&= 4 \int_0^{2\pi} \sin\frac{\theta}{2}\, d\theta && \text{\color{red}$\sin\frac{\theta}{2} \geq 0$ for $0 \leq \theta \leq 2\pi$}\\[2mm]
&= 8\left[-\cos\frac{\theta}{2} \right]_0^{2\pi} \\[2mm]
&= 8(1 + 1) \\[1mm]
&= 16
\end{aligned}
$$

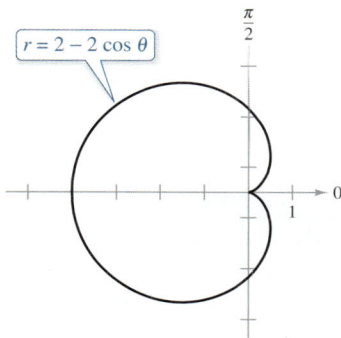

$r = 2 - 2 \cos \theta$

Figure 10.55

Using Figure 10.55, you can determine the reasonableness of this answer by comparing it with the circumference of a circle. For example, a circle of radius $\frac{5}{2}$ has a circumference of

$$5\pi \approx 15.7.$$

Note that in the fifth step of the solution, it is legitimate to write

$$\sqrt{2\sin^2 \frac{\theta}{2}} = \sqrt{2}\sin\frac{\theta}{2}$$

rather than

$$\sqrt{2\sin^2 \frac{\theta}{2}} = \sqrt{2}\left|\sin\frac{\theta}{2}\right|$$

because $\sin(\theta/2) \geq 0$ for $0 \leq \theta \leq 2\pi$.

Area of a Surface of Revolution

The polar coordinate versions of the formulas for the area of a surface of revolution can be obtained from the parametric versions given in Theorem 10.9, using the equations $x = r\cos\theta$ and $y = r\sin\theta$.

· · · · · · · · · · · · · · · · · ·▷

· · REMARK   When using Theorem 10.15, check to see that the graph of $r = f(\theta)$ is traced only once on the interval $\alpha \leq \theta \leq \beta$. For example, the circle $r = \cos\theta$ is traced only once on the interval $0 \leq \theta \leq \pi$.

THEOREM 10.15   Area of a Surface of Revolution

Let f be a function whose derivative is continuous on an interval $\alpha \leq \theta \leq \beta$. The area of the surface formed by revolving the graph of $r = f(\theta)$ from $\theta = \alpha$ to $\theta = \beta$ about the indicated line is as follows.

1. $S = 2\pi \displaystyle\int_\alpha^\beta f(\theta) \sin\theta \sqrt{[f(\theta)]^2 + [f'(\theta)]^2}\, d\theta$ About the polar axis

2. $S = 2\pi \displaystyle\int_\alpha^\beta f(\theta) \cos\theta \sqrt{[f(\theta)]^2 + [f'(\theta)]^2}\, d\theta$ About the line $\theta = \dfrac{\pi}{2}$

EXAMPLE 5   **Finding the Area of a Surface of Revolution**

Find the area of the surface formed by revolving the circle $r = f(\theta) = \cos\theta$ about the line $\theta = \pi/2$, as shown in Figure 10.56.

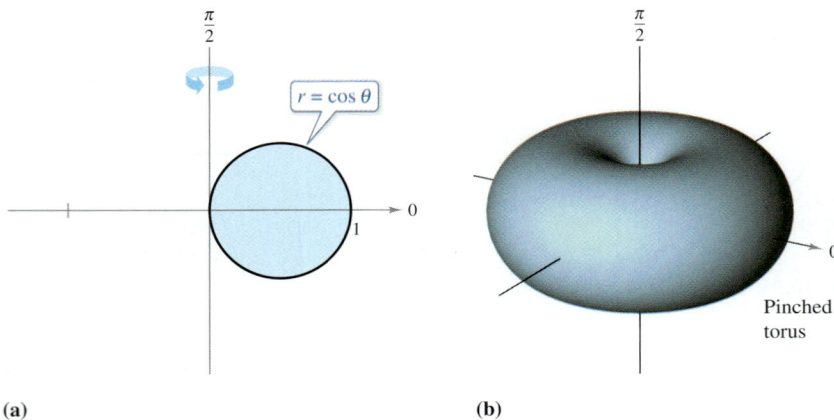

(a)   **(b)**   Pinched torus

Figure 10.56

Solution   Use the second formula in Theorem 10.15 with $f'(\theta) = -\sin\theta$. Because the circle is traced once as θ increases from 0 to π, you have

$$S = 2\pi \int_\alpha^\beta f(\theta) \cos\theta \sqrt{[f(\theta)]^2 + [f'(\theta)]^2}\, d\theta \qquad \text{Formula for area of a surface of revolution}$$

$$= 2\pi \int_0^\pi \cos\theta(\cos\theta)\sqrt{\cos^2\theta + \sin^2\theta}\, d\theta$$

$$= 2\pi \int_0^\pi \cos^2\theta\, d\theta \qquad \text{Trigonometric identity}$$

$$= \pi \int_0^\pi (1 + \cos 2\theta)\, d\theta \qquad \text{Trigonometric identity}$$

$$= \pi \left[\theta + \frac{\sin 2\theta}{2} \right]_0^\pi$$

$$= \pi^2.$$

10.5 Exercises

See **CalcChat.com** for tutorial help and worked-out solutions to odd-numbered exercises.

Area of a Polar Region In Exercises 1–4, write an integral that represents the area of the shaded region of the figure. Do not evaluate the integral.

1. $r = 4 \sin \theta$

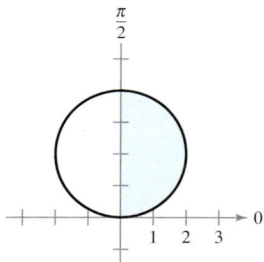

2. $r = \cos 2\theta$

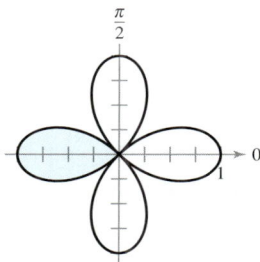

3. $r = 3 - 2 \sin \theta$

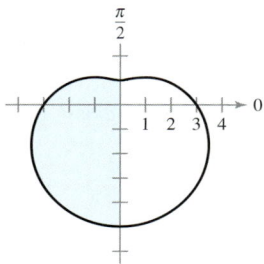

4. $r = 1 - \cos 2\theta$

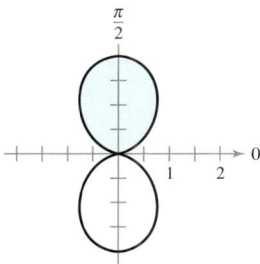

Finding the Area of a Polar Region In Exercises 5–16, find the area of the region.

5. Interior of $r = 6 \sin \theta$

6. Interior of $r = 3 \cos \theta$

7. One petal of $r = 2 \cos 3\theta$

8. One petal of $r = 4 \sin 3\theta$

9. One petal of $r = \sin 2\theta$

10. One petal of $r = \cos 5\theta$

11. Interior of $r = 1 - \sin \theta$

12. Interior of $r = 1 - \sin \theta$ (above the polar axis)

13. Interior of $r = 5 + 2 \sin \theta$

14. Interior of $r = 4 - 4 \cos \theta$

15. Interior of $r^2 = 4 \cos 2\theta$

16. Interior of $r^2 = 6 \sin 2\theta$

Finding the Area of a Polar Region In Exercises 17–24, use a graphing utility to graph the polar equation. Find the area of the given region analytically.

17. Inner loop of $r = 1 + 2 \cos \theta$

18. Inner loop of $r = 2 - 4 \cos \theta$

19. Inner loop of $r = 1 + 2 \sin \theta$

20. Inner loop of $r = 4 - 6 \sin \theta$

21. Between the loops of $r = 1 + 2 \cos \theta$

22. Between the loops of $r = 2(1 + 2 \sin \theta)$

23. Between the loops of $r = 3 - 6 \sin \theta$

24. Between the loops of $r = \frac{1}{2} + \cos \theta$

Finding Points of Intersection In Exercises 25–32, find the points of intersection of the graphs of the equations.

25. $r = 1 + \cos \theta$
$r = 1 - \cos \theta$

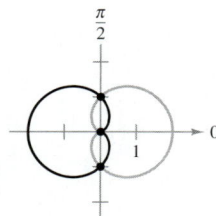

26. $r = 3(1 + \sin \theta)$
$r = 3(1 - \sin \theta)$

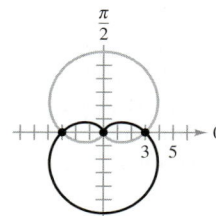

27. $r = 1 + \cos \theta$
$r = 1 - \sin \theta$

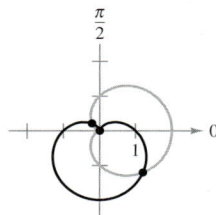

28. $r = 2 - 3 \cos \theta$
$r = \cos \theta$

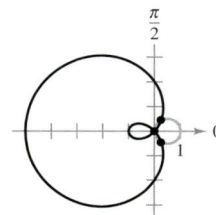

29. $r = 4 - 5 \sin \theta$
$r = 3 \sin \theta$

30. $r = 3 + \sin \theta$
$r = 2 \csc \theta$

31. $r = \frac{\theta}{2}$
$r = 2$

32. $\theta = \frac{\pi}{4}$
$r = 2$

Writing In Exercises 33 and 34, use a graphing utility to graph the polar equations and approximate the points of intersection of the graphs. Watch the graphs as they are traced in the viewing window. Explain why the pole is not a point of intersection obtained by solving the equations simultaneously.

33. $r = \cos \theta$
$r = 2 - 3 \sin \theta$

34. $r = 4 \sin \theta$
$r = 2(1 + \sin \theta)$

Finding the Area of a Polar Region Between Two Curves In Exercises 35–42, use a graphing utility to graph the polar equations. Find the area of the given region analytically.

35. Common interior of $r = 4 \sin 2\theta$ and $r = 2$

36. Common interior of $r = 2(1 + \cos \theta)$ and $r = 2(1 - \cos \theta)$

37. Common interior of $r = 3 - 2 \sin \theta$ and $r = -3 + 2 \sin \theta$

38. Common interior of $r = 5 - 3 \sin \theta$ and $r = 5 - 3 \cos \theta$

39. Common interior of $r = 4 \sin \theta$ and $r = 2$

40. Common interior of $r = 2 \cos \theta$ and $r = 2 \sin \theta$

41. Inside $r = 2 \cos \theta$ and outside $r = 1$

42. Inside $r = 3 \sin \theta$ and outside $r = 1 + \sin \theta$

Finding the Area of a Polar Region Between Two Curves In Exercises 43–46, find the area of the region.

43. Inside $r = a(1 + \cos \theta)$ and outside $r = a \cos \theta$

44. Inside $r = 2a \cos \theta$ and outside $r = a$

45. Common interior of $r = a(1 + \cos \theta)$ and $r = a \sin \theta$

46. Common interior of $r = a \cos \theta$ and $r = a \sin \theta$, where $a > 0$

47. Antenna Radiation

The radiation from a transmitting antenna is not uniform in all directions. The intensity from a particular antenna is modeled by $r = a \cos^2 \theta$.

(a) Convert the polar equation to rectangular form.

(b) Use a graphing utility to graph the model for $a = 4$ and $a = 6$.

(c) Find the area of the geographical region between the two curves in part (b).

48. Area The area inside one or more of the three interlocking circles

$$r = 2a \cos \theta, \quad r = 2a \sin \theta, \quad \text{and} \quad r = a$$

is divided into seven regions. Find the area of each region.

49. Conjecture Find the area of the region enclosed by

$$r = a \cos(n\theta)$$

for $n = 1, 2, 3, \ldots$. Use the results to make a conjecture about the area enclosed by the function when n is even and when n is odd.

50. Area Sketch the strophoid

$$r = \sec \theta - 2 \cos \theta, \quad -\frac{\pi}{2} < \theta < \frac{\pi}{2}.$$

Convert this equation to rectangular coordinates. Find the area enclosed by the loop.

Finding the Arc Length of a Polar Curve In Exercises 51–56, find the length of the curve over the given interval.

Polar Equation	Interval
51. $r = 8$	$0 \leq \theta \leq 2\pi$
52. $r = a$	$0 \leq \theta \leq 2\pi$
53. $r = 4 \sin \theta$	$0 \leq \theta \leq \pi$
54. $r = 2a \cos \theta$	$-\dfrac{\pi}{2} \leq \theta \leq \dfrac{\pi}{2}$
55. $r = 1 + \sin \theta$	$0 \leq \theta \leq 2\pi$
56. $r = 8(1 + \cos \theta)$	$0 \leq \theta \leq 2\pi$

Finding the Arc Length of a Polar Curve In Exercises 57–62, use a graphing utility to graph the polar equation over the given interval. Use the integration capabilities of the graphing utility to approximate the length of the curve accurate to two decimal places.

Polar Equation	Interval
57. $r = 2\theta$	$0 \leq \theta \leq \dfrac{\pi}{2}$
58. $r = \sec \theta$	$0 \leq \theta \leq \dfrac{\pi}{3}$
59. $r = \dfrac{1}{\theta}$	$\pi \leq \theta \leq 2\pi$
60. $r = e^{\theta}$	$0 \leq \theta \leq \pi$
61. $r = \sin(3 \cos \theta)$	$0 \leq \theta \leq \pi$
62. $r = 2 \sin(2 \cos \theta)$	$0 \leq \theta \leq \pi$

Finding the Area of a Surface of Revolution In Exercises 63–66, find the area of the surface formed by revolving the curve about the given line.

Polar Equation	Interval	Axis of Revolution
63. $r = 6 \cos \theta$	$0 \leq \theta \leq \dfrac{\pi}{2}$	Polar axis
64. $r = a \cos \theta$	$0 \leq \theta \leq \dfrac{\pi}{2}$	$\theta = \dfrac{\pi}{2}$
65. $r = e^{a\theta}$	$0 \leq \theta \leq \dfrac{\pi}{2}$	$\theta = \dfrac{\pi}{2}$
66. $r = a(1 + \cos \theta)$	$0 \leq \theta \leq \pi$	Polar axis

Finding the Area of a Surface of Revolution In Exercises 67 and 68, use the integration capabilities of a graphing utility to approximate, to two decimal places, the area of the surface formed by revolving the curve about the polar axis.

67. $r = 4 \cos 2\theta, \quad 0 \leq \theta \leq \dfrac{\pi}{4}$

68. $r = \theta, \quad 0 \leq \theta \leq \pi$

WRITING ABOUT CONCEPTS

69. Points of Intersection Explain why finding points of intersection of polar graphs may require further analysis beyond solving two equations simultaneously.

70. Area of a Surface of Revolution Give the integral formulas for the area of the surface of revolution formed when the graph of $r = f(\theta)$ is revolved about

(a) the polar axis.

(b) the line $\theta = \pi/2$.

71. Area of a Region For each polar equation, sketch its graph, determine the interval that traces the graph only once, and find the area of the region bounded by the graph using a geometric formula and integration.

(a) $r = 10 \cos \theta$ (b) $r = 5 \sin \theta$

72. **HOW DO YOU SEE IT?** Which graph, traced out only once, has a larger arc length? Explain your reasoning.

(a)

(b)

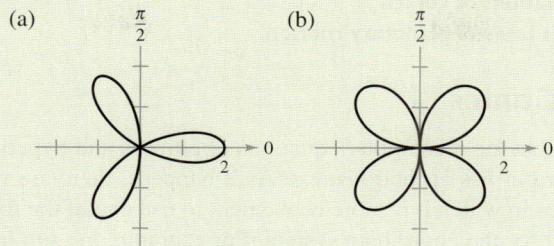

73. Surface Area of a Torus Find the surface area of the torus generated by revolving the circle given by $r = 2$ about the line $r = 5 \sec \theta$.

74. Surface Area of a Torus Find the surface area of the torus generated by revolving the circle given by $r = a$ about the line $r = b \sec \theta$, where $0 < a < b$.

75. Approximating Area Consider the circle $r = 8 \cos \theta$.

(a) Find the area of the circle.

(b) Complete the table giving the areas A of the sectors of the circle between $\theta = 0$ and the values of θ in the table.

θ	0.2	0.4	0.6	0.8	1.0	1.2	1.4
A							

(c) Use the table in part (b) to approximate the values of θ for which the sector of the circle composes $\frac{1}{4}$, $\frac{1}{2}$, and $\frac{3}{4}$ of the total area of the circle.

(d) Use a graphing utility to approximate, to two decimal places, the angles θ for which the sector of the circle composes $\frac{1}{4}$, $\frac{1}{2}$, and $\frac{3}{4}$ of the total area of the circle.

(e) Do the results of part (d) depend on the radius of the circle? Explain.

76. Approximating Area Consider the circle $r = 3 \sin \theta$.

(a) Find the area of the circle.

(b) Complete the table giving the areas A of the sectors of the circle between $\theta = 0$ and the values of θ in the table.

θ	0.2	0.4	0.6	0.8	1.0	1.2	1.4
A							

(c) Use the table in part (b) to approximate the values of θ for which the sector of the circle composes $\frac{1}{8}$, $\frac{1}{4}$, and $\frac{1}{2}$ of the total area of the circle.

(d) Use a graphing utility to approximate, to two decimal places, the angles θ for which the sector of the circle composes $\frac{1}{8}$, $\frac{1}{4}$, and $\frac{1}{2}$ of the total area of the circle.

77. Conic What conic section does the polar equation $r = a \sin \theta + b \cos \theta$ represent?

78. Area Find the area of the circle given by

$$r = \sin \theta + \cos \theta.$$

Check your result by converting the polar equation to rectangular form, then using the formula for the area of a circle.

79. Spiral of Archimedes The curve represented by the equation $r = a\theta$, where a is a constant, is called the spiral of Archimedes.

(a) Use a graphing utility to graph $r = \theta$, where $\theta \geq 0$. What happens to the graph of $r = a\theta$ as a increases? What happens if $\theta \leq 0$?

(b) Determine the points on the spiral $r = a\theta$ ($a > 0$, $\theta \geq 0$), where the curve crosses the polar axis.

(c) Find the length of $r = \theta$ over the interval $0 \leq \theta \leq 2\pi$.

(d) Find the area under the curve $r = \theta$ for $0 \leq \theta \leq 2\pi$.

80. Logarithmic Spiral The curve represented by the equation $r = ae^{b\theta}$, where a and b are constants, is called a **logarithmic spiral.** The figure shows the graph of $r = e^{\theta/6}$, $-2\pi \leq \theta \leq 2\pi$. Find the area of the shaded region.

Figure for 80

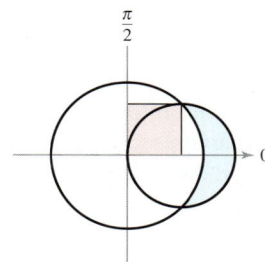

Figure for 81

81. Area The larger circle in the figure is the graph of $r = 1$. Find the polar equation of the smaller circle such that the shaded regions are equal.

82. Folium of Descartes A curve called the **folium of Descartes** can be represented by the parametric equations

$$x = \frac{3t}{1 + t^3} \quad \text{and} \quad y = \frac{3t^2}{1 + t^3}.$$

(a) Convert the parametric equations to polar form.

(b) Sketch the graph of the polar equation from part (a).

(c) Use a graphing utility to approximate the area enclosed by the loop of the curve.

True or False? In Exercises 83 and 84, determine whether the statement is true or false. If it is false, explain why or give an example that shows it is false.

83. If $f(\theta) > 0$ for all θ and $g(\theta) < 0$ for all θ, then the graphs of $r = f(\theta)$ and $r = g(\theta)$ do not intersect.

84. If $f(\theta) = g(\theta)$ for $\theta = 0$, $\pi/2$, and $3\pi/2$, then the graphs of $r = f(\theta)$ and $r = g(\theta)$ have at least four points of intersection.

85. Arc Length in Polar Form Use the formula for the arc length of a curve in parametric form to derive the formula for the arc length of a polar curve.

10.6 Polar Equations of Conics and Kepler's Laws

■ Analyze and write polar equations of conics.
■ Understand and use Kepler's Laws of planetary motion.

Polar Equations of Conics

In this chapter, you have seen that the rectangular equations of ellipses and hyperbolas take simple forms when the origin lies at their *centers*. As it happens, there are many important applications of conics in which it is more convenient to use one of the foci as the reference point (the origin) for the coordinate system. For example, the sun lies at a focus of Earth's orbit. Similarly, the light source of a parabolic reflector lies at its focus. In this section, you will see that polar equations of conics take simpler forms when one of the foci lies at the pole.

The next theorem uses the concept of *eccentricity*, as defined in Section 10.1, to classify the three basic types of conics.

THEOREM 10.16   Classification of Conics by Eccentricity

Let F be a fixed point (*focus*) and let D be a fixed line (*directrix*) in the plane. Let P be another point in the plane and let e (*eccentricity*) be the ratio of the distance between P and F to the distance between P and D. The collection of all points P with a given eccentricity is a conic.

1. The conic is an ellipse for $0 < e < 1$.
2. The conic is a parabola for $e = 1$.
3. The conic is a hyperbola for $e > 1$.

A proof of this theorem is given in Appendix A.
See LarsonCalculus.com for Bruce Edwards's video of this proof.

In Figure 10.57, note that for each type of conic, the pole corresponds to the fixed point (focus) given in the definition.

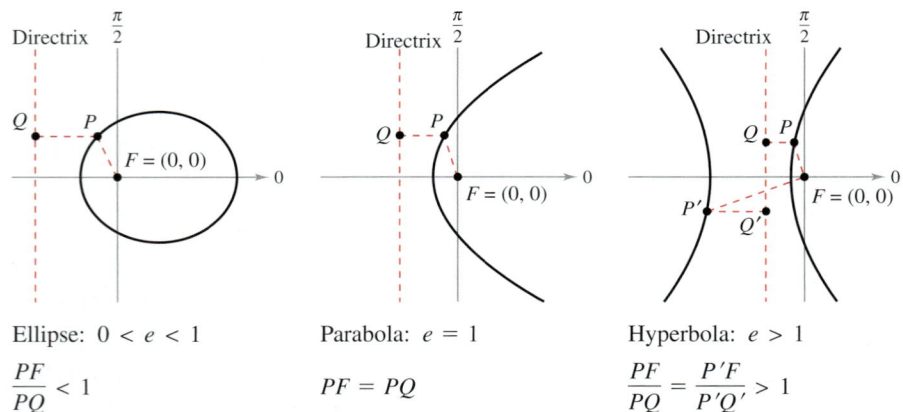

Ellipse: $0 < e < 1$

$$\frac{PF}{PQ} < 1$$

Parabola: $e = 1$

$$PF = PQ$$

Hyperbola: $e > 1$

$$\frac{PF}{PQ} = \frac{P'F}{P'Q'} > 1$$

Figure 10.57

The benefit of locating a focus of a conic at the pole is that the equation of the conic becomes simpler, as seen in the proof of the next theorem.

THEOREM 10.17 Polar Equations of Conics

The graph of a polar equation of the form

$$r = \frac{ed}{1 \pm e \cos \theta} \quad \text{or} \quad r = \frac{ed}{1 \pm e \sin \theta}$$

is a conic, where $e > 0$ is the eccentricity and $|d|$ is the distance between the focus at the pole and its corresponding directrix.

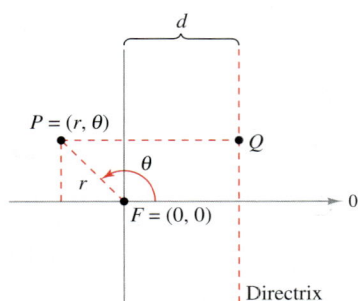

Figure 10.58

Proof This is a proof for $r = ed/(1 + e \cos \theta)$ with $d > 0$. In Figure 10.58, consider a vertical directrix d units to the right of the focus $F = (0, 0)$. If $P = (r, \theta)$ is a point on the graph of $r = ed/(1 + e \cos \theta)$, then the distance between P and the directrix can be shown to be

$$PQ = |d - x| = |d - r \cos \theta| = \left| \frac{r(1 + e \cos \theta)}{e} - r \cos \theta \right| = \left| \frac{r}{e} \right|.$$

Because the distance between P and the pole is simply $PF = |r|$, the ratio of PF to PQ is

$$\frac{PF}{PQ} = \frac{|r|}{|r/e|} = |e| = e$$

and, by Theorem 10.16, the graph of the equation must be a conic. The proofs of the other cases are similar.

See LarsonCalculus.com for Bruce Edwards's video of this proof.

The four types of equations indicated in Theorem 10.17 can be classified as follows, where $d > 0$.

a. Horizontal directrix above the pole: $r = \dfrac{ed}{1 + e \sin \theta}$

b. Horizontal directrix below the pole: $r = \dfrac{ed}{1 - e \sin \theta}$

c. Vertical directrix to the right of the pole: $r = \dfrac{ed}{1 + e \cos \theta}$

d. Vertical directrix to the left of the pole: $r = \dfrac{ed}{1 - e \cos \theta}$

Figure 10.59 illustrates these four possibilities for a parabola. Note that for convenience, the equation for the directrix is shown in rectangular form.

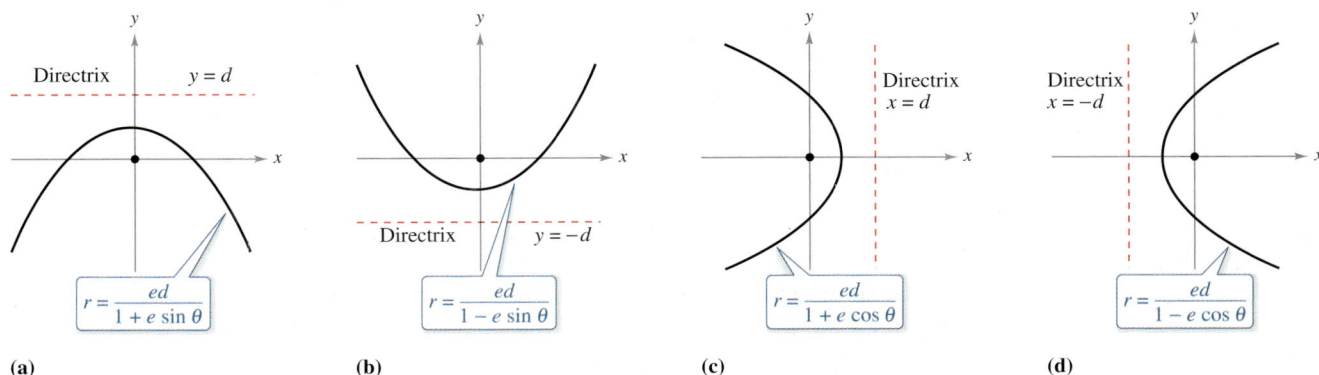

(a) (b) (c) (d)

The four types of polar equations for a parabola

Figure 10.59

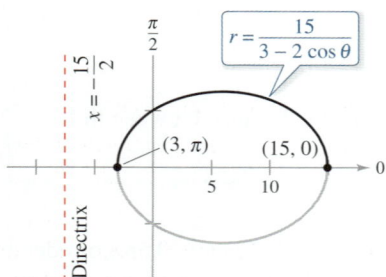

$r = \dfrac{15}{3 - 2\cos\theta}$

The graph of the conic is an ellipse with $e = \frac{2}{3}$.
Figure 10.60

EXAMPLE 1 **Determining a Conic from Its Equation**

Sketch the graph of the conic $r = \dfrac{15}{3 - 2\cos\theta}$.

Solution To determine the type of conic, rewrite the equation as

$$r = \frac{15}{3 - 2\cos\theta} \qquad \text{Write original equation.}$$

$$= \frac{5}{1 - (2/3)\cos\theta}. \qquad \begin{array}{l}\text{Divide numerator and}\\\text{denominator by 3.}\end{array}$$

So, the graph is an ellipse with $e = \frac{2}{3}$. You can sketch the upper half of the ellipse by plotting points from $\theta = 0$ to $\theta = \pi$, as shown in Figure 10.60. Then, using symmetry with respect to the polar axis, you can sketch the lower half.

For the ellipse in Figure 10.60, the major axis is horizontal and the vertices lie at $(15, 0)$ and $(3, \pi)$. So, the length of the *major* axis is $2a = 18$. To find the length of the minor axis, you can use the equations $e = c/a$ and $b^2 = a^2 - c^2$ to conclude that

$$b^2 = a^2 - c^2 = a^2 - (ea)^2 = a^2(1 - e^2). \qquad \text{Ellipse}$$

Because $e = \frac{2}{3}$, you have

$$b^2 = 9^2\left[1 - \left(\tfrac{2}{3}\right)^2\right] = 45$$

which implies that $b = \sqrt{45} = 3\sqrt{5}$. So, the length of the minor axis is $2b = 6\sqrt{5}$. A similar analysis for hyperbolas yields

$$b^2 = c^2 - a^2 = (ea)^2 - a^2 = a^2(e^2 - 1). \qquad \text{Hyperbola}$$

EXAMPLE 2 **Sketching a Conic from Its Polar Equation**

• • • ▷ *See LarsonCalculus.com for an interactive version of this type of example.*

Sketch the graph of the polar equation $r = \dfrac{32}{3 + 5\sin\theta}$.

Solution Dividing the numerator and denominator by 3 produces

$$r = \frac{32/3}{1 + (5/3)\sin\theta}.$$

Because $e = \frac{5}{3} > 1$, the graph is a hyperbola. Because $d = \frac{32}{5}$, the directrix is the line $y = \frac{32}{5}$. The transverse axis of the hyperbola lies on the line $\theta = \pi/2$, and the vertices occur at

$$(r, \theta) = \left(4, \frac{\pi}{2}\right) \quad \text{and} \quad (r, \theta) = \left(-16, \frac{3\pi}{2}\right).$$

Because the length of the transverse axis is 12, you can see that $a = 6$. To find b, write

$$b^2 = a^2(e^2 - 1) = 6^2\left[\left(\frac{5}{3}\right)^2 - 1\right] = 64.$$

Therefore, $b = 8$. Finally, you can use a and b to determine the asymptotes of the hyperbola and obtain the sketch shown in Figure 10.61.

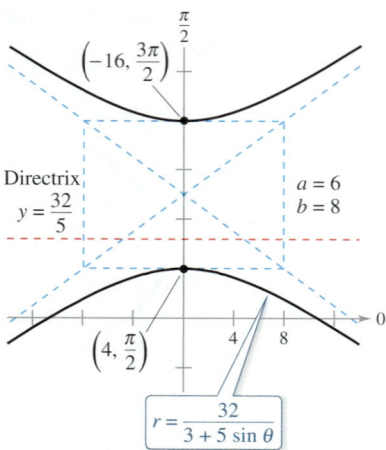

Directrix $y = \dfrac{32}{5}$

$a = 6$
$b = 8$

$r = \dfrac{32}{3 + 5\sin\theta}$

The graph of the conic is a hyperbola with $e = \frac{5}{3}$.
Figure 10.61

Kepler's Laws

Kepler's Laws, named after the German astronomer Johannes Kepler, can be used to describe the orbits of the planets about the sun.

1. Each planet moves in an elliptical orbit with the sun as a focus.
2. A ray from the sun to the planet sweeps out equal areas of the ellipse in equal times.
3. The square of the period is proportional to the cube of the mean distance between the planet and the sun.*

Although Kepler derived these laws empirically, they were later validated by Newton. In fact, Newton was able to show that each law can be deduced from a set of universal laws of motion and gravitation that govern the movement of all heavenly bodies, including comets and satellites. This is shown in the next example, involving the comet named after the English mathematician and physicist Edmund Halley (1656–1742).

EXAMPLE 3 Halley's Comet

Halley's comet has an elliptical orbit with the sun at one focus and has an eccentricity of $e \approx 0.967$. The length of the major axis of the orbit is approximately 35.88 astronomical units (AU). (An astronomical unit is defined as the mean distance between Earth and the sun, 93 million miles.) Find a polar equation for the orbit. How close does Halley's comet come to the sun?

Solution Using a vertical axis, you can choose an equation of the form

$$r = \frac{ed}{(1 + e \sin \theta)}.$$

Because the vertices of the ellipse occur when $\theta = \pi/2$ and $\theta = 3\pi/2$, you can determine the length of the major axis to be the sum of the r-values of the vertices, as shown in Figure 10.62. That is,

$$2a = \frac{0.967d}{1 + 0.967} + \frac{0.967d}{1 - 0.967}$$

$$35.88 \approx 29.79d. \qquad \text{\textcolor{red}{$2a \approx 35.88$}}$$

So, $d \approx 1.204$ and

$$ed \approx (0.967)(1.204) \approx 1.164.$$

Using this value in the equation produces

$$r = \frac{1.164}{1 + 0.967 \sin \theta}$$

where r is measured in astronomical units. To find the closest point to the sun (the focus), you can write

$$c = ea \approx (0.967)(17.94) \approx 17.35.$$

Because c is the distance between the focus and the center, the closest point is

$$a - c \approx 17.94 - 17.35$$
$$\approx 0.59 \text{ AU}$$
$$\approx 55,000,000 \text{ miles.}$$

JOHANNES KEPLER (1571–1630)

Kepler formulated his three laws from the extensive data recorded by Danish astronomer Tycho Brahe, and from direct observation of the orbit of Mars.
See LarsonCalculus.com to read more of this biography.

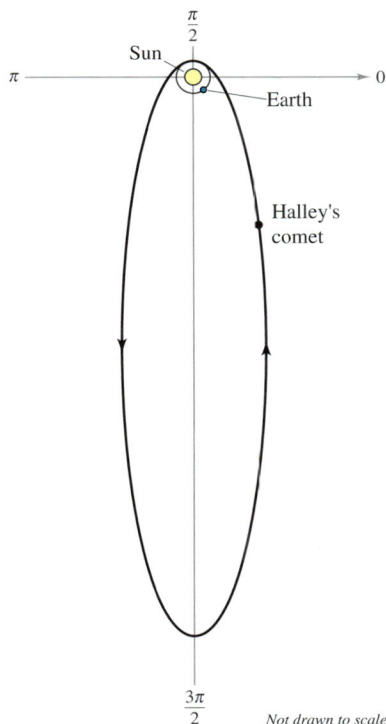

Figure 10.62

Not drawn to scale

---

* If Earth is used as a reference with a period of 1 year and a distance of 1 astronomical unit, then the proportionality constant is 1. For example, because Mars has a mean distance to the sun of $D \approx 1.524$ AU, its period P is $D^3 = P^2$. So, the period for Mars is $P \approx 1.88$.

Kepler's Second Law states that as a planet moves about the sun, a ray from the sun to the planet sweeps out equal areas in equal times. This law can also be applied to comets or asteroids with elliptical orbits. For example, Figure 10.63 shows the orbit of the asteroid Apollo about the sun. Applying Kepler's Second Law to this asteroid, you know that the closer it is to the sun, the greater its velocity, because a short ray must be moving quickly to sweep out as much area as a long ray.

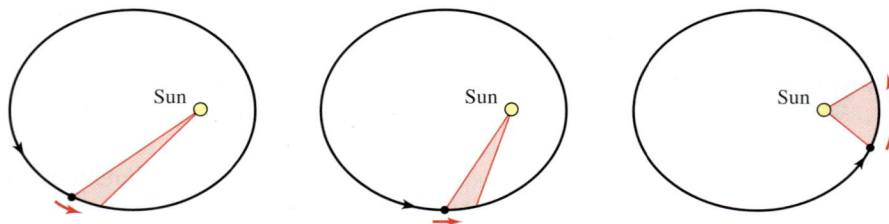

A ray from the sun to the asteroid Apollo sweeps out equal areas in equal times.
Figure 10.63

EXAMPLE 4 **The Asteroid Apollo**

The asteroid Apollo has a period of 661 Earth days, and its orbit is approximated by the ellipse

$$r = \frac{1}{1 + (5/9) \cos \theta} = \frac{9}{9 + 5 \cos \theta}$$

where r is measured in astronomical units. How long does it take Apollo to move from the position $\theta = -\pi/2$ to $\theta = \pi/2$, as shown in Figure 10.64?

Solution Begin by finding the area swept out as θ increases from $-\pi/2$ to $\pi/2$.

$$A = \frac{1}{2} \int_\alpha^\beta r^2 \, d\theta \qquad \text{Formula for area of a polar graph}$$

$$= \frac{1}{2} \int_{-\pi/2}^{\pi/2} \left(\frac{9}{9 + 5 \cos \theta} \right)^2 d\theta$$

Using the substitution $u = \tan(\theta/2)$, as discussed in Section 8.6, you obtain

$$A = \frac{81}{112} \left[\frac{-5 \sin \theta}{9 + 5 \cos \theta} + \frac{18}{\sqrt{56}} \arctan \frac{\sqrt{56} \tan(\theta/2)}{14} \right]_{-\pi/2}^{\pi/2} \approx 0.90429.$$

Because the major axis of the ellipse has length $2a = 81/28$ and the eccentricity is $e = 5/9$, you can determine that

$$b = a\sqrt{1 - e^2} = \frac{9}{\sqrt{56}}.$$

So, the area of the ellipse is

$$\text{Area of ellipse} = \pi ab = \pi \left(\frac{81}{56} \right) \left(\frac{9}{\sqrt{56}} \right) \approx 5.46507.$$

Because the time required to complete the orbit is 661 days, you can apply Kepler's Second Law to conclude that the time t required to move from the position $\theta = -\pi/2$ to $\theta = \pi/2$ is

$$\frac{t}{661} = \frac{\text{area of elliptical segment}}{\text{area of ellipse}} \approx \frac{0.90429}{5.46507}$$

which implies that $t \approx 109$ days.

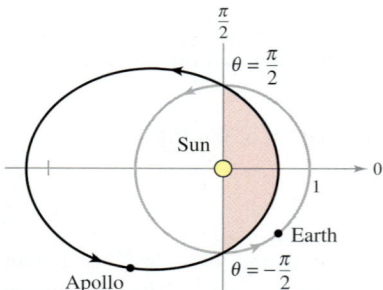

Apollo

$\theta = \frac{\pi}{2}$

Sun

Earth

$\theta = -\frac{\pi}{2}$

Figure 10.64

10.6 Exercises

Graphical Reasoning In Exercises 1–4, use a graphing utility to graph the polar equation when (a) $e = 1$, (b) $e = 0.5$, and (c) $e = 1.5$. Identify the conic.

1. $r = \dfrac{2e}{1 + e \cos \theta}$ **2.** $r = \dfrac{2e}{1 - e \cos \theta}$

3. $r = \dfrac{2e}{1 - e \sin \theta}$ **4.** $r = \dfrac{2e}{1 + e \sin \theta}$

5. Writing Consider the polar equation

$$r = \frac{4}{1 + e \sin \theta}.$$

(a) Use a graphing utility to graph the equation for $e = 0.1$, $e = 0.25$, $e = 0.5$, $e = 0.75$, and $e = 0.9$. Identify the conic and discuss the change in its shape as $e \to 1^-$ and $e \to 0^+$.

(b) Use a graphing utility to graph the equation for $e = 1$. Identify the conic.

(c) Use a graphing utility to graph the equation for $e = 1.1$, $e = 1.5$, and $e = 2$. Identify the conic and discuss the change in its shape as $e \to 1^+$ and $e \to \infty$.

6. Writing Consider the polar equation

$$r = \frac{4}{1 - 0.4 \cos \theta}.$$

(a) Identify the conic without graphing the equation.

(b) Without graphing the following polar equations, describe how each differs from the polar equation above.

$$r = \frac{4}{1 + 0.4 \cos \theta}, \quad r = \frac{4}{1 - 0.4 \sin \theta}$$

(c) Verify the results of part (b) graphically.

Matching In Exercises 7–12, match the polar equation with the correct graph. [The graphs are labeled (a), (b), (c), (d), (e), and (f).]

(a)

(b)

(c)

(d)

(e)

(f)

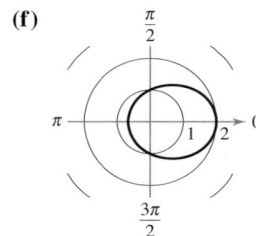

7. $r = \dfrac{6}{1 - \cos \theta}$ **8.** $r = \dfrac{2}{2 - \cos \theta}$

9. $r = \dfrac{3}{1 - 2 \sin \theta}$ **10.** $r = \dfrac{2}{1 + \sin \theta}$

11. $r = \dfrac{6}{2 - \sin \theta}$ **12.** $r = \dfrac{2}{2 + 3 \cos \theta}$

Sketching and Identifying a Conic In Exercises 13–22, find the eccentricity and the distance from the pole to the directrix of the conic. Then sketch and identify the graph. Use a graphing utility to confirm your results.

13. $r = \dfrac{1}{1 - \cos \theta}$ **14.** $r = \dfrac{6}{3 - 2 \cos \theta}$

15. $r = \dfrac{3}{2 + 6 \sin \theta}$ **16.** $r = \dfrac{4}{1 + \cos \theta}$

17. $r = \dfrac{5}{-1 + 2 \cos \theta}$ **18.** $r = \dfrac{10}{5 + 4 \sin \theta}$

19. $r = \dfrac{6}{2 + \cos \theta}$ **20.** $r = \dfrac{-6}{3 + 7 \sin \theta}$

21. $r = \dfrac{300}{-12 + 6 \sin \theta}$ **22.** $r = \dfrac{1}{1 + \sin \theta}$

Identifying a Conic In Exercises 23–26, use a graphing utility to graph the polar equation. Identify the graph and find its eccentricity.

23. $r = \dfrac{3}{-4 + 2 \sin \theta}$ **24.** $r = \dfrac{-15}{2 + 8 \sin \theta}$

25. $r = \dfrac{-10}{1 - \cos \theta}$ **26.** $r = \dfrac{6}{6 + 7 \cos \theta}$

Comparing Graphs In Exercises 27–30, use a graphing utility to graph the conic. Describe how the graph differs from the graph in the indicated exercise.

27. $r = \dfrac{4}{1 + \cos(\theta - \pi/3)}$ (See Exercise 16.)

28. $r = \dfrac{10}{5 + 4 \sin(\theta - \pi/4)}$ (See Exercise 18.)

29. $r = \dfrac{6}{2 + \cos(\theta + \pi/6)}$ (See Exercise 19.)

30. $r = \dfrac{-6}{3 + 7 \sin(\theta + 2\pi/3)}$ (See Exercise 20.)

31. Rotated Ellipse Write the equation for the ellipse rotated $\pi/6$ radian clockwise from the ellipse

$$r = \frac{8}{8 + 5\cos\theta}.$$

32. Rotated Parabola Write the equation for the parabola rotated $\pi/4$ radian counterclockwise from the parabola

$$r = \frac{9}{1 + \sin\theta}.$$

Finding a Polar Equation In Exercises 33–44, find a polar equation for the conic with its focus at the pole. (For convenience, the equation for the directrix is given in rectangular form.)

Conic	Eccentricity	Directrix
33. Parabola	$e = 1$	$x = -3$
34. Parabola	$e = 1$	$y = 4$
35. Ellipse	$e = \frac{1}{2}$	$y = 1$
36. Ellipse	$e = \frac{3}{4}$	$y = -2$
37. Hyperbola	$e = 2$	$x = 1$
38. Hyperbola	$e = \frac{3}{2}$	$x = -1$

Conic	Vertex or Vertices
39. Parabola	$\left(1, -\frac{\pi}{2}\right)$
40. Parabola	$(5, \pi)$
41. Ellipse	$(2, 0), (8, \pi)$
42. Ellipse	$\left(2, \frac{\pi}{2}\right), \left(4, \frac{3\pi}{2}\right)$
43. Hyperbola	$\left(1, \frac{3\pi}{2}\right), \left(9, \frac{3\pi}{2}\right)$
44. Hyperbola	$(2, 0), (10, 0)$

45. Finding a Polar Equation Find a polar equation for the ellipse with focus $(0, 0)$, eccentricity $\frac{1}{2}$, and a directrix at $r = 4\sec\theta$.

46. Finding a Polar Equation Find a polar equation for the hyperbola with focus $(0, 0)$, eccentricity 2, and a directrix at $r = -8\csc\theta$.

WRITING ABOUT CONCEPTS

47. Eccentricity Classify the conics by their eccentricities.

48. Identifying Conics Identify each conic.

(a) $r = \dfrac{5}{1 - 2\cos\theta}$ (b) $r = \dfrac{5}{10 - \sin\theta}$

(c) $r = \dfrac{5}{3 - 3\cos\theta}$ (d) $r = \dfrac{5}{1 - 3\sin(\theta - \pi/4)}$

49. Distance Describe what happens to the distance between the directrix and the center of an ellipse when the foci remain fixed and e approaches 0.

---

50. HOW DO YOU SEE IT? Identify the conic in the graph and give the possible values for the eccentricity.

(a)

(b)

(c)

(d)

51. Ellipse Show that the polar equation for $\dfrac{x^2}{a^2} + \dfrac{y^2}{b^2} = 1$ is

$$r^2 = \frac{b^2}{1 - e^2\cos^2\theta}. \qquad \text{Ellipse}$$

52. Hyperbola Show that the polar equation for $\dfrac{x^2}{a^2} - \dfrac{y^2}{b^2} = 1$ is

$$r^2 = \frac{-b^2}{1 - e^2\cos^2\theta}. \qquad \text{Hyperbola}$$

Finding a Polar Equation In Exercises 53–56, use the results of Exercises 51 and 52 to write the polar form of the equation of the conic.

53. Ellipse: focus at $(4, 0)$; vertices at $(5, 0)$, $(5, \pi)$

54. Hyperbola: focus at $(5, 0)$; vertices at $(4, 0)$, $(4, \pi)$

55. $\dfrac{x^2}{9} - \dfrac{y^2}{16} = 1$

56. $\dfrac{x^2}{4} + y^2 = 1$

Area of a Region In Exercises 57–60, use the integration capabilities of a graphing utility to approximate, to two decimal places, the area of the region bounded by the graph of the polar equation.

57. $r = \dfrac{3}{2 - \cos\theta}$ **58.** $r = \dfrac{9}{4 + \cos\theta}$

59. $r = \dfrac{2}{3 - 2\sin\theta}$ **60.** $r = \dfrac{3}{6 + 5\sin\theta}$

61. Explorer 18 On November 27, 1963, the United States launched Explorer 18. Its low and high points above the surface of Earth were approximately 119 miles and 123,000 miles (see figure). The center of Earth is a focus of the orbit. Find the polar equation for the orbit and find the distance between the surface of Earth and the satellite when $\theta = 60°$. (Assume that the radius of Earth is 4000 miles.)

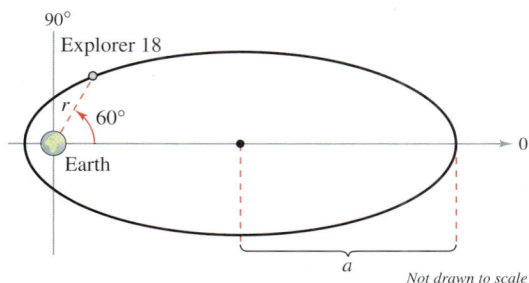

Not drawn to scale

62. Planetary Motion The planets travel in elliptical orbits with the sun as a focus, as shown in the figure.

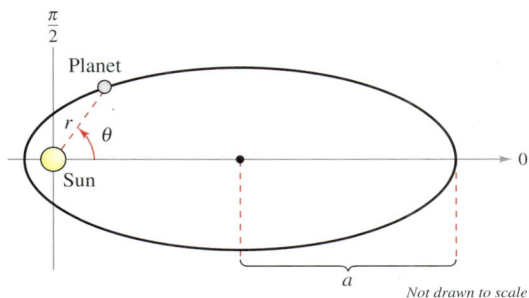

Not drawn to scale

(a) Show that the polar equation of the orbit is given by

$$r = \frac{(1 - e^2)a}{1 - e \cos \theta}$$

where e is the eccentricity.

(b) Show that the minimum distance (*perihelion*) from the sun to the planet is $r = a(1 - e)$ and the maximum distance (*aphelion*) is $r = a(1 + e)$.

Planetary Motion In Exercises 63–66, use Exercise 62 to find the polar equation of the elliptical orbit of the planet, and the perihelion and aphelion distances.

63. Earth $a = 1.496 \times 10^8$ kilometers
                   $e = 0.0167$

64. Saturn $a = 1.427 \times 10^9$ kilometers
                   $e = 0.0542$

65. Neptune $a = 4.498 \times 10^9$ kilometers
                   $e = 0.0086$

66. Mercury $a = 5.791 \times 10^7$ kilometers
                   $e = 0.2056$

NASA

• • 67. Planetary Motion • • • • • • • • • • • • • • • •

In Exercise 65, the polar equation for the elliptical orbit of Neptune was found. Use the equation and a computer algebra system to perform each of the following.

(a) Approximate the area swept out by a ray from the sun to the planet as θ increases from 0 to $\pi/9$. Use this result to determine the number of years required for the planet to move through this arc when the period of one revolution around the sun is 165 years.

(b) By trial and error, approximate the angle α such that the area swept out by a ray from the sun to the planet as θ increases from π to α equals the area found in part (a) (see figure). Does the ray sweep through a larger or smaller angle than in part (a) to generate the same area? Why is this the case?

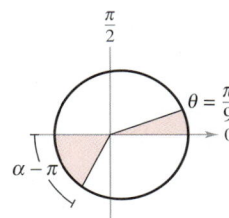

(c) Approximate the distances the planet traveled in parts (a) and (b). Use these distances to approximate the average number of kilometers per year the planet traveled in the two cases.

68. Comet Hale-Bopp The comet Hale-Bopp has an elliptical orbit with the sun at one focus and has an eccentricity of $e \approx 0.995$. The length of the major axis of the orbit is approximately 500 astronomical units.

(a) Find the length of its minor axis.

(b) Find a polar equation for the orbit.

(c) Find the perihelion and aphelion distances.

Eccentricity In Exercises 69 and 70, let r_0 represent the distance from a focus to the nearest vertex, and let r_1 represent the distance from the focus to the farthest vertex.

69. Show that the eccentricity of an ellipse can be written as $e = \dfrac{r_1 - r_0}{r_1 + r_0}$. Then show that $\dfrac{r_1}{r_0} = \dfrac{1 + e}{1 - e}$.

70. Show that the eccentricity of a hyperbola can be written as $e = \dfrac{r_1 + r_0}{r_1 - r_0}$. Then show that $\dfrac{r_1}{r_0} = \dfrac{e + 1}{e - 1}$.

Matching In Exercises 1–6, match the equation with the correct graph. [The graphs are labeled (a), (b), (c), (d), (e), and (f).]

(a)

(b)

(c)

(d)

(e)

(f)

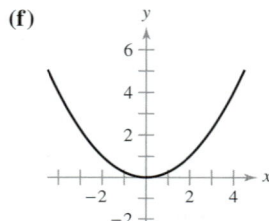

1. $4x^2 + y^2 = 4$

2. $4x^2 - y^2 = 4$

3. $y^2 = -4x$

4. $y^2 - 4x^2 = 4$

5. $x^2 + 4y^2 = 4$

6. $x^2 = 4y$

Identifying a Conic In Exercises 7–14, identify the conic, analyze the equation (center, radius, vertices, foci, eccentricity, directrix, and asymptotes, if possible), and sketch its graph. Use a graphing utility to confirm your results.

7. $16x^2 + 16y^2 - 16x + 24y - 3 = 0$

8. $y^2 - 12y - 8x + 20 = 0$

9. $3x^2 - 2y^2 + 24x + 12y + 24 = 0$

10. $5x^2 + y^2 - 20x + 19 = 0$

11. $3x^2 + 2y^2 - 12x + 12y + 29 = 0$

12. $12x^2 - 12y^2 - 12x + 24y - 45 = 0$

13. $x^2 - 6x - 8y + 1 = 0$

14. $9x^2 + 25y^2 + 18x - 100y - 116 = 0$

Finding an Equation of a Parabola In Exercises 15 and 16, find an equation of the parabola.

15. Vertex: $(0, 2)$
    Directrix: $x = -3$

16. Vertex: $(2, 6)$
    Focus: $(2, 4)$

Finding an Equation of an Ellipse In Exercises 17–20, find an equation of the ellipse.

17. Center: $(0, 0)$
    Focus: $(5, 0)$
    Vertex: $(7, 0)$

18. Center: $(0, 0)$
    Major axis: vertical
    Points on the ellipse:
    $(1, 2), (2, 0)$

19. Vertices: $(3, 1), (3, 7)$
    Eccentricity: $\frac{2}{3}$

20. Foci: $(0, \pm 7)$
    Major axis length: 20

Finding an Equation of a Hyperbola In Exercises 21–24, find an equation of the hyperbola.

21. Vertices: $(0, \pm 8)$
    Asymptotes: $y = \pm 2x$

22. Vertices: $(\pm 2, 0)$
    Asymptotes: $y = \pm 32x$

23. Vertices: $(\pm 7, -1)$
    Foci: $(\pm 9, -1)$

24. Center: $(0, 0)$
    Vertex: $(0, 3)$
    Focus: $(0, 6)$

25. **Satellite Antenna** A cross section of a large parabolic antenna is modeled by the graph of

$$y = \frac{x^2}{200}, \quad -100 \le x \le 100.$$

The receiving and transmitting equipment is positioned at the focus.

(a) Find the coordinates of the focus.

(b) Find the surface area of the antenna.

26. **Using an Ellipse** Consider the ellipse $\dfrac{x^2}{25} + \dfrac{y^2}{9} = 1$.

(a) Find the area of the region bounded by the ellipse.

(b) Find the volume of the solid generated by revolving the region about its major axis.

Using Parametric Equations In Exercises 27–34, sketch the curve represented by the parametric equations (indicate the orientation of the curve), and write the corresponding rectangular equation by eliminating the parameter.

27. $x = 1 + 8t, \ y = 3 - 4t$

28. $x = t - 6, \ y = t^2$

29. $x = e^t - 1, \ y = e^{3t}$

30. $x = e^{4t}, \ y = t + 4$

31. $x = 6 \cos \theta, \ y = 6 \sin \theta$

32. $x = 2 + 5 \cos t, \ y = 3 + 2 \sin t$

33. $x = 2 + \sec \theta, \ y = 3 + \tan \theta$

34. $x = 5 \sin^3 \theta, \ y = 5 \cos^3 \theta$

Finding Parametric Equations In Exercises 35 and 36, find two different sets of parametric equations for the rectangular equation.

35. $y = 4x + 3$

36. $y = x^2 - 2$

37. Rotary Engine   The rotary engine was developed by Felix Wankel in the 1950s. It features a rotor that is a modified equilateral triangle. The rotor moves in a chamber that, in two dimensions, is an epitrochoid. Use a graphing utility to graph the chamber modeled by the parametric equations

$$x = \cos 3\theta + 5 \cos \theta$$

and

$$y = \sin 3\theta + 5 \sin \theta.$$

38. Serpentine Curve   Consider the parametric equations $x = 2 \cot \theta$ and $y = 4 \sin \theta \cos \theta, 0 < \theta < \pi$.

(a) Use a graphing utility to graph the curve.

(b) Eliminate the parameter to show that the rectangular equation of the serpentine curve is $(4 + x^2)y = 8x$.

Finding Slope and Concavity   In Exercises 39–46, find dy/dx and d^2y/dx^2, and find the slope and concavity (if possible) at the given value of the parameter.

Parametric Equations	Parameter
39. $x = 2 + 5t, \quad y = 1 - 4t$	$t = 3$
40. $x = t - 6, \quad y = t^2$	$t = 5$
41. $x = \dfrac{1}{t}, \quad y = 2t + 3$	$t = -1$
42. $x = \dfrac{1}{t}, \quad y = t^2$	$t = -2$
43. $x = 5 + \cos \theta, \quad y = 3 + 4 \sin \theta$	$\theta = \dfrac{\pi}{6}$
44. $x = 10 \cos \theta, \quad y = 10 \sin \theta$	$\theta = \dfrac{\pi}{4}$
45. $x = \cos^3 \theta, \quad y = 4 \sin^3 \theta$	$\theta = \dfrac{\pi}{3}$
46. $x = e^t, \quad y = e^{-t}$	$t = 1$

Finding an Equation of a Tangent Line   In Exercises 47 and 48, (a) use a graphing utility to graph the curve represented by the parametric equations, (b) use a graphing utility to find $dx/d\theta$, $dy/d\theta$, and dy/dx at the given value of the parameter, (c) find an equation of the tangent line to the curve at the given value of the parameter, and (d) use a graphing utility to graph the curve and the tangent line from part (c).

Parametric Equations	Parameter
47. $x = \cot \theta, \quad y = \sin 2\theta$	$\theta = \dfrac{\pi}{6}$
48. $x = \dfrac{1}{4} \tan \theta, \quad y = 6 \sin \theta$	$\theta = \dfrac{\pi}{3}$

Horizontal and Vertical Tangency   In Exercises 49–52, find all points (if any) of horizontal and vertical tangency to the curve. Use a graphing utility to confirm your results.

49. $x = 5 - t, \quad y = 2t^2$

50. $x = t + 2, \quad y = t^3 - 2t$

51. $x = 2 + 2 \sin \theta, \quad y = 1 + \cos \theta$

52. $x = 2 - 2 \cos \theta, \quad y = 2 \sin 2\theta$

Arc Length   In Exercises 53 and 54, find the arc length of the curve on the given interval.

Parametric Equations	Interval
53. $x = t^2 + 1, \quad y = 4t^3 + 3$	$0 \le t \le 2$
54. $x = 6 \cos \theta, \quad y = 6 \sin \theta$	$0 \le \theta \le \pi$

Surface Area   In Exercises 55 and 56, find the area of the surface generated by revolving the curve about (a) the x-axis and (b) the y-axis.

55. $x = t, \quad y = 3t, \quad 0 \le t \le 2$

56. $x = 2 \cos \theta, \quad y = 2 \sin \theta, \quad 0 \le \theta \le \dfrac{\pi}{2}$

Area   In Exercises 57 and 58, find the area of the region.

57. $x = 3 \sin \theta$
$y = 2 \cos \theta$
$-\dfrac{\pi}{2} \le \theta \le \dfrac{\pi}{2}$

58. $x = 2 \cos \theta$
$y = \sin \theta$
$0 \le \theta \le \pi$

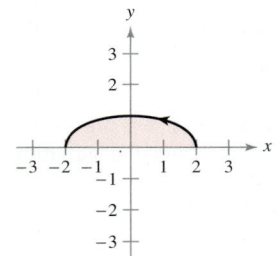

Polar-to-Rectangular Conversion   In Exercises 59–62, plot the point in polar coordinates and find the corresponding rectangular coordinates of the point.

59. $\left(5, \dfrac{3\pi}{2}\right)$

60. $\left(-6, \dfrac{7\pi}{6}\right)$

61. $\left(\sqrt{3}, 1.56\right)$

62. $(-2, -2.45)$

Rectangular-to-Polar Conversion   In Exercises 63–66, the rectangular coordinates of a point are given. Plot the point and find *two* sets of polar coordinates of the point for $0 \le \theta < 2\pi$.

63. $(4, -4)$

64. $(0, -7)$

65. $(-1, 3)$

66. $\left(-\sqrt{3}, -\sqrt{3}\right)$

Rectangular-to-Polar Conversion   In Exercises 67–72, convert the rectangular equation to polar form and sketch its graph.

67. $x^2 + y^2 = 25$

68. $x^2 - y^2 = 4$

69. $y = 9$

70. $x = 6$

71. $x^2 = 4y$

72. $x^2 + y^2 - 4x = 0$

Polar-to-Rectangular Conversion In Exercises 73–78, convert the polar equation to rectangular form and sketch its graph.

73. $r = 3 \cos \theta$

74. $r = 10$

75. $r = 6 \sin \theta$

76. $r = 3 \csc \theta$

77. $r = -2 \sec \theta \tan \theta$

78. $\theta = \dfrac{3\pi}{4}$

Graphing a Polar Equation In Exercises 79–82, use a graphing utility to graph the polar equation.

79. $r = \dfrac{3}{\cos(\theta - \pi/4)}$

80. $r = 2 \sin \theta \, \cos^2 \theta$

81. $r = 4 \cos 2\theta \sec \theta$

82. $r = 4(\sec \theta - \cos \theta)$

Horizontal and Vertical Tangency In Exercises 83 and 84, find the points of horizontal and vertical tangency (if any) to the polar curve.

83. $r = 1 - \cos \theta$

84. $r = 3 \tan \theta$

Tangent Lines at the Pole In Exercises 85 and 86, sketch a graph of the polar equation and find the tangents at the pole.

85. $r = 4 \sin 3\theta$

86. $r = 3 \cos 4\theta$

Sketching a Polar Graph In Exercises 87–96, sketch a graph of the polar equation.

87. $r = 6$

88. $\theta = \dfrac{\pi}{10}$

89. $r = -\sec \theta$

90. $r = 5 \csc \theta$

91. $r^2 = 4 \sin^2 2\theta$

92. $r = 3 - 4 \cos \theta$

93. $r = 4 - 3 \cos \theta$

94. $r = 4\theta$

95. $r = -3 \cos 2\theta$

96. $r = \cos 5\theta$

Finding the Area of a Polar Region In Exercises 97–102, find the area of the region.

97. One petal of $r = 3 \cos 5\theta$

98. One petal of $r = 2 \sin 6\theta$

99. Interior of $r = 2 + \cos \theta$

100. Interior of $r = 5(1 - \sin \theta)$

101. Interior of $r^2 = 4 \sin 2\theta$

102. Common interior of $r = 4 \cos \theta$ and $r = 2$

Finding the Area of a Polar Region In Exercises 103–106, use a graphing utility to graph the polar equation. Find the area of the given region analytically.

103. Inner loop of $r = 3 - 6 \cos \theta$

104. Inner loop of $r = 2 + 4 \sin \theta$

105. Between the loops of $r = 3 - 6 \cos \theta$

106. Between the loops of $r = 2 + 4 \sin \theta$

Finding Points of Intersection In Exercises 107 and 108, find the points of intersection of the graphs of the equations.

107. $r = 1 - \cos \theta$
$r = 1 + \sin \theta$

108. $r = 1 + \sin \theta$
$r = 3 \sin \theta$

Finding the Arc Length of a Polar Curve In Exercises 109 and 110, find the length of the curve over the given interval.

Polar Equation	Interval
109. $r = 5 \cos \theta$	$\dfrac{\pi}{2} \leq \theta \leq \pi$
110. $r = 3(1 - \cos \theta)$	$0 \leq \theta \leq \pi$

Finding the Area of a Surface of Revolution In Exercises 111 and 112, write an integral that represents the area of the surface formed by revolving the curve about the given line. Use the integration capabilities of a graphing utility to approximate the integral accurate to two decimal places.

Polar Equation	Interval	Axis of Revolution
111. $r = 1 + 4 \cos \theta$	$0 \leq \theta \leq \dfrac{\pi}{2}$	Polar axis
112. $r = 2 \sin \theta$	$0 \leq \theta \leq \dfrac{\pi}{2}$	$\theta = \dfrac{\pi}{2}$

Sketching and Identifying a Conic In Exercises 113–118, find the eccentricity and the distance from the pole to the directrix of the conic. Then sketch and identify the graph. Use a graphing utility to confirm your results.

113. $r = \dfrac{6}{1 - \sin \theta}$

114. $r = \dfrac{2}{1 + \cos \theta}$

115. $r = \dfrac{6}{3 + 2 \cos \theta}$

116. $r = \dfrac{4}{5 - 3 \sin \theta}$

117. $r = \dfrac{4}{2 - 3 \sin \theta}$

118. $r = \dfrac{8}{2 - 5 \cos \theta}$

Finding a Polar Equation In Exercises 119–124, find a polar equation for the conic with its focus at the pole. (For convenience, the equation for the directrix is given in rectangular form.)

Conic	Eccentricity	Directrix
119. Parabola	$e = 1$	$x = 4$
120. Ellipse	$e = \dfrac{3}{4}$	$y = -2$
121. Hyperbola	$e = 3$	$y = 3$

Conic	Vertex or Vertices
122. Parabola	$\left(2, \dfrac{\pi}{2}\right)$
123. Ellipse	$(5, 0), (1, \pi)$
124. Hyperbola	$(1, 0), (7, 0)$

P.S. Problem Solving

1. Using a Parabola Consider the parabola $x^2 = 4y$ and the focal chord $y = \frac{3}{4}x + 1$.

(a) Sketch the graph of the parabola and the focal chord.

(b) Show that the tangent lines to the parabola at the endpoints of the focal chord intersect at right angles.

(c) Show that the tangent lines to the parabola at the endpoints of the focal chord intersect on the directrix of the parabola.

2. Using a Parabola Consider the parabola $x^2 = 4py$ and one of its focal chords.

(a) Show that the tangent lines to the parabola at the endpoints of the focal chord intersect at right angles.

(b) Show that the tangent lines to the parabola at the endpoints of the focal chord intersect on the directrix of the parabola.

3. Proof Prove Theorem 10.2, Reflective Property of a Parabola, as shown in the figure.

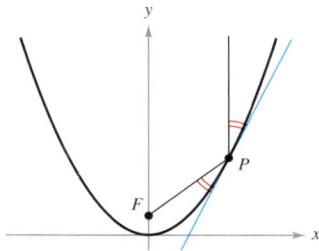

4. Flight Paths An air traffic controller spots two planes at the same altitude flying toward each other (see figure). Their flight paths are $20°$ and $315°$. One plane is 150 miles from point P with a speed of 375 miles per hour. The other is 190 miles from point P with a speed of 450 miles per hour.

(a) Find parametric equations for the path of each plane where t is the time in hours, with $t = 0$ corresponding to the time at which the air traffic controller spots the planes.

(b) Use the result of part (a) to write the distance between the planes as a function of t.

(c) Use a graphing utility to graph the function in part (b). When will the distance between the planes be minimum? If the planes must keep a separation of at least 3 miles, is the requirement met?

5. Strophoid The curve given by the parametric equations

$$x(t) = \frac{1 - t^2}{1 + t^2} \quad \text{and} \quad y(t) = \frac{t(1 - t^2)}{1 + t^2}$$

is called a **strophoid.**

(a) Find a rectangular equation of the strophoid.

(b) Find a polar equation of the strophoid.

(c) Sketch a graph of the strophoid.

(d) Find the equations of the two tangent lines at the origin.

(e) Find the points on the graph at which the tangent lines are horizontal.

6. Finding a Rectangular Equation Find a rectangular equation of the portion of the cycloid given by the parametric equations $x = a(\theta - \sin\theta)$ and $y = a(1 - \cos\theta)$, $0 \le \theta \le \pi$, as shown in the figure.

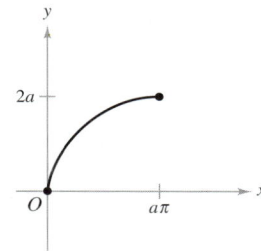

7. Cornu Spiral Consider the **cornu spiral** given by

$$x(t) = \int_0^t \cos\left(\frac{\pi u^2}{2}\right) du \quad \text{and} \quad y(t) = \int_0^t \sin\left(\frac{\pi u^2}{2}\right) du.$$

(a) Use a graphing utility to graph the spiral over the interval $-\pi \le t \le \pi$.

(b) Show that the cornu spiral is symmetric with respect to the origin.

(c) Find the length of the cornu spiral from $t = 0$ to $t = a$. What is the length of the spiral from $t = -\pi$ to $t = \pi$?

8. Using an Ellipse Consider the region bounded by the ellipse $x^2/a^2 + y^2/b^2 = 1$, with eccentricity $e = c/a$.

(a) Show that the area of the region is πab.

(b) Show that the solid (oblate spheroid) generated by revolving the region about the minor axis of the ellipse has a volume of $V = 4\pi^2 b/3$ and a surface area of

$$S = 2\pi a^2 + \pi\left(\frac{b^2}{e}\right)\ln\left(\frac{1 + e}{1 - e}\right).$$

(c) Show that the solid (prolate spheroid) generated by revolving the region about the major axis of the ellipse has a volume of $V = 4\pi ab^2/3$ and a surface area of

$$S = 2\pi b^2 + 2\pi\left(\frac{ab}{e}\right)\arcsin e.$$

9. Area   Let a and b be positive constants. Find the area of the region in the first quadrant bounded by the graph of the polar equation

$$r = \frac{ab}{(a \sin \theta + b \cos \theta)}, \quad 0 \le \theta \le \frac{\pi}{2}.$$

10. Using a Right Triangle   Consider the right triangle shown in the figure.

(a) Show that the area of the triangle is $A(\alpha) = \dfrac{1}{2}\displaystyle\int_0^{\alpha} \sec^2 \theta \, d\theta.$

(b) Show that $\tan \alpha = \displaystyle\int_0^{\alpha} \sec^2 \theta \, d\theta.$

(c) Use part (b) to derive the formula for the derivative of the tangent function.

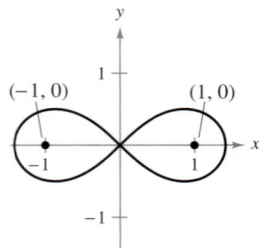

Figure for 10   Figure for 11

11. Finding a Polar Equation   Determine the polar equation of the set of all points (r, θ), the product of whose distances from the points $(1, 0)$ and $(-1, 0)$ is equal to 1, as shown in the figure.

12. Arc Length   A particle is moving along the path described by the parametric equations $x = 1/t$ and $y = (\sin t)/t$, for $1 \le t < \infty$, as shown in the figure. Find the length of this path.

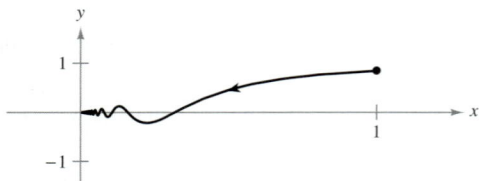

13. Finding a Polar Equation   Four dogs are located at the corners of a square with sides of length d. The dogs all move counterclockwise at the same speed directly toward the next dog, as shown in the figure. Find the polar equation of a dog's path as it spirals toward the center of the square.

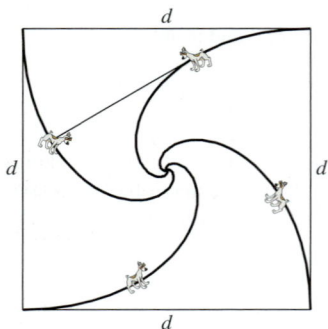

14. Using a Hyperbola   Consider the hyperbola

$$\frac{x^2}{a^2} - \frac{y^2}{b^2} = 1$$

with foci F_1 and F_2, as shown in the figure. Let T be the tangent line at a point M on the hyperbola. Show that incoming rays of light aimed at one focus are reflected by a hyperbolic mirror toward the other focus.

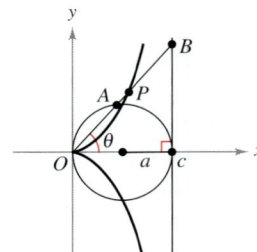

Figure for 14   Figure for 15

15. Cissoid of Diocles   Consider a circle of radius a tangent to the y-axis and the line $x = 2a$, as shown in the figure. Let A be the point where the segment OB intersects the circle. The **cissoid of Diocles** consists of all points P such that $OP = AB$.

(a) Find a polar equation of the cissoid.

(b) Find a set of parametric equations for the cissoid that does not contain trigonometric functions.

(c) Find a rectangular equation of the cissoid.

16. Butterfly Curve   Use a graphing utility to graph the curve shown below. The curve is given by

$$r = e^{\cos \theta} - 2 \cos 4\theta + \sin^5 \frac{\theta}{12}.$$

Over what interval must θ vary to produce the curve?

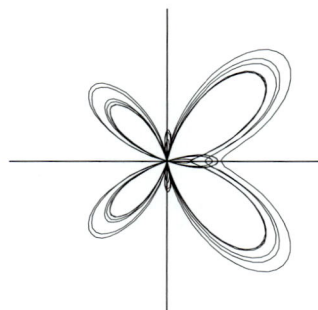

■ **FOR FURTHER INFORMATION**   For more information on this curve, see the article "A Study in Step Size" by Temple H. Fay in *Mathematics Magazine*. To view this article, go to *MathArticles.com*.

17. Graphing Polar Equations   Use a graphing utility to graph the polar equation $r = \cos 5\theta + n \cos \theta$ for $0 \le \theta < \pi$ and for the integers $n = -5$ to $n = 5$. What values of n produce the "heart" portion of the curve? What values of n produce the "bell" portion? (This curve, created by Michael W. Chamberlin, appeared in *The College Mathematics Journal*.)

11 Vectors and the Geometry of Space

Geography (*Exercise 45, p. 803*)

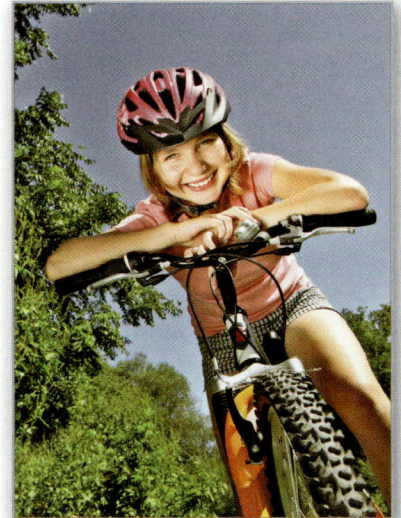

Torque (*Exercise 29, p. 781*)

Work (*Exercise 64, p. 774*)

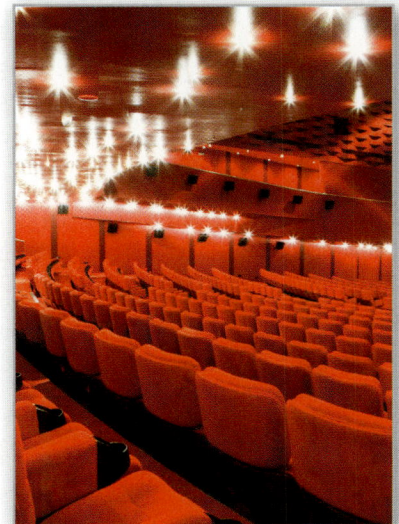

Auditorium Lights
(*Exercise 101, p. 765*)

Navigation (*Exercise 84, p. 757*)

11.1 Vectors in the Plane

■ Write the component form of a vector.
■ Perform vector operations and interpret the results geometrically.
■ Write a vector as a linear combination of standard unit vectors.

Component Form of a Vector

Many quantities in geometry and physics, such as area, volume, temperature, mass, and time, can be characterized by a single real number that is scaled to appropriate units of measure. These are called **scalar quantities,** and the real number associated with each is called a **scalar.**

Q
Terminal
point

P
$\overrightarrow{PQ}$

Initial
point

A directed line segment
Figure 11.1

Other quantities, such as force, velocity, and acceleration, involve both magnitude and direction and cannot be characterized completely by a single real number. A **directed line segment** is used to represent such a quantity, as shown in Figure 11.1. The directed line segment $\overrightarrow{PQ}$ has **initial point** P and **terminal point** Q, and its **length** (or **magnitude**) is denoted by $\|\overrightarrow{PQ}\|$. Directed line segments that have the same length and direction are **equivalent,** as shown in Figure 11.2. The set of all directed line segments that are equivalent to a given directed line segment $\overrightarrow{PQ}$ is a **vector in the plane** and is denoted by

$$\mathbf{v} = \overrightarrow{PQ}.$$

In typeset material, vectors are usually denoted by lowercase, boldface letters such as **u**, **v**, and **w**. When written by hand, however, vectors are often denoted by letters with arrows above them, such as $\vec{u}$, $\vec{v}$, and $\vec{w}$.

Equivalent directed line segments
Figure 11.2

Be sure you understand that a vector represents a *set* of directed line segments (each having the same length and direction). In practice, however, it is common not to distinguish between a vector and one of its representatives.

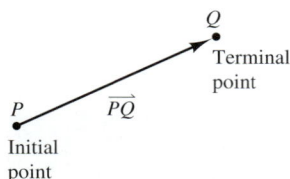

EXAMPLE 1 Vector Representation: Directed Line Segments

Let **v** be represented by the directed line segment from $(0, 0)$ to $(3, 2)$, and let **u** be represented by the directed line segment from $(1, 2)$ to $(4, 4)$. Show that **v** and **u** are equivalent.

Solution Let $P(0, 0)$ and $Q(3, 2)$ be the initial and terminal points of **v**, and let $R(1, 2)$ and $S(4, 4)$ be the initial and terminal points of **u**, as shown in Figure 11.3. You can use the Distance Formula to show that $\overrightarrow{PQ}$ and $\overrightarrow{RS}$ have the *same length.*

$$\|\overrightarrow{PQ}\| = \sqrt{(3 - 0)^2 + (2 - 0)^2} = \sqrt{13}$$
$$\|\overrightarrow{RS}\| = \sqrt{(4 - 1)^2 + (4 - 2)^2} = \sqrt{13}$$

Both line segments have the *same direction,* because they both are directed toward the upper right on lines having the same slope.

$$\text{Slope of } \overrightarrow{PQ} = \frac{2 - 0}{3 - 0} = \frac{2}{3}$$

and

$$\text{Slope of } \overrightarrow{RS} = \frac{4 - 2}{4 - 1} = \frac{2}{3}$$

Because $\overrightarrow{PQ}$ and $\overrightarrow{RS}$ have the same length and direction, you can conclude that the two vectors are equivalent. That is, **v** and **u** are equivalent.

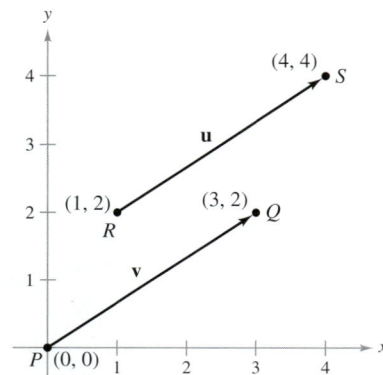

The vectors **u** and **v** are equivalent.
Figure 11.3

The directed line segment whose initial point is the origin is often the most convenient representative of a set of equivalent directed line segments such as those shown in Figure 11.3. This representation of $\mathbf{v}$ is said to be in **standard position.** A directed line segment whose initial point is the origin can be uniquely represented by the coordinates of its terminal point $Q(v_1, v_2)$, as shown in Figure 11.4.

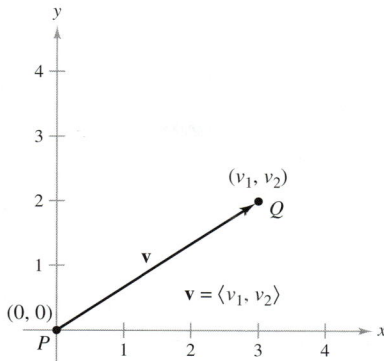

A vector in standard position
Figure 11.4

Definition of Component Form of a Vector in the Plane

If $\mathbf{v}$ is a vector in the plane whose initial point is the origin and whose terminal point is (v_1, v_2), then the **component form of v** is

$$\mathbf{v} = \langle v_1, v_2 \rangle.$$

The coordinates v_1 and v_2 are called the **components of v.** If both the initial point and the terminal point lie at the origin, then $\mathbf{v}$ is called the **zero vector** and is denoted by $\mathbf{0} = \langle 0, 0 \rangle$.

This definition implies that two vectors $\mathbf{u} = \langle u_1, u_2 \rangle$ and $\mathbf{v} = \langle v_1, v_2 \rangle$ are **equal** if and only if $u_1 = v_1$ and $u_2 = v_2$.

The procedures listed below can be used to convert directed line segments to component form or vice versa.

1. If $P(p_1, p_2)$ and $Q(q_1, q_2)$ are the initial and terminal points of a directed line segment, then the component form of the vector $\mathbf{v}$ represented by $\overrightarrow{PQ}$ is

$$\langle v_1, v_2 \rangle = \langle q_1 - p_1, q_2 - p_2 \rangle.$$

Moreover, from the Distance Formula, you can see that the **length** (or **magnitude**) of $\mathbf{v}$ is

$$\|\mathbf{v}\| = \sqrt{(q_1 - p_1)^2 + (q_2 - p_2)^2} \qquad \text{Length of a vector}$$
$$= \sqrt{v_1^2 + v_2^2}.$$

2. If $\mathbf{v} = \langle v_1, v_2 \rangle$, then $\mathbf{v}$ can be represented by the directed line segment, in standard position, from $P(0, 0)$ to $Q(v_1, v_2)$.

The length of $\mathbf{v}$ is also called the **norm of v.** If $\|\mathbf{v}\| = 1$, then $\mathbf{v}$ is a **unit vector.** Moreover, $\|\mathbf{v}\| = 0$ if and only if $\mathbf{v}$ is the zero vector $\mathbf{0}$.

EXAMPLE 2   **Component Form and Length of a Vector**

Find the component form and length of the vector $\mathbf{v}$ that has initial point $(3, -7)$ and terminal point $(-2, 5)$.

Solution   Let $P(3, -7) = (p_1, p_2)$ and $Q(-2, 5) = (q_1, q_2)$. Then the components of $\mathbf{v} = \langle v_1, v_2 \rangle$ are

$$v_1 = q_1 - p_1 = -2 - 3 = -5$$

and

$$v_2 = q_2 - p_2 = 5 - (-7) = 12.$$

So, as shown in Figure 11.5, $\mathbf{v} = \langle -5, 12 \rangle$, and the length of $\mathbf{v}$ is

$$\|\mathbf{v}\| = \sqrt{(-5)^2 + 12^2}$$
$$= \sqrt{169}$$
$$= 13.$$

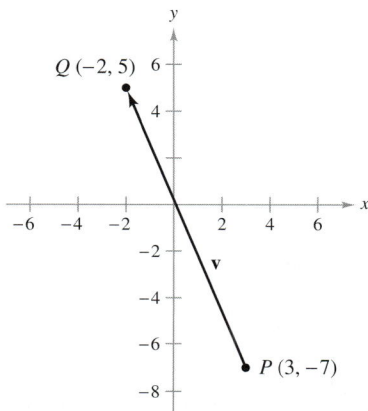

Component form of $\mathbf{v}$: $\mathbf{v} = \langle -5, 12 \rangle$
Figure 11.5

Vector Operations

Definitions of Vector Addition and Scalar Multiplication

Let $\mathbf{u} = \langle u_1, u_2 \rangle$ and $\mathbf{v} = \langle v_1, v_2 \rangle$ be vectors and let c be a scalar.

1. The **vector sum** of $\mathbf{u}$ and $\mathbf{v}$ is the vector $\mathbf{u} + \mathbf{v} = \langle u_1 + v_1, u_2 + v_2 \rangle$.

2. The **scalar multiple** of c and $\mathbf{u}$ is the vector

$$c\mathbf{u} = \langle cu_1, cu_2 \rangle.$$

3. The **negative** of $\mathbf{v}$ is the vector

$$-\mathbf{v} = (-1)\mathbf{v} = \langle -v_1, -v_2 \rangle.$$

4. The **difference** of $\mathbf{u}$ and $\mathbf{v}$ is

$$\mathbf{u} - \mathbf{v} = \mathbf{u} + (-\mathbf{v}) = \langle u_1 - v_1, u_2 - v_2 \rangle.$$

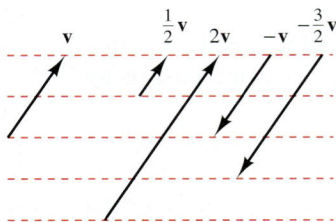

The scalar multiplication of $\mathbf{v}$
Figure 11.6

Geometrically, the scalar multiple of a vector $\mathbf{v}$ and a scalar c is the vector that is $|c|$ times as long as $\mathbf{v}$, as shown in Figure 11.6. If c is positive, then $c\mathbf{v}$ has the same direction as $\mathbf{v}$. If c is negative, then $c\mathbf{v}$ has the opposite direction.

The sum of two vectors can be represented geometrically by positioning the vectors (without changing their magnitudes or directions) so that the initial point of one coincides with the terminal point of the other, as shown in Figure 11.7. The vector $\mathbf{u} + \mathbf{v}$, called the **resultant vector,** is the diagonal of a parallelogram having $\mathbf{u}$ and $\mathbf{v}$ as its adjacent sides.

To find $\mathbf{u} + \mathbf{v}$,   (1) move the initial point of $\mathbf{v}$ to the terminal point of $\mathbf{u}$, or   (2) move the initial point of $\mathbf{u}$ to the terminal point of $\mathbf{v}$.

Figure 11.7

Figure 11.8 shows the equivalence of the geometric and algebraic definitions of vector addition and scalar multiplication, and presents (at far right) a geometric interpretation of $\mathbf{u} - \mathbf{v}$.

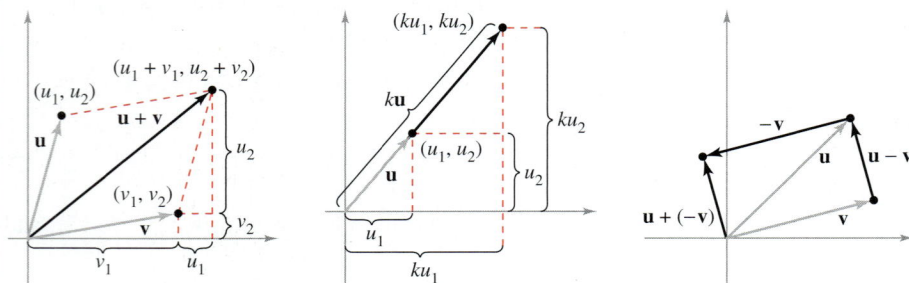

Vector addition   Scalar multiplication   Vector subtraction
Figure 11.8

WILLIAM ROWAN HAMILTON (1805–1865)

Some of the earliest work with vectors was done by the Irish mathematician William Rowan Hamilton. Hamilton spent many years developing a system of vector-like quantities called *quaternions*. It wasn't until the latter half of the nineteenth century that the Scottish physicist James Maxwell (1831–1879) restructured Hamilton's quaternions in a form useful for representing physical quantities such as force, velocity, and acceleration.
See LarsonCalculus.com to read more of this biography.

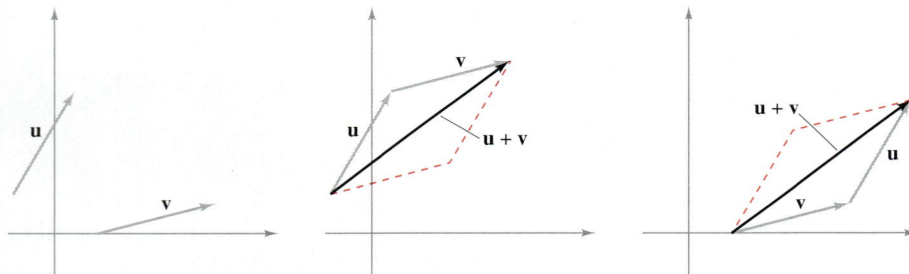

EXAMPLE 3 **Vector Operations**

For $\mathbf{v} = \langle -2, 5 \rangle$ and $\mathbf{w} = \langle 3, 4 \rangle$, find each of the vectors.

a. $\frac{1}{2}\mathbf{v}$ **b.** $\mathbf{w} - \mathbf{v}$ **c.** $\mathbf{v} + 2\mathbf{w}$

Solution

a. $\frac{1}{2}\mathbf{v} = \left\langle \frac{1}{2}(-2), \frac{1}{2}(5) \right\rangle = \left\langle -1, \frac{5}{2} \right\rangle$

b. $\mathbf{w} - \mathbf{v} = \langle w_1 - v_1, w_2 - v_2 \rangle = \langle 3 - (-2), 4 - 5 \rangle = \langle 5, -1 \rangle$

c. Using $2\mathbf{w} = \langle 6, 8 \rangle$, you have

$$\begin{aligned} \mathbf{v} + 2\mathbf{w} &= \langle -2, 5 \rangle + \langle 6, 8 \rangle \\ &= \langle -2 + 6, 5 + 8 \rangle \\ &= \langle 4, 13 \rangle. \end{aligned}$$

Vector addition and scalar multiplication share many properties of ordinary arithmetic, as shown in the next theorem.

EMMY NOETHER (1882–1935)

One person who contributed to our knowledge of axiomatic systems was the German mathematician Emmy Noether. Noether is generally recognized as the leading woman mathematician in recent history.

> **THEOREM 11.1** **Properties of Vector Operations**
>
> Let $\mathbf{u}$, $\mathbf{v}$, and $\mathbf{w}$ be vectors in the plane, and let c and d be scalars.
>
> **1.** $\mathbf{u} + \mathbf{v} = \mathbf{v} + \mathbf{u}$ Commutative Property
> **2.** $(\mathbf{u} + \mathbf{v}) + \mathbf{w} = \mathbf{u} + (\mathbf{v} + \mathbf{w})$ Associative Property
> **3.** $\mathbf{u} + \mathbf{0} = \mathbf{u}$ Additive Identity Property
> **4.** $\mathbf{u} + (-\mathbf{u}) = \mathbf{0}$ Additive Inverse Property
> **5.** $c(d\mathbf{u}) = (cd)\mathbf{u}$
> **6.** $(c + d)\mathbf{u} = c\mathbf{u} + d\mathbf{u}$ Distributive Property
> **7.** $c(\mathbf{u} + \mathbf{v}) = c\mathbf{u} + c\mathbf{v}$ Distributive Property
> **8.** $1(\mathbf{u}) = \mathbf{u}, \ 0(\mathbf{u}) = \mathbf{0}$

Proof The proof of the *Associative Property* of vector addition uses the Associative Property of addition of real numbers.

$$\begin{aligned} (\mathbf{u} + \mathbf{v}) + \mathbf{w} &= [\langle u_1, u_2 \rangle + \langle v_1, v_2 \rangle] + \langle w_1, w_2 \rangle \\ &= \langle u_1 + v_1, u_2 + v_2 \rangle + \langle w_1, w_2 \rangle \\ &= \langle (u_1 + v_1) + w_1, (u_2 + v_2) + w_2 \rangle \\ &= \langle u_1 + (v_1 + w_1), u_2 + (v_2 + w_2) \rangle \\ &= \langle u_1, u_2 \rangle + \langle v_1 + w_1, v_2 + w_2 \rangle \\ &= \mathbf{u} + (\mathbf{v} + \mathbf{w}) \end{aligned}$$

The other properties can be proved in a similar manner.

See LarsonCalculus.com for Bruce Edwards's video of this proof.

Any set of vectors (with an accompanying set of scalars) that satisfies the eight properties listed in Theorem 11.1 is a **vector space.*** The eight properties are the *vector space axioms*. So, this theorem states that the set of vectors in the plane (with the set of real numbers) forms a vector space.

■ **FOR FURTHER INFORMATION**
For more information on Emmy Noether, see the article "Emmy Noether, Greatest Woman Mathematician" by Clark Kimberling in *Mathematics Teacher*. To view this article, go to *MathArticles.com*.

---

* For more information about vector spaces, see *Elementary Linear Algebra*, Seventh Edition, by Ron Larson (Boston, Massachusetts: Brooks/Cole, Cengage Learning, 2013).

> **THEOREM 11.2 Length of a Scalar Multiple**
>
> Let $\mathbf{v}$ be a vector and let c be a scalar. Then
>
> $$\|c\mathbf{v}\| = |c|\,\|\mathbf{v}\|. \qquad \textcolor{red}{|c| \text{ is the absolute value of } c.}$$

Proof Because $c\mathbf{v} = \langle cv_1, cv_2 \rangle$, it follows that

$$
\begin{aligned}
\|c\mathbf{v}\| &= \|\langle cv_1, cv_2 \rangle\| \\
&= \sqrt{(cv_1)^2 + (cv_2)^2} \\
&= \sqrt{c^2 v_1^2 + c^2 v_2^2} \\
&= \sqrt{c^2(v_1^2 + v_2^2)} \\
&= |c|\sqrt{v_1^2 + v_2^2} \\
&= |c|\,\|\mathbf{v}\|.
\end{aligned}
$$

See LarsonCalculus.com for Bruce Edwards's video of this proof.

In many applications of vectors, it is useful to find a unit vector that has the same direction as a given vector. The next theorem gives a procedure for doing this.

> **THEOREM 11.3 Unit Vector in the Direction of v**
>
> If $\mathbf{v}$ is a nonzero vector in the plane, then the vector
>
> $$\mathbf{u} = \frac{\mathbf{v}}{\|\mathbf{v}\|} = \frac{1}{\|\mathbf{v}\|}\mathbf{v}$$
>
> has length 1 and the same direction as $\mathbf{v}$.

Proof Because $1/\|\mathbf{v}\|$ is positive and $\mathbf{u} = (1/\|\mathbf{v}\|)\mathbf{v}$, you can conclude that $\mathbf{u}$ has the same direction as $\mathbf{v}$. To see that $\|\mathbf{u}\| = 1$, note that

$$\|\mathbf{u}\| = \left\| \left(\frac{1}{\|\mathbf{v}\|} \right) \mathbf{v} \right\| = \left| \frac{1}{\|\mathbf{v}\|} \right| \|\mathbf{v}\| = \frac{1}{\|\mathbf{v}\|}\|\mathbf{v}\| = 1.$$

So, $\mathbf{u}$ has length 1 and the same direction as $\mathbf{v}$.

See LarsonCalculus.com for Bruce Edwards's video of this proof.

In Theorem 11.3, $\mathbf{u}$ is called a **unit vector in the direction of v.** The process of multiplying $\mathbf{v}$ by $1/\|\mathbf{v}\|$ to get a unit vector is called **normalization of v.**

EXAMPLE 4 **Finding a Unit Vector**

Find a unit vector in the direction of $\mathbf{v} = \langle -2, 5 \rangle$ and verify that it has length 1.

Solution From Theorem 11.3, the unit vector in the direction of $\mathbf{v}$ is

$$\frac{\mathbf{v}}{\|\mathbf{v}\|} = \frac{\langle -2, 5 \rangle}{\sqrt{(-2)^2 + (5)^2}} = \frac{1}{\sqrt{29}}\langle -2, 5 \rangle = \left\langle \frac{-2}{\sqrt{29}}, \frac{5}{\sqrt{29}} \right\rangle.$$

This vector has length 1, because

$$\sqrt{\left(\frac{-2}{\sqrt{29}}\right)^2 + \left(\frac{5}{\sqrt{29}}\right)^2} = \sqrt{\frac{4}{29} + \frac{25}{29}} = \sqrt{\frac{29}{29}} = 1.$$

Generally, the length of the sum of two vectors is not equal to the sum of their lengths. To see this, consider the vectors **u** and **v** as shown in Figure 11.9. With **u** and **v** as two sides of a triangle, the length of the third side is $\|\mathbf{u}+\mathbf{v}\|$, and

$$\|\mathbf{u}+\mathbf{v}\| \le \|\mathbf{u}\| + \|\mathbf{v}\|.$$

Equality occurs only when the vectors **u** and **v** have the *same direction*. This result is called the **triangle inequality** for vectors. (You are asked to prove this in Exercise 77, Section 11.3.)

Standard Unit Vectors

The unit vectors $\langle 1,0\rangle$ and $\langle 0,1\rangle$ are called the **standard unit vectors** in the plane and are denoted by

$$\mathbf{i}=\langle 1,0\rangle \quad \text{and} \quad \mathbf{j}=\langle 0,1\rangle \qquad \text{Standard unit vectors}$$

as shown in Figure 11.10. These vectors can be used to represent any vector uniquely, as follows.

$$\mathbf{v}=\langle v_1,v_2\rangle = \langle v_1,0\rangle + \langle 0,v_2\rangle = v_1\langle 1,0\rangle + v_2\langle 0,1\rangle = v_1\mathbf{i}+v_2\mathbf{j}$$

The vector $\mathbf{v}=v_1\mathbf{i}+v_2\mathbf{j}$ is called a **linear combination** of **i** and **j**. The scalars v_1 and v_2 are called the **horizontal** and **vertical components of v.**

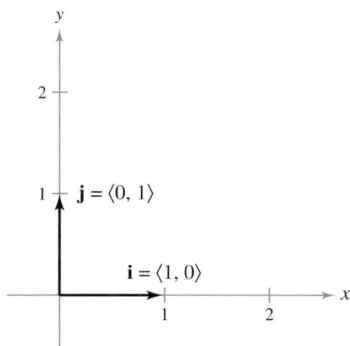

EXAMPLE 5 **Writing a Linear Combination of Unit Vectors**

Let **u** be the vector with initial point $(2,-5)$ and terminal point $(-1,3)$, and let $\mathbf{v}=2\mathbf{i}-\mathbf{j}$. Write each vector as a linear combination of **i** and **j**.

a. u   **b. w** $=2\mathbf{u}-3\mathbf{v}$

Solution

a. $\mathbf{u}=\langle q_1-p_1,q_2-p_2\rangle = \langle -1-2,3-(-5)\rangle = \langle -3,8\rangle = -3\mathbf{i}+8\mathbf{j}$
b. $\mathbf{w}=2\mathbf{u}-3\mathbf{v}=2(-3\mathbf{i}+8\mathbf{j})-3(2\mathbf{i}-\mathbf{j})=-6\mathbf{i}+16\mathbf{j}-6\mathbf{i}+3\mathbf{j}=-12\mathbf{i}+19\mathbf{j}$

If **u** is a unit vector and θ is the angle (measured counterclockwise) from the positive x-axis to **u**, then the terminal point of **u** lies on the unit circle, and you have

$$\mathbf{u}=\langle \cos\theta,\sin\theta\rangle = \cos\theta\mathbf{i}+\sin\theta\mathbf{j} \qquad \text{Unit vector}$$

as shown in Figure 11.11. Moreover, it follows that any other nonzero vector **v** making an angle θ with the positive x-axis has the same direction as **u**, and you can write

$$\mathbf{v}=\|\mathbf{v}\|\langle \cos\theta,\sin\theta\rangle = \|\mathbf{v}\|\cos\theta\mathbf{i}+\|\mathbf{v}\|\sin\theta\mathbf{j}.$$

EXAMPLE 6 **Writing a Vector of Given Magnitude and Direction**

The vector **v** has a magnitude of 3 and makes an angle of $30°=\pi/6$ with the positive x-axis. Write **v** as a linear combination of the unit vectors **i** and **j**.

Solution Because the angle between **v** and the positive x-axis is $\theta=\pi/6$, you can write

$$\mathbf{v}=\|\mathbf{v}\|\cos\theta\mathbf{i}+\|\mathbf{v}\|\sin\theta\mathbf{j}=3\cos\frac{\pi}{6}\mathbf{i}+3\sin\frac{\pi}{6}\mathbf{j}=\frac{3\sqrt{3}}{2}\mathbf{i}+\frac{3}{2}\mathbf{j}.$$

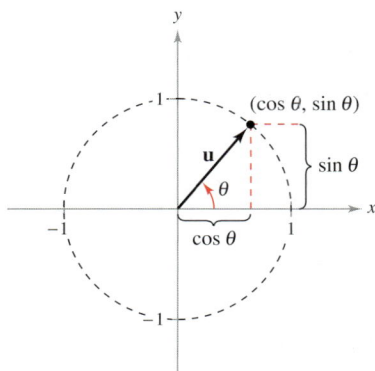

Triangle inequality
Figure 11.9

Standard unit vectors **i** and **j**
Figure 11.10

The angle θ from the positive x-axis to the vector **u**
Figure 11.11

Vectors have many applications in physics and engineering. One example is force. A vector can be used to represent force, because force has both magnitude and direction. If two or more forces are acting on an object, then the **resultant force** on the object is the vector sum of the vector forces.

EXAMPLE 7 **Finding the Resultant Force**

Two tugboats are pushing an ocean liner, as shown in Figure 11.12. Each boat is exerting a force of 400 pounds. What is the resultant force on the ocean liner?

Solution Using Figure 11.12, you can represent the forces exerted by the first and second tugboats as

$$\mathbf{F}_1 = 400\langle \cos 20°, \sin 20° \rangle = 400 \cos(20°)\mathbf{i} + 400 \sin(20°)\mathbf{j}$$
$$\mathbf{F}_2 = 400\langle \cos(-20°), \sin(-20°) \rangle = 400 \cos(20°)\mathbf{i} - 400 \sin(20°)\mathbf{j}.$$

The resultant force on the ocean liner is

$$\mathbf{F} = \mathbf{F}_1 + \mathbf{F}_2$$
$$= [400 \cos(20°)\mathbf{i} + 400 \sin(20°)\mathbf{j}] + [400 \cos(20°)\mathbf{i} - 400 \sin(20°)\mathbf{j}]$$
$$= 800 \cos(20°)\mathbf{i}$$
$$\approx 752\mathbf{i}.$$

So, the resultant force on the ocean liner is approximately 752 pounds in the direction of the positive x-axis.

The resultant force on the ocean liner that is exerted by the two tugboats
Figure 11.12

In surveying and navigation, a **bearing** is a direction that measures the acute angle that a path or line of sight makes with a fixed north-south line. In air navigation, bearings are measured in degrees clockwise from north.

EXAMPLE 8 **Finding a Velocity**

⋅ ⋅ ⋅▷ *See LarsonCalculus.com for an interactive version of this type of example.*

An airplane is traveling at a fixed altitude with a negligible wind factor. The airplane is traveling at a speed of 500 miles per hour with a bearing of 330°, as shown in Figure 11.13(a). As the airplane reaches a certain point, it encounters wind with a velocity of 70 miles per hour in the direction N 45° E (45° east of north), as shown in Figure 11.13(b). What are the resultant speed and direction of the airplane?

Solution Using Figure 11.13(a), represent the velocity of the airplane (alone) as

$$\mathbf{v}_1 = 500 \cos(120°)\mathbf{i} + 500 \sin(120°)\mathbf{j}.$$

The velocity of the wind is represented by the vector

$$\mathbf{v}_2 = 70 \cos(45°)\mathbf{i} + 70 \sin(45°)\mathbf{j}.$$

The resultant velocity of the airplane (in the wind) is

$$\mathbf{v} = \mathbf{v}_1 + \mathbf{v}_2$$
$$= 500 \cos(120°)\mathbf{i} + 500 \sin(120°)\mathbf{j} + 70 \cos(45°)\mathbf{i} + 70 \sin(45°)\mathbf{j}$$
$$\approx -200.5\mathbf{i} + 482.5\mathbf{j}.$$

To find the resultant speed and direction, write $\mathbf{v} = \|\mathbf{v}\|(\cos \theta \, \mathbf{i} + \sin \theta \, \mathbf{j})$. Because $\|\mathbf{v}\| \approx \sqrt{(-200.5)^2 + (482.5)^2} \approx 522.5$, you can write

$$\mathbf{v} \approx 522.5\left(\frac{-200.5}{522.5}\mathbf{i} + \frac{482.5}{522.5}\mathbf{j} \right) \approx 522.5\,[\cos(112.6°)\mathbf{i} + \sin(112.6°)\mathbf{j}].$$

The new speed of the airplane, as altered by the wind, is approximately 522.5 miles per hour in a path that makes an angle of 112.6° with the positive x-axis.

(a) Direction without wind

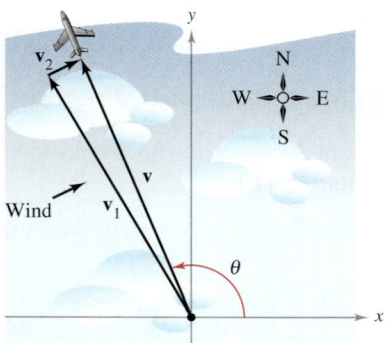

(b) Direction with wind
Figure 11.13

11.1 Exercises

See **CalcChat.com** for tutorial help and worked-out solutions to odd-numbered exercises.

Sketching a Vector In Exercises 1–4, (a) find the component form of the vector v and (b) sketch the vector with its initial point at the origin.

1.

2.

3.

4.
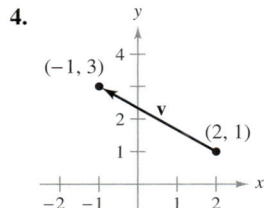

Equivalent Vectors In Exercises 5–8, find the vectors u and v whose initial and terminal points are given. Show that u and v are equivalent.

5. u: $(3, 2)$, $(5, 6)$
   v: $(1, 4)$, $(3, 8)$

6. u: $(-4, 0)$, $(1, 8)$
   v: $(2, -1)$, $(7, 7)$

7. u: $(0, 3)$, $(6, -2)$
   v: $(3, 10)$, $(9, 5)$

8. u: $(-4, -1)$, $(11, -4)$
   v: $(10, 13)$, $(25, 10)$

Writing a Vector in Different Forms In Exercises 9–16, the initial and terminal points of a vector v are given. (a) Sketch the given directed line segment, (b) write the vector in component form, (c) write the vector as the linear combination of the standard unit vectors i and j, and (d) sketch the vector with its initial point at the origin.

	Terminal		Terminal
Initial Point	**Point**	**Initial Point**	**Point**
9. $(2, 0)$	$(5, 5)$	10. $(4, -6)$	$(3, 6)$
11. $(8, 3)$	$(6, -1)$	12. $(0, -4)$	$(-5, -1)$
13. $(6, 2)$	$(6, 6)$	14. $(7, -1)$	$(-3, -1)$
15. $\left(\frac{3}{2}, \frac{4}{3}\right)$	$\left(\frac{1}{2}, 3\right)$	16. $(0.12, 0.60)$	$(0.84, 1.25)$

Sketching Scalar Multiples In Exercises 17 and 18, sketch each scalar multiple of v.

17. $\mathbf{v} = \langle 3, 5 \rangle$
    (a) $2\mathbf{v}$ (b) $-3\mathbf{v}$ (c) $\frac{7}{2}\mathbf{v}$ (d) $\frac{2}{3}\mathbf{v}$

18. $\mathbf{v} = \langle -2, 3 \rangle$
    (a) $4\mathbf{v}$ (b) $-\frac{1}{2}\mathbf{v}$ (c) $0\mathbf{v}$ (d) $-6\mathbf{v}$

Using Vector Operations In Exercises 19 and 20, find (a) $\frac{2}{3}\mathbf{u}$, (b) $3\mathbf{v}$, (c) $\mathbf{v} - \mathbf{u}$, and (d) $2\mathbf{u} + 5\mathbf{v}$.

19. $\mathbf{u} = \langle 4, 9 \rangle$, $\mathbf{v} = \langle 2, -5 \rangle$
20. $\mathbf{u} = \langle -3, -8 \rangle$, $\mathbf{v} = \langle 8, 25 \rangle$

Sketching a Vector In Exercises 21–26, use the figure to sketch a graph of the vector. To print an enlarged copy of the graph, go to *MathGraphs.com*.

21. $-\mathbf{u}$
22. $2\mathbf{u}$
23. $-\mathbf{v}$
24. $\frac{1}{2}\mathbf{v}$
25. $\mathbf{u} - \mathbf{v}$
26. $\mathbf{u} + 2\mathbf{v}$

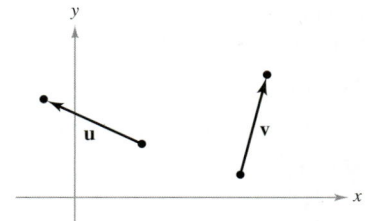

Finding a Terminal Point In Exercises 27 and 28, the vector v and its initial point are given. Find the terminal point.

27. $\mathbf{v} = \langle -1, 3 \rangle$; Initial point: $(4, 2)$
28. $\mathbf{v} = \langle 4, -9 \rangle$; Initial point: $(5, 3)$

Finding a Magnitude of a Vector In Exercises 29–34, find the magnitude of v.

29. $\mathbf{v} = 7\mathbf{i}$
30. $\mathbf{v} = -3\mathbf{i}$
31. $\mathbf{v} = \langle 4, 3 \rangle$
32. $\mathbf{v} = \langle 12, -5 \rangle$
33. $\mathbf{v} = 6\mathbf{i} - 5\mathbf{j}$
34. $\mathbf{v} = -10\mathbf{i} + 3\mathbf{j}$

Finding a Unit Vector In Exercises 35–38, find the unit vector in the direction of v and verify that it has length 1.

35. $\mathbf{v} = \langle 3, 12 \rangle$
36. $\mathbf{v} = \langle -5, 15 \rangle$
37. $\mathbf{v} = \left\langle \frac{3}{2}, \frac{5}{2} \right\rangle$
38. $\mathbf{v} = \langle -6.2, 3.4 \rangle$

Finding Magnitudes In Exercises 39–42, find the following.

(a) $\|\mathbf{u}\|$ (b) $\|\mathbf{v}\|$ (c) $\|\mathbf{u} + \mathbf{v}\|$

(d) $\left\| \dfrac{\mathbf{u}}{\|\mathbf{u}\|} \right\|$ (e) $\left\| \dfrac{\mathbf{v}}{\|\mathbf{v}\|} \right\|$ (f) $\left\| \dfrac{\mathbf{u} + \mathbf{v}}{\|\mathbf{u} + \mathbf{v}\|} \right\|$

39. $\mathbf{u} = \langle 1, -1 \rangle$, $\mathbf{v} = \langle -1, 2 \rangle$
40. $\mathbf{u} = \langle 0, 1 \rangle$, $\mathbf{v} = \langle 3, -3 \rangle$
41. $\mathbf{u} = \left\langle 1, \frac{1}{2} \right\rangle$, $\mathbf{v} = \langle 2, 3 \rangle$
42. $\mathbf{u} = \langle 2, -4 \rangle$, $\mathbf{v} = \langle 5, 5 \rangle$

Using the Triangle Inequality In Exercises 43 and 44, sketch a graph of u, v, and u + v. Then demonstrate the triangle inequality using the vectors u and v.

43. $\mathbf{u} = \langle 2, 1 \rangle$, $\mathbf{v} = \langle 5, 4 \rangle$
44. $\mathbf{u} = \langle -3, 2 \rangle$, $\mathbf{v} = \langle 1, -2 \rangle$

Finding a Vector In Exercises 45–48, find the vector v with the given magnitude and the same direction as u.

	Magnitude	Direction
45.	$\|\mathbf{v}\| = 6$	$\mathbf{u} = \langle 0, 3 \rangle$
46.	$\|\mathbf{v}\| = 4$	$\mathbf{u} = \langle 1, 1 \rangle$
47.	$\|\mathbf{v}\| = 5$	$\mathbf{u} = \langle -1, 2 \rangle$
48.	$\|\mathbf{v}\| = 2$	$\mathbf{u} = \left\langle \sqrt{3}, 3 \right\rangle$

Finding a Vector In Exercises 49–52, find the component form of v given its magnitude and the angle it makes with the positive x-axis.

49. $\|\mathbf{v}\| = 3$, $\theta = 0°$ **50.** $\|\mathbf{v}\| = 5$, $\theta = 120°$

51. $\|\mathbf{v}\| = 2$, $\theta = 150°$ **52.** $\|\mathbf{v}\| = 4$, $\theta = 3.5°$

Finding a Vector In Exercises 53–56, find the component form of u + v given the lengths of u and v and the angles that u and v make with the positive x-axis.

53. $\|\mathbf{u}\| = 1$, $\theta_\mathbf{u} = 0°$
    54. $\|\mathbf{u}\| = 4$, $\theta_\mathbf{u} = 0°$
$\|\mathbf{v}\| = 3$, $\theta_\mathbf{v} = 45°$
       $\|\mathbf{v}\| = 2$, $\theta_\mathbf{v} = 60°$

55. $\|\mathbf{u}\| = 2$, $\theta_\mathbf{u} = 4$
    56. $\|\mathbf{u}\| = 5$, $\theta_\mathbf{u} = -0.5$
$\|\mathbf{v}\| = 1$, $\theta_\mathbf{v} = 2$
       $\|\mathbf{v}\| = 5$, $\theta_\mathbf{v} = 0.5$

WRITING ABOUT CONCEPTS

57. Scalar and Vector In your own words, state the difference between a scalar and a vector. Give examples of each.

58. Scalar or Vector Identify the quantity as a scalar or as a vector. Explain your reasoning.

(a) The muzzle velocity of a gun

(b) The price of a company's stock

(c) The air temperature in a room

(d) The weight of a car

---

59. Using a Parallelogram Three vertices of a parallelogram are $(1, 2), (3, 1),$ and $(8, 4)$. Find the three possible fourth vertices (see figure).

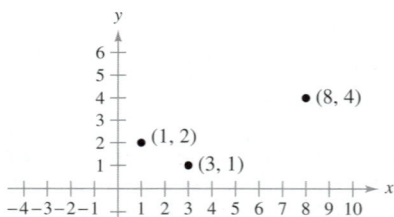

60. HOW DO YOU SEE IT? Use the figure to determine whether each statement is true or false. Justify your answer.

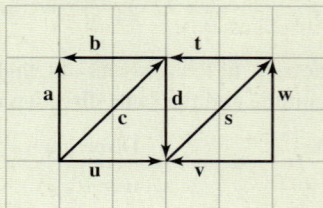

(a) $\mathbf{a} = -\mathbf{d}$ (b) $\mathbf{c} = \mathbf{s}$

(c) $\mathbf{a} + \mathbf{u} = \mathbf{c}$ (d) $\mathbf{v} + \mathbf{w} = -\mathbf{s}$

(e) $\mathbf{a} + \mathbf{d} = 0$ (f) $\mathbf{u} - \mathbf{v} = -2(\mathbf{b} + \mathbf{t})$

Finding Values In Exercises 61–66, find a and b such that $\mathbf{v} = a\mathbf{u} + b\mathbf{w}$, where $\mathbf{u} = \langle 1, 2 \rangle$ and $\mathbf{w} = \langle 1, -1 \rangle$.

61. $\mathbf{v} = \langle 2, 1 \rangle$ **62.** $\mathbf{v} = \langle 0, 3 \rangle$

63. $\mathbf{v} = \langle 3, 0 \rangle$ **64.** $\mathbf{v} = \langle 3, 3 \rangle$

65. $\mathbf{v} = \langle 1, 1 \rangle$ **66.** $\mathbf{v} = \langle -1, 7 \rangle$

Finding Unit Vectors In Exercises 67–72, find a unit vector (a) parallel to and (b) perpendicular to the graph of f at the given point. Then sketch the graph of f and sketch the vectors at the given point.

67. $f(x) = x^2$, $(3, 9)$ **68.** $f(x) = -x^2 + 5$, $(1, 4)$

69. $f(x) = x^3$, $(1, 1)$ **70.** $f(x) = x^3$, $(-2, -8)$

71. $f(x) = \sqrt{25 - x^2}$, $(3, 4)$

72. $f(x) = \tan x$, $\left(\dfrac{\pi}{4}, 1 \right)$

Finding a Vector In Exercises 73 and 74, find the component form of v given the magnitudes of u and u + v and the angles that u and u + v make with the positive x-axis.

73. $\|\mathbf{u}\| = 1$, $\theta = 45°$
    74. $\|\mathbf{u}\| = 4$, $\theta = 30°$
$\|\mathbf{u} + \mathbf{v}\| = \sqrt{2}$, $\theta = 90°$
     $\|\mathbf{u} + \mathbf{v}\| = 6$, $\theta = 120°$

75. Resultant Force Forces with magnitudes of 500 pounds and 200 pounds act on a machine part at angles of 30° and $-45°$, respectively, with the x-axis (see figure). Find the direction and magnitude of the resultant force.

Figure for 75   Figure for 76

76. Numerical and Graphical Analysis Forces with magnitudes of 180 newtons and 275 newtons act on a hook (see figure). The angle between the two forces is θ degrees.

(a) When $\theta = 30°$, find the direction and magnitude of the resultant force.

(b) Write the magnitude M and direction α of the resultant force as functions of θ, where $0° \le \theta \le 180°$.

(c) Use a graphing utility to complete the table.

θ	0°	30°	60°	90°	120°	150°	180°
M							
α							

(d) Use a graphing utility to graph the two functions M and α.

(e) Explain why one of the functions decreases for increasing values of θ, whereas the other does not.

77. Resultant Force   Three forces with magnitudes of 75 pounds, 100 pounds, and 125 pounds act on an object at angles of 30°, 45°, and 120°, respectively, with the positive x-axis. Find the direction and magnitude of the resultant force.

78. Resultant Force   Three forces with magnitudes of 400 newtons, 280 newtons, and 350 newtons act on an object at angles of $-30°$, 45°, and 135°, respectively, with the positive x-axis. Find the direction and magnitude of the resultant force.

79. Think About It   Consider two forces of equal magnitude acting on a point.

(a) When the magnitude of the resultant is the sum of the magnitudes of the two forces, make a conjecture about the angle between the forces.

(b) When the resultant of the forces is 0, make a conjecture about the angle between the forces.

(c) Can the magnitude of the resultant be greater than the sum of the magnitudes of the two forces? Explain.

80. Cable Tension   Determine the tension in each cable supporting the given load for each figure.

(a)

(b)

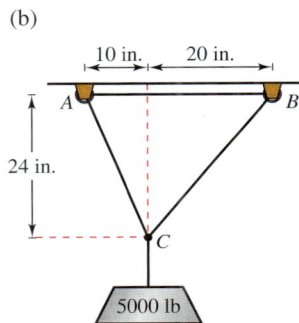

81. Projectile Motion   A gun with a muzzle velocity of 1200 feet per second is fired at an angle of 6° above the horizontal. Find the vertical and horizontal components of the velocity.

82. Shared Load   To carry a 100-pound cylindrical weight, two workers lift on the ends of short ropes tied to an eyelet on the top center of the cylinder. One rope makes a 20° angle away from the vertical and the other makes a 30° angle (see figure).

(a) Find each rope's tension when the resultant force is vertical.

(b) Find the vertical component of each worker's force.

Figure for 82

Figure for 83

83. Navigation   A plane is flying with a bearing of 302°. Its speed with respect to the air is 900 kilometers per hour. The wind at the plane's altitude is from the southwest at 100 kilometers per hour (see figure). What is the true direction of the plane, and what is its speed with respect to the ground?

84. Navigation

A plane flies at a constant groundspeed of 400 miles per hour due east and encounters a 50-mile-per-hour wind from the northwest. Find the airspeed and compass direction that will allow the plane to maintain its groundspeed and eastward direction.

True or False?   **In Exercises 85–90, determine whether the statement is true or false. If it is false, explain why or give an example that shows it is false.**

85. If $\mathbf{u}$ and $\mathbf{v}$ have the same magnitude and direction, then $\mathbf{u}$ and $\mathbf{v}$ are equivalent.

86. If $\mathbf{u}$ is a unit vector in the direction of $\mathbf{v}$, then $\mathbf{v} = \|\mathbf{v}\|\,\mathbf{u}$.

87. If $\mathbf{u} = a\mathbf{i} + b\mathbf{j}$ is a unit vector, then $a^2 + b^2 = 1$.

88. If $\mathbf{v} = a\mathbf{i} + b\mathbf{j} = \mathbf{0}$, then $a = -b$.

89. If $a = b$, then $\|a\mathbf{i} + b\mathbf{j}\| = \sqrt{2}\,a$.

90. If $\mathbf{u}$ and $\mathbf{v}$ have the same magnitude but opposite directions, then $\mathbf{u} + \mathbf{v} = \mathbf{0}$.

91. Proof   Prove that

$$\mathbf{u} = (\cos\theta)\mathbf{i} - (\sin\theta)\mathbf{j} \quad \text{and} \quad \mathbf{v} = (\sin\theta)\mathbf{i} + (\cos\theta)\mathbf{j}$$

are unit vectors for any angle θ.

92. Geometry   Using vectors, prove that the line segment joining the midpoints of two sides of a triangle is parallel to, and one-half the length of, the third side.

93. Geometry   Using vectors, prove that the diagonals of a parallelogram bisect each other.

94. Proof   Prove that the vector $\mathbf{w} = \|\mathbf{u}\|\mathbf{v} + \|\mathbf{v}\|\mathbf{u}$ bisects the angle between $\mathbf{u}$ and $\mathbf{v}$.

95. Using a Vector   Consider the vector $\mathbf{u} = \langle x, y \rangle$. Describe the set of all points (x, y) such that $\|\mathbf{u}\| = 5$.

PUTNAM EXAM CHALLENGE

96. A coast artillery gun can fire at any angle of elevation between 0° and 90° in a fixed vertical plane. If air resistance is neglected and the muzzle velocity is constant ($= v_0$), determine the set H of points in the plane and above the horizontal which can be hit.

11.2 Space Coordinates and Vectors in Space

■ Understand the three-dimensional rectangular coordinate system.
■ Analyze vectors in space.

Coordinates in Space

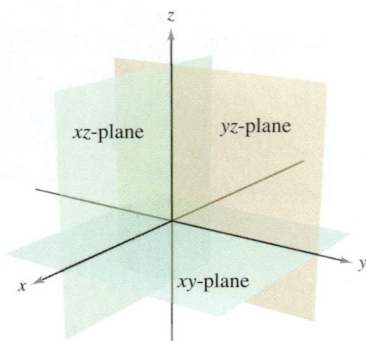

The three-dimensional coordinate
system
Figure 11.14

Up to this point in the text, you have been primarily concerned with the two-dimensional coordinate system. Much of the remaining part of your study of calculus will involve the three-dimensional coordinate system.

Before extending the concept of a vector to three dimensions, you must be able to identify points in the **three-dimensional coordinate system.** You can construct this system by passing a z-axis perpendicular to both the x- and y-axes at the origin, as shown in Figure 11.14. Taken as pairs, the axes determine three **coordinate planes:** the **xy-plane,** the **xz-plane,** and the **yz-plane.** These three coordinate planes separate three-space into eight **octants.** The first octant is the one for which all three coordinates are positive. In this three-dimensional system, a point P in space is determined by an ordered triple (x, y, z), where x, y, and z are as follows.

x = directed distance from yz-plane to P

y = directed distance from xz-plane to P

z = directed distance from xy-plane to P

Several points are shown in Figure 11.15.

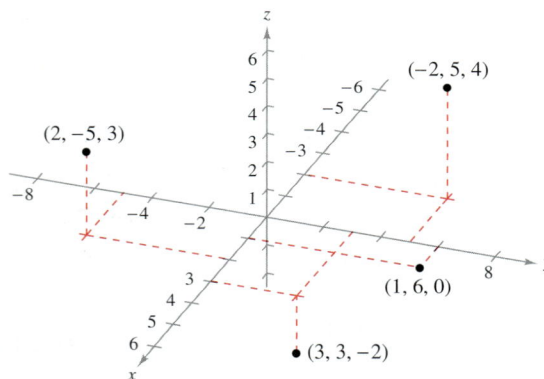

Points in the three-dimensional coordinate system are
represented by ordered triples.
Figure 11.15

• • **REMARK** The three-
• dimensional rotatable graphs
• that are available at
• *LarsonCalculus.com* can help
• you visualize points or objects
• in a three-dimensional
• coordinate system.

A three-dimensional coordinate system can have either a **right-handed** or a **left-handed** orientation. To determine the orientation of a system, imagine that you are standing at the origin, with your arms pointing in the direction of the positive x- and y-axes, and with the z-axis pointing up, as shown in Figure 11.16. The system is right-handed or left-handed depending on which hand points along the x-axis. In this text, you will work exclusively with the right-handed system.

Right-handed   Left-handed
system   system
Figure 11.16

Many of the formulas established for the two-dimensional coordinate system can be extended to three dimensions. For example, to find the distance between two points in space, you can use the Pythagorean Theorem twice, as shown in Figure 11.17. By doing this, you will obtain the formula for the distance between the points (x_1, y_1, z_1) and (x_2, y_2, z_2).

$$d = \sqrt{(x_2 - x_1)^2 + (y_2 - y_1)^2 + (z_2 - z_1)^2} \qquad \text{Distance Formula}$$

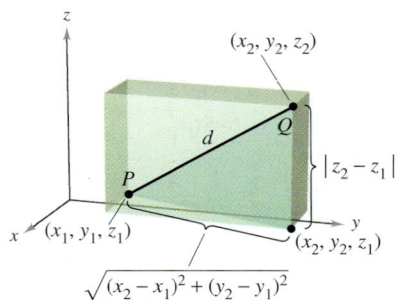

The distance between two points in space

Figure 11.17

EXAMPLE 1 **Finding the Distance Between Two Points in Space**

Find the distance between the points $(2, -1, 3)$ and $(1, 0, -2)$.

Solution

$$\begin{aligned} d &= \sqrt{(1 - 2)^2 + (0 + 1)^2 + (-2 - 3)^2} \qquad \text{Distance Formula} \\ &= \sqrt{1 + 1 + 25} \\ &= \sqrt{27} \\ &= 3\sqrt{3} \end{aligned}$$

A **sphere** with center at (x_0, y_0, z_0) and radius r is defined to be the set of all points (x, y, z) such that the distance between (x, y, z) and (x_0, y_0, z_0) is r. You can use the Distance Formula to find the **standard equation of a sphere** of radius r, centered at (x_0, y_0, z_0). If (x, y, z) is an arbitrary point on the sphere, then the equation of the sphere is

$$(x - x_0)^2 + (y - y_0)^2 + (z - z_0)^2 = r^2 \qquad \text{Equation of sphere}$$

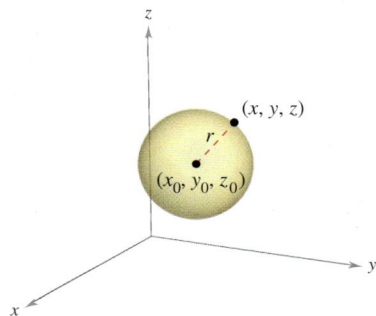

Figure 11.18

as shown in Figure 11.18. Moreover, the midpoint of the line segment joining the points (x_1, y_1, z_1) and (x_2, y_2, z_2) has coordinates

$$\left(\frac{x_1 + x_2}{2}, \frac{y_1 + y_2}{2}, \frac{z_1 + z_2}{2} \right). \qquad \text{Midpoint Formula}$$

EXAMPLE 2 **Finding the Equation of a Sphere**

Find the standard equation of the sphere that has the points

$$(5, -2, 3) \quad \text{and} \quad (0, 4, -3)$$

as endpoints of a diameter.

Solution Using the Midpoint Formula, the center of the sphere is

$$\left(\frac{5 + 0}{2}, \frac{-2 + 4}{2}, \frac{3 - 3}{2} \right) = \left(\frac{5}{2}, 1, 0 \right). \qquad \text{Midpoint Formula}$$

By the Distance Formula, the radius is

$$r = \sqrt{\left(0 - \frac{5}{2} \right)^2 + (4 - 1)^2 + (-3 - 0)^2} = \sqrt{\frac{97}{4}} = \frac{\sqrt{97}}{2}.$$

Therefore, the standard equation of the sphere is

$$\left(x - \frac{5}{2} \right)^2 + (y - 1)^2 + z^2 = \frac{97}{4}. \qquad \text{Equation of sphere}$$

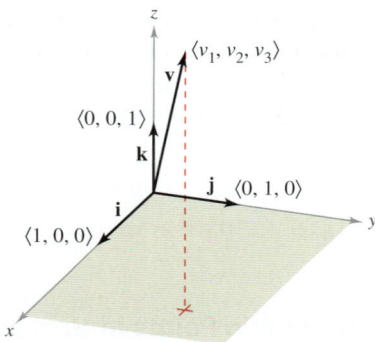

The standard unit vectors in space
Figure 11.19

Vectors in Space

In space, vectors are denoted by ordered triples $\mathbf{v} = \langle v_1, v_2, v_3 \rangle$. The **zero vector** is denoted by $\mathbf{0} = \langle 0, 0, 0 \rangle$. Using the unit vectors

$$\mathbf{i} = \langle 1, 0, 0 \rangle, \quad \mathbf{j} = \langle 0, 1, 0 \rangle, \quad \text{and} \quad \mathbf{k} = \langle 0, 0, 1 \rangle$$

the **standard unit vector notation** for $\mathbf{v}$ is

$$\mathbf{v} = v_1\mathbf{i} + v_2\mathbf{j} + v_3\mathbf{k}$$

as shown in Figure 11.19. If $\mathbf{v}$ is represented by the directed line segment from $P(p_1, p_2, p_3)$ to $Q(q_1, q_2, q_3)$, as shown in Figure 11.20, then the component form of $\mathbf{v}$ is written by subtracting the coordinates of the initial point from the coordinates of the terminal point, as follows.

$$\mathbf{v} = \langle v_1, v_2, v_3 \rangle = \langle q_1 - p_1, q_2 - p_2, q_3 - p_3 \rangle$$

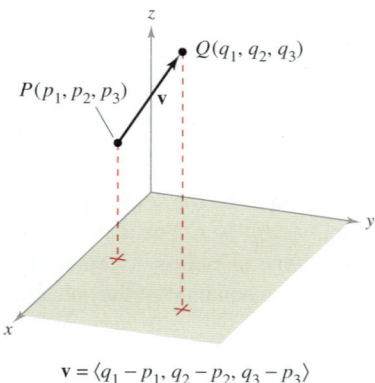

$$\mathbf{v} = \langle q_1 - p_1, q_2 - p_2, q_3 - p_3 \rangle$$

Figure 11.20

Vectors in Space

Let $\mathbf{u} = \langle u_1, u_2, u_3 \rangle$ and $\mathbf{v} = \langle v_1, v_2, v_3 \rangle$ be vectors in space and let c be a scalar.

1. *Equality of Vectors:* $\mathbf{u} = \mathbf{v}$ if and only if $u_1 = v_1$, $u_2 = v_2$, and $u_3 = v_3$.
2. *Component Form:* If $\mathbf{v}$ is represented by the directed line segment from $P(p_1, p_2, p_3)$ to $Q(q_1, q_2, q_3)$, then
$$\mathbf{v} = \langle v_1, v_2, v_3 \rangle = \langle q_1 - p_1, q_2 - p_2, q_3 - p_3 \rangle.$$
3. *Length:* $\|\mathbf{v}\| = \sqrt{v_1^2 + v_2^2 + v_3^2}$
4. *Unit Vector in the Direction of $\mathbf{v}$:* $\dfrac{\mathbf{v}}{\|\mathbf{v}\|} = \left(\dfrac{1}{\|\mathbf{v}\|}\right)\langle v_1, v_2, v_3 \rangle, \quad \mathbf{v} \neq \mathbf{0}$
5. *Vector Addition:* $\mathbf{v} + \mathbf{u} = \langle v_1 + u_1, v_2 + u_2, v_3 + u_3 \rangle$
6. *Scalar Multiplication:* $c\mathbf{v} = \langle cv_1, cv_2, cv_3 \rangle$

Note that the properties of vector operations listed in Theorem 11.1 (see Section 11.1) are also valid for vectors in space.

EXAMPLE 3 Finding the Component Form of a Vector in Space

· · · ▷ *See LarsonCalculus.com for an interactive version of this type of example.*

Find the component form and magnitude of the vector $\mathbf{v}$ having initial point $(-2, 3, 1)$ and terminal point $(0, -4, 4)$. Then find a unit vector in the direction of $\mathbf{v}$.

Solution The component form of $\mathbf{v}$ is

$$\mathbf{v} = \langle q_1 - p_1, q_2 - p_2, q_3 - p_3 \rangle = \langle 0 - (-2), -4 - 3, 4 - 1 \rangle = \langle 2, -7, 3 \rangle$$

which implies that its magnitude is

$$\|\mathbf{v}\| = \sqrt{2^2 + (-7)^2 + 3^2} = \sqrt{62}.$$

The unit vector in the direction of $\mathbf{v}$ is

$$\begin{aligned}
\mathbf{u} &= \frac{\mathbf{v}}{\|\mathbf{v}\|} \\
&= \frac{1}{\sqrt{62}}\langle 2, -7, 3 \rangle \\
&= \left\langle \frac{2}{\sqrt{62}}, \frac{-7}{\sqrt{62}}, \frac{3}{\sqrt{62}} \right\rangle.
\end{aligned}$$

Recall from the definition of scalar multiplication that positive scalar multiples of a nonzero vector **v** have the same direction as **v**, whereas negative multiples have the direction opposite of **v**. In general, two nonzero vectors **u** and **v** are **parallel** when there is some scalar c such that $\mathbf{u} = c\mathbf{v}$. For example, in Figure 11.21, the vectors **u**, **v**, and **w** are parallel because

$$\mathbf{u} = 2\mathbf{v} \quad \text{and} \quad \mathbf{w} = -\mathbf{v}.$$

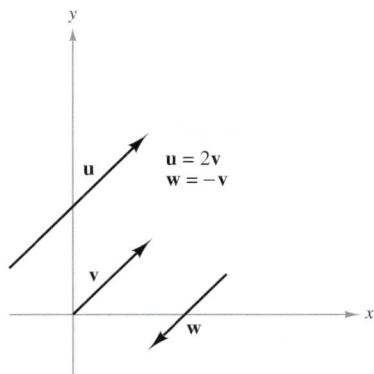

Parallel vectors
Figure 11.21

Definition of Parallel Vectors

Two nonzero vectors **u** and **v** are **parallel** when there is some scalar c such that $\mathbf{u} = c\mathbf{v}$.

EXAMPLE 4   **Parallel Vectors**

Vector **w** has initial point $(2, -1, 3)$ and terminal point $(-4, 7, 5)$. Which of the following vectors is parallel to **w**?

a. $\mathbf{u} = \langle 3, -4, -1 \rangle$

b. $\mathbf{v} = \langle 12, -16, 4 \rangle$

Solution   Begin by writing **w** in component form.

$$\mathbf{w} = \langle -4 - 2, 7 - (-1), 5 - 3 \rangle = \langle -6, 8, 2 \rangle$$

a. Because $\mathbf{u} = \langle 3, -4, -1 \rangle = -\frac{1}{2}\langle -6, 8, 2 \rangle = -\frac{1}{2}\mathbf{w}$, you can conclude that **u** is parallel to **w**.

b. In this case, you want to find a scalar c such that

$$\langle 12, -16, 4 \rangle = c\langle -6, 8, 2 \rangle.$$

To find c, equate the corresponding components and solve as shown.

$$12 = -6c \implies c = -2$$
$$-16 = 8c \implies c = -2$$
$$4 = 2c \implies c = 2$$

Note that $c = -2$ for the first two components and $c = 2$ for the third component. This means that the equation $\langle 12, -16, 4 \rangle = c\langle -6, 8, 2 \rangle$ has no solution, and the vectors are not parallel.

EXAMPLE 5   **Using Vectors to Determine Collinear Points**

Determine whether the points

$$P(1, -2, 3), \quad Q(2, 1, 0), \quad \text{and} \quad R(4, 7, -6)$$

are collinear.

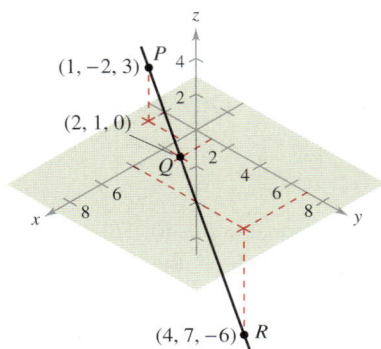

The points P, Q, and R lie on the same line.
Figure 11.22

Solution   The component forms of $\overrightarrow{PQ}$ and $\overrightarrow{PR}$ are

$$\overrightarrow{PQ} = \langle 2 - 1, 1 - (-2), 0 - 3 \rangle = \langle 1, 3, -3 \rangle$$

and

$$\overrightarrow{PR} = \langle 4 - 1, 7 - (-2), -6 - 3 \rangle = \langle 3, 9, -9 \rangle.$$

These two vectors have a common initial point. So, P, Q, and R lie on the same line if and only if $\overrightarrow{PQ}$ and $\overrightarrow{PR}$ are parallel—which they are because $\overrightarrow{PR} = 3\,\overrightarrow{PQ}$, as shown in Figure 11.22.

Standard Unit Vector Notation

a. Write the vector $\mathbf{v} = 4\mathbf{i} - 5\mathbf{k}$ in component form.

b. Find the terminal point of the vector $\mathbf{v} = 7\mathbf{i} - \mathbf{j} + 3\mathbf{k}$, given that the initial point is $P(-2, 3, 5)$.

c. Find the magnitude of the vector $\mathbf{v} = -6\mathbf{i} + 2\mathbf{j} - 3\mathbf{k}$. Then find a unit vector in the direction of $\mathbf{v}$.

Solution

a. Because $\mathbf{j}$ is missing, its component is 0 and

$$\mathbf{v} = 4\mathbf{i} - 5\mathbf{k} = \langle 4, 0, -5 \rangle.$$

b. You need to find $Q(q_1, q_2, q_3)$ such that

$$\mathbf{v} = \overrightarrow{PQ} = 7\mathbf{i} - \mathbf{j} + 3\mathbf{k}.$$

This implies that $q_1 - (-2) = 7$, $q_2 - 3 = -1$, and $q_3 - 5 = 3$. The solution of these three equations is $q_1 = 5$, $q_2 = 2$, and $q_3 = 8$. Therefore, Q is $(5, 2, 8)$.

c. Note that $v_1 = -6$, $v_2 = 2$, and $v_3 = -3$. So, the magnitude of $\mathbf{v}$ is

$$\|\mathbf{v}\| = \sqrt{(-6)^2 + 2^2 + (-3)^2} = \sqrt{49} = 7.$$

The unit vector in the direction of $\mathbf{v}$ is

$$\tfrac{1}{7}(-6\mathbf{i} + 2\mathbf{j} - 3\mathbf{k}) = -\tfrac{6}{7}\mathbf{i} + \tfrac{2}{7}\mathbf{j} - \tfrac{3}{7}\mathbf{k}.$$

Measuring Force

A television camera weighing 120 pounds is supported by a tripod, as shown in Figure 11.23. Represent the force exerted on each leg of the tripod as a vector.

Solution Let the vectors $\mathbf{F}_1$, $\mathbf{F}_2$, and $\mathbf{F}_3$ represent the forces exerted on the three legs. From Figure 11.23, you can determine the directions of $\mathbf{F}_1$, $\mathbf{F}_2$, and $\mathbf{F}_3$ to be as follows.

$$\overrightarrow{PQ_1} = \langle 0 - 0, -1 - 0, 0 - 4 \rangle = \langle 0, -1, -4 \rangle$$

$$\overrightarrow{PQ_2} = \left\langle \frac{\sqrt{3}}{2} - 0, \frac{1}{2} - 0, 0 - 4 \right\rangle = \left\langle \frac{\sqrt{3}}{2}, \frac{1}{2}, -4 \right\rangle$$

$$\overrightarrow{PQ_3} = \left\langle -\frac{\sqrt{3}}{2} - 0, \frac{1}{2} - 0, 0 - 4 \right\rangle = \left\langle -\frac{\sqrt{3}}{2}, \frac{1}{2}, -4 \right\rangle$$

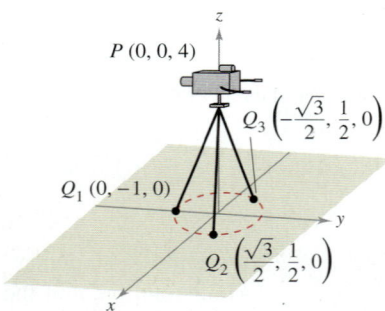

$P(0, 0, 4)$

$Q_3 \left(-\frac{\sqrt{3}}{2}, \frac{1}{2}, 0 \right)$

$Q_1 (0, -1, 0)$

$Q_2 \left(\frac{\sqrt{3}}{2}, \frac{1}{2}, 0 \right)$

Figure 11.23

Because each leg has the same length, and the total force is distributed equally among the three legs, you know that $\|\mathbf{F}_1\| = \|\mathbf{F}_2\| = \|\mathbf{F}_3\|$. So, there exists a constant c such that

$$\mathbf{F}_1 = c\langle 0, -1, -4 \rangle, \quad \mathbf{F}_2 = c\left\langle \frac{\sqrt{3}}{2}, \frac{1}{2}, -4 \right\rangle, \quad \text{and} \quad \mathbf{F}_3 = c\left\langle -\frac{\sqrt{3}}{2}, \frac{1}{2}, -4 \right\rangle.$$

Let the total force exerted by the object be given by $\mathbf{F} = \langle 0, 0, -120 \rangle$. Then, using the fact that

$$\mathbf{F} = \mathbf{F}_1 + \mathbf{F}_2 + \mathbf{F}_3$$

you can conclude that $\mathbf{F}_1$, $\mathbf{F}_2$, and $\mathbf{F}_3$ all have a vertical component of -40. This implies that $c(-4) = -40$ and $c = 10$. Therefore, the forces exerted on the legs can be represented by

$$\mathbf{F}_1 = \langle 0, -10, -40 \rangle,$$
$$\mathbf{F}_2 = \langle 5\sqrt{3}, 5, -40 \rangle,$$

and

$$\mathbf{F}_3 = \langle -5\sqrt{3}, 5, -40 \rangle.$$

Plotting Points In Exercises 1–4, plot the points in the same three-dimensional coordinate system.

1. (a) $(2, 1, 3)$ (b) $(-1, 2, 1)$
2. (a) $(3, -2, 5)$ (b) $\left(\frac{3}{2}, 4, -2\right)$
3. (a) $(5, -2, 2)$ (b) $(5, -2, -2)$
4. (a) $(0, 4, -5)$ (b) $(4, 0, 5)$

Finding Coordinates of a Point In Exercises 5–8, find the coordinates of the point.

5. The point is located three units behind the yz-plane, four units to the right of the xz-plane, and five units above the xy-plane.

6. The point is located seven units in front of the yz-plane, two units to the left of the xz-plane, and one unit below the xy-plane.

7. The point is located on the x-axis, 12 units in front of the yz-plane.

8. The point is located in the yz-plane, three units to the right of the xz-plane, and two units above the xy-plane.

9. **Think About It** What is the z-coordinate of any point in the xy-plane?

10. **Think About It** What is the x-coordinate of any point in the yz-plane?

Using the Three-Dimensional Coordinate System In Exercises 11–22, determine the location of a point (x, y, z) that satisfies the condition(s).

11. $z = 6$ 12. $y = 2$
13. $x = -3$ 14. $z = -\frac{5}{2}$
15. $y < 0$ 16. $x > 0$
17. $|y| \le 3$ 18. $|x| > 4$
19. $xy > 0, \quad z = -3$ 20. $xy < 0, \quad z = 4$
21. $xyz < 0$ 22. $xyz > 0$

Finding the Distance Between Two Points in Space In Exercises 23–26, find the distance between the points.

23. $(0, 0, 0), (-4, 2, 7)$ 24. $(-2, 3, 2), (2, -5, -2)$
25. $(1, -2, 4), (6, -2, -2)$ 26. $(2, 2, 3), (4, -5, 6)$

Classifying a Triangle In Exercises 27–30, find the lengths of the sides of the triangle with the indicated vertices, and determine whether the triangle is a right triangle, an isosceles triangle, or neither.

27. $(0, 0, 4), (2, 6, 7), (6, 4, -8)$
28. $(3, 4, 1), (0, 6, 2), (3, 5, 6)$
29. $(-1, 0, -2), (-1, 5, 2), (-3, -1, 1)$
30. $(4, -1, -1), (2, 0, -4), (3, 5, -1)$

31. **Think About It** The triangle in Exercise 27 is translated five units upward along the z-axis. Determine the coordinates of the translated triangle.

32. **Think About It** The triangle in Exercise 28 is translated three units to the right along the y-axis. Determine the coordinates of the translated triangle.

Finding the Midpoint In Exercises 33–36, find the coordinates of the midpoint of the line segment joining the points.

33. $(3, 4, 6), (1, 8, 0)$ 34. $(7, 2, 2), (-5, -2, -3)$
35. $(5, -9, 7), (-2, 3, 3)$ 36. $(4, 0, -6), (8, 8, 20)$

Finding the Equation of a Sphere In Exercises 37–40, find the standard equation of the sphere.

37. Center: $(0, 2, 5)$ 38. Center: $(4, -1, 1)$
    Radius: 2   Radius: 5

39. Endpoints of a diameter: $(2, 0, 0), (0, 6, 0)$

40. Center: $(-3, 2, 4)$, tangent to the yz-plane

Finding the Equation of a Sphere In Exercises 41–44, complete the square to write the equation of the sphere in standard form. Find the center and radius.

41. $x^2 + y^2 + z^2 - 2x + 6y + 8z + 1 = 0$
42. $x^2 + y^2 + z^2 + 9x - 2y + 10z + 19 = 0$
43. $9x^2 + 9y^2 + 9z^2 - 6x + 18y + 1 = 0$
44. $4x^2 + 4y^2 + 4z^2 - 24x - 4y + 8z - 23 = 0$

Finding the Component Form of a Vector in Space In Exercises 45–48, (a) find the component form of the vector v, (b) write the vector using standard unit vector notation, and (c) sketch the vector with its initial point at the origin.

45.

46.

47.

48.
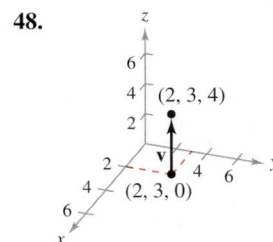

Finding the Component Form of a Vector in Space In Exercises 49 and 50, find the component form and magnitude of the vector v with the given initial and terminal points. Then find a unit vector in the direction of v.

49. Initial point: $(3, 2, 0)$

   Terminal point: $(4, 1, 6)$

50. Initial point: $(1, -2, 4)$

   Terminal point: $(2, 4, -2)$

Writing a Vector in Different Forms In Exercises 51 and 52, the initial and terminal points of a vector v are given. (a) Sketch the directed line segment, (b) find the component form of the vector, (c) write the vector using standard unit vector notation, and (d) sketch the vector with its initial point at the origin.

51. Initial point: $(-1, 2, 3)$

   Terminal point: $(3, 3, 4)$

52. Initial point: $(2, -1, -2)$

   Terminal point: $(-4, 3, 7)$

Finding a Terminal Point In Exercises 53 and 54, the vector v and its initial point are given. Find the terminal point.

53. $\mathbf{v} = \langle 3, -5, 6 \rangle$

   Initial point: $(0, 6, 2)$

54. $\mathbf{v} = \left\langle 1, -\frac{2}{3}, \frac{1}{2} \right\rangle$

   Initial point: $\left(0, 2, \frac{5}{2}\right)$

Finding Scalar Multiples In Exercises 55 and 56, find each scalar multiple of v and sketch its graph.

55. $\mathbf{v} = \langle 1, 2, 2 \rangle$

   (a) $2\mathbf{v}$ (b) $-\mathbf{v}$

   (c) $\frac{3}{2}\mathbf{v}$ (d) $0\mathbf{v}$

56. $\mathbf{v} = \langle 2, -2, 1 \rangle$

   (a) $-\mathbf{v}$ (b) $2\mathbf{v}$

   (c) $\frac{1}{2}\mathbf{v}$ (d) $\frac{5}{2}\mathbf{v}$

Finding a Vector In Exercises 57–60, find the vector z, given that $\mathbf{u} = \langle 1, 2, 3 \rangle$, $\mathbf{v} = \langle 2, 2, -1 \rangle$, and $\mathbf{w} = \langle 4, 0, -4 \rangle$.

57. $\mathbf{z} = \mathbf{u} - \mathbf{v} + 2\mathbf{w}$

58. $\mathbf{z} = 5\mathbf{u} - 3\mathbf{v} - \frac{1}{2}\mathbf{w}$

59. $2\mathbf{z} - 3\mathbf{u} = \mathbf{w}$

60. $2\mathbf{u} + \mathbf{v} - \mathbf{w} + 3\mathbf{z} = 0$

Parallel Vectors In Exercises 61–64, determine which of the vectors is (are) parallel to z. Use a graphing utility to confirm your results.

61. $\mathbf{z} = \langle 3, 2, -5 \rangle$

   (a) $\langle -6, -4, 10 \rangle$

   (b) $\left\langle 2, \frac{4}{3}, -\frac{10}{3} \right\rangle$

   (c) $\langle 6, 4, 10 \rangle$

   (d) $\langle 1, -4, 2 \rangle$

62. $\mathbf{z} = \frac{1}{2}\mathbf{i} - \frac{2}{3}\mathbf{j} + \frac{3}{4}\mathbf{k}$

   (a) $6\mathbf{i} - 4\mathbf{j} + 9\mathbf{k}$

   (b) $-\mathbf{i} + \frac{4}{3}\mathbf{j} - \frac{3}{2}\mathbf{k}$

   (c) $12\mathbf{i} + 9\mathbf{k}$

   (d) $\frac{3}{4}\mathbf{i} - \mathbf{j} + \frac{9}{8}\mathbf{k}$

63. z has initial point $(1, -1, 3)$ and terminal point $(-2, 3, 5)$.

   (a) $-6\mathbf{i} + 8\mathbf{j} + 4\mathbf{k}$ (b) $4\mathbf{j} + 2\mathbf{k}$

64. z has initial point $(5, 4, 1)$ and terminal point $(-2, -4, 4)$.

   (a) $\langle 7, 6, 2 \rangle$ (b) $\langle 14, 16, -6 \rangle$

Using Vectors to Determine Collinear Points In Exercises 65–68, use vectors to determine whether the points are collinear.

65. $(0, -2, -5), (3, 4, 4), (2, 2, 1)$

66. $(4, -2, 7), (-2, 0, 3), (7, -3, 9)$

67. $(1, 2, 4), (2, 5, 0), (0, 1, 5)$

68. $(0, 0, 0), (1, 3, -2), (2, -6, 4)$

Verifying a Parallelogram In Exercises 69 and 70, use vectors to show that the points form the vertices of a parallelogram.

69. $(2, 9, 1), (3, 11, 4), (0, 10, 2), (1, 12, 5)$

70. $(1, 1, 3), (9, -1, -2), (11, 2, -9), (3, 4, -4)$

Finding the Magnitude In Exercises 71–76, find the magnitude of v.

71. $\mathbf{v} = \langle 0, 0, 0 \rangle$

72. $\mathbf{v} = \langle 1, 0, 3 \rangle$

73. $\mathbf{v} = 3\mathbf{j} - 5\mathbf{k}$

74. $\mathbf{v} = 2\mathbf{i} + 5\mathbf{j} - \mathbf{k}$

75. $\mathbf{v} = \mathbf{i} - 2\mathbf{j} - 3\mathbf{k}$

76. $\mathbf{v} = -4\mathbf{i} + 3\mathbf{j} + 7\mathbf{k}$

Finding Unit Vectors In Exercises 77–80, find a unit vector (a) in the direction of v and (b) in the direction opposite of v.

77. $\mathbf{v} = \langle 2, -1, 2 \rangle$

78. $\mathbf{v} = \langle 6, 0, 8 \rangle$

79. $\mathbf{v} = 4\mathbf{i} - 5\mathbf{j} + 3\mathbf{k}$

80. $\mathbf{v} = 5\mathbf{i} + 3\mathbf{j} - \mathbf{k}$

81. Using Vectors Consider the two nonzero vectors u and v, and let s and t be real numbers. Describe the geometric figure generated by the terminal points of the three vectors $t\mathbf{v}$, $\mathbf{u} + t\mathbf{v}$, and $s\mathbf{u} + t\mathbf{v}$.

82. HOW DO YOU SEE IT? Determine (x, y, z) for each figure. Then find the component form of the vector from the point on the x-axis to the point (x, y, z).

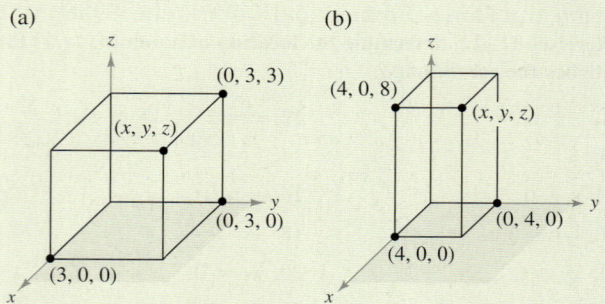

(a)

(b)

Finding a Vector In Exercises 83–86, find the vector v with the given magnitude and the same direction as u.

	Magnitude	Direction
83.	$\|\mathbf{v}\| = 10$	$\mathbf{u} = \langle 0, 3, 3 \rangle$
84.	$\|\mathbf{v}\| = 3$	$\mathbf{u} = \langle 1, 1, 1 \rangle$
85.	$\|\mathbf{v}\| = \frac{3}{2}$	$\mathbf{u} = \langle 2, -2, 1 \rangle$
86.	$\|\mathbf{v}\| = 7$	$\mathbf{u} = \langle -4, 6, 2 \rangle$

Sketching a Vector In Exercises 87 and 88, sketch the vector v and write its component form.

87. v lies in the yz-plane, has magnitude 2, and makes an angle of $30°$ with the positive y-axis.

88. $\mathbf{v}$ lies in the xz-plane, has magnitude 5, and makes an angle of 45° with the positive z-axis.

Finding a Point Using Vectors   In Exercises 89 and 90, use vectors to find the point that lies two-thirds of the way from P to Q.

89. $P(4, 3, 0), Q(1, -3, 3)$

90. $P(1, 2, 5), Q(6, 8, 2)$

91. Using Vectors Let $\mathbf{u} = \mathbf{i} + \mathbf{j}$, $\mathbf{v} = \mathbf{j} + \mathbf{k}$, and $\mathbf{w} = a\mathbf{u} + b\mathbf{v}$.

(a) Sketch $\mathbf{u}$ and $\mathbf{v}$.

(b) If $\mathbf{w} = \mathbf{0}$, show that a and b must both be zero.

(c) Find a and b such that $\mathbf{w} = \mathbf{i} + 2\mathbf{j} + \mathbf{k}$.

(d) Show that no choice of a and b yields $\mathbf{w} = \mathbf{i} + 2\mathbf{j} + 3\mathbf{k}$.

92. Writing The initial and terminal points of the vector $\mathbf{v}$ are (x_1, y_1, z_1) and (x, y, z). Describe the set of all points (x, y, z) such that $\|\mathbf{v}\| = 4$.

WRITING ABOUT CONCEPTS

93. Describing Coordinates A point in the three-dimensional coordinate system has coordinates (x_0, y_0, z_0). Describe what each coordinate measures.

94. Distance Formula Give the formula for the distance between the points (x_1, y_1, z_1) and (x_2, y_2, z_2).

95. Standard Equation of a Sphere Give the standard equation of a sphere of radius r, centered at (x_0, y_0, z_0).

96. Parallel Vectors State the definition of parallel vectors.

97. Using a Triangle and Vectors Let A, B, and C be vertices of a triangle. Find $\overrightarrow{AB} + \overrightarrow{BC} + \overrightarrow{CA}$.

98. Using Vectors Let $\mathbf{r} = \langle x, y, z \rangle$ and $\mathbf{r}_0 = \langle 1, 1, 1 \rangle$. Describe the set of all points (x, y, z) such that $\|\mathbf{r} - \mathbf{r}_0\| = 2$.

99. Diagonal of a Cube Find the component form of the unit vector $\mathbf{v}$ in the direction of the diagonal of the cube shown in the figure.

Figure for 99

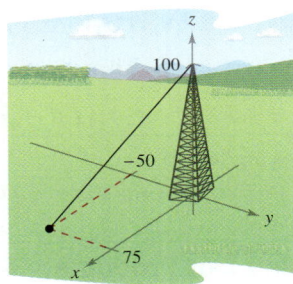

Figure for 100

100. Tower Guy Wire The guy wire supporting a 100-foot tower has a tension of 550 pounds. Using the distances shown in the figure, write the component form of the vector $\mathbf{F}$ representing the tension in the wire.

Losevsky Photo and Video/Shutterstock.com

101. Auditorium Lights

The lights in an auditorium are 24-pound discs of radius 18 inches. Each disc is supported by three equally spaced cables that are L inches long (see figure).

(a) Write the tension T in each cable as a function of L. Determine the domain of the function.

(b) Use a graphing utility and the function in part (a) to complete the table.

L	20	25	30	35	40	45	50
T							

(c) Use a graphing utility to graph the function in part (a). Determine the asymptotes of the graph.

(d) Confirm the asymptotes of the graph in part (c) analytically.

(e) Determine the minimum length of each cable when a cable is designed to carry a maximum load of 10 pounds.

102. Think About It Suppose the length of each cable in Exercise 101 has a fixed length $L = a$, and the radius of each disc is r_0 inches. Make a conjecture about the limit $\lim_{r_0 \to a^-} T$ and give a reason for your answer.

103. Load Supports Find the tension in each of the supporting cables in the figure when the weight of the crate is 500 newtons.

Figure for 103

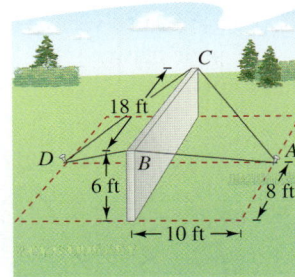

Figure for 104

104. Construction A precast concrete wall is temporarily kept in its vertical position by ropes (see figure). Find the total force exerted on the pin at position A. The tensions in AB and AC are 420 pounds and 650 pounds.

105. Geometry Write an equation whose graph consists of the set of points $P(x, y, z)$ that are twice as far from $A(0, -1, 1)$ as from $B(1, 2, 0)$. Describe the geometric figure represented by the equation.

11.3 The Dot Product of Two Vectors

- Use properties of the dot product of two vectors.
- Find the angle between two vectors using the dot product.
- Find the direction cosines of a vector in space.
- Find the projection of a vector onto another vector.
- Use vectors to find the work done by a constant force.

The Dot Product

So far, you have studied two operations with vectors—vector addition and multiplication by a scalar—each of which yields another vector. In this section, you will study a third vector operation, the **dot product.** This product yields a scalar, rather than a vector.

REMARK Because the dot product of two vectors yields a scalar, it is also called the *scalar product* (or *inner product*) of the two vectors.

Definition of Dot Product

The **dot product** of $\mathbf{u} = \langle u_1, u_2 \rangle$ and $\mathbf{v} = \langle v_1, v_2 \rangle$ is

$$\mathbf{u} \cdot \mathbf{v} = u_1 v_1 + u_2 v_2.$$

The **dot product** of $\mathbf{u} = \langle u_1, u_2, u_3 \rangle$ and $\mathbf{v} = \langle v_1, v_2, v_3 \rangle$ is

$$\mathbf{u} \cdot \mathbf{v} = u_1 v_1 + u_2 v_2 + u_3 v_3.$$

Exploration

Interpreting a Dot Product
Several vectors are shown below on the unit circle. Find the dot products of several pairs of vectors. Then find the angle between each pair that you used. Make a conjecture about the relationship between the dot product of two vectors and the angle between the vectors.

THEOREM 11.4 Properties of the Dot Product

Let **u**, **v**, and **w** be vectors in the plane or in space and let c be a scalar.

1. $\mathbf{u} \cdot \mathbf{v} = \mathbf{v} \cdot \mathbf{u}$ Commutative Property
2. $\mathbf{u} \cdot (\mathbf{v} + \mathbf{w}) = \mathbf{u} \cdot \mathbf{v} + \mathbf{u} \cdot \mathbf{w}$ Distributive Property
3. $c(\mathbf{u} \cdot \mathbf{v}) = c\mathbf{u} \cdot \mathbf{v} = \mathbf{u} \cdot c\mathbf{v}$
4. $\mathbf{0} \cdot \mathbf{v} = 0$
5. $\mathbf{v} \cdot \mathbf{v} = \|\mathbf{v}\|^2$

Proof To prove the first property, let $\mathbf{u} = \langle u_1, u_2, u_3 \rangle$ and $\mathbf{v} = \langle v_1, v_2, v_3 \rangle$. Then

$$\mathbf{u} \cdot \mathbf{v} = u_1 v_1 + u_2 v_2 + u_3 v_3 = v_1 u_1 + v_2 u_2 + v_3 u_3 = \mathbf{v} \cdot \mathbf{u}.$$

For the fifth property, let $\mathbf{v} = \langle v_1, v_2, v_3 \rangle$. Then

$$\mathbf{v} \cdot \mathbf{v} = v_1^2 + v_2^2 + v_3^2 = \left(\sqrt{v_1^2 + v_2^2 + v_3^2} \right)^2 = \|\mathbf{v}\|^2.$$

Proofs of the other properties are left to you.
See LarsonCalculus.com for Bruce Edwards's video of this proof.

EXAMPLE 1   Finding Dot Products

Let $\mathbf{u} = \langle 2, -2 \rangle$, $\mathbf{v} = \langle 5, 8 \rangle$, and $\mathbf{w} = \langle -4, 3 \rangle$.

a. $\mathbf{u} \cdot \mathbf{v} = \langle 2, -2 \rangle \cdot \langle 5, 8 \rangle = 2(5) + (-2)(8) = -6$

b. $(\mathbf{u} \cdot \mathbf{v})\mathbf{w} = -6\langle -4, 3 \rangle = \langle 24, -18 \rangle$

c. $\mathbf{u} \cdot (2\mathbf{v}) = 2(\mathbf{u} \cdot \mathbf{v}) = 2(-6) = -12$

d. $\|\mathbf{w}\|^2 = \mathbf{w} \cdot \mathbf{w} = \langle -4, 3 \rangle \cdot \langle -4, 3 \rangle = (-4)(-4) + (3)(3) = 25$

Notice that the result of part (b) is a *vector* quantity, whereas the results of the other three parts are *scalar* quantities.

Angle Between Two Vectors

The **angle between two nonzero vectors** is the angle θ, $0 \leq \theta \leq \pi$, between their respective standard position vectors, as shown in Figure 11.24. The next theorem shows how to find this angle using the dot product. (Note that the angle between the zero vector and another vector is not defined here.)

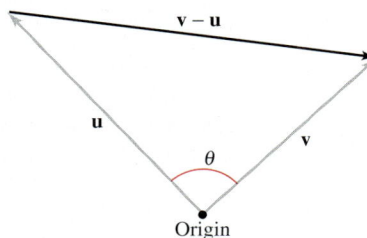

The angle between two vectors
Figure 11.24

> ### THEOREM 11.5   Angle Between Two Vectors
>
> If θ is the angle between two nonzero vectors $\mathbf{u}$ and $\mathbf{v}$, where $0 \leq \theta \leq \pi$, then
>
> $$\cos \theta = \frac{\mathbf{u} \cdot \mathbf{v}}{\|\mathbf{u}\| \|\mathbf{v}\|}.$$

Proof   Consider the triangle determined by vectors $\mathbf{u}$, $\mathbf{v}$, and $\mathbf{v} - \mathbf{u}$, as shown in Figure 11.24. By the Law of Cosines, you can write

$$\|\mathbf{v} - \mathbf{u}\|^2 = \|\mathbf{u}\|^2 + \|\mathbf{v}\|^2 - 2\|\mathbf{u}\| \|\mathbf{v}\| \cos \theta.$$

Using the properties of the dot product, the left side can be rewritten as

$$\begin{aligned} \|\mathbf{v} - \mathbf{u}\|^2 &= (\mathbf{v} - \mathbf{u}) \cdot (\mathbf{v} - \mathbf{u}) \\ &= (\mathbf{v} - \mathbf{u}) \cdot \mathbf{v} - (\mathbf{v} - \mathbf{u}) \cdot \mathbf{u} \\ &= \mathbf{v} \cdot \mathbf{v} - \mathbf{u} \cdot \mathbf{v} - \mathbf{v} \cdot \mathbf{u} + \mathbf{u} \cdot \mathbf{u} \\ &= \|\mathbf{v}\|^2 - 2\mathbf{u} \cdot \mathbf{v} + \|\mathbf{u}\|^2 \end{aligned}$$

and substitution back into the Law of Cosines yields

$$\begin{aligned} \|\mathbf{v}\|^2 - 2\mathbf{u} \cdot \mathbf{v} + \|\mathbf{u}\|^2 &= \|\mathbf{u}\|^2 + \|\mathbf{v}\|^2 - 2\|\mathbf{u}\| \|\mathbf{v}\| \cos \theta \\ -2\mathbf{u} \cdot \mathbf{v} &= -2\|\mathbf{u}\| \|\mathbf{v}\| \cos \theta \\ \cos \theta &= \frac{\mathbf{u} \cdot \mathbf{v}}{\|\mathbf{u}\| \|\mathbf{v}\|}. \end{aligned}$$

See LarsonCalculus.com for Bruce Edwards's video of this proof.

Note in Theorem 11.5 that because $\|\mathbf{u}\|$ and $\|\mathbf{v}\|$ are always positive, $\mathbf{u} \cdot \mathbf{v}$ and $\cos \theta$ will always have the same sign. Figure 11.25 shows the possible orientations of two vectors.

Opposite direction	$\mathbf{u} \cdot \mathbf{v} < 0$	$\mathbf{u} \cdot \mathbf{v} = 0$	$\mathbf{u} \cdot \mathbf{v} > 0$	Same direction
$\theta = \pi$	$\pi/2 < \theta < \pi$	$\theta = \pi/2$	$0 < \theta < \pi/2$	$\theta = 0$
$\cos \theta = -1$	$-1 < \cos \theta < 0$	$\cos \theta = 0$	$0 < \cos \theta < 1$	$\cos \theta = 1$

Figure 11.25

From Theorem 11.5, you can see that two nonzero vectors meet at a right angle if and only if their dot product is zero. Two such vectors are said to be **orthogonal.**

Definition of Orthogonal Vectors

The vectors $\mathbf{u}$ and $\mathbf{v}$ are orthogonal when $\mathbf{u} \cdot \mathbf{v} = 0$.

REMARK The terms "perpendicular," "orthogonal," and "normal" all mean essentially the same thing—meeting at right angles. It is common, however, to say that two vectors are *orthogonal*, two lines or planes are *perpendicular*, and a vector is *normal* to a line or plane.

From this definition, it follows that the zero vector is orthogonal to every vector $\mathbf{u}$, because $\mathbf{0} \cdot \mathbf{u} = 0$. Moreover, for $0 \le \theta \le \pi$, you know that $\cos\theta = 0$ if and only if $\theta = \pi/2$. So, you can use Theorem 11.5 to conclude that two *nonzero* vectors are orthogonal if and only if the angle between them is $\pi/2$.

EXAMPLE 2 **Finding the Angle Between Two Vectors**

⋯▷ *See LarsonCalculus.com for an interactive version of this type of example.*

For $\mathbf{u} = \langle 3, -1, 2 \rangle$, $\mathbf{v} = \langle -4, 0, 2 \rangle$, $\mathbf{w} = \langle 1, -1, -2 \rangle$, and $\mathbf{z} = \langle 2, 0, -1 \rangle$, find the angle between each pair of vectors.

a. $\mathbf{u}$ and $\mathbf{v}$ **b.** $\mathbf{u}$ and $\mathbf{w}$ **c.** $\mathbf{v}$ and $\mathbf{z}$

Solution

a. $\cos\theta = \dfrac{\mathbf{u} \cdot \mathbf{v}}{\|\mathbf{u}\|\,\|\mathbf{v}\|} = \dfrac{-12 + 0 + 4}{\sqrt{14}\sqrt{20}} = \dfrac{-8}{2\sqrt{14}\sqrt{5}} = \dfrac{-4}{\sqrt{70}}$

Because $\mathbf{u} \cdot \mathbf{v} < 0$, $\theta = \arccos\dfrac{-4}{\sqrt{70}} \approx 2.069$ radians.

REMARK The angle between $\mathbf{u}$ and $\mathbf{v}$ in Example 3(a) can also be written as approximately $118.561°$.

b. $\cos\theta = \dfrac{\mathbf{u} \cdot \mathbf{w}}{\|\mathbf{u}\|\,\|\mathbf{w}\|} = \dfrac{3 + 1 - 4}{\sqrt{14}\sqrt{6}} = \dfrac{0}{\sqrt{84}} = 0$

Because $\mathbf{u} \cdot \mathbf{w} = 0$, $\mathbf{u}$ and $\mathbf{w}$ are *orthogonal*. So, $\theta = \pi/2$.

c. $\cos\theta = \dfrac{\mathbf{v} \cdot \mathbf{z}}{\|\mathbf{v}\|\,\|\mathbf{z}\|} = \dfrac{-8 + 0 - 2}{\sqrt{20}\sqrt{5}} = \dfrac{-10}{\sqrt{100}} = -1$

Consequently, $\theta = \pi$. Note that $\mathbf{v}$ and $\mathbf{z}$ are parallel, with $\mathbf{v} = -2\mathbf{z}$. ◼

When the angle between two vectors is known, rewriting Theorem 11.5 in the form

$$\mathbf{u} \cdot \mathbf{v} = \|\mathbf{u}\|\,\|\mathbf{v}\| \cos\theta \qquad \text{\color{red}Alternative form of dot product}$$

produces an alternative way to calculate the dot product.

EXAMPLE 3 **Alternative Form of the Dot Product**

Given that $\|\mathbf{u}\| = 10$, $\|\mathbf{v}\| = 7$, and the angle between $\mathbf{u}$ and $\mathbf{v}$ is $\pi/4$, find $\mathbf{u} \cdot \mathbf{v}$.

Solution Use the alternative form of the dot product as shown.

$$\mathbf{u} \cdot \mathbf{v} = \|\mathbf{u}\|\,\|\mathbf{v}\| \cos\theta = (10)(7)\cos\frac{\pi}{4} = 35\sqrt{2}$$

◼

Direction Cosines

For a vector in the plane, you have seen that it is convenient to measure direction in terms of the angle, measured counterclockwise, *from* the positive *x*-axis *to* the vector. In space, it is more convenient to measure direction in terms of the angles *between* the nonzero vector **v** and the three unit vectors **i**, **j**, and **k**, as shown in Figure 11.26. The angles α, β, and γ are the **direction angles of v**, and $\cos \alpha$, $\cos \beta$, and $\cos \gamma$ are the **direction cosines of v**. Because

$$\mathbf{v} \cdot \mathbf{i} = \|\mathbf{v}\| \|\mathbf{i}\| \cos \alpha = \|\mathbf{v}\| \cos \alpha$$

and

$$\mathbf{v} \cdot \mathbf{i} = \langle v_1, v_2, v_3 \rangle \cdot \langle 1, 0, 0 \rangle = v_1$$

it follows that $\cos \alpha = v_1 / \|\mathbf{v}\|$. By similar reasoning with the unit vectors **j** and **k**, you have

$$\cos \alpha = \frac{v_1}{\|\mathbf{v}\|}$$ α is the angle between **v** and **i**.

$$\cos \beta = \frac{v_2}{\|\mathbf{v}\|}$$ β is the angle between **v** and **j**.

$$\cos \gamma = \frac{v_3}{\|\mathbf{v}\|}.$$ γ is the angle between **v** and **k**.

Consequently, any nonzero vector **v** in space has the normalized form

$$\frac{\mathbf{v}}{\|\mathbf{v}\|} = \frac{v_1}{\|\mathbf{v}\|}\mathbf{i} + \frac{v_2}{\|\mathbf{v}\|}\mathbf{j} + \frac{v_3}{\|\mathbf{v}\|}\mathbf{k} = \cos \alpha \mathbf{i} + \cos \beta \mathbf{j} + \cos \gamma \mathbf{k}$$

and because $\mathbf{v}/\|\mathbf{v}\|$ is a unit vector, it follows that

$$\cos^2 \alpha + \cos^2 \beta + \cos^2 \gamma = 1.$$

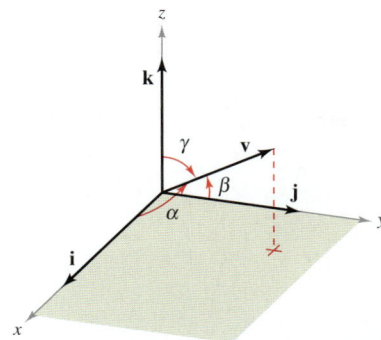
Direction angles
Figure 11.26

REMARK Recall that α, β, and γ are the Greek letters alpha, beta, and gamma, respectively.

EXAMPLE 4   Finding Direction Angles

Find the direction cosines and angles for the vector $\mathbf{v} = 2\mathbf{i} + 3\mathbf{j} + 4\mathbf{k}$, and show that $\cos^2 \alpha + \cos^2 \beta + \cos^2 \gamma = 1$.

Solution   Because $\|\mathbf{v}\| = \sqrt{2^2 + 3^2 + 4^2} = \sqrt{29}$, you can write the following.

$$\cos \alpha = \frac{v_1}{\|\mathbf{v}\|} = \frac{2}{\sqrt{29}} \implies \alpha \approx 68.2°$$ Angle between **v** and **i**

$$\cos \beta = \frac{v_2}{\|\mathbf{v}\|} = \frac{3}{\sqrt{29}} \implies \beta \approx 56.1°$$ Angle between **v** and **j**

$$\cos \gamma = \frac{v_3}{\|\mathbf{v}\|} = \frac{4}{\sqrt{29}} \implies \gamma \approx 42.0°$$ Angle between **v** and **k**

Furthermore, the sum of the squares of the direction cosines is

$$\cos^2 \alpha + \cos^2 \beta + \cos^2 \gamma = \frac{4}{29} + \frac{9}{29} + \frac{16}{29}$$

$$= \frac{29}{29}$$

$$= 1.$$

See Figure 11.27.

α = angle between **v** and **i**
β = angle between **v** and **j**
γ = angle between **v** and **k**

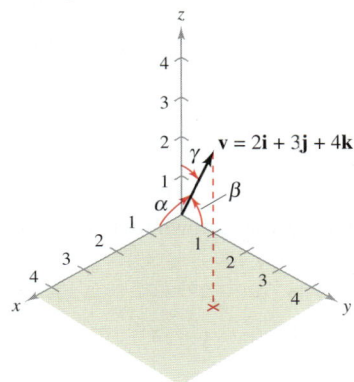

The direction angles of **v**
Figure 11.27

Projections and Vector Components

The force due to gravity pulls the boat against the ramp and down the ramp.
Figure 11.28

You have already seen applications in which two vectors are added to produce a resultant vector. Many applications in physics and engineering pose the reverse problem—decomposing a vector into the sum of two **vector components.** The following physical example enables you to see the usefulness of this procedure.

Consider a boat on an inclined ramp, as shown in Figure 11.28. The force $\mathbf{F}$ due to gravity pulls the boat *down* the ramp and *against* the ramp. These two forces, $\mathbf{w}_1$ and $\mathbf{w}_2$, are orthogonal—they are called the vector components of $\mathbf{F}$.

$$\mathbf{F} = \mathbf{w}_1 + \mathbf{w}_2 \qquad \text{Vector components of } \mathbf{F}$$

The forces $\mathbf{w}_1$ and $\mathbf{w}_2$ help you analyze the effect of gravity on the boat. For example, $\mathbf{w}_1$ indicates the force necessary to keep the boat from rolling down the ramp, whereas $\mathbf{w}_2$ indicates the force that the tires must withstand.

Definitions of Projection and Vector Components

Let $\mathbf{u}$ and $\mathbf{v}$ be nonzero vectors. Moreover, let

$$\mathbf{u} = \mathbf{w}_1 + \mathbf{w}_2$$

where $\mathbf{w}_1$ is parallel to $\mathbf{v}$ and $\mathbf{w}_2$ is orthogonal to $\mathbf{v}$, as shown in Figure 11.29.

1. $\mathbf{w}_1$ is called the **projection of u onto v** or the **vector component of u along v**, and is denoted by $\mathbf{w}_1 = \text{proj}_\mathbf{v}\mathbf{u}$.
2. $\mathbf{w}_2 = \mathbf{u} - \mathbf{w}_1$ is called the **vector component of u orthogonal to v.**

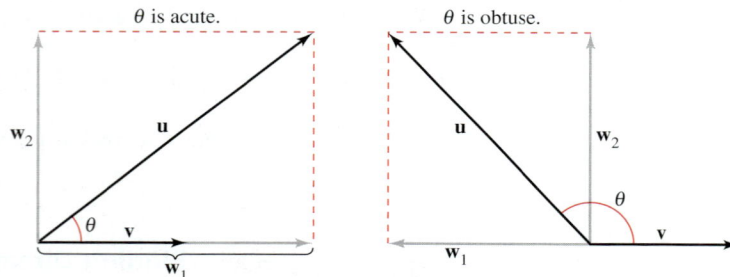

$\mathbf{w}_1 = \text{proj}_\mathbf{v}\mathbf{u} = $ projection of $\mathbf{u}$ onto $\mathbf{v} = $ vector component of $\mathbf{u}$ along $\mathbf{v}$
$\mathbf{w}_2 = $ vector component of $\mathbf{u}$ orthogonal to $\mathbf{v}$
Figure 11.29

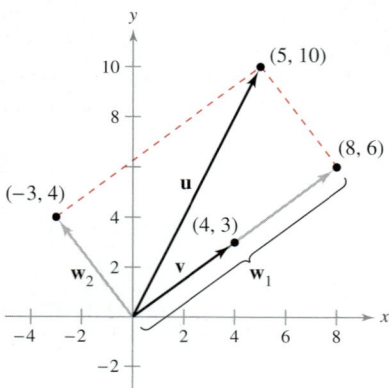

$\mathbf{u} = \mathbf{w}_1 + \mathbf{w}_2$
Figure 11.30

EXAMPLE 5   **Finding a Vector Component of u Orthogonal to v**

Find the vector component of $\mathbf{u} = \langle 5, 10 \rangle$ that is orthogonal to $\mathbf{v} = \langle 4, 3 \rangle$, given that

$$\mathbf{w}_1 = \text{proj}_\mathbf{v}\mathbf{u} = \langle 8, 6 \rangle$$

and

$$\mathbf{u} = \langle 5, 10 \rangle = \mathbf{w}_1 + \mathbf{w}_2.$$

Solution   Because $\mathbf{u} = \mathbf{w}_1 + \mathbf{w}_2$, where $\mathbf{w}_1$ is parallel to $\mathbf{v}$, it follows that $\mathbf{w}_2$ is the vector component of $\mathbf{u}$ orthogonal to $\mathbf{v}$. So, you have

$$\begin{aligned} \mathbf{w}_2 &= \mathbf{u} - \mathbf{w}_1 \\ &= \langle 5, 10 \rangle - \langle 8, 6 \rangle \\ &= \langle -3, 4 \rangle. \end{aligned}$$

Check to see that $\mathbf{w}_2$ is orthogonal to $\mathbf{v}$, as shown in Figure 11.30.

From Example 5, you can see that it is easy to find the vector component $\mathbf{w}_2$ once you have found the projection, $\mathbf{w}_1$, of $\mathbf{u}$ onto $\mathbf{v}$. To find this projection, use the dot product in the next theorem, which you will prove in Exercise 78.

> **THEOREM 11.6   Projection Using the Dot Product**
>
> If $\mathbf{u}$ and $\mathbf{v}$ are nonzero vectors, then the projection of $\mathbf{u}$ onto $\mathbf{v}$ is
>
> $$\text{proj}_\mathbf{v}\mathbf{u} = \left(\frac{\mathbf{u} \cdot \mathbf{v}}{\|\mathbf{v}\|^2}\right)\mathbf{v}.$$

•• **REMARK** Note the distinction between the terms "component" and "vector component." For example, using the standard unit vectors with $\mathbf{u} = u_1\mathbf{i} + u_2\mathbf{j}$, u_1 is the *component* of $\mathbf{u}$ in the direction of $\mathbf{i}$ and $u_1\mathbf{i}$ is the *vector component* in the direction of $\mathbf{i}$.

The projection of $\mathbf{u}$ onto $\mathbf{v}$ can be written as a scalar multiple of a unit vector in the direction of $\mathbf{v}$. That is,

$$\left(\frac{\mathbf{u} \cdot \mathbf{v}}{\|\mathbf{v}\|^2}\right)\mathbf{v} = \left(\frac{\mathbf{u} \cdot \mathbf{v}}{\|\mathbf{v}\|}\right)\frac{\mathbf{v}}{\|\mathbf{v}\|} = (k)\frac{\mathbf{v}}{\|\mathbf{v}\|}.$$

The scalar k is called the **component of $\mathbf{u}$ in the direction of $\mathbf{v}$.** So,

$$k = \frac{\mathbf{u} \cdot \mathbf{v}}{\|\mathbf{v}\|} = \|\mathbf{u}\| \cos\theta.$$

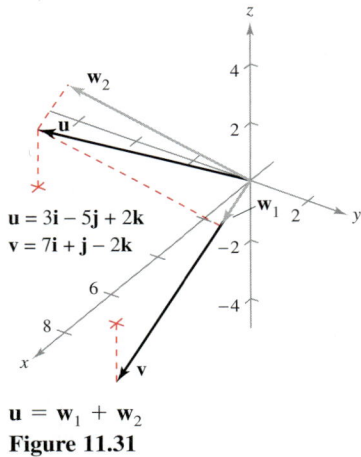

$\mathbf{u} = 3\mathbf{i} - 5\mathbf{j} + 2\mathbf{k}$
$\mathbf{v} = 7\mathbf{i} + \mathbf{j} - 2\mathbf{k}$

$\mathbf{u} = \mathbf{w}_1 + \mathbf{w}_2$
Figure 11.31

EXAMPLE 6   Decomposing a Vector into Vector Components

Find the projection of $\mathbf{u}$ onto $\mathbf{v}$ and the vector component of $\mathbf{u}$ orthogonal to $\mathbf{v}$ for

$$\mathbf{u} = 3\mathbf{i} - 5\mathbf{j} + 2\mathbf{k} \quad \text{and} \quad \mathbf{v} = 7\mathbf{i} + \mathbf{j} - 2\mathbf{k}.$$

Solution   The projection of $\mathbf{u}$ onto $\mathbf{v}$ is

$$\mathbf{w}_1 = \text{proj}_\mathbf{v}\mathbf{u} = \left(\frac{\mathbf{u} \cdot \mathbf{v}}{\|\mathbf{v}\|^2}\right)\mathbf{v} = \left(\frac{12}{54}\right)(7\mathbf{i} + \mathbf{j} - 2\mathbf{k}) = \frac{14}{9}\mathbf{i} + \frac{2}{9}\mathbf{j} - \frac{4}{9}\mathbf{k}.$$

The vector component of $\mathbf{u}$ orthogonal to $\mathbf{v}$ is the vector

$$\mathbf{w}_2 = \mathbf{u} - \mathbf{w}_1 = (3\mathbf{i} - 5\mathbf{j} + 2\mathbf{k}) - \left(\frac{14}{9}\mathbf{i} + \frac{2}{9}\mathbf{j} - \frac{4}{9}\mathbf{k}\right) = \frac{13}{9}\mathbf{i} - \frac{47}{9}\mathbf{j} + \frac{22}{9}\mathbf{k}.$$

See Figure 11.31.

EXAMPLE 7   Finding a Force

A 600-pound boat sits on a ramp inclined at $30°$, as shown in Figure 11.32. What force is required to keep the boat from rolling down the ramp?

Solution   Because the force due to gravity is vertical and downward, you can represent the gravitational force by the vector $\mathbf{F} = -600\mathbf{j}$. To find the force required to keep the boat from rolling down the ramp, project $\mathbf{F}$ onto a unit vector $\mathbf{v}$ in the direction of the ramp, as follows.

$$\mathbf{v} = \cos 30°\mathbf{i} + \sin 30°\mathbf{j} = \frac{\sqrt{3}}{2}\mathbf{i} + \frac{1}{2}\mathbf{j} \qquad \text{Unit vector along ramp}$$

Therefore, the projection of $\mathbf{F}$ onto $\mathbf{v}$ is

$$\mathbf{w}_1 = \text{proj}_\mathbf{v}\mathbf{F} = \left(\frac{\mathbf{F} \cdot \mathbf{v}}{\|\mathbf{v}\|^2}\right)\mathbf{v} = (\mathbf{F} \cdot \mathbf{v})\mathbf{v} = (-600)\left(\frac{1}{2}\right)\mathbf{v} = -300\left(\frac{\sqrt{3}}{2}\mathbf{i} + \frac{1}{2}\mathbf{j}\right).$$

The magnitude of this force is 300, and therefore a force of 300 pounds is required to keep the boat from rolling down the ramp.

$\mathbf{w}_1 = \text{proj}_\mathbf{v}(\mathbf{F})$
Figure 11.32

Work

The work W done by the constant force $\mathbf{F}$ acting along the line of motion of an object is given by

$$W = (\text{magnitude of force})(\text{distance}) = \|\mathbf{F}\|\,\|\overrightarrow{PQ}\|$$

as shown in Figure 11.33(a). When the constant force $\mathbf{F}$ is not directed along the line of motion, you can see from Figure 11.33(b) that the work W done by the force is

$$W = \|\text{proj}_{\overrightarrow{PQ}}\mathbf{F}\|\,\|\overrightarrow{PQ}\| = (\cos\theta)\|\mathbf{F}\|\,\|\overrightarrow{PQ}\| = \mathbf{F}\cdot\overrightarrow{PQ}.$$

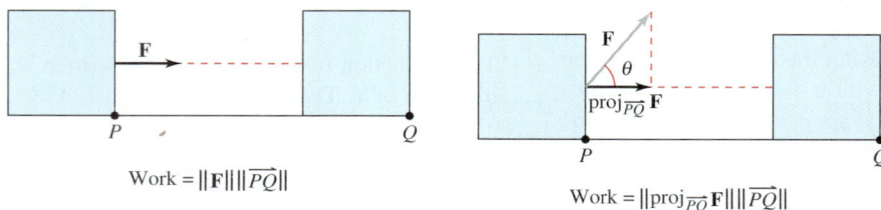

$$\text{Work} = \|\mathbf{F}\|\,\|\overrightarrow{PQ}\|$$

(a) Force acts along the line of motion.

$$\text{Work} = \|\text{proj}_{\overrightarrow{PQ}}\mathbf{F}\|\,\|\overrightarrow{PQ}\|$$

(b) Force acts at angle θ with the line of motion.

Figure 11.33

This notion of work is summarized in the next definition.

Definition of Work

The work W done by a constant force $\mathbf{F}$ as its point of application moves along the vector $\overrightarrow{PQ}$ is one of the following.

1. $W = \|\text{proj}_{\overrightarrow{PQ}}\mathbf{F}\|\,\|\overrightarrow{PQ}\|$ Projection form

2. $W = \mathbf{F}\cdot\overrightarrow{PQ}$ Dot product form

EXAMPLE 8 **Finding Work**

To close a sliding door, a person pulls on a rope with a constant force of 50 pounds at a constant angle of 60°, as shown in Figure 11.34. Find the work done in moving the door 12 feet to its closed position.

Figure 11.34

Solution Using a projection, you can calculate the work as follows.

$$W = \|\text{proj}_{\overrightarrow{PQ}}\mathbf{F}\|\,\|\overrightarrow{PQ}\| = \cos(60°)\|\mathbf{F}\|\,\|\overrightarrow{PQ}\| = \frac{1}{2}(50)(12) = 300 \text{ foot-pounds}$$

11.3 Exercises

See **CalcChat.com** for tutorial help and worked-out solutions to odd-numbered exercises.

Finding Dot Products In Exercises 1–8, find (a) $\mathbf{u} \cdot \mathbf{v}$, (b) $\mathbf{u} \cdot \mathbf{u}$, (c) $\|\mathbf{u}\|^2$, (d) $(\mathbf{u} \cdot \mathbf{v})\mathbf{v}$, and (e) $\mathbf{u} \cdot (2\mathbf{v})$.

1. $\mathbf{u} = \langle 3, 4 \rangle$, $\mathbf{v} = \langle -1, 5 \rangle$ 2. $\mathbf{u} = \langle 4, 10 \rangle$, $\mathbf{v} = \langle -2, 3 \rangle$

3. $\mathbf{u} = \langle 6, -4 \rangle$, $\mathbf{v} = \langle -3, 2 \rangle$ 4. $\mathbf{u} = \langle -4, 8 \rangle$, $\mathbf{v} = \langle 7, 5 \rangle$

5. $\mathbf{u} = \langle 2, -3, 4 \rangle$, $\mathbf{v} = \langle 0, 6, 5 \rangle$ 6. $\mathbf{u} = \mathbf{i}$, $\mathbf{v} = \mathbf{i}$

7. $\mathbf{u} = 2\mathbf{i} - \mathbf{j} + \mathbf{k}$ 8. $\mathbf{u} = 2\mathbf{i} + \mathbf{j} - 2\mathbf{k}$

   $\mathbf{v} = \mathbf{i} - \mathbf{k}$ $\mathbf{v} = \mathbf{i} - 3\mathbf{j} + 2\mathbf{k}$

Finding the Angle Between Two Vectors In Exercises 9–16, find the angle θ between the vectors (a) in radians and (b) in degrees.

9. $\mathbf{u} = \langle 1, 1 \rangle$, $\mathbf{v} = \langle 2, -2 \rangle$ 10. $\mathbf{u} = \langle 3, 1 \rangle$, $\mathbf{v} = \langle 2, -1 \rangle$

11. $\mathbf{u} = 3\mathbf{i} + \mathbf{j}$, $\mathbf{v} = -2\mathbf{i} + 4\mathbf{j}$

12. $\mathbf{u} = \cos\left(\dfrac{\pi}{6}\right)\mathbf{i} + \sin\left(\dfrac{\pi}{6}\right)\mathbf{j}$, $\mathbf{v} = \cos\left(\dfrac{3\pi}{4}\right)\mathbf{i} + \sin\left(\dfrac{3\pi}{4}\right)\mathbf{j}$

13. $\mathbf{u} = \langle 1, 1, 1 \rangle$ 14. $\mathbf{u} = 3\mathbf{i} + 2\mathbf{j} + \mathbf{k}$

    $\mathbf{v} = \langle 2, 1, -1 \rangle$ $\mathbf{v} = 2\mathbf{i} - 3\mathbf{j}$

15. $\mathbf{u} = 3\mathbf{i} + 4\mathbf{j}$ 16. $\mathbf{u} = 2\mathbf{i} - 3\mathbf{j} + \mathbf{k}$

    $\mathbf{v} = -2\mathbf{j} + 3\mathbf{k}$ $\mathbf{v} = \mathbf{i} - 2\mathbf{j} + \mathbf{k}$

Alternative Form of Dot Product In Exercises 17 and 18, use the alternative form of the dot product to find $\mathbf{u} \cdot \mathbf{v}$.

17. $\|\mathbf{u}\| = 8$, $\|\mathbf{v}\| = 5$, and the angle between $\mathbf{u}$ and $\mathbf{v}$ is $\pi/3$.

18. $\|\mathbf{u}\| = 40$, $\|\mathbf{v}\| = 25$, and the angle between $\mathbf{u}$ and $\mathbf{v}$ is $5\pi/6$.

Comparing Vectors In Exercises 19–24, determine whether u and v are orthogonal, parallel, or neither.

19. $\mathbf{u} = \langle 4, 3 \rangle$, $\mathbf{v} = \left\langle \frac{1}{2}, -\frac{2}{3} \right\rangle$ 20. $\mathbf{u} = -\frac{1}{3}(\mathbf{i} - 2\mathbf{j})$, $\mathbf{v} = 2\mathbf{i} - 4\mathbf{j}$

21. $\mathbf{u} = \mathbf{j} + 6\mathbf{k}$ 22. $\mathbf{u} = -2\mathbf{i} + 3\mathbf{j} - \mathbf{k}$

    $\mathbf{v} = \mathbf{i} - 2\mathbf{j} - \mathbf{k}$ $\mathbf{v} = 2\mathbf{i} + \mathbf{j} - \mathbf{k}$

23. $\mathbf{u} = \langle 2, -3, 1 \rangle$ 24. $\mathbf{u} = \langle \cos\theta, \sin\theta, -1 \rangle$

    $\mathbf{v} = \langle -1, -1, -1 \rangle$ $\mathbf{v} = \langle \sin\theta, -\cos\theta, 0 \rangle$

Classifying a Triangle In Exercises 25–28, the vertices of a triangle are given. Determine whether the triangle is an acute triangle, an obtuse triangle, or a right triangle. Explain your reasoning.

25. $(1, 2, 0)$, $(0, 0, 0)$, $(-2, 1, 0)$

26. $(-3, 0, 0)$, $(0, 0, 0)$, $(1, 2, 3)$

27. $(2, 0, 1)$, $(0, 1, 2)$, $(-0.5, 1.5, 0)$

28. $(2, -7, 3)$, $(-1, 5, 8)$, $(4, 6, -1)$

Finding Direction Angles In Exercises 29–34, find the direction cosines and angles of u, and demonstrate that the sum of the squares of the direction cosines is 1.

29. $\mathbf{u} = \mathbf{i} + 2\mathbf{j} + 2\mathbf{k}$ 30. $\mathbf{u} = 5\mathbf{i} + 3\mathbf{j} - \mathbf{k}$

31. $\mathbf{u} = 3\mathbf{i} + 2\mathbf{j} - 2\mathbf{k}$ 32. $\mathbf{u} = -4\mathbf{i} + 3\mathbf{j} + 5\mathbf{k}$

33. $\mathbf{u} = \langle 0, 6, -4 \rangle$ 34. $\mathbf{u} = \langle -1, 5, 2 \rangle$

Finding the Projection of u onto v In Exercises 35–42, (a) find the projection of u onto v, and (b) find the vector component of u orthogonal to v.

35. $\mathbf{u} = \langle 6, 7 \rangle$, $\mathbf{v} = \langle 1, 4 \rangle$ 36. $\mathbf{u} = \langle 9, 7 \rangle$, $\mathbf{v} = \langle 1, 3 \rangle$

37. $\mathbf{u} = 2\mathbf{i} + 3\mathbf{j}$, $\mathbf{v} = 5\mathbf{i} + \mathbf{j}$

38. $\mathbf{u} = 2\mathbf{i} - 3\mathbf{j}$, $\mathbf{v} = 3\mathbf{i} + 2\mathbf{j}$

39. $\mathbf{u} = \langle 0, 3, 3 \rangle$, $\mathbf{v} = \langle -1, 1, 1 \rangle$

40. $\mathbf{u} = \langle 8, 2, 0 \rangle$, $\mathbf{v} = \langle 2, 1, -1 \rangle$

41. $\mathbf{u} = 2\mathbf{i} + \mathbf{j} + 2\mathbf{k}$, $\mathbf{v} = 3\mathbf{j} + 4\mathbf{k}$

42. $\mathbf{u} = \mathbf{i} + 4\mathbf{k}$, $\mathbf{v} = 3\mathbf{i} + 2\mathbf{k}$

WRITING ABOUT CONCEPTS

43. **Dot Product** Define the dot product of vectors **u** and **v**.

44. **Orthogonal Vectors** State the definition of orthogonal vectors. When vectors are neither parallel nor orthogonal, how do you find the angle between them? Explain.

45. **Using Vectors** Determine which of the following are defined for nonzero vectors **u**, **v**, and **w**. Explain your reasoning.

    (a) $\mathbf{u} \cdot (\mathbf{v} + \mathbf{w})$ (b) $(\mathbf{u} \cdot \mathbf{v})\mathbf{w}$

    (c) $\mathbf{u} \cdot \mathbf{v} + \mathbf{w}$ (d) $\|\mathbf{u}\| \cdot (\mathbf{v} + \mathbf{w})$

46. **Direction Cosines** Describe direction cosines and direction angles of a vector **v**.

47. **Projection** Give a geometric description of the projection of **u** onto **v**.

48. **Projection** What can be said about the vectors **u** and **v** when (a) the projection of **u** onto **v** equals **u** and (b) the projection of **u** onto **v** equals **0**?

49. **Projection** When the projection of **u** onto **v** has the same magnitude as the projection of **v** onto **u**, can you conclude that $\|\mathbf{u}\| = \|\mathbf{v}\|$? Explain.

50. **HOW DO YOU SEE IT?** What is known about θ, the angle between two nonzero vectors **u** and **v**, when

    (a) $\mathbf{u} \cdot \mathbf{v} = 0$?   (b) $\mathbf{u} \cdot \mathbf{v} > 0$?   (c) $\mathbf{u} \cdot \mathbf{v} < 0$?

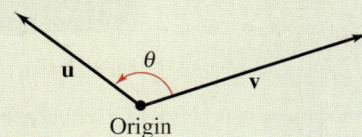

51. Revenue The vector $\mathbf{u} = \langle 3240, 1450, 2235 \rangle$ gives the numbers of hamburgers, chicken sandwiches, and cheeseburgers, respectively, sold at a fast-food restaurant in one week. The vector $\mathbf{v} = \langle 2.25, 2.95, 2.65 \rangle$ gives the prices (in dollars) per unit for the three food items. Find the dot product $\mathbf{u} \cdot \mathbf{v}$, and explain what information it gives.

52. Revenue Repeat Exercise 51 after increasing prices by 4%. Identify the vector operation used to increase prices by 4%.

Orthogonal Vectors In Exercises 53–56, find two vectors in opposite directions that are orthogonal to the vector u. (The answers are not unique.)

53. $\mathbf{u} = -\frac{1}{4}\mathbf{i} + \frac{3}{2}\mathbf{j}$ **54.** $\mathbf{u} = 9\mathbf{i} - 4\mathbf{j}$

55. $\mathbf{u} = \langle 3, 1, -2 \rangle$ **56.** $\mathbf{u} = \langle 4, -3, 6 \rangle$

57. Finding an Angle Find the angle between a cube's diagonal and one of its edges.

58. Finding an Angle Find the angle between the diagonal of a cube and the diagonal of one of its sides.

59. Braking Load A 48,000-pound truck is parked on a 10° slope (see figure). Assume the only force to overcome is that due to gravity. Find (a) the force required to keep the truck from rolling down the hill and (b) the force perpendicular to the hill.

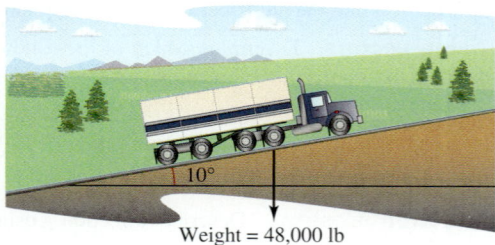

Weight = 48,000 lb

60. Braking Load A 5400-pound sport utility vehicle is parked on an 18° slope. Assume the only force to overcome is that due to gravity. Find (a) the force required to keep the vehicle from rolling down the hill and (b) the force perpendicular to the hill.

61. Work An object is pulled 10 feet across a floor, using a force of 85 pounds. The direction of the force is 60° above the horizontal (see figure). Find the work done.

85 lb

60°

← 10 ft →

Not drawn to scale

Figure for 61

20°

Figure for 62

62. Work A toy wagon is pulled by exerting a force of 25 pounds on a handle that makes a 20° angle with the horizontal (see figure). Find the work done in pulling the wagon 50 feet.

Ziva_K/iStockphoto.com

63. Work A car is towed using a force of 1600 newtons. The chain used to pull the car makes a 25° angle with the horizontal. Find the work done in towing the car 2 kilometers.

64. Work

A sled is pulled by exerting a force of 100 newtons on a rope that makes a 25° angle with the horizontal. Find the work done in pulling the sled 40 meters.

True or False? In Exercises 65 and 66, determine whether the statement is true or false. If it is false, explain why or give an example that shows it is false.

65. If $\mathbf{u} \cdot \mathbf{v} = \mathbf{u} \cdot \mathbf{w}$ and $\mathbf{u} \neq \mathbf{0}$, then $\mathbf{v} = \mathbf{w}$.

66. If $\mathbf{u}$ and $\mathbf{v}$ are orthogonal to $\mathbf{w}$, then $\mathbf{u} + \mathbf{v}$ is orthogonal to $\mathbf{w}$.

Using Points of Intersection In Exercises 67–70, (a) find all points of intersection of the graphs of the two equations, (b) find the unit tangent vectors to each curve at their points of intersection, and (c) find the angles ($0° \leq \theta \leq 90°$) between the curves at their points of intersection.

67. $y = x^2, \quad y = x^{1/3}$ **68.** $y = x^3, \quad y = x^{1/3}$

69. $y = 1 - x^2, \quad y = x^2 - 1$ **70.** $(y + 1)^2 = x, \quad y = x^3 - 1$

71. Proof Use vectors to prove that the diagonals of a rhombus are perpendicular.

72. Proof Use vectors to prove that a parallelogram is a rectangle if and only if its diagonals are equal in length.

73. Bond Angle Consider a regular tetrahedron with vertices $(0, 0, 0)$, $(k, k, 0)$, $(k, 0, k)$, and $(0, k, k)$, where k is a positive real number.

(a) Sketch the graph of the tetrahedron.

(b) Find the length of each edge.

(c) Find the angle between any two edges.

(d) Find the angle between the line segments from the centroid $(k/2, k/2, k/2)$ to two vertices. This is the bond angle for a molecule such as CH_4 or $PbCl_4$, where the structure of the molecule is a tetrahedron.

74. Proof Consider the vectors $\mathbf{u} = \langle \cos \alpha, \sin \alpha, 0 \rangle$ and $\mathbf{v} = \langle \cos \beta, \sin \beta, 0 \rangle$, where $\alpha > \beta$. Find the dot product of the vectors and use the result to prove the identity

$$\cos(\alpha - \beta) = \cos \alpha \cos \beta + \sin \alpha \sin \beta.$$

75. Proof Prove that $\|\mathbf{u} - \mathbf{v}\|^2 = \|\mathbf{u}\|^2 + \|\mathbf{v}\|^2 - 2\mathbf{u} \cdot \mathbf{v}$.

76. Proof Prove the **Cauchy-Schwarz Inequality**,

$$|\mathbf{u} \cdot \mathbf{v}| \leq \|\mathbf{u}\| \|\mathbf{v}\|.$$

77. Proof Prove the triangle inequality $\|\mathbf{u} + \mathbf{v}\| \leq \|\mathbf{u}\| + \|\mathbf{v}\|$.

78. Proof Prove Theorem 11.6.

11.4 The Cross Product of Two Vectors in Space

■ Find the cross product of two vectors in space.
■ Use the triple scalar product of three vectors in space.

The Cross Product

Exploration

Geometric Property of the Cross Product   Three pairs of vectors are shown below. Use the definition to find the cross product of each pair. Sketch all three vectors in a three-dimensional system. Describe any relationships among the three vectors. Use your description to write a conjecture about **u**, **v**, and **u** × **v**.

a. $\mathbf{u} = \langle 3, 0, 3 \rangle$, $\mathbf{v} = \langle 3, 0, -3 \rangle$

b. $\mathbf{u} = \langle 0, 3, 3 \rangle$, $\mathbf{v} = \langle 0, -3, 3 \rangle$

c. $\mathbf{u} = \langle 3, 3, 0 \rangle$, $\mathbf{v} = \langle 3, -3, 0 \rangle$

Many applications in physics, engineering, and geometry involve finding a vector in space that is orthogonal to two given vectors. In this section, you will study a product that will yield such a vector. It is called the **cross product,** and it is most conveniently defined and calculated using the standard unit vector form. Because the cross product yields a vector, it is also called the **vector product.**

Definition of Cross Product of Two Vectors in Space

Let

$$\mathbf{u} = u_1\mathbf{i} + u_2\mathbf{j} + u_3\mathbf{k} \quad \text{and} \quad \mathbf{v} = v_1\mathbf{i} + v_2\mathbf{j} + v_3\mathbf{k}$$

be vectors in space. The **cross product** of **u** and **v** is the vector

$$\mathbf{u} \times \mathbf{v} = (u_2v_3 - u_3v_2)\mathbf{i} - (u_1v_3 - u_3v_1)\mathbf{j} + (u_1v_2 - u_2v_1)\mathbf{k}.$$

It is important to note that this definition applies only to three-dimensional vectors. The cross product is not defined for two-dimensional vectors.

A convenient way to calculate $\mathbf{u} \times \mathbf{v}$ is to use the *determinant form* with cofactor expansion shown below. (This 3×3 determinant form is used simply to help remember the formula for the cross product—it is technically not a determinant because not all the entries of the corresponding matrix are real numbers.)

$$\mathbf{u} \times \mathbf{v} = \begin{vmatrix} \mathbf{i} & \mathbf{j} & \mathbf{k} \\ u_1 & u_2 & u_3 \\ v_1 & v_2 & v_3 \end{vmatrix} \qquad \longleftarrow \text{Put "}\mathbf{u}\text{" in Row 2.} \\ \longleftarrow \text{Put "}\mathbf{v}\text{" in Row 3.}$$

$$= \begin{vmatrix} \mathbf{i} & \mathbf{j} & \mathbf{k} \\ u_1 & u_2 & u_3 \\ v_1 & v_2 & v_3 \end{vmatrix}\mathbf{i} - \begin{vmatrix} \mathbf{i} & \mathbf{j} & \mathbf{k} \\ u_1 & u_2 & u_3 \\ v_1 & v_2 & v_3 \end{vmatrix}\mathbf{j} + \begin{vmatrix} \mathbf{i} & \mathbf{j} & \mathbf{k} \\ u_1 & u_2 & u_3 \\ v_1 & v_2 & v_3 \end{vmatrix}\mathbf{k}$$

$$= \begin{vmatrix} u_2 & u_3 \\ v_2 & v_3 \end{vmatrix}\mathbf{i} - \begin{vmatrix} u_1 & u_3 \\ v_1 & v_3 \end{vmatrix}\mathbf{j} + \begin{vmatrix} u_1 & u_2 \\ v_1 & v_2 \end{vmatrix}\mathbf{k}$$

$$= (u_2v_3 - u_3v_2)\mathbf{i} - (u_1v_3 - u_3v_1)\mathbf{j} + (u_1v_2 - u_2v_1)\mathbf{k}$$

Note the minus sign in front of the **j**-component. Each of the three 2×2 determinants can be evaluated by using the diagonal pattern

$$\begin{vmatrix} a & b \\ c & d \end{vmatrix} = ad - bc.$$

Here are a couple of examples.

$$\begin{vmatrix} 2 & 4 \\ 3 & -1 \end{vmatrix} = (2)(-1) - (4)(3) = -2 - 12 = -14$$

and

$$\begin{vmatrix} 4 & 0 \\ -6 & 3 \end{vmatrix} = (4)(3) - (0)(-6) = 12$$

NOTATION FOR DOT AND CROSS PRODUCTS

The notation for the dot product and cross product of vectors was first introduced by the American physicist Josiah Willard Gibbs (1839–1903). In the early 1880s, Gibbs built a system to represent physical quantities called "vector analysis." The system was a departure from Hamilton's theory of quaternions.

EXAMPLE 1 **Finding the Cross Product**

For $\mathbf{u} = \mathbf{i} - 2\mathbf{j} + \mathbf{k}$ and $\mathbf{v} = 3\mathbf{i} + \mathbf{j} - 2\mathbf{k}$, find each of the following.

a. $\mathbf{u} \times \mathbf{v}$ **b.** $\mathbf{v} \times \mathbf{u}$ **c.** $\mathbf{v} \times \mathbf{v}$

Solution

a. $\mathbf{u} \times \mathbf{v} = \begin{vmatrix} \mathbf{i} & \mathbf{j} & \mathbf{k} \\ 1 & -2 & 1 \\ 3 & 1 & -2 \end{vmatrix}$

$= \begin{vmatrix} -2 & 1 \\ 1 & -2 \end{vmatrix} \mathbf{i} - \begin{vmatrix} 1 & 1 \\ 3 & -2 \end{vmatrix} \mathbf{j} + \begin{vmatrix} 1 & -2 \\ 3 & 1 \end{vmatrix} \mathbf{k}$

$= (4 - 1)\mathbf{i} - (-2 - 3)\mathbf{j} + (1 + 6)\mathbf{k}$

$= 3\mathbf{i} + 5\mathbf{j} + 7\mathbf{k}$

b. $\mathbf{v} \times \mathbf{u} = \begin{vmatrix} \mathbf{i} & \mathbf{j} & \mathbf{k} \\ 3 & 1 & -2 \\ 1 & -2 & 1 \end{vmatrix}$

$= \begin{vmatrix} 1 & -2 \\ -2 & 1 \end{vmatrix} \mathbf{i} - \begin{vmatrix} 3 & -2 \\ 1 & 1 \end{vmatrix} \mathbf{j} + \begin{vmatrix} 3 & 1 \\ 1 & -2 \end{vmatrix} \mathbf{k}$

$= (1 - 4)\mathbf{i} - (3 + 2)\mathbf{j} + (-6 - 1)\mathbf{k}$

$= -3\mathbf{i} - 5\mathbf{j} - 7\mathbf{k}$

REMARK Note that this result is the negative of that in part (a).

c. $\mathbf{v} \times \mathbf{v} = \begin{vmatrix} \mathbf{i} & \mathbf{j} & \mathbf{k} \\ 3 & 1 & -2 \\ 3 & 1 & -2 \end{vmatrix} = \mathbf{0}$

The results obtained in Example 1 suggest some interesting *algebraic* properties of the cross product. For instance, $\mathbf{u} \times \mathbf{v} = -(\mathbf{v} \times \mathbf{u})$, and $\mathbf{v} \times \mathbf{v} = \mathbf{0}$. These properties, and several others, are summarized in the next theorem.

THEOREM 11.7 **Algebraic Properties of the Cross Product**

Let $\mathbf{u}$, $\mathbf{v}$, and $\mathbf{w}$ be vectors in space, and let c be a scalar.

1. $\mathbf{u} \times \mathbf{v} = -(\mathbf{v} \times \mathbf{u})$
2. $\mathbf{u} \times (\mathbf{v} + \mathbf{w}) = (\mathbf{u} \times \mathbf{v}) + (\mathbf{u} \times \mathbf{w})$
3. $c(\mathbf{u} \times \mathbf{v}) = (c\mathbf{u}) \times \mathbf{v} = \mathbf{u} \times (c\mathbf{v})$
4. $\mathbf{u} \times \mathbf{0} = \mathbf{0} \times \mathbf{u} = \mathbf{0}$
5. $\mathbf{u} \times \mathbf{u} = \mathbf{0}$
6. $\mathbf{u} \cdot (\mathbf{v} \times \mathbf{w}) = (\mathbf{u} \times \mathbf{v}) \cdot \mathbf{w}$

Proof To prove Property 1, let $\mathbf{u} = u_1\mathbf{i} + u_2\mathbf{j} + u_3\mathbf{k}$ and $\mathbf{v} = v_1\mathbf{i} + v_2\mathbf{j} + v_3\mathbf{k}$. Then,

$$\mathbf{u} \times \mathbf{v} = (u_2v_3 - u_3v_2)\mathbf{i} - (u_1v_3 - u_3v_1)\mathbf{j} + (u_1v_2 - u_2v_1)\mathbf{k}$$

and

$$\mathbf{v} \times \mathbf{u} = (v_2u_3 - v_3u_2)\mathbf{i} - (v_1u_3 - v_3u_1)\mathbf{j} + (v_1u_2 - v_2u_1)\mathbf{k}$$

which implies that $\mathbf{u} \times \mathbf{v} = -(\mathbf{v} \times \mathbf{u})$. Proofs of Properties 2, 3, 5, and 6 are left as exercises (see Exercises 51–54).

See LarsonCalculus.com for Bruce Edwards's video of this proof.

Note that Property 1 of Theorem 11.7 indicates that the cross product is *not commutative*. In particular, this property indicates that the vectors $\mathbf{u} \times \mathbf{v}$ and $\mathbf{v} \times \mathbf{u}$ have equal lengths but opposite directions. The next theorem lists some other *geometric* properties of the cross product of two vectors.

> **THEOREM 11.8 Geometric Properties of the Cross Product**
>
> Let $\mathbf{u}$ and $\mathbf{v}$ be nonzero vectors in space, and let θ be the angle between $\mathbf{u}$ and $\mathbf{v}$.
>
> 1. $\mathbf{u} \times \mathbf{v}$ is orthogonal to both $\mathbf{u}$ and $\mathbf{v}$.
> 2. $\|\mathbf{u} \times \mathbf{v}\| = \|\mathbf{u}\|\,\|\mathbf{v}\| \sin \theta$
> 3. $\mathbf{u} \times \mathbf{v} = \mathbf{0}$ if and only if $\mathbf{u}$ and $\mathbf{v}$ are scalar multiples of each other.
> 4. $\|\mathbf{u} \times \mathbf{v}\| = $ area of parallelogram having $\mathbf{u}$ and $\mathbf{v}$ as adjacent sides.

REMARK It follows from Properties 1 and 2 in Theorem 11.8 that if $\mathbf{n}$ is a unit vector orthogonal to both $\mathbf{u}$ and $\mathbf{v}$, then

$$\mathbf{u} \times \mathbf{v} = \pm(\|\mathbf{u}\|\,\|\mathbf{v}\| \sin \theta)\mathbf{n}.$$

Proof To prove Property 2, note because $\cos \theta = (\mathbf{u} \cdot \mathbf{v})/(\|\mathbf{u}\|\,\|\mathbf{v}\|)$, it follows that

$$\|\mathbf{u}\|\,\|\mathbf{v}\| \sin \theta = \|\mathbf{u}\|\,\|\mathbf{v}\|\sqrt{1 - \cos^2 \theta}$$
$$= \|\mathbf{u}\|\,\|\mathbf{v}\|\sqrt{1 - \frac{(\mathbf{u} \cdot \mathbf{v})^2}{\|\mathbf{u}\|^2\|\mathbf{v}\|^2}}$$
$$= \sqrt{\|\mathbf{u}\|^2\|\mathbf{v}\|^2 - (\mathbf{u} \cdot \mathbf{v})^2}$$
$$= \sqrt{(u_1^2 + u_2^2 + u_3^2)(v_1^2 + v_2^2 + v_3^2) - (u_1v_1 + u_2v_2 + u_3v_3)^2}$$
$$= \sqrt{(u_2v_3 - u_3v_2)^2 + (u_1v_3 - u_3v_1)^2 + (u_1v_2 - u_2v_1)^2}$$
$$= \|\mathbf{u} \times \mathbf{v}\|.$$

To prove Property 4, refer to Figure 11.35, which is a parallelogram having $\mathbf{v}$ and $\mathbf{u}$ as adjacent sides. Because the height of the parallelogram is $\|\mathbf{v}\| \sin \theta$, the area is

$$\text{Area} = (\text{base})(\text{height})$$
$$= \|\mathbf{u}\|\,\|\mathbf{v}\| \sin \theta$$
$$= \|\mathbf{u} \times \mathbf{v}\|.$$

Proofs of Properties 1 and 3 are left as exercises (see Exercises 55 and 56).
See LarsonCalculus.com for Bruce Edwards's video of this proof.

The vectors $\mathbf{u}$ and $\mathbf{v}$ form adjacent sides of a parallelogram.
Figure 11.35

Both $\mathbf{u} \times \mathbf{v}$ and $\mathbf{v} \times \mathbf{u}$ are perpendicular to the plane determined by $\mathbf{u}$ and $\mathbf{v}$. One way to remember the orientations of the vectors $\mathbf{u}$, $\mathbf{v}$, and $\mathbf{u} \times \mathbf{v}$ is to compare them with the unit vectors $\mathbf{i}$, $\mathbf{j}$, and $\mathbf{k} = \mathbf{i} \times \mathbf{j}$, as shown in Figure 11.36. The three vectors $\mathbf{u}$, $\mathbf{v}$, and $\mathbf{u} \times \mathbf{v}$ form a *right-handed system,* whereas the three vectors $\mathbf{u}$, $\mathbf{v}$, and $\mathbf{v} \times \mathbf{u}$ form a *left-handed system.*

Right-handed systems
Figure 11.36

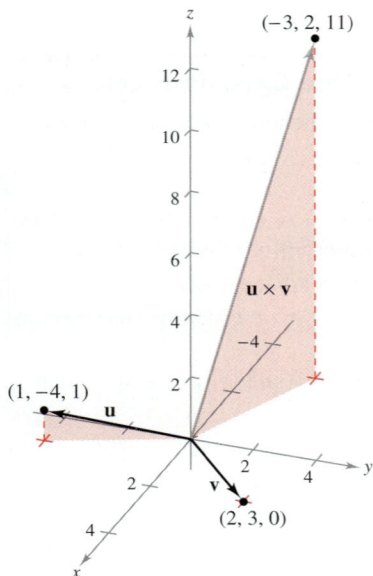

The vector $\mathbf{u} \times \mathbf{v}$ is orthogonal to both **u** and **v**.
Figure 11.37

EXAMPLE 2 Using the Cross Product

· · · ▷ *See LarsonCalculus.com for an interactive version of this type of example.*

Find a unit vector that is orthogonal to both

$$\mathbf{u} = \mathbf{i} - 4\mathbf{j} + \mathbf{k}$$

and

$$\mathbf{v} = 2\mathbf{i} + 3\mathbf{j}.$$

Solution The cross product $\mathbf{u} \times \mathbf{v}$, as shown in Figure 11.37, is orthogonal to both **u** and **v**.

$$\mathbf{u} \times \mathbf{v} = \begin{vmatrix} \mathbf{i} & \mathbf{j} & \mathbf{k} \\ 1 & -4 & 1 \\ 2 & 3 & 0 \end{vmatrix} \qquad \text{Cross product}$$

$$= -3\mathbf{i} + 2\mathbf{j} + 11\mathbf{k}$$

Because

$$\|\mathbf{u} \times \mathbf{v}\| = \sqrt{(-3)^2 + 2^2 + 11^2} = \sqrt{134}$$

a unit vector orthogonal to both **u** and **v** is

$$\frac{\mathbf{u} \times \mathbf{v}}{\|\mathbf{u} \times \mathbf{v}\|} = -\frac{3}{\sqrt{134}}\mathbf{i} + \frac{2}{\sqrt{134}}\mathbf{j} + \frac{11}{\sqrt{134}}\mathbf{k}.$$

In Example 2, note that you could have used the cross product $\mathbf{v} \times \mathbf{u}$ to form a unit vector that is orthogonal to both **u** and **v**. With that choice, you would have obtained the negative of the unit vector found in the example.

EXAMPLE 3 Geometric Application of the Cross Product

The vertices of a quadrilateral are listed below. Show that the quadrilateral is a parallelogram, and find its area.

$$A = (5, 2, 0) \qquad B = (2, 6, 1)$$
$$C = (2, 4, 7) \qquad D = (5, 0, 6)$$

Solution From Figure 11.38, you can see that the sides of the quadrilateral correspond to the following four vectors.

$$\overrightarrow{AB} = -3\mathbf{i} + 4\mathbf{j} + \mathbf{k} \qquad \overrightarrow{CD} = 3\mathbf{i} - 4\mathbf{j} - \mathbf{k} = -\overrightarrow{AB}$$
$$\overrightarrow{AD} = 0\mathbf{i} - 2\mathbf{j} + 6\mathbf{k} \qquad \overrightarrow{CB} = 0\mathbf{i} + 2\mathbf{j} - 6\mathbf{k} = -\overrightarrow{AD}$$

So, $\overrightarrow{AB}$ is parallel to $\overrightarrow{CD}$ and $\overrightarrow{AD}$ is parallel to $\overrightarrow{CB}$, and you can conclude that the quadrilateral is a parallelogram with $\overrightarrow{AB}$ and $\overrightarrow{AD}$ as adjacent sides. Moreover, because

$$\overrightarrow{AB} \times \overrightarrow{AD} = \begin{vmatrix} \mathbf{i} & \mathbf{j} & \mathbf{k} \\ -3 & 4 & 1 \\ 0 & -2 & 6 \end{vmatrix} \qquad \text{Cross product}$$

$$= 26\mathbf{i} + 18\mathbf{j} + 6\mathbf{k}$$

the area of the parallelogram is

$$\|\overrightarrow{AB} \times \overrightarrow{AD}\| = \sqrt{1036} \approx 32.19.$$

Is the parallelogram a rectangle? You can determine whether it is by finding the angle between the vectors $\overrightarrow{AB}$ and $\overrightarrow{AD}$.

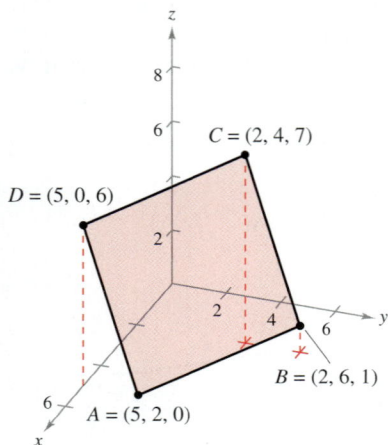

The area of the parallelogram is approximately 32.19.
Figure 11.38

In physics, the cross product can be used to measure **torque**—the **moment M of a force F about a point** *P*, as shown in Figure 11.39. If the point of application of the force is *Q*, then the moment of **F** about *P* is

$$\mathbf{M} = \overrightarrow{PQ} \times \mathbf{F}. \qquad \text{\color{red}{Moment of \textbf{F} about } P}$$

The magnitude of the moment **M** measures the tendency of the vector $\overrightarrow{PQ}$ to rotate counterclockwise (using the right-hand rule) about an axis directed along the vector **M**.

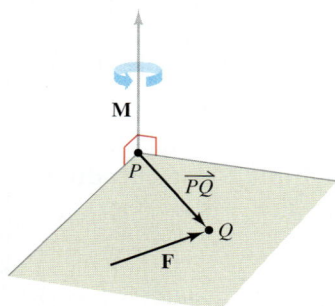

The moment of **F** about *P*
Figure 11.39

EXAMPLE 4   An Application of the Cross Product

A vertical force of 50 pounds is applied to the end of a one-foot lever that is attached to an axle at point *P*, as shown in Figure 11.40. Find the moment of this force about the point *P* when $\theta = 60°$.

Solution   Represent the 50-pound force as

$$\mathbf{F} = -50\mathbf{k}$$

and the lever as

$$\overrightarrow{PQ} = \cos(60°)\mathbf{j} + \sin(60°)\mathbf{k} = \frac{1}{2}\mathbf{j} + \frac{\sqrt{3}}{2}\mathbf{k}.$$

The moment of **F** about *P* is

$$\mathbf{M} = \overrightarrow{PQ} \times \mathbf{F} = \begin{vmatrix} \mathbf{i} & \mathbf{j} & \mathbf{k} \\ 0 & \dfrac{1}{2} & \dfrac{\sqrt{3}}{2} \\ 0 & 0 & -50 \end{vmatrix} = -25\mathbf{i}. \qquad \text{\color{red}{Moment of \textbf{F} about } P}$$

The magnitude of this moment is 25 foot-pounds.

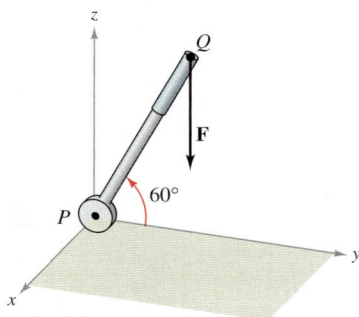

A vertical force of 50 pounds is applied at point *Q*.
Figure 11.40

In Example 4, note that the moment (the tendency of the lever to rotate about its axle) is dependent on the angle θ. When $\theta = \pi/2$, the moment is 0. The moment is greatest when $\theta = 0$.

The Triple Scalar Product

For vectors **u**, **v**, and **w** in space, the dot product of **u** and $\mathbf{v} \times \mathbf{w}$

$$\mathbf{u} \cdot (\mathbf{v} \times \mathbf{w})$$

is called the **triple scalar product,** as defined in Theorem 11.9. The proof of this theorem is left as an exercise (see Exercise 59).

THEOREM 11.9   The Triple Scalar Product

For $\mathbf{u} = u_1\mathbf{i} + u_2\mathbf{j} + u_3\mathbf{k}$, $\mathbf{v} = v_1\mathbf{i} + v_2\mathbf{j} + v_3\mathbf{k}$, and $\mathbf{w} = w_1\mathbf{i} + w_2\mathbf{j} + w_3\mathbf{k}$, the triple scalar product is

$$\mathbf{u} \cdot (\mathbf{v} \times \mathbf{w}) = \begin{vmatrix} u_1 & u_2 & u_3 \\ v_1 & v_2 & v_3 \\ w_1 & w_2 & w_3 \end{vmatrix}.$$

Note that the value of a determinant is multiplied by -1 when two rows are interchanged. After two such interchanges, the value of the determinant will be unchanged. So, the following triple scalar products are equivalent.

$$\mathbf{u} \cdot (\mathbf{v} \times \mathbf{w}) = \mathbf{v} \cdot (\mathbf{w} \times \mathbf{u}) = \mathbf{w} \cdot (\mathbf{u} \times \mathbf{v})$$

$\mathbf{v} \times \mathbf{w}$

$\mathbf{u}$

$\mathbf{w}$ $\mathbf{v}$

$\|\text{proj}_{\mathbf{v} \times \mathbf{w}}\mathbf{u}\|$

Area of base $= \|\mathbf{v} \times \mathbf{w}\|$
Volume of parallelepiped $= |\mathbf{u} \cdot (\mathbf{v} \times \mathbf{w})|$
Figure 11.41

If the vectors **u**, **v**, and **w** do not lie in the same plane, then the triple scalar product $\mathbf{u} \cdot (\mathbf{v} \times \mathbf{w})$ can be used to determine the volume of the parallelepiped (a polyhedron, all of whose faces are parallelograms) with **u**, **v**, and **w** as adjacent edges, as shown in Figure 11.41. This is established in the next theorem.

> **THEOREM 11.10   Geometric Property of the Triple Scalar Product**
>
> The volume V of a parallelepiped with vectors **u**, **v**, and **w** as adjacent edges is
>
> $$V = |\mathbf{u} \cdot (\mathbf{v} \times \mathbf{w})|.$$

Proof   In Figure 11.41, note that the area of the base is $\|\mathbf{v} \times \mathbf{w}\|$ and the height of the parallelpiped is $\|\text{proj}_{\mathbf{v} \times \mathbf{w}}\mathbf{u}\|$. Therefore, the volume is

$$V = (\text{height})(\text{area of base})$$
$$= \|\text{proj}_{\mathbf{v} \times \mathbf{w}}\mathbf{u}\| \|\mathbf{v} \times \mathbf{w}\|$$
$$= \left| \frac{\mathbf{u} \cdot (\mathbf{v} \times \mathbf{w})}{\|\mathbf{v} \times \mathbf{w}\|} \right| \|\mathbf{v} \times \mathbf{w}\|$$
$$= |\mathbf{u} \cdot (\mathbf{v} \times \mathbf{w})|.$$

See LarsonCalculus.com for Bruce Edwards's video of this proof.

EXAMPLE 5   **Volume by the Triple Scalar Product**

Find the volume of the parallelepiped shown in Figure 11.42 having

$$\mathbf{u} = 3\mathbf{i} - 5\mathbf{j} + \mathbf{k}$$
$$\mathbf{v} = 2\mathbf{j} - 2\mathbf{k}$$

and

$$\mathbf{w} = 3\mathbf{i} + \mathbf{j} + \mathbf{k}$$

as adjacent edges.

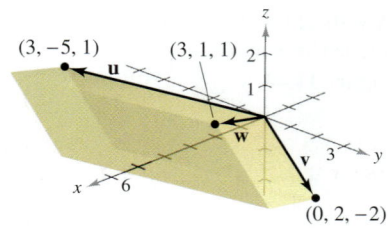

The parallelepiped has a volume of 36.
Figure 11.42

Solution   By Theorem 11.10, you have

$$V = |\mathbf{u} \cdot (\mathbf{v} \times \mathbf{w})| \qquad \text{\color{red}Triple scalar product}$$

$$= \begin{vmatrix} 3 & -5 & 1 \\ 0 & 2 & -2 \\ 3 & 1 & 1 \end{vmatrix}$$

$$= 3 \begin{vmatrix} 2 & -2 \\ 1 & 1 \end{vmatrix} - (-5) \begin{vmatrix} 0 & -2 \\ 3 & 1 \end{vmatrix} + (1) \begin{vmatrix} 0 & 2 \\ 3 & 1 \end{vmatrix}$$

$$= 3(4) + 5(6) + 1(-6)$$

$$= 36.$$

A natural consequence of Theorem 11.10 is that the volume of the parallelepiped is 0 if and only if the three vectors are coplanar. That is, when the vectors $\mathbf{u} = \langle u_1, u_2, u_3 \rangle$, $\mathbf{v} = \langle v_1, v_2, v_3 \rangle$, and $\mathbf{w} = \langle w_1, w_2, w_3 \rangle$ have the same initial point, they lie in the same plane if and only if

$$\mathbf{u} \cdot (\mathbf{v} \times \mathbf{w}) = \begin{vmatrix} u_1 & u_2 & u_3 \\ v_1 & v_2 & v_3 \\ w_1 & w_2 & w_3 \end{vmatrix} = 0.$$

11.4 Exercises

Cross Product of Unit Vectors In Exercises 1–6, find the cross product of the unit vectors and sketch your result.

1. $\mathbf{j} \times \mathbf{i}$

2. $\mathbf{i} \times \mathbf{j}$

3. $\mathbf{j} \times \mathbf{k}$

4. $\mathbf{k} \times \mathbf{j}$

5. $\mathbf{i} \times \mathbf{k}$

6. $\mathbf{k} \times \mathbf{i}$

Finding Cross Products In Exercises 7–10, find (a) $\mathbf{u} \times \mathbf{v}$, (b) $\mathbf{v} \times \mathbf{u}$, and (c) $\mathbf{v} \times \mathbf{v}$.

7. $\mathbf{u} = -2\mathbf{i} + 4\mathbf{j}$
   $\mathbf{v} = 3\mathbf{i} + 2\mathbf{j} + 5\mathbf{k}$

8. $\mathbf{u} = 3\mathbf{i} + 5\mathbf{k}$
   $\mathbf{v} = 2\mathbf{i} + 3\mathbf{j} - 2\mathbf{k}$

9. $\mathbf{u} = \langle 7, 3, 2 \rangle$
   $\mathbf{v} = \langle 1, -1, 5 \rangle$

10. $\mathbf{u} = \langle 3, -2, -2 \rangle$
    $\mathbf{v} = \langle 1, 5, 1 \rangle$

Finding a Cross Product In Exercises 11–16, find $\mathbf{u} \times \mathbf{v}$ and show that it is orthogonal to both $\mathbf{u}$ and $\mathbf{v}$.

11. $\mathbf{u} = \langle 12, -3, 0 \rangle$
    $\mathbf{v} = \langle -2, 5, 0 \rangle$

12. $\mathbf{u} = \langle -1, 1, 2 \rangle$
    $\mathbf{v} = \langle 0, 1, 0 \rangle$

13. $\mathbf{u} = \langle 2, -3, 1 \rangle$
    $\mathbf{v} = \langle 1, -2, 1 \rangle$

14. $\mathbf{u} = \langle -10, 0, 6 \rangle$
    $\mathbf{v} = \langle 5, -3, 0 \rangle$

15. $\mathbf{u} = \mathbf{i} + \mathbf{j} + \mathbf{k}$
    $\mathbf{v} = 2\mathbf{i} + \mathbf{j} - \mathbf{k}$

16. $\mathbf{u} = \mathbf{i} + 6\mathbf{j}$
    $\mathbf{v} = -2\mathbf{i} + \mathbf{j} + \mathbf{k}$

Finding a Unit Vector In Exercises 17–20, find a unit vector that is orthogonal to both $\mathbf{u}$ and $\mathbf{v}$.

17. $\mathbf{u} = \langle 4, -3, 1 \rangle$
    $\mathbf{v} = \langle 2, 5, 3 \rangle$

18. $\mathbf{u} = \langle -8, -6, 4 \rangle$
    $\mathbf{v} = \langle 10, -12, -2 \rangle$

19. $\mathbf{u} = -3\mathbf{i} + 2\mathbf{j} - 5\mathbf{k}$
    $\mathbf{v} = \mathbf{i} - \mathbf{j} + 4\mathbf{k}$

20. $\mathbf{u} = 2\mathbf{k}$
    $\mathbf{v} = 4\mathbf{i} + 6\mathbf{k}$

Area In Exercises 21–24, find the area of the parallelogram that has the given vectors as adjacent sides. Use a computer algebra system or a graphing utility to verify your result.

21. $\mathbf{u} = \mathbf{j}$
    $\mathbf{v} = \mathbf{j} + \mathbf{k}$

22. $\mathbf{u} = \mathbf{i} + \mathbf{j} + \mathbf{k}$
    $\mathbf{v} = \mathbf{j} + \mathbf{k}$

23. $\mathbf{u} = \langle 3, 2, -1 \rangle$
    $\mathbf{v} = \langle 1, 2, 3 \rangle$

24. $\mathbf{u} = \langle 2, -1, 0 \rangle$
    $\mathbf{v} = \langle -1, 2, 0 \rangle$

Area In Exercises 25 and 26, verify that the points are the vertices of a parallelogram, and find its area.

25. $A(0, 3, 2)$, $B(1, 5, 5)$, $C(6, 9, 5)$, $D(5, 7, 2)$

26. $A(2, -3, 1)$, $B(6, 5, -1)$, $C(7, 2, 2)$, $D(3, -6, 4)$

Area In Exercises 27 and 28, find the area of the triangle with the given vertices. (*Hint:* $\frac{1}{2} \| \mathbf{u} \times \mathbf{v} \|$ is the area of the triangle having $\mathbf{u}$ and $\mathbf{v}$ as adjacent sides.)

27. $A(0, 0, 0)$, $B(1, 0, 3)$, $C(-3, 2, 0)$

28. $A(2, -3, 4)$, $B(0, 1, 2)$, $C(-1, 2, 0)$

29. Torque

A child applies the brakes on a bicycle by applying a downward force of 20 pounds on the pedal when the crank makes a 40° angle with the horizontal (see figure). The crank is 6 inches in length. Find the torque at P.

30. Torque Both the magnitude and the direction of the force on a crankshaft change as the crankshaft rotates. Find the torque on the crankshaft using the position and data shown in the figure.

Figure for 30

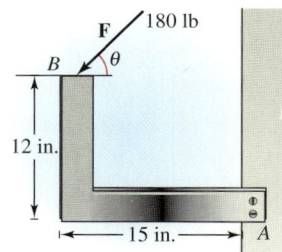

Figure for 31

31. Optimization A force of 180 pounds acts on the bracket shown in the figure.

(a) Determine the vector $\overrightarrow{AB}$ and the vector $\mathbf{F}$ representing the force. ($\mathbf{F}$ will be in terms of θ.)

(b) Find the magnitude of the moment about A by evaluating $\| \overrightarrow{AB} \times \mathbf{F} \|$.

(c) Use the result of part (b) to determine the magnitude of the moment when $\theta = 30°$.

(d) Use the result of part (b) to determine the angle θ when the magnitude of the moment is maximum. At that angle, what is the relationship between the vectors $\mathbf{F}$ and $\overrightarrow{AB}$? Is it what you expected? Why or why not?

(e) Use a graphing utility to graph the function for the magnitude of the moment about A for $0° \leq \theta \leq 180°$. Find the zero of the function in the given domain. Interpret the meaning of the zero in the context of the problem.

32. Optimization A force of 56 pounds acts on the pipe wrench shown in the figure.

(a) Find the magnitude of the moment about O by evaluating $\|\overrightarrow{OA} \times \mathbf{F}\|$. Use a graphing utility to graph the resulting function of θ.

(b) Use the result of part (a) to determine the magnitude of the moment when $\theta = 45°$.

(c) Use the result of part (a) to determine the angle θ when the magnitude of the moment is maximum. Is the answer what you expected? Why or why not?

Finding a Triple Scalar Product In Exercises 33–36, find $\mathbf{u} \cdot (\mathbf{v} \times \mathbf{w})$.

33. $\mathbf{u} = \mathbf{i}$
$\mathbf{v} = \mathbf{j}$
$\mathbf{w} = \mathbf{k}$

34. $\mathbf{u} = \langle 1, 1, 1 \rangle$
$\mathbf{v} = \langle 2, 1, 0 \rangle$
$\mathbf{w} = \langle 0, 0, 1 \rangle$

35. $\mathbf{u} = \langle 2, 0, 1 \rangle$
$\mathbf{v} = \langle 0, 3, 0 \rangle$
$\mathbf{w} = \langle 0, 0, 1 \rangle$

36. $\mathbf{u} = \langle 2, 0, 0 \rangle$
$\mathbf{v} = \langle 1, 1, 1 \rangle$
$\mathbf{w} = \langle 0, 2, 2 \rangle$

Volume In Exercises 37 and 38, use the triple scalar product to find the volume of the parallelepiped having adjacent edges $\mathbf{u}$, $\mathbf{v}$, and $\mathbf{w}$.

37. $\mathbf{u} = \mathbf{i} + \mathbf{j}$
$\mathbf{v} = \mathbf{j} + \mathbf{k}$
$\mathbf{w} = \mathbf{i} + \mathbf{k}$

38. $\mathbf{u} = \langle 1, 3, 1 \rangle$
$\mathbf{v} = \langle 0, 6, 6 \rangle$
$\mathbf{w} = \langle -4, 0, -4 \rangle$

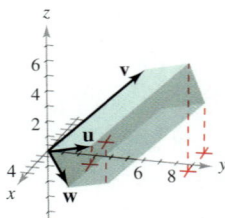

Volume In Exercises 39 and 40, find the volume of the parallelepiped with the given vertices.

39. $(0, 0, 0)$, $(3, 0, 0)$, $(0, 5, 1)$, $(2, 0, 5)$
$(3, 5, 1)$, $(5, 0, 5)$, $(2, 5, 6)$, $(5, 5, 6)$

40. $(0, 0, 0)$, $(0, 4, 0)$, $(-3, 0, 0)$, $(-1, 1, 5)$
$(-3, 4, 0)$, $(-1, 5, 5)$, $(-4, 1, 5)$, $(-4, 5, 5)$

41. Comparing Dot Products Identify the dot products that are equal. Explain your reasoning. (Assume $\mathbf{u}$, $\mathbf{v}$, and $\mathbf{w}$ are nonzero vectors.)

(a) $\mathbf{u} \cdot (\mathbf{v} \times \mathbf{w})$
(b) $(\mathbf{v} \times \mathbf{w}) \cdot \mathbf{u}$
(c) $(\mathbf{u} \times \mathbf{v}) \cdot \mathbf{w}$
(d) $(\mathbf{u} \times -\mathbf{w}) \cdot \mathbf{v}$
(e) $\mathbf{u} \cdot (\mathbf{w} \times \mathbf{v})$
(f) $\mathbf{w} \cdot (\mathbf{v} \times \mathbf{u})$
(g) $(-\mathbf{u} \times \mathbf{v}) \cdot \mathbf{w}$
(h) $(\mathbf{w} \times \mathbf{u}) \cdot \mathbf{v}$

42. Using Dot and Cross Products When $\mathbf{u} \times \mathbf{v} = \mathbf{0}$ and $\mathbf{u} \cdot \mathbf{v} = 0$, what can you conclude about $\mathbf{u}$ and $\mathbf{v}$?

WRITING ABOUT CONCEPTS

43. Cross Product Define the cross product of vectors $\mathbf{u}$ and $\mathbf{v}$.

44. Cross Product State the geometric properties of the cross product.

45. Magnitude When the magnitudes of two vectors are doubled, how will the magnitude of the cross product of the vectors change? Explain.

46. HOW DO YOU SEE IT? The vertices of a triangle in space are (x_1, y_1, z_1), (x_2, y_2, z_2), and (x_3, y_3, z_3). Explain how to find a vector perpendicular to the triangle.

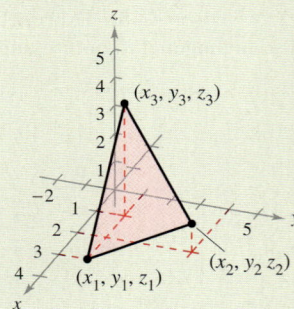

True or False? In Exercises 47–50, determine whether the statement is true or false. If it is false, explain why or give an example that shows it is false.

47. It is possible to find the cross product of two vectors in a two-dimensional coordinate system.

48. If $\mathbf{u}$ and $\mathbf{v}$ are vectors in space that are nonzero and nonparallel, then $\mathbf{u} \times \mathbf{v} = \mathbf{v} \times \mathbf{u}$.

49. If $\mathbf{u} \neq \mathbf{0}$ and $\mathbf{u} \times \mathbf{v} = \mathbf{u} \times \mathbf{w}$, then $\mathbf{v} = \mathbf{w}$.

50. If $\mathbf{u} \neq \mathbf{0}$, $\mathbf{u} \cdot \mathbf{v} = \mathbf{u} \cdot \mathbf{w}$, and $\mathbf{u} \times \mathbf{v} = \mathbf{u} \times \mathbf{w}$, then $\mathbf{v} = \mathbf{w}$.

Proof In Exercises 51–56, prove the property of the cross product.

51. $\mathbf{u} \times (\mathbf{v} + \mathbf{w}) = (\mathbf{u} \times \mathbf{v}) + (\mathbf{u} \times \mathbf{w})$

52. $c(\mathbf{u} \times \mathbf{v}) = (c\mathbf{u}) \times \mathbf{v} = \mathbf{u} \times (c\mathbf{v})$

53. $\mathbf{u} \times \mathbf{u} = \mathbf{0}$ **54.** $\mathbf{u} \cdot (\mathbf{v} \times \mathbf{w}) = (\mathbf{u} \times \mathbf{v}) \cdot \mathbf{w}$

55. $\mathbf{u} \times \mathbf{v}$ is orthogonal to both $\mathbf{u}$ and $\mathbf{v}$.

56. $\mathbf{u} \times \mathbf{v} = \mathbf{0}$ if and only if $\mathbf{u}$ and $\mathbf{v}$ are scalar multiples of each other.

57. Proof Prove that $\|\mathbf{u} \times \mathbf{v}\| = \|\mathbf{u}\| \, \|\mathbf{v}\|$ if $\mathbf{u}$ and $\mathbf{v}$ are orthogonal.

58. Proof Prove that $\mathbf{u} \times (\mathbf{v} \times \mathbf{w}) = (\mathbf{u} \cdot \mathbf{w})\mathbf{v} - (\mathbf{u} \cdot \mathbf{v})\mathbf{w}$.

59. Proof Prove Theorem 11.9.

11.5   Lines and Planes in Space

■ Write a set of parametric equations for a line in space.
■ Write a linear equation to represent a plane in space.
■ Sketch the plane given by a linear equation.
■ Find the distances between points, planes, and lines in space.

Lines in Space

In the plane, *slope* is used to determine the equation of a line. In space, it is more convenient to use *vectors* to determine the equation of a line.

In Figure 11.43, consider the line L through the point $P(x_1, y_1, z_1)$ and parallel to the vector $\mathbf{v} = \langle a, b, c \rangle$. The vector $\mathbf{v}$ is a **direction vector** for the line L, and a, b, and c are **direction numbers.** One way of describing the line L is to say that it consists of all points $Q(x, y, z)$ for which the vector $\overrightarrow{PQ}$ is parallel to $\mathbf{v}$. This means that $\overrightarrow{PQ}$ is a scalar multiple of $\mathbf{v}$ and you can write $\overrightarrow{PQ} = t\mathbf{v}$, where t is a scalar (a real number).

$$\overrightarrow{PQ} = \langle x - x_1, y - y_1, z - z_1 \rangle = \langle at, bt, ct \rangle = t\mathbf{v}$$

By equating corresponding components, you can obtain **parametric equations** of a line in space.

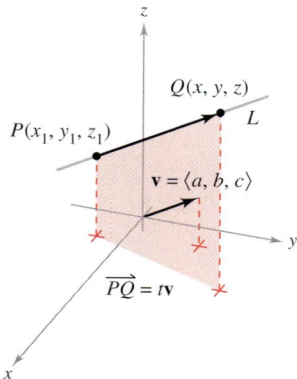

Line L and its direction vector $\mathbf{v}$
Figure 11.43

THEOREM 11.11   Parametric Equations of a Line in Space

A line L parallel to the vector $\mathbf{v} = \langle a, b, c \rangle$ and passing through the point $P(x_1, y_1, z_1)$ is represented by the **parametric equations**

$$x = x_1 + at, \quad y = y_1 + bt, \quad \text{and} \quad z = z_1 + ct.$$

If the direction numbers a, b, and c are all nonzero, then you can eliminate the parameter t to obtain **symmetric equations** of the line.

$$\frac{x - x_1}{a} = \frac{y - y_1}{b} = \frac{z - z_1}{c} \qquad \text{Symmetric equations}$$

EXAMPLE 1   Finding Parametric and Symmetric Equations

Find parametric and symmetric equations of the line L that passes through the point $(1, -2, 4)$ and is parallel to $\mathbf{v} = \langle 2, 4, -4 \rangle$, as shown in Figure 11.44.

Solution   To find a set of parametric equations of the line, use the coordinates $x_1 = 1$, $y_1 = -2$, and $z_1 = 4$ and direction numbers $a = 2$, $b = 4$, and $c = -4$.

$$x = 1 + 2t, \quad y = -2 + 4t, \quad z = 4 - 4t \qquad \text{Parametric equations}$$

Because a, b, and c are all nonzero, a set of symmetric equations is

$$\frac{x - 1}{2} = \frac{y + 2}{4} = \frac{z - 4}{-4}. \qquad \text{Symmetric equations}$$

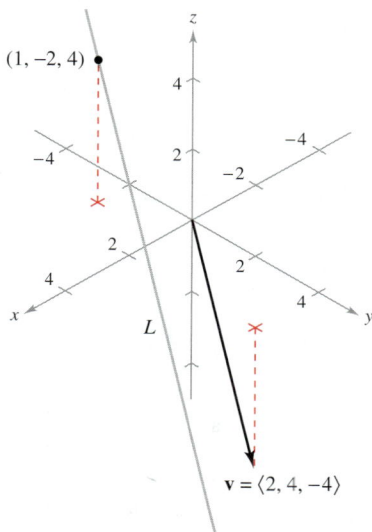

The vector $\mathbf{v}$ is parallel to the line L.
Figure 11.44

Neither parametric equations nor symmetric equations of a given line are unique. For instance, in Example 1, by letting $t = 1$ in the parametric equations, you would obtain the point $(3, 2, 0)$. Using this point with the direction numbers $a = 2$, $b = 4$, and $c = -4$ would produce a different set of parametric equations

$$x = 3 + 2t, \quad y = 2 + 4t, \quad \text{and} \quad z = -4t.$$

| EXAMPLE 2 | **Parametric Equations of a Line Through Two Points** |

⋯▷ See LarsonCalculus.com for an interactive version of this type of example.

Find a set of parametric equations of the line that passes through the points

$$(-2, 1, 0) \quad \text{and} \quad (1, 3, 5).$$

Solution Begin by using the points $P(-2, 1, 0)$ and $Q(1, 3, 5)$ to find a direction vector for the line passing through P and Q.

$$\mathbf{v} = \overrightarrow{PQ} = \langle 1 - (-2), 3 - 1, 5 - 0 \rangle = \langle 3, 2, 5 \rangle = \langle a, b, c \rangle$$

Using the direction numbers $a = 3$, $b = 2$, and $c = 5$ with the point $P(-2, 1, 0)$, you can obtain the parametric equations

$$x = -2 + 3t, \quad y = 1 + 2t, \quad \text{and} \quad z = 5t. \qquad \blacksquare$$

REMARK As t varies over all real numbers, the parametric equations in Example 2 determine the points (x, y, z) on the line. In particular, note that $t = 0$ and $t = 1$ give the original points $(-2, 1, 0)$ and $(1, 3, 5)$.

Planes in Space

You have seen how an equation of a line in space can be obtained from a point on the line and a vector *parallel* to it. You will now see that an equation of a plane in space can be obtained from a point in the plane and a vector *normal* (perpendicular) to the plane.

Consider the plane containing the point $P(x_1, y_1, z_1)$ having a nonzero normal vector

$$\mathbf{n} = \langle a, b, c \rangle$$

as shown in Figure 11.45. This plane consists of all points $Q(x, y, z)$ for which vector $\overrightarrow{PQ}$ is orthogonal to $\mathbf{n}$. Using the dot product, you can write the following.

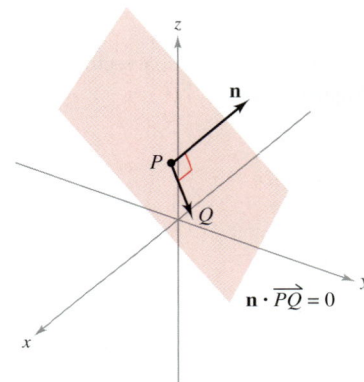

The normal vector $\mathbf{n}$ is orthogonal to each vector $\overrightarrow{PQ}$ in the plane.
Figure 11.45

$$\mathbf{n} \cdot \overrightarrow{PQ} = 0$$
$$\langle a, b, c \rangle \cdot \langle x - x_1, y - y_1, z - z_1 \rangle = 0$$
$$a(x - x_1) + b(y - y_1) + c(z - z_1) = 0$$

The third equation of the plane is said to be in **standard form**.

THEOREM 11.12 **Standard Equation of a Plane in Space**

The plane containing the point (x_1, y_1, z_1) and having normal vector

$$\mathbf{n} = \langle a, b, c \rangle$$

can be represented by the **standard form** of the equation of a plane

$$a(x - x_1) + b(y - y_1) + c(z - z_1) = 0.$$

By regrouping terms, you obtain the **general form** of the equation of a plane in space.

$$ax + by + cz + d = 0 \qquad \text{General form of equation of plane}$$

Given the general form of the equation of a plane, it is easy to find a normal vector to the plane. Simply use the coefficients of x, y, and z and write

$$\mathbf{n} = \langle a, b, c \rangle.$$

EXAMPLE 3 Finding an Equation of a Plane in Three-Space

Find the general equation of the plane containing the points

$$(2, 1, 1), \quad (0, 4, 1), \quad \text{and} \quad (-2, 1, 4).$$

Solution To apply Theorem 11.12, you need a point in the plane and a vector that is normal to the plane. There are three choices for the point, but no normal vector is given. To obtain a normal vector, use the cross product of vectors $\mathbf{u}$ and $\mathbf{v}$ extending from the point $(2, 1, 1)$ to the points $(0, 4, 1)$ and $(-2, 1, 4)$, as shown in Figure 11.46. The component forms of $\mathbf{u}$ and $\mathbf{v}$ are

$$\mathbf{u} = \langle 0 - 2, 4 - 1, 1 - 1 \rangle = \langle -2, 3, 0 \rangle$$
$$\mathbf{v} = \langle -2 - 2, 1 - 1, 4 - 1 \rangle = \langle -4, 0, 3 \rangle$$

and it follows that

$$\mathbf{n} = \mathbf{u} \times \mathbf{v}$$
$$= \begin{vmatrix} \mathbf{i} & \mathbf{j} & \mathbf{k} \\ -2 & 3 & 0 \\ -4 & 0 & 3 \end{vmatrix}$$
$$= 9\mathbf{i} + 6\mathbf{j} + 12\mathbf{k}$$
$$= \langle a, b, c \rangle$$

is normal to the given plane. Using the direction numbers for $\mathbf{n}$ and the point $(x_1, y_1, z_1) = (2, 1, 1)$, you can determine an equation of the plane to be

$a(x - x_1) + b(y - y_1) + c(z - z_1) = 0$	
$9(x - 2) + 6(y - 1) + 12(z - 1) = 0$	Standard form
$9x + 6y + 12z - 36 = 0$	General form
$3x + 2y + 4z - 12 = 0.$	Simplified general form

REMARK In Example 3, check to see that each of the three original points satisfies the equation

$$3x + 2y + 4z - 12 = 0.$$

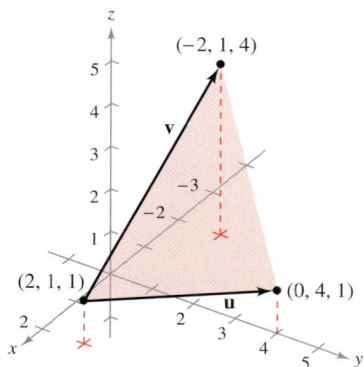

A plane determined by $\mathbf{u}$ and $\mathbf{v}$
Figure 11.46

Two distinct planes in three-space either are parallel or intersect in a line. For two planes that intersect, you can determine the angle $(0 \le \theta \le \pi/2)$ between them from the angle between their normal vectors, as shown in Figure 11.47. Specifically, if vectors $\mathbf{n}_1$ and $\mathbf{n}_2$ are normal to two intersecting planes, then the angle θ between the normal vectors is equal to the angle between the two planes and is

$$\cos \theta = \frac{|\mathbf{n}_1 \cdot \mathbf{n}_2|}{\|\mathbf{n}_1\| \|\mathbf{n}_2\|}. \qquad \text{Angle between two planes}$$

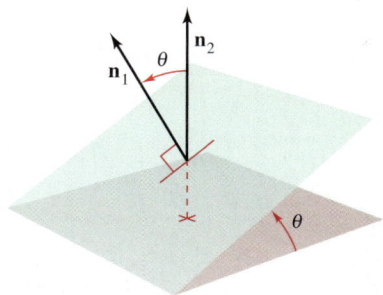

The angle θ between two planes
Figure 11.47

Consequently, two planes with normal vectors $\mathbf{n}_1$ and $\mathbf{n}_2$ are

1. *perpendicular* when $\mathbf{n}_1 \cdot \mathbf{n}_2 = 0$.
2. *parallel* when $\mathbf{n}_1$ is a scalar multiple of $\mathbf{n}_2$.

> **EXAMPLE 4** **Finding the Line of Intersection of Two Planes**

Find the angle between the two planes

$$x - 2y + z = 0 \quad \text{and} \quad 2x + 3y - 2z = 0.$$

Then find parametric equations of their line of intersection (see Figure 11.48).

•• **REMARK** The three-dimensional rotatable graphs that are available at *LarsonCalculus.com* can help you visualize surfaces such as those shown in Figure 11.48. If you have access to these graphs, you should use them to help your spatial intuition when studying this section and other sections in the text that deal with vectors, curves, or surfaces in space.

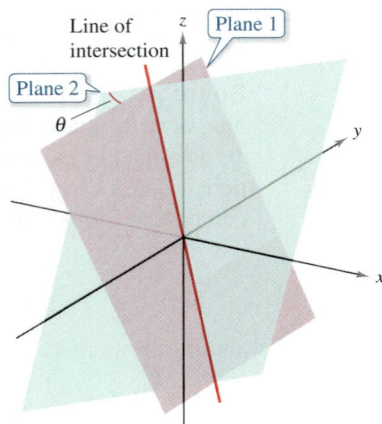

Figure 11.48

Solution Normal vectors for the planes are $\mathbf{n}_1 = \langle 1, -2, 1 \rangle$ and $\mathbf{n}_2 = \langle 2, 3, -2 \rangle$. Consequently, the angle between the two planes is determined as follows.

$$\cos \theta = \frac{|\mathbf{n}_1 \cdot \mathbf{n}_2|}{\|\mathbf{n}_1\| \|\mathbf{n}_2\|} = \frac{|-6|}{\sqrt{6}\sqrt{17}} = \frac{6}{\sqrt{102}} \approx 0.59409$$

This implies that the angle between the two planes is $\theta \approx 53.55°$. You can find the line of intersection of the two planes by simultaneously solving the two linear equations representing the planes. One way to do this is to multiply the first equation by -2 and add the result to the second equation.

$$\begin{array}{lll} x - 2y + z = 0 & \Longrightarrow & -2x + 4y - 2z = 0 \\ 2x + 3y - 2z = 0 & & \underline{2x + 3y - 2z = 0} \\ & & 7y - 4z = 0 \quad \Longrightarrow \quad y = \dfrac{4z}{7} \end{array}$$

Substituting $y = 4z/7$ back into one of the original equations, you can determine that $x = z/7$. Finally, by letting $t = z/7$, you obtain the parametric equations

$$x = t, \quad y = 4t, \quad \text{and} \quad z = 7t \qquad \textcolor{red}{\text{Line of intersection}}$$

which indicate that 1, 4, and 7 are direction numbers for the line of intersection. ∎

Note that the direction numbers in Example 4 can be obtained from the cross product of the two normal vectors as follows.

$$\mathbf{n}_1 \times \mathbf{n}_2 = \begin{vmatrix} \mathbf{i} & \mathbf{j} & \mathbf{k} \\ 1 & -2 & 1 \\ 2 & 3 & -2 \end{vmatrix}$$

$$= \begin{vmatrix} -2 & 1 \\ 3 & -2 \end{vmatrix} \mathbf{i} - \begin{vmatrix} 1 & 1 \\ 2 & -2 \end{vmatrix} \mathbf{j} + \begin{vmatrix} 1 & -2 \\ 2 & 3 \end{vmatrix} \mathbf{k}$$

$$= \mathbf{i} + 4\mathbf{j} + 7\mathbf{k}$$

This means that the line of intersection of the two planes is parallel to the cross product of their normal vectors.

Sketching Planes in Space

If a plane in space intersects one of the coordinate planes, then the line of intersection is called the **trace** of the given plane in the coordinate plane. To sketch a plane in space, it is helpful to find its points of intersection with the coordinate axes and its traces in the coordinate planes. For example, consider the plane

$$3x + 2y + 4z = 12.$$ Equation of plane

You can find the xy-trace by letting $z = 0$ and sketching the line

$$3x + 2y = 12$$ xy-trace

in the xy-plane. This line intersects the x-axis at $(4, 0, 0)$ and the y-axis at $(0, 6, 0)$. In Figure 11.49, this process is continued by finding the yz-trace and the xz-trace, and then shading the triangular region lying in the first octant.

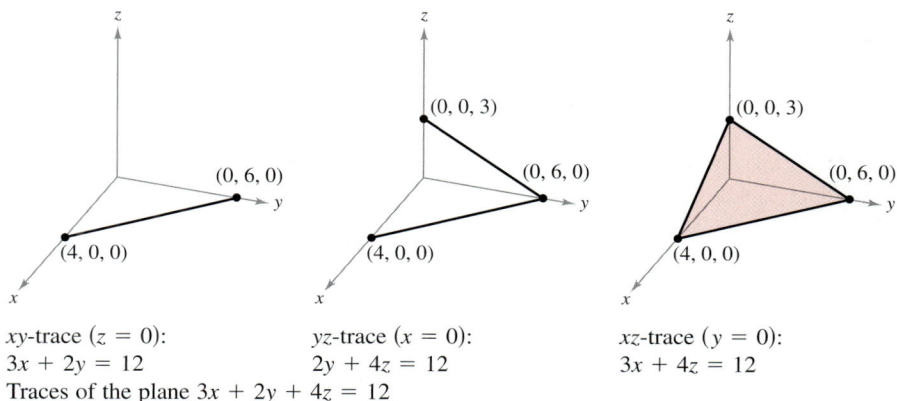

xy-trace ($z = 0$): $3x + 2y = 12$
yz-trace ($x = 0$): $2y + 4z = 12$
xz-trace ($y = 0$): $3x + 4z = 12$
Traces of the plane $3x + 2y + 4z = 12$
Figure 11.49

If an equation of a plane has a missing variable, such as

$$2x + z = 1$$

then the plane must be *parallel to the axis* represented by the missing variable, as shown in Figure 11.50. If two variables are missing from an equation of a plane, such as

$$ax + d = 0$$

then it is *parallel to the coordinate plane* represented by the missing variables, as shown in Figure 11.51.

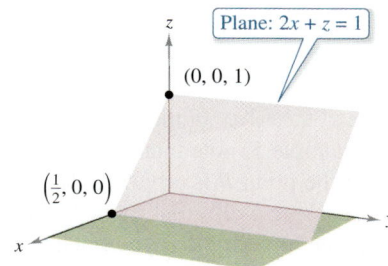

Plane $2x + z = 1$ is parallel to the y-axis.
Figure 11.50

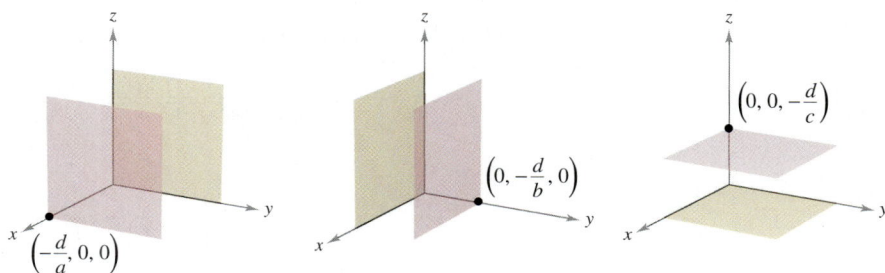

Plane $ax + d = 0$ is parallel to the yz-plane.
Plane $by + d = 0$ is parallel to the xz-plane.
Plane $cz + d = 0$ is parallel to the xy-plane.
Figure 11.51

Distances Between Points, Planes, and Lines

Consider two types of problems involving distance in space: (1) finding the distance between a point and a plane, and (2) finding the distance between a point and a line. The solutions of these problems illustrate the versatility and usefulness of vectors in coordinate geometry: the first problem uses the *dot product* of two vectors, and the second problem uses the *cross product*.

The distance D between a point Q and a plane is the length of the shortest line segment connecting Q to the plane, as shown in Figure 11.52. For *any* point P in the plane, you can find this distance by projecting the vector $\overrightarrow{PQ}$ onto the normal vector $\mathbf{n}$. The length of this projection is the desired distance.

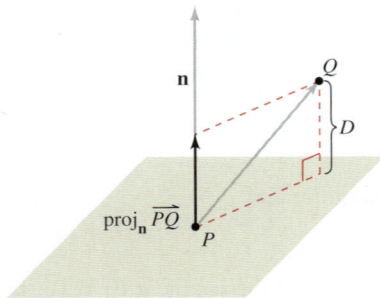

$D = \|\operatorname{proj}_\mathbf{n} \overrightarrow{PQ}\|$

The distance between a point and a plane
Figure 11.52

> **THEOREM 11.13 Distance Between a Point and a Plane**
>
> The distance between a plane and a point Q (not in the plane) is
>
> $$D = \|\operatorname{proj}_\mathbf{n}\overrightarrow{PQ}\| = \frac{|\overrightarrow{PQ} \cdot \mathbf{n}|}{\|\mathbf{n}\|}$$
>
> where P is a point in the plane and $\mathbf{n}$ is normal to the plane.

To find a point in the plane $ax + by + cz + d = 0$, where $a \neq 0$, let $y = 0$ and $z = 0$. Then, from the equation $ax + d = 0$, you can conclude that the point

$$\left(-\frac{d}{a}, 0, 0\right)$$

lies in the plane.

EXAMPLE 5 Finding the Distance Between a Point and a Plane

Find the distance between the point $Q(1, 5, -4)$ and the plane $3x - y + 2z = 6$.

Solution You know that $\mathbf{n} = \langle 3, -1, 2 \rangle$ is normal to the plane. To find a point in the plane, let $y = 0$ and $z = 0$, and obtain the point $P(2, 0, 0)$. The vector from P to Q is

$$\overrightarrow{PQ} = \langle 1 - 2, 5 - 0, -4 - 0 \rangle$$
$$= \langle -1, 5, -4 \rangle.$$

Using the Distance Formula given in Theorem 11.13 produces

$$D = \frac{|\overrightarrow{PQ} \cdot \mathbf{n}|}{\|\mathbf{n}\|} = \frac{|\langle -1, 5, -4 \rangle \cdot \langle 3, -1, 2 \rangle|}{\sqrt{9 + 1 + 4}} = \frac{|-3 - 5 - 8|}{\sqrt{14}} = \frac{16}{\sqrt{14}} \approx 4.28.$$

REMARK In the solution to Example 5, note that the choice of the point P is arbitrary. Try choosing a different point in the plane to verify that you obtain the same distance.

From Theorem 11.13, you can determine that the distance between the point $Q(x_0, y_0, z_0)$ and the plane $ax + by + cz + d = 0$ is

$$D = \frac{|a(x_0 - x_1) + b(y_0 - y_1) + c(z_0 - z_1)|}{\sqrt{a^2 + b^2 + c^2}}$$

or

$$D = \frac{|ax_0 + by_0 + cz_0 + d|}{\sqrt{a^2 + b^2 + c^2}} \qquad \text{Distance between a point and a plane}$$

where $P(x_1, y_1, z_1)$ is a point in the plane and $d = -(ax_1 + by_1 + cz_1)$.

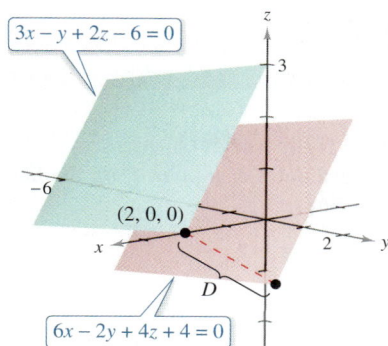

$3x - y + 2z - 6 = 0$

$6x - 2y + 4z + 4 = 0$

$(2, 0, 0)$

The distance between the parallel planes is approximately 2.14.
Figure 11.53

EXAMPLE 6   Finding the Distance Between Two Parallel Planes

Two parallel planes, $3x - y + 2z - 6 = 0$ and $6x - 2y + 4z + 4 = 0$, are shown in Figure 11.53. To find the distance between the planes, choose a point in the first plane, such as $(x_0, y_0, z_0) = (2, 0, 0)$. Then, from the second plane, you can determine that $a = 6$, $b = -2$, $c = 4$, and $d = 4$, and conclude that the distance is

$$D = \frac{|ax_0 + by_0 + cz_0 + d|}{\sqrt{a^2 + b^2 + c^2}}$$

$$= \frac{|6(2) + (-2)(0) + (4)(0) + 4|}{\sqrt{6^2 + (-2)^2 + 4^2}}$$

$$= \frac{16}{\sqrt{56}} = \frac{8}{\sqrt{14}} \approx 2.14.$$

The formula for the distance between a point and a line in space resembles that for the distance between a point and a plane—except that you replace the dot product with the length of the cross product and the normal vector **n** with a direction vector for the line.

THEOREM 11.14   Distance Between a Point and a Line in Space

The distance between a point Q and a line in space is

$$D = \frac{\|\overrightarrow{PQ} \times \mathbf{u}\|}{\|\mathbf{u}\|}$$

where **u** is a direction vector for the line and P is a point on the line.

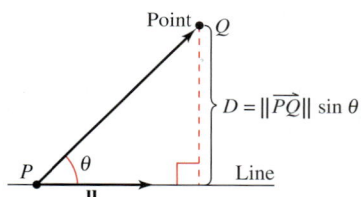

$D = \|\overrightarrow{PQ}\| \sin \theta$

The distance between a point and a line
Figure 11.54

Proof   In Figure 11.54, let D be the distance between the point Q and the line. Then $D = \|\overrightarrow{PQ}\| \sin \theta$, where θ is the angle between **u** and $\overrightarrow{PQ}$. By Property 2 of Theorem 11.8, you have $\|\mathbf{u}\| \|\overrightarrow{PQ}\| \sin \theta = \|\mathbf{u} \times \overrightarrow{PQ}\| = \|\overrightarrow{PQ} \times \mathbf{u}\|$. Consequently,

$$D = \|\overrightarrow{PQ}\| \sin \theta = \frac{\|\overrightarrow{PQ} \times \mathbf{u}\|}{\|\mathbf{u}\|}.$$

See LarsonCalculus.com for Bruce Edwards's video of this proof.

EXAMPLE 7   Finding the Distance Between a Point and a Line

Find the distance between the point $Q(3, -1, 4)$ and the line

$$x = -2 + 3t, \quad y = -2t, \quad \text{and} \quad z = 1 + 4t.$$

Solution   Using the direction numbers 3, -2, and 4, a direction vector for the line is $\mathbf{u} = \langle 3, -2, 4 \rangle$. To find a point on the line, let $t = 0$ and obtain $P = (-2, 0, 1)$. So,

$$\overrightarrow{PQ} = \langle 3 - (-2), -1 - 0, 4 - 1 \rangle = \langle 5, -1, 3 \rangle$$

and you can form the cross product

$$\overrightarrow{PQ} \times \mathbf{u} = \begin{vmatrix} \mathbf{i} & \mathbf{j} & \mathbf{k} \\ 5 & -1 & 3 \\ 3 & -2 & 4 \end{vmatrix} = 2\mathbf{i} - 11\mathbf{j} - 7\mathbf{k} = \langle 2, -11, -7 \rangle.$$

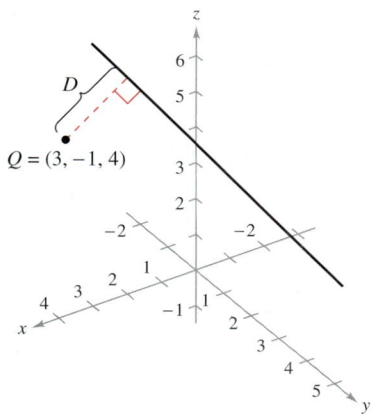

$Q = (3, -1, 4)$

The distance between the point Q and the line is $\sqrt{6} \approx 2.45$.
Figure 11.55

Finally, using Theorem 11.14, you can find the distance to be

$$D = \frac{\|\overrightarrow{PQ} \times \mathbf{u}\|}{\|\mathbf{u}\|} = \frac{\sqrt{174}}{\sqrt{29}} = \sqrt{6} \approx 2.45. \qquad \text{See Figure 11.55.}$$

11.5 Exercises

See **CalcChat.com** for tutorial help and worked-out solutions to odd-numbered exercises.

Checking Points on a Line In Exercises 1 and 2, determine whether each point lies on the line.

1. $x = -2 + t$, $y = 3t$, $z = 4 + t$

   (a) $(0, 6, 6)$ (b) $(2, 3, 5)$

2. $\dfrac{x - 3}{2} = \dfrac{y - 7}{8} = z + 2$

   (a) $(7, 23, 0)$ (b) $(1, -1, -3)$

Finding Parametric and Symmetric Equations In Exercises 3–8, find sets of (a) parametric equations and (b) symmetric equations of the line through the point parallel to the given vector or line (if possible). (For each line, write the direction numbers as integers.)

Point	Parallel to
3. $(0, 0, 0)$	$\mathbf{v} = \langle 3, 1, 5 \rangle$
4. $(0, 0, 0)$	$\mathbf{v} = \left\langle -2, \frac{5}{2}, 1 \right\rangle$
5. $(-2, 0, 3)$	$\mathbf{v} = 2\mathbf{i} + 4\mathbf{j} - 2\mathbf{k}$
6. $(-3, 0, 2)$	$\mathbf{v} = 6\mathbf{j} + 3\mathbf{k}$
7. $(1, 0, 1)$	$x = 3 + 3t$, $y = 5 - 2t$, $z = -7 + t$
8. $(-3, 5, 4)$	$\dfrac{x - 1}{3} = \dfrac{y + 1}{-2} = z - 3$

Finding Parametric and Symmetric Equations In Exercises 9–12, find sets of (a) parametric equations and (b) symmetric equations of the line through the two points (if possible). (For each line, write the direction numbers as integers.)

9. $(5, -3, -2)$, $\left(-\frac{2}{3}, \frac{2}{3}, 1\right)$ **10.** $(0, 4, 3)$, $(-1, 2, 5)$

11. $(7, -2, 6)$, $(-3, 0, 6)$ **12.** $(0, 0, 25)$, $(10, 10, 0)$

Finding Parametric Equations In Exercises 13–20, find a set of parametric equations of the line.

13. The line passes through the point $(2, 3, 4)$ and is parallel to the xz-plane and the yz-plane.

14. The line passes through the point $(-4, 5, 2)$ and is parallel to the xy-plane and the yz-plane.

15. The line passes through the point $(2, 3, 4)$ and is perpendicular to the plane given by $3x + 2y - z = 6$.

16. The line passes through the point $(-4, 5, 2)$ and is perpendicular to the plane given by $-x + 2y + z = 5$.

17. The line passes through the point $(5, -3, -4)$ and is parallel to $\mathbf{v} = \langle 2, -1, 3 \rangle$.

18. The line passes through the point $(-1, 4, -3)$ and is parallel to $\mathbf{v} = 5\mathbf{i} - \mathbf{j}$.

19. The line passes through the point $(2, 1, 2)$ and is parallel to the line $x = -t$, $y = 1 + t$, $z = -2 + t$.

20. The line passes through the point $(-6, 0, 8)$ and is parallel to the line $x = 5 - 2t$, $y = -4 + 2t$, $z = 0$.

Using Parametric and Symmetric Equations In Exercises 21–24, find the coordinates of a point P on the line and a vector $\mathbf{v}$ parallel to the line.

21. $x = 3 - t$, $y = -1 + 2t$, $z = -2$

22. $x = 4t$, $y = 5 - t$, $z = 4 + 3t$

23. $\dfrac{x - 7}{4} = \dfrac{y + 6}{2} = z + 2$ **24.** $\dfrac{x + 3}{5} = \dfrac{y}{8} = \dfrac{z - 3}{6}$

Determining Parallel Lines In Exercises 25–28, determine whether any of the lines are parallel or identical.

25. L_1: $x = 6 - 3t$, $y = -2 + 2t$, $z = 5 + 4t$

    L_2: $x = 6t$, $y = 2 - 4t$, $z = 13 - 8t$

    L_3: $x = 10 - 6t$, $y = 3 + 4t$, $z = 7 + 8t$

    L_4: $x = -4 + 6t$, $y = 3 + 4t$, $z = 5 - 6t$

26. L_1: $x = 3 + 2t$, $y = -6t$, $z = 1 - 2t$

    L_2: $x = 1 + 2t$, $y = -1 - t$, $z = 3t$

    L_3: $x = -1 + 2t$, $y = 3 - 10t$, $z = 1 - 4t$

    L_4: $x = 5 + 2t$, $y = 1 - t$, $z = 8 + 3t$

27. L_1: $\dfrac{x - 8}{4} = \dfrac{y + 5}{-2} = \dfrac{z + 9}{3}$

    L_2: $\dfrac{x + 7}{2} = \dfrac{y - 4}{1} = \dfrac{z + 6}{5}$

    L_3: $\dfrac{x + 4}{-8} = \dfrac{y - 1}{4} = \dfrac{z + 18}{-6}$

    L_4: $\dfrac{x - 2}{-2} = \dfrac{y + 3}{1} = \dfrac{z - 4}{1.5}$

28. L_1: $\dfrac{x - 3}{2} = \dfrac{y - 2}{1} = \dfrac{z + 2}{2}$

    L_2: $\dfrac{x - 1}{4} = \dfrac{y - 1}{2} = \dfrac{z + 3}{4}$

    L_3: $\dfrac{x + 2}{1} = \dfrac{y - 1}{0.5} = \dfrac{z - 3}{1}$

    L_4: $\dfrac{x - 3}{2} = \dfrac{y + 1}{4} = \dfrac{z - 2}{-1}$

Finding a Point of Intersection In Exercises 29–32, determine whether the lines intersect, and if so, find the point of intersection and the cosine of the angle of intersection.

29. $x = 4t + 2$, $y = 3$, $z = -t + 1$

    $x = 2s + 2$, $y = 2s + 3$, $z = s + 1$

30. $x = -3t + 1$, $y = 4t + 1$, $z = 2t + 4$

    $x = 3s + 1$, $y = 2s + 4$, $z = -s + 1$

31. $\dfrac{x}{3} = \dfrac{y - 2}{-1} = z + 1$, $\dfrac{x - 1}{4} = y + 2 = \dfrac{z + 3}{-3}$

32. $\dfrac{x - 2}{-3} = \dfrac{y - 2}{6} = z - 3$, $\dfrac{x - 3}{2} = y + 5 = \dfrac{z + 2}{4}$

Checking Points on a Plane   In Exercises 33 and 34, determine whether the plane passes through each point.

33. $x + 2y - 4z - 1 = 0$

   (a) $(-7, 2, -1)$ (b) $(5, 2, 2)$

34. $2x + y + 3z - 6 = 0$

   (a) $(3, 6, -2)$ (b) $(-1, 5, -1)$

Finding an Equation of a Plane   In Exercises 35–40, find an equation of the plane passing through the point perpendicular to the given vector or line.

Point	Perpendicular to
35. $(1, 3, -7)$	$\mathbf{n} = \mathbf{j}$
36. $(0, -1, 4)$	$\mathbf{n} = \mathbf{k}$
37. $(3, 2, 2)$	$\mathbf{n} = 2\mathbf{i} + 3\mathbf{j} - \mathbf{k}$
38. $(0, 0, 0)$	$\mathbf{n} = -3\mathbf{i} + 2\mathbf{k}$
39. $(-1, 4, 0)$	$x = -1 + 2t, y = 5 - t, z = 3 - 2t$
40. $(3, 2, 2)$	$\dfrac{x - 1}{4} = y + 2 = \dfrac{z + 3}{-3}$

Finding an Equation of a Plane   In Exercises 41–52, find an equation of the plane.

41. The plane passes through $(0, 0, 0)$, $(2, 0, 3)$, and $(-3, -1, 5)$.

42. The plane passes through $(3, -1, 2)$, $(2, 1, 5)$, and $(1, -2, -2)$.

43. The plane passes through $(1, 2, 3)$, $(3, 2, 1)$, and $(-1, -2, 2)$.

44. The plane passes through the point $(1, 2, 3)$ and is parallel to the yz-plane.

45. The plane passes through the point $(1, 2, 3)$ and is parallel to the xy-plane.

46. The plane contains the y-axis and makes an angle of $\pi/6$ with the positive x-axis.

47. The plane contains the lines given by

$$\frac{x - 1}{-2} = y - 4 = z$$

and

$$\frac{x - 2}{-3} = \frac{y - 1}{4} = \frac{z - 2}{-1}.$$

48. The plane passes through the point $(2, 2, 1)$ and contains the line given by

$$\frac{x}{2} = \frac{y - 4}{-1} = z.$$

49. The plane passes through the points $(2, 2, 1)$ and $(-1, 1, -1)$ and is perpendicular to the plane $2x - 3y + z = 3$.

50. The plane passes through the points $(3, 2, 1)$ and $(3, 1, -5)$ and is perpendicular to the plane $6x + 7y + 2z = 10$.

51. The plane passes through the points $(1, -2, -1)$ and $(2, 5, 6)$ and is parallel to the x-axis.

52. The plane passes through the points $(4, 2, 1)$ and $(-3, 5, 7)$ and is parallel to the z-axis.

Finding an Equation of a Plane   In Exercises 53–56, find an equation of the plane that contains all the points that are equidistant from the given points.

53. $(2, 2, 0)$, $(0, 2, 2)$ **54.** $(1, 0, 2)$, $(2, 0, 1)$

55. $(-3, 1, 2)$, $(6, -2, 4)$ **56.** $(-5, 1, -3)$, $(2, -1, 6)$

Comparing Planes   In Exercises 57–62, determine whether the planes are parallel, orthogonal, or neither. If they are neither parallel nor orthogonal, find the angle of intersection.

57. $5x - 3y + z = 4$ **58.** $3x + y - 4z = 3$

   $x + 4y + 7z = 1$ $-9x - 3y + 12z = 4$

59. $x - 3y + 6z = 4$ **60.** $3x + 2y - z = 7$

   $5x + y - z = 4$ $x - 4y + 2z = 0$

61. $x - 5y - z = 1$ **62.** $2x - z = 1$

   $5x - 25y - 5z = -3$ $4x + y + 8z = 10$

Sketching a Graph of a Plane   In Exercises 63–70, sketch a graph of the plane and label any intercepts.

63. $4x + 2y + 6z = 12$ **64.** $3x + 6y + 2z = 6$

65. $2x - y + 3z = 4$ **66.** $2x - y + z = 4$

67. $x + z = 6$ **68.** $2x + y = 8$

69. $x = 5$ **70.** $z = 8$

Parallel Planes   In Exercises 71–74, determine whether any of the planes are parallel or identical.

71. $P_1: -5x + 2y - 8z = 6$ **72.** $P_1: 2x - y + 3z = 8$

   $P_2: 15x - 6y + 24z = 17$ $P_2: 3x - 5y - 2z = 6$

   $P_3: 6x - 4y + 4z = 9$ $P_3: 8x - 4y + 12z = 5$

   $P_4: 3x - 2y - 2z = 4$ $P_4: -4x - 2y + 6z = 11$

73. $P_1: 3x - 2y + 5z = 10$

   $P_2: -6x + 4y - 10z = 5$

   $P_3: -3x + 2y + 5z = 8$

   $P_4: 75x - 50y + 125z = 250$

74. $P_1: -60x + 90y + 30z = 27$

   $P_2: 6x - 9y - 3z = 2$

   $P_3: -20x + 30y + 10z = 9$

   $P_4: 12x - 18y + 6z = 5$

Intersection of Planes   In Exercises 75 and 76, (a) find the angle between the two planes, and (b) find a set of parametric equations for the line of intersection of the planes.

75. $3x + 2y - z = 7$ **76.** $6x - 3y + z = 5$

   $x - 4y + 2z = 0$ $-x + y + 5z = 5$

Intersection of a Plane and a Line   In Exercises 77–80, find the point(s) of intersection (if any) of the plane and the line. Also, determine whether the line lies in the plane.

77. $2x - 2y + z = 12,\quad x - \dfrac{1}{2} = \dfrac{y + (3/2)}{-1} = \dfrac{z + 1}{2}$

78. $2x + 3y = -5,\quad \dfrac{x - 1}{4} = \dfrac{y}{2} = \dfrac{z - 3}{6}$

79. $2x + 3y = 10$, $\dfrac{x-1}{3} = \dfrac{y+1}{-2} = z - 3$

80. $5x + 3y = 17$, $\dfrac{x-4}{2} = \dfrac{y+1}{-3} = \dfrac{z+2}{5}$

Finding the Distance Between a Point and a Plane In Exercises 81–84, find the distance between the point and the plane.

81. $(0, 0, 0)$
$2x + 3y + z = 12$

82. $(0, 0, 0)$
$5x + y - z = 9$

83. $(2, 8, 4)$
$2x + y + z = 5$

84. $(1, 3, -1)$
$3x - 4y + 5z = 6$

Finding the Distance Between Two Parallel Planes In Exercises 85–88, verify that the two planes are parallel, and find the distance between the planes.

85. $x - 3y + 4z = 10$
$x - 3y + 4z = 6$

86. $4x - 4y + 9z = 7$
$4x - 4y + 9z = 18$

87. $-3x + 6y + 7z = 1$
$6x - 12y - 14z = 25$

88. $2x - 4z = 4$
$2x - 4z = 10$

Finding the Distance Between a Point and a Line In Exercises 89–92, find the distance between the point and the line given by the set of parametric equations.

89. $(1, 5, -2)$; $\quad x = 4t - 2$, $\quad y = 3$, $\quad z = -t + 1$

90. $(1, -2, 4)$; $\quad x = 2t$, $\quad y = t - 3$, $\quad z = 2t + 2$

91. $(-2, 1, 3)$; $\quad x = 1 - t$, $\quad y = 2 + t$, $\quad z = -2t$

92. $(4, -1, 5)$; $\quad x = 3$, $\quad y = 1 + 3t$, $\quad z = 1 + t$

Finding the Distance Between Two Parallel Lines In Exercises 93 and 94, verify that the lines are parallel, and find the distance between them.

93. L_1: $x = 2 - t$, $\quad y = 3 + 2t$, $\quad z = 4 + t$

L_2: $x = 3t$, $\quad y = 1 - 6t$, $\quad z = 4 - 3t$

94. L_1: $x = 3 + 6t$, $\quad y = -2 + 9t$, $\quad z = 1 - 12t$

L_2: $x = -1 + 4t$, $\quad y = 3 + 6t$, $\quad z = -8t$

WRITING ABOUT CONCEPTS

95. Parametric and Symmetric Equations Give the parametric equations and the symmetric equations of a line in space. Describe what is required to find these equations.

96. Standard Equation of a Plane in Space Give the standard equation of a plane in space. Describe what is required to find this equation.

97. Intersection of Two Planes Describe a method of finding the line of intersection of two planes.

98. Parallel and Perpendicular Planes Describe a method for determining when two planes, $a_1x + b_1y + c_1z + d_1 = 0$ and $a_2x + b_2y + c_2z + d_2 = 0$, are (a) parallel and (b) perpendicular. Explain your reasoning.

WRITING ABOUT CONCEPTS **(continued)**

99. Normal Vector Let L_1 and L_2 be nonparallel lines that do not intersect. Is it possible to find a nonzero vector **v** such that **v** is normal to both L_1 and L_2? Explain your reasoning.

100. HOW DO YOU SEE IT? Match the general equation with its graph. Then state what axis or plane the equation is parallel to.

(a) $ax + by + d = 0$
(b) $ax + d = 0$
(c) $cz + d = 0$
(d) $ax + cz + d = 0$

(i)

(ii)

(iii)

(iv)

101. Modeling Data Personal consumption expenditures (in billions of dollars) for several types of recreation from 2005 through 2010 are shown in the table, where x is the expenditures on amusement parks and campgrounds, y is the expenditures on live entertainment (excluding sports), and z is the expenditures on spectator sports. (*Source: U.S. Bureau of Economic Analysis*)

Year	2005	2006	2007	2008	2009	2010
x	36.4	39.0	42.4	44.7	43.0	45.2
y	15.3	16.6	17.4	17.5	17.0	17.3
z	16.4	18.1	20.0	20.5	20.1	21.4

A model for the data is given by

$0.46x + 0.30y - z = 4.94$.

(a) Complete a fourth row in the table using the model to approximate z for the given values of x and y. Compare the approximations with the actual values of z.

(b) According to this model, increases in expenditures on recreation types x and y would correspond to what kind of change in expenditures on recreation type z?

102. Mechanical Design The figure shows a chute at the top of a grain elevator of a combine that funnels the grain into a bin. Find the angle between two adjacent sides.

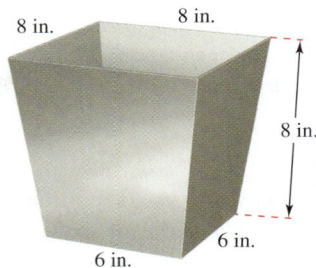

8 in. 8 in. 8 in. 6 in. 6 in.

103. Distance Two insects are crawling along different lines in three-space. At time t (in minutes), the first insect is at the point (x, y, z) on the line $x = 6 + t$, $y = 8 - t$, $z = 3 + t$. Also, at time t, the second insect is at the point (x, y, z) on the line $x = 1 + t$, $y = 2 + t$, $z = 2t$. Assume that distances are given in inches.

(a) Find the distance between the two insects at time $t = 0$.

(b) Use a graphing utility to graph the distance between the insects from $t = 0$ to $t = 10$.

(c) Using the graph from part (b), what can you conclude about the distance between the insects?

(d) How close to each other do the insects get?

104. Finding an Equation of a Sphere Find the standard equation of the sphere with center $(-3, 2, 4)$ that is tangent to the plane given by $2x + 4y - 3z = 8$.

105. Finding a Point of Intersection Find the point of intersection of the plane $3x - y + 4z = 7$ and the line through $(5, 4, -3)$ that is perpendicular to this plane.

106. Finding the Distance Between a Plane and a Line Show that the plane $2x - y - 3z = 4$ is parallel to the line $x = -2 + 2t$, $y = -1 + 4t$, $z = 4$, and find the distance between them.

107. Finding a Point of Intersection Find the point of intersection of the line through $(1, -3, 1)$ and $(3, -4, 2)$ and the plane given by $x - y + z = 2$.

108. Finding Parametric Equations Find a set of parametric equations for the line passing through the point $(1, 0, 2)$ that is parallel to the plane given by $x + y + z = 5$ and perpendicular to the line $x = t$, $y = 1 + t$, $z = 1 + t$.

True or False? In Exercises 109–114, determine whether the statement is true or false. If it is false, explain why or give an example that shows it is false.

109. If $\mathbf{v} = a_1\mathbf{i} + b_1\mathbf{j} + c_1\mathbf{k}$ is any vector in the plane given by $a_2x + b_2y + c_2z + d_2 = 0$, then $a_1a_2 + b_1b_2 + c_1c_2 = 0$.

110. Every two lines in space are either intersecting or parallel.

111. Two planes in space are either intersecting or parallel.

112. If two lines L_1 and L_2 are parallel to a plane P, then L_1 and L_2 are parallel.

113. Two planes perpendicular to a third plane in space are parallel.

114. A plane and a line in space are either intersecting or parallel.

SECTION PROJECT

Distances in Space

You have learned two distance formulas in this section—one for the distance between a point and a plane, and one for the distance between a point and a line. In this project, you will study a third distance problem—the distance between two skew lines. Two lines in space are *skew* if they are neither parallel nor intersecting (see figure).

(a) Consider the following two lines in space.

L_1: $x = 4 + 5t$, $y = 5 + 5t$, $z = 1 - 4t$

L_2: $x = 4 + s$, $y = -6 + 8s$, $z = 7 - 3s$

   (i) Show that these lines are not parallel.

   (ii) Show that these lines do not intersect, and therefore are skew lines.

   (iii) Show that the two lines lie in parallel planes.

   (iv) Find the distance between the parallel planes from part (iii). This is the distance between the original skew lines.

(b) Use the procedure in part (a) to find the distance between the lines.

L_1: $x = 2t$, $y = 4t$, $z = 6t$

L_2: $x = 1 - s$, $y = 4 + s$, $z = -1 + s$

(c) Use the procedure in part (a) to find the distance between the lines.

L_1: $x = 3t$, $y = 2 - t$, $z = -1 + t$

L_2: $x = 1 + 4s$, $y = -2 + s$, $z = -3 - 3s$

(d) Develop a formula for finding the distance between the skew lines.

L_1: $x = x_1 + a_1t$, $y = y_1 + b_1t$, $z = z_1 + c_1t$

L_2: $x = x_2 + a_2s$, $y = y_2 + b_2s$, $z = z_2 + c_2s$

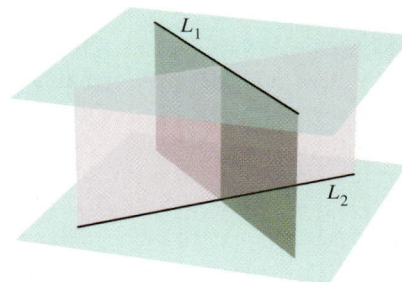

L_1 L_2

11.6 Surfaces in Space

■ Recognize and write equations of cylindrical surfaces.
■ Recognize and write equations of quadric surfaces.
■ Recognize and write equations of surfaces of revolution.

Cylindrical Surfaces

The first five sections of this chapter contained the vector portion of the preliminary work necessary to study vector calculus and the calculus of space. In this and the next section, you will study surfaces in space and alternative coordinate systems for space. You have already studied two special types of surfaces.

1. Spheres: $(x - x_0)^2 + (y - y_0)^2 + (z - z_0)^2 = r^2$ Section 11.2
2. Planes: $ax + by + cz + d = 0$ Section 11.5

A third type of surface in space is a **cylindrical surface,** or simply a **cylinder.** To define a cylinder, consider the familiar right circular cylinder shown in Figure 11.56. The cylinder was generated by a vertical line moving around the circle $x^2 + y^2 = a^2$ in the xy-plane. This circle is a **generating curve** for the cylinder, as indicated in the next definition.

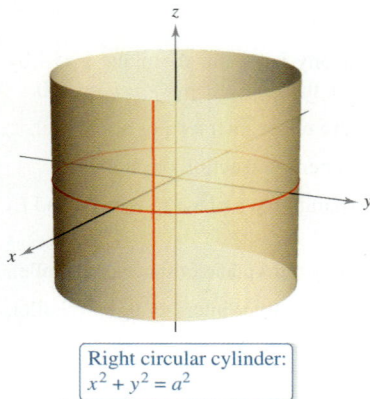

Right circular cylinder:
$x^2 + y^2 = a^2$

Rulings are parallel to z-axis
Figure 11.56

Definition of a Cylinder

Let C be a curve in a plane and let L be a line not in a parallel plane. The set of all lines parallel to L and intersecting C is a **cylinder.** The curve C is the **generating curve** (or **directrix**) of the cylinder, and the parallel lines are **rulings.**

Without loss of generality, you can assume that C lies in one of the three coordinate planes. Moreover, this text restricts the discussion to *right* cylinders— cylinders whose rulings are perpendicular to the coordinate plane containing C, as shown in Figure 11.57. Note that the rulings intersect C and are parallel to the line L.

For the right circular cylinder shown in Figure 11.56, the equation of the generating curve in the xy-plane is

$$x^2 + y^2 = a^2.$$

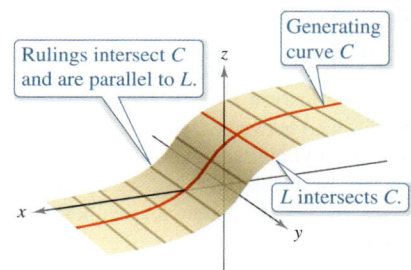

Right cylinder: A cylinder whose rulings are perpendicular to the coordinate plane containing C
Figure 11.57

To find an equation of the cylinder, note that you can generate any one of the rulings by fixing the values of x and y and then allowing z to take on all real values. In this sense, the value of z is arbitrary and is, therefore, not included in the equation. In other words, the equation of this cylinder is simply the equation of its generating curve.

$$x^2 + y^2 = a^2$$ Equation of cylinder in space

Equations of Cylinders

The equation of a cylinder whose rulings are parallel to one of the coordinate axes contains only the variables corresponding to the other two axes.

| EXAMPLE 1 | **Sketching a Cylinder** |

Sketch the surface represented by each equation.

a. $z = y^2$ **b.** $z = \sin x, \quad 0 \le x \le 2\pi$

Solution

a. The graph is a cylinder whose generating curve, $z = y^2$, is a parabola in the yz-plane. The rulings of the cylinder are parallel to the x-axis, as shown in Figure 11.58(a).

b. The graph is a cylinder generated by the sine curve in the xz-plane. The rulings are parallel to the y-axis, as shown in Figure 11.58(b).

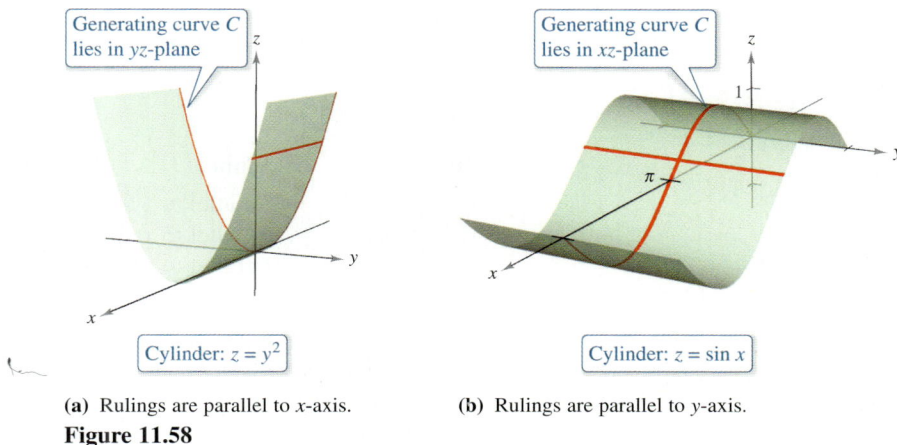

Generating curve C lies in yz-plane

Cylinder: $z = y^2$

(a) Rulings are parallel to x-axis.

Generating curve C lies in xz-plane

Cylinder: $z = \sin x$

(b) Rulings are parallel to y-axis.

Figure 11.58

Quadric Surfaces

The fourth basic type of surface in space is a **quadric surface.** Quadric surfaces are the three-dimensional analogs of conic sections.

Quadric Surface

The equation of a **quadric surface** in space is a second-degree equation in three variables. The **general form** of the equation is

$$Ax^2 + By^2 + Cz^2 + Dxy + Exz + Fyz + Gx + Hy + Iz + J = 0.$$

There are six basic types of quadric surfaces: **ellipsoid, hyperboloid of one sheet, hyperboloid of two sheets, elliptic cone, elliptic paraboloid, and hyperbolic paraboloid.**

The intersection of a surface with a plane is called the **trace of the surface** in the plane. To visualize a surface in space, it is helpful to determine its traces in some well-chosen planes. The traces of quadric surfaces are conics. These traces, together with the **standard form** of the equation of each quadric surface, are shown in the table on the next two pages.

In the table on the next two pages, only one of several orientations of each quadric surface is shown. When the surface is oriented along a different axis, its standard equation will change accordingly, as illustrated in Examples 2 and 3. The fact that the two types of paraboloids have one variable raised to the first power can be helpful in classifying quadric surfaces. The other four types of basic quadric surfaces have equations that are of *second degree* in all three variables.

Ellipsoid

$$\frac{x^2}{a^2} + \frac{y^2}{b^2} + \frac{z^2}{c^2} = 1$$

Trace	Plane
Ellipse	Parallel to xy-plane
Ellipse	Parallel to xz-plane
Ellipse	Parallel to yz-plane

The surface is a sphere when $a = b = c \neq 0$.

Hyperboloid of One Sheet

$$\frac{x^2}{a^2} + \frac{y^2}{b^2} - \frac{z^2}{c^2} = 1$$

Trace	Plane
Ellipse	Parallel to xy-plane
Hyperbola	Parallel to xz-plane
Hyperbola	Parallel to yz-plane

The axis of the hyperboloid corresponds to the variable whose coefficient is negative.

Hyperboloid of Two Sheets

$$\frac{z^2}{c^2} - \frac{x^2}{a^2} - \frac{y^2}{b^2} = 1$$

Trace	Plane
Ellipse	Parallel to xy-plane
Hyperbola	Parallel to xz-plane
Hyperbola	Parallel to yz-plane

The axis of the hyperboloid corresponds to the variable whose coefficient is positive. There is no trace in the coordinate plane perpendicular to this axis.

Elliptic Cone

$$\frac{x^2}{a^2} + \frac{y^2}{b^2} - \frac{z^2}{c^2} = 0$$

Trace	Plane
Ellipse	Parallel to xy-plane
Hyperbola	Parallel to xz-plane
Hyperbola	Parallel to yz-plane

The axis of the cone corresponds to the variable whose coefficient is negative. The traces in the coordinate planes parallel to this axis are intersecting lines.

Elliptic Paraboloid

$$z = \frac{x^2}{a^2} + \frac{y^2}{b^2}$$

Trace	Plane
Ellipse	Parallel to xy-plane
Parabola	Parallel to xz-plane
Parabola	Parallel to yz-plane

The axis of the paraboloid corresponds to the variable raised to the first power.

Hyperbolic Paraboloid

$$z = \frac{y^2}{b^2} - \frac{x^2}{a^2}$$

Trace	Plane
Hyperbola	Parallel to xy-plane
Parabola	Parallel to xz-plane
Parabola	Parallel to yz-plane

The axis of the paraboloid corresponds to the variable raised to the first power.

To classify a quadric surface, begin by writing the equation of the surface in standard form. Then, determine several traces taken in the coordinate planes *or* taken in planes that are parallel to the coordinate planes.

EXAMPLE 2 Sketching a Quadric Surface

Classify and sketch the surface

$$4x^2 - 3y^2 + 12z^2 + 12 = 0.$$

Solution Begin by writing the equation in standard form.

$4x^2 - 3y^2 + 12z^2 + 12 = 0$	Write original equation.
$\dfrac{x^2}{-3} + \dfrac{y^2}{4} - z^2 - 1 = 0$	Divide by -12.
$\dfrac{y^2}{4} - \dfrac{x^2}{3} - \dfrac{z^2}{1} = 1$	Standard form

From the table on pages 796 and 797, you can conclude that the surface is a hyperboloid of two sheets with the y-axis as its axis. To sketch the graph of this surface, it helps to find the traces in the coordinate planes.

xy-trace $(z = 0)$:	$\dfrac{y^2}{4} - \dfrac{x^2}{3} = 1$	Hyperbola
xz-trace $(y = 0)$:	$\dfrac{x^2}{3} + \dfrac{z^2}{1} = -1$	No trace
yz-trace $(x = 0)$:	$\dfrac{y^2}{4} - \dfrac{z^2}{1} = 1$	Hyperbola

The graph is shown in Figure 11.59.

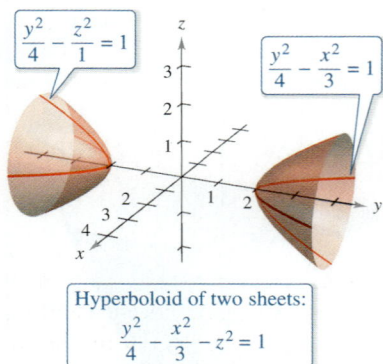

$\dfrac{y^2}{4} - \dfrac{z^2}{1} = 1$

$\dfrac{y^2}{4} - \dfrac{x^2}{3} = 1$

Hyperboloid of two sheets:
$\dfrac{y^2}{4} - \dfrac{x^2}{3} - z^2 = 1$

Figure 11.59

EXAMPLE 3 Sketching a Quadric Surface

Classify and sketch the surface

$$x - y^2 - 4z^2 = 0.$$

Solution Because x is raised only to the first power, the surface is a paraboloid. The axis of the paraboloid is the x-axis. In standard form, the equation is

$$x = y^2 + 4z^2.\qquad \text{Standard form}$$

Some convenient traces are listed below.

xy-trace $(z = 0)$:	$x = y^2$	Parabola
xz-trace $(y = 0)$:	$x = 4z^2$	Parabola
parallel to yz-plane $(x = 4)$:	$\dfrac{y^2}{4} + \dfrac{z^2}{1} = 1$	Ellipse

The surface is an *elliptic* paraboloid, as shown in Figure 11.60.

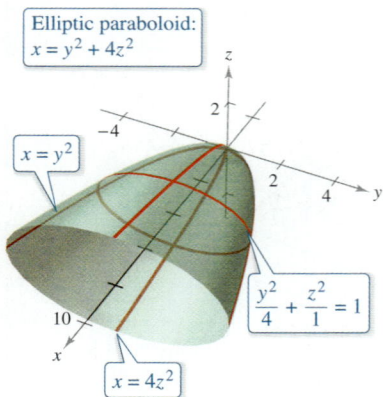

Elliptic paraboloid:
$x = y^2 + 4z^2$

$x = y^2$

$\dfrac{y^2}{4} + \dfrac{z^2}{1} = 1$

$x = 4z^2$

Figure 11.60

Some second-degree equations in x, y, and z do not represent any of the basic types of quadric surfaces. For example, the graph of

$$x^2 + y^2 + z^2 = 0 \qquad \text{Single point}$$

is a single point, and the graph of

$$x^2 + y^2 = 1 \qquad \text{Right circular cylinder}$$

is a right circular cylinder.

For a quadric surface not centered at the origin, you can form the standard equation by completing the square, as demonstrated in Example 4.

EXAMPLE 4   A Quadric Surface Not Centered at the Origin

•••▷ *See LarsonCalculus.com for an interactive version of this type of example.*

Classify and sketch the surface

$$x^2 + 2y^2 + z^2 - 4x + 4y - 2z + 3 = 0.$$

Solution   Begin by grouping terms and factoring where possible.

$$x^2 - 4x + 2(y^2 + 2y) + z^2 - 2z = -3$$

Next, complete the square for each variable and write the equation in standard form.

$$(x^2 - 4x + \ \ ) + 2(y^2 + 2y + \ \ ) + (z^2 - 2z + \ \ ) = -3$$
$$(x^2 - 4x + 4) + 2(y^2 + 2y + 1) + (z^2 - 2z + 1) = -3 + 4 + 2 + 1$$
$$(x - 2)^2 + 2(y + 1)^2 + (z - 1)^2 = 4$$
$$\frac{(x - 2)^2}{4} + \frac{(y + 1)^2}{2} + \frac{(z - 1)^2}{4} = 1$$

From this equation, you can see that the quadric surface is an ellipsoid that is centered at $(2, -1, 1)$. Its graph is shown in Figure 11.61.

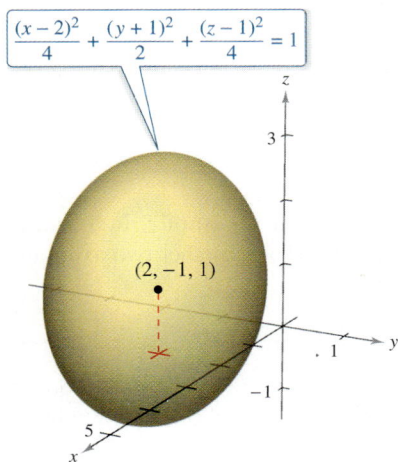

$$\frac{(x - 2)^2}{4} + \frac{(y + 1)^2}{2} + \frac{(z - 1)^2}{4} = 1$$

$(2, -1, 1)$

An ellipsoid centered at $(2, -1, 1)$
Figure 11.61

▷ **TECHNOLOGY**   A 3-D graphing utility can help you visualize a surface in space.* Such a graphing utility may create a three-dimensional graph by sketching several traces of the surface and then applying a "hidden-line" routine that blocks out portions of the surface that lie behind other portions of the surface. Two examples of figures that were generated by *Mathematica* are shown below.

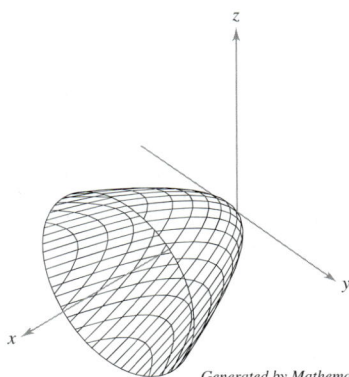

Generated by Mathematica

Elliptic paraboloid
$$x = \frac{y^2}{2} + \frac{z^2}{2}$$

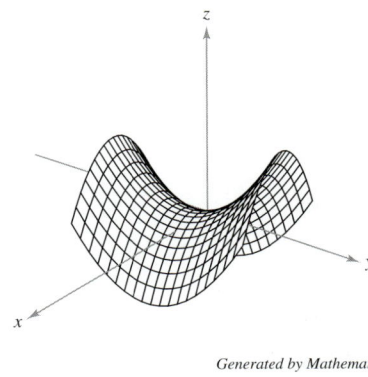

Generated by Mathematica

Hyperbolic paraboloid
$$z = \frac{y^2}{16} - \frac{x^2}{16}$$

Using a graphing utility to graph a surface in space requires practice. For one thing, you must know enough about the surface to be able to specify a *viewing window* that gives a representative view of the surface. Also, you can often improve the view of a surface by rotating the axes. For instance, note that the elliptic paraboloid in the figure is seen from a line of sight that is "higher" than the line of sight used to view the hyperbolic paraboloid.

---

* Some 3-D graphing utilities require surfaces to be entered with parametric equations. For a discussion of this technique, see Section 15.5.

Surfaces of Revolution

Circular cross section

z

Generating curve $y = r(z)$

$(0, 0, z)$

$(0, r(z), z)$

(x, y, z)

$r(z)$

y

x

Figure 11.62

The fifth special type of surface you will study is a **surface of revolution.** In Section 7.4, you studied a method for finding the *area* of such a surface. You will now look at a procedure for finding its *equation.* Consider the graph of the **radius function**

$$y = r(z) \qquad \text{\color{red}Generating curve}$$

in the *yz*-plane. When this graph is revolved about the *z*-axis, it forms a surface of revolution, as shown in Figure 11.62. The trace of the surface in the plane $z = z_0$ is a circle whose radius is $r(z_0)$ and whose equation is

$$x^2 + y^2 = [r(z_0)]^2. \qquad \text{\color{red}Circular trace in plane: } z = z_0$$

Replacing z_0 with z produces an equation that is valid for all values of z. In a similar manner, you can obtain equations for surfaces of revolution for the other two axes, and the results are summarized as follows.

Surface of Revolution

If the graph of a radius function r is revolved about one of the coordinate axes, then the equation of the resulting surface of revolution has one of the forms listed below.

1. Revolved about the *x*-axis: $y^2 + z^2 = [r(x)]^2$
2. Revolved about the *y*-axis: $x^2 + z^2 = [r(y)]^2$
3. Revolved about the *z*-axis: $x^2 + y^2 = [r(z)]^2$

EXAMPLE 5 **Finding an Equation for a Surface of Revolution**

Find an equation for the surface of revolution formed by revolving (a) the graph of $y = 1/z$ about the *z*-axis and (b) the graph of $9x^2 = y^3$ about the *y*-axis.

Solution

a. An equation for the surface of revolution formed by revolving the graph of

$$y = \frac{1}{z} \qquad \text{\color{red}Radius function}$$

about the *z*-axis is

$$x^2 + y^2 = [r(z)]^2 \qquad \text{\color{red}Revolved about the } z\text{-axis}$$

$$x^2 + y^2 = \left(\frac{1}{z}\right)^2. \qquad \text{\color{red}Substitute } 1/z \text{ for } r(z).$$

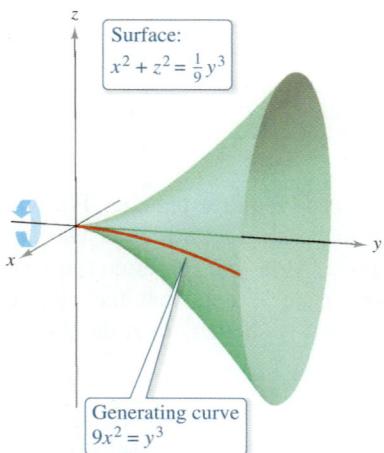

z

Surface: $x^2 + z^2 = \frac{1}{9}y^3$

y

x

Generating curve $9x^2 = y^3$

Figure 11.63

b. To find an equation for the surface formed by revolving the graph of $9x^2 = y^3$ about the *y*-axis, solve for x in terms of y to obtain

$$x = \frac{1}{3}y^{3/2} = r(y). \qquad \text{\color{red}Radius function}$$

So, the equation for this surface is

$$x^2 + z^2 = [r(y)]^2 \qquad \text{\color{red}Revolved about the } y\text{-axis}$$

$$x^2 + z^2 = \left(\frac{1}{3}y^{3/2}\right)^2 \qquad \text{\color{red}Substitute } \tfrac{1}{3}y^{3/2} \text{ for } r(y).$$

$$x^2 + z^2 = \frac{1}{9}y^3. \qquad \text{\color{red}Equation of surface}$$

The graph is shown in Figure 11.63.

The generating curve for a surface of revolution is not unique. For instance, the surface

$$x^2 + z^2 = e^{-2y}$$

can be formed by revolving either the graph of

$$x = e^{-y}$$

about the y-axis or the graph of

$$z = e^{-y}$$

about the y-axis, as shown in Figure 11.64.

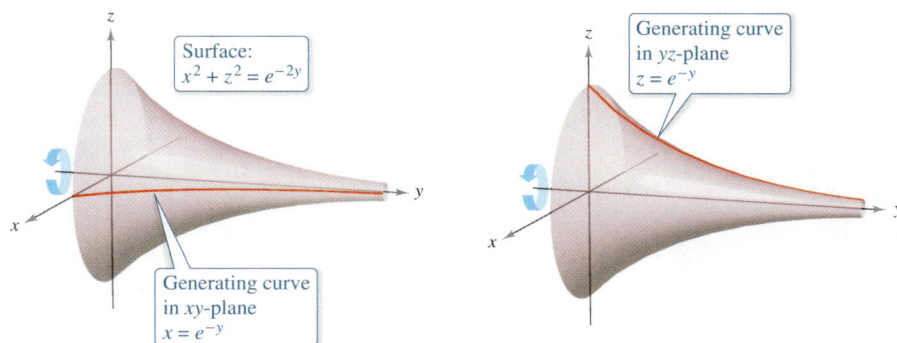

Figure 11.64

EXAMPLE 6 Finding a Generating Curve

Find a generating curve and the axis of revolution for the surface

$$x^2 + 3y^2 + z^2 = 9.$$

Solution The equation has one of the forms listed below.

$$x^2 + y^2 = [r(z)]^2 \qquad \text{Revolved about } z\text{-axis}$$
$$y^2 + z^2 = [r(x)]^2 \qquad \text{Revolved about } x\text{-axis}$$
$$x^2 + z^2 = [r(y)]^2 \qquad \text{Revolved about } y\text{-axis}$$

Because the coefficients of x^2 and z^2 are equal, you should choose the third form and write

$$x^2 + z^2 = 9 - 3y^2.$$

The y-axis is the axis of revolution. You can choose a generating curve from either of the traces

$$x^2 = 9 - 3y^2 \qquad \text{Trace in } xy\text{-plane}$$

or

$$z^2 = 9 - 3y^2. \qquad \text{Trace in } yz\text{-plane}$$

For instance, using the first trace, the generating curve is the semiellipse

$$x = \sqrt{9 - 3y^2}. \qquad \text{Generating curve}$$

The graph of this surface is shown in Figure 11.65.

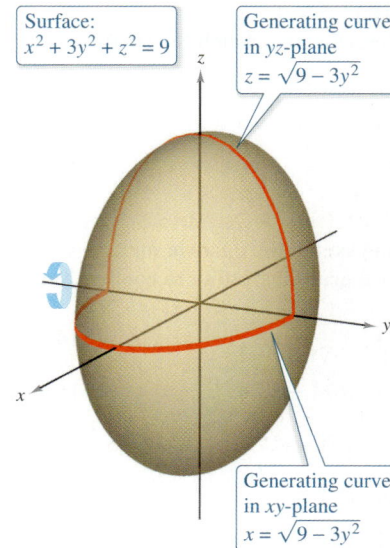

Figure 11.65

11.6 Exercises

See **CalcChat.com** for tutorial help and worked-out solutions to odd-numbered exercises.

Matching In Exercises 1–6, match the equation with its graph. [The graphs are labeled (a), (b), (c), (d), (e), and (f).]

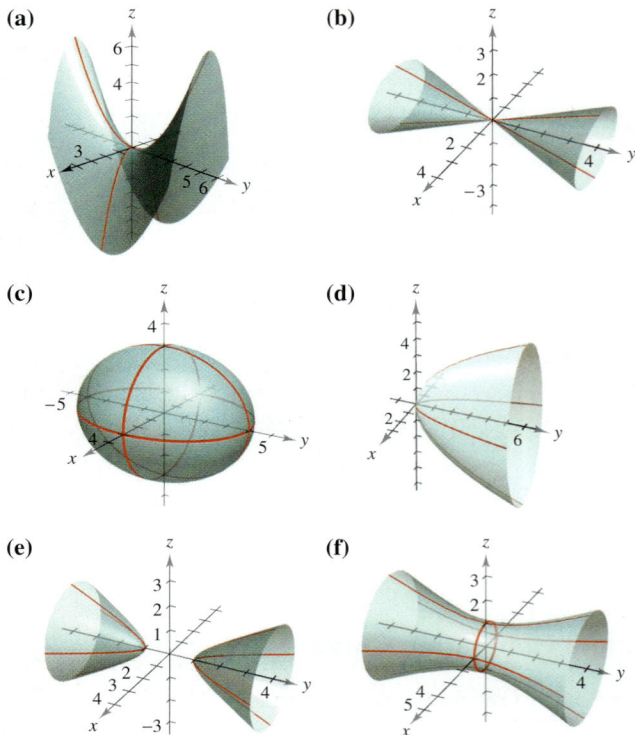

(a)

(b)

(c)

(d)

(e)

(f)

1. $\dfrac{x^2}{9} + \dfrac{y^2}{16} + \dfrac{z^2}{9} = 1$

2. $15x^2 - 4y^2 + 15z^2 = -4$

3. $4x^2 - y^2 + 4z^2 = 4$

4. $y^2 = 4x^2 + 9z^2$

5. $4x^2 - 4y + z^2 = 0$

6. $4x^2 - y^2 + 4z = 0$

Sketching a Surface in Space In Exercises 7–12, describe and sketch the surface.

7. $y = 5$

8. $z = 2$

9. $y^2 + z^2 = 9$

10. $y^2 + z = 6$

11. $4x^2 + y^2 = 4$

12. $y^2 - z^2 = 16$

Sketching a Quadric Surface In Exercises 13–24, classify and sketch the quadric surface. Use a computer algebra system or a graphing utility to confirm your sketch.

13. $x^2 + \dfrac{y^2}{4} + z^2 = 1$

14. $\dfrac{x^2}{16} + \dfrac{y^2}{25} + \dfrac{z^2}{25} = 1$

15. $16x^2 - y^2 + 16z^2 = 4$

16. $-8x^2 + 18y^2 + 18z^2 = 2$

17. $4x^2 - y^2 - z^2 = 1$

18. $z^2 - x^2 - \dfrac{y^2}{4} = 1$

19. $x^2 - y + z^2 = 0$

20. $z = x^2 + 4y^2$

21. $x^2 - y^2 + z = 0$

22. $3z = -y^2 + x^2$

23. $z^2 = x^2 + \dfrac{y^2}{9}$

24. $x^2 = 2y^2 + 2z^2$

WRITING ABOUT CONCEPTS

25. Cylinder State the definition of a cylinder.

26. Trace of a Surface What is meant by the trace of a surface? How do you find a trace?

27. Quadric Surfaces Identify the six quadric surfaces and give the standard form of each.

28. Classifying an Equation What does the equation $z = x^2$ represent in the xz-plane? What does it represent in three-space?

29. Classifying an Equation What does the equation $4x^2 + 6y^2 - 3z^2 = 12$ represent in the xy-plane? What does it represent in three-space?

30. **HOW DO YOU SEE IT?** The four figures are graphs of the quadric surface $z = x^2 + y^2$. Match each of the four graphs with the point in space from which the paraboloid is viewed. The four points are $(0, 0, 20)$, $(0, 20, 0)$, $(20, 0, 0)$, and $(10, 10, 20)$.

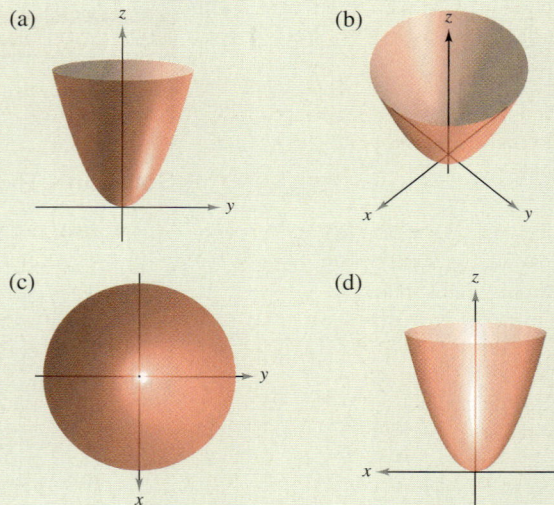

(a)

(b)

(c)

(d)

Finding an Equation of a Surface of Revolution In Exercises 31–36, find an equation for the surface of revolution formed by revolving the curve in the indicated coordinate plane about the given axis.

Equation of Curve	Coordinate Plane	Axis of Revolution
31. $z^2 = 4y$	yz-plane	y-axis
32. $z = 3y$	yz-plane	y-axis
33. $z = 2y$	yz-plane	z-axis

Equation of Curve	Coordinate Plane	Axis of Revolution
34. $2z = \sqrt{4 - x^2}$	xz-plane	x-axis
35. $xy = 2$	xy-plane	x-axis
36. $z = \ln y$	yz-plane	z-axis

Finding a Generating Curve   In Exercises 37 and 38, find an equation of a generating curve given the equation of its surface of revolution.

37. $x^2 + y^2 - 2z = 0$ **38.** $x^2 + z^2 = \cos^2 y$

Finding the Volume of a Solid   In Exercises 39 and 40, use the shell method to find the volume of the solid below the surface of revolution and above the xy-plane.

39. The curve $z = 4x - x^2$ in the xz-plane is revolved about the z-axis.

40. The curve $z = \sin y$ $(0 \leq y \leq \pi)$ in the yz-plane is revolved about the z-axis.

Analyzing a Trace   In Exercises 41 and 42, analyze the trace when the surface

$$z = \tfrac{1}{2}x^2 + \tfrac{1}{4}y^2$$

is intersected by the indicated planes.

41. Find the lengths of the major and minor axes and the coordinates of the foci of the ellipse generated when the surface is intersected by the planes given by

   (a) $z = 2$ and   (b) $z = 8$.

42. Find the coordinates of the focus of the parabola formed when the surface is intersected by the planes given by

   (a) $y = 4$ and   (b) $x = 2$.

Finding an Equation of a Surface   In Exercises 43 and 44, find an equation of the surface satisfying the conditions, and identify the surface.

43. The set of all points equidistant from the point $(0, 2, 0)$ and the plane $y = -2$

44. The set of all points equidistant from the point $(0, 0, 4)$ and the xy-plane

45. Geography

Because of the forces caused by its rotation, Earth is an oblate ellipsoid rather than a sphere. The equatorial radius is 3963 miles and the polar radius is 3950 miles. Find an equation of the ellipsoid. (Assume that the center of Earth is at the origin and that the trace formed by the plane $z = 0$ corresponds to the equator.)

46. Machine Design   The top of a rubber bushing designed to absorb vibrations in an automobile is the surface of revolution generated by revolving the curve

$$z = \tfrac{1}{2}y^2 + 1$$

for $0 \leq y \leq 2$ in the yz-plane about the z-axis.

   (a) Find an equation for the surface of revolution.

   (b) All measurements are in centimeters and the bushing is set on the xy-plane. Use the shell method to find its volume.

   (c) The bushing has a hole of diameter 1 centimeter through its center and parallel to the axis of revolution. Find the volume of the rubber bushing.

47. Using a Hyperbolic Paraboloid   Determine the intersection of the hyperbolic paraboloid

$$z = \frac{y^2}{b^2} - \frac{x^2}{a^2}$$

with the plane $bx + ay - z = 0$. (Assume $a, b > 0$.)

48. Intersection of Surfaces   Explain why the curve of intersection of the surfaces

$$x^2 + 3y^2 - 2z^2 + 2y = 4$$

and

$$2x^2 + 6y^2 - 4z^2 - 3x = 2$$

lies in a plane.

True or False?   In Exercises 49–52, determine whether the statement is true or false. If it is false, explain why or give an example that shows it is false.

49. A sphere is an ellipsoid.

50. The generating curve for a surface of revolution is unique.

51. All traces of an ellipsoid are ellipses.

52. All traces of a hyperboloid of one sheet are hyperboloids.

53. Think About It   Three types of classic "topological" surfaces are shown below. The sphere and torus have both an "inside" and an "outside." Does the Klein bottle have both an inside and an outside? Explain.

Sphere   Torus

Klein bottle   Klein bottle

11.7 Cylindrical and Spherical Coordinates

■ Use cylindrical coordinates to represent surfaces in space.
■ Use spherical coordinates to represent surfaces in space.

Cylindrical Coordinates

You have already seen that some two-dimensional graphs are easier to represent in polar coordinates than in rectangular coordinates. A similar situation exists for surfaces in space. In this section, you will study two alternative space-coordinate systems. The first, the **cylindrical coordinate system,** is an extension of polar coordinates in the plane to three-dimensional space.

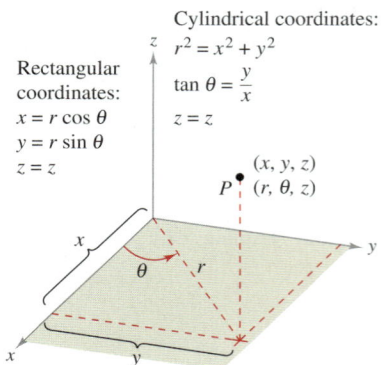

Cylindrical coordinates:
$r^2 = x^2 + y^2$
$\tan \theta = \dfrac{y}{x}$
$z = z$

Rectangular coordinates:
$x = r \cos \theta$
$y = r \sin \theta$
$z = z$

Figure 11.66

The Cylindrical Coordinate System

In a **cylindrical coordinate system,** a point P in space is represented by an ordered triple (r, θ, z).

1. (r, θ) is a polar representation of the projection of P in the xy-plane.
2. z is the directed distance from (r, θ) to P.

To convert from rectangular to cylindrical coordinates (or vice versa), use the conversion guidelines for polar coordinates listed below and illustrated in Figure 11.66.

Cylindrical to rectangular:

$$x = r \cos \theta, \qquad y = r \sin \theta, \qquad z = z$$

Rectangular to cylindrical:

$$r^2 = x^2 + y^2, \qquad \tan \theta = \frac{y}{x}, \qquad z = z$$

The point $(0, 0, 0)$ is called the **pole.** Moreover, because the representation of a point in the polar coordinate system is not unique, it follows that the representation in the cylindrical coordinate system is also not unique.

$(x, y, z) = (-2\sqrt{3}, 2, 3)$

$(r, \theta, z) = \left(4, \dfrac{5\pi}{6}, 3\right)$

Figure 11.67

EXAMPLE 1 **Cylindrical-to-Rectangular Conversion**

Convert the point $(r, \theta, z) = (4, 5\pi/6, 3)$ to rectangular coordinates.

Solution Using the cylindrical-to-rectangular conversion equations produces

$$x = 4 \cos \frac{5\pi}{6} = 4\left(-\frac{\sqrt{3}}{2}\right) = -2\sqrt{3}$$

$$y = 4 \sin \frac{5\pi}{6} = 4\left(\frac{1}{2}\right) = 2$$

$$z = 3.$$

So, in rectangular coordinates, the point is $(x, y, z) = \left(-2\sqrt{3}, 2, 3\right)$, as shown in Figure 11.67.

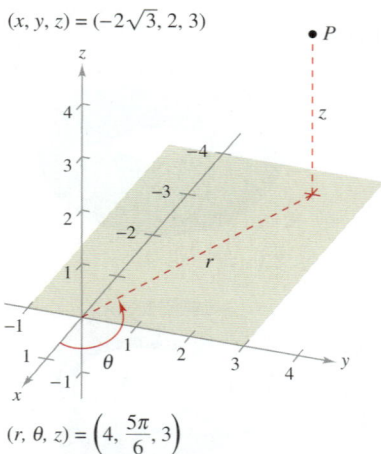

EXAMPLE 2   **Rectangular-to-Cylindrical Conversion**

Convert the point

$$(x, y, z) = \left(1, \sqrt{3}, 2\right)$$

to cylindrical coordinates.

Solution   Use the rectangular-to-cylindrical conversion equations.

$$r = \pm\sqrt{1 + 3} = \pm 2$$

$$\tan \theta = \sqrt{3} \quad \Longrightarrow \quad \theta = \arctan\left(\sqrt{3}\right) + n\pi = \frac{\pi}{3} + n\pi$$

$$z = 2$$

You have two choices for r and infinitely many choices for θ. As shown in Figure 11.68, two convenient representations of the point are

$$\left(2, \frac{\pi}{3}, 2\right) \qquad \text{\textit{\color{red}{r > 0 and θ in Quadrant I}}}$$

and

$$\left(-2, \frac{4\pi}{3}, 2\right). \qquad \text{\textit{\color{red}{r < 0 and θ in Quadrant III}}}$$

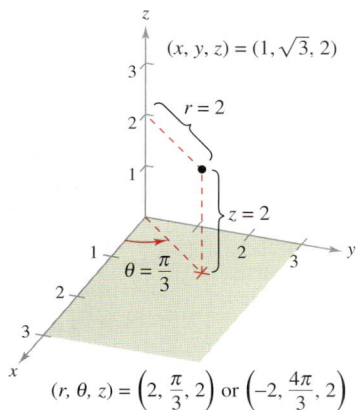

$(x, y, z) = (1, \sqrt{3}, 2)$

$r = 2$

$z = 2$

$\theta = \dfrac{\pi}{3}$

$(r, \theta, z) = \left(2, \dfrac{\pi}{3}, 2\right)$ or $\left(-2, \dfrac{4\pi}{3}, 2\right)$

Figure 11.68

Cylindrical coordinates are especially convenient for representing cylindrical surfaces and surfaces of revolution with the z-axis as the axis of symmetry, as shown in Figure 11.69.

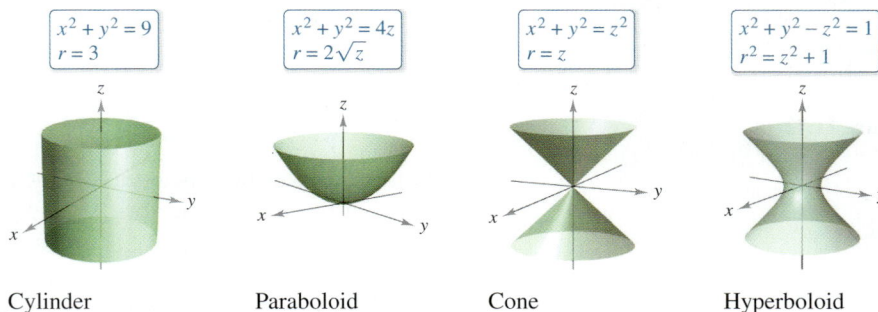

$x^2 + y^2 = 9$
$r = 3$

$x^2 + y^2 = 4z$
$r = 2\sqrt{z}$

$x^2 + y^2 = z^2$
$r = z$

$x^2 + y^2 - z^2 = 1$
$r^2 = z^2 + 1$

Cylinder   Paraboloid   Cone   Hyperboloid
Figure 11.69

Vertical planes containing the z-axis and horizontal planes also have simple cylindrical coordinate equations, as shown in Figure 11.70.

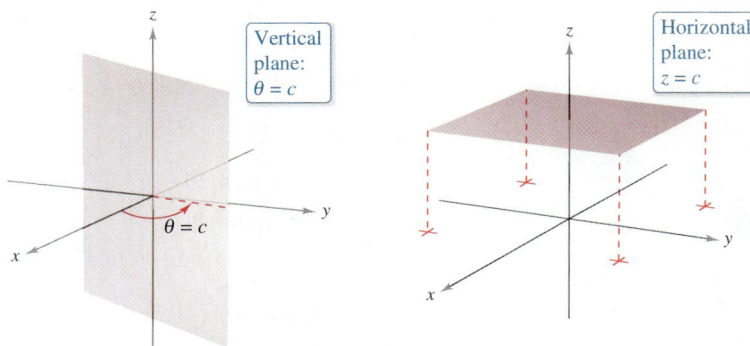

Vertical plane:
$\theta = c$

$\theta = c$

Horizontal plane:
$z = c$

Figure 11.70

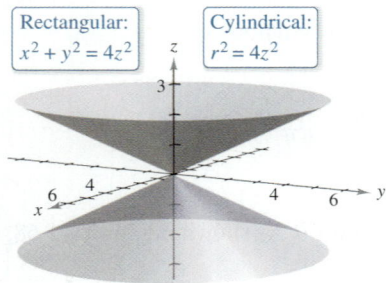

Rectangular:
$x^2 + y^2 = 4z^2$

Cylindrical:
$r^2 = 4z^2$

Figure 11.71

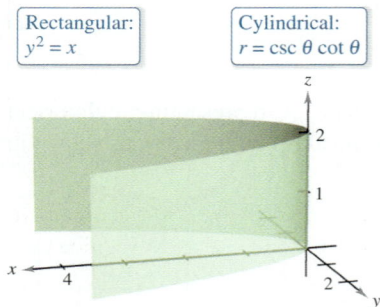

Rectangular:
$y^2 = x$

Cylindrical:
$r = \csc \theta \cot \theta$

Figure 11.72

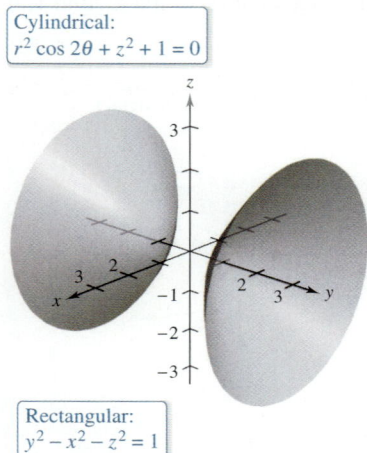

Cylindrical:
$r^2 \cos 2\theta + z^2 + 1 = 0$

Rectangular:
$y^2 - x^2 - z^2 = 1$

Figure 11.73

EXAMPLE 3   **Rectangular-to-Cylindrical Conversion**

Find an equation in cylindrical coordinates for the surface represented by each rectangular equation.

a. $x^2 + y^2 = 4z^2$

b. $y^2 = x$

Solution

a. From Section 11.6, you know that the graph of

$$x^2 + y^2 = 4z^2$$

is an elliptic cone with its axis along the z-axis, as shown in Figure 11.71. When you replace $x^2 + y^2$ with r^2, the equation in cylindrical coordinates is

$$x^2 + y^2 = 4z^2 \qquad \text{Rectangular equation}$$
$$r^2 = 4z^2. \qquad \text{Cylindrical equation}$$

b. The graph of the surface

$$y^2 = x$$

is a parabolic cylinder with rulings parallel to the z-axis, as shown in Figure 11.72. To obtain the equation in cylindrical coordinates, replace y^2 with $r^2 \sin^2 \theta$ and x with $r \cos \theta$, as shown.

$$y^2 = x \qquad \text{Rectangular equation}$$
$$r^2 \sin^2 \theta = r \cos \theta \qquad \text{Substitute } r \sin \theta \text{ for } y \text{ and } r \cos \theta \text{ for } x.$$
$$r(r \sin^2 \theta - \cos \theta) = 0 \qquad \text{Collect terms and factor.}$$
$$r \sin^2 \theta - \cos \theta = 0 \qquad \text{Divide each side by } r.$$
$$r = \frac{\cos \theta}{\sin^2 \theta} \qquad \text{Solve for } r.$$
$$r = \csc \theta \cot \theta \qquad \text{Cylindrical equation}$$

Note that this equation includes a point for which $r = 0$, so nothing was lost by dividing each side by the factor r. ∎

Converting from cylindrical coordinates to rectangular coordinates is less straightforward than converting from rectangular coordinates to cylindrical coordinates, as demonstrated in Example 4.

EXAMPLE 4   **Cylindrical-to-Rectangular Conversion**

Find an equation in rectangular coordinates for the surface represented by the cylindrical equation

$$r^2 \cos 2\theta + z^2 + 1 = 0.$$

Solution

$$r^2 \cos 2\theta + z^2 + 1 = 0 \qquad \text{Cylindrical equation}$$
$$r^2(\cos^2 \theta - \sin^2 \theta) + z^2 + 1 = 0 \qquad \text{Trigonometric identity}$$
$$r^2 \cos^2 \theta - r^2 \sin^2 \theta + z^2 = -1$$
$$x^2 - y^2 + z^2 = -1 \qquad \text{Replace } r \cos \theta \text{ with } x \text{ and } r \sin \theta \text{ with } y.$$
$$y^2 - x^2 - z^2 = 1 \qquad \text{Rectangular equation}$$

This is a hyperboloid of two sheets whose axis lies along the y-axis, as shown in Figure 11.73. ∎

Spherical Coordinates

In the **spherical coordinate system,** each point is represented by an ordered triple: the first coordinate is a distance, and the second and third coordinates are angles. This system is similar to the latitude-longitude system used to identify points on the surface of Earth. For example, the point on the surface of Earth whose latitude is 40° North (of the equator) and whose longitude is 80° West (of the prime meridian) is shown in Figure 11.74. Assuming that Earth is spherical and has a radius of 4000 miles, you would label this point as

$$(4000, -80°, 50°).$$

Radius   80° clockwise from prime meridian   50° down from North Pole

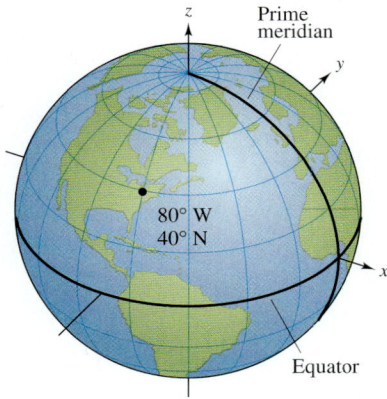

Figure 11.74

The Spherical Coordinate System

In a **spherical coordinate system,** a point P in space is represented by an ordered triple (ρ, θ, ϕ), where ρ is the lowercase Greek letter *rho* and ϕ is the lowercase Greek letter *phi.*

1. ρ is the distance between P and the origin, $\rho \geq 0$.
2. θ is the same angle used in cylindrical coordinates for $r \geq 0$.
3. ϕ is the angle *between* the positive z-axis and the line segment $\overrightarrow{OP}$, $0 \leq \phi \leq \pi$.

Note that the first and third coordinates, ρ and ϕ, are nonnegative.

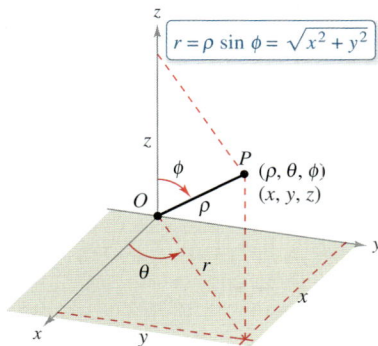

Spherical coordinates
Figure 11.75

The relationship between rectangular and spherical coordinates is illustrated in Figure 11.75. To convert from one system to the other, use the conversion guidelines listed below.

Spherical to rectangular:

$$x = \rho \sin \phi \cos \theta, \quad y = \rho \sin \phi \sin \theta, \quad z = \rho \cos \phi$$

Rectangular to spherical:

$$\rho^2 = x^2 + y^2 + z^2, \quad \tan \theta = \frac{y}{x}, \quad \phi = \arccos\left(\frac{z}{\sqrt{x^2 + y^2 + z^2}}\right)$$

To change coordinates between the cylindrical and spherical systems, use the conversion guidelines listed below.

Spherical to cylindrical $(r \geq 0)$:

$$r^2 = \rho^2 \sin^2 \phi, \quad \theta = \theta, \quad z = \rho \cos \phi$$

Cylindrical to spherical $(r \geq 0)$:

$$\rho = \sqrt{r^2 + z^2}, \quad \theta = \theta, \quad \phi = \arccos\left(\frac{z}{\sqrt{r^2 + z^2}}\right)$$

The spherical coordinate system is useful primarily for surfaces in space that have a *point* or *center* of symmetry. For example, Figure 11.76 shows three surfaces with simple spherical equations.

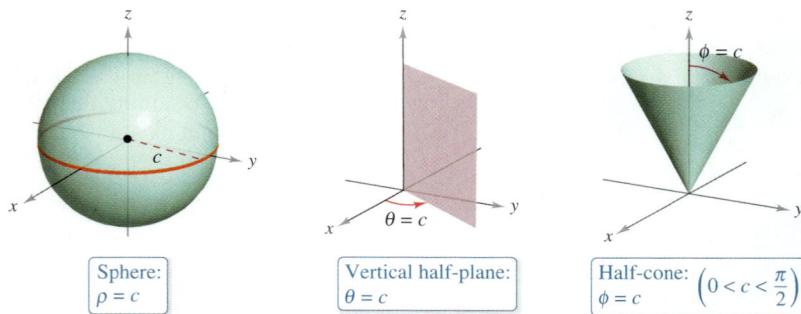

Sphere:
$\rho = c$

Vertical half-plane:
$\theta = c$

Half-cone:
$\phi = c$ $\left(0 < c < \dfrac{\pi}{2}\right)$

Figure 11.76

EXAMPLE 5   Rectangular-to-Spherical Conversion

⋯▷ *See LarsonCalculus.com for an interactive version of this type of example.*

Find an equation in spherical coordinates for the surface represented by each rectangular equation.

a. Cone: $x^2 + y^2 = z^2$ **b.** Sphere: $x^2 + y^2 + z^2 - 4z = 0$

Solution

a. Use the spherical-to-rectangular equations

$$x = \rho \sin \phi \cos \theta, \quad y = \rho \sin \phi \sin \theta, \quad \text{and} \quad z = \rho \cos \phi$$

and substitute in the rectangular equation as shown.

$$x^2 + y^2 = z^2$$
$$\rho^2 \sin^2 \phi \cos^2 \theta + \rho^2 \sin^2 \phi \sin^2 \theta = \rho^2 \cos^2 \phi$$
$$\rho^2 \sin^2 \phi (\cos^2 \theta + \sin^2 \theta) = \rho^2 \cos^2 \phi$$
$$\rho^2 \sin^2 \phi = \rho^2 \cos^2 \phi$$
$$\frac{\sin^2 \phi}{\cos^2 \phi} = 1 \qquad \rho \geq 0$$
$$\tan^2 \phi = 1$$
$$\tan \phi = \pm 1$$

So, you can conclude that

$$\phi = \frac{\pi}{4} \quad \text{or} \quad \phi = \frac{3\pi}{4}.$$

The equation $\phi = \pi/4$ represents the *upper* half-cone, and the equation $\phi = 3\pi/4$ represents the *lower* half-cone.

b. Because $\rho^2 = x^2 + y^2 + z^2$ and $z = \rho \cos \phi$, the rectangular equation has the following spherical form.

$$\rho^2 - 4\rho \cos \phi = 0 \implies \rho(\rho - 4 \cos \phi) = 0$$

Temporarily discarding the possibility that $\rho = 0$, you have the spherical equation

$$\rho - 4 \cos \phi = 0 \quad \text{or} \quad \rho = 4 \cos \phi.$$

Note that the solution set for this equation includes a point for which $\rho = 0$, so nothing is lost by discarding the factor ρ. The sphere represented by the equation $\rho = 4 \cos \phi$ is shown in Figure 11.77.

Rectangular:
$x^2 + y^2 + z^2 - 4z = 0$

Spherical:
$\rho = 4 \cos \phi$

Figure 11.77

11.7 Exercises

See **CalcChat.com** for tutorial help and worked-out solutions to odd-numbered exercises.

Cylindrical-to-Rectangular Conversion In Exercises 1–6, convert the point from cylindrical coordinates to rectangular coordinates.

1. $(-7, 0, 5)$

2. $(2, -\pi, -4)$

3. $\left(3, \dfrac{\pi}{4}, 1\right)$

4. $\left(6, -\dfrac{\pi}{4}, 2\right)$

5. $\left(4, \dfrac{7\pi}{6}, 3\right)$

6. $\left(-0.5, \dfrac{4\pi}{3}, 8\right)$

Rectangular-to-Cylindrical Conversion In Exercises 7–12, convert the point from rectangular coordinates to cylindrical coordinates.

7. $(0, 5, 1)$

8. $\left(2\sqrt{2}, -2\sqrt{2}, 4\right)$

9. $(2, -2, -4)$

10. $(3, -3, 7)$

11. $\left(1, \sqrt{3}, 4\right)$

12. $\left(2\sqrt{3}, -2, 6\right)$

Rectangular-to-Cylindrical Conversion In Exercises 13–20, find an equation in cylindrical coordinates for the equation given in rectangular coordinates.

13. $z = 4$

14. $x = 9$

15. $x^2 + y^2 + z^2 = 17$

16. $z = x^2 + y^2 - 11$

17. $y = x^2$

18. $x^2 + y^2 = 8x$

19. $y^2 = 10 - z^2$

20. $x^2 + y^2 + z^2 - 3z = 0$

Cylindrical-to-Rectangular Conversion In Exercises 21–28, find an equation in rectangular coordinates for the equation given in cylindrical coordinates, and sketch its graph.

21. $r = 3$

22. $z = 2$

23. $\theta = \dfrac{\pi}{6}$

24. $r = \dfrac{1}{2}z$

25. $r^2 + z^2 = 5$

26. $z = r^2 \cos^2 \theta$

27. $r = 2 \sin \theta$

28. $r = 2 \cos \theta$

Rectangular-to-Spherical Conversion In Exercises 29–34, convert the point from rectangular coordinates to spherical coordinates.

29. $(4, 0, 0)$

30. $(-4, 0, 0)$

31. $\left(-2, 2\sqrt{3}, 4\right)$

32. $\left(2, 2, 4\sqrt{2}\right)$

33. $\left(\sqrt{3}, 1, 2\sqrt{3}\right)$

34. $(-1, 2, 1)$

Spherical-to-Rectangular Conversion In Exercises 35–40, convert the point from spherical coordinates to rectangular coordinates.

35. $\left(4, \dfrac{\pi}{6}, \dfrac{\pi}{4}\right)$

36. $\left(12, \dfrac{3\pi}{4}, \dfrac{\pi}{9}\right)$

37. $\left(12, -\dfrac{\pi}{4}, 0\right)$

38. $\left(9, \dfrac{\pi}{4}, \pi\right)$

39. $\left(5, \dfrac{\pi}{4}, \dfrac{3\pi}{4}\right)$

40. $\left(6, \pi, \dfrac{\pi}{2}\right)$

Rectangular-to-Spherical Conversion In Exercises 41–48, find an equation in spherical coordinates for the equation given in rectangular coordinates.

41. $y = 2$

42. $z = 6$

43. $x^2 + y^2 + z^2 = 49$

44. $x^2 + y^2 - 3z^2 = 0$

45. $x^2 + y^2 = 16$

46. $x = 13$

47. $x^2 + y^2 = 2z^2$

48. $x^2 + y^2 + z^2 - 9z = 0$

Spherical-to-Rectangular Conversion In Exercises 49–56, find an equation in rectangular coordinates for the equation given in spherical coordinates, and sketch its graph.

49. $\rho = 5$

50. $\theta = \dfrac{3\pi}{4}$

51. $\phi = \dfrac{\pi}{6}$

52. $\phi = \dfrac{\pi}{2}$

53. $\rho = 4 \cos \phi$

54. $\rho = 2 \sec \phi$

55. $\rho = \csc \phi$

56. $\rho = 4 \csc \phi \sec \theta$

Matching In Exercises 57–62, match the equation (written in terms of cylindrical or spherical coordinates) with its graph. [The graphs are labeled (a), (b), (c), (d), (e), and (f).]

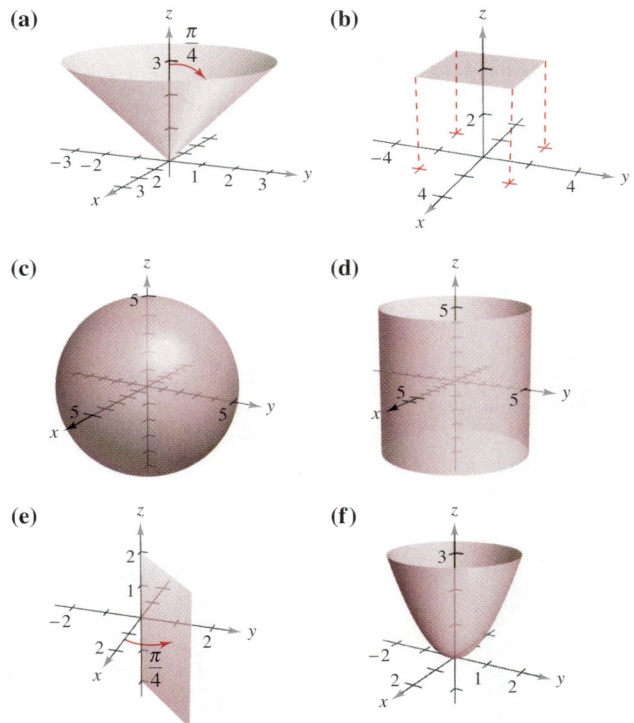

(a)

(b)

(c)

(d)

(e)

(f)

57. $r = 5$

58. $\theta = \dfrac{\pi}{4}$

59. $\rho = 5$

60. $\phi = \dfrac{\pi}{4}$

61. $r^2 = z$

62. $\rho = 4 \sec \phi$

Cylindrical-to-Spherical Conversion In Exercises 63–70, convert the point from cylindrical coordinates to spherical coordinates.

63. $\left(4, \dfrac{\pi}{4}, 0\right)$ 64. $\left(3, -\dfrac{\pi}{4}, 0\right)$

65. $\left(4, \dfrac{\pi}{2}, 4\right)$ 66. $\left(2, \dfrac{2\pi}{3}, -2\right)$

67. $\left(4, -\dfrac{\pi}{6}, 6\right)$ 68. $\left(-4, \dfrac{\pi}{3}, 4\right)$

69. $(12, \pi, 5)$ 70. $\left(4, \dfrac{\pi}{2}, 3\right)$

Spherical-to-Cylindrical Conversion In Exercises 71–78, convert the point from spherical coordinates to cylindrical coordinates.

71. $\left(10, \dfrac{\pi}{6}, \dfrac{\pi}{2}\right)$ 72. $\left(4, \dfrac{\pi}{18}, \dfrac{\pi}{2}\right)$

73. $\left(36, \pi, \dfrac{\pi}{2}\right)$ 74. $\left(18, \dfrac{\pi}{3}, \dfrac{\pi}{3}\right)$

75. $\left(6, -\dfrac{\pi}{6}, \dfrac{\pi}{3}\right)$ 76. $\left(5, -\dfrac{5\pi}{6}, \pi\right)$

77. $\left(8, \dfrac{7\pi}{6}, \dfrac{\pi}{6}\right)$ 78. $\left(7, \dfrac{\pi}{4}, \dfrac{3\pi}{4}\right)$

WRITING ABOUT CONCEPTS

79. **Rectangular and Cylindrical Coordinates** Give the equations for the coordinate conversion from rectangular to cylindrical coordinates and vice versa.

80. **Spherical Coordinates** Explain why in spherical coordinates the graph of $\theta = c$ is a half-plane and not an entire plane.

81. **Rectangular and Spherical Coordinates** Give the equations for the coordinate conversion from rectangular to spherical coordinates and vice versa.

---

82. **HOW DO YOU SEE IT?** Identify the surface graphed and match the graph with its rectangular equation. Then find an equation in cylindrical coordinates for the equation given in rectangular coordinates.

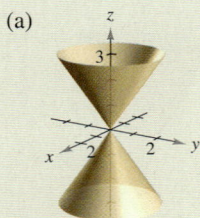

(a)

(b)

(i) $x^2 + y^2 = \dfrac{4}{9}z^2$ (ii) $x^2 + y^2 - z^2 = 2$

Converting a Rectangular Equation In Exercises 83–90, convert the rectangular equation to an equation in (a) cylindrical coordinates and (b) spherical coordinates.

83. $x^2 + y^2 + z^2 = 25$ 84. $4(x^2 + y^2) = z^2$

85. $x^2 + y^2 + z^2 - 2z = 0$ 86. $x^2 + y^2 = z$

87. $x^2 + y^2 = 4y$ 88. $x^2 + y^2 = 36$

89. $x^2 - y^2 = 9$ 90. $y = 4$

Sketching a Solid In Exercises 91–94, sketch the solid that has the given description in cylindrical coordinates.

91. $0 \le \theta \le \pi/2, 0 \le r \le 2, 0 \le z \le 4$

92. $-\pi/2 \le \theta \le \pi/2, 0 \le r \le 3, 0 \le z \le r\cos\theta$

93. $0 \le \theta \le 2\pi, 0 \le r \le a, r \le z \le a$

94. $0 \le \theta \le 2\pi, 2 \le r \le 4, z^2 \le -r^2 + 6r - 8$

Sketching a Solid In Exercises 95–98, sketch the solid that has the given description in spherical coordinates.

95. $0 \le \theta \le 2\pi, 0 \le \phi \le \pi/6, 0 \le \rho \le a\sec\phi$

96. $0 \le \theta \le 2\pi, \pi/4 \le \phi \le \pi/2, 0 \le \rho \le 1$

97. $0 \le \theta \le \pi/2, 0 \le \phi \le \pi/2, 0 \le \rho \le 2$

98. $0 \le \theta \le \pi, 0 \le \phi \le \pi/2, 1 \le \rho \le 3$

Think About It In Exercises 99–104, find inequalities that describe the solid, and state the coordinate system used. Position the solid on the coordinate system such that the inequalities are as simple as possible.

99. A cube with each edge 10 centimeters long

100. A cylindrical shell 8 meters long with an inside diameter of 0.75 meter and an outside diameter of 1.25 meters

101. A spherical shell with inside and outside radii of 4 inches and 6 inches, respectively

102. The solid that remains after a hole 1 inch in diameter is drilled through the center of a sphere 6 inches in diameter

103. The solid inside both $x^2 + y^2 + z^2 = 9$ and $\left(x - \dfrac{3}{2}\right)^2 + y^2 = \dfrac{9}{4}$

104. The solid between the spheres $x^2 + y^2 + z^2 = 4$ and $x^2 + y^2 + z^2 = 9$, and inside the cone $z^2 = x^2 + y^2$

True or False? In Exercises 105–108, determine whether the statement is true or false. If it is false, explain why or give an example that shows it is false.

105. In cylindrical coordinates, the equation $r = z$ is a cylinder.

106. The equations $\rho = 2$ and $x^2 + y^2 + z^2 = 4$ represent the same surface.

107. The cylindrical coordinates of a point (x, y, z) are unique.

108. The spherical coordinates of a point (x, y, z) are unique.

109. **Intersection of Surfaces** Identify the curve of intersection of the surfaces (in cylindrical coordinates) $z = \sin\theta$ and $r = 1$.

110. **Intersection of Surfaces** Identify the curve of intersection of the surfaces (in spherical coordinates) $\rho = 2\sec\phi$ and $\rho = 4$.

Review Exercises

See CalcChat.com for tutorial help and worked-out solutions to odd-numbered exercises.

Writing Vectors in Different Forms In Exercises 1 and 2, let $\mathbf{u} = \overrightarrow{PQ}$ and $\mathbf{v} = \overrightarrow{PR}$, and (a) write u and v in component form, (b) write u and v as the linear combination of the standard unit vectors i and j, (c) find the magnitudes of u and v, and (d) find 2u + v.

1. $P = (1, 2), Q = (4, 1), R = (5, 4)$

2. $P = (-2, -1), Q = (5, -1), R = (2, 4)$

Finding a Vector In Exercises 3 and 4, find the component form of v given its magnitude and the angle it makes with the positive x-axis.

3. $\|\mathbf{v}\| = 8, \ \theta = 60°$ 4. $\|\mathbf{v}\| = \frac{1}{2}, \ \theta = 225°$

5. **Finding Coordinates of a Point** Find the coordinates of the point located in the xy-plane, four units to the right of the xz-plane, and five units behind the yz-plane.

6. **Finding Coordinates of a Point** Find the coordinates of the point located on the y-axis and seven units to the left of the xz-plane.

Finding the Distance Between Two Points in Space In Exercises 7 and 8, find the distance between the points.

7. $(1, 6, 3), (-2, 3, 5)$

8. $(-2, 1, -5), (4, -1, -1)$

Finding the Equation of a Sphere In Exercises 9 and 10, find the standard equation of the sphere.

9. Center: $(3, -2, 6)$; Diameter: 15

10. Endpoints of a diameter: $(0, 0, 4), (4, 6, 0)$

Finding the Equation of a Sphere In Exercises 11 and 12, complete the square to write the equation of the sphere in standard form. Find the center and radius.

11. $x^2 + y^2 + z^2 - 4x - 6y + 4 = 0$

12. $x^2 + y^2 + z^2 - 10x + 6y - 4z + 34 = 0$

Writing a Vector in Different Forms In Exercises 13 and 14, the initial and terminal points of a vector are given. (a) Sketch the directed line segment, (b) find the component form of the vector, (c) write the vector using standard unit vector notation, and (d) sketch the vector with its initial point at the origin.

13. Initial point: $(2, -1, 3)$ 14. Initial point: $(6, 2, 0)$

Terminal point: $(4, 4, -7)$ Terminal point: $(3, -3, 8)$

Using Vectors to Determine Collinear Points In Exercises 15 and 16, use vectors to determine whether the points are collinear.

15. $(3, 4, -1), (-1, 6, 9), (5, 3, -6)$

16. $(5, -4, 7), (8, -5, 5), (11, 6, 3)$

17. **Finding a Unit Vector** Find a unit vector in the direction of $\mathbf{u} = \langle 2, 3, 5 \rangle$.

18. **Finding a Vector** Find the vector v of magnitude 8 in the direction $\langle 6, -3, 2 \rangle$.

Finding Dot Products In Exercises 19 and 20, let $\mathbf{u} = \overrightarrow{PQ}$ and $\mathbf{v} = \overrightarrow{PR}$, and find (a) the component forms of u and v, (b) u · v, and (c) v · v.

19. $P = (5, 0, 0), \ Q = (4, 4, 0), R = (2, 0, 6)$

20. $P = (2, -1, 3), \ Q = (0, 5, 1), \ R = (5, 5, 0)$

Finding the Angle Between Two Vectors In Exercises 21–24, find the angle θ between the vectors (a) in radians and (b) in degrees.

21. $\mathbf{u} = 5[\cos(3\pi/4)\mathbf{i} + \sin(3\pi/4)\mathbf{j}]$

    $\mathbf{v} = 2[\cos(2\pi/3)\mathbf{i} + \sin(2\pi/3)\mathbf{j}]$

22. $\mathbf{u} = 6\mathbf{i} + 2\mathbf{j} - 3\mathbf{k}, \quad \mathbf{v} = -\mathbf{i} + 5\mathbf{j}$

23. $\mathbf{u} = \langle 10, -5, 15 \rangle, \quad \mathbf{v} = \langle -2, 1, -3 \rangle$

24. $\mathbf{u} = \langle 1, 0, -3 \rangle, \quad \mathbf{v} = \langle 2, -2, 1 \rangle$

Comparing Vectors In Exercises 25 and 26, determine whether u and v are orthogonal, parallel, or neither.

25. $\mathbf{u} = \langle 7, -2, 3 \rangle$ 26. $\mathbf{u} = \langle -4, 3, -6 \rangle$

    $\mathbf{v} = \langle -1, 4, 5 \rangle$ $\mathbf{v} = \langle 16, -12, 24 \rangle$

Finding the Projection of u onto v In Exercises 27–30, find the projection of u onto v.

27. $\mathbf{u} = \langle 7, 9 \rangle, \quad \mathbf{v} = \langle 1, 5 \rangle$

28. $\mathbf{u} = 4\mathbf{i} + 2\mathbf{j}, \quad \mathbf{v} = 3\mathbf{i} + 4\mathbf{j}$

29. $\mathbf{u} = \langle 1, -1, 1 \rangle, \quad \mathbf{v} = \langle 2, 0, 2 \rangle$

30. $\mathbf{u} = 5\mathbf{i} + \mathbf{j} + 3\mathbf{k}, \quad \mathbf{v} = 2\mathbf{i} + 3\mathbf{j} + \mathbf{k}$

31. **Orthogonal Vectors** Find two vectors in opposite directions that are orthogonal to the vector $\mathbf{u} = \langle 5, 6, -3 \rangle$.

32. **Work** An object is pulled 8 feet across a floor using a force of 75 pounds. The direction of the force is 30° above the horizontal. Find the work done.

Finding Cross Products In Exercises 33–36, find (a) u × v, (b) v × u, and (c) v × v.

33. $\mathbf{u} = 4\mathbf{i} + 3\mathbf{j} + 6\mathbf{k}$ 34. $\mathbf{u} = 6\mathbf{i} - 5\mathbf{j} + 2\mathbf{k}$

    $\mathbf{v} = 5\mathbf{i} + 2\mathbf{j} + \mathbf{k}$ $\mathbf{v} = -4\mathbf{i} + 2\mathbf{j} + 3\mathbf{k}$

35. $\mathbf{u} = \langle 2, -4, -4 \rangle$ 36. $\mathbf{u} = \langle 0, 2, 1 \rangle$

    $\mathbf{v} = \langle 1, 1, 3 \rangle$ $\mathbf{v} = \langle 1, -3, 4 \rangle$

37. **Finding a Unit Vector** Find a unit vector that is orthogonal to both $\mathbf{u} = \langle 2, -10, 8 \rangle$ and $\mathbf{v} = \langle 4, 6, -8 \rangle$.

38. **Area** Find the area of the parallelogram that has the vectors $\mathbf{u} = \langle 3, -1, 5 \rangle$ and $\mathbf{v} = \langle 2, -4, 1 \rangle$ as adjacent sides.

39. Torque The specifications for a tractor state that the torque on a bolt with head size $\frac{7}{8}$ inch cannot exceed 200 foot-pounds. Determine the maximum force $\|\mathbf{F}\|$ that can be applied to the wrench in the figure.

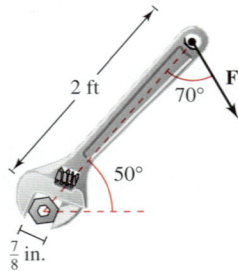

40. Volume Use the triple scalar product to find the volume of the parallelepiped having adjacent edges $\mathbf{u} = 2\mathbf{i} + \mathbf{j}$, $\mathbf{v} = 2\mathbf{j} + \mathbf{k}$, and $\mathbf{w} = -\mathbf{j} + 2\mathbf{k}$.

Finding Parametric and Symmetric Equations In Exercises 41 and 42, find sets of (a) **parametric equations and** (b) **symmetric equations of the line through the two points.** (For each line, write the direction numbers as integers.)

41. $(3, 0, 2)$, $(9, 11, 6)$ **42.** $(-1, 4, 3)$, $(8, 10, 5)$

Finding Parametric Equations In Exercises 43–46, find a set of parametric equations of the line.

43. The line passes through the point $(1, 2, 3)$ and is perpendicular to the xz-plane.

44. The line passes through the point $(1, 2, 3)$ and is parallel to the line given by $x = y = z$.

45. The line is the intersection of the planes $3x - 3y - 7z = -4$ and $x - y + 2z = 3$.

46. The line passes through the point $(0, 1, 4)$ and is perpendicular to $\mathbf{u} = \langle 2, -5, 1 \rangle$ and $\mathbf{v} = \langle -3, 1, 4 \rangle$.

Finding an Equation of a Plane In Exercises 47–50, find an equation of the plane.

47. The plane passes through $(-3, -4, 2)$, $(-3, 4, 1)$, and $(1, 1, -2)$.

48. The plane passes through the point $(-2, 3, 1)$ and is perpendicular to $\mathbf{n} = 3\mathbf{i} - \mathbf{j} + \mathbf{k}$.

49. The plane contains the lines given by

$$\frac{x - 1}{-2} = y = z + 1 \quad \text{and} \quad \frac{x + 1}{-2} = y - 1 = z - 2.$$

50. The plane passes through the points $(5, 1, 3)$ and $(2, -2, 1)$ and is perpendicular to the plane $2x + y - z = 4$.

51. Distance Find the distance between the point $(1, 0, 2)$ and the plane $2x - 3y + 6z = 6$.

52. Distance Find the distance between the point $(3, -2, 4)$ and the plane $2x - 5y + z = 10$.

53. Distance Find the distance between the planes $5x - 3y + z = 2$ and $5x - 3y + z = -3$.

54. Distance Find the distance between the point $(-5, 1, 3)$ and the line given by $x = 1 + t$, $y = 3 - 2t$, and $z = 5 - t$.

Sketching a Surface in Space In Exercises 55–64, describe and sketch the surface.

55. $x + 2y + 3z = 6$

56. $y = z^2$

57. $y = \frac{1}{2}z$

58. $y = \cos z$

59. $\dfrac{x^2}{16} + \dfrac{y^2}{9} + z^2 = 1$

60. $16x^2 + 16y^2 - 9z^2 = 0$

61. $\dfrac{x^2}{16} - \dfrac{y^2}{9} + z^2 = -1$

62. $\dfrac{x^2}{25} + \dfrac{y^2}{4} - \dfrac{z^2}{100} = 1$

63. $x^2 + z^2 = 4$

64. $y^2 + z^2 = 16$

65. Surface of Revolution Find an equation for the surface of revolution formed by revolving the curve $z^2 = 2y$ in the yz-plane about the y-axis.

66. Surface of Revolution Find an equation for the surface of revolution formed by revolving the curve $2x + 3z = 1$ in the xz-plane about the x-axis.

Converting Rectangular Coordinates In Exercises 67 and 68, convert the point from rectangular coordinates to (a) **cylindrical coordinates and** (b) **spherical coordinates.**

67. $\left(-2\sqrt{2}, 2\sqrt{2}, 2\right)$ **68.** $\left(\dfrac{\sqrt{3}}{4}, \dfrac{3}{4}, \dfrac{3\sqrt{3}}{2}\right)$

Cylindrical-to-Spherical Conversion In Exercises 69 and 70, convert the point from cylindrical coordinates to spherical coordinates.

69. $\left(100, -\dfrac{\pi}{6}, 50\right)$ **70.** $\left(81, -\dfrac{5\pi}{6}, 27\sqrt{3}\right)$

Spherical-to-Cylindrical Conversion In Exercises 71 and 72, convert the point from spherical coordinates to cylindrical coordinates.

71. $\left(25, -\dfrac{\pi}{4}, \dfrac{3\pi}{4}\right)$ **72.** $\left(12, -\dfrac{\pi}{2}, \dfrac{2\pi}{3}\right)$

Converting a Rectangular Equation In Exercises 73 and 74, convert the rectangular equation to an equation in (a) **cylindrical coordinates and** (b) **spherical coordinates.**

73. $x^2 - y^2 = 2z$

74. $x^2 + y^2 + z^2 = 16$

Cylindrical-to-Rectangular Conversion In Exercises 75 and 76, find an equation in rectangular coordinates for the equation given in cylindrical coordinates, and sketch its graph.

75. $r = 5 \cos \theta$

76. $z = 4$

Spherical-to-Rectangular Conversion In Exercises 77 and 78, find an equation in rectangular coordinates for the equation given in spherical coordinates, and sketch its graph.

77. $\theta = \dfrac{\pi}{4}$

78. $\rho = 3 \cos \phi$

P.S. Problem Solving

See **CalcChat.com** for tutorial help and
worked-out solutions to odd-numbered exercises.

1. Proof Using vectors, prove the Law of Sines: If **a**, **b**, and **c** are the three sides of the triangle shown in the figure, then

$$\frac{\sin A}{\|\mathbf{a}\|} = \frac{\sin B}{\|\mathbf{b}\|} = \frac{\sin C}{\|\mathbf{c}\|}.$$

2. Using an Equation Consider the function

$$f(x) = \int_0^x \sqrt{t^4 + 1} \, dt.$$

(a) Use a graphing utility to graph the function on the interval $-2 \le x \le 2$.

(b) Find a unit vector parallel to the graph of f at the point $(0, 0)$.

(c) Find a unit vector perpendicular to the graph of f at the point $(0, 0)$.

(d) Find the parametric equations of the tangent line to the graph of f at the point $(0, 0)$.

3. Proof Using vectors, prove that the line segments joining the midpoints of the sides of a parallelogram form a parallelogram (see figure).

4. Proof Using vectors, prove that the diagonals of a rhombus are perpendicular (see figure).

5. Distance

(a) Find the shortest distance between the point $Q(2, 0, 0)$ and the line determined by the points $P_1(0, 0, 1)$ and $P_2(0, 1, 2)$.

(b) Find the shortest distance between the point $Q(2, 0, 0)$ and the line segment joining the points $P_1(0, 0, 1)$ and $P_2(0, 1, 2)$.

6. Orthogonal Vectors Let P_0 be a point in the plane with normal vector **n**. Describe the set of points P in the plane for which $(\mathbf{n} + \overrightarrow{PP_0})$ is orthogonal to $(\mathbf{n} - \overrightarrow{PP_0})$.

7. Volume

(a) Find the volume of the solid bounded below by the paraboloid $z = x^2 + y^2$ and above by the plane $z = 1$.

(b) Find the volume of the solid bounded below by the elliptic paraboloid

$$z = \frac{x^2}{a^2} + \frac{y^2}{b^2}$$

and above by the plane $z = k$, where $k > 0$.

(c) Show that the volume of the solid in part (b) is equal to one-half the product of the area of the base times the altitude, as shown in the figure.

8. Volume

(a) Use the disk method to find the volume of the sphere $x^2 + y^2 + z^2 = r^2$.

(b) Find the volume of the ellipsoid $\dfrac{x^2}{a^2} + \dfrac{y^2}{b^2} + \dfrac{z^2}{c^2} = 1$.

9. Proof Prove the following property of the cross product.

$$(\mathbf{u} \times \mathbf{v}) \times (\mathbf{w} \times \mathbf{z}) = (\mathbf{u} \times \mathbf{v} \cdot \mathbf{z})\mathbf{w} - (\mathbf{u} \times \mathbf{v} \cdot \mathbf{w})\mathbf{z}$$

10. Using Parametric Equations Consider the line given by the parametric equations

$$x = -t + 3, \quad y = \tfrac{1}{2}t + 1, \quad z = 2t - 1$$

and the point $(4, 3, s)$ for any real number s.

(a) Write the distance between the point and the line as a function of s.

(b) Use a graphing utility to graph the function in part (a). Use the graph to find the value of s such that the distance between the point and the line is minimum.

(c) Use the *zoom* feature of a graphing utility to zoom out several times on the graph in part (b). Does it appear that the graph has slant asymptotes? Explain. If it appears to have slant asymptotes, find them.

11. Sketching Graphs Sketch the graph of each equation given in spherical coordinates.

(a) $\rho = 2 \sin \phi$ (b) $\rho = 2 \cos \phi$

12. Sketching Graphs Sketch the graph of each equation given in cylindrical coordinates.

(a) $r = 2 \cos \theta$ (b) $z = r^2 \cos 2\theta$

13. Tetherball A tetherball weighing 1 pound is pulled outward from the pole by a horizontal force **u** until the rope makes an angle of θ degrees with the pole (see figure).

(a) Determine the resulting tension in the rope and the magnitude of **u** when $\theta = 30°$.

(b) Write the tension T in the rope and the magnitude of **u** as functions of θ. Determine the domains of the functions.

(c) Use a graphing utility to complete the table.

θ	0°	10°	20°	30°	40°	50°	60°
T							
$\|\mathbf{u}\|$							

(d) Use a graphing utility to graph the two functions for $0° \leq \theta \leq 60°$.

(e) Compare T and $\|\mathbf{u}\|$ as θ increases.

(f) Find (if possible) $\lim\limits_{\theta \to \pi/2^-} T$ and $\lim\limits_{\theta \to \pi/2^-} \|\mathbf{u}\|$. Are the results what you expected? Explain.

Figure for 13

Figure for 14

14. Towing A loaded barge is being towed by two tugboats, and the magnitude of the resultant is 6000 pounds directed along the axis of the barge (see figure). Each towline makes an angle of θ degrees with the axis of the barge.

(a) Find the tension in the towlines when $\theta = 20°$.

(b) Write the tension T of each line as a function of θ. Determine the domain of the function.

(c) Use a graphing utility to complete the table.

θ	10°	20°	30°	40°	50°	60°
T						

(d) Use a graphing utility to graph the tension function.

(e) Explain why the tension increases as θ increases.

15. Proof Consider the vectors $\mathbf{u} = \langle \cos \alpha, \sin \alpha, 0 \rangle$ and $\mathbf{v} = \langle \cos \beta, \sin \beta, 0 \rangle$, where $\alpha > \beta$. Find the cross product of the vectors and use the result to prove the identity

$$\sin(\alpha - \beta) = \sin \alpha \cos \beta - \cos \alpha \sin \beta.$$

16. Latitude-Longitude System Los Angeles is located at 34.05° North latitude and 118.24° West longitude, and Rio de Janeiro, Brazil, is located at 22.90° South latitude and 43.23° West longitude (see figure). Assume that Earth is spherical and has a radius of 4000 miles.

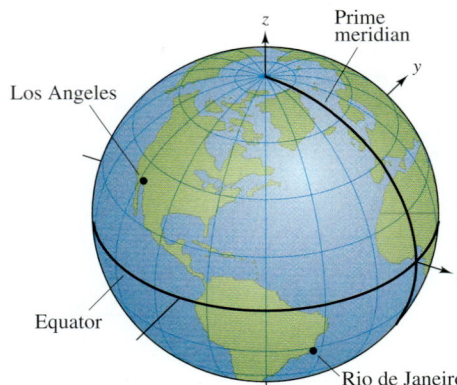

(a) Find the spherical coordinates for the location of each city.

(b) Find the rectangular coordinates for the location of each city.

(c) Find the angle (in radians) between the vectors from the center of Earth to the two cities.

(d) Find the great-circle distance s between the cities. (*Hint:* $s = r\theta$)

(e) Repeat parts (a)–(d) for the cities of Boston, located at 42.36° North latitude and 71.06° West longitude, and Honolulu, located at 21.31° North latitude and 157.86° West longitude.

17. Distance Between a Point and a Plane Consider the plane that passes through the points P, R, and S. Show that the distance from a point Q to this plane is

$$\text{Distance} = \frac{|\mathbf{u} \cdot (\mathbf{v} \times \mathbf{w})|}{\|\mathbf{u} \times \mathbf{v}\|}$$

where $\mathbf{u} = \overrightarrow{PR}$, $\mathbf{v} = \overrightarrow{PS}$, and $\mathbf{w} = \overrightarrow{PQ}$.

18. Distance Between Parallel Planes Show that the distance between the parallel planes

$$ax + by + cz + d_1 = 0 \quad \text{and} \quad ax + by + cz + d_2 = 0$$

is

$$\text{Distance} = \frac{|d_1 - d_2|}{\sqrt{a^2 + b^2 + c^2}}.$$

19. Intersection of Planes Show that the curve of intersection of the plane $z = 2y$ and the cylinder $x^2 + y^2 = 1$ is an ellipse.

20. Vector Algebra Read the article "Tooth Tables: Solution of a Dental Problem by Vector Algebra" by Gary Hosler Meisters in *Mathematics Magazine*. (To view this article, go to *MathArticles.com*.) Then write a paragraph explaining how vectors and vector algebra can be used in the construction of dental inlays.

12 Vector-Valued Functions

Speed *(Exercise 68, p. 861)*

Air Traffic Control
(Exercise 65, p. 850)

Football *(Exercise 32, p. 839)*

Shot-Put Throw
(Exercise 42, p. 839)

Playground Slide *(Exercise 81, p. 823)*

Clockwise from top left, 06photo/Shutterstock.com; Elena Aliaga/Shutterstock.com;
Jamie Roach/Shutterstock.com; Jack.Q/Shutterstock.com; Nicholas Moore/Shutterstock.com

12.1 Vector-Valued Functions

■ Analyze and sketch a space curve given by a vector-valued function.
■ Extend the concepts of limits and continuity to vector-valued functions.

Space Curves and Vector-Valued Functions

In Section 10.2, a *plane curve* was defined as the set of ordered pairs $(f(t), g(t))$ together with their defining parametric equations

$$x = f(t) \quad \text{and} \quad y = g(t)$$

where f and g are continuous functions of t on an interval I. This definition can be extended naturally to three-dimensional space. A **space curve** C is the set of all ordered triples $(f(t), g(t), h(t))$ together with their defining parametric equations

$$x = f(t), \quad y = g(t), \quad \text{and} \quad z = h(t)$$

where f, g, and h are continuous functions of t on an interval I.

Before looking at examples of space curves, a new type of function, called a **vector-valued function,** is introduced. This type of function maps real numbers to vectors.

Definition of Vector-Valued Function

A function of the form

$$\mathbf{r}(t) = f(t)\mathbf{i} + g(t)\mathbf{j} \qquad \text{Plane}$$

or

$$\mathbf{r}(t) = f(t)\mathbf{i} + g(t)\mathbf{j} + h(t)\mathbf{k} \qquad \text{Space}$$

is a **vector-valued function,** where the **component functions** f, g, and h are real-valued functions of the parameter t. Vector-valued functions are sometimes denoted as

$$\mathbf{r}(t) = \langle f(t), g(t) \rangle \qquad \text{Plane}$$

or

$$\mathbf{r}(t) = \langle f(t), g(t), h(t) \rangle. \qquad \text{Space}$$

Technically, a curve in a plane or in space consists of a collection of points and the defining parametric equations. Two different curves can have the same graph. For instance, each of the curves

$$\mathbf{r}(t) = \sin t\,\mathbf{i} + \cos t\,\mathbf{j} \quad \text{and} \quad \mathbf{r}(t) = \sin t^2\,\mathbf{i} + \cos t^2\,\mathbf{j}$$

has the unit circle as its graph, but these equations do not represent the same curve—because the circle is traced out in different ways on the graphs.

Be sure you see the distinction between the vector-valued function $\mathbf{r}$ and the real-valued functions f, g, and h. All are functions of the real variable t, but $\mathbf{r}(t)$ is a vector, whereas $f(t)$, $g(t)$, and $h(t)$ are real numbers (for each specific value of t).

Vector-valued functions serve dual roles in the representation of curves. By letting the parameter t represent time, you can use a vector-valued function to represent *motion* along a curve. Or, in the more general case, you can use a vector-valued function to *trace the graph* of a curve. In either case, the terminal point of the position vector $\mathbf{r}(t)$ coincides with the point (x, y) or (x, y, z) on the curve given by the parametric equations, as shown in Figure 12.1. The arrowhead on the curve indicates the curve's *orientation* by pointing in the direction of increasing values of t.

Curve in a plane

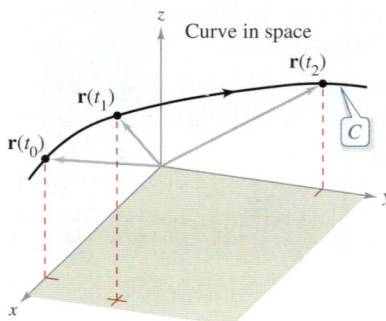

Curve in space

Curve C is traced out by the terminal point of position vector $\mathbf{r}(t)$.
Figure 12.1

Unless stated otherwise, the **domain** of a vector-valued function **r** is considered to be the intersection of the domains of the component functions f, g, and h. For instance, the domain of $\mathbf{r}(t) = \ln t\,\mathbf{i} + \sqrt{1-t}\,\mathbf{j} + t\mathbf{k}$ is the interval $(0, 1]$.

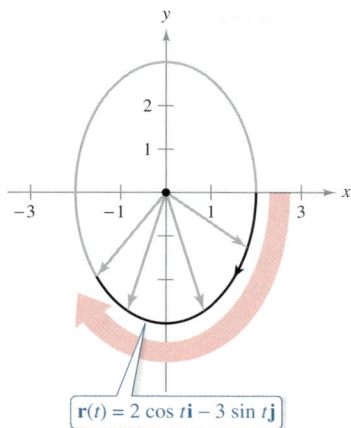

$\mathbf{r}(t) = 2\cos t\mathbf{i} - 3\sin t\mathbf{j}$

The ellipse is traced clockwise as t increases from 0 to 2π.
Figure 12.2

EXAMPLE 1   Sketching a Plane Curve

Sketch the plane curve represented by the vector-valued function

$$\mathbf{r}(t) = 2\cos t\mathbf{i} - 3\sin t\mathbf{j}, \quad 0 \le t \le 2\pi. \qquad \text{\color{red}Vector-valued function}$$

Solution   From the position vector $\mathbf{r}(t)$, you can write the parametric equations

$$x = 2\cos t \quad \text{and} \quad y = -3\sin t.$$

Solving for $\cos t$ and $\sin t$ and using the identity $\cos^2 t + \sin^2 t = 1$ produces the rectangular equation

$$\frac{x^2}{2^2} + \frac{y^2}{3^2} = 1. \qquad \text{\color{red}Rectangular equation}$$

The graph of this rectangular equation is the ellipse shown in Figure 12.2. The curve has a *clockwise* orientation. That is, as t increases from 0 to 2π, the position vector $\mathbf{r}(t)$ moves clockwise, and its terminal point traces the ellipse.

EXAMPLE 2   Sketching a Space Curve

$\cdots\cdots\triangleright$ *See LarsonCalculus.com for an interactive version of this type of example.*

Sketch the space curve represented by the vector-valued function

$$\mathbf{r}(t) = 4\cos t\mathbf{i} + 4\sin t\mathbf{j} + t\mathbf{k}, \quad 0 \le t \le 4\pi. \qquad \text{\color{red}Vector-valued function}$$

Solution   From the first two parametric equations

$$x = 4\cos t \quad \text{and} \quad y = 4\sin t$$

you can obtain

$$x^2 + y^2 = 16. \qquad \text{\color{red}Rectangular equation}$$

This means that the curve lies on a right circular cylinder of radius 4, centered about the z-axis. To locate the curve on this cylinder, you can use the third parametric equation

$$z = t.$$

In Figure 12.3, note that as t increases from 0 to 4π, the point (x, y, z) spirals up the cylinder to produce a **helix**. A real-life example of a helix is shown in the drawing at the left.

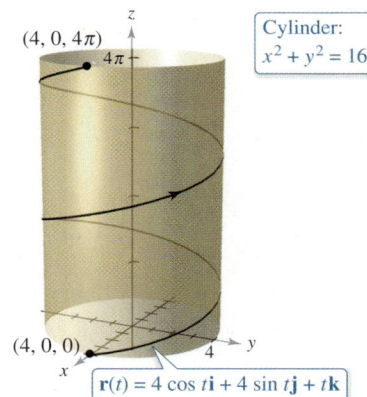

Cylinder:
$x^2 + y^2 = 16$

$\mathbf{r}(t) = 4\cos t\mathbf{i} + 4\sin t\mathbf{j} + t\mathbf{k}$

As t increases from 0 to 4π, two spirals on the helix are traced out.
Figure 12.3

In 1953, Francis Crick and James D. Watson discovered the double helix structure of DNA.

In Examples 1 and 2, you were given a vector-valued function and were asked to sketch the corresponding curve. The next two examples address the reverse problem—finding a vector-valued function to represent a given graph. Of course, when the graph is described parametrically, representation by a vector-valued function is straightforward. For instance, to represent the line in space given by $x = 2 + t$, $y = 3t$, and $z = 4 - t$, you can simply use the vector-valued function

$$\mathbf{r}(t) = (2 + t)\mathbf{i} + 3t\mathbf{j} + (4 - t)\mathbf{k}.$$

When a set of parametric equations for the graph is not given, the problem of representing the graph by a vector-valued function boils down to finding a set of parametric equations.

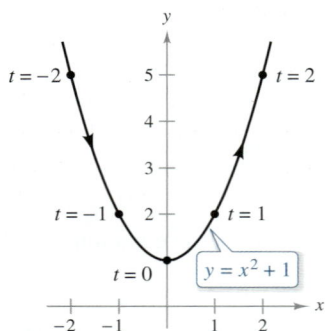

There are many ways to parametrize this graph. One way is to let $x = t$.

Figure 12.4

EXAMPLE 3 **Representing a Graph: Vector-Valued Function**

Represent the parabola

$$y = x^2 + 1$$

by a vector-valued function.

Solution Although there are many ways to choose the parameter t, a natural choice is to let $x = t$. Then $y = t^2 + 1$ and you have

$$\mathbf{r}(t) = t\mathbf{i} + (t^2 + 1)\mathbf{j}. \qquad \text{Vector-valued function}$$

Note in Figure 12.4 the orientation produced by this particular choice of parameter. Had you chosen $x = -t$ as the parameter, the curve would have been oriented in the opposite direction.

EXAMPLE 4 **Representing a Graph: Vector-Valued Function**

Sketch the space curve C represented by the intersection of the semiellipsoid

$$\frac{x^2}{12} + \frac{y^2}{24} + \frac{z^2}{4} = 1, \quad z \geq 0$$

and the parabolic cylinder $y = x^2$. Then find a vector-valued function to represent the graph.

Solution The intersection of the two surfaces is shown in Figure 12.5. As in Example 3, a natural choice of parameter is $x = t$. For this choice, you can use the given equation $y = x^2$ to obtain $y = t^2$. Then it follows that

$$\frac{z^2}{4} = 1 - \frac{x^2}{12} - \frac{y^2}{24} = 1 - \frac{t^2}{12} - \frac{t^4}{24} = \frac{24 - 2t^2 - t^4}{24} = \frac{(6 + t^2)(4 - t^2)}{24}.$$

Because the curve lies above the xy-plane, you should choose the positive square root for z and obtain the parametric equations

$$x = t, \quad y = t^2, \quad \text{and} \quad z = \sqrt{\frac{(6 + t^2)(4 - t^2)}{6}}.$$

The resulting vector-valued function is

$$\mathbf{r}(t) = t\mathbf{i} + t^2\mathbf{j} + \sqrt{\frac{(6 + t^2)(4 - t^2)}{6}}\,\mathbf{k}, \quad -2 \leq t \leq 2. \qquad \text{Vector-valued function}$$

(Note that the $\mathbf{k}$-component of $\mathbf{r}(t)$ implies $-2 \leq t \leq 2$.) From the points $(-2, 4, 0)$ and $(2, 4, 0)$ shown in Figure 12.5, you can see that the curve is traced as t increases from -2 to 2.

• • **REMARK** Curves in space can be specified in various ways. For instance, the curve in Example 4 is described as the intersection of two surfaces in space.

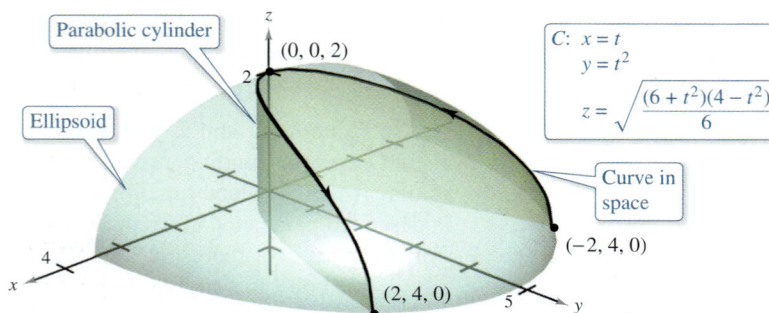

The curve C is the intersection of the semiellipsoid and the parabolic cylinder.

Figure 12.5

Limits and Continuity

Many techniques and definitions used in the calculus of real-valued functions can be applied to vector-valued functions. For instance, you can add and subtract vector-valued functions, multiply a vector-valued function by a scalar, take the limit of a vector-valued function, differentiate a vector-valued function, and so on. The basic approach is to capitalize on the linearity of vector operations by extending the definitions on a component-by-component basis. For example, to add two vector-valued functions (in the plane), you can write

$$\mathbf{r}_1(t) + \mathbf{r}_2(t) = [f_1(t)\mathbf{i} + g_1(t)\mathbf{j}] + [f_2(t)\mathbf{i} + g_2(t)\mathbf{j}] \qquad \text{Sum}$$
$$= [f_1(t) + f_2(t)]\mathbf{i} + [g_1(t) + g_2(t)]\mathbf{j}.$$

To subtract two vector-valued functions, you can write

$$\mathbf{r}_1(t) - \mathbf{r}_2(t) = [f_1(t)\mathbf{i} + g_1(t)\mathbf{j}] - [f_2(t)\mathbf{i} + g_2(t)\mathbf{j}] \qquad \text{Difference}$$
$$= [f_1(t) - f_2(t)]\mathbf{i} + [g_1(t) - g_2(t)]\mathbf{j}.$$

Similarly, to multiply a vector-valued function by a scalar, you can write

$$c\mathbf{r}(t) = c[f_1(t)\mathbf{i} + g_1(t)\mathbf{j}] \qquad \text{Scalar multiplication}$$
$$= cf_1(t)\mathbf{i} + cg_1(t)\mathbf{j}.$$

To divide a vector-valued function by a scalar, you can write

$$\frac{\mathbf{r}(t)}{c} = \frac{[f_1(t)\mathbf{i} + g_1(t)\mathbf{j}]}{c}, \quad c \neq 0 \qquad \text{Scalar division}$$
$$= \frac{f_1(t)}{c}\mathbf{i} + \frac{g_1(t)}{c}\mathbf{j}.$$

This component-by-component extension of operations with real-valued functions to vector-valued functions is further illustrated in the definition of the limit of a vector-valued function.

Definition of the Limit of a Vector-Valued Function

1. If $\mathbf{r}$ is a vector-valued function such that $\mathbf{r}(t) = f(t)\mathbf{i} + g(t)\mathbf{j}$, then

$$\lim_{t \to a} \mathbf{r}(t) = \left[\lim_{t \to a} f(t)\right]\mathbf{i} + \left[\lim_{t \to a} g(t)\right]\mathbf{j} \qquad \text{Plane}$$

provided f and g have limits as $t \to a$.

2. If $\mathbf{r}$ is a vector-valued function such that $\mathbf{r}(t) = f(t)\mathbf{i} + g(t)\mathbf{j} + h(t)\mathbf{k}$, then

$$\lim_{t \to a} \mathbf{r}(t) = \left[\lim_{t \to a} f(t)\right]\mathbf{i} + \left[\lim_{t \to a} g(t)\right]\mathbf{j} + \left[\lim_{t \to a} h(t)\right]\mathbf{k} \qquad \text{Space}$$

provided f, g, and h have limits as $t \to a$.

If $\mathbf{r}(t)$ approaches the vector $\mathbf{L}$ as $t \to a$, then the length of the vector $\mathbf{r}(t) - \mathbf{L}$ approaches 0. That is,

$$\|\mathbf{r}(t) - \mathbf{L}\| \to 0 \quad \text{as} \quad t \to a.$$

This is illustrated graphically in Figure 12.6. With this definition of the limit of a vector-valued function, you can develop vector versions of most of the limit theorems given in Chapter 1. For example, the limit of the sum of two vector-valued functions is the sum of their individual limits. Also, you can use the orientation of the curve $\mathbf{r}(t)$ to define one-sided limits of vector-valued functions. The next definition extends the notion of continuity to vector-valued functions.

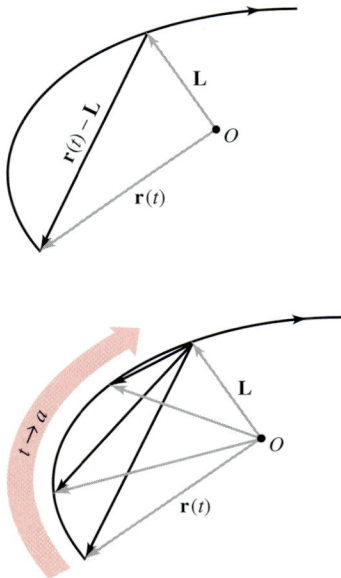

As t approaches a, $\mathbf{r}(t)$ approaches the limit $\mathbf{L}$. For the limit $\mathbf{L}$ to exist, it is not necessary that $\mathbf{r}(a)$ be defined or that $\mathbf{r}(a)$ be equal to $\mathbf{L}$.

Figure 12.6

> **Definition of Continuity of a Vector-Valued Function**
>
> A vector-valued function **r** is **continuous at the point** given by $t = a$ when the limit of $\mathbf{r}(t)$ exists as $t \to a$ and
>
> $$\lim_{t \to a} \mathbf{r}(t) = \mathbf{r}(a).$$
>
> A vector-valued function **r** is **continuous on an interval** *I* when it is continuous at every point in the interval.

From this definition, it follows that a vector-valued function is continuous at $t = a$ if and only if each of its component functions is continuous at $t = a$.

EXAMPLE 5   Continuity of a Vector-Valued Function

Discuss the continuity of the vector-valued function

$$\mathbf{r}(t) = t\mathbf{i} + a\mathbf{j} + (a^2 - t^2)\mathbf{k} \qquad \text{\textcolor{red}{a is a constant.}}$$

at $t = 0$.

Solution   As t approaches 0, the limit is

$$\lim_{t \to 0} \mathbf{r}(t) = \left[\lim_{t \to 0} t\right]\mathbf{i} + \left[\lim_{t \to 0} a\right]\mathbf{j} + \left[\lim_{t \to 0} (a^2 - t^2)\right]\mathbf{k}$$
$$= 0\mathbf{i} + a\mathbf{j} + a^2\mathbf{k}$$
$$= a\mathbf{j} + a^2\mathbf{k}.$$

Because

$$\mathbf{r}(0) = (0)\mathbf{i} + (a)\mathbf{j} + (a^2)\mathbf{k}$$
$$= a\mathbf{j} + a^2\mathbf{k}$$

you can conclude that **r** is continuous at $t = 0$. By similar reasoning, you can conclude that the vector-valued function **r** is continuous at all real-number values of t.   ■

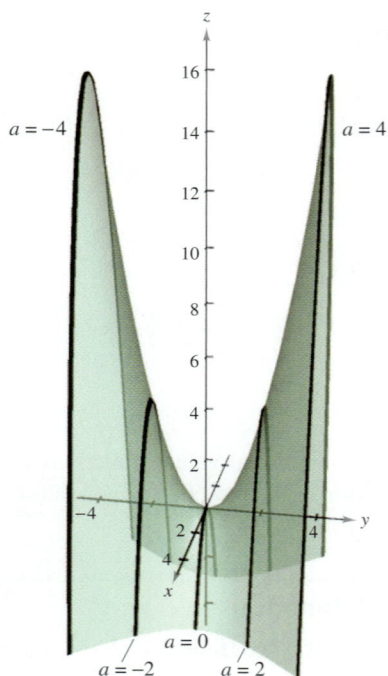

For each value of a, the curve represented by the vector-valued function $\mathbf{r}(t) = t\mathbf{i} + a\mathbf{j} + (a^2 - t^2)\mathbf{k}$ is a parabola.
Figure 12.7

For each value of a, the curve represented by the vector-valued function in Example 5

$$\mathbf{r}(t) = t\mathbf{i} + a\mathbf{j} + (a^2 - t^2)\mathbf{k} \qquad \text{\textcolor{red}{a is a constant.}}$$

is a parabola. You can think of each parabola as the intersection of the vertical plane $y = a$ and the hyperbolic paraboloid

$$y^2 - x^2 = z$$

as shown in Figure 12.7.

EXAMPLE 6   Continuity of a Vector-Valued Function

Determine the interval(s) on which the vector-valued function

$$\mathbf{r}(t) = t\mathbf{i} + \sqrt{t + 1}\,\mathbf{j} + (t^2 + 1)\mathbf{k}$$

is continuous.

Solution   The component functions are $f(t) = t$, $g(t) = \sqrt{t + 1}$, and $h(t) = (t^2 + 1)$. Both f and h are continuous for all real-number values of t. The function g, however, is continuous only for $t \geq -1$. So, **r** is continuous on the interval $[-1, \infty)$.   ■

▷ **TECHNOLOGY** Almost any type of three-dimensional sketch is difficult to do by hand, but sketching curves in space is especially difficult. The problem is trying to create the illusion of three dimensions. Graphing utilities use a variety of techniques to add "three-dimensionality" to graphs of space curves: one way is to show the curve on a surface, as in Figure 12.7.

12.1 Exercises

See **CalcChat.com** for tutorial help and worked-out solutions to odd-numbered exercises.

Finding the Domain In Exercises 1–8, find the domain of the vector-valued function.

1. $\mathbf{r}(t) = \dfrac{1}{t+1}\mathbf{i} + \dfrac{t}{2}\mathbf{j} - 3t\mathbf{k}$

2. $\mathbf{r}(t) = \sqrt{4-t^2}\,\mathbf{i} + t^2\mathbf{j} - 6t\mathbf{k}$

3. $\mathbf{r}(t) = \ln t\,\mathbf{i} - e^t\mathbf{j} - t\mathbf{k}$

4. $\mathbf{r}(t) = \sin t\,\mathbf{i} + 4\cos t\,\mathbf{j} + t\mathbf{k}$

5. $\mathbf{r}(t) = \mathbf{F}(t) + \mathbf{G}(t)$, where

$\mathbf{F}(t) = \cos t\,\mathbf{i} - \sin t\,\mathbf{j} + \sqrt{t}\,\mathbf{k}, \quad \mathbf{G}(t) = \cos t\,\mathbf{i} + \sin t\,\mathbf{j}$

6. $\mathbf{r}(t) = \mathbf{F}(t) - \mathbf{G}(t)$, where

$\mathbf{F}(t) = \ln t\,\mathbf{i} + 5t\mathbf{j} - 3t^2\mathbf{k}, \quad \mathbf{G}(t) = \mathbf{i} + 4t\mathbf{j} - 3t^2\mathbf{k}$

7. $\mathbf{r}(t) = \mathbf{F}(t) \times \mathbf{G}(t)$, where

$\mathbf{F}(t) = \sin t\,\mathbf{i} + \cos t\,\mathbf{j}, \quad \mathbf{G}(t) = \sin t\,\mathbf{j} + \cos t\,\mathbf{k}$

8. $\mathbf{r}(t) = \mathbf{F}(t) \times \mathbf{G}(t)$, where

$\mathbf{F}(t) = t^3\mathbf{i} - t\mathbf{j} + t\mathbf{k}, \quad \mathbf{G}(t) = \sqrt[3]{t}\,\mathbf{i} + \dfrac{1}{t+1}\mathbf{j} + (t+2)\mathbf{k}$

Evaluating a Function In Exercises 9–12, evaluate (if possible) the vector-valued function at each given value of t.

9. $\mathbf{r}(t) = \frac{1}{2}t^2\mathbf{i} - (t-1)\mathbf{j}$

  (a) $\mathbf{r}(1)$ (b) $\mathbf{r}(0)$ (c) $\mathbf{r}(s+1)$

  (d) $\mathbf{r}(2 + \Delta t) - \mathbf{r}(2)$

10. $\mathbf{r}(t) = \cos t\,\mathbf{i} + 2\sin t\,\mathbf{j}$

  (a) $\mathbf{r}(0)$ (b) $\mathbf{r}(\pi/4)$ (c) $\mathbf{r}(\theta - \pi)$

  (d) $\mathbf{r}(\pi/6 + \Delta t) - \mathbf{r}(\pi/6)$

11. $\mathbf{r}(t) = \ln t\,\mathbf{i} + \dfrac{1}{t}\mathbf{j} + 3t\mathbf{k}$

  (a) $\mathbf{r}(2)$ (b) $\mathbf{r}(-3)$ (c) $\mathbf{r}(t-4)$

  (d) $\mathbf{r}(1 + \Delta t) - \mathbf{r}(1)$

12. $\mathbf{r}(t) = \sqrt{t}\,\mathbf{i} + t^{3/2}\mathbf{j} + e^{-t/4}\mathbf{k}$

  (a) $\mathbf{r}(0)$ (b) $\mathbf{r}(4)$ (c) $\mathbf{r}(c+2)$

  (d) $\mathbf{r}(9 + \Delta t) - \mathbf{r}(9)$

Writing a Vector-Valued Function In Exercises 13–16, represent the line segment from P to Q by a vector-valued function and by a set of parametric equations.

13. $P(0, 0, 0), Q(3, 1, 2)$

14. $P(0, 2, -1), Q(4, 7, 2)$

15. $P(-2, 5, -3), Q(-1, 4, 9)$

16. $P(1, -6, 8), Q(-3, -2, 5)$

Think About It In Exercises 17 and 18, find $\mathbf{r}(t) \cdot \mathbf{u}(t)$. Is the result a vector-valued function? Explain.

17. $\mathbf{r}(t) = (3t - 1)\mathbf{i} + \frac{1}{4}t^3\mathbf{j} + 4\mathbf{k}, \quad \mathbf{u}(t) = t^2\mathbf{i} - 8\mathbf{j} + t^3\mathbf{k}$

18. $\mathbf{r}(t) = \langle 3\cos t, 2\sin t, t - 2 \rangle, \quad \mathbf{u}(t) = \langle 4\sin t, -6\cos t, t^2 \rangle$

Matching In Exercises 19–22, match the equation with its graph. [The graphs are labeled (a), (b), (c), and (d).]

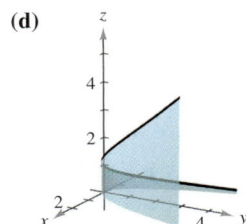

(a)

(b)

(c)

(d)

19. $\mathbf{r}(t) = t\mathbf{i} + 2t\mathbf{j} + t^2\mathbf{k}, \quad -2 \le t \le 2$

20. $\mathbf{r}(t) = \cos(\pi t)\mathbf{i} + \sin(\pi t)\mathbf{j} + t^2\mathbf{k}, \quad -1 \le t \le 1$

21. $\mathbf{r}(t) = t\mathbf{i} + t^2\mathbf{j} + e^{0.75t}\mathbf{k}, \quad -2 \le t \le 2$

22. $\mathbf{r}(t) = t\mathbf{i} + \ln t\,\mathbf{j} + \dfrac{2t}{3}\mathbf{k}, \quad 0.1 \le t \le 5$

Sketching a Curve In Exercises 23–38, sketch the curve represented by the vector-valued function and give the orientation of the curve.

23. $\mathbf{r}(t) = \dfrac{t}{4}\mathbf{i} + (t-1)\mathbf{j}$ **24.** $\mathbf{r}(t) = (5-t)\mathbf{i} + \sqrt{t}\,\mathbf{j}$

25. $\mathbf{r}(t) = t^3\mathbf{i} + t^2\mathbf{j}$ **26.** $\mathbf{r}(t) = (t^2 + t)\mathbf{i} + (t^2 - t)\mathbf{j}$

27. $\mathbf{r}(\theta) = \cos\theta\mathbf{i} + 3\sin\theta\mathbf{j}$

28. $\mathbf{r}(t) = 2\cos t\,\mathbf{i} + 2\sin t\,\mathbf{j}$

29. $\mathbf{r}(\theta) = 3\sec\theta\mathbf{i} + 2\tan\theta\mathbf{j}$

30. $\mathbf{r}(t) = 2\cos^3 t\,\mathbf{i} + 2\sin^3 t\,\mathbf{j}$

31. $\mathbf{r}(t) = (-t + 1)\mathbf{i} + (4t + 2)\mathbf{j} + (2t + 3)\mathbf{k}$

32. $\mathbf{r}(t) = t\mathbf{i} + (2t - 5)\mathbf{j} + 3t\mathbf{k}$

33. $\mathbf{r}(t) = 2\cos t\,\mathbf{i} + 2\sin t\,\mathbf{j} + t\mathbf{k}$

34. $\mathbf{r}(t) = t\mathbf{i} + 3\cos t\,\mathbf{j} + 3\sin t\,\mathbf{k}$

35. $\mathbf{r}(t) = 2\sin t\,\mathbf{i} + 2\cos t\,\mathbf{j} + e^{-t}\mathbf{k}$

36. $\mathbf{r}(t) = t^2\mathbf{i} + 2t\mathbf{j} + \frac{3}{2}t\mathbf{k}$

37. $\mathbf{r}(t) = \left\langle t, t^2, \frac{2}{3}t^3 \right\rangle$

38. $\mathbf{r}(t) = \langle \cos t + t\sin t, \sin t - t\cos t, t \rangle$

Identifying a Common Curve In Exercises 39–42, use a computer algebra system to graph the vector-valued function and identify the common curve.

39. $\mathbf{r}(t) = -\dfrac{1}{2}t^2\mathbf{i} + t\mathbf{j} - \dfrac{\sqrt{3}}{2}t^2\mathbf{k}$

40. $\mathbf{r}(t) = t\mathbf{i} - \dfrac{\sqrt{3}}{2}t^2\mathbf{j} + \dfrac{1}{2}t^2\mathbf{k}$

41. $\mathbf{r}(t) = \sin t\mathbf{i} + \left(\dfrac{\sqrt{3}}{2}\cos t - \dfrac{1}{2}t\right)\mathbf{j} + \left(\dfrac{1}{2}\cos t + \dfrac{\sqrt{3}}{2}\right)\mathbf{k}$

42. $\mathbf{r}(t) = -\sqrt{2}\sin t\mathbf{i} + 2\cos t\mathbf{j} + \sqrt{2}\sin t\mathbf{k}$

⚡ **Think About It** In Exercises 43 and 44, use a computer algebra system to graph the vector-valued function **r**(t). For each **u**(t), make a conjecture about the transformation (if any) of the graph of **r**(t). Use a computer algebra system to verify your conjecture.

43. $\mathbf{r}(t) = 2\cos t\mathbf{i} + 2\sin t\mathbf{j} + \frac{1}{2}t\mathbf{k}$

(a) $\mathbf{u}(t) = 2(\cos t - 1)\mathbf{i} + 2\sin t\mathbf{j} + \frac{1}{2}t\mathbf{k}$

(b) $\mathbf{u}(t) = 2\cos t\mathbf{i} + 2\sin t\mathbf{j} + 2t\mathbf{k}$

(c) $\mathbf{u}(t) = 2\cos(-t)\mathbf{i} + 2\sin(-t)\mathbf{j} + \frac{1}{2}(-t)\mathbf{k}$

(d) $\mathbf{u}(t) = \frac{1}{2}t\mathbf{i} + 2\sin t\mathbf{j} + 2\cos t\mathbf{k}$

(e) $\mathbf{u}(t) = 6\cos t\mathbf{i} + 6\sin t\mathbf{j} + \frac{1}{2}t\mathbf{k}$

44. $\mathbf{r}(t) = t\mathbf{i} + t^2\mathbf{j} + \frac{1}{2}t^3\mathbf{k}$

(a) $\mathbf{u}(t) = t\mathbf{i} + (t^2 - 2)\mathbf{j} + \frac{1}{2}t^3\mathbf{k}$

(b) $\mathbf{u}(t) = t^2\mathbf{i} + t\mathbf{j} + \frac{1}{2}t^3\mathbf{k}$

(c) $\mathbf{u}(t) = t\mathbf{i} + t^2\mathbf{j} + \left(\frac{1}{2}t^3 + 4\right)\mathbf{k}$

(d) $\mathbf{u}(t) = t\mathbf{i} + t^2\mathbf{j} + \frac{1}{8}t^3\mathbf{k}$

(e) $\mathbf{u}(t) = (-t)\mathbf{i} + (-t)^2\mathbf{j} + \frac{1}{2}(-t)^3\mathbf{k}$

Representing a Graph by a Vector-Valued Function In Exercises 45–52, represent the plane curve by a vector-valued function. (There are many correct answers.)

45. $y = x + 5$

46. $2x - 3y + 5 = 0$

47. $y = (x - 2)^2$

48. $y = 4 - x^2$

49. $x^2 + y^2 = 25$

50. $(x - 2)^2 + y^2 = 4$

51. $\dfrac{x^2}{16} - \dfrac{y^2}{4} = 1$

52. $\dfrac{x^2}{9} + \dfrac{y^2}{16} = 1$

Representing a Graph by a Vector-Valued Function In Exercises 53–60, sketch the space curve represented by the intersection of the surfaces. Then represent the curve by a vector-valued function using the given parameter.

Surfaces	Parameter
53. $z = x^2 + y^2$, $x + y = 0$	$x = t$
54. $z = x^2 + y^2$, $z = 4$	$x = 2\cos t$
55. $x^2 + y^2 = 4$, $z = x^2$	$x = 2\sin t$
56. $4x^2 + 4y^2 + z^2 = 16$, $x = z^2$	$z = t$
57. $x^2 + y^2 + z^2 = 4$, $x + z = 2$	$x = 1 + \sin t$
58. $x^2 + y^2 + z^2 = 10$, $x + y = 4$	$x = 2 + \sin t$
59. $x^2 + z^2 = 4$, $y^2 + z^2 = 4$	$x = t$ (first octant)
60. $x^2 + y^2 + z^2 = 16$, $xy = 4$	$x = t$ (first octant)

61. Sketching a Curve Show that the vector-valued function $\mathbf{r}(t) = t\mathbf{i} + 2t\cos t\mathbf{j} + 2t\sin t\mathbf{k}$ lies on the cone $4x^2 = y^2 + z^2$. Sketch the curve.

62. Sketching a Curve Show that the vector-valued function $\mathbf{r}(t) = e^{-t}\cos t\mathbf{i} + e^{-t}\sin t\mathbf{j} + e^{-t}\mathbf{k}$ lies on the cone $z^2 = x^2 + y^2$. Sketch the curve.

Finding a Limit In Exercises 63–68, find the limit (if it exists).

63. $\displaystyle\lim_{t\to\pi} (t\mathbf{i} + \cos t\mathbf{j} + \sin t\mathbf{k})$

64. $\displaystyle\lim_{t\to2} \left(3t\mathbf{i} + \dfrac{2}{t^2 - 1}\mathbf{j} + \dfrac{1}{t}\mathbf{k}\right)$

65. $\displaystyle\lim_{t\to0} \left(t^2\mathbf{i} + 3t\mathbf{j} + \dfrac{1 - \cos t}{t}\mathbf{k}\right)$

66. $\displaystyle\lim_{t\to1} \left(\sqrt{t}\,\mathbf{i} + \dfrac{\ln t}{t^2 - 1}\mathbf{j} + \dfrac{1}{t - 1}\mathbf{k}\right)$

67. $\displaystyle\lim_{t\to0} \left(e^t\mathbf{i} + \dfrac{\sin t}{t}\mathbf{j} + e^{-t}\mathbf{k}\right)$

68. $\displaystyle\lim_{t\to\infty} \left(e^{-t}\mathbf{i} + \dfrac{1}{t}\mathbf{j} + \dfrac{t}{t^2 + 1}\mathbf{k}\right)$

Continuity of a Vector-Valued Function In Exercises 69–74, determine the interval(s) on which the vector-valued function is continuous.

69. $\mathbf{r}(t) = t\mathbf{i} + \dfrac{1}{t}\mathbf{j}$

70. $\mathbf{r}(t) = \sqrt{t}\,\mathbf{i} + \sqrt{t - 1}\,\mathbf{j}$

71. $\mathbf{r}(t) = t\mathbf{i} + \arcsin t\mathbf{j} + (t - 1)\mathbf{k}$

72. $\mathbf{r}(t) = 2e^{-t}\mathbf{i} + e^{-t}\mathbf{j} + \ln(t - 1)\mathbf{k}$

73. $\mathbf{r}(t) = \langle e^{-t}, t^2, \tan t\rangle$ **74.** $\mathbf{r}(t) = \langle 8, \sqrt{t}, \sqrt[3]{t}\rangle$

WRITING ABOUT CONCEPTS

Writing a Transformation In Exercises 75–78, consider the vector-valued function

$\mathbf{r}(t) = t^2\mathbf{i} + (t - 3)\mathbf{j} + t\mathbf{k}.$

Write a vector-valued function **s**(t) that is the specified transformation of **r**.

75. A vertical translation three units upward

76. A vertical translation four units downward

77. A horizontal translation two units in the direction of the negative x-axis

78. A horizontal translation five units in the direction of the positive y-axis

79. Continuity of a Vector-Valued Function State the definition of continuity of a vector-valued function. Give an example of a vector-valued function that is defined but not continuous at $t = 2$.

---

80. Comparing Functions Which of the following vector-valued functions represent the same graph?

(a) $\mathbf{r}(t) = (-3\cos t + 1)\mathbf{i} + (5\sin t + 2)\mathbf{j} + 4\mathbf{k}$

(b) $\mathbf{r}(t) = 4\mathbf{i} + (-3\cos t + 1)\mathbf{j} + (5\sin t + 2)\mathbf{k}$

(c) $\mathbf{r}(t) = (3\cos t - 1)\mathbf{i} + (-5\sin t - 2)\mathbf{j} + 4\mathbf{k}$

(d) $\mathbf{r}(t) = (-3\cos 2t + 1)\mathbf{i} + (5\sin 2t + 2)\mathbf{j} + 4\mathbf{k}$

81. Playground Slide

The outer edge of a playground slide is in the shape of a helix of radius 1.5 meters. The slide has a height of 2 meters and makes one complete revolution from top to bottom. Find a vector-valued function for the helix. Use a computer algebra system to graph your function. (There are many correct answers.)

82. HOW DO YOU SEE IT? The four figures below are graphs of the vector-valued function $\mathbf{r}(t) = 4\cos t\,\mathbf{i} + 4\sin t\,\mathbf{j} + (t/4)\mathbf{k}$. Match each of the four graphs with the point in space from which the helix is viewed. The four points are $(0, 0, 20)$, $(20, 0, 0)$, $(-20, 0, 0)$, and $(10, 20, 10)$.

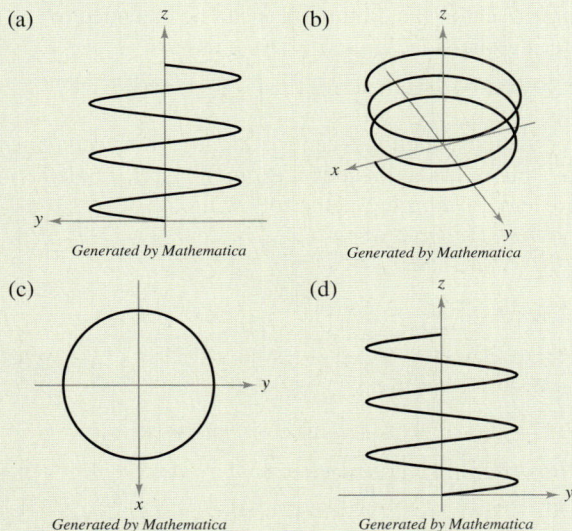

(a)

Generated by Mathematica

(b)

Generated by Mathematica

(c)

Generated by Mathematica

(d)

Generated by Mathematica

83. Proof Let $\mathbf{r}(t)$ and $\mathbf{u}(t)$ be vector-valued functions whose limits exist as $t \to c$. Prove that

$$\lim_{t\to c}\left[\mathbf{r}(t) \times \mathbf{u}(t)\right] = \lim_{t\to c}\mathbf{r}(t) \times \lim_{t\to c}\mathbf{u}(t).$$

84. Proof Let $\mathbf{r}(t)$ and $\mathbf{u}(t)$ be vector-valued functions whose limits exist as $t \to c$. Prove that

$$\lim_{t\to c}\left[\mathbf{r}(t) \cdot \mathbf{u}(t)\right] = \lim_{t\to c}\mathbf{r}(t) \cdot \lim_{t\to c}\mathbf{u}(t).$$

85. Proof Prove that if $\mathbf{r}$ is a vector-valued function that is continuous at c, then $\|\mathbf{r}\|$ is continuous at c.

86. Verifying a Converse Verify that the converse of Exercise 85 is not true by finding a vector-valued function $\mathbf{r}$ such that $\|\mathbf{r}\|$ is continuous at c but $\mathbf{r}$ is not continuous at c.

Particle Motion In Exercises 87 and 88, two particles travel along the space curves $\mathbf{r}(t)$ and $\mathbf{u}(t)$. A collision will occur at the point of intersection P when both particles are at P at the same time. Do the particles collide? Do their paths intersect?

87. $\mathbf{r}(t) = t^2\mathbf{i} + (9t - 20)\mathbf{j} + t^2\mathbf{k}$

$\mathbf{u}(t) = (3t + 4)\mathbf{i} + t^2\mathbf{j} + (5t - 4)\mathbf{k}$

88. $\mathbf{r}(t) = t\mathbf{i} + t^2\mathbf{j} + t^3\mathbf{k}$

$\mathbf{u}(t) = (-2t + 3)\mathbf{i} + 8t\mathbf{j} + (12t + 2)\mathbf{k}$

Think About It In Exercises 89 and 90, two particles travel along the space curves $\mathbf{r}(t)$ and $\mathbf{u}(t)$.

89. If $\mathbf{r}(t)$ and $\mathbf{u}(t)$ intersect, will the particles collide?

90. If the particles collide, do their paths $\mathbf{r}(t)$ and $\mathbf{u}(t)$ intersect?

True or False? In Exercises 91–94, determine whether the statement is true or false. If it is false, explain why or give an example that shows it is false.

91. If f, g, and h are first-degree polynomial functions, then the curve given by $x = f(t)$, $y = g(t)$, and $z = h(t)$ is a line.

92. If the curve given by $x = f(t)$, $y = g(t)$, and $z = h(t)$ is a line, then f, g, and h are first-degree polynomial functions of t.

93. Two particles travel along the space curves $\mathbf{r}(t)$ and $\mathbf{u}(t)$. The intersection of their paths depends only on the curves traced out by $\mathbf{r}(t)$ and $\mathbf{u}(t)$, while collision depends on the parametrizations.

94. The vector-valued function $\mathbf{r}(t) = t^2\mathbf{i} + t\sin t\,\mathbf{j} + t\cos t\,\mathbf{k}$ lies on the paraboloid $x = y^2 + z^2$.

SECTION PROJECT

Witch of Agnesi

In Section 3.5, you studied a famous curve called the **Witch of Agnesi**. In this project, you will take a closer look at this function.

Consider a circle of radius a centered on the y-axis at $(0, a)$. Let A be a point on the horizontal line $y = 2a$, let O be the origin, and let B be the point where the segment OA intersects the circle. A point P is on the Witch of Agnesi when P lies on the horizontal line through B and on the vertical line through A.

(a) Show that the point A is traced out by the vector-valued function $\mathbf{r}_A(\theta) = 2a\cot\theta\,\mathbf{i} + 2a\mathbf{j}$ for $0 < \theta < \pi$, where θ is the angle that OA makes with the positive x-axis.

(b) Show that the point B is traced out by the vector-valued function $\mathbf{r}_B(\theta) = a\sin 2\theta\,\mathbf{i} + a(1 - \cos 2\theta)\mathbf{j}$ for $0 < \theta < \pi$.

(c) Combine the results of parts (a) and (b) to find the vector-valued function $\mathbf{r}(\theta)$ for the Witch of Agnesi. Use a graphing utility to graph this curve for $a = 1$.

(d) Describe the limits $\lim_{\theta\to 0^+}\mathbf{r}(\theta)$ and $\lim_{\theta\to\pi^-}\mathbf{r}(\theta)$.

(e) Eliminate the parameter θ and determine the rectangular equation of the Witch of Agnesi. Use a graphing utility to graph this function for $a = 1$ and compare your graph with that obtained in part (c).

12.2 Differentiation and Integration of Vector-Valued Functions

■ Differentiate a vector-valued function.
■ Integrate a vector-valued function.

Differentiation of Vector-Valued Functions

In Sections 12.3–12.5, you will study several important applications involving the calculus of vector-valued functions. In preparation for that study, this section is devoted to the mechanics of differentiation and integration of vector-valued functions.

The definition of the derivative of a vector-valued function parallels the definition for real-valued functions.

> **Definition of the Derivative of a Vector-Valued Function**
>
> The **derivative of a vector-valued function r** is
>
> $$\mathbf{r}'(t) = \lim_{\Delta t \to 0} \frac{\mathbf{r}(t + \Delta t) - \mathbf{r}(t)}{\Delta t}$$
>
> for all t for which the limit exists. If $\mathbf{r}'(t)$ exists, then **r** is **differentiable at t**. If $\mathbf{r}'(t)$ exists for all t in an open interval I, then **r** is **differentiable on the interval I**. Differentiability of vector-valued functions can be extended to closed intervals by considering one-sided limits.

• •REMARK In addition to $\mathbf{r}'(t)$, other notations for the derivative of a vector-valued function are

$$\frac{d}{dt}[\mathbf{r}(t)], \quad \frac{d\mathbf{r}}{dt}, \quad \text{and} \quad D_t[\mathbf{r}(t)].$$

Differentiation of vector-valued functions can be done on a *component-by-component basis*. To see why this is true, consider the function $\mathbf{r}(t) = f(t)\mathbf{i} + g(t)\mathbf{j}$. Applying the definition of the derivative produces the following.

$$\mathbf{r}'(t) = \lim_{\Delta t \to 0} \frac{\mathbf{r}(t + \Delta t) - \mathbf{r}(t)}{\Delta t}$$

$$= \lim_{\Delta t \to 0} \frac{f(t + \Delta t)\mathbf{i} + g(t + \Delta t)\mathbf{j} - f(t)\mathbf{i} - g(t)\mathbf{j}}{\Delta t}$$

$$= \lim_{\Delta t \to 0} \left\{ \left[\frac{f(t + \Delta t) - f(t)}{\Delta t} \right]\mathbf{i} + \left[\frac{g(t + \Delta t) - g(t)}{\Delta t} \right]\mathbf{j} \right\}$$

$$= \left\{ \lim_{\Delta t \to 0} \left[\frac{f(t + \Delta t) - f(t)}{\Delta t} \right] \right\}\mathbf{i} + \left\{ \lim_{\Delta t \to 0} \left[\frac{g(t + \Delta t) - g(t)}{\Delta t} \right] \right\}\mathbf{j}$$

$$= f'(t)\mathbf{i} + g'(t)\mathbf{j}$$

This important result is listed in the theorem shown below. Note that the derivative of the vector-valued function **r** is itself a vector-valued function. You can see from Figure 12.8 that $\mathbf{r}'(t)$ is a vector tangent to the curve given by $\mathbf{r}(t)$ and pointing in the direction of increasing t-values.

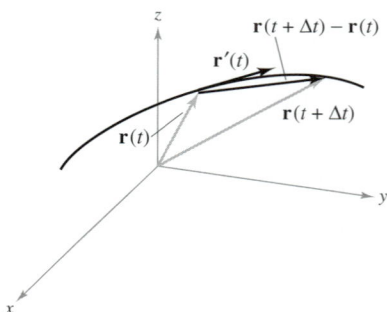

Figure 12.8

> **THEOREM 12.1 Differentiation of Vector-Valued Functions**
>
> **1.** If $\mathbf{r}(t) = f(t)\mathbf{i} + g(t)\mathbf{j}$, where f and g are differentiable functions of t, then
>
> $$\mathbf{r}'(t) = f'(t)\mathbf{i} + g'(t)\mathbf{j}. \qquad \text{Plane}$$
>
> **2.** If $\mathbf{r}(t) = f(t)\mathbf{i} + g(t)\mathbf{j} + h(t)\mathbf{k}$, where f, g, and h are differentiable functions of t, then
>
> $$\mathbf{r}'(t) = f'(t)\mathbf{i} + g'(t)\mathbf{j} + h'(t)\mathbf{k}. \qquad \text{Space}$$

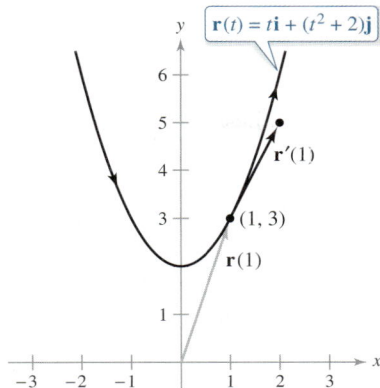

Figure 12.9

EXAMPLE 1 **Differentiation of a Vector-Valued Function**

⋮ ▷ *See LarsonCalculus.com for an interactive version of this type of example.*

For the vector-valued function

$$\mathbf{r}(t) = t\mathbf{i} + (t^2 + 2)\mathbf{j}$$

find $\mathbf{r}'(t)$. Then sketch the plane curve represented by $\mathbf{r}(t)$ and the graphs of $\mathbf{r}(1)$ and $\mathbf{r}'(1)$.

Solution Differentiate on a component-by-component basis to obtain

$$\mathbf{r}'(t) = \mathbf{i} + 2t\mathbf{j}. \qquad \text{Derivative}$$

From the position vector $\mathbf{r}(t)$, you can write the parametric equations $x = t$ and $y = t^2 + 2$. The corresponding rectangular equation is $y = x^2 + 2$. When $t = 1$,

$$\mathbf{r}(1) = \mathbf{i} + 3\mathbf{j}$$

and

$$\mathbf{r}'(1) = \mathbf{i} + 2\mathbf{j}.$$

In Figure 12.9, $\mathbf{r}(1)$ is drawn starting at the origin, and $\mathbf{r}'(1)$ is drawn starting at the terminal point of $\mathbf{r}(1)$. ∎

Higher-order derivatives of vector-valued functions are obtained by successive differentiation of each component function.

EXAMPLE 2 **Higher-Order Differentiation**

For the vector-valued function

$$\mathbf{r}(t) = \cos t\mathbf{i} + \sin t\mathbf{j} + 2t\mathbf{k}$$

find each of the following.

a. $\mathbf{r}'(t)$
b. $\mathbf{r}''(t)$
c. $\mathbf{r}'(t) \cdot \mathbf{r}''(t)$
d. $\mathbf{r}'(t) \times \mathbf{r}''(t)$

Solution

a. $\mathbf{r}'(t) = -\sin t\mathbf{i} + \cos t\mathbf{j} + 2\mathbf{k}$ ⟶ First derivative

b. $\mathbf{r}''(t) = -\cos t\mathbf{i} - \sin t\mathbf{j} + 0\mathbf{k}$

$\qquad = -\cos t\mathbf{i} - \sin t\mathbf{j}$ ⟶ Second derivative

c. $\mathbf{r}'(t) \cdot \mathbf{r}''(t) = \sin t \cos t - \sin t \cos t = 0$ ⟶ Dot product

d. $\mathbf{r}'(t) \times \mathbf{r}''(t) = \begin{vmatrix} \mathbf{i} & \mathbf{j} & \mathbf{k} \\ -\sin t & \cos t & 2 \\ -\cos t & -\sin t & 0 \end{vmatrix}$ ⟶ Cross product

$$= \begin{vmatrix} \cos t & 2 \\ -\sin t & 0 \end{vmatrix}\mathbf{i} - \begin{vmatrix} -\sin t & 2 \\ -\cos t & 0 \end{vmatrix}\mathbf{j} + \begin{vmatrix} -\sin t & \cos t \\ -\cos t & -\sin t \end{vmatrix}\mathbf{k}$$

$$= 2 \sin t\mathbf{i} - 2 \cos t\mathbf{j} + \mathbf{k} \qquad \blacksquare$$

In Example 2(c), note that the dot product is a real-valued function, not a vector-valued function.

The parametrization of the curve represented by the vector-valued function

$$\mathbf{r}(t) = f(t)\mathbf{i} + g(t)\mathbf{j} + h(t)\mathbf{k}$$

is **smooth on an open interval** I when f', g', and h' are continuous on I and $\mathbf{r}'(t) \neq \mathbf{0}$ for any value of t in the interval I.

EXAMPLE 3   Finding Intervals on Which a Curve Is Smooth

Find the intervals on which the epicycloid C given by

$$\mathbf{r}(t) = (5 \cos t - \cos 5t)\mathbf{i} + (5 \sin t - \sin 5t)\mathbf{j}, \quad 0 \leq t \leq 2\pi$$

is smooth.

Solution   The derivative of $\mathbf{r}$ is

$$\mathbf{r}'(t) = (-5 \sin t + 5 \sin 5t)\mathbf{i} + (5 \cos t - 5 \cos 5t)\mathbf{j}.$$

In the interval $[0, 2\pi]$, the only values of t for which

$$\mathbf{r}'(t) = 0\mathbf{i} + 0\mathbf{j}$$

are $t = 0$, $\pi/2$, π, $3\pi/2$, and 2π. Therefore, you can conclude that C is smooth on the intervals

$$\left(0, \frac{\pi}{2}\right), \quad \left(\frac{\pi}{2}, \pi\right), \quad \left(\pi, \frac{3\pi}{2}\right), \quad \text{and} \quad \left(\frac{3\pi}{2}, 2\pi\right)$$

as shown in Figure 12.10.

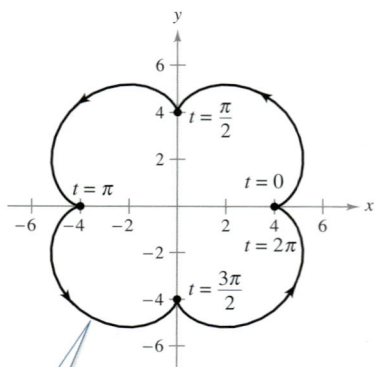

$\mathbf{r}(t) = (5 \cos t - \cos 5t)\mathbf{i} + (5 \sin t - \sin 5t)\mathbf{j}$

The epicycloid is not smooth at the points where it intersects the axes.

Figure 12.10

In Figure 12.10, note that the curve is not smooth at points at which the curve makes abrupt changes in direction. Such points are called **cusps** or **nodes.**

Most of the differentiation rules in Chapter 2 have counterparts for vector-valued functions, and several of these are listed in the next theorem. Note that the theorem contains three versions of "product rules." Property 3 gives the derivative of the product of a real-valued function w and a vector-valued function $\mathbf{r}$, Property 4 gives the derivative of the dot product of two vector-valued functions, and Property 5 gives the derivative of the cross product of two vector-valued functions (in space).

THEOREM 12.2   Properties of the Derivative

Let $\mathbf{r}$ and $\mathbf{u}$ be differentiable vector-valued functions of t, let w be a differentiable real-valued function of t, and let c be a scalar.

1. $\dfrac{d}{dt}[c\mathbf{r}(t)] = c\mathbf{r}'(t)$

2. $\dfrac{d}{dt}[\mathbf{r}(t) \pm \mathbf{u}(t)] = \mathbf{r}'(t) \pm \mathbf{u}'(t)$

3. $\dfrac{d}{dt}[w(t)\mathbf{r}(t)] = w(t)\mathbf{r}'(t) + w'(t)\mathbf{r}(t)$

4. $\dfrac{d}{dt}[\mathbf{r}(t) \cdot \mathbf{u}(t)] = \mathbf{r}(t) \cdot \mathbf{u}'(t) + \mathbf{r}'(t) \cdot \mathbf{u}(t)$

5. $\dfrac{d}{dt}[\mathbf{r}(t) \times \mathbf{u}(t)] = \mathbf{r}(t) \times \mathbf{u}'(t) + \mathbf{r}'(t) \times \mathbf{u}(t)$

6. $\dfrac{d}{dt}[\mathbf{r}(w(t))] = \mathbf{r}'(w(t))w'(t)$

7. If $\mathbf{r}(t) \cdot \mathbf{r}(t) = c$, then $\mathbf{r}(t) \cdot \mathbf{r}'(t) = 0$.

•• REMARK   Note that Property 5 applies only to three-dimensional vector-valued functions because the cross product is not defined for two-dimensional vectors.

Proof   To prove Property 4, let

$$\mathbf{r}(t) = f_1(t)\mathbf{i} + g_1(t)\mathbf{j} \quad \text{and} \quad \mathbf{u}(t) = f_2(t)\mathbf{i} + g_2(t)\mathbf{j}$$

where $f_1, f_2, g_1,$ and g_2 are differentiable functions of t. Then

$$\mathbf{r}(t) \cdot \mathbf{u}(t) = f_1(t)f_2(t) + g_1(t)g_2(t)$$

and it follows that

$$\frac{d}{dt}[\mathbf{r}(t) \cdot \mathbf{u}(t)] = f_1(t)f_2'(t) + f_1'(t)f_2(t) + g_1(t)g_2'(t) + g_1'(t)g_2(t)$$

$$= [f_1(t)f_2'(t) + g_1(t)g_2'(t)] + [f_1'(t)f_2(t) + g_1'(t)g_2(t)]$$

$$= \mathbf{r}(t) \cdot \mathbf{u}'(t) + \mathbf{r}'(t) \cdot \mathbf{u}(t).$$

See LarsonCalculus.com for Bruce Edwards's video of this proof.

Proofs of the other properties are left as exercises (see Exercises 67–71 and Exercise 74).

> **Exploration**
>
> Let $\mathbf{r}(t) = \cos t\mathbf{i} + \sin t\mathbf{j}$. Sketch the graph of $\mathbf{r}(t)$. Explain why the graph is a circle of radius 1 centered at the origin. Calculate $\mathbf{r}(\pi/4)$ and $\mathbf{r}'(\pi/4)$. Position the vector $\mathbf{r}'(\pi/4)$ so that its initial point is at the terminal point of $\mathbf{r}(\pi/4)$. What do you observe? Show that $\mathbf{r}(t) \cdot \mathbf{r}(t)$ is constant and that $\mathbf{r}(t) \cdot \mathbf{r}'(t) = 0$ for all t. How does this example relate to Property 7 of Theorem 12.2?

EXAMPLE 4   **Using Properties of the Derivative**

For $\mathbf{r}(t) = \dfrac{1}{t}\mathbf{i} - \mathbf{j} + \ln t\mathbf{k}$ and $\mathbf{u}(t) = t^2\mathbf{i} - 2t\mathbf{j} + \mathbf{k}$, find

a. $\dfrac{d}{dt}[\mathbf{r}(t) \cdot \mathbf{u}(t)]$ and   **b.** $\dfrac{d}{dt}[\mathbf{u}(t) \times \mathbf{u}'(t)]$.

Solution

a. Because $\mathbf{r}'(t) = -\dfrac{1}{t^2}\mathbf{i} + \dfrac{1}{t}\mathbf{k}$ and $\mathbf{u}'(t) = 2t\mathbf{i} - 2\mathbf{j}$, you have

$$\frac{d}{dt}[\mathbf{r}(t) \cdot \mathbf{u}(t)]$$

$$= \mathbf{r}(t) \cdot \mathbf{u}'(t) + \mathbf{r}'(t) \cdot \mathbf{u}(t)$$

$$= \left(\frac{1}{t}\mathbf{i} - \mathbf{j} + \ln t\mathbf{k}\right) \cdot (2t\mathbf{i} - 2\mathbf{j}) + \left(-\frac{1}{t^2}\mathbf{i} + \frac{1}{t}\mathbf{k}\right) \cdot (t^2\mathbf{i} - 2t\mathbf{j} + \mathbf{k})$$

$$= 2 + 2 + (-1) + \frac{1}{t}$$

$$= 3 + \frac{1}{t}.$$

b. Because $\mathbf{u}'(t) = 2t\mathbf{i} - 2\mathbf{j}$ and $\mathbf{u}''(t) = 2\mathbf{i}$, you have

$$\frac{d}{dt}[\mathbf{u}(t) \times \mathbf{u}'(t)] = [\mathbf{u}(t) \times \mathbf{u}''(t)] + [\mathbf{u}'(t) \times \mathbf{u}'(t)]$$

$$= \begin{vmatrix} \mathbf{i} & \mathbf{j} & \mathbf{k} \\ t^2 & -2t & 1 \\ 2 & 0 & 0 \end{vmatrix} + \mathbf{0}$$

$$= \begin{vmatrix} -2t & 1 \\ 0 & 0 \end{vmatrix}\mathbf{i} - \begin{vmatrix} t^2 & 1 \\ 2 & 0 \end{vmatrix}\mathbf{j} + \begin{vmatrix} t^2 & -2t \\ 2 & 0 \end{vmatrix}\mathbf{k}$$

$$= 0\mathbf{i} - (-2)\mathbf{j} + 4t\mathbf{k}$$

$$= 2\mathbf{j} + 4t\mathbf{k}.$$

Try reworking parts (a) and (b) in Example 4 by first forming the dot and cross products and then differentiating to see that you obtain the same results.

Integration of Vector-Valued Functions

The next definition is a consequence of the definition of the derivative of a vector-valued function.

Definition of Integration of Vector-Valued Functions

1. If $\mathbf{r}(t) = f(t)\mathbf{i} + g(t)\mathbf{j}$, where f and g are continuous on $[a, b]$, then the **indefinite integral (antiderivative)** of $\mathbf{r}$ is

$$\int \mathbf{r}(t)\,dt = \left[\int f(t)\,dt\right]\mathbf{i} + \left[\int g(t)\,dt\right]\mathbf{j} \qquad \text{Plane}$$

and its **definite integral** over the interval $a \le t \le b$ is

$$\int_a^b \mathbf{r}(t)\,dt = \left[\int_a^b f(t)\,dt\right]\mathbf{i} + \left[\int_a^b g(t)\,dt\right]\mathbf{j}.$$

2. If $\mathbf{r}(t) = f(t)\mathbf{i} + g(t)\mathbf{j} + h(t)\mathbf{k}$, where f, g, and h are continuous on $[a, b]$, then the **indefinite integral (antiderivative)** of $\mathbf{r}$ is

$$\int \mathbf{r}(t)\,dt = \left[\int f(t)\,dt\right]\mathbf{i} + \left[\int g(t)\,dt\right]\mathbf{j} + \left[\int h(t)\,dt\right]\mathbf{k} \qquad \text{Space}$$

and its **definite integral** over the interval $a \le t \le b$ is

$$\int_a^b \mathbf{r}(t)\,dt = \left[\int_a^b f(t)\,dt\right]\mathbf{i} + \left[\int_a^b g(t)\,dt\right]\mathbf{j} + \left[\int_a^b h(t)\,dt\right]\mathbf{k}.$$

The antiderivative of a vector-valued function is a family of vector-valued functions all differing by a constant vector $\mathbf{C}$. For instance, if $\mathbf{r}(t)$ is a three-dimensional vector-valued function, then for the indefinite integral $\int \mathbf{r}(t)\,dt$, you obtain three constants of integration

$$\int f(t)\,dt = F(t) + C_1, \quad \int g(t)\,dt = G(t) + C_2, \quad \int h(t)\,dt = H(t) + C_3$$

where $F'(t) = f(t)$, $G'(t) = g(t)$, and $H'(t) = h(t)$. These three *scalar* constants produce one *vector* constant of integration

$$\int \mathbf{r}(t)\,dt = [F(t) + C_1]\mathbf{i} + [G(t) + C_2]\mathbf{j} + [H(t) + C_3]\mathbf{k}$$
$$= [F(t)\mathbf{i} + G(t)\mathbf{j} + H(t)\mathbf{k}] + [C_1\mathbf{i} + C_2\mathbf{j} + C_3\mathbf{k}]$$
$$= \mathbf{R}(t) + \mathbf{C}$$

where $\mathbf{R}'(t) = \mathbf{r}(t)$.

EXAMPLE 5 **Integrating a Vector-Valued Function**

Find the indefinite integral

$$\int (t\,\mathbf{i} + 3\mathbf{j})\,dt.$$

Solution Integrating on a component-by-component basis produces

$$\int (t\,\mathbf{i} + 3\mathbf{j})\,dt = \frac{t^2}{2}\mathbf{i} + 3t\mathbf{j} + \mathbf{C}.$$

Example 6 shows how to evaluate the definite integral of a vector-valued function.

EXAMPLE 6   **Definite Integral of a Vector-Valued Function**

Evaluate the integral

$$\int_0^1 \mathbf{r}(t)\, dt = \int_0^1 \left(\sqrt[3]{t}\,\mathbf{i} + \frac{1}{t+1}\mathbf{j} + e^{-t}\,\mathbf{k} \right) dt.$$

Solution

$$\int_0^1 \mathbf{r}(t)\, dt = \left(\int_0^1 t^{1/3}\, dt \right)\mathbf{i} + \left(\int_0^1 \frac{1}{t+1}\, dt \right)\mathbf{j} + \left(\int_0^1 e^{-t}\, dt \right)\mathbf{k}$$

$$= \left[\left(\frac{3}{4}\right) t^{4/3} \right]_0^1 \mathbf{i} + \left[\ln|t+1| \right]_0^1 \mathbf{j} + \left[-e^{-t} \right]_0^1 \mathbf{k}$$

$$= \frac{3}{4}\mathbf{i} + (\ln 2)\mathbf{j} + \left(1 - \frac{1}{e} \right)\mathbf{k}$$

As with real-valued functions, you can narrow the family of antiderivatives of a vector-valued function $\mathbf{r}'$ down to a single antiderivative by imposing an initial condition on the vector-valued function $\mathbf{r}$. This is demonstrated in the next example.

EXAMPLE 7   **The Antiderivative of a Vector-Valued Function**

Find the antiderivative of

$$\mathbf{r}'(t) = \cos 2t\,\mathbf{i} - 2\sin t\,\mathbf{j} + \frac{1}{1+t^2}\mathbf{k}$$

that satisfies the initial condition

$$\mathbf{r}(0) = 3\mathbf{i} - 2\mathbf{j} + \mathbf{k}.$$

Solution

$$\mathbf{r}(t) = \int \mathbf{r}'(t)\, dt$$

$$= \left(\int \cos 2t\, dt \right)\mathbf{i} + \left(\int -2\sin t\, dt \right)\mathbf{j} + \left(\int \frac{1}{1+t^2}\, dt \right)\mathbf{k}$$

$$= \left(\frac{1}{2}\sin 2t + C_1 \right)\mathbf{i} + (2\cos t + C_2)\mathbf{j} + (\arctan t + C_3)\mathbf{k}$$

Letting $t = 0$, you can write

$$\mathbf{r}(0) = (0 + C_1)\mathbf{i} + (2 + C_2)\mathbf{j} + (0 + C_3)\mathbf{k}.$$

Using the fact that $\mathbf{r}(0) = 3\mathbf{i} - 2\mathbf{j} + \mathbf{k}$, you have

$$(0 + C_1)\mathbf{i} + (2 + C_2)\mathbf{j} + (0 + C_3)\mathbf{k} = 3\mathbf{i} - 2\mathbf{j} + \mathbf{k}.$$

Equating corresponding components produces

$$C_1 = 3, \quad 2 + C_2 = -2, \quad \text{and} \quad C_3 = 1.$$

So, the antiderivative that satisfies the initial condition is

$$\mathbf{r}(t) = \left(\frac{1}{2}\sin 2t + 3 \right)\mathbf{i} + (2\cos t - 4)\mathbf{j} + (\arctan t + 1)\mathbf{k}.$$

12.2 Exercises

Differentiation of Vector-Valued Functions In Exercises 1–6, find $r'(t)$, $r(t_0)$, and $r'(t_0)$ for the given value of t_0. Then sketch the plane curve represented by the vector-valued function, and sketch the vectors $r(t_0)$ and $r'(t_0)$. Position the vectors such that the initial point of $r(t_0)$ is at the origin and the initial point of $r'(t_0)$ is at the terminal point of $r(t_0)$. What is the relationship between $r'(t_0)$ and the curve?

1. $r(t) = t^2 i + tj$, $t_0 = 2$

2. $r(t) = (1 + t)i + t^3 j$, $t_0 = 1$

3. $r(t) = \cos t i + \sin t j$, $t_0 = \dfrac{\pi}{2}$

4. $r(t) = 3 \sin t i + 4 \cos t j$, $t_0 = \dfrac{\pi}{2}$

5. $r(t) = \langle e^t, e^{2t} \rangle$, $t_0 = 0$

6. $r(t) = \langle e^{-t}, e^t \rangle$, $t_0 = 0$

Differentiation of Vector-Valued Functions In Exercises 7 and 8, find $r'(t)$, $r(t_0)$, and $r'(t_0)$ for the given value of t_0. Then sketch the space curve represented by the vector-valued function, and sketch the vectors $r(t_0)$ and $r'(t_0)$.

7. $r(t) = 2 \cos t i + 2 \sin t j + t k$, $t_0 = \dfrac{3\pi}{2}$

8. $r(t) = t i + t^2 j + \frac{3}{2} k$, $t_0 = 2$

Finding a Derivative In Exercises 9–20, find $r'(t)$.

9. $r(t) = t^3 i - 3t j$

10. $r(t) = \sqrt{t} i + (1 - t^3) j$

11. $r(t) = \langle 2 \cos t, 5 \sin t \rangle$

12. $r(t) = \langle t \cos t, -2 \sin t \rangle$

13. $r(t) = 6t i - 7t^2 j + t^3 k$

14. $r(t) = \dfrac{1}{t} i + 16t j + \dfrac{t^2}{2} k$

15. $r(t) = a \cos^3 t i + a \sin^3 t j + k$

16. $r(t) = 4\sqrt{t} i + t^2 \sqrt{t} j + \ln t^2 k$

17. $r(t) = e^{-t} i + 4j + 5te^t k$

18. $r(t) = \langle t^3, \cos 3t, \sin 3t \rangle$

19. $r(t) = \langle t \sin t, t \cos t, t \rangle$

20. $r(t) = \langle \arcsin t, \arccos t, 0 \rangle$

Higher-Order Differentiation In Exercises 21–24, find (a) $r'(t)$, (b) $r''(t)$, and (c) $r'(t) \cdot r''(t)$.

21. $r(t) = t^3 i + \frac{1}{2} t^2 j$

22. $r(t) = (t^2 + t)i + (t^2 - t)j$

23. $r(t) = 4 \cos t i + 4 \sin t j$

24. $r(t) = 8 \cos t i + 3 \sin t j$

Higher-Order Differentiation In Exercises 25–28, find (a) $r'(t)$, (b) $r''(t)$, (c) $r'(t) \cdot r''(t)$, and (d) $r'(t) \times r''(t)$.

25. $r(t) = \frac{1}{2} t^2 i - tj + \frac{1}{6} t^3 k$

26. $r(t) = t^3 i + (2t^2 + 3)j + (3t - 5)k$

27. $r(t) = \langle \cos t + t \sin t, \sin t - t \cos t, t \rangle$

28. $r(t) = \langle e^{-t}, t^2, \tan t \rangle$

Finding Intervals on Which a Curve Is Smooth In Exercises 29–38, find the open interval(s) on which the curve given by the vector-valued function is smooth.

29. $r(t) = t^2 i + t^3 j$

30. $r(t) = \dfrac{1}{t - 1} i + 3t j$

31. $r(\theta) = 2 \cos^3 \theta i + 3 \sin^3 \theta j$

32. $r(\theta) = (\theta + \sin \theta)i + (1 - \cos \theta)j$

33. $r(\theta) = (\theta - 2 \sin \theta)i + (1 - 2 \cos \theta)j$

34. $r(t) = \dfrac{2t}{8 + t^3} i + \dfrac{2t^2}{8 + t^3} j$ **35.** $r(t) = (t - 1)i + \dfrac{1}{t} j - t^2 k$

36. $r(t) = e^t i - e^{-t} j + 3t k$

37. $r(t) = t i - 3t j + \tan t k$

38. $r(t) = \sqrt{t} i + (t^2 - 1)j + \frac{1}{4} t k$

Using Properties of the Derivative In Exercises 39 and 40, use the properties of the derivative to find the following.

(a) $r'(t)$ (b) $\dfrac{d}{dt} [3r(t) - u(t)]$ (c) $\dfrac{d}{dt} (5t)u(t)$

(d) $\dfrac{d}{dt} [r(t) \cdot u(t)]$ (e) $\dfrac{d}{dt} [r(t) \times u(t)]$ (f) $\dfrac{d}{dt} r(2t)$

39. $r(t) = t i + 3t j + t^2 k$, $\quad u(t) = 4t i + t^2 j + t^3 k$

40. $r(t) = t i + 2 \sin t j + 2 \cos t k$

$u(t) = \dfrac{1}{t} i + 2 \sin t j + 2 \cos t k$

Using Two Methods In Exercises 41 and 42, find (a) $\dfrac{d}{dt} [r(t) \cdot u(t)]$ and (b) $\dfrac{d}{dt} [r(t) \times u(t)]$ in two different ways.

(i) Find the product first, then differentiate.
(ii) Apply the properties of Theorem 12.2.

41. $r(t) = t i + 2t^2 j + t^3 k$, $\quad u(t) = t^4 k$

42. $r(t) = \cos t i + \sin t j + t k$, $\quad u(t) = j + t k$

Finding an Indefinite Integral In Exercises 43–50, find the indefinite integral.

43. $\displaystyle\int (2t i + j + k) \, dt$ **44.** $\displaystyle\int \left(4t^3 i + 6t j - 4\sqrt{t} k \right) dt$

45. $\displaystyle\int \left(\dfrac{1}{t} i + j - t^{3/2} k \right) dt$ **46.** $\displaystyle\int \left(\ln t i + \dfrac{1}{t} j + k \right) dt$

47. $\displaystyle\int \left[(2t - 1)i + 4t^3 j + 3\sqrt{t} k \right] dt$

48. $\displaystyle\int (e^t i + \sin t j + \cos t k) \, dt$

49. $\displaystyle\int \left(\sec^2 t i + \dfrac{1}{1 + t^2} j \right) dt$

50. $\displaystyle\int (e^{-t} \sin t i + e^{-t} \cos t j) \, dt$

Evaluating a Definite Integral In Exercises 51–56, evaluate the definite integral.

51. $\int_0^1 (8t\mathbf{i} + t\mathbf{j} - \mathbf{k})\, dt$ **52.** $\int_{-1}^1 \left(t\mathbf{i} + t^3\mathbf{j} + \sqrt[3]{t}\,\mathbf{k}\right) dt$

53. $\int_0^{\pi/2} [(a\cos t)\mathbf{i} + (a\sin t)\mathbf{j} + \mathbf{k}]\, dt$

54. $\int_0^{\pi/4} [(\sec t \tan t)\mathbf{i} + (\tan t)\mathbf{j} + (2\sin t \cos t)\mathbf{k}]\, dt$

55. $\int_0^2 (t\mathbf{i} + e^t\mathbf{j} - te^t\mathbf{k})\, dt$ **56.** $\int_0^3 \|t\mathbf{i} + t^2\mathbf{j}\|\, dt$

Finding an Antiderivative In Exercises 57–62, find $\mathbf{r}(t)$ that satisifies the initial condition(s).

57. $\mathbf{r}'(t) = 4e^{2t}\mathbf{i} + 3e^t\mathbf{j}, \quad \mathbf{r}(0) = 2\mathbf{i}$

58. $\mathbf{r}'(t) = 3t^2\mathbf{j} + 6\sqrt{t}\,\mathbf{k}, \quad \mathbf{r}(0) = \mathbf{i} + 2\mathbf{j}$

59. $\mathbf{r}''(t) = -32\mathbf{j}, \quad \mathbf{r}'(0) = 600\sqrt{3}\,\mathbf{i} + 600\mathbf{j}, \quad \mathbf{r}(0) = \mathbf{0}$

60. $\mathbf{r}''(t) = -4\cos t\mathbf{j} - 3\sin t\mathbf{k}, \quad \mathbf{r}'(0) = 3\mathbf{k}, \quad \mathbf{r}(0) = 4\mathbf{j}$

61. $\mathbf{r}'(t) = te^{-t^2}\mathbf{i} - e^{-t}\mathbf{j} + \mathbf{k}, \quad \mathbf{r}(0) = \frac{1}{2}\mathbf{i} - \mathbf{j} + \mathbf{k}$

62. $\mathbf{r}'(t) = \frac{1}{1+t^2}\mathbf{i} + \frac{1}{t^2}\mathbf{j} + \frac{1}{t}\mathbf{k}, \quad \mathbf{r}(1) = 2\mathbf{i}$

WRITING ABOUT CONCEPTS

63. Differentiation State the definition of the derivative of a vector-valued function. Describe how to find the derivative of a vector-valued function and give its geometric interpretation.

64. Integration How do you find the integral of a vector-valued function?

65. Using a Derivative The three components of the derivative of the vector-valued function $\mathbf{u}$ are positive at $t = t_0$. Describe the behavior of $\mathbf{u}$ at $t = t_0$.

66. Using a Derivative The z-component of the derivative of the vector-valued function $\mathbf{u}$ is 0 for t in the domain of the function. What does this imply about the graph of $\mathbf{u}$?

Proof In Exercises 67–74, prove the property. In each case, assume r, u, and v are differentiable vector-valued functions of t in space, w is a differentiable real-valued function of t, and c is a scalar.

67. $\frac{d}{dt}[c\mathbf{r}(t)] = c\mathbf{r}'(t)$

68. $\frac{d}{dt}[\mathbf{r}(t) \pm \mathbf{u}(t)] = \mathbf{r}'(t) \pm \mathbf{u}'(t)$

69. $\frac{d}{dt}[w(t)\mathbf{r}(t)] = w(t)\mathbf{r}'(t) + w'(t)\mathbf{r}(t)$

70. $\frac{d}{dt}[\mathbf{r}(t) \times \mathbf{u}(t)] = \mathbf{r}(t) \times \mathbf{u}'(t) + \mathbf{r}'(t) \times \mathbf{u}(t)$

71. $\frac{d}{dt}[\mathbf{r}(w(t))] = \mathbf{r}'(w(t))w'(t)$

72. $\frac{d}{dt}[\mathbf{r}(t) \times \mathbf{r}'(t)] = \mathbf{r}(t) \times \mathbf{r}''(t)$

73. $\frac{d}{dt}\{\mathbf{r}(t) \cdot [\mathbf{u}(t) \times \mathbf{v}(t)]\} = \mathbf{r}'(t) \cdot [\mathbf{u}(t) \times \mathbf{v}(t)] + \mathbf{r}(t) \cdot [\mathbf{u}'(t) \times \mathbf{v}(t)] + \mathbf{r}(t) \cdot [\mathbf{u}(t) \times \mathbf{v}'(t)]$

74. If $\mathbf{r}(t) \cdot \mathbf{r}(t)$ is a constant, then $\mathbf{r}(t) \cdot \mathbf{r}'(t) = 0$.

75. Particle Motion A particle moves in the xy-plane along the curve represented by the vector-valued function $\mathbf{r}(t) = (t - \sin t)\mathbf{i} + (1 - \cos t)\mathbf{j}$.

(a) Use a graphing utility to graph $\mathbf{r}$. Describe the curve.

(b) Find the minimum and maximum values of $\|\mathbf{r}'\|$ and $\|\mathbf{r}''\|$.

76. Particle Motion A particle moves in the yz-plane along the curve represented by the vector-valued function $\mathbf{r}(t) = (2\cos t)\mathbf{j} + (3\sin t)\mathbf{k}$.

(a) Describe the curve.

(b) Find the minimum and maximum values of $\|\mathbf{r}'\|$ and $\|\mathbf{r}''\|$.

77. Perpendicular Vectors Consider the vector-valued function $\mathbf{r}(t) = (e^t \sin t)\mathbf{i} + (e^t \cos t)\mathbf{j}$. Show that $\mathbf{r}(t)$ and $\mathbf{r}''(t)$ are always perpendicular to each other.

78. HOW DO YOU SEE IT? The graph shows a vector-valued function $\mathbf{r}(t)$ for $0 \le t \le 2\pi$ and its derivative $\mathbf{r}'(t)$ for several values of t.

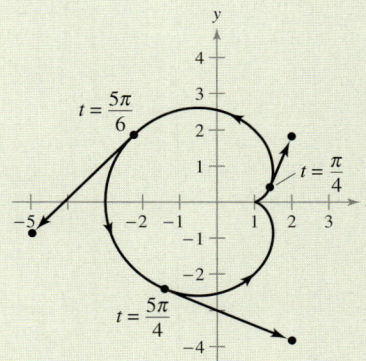

(a) For each derivative shown in the graph, determine whether each component is positive or negative.

(b) Is the curve smooth on the interval $[0, 2\pi]$? Explain.

True or False? In Exercises 79–82, determine whether the statement is true or false. If it is false, explain why or give an example that shows it is false.

79. If a particle moves along a sphere centered at the origin, then its derivative vector is always tangent to the sphere.

80. The definite integral of a vector-valued function is a real number.

81. $\frac{d}{dt}[\|\mathbf{r}(t)\|] = \|\mathbf{r}'(t)\|$

82. If $\mathbf{r}$ and $\mathbf{u}$ are differentiable vector-valued functions of t, then

$$\frac{d}{dt}[\mathbf{r}(t) \cdot \mathbf{u}(t)] = \mathbf{r}'(t) \cdot \mathbf{u}'(t).$$

12.3 Velocity and Acceleration

■ Describe the velocity and acceleration associated with a vector-valued function.
■ Use a vector-valued function to analyze projectile motion.

Velocity and Acceleration

You are now ready to combine your study of parametric equations, curves, vectors, and vector-valued functions to form a model for motion along a curve. You will begin by looking at the motion of an object in the plane. (The motion of an object in space can be developed similarly.)

As an object moves along a curve in the plane, the coordinates x and y of its center of mass are each functions of time t. Rather than using the letters f and g to represent these two functions, it is convenient to write $x = x(t)$ and $y = y(t)$. So, the position vector $\mathbf{r}(t)$ takes the form

$$\mathbf{r}(t) = x(t)\mathbf{i} + y(t)\mathbf{j}. \qquad \text{Position vector}$$

The beauty of this vector model for representing motion is that you can use the first and second derivatives of the vector-valued function $\mathbf{r}$ to find the object's velocity and acceleration. (Recall from the preceding chapter that velocity and acceleration are both vector quantities having magnitude and direction.) To find the velocity and acceleration vectors at a given time t, consider a point $Q(x(t + \Delta t), y(t + \Delta t))$ that is approaching the point $P(x(t), y(t))$ along the curve C given by $\mathbf{r}(t) = x(t)\mathbf{i} + y(t)\mathbf{j}$, as shown in Figure 12.11. As $\Delta t \to 0$, the direction of the vector $\overrightarrow{PQ}$ (denoted by $\Delta\mathbf{r}$) approaches the *direction of motion* at time t.

$$\Delta\mathbf{r} = \mathbf{r}(t + \Delta t) - \mathbf{r}(t)$$
$$\frac{\Delta\mathbf{r}}{\Delta t} = \frac{\mathbf{r}(t + \Delta t) - \mathbf{r}(t)}{\Delta t}$$
$$\lim_{\Delta t \to 0} \frac{\Delta\mathbf{r}}{\Delta t} = \lim_{\Delta t \to 0} \frac{\mathbf{r}(t + \Delta t) - \mathbf{r}(t)}{\Delta t}$$

If this limit exists, it is defined as the **velocity vector** or **tangent vector** to the curve at point P. Note that this is the same limit used to define $\mathbf{r}'(t)$. So, the direction of $\mathbf{r}'(t)$ gives the direction of motion at time t. Moreover, the magnitude of the vector $\mathbf{r}'(t)$

$$\|\mathbf{r}'(t)\| = \|x'(t)\mathbf{i} + y'(t)\mathbf{j}\| = \sqrt{[x'(t)]^2 + [y'(t)]^2}$$

gives the **speed** of the object at time t. Similarly, you can use $\mathbf{r}''(t)$ to find acceleration, as indicated in the definitions at the top of the next page.

Exploration

Exploring Velocity Consider the circle given by

$$\mathbf{r}(t) = (\cos \omega t)\mathbf{i} + (\sin \omega t)\mathbf{j}.$$

(The symbol ω is the Greek letter omega.) Use a graphing utility in *parametric* mode to graph this circle for several values of ω. How does ω affect the velocity of the terminal point as it traces out the curve? For a given value of ω, does the speed appear constant? Does the acceleration appear constant? Explain your reasoning.

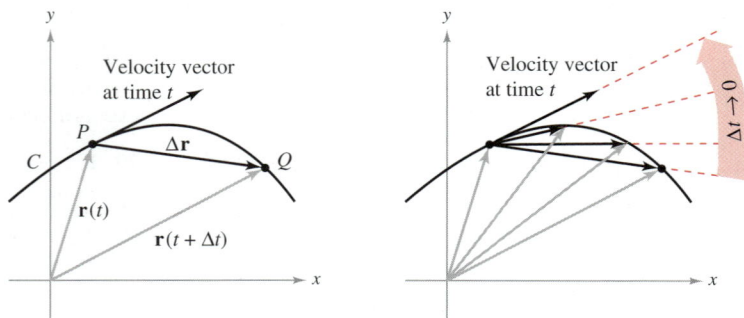

As $\Delta t \to 0$, $\dfrac{\Delta\mathbf{r}}{\Delta t}$ approaches the velocity vector.

Figure 12.11

Definitions of Velocity and Acceleration

If x and y are twice-differentiable functions of t, and $\mathbf{r}$ is a vector-valued function given by $\mathbf{r}(t) = x(t)\mathbf{i} + y(t)\mathbf{j}$, then the velocity vector, acceleration vector, and speed at time t are as follows.

$$\textbf{Velocity} = \mathbf{v}(t) \quad = \mathbf{r}\,'(t) \quad = x\,'(t)\mathbf{i} + y\,'(t)\mathbf{j}$$

$$\textbf{Acceleration} = \mathbf{a}(t) \quad = \mathbf{r}\,''(t) \quad = x''(t)\mathbf{i} + y''(t)\mathbf{j}$$

$$\textbf{Speed} = \|\mathbf{v}(t)\| = \|\mathbf{r}\,'(t)\| = \sqrt{[x\,'(t)]^2 + [y\,'(t)]^2}$$

For motion along a space curve, the definitions are similar. That is, for $\mathbf{r}(t) = x(t)\mathbf{i} + y(t)\mathbf{j} + z(t)\mathbf{k}$, you have

$$\textbf{Velocity} = \mathbf{v}(t) \quad = \mathbf{r}'(t) \quad = x\,'(t)\mathbf{i} + y\,'(t)\mathbf{j} + z\,'(t)\mathbf{k}$$

$$\textbf{Acceleration} = \mathbf{a}(t) \quad = \mathbf{r}''(t) \quad = x''(t)\mathbf{i} + y''(t)\mathbf{j} + z''(t)\mathbf{k}$$

$$\textbf{Speed} = \|\mathbf{v}(t)\| = \|\mathbf{r}\,'(t)\| = \sqrt{[x\,'(t)]^2 + [y\,'(t)]^2 + [z\,'(t)]^2}.$$

EXAMPLE 1   **Velocity and Acceleration Along a Plane Curve**

Find the velocity vector, speed, and acceleration vector of a particle that moves along the plane curve C described by

$$\mathbf{r}(t) = 2\sin\frac{t}{2}\mathbf{i} + 2\cos\frac{t}{2}\mathbf{j}. \qquad \text{Position vector}$$

Solution

The velocity vector is

$$\mathbf{v}(t) = \mathbf{r}\,'(t) = \cos\frac{t}{2}\mathbf{i} - \sin\frac{t}{2}\mathbf{j}. \qquad \text{Velocity vector}$$

The speed (at any time) is

$$\|\mathbf{r}\,'(t)\| = \sqrt{\cos^2\frac{t}{2} + \sin^2\frac{t}{2}} = 1. \qquad \text{Speed}$$

The acceleration vector is

$$\mathbf{a}(t) = \mathbf{r}''(t) = -\frac{1}{2}\sin\frac{t}{2}\mathbf{i} - \frac{1}{2}\cos\frac{t}{2}\mathbf{j}. \qquad \text{Acceleration vector}$$

The parametric equations for the curve in Example 1 are

$$x = 2\sin\frac{t}{2} \quad \text{and} \quad y = 2\cos\frac{t}{2}.$$

By eliminating the parameter t, you obtain the rectangular equation

$$x^2 + y^2 = 4. \qquad \text{Rectangular equation}$$

So, the curve is a circle of radius 2 centered at the origin, as shown in Figure 12.12. Because the velocity vector

$$\mathbf{v}(t) = \cos\frac{t}{2}\mathbf{i} - \sin\frac{t}{2}\mathbf{j}$$

has a constant magnitude but a changing direction as t increases, the particle moves around the circle at a constant speed.

REMARK In Example 1, note that the velocity and acceleration vectors are orthogonal at any point in time. This is characteristic of motion at a constant speed. (See Exercise 53.)

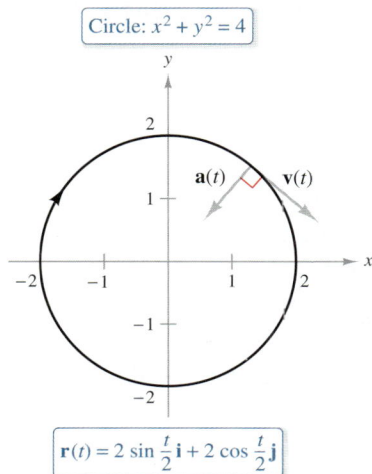

Circle: $x^2 + y^2 = 4$

$\mathbf{r}(t) = 2\sin\dfrac{t}{2}\mathbf{i} + 2\cos\dfrac{t}{2}\mathbf{j}$

The particle moves around the circle at a constant speed.

Figure 12.12

$$\boxed{\mathbf{r}(t) = (t^2 - 4)\mathbf{i} + t\mathbf{j}}$$

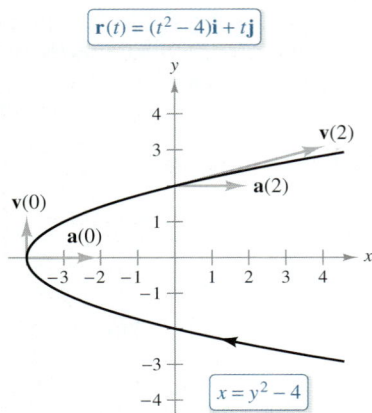

At each point on the curve, the acceleration vector points to the right.
Figure 12.13

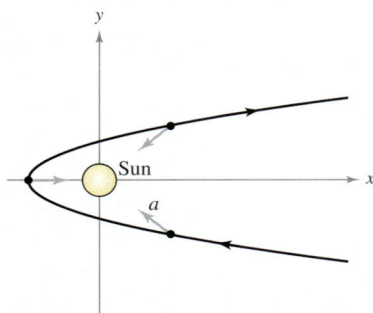

At each point in the comet's orbit, the acceleration vector points toward the sun.
Figure 12.14

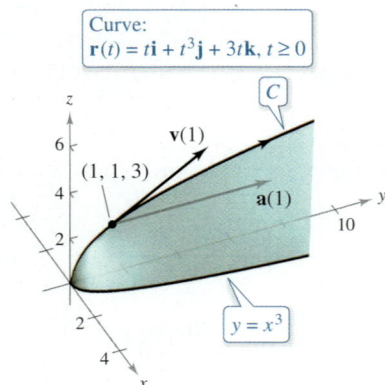

Figure 12.15

EXAMPLE 2 Velocity and Acceleration Vectors in the Plane

Sketch the path of an object moving along the plane curve given by

$$\mathbf{r}(t) = (t^2 - 4)\mathbf{i} + t\mathbf{j} \qquad \text{Position vector}$$

and find the velocity and acceleration vectors when $t = 0$ and $t = 2$.

Solution Using the parametric equations $x = t^2 - 4$ and $y = t$, you can determine that the curve is a parabola given by

$$x = y^2 - 4 \qquad \text{Rectangular equation}$$

as shown in Figure 12.13. The velocity vector (at any time) is

$$\mathbf{v}(t) = \mathbf{r}'(t) = 2t\mathbf{i} + \mathbf{j} \qquad \text{Velocity vector}$$

and the acceleration vector (at any time) is

$$\mathbf{a}(t) = \mathbf{r}''(t) = 2\mathbf{i}. \qquad \text{Acceleration vector}$$

When $t = 0$, the velocity and acceleration vectors are

$$\mathbf{v}(0) = 2(0)\mathbf{i} + \mathbf{j} = \mathbf{j} \quad \text{and} \quad \mathbf{a}(0) = 2\mathbf{i}.$$

When $t = 2$, the velocity and acceleration vectors are

$$\mathbf{v}(2) = 2(2)\mathbf{i} + \mathbf{j} = 4\mathbf{i} + \mathbf{j} \quad \text{and} \quad \mathbf{a}(2) = 2\mathbf{i}.$$

For the object moving along the path shown in Figure 12.13, note that the acceleration vector is constant (it has a magnitude of 2 and points to the right). This implies that the speed of the object is decreasing as the object moves toward the vertex of the parabola, and the speed is increasing as the object moves away from the vertex of the parabola.

This type of motion is *not* characteristic of comets that travel on parabolic paths through our solar system. For such comets, the acceleration vector always points to the origin (the sun), which implies that the comet's speed increases as it approaches the vertex of the path and decreases as it moves away from the vertex. (See Figure 12.14.)

EXAMPLE 3 Velocity and Acceleration Vectors in Space

•••▷ *See LarsonCalculus.com for an interactive version of this type of example.*

Sketch the path of an object moving along the space curve C given by

$$\mathbf{r}(t) = t\mathbf{i} + t^3\mathbf{j} + 3t\mathbf{k}, \quad t \geq 0 \qquad \text{Position vector}$$

and find the velocity and acceleration vectors when $t = 1$.

Solution Using the parametric equations $x = t$ and $y = t^3$, you can determine that the path of the object lies on the cubic cylinder given by

$$y = x^3. \qquad \text{Rectangular equation}$$

Moreover, because $z = 3t$, the object starts at $(0, 0, 0)$ and moves upward as t increases, as shown in Figure 12.15. Because $\mathbf{r}(t) = t\mathbf{i} + t^3\mathbf{j} + 3t\mathbf{k}$, you have

$$\mathbf{v}(t) = \mathbf{r}'(t) = \mathbf{i} + 3t^2\mathbf{j} + 3\mathbf{k} \qquad \text{Velocity vector}$$

and

$$\mathbf{a}(t) = \mathbf{r}''(t) = 6t\mathbf{j}. \qquad \text{Acceleration vector}$$

When $t = 1$, the velocity and acceleration vectors are

$$\mathbf{v}(1) = \mathbf{r}'(1) = \mathbf{i} + 3\mathbf{j} + 3\mathbf{k} \quad \text{and} \quad \mathbf{a}(1) = \mathbf{r}''(1) = 6\mathbf{j}.$$

So far in this section, you have concentrated on finding the velocity and acceleration by differentiating the position vector. Many practical applications involve the reverse problem—finding the position vector for a given velocity or acceleration. This is demonstrated in the next example.

EXAMPLE 4 **Finding a Position Vector by Integration**

An object starts from rest at the point $(1, 2, 0)$ and moves with an acceleration of

$$\mathbf{a}(t) = \mathbf{j} + 2\mathbf{k} \qquad \text{Acceleration vector}$$

where $\|\mathbf{a}(t)\|$ is measured in feet per second per second. Find the location of the object after $t = 2$ seconds.

Solution From the description of the object's motion, you can deduce the following *initial conditions*. Because the object starts from rest, you have

$$\mathbf{v}(0) = \mathbf{0}.$$

Moreover, because the object starts at the point $(x, y, z) = (1, 2, 0)$, you have

$$\mathbf{r}(0) = x(0)\mathbf{i} + y(0)\mathbf{j} + z(0)\mathbf{k} = 1\mathbf{i} + 2\mathbf{j} + 0\mathbf{k} = \mathbf{i} + 2\mathbf{j}.$$

To find the position vector, you should integrate twice, each time using one of the initial conditions to solve for the constant of integration. The velocity vector is

$$\mathbf{v}(t) = \int \mathbf{a}(t)\, dt$$
$$= \int (\mathbf{j} + 2\mathbf{k})\, dt$$
$$= t\mathbf{j} + 2t\mathbf{k} + \mathbf{C}$$

where $\mathbf{C} = C_1\mathbf{i} + C_2\mathbf{j} + C_3\mathbf{k}$. Letting $t = 0$ and applying the initial condition $\mathbf{v}(0) = \mathbf{0}$, you obtain

$$\mathbf{v}(0) = C_1\mathbf{i} + C_2\mathbf{j} + C_3\mathbf{k} = \mathbf{0} \implies C_1 = C_2 = C_3 = 0.$$

So, the *velocity* at any time t is

$$\mathbf{v}(t) = t\mathbf{j} + 2t\mathbf{k}. \qquad \text{Velocity vector}$$

Integrating once more produces

$$\mathbf{r}(t) = \int \mathbf{v}(t)\, dt$$
$$= \int (t\mathbf{j} + 2t\mathbf{k})\, dt$$
$$= \frac{t^2}{2}\mathbf{j} + t^2\mathbf{k} + \mathbf{C}$$

where $\mathbf{C} = C_4\mathbf{i} + C_5\mathbf{j} + C_6\mathbf{k}$. Letting $t = 0$ and applying the initial condition $\mathbf{r}(0) = \mathbf{i} + 2\mathbf{j}$, you have

$$\mathbf{r}(0) = C_4\mathbf{i} + C_5\mathbf{j} + C_6\mathbf{k} = \mathbf{i} + 2\mathbf{j} \implies C_4 = 1, C_5 = 2, C_6 = 0.$$

So, the *position* vector is

$$\mathbf{r}(t) = \mathbf{i} + \left(\frac{t^2}{2} + 2\right)\mathbf{j} + t^2\mathbf{k}. \qquad \text{Position vector}$$

The location of the object after $t = 2$ seconds is given by

$$\mathbf{r}(2) = \mathbf{i} + 4\mathbf{j} + 4\mathbf{k}$$

as shown in Figure 12.16.

Curve:
$\mathbf{r}(t) = \mathbf{i} + \left(\frac{t^2}{2} + 2\right)\mathbf{j} + t^2\mathbf{k}$

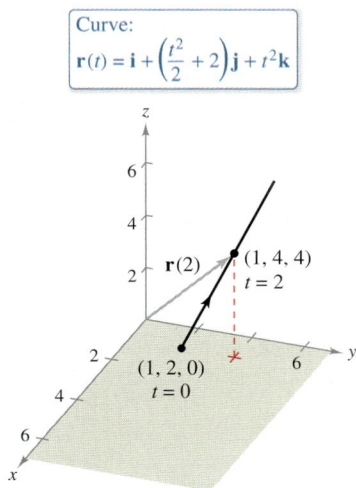

The object takes 2 seconds to move from point $(1, 2, 0)$ to point $(1, 4, 4)$ along the curve.
Figure 12.16

Projectile Motion

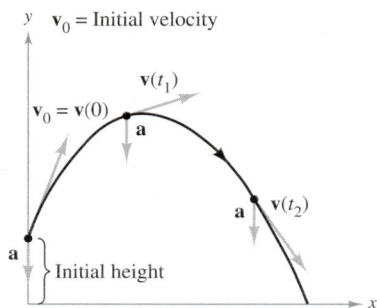

$\mathbf{v}_0 =$ Initial velocity

Figure 12.17

You now have the machinery to derive the parametric equations for the path of a projectile. Assume that gravity is the only force acting on the projectile after it is launched. So, the motion occurs in a vertical plane, which can be represented by the xy-coordinate system with the origin as a point on Earth's surface, as shown in Figure 12.17. For a projectile of mass m, the force due to gravity is

$$\mathbf{F} = -mg\mathbf{j} \qquad \text{Force due to gravity}$$

where the acceleration due to gravity is $g = 32$ feet per second per second, or 9.81 meters per second per second. By **Newton's Second Law of Motion**, this same force produces an acceleration $\mathbf{a} = \mathbf{a}(t)$ and satisfies the equation $\mathbf{F} = m\mathbf{a}$. Consequently, the acceleration of the projectile is given by $m\mathbf{a} = -mg\mathbf{j}$, which implies that

$$\mathbf{a} = -g\mathbf{j}. \qquad \text{Acceleration of projectile}$$

EXAMPLE 5 **Derivation of the Position Vector for a Projectile**

A projectile of mass m is launched from an initial position $\mathbf{r}_0$ with an initial velocity $\mathbf{v}_0$. Find its position vector as a function of time.

Solution Begin with the acceleration $\mathbf{a}(t) = -g\mathbf{j}$ and integrate twice.

$$\mathbf{v}(t) = \int \mathbf{a}(t)\, dt = \int -g\mathbf{j}\, dt = -gt\mathbf{j} + \mathbf{C}_1$$

$$\mathbf{r}(t) = \int \mathbf{v}(t)\, dt = \int (-gt\mathbf{j} + \mathbf{C}_1)\, dt = -\frac{1}{2}gt^2\mathbf{j} + \mathbf{C}_1 t + \mathbf{C}_2$$

You can use the facts that $\mathbf{v}(0) = \mathbf{v}_0$ and $\mathbf{r}(0) = \mathbf{r}_0$ to solve for the constant vectors $\mathbf{C}_1$ and $\mathbf{C}_2$. Doing this produces

$$\mathbf{C}_1 = \mathbf{v}_0 \quad \text{and} \quad \mathbf{C}_2 = \mathbf{r}_0.$$

Therefore, the position vector is

$$\mathbf{r}(t) = -\frac{1}{2}gt^2\mathbf{j} + t\mathbf{v}_0 + \mathbf{r}_0. \qquad \text{Position vector}$$

In many projectile problems, the constant vectors $\mathbf{r}_0$ and $\mathbf{v}_0$ are not given explicitly. Often you are given the initial height h, the initial speed v_0, and the angle θ at which the projectile is launched, as shown in Figure 12.18. From the given height, you can deduce that $\mathbf{r}_0 = h\mathbf{j}$. Because the speed gives the magnitude of the initial velocity, it follows that $v_0 = \|\mathbf{v}_0\|$ and you can write

$$\begin{aligned}
\mathbf{v}_0 &= x\mathbf{i} + y\mathbf{j} \\
&= (\|\mathbf{v}_0\| \cos\theta)\mathbf{i} + (\|\mathbf{v}_0\| \sin\theta)\mathbf{j} \\
&= v_0 \cos\theta\mathbf{i} + v_0 \sin\theta\mathbf{j}.
\end{aligned}$$

So, the position vector can be written in the form

$$\boxed{\mathbf{r}(t) = -\frac{1}{2}gt^2\mathbf{j} + t\mathbf{v}_0 + \mathbf{r}_0} \qquad \text{Position vector}$$

$$= -\frac{1}{2}gt^2\mathbf{j} + tv_0 \cos\theta\mathbf{i} + tv_0 \sin\theta\mathbf{j} + h\mathbf{j}$$

$$= (v_0 \cos\theta)t\mathbf{i} + \left[h + (v_0 \sin\theta)t - \frac{1}{2}gt^2 \right]\mathbf{j}.$$

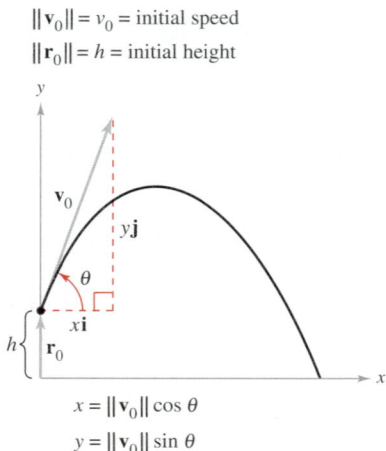

$\|\mathbf{v}_0\| = v_0 =$ initial speed
$\|\mathbf{r}_0\| = h =$ initial height

$x = \|\mathbf{v}_0\| \cos\theta$
$y = \|\mathbf{v}_0\| \sin\theta$

Figure 12.18

THEOREM 12.3 Position Vector for a Projectile

Neglecting air resistance, the path of a projectile launched from an initial height h with initial speed v_0 and angle of elevation θ is described by the vector function

$$\mathbf{r}(t) = (v_0 \cos \theta)t\mathbf{i} + \left[h + (v_0 \sin \theta)t - \tfrac{1}{2}gt^2\right]\mathbf{j}$$

where g is the acceleration due to gravity.

EXAMPLE 6 **Describing the Path of a Baseball**

A baseball is hit 3 feet above ground level at 100 feet per second and at an angle of 45° with respect to the ground, as shown in Figure 12.19. Find the maximum height reached by the baseball. Will it clear a 10-foot-high fence located 300 feet from home plate?

Figure 12.19

Solution You are given

$$h = 3, \quad v_0 = 100, \quad \text{and} \quad \theta = 45°.$$

So, using Theorem 12.3 with $g = 32$ feet per second per second produces

$$\mathbf{r}(t) = \left(100 \cos \frac{\pi}{4}\right)t\mathbf{i} + \left[3 + \left(100 \sin \frac{\pi}{4}\right)t - 16t^2\right]\mathbf{j}$$

$$= \left(50\sqrt{2}t\right)\mathbf{i} + \left(3 + 50\sqrt{2}t - 16t^2\right)\mathbf{j}.$$

The velocity vector is

$$\mathbf{v}(t) = \mathbf{r}'(t) = 50\sqrt{2}\mathbf{i} + \left(50\sqrt{2} - 32t\right)\mathbf{j}.$$

The maximum height occurs when

$$y'(t) = 50\sqrt{2} - 32t$$

is equal to 0, which implies that

$$t = \frac{25\sqrt{2}}{16} \approx 2.21 \text{ seconds}.$$

So, the maximum height reached by the ball is

$$y = 3 + 50\sqrt{2}\left(\frac{25\sqrt{2}}{16}\right) - 16\left(\frac{25\sqrt{2}}{16}\right)^2$$

$$= \frac{649}{8}$$

$$\approx 81 \text{ feet.} \qquad \textcolor{red}{\text{Maximum height when } t \approx 2.21 \text{ seconds}}$$

The ball is 300 feet from where it was hit when

$$300 = x(t) \quad \Longrightarrow \quad 300 = 50\sqrt{2}t.$$

Solving this equation for t produces $t = 3\sqrt{2} \approx 4.24$ seconds. At this time, the height of the ball is

$$y = 3 + 50\sqrt{2}\left(3\sqrt{2}\right) - 16\left(3\sqrt{2}\right)^2$$

$$= 303 - 288$$

$$= 15 \text{ feet.} \qquad \textcolor{red}{\text{Height when } t \approx 4.24 \text{ seconds}}$$

Therefore, the ball clears the 10-foot fence for a home run.

Finding Velocity and Acceleration Along a Plane Curve
In Exercises 1–8, the position vector r describes the path of an object moving in the xy-plane.

(a) Find the velocity vector, speed, and acceleration vector of the object.

(b) Evaluate the velocity vector and acceleration vector of the object at the given point.

(c) Sketch a graph of the path, and sketch the velocity and acceleration vectors at the given point.

Position Vector	Point
1. $\mathbf{r}(t) = 3t\mathbf{i} + (t-1)\mathbf{j}$	$(3, 0)$
2. $\mathbf{r}(t) = t\mathbf{i} + (-t^2 + 4)\mathbf{j}$	$(1, 3)$
3. $\mathbf{r}(t) = t^2\mathbf{i} + t\mathbf{j}$	$(4, 2)$
4. $\mathbf{r}(t) = \left(\frac{1}{4}t^3 + 1\right)\mathbf{i} + t\mathbf{j}$	$(3, 2)$
5. $\mathbf{r}(t) = 2\cos t\mathbf{i} + 2\sin t\mathbf{j}$	$(\sqrt{2}, \sqrt{2})$
6. $\mathbf{r}(t) = 3\cos t\mathbf{i} + 2\sin t\mathbf{j}$	$(3, 0)$
7. $\mathbf{r}(t) = \langle t - \sin t, 1 - \cos t \rangle$	$(\pi, 2)$
8. $\mathbf{r}(t) = \langle e^{-t}, e^t \rangle$	$(1, 1)$

Finding Velocity and Acceleration Vectors In Exercises 9–18, the position vector r describes the path of an object moving in space.

(a) Find the velocity vector, speed, and acceleration vector of the object.

(b) Evaluate the velocity vector and acceleration vector of the object at the given value of t.

Position Vector	Time
9. $\mathbf{r}(t) = t\mathbf{i} + 5t\mathbf{j} + 3t\mathbf{k}$	$t = 1$
10. $\mathbf{r}(t) = 4t\mathbf{i} + 4t\mathbf{j} + 2t\mathbf{k}$	$t = 3$
11. $\mathbf{r}(t) = t\mathbf{i} + t^2\mathbf{j} + \frac{1}{2}t^2\mathbf{k}$	$t = 4$
12. $\mathbf{r}(t) = 3t\mathbf{i} + t\mathbf{j} + \frac{1}{4}t^2\mathbf{k}$	$t = 2$
13. $\mathbf{r}(t) = t\mathbf{i} + t\mathbf{j} + \sqrt{9 - t^2}\,\mathbf{k}$	$t = 0$
14. $\mathbf{r}(t) = t^2\mathbf{i} + t\mathbf{j} + 2t^{3/2}\mathbf{k}$	$t = 4$
15. $\mathbf{r}(t) = \langle 4t, 3\cos t, 3\sin t \rangle$	$t = \pi$
16. $\mathbf{r}(t) = \langle 2\cos t, 2\sin t, t^2 \rangle$	$t = \frac{\pi}{4}$
17. $\mathbf{r}(t) = \langle e^t \cos t, e^t \sin t, e^t \rangle$	$t = 0$
18. $\mathbf{r}(t) = \left\langle \ln t, \frac{1}{t}, t^4 \right\rangle$	$t = 2$

Finding a Position Vector by Integration In Exercises 19–24, use the given acceleration vector to find the velocity and position vectors. Then find the position at time $t = 2$.

19. $\mathbf{a}(t) = \mathbf{i} + \mathbf{j} + \mathbf{k}, \quad \mathbf{v}(0) = \mathbf{0}, \quad \mathbf{r}(0) = \mathbf{0}$

20. $\mathbf{a}(t) = 2\mathbf{i} + 3\mathbf{k}, \quad \mathbf{v}(0) = 4\mathbf{j}, \quad \mathbf{r}(0) = \mathbf{0}$

21. $\mathbf{a}(t) = t\mathbf{j} + t\mathbf{k}, \quad \mathbf{v}(1) = 5\mathbf{j}, \quad \mathbf{r}(1) = \mathbf{0}$

22. $\mathbf{a}(t) = -32\mathbf{k}, \quad \mathbf{v}(0) = 3\mathbf{i} - 2\mathbf{j} + \mathbf{k}, \quad \mathbf{r}(0) = 5\mathbf{j} + 2\mathbf{k}$

23. $\mathbf{a}(t) = -\cos t\mathbf{i} - \sin t\mathbf{j}, \quad \mathbf{v}(0) = \mathbf{j} + \mathbf{k}, \quad \mathbf{r}(0) = \mathbf{i}$

24. $\mathbf{a}(t) = e^t\mathbf{i} - 8\mathbf{k}, \quad \mathbf{v}(0) = 2\mathbf{i} + 3\mathbf{j} + \mathbf{k}, \quad \mathbf{r}(0) = \mathbf{0}$

Projectile Motion In Exercises 25–38, use the model for projectile motion, assuming there is no air resistance.

25. A baseball is hit from a height of 2.5 feet above the ground with an initial velocity of 140 feet per second and at an angle of 22° above the horizontal. Find the maximum height reached by the baseball. Determine whether it will clear a 10-foot-high fence located 375 feet from home plate.

26. Determine the maximum height and range of a projectile fired at a height of 3 feet above the ground with an initial velocity of 900 feet per second and at an angle of 45° above the horizontal.

27. A baseball, hit 3 feet above the ground, leaves the bat at an angle of 45° and is caught by an outfielder 3 feet above the ground and 300 feet from home plate. What is the initial speed of the ball, and how high does it rise?

28. A baseball player at second base throws a ball 90 feet to the player at first base. The ball is released at a point 5 feet above the ground with an initial velocity of 50 miles per hour and at an angle of 15° above the horizontal. At what height does the player at first base catch the ball?

29. Eliminate the parameter t from the position vector for the motion of a projectile to show that the rectangular equation is

$$y = -\frac{16\sec^2 \theta}{v_0^2}x^2 + (\tan \theta)x + h.$$

30. The path of a ball is given by the rectangular equation

$$y = x - 0.005x^2.$$

Use the result of Exercise 29 to find the position vector. Then find the speed and direction of the ball at the point at which it has traveled 60 feet horizontally.

31. The Rogers Centre in Toronto, Ontario, has a center field fence that is 10 feet high and 400 feet from home plate. A ball is hit 3 feet above the ground and leaves the bat at a speed of 100 miles per hour.

(a) The ball leaves the bat at an angle of $\theta = \theta_0$ with the horizontal. Write the vector-valued function for the path of the ball.

(b) Use a graphing utility to graph the vector-valued function for $\theta_0 = 10°$, $\theta_0 = 15°$, $\theta_0 = 20°$, and $\theta_0 = 25°$. Use the graphs to approximate the minimum angle required for the hit to be a home run.

(c) Determine analytically the minimum angle required for the hit to be a home run.

32. Football

The quarterback of a football team releases a pass at a height of 7 feet above the playing field, and the football is caught by a receiver 30 yards directly downfield at a height of 4 feet. The pass is released at an angle of 35° with the horizontal.

(a) Find the speed of the football when it is released.

(b) Find the maximum height of the football.

(c) Find the time the receiver has to reach the proper position after the quarterback releases the football.

33. A bale ejector consists of two variable-speed belts at the end of a baler. Its purpose is to toss bales into a trailing wagon. In loading the back of a wagon, a bale must be thrown to a position 8 feet above and 16 feet behind the ejector.

(a) Find the minimum initial speed of the bale and the corresponding angle at which it must be ejected from the baler.

(b) The ejector has a fixed angle of 45°. Find the initial speed required.

34. A bomber is flying at an altitude of 30,000 feet at a speed of 540 miles per hour (see figure). When should the bomb be released for it to hit the target? (Give your answer in terms of the angle of depression from the plane to the target.) What is the speed of the bomb at the time of impact?

35. A shot fired from a gun with a muzzle velocity of 1200 feet per second is to hit a target 3000 feet away. Determine the minimum angle of elevation of the gun.

36. A projectile is fired from ground level at an angle of 12° with the horizontal. The projectile is to have a range of 200 feet. Find the minimum initial velocity necessary.

37. Use a graphing utility to graph the paths of a projectile for the given values of θ and v_0. For each case, use the graph to approximate the maximum height and range of the projectile. (Assume that the projectile is launched from ground level.)

(a) $\theta = 10°$, $v_0 = 66$ ft/sec

(b) $\theta = 10°$, $v_0 = 146$ ft/sec

(c) $\theta = 45°$, $v_0 = 66$ ft/sec

(d) $\theta = 45°$, $v_0 = 146$ ft/sec

(e) $\theta = 60°$, $v_0 = 66$ ft/sec

(f) $\theta = 60°$, $v_0 = 146$ ft/sec

38. Find the angles at which an object must be thrown to obtain (a) the maximum range and (b) the maximum height.

Projectile Motion In Exercises 39 and 40, use the model for projectile motion, assuming there is no air resistance. [$g = -9.8$ meters per second per second]

39. Determine the maximum height and range of a projectile fired at a height of 1.5 meters above the ground with an initial velocity of 100 meters per second and at an angle of 30° above the horizontal.

40. A projectile is fired from ground level at an angle of 8° with the horizontal. The projectile is to have a range of 50 meters. Find the minimum initial velocity necessary.

41. **Shot-Put Throw** The path of a shot thrown at an angle θ is

$$\mathbf{r}(t) = (v_0 \cos \theta)t\,\mathbf{i} + \left[h + (v_0 \sin \theta)t - \frac{1}{2}gt^2 \right]\mathbf{j}$$

where v_0 is the initial speed, h is the initial height, t is the time in seconds, and g is the acceleration due to gravity. Verify that the shot will remain in the air for a total of

$$t = \frac{v_0 \sin \theta + \sqrt{v_0^2 \sin^2 \theta + 2gh}}{g} \text{ seconds}$$

and will travel a horizontal distance of

$$\frac{v_0^2 \cos \theta}{g} \left(\sin \theta + \sqrt{\sin^2 \theta + \frac{2gh}{v_0^2}} \right) \text{ feet.}$$

42. Shot-Put Throw

A shot is thrown from a height of $h = 6$ feet with an initial speed of $v_0 = 45$ feet per second and at an angle of $\theta = 42.5°$ with the horizontal. Use the result of Exercise 41 to find the total time of travel and the total horizontal distance traveled.

Cycloidal Motion In Exercises 43 and 44, consider the motion of a point (or particle) on the circumference of a rolling circle. As the circle rolls, it generates the cycloid

$$\mathbf{r}(t) = b(\omega t - \sin \omega t)\mathbf{i} + b(1 - \cos \omega t)\mathbf{j}$$

where ω is the constant angular velocity of the circle and b is the radius of the circle.

43. Find the velocity and acceleration vectors of the particle. Use the results to determine the times at which the speed of the particle will be (a) zero and (b) maximized.

44. Find the maximum speed of a point on the circumference of an automobile tire of radius 1 foot when the automobile is traveling at 60 miles per hour. Compare this speed with the speed of the automobile.

Circular Motion In Exercises 45–48, consider a particle moving on a circular path of radius b described by $\mathbf{r}(t) = b \cos \omega t \,\mathbf{i} + b \sin \omega t \,\mathbf{j}$, where $\omega = du/dt$ is the constant angular velocity.

45. Find the velocity vector and show that it is orthogonal to $\mathbf{r}(t)$.

46. (a) Show that the speed of the particle is $b\omega$.

(b) Use a graphing utility in *parametric* mode to graph the circle for $b = 6$. Try different values of ω. Does the graphing utility draw the circle faster for greater values of ω?

47. Find the acceleration vector and show that its direction is always toward the center of the circle.

48. Show that the magnitude of the acceleration vector is $b\omega^2$.

Circular Motion In Exercises 49 and 50, use the results of Exercises 45–48.

49. A stone weighing 1 pound is attached to a two-foot string and is whirled horizontally (see figure). The string will break under a force of 10 pounds. Find the maximum speed the stone can attain without breaking the string. $\left(\text{Use } \mathbf{F} = m\mathbf{a}, \text{ where } m = \frac{1}{32}.\right)$

Figure for 49

Figure for 50

50. A 3400-pound automobile is negotiating a circular interchange of radius 300 feet at 30 miles per hour (see figure). Assuming the roadway is level, find the force between the tires and the road such that the car stays on the circular path and does not skid. (Use $\mathbf{F} = m\mathbf{a}$, where $m = 3400/32$.) Find the angle at which the roadway should be banked so that no lateral frictional force is exerted on the tires of the automobile.

WRITING ABOUT CONCEPTS

51. Velocity and Speed In your own words, explain the difference between the velocity of an object and its speed.

52. Particle Motion Consider a particle that is moving on the path $\mathbf{r}_1(t) = x(t)\mathbf{i} + y(t)\mathbf{j} + z(t)\mathbf{k}$.

(a) Discuss any changes in the position, velocity, or acceleration of the particle when its position is given by the vector-valued function $\mathbf{r}_2(t) = \mathbf{r}_1(2t)$.

(b) Generalize the results for the vector-valued function $\mathbf{r}_3(t) = \mathbf{r}_1(\omega t)$.

---

53. Proof Prove that when an object is traveling at a constant speed, its velocity and acceleration vectors are orthogonal.

54. Proof Prove that an object moving in a straight line at a constant speed has an acceleration of 0.

55. Investigation A particle moves on an elliptical path given by the vector-valued function $\mathbf{r}(t) = 6 \cos t \,\mathbf{i} + 3 \sin t \,\mathbf{j}$.

(a) Find $\mathbf{v}(t)$, $\|\mathbf{v}(t)\|$, and $\mathbf{a}(t)$.

(b) Use a graphing utility to complete the table.

t	0	$\dfrac{\pi}{4}$	$\dfrac{\pi}{2}$	$\dfrac{2\pi}{3}$	π
Speed					

(c) Graph the elliptical path and the velocity and acceleration vectors at the values of t given in the table in part (b).

(d) Use the results of parts (b) and (c) to describe the geometric relationship between the velocity and acceleration vectors when the speed of the particle is increasing, and when it is decreasing.

56. Particle Motion Consider a particle moving on an elliptical path described by $\mathbf{r}(t) = a \cos \omega t \,\mathbf{i} + b \sin \omega t \,\mathbf{j}$, where $\omega = d\theta/dt$ is the constant angular velocity.

(a) Find the velocity vector. What is the speed of the particle?

(b) Find the acceleration vector and show that its direction is always toward the center of the ellipse.

57. Path of an Object When $t = 0$, an object is at the point $(0, 1)$ and has a velocity vector $\mathbf{v}(0) = -\mathbf{i}$. It moves with an acceleration of $\mathbf{a}(t) = \sin t \,\mathbf{i} - \cos t \,\mathbf{j}$. Show that the path of the object is a circle.

58. **HOW DO YOU SEE IT?** The graph shows the path of a projectile and the velocity and acceleration vectors at times t_1 and t_2. Classify the angle between the velocity vector and the acceleration vector at times t_1 and t_2. Is the speed increasing or decreasing at times t_1 and t_2? Explain your reasoning.

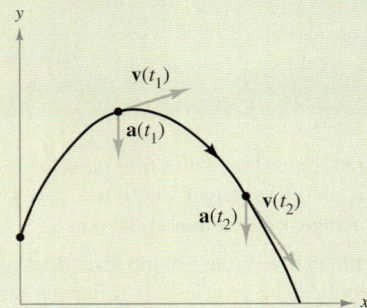

True or False? In Exercises 59–62, determine whether the statement is true or false. If it is false, explain why or give an example that shows it is false.

59. The acceleration of an object is the derivative of the speed.

60. The velocity of an object is the derivative of the position.

61. The velocity vector points in the direction of motion.

62. If a particle moves along a straight line, then the velocity and acceleration vectors are orthogonal.

12.4 Tangent Vectors and Normal Vectors

■ Find a unit tangent vector and a principal unit normal vector at a point on a space curve.
■ Find the tangential and normal components of acceleration.

Tangent Vectors and Normal Vectors

In the preceding section, you learned that the velocity vector points in the direction of motion. This observation leads to the next definition, which applies to any smooth curve—not just to those for which the parameter represents time.

Definition of Unit Tangent Vector

Let C be a smooth curve represented by $\mathbf{r}$ on an open interval I. The **unit tangent vector** $\mathbf{T}(t)$ at t is defined as

$$\mathbf{T}(t) = \frac{\mathbf{r}'(t)}{\|\mathbf{r}'(t)\|}, \quad \mathbf{r}'(t) \neq \mathbf{0}.$$

Recall that a curve is *smooth* on an interval when $\mathbf{r}'$ is continuous and nonzero on the interval. So, "smoothness" is sufficient to guarantee that a curve has a unit tangent vector.

EXAMPLE 1 **Finding the Unit Tangent Vector**

Find the unit tangent vector to the curve given by

$$\mathbf{r}(t) = t\mathbf{i} + t^2\mathbf{j}$$

when $t = 1$.

Solution The derivative of $\mathbf{r}(t)$ is

$$\mathbf{r}'(t) = \mathbf{i} + 2t\mathbf{j}. \qquad \text{Derivative of } \mathbf{r}(t)$$

So, the unit tangent vector is

$$\mathbf{T}(t) = \frac{\mathbf{r}'(t)}{\|\mathbf{r}'(t)\|} \qquad \text{Definition of } \mathbf{T}(t)$$

$$= \frac{1}{\sqrt{1 + 4t^2}}(\mathbf{i} + 2t\mathbf{j}). \qquad \text{Substitute for } \mathbf{r}'(t).$$

When $t = 1$, the unit tangent vector is

$$\mathbf{T}(1) = \frac{1}{\sqrt{5}}(\mathbf{i} + 2\mathbf{j})$$

as shown in Figure 12.20.

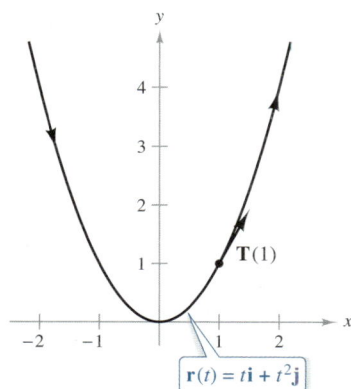

$$\mathbf{r}(t) = t\mathbf{i} + t^2\mathbf{j}$$

The direction of the unit tangent vector depends on the orientation of the curve.
Figure 12.20

In Example 1, note that the direction of the unit tangent vector depends on the orientation of the curve. For the parabola described by

$$\mathbf{r}(t) = -(t - 2)\mathbf{i} + (t - 2)^2\mathbf{j}$$

$\mathbf{T}(1)$ would still represent the unit tangent vector at the point $(1, 1)$, but it would point in the opposite direction. Try verifying this.

The **tangent line to a curve** at a point is the line that passes through the point and is parallel to the unit tangent vector. In Example 2, the unit tangent vector is used to find the tangent line at a point on a helix.

EXAMPLE 2 **Finding the Tangent Line at a Point on a Curve**

Find $\mathbf{T}(t)$ and then find a set of parametric equations for the tangent line to the helix given by

$$\mathbf{r}(t) = 2 \cos t \mathbf{i} + 2 \sin t \mathbf{j} + t \mathbf{k}$$

at the point $\left(\sqrt{2}, \sqrt{2}, \dfrac{\pi}{4} \right)$.

Solution The derivative of $\mathbf{r}(t)$ is

$$\mathbf{r}'(t) = -2 \sin t \mathbf{i} + 2 \cos t \mathbf{j} + \mathbf{k}$$

which implies that $\|\mathbf{r}'(t)\| = \sqrt{4 \sin^2 t + 4 \cos^2 t + 1} = \sqrt{5}$. Therefore, the unit tangent vector is

$$
\begin{aligned}
\mathbf{T}(t) &= \frac{\mathbf{r}'(t)}{\|\mathbf{r}'(t)\|} \\
&= \frac{1}{\sqrt{5}}(-2 \sin t \mathbf{i} + 2 \cos t \mathbf{j} + \mathbf{k}). \qquad \text{\color{red} Unit tangent vector}
\end{aligned}
$$

At the point $\left(\sqrt{2}, \sqrt{2}, \pi/4 \right)$, $t = \pi/4$ and the unit tangent vector is

$$
\begin{aligned}
\mathbf{T}\!\left(\frac{\pi}{4}\right) &= \frac{1}{\sqrt{5}}\left(-2\frac{\sqrt{2}}{2}\mathbf{i} + 2\frac{\sqrt{2}}{2}\mathbf{j} + \mathbf{k}\right) \\
&= \frac{1}{\sqrt{5}}(-\sqrt{2}\,\mathbf{i} + \sqrt{2}\,\mathbf{j} + \mathbf{k}).
\end{aligned}
$$

Using the direction numbers $a = -\sqrt{2}$, $b = \sqrt{2}$, and $c = 1$, and the point $(x_1, y_1, z_1) = \left(\sqrt{2}, \sqrt{2}, \pi/4 \right)$, you can obtain the parametric equations (given with parameter s) listed below.

$$
\begin{aligned}
x &= x_1 + as = \sqrt{2} - \sqrt{2}s \\
y &= y_1 + bs = \sqrt{2} + \sqrt{2}s \\
z &= z_1 + cs = \frac{\pi}{4} + s
\end{aligned}
$$

This tangent line is shown in Figure 12.21. ■

Curve:
$\mathbf{r}(t) = 2 \cos t \mathbf{i} + 2 \sin t \mathbf{j} + t \mathbf{k}$

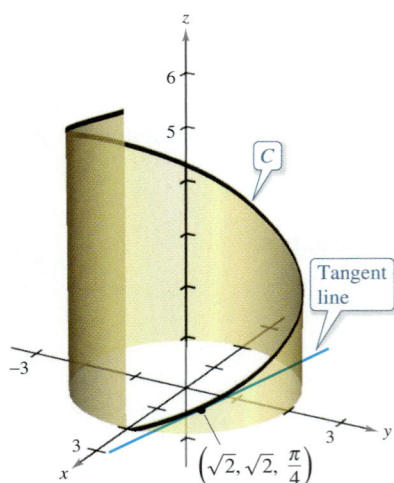

The tangent line to a curve at a point is determined by the unit tangent vector at the point.
Figure 12.21

In Example 2, there are infinitely many vectors that are orthogonal to the tangent vector $\mathbf{T}(t)$. One of these is the vector $\mathbf{T}'(t)$. This follows from Property 7 of Theorem 12.2. That is,

$$\mathbf{T}(t) \cdot \mathbf{T}(t) = \|\mathbf{T}(t)\|^2 = 1 \quad \Longrightarrow \quad \mathbf{T}(t) \cdot \mathbf{T}'(t) = 0.$$

By normalizing the vector $\mathbf{T}'(t)$, you obtain a special vector called the **principal unit normal vector,** as indicated in the next definition.

Definition of Principal Unit Normal Vector

Let C be a smooth curve represented by $\mathbf{r}$ on an open interval I. If $\mathbf{T}'(t) \neq \mathbf{0}$, then the **principal unit normal vector** at t is defined as

$$\mathbf{N}(t) = \frac{\mathbf{T}'(t)}{\|\mathbf{T}'(t)\|}.$$

EXAMPLE 3   **Finding the Principal Unit Normal Vector**

Find $\mathbf{N}(t)$ and $\mathbf{N}(1)$ for the curve represented by $\mathbf{r}(t) = 3t\mathbf{i} + 2t^2\mathbf{j}$.

Solution   By differentiating, you obtain

$$\mathbf{r}'(t) = 3\mathbf{i} + 4t\mathbf{j}$$

which implies that

$$\|\mathbf{r}'(t)\| = \sqrt{9 + 16t^2}.$$

So, the unit tangent vector is

$$\mathbf{T}(t) = \frac{\mathbf{r}'(t)}{\|\mathbf{r}'(t)\|}$$

$$= \frac{1}{\sqrt{9 + 16t^2}}(3\mathbf{i} + 4t\mathbf{j}). \qquad \text{Unit tangent vector}$$

Using Theorem 12.2, differentiate $\mathbf{T}(t)$ with respect to t to obtain

$$\mathbf{T}'(t) = \frac{1}{\sqrt{9 + 16t^2}}(4\mathbf{j}) - \frac{16t}{(9 + 16t^2)^{3/2}}(3\mathbf{i} + 4t\mathbf{j})$$

$$= \frac{12}{(9 + 16t^2)^{3/2}}(-4t\mathbf{i} + 3\mathbf{j})$$

which implies that

$$\|\mathbf{T}'(t)\| = 12\sqrt{\frac{9 + 16t^2}{(9 + 16t^2)^3}} = \frac{12}{9 + 16t^2}.$$

Therefore, the principal unit normal vector is

$$\mathbf{N}(t) = \frac{\mathbf{T}'(t)}{\|\mathbf{T}'(t)\|}$$

$$= \frac{1}{\sqrt{9 + 16t^2}}(-4t\mathbf{i} + 3\mathbf{j}). \qquad \text{Principal unit normal vector}$$

When $t = 1$, the principal unit normal vector is

$$\mathbf{N}(1) = \frac{1}{5}(-4\mathbf{i} + 3\mathbf{j})$$

as shown in Figure 12.22.

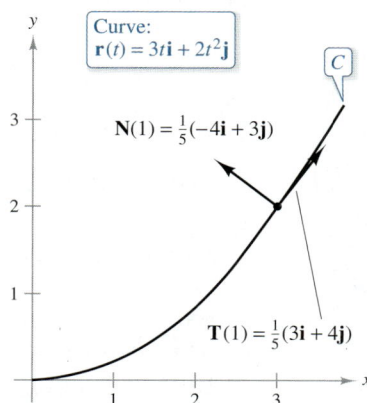

Curve:
$\mathbf{r}(t) = 3t\mathbf{i} + 2t^2\mathbf{j}$

$\mathbf{N}(1) = \frac{1}{5}(-4\mathbf{i} + 3\mathbf{j})$

$\mathbf{T}(1) = \frac{1}{5}(3\mathbf{i} + 4\mathbf{j})$

The principal unit normal vector points toward the concave side of the curve.
Figure 12.22

The principal unit normal vector can be difficult to evaluate algebraically. For plane curves, you can simplify the algebra by finding

$$\mathbf{T}(t) = x(t)\mathbf{i} + y(t)\mathbf{j}$$ Unit tangent vector

and observing that $\mathbf{N}(t)$ must be either

$$\mathbf{N}_1(t) = y(t)\mathbf{i} - x(t)\mathbf{j} \quad \text{or} \quad \mathbf{N}_2(t) = -y(t)\mathbf{i} + x(t)\mathbf{j}.$$

Because $\sqrt{[x(t)]^2 + [y(t)]^2} = 1$, it follows that both $\mathbf{N}_1(t)$ and $\mathbf{N}_2(t)$ are unit normal vectors. The *principal* unit normal vector $\mathbf{N}$ is the one that points toward the concave side of the curve, as shown in Figure 12.22 (see Exercise 76). This also holds for curves in space. That is, for an object moving along a curve C in space, the vector $\mathbf{T}(t)$ points in the direction the object is moving, whereas the vector $\mathbf{N}(t)$ is orthogonal to $\mathbf{T}(t)$ and points in the direction in which the object is turning, as shown in Figure 12.23.

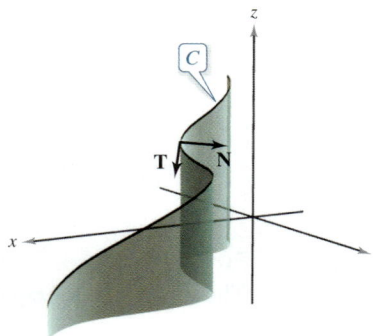

At any point on a curve, a unit normal vector is orthogonal to the unit tangent vector. The *principal* unit normal vector points in the direction in which the curve is turning.
Figure 12.23

Helix:
$\mathbf{r}(t) = 2\cos t\mathbf{i} + 2\sin t\mathbf{j} + t\mathbf{k}$

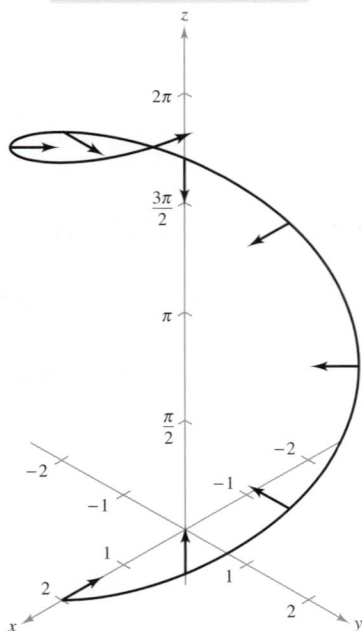

$\mathbf{N}(t)$ is horizontal and points toward the z-axis.
Figure 12.24

EXAMPLE 4   **Finding the Principal Unit Normal Vector**

Find the principal unit normal vector for the helix $\mathbf{r}(t) = 2\cos t\mathbf{i} + 2\sin t\mathbf{j} + t\mathbf{k}$.

Solution   From Example 2, you know that the unit tangent vector is

$$\mathbf{T}(t) = \frac{1}{\sqrt{5}}(-2\sin t\mathbf{i} + 2\cos t\mathbf{j} + \mathbf{k}).$$ Unit tangent vector

So, $\mathbf{T}'(t)$ is given by

$$\mathbf{T}'(t) = \frac{1}{\sqrt{5}}(-2\cos t\mathbf{i} - 2\sin t\mathbf{j}).$$

Because $\|\mathbf{T}'(t)\| = 2/\sqrt{5}$, it follows that the principal unit normal vector is

$$\mathbf{N}(t) = \frac{\mathbf{T}'(t)}{\|\mathbf{T}'(t)\|}$$
$$= \frac{1}{2}(-2\cos t\mathbf{i} - 2\sin t\mathbf{j})$$
$$= -\cos t\mathbf{i} - \sin t\mathbf{j}.$$ Principal unit normal vector

Note that this vector is horizontal and points toward the z-axis, as shown in Figure 12.24.

Tangential and Normal Components of Acceleration

In the preceding section, you considered the problem of describing the motion of an object along a curve. You saw that for an object traveling at a *constant speed,* the velocity and acceleration vectors are perpendicular. This seems reasonable, because the speed would not be constant if any acceleration were acting in the direction of motion. You can verify this observation by noting that

$$\mathbf{r}''(t) \cdot \mathbf{r}'(t) = 0$$

when $\|\mathbf{r}'(t)\|$ is a constant. (See Property 7 of Theorem 12.2.)

For an object traveling at a *variable speed,* however, the velocity and acceleration vectors are not necessarily perpendicular. For instance, you saw that the acceleration vector for a projectile always points down, regardless of the direction of motion.

In general, part of the acceleration (the tangential component) acts in the line of motion, and part of it (the normal component) acts perpendicular to the line of motion. In order to determine these two components, you can use the unit vectors $\mathbf{T}(t)$ and $\mathbf{N}(t)$, which serve in much the same way as do $\mathbf{i}$ and $\mathbf{j}$ in representing vectors in the plane. The next theorem states that the acceleration vector lies in the plane determined by $\mathbf{T}(t)$ and $\mathbf{N}(t)$.

THEOREM 12.4   Acceleration Vector

If $\mathbf{r}(t)$ is the position vector for a smooth curve C and $\mathbf{N}(t)$ exists, then the acceleration vector $\mathbf{a}(t)$ lies in the plane determined by $\mathbf{T}(t)$ and $\mathbf{N}(t)$.

Proof   To simplify the notation, write $\mathbf{T}$ for $\mathbf{T}(t)$, $\mathbf{T}'$ for $\mathbf{T}'(t)$, and so on. Because $\mathbf{T} = \mathbf{r}'/\|\mathbf{r}'\| = \mathbf{v}/\|\mathbf{v}\|$, it follows that

$$\mathbf{v} = \|\mathbf{v}\|\mathbf{T}.$$

By differentiating, you obtain

$$\mathbf{a} = \mathbf{v}' \qquad\qquad\qquad\qquad \text{\color{red}{Product Rule}}$$

$$= \frac{d}{dt}[\|\mathbf{v}\|]\mathbf{T} + \|\mathbf{v}\|\mathbf{T}'$$

$$= \frac{d}{dt}[\|\mathbf{v}\|]\mathbf{T} + \|\mathbf{v}\|\mathbf{T}'\left(\frac{\|\mathbf{T}'\|}{\|\mathbf{T}'\|}\right)$$

$$= \frac{d}{dt}[\|\mathbf{v}\|]\mathbf{T} + \|\mathbf{v}\|\,\|\mathbf{T}'\|\,\mathbf{N}. \qquad \text{\color{red}{$\mathbf{N} = \mathbf{T}'/\|\mathbf{T}'\|$}}$$

Because $\mathbf{a}$ is written as a linear combination of $\mathbf{T}$ and $\mathbf{N}$, it follows that $\mathbf{a}$ lies in the plane determined by $\mathbf{T}$ and $\mathbf{N}$.

See LarsonCalculus.com for Bruce Edwards's video of this proof.

The coefficients of $\mathbf{T}$ and $\mathbf{N}$ in the proof of Theorem 12.4 are called the **tangential and normal components of acceleration** and are denoted by

$$a_{\mathbf{T}} = \frac{d}{dt}[\|\mathbf{v}\|]$$

and $a_{\mathbf{N}} = \|\mathbf{v}\|\,\|\mathbf{T}'\|$. So, you can write

$$\mathbf{a}(t) = a_{\mathbf{T}}\mathbf{T}(t) + a_{\mathbf{N}}\mathbf{N}(t).$$

The next theorem lists some convenient formulas for $a_{\mathbf{N}}$ and $a_{\mathbf{T}}$.

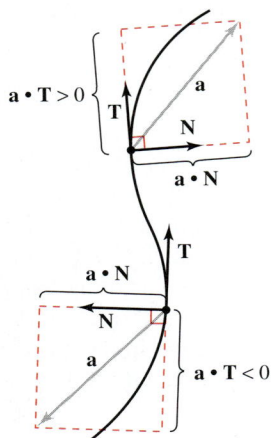

The tangential and normal components of acceleration are obtained by projecting **a** onto **T** and **N**.
Figure 12.25

> **THEOREM 12.5** **Tangential and Normal Components of Acceleration**
>
> If $\mathbf{r}(t)$ is the position vector for a smooth curve C [for which $\mathbf{N}(t)$ exists], then the tangential and normal components of acceleration are as follows.
>
> $$a_{\mathbf{T}} = \frac{d}{dt}[\|\mathbf{v}\|] = \mathbf{a} \cdot \mathbf{T} = \frac{\mathbf{v} \cdot \mathbf{a}}{\|\mathbf{v}\|}$$
>
> $$a_{\mathbf{N}} = \|\mathbf{v}\|\,\|\mathbf{T}'\| = \mathbf{a} \cdot \mathbf{N} = \frac{\|\mathbf{v} \times \mathbf{a}\|}{\|\mathbf{v}\|} = \sqrt{\|\mathbf{a}\|^2 - a_{\mathbf{T}}^2}$$
>
> Note that $a_{\mathbf{N}} \geq 0$. The normal component of acceleration is also called the **centripetal component of acceleration.**

Proof   Note that a lies in the plane of **T** and **N**. So, you can use Figure 12.25 to conclude that, for any time t, the components of the projection of the acceleration vector onto **T** and onto **N** are given by $a_{\mathbf{T}} = \mathbf{a} \cdot \mathbf{T}$ and $a_{\mathbf{N}} = \mathbf{a} \cdot \mathbf{N}$, respectively. Moreover, because $\mathbf{a} = \mathbf{v}'$ and $\mathbf{T} = \mathbf{v}/\|\mathbf{v}\|$, you have

$$a_{\mathbf{T}} = \mathbf{a} \cdot \mathbf{T} = \mathbf{T} \cdot \mathbf{a} = \frac{\mathbf{v}}{\|\mathbf{v}\|} \cdot \mathbf{a} = \frac{\mathbf{v} \cdot \mathbf{a}}{\|\mathbf{v}\|}.$$

In Exercises 78 and 79, you are asked to prove the other parts of the theorem.
See LarsonCalculus.com for Bruce Edwards's video of this proof.

EXAMPLE 5   **Tangential and Normal Components of Acceleration**

⋮ ⋯▷ *See LarsonCalculus.com for an interactive version of this type of example.*

Find the tangential and normal components of acceleration for the position vector given by $\mathbf{r}(t) = 3t\mathbf{i} - t\mathbf{j} + t^2\mathbf{k}$.

Solution   Begin by finding the velocity, speed, and acceleration.

$$\mathbf{v}(t) = \mathbf{r}'(t) = 3\mathbf{i} - \mathbf{j} + 2t\mathbf{k} \qquad \text{Velocity vector}$$

$$\|\mathbf{v}(t)\| = \sqrt{9 + 1 + 4t^2} = \sqrt{10 + 4t^2} \qquad \text{Speed}$$

$$\mathbf{a}(t) = \mathbf{r}''(t) = 2\mathbf{k} \qquad \text{Acceleration vector}$$

By Theorem 12.5, the tangential component of acceleration is

$$a_{\mathbf{T}} = \frac{\mathbf{v} \cdot \mathbf{a}}{\|\mathbf{v}\|} = \frac{4t}{\sqrt{10 + 4t^2}} \qquad \text{Tangential component of acceleration}$$

and because

$$\mathbf{v} \times \mathbf{a} = \begin{vmatrix} \mathbf{i} & \mathbf{j} & \mathbf{k} \\ 3 & -1 & 2t \\ 0 & 0 & 2 \end{vmatrix} = -2\mathbf{i} - 6\mathbf{j}$$

the normal component of acceleration is

$$a_{\mathbf{N}} = \frac{\|\mathbf{v} \times \mathbf{a}\|}{\|\mathbf{v}\|} = \frac{\sqrt{4 + 36}}{\sqrt{10 + 4t^2}} = \frac{2\sqrt{10}}{\sqrt{10 + 4t^2}}. \qquad \text{Normal component of acceleration}$$

In Example 5, you could have used the alternative formula for $a_{\mathbf{N}}$ as follows.

$$a_{\mathbf{N}} = \sqrt{\|\mathbf{a}\|^2 - a_{\mathbf{T}}^2} = \sqrt{(2)^2 - \frac{16t^2}{10 + 4t^2}} = \frac{2\sqrt{10}}{\sqrt{10 + 4t^2}}$$

EXAMPLE 6 Finding $a_{\mathbf{T}}$ and $a_{\mathbf{N}}$ for a Circular Helix

Find the tangential and normal components of acceleration for the helix given by

$$\mathbf{r}(t) = b \cos t\,\mathbf{i} + b \sin t\,\mathbf{j} + ct\mathbf{k}, \quad b > 0.$$

Solution

$$\mathbf{v}(t) = \mathbf{r}'(t) = -b \sin t\,\mathbf{i} + b \cos t\,\mathbf{j} + c\mathbf{k} \qquad \text{Velocity vector}$$

$$\|\mathbf{v}(t)\| = \sqrt{b^2 \sin^2 t + b^2 \cos^2 t + c^2} = \sqrt{b^2 + c^2} \qquad \text{Speed}$$

$$\mathbf{a}(t) = \mathbf{r}''(t) = -b \cos t\,\mathbf{i} - b \sin t\,\mathbf{j} \qquad \text{Acceleration vector}$$

By Theorem 12.5, the tangential component of acceleration is

$$a_{\mathbf{T}} = \frac{\mathbf{v} \cdot \mathbf{a}}{\|\mathbf{v}\|} = \frac{b^2 \sin t \cos t - b^2 \sin t \cos t + 0}{\sqrt{b^2 + c^2}} = 0. \qquad \begin{array}{l}\text{Tangential component}\\\text{of acceleration}\end{array}$$

Moreover, because

$$\|\mathbf{a}\| = \sqrt{b^2 \cos^2 t + b^2 \sin^2 t} = b$$

you can use the alternative formula for the normal component of acceleration to obtain

$$a_{\mathbf{N}} = \sqrt{\|\mathbf{a}\|^2 - a_{\mathbf{T}}^2} = \sqrt{b^2 - 0^2} = b. \qquad \begin{array}{l}\text{Normal component}\\\text{of acceleration}\end{array}$$

Note that the normal component of acceleration is equal to the magnitude of the acceleration. In other words, because the speed is constant, the acceleration is perpendicular to the velocity. See Figure 12.26.

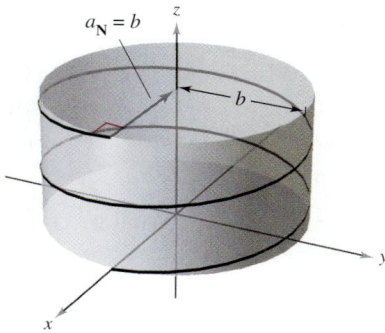

$a_{\mathbf{N}} = b$

The normal component of acceleration is equal to the radius of the cylinder around which the helix is spiraling.
Figure 12.26

EXAMPLE 7 Projectile Motion

The position vector for the projectile shown in Figure 12.27 is

$$\mathbf{r}(t) = \left(50\sqrt{2}\,t\right)\mathbf{i} + \left(50\sqrt{2}\,t - 16t^2\right)\mathbf{j}. \qquad \text{Position vector}$$

Find the tangential components of acceleration when $t = 0$, 1, and $25\sqrt{2}/16$.

Solution

$$\mathbf{v}(t) = 50\sqrt{2}\,\mathbf{i} + \left(50\sqrt{2} - 32t\right)\mathbf{j} \qquad \text{Velocity vector}$$

$$\|\mathbf{v}(t)\| = 2\sqrt{50^2 - 16(50)\sqrt{2}\,t + 16^2 t^2} \qquad \text{Speed}$$

$$\mathbf{a}(t) = -32\mathbf{j} \qquad \text{Acceleration vector}$$

The tangential component of acceleration is

$$a_{\mathbf{T}}(t) = \frac{\mathbf{v}(t) \cdot \mathbf{a}(t)}{\|\mathbf{v}(t)\|} = \frac{-32\left(50\sqrt{2} - 32t\right)}{2\sqrt{50^2 - 16(50)\sqrt{2}\,t + 16^2 t^2}}. \qquad \begin{array}{l}\text{Tangential component}\\\text{of acceleration}\end{array}$$

At the specified times, you have

$$a_{\mathbf{T}}(0) = \frac{-32\left(50\sqrt{2}\right)}{100} = -16\sqrt{2} \approx -22.6$$

$$a_{\mathbf{T}}(1) = \frac{-32\left(50\sqrt{2} - 32\right)}{2\sqrt{50^2 - 16(50)\sqrt{2} + 16^2}} \approx -15.4$$

$$a_{\mathbf{T}}\left(\frac{25\sqrt{2}}{16}\right) = \frac{-32\left(50\sqrt{2} - 50\sqrt{2}\right)}{50\sqrt{2}} = 0.$$

You can see from Figure 12.27 that at the maximum height, when $t = 25\sqrt{2}/16$, the tangential component is 0. This is reasonable because the direction of motion is horizontal at the point and the tangential component of the acceleration is equal to the horizontal component of the acceleration.

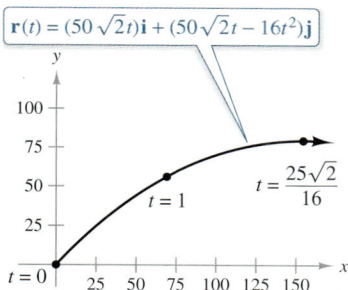

$\mathbf{r}(t) = (50\sqrt{2}\,t)\mathbf{i} + (50\sqrt{2}\,t - 16t^2)\mathbf{j}$

$t = \dfrac{25\sqrt{2}}{16}$

$t = 1$

$t = 0$

The path of a projectile
Figure 12.27

12.4 Exercises

See **CalcChat.com** for tutorial help and worked-out solutions to odd-numbered exercises.

Finding the Unit Tangent Vector In Exercises 1–6, find the unit tangent vector to the curve at the specified value of the parameter.

1. $\mathbf{r}(t) = t^2\mathbf{i} + 2t\mathbf{j}, \quad t = 1$ 2. $\mathbf{r}(t) = t^3\mathbf{i} + 2t^2\mathbf{j}, \quad t = 1$

3. $\mathbf{r}(t) = 4\cos t\mathbf{i} + 4\sin t\mathbf{j}, \quad t = \dfrac{\pi}{4}$

4. $\mathbf{r}(t) = 6\cos t\mathbf{i} + 2\sin t\mathbf{j}, \quad t = \dfrac{\pi}{3}$

5. $\mathbf{r}(t) = 3t\mathbf{i} - \ln t\mathbf{j}, \quad t = e$ 6. $\mathbf{r}(t) = e^t\cos t\mathbf{i} + e^t\mathbf{j}, \quad t = 0$

Finding a Tangent Line In Exercises 7–12, find the unit tangent vector $\mathbf{T}(t)$ and find a set of parametric equations for the line tangent to the space curve at point P.

7. $\mathbf{r}(t) = t\mathbf{i} + t^2\mathbf{j} + t\mathbf{k}, \quad P(0, 0, 0)$
8. $\mathbf{r}(t) = t^2\mathbf{i} + t\mathbf{j} + \frac{4}{3}\mathbf{k}, \quad P\left(1, 1, \frac{4}{3}\right)$
9. $\mathbf{r}(t) = 3\cos t\mathbf{i} + 3\sin t\mathbf{j} + t\mathbf{k}, \quad P(3, 0, 0)$
10. $\mathbf{r}(t) = \langle t, t, \sqrt{4 - t^2}\rangle, \quad P(1, 1, \sqrt{3})$
11. $\mathbf{r}(t) = \langle 2\cos t, 2\sin t, 4\rangle, \quad P(\sqrt{2}, \sqrt{2}, 4)$
12. $\mathbf{r}(t) = \langle 2\sin t, 2\cos t, 4\sin^2 t\rangle, \quad P(1, \sqrt{3}, 1)$

Finding the Principal Unit Normal Vector In Exercises 13–20, find the principal unit normal vector to the curve at the specified value of the parameter.

13. $\mathbf{r}(t) = t\mathbf{i} + \frac{1}{2}t^2\mathbf{j}, \quad t = 2$
14. $\mathbf{r}(t) = t\mathbf{i} + \dfrac{6}{t}\mathbf{j}, \quad t = 3$
15. $\mathbf{r}(t) = \ln t\mathbf{i} + (t + 1)\mathbf{j}, \quad t = 2$
16. $\mathbf{r}(t) = \pi\cos t\mathbf{i} + \pi\sin t\mathbf{j}, \quad t = \dfrac{\pi}{6}$
17. $\mathbf{r}(t) = t\mathbf{i} + t^2\mathbf{j} + \ln t\mathbf{k}, \quad t = 1$
18. $\mathbf{r}(t) = \sqrt{2}t\mathbf{i} + e^t\mathbf{j} + e^{-t}\mathbf{k}, \quad t = 0$
19. $\mathbf{r}(t) = 6\cos t\mathbf{i} + 6\sin t\mathbf{j} + \mathbf{k}, \quad t = \dfrac{3\pi}{4}$
20. $\mathbf{r}(t) = \cos 3t\mathbf{i} + 2\sin 3t\mathbf{j} + \mathbf{k}, \quad t = \pi$

Finding Tangential and Normal Components of Acceleration In Exercises 21–28, find $\mathbf{T}(t)$, $\mathbf{N}(t)$, $a_{\mathbf{T}}$, and $a_{\mathbf{N}}$ at the given time t for the plane curve $\mathbf{r}(t)$.

21. $\mathbf{r}(t) = t\mathbf{i} + \dfrac{1}{t}\mathbf{j}, \quad t = 1$ 22. $\mathbf{r}(t) = t^2\mathbf{i} + 2t\mathbf{j}, \quad t = 1$

23. $\mathbf{r}(t) = (t - t^3)\mathbf{i} + 2t^2\mathbf{j}, \quad t = 1$
24. $\mathbf{r}(t) = (t^3 - 4t)\mathbf{i} + (t^2 - 1)\mathbf{j}, \quad t = 0$
25. $\mathbf{r}(t) = e^t\mathbf{i} + e^{-2t}\mathbf{j}, \quad t = 0$
26. $\mathbf{r}(t) = e^t\mathbf{i} + e^{-t}\mathbf{j} + t\mathbf{k}, \quad t = 0$
27. $\mathbf{r}(t) = e^t\cos t\mathbf{i} + e^t\sin t\mathbf{j}, \quad t = \dfrac{\pi}{2}$
28. $\mathbf{r}(t) = 4\cos 3t\mathbf{i} + 4\sin 3t\mathbf{j}, \quad t = \pi$

Circular Motion In Exercises 29–32, consider an object moving according to the position vector

$\mathbf{r}(t) = a\cos \omega t\,\mathbf{i} + a\sin \omega t\,\mathbf{j}.$

29. Find $\mathbf{T}(t)$, $\mathbf{N}(t)$, $a_{\mathbf{T}}$, and $a_{\mathbf{N}}$.
30. Determine the directions of $\mathbf{T}$ and $\mathbf{N}$ relative to the position vector $\mathbf{r}$.
31. Determine the speed of the object at any time t and explain its value relative to the value of $a_{\mathbf{T}}$.
32. When the angular velocity ω is halved, by what factor is $a_{\mathbf{N}}$ changed?

Sketching a Graph and Vectors In Exercises 33–36, sketch the graph of the plane curve given by the vector-valued function, and, at the point on the curve determined by $\mathbf{r}(t_0)$, sketch the vectors $\mathbf{T}$ and $\mathbf{N}$. Note that $\mathbf{N}$ points toward the concave side of the curve.

Vector-Valued Function	Time
33. $\mathbf{r}(t) = t\mathbf{i} + \dfrac{1}{t}\mathbf{j}$	$t_0 = 2$
34. $\mathbf{r}(t) = t^3\mathbf{i} + t\mathbf{j}$	$t_0 = 1$
35. $\mathbf{r}(t) = (2t + 1)\mathbf{i} - t^2\mathbf{j}$	$t_0 = 2$
36. $\mathbf{r}(t) = 2\cos t\mathbf{i} + 2\sin t\mathbf{j}$	$t_0 = \dfrac{\pi}{4}$

Finding Vectors In Exercises 37–42, find $\mathbf{T}(t)$, $\mathbf{N}(t)$, $a_{\mathbf{T}}$, and $a_{\mathbf{N}}$ at the given time t for the space curve $\mathbf{r}(t)$. [*Hint:* Find $\mathbf{a}(t)$, $\mathbf{T}(t)$, $a_{\mathbf{T}}$, and $a_{\mathbf{N}}$. Solve for $\mathbf{N}$ in the equation $\mathbf{a}(t) = a_{\mathbf{T}}\mathbf{T} + a_{\mathbf{N}}\mathbf{N}$.]

Vector-Valued Function	Time
37. $\mathbf{r}(t) = t\mathbf{i} + 2t\mathbf{j} - 3t\mathbf{k}$	$t = 1$
38. $\mathbf{r}(t) = \cos t\mathbf{i} + \sin t\mathbf{j} + 2t\mathbf{k}$	$t = \dfrac{\pi}{3}$
39. $\mathbf{r}(t) = t\mathbf{i} + t^2\mathbf{j} + \dfrac{t^2}{2}\mathbf{k}$	$t = 1$
40. $\mathbf{r}(t) = (2t - 1)\mathbf{i} + t^2\mathbf{j} - 4t\mathbf{k}$	$t = 2$
41. $\mathbf{r}(t) = e^t\sin t\mathbf{i} + e^t\cos t\mathbf{j} + e^t\mathbf{k}$	$t = 0$
42. $\mathbf{r}(t) = e^t\mathbf{i} + 2t\mathbf{j} + e^{-t}\mathbf{k}$	$t = 0$

WRITING ABOUT CONCEPTS

43. **Definitions** Define the unit tangent vector, the principal unit normal vector, and the tangential and normal components of acceleration.

44. **Unit Tangent Vector** How is the unit tangent vector related to the orientation of a curve? Explain.

45. **Acceleration** Describe the motion of a particle when the normal component of acceleration is 0.

46. **Acceleration** Describe the motion of a particle when the tangential component of acceleration is 0.

47. Finding Vectors An object moves along the path given by

$$\mathbf{r}(t) = 3t\mathbf{i} + 4t\mathbf{j}.$$

Find $\mathbf{v}(t)$, $\mathbf{a}(t)$, $\mathbf{T}(t)$, and $\mathbf{N}(t)$ (if it exists). What is the form of the path? Is the speed of the object constant or changing?

48. HOW DO YOU SEE IT? The figures show the paths of two particles.

(a) Which vector, **s** or **t**, represents the unit tangent vector?

(b) Which vector, **y** or **z**, represents the principal unit normal vector? Explain.

49. Cycloidal Motion The figure shows the path of a particle modeled by the vector-valued function

$$\mathbf{r}(t) = \langle \pi t - \sin \pi t, 1 - \cos \pi t \rangle.$$

The figure also shows the vectors $\mathbf{v}(t)/\|\mathbf{v}(t)\|$ and $\mathbf{a}(t)/\|\mathbf{a}(t)\|$ at the indicated values of t.

(a) Find $a_{\mathbf{T}}$ and $a_{\mathbf{N}}$ at $t = \frac{1}{2}$, $t = 1$, and $t = \frac{3}{2}$.

(b) Determine whether the speed of the particle is increasing or decreasing at each of the indicated values of t. Give reasons for your answers.

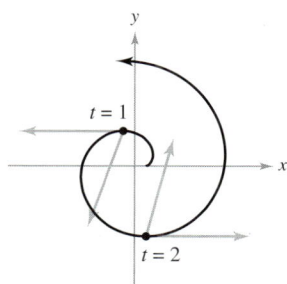

Figure for 49   Figure for 50

50. Motion Along an Involute of a Circle The figure shows a particle moving along a path modeled by

$$\mathbf{r}(t) = \langle \cos \pi t + \pi t \sin \pi t, \sin \pi t - \pi t \cos \pi t \rangle.$$

The figure also shows the vectors $\mathbf{v}(t)$ and $\mathbf{a}(t)$ for $t = 1$ and $t = 2$.

(a) Find $a_{\mathbf{T}}$ and $a_{\mathbf{N}}$ at $t = 1$ and $t = 2$.

(b) Determine whether the speed of the particle is increasing or decreasing at each of the indicated values of t. Give reasons for your answers.

Finding a Binormal Vector In Exercises 51–56, find the vectors **T** and **N**, and the binormal vector $\mathbf{B} = \mathbf{T} \times \mathbf{N}$, for the vector-valued function $\mathbf{r}(t)$ at the given value of t.

51. $\mathbf{r}(t) = 2 \cos t\mathbf{i} + 2 \sin t\mathbf{j} + \frac{t}{2}\mathbf{k}$, $\quad t_0 = \frac{\pi}{2}$

52. $\mathbf{r}(t) = t\mathbf{i} + t^2\mathbf{j} + \frac{t^3}{3}\mathbf{k}$, $\quad t_0 = 1$

53. $\mathbf{r}(t) = \mathbf{i} + \sin t\mathbf{j} + \cos t\mathbf{k}$, $\quad t_0 = \frac{\pi}{4}$

54. $\mathbf{r}(t) = 2e^t\mathbf{i} + e^t \cos t\mathbf{j} + e^t \sin t\mathbf{k}$, $\quad t_0 = 0$

55. $\mathbf{r}(t) = 4 \sin t\mathbf{i} + 4 \cos t\mathbf{j} + 2t\mathbf{k}$, $\quad t_0 = \frac{\pi}{3}$

56. $\mathbf{r}(t) = 3 \cos 2t\mathbf{i} + 3 \sin 2t\mathbf{j} + t\mathbf{k}$, $\quad t_0 = \frac{\pi}{4}$

Alternative Formula for the Principal Unit Normal Vector In Exercises 57–60, use the vector-valued function $\mathbf{r}(t)$ to find the principal unit normal vector $\mathbf{N}(t)$ using the alternative formula

$$\mathbf{N} = \frac{(\mathbf{v} \cdot \mathbf{v})\mathbf{a} - (\mathbf{v} \cdot \mathbf{a})\mathbf{v}}{\|(\mathbf{v} \cdot \mathbf{v})\mathbf{a} - (\mathbf{v} \cdot \mathbf{a})\mathbf{v}\|}.$$

57. $\mathbf{r}(t) = 3t\mathbf{i} + 2t^2\mathbf{j}$

58. $\mathbf{r}(t) = 3 \cos 2t\mathbf{i} + 3 \sin 2t\mathbf{j}$

59. $\mathbf{r}(t) = 2t\mathbf{i} + 4t\mathbf{j} + t^2\mathbf{k}$

60. $\mathbf{r}(t) = 5 \cos t\mathbf{i} + 5 \sin t\mathbf{j} + 3t\mathbf{k}$

61. Projectile Motion Find the tangential and normal components of acceleration for a projectile fired at an angle θ with the horizontal at an initial speed of v_0. What are the components when the projectile is at its maximum height?

62. Projectile Motion Use your results from Exercise 61 to find the tangential and normal components of acceleration for a projectile fired at an angle of $45°$ with the horizontal at an initial speed of 150 feet per second. What are the components when the projectile is at its maximum height?

63. Projectile Motion A projectile is launched with an initial velocity of 120 feet per second at a height of 5 feet and at an angle of $30°$ with the horizontal.

(a) Determine the vector-valued function for the path of the projectile.

(b) Use a graphing utility to graph the path and approximate the maximum height and range of the projectile.

(c) Find $\mathbf{v}(t)$, $\|\mathbf{v}(t)\|$, and $\mathbf{a}(t)$.

(d) Use a graphing utility to complete the table.

t	0.5	1.0	1.5	2.0	2.5	3.0
Speed						

(e) Use a graphing utility to graph the scalar functions $a_{\mathbf{T}}$ and $a_{\mathbf{N}}$. How is the speed of the projectile changing when $a_{\mathbf{T}}$ and $a_{\mathbf{N}}$ have opposite signs?

64. Projectile Motion A projectile is launched with an initial velocity of 220 feet per second at a height of 4 feet and at an angle of 45° with the horizontal.

(a) Determine the vector-valued function for the path of the projectile.

(b) Use a graphing utility to graph the path and approximate the maximum height and range of the projectile.

(c) Find $\mathbf{v}(t)$, $\|\mathbf{v}(t)\|$, and $\mathbf{a}(t)$.

(d) Use a graphing utility to complete the table.

t	0.5	1.0	1.5	2.0	2.5	3.0
Speed						

65. Air Traffic Control

Because of a storm, ground controllers instruct the pilot of a plane flying at an altitude of 4 miles to make a 90° turn and climb to an altitude of 4.2 miles. The model for the path of the plane during this maneuver is

$$\mathbf{r}(t) = \langle 10 \cos 10\pi t, 10 \sin 10\pi t, 4 + 4t \rangle, \quad 0 \le t \le \tfrac{1}{20}$$

where t is the time in hours and $\mathbf{r}$ is the distance in miles.

(a) Determine the speed of the plane.

(b) Calculate $a_\mathbf{T}$ and $a_\mathbf{N}$. Why is one of these equal to 0?

66. Projectile Motion A plane flying at an altitude of 36,000 feet at a speed of 600 miles per hour releases a bomb. Find the tangential and normal components of acceleration acting on the bomb.

67. Centripetal Acceleration An object is spinning at a constant speed on the end of a string, according to the position vector given in Exercises 29–32.

(a) When the angular velocity ω is doubled, how is the centripetal component of acceleration changed?

(b) When the angular velocity is unchanged but the length of the string is halved, how is the centripetal component of acceleration changed?

68. Centripetal Force An object of mass m moves at a constant speed v in a circular path of radius r. The force required to produce the centripetal component of acceleration is called the *centripetal force* and is given by $F = mv^2/r$. Newton's Law of Universal Gravitation is given by $F = GMm/d^2$, where d is the distance between the centers of the two bodies of masses M and m, and G is a gravitational constant. Use this law to show that the speed required for circular motion is $v = \sqrt{GM/r}$.

Orbital Speed In Exercises 69–72, use the result of Exercise 68 to find the speed necessary for the given circular orbit around Earth. Let $GM = 9.56 \times 10^4$ cubic miles per second per second, and assume the radius of Earth is 4000 miles.

69. The orbit of the International Space Station 255 miles above the surface of Earth

70. The orbit of the Hubble telescope 360 miles above the surface of Earth

71. The orbit of a heat capacity mapping satellite 385 miles above the surface of Earth

72. The orbit of a communications satellite r miles above the surface of Earth that is in geosynchronous orbit. [The satellite completes one orbit per sidereal day (approximately 23 hours, 56 minutes), and therefore appears to remain stationary above a point on Earth.]

True or False? In Exercises 73 and 74, determine whether the statement is true or false. If it is false, explain why or give an example that shows it is false.

73. If a car's speedometer is constant, then the car cannot be accelerating.

74. If $a_\mathbf{N} = 0$ for a moving object, then the object is moving in a straight line.

75. Motion of a Particle A particle moves along a path modeled by

$$\mathbf{r}(t) = \cosh(bt)\mathbf{i} + \sinh(bt)\mathbf{j}$$

where b is a positive constant.

(a) Show that the path of the particle is a hyperbola.

(b) Show that $\mathbf{a}(t) = b^2\,\mathbf{r}(t)$.

76. Proof Prove that the principal unit normal vector $\mathbf{N}$ points toward the concave side of a plane curve.

77. Proof Prove that the vector $\mathbf{T}'(t)$ is 0 for an object moving in a straight line.

78. Proof Prove that $a_\mathbf{N} = \dfrac{\|\mathbf{v} \times \mathbf{a}\|}{\|\mathbf{v}\|}$.

79. Proof Prove that $a_\mathbf{N} = \sqrt{\|\mathbf{a}\|^2 - a_\mathbf{T}^2}$.

PUTNAM EXAM CHALLENGE

80. A particle of unit mass moves on a straight line under the action of a force which is a function $f(v)$ of the velocity v of the particle, but the form of this function is not known. A motion is observed, and the distance x covered in time t is found to be connected with t by the formula $x = at + bt^2 + ct^3$, where a, b, and c have numerical values determined by observation of the motion. Find the function $f(v)$ for the range of v covered by the experiment.

This problem was composed by the Committee on the Putnam Prize Competition.
© The Mathematical Association of America. All rights reserved.

Elena Aliaga/Shutterstock.com

12.5 Arc Length and Curvature

■ Find the arc length of a space curve.
■ Use the arc length parameter to describe a plane curve or space curve.
■ Find the curvature of a curve at a point on the curve.
■ Use a vector-valued function to find frictional force.

Arc Length

In Section 10.3, you saw that the arc length of a smooth *plane* curve C given by the parametric equations $x = x(t)$ and $y = y(t)$, $a \le t \le b$, is

$$s = \int_a^b \sqrt{[x'(t)]^2 + [y'(t)]^2}\, dt.$$

In vector form, where C is given by $\mathbf{r}(t) = x(t)\mathbf{i} + y(t)\mathbf{j}$, you can rewrite this equation for arc length as

$$s = \int_a^b \|\mathbf{r}'(t)\|\, dt.$$

The formula for the arc length of a plane curve has a natural extension to a smooth curve in *space,* as stated in the next theorem.

> **THEOREM 12.6 Arc Length of a Space Curve**
>
> If C is a smooth curve given by $\mathbf{r}(t) = x(t)\mathbf{i} + y(t)\mathbf{j} + z(t)\mathbf{k}$ on an interval $[a, b]$, then the arc length of C on the interval is
>
> $$s = \int_a^b \sqrt{[x'(t)]^2 + [y'(t)]^2 + [z'(t)]^2}\, dt = \int_a^b \|\mathbf{r}'(t)\|\, dt.$$

See LarsonCalculus.com for an interactive version of this type of example.

EXAMPLE 1 Finding the Arc Length of a Curve in Space

Find the arc length of the curve given by

$$\mathbf{r}(t) = t\mathbf{i} + \frac{4}{3}t^{3/2}\mathbf{j} + \frac{1}{2}t^2\mathbf{k}$$

from $t = 0$ to $t = 2$, as shown in Figure 12.28.

Solution Using $x(t) = t$, $y(t) = \frac{4}{3}t^{3/2}$, and $z(t) = \frac{1}{2}t^2$, you obtain $x'(t) = 1$, $y'(t) = 2t^{1/2}$, and $z'(t) = t$. So, the arc length from $t = 0$ to $t = 2$ is given by

$$s = \int_0^2 \sqrt{[x'(t)]^2 + [y'(t)]^2 + [z'(t)]^2}\, dt \quad \text{Formula for arc length}$$

$$= \int_0^2 \sqrt{1 + 4t + t^2}\, dt$$

$$= \int_0^2 \sqrt{(t+2)^2 - 3}\, dt \quad \text{Integration tables (Appendix B), Formula 26}$$

$$= \left[\frac{t+2}{2}\sqrt{(t+2)^2 - 3} - \frac{3}{2}\ln\left|(t+2) + \sqrt{(t+2)^2 - 3}\right|\right]_0^2$$

$$= 2\sqrt{13} - \frac{3}{2}\ln\left(4 + \sqrt{13}\right) - 1 + \frac{3}{2}\ln 3$$

$$\approx 4.816.$$

Exploration

Arc Length Formula The formula for the arc length of a space curve is given in terms of the parametric equations used to represent the curve. Does this mean that the arc length of the curve depends on the parameter being used? Would you want this to be true? Explain your reasoning.

Here is a different parametric representation of the curve in Example 1.

$$\mathbf{r}(t) = t^2\mathbf{i} + \frac{4}{3}t^3\mathbf{j} + \frac{1}{2}t^4\mathbf{k}$$

Find the arc length from $t = 0$ to $t = \sqrt{2}$ and compare the result with that found in Example 1.

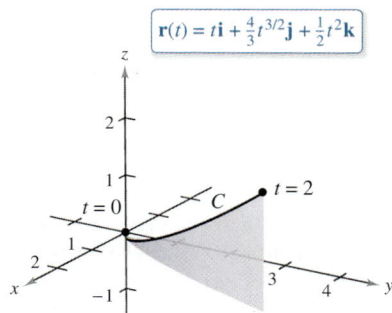

As t increases from 0 to 2, the vector $\mathbf{r}(t)$ traces out a curve.
Figure 12.28

Curve:
$\mathbf{r}(t) = b \cos t\mathbf{i} + b \sin t\mathbf{j} + \sqrt{1 - b^2}\, t\mathbf{k}$

One turn of a helix
Figure 12.29

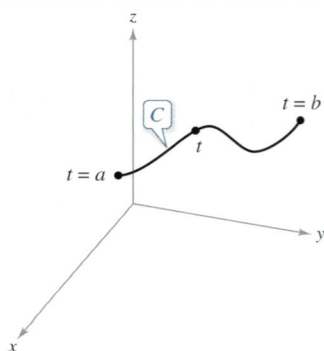

Figure 12.30

EXAMPLE 2 **Finding the Arc Length of a Helix**

Find the length of one turn of the helix given by

$$\mathbf{r}(t) = b \cos t\mathbf{i} + b \sin t\mathbf{j} + \sqrt{1 - b^2}\, t\mathbf{k}$$

as shown in Figure 12.29.

Solution Begin by finding the derivative.

$$\mathbf{r}'(t) = -b \sin t\mathbf{i} + b \cos t\mathbf{j} + \sqrt{1 - b^2}\,\mathbf{k} \qquad \text{Derivative}$$

Now, using the formula for arc length, you can find the length of one turn of the helix by integrating $\|\mathbf{r}'(t)\|$ from 0 to 2π.

$$s = \int_0^{2\pi} \|\mathbf{r}'(t)\|\, dt \qquad \text{Formula for arc length}$$

$$= \int_0^{2\pi} \sqrt{b^2(\sin^2 t + \cos^2 t) + (1 - b^2)}\, dt$$

$$= \int_0^{2\pi} dt$$

$$= t \Big]_0^{2\pi}$$

$$= 2\pi$$

So, the length is 2π units. ■

Arc Length Parameter

You have seen that curves can be represented by vector-valued functions in different ways, depending on the choice of parameter. For *motion* along a curve, the convenient parameter is time t. For studying the *geometric properties* of a curve, however, the convenient parameter is often arc length s.

$s(t) = \int_a^t \sqrt{[x'(u)]^2 + [y'(u)]^2 + [z'(u)]^2}\, du$

Definition of Arc Length Function

Let C be a smooth curve given by $\mathbf{r}(t)$ defined on the closed interval $[a, b]$. For $a \le t \le b$, the **arc length function** is

$$s(t) = \int_a^t \|\mathbf{r}'(u)\|\, du = \int_a^t \sqrt{[x'(u)]^2 + [y'(u)]^2 + [z'(u)]^2}\, du.$$

The arc length s is called the **arc length parameter.** (See Figure 12.30.)

Note that the arc length function s is *nonnegative*. It measures the distance along C from the initial point $(x(a), y(a), z(a))$ to the point $(x(t), y(t), z(t))$.

Using the definition of the arc length function and the Second Fundamental Theorem of Calculus, you can conclude that

$$\frac{ds}{dt} = \|\mathbf{r}'(t)\|. \qquad \text{Derivative of arc length function}$$

In differential form, you can write

$$ds = \|\mathbf{r}'(t)\|\, dt.$$

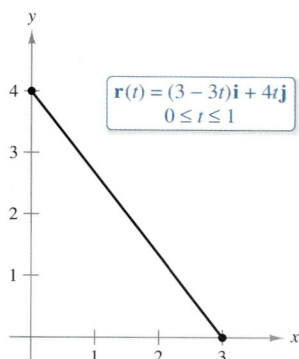

$r(t) = (3 - 3t)\mathbf{i} + 4t\mathbf{j}$
$0 \le t \le 1$

The line segment from $(3, 0)$ to $(0, 4)$ can be parametrized using the arc length parameter s.

Figure 12.31

EXAMPLE 3 **Finding the Arc Length Function for a Line**

Find the arc length function $s(t)$ for the line segment given by

$$\mathbf{r}(t) = (3 - 3t)\mathbf{i} + 4t\mathbf{j}, \quad 0 \le t \le 1$$

and write $\mathbf{r}$ as a function of the parameter s. (See Figure 12.31.)

Solution Because $\mathbf{r}'(t) = -3\mathbf{i} + 4\mathbf{j}$ and

$$\|\mathbf{r}'(t)\| = \sqrt{(-3)^2 + 4^2} = 5$$

you have

$$
\begin{aligned}
s(t) &= \int_0^t \|\mathbf{r}'(u)\| \, du \\
&= \int_0^t 5 \, du \\
&= 5t.
\end{aligned}
$$

Using $s = 5t$ (or $t = s/5$), you can rewrite $\mathbf{r}$ using the arc length parameter as follows.

$$\mathbf{r}(s) = \left(3 - \frac{3}{5}s\right)\mathbf{i} + \frac{4}{5}s\mathbf{j}, \quad 0 \le s \le 5$$

One of the advantages of writing a vector-valued function in terms of the arc length parameter is that $\|\mathbf{r}'(s)\| = 1$. For instance, in Example 3, you have

$$\|\mathbf{r}'(s)\| = \sqrt{\left(-\frac{3}{5}\right)^2 + \left(\frac{4}{5}\right)^2} = 1.$$

So, for a smooth curve C represented by $\mathbf{r}(s)$, where s is the arc length parameter, the arc length between a and b is

$$
\begin{aligned}
\text{Length of arc} &= \int_a^b \|\mathbf{r}'(s)\| \, ds \\
&= \int_a^b ds \\
&= b - a \\
&= \text{length of interval.}
\end{aligned}
$$

Furthermore, if t is *any* parameter such that $\|\mathbf{r}'(t)\| = 1$, then t must be the arc length parameter. These results are summarized in the next theorem, which is stated without proof.

THEOREM 12.7 Arc Length Parameter

If C is a smooth curve given by

$$\mathbf{r}(s) = x(s)\mathbf{i} + y(s)\mathbf{j} \qquad \text{\color{#C00}Plane curve}$$

or

$$\mathbf{r}(s) = x(s)\mathbf{i} + y(s)\mathbf{j} + z(s)\mathbf{k} \qquad \text{\color{#C00}Space curve}$$

where s is the arc length parameter, then

$$\|\mathbf{r}'(s)\| = 1.$$

Moreover, if t is *any* parameter for the vector-valued function $\mathbf{r}$ such that $\|\mathbf{r}'(t)\| = 1$, then t must be the arc length parameter.

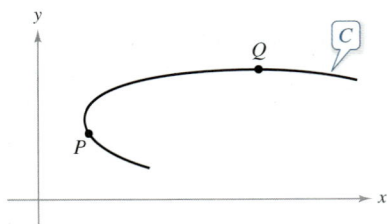

Curvature at P is greater than at Q.
Figure 12.32

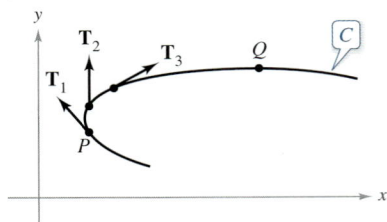

The magnitude of the rate of change of **T** with respect to the arc length is the curvature of a curve.
Figure 12.33

Curvature

An important use of the arc length parameter is to find **curvature**—the measure of how sharply a curve bends. For instance, in Figure 12.32, the curve bends more sharply at P than at Q, and you can say that the curvature is greater at P than at Q. You can calculate curvature by calculating the magnitude of the rate of change of the unit tangent vector **T** with respect to the arc length s, as shown in Figure 12.33.

Definition of Curvature

Let C be a smooth curve (in the plane *or* in space) given by $\mathbf{r}(s)$, where s is the arc length parameter. The **curvature** K at s is

$$K = \left\| \frac{d\mathbf{T}}{ds} \right\| = \|\mathbf{T}'(s)\|.$$

A circle has the same curvature at any point. Moreover, the curvature and the radius of the circle are inversely related. That is, a circle with a large radius has a small curvature, and a circle with a small radius has a large curvature. This inverse relationship is made explicit in the next example.

EXAMPLE 4   **Finding the Curvature of a Circle**

Show that the curvature of a circle of radius r is

$$K = \frac{1}{r}.$$

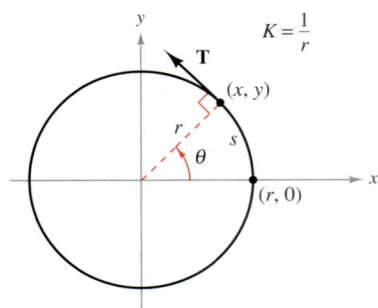

The curvature of a circle is constant.
Figure 12.34

Solution   Without loss of generality, you can consider the circle to be centered at the origin. Let (x, y) be any point on the circle and let s be the length of the arc from $(r, 0)$ to (x, y), as shown in Figure 12.34. By letting θ be the central angle of the circle, you can represent the circle by

$$\mathbf{r}(\theta) = r \cos \theta \mathbf{i} + r \sin \theta \mathbf{j}. \qquad \text{\textcolor{red}{θ is the parameter.}}$$

Using the formula for the length of a circular arc $s = r\theta$, you can rewrite $\mathbf{r}(\theta)$ in terms of the arc length parameter as follows.

$$\mathbf{r}(s) = r \cos \frac{s}{r} \mathbf{i} + r \sin \frac{s}{r} \mathbf{j} \qquad \text{\textcolor{red}{Arc length s is the parameter.}}$$

So, $\mathbf{r}'(s) = -\sin \frac{s}{r} \mathbf{i} + \cos \frac{s}{r} \mathbf{j}$, and it follows that $\|\mathbf{r}'(s)\| = 1$, which implies that the unit tangent vector is

$$\mathbf{T}(s) = \frac{\mathbf{r}'(s)}{\|\mathbf{r}'(s)\|} = -\sin \frac{s}{r} \mathbf{i} + \cos \frac{s}{r} \mathbf{j}$$

and the curvature is

$$K = \|\mathbf{T}'(s)\| = \left\| -\frac{1}{r} \cos \frac{s}{r} \mathbf{i} - \frac{1}{r} \sin \frac{s}{r} \mathbf{j} \right\| = \frac{1}{r}$$

at every point on the circle.

Because a straight line doesn't curve, you would expect its curvature to be 0. Try checking this by finding the curvature of the line given by

$$\mathbf{r}(s) = \left(3 - \frac{3}{5} s \right) \mathbf{i} + \frac{4}{5} s \mathbf{j}.$$

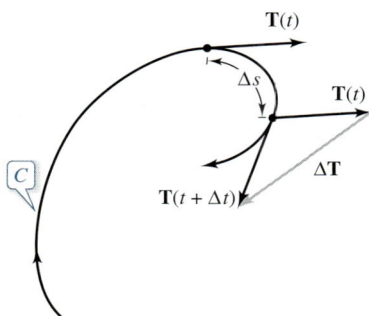

In Example 4, the curvature was found by applying the definition directly. This requires that the curve be written in terms of the arc length parameter s. The next theorem gives two other formulas for finding the curvature of a curve written in terms of an arbitrary parameter t. The proof of this theorem is left as an exercise [see Exercise 84, parts (a) and (b)].

THEOREM 12.8 Formulas for Curvature

If C is a smooth curve given by $\mathbf{r}(t)$, then the curvature K of C at t is

$$K = \frac{\|\mathbf{T}'(t)\|}{\|\mathbf{r}'(t)\|} = \frac{\|\mathbf{r}'(t) \times \mathbf{r}''(t)\|}{\|\mathbf{r}'(t)\|^3}.$$

Because $\|\mathbf{r}'(t)\| = ds/dt$, the first formula implies that curvature is the ratio of the rate of change in the tangent vector $\mathbf{T}$ to the rate of change in arc length. To see that this is reasonable, let Δt be a "small number." Then,

$$\frac{\mathbf{T}'(t)}{ds/dt} \approx \frac{[\mathbf{T}(t + \Delta t) - \mathbf{T}(t)]/\Delta t}{[s(t + \Delta t) - s(t)]/\Delta t} = \frac{\mathbf{T}(t + \Delta t) - \mathbf{T}(t)}{s(t + \Delta t) - s(t)} = \frac{\Delta \mathbf{T}}{\Delta s}.$$

In other words, for a given Δs, the greater the length of $\Delta \mathbf{T}$, the more the curve bends at t, as shown in Figure 12.35.

Figure 12.35

EXAMPLE 5 **Finding the Curvature of a Space Curve**

Find the curvature of the curve given by

$$\mathbf{r}(t) = 2t\mathbf{i} + t^2\mathbf{j} - \frac{1}{3}t^3\mathbf{k}.$$

Solution It is not apparent whether this parameter represents arc length, so you should use the formula $K = \|\mathbf{T}'(t)\|/\|\mathbf{r}'(t)\|$.

$$\mathbf{r}'(t) = 2\mathbf{i} + 2t\mathbf{j} - t^2\mathbf{k}$$

$$\|\mathbf{r}'(t)\| = \sqrt{4 + 4t^2 + t^4} \qquad \text{Length of } \mathbf{r}'(t)$$

$$= t^2 + 2$$

$$\mathbf{T}(t) = \frac{\mathbf{r}'(t)}{\|\mathbf{r}'(t)\|}$$

$$= \frac{2\mathbf{i} + 2t\mathbf{j} - t^2\mathbf{k}}{t^2 + 2}$$

$$\mathbf{T}'(t) = \frac{(t^2 + 2)(2\mathbf{j} - 2t\mathbf{k}) - (2t)(2\mathbf{i} + 2t\mathbf{j} - t^2\mathbf{k})}{(t^2 + 2)^2}$$

$$= \frac{-4t\mathbf{i} + (4 - 2t^2)\mathbf{j} - 4t\mathbf{k}}{(t^2 + 2)^2}$$

$$\|\mathbf{T}'(t)\| = \frac{\sqrt{16t^2 + 16 - 16t^2 + 4t^4 + 16t^2}}{(t^2 + 2)^2}$$

$$= \frac{2(t^2 + 2)}{(t^2 + 2)^2}$$

$$= \frac{2}{t^2 + 2} \qquad \text{Length of } \mathbf{T}'(t)$$

Therefore,

$$K = \frac{\|\mathbf{T}'(t)\|}{\|\mathbf{r}'(t)\|} = \frac{2}{(t^2 + 2)^2}. \qquad \text{Curvature}$$

The next theorem presents a formula for calculating the curvature of a plane curve given by $y = f(x)$.

THEOREM 12.9 **Curvature in Rectangular Coordinates**

If C is the graph of a twice-differentiable function given by $y = f(x)$, then the curvature K at the point (x, y) is

$$K = \frac{|y''|}{[1 + (y')^2]^{3/2}}.$$

Proof By representing the curve C by $\mathbf{r}(x) = x\mathbf{i} + f(x)\mathbf{j} + 0\mathbf{k}$ (where x is the parameter), you obtain $\mathbf{r}'(x) = \mathbf{i} + f'(x)\mathbf{j}$,

$$\|\mathbf{r}'(x)\| = \sqrt{1 + [f'(x)]^2}$$

and $\mathbf{r}''(x) = f''(x)\mathbf{j}$. Because $\mathbf{r}'(x) \times \mathbf{r}''(x) = f''(x)\mathbf{k}$, it follows that the curvature is

$$K = \frac{\|\mathbf{r}'(x) \times \mathbf{r}''(x)\|}{\|\mathbf{r}'(x)\|^3}$$

$$= \frac{|f''(x)|}{\{1 + [f'(x)]^2\}^{3/2}}$$

$$= \frac{|y''|}{[1 + (y')^2]^{3/2}}.$$

See LarsonCalculus.com for Bruce Edwards's video of this proof.

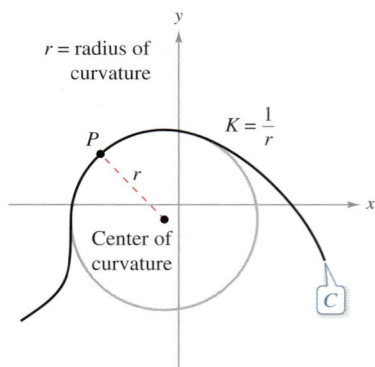

Let C be a curve with curvature K at point P. The circle passing through point P with radius $r = 1/K$ is called the **circle of curvature** when the circle lies on the concave side of the curve and shares a common tangent line with the curve at point P. The radius is called the **radius of curvature** at P, and the center of the circle is called the **center of curvature.**

The circle of curvature gives you a nice way to estimate the curvature K at a point P on a curve graphically. Using a compass, you can sketch a circle that lies against the concave side of the curve at point P, as shown in Figure 12.36. If the circle has a radius of r, then you can estimate the curvature to be $K = 1/r$.

The circle of curvature
Figure 12.36

EXAMPLE 6 **Finding Curvature in Rectangular Coordinates**

Find the curvature of the parabola given by $y = x - \frac{1}{4}x^2$ at $x = 2$. Sketch the circle of curvature at $(2, 1)$.

Solution The curvature at $x = 2$ is as follows.

$$y' = 1 - \frac{x}{2} \qquad\qquad y' = 0$$

$$y'' = -\frac{1}{2} \qquad\qquad y'' = -\frac{1}{2}$$

$$K = \frac{|y''|}{[1 + (y')^2]^{3/2}} \qquad K = \frac{1}{2}$$

Because the curvature at $P(2, 1)$ is $\frac{1}{2}$, it follows that the radius of the circle of curvature at that point is 2. So, the center of curvature is $(2, -1)$, as shown in Figure 12.37. [In the figure, note that the curve has the greatest curvature at P. Try showing that the curvature at $Q(4, 0)$ is $1/2^{5/2} \approx 0.177$.]

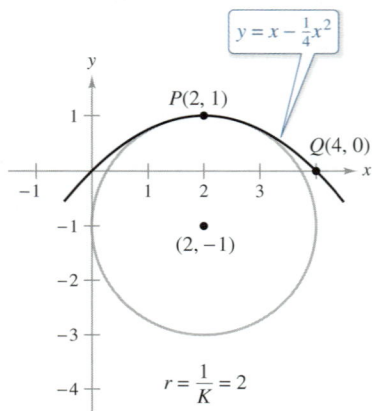

The circle of curvature
Figure 12.37

The amount of thrust felt by passengers in a car that is turning depends on two things—the speed of the car and the sharpness of the turn.

Figure 12.38

Arc length and curvature are closely related to the tangential and normal components of acceleration. The tangential component of acceleration is the rate of change of the speed, which in turn is the rate of change of the arc length. This component is negative as a moving object slows down and positive as it speeds up—regardless of whether the object is turning or traveling in a straight line. So, the tangential component is solely a function of the arc length and is independent of the curvature.

On the other hand, the normal component of acceleration is a function of *both* speed and curvature. This component measures the acceleration acting perpendicular to the direction of motion. To see why the normal component is affected by both speed and curvature, imagine that you are driving a car around a turn, as shown in Figure 12.38. When your speed is high and the turn is sharp, you feel yourself thrown against the car door. By lowering your speed *or* taking a more gentle turn, you are able to lessen this sideways thrust.

The next theorem explicitly states the relationships among speed, curvature, and the components of acceleration.

> **REMARK** Note that Theorem 12.10 gives additional formulas for $a_\mathbf{T}$ and $a_\mathbf{N}$.

THEOREM 12.10   Acceleration, Speed, and Curvature

If $\mathbf{r}(t)$ is the position vector for a smooth curve C, then the acceleration vector is given by

$$\mathbf{a}(t) = \frac{d^2s}{dt^2}\mathbf{T} + K\left(\frac{ds}{dt}\right)^2\mathbf{N}$$

where K is the curvature of C and ds/dt is the speed.

Proof For the position vector $\mathbf{r}(t)$, you have

$$\mathbf{a}(t) = a_\mathbf{T}\mathbf{T} + a_\mathbf{N}\mathbf{N}$$

$$= \frac{d}{dt}[\|\mathbf{v}\|]\mathbf{T} + \|\mathbf{v}\|\,\|\mathbf{T}'\|\mathbf{N}$$

$$= \frac{d^2s}{dt^2}\mathbf{T} + \frac{ds}{dt}(\|\mathbf{v}\|K)\mathbf{N}$$

$$= \frac{d^2s}{dt^2}\mathbf{T} + K\left(\frac{ds}{dt}\right)^2\mathbf{N}.$$

See LarsonCalculus.com for Bruce Edwards's video of this proof.

EXAMPLE 7   **Tangential and Normal Components of Acceleration**

Find $a_\mathbf{T}$ and $a_\mathbf{N}$ for the curve given by

$$\mathbf{r}(t) = 2t\,\mathbf{i} + t^2\mathbf{j} - \tfrac{1}{3}t^3\mathbf{k}.$$

Solution From Example 5, you know that

$$\frac{ds}{dt} = \|\mathbf{r}'(t)\| = t^2 + 2 \quad \text{and} \quad K = \frac{2}{(t^2 + 2)^2}.$$

Therefore,

$$a_\mathbf{T} = \frac{d^2s}{dt^2} = 2t \qquad\qquad \text{Tangential component}$$

and

$$a_\mathbf{N} = K\left(\frac{ds}{dt}\right)^2 = \frac{2}{(t^2 + 2)^2}(t^2 + 2)^2 = 2. \qquad \text{Normal component}$$

Application

There are many applications in physics and engineering dynamics that involve the relationships among speed, arc length, curvature, and acceleration. One such application concerns frictional force.

A moving object with mass m is in contact with a stationary object. The total force required to produce an acceleration $\mathbf{a}$ along a given path is

$$\mathbf{F} = m\mathbf{a}$$
$$= m\left(\frac{d^2s}{dt^2}\right)\mathbf{T} + mK\left(\frac{ds}{dt}\right)^2\mathbf{N}$$
$$= ma_\mathbf{T}\mathbf{T} + ma_\mathbf{N}\mathbf{N}.$$

The portion of this total force that is supplied by the stationary object is called the **force of friction.** For example, when a car moving with constant speed is rounding a turn, the roadway exerts a frictional force that keeps the car from sliding off the road. If the car is not sliding, the frictional force is perpendicular to the direction of motion and has magnitude equal to the normal component of acceleration, as shown in Figure 12.39. The potential frictional force of a road around a turn can be increased by banking the roadway.

Force of friction

The force of friction is perpendicular to the direction of motion.
Figure 12.39

60 km/h

12 m

Figure 12.40

EXAMPLE 8 **Frictional Force**

A 360-kilogram go-cart is driven at a speed of 60 kilometers per hour around a circular racetrack of radius 12 meters, as shown in Figure 12.40. To keep the cart from skidding off course, what frictional force must the track surface exert on the tires?

Solution The frictional force must equal the mass times the normal component of acceleration. For this circular path, you know that the curvature is

$$K = \frac{1}{12}. \qquad \text{Curvature of circular racetrack}$$

Therefore, the frictional force is

$$ma_\mathbf{N} = mK\left(\frac{ds}{dt}\right)^2$$
$$= (360 \text{ kg})\left(\frac{1}{12 \text{ m}}\right)\left(\frac{60,000 \text{ m}}{3600 \text{ sec}}\right)^2$$
$$\approx 8333 \text{ (kg)(m)/sec}^2.$$

SUMMARY OF VELOCITY, ACCELERATION, AND CURVATURE

Unless noted otherwise, let C be a curve (in the plane or in space) given by the position vector

$$\mathbf{r}(t) = x(t)\mathbf{i} + y(t)\mathbf{j}$$ Curve in the plane

or

$$\mathbf{r}(t) = x(t)\mathbf{i} + y(t)\mathbf{j} + z(t)\mathbf{k}$$ Curve in space

where x, y, and z are twice-differentiable functions of t.

Velocity vector, speed, and acceleration vector

$$\mathbf{v}(t) = \mathbf{r}'(t)$$ Velocity vector

$$\|\mathbf{v}(t)\| = \frac{ds}{dt} = \|\mathbf{r}'(t)\|$$ Speed

$$\mathbf{a}(t) = \mathbf{r}''(t)$$ Acceleration vector

$$= a_{\mathbf{T}}\mathbf{T}(t) + a_{\mathbf{N}}\mathbf{N}(t)$$

$$= \frac{d^2s}{dt^2}\mathbf{T}(t) + K\left(\frac{ds}{dt}\right)^2\mathbf{N}(t)$$ K is curvature and $\frac{ds}{dt}$ is speed.

Unit tangent vector and principal unit normal vector

$$\mathbf{T}(t) = \frac{\mathbf{r}'(t)}{\|\mathbf{r}'(t)\|}$$ Unit tangent vector

$$\mathbf{N}(t) = \frac{\mathbf{T}'(t)}{\|\mathbf{T}'(t)\|}$$ Principal unit normal vector

Components of acceleration

$$a_{\mathbf{T}} = \mathbf{a} \cdot \mathbf{T} = \frac{\mathbf{v} \cdot \mathbf{a}}{\|\mathbf{v}\|} = \frac{d^2s}{dt^2}$$ Tangential component of acceleration

$$a_{\mathbf{N}} = \mathbf{a} \cdot \mathbf{N}$$ Normal component of acceleration

$$= \frac{\|\mathbf{v} \times \mathbf{a}\|}{\|\mathbf{v}\|}$$

$$= \sqrt{\|\mathbf{a}\|^2 - a_{\mathbf{T}}^2}$$

$$= K\left(\frac{ds}{dt}\right)^2$$ K is curvature and $\frac{ds}{dt}$ is speed.

Formulas for curvature in the plane

$$K = \frac{|y''|}{[1 + (y')^2]^{3/2}}$$ C given by $y = f(x)$

$$K = \frac{|x'y'' - y'x''|}{[(x')^2 + (y')^2]^{3/2}}$$ C given by $x = x(t), y = y(t)$

Formulas for curvature in the plane or in space

$$K = \|\mathbf{T}'(s)\| = \|\mathbf{r}''(s)\|$$ s is arc length parameter.

$$K = \frac{\|\mathbf{T}'(t)\|}{\|\mathbf{r}'(t)\|} = \frac{\|\mathbf{r}'(t) \times \mathbf{r}''(t)\|}{\|\mathbf{r}'(t)\|^3}$$ t is general parameter.

$$K = \frac{\mathbf{a}(t) \cdot \mathbf{N}(t)}{\|\mathbf{v}(t)\|^2}$$

Cross product formulas apply only to curves in space.

Finding the Arc Length of a Plane Curve In Exercises 1–6, sketch the plane curve and find its length over the given interval.

1. $\mathbf{r}(t) = 3t\mathbf{i} - t\mathbf{j}$, $[0, 3]$

2. $\mathbf{r}(t) = t\mathbf{i} + t^2\mathbf{j}$, $[0, 4]$

3. $\mathbf{r}(t) = t^3\mathbf{i} + t^2\mathbf{j}$, $[0, 1]$

4. $\mathbf{r}(t) = (t + 1)\mathbf{i} + t^2\mathbf{j}$, $[0, 6]$

5. $\mathbf{r}(t) = a\cos^3 t\mathbf{i} + a\sin^3 t\mathbf{j}$, $[0, 2\pi]$

6. $\mathbf{r}(t) = a\cos t\mathbf{i} + a\sin t\mathbf{j}$, $[0, 2\pi]$

7. **Projectile Motion** A baseball is hit 3 feet above the ground at 100 feet per second and at an angle of 45° with respect to the ground.

 (a) Find the vector-valued function for the path of the baseball.

 (b) Find the maximum height.

 (c) Find the range.

 (d) Find the arc length of the trajectory.

8. **Projectile Motion** Repeat Exercise 7 for a baseball that is hit 4 feet above the ground at 80 feet per second and at an angle of 30° with respect to the ground.

Finding the Arc Length of a Curve in Space In Exercises 9–14, sketch the space curve and find its length over the given interval.

Vector-Valued Function	Interval
9. $\mathbf{r}(t) = -t\mathbf{i} + 4t\mathbf{j} + 3t\mathbf{k}$	$[0, 1]$
10. $\mathbf{r}(t) = \mathbf{i} + t^2\mathbf{j} + t^3\mathbf{k}$	$[0, 2]$
11. $\mathbf{r}(t) = \langle 4t, -\cos t, \sin t \rangle$	$\left[0, \dfrac{3\pi}{2}\right]$
12. $\mathbf{r}(t) = \langle 2\sin t, 5t, 2\cos t \rangle$	$[0, \pi]$
13. $\mathbf{r}(t) = a\cos t\mathbf{i} + a\sin t\mathbf{j} + bt\mathbf{k}$	$[0, 2\pi]$
14. $\mathbf{r}(t) = \langle \cos t + t\sin t, \sin t - t\cos t, t^2 \rangle$	$\left[0, \dfrac{\pi}{2}\right]$

15. **Investigation** Consider the graph of the vector-valued function $\mathbf{r}(t) = t\mathbf{i} + (4 - t^2)\mathbf{j} + t^3\mathbf{k}$ on the interval $[0, 2]$.

 (a) Approximate the length of the curve by finding the length of the line segment connecting its endpoints.

 (b) Approximate the length of the curve by summing the lengths of the line segments connecting the terminal points of the vectors $\mathbf{r}(0)$, $\mathbf{r}(0.5)$, $\mathbf{r}(1)$, $\mathbf{r}(1.5)$, and $\mathbf{r}(2)$.

 (c) Describe how you could obtain a more accurate approximation by continuing the processes in parts (a) and (b).

 (d) Use the integration capabilities of a graphing utility to approximate the length of the curve. Compare this result with the answers in parts (a) and (b).

16. **Investigation** Repeat Exercise 15 for the vector-valued function $\mathbf{r}(t) = 6\cos(\pi t/4)\mathbf{i} + 2\sin(\pi t/4)\mathbf{j} + t\mathbf{k}$.

17. **Investigation** Consider the helix represented by the vector-valued function $\mathbf{r}(t) = \langle 2\cos t, 2\sin t, t \rangle$.

 (a) Write the length of the arc s on the helix as a function of t by evaluating the integral

$$s = \int_0^t \sqrt{[x'(u)]^2 + [y'(u)]^2 + [z'(u)]^2}\, du.$$

 (b) Solve for t in the relationship derived in part (a), and substitute the result into the original set of parametric equations. This yields a parametrization of the curve in terms of the arc length parameter s.

 (c) Find the coordinates of the point on the helix for arc lengths $s = \sqrt{5}$ and $s = 4$.

 (d) Verify that $\|\mathbf{r}'(s)\| = 1$.

18. **Investigation** Repeat Exercise 17 for the curve represented by the vector-valued function

$$\mathbf{r}(t) = \left\langle 4(\sin t - t\cos t), 4(\cos t + t\sin t), \tfrac{3}{2}t^2 \right\rangle.$$

Finding Curvature In Exercises 19–22, find the curvature K of the curve, where s is the arc length parameter.

19. $\mathbf{r}(s) = \left(1 + \dfrac{\sqrt{2}}{2}s\right)\mathbf{i} + \left(1 - \dfrac{\sqrt{2}}{2}s\right)\mathbf{j}$

20. $\mathbf{r}(s) = (3 + s)\mathbf{i} + \mathbf{j}$

21. Helix in Exercise 17: $\mathbf{r}(t) = \langle 2\cos t, 2\sin t, t \rangle$

22. Curve in Exercise 18:

$$\mathbf{r}(t) = \left\langle 4(\sin t - t\cos t), 4(\cos t + t\sin t), \tfrac{3}{2}t^2 \right\rangle$$

Finding Curvature In Exercises 23–28, find the curvature K of the plane curve at the given value of the parameter.

23. $\mathbf{r}(t) = 4t\mathbf{i} - 2t\mathbf{j}$, $t = 1$ 24. $\mathbf{r}(t) = t^2\mathbf{i} + \mathbf{j}$, $t = 2$

25. $\mathbf{r}(t) = t\mathbf{i} + \dfrac{1}{t}\mathbf{j}$, $t = 1$ 26. $\mathbf{r}(t) = t\mathbf{i} + \dfrac{1}{9}t^3\mathbf{j}$, $t = 2$

27. $\mathbf{r}(t) = \langle t, \sin t \rangle$, $t = \dfrac{\pi}{2}$

28. $\mathbf{r}(t) = \langle 5\cos t, 4\sin t \rangle$, $t = \dfrac{\pi}{3}$

Finding Curvature In Exercises 29–36, find the curvature K of the curve.

29. $\mathbf{r}(t) = 4\cos 2\pi t\mathbf{i} + 4\sin 2\pi t\mathbf{j}$

30. $\mathbf{r}(t) = 2\cos \pi t\mathbf{i} + \sin \pi t\mathbf{j}$

31. $\mathbf{r}(t) = a\cos \omega t\mathbf{i} + a\sin \omega t\mathbf{j}$

32. $\mathbf{r}(t) = a\cos \omega t\mathbf{i} + b\sin \omega t\mathbf{j}$

33. $\mathbf{r}(t) = t\mathbf{i} + t^2\mathbf{j} + \dfrac{t^2}{2}\mathbf{k}$ 34. $\mathbf{r}(t) = 2t^2\mathbf{i} + t\mathbf{j} + \dfrac{1}{2}t^2\mathbf{k}$

35. $\mathbf{r}(t) = 4t\mathbf{i} + 3\cos t\mathbf{j} + 3\sin t\mathbf{k}$

36. $\mathbf{r}(t) = e^{2t}\mathbf{i} + e^{2t}\cos t\mathbf{j} + e^{2t}\sin t\mathbf{k}$

Finding Curvature   In Exercises 37–40, find the curvature K of the curve at the point P.

37. $\mathbf{r}(t) = 3t\mathbf{i} + 2t^2\mathbf{j}, \quad P(-3, 2)$

38. $\mathbf{r}(t) = e^t\mathbf{i} + 4t\mathbf{j}, \quad P(1, 0)$

39. $\mathbf{r}(t) = t\mathbf{i} + t^2\mathbf{j} + \dfrac{t^3}{4}\mathbf{k}, \quad P(2, 4, 2)$

40. $\mathbf{r}(t) = e^t \cos t\mathbf{i} + e^t \sin t\mathbf{j} + e^t\mathbf{k}, \quad P(1, 0, 1)$

Finding Curvature in Rectangular Coordinates   In Exercises 41–48, find the curvature and radius of curvature of the plane curve at the given value of x.

41. $y = 3x - 2, \quad x = a$

42. $y = 2x + \dfrac{4}{x}, \quad x = 1$

43. $y = 2x^2 + 3, \quad x = -1$

44. $y = \tfrac{3}{4}\sqrt{16 - x^2}, \quad x = 0$

45. $y = \cos 2x, \quad x = 2\pi$

46. $y = e^{3x}, \quad x = 0$

47. $y = x^3, \quad x = 2$

48. $y = x^n, \quad x = 1, \quad n \geq 2$

Maximum Curvature   In Exercises 49–54, (a) find the point on the curve at which the curvature K is a maximum, and (b) find the limit of K as $x \to \infty$.

49. $y = (x - 1)^2 + 3$

50. $y = x^3$

51. $y = x^{2/3}$

52. $y = \dfrac{1}{x}$

53. $y = \ln x$

54. $y = e^x$

Curvature   In Exercises 55–58, find all points on the graph of the function such that the curvature is zero.

55. $y = 1 - x^3$

56. $y = (x - 1)^3 + 3$

57. $y = \cos x$

58. $y = \sin x$

WRITING ABOUT CONCEPTS

59. Arc Length   Give the formula for the arc length of a smooth curve in space.

60. Curvature   Give the formulas for curvature in the plane and in space.

61. Curvature   Describe the graph of a vector-valued function for which the curvature is 0 for all values of t in its domain.

62. Curvature   Given a twice-differentiable function $y = f(x)$, determine its curvature at a relative extremum. Can the curvature ever be greater than it is at a relative extremum? Why or why not?

---

63. Investigation   Consider the function $f(x) = x^4 - x^2$.

(a) Use a computer algebra system to find the curvature K of the curve as a function of x.

(b) Use the result of part (a) to find the circles of curvature to the graph of f when $x = 0$ and $x = 1$. Use a computer algebra system to graph the function and the two circles of curvature.

(c) Graph the function $K(x)$ and compare it with the graph of $f(x)$. For example, do the extrema of f and K occur at the same critical numbers? Explain your reasoning.

06photo/Shutterstock.com

64. Motion of a Particle   A particle moves along the plane curve C described by $\mathbf{r}(t) = t\mathbf{i} + t^2\mathbf{j}$.

(a) Find the length of C on the interval $0 \leq t \leq 2$.

(b) Find the curvature K of the plane curve at $t = 0$, $t = 1$, and $t = 2$.

(c) Describe the curvature of C as t changes from $t = 0$ to $t = 2$.

65. Investigation   Find all a and b such that the two curves given by

$$y_1 = ax(b - x) \quad \text{and} \quad y_2 = \frac{x}{x + 2}$$

intersect at only one point and have a common tangent line and equal curvature at that point. Sketch a graph for each set of values of a and b.

66. **HOW DO YOU SEE IT?**   Using the graph of the ellipse, at what point(s) is the curvature the least and the greatest?

$x^2 + 4y^2 = 4$

67. Sphere and Paraboloid   A sphere of radius 4 is dropped into the paraboloid given by $z = x^2 + y^2$.

(a) How close will the sphere come to the vertex of the paraboloid?

(b) What is the radius of the largest sphere that will touch the vertex?

68. Speed

The smaller the curvature of a bend in a road, the faster a car can travel. Assume that the maximum speed around a turn is inversely proportional to the square root of the curvature. A car moving on the path $y = \tfrac{1}{3}x^3$, where x and y are measured in miles, can safely go 30 miles per hour at $\left(1, \tfrac{1}{3}\right)$. How fast can it go at $\left(\tfrac{3}{2}, \tfrac{9}{8}\right)$?

69. Center of Curvature   Let C be a curve given by $y = f(x)$. Let K be the curvature $(K \neq 0)$ at the point $P(x_0, y_0)$ and let

$$z = \frac{1 + f'(x_0)^2}{f''(x_0)}.$$

Show that the coordinates (α, β) of the center of curvature at P are $(\alpha, \beta) = (x_0 - f'(x_0)z, y_0 + z)$.

70. Center of Curvature Use the result of Exercise 69 to find the center of curvature for the curve at the given point.

(a) $y = e^x$, $(0, 1)$ (b) $y = \dfrac{x^2}{2}, \left(1, \dfrac{1}{2}\right)$ (c) $y = x^2$, $(0, 0)$

71. Curvature A curve C is given by the polar equation $r = f(\theta)$. Show that the curvature K at the point (r, θ) is

$$K = \frac{|2(r')^2 - rr'' + r^2|}{[(r')^2 + r^2]^{3/2}}.$$

[*Hint:* Represent the curve by $\mathbf{r}(\theta) = r \cos \theta \mathbf{i} + r \sin \theta \mathbf{j}$.]

72. Curvature Use the result of Exercise 71 to find the curvature of each polar curve.

(a) $r = 1 + \sin \theta$ (b) $r = \theta$

(c) $r = a \sin \theta$ (d) $r = e^\theta$

73. Curvature Given the polar curve $r = e^{a\theta}$, $a > 0$, find the curvature K and determine the limit of K as (a) $\theta \to \infty$ and (b) $a \to \infty$.

74. Curvature at the Pole Show that the formula for the curvature of a polar curve $r = f(\theta)$ given in Exercise 71 reduces to $K = 2/|r'|$ for the curvature *at the pole*.

Curvature at the Pole In Exercises 75 and 76, use the result of Exercise 74 to find the curvature of the rose curve at the pole.

75. $r = 4 \sin 2\theta$ **76.** $r = 6 \cos 3\theta$

77. Proof For a smooth curve given by the parametric equations $x = f(t)$ and $y = g(t)$, prove that the curvature is given by

$$K = \frac{|f'(t)g''(t) - g'(t)f''(t)|}{\{[f'(t)]^2 + [g'(t)]^2\}^{3/2}}.$$

78. Horizontal Asymptotes Use the result of Exercise 77 to find the curvature K of the curve represented by the parametric equations $x(t) = t^3$ and $y(t) = \frac{1}{2}t^2$. Use a graphing utility to graph K and determine any horizontal asymptotes. Interpret the asymptotes in the context of the problem.

79. Curvature of a Cycloid Use the result of Exercise 77 to find the curvature K of the cycloid represented by the parametric equations

$$x(\theta) = a(\theta - \sin \theta) \quad \text{and} \quad y(\theta) = a(1 - \cos \theta).$$

What are the minimum and maximum values of K?

80. Tangential and Normal Components of Acceleration Use Theorem 12.10 to find a_T and a_N for each curve given by the vector-valued function.

(a) $\mathbf{r}(t) = 3t^2\mathbf{i} + (3t - t^3)\mathbf{j}$

(b) $\mathbf{r}(t) = t\mathbf{i} + t^2\mathbf{j} + \frac{1}{2}t^2\mathbf{k}$

81. Frictional Force A 5500-pound vehicle is driven at a speed of 30 miles per hour on a circular interchange of radius 100 feet. To keep the vehicle from skidding off course, what frictional force must the road surface exert on the tires?

82. Frictional Force A 6400-pound vehicle is driven at a speed of 35 miles per hour on a circular interchange of radius 250 feet. To keep the vehicle from skidding off course, what frictional force must the road surface exert on the tires?

83. Curvature Verify that the curvature at any point (x, y) on the graph of $y = \cosh x$ is $1/y^2$.

84. Formulas for Curvature Use the definition of curvature in space, $K = \|\mathbf{T}'(s)\| = \|\mathbf{r}''(s)\|$, to verify each formula.

(a) $K = \dfrac{\|\mathbf{T}'(t)\|}{\|\mathbf{r}'(t)\|}$

(b) $K = \dfrac{\|\mathbf{r}'(t) \times \mathbf{r}''(t)\|}{\|\mathbf{r}'(t)\|^3}$

(c) $K = \dfrac{\mathbf{a}(t) \cdot \mathbf{N}(t)}{\|\mathbf{v}(t)\|^2}$

True or False? In Exercises 85–88, determine whether the statement is true or false. If it is false, explain why or give an example that shows it is false.

85. The arc length of a space curve depends on the parametrization.

86. The curvature of a circle is the same as its radius.

87. The curvature of a line is 0.

88. The normal component of acceleration is a function of both speed and curvature.

Kepler's Laws In Exercises 89–96, you are asked to verify Kepler's Laws of Planetary Motion. For these exercises, assume that each planet moves in an orbit given by the vector-valued function **r**. Let $r = \|\mathbf{r}\|$, let G represent the universal gravitational constant, let M represent the mass of the sun, and let m represent the mass of the planet.

89. Prove that $\mathbf{r} \cdot \mathbf{r}' = r\dfrac{dr}{dt}$.

90. Using Newton's Second Law of Motion, $\mathbf{F} = m\mathbf{a}$, and Newton's Second Law of Gravitation

$$\mathbf{F} = -\frac{GmM}{r^3}\mathbf{r}$$

show that $\mathbf{a}$ and $\mathbf{r}$ are parallel, and that $\mathbf{r}(t) \times \mathbf{r}'(t) = \mathbf{L}$ is a constant vector. So, $\mathbf{r}(t)$ moves *in a fixed plane*, orthogonal to $\mathbf{L}$.

91. Prove that $\dfrac{d}{dt}\left[\dfrac{\mathbf{r}}{r}\right] = \dfrac{1}{r^3}\{[\mathbf{r} \times \mathbf{r}'] \times \mathbf{r}\}$.

92. Show that $\dfrac{\mathbf{r}'}{GM} \times \mathbf{L} - \dfrac{\mathbf{r}}{r} = \mathbf{e}$ is a constant vector.

93. Prove Kepler's First Law: Each planet moves in an elliptical orbit with the sun as a focus.

94. Assume that the elliptical orbit

$$r = \frac{ed}{1 + e \cos \theta}$$

is in the xy-plane, with $\mathbf{L}$ along the z-axis. Prove that

$$\|\mathbf{L}\| = r^2\frac{d\theta}{dt}.$$

95. Prove Kepler's Second Law: Each ray from the sun to a planet sweeps out equal areas of the ellipse in equal times.

96. Prove Kepler's Third Law: The square of the period of a planet's orbit is proportional to the cube of the mean distance between the planet and the sun.

Review Exercises

See **CalcChat.com** for tutorial help and worked-out solutions to odd-numbered exercises.

Domain and Continuity In Exercises 1–4, (a) find the domain of r, and (b) determine the values (if any) of t for which the function is continuous.

1. $\mathbf{r}(t) = \tan t\,\mathbf{i} + \mathbf{j} + t\,\mathbf{k}$ **2.** $\mathbf{r}(t) = \sqrt{t}\,\mathbf{i} + \dfrac{1}{t-4}\mathbf{j} + \mathbf{k}$

3. $\mathbf{r}(t) = \ln t\,\mathbf{i} + t\,\mathbf{j} + t\,\mathbf{k}$

4. $\mathbf{r}(t) = (2t+1)\mathbf{i} + t^2\mathbf{j} + t\,\mathbf{k}$

Evaluating a Function In Exercises 5 and 6, evaluate (if possible) the vector-valued function at each given value of t.

5. $\mathbf{r}(t) = (2t+1)\mathbf{i} + t^2\mathbf{j} - \sqrt{t+2}\,\mathbf{k}$

   (a) $\mathbf{r}(0)$ (b) $\mathbf{r}(-2)$ (c) $\mathbf{r}(c-1)$

   (d) $\mathbf{r}(1+\Delta t) - \mathbf{r}(1)$

6. $\mathbf{r}(t) = 3\cos t\,\mathbf{i} + (1-\sin t)\mathbf{j} - t\,\mathbf{k}$

   (a) $\mathbf{r}(0)$ (b) $\mathbf{r}\!\left(\dfrac{\pi}{2}\right)$ (c) $\mathbf{r}(s-\pi)$

   (d) $\mathbf{r}(\pi + \Delta t) - \mathbf{r}(\pi)$

Writing a Vector-Valued Function In Exercises 7 and 8, represent the line segment from P to Q by a vector-valued function and by a set of parametric equations.

7. $P(3, 0, 5),\quad Q(2, -2, 3)$

8. $P(-2, -3, 8),\quad Q(5, 1, -2)$

Sketching a Curve In Exercises 9–12, sketch the curve represented by the vector-valued function and give the orientation of the curve.

9. $\mathbf{r}(t) = \langle \pi\cos t, \pi\sin t\rangle$ **10.** $\mathbf{r}(t) = \langle t+2, t^2-1\rangle$

11. $\mathbf{r}(t) = (t+1)\mathbf{i} + (3t-1)\mathbf{j} + 2t\,\mathbf{k}$

12. $\mathbf{r}(t) = 2\cos t\,\mathbf{i} + t\,\mathbf{j} + 2\sin t\,\mathbf{k}$

Representing a Graph by a Vector-Valued Function In Exercises 13 and 14, represent the plane curve by a vector-valued function. (There are many correct answers.)

13. $3x + 4y - 12 = 0$ **14.** $y = 9 - x^2$

Representing a Graph by a Vector-Valued Function In Exercises 15 and 16, sketch the space curve represented by the intersection of the surfaces. Use the parameter $x = t$ to find a vector-valued function for the space curve.

15. $z = x^2 + y^2,\quad x + y = 0$

16. $x^2 + z^2 = 4,\quad x - y = 0$

Finding a Limit In Exercises 17 and 18, find the limit.

17. $\lim\limits_{t\to 4^-}\left(t\,\mathbf{i} + \sqrt{4-t}\,\mathbf{j} + \mathbf{k}\right)$

18. $\lim\limits_{t\to 0}\left(\dfrac{\sin 2t}{t}\mathbf{i} + e^{-t}\mathbf{j} + e^{t}\mathbf{k}\right)$

Higher-Order Differentiation In Exercises 19 and 20, find (a) $\mathbf{r}'(t)$, (b) $\mathbf{r}''(t)$, and (c) $\mathbf{r}'(t) \cdot \mathbf{r}''(t)$.

19. $\mathbf{r}(t) = (t^2 + 4t)\mathbf{i} - 3t^2\mathbf{j}$

20. $\mathbf{r}(t) = 5\cos t\,\mathbf{i} + 2\sin t\,\mathbf{j}$

Higher-Order Differentiation In Exercises 21 and 22, find (a) $\mathbf{r}'(t)$, (b) $\mathbf{r}''(t)$, (c) $\mathbf{r}'(t) \cdot \mathbf{r}''(t)$, and (d) $\mathbf{r}'(t) \times \mathbf{r}''(t)$.

21. $\mathbf{r}(t) = 2t^3\mathbf{i} + 4t\,\mathbf{j} - t^2\mathbf{k}$

22. $\mathbf{r}(t) = (4t+3)\mathbf{i} + t^2\mathbf{j} + (2t^2+4)\mathbf{k}$

Using Properties of the Derivative In Exercises 23 and 24, use the properties of the derivative to find the following.

  (a) $\mathbf{r}'(t)$ (b) $\dfrac{d}{dt}[\mathbf{u}(t) - 2\mathbf{r}(t)]$ (c) $\dfrac{d}{dt}(3t)\mathbf{r}(t)$

  (d) $\dfrac{d}{dt}[\mathbf{r}(t) \cdot \mathbf{u}(t)]$ (e) $\dfrac{d}{dt}[\mathbf{r}(t) \times \mathbf{u}(t)]$ (f) $\dfrac{d}{dt}\mathbf{u}(2t)$

23. $\mathbf{r}(t) = 3t\,\mathbf{i} + (t-1)\mathbf{j},\quad \mathbf{u}(t) = t\,\mathbf{i} + t^2\mathbf{j} + \tfrac{2}{3}t^3\mathbf{k}$

24. $\mathbf{r}(t) = \sin t\,\mathbf{i} + \cos t\,\mathbf{j} + t\,\mathbf{k},\quad \mathbf{u}(t) = \sin t\,\mathbf{i} + \cos t\,\mathbf{j} + \dfrac{1}{t}\mathbf{k}$

Finding an Indefinite Integral In Exercises 25–28, find the indefinite integral.

25. $\displaystyle\int (\mathbf{i} + 3\mathbf{j} + 4t\mathbf{k})\,dt$

26. $\displaystyle\int (t^2\mathbf{i} + 5t\mathbf{j} + 8t^3\mathbf{k})\,dt$

27. $\displaystyle\int \left(3\sqrt{t}\,\mathbf{i} + \dfrac{2}{t}\mathbf{j} + \mathbf{k}\right)dt$

28. $\displaystyle\int (\sin t\,\mathbf{i} + \cos t\,\mathbf{j} + e^{2t}\mathbf{k})\,dt$

Evaluating a Definite Integral In Exercises 29–32, evaluate the definite integral.

29. $\displaystyle\int_{-2}^{2} (3t\,\mathbf{i} + 2t^2\mathbf{j} - t^3\mathbf{k})\,dt$

30. $\displaystyle\int_{0}^{1} (t\,\mathbf{i} + \sqrt{t}\,\mathbf{j} + 4t\mathbf{k})\,dt$

31. $\displaystyle\int_{0}^{2} (e^{t/2}\,\mathbf{i} - 3t^2\mathbf{j} - \mathbf{k})\,dt$

32. $\displaystyle\int_{0}^{\pi/3} (2\cos t\,\mathbf{i} + \sin t\,\mathbf{j} + 3\mathbf{k})\,dt$

Finding an Antiderivative In Exercises 33 and 34, find $\mathbf{r}(t)$ that satisfies the initial condition(s).

33. $\mathbf{r}'(t) = 2t\,\mathbf{i} + e^{t}\mathbf{j} + e^{-t}\mathbf{k},\quad \mathbf{r}(0) = \mathbf{i} + 3\mathbf{j} - 5\mathbf{k}$

34. $\mathbf{r}'(t) = \sec t\,\mathbf{i} + \tan t\,\mathbf{j} + t^2\mathbf{k},\quad \mathbf{r}(0) = 3\mathbf{k}$

Finding Velocity and Acceleration Vectors In Exercises 35–38, the position vector r describes the path of an object moving in space.

(a) Find the velocity vector, speed, and acceleration vector of the object.

(b) Evaluate the velocity vector and acceleration vector of the object at the given value of t.

Position Vector	Time
35. $\mathbf{r}(t) = 4t\mathbf{i} + t^3\mathbf{j} - t\mathbf{k}$	$t = 1$
36. $\mathbf{r}(t) = \sqrt{t}\,\mathbf{i} + 5t\mathbf{j} + 2t^2\mathbf{k}$	$t = 4$
37. $\mathbf{r}(t) = \langle \cos^3 t, \sin^3 t, 3t \rangle$	$t = \pi$
38. $\mathbf{r}(t) = \langle t, -\tan t, e^t \rangle$	$t = 0$

Projectile Motion In Exercises 39–42, use the model for projectile motion, assuming there is no air resistance. $[a(t) = -32$ feet per second per second or $a(t) = -9.8$ meters per second per second]

39. A projectile is fired from ground level with an initial velocity of 84 feet per second at an angle of $30°$ with the horizontal. Find the range of the projectile.

40. A baseball is hit from a height of 3.5 feet above the ground with an initial velocity of 120 feet per second and at an angle of $30°$ above the horizontal. Find the maximum height reached by the baseball. Determine whether it will clear an 8-foot-high fence located 375 feet from home plate.

41. A projectile is fired from ground level at an angle of $20°$ with the horizontal. The projectile has a range of 95 meters. Find the minimum initial velocity.

42. Use a graphing utility to graph the paths of a projectile for $v_0 = 20$ meters per second, $h = 0$ and (a) $\theta = 30°$, (b) $\theta = 45°$, and (c) $\theta = 60°$. Use the graphs to approximate the maximum height and range of the projectile for each case.

Finding the Unit Tangent Vector In Exercises 43 and 44, find the unit tangent vector to the curve at the specified value of the parameter.

43. $\mathbf{r}(t) = 3t\mathbf{i} + 3t^3\mathbf{j}, \quad t = 1$

44. $\mathbf{r}(t) = 2 \sin t\mathbf{i} + 4 \cos t\mathbf{j}, \quad t = \dfrac{\pi}{6}$

Finding a Tangent Line In Exercises 45 and 46, find the unit tangent vector $\mathbf{T}(t)$ and find a set of parametric equations for the line tangent to the space curve at point P.

45. $\mathbf{r}(t) = 2 \cos t\mathbf{i} + 2 \sin t\mathbf{j} + t\mathbf{k}, \quad P\left(1, \sqrt{3}, \dfrac{\pi}{3}\right)$

46. $\mathbf{r}(t) = t\mathbf{i} + t^2\mathbf{j} + \frac{2}{3}t^3\mathbf{k}, \quad P\left(2, 4, \dfrac{16}{3}\right)$

Finding the Principal Unit Normal Vector In Exercises 47–50, find the principal unit normal vector to the curve at the specified value of the parameter.

47. $\mathbf{r}(t) = 2t\mathbf{i} + 3t^2\mathbf{j}, \quad t = 1$ **48.** $\mathbf{r}(t) = t\mathbf{i} + \ln t\mathbf{j}, \quad t = 2$

49. $\mathbf{r}(t) = 3 \cos 2t\mathbf{i} + 3 \sin 2t\mathbf{j} + 3\mathbf{k}, \quad t = \dfrac{\pi}{4}$

50. $\mathbf{r}(t) = 4 \cos t\mathbf{i} + 4 \sin t\mathbf{j} + \mathbf{k}, \quad t = \dfrac{2\pi}{3}$

Finding Tangential and Normal Components of Acceleration In Exercises 51 and 52, find $\mathbf{T}(t)$, $\mathbf{N}(t)$, $a_\mathbf{T}$, and $a_\mathbf{N}$ at the given time t for the plane curve $\mathbf{r}(t)$.

51. $\mathbf{r}(t) = \dfrac{3}{t}\mathbf{i} - 6t\mathbf{j}, \quad t = 3$

52. $\mathbf{r}(t) = 3 \cos 2t\mathbf{i} + 3 \sin 2t\mathbf{j}, \quad t = \dfrac{\pi}{6}$

Finding the Arc Length of a Plane Curve In Exercises 53–56, sketch the plane curve and find its length over the given interval.

	Vector-Valued Function	Interval
53.	$\mathbf{r}(t) = 2t\mathbf{i} - 3t\mathbf{j}$	$[0, 5]$
54.	$\mathbf{r}(t) = t^2\mathbf{i} + 2t\mathbf{k}$	$[0, 3]$
55.	$\mathbf{r}(t) = 10 \cos^3 t\mathbf{i} + 10 \sin^3 t\mathbf{j}$	$[0, 2\pi]$
56.	$\mathbf{r}(t) = 10 \cos t\mathbf{i} + 10 \sin t\mathbf{j}$	$[0, 2\pi]$

Finding the Arc Length of a Curve in Space In Exercises 57–60, sketch the space curve and find its length over the given interval.

	Vector-Valued Function	Interval
57.	$\mathbf{r}(t) = -3t\mathbf{i} + 2t\mathbf{j} + 4t\mathbf{k}$	$[0, 3]$
58.	$\mathbf{r}(t) = t\mathbf{i} + t^2\mathbf{j} + 2t\mathbf{k}$	$[0, 2]$
59.	$\mathbf{r}(t) = \langle 8 \cos t, 8 \sin t, t \rangle$	$\left[0, \dfrac{\pi}{2}\right]$
60.	$\mathbf{r}(t) = \langle 2(\sin t - t \cos t), 2(\cos t + t \sin t), t \rangle$	$\left[0, \dfrac{\pi}{2}\right]$

Finding Curvature In Exercises 61–64, find the curvature K of the curve.

61. $\mathbf{r}(t) = 3t\mathbf{i} + 2t\mathbf{j}$ **62.** $\mathbf{r}(t) = 2\sqrt{t}\,\mathbf{i} + 3t\mathbf{j}$

63. $\mathbf{r}(t) = 2t\mathbf{i} + \frac{1}{2}t^2\mathbf{j} + t^2\mathbf{k}$

64. $\mathbf{r}(t) = 2t\mathbf{i} + 5 \cos t\mathbf{j} + 5 \sin t\mathbf{k}$

Finding Curvature In Exercises 65 and 66, find the curvature K of the curve at the point P.

65. $\mathbf{r}(t) = \frac{1}{2}t^2\mathbf{i} + t\mathbf{j} + \frac{1}{3}t^3\mathbf{k}, \quad P\left(\frac{1}{2}, 1, \frac{1}{3}\right)$

66. $\mathbf{r}(t) = 4 \cos t\mathbf{i} + 3 \sin t\mathbf{j} + t\mathbf{k}, \quad P(-4, 0, \pi)$

Finding Curvature in Rectangular Coordinates In Exercises 67–70, find the curvature and radius of curvature of the plane curve at the given value of x.

67. $y = \frac{1}{2}x^2 + 2, \quad x = 4$ **68.** $y = e^{-x/2}, \quad x = 0$

69. $y = \ln x, \quad x = 1$ **70.** $y = \tan x, \quad x = \dfrac{\pi}{4}$

71. Frictional Force A 7200-pound vehicle is driven at a speed of 25 miles per hour on a circular interchange of radius 150 feet. To keep the vehicle from skidding off course, what frictional force must the road surface exert on the tires?

P.S. Problem Solving

See **CalcChat.com** for tutorial help and worked-out solutions to odd-numbered exercises.

1. Cornu Spiral The **cornu spiral** is given by

$$x(t) = \int_0^t \cos\left(\frac{\pi u^2}{2}\right) du \quad \text{and} \quad y(t) = \int_0^t \sin\left(\frac{\pi u^2}{2}\right) du.$$

The spiral shown in the figure was plotted over the interval $-\pi \le t \le \pi$.

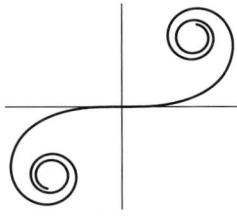

Generated by Mathematica

(a) Find the arc length of this curve from $t = 0$ to $t = a$.

(b) Find the curvature of the graph when $t = a$.

(c) The cornu spiral was discovered by James Bernoulli. He found that the spiral has an amazing relationship between curvature and arc length. What is this relationship?

2. Radius of Curvature Let T be the tangent line at the point $P(x, y)$ to the graph of the curve $x^{2/3} + y^{2/3} = a^{2/3}$, $a > 0$, as shown in the figure. Show that the radius of curvature at P is three times the distance from the origin to the tangent line T.

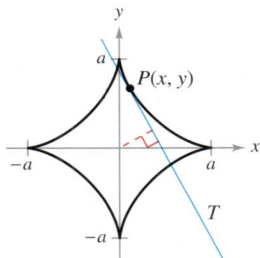

3. Projectile Motion A bomber is flying horizontally at an altitude of 3200 feet with a velocity of 400 feet per second when it releases a bomb. A projectile is launched 5 seconds later from a cannon at a site facing the bomber and 5000 feet from the point that was directly beneath the bomber when the bomb was released, as shown in the figure. The projectile is to intercept the bomb at an altitude of 1600 feet. Determine the required initial speed and angle of inclination of the projectile. (Ignore air resistance.)

4. Projectile Motion Repeat Exercise 3 for the case in which the bomber is facing *away* from the launch site, as shown in the figure.

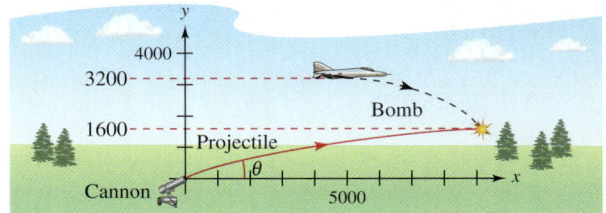

5. Cycloid Consider one arch of the cycloid

$$\mathbf{r}(\theta) = (\theta - \sin\theta)\mathbf{i} + (1 - \cos\theta)\mathbf{j}, \quad 0 \le \theta \le 2\pi$$

as shown in the figure. Let $s(\theta)$ be the arc length from the highest point on the arch to the point $(x(\theta), y(\theta))$, and let $\rho(\theta) = 1/K$ be the radius of curvature at the point $(x(\theta), y(\theta))$. Show that s and ρ are related by the equation $s^2 + \rho^2 = 16$. (This equation is called a *natural equation* for the curve.)

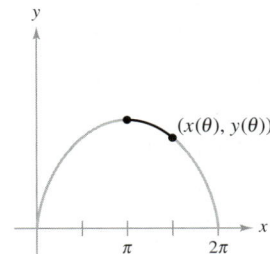

6. Cardioid Consider the cardioid

$$r = 1 - \cos\theta, \quad 0 \le \theta \le 2\pi$$

as shown in the figure. Let $s(\theta)$ be the arc length from the point $(2, \pi)$ on the cardioid to the point (r, θ), and let $p(\theta) = 1/K$ be the radius of curvature at the point (r, θ). Show that s and ρ are related by the equation $s^2 + 9\rho^2 = 16$. (This equation is called a *natural equation* for the curve.)

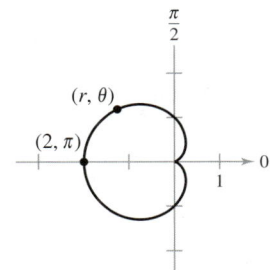

7. Proof If $\mathbf{r}(t)$ is a nonzero differentiable function of t, prove that

$$\frac{d}{dt}(\|\mathbf{r}(t)\|) = \frac{1}{\|\mathbf{r}(t)\|}\mathbf{r}(t) \cdot \mathbf{r}'(t).$$

8. Satellite A communications satellite moves in a circular orbit around Earth at a distance of 42,000 kilometers from the center of Earth. The angular velocity

$$\frac{d\theta}{dt} = \omega = \frac{\pi}{12} \text{ radian per hour}$$

is constant.

(a) Use polar coordinates to show that the acceleration vector is given by

$$\mathbf{a} = \frac{d^2\mathbf{r}}{dt^2} = \left[\frac{d^2r}{dt^2} - r\left(\frac{d\theta}{dt}\right)^2\right]\mathbf{u}_r + \left[r\frac{d^2\theta}{dt^2} + 2\frac{dr}{dt}\frac{d\theta}{dt}\right]\mathbf{u}_\theta$$

where $\mathbf{u}_r = \cos\theta\mathbf{i} + \sin\theta\mathbf{j}$ is the unit vector in the radial direction and $\mathbf{u}_\theta = -\sin\theta\mathbf{i} + \cos\theta\mathbf{j}$.

(b) Find the radial and angular components of acceleration for the satellite.

Binormal Vector **In Exercises 9–11, use the binormal vector defined by the equation $\mathbf{B} = \mathbf{T} \times \mathbf{N}$.**

9. Find the unit tangent, unit normal, and binormal vectors for the helix

$$\mathbf{r}(t) = 4\cos t\mathbf{i} + 4\sin t\mathbf{j} + 3t\mathbf{k}$$

at $t = \pi/2$. Sketch the helix together with these three mutually orthogonal unit vectors.

10. Find the unit tangent, unit normal, and binormal vectors for the curve

$$\mathbf{r}(t) = \cos t\mathbf{i} + \sin t\mathbf{j} - \mathbf{k}$$

at $t = \pi/4$. Sketch the curve together with these three mutually orthogonal unit vectors.

11. (a) Prove that there exists a scalar τ, called the **torsion**, such that $d\mathbf{B}/ds = -\tau\mathbf{N}$.

(b) Prove that $\dfrac{d\mathbf{N}}{ds} = -K\mathbf{T} + \tau\mathbf{B}$.

(The three equations $d\mathbf{T}/ds = K\mathbf{N}$, $d\mathbf{N}/ds = -K\mathbf{T} + \tau\mathbf{B}$, and $d\mathbf{B}/ds = -\tau\mathbf{N}$ are called the *Frenet-Serret formulas.*)

12. Exit Ramp A highway has an exit ramp that begins at the origin of a coordinate system and follows the curve

$$y = \frac{1}{32}x^{5/2}$$

to the point $(4, 1)$ (see figure). Then it follows a circular path whose curvature is that given by the curve at $(4, 1)$. What is the radius of the circular arc? Explain why the curve and the circular arc should have the same curvature at $(4, 1)$.

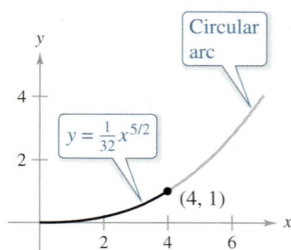

13. Arc Length and Curvature Consider the vector-valued function

$$\mathbf{r}(t) = \langle t\cos\pi t, t\sin\pi t\rangle, \quad 0 \le t \le 2.$$

(a) Use a graphing utility to graph the function.

(b) Find the length of the arc in part (a).

(c) Find the curvature K as a function of t. Find the curvatures for t-values of 0, 1, and 2.

(d) Use a graphing utility to graph the function K.

(e) Find (if possible) $\lim_{t\to\infty} K$.

(f) Using the result of part (e), make a conjecture about the graph of $\mathbf{r}$ as $t \to \infty$.

14. Ferris Wheel You want to toss an object to a friend who is riding a Ferris wheel (see figure). The following parametric equations give the path of the friend $\mathbf{r}_1(t)$ and the path of the object $\mathbf{r}_2(t)$. Distance is measured in meters and time is measured in seconds.

$$\mathbf{r}_1(t) = 15\left(\sin\frac{\pi t}{10}\right)\mathbf{i} + \left(16 - 15\cos\frac{\pi t}{10}\right)\mathbf{j}$$

$$\mathbf{r}_2(t) = [22 - 8.03(t - t_0)]\mathbf{i} +$$
$$[1 + 11.47(t - t_0) - 4.9(t - t_0)^2]\mathbf{j}$$

(a) Locate your friend's position on the Ferris wheel at time $t = 0$.

(b) Determine the number of revolutions per minute of the Ferris wheel.

(c) What are the speed and angle of inclination (in degrees) at which the object is thrown at time $t = t_0$?

(d) Use a graphing utility to graph the vector-valued functions using a value of t_0 that allows your friend to be within reach of the object. (Do this by trial and error.) Explain the significance of t_0.

(e) Find the approximate time your friend should be able to catch the object. Approximate the speeds of your friend and the object at that time.

13

Functions of Several Variables

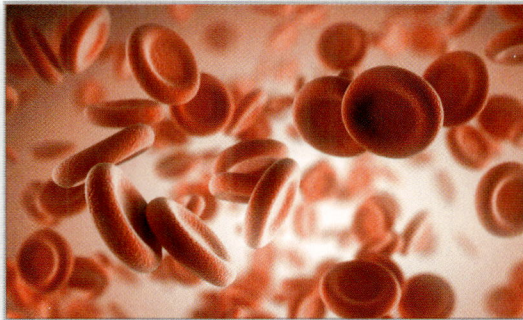
Hardy-Weinberg Law *(Exercise 15, p. 949)*

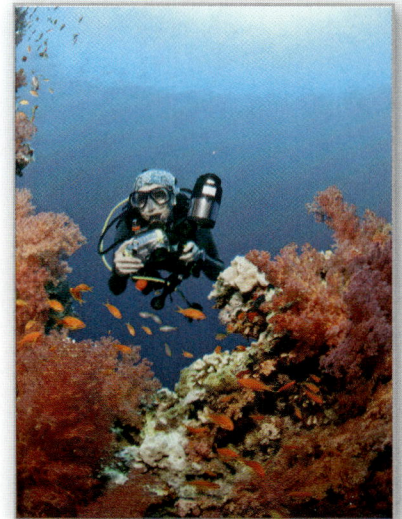
Ocean Floor
(Exercise 74, p. 926)

Wind Chill *(Exercise 31, p. 906)*

Marginal Costs
(Exercise 110, p. 898)

Forestry *(Exercise 75, p. 878)*

Clockwise from top left, Sashkin/Shutterstock.com; Brandelet/Shutterstock.com;
Amy Walters/Shutterstock.com; Val Thoermer/Shutterstock.com; Roca/Shutterstock.com

13.1 Introduction to Functions of Several Variables

- ■ Understand the notation for a function of several variables.
- ■ Sketch the graph of a function of two variables.
- ■ Sketch level curves for a function of two variables.
- ■ Sketch level surfaces for a function of three variables.
- ■ Use computer graphics to graph a function of two variables.

Functions of Several Variables

So far in this text, you have dealt only with functions of a single (independent) variable. Many familiar quantities, however, are functions of two or more variables. Here are three examples.

1. The work done by a force, $W = FD$, is a function of two variables.
2. The volume of a right circular cylinder, $V = \pi r^2 h$, is a function of two variables.
3. The volume of a rectangular solid, $V = lwh$, is a function of three variables.

The notation for a function of two or more variables is similar to that for a function of a single variable. Here are two examples.

$$z = f(\underbrace{x, y}_{\text{2 variables}}) = x^2 + xy \qquad \text{Function of two variables}$$

and

$$w = f(\underbrace{x, y, z}_{\text{3 variables}}) = x + 2y - 3z \qquad \text{Function of three variables}$$

Exploration

Without using a graphing utility, describe the graph of each function of two variables.

a. $z = x^2 + y^2$

b. $z = x + y$

c. $z = x^2 + y$

d. $z = \sqrt{x^2 + y^2}$

e. $z = \sqrt{1 - x^2 + y^2}$

MARY FAIRFAX SOMERVILLE (1780–1872)

Somerville was interested in the problem of creating geometric models for functions of several variables. Her most well-known book, *The Mechanics of the Heavens,* was published in 1831. *See LarsonCalculus.com to read more of this biography.*

Definition of a Function of Two Variables

Let D be a set of ordered pairs of real numbers. If to each ordered pair (x, y) in D there corresponds a unique real number $f(x, y)$, then f is a **function of x and y.** The set D is the **domain** of f, and the corresponding set of values for $f(x, y)$ is the **range** of f. For the function

$$z = f(x, y)$$

x and y are called the **independent variables** and z is called the **dependent variable.**

Similar definitions can be given for functions of three, four, or n variables, where the domains consist of ordered triples (x_1, x_2, x_3), quadruples (x_1, x_2, x_3, x_4), and n-tuples $(x_1, x_2, \ldots, x_n)$. In all cases, the range is a set of real numbers. In this chapter, you will study only functions of two or three variables.

As with functions of one variable, the most common way to describe a function of several variables is with an *equation,* and unless it is otherwise restricted, you can assume that the domain is the set of all points for which the equation is defined. For instance, the domain of the function

$$f(x, y) = x^2 + y^2$$

is the entire xy-plane. Similarly, the domain of

$$f(x, y) = \ln xy$$

is the set of all points (x, y) in the plane for which $xy > 0$. This consists of all points in the first and third quadrants.

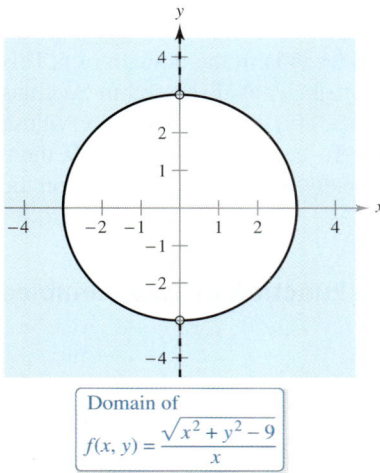

Domain of
$f(x, y) = \dfrac{\sqrt{x^2 + y^2 - 9}}{x}$

Figure 13.1

EXAMPLE 1 **Domains of Functions of Several Variables**

Find the domain of each function.

a. $f(x, y) = \dfrac{\sqrt{x^2 + y^2 - 9}}{x}$ **b.** $g(x, y, z) = \dfrac{x}{\sqrt{9 - x^2 - y^2 - z^2}}$

Solution

a. The function f is defined for all points (x, y) such that $x \neq 0$ and

$$x^2 + y^2 \geq 9.$$

So, the domain is the set of all points lying on or outside the circle $x^2 + y^2 = 9$, *except* those points on the y-axis, as shown in Figure 13.1.

b. The function g is defined for all points (x, y, z) such that

$$x^2 + y^2 + z^2 < 9.$$

Consequently, the domain is the set of all points (x, y, z) lying inside a sphere of radius 3 that is centered at the origin. ∎

Functions of several variables can be combined in the same ways as functions of single variables. For instance, you can form the sum, difference, product, and quotient of two functions of two variables as follows.

$$(f \pm g)(x, y) = f(x, y) \pm g(x, y) \qquad \text{Sum or difference}$$
$$(fg)(x, y) = f(x, y)g(x, y) \qquad \text{Product}$$
$$\frac{f}{g}(x, y) = \frac{f(x, y)}{g(x, y)}, \quad g(x, y) \neq 0 \qquad \text{Quotient}$$

You cannot form the composite of two functions of several variables. You can, however, form the **composite** function $(g \circ h)(x, y)$, where g is a function of a single variable and h is a function of two variables.

$$(g \circ h)(x, y) = g(h(x, y)) \qquad \text{Composition}$$

The domain of this composite function consists of all (x, y) in the domain of h such that $h(x, y)$ is in the domain of g. For example, the function

$$f(x, y) = \sqrt{16 - 4x^2 - y^2}$$

can be viewed as the composite of the function of two variables given by

$$h(x, y) = 16 - 4x^2 - y^2$$

and the function of a single variable given by

$$g(u) = \sqrt{u}.$$

The domain of this function is the set of all points lying on or inside the ellipse $4x^2 + y^2 = 16$.

A function that can be written as a sum of functions of the form $cx^m y^n$ (where c is a real number and m and n are nonnegative integers) is called a **polynomial function** of two variables. For instance, the functions

$$f(x, y) = x^2 + y^2 - 2xy + x + 2 \quad \text{and} \quad g(x, y) = 3xy^2 + x - 2$$

are polynomial functions of two variables. A **rational function** is the quotient of two polynomial functions. Similar terminology is used for functions of more than two variables.

The Graph of a Function of Two Variables

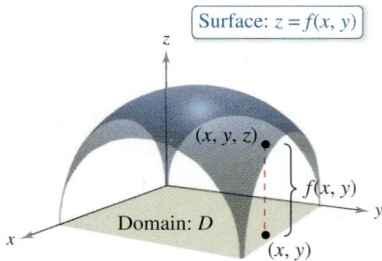

Surface: $z = f(x, y)$

Domain: D

Figure 13.2

As with functions of a single variable, you can learn a lot about the behavior of a function of two variables by sketching its graph. The **graph** of a function f of two variables is the set of all points (x, y, z) for which $z = f(x, y)$ and (x, y) is in the domain of f. This graph can be interpreted geometrically as a *surface in space,* as discussed in Sections 11.5 and 11.6. In Figure 13.2, note that the graph of $z = f(x, y)$ is a surface whose projection onto the xy-plane is D, the domain of f. To each point (x, y) in D there corresponds a point (x, y, z) on the surface, and, conversely, to each point (x, y, z) on the surface there corresponds a point (x, y) in D.

EXAMPLE 2   Describing the Graph of a Function of Two Variables

What is the range of

$$f(x, y) = \sqrt{16 - 4x^2 - y^2}?$$

Describe the graph of f.

Solution   The domain D implied by the equation of f is the set of all points (x, y) such that

$$16 - 4x^2 - y^2 \geq 0.$$

So, D is the set of all points lying on or inside the ellipse

$$\frac{x^2}{4} + \frac{y^2}{16} = 1. \qquad \text{Ellipse in the } xy\text{-plane}$$

The range of f is all values $z = f(x, y)$ such that $0 \leq z \leq \sqrt{16}$, or

$$0 \leq z \leq 4. \qquad \text{Range of } f$$

A point (x, y, z) is on the graph of f if and only if

$$z = \sqrt{16 - 4x^2 - y^2}$$
$$z^2 = 16 - 4x^2 - y^2$$
$$4x^2 + y^2 + z^2 = 16$$
$$\frac{x^2}{4} + \frac{y^2}{16} + \frac{z^2}{16} = 1, \qquad 0 \leq z \leq 4.$$

Surface: $z = \sqrt{16 - 4x^2 - y^2}$

Trace in plane $z = 2$

Range

Domain

The graph of $f(x, y) = \sqrt{16 - 4x^2 - y^2}$ is the upper half of an ellipsoid.

Figure 13.3

From Section 11.6, you know that the graph of f is the upper half of an ellipsoid, as shown in Figure 13.3.

To sketch a surface in space *by hand,* it helps to use traces in planes parallel to the coordinate planes, as shown in Figure 13.3. For example, to find the trace of the surface in the plane $z = 2$, substitute $z = 2$ in the equation $z = \sqrt{16 - 4x^2 - y^2}$ and obtain

$$2 = \sqrt{16 - 4x^2 - y^2} \implies \frac{x^2}{3} + \frac{y^2}{12} = 1.$$

So, the trace is an ellipse centered at the point $(0, 0, 2)$ with major and minor axes of lengths

$$4\sqrt{3} \quad \text{and} \quad 2\sqrt{3}.$$

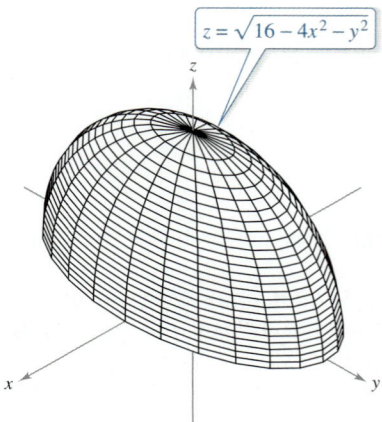

$z = \sqrt{16 - 4x^2 - y^2}$

Figure 13.4

Traces are also used with most three-dimensional graphing utilities. For instance, Figure 13.4 shows a computer-generated version of the surface given in Example 2. For this graph, the computer took 25 traces parallel to the xy-plane and 12 traces in vertical planes.

If you have access to a three-dimensional graphing utility, use it to graph several surfaces.

Level Curves

A second way to visualize a function of two variables is to use a **scalar field** in which the scalar

$$z = f(x, y)$$

is assigned to the point (x, y). A scalar field can be characterized by **level curves** (or **contour lines**) along which the value of $f(x, y)$ is constant. For instance, the weather map in Figure 13.5 shows level curves of equal pressure called **isobars.** In weather maps for which the level curves represent points of equal temperature, the level curves are called **isotherms,** as shown in Figure 13.6. Another common use of level curves is in representing electric potential fields. In this type of map, the level curves are called **equipotential lines.**

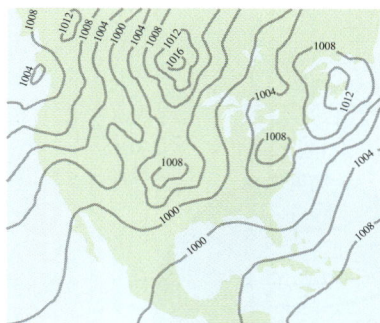

Level curves show the lines of equal pressure (isobars), measured in millibars.

Figure 13.5

Level curves show the lines of equal temperature (isotherms), measured in degrees Fahrenheit.

Figure 13.6

Contour maps are commonly used to show regions on Earth's surface, with the level curves representing the height above sea level. This type of map is called a **topographic map.** For example, the mountain shown in Figure 13.7 is represented by the topographic map in Figure 13.8.

Figure 13.7

Figure 13.8

A contour map depicts the variation of z with respect to x and y by the spacing between level curves. Much space between level curves indicates that z is changing slowly, whereas little space indicates a rapid change in z. Furthermore, to produce a good three-dimensional illusion in a contour map, it is important to choose c-values that are *evenly spaced.*

EXAMPLE 3 Sketching a Contour Map

The hemisphere

$$f(x, y) = \sqrt{64 - x^2 - y^2}$$

is shown in Figure 13.9. Sketch a contour map of this surface using level curves corresponding to $c = 0, 1, 2, \ldots, 8$.

Solution For each value of c, the equation $f(x, y) = c$ is a circle (or point) in the xy-plane. For example, when $c_1 = 0$, the level curve is

$$x^2 + y^2 = 64 \qquad \text{Circle of radius 8}$$

which is a circle of radius 8. Figure 13.10 shows the nine level curves for the hemisphere.

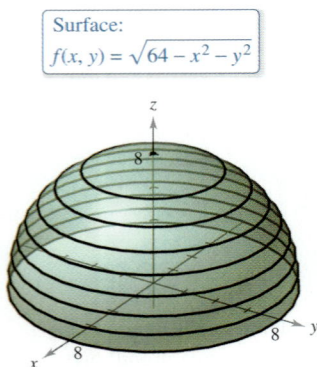

Surface:
$f(x, y) = \sqrt{64 - x^2 - y^2}$

Hemisphere
Figure 13.9

Contour map
Figure 13.10

EXAMPLE 4 Sketching a Contour Map

$\cdots\triangleright$ *See LarsonCalculus.com for an interactive version of this type of example.*

The hyperbolic paraboloid

$$z = y^2 - x^2$$

is shown in Figure 13.11. Sketch a contour map of this surface.

Solution For each value of c, let $f(x, y) = c$ and sketch the resulting level curve in the xy-plane. For this function, each of the level curves ($c \neq 0$) is a hyperbola whose asymptotes are the lines $y = \pm x$. When $c < 0$, the transverse axis is horizontal. For instance, the level curve for $c = -4$ is

$$\frac{x^2}{2^2} - \frac{y^2}{2^2} = 1.$$

When $c > 0$, the transverse axis is vertical. For instance, the level curve for $c = 4$ is

$$\frac{y^2}{2^2} - \frac{x^2}{2^2} = 1.$$

When $c = 0$, the level curve is the degenerate conic representing the intersecting asymptotes, as shown in Figure 13.12.

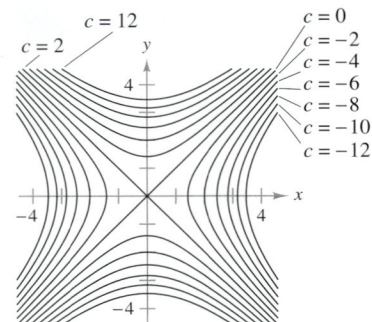

Surface:
$z = y^2 - x^2$

Hyperbolic paraboloid
Figure 13.11

Hyperbolic level curves (at increments of 2)
Figure 13.12

One example of a function of two variables used in economics is the **Cobb-Douglas production function.** This function is used as a model to represent the numbers of units produced by varying amounts of labor and capital. If x measures the units of labor and y measures the units of capital, then the number of units produced is

$$f(x, y) = Cx^a y^{1-a}$$

where C and a are constants with $0 < a < 1$.

EXAMPLE 5   **The Cobb-Douglas Production Function**

A toy manufacturer estimates a production function to be

$$f(x, y) = 100x^{0.6} y^{0.4}$$

where x is the number of units of labor and y is the number of units of capital. Compare the production level when $x = 1000$ and $y = 500$ with the production level when $x = 2000$ and $y = 1000$.

Solution   When $x = 1000$ and $y = 500$, the production level is

$$f(1000, 500) = 100(1000^{0.6})(500^{0.4}) \approx 75{,}786.$$

When $x = 2000$ and $y = 1000$, the production level is

$$f(2000, 1000) = 100(2000^{0.6})(1000^{0.4}) = 151{,}572.$$

The level curves of $z = f(x, y)$ are shown in Figure 13.13. Note that by doubling both x and y, you double the production level (see Exercise 79).

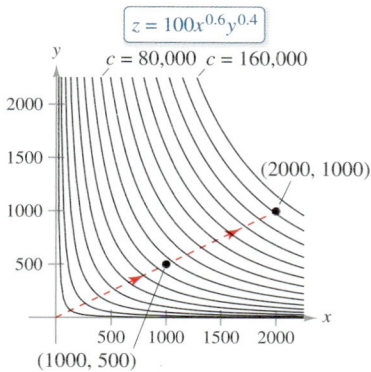

$z = 100x^{0.6}y^{0.4}$
$c = 80{,}000 \quad c = 160{,}000$
$(2000, 1000)$
$(1000, 500)$

Level curves (at increments of 10,000)
Figure 13.13

Level Surfaces

The concept of a level curve can be extended by one dimension to define a **level surface.** If f is a function of three variables and c is a constant, then the graph of the equation

$$f(x, y, z) = c$$

is a **level surface** of the function f, as shown in Figure 13.14.

$f(x, y, z) = c_3$
$f(x, y, z) = c_2$
$f(x, y, z) = c_1$

Level surfaces of f
Figure 13.14

| EXAMPLE 6 | **Level Surfaces** |

Describe the level surfaces of

$$f(x, y, z) = 4x^2 + y^2 + z^2.$$

Solution Each level surface has an equation of the form

$$4x^2 + y^2 + z^2 = c. \qquad \text{Equation of level surface}$$

So, the level surfaces are ellipsoids (whose cross sections parallel to the yz-plane are circles). As c increases, the radii of the circular cross sections increase according to the square root of c. For example, the level surfaces corresponding to the values $c = 0$, $c = 4$, and $c = 16$ are as follows.

$$4x^2 + y^2 + z^2 = 0 \qquad \text{Level surface for } c = 0 \text{ (single point)}$$

$$\frac{x^2}{1} + \frac{y^2}{4} + \frac{z^2}{4} = 1 \qquad \text{Level surface for } c = 4 \text{ (ellipsoid)}$$

$$\frac{x^2}{4} + \frac{y^2}{16} + \frac{z^2}{16} = 1 \qquad \text{Level surface for } c = 16 \text{ (ellipsoid)}$$

These level surfaces are shown in Figure 13.15.

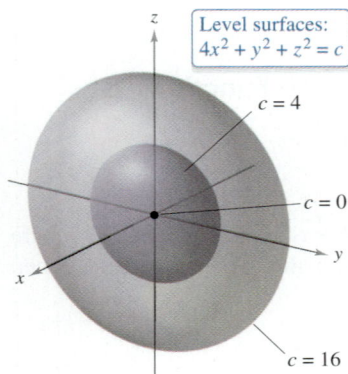

Level surfaces: $4x^2 + y^2 + z^2 = c$

Figure 13.15

If the function in Example 6 represented the *temperature* at the point (x, y, z), then the level surfaces shown in Figure 13.15 would be called **isothermal surfaces.**

Computer Graphics

The problem of sketching the graph of a surface in space can be simplified by using a computer. Although there are several types of three-dimensional graphing utilities, most use some form of trace analysis to give the illusion of three dimensions. To use such a graphing utility, you usually need to enter the equation of the surface and the region in the xy-plane over which the surface is to be plotted. (You might also need to enter the number of traces to be taken.) For instance, to graph the surface

$$f(x, y) = (x^2 + y^2)e^{1-x^2-y^2}$$

you might choose the following bounds for x, y, and z.

$$-3 \le x \le 3 \qquad \text{Bounds for } x$$
$$-3 \le y \le 3 \qquad \text{Bounds for } y$$
$$0 \le z \le 3 \qquad \text{Bounds for } z$$

Figure 13.16 shows a computer-generated graph of this surface using 26 traces taken parallel to the yz-plane. To heighten the three-dimensional effect, the program uses a "hidden line" routine. That is, it begins by plotting the traces in the foreground (those corresponding to the largest x-values), and then, as each new trace is plotted, the program determines whether all or only part of the next trace should be shown.

The graphs on the next page show a variety of surfaces that were plotted by computer. If you have access to a computer drawing program, use it to reproduce these surfaces. Remember also that the three-dimensional graphics in this text can be viewed and rotated. These rotatable graphs are available at *LarsonCalculus.com*.

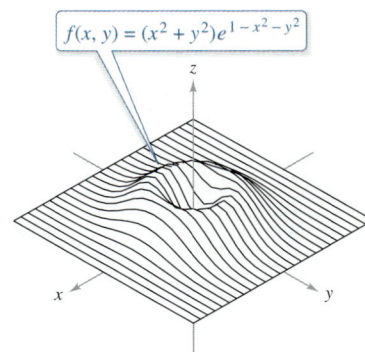

$f(x, y) = (x^2 + y^2)e^{1-x^2-y^2}$

Figure 13.16

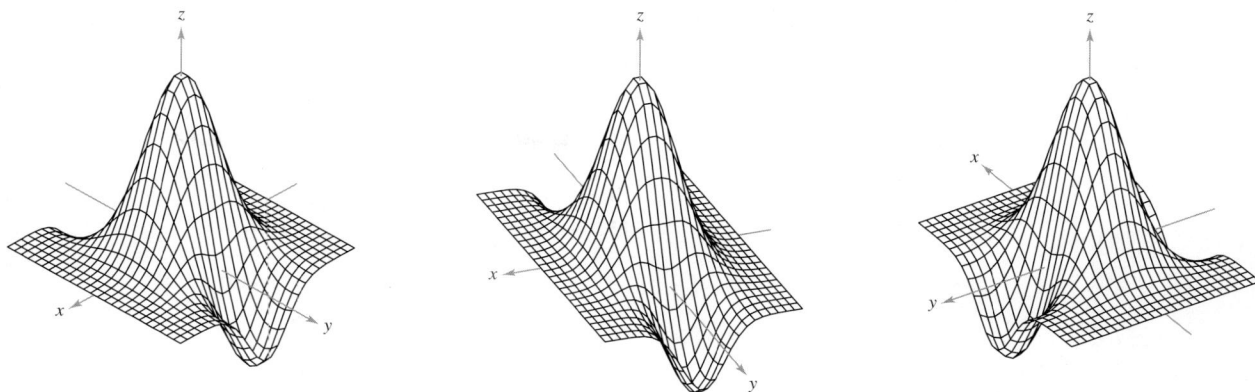

Three different views of the graph of $f(x, y) = (2 - y^2 + x^2)e^{1-x^2-(y^2/4)}$

Single traces

Double traces

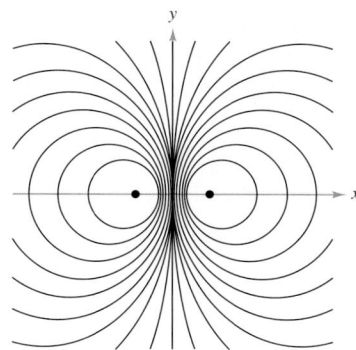

Level curves

Traces and level curves of the graph of $f(x, y) = \dfrac{-4x}{x^2 + y^2 + 1}$

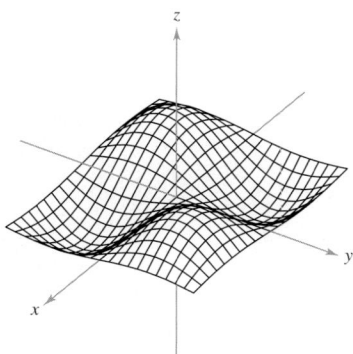

$f(x, y) = \sin x \sin y$

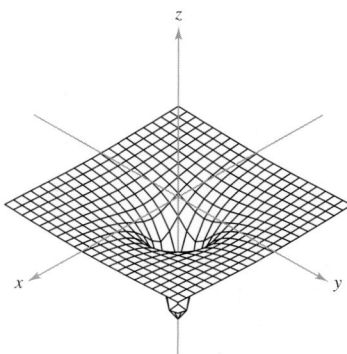

$f(x, y) = -\dfrac{1}{\sqrt{x^2 + y^2}}$

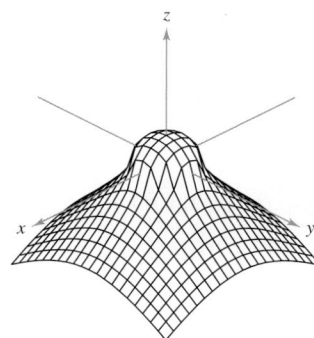

$f(x, y) = \dfrac{1 - x^2 - y^2}{\sqrt{|1 - x^2 - y^2|}}$

Determining Whether a Graph Is a Function In Exercises 1 and 2, use the graph to determine whether z is a function of x and y. Explain.

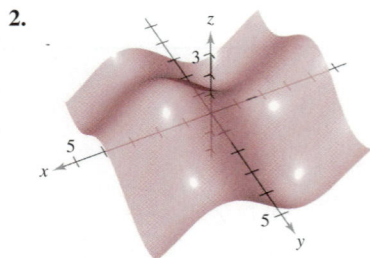

1.

2.

Determining Whether an Equation Is a Function In Exercises 3–6, determine whether z is a function of x and y.

3. $x^2z + 3y^2 - xy = 10$ **4.** $xz^2 + 2xy - y^2 = 4$

5. $\dfrac{x^2}{4} + \dfrac{y^2}{9} + z^2 = 1$ **6.** $z + x \ln y - 8yz = 0$

Evaluating a Function In Exercises 7–18, find and simplify the function values.

7. $f(x, y) = xy$

(a) $(3, 2)$ (b) $(-1, 4)$ (c) $(30, 5)$

(d) $(5, y)$ (e) $(x, 2)$ (f) $(5, t)$

8. $f(x, y) = 4 - x^2 - 4y^2$

(a) $(0, 0)$ (b) $(0, 1)$ (c) $(2, 3)$

(d) $(1, y)$ (e) $(x, 0)$ (f) $(t, 1)$

9. $f(x, y) = xe^y$

(a) $(5, 0)$ (b) $(3, 2)$ (c) $(2, -1)$

(d) $(5, y)$ (e) $(x, 2)$ (f) (t, t)

10. $g(x, y) = \ln|x + y|$

(a) $(1, 0)$ (b) $(0, -1)$ (c) $(0, e)$

(d) $(1, 1)$ (e) $(e, e/2)$ (f) $(2, 5)$

11. $h(x, y, z) = \dfrac{xy}{z}$

(a) $(2, 3, 9)$ (b) $(1, 0, 1)$ (c) $(-2, 3, 4)$ (d) $(5, 4, -6)$

12. $f(x, y, z) = \sqrt{x + y + z}$

(a) $(0, 5, 4)$ (b) $(6, 8, -3)$

(c) $(4, 6, 2)$ (d) $(10, -4, -3)$

13. $f(x, y) = x \sin y$

(a) $(2, \pi/4)$ (b) $(3, 1)$ (c) $(-3, \pi/3)$ (d) $(4, \pi/2)$

14. $V(r, h) = \pi r^2 h$

(a) $(3, 10)$ (b) $(5, 2)$ (c) $(4, 8)$ (d) $(6, 4)$

15. $g(x, y) = \displaystyle\int_x^y (2t - 3)\, dt$

(a) $(4, 0)$ (b) $(4, 1)$ (c) $\left(4, \tfrac{3}{2}\right)$ (d) $\left(\tfrac{3}{2}, 0\right)$

16. $g(x, y) = \displaystyle\int_x^y \dfrac{1}{t}\, dt$

(a) $(4, 1)$ (b) $(6, 3)$ (c) $(2, 5)$ (d) $\left(\tfrac{1}{2}, 7\right)$

17. $f(x, y) = 2x + y^2$

(a) $\dfrac{f(x + \Delta x, y) - f(x, y)}{\Delta x}$ (b) $\dfrac{f(x, y + \Delta y) - f(x, y)}{\Delta y}$

18. $f(x, y) = 3x^2 - 2y$

(a) $\dfrac{f(x + \Delta x, y) - f(x, y)}{\Delta x}$ (b) $\dfrac{f(x, y + \Delta y) - f(x, y)}{\Delta y}$

Finding the Domain and Range of a Function In Exercises 19–30, find the domain and range of the function.

19. $f(x, y) = x^2 + y^2$ **20.** $f(x, y) = e^{xy}$

21. $g(x, y) = x\sqrt{y}$ **22.** $g(x, y) = \dfrac{y}{\sqrt{x}}$

23. $z = \dfrac{x + y}{xy}$ **24.** $z = \dfrac{xy}{x - y}$

25. $f(x, y) = \sqrt{4 - x^2 - y^2}$ **26.** $f(x, y) = \sqrt{4 - x^2 - 4y^2}$

27. $f(x, y) = \arccos(x + y)$ **28.** $f(x, y) = \arcsin(y/x)$

29. $f(x, y) = \ln(4 - x - y)$ **30.** $f(x, y) = \ln(xy - 6)$

31. **Think About It** The graphs labeled (a), (b), (c), and (d) are graphs of the function $f(x, y) = -4x/(x^2 + y^2 + 1)$. Match the four graphs with the points in space from which the surface is viewed. The four points are $(20, 15, 25)$, $(-15, 10, 20)$, $(20, 20, 0)$, and $(20, 0, 0)$.

(a)

Generated by Maple

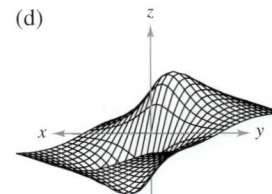

(b)

Generated by Maple

(c)

Generated by Maple

(d)

Generated by Maple

32. Think About It Use the function given in Exercise 31.

(a) Find the domain and range of the function.

(b) Identify the points in the xy-plane at which the function value is 0.

(c) Does the surface pass through all the octants of the rectangular coordinate system? Give reasons for your answer.

Sketching a Surface In Exercises 33–40, sketch the surface given by the function.

33. $f(x, y) = 4$

34. $f(x, y) = 6 - 2x - 3y$

35. $f(x, y) = y^2$

36. $g(x, y) = \frac{1}{2}y$

37. $z = -x^2 - y^2$

38. $z = \frac{1}{2}\sqrt{x^2 + y^2}$

39. $f(x, y) = e^{-x}$

40. $f(x, y) = \begin{cases} xy, & x \geq 0, y \geq 0 \\ 0, & x < 0 \text{ or } y < 0 \end{cases}$

Graphing a Function Using Technology In Exercises 41–44, use a computer algebra system to graph the function.

41. $z = y^2 - x^2 + 1$

42. $z = \frac{1}{12}\sqrt{144 - 16x^2 - 9y^2}$

43. $f(x, y) = x^2 e^{(-xy/2)}$

44. $f(x, y) = x \sin y$

Matching In Exercises 45–48, match the graph of the surface with one of the contour maps. [The contour maps are labeled (a), (b), (c), and (d).]

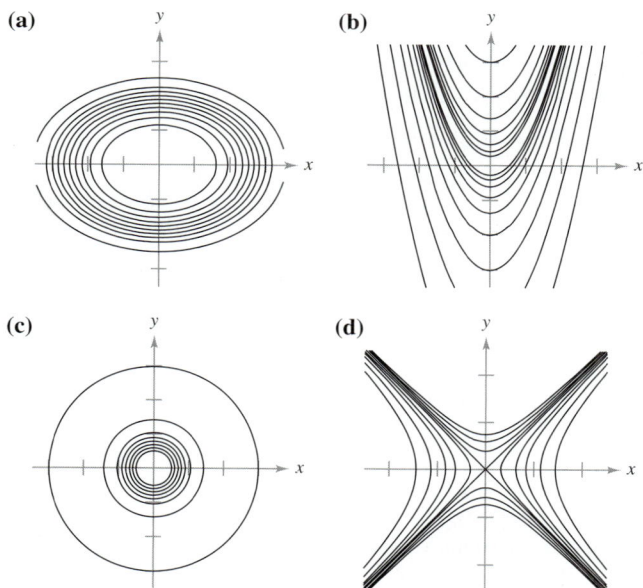

(a)

(b)

(c)

(d)

45. $f(x, y) = e^{1-x^2-y^2}$

46. $f(x, y) = e^{1-x^2+y^2}$

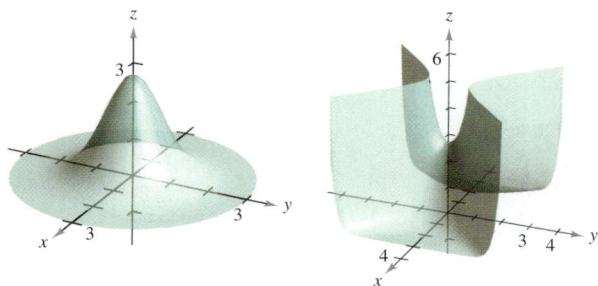

47. $f(x, y) = \ln|y - x^2|$

48. $f(x, y) = \cos\left(\dfrac{x^2 + 2y^2}{4}\right)$

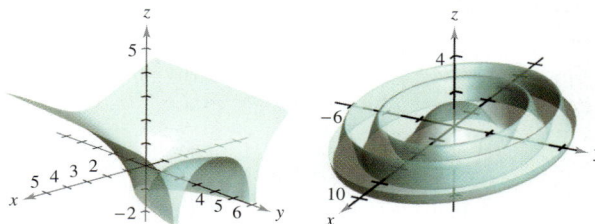

Sketching a Contour Map In Exercises 49–56, describe the level curves of the function. Sketch a contour map of the surface using level curves for the given c-values.

49. $z = x + y$, $c = -1, 0, 2, 4$

50. $z = 6 - 2x - 3y$, $c = 0, 2, 4, 6, 8, 10$

51. $z = x^2 + 4y^2$, $c = 0, 1, 2, 3, 4$

52. $f(x, y) = \sqrt{9 - x^2 - y^2}$, $c = 0, 1, 2, 3$

53. $f(x, y) = xy$, $c = \pm1, \pm2, \ldots, \pm6$

54. $f(x, y) = e^{xy/2}$, $c = 2, 3, 4, \frac{1}{2}, \frac{1}{3}, \frac{1}{4}$

55. $f(x, y) = x/(x^2 + y^2)$, $c = \pm\frac{1}{2}, \pm1, \pm\frac{3}{2}, \pm2$

56. $f(x, y) = \ln(x - y)$, $c = 0, \pm\frac{1}{2}, \pm1, \pm\frac{3}{2}, \pm2$

Graphing Level Curves In Exercises 57–60, use a graphing utility to graph six level curves of the function.

57. $f(x, y) = x^2 - y^2 + 2$

58. $f(x, y) = |xy|$

59. $g(x, y) = \dfrac{8}{1 + x^2 + y^2}$

60. $h(x, y) = 3\sin(|x| + |y|)$

WRITING ABOUT CONCEPTS

61. Function of Two Variables What is a graph of a function of two variables? How is it interpreted geometrically? Describe level curves.

62. Using Level Curves All of the level curves of the surface given by $z = f(x, y)$ are concentric circles. Does this imply that the graph of f is a hemisphere? Illustrate your answer with an example.

63. Creating a Function Construct a function whose level curves are lines passing through the origin.

64. Conjecture Consider the function $f(x, y) = xy$, for $x \geq 0$ and $y \geq 0$.

(a) Sketch the graph of the surface given by f.

(b) Make a conjecture about the relationship between the graphs of f and $g(x, y) = f(x, y) - 3$. Explain your reasoning.

(c) Make a conjecture about the relationship between the graphs of f and $g(x, y) = -f(x, y)$. Explain your reasoning.

(d) Make a conjecture about the relationship between the graphs of f and $g(x, y) = \frac{1}{2}f(x, y)$. Explain your reasoning.

(e) On the surface in part (a), sketch the graph of $z = f(x, x)$.

Writing In Exercises 65 and 66, use the graphs of the level curves (*c*-values evenly spaced) of the function f to write a description of a possible graph of f. Is the graph of f unique? Explain.

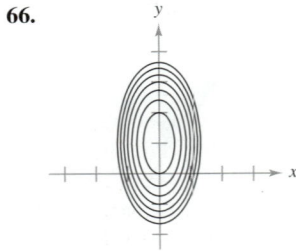

65.

66.

67. Investment In 2012, an investment of $1000 was made in a bond earning 6% compounded annually. Assume that the buyer pays tax at rate R and the annual rate of inflation is I. In the year 2022, the value V of the investment in constant 2012 dollars is

$$V(I, R) = 1000\left[\frac{1 + 0.06(1 - R)}{1 + I}\right]^{10}.$$

Use this function of two variables to complete the table.

	Inflation Rate		
Tax Rate	0	0.03	0.05
0			
0.28			
0.35			

68. Investment A principal of $5000 is deposited in a savings account that earns interest at a rate of r (written as a decimal), compounded continuously. The amount $A(r, t)$ after t years is

$$A(r, t) = 5000e^{rt}.$$

Use this function of two variables to complete the table.

	Number of Years			
Rate	5	10	15	20
0.02				
0.03				
0.04				
0.05				

Sketching a Level Surface In Exercises 69–74, sketch the graph of the level surface $f(x, y, z) = c$ at the given value of c.

69. $f(x, y, z) = x - y + z, \quad c = 1$

70. $f(x, y, z) = 4x + y + 2z, \quad c = 4$

71. $f(x, y, z) = x^2 + y^2 + z^2, \quad c = 9$

72. $f(x, y, z) = x^2 + \frac{1}{4}y^2 - z, \quad c = 1$

73. $f(x, y, z) = 4x^2 + 4y^2 - z^2, \quad c = 0$

74. $f(x, y, z) = \sin x - z, \quad c = 0$

75. Forestry

The **Doyle Log Rule** is one of several methods used to determine the lumber yield of a log (in board-feet) in terms of its diameter d (in inches) and its length L (in feet). The number of board-feet is

$$N(d, L) = \left(\frac{d - 4}{4}\right)^2 L.$$

(a) Find the number of board-feet of lumber in a log 22 inches in diameter and 12 feet in length.

(b) Find $N(30, 12)$.

76. Queuing Model The average length of time that a customer waits in line for service is

$$W(x, y) = \frac{1}{x - y}, \quad x > y$$

where y is the average arrival rate, written as the number of customers per unit of time, and x is the average service rate, written in the same units. Evaluate each of the following.

(a) $W(15, 9)$ (b) $W(15, 13)$

(c) $W(12, 7)$ (d) $W(5, 2)$

77. Temperature Distribution The temperature T (in degrees Celsius) at any point (x, y) in a circular steel plate of radius 10 meters is

$$T = 600 - 0.75x^2 - 0.75y^2$$

where x and y are measured in meters. Sketch some of the isothermal curves.

78. Electric Potential The electric potential V at any point (x, y) is

$$V(x, y) = \frac{5}{\sqrt{25 + x^2 + y^2}}.$$

Sketch the equipotential curves for $V = \frac{1}{2}$, $V = \frac{1}{3}$, and $V = \frac{1}{4}$.

79. Cobb-Douglas Production Function Use the Cobb-Douglas production function (see Example 5) to show that when the number of units of labor and the number of units of capital are doubled, the production level is also doubled.

80. Cobb-Douglas Production Function Show that the Cobb-Douglas production function

$$z = Cx^a y^{1-a}$$

can be rewritten as

$$\ln \frac{z}{y} = \ln C + a \ln \frac{x}{y}.$$

81. Ideal Gas Law According to the Ideal Gas Law,

$$PV = kT$$

where P is pressure, V is volume, T is temperature (in kelvins), and k is a constant of proportionality. A tank contains 2000 cubic inches of nitrogen at a pressure of 26 pounds per square inch and a temperature of 300 K.

(a) Determine k.

(b) Write P as a function of V and T and describe the level curves.

82. Modeling Data The table shows the net sales x (in billions of dollars), the total assets y (in billions of dollars), and the shareholder's equity z (in billions of dollars) for Apple for the years 2006 through 2011. *(Source: Apple Inc.)*

Year	2006	2007	2008	2009	2010	2011
x	19.3	24.6	37.5	42.9	65.2	108.2
y	17.2	24.9	36.2	47.5	75.2	116.4
z	10.0	14.5	22.3	31.6	47.8	76.6

A model for these data is

$$z = f(x, y) = 0.035x + 0.640y - 1.77.$$

(a) Use a graphing utility and the model to approximate z for the given values of x and y.

(b) Which of the two variables in this model has the greater influence on shareholder's equity? Explain.

(c) Simplify the expression for $f(x, 150)$ and interpret its meaning in the context of the problem.

83. Meteorology Meteorologists measure the atmospheric pressure in millibars. From these observations, they create weather maps on which the curves of equal atmospheric pressure (isobars), are drawn (see figure). On the map, the closer the isobars, the higher the wind speed. Match points A, B, and C with (a) highest pressure, (b) lowest pressure, and (c) highest wind velocity.

Figure for 83

Figure for 84

84. Acid Rain The acidity of rainwater is measured in units called pH. A pH of 7 is neutral, smaller values are increasingly acidic, and larger values are increasingly alkaline. The map shows curves of equal pH and gives evidence that downwind of heavily industrialized areas, the acidity has been increasing. Using the level curves on the map, determine the direction of the prevailing winds in the northeastern United States.

NASA

85. Construction Cost A rectangular box with an open top has a length of x feet, a width of y feet, and a height of z feet. It costs \$1.20 per square foot to build the base and \$0.75 per square foot to build the sides. Write the cost C of constructing the box as a function of x, y, and z.

86. HOW DO YOU SEE IT? The contour map shown in the figure was computer generated using data collected by satellite instrumentation. Color is used to show the "ozone hole" in Earth's atmosphere. The purple and blue areas represent the lowest levels of ozone, and the green areas represent the highest levels. *(Source: National Aeronautics and Space Administration)*

(a) Do the level curves correspond to equally spaced ozone levels? Explain.

(b) Describe how to obtain a more detailed contour map.

True or False? In Exercises 87–90, determine whether the statement is true or false. If it is false, explain why or give an example that shows it is false.

87. If $f(x_0, y_0) = f(x_1, y_1)$, then $x_0 = x_1$ and $y_0 = y_1$.

88. If f is a function, then $f(ax, ay) = a^2 f(x, y)$.

89. A vertical line can intersect the graph of $z = f(x, y)$ at most once.

90. Two different level curves of the graph of $z = f(x, y)$ can intersect.

PUTNAM EXAM CHALLENGE

91. Let $f \colon \mathbb{R}^2 \to \mathbb{R}$ be a function such that

$$f(x, y) + f(y, z) + f(z, x) = 0$$

for all real numbers x, y, and z. Prove that there exists a function $g \colon \mathbb{R} \to \mathbb{R}$ such that

$$f(x, y) = g(x) - g(y)$$

for all real numbers x and y.

13.2 Limits and Continuity

- Understand the definition of a neighborhood in the plane.
- Understand and use the definition of the limit of a function of two variables.
- Extend the concept of continuity to a function of two variables.
- Extend the concept of continuity to a function of three variables.

Neighborhoods in the Plane

In this section, you will study limits and continuity involving functions of two or three variables. The section begins with functions of two variables. At the end of the section, the concepts are extended to functions of three variables.

Your study of the limit of a function of two variables begins by defining a two-dimensional analog to an interval on the real number line. Using the formula for the distance between two points

$$(x, y) \quad \text{and} \quad (x_0, y_0)$$

in the plane, you can define the **δ-neighborhood** about (x_0, y_0) to be the **disk** centered at (x_0, y_0) with radius $\delta > 0$

$$\left\{ (x, y): \ \sqrt{(x - x_0)^2 + (y - y_0)^2} < \delta \right\} \qquad \text{Open disk}$$

as shown in Figure 13.17. When this formula contains the *less than* inequality sign, $<$, the disk is called **open,** and when it contains the *less than or equal to* inequality sign, $\leq$, the disk is called **closed.** This corresponds to the use of $<$ and $\leq$ to define open and closed intervals.

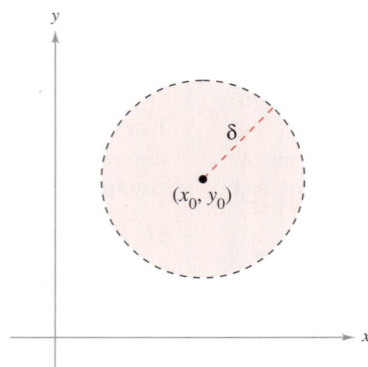

An open disk

Figure 13.17

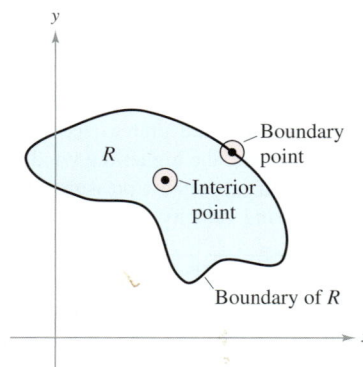

The boundary and interior points of a region R

Figure 13.18

A point (x_0, y_0) in a plane region R is an **interior point** of R if there exists a δ-neighborhood about (x_0, y_0) that lies entirely in R, as shown in Figure 13.18. If every point in R is an interior point, then R is an **open region.** A point (x_0, y_0) is a **boundary point** of R if every open disk centered at (x_0, y_0) contains points inside R *and* points outside R. By definition, a region must contain its interior points, but it need not contain its boundary points. If a region contains all its boundary points, then the region is **closed.** A region that contains some but not all of its boundary points is neither open nor closed.

■ **FOR FURTHER INFORMATION** For more information on Sonya Kovalevsky, see the article "S. Kovalevsky: A Mathematical Lesson" by Karen D. Rappaport in *The American Mathematical Monthly.* To view this article, go to *MathArticles.com.*

SONYA KOVALEVSKY (1850–1891)

Much of the terminology used to define limits and continuity of a function of two or three variables was introduced by the German mathematician Karl Weierstrass (1815–1897). Weierstrass's rigorous approach to limits and other topics in calculus gained him the reputation as the "father of modern analysis." Weierstrass was a gifted teacher. One of his best-known students was the Russian mathematician Sonya Kovalevsky, who applied many of Weierstrass's techniques to problems in mathematical physics and became one of the first women to gain acceptance as a research mathematician.

Limit of a Function of Two Variables

> ### Definition of the Limit of a Function of Two Variables
>
> Let f be a function of two variables defined, except possibly at (x_0, y_0), on an open disk centered at (x_0, y_0), and let L be a real number. Then
>
> $$\lim_{(x,\, y) \to (x_0,\, y_0)} f(x, y) = L$$
>
> if for each $\varepsilon > 0$ there corresponds a $\delta > 0$ such that
>
> $$|f(x, y) - L| < \varepsilon \quad \text{whenever} \quad 0 < \sqrt{(x - x_0)^2 + (y - y_0)^2} < \delta.$$

Graphically, the definition of the limit of a function of two variables implies that for any point $(x, y) \neq (x_0, y_0)$ in the disk of radius δ, the value $f(x, y)$ lies between $L + \varepsilon$ and $L - \varepsilon$, as shown in Figure 13.19.

The definition of the limit of a function of two variables is similar to the definition of the limit of a function of a single variable, yet there is a critical difference. To determine whether a function of a single variable has a limit, you need only test the approach from two directions—from the right and from the left. When the function approaches the same limit from the right and from the left, you can conclude that the limit exists. For a function of two variables, however, the statement

$$(x, y) \to (x_0, y_0)$$

means that the point (x, y) is allowed to approach (x_0, y_0) from any direction. If the value of

$$\lim_{(x,\, y) \to (x_0,\, y_0)} f(x, y)$$

is not the same for all possible approaches, or **paths,** to (x_0, y_0), then the limit does not exist.

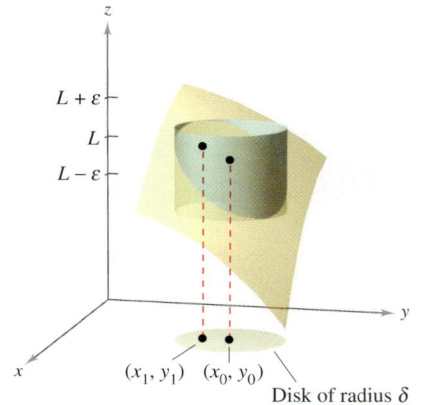

For any (x, y) in the disk of radius δ, the value $f(x, y)$ lies between $L + \varepsilon$ and $L - \varepsilon$.

Figure 13.19

EXAMPLE 1   **Verifying a Limit by the Definition**

Show that $\displaystyle\lim_{(x,\, y) \to (a,\, b)} x = a$.

Solution   Let $f(x, y) = x$ and $L = a$. You need to show that for each $\varepsilon > 0$, there exists a δ-neighborhood about (a, b) such that

$$|f(x, y) - L| = |x - a| < \varepsilon$$

whenever $(x, y) \neq (a, b)$ lies in the neighborhood. You can first observe that from

$$0 < \sqrt{(x - a)^2 + (y - b)^2} < \delta$$

it follows that

$$
\begin{aligned}
|f(x, y) - a| &= |x - a| \\
&= \sqrt{(x - a)^2} \\
&\leq \sqrt{(x - a)^2 + (y - b)^2} \\
&< \delta.
\end{aligned}
$$

So, you can choose $\delta = \varepsilon$, and the limit is verified.

Limits of functions of several variables have the same properties regarding sums, differences, products, and quotients as do limits of functions of single variables. (See Theorem 1.2 in Section 1.3.) Some of these properties are used in the next example.

EXAMPLE 2 Verifying a Limit

Evaluate

$$\lim_{(x, y)\to(1, 2)} \frac{5x^2y}{x^2 + y^2}.$$

Solution By using the properties of limits of products and sums, you obtain

$$\lim_{(x, y)\to(1, 2)} 5x^2y = 5(1^2)(2) = 10$$

and

$$\lim_{(x, y)\to(1, 2)} (x^2 + y^2) = (1^2 + 2^2) = 5.$$

Because the limit of a quotient is equal to the quotient of the limits (and the denominator is not 0), you have

$$\lim_{(x, y)\to(1, 2)} \frac{5x^2y}{x^2 + y^2} = \frac{10}{5} = 2.$$

EXAMPLE 3 Verifying a Limit

Evaluate $\lim\limits_{(x, y)\to(0, 0)} \dfrac{5x^2y}{x^2 + y^2}.$

Solution In this case, the limits of the numerator and of the denominator are both 0, and so you cannot determine the existence (or nonexistence) of a limit by taking the limits of the numerator and denominator separately and then dividing. From the graph of f in Figure 13.20, however, it seems reasonable that the limit might be 0. So, you can try applying the definition to $L = 0$. First, note that

$$|y| \le \sqrt{x^2 + y^2}$$

and

$$\frac{x^2}{x^2 + y^2} \le 1.$$

Then, in a δ-neighborhood about $(0, 0)$, you have

$$0 < \sqrt{x^2 + y^2} < \delta$$

and it follows that, for $(x, y) \ne (0, 0)$,

$$|f(x, y) - 0| = \left| \frac{5x^2y}{x^2 + y^2} \right|$$

$$= 5|y|\left(\frac{x^2}{x^2 + y^2}\right)$$

$$\le 5|y|$$

$$\le 5\sqrt{x^2 + y^2}$$

$$< 5\delta.$$

So, you can choose $\delta = \varepsilon/5$ and conclude that

$$\lim_{(x, y)\to(0, 0)} \frac{5x^2y}{x^2 + y^2} = 0.$$

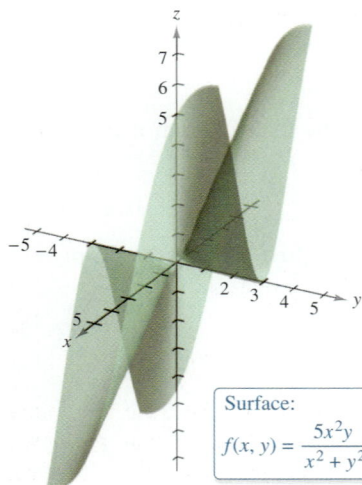

Surface:

$$f(x, y) = \frac{5x^2y}{x^2 + y^2}$$

Figure 13.20

For some functions, it is easy to recognize that a limit does not exist. For instance, it is clear that the limit

$$\lim_{(x, y)\to(0, 0)} \frac{1}{x^2 + y^2}$$

does not exist because the values of $f(x, y)$ increase without bound as (x, y) approaches $(0, 0)$ along *any* path (see Figure 13.21).

For other functions, it is not so easy to recognize that a limit does not exist. For instance, the next example describes a limit that does not exist because the function approaches different values along different paths.

$\lim_{(x, y)\to(0, 0)} \dfrac{1}{x^2 + y^2}$ does not exist.

Figure 13.21

EXAMPLE 4   A Limit That Does Not Exist

▷ *See LarsonCalculus.com for an interactive version of this type of example.*

Show that the limit does not exist.

$$\lim_{(x, y)\to(0, 0)} \left(\frac{x^2 - y^2}{x^2 + y^2}\right)^2$$

Solution   The domain of the function

$$f(x, y) = \left(\frac{x^2 - y^2}{x^2 + y^2}\right)^2$$

consists of all points in the xy-plane except for the point $(0, 0)$. To show that the limit as (x, y) approaches $(0, 0)$ does not exist, consider approaching $(0, 0)$ along two different "paths," as shown in Figure 13.22. Along the x-axis, every point is of the form

$$(x, 0)$$

and the limit along this approach is

$$\lim_{(x, 0)\to(0, 0)} \left(\frac{x^2 - 0^2}{x^2 + 0^2}\right)^2 = \lim_{(x, 0)\to(0, 0)} 1^2 = 1. \qquad \text{Limit along } x\text{-axis}$$

However, when (x, y) approaches $(0, 0)$ along the line $y = x$, you obtain

$$\lim_{(x, x)\to(0, 0)} \left(\frac{x^2 - x^2}{x^2 + x^2}\right)^2 = \lim_{(x, x)\to(0, 0)} \left(\frac{0}{2x^2}\right)^2 = 0. \qquad \text{Limit along line } y = x$$

This means that in any open disk centered at $(0, 0)$, there are points (x, y) at which f takes on the value 1, and other points at which f takes on the value 0. For instance,

$$f(x, y) = 1$$

at $(1, 0)$, $(0.1, 0)$, $(0.01, 0)$, and $(0.001, 0)$, and

$$f(x, y) = 0$$

at $(1, 1)$, $(0.1, 0.1)$, $(0.01, 0.01)$, and $(0.001, 0.001)$. So, f does not have a limit as (x, y) approaches $(0, 0)$.

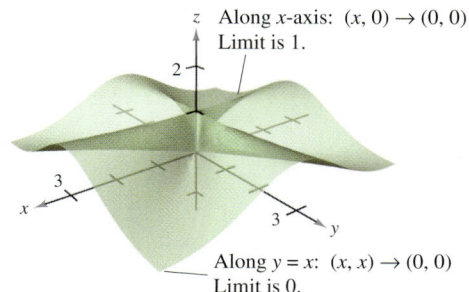

Along x-axis: $(x, 0) \to (0, 0)$
Limit is 1.

Along $y = x$: $(x, x) \to (0, 0)$
Limit is 0.

$\lim_{(x, y)\to(0, 0)} \left(\dfrac{x^2 - y^2}{x^2 + y^2}\right)^2$ does not exist.

Figure 13.22

In Example 4, you could conclude that the limit does not exist because you found two approaches that produced different limits. Be sure you understand that when two approaches produce the same limit, you *cannot* conclude that the limit exists. To form such a conclusion, you must show that the limit is the same along *all* possible approaches.

Continuity of a Function of Two Variables

Notice in Example 2 that the limit of $f(x, y) = 5x^2y/(x^2 + y^2)$ as $(x, y) \to (1, 2)$ can be evaluated by direct substitution. That is, the limit is $f(1, 2) = 2$. In such cases, the function f is said to be **continuous** at the point $(1, 2)$.

• • • • • • • • • • • • • • • ▷

• • REMARK This definition of continuity can be extended to *boundary points* of the open region R by considering a special type of limit in which (x, y) is allowed to approach (x_0, y_0) along paths lying in the region R. This notion is similar to that of one-sided limits, as discussed in Chapter 1.

Definition of Continuity of a Function of Two Variables

A function f of two variables is **continuous at a point** (x_0, y_0) in an open region R if $f(x_0, y_0)$ is equal to the limit of $f(x, y)$ as (x, y) approaches (x_0, y_0). That is,

$$\lim_{(x, y)\to(x_0, y_0)} f(x, y) = f(x_0, y_0).$$

The function f is **continuous in the open region** R if it is continuous at every point in R.

In Example 3, it was shown that the function

$$f(x, y) = \frac{5x^2y}{x^2 + y^2}$$

is not continuous at $(0, 0)$. Because the limit at this point exists, however, you can remove the discontinuity by defining f at $(0, 0)$ as being equal to its limit there. Such a discontinuity is called **removable**. In Example 4, the function

$$f(x, y) = \left(\frac{x^2 - y^2}{x^2 + y^2}\right)^2$$

was also shown not to be continuous at $(0, 0)$, but this discontinuity is **nonremovable.**

THEOREM 13.1 Continuous Functions of Two Variables

If k is a real number and f and g are continuous at (x_0, y_0), then the following functions are also continuous at (x_0, y_0).

1. Scalar multiple: kf **2.** Sum or difference: $f \pm g$

3. Product: fg **4.** Quotient: f/g, $g(x_0, y_0) \neq 0$

Theorem 13.1 establishes the continuity of *polynomial* and *rational* functions at every point in their domains. Furthermore, the continuity of other types of functions can be extended naturally from one to two variables. For instance, the functions whose graphs are shown in Figures 13.23 and 13.24 are continuous at every point in the plane.

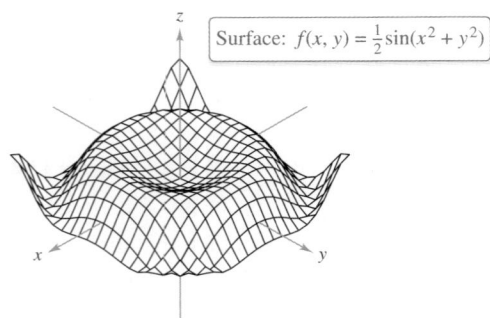

Surface: $f(x, y) = \frac{1}{2}\sin(x^2 + y^2)$

The function f is continuous at every point in the plane.
Figure 13.23

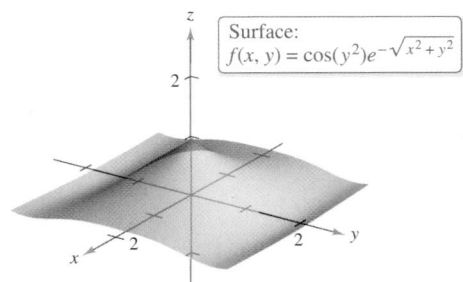

Surface:
$f(x, y) = \cos(y^2)e^{-\sqrt{x^2 + y^2}}$

The function f is continuous at every point in the plane.
Figure 13.24

The next theorem states conditions under which a composite function is continuous.

> **THEOREM 13.2   Continuity of a Composite Function**
>
> If h is continuous at (x_0, y_0) and g is continuous at $h(x_0, y_0)$, then the composite function given by $(g \circ h)(x, y) = g(h(x, y))$ is continuous at (x_0, y_0). That is,
>
> $$\lim_{(x, y) \to (x_0, y_0)} g(h(x, y)) = g(h(x_0, y_0)).$$

Note in Theorem 13.2 that h is a function of two variables and g is a function of one variable.

EXAMPLE 5   Testing for Continuity

Discuss the continuity of each function.

a. $f(x, y) = \dfrac{x - 2y}{x^2 + y^2}$ **b.** $g(x, y) = \dfrac{2}{y - x^2}$

Solution

a. Because a rational function is continuous at every point in its domain, you can conclude that f is continuous at each point in the xy-plane except at $(0, 0)$, as shown in Figure 13.25.

b. The function

$$g(x, y) = \frac{2}{y - x^2}$$

is continuous except at the points at which the denominator is 0, which is given by the equation

$$y - x^2 = 0.$$

So, you can conclude that the function is continuous at all points except those lying on the parabola $y = x^2$. Inside this parabola, you have $y > x^2$, and the surface represented by the function lies above the xy-plane, as shown in Figure 13.26. Outside the parabola, $y < x^2$, and the surface lies below the xy-plane.

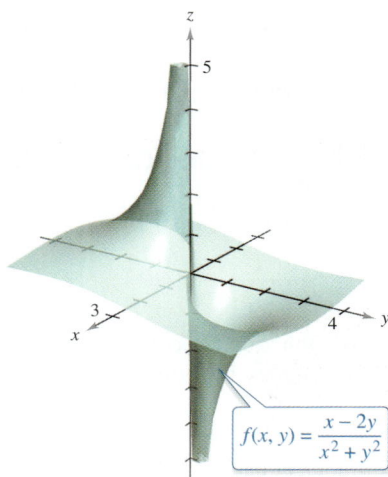

$$f(x, y) = \frac{x - 2y}{x^2 + y^2}$$

The function f is not continuous at $(0, 0)$.

Figure 13.25

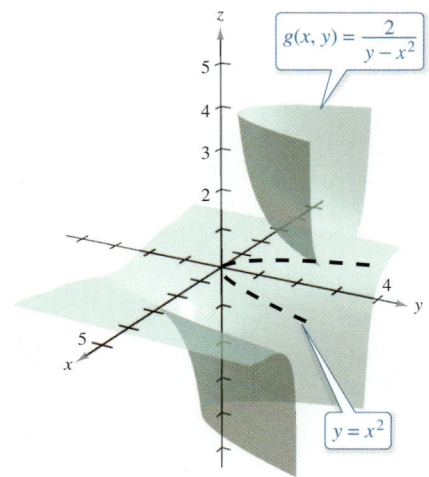

$$g(x, y) = \frac{2}{y - x^2}$$

$$y = x^2$$

The function g is not continuous on the parabola $y = x^2$.

Figure 13.26

Continuity of a Function of Three Variables

The preceding definitions of limits and continuity can be extended to functions of three variables by considering points (x, y, z) within the *open sphere*

$$(x - x_0)^2 + (y - y_0)^2 + (z - z_0)^2 < \delta^2. \qquad \text{Open sphere}$$

The radius of this sphere is δ, and the sphere is centered at (x_0, y_0, z_0), as shown in Figure 13.27.

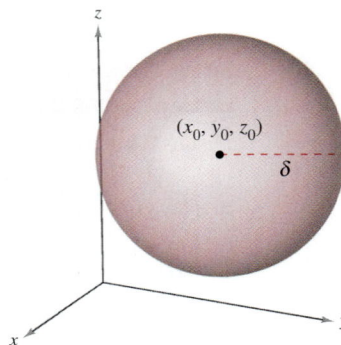

Open sphere in space
Figure 13.27

A point (x_0, y_0, z_0) in a region R in space is an **interior point** of R if there exists a δ-sphere about (x_0, y_0, z_0) that lies entirely in R. If every point in R is an interior point, then R is called **open.**

Definition of Continuity of a Function of Three Variables

A function f of three variables is **continuous at a point (x_0, y_0, z_0)** in an open region R if $f(x_0, y_0, z_0)$ is defined and is equal to the limit of $f(x, y, z)$ as (x, y, z) approaches (x_0, y_0, z_0). That is,

$$\lim_{(x, y, z) \to (x_0, y_0, z_0)} f(x, y, z) = f(x_0, y_0, z_0).$$

The function f is **continuous in the open region R** if it is continuous at every point in R.

EXAMPLE 6   **Testing Continuity of a Function of Three Variables**

Discuss the continuity of

$$f(x, y, z) = \frac{1}{x^2 + y^2 - z}.$$

Solution   The function f is continuous except at the points at which the denominator is 0, which are given by the equation

$$x^2 + y^2 - z = 0.$$

So, f is continuous at each point in space except at the points on the paraboloid

$$z = x^2 + y^2.$$

13.2 Exercises

See **CalcChat.com** for tutorial help and worked-out solutions to odd-numbered exercises.

Verifying a Limit by the Definition In Exercises 1–4, use the definition of the limit of a function of two variables to verify the limit.

1. $\displaystyle\lim_{(x,\,y)\to(1,\,0)} x = 1$

2. $\displaystyle\lim_{(x,\,y)\to(4,\,-1)} x = 4$

3. $\displaystyle\lim_{(x,\,y)\to(1,\,-3)} y = -3$

4. $\displaystyle\lim_{(x,\,y)\to(a,\,b)} y = b$

Using Properties of Limits In Exercises 5–8, find the indicated limit by using the limits

$$\lim_{(x,\,y)\to(a,\,b)} f(x,y) = 4 \quad \text{and} \quad \lim_{(x,\,y)\to(a,\,b)} g(x,y) = 3.$$

5. $\displaystyle\lim_{(x,\,y)\to(a,\,b)} [f(x,y) - g(x,y)]$

6. $\displaystyle\lim_{(x,\,y)\to(a,\,b)} \left[\frac{5f(x,y)}{g(x,y)}\right]$

7. $\displaystyle\lim_{(x,\,y)\to(a,\,b)} [f(x,y)g(x,y)]$

8. $\displaystyle\lim_{(x,\,y)\to(a,\,b)} \left[\frac{f(x,y) + g(x,y)}{f(x,y)}\right]$

Limit and Continuity In Exercises 9–22, find the limit and discuss the continuity of the function.

9. $\displaystyle\lim_{(x,\,y)\to(2,\,1)} (2x^2 + y)$

10. $\displaystyle\lim_{(x,\,y)\to(0,\,0)} (x + 4y + 1)$

11. $\displaystyle\lim_{(x,\,y)\to(1,\,2)} e^{xy}$

12. $\displaystyle\lim_{(x,\,y)\to(2,\,4)} \frac{x + y}{x^2 + 1}$

13. $\displaystyle\lim_{(x,\,y)\to(0,\,2)} \frac{x}{y}$

14. $\displaystyle\lim_{(x,\,y)\to(-1,\,2)} \frac{x + y}{x - y}$

15. $\displaystyle\lim_{(x,\,y)\to(1,\,1)} \frac{xy}{x^2 + y^2}$

16. $\displaystyle\lim_{(x,\,y)\to(1,\,1)} \frac{x}{\sqrt{x + y}}$

17. $\displaystyle\lim_{(x,\,y)\to(\pi/4,\,2)} y \cos xy$

18. $\displaystyle\lim_{(x,\,y)\to(2\pi,\,4)} \sin\frac{x}{y}$

19. $\displaystyle\lim_{(x,\,y)\to(0,\,1)} \frac{\arcsin xy}{1 - xy}$

20. $\displaystyle\lim_{(x,\,y)\to(0,\,1)} \frac{\arccos(x/y)}{1 + xy}$

21. $\displaystyle\lim_{(x,\,y,\,z)\to(1,\,3,\,4)} \sqrt{x + y + z}$

22. $\displaystyle\lim_{(x,\,y,\,z)\to(-2,\,1,\,0)} xe^{yz}$

Finding a Limit In Exercises 23–34, find the limit (if it exists). If the limit does not exist, explain why.

23. $\displaystyle\lim_{(x,\,y)\to(1,\,1)} \frac{xy - 1}{1 + xy}$

24. $\displaystyle\lim_{(x,\,y)\to(1,\,-1)} \frac{x^2y}{1 + xy^2}$

25. $\displaystyle\lim_{(x,\,y)\to(0,\,0)} \frac{1}{x + y}$

26. $\displaystyle\lim_{(x,\,y)\to(0,\,0)} \frac{1}{x^2y^2}$

27. $\displaystyle\lim_{(x,\,y)\to(0,\,0)} \frac{x - y}{\sqrt{x} - \sqrt{y}}$

28. $\displaystyle\lim_{(x,\,y)\to(2,\,1)} \frac{x - y - 1}{\sqrt{x - y} - 1}$

29. $\displaystyle\lim_{(x,\,y)\to(0,\,0)} \frac{x + y}{x^2 + y}$

30. $\displaystyle\lim_{(x,\,y)\to(0,\,0)} \frac{x}{x^2 - y^2}$

31. $\displaystyle\lim_{(x,\,y)\to(0,\,0)} \frac{x^2}{(x^2 + 1)(y^2 + 1)}$

32. $\displaystyle\lim_{(x,\,y)\to(0,\,0)} \ln(x^2 + y^2)$

33. $\displaystyle\lim_{(x,\,y,\,z)\to(0,\,0,\,0)} \frac{xy + yz + xz}{x^2 + y^2 + z^2}$

34. $\displaystyle\lim_{(x,\,y,\,z)\to(0,\,0,\,0)} \frac{xy + yz^2 + xz^2}{x^2 + y^2 + z^2}$

Continuity In Exercises 35 and 36, discuss the continuity of the function and evaluate the limit of $f(x,y)$ (if it exists) as $(x,y) \to (0,0)$.

35. $f(x,y) = e^{xy}$

36. $f(x,y) = 1 - \dfrac{\cos(x^2 + y^2)}{x^2 + y^2}$

Limit and Continuity In Exercises 37–40, use a graphing utility to make a table showing the values of $f(x,y)$ at the given points for each path. Use the result to make a conjecture about the limit of $f(x,y)$ as $(x,y) \to (0,0)$. Determine analytically whether the limit exists and discuss the continuity of the function.

37. $f(x,y) = \dfrac{xy}{x^2 + y^2}$

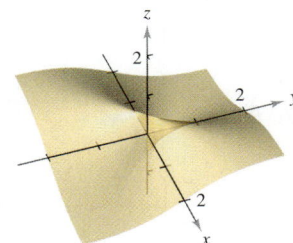

Path: $y = 0$

Points: $(1, 0), (0.5, 0),$ $(0.1, 0), (0.01, 0),$ $(0.001, 0)$

Path: $y = x$

Points: $(1, 1), (0.5, 0.5),$ $(0.1, 0.1), (0.01, 0.01),$ $(0.001, 0.001)$

888   **Chapter 13**   Functions of Several Variables

38. $f(x, y) = -\dfrac{xy^2}{x^2 + y^4}$

Path: $x = y^2$

Points: $(1, 1)$, $(0.25, 0.5)$,
$(0.01, 0.1)$,
$(0.0001, 0.01)$,
$(0.000001, 0.001)$

Path: $x = -y^2$

Points: $(-1, 1)$, $(-0.25, 0.5)$,
$(-0.01, 0.1)$,
$(-0.0001, 0.01)$,
$(-0.000001, 0.001)$

39. $f(x, y) = \dfrac{y}{x^2 + y^2}$

Path: $y = 0$

Points: $(1, 0)$, $(0.5, 0)$,
$(0.1, 0)$, $(0.01, 0)$,
$(0.001, 0)$

Path: $y = x$

Points: $(1, 1)$, $(0.5, 0.5)$,
$(0.1, 0.1)$,
$(0.01, 0.01)$,
$(0.001, 0.001)$

40. $f(x, y) = \dfrac{2x - y^2}{2x^2 + y}$

Path: $y = 0$

Points: $(1, 0)$, $(0.25, 0)$,
$(0.01, 0)$,
$(0.001, 0)$,
$(0.000001, 0)$

Path: $y = x$

Points: $(1, 1)$, $(0.25, 0.25)$,
$(0.01, 0.01)$,
$(0.001, 0.001)$,
$(0.0001, 0.0001)$

Comparing Continuity   In Exercises 41 and 42, discuss the continuity of the functions f and g. Explain any differences.

41. $f(x, y) = \begin{cases} \dfrac{x^4 - y^4}{x^2 + y^2}, & (x, y) \neq (0, 0) \\ 0, & (x, y) = (0, 0) \end{cases}$

$g(x, y) = \begin{cases} \dfrac{x^4 - y^4}{x^2 + y^2}, & (x, y) \neq (0, 0) \\ 1, & (x, y) = (0, 0) \end{cases}$

42. $f(x, y) = \begin{cases} \dfrac{x^2 + 2xy^2 + y^2}{x^2 + y^2}, & (x, y) \neq (0, 0) \\ 0, & (x, y) = (0, 0) \end{cases}$

$g(x, y) = \begin{cases} \dfrac{x^2 + 2xy^2 + y^2}{x^2 + y^2}, & (x, y) \neq (0, 0) \\ 1, & (x, y) = (0, 0) \end{cases}$

Finding a Limit Using Polar Coordinates   In Exercises 43–48, use polar coordinates to find the limit. [*Hint:* Let $x = r \cos \theta$ and $y = r \sin \theta$, and note that $(x, y) \to (0, 0)$ implies $r \to 0$.]

43. $\displaystyle \lim_{(x, y) \to (0, 0)} \dfrac{xy^2}{x^2 + y^2}$

44. $\displaystyle \lim_{(x, y) \to (0, 0)} \dfrac{x^3 + y^3}{x^2 + y^2}$

45. $\displaystyle \lim_{(x, y) \to (0, 0)} \dfrac{x^2 y^2}{x^2 + y^2}$

46. $\displaystyle \lim_{(x, y) \to (0, 0)} \dfrac{x^2 - y^2}{\sqrt{x^2 + y^2}}$

47. $\displaystyle \lim_{(x, y) \to (0, 0)} \cos(x^2 + y^2)$

48. $\displaystyle \lim_{(x, y) \to (0, 0)} \sin \sqrt{x^2 + y^2}$

Finding a Limit Using Polar Coordinates   In Exercises 49–52, use polar coordinates and L'Hôpital's Rule to find the limit.

49. $\displaystyle \lim_{(x, y) \to (0, 0)} \dfrac{\sin \sqrt{x^2 + y^2}}{\sqrt{x^2 + y^2}}$

50. $\displaystyle \lim_{(x, y) \to (0, 0)} \dfrac{\sin(x^2 + y^2)}{x^2 + y^2}$

51. $\displaystyle \lim_{(x, y) \to (0, 0)} \dfrac{1 - \cos(x^2 + y^2)}{x^2 + y^2}$

52. $\displaystyle \lim_{(x, y) \to (0, 0)} (x^2 + y^2) \ln (x^2 + y^2)$

Continuity   In Exercises 53–58, discuss the continuity of the function.

53. $f(x, y, z) = \dfrac{1}{\sqrt{x^2 + y^2 + z^2}}$

54. $f(x, y, z) = \dfrac{z}{x^2 + y^2 - 4}$

55. $f(x, y, z) = \dfrac{\sin z}{e^x + e^y}$

56. $f(x, y, z) = xy \sin z$

57. $f(x, y) = \begin{cases} \dfrac{\sin xy}{xy}, & xy \neq 0 \\ 1, & xy = 0 \end{cases}$

58. $f(x, y) = \begin{cases} \dfrac{\sin(x^2 - y^2)}{x^2 - y^2}, & x^2 \neq y^2 \\ 1, & x^2 = y^2 \end{cases}$

Continuity of a Composite Function   In Exercises 59–62, discuss the continuity of the composite function $f \circ g$.

59. $f(t) = t^2$

$g(x, y) = 2x - 3y$

60. $f(t) = \dfrac{1}{t}$

$g(x, y) = x^2 + y^2$

61. $f(t) = \dfrac{1}{t}$

$g(x, y) = 2x - 3y$

62. $f(t) = \dfrac{1}{1 - t}$

$g(x, y) = x^2 + y^2$

Finding a Limit   In Exercises 63–68, find each limit.

(a) $\displaystyle \lim_{\Delta x \to 0} \dfrac{f(x + \Delta x, y) - f(x, y)}{\Delta x}$

(b) $\displaystyle \lim_{\Delta y \to 0} \dfrac{f(x, y + \Delta y) - f(x, y)}{\Delta y}$

63. $f(x, y) = x^2 - 4y$

64. $f(x, y) = x^2 + y^2$

65. $f(x, y) = \dfrac{x}{y}$

66. $f(x, y) = \dfrac{1}{x + y}$

67. $f(x, y) = 3x + xy - 2y$

68. $f(x, y) = \sqrt{y} \, (y + 1)$

True or False? In Exercises 69–72, determine whether the statement is true or false. If it is false, explain why or give an example that shows it is false.

69. If $\lim_{(x,\,y)\to(0,\,0)} f(x, y) = 0$, then $\lim_{x\to 0} f(x, 0) = 0$.

70. If $\lim_{(x,\,y)\to(0,\,0)} f(0, y) = 0$, then $\lim_{(x,\,y)\to(0,\,0)} f(x, y) = 0$.

71. If f is continuous for all nonzero x and y, and $f(0, 0) = 0$, then $\lim_{(x,\,y)\to(0,\,0)} f(x, y) = 0$.

72. If g and h are continuous functions of x and y, and $f(x, y) = g(x) + h(y)$, then f is continuous.

73. Limit Consider $\lim_{(x,\,y)\to(0,\,0)} \dfrac{x^2 + y^2}{xy}$ (see figure).

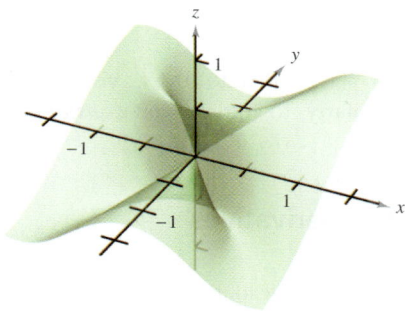

(a) Determine (if possible) the limit along any line of the form $y = ax$.

(b) Determine (if possible) the limit along the parabola $y = x^2$.

(c) Does the limit exist? Explain.

74. Limit Consider $\lim_{(x,\,y)\to(0,\,0)} \dfrac{x^2 y}{x^4 + y^2}$ (see figure).

(a) Determine (if possible) the limit along any line of the form $y = ax$.

(b) Determine (if possible) the limit along the parabola $y = x^2$.

(c) Does the limit exist? Explain.

Finding a Limit Using Spherical Coordinates In Exercises 75 and 76, use spherical coordinates to find the limit. [*Hint:* Let $x = \rho \sin \phi \cos \theta$, $y = \rho \sin \phi \sin \theta$, and $z = \rho \cos \phi$, and note that $(x, y, z) \to (0, 0, 0)$ implies $\rho \to 0^+$.]

75. $\lim_{(x,\,y,\,z)\to(0,\,0,\,0)} \dfrac{xyz}{x^2 + y^2 + z^2}$

76. $\lim_{(x,\,y,\,z)\to(0,\,0,\,0)} \tan^{-1}\left[\dfrac{1}{x^2 + y^2 + z^2}\right]$

77. Finding a Limit Find the following limit.
$$\lim_{(x,\,y)\to(0,\,1)} \tan^{-1}\left[\frac{x^2 + 1}{x^2 + (y - 1)^2}\right]$$

78. Continuity For the function
$$f(x, y) = xy\left(\frac{x^2 - y^2}{x^2 + y^2}\right)$$
define $f(0, 0)$ such that f is continuous at the origin.

79. Proof Prove that
$$\lim_{(x,\,y)\to(a,\,b)} [f(x, y) + g(x, y)] = L_1 + L_2$$
where $f(x, y)$ approaches L_1 and $g(x, y)$ approaches L_2 as $(x, y) \to (a, b)$.

80. Proof Prove that if f is continuous and $f(a, b) < 0$, then there exists a δ-neighborhood about (a, b) such that $f(x, y) < 0$ for every point (x, y) in the neighborhood.

WRITING ABOUT CONCEPTS

81. Limit Define the limit of a function of two variables. Describe a method for showing that
$$\lim_{(x,\,y)\to(x_0,\,y_0)} f(x, y)$$
does not exist.

82. Continuity State the definition of continuity of a function of two variables.

83. Limits and Function Values

(a) If $f(2, 3) = 4$, can you conclude anything about $\lim_{(x,\,y)\to(2,\,3)} f(x, y)$? Give reasons for your answer.

(b) If $\lim_{(x,\,y)\to(2,\,3)} f(x, y) = 4$, can you conclude anything about $f(2, 3)$? Give reasons for your answer.

84. HOW DO YOU SEE IT? The figure shows the graph of $f(x, y) = \ln x^2 + y^2$. From the graph, does it appear that the limit at each point exists?

(a) $(-1, -1)$ (b) $(0, 3)$
(c) $(0, 0)$ (d) $(2, 0)$

13.3 Partial Derivatives

- Find and use partial derivatives of a function of two variables.
- Find and use partial derivatives of a function of three or more variables.
- Find higher-order partial derivatives of a function of two or three variables.

Partial Derivatives of a Function of Two Variables

In applications of functions of several variables, the question often arises, "How will the value of a function be affected by a change in one of its independent variables?" You can answer this by considering the independent variables one at a time. For example, to determine the effect of a catalyst in an experiment, a chemist could conduct the experiment several times using varying amounts of the catalyst, while keeping constant other variables such as temperature and pressure. You can use a similar procedure to determine the rate of change of a function f with respect to one of its several independent variables. This process is called **partial differentiation,** and the result is referred to as the **partial derivative** of f with respect to the chosen independent variable.

JEAN LE ROND D'ALEMBERT (1717–1783)

The introduction of partial derivatives followed Newton's and Leibniz's work in calculus by several years. Between 1730 and 1760, Leonhard Euler and Jean Le Rond d'Alembert separately published several papers on dynamics, in which they established much of the theory of partial derivatives. These papers used functions of two or more variables to study problems involving equilibrium, fluid motion, and vibrating strings.
See LarsonCalculus.com to read more of this biography.

Definition of Partial Derivatives of a Function of Two Variables

If $z = f(x, y)$, then the **first partial derivatives** of f with respect to x and y are the functions f_x and f_y defined by

$$f_x(x, y) = \lim_{\Delta x \to 0} \frac{f(x + \Delta x, y) - f(x, y)}{\Delta x} \qquad \text{Partial derivative with respect to } x$$

and

$$f_y(x, y) = \lim_{\Delta y \to 0} \frac{f(x, y + \Delta y) - f(x, y)}{\Delta y} \qquad \text{Partial derivative with respect to } y$$

provided the limits exist.

This definition indicates that if $z = f(x, y)$, then to find f_x, you *consider y constant* and differentiate with respect to x. Similarly, to find f_y, you *consider x constant* and differentiate with respect to y.

EXAMPLE 1 Finding Partial Derivatives

a. To find f_x for $f(x, y) = 3x - x^2y^2 + 2x^3y$, consider y to be constant and differentiate with respect to x.

$$f_x(x, y) = 3 - 2xy^2 + 6x^2y \qquad \text{Partial derivative with respect to } x$$

To find f_y, consider x to be constant and differentiate with respect to y.

$$f_y(x, y) = -2x^2y + 2x^3 \qquad \text{Partial derivative with respect to } y$$

b. To find f_x for $f(x, y) = (\ln x)(\sin x^2y)$, consider y to be constant and differentiate with respect to x.

$$f_x(x, y) = (\ln x)(\cos x^2y)(2xy) + \frac{\sin x^2y}{x} \qquad \text{Partial derivative with respect to } x$$

To find f_y, consider x to be constant and differentiate with respect to y.

$$f_y(x, y) = (\ln x)(\cos x^2y)(x^2) \qquad \text{Partial derivative with respect to } y$$

• • REMARK   The notation $\partial z / \partial x$ is read as "the partial derivative of z with respect to x," and $\partial z / \partial y$ is read as "the partial derivative of z with respect to y."

Notation for First Partial Derivatives

For $z = f(x, y)$, the partial derivatives f_x and f_y are denoted by

$$\frac{\partial}{\partial x} f(x, y) = f_x(x, y) = z_x = \frac{\partial z}{\partial x} \qquad \text{Partial derivative with respect to } x$$

and

$$\frac{\partial}{\partial y} f(x, y) = f_y(x, y) = z_y = \frac{\partial z}{\partial y}. \qquad \text{Partial derivative with respect to } y$$

The first partials evaluated at the point (a, b) are denoted by

$$\left. \frac{\partial z}{\partial x} \right|_{(a, b)} = f_x(a, b)$$

and

$$\left. \frac{\partial z}{\partial y} \right|_{(a, b)} = f_y(a, b).$$

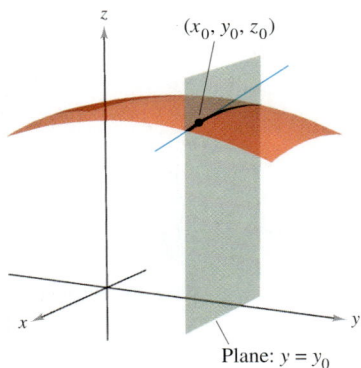

$\dfrac{\partial f}{\partial x}$ = slope in x-direction

Figure 13.28

EXAMPLE 2   **Finding and Evaluating Partial Derivatives**

For $f(x, y) = xe^{x^2 y}$, find f_x and f_y, and evaluate each at the point $(1, \ln 2)$.

Solution   Because

$$f_x(x, y) = xe^{x^2 y}(2xy) + e^{x^2 y} \qquad \text{Partial derivative with respect to } x$$

the partial derivative of f with respect to x at $(1, \ln 2)$ is

$$f_x(1, \ln 2) = e^{\ln 2}(2 \ln 2) + e^{\ln 2}$$
$$= 4 \ln 2 + 2.$$

Because

$$f_y(x, y) = xe^{x^2 y}(x^2)$$
$$= x^3 e^{x^2 y} \qquad \text{Partial derivative with respect to } y$$

the partial derivative of f with respect to y at $(1, \ln 2)$ is

$$f_y(1, \ln 2) = e^{\ln 2}$$
$$= 2.$$

The partial derivatives of a function of two variables, $z = f(x, y)$, have a useful geometric interpretation. If $y = y_0$, then $z = f(x, y_0)$ represents the curve formed by intersecting the surface $z = f(x, y)$ with the plane $y = y_0$, as shown in Figure 13.28. Therefore,

$$f_x(x_0, y_0) = \lim_{\Delta x \to 0} \frac{f(x_0 + \Delta x, y_0) - f(x_0, y_0)}{\Delta x}$$

represents the slope of this curve at the point $(x_0, y_0, f(x_0, y_0))$. Note that both the curve and the tangent line lie in the plane $y = y_0$. Similarly,

$$f_y(x_0, y_0) = \lim_{\Delta y \to 0} \frac{f(x_0, y_0 + \Delta y) - f(x_0, y_0)}{\Delta y}$$

represents the slope of the curve given by the intersection of $z = f(x, y)$ and the plane $x = x_0$ at $(x_0, y_0, f(x_0, y_0))$, as shown in Figure 13.29.

Informally, the values of $\partial f / \partial x$ and $\partial f / \partial y$ at the point (x_0, y_0, z_0) denote the **slopes of the surface in the x- and y-directions,** respectively.

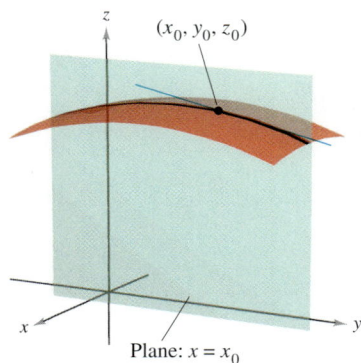

$\dfrac{\partial f}{\partial y}$ = slope in y-direction

Figure 13.29

EXAMPLE 3   **Finding the Slopes of a Surface**

•••▷ *See LarsonCalculus.com for an interactive version of this type of example.*

Find the slopes in the x-direction and in the y-direction of the surface

$$f(x, y) = -\frac{x^2}{2} - y^2 + \frac{25}{8}$$

at the point $\left(\frac{1}{2}, 1, 2\right)$.

Solution   The partial derivatives of f with respect to x and y are

$$f_x(x, y) = -x \quad \text{and} \quad f_y(x, y) = -2y. \qquad \text{Partial derivatives}$$

So, in the x-direction, the slope is

$$f_x\left(\frac{1}{2}, 1\right) = -\frac{1}{2} \qquad \text{Figure 13.30}$$

and in the y-direction, the slope is

$$f_y\left(\frac{1}{2}, 1\right) = -2. \qquad \text{Figure 13.31}$$

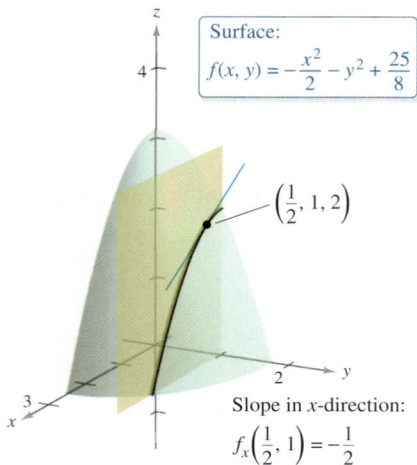

Surface:
$$f(x, y) = -\frac{x^2}{2} - y^2 + \frac{25}{8}$$

Slope in x-direction:
$$f_x\left(\frac{1}{2}, 1\right) = -\frac{1}{2}$$

Figure 13.30

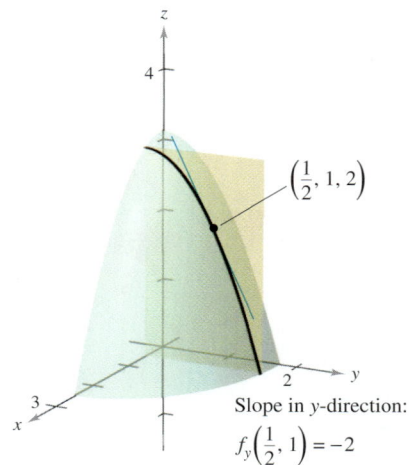

Slope in y-direction:
$$f_y\left(\frac{1}{2}, 1\right) = -2$$

Figure 13.31

EXAMPLE 4   **Finding the Slopes of a Surface**

Find the slopes of the surface

$$f(x, y) = 1 - (x - 1)^2 - (y - 2)^2$$

at the point $(1, 2, 1)$ in the x-direction and in the y-direction.

Solution   The partial derivatives of f with respect to x and y are

$$f_x(x, y) = -2(x - 1) \quad \text{and} \quad f_y(x, y) = -2(y - 2). \qquad \text{Partial derivatives}$$

So, at the point $(1, 2, 1)$, the slope in the x-direction is

$$f_x(1, 2) = -2(1 - 1) = 0$$

and the slope in the y-direction is

$$f_y(1, 2) = -2(2 - 2) = 0$$

as shown in Figure 13.32.

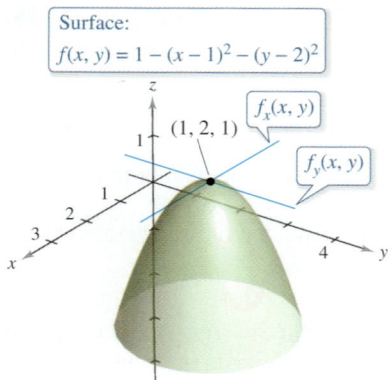

Surface:
$$f(x, y) = 1 - (x - 1)^2 - (y - 2)^2$$

Figure 13.32

No matter how many variables are involved, partial derivatives can be interpreted as *rates of change.*

EXAMPLE 5 **Using Partial Derivatives to Find Rates of Change**

The area of a parallelogram with adjacent sides a and b and included angle θ is given by $A = ab \sin \theta$, as shown in Figure 13.33.

a. Find the rate of change of A with respect to a for $a = 10$, $b = 20$, and $\theta = \dfrac{\pi}{6}$.

b. Find the rate of change of A with respect to θ for $a = 10$, $b = 20$, and $\theta = \dfrac{\pi}{6}$.

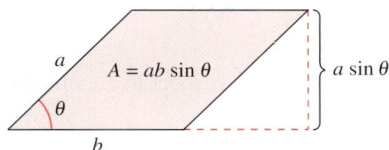

The area of the parallelogram is $ab \sin \theta$.

Figure 13.33

Solution

a. To find the rate of change of the area with respect to a, hold b and θ constant and differentiate with respect to a to obtain

$$\frac{\partial A}{\partial a} = b \sin \theta. \qquad \text{Find partial derivative with respect to } a.$$

For $a = 10$, $b = 20$, and $\theta = \pi/6$, the rate of change of the area with respect to a is

$$\frac{\partial A}{\partial a} = 20 \sin \frac{\pi}{6} = 10. \qquad \text{Substitute for } b \text{ and } \theta.$$

b. To find the rate of change of the area with respect to θ, hold a and b constant and differentiate with respect to θ to obtain

$$\frac{\partial A}{\partial \theta} = ab \cos \theta. \qquad \text{Find partial derivative with respect to } \theta.$$

For $a = 10$, $b = 20$, and $\theta = \pi/6$, the rate of change of the area with respect to θ is

$$\frac{\partial A}{\partial \theta} = 200 \cos \frac{\pi}{6} = 100\sqrt{3}. \qquad \text{Substitute for } a, b, \text{ and } \theta.$$

Partial Derivatives of a Function of Three or More Variables

The concept of a partial derivative can be extended naturally to functions of three or more variables. For instance, if $w = f(x, y, z)$, then there are three partial derivatives, each of which is formed by holding two of the variables constant. That is, to define the partial derivative of w with respect to x, consider y and z to be constant and differentiate with respect to x. A similar process is used to find the derivatives of w with respect to y and with respect to z.

$$\frac{\partial w}{\partial x} = f_x(x, y, z) = \lim_{\Delta x \to 0} \frac{f(x + \Delta x, y, z) - f(x, y, z)}{\Delta x}$$

$$\frac{\partial w}{\partial y} = f_y(x, y, z) = \lim_{\Delta y \to 0} \frac{f(x, y + \Delta y, z) - f(x, y, z)}{\Delta y}$$

$$\frac{\partial w}{\partial z} = f_z(x, y, z) = \lim_{\Delta z \to 0} \frac{f(x, y, z + \Delta z) - f(x, y, z)}{\Delta z}$$

In general, if $w = f(x_1, x_2, \ldots, x_n)$, then there are n partial derivatives denoted by

$$\frac{\partial w}{\partial x_k} = f_{x_k}(x_1, x_2, \ldots, x_n), \quad k = 1, 2, \ldots, n.$$

To find the partial derivative with respect to one of the variables, hold the other variables constant and differentiate with respect to the given variable.

EXAMPLE 6   **Finding Partial Derivatives**

a. To find the partial derivative of $f(x, y, z) = xy + yz^2 + xz$ with respect to z, consider x and y to be constant and obtain

$$\frac{\partial}{\partial z}[xy + yz^2 + xz] = 2yz + x.$$

b. To find the partial derivative of $f(x, y, z) = z \sin(xy^2 + 2z)$ with respect to z, consider x and y to be constant. Then, using the Product Rule, you obtain

$$\frac{\partial}{\partial z}[z \sin(xy^2 + 2z)] = (z)\frac{\partial}{\partial z}[\sin(xy^2 + 2z)] + \sin(xy^2 + 2z)\frac{\partial}{\partial z}[z]$$

$$= (z)[\cos(xy^2 + 2z)](2) + \sin(xy^2 + 2z)$$

$$= 2z \cos(xy^2 + 2z) + \sin(xy^2 + 2z).$$

c. To find the partial derivative of

$$f(x, y, z, w) = \frac{x + y + z}{w}$$

with respect to w, consider x, y, and z to be constant and obtain

$$\frac{\partial}{\partial w}\left[\frac{x + y + z}{w}\right] = -\frac{x + y + z}{w^2}.$$

Higher-Order Partial Derivatives

As is true for ordinary derivatives, it is possible to take second, third, and higher-order partial derivatives of a function of several variables, provided such derivatives exist. Higher-order derivatives are denoted by the order in which the differentiation occurs. For instance, the function $z = f(x, y)$ has the following second partial derivatives.

1. Differentiate twice with respect to x:

$$\frac{\partial}{\partial x}\left(\frac{\partial f}{\partial x}\right) = \frac{\partial^2 f}{\partial x^2} = f_{xx}.$$

2. Differentiate twice with respect to y:

$$\frac{\partial}{\partial y}\left(\frac{\partial f}{\partial y}\right) = \frac{\partial^2 f}{\partial y^2} = f_{yy}.$$

3. Differentiate first with respect to x and then with respect to y:

$$\frac{\partial}{\partial y}\left(\frac{\partial f}{\partial x}\right) = \frac{\partial^2 f}{\partial y \partial x} = f_{xy}.$$

4. Differentiate first with respect to y and then with respect to x:

$$\frac{\partial}{\partial x}\left(\frac{\partial f}{\partial y}\right) = \frac{\partial^2 f}{\partial x \partial y} = f_{yx}.$$

• • REMARK Note that the two types of notation for mixed partials have different conventions for indicating the order of differentiation.

$$\frac{\partial}{\partial y}\left(\frac{\partial f}{\partial x}\right) = \frac{\partial^2 f}{\partial y \partial x} \quad \text{Right-to-left order}$$

$$(f_x)_y = f_{xy} \quad \text{Left-to-right order}$$

You can remember the order by observing that in both notations you differentiate first with respect to the variable "nearest" f.

The third and fourth cases are called **mixed partial derivatives.**

EXAMPLE 7   **Finding Second Partial Derivatives**

Find the second partial derivatives of

$$f(x, y) = 3xy^2 - 2y + 5x^2y^2$$

and determine the value of $f_{xy}(-1, 2)$.

Solution   Begin by finding the first partial derivatives with respect to x and y.

$$f_x(x, y) = 3y^2 + 10xy^2 \quad \text{and} \quad f_y(x, y) = 6xy - 2 + 10x^2y$$

Then, differentiate each of these with respect to x and y.

$$f_{xx}(x, y) = 10y^2 \qquad \text{and} \quad f_{yy}(x, y) = 6x + 10x^2$$
$$f_{xy}(x, y) = 6y + 20xy \quad \text{and} \quad f_{yx}(x, y) = 6y + 20xy$$

At $(-1, 2)$, the value of f_{xy} is

$$f_{xy}(-1, 2) = 12 - 40 = -28.$$

Notice in Example 7 that the two mixed partials are equal. Sufficient conditions for this occurrence are given in Theorem 13.3.

THEOREM 13.3   Equality of Mixed Partial Derivatives

If f is a function of x and y such that f_{xy} and f_{yx} are continuous on an open disk R, then, for every (x, y) in R,

$$f_{xy}(x, y) = f_{yx}(x, y).$$

Theorem 13.3 also applies to a function f of *three or more variables* so long as all second partial derivatives are continuous. For example, if

$$w = f(x, y, z) \qquad \text{Function of three variables}$$

and all the second partial derivatives are continuous in an open region R, then at each point in R, the order of differentiation in the mixed second partial derivatives is irrelevant. If the third partial derivatives of f are also continuous, then the order of differentiation of the mixed third partial derivatives is irrelevant.

EXAMPLE 8   **Finding Higher-Order Partial Derivatives**

Show that $f_{xz} = f_{zx}$ and $f_{xzz} = f_{zxz} = f_{zzx}$ for the function

$$f(x, y, z) = ye^x + x \ln z.$$

Solution

First partials:

$$f_x(x, y, z) = ye^x + \ln z, \quad f_z(x, y, z) = \frac{x}{z}$$

Second partials (note that the first two are equal):

$$f_{xz}(x, y, z) = \frac{1}{z}, \quad f_{zx}(x, y, z) = \frac{1}{z}, \quad f_{zz}(x, y, z) = -\frac{x}{z^2}$$

Third partials (note that all three are equal):

$$f_{xzz}(x, y, z) = -\frac{1}{z^2}, \quad f_{zxz}(x, y, z) = -\frac{1}{z^2}, \quad f_{zzx}(x, y, z) = -\frac{1}{z^2}.$$

13.3 Exercises

See **CalcChat.com** for tutorial help and worked-out solutions to odd-numbered exercises.

Examining a Partial Derivative In Exercises 1–6, explain whether the Quotient Rule should be used to find the partial derivative. Do not differentiate.

1. $\dfrac{\partial}{\partial x}\left(\dfrac{x^2 y}{y^2 - 3}\right)$

2. $\dfrac{\partial}{\partial y}\left(\dfrac{x^2 y}{y^2 - 3}\right)$

3. $\dfrac{\partial}{\partial y}\left(\dfrac{x - y}{x^2 + 1}\right)$

4. $\dfrac{\partial}{\partial x}\left(\dfrac{x - y}{x^2 + 1}\right)$

5. $\dfrac{\partial}{\partial x}\left(\dfrac{xy}{x^2 + 1}\right)$

6. $\dfrac{\partial}{\partial y}\left(\dfrac{xy}{x^2 + 1}\right)$

Finding Partial Derivatives In Exercises 7–38, find both first partial derivatives.

7. $f(x, y) = 2x - 5y + 3$

8. $f(x, y) = x^2 - 2y^2 + 4$

9. $f(x, y) = x^2 y^3$

10. $f(x, y) = 4x^3 y^{-2}$

11. $z = x\sqrt{y}$

12. $z = 2y^2\sqrt{x}$

13. $z = x^2 - 4xy + 3y^2$

14. $z = y^3 - 2xy^2 - 1$

15. $z = e^{xy}$

16. $z = e^{x/y}$

17. $z = x^2 e^{2y}$

18. $z = y e^{y/x}$

19. $z = \ln\dfrac{x}{y}$

20. $z = \ln\sqrt{xy}$

21. $z = \ln(x^2 + y^2)$

22. $z = \ln\dfrac{x + y}{x - y}$

23. $z = \dfrac{x^2}{2y} + \dfrac{3y^2}{x}$

24. $z = \dfrac{xy}{x^2 + y^2}$

25. $h(x, y) = e^{-(x^2 + y^2)}$

26. $g(x, y) = \ln\sqrt{x^2 + y^2}$

27. $f(x, y) = \sqrt{x^2 + y^2}$

28. $f(x, y) = \sqrt{2x + y^3}$

29. $z = \cos xy$

30. $z = \sin(x + 2y)$

31. $z = \tan(2x - y)$

32. $z = \sin 5x \cos 5y$

33. $z = e^y \sin xy$

34. $z = \cos(x^2 + y^2)$

35. $z = \sinh(2x + 3y)$

36. $z = \cosh xy^2$

37. $f(x, y) = \displaystyle\int_x^y (t^2 - 1)\, dt$

38. $f(x, y) = \displaystyle\int_x^y (2t + 1)\, dt + \int_y^x (2t - 1)\, dt$

Finding Partial Derivatives In Exercises 39–42, use the limit definition of partial derivatives to find $f_x(x, y)$ and $f_y(x, y)$.

39. $f(x, y) = 3x + 2y$

40. $f(x, y) = x^2 - 2xy + y^2$

41. $f(x, y) = \sqrt{x + y}$

42. $f(x, y) = \dfrac{1}{x + y}$

Evaluating Partial Derivatives In Exercises 43–50, evaluate f_x and f_y at the given point.

43. $f(x, y) = e^y \sin x$, $(\pi, 0)$

44. $f(x, y) = e^{-x} \cos y$, $(0, 0)$

45. $f(x, y) = \cos(2x - y)$, $\left(\dfrac{\pi}{4}, \dfrac{\pi}{3}\right)$

46. $f(x, y) = \sin xy$, $\left(2, \dfrac{\pi}{4}\right)$

47. $f(x, y) = \arctan\dfrac{y}{x}$, $(2, -2)$

48. $f(x, y) = \arccos xy$, $(1, 1)$

49. $f(x, y) = \dfrac{xy}{x - y}$, $(2, -2)$

50. $f(x, y) = \dfrac{2xy}{\sqrt{4x^2 + 5y^2}}$, $(1, 1)$

Finding the Slopes of a Surface In Exercises 51 and 52, find the slopes of the surface in the x- and y-directions at the given point.

51. $g(x, y) = 4 - x^2 - y^2$
$(1, 1, 2)$

52. $h(x, y) = x^2 - y^2$
$(-2, 1, 3)$

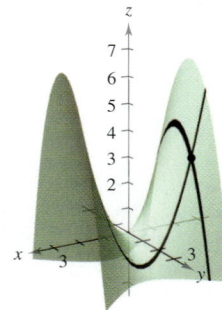

Finding Partial Derivatives In Exercises 53–58, find the first partial derivatives with respect to x, y, and z.

53. $H(x, y, z) = \sin(x + 2y + 3z)$

54. $f(x, y, z) = 3x^2 y - 5xyz + 10yz^2$

55. $w = \sqrt{x^2 + y^2 + z^2}$

56. $w = \dfrac{7xz}{x + y}$

57. $F(x, y, z) = \ln\sqrt{x^2 + y^2 + z^2}$

58. $G(x, y, z) = \dfrac{1}{\sqrt{1 - x^2 - y^2 - z^2}}$

Evaluating Partial Derivatives In Exercises 59–64, evaluate $f_x, f_y,$ and f_z at the given point.

59. $f(x, y, z) = x^3 yz^2$, $(1, 1, 1)$

60. $f(x, y, z) = x^2 y^3 + 2xyz - 3yz$, $(-2, 1, 2)$

61. $f(x, y, z) = \dfrac{x}{yz}$, $(1, -1, -1)$

62. $f(x, y, z) = \dfrac{xy}{x + y + z}$, $(3, 1, -1)$

63. $f(x, y, z) = z \sin(x + y)$, $\left(0, \dfrac{\pi}{2}, -4\right)$

64. $f(x, y, z) = \sqrt{3x^2 + y^2 - 2z^2}$, $(1, -2, 1)$

Using First Partial Derivatives In Exercises 65–72, for $f(x, y)$, find all values of x and y such that $f_x(x, y) = 0$ and $f_y(x, y) = 0$ simultaneously.

65. $f(x, y) = x^2 + xy + y^2 - 2x + 2y$

66. $f(x, y) = x^2 - xy + y^2 - 5x + y$

67. $f(x, y) = x^2 + 4xy + y^2 - 4x + 16y + 3$

68. $f(x, y) = x^2 - xy + y^2$

69. $f(x, y) = \dfrac{1}{x} + \dfrac{1}{y} + xy$

70. $f(x, y) = 3x^3 - 12xy + y^3$

71. $f(x, y) = e^{x^2 + xy + y^2}$

72. $f(x, y) = \ln(x^2 + y^2 + 1)$

Finding Second Partial Derivatives In Exercises 73–82, find the four second partial derivatives. Observe that the second mixed partials are equal.

73. $z = 3xy^2$

74. $z = x^2 + 3y^2$

75. $z = x^2 - 2xy + 3y^2$

76. $z = x^4 - 3x^2y^2 + y^4$

77. $z = \sqrt{x^2 + y^2}$

78. $z = \ln(x - y)$

79. $z = e^x \tan y$

80. $z = 2xe^y - 3ye^{-x}$

81. $z = \cos xy$

82. $z = \arctan \dfrac{y}{x}$

Finding Partial Derivatives Using Technology In Exercises 83–86, use a computer algebra system to find the first and second partial derivatives of the function. Determine whether there exist values of x and y such that $f_x(x, y) = 0$ and $f_y(x, y) = 0$ simultaneously.

83. $f(x, y) = x \sec y$

84. $f(x, y) = \sqrt{25 - x^2 - y^2}$

85. $f(x, y) = \ln \dfrac{x}{x^2 + y^2}$

86. $f(x, y) = \dfrac{xy}{x - y}$

Comparing Mixed Partial Derivatives In Exercises 87–90, show that the mixed partial derivatives $f_{xyy}, f_{yxy},$ and f_{yyx} are equal.

87. $f(x, y, z) = xyz$

88. $f(x, y, z) = x^2 - 3xy + 4yz + z^3$

89. $f(x, y, z) = e^{-x} \sin yz$

90. $f(x, y, z) = \dfrac{2z}{x + y}$

Laplace's Equation In Exercises 91–94, show that the function satisfies Laplace's equation $\partial^2 z/\partial x^2 + \partial^2 z/\partial y^2 = 0$.

91. $z = 5xy$

92. $z = \frac{1}{2}(e^y - e^{-y})\sin x$

93. $z = e^x \sin y$

94. $z = \arctan \dfrac{y}{x}$

Wave Equation In Exercises 95–98, show that the function satisfies the wave equation $\partial^2 z/\partial t^2 = c^2(\partial^2 z/\partial x^2)$.

95. $z = \sin(x - ct)$

96. $z = \cos(4x + 4ct)$

97. $z = \ln(x + ct)$

98. $z = \sin \omega ct \sin \omega x$

Heat Equation In Exercises 99 and 100, show that the function satisfies the heat equation $\partial z/\partial t = c^2(\partial^2 z/\partial x^2)$.

99. $z = e^{-t} \cos \dfrac{x}{c}$

100. $z = e^{-t} \sin \dfrac{x}{c}$

Using First Partial Derivatives In Exercises 101 and 102, determine whether there exists a function $f(x, y)$ with the given partial derivatives. Explain your reasoning. If such a function exists, give an example.

101. $f_x(x, y) = -3 \sin(3x - 2y)$, $f_y(x, y) = 2 \sin(3x - 2y)$

102. $f_x(x, y) = 2x + y$, $f_y(x, y) = x - 4y$

WRITING ABOUT CONCEPTS

103. First Partial Derivatives Let f be a function of two variables x and y. Describe the procedure for finding the first partial derivatives.

104. First Partial Derivatives Sketch a surface representing a function f of two variables x and y. Use the sketch to give geometric interpretations of $\partial f/\partial x$ and $\partial f/\partial y$.

105. Sketching a Graph Sketch the graph of a function $z = f(x, y)$ whose derivative f_x is always negative and whose derivative f_y is always positive.

106. Sketching a Graph Sketch the graph of a function $z = f(x, y)$ whose derivatives f_x and f_y are always positive.

107. Mixed Partial Derivatives If f is a function of x and y such that f_{xy} and f_{yx} are continuous, what is the relationship between the mixed partial derivatives? Explain.

---

108. HOW DO YOU SEE IT? Use the graph of the surface to determine the sign of each partial derivative. Explain your reasoning.

(a) $f_x(4, 1)$ (b) $f_y(4, 1)$

(c) $f_x(-1, -2)$ (d) $f_y(-1, -2)$

---

109. Marginal Revenue A pharmaceutical corporation has two plants that produce the same over-the-counter medicine. If x_1 and x_2 are the numbers of units produced at plant 1 and plant 2, respectively, then the total revenue for the product is given by $R = 200x_1 + 200x_2 - 4x_1^2 - 8x_1x_2 - 4x_2^2$. When $x_1 = 4$ and $x_2 = 12$, find (a) the marginal revenue for plant 1, $\partial R/\partial x_1$, and (b) the marginal revenue for plant 2, $\partial R/\partial x_2$.

110. Marginal Costs

A company manufactures two types of wood-burning stoves: a freestanding model and a fireplace-insert model. The cost function for producing x freestanding and y fireplace-insert stoves is

$$C = 32\sqrt{xy} + 175x + 205y + 1050.$$

(a) Find the marginal costs ($\partial C/\partial x$ and $\partial C/\partial y$) when $x = 80$ and $y = 20$.

(b) When additional production is required, which model of stove results in the cost increasing at a higher rate? How can this be determined from the cost model?

111. Psychology Early in the twentieth century, an intelligence test called the *Stanford-Binet Test* (more commonly known as the IQ test) was developed. In this test, an individual's mental age M is divided by the individual's chronological age C and the quotient is multiplied by 100. The result is the individual's IQ.

$$IQ(M, C) = \frac{M}{C} \times 100$$

Find the partial derivatives of IQ with respect to M and with respect to C. Evaluate the partial derivatives at the point $(12, 10)$ and interpret the result. *(Source: Adapted from Bernstein/Clark-Stewart/Roy/Wickens, Psychology, Fourth Edition)*

112. Marginal Productivity Consider the Cobb-Douglas production function $f(x, y) = 200x^{0.7}y^{0.3}$. When $x = 1000$ and $y = 500$, find (a) the marginal productivity of labor, $\partial f/\partial x$, and (b) the marginal productivity of capital, $\partial f/\partial y$.

113. Think About It Let N be the number of applicants to a university, p the charge for food and housing at the university, and t the tuition. Suppose that N is a function of p and t such that $\partial N/\partial p < 0$ and $\partial N/\partial t < 0$. What information is gained by noticing that both partials are negative?

114. Investment The value of an investment of $1000 earning 6% compounded annually is

$$V(I, R) = 1000\left[\frac{1 + 0.06(1 - R)}{1 + I}\right]^{10}$$

where I is the annual rate of inflation and R is the tax rate for the person making the investment. Calculate $V_I(0.03, 0.28)$ and $V_R(0.03, 0.28)$. Determine whether the tax rate or the rate of inflation is the greater "negative" factor in the growth of the investment.

115. Temperature Distribution The temperature at any point (x, y) in a steel plate is $T = 500 - 0.6x^2 - 1.5y^2$, where x and y are measured in meters. At the point $(2, 3)$, find the rates of change of the temperature with respect to the distances moved along the plate in the directions of the x- and y-axes.

116. Apparent Temperature A measure of how hot weather feels to an average person is the Apparent Temperature Index. A model for this index is

$$A = 0.885t - 22.4h + 1.20th - 0.544$$

where A is the apparent temperature in degrees Celsius, t is the air temperature, and h is the relative humidity in decimal form. *(Source: The UMAP Journal)*

(a) Find $\dfrac{\partial A}{\partial t}$ and $\dfrac{\partial A}{\partial h}$ when $t = 30°$ and $h = 0.80$.

(b) Which has a greater effect on A, air temperature or humidity? Explain.

117. Ideal Gas Law The Ideal Gas Law states that $PV = nRT$, where P is pressure, V is volume, n is the number of moles of gas, R is a fixed constant (the gas constant), and T is absolute temperature. Show that

$$\frac{\partial T}{\partial P} \cdot \frac{\partial P}{\partial V} \cdot \frac{\partial V}{\partial T} = -1.$$

118. Marginal Utility The utility function $U = f(x, y)$ is a measure of the utility (or satisfaction) derived by a person from the consumption of two products x and y. The utility function for two products is

$$U = -5x^2 + xy - 3y^2.$$

(a) Determine the marginal utility of product x.

(b) Determine the marginal utility of product y.

(c) When $x = 2$ and $y = 3$, should a person consume one more unit of product x or one more unit of product y? Explain your reasoning.

(d) Use a computer algebra system to graph the function. Interpret the marginal utilities of products x and y graphically.

119. Modeling Data The expenditures (in billions of dollars) for different types of recreation in the United States from 2005 through 2010 are shown in the table. Expenditures on amusement parks and campgrounds, live entertainment (excluding sports), and spectator sports are represented by the variables x, y, and z. *(Source: U.S. Bureau of Economic Analysis)*

Year	2005	2006	2007	2008	2009	2010
x	36.4	39.0	42.4	44.7	43.0	45.2
y	15.3	16.6	17.4	17.5	17.0	17.3
z	16.4	18.1	20.0	20.5	20.1	21.4

A model for the data is given by

$$z = 0.461x + 0.301y - 494.$$

(a) Find $\dfrac{\partial z}{\partial x}$ and $\dfrac{\partial z}{\partial y}$.

(b) Interpret the partial derivatives in the context of the problem.

Amy Walters/Shutterstock.com

120. Modeling Data The table shows the public medical expenditures (in billions of dollars) for workers' compensation x, Medicaid y, and Medicare z from 2005 through 2010. (*Source: Centers for Medicare and Medicaid Services*)

Year	2005	2006	2007	2008	2009	2010
x	41.2	41.6	41.2	40.1	36.7	37.2
y	309.5	306.8	326.4	343.8	374.4	401.4
z	338.8	403.1	432.3	466.9	499.8	524.6

A model for the data is given by

$z = 11.734x^2 - 0.028y^2 - 888.24x + 23.09y + 12{,}573.9.$

(a) Find $\dfrac{\partial^2 z}{\partial x^2}$ and $\dfrac{\partial^2 z}{\partial y^2}$.

(b) Determine the concavity of traces parallel to the xz-plane. Interpret the result in the context of the problem.

(c) Determine the concavity of traces parallel to the yz-plane. Interpret the result in the context of the problem.

True or False? **In Exercises 121–124, determine whether the statement is true or false. If it is false, explain why or give an example that shows it is false.**

121. If $z = f(x, y)$ and $\dfrac{\partial z}{\partial x} = \dfrac{\partial z}{\partial y}$, then $z = c(x + y)$.

122. If $z = f(x)g(y)$, then $\dfrac{\partial z}{\partial x} + \dfrac{\partial z}{\partial y} = f'(x)g(y) + f(x)g'(y)$.

123. If $z = e^{xy}$, then $\dfrac{\partial^2 z}{\partial y \partial x} = (xy + 1)e^{xy}$.

124. If a cylindrical surface $z = f(x, y)$ has rulings parallel to the y-axis, then $\partial z/\partial y = 0$.

125. Using a Function Consider the function defined by

$$f(x, y) = \begin{cases} \dfrac{xy(x^2 - y^2)}{x^2 + y^2}, & (x, y) \neq (0, 0) \\ 0, & (x, y) = (0, 0) \end{cases}.$$

(a) Find $f_x(x, y)$ and $f_y(x, y)$ for $(x, y) \neq (0, 0)$.

(b) Use the definition of partial derivatives to find $f_x(0, 0)$ and $f_y(0, 0)$.

$$\left[\textit{Hint: } f_x(0, 0) = \lim_{\Delta x \to 0} \frac{f(\Delta x, 0) - f(0, 0)}{\Delta x} \right]$$

(c) Use the definition of partial derivatives to find $f_{xy}(0, 0)$ and $f_{yx}(0, 0)$.

(d) Using Theorem 13.3 and the result of part (c), what can be said about f_{xy} or f_{yx}?

126. Using a Function Consider the function

$f(x, y) = (x^3 + y^3)^{1/3}.$

(a) Find $f_x(0, 0)$ and $f_y(0, 0)$.

(b) Determine the points (if any) at which $f_x(x, y)$ or $f_y(x, y)$ fails to exist.

127. Using a Function Consider the function

$f(x, y) = (x^2 + y^2)^{2/3}.$

Show that

$$f_x(x, y) = \begin{cases} \dfrac{4x}{3(x^2 + y^2)^{1/3}}, & (x, y) \neq (0, 0) \\ 0, & (x, y) = (0, 0) \end{cases}.$$

■ **FOR FURTHER INFORMATION** For more information about this problem, see the article "A Classroom Note on a Naturally Occurring Piecewise Defined Function" by Don Cohen in *Mathematics and Computer Education.*

SECTION PROJECT

Moiré Fringes

Read the article "Moiré Fringes and the Conic Sections" by Mike Cullen in *The College Mathematics Journal.* The article describes how two families of level curves given by

$$f(x, y) = a \quad \text{and} \quad g(x, y) = b$$

can form Moiré patterns. After reading the article, write a paper explaining how the expression

$$\frac{\partial f}{\partial x} \cdot \frac{\partial g}{\partial x} + \frac{\partial f}{\partial y} \cdot \frac{\partial g}{\partial y}$$

is related to the Moiré patterns formed by intersecting the two families of level curves. Use one of the following patterns as an example in your paper.

13.4 Differentials

■ Understand the concepts of increments and differentials.
■ Extend the concept of differentiability to a function of two variables.
■ Use a differential as an approximation.

Increments and Differentials

In this section, the concepts of increments and differentials are generalized to functions of two or more variables. Recall from Section 3.9 that for $y = f(x)$, the differential of y was defined as

$$dy = f'(x)\, dx.$$

Similar terminology is used for a function of two variables, $z = f(x, y)$. That is, Δx and Δy are the **increments of x and y,** and the **increment of z** is

$$\Delta z = f(x + \Delta x, y + \Delta y) - f(x, y). \qquad \text{Increment of } z$$

Definition of Total Differential

If $z = f(x, y)$ and Δx and Δy are increments of x and y, then the **differentials** of the independent variables x and y are

$$dx = \Delta x \quad \text{and} \quad dy = \Delta y$$

and the **total differential** of the dependent variable z is

$$dz = \frac{\partial z}{\partial x}\, dx + \frac{\partial z}{\partial y}\, dy = f_x(x, y)\, dx + f_y(x, y)\, dy.$$

This definition can be extended to a function of three or more variables. For instance, if $w = f(x, y, z, u)$, then $dx = \Delta x$, $dy = \Delta y$, $dz = \Delta z$, $du = \Delta u$, and the total differential of w is

$$dw = \frac{\partial w}{\partial x}\, dx + \frac{\partial w}{\partial y}\, dy + \frac{\partial w}{\partial z}\, dz + \frac{\partial w}{\partial u}\, du.$$

EXAMPLE 1   **Finding the Total Differential**

Find the total differential for each function.

a. $z = 2x \sin y - 3x^2 y^2$ **b.** $w = x^2 + y^2 + z^2$

Solution

a. The total differential dz for $z = 2x \sin y - 3x^2 y^2$ is

$$dz = \frac{\partial z}{\partial x}\, dx + \frac{\partial z}{\partial y}\, dy \qquad \text{Total differential } dz$$

$$= (2 \sin y - 6xy^2)\, dx + (2x \cos y - 6x^2 y)\, dy.$$

b. The total differential dw for $w = x^2 + y^2 + z^2$ is

$$dw = \frac{\partial w}{\partial x}\, dx + \frac{\partial w}{\partial y}\, dy + \frac{\partial w}{\partial z}\, dz \qquad \text{Total differential } dw$$

$$= 2x\, dx + 2y\, dy + 2z\, dz.$$

Differentiability

In Section 3.9, you learned that for a *differentiable* function given by $y = f(x)$, you can use the differential $dy = f'(x)\,dx$ as an approximation (for small Δx) of the value $\Delta y = f(x + \Delta x) - f(x)$. When a similar approximation is possible for a function of two variables, the function is said to be **differentiable.** This is stated explicitly in the next definition.

Definition of Differentiability

A function f given by $z = f(x, y)$ is **differentiable** at (x_0, y_0) if Δz can be written in the form

$$\Delta z = f_x(x_0, y_0)\,\Delta x + f_y(x_0, y_0)\,\Delta y + \varepsilon_1 \Delta x + \varepsilon_2 \Delta y$$

where both ε_1 and $\varepsilon_2 \to 0$ as

$$(\Delta x, \Delta y) \to (0, 0).$$

The function f is **differentiable in a region R** if it is differentiable at each point in R.

EXAMPLE 2   **Showing that a Function Is Differentiable**

Show that the function

$$f(x, y) = x^2 + 3y$$

is differentiable at every point in the plane.

Solution   Letting $z = f(x, y)$, the increment of z at an arbitrary point (x, y) in the plane is

$$\begin{aligned}
\Delta z &= f(x + \Delta x, y + \Delta y) - f(x, y) \qquad \text{\textcolor{red}{Increment of } } z\\
&= (x^2 + 2x\Delta x + \Delta x^2) + 3(y + \Delta y) - (x^2 + 3y)\\
&= 2x\Delta x + \Delta x^2 + 3\Delta y\\
&= 2x(\Delta x) + 3(\Delta y) + \Delta x(\Delta x) + 0(\Delta y)\\
&= f_x(x, y)\,\Delta x + f_y(x, y)\,\Delta y + \varepsilon_1 \Delta x + \varepsilon_2 \Delta y
\end{aligned}$$

where $\varepsilon_1 = \Delta x$ and $\varepsilon_2 = 0$. Because $\varepsilon_1 \to 0$ and $\varepsilon_2 \to 0$ as $(\Delta x, \Delta y) \to (0, 0)$, it follows that f is differentiable at every point in the plane. The graph of f is shown in Figure 13.34.

Figure 13.34

Be sure you see that the term "differentiable" is used differently for functions of two variables than for functions of one variable. A function of one variable is differentiable at a point when its derivative exists at the point. For a function of two variables, however, the existence of the partial derivatives f_x and f_y does not guarantee that the function is differentiable (see Example 5). The next theorem gives a *sufficient* condition for differentiability of a function of two variables.

THEOREM 13.4   Sufficient Condition for Differentiability

If f is a function of x and y, where f_x and f_y are continuous in an open region R, then f is differentiable on R.

A proof of Theorem 13.4 is given in Appendix A.

See LarsonCalculus.com for Bruce Edwards's video of this proof.

Approximation by Differentials

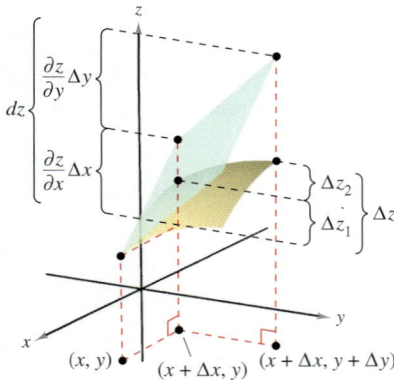

The exact change in z is Δz. This change can be approximated by the differential dz.

Figure 13.35

Theorem 13.4 tells you that you can choose $(x + \Delta x, y + \Delta y)$ close enough to (x, y) to make $\varepsilon_1 \Delta x$ and $\varepsilon_2 \Delta y$ insignificant. In other words, for small Δx and Δy, you can use the approximation

$$\Delta z \approx dz.$$

This approximation is illustrated graphically in Figure 13.35. Recall that the partial derivatives $\partial z/\partial x$ and $\partial z/\partial y$ can be interpreted as the slopes of the surface in the x- and y-directions. This means that

$$dz = \frac{\partial z}{\partial x} \Delta x + \frac{\partial z}{\partial y} \Delta y$$

represents the change in height of a plane that is tangent to the surface at the point $(x, y, f(x, y))$. Because a plane in space is represented by a linear equation in the variables x, y, and z, the approximation of Δz by dz is called a **linear approximation.** You will learn more about this geometric interpretation in Section 13.7.

EXAMPLE 3   **Using a Differential as an Approximation**

⋯▷ *See LarsonCalculus.com for an interactive version of this type of example.*

Use the differential dz to approximate the change in $z = \sqrt{4 - x^2 - y^2}$ as (x, y) moves from the point $(1, 1)$ to the point $(1.01, 0.97)$. Compare this approximation with the exact change in z.

Solution   Letting $(x, y) = (1, 1)$ and $(x + \Delta x, y + \Delta y) = (1.01, 0.97)$ produces

$$dx = \Delta x = 0.01 \quad \text{and} \quad dy = \Delta y = -0.03.$$

So, the change in z can be approximated by

$$\Delta z \approx dz = \frac{\partial z}{\partial x} dx + \frac{\partial z}{\partial y} dy = \frac{-x}{\sqrt{4 - x^2 - y^2}} \Delta x + \frac{-y}{\sqrt{4 - x^2 - y^2}} \Delta y.$$

When $x = 1$ and $y = 1$, you have

$$\Delta z \approx -\frac{1}{\sqrt{2}}(0.01) - \frac{1}{\sqrt{2}}(-0.03) = \frac{0.02}{\sqrt{2}} = \sqrt{2}(0.01) \approx 0.0141.$$

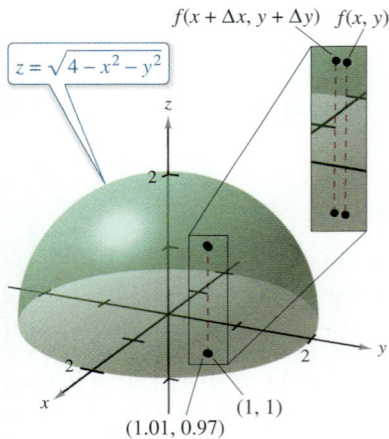

As (x, y) moves from $(1, 1)$ to the point $(1.01, 0.97)$, the value of $f(x, y)$ changes by about 0.0137.

Figure 13.36

In Figure 13.36, you can see that the exact change corresponds to the difference in the heights of two points on the surface of a hemisphere. This difference is given by

$$\Delta z = f(1.01, 0.97) - f(1, 1)$$
$$= \sqrt{4 - (1.01)^2 - (0.97)^2} - \sqrt{4 - 1^2 - 1^2}$$
$$\approx 0.0137.$$

A function of three variables $w = f(x, y, z)$ is **differentiable** at (x, y, z) provided that

$$\Delta w = f(x + \Delta x, y + \Delta y, z + \Delta z) - f(x, y, z)$$

can be written in the form

$$\Delta w = f_x \Delta x + f_y \Delta y + f_z \Delta z + \varepsilon_1 \Delta x + \varepsilon_2 \Delta y + \varepsilon_3 \Delta z$$

where ε_1, ε_2, and $\varepsilon_3 \to 0$ as $(\Delta x, \Delta y, \Delta z) \to (0, 0, 0)$. With this definition of differentiability, Theorem 13.4 has the following extension for functions of three variables: If f is a function of x, y, and z, where $f, f_x, f_y,$ and f_z are continuous in an open region R, then f is differentiable on R.

In Section 3.9, you used differentials to approximate the propagated error introduced by an error in measurement. This application of differentials is further illustrated in Example 4.

Volume $= xyz$
Figure 13.37

EXAMPLE 4   **Error Analysis**

The possible error involved in measuring each dimension of a rectangular box is ± 0.1 millimeter. The dimensions of the box are $x = 50$ centimeters, $y = 20$ centimeters, and $z = 15$ centimeters, as shown in Figure 13.37. Use dV to estimate the propagated error and the relative error in the calculated volume of the box.

Solution   The volume of the box is $V = xyz$, and so

$$dV = \frac{\partial V}{\partial x}\, dx + \frac{\partial V}{\partial y}\, dy + \frac{\partial V}{\partial z}\, dz$$
$$= yz\, dx + xz\, dy + xy\, dz.$$

Using 0.1 millimeter $= 0.01$ centimeter, you have

$$dx = dy = dz = \pm 0.01$$

and the propagated error is approximately

$$dV = (20)(15)(\pm 0.01) + (50)(15)(\pm 0.01) + (50)(20)(\pm 0.01)$$
$$= 300(\pm 0.01) + 750(\pm 0.01) + 1000(\pm 0.01)$$
$$= 2050(\pm 0.01)$$
$$= \pm 20.5 \text{ cubic centimeters.}$$

Because the measured volume is

$$V = (50)(20)(15) = 15{,}000 \text{ cubic centimeters,}$$

the relative error, $\Delta V / V$, is approximately

$$\frac{\Delta V}{V} \approx \frac{dV}{V} = \frac{20.5}{15{,}000} \approx 0.14\%.$$

As is true for a function of a single variable, when a function in two or more variables is differentiable at a point, it is also continuous there.

THEOREM 13.5   Differentiability Implies Continuity

If a function of x and y is differentiable at (x_0, y_0), then it is continuous at (x_0, y_0).

Proof   Let f be differentiable at (x_0, y_0), where $z = f(x, y)$. Then

$$\Delta z = [f_x(x_0, y_0) + \varepsilon_1]\Delta x + [f_y(x_0, y_0) + \varepsilon_2]\Delta y$$

where both ε_1 and $\varepsilon_2 \to 0$ as $(\Delta x, \Delta y) \to (0, 0)$. However, by definition, you know that Δz is

$$\Delta z = f(x_0 + \Delta x, y_0 + \Delta y) - f(x_0, y_0).$$

Letting $x = x_0 + \Delta x$ and $y = y_0 + \Delta y$ produces

$$f(x, y) - f(x_0, y_0) = [f_x(x_0, y_0) + \varepsilon_1]\Delta x + [f_y(x_0, y_0) + \varepsilon_2]\Delta y$$
$$= [f_x(x_0, y_0) + \varepsilon_1](x - x_0) + [f_y(x_0, y_0) + \varepsilon_2](y - y_0).$$

Taking the limit as $(x, y) \to (x_0, y_0)$, you have

$$\lim_{(x, y) \to (x_0, y_0)} f(x, y) = f(x_0, y_0)$$

which means that f is continuous at (x_0, y_0).

See LarsonCalculus.com for Bruce Edwards's video of this proof.

Remember that the existence of f_x and f_y is not sufficient to guarantee differentiability, as illustrated in the next example.

EXAMPLE 5 **A Function That Is Not Differentiable**

For the function

$$f(x, y) = \begin{cases} \dfrac{-3xy}{x^2 + y^2}, & (x, y) \neq (0, 0) \\ 0, & (x, y) = (0, 0) \end{cases}$$

show that $f_x(0, 0)$ and $f_y(0, 0)$ both exist, but that f is not differentiable at $(0, 0)$.

Solution You can show that f is not differentiable at $(0, 0)$ by showing that it is not continuous at this point. To see that f is not continuous at $(0, 0)$, look at the values of $f(x, y)$ along two different approaches to $(0, 0)$, as shown in Figure 13.38. Along the line $y = x$, the limit is

$$\lim_{(x, x) \to (0, 0)} f(x, y) = \lim_{(x, x) \to (0, 0)} \frac{-3x^2}{2x^2} = -\frac{3}{2}$$

whereas along $y = -x$, you have

$$\lim_{(x, -x) \to (0, 0)} f(x, y) = \lim_{(x, -x) \to (0, 0)} \frac{3x^2}{2x^2} = \frac{3}{2}.$$

So, the limit of $f(x, y)$ as $(x, y) \to (0, 0)$ does not exist, and you can conclude that f is not continuous at $(0, 0)$. Therefore, by Theorem 13.5, you know that f is not differentiable at $(0, 0)$. On the other hand, by the definition of the partial derivatives f_x and f_y, you have

$$f_x(0, 0) = \lim_{\Delta x \to 0} \frac{f(\Delta x, 0) - f(0, 0)}{\Delta x} = \lim_{\Delta x \to 0} \frac{0 - 0}{\Delta x} = 0$$

and

$$f_y(0, 0) = \lim_{\Delta y \to 0} \frac{f(0, \Delta y) - f(0, 0)}{\Delta y} = \lim_{\Delta y \to 0} \frac{0 - 0}{\Delta y} = 0.$$

So, the partial derivatives at $(0, 0)$ exist.

Along the line $y = -x$, $f(x, y)$ approaches $3/2$.

$$f(x, y) = \begin{cases} \dfrac{-3xy}{x^2 + y^2}, & (x, y) \neq (0, 0) \\ 0, & (x, y) = (0, 0) \end{cases}$$

$(0, 0, 0)$

Along the line $y = x$, $f(x, y)$ approaches $-3/2$.

Figure 13.38

▷ **TECHNOLOGY** A graphing utility can be used to graph piecewise-defined functions like the one given in Example 5. For instance, the graph shown at the left was generated by *Mathematica*.

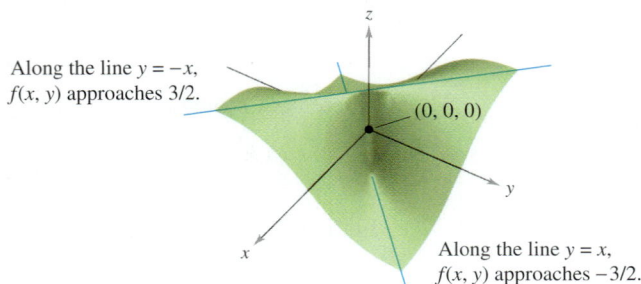

Generated by Mathematica

13.4 Exercises

See **CalcChat.com** for tutorial help and worked-out solutions to odd-numbered exercises.

Finding a Total Differential   In Exercises 1–10, find the total differential.

1. $z = 2x^2y^3$

2. $z = 2x^4y - 8x^2y^3$

3. $z = \dfrac{-1}{x^2 + y^2}$

4. $w = \dfrac{x + y}{z - 3y}$

5. $z = x\cos y - y\cos x$

6. $z = \frac{1}{2}(e^{x^2+y^2} - e^{-x^2-y^2})$

7. $z = e^x \sin y$

8. $w = e^y \cos x + z^2$

9. $w = 2z^3y \sin x$

10. $w = x^2yz^2 + \sin yz$

Using a Differential as an Approximation   In Exercises 11–16, (a) evaluate $f(2, 1)$ and $f(2.1, 1.05)$ and calculate Δz, and (b) use the total differential dz to approximate Δz.

11. $f(x, y) = 2x - 3y$

12. $f(x, y) = x^2 + y^2$

13. $f(x, y) = 16 - x^2 - y^2$

14. $f(x, y) = \dfrac{y}{x}$

15. $f(x, y) = ye^x$

16. $f(x, y) = x\cos y$

Approximating an Expression   In Exercises 17–20, find $z = f(x, y)$ and use the total differential to approximate the quantity.

17. $(2.01)^2(9.02) - 2^2 \cdot 9$

18. $\dfrac{1 - (3.05)^2}{(5.95)^2} - \dfrac{1 - 3^2}{6^2}$

19. $\sqrt{(5.05)^2 + (3.1)^2} - \sqrt{5^2 + 3^2}$

20. $\sin[(1.05)^2 + (0.95)^2] - \sin(1^2 + 1^2)$

WRITING ABOUT CONCEPTS

21. Approximation   Describe the change in accuracy of dz as an approximation of Δz as Δx and Δy increase.

22. Linear Approximation   What is meant by a linear approximation of $z = f(x, y)$ at the point $P(x_0, y_0)$?

23. Using Differentials   When using differentials, what is meant by the terms *propagated error* and *relative error*?

24. HOW DO YOU SEE IT?   Which point has a greater differential, $(2, 2)$ or $\left(\frac{1}{2}, \frac{1}{2}\right)$? Explain. (Assume that dx and dy are the same for both points.)

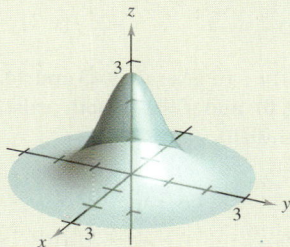

25. Area   The area of the shaded rectangle in the figure is $A = lh$. The possible errors in the length and height are Δl and Δh, respectively. Find dA and identify the regions in the figure whose areas are given by the terms of dA. What region represents the difference between ΔA and dA?

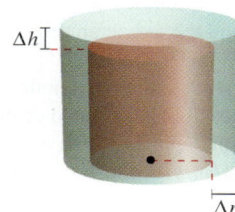

Figure for 25   Figure for 26

26. Volume   The volume of the red right circular cylinder in the figure is $V = \pi r^2 h$. The possible errors in the radius and the height are Δr and Δh, respectively. Find dV and identify the solids in the figure whose volumes are given by the terms of dV. What solid represents the difference between ΔV and dV?

27. Numerical Analysis   A right circular cone of height $h = 8$ and radius $r = 4$ is constructed, and in the process, errors Δr and Δh are made in the radius and height, respectively. Complete the table to show the relationship between ΔV and dV for the indicated errors.

Δr	Δh	dV or dS	ΔV or ΔS	$\Delta V - dV$ or $\Delta S - dS$
0.1	0.1			
0.1	-0.1			
0.001	0.002			
-0.0001	0.0002			

Table for Exercises 27 and 28

28. Numerical Analysis   The height and radius of a right circular cone are measured as $h = 16$ meters and $r = 6$ meters. In the process of measuring, errors Δr and Δh are made. Let S be the lateral surface area of the cone. Complete the table above to show the relationship between ΔS and dS for the indicated errors.

29. Volume   The possible error involved in measuring each dimension of a rectangular box is ± 0.02 inch. The dimensions of the box are 8 inches by 5 inches by 12 inches. Approximate the propagated error and the relative error in the calculated volume of the box.

30. Volume   The possible error involved in measuring each dimension of a right circular cylinder is ± 0.05 centimeter. The radius is 3 centimeters and the height is 10 centimeters. Approximate the propagated error and the relative error in the calculated volume of the cylinder.

31. Wind Chill

The formula for wind chill C (in degrees Fahrenheit) is given by

$$C = 35.74 + 0.6215T - 35.75v^{0.16} + 0.4275Tv^{0.16}$$

where v is the wind speed in miles per hour and T is the temperature in degrees Fahrenheit. The wind speed is 23 ± 3 miles per hour and the temperature is $8° \pm 1°$. Use dC to estimate the maximum possible propagated error and relative error in calculating the wind chill. *(Source: National Oceanic and Atmospheric Administration)*

32. Resistance The total resistance R (in ohms) of two resistors connected in parallel is given by

$$\frac{1}{R} = \frac{1}{R_1} + \frac{1}{R_2}.$$

Approximate the change in R as R_1 is increased from 10 ohms to 10.5 ohms and R_2 is decreased from 15 ohms to 13 ohms.

33. Power Electrical power P is given by

$$P = \frac{E^2}{R}$$

where E is voltage and R is resistance. Approximate the maximum percent error in calculating power when 120 volts is applied to a 2000-ohm resistor and the possible percent errors in measuring E and R are 3% and 4%, respectively.

34. Acceleration The centripetal acceleration of a particle moving in a circle is $a = v^2/r$, where v is the velocity and r is the radius of the circle. Approximate the maximum percent error in measuring the acceleration due to errors of 3% in v and 2% in r.

35. Volume A trough is 16 feet long (see figure). Its cross sections are isosceles triangles with each of the two equal sides measuring 18 inches. The angle between the two equal sides is θ.

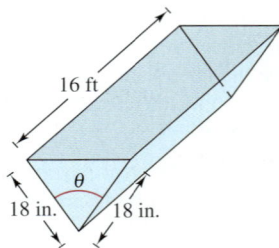

Not drawn to scale

(a) Write the volume of the trough as a function of θ and determine the value of θ such that the volume is a maximum.

(b) The maximum error in the linear measurements is one-half inch and the maximum error in the angle measure is 2°. Approximate the change in the maximum volume.

36. Sports A baseball player in center field is playing approximately 330 feet from a television camera that is behind home plate. A batter hits a fly ball that goes to the wall 420 feet from the camera (see figure).

(a) The camera turns 9° to follow the play. Approximate the number of feet that the center fielder has to run to make the catch.

(b) The position of the center fielder could be in error by as much as 6 feet and the maximum error in measuring the rotation of the camera is 1°. Approximate the maximum possible error in the result of part (a).

37. Inductance The inductance L (in microhenrys) of a straight nonmagnetic wire in free space is

$$L = 0.00021\left(\ln \frac{2h}{r} - 0.75\right)$$

where h is the length of the wire in millimeters and r is the radius of a circular cross section. Approximate L when $r = 2 \pm \frac{1}{16}$ millimeters and $h = 100 \pm \frac{1}{100}$ millimeters.

38. Pendulum The period T of a pendulum of length L is

$$T = \frac{2\pi\sqrt{L}}{\sqrt{g}}$$

where g is the acceleration due to gravity. A pendulum is moved from the Canal Zone, where $g = 32.09$ feet per second per second, to Greenland, where $g = 32.23$ feet per second per second. Because of the change in temperature, the length of the pendulum changes from 2.5 feet to 2.48 feet. Approximate the change in the period of the pendulum.

Differentiability In Exercises 39–42, show that the function is differentiable by finding values of ε_1 and ε_2 as designated in the definition of differentiability, and verify that both ε_1 and ε_2 approach 0 as $(\Delta x, \Delta y) \to (0, 0)$.

39. $f(x, y) = x^2 - 2x + y$ **40.** $f(x, y) = x^2 + y^2$

41. $f(x, y) = x^2y$ **42.** $f(x, y) = 5x - 10y + y^3$

Differentiability In Exercises 43 and 44, use the function to show that $f_x(0, 0)$ and $f_y(0, 0)$ both exist, but that f is not differentiable at $(0, 0)$.

43. $f(x, y) = \begin{cases} \dfrac{3x^2y}{x^4 + y^2}, & (x, y) \neq (0, 0) \\ 0, & (x, y) = (0, 0) \end{cases}$

44. $f(x, y) = \begin{cases} \dfrac{5x^2y}{x^3 + y^3}, & (x, y) \neq (0, 0) \\ 0, & (x, y) = (0, 0) \end{cases}$

13.5   Chain Rules for Functions of Several Variables

■ Use the Chain Rules for functions of several variables.
■ Find partial derivatives implicitly.

Chain Rules for Functions of Several Variables

Your work with differentials in the preceding section provides the basis for the extension of the Chain Rule to functions of two variables. There are two cases—the first case involves w as a function of x and y, where x and y are functions of a single independent variable t, as shown in Theorem 13.6.

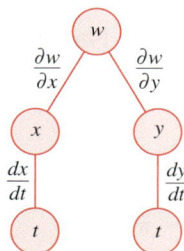

Chain Rule: one independent variable
w is a function of x and y, which are each functions of t. This diagram represents the derivative of w with respect to t.
Figure 13.39

THEOREM 13.6   Chain Rule: One Independent Variable

Let $w = f(x, y)$, where f is a differentiable function of x and y. If $x = g(t)$ and $y = h(t)$, where g and h are differentiable functions of t, then w is a differentiable function of t, and

$$\frac{dw}{dt} = \frac{\partial w}{\partial x}\frac{dx}{dt} + \frac{\partial w}{\partial y}\frac{dy}{dt}.$$

The Chain Rule is shown schematically in Figure 13.39.
A proof of Theorem 13.6 is given in Appendix A.
See LarsonCalculus.com for Bruce Edwards's video of this proof.

EXAMPLE 1   **Chain Rule: One Independent Variable**

Let $w = x^2 y - y^2$, where $x = \sin t$ and $y = e^t$. Find dw/dt when $t = 0$.

Solution   By the Chain Rule for one independent variable, you have

$$\begin{aligned}
\frac{dw}{dt} &= \frac{\partial w}{\partial x}\frac{dx}{dt} + \frac{\partial w}{\partial y}\frac{dy}{dt} \\
&= 2xy(\cos t) + (x^2 - 2y)e^t \\
&= 2(\sin t)(e^t)(\cos t) + (\sin^2 t - 2e^t)e^t \\
&= 2e^t \sin t \cos t + e^t \sin^2 t - 2e^{2t}.
\end{aligned}$$

When $t = 0$, it follows that

$$\frac{dw}{dt} = -2.$$

The Chain Rules presented in this section provide alternative techniques for solving many problems in single-variable calculus. For instance, in Example 1, you could have used single-variable techniques to find dw/dt by first writing w as a function of t,

$$\begin{aligned}
w &= x^2 y - y^2 \\
&= (\sin t)^2(e^t) - (e^t)^2 \\
&= e^t \sin^2 t - e^{2t}
\end{aligned}$$

and then differentiating as usual.

$$\frac{dw}{dt} = 2e^t \sin t \cos t + e^t \sin^2 t - 2e^{2t}$$

The Chain Rule in Theorem 13.6 can be extended to any number of variables. For example, if each x_i is a differentiable function of a single variable t, then for

$$w = f(x_1, x_2, \ldots, x_n)$$

you have

$$\frac{dw}{dt} = \frac{\partial w}{\partial x_1}\frac{dx_1}{dt} + \frac{\partial w}{\partial x_2}\frac{dx_2}{dt} + \cdots + \frac{\partial w}{\partial x_n}\frac{dx_n}{dt}.$$

EXAMPLE 2   An Application of a Chain Rule to Related Rates

Two objects are traveling in elliptical paths given by the following parametric equations.

$$x_1 = 4\cos t \quad \text{and} \quad y_1 = 2\sin t \qquad \text{First object}$$
$$x_2 = 2\sin 2t \quad \text{and} \quad y_2 = 3\cos 2t \qquad \text{Second object}$$

At what rate is the distance between the two objects changing when $t = \pi$?

Solution   From Figure 13.40, you can see that the distance s between the two objects is given by

$$s = \sqrt{(x_2 - x_1)^2 + (y_2 - y_1)^2}$$

and that when $t = \pi$, you have $x_1 = -4$, $y_1 = 0$, $x_2 = 0$, $y_2 = 3$, and

$$s = \sqrt{(0 + 4)^2 + (3 - 0)^2} = 5.$$

When $t = \pi$, the partial derivatives of s are as follows.

$$\frac{\partial s}{\partial x_1} = \frac{-(x_2 - x_1)}{\sqrt{(x_2 - x_1)^2 + (y_2 - y_1)^2}} = -\frac{1}{5}(0 + 4) = -\frac{4}{5}$$

$$\frac{\partial s}{\partial y_1} = \frac{-(y_2 - y_1)}{\sqrt{(x_2 - x_1)^2 + (y_2 - y_1)^2}} = -\frac{1}{5}(3 - 0) = -\frac{3}{5}$$

$$\frac{\partial s}{\partial x_2} = \frac{(x_2 - x_1)}{\sqrt{(x_2 - x_1)^2 + (y_2 - y_1)^2}} = \frac{1}{5}(0 + 4) = \frac{4}{5}$$

$$\frac{\partial s}{\partial y_2} = \frac{(y_2 - y_1)}{\sqrt{(x_2 - x_1)^2 + (y_2 - y_1)^2}} = \frac{1}{5}(3 - 0) = \frac{3}{5}$$

When $t = \pi$, the derivatives of x_1, y_1, x_2, and y_2 are

$$\frac{dx_1}{dt} = -4\sin t = 0$$

$$\frac{dy_1}{dt} = 2\cos t = -2$$

$$\frac{dx_2}{dt} = 4\cos 2t = 4$$

$$\frac{dy_2}{dt} = -6\sin 2t = 0.$$

So, using the appropriate Chain Rule, you know that the distance is changing at a rate of

$$\frac{ds}{dt} = \frac{\partial s}{\partial x_1}\frac{dx_1}{dt} + \frac{\partial s}{\partial y_1}\frac{dy_1}{dt} + \frac{\partial s}{\partial x_2}\frac{dx_2}{dt} + \frac{\partial s}{\partial y_2}\frac{dy_2}{dt}$$

$$= \left(-\frac{4}{5}\right)(0) + \left(-\frac{3}{5}\right)(-2) + \left(\frac{4}{5}\right)(4) + \left(\frac{3}{5}\right)(0)$$

$$= \frac{22}{5}.$$

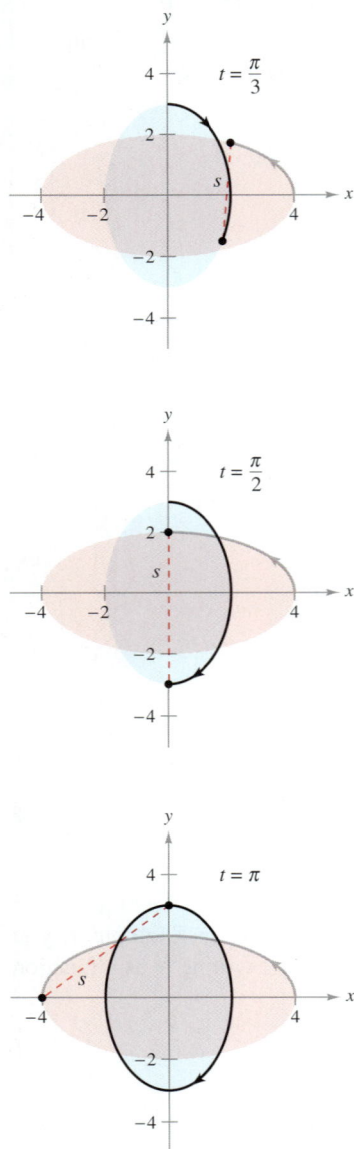

Paths of two objects traveling in elliptical orbits
Figure 13.40

In Example 2, note that s is the function of four *intermediate* variables, x_1, y_1, x_2, and y_2, each of which is a function of a single variable t. Another type of composite function is one in which the intermediate variables are themselves functions of more than one variable. For instance, for $w = f(x, y)$, where $x = g(s, t)$ and $y = h(s, t)$, it follows that w is a function of s and t, and you can consider the partial derivatives of w with respect to s and t. One way to find these partial derivatives is to write w as a function of s and t explicitly by substituting the equations $x = g(s, t)$ and $y = h(s, t)$ into the equation $w = f(x, y)$. Then you can find the partial derivatives in the usual way, as demonstrated in the next example.

EXAMPLE 3   Finding Partial Derivatives by Substitution

Find $\partial w/\partial s$ and $\partial w/\partial t$ for $w = 2xy$, where $x = s^2 + t^2$ and $y = s/t$.

Solution   Begin by substituting $x = s^2 + t^2$ and $y = s/t$ into the equation $w = 2xy$ to obtain

$$w = 2xy = 2(s^2 + t^2)\left(\frac{s}{t}\right) = 2\left(\frac{s^3}{t} + st\right).$$

Then, to find $\partial w/\partial s$, hold t constant and differentiate with respect to s.

$$\frac{\partial w}{\partial s} = 2\left(\frac{3s^2}{t} + t\right)$$

$$= \frac{6s^2 + 2t^2}{t}$$

Similarly, to find $\partial w/\partial t$, hold s constant and differentiate with respect to t to obtain

$$\frac{\partial w}{\partial t} = 2\left(-\frac{s^3}{t^2} + s\right)$$

$$= 2\left(\frac{-s^3 + st^2}{t^2}\right)$$

$$= \frac{2st^2 - 2s^3}{t^2}.$$

Theorem 13.7 gives an alternative method for finding the partial derivatives in Example 3, without explicitly writing w as a function of s and t.

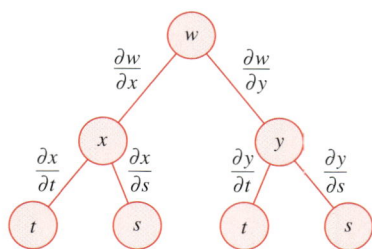

Chain Rule: two independent variables
Figure 13.41

THEOREM 13.7   Chain Rule: Two Independent Variables

Let $w = f(x, y)$, where f is a differentiable function of x and y. If $x = g(s, t)$ and $y = h(s, t)$ such that the first partials $\partial x/\partial s$, $\partial x/\partial t$, $\partial y/\partial s$, and $\partial y/\partial t$ all exist, then $\partial w/\partial s$ and $\partial w/\partial t$ exist and are given by

$$\frac{\partial w}{\partial s} = \frac{\partial w}{\partial x}\frac{\partial x}{\partial s} + \frac{\partial w}{\partial y}\frac{\partial y}{\partial s}$$

and

$$\frac{\partial w}{\partial t} = \frac{\partial w}{\partial x}\frac{\partial x}{\partial t} + \frac{\partial w}{\partial y}\frac{\partial y}{\partial t}.$$

The Chain Rule is shown schematically in Figure 13.41.

Proof   To obtain $\partial w/\partial s$, hold t constant and apply Theorem 13.6 to obtain the desired result. Similarly, for $\partial w/\partial t$, hold s constant and apply Theorem 13.6.

See LarsonCalculus.com for Bruce Edwards's video of this proof.

> **EXAMPLE 4**   **The Chain Rule with Two Independent Variables**
>
> •••▷ *See LarsonCalculus.com for an interactive version of this type of example.*

Use the Chain Rule to find $\partial w/\partial s$ and $\partial w/\partial t$ for

$$w = 2xy$$

where $x = s^2 + t^2$ and $y = s/t$.

Solution   Note that these same partials were found in Example 3. This time, using Theorem 13.7, you can hold t constant and differentiate with respect to s to obtain

$$\frac{\partial w}{\partial s} = \frac{\partial w}{\partial x}\frac{\partial x}{\partial s} + \frac{\partial w}{\partial y}\frac{\partial y}{\partial s}$$

$$= 2y(2s) + 2x\left(\frac{1}{t}\right)$$

$$= 2\left(\frac{s}{t}\right)(2s) + 2(s^2 + t^2)\left(\frac{1}{t}\right) \qquad \text{Substitute } \frac{s}{t} \text{ for } y \text{ and } s^2 + t^2 \text{ for } x.$$

$$= \frac{4s^2}{t} + \frac{2s^2 + 2t^2}{t}$$

$$= \frac{6s^2 + 2t^2}{t}.$$

Similarly, holding s constant gives

$$\frac{\partial w}{\partial t} = \frac{\partial w}{\partial x}\frac{\partial x}{\partial t} + \frac{\partial w}{\partial y}\frac{\partial y}{\partial t}$$

$$= 2y(2t) + 2x\left(\frac{-s}{t^2}\right)$$

$$= 2\left(\frac{s}{t}\right)(2t) + 2(s^2 + t^2)\left(\frac{-s}{t^2}\right) \qquad \text{Substitute } \frac{s}{t} \text{ for } y \text{ and } s^2 + t^2 \text{ for } x.$$

$$= 4s - \frac{2s^3 + 2st^2}{t^2}$$

$$= \frac{4st^2 - 2s^3 - 2st^2}{t^2}$$

$$= \frac{2st^2 - 2s^3}{t^2}.$$

The Chain Rule in Theorem 13.7 can also be extended to any number of variables. For example, if w is a differentiable function of the n variables

$$x_1, x_2, \ldots, x_n$$

where each x_i is a differentiable function of the m variables $t_1, t_2, \ldots, t_m$, then for

$$w = f(x_1, x_2, \ldots, x_n)$$

you obtain the following.

$$\frac{\partial w}{\partial t_1} = \frac{\partial w}{\partial x_1}\frac{\partial x_1}{\partial t_1} + \frac{\partial w}{\partial x_2}\frac{\partial x_2}{\partial t_1} + \cdots + \frac{\partial w}{\partial x_n}\frac{\partial x_n}{\partial t_1}$$

$$\frac{\partial w}{\partial t_2} = \frac{\partial w}{\partial x_1}\frac{\partial x_1}{\partial t_2} + \frac{\partial w}{\partial x_2}\frac{\partial x_2}{\partial t_2} + \cdots + \frac{\partial w}{\partial x_n}\frac{\partial x_n}{\partial t_2}$$

$$\vdots$$

$$\frac{\partial w}{\partial t_m} = \frac{\partial w}{\partial x_1}\frac{\partial x_1}{\partial t_m} + \frac{\partial w}{\partial x_2}\frac{\partial x_2}{\partial t_m} + \cdots + \frac{\partial w}{\partial x_n}\frac{\partial x_n}{\partial t_m}$$

EXAMPLE 5   **The Chain Rule for a Function of Three Variables**

Find $\partial w/\partial s$ and $\partial w/\partial t$ when $s = 1$ and $t = 2\pi$ for

$$w = xy + yz + xz$$

where $x = s \cos t$, $y = s \sin t$, and $z = t$.

Solution   By extending the result of Theorem 13.7, you have

$$\frac{\partial w}{\partial s} = \frac{\partial w}{\partial x}\frac{\partial x}{\partial s} + \frac{\partial w}{\partial y}\frac{\partial y}{\partial s} + \frac{\partial w}{\partial z}\frac{\partial z}{\partial s}$$
$$= (y + z)(\cos t) + (x + z)(\sin t) + (y + x)(0)$$
$$= (y + z)(\cos t) + (x + z)(\sin t).$$

When $s = 1$ and $t = 2\pi$, you have $x = 1$, $y = 0$, and $z = 2\pi$. So,

$$\frac{\partial w}{\partial s} = (0 + 2\pi)(1) + (1 + 2\pi)(0) = 2\pi.$$

Furthermore,

$$\frac{\partial w}{\partial t} = \frac{\partial w}{\partial x}\frac{\partial x}{\partial t} + \frac{\partial w}{\partial y}\frac{\partial y}{\partial t} + \frac{\partial w}{\partial z}\frac{\partial z}{\partial t}$$
$$= (y + z)(-s \sin t) + (x + z)(s \cos t) + (y + x)(1)$$

and for $s = 1$ and $t = 2\pi$, it follows that

$$\frac{\partial w}{\partial t} = (0 + 2\pi)(0) + (1 + 2\pi)(1) + (0 + 1)(1)$$
$$= 2 + 2\pi.$$

Implicit Partial Differentiation

This section concludes with an application of the Chain Rule to determine the derivative of a function defined *implicitly*. Let x and y be related by the equation $F(x, y) = 0$, where $y = f(x)$ is a differentiable function of x. To find dy/dx, you could use the techniques discussed in Section 2.5. You will see, however, that the Chain Rule provides a convenient alternative. Consider the function

$$w = F(x, y) = F(x, f(x)).$$

You can apply Theorem 13.6 to obtain

$$\frac{dw}{dx} = F_x(x, y)\frac{dx}{dx} + F_y(x, y)\frac{dy}{dx}.$$

Because $w = F(x, y) = 0$ for all x in the domain of f, you know that

$$\frac{dw}{dx} = 0$$

and you have

$$F_x(x, y)\frac{dx}{dx} + F_y(x, y)\frac{dy}{dx} = 0.$$

Now, if $F_y(x, y) \neq 0$, you can use the fact that $dx/dx = 1$ to conclude that

$$\frac{dy}{dx} = -\frac{F_x(x, y)}{F_y(x, y)}.$$

A similar procedure can be used to find the partial derivatives of functions of several variables that are defined implicitly.

THEOREM 13.8 Chain Rule: Implicit Differentiation

If the equation $F(x, y) = 0$ defines y implicitly as a differentiable function of x, then

$$\frac{dy}{dx} = -\frac{F_x(x, y)}{F_y(x, y)}, \quad F_y(x, y) \neq 0.$$

If the equation $F(x, y, z) = 0$ defines z implicitly as a differentiable function of x and y, then

$$\frac{\partial z}{\partial x} = -\frac{F_x(x, y, z)}{F_z(x, y, z)} \quad \text{and} \quad \frac{\partial z}{\partial y} = -\frac{F_y(x, y, z)}{F_z(x, y, z)}, \quad F_z(x, y, z) \neq 0.$$

This theorem can be extended to differentiable functions defined implicitly with any number of variables.

EXAMPLE 6 Finding a Derivative Implicitly

Find dy/dx for

$$y^3 + y^2 - 5y - x^2 + 4 = 0.$$

Solution Begin by letting

$$F(x, y) = y^3 + y^2 - 5y - x^2 + 4.$$

Then

$$F_x(x, y) = -2x \quad \text{and} \quad F_y(x, y) = 3y^2 + 2y - 5.$$

Using Theorem 13.8, you have

$$\frac{dy}{dx} = -\frac{F_x(x, y)}{F_y(x, y)} = \frac{-(-2x)}{3y^2 + 2y - 5} = \frac{2x}{3y^2 + 2y - 5}.$$

· · REMARK Compare the solution to Example 6 with the solution to Example 2 in Section 2.5.

EXAMPLE 7 Finding Partial Derivatives Implicitly

Find $\partial z/\partial x$ and $\partial z/\partial y$ for

$$3x^2z - x^2y^2 + 2z^3 + 3yz - 5 = 0.$$

Solution Begin by letting

$$F(x, y, z) = 3x^2z - x^2y^2 + 2z^3 + 3yz - 5.$$

Then

$$F_x(x, y, z) = 6xz - 2xy^2$$
$$F_y(x, y, z) = -2x^2y + 3z$$

and

$$F_z(x, y, z) = 3x^2 + 6z^2 + 3y.$$

Using Theorem 13.8, you have

$$\frac{\partial z}{\partial x} = -\frac{F_x(x, y, z)}{F_z(x, y, z)} = \frac{2xy^2 - 6xz}{3x^2 + 6z^2 + 3y}$$

and

$$\frac{\partial z}{\partial y} = -\frac{F_y(x, y, z)}{F_z(x, y, z)} = \frac{2x^2y - 3z}{3x^2 + 6z^2 + 3y}.$$

13.5 Exercises

See **CalcChat.com** for tutorial help and worked-out solutions to odd-numbered exercises.

Using the Chain Rule In Exercises 1–4, find dw/dt using the appropriate Chain Rule. Evaluate dw/dt at the given value of t.

Function	Value
1. $w = x^2 + y^2$	$t = 2$
$\quad x = 2t, y = 3t$	
2. $w = \sqrt{x^2 + y^2}$	$t = 0$
$\quad x = \cos t, y = e^t$	
3. $w = x \sin y$	$t = 0$
$\quad x = e^t, \quad y = \pi - t$	
4. $w = \ln \dfrac{y}{x}$	$t = \dfrac{\pi}{4}$
$\quad x = \cos t, y = \sin t$	

Using Different Methods In Exercises 5–10, find dw/dt (a) by using the appropriate Chain Rule and (b) by converting w to a function of t before differentiating.

5. $w = xy, \quad x = e^t, \quad y = e^{-2t}$

6. $w = \cos(x - y), \quad x = t^2, \quad y = 1$

7. $w = x^2 + y^2 + z^2, \quad x = \cos t, \quad y = \sin t, \quad z = e^t$

8. $w = xy \cos z, \quad x = t, \quad y = t^2, \quad z = \arccos t$

9. $w = xy + xz + yz, \quad x = t - 1, \quad y = t^2 - 1, \quad z = t$

10. $w = xy^2 + x^2z + yz^2, \quad x = t^2, \quad y = 2t, \quad z = 2$

Projectile Motion In Exercises 11 and 12, the parametric equations for the paths of two projectiles are given. At what rate is the distance between the two objects changing at the given value of t?

11. $x_1 = 10 \cos 2t, \ y_1 = 6 \sin 2t$ First object

$\quad x_2 = 7 \cos t, \ y_2 = 4 \sin t$ Second object

$\quad t = \pi/2$

12. $x_1 = 48\sqrt{2}\,t, \ y_1 = 48\sqrt{2}\,t - 16t^2$ First object

$\quad x_2 = 48\sqrt{3}\,t, \ y_2 = 48t - 16t^2$ Second object

$\quad t = 1$

Finding Partial Derivatives In Exercises 13–16, find $\partial w/\partial s$ and $\partial w/\partial t$ using the appropriate Chain Rule. Evaluate each partial derivative at the given values of s and t.

Function	Values
13. $w = x^2 + y^2$	$s = 1, \quad t = 0$
$\quad x = s + t, \quad y = s - t$	
14. $w = y^3 - 3x^2y$	$s = -1, \quad t = 2$
$\quad x = e^s, \quad y = e^t$	
15. $w = \sin(2x + 3y)$	$s = 0, \quad t = \dfrac{\pi}{2}$
$\quad x = s + t, \quad y = s - t$	
16. $w = x^2 - y^2$	$s = 3, \quad t = \dfrac{\pi}{4}$
$\quad x = s \cos t, \quad y = s \sin t$	

Using Different Methods In Exercises 17–20, find $\partial w/\partial s$ and $\partial w/\partial t$ (a) by using the appropriate Chain Rule and (b) by converting w to a function of s and t before differentiating.

17. $w = xyz, \quad x = s + t, \quad y = s - t, \quad z = st^2$

18. $w = x^2 + y^2 + z^2, \quad x = t \sin s, \quad y = t \cos s, \quad z = st^2$

19. $w = ze^{xy}, \quad x = s - t, \quad y = s + t, \quad z = st$

20. $w = x \cos yz, \quad x = s^2, \quad y = t^2, \quad z = s - 2t$

Finding a Derivative Implicitly In Exercises 21–24, differentiate implicitly to find dy/dx.

21. $x^2 - xy + y^2 - x + y = 0$

22. $\sec xy + \tan xy + 5 = 0$

23. $\ln\sqrt{x^2 + y^2} + x + y = 4$

24. $\dfrac{x}{x^2 + y^2} - y^2 = 6$

Finding Partial Derivatives Implicitly In Exercises 25–32, differentiate implicitly to find the first partial derivatives of z.

25. $x^2 + y^2 + z^2 = 1$ **26.** $xz + yz + xy = 0$

27. $x^2 + 2yz + z^2 = 1$ **28.** $x + \sin(y + z) = 0$

29. $\tan(x + y) + \tan(y + z) = 1$

30. $z = e^x \sin(y + z)$

31. $e^{xz} + xy = 0$

32. $x \ln y + y^2z + z^2 = 8$

Finding Partial Derivatives Implicitly In Exercises 33–36, differentiate implicitly to find the first partial derivatives of w.

33. $xy + yz - wz + wx = 5$

34. $x^2 + y^2 + z^2 - 5yw + 10w^2 = 2$

35. $\cos xy + \sin yz + wz = 20$

36. $w - \sqrt{x - y} - \sqrt{y - z} = 0$

Homogeneous Functions A function f is *homogeneous of degree n* when $f(tx, ty) = t^n f(x, y)$. In Exercises 37–40, (a) show that the function is homogeneous and determine n, and (b) show that $xf_x(x, y) + yf_y(x, y) = nf(x, y)$.

37. $f(x, y) = \dfrac{xy}{\sqrt{x^2 + y^2}}$

38. $f(x, y) = x^3 - 3xy^2 + y^3$

39. $f(x, y) = e^{x/y}$

40. $f(x, y) = \dfrac{x^2}{\sqrt{x^2 + y^2}}$

41. Using a Table of Values Let $w = f(x, y)$, $x = g(t)$, and $y = h(t)$, where f, g, and h are differentiable. Use the appropriate Chain Rule to find dw/dt when $t = 2$, given the following table of values.

$g(2)$	$h(2)$	$g'(2)$	$h'(2)$	$f_x(4, 3)$	$f_y(4, 3)$
4	3	-1	6	-5	7

42. Using a Table of Values Let $w = f(x, y)$, $x = g(s, t)$, and $y = h(s, t)$, where f, g, and h are differentiable. Use the appropriate Chain Rule to find $w_s(1, 2)$ and $w_t(1, 2)$, given the following table of values.

$g(1, 2)$	$h(1, 2)$	$g_s(1, 2)$	$h_s(1, 2)$
4	3	-3	5

$g_t(1, 2)$	$h_t(1, 2)$	$f_x(4, 3)$	$f_y(4, 3)$
-2	8	-5	7

43. Chain Rule Let $w = f(x, y)$ be a function in which x and y are functions of a single variable t. Give the Chain Rule for finding dw/dt.

44. Chain Rule Let $w = f(x, y)$ be a function in which x and y are functions of two variables s and t. Give the Chain Rule for finding $\partial w/\partial s$ and $\partial w/\partial t$.

45. Implicit Differentiation For $f(x, y) = 0$, give the rule for finding dy/dx implicitly. For $f(x, y, z) = 0$, give the rule for finding $\partial z/\partial x$ and $\partial z/\partial y$ implicitly.

46. HOW DO YOU SEE IT? The graph of the function $w = f(x, y)$ is shown below.

(a) Assume that x and y are functions of a single variable r. Give the chain rule for finding dw/dr.

(b) Assume that x and y are functions of two variables r and θ. Give the chain rule for finding $\partial w/\partial r$ and $\partial w/\partial \theta$.

47. Volume and Surface Area The radius of a right circular cylinder is increasing at a rate of 6 inches per minute, and the height is decreasing at a rate of 4 inches per minute. What are the rates of change of the volume and surface area when the radius is 12 inches and the height is 36 inches?

48. Ideal Gas Law The Ideal Gas Law is $pV = mRT$, where p is the pressure, V is the volume, m is the constant mass, R is a constant, T is the temperature, and p and V are functions of time. Find dT/dt, the rate at which the temperature changes with respect to time.

49. Moment of Inertia An annular cylinder has an inside radius of r_1 and an outside radius of r_2 (see figure). Its moment of inertia is $I = \frac{1}{2}m(r_1^2 + r_2^2)$, where m is the mass. The two radii are increasing at a rate of 2 centimeters per second. Find the rate at which I is changing at the instant the radii are 6 centimeters and 8 centimeters. (Assume mass is a constant.)

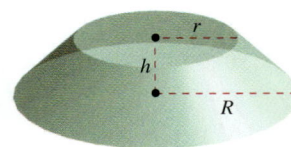

Figure for 49   Figure for 50

50. Volume and Surface Area The two radii of the frustum of a right circular cone are increasing at a rate of 4 centimeters per minute, and the height is increasing at a rate of 12 centimeters per minute (see figure). Find the rates at which the volume and surface area are changing when the two radii are 15 centimeters and 25 centimeters, and the height is 10 centimeters.

51. Using the Chain Rule Show that

$$\frac{\partial w}{\partial u} + \frac{\partial w}{\partial v} = 0$$

for $w = f(x, y)$, $x = u - v$, and $y = v - u$.

52. Using the Chain Rule Demonstrate the result of Exercise 51 for

$$w = (x - y) \sin(y - x).$$

53. Cauchy-Riemann Equations Given the functions $u(x, y)$ and $v(x, y)$, verify that the **Cauchy-Riemann differential equations**

$$\frac{\partial u}{\partial x} = \frac{\partial v}{\partial y} \quad \text{and} \quad \frac{\partial u}{\partial y} = -\frac{\partial v}{\partial x}$$

can be written in polar coordinate form as

$$\frac{\partial u}{\partial r} = \frac{1}{r} \cdot \frac{\partial v}{\partial \theta} \quad \text{and} \quad \frac{\partial v}{\partial r} = -\frac{1}{r} \cdot \frac{\partial u}{\partial \theta}.$$

54. Cauchy-Riemann Equations Demonstrate the result of Exercise 53 for the functions

$$u = \ln \sqrt{x^2 + y^2} \quad \text{and} \quad v = \arctan \frac{y}{x}.$$

55. Homogeneous Function Show that if $f(x, y)$ is homogeneous of degree n, then

$$xf_x(x, y) + yf_y(x, y) = nf(x, y).$$

[*Hint*: Let $g(t) = f(tx, ty) = t^n f(x, y)$. Find $g'(t)$ and then let $t = 1$.]

13.6 Directional Derivatives and Gradients

- Find and use directional derivatives of a function of two variables.
- Find the gradient of a function of two variables.
- Use the gradient of a function of two variables in applications.
- Find directional derivatives and gradients of functions of three variables.

Directional Derivative

You are standing on the hillside represented by $z = f(x, y)$ in Figure 13.42 and want to determine the hill's incline toward the z-axis. You already know how to determine the slopes in two different directions—the slope in the y-direction is given by the partial derivative $f_y(x, y)$, and the slope in the x-direction is given by the partial derivative $f_x(x, y)$. In this section, you will see that these two partial derivatives can be used to find the slope in *any* direction.

To determine the slope at a point on a surface, you will define a new type of derivative called a **directional derivative.**

Begin by letting $z = f(x, y)$ be a *surface* and $P(x_0, y_0)$ be a *point* in the domain of f, as shown in Figure 13.43. The "direction" of the directional derivative is given by a unit vector

$$\mathbf{u} = \cos \theta \mathbf{i} + \sin \theta \mathbf{j}$$

where θ is the angle the vector makes with the positive x-axis. To find the desired slope, reduce the problem to two dimensions by intersecting the surface with a vertical plane passing through the point P and parallel to $\mathbf{u}$, as shown in Figure 13.44. This vertical plane intersects the surface to form a curve C. The slope of the surface at $(x_0, y_0, f(x_0, y_0))$ in the direction of $\mathbf{u}$ is defined as the slope of the curve C at that point.

Informally, you can write the slope of the curve C as a limit that looks much like those used in single-variable calculus. The vertical plane used to form C intersects the xy-plane in a line L, represented by the parametric equations

$$x = x_0 + t \cos \theta$$

and

$$y = y_0 + t \sin \theta$$

so that for any value of t, the point $Q(x, y)$ lies on the line L. For each of the points P and Q, there is a corresponding point on the surface.

$(x_0, y_0, f(x_0, y_0))$ Point above P
$(x, y, f(x, y))$ Point above Q

Moreover, because the distance between P and Q is

$$\sqrt{(x - x_0)^2 + (y - y_0)^2} = \sqrt{(t \cos \theta)^2 + (t \sin \theta)^2}$$
$$= |t|$$

you can write the slope of the secant line through $(x_0, y_0, f(x_0, y_0))$ and $(x, y, f(x, y))$ as

$$\frac{f(x, y) - f(x_0, y_0)}{t} = \frac{f(x_0 + t \cos \theta, y_0 + t \sin \theta) - f(x_0, y_0)}{t}.$$

Finally, by letting t approach 0, you arrive at the definition on the next page.

Figure 13.42

Figure 13.43

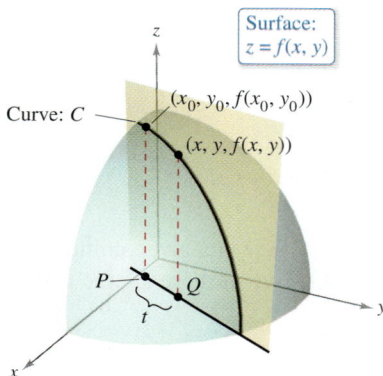
Figure 13.44

Definition of Directional Derivative

Let f be a function of two variables x and y and let $\mathbf{u} = \cos \theta \mathbf{i} + \sin \theta \mathbf{j}$ be a unit vector. Then the **directional derivative of f in the direction of u,** denoted by $D_{\mathbf{u}} f$, is

$$D_{\mathbf{u}} f(x, y) = \lim_{t \to 0} \frac{f(x + t \cos \theta, y + t \sin \theta) - f(x, y)}{t}$$

provided this limit exists.

Calculating directional derivatives by this definition is similar to finding the derivative of a function of one variable by the limit process (given in Section 2.1). A simpler "working" formula for finding directional derivatives involves the partial derivatives f_x and f_y.

THEOREM 13.9   Directional Derivative

If f is a differentiable function of x and y, then the directional derivative of f in the direction of the unit vector $\mathbf{u} = \cos \theta \mathbf{i} + \sin \theta \mathbf{j}$ is

$$D_{\mathbf{u}} f(x, y) = f_x(x, y) \cos \theta + f_y(x, y) \sin \theta.$$

Proof   For a fixed point (x_0, y_0), let

$$x = x_0 + t \cos \theta \quad \text{and} \quad y = y_0 + t \sin \theta.$$

Then, let $g(t) = f(x, y)$. Because f is differentiable, you can apply the Chain Rule given in Theorem 13.6 to obtain

$$g'(t) = f_x(x, y) x'(t) + f_y(x, y) y'(t) = f_x(x, y) \cos \theta + f_y(x, y) \sin \theta.$$

If $t = 0$, then $x = x_0$ and $y = y_0$, so

$$g'(0) = f_x(x_0, y_0) \cos \theta + f_y(x_0, y_0) \sin \theta.$$

By the definition of $g'(t)$, it is also true that

$$g'(0) = \lim_{t \to 0} \frac{g(t) - g(0)}{t}$$

$$= \lim_{t \to 0} \frac{f(x_0 + t \cos \theta, y_0 + t \sin \theta) - f(x_0, y_0)}{t}.$$

Consequently, $D_{\mathbf{u}} f(x_0, y_0) = f_x(x_0, y_0) \cos \theta + f_y(x_0, y_0) \sin \theta.$

See LarsonCalculus.com for Bruce Edwards's video of this proof.

There are infinitely many directional derivatives of a surface at a given point—one for each direction specified by $\mathbf{u}$, as shown in Figure 13.45. Two of these are the partial derivatives f_x and f_y.

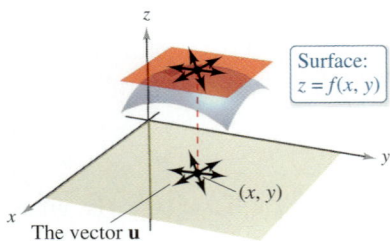
The vector **u**

Figure 13.45

1. Direction of positive x-axis $(\theta = 0)$: $\mathbf{u} = \cos 0 \mathbf{i} + \sin 0 \mathbf{j} = \mathbf{i}$

$$D_{\mathbf{i}} f(x, y) = f_x(x, y) \cos 0 + f_y(x, y) \sin 0 = f_x(x, y)$$

2. Direction of positive y-axis $\left(\theta = \dfrac{\pi}{2}\right)$: $\mathbf{u} = \cos \dfrac{\pi}{2} \mathbf{i} + \sin \dfrac{\pi}{2} \mathbf{j} = \mathbf{j}$

$$D_{\mathbf{j}} f(x, y) = f_x(x, y) \cos \frac{\pi}{2} + f_y(x, y) \sin \frac{\pi}{2} = f_y(x, y)$$

Finding a Directional Derivative

Find the directional derivative of

$$f(x, y) = 4 - x^2 - \frac{1}{4}y^2 \qquad \text{Surface}$$

at $(1, 2)$ in the direction of

$$\mathbf{u} = \left(\cos\frac{\pi}{3}\right)\mathbf{i} + \left(\sin\frac{\pi}{3}\right)\mathbf{j}. \qquad \text{Direction}$$

Solution   Because $f_x(x, y) = -2x$ and $f_y(x, y) = -y/2$ are continuous, f is differentiable, and you can apply Theorem 13.9.

$$D_{\mathbf{u}}f(x, y) = f_x(x, y)\cos\theta + f_y(x, y)\sin\theta = (-2x)\cos\theta + \left(-\frac{y}{2}\right)\sin\theta$$

Evaluating at $\theta = \pi/3$, $x = 1$, and $y = 2$ produces

$$D_{\mathbf{u}}f(1, 2) = (-2)\left(\frac{1}{2}\right) + (-1)\left(\frac{\sqrt{3}}{2}\right)$$

$$= -1 - \frac{\sqrt{3}}{2}$$

$$\approx -1.866. \qquad \text{See Figure 13.46.}$$

Note in Figure 13.46 that you can interpret the directional derivative as giving the slope of the surface at the point $(1, 2, 2)$ in the direction of the unit vector $\mathbf{u}$.

Surface:
$$f(x, y) = 4 - x^2 - \frac{1}{4}y^2$$

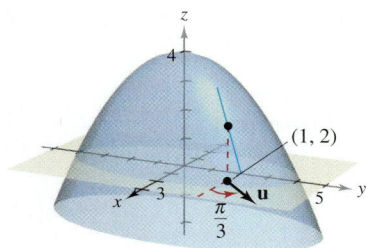

Figure 13.46

You have been specifying direction by a unit vector $\mathbf{u}$. When the direction is given by a vector whose length is not 1, you must normalize the vector before applying the formula in Theorem 13.9.

Finding a Directional Derivative

$\bullet \cdots \triangleright$ *See LarsonCalculus.com for an interactive version of this type of example.*

Find the directional derivative of

$$f(x, y) = x^2 \sin 2y \qquad \text{Surface}$$

at $(1, \pi/2)$ in the direction of

$$\mathbf{v} = 3\mathbf{i} - 4\mathbf{j}. \qquad \text{Direction}$$

Solution   Because $f_x(x, y) = 2x\sin 2y$ and $f_y(x, y) = 2x^2\cos 2y$ are continuous, f is differentiable, and you can apply Theorem 13.9. Begin by finding a unit vector in the direction of $\mathbf{v}$.

$$\mathbf{u} = \frac{\mathbf{v}}{\|\mathbf{v}\|} = \frac{3}{5}\mathbf{i} - \frac{4}{5}\mathbf{j} = \cos\theta\,\mathbf{i} + \sin\theta\,\mathbf{j}$$

Using this unit vector, you have

$$D_{\mathbf{u}}f(x, y) = (2x\sin 2y)(\cos\theta) + (2x^2\cos 2y)(\sin\theta)$$

$$D_{\mathbf{u}}f\left(1, \frac{\pi}{2}\right) = (2\sin\pi)\left(\frac{3}{5}\right) + (2\cos\pi)\left(-\frac{4}{5}\right)$$

$$= (0)\left(\frac{3}{5}\right) + (-2)\left(-\frac{4}{5}\right)$$

$$= \frac{8}{5}. \qquad \text{See Figure 13.47.}$$

Surface:
$$f(x, y) = x^2 \sin 2y$$

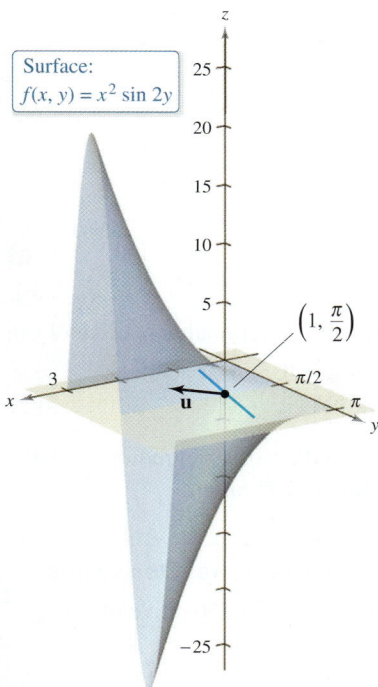

Figure 13.47

The Gradient of a Function of Two Variables

The **gradient** of a function of two variables is a vector-valued function of two variables. This function has many important uses, some of which are described later in this section.

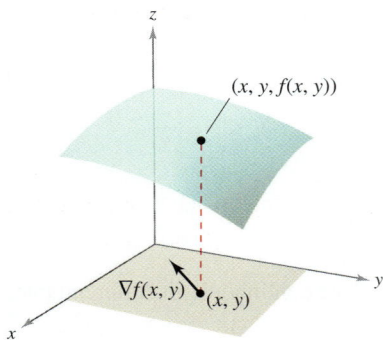

The gradient of f is a vector in the xy-plane.

Figure 13.48

Definition of Gradient of a Function of Two Variables

Let $z = f(x, y)$ be a function of x and y such that f_x and f_y exist. Then the **gradient of f,** denoted by $\nabla f(x, y)$, is the vector

$$\nabla f(x, y) = f_x(x, y)\mathbf{i} + f_y(x, y)\mathbf{j}.$$

(The symbol ∇f is read as "del f.") Another notation for the gradient is **grad** $f(x, y)$. In Figure 13.48, note that for each (x, y), the gradient $\nabla f(x, y)$ is a vector in the plane (not a vector in space).

Notice that no value is assigned to the symbol ∇ by itself. It is an operator in the same sense that d/dx is an operator. When ∇ operates on $f(x, y)$, it produces the vector $\nabla f(x, y)$.

EXAMPLE 3   **Finding the Gradient of a Function**

Find the gradient of

$$f(x, y) = y \ln x + xy^2$$

at the point $(1, 2)$.

Solution   Using

$$f_x(x, y) = \frac{y}{x} + y^2 \quad \text{and} \quad f_y(x, y) = \ln x + 2xy$$

you have

$$\begin{aligned} \nabla f(x, y) &= f_x(x, y)\mathbf{i} + f_y(x, y)\mathbf{j} \\ &= \left(\frac{y}{x} + y^2\right)\mathbf{i} + (\ln x + 2xy)\mathbf{j}. \end{aligned}$$

At the point $(1, 2)$, the gradient is

$$\begin{aligned} \nabla f(1, 2) &= \left(\frac{2}{1} + 2^2\right)\mathbf{i} + [\ln 1 + 2(1)(2)]\mathbf{j} \\ &= 6\mathbf{i} + 4\mathbf{j}. \end{aligned}$$

Because the gradient of f is a vector, you can write the directional derivative of f in the direction of $\mathbf{u}$ as

$$D_{\mathbf{u}}f(x, y) = [f_x(x, y)\mathbf{i} + f_y(x, y)\mathbf{j}] \cdot [\cos\theta\mathbf{i} + \sin\theta\mathbf{j}].$$

In other words, the directional derivative is the dot product of the gradient and the direction vector. This useful result is summarized in the next theorem.

THEOREM 13.10   **Alternative Form of the Directional Derivative**

If f is a differentiable function of x and y, then the directional derivative of f in the direction of the unit vector $\mathbf{u}$ is

$$D_{\mathbf{u}}f(x, y) = \nabla f(x, y) \cdot \mathbf{u}.$$

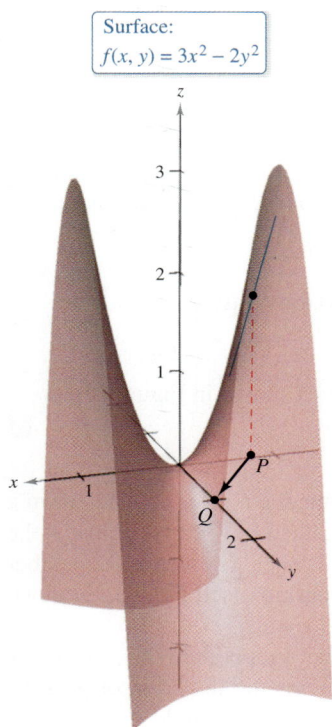

Surface:
$f(x, y) = 3x^2 - 2y^2$

Figure 13.49

⊳

.....
REMARK Property 2 of Theorem 13.11 says that at the point (x, y), f increases most rapidly in the direction of the gradient, $\nabla f(x, y)$.

| EXAMPLE 4 | **Using $\nabla f(x, y)$ to Find a Directional Derivative** |

Find the directional derivative of

$$f(x, y) = 3x^2 - 2y^2$$

at $\left(-\frac{3}{4}, 0\right)$ in the direction from $P\left(-\frac{3}{4}, 0\right)$ to $Q(0, 1)$.

Solution Because the partials of f are continuous, f is differentiable and you can apply Theorem 13.10. A vector in the specified direction is

$$\overrightarrow{PQ} = \left(0 + \frac{3}{4}\right)\mathbf{i} + (1 - 0)\mathbf{j}$$

$$= \frac{3}{4}\mathbf{i} + \mathbf{j}$$

and a unit vector in this direction is

$$\mathbf{u} = \frac{\overrightarrow{PQ}}{\|\overrightarrow{PQ}\|} = \frac{3}{5}\mathbf{i} + \frac{4}{5}\mathbf{j}. \qquad \text{Unit vector in direction of } \overrightarrow{PQ}$$

Because

$$\nabla f(x, y) = f_x(x, y)\mathbf{i} + f_y(x, y)\mathbf{j} = 6x\mathbf{i} - 4y\mathbf{j}$$

the gradient at $\left(-\frac{3}{4}, 0\right)$ is

$$\nabla f\left(-\frac{3}{4}, 0\right) = -\frac{9}{2}\mathbf{i} + 0\mathbf{j}. \qquad \text{Gradient at } \left(-\frac{3}{4}, 0\right)$$

Consequently, at $\left(-\frac{3}{4}, 0\right)$, the directional derivative is

$$D_{\mathbf{u}}f\left(-\frac{3}{4}, 0\right) = \nabla f\left(-\frac{3}{4}, 0\right) \cdot \mathbf{u}$$

$$= \left(-\frac{9}{2}\mathbf{i} + 0\mathbf{j}\right) \cdot \left(\frac{3}{5}\mathbf{i} + \frac{4}{5}\mathbf{j}\right)$$

$$= -\frac{27}{10}. \qquad \text{Directional derivative at } \left(-\frac{3}{4}, 0\right)$$

See Figure 13.49.

Applications of the Gradient

You have already seen that there are many directional derivatives at the point (x, y) on a surface. In many applications, you may want to know in which direction to move so that $f(x, y)$ increases most rapidly. This direction is called the direction of *steepest ascent*, and it is given by the gradient, as stated in the next theorem.

THEOREM 13.11 **Properties of the Gradient**

Let f be differentiable at the point (x, y).

1. If $\nabla f(x, y) = \mathbf{0}$, then $D_{\mathbf{u}}f(x, y) = 0$ for all $\mathbf{u}$.
2. The direction of *maximum* increase of f is given by $\nabla f(x, y)$. The maximum value of $D_{\mathbf{u}}f(x, y)$ is

   $$\|\nabla f(x, y)\|. \qquad \text{Maximum value of } D_{\mathbf{u}}f(x, y)$$

3. The direction of *minimum* increase of f is given by $-\nabla f(x, y)$. The minimum value of $D_{\mathbf{u}}f(x, y)$ is

   $$-\|\nabla f(x, y)\|. \qquad \text{Minimum value of } D_{\mathbf{u}}f(x, y)$$

Proof If $\nabla f(x, y) = \mathbf{0}$, then for any direction (any $\mathbf{u}$), you have

$$D_\mathbf{u} f(x, y) = \nabla f(x, y) \cdot \mathbf{u}$$
$$= (0\mathbf{i} + 0\mathbf{j}) \cdot (\cos \theta \mathbf{i} + \sin \theta \mathbf{j})$$
$$= 0.$$

If $\nabla f(x, y) \neq \mathbf{0}$, then let ϕ be the angle between $\nabla f(x, y)$ and a unit vector $\mathbf{u}$. Using the dot product, you can apply Theorem 11.5 to conclude that

$$D_\mathbf{u} f(x, y) = \nabla f(x, y) \cdot \mathbf{u}$$
$$= \|\nabla f(x, y)\| \, \|\mathbf{u}\| \cos \phi$$
$$= \|\nabla f(x, y)\| \cos \phi$$

and it follows that the maximum value of $D_\mathbf{u} f(x, y)$ will occur when

$$\cos \phi = 1.$$

So, $\phi = 0$, and the maximum value of the directional derivative occurs when $\mathbf{u}$ has the same direction as $\nabla f(x, y)$. Moreover, this largest value of $D_\mathbf{u} f(x, y)$ is precisely

$$\|\nabla f(x, y)\| \cos \phi = \|\nabla f(x, y)\|.$$

Similarly, the minimum value of $D_\mathbf{u} f(x, y)$ can be obtained by letting

$$\phi = \pi$$

so that $\mathbf{u}$ points in the direction opposite that of $\nabla f(x, y)$, as shown in Figure 13.50.

See LarsonCalculus.com for Bruce Edwards's video of this proof.

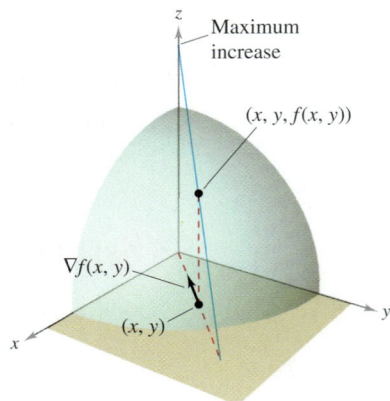

The gradient of f is a vector in the xy-plane that points in the direction of maximum increase on the surface given by $z = f(x, y)$.
Figure 13.50

To visualize one of the properties of the gradient, imagine a skier coming down a mountainside. If $f(x, y)$ denotes the altitude of the skier, then $-\nabla f(x, y)$ indicates the *compass direction* the skier should take to ski the path of steepest descent. (Remember that the gradient indicates direction in the xy-plane and does not itself point up or down the mountainside.)

As another illustration of the gradient, consider the temperature $T(x, y)$ at any point (x, y) on a flat metal plate. In this case, $\nabla T(x, y)$ gives the direction of greatest temperature increase at the point (x, y), as illustrated in the next example.

EXAMPLE 5   **Finding the Direction of Maximum Increase**

The temperature in degrees Celsius on the surface of a metal plate is

$$T(x, y) = 20 - 4x^2 - y^2$$

where x and y are measured in centimeters. In what direction from $(2, -3)$ does the temperature increase most rapidly? What is this rate of increase?

Solution   The gradient is

$$\nabla T(x, y) = T_x(x, y)\mathbf{i} + T_y(x, y)\mathbf{j}$$
$$= -8x\mathbf{i} - 2y\mathbf{j}.$$

It follows that the direction of maximum increase is given by

$$\nabla T(2, -3) = -16\mathbf{i} + 6\mathbf{j}$$

as shown in Figure 13.51, and the rate of increase is

$$\|\nabla T(2, -3)\| = \sqrt{256 + 36}$$
$$= \sqrt{292}$$
$$\approx 17.09° \text{ per centimeter.}$$

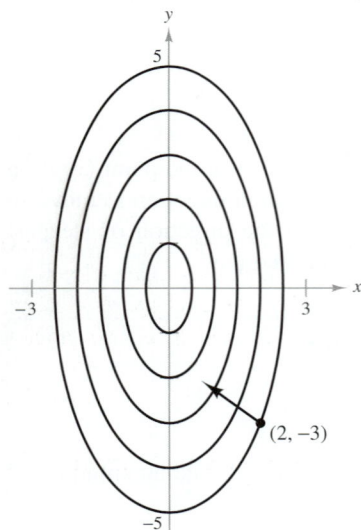

Level curves:
$T(x, y) = 20 - 4x^2 - y^2$

The direction of most rapid increase in temperature at $(2, -3)$ is given by $-16\mathbf{i} + 6\mathbf{j}$.
Figure 13.51

The solution presented in Example 5 can be misleading. Although the gradient points in the direction of maximum temperature increase, it does not necessarily point toward the hottest spot on the plate. In other words, the gradient provides a local solution to finding an increase relative to the temperature at the point $(2, -3)$. *Once you leave that position, the direction of maximum increase may change.*

EXAMPLE 6   **Finding the Path of a Heat-Seeking Particle**

A heat-seeking particle is located at the point $(2, -3)$ on a metal plate whose temperature at (x, y) is

$$T(x, y) = 20 - 4x^2 - y^2.$$

Find the path of the particle as it continuously moves in the direction of maximum temperature increase.

Solution   Let the path be represented by the position vector

$$\mathbf{r}(t) = x(t)\mathbf{i} + y(t)\mathbf{j}.$$

A tangent vector at each point $(x(t), y(t))$ is given by

$$\mathbf{r}'(t) = \frac{dx}{dt}\mathbf{i} + \frac{dy}{dt}\mathbf{j}.$$

Because the particle seeks maximum temperature increase, the directions of $\mathbf{r}'(t)$ and $\nabla T(x, y) = -8x\mathbf{i} - 2y\mathbf{j}$ are the same at each point on the path. So,

$$-8x = k\frac{dx}{dt} \quad \text{and} \quad -2y = k\frac{dy}{dt}$$

where k depends on t. By solving each equation for dt/k and equating the results, you obtain

$$\frac{dx}{-8x} = \frac{dy}{-2y}.$$

The solution of this differential equation is $x = Cy^4$. Because the particle starts at the point $(2, -3)$, you can determine that $C = 2/81$. So, the path of the heat-seeking particle is

$$x = \frac{2}{81}y^4.$$

The path is shown in Figure 13.52.

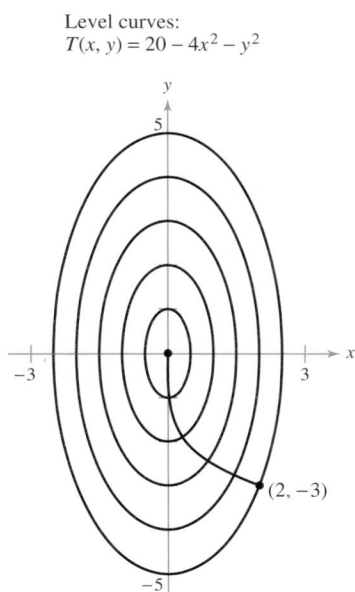

Level curves:
$T(x, y) = 20 - 4x^2 - y^2$

$(2, -3)$

Path followed by a heat-seeking particle
Figure 13.52

In Figure 13.52, the path of the particle (determined by the gradient at each point) appears to be orthogonal to each of the level curves. This becomes clear when you consider that the temperature $T(x, y)$ is constant along a given level curve. So, at any point (x, y) on the curve, the rate of change of T in the direction of a unit tangent vector $\mathbf{u}$ is 0, and you can write

$$\nabla f(x, y) \cdot \mathbf{u} = D_{\mathbf{u}} T(x, y) = 0. \qquad \text{u is a unit tangent vector.}$$

Because the dot product of $\nabla f(x, y)$ and $\mathbf{u}$ is 0, you can conclude that they must be orthogonal. This result is stated in the next theorem.

THEOREM 13.12   Gradient Is Normal to Level Curves

If f is differentiable at (x_0, y_0) and $\nabla f(x_0, y_0) \neq \mathbf{0}$, then $\nabla f(x_0, y_0)$ is normal to the level curve through (x_0, y_0).

EXAMPLE 7   Finding a Normal Vector to a Level Curve

Sketch the level curve corresponding to $c = 0$ for the function given by

$$f(x, y) = y - \sin x$$

and find a normal vector at several points on the curve.

Solution   The level curve for $c = 0$ is given by

$$0 = y - \sin x$$

or

$$y = \sin x$$

as shown in Figure 13.53(a). Because the gradient vector of f at (x, y) is

$$\nabla f(x, y) = f_x(x, y)\mathbf{i} + f_y(x, y)\mathbf{j}$$
$$= -\cos x\mathbf{i} + \mathbf{j}$$

you can use Theorem 13.12 to conclude that $\nabla f(x, y)$ is normal to the level curve at the point (x, y). Some gradient vectors are

$$\nabla f(-\pi, 0) = \mathbf{i} + \mathbf{j}$$
$$\nabla f\left(-\frac{2\pi}{3}, -\frac{\sqrt{3}}{2}\right) = \frac{1}{2}\mathbf{i} + \mathbf{j}$$
$$\nabla f\left(-\frac{\pi}{2}, -1\right) = \mathbf{j}$$
$$\nabla f\left(-\frac{\pi}{3}, -\frac{\sqrt{3}}{2}\right) = -\frac{1}{2}\mathbf{i} + \mathbf{j}$$
$$\nabla f(0, 0) = -\mathbf{i} + \mathbf{j}$$
$$\nabla f\left(\frac{\pi}{3}, \frac{\sqrt{3}}{2}\right) = -\frac{1}{2}\mathbf{i} + \mathbf{j}$$
$$\nabla f\left(\frac{\pi}{2}, 1\right) = \mathbf{j}$$
$$\nabla f\left(\frac{2\pi}{3}, \frac{\sqrt{3}}{2}\right) = \frac{1}{2}\mathbf{i} + \mathbf{j}$$

and

$$\nabla f(\pi, 0) = \mathbf{i} + \mathbf{j}.$$

These are shown in Figure 13.53(b).

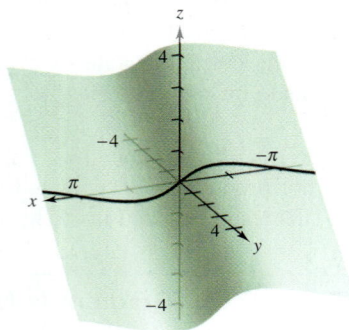

(a) The surface is given by $f(x, y) = y - \sin x$.

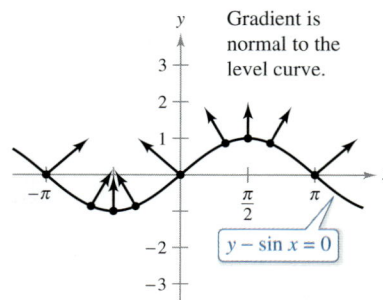

(b) The level curve is given by $f(x, y) = 0.$

Figure 13.53

Functions of Three Variables

The definitions of the directional derivative and the gradient can be extended naturally to functions of three or more variables. As often happens, some of the geometric interpretation is lost in the generalization from functions of two variables to those of three variables. For example, you cannot interpret the directional derivative of a function of three variables as representing slope.

The definitions and properties of the directional derivative and the gradient of a function of three variables are listed below.

Directional Derivative and Gradient for Three Variables

Let f be a function of x, y, and z, with continuous first partial derivatives. The **directional derivative of f** in the direction of a unit vector

$$\mathbf{u} = a\mathbf{i} + b\mathbf{j} + c\mathbf{k}$$

is given by

$$D_{\mathbf{u}} f(x, y, z) = af_x(x, y, z) + bf_y(x, y, z) + cf_z(x, y, z).$$

The **gradient of f** is defined as

$$\nabla f(x, y, z) = f_x(x, y, z)\mathbf{i} + f_y(x, y, z)\mathbf{j} + f_z(x, y, z)\mathbf{k}.$$

Properties of the gradient are as follows.

1. $D_{\mathbf{u}} f(x, y, z) = \nabla f(x, y, z) \cdot \mathbf{u}$
2. If $\nabla f(x, y, z) = \mathbf{0}$, then $D_{\mathbf{u}} f(x, y, z) = 0$ for all $\mathbf{u}$.
3. The direction of *maximum* increase of f is given by $\nabla f(x, y, z)$. The maximum value of $D_{\mathbf{u}} f(x, y, z)$ is

   $$\|\nabla f(x, y, z)\|. \qquad \text{\textcolor{red}{Maximum value of } } D_{\mathbf{u}} f(x, y, z)$$

4. The direction of *minimum* increase of f is given by $-\nabla f(x, y, z)$. The minimum value of $D_{\mathbf{u}} f(x, y, z)$ is

   $$-\|\nabla f(x, y, z)\|. \qquad \text{\textcolor{red}{Minimum value of } } D_{\mathbf{u}} f(x, y, z)$$

You can generalize Theorem 13.12 to functions of three variables. Under suitable hypotheses,

$$\nabla f(x_0, y_0, z_0)$$

is normal to the level surface through (x_0, y_0, z_0).

EXAMPLE 8 **Finding the Gradient of a Function**

Find $\nabla f(x, y, z)$ for the function

$$f(x, y, z) = x^2 + y^2 - 4z$$

and find the direction of maximum increase of f at the point $(2, -1, 1)$.

Solution The gradient vector is

$$\nabla f(x, y, z) = f_x(x, y, z)\mathbf{i} + f_y(x, y, z)\mathbf{j} + f_z(x, y, z)\mathbf{k}$$
$$= 2x\mathbf{i} + 2y\mathbf{j} - 4\mathbf{k}.$$

So, it follows that the direction of maximum increase at $(2, -1, 1)$ is

$$\nabla f(2, -1, 1) = 4\mathbf{i} - 2\mathbf{j} - 4\mathbf{k}. \qquad \text{\textcolor{red}{See Figure 13.54.}}$$

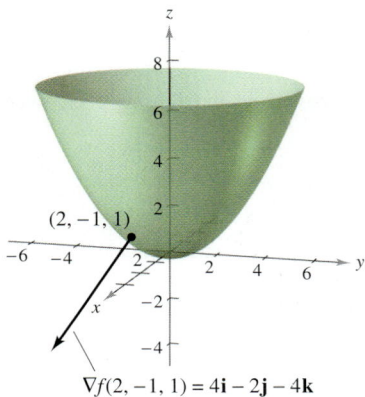

$\nabla f(2, -1, 1) = 4\mathbf{i} - 2\mathbf{j} - 4\mathbf{k}$

Level surface and gradient vector at $(2, -1, 1)$ for $f(x, y, z) = x^2 + y^2 - 4z$
Figure 13.54

13.6 Exercises

See **CalcChat.com** for tutorial help and worked-out solutions to odd-numbered exercises.

Finding a Directional Derivative In Exercises 1–4, use Theorem 13.9 to find the directional derivative of the function at P in the direction of the unit vector $\mathbf{u} = \cos\theta\mathbf{i} + \sin\theta\mathbf{j}$.

1. $f(x, y) = x^2 + y^2$, $\quad P(1, -2)$, $\quad \theta = \dfrac{\pi}{4}$

2. $f(x, y) = \dfrac{y}{x + y}$, $\quad P(3, 0)$, $\quad \theta = -\dfrac{\pi}{6}$

3. $f(x, y) = \sin(2x + y)$, $\quad P(0, 0)$, $\quad \theta = \dfrac{\pi}{3}$

4. $g(x, y) = xe^y$, $\quad P(0, 2)$, $\quad \theta = \dfrac{2\pi}{3}$

Finding a Directional Derivative In Exercises 5–8, use Theorem 13.9 to find the directional derivative of the function at P in the direction of $\mathbf{v}$.

5. $f(x, y) = 3x - 4xy + 9y$, $\quad P(1, 2)$, $\quad \mathbf{v} = \frac{3}{5}\mathbf{i} + \frac{4}{5}\mathbf{j}$

6. $f(x, y) = x^3 - y^3$, $\quad P(4, 3)$, $\quad \mathbf{v} = \dfrac{\sqrt{2}}{2}(\mathbf{i} + \mathbf{j})$

7. $g(x, y) = \sqrt{x^2 + y^2}$, $\quad P(3, 4)$, $\quad \mathbf{v} = 3\mathbf{i} - 4\mathbf{j}$

8. $h(x, y) = e^{-(x^2 + y^2)}$, $\quad P(0, 0)$, $\quad \mathbf{v} = \mathbf{i} + \mathbf{j}$

Finding a Directional Derivative In Exercises 9–12, use Theorem 13.9 to find the directional derivative of the function at P in the direction of Q.

9. $f(x, y) = x^2 + 3y^2$, $\quad P(1, 1)$, $\quad Q(4, 5)$

10. $f(x, y) = \cos(x + y)$, $\quad P(0, \pi)$, $\quad Q\left(\dfrac{\pi}{2}, 0\right)$

11. $f(x, y) = e^y \sin x$, $\quad P(0, 0)$, $\quad Q(2, 1)$

12. $f(x, y) = \sin 2x \cos y$, $\quad P(\pi, 0)$, $\quad Q\left(\dfrac{\pi}{2}, \pi\right)$

Finding the Gradient of a Function In Exercises 13–18, find the gradient of the function at the given point.

13. $f(x, y) = 3x + 5y^2 + 1$, $\quad (2, 1)$

14. $g(x, y) = 2xe^{y/x}$, $\quad (2, 0)$

15. $z = \ln(x^2 - y)$, $\quad (2, 3)$

16. $z = \cos(x^2 + y^2)$, $\quad (3, -4)$

17. $w = 3x^2 - 5y^2 + 2z^2$, $\quad (1, 1, -2)$

18. $w = x\tan(y + z)$, $\quad (4, 3, -1)$

Finding a Directional Derivative In Exercises 19–22, use the gradient to find the directional derivative of the function at P in the direction of $\mathbf{v}$.

19. $f(x, y) = xy$, $\quad P(0, -2)$, $\quad \mathbf{v} = \frac{1}{2}(\mathbf{i} + \sqrt{3}\mathbf{j})$

20. $h(x, y) = e^x \sin y$, $\quad P\left(1, \dfrac{\pi}{2}\right)$, $\quad \mathbf{v} = -\mathbf{i}$

21. $f(x, y, z) = x^2 + y^2 + z^2$, $\quad P(1, 1, 1)$, $\quad \mathbf{v} = \dfrac{\sqrt{3}}{3}(\mathbf{i} - \mathbf{j} + \mathbf{k})$

22. $f(x, y, z) = xy + yz + xz$, $\quad P(1, 2, -1)$, $\quad \mathbf{v} = 2\mathbf{i} + \mathbf{j} - \mathbf{k}$

Finding a Directional Derivative Using the Gradient In Exercises 23–26, use the gradient to find the directional derivative of the function at P in the direction of Q.

23. $g(x, y) = x^2 + y^2 + 1$, $\quad P(1, 2)$, $\quad Q(2, 3)$

24. $f(x, y) = 3x^2 - y^2 + 4$, $\quad P(-1, 4)$, $\quad Q(3, 6)$

25. $g(x, y, z) = xye^z$, $\quad P(2, 4, 0)$, $\quad Q(0, 0, 0)$

26. $h(x, y, z) = \ln(x + y + z)$, $\quad P(1, 0, 0)$, $\quad Q(4, 3, 1)$

Using Properties of the Gradient In Exercises 27–36, find the gradient of the function and the maximum value of the directional derivative at the given point.

Function	Point
27. $f(x, y) = x^2 + 2xy$	$(1, 0)$
28. $f(x, y) = \dfrac{x + y}{y + 1}$	$(0, 1)$
29. $h(x, y) = x\tan y$	$\left(2, \dfrac{\pi}{4}\right)$
30. $h(x, y) = y\cos(x - y)$	$\left(0, \dfrac{\pi}{3}\right)$
31. $g(x, y) = ye^{-x}$	$(0, 5)$
32. $g(x, y) = \ln\sqrt[3]{x^2 + y^2}$	$(1, 2)$
33. $f(x, y, z) = \sqrt{x^2 + y^2 + z^2}$	$(1, 4, 2)$
34. $w = \dfrac{1}{\sqrt{1 - x^2 - y^2 - z^2}}$	$(0, 0, 0)$
35. $w = xy^2z^2$	$(2, 1, 1)$
36. $f(x, y, z) = xe^{yz}$	$(2, 0, -4)$

Using a Function In Exercises 37–42, consider the function $f(x, y) = 3 - \dfrac{x}{3} - \dfrac{y}{2}$.

37. Sketch the graph of f in the first octant and plot the point $(3, 2, 1)$ on the surface.

38. Find $D_{\mathbf{u}}f(3, 2)$, where $\mathbf{u} = \cos\theta\mathbf{i} + \sin\theta\mathbf{j}$, using each given value of θ.

(a) $\theta = \dfrac{\pi}{4}$ (b) $\theta = \dfrac{2\pi}{3}$ (c) $\theta = \dfrac{4\pi}{3}$ (d) $\theta = -\dfrac{\pi}{6}$

39. Find $D_{\mathbf{u}}f(3, 2)$, where $\mathbf{u} = \dfrac{\mathbf{v}}{\|\mathbf{v}\|}$, using each given vector $\mathbf{v}$.

(a) $\mathbf{v} = \mathbf{i} + \mathbf{j}$ (b) $\mathbf{v} = -3\mathbf{i} - 4\mathbf{j}$

(c) $\mathbf{v}$ is the vector from $(1, 2)$ to $(-2, 6)$.

(d) $\mathbf{v}$ is the vector from $(3, 2)$ to $(4, 5)$.

40. Find $\nabla f(x, y)$.

41. Find the maximum value of the directional derivative at $(3, 2)$.

42. Find a unit vector $\mathbf{u}$ orthogonal to $\nabla f(3, 2)$ and calculate $D_{\mathbf{u}}f(3, 2)$. Discuss the geometric meaning of the result.

Investigation   In Exercises 43 and 44, (a) use the graph to estimate the components of the vector in the direction of the maximum rate of increase in the function at the given point. (b) Find the gradient at the point and compare it with your estimate in part (a). (c) In what direction would the function be decreasing at the greatest rate? Explain.

43. $f(x, y) = \frac{1}{10}(x^2 - 3xy + y^2)$ **44.** $f(x, y) = \frac{1}{2}y\sqrt{x}$

   $(1, 2)$ $(1, 2)$

Generated by Maple   *Generated by Maple*

45. Investigation   Consider the function

$$f(x, y) = x^2 - y^2$$

at the point $(4, -3, 7)$.

(a) Use a computer algebra system to graph the surface represented by the function.

(b) Determine the directional derivative $D_{\mathbf{u}}f(4, -3)$ as a function of θ, where $\mathbf{u} = \cos\theta\,\mathbf{i} + \sin\theta\,\mathbf{j}$. Use a computer algebra system to graph the function on the interval $[0, 2\pi)$.

(c) Approximate the zeros of the function in part (b) and interpret each in the context of the problem.

(d) Approximate the critical numbers of the function in part (b) and interpret each in the context of the problem.

(e) Find $\|\nabla f(4, -3)\|$ and explain its relationship to your answers in part (d).

(f) Use a computer algebra system to graph the level curve of the function f at the level $c = 7$. On this curve, graph the vector in the direction of $\nabla f(4, -3)$, and state its relationship to the level curve.

46. Investigation   Consider the function

$$f(x, y) = \frac{8y}{1 + x^2 + y^2}.$$

(a) Analytically verify that the level curve of $f(x, y)$ at the level $c = 2$ is a circle.

(b) At the point $\left(\sqrt{3}, 2\right)$ on the level curve for which $c = 2$, sketch the vector showing the direction of the greatest rate of increase of the function. (To print an enlarged copy of the graph, go to *MathGraphs.com*.)

(c) At the point $\left(\sqrt{3}, 2\right)$ on the level curve, sketch a vector such that the directional derivative is 0.

(d) Use a computer algebra system to graph the surface to verify your answers in parts (a)–(c).

Finding a Normal Vector   In Exercises 47–50, find a normal vector to the level curve $f(x, y) = c$ at P.

47. $f(x, y) = 6 - 2x - 3y$ **48.** $f(x, y) = x^2 + y^2$

   $c = 6$, $P(0, 0)$ $c = 25$, $P(3, 4)$

49. $f(x, y) = xy$ **50.** $f(x, y) = \dfrac{x}{x^2 + y^2}$

   $c = -3$, $P(-1, 3)$ $c = \frac{1}{2}$, $P(1, 1)$

Using a Function   In Exercises 51–54, (a) find the gradient of the function at P, (b) find a unit normal vector to the level curve $f(x, y) = c$ at P, (c) find the tangent line to the level curve $f(x, y) = c$ at P, and (d) sketch the level curve, the unit normal vector, and the tangent line in the xy-plane.

51. $f(x, y) = 4x^2 - y$ **52.** $f(x, y) = x - y^2$

   $c = 6$, $P(2, 10)$ $c = 3$, $P(4, -1)$

53. $f(x, y) = 3x^2 - 2y^2$ **54.** $f(x, y) = 9x^2 + 4y^2$

   $c = 1$, $P(1, 1)$ $c = 40$, $P(2, -1)$

WRITING ABOUT CONCEPTS

55. Directional Derivative   Define the derivative of the function $z = f(x, y)$ in the direction $\mathbf{u} = \cos\theta\,\mathbf{i} + \sin\theta\,\mathbf{j}$.

56. Directional Derivative   Write a paragraph describing the directional derivative of the function f in the direction $\mathbf{u} = \cos\theta\,\mathbf{i} + \sin\theta\,\mathbf{j}$ when (a) $\theta = 0°$ and (b) $\theta = 90°$.

57. Gradient   Define the gradient of a function of two variables. State the properties of the gradient.

58. Sketching a Graph and a Vector   Sketch the graph of a surface and select a point P on the surface. Sketch a vector in the xy-plane giving the direction of steepest ascent on the surface at P.

59. Gradient and Level Curves   Describe the relationship of the gradient to the level curves of a surface given by $z = f(x, y)$.

60. Using a Function   Consider the function

$$f(x, y) = 9 - x^2 - y^2.$$

(a) Sketch the graph of f in the first octant and plot the point $(1, 2, 4)$ on the surface.

(b) Find $D_{\mathbf{u}}f(1, 2)$, where $\mathbf{u} = \cos\theta\,\mathbf{i} + \sin\theta\,\mathbf{j}$, for $\theta = -\pi/4$.

(c) Repeat part (b) for $\theta = \pi/3$.

(d) Find $\nabla f(1, 2)$ and $\|\nabla f(1, 2)\|$.

(e) Find a unit vector $\mathbf{u}$ orthogonal to $\nabla f(1, 2)$ and calculate $D_{\mathbf{u}}f(1, 2)$. Discuss the geometric meaning of the result.

61. Topography   The surface of a mountain is modeled by the equation

$$h(x, y) = 5000 - 0.001x^2 - 0.004y^2.$$

A mountain climber is at the point $(500, 300, 4390)$. In what direction should the climber move in order to ascend at the greatest rate?

62. **HOW DO YOU SEE IT?** The figure shows a topographic map carried by a group of hikers. Sketch the paths of steepest descent when the hikers start at point A and when they start at point B. (To print an enlarged copy of the graph, go to *MathGraphs.com*.)

63. Temperature Distribution The temperature at the point (x, y) on a metal plate is

$$T = \frac{x}{x^2 + y^2}.$$

Find the direction of greatest increase in heat from the point $(3, 4)$.

64. Temperature The temperature at the point (x, y) on a metal plate is modeled by $T(x, y) = 400e^{-(x^2+y)/2}$, $x \geq 0$, $y \geq 0$.

(a) Use a computer algebra system to graph the temperature distribution function.

(b) Find the directions of no change in heat on the plate from the point $(3, 5)$.

(c) Find the direction of greatest increase in heat from the point $(3, 5)$.

Finding the Direction of Maximum Increase In Exercises 65 and 66, the temperature in degrees Celsius on the surface of a metal plate is given by $T(x, y)$, where x and y are measured in centimeters. Find the direction from point P where the temperature increases most rapidly and the rate of increase.

65. $T(x, y) = 80 - 3x^2 - y^2$, $P(-1, 5)$

66. $T(x, y) = 50 - x^2 - 4y^2$, $P(2, -1)$

Heat-Seeking Path In Exercises 67 and 68, find the path of a heat-seeking particle placed at point P on a metal plate with a temperature field $T(x, y)$.

67. $T(x, y) = 400 - 2x^2 - y^2$, $P(10, 10)$

68. $T(x, y) = 100 - x^2 - 2y^2$, $P(4, 3)$

True or False? In Exercises 69–72, determine whether the statement is true or false. If it is false, explain why or give an example that shows it is false.

69. If $f(x, y) = \sqrt{1 - x^2 - y^2}$, then $D_\mathbf{u} f(0, 0) = 0$ for any unit vector $\mathbf{u}$.

70. If $f(x, y) = x + y$, then $-1 \leq D_\mathbf{u} f(x, y) \leq 1$.

71. If $D_\mathbf{u} f(x, y)$ exists, then $D_\mathbf{u} f(x, y) = -D_{-\mathbf{u}} f(x, y)$.

72. If $D_\mathbf{u} f(x_0, y_0) = c$ for any unit vector $\mathbf{u}$, then $c = 0$.

73. Finding a Function Find a function f such that

$$\nabla f = e^x \cos y \, \mathbf{i} - e^x \sin y \, \mathbf{j} + z \, \mathbf{k}.$$

74. Ocean Floor

A team of oceanographers is mapping the ocean floor to assist in the recovery of a sunken ship. Using sonar, they develop the model

$$D = 250 + 30x^2 + 50 \sin \frac{\pi y}{2}, \quad 0 \leq x \leq 2, 0 \leq y \leq 2$$

where D is the depth in meters, and x and y are the distances in kilometers.

(a) Use a computer algebra system to graph the surface.

(b) Because the graph in part (a) is showing depth, it is not a map of the ocean floor. How could the model be changed so that the graph of the ocean floor could be obtained?

(c) What is the depth of the ship if it is located at the coordinates $x = 1$ and $y = 0.5$?

(d) Determine the steepness of the ocean floor in the positive x-direction from the position of the ship.

(e) Determine the steepness of the ocean floor in the positive y-direction from the position of the ship.

(f) Determine the direction of the greatest rate of change of depth from the position of the ship.

75. Using a Function Consider the function

$$f(x, y) = \sqrt[3]{xy}.$$

(a) Show that f is continuous at the origin.

(b) Show that f_x and f_y exist at the origin, but that the directional derivatives at the origin in all other directions do not exist.

(c) Use a computer algebra system to graph f near the origin to verify your answers in parts (a) and (b). Explain.

76. Directional Derivative Consider the function

$$f(x, y) = \begin{cases} \dfrac{4xy}{x^2 + y^2}, & (x, y) \neq (0, 0) \\ 0, & (x, y) = (0, 0) \end{cases}$$

and the unit vector

$$\mathbf{u} = \frac{1}{\sqrt{2}}(\mathbf{i} + \mathbf{j}).$$

Does the directional derivative of f at $P(0, 0)$ in the direction of $\mathbf{u}$ exist? If $f(0, 0)$ were defined as 2 instead of 0, would the directional derivative exist?

13.7 Tangent Planes and Normal Lines

- Find equations of tangent planes and normal lines to surfaces.
- Find the angle of inclination of a plane in space.
- Compare the gradients $\nabla f(x, y)$ and $\nabla F(x, y, z)$.

Tangent Plane and Normal Line to a Surface

So far, you have represented surfaces in space primarily by equations of the form

$$z = f(x, y). \qquad \text{Equation of a surface } S$$

In the development to follow, however, it is convenient to use the more general representation $F(x, y, z) = 0$. For a surface S given by $z = f(x, y)$, you can convert to the general form by defining F as

$$F(x, y, z) = f(x, y) - z.$$

Because $f(x, y) - z = 0$, you can consider S to be the level surface of F given by

$$F(x, y, z) = 0. \qquad \text{Alternative equation of surface } S$$

EXAMPLE 1 Writing an Equation of a Surface

For the function

$$F(x, y, z) = x^2 + y^2 + z^2 - 4$$

describe the level surface given by

$$F(x, y, z) = 0.$$

Solution The level surface given by $F(x, y, z) = 0$ can be written as

$$x^2 + y^2 + z^2 = 4$$

which is a sphere of radius 2 whose center is at the origin. ∎

You have seen many examples of the usefulness of normal lines in applications involving curves. Normal lines are equally important in analyzing surfaces and solids. For example, consider the collision of two billiard balls. When a stationary ball is struck at a point P on its surface, it moves along the **line of impact** determined by P and the center of the ball. The impact can occur in *two* ways. When the cue ball is moving along the line of impact, it stops dead and imparts all of its momentum to the stationary ball, as shown in Figure 13.55. When the cue ball is not moving along the line of impact, it is deflected to one side or the other and retains part of its momentum. The part of the momentum that is transferred to the stationary ball occurs along the line of impact, *regardless* of the direction of the cue ball, as shown in Figure 13.56. This line of impact is called the **normal line** to the surface of the ball at the point P.

Figure 13.55

Figure 13.56

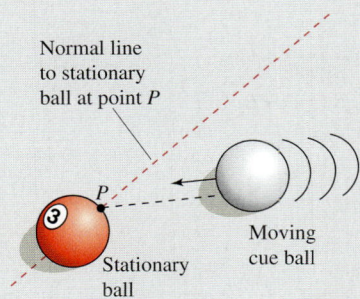

In the process of finding a normal line to a surface, you are also able to solve the problem of finding a **tangent plane** to the surface. Let S be a surface given by

$$F(x, y, z) = 0$$

and let $P(x_0, y_0, z_0)$ be a point on S. Let C be a curve on S through P that is defined by the vector-valued function

$$\mathbf{r}(t) = x(t)\mathbf{i} + y(t)\mathbf{j} + z(t)\mathbf{k}.$$

Then, for all t,

$$F(x(t), \ y(t), \ z(t)) = 0.$$

If F is differentiable and $x'(t)$, $y'(t)$, and $z'(t)$ all exist, then it follows from the Chain Rule that

$$
\begin{aligned}
0 &= F'(t) \\
&= F_x(x, y, z)x'(t) + F_y(x, y, z)y'(t) + F_z(x, y, z)z'(t).
\end{aligned}
$$

At (x_0, y_0, z_0), the equivalent vector form is

$$0 = \underbrace{\nabla F(x_0, y_0, z_0)}_{\text{Gradient}} \cdot \underbrace{\mathbf{r}'(t_0)}_{\substack{\text{Tangent} \\ \text{vector}}}.$$

Surface S:
$F(x, y, z) = 0$

$P(x_0, y_0, z_0)$

Tangent plane to surface S at P
Figure 13.57

This result means that the gradient at P is orthogonal to the tangent vector of every curve on S through P. So, all tangent lines on S lie in a plane that is normal to $\nabla F(x_0, y_0, z_0)$ and contains P, as shown in Figure 13.57.

Definitions of Tangent Plane and Normal Line

Let F be differentiable at the point $P(x_0, y_0, z_0)$ on the surface S given by $F(x, y, z) = 0$ such that

$$\nabla F(x_0, y_0, z_0) \neq \mathbf{0}.$$

1. The plane through P that is normal to $\nabla F(x_0, y_0, z_0)$ is called the **tangent plane to S at P.**
2. The line through P having the direction of $\nabla F(x_0, y_0, z_0)$ is called the **normal line to S at P.**

> **REMARK** In the remainder of this section, assume $\nabla F(x_0, y_0, z_0)$ to be nonzero unless stated otherwise.

To find an equation for the tangent plane to S at (x_0, y_0, z_0), let (x, y, z) be an arbitrary point in the tangent plane. Then the vector

$$\mathbf{v} = (x - x_0)\mathbf{i} + (y - y_0)\mathbf{j} + (z - z_0)\mathbf{k}$$

lies in the tangent plane. Because $\nabla F(x_0, y_0, z_0)$ is normal to the tangent plane at (x_0, y_0, z_0), it must be orthogonal to every vector in the tangent plane, and you have

$$\nabla F(x_0, y_0, z_0) \cdot \mathbf{v} = 0$$

which leads to the next theorem.

THEOREM 13.13 Equation of Tangent Plane

If F is differentiable at (x_0, y_0, z_0), then an equation of the tangent plane to the surface given by $F(x, y, z) = 0$ at (x_0, y_0, z_0) is

$$F_x(x_0, y_0, z_0)(x - x_0) + F_y(x_0, y_0, z_0)(y - y_0) + F_z(x_0, y_0, z_0)(z - z_0) = 0.$$

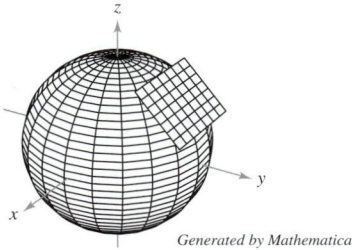
EXAMPLE 2   **Finding an Equation of a Tangent Plane**

Find an equation of the tangent plane to the hyperboloid

$$z^2 - 2x^2 - 2y^2 = 12$$

at the point $(1, -1, 4)$.

Solution   Begin by writing the equation of the surface as

$$z^2 - 2x^2 - 2y^2 - 12 = 0.$$

Then, considering

$$F(x, y, z) = z^2 - 2x^2 - 2y^2 - 12$$

you have

$$F_x(x, y, z) = -4x, \quad F_y(x, y, z) = -4y, \quad \text{and} \quad F_z(x, y, z) = 2z.$$

At the point $(1, -1, 4)$, the partial derivatives are

$$F_x(1, -1, 4) = -4, \quad F_y(1, -1, 4) = 4, \quad \text{and} \quad F_z(1, -1, 4) = 8.$$

So, an equation of the tangent plane at $(1, -1, 4)$ is

$$-4(x - 1) + 4(y + 1) + 8(z - 4) = 0$$
$$-4x + 4 + 4y + 4 + 8z - 32 = 0$$
$$-4x + 4y + 8z - 24 = 0$$
$$x - y - 2z + 6 = 0.$$

Figure 13.58 shows a portion of the hyperboloid and the tangent plane.

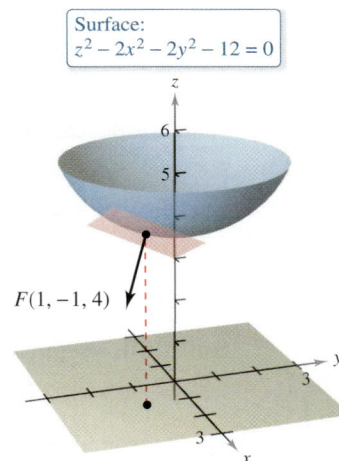

Surface:
$z^2 - 2x^2 - 2y^2 - 12 = 0$

$F(1, -1, 4)$

Tangent plane to surface
Figure 13.58

To find the equation of the tangent plane at a point on a surface given by $z = f(x, y)$, you can define the function F by

$$F(x, y, z) = f(x, y) - z.$$

Then S is given by the level surface $F(x, y, z) = 0$, and by Theorem 13.13, an equation of the tangent plane to S at the point (x_0, y_0, z_0) is

$$f_x(x_0, y_0)(x - x_0) + f_y(x_0, y_0)(y - y_0) - (z - z_0) = 0.$$

EXAMPLE 3   Finding an Equation of the Tangent Plane

Find the equation of the tangent plane to the paraboloid

$$z = 1 - \frac{1}{10}(x^2 + 4y^2)$$

at the point $\left(1, 1, \frac{1}{2}\right)$.

Solution   From $z = f(x, y) = 1 - \frac{1}{10}(x^2 + 4y^2)$, you obtain

$$f_x(x, y) = -\frac{x}{5} \implies f_x(1, 1) = -\frac{1}{5}$$

and

$$f_y(x, y) = -\frac{4y}{5} \implies f_y(1, 1) = -\frac{4}{5}.$$

So, an equation of the tangent plane at $\left(1, 1, \frac{1}{2}\right)$ is

$$f_x(1, 1)(x - 1) + f_y(1, 1)(y - 1) - \left(z - \frac{1}{2}\right) = 0$$

$$-\frac{1}{5}(x - 1) - \frac{4}{5}(y - 1) - \left(z - \frac{1}{2}\right) = 0$$

$$-\frac{1}{5}x - \frac{4}{5}y - z + \frac{3}{2} = 0.$$

This tangent plane is shown in Figure 13.59.

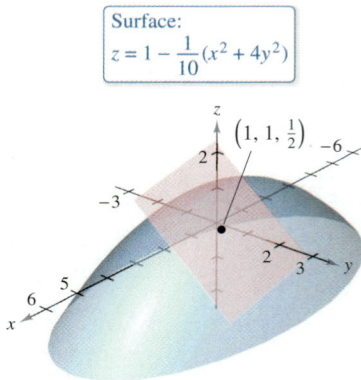

Surface:
$z = 1 - \frac{1}{10}(x^2 + 4y^2)$

Figure 13.59

The gradient $\nabla F(x, y, z)$ provides a convenient way to find equations of normal lines, as shown in Example 4.

EXAMPLE 4   Finding an Equation of a Normal Line to a Surface

⋯▷ *See LarsonCalculus.com for an interactive version of this type of example.*

Find a set of symmetric equations for the normal line to the surface

$$xyz = 12$$

at the point $(2, -2, -3)$.

Solution   Begin by letting

$$F(x, y, z) = xyz - 12.$$

Then, the gradient is given by

$$\nabla F(x, y, z) = F_x(x, y, z)\mathbf{i} + F_y(x, y, z)\mathbf{j} + F_z(x, y, z)\mathbf{k}$$
$$= yz\mathbf{i} + xz\mathbf{j} + xy\mathbf{k}$$

and at the point $(2, -2, -3)$, you have

$$\nabla F(2, -2, -3) = (-2)(-3)\mathbf{i} + (2)(-3)\mathbf{j} + (2)(-2)\mathbf{k}$$
$$= 6\mathbf{i} - 6\mathbf{j} - 4\mathbf{k}.$$

The normal line at $(2, -2, -3)$ has direction numbers 6, -6, and -4, and the corresponding set of symmetric equations is

$$\frac{x - 2}{6} = \frac{y + 2}{-6} = \frac{z + 3}{-4}.$$

See Figure 13.60.

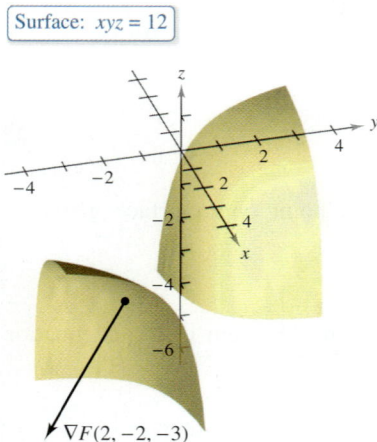

Surface: $xyz = 12$

$\nabla F(2, -2, -3)$

Figure 13.60

Knowing that the gradient $\nabla F(x, y, z)$ is normal to the surface given by $F(x, y, z) = 0$ allows you to solve a variety of problems dealing with surfaces and curves in space.

EXAMPLE 5 **Finding the Equation of a Tangent Line to a Curve**

Describe the tangent line to the curve of intersection of the ellipsoid

$$x^2 + 2y^2 + 2z^2 = 20 \qquad \text{Ellipsoid}$$

and the paraboloid

$$x^2 + y^2 + z = 4 \qquad \text{Paraboloid}$$

at the point $(0, 1, 3)$, as shown in Figure 13.61.

Solution   Begin by finding the gradients to both surfaces at the point $(0, 1, 3)$.

Ellipsoid

$$F(x, y, z) = x^2 + 2y^2 + 2z^2 - 20$$
$$\nabla F(x, y, z) = 2x\mathbf{i} + 4y\mathbf{j} + 4z\mathbf{k}$$
$$\nabla F(0, 1, 3) = 4\mathbf{j} + 12\mathbf{k}$$

Paraboloid

$$G(x, y, z) = x^2 + y^2 + z - 4$$
$$\nabla G(x, y, z) = 2x\mathbf{i} + 2y\mathbf{j} + \mathbf{k}$$
$$\nabla G(0, 1, 3) = 2\mathbf{j} + \mathbf{k}$$

The cross product of these two gradients is a vector that is tangent to both surfaces at the point $(0, 1, 3)$.

$$\nabla F(0, 1, 3) \times \nabla G(0, 1, 3) = \begin{vmatrix} \mathbf{i} & \mathbf{j} & \mathbf{k} \\ 0 & 4 & 12 \\ 0 & 2 & 1 \end{vmatrix} = -20\mathbf{i}$$

So, the tangent line to the curve of intersection of the two surfaces at the point $(0, 1, 3)$ is a line that is parallel to the x-axis and passes through the point $(0, 1, 3)$. ■

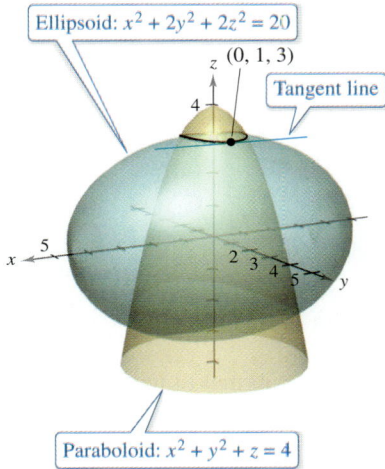

Ellipsoid: $x^2 + 2y^2 + 2z^2 = 20$

Tangent line

Paraboloid: $x^2 + y^2 + z = 4$

Figure 13.61

The Angle of Inclination of a Plane

Another use of the gradient

$$\nabla F(x, y, z)$$

is to determine the angle of inclination of the tangent plane to a surface. The **angle of inclination** of a plane is defined as the angle θ $(0 \le \theta \le \pi/2)$ between the given plane and the xy-plane, as shown in Figure 13.62. (The angle of inclination of a horizontal plane is defined as zero.) Because the vector $\mathbf{k}$ is normal to the xy-plane, you can use the formula for the cosine of the angle between two planes (given in Section 11.5) to conclude that the angle of inclination of a plane with normal vector $\mathbf{n}$ is

$$\cos \theta = \frac{|\mathbf{n} \cdot \mathbf{k}|}{\|\mathbf{n}\| \, \|\mathbf{k}\|} = \frac{|\mathbf{n} \cdot \mathbf{k}|}{\|\mathbf{n}\|}. \qquad \text{Angle of inclination of a plane}$$

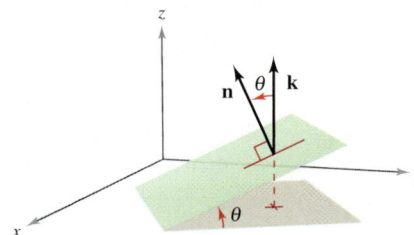

The angle of inclination
Figure 13.62

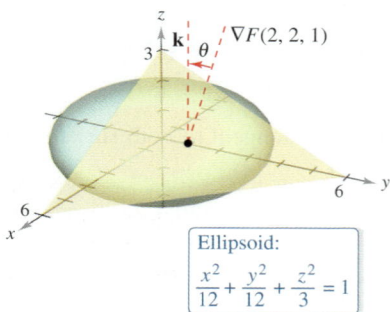

Figure 13.63

| EXAMPLE 6 | **Finding the Angle of Inclination of a Tangent Plane** |

Find the angle of inclination of the tangent plane to the ellipsoid

$$\frac{x^2}{12} + \frac{y^2}{12} + \frac{z^2}{3} = 1$$

at the point $(2, 2, 1)$.

Solution   Begin by letting

$$F(x, y, z) = \frac{x^2}{12} + \frac{y^2}{12} + \frac{z^2}{3} - 1.$$

Then, the gradient of F at the point $(2, 2, 1)$ is

$$\nabla F(x, y, z) = \frac{x}{6}\mathbf{i} + \frac{y}{6}\mathbf{j} + \frac{2z}{3}\mathbf{k}$$

$$\nabla F(2, 2, 1) = \frac{1}{3}\mathbf{i} + \frac{1}{3}\mathbf{j} + \frac{2}{3}\mathbf{k}.$$

Because $\nabla F(2, 2, 1)$ is normal to the tangent plane and $\mathbf{k}$ is normal to the xy-plane, it follows that the angle of inclination of the tangent plane is

$$\cos \theta = \frac{|\nabla F(2, 2, 1) \cdot \mathbf{k}|}{\|\nabla F(2, 2, 1)\|} = \frac{2/3}{\sqrt{(1/3)^2 + (1/3)^2 + (2/3)^2}} = \sqrt{\frac{2}{3}}$$

which implies that

$$\theta = \arccos \sqrt{\frac{2}{3}} \approx 35.3°,$$

as shown in Figure 13.63.   ∎

A special case of the procedure shown in Example 6 is worth noting. The angle of inclination θ of the tangent plane to the surface $z = f(x, y)$ at (x_0, y_0, z_0) is

$$\cos \theta = \frac{1}{\sqrt{[f_x(x_0, y_0)]^2 + [f_y(x_0, y_0)]^2 + 1}}.$$

Alternative formula for angle of inclination (See Exercise 67.)

A Comparison of the Gradients $\nabla f(x, y)$ and $\nabla F(x, y, z)$

This section concludes with a comparison of the gradients $\nabla f(x, y)$ and $\nabla F(x, y, z)$. In the preceding section, you saw that the gradient of a function f of two variables is normal to the level curves of f. Specifically, Theorem 13.12 states that if f is differentiable at (x_0, y_0) and $\nabla f(x_0, y_0) \neq \mathbf{0}$, then $\nabla f(x_0, y_0)$ is normal to the level curve through (x_0, y_0). Having developed normal lines to surfaces, you can now extend this result to a function of three variables. The proof of Theorem 13.14 is left as an exercise (see Exercise 68).

> **THEOREM 13.14   Gradient Is Normal to Level Surfaces**
>
> If F is differentiable at (x_0, y_0, z_0) and
>
> $$\nabla F(x_0, y_0, z_0) \neq \mathbf{0}$$
>
> then $\nabla F(x_0, y_0, z_0)$ is normal to the level surface through (x_0, y_0, z_0).

When working with the gradients $\nabla f(x, y)$ and $\nabla F(x, y, z)$, be sure you remember that $\nabla f(x, y)$ is a vector in the xy-plane and $\nabla F(x, y, z)$ is a vector in space.

13.7 Exercises

See **CalcChat.com** for tutorial help and worked-out solutions to odd-numbered exercises.

Describing a Surface   In Exercises 1–4, describe the level surface $F(x, y, z) = 0$.

1. $F(x, y, z) = 3x - 5y + 3z - 15$

2. $F(x, y, z) = x^2 + y^2 + z^2 - 25$

3. $F(x, y, z) = 4x^2 + 9y^2 - 4z^2$

4. $F(x, y, z) = 16x^2 - 9y^2 + 36z$

Finding a Unit Normal Vector   In Exercises 5–8, find a unit normal vector to the surface at the given point. [*Hint:* Normalize the gradient vector $\nabla F(x, y, z)$.]

Surface	Point
5. $3x + 4y + 12z = 0$	$(0, 0, 0)$
6. $x^2 + y^2 + z^2 = 6$	$(1, 1, 2)$
7. $x^2 + 3y + z^3 = 9$	$(2, -1, 2)$
8. $x^2y^3 - y^2z + 2xz^3 = 4$	$(-1, 1, -1)$

Finding an Equation of a Tangent Plane   In Exercises 9–20, find an equation of the tangent plane to the surface at the given point.

9. $z = x^2 + y^2 + 3$

    $(2, 1, 8)$

10. $f(x, y) = \dfrac{y}{x}$

    $(1, 2, 2)$

$(2, 1, 8)$

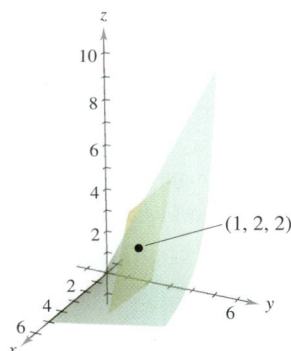

$(1, 2, 2)$

11. $z = \sqrt{x^2 + y^2}$, $(3, 4, 5)$

12. $g(x, y) = \arctan \dfrac{y}{x}$, $(1, 0, 0)$

13. $g(x, y) = x^2 + y^2$, $(1, -1, 2)$

14. $f(x, y) = x^2 - 2xy + y^2$, $(1, 2, 1)$

15. $h(x, y) = \ln \sqrt{x^2 + y^2}$, $(3, 4, \ln 5)$

16. $h(x, y) = \cos y$, $\left(5, \dfrac{\pi}{4}, \dfrac{\sqrt{2}}{2}\right)$

17. $x^2 + 4y^2 + z^2 = 36$, $(2, -2, 4)$

18. $x^2 + 2z^2 = y^2$, $(1, 3, -2)$

19. $xy^2 + 3x - z^2 = 8$, $(1, -3, 2)$

20. $z = e^x(\sin y + 1)$, $\left(0, \dfrac{\pi}{2}, 2\right)$

Finding an Equation of a Tangent Plane and a Normal Line   In Exercises 21–30, find an equation of the tangent plane and find a set of symmetric equations for the normal line to the surface at the given point.

21. $x + y + z = 9$, $(3, 3, 3)$

22. $x^2 + y^2 + z^2 = 9$, $(1, 2, 2)$

23. $x^2 + y^2 + z = 9$, $(1, 2, 4)$

24. $z = 16 - x^2 - y^2$, $(2, 2, 8)$

25. $z = x^2 - y^2$, $(3, 2, 5)$

26. $xy - z = 0$, $(-2, -3, 6)$

27. $xyz = 10$, $(1, 2, 5)$

28. $z = ye^{2xy}$, $(0, 2, 2)$

29. $z = \arctan \dfrac{y}{x}$, $\left(1, 1, \dfrac{\pi}{4}\right)$

30. $y \ln xz^2 = 2$, $(e, 2, 1)$

Finding the Equation of a Tangent Line to a Curve   In Exercises 31–36, (a) find a set of symmetric equations for the tangent line to the curve of intersection of the surfaces at the given point, and (b) find the cosine of the angle between the gradient vectors at this point. State whether the surfaces are orthogonal at the point of intersection.

31. $x^2 + y^2 = 2$, $z = x$, $(1, 1, 1)$

32. $z = x^2 + y^2$, $z = 4 - y$, $(2, -1, 5)$

33. $x^2 + z^2 = 25$, $y^2 + z^2 = 25$, $(3, 3, 4)$

34. $z = \sqrt{x^2 + y^2}$, $5x - 2y + 3z = 22$, $(3, 4, 5)$

35. $x^2 + y^2 + z^2 = 14$, $x - y - z = 0$, $(3, 1, 2)$

36. $z = x^2 + y^2$, $x + y + 6z = 33$, $(1, 2, 5)$

Finding the Angle of Inclination of a Tangent Plane   In Exercises 37–40, find the angle of inclination θ of the tangent plane to the surface at the given point.

37. $3x^2 + 2y^2 - z = 15$, $(2, 2, 5)$

38. $2xy - z^3 = 0$, $(2, 2, 2)$

39. $x^2 - y^2 + z = 0$, $(1, 2, 3)$

40. $x^2 + y^2 = 5$, $(2, 1, 3)$

Horizontal Tangent Plane   In Exercises 41–46, find the point(s) on the surface at which the tangent plane is horizontal.

41. $z = 3 - x^2 - y^2 + 6y$

42. $z = 3x^2 + 2y^2 - 3x + 4y - 5$

43. $z = x^2 - xy + y^2 - 2x - 2y$

44. $z = 4x^2 + 4xy - 2y^2 + 8x - 5y - 4$

45. $z = 5xy$

46. $z = xy + \dfrac{1}{x} + \dfrac{1}{y}$

Tangent Surfaces In Exercises 47 and 48, show that the surfaces are tangent to each other at the given point by showing that the surfaces have the same tangent plane at this point.

47. $x^2 + 2y^2 + 3z^2 = 3$, $x^2 + y^2 + z^2 + 6x - 10y + 14 = 0$,

$(-1, 1, 0)$

48. $x^2 + y^2 + z^2 - 8x - 12y + 4z + 42 = 0$,

$x^2 + y^2 + 2z = 7$, $(2, 3, -3)$

Perpendicular Tangent Planes In Exercises 49 and 50, (a) show that the surfaces intersect at the given point, and (b) show that the surfaces have perpendicular tangent planes at this point.

49. $z = 2xy^2$, $8x^2 - 5y^2 - 8z = -13$, $(1, 1, 2)$

50. $x^2 + y^2 + z^2 + 2x - 4y - 4z - 12 = 0$,

$4x^2 + y^2 + 16z^2 = 24$, $(1, -2, 1)$

51. Using an Ellipsoid Find a point on the ellipsoid $x^2 + 4y^2 + z^2 = 9$ where the tangent plane is perpendicular to the line with parametric equations

$x = 2 - 4t$, $y = 1 + 8t$, and $z = 3 - 2t$.

52. Using a Hyperboloid Find a point on the hyperboloid $x^2 + 4y^2 - z^2 = 1$ where the tangent plane is parallel to the plane $x + 4y - z = 0$.

> **WRITING ABOUT CONCEPTS**
>
> **53. Tangent Plane** Give the standard form of the equation of the tangent plane to a surface given by $F(x, y, z) = 0$ at (x_0, y_0, z_0).
>
> **54. Normal Lines** For some surfaces, the normal lines at any point pass through the same geometric object. What is the common geometric object for a sphere? What is the common geometric object for a right circular cylinder? Explain.
>
> **55. Tangent Plane** Discuss the relationship between the tangent plane to a surface and approximation by differentials.

56. **HOW DO YOU SEE IT?** The graph shows the ellipsoid $x^2 + 4y^2 + z^2 = 16$. Use the graph to determine the equation of the tangent plane at each of the given points.

(a) $(4, 0, 0)$ (b) $(0, -2, 0)$ (c) $(0, 0, -4)$

57. Investigation Consider the function

$$f(x, y) = \frac{4xy}{(x^2 + 1)(y^2 + 1)}$$

on the intervals $-2 \le x \le 2$ and $0 \le y \le 3$.

(a) Find a set of parametric equations of the normal line and an equation of the tangent plane to the surface at the point $(1, 1, 1)$.

(b) Repeat part (a) for the point $\left(-1, 2, -\frac{4}{5}\right)$.

(c) Use a computer algebra system to graph the surface, the normal lines, and the tangent planes found in parts (a) and (b).

58. Investigation Consider the function

$$f(x, y) = \frac{\sin y}{x}$$

on the intervals $-3 \le x \le 3$ and $0 \le y \le 2\pi$.

(a) Find a set of parametric equations of the normal line and an equation of the tangent plane to the surface at the point

$\left(2, \frac{\pi}{2}, \frac{1}{2}\right)$.

(b) Repeat part (a) for the point $\left(-\frac{2}{3}, \frac{3\pi}{2}, \frac{3}{2}\right)$.

(c) Use a computer algebra system to graph the surface, the normal lines, and the tangent planes found in parts (a) and (b).

59. Using Functions Consider the functions

$$f(x, y) = 6 - x^2 - \frac{y^2}{4} \quad \text{and} \quad g(x, y) = 2x + y.$$

(a) Find a set of parametric equations of the tangent line to the curve of intersection of the surfaces at the point $(1, 2, 4)$, and find the angle between the gradient vectors.

(b) Use a computer algebra system to graph the surfaces. Graph the tangent line found in part (a).

60. Using Functions Consider the functions

$$f(x, y) = \sqrt{16 - x^2 - y^2 + 2x - 4y}$$

and

$$g(x, y) = \frac{\sqrt{2}}{2}\sqrt{1 - 3x^2 + y^2 + 6x + 4y}.$$

(a) Use a computer algebra system to graph the first-octant portion of the surfaces represented by f and g.

(b) Find one first-octant point on the curve of intersection and show that the surfaces are orthogonal at this point.

(c) These surfaces are orthogonal along the curve of intersection. Does part (b) prove this fact? Explain.

Writing a Tangent Plane In Exercises 61 and 62, show that the tangent plane to the quadric surface at the point (x_0, y_0, z_0) can be written in the given form.

61. Ellipsoid: $\dfrac{x^2}{a^2} + \dfrac{y^2}{b^2} + \dfrac{z^2}{c^2} = 1$

Plane: $\dfrac{x_0 x}{a^2} + \dfrac{y_0 y}{b^2} + \dfrac{z_0 z}{c^2} = 1$

62. Hyperboloid: $\dfrac{x^2}{a^2} + \dfrac{y^2}{b^2} - \dfrac{z^2}{c^2} = 1$

Plane: $\dfrac{x_0 x}{a^2} + \dfrac{y_0 y}{b^2} - \dfrac{z_0 z}{c^2} = 1$

63. Tangent Planes of a Cone Show that any tangent plane to the cone

$$z^2 = a^2 x^2 + b^2 y^2$$

passes through the origin.

64. Tangent Planes Let f be a differentiable function and consider the surface

$$z = xf\left(\frac{y}{x}\right).$$

Show that the tangent plane at any point $P(x_0, y_0, z_0)$ on the surface passes through the origin.

65. Approximation Consider the following approximations for a function $f(x, y)$ centered at $(0, 0)$.

Linear Approximation:

$$P_1(x, y) = f(0, 0) + f_x(0, 0)x + f_y(0, 0)y$$

Quadratic Approximation:

$$P_2(x, y) = f(0, 0) + f_x(0, 0)x + f_y(0, 0)y +$$
$$\tfrac{1}{2}f_{xx}(0, 0)x^2 + f_{xy}(0, 0)xy + \tfrac{1}{2}f_{yy}(0, 0)y^2$$

[Note that the linear approximation is the tangent plane to the surface at $(0, 0, f(0, 0))$.]

(a) Find the linear approximation of $f(x, y) = e^{(x-y)}$ centered at $(0, 0)$.

(b) Find the quadratic approximation of $f(x, y) = e^{(x-y)}$ centered at $(0, 0)$.

(c) When $x = 0$ in the quadratic approximation, you obtain the second-degree Taylor polynomial for what function? Answer the same question for $y = 0$.

(d) Complete the table.

x	y	$f(x, y)$	$P_1(x, y)$	$P_2(x, y)$
0	0			
0	0.1			
0.2	0.1			
0.2	0.5			
1	0.5			

(e) Use a computer algebra system to graph the surfaces $z = f(x, y)$, $z = P_1(x, y)$, and $z = P_2(x, y)$.

66. Approximation Repeat Exercise 65 for the function $f(x, y) = \cos(x + y)$.

67. Proof Prove that the angle of inclination θ of the tangent plane to the surface $z = f(x, y)$ at the point (x_0, y_0, z_0) is given by

$$\cos\theta = \frac{1}{\sqrt{[f_x(x_0, y_0)]^2 + [f_y(x_0, y_0)]^2 + 1}}.$$

68. Proof Prove Theorem 13.14.

SECTION PROJECT

Wildflowers

The diversity of wildflowers in a meadow can be measured by counting the numbers of daisies, buttercups, shooting stars, and so on. When there are n types of wildflowers, each with a proportion p_i of the total population, it follows that

$$p_1 + p_2 + \cdots + p_n = 1.$$

The measure of diversity of the population is defined as

$$H = -\sum_{i=1}^{n} p_i \log_2 p_i.$$

In this definition, it is understood that $p_i \log_2 p_i = 0$ when $p_i = 0$. The tables show proportions of wildflowers in a meadow in May, June, August, and September.

May

Flower type	1	2	3	4
Proportion	$\frac{5}{16}$	$\frac{5}{16}$	$\frac{5}{16}$	$\frac{1}{16}$

June

Flower type	1	2	3	4
Proportion	$\frac{1}{4}$	$\frac{1}{4}$	$\frac{1}{4}$	$\frac{1}{4}$

August

Flower type	1	2	3	4
Proportion	$\frac{1}{4}$	0	$\frac{1}{4}$	$\frac{1}{2}$

September

Flower type	1	2	3	4
Proportion	0	0	0	1

(a) Determine the wildflower diversity for each month. How would you interpret September's diversity? Which month had the greatest diversity?

(b) When the meadow contains 10 types of wildflowers in roughly equal proportions, is the diversity of the population greater than or less than the diversity of a similar distribution of 4 types of flowers? What type of distribution (of 10 types of wildflowers) would produce maximum diversity?

(c) Let H_n represent the maximum diversity of n types of wildflowers. Does H_n approach a limit as n approaches ∞?

■ **FOR FURTHER INFORMATION** Biologists use the concept of diversity to measure the proportions of different types of organisms within an environment. For more information on this technique, see the article "Information Theory and Biological Diversity" by Steven Kolmes and Kevin Mitchell in the *UMAP Modules*.

13.8 Extrema of Functions of Two Variables

■ Find absolute and relative extrema of a function of two variables.
■ Use the Second Partials Test to find relative extrema of a function of two variables.

Absolute Extrema and Relative Extrema

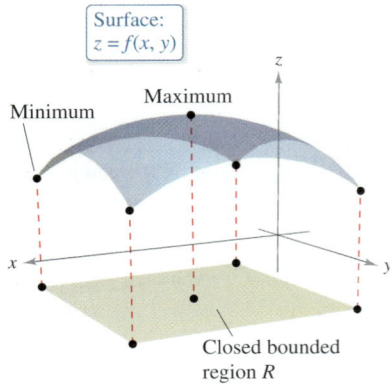

Surface:
$z = f(x, y)$

Minimum
Maximum

Closed bounded
region R

R contains point(s) at which $f(x, y)$ is a minimum and point(s) at which $f(x, y)$ is a maximum.
Figure 13.64

In Chapter 3, you studied techniques for finding the extreme values of a function of a single variable. In this section, you will extend these techniques to functions of two variables. For example, in Theorem 13.15 below, the Extreme Value Theorem for a function of a single variable is extended to a function of two variables.

Consider the continuous function f of two variables, defined on a closed bounded region R. The values $f(a, b)$ and $f(c, d)$ such that

$$f(a, b) \leq f(x, y) \leq f(c, d) \qquad (a, b) \text{ and } (c, d) \text{ are in } R.$$

for all (x, y) in R are called the **minimum** and **maximum** of f in the region R, as shown in Figure 13.64. Recall from Section 13.2 that a region in the plane is *closed* when it contains all of its boundary points. The Extreme Value Theorem deals with a region in the plane that is both closed and *bounded*. A region in the plane is **bounded** when it is a subregion of a closed disk in the plane.

THEOREM 13.15   Extreme Value Theorem

Let f be a continuous function of two variables x and y defined on a closed bounded region R in the xy-plane.

1. There is at least one point in R at which f takes on a minimum value.
2. There is at least one point in R at which f takes on a maximum value.

A minimum is also called an **absolute minimum** and a maximum is also called an **absolute maximum.** As in single-variable calculus, there is a distinction made between absolute extrema and **relative extrema.**

Definition of Relative Extrema

Let f be a function defined on a region R containing (x_0, y_0).

1. The function f has a **relative minimum** at (x_0, y_0) if

$$f(x, y) \geq f(x_0, y_0)$$

    for all (x, y) in an *open* disk containing (x_0, y_0).
2. The function f has a **relative maximum** at (x_0, y_0) if

$$f(x, y) \leq f(x_0, y_0)$$

    for all (x, y) in an *open* disk containing (x_0, y_0).

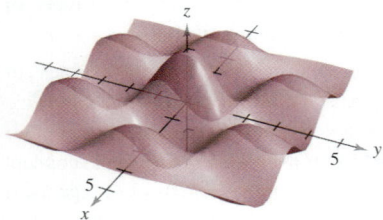

Relative extrema
Figure 13.65

To say that f has a relative maximum at (x_0, y_0) means that the point (x_0, y_0, z_0) is at least as high as all nearby points on the graph of

$$z = f(x, y).$$

Similarly, f has a relative minimum at (x_0, y_0) when (x_0, y_0, z_0) is at least as low as all nearby points on the graph. (See Figure 13.65.)

To locate relative extrema of f, you can investigate the points at which the gradient of f is **0** or the points at which one of the partial derivatives does not exist. Such points are called **critical points** of f.

Definition of Critical Point

Let f be defined on an open region R containing (x_0, y_0). The point (x_0, y_0) is a **critical point** of f if one of the following is true.

1. $f_x(x_0, y_0) = 0$ and $f_y(x_0, y_0) = 0$
2. $f_x(x_0, y_0)$ or $f_y(x_0, y_0)$ does not exist.

Recall from Theorem 13.11 that if f is differentiable and

$$\nabla f(x_0, y_0) = f_x(x_0, y_0)\mathbf{i} + f_y(x_0, y_0)\mathbf{j} = 0\mathbf{i} + 0\mathbf{j}$$

then every directional derivative at (x_0, y_0) must be 0. This implies that the function has a horizontal tangent plane at the point (x_0, y_0), as shown in Figure 13.66. It appears that such a point is a likely location of a relative extremum. This is confirmed by Theorem 13.16.

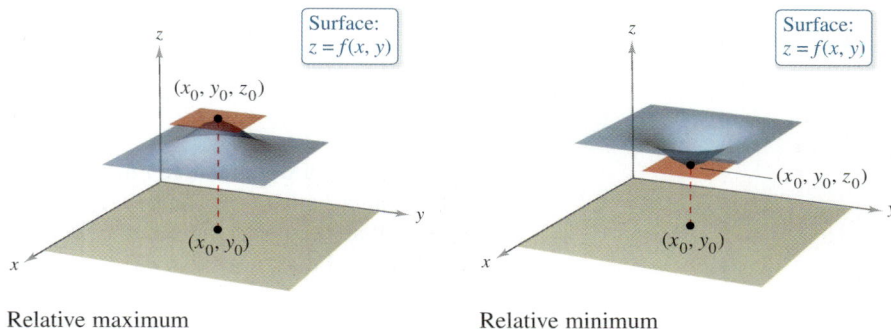

Relative maximum

Relative minimum

Figure 13.66

THEOREM 13.16 Relative Extrema Occur Only at Critical Points

If f has a relative extremum at (x_0, y_0) on an open region R, then (x_0, y_0) is a critical point of f.

Exploration

Use a graphing utility to graph $z = x^3 - 3xy + y^3$ using the bounds $0 \le x \le 3$, $0 \le y \le 3$, and $-3 \le z \le 3$. This view makes it appear as though the surface has an absolute minimum. But does it?

Determine the relative extrema of

$$f(x, y) = 2x^2 + y^2 + 8x - 6y + 20.$$

Solution Begin by finding the critical points of f. Because

$$f_x(x, y) = 4x + 8 \qquad \text{Partial with respect to } x$$

and

$$f_y(x, y) = 2y - 6 \qquad \text{Partial with respect to } y$$

are defined for all x and y, the only critical points are those for which both first partial derivatives are 0. To locate these points, set $f_x(x, y)$ and $f_y(x, y)$ equal to 0, and solve the equations

$$4x + 8 = 0 \quad \text{and} \quad 2y - 6 = 0$$

to obtain the critical point $(-2, 3)$. By completing the square for f, you can see that for all $(x, y) \neq (-2, 3)$

$$f(x, y) = 2(x + 2)^2 + (y - 3)^2 + 3 > 3.$$

So, a relative *minimum* of f occurs at $(-2, 3)$. The value of the relative minimum is $f(-2, 3) = 3$, as shown in Figure 13.67.

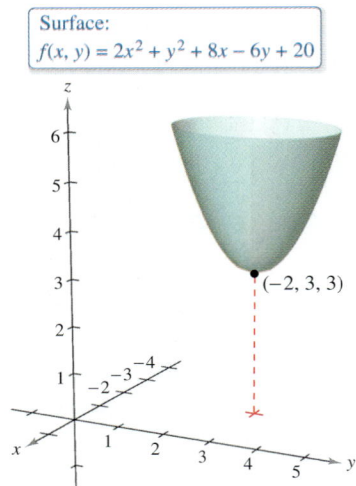

Surface:
$f(x, y) = 2x^2 + y^2 + 8x - 6y + 20$

The function $z = f(x, y)$ has a relative minimum at $(-2, 3)$.
Figure 13.67

Example 1 shows a relative minimum occurring at one type of critical point—the type for which both $f_x(x, y)$ and $f_y(x, y)$ are 0. The next example concerns a relative maximum that occurs at the other type of critical point—the type for which either $f_x(x, y)$ or $f_y(x, y)$ does not exist.

EXAMPLE 2 **Finding a Relative Extremum**

Determine the relative extrema of

$$f(x, y) = 1 - (x^2 + y^2)^{1/3}.$$

Solution Because

$$f_x(x, y) = -\frac{2x}{3(x^2 + y^2)^{2/3}} \qquad \text{Partial with respect to } x$$

and

$$f_y(x, y) = -\frac{2y}{3(x^2 + y^2)^{2/3}} \qquad \text{Partial with respect to } y$$

it follows that both partial derivatives exist for all points in the xy-plane except for $(0, 0)$. Moreover, because the partial derivatives cannot both be 0 unless both x and y are 0, you can conclude that $(0, 0)$ is the only critical point. In Figure 13.68, note that $f(0, 0)$ is 1. For all other (x, y), it is clear that

$$f(x, y) = 1 - (x^2 + y^2)^{1/3} < 1.$$

So, f has a relative *maximum* at $(0, 0)$.

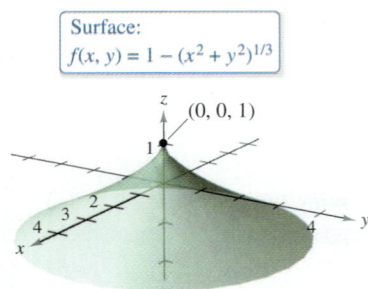

Surface:
$f(x, y) = 1 - (x^2 + y^2)^{1/3}$

$f_x(x, y)$ and $f_y(x, y)$ are undefined at $(0, 0)$.
Figure 13.68

In Example 2, $f_x(x, y) = 0$ for every point on the y-axis other than $(0, 0)$. However, because $f_y(x, y)$ is nonzero, these are not critical points. Remember that *one* of the partials must not exist or *both* must be 0 in order to yield a critical point.

The Second Partials Test

Theorem 13.16 tells you that to find relative extrema, you need only examine values of $f(x, y)$ at critical points. However, as is true for a function of one variable, the critical points of a function of two variables do not always yield relative maxima or minima. Some critical points yield **saddle points,** which are neither relative maxima nor relative minima.

As an example of a critical point that does not yield a relative extremum, consider the hyperbolic paraboloid

$$f(x, y) = y^2 - x^2$$

as shown in Figure 13.69. At the point $(0, 0)$, both partial derivatives

$$f_x(x, y) = -2x \quad \text{and} \quad f_y(x, y) = 2y$$

are 0. The function f does not, however, have a relative extremum at this point because in any open disk centered at $(0, 0)$, the function takes on both negative values (along the x-axis) *and* positive values (along the y-axis). So, the point $(0, 0, 0)$ is a saddle point of the surface. (The term "saddle point" comes from the fact that surfaces such as the one shown in Figure 13.69 resemble saddles.)

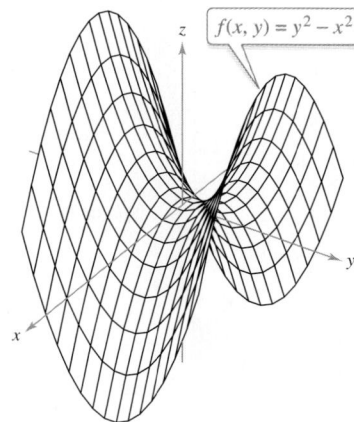

$f(x, y) = y^2 - x^2$

Saddle point at $(0, 0, 0)$:
$f_x(0, 0) = f_y(0, 0) = 0$
Figure 13.69

For the functions in Examples 1 and 2, it was relatively easy to determine the relative extrema, because each function was either given, or able to be written, in completed square form. For more complicated functions, algebraic arguments are less convenient and it is better to rely on the analytic means presented in the following Second Partials Test. This is the two-variable counterpart of the Second Derivative Test for functions of one variable. The proof of this theorem is best left to a course in advanced calculus.

THEOREM 13.17   Second Partials Test

Let f have continuous second partial derivatives on an open region containing a point (a, b) for which

$$f_x(a, b) = 0 \quad \text{and} \quad f_y(a, b) = 0.$$

To test for relative extrema of f, consider the quantity

$$d = f_{xx}(a, b)f_{yy}(a, b) - [f_{xy}(a, b)]^2.$$

1. If $d > 0$ and $f_{xx}(a, b) > 0$, then f has a **relative minimum** at (a, b).
2. If $d > 0$ and $f_{xx}(a, b) < 0$, then f has a **relative maximum** at (a, b).
3. If $d < 0$, then $(a, b, f(a, b))$ is a **saddle point.**
4. The test is inconclusive if $d = 0$.

• • REMARK If $d > 0$, then $f_{xx}(a, b)$ and $f_{yy}(a, b)$ must have the same sign. This means that $f_{xx}(a, b)$ can be replaced by $f_{yy}(a, b)$ in the first two parts of the test.

A convenient device for remembering the formula for d in the Second Partials Test is given by the 2×2 determinant

$$d = \begin{vmatrix} f_{xx}(a, b) & f_{xy}(a, b) \\ f_{yx}(a, b) & f_{yy}(a, b) \end{vmatrix}$$

where $f_{xy}(a, b) = f_{yx}(a, b)$ by Theorem 13.3.

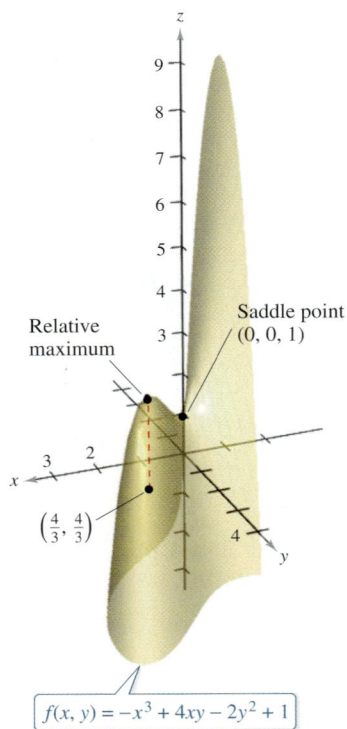

Relative maximum

Saddle point
$(0, 0, 1)$

$\left(\frac{4}{3}, \frac{4}{3}\right)$

$f(x, y) = -x^3 + 4xy - 2y^2 + 1$

Figure 13.70

EXAMPLE 3   **Using the Second Partials Test**

Find the relative extrema of $f(x, y) = -x^3 + 4xy - 2y^2 + 1$.

Solution   Begin by finding the critical points of f. Because

$$f_x(x, y) = -3x^2 + 4y \quad \text{and} \quad f_y(x, y) = 4x - 4y$$

exist for all x and y, the only critical points are those for which both first partial derivatives are 0. To locate these points, set $f_x(x, y)$ and $f_y(x, y)$ equal to 0 to obtain

$$-3x^2 + 4y = 0 \quad \text{and} \quad 4x - 4y = 0.$$

From the second equation, you know that $x = y$, and, by substitution into the first equation, you obtain two solutions: $y = x = 0$ and $y = x = \frac{4}{3}$. Because

$$f_{xx}(x, y) = -6x, \quad f_{yy}(x, y) = -4, \quad \text{and} \quad f_{xy}(x, y) = 4$$

it follows that, for the critical point $(0, 0)$,

$$d = f_{xx}(0, 0)f_{yy}(0, 0) - [f_{xy}(0, 0)]^2 = 0 - 16 < 0$$

and, by the Second Partials Test, you can conclude that $(0, 0, 1)$ is a saddle point of f. Furthermore, for the critical point $\left(\frac{4}{3}, \frac{4}{3}\right)$,

$$d = f_{xx}\left(\frac{4}{3}, \frac{4}{3}\right)f_{yy}\left(\frac{4}{3}, \frac{4}{3}\right) - \left[f_{xy}\left(\frac{4}{3}, \frac{4}{3}\right)\right]^2$$
$$= -8(-4) - 16$$
$$= 16$$
$$> 0$$

and because $f_{xx}\left(\frac{4}{3}, \frac{4}{3}\right) = -8 < 0$, you can conclude that f has a relative maximum at $\left(\frac{4}{3}, \frac{4}{3}\right)$, as shown in Figure 13.70.

The Second Partials Test can fail to find relative extrema in two ways. If either of the first partial derivatives does not exist, you cannot use the test. Also, if

$$d = f_{xx}(a, b)f_{yy}(a, b) - [f_{xy}(a, b)]^2 = 0$$

the test fails. In such cases, you can try a sketch or some other approach, as demonstrated in the next example.

EXAMPLE 4   **Failure of the Second Partials Test**

Find the relative extrema of $f(x, y) = x^2y^2$.

Solution   Because $f_x(x, y) = 2xy^2$ and $f_y(x, y) = 2x^2y$, you know that both partial derivatives are 0 when $x = 0$ or $y = 0$. That is, every point along the x- or y-axis is a critical point. Moreover, because

$$f_{xx}(x, y) = 2y^2, \quad f_{yy}(x, y) = 2x^2, \quad \text{and} \quad f_{xy}(x, y) = 4xy$$

you know that

$$d = f_{xx}(x, y)f_{yy}(x, y) - [f_{xy}(x, y)]^2$$
$$= 4x^2y^2 - 16x^2y^2$$
$$= -12x^2y^2$$

which is 0 when either $x = 0$ or $y = 0$. So, the Second Partials Test fails. However, because $f(x, y) = 0$ for every point along the x- or y-axis and $f(x, y) = x^2y^2 > 0$ for all other points, you can conclude that each of these critical points yields an absolute minimum, as shown in Figure 13.71.

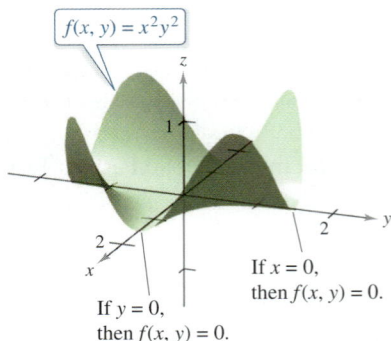

$f(x, y) = x^2y^2$

If $x = 0$, then $f(x, y) = 0$.

If $y = 0$, then $f(x, y) = 0$.

Figure 13.71

Absolute extrema of a function can occur in two ways. First, some relative extrema also happen to be absolute extrema. For instance, in Example 1, $f(-2, 3)$ is an absolute minimum of the function. (On the other hand, the relative maximum found in Example 3 is not an absolute maximum of the function.) Second, absolute extrema can occur at a boundary point of the domain. This is illustrated in Example 5.

> **EXAMPLE 5**   **Finding Absolute Extrema**

Find the absolute extrema of the function

$$f(x, y) = \sin xy$$

on the closed region given by

$$0 \leq x \leq \pi \quad \text{and} \quad 0 \leq y \leq 1.$$

Solution   From the partial derivatives

$$f_x(x, y) = y \cos xy \quad \text{and} \quad f_y(x, y) = x \cos xy$$

you can see that each point lying on the hyperbola $xy = \pi/2$ is a critical point. These points each yield the value

$$f(x, y) = \sin \frac{\pi}{2} = 1$$

which you know is the absolute maximum, as shown in Figure 13.72. The only other critical point of *f lying in the given region* is $(0, 0)$. It yields an absolute minimum of 0, because

$$0 \leq xy \leq \pi$$

implies that

$$0 \leq \sin xy \leq 1.$$

To locate other absolute extrema, you should consider the four boundaries of the region formed by taking traces with the vertical planes $x = 0$, $x = \pi$, $y = 0$, and $y = 1$. In doing this, you will find that $\sin xy = 0$ at all points on the x-axis, at all points on the y-axis, and at the point $(\pi, 1)$. Each of these points yields an absolute minimum for the surface, as shown in Figure 13.72.

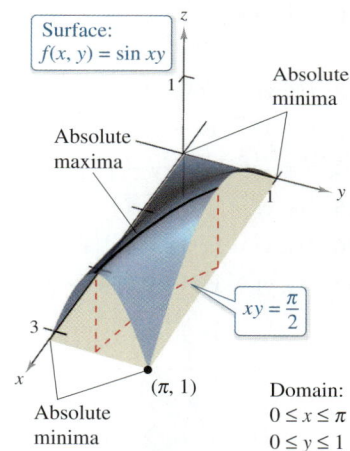

Figure 13.72

The concepts of relative extrema and critical points can be extended to functions of three or more variables. When all first partial derivatives of

$$w = f(x_1, x_2, x_3, \ldots, x_n)$$

exist, it can be shown that a relative maximum or minimum can occur at $(x_1, x_2, x_3, \ldots, x_n)$ only when every first partial derivative is 0 at that point. This means that the critical points are obtained by solving the following system of equations.

$$f_{x_1}(x_1, x_2, x_3, \ldots, x_n) = 0$$
$$f_{x_2}(x_1, x_2, x_3, \ldots, x_n) = 0$$
$$\vdots$$
$$f_{x_n}(x_1, x_2, x_3, \ldots, x_n) = 0$$

The extension of Theorem 13.17 to three or more variables is also possible, although you will not consider such an extension in this text.

13.8 Exercises

See **CalcChat.com** for tutorial help and worked-out solutions to odd-numbered exercises.

Finding Relative Extrema In Exercises 1–6, identify any extrema of the function by recognizing its given form or its form after completing the square. Verify your results by using the partial derivatives to locate any critical points and test for relative extrema.

1. $g(x, y) = (x - 1)^2 + (y - 3)^2$

2. $g(x, y) = 5 - (x - 3)^2 - (y + 2)^2$

3. $f(x, y) = \sqrt{x^2 + y^2 + 1}$

4. $f(x, y) = \sqrt{25 - (x - 2)^2 - y^2}$

5. $f(x, y) = x^2 + y^2 + 2x - 6y + 6$

6. $f(x, y) = -x^2 - y^2 + 10x + 12y - 64$

Using the Second Partials Test In Exercises 7–20, examine the function for relative extrema and saddle points.

7. $h(x, y) = 80x + 80y - x^2 - y^2$

8. $g(x, y) = x^2 - y^2 - x - y$

9. $g(x, y) = xy$

10. $h(x, y) = x^2 - 3xy - y^2$

11. $f(x, y) = -3x^2 - 2y^2 + 3x - 4y + 5$

12. $f(x, y) = 2x^2 + 2xy + y^2 + 2x - 3$

13. $z = x^2 + xy + \frac{1}{2}y^2 - 2x + y$

14. $z = -5x^2 + 4xy - y^2 + 16x + 10$

15. $f(x, y) = \sqrt{x^2 + y^2}$

16. $h(x, y) = (x^2 + y^2)^{1/3} + 2$

17. $f(x, y) = x^2 - xy - y^2 - 3x - y$

18. $f(x, y) = 2xy - \frac{1}{2}(x^4 + y^4) + 1$

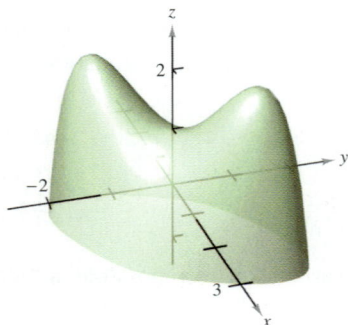

19. $z = e^{-x} \sin y$

20. $z = \left(\frac{1}{2} - x^2 + y^2\right)e^{1 - x^2 - y^2}$

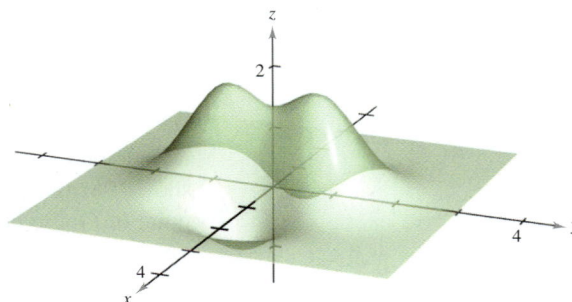

Finding Relative Extrema and Saddle Points Using Technology In Exercises 21–24, use a computer algebra system to graph the surface and locate any relative extrema and saddle points.

21. $z = \dfrac{-4x}{x^2 + y^2 + 1}$

22. $f(x, y) = y^3 - 3yx^2 - 3y^2 - 3x^2 + 1$

23. $z = (x^2 + 4y^2)e^{1 - x^2 - y^2}$

24. $z = e^{xy}$

Finding Relative Extrema In Exercises 25 and 26, examine the function for extrema without using the derivative tests, and use a computer algebra system to graph the surface. (*Hint:* By observation, determine if it is possible for z to be negative. When is z equal to 0?)

25. $z = \dfrac{(x - y)^4}{x^2 + y^2}$

26. $z = \dfrac{(x^2 - y^2)^2}{x^2 + y^2}$

Think About It In Exercises 27–30, determine whether there is a relative maximum, a relative minimum, a saddle point, or insufficient information to determine the nature of the function $f(x, y)$ at the critical point (x_0, y_0).

27. $f_{xx}(x_0, y_0) = 9, \quad f_{yy}(x_0, y_0) = 4, \quad f_{xy}(x_0, y_0) = 6$

28. $f_{xx}(x_0, y_0) = -3, \quad f_{yy}(x_0, y_0) = -8, \quad f_{xy}(x_0, y_0) = 2$

29. $f_{xx}(x_0, y_0) = -9, \quad f_{yy}(x_0, y_0) = 6, \quad f_{xy}(x_0, y_0) = 10$

30. $f_{xx}(x_0, y_0) = 25, \quad f_{yy}(x_0, y_0) = 8, \quad f_{xy}(x_0, y_0) = 10$

31. Using the Second Partials Test A function f has continuous second partial derivatives on an open region containing the critical point $(3, 7)$. The function has a minimum at $(3, 7)$, and $d > 0$ for the Second Partials Test. Determine the interval for $f_{xy}(3, 7)$ when $f_{xx}(3, 7) = 2$ and $f_{yy}(3, 7) = 8$.

32. Using the Second Partials Test A function f has continuous second partial derivatives on an open region containing the critical point (a, b). If $f_{xx}(a, b)$ and $f_{yy}(a, b)$ have opposite signs, what is implied? Explain.

Finding Relative Extrema and Saddle Points In Exercises 33–38, (a) find the critical points, (b) test for relative extrema, (c) list the critical points for which the Second Partials Test fails, and (d) use a computer algebra system to graph the function, labeling any extrema and saddle points.

33. $f(x, y) = x^3 + y^3$

34. $f(x, y) = x^3 + y^3 - 6x^2 + 9y^2 + 12x + 27y + 19$

35. $f(x, y) = (x - 1)^2(y + 4)^2$

36. $f(x, y) = \sqrt{(x - 1)^2 + (y + 2)^2}$

37. $f(x, y) = x^{2/3} + y^{2/3}$

38. $f(x, y) = (x^2 + y^2)^{2/3}$

Examining a Function In Exercises 39 and 40, find the critical points of the function and, from the form of the function, determine whether a relative maximum or a relative minimum occurs at each point.

39. $f(x, y, z) = x^2 + (y - 3)^2 + (z + 1)^2$

40. $f(x, y, z) = 9 - [x(y - 1)(z + 2)]^2$

Finding Absolute Extrema In Exercises 41–48, find the absolute extrema of the function over the region R. (In each case, R contains the boundaries.) Use a computer algebra system to confirm your results.

41. $f(x, y) = x^2 - 4xy + 5$

$R = \{(x, y): 1 \le x \le 4, \ 0 \le y \le 2\}$

42. $f(x, y) = x^2 + xy, \quad R = \{(x, y): |x| \le 2, \ |y| \le 1\}$

43. $f(x, y) = 12 - 3x - 2y$

R: The triangular region in the xy-plane with vertices $(2, 0)$, $(0, 1)$, and $(1, 2)$

44. $f(x, y) = (2x - y)^2$

R: The triangular region in the xy-plane with vertices $(2, 0)$, $(0, 1)$, and $(1, 2)$

45. $f(x, y) = 3x^2 + 2y^2 - 4y$

R: The region in the xy-plane bounded by the graphs of $y = x^2$ and $y = 4$

46. $f(x, y) = 2x - 2xy + y^2$

R: The region in the xy-plane bounded by the graphs of $y = x^2$ and $y = 1$

47. $f(x, y) = x^2 + 2xy + y^2, \quad R = \{(x, y): |x| \le 2, \ |y| \le 1\}$

48. $f(x, y) = \dfrac{4xy}{(x^2 + 1)(y^2 + 1)}$

$R = \{(x, y): 0 \le x \le 1, \ 0 \le y \le 1\}$

WRITING ABOUT CONCEPTS

49. Defining Terms Define each of the following for a function of two variables.

(a) Relative minimum   (b) Relative maximum

(c) Critical point   (d) Saddle point

Sketching a Graph In Exercises 50–52, sketch the graph of an arbitrary function f satisfying the given conditions. State whether the function has any extrema or saddle points. (There are many correct answers.)

50. All of the first and second partial derivatives of f are 0.

51. $f_x(x, y) > 0$ and $f_y(x, y) < 0$ for all (x, y).

52. $f_x(0, 0) = 0, \ f_y(0, 0) = 0$

$$f_x(x, y) \begin{cases} < 0, & x < 0 \\ > 0, & x > 0 \end{cases}, \quad f_y(x, y) \begin{cases} > 0, & y < 0 \\ < 0, & y > 0 \end{cases}$$

$f_{xx}(x, y) > 0, f_{yy}(x, y) < 0,$ and $f_{xy}(x, y) = 0$ for all (x, y).

53. Comparing Functions Consider the functions

$$f(x, y) = x^2 - y^2 \quad \text{and} \quad g(x, y) = x^2 + y^2.$$

(a) Show that both functions have a critical point at $(0, 0)$.

(b) Explain how f and g behave differently at this critical point.

54. HOW DO YOU SEE IT? The figure shows the level curves for an unknown function $f(x, y)$. What, if any, information can be given about f at the points A, B, C, and D? Explain your reasoning.

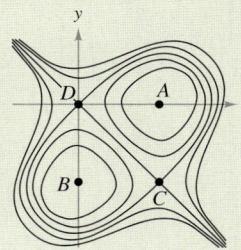

True or False? In Exercises 55–58, determine whether the statement is true or false. If it is false, explain why or give an example that shows it is false.

55. If f has a relative maximum at (x_0, y_0, z_0), then

$$f_x(x_0, y_0) = f_y(x_0, y_0) = 0.$$

56. If $f_x(x_0, y_0) = f_y(x_0, y_0) = 0$, then f has a relative maximum at (x_0, y_0, z_0).

57. Between any two relative minima of f, there must be at least one relative maximum of f.

58. If f is continuous for all x and y and has two relative minima, then f must have at least one relative maximum.

13.9 Applications of Extrema

■ Solve optimization problems involving functions of several variables.
■ Use the method of least squares.

Applied Optimization Problems

In this section, you will survey a few of the many applications of extrema of functions of two (or more) variables.

EXAMPLE 1   **Finding Maximum Volume**

⋯▷ *See LarsonCalculus.com for an interactive version of this type of example.*

A rectangular box is resting on the xy-plane with one vertex at the origin. The opposite vertex lies in the plane

$$6x + 4y + 3z = 24$$

as shown in Figure 13.73. Find the maximum volume of such a box.

Solution   Let x, y, and z represent the length, width, and height of the box. Because one vertex of the box lies in the plane $6x + 4y + 3z = 24$, you know that $z = \frac{1}{3}(24 - 6x - 4y)$, and you can write the volume xyz of the box as a function of two variables.

$$V(x, y) = (x)(y)\left[\tfrac{1}{3}(24 - 6x - 4y)\right]$$
$$= \tfrac{1}{3}(24xy - 6x^2y - 4xy^2)$$

Next, find the first partial derivatives of V.

$$V_x(x, y) = \frac{1}{3}(24y - 12xy - 4y^2) = \frac{y}{3}(24 - 12x - 4y)$$

$$V_y(x, y) = \frac{1}{3}(24x - 6x^2 - 8xy) = \frac{x}{3}(24 - 6x - 8y)$$

Note that the first partial derivatives are defined for all x and y. So, by setting $V_x(x, y)$ and $V_y(x, y)$ equal to 0 and solving the equations $\frac{1}{3}y(24 - 12x - 4y) = 0$ and $\frac{1}{3}x(24 - 6x - 8y) = 0$, you obtain the critical points $(0, 0)$ and $\left(\frac{4}{3}, 2\right)$. At $(0, 0)$, the volume is 0, so that point does not yield a maximum volume. At the point $\left(\frac{4}{3}, 2\right)$, you can apply the Second Partials Test.

$$V_{xx}(x, y) = -4y$$

$$V_{yy}(x, y) = \frac{-8x}{3}$$

$$V_{xy}(x, y) = \frac{1}{3}(24 - 12x - 8y)$$

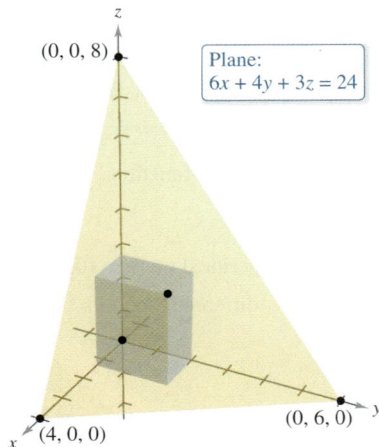

(0, 0, 8)

Plane:
$6x + 4y + 3z = 24$

(0, 6, 0)

(4, 0, 0)

Figure 13.73

Because

$$V_{xx}\left(\tfrac{4}{3}, 2\right)V_{yy}\left(\tfrac{4}{3}, 2\right) - \left[V_{xy}\left(\tfrac{4}{3}, 2\right)\right]^2 = (-8)\left(-\tfrac{32}{9}\right) - \left(-\tfrac{8}{3}\right)^2 = \tfrac{64}{3} > 0$$

and

$$V_{xx}\left(\tfrac{4}{3}, 2\right) = -8 < 0$$

you can conclude from the Second Partials Test that the maximum volume is

$$V\left(\tfrac{4}{3}, 2\right) = \tfrac{1}{3}\left[24\left(\tfrac{4}{3}\right)(2) - 6\left(\tfrac{4}{3}\right)^2(2) - 4\left(\tfrac{4}{3}\right)(2^2)\right] = \tfrac{64}{9} \text{ cubic units.}$$

Note that the volume is 0 at the boundary points of the triangular domain of V.   ■

⋯**REMARK** In many applied problems, the domain of the function to be optimized is a closed bounded region. To find minimum or maximum points, you must not only test critical points, but also consider the values of the function at points on the boundary.

Applications of extrema in economics and business often involve more than one independent variable. For instance, a company may produce several models of one type of product. The price per unit and profit per unit are usually different for each model. Moreover, the demand for each model is often a function of the prices of the other models (as well as its own price). The next example illustrates an application involving two products.

EXAMPLE 2   **Finding the Maximum Profit**

An electronics manufacturer determines that the profit P (in dollars) obtained by producing and selling x units of an LCD television and y units of a plasma television is approximated by the model

$$P(x, y) = 8x + 10y - (0.001)(x^2 + xy + y^2) - 10,000.$$

Find the production level that produces a maximum profit. What is the maximum profit?

Solution   The partial derivatives of the profit function are

$$P_x(x, y) = 8 - (0.001)(2x + y)$$

and

$$P_y(x, y) = 10 - (0.001)(x + 2y).$$

By setting these partial derivatives equal to 0, you obtain the following system of equations.

$$8 - (0.001)(2x + y) = 0$$
$$10 - (0.001)(x + 2y) = 0$$

After simplifying, this system of linear equations can be written as

$$2x + y = 8000$$
$$x + 2y = 10,000.$$

Solving this system produces $x = 2000$ and $y = 4000$. The second partial derivatives of P are

$$P_{xx}(2000, 4000) = -0.002$$
$$P_{yy}(2000, 4000) = -0.002$$
$$P_{xy}(2000, 4000) = -0.001.$$

Because $P_{xx} < 0$ and

$$P_{xx}(2000, 4000)P_{yy}(2000, 4000) - [P_{xy}(2000, 4000)]^2 = (-0.002)^2 - (-0.001)^2$$

is greater than 0, you can conclude that the production level of $x = 2000$ units and $y = 4000$ units yields a *maximum* profit. The maximum profit is

$$P(2000, 4000)$$
$$= 8(2000) + 10(4000) - (0.001)[2000^2 + 2000(4000) + (4000^2)] - 10,000$$
$$= \$18,000. \qquad \blacksquare$$

In Example 2, it was assumed that the manufacturing plant is able to produce the required number of units to yield a maximum profit. In actual practice, the production would be bounded by physical constraints. You will study such constrained optimization problems in the next section.

■ **FOR FURTHER INFORMATION**   For more information on the use of mathematics in economics, see the article "Mathematical Methods of Economics" by Joel Franklin in *The American Mathematical Monthly*. To view this article, go to *MathArticles.com*.

The Method of Least Squares

Many of the examples in this text have involved **mathematical models.** For instance, Example 2 involves a quadratic model for profit. There are several ways to develop such models; one is called the **method of least squares.**

In constructing a model to represent a particular phenomenon, the goals are simplicity and accuracy. Of course, these goals often conflict. For instance, a simple linear model for the points in Figure 13.74 is

$$y = 1.9x - 5.$$

However, Figure 13.75 shows that by choosing the slightly more complicated quadratic model

$$y = 0.20x^2 - 0.7x + 1$$

you can achieve greater accuracy.

Figure 13.74

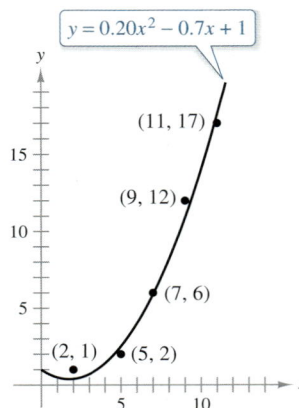

Figure 13.75

As a measure of how well the model $y = f(x)$ fits the collection of points

$$\{(x_1, y_1), (x_2, y_2), (x_3, y_3), \ldots, (x_n, y_n)\}$$

you can add the squares of the differences between the actual y-values and the values given by the model to obtain the **sum of the squared errors**

$$S = \sum_{i=1}^{n} [f(x_i) - y_i]^2. \qquad \text{Sum of the squared errors}$$

Graphically, S can be interpreted as the sum of the squares of the vertical distances between the graph of f and the given points in the plane, as shown in Figure 13.76. If the model is perfect, then $S = 0$. However, when perfection is not feasible, you can settle for a model that minimizes S. For instance, the sum of the squared errors for the linear model in Figure 13.74 is

$$S = 17.6.$$

Statisticians call the *linear model* that minimizes S the **least squares regression line.** The proof that this line actually minimizes S involves the minimizing of a function of two variables.

• • REMARK A method for finding the least squares regression quadratic for a collection of data is described in Exercise 31.

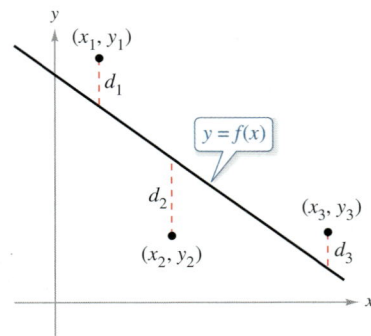

Sum of the squared errors:
$$S = d_1^2 + d_2^2 + d_3^2$$
Figure 13.76

THEOREM 13.18   Least Squares Regression Line

The **least squares regression line** for $\{(x_1, y_1), (x_2, y_2), \ldots, (x_n, y_n)\}$ is given by $f(x) = ax + b$, where

$$a = \frac{n\displaystyle\sum_{i=1}^{n} x_i y_i - \sum_{i=1}^{n} x_i \sum_{i=1}^{n} y_i}{n\displaystyle\sum_{i=1}^{n} x_i^2 - \left(\sum_{i=1}^{n} x_i\right)^2} \quad \text{and} \quad b = \frac{1}{n}\left(\sum_{i=1}^{n} y_i - a\sum_{i=1}^{n} x_i\right).$$

Proof   Let $S(a, b)$ represent the sum of the squared errors for the model

$$f(x) = ax + b$$

and the given set of points. That is,

$$S(a, b) = \sum_{i=1}^{n} [f(x_i) - y_i]^2$$

$$= \sum_{i=1}^{n} (ax_i + b - y_i)^2$$

where the points (x_i, y_i) represent constants. Because S is a function of a and b, you can use the methods discussed in the preceding section to find the minimum value of S. Specifically, the first partial derivatives of S are

$$S_a(a, b) = \sum_{i=1}^{n} 2x_i(ax_i + b - y_i)$$

$$= 2a\sum_{i=1}^{n} x_i^2 + 2b\sum_{i=1}^{n} x_i - 2\sum_{i=1}^{n} x_i y_i$$

and

$$S_b(a, b) = \sum_{i=1}^{n} 2(ax_i + b - y_i)$$

$$= 2a\sum_{i=1}^{n} x_i + 2nb - 2\sum_{i=1}^{n} y_i.$$

By setting these two partial derivatives equal to 0, you obtain the values of a and b that are listed in the theorem. It is left to you to apply the Second Partials Test (see Exercise 41) to verify that these values of a and b yield a minimum.

See LarsonCalculus.com for Bruce Edwards's video of this proof.

If the x-values are symmetrically spaced about the y-axis, then $\Sigma x_i = 0$ and the formulas for a and b simplify to

$$a = \frac{\displaystyle\sum_{i=1}^{n} x_i y_i}{\displaystyle\sum_{i=1}^{n} x_i^2}$$

and

$$b = \frac{1}{n}\sum_{i=1}^{n} y_i.$$

This simplification is often possible with a translation of the x-values. For instance, given that the x-values in a data collection consist of the years 2009, 2010, 2011, 2012, and 2013, you could let 2011 be represented by 0.

EXAMPLE 3 **Finding the Least Squares Regression Line**

Find the least squares regression line for the points

$$(-3, 0), (-1, 1), (0, 2), \text{ and } (2, 3).$$

Solution The table shows the calculations involved in finding the least squares regression line using $n = 4$.

x	y	xy	x^2
-3	0	0	9
-1	1	-1	1
0	2	0	0
2	3	6	4
$\displaystyle\sum_{i=1}^{n} x_i = -2$	$\displaystyle\sum_{i=1}^{n} y_i = 6$	$\displaystyle\sum_{i=1}^{n} x_i y_i = 5$	$\displaystyle\sum_{i=1}^{n} x_i^2 = 14$

▷ **TECHNOLOGY** Many calculators have "built-in" least squares regression programs. If your calculator has such a program, use it to duplicate the results of Example 3.

Applying Theorem 13.18 produces

$$a = \frac{n\displaystyle\sum_{i=1}^{n} x_i y_i - \displaystyle\sum_{i=1}^{n} x_i \displaystyle\sum_{i=1}^{n} y_i}{n\displaystyle\sum_{i=1}^{n} x_i^2 - \left(\displaystyle\sum_{i=1}^{n} x_i\right)^2}$$

$$= \frac{4(5) - (-2)(6)}{4(14) - (-2)^2}$$

$$= \frac{8}{13}$$

and

$$b = \frac{1}{n}\left(\displaystyle\sum_{i=1}^{n} y_i - a\displaystyle\sum_{i=1}^{n} x_i\right)$$

$$= \frac{1}{4}\left[6 - \frac{8}{13}(-2)\right]$$

$$= \frac{47}{26}.$$

The least squares regression line is

$$f(x) = \frac{8}{13}x + \frac{47}{26}$$

as shown in Figure 13.77.

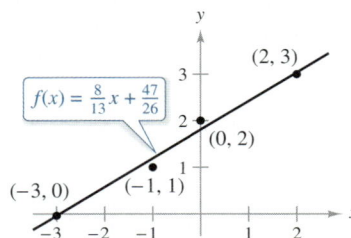

Least squares regression line
Figure 13.77

13.9 Exercises

Finding Minimum Distance   In Exercises 1 and 2, find the minimum distance from the point to the plane $x - y + z = 3$. (*Hint:* To simplify the computations, minimize the square of the distance.)

1. $(0, 0, 0)$ **2.** $(1, 2, 3)$

Finding Minimum Distance   In Exercises 3 and 4, find the minimum distance from the point to the surface $z = \sqrt{1 - 2x - 2y}$. (*Hint:* To simplify the computations, minimize the square of the distance.)

3. $(-2, -2, 0)$ **4.** $(-4, 1, 0)$

Finding Positive Numbers   In Exercises 5–8, find three positive integers $x, y,$ and z that satisfy the given conditions.

5. The product is 27, and the sum is a minimum.

6. The sum is 32, and $P = xy^2z$ is a maximum.

7. The sum is 30, and the sum of the squares is a minimum.

8. The product is 1, and the sum of the squares is a minimum.

9. Cost   A home improvement contractor is painting the walls and ceiling of a rectangular room. The volume of the room is 668.25 cubic feet. The cost of wall paint is \$0.06 per square foot and the cost of ceiling paint is \$0.11 per square foot. Find the room dimensions that result in a minimum cost for the paint. What is the minimum cost for the paint?

10. Maximum Volume   The material for constructing the base of an open box costs 1.5 times as much per unit area as the material for constructing the sides. For a fixed amount of money C, find the dimensions of the box of largest volume that can be made.

11. Volume and Surface Area   Show that a rectangular box of given volume and minimum surface area is a cube.

12. Maximum Volume   Show that the rectangular box of maximum volume inscribed in a sphere of radius r is a cube.

13. Maximum Revenue   A company manufactures running shoes and basketball shoes. The total revenue from x_1 units of running shoes and x_2 units of basketball shoes is

$$R = -5x_1^2 - 8x_2^2 - 2x_1x_2 + 42x_1 + 102x_2$$

where x_1 and x_2 are in thousands of units. Find x_1 and x_2 so as to maximize the revenue.

14. Maximum Profit   A corporation manufactures candles at two locations. The cost of producing x_1 units at location 1 is

$$C_1 = 0.02x_1^2 + 4x_1 + 500$$

and the cost of producing x_2 units at location 2 is

$$C_2 = 0.05x_2^2 + 4x_2 + 275.$$

The candles sell for \$15 per unit. Find the quantity that should be produced at each location to maximize the profit $P = 15(x_1 + x_2) - C_1 - C_2$.

Sashkin/Shutterstock.com

15. Hardy-Weinberg Law

Common blood types are determined genetically by three alleles A, B, and O. (An allele is any of a group of possible mutational forms of a gene.) A person whose blood type is AA, BB, or OO is homozygous. A person whose blood type is AB, AO, or BO is heterozygous. The Hardy-Weinberg Law states that the proportion P of heterozygous individuals in any given population is

$$P(p, q, r) = 2pq + 2pr + 2qr$$

where p represents the percent of allele A in the population, q represents the percent of allele B in the population, and r represents the percent of allele O in the population. Use the fact that

$$p + q + r = 1$$

to show that the maximum proportion of heterozygous individuals in any population is $\frac{2}{3}$.

16. Shannon Diversity Index   One way to measure species diversity is to use the Shannon diversity index H. If a habitat consists of three species, A, B, and C, then its Shannon diversity index is

$$H = -x \ln x - y \ln y - z \ln z$$

where x is the percent of species A in the habitat, y is the percent of species B in the habitat, and z is the percent of species C in the habitat. Use the fact that

$$x + y + z = 1$$

to show that the maximum value of H occurs when $x = y = z = \frac{1}{3}$. What is the maximum value of H?

17. Minimum Cost   A water line is to be built from point P to point S and must pass through regions where construction costs differ (see figure). The cost per kilometer (in dollars) is $3k$ from P to Q, $2k$ from Q to R, and k from R to S. Find x and y such that the total cost C will be minimized.

Figure for 17   Figure for 18

18. Area   A trough with trapezoidal cross sections is formed by turning up the edges of a 30-inch-wide sheet of aluminum (see figure). Find the cross section of maximum area.

19. Applied Optimization Problems In your own words, state the problem-solving strategy for applied minimum and maximum problems.

20. Method of Least Squares In your own words, describe the method of least squares for finding mathematical models.

Finding the Least Squares Regression Line In Exercises 21–24, (a) find the least squares regression line, and (b) calculate S, the sum of the squared errors. Use the regression capabilities of a graphing utility to verify your results.

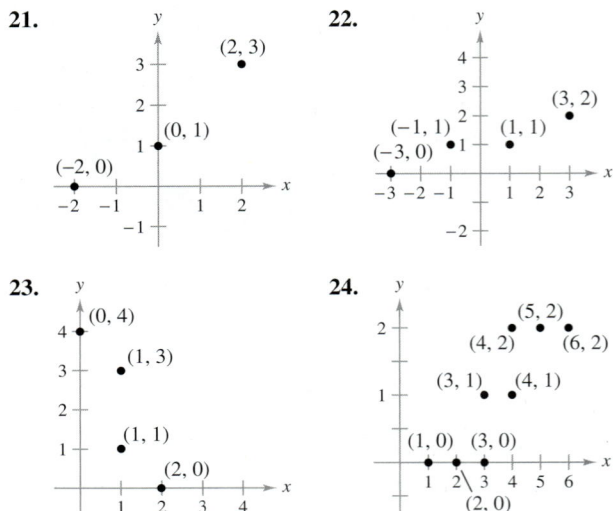

21.

22.

23.

24.

Finding the Least Squares Regression Line In Exercises 25–28, find the least squares regression line for the points. Use the regression capabilities of a graphing utility to verify your results. Use the graphing utility to plot the points and graph the regression line.

25. (0, 0), (1, 1), (3, 4), (4, 2), (5, 5)

26. (1, 0), (3, 3), (5, 6)

27. (0, 6), (4, 3), (5, 0), (8, −4), (10, −5)

28. (6, 4), (1, 2), (3, 3), (8, 6), (11, 8), (13, 8)

29. Modeling Data The ages x (in years) and systolic blood pressures y of seven men are shown in the table.

Age, x	16	25	39	45	49	64	70
Systolic Blood Pressure, y	109	122	150	165	159	183	199

(a) Use the regression capabilities of a graphing utility to find the least squares regression line for the data.

(b) Use the model to approximate the change in systolic blood pressure for each one-year increase in age.

30. Modeling Data The table shows the gross income tax collections (in billions of dollars) by the Internal Revenue Service for individuals x and businesses y. (*Source: U.S. Internal Revenue Service*)

Year	1975	1980	1985	1990
Individual, x	156	288	397	540
Business, y	46	72	77	110

Year	1995	2000	2005	2010
Individual, x	676	1137	1108	1164
Business, y	174	236	307	278

(a) Use the regression capabilities of a graphing utility to find the least squares regression line for the data.

(b) Use the model to estimate the business income taxes collected when the individual income taxes collected is $1300 billion.

31. Method of Least Squares Find a system of equations whose solution yields the coefficients a, b, and c for the least squares regression quadratic

$$y = ax^2 + bx + c$$

for the points $(x_1, y_1), (x_2, y_2), \ldots, (x_n, y_n)$ by minimizing the sum

$$S(a, b, c) = \sum_{i=1}^{n} (y_i - ax_i^2 - bx_i - c)^2.$$

32. HOW DO YOU SEE IT? Match the regression equation with the appropriate graph. Explain your reasoning. (Note that the x- and y-axes are broken.)

(a) $y = 0.22x - 7.5$ (b) $y = -0.35x + 11.5$
(c) $y = 0.09x + 19.8$ (d) $y = -1.29x + 89.8$

(i)

(ii)

(iii)

(iv)

Finding the Least Squares Regression Quadratic In Exercises 33–36, use the result of Exercise 31 to find the least squares regression quadratic for the given points. Use the regression capabilities of a graphing utility to confirm your results. Use the graphing utility to plot the points and graph the least squares regression quadratic.

33. $(-2, 0), (-1, 0), (0, 1), (1, 2), (2, 5)$

34. $(-4, 5), (-2, 6), (2, 6), (4, 2)$

35. $(0, 0), (2, 2), (3, 6), (4, 12)$ **36.** $(0, 10), (1, 9), (2, 6), (3, 0)$

37. Modeling Data After a new turbocharger for an automobile engine was developed, the following experimental data were obtained for speed y in miles per hour at two-second time intervals x.

Time, x	0	2	4	6	8	10
Speed, y	0	15	30	50	65	70

(a) Find a least squares regression quadratic for the data. Use a graphing utility to confirm your results.

(b) Use a graphing utility to plot the points and graph the model.

38. Modeling Data The table shows the world populations y (in billions) for five different years. Let $x = 3$ represent the year 2003. (*Source: U.S. Census Bureau, International Data Base*)

Year, x	2003	2005	2007	2009	2011
Population, y	6.3	6.5	6.6	6.8	6.9

(a) Use the regression capabilities of a graphing utility to find the least squares regression line for the data.

(b) Use the regression capabilities of a graphing utility to find the least squares regression quadratic for the data.

(c) Use a graphing utility to plot the data and graph the models.

(d) Use both models to forecast the world population for the year 2020. How do the two models differ as you extrapolate into the future?

39. Modeling Data A meteorologist measures the atmospheric pressure P (in kilograms per square meter) at altitude h (in kilometers). The data are shown below.

Altitude, h	0	5	10	15	20
Pressure, P	10,332	5583	2376	1240	517

(a) Use the regression capabilities of a graphing utility to find a least squares regression line for the points $(h, \ln P)$.

(b) The result in part (a) is an equation of the form $\ln P = ah + b$. Write this logarithmic form in exponential form.

(c) Use a graphing utility to plot the original data and graph the exponential model in part (b).

(d) If your graphing utility can fit logarithmic models to data, then use it to verify the result in part (b).

40. Modeling Data The endpoints of the interval over which distinct vision is possible are called the near point and far point of the eye. With increasing age, these points normally change. The table shows the approximate near points y (in inches) for various ages x (in years). (*Source: Ophthalmology & Physiological Optics*)

Age, x	16	32	44	50	60
Near Point, y	3.0	4.7	9.8	19.7	39.4

(a) Find a rational model for the data by taking the reciprocals of the near points to generate the points $(x, 1/y)$. Use the regression capabilities of a graphing utility to find a least squares regression line for the revised data. The resulting line has the form $1/y = ax + b$. Solve for y.

(b) Use a graphing utility to plot the data and graph the model.

(c) Do you think the model can be used to predict the near point for a person who is 70 years old? Explain.

41. Using the Second Partials Test Use the Second Partials Test to verify that the formulas for a and b given in Theorem 13.18 yield a minimum.

$$\left[\text{Hint: Use the fact that } n \sum_{i=1}^{n} x_i^2 \geq \left(\sum_{i=1}^{n} x_i \right)^2. \right]$$

SECTION PROJECT

Building a Pipeline

An oil company wishes to construct a pipeline from its offshore facility A to its refinery B. The offshore facility is 2 miles from shore, and the refinery is 1 mile inland. Furthermore, A and B are 5 miles apart, as shown in the figure.

The cost of building the pipeline is \$3 million per mile in the water and \$4 million per mile on land. So, the cost of the pipeline depends on the location of point P, where it meets the shore. What would be the most economical route of the pipeline?

Imagine that you are to write a report to the oil company about this problem. Let x be the distance shown in the figure. Determine the cost of building the pipeline from A to P, and the cost of building it from P to B. Analyze some sample pipeline routes and their corresponding costs. For instance, what is the cost of the most direct route? Then use calculus to determine the route of the pipeline that minimizes the cost. Explain all steps of your development and include any relevant graphs.

13.10 Lagrange Multipliers

■ Understand the Method of Lagrange Multipliers.
■ Use Lagrange multipliers to solve constrained optimization problems.
■ Use the Method of Lagrange Multipliers with two constraints.

Lagrange Multipliers

> **LAGRANGE MULTIPLIERS**
>
> The Method of Lagrange Multipliers is named after the French mathematician Joseph-Louis Lagrange. Lagrange first introduced the method in his famous paper on mechanics, written when he was just 19 years old.

Many optimization problems have restrictions, or **constraints,** on the values that can be used to produce the optimal solution. Such constraints tend to complicate optimization problems because the optimal solution can occur at a boundary point of the domain. In this section, you will study an ingenious technique for solving such problems. It is called the **Method of Lagrange Multipliers.**

To see how this technique works, consider the problem of finding the rectangle of maximum area that can be inscribed in the ellipse

$$\frac{x^2}{3^2} + \frac{y^2}{4^2} = 1.$$

Let (x, y) be the vertex of the rectangle in the first quadrant, as shown in Figure 13.78. Because the rectangle has sides of lengths $2x$ and $2y$, its area is given by

$$f(x, y) = 4xy. \qquad \text{\textcolor{red}{Objective function}}$$

You want to find x and y such that $f(x, y)$ is a maximum. Your choice of (x, y) is restricted to first-quadrant points that lie on the ellipse

$$\frac{x^2}{3^2} + \frac{y^2}{4^2} = 1. \qquad \text{\textcolor{red}{Constraint}}$$

Now, consider the constraint equation to be a fixed level curve of

$$g(x, y) = \frac{x^2}{3^2} + \frac{y^2}{4^2}.$$

The level curves of f represent a family of hyperbolas

$$f(x, y) = 4xy = k.$$

In this family, the level curves that meet the constraint correspond to the hyperbolas that intersect the ellipse. Moreover, to maximize $f(x, y)$, you want to find the hyperbola that just barely satisfies the constraint. The level curve that does this is the one that is *tangent* to the ellipse, as shown in Figure 13.79.

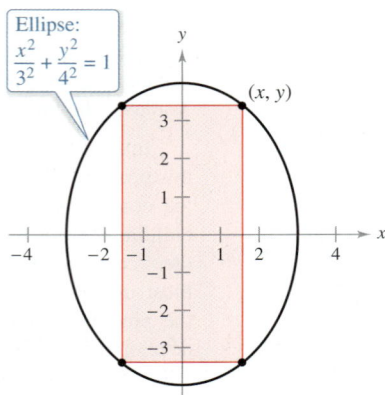

Objective function: $f(x, y) = 4xy$

Figure 13.78

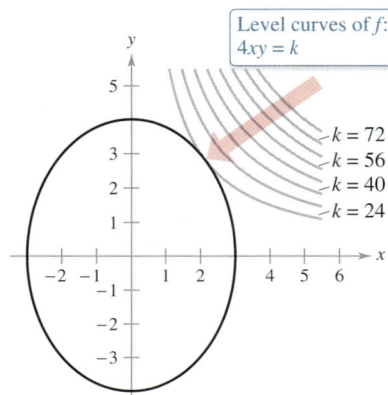

Constraint: $g(x, y) = \dfrac{x^2}{3^2} + \dfrac{y^2}{4^2} = 1$

Figure 13.79

To find the appropriate hyperbola, use the fact that two curves are tangent at a point if and only if their gradient vectors are parallel. This means that $\nabla f(x, y)$ must be a scalar multiple of $\nabla g(x, y)$ at the point of tangency. In the context of constrained optimization problems, this scalar is denoted by λ (the lowercase Greek letter lambda).

$$\nabla f(x, y) = \lambda \nabla g(x, y)$$

The scalar λ is called a **Lagrange multiplier.** Theorem 13.19 gives the necessary conditions for the existence of such multipliers.

· · · · · · · · · · · · · · · · ·▷

· · REMARK Lagrange's Theorem can be shown to be true for functions of three variables, using a similar argument with level surfaces and Theorem 13.14.

THEOREM 13.19   Lagrange's Theorem

Let f and g have continuous first partial derivatives such that f has an extremum at a point (x_0, y_0) on the smooth constraint curve $g(x, y) = c$. If $\nabla g(x_0, y_0) \neq \mathbf{0}$, then there is a real number λ such that

$$\nabla f(x_0, y_0) = \lambda \nabla g(x_0, y_0).$$

Proof   To begin, represent the smooth curve given by $g(x, y) = c$ by the vector-valued function

$$\mathbf{r}(t) = x(t)\mathbf{i} + y(t)\mathbf{j}, \quad \mathbf{r}'(t) \neq \mathbf{0}$$

where x' and y' are continuous on an open interval I. Define the function h as $h(t) = f(x(t), y(t))$. Then, because $f(x_0, y_0)$ is an extreme value of f, you know that

$$h(t_0) = f(x(t_0), y(t_0)) = f(x_0, y_0)$$

is an extreme value of h. This implies that $h'(t_0) = 0$, and, by the Chain Rule,

$$h'(t_0) = f_x(x_0, y_0)x'(t_0) + f_y(x_0, y_0)y'(t_0) = \nabla f(x_0, y_0) \cdot \mathbf{r}'(t_0) = 0.$$

So, $\nabla f(x_0, y_0)$ is orthogonal to $\mathbf{r}'(t_0)$. Moreover, by Theorem 13.12, $\nabla g(x_0, y_0)$ is also orthogonal to $\mathbf{r}'(t_0)$. Consequently, the gradients $\nabla f(x_0, y_0)$ and $\nabla g(x_0, y_0)$ are parallel, and there must exist a scalar λ such that

$$\nabla f(x_0, y_0) = \lambda \nabla g(x_0, y_0).$$

See LarsonCalculus.com for Bruce Edwards's video of this proof.

The Method of Lagrange Multipliers uses Theorem 13.19 to find the extreme values of a function f subject to a constraint.

· · · · · · · · · · · · · · · · ·▷

· · REMARK   As you will see in Examples 1 and 2, the Method of Lagrange Multipliers requires solving systems of nonlinear equations. This often can require some tricky algebraic manipulation.

Method of Lagrange Multipliers

Let f and g satisfy the hypothesis of Lagrange's Theorem, and let f have a minimum or maximum subject to the constraint $g(x, y) = c$. To find the minimum or maximum of f, use these steps.

1. Simultaneously solve the equations $\nabla f(x, y) = \lambda \nabla g(x, y)$ and $g(x, y) = c$ by solving the following system of equations.

$$f_x(x, y) = \lambda g_x(x, y)$$
$$f_y(x, y) = \lambda g_y(x, y)$$
$$g(x, y) = c$$

2. Evaluate f at each solution point obtained in the first step. The greatest value yields the maximum of f subject to the constraint $g(x, y) = c$, and the least value yields the minimum of f subject to the constraint $g(x, y) = c$.

Constrained Optimization Problems

In the problem at the beginning of this section, you wanted to maximize the area of a rectangle that is inscribed in an ellipse. Example 1 shows how to use Lagrange multipliers to solve this problem.

EXAMPLE 1 Using a Lagrange Multiplier with One Constraint

Find the maximum value of $f(x, y) = 4xy$, where $x > 0$ and $y > 0$, subject to the constraint $(x^2/3^2) + (y^2/4^2) = 1$.

Solution To begin, let

$$g(x, y) = \frac{x^2}{3^2} + \frac{y^2}{4^2} = 1.$$

By equating $\nabla f(x, y) = 4y\mathbf{i} + 4x\mathbf{j}$ and $\lambda \nabla g(x, y) = (2\lambda x/9)\mathbf{i} + (\lambda y/8)\mathbf{j}$, you can obtain the following system of equations.

$$4y = \frac{2}{9}\lambda x \qquad f_x(x, y) = \lambda g_x(x, y)$$

$$4x = \frac{1}{8}\lambda y \qquad f_y(x, y) = \lambda g_y(x, y)$$

$$\frac{x^2}{3^2} + \frac{y^2}{4^2} = 1 \qquad \text{Constraint}$$

From the first equation, you obtain $\lambda = 18y/x$, and substitution into the second equation produces

$$4x = \frac{1}{8}\left(\frac{18y}{x}\right)y \quad \Longrightarrow \quad x^2 = \frac{9}{16}y^2.$$

Substituting this value for x^2 into the third equation produces

$$\frac{1}{9}\left(\frac{9}{16}y^2\right) + \frac{1}{16}y^2 = 1 \quad \Longrightarrow \quad y^2 = 8.$$

So, $y = \pm 2\sqrt{2}$. Because it is required that $y > 0$, choose the positive value and find that

$$x^2 = \frac{9}{16}y^2$$

$$= \frac{9}{16}(8)$$

$$= \frac{9}{2}$$

$$x = \frac{3}{\sqrt{2}}.$$

So, the maximum value of f is

$$f\left(\frac{3}{\sqrt{2}}, 2\sqrt{2}\right) = 4xy = 4\left(\frac{3}{\sqrt{2}}\right)\left(2\sqrt{2}\right) = 24.$$

Note that writing the constraint as

$$g(x, y) = \frac{x^2}{3^2} + \frac{y^2}{4^2} = 1 \quad \text{or} \quad g(x, y) = \frac{x^2}{3^2} + \frac{y^2}{4^2} - 1 = 0$$

does not affect the solution—the constant is eliminated when you form ∇g.

REMARK Example 1 can also be solved using the techniques you learned in Chapter 3. To see how, try to find the maximum value of $A = 4xy$ given that

$$\frac{x^2}{3^2} + \frac{y^2}{4^2} = 1.$$

To begin, solve the second equation for y to obtain

$$y = \tfrac{4}{3}\sqrt{9 - x^2}.$$

Then substitute into the first equation to obtain

$$A = 4x\left(\tfrac{4}{3}\sqrt{9 - x^2}\right).$$

Finally, use the techniques of Chapter 3 to maximize A.

EXAMPLE 2   **A Business Application**

The Cobb-Douglas production function (see Example 5, Section 13.1) for a software manufacturer is given by

$$f(x, y) = 100x^{3/4}y^{1/4} \qquad \text{Objective function}$$

where x represents the units of labor (at \$150 per unit) and y represents the units of capital (at \$250 per unit). The total cost of labor and capital is limited to \$50,000. Find the maximum production level for this manufacturer.

Solution   The gradient of f is

$$\nabla f(x, y) = 75x^{-1/4}y^{1/4}\,\mathbf{i} + 25x^{3/4}y^{-3/4}\,\mathbf{j}.$$

The limit on the cost of labor and capital produces the constraint

$$g(x, y) = 150x + 250y = 50{,}000. \qquad \text{Constraint}$$

So, $\lambda \nabla g(x, y) = 150\lambda\mathbf{i} + 250\lambda\mathbf{j}$. This gives rise to the following system of equations.

$$75x^{-1/4}y^{1/4} = 150\lambda \qquad f_x(x, y) = \lambda g_x(x, y)$$
$$25x^{3/4}y^{-3/4} = 250\lambda \qquad f_y(x, y) = \lambda g_y(x, y)$$
$$150x + 250y = 50{,}000 \qquad \text{Constraint}$$

By solving for λ in the first equation

$$\lambda = \frac{75x^{-1/4}y^{1/4}}{150} = \frac{x^{-1/4}y^{1/4}}{2}$$

and substituting into the second equation, you obtain

$$25x^{3/4}y^{-3/4} = 250\left(\frac{x^{-1/4}y^{1/4}}{2}\right)$$
$$25x = 125y \qquad \text{Multiply by } x^{1/4}y^{3/4}.$$
$$x = 5y.$$

By substituting this value for x in the third equation, you have

$$150(5y) + 250y = 50{,}000$$
$$1000y = 50{,}000$$
$$y = 50 \text{ units of capital.}$$

This means that the value of x is

$$x = 5(50)$$
$$= 250 \text{ units of labor.}$$

So, the maximum production level is

$$f(250, 50) = 100(250)^{3/4}(50)^{1/4}$$
$$\approx 16{,}719 \text{ product units.}$$

■ FOR FURTHER INFORMATION For more information on the use of Lagrange multipliers in economics, see the article "Lagrange Multiplier Problems in Economics" by John V. Baxley and John C. Moorhouse in *The American Mathematical Monthly.* To view this article, go to *MathArticles.com.*

Economists call the Lagrange multiplier obtained in a production function the **marginal productivity of money.** For instance, in Example 2, the marginal productivity of money at $x = 250$ and $y = 50$ is

$$\lambda = \frac{x^{-1/4}y^{1/4}}{2} = \frac{(250)^{-1/4}(50)^{1/4}}{2} \approx 0.334$$

which means that for each additional dollar spent on production, an additional 0.334 unit of the product can be produced.

EXAMPLE 3 **Lagrange Multipliers and Three Variables**

• • • ▷ *See LarsonCalculus.com for an interactive version of this type of example.*

Find the minimum value of

$$f(x, y, z) = 2x^2 + y^2 + 3z^2 \qquad \text{Objective function}$$

subject to the constraint $2x - 3y - 4z = 49$.

Solution Let $g(x, y, z) = 2x - 3y - 4z = 49$. Then, because

$$\nabla f(x, y, z) = 4x\mathbf{i} + 2y\mathbf{j} + 6z\mathbf{k}$$

and

$$\lambda \nabla g(x, y, z) = 2\lambda \mathbf{i} - 3\lambda \mathbf{j} - 4\lambda \mathbf{k}$$

you obtain the following system of equations.

$$4x = 2\lambda \qquad\qquad f_x(x, y, z) = \lambda g_x(x, y, z)$$
$$2y = -3\lambda \qquad\qquad f_y(x, y, z) = \lambda g_y(x, y, z)$$
$$6z = -4\lambda \qquad\qquad f_z(x, y, z) = \lambda g_z(x, y, z)$$
$$2x - 3y - 4z = 49 \qquad\qquad \text{Constraint}$$

The solution of this system is $x = 3$, $y = -9$, and $z = -4$. So, the optimum value of f is

$$f(3, -9, -4) = 2(3)^2 + (-9)^2 + 3(-4)^2$$
$$= 147.$$

From the original function and constraint, it is clear that $f(x, y, z)$ has no maximum. So, the optimum value of f determined above is a minimum. ■

A graphical interpretation of constrained optimization problems in two variables was given at the beginning of this section. In three variables, the interpretation is similar, except that level surfaces are used instead of level curves. For instance, in Example 3, the level surfaces of f are ellipsoids centered at the origin, and the constraint

$$2x - 3y - 4z = 49$$

is a plane. The minimum value of f is represented by the ellipsoid that is tangent to the constraint plane, as shown in Figure 13.80.

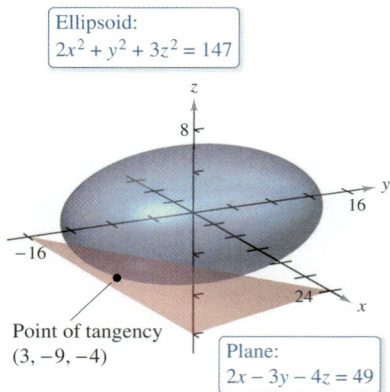

Ellipsoid:
$2x^2 + y^2 + 3z^2 = 147$

Point of tangency
$(3, -9, -4)$

Plane:
$2x - 3y - 4z = 49$

Figure 13.80

EXAMPLE 4 **Optimization Inside a Region**

Find the extreme values of

$$f(x, y) = x^2 + 2y^2 - 2x + 3 \qquad \text{Objective function}$$

subject to the constraint $x^2 + y^2 \le 10$.

Solution To solve this problem, you can break the constraint into two cases.

a. For points *on the circle* $x^2 + y^2 = 10$, you can use Lagrange multipliers to find that the maximum value of $f(x, y)$ is 24—this value occurs at $(-1, 3)$ and at $(-1, -3)$. In a similar way, you can determine that the minimum value of $f(x, y)$ is approximately 6.675—this value occurs at $\left(\sqrt{10}, 0\right)$.

b. For points *inside the circle,* you can use the techniques discussed in Section 13.8 to conclude that the function has a relative minimum of 2 at the point $(1, 0)$.

By combining these two results, you can conclude that f has a maximum of 24 at $(-1, \pm 3)$ and a minimum of 2 at $(1, 0)$, as shown in Figure 13.81. ■

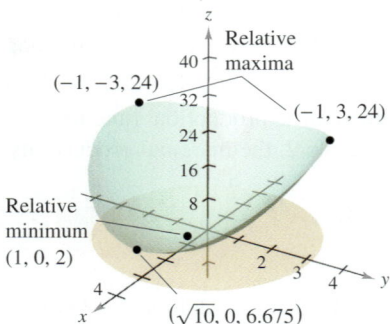

Relative maxima
$(-1, -3, 24)$
$(-1, 3, 24)$

Relative minimum
$(1, 0, 2)$

$\left(\sqrt{10}, 0, 6.675\right)$

Figure 13.81

The Method of Lagrange Multipliers with Two Constraints

For optimization problems involving *two* constraint functions g and h, you can introduce a second Lagrange multiplier, μ (the lowercase Greek letter mu), and then solve the equation

$$\nabla f = \lambda \nabla g + \mu \nabla h$$

where the gradient vectors are not parallel, as illustrated in Example 5.

EXAMPLE 5   Optimization with Two Constraints

Let $T(x, y, z) = 20 + 2x + 2y + z^2$ represent the temperature at each point on the sphere

$$x^2 + y^2 + z^2 = 11.$$

Find the extreme temperatures on the curve formed by the intersection of the plane $x + y + z = 3$ and the sphere.

Solution   The two constraints are

$$g(x, y, z) = x^2 + y^2 + z^2 = 11 \quad \text{and} \quad h(x, y, z) = x + y + z = 3.$$

Using

$$\nabla T(x, y, z) = 2\mathbf{i} + 2\mathbf{j} + 2z\mathbf{k}$$
$$\lambda \nabla g(x, y, z) = 2\lambda x\mathbf{i} + 2\lambda y\mathbf{j} + 2\lambda z\mathbf{k}$$

and

$$\mu \nabla h(x, y, z) = \mu\mathbf{i} + \mu\mathbf{j} + \mu\mathbf{k}$$

you can write the following system of equations.

$$2 = 2\lambda x + \mu \qquad \textcolor{red}{T_x(x, y, z) = \lambda g_x(x, y, z) + \mu h_x(x, y, z)}$$
$$2 = 2\lambda y + \mu \qquad \textcolor{red}{T_y(x, y, z) = \lambda g_y(x, y, z) + \mu h_y(x, y, z)}$$
$$2z = 2\lambda z + \mu \qquad \textcolor{red}{T_z(x, y, z) = \lambda g_z(x, y, z) + \mu h_z(x, y, z)}$$
$$x^2 + y^2 + z^2 = 11 \qquad \textcolor{red}{\text{Constraint 1}}$$
$$x + y + z = 3 \qquad \textcolor{red}{\text{Constraint 2}}$$

By subtracting the second equation from the first, you can obtain the following system.

$$\lambda(x - y) = 0$$
$$2z(1 - \lambda) - \mu = 0$$
$$x^2 + y^2 + z^2 = 11$$
$$x + y + z = 3$$

> **• • REMARK**   The systems of equations that arise when the Method of Lagrange Multipliers is used are not, in general, linear systems, and finding the solutions often requires ingenuity.

From the first equation, you can conclude that $\lambda = 0$ or $x = y$. For $\lambda = 0$, you can show that the critical points are $(3, -1, 1)$ and $(-1, 3, 1)$. (Try doing this—it takes a little work.) For $\lambda \neq 0$, then $x = y$ and you can show that the critical points occur when $x = y = (3 \pm 2\sqrt{3})/3$ and $z = (3 \mp 4\sqrt{3})/3$. Finally, to find the optimal solutions, compare the temperatures at the four critical points.

$$T(3, -1, 1) = T(-1, 3, 1) = 25$$
$$T\left(\frac{3 - 2\sqrt{3}}{3}, \frac{3 - 2\sqrt{3}}{3}, \frac{3 + 4\sqrt{3}}{3}\right) = \frac{91}{3} \approx 30.33$$
$$T\left(\frac{3 + 2\sqrt{3}}{3}, \frac{3 + 2\sqrt{3}}{3}, \frac{3 - 4\sqrt{3}}{3}\right) = \frac{91}{3} \approx 30.33$$

So, $T = 25$ is the minimum temperature and $T = \frac{91}{3}$ is the maximum temperature on the curve.

13.10 Exercises

See **CalcChat.com** for tutorial help and worked-out solutions to odd-numbered exercises.

Using Lagrange Multipliers In Exercises 1–8, use Lagrange multipliers to find the indicated extrema, assuming that x and y are positive.

1. Maximize: $f(x, y) = xy$

   Constraint: $x + y = 10$

2. Minimize: $f(x, y) = 2x + y$

   Constraint: $xy = 32$

3. Minimize $f(x, y) = x^2 + y^2$

   Constraint: $x + 2y - 5 = 0$

4. Maximize $f(x, y) = x^2 - y^2$

   Constraint: $2y - x^2 = 0$

5. Maximize $f(x, y) = 2x + 2xy + y$

   Constraint: $2x + y = 100$

6. Minimize $f(x, y) = 3x + y + 10$

   Constraint: $x^2 y = 6$

7. Maximize $f(x, y) = \sqrt{6 - x^2 - y^2}$

   Constraint: $x + y - 2 = 0$

8. Minimize $f(x, y) = \sqrt{x^2 + y^2}$

   Constraint: $2x + 4y - 15 = 0$

Using Lagrange Multipliers In Exercises 9–12, use Lagrange multipliers to find the indicated extrema, assuming that x, y, and z are positive.

9. Minimize $f(x, y, z) = x^2 + y^2 + z^2$

   Constraint: $x + y + z - 9 = 0$

10. Maximize $f(x, y, z) = xyz$

    Constraint: $x + y + z - 3 = 0$

11. Minimize $f(x, y, z) = x^2 + y^2 + z^2$

    Constraint: $x + y + z = 1$

12. Maximize $f(x, y, z) = x + y + z$

    Constraint: $x^2 + y^2 + z^2 = 1$

Using Lagrange Multipliers In Exercises 13 and 14, use Lagrange multipliers to find any extrema of the function subject to the constraint $x^2 + y^2 \leq 1$.

13. $f(x, y) = x^2 + 3xy + y^2$ 14. $f(x, y) = e^{-xy/4}$

Using Lagrange Multipliers In Exercises 15 and 16, use Lagrange multipliers to find the indicated extrema of f subject to two constraints. In each case, assume that x, y, and z are nonnegative.

15. Maximize $f(x, y, z) = xyz$

    Constraints: $x + y + z = 32$, $x - y + z = 0$

16. Minimize $f(x, y, z) = x^2 + y^2 + z^2$

    Constraints: $x + 2z = 6$, $x + y = 12$

Finding Minimum Distance In Exercises 17–26, use Lagrange multipliers to find the minimum distance from the curve or surface to the indicated point. (*Hint:* To simplify the computations, minimize the square of the distance.)

Curve	Point
17. Line: $x + y = 1$	$(0, 0)$
18. Line: $2x + 3y = -1$	$(0, 0)$
19. Line: $x - y = 4$	$(0, 2)$
20. Line: $x + 4y = 3$	$(1, 0)$
21. Parabola: $y = x^2$	$(0, 3)$
22. Parabola: $y = x^2$	$(-3, 0)$
23. Circle: $x^2 + (y - 1)^2 = 9$	$(4, 4)$
24. Circle: $(x - 4)^2 + y^2 = 4$	$(0, 10)$

Surface	Point
25. Plane: $x + y + z = 1$	$(2, 1, 1)$
26. Cone: $z = \sqrt{x^2 + y^2}$	$(4, 0, 0)$

Intersection of Surfaces In Exercises 27 and 28, find the highest point on the curve of intersection of the surfaces.

27. Cone: $x^2 + y^2 - z^2 = 0$

    Plane: $x + 2z = 4$

28. Sphere: $x^2 + y^2 + z^2 = 36$

    Plane: $2x + y - z = 2$

WRITING ABOUT CONCEPTS

29. **Constrained Optimization Problems** Explain what is meant by constrained optimization problems.

30. **Method of Lagrange Multipliers** Explain the Method of Lagrange Multipliers for solving constrained optimization problems.

Using Lagrange Multipliers In Exercises 31–38, use Lagrange multipliers to solve the indicated exercise in Section 13.9.

31. Exercise 1	32. Exercise 2
33. Exercise 5	34. Exercise 6
35. Exercise 9	36. Exercise 10
37. Exercise 15	38. Exercise 16

39. **Maximum Volume** Use Lagrange multipliers to find the dimensions of a rectangular box of maximum volume that can be inscribed (with edges parallel to the coordinate axes) in the ellipsoid

$$\frac{x^2}{a^2} + \frac{y^2}{b^2} + \frac{z^2}{c^2} = 1.$$

40. HOW DO YOU SEE IT? The graphs show the constraint and several level curves of the objective function. Use the graph to approximate the indicated extrema.

(a) Maximize $z = xy$

Constraint: $2x + y = 4$

(b) Minimize $z = x^2 + y^2$

Constraint: $x + y - 4 = 0$

41. Minimum Cost A cargo container (in the shape of a rectangular solid) must have a volume of 480 cubic feet. The bottom will cost $5 per square foot to construct and the sides and the top will cost $3 per square foot to construct. Use Lagrange multipliers to find the dimensions of the container of this size that has minimum cost.

42. Geometric and Arithmetic Means

(a) Use Lagrange multipliers to prove that the product of three positive numbers x, y, and z, whose sum has the constant value S, is a maximum when the three numbers are equal. Use this result to prove that

$$\sqrt[3]{xyz} \le \frac{x + y + z}{3}.$$

(b) Generalize the result of part (a) to prove that the product

$$x_1 x_2 x_3 \cdot \cdot \cdot x_n$$

is a maximum when

$$x_1 = x_2 = x_3 = \cdot \cdot \cdot = x_n, \quad \sum_{i=1}^{n} x_i = S, \text{ and all } x_i \ge 0.$$

Then prove that

$$\sqrt[n]{x_1 x_2 x_3 \cdot \cdot \cdot x_n} \le \frac{x_1 + x_2 + x_3 + \cdot \cdot \cdot + x_n}{n}.$$

This shows that the geometric mean is never greater than the arithmetic mean.

43. Minimum Surface Area Use Lagrange multipliers to find the dimensions of a right circular cylinder with volume V_0 cubic units and minimum surface area.

44. Temperature Let $T(x, y, z) = 100 + x^2 + y^2$ represent the temperature at each point on the sphere

$$x^2 + y^2 + z^2 = 50.$$

Find the maximum temperature on the curve formed by the intersection of the sphere and the plane $x - z = 0$.

45. Refraction of Light When light waves traveling in a transparent medium strike the surface of a second transparent medium, they tend to "bend" in order to follow the path of minimum time. This tendency is called *refraction* and is described by **Snell's Law of Refraction,**

$$\frac{\sin \theta_1}{v_1} = \frac{\sin \theta_2}{v_2}$$

where θ_1 and θ_2 are the magnitudes of the angles shown in the figure, and v_1 and v_2 are the velocities of light in the two media. Use Lagrange multipliers to derive this law using $x + y = a$.

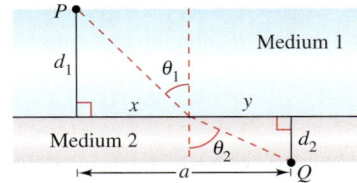

46. Area and Perimeter A semicircle is on top of a rectangle (see figure). When the area is fixed and the perimeter is a minimum, or when the perimeter is fixed and the area is a maximum, use Lagrange multipliers to verify that the length of the rectangle is twice its height.

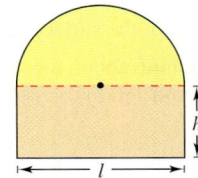

Production Level In Exercises 47 and 48, find the maximum production level P when the total cost of labor (at $72 per unit) and capital (at $60 per unit) is limited to $250,000, where x is the number of units of labor and y is the number of units of capital.

47. $P(x, y) = 100x^{0.25}y^{0.75}$

48. $P(x, y) = 100x^{0.4}y^{0.6}$

Cost In Exercises 49 and 50, find the minimum cost of producing 50,000 units of a product, where x is the number of units of labor (at $72 per unit) and y is the number of units of capital (at $60 per unit).

49. $P(x, y) = 100x^{0.25}y^{0.75}$

50. $P(x, y) = 100x^{0.6}y^{0.4}$

PUTNAM EXAM CHALLENGE

51. A can buoy is to be made of three pieces, namely, a cylinder and two equal cones, the altitude of each cone being equal to the altitude of the cylinder. For a given area of surface, what shape will have the greatest volume?

Evaluating a Function In Exercises 1 and 2, find and simplify the function values.

1. $f(x, y) = 3x^2y$

   (a) $(1, 3)$ (b) $(-1, 1)$ (c) $(-4, 0)$ (d) $(x, 2)$

2. $f(x, y) = 6 - 4x - 2y^2$

   (a) $(0, 2)$ (b) $(5, 0)$ (c) $(-1, -2)$ (d) $(-3, y)$

Finding the Domain and Range of a Function In Exercises 3 and 4, find the domain and range of the function.

3. $f(x, y) = \dfrac{\sqrt{x}}{y}$

4. $f(x, y) = \sqrt{36 - x^2 - y^2}$

Sketching a Contour Map In Exercises 5 and 6, describe the level curves of the function. Sketch a contour map of the surface using level curves for the given c-values.

5. $z = 3 - 2x + y$, $c = 0, 2, 4, 6, 8$

6. $z = 2x^2 + y^2$, $c = 1, 2, 3, 4, 5$

7. **Conjecture** Consider the function $f(x, y) = x^2 + y^2$.

   (a) Sketch the graph of the surface given by f.

   (b) Make a conjecture about the relationship between the graphs of f and $g(x, y) = f(x, y) + 2$. Explain your reasoning.

   (c) Make a conjecture about the relationship between the graphs of f and $g(x, y) = f(x, y - 2)$. Explain your reasoning.

   (d) On the surface in part (a), sketch the graphs of $z = f(1, y)$ and $z = f(x, 1)$.

8. **Investment** A principal of \$2000 is deposited in a savings account that earns interest at a rate of r (written as a decimal) compounded continuously. The amount $A(r, t)$ after t years is

$$A(r, t) = 2000e^{rt}.$$

Use this function of two variables to complete the table.

Rate	Number of Years			
	5	10	15	20
0.02				
0.04				
0.06				
0.07				

Sketching a Level Surface In Exercises 9 and 10, sketch the graph of the level surface $f(x, y, z) = c$ at the given value of c.

9. $f(x, y, z) = x^2 - y + z^2$, $c = 2$

10. $f(x, y, z) = 4x^2 - y^2 + 4z^2$, $c = 0$

Limit and Continuity In Exercises 11–14, find the limit (if it exists) and discuss the continuity of the function.

11. $\displaystyle\lim_{(x, y)\to(1, 1)} \dfrac{xy}{x^2 + y^2}$

12. $\displaystyle\lim_{(x, y)\to(1, 1)} \dfrac{xy}{x^2 - y^2}$

13. $\displaystyle\lim_{(x, y)\to(0, 0)} \dfrac{y + xe^{-y^2}}{1 + x^2}$

14. $\displaystyle\lim_{(x, y)\to(0, 0)} \dfrac{x^2y}{x^4 + y^2}$

Finding Partial Derivatives In Exercises 15–22, find all first partial derivatives.

15. $f(x, y) = 5x^3 + 7y - 3$

16. $f(x, y) = 4x^2 - 2xy + y^2$

17. $f(x, y) = e^x \cos y$

18. $f(x, y) = \dfrac{xy}{x + y}$

19. $f(x, y) = y^3e^{4x}$

20. $z = \ln(x^2 + y^2 + 1)$

21. $f(x, y, z) = 2xz^2 + 6xyz - 5xy^3$

22. $w = \sqrt{x^2 - y^2 - z^2}$

Finding Second Partial Derivatives In Exercises 23–26, find the four second partial derivatives. Observe that the second mixed partials are equal.

23. $f(x, y) = 3x^2 - xy + 2y^3$

24. $h(x, y) = \dfrac{x}{x + y}$

25. $h(x, y) = x \sin y + y \cos x$

26. $g(x, y) = \cos(x - 2y)$

27. **Finding the Slopes of a Surface** Find the slopes of the surface $z = x^2 \ln(y + 1)$ in the x- and y-directions at the point $(2, 0, 0)$.

28. **Marginal Revenue** A company has two plants that produce the same lawn mower. If x_1 and x_2 are the numbers of units produced at plant 1 and plant 2, respectively, then the total revenue for the product is given by

$$R = 300x_1 + 300x_2 - 5x_1^2 - 10x_1x_2 - 5x_2^2.$$

When $x_1 = 5$ and $x_2 = 8$, find (a) the marginal revenue for plant 1, $\partial R/\partial x_1$, and (b) the marginal revenue for plant 2, $\partial R/\partial x_2$.

Finding a Total Differential In Exercises 29–32, find the total differential.

29. $z = x \sin xy$

30. $z = 5x^4y^3$

31. $w = 3xy^2 - 2x^3yz^2$

32. $w = \dfrac{3x + 4y}{y + 3z}$

Using a Differential as an Approximation In Exercises 33 and 34, (a) evaluate $f(2, 1)$ and $f(2.1, 1.05)$ and calculate Δz, and (b) use the total differential dz to approximate Δz.

33. $f(x, y) = 4x + 2y$

34. $f(x, y) = 36 - x^2 - y^2$

35. **Volume** The possible error involved in measuring each dimension of a right circular cone is $\pm\frac{1}{8}$ inch. The radius is 2 inches and the height is 5 inches. Approximate the propagated error and the relative error in the calculated volume of the cone.

36. Lateral Surface Area   Approximate the propagated error and the relative error in the computation of the lateral surface area of the cone in Exercise 35. $\left(\text{The lateral surface area is given by } A = \pi r \sqrt{r^2 + h^2}.\right)$

Using Different Methods   In Exercises 37 and 38, find dw/dt **(a)** by using the appropriate Chain Rule, and **(b)** by converting w to a function of t before differentiating.

37. $w = \ln(x^2 + y), \quad x = 2t, \quad y = 4 - t$

38. $w = y^2 - x, \quad x = \cos t, \quad y = \sin t$

Using Different Methods   In Exercises 39 and 40, find $\partial w/\partial r$ and $\partial w/\partial t$ **(a)** by using the appropriate Chain Rule and **(b)** by converting w to a function of r and t before differentiating.

39. $w = \dfrac{xy}{z}, \quad x = 2r + t, \quad y = rt, \quad z = 2r - t$

40. $w = x^2 + y^2 + z^2, \quad x = r \cos t, \quad y = r \sin t, \quad z = t$

Finding Partial Derivatives Implicitly   In Exercises 41 and 42, differentiate implicitly to find the first partial derivatives of z.

41. $x^2 + xy + y^2 + yz + z^2 = 0$ **42.** $xz^2 - y \sin z = 0$

Finding a Directional Derivative   In Exercises 43 and 44, use Theorem 13.9 to find the directional derivative of the function at P in the direction of $\mathbf{v}$.

43. $f(x, y) = x^2 y, \quad P(-5, 5), \quad \mathbf{v} = 3\mathbf{i} - 4\mathbf{j}$

44. $f(x, y) = \frac{1}{4}y^2 - x^2, \quad P(1, 4), \quad \mathbf{v} = 2\mathbf{i} + \mathbf{j}$

Finding a Directional Derivative   In Exercises 45 and 46, use the gradient to find the directional derivative of the function at P in the direction of $\mathbf{v}$.

45. $w = y^2 + xz, \quad P(1, 2, 2), \quad \mathbf{v} = 2\mathbf{i} - \mathbf{j} + 2\mathbf{k}$

46. $w = 5x^2 + 2xy - 3y^2 z, \quad P(1, 0, 1), \quad \mathbf{v} = \mathbf{i} + \mathbf{j} - \mathbf{k}$

Using Properties of the Gradient   In Exercises 47–50, find the gradient of the function and the maximum value of the directional derivative at the given point.

47. $z = x^2 y, \quad (2, 1)$

48. $z = e^{-x} \cos y, \quad \left(0, \dfrac{\pi}{4}\right)$

49. $z = \dfrac{y}{x^2 + y^2}, \quad (1, 1)$

50. $z = \dfrac{x^2}{x - y}, \quad (2, 1)$

Using a Function   In Exercises 51 and 52, **(a)** find the gradient of the function at P, **(b)** find a unit normal vector to the level curve $f(x, y) = c$ at P, **(c)** find the tangent line to the level curve $f(x, y) = c$ at P, and **(d)** sketch the level curve, the unit normal vector, and the tangent line in the xy-plane.

51. $f(x, y) = 9x^2 - 4y^2$ **52.** $f(x, y) = 4y \sin x - y$

$\quad c = 65, \; P(3, 2)$ $\quad c = 3, \; P\left(\dfrac{\pi}{2}, 1\right)$

Finding an Equation of a Tangent Plane   In Exercises 53–56, find an equation of the tangent plane to the surface at the given point.

53. $z = x^2 + y^2 + 2, \quad (1, 3, 12)$

54. $9x^2 + y^2 + 4z^2 = 25, \quad (0, -3, 2)$

55. $z = -9 + 4x - 6y - x^2 - y^2, \quad (2, -3, 4)$

56. $f(x, y) = \sqrt{25 - y^2}, \quad (2, 3, 4)$

Finding an Equation of a Tangent Plane and a Normal Line   In Exercises 57 and 58, find an equation of the tangent plane and find a set of symmetric equations for the normal line to the surface at the given point.

57. $f(x, y) = x^2 y, \quad (2, 1, 4)$

58. $z = \sqrt{9 - x^2 - y^2}, \quad (1, 2, 2)$

59. Angle of Inclination   Find the angle of inclination θ of the tangent plane to the surface $x^2 + y^2 + z^2 = 14$ at the point $(2, 1, 3)$.

60. Approximation   Consider the following approximations for a function $f(x, y)$ centered at $(0, 0)$.

Linear approximation:

$$P_1(x, y) = f(0, 0) + f_x(0, 0)x + f_y(0, 0)y$$

Quadratic approximation:

$$P_2(x, y) = f(0, 0) + f_x(0, 0)x + f_y(0, 0)y +$$
$$\tfrac{1}{2}f_{xx}(0, 0)x^2 + f_{xy}(0, 0)xy + \tfrac{1}{2}f_{yy}(0, 0)y^2$$

[Note that the linear approximation is the tangent plane to the surface at $(0, 0, f(0, 0))$.]

(a) Find the linear approximation of

$$f(x, y) = \cos x + \sin y$$

centered at $(0, 0)$.

(b) Find the quadratic approximation of

$$f(x, y) = \cos x + \sin y$$

centered at $(0, 0)$.

(c) When $y = 0$ in the quadratic approximation, you obtain the second-degree Taylor polynomial for what function?

(d) Complete the table.

x	y	$f(x, y)$	$P_1(x, y)$	$P_2(x, y)$
0	0			
0	0.1			
0.2	0.1			
0.5	0.3			
1	0.5			

(e) Use a computer algebra system to graph the surfaces $z = f(x, y)$, $z = P_1(x, y)$, and $z = P_2(x, y)$. How does the accuracy of the approximations change as the distance from $(0, 0)$ increases?

Using the Second Partials Test In Exercises 61–66, examine the function for relative extrema and saddle points.

61. $f(x, y) = -x^2 - 4y^2 + 8x - 8y - 11$

62. $f(x, y) = x^2 - y^2 - 16x - 16y$

63. $f(x, y) = 2x^2 + 6xy + 9y^2 + 8x + 14$

64. $f(x, y) = x^2 + 3xy + y^2 - 5x$

65. $f(x, y) = xy + \dfrac{1}{x} + \dfrac{1}{y}$

66. $f(x, y) = -8x^2 + 4xy - y^2 + 12x + 7$

67. Finding Minimum Distance Find the minimum distance from the point $(2, 1, 4)$ to the surface $x + y + z = 4$. (*Hint:* To simplify the computations, minimize the square of the distance.)

68. Finding Positive Numbers Find three positive integers, x, y, and z, such that the product is 64 and the sum is a minimum.

69. Maximum Revenue A company manufactures two types of bicycles, a racing bicycle and a mountain bicycle. The total revenue from x_1 units of racing bicycles and x_2 units of mountain bicycles is

$$R = -6x_1^2 - 10x_2^2 - 2x_1x_2 + 32x_1 + 84x_2$$

where x_1 and x_2 are in thousands of units. Find x_1 and x_2 so as to maximize the revenue.

70. Maximum Profit A corporation manufactures digital cameras at two locations. The cost of producing x_1 units at location 1 is $C_1 = 0.05x_1^2 + 15x_1 + 5400$ and the cost of producing x_2 units at location 2 is $C_2 = 0.03x_2^2 + 15x_2 + 6100$. The digital cameras sell for $180 per unit. Find the quantity that should be produced at each location to maximize the profit $P = 180(x_1 + x_2) - C_1 - C_2$.

Finding the Least Squares Regression Line In Exercises 71 and 72, find the least squares regression line for the points. Use the regression capabilities of a graphing utility to verify your results. Use the graphing utility to plot the points and graph the regression line.

71. $(0, 4)$, $(1, 5)$, $(3, 6)$, $(6, 8)$, $(8, 10)$

72. $(0, 10)$, $(2, 8)$, $(4, 7)$, $(7, 5)$, $(9, 3)$, $(12, 0)$

73. Modeling Data An agronomist used four test plots to determine the relationship between the wheat yield y (in bushels per acre) and the amount of fertilizer x (in pounds per acre). The results are shown in the table.

Fertilizer, x	100	150	200	250
Yield, y	35	44	50	56

(a) Use the regression capabilities of a graphing utility to find the least squares regression line for the data.

(b) Use the model to approximate the wheat yield for a fertilizer application of 175 pounds per acre.

74. Modeling Data The data in the table show the yield y (in milligrams) of a chemical reaction after t minutes.

Minutes, t	1	2	3	4
Yield, y	1.2	7.1	9.9	13.1

Minutes, t	5	6	7	8
Yield, y	15.5	16.0	17.9	18.0

(a) Use the regression capabilities of a graphing utility to find the least squares regression line for the data. Then use the graphing utility to plot the data and graph the model.

(b) Use a graphing utility to plot the points $(\ln t, y)$. Do these points appear to follow a linear pattern more closely than the plot of the given data in part (a)?

(c) Use the regression capabilities of a graphing utility to find the least squares regression line for the points $(\ln t, y)$ and obtain the logarithmic model $y = a + b \ln t$.

(d) Use a graphing utility to plot the original data and graph the linear and logarithmic models. Which is a better model? Explain.

Using Lagrange Multipliers In Exercises 75–80, use Lagrange multipliers to find the indicated extrema, assuming that x and y are positive.

75. Minimize: $f(x, y) = x^2 + y^2$

Constraint: $x + y - 8 = 0$

76. Maximize: $f(x, y) = xy$

Constraint: $x + 3y - 6 = 0$

77. Maximize: $f(x, y) = 2x + 3xy + y$

Constraint: $x + 2y = 29$

78. Minimize: $f(x, y) = x^2 - y^2$

Constraint: $x - 2y + 6 = 0$

79. Maximize: $f(x, y) = 2xy$

Constraint: $2x + y = 12$

80. Minimize: $f(x, y) = 3x^2 - y^2$

Constraint: $2x - 2y + 5 = 0$

81. Minimum Cost A water line is to be built from point P to point S and must pass through regions where construction costs differ (see figure). The cost per kilometer in dollars is $3k$ from P to Q, $2k$ from Q to R, and k from R to S. For simplicity, let $k = 1$. Use Lagrange multipliers to find x, y, and z such that the total cost C will be minimized.

P.S. Problem Solving

See **CalcChat.com** for tutorial help and worked-out solutions to odd-numbered exercises.

1. **Area**   **Heron's Formula** states that the area of a triangle with sides of lengths a, b, and c is given by

$$A = \sqrt{s(s-a)(s-b)(s-c)}$$

where $s = \dfrac{a+b+c}{2}$, as shown in the figure.

(a) Use Heron's Formula to find the area of the triangle with vertices $(0, 0)$, $(3, 4)$, and $(6, 0)$.

(b) Show that among all triangles having a fixed perimeter, the triangle with the largest area is an equilateral triangle.

(c) Show that among all triangles having a fixed area, the triangle with the smallest perimeter is an equilateral triangle.

Figure for 1   Figure for 2

2. **Minimizing Material**   An industrial container is in the shape of a cylinder with hemispherical ends, as shown in the figure. The container must hold 1000 liters of fluid. Determine the radius r and length h that minimize the amount of material used in the construction of the tank.

3. **Tangent Plane**   Let $P(x_0, y_0, z_0)$ be a point in the first octant on the surface $xyz = 1$.

(a) Find the equation of the tangent plane to the surface at the point P.

(b) Show that the volume of the tetrahedron formed by the three coordinate planes and the tangent plane is constant, independent of the point of tangency (see figure).

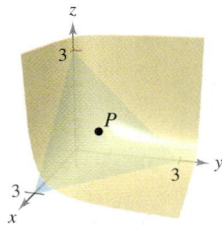

4. **Using Functions**   Use a graphing utility to graph the functions

$$f(x) = \sqrt[3]{x^3 - 1} \quad \text{and} \quad g(x) = x$$

in the same viewing window.

(a) Show that

$$\lim_{x \to \infty} [f(x) - g(x)] = 0 \quad \text{and} \quad \lim_{x \to -\infty} [f(x) - g(x)] = 0.$$

(b) Find the point on the graph of f that is farthest from the graph of g.

5. **Finding Maximum and Minimum Values**

(a) Let $f(x, y) = x - y$ and $g(x, y) = x^2 + y^2 = 4$. Graph various level curves of f and the constraint g in the xy-plane. Use the graph to determine the maximum value of f subject to the constraint $g = 4$. Then verify your answer using Lagrange multipliers.

(b) Let $f(x, y) = x - y$ and $g(x, y) = x^2 + y^2 = 0$. Find the maximum and minimum values of f subject to the constraint $g = 0$. Does the Method of Lagrange Multipliers work in this case? Explain.

6. **Minimizing Costs**   A heated storage room has the shape of a rectangular prism and has a volume of 1000 cubic feet, as shown in the figure. Because warm air rises, the heat loss per unit of area through the ceiling is five times as great as the heat loss through the floor. The heat loss through the four walls is three times as great as the heat loss through the floor. Determine the room dimensions that will minimize heat loss and therefore minimize heating costs.

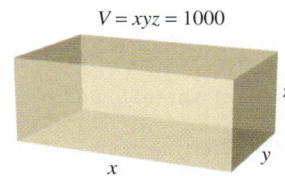

$$V = xyz = 1000$$

7. **Minimizing Costs**   Repeat Exercise 6 assuming that the heat loss through the walls and ceiling remain the same, but the floor is insulated so that there is no heat loss through the floor.

8. **Temperature**   Consider a circular plate of radius 1 given by $x^2 + y^2 \le 1$, as shown in the figure. The temperature at any point $P(x, y)$ on the plate is $T(x, y) = 2x^2 + y^2 - y + 10$.

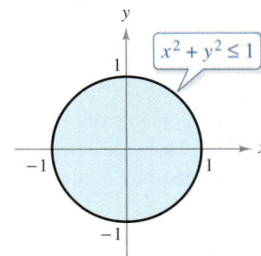

$$x^2 + y^2 \le 1$$

(a) Sketch the isotherm $T(x, y) = 10$. To print an enlarged copy of the graph, go to *MathGraphs.com*.

(b) Find the hottest and coldest points on the plate.

9. **Cobb-Douglas Production Function**   Consider the Cobb-Douglas production function

$$f(x, y) = Cx^a y^{1-a}, \quad 0 < a < 1.$$

(a) Show that f satisfies the equation $x \dfrac{\partial f}{\partial x} + y \dfrac{\partial f}{\partial y} = f$.

(b) Show that $f(tx, ty) = tf(x, y)$.

10. Minimizing Area   Consider the ellipse

$$\frac{x^2}{a^2} + \frac{y^2}{b^2} = 1$$

that encloses the circle $x^2 + y^2 = 2x$. Find values of a and b that minimize the area of the ellipse.

11. Projectile Motion   A projectile is launched at an angle of $45°$ with the horizontal and with an initial velocity of 64 feet per second. A television camera is located in the plane of the path of the projectile 50 feet behind the launch site (see figure).

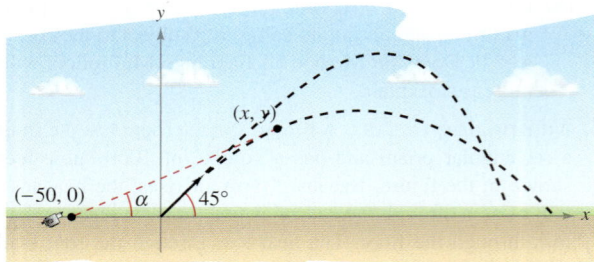

(a) Find parametric equations for the path of the projectile in terms of the parameter t representing time.

(b) Write the angle α that the camera makes with the horizontal in terms of x and y and in terms of t.

(c) Use the results of part (b) to find $\dfrac{d\alpha}{dt}$.

(d) Use a graphing utility to graph α in terms of t. Is the graph symmetric to the axis of the parabolic arch of the projectile? At what time is the rate of change of α greatest?

(e) At what time is the angle α maximum? Does this occur when the projectile is at its greatest height?

12. Distance   Consider the distance d between the launch site and the projectile in Exercise 11.

(a) Write the distance d in terms of x and y and in terms of the parameter t.

(b) Use the results of part (a) to find the rate of change of d.

(c) Find the rate of change of the distance when $t = 2$.

(d) When is the rate of change of d minimum during the flight of the projectile? Does this occur at the time when the projectile reaches its maximum height?

13. Finding Extrema and Saddle Points Using Technology   Consider the function

$$f(x, y) = (\alpha x^2 + \beta y^2)e^{-(x^2+y^2)}, \quad 0 < |\alpha| < \beta.$$

(a) Use a computer algebra system to graph the function for $\alpha = 1$ and $\beta = 2$, and identify any extrema or saddle points.

(b) Use a computer algebra system to graph the function for $\alpha = -1$ and $\beta = 2$, and identify any extrema or saddle points.

(c) Generalize the results in parts (a) and (b) for the function f.

14. Proof   Prove that if f is a differentiable function such that $\nabla f(x_0, y_0) = \mathbf{0}$, then the tangent plane at (x_0, y_0) is horizontal.

15. Area   The figure shows a rectangle that is approximately $l = 6$ centimeters long and $h = 1$ centimeter high.

(a) Draw a rectangular strip along the rectangular region showing a small increase in length.

(b) Draw a rectangular strip along the rectangular region showing a small increase in height.

(c) Use the results in parts (a) and (b) to identify the measurement that has more effect on the area A of the rectangle.

(d) Verify your answer in part (c) analytically by comparing the value of dA when $dl = 0.01$ and when $dh = 0.01$.

16. Tangent Planes   Let f be a differentiable function of one variable. Show that all tangent planes to the surface $z = yf(x/y)$ intersect in a common point.

17. Wave Equation   Show that

$$u(x, t) = \frac{1}{2}[\sin(x - t) + \sin(x + t)]$$

is a solution to the one-dimensional wave equation

$$\frac{\partial^2 u}{\partial t^2} = \frac{\partial^2 u}{\partial x^2}.$$

18. Wave Equation   Show that

$$u(x, t) = \frac{1}{2}[f(x - ct) + f(x + ct)]$$

is a solution to the one-dimensional wave equation

$$\frac{\partial^2 u}{\partial t^2} = c^2 \frac{\partial^2 u}{\partial x^2}.$$

(This equation describes the small transverse vibration of an elastic string such as those on certain musical instruments.)

19. Verifying Equations   Consider the function $w = f(x, y)$, where $x = r\cos\theta$ and $y = r\sin\theta$. Verify each of the following.

(a) $\dfrac{\partial w}{\partial x} = \dfrac{\partial w}{\partial r}\cos\theta - \dfrac{\partial w}{\partial \theta}\dfrac{\sin\theta}{r}$

$\dfrac{\partial w}{\partial y} = \dfrac{\partial w}{\partial r}\sin\theta + \dfrac{\partial w}{\partial \theta}\dfrac{\cos\theta}{r}$

(b) $\left(\dfrac{\partial w}{\partial x}\right)^2 + \left(\dfrac{\partial w}{\partial y}\right)^2 = \left(\dfrac{\partial w}{\partial r}\right)^2 + \left(\dfrac{1}{r^2}\right)\left(\dfrac{\partial w}{\partial \theta}\right)^2$

20. Using a Function   Demonstrate the result of Exercise 19(b) for

$$w = \arctan\frac{y}{x}.$$

21. Laplace's Equation   Rewrite Laplace's equation

$$\frac{\partial^2 u}{\partial x^2} + \frac{\partial^2 u}{\partial y^2} + \frac{\partial^2 u}{\partial z^2} = 0$$

in cylindrical coordinates.

14 Multiple Integration

Modeling Data *(Exercise 34, p. 1008)*

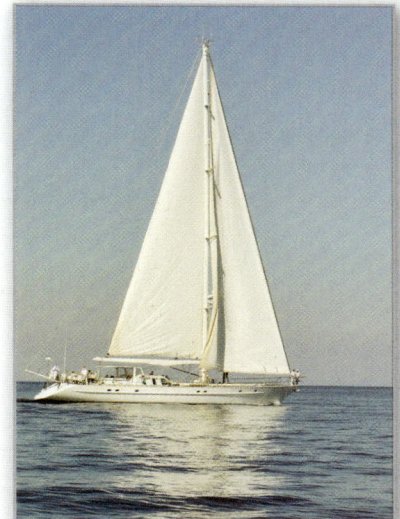

Center of Pressure on a Sail
(Section Project, p. 1001)

Glacier *(Exercise 60, p. 993)*

Population
(Exercise 57, p. 992)

Average Production *(Exercise 57, p. 984)*

14.1 Iterated Integrals and Area in the Plane

■ Evaluate an iterated integral.
■ Use an iterated integral to find the area of a plane region.

Iterated Integrals

In Chapters 14 and 15, you will study several applications of integration involving functions of several variables. Chapter 14 is like Chapter 7 in that it surveys the use of integration to find plane areas, volumes, surface areas, moments, and centers of mass.

In Chapter 13, you saw that it is meaningful to differentiate functions of several variables with respect to one variable while holding the other variables constant. You can *integrate* functions of several variables by a similar procedure. For example, consider the partial derivative $f_x(x, y) = 2xy$. By considering y constant, you can integrate with respect to x to obtain

$$f(x, y) = \int f_x(x, y)\, dx \qquad \text{Integrate with respect to } x.$$

$$= \int 2xy\, dx \qquad \text{Hold } y \text{ constant.}$$

$$= y \int 2x\, dx \qquad \text{Factor out constant } y.$$

$$= y(x^2) + C(y) \qquad \text{Antiderivative of } 2x \text{ is } x^2.$$

$$= x^2 y + C(y). \qquad C(y) \text{ is a function of } y.$$

The "constant" of integration, $C(y)$, is a function of y. In other words, by integrating with respect to x, you are able to recover $f(x, y)$ only partially. The total recovery of a function of x and y from its partial derivatives is a topic you will study in Chapter 15. For now, you will focus on extending definite integrals to functions of several variables. For instance, by considering y constant, you can apply the Fundamental Theorem of Calculus to evaluate

$$\int_1^{2y} 2xy\, dx = x^2 y \Big]_1^{2y} = (2y)^2 y - (1)^2 y = 4y^3 - y.$$

x is the variable
of integration
and y is fixed.

Replace x by
the limits of
integration.

The result is
a function
of y.

Similarly, you can integrate with respect to y by holding x fixed. Both procedures are summarized as follows.

$$\int_{h_1(y)}^{h_2(y)} f_x(x, y)\, dx = f(x, y) \Big]_{h_1(y)}^{h_2(y)} = f(h_2(y), y) - f(h_1(y), y) \qquad \text{With respect to } x$$

$$\int_{g_1(x)}^{g_2(x)} f_y(x, y)\, dy = f(x, y) \Big]_{g_1(x)}^{g_2(x)} = f(x, g_2(x)) - f(x, g_1(x)) \qquad \text{With respect to } y$$

Note that the variable of integration cannot appear in either limit of integration. For instance, it makes no sense to write

$$\int_0^x y\, dx.$$

EXAMPLE 1   **Integrating with Respect to *y***

Evaluate $\int_{1}^{x} (2x^2y^{-2} + 2y) \, dy$.

Solution   Considering *x* to be constant and integrating with respect to *y* produces

$$\int_{1}^{x} (2x^2y^{-2} + 2y) \, dy = \left[\frac{-2x^2}{y} + y^2 \right]_{1}^{x} \qquad \text{Integrate with respect to } y.$$

$$= \left(\frac{-2x^2}{x} + x^2 \right) - \left(\frac{-2x^2}{1} + 1 \right)$$

$$= 3x^2 - 2x - 1.$$

Notice in Example 1 that the integral defines a function of *x* and can *itself* be integrated, as shown in the next example.

EXAMPLE 2   **The Integral of an Integral**

Evaluate $\int_{1}^{2} \left[\int_{1}^{x} (2x^2y^{-2} + 2y) \, dy \right] dx$.

Solution   Using the result of Example 1, you have

$$\int_{1}^{2} \left[\int_{1}^{x} (2x^2y^{-2} + 2y) \, dy \right] dx = \int_{1}^{2} (3x^2 - 2x - 1) \, dx$$

$$= \left[x^3 - x^2 - x \right]_{1}^{2} \qquad \text{Integrate with respect to } x.$$

$$= 2 - (-1)$$

$$= 3.$$

The integral in Example 2 is an **iterated integral.** The brackets used in Example 2 are normally not written. Instead, iterated integrals are usually written simply as

$$\int_{a}^{b} \int_{g_1(x)}^{g_2(x)} f(x, y) \, dy \, dx \quad \text{and} \quad \int_{c}^{d} \int_{h_1(y)}^{h_2(y)} f(x, y) \, dx \, dy.$$

The **inside limits of integration** can be variable with respect to the outer variable of integration. However, the **outside limits of integration** *must be* constant with respect to both variables of integration. After performing the inside integration, you obtain a "standard" definite integral, and the second integration produces a real number. The limits of integration for an iterated integral identify two sets of boundary intervals for the variables. For instance, in Example 2, the outside limits indicate that *x* lies in the interval $1 \le x \le 2$ and the inside limits indicate that *y* lies in the interval $1 \le y \le x$. Together, these two intervals determine the **region of integration R** of the iterated integral, as shown in Figure 14.1.

Because an iterated integral is just a special type of definite integral—one in which the integrand is also an integral—you can use the properties of definite integrals to evaluate iterated integrals.

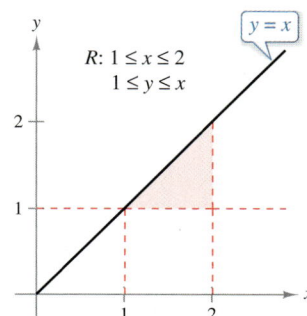

The region of integration for
$$\int_{1}^{2} \int_{1}^{x} f(x, y) \, dy \, dx$$

Figure 14.1

Area of a Plane Region

In the remainder of this section, you will take a new look at an old problem—that of finding the area of a plane region. Consider the plane region R bounded by $a \le x \le b$ and $g_1(x) \le y \le g_2(x)$, as shown in Figure 14.2. The area of R is

$$\int_a^b [g_2(x) - g_1(x)] \, dx. \qquad \text{Area of } R$$

Using the Fundamental Theorem of Calculus, you can rewrite the integrand $g_2(x) - g_1(x)$ as a definite integral. Specifically, consider x to be fixed and let y vary from $g_1(x)$ to $g_2(x)$, and you can write

$$\int_{g_1(x)}^{g_2(x)} dy = y \Big]_{g_1(x)}^{g_2(x)} = g_2(x) - g_1(x).$$

Combining these two integrals, you can write the area of the region R as an iterated integral

$$\int_a^b \int_{g_1(x)}^{g_2(x)} dy \, dx = \int_a^b y \Big]_{g_1(x)}^{g_2(x)} dx = \int_a^b [g_2(x) - g_1(x)] \, dx. \qquad \text{Area of } R$$

Placing a representative rectangle in the region R helps determine both the order and the limits of integration. A vertical rectangle implies the order $dy \, dx$, with the inside limits corresponding to the upper and lower bounds of the rectangle, as shown in Figure 14.2. This type of region is **vertically simple,** because the outside limits of integration represent the vertical lines

$$x = a$$

and

$$x = b.$$

Similarly, a horizontal rectangle implies the order $dx \, dy$, with the inside limits determined by the left and right bounds of the rectangle, as shown in Figure 14.3. This type of region is **horizontally simple,** because the outside limits represent the horizontal lines

$$y = c$$

and

$$y = d.$$

The iterated integrals used for these two types of simple regions are summarized as follows.

Region is bounded by
$a \le x \le b$ and
$g_1(x) \le y \le g_2(x)$

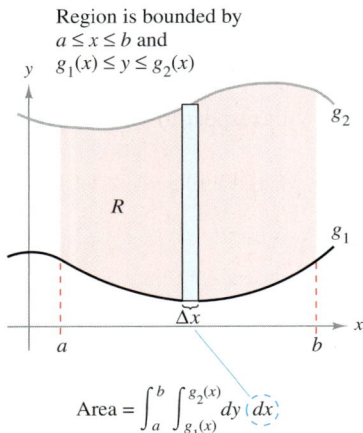

$$\text{Area} = \int_a^b \int_{g_1(x)}^{g_2(x)} dy \, dx$$

Vertically simple region
Figure 14.2

Region is bounded by
$c \le y \le d$ and
$h_1(y) \le x \le h_2(y)$

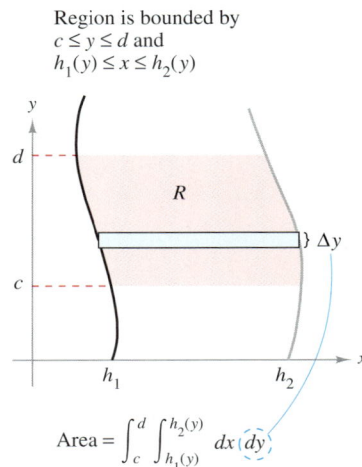

$$\text{Area} = \int_c^d \int_{h_1(y)}^{h_2(y)} dx \, dy$$

Horizontally simple region
Figure 14.3

> **REMARK** Be sure you see that the orders of integration of these two integrals are different—the order $dy \, dx$ corresponds to a vertically simple region, and the order $dx \, dy$ corresponds to a horizontally simple region.

Area of a Region in the Plane

1. If R is defined by $a \le x \le b$ and $g_1(x) \le y \le g_2(x)$, where g_1 and g_2 are continuous on $[a, b]$, then the area of R is

$$A = \int_a^b \int_{g_1(x)}^{g_2(x)} dy \, dx. \qquad \text{Figure 14.2 (vertically simple)}$$

2. If R is defined by $c \le y \le d$ and $h_1(y) \le x \le h_2(y)$, where h_1 and h_2 are continuous on $[c, d]$, then the area of R is

$$A = \int_c^d \int_{h_1(y)}^{h_2(y)} dx \, dy. \qquad \text{Figure 14.3 (horizontally simple)}$$

If all four limits of integration happen to be constants, then the region of integration is rectangular, as shown in Example 3.

EXAMPLE 3   **The Area of a Rectangular Region**

Use an iterated integral to represent the area of the rectangle shown in Figure 14.4.

Solution   The region shown in Figure 14.4 is both vertically simple and horizontally simple, so you can use either order of integration. By choosing the order $dy\,dx$, you obtain the following.

$$\int_a^b \int_c^d dy\,dx = \int_a^b y \Big]_c^d dx \qquad \text{Integrate with respect to } y.$$

$$= \int_a^b (d - c)\,dx$$

$$= \Big[(d - c)x\Big]_a^b \qquad \text{Integrate with respect to } x.$$

$$= (d - c)(b - a)$$

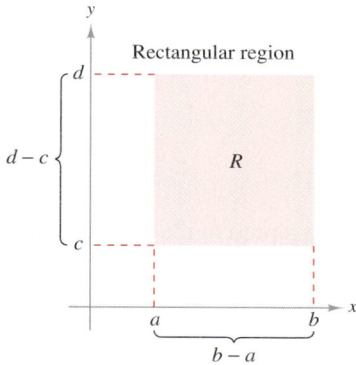

Figure 14.4

Notice that this answer is consistent with what you know from geometry.

EXAMPLE 4   **Finding Area by an Iterated Integral**

Use an iterated integral to find the area of the region bounded by the graphs of

$$f(x) = \sin x \qquad \text{Sine curve forms upper boundary.}$$

and

$$g(x) = \cos x \qquad \text{Cosine curve forms lower boundary.}$$

between $x = \pi/4$ and $x = 5\pi/4$.

Solution   Because f and g are given as functions of x, a vertical representative rectangle is convenient, and you can choose $dy\,dx$ as the order of integration, as shown in Figure 14.5. The outside limits of integration are

$$\frac{\pi}{4} \le x \le \frac{5\pi}{4}.$$

Moreover, because the rectangle is bounded above by $f(x) = \sin x$ and below by $g(x) = \cos x$, you have

$$\text{Area of } R = \int_{\pi/4}^{5\pi/4} \int_{\cos x}^{\sin x} dy\,dx$$

$$= \int_{\pi/4}^{5\pi/4} y \Big]_{\cos x}^{\sin x} dx \qquad \text{Integrate with respect to } y.$$

$$= \int_{\pi/4}^{5\pi/4} (\sin x - \cos x)\,dx$$

$$= \Big[-\cos x - \sin x \Big]_{\pi/4}^{5\pi/4} \qquad \text{Integrate with respect to } x.$$

$$= 2\sqrt{2}.$$

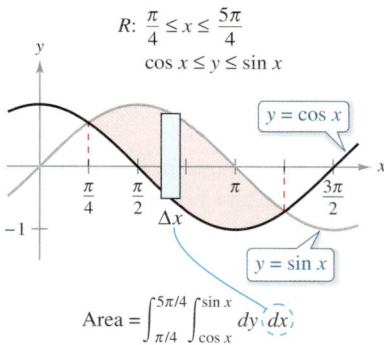

Figure 14.5

The region of integration of an iterated integral need not have any straight lines as boundaries. For instance, the region of integration shown in Figure 14.5 is *vertically simple* even though it has no vertical lines as left and right boundaries. The quality that makes the region vertically simple is that it is bounded above and below by the graphs of *functions of x*.

One order of integration will often produce a simpler integration problem than the other order. For instance, try reworking Example 4 with the order $dx\,dy$—you may be surprised to see that the task is formidable. However, if you succeed, you will see that the answer is the same. In other words, the order of integration affects the ease of integration, but not the value of the integral.

EXAMPLE 5   **Comparing Different Orders of Integration**

⋯▷ *See LarsonCalculus.com for an interactive version of this type of example.*

Sketch the region whose area is represented by the integral

$$\int_0^2 \int_{y^2}^4 dx\,dy.$$

Then find another iterated integral using the order $dy\,dx$ to represent the same area and show that both integrals yield the same value.

Solution   From the given limits of integration, you know that

$$y^2 \le x \le 4 \qquad \text{\textcolor{red}{Inner limits of integration}}$$

which means that the region R is bounded on the left by the parabola $x = y^2$ and on the right by the line $x = 4$. Furthermore, because

$$0 \le y \le 2 \qquad \text{\textcolor{red}{Outer limits of integration}}$$

you know that R is bounded below by the x-axis, as shown in Figure 14.6(a). The value of this integral is

$$\int_0^2 \int_{y^2}^4 dx\,dy = \int_0^2 x \Big]_{y^2}^4 dy \qquad \text{\textcolor{red}{Integrate with respect to } x.}$$

$$= \int_0^2 (4 - y^2)\,dy$$

$$= \left[4y - \frac{y^3}{3} \right]_0^2 \qquad \text{\textcolor{red}{Integrate with respect to } y.}$$

$$= \frac{16}{3}.$$

To change the order of integration to $dy\,dx$, place a vertical rectangle in the region, as shown in Figure 14.6(b). From this, you can see that the constant bounds $0 \le x \le 4$ serve as the outer limits of integration. By solving for y in the equation $x = y^2$, you can conclude that the inner bounds are $0 \le y \le \sqrt{x}$. So, the area of the region can also be represented by

$$\int_0^4 \int_0^{\sqrt{x}} dy\,dx.$$

By evaluating this integral, you can see that it has the same value as the original integral.

$$\int_0^4 \int_0^{\sqrt{x}} dy\,dx = \int_0^4 y \Big]_0^{\sqrt{x}} dx \qquad \text{\textcolor{red}{Integrate with respect to } y.}$$

$$= \int_0^4 \sqrt{x}\,dx$$

$$= \frac{2}{3} x^{3/2} \Big]_0^4 \qquad \text{\textcolor{red}{Integrate with respect to } x.}$$

$$= \frac{16}{3}$$

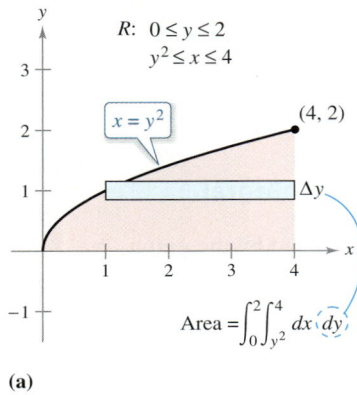

R: $0 \le y \le 2$
$y^2 \le x \le 4$

$x = y^2$ (4, 2)

Δy

Area $= \int_0^2 \int_{y^2}^4 dx\,dy$

(a)

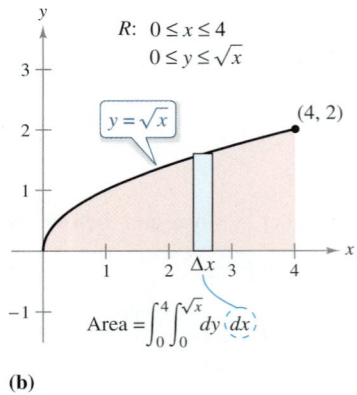

R: $0 \le x \le 4$
$0 \le y \le \sqrt{x}$

$y = \sqrt{x}$ (4, 2)

Δx

Area $= \int_0^4 \int_0^{\sqrt{x}} dy\,dx$

(b)

Figure 14.6

Sometimes it is not possible to calculate the area of a region with a single iterated integral. In these cases, you can divide the region into subregions such that the area of each subregion can be calculated by an iterated integral. The total area is then the sum of the iterated integrals.

▷ **TECHNOLOGY**   Some computer software can perform symbolic integration for integrals such as those in Example 6. If you have access to such software, use it to evaluate the integrals in the exercises and examples given in this section.

EXAMPLE 6   An Area Represented by Two Iterated Integrals

Find the area of the region R that lies below the parabola

$$y = 4x - x^2 \qquad \text{Parabola forms upper boundary.}$$

above the x-axis, and above the line

$$y = -3x + 6. \qquad \text{Line and } x\text{-axis form lower boundary.}$$

Solution   Begin by dividing R into the two subregions R_1 and R_2 shown in Figure 14.7.

•• **REMARK**   In Examples 3 through 6, be sure you see the benefit of sketching the region of integration. You should develop the habit of making sketches to help you determine the limits of integration for all iterated integrals in this chapter.
•••••••••••••••••▷

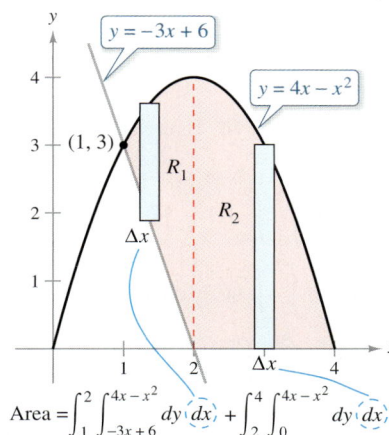

$$\text{Area} = \int_1^2 \int_{-3x+6}^{4x-x^2} dy\, dx + \int_2^4 \int_0^{4x-x^2} dy\, dx$$

Figure 14.7

In both regions, it is convenient to use vertical rectangles, and you have

$$\begin{aligned}
\text{Area} &= \int_1^2 \int_{-3x+6}^{4x-x^2} dy\, dx + \int_2^4 \int_0^{4x-x^2} dy\, dx \\[2mm]
&= \int_1^2 (4x - x^2 + 3x - 6)\, dx + \int_2^4 (4x - x^2)\, dx \\[2mm]
&= \left[\frac{7x^2}{2} - \frac{x^3}{3} - 6x \right]_1^2 + \left[2x^2 - \frac{x^3}{3} \right]_2^4 \\[2mm]
&= \left(14 - \frac{8}{3} - 12 - \frac{7}{2} + \frac{1}{3} + 6 \right) + \left(32 - \frac{64}{3} - 8 + \frac{8}{3} \right) \\[2mm]
&= \frac{15}{2}.
\end{aligned}$$

The area of the region is $15/2$ square units. Try checking this using the procedure for finding the area between two curves, as presented in Section 7.1. ■

At this point, you may be wondering why you would need iterated integrals. After all, you already know how to use conventional integration to find the area of a region in the plane. (For instance, compare the solution of Example 4 in this section with that given in Example 3 in Section 7.1.) The need for iterated integrals will become clear in the next section. In this section, primary attention is given to procedures for finding the limits of integration of the region of an iterated integral, and the following exercise set is designed to develop skill in this important procedure.

14.1 Exercises

See **CalcChat.com** for tutorial help and worked-out solutions to odd-numbered exercises.

Evaluating an Integral In Exercises 1–10, evaluate the integral.

1. $\displaystyle\int_0^x (x + 2y)\, dy$

2. $\displaystyle\int_x^{x^2} \frac{y}{x}\, dy$

3. $\displaystyle\int_1^{2y} \frac{y}{x}\, dx, \quad y > 0$

4. $\displaystyle\int_0^{\cos y} y\, dx$

5. $\displaystyle\int_0^{\sqrt{4-x^2}} x^2 y\, dy$

6. $\displaystyle\int_{x^3}^{\sqrt{x}} (x^2 + 3y^2)\, dy$

7. $\displaystyle\int_{e^y}^y \frac{y\ln x}{x}\, dx, \quad y > 0$

8. $\displaystyle\int_{-\sqrt{1-y^2}}^{\sqrt{1-y^2}} (x^2 + y^2)\, dx$

9. $\displaystyle\int_0^{x^3} ye^{-y/x}\, dy$

10. $\displaystyle\int_y^{\pi/2} \sin^3 x \cos y\, dx$

Evaluating an Iterated Integral In Exercises 11–30, evaluate the iterated integral.

11. $\displaystyle\int_0^1 \int_0^2 (x + y)\, dy\, dx$

12. $\displaystyle\int_{-1}^1 \int_{-2}^2 (x^2 - y^2)\, dy\, dx$

13. $\displaystyle\int_1^2 \int_0^4 (x^2 - 2y^2)\, dx\, dy$

14. $\displaystyle\int_{-1}^2 \int_1^3 (x + y^2)\, dx\, dy$

15. $\displaystyle\int_0^{\pi/2} \int_0^1 y\cos x\, dy\, dx$

16. $\displaystyle\int_0^{\ln 4} \int_0^{\ln 3} e^{x+y}\, dy\, dx$

17. $\displaystyle\int_0^{\pi} \int_0^{\sin x} (1 + \cos x)\, dy\, dx$

18. $\displaystyle\int_1^4 \int_1^{\sqrt{x}} 2ye^{-x}\, dy\, dx$

19. $\displaystyle\int_0^1 \int_0^x \sqrt{1 - x^2}\, dy\, dx$

20. $\displaystyle\int_{-4}^4 \int_0^{x^2} \sqrt{64 - x^3}\, dy\, dx$

21. $\displaystyle\int_{-1}^5 \int_0^{3y} \left(3 + x^2 + \frac{1}{4}y^2\right) dx\, dy$

22. $\displaystyle\int_0^2 \int_y^{2y} (10 + 2x^2 + 2y^2)\, dx\, dy$

23. $\displaystyle\int_0^1 \int_0^{\sqrt{1-y^2}} (x + y)\, dx\, dy$

24. $\displaystyle\int_0^2 \int_{3y^2-6y}^{2y-y^2} 3y\, dx\, dy$

25. $\displaystyle\int_0^2 \int_0^{\sqrt{4-y^2}} \frac{2}{\sqrt{4 - y^2}}\, dx\, dy$

26. $\displaystyle\int_1^3 \int_0^y \frac{4}{x^2 + y^2}\, dx\, dy$

27. $\displaystyle\int_0^{\pi/2} \int_0^{2\cos\theta} r\, dr\, d\theta$

28. $\displaystyle\int_0^{\pi/4} \int_{\sqrt{3}}^{\sqrt{3}\cos\theta} r\, dr\, d\theta$

29. $\displaystyle\int_0^{\pi/2} \int_0^{\sin\theta} \theta r\, dr\, d\theta$

30. $\displaystyle\int_0^{\pi/4} \int_0^{\cos\theta} 3r^2 \sin\theta\, dr\, d\theta$

Evaluating an Improper Iterated Integral In Exercises 31–34, evaluate the improper iterated integral.

31. $\displaystyle\int_1^{\infty} \int_0^{1/x} y\, dy\, dx$

32. $\displaystyle\int_0^3 \int_0^{\infty} \frac{x^2}{1 + y^2}\, dy\, dx$

33. $\displaystyle\int_1^{\infty} \int_1^{\infty} \frac{1}{xy}\, dx\, dy$

34. $\displaystyle\int_0^{\infty} \int_0^{\infty} xye^{-(x^2+y^2)}\, dx\, dy$

Finding the Area of a Region In Exercises 35–38, use an iterated integral to find the area of the region.

35.

36.

37.

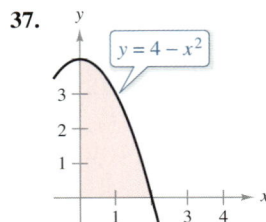

$y = 4 - x^2$

38.

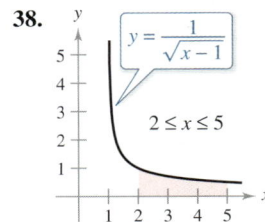

$y = \dfrac{1}{\sqrt{x-1}}$

$2 \le x \le 5$

Finding the Area of a Region In Exercises 39–44, use an iterated integral to find the area of the region bounded by the graphs of the equations.

39. $\sqrt{x} + \sqrt{y} = 2, \quad x = 0, \quad y = 0$

40. $y = x^{3/2}, \quad y = 2x$

41. $2x - 3y = 0, \quad x + y = 5, \quad y = 0$

42. $\dfrac{x^2}{a^2} + \dfrac{y^2}{b^2} = 1$

43. $y = 4 - x^2, \quad y = x + 2$

44. $y = x, \quad y = 2x, \quad x = 2$

Switching the Order of Integration In Exercises 45–52, sketch the region R of integration and switch the order of integration.

45. $\displaystyle\int_0^4 \int_0^y f(x, y)\, dx\, dy$

46. $\displaystyle\int_0^4 \int_{\sqrt{y}}^2 f(x, y)\, dx\, dy$

47. $\displaystyle\int_{-2}^2 \int_0^{\sqrt{4-x^2}} f(x, y)\, dy\, dx$

48. $\displaystyle\int_0^2 \int_0^{4-x^2} f(x, y)\, dy\, dx$

49. $\displaystyle\int_1^{10} \int_0^{\ln y} f(x, y)\, dx\, dy$

50. $\displaystyle\int_{-1}^2 \int_0^{e^{-x}} f(x, y)\, dy\, dx$

51. $\displaystyle\int_{-1}^1 \int_{x^2}^1 f(x, y)\, dy\, dx$

52. $\displaystyle\int_{-\pi/2}^{\pi/2} \int_0^{\cos x} f(x, y)\, dy\, dx$

Switching the Order of Integration In Exercises 53–62, sketch the region R whose area is given by the iterated integral. Then switch the order of integration and show that both orders yield the same area.

53. $\displaystyle\int_0^1 \int_0^2 dy\, dx$

54. $\displaystyle\int_1^2 \int_2^4 dx\, dy$

55. $\displaystyle\int_0^1 \int_{-\sqrt{1-y^2}}^{\sqrt{1-y^2}} dx\, dy$

56. $\displaystyle\int_{-2}^2 \int_{-\sqrt{4-x^2}}^{\sqrt{4-x^2}} dy\, dx$

57. $\displaystyle\int_0^2\int_0^x dy\,dx + \int_2^4\int_0^{4-x} dy\,dx$

58. $\displaystyle\int_0^4\int_0^{x/2} dy\,dx + \int_4^6\int_0^{6-x} dy\,dx$

59. $\displaystyle\int_0^2\int_{x/2}^1 dy\,dx$

60. $\displaystyle\int_0^9\int_{\sqrt{x}}^3 dy\,dx$

61. $\displaystyle\int_0^1\int_{y^2}^{\sqrt[3]{y}} dx\,dy$

62. $\displaystyle\int_{-2}^2\int_0^{4-y^2} dx\,dy$

63. Think About It   Give a geometric argument for the equality. Verify the equality analytically.

$$\int_0^5\int_x^{\sqrt{50-x^2}} x^2y^2\,dy\,dx =$$

$$\int_0^5\int_0^y x^2y^2\,dx\,dy + \int_5^{5\sqrt{2}}\int_0^{\sqrt{50-y^2}} x^2y^2\,dx\,dy$$

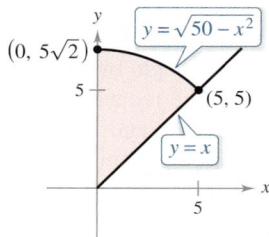
$(0, 5\sqrt{2})$ $y=\sqrt{50-x^2}$ $(5,5)$ $y=x$

64. HOW DO YOU SEE IT?   Complete the iterated integrals so that each one represents the area of the region R (see figure).

(a) Area $= \displaystyle\int\int dx\,dy$ (b) Area $= \displaystyle\int\int dy\,dx$

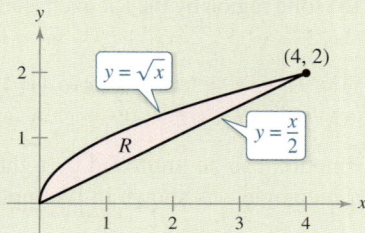
$y=\sqrt{x}$ $(4,2)$ R $y=\dfrac{x}{2}$

Switching the Order of Integration   In Exercises 65–70, sketch the region of integration. Then evaluate the iterated integral. (*Hint:* Note that it is necessary to switch the order of integration.)

65. $\displaystyle\int_0^2\int_x^2 x\sqrt{1+y^3}\,dy\,dx$

66. $\displaystyle\int_0^4\int_{\sqrt{x}}^2 \frac{3}{2+y^3}\,dy\,dx$

67. $\displaystyle\int_0^1\int_{2x}^2 4e^{y^2}\,dy\,dx$

68. $\displaystyle\int_0^2\int_x^2 e^{-y^2}\,dy\,dx$

69. $\displaystyle\int_0^1\int_y^1 \sin x^2\,dx\,dy$

70. $\displaystyle\int_0^2\int_{y^2}^4 \sqrt{x}\sin x\,dx\,dy$

Evaluating an Iterated Integral Using Technology   In Exercises 71–78, use a computer algebra system to evaluate the iterated integral.

71. $\displaystyle\int_0^2\int_{x^2}^{2x} (x^3+3y^2)\,dy\,dx$

72. $\displaystyle\int_0^1\int_y^{2y} \sin(x+y)\,dx\,dy$

73. $\displaystyle\int_0^4\int_0^y \frac{2}{(x+1)(y+1)}\,dx\,dy$

74. $\displaystyle\int_0^a\int_0^{a-x} (x^2+y^2)\,dy\,dx$

75. $\displaystyle\int_0^2\int_0^{4-x^2} e^{xy}\,dy\,dx$

76. $\displaystyle\int_0^2\int_x^2 \sqrt{16-x^3-y^3}\,dy\,dx$

77. $\displaystyle\int_0^{2\pi}\int_0^{1+\cos\theta} 6r^2\cos\theta\,dr\,d\theta$

78. $\displaystyle\int_0^{\pi/2}\int_0^{1+\sin\theta} 15\theta r\,dr\,d\theta$

Comparing Different Orders of Integration Using Technology   In Exercises 79 and 80, (a) sketch the region of integration, (b) switch the order of integration, and (c) use a computer algebra system to show that both orders yield the same value.

79. $\displaystyle\int_0^2\int_{y^3}^{4\sqrt{2y}} (x^2y-xy^2)\,dx\,dy$

80. $\displaystyle\int_0^2\int_{\sqrt{4-x^2}}^{4-x^2/4} \frac{xy}{x^2+y^2+1}\,dy\,dx$

WRITING ABOUT CONCEPTS

81. Iterated Integral   Explain what is meant by an iterated integral. How is it evaluated?

82. Vertically Simple and Horizontally Simple   Describe regions that are vertically simple and regions that are horizontally simple.

83. Region of Integration   Give a geometric description of the region of integration when the inside and outside limits of integration are constants.

84. Order of Integration   Explain why it is sometimes an advantage to change the order of integration.

True or False?   In Exercises 85 and 86, determine whether the statement is true or false. If it is false, explain why or give an example that shows it is false.

85. $\displaystyle\int_a^b\int_c^d f(x,y)\,dy\,dx = \int_c^d\int_a^b f(x,y)\,dx\,dy$

86. $\displaystyle\int_0^1\int_0^x f(x,y)\,dy\,dx = \int_0^1\int_0^y f(x,y)\,dx\,dy$

14.2 Double Integrals and Volume

- Use a double integral to represent the volume of a solid region and use properties of double integrals.
- Evaluate a double integral as an iterated integral.
- Find the average value of a function over a region.

Double Integrals and Volume of a Solid Region

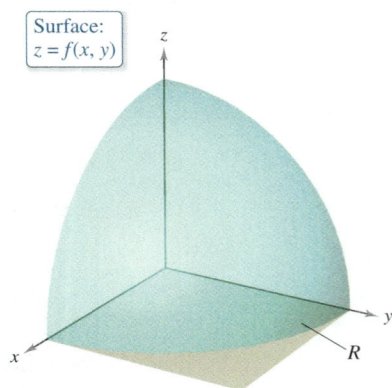

Surface:
$z = f(x, y)$

Figure 14.8

You already know that a definite integral over an *interval* uses a limit process to assign measures to quantities such as area, volume, arc length, and mass. In this section, you will use a similar process to define the **double integral** of a function of two variables over a *region in the plane*.

Consider a continuous function f such that $f(x, y) \geq 0$ for all (x, y) in a region R in the xy-plane. The goal is to find the volume of the solid region lying between the surface given by

$$z = f(x, y) \qquad \text{Surface lying above the } xy\text{-plane}$$

and the xy-plane, as shown in Figure 14.8. You can begin by superimposing a rectangular grid over the region, as shown in Figure 14.9. The rectangles lying entirely within R form an **inner partition** Δ, whose **norm** $\|\Delta\|$ is defined as the length of the longest diagonal of the n rectangles. Next, choose a point (x_i, y_i) in each rectangle and form the rectangular prism whose height is

$$f(x_i, y_i) \qquad \text{Height of } i\text{th prism}$$

as shown in Figure 14.10. Because the area of the ith rectangle is

$$\Delta A_i \qquad \text{Area of } i\text{th rectangle}$$

it follows that the volume of the ith prism is

$$f(x_i, y_i)\, \Delta A_i \qquad \text{Volume of } i\text{th prism}$$

and you can approximate the volume of the solid region by the Riemann sum of the volumes of all n prisms,

$$\sum_{i=1}^{n} f(x_i, y_i)\, \Delta A_i \qquad \text{Riemann sum}$$

as shown in Figure 14.11. This approximation can be improved by tightening the mesh of the grid to form smaller and smaller rectangles, as shown in Example 1.

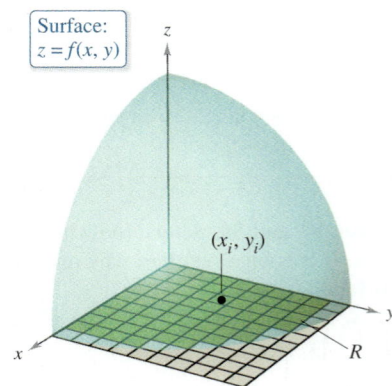

Surface:
$z = f(x, y)$

The rectangles lying within R form an inner partition of R.
Figure 14.9

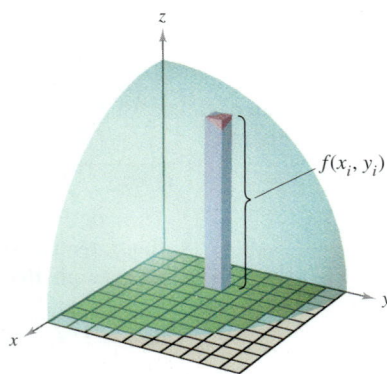

Rectangular prism whose base has an area of ΔA_i and whose height is $f(x_i, y_i)$
Figure 14.10

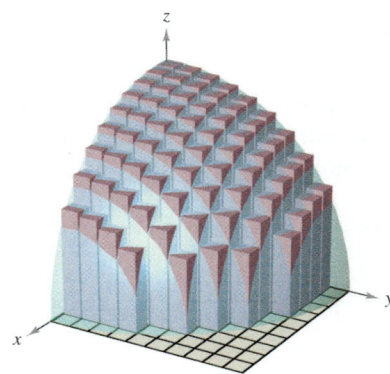

Volume approximated by rectangular prisms
Figure 14.11

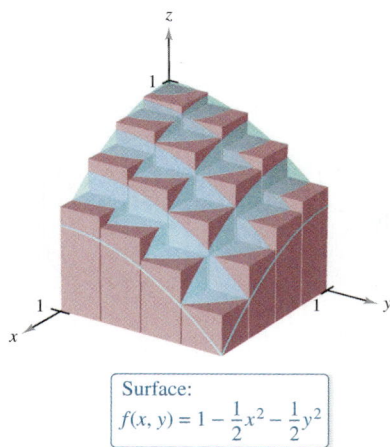

Surface:
$$f(x, y) = 1 - \frac{1}{2}x^2 - \frac{1}{2}y^2$$

Figure 14.12

EXAMPLE 1   **Approximating the Volume of a Solid**

Approximate the volume of the solid lying between the paraboloid

$$f(x, y) = 1 - \frac{1}{2}x^2 - \frac{1}{2}y^2$$

and the square region R given by $0 \le x \le 1$, $0 \le y \le 1$. Use a partition made up of squares whose sides have a length of $\frac{1}{4}$.

Solution   Begin by forming the specified partition of R. For this partition, it is convenient to choose the centers of the subregions as the points at which to evaluate $f(x, y)$.

$\left(\frac{1}{8}, \frac{1}{8}\right)$	$\left(\frac{1}{8}, \frac{3}{8}\right)$	$\left(\frac{1}{8}, \frac{5}{8}\right)$	$\left(\frac{1}{8}, \frac{7}{8}\right)$
$\left(\frac{3}{8}, \frac{1}{8}\right)$	$\left(\frac{3}{8}, \frac{3}{8}\right)$	$\left(\frac{3}{8}, \frac{5}{8}\right)$	$\left(\frac{3}{8}, \frac{7}{8}\right)$
$\left(\frac{5}{8}, \frac{1}{8}\right)$	$\left(\frac{5}{8}, \frac{3}{8}\right)$	$\left(\frac{5}{8}, \frac{5}{8}\right)$	$\left(\frac{5}{8}, \frac{7}{8}\right)$
$\left(\frac{7}{8}, \frac{1}{8}\right)$	$\left(\frac{7}{8}, \frac{3}{8}\right)$	$\left(\frac{7}{8}, \frac{5}{8}\right)$	$\left(\frac{7}{8}, \frac{7}{8}\right)$

Because the area of each square is $\Delta A_i = \frac{1}{16}$, you can approximate the volume by the sum

$$\sum_{i=1}^{16} f(x_i, y_i)\,\Delta A_i = \sum_{i=1}^{16}\left(1 - \frac{1}{2}x_i^2 - \frac{1}{2}y_i^2\right)\left(\frac{1}{16}\right) \approx 0.672.$$

This approximation is shown graphically in Figure 14.12. The exact volume of the solid is $\frac{2}{3}$ (see Example 2). You can obtain a better approximation by using a finer partition. For example, with a partition of squares with sides of length $\frac{1}{10}$, the approximation is 0.668.

▷ **TECHNOLOGY**   Some three-dimensional graphing utilities are capable of graphing figures such as that shown in Figure 14.12. For instance, the graph shown at the right was drawn with a computer program. In this graph, note that each of the rectangular prisms lies within the solid region.

In Example 1, note that by using finer partitions, you obtain better approximations of the volume. This observation suggests that you could obtain the exact volume by taking a limit. That is,

$$\text{Volume} = \lim_{\|\Delta\| \to 0} \sum_{i=1}^{n} f(x_i, y_i)\,\Delta A_i.$$

The precise meaning of this limit is that the limit is equal to L if for every $\varepsilon > 0$, there exists a $\delta > 0$ such that

$$\left| L - \sum_{i=1}^{n} f(x_i, y_i)\,\Delta A_i \right| < \varepsilon$$

for all partitions Δ of the plane region R (that satisfy $\|\Delta\| < \delta$) and for all possible choices of x_i and y_i in the ith region.

Using the limit of a Riemann sum to define volume is a special case of using the limit to define a **double integral**. The general case, however, does not require that the function be positive or continuous.

Exploration

The entries in the table represent the depths (in 10-yard units) of earth at the centers of the squares in the figure below.

y \ x	1	2	3
1	10	9	7
2	7	7	4
3	5	5	4
4	4	5	3

Approximate the number of cubic yards of earth in the first octant. (This exploration was submitted by Robert Vojack, Ridgewood High School, Ridgewood, NJ.)

Definition of Double Integral

If f is defined on a closed, bounded region R in the xy-plane, then the **double integral of f over R** is

$$\int_R \int f(x, y)\, dA = \lim_{\|\Delta\| \to 0} \sum_{i=1}^{n} f(x_i, y_i)\, \Delta A_i$$

provided the limit exists. If the limit exists, then f is **integrable** over R.

Having defined a double integral, you will see that a definite integral is occasionally referred to as a **single integral.**

Sufficient conditions for the double integral of f on the region R to exist are that R can be written as a union of a finite number of nonoverlapping subregions (see figure at the right) that are vertically or horizontally simple *and* that f is continuous on the region R. This means that the intersection of two nonoverlapping regions is a set that has an area of 0. In Figure 14.13, the area of the line segment common to R_1 and R_2 is 0.

A double integral can be used to find the volume of a solid region that lies between the xy-plane and the surface given by $z = f(x, y)$.

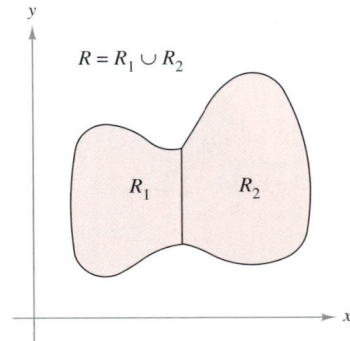

The two regions R_1 and R_2 are nonoverlapping.

Figure 14.13

Volume of a Solid Region

If f is integrable over a plane region R and $f(x, y) \geq 0$ for all (x, y) in R, then the volume of the solid region that lies above R and below the graph of f is

$$V = \int_R \int f(x, y)\, dA.$$

Double integrals share many properties of single integrals.

THEOREM 14.1 Properties of Double Integrals

Let f and g be continuous over a closed, bounded plane region R, and let c be a constant.

1. $\int_R \int cf(x, y)\, dA = c \int_R \int f(x, y)\, dA$

2. $\int_R \int [f(x, y) \pm g(x, y)]\, dA = \int_R \int f(x, y)\, dA \pm \int_R \int g(x, y)\, dA$

3. $\int_R \int f(x, y)\, dA \geq 0, \quad$ if $f(x, y) \geq 0$

4. $\int_R \int f(x, y)\, dA \geq \int_R \int g(x, y)\, dA, \quad$ if $f(x, y) \geq g(x, y)$

5. $\int_R \int f(x, y)\, dA = \int_{R_1} \int f(x, y)\, dA + \int_{R_2} \int f(x, y)\, dA$, where R is the union of two nonoverlapping subregions R_1 and R_2.

Evaluation of Double Integrals

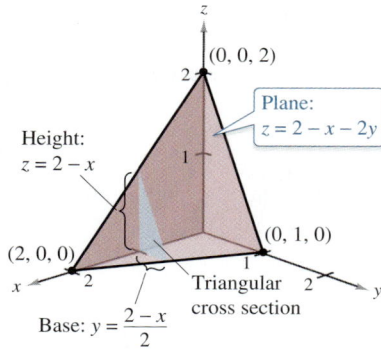

Figure 14.14

Normally, the first step in evaluating a double integral is to rewrite it as an iterated integral. To show how this is done, a geometric model of a double integral is used as the volume of a solid.

Consider the solid region bounded by the plane $z = f(x, y) = 2 - x - 2y$ and the three coordinate planes, as shown in Figure 14.14. Each vertical cross section taken parallel to the yz-plane is a triangular region whose base has a length of $y = (2 - x)/2$ and whose height is $z = 2 - x$. This implies that for a fixed value of x, the area of the triangular cross section is

$$A(x) = \frac{1}{2}(\text{base})(\text{height}) = \frac{1}{2}\left(\frac{2-x}{2}\right)(2-x) = \frac{(2-x)^2}{4}.$$

By the formula for the volume of a solid with known cross sections (Section 7.2), the volume of the solid is

$$\text{Volume} = \int_a^b A(x)\, dx$$

$$= \int_0^2 \frac{(2-x)^2}{4}\, dx$$

$$= -\frac{(2-x)^3}{12}\Big]_0^2$$

$$= \frac{2}{3}.$$

This procedure works no matter how $A(x)$ is obtained. In particular, you can find $A(x)$ by integration, as shown in Figure 14.15. That is, you consider x to be constant, and integrate $z = 2 - x - 2y$ from 0 to $(2 - x)/2$ to obtain

$$A(x) = \int_0^{(2-x)/2} (2 - x - 2y)\, dy$$

$$= \left[(2-x)y - y^2\right]_0^{(2-x)/2}$$

$$= \frac{(2-x)^2}{4}.$$

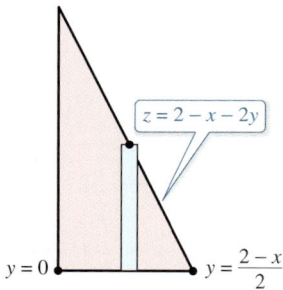

Triangular cross section
Figure 14.15

Combining these results, you have the *iterated integral*

$$\text{Volume} = \int_R\!\!\int f(x, y)\, dA = \int_0^2 \int_0^{(2-x)/2} (2 - x - 2y)\, dy\, dx.$$

To understand this procedure better, it helps to imagine the integration as two sweeping motions. For the inner integration, a vertical line sweeps out the area of a cross section. For the outer integration, the triangular cross section sweeps out the volume, as shown in Figure 14.16.

Integrate with respect to y to obtain the area of the cross section.   Integrate with respect to x to obtain the volume of the solid.
Figure 14.16

The next theorem was proved by the Italian mathematician Guido Fubini (1879–1943). The theorem states that if R is a vertically or horizontally simple region and f is continuous on R, then the double integral of f on R is equal to an iterated integral.

THEOREM 14.2   Fubini's Theorem

Let f be continuous on a plane region R.

1. If R is defined by $a \le x \le b$ and $g_1(x) \le y \le g_2(x)$, where g_1 and g_2 are continuous on $[a, b]$, then

$$\int_R \int f(x, y)\, dA = \int_a^b \int_{g_1(x)}^{g_2(x)} f(x, y)\, dy\, dx.$$

2. If R is defined by $c \le y \le d$ and $h_1(y) \le x \le h_2(y)$, where h_1 and h_2 are continuous on $[c, d]$, then

$$\int_R \int f(x, y)\, dA = \int_c^d \int_{h_1(y)}^{h_2(y)} f(x, y)\, dx\, dy.$$

EXAMPLE 2   Evaluating a Double Integral as an Iterated Integral

Evaluate

$$\int_R \int \left(1 - \frac{1}{2}x^2 - \frac{1}{2}y^2\right) dA$$

where R is the region given by

$$0 \le x \le 1, \quad 0 \le y \le 1.$$

Solution Because the region R is a square, it is both vertically and horizontally simple, and you can use either order of integration. Choose $dy\, dx$ by placing a vertical representative rectangle in the region (see the figure at the right). This produces the following.

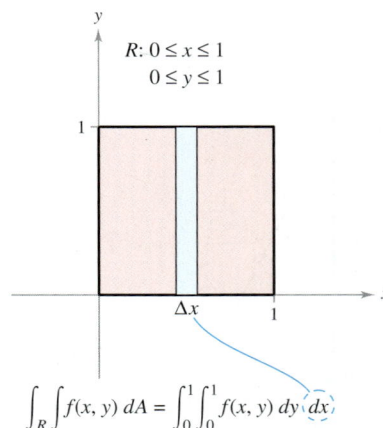

$R: 0 \le x \le 1$
$0 \le y \le 1$

$$\int_R \int f(x, y)\, dA = \int_0^1 \int_0^1 f(x, y)\, dy\, dx$$

$$\int_R \int \left(1 - \frac{1}{2}x^2 - \frac{1}{2}y^2\right) dA = \int_0^1 \int_0^1 \left(1 - \frac{1}{2}x^2 - \frac{1}{2}y^2\right) dy\, dx$$

$$= \int_0^1 \left[\left(1 - \frac{1}{2}x^2\right)y - \frac{y^3}{6}\right]_0^1 dx$$

$$= \int_0^1 \left(\frac{5}{6} - \frac{1}{2}x^2\right) dx$$

$$= \left[\frac{5}{6}x - \frac{x^3}{6}\right]_0^1$$

$$= \frac{2}{3}$$

The double integral evaluated in Example 2 represents the volume of the solid region approximated in Example 1. Note that the approximation obtained in Example 1 is quite good $\left(0.672 \text{ vs. } \frac{2}{3}\right)$, even though you used a partition consisting of only 16 squares. The error resulted because the centers of the square subregions were used as the points in the approximation. This is comparable to the Midpoint Rule approximation of a single integral.

The difficulty of evaluating a single integral $\int_a^b f(x)\,dx$ usually depends on the function f, not on the interval $[a, b]$. This is a major difference between single and double integrals. In the next example, you will integrate a function similar to the one in Examples 1 and 2. Notice that a change in the region R produces a much more difficult integration problem.

Finding Volume by a Double Integral

Find the volume of the solid region bounded by the paraboloid $z = 4 - x^2 - 2y^2$ and the xy-plane, as shown in Figure 14.17(a).

Solution   By letting $z = 0$, you can see that the base of the region in the xy-plane is the ellipse $x^2 + 2y^2 = 4$, as shown in Figure 14.17(b). This plane region is both vertically and horizontally simple, so the order $dy\,dx$ is appropriate.

Variable bounds for y: $\quad -\sqrt{\dfrac{(4 - x^2)}{2}} \le y \le \sqrt{\dfrac{(4 - x^2)}{2}}$

Constant bounds for x: $\quad -2 \le x \le 2$

The volume is

$$
\begin{aligned}
V &= \int_{-2}^{2}\int_{-\sqrt{(4-x^2)/2}}^{\sqrt{(4-x^2)/2}} (4 - x^2 - 2y^2)\,dy\,dx && \text{\color{red}See Figure 14.17(b).}\\[2mm]
&= \int_{-2}^{2}\left[(4 - x^2)y - \frac{2y^3}{3}\right]_{-\sqrt{(4-x^2)/2}}^{\sqrt{(4-x^2)/2}} dx \\[2mm]
&= \frac{4}{3\sqrt{2}}\int_{-2}^{2}(4 - x^2)^{3/2}\,dx \\[2mm]
&= \frac{4}{3\sqrt{2}}\int_{-\pi/2}^{\pi/2} 16\cos^4\theta\,d\theta && \text{\color{red}$x = 2\sin\theta$}\\[2mm]
&= \frac{64}{3\sqrt{2}}(2)\int_{0}^{\pi/2} \cos^4\theta\,d\theta \\[2mm]
&= \frac{128}{3\sqrt{2}}\left(\frac{3\pi}{16}\right) && \text{\color{red}Wallis's Formula}\\[2mm]
&= 4\sqrt{2}\,\pi.
\end{aligned}
$$

· · REMARK   In Example 3, note the usefulness of Wallis's Formula to evaluate $\int_0^{\pi/2}\cos^n\theta\,d\theta$. You may want to review this formula in Section 8.3.

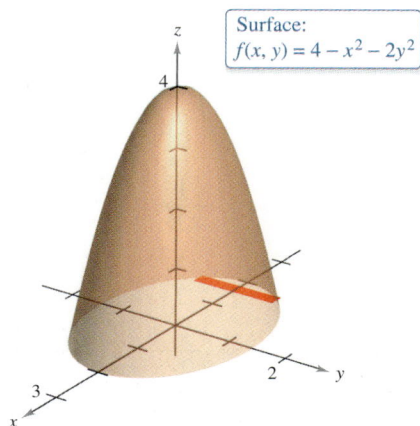

$$\boxed{\text{Surface:}\ f(x, y) = 4 - x^2 - 2y^2}$$

(a)

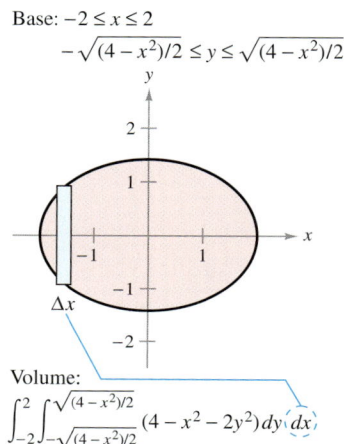

Base: $-2 \le x \le 2$
$-\sqrt{(4-x^2)/2} \le y \le \sqrt{(4-x^2)/2}$

Volume:
$$\int_{-2}^{2}\int_{-\sqrt{(4-x^2)/2}}^{\sqrt{(4-x^2)/2}} (4 - x^2 - 2y^2)\,dy\,dx$$

(b)

Figure 14.17

In Examples 2 and 3, the problems could be solved with either order of integration because the regions were both vertically and horizontally simple. Moreover, had you used the order $dx\,dy$, you would have obtained integrals of comparable difficulty. There are, however, some occasions in which one order of integration is much more convenient than the other. Example 4 shows such a case.

EXAMPLE 4   **Comparing Different Orders of Integration**

▷ *See LarsonCalculus.com for an interactive version of this type of example.*

Find the volume of the solid region bounded by the surface

$$f(x, y) = e^{-x^2} \qquad \text{Surface}$$

and the planes $z = 0$, $y = 0$, $y = x$, and $x = 1$, as shown in Figure 14.18.

Solution   The base of the solid region in the xy-plane is bounded by the lines $y = 0$, $x = 1$, and $y = x$. The two possible orders of integration are shown in Figure 14.19.

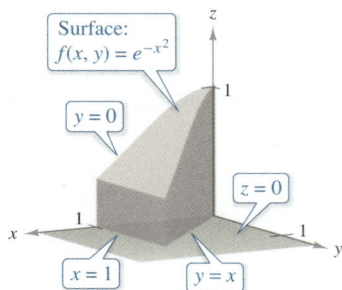

Surface:
$f(x, y) = e^{-x^2}$
$y = 0$
$z = 0$
$x = 1$
$y = x$

Base is bounded by $y = 0$, $y = x$, and $x = 1$.
Figure 14.18

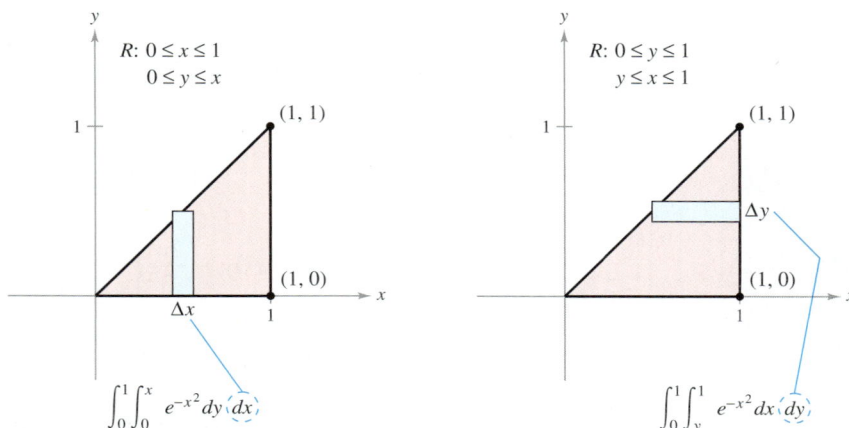

$R: 0 \le x \le 1$
$0 \le y \le x$
$(1, 1)$
$(1, 0)$
Δx
$\int_0^1 \int_0^x e^{-x^2}\,dy\,dx$

$R: 0 \le y \le 1$
$y \le x \le 1$
$(1, 1)$
Δy
$(1, 0)$
$\int_0^1 \int_y^1 e^{-x^2}\,dx\,dy$

Figure 14.19

By setting up the corresponding iterated integrals, you can see that the order $dx\,dy$ requires the antiderivative

$$\int e^{-x^2}\,dx$$

which is not an elementary function. On the other hand, the order $dy\,dx$ produces

$$\int_0^1 \int_0^x e^{-x^2}\,dy\,dx = \int_0^1 e^{-x^2}y \Big]_0^x dx$$

$$= \int_0^1 xe^{-x^2}\,dx$$

$$= -\frac{1}{2}e^{-x^2}\Big]_0^1$$

$$= -\frac{1}{2}\left(\frac{1}{e} - 1\right)$$

$$= \frac{e - 1}{2e}$$

$$\approx 0.316.$$

▷ **TECHNOLOGY** Try using a symbolic integration utility to evaluate the integral in Example 4.

EXAMPLE 5 **Volume of a Region Bounded by Two Surfaces**

Find the volume of the solid region bounded above by the paraboloid

$$z = 1 - x^2 - y^2 \qquad \text{Paraboloid}$$

and below by the plane

$$z = 1 - y \qquad \text{Plane}$$

as shown in Figure 14.20.

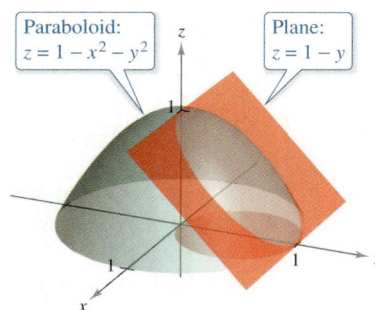

Figure 14.20

Solution Equating z-values, you can determine that the intersection of the two surfaces occurs on the right circular cylinder given by

$$1 - y = 1 - x^2 - y^2 \quad \Longrightarrow \quad x^2 = y - y^2.$$

So, the region R in the xy-plane is a circle, as shown in Figure 14.21. Because the volume of the solid region is the difference between the volume under the paraboloid and the volume under the plane, you have

$$\text{Volume} = (\text{volume under paraboloid}) - (\text{volume under plane})$$

$$= \int_0^1 \int_{-\sqrt{y-y^2}}^{\sqrt{y-y^2}} (1 - x^2 - y^2)\, dx\, dy - \int_0^1 \int_{-\sqrt{y-y^2}}^{\sqrt{y-y^2}} (1 - y)\, dx\, dy$$

$$= \int_0^1 \int_{-\sqrt{y-y^2}}^{\sqrt{y-y^2}} (y - y^2 - x^2)\, dx\, dy$$

$$= \int_0^1 \left[(y - y^2)x - \frac{x^3}{3} \right]_{-\sqrt{y-y^2}}^{\sqrt{y-y^2}}\, dy$$

$$= \frac{4}{3}\int_0^1 (y - y^2)^{3/2}\, dy$$

$$= \left(\frac{4}{3}\right)\left(\frac{1}{8}\right)\int_0^1 [1 - (2y - 1)^2]^{3/2}\, dy$$

$$= \frac{1}{6}\int_{-\pi/2}^{\pi/2} \frac{\cos^4 \theta}{2}\, d\theta \qquad 2y - 1 = \sin \theta$$

$$= \frac{1}{6}\int_0^{\pi/2} \cos^4 \theta\, d\theta$$

$$= \left(\frac{1}{6}\right)\left(\frac{3\pi}{16}\right) \qquad \text{Wallis's Formula}$$

$$= \frac{\pi}{32}.$$

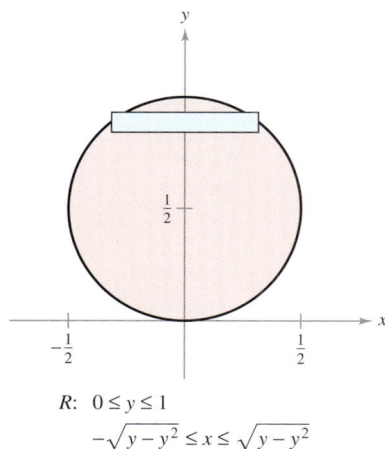

$R:\ 0 \le y \le 1$

$-\sqrt{y-y^2} \le x \le \sqrt{y-y^2}$

Figure 14.21

Average Value of a Function

Recall from Section 4.4 that for a function f in one variable, the average value of f on the interval $[a, b]$ is

$$\frac{1}{b-a}\int_a^b f(x)\, dx.$$

Given a function f in two variables, you can find the average value of f over the plane region R as shown in the following definition.

Definition of the Average Value of a Function Over a Region

If f is integrable over the plane region R, then the **average value** of f over R is

$$\text{Average value} = \frac{1}{A}\int_R\int f(x, y)\, dA$$

where A is the area of R.

EXAMPLE 6   Finding the Average Value of a Function

Find the average value of

$$f(x, y) = \frac{1}{2}xy$$

over the plane region R, where R is a rectangle with vertices

$$(0, 0),\ (4, 0),\ (4, 3),\ \text{and}\ (0, 3).$$

Solution   The area of the rectangular region R is

$$A = (4)(3) = 12$$

as shown in Figure 14.22. The bounds for x are

$$0 \le x \le 4$$

and the bounds for y are

$$0 \le y \le 3.$$

So, the average value is

$$\text{Average value} = \frac{1}{A}\int_R\int f(x, y)\, dA$$

$$= \frac{1}{12}\int_0^4\int_0^3 \frac{1}{2}xy\, dy\, dx$$

$$= \frac{1}{12}\int_0^4 \frac{1}{4}xy^2\Big]_0^3\, dx$$

$$= \left(\frac{1}{12}\right)\left(\frac{9}{4}\right)\int_0^4 x\, dx$$

$$= \frac{3}{16}\left[\frac{1}{2}x^2\right]_0^4$$

$$= \left(\frac{3}{16}\right)(8)$$

$$= \frac{3}{2}.$$

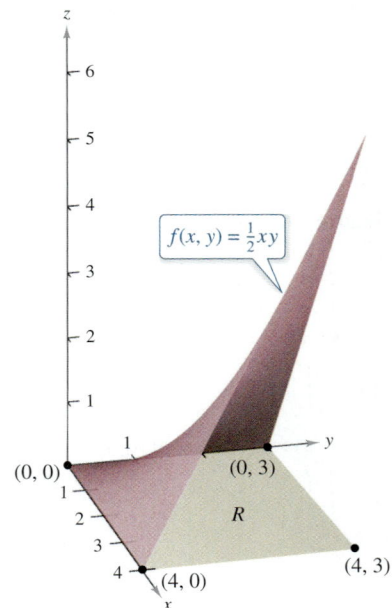

Figure 14.22

14.2 Exercises

See **CalcChat.com** for tutorial help and worked-out solutions to odd-numbered exercises.

Approximation In Exercises 1–4, approximate the integral $\int_R \int f(x, y)\, dA$ by dividing the rectangle R with vertices $(0, 0)$, $(4, 0)$, $(4, 2)$, and $(0, 2)$ into eight equal squares and finding the sum $\sum_{i=1}^{8} f(x_i, y_i)\, \Delta A_i$, where (x_i, y_i) is the center of the ith square. Evaluate the iterated integral and compare it with the approximation.

1. $\int_0^4 \int_0^2 (x + y)\, dy\, dx$ **2.** $\frac{1}{2}\int_0^4 \int_0^2 x^2 y\, dy\, dx$

3. $\int_0^4 \int_0^2 (x^2 + y^2)\, dy\, dx$ **4.** $\int_0^4 \int_0^2 \frac{1}{(x + 1)(y + 1)}\, dy\, dx$

Evaluating a Double Integral In Exercises 5–10, sketch the region R and evaluate the iterated integral $\int_R \int f(x, y)\, dA$.

5. $\int_0^2 \int_0^1 (1 + 2x + 2y)\, dy\, dx$ **6.** $\int_0^\pi \int_0^{\pi/2} \sin^2 x \cos^2 y\, dy\, dx$

7. $\int_0^6 \int_{y/2}^3 (x + y)\, dx\, dy$ **8.** $\int_0^4 \int_{\frac{1}{2}y}^{\sqrt{y}} x^2 y^2\, dx\, dy$

9. $\int_{-a}^a \int_{-\sqrt{a^2-x^2}}^{\sqrt{a^2-x^2}} (x + y)\, dy\, dx$

10. $\int_0^1 \int_{y-1}^0 e^{x+y}\, dx\, dy + \int_0^1 \int_0^{1-y} e^{x+y}\, dx\, dy$

Evaluating a Double Integral In Exercises 11–18, set up integrals for both orders of integration. Use the more convenient order to evaluate the integral over the region R.

11. $\int_R \int xy\, dA$

   R: rectangle with vertices $(0, 0)$, $(0, 5)$, $(3, 5)$, $(3, 0)$

12. $\int_R \int \sin x \sin y\, dA$

   R: rectangle with vertices $(-\pi, 0)$, $(\pi, 0)$, $(\pi, \pi/2)$, $(-\pi, \pi/2)$

13. $\int_R \int \frac{y}{x^2 + y^2}\, dA$

   R: trapezoid bounded by $y = x$, $y = 2x$, $x = 1$, $x = 2$

14. $\int_R \int xe^y\, dA$

   R: triangle bounded by $y = 4 - x$, $y = 0$, $x = 0$

15. $\int_R \int -2y\, dA$

   R: region bounded by $y = 4 - x^2$, $y = 4 - x$

16. $\int_R \int \frac{y}{1 + x^2}\, dA$

   R: region bounded by $y = 0$, $y = \sqrt{x}$, $x = 4$

17. $\int_R \int x\, dA$

   R: sector of a circle in the first quadrant bounded by $y = \sqrt{25 - x^2}$, $3x - 4y = 0$, $y = 0$

18. $\int_R \int (x^2 + y^2)\, dA$

   R: semicircle bounded by $y = \sqrt{4 - x^2}$, $y = 0$

Finding Volume In Exercises 19–26, use a double integral to find the volume of the indicated solid.

19.

$z = \frac{y}{2}$

$0 \le x \le 4$
$0 \le y \le 2$

20.

$z = 6 - 2y$

$0 \le x \le 4$
$0 \le y \le 2$

21.

$z = 4 - x - y$

$y = x$ $y = 2$

22.

$2x + 3y + 4z = 12$

23.

$z = 1 - xy$

$y = x$ $y = 1$

24.

$z = 4 - y^2$

$y = x$ $y = 2$

25. Improper integral

$z = \dfrac{1}{(x + 1)^2(y + 1)^2}$

$0 \le x < \infty$
$0 \le y < \infty$

26. Improper integral

$z = e^{-(x+y)/2}$

$0 \le x < \infty$
$0 \le y < \infty$

Finding Volume   In Exercises 27–32, set up and evaluate a double integral to find the volume of the solid bounded by the graphs of the equations.

27. $z = xy$, $z = 0$, $y = x$, $x = 1$, first octant

28. $z = 0$, $z = x^2$, $x = 0$, $x = 2$, $y = 0$, $y = 4$

29. $x^2 + z^2 = 1$, $y^2 + z^2 = 1$, first octant

30. $y = 4 - x^2$, $z = 4 - x^2$, first octant

31. $z = x + y$, $x^2 + y^2 = 4$, first octant

32. $z = \dfrac{1}{1 + y^2}$, $x = 0$, $x = 2$, $y \geq 0$

Volume of a Region Bounded by Two Surfaces   In Exercises 33–38, set up a double integral to find the volume of the solid region bounded by the graphs of the equations. Do not evaluate the integral.

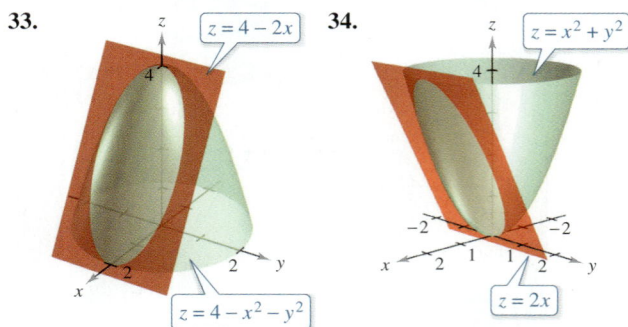

33.

$z = 4 - 2x$

$z = 4 - x^2 - y^2$

34.

$z = x^2 + y^2$

$z = 2x$

35. $z = x^2 + y^2$, $x^2 + y^2 = 4$, $z = 0$

36. $z = \sin^2 x$, $z = 0$, $0 \leq x \leq \pi$, $0 \leq y \leq 5$

37. $z = x^2 + 2y^2$, $z = 4y$

38. $z = x^2 + y^2$, $z = 18 - x^2 - y^2$

Finding Volume Using Technology   In Exercises 39–42, use a computer algebra system to find the volume of the solid bounded by the graphs of the equations.

39. $z = 9 - x^2 - y^2$, $z = 0$

40. $x^2 = 9 - y$, $z^2 = 9 - y$, first octant

41. $z = \dfrac{2}{1 + x^2 + y^2}$, $z = 0$, $y = 0$, $x = 0$, $y = -0.5x + 1$

42. $z = \ln(1 + x + y)$, $z = 0$, $y = 0$, $x = 0$, $x = 4 - \sqrt{y}$

43. Proof   Let f be a continuous function such that $0 \leq f(x, y) \leq 1$ over a region R of area 1. Prove that $0 \leq \int_R \int f(x, y)\, dA \leq 1$.

44. Finding Volume   Find the volume of the solid in the first octant bounded by the coordinate planes and the plane $(x/a) + (y/b) + (z/c) = 1$, where $a > 0$, $b > 0$, and $c > 0$.

Evaluating an Iterated Integral   In Exercises 45–50, sketch the region of integration. Then evaluate the iterated integral, switching the order of integration if necessary.

45. $\displaystyle \int_0^1 \int_{y/2}^{1/2} e^{-x^2}\, dx\, dy$

46. $\displaystyle \int_0^{\ln 10} \int_{e^x}^{10} \frac{1}{\ln y}\, dy\, dx$

47. $\displaystyle \int_{-2}^{2} \int_{-\sqrt{4 - x^2}}^{\sqrt{4 - x^2}} \sqrt{4 - y^2}\, dy\, dx$

48. $\displaystyle \int_0^3 \int_{y/3}^1 \frac{1}{1 + x^4}\, dx\, dy$

49. $\displaystyle \int_0^1 \int_0^{\arccos y} \sin x \sqrt{1 + \sin^2 x}\, dx\, dy$

50. $\displaystyle \int_0^2 \int_{(1/2)x^2}^2 \sqrt{y}\, \cos y\, dy\, dx$

Average Value   In Exercises 51–56, find the average value of $f(x, y)$ over the plane region R.

51. $f(x, y) = x$

R: rectangle with vertices $(0, 0)$, $(4, 0)$, $(4, 2)$, $(0, 2)$

52. $f(x, y) = 2xy$

R: rectangle with vertices $(0, 0)$, $(5, 0)$, $(5, 3)$, $(0, 3)$

53. $f(x, y) = x^2 + y^2$

R: square with vertices $(0, 0)$, $(2, 0)$, $(2, 2)$, $(0, 2)$

54. $f(x, y) = \dfrac{1}{x + y}$, R: triangle with vertices $(0, 0)$, $(1, 0)$, $(1, 1)$

55. $f(x, y) = e^{x+y}$, R: triangle with vertices $(0, 0)$, $(0, 1)$, $(1, 1)$

56. $f(x, y) = \sin(x + y)$

R: rectangle with vertices $(0, 0)$, $(\pi, 0)$, (π, π), $(0, \pi)$

57. Average Production

The Cobb-Douglas production function for an automobile manufacturer is

$f(x, y) = 100x^{0.6}y^{0.4}$

where x is the number of units of labor and y is the number of units of capital. Estimate the average production level when the number of units of labor x varies between 200 and 250 and the number of units of capital y varies between 300 and 325.

58. Average Temperature   The temperature in degrees Celsius on the surface of a metal plate is

$T(x, y) = 20 - 4x^2 - y^2$

where x and y are measured in centimeters. Estimate the average temperature when x varies between 0 and 2 centimeters and y varies between 0 and 4 centimeters.

WRITING ABOUT CONCEPTS

59. Double Integral   State the definition of a double integral. When the integrand is a nonnegative function over the region of integration, give the geometric interpretation of a double integral.

60. Volume   Let R be a region in the xy-plane whose area is B. When $f(x, y) = k$ for every point (x, y) in R, what is the value of $\int_R \int f(x, y)\, dA$? Explain.

WRITING ABOUT CONCEPTS (continued)

61. Volume Let the plane region R be a unit circle and let the maximum value of f on R be 6. Is the greatest possible value of $\int_R \int f(x, y)\, dy\, dx$ equal to 6? Why or why not? If not, what is the greatest possible value?

62. Comparing Iterated Integrals The following iterated integrals represent the solution to the same problem. Which iterated integral is easier to evaluate? Explain your reasoning.

$$\int_0^4 \int_{x/2}^2 \sin y^2\, dy\, dx = \int_0^2 \int_0^{2y} \sin y^2\, dx\, dy$$

Probability A joint density function of the continuous random variables x and y is a function $f(x, y)$ satisfying the following properties.

(a) $f(x, y) \geq 0$ for all (x, y) (b) $\int_{-\infty}^{\infty} \int_{-\infty}^{\infty} f(x, y)\, dA = 1$

(c) $P[(x, y) \in R] = \int_R \int f(x, y)\, dA$

In Exercises 63–66, show that the function is a joint density function and find the required probability.

63. $f(x, y) = \begin{cases} \frac{1}{10}, & 0 \leq x \leq 5,\ 0 \leq y \leq 2 \\ 0, & \text{elsewhere} \end{cases}$

$P(0 \leq x \leq 2,\ 1 \leq y \leq 2)$

64. $f(x, y) = \begin{cases} \frac{1}{4}xy, & 0 \leq x \leq 2,\ 0 \leq y \leq 2 \\ 0, & \text{elsewhere} \end{cases}$

$P(0 \leq x \leq 1,\ 1 \leq y \leq 2)$

65. $f(x, y) = \begin{cases} \frac{1}{27}(9 - x - y), & 0 \leq x \leq 3,\ 3 \leq y \leq 6 \\ 0, & \text{elsewhere} \end{cases}$

$P(0 \leq x \leq 1,\ 4 \leq y \leq 6)$

66. $f(x, y) = \begin{cases} e^{-x-y}, & x \geq 0,\ y \geq 0 \\ 0, & \text{elsewhere} \end{cases}$

$P(0 \leq x \leq 1,\ x \leq y \leq 1)$

67. Approximation The table shows values of a function f over a square region R. Divide the region into 16 equal squares and select (x_i, y_i) to be the point in the ith square closest to the origin. Compare this approximation with that obtained by using the point in the ith square farthest from the origin.

$$\int_0^4 \int_0^4 f(x, y)\, dy\, dx$$

x \ y	0	1	2	3	4
0	32	31	28	23	16
1	31	30	27	22	15
2	28	27	24	19	12
3	23	22	19	14	7
4	16	15	12	7	0

68. HOW DO YOU SEE IT? The figure below shows Erie County, New York. Let $f(x, y)$ represent the total annual snowfall at the point (x, y) in the county, where R is the county. Interpret each of the following.

(a) $\int_R \int f(x, y)\, dA$

(b) $\dfrac{\int_R \int f(x, y)\, dA}{\int_R \int dA}$

True or False? In Exercises 69 and 70, determine whether the statement is true or false. If it is false, explain why or give an example that shows it is false.

69. The volume of the sphere $x^2 + y^2 + z^2 = 1$ is given by the integral

$$V = 8 \int_0^1 \int_0^1 \sqrt{1 - x^2 - y^2}\, dx\, dy.$$

70. If $f(x, y) \leq g(x, y)$ for all (x, y) in R, and both f and g are continuous over R, then $\int_R \int f(x, y)\, dA \leq \int_R \int g(x, y)\, dA$.

71. Maximizing a Double Integral Determine the region R in the xy-plane that maximizes the value of

$$\int_R \int (9 - x^2 - y^2)\, dA.$$

72. Minimizing a Double Integral Determine the region R in the xy-plane that minimizes the value of

$$\int_R \int (x^2 + y^2 - 4)\, dA.$$

73. Average Value Let

$$f(x) = \int_1^x e^{t^2}\, dt.$$

Find the average value of f on the interval $[0, 1]$.

74. Using Geometry Use a geometric argument to show that

$$\int_0^3 \int_0^{\sqrt{9-y^2}} \sqrt{9 - x^2 - y^2}\, dx\, dy = \frac{9\pi}{2}.$$

PUTNAM EXAM CHALLENGE

75. Evaluate $\int_0^a \int_0^b e^{\max\{b^2 x^2,\, a^2 y^2\}}\, dy\, dx$, where a and b are positive.

76. Show that if $\lambda > \frac{1}{2}$ there does not exist a real-valued function u such that for all x in the closed interval $0 \leq x \leq 1$, $u(x) = 1 + \lambda \int_x^1 u(y)u(y - x)\, dy$.

These problems were composed by the Committee on the Putnam Prize Competition.
© The Mathematical Association of America. All rights reserved.

14.3 Change of Variables: Polar Coordinates

■ **Write and evaluate double integrals in polar coordinates.**

Double Integrals in Polar Coordinates

Some double integrals are *much* easier to evaluate in polar form than in rectangular form. This is especially true for regions such as circles, cardioids, and rose curves, and for integrands that involve $x^2 + y^2$.

In Section 10.4, you learned that the polar coordinates (r, θ) of a point are related to the rectangular coordinates (x, y) of the point as follows.

$$x = r \cos \theta \quad \text{and} \quad y = r \sin \theta$$

$$r^2 = x^2 + y^2 \quad \text{and} \quad \tan \theta = \frac{y}{x}$$

EXAMPLE 1   **Using Polar Coordinates to Describe a Region**

Use polar coordinates to describe each region shown in Figure 14.23.

(a)

(b)

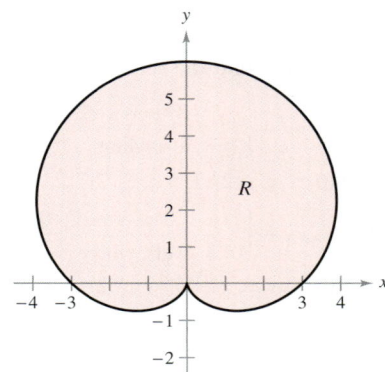
(c)

Figure 14.23

Solution

a. The region R is a quarter circle of radius 2. It can be described in polar coordinates as

$$R = \{(r, \theta): 0 \le r \le 2, \quad 0 \le \theta \le \pi/2\}.$$

b. The region R consists of all points between concentric circles of radii 1 and 3. It can be described in polar coordinates as

$$R = \{(r, \theta): 1 \le r \le 3, \quad 0 \le \theta \le 2\pi\}.$$

c. The region R is a cardioid with $a = b = 3$. It can be described in polar coordinates as

$$R = \{(r, \theta): 0 \le r \le 3 + 3 \sin \theta, 0 \le \theta \le 2\pi\}.$$ ■

The regions in Example 1 are special cases of **polar sectors**

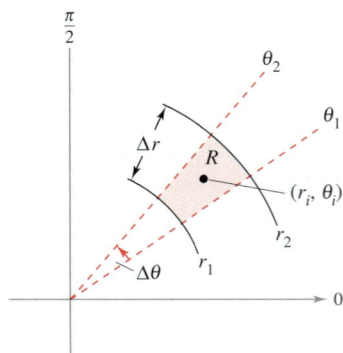

Polar sector
Figure 14.24

$$\boxed{R = \{(r, \theta): r_1 \le r \le r_2, \quad \theta_1 \le \theta \le \theta_2\}} \qquad \text{Polar sector}$$

as shown in Figure 14.24.

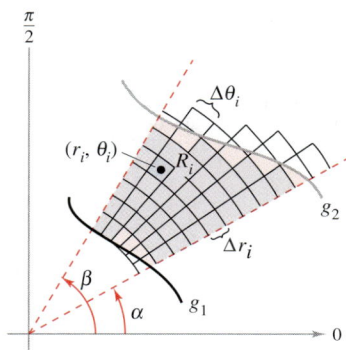

Polar grid superimposed over region R
Figure 14.25

To define a double integral of a continuous function $z = f(x, y)$ in polar coordinates, consider a region R bounded by the graphs of

$$r = g_1(\theta) \quad \text{and} \quad r = g_2(\theta)$$

and the lines $\theta = \alpha$ and $\theta = \beta$. Instead of partitioning R into small rectangles, use a partition of small polar sectors. On R, superimpose a polar grid made of rays and circular arcs, as shown in Figure 14.25. The polar sectors R_i lying entirely within R form an **inner polar partition** Δ, whose **norm** $\|\Delta\|$ is the length of the longest diagonal of the n polar sectors.

Consider a specific polar sector R_i, as shown in Figure 14.26. It can be shown (see Exercise 70) that the area of R_i is

$$\Delta A_i = r_i \, \Delta r_i \, \Delta \theta_i \qquad \text{Area of } R_i$$

where $\Delta r_i = r_2 - r_1$ and $\Delta \theta_i = \theta_2 - \theta_1$. This implies that the volume of the solid of height $f(r_i \cos \theta_i, r_i \sin \theta_i)$ above R_i is approximately

$$f(r_i \cos \theta_i, r_i \sin \theta_i) r_i \, \Delta r_i \, \Delta \theta_i$$

and you have

$$\int_R \int f(x, y) \, dA \approx \sum_{i=1}^{n} f(r_i \cos \theta_i, r_i \sin \theta_i) r_i \, \Delta r_i \, \Delta \theta_i.$$

The sum on the right can be interpreted as a Riemann sum for

$$f(r \cos \theta, r \sin \theta) r.$$

The region R corresponds to a *horizontally simple* region S in the $r\theta$-plane, as shown in Figure 14.27. The polar sectors R_i correspond to rectangles S_i, and the area ΔA_i of S_i is $\Delta r_i \, \Delta \theta_i$. So, the right-hand side of the equation corresponds to the double integral

$$\int_S \int f(r \cos \theta, r \sin \theta) r \, dA.$$

From this, you can apply Theorem 14.2 to write

$$\int_R \int f(x, y) \, dA = \int_S \int f(r \cos \theta, r \sin \theta) r \, dA$$

$$= \int_\alpha^\beta \int_{g_1(\theta)}^{g_2(\theta)} f(r \cos \theta, r \sin \theta) r \, dr \, d\theta.$$

This suggests the theorem on the next page, the proof of which is discussed in Section 14.8.

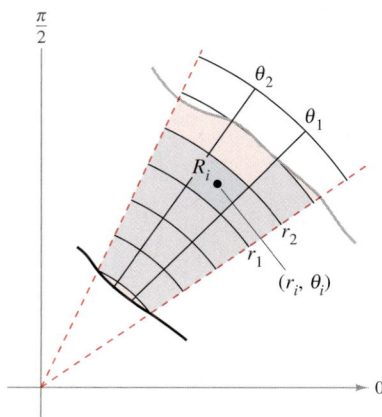

The polar sector R_i is the set of all points (r, θ) such that $r_1 \leq r \leq r_2$ and $\theta_1 \leq \theta \leq \theta_2$.
Figure 14.26

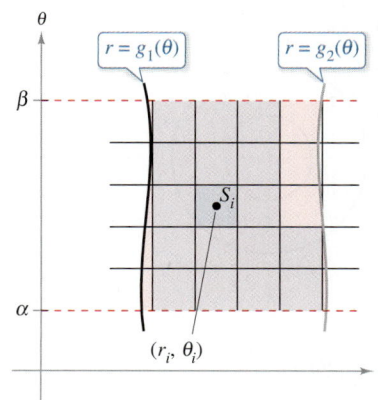

Horizontally simple region S
Figure 14.27

THEOREM 14.3 Change of Variables to Polar Form

Let R be a plane region consisting of all points $(x, y) = (r \cos \theta, r \sin \theta)$ satisfying the conditions $0 \le g_1(\theta) \le r \le g_2(\theta)$, $\alpha \le \theta \le \beta$, where $0 \le (\beta - \alpha) \le 2\pi$. If g_1 and g_2 are continuous on $[\alpha, \beta]$ and f is continuous on R, then

$$\iint_R f(x, y)\, dA = \int_\alpha^\beta \int_{g_1(\theta)}^{g_2(\theta)} f(r \cos \theta, r \sin \theta)\, r\, dr\, d\theta.$$

Exploration

Volume of a Paraboloid Sector In the Exploration on page 979, you were asked to summarize the different ways you know of finding the volume of the solid bounded by the paraboloid

$$z = a^2 - x^2 - y^2, a > 0$$

and the xy-plane. You now know another way. Use it to find the volume of the solid.

If $z = f(x, y)$ is nonnegative on R, then the integral in Theorem 14.3 can be interpreted as the volume of the solid region between the graph of f and the region R. When using the integral in Theorem 14.3, be certain not to omit the extra factor of r in the integrand.

The region R is restricted to two basic types, **r-simple** regions and **θ-simple** regions, as shown in Figure 14.28.

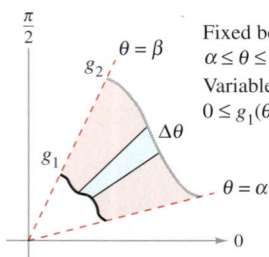

Fixed bounds for θ:
$\alpha \le \theta \le \beta$
Variable bounds for r:
$0 \le g_1(\theta) \le r \le g_2(\theta)$

r-Simple region

Variable bounds for θ:
$0 \le h_1(r) \le \theta \le h_2(r)$
Fixed bounds for r:
$r_1 \le r \le r_2$

θ-Simple region

Figure 14.28

EXAMPLE 2 **Evaluating a Double Polar Integral**

Let R be the annular region lying between the two circles $x^2 + y^2 = 1$ and $x^2 + y^2 = 5$. Evaluate the integral

$$\iint_R (x^2 + y)\, dA.$$

Solution The polar boundaries are $1 \le r \le \sqrt{5}$ and $0 \le \theta \le 2\pi$, as shown in Figure 14.29. Furthermore, $x^2 = (r \cos \theta)^2$ and $y = r \sin \theta$. So, you have

$$\iint_R (x^2 + y)\, dA = \int_0^{2\pi} \int_1^{\sqrt{5}} (r^2 \cos^2 \theta + r \sin \theta) r\, dr\, d\theta$$

$$= \int_0^{2\pi} \int_1^{\sqrt{5}} (r^3 \cos^2 \theta + r^2 \sin \theta)\, dr\, d\theta$$

$$= \int_0^{2\pi} \left(\frac{r^4}{4} \cos^2 \theta + \frac{r^3}{3} \sin \theta \right) \Big]_1^{\sqrt{5}} d\theta$$

$$= \int_0^{2\pi} \left(6 \cos^2 \theta + \frac{5\sqrt{5} - 1}{3} \sin \theta \right) d\theta$$

$$= \int_0^{2\pi} \left(3 + 3 \cos 2\theta + \frac{5\sqrt{5} - 1}{3} \sin \theta \right) d\theta$$

$$= \left(3\theta + \frac{3 \sin 2\theta}{2} - \frac{5\sqrt{5} - 1}{3} \cos \theta \right) \Big]_0^{2\pi}$$

$$= 6\pi.$$

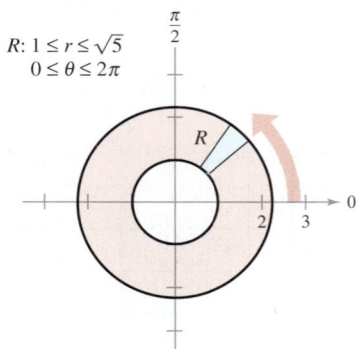

$R: 1 \le r \le \sqrt{5}$
$0 \le \theta \le 2\pi$

r-Simple region
Figure 14.29

In Example 2, be sure to notice the extra factor of r in the integrand. This comes from the formula for the area of a polar sector. In differential notation, you can write

$$dA = r \, dr \, d\theta$$

which indicates that the area of a polar sector increases as you move away from the origin.

Surface: $z = \sqrt{16 - x^2 - y^2}$

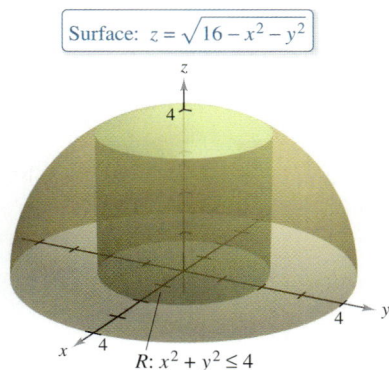

Figure 14.30

$R: x^2 + y^2 \leq 4$

EXAMPLE 3   Change of Variables to Polar Coordinates

Use polar coordinates to find the volume of the solid region bounded above by the hemisphere

$$z = \sqrt{16 - x^2 - y^2}$$ Hemisphere forms upper surface.

and below by the circular region R given by

$$x^2 + y^2 \leq 4$$ Circular region forms lower surface.

as shown in Figure 14.30.

Solution   In Figure 14.30, you can see that R has the bounds

$$-\sqrt{4 - y^2} \leq x \leq \sqrt{4 - y^2}, \quad -2 \leq y \leq 2$$

and that $0 \leq z \leq \sqrt{16 - x^2 - y^2}$. In polar coordinates, the bounds are

$$0 \leq r \leq 2 \quad \text{and} \quad 0 \leq \theta \leq 2\pi$$

with height $z = \sqrt{16 - x^2 - y^2} = \sqrt{16 - r^2}$. Consequently, the volume V is

$$
\begin{aligned}
V &= \iint_R f(x, y) \, dA \\
&= \int_0^{2\pi} \int_0^2 \sqrt{16 - r^2} \, r \, dr \, d\theta \\
&= -\frac{1}{3} \int_0^{2\pi} (16 - r^2)^{3/2} \Big]_0^2 \, d\theta \\
&= -\frac{1}{3} \int_0^{2\pi} \left(24\sqrt{3} - 64 \right) d\theta \\
&= -\frac{8}{3} (3\sqrt{3} - 8) \theta \Big]_0^{2\pi} \\
&= \frac{16\pi}{3} (8 - 3\sqrt{3}) \\
&\approx 46.979.
\end{aligned}
$$

■

• • • • • • • • • • • • • • • • • • • ▷

·· REMARK   To see the benefit of polar coordinates in Example 3, you should try to evaluate the corresponding rectangular iterated integral

$$\int_{-2}^{2} \int_{-\sqrt{4-y^2}}^{\sqrt{4-y^2}} \sqrt{16 - x^2 - y^2} \, dx \, dy.$$

▷ **TECHNOLOGY**   Any computer algebra system that can evaluate double integrals in rectangular coordinates can also evaluate double integrals in polar coordinates. The reason this is true is that once you have formed the iterated integral, its value is not changed by using different variables. In other words, if you use a computer algebra system to evaluate

$$\int_0^{2\pi} \int_0^2 \sqrt{16 - x^2} \, x \, dx \, dy$$

you should obtain the same value as that obtained in Example 3.

Just as with rectangular coordinates, the double integral

$$\iint_R dA$$

can be used to find the area of a region in the plane.

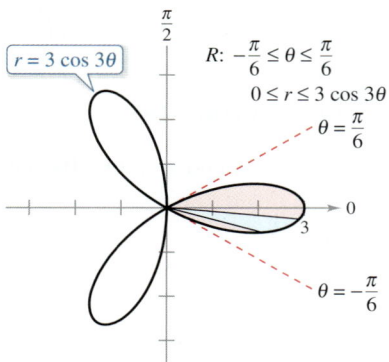

$r = 3 \cos 3\theta$

$R: -\dfrac{\pi}{6} \le \theta \le \dfrac{\pi}{6}$

$0 \le r \le 3 \cos 3\theta$

$\theta = \dfrac{\pi}{6}$

$\theta = -\dfrac{\pi}{6}$

Figure 14.31

EXAMPLE 4   Finding Areas of Polar Regions

••••▷ *See LarsonCalculus.com for an interactive version of this type of example.*

To use a double integral to find the area enclosed by the graph of $r = 3 \cos 3\theta$, let R be one petal of the curve shown in Figure 14.31. This region is r-simple, and the boundaries are $-\pi/6 \le \theta \le \pi/6$ and $0 \le r \le 3 \cos 3\theta$. So, the area of one petal is

$$\frac{1}{3}A = \int_R \int dA = \int_{-\pi/6}^{\pi/6} \int_0^{3\cos 3\theta} r\, dr\, d\theta$$

$$= \int_{-\pi/6}^{\pi/6} \frac{r^2}{2} \Big]_0^{3\cos 3\theta} d\theta$$

$$= \frac{9}{2} \int_{-\pi/6}^{\pi/6} \cos^2 3\theta\, d\theta$$

$$= \frac{9}{4} \int_{-\pi/6}^{\pi/6} (1 + \cos 6\theta)\, d\theta$$

$$= \frac{9}{4} \left[\theta + \frac{1}{6} \sin 6\theta \right]_{-\pi/6}^{\pi/6}$$

$$= \frac{3\pi}{4}.$$

So, the total area is $A = 9\pi/4$. ◼

As illustrated in Example 4, the area of a region in the plane can be represented by

$$A = \int_\alpha^\beta \int_{g_1(\theta)}^{g_2(\theta)} r\, dr\, d\theta.$$

For $g_1(\theta) = 0$, you obtain

$$A = \int_\alpha^\beta \int_0^{g_2(\theta)} r\, dr\, d\theta = \int_\alpha^\beta \frac{r^2}{2} \Big]_0^{g_2(\theta)} d\theta = \int_\alpha^\beta \frac{1}{2}(g_2(\theta))^2 \, d\theta$$

which agrees with Theorem 10.13.

So far in this section, all of the examples of iterated integrals in polar form have been of the form

$$\int_\alpha^\beta \int_{g_1(\theta)}^{g_2(\theta)} f(r \cos\theta, r \sin\theta) r\, dr\, d\theta$$

in which the order of integration is with respect to r first. Sometimes you can obtain a simpler integration problem by switching the order of integration.

EXAMPLE 5   Changing the Order of Integration

Find the area of the region bounded above by the spiral $r = \pi/(3\theta)$ and below by the polar axis, between $r = 1$ and $r = 2$.

Solution   The region is shown in Figure 14.32. The polar boundaries for the region are

$$1 \le r \le 2 \quad \text{and} \quad 0 \le \theta \le \frac{\pi}{3r}.$$

So, the area of the region can be evaluated as follows.

$$A = \int_1^2 \int_0^{\pi/(3r)} r\, d\theta\, dr = \int_1^2 r\theta \Big]_0^{\pi/(3r)} dr = \int_1^2 \frac{\pi}{3}\, dr = \frac{\pi r}{3} \Big]_1^2 = \frac{\pi}{3} \quad ◼$$

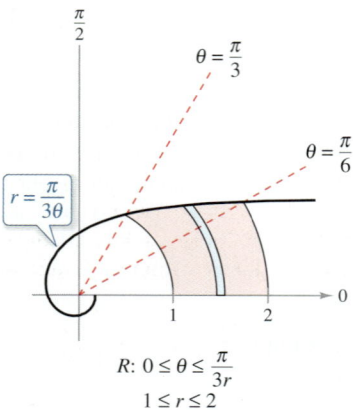

$\theta = \dfrac{\pi}{3}$

$\theta = \dfrac{\pi}{6}$

$r = \dfrac{\pi}{3\theta}$

$R: 0 \le \theta \le \dfrac{\pi}{3r}$

$1 \le r \le 2$

θ-Simple region
Figure 14.32

Choosing a Coordinate System In Exercises 1–4, the region R for the integral $\int_R \int f(x, y) \, dA$ is shown. State whether you would use rectangular or polar coordinates to evaluate the integral.

1.

2.

3.

4.

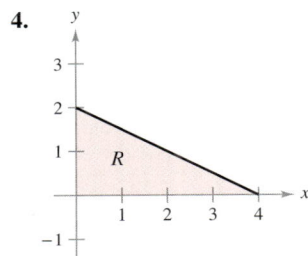

Describing a Region In Exercises 5–8, use polar coordinates to describe the region shown.

5.

6.

7.

8.

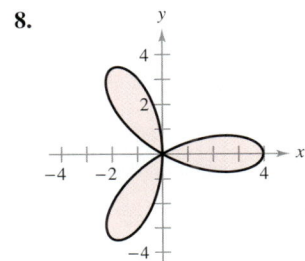

Evaluating a Double Integral In Exercises 9–16, evaluate the double integral $\int_R \int f(r, \theta) \, dA$, and sketch the region R.

9. $\displaystyle \int_0^\pi \int_0^{\cos \theta} r \, dr \, d\theta$

10. $\displaystyle \int_0^\pi \int_0^{\sin \theta} r^2 \, dr \, d\theta$

11. $\displaystyle \int_0^{2\pi} \int_0^6 3r^2 \sin \theta \, dr \, d\theta$

12. $\displaystyle \int_0^{\pi/4} \int_0^4 r^2 \sin \theta \cos \theta \, dr \, d\theta$

13. $\displaystyle \int_0^{\pi/2} \int_2^3 \sqrt{9 - r^2} \, r \, dr \, d\theta$

14. $\displaystyle \int_0^{\pi/2} \int_0^3 re^{-r^2} \, dr \, d\theta$

15. $\displaystyle \int_0^{\pi/2} \int_0^{1+\sin \theta} \theta r \, dr \, d\theta$

16. $\displaystyle \int_0^{\pi/2} \int_0^{1-\cos \theta} (\sin \theta) r \, dr \, d\theta$

Converting to Polar Coordinates In Exercises 17–26, evaluate the iterated integral by converting to polar coordinates.

17. $\displaystyle \int_0^a \int_0^{\sqrt{a^2 - y^2}} y \, dx \, dy$

18. $\displaystyle \int_0^a \int_0^{\sqrt{a^2 - x^2}} x \, dy \, dx$

19. $\displaystyle \int_{-2}^2 \int_0^{\sqrt{4 - x^2}} (x^2 + y^2) \, dy \, dx$

20. $\displaystyle \int_0^1 \int_{-\sqrt{x - x^2}}^{\sqrt{x - x^2}} (x^2 + y^2) \, dy \, dx$

21. $\displaystyle \int_0^3 \int_0^{\sqrt{9 - x^2}} (x^2 + y^2)^{3/2} \, dy \, dx$

22. $\displaystyle \int_0^2 \int_y^{\sqrt{8 - y^2}} \sqrt{x^2 + y^2} \, dx \, dy$

23. $\displaystyle \int_0^2 \int_0^{\sqrt{2x - x^2}} xy \, dy \, dx$

24. $\displaystyle \int_0^4 \int_0^{\sqrt{4y - y^2}} x^2 \, dx \, dy$

25. $\displaystyle \int_{-1}^1 \int_0^{\sqrt{1 - x^2}} \cos(x^2 + y^2) \, dy \, dx$

26. $\displaystyle \int_0^2 \int_0^{\sqrt{4 - x^2}} \sin \sqrt{x^2 + y^2} \, dy \, dx$

Converting to Polar Coordinates In Exercises 27 and 28, combine the sum of the two iterated integrals into a single iterated integral by converting to polar coordinates. Evaluate the resulting iterated integral.

27. $\displaystyle \int_0^2 \int_0^x \sqrt{x^2 + y^2} \, dy \, dx + \int_2^{2\sqrt{2}} \int_0^{\sqrt{8 - x^2}} \sqrt{x^2 + y^2} \, dy \, dx$

28. $\displaystyle \int_0^{5\sqrt{2}/2} \int_0^x xy \, dy \, dx + \int_{5\sqrt{2}/2}^5 \int_0^{\sqrt{25 - x^2}} xy \, dy \, dx$

Converting to Polar Coordinates In Exercises 29–32, use polar coordinates to set up and evaluate the double integral $\int_R\int f(x, y)\, dA$.

29. $f(x, y) = x + y$

  $R: x^2 + y^2 \le 4, x \ge 0, y \ge 0$

30. $f(x, y) = e^{-(x^2+y^2)/2}$

  $R: x^2 + y^2 \le 25, x \ge 0$

31. $f(x, y) = \arctan \dfrac{y}{x}$

  $R: x^2 + y^2 \ge 1, x^2 + y^2 \le 4, 0 \le y \le x$

32. $f(x, y) = 9 - x^2 - y^2$

  $R: x^2 + y^2 \le 9, x \ge 0, y \ge 0$

Volume In Exercises 33–38, use a double integral in polar coordinates to find the volume of the solid bounded by the graphs of the equations.

33. $z = xy, x^2 + y^2 = 1$, first octant

34. $z = x^2 + y^2 + 3, z = 0, x^2 + y^2 = 1$

35. $z = \sqrt{x^2 + y^2}, z = 0, x^2 + y^2 = 25$

36. $z = \ln(x^2 + y^2), z = 0, x^2 + y^2 \ge 1, x^2 + y^2 \le 4$

37. Inside the hemisphere $z = \sqrt{16 - x^2 - y^2}$ and inside the cylinder $x^2 + y^2 - 4x = 0$

38. Inside the hemisphere $z = \sqrt{16 - x^2 - y^2}$ and outside the cylinder $x^2 + y^2 = 1$

39. Volume Find a such that the volume inside the hemisphere

$$z = \sqrt{16 - x^2 - y^2}$$

and outside the cylinder

$$x^2 + y^2 = a^2$$

is one-half the volume of the hemisphere.

40. Volume Use a double integral in polar coordinates to find the volume of a sphere of radius a.

Area In Exercises 41–46, use a double integral to find the area of the shaded region.

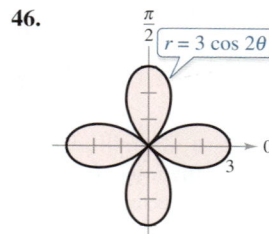

41.

$r = 6 \cos \theta$

42.

$r = 2$ $r = 4$

43.

$r = 1 + \cos \theta$

44.

$r = 2 + \sin \theta$

45.

$r = 2 \sin 3\theta$

46.

$r = 3 \cos 2\theta$

Area In Exercises 47–52, sketch a graph of the region bounded by the graphs of the equations. Then use a double integral to find the area of the region.

47. Inside the circle $r = 2 \cos \theta$ and outside the circle $r = 1$

48. Inside the cardioid $r = 2 + 2 \cos \theta$ and outside the circle $r = 1$

49. Inside the circle $r = 3 \cos \theta$ and outside the cardioid $r = 1 + \cos \theta$

50. Inside the cardioid $r = 1 + \cos \theta$ and outside the circle $r = 3 \cos \theta$

51. Inside the rose curve $r = 4 \sin 3\theta$ and outside the circle $r = 2$

52. Inside the circle $r = 2$ and outside the cardioid $r = 2 - 2 \cos \theta$

WRITING ABOUT CONCEPTS

53. Polar Coordinates Describe the partition of the region R of integration in the xy-plane when polar coordinates are used to evaluate a double integral.

54. Converting Coordinates Explain how to change from rectangular coordinates to polar coordinates in a double integral.

55. Describing Regions In your own words, describe r-simple regions and θ-simple regions.

56. Comparing Integrals Let R be the region bounded by the circle $x^2 + y^2 = 9$.

  (a) Set up the integral $\int_R\int f(x, y)\, dA$.

  (b) Convert the integral in part (a) to polar coordinates.

  (c) Which integral would you choose to evaluate? Why?

57. Population

The population density of a city is approximated by the model

$$f(x, y) = 4000e^{-0.01(x^2+y^2)}$$

for the region $x^2 + y^2 \le 49$, where x and y are measured in miles. Integrate the density function over the indicated circular region to approximate the population of the city.

58. HOW DO YOU SEE IT? Each figure shows a region of integration for the double integral $\int_R\int f(x, y)\, dA$. For each region, state whether horizontal representative elements, vertical representative elements, or polar sectors would yield the easiest method for obtaining the limits of integration. Explain your reasoning.

(a)

(b)

(c)
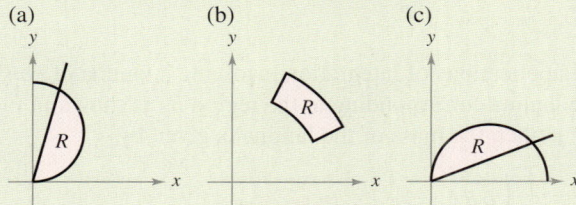

59. Volume Determine the diameter of a hole that is drilled vertically through the center of the solid bounded by the graphs of the equations $z = 25e^{-(x^2+y^2)/4}$, $z = 0$, and $x^2 + y^2 = 16$ when one-tenth of the volume of the solid is removed.

60. Glacier

Horizontal cross sections of a piece of ice that broke from a glacier are in the shape of a quarter of a circle with a radius of approximately 50 feet. The base is divided into 20 subregions, as shown in the figure. At the center of each subregion, the height of the ice is measured, yielding the following points in cylindrical coordinates.

$\left(5, \frac{\pi}{16}, 7\right), \left(15, \frac{\pi}{16}, 8\right), \left(25, \frac{\pi}{16}, 10\right), \left(35, \frac{\pi}{16}, 12\right), \left(45, \frac{\pi}{16}, 9\right),$
$\left(5, \frac{3\pi}{16}, 9\right), \left(15, \frac{3\pi}{16}, 10\right), \left(25, \frac{3\pi}{16}, 14\right), \left(35, \frac{3\pi}{16}, 15\right), \left(45, \frac{3\pi}{16}, 10\right),$
$\left(5, \frac{5\pi}{16}, 9\right), \left(15, \frac{5\pi}{16}, 11\right), \left(25, \frac{5\pi}{16}, 15\right), \left(35, \frac{5\pi}{16}, 18\right), \left(45, \frac{5\pi}{16}, 14\right),$
$\left(5, \frac{7\pi}{16}, 5\right), \left(15, \frac{7\pi}{16}, 8\right), \left(25, \frac{7\pi}{16}, 11\right), \left(35, \frac{7\pi}{16}, 16\right), \left(45, \frac{7\pi}{16}, 12\right)$

(a) Approximate the volume of the solid.

(b) Ice weighs approximately 57 pounds per cubic foot. Approximate the weight of the solid.

(c) There are 7.48 gallons of water per cubic foot. Approximate the number of gallons of water in the solid.

Approximation In Exercises 61 and 62, use a computer algebra system to approximate the iterated integral.

61. $\int_{\pi/4}^{\pi/2} \int_0^5 r\sqrt{1 + r^3}\, \sin\sqrt{\theta}\, dr\, d\theta$

62. $\int_0^{\pi/4} \int_0^4 5re^{\sqrt{r\theta}}\, dr\, d\theta$

True or False? In Exercises 63 and 64, determine whether the statement is true or false. If it is false, explain why or give an example that shows it is false.

63. If $\int_R\int f(r, \theta)\, dA > 0$, then $f(r, \theta) > 0$ for all (r, θ) in R.

64. If $f(r, \theta)$ is a constant function and the area of the region S is twice that of the region R, then
$$2\int_R\int f(r, \theta)\, dA = \int_S\int f(r, \theta)\, dA.$$

65. Probability The value of the integral
$$I = \int_{-\infty}^{\infty} e^{-x^2/2}\, dx$$
is required in the development of the normal probability density function.

(a) Use polar coordinates to evaluate the improper integral.
$$I^2 = \left(\int_{-\infty}^{\infty} e^{-x^2/2}\, dx\right)\left(\int_{-\infty}^{\infty} e^{-y^2/2}\, dy\right)$$
$$= \int_{-\infty}^{\infty}\int_{-\infty}^{\infty} e^{-(x^2+y^2)/2}\, dA$$

(b) Use the result of part (a) to determine I.

FOR FURTHER INFORMATION For more information on this problem, see the article "Integrating e^{-x^2} Without Polar Coordinates" by William Dunham in *Mathematics Teacher*. To view this article, go to *MathArticles.com*.

66. Evaluating Integrals Use the result of Exercise 65 and a change of variables to evaluate each integral. No integration is required.

(a) $\int_{-\infty}^{\infty} e^{-x^2}\, dx$

(b) $\int_{-\infty}^{\infty} e^{-4x^2}\, dx$

67. Think About It Consider the region bounded by the graphs of $y = 2$, $y = 4$, $y = x$, and $y = \sqrt{3}x$ and the double integral $\int_R\int f\, dA$. Determine the limits of integration when the region R is divided into (a) horizontal representative elements, (b) vertical representative elements, and (c) polar sectors.

68. Think About It Repeat Exercise 67 for a region R bounded by the graph of the equation $(x - 2)^2 + y^2 = 4$.

69. Probability Find k such that the function
$$f(x, y) = \begin{cases} ke^{-(x^2+y^2)}, & x \geq 0, y \geq 0 \\ 0, & \text{elsewhere} \end{cases}$$
is a probability density function.

70. Area Show that the area A of the polar sector R (see figure) is $A = r\Delta r\Delta\theta$, where $r = (r_1 + r_2)/2$ is the average radius of R.

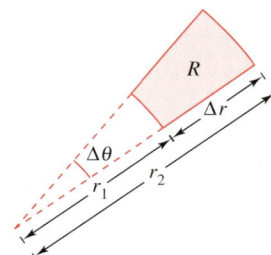

14.4 Center of Mass and Moments of Inertia

■ Find the mass of a planar lamina using a double integral.
■ Find the center of mass of a planar lamina using double integrals.
■ Find moments of inertia using double integrals.

Mass

Section 7.6 discussed several applications of integration involving a lamina of *constant* density ρ. For example, if the lamina corresponding to the region R, as shown in Figure 14.33, has a constant density ρ, then the mass of the lamina is given by

$$\text{Mass} = \rho A = \rho \int_R \int dA = \int_R \int \rho \, dA. \qquad \text{Constant density}$$

If not otherwise stated, a lamina is assumed to have a constant density. In this section, however, you will extend the definition of the term *lamina* to include thin plates of *variable* density. Double integrals can be used to find the mass of a lamina of variable density, where the density at (x, y) is given by the **density function ρ.**

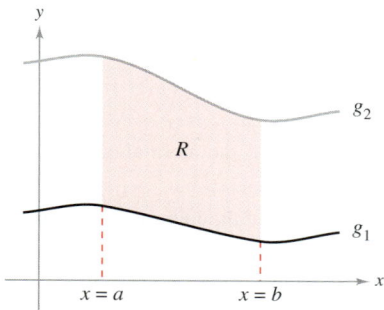

Lamina of constant density ρ
Figure 14.33

Definition of Mass of a Planar Lamina of Variable Density

If ρ is a continuous density function on the lamina corresponding to a plane region R, then the mass m of the lamina is given by

$$m = \int_R \int \rho(x, y) \, dA. \qquad \text{Variable density}$$

Density is normally expressed as mass per unit volume. For a planar lamina, however, density is mass per unit surface area.

EXAMPLE 1   Finding the Mass of a Planar Lamina

Find the mass of the triangular lamina with vertices $(0, 0)$, $(0, 3)$, and $(2, 3)$, given that the density at (x, y) is $\rho(x, y) = 2x + y$.

Solution   As shown in Figure 14.34, region R has the boundaries $x = 0$, $y = 3$, and $y = 3x/2$ (or $x = 2y/3$). Therefore, the mass of the lamina is

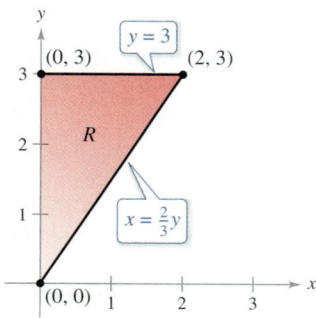

$$m = \int_R \int (2x + y) \, dA$$
$$= \int_0^3 \int_0^{2y/3} (2x + y) \, dx \, dy$$
$$= \int_0^3 \left[x^2 + xy \right]_0^{2y/3} dy$$
$$= \frac{10}{9} \int_0^3 y^2 \, dy$$
$$= \frac{10}{9} \left[\frac{y^3}{3} \right]_0^3$$
$$= 10.$$

Lamina of variable density
$\rho(x, y) = 2x + y$
Figure 14.34

In Figure 14.34, note that the planar lamina is shaded so that the darkest shading corresponds to the densest part.

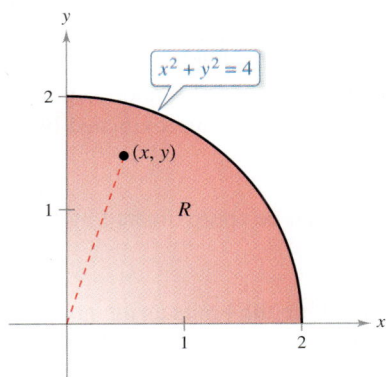

Density at (x, y): $\rho(x, y) = k\sqrt{x^2 + y^2}$

Figure 14.35

EXAMPLE 2   **Finding Mass by Polar Coordinates**

Find the mass of the lamina corresponding to the first-quadrant portion of the circle

$$x^2 + y^2 = 4$$

where the density at the point (x, y) is proportional to the distance between the point and the origin, as shown in Figure 14.35.

Solution   At any point (x, y), the density of the lamina is

$$\rho(x, y) = k\sqrt{(x - 0)^2 + (y - 0)^2}$$
$$= k\sqrt{x^2 + y^2}.$$

Because $0 \leq x \leq 2$ and $0 \leq y \leq \sqrt{4 - x^2}$, the mass is given by

$$m = \int_R \int k\sqrt{x^2 + y^2}\, dA$$
$$= \int_0^2 \int_0^{\sqrt{4-x^2}} k\sqrt{x^2 + y^2}\, dy\, dx.$$

To simplify the integration, you can convert to polar coordinates, using the bounds

$$0 \leq \theta \leq \pi/2 \quad \text{and} \quad 0 \leq r \leq 2.$$

So, the mass is

$$m = \int_R \int k\sqrt{x^2 + y^2}\, dA$$
$$= \int_0^{\pi/2} \int_0^2 k\sqrt{r^2}\, r\, dr\, d\theta$$
$$= \int_0^{\pi/2} \int_0^2 kr^2\, dr\, d\theta$$
$$= \int_0^{\pi/2} \frac{kr^3}{3} \Big]_0^2 \, d\theta$$
$$= \frac{8k}{3} \int_0^{\pi/2} d\theta$$
$$= \frac{8k}{3} \Big[\theta \Big]_0^{\pi/2}$$
$$= \frac{4\pi k}{3}.$$

▷ **TECHNOLOGY** On many occasions, this text has mentioned the benefits of computer programs that perform symbolic integration. Even if you use such a program regularly, you should remember that its greatest benefit comes only in the hands of a knowledgeable user. For instance, notice how much simpler the integral in Example 2 becomes when it is converted to polar form.

Rectangular Form

$$\int_0^2 \int_0^{\sqrt{4-x^2}} k\sqrt{x^2 + y^2}\, dy\, dx$$

Polar Form

$$\int_0^{\pi/2} \int_0^2 kr^2\, dr\, d\theta$$

If you have access to software that performs symbolic integration, use it to evaluate both integrals. Some software programs cannot handle the first integral, but any program that can handle double integrals can evaluate the second integral.

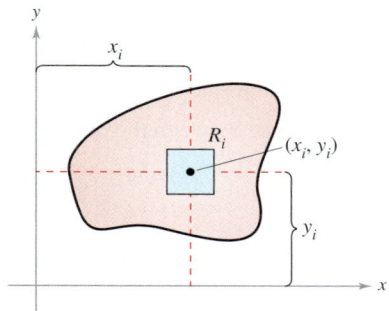

$M_x = (\text{mass})(y_i)$
$M_y = (\text{mass})(x_i)$

Figure 14.36

Moments and Center of Mass

For a lamina of variable density, moments of mass are defined in a manner similar to that used for the uniform density case. For a partition Δ of a lamina corresponding to a plane region R, consider the ith rectangle R_i of one area ΔA_i, as shown in Figure 14.36. Assume that the mass of R_i is concentrated at one of its interior points (x_i, y_i). The moment of mass of R_i with respect to the x-axis can be approximated by

$$(\text{Mass})(y_i) \approx [\rho(x_i, y_i)\,\Delta A_i](y_i).$$

Similarly, the moment of mass with respect to the y-axis can be approximated by

$$(\text{Mass})(x_i) \approx [\rho(x_i, y_i)\,\Delta A_i](x_i).$$

By forming the Riemann sum of all such products and taking the limits as the norm of Δ approaches 0, you obtain the following definitions of moments of mass with respect to the x- and y-axes.

Moments and Center of Mass of a Variable Density Planar Lamina

Let ρ be a continuous density function on the planar lamina R. The **moments of mass** with respect to the x- and y-axes are

$$M_x = \int_R \int y\rho(x, y)\, dA$$

and

$$M_y = \int_R \int x\rho(x, y)\, dA.$$

If m is the mass of the lamina, then the **center of mass** is

$$(\bar{x}, \bar{y}) = \left(\frac{M_y}{m}, \frac{M_x}{m} \right).$$

If R represents a simple plane region rather than a lamina, then the point $(\bar{x}, \bar{y})$ is called the **centroid** of the region.

For some planar laminas with a constant density ρ, you can determine the center of mass (or one of its coordinates) using symmetry rather than using integration. For instance, consider the laminas of constant density shown in Figure 14.37. Using symmetry, you can see that $\bar{y} = 0$ for the first lamina and $\bar{x} = 0$ for the second lamina.

$R: 0 \le x \le 1$
$-\sqrt{1 - x^2} \le y \le \sqrt{1 - x^2}$

$R: -\sqrt{1 - y^2} \le x \le \sqrt{1 - y^2}$
$0 \le y \le 1$

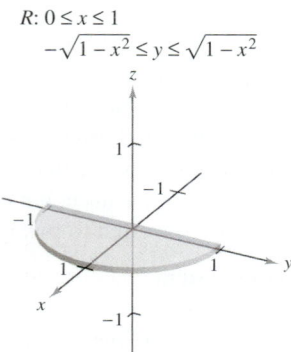

Lamina of constant density that is symmetric with respect to the x-axis

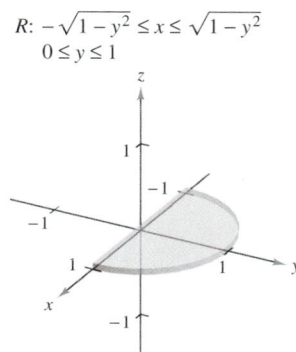

Lamina of constant density that is symmetric with respect to the y-axis

Figure 14.37

EXAMPLE 3 **Finding the Center of Mass**

⋯▷ *See LarsonCalculus.com for an interactive version of this type of example.*

Find the center of mass of the lamina corresponding to the parabolic region

$$0 \le y \le 4 - x^2 \qquad \text{Parabolic region}$$

where the density at the point (x, y) is proportional to the distance between (x, y) and the x-axis, as shown in Figure 14.38.

Solution The lamina is symmetric with respect to the y-axis and $\rho(x, y) = ky$. So, the center of mass lies on the y-axis and $\bar{x} = 0$. To find $\bar{y}$, first find the mass of the lamina.

$$
\begin{aligned}
\text{Mass} &= \int_{-2}^{2} \int_{0}^{4-x^2} ky \, dy \, dx \\
&= \frac{k}{2} \int_{-2}^{2} y^2 \Big]_{0}^{4-x^2} dx \\
&= \frac{k}{2} \int_{-2}^{2} (16 - 8x^2 + x^4) \, dx \\
&= \frac{k}{2} \left[16x - \frac{8x^3}{3} + \frac{x^5}{5} \right]_{-2}^{2} \\
&= k \left(32 - \frac{64}{3} + \frac{32}{5} \right) \\
&= \frac{256k}{15}
\end{aligned}
$$

Next, find the moment about the x-axis.

$$
\begin{aligned}
M_x &= \int_{-2}^{2} \int_{0}^{4-x^2} (y)(ky) \, dy \, dx \\
&= \frac{k}{3} \int_{-2}^{2} y^3 \Big]_{0}^{4-x^2} dx \\
&= \frac{k}{3} \int_{-2}^{2} (64 - 48x^2 + 12x^4 - x^6) \, dx \\
&= \frac{k}{3} \left[64x - 16x^3 + \frac{12x^5}{5} - \frac{x^7}{7} \right]_{-2}^{2} \\
&= \frac{4096k}{105}
\end{aligned}
$$

So,

$$\bar{y} = \frac{M_x}{m} = \frac{4096k/105}{256k/15} = \frac{16}{7}$$

and the center of mass is $\left(0, \frac{16}{7} \right)$. ◼

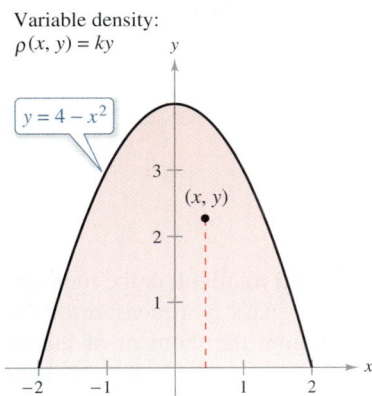

Although you can think of the moments M_x and M_y as measuring the tendency to rotate about the x- or y-axis, the calculation of moments is usually an intermediate step toward a more tangible goal. The use of the moments M_x and M_y is typical—to find the center of mass. Determination of the center of mass is useful in a variety of applications that allow you to treat a lamina as if its mass were concentrated at just one point. Intuitively, you can think of the center of mass as the balancing point of the lamina. For instance, the lamina in Example 3 should balance on the point of a pencil placed at $\left(0, \frac{16}{7} \right)$, as shown in Figure 14.39.

Variable density:
$\rho(x, y) = ky$

$y = 4 - x^2$

(x, y)

Parabolic region of variable density
Figure 14.38

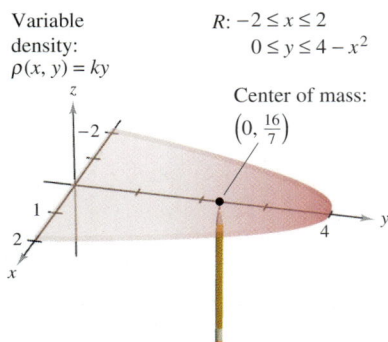

Variable density:
$\rho(x, y) = ky$

$R: -2 \le x \le 2$
$0 \le y \le 4 - x^2$

Center of mass:
$\left(0, \frac{16}{7} \right)$

Figure 14.39

Moments of Inertia

The moments of M_x and M_y used in determining the center of mass of a lamina are sometimes called the **first moments** about the x- and y-axes. In each case, the moment is the product of a mass times a distance.

$$M_x = \int_R \int (y)\rho(x, y)\, dA \qquad M_y = \int_R \int (x)\rho(x, y)\, dA$$

Distance to x-axis   Mass   Distance to y-axis   Mass

You will now look at another type of moment—the **second moment,** or the **moment of inertia** of a lamina about a line. In the same way that mass is a measure of the tendency of matter to resist a change in straight-line motion, the moment of inertia about a line is a *measure of the tendency of matter to resist a change in rotational motion.* For example, when a particle of mass m is a distance d from a fixed line, its moment of inertia about the line is defined as

$$I = md^2 = (\text{mass})(\text{distance})^2.$$

As with moments of mass, you can generalize this concept to obtain the moments of inertia about the x- and y-axes of a lamina of variable density. These second moments are denoted by I_x and I_y, and in each case the moment is the product of a mass times the square of a distance.

$$I_x = \int_R \int (y^2)\rho(x, y)\, dA \qquad I_y = \int_R \int (x^2)\rho(x, y)\, dA$$

Square of distance to x-axis   Mass   Square of distance to y-axis   Mass

The sum of the moments I_x and I_y is called the **polar moment of inertia** and is denoted by I_0. For a lamina in the xy-plane, I_0 represents the moment of inertia of the lamina about the z-axis. The term "polar moment of inertia" stems from the fact that the square of the polar distance r is used in the calculation.

$$I_0 = \int_R \int (x^2 + y^2)\rho(x, y)\, dA = \int_R \int r^2\rho(x, y)\, dA$$

EXAMPLE 4   **Finding the Moment of Inertia**

Find the moment of inertia about the x-axis of the lamina in Example 3.

Solution   From the definition of moment of inertia, you have

$$I_x = \int_{-2}^{2} \int_{0}^{4-x^2} y^2(ky)\, dy\, dx$$

$$= \frac{k}{4}\int_{-2}^{2} y^4\Big]_{0}^{4-x^2} dx$$

$$= \frac{k}{4}\int_{-2}^{2} (256 - 256x^2 + 96x^4 - 16x^6 + x^8)\, dx$$

$$= \frac{k}{4}\left[256x - \frac{256x^3}{3} + \frac{96x^5}{5} - \frac{16x^7}{7} + \frac{x^9}{9}\right]_{-2}^{2}$$

$$= \frac{32{,}768k}{315}.$$

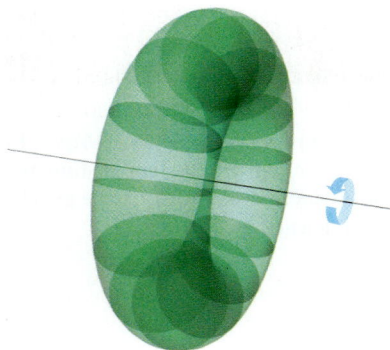

Planar lamina revolving at ω radians per second
Figure 14.40

The moment of inertia I of a revolving lamina can be used to measure its kinetic energy. For example, suppose a planar lamina is revolving about a line with an **angular speed** of ω radians per second, as shown in Figure 14.40. The kinetic energy E of the revolving lamina is

$$E = \frac{1}{2} I\omega^2. \qquad \text{Kinetic energy for rotational motion}$$

On the other hand, the kinetic energy E of a mass m moving in a straight line at a velocity v is

$$E = \frac{1}{2} mv^2. \qquad \text{Kinetic energy for linear motion}$$

So, the kinetic energy of a mass moving in a straight line is proportional to its mass, but the kinetic energy of a mass revolving about an axis is proportional to its moment of inertia.

The **radius of gyration** $\bar{\bar{r}}$ of a revolving mass m with moment of inertia I is defined as

$$\bar{\bar{r}} = \sqrt{\frac{I}{m}}. \qquad \text{Radius of gyration}$$

If the entire mass were located at a distance $\bar{\bar{r}}$ from its axis of revolution, it would have the same moment of inertia and, consequently, the same kinetic energy. For instance, the radius of gyration of the lamina in Example 4 about the x-axis is

$$\bar{\bar{y}} = \sqrt{\frac{I_x}{m}} = \sqrt{\frac{32{,}768k/315}{256k/15}} = \sqrt{\frac{128}{21}} \approx 2.469.$$

EXAMPLE 5 **Finding the Radius of Gyration**

Find the radius of gyration about the y-axis for the lamina corresponding to the region $R: 0 \le y \le \sin x$, $0 \le x \le \pi$, where the density at (x, y) is given by $\rho(x, y) = x$.

Solution The region R is shown in Figure 14.41. By integrating $\rho(x, y) = x$ over the region R, you can determine that the mass of the region is π. The moment of inertia about the y-axis is

$$I_y = \int_0^{\pi} \int_0^{\sin x} x^3 \, dy \, dx$$

$$= \int_0^{\pi} x^3 y \Big]_0^{\sin x} dx$$

$$= \int_0^{\pi} x^3 \sin x \, dx$$

$$= \left[(3x^2 - 6)(\sin x) - (x^3 - 6x)(\cos x) \right]_0^{\pi}$$

$$= \pi^3 - 6\pi.$$

So, the radius of gyration about the y-axis is

$$\bar{\bar{x}} = \sqrt{\frac{I_y}{m}}$$

$$= \sqrt{\frac{\pi^3 - 6\pi}{\pi}}$$

$$= \sqrt{\pi^2 - 6}$$

$$\approx 1.967.$$

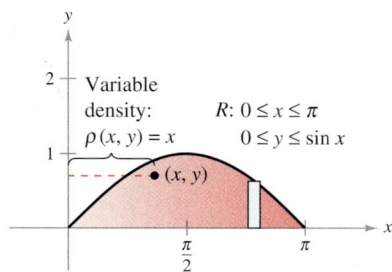

Figure 14.41

14.4 Exercises

See **CalcChat.com** for tutorial help and worked-out solutions to odd-numbered exercises.

Finding the Mass of a Lamina In Exercises 1–4, find the mass of the lamina described by the inequalities, given that its density is $\rho(x, y) = xy$. (*Hint:* Some of the integrals are simpler in polar coordinates.)

1. $0 \le x \le 2, \ 0 \le y \le 2$

2. $0 \le x \le 3, \ 0 \le y \le 9 - x^2$

3. $0 \le x \le 1, \ 0 \le y \le \sqrt{1 - x^2}$

4. $x \ge 0, \ 3 \le y \le 3 + \sqrt{9 - x^2}$

Finding the Center of Mass In Exercises 5–8, find the mass and center of mass of the lamina for each density.

5. R: square with vertices $(0, 0), (a, 0), (0, a), (a, a)$

   (a) $\rho = k$ (b) $\rho = ky$ (c) $\rho = kx$

6. R: rectangle with vertices $(0, 0), (a, 0), (0, b), (a, b)$

   (a) $\rho = kxy$ (b) $\rho = k(x^2 + y^2)$

7. R: triangle with vertices $(0, 0), (0, a), (a, a)$

   (a) $\rho = k$ (b) $\rho = ky$ (c) $\rho = kx$

8. R: triangle with vertices $(0, 0), (a/2, a), (a, 0)$

   (a) $\rho = k$ (b) $\rho = kxy$

9. **Translations in the Plane** Translate the lamina in Exercise 5 to the right five units and determine the resulting center of mass.

10. **Conjecture** Use the result of Exercise 9 to make a conjecture about the change in the center of mass when a lamina of constant density is translated c units horizontally or d units vertically. Is the conjecture true when the density is not constant? Explain.

Finding the Center of Mass In Exercises 11–22, find the mass and center of mass of the lamina bounded by the graphs of the equations for the given density. (*Hint:* Some of the integrals are simpler in polar coordinates.)

11. $y = \sqrt{x}, \ y = 0, \ x = 1, \ \rho = ky$

12. $y = x^2, \ y = 0, \ x = 2, \ \rho = kxy$

13. $y = 4/x, \ y = 0, \ x = 1, \ x = 4, \ \rho = kx^2$

14. $y = \dfrac{1}{1 + x^2}, \ y = 0, \ x = -1, \ x = 1, \ \rho = k$

15. $y = e^x, \ y = 0, \ x = 0, \ x = 1, \ \rho = k$

16. $y = e^{-x}, \ y = 0, \ x = 0, \ x = 1, \ \rho = ky^2$

17. $y = 4 - x^2, \ y = 0, \ \rho = ky$

18. $x = 9 - y^2, \ x = 0, \ \rho = kx$

19. $y = \sin \dfrac{\pi x}{L}, \ y = 0, \ x = 0, \ x = L, \ \rho = k$

20. $y = \cos \dfrac{\pi x}{L}, \ y = 0, \ x = 0, \ x = \dfrac{L}{2}, \ \rho = ky$

21. $y = \sqrt{a^2 - x^2}, \ 0 \le y \le x, \ \rho = k$

22. $x^2 + y^2 = a^2, \ x \ge 0, y \ge 0, \rho = k(x^2 + y^2)$

Finding the Center of Mass Using Technology In Exercises 23–26, use a computer algebra system to find the mass and center of mass of the lamina bounded by the graphs of the equations for the given density.

23. $y = e^{-x}, \ y = 0, \ x = 0, \ x = 2, \ \rho = kxy$

24. $y = \ln x, \ y = 0, \ x = 1, \ x = e, \ \rho = \dfrac{k}{x}$

25. $r = 2 \cos 3\theta, \ -\dfrac{\pi}{6} \le \theta \le \dfrac{\pi}{6}, \ \rho = k$

26. $r = 1 + \cos \theta, \ \rho = k$

Finding the Radius of Gyration About Each Axis In Exercises 27–32, verify the given moment(s) of inertia and find $\bar{x}$ and $\bar{y}$. Assume that each lamina has a density of $\rho = 1$ gram per square centimeter. (These regions are common shapes used in engineering.)

27. Rectangle

$I_x = \frac{1}{3}bh^3$
$I_y = \frac{1}{3}b^3h$

28. Right triangle

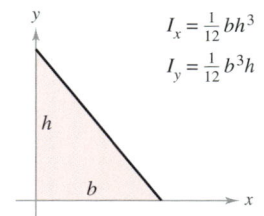

$I_x = \frac{1}{12}bh^3$
$I_y = \frac{1}{12}b^3h$

29. Circle

$I_0 = \frac{1}{2}\pi a^4$

30. Semicircle

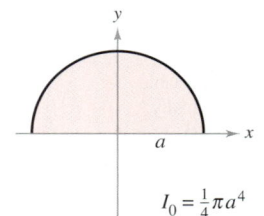

$I_0 = \frac{1}{4}\pi a^4$

31. Quarter circle

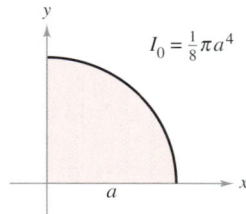

$I_0 = \frac{1}{8}\pi a^4$

32. Ellipse

$I_0 = \frac{1}{4}\pi ab(a^2 + b^2)$

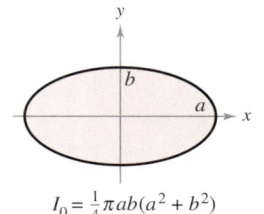

Finding Moments of Inertia and Radii of Gyration In Exercises 33–36, find $I_x, I_y, I_0, \bar{x}$, and $\bar{y}$ for the lamina bounded by the graphs of the equations.

33. $y = 4 - x^2, \ y = 0, \ x > 0, \ \rho = kx$

34. $y = x, \ y = x^2, \ \rho = kxy$

35. $y = \sqrt{x}, \ y = 0, \ x = 4, \ \rho = kxy$

36. $y = x^2, \ y^2 = x, \ \rho = kx$

Finding a Moment of Inertia Using Technology In Exercises 37–40, set up the double integral required to find the moment of inertia I, about the given line, of the lamina bounded by the graphs of the equations. Use a computer algebra system to evaluate the double integral.

37. $x^2 + y^2 = b^2$, $\rho = k$, line: $x = a$ $(a > b)$

38. $y = \sqrt{x}$, $y = 0$, $x = 4$, $\rho = kx$, line: $x = 6$

39. $y = \sqrt{a^2 - x^2}$, $y = 0$, $\rho = ky$, line: $y = a$

40. $y = 4 - x^2$, $y = 0$, $\rho = k$, line: $y = 2$

Hydraulics In Exercises 41–44, determine the location of the horizontal axis y_a at which a vertical gate in a dam is to be hinged so that there is no moment causing rotation under the indicated loading (see figure). The model for y_a is

$$y_a = \bar{y} - \frac{I_{\bar{y}}}{hA}$$

where $\bar{y}$ is the y-coordinate of the centroid of the gate, $I_{\bar{y}}$ is the moment of inertia of the gate about the line $y = \bar{y}$, h is the depth of the centroid below the surface, and A is the area of the gate.

41.

42.

43.

44.
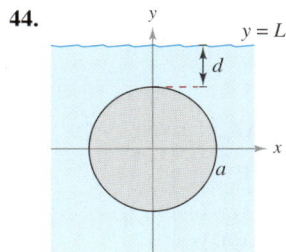

WRITING ABOUT CONCEPTS

45. Moments and Center of Mass Give the formulas for finding the moments and center of mass of a variable density planar lamina.

46. Moments of Inertia Give the formulas for finding the moments of inertia about the x- and y-axes for a variable density planar lamina.

WRITING ABOUT CONCEPTS   (continued)

47. Radius of Gyration In your own words, describe what the radius of gyration measures.

---

48. HOW DO YOU SEE IT? The center of mass of the lamina of constant density shown in the figure is $\left(2, \frac{8}{5}\right)$. Make a conjecture about how the center of mass $(\bar{x}, \bar{y})$ changes for each given nonconstant density $\rho(x, y)$. Explain. (Make your conjecture *without* performing any calculations.)

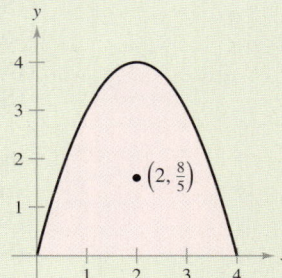

(a) $\rho(x, y) = ky$ (b) $\rho(x, y) = k|2 - x|$

(c) $\rho(x, y) = kxy$ (d) $\rho(x, y) = k(4 - x)(4 - y)$

---

49. Proof Prove the following Theorem of Pappus: Let R be a region in a plane and let L be a line in the same plane such that L does not intersect the interior of R. If r is the distance between the centroid of R and the line, then the volume V of the solid of revolution formed by revolving R about the line is given by $V = 2\pi rA$, where A is the area of R.

SECTION PROJECT

Center of Pressure on a Sail

The center of pressure on a sail is the point (x_p, y_p) at which the total aerodynamic force may be assumed to act. If the sail is represented by a plane region R, then the center of pressure is

$$x_p = \frac{\int_R \int xy \, dA}{\int_R \int y \, dA} \quad \text{and} \quad y_p = \frac{\int_R \int y^2 \, dA}{\int_R \int y \, dA}.$$

Consider a triangular sail with vertices at $(0, 0)$, $(2, 1)$, and $(0, 5)$. Verify the value of each integral.

(a) $\displaystyle\int_R \int y \, dA = 10$

(b) $\displaystyle\int_R \int xy \, dA = \frac{35}{6}$

(c) $\displaystyle\int_R \int y^2 \, dA = \frac{155}{6}$

Calculate the coordinates (x_p, y_p) of the center of pressure. Sketch a graph of the sail and indicate the location of the center of pressure.

14.5 Surface Area

■ Use a double integral to find the area of a surface.

Surface Area

At this point, you know a great deal about the solid region lying between a surface and a closed and bounded region R in the xy-plane, as shown in Figure 14.42. For example, you know how to find the extrema of f on R (Section 13.8), the area of the base R of the solid (Section 14.1), the volume of the solid (Section 14.2), and the centroid of the base R (Section 14.4).

Surface:
$z = f(x, y)$

Region R in xy-plane

Figure 14.42

In this section, you will learn how to find the upper **surface area** of the solid. Later, you will learn how to find the centroid of the solid (Section 14.6) and the lateral surface area (Section 15.2).

To begin, consider a surface S given by

$$z = f(x, y) \qquad \text{Surface defined over a region } R$$

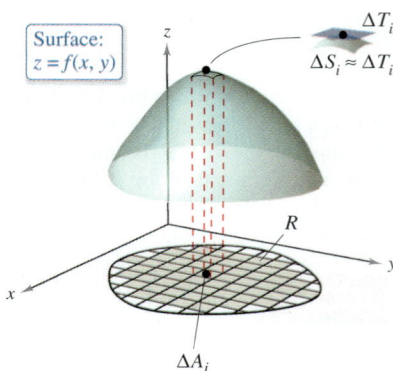

Surface:
$z = f(x, y)$

ΔT_i

$\Delta S_i \approx \Delta T_i$

R

ΔA_i

Figure 14.43

defined over a region R. Assume that R is closed and bounded and that f has continuous first partial derivatives. To find the surface area, construct an inner partition of R consisting of n rectangles, where the area of the ith rectangle R_i is $\Delta A_i = \Delta x_i \Delta y_i$, as shown in Figure 14.43. In each R_i, let (x_i, y_i) be the point that is closest to the origin. At the point $(x_i, y_i, z_i) = (x_i, y_i, f(x_i, y_i))$ on the surface S, construct a tangent plane T_i. The area of the portion of the tangent plane that lies directly above R_i is approximately equal to the area of the surface lying directly above R_i. That is, $\Delta T_i \approx \Delta S_i$. So, the surface area of S is approximated by

$$\sum_{i=1}^{n} \Delta S_i \approx \sum_{i=1}^{n} \Delta T_i.$$

To find the area of the parallelogram ΔT_i, note that its sides are given by the vectors

$$\mathbf{u} = \Delta x_i \mathbf{i} + f_x(x_i, y_i) \Delta x_i \mathbf{k}$$

and

$$\mathbf{v} = \Delta y_i \mathbf{j} + f_y(x_i, y_i) \Delta y_i \mathbf{k}.$$

From Theorem 11.8, the area of ΔT_i is given by $\|\mathbf{u} \times \mathbf{v}\|$, where

$$\mathbf{u} \times \mathbf{v} = \begin{vmatrix} \mathbf{i} & \mathbf{j} & \mathbf{k} \\ \Delta x_i & 0 & f_x(x_i, y_i) \Delta x_i \\ 0 & \Delta y_i & f_y(x_i, y_i) \Delta y_i \end{vmatrix}$$

$$= -f_x(x_i, y_i) \Delta x_i \Delta y_i \mathbf{i} - f_y(x_i, y_i) \Delta x_i \Delta y_i \mathbf{j} + \Delta x_i \Delta y_i \mathbf{k}$$

$$= (-f_x(x_i, y_i) \mathbf{i} - f_y(x_i, y_i) \mathbf{j} + \mathbf{k}) \Delta A_i.$$

So, the area of ΔT_i is $\|\mathbf{u} \times \mathbf{v}\| = \sqrt{[f_x(x_i, y_i)]^2 + [f_y(x_i, y_i)]^2 + 1} \, \Delta A_i$, and

$$\text{Surface area of } S \approx \sum_{i=1}^{n} \Delta S_i$$

$$\approx \sum_{i=1}^{n} \sqrt{1 + [f_x(x_i, y_i)]^2 + [f_y(x_i, y_i)]^2} \, \Delta A_i.$$

This suggests the definition of surface area on the next page.

Definition of Surface Area

If f and its first partial derivatives are continuous on the closed region R in the xy-plane, then the **area of the surface S** given by $z = f(x, y)$ over R is defined as

$$\text{Surface area} = \iint_R dS$$

$$= \iint_R \sqrt{1 + [f_x(x, y)]^2 + [f_y(x, y)]^2} \, dA.$$

As an aid to remembering the double integral for surface area, it is helpful to note its similarity to the integral for arc length.

Length on x-axis: $\displaystyle\int_a^b dx$

Arc length in xy-plane: $\displaystyle\int_a^b ds = \int_a^b \sqrt{1 + [f'(x)]^2} \, dx$

Area in xy-plane: $\displaystyle\iint_R dA$

Surface area in space: $\displaystyle\iint_R dS = \iint_R \sqrt{1 + [f_x(x, y)]^2 + [f_y(x, y)]^2} \, dA$

Like integrals for arc length, integrals for surface area are often very difficult to evaluate. However, one type that is easily evaluated is demonstrated in the next example.

EXAMPLE 1   **The Surface Area of a Plane Region**

Find the surface area of the portion of the plane

$$z = 2 - x - y$$

that lies above the circle $x^2 + y^2 \leq 1$ in the first quadrant, as shown in Figure 14.44.

Solution   Because $f_x(x, y) = -1$ and $f_y(x, y) = -1$, the surface area is given by

$$S = \iint_R \sqrt{1 + [f_x(x, y)]^2 + [f_y(x, y)]^2} \, dA \qquad \text{\color{red}Formula for surface area}$$

$$= \iint_R \sqrt{1 + (-1)^2 + (-1)^2} \, dA \qquad \text{\color{red}Substitute.}$$

$$= \iint_R \sqrt{3} \, dA$$

$$= \sqrt{3} \iint_R dA.$$

Note that the last integral is $\sqrt{3}$ times the area of the region R. R is a quarter circle of radius 1, with an area of $\frac{1}{4}\pi(1^2)$ or $\pi/4$. So, the area of S is

$$S = \sqrt{3} \, (\text{area of } R)$$

$$= \sqrt{3}\left(\frac{\pi}{4}\right)$$

$$= \frac{\sqrt{3}\,\pi}{4}.$$

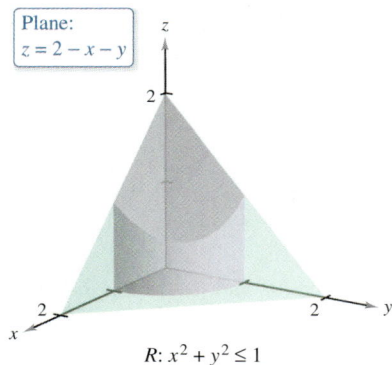

Plane:
$z = 2 - x - y$

$R: x^2 + y^2 \leq 1$

Figure 14.44

Figure 14.45

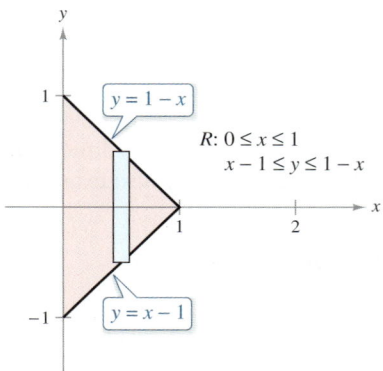

Figure 14.46

EXAMPLE 2 **Finding Surface Area**

⋅⋅⋅▷ *See LarsonCalculus.com for an interactive version of this type of example.*

Find the area of the portion of the surface $f(x, y) = 1 - x^2 + y$ that lies above the triangular region with vertices $(1, 0, 0)$, $(0, -1, 0)$, and $(0, 1, 0)$, as shown in Figure 14.45.

Solution Because $f_x(x, y) = -2x$ and $f_y(x, y) = 1$, you have

$$S = \iint_R \sqrt{1 + [f_x(x, y)]^2 + [f_y(x, y)]^2}\, dA = \iint_R \sqrt{1 + 4x^2 + 1}\, dA.$$

In Figure 14.46, you can see that the bounds for R are $0 \le x \le 1$ and $x - 1 \le y \le 1 - x$. So, the integral becomes

$$S = \int_0^1 \int_{x-1}^{1-x} \sqrt{2 + 4x^2}\, dy\, dx$$

$$= \int_0^1 y\sqrt{2 + 4x^2}\, \bigg]_{x-1}^{1-x} dx$$

$$= \int_0^1 \left[(1 - x)\sqrt{2 + 4x^2} - (x - 1)\sqrt{2 + 4x^2} \right] dx$$

$$= \int_0^1 \left(2\sqrt{2 + 4x^2} - 2x\sqrt{2 + 4x^2} \right) dx \quad \text{\color{red}Integration tables (Appendix B), Formula 26 and Power Rule}$$

$$= \left[x\sqrt{2 + 4x^2} + \ln\left(2x + \sqrt{2 + 4x^2}\right) - \frac{(2 + 4x^2)^{3/2}}{6} \right]_0^1$$

$$= \sqrt{6} + \ln\left(2 + \sqrt{6}\right) - \sqrt{6} - \ln\sqrt{2} + \frac{1}{3}\sqrt{2}$$

$$\approx 1.618.$$

EXAMPLE 3 **Change of Variables to Polar Coordinates**

Find the surface area of the paraboloid $z = 1 + x^2 + y^2$ that lies above the unit circle, as shown in Figure 14.47.

Solution Because $f_x(x, y) = 2x$ and $f_y(x, y) = 2y$, you have

$$S = \iint_R \sqrt{1 + [f_x(x, y)]^2 + [f_y(x, y)]^2}\, dA = \iint_R \sqrt{1 + 4x^2 + 4y^2}\, dA.$$

You can convert to polar coordinates by letting $x = r\cos\theta$ and $y = r\sin\theta$. Then, because the region R is bounded by $0 \le r \le 1$ and $0 \le \theta \le 2\pi$, you have

$$S = \int_0^{2\pi} \int_0^1 \sqrt{1 + 4r^2}\, r\, dr\, d\theta$$

$$= \int_0^{2\pi} \frac{1}{12}(1 + 4r^2)^{3/2} \bigg]_0^1 d\theta$$

$$= \int_0^{2\pi} \frac{5\sqrt{5} - 1}{12}\, d\theta$$

$$= \frac{5\sqrt{5} - 1}{12}\, \theta \bigg]_0^{2\pi}$$

$$= \frac{\pi\left(5\sqrt{5} - 1\right)}{6}$$

$$\approx 5.33.$$

Figure 14.47

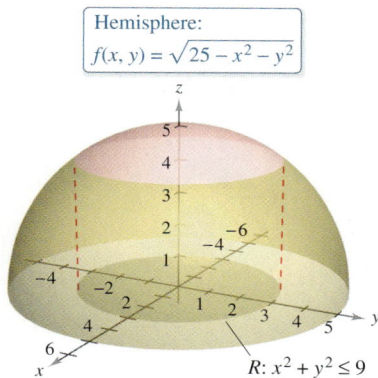

Hemisphere:
$f(x, y) = \sqrt{25 - x^2 - y^2}$

$R: x^2 + y^2 \leq 9$

Figure 14.48

EXAMPLE 4 **Finding Surface Area**

Find the surface area S of the portion of the hemisphere

$$f(x, y) = \sqrt{25 - x^2 - y^2} \qquad \text{Hemisphere}$$

that lies above the region R bounded by the circle $x^2 + y^2 \leq 9$, as shown in Figure 14.48.

Solution The first partial derivatives of f are

$$f_x(x, y) = \frac{-x}{\sqrt{25 - x^2 - y^2}}$$

and

$$f_y(x, y) = \frac{-y}{\sqrt{25 - x^2 - y^2}}$$

and, from the formula for surface area, you have

$$dS = \sqrt{1 + [f_x(x, y)]^2 + [f_y(x, y)]^2}\, dA$$

$$= \sqrt{1 + \left(\frac{-x}{\sqrt{25 - x^2 - y^2}}\right)^2 + \left(\frac{-y}{\sqrt{25 - x^2 - y^2}}\right)^2}\, dA$$

$$= \frac{5}{\sqrt{25 - x^2 - y^2}}\, dA.$$

So, the surface area is

$$S = \int\!\!\int_R \frac{5}{\sqrt{25 - x^2 - y^2}}\, dA.$$

You can convert to polar coordinates by letting $x = r\cos\theta$ and $y = r\sin\theta$. Then, because the region R is bounded by $0 \leq r \leq 3$ and $0 \leq \theta \leq 2\pi$, you obtain

$$S = \int_0^{2\pi}\int_0^3 \frac{5}{\sqrt{25 - r^2}}\, r\, dr\, d\theta$$

$$= 5\int_0^{2\pi} -\sqrt{25 - r^2}\,\Big]_0^3\, d\theta$$

$$= 5\int_0^{2\pi} d\theta$$

$$= 10\pi.$$

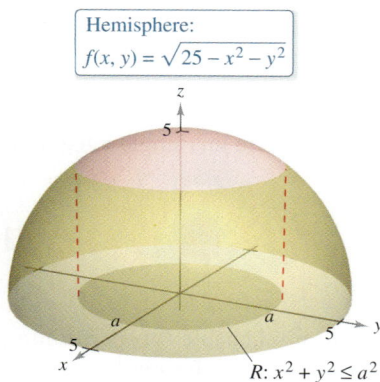

Hemisphere:
$f(x, y) = \sqrt{25 - x^2 - y^2}$

$R: x^2 + y^2 \leq a^2$

Figure 14.49

The procedure used in Example 4 can be extended to find the surface area of a sphere by using the region R bounded by the circle $x^2 + y^2 \leq a^2$, where $0 < a < 5$, as shown in Figure 14.49. The surface area of the portion of the hemisphere

$$f(x, y) = \sqrt{25 - x^2 - y^2}$$

lying above the circular region can be shown to be

$$S = \int\!\!\int_R \frac{5}{\sqrt{25 - x^2 - y^2}}\, dA$$

$$= \int_0^{2\pi}\int_0^a \frac{5}{\sqrt{25 - r^2}}\, r\, dr\, d\theta$$

$$= 10\pi\left(5 - \sqrt{25 - a^2}\right).$$

By taking the limit as a approaches 5 and doubling the result, you obtain a total area of 100π. (The surface area of a sphere of radius r is $S = 4\pi r^2$.)

You can use Simpson's Rule or the Trapezoidal Rule to approximate the value of a double integral, *provided* you can get through the first integration. This is demonstrated in the next example.

EXAMPLE 5 **Approximating Surface Area by Simpson's Rule**

Paraboloid:
$f(x, y) = 2 - x^2 - y^2$

Figure 14.50

Find the area of the surface of the paraboloid

$$f(x, y) = 2 - x^2 - y^2 \qquad \text{Paraboloid}$$

that lies above the square region bounded by

$$-1 \le x \le 1 \quad \text{and} \quad -1 \le y \le 1$$

as shown in Figure 14.50.

Solution Using the partial derivatives

$$f_x(x, y) = -2x \quad \text{and} \quad f_y(x, y) = -2y$$

you have a surface area of

$$S = \iint_R \sqrt{1 + [f_x(x, y)]^2 + [f_y(x, y)]^2} \, dA$$

$$= \iint_R \sqrt{1 + (-2x)^2 + (-2y)^2} \, dA$$

$$= \iint_R \sqrt{1 + 4x^2 + 4y^2} \, dA.$$

In polar coordinates, the line $x = 1$ is given by

$$r \cos \theta = 1 \quad \text{or} \quad r = \sec \theta$$

and you can determine from Figure 14.51 that one-fourth of the region R is bounded by

$$0 \le r \le \sec \theta \quad \text{and} \quad -\frac{\pi}{4} \le \theta \le \frac{\pi}{4}.$$

Letting $x = r \cos \theta$ and $y = r \sin \theta$ produces

$$\frac{1}{4} S = \frac{1}{4} \iint_R \sqrt{1 + 4x^2 + 4y^2} \, dA$$

$$= \int_{-\pi/4}^{\pi/4} \int_0^{\sec \theta} \sqrt{1 + 4r^2} \, r \, dr \, d\theta$$

$$= \int_{-\pi/4}^{\pi/4} \frac{1}{12} (1 + 4r^2)^{3/2} \bigg]_0^{\sec \theta} \, d\theta$$

$$= \frac{1}{12} \int_{-\pi/4}^{\pi/4} [(1 + 4 \sec^2 \theta)^{3/2} - 1] \, d\theta.$$

One-fourth of the region R is bounded by $0 \le r \le \sec \theta$ and $-\frac{\pi}{4} \le \theta \le \frac{\pi}{4}$.

Figure 14.51

Finally, using Simpson's Rule with $n = 10$, you can approximate this single integral to be

$$S = \frac{1}{3} \int_{-\pi/4}^{\pi/4} [(1 + 4 \sec^2 \theta)^{3/2} - 1] \, d\theta \approx 7.450.$$

▷ **TECHNOLOGY** Most computer programs that are capable of performing symbolic integration for multiple integrals are also capable of performing numerical approximation techniques. If you have access to such software, use it to approximate the value of the integral in Example 5.

14.5 Exercises

See **CalcChat.com** for tutorial help and worked-out solutions to odd-numbered exercises.

Finding Surface Area In Exercises 1–14, find the area of the surface given by $z = f(x, y)$ over the region R. (*Hint:* Some of the integrals are simpler in polar coordinates.)

1. $f(x, y) = 2x + 2y$

R: triangle with vertices $(0, 0)$, $(4, 0)$, $(0, 4)$

2. $f(x, y) = 15 + 2x - 3y$

R: square with vertices $(0, 0)$, $(3, 0)$, $(0, 3)$, $(3, 3)$

3. $f(x, y) = 7 + 2x + 2y$, $R = \{(x, y): x^2 + y^2 \le 4\}$

4. $f(x, y) = 12 + 2x - 3y$, $R = \{(x, y): x^2 + y^2 \le 9\}$

5. $f(x, y) = 9 - x^2$

R: square with vertices $(0, 0)$, $(2, 0)$, $(0, 2)$, $(2, 2)$

6. $f(x, y) = y^2$

R: square with vertices $(0, 0)$, $(3, 0)$, $(0, 3)$, $(3, 3)$

7. $f(x, y) = 3 + x^{3/2}$

R: rectangle with vertices $(0, 0)$, $(0, 4)$, $(3, 4)$, $(3, 0)$

8. $f(x, y) = 2 + \frac{2}{3}y^{3/2}$

$R = \{(x, y): 0 \le x \le 2, 0 \le y \le 2 - x\}$

9. $f(x, y) = \ln|\sec x|$

$R = \left\{(x, y): 0 \le x \le \dfrac{\pi}{4}, 0 \le y \le \tan x\right\}$

10. $f(x, y) = 13 + x^2 - y^2$, $R = \{(x, y): x^2 + y^2 \le 4\}$

11. $f(x, y) = \sqrt{x^2 + y^2}$, $R = \{(x, y): 0 \le f(x, y) \le 1\}$

12. $f(x, y) = xy$, $R = \{(x, y): x^2 + y^2 \le 16\}$

13. $f(x, y) = \sqrt{a^2 - x^2 - y^2}$

$R = \{(x, y): x^2 + y^2 \le b^2, 0 < b < a\}$

14. $f(x, y) = \sqrt{a^2 - x^2 - y^2}$

$R = \{(x, y): x^2 + y^2 \le a^2\}$

Finding Surface Area In Exercises 15–18, find the area of the surface.

15. The portion of the plane $z = 24 - 3x - 2y$ in the first octant

16. The portion of the paraboloid $z = 16 - x^2 - y^2$ in the first octant

17. The portion of the sphere $x^2 + y^2 + z^2 = 25$ inside the cylinder $x^2 + y^2 = 9$

18. The portion of the cone $z = 2\sqrt{x^2 + y^2}$ inside the cylinder $x^2 + y^2 = 4$

Finding Surface Area Using Technology In Exercises 19–24, write a double integral that represents the surface area of $z = f(x, y)$ over the region R. Use a computer algebra system to evaluate the double integral.

19. $f(x, y) = 2y + x^2$, R: triangle with vertices $(0, 0)$, $(1, 0)$, $(1, 1)$

20. $f(x, y) = 2x + y^2$, R: triangle with vertices $(0, 0)$, $(2, 0)$, $(2, 2)$

21. $f(x, y) = 9 - x^2 - y^2$, $R = \{(x, y): 0 \le f(x, y)\}$

22. $f(x, y) = x^2 + y^2$, $R = \{(x, y): 0 \le f(x, y) \le 16\}$

23. $f(x, y) = 4 - x^2 - y^2$

$R = \{(x, y): 0 \le x \le 1, 0 \le y \le 1\}$

24. $f(x, y) = \frac{2}{3}x^{3/2} + \cos x$

$R = \{(x, y): 0 \le x \le 1, 0 \le y \le 1\}$

Setting Up a Double Integral In Exercises 25–28, set up a double integral that gives the area of the surface on the graph of f over the region R.

25. $f(x, y) = e^{xy}$, $R = \{(x, y): 0 \le x \le 4, 0 \le y \le 10\}$

26. $f(x, y) = x^2 - 3xy - y^2$

$R = \{(x, y): 0 \le x \le 4, 0 \le y \le x\}$

27. $f(x, y) = e^{-x} \sin y$, $R = \{(x, y): x^2 + y^2 \le 4\}$

28. $f(x, y) = \cos(x^2 + y^2)$, $R = \left\{(x, y): x^2 + y^2 \le \dfrac{\pi}{2}\right\}$

WRITING ABOUT CONCEPTS

29. Surface Area State the double integral definition of the area of a surface S given by $z = f(x, y)$ over the region R in the xy-plane.

30. Surface Area Answer each question about the surface area S on a surface given by a positive function $z = f(x, y)$ over a region R in the xy-plane. Explain each answer.

(a) Is it possible for S to equal the area of R?

(b) Can S be greater than the area of R?

(c) Can S be less than the area of R?

31. Surface Area Will the surface area of the graph of a function $z = f(x, y)$ over a region R increase when the graph is shifted k units vertically? Why or why not?

---

32. HOW DO YOU SEE IT? Consider the surface $f(x, y) = x^2 + y^2$ (see figure) and the surface area of f over each region R. Without integrating, order the surface areas from least to greatest. Explain.

(a) R: rectangle with vertices $(0, 0)$, $(2, 0)$, $(2, 2)$, $(0, 2)$

(b) R: triangle with vertices $(0, 0)$, $(2, 0)$, $(0, 2)$

(c) $R = \{(x, y): x^2 + y^2 \le 4$, first quadrant only$\}$

33. Product Design A company produces a spherical object of radius 25 centimeters. A hole of radius 4 centimeters is drilled through the center of the object.

(a) Find the volume of the object.

(b) Find the outer surface area of the object.

34. Modeling Data

A company builds a warehouse with dimensions 30 feet by 50 feet. The symmetrical shape and selected heights of the roof are shown in the figure.

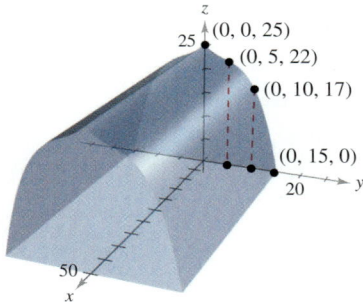

(a) Use the regression capabilities of a graphing utility to find a model of the form

$$z = ay^3 + by^2 + cy + d$$

for the roof line.

(b) Use the numerical integration capabilities of a graphing utility and the model in part (a) to approximate the volume of storage space in the warehouse.

(c) Use the numerical integration utility and the model in part (a) to approximate the surface area of the roof.

(d) Approximate the arc length of the roof line and find the surface area of the roof by multiplying the arc length by the length of the warehouse. Compare the results and the integrations with those found in part (c).

35. Surface Area Find the surface area of the solid of intersection of the cylinders $x^2 + z^2 = 1$ and $y^2 + z^2 = 1$ (see figure).

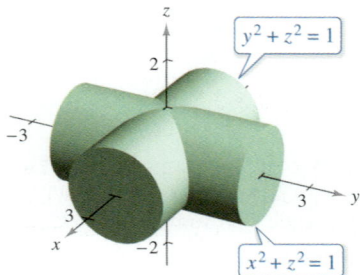

36. Surface Area Show that the surface area of the cone $z = k\sqrt{x^2 + y^2}, k > 0$, over the circular region $x^2 + y^2 \le r^2$ in the xy-plane is $\pi r^2 \sqrt{k^2 + 1}$ (see figure).

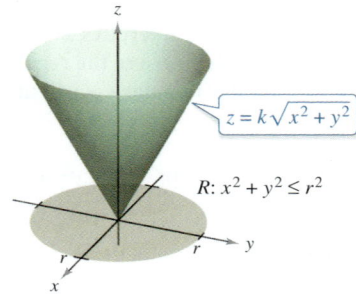

SECTION PROJECT

Capillary Action

A well-known property of liquids is that they will rise in narrow vertical channels—this property is called "capillary action." The figure shows two plates, that form a narrow wedge in a container of liquid. The upper surface of the liquid follows a hyperbolic shape given by

$$z = \frac{k}{\sqrt{x^2 + y^2}}$$

where x, y, and z are measured in inches. The constant k depends on the angle of the wedge, the type of liquid, and the material that comprises the flat plates.

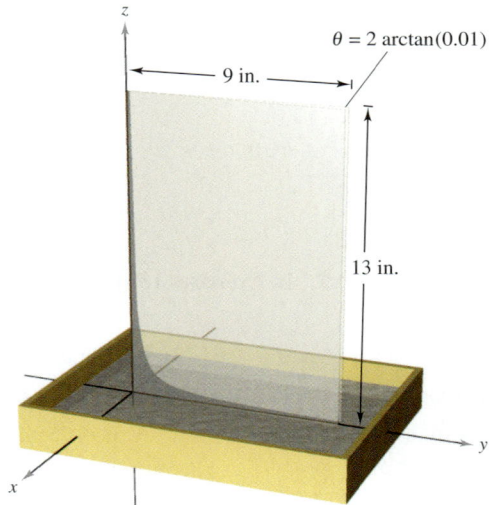

(a) Find the volume of the liquid that has risen in the wedge. (Assume $k = 1$.)

(b) Find the horizontal surface area of the liquid that has risen in the wedge.

Adaptation of Capillary Action problem from "Capillary Phenomena" by Thomas B. Greenslade, Jr., *Physics Teacher,* May 1992. By permission of the author.

14.6 Triple Integrals and Applications

■ Use a triple integral to find the volume of a solid region.
■ Find the center of mass and moments of inertia of a solid region.

Triple Integrals

The procedure used to define a **triple integral** follows that used for double integrals. Consider a function f of three variables that is continuous over a bounded solid region Q. Then, encompass Q with a network of boxes and form the **inner partition** consisting of all boxes lying entirely within Q, as shown in Figure 14.52. The volume of the ith box is

$$\Delta V_i = \Delta x_i \Delta y_i \Delta z_i. \qquad \text{Volume of } i\text{th box}$$

The **norm** $\|\Delta\|$ of the partition is the length of the longest diagonal of the n boxes in the partition. Choose a point (x_i, y_i, z_i) in each box and form the Riemann sum

$$\sum_{i=1}^{n} f(x_i, y_i, z_i)\, \Delta V_i.$$

Taking the limit as $\|\Delta\| \to 0$ leads to the following definition.

Solid region Q

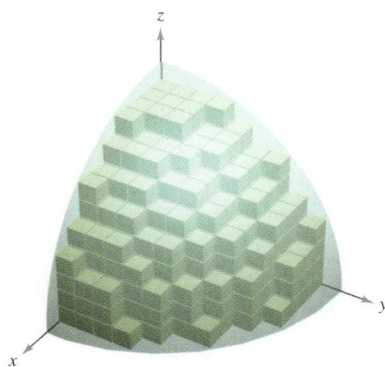

Definition of Triple Integral

If f is continuous over a bounded solid region Q, then the **triple integral of f over Q** is defined as

$$\iiint\limits_{Q} f(x, y, z)\, dV = \lim_{\|\Delta\|\to 0} \sum_{i=1}^{n} f(x_i, y_i, z_i)\, \Delta V_i$$

provided the limit exists. The **volume** of the solid region Q is given by

$$\text{Volume of } Q = \iiint\limits_{Q} dV.$$

Volume of $Q \approx \sum_{i=1}^{n} \Delta V_i$

Figure 14.52

Some of the properties of double integrals in Theorem 14.1 can be restated in terms of triple integrals.

1. $\displaystyle\iiint\limits_{Q} cf(x, y, z)\, dV = c\iiint\limits_{Q} f(x, y, z)\, dV$

2. $\displaystyle\iiint\limits_{Q} [f(x, y, z) \pm g(x, y, z)]\, dV = \iiint\limits_{Q} f(x, y, z)\, dV \pm \iiint\limits_{Q} g(x, y, z)\, dV$

3. $\displaystyle\iiint\limits_{Q} f(x, y, z)\, dV = \iiint\limits_{Q_1} f(x, y, z)\, dV + \iiint\limits_{Q_2} f(x, y, z)\, dV$

In the properties above, Q is the union of two nonoverlapping solid subregions Q_1 and Q_2. If the solid region Q is simple, then the triple integral $\iiint f(x, y, z)\, dV$ can be evaluated with an iterated integral using one of the six possible orders of integration:

$$dx\, dy\, dz \quad dy\, dx\, dz \quad dz\, dx\, dy$$
$$dx\, dz\, dy \quad dy\, dz\, dx \quad dz\, dy\, dx.$$

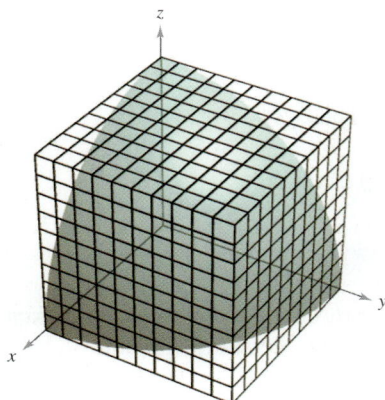

Exploration

Volume of a Paraboloid Sector In the Explorations on pages 979 and 998, you were asked to summarize the different ways you know of finding the volume of the solid bounded by the paraboloid

$$z = a^2 - x^2 - y^2, \quad a > 0$$

and the *xy*-plane. You now know one more way. Use it to find the volume of the solid.

The following version of Fubini's Theorem describes a region that is considered simple with respect to the order *dz dy dx*. Similar descriptions can be given for the other five orders.

THEOREM 14.4   Evaluation by Iterated Integrals

Let *f* be continuous on a solid region *Q* defined by

$$a \le x \le b,$$
$$h_1(x) \le y \le h_2(x),$$
$$g_1(x, y) \le z \le g_2(x, y)$$

where h_1, h_2, g_1, and g_2 are continuous functions. Then,

$$\iiint\limits_{Q} f(x, y, z)\, dV = \int_a^b \int_{h_1(x)}^{h_2(x)} \int_{g_1(x,y)}^{g_2(x,y)} f(x, y, z)\, dz\, dy\, dx.$$

To evaluate a triple iterated integral in the order *dz dy dx*, hold *both x* and *y* constant for the innermost integration. Then, hold *x* constant for the second integration.

EXAMPLE 1   Evaluating a Triple Iterated Integral

Evaluate the triple iterated integral

$$\int_0^2 \int_0^x \int_0^{x+y} e^x(y + 2z)\, dz\, dy\, dx.$$

Solution For the first integration, hold *x* and *y* constant and integrate with respect to *z*.

$$\int_0^2 \int_0^x \int_0^{x+y} e^x(y + 2z)\, dz\, dy\, dx = \int_0^2 \int_0^x e^x(yz + z^2) \Big]_0^{x+y} dy\, dx$$

$$= \int_0^2 \int_0^x e^x(x^2 + 3xy + 2y^2)\, dy\, dx$$

For the second integration, hold *x* constant and integrate with respect to *y*.

$$\int_0^2 \int_0^x e^x(x^2 + 3xy + 2y^2)\, dy\, dx = \int_0^2 \left[e^x\left(x^2 y + \frac{3xy^2}{2} + \frac{2y^3}{3}\right) \right]_0^x dx$$

$$= \frac{19}{6} \int_0^2 x^3 e^x\, dx$$

Finally, integrate with respect to *x*.

$$\frac{19}{6} \int_0^2 x^3 e^x\, dx = \frac{19}{6} \left[e^x(x^3 - 3x^2 + 6x - 6) \right]_0^2$$

$$= 19\left(\frac{e^2}{3} + 1\right)$$

$$\approx 65.797$$

Example 1 demonstrates the integration order *dz dy dx*. For other orders, you can follow a similar procedure. For instance, to evaluate a triple iterated integral in the order *dx dy dz*, hold both *y* and *z* constant for the innermost integration and integrate with respect to *x*. Then, for the second integration, hold *z* constant and integrate with respect to *y*. Finally, for the third integration, integrate with respect to *z*.

To find the limits for a particular order of integration, it is generally advisable first to determine the innermost limits, which may be functions of the outer two variables. Then, by projecting the solid Q onto the coordinate plane of the outer two variables, you can determine their limits of integration by the methods used for double integrals. For instance, to evaluate

$$\iiint\limits_Q f(x, y, z)\, dz\, dy\, dx$$

first determine the limits for z; the integral then has the form

$$\iint\left[\int_{g_1(x, y)}^{g_2(x, y)} f(x, y, z)\, dz\right] dy\, dx.$$

By projecting the solid Q onto the xy-plane, you can determine the limits for x and y as you did for double integrals, as shown in Figure 14.53.

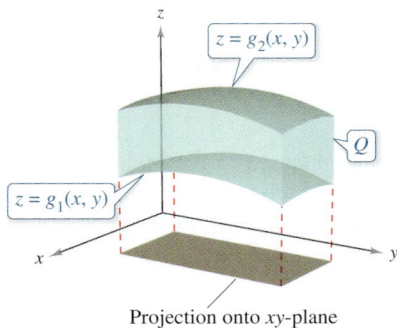

$z = g_2(x, y)$

Q

$z = g_1(x, y)$

Projection onto xy-plane

Solid region Q lies between two surfaces.
Figure 14.53

EXAMPLE 2   Using a Triple Integral to Find Volume

Find the volume of the ellipsoid given by $4x^2 + 4y^2 + z^2 = 16$.

Solution   Because x, y, and z play similar roles in the equation, the order of integration is probably immaterial, and you can arbitrarily choose $dz\, dy\, dx$. Moreover, you can simplify the calculation by considering only the portion of the ellipsoid lying in the first octant, as shown in Figure 14.54. From the order $dz\, dy\, dx$, you first determine the bounds for z.

$$0 \le z \le 2\sqrt{4 - x^2 - y^2}$$

In Figure 14.55, you can see that the boundaries for x and y are

$$0 \le x \le 2 \quad \text{and} \quad 0 \le y \le \sqrt{4 - x^2}.$$

So, the volume of the ellipsoid is

$$V = \iiint\limits_Q dV$$

$$= 8\int_0^2 \int_0^{\sqrt{4-x^2}} \int_0^{2\sqrt{4-x^2-y^2}} dz\, dy\, dx$$

$$= 8\int_0^2 \int_0^{\sqrt{4-x^2}} z\Big]_0^{2\sqrt{4-x^2-y^2}} dy\, dx$$

$$= 16\int_0^2 \int_0^{\sqrt{4-x^2}} \sqrt{(4 - x^2) - y^2}\, dy\, dx \qquad \begin{array}{l}\text{Integration tables (Appendix B)}\\ \text{Formula 37}\end{array}$$

$$= 8\int_0^2 \left[y\sqrt{4 - x^2 - y^2} + (4 - x^2)\arcsin\!\left(\frac{y}{\sqrt{4 - x^2}}\right)\right]_0^{\sqrt{4-x^2}} dx$$

$$= 8\int_0^2 [0 + (4 - x^2)\arcsin(1) - 0 - 0]\, dx$$

$$= 8\int_0^2 (4 - x^2)\left(\frac{\pi}{2}\right) dx$$

$$= 4\pi\left[4x - \frac{x^3}{3}\right]_0^2$$

$$= \frac{64\pi}{3}.$$

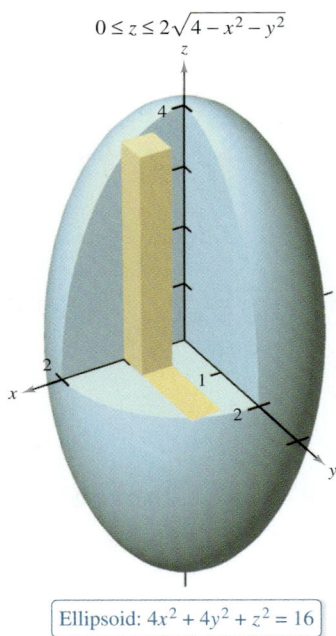

$0 \le z \le 2\sqrt{4 - x^2 - y^2}$

Ellipsoid: $4x^2 + 4y^2 + z^2 = 16$

Figure 14.54

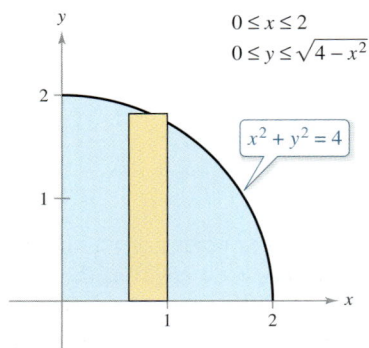

$0 \le x \le 2$
$0 \le y \le \sqrt{4 - x^2}$

$x^2 + y^2 = 4$

Figure 14.55

Example 2 is unusual in that all six possible orders of integration produce integrals of comparable difficulty. Try setting up some other possible orders of integration to find the volume of the elipsoid. For instance, the order $dx\,dy\,dz$ yields the integral

$$V = 8 \int_0^4 \int_0^{\sqrt{16-z^2}/2} \int_0^{\sqrt{16-4y^2-z^2}/2} dx\,dy\,dz.$$

The evaluation of this integral yields the same volume obtained in Example 2. This is always the case—the order of integration does not affect the value of the integral. However, the order of integration often does affect the complexity of the integral. In Example 3, the given order of integration is not convenient, so you can change the order to simplify the problem.

EXAMPLE 3 **Changing the Order of Integration**

Evaluate $\displaystyle\int_0^{\sqrt{\pi/2}} \int_x^{\sqrt{\pi/2}} \int_1^3 \sin(y^2)\,dz\,dy\,dx.$

Solution Note that after one integration in the given order, you would encounter the integral $2\int \sin(y^2)\,dy$, which is not an elementary function. To avoid this problem, change the order of integration to $dz\,dx\,dy$, so that y is the outer variable. From Figure 14.56, you can see that the solid region Q is

$$0 \le x \le \sqrt{\frac{\pi}{2}}$$

$$x \le y \le \sqrt{\frac{\pi}{2}}$$

$$1 \le z \le 3$$

and the projection of Q in the xy-plane yields the bounds

$$0 \le y \le \sqrt{\frac{\pi}{2}}$$

and

$$0 \le x \le y.$$

So, evaluating the triple integral using the order $dz\,dx\,dy$ produces

$$\int_0^{\sqrt{\pi/2}} \int_0^y \int_1^3 \sin(y^2)\,dz\,dx\,dy = \int_0^{\sqrt{\pi/2}} \int_0^y z \sin(y^2)\Big]_1^3 dx\,dy$$

$$= 2 \int_0^{\sqrt{\pi/2}} \int_0^y \sin(y^2)\,dx\,dy$$

$$= 2 \int_0^{\sqrt{\pi/2}} x \sin(y^2)\Big]_0^y dy$$

$$= 2 \int_0^{\sqrt{\pi/2}} y \sin(y^2)\,dy$$

$$= -\cos(y^2)\Big]_0^{\sqrt{\pi/2}}$$

$$= 1.$$

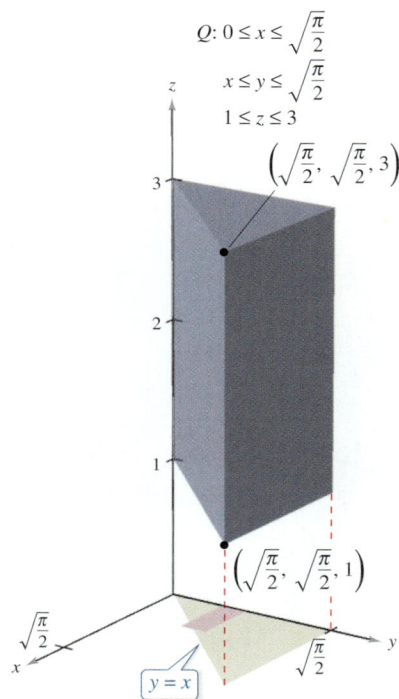

$Q: 0 \le x \le \sqrt{\dfrac{\pi}{2}}$

$\quad x \le y \le \sqrt{\dfrac{\pi}{2}}$

$\quad 1 \le z \le 3$

$\left(\sqrt{\dfrac{\pi}{2}},\ \sqrt{\dfrac{\pi}{2}},\ 3\right)$

$\left(\sqrt{\dfrac{\pi}{2}},\ \sqrt{\dfrac{\pi}{2}},\ 1\right)$

$y = x$

Figure 14.56

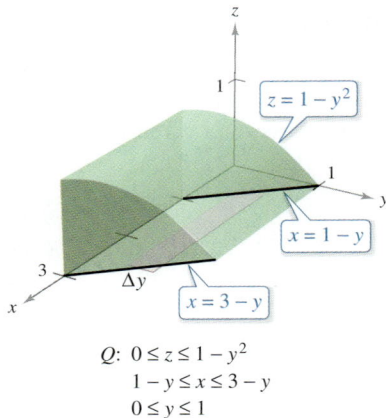

$z = 1 - y^2$

$x = 1 - y$

$x = 3 - y$

Q: $0 \le z \le 1 - y^2$
$1 - y \le x \le 3 - y$
$0 \le y \le 1$

Figure 14.57

EXAMPLE 4   **Determining the Limits of Integration**

Set up a triple integral for the volume of each solid region.

a. The region in the first octant bounded above by the cylinder $z = 1 - y^2$ and lying between the vertical planes $x + y = 1$ and $x + y = 3$

b. The upper hemisphere $z = \sqrt{1 - x^2 - y^2}$

c. The region bounded below by the paraboloid $z = x^2 + y^2$ and above by the sphere $x^2 + y^2 + z^2 = 6$

Solution

a. In Figure 14.57, note that the solid is bounded below by the xy-plane ($z = 0$) and above by the cylinder $z = 1 - y^2$. So,

$$0 \le z \le 1 - y^2. \qquad \text{Bounds for } z$$

Projecting the region onto the xy-plane produces a parallelogram. Because two sides of the parallelogram are parallel to the x-axis, you have the following bounds:

$$1 - y \le x \le 3 - y \quad \text{and} \quad 0 \le y \le 1.$$

So, the volume of the region is given by

$$V = \iiint_Q dV = \int_0^1 \int_{1-y}^{3-y} \int_0^{1-y^2} dz\, dx\, dy.$$

b. For the upper hemisphere $z = \sqrt{1 - x^2 - y^2}$, you have

$$0 \le z \le \sqrt{1 - x^2 - y^2}. \qquad \text{Bounds for } z$$

In Figure 14.58, note that the projection of the hemisphere onto the xy-plane is the circle

$$x^2 + y^2 = 1$$

and you can use either order $dx\, dy$ or $dy\, dx$. Choosing the first produces

$$-\sqrt{1-y^2} \le x \le \sqrt{1-y^2} \quad \text{and} \quad -1 \le y \le 1$$

which implies that the volume of the region is given by

$$V = \iiint_Q dV = \int_{-1}^{1} \int_{-\sqrt{1-y^2}}^{\sqrt{1-y^2}} \int_0^{\sqrt{1-x^2-y^2}} dz\, dx\, dy.$$

c. For the region bounded below by the paraboloid $z = x^2 + y^2$ and above by the sphere $x^2 + y^2 + z^2 = 6$, you have

$$x^2 + y^2 \le z \le \sqrt{6 - x^2 - y^2}. \qquad \text{Bounds for } z$$

The sphere and the paraboloid intersect at $z = 2$. Moreover, you can see in Figure 14.59 that the projection of the solid region onto the xy-plane is the circle

$$x^2 + y^2 = 2.$$

Using the order $dy\, dx$ produces

$$-\sqrt{2 - x^2} \le y \le \sqrt{2 - x^2} \quad \text{and} \quad -\sqrt{2} \le x \le \sqrt{2}$$

which implies that the volume of the region is given by

$$V = \iiint_Q dV = \int_{-\sqrt{2}}^{\sqrt{2}} \int_{-\sqrt{2-x^2}}^{\sqrt{2-x^2}} \int_{x^2+y^2}^{\sqrt{6-x^2-y^2}} dz\, dy\, dx. \qquad \blacksquare$$

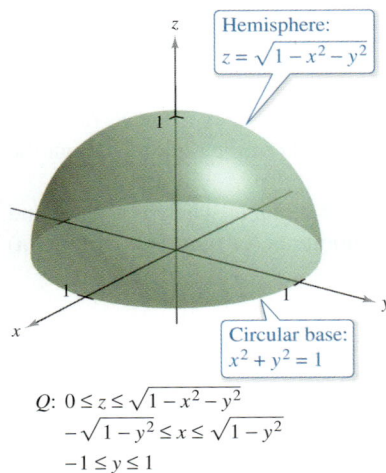

Hemisphere:
$z = \sqrt{1 - x^2 - y^2}$

Circular base:
$x^2 + y^2 = 1$

Q: $0 \le z \le \sqrt{1 - x^2 - y^2}$
$-\sqrt{1-y^2} \le x \le \sqrt{1-y^2}$
$-1 \le y \le 1$

Figure 14.58

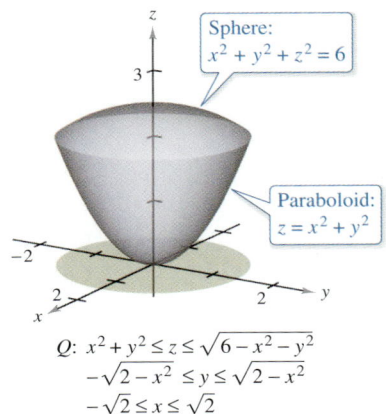

Sphere:
$x^2 + y^2 + z^2 = 6$

Paraboloid:
$z = x^2 + y^2$

Q: $x^2 + y^2 \le z \le \sqrt{6 - x^2 - y^2}$
$-\sqrt{2-x^2} \le y \le \sqrt{2-x^2}$
$-\sqrt{2} \le x \le \sqrt{2}$

Figure 14.59

Center of Mass and Moments of Inertia

In the remainder of this section, two important engineering applications of triple integrals are discussed. Consider a solid region Q whose density is given by the **density function** ρ. The **center of mass** of a solid region Q of mass m is given by $(\bar{x}, \bar{y}, \bar{z})$, where

$$m = \iiint_Q \rho(x, y, z)\, dV \qquad \text{Mass of the solid}$$

$$M_{yz} = \iiint_Q x\rho(x, y, z)\, dV \qquad \text{First moment about } yz\text{-plane}$$

$$M_{xz} = \iiint_Q y\rho(x, y, z)\, dV \qquad \text{First moment about } xz\text{-plane}$$

$$M_{xy} = \iiint_Q z\rho(x, y, z)\, dV \qquad \text{First moment about } xy\text{-plane}$$

and

$$\bar{x} = \frac{M_{yz}}{m}, \quad \bar{y} = \frac{M_{xz}}{m}, \quad \bar{z} = \frac{M_{xy}}{m}.$$

The quantities M_{yz}, M_{xz}, and M_{xy} are called the **first moments** of the region Q about the yz-, xz-, and xy-planes, respectively.

The first moments for solid regions are taken about a plane, whereas the second moments for solids are taken about a line. The **second moments** (or **moments of inertia**) about the x-, y-, and z-axes are

$$I_x = \iiint_Q (y^2 + z^2)\rho(x, y, z)\, dV \qquad \text{Moment of inertia about } x\text{-axis}$$

$$I_y = \iiint_Q (x^2 + z^2)\rho(x, y, z)\, dV \qquad \text{Moment of inertia about } y\text{-axis}$$

and

$$I_z = \iiint_Q (x^2 + y^2)\rho(x, y, z)\, dV. \qquad \text{Moment of inertia about } z\text{-axis}$$

> ····▷
> **REMARK** In engineering and physics, the moment of inertia of a mass is used to find the time required for the mass to reach a given speed of rotation about an axis, as shown in Figure 14.60. The greater the moment of inertia, the longer a force must be applied for the mass to reach the given speed.

Figure 14.60

For problems requiring the calculation of all three moments, considerable effort can be saved by applying the additive property of triple integrals and writing

$$I_x = I_{xz} + I_{xy}, \quad I_y = I_{yz} + I_{xy}, \quad \text{and} \quad I_z = I_{yz} + I_{xz}$$

where I_{xy}, I_{xz}, and I_{yz} are

$$I_{xy} = \iiint_Q z^2\rho(x, y, z)\, dV$$

$$I_{xz} = \iiint_Q y^2\rho(x, y, z)\, dV$$

and

$$I_{yz} = \iiint_Q x^2\rho(x, y, z)\, dV.$$

Finding the Center of Mass of a Solid Region

• • • ▷ *See LarsonCalculus.com for an interactive version of this type of example.*

Find the center of mass of the unit cube shown in Figure 14.61, given that the density at the point (x, y, z) is proportional to the square of its distance from the origin.

Solution Because the density at (x, y, z) is proportional to the square of the distance between $(0, 0, 0)$ and (x, y, z), you have

$$\rho(x, y, z) = k(x^2 + y^2 + z^2).$$

You can use this density function to find the mass of the cube. Because of the symmetry of the region, any order of integration will produce an integral of comparable difficulty.

$$
\begin{aligned}
m &= \int_0^1 \int_0^1 \int_0^1 k(x^2 + y^2 + z^2)\, dz\, dy\, dx \\
&= k \int_0^1 \int_0^1 \left[(x^2 + y^2)z + \frac{z^3}{3} \right]_0^1 dy\, dx \\
&= k \int_0^1 \int_0^1 \left(x^2 + y^2 + \frac{1}{3} \right) dy\, dx \\
&= k \int_0^1 \left[\left(x^2 + \frac{1}{3} \right)y + \frac{y^3}{3} \right]_0^1 dx \\
&= k \int_0^1 \left(x^2 + \frac{2}{3} \right) dx \\
&= k \left[\frac{x^3}{3} + \frac{2x}{3} \right]_0^1 \\
&= k
\end{aligned}
$$

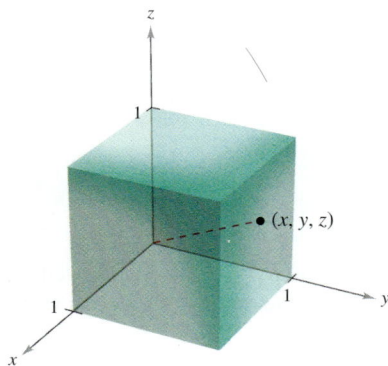

Variable density:
$\rho(x, y, z) = k(x^2 + y^2 + z^2)$
Figure 14.61

The first moment about the yz-plane is

$$
\begin{aligned}
M_{yz} &= k \int_0^1 \int_0^1 \int_0^1 x(x^2 + y^2 + z^2)\, dz\, dy\, dx \\
&= k \int_0^1 x \left[\int_0^1 \int_0^1 (x^2 + y^2 + z^2)\, dz\, dy \right] dx.
\end{aligned}
$$

Note that x can be factored out of the two inner integrals, because it is constant with respect to y and z. After factoring, the two inner integrals are the same as for the mass m. Therefore, you have

$$
\begin{aligned}
M_{yz} &= k \int_0^1 x \left(x^2 + \frac{2}{3} \right) dx \\
&= k \left[\frac{x^4}{4} + \frac{x^2}{3} \right]_0^1 \\
&= \frac{7k}{12}.
\end{aligned}
$$

So,

$$\bar{x} = \frac{M_{yz}}{m} = \frac{7k/12}{k} = \frac{7}{12}.$$

Finally, from the nature of ρ and the symmetry of x, y, and z in this solid region, you have $\bar{x} = \bar{y} = \bar{z}$, and the center of mass is $\left(\frac{7}{12}, \frac{7}{12}, \frac{7}{12} \right)$.

EXAMPLE 6   **Moments of Inertia for a Solid Region**

Find the moments of inertia about the x- and y-axes for the solid region lying between the hemisphere

$$z = \sqrt{4 - x^2 - y^2}$$

and the xy-plane, given that the density at (x, y, z) is proportional to the distance between (x, y, z) and the xy-plane.

$0 \le z \le \sqrt{4 - x^2 - y^2}$
$-\sqrt{4-x^2} \le y \le \sqrt{4-x^2}$
$-2 \le x \le 2$

Hemisphere:
$z = \sqrt{4 - x^2 - y^2}$

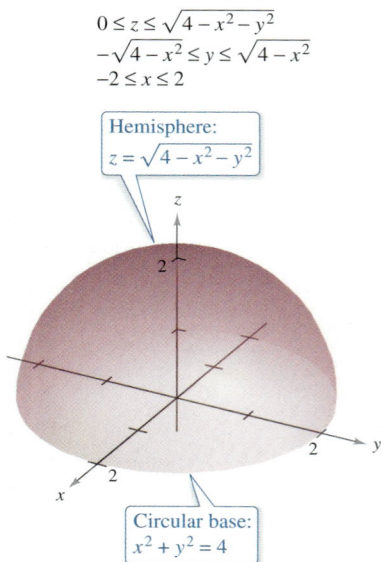

Circular base:
$x^2 + y^2 = 4$

Variable density: $\rho(x, y, z) = kz$
Figure 14.62

Solution   The density of the region is given by

$$\rho(x, y, z) = kz.$$

Considering the symmetry of this problem, you know that $I_x = I_y$, and you need to compute only one moment, say I_x. From Figure 14.62, choose the order $dz\,dy\,dx$ and write

$$I_x = \iiint_Q (y^2 + z^2)\rho(x, y, z)\, dV$$

$$= \int_{-2}^{2}\int_{-\sqrt{4-x^2}}^{\sqrt{4-x^2}}\int_{0}^{\sqrt{4-x^2-y^2}} (y^2 + z^2)(kz)\, dz\, dy\, dx$$

$$= k\int_{-2}^{2}\int_{-\sqrt{4-x^2}}^{\sqrt{4-x^2}} \left[\frac{y^2 z^2}{2} + \frac{z^4}{4}\right]_{0}^{\sqrt{4-x^2-y^2}} dy\, dx$$

$$= k\int_{-2}^{2}\int_{-\sqrt{4-x^2}}^{\sqrt{4-x^2}} \left[\frac{y^2(4 - x^2 - y^2)}{2} + \frac{(4 - x^2 - y^2)^2}{4}\right] dy\, dx$$

$$= \frac{k}{4}\int_{-2}^{2}\int_{-\sqrt{4-x^2}}^{\sqrt{4-x^2}} [(4 - x^2)^2 - y^4]\, dy\, dx$$

$$= \frac{k}{4}\int_{-2}^{2} \left[(4 - x^2)^2 y - \frac{y^5}{5}\right]_{-\sqrt{4-x^2}}^{\sqrt{4-x^2}} dx$$

$$= \frac{k}{4}\int_{-2}^{2} \frac{8}{5}(4 - x^2)^{5/2}\, dx$$

$$= \frac{4k}{5}\int_{0}^{2} (4 - x^2)^{5/2}\, dx \qquad x = 2\sin\theta$$

$$= \frac{4k}{5}\int_{0}^{\pi/2} 64\cos^6\theta\, d\theta$$

$$= \left(\frac{256k}{5}\right)\left(\frac{5\pi}{32}\right) \qquad \text{Wallis's Formula}$$

$$= 8k\pi.$$

So, $I_x = 8k\pi = I_y$.

In Example 6, notice that the moments of inertia about the x- and y-axes are equal to each other. The moment about the z-axis, however, is different. Does it seem that the moment of inertia about the z-axis should be less than or greater than the moments calculated in Example 6? By performing the calculations, you can determine that

$$I_z = \frac{16}{3}k\pi.$$

This tells you that the solid shown in Figure 14.62 has a greater resistance to rotation about the x- or y-axis than about the z-axis.

14.6 Exercises

See **CalcChat.com** for tutorial help and worked-out solutions to odd-numbered exercises.

Evaluating a Triple Iterated Integral In Exercises 1–8, evaluate the triple iterated integral.

1. $\displaystyle\int_0^3\int_0^2\int_0^1 (x + y + z)\, dx\, dz\, dy$

2. $\displaystyle\int_{-1}^1\int_{-1}^1\int_{-1}^1 x^2 y^2 z^2\, dx\, dy\, dz$

3. $\displaystyle\int_0^1\int_0^x\int_0^{xy} x\, dz\, dy\, dx$

4. $\displaystyle\int_0^9\int_0^{y/3}\int_0^{\sqrt{y^2 - 9x^2}} z\, dz\, dx\, dy$

5. $\displaystyle\int_1^4\int_0^1\int_0^x 2ze^{-x^2}\, dy\, dx\, dz$

6. $\displaystyle\int_1^4\int_1^{e^2}\int_0^{1/xz} \ln z\, dy\, dz\, dx$

7. $\displaystyle\int_0^4\int_0^{\pi/2}\int_0^{1-x} x\cos y\, dz\, dy\, dx$

8. $\displaystyle\int_0^{\pi/2}\int_0^{y/2}\int_0^{1/y} \sin y\, dz\, dx\, dy$

Approximating a Triple Iterated Integral Using Technology In Exercises 9 and 10, use a computer algebra system to approximate the triple iterated integral.

9. $\displaystyle\int_0^3\int_{-\sqrt{9-y^2}}^{\sqrt{9-y^2}}\int_0^{y^2} y\, dz\, dx\, dy$

10. $\displaystyle\int_0^3\int_0^{2-(2y/3)}\int_0^{6-2y-3z} ze^{-x^2y^2}\, dx\, dz\, dy$

Setting Up a Triple Integral In Exercises 11–16, set up a triple integral for the volume of the solid.

11. The solid in the first octant bounded by the coordinate planes and the plane $z = 5 - x - y$

12. The solid bounded by $z = 9 - x^2$, $z = 0$, $y = 0$, and $y = 2x$

13. The solid bounded by $z = 6 - x^2 - y^2$ and $z = 0$

14. The solid bounded by $z = \sqrt{16 - x^2 - y^2}$ and $z = 0$

15. The solid that is the common interior below the sphere $x^2 + y^2 + z^2 = 80$ and above the paraboloid $z = \frac{1}{2}(x^2 + y^2)$

16. The solid bounded above by the cylinder $z = 4 - x^2$ and below by the paraboloid $z = x^2 + 3y^2$

Volume In Exercises 17–20, use a triple integral to find the volume of the solid shown in the figure.

17.

18.

19.

20.

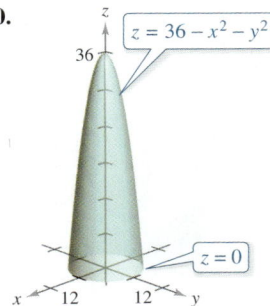

Volume In Exercises 21–24, use a triple integral to find the volume of the solid bounded by the graphs of the equations.

21. $z = 4 - x^2$, $y = 4 - x^2$, first octant

22. $z = 9 - x^3$, $y = -x^2 + 2$, $y = 0$, $z = 0$, $x \ge 0$

23. $z = 2 - y$, $z = 4 - y^2$, $x = 0$, $x = 3$, $y = 0$

24. $z = x$, $y = x + 2$, $y = x^2$, first octant

Changing the Order of Integration In Exercises 25–30, sketch the solid whose volume is given by the iterated integral and rewrite the integral using the indicated order of integration.

25. $\displaystyle\int_0^1\int_{-1}^0\int_0^{y^2} dz\, dy\, dx$

Rewrite using the order $dy\, dz\, dx$.

26. $\displaystyle\int_{-1}^1\int_{y^2}^1\int_0^{1-x} dz\, dx\, dy$

Rewrite using the order $dx\, dz\, dy$.

27. $\displaystyle\int_0^4\int_0^{(4-x)/2}\int_0^{(12-3x-6y)/4} dz\, dy\, dx$

Rewrite using the order $dy\, dx\, dz$.

28. $\displaystyle\int_0^3\int_0^{\sqrt{9-x^2}}\int_0^{6-x-y} dz\, dy\, dx$

Rewrite using the order $dz\, dx\, dy$.

29. $\displaystyle\int_0^1\int_y^1\int_0^{\sqrt{1-y^2}} dz\, dx\, dy$

Rewrite using the order $dz\, dy\, dx$.

30. $\displaystyle\int_0^2\int_{2x}^4\int_0^{\sqrt{y^2-4x^2}} dz\, dy\, dx$

Rewrite using the order $dx\, dy\, dz$.

Orders of Integration In Exercises 31–34, list the six possible orders of integration for the triple integral over the solid region Q, $\iiint\limits_Q xyz\, dV$.

31. $Q = \{(x, y, z)\colon 0 \le x \le 1, 0 \le y \le x, 0 \le z \le 3\}$

32. $Q = \{(x, y, z)\colon 0 \le x \le 2, x^2 \le y \le 4, 0 \le z \le 2 - x\}$

33. $Q = \{(x, y, z)\colon x^2 + y^2 \le 9, 0 \le z \le 4\}$

34. $Q = \{(x, y, z)\colon 0 \le x \le 1, y \le 1 - x^2, 0 \le z \le 6\}$

Orders of Integration In Exercises 35 and 36, the figure shows the region of integration for the given integral. Rewrite the integral as an equivalent iterated integral in the five other orders.

35. $\displaystyle\int_0^1 \int_0^{1-y^2} \int_0^{1-y} dz\, dx\, dy$ **36.** $\displaystyle\int_0^3 \int_0^x \int_0^{9-x^2} dz\, dy\, dx$

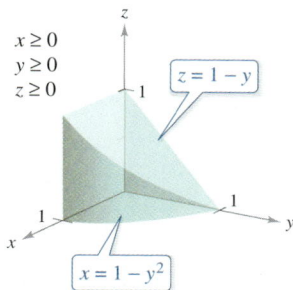

$x \geq 0$
$y \geq 0$
$z \geq 0$
$z = 1 - y$
$x = 1 - y^2$

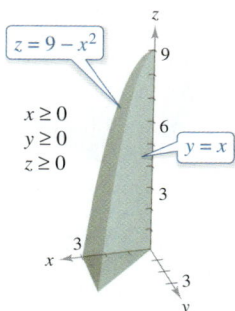

$z = 9 - x^2$
$x \geq 0$
$y \geq 0$
$z \geq 0$
$y = x$

Mass and Center of Mass In Exercises 37–40, find the mass and the indicated coordinates of the center of mass of the solid region Q of density ρ bounded by the graphs of the equations.

37. Find $\bar{x}$ using $\rho(x, y, z) = k$.

Q: $2x + 3y + 6z = 12$, $x = 0$, $y = 0$, $z = 0$

38. Find $\bar{y}$ using $\rho(x, y, z) = ky$.

Q: $3x + 3y + 5z = 15$, $x = 0$, $y = 0$, $z = 0$

39. Find $\bar{z}$ using $\rho(x, y, z) = kx$.

Q: $z = 4 - x$, $z = 0$, $y = 0$, $y = 4$, $x = 0$

40. Find $\bar{y}$ using $\rho(x, y, z) = k$.

Q: $\dfrac{x}{a} + \dfrac{y}{b} + \dfrac{z}{c} = 1$ $(a, b, c > 0)$, $x = 0$, $y = 0$, $z = 0$

Mass and Center of Mass In Exercises 41 and 42, set up the triple integrals for finding the mass and the center of mass of the solid of density ρ bounded by the graphs of the equations.

41. $x = 0$, $x = b$, $y = 0$, $y = b$, $z = 0$, $z = b$

$\rho(x, y, z) = kxy$

42. $x = 0$, $x = a$, $y = 0$, $y = b$, $z = 0$, $z = c$

$\rho(x, y, z) = kz$

Think About It The center of mass of a solid of constant density is shown in the figure. In Exercises 43–46, make a conjecture about how the center of mass $(\bar{x}, \bar{y}, \bar{z})$ will change for the nonconstant density $\rho(x, y, z)$. Explain.

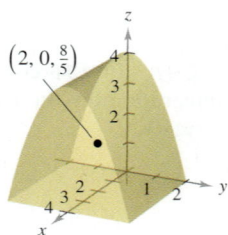

$\left(2, 0, \frac{8}{5}\right)$

43. $\rho(x, y, z) = kx$ **44.** $\rho(x, y, z) = kz$

45. $\rho(x, y, z) = k(y + 2)$ **46.** $\rho(x, y, z) = kxz^2(y + 2)^2$

Centroid In Exercises 47–52, find the centroid of the solid region bounded by the graphs of the equations or described by the figure. Use a computer algebra system to evaluate the triple integrals. (Assume uniform density and find the center of mass.)

47. $z = \dfrac{h}{r}\sqrt{x^2 + y^2}$, $z = h$

48. $y = \sqrt{9 - x^2}$, $z = y$, $z = 0$

49. $z = \sqrt{16 - x^2 - y^2}$, $z = 0$

50. $z = \dfrac{1}{y^2 + 1}$, $z = 0$, $x = -2$, $x = 2$, $y = 0$, $y = 1$

51.

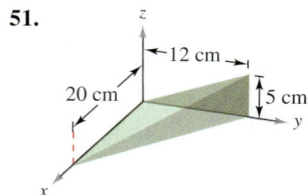

12 cm
20 cm
5 cm

52.

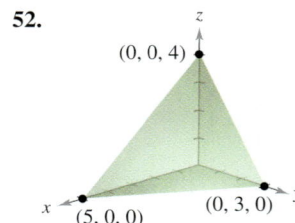

$(0, 0, 4)$
$(5, 0, 0)$
$(0, 3, 0)$

Moments of Inertia In Exercises 53–56, find I_x, I_y, and I_z for the solid of given density. Use a computer algebra system to evaluate the triple integrals.

53. (a) $\rho = k$
(b) $\rho = kxyz$

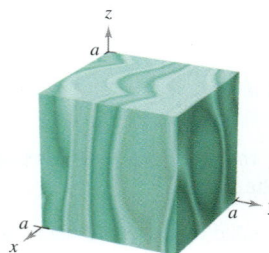

a

54. (a) $\rho(x, y, z) = k$
(b) $\rho(x, y, z) = k(x^2 + y^2)$

$\frac{a}{2}$

55. (a) $\rho(x, y, z) = k$
(b) $\rho = ky$

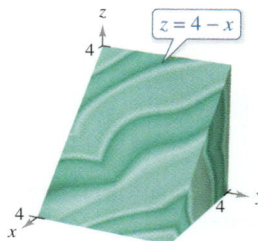

$z = 4 - x$

56. (a) $\rho = kz$
(b) $\rho = k(4 - z)$

$z = 4 - y^2$

Moments of Inertia In Exercises 57 and 58, verify the moments of inertia for the solid of uniform density. Use a computer algebra system to evaluate the triple integrals.

57. $I_x = \dfrac{1}{12}m(3a^2 + L^2)$

$I_y = \dfrac{1}{2}ma^2$

$I_z = \dfrac{1}{12}m(3a^2 + L^2)$

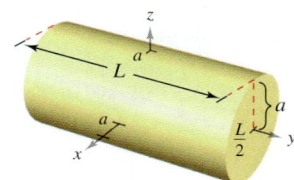

L
$\frac{L}{2}$
a

58. $I_x = \frac{1}{12}m(a^2 + b^2)$

$I_y = \frac{1}{12}m(b^2 + c^2)$

$I_z = \frac{1}{12}m(a^2 + c^2)$

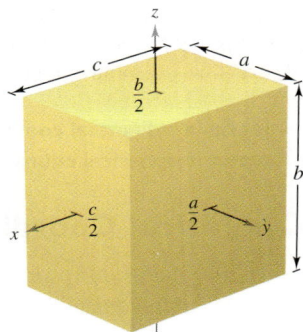

Moments of Inertia   In Exercises 59 and 60, set up a triple integral that gives the moment of inertia about the z-axis of the solid region Q of density ρ.

59. $Q = \{(x, y, z): -1 \le x \le 1, -1 \le y \le 1, 0 \le z \le 1 - x\}$

$\rho = \sqrt{x^2 + y^2 + z^2}$

60. $Q = \{(x, y, z): x^2 + y^2 \le 1, 0 \le z \le 4 - x^2 - y^2\}$

$\rho = kx^2$

Setting Up Triple Integrals   In Exercises 61 and 62, using the description of the solid region, set up the integral for (a) the mass, (b) the center of mass, and (c) the moment of inertia about the z-axis.

61. The solid bounded by $z = 4 - x^2 - y^2$ and $z = 0$ with density function $\rho = kz$

62. The solid in the first octant bounded by the coordinate planes and $x^2 + y^2 + z^2 = 25$ with density function $\rho = kxy$

Average Value   In Exercises 63–66, find the average value of the function over the given solid. The average value of a continuous function $f(x, y, z)$ over a solid region Q is

$$\frac{1}{V} \iiint_Q f(x, y, z)\, dV$$

where V is the volume of the solid region Q.

63. $f(x, y, z) = z^2 + 4$ over the cube in the first octant bounded by the coordinate planes and the planes $x = 1$, $y = 1$, and $z = 1$

64. $f(x, y, z) = xyz$ over the cube in the first octant bounded by the coordinate planes and the planes $x = 4$, $y = 4$, and $z = 4$

65. $f(x, y, z) = x + y + z$ over the tetrahedron in the first octant with vertices $(0, 0, 0)$, $(2, 0, 0)$, $(0, 2, 0)$, and $(0, 0, 2)$

66. $f(x, y, z) = x + y$ over the solid bounded by the sphere $x^2 + y^2 + z^2 = 3$

WRITING ABOUT CONCEPTS

67. Triple Integral   Define a triple integral and describe a method of evaluating a triple integral.

68. Moment of Inertia   Determine whether the moment of inertia about the y-axis of the cylinder in Exercise 57 will increase or decrease for the nonconstant density $\rho(x, y, z) = \sqrt{x^2 + z^2}$ and $a = 4$.

WRITING ABOUT CONCEPTS   (continued)

69. Think About It   Which of the integrals below is equal to $\int_1^3 \int_0^2 \int_{-1}^1 f(x, y, z)\, dz\, dy\, dx$? Explain.

(a) $\int_1^3 \int_0^2 \int_{-1}^1 f(x, y, z)\, dz\, dx\, dy$

(b) $\int_{-1}^1 \int_0^2 \int_1^3 f(x, y, z)\, dx\, dy\, dz$

(c) $\int_0^2 \int_1^3 \int_{-1}^1 f(x, y, z)\, dy\, dx\, dz$

---

70.   HOW DO YOU SEE IT?   Consider two solids, solid A and solid B, of equal weight as shown below.

Solid A   Solid B

(a) Because the solids have the same weight, which has the greater density? Explain.

(b) Which solid has the greater moment of inertia?

(c) The solids are rolled down an inclined plane. They are started at the same time and at the same height. Which will reach the bottom first? Explain.

71. Maximizing a Triple Integral   Find the solid region Q where the triple integral

$$\iiint_Q (1 - 2x^2 - y^2 - 3z^2)\, dV$$

is a maximum. Use a computer algebra system to approximate the maximum value. What is the exact maximum value?

72. Finding a Value   Solve for a in the triple integral.

$$\int_0^1 \int_0^{3-a-y^2} \int_a^{4-x-y^2} dz\, dx\, dy = \frac{14}{15}$$

PUTNAM EXAM CHALLENGE

73. Evaluate

$$\lim_{n \to \infty} \int_0^1 \int_0^1 \cdots \int_0^1 \cos^2\left\{\frac{\pi}{2n}(x_1 + x_2 + \cdots + x_n)\right\} dx_1\, dx_2 \cdots dx_n.$$

This problem was composed by the Committee on the Putnam Prize Competition.
© The Mathematical Association of America. All rights reserved.

14.7 Triple Integrals in Other Coordinates

■ Write and evaluate a triple integral in cylindrical coordinates.
■ Write and evaluate a triple integral in spherical coordinates.

Triple Integrals in Cylindrical Coordinates

Many common solid regions, such as spheres, ellipsoids, cones, and paraboloids, can yield difficult triple integrals in rectangular coordinates. In fact, it is precisely this difficulty that led to the introduction of nonrectangular coordinate systems. In this section, you will learn how to use *cylindrical* and *spherical* coordinates to evaluate triple integrals.

Recall from Section 11.7 that the rectangular conversion equations for cylindrical coordinates are

$$x = r \cos \theta$$
$$y = r \sin \theta$$
$$z = z.$$

An easy way to remember these conversions is to note that the equations for x and y are the same as in polar coordinates and z is unchanged.

In this coordinate system, the simplest solid region is a cylindrical block determined by

$$r_1 \leq r \leq r_2$$
$$\theta_1 \leq \theta \leq \theta_2$$

and

$$z_1 \leq z \leq z_2$$

as shown in Figure 14.63.

To obtain the cylindrical coordinate form of a triple integral, consider a solid region Q whose projection R onto the xy-plane can be described in polar coordinates. That is,

$$Q = \{(x, y, z): (x, y) \text{ is in } R, \quad h_1(x, y) \leq z \leq h_2(x, y)\}$$

and

$$R = \{(r, \theta): \theta_1 \leq \theta \leq \theta_2, \quad g_1(\theta) \leq r \leq g_2(\theta)\}.$$

If f is a continuous function on the solid Q, then you can write the triple integral of f over Q as

$$\iiint_Q f(x, y, z) \, dV = \iint_R \left[\int_{h_1(x, y)}^{h_2(x, y)} f(x, y, z) \, dz \right] dA$$

where the double integral over R is evaluated in polar coordinates. That is, R is a plane region that is either r-simple or θ-simple. If R is r-simple, then the iterated form of the triple integral in cylindrical form is

$$\iiint_Q f(x, y, z) \, dV = \int_{\theta_1}^{\theta_2} \int_{g_1(\theta)}^{g_2(\theta)} \int_{h_1(r \cos \theta, \, r \sin \theta)}^{h_2(r \cos \theta, \, r \sin \theta)} f(r \cos \theta, r \sin \theta, z) r \, dz \, dr \, d\theta.$$

This is only one of six possible orders of integration. The other five are $dz \, d\theta \, dr$, $dr \, dz \, d\theta$, $dr \, d\theta \, dz$, $d\theta \, dz \, dr$, and $d\theta \, dr \, dz$.

PIERRE SIMON DE LAPLACE (1749–1827)

One of the first to use a cylindrical coordinate system was the French mathematician Pierre Simon de Laplace. Laplace has been called the "Newton of France," and he published many important works in mechanics, differential equations, and probability. *See LarsonCalculus.com to read more of this biography.*

Volume of cylindrical block:
$\Delta V_i = r_i \Delta r_i \Delta \theta_i \Delta z_i$
Figure 14.63

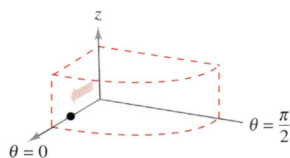
Integrate with respect to r.

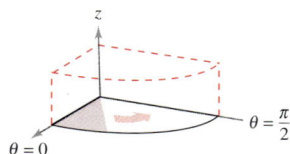
Integrate with respect to θ.

Integrate with respect to z.
Figure 14.64

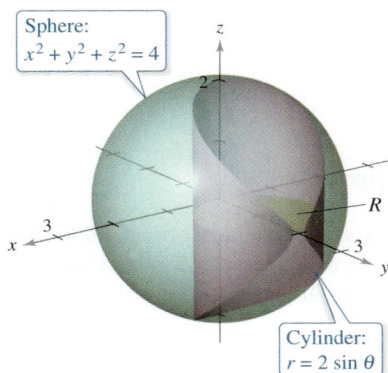
Sphere:
$x^2 + y^2 + z^2 = 4$
Cylinder:
$r = 2 \sin \theta$

Figure 14.65

To visualize a particular order of integration, it helps to view the iterated integral in terms of three sweeping motions—each adding another dimension to the solid. For instance, in the order $dr\, d\theta\, dz$, the first integration occurs in the r-direction as a point sweeps out a ray. Then, as θ increases, the line sweeps out a sector. Finally, as z increases, the sector sweeps out a solid wedge, as shown in Figure 14.64.

Exploration

Volume of a Paraboloid Sector   In the Explorations on pages 979, 998, and 1010, you were asked to summarize the different ways you know of finding the volume of the solid bounded by the paraboloid

$$z = a^2 - x^2 - y^2, \quad a > 0$$

and the xy-plane. You now know one more way. Use it to find the volume of the solid. Compare the different methods. What are the advantages and disadvantages of each?

EXAMPLE 1   **Finding Volume in Cylindrical Coordinates**

Find the volume of the solid region Q cut from the sphere $x^2 + y^2 + z^2 = 4$ by the cylinder $r = 2 \sin \theta$, as shown in Figure 14.65.

Solution   Because $x^2 + y^2 + z^2 = r^2 + z^2 = 4$, the bounds on z are

$$-\sqrt{4 - r^2} \le z \le \sqrt{4 - r^2}.$$

Let R be the circular projection of the solid onto the $r\theta$-plane. Then the bounds on R are

$$0 \le r \le 2 \sin \theta \quad \text{and} \quad 0 \le \theta \le \pi.$$

So, the volume of Q is

$$
\begin{aligned}
V &= \int_0^\pi \int_0^{2\sin\theta} \int_{-\sqrt{4-r^2}}^{\sqrt{4-r^2}} r\, dz\, dr\, d\theta \\
&= 2\int_0^{\pi/2} \int_0^{2\sin\theta} \int_{-\sqrt{4-r^2}}^{\sqrt{4-r^2}} r\, dz\, dr\, d\theta \\
&= 2\int_0^{\pi/2} \int_0^{2\sin\theta} 2r\sqrt{4-r^2}\, dr\, d\theta \\
&= 2\int_0^{\pi/2} -\frac{2}{3}(4-r^2)^{3/2}\Big]_0^{2\sin\theta} d\theta \\
&= \frac{4}{3}\int_0^{\pi/2} (8 - 8\cos^3\theta)\, d\theta \\
&= \frac{32}{3}\int_0^{\pi/2} [1 - (\cos\theta)(1 - \sin^2\theta)]\, d\theta \\
&= \frac{32}{3}\left[\theta - \sin\theta + \frac{\sin^3\theta}{3}\right]_0^{\pi/2} \\
&= \frac{16}{9}(3\pi - 4) \\
&\approx 9.644.
\end{aligned}
$$

$0 \le z \le \sqrt{16 - 4r^2}$

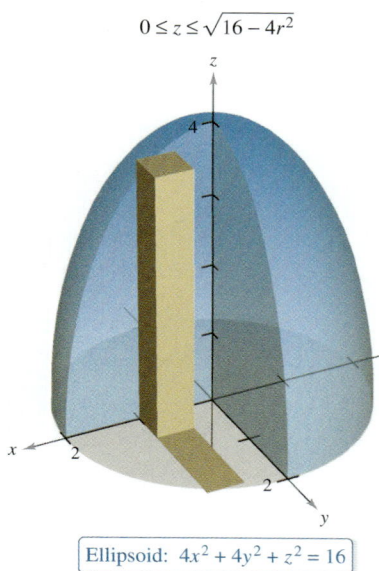

Ellipsoid: $4x^2 + 4y^2 + z^2 = 16$

Figure 14.66

EXAMPLE 2 **Finding Mass in Cylindrical Coordinates**

Find the mass of the ellipsoid Q given by $4x^2 + 4y^2 + z^2 = 16$, lying above the xy-plane. The density at a point in the solid is proportional to the distance between the point and the xy-plane.

Solution The density function is $\rho(r, \theta, z) = kz$. The bounds on z are

$$0 \le z \le \sqrt{16 - 4x^2 - 4y^2} = \sqrt{16 - 4r^2}$$

where $0 \le r \le 2$ and $0 \le \theta \le 2\pi$, as shown in Figure 14.66. The mass of the solid is

$$m = \int_0^{2\pi} \int_0^2 \int_0^{\sqrt{16-4r^2}} kzr \, dz \, dr \, d\theta$$

$$= \frac{k}{2} \int_0^{2\pi} \int_0^2 z^2 r \Big]_0^{\sqrt{16-4r^2}} dr \, d\theta$$

$$= \frac{k}{2} \int_0^{2\pi} \int_0^2 (16r - 4r^3) \, dr \, d\theta$$

$$= \frac{k}{2} \int_0^{2\pi} \Big[8r^2 - r^4 \Big]_0^2 d\theta$$

$$= 8k \int_0^{2\pi} d\theta$$

$$= 16\pi k.$$

Integration in cylindrical coordinates is useful when factors involving $x^2 + y^2$ appear in the integrand, as illustrated in Example 3.

EXAMPLE 3 **Finding a Moment of Inertia**

Find the moment of inertia about the axis of symmetry of the solid Q bounded by the paraboloid $z = x^2 + y^2$ and the plane $z = 4$, as shown in Figure 14.67. The density at each point is proportional to the distance between the point and the z-axis.

Solution Because the z-axis is the axis of symmetry and $\rho(x, y, z) = k\sqrt{x^2 + y^2}$, it follows that

$$I_z = \iiint_Q k(x^2 + y^2)\sqrt{x^2 + y^2} \, dV.$$

In cylindrical coordinates, $0 \le r \le \sqrt{x^2 + y^2} = \sqrt{z}$. So, you have

$$I_z = k \int_0^4 \int_0^{2\pi} \int_0^{\sqrt{z}} r^2(r)r \, dr \, d\theta \, dz$$

$$= k \int_0^4 \int_0^{2\pi} \frac{r^5}{5} \Big]_0^{\sqrt{z}} d\theta \, dz$$

$$= k \int_0^4 \int_0^{2\pi} \frac{z^{5/2}}{5} d\theta \, dz$$

$$= \frac{k}{5} \int_0^4 z^{5/2}(2\pi) \, dz$$

$$= \frac{2\pi k}{5} \Big[\frac{2}{7} z^{7/2} \Big]_0^4$$

$$= \frac{512k\pi}{35}.$$

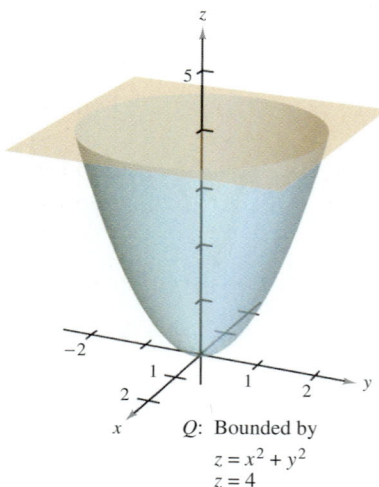

Q: Bounded by
$z = x^2 + y^2$
$z = 4$

Figure 14.67

Triple Integrals in Spherical Coordinates

Triple integrals involving spheres or cones are often easier to evaluate by converting to spherical coordinates. Recall from Section 11.7 that the rectangular conversion equations for spherical coordinates are

$$x = \rho \sin \phi \cos \theta$$
$$y = \rho \sin \phi \sin \theta$$
$$z = \rho \cos \phi.$$

REMARK   The Greek letter ρ used in spherical coordinates is not related to density. Rather, it is the three-dimensional analog of the r used in polar coordinates. For problems involving spherical coordinates and a density function, this text uses a different symbol to denote density.

In this coordinate system, the simplest region is a spherical block determined by

$$\{(\rho, \theta, \phi): \rho_1 \le \rho \le \rho_2, \quad \theta_1 \le \theta \le \theta_2, \quad \phi_1 \le \phi \le \phi_2\}$$

where $\rho_1 \ge 0$, $\theta_2 - \theta_1 \le 2\pi$, and $0 \le \phi_1 \le \phi_2 \le \pi$, as shown in Figure 14.68. If (ρ, θ, ϕ) is a point in the interior of such a block, then the volume of the block can be approximated by $\Delta V \approx \rho^2 \sin \phi \, \Delta\rho \, \Delta\phi \, \Delta\theta$. (See Exercise 8 in the Problem Solving exercises at the end of this chapter.)

Using the usual process involving an inner partition, summation, and a limit, you can develop a triple integral in spherical coordinates for a continuous function f defined on the solid region Q. This formula, shown below, can be modified for different orders of integration and generalized to include regions with variable boundaries.

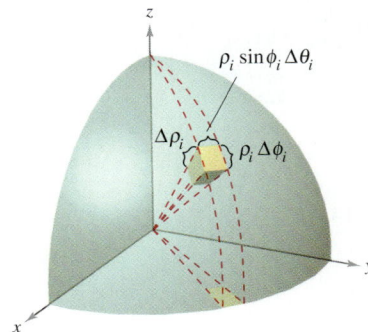

Spherical block: $\Delta V_i \approx \rho_i^2 \sin \phi_i \, \Delta\rho_i \, \Delta\phi_i \, \Delta\theta_i$
Figure 14.68

$$\iiint_Q f(x, y, z) \, dV = \int_{\theta_1}^{\theta_2} \int_{\phi_1}^{\phi_2} \int_{\rho_1}^{\rho_2} f(\rho \sin \phi \cos \theta, \rho \sin \phi \sin \theta, \rho \cos \phi)\rho^2 \sin \phi \, d\rho \, d\phi \, d\theta$$

Like triple integrals in cylindrical coordinates, triple integrals in spherical coordinates are evaluated with iterated integrals. As with cylindrical coordinates, you can visualize a particular order of integration by viewing the iterated integral in terms of three sweeping motions—each adding another dimension to the solid. For instance, the iterated integral

$$\int_0^{2\pi} \int_0^{\pi/4} \int_0^3 \rho^2 \sin \phi \, d\rho \, d\phi \, d\theta$$

(which is used in Example 4) is illustrated in Figure 14.69.

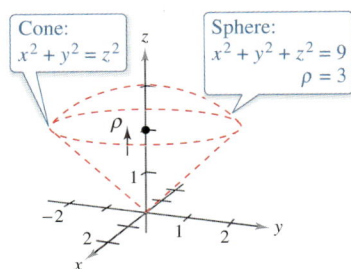

ρ varies from 0 to 3 with ϕ and θ held constant.

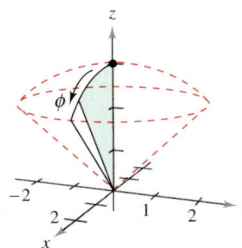

ϕ varies from 0 to $\pi/4$ with θ held constant.

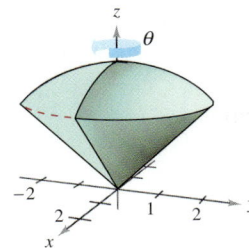

θ varies from 0 to 2π.

Figure 14.69

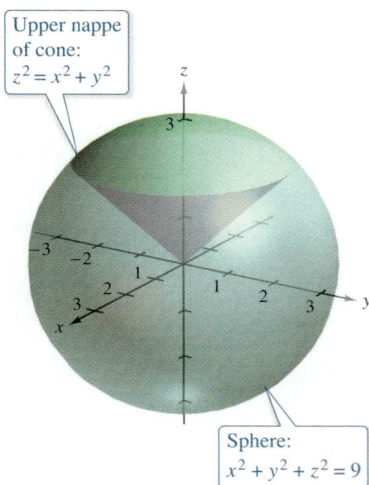

Upper nappe
of cone:
$z^2 = x^2 + y^2$

Sphere:
$x^2 + y^2 + z^2 = 9$

Figure 14.70

EXAMPLE 4   **Finding Volume in Spherical Coordinates**

Find the volume of the solid region Q bounded below by the upper nappe of the cone $z^2 = x^2 + y^2$ and above by the sphere $x^2 + y^2 + z^2 = 9$, as shown in Figure 14.70.

Solution   In spherical coordinates, the equation of the sphere is

$$\rho^2 = x^2 + y^2 + z^2 = 9 \quad \Longrightarrow \quad \rho = 3.$$

Furthermore, the sphere and cone intersect when

$$(x^2 + y^2) + z^2 = (z^2) + z^2 = 9 \quad \Longrightarrow \quad z = \frac{3}{\sqrt{2}}$$

and, because $z = \rho \cos \phi$, it follows that

$$\left(\frac{3}{\sqrt{2}}\right)\left(\frac{1}{3}\right) = \cos \phi \quad \Longrightarrow \quad \phi = \frac{\pi}{4}.$$

Consequently, you can use the integration order $d\rho\, d\phi\, d\theta$, where $0 \le \rho \le 3$, $0 \le \phi \le \pi/4$, and $0 \le \theta \le 2\pi$. The volume is

$$\iiint\limits_{Q} dV = \int_0^{2\pi} \int_0^{\pi/4} \int_0^3 \rho^2 \sin \phi \, d\rho \, d\phi \, d\theta$$

$$= \int_0^{2\pi} \int_0^{\pi/4} 9 \sin \phi \, d\phi \, d\theta$$

$$= 9 \int_0^{2\pi} -\cos \phi \Big]_0^{\pi/4} d\theta$$

$$= 9 \int_0^{2\pi} \left(1 - \frac{\sqrt{2}}{2}\right) d\theta$$

$$= 9\pi\left(2 - \sqrt{2}\right)$$

$$\approx 16.563.$$

EXAMPLE 5   **Finding the Center of Mass of a Solid Region**

•••▷ *See LarsonCalculus.com for an interactive version of this type of example.*

Find the center of mass of the solid region Q of uniform density, bounded below by the upper nappe of the cone $z^2 = x^2 + y^2$ and above by the sphere $x^2 + y^2 + z^2 = 9$.

Solution   Because the density is uniform, you can consider the density at the point (x, y, z) to be k. By symmetry, the center of mass lies on the z-axis, and you need only calculate $\bar{z} = M_{xy}/m$, where $m = kV = 9k\pi\left(2 - \sqrt{2}\right)$ from Example 4. Because $z = \rho \cos \phi$, it follows that

$$M_{xy} = \iiint\limits_{Q} kz \, dV = k \int_0^3 \int_0^{2\pi} \int_0^{\pi/4} (\rho \cos \phi)\rho^2 \sin \phi \, d\phi \, d\theta \, d\rho$$

$$= k \int_0^3 \int_0^{2\pi} \rho^3 \frac{\sin^2 \phi}{2} \Big]_0^{\pi/4} d\theta \, d\rho$$

$$= \frac{k}{4} \int_0^3 \int_0^{2\pi} \rho^3 \, d\theta \, d\rho = \frac{k\pi}{2} \int_0^3 \rho^3 \, d\rho = \frac{81k\pi}{8}.$$

So,

$$\bar{z} = \frac{M_{xy}}{m} = \frac{81k\pi/8}{9k\pi\left(2 - \sqrt{2}\right)} = \frac{9\left(2 + \sqrt{2}\right)}{16} \approx 1.920$$

and the center of mass is approximately $(0, 0, 1.92)$.

14.7 Exercises

See **CalcChat.com** for tutorial help and worked-out solutions to odd-numbered exercises.

Evaluating an Iterated Integral In Exercises 1–6, evaluate the iterated integral.

1. $\displaystyle\int_{-1}^{5}\int_{0}^{\pi/2}\int_{0}^{3} r\cos\theta \, dr \, d\theta \, dz$ **2.** $\displaystyle\int_{0}^{\pi/4}\int_{0}^{6}\int_{0}^{6-r} rz \, dz \, dr \, d\theta$

3. $\displaystyle\int_{0}^{\pi/2}\int_{0}^{2\cos^2\theta}\int_{0}^{4-r^2} r\sin\theta \, dz \, dr \, d\theta$

4. $\displaystyle\int_{0}^{\pi/2}\int_{0}^{\pi}\int_{0}^{2} e^{-\rho^3}\rho^2 \, d\rho \, d\theta \, d\phi$

5. $\displaystyle\int_{0}^{2\pi}\int_{0}^{\pi/4}\int_{0}^{\cos\phi} \rho^2\sin\phi \, d\rho \, d\phi \, d\theta$

6. $\displaystyle\int_{0}^{\pi/4}\int_{0}^{\pi/4}\int_{0}^{\cos\theta} \rho^2\sin\phi\cos\phi \, d\rho \, d\theta \, d\phi$

Approximating an Iterated Integral Using Technology In Exercises 7 and 8, use a computer algebra system to approximate the iterated integral.

7. $\displaystyle\int_{0}^{4}\int_{0}^{z}\int_{0}^{\pi/2} re^r \, d\theta \, dr \, dz$

8. $\displaystyle\int_{0}^{\pi/2}\int_{0}^{\pi}\int_{0}^{\sin\theta} (2\cos\phi)\rho^2 \, d\rho \, d\theta \, d\phi$

Volume In Exercises 9–12, sketch the solid region whose volume is given by the iterated integral, and evaluate the iterated integral.

9. $\displaystyle\int_{0}^{\pi/2}\int_{0}^{3}\int_{0}^{e^{-r^2}} r \, dz \, dr \, d\theta$ **10.** $\displaystyle\int_{0}^{2\pi}\int_{0}^{\sqrt{5}}\int_{0}^{5-r^2} r \, dz \, dr \, d\theta$

11. $\displaystyle\int_{0}^{2\pi}\int_{\pi/6}^{\pi/2}\int_{0}^{4} \rho^2\sin\phi \, d\rho \, d\phi \, d\theta$

12. $\displaystyle\int_{0}^{2\pi}\int_{0}^{\pi}\int_{2}^{5} \rho^2\sin\phi \, d\rho \, d\phi \, d\theta$

Converting Coordinates In Exercises 13–16, convert the integral from rectangular coordinates to both cylindrical and spherical coordinates, and evaluate the simplest iterated integral.

13. $\displaystyle\int_{-2}^{2}\int_{-\sqrt{4-x^2}}^{\sqrt{4-x^2}}\int_{x^2+y^2}^{4} x \, dz \, dy \, dx$

14. $\displaystyle\int_{0}^{2}\int_{0}^{\sqrt{4-x^2}}\int_{0}^{\sqrt{16-x^2-y^2}} \sqrt{x^2+y^2} \, dz \, dy \, dx$

15. $\displaystyle\int_{-a}^{a}\int_{-\sqrt{a^2-x^2}}^{\sqrt{a^2-x^2}}\int_{a}^{a+\sqrt{a^2-x^2-y^2}} x \, dz \, dy \, dx$

16. $\displaystyle\int_{0}^{3}\int_{0}^{\sqrt{9-x^2}}\int_{0}^{\sqrt{9-x^2-y^2}} \sqrt{x^2+y^2+z^2} \, dz \, dy \, dx$

Volume In Exercises 17–22, use cylindrical coordinates to find the volume of the solid.

17. Solid inside both $x^2 + y^2 + z^2 = a^2$

and $(x - a/2)^2 + y^2 = (a/2)^2$

18. Solid inside $x^2 + y^2 + z^2 = 16$ and outside $z = \sqrt{x^2 + y^2}$

19. Solid bounded above by $z = 2x$ and below by $z = 2x^2 + 2y^2$

20. Solid bounded above by $z = 2 - x^2 - y^2$ and below by $z = x^2 + y^2$

21. Solid bounded by the graphs of the sphere $r^2 + z^2 = a^2$ and the cylinder $r = a\cos\theta$

22. Solid inside the sphere $x^2 + y^2 + z^2 = 4$ and above the upper nappe of the cone $z^2 = x^2 + y^2$

Mass In Exercises 23 and 24, use cylindrical coordinates to find the mass of the solid Q of density ρ.

23. $Q = \{(x, y, z): 0 \le z \le 9 - x - 2y, x^2 + y^2 \le 4\}$

$\rho(x, y, z) = k\sqrt{x^2 + y^2}$

24. $Q = \{(x, y, z): 0 \le z \le 12e^{-(x^2+y^2)}, x^2 + y^2 \le 4, x \ge 0, y \ge 0\}$

$\rho(x, y, z) = k$

Using Cylindrical Coordinates In Exercises 25–30, use cylindrical coordinates to find the indicated characteristic of the cone shown in the figure.

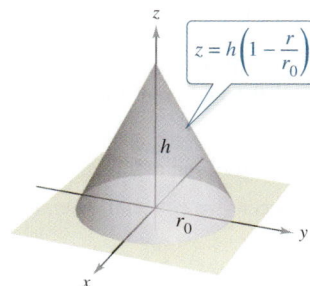

$z = h\left(1 - \dfrac{r}{r_0}\right)$

25. Find the volume of the cone.

26. Find the centroid of the cone.

27. Find the center of mass of the cone, assuming that its density at any point is proportional to the distance between the point and the axis of the cone. Use a computer algebra system to evaluate the triple integral.

28. Find the center of mass of the cone, assuming that its density at any point is proportional to the distance between the point and the base. Use a computer algebra system to evaluate the triple integral.

29. Assume that the cone has uniform density and show that the moment of inertia about the z-axis is

$I_z = \dfrac{3}{10}mr_0^2.$

30. Assume that the density of the cone is $\rho(x, y, z) = k\sqrt{x^2 + y^2}$ and find the moment of inertia about the z-axis.

Moment of Inertia In Exercises 31 and 32, use cylindrical coordinates to verify the given formula for the moment of inertia of the solid of uniform density.

31. Cylindrical shell: $I_z = \dfrac{1}{2}m(a^2 + b^2)$

$0 < a \le r \le b, \quad 0 \le z \le h$

32. Right circular cylinder: $I_z = \frac{3}{2}ma^2$

$r = 2a \sin\theta, \quad 0 \le z \le h$

Use a computer algebra system to evaluate the triple integral.

Volume In Exercises 33–36, use spherical coordinates to find the volume of the solid.

33. Solid inside $x^2 + y^2 + z^2 = 9$, outside $z = \sqrt{x^2 + y^2}$, and above the xy-plane

34. Solid bounded above by $x^2 + y^2 + z^2 = z$ and below by $z = \sqrt{x^2 + y^2}$

35. The torus given by $\rho = 4\sin\phi$ (Use a computer algebra system to evaluate the triple integral.)

36. The solid between the spheres

$x^2 + y^2 + z^2 = a^2$ and $x^2 + y^2 + z^2 = b^2, b > a$,

and inside the cone $z^2 = x^2 + y^2$

Mass In Exercises 37 and 38, use spherical coordinates to find the mass of the sphere $x^2 + y^2 + z^2 = a^2$ with the given density.

37. The density at any point is proportional to the distance between the point and the origin.

38. The density at any point is proportional to the distance of the point from the z-axis.

Center of Mass In Exercises 39 and 40, use spherical coordinates to find the center of mass of the solid of uniform density.

39. Hemispherical solid of radius r

40. Solid lying between two concentric hemispheres of radii r and R, where $r < R$

Moment of Inertia In Exercises 41 and 42, use spherical coordinates to find the moment of inertia about the z-axis of the solid of uniform density.

41. Solid bounded by the hemisphere $\rho = \cos\phi, \frac{\pi}{4} \le \phi \le \frac{\pi}{2}$, and the cone $\phi = \frac{\pi}{4}$

42. Solid lying between two concentric hemispheres of radii r and R, where $r < R$

WRITING ABOUT CONCEPTS

43. Coordinate Conversion Give the equations for conversion from rectangular to cylindrical coordinates and vice versa.

44. Coordinate Conversion Give the equations for conversion from rectangular to spherical coordinates and vice versa.

45. Cylindrical Form Give the iterated form of the triple integral $\iiint_Q f(x, y, z)\, dV$ in cylindrical form.

46. Spherical Form Give the iterated form of the triple integral $\iiint_Q f(x, y, z)\, dV$ in spherical form.

WRITING ABOUT CONCEPTS (continued)

47. Using Coordinates Describe the surface whose equation is a coordinate equal to a constant for each of the coordinates in (a) the cylindrical coordinate system and (b) the spherical coordinate system.

48. HOW DO YOU SEE IT? The solid is bounded below by the upper nappe of a cone and above by a sphere (see figure). Would it be easier to use cylindrical coordinates or spherical coordinates to find the volume of the solid? Explain.

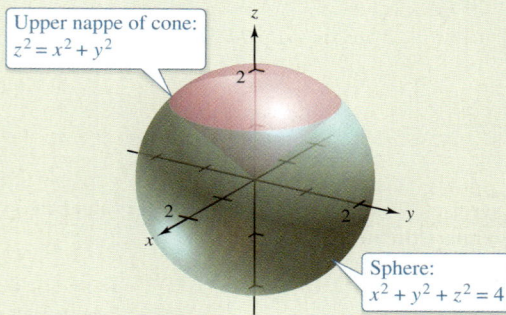

Upper nappe of cone: $z^2 = x^2 + y^2$

Sphere: $x^2 + y^2 + z^2 = 4$

PUTNAM EXAM CHALLENGE

49. Find the volume of the region of points (x, y, z) such that $(x^2 + y^2 + z^2 + 8)^2 \le 36(x^2 + y^2)$.

This problem was composed by the Committee on the Putnam Prize Competition.
© The Mathematical Association of America. All rights reserved.

SECTION PROJECT

Wrinkled and Bumpy Spheres

In parts (a) and (b), find the volume of the wrinkled sphere or bumpy sphere. These solids are used as models for tumors.

(a) Wrinkled sphere

$\rho = 1 + 0.2\sin 8\theta \sin\phi$

$0 \le \theta \le 2\pi, 0 \le \phi \le \pi$

(b) Bumpy sphere

$\rho = 1 + 0.2\sin 8\theta \sin 4\phi$

$0 \le \theta \le 2\pi, 0 \le \phi \le \pi$

Generated by Maple

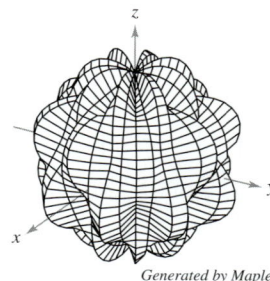
Generated by Maple

FOR FURTHER INFORMATION For more information on these types of spheres, see the article "Heat Therapy for Tumors" by Leah Edelstein-Keshet in *The UMAP Journal*.

14.8 Change of Variables: Jacobians

■ Understand the concept of a Jacobian.
■ Use a Jacobian to change variables in a double integral.

Jacobians

For the single integral

$$\int_a^b f(x)\,dx$$

you can change variables by letting $x = g(u)$, so that $dx = g'(u)\,du$, and obtain

$$\int_a^b f(x)\,dx = \int_c^d f(g(u))g'(u)\,du$$

where $a = g(c)$ and $b = g(d)$. Note that the change of variables process introduces an additional factor $g'(u)$ into the integrand. This also occurs in the case of double integrals

$$\iint_R f(x, y)\,dA = \iint_S f(g(u, v), h(u, v))\underbrace{\left| \frac{\partial x}{\partial u}\frac{\partial y}{\partial v} - \frac{\partial y}{\partial u}\frac{\partial x}{\partial v} \right|}_{\text{Jacobian}}\,du\,dv$$

where the change of variables

$$x = g(u, v) \quad \text{and} \quad y = h(u, v)$$

introduces a factor called the Jacobian of x and y with respect to u and v. In defining the Jacobian, it is convenient to use the determinant notation shown below.

Definition of the Jacobian

If $x = g(u, v)$ and $y = h(u, v)$, then the **Jacobian** of x and y with respect to u and v, denoted by $\partial(x, y)/\partial(u, v)$, is

$$\frac{\partial(x, y)}{\partial(u, v)} = \begin{vmatrix} \dfrac{\partial x}{\partial u} & \dfrac{\partial x}{\partial v} \\[2mm] \dfrac{\partial y}{\partial u} & \dfrac{\partial y}{\partial v} \end{vmatrix} = \frac{\partial x}{\partial u}\frac{\partial y}{\partial v} - \frac{\partial y}{\partial u}\frac{\partial x}{\partial v}.$$

EXAMPLE 1   **The Jacobian for Rectangular-to-Polar Conversion**

Find the Jacobian for the change of variables defined by

$$x = r\cos\theta \quad \text{and} \quad y = r\sin\theta.$$

Solution   From the definition of the Jacobian, you obtain

$$\frac{\partial(x, y)}{\partial(r, \theta)} = \begin{vmatrix} \dfrac{\partial x}{\partial r} & \dfrac{\partial x}{\partial \theta} \\[2mm] \dfrac{\partial y}{\partial r} & \dfrac{\partial y}{\partial \theta} \end{vmatrix}$$

$$= \begin{vmatrix} \cos\theta & -r\sin\theta \\ \sin\theta & r\cos\theta \end{vmatrix}$$

$$= r\cos^2\theta + r\sin^2\theta$$

$$= r.$$

CARL GUSTAV JACOBI (1804–1851)

The Jacobian is named after the German mathematician Carl Gustav Jacobi. Jacobi is known for his work in many areas of mathematics, but his interest in integration stemmed from the problem of finding the circumference of an ellipse.
See LarsonCalculus.com to read more of this biography.

$T(r, \theta) = (r \cos \theta, r \sin \theta)$

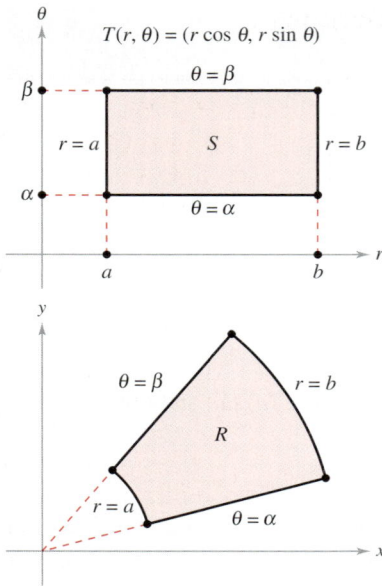

S in the region in the $r\theta$-plane that corresponds to R in the xy-plane.
Figure 14.71

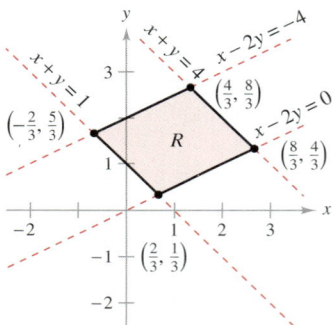

Example 1 points out that the change of variables from rectangular to polar coordinates for a double integral can be written as

$$\int_R \int f(x, y)\, dA = \int_S \int f(r \cos \theta, r \sin \theta) r\, dr\, d\theta, \quad r > 0$$

$$= \int_S \int f(r \cos \theta, r \sin \theta) \left| \frac{\partial(x, y)}{\partial(r, \theta)} \right| dr\, d\theta$$

where S is the region in the $r\theta$-plane that corresponds to the region R in the xy-plane, as shown in Figure 14.71. This formula is similar to that found in Theorem 14.3 on page 988.

In general, a change of variables is given by a one-to-one **transformation** T from a region S in the uv-plane to a region R in the xy-plane, to be given by

$$T(u, v) = (x, y) = (g(u, v), h(u, v))$$

where g and h have continuous first partial derivatives in the region S. Note that the point (u, v) lies in S and the point (x, y) lies in R. In most cases, you are hunting for a transformation in which the region S is simpler than the region R.

EXAMPLE 2 **Finding a Change of Variables to Simplify a Region**

Let R be the region bounded by the lines

$$x - 2y = 0, \quad x - 2y = -4, \quad x + y = 4, \quad \text{and} \quad x + y = 1$$

as shown in Figure 14.72. Find a transformation T from a region S to R such that S is a rectangular region (with sides parallel to the u- or v-axis).

Solution To begin, let $u = x + y$ and $v = x - 2y$. Solving this system of equations for x and y produces $T(u, v) = (x, y)$, where

$$x = \frac{1}{3}(2u + v) \quad \text{and} \quad y = \frac{1}{3}(u - v).$$

The four boundaries for R in the xy-plane give rise to the following bounds for S in the uv-plane.

Region R in the xy-plane
Figure 14.72

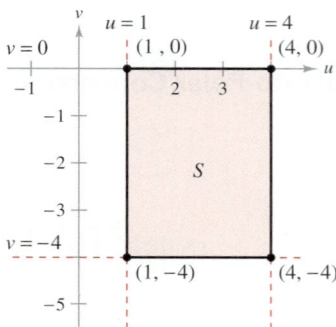

Bounds in the xy-Plane		**Bounds in the uv-Plane**
$x + y = 1$	$\Rightarrow$	$u = 1$
$x + y = 4$	$\Rightarrow$	$u = 4$
$x - 2y = 0$	$\Rightarrow$	$v = 0$
$x - 2y = -4$	$\Rightarrow$	$v = -4$

The region S is shown in Figure 14.73. Note that the transformation

$$T(u, v) = (x, y) = \left(\frac{1}{3}[2u + v], \frac{1}{3}[u - v] \right)$$

maps the vertices of the region S onto the vertices of the region R. For instance,

$$T(1, 0) = \left(\frac{1}{3}[2(1) + 0], \frac{1}{3}[1 - 0] \right) = \left(\frac{2}{3}, \frac{1}{3} \right)$$

$$T(4, 0) = \left(\frac{1}{3}[2(4) + 0], \frac{1}{3}[4 - 0] \right) = \left(\frac{8}{3}, \frac{4}{3} \right)$$

$$T(4, -4) = \left(\frac{1}{3}[2(4) - 4], \frac{1}{3}[4 - (-4)] \right) = \left(\frac{4}{3}, \frac{8}{3} \right)$$

$$T(1, -4) = \left(\frac{1}{3}[2(1) - 4], \frac{1}{3}[1 - (-4)] \right) = \left(-\frac{2}{3}, \frac{5}{3} \right).$$

Region S in the uv-plane
Figure 14.73

Change of Variables for Double Integrals

> **THEOREM 14.5** Change of Variables for Double Integrals
>
> Let R be a vertically or horizontally simple region in the xy-plane, and let S be a vertically or horizontally simple region in the uv-plane. Let T from S to R be given by $T(u, v) = (x, y) = (g(u, v), h(u, v))$, where g and h have continuous first partial derivatives. Assume that T is one-to-one except possibly on the boundary of S. If f is continuous on R, and $\partial(x, y)/\partial(u, v)$ is nonzero on S, then
>
> $$\int_R \int f(x, y)\, dx\, dy = \int_S \int f(g(u, v), h(u, v)) \left| \frac{\partial(x, y)}{\partial(u, v)} \right| du\, dv.$$

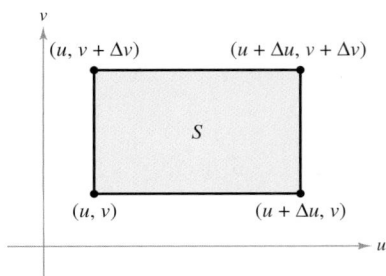

$(u, v + \Delta v)$ $(u + \Delta u, v + \Delta v)$

S

(u, v) $(u + \Delta u, v)$

Area of $S = \Delta u\, \Delta v$
$\Delta u > 0,\ \Delta v > 0$
Figure 14.74

Proof Consider the case in which S is a rectangular region in the uv-plane with vertices (u, v), $(u + \Delta u, v)$, $(u + \Delta u, v + \Delta v)$, and $(u, v + \Delta v)$, as shown in Figure 14.74. The images of these vertices in the xy-plane are shown in Figure 14.75. If Δu and Δv are small, then the continuity of g and h implies that R is approximately a parallelogram determined by the vectors $\overrightarrow{MN}$ and $\overrightarrow{MQ}$. So, the area of R is

$$\Delta A \approx \left\| \overrightarrow{MN} \times \overrightarrow{MQ} \right\|.$$

Moreover, for small Δu and Δv, the partial derivatives of g and h with respect to u can be approximated by

$$g_u(u, v) \approx \frac{g(u + \Delta u, v) - g(u, v)}{\Delta u} \quad \text{and} \quad h_u(u, v) \approx \frac{h(u + \Delta u, v) - h(u, v)}{\Delta u}.$$

Consequently,

$$\overrightarrow{MN} = [g(u + \Delta u, v) - g(u, v)]\mathbf{i} + [h(u + \Delta u, v) - h(u, v)]\mathbf{j}$$
$$\approx [g_u(u, v)\, \Delta u]\mathbf{i} + [h_u(u, v)\, \Delta u]\mathbf{j}$$
$$= \frac{\partial x}{\partial u}\, \Delta u\, \mathbf{i} + \frac{\partial y}{\partial u}\, \Delta u\, \mathbf{j}.$$

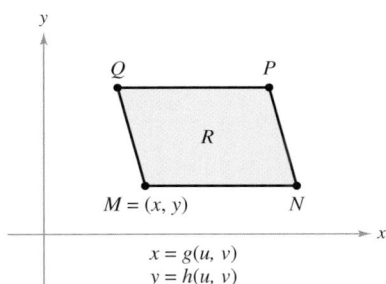

Q P

R

$M = (x, y)$ N

$x = g(u, v)$
$y = h(u, v)$

The vertices in the xy-plane are
$M(g(u, v), h(u, v))$,
$N(g(u + \Delta u, v), h(u + \Delta u, v))$,
$P(g(u + \Delta u, v + \Delta v),$
$h(u + \Delta u, v + \Delta v))$, and
$Q(g(u, v + \Delta v), h(u, v + \Delta v))$.
Figure 14.75

Similarly, you can approximate $\overrightarrow{MQ}$ by $\dfrac{\partial x}{\partial v}\, \Delta v\, \mathbf{i} + \dfrac{\partial y}{\partial v}\, \Delta v\, \mathbf{j}$, which implies that

$$\overrightarrow{MN} \times \overrightarrow{MQ} \approx \begin{vmatrix} \mathbf{i} & \mathbf{j} & \mathbf{k} \\ \dfrac{\partial x}{\partial u}\, \Delta u & \dfrac{\partial y}{\partial u}\, \Delta u & 0 \\ \dfrac{\partial x}{\partial v}\, \Delta v & \dfrac{\partial y}{\partial v}\, \Delta v & 0 \end{vmatrix} = \begin{vmatrix} \dfrac{\partial x}{\partial u} & \dfrac{\partial y}{\partial u} \\ \dfrac{\partial x}{\partial v} & \dfrac{\partial y}{\partial v} \end{vmatrix} \Delta u\, \Delta v\, \mathbf{k}.$$

It follows that, in Jacobian notation,

$$\Delta A \approx \left\| \overrightarrow{MN} \times \overrightarrow{MQ} \right\| \approx \left| \frac{\partial(x, y)}{\partial(u, v)} \right| \Delta u\, \Delta v.$$

Because this approximation improves as Δu and Δv approach 0, the limiting case can be written as

$$dA \approx \left\| \overrightarrow{MN} \times \overrightarrow{MQ} \right\| \approx \left| \frac{\partial(x, y)}{\partial(u, v)} \right| du\, dv.$$

So,

$$\int_R \int f(x, y)\, dx\, dy = \int_S \int f(g(u, v), h(u, v)) \left| \frac{\partial(x, y)}{\partial(u, v)} \right| du\, dv.$$

See LarsonCalculus.com for Bruce Edwards's video of this proof.

The next two examples show how a change of variables can simplify the integration process. The simplification can occur in various ways. You can make a change of variables to simplify either the *region R* or the *integrand f(x, y)*, or both.

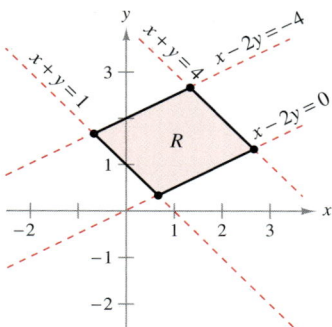

Figure 14.76

EXAMPLE 3   **Using a Change of Variables to Simplify a Region**

$\cdots\triangleright$ *See LarsonCalculus.com for an interactive version of this type of example.*

Let R be the region bounded by the lines

$$x - 2y = 0, \quad x - 2y = -4, \quad x + y = 4, \quad \text{and} \quad x + y = 1$$

as shown in Figure 14.76. Evaluate the double integral

$$\iint_R 3xy \, dA.$$

Solution   From Example 2, you can use the following change of variables.

$$x = \frac{1}{3}(2u + v) \quad \text{and} \quad y = \frac{1}{3}(u - v)$$

The partial derivatives of x and y are

$$\frac{\partial x}{\partial u} = \frac{2}{3}, \quad \frac{\partial x}{\partial v} = \frac{1}{3}, \quad \frac{\partial y}{\partial u} = \frac{1}{3}, \quad \text{and} \quad \frac{\partial y}{\partial v} = -\frac{1}{3}$$

which implies that the Jacobian is

$$\begin{aligned}
\frac{\partial(x, y)}{\partial(u, v)} &= \begin{vmatrix} \dfrac{\partial x}{\partial u} & \dfrac{\partial x}{\partial v} \\[2mm] \dfrac{\partial y}{\partial u} & \dfrac{\partial y}{\partial v} \end{vmatrix} \\[3mm]
&= \begin{vmatrix} \dfrac{2}{3} & \dfrac{1}{3} \\[2mm] \dfrac{1}{3} & -\dfrac{1}{3} \end{vmatrix} \\[3mm]
&= -\frac{2}{9} - \frac{1}{9} \\[2mm]
&= -\frac{1}{3}.
\end{aligned}$$

So, by Theorem 14.5, you obtain

$$\begin{aligned}
\iint_R 3xy \, dA &= \iint_S 3\left[\frac{1}{3}(2u + v)\frac{1}{3}(u - v)\right]\left|\frac{\partial(x, y)}{\partial(u, v)}\right| dv \, du \\[2mm]
&= \int_1^4 \int_{-4}^0 \frac{1}{9}(2u^2 - uv - v^2) \, dv \, du \\[2mm]
&= \frac{1}{9}\int_1^4 \left[2u^2v - \frac{uv^2}{2} - \frac{v^3}{3}\right]_{-4}^0 du \\[2mm]
&= \frac{1}{9}\int_1^4 \left(8u^2 + 8u - \frac{64}{3}\right) du \\[2mm]
&= \frac{1}{9}\left[\frac{8u^3}{3} + 4u^2 - \frac{64}{3}u\right]_1^4 \\[2mm]
&= \frac{164}{9}.
\end{aligned}$$

EXAMPLE 4 **Change of Variables: Simplifying an Integrand**

Let R be the region bounded by the square with vertices $(0, 1)$, $(1, 2)$, $(2, 1)$, and $(1, 0)$. Evaluate the integral

$$\int_R \int (x + y)^2 \sin^2(x - y) \, dA.$$

Solution Note that the sides of R lie on the lines $x + y = 1$, $x - y = 1$, $x + y = 3$, and $x - y = -1$, as shown in Figure 14.77. Letting $u = x + y$ and $v = x - y$, you can determine the bounds for region S in the uv-plane to be

$$1 \le u \le 3 \quad \text{and} \quad -1 \le v \le 1$$

as shown in Figure 14.78. Solving for x and y in terms of u and v produces

$$x = \frac{1}{2}(u + v) \quad \text{and} \quad y = \frac{1}{2}(u - v).$$

The partial derivatives of x and y are

$$\frac{\partial x}{\partial u} = \frac{1}{2}, \quad \frac{\partial x}{\partial v} = \frac{1}{2}, \quad \frac{\partial y}{\partial u} = \frac{1}{2}, \quad \text{and} \quad \frac{\partial y}{\partial v} = -\frac{1}{2}$$

which implies that the Jacobian is

$$\frac{\partial(x, y)}{\partial(u, v)} = \begin{vmatrix} \dfrac{\partial x}{\partial u} & \dfrac{\partial x}{\partial v} \\ \dfrac{\partial y}{\partial u} & \dfrac{\partial y}{\partial v} \end{vmatrix} = \begin{vmatrix} \dfrac{1}{2} & \dfrac{1}{2} \\ \dfrac{1}{2} & -\dfrac{1}{2} \end{vmatrix} = -\frac{1}{4} - \frac{1}{4} = -\frac{1}{2}.$$

By Theorem 14.5, it follows that

$$\int_R \int (x + y)^2 \sin^2(x - y) \, dA = \int_{-1}^{1} \int_{1}^{3} u^2 \sin^2 v \left(\frac{1}{2}\right) du \, dv$$

$$= \frac{1}{2} \int_{-1}^{1} (\sin^2 v) \frac{u^3}{3} \Big]_{1}^{3} dv$$

$$= \frac{13}{3} \int_{-1}^{1} \sin^2 v \, dv$$

$$= \frac{13}{6} \int_{-1}^{1} (1 - \cos 2v) \, dv$$

$$= \frac{13}{6} \left[v - \frac{1}{2} \sin 2v \right]_{-1}^{1}$$

$$= \frac{13}{6} \left[2 - \frac{1}{2} \sin 2 + \frac{1}{2} \sin(-2) \right]$$

$$= \frac{13}{6} (2 - \sin 2)$$

$$\approx 2.363.$$

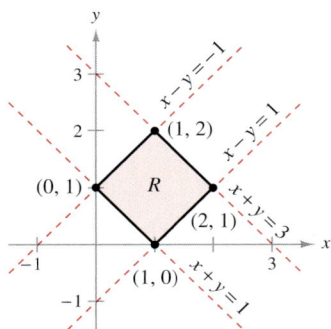

Region R in the xy-plane
Figure 14.77

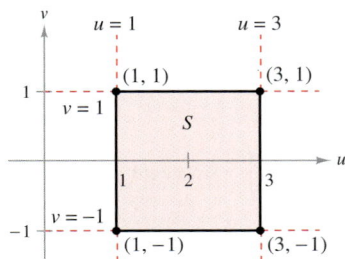

Region S in the uv-plane
Figure 14.78

In each of the change of variables examples in this section, the region S has been a rectangle with sides parallel to the u- or v-axis. Occasionally, a change of variables can be used for other types of regions. For instance, letting $T(u, v) = \left(x, \frac{1}{2}y\right)$ changes the circular region $u^2 + v^2 = 1$ to the elliptical region

$$x^2 + \frac{y^2}{4} = 1.$$

14.8 Exercises

See **CalcChat.com** for tutorial help and worked-out solutions to odd-numbered exercises.

Finding a Jacobian In Exercises 1–8, find the Jacobian $\partial(x, y)/\partial(u, v)$ for the indicated change of variables.

1. $x = -\frac{1}{2}(u - v), y = \frac{1}{2}(u + v)$

2. $x = au + bv, y = cu + dv$

3. $x = u - v^2, y = u + v$

4. $x = uv - 2u, y = uv$

5. $x = u \cos \theta - v \sin \theta, y = u \sin \theta + v \cos \theta$

6. $x = u + a, y = v + a$

7. $x = e^u \sin v, y = e^u \cos v$

8. $x = u/v, y = u + v$

Using a Transformation In Exercises 9–12, sketch the image S in the uv-plane of the region R in the xy-plane using the given transformations.

9. $x = 3u + 2v$

$y = 3v$

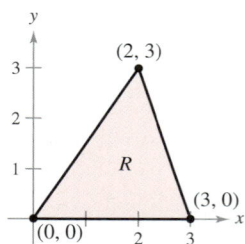

10. $x = \frac{1}{3}(4u - v)$

$y = \frac{1}{3}(u - v)$

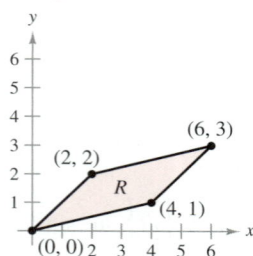

11. $x = \frac{1}{2}(u + v)$

$y = \frac{1}{2}(u - v)$

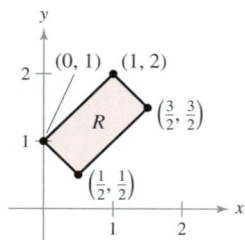

12. $x = \frac{1}{3}(v - u)$

$y = \frac{1}{3}(2v + u)$

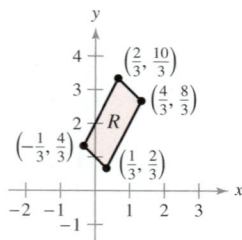

Verifying a Change of Variables In Exercises 13 and 14, verify the result of the indicated example by setting up the integral using $dy\, dx$ or $dx\, dy$ for dA. Then use a computer algebra system to evaluate the integral.

13. Example 3

14. Example 4

Evaluating a Double Integral Using a Change of Variables In Exercises 15–20, use the indicated change of variables to evaluate the double integral.

15. $\displaystyle\int_R\!\!\int 4(x^2 + y^2)\, dA$

$x = \frac{1}{2}(u + v)$

$y = \frac{1}{2}(u - v)$

16. $\displaystyle\int_R\!\!\int 60xy\, dA$

$x = \frac{1}{2}(u + v)$

$y = -\frac{1}{2}(u - v)$

Figure for 15

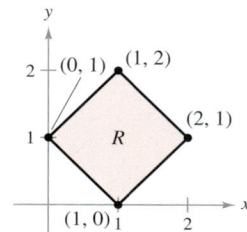

Figure for 16

17. $\displaystyle\int_R\!\!\int y(x - y)\, dA$

$x = u + v$

$y = u$

18. $\displaystyle\int_R\!\!\int 4(x + y)e^{x-y}\, dA$

$x = \frac{1}{2}(u + v)$

$y = \frac{1}{2}(u - v)$

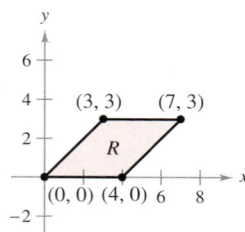

19. $\displaystyle\int_R\!\!\int e^{-xy/2}\, dA$

$x = \sqrt{\dfrac{v}{u}}, \quad y = \sqrt{uv}$

20. $\displaystyle\int_R\!\!\int y \sin xy\, dA$

$x = \dfrac{u}{v}, \quad y = v$

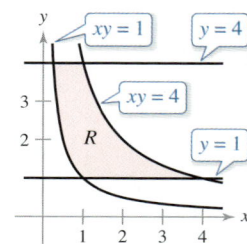

Finding Volume Using a Change of Variables In Exercises 21–28, use a change of variables to find the volume of the solid region lying below the surface $z = f(x, y)$ and above the plane region R.

21. $f(x, y) = 48xy$

R: region bounded by the square with vertices $(1, 0), (0, 1), (1, 2), (2, 1)$

22. $f(x, y) = (3x + 2y)^2 \sqrt{2y - x}$

R: region bounded by the parallelogram with vertices $(0, 0), (-2, 3), (2, 5), (4, 2)$

23. $f(x, y) = (x + y)e^{x-y}$

R: region bounded by the square with vertices $(4, 0), (6, 2), (4, 4), (2, 2)$

24. $f(x, y) = (x + y)^2 \sin^2(x - y)$

R: region bounded by the square with vertices $(\pi, 0), (3\pi/2, \pi/2), (\pi, \pi), (\pi/2, \pi/2)$

25. $f(x, y) = \sqrt{(x - y)(x + 4y)}$

   R: region bounded by the parallelogram with vertices $(0, 0)$, $(1, 1)$, $(5, 0)$, $(4, -1)$

26. $f(x, y) = (3x + 2y)(2y - x)^{3/2}$

   R: region bounded by the parallelogram with vertices $(0, 0)$, $(-2, 3)$, $(2, 5)$, $(4, 2)$

27. $f(x, y) = \sqrt{x + y}$

   R: region bounded by the triangle with vertices $(0, 0)$, $(a, 0)$, $(0, a)$, where $a > 0$

28. $f(x, y) = \dfrac{xy}{1 + x^2y^2}$

   R: region bounded by the graphs of $xy = 1$, $xy = 4$, $x = 1$, $x = 4$ (*Hint:* Let $x = u$, $y = v/u$.)

29. Using a Transformation   The substitutions $u = 2x - y$ and $v = x + y$ make the region R (see figure) into a simpler region S in the uv-plane. Determine the total number of sides of S that are parallel to either the u-axis or the v-axis.

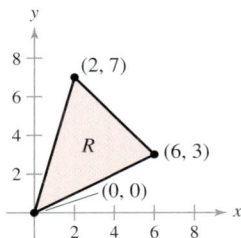

30. HOW DO YOU SEE IT?   The region R is transformed into a simpler region S (see figure). Which substitution can be used to make the transformation?

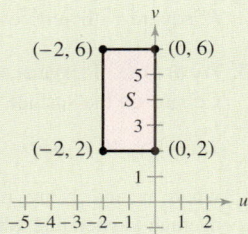

   (a) $u = 3y - x$, $v = y - x$ (b) $u = y - x$, $v = 3y - x$

31. Using an Ellipse   Consider the region R in the xy-plane bounded by the ellipse

$$\frac{x^2}{a^2} + \frac{y^2}{b^2} = 1$$

   and the transformations $x = au$ and $y = bv$.

   (a) Sketch the graph of the region R and its image S under the given transformation.

   (b) Find $\dfrac{\partial(x, y)}{\partial(u, v)}$.

   (c) Find the area of the ellipse.

32. Volume   Use the result of Exercise 31 to find the volume of each dome-shaped solid lying below the surface $z = f(x, y)$ and above the elliptical region R. (*Hint:* After making the change of variables given by the results in Exercise 31, make a second change of variables to polar coordinates.)

   (a) $f(x, y) = 16 - x^2 - y^2$

   $$R: \frac{x^2}{16} + \frac{y^2}{9} \le 1$$

   (b) $f(x, y) = A \cos\left(\dfrac{\pi}{2}\sqrt{\dfrac{x^2}{a^2} + \dfrac{y^2}{b^2}}\right)$

   $$R: \frac{x^2}{a^2} + \frac{y^2}{b^2} \le 1$$

WRITING ABOUT CONCEPTS

33. Jacobian   State the definition of the Jacobian.

34. Change of Variables   Describe how to use the Jacobian to change variables in double integrals.

Finding a Jacobian   In Exercises 35–40, find the Jacobian

$$\frac{\partial(x, y, z)}{\partial(u, v, w)}$$

for the indicated change of variables. If

$$x = f(u, v, w), \quad y = g(u, v, w), \quad \text{and} \quad z = h(u, v, w)$$

then the Jacobian of x, y, and z with respect to u, v, and w is

$$\frac{\partial(x, y, z)}{\partial(u, v, w)} = \begin{vmatrix} \dfrac{\partial x}{\partial u} & \dfrac{\partial x}{\partial v} & \dfrac{\partial x}{\partial w} \\[2mm] \dfrac{\partial y}{\partial u} & \dfrac{\partial y}{\partial v} & \dfrac{\partial y}{\partial w} \\[2mm] \dfrac{\partial z}{\partial u} & \dfrac{\partial z}{\partial v} & \dfrac{\partial z}{\partial w} \end{vmatrix}.$$

35. $x = u(1 - v)$, $y = uv(1 - w)$, $z = uvw$

36. $x = 4u - v$, $y = 4v - w$, $z = u + w$

37. $x = \frac{1}{2}(u + v)$, $y = \frac{1}{2}(u - v)$, $z = 2uvw$

38. $x = u - v + w$, $y = 2uv$, $z = u + v + w$

39. Spherical Coordinates

   $x = \rho \sin\phi \cos\theta$, $y = \rho \sin\phi \sin\theta$, $z = \rho \cos\phi$

40. Cylindrical Coordinates

   $x = r \cos\theta$, $y = r \sin\theta$, $z = z$

PUTNAM EXAM CHALLENGE

41. Let A be the area of the region in the first quadrant bounded by the line $y = \frac{1}{2}x$, the x-axis, and the ellipse $\frac{1}{9}x^2 + y^2 = 1$. Find the positive number m such that A is equal to the area of the region in the first quadrant bounded by the line $y = mx$, the y-axis, and the ellipse $\frac{1}{9}x^2 + y^2 = 1$.

This problem was composed by the Committee on the Putnam Prize Competition.
© The Mathematical Association of America. All rights reserved.

Evaluating an Integral In Exercises 1 and 2, evaluate the integral.

1. $\int_0^{2x} xy^3\, dy$

2. $\int_y^{2y} (x^2 + y^2)\, dx$

Evaluating an Iterated Integral In Exercises 3–6, evaluate the iterated integral.

3. $\int_0^1 \int_0^{1+x} (3x + 2y)\, dy\, dx$

4. $\int_0^2 \int_{x^2}^{2x} (x^2 + 2y)\, dy\, dx$

5. $\int_0^3 \int_0^{\sqrt{9-x^2}} 4x\, dy\, dx$

6. $\int_0^1 \int_0^{2y} (9 + 3x^2 + 3y^2)\, dx\, dy$

Finding the Area of a Region In Exercises 7–10, use an iterated integral to find the area of the region bounded by the graphs of the equations.

7. $x + 3y = 3,\ x = 0,\ y = 0$

8. $y = 6x - x^2,\ y = x^2 - 2x$

9. $y = x,\ y = 2x + 2,\ x = 0,\ x = 4$

10. $x = y^2 + 1,\ x = 0,\ y = 0,\ y = 2$

Switching the Order of Integration In Exercises 11–14, sketch the region R whose area is given by the iterated integral. Then switch the order of integration and show that both orders yield the same area.

11. $\int_2^4 \int_1^5 dx\, dy$

12. $\int_0^3 \int_0^x dy\, dx + \int_3^6 \int_0^{6-x} dy\, dx$

13. $\int_0^4 \int_{2x}^8 dy\, dx$

14. $\int_{-3}^3 \int_0^{9-y^2} dx\, dy$

Evaluating a Double Integral In Exercises 15 and 16, set up integrals for both orders of integration. Use the more convenient order to evaluate the integral over the region R.

15. $\int\int_R 4xy\, dA$

R: rectangle with vertices $(0, 0),\ (0, 4),\ (2, 4),\ (2, 0)$

16. $\int\int_R 6x^2\, dA$

R: region bounded by $y = 0,\ y = \sqrt{x},\ x = 1$

Finding Volume In Exercises 17–20, use a double integral to find the volume of the indicated solid.

17.

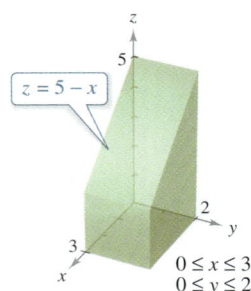

$z = 5 - x$
$0 \le x \le 3$
$0 \le y \le 2$

18.

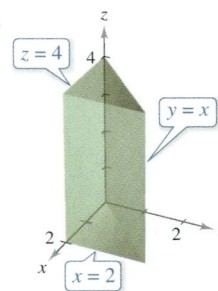

$z = 4$
$y = x$
$x = 2$

19.

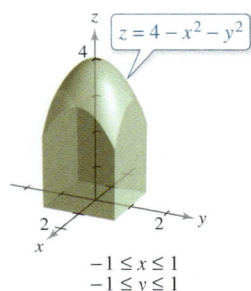

$z = 4 - x^2 - y^2$
$-1 \le x \le 1$
$-1 \le y \le 1$

20.

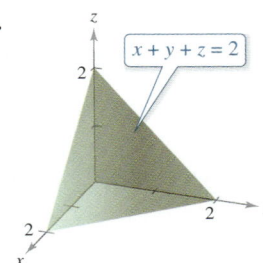

$x + y + z = 2$

Average Value In Exercises 21 and 22, find the average value of $f(x, y)$ over the plane region R.

21. $f(x) = 16 - x^2 - y^2$

R: rectangle with vertices $(2, 2),\ (-2, 2),\ (-2, -2),\ (2, -2)$

22. $f(x) = 2x^2 + y^2$

R: square with vertices $(0, 0),\ (3, 0),\ (3, 3),\ (0, 3)$

23. **Average Temperature** The temperature in degrees Celsius on the surface of a metal plate is

$T(x, y) = 40 - 6x^2 - y^2$

where x and y are measured in centimeters. Estimate the average temperature when x varies between 0 and 3 centimeters and y varies between 0 and 5 centimeters.

24. **Average Profit** A firm's profit P from marketing two soft drinks is

$P = 192x + 576y - x^2 - 5y^2 - 2xy - 5000$

where x and y represent the numbers of units of the two soft drinks. Estimate the average weekly profit when x varies between 40 and 50 units and y varies between 45 and 60 units.

Converting to Polar Coordinates In Exercises 25 and 26, evaluate the iterated integral by converting to polar coordinates.

25. $\int_0^h \int_0^x \sqrt{x^2 + y^2}\, dy\, dx$

26. $\int_0^4 \int_0^{\sqrt{16-y^2}} (x^2 + y^2)\, dx\, dy$

Volume In Exercises 27 and 28, use a double integral in polar coordinates to find the volume of the solid bounded by the graphs of the equations.

27. $z = xy^2$, $x^2 + y^2 = 9$, first octant

28. $z = \sqrt{25 - x^2 - y^2}$, $z = 0$, $x^2 + y^2 = 16$

Area In Exercises 29 and 30, use a double integral to find the area of the shaded region.

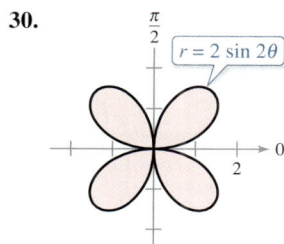

29.

$r = 2 + \cos\theta$

30.

$r = 2\sin 2\theta$

Area In Exercises 31 and 32, sketch a graph of the region bounded by the graphs of the equations. Then use a double integral to find the area of the region.

31. Inside the cardioid $r = 2 + 2\cos\theta$ and outside the circle $r = 3$

32. Inside the circle $r = 3\sin\theta$ and outside the cardioid $r = 1 + \sin\theta$

33. Area and Volume Consider the region R in the xy-plane bounded by the graph of the equation

$$(x^2 + y^2)^2 = 9(x^2 - y^2).$$

(a) Convert the equation to polar coordinates. Use a graphing utility to graph the equation.

(b) Use a double integral to find the area of the region R.

(c) Use a computer algebra system to determine the volume of the solid over the region R and beneath the hemisphere $z = \sqrt{9 - x^2 - y^2}$.

34. Converting to Polar Coordinates Combine the sum of the two iterated integrals into a single iterated integral by converting to polar coordinates. Evaluate the resulting iterated integral.

$$\int_0^{8/\sqrt{13}} \int_0^{3x/2} xy\, dy\, dx + \int_{8/\sqrt{13}}^4 \int_0^{\sqrt{16-x^2}} xy\, dy\, dx$$

Finding the Center of Mass In Exercises 35–38, find the mass and center of mass of the lamina bounded by the graphs of the equations for the given density. (*Hint:* Some of the integrals are simpler in polar coordinates.)

35. $y = x^3$, $y = 0$, $x = 2$, $\rho = kx$

36. $y = \dfrac{2}{x}$, $y = 0$, $x = 1$, $x = 2$, $\rho = ky$

37. $y = 2x$, $y = 2x^3$, $x \ge 0$, $y \ge 0$, $\rho = kxy$

38. $y = 6 - x$, $y = 0$, $x = 0$, $\rho = kx^2$

Finding Moments of Inertia and Radii of Gyration In Exercises 39 and 40, find I_x, I_y, I_0, $\bar{\bar{x}}$, and $\bar{\bar{y}}$ for the lamina bounded by the graphs of the equations.

39. $y = 0$, $y = b$, $x = 0$, $x = a$, $\rho = kx$

40. $y = 4 - x^2$, $y = 0$, $x > 0$, $\rho = ky$

Finding Surface Area In Exercises 41–44, find the area of the surface given by $z = f(x, y)$ over the region R. (*Hint:* Some of the integrals are simpler in polar coordinates.)

41. $f(x, y) = 25 - x^2 - y^2$

$R = \{(x, y):\ x^2 + y^2 \le 25\}$

42. $f(x, y) = 8 + 4x - 5y$

$R = \{(x, y):\ x^2 + y^2 \le 1\}$

43. $f(x, y) = 9 - y^2$

R: triangle bounded by the graphs of the equations $y = x$, $y = -x$, and $y = 3$

44. $f(x, y) = 4 - x^2$

R: triangle bounded by the graphs of the equations $y = x$, $y = -x$, and $y = 2$

45. Building Design A new auditorium is built with a foundation in the shape of one-fourth of a circle of radius 50 feet. So, it forms a region R bounded by the graph of

$$x^2 + y^2 = 50^2$$

with $x \ge 0$ and $y \ge 0$. The following equations are models for the floor and ceiling.

Floor: $z = \dfrac{x + y}{5}$

Ceiling: $z = 20 + \dfrac{xy}{100}$

(a) Calculate the volume of the room, which is needed to determine the heating and cooling requirements.

(b) Find the surface area of the ceiling.

46. Surface Area The roof over the stage of an open air theater at a theme park is modeled by

$$f(x, y) = 25\left[1 + e^{-(x^2 + y^2)/1000} \cos^2\left(\frac{x^2 + y^2}{1000}\right)\right]$$

where the stage is a semicircle bounded by the graphs of $y = \sqrt{50^2 - x^2}$ and $y = 0$.

(a) Use a computer algebra system to graph the surface.

(b) Use a computer algebra system to approximate the number of square feet of roofing required to cover the surface.

Evaluating a Triple Iterated Integral In Exercises 47–50, evaluate the triple iterated integral.

47. $\displaystyle \int_0^4 \int_0^1 \int_0^2 (2x + y + 4z)\, dy\, dz\, dx$

48. $\displaystyle \int_0^2 \int_0^y \int_0^{xy} y\, dz\, dx\, dy$

49. $\displaystyle\int_0^a \int_0^b \int_0^c (x^2 + y^2 + z^2)\, dx\, dy\, dz$

50. $\displaystyle\int_0^3 \int_{\pi/2}^{\pi} \int_2^5 z \sin x\, dy\, dx\, dz$

Approximating a Triple Iterated Integral Using Technology In Exercises 51 and 52, use a computer algebra system to approximate the iterated integral.

51. $\displaystyle\int_{-1}^1 \int_{-\sqrt{1-x^2}}^{\sqrt{1-x^2}} \int_{-\sqrt{1-x^2-y^2}}^{\sqrt{1-x^2-y^2}} (x^2 + y^2)\, dz\, dy\, dx$

52. $\displaystyle\int_0^2 \int_0^{\sqrt{4-x^2}} \int_0^{\sqrt{4-x^2-y^2}} xyz\, dz\, dy\, dx$

Volume In Exercises 53 and 54, use a triple integral to find the volume of the solid bounded by the graphs of the equations.

53. $z = xy,\ z = 0,\ 0 \le x \le 3,\ 0 \le y \le 4$

54. $z = 8 - x - y,\ z = 0,\ y = x,\ y = 3,\ x = 0$

Changing the Order of Integration In Exercises 55 and 56, sketch the solid whose volume is given by the iterated integral and rewrite the integral using the indicated order of integration.

55. $\displaystyle\int_0^1 \int_0^y \int_0^{\sqrt{1-x^2}} dz\, dx\, dy$

Rewrite using the order $dz\, dy\, dx$.

56. $\displaystyle\int_0^6 \int_0^{6-x} \int_0^{6-x-y} dz\, dy\, dx$

Rewrite using the order $dy\, dx\, dz$.

Mass and Center of Mass In Exercises 57 and 58, find the mass and the indicated coordinates of the center of mass of the solid region Q of density ρ bounded by the graphs of the equations.

57. Find $\bar{x}$ using $\rho(x, y, z) = k$.

Q: $x + y + z = 10,\ x = 0,\ y = 0,\ z = 0$

58. Find $\bar{y}$ using $\rho(x, y, z) = kx$.

Q: $z = 5 - y,\ z = 0,\ y = 0,\ x = 0,\ x = 5$

Evaluating an Iterated Integral In Exercises 59–62, evaluate the iterated integral.

59. $\displaystyle\int_0^3 \int_0^{\pi/3} \int_0^4 r \cos \theta\, dr\, d\theta\, dz$

60. $\displaystyle\int_0^{\pi/2} \int_0^3 \int_0^{4-z} z\, dr\, dz\, d\theta$

61. $\displaystyle\int_0^{\pi/2} \int_0^{\pi/2} \int_0^2 \rho^2\, d\rho\, d\theta\, d\phi$

62. $\displaystyle\int_0^{\pi/4} \int_0^{\pi/4} \int_0^{\cos \phi} \cos \theta\, d\rho\, d\phi\, d\theta$

Approximating an Iterated Integral Using Technology In Exercises 63 and 64, use a computer algebra system to approximate the iterated integral.

63. $\displaystyle\int_0^{\pi} \int_0^2 \int_0^3 \sqrt{z^2 + 4}\, dz\, dr\, d\theta$

64. $\displaystyle\int_0^{\pi/2} \int_0^{\pi/2} \int_0^{\cos \phi} \rho^2 \cos \theta\, d\rho\, d\theta\, d\phi$

65. Volume Use cylindrical coordinates to find the volume of the solid bounded above by $z = 8 - x^2 - y^2$ and below by $z = x^2 + y^2$.

66. Volume Use spherical coordinates to find the volume of the solid bounded above by $x^2 + y^2 + z^2 = 36$ and below by $z = \sqrt{x^2 + y^2}$.

Finding a Jacobian In Exercises 67–70, find the Jacobian $\partial(x, y)/\partial(u, v)$ for the indicated change of variables.

67. $x = u + 3v,\ y = 2u - 3v$

68. $x = u^2 + v^2,\ y = u^2 - v^2$

69. $x = u \sin \theta + v \cos \theta,\ y = u \cos \theta + v \sin \theta$

70. $x = uv,\ y = \dfrac{v}{u}$

Evaluating a Double Integral Using a Change of Variables In Exercises 71–74, use the indicated change of variables to evaluate the double integral.

71. $\displaystyle\iint_R \ln(x + y)\, dA$

$x = \dfrac{1}{2}(u + v),\ y = \dfrac{1}{2}(u - v)$

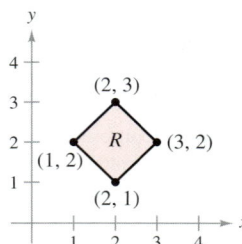

72. $\displaystyle\iint_R 16xy\, dA$

$x = \dfrac{1}{4}(u + v),\ y = \dfrac{1}{2}(v - u)$

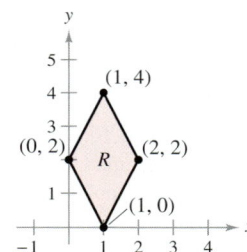

73. $\displaystyle\iint_R (xy + x^2)\, dA$

$x = u,\ y = \dfrac{1}{3}(u - v)$

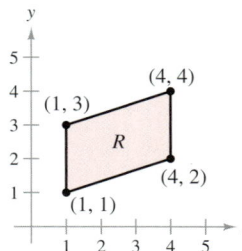

74. $\displaystyle\iint_R \dfrac{x}{1 + x^2 y^2}\, dA$

$x = u,\ y = \dfrac{v}{u}$

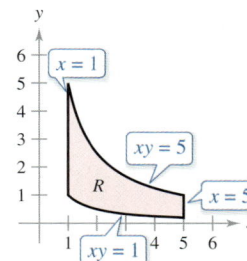

P.S. Problem Solving

See **CalcChat.com** for tutorial help and worked-out solutions to odd-numbered exercises.

1. Volume   Find the volume of the solid of intersection of the three cylinders $x^2 + z^2 = 1$, $y^2 + z^2 = 1$, and $x^2 + y^2 = 1$ (see figure).

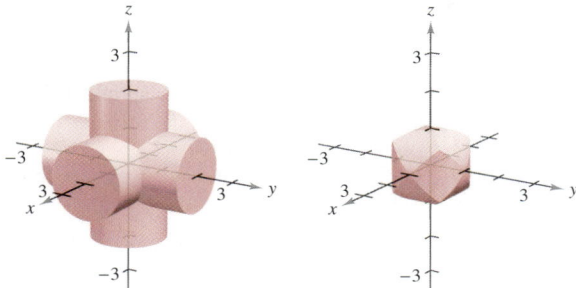

2. Surface Area   Let a, b, c, and d be positive real numbers. The first octant of the plane $ax + by + cz = d$ is shown in the figure. Show that the surface area of this portion of the plane is equal to

$$\frac{A(R)}{c}\sqrt{a^2 + b^2 + c^2}$$

where $A(R)$ is the area of the triangular region R in the xy-plane, as shown in the figure.

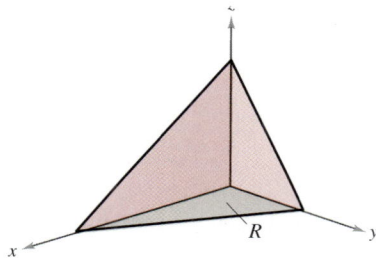

3. Using a Change of Variables   The figure shows the region R bounded by the curves

$$y = \sqrt{x}, \; y = \sqrt{2x}, \; y = \frac{x^2}{3}, \text{ and } y = \frac{x^2}{4}.$$

Use the change of variables $x = u^{1/3}v^{2/3}$ and $y = u^{2/3}v^{1/3}$ to find the area of the region R.

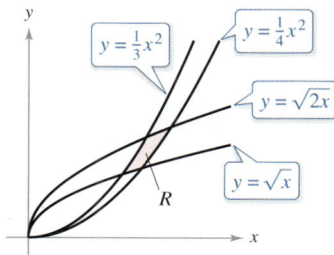

4. Proof   Prove that $\displaystyle\lim_{n\to\infty} \int_0^1 \int_0^1 x^n y^n \, dx \, dy = 0$.

5. Deriving a Sum   Derive Euler's famous result that was mentioned in Section 9.3,

$$\sum_{n=1}^{\infty} \frac{1}{n^2} = \frac{\pi^2}{6}$$

by completing each step.

(a) Prove that $\displaystyle\int \frac{dv}{2 - u^2 + v^2} = \frac{1}{\sqrt{2 - u^2}} \arctan \frac{v}{\sqrt{2 - u^2}} + C.$

(b) Prove that

$$I_1 = \int_0^{\sqrt{2}/2} \int_{-u}^{u} \frac{2}{2 - u^2 + v^2} \, dv \, du = \frac{\pi^2}{18}$$

by using the substitution $u = \sqrt{2} \sin \theta$.

(c) Prove that

$$I_2 = \int_{\sqrt{2}/2}^{\sqrt{2}} \int_{u - \sqrt{2}}^{-u + \sqrt{2}} \frac{2}{2 - u^2 + v^2} \, dv \, du$$

$$= 4 \int_{\pi/6}^{\pi/2} \arctan \frac{1 - \sin \theta}{\cos \theta} \, d\theta$$

by using the substitution $u = \sqrt{2} \sin \theta$.

(d) Prove the trigonometric identity

$$\frac{1 - \sin \theta}{\cos \theta} = \tan\left(\frac{(\pi/2) - \theta}{2}\right).$$

(e) Prove that $\displaystyle I_2 = \int_{\sqrt{2}/2}^{\sqrt{2}} \int_{u - \sqrt{2}}^{-u + \sqrt{2}} \frac{2}{2 - u^2 + v^2} \, dv \, du = \frac{\pi^2}{9}.$

(f) Use the formula for the sum of an infinite geometric series to verify that

$$\sum_{n=1}^{\infty} \frac{1}{n^2} = \int_0^1 \int_0^1 \frac{1}{1 - xy} \, dx \, dy.$$

(g) Use the change of variables

$$u = \frac{x + y}{\sqrt{2}} \quad \text{and} \quad v = \frac{y - x}{\sqrt{2}}$$

to prove that

$$\sum_{n=1}^{\infty} \frac{1}{n^2} = \int_0^1 \int_0^1 \frac{1}{1 - xy} \, dx \, dy = I_1 + I_2 = \frac{\pi^2}{6}.$$

6. Evaluating a Double Integral   Evaluate the integral

$$\int_0^{\infty} \int_0^{\infty} \frac{1}{(1 + x^2 + y^2)^2} \, dx \, dy.$$

7. Evaluating Double Integrals   Evaluate the integrals

$$\int_0^1 \int_0^1 \frac{x - y}{(x + y)^3} \, dx \, dy \quad \text{and} \quad \int_0^1 \int_0^1 \frac{x - y}{(x + y)^3} \, dy \, dx.$$

Are the results the same? Why or why not?

8. Volume   Show that the volume of a spherical block can be approximated by $\Delta V \approx \rho^2 \sin \phi \, \Delta\rho \, \Delta\phi \, \Delta\theta$.

Evaluating an Integral In Exercises 9 and 10, evaluate the integral. (*Hint:* See Exercise 65 in Section 14.3.)

9. $\displaystyle\int_0^\infty x^2 e^{-x^2}\, dx$

10. $\displaystyle\int_0^1 \sqrt{\ln\frac{1}{x}}\, dx$

11. Joint Density Function Consider the function

$$f(x, y) = \begin{cases} ke^{-(x+y)/a}, & x \ge 0,\ y \ge 0 \\ 0, & \text{elsewhere.} \end{cases}$$

Find the relationship between the positive constants a and k such that f is a joint density function of the continuous random variables x and y.

12. Volume Find the volume of the solid generated by revolving the region in the first quadrant bounded by $y = e^{-x^2}$ about the y-axis. Use this result to find

$$\int_{-\infty}^{\infty} e^{-x^2}\, dx.$$

13. Volume and Surface Area From 1963 to 1986, the volume of the Great Salt Lake approximately tripled while its top surface area approximately doubled. Read the article "Relations between Surface Area and Volume in Lakes" by Daniel Cass and Gerald Wildenberg in *The College Mathematics Journal*. Then give examples of solids that have "water levels" a and b such that $V(b) = 3V(a)$ and $A(b) = 2A(a)$ (see figure), where V is volume and A is area.

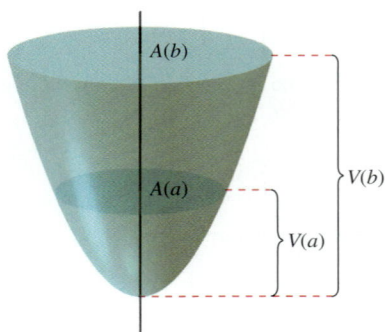

14. Proof The angle between a plane P and the xy-plane is θ, where $0 \le \theta < \pi/2$. The projection of a rectangular region in P onto the xy-plane is a rectangle whose sides have lengths Δx and Δy, as shown in the figure. Prove that the area of the rectangular region in P is $\sec\theta\,\Delta x\,\Delta y$.

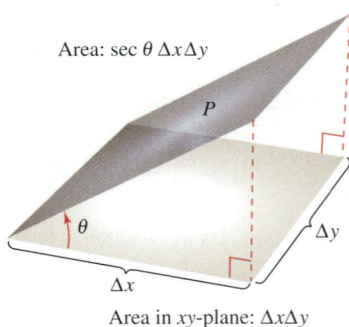

Area: $\sec\theta\,\Delta x\Delta y$

Area in xy-plane: $\Delta x\Delta y$

15. Surface Area Use the result of Exercise 14 to order the planes in ascending order of their surface areas for a fixed region R in the xy-plane. Explain your ordering without doing any calculations.

(a) $z_1 = 2 + x$

(b) $z_2 = 5$

(c) $z_3 = 10 - 5x + 9y$

(d) $z_4 = 3 + x - 2y$

16. Sprinkler Consider a circular lawn with a radius of 10 feet, as shown in the figure. Assume that a sprinkler distributes water in a radial fashion according to the formula

$$f(r) = \frac{r}{16} - \frac{r^2}{160}$$

(measured in cubic feet of water per hour per square foot of lawn), where r is the distance in feet from the sprinkler. Find the amount of water that is distributed in 1 hour in the following two annular regions.

$$A = \{(r, \theta)\colon 4 \le r \le 5,\, 0 \le \theta \le 2\pi\}$$

$$B = \{(r, \theta)\colon 9 \le r \le 10,\, 0 \le \theta \le 2\pi\}$$

Is the distribution of water uniform? Determine the amount of water the entire lawn receives in 1 hour.

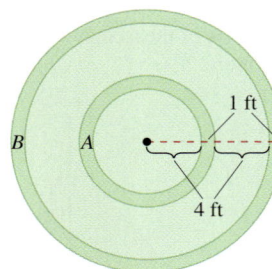

17. Changing the Order of Integration Sketch the solid whose volume is given by the sum of the iterated integrals

$$\int_0^6 \int_{z/2}^3 \int_{z/2}^y dx\, dy\, dz + \int_0^6 \int_3^{(12-z)/2} \int_{z/2}^{6-y} dx\, dy\, dz.$$

Then write the volume as a single iterated integral in the order $dy\, dz\, dx$.

18. Volume The figure shows a solid bounded below by the plane $z = 2$ and above by the sphere $x^2 + y^2 + z^2 = 8$.

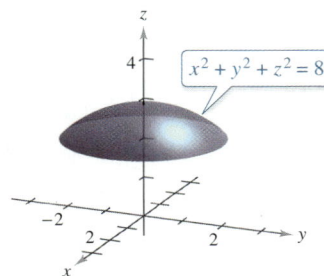

$x^2 + y^2 + z^2 = 8$

(a) Find the volume of the solid using cylindrical coordinates.

(b) Find the volume of the solid using spherical coordinates.

15 Vector Analysis

Work (Exercise 39, p. 1073)

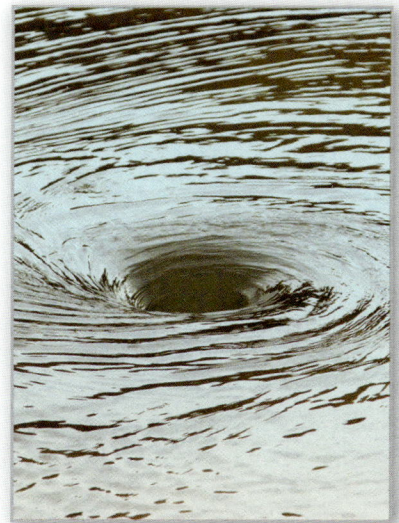
An Application of Curl
(Example 3, p. 1118)

Finding the Mass of a Spring (Example 5, p. 1055)

Building Design
(Exercise 72, p. 1064)

Earth's Magnetic Field (Exercise 83, p. 1050)

15.1 Vector Fields

■ Understand the concept of a vector field.
■ Determine whether a vector field is conservative.
■ Find the curl of a vector field.
■ Find the divergence of a vector field.

Vector Fields

In Chapter 12, you studied vector-valued functions—functions that assign a vector to a *real number.* There you saw that vector-valued functions of real numbers are useful in representing curves and motion along a curve. In this chapter, you will study two other types of vector-valued functions—functions that assign a vector to a *point in the plane* or a *point in space.* Such functions are called **vector fields,** and they are useful in representing various types of **force fields** and **velocity fields.**

> **Definition of Vector Field**
>
> A **vector field over a plane region R** is a function $\mathbf{F}$ that assigns a vector $\mathbf{F}(x, y)$ to each point in R.
>
> A **vector field over a solid region Q in space** is a function $\mathbf{F}$ that assigns a vector $\mathbf{F}(x, y, z)$ to each point in Q.

Although a vector field consists of infinitely many vectors, you can get a good idea of what the vector field looks like by sketching several representative vectors $\mathbf{F}(x, y)$ whose initial points are (x, y).

The *gradient* is one example of a vector field. For instance, if

$$f(x, y) = x^2 y + 3xy^3$$

then the gradient of f

$$\nabla f(x, y) = f_x(x, y)\mathbf{i} + f_y(x, y)\mathbf{j}$$
$$= (2xy + 3y^3)\mathbf{i} + (x^2 + 9xy^2)\mathbf{j} \qquad \text{Vector field in the plane}$$

is a vector field in the plane. From Chapter 13, the graphical interpretation of this field is a family of vectors, each of which points in the direction of maximum increase along the surface given by $z = f(x, y)$.

Similarly, if

$$f(x, y, z) = x^2 + y^2 + z^2$$

then the gradient of f

$$\nabla f(x, y, z) = f_x(x, y, z)\mathbf{i} + f_y(x, y, z)\mathbf{j} + f_z(x, y, z)\mathbf{k}$$
$$= 2x\mathbf{i} + 2y\mathbf{j} + 2z\mathbf{k} \qquad \text{Vector field in space}$$

is a vector field in space. Note that the component functions for this particular vector field are $2x$, $2y$, and $2z$.

A vector field

$$\mathbf{F}(x, y, z) = M(x, y, z)\mathbf{i} + N(x, y, z)\mathbf{j} + P(x, y, z)\mathbf{k}$$

is **continuous** at a point if and only if each of its component functions M, N, and P is continuous at that point.

Velocity field

Rotating wheel
Figure 15.1

Air flow vector field
Figure 15.2

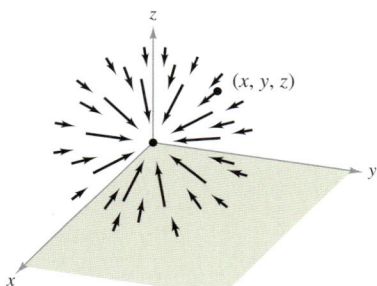

m_1 is located at (x, y, z).
m_2 is located at $(0, 0, 0)$.

Gravitational force field
Figure 15.3

Some common *physical* examples of vector fields are **velocity fields, gravitational fields,** and **electric force fields.**

1. *Velocity fields* describe the motions of systems of particles in the plane or in space. For instance, Figure 15.1 shows the vector field determined by a wheel rotating on an axle. Notice that the velocity vectors are determined by the locations of their initial points—the farther a point is from the axle, the greater its velocity. Velocity fields are also determined by the flow of liquids through a container or by the flow of air currents around a moving object, as shown in Figure 15.2.

2. *Gravitational fields* are defined by **Newton's Law of Gravitation,** which states that the force of attraction exerted on a particle of mass m_1 located at (x, y, z) by a particle of mass m_2 located at $(0, 0, 0)$ is

$$\mathbf{F}(x, y, z) = \frac{-Gm_1m_2}{x^2 + y^2 + z^2}\,\mathbf{u}$$

where G is the gravitational constant and $\mathbf{u}$ is the unit vector in the direction from the origin to (x, y, z). In Figure 15.3, you can see that the gravitational field $\mathbf{F}$ has the properties that $\mathbf{F}(x, y, z)$ always points toward the origin, and that the magnitude of $\mathbf{F}(x, y, z)$ is the same at all points equidistant from the origin. A vector field with these two properties is called a **central force field.** Using the position vector

$$\mathbf{r} = x\mathbf{i} + y\mathbf{j} + z\mathbf{k}$$

for the point (x, y, z), you can write the gravitational field $\mathbf{F}$ as

$$\mathbf{F}(x, y, z) = \frac{-Gm_1m_2}{\|\mathbf{r}\|^2}\left(\frac{\mathbf{r}}{\|\mathbf{r}\|}\right) = \frac{-Gm_1m_2}{\|\mathbf{r}\|^2}\,\mathbf{u}.$$

3. *Electric force fields* are defined by **Coulomb's Law,** which states that the force exerted on a particle with electric charge q_1 located at (x, y, z) by a particle with electric charge q_2 located at $(0, 0, 0)$ is

$$\mathbf{F}(x, y, z) = \frac{cq_1q_2}{\|\mathbf{r}\|^2}\,\mathbf{u}$$

where $\mathbf{r} = x\mathbf{i} + y\mathbf{j} + z\mathbf{k}$, $\mathbf{u} = \mathbf{r}/\|\mathbf{r}\|$, and c is a constant that depends on the choice of units for $\|\mathbf{r}\|$, q_1, and q_2.

Note that an electric force field has the same form as a gravitational field. That is,

$$\mathbf{F}(x, y, z) = \frac{k}{\|\mathbf{r}\|^2}\,\mathbf{u}.$$

Such a force field is called an **inverse square field.**

Definition of Inverse Square Field

Let $\mathbf{r}(t) = x(t)\mathbf{i} + y(t)\mathbf{j} + z(t)\mathbf{k}$ be a position vector. The vector field $\mathbf{F}$ is an **inverse square field** if

$$\mathbf{F}(x, y, z) = \frac{k}{\|\mathbf{r}\|^2}\,\mathbf{u}$$

where k is a real number and

$$\mathbf{u} = \frac{\mathbf{r}}{\|\mathbf{r}\|}$$

is a unit vector in the direction of $\mathbf{r}$.

Because vector fields consist of infinitely many vectors, it is not possible to create a sketch of the entire field. Instead, when you sketch a vector field, your goal is to sketch representative vectors that help you visualize the field.

EXAMPLE 1   Sketching a Vector Field

Sketch some vectors in the vector field

$$\mathbf{F}(x, y) = -y\mathbf{i} + x\mathbf{j}.$$

Solution   You could plot vectors at several random points in the plane. It is more enlightening, however, to plot vectors of equal magnitude. This corresponds to finding level curves in scalar fields. In this case, vectors of equal magnitude lie on circles.

$$\|\mathbf{F}\| = c \qquad \text{Vectors of length } c$$
$$\sqrt{x^2 + y^2} = c$$
$$x^2 + y^2 = c^2 \qquad \text{Equation of circle}$$

To begin making the sketch, choose a value for c and plot several vectors on the resulting circle. For instance, the following vectors occur on the unit circle.

Point	Vector
$(1, 0)$	$\mathbf{F}(1, 0) = \mathbf{j}$
$(0, 1)$	$\mathbf{F}(0, 1) = -\mathbf{i}$
$(-1, 0)$	$\mathbf{F}(-1, 0) = -\mathbf{j}$
$(0, -1)$	$\mathbf{F}(0, -1) = \mathbf{i}$

These and several other vectors in the vector field are shown in Figure 15.4. Note in the figure that this vector field is similar to that given by the rotating wheel shown in Figure 15.1.

Vector field:
$\mathbf{F}(x, y) = -y\mathbf{i} + x\mathbf{j}$

Figure 15.4

EXAMPLE 2   Sketching a Vector Field

Sketch some vectors in the vector field

$$\mathbf{F}(x, y) = 2x\mathbf{i} + y\mathbf{j}.$$

Solution   For this vector field, vectors of equal length lie on ellipses given by

$$\|\mathbf{F}\| = c$$
$$\sqrt{(2x)^2 + (y)^2} = c$$

which implies that

$$4x^2 + y^2 = c^2. \qquad \text{Equation of ellipse}$$

For $c = 1$, sketch several vectors $2x\mathbf{i} + y\mathbf{j}$ of magnitude 1 at points on the ellipse given by

$$4x^2 + y^2 = 1.$$

For $c = 2$, sketch several vectors $2x\mathbf{i} + y\mathbf{j}$ of magnitude 2 at points on the ellipse given by

$$4x^2 + y^2 = 4.$$

These vectors are shown in Figure 15.5.

Vector field:
$\mathbf{F}(x, y) = 2x\mathbf{i} + y\mathbf{j}$

Figure 15.5

▷ **TECHNOLOGY** A computer algebra system can be used to graph vectors in a vector field. If you have access to a computer algebra system, use it to graph several representative vectors for the vector field in Example 2.

EXAMPLE 3   **Sketching a Velocity Field**

Sketch some vectors in the velocity field

$$\mathbf{v}(x, y, z) = (16 - x^2 - y^2)\mathbf{k}$$

where $x^2 + y^2 \leq 16$.

Solution   You can imagine that $\mathbf{v}$ describes the velocity of a liquid flowing through a tube of radius 4. Vectors near the z-axis are longer than those near the edge of the tube. For instance, at the point $(0, 0, 0)$, the velocity vector is $\mathbf{v}(0, 0, 0) = 16\mathbf{k}$, whereas at the point $(0, 3, 0)$, the velocity vector is $\mathbf{v}(0, 3, 0) = 7\mathbf{k}$. Figure 15.6 shows these and several other vectors for the velocity field. From the figure, you can see that the speed of the liquid is greater near the center of the tube than near the edges of the tube.

Velocity field:
$\mathbf{v}(x, y, z) = (16 - x^2 - y^2)\mathbf{k}$

Figure 15.6

Conservative Vector Fields

Notice in Figure 15.5 that all the vectors appear to be normal to the level curve from which they emanate. Because this is a property of gradients, it is natural to ask whether the vector field

$$\mathbf{F}(x, y) = 2x\mathbf{i} + y\mathbf{j}$$

is the *gradient* of some differentiable function f. The answer is that some vector fields can be represented as the gradients of differentiable functions and some cannot—those that can are called **conservative** vector fields.

Definition of Conservative Vector Field

A vector field $\mathbf{F}$ is called **conservative** when there exists a differentiable function f such that $\mathbf{F} = \nabla f$. The function f is called the **potential function** for $\mathbf{F}$.

EXAMPLE 4   **Conservative Vector Fields**

a. The vector field given by $\mathbf{F}(x, y) = 2x\mathbf{i} + y\mathbf{j}$ is conservative. To see this, consider the potential function $f(x, y) = x^2 + \frac{1}{2}y^2$. Because

$$\nabla f = 2x\mathbf{i} + y\mathbf{j} = \mathbf{F}$$

it follows that $\mathbf{F}$ is conservative.

b. Every inverse square field is conservative. To see this, let

$$\mathbf{F}(x, y, z) = \frac{k}{\|\mathbf{r}\|^2}\mathbf{u} \quad \text{and} \quad f(x, y, z) = \frac{-k}{\sqrt{x^2 + y^2 + z^2}}$$

where $\mathbf{u} = \mathbf{r}/\|\mathbf{r}\|$. Because

$$
\begin{aligned}
\nabla f &= \frac{kx}{(x^2 + y^2 + z^2)^{3/2}}\mathbf{i} + \frac{ky}{(x^2 + y^2 + z^2)^{3/2}}\mathbf{j} + \frac{kz}{(x^2 + y^2 + z^2)^{3/2}}\mathbf{k} \\
&= \frac{k}{x^2 + y^2 + z^2}\left(\frac{x\mathbf{i} + y\mathbf{j} + z\mathbf{k}}{\sqrt{x^2 + y^2 + z^2}}\right) \\
&= \frac{k}{\|\mathbf{r}\|^2}\frac{\mathbf{r}}{\|\mathbf{r}\|} \\
&= \frac{k}{\|\mathbf{r}\|^2}\mathbf{u}
\end{aligned}
$$

it follows that $\mathbf{F}$ is conservative.

As can be seen in Example 4(b), many important vector fields, including gravitational fields and electric force fields, are conservative. Most of the terminology in this chapter comes from physics. For example, the term "conservative" is derived from the classic physical law regarding the conservation of energy. This law states that the sum of the kinetic energy and the potential energy of a particle moving in a conservative force field is constant. (The kinetic energy of a particle is the energy due to its motion, and the potential energy is the energy due to its position in the force field.)

The next theorem gives a necessary and sufficient condition for a vector field *in the plane* to be conservative.

▷

• REMARK Theorem 15.1 is valid on simply connected domains. A plane region *R* is simply connected when every simple closed curve in *R* encloses only points that are in *R*. (See Figure 15.26 in Section 15.4.)

THEOREM 15.1 Test for Conservative Vector Field in the Plane

Let *M* and *N* have continuous first partial derivatives on an open disk *R*. The vector field $\mathbf{F}(x, y) = M\mathbf{i} + N\mathbf{j}$ is conservative if and only if

$$\frac{\partial N}{\partial x} = \frac{\partial M}{\partial y}.$$

Proof To prove that the given condition is necessary for **F** to be conservative, suppose there exists a potential function *f* such that

$$\mathbf{F}(x, y) = \nabla f(x, y) = M\mathbf{i} + N\mathbf{j}.$$

Then you have

$$f_x(x, y) = M \implies f_{xy}(x, y) = \frac{\partial M}{\partial y}$$

$$f_y(x, y) = N \implies f_{yx}(x, y) = \frac{\partial N}{\partial x}$$

and, by the equivalence of the mixed partials f_{xy} and f_{yx}, you can conclude that $\partial N/\partial x = \partial M/\partial y$ for all (x, y) in *R*. The sufficiency of this condition is proved in Section 15.4.

See LarsonCalculus.com for Bruce Edwards's video of this proof.

EXAMPLE 5   **Testing for Conservative Vector Fields in the Plane**

Decide whether the vector field given by **F** is conservative.

a. $\mathbf{F}(x, y) = x^2 y\mathbf{i} + xy\mathbf{j}$ **b.** $\mathbf{F}(x, y) = 2x\mathbf{i} + y\mathbf{j}$

Solution

a. The vector field

$$\mathbf{F}(x, y) = x^2 y\mathbf{i} + xy\mathbf{j}$$

is not conservative because

$$\frac{\partial M}{\partial y} = \frac{\partial}{\partial y}[x^2 y] = x^2 \quad \text{and} \quad \frac{\partial N}{\partial x} = \frac{\partial}{\partial x}[xy] = y.$$

b. The vector field

$$\mathbf{F}(x, y) = 2x\mathbf{i} + y\mathbf{j}$$

is conservative because

$$\frac{\partial M}{\partial y} = \frac{\partial}{\partial y}[2x] = 0 \quad \text{and} \quad \frac{\partial N}{\partial x} = \frac{\partial}{\partial x}[y] = 0.$$

Theorem 15.1 tells you whether a vector field is conservative. It does not tell you how to find a potential function of $\mathbf{F}$. The problem is comparable to antidifferentiation. Sometimes you will be able to find a potential function by simple inspection. For instance, in Example 4, you observed that

$$f(x, y) = x^2 + \frac{1}{2}y^2$$

has the property that

$$\nabla f(x, y) = 2x\mathbf{i} + y\mathbf{j}.$$

EXAMPLE 6   **Finding a Potential Function for $F(x, y)$**

Find a potential function for

$$\mathbf{F}(x, y) = 2xy\mathbf{i} + (x^2 - y)\mathbf{j}.$$

Solution   From Theorem 15.1, it follows that $\mathbf{F}$ is conservative because

$$\frac{\partial}{\partial y}[2xy] = 2x \quad \text{and} \quad \frac{\partial}{\partial x}[x^2 - y] = 2x.$$

If f is a function whose gradient is equal to $\mathbf{F}(x, y)$, then

$$\nabla f(x, y) = 2xy\mathbf{i} + (x^2 - y)\mathbf{j}$$

which implies that

$$f_x(x, y) = 2xy$$

and

$$f_y(x, y) = x^2 - y.$$

To reconstruct the function f from these two partial derivatives, integrate $f_x(x, y)$ with respect to x

$$f(x, y) = \int f_x(x, y)\, dx = \int 2xy\, dx = x^2y + g(y)$$

and integrate $f_y(x, y)$ with respect to y

$$f(x, y) = \int f_y(x, y)\, dy = \int (x^2 - y)\, dy = x^2y - \frac{y^2}{2} + h(x).$$

Notice that $g(y)$ is constant with respect to x and $h(x)$ is constant with respect to y. To find a single expression that represents $f(x, y)$, let

$$g(y) = -\frac{y^2}{2} \quad \text{and} \quad h(x) = K.$$

Then, you can write

$$f(x, y) = x^2y + g(y) + K$$
$$= x^2y - \frac{y^2}{2} + K.$$

You can check this result by forming the gradient of f. You will see that it is equal to the original function $\mathbf{F}$.

Notice that the solution to Example 6 is comparable to that given by an indefinite integral. That is, the solution represents a family of potential functions, any two of which differ by a constant. To find a unique solution, you would have to be given an initial condition that is satisfied by the potential function.

Curl of a Vector Field

Theorem 15.1 has a counterpart for vector fields in space. Before stating that result, the definition of the **curl of a vector field** in space is given.

Definition of Curl of a Vector Field

The curl of $\mathbf{F}(x, y, z) = M\mathbf{i} + N\mathbf{j} + P\mathbf{k}$ is

$$\text{curl } \mathbf{F}(x, y, z) = \nabla \times \mathbf{F}(x, y, z)$$

$$= \left(\frac{\partial P}{\partial y} - \frac{\partial N}{\partial z} \right)\mathbf{i} - \left(\frac{\partial P}{\partial x} - \frac{\partial M}{\partial z} \right)\mathbf{j} + \left(\frac{\partial N}{\partial x} - \frac{\partial M}{\partial y} \right)\mathbf{k}.$$

If curl $\mathbf{F} = \mathbf{0}$, then $\mathbf{F}$ is said to be **irrotational.**

The cross product notation used for curl comes from viewing the gradient ∇f as the result of the **differential operator** ∇ acting on the function f. In this context, you can use the following determinant form as an aid in remembering the formula for curl.

$$\text{curl } \mathbf{F}(x, y, z) = \nabla \times \mathbf{F}(x, y, z)$$

$$= \begin{vmatrix} \mathbf{i} & \mathbf{j} & \mathbf{k} \\ \dfrac{\partial}{\partial x} & \dfrac{\partial}{\partial y} & \dfrac{\partial}{\partial z} \\ M & N & P \end{vmatrix}$$

$$= \left(\frac{\partial P}{\partial y} - \frac{\partial N}{\partial z} \right)\mathbf{i} - \left(\frac{\partial P}{\partial x} - \frac{\partial M}{\partial z} \right)\mathbf{j} + \left(\frac{\partial N}{\partial x} - \frac{\partial M}{\partial y} \right)\mathbf{k}$$

EXAMPLE 7 Finding the Curl of a Vector Field

▷ *See LarsonCalculus.com for an interactive version of this type of example.*

Find curl $\mathbf{F}$ of the vector field

$$\mathbf{F}(x, y, z) = 2xy\mathbf{i} + (x^2 + z^2)\mathbf{j} + 2yz\mathbf{k}.$$

Is $\mathbf{F}$ irrotational?

Solution The curl of $\mathbf{F}$ is

$$\text{curl } \mathbf{F}(x, y, z) = \nabla \times \mathbf{F}(x, y, z)$$

$$= \begin{vmatrix} \mathbf{i} & \mathbf{j} & \mathbf{k} \\ \dfrac{\partial}{\partial x} & \dfrac{\partial}{\partial y} & \dfrac{\partial}{\partial z} \\ 2xy & x^2 + z^2 & 2yz \end{vmatrix}$$

$$= \begin{vmatrix} \dfrac{\partial}{\partial y} & \dfrac{\partial}{\partial z} \\ x^2 + z^2 & 2yz \end{vmatrix}\mathbf{i} - \begin{vmatrix} \dfrac{\partial}{\partial x} & \dfrac{\partial}{\partial z} \\ 2xy & 2yz \end{vmatrix}\mathbf{j} + \begin{vmatrix} \dfrac{\partial}{\partial x} & \dfrac{\partial}{\partial y} \\ 2xy & x^2 + z^2 \end{vmatrix}\mathbf{k}$$

$$= (2z - 2z)\mathbf{i} - (0 - 0)\mathbf{j} + (2x - 2x)\mathbf{k}$$

$$= \mathbf{0}.$$

Because curl $\mathbf{F} = \mathbf{0}$, $\mathbf{F}$ is irrotational.

▷ **TECHNOLOGY** Some computer algebra systems have a command that can be used to find the curl of a vector field. If you have access to a computer algebra system that has such a command, use it to find the curl of the vector field in Example 7.

Later in this chapter, you will assign a physical interpretation to the curl of a vector field. But for now, the primary use of curl is shown in the following test for conservative vector fields in space. The test states that for a vector field in space, the curl is **0** at every point in its domain if and only if **F** is conservative. The proof is similar to that given for Theorem 15.1.

⊳

• • REMARK Theorem 15.2 is valid for *simply connected* domains in space. A simply connected domain in space is a domain *D* for which every simple closed curve in *D* can be shrunk to a point in *D* without leaving *D*. (See Section 15.4.)

THEOREM 15.2   Test for Conservative Vector Field in Space

Suppose that *M*, *N*, and *P* have continuous first partial derivatives in an open sphere *Q* in space. The vector field

$$\mathbf{F}(x, y, z) = M\mathbf{i} + N\mathbf{j} + P\mathbf{k}$$

is conservative if and only if

$$\text{curl } \mathbf{F}(x, y, z) = \mathbf{0}.$$

That is, **F** is conservative if and only if

$$\frac{\partial P}{\partial y} = \frac{\partial N}{\partial z}, \quad \frac{\partial P}{\partial x} = \frac{\partial M}{\partial z}, \quad \text{and} \quad \frac{\partial N}{\partial x} = \frac{\partial M}{\partial y}.$$

From Theorem 15.2, you can see that the vector field given in Example 7 is conservative because curl $\mathbf{F}(x, y, z) = \mathbf{0}$. Try showing that the vector field

$$\mathbf{F}(x, y, z) = x^3 y^2 z \mathbf{i} + x^2 z \mathbf{j} + x^2 y \mathbf{k}$$

is not conservative—you can do this by showing that its curl is

$$\text{curl } \mathbf{F}(x, y, z) = (x^3 y^2 - 2xy)\mathbf{j} + (2xz - 2x^3 yz)\mathbf{k} \neq \mathbf{0}.$$

For vector fields in space that pass the test for being conservative, you can find a potential function by following the same pattern used in the plane (as demonstrated in Example 6).

⊳

• • REMARK Examples 6 and 8 are illustrations of a type of problem called *recovering a function from its gradient*. If you go on to take a course in differential equations, you will study other methods for solving this type of problem. One popular method gives an interplay between successive "partial integrations" and partial differentiations.

EXAMPLE 8   **Finding a Potential Function for F(x, y, z)**

Find a potential function for

$$\mathbf{F}(x, y, z) = 2xy\mathbf{i} + (x^2 + z^2)\mathbf{j} + 2yz\mathbf{k}.$$

Solution   From Example 7, you know that the vector field given by **F** is conservative. If *f* is a function such that $\mathbf{F}(x, y, z) = \nabla f(x, y, z)$, then

$$f_x(x, y, z) = 2xy, \quad f_y(x, y, z) = x^2 + z^2, \quad \text{and} \quad f_z(x, y, z) = 2yz$$

and integrating with respect to *x*, *y*, and *z* separately produces

$$f(x, y, z) = \int M \, dx = \int 2xy \, dx = x^2 y + g(y, z)$$

$$f(x, y, z) = \int N \, dy = \int (x^2 + z^2) \, dy = x^2 y + yz^2 + h(x, z)$$

$$f(x, y, z) = \int P \, dz = \int 2yz \, dz = yz^2 + k(x, y).$$

Comparing these three versions of $f(x, y, z)$, you can conclude that

$$g(y, z) = yz^2 + K, \quad h(x, z) = K, \quad \text{and} \quad k(x, y) = x^2 y + K.$$

So, $f(x, y, z)$ is given by

$$f(x, y, z) = x^2 y + yz^2 + K. \qquad \blacksquare$$

Divergence of a Vector Field

You have seen that the curl of a vector field **F** is itself a vector field. Another important function defined on a vector field is **divergence,** which is a scalar function.

> ### Definition of Divergence of a Vector Field
>
> The **divergence** of $\mathbf{F}(x, y) = M\mathbf{i} + N\mathbf{j}$ is
>
> $$\operatorname{div} \mathbf{F}(x, y) = \nabla \cdot \mathbf{F}(x, y) = \frac{\partial M}{\partial x} + \frac{\partial N}{\partial y}. \qquad \text{Plane}$$
>
> The **divergence** of $\mathbf{F}(x, y, z) = M\mathbf{i} + N\mathbf{j} + P\mathbf{k}$ is
>
> $$\operatorname{div} \mathbf{F}(x, y, z) = \nabla \cdot \mathbf{F}(x, y, z) = \frac{\partial M}{\partial x} + \frac{\partial N}{\partial y} + \frac{\partial P}{\partial z}. \qquad \text{Space}$$
>
> If div $\mathbf{F} = 0$, then $\mathbf{F}$ is said to be **divergence free.**

The dot product notation used for divergence comes from considering ∇ as a **differential operator,** as follows.

$$\nabla \cdot \mathbf{F}(x, y, z) = \left[\left(\frac{\partial}{\partial x} \right)\mathbf{i} + \left(\frac{\partial}{\partial y} \right)\mathbf{j} + \left(\frac{\partial}{\partial z} \right)\mathbf{k} \right] \cdot (M\mathbf{i} + N\mathbf{j} + P\mathbf{k})$$

$$= \frac{\partial M}{\partial x} + \frac{\partial N}{\partial y} + \frac{\partial P}{\partial z}$$

▷ **TECHNOLOGY** Some computer algebra systems have a command that can be used to find the divergence of a vector field. If you have access to a computer algebra system that has such a command, use it to find the divergence of the vector field in Example 9.

EXAMPLE 9 **Finding the Divergence of a Vector Field**

Find the divergence at $(2, 1, -1)$ for the vector field

$$\mathbf{F}(x, y, z) = x^3 y^2 z\mathbf{i} + x^2 z\mathbf{j} + x^2 y\mathbf{k}.$$

Solution The divergence of **F** is

$$\operatorname{div} \mathbf{F}(x, y, z) = \frac{\partial}{\partial x}[x^3 y^2 z] + \frac{\partial}{\partial y}[x^2 z] + \frac{\partial}{\partial z}[x^2 y] = 3x^2 y^2 z.$$

At the point $(2, 1, -1)$, the divergence is

$$\operatorname{div} \mathbf{F}(2, 1, -1) = 3(2^2)(1^2)(-1) = -12.$$

Divergence can be viewed as a type of derivative of **F** in that, for vector fields representing velocities of moving particles, the divergence measures the rate of particle flow per unit volume at a point. In hydrodynamics (the study of fluid motion), a velocity field that is divergence free is called **incompressible.** In the study of electricity and magnetism, a vector field that is divergence free is called **solenoidal.**

There are many important properties of the divergence and curl of a vector field **F** [see Exercise 77(a)–(g)]. One that is used often is described in Theorem 15.3. You are asked to prove this theorem in Exercise 77(h).

> ### THEOREM 15.3 Divergence and Curl
>
> If $\mathbf{F}(x, y, z) = M\mathbf{i} + N\mathbf{j} + P\mathbf{k}$ is a vector field and M, N, and P have continuous second partial derivatives, then
>
> $$\operatorname{div}(\operatorname{curl} \mathbf{F}) = 0.$$

15.1 Exercises

See **CalcChat.com** for tutorial help and worked-out solutions to odd-numbered exercises.

Matching   In Exercises 1–4, match the vector field with its graph. [The graphs are labeled (a), (b), (c), and (d).]

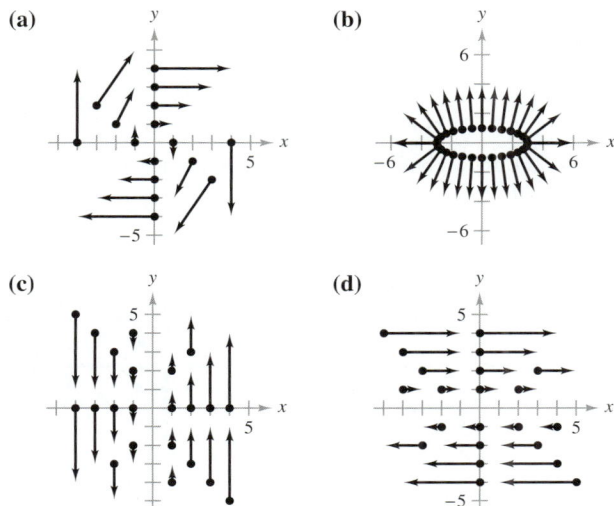

(a)

(b)

(c)

(d)

1. $\mathbf{F}(x, y) = y\mathbf{i}$

2. $\mathbf{F}(x, y) = x\mathbf{j}$

3. $\mathbf{F}(x, y) = y\mathbf{i} - x\mathbf{j}$

4. $\mathbf{F}(x, y) = x\mathbf{i} + 3y\mathbf{j}$

Sketching a Vector Field   In Exercises 5–10, find $\|\mathbf{F}\|$ and sketch several representative vectors in the vector field.

5. $\mathbf{F}(x, y) = \mathbf{i} + \mathbf{j}$

6. $\mathbf{F}(x, y) = y\mathbf{i} - 2x\mathbf{j}$

7. $\mathbf{F}(x, y, z) = 3y\mathbf{j}$

8. $\mathbf{F}(x, y) = y\mathbf{i} + x\mathbf{j}$

9. $\mathbf{F}(x, y, z) = \mathbf{i} + \mathbf{j} + \mathbf{k}$

10. $\mathbf{F}(x, y, z) = x\mathbf{i} + y\mathbf{j} + z\mathbf{k}$

Graphing a Vector Field   In Exercises 11–14, use a computer algebra system to graph several representative vectors in the vector field.

11. $\mathbf{F}(x, y) = \frac{1}{8}(2xy\mathbf{i} + y^2\mathbf{j})$

12. $\mathbf{F}(x, y) = \langle 2y - x, 2y + x \rangle$

13. $\mathbf{F}(x, y, z) = \dfrac{x\mathbf{i} + y\mathbf{j} + z\mathbf{k}}{\sqrt{x^2 + y^2 + z^2}}$

14. $\mathbf{F}(x, y, z) = \langle x, -y, z \rangle$

Finding a Conservative Vector Field   In Exercises 15–24, find the conservative vector field for the potential function by finding its gradient.

15. $f(x, y) = x^2 + 2y^2$

16. $f(x, y) = x^2 - \frac{1}{4}y^2$

17. $g(x, y) = 5x^2 + 3xy + y^2$

18. $g(x, y) = \sin 3x \cos 4y$

19. $f(x, y, z) = 6xyz$

20. $f(x, y, z) = \sqrt{x^2 + 4y^2 + z^2}$

21. $g(x, y, z) = z + ye^{x^2}$

22. $g(x, y, z) = \dfrac{y}{z} + \dfrac{z}{x} - \dfrac{xz}{y}$

23. $h(x, y, z) = xy \ln(x + y)$

24. $h(x, y, z) = x \arcsin yz$

Testing for a Conservative Vector Field   In Exercises 25–32, determine whether the vector field is conservative.

25. $\mathbf{F}(x, y) = xy^2\mathbf{i} + x^2y\mathbf{j}$

26. $\mathbf{F}(x, y) = \dfrac{1}{x^2}(y\mathbf{i} - x\mathbf{j})$

27. $\mathbf{F}(x, y) = \sin y\mathbf{i} + x \cos y\mathbf{j}$

28. $\mathbf{F}(x, y) = 5y^2(y\mathbf{i} + 3x\mathbf{j})$

29. $\mathbf{F}(x, y) = \dfrac{1}{xy}(y\mathbf{i} - x\mathbf{j})$

30. $\mathbf{F}(x, y) = \dfrac{2}{y^2}e^{2x/y}(y\mathbf{i} - x\mathbf{j})$

31. $\mathbf{F}(x, y) = \dfrac{1}{\sqrt{x^2 + y^2}}(\mathbf{i} + \mathbf{j})$

32. $\mathbf{F}(x, y) = \dfrac{1}{\sqrt{1 + xy}}(y\mathbf{i} + x\mathbf{j})$

Finding a Potential Function   In Exercises 33–42, determine whether the vector field is conservative. If it is, find a potential function for the vector field.

33. $\mathbf{F}(x, y) = y\mathbf{i} + x\mathbf{j}$

34. $\mathbf{F}(x, y) = 3x^2y^2\mathbf{i} + 2x^3y\mathbf{j}$

35. $\mathbf{F}(x, y) = 2xy\mathbf{i} + x^2\mathbf{j}$

36. $\mathbf{F}(x, y) = xe^{x^2y}(2y\mathbf{i} + x\mathbf{j})$

37. $\mathbf{F}(x, y) = 15y^3\mathbf{i} - 5xy^2\mathbf{j}$

38. $\mathbf{F}(x, y) = \dfrac{1}{y^2}(y\mathbf{i} - 2x\mathbf{j})$

39. $\mathbf{F}(x, y) = \dfrac{2y}{x}\mathbf{i} - \dfrac{x^2}{y^2}\mathbf{j}$

40. $\mathbf{F}(x, y) = \dfrac{x\mathbf{i} + y\mathbf{j}}{x^2 + y^2}$

41. $\mathbf{F}(x, y) = e^x(\cos y\mathbf{i} - \sin y\mathbf{j})$

42. $\mathbf{F}(x, y) = \dfrac{2x\mathbf{i} + 2y\mathbf{j}}{(x^2 + y^2)^2}$

Finding the Curl of a Vector Field   In Exercises 43–46, find curl $\mathbf{F}$ for the vector field at the given point.

43. $\mathbf{F}(x, y, z) = xyz\mathbf{i} + xyz\mathbf{j} + xyz\mathbf{k}$; (2, 1, 3)

44. $\mathbf{F}(x, y, z) = x^2z\mathbf{i} - 2xz\mathbf{j} + yz\mathbf{k}$; (2, -1, 3)

45. $\mathbf{F}(x, y, z) = e^x \sin y\mathbf{i} - e^x \cos y\mathbf{j}$; (0, 0, 1)

46. $\mathbf{F}(x, y, z) = e^{-xyz}(\mathbf{i} + \mathbf{j} + \mathbf{k})$; (3, 2, 0)

Finding the Curl of a Vector Field   In Exercises 47–50, use a computer algebra system to find the curl $\mathbf{F}$ for the vector field.

47. $\mathbf{F}(x, y, z) = \arctan\left(\dfrac{x}{y}\right)\mathbf{i} + \ln\sqrt{x^2 + y^2}\mathbf{j} + \mathbf{k}$

48. $\mathbf{F}(x, y, z) = \dfrac{yz}{y - z}\mathbf{i} + \dfrac{xz}{x - z}\mathbf{j} + \dfrac{xy}{x - y}\mathbf{k}$

49. $\mathbf{F}(x, y, z) = \sin(x - y)\mathbf{i} + \sin(y - z)\mathbf{j} + \sin(z - x)\mathbf{k}$

50. $\mathbf{F}(x, y, z) = \sqrt{x^2 + y^2 + z^2}\,(\mathbf{i} + \mathbf{j} + \mathbf{k})$

Finding a Potential Function   In Exercises 51–56, determine whether the vector field $\mathbf{F}$ is conservative. If it is, find a potential function for the vector field.

51. $\mathbf{F}(x, y, z) = xy^2z^2\mathbf{i} + x^2yz^2\mathbf{j} + x^2y^2z\mathbf{k}$

52. $\mathbf{F}(x, y, z) = y^2z^3\mathbf{i} + 2xyz^3\mathbf{j} + 3xy^2z^2\mathbf{k}$

53. $\mathbf{F}(x, y, z) = \sin z\mathbf{i} + \sin x\mathbf{j} + \sin y\mathbf{k}$

54. $\mathbf{F}(x, y, z) = ye^z\mathbf{i} + ze^x\mathbf{j} + xe^y\mathbf{k}$

55. $\mathbf{F}(x, y, z) = \dfrac{z}{y}\mathbf{i} - \dfrac{xz}{y^2}\mathbf{j} + \dfrac{x}{y}\mathbf{k}$

56. $\mathbf{F}(x, y, z) = \dfrac{x}{x^2 + y^2}\mathbf{i} + \dfrac{y}{x^2 + y^2}\mathbf{j} + \mathbf{k}$

Finding the Divergence of a Vector Field In Exercises 57–60, find the divergence of the vector field F.

57. $\mathbf{F}(x, y) = x^2\mathbf{i} + 2y^2\mathbf{j}$

58. $\mathbf{F}(x, y) = xe^x\mathbf{i} + ye^y\mathbf{j}$

59. $\mathbf{F}(x, y, z) = \sin x\mathbf{i} + \cos y\mathbf{j} + z^2\mathbf{k}$

60. $\mathbf{F}(x, y, z) = \ln(x^2 + y^2)\mathbf{i} + xy\mathbf{j} + \ln(y^2 + z^2)\mathbf{k}$

Finding the Divergence of a Vector Field In Exercises 61–64, find the divergence of the vector field F at the given point.

61. $\mathbf{F}(x, y, z) = xyz\mathbf{i} + xy\mathbf{j} + z\mathbf{k}$; (2, 1, 1)

62. $\mathbf{F}(x, y, z) = x^2z\mathbf{i} - 2xz\mathbf{j} + yz\mathbf{k}$; (2, −1, 3)

63. $\mathbf{F}(x, y, z) = e^x \sin y\mathbf{i} - e^x \cos y\mathbf{j} + z^2\mathbf{k}$; (3, 0, 0)

64. $\mathbf{F}(x, y, z) = \ln(xyz)(\mathbf{i} + \mathbf{j} + \mathbf{k})$; (3, 2, 1)

WRITING ABOUT CONCEPTS

65. **Vector Field** Define a vector field in the plane and in space. Give some physical examples of vector fields.

66. **Conservative Vector Field** What is a conservative vector field? How do you test for it in the plane and in space?

67. **Curl** Define the curl of a vector field.

68. **Divergence** Define the divergence of a vector field in the plane and in space.

Curl of a Cross Product In Exercises 69 and 70, find curl $(\mathbf{F} \times \mathbf{G}) = \nabla \times (\mathbf{F} \times \mathbf{G})$.

69. $\mathbf{F}(x, y, z) = \mathbf{i} + 3x\mathbf{j} + 2y\mathbf{k}$ 70. $\mathbf{F}(x, y, z) = x\mathbf{i} - z\mathbf{k}$
    $\mathbf{G}(x, y, z) = x\mathbf{i} - y\mathbf{j} + z\mathbf{k}$ $\mathbf{G}(x, y, z) = x^2\mathbf{i} + y\mathbf{j} + z^2\mathbf{k}$

Curl of the Curl of a Vector Field In Exercises 71 and 72, find curl(curl F) $= \nabla \times (\nabla \times \mathbf{F})$.

71. $\mathbf{F}(x, y, z) = xyz\mathbf{i} + y\mathbf{j} + z\mathbf{k}$

72. $\mathbf{F}(x, y, z) = x^2z\mathbf{i} - 2xz\mathbf{j} + yz\mathbf{k}$

Divergence of a Cross Product In Exercises 73 and 74, find div$(\mathbf{F} \times \mathbf{G}) = \nabla \cdot (\mathbf{F} \times \mathbf{G})$.

73. $\mathbf{F}(x, y, z) = \mathbf{i} + 3x\mathbf{j} + 2y\mathbf{k}$ 74. $\mathbf{F}(x, y, z) = x\mathbf{i} - z\mathbf{k}$
    $\mathbf{G}(x, y, z) = x\mathbf{i} - y\mathbf{j} + z\mathbf{k}$ $\mathbf{G}(x, y, z) = x^2\mathbf{i} + y\mathbf{j} + z^2\mathbf{k}$

Divergence of the Curl of a Vector Field In Exercises 75 and 76, find div(curl F) $= \nabla \cdot (\nabla \times \mathbf{F})$.

75. $\mathbf{F}(x, y, z) = xyz\mathbf{i} + y\mathbf{j} + z\mathbf{k}$

76. $\mathbf{F}(x, y, z) = x^2z\mathbf{i} - 2xz\mathbf{j} + yz\mathbf{k}$

77. **Proof** In parts (a)–(h), prove the property for vector fields F and G and scalar function f. (Assume that the required partial derivatives are continuous.)

(a) curl$(\mathbf{F} + \mathbf{G})$ = curl F + curl G

(b) curl$(\nabla f) = \nabla \times (\nabla f) = \mathbf{0}$

(c) div$(\mathbf{F} + \mathbf{G})$ = div F + div G

(d) div$(\mathbf{F} \times \mathbf{G})$ = (curl F) $\cdot$ G − F $\cdot$ (curl G)

(e) $\nabla \times [\nabla f + (\nabla \times \mathbf{F})] = \nabla \times (\nabla \times \mathbf{F})$

(f) $\nabla \times (f\mathbf{F}) = f(\nabla \times \mathbf{F}) + (\nabla f) \times \mathbf{F}$

(g) div$(f\mathbf{F}) = f$ div F + $\nabla f \cdot \mathbf{F}$

(h) div(curl F) = 0   (Theorem 15.3)

78. **HOW DO YOU SEE IT?** Several representative vectors in the vector fields

$$\mathbf{F}(x, y) = \frac{x\mathbf{i} + y\mathbf{j}}{\sqrt{x^2 + y^2}} \quad \text{and} \quad \mathbf{G}(x, y) = \frac{x\mathbf{i} - y\mathbf{j}}{\sqrt{x^2 + y^2}}$$

are shown below. Explain any similarities or differences in the vector fields.

Vector field:
$\mathbf{F}(x, y) = \dfrac{x\mathbf{i} + y\mathbf{j}}{\sqrt{x^2 + y^2}}$

Vector field:
$\mathbf{G}(x, y) = \dfrac{x\mathbf{i} - y\mathbf{j}}{\sqrt{x^2 + y^2}}$

True or False? In Exercises 79–82, determine whether the statement is true or false. If it is false, explain why or give an example that shows it is false.

79. If $\mathbf{F}(x, y) = 4x\mathbf{i} - y^2\mathbf{j}$, then $\|\mathbf{F}(x, y)\| \to 0$ as $(x, y) \to (0, 0)$.

80. If $\mathbf{F}(x, y) = 4x\mathbf{i} - y^2\mathbf{j}$ and (x, y) is on the positive y-axis, then the vector points in the negative y-direction.

81. If f is a scalar field, then curl f is a meaningful expression.

82. If F is a vector field and curl F $= \mathbf{0}$, then F is irrotational but not conservative.

83. **Earth's Magnetic Field**

A cross section of Earth's magnetic field can be represented as a vector field in which the center of Earth is located at the origin and the positive y-axis points in the direction of the magnetic north pole. The equation for this field is

$$\mathbf{F}(x, y) = M(x, y)\mathbf{i} + N(x, y)\mathbf{j}$$
$$= \frac{m}{(x^2 + y^2)^{5/2}}[3xy\mathbf{i} + (2y^2 - x^2)\mathbf{j}]$$

where m is the magnetic moment of Earth. Show that this vector field is conservative.

15.2 Line Integrals

■ Understand and use the concept of a piecewise smooth curve.
■ Write and evaluate a line integral.
■ Write and evaluate a line integral of a vector field.
■ Write and evaluate a line integral in differential form.

Piecewise Smooth Curves

A classic property of gravitational fields is that, subject to certain physical constraints, the work done by gravity on an object moving between two points in the field is independent of the path taken by the object. One of the constraints is that the **path** must be a piecewise smooth curve. Recall that a plane curve C given by

$$\mathbf{r}(t) = x(t)\mathbf{i} + y(t)\mathbf{j}, \quad a \le t \le b$$

is **smooth** when

$$\frac{dx}{dt} \quad \text{and} \quad \frac{dy}{dt}$$

are continuous on $[a, b]$ and not simultaneously 0 on (a, b). Similarly, a space curve C given by

$$\mathbf{r}(t) = x(t)\mathbf{i} + y(t)\mathbf{j} + z(t)\mathbf{k}, \quad a \le t \le b$$

is **smooth** when

$$\frac{dx}{dt}, \quad \frac{dy}{dt}, \quad \text{and} \quad \frac{dz}{dt}$$

are continuous on $[a, b]$ and not simultaneously 0 on (a, b). A curve C is **piecewise smooth** when the interval $[a, b]$ can be partitioned into a finite number of subintervals, on each of which C is smooth.

JOSIAH WILLARD GIBBS (1839–1903)

Many physicists and mathematicians have contributed to the theory and applications described in this chapter—Newton, Gauss, Laplace, Hamilton, and Maxwell, among others. However, the use of vector analysis to describe these results is attributed primarily to the American mathematical physicist Josiah Willard Gibbs.
See LarsonCalculus.com to read more of this biography.

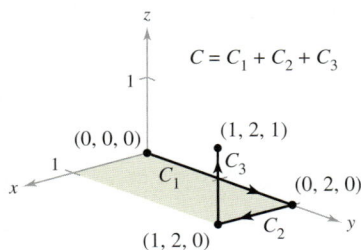

EXAMPLE 1   **Finding a Piecewise Smooth Parametrization**

Find a piecewise smooth parametrization of the graph of C shown in Figure 15.7.

Solution   Because C consists of three line segments C_1, C_2, and C_3, you can construct a smooth parametrization for each segment and piece them together by making the last t-value in C_i correspond to the first t-value in C_{i+1}.

C_1: $x(t) = 0$, $y(t) = 2t$, $z(t) = 0$, $0 \le t \le 1$
C_2: $x(t) = t - 1$, $y(t) = 2$, $z(t) = 0$, $1 \le t \le 2$
C_3: $x(t) = 1$, $y(t) = 2$, $z(t) = t - 2$, $2 \le t \le 3$

So, C is given by

$$\mathbf{r}(t) = \begin{cases} 2t\mathbf{j}, & 0 \le t \le 1 \\ (t-1)\mathbf{i} + 2\mathbf{j}, & 1 \le t \le 2 \\ \mathbf{i} + 2\mathbf{j} + (t-2)\mathbf{k}, & 2 \le t \le 3 \end{cases}.$$

Because C_1, C_2, and C_3 are smooth, it follows that C is piecewise smooth.

Figure 15.7

Recall that parametrization of a curve induces an **orientation** to the curve. For instance, in Example 1, the curve is oriented such that the positive direction is from $(0, 0, 0)$, following the curve to $(1, 2, 1)$. Try finding a parametrization that induces the opposite orientation.

Line Integrals

Up to this point in the text, you have studied various types of integrals. For a single integral

$$\int_a^b f(x)\, dx \qquad\qquad \text{Integrate over interval } [a, b].$$

you integrated over the interval $[a, b]$. Similarly, for a double integral

$$\int_R\!\!\int f(x, y)\, dA \qquad\qquad \text{Integrate over region } R.$$

you integrated over the region R in the plane. In this section, you will study a new type of integral called a **line integral**

$$\int_C f(x, y)\, ds \qquad\qquad \text{Integrate over curve } C.$$

for which you integrate over a piecewise smooth curve C. (The terminology is somewhat unfortunate—this type of integral might be better described as a "curve integral.")

To introduce the concept of a line integral, consider the mass of a wire of finite length, given by a curve C in space. The density (mass per unit length) of the wire at the point (x, y, z) is given by $f(x, y, z)$. Partition the curve C by the points

$$P_0, P_1, \ldots, P_n$$

producing n subarcs, as shown in Figure 15.8. The length of the ith subarc is given by Δs_i. Next, choose a point (x_i, y_i, z_i) in each subarc. If the length of each subarc is small, then the total mass of the wire can be approximated by the sum

$$\text{Mass of wire} \approx \sum_{i=1}^n f(x_i, y_i, z_i)\, \Delta s_i.$$

By letting $\|\Delta\|$ denote the length of the longest subarc and letting $\|\Delta\|$ approach 0, it seems reasonable that the limit of this sum approaches the mass of the wire. This leads to the next definition.

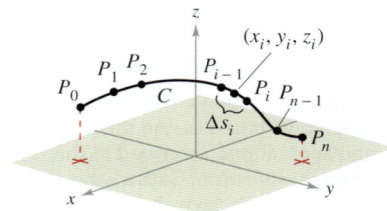

Partitioning of curve C
Figure 15.8

Definition of Line Integral

If f is defined in a region containing a smooth curve C of finite length, then the **line integral of f along C** is given by

$$\int_C f(x, y)\, ds = \lim_{\|\Delta\| \to 0} \sum_{i=1}^n f(x_i, y_i)\, \Delta s_i \qquad \text{Plane}$$

or

$$\int_C f(x, y, z)\, ds = \lim_{\|\Delta\| \to 0} \sum_{i=1}^n f(x_i, y_i, z_i)\, \Delta s_i \qquad \text{Space}$$

provided this limit exists.

As with the integrals discussed in Chapter 14, evaluation of a line integral is best accomplished by converting it to a definite integral. It can be shown that if f is *continuous*, then the limit given above exists and is the same for all smooth parametrizations of C.

To evaluate a line integral over a plane curve C given by $\mathbf{r}(t) = x(t)\mathbf{i} + y(t)\mathbf{j}$, use the fact that

$$ds = \|\mathbf{r}'(t)\| \, dt = \sqrt{[x'(t)]^2 + [y'(t)]^2} \, dt.$$

A similar formula holds for a space curve, as indicated in Theorem 15.4.

THEOREM 15.4   Evaluation of a Line Integral as a Definite Integral

Let f be continuous in a region containing a smooth curve C. If C is given by $\mathbf{r}(t) = x(t)\mathbf{i} + y(t)\mathbf{j}$, where $a \le t \le b$, then

$$\int_C f(x, y) \, ds = \int_a^b f(x(t), y(t)) \sqrt{[x'(t)]^2 + [y'(t)]^2} \, dt.$$

If C is given by $\mathbf{r}(t) = x(t)\mathbf{i} + y(t)\mathbf{j} + z(t)\mathbf{k}$, where $a \le t \le b$, then

$$\int_C f(x, y, z) \, ds = \int_a^b f(x(t), y(t), z(t)) \sqrt{[x'(t)]^2 + [y'(t)]^2 + [z'(t)]^2} \, dt.$$

Note that if $f(x, y, z) = 1$, then the line integral gives the arc length of the curve C, as defined in Section 12.5. That is,

$$\int_C 1 \, ds = \int_a^b \|\mathbf{r}'(t)\| \, dt = \text{length of curve } C.$$

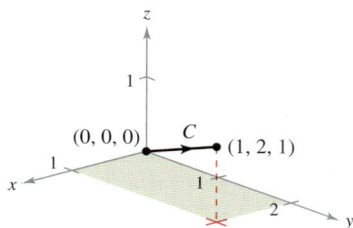

Figure 15.9

EXAMPLE 2   Evaluating a Line Integral

Evaluate

$$\int_C (x^2 - y + 3z) \, ds$$

where C is the line segment shown in Figure 15.9.

Solution   Begin by writing a parametric form of the equation of the line segment:

$$x = t, \quad y = 2t, \quad \text{and} \quad z = t, \quad 0 \le t \le 1.$$

Therefore, $x'(t) = 1$, $y'(t) = 2$, and $z'(t) = 1$, which implies that

$$\sqrt{[x'(t)]^2 + [y'(t)]^2 + [z'(t)]^2} = \sqrt{1^2 + 2^2 + 1^2} = \sqrt{6}.$$

So, the line integral takes the following form.

$$\begin{aligned}
\int_C (x^2 - y + 3z) \, ds &= \int_0^1 (t^2 - 2t + 3t) \sqrt{6} \, dt \\
&= \sqrt{6} \int_0^1 (t^2 + t) \, dt \\
&= \sqrt{6} \left[\frac{t^3}{3} + \frac{t^2}{2} \right]_0^1 \\
&= \frac{5\sqrt{6}}{6}
\end{aligned}$$

The value of the line integral in Example 2 does not depend on the parametrization of the line segment C; any smooth parametrization will produce the same value. To convince yourself of this, try some other parametrizations, such as $x = 1 + 2t$, $y = 2 + 4t$, and $z = 1 + 2t$, $-\frac{1}{2} \le t \le 0$, or $x = -t$, $y = -2t$, and $z = -t$, $-1 \le t \le 0$.

Let C be a path composed of smooth curves $C_1, C_2, \ldots, C_n$. If f is continuous on C, then it can be shown that

$$\int_C f(x, y) \, ds = \int_{C_1} f(x, y) \, ds + \int_{C_2} f(x, y) \, ds + \cdots + \int_{C_n} f(x, y) \, ds.$$

This property is used in Example 3.

EXAMPLE 3   **Evaluating a Line Integral Over a Path**

Evaluate

$$\int_C x \, ds$$

where C is the piecewise smooth curve shown in Figure 15.10.

Solution   Begin by integrating up the line $y = x$, using the following parametrization.

$$C_1 \colon x = t, \; y = t, \quad 0 \le t \le 1$$

For this curve, $\mathbf{r}(t) = t\mathbf{i} + t\mathbf{j}$, which implies that $x'(t) = 1$ and $y'(t) = 1$. So,

$$\sqrt{[x'(t)]^2 + [y'(t)]^2} = \sqrt{2}$$

and you have

$$\int_{C_1} x \, ds = \int_0^1 t\sqrt{2} \, dt = \frac{\sqrt{2}}{2} t^2 \Big]_0^1 = \frac{\sqrt{2}}{2}.$$

Next, integrate down the parabola $y = x^2$, using the parametrization

$$C_2 \colon x = 1 - t, \quad y = (1 - t)^2, \quad 0 \le t \le 1.$$

For this curve,

$$\mathbf{r}(t) = (1 - t)\mathbf{i} + (1 - t)^2\mathbf{j}$$

which implies that $x'(t) = -1$ and $y'(t) = -2(1 - t)$. So,

$$\sqrt{[x'(t)]^2 + [y'(t)]^2} = \sqrt{1 + 4(1 - t)^2}$$

and you have

$$\int_{C_2} x \, ds = \int_0^1 (1 - t)\sqrt{1 + 4(1 - t)^2} \, dt$$

$$= -\frac{1}{8}\left[\frac{2}{3}[1 + 4(1 - t)^2]^{3/2}\right]_0^1$$

$$= \frac{1}{12}(5^{3/2} - 1).$$

Consequently,

$$\int_C x \, ds = \int_{C_1} x \, ds + \int_{C_2} x \, ds = \frac{\sqrt{2}}{2} + \frac{1}{12}(5^{3/2} - 1) \approx 1.56.$$

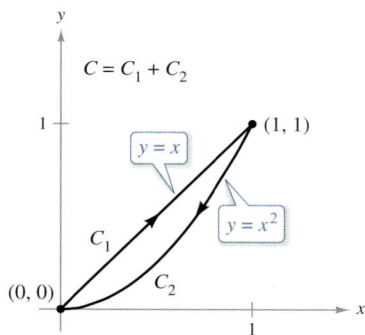

Figure 15.10

For parametrizations given by $\mathbf{r}(t) = x(t)\mathbf{i} + y(t)\mathbf{j} + z(t)\mathbf{k}$, it is helpful to remember the form of ds as

$$ds = \|\mathbf{r}'(t)\| \, dt = \sqrt{[x'(t)]^2 + [y'(t)]^2 + [z'(t)]^2} \, dt.$$

This is demonstrated in Example 4.

EXAMPLE 4 **Evaluating a Line Integral**

Evaluate $\displaystyle\int_C (x + 2)\, ds$, where C is the curve represented by

$$\mathbf{r}(t) = t\mathbf{i} + \frac{4}{3}t^{3/2}\mathbf{j} + \frac{1}{2}t^2\mathbf{k}, \quad 0 \le t \le 2.$$

Solution Because $\mathbf{r}'(t) = \mathbf{i} + 2t^{1/2}\mathbf{j} + t\mathbf{k}$ and

$$\|\mathbf{r}'(t)\| = \sqrt{[x'(t)]^2 + [y'(t)]^2 + [z'(t)]^2} = \sqrt{1 + 4t + t^2}$$

it follows that

$$\begin{aligned}
\int_C (x + 2)\, ds &= \int_0^2 (t + 2)\sqrt{1 + 4t + t^2}\, dt \\
&= \frac{1}{2}\int_0^2 2(t + 2)(1 + 4t + t^2)^{1/2}\, dt \\
&= \frac{1}{3}\Big[(1 + 4t + t^2)^{3/2}\Big]_0^2 \\
&= \frac{1}{3}\big(13\sqrt{13} - 1\big) \\
&\approx 15.29.
\end{aligned}$$

The next example shows how a line integral can be used to find the mass of a spring whose density varies. In Figure 15.11, note that the density of this spring increases as the spring spirals up the z-axis.

EXAMPLE 5 **Finding the Mass of a Spring**

Find the mass of a spring in the shape of the circular helix

$$\mathbf{r}(t) = \frac{1}{\sqrt{2}}(\cos t\mathbf{i} + \sin t\mathbf{j} + t\mathbf{k})$$

where $0 \le t \le 6\pi$ and the density of the spring is

$$\rho(x, y, z) = 1 + z$$

as shown in Figure 15.11.

Solution Because

$$\|\mathbf{r}'(t)\| = \frac{1}{\sqrt{2}}\sqrt{(-\sin t)^2 + (\cos t)^2 + (1)^2} = 1$$

it follows that the mass of the spring is

$$\begin{aligned}
\text{Mass} &= \int_C (1 + z)\, ds \\
&= \int_0^{6\pi}\left(1 + \frac{t}{\sqrt{2}}\right)dt \\
&= \left[t + \frac{t^2}{2\sqrt{2}}\right]_0^{6\pi} \\
&= 6\pi\left(1 + \frac{3\pi}{\sqrt{2}}\right) \\
&\approx 144.47.
\end{aligned}$$

Density:
$\rho(x, y, z) = 1 + z$

$\mathbf{r}(t) = \dfrac{1}{\sqrt{2}}(\cos t\mathbf{i} + \sin t\mathbf{j} + t\mathbf{k})$

Figure 15.11

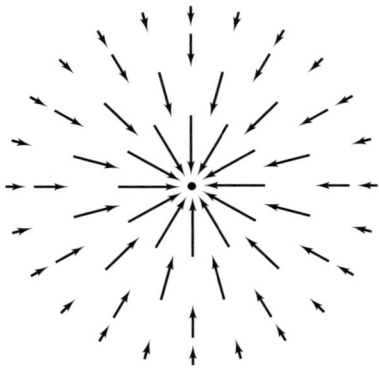

Inverse square force field **F**

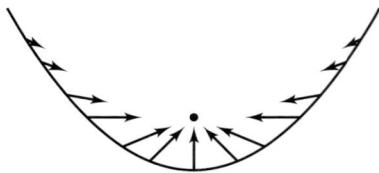

Vectors along a parabolic path in the force field **F**

Figure 15.12

Line Integrals of Vector Fields

One of the most important physical applications of line integrals is that of finding the **work** done on an object moving in a force field. For example, Figure 15.12 shows an inverse square force field similar to the gravitational field of the sun. Note that the magnitude of the force along a circular path about the center is constant, whereas the magnitude of the force along a parabolic path varies from point to point.

To see how a line integral can be used to find work done in a force field **F**, consider an object moving along a path C in the field, as shown in Figure 15.13. To determine the work done by the force, you need consider only that part of the force that is acting in the same direction as that in which the object is moving (or the opposite direction). This means that at each point on C, you can consider the projection $\mathbf{F} \cdot \mathbf{T}$ of the force vector **F** onto the unit tangent vector **T**. On a small subarc of length Δs_i, the increment of work is

$$\Delta W_i = (\text{force})(\text{distance})$$
$$\approx [\mathbf{F}(x_i, y_i, z_i) \cdot \mathbf{T}(x_i, y_i, z_i)] \, \Delta s_i$$

where (x_i, y_i, z_i) is a point in the ith subarc. Consequently, the total work done is given by the integral

$$W = \int_C \mathbf{F}(x, y, z) \cdot \mathbf{T}(x, y, z) \, ds.$$

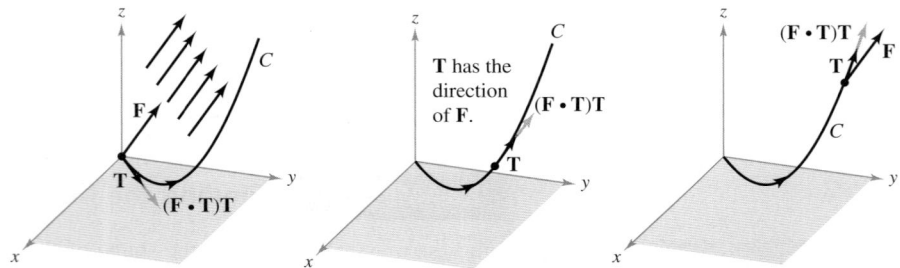

At each point on C, the force in the direction of motion is $(\mathbf{F} \cdot \mathbf{T})\mathbf{T}$.

Figure 15.13

This line integral appears in other contexts and is the basis of the definition of the **line integral of a vector field** shown below. Note in the definition that

$$\mathbf{F} \cdot \mathbf{T} \, ds = \mathbf{F} \cdot \frac{\mathbf{r}'(t)}{\|\mathbf{r}'(t)\|} \|\mathbf{r}'(t)\| \, dt$$
$$= \mathbf{F} \cdot \mathbf{r}'(t) \, dt$$
$$= \mathbf{F} \cdot d\mathbf{r}.$$

Definition of the Line Integral of a Vector Field

Let **F** be a continuous vector field defined on a smooth curve C given by

$$\mathbf{r}(t), \quad a \le t \le b.$$

The **line integral** of **F** on C is given by

$$\int_C \mathbf{F} \cdot d\mathbf{r} = \int_C \mathbf{F} \cdot \mathbf{T} \, ds$$
$$= \int_a^b \mathbf{F}(x(t), y(t), z(t)) \cdot \mathbf{r}'(t) \, dt.$$

EXAMPLE 6   Work Done by a Force

•••▷ *See LarsonCalculus.com for an interactive version of this type of example.*

Find the work done by the force field

$$\mathbf{F}(x, y, z) = -\frac{1}{2}x\mathbf{i} - \frac{1}{2}y\mathbf{j} + \frac{1}{4}\mathbf{k} \qquad \text{Force field } \mathbf{F}$$

on a particle as it moves along the helix given by

$$\mathbf{r}(t) = \cos t\,\mathbf{i} + \sin t\,\mathbf{j} + t\mathbf{k} \qquad \text{Space curve } C$$

from the point $(1, 0, 0)$ to $(-1, 0, 3\pi)$, as shown in Figure 15.14.

Solution   Because

$$\begin{aligned}\mathbf{r}(t) &= x(t)\mathbf{i} + y(t)\mathbf{j} + z(t)\mathbf{k}\\ &= \cos t\,\mathbf{i} + \sin t\,\mathbf{j} + t\mathbf{k}\end{aligned}$$

it follows that

$$x(t) = \cos t, \quad y(t) = \sin t, \quad \text{and} \quad z(t) = t.$$

So, the force field can be written as

$$\mathbf{F}(x(t), y(t), z(t)) = -\frac{1}{2}\cos t\,\mathbf{i} - \frac{1}{2}\sin t\,\mathbf{j} + \frac{1}{4}\mathbf{k}.$$

To find the work done by the force field in moving a particle along the curve C, use the fact that

$$\mathbf{r}'(t) = -\sin t\,\mathbf{i} + \cos t\,\mathbf{j} + \mathbf{k}$$

and write the following.

$$\begin{aligned}W &= \int_C \mathbf{F} \cdot d\mathbf{r}\\ &= \int_a^b \mathbf{F}(x(t), y(t), z(t)) \cdot \mathbf{r}'(t)\, dt\\ &= \int_0^{3\pi} \left(-\frac{1}{2}\cos t\,\mathbf{i} - \frac{1}{2}\sin t\,\mathbf{j} + \frac{1}{4}\mathbf{k}\right) \cdot (-\sin t\,\mathbf{i} + \cos t\,\mathbf{j} + \mathbf{k})\, dt\\ &= \int_0^{3\pi} \left(\frac{1}{2}\sin t\cos t - \frac{1}{2}\sin t\cos t + \frac{1}{4}\right) dt\\ &= \int_0^{3\pi} \frac{1}{4}\, dt\\ &= \frac{1}{4}t\Big]_0^{3\pi}\\ &= \frac{3\pi}{4}\end{aligned}$$

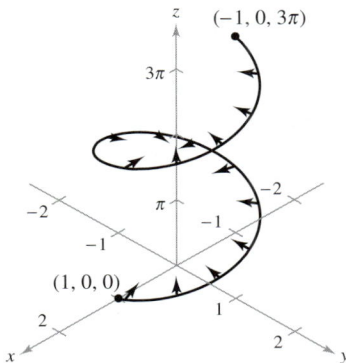

Figure 15.14

In Example 6, note that the x- and y-components of the force field end up contributing nothing to the total work. This occurs because *in this particular example*, the z-component of the force field is the only portion of the force that is acting in the same (or opposite) direction in which the particle is moving (see Figure 15.15).

▷ **TECHNOLOGY**   The computer-generated view of the force field in Example 6 shown in Figure 15.15 indicates that each vector in the force field points toward the z-axis.

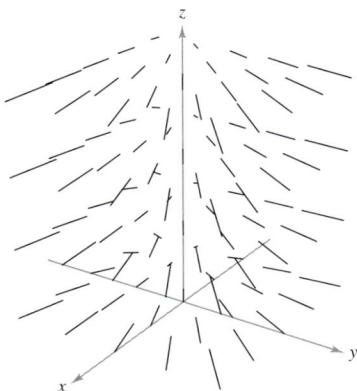

Generated by Mathematica

Figure 15.15

For line integrals of vector functions, the orientation of the curve C is important. If the orientation of the curve is reversed, the unit tangent vector $\mathbf{T}(t)$ is changed to $-\mathbf{T}(t)$, and you obtain

$$\int_{-C} \mathbf{F} \cdot d\mathbf{r} = -\int_{C} \mathbf{F} \cdot d\mathbf{r}.$$

EXAMPLE 7 Orientation and Parametrization of a Curve

C_1: $\mathbf{r}_1(t) = (4 - t)\mathbf{i} + (4t - t^2)\mathbf{j}$
C_2: $\mathbf{r}_2(t) = t\mathbf{i} + (4t - t^2)\mathbf{j}$

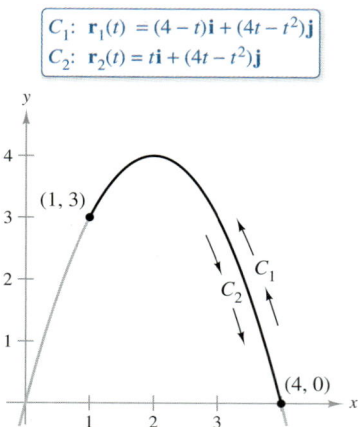

Figure 15.16

Let $\mathbf{F}(x, y) = y\mathbf{i} + x^2\mathbf{j}$ and evaluate the line integral

$$\int_{C} \mathbf{F} \cdot d\mathbf{r}$$

for each parabolic curve shown in Figure 15.16.

a. C_1: $\mathbf{r}_1(t) = (4 - t)\mathbf{i} + (4t - t^2)\mathbf{j}, \quad 0 \le t \le 3$

b. C_2: $\mathbf{r}_2(t) = t\mathbf{i} + (4t - t^2)\mathbf{j}, \quad 1 \le t \le 4$

Solution

a. Because $\mathbf{r}_1'(t) = -\mathbf{i} + (4 - 2t)\mathbf{j}$ and

$$\mathbf{F}(x(t), y(t)) = (4t - t^2)\mathbf{i} + (4 - t)^2\mathbf{j}$$

the line integral is

$$\int_{C_1} \mathbf{F} \cdot d\mathbf{r} = \int_0^3 [(4t - t^2)\mathbf{i} + (4 - t)^2\mathbf{j}] \cdot [-\mathbf{i} + (4 - 2t)\mathbf{j}] \, dt$$

$$= \int_0^3 (-4t + t^2 + 64 - 64t + 20t^2 - 2t^3) \, dt$$

$$= \int_0^3 (-2t^3 + 21t^2 - 68t + 64) \, dt$$

$$= \left[-\frac{t^4}{2} + 7t^3 - 34t^2 + 64t \right]_0^3$$

$$= \frac{69}{2}.$$

b. Because $\mathbf{r}_2'(t) = \mathbf{i} + (4 - 2t)\mathbf{j}$ and

$$\mathbf{F}(x(t), y(t)) = (4t - t^2)\mathbf{i} + t^2\mathbf{j}$$

the line integral is

$$\int_{C_2} \mathbf{F} \cdot d\mathbf{r} = \int_1^4 [(4t - t^2)\mathbf{i} + t^2\mathbf{j}] \cdot [\mathbf{i} + (4 - 2t)\mathbf{j}] \, dt$$

$$= \int_1^4 (4t - t^2 + 4t^2 - 2t^3) \, dt$$

$$= \int_1^4 (-2t^3 + 3t^2 + 4t) \, dt$$

$$= \left[-\frac{t^4}{2} + t^3 + 2t^2 \right]_1^4$$

$$= -\frac{69}{2}.$$

The answer in part (b) is the negative of that in part (a) because C_1 and C_2 represent opposite orientations of the same parabolic segment.

•• **REMARK** Although the value of the line integral in Example 7 depends on the orientation of C, it does not depend on the parametrization of C. To see this, let C_3 be represented by

$$\mathbf{r}_3 = (t + 2)\mathbf{i} + (4 - t^2)\mathbf{j}$$

where $-1 \le t \le 2$. The graph of this curve is the same parabolic segment shown in Figure 15.16. Does the value of the line integral over C_3 agree with the value over C_1 or C_2? Why or why not?

Line Integrals in Differential Form

A second commonly used form of line integrals is derived from the vector field notation used in Section 15.1. If $\mathbf{F}$ is a vector field of the form $\mathbf{F}(x, y) = M\mathbf{i} + N\mathbf{j}$, and C is given by $\mathbf{r}(t) = x(t)\mathbf{i} + y(t)\mathbf{j}$, then $\mathbf{F} \cdot d\mathbf{r}$ is often written as $M\,dx + N\,dy$.

$$
\begin{aligned}
\int_C \mathbf{F} \cdot d\mathbf{r} &= \int_C \mathbf{F} \cdot \frac{d\mathbf{r}}{dt}\,dt \\
&= \int_a^b (M\mathbf{i} + N\mathbf{j}) \cdot (x'(t)\mathbf{i} + y'(t)\mathbf{j})\,dt \\
&= \int_a^b \left(M\frac{dx}{dt} + N\frac{dy}{dt} \right) dt \\
&= \int_C (M\,dx + N\,dy)
\end{aligned}
$$

This **differential form** can be extended to three variables.

▷

• • REMARK The parentheses are often omitted from this differential form, as shown below.

$$\int_C M\,dx + N\,dy$$

In three variables, the differential form is

$$\int_C M\,dx + N\,dy + P\,dz.$$

EXAMPLE 8   **Evaluating a Line Integral in Differential Form**

Let C be the circle of radius 3 given by

$$\mathbf{r}(t) = 3\cos t\,\mathbf{i} + 3\sin t\,\mathbf{j}, \quad 0 \le t \le 2\pi$$

as shown in Figure 15.17. Evaluate the line integral

$$\int_C y^3\,dx + (x^3 + 3xy^2)\,dy.$$

Solution   Because $x = 3\cos t$ and $y = 3\sin t$, you have $dx = -3\sin t\,dt$ and $dy = 3\cos t\,dt$. So, the line integral is

$$
\begin{aligned}
&\int_C M\,dx + N\,dy \\
&= \int_C y^3\,dx + (x^3 + 3xy^2)\,dy \\
&= \int_0^{2\pi} [(27\sin^3 t)(-3\sin t) + (27\cos^3 t + 81\cos t\sin^2 t)(3\cos t)]\,dt \\
&= 81\int_0^{2\pi} (\cos^4 t - \sin^4 t + 3\cos^2 t\sin^2 t)\,dt \\
&= 81\int_0^{2\pi} \left(\cos^2 t - \sin^2 t + \frac{3}{4}\sin^2 2t \right) dt \\
&= 81\int_0^{2\pi} \left[\cos 2t + \frac{3}{4}\left(\frac{1 - \cos 4t}{2} \right) \right] dt \\
&= 81\left[\frac{\sin 2t}{2} + \frac{3}{8}t - \frac{3\sin 4t}{32} \right]_0^{2\pi} \\
&= \frac{243\pi}{4}.
\end{aligned}
$$

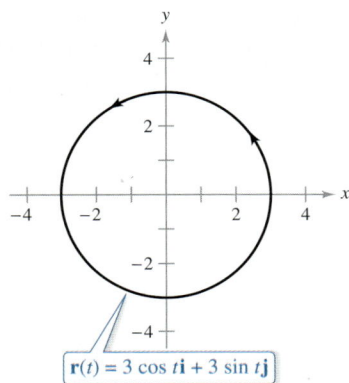

$\mathbf{r}(t) = 3\cos t\,\mathbf{i} + 3\sin t\,\mathbf{j}$

Figure 15.17

The orientation of C affects the value of the differential form of a line integral. Specifically, if $-C$ has the orientation opposite to that of C, then

$$\int_{-C} M\,dx + N\,dy = -\int_C M\,dx + N\,dy.$$

So, of the three line integral forms presented in this section, the orientation of C does not affect the form $\int_C f(x, y)\,ds$, but it does affect the vector form and the differential form.

For curves represented by $y = g(x)$, $a \le x \le b$, you can let $x = t$ and obtain the parametric form

$$x = t \quad \text{and} \quad y = g(t), \quad a \le t \le b.$$

Because $dx = dt$ for this form, you have the option of evaluating the line integral in the variable x or the variable t. This is demonstrated in Example 9.

EXAMPLE 9 **Evaluating a Line Integral in Differential Form**

Evaluate

$$\int_C y \, dx + x^2 \, dy$$

where C is the parabolic arc given by $y = 4x - x^2$ from $(4, 0)$ to $(1, 3)$, as shown in Figure 15.18.

Solution Rather than converting to the parameter t, you can simply retain the variable x and write

$$y = 4x - x^2 \quad \Longrightarrow \quad dy = (4 - 2x) \, dx.$$

Then, in the direction from $(4, 0)$ to $(1, 3)$, the line integral is

$$\int_C y \, dx + x^2 \, dy = \int_4^1 \left[(4x - x^2) \, dx + x^2(4 - 2x) \, dx \right]$$

$$= \int_4^1 (4x + 3x^2 - 2x^3) \, dx$$

$$= \left[2x^2 + x^3 - \frac{x^4}{2} \right]_4^1$$

$$= \frac{69}{2}. \qquad \text{See Example 7.}$$

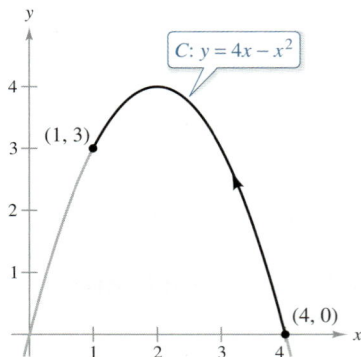
$C: y = 4x - x^2$

Figure 15.18

Exploration

Finding Lateral Surface Area The figure below shows a piece of tin that has been cut from a circular cylinder. The base of the circular cylinder is modeled by $x^2 + y^2 = 9$. At any point (x, y) on the base, the height of the object is

$$f(x, y) = 1 + \cos \frac{\pi x}{4}.$$

Explain how to use a line integral to find the surface area of the piece of tin.

$1 + \cos \dfrac{\pi x}{4}$
$x^2 + y^2 = 9$

15.2 Exercises

See **CalcChat.com** for tutorial help and worked-out solutions to odd-numbered exercises.

Finding a Piecewise Smooth Parametrization In Exercises 1–6, find a piecewise smooth parametrization of the path C. (There is more than one correct answer.)

1.

2.

3.

4.

5.

6.

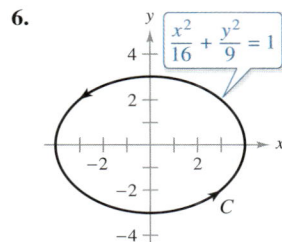

Evaluating a Line Integral In Exercises 7–10, evaluate the line integral along the given path.

7. $\displaystyle\int_C xy \, ds$

$C: \mathbf{r}(t) = 4t\mathbf{i} + 3t\mathbf{j}$

$0 \le t \le 1$

8. $\displaystyle\int_C 3(x - y) \, ds$

$C: \mathbf{r}(t) = t\mathbf{i} + (2 - t)\mathbf{j}$

$0 \le t \le 2$

9. $\displaystyle\int_C (x^2 + y^2 + z^2) \, ds$

$C: \mathbf{r}(t) = \sin t\mathbf{i} + \cos t\mathbf{j} + 2\mathbf{k}$

$0 \le t \le \pi/2$

10. $\displaystyle\int_C 2xyz \, ds$

$C: \mathbf{r}(t) = 12t\mathbf{i} + 5t\mathbf{j} + 84t\mathbf{k}$

$0 \le t \le 1$

Evaluating a Line Integral In Exercises 11–14, (a) find a parametrization of the path C, and (b) evaluate

$$\int_C (x^2 + y^2) \, ds$$

along C.

11. C: line segment from $(0, 0)$ to $(1, 1)$

12. C: line segment from $(0, 0)$ to $(2, 4)$

13. C: counterclockwise around the circle $x^2 + y^2 = 1$ from $(1, 0)$ to $(0, 1)$

14. C: counterclockwise around the circle $x^2 + y^2 = 4$ from $(2, 0)$ to $(0, 2)$

Evaluating a Line Integral In Exercises 15–18, (a) find a parametrization of the path C, and (b) evaluate

$$\int_C \left(x + 4\sqrt{y}\right) ds$$

along C.

15. C: x-axis from $x = 0$ to $x = 1$

16. C: y-axis from $y = 1$ to $y = 9$

17. C: counterclockwise around the triangle with vertices $(0, 0)$, $(1, 0)$, and $(0, 1)$

18. C: counterclockwise around the square with vertices $(0, 0)$, $(2, 0)$, $(2, 2)$, and $(0, 2)$

Finding a Parametrization and Evaluating a Line Integral In Exercises 19 and 20, (a) find a piecewise smooth parametrization of the path C shown in the figure, and (b) evaluate

$$\int_C (2x + y^2 - z) \, ds$$

along C.

19.

20.

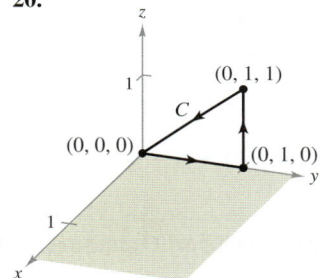

Mass In Exercises 21 and 22, find the total mass of two turns of a spring with density ρ in the shape of the circular helix

$\mathbf{r}(t) = 2\cos t\mathbf{i} + 2\sin t\mathbf{j} + t\mathbf{k}, \ 0 \le t \le 4\pi.$

21. $\rho(x, y, z) = \frac{1}{2}(x^2 + y^2 + z^2)$

22. $\rho(x, y, z) = z$

Mass In Exercises 23–26, find the total mass of the wire with density ρ.

23. $\mathbf{r}(t) = \cos t\mathbf{i} + \sin t\mathbf{j}, \ \rho(x, y) = x + y + 2, \ 0 \le t \le \pi$

24. $\mathbf{r}(t) = t^2\mathbf{i} + 2t\mathbf{j}, \ \rho(x, y) = \frac{3}{4}y, \ 0 \le t \le 1$

25. $\mathbf{r}(t) = t^2\mathbf{i} + 2t\mathbf{j} + t\mathbf{k}, \ \rho(x, y, z) = kz \ (k > 0), \ 1 \le t \le 3$

26. $\mathbf{r}(t) = 2\cos t\mathbf{i} + 2\sin t\mathbf{j} + 3t\mathbf{k}, \ \rho(x, y, z) = k + z$

$(k > 0), \ 0 \le t \le 2\pi$

Evaluating a Line Integral of a Vector Field In Exercises 27–32, evaluate

$$\int_C \mathbf{F} \cdot d\mathbf{r}$$

where C is represented by $\mathbf{r}(t)$.

27. $\mathbf{F}(x, y) = x\mathbf{i} + y\mathbf{j}$

C: $\mathbf{r}(t) = t\mathbf{i} + t\mathbf{j}, \quad 0 \le t \le 1$

28. $\mathbf{F}(x, y) = xy\mathbf{i} + y\mathbf{j}$

C: $\mathbf{r}(t) = 4 \cos t\mathbf{i} + 4 \sin t\mathbf{j}, \quad 0 \le t \le \pi/2$

29. $\mathbf{F}(x, y) = 3x\mathbf{i} + 4y\mathbf{j}$

C: $\mathbf{r}(t) = \cos t\mathbf{i} + \sin t\mathbf{j}, \quad 0 \le t \le \pi/2$

30. $\mathbf{F}(x, y) = 3x\mathbf{i} + 4y\mathbf{j}$

C: $\mathbf{r}(t) = t\mathbf{i} + \sqrt{4 - t^2}\mathbf{j}, \quad -2 \le t \le 2$

31. $\mathbf{F}(x, y, z) = xy\mathbf{i} + xz\mathbf{j} + yz\mathbf{k}$

C: $\mathbf{r}(t) = t\mathbf{i} + t^2\mathbf{j} + 2t\mathbf{k}, \quad 0 \le t \le 1$

32. $\mathbf{F}(x, y, z) = x^2\mathbf{i} + y^2\mathbf{j} + z^2\mathbf{k}$

C: $\mathbf{r}(t) = 2 \sin t\mathbf{i} + 2 \cos t\mathbf{j} + \frac{1}{2}t^2\mathbf{k}, \quad 0 \le t \le \pi$

Evaluating a Line Integral of a Vector Field In Exercises 33 and 34, use a computer algebra system to evaluate the integral

$$\int_C \mathbf{F} \cdot d\mathbf{r}$$

where C is represented by $\mathbf{r}(t)$.

33. $\mathbf{F}(x, y, z) = x^2z\mathbf{i} + 6y\mathbf{j} + yz^2\mathbf{k}$

C: $\mathbf{r}(t) = t\mathbf{i} + t^2\mathbf{j} + \ln t\mathbf{k}, \quad 1 \le t \le 3$

34. $\mathbf{F}(x, y, z) = \dfrac{x\mathbf{i} + y\mathbf{j} + z\mathbf{k}}{\sqrt{x^2 + y^2 + z^2}}$

C: $\mathbf{r}(t) = t\mathbf{i} + t\mathbf{j} + e^t\mathbf{k}, \quad 0 \le t \le 2$

Work In Exercises 35–40, find the work done by the force field $\mathbf{F}$ on a particle moving along the given path.

35. $\mathbf{F}(x, y) = x\mathbf{i} + 2y\mathbf{j}$

C: $x = t, y = t^3$ from $(0, 0)$ to $(2, 8)$

Figure for 35

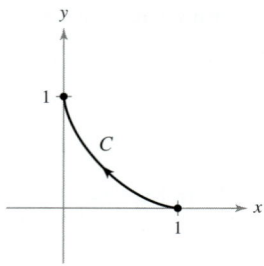
Figure for 36

36. $\mathbf{F}(x, y) = x^2\mathbf{i} - xy\mathbf{j}$

C: $x = \cos^3 t, y = \sin^3 t$ from $(1, 0)$ to $(0, 1)$

37. $\mathbf{F}(x, y) = x\mathbf{i} + y\mathbf{j}$

C: counterclockwise around the triangle with vertices $(0, 0)$, $(1, 0)$, and $(0, 1)$ (*Hint:* See Exercise 17a.)

Figure for 37

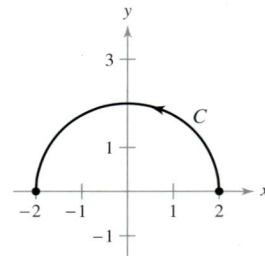
Figure for 38

38. $\mathbf{F}(x, y) = -y\mathbf{i} - x\mathbf{j}$

C: counterclockwise along the semicircle $y = \sqrt{4 - x^2}$ from $(2, 0)$ to $(-2, 0)$

39. $\mathbf{F}(x, y, z) = x\mathbf{i} + y\mathbf{j} - 5z\mathbf{k}$

C: $\mathbf{r}(t) = 2 \cos t\mathbf{i} + 2 \sin t\mathbf{j} + t\mathbf{k}, \quad 0 \le t \le 2\pi$

Figure for 39

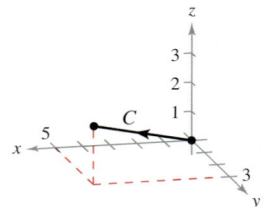
Figure for 40

40. $\mathbf{F}(x, y, z) = yz\mathbf{i} + xz\mathbf{j} + xy\mathbf{k}$

C: line from $(0, 0, 0)$ to $(5, 3, 2)$

Work In Exercises 41–44, determine whether the work done along the path C is positive, negative, or zero. Explain.

41.

42.

43.

44.

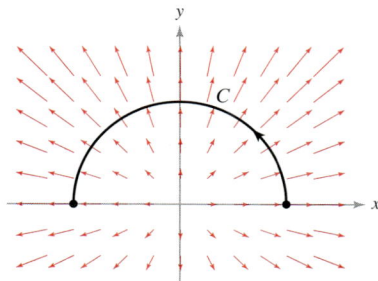

Evaluating a Line Integral of a Vector Field In Exercises 45 and 46, evaluate $\int_C \mathbf{F} \cdot d\mathbf{r}$ for each curve. Discuss the orientation of the curve and its effect on the value of the integral.

45. $\mathbf{F}(x, y) = x^2\mathbf{i} + xy\mathbf{j}$

  (a) $\mathbf{r}_1(t) = 2t\mathbf{i} + (t - 1)\mathbf{j}, \quad 1 \le t \le 3$

  (b) $\mathbf{r}_2(t) = 2(3 - t)\mathbf{i} + (2 - t)\mathbf{j}, \quad 0 \le t \le 2$

46. $\mathbf{F}(x, y) = x^2y\mathbf{i} + xy^{3/2}\mathbf{j}$

  (a) $\mathbf{r}_1(t) = (t + 1)\mathbf{i} + t^2\mathbf{j}, \quad 0 \le t \le 2$

  (b) $\mathbf{r}_2(t) = (1 + 2\cos t)\mathbf{i} + (4\cos^2 t)\mathbf{j}, \quad 0 \le t \le \pi/2$

Demonstrate a Property In Exercises 47–50, demonstrate the property that

$$\int_C \mathbf{F} \cdot d\mathbf{r} = 0$$

regardless of the initial and terminal points of C, where the tangent vector $\mathbf{r}'(t)$ is orthogonal to the force field $\mathbf{F}$.

47. $\mathbf{F}(x, y) = y\mathbf{i} - x\mathbf{j}$ **48.** $\mathbf{F}(x, y) = -3y\mathbf{i} + x\mathbf{j}$

  $C: \mathbf{r}(t) = t\mathbf{i} - 2t\mathbf{j}$ $C: \mathbf{r}(t) = t\mathbf{i} - t^3\mathbf{j}$

49. $\mathbf{F}(x, y) = (x^3 - 2x^2)\mathbf{i} + \left(x - \dfrac{y}{2}\right)\mathbf{j}$

  $C: \mathbf{r}(t) = t\mathbf{i} + t^2\mathbf{j}$

50. $\mathbf{F}(x, y) = x\mathbf{i} + y\mathbf{j}$

  $C: \mathbf{r}(t) = 3\sin t\mathbf{i} + 3\cos t\mathbf{j}$

Evaluating a Line Integral in Differential Form In Exercises 51–54, evaluate the line integral along the path C given by $x = 2t, y = 10t$, where $0 \le t \le 1$.

51. $\displaystyle\int_C (x + 3y^2)\, dy$ **52.** $\displaystyle\int_C (x + 3y^2)\, dx$

53. $\displaystyle\int_C xy\, dx + y\, dy$ **54.** $\displaystyle\int_C (3y - x)\, dx + y^2\, dy$

Evaluating a Line Integral in Differential Form In Exercises 55–62, evaluate the integral

$$\int_C (2x - y)\, dx + (x + 3y)\, dy$$

along the path C.

55. C: x-axis from $x = 0$ to $x = 5$

56. C: y-axis from $y = 0$ to $y = 2$

57. C: line segments from $(0, 0)$ to $(3, 0)$ and $(3, 0)$ to $(3, 3)$

58. C: line segments from $(0, 0)$ to $(0, -3)$ and $(0, -3)$ to $(2, -3)$

59. C: arc on $y = 1 - x^2$ from $(0, 1)$ to $(1, 0)$

60. C: arc on $y = x^{3/2}$ from $(0, 0)$ to $(4, 8)$

61. C: parabolic path $x = t, y = 2t^2$ from $(0, 0)$ to $(2, 8)$

62. C: elliptic path $x = 4\sin t, y = 3\cos t$ from $(0, 3)$ to $(4, 0)$

Lateral Surface Area In Exercises 63–70, find the area of the lateral surface (see figure) over the curve C in the xy-plane and under the surface $z = f(x, y)$, where

$$\text{Lateral surface area} = \int_C f(x, y)\, ds.$$

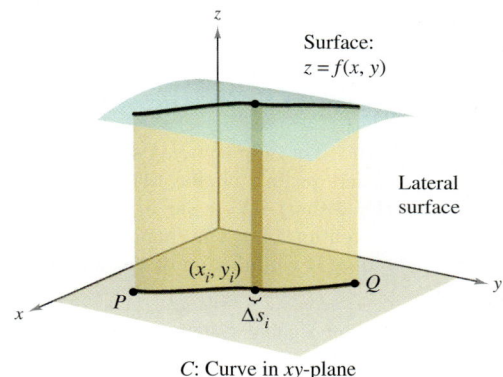

C: Curve in xy-plane

63. $f(x, y) = h$, C: line from $(0, 0)$ to $(3, 4)$

64. $f(x, y) = y$, C: line from $(0, 0)$ to $(4, 4)$

65. $f(x, y) = xy$, C: $x^2 + y^2 = 1$ from $(1, 0)$ to $(0, 1)$

66. $f(x, y) = x + y$, C: $x^2 + y^2 = 1$ from $(1, 0)$ to $(0, 1)$

67. $f(x, y) = h$, C: $y = 1 - x^2$ from $(1, 0)$ to $(0, 1)$

68. $f(x, y) = y + 1$, C: $y = 1 - x^2$ from $(1, 0)$ to $(0, 1)$

69. $f(x, y) = xy$, C: $y = 1 - x^2$ from $(1, 0)$ to $(0, 1)$

70. $f(x, y) = x^2 - y^2 + 4$, C: $x^2 + y^2 = 4$

71. Engine Design A tractor engine has a steel component with a circular base modeled by the vector-valued function $\mathbf{r}(t) = 2\cos t\mathbf{i} + 2\sin t\mathbf{j}$. Its height is given by $z = 1 + y^2$. (All measurements of the component are in centimeters.)

  (a) Find the lateral surface area of the component.

  (b) The component is in the form of a shell of thickness 0.2 centimeter. Use the result of part (a) to approximate the amount of steel used in its manufacture.

  (c) Draw a sketch of the component.

72. Building Design

The ceiling of a building has a height above the floor given by $z = 20 + \frac{1}{4}x$. One of the walls follows a path modeled by $y = x^{3/2}$. Find the surface area of the wall for $0 \le x \le 40$. (All measurements are in feet.)

Moments of Inertia Consider a wire of density $\rho(x, y)$ given by the space curve

$$C: \mathbf{r}(t) = x(t)\mathbf{i} + y(t)\mathbf{j}, \quad 0 \le t \le b.$$

The moments of inertia about the x- and y-axes are given by

$$I_x = \int_C y^2 \rho(x, y)\, ds \quad \text{and} \quad I_y = \int_C x^2 \rho(x, y)\, ds.$$

In Exercises 73 and 74, find the moments of inertia for the wire of density ρ.

73. A wire lies along $\mathbf{r}(t) = a \cos t\mathbf{i} + a \sin t\mathbf{j}$, where $0 \le t \le 2\pi$ and $a > 0$, with density $\rho(x, y) = 1$.

74. A wire lies along $\mathbf{r}(t) = a \cos t\mathbf{i} + a \sin t\mathbf{j}$, where $0 \le t \le 2\pi$ and $a > 0$, with density $\rho(x, y) = y$.

75. Investigation The top outer edge of a solid with vertical sides and resting on the xy-plane is modeled by $\mathbf{r}(t) = 3 \cos t\mathbf{i} + 3 \sin t\mathbf{j} + (1 + \sin^2 2t)\mathbf{k}$, where all measurements are in centimeters. The intersection of the plane $y = b\,(-3 < b < 3)$ with the top of the solid is a horizontal line.

(a) Use a computer algebra system to graph the solid.

(b) Use a computer algebra system to approximate the lateral surface area of the solid.

(c) Find (if possible) the volume of the solid.

76. Work A particle moves along the path $y = x^2$ from the point $(0, 0)$ to the point $(1, 1)$. The force field $\mathbf{F}$ is measured at five points along the path, and the results are shown in the table. Use Simpson's Rule or a graphing utility to approximate the work done by the force field.

(x, y)	$(0, 0)$	$\left(\frac{1}{4}, \frac{1}{16}\right)$	$\left(\frac{1}{2}, \frac{1}{4}\right)$	$\left(\frac{3}{4}, \frac{9}{16}\right)$	$(1, 1)$
$\mathbf{F}(x, y)$	$\langle 5, 0 \rangle$	$\langle 3.5, 1 \rangle$	$\langle 2, 2 \rangle$	$\langle 1.5, 3 \rangle$	$\langle 1, 5 \rangle$

77. Work Find the work done by a person weighing 175 pounds walking exactly one revolution up a circular helical staircase of radius 3 feet when the person rises 10 feet.

78. Investigation Determine the value of c such that the work done by the force field $\mathbf{F}(x, y) = 15[(4 - x^2 y)\mathbf{i} - xy\mathbf{j}]$ on an object moving along the parabolic path $y = c(1 - x^2)$ between the points $(-1, 0)$ and $(1, 0)$ is a minimum. Compare the result with the work required to move the object along the straight-line path connecting the points.

WRITING ABOUT CONCEPTS

79. Line Integral Define a line integral of a function f along a smooth curve C in the plane and in space. How do you evaluate the line integral as a definite integral?

80. Line Integral of a Vector Field Define a line integral of a continuous vector field $\mathbf{F}$ on a smooth curve C. How do you evaluate the line integral as a definite integral?

81. Ordering Surfaces Order the surfaces in ascending order of the lateral surface area under the surface and over the curve $y = \sqrt{x}$ from $(0, 0)$ to $(4, 2)$ in the xy-plane. Explain your ordering without doing any calculations.

(a) $z_1 = 2 + x$ (b) $z_2 = 5 + x$

(c) $z_3 = 2$ (d) $z_4 = 10 + x + 2y$

82. HOW DO YOU SEE IT? For each of the following, determine whether the work done in moving an object from the first to the second point through the force field shown in the figure is positive, negative, or zero. Explain your answer.

(a) From $(-3, -3)$ to $(3, 3)$

(b) From $(-3, 0)$ to $(0, 3)$

(c) From $(5, 0)$ to $(0, 3)$

True or False? In Exercises 83–86, determine whether the statement is true or false. If it is false, explain why or give an example that shows it is false.

83. If C is given by $x(t) = t$, $y(t) = t$, where $0 \le t \le 1$, then

$$\int_C xy\, ds = \int_0^1 t^2\, dt.$$

84. If $C_2 = -C_1$, then $\displaystyle\int_{C_1} f(x, y)\, ds + \int_{C_2} f(x, y)\, ds = 0$.

85. The vector functions $\mathbf{r}_1 = t\mathbf{i} + t^2\mathbf{j}$, where $0 \le t \le 1$, and $\mathbf{r}_2 = (1 - t)\mathbf{i} + (1 - t)^2\mathbf{j}$, where $0 \le t \le 1$, define the same curve.

86. If $\displaystyle\int_C \mathbf{F} \cdot \mathbf{T}\, ds = 0$, then $\mathbf{F}$ and $\mathbf{T}$ are orthogonal.

87. Work Consider a particle that moves through the force field $\mathbf{F}(x, y) = (y - x)\mathbf{i} + xy\mathbf{j}$ from the point $(0, 0)$ to the point $(0, 1)$ along the curve $x = kt(1 - t)$, $y = t$. Find the value of k such that the work done by the force field is 1.

15.3 Conservative Vector Fields and Independence of Path

- Understand and use the Fundamental Theorem of Line Integrals.
- Understand the concept of independence of path.
- Understand the concept of conservation of energy.

Fundamental Theorem of Line Integrals

The discussion at the beginning of Section 15.2 pointed out that in a gravitational field the work done by gravity on an object moving between two points in the field is independent of the path taken by the object. In this section, you will study an important generalization of this result—it is called the **Fundamental Theorem of Line Integrals.** To begin, an example is presented in which the line integral of a *conservative vector field* is evaluated over three different paths.

EXAMPLE 1   **Line Integral of a Conservative Vector Field**

Find the work done by the force field

$$\mathbf{F}(x, y) = \frac{1}{2}xy\mathbf{i} + \frac{1}{4}x^2\mathbf{j}$$

on a particle that moves from $(0, 0)$ to $(1, 1)$ along each path, as shown in Figure 15.19.

a. C_1: $y = x$ **b.** C_2: $x = y^2$ **c.** C_3: $y = x^3$

Solution   Note that $\mathbf{F}$ is conservative because the first partial derivatives are equal.

$$\frac{\partial}{\partial y}\left[\frac{1}{2}xy\right] = \frac{1}{2}x \quad \text{and} \quad \frac{\partial}{\partial x}\left[\frac{1}{4}x^2\right] = \frac{1}{2}x$$

a. Let $\mathbf{r}(t) = t\mathbf{i} + t\mathbf{j}$ for $0 \le t \le 1$, so that

$$d\mathbf{r} = (\mathbf{i} + \mathbf{j})\, dt \quad \text{and} \quad \mathbf{F}(x, y) = \frac{1}{2}t^2\mathbf{i} + \frac{1}{4}t^2\mathbf{j}.$$

Then, the work done is

$$W = \int_{C_1} \mathbf{F} \cdot d\mathbf{r} = \int_0^1 \frac{3}{4}t^2 \, dt = \frac{1}{4}t^3 \Big]_0^1 = \frac{1}{4}.$$

b. Let $\mathbf{r}(t) = t\mathbf{i} + \sqrt{t}\,\mathbf{j}$ for $0 \le t \le 1$, so that

$$d\mathbf{r} = \left(\mathbf{i} + \frac{1}{2\sqrt{t}}\mathbf{j}\right) dt \quad \text{and} \quad \mathbf{F}(x, y) = \frac{1}{2}t^{3/2}\mathbf{i} + \frac{1}{4}t^2\mathbf{j}.$$

Then, the work done is

$$W = \int_{C_2} \mathbf{F} \cdot d\mathbf{r} = \int_0^1 \frac{5}{8}t^{3/2} \, dt = \frac{1}{4}t^{5/2} \Big]_0^1 = \frac{1}{4}.$$

c. Let $\mathbf{r}(t) = \frac{1}{2}t\mathbf{i} + \frac{1}{8}t^3\mathbf{j}$ for $0 \le t \le 2$, so that

$$d\mathbf{r} = \left(\frac{1}{2}\mathbf{i} + \frac{3}{8}t^2\mathbf{j}\right) dt \quad \text{and} \quad \mathbf{F}(x, y) = \frac{1}{32}t^4\mathbf{i} + \frac{1}{16}t^2\mathbf{j}.$$

Then, the work done is

$$W = \int_{C_3} \mathbf{F} \cdot d\mathbf{r} = \int_0^2 \frac{5}{128}t^4 \, dt = \frac{1}{128}t^5 \Big]_0^2 = \frac{1}{4}.$$

So, the work done by the conservative vector field $\mathbf{F}$ is the same for each path.

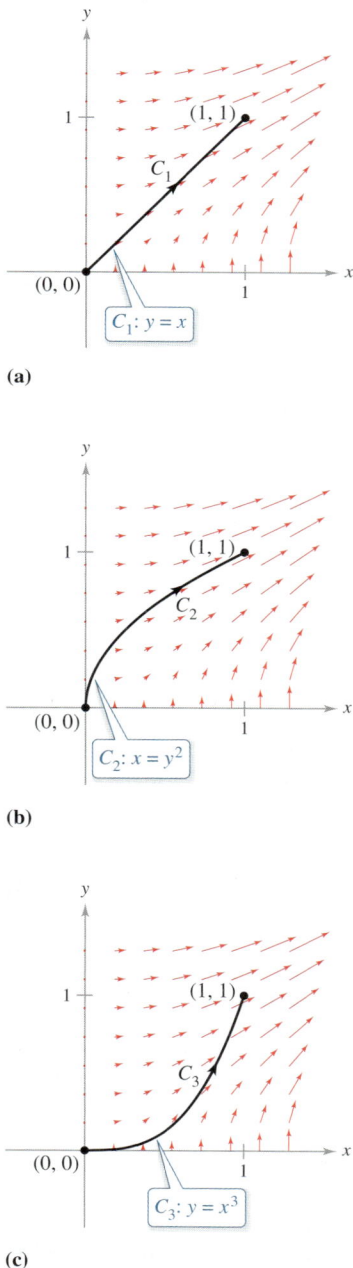

(a)

C_1: $y = x$

(b)

C_2: $x = y^2$

(c)

C_3: $y = x^3$

Figure 15.19

In Example 1, note that the vector field $F(x, y) = \frac{1}{2}xy\mathbf{i} + \frac{1}{4}x^2\mathbf{j}$ is conservative because $F(x, y) = \nabla f(x, y)$, where $f(x, y) = \frac{1}{4}x^2y$. In such cases, the next theorem states that the value of $\int_C F \cdot d\mathbf{r}$ is given by

$$\int_C F \cdot d\mathbf{r} = f(x(1), y(1)) - f(x(0), y(0))$$

$$= \frac{1}{4} - 0$$

$$= \frac{1}{4}.$$

· · · · · · · · · · · · · · · ▷

REMARK Notice how the Fundamental Theorem of Line Integrals is similar to the Fundamental Theorem of Calculus (Section 4.4), which states that

$$\int_a^b f(x)\, dx = F(b) - F(a)$$

where $F'(x) = f(x)$.

> **THEOREM 15.5   Fundamental Theorem of Line Integrals**
>
> Let C be a piecewise smooth curve lying in an open region R and given by
>
> $$\mathbf{r}(t) = x(t)\mathbf{i} + y(t)\mathbf{j}, \quad a \le t \le b.$$
>
> If $F(x, y) = M\mathbf{i} + N\mathbf{j}$ is conservative in R, and M and N are continuous in R, then
>
> $$\int_C F \cdot d\mathbf{r} = \int_C \nabla f \cdot d\mathbf{r} = f(x(b), y(b)) - f(x(a), y(a))$$
>
> where f is a potential function of F. That is, $F(x, y) = \nabla f(x, y)$.

Proof   A proof is provided only for a smooth curve. For piecewise smooth curves, the procedure is carried out separately on each smooth portion. Because

$$F(x, y) = \nabla f(x, y) = f_x(x, y)\mathbf{i} + f_y(x, y)\mathbf{j}$$

it follows that

$$\int_C F \cdot d\mathbf{r} = \int_a^b F \cdot \frac{d\mathbf{r}}{dt}\, dt$$

$$= \int_a^b \left[f_x(x, y) \frac{dx}{dt} + f_y(x, y) \frac{dy}{dt} \right] dt$$

and, by the Chain Rule (Theorem 13.6), you have

$$\int_C F \cdot d\mathbf{r} = \int_a^b \frac{d}{dt}[f(x(t), y(t))]\, dt$$

$$= f(x(b), y(b)) - f(x(a), y(a)).$$

The last step is an application of the Fundamental Theorem of Calculus.

See LarsonCalculus.com for Bruce Edwards's video of this proof.

In space, the Fundamental Theorem of Line Integrals takes the following form. Let C be a piecewise smooth curve lying in an open region Q and given by

$$\mathbf{r}(t) = x(t)\mathbf{i} + y(t)\mathbf{j} + z(t)\mathbf{k}, \quad a \le t \le b.$$

If $F(x, y, z) = M\mathbf{i} + N\mathbf{j} + P\mathbf{k}$ is conservative and M, N, and P are continuous, then

$$\int_C F \cdot d\mathbf{r} = \int_C \nabla f \cdot d\mathbf{r} = f(x(b), y(b), z(b)) - f(x(a), y(a), z(a))$$

where $F(x, y, z) = \nabla f(x, y, z)$.

The Fundamental Theorem of Line Integrals states that if the vector field F is conservative, then the line integral between any two points is simply the difference in the values of the *potential* function f at these points.

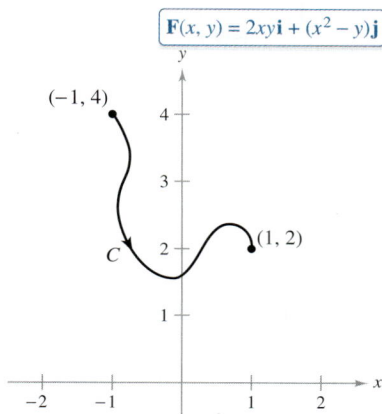

$F(x, y) = 2xy\mathbf{i} + (x^2 - y)\mathbf{j}$

Using the Fundamental Theorem of
Line Integrals, $\int_C \mathbf{F} \cdot d\mathbf{r}$
Figure 15.20

EXAMPLE 2 **Using the Fundamental Theorem of Line Integrals**

Evaluate $\int_C \mathbf{F} \cdot d\mathbf{r}$, where C is a piecewise smooth curve from $(-1, 4)$ to $(1, 2)$ and

$$\mathbf{F}(x, y) = 2xy\mathbf{i} + (x^2 - y)\mathbf{j}$$

as shown in Figure 15.20.

Solution From Example 6 in Section 15.1, you know that $\mathbf{F}$ is the gradient of f, where

$$f(x, y) = x^2y - \frac{y^2}{2} + K.$$

Consequently, $\mathbf{F}$ is conservative, and by the Fundamental Theorem of Line Integrals, it follows that

$$\int_C \mathbf{F} \cdot d\mathbf{r} = f(1, 2) - f(-1, 4)$$
$$= \left[1^2(2) - \frac{2^2}{2}\right] - \left[(-1)^2(4) - \frac{4^2}{2}\right]$$
$$= 4.$$

Note that it is unnecessary to include a constant K as part of f, because it is canceled by subtraction.

EXAMPLE 3 **Using the Fundamental Theorem of Line Integrals**

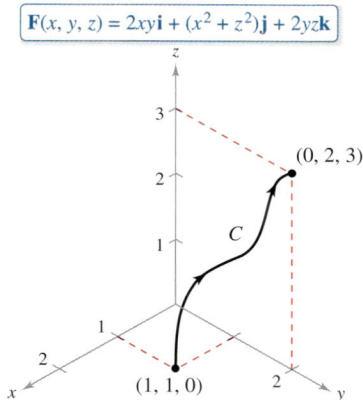

$F(x, y, z) = 2xy\mathbf{i} + (x^2 + z^2)\mathbf{j} + 2yz\mathbf{k}$

Using the Fundamental Theorem of
Line Integrals, $\int_C \mathbf{F} \cdot d\mathbf{r}$
Figure 15.21

Evaluate $\int_C \mathbf{F} \cdot d\mathbf{r}$, where C is a piecewise smooth curve from $(1, 1, 0)$ to $(0, 2, 3)$ and

$$\mathbf{F}(x, y, z) = 2xy\mathbf{i} + (x^2 + z^2)\mathbf{j} + 2yz\mathbf{k}$$

as shown in Figure 15.21.

Solution From Example 8 in Section 15.1, you know that $\mathbf{F}$ is the gradient of f, where

$$f(x, y, z) = x^2y + yz^2 + K.$$

Consequently, $\mathbf{F}$ is conservative, and by the Fundamental Theorem of Line Integrals, it follows that

$$\int_C \mathbf{F} \cdot d\mathbf{r} = f(0, 2, 3) - f(1, 1, 0)$$
$$= [(0)^2(2) + (2)(3)^2] - [(1)^2(1) + (1)(0)^2]$$
$$= 17.$$

In Examples 2 and 3, be sure you see that the value of the line integral is the same for any smooth curve C that has the given initial and terminal points. For instance, in Example 3, try evaluating the line integral for the curve given by

$$\mathbf{r}(t) = (1 - t)\mathbf{i} + (1 + t)\mathbf{j} + 3t\mathbf{k}.$$

You should obtain

$$\int_C \mathbf{F} \cdot d\mathbf{r} = \int_0^1 (30t^2 + 16t - 1)\, dt$$
$$= 17.$$

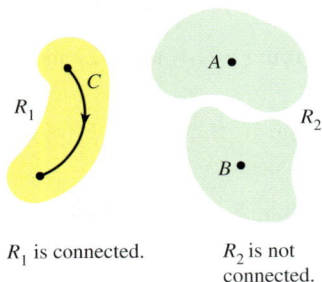

R_1 is connected. R_2 is not connected.

Figure 15.22

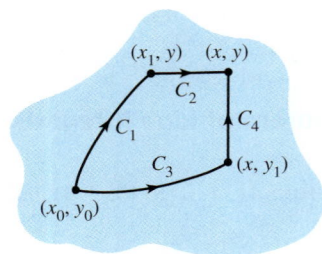

Figure 15.23

Independence of Path

From the Fundamental Theorem of Line Integrals, it is clear that if $\mathbf{F}$ is continuous and conservative in an open region R, then the value of $\int_C \mathbf{F} \cdot d\mathbf{r}$ is the same for every piecewise smooth curve C from one fixed point in R to another fixed point in R. This result is described by saying that the line integral $\int_C \mathbf{F} \cdot d\mathbf{r}$ is **independent of path** in the region R.

A region in the plane (or in space) is **connected** when any two points in the region can be joined by a piecewise smooth curve lying entirely within the region, as shown in Figure 15.22. In open regions that are *connected,* the path independence of $\int_C \mathbf{F} \cdot d\mathbf{r}$ is equivalent to the condition that $\mathbf{F}$ is conservative.

THEOREM 15.6 Independence of Path and Conservative Vector Fields

If $\mathbf{F}$ is continuous on an open connected region, then the line integral

$$\int_C \mathbf{F} \cdot d\mathbf{r}$$

is independent of path if and only if $\mathbf{F}$ is conservative.

Proof If $\mathbf{F}$ is conservative, then, by the Fundamental Theorem of Line Integrals, the line integral is independent of path. Now establish the converse for a plane region R. Let $\mathbf{F}(x, y) = M\mathbf{i} + N\mathbf{j}$, and let (x_0, y_0) be a fixed point in R. For any point (x, y) in R, choose a piecewise smooth curve C running from (x_0, y_0) to (x, y), and define f by

$$f(x, y) = \int_C \mathbf{F} \cdot d\mathbf{r} = \int_C M\,dx + N\,dy.$$

The existence of C in R is guaranteed by the fact that R is connected. You can show that f is a potential function of $\mathbf{F}$ by considering two different paths between (x_0, y_0) and (x, y). For the *first* path, choose (x_1, y) in R such that $x \neq x_1$. This is possible because R is open. Then choose C_1 and C_2, as shown in Figure 15.23. Using the independence of path, it follows that

$$f(x, y) = \int_C M\,dx + N\,dy$$
$$= \int_{C_1} M\,dx + N\,dy + \int_{C_2} M\,dx + N\,dy.$$

Because the first integral does not depend on x, and because $dy = 0$ in the second integral, you have

$$f(x, y) = g(y) + \int_{C_2} M\,dx$$

and it follows that the partial derivative of f with respect to x is $f_x(x, y) = M$. For the *second* path, choose a point (x, y_1). Using reasoning similar to that used for the first path, you can conclude that $f_y(x, y) = N$. Therefore,

$$\nabla f(x, y) = f_x(x, y)\mathbf{i} + f_y(x, y)\mathbf{j}$$
$$= M\mathbf{i} + N\mathbf{j}$$
$$= \mathbf{F}(x, y)$$

and it follows that $\mathbf{F}$ is conservative.

See LarsonCalculus.com for Bruce Edwards's video of this proof.

EXAMPLE 4 **Finding Work in a Conservative Force Field**

For the force field given by

$$F(x, y, z) = e^x \cos y\mathbf{i} - e^x \sin y\mathbf{j} + 2\mathbf{k}$$

show that $\int_C \mathbf{F} \cdot d\mathbf{r}$ is independent of path, and calculate the work done by $\mathbf{F}$ on an object moving along a curve C from $(0, \pi/2, 1)$ to $(1, \pi, 3)$.

Solution Writing the force field in the form $\mathbf{F}(x, y, z) = M\mathbf{i} + N\mathbf{j} + P\mathbf{k}$, you have $M = e^x \cos y$, $N = -e^x \sin y$, and $P = 2$, and it follows that

$$\frac{\partial P}{\partial y} = 0 = \frac{\partial N}{\partial z}$$

$$\frac{\partial P}{\partial x} = 0 = \frac{\partial M}{\partial z}$$

and

$$\frac{\partial N}{\partial x} = -e^x \sin y = \frac{\partial M}{\partial y}.$$

So, $\mathbf{F}$ is conservative. If f is a potential function of $\mathbf{F}$, then

$$f_x(x, y, z) = e^x \cos y$$
$$f_y(x, y, z) = -e^x \sin y$$

and

$$f_z(x, y, z) = 2.$$

By integrating with respect to x, y, and z separately, you obtain

$$f(x, y, z) = \int f_x(x, y, z)\, dx = \int e^x \cos y\, dx = e^x \cos y + g(y, z)$$

$$f(x, y, z) = \int f_y(x, y, z)\, dy = \int -e^x \sin y\, dy = e^x \cos y + h(x, z)$$

and

$$f(x, y, z) = \int f_z(x, y, z)\, dz = \int 2\, dz = 2z + k(x, y).$$

By comparing these three versions of $f(x, y, z)$, you can conclude that

$$f(x, y, z) = e^x \cos y + 2z + K.$$

Therefore, the work done by $\mathbf{F}$ along *any* curve C from $(0, \pi/2, 1)$ to $(1, \pi, 3)$ is

$$W = \int_C \mathbf{F} \cdot d\mathbf{r}$$

$$= \left[e^x \cos y + 2z \right]_{(0, \pi/2, 1)}^{(1, \pi, 3)}$$

$$= (-e + 6) - (0 + 2)$$

$$= 4 - e.$$

For the object in Example 4, how much work is done when the object moves on a curve from $(0, \pi/2, 1)$ to $(1, \pi, 3)$ and then back to the starting point $(0, \pi/2, 1)$? The Fundamental Theorem of Line Integrals states that there is zero work done. Remember that, by definition, work can be negative. So, by the time the object gets back to its starting point, the amount of work that registers positively is canceled out by the amount of work that registers negatively.

A curve C given by $\mathbf{r}(t)$ for $a \le t \le b$ is **closed** when $\mathbf{r}(a) = \mathbf{r}(b)$. By the Fundamental Theorem of Line Integrals, you can conclude that if $\mathbf{F}$ is continuous and conservative on an open region R, then the line integral over every closed curve C is 0.

> **THEOREM 15.7 Equivalent Conditions**
>
> Let $\mathbf{F}(x, y, z) = M\mathbf{i} + N\mathbf{j} + P\mathbf{k}$ have continuous first partial derivatives in an open connected region R, and let C be a piecewise smooth curve in R. The conditions listed below are equivalent.
>
> 1. $\mathbf{F}$ is conservative. That is, $\mathbf{F} = \nabla f$ for some function f.
> 2. $\displaystyle\int_C \mathbf{F} \cdot d\mathbf{r}$ is independent of path.
> 3. $\displaystyle\int_C \mathbf{F} \cdot d\mathbf{r} = 0$ for every *closed* curve C in R.

REMARK Theorem 15.7 gives you options for evaluating a line integral involving a conservative vector field. You can use a potential function, or it might be more convenient to choose a particularly simple path, such as a straight line.

EXAMPLE 5 Evaluating a Line Integral

> See LarsonCalculus.com for an interactive version of this type of example.

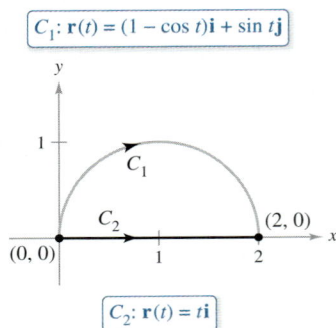

Evaluate $\displaystyle\int_{C_1} \mathbf{F} \cdot d\mathbf{r}$, where

$$\mathbf{F}(x, y) = (y^3 + 1)\mathbf{i} + (3xy^2 + 1)\mathbf{j}$$

and C_1 is the semicircular path from $(0, 0)$ to $(2, 0)$, as shown in Figure 15.24.

Solution You have the following three options.

a. You can use the method presented in Section 15.2 to evaluate the line integral along the *given curve*. To do this, you can use the parametrization $\mathbf{r}(t) = (1 - \cos t)\mathbf{i} + \sin t\mathbf{j}$, where $0 \le t \le \pi$. For this parametrization, it follows that

$$d\mathbf{r} = \mathbf{r}'(t)\, dt = (\sin t\mathbf{i} + \cos t\mathbf{j})\, dt$$

and

$$\int_{C_1} \mathbf{F} \cdot d\mathbf{r} = \int_0^\pi (\sin t + \sin^4 t + \cos t + 3\sin^2 t \cos t - 3\sin^2 t \cos^2 t)\, dt.$$

This integral should dampen your enthusiasm for this option.

b. You can try to find a *potential function* and evaluate the line integral by the Fundamental Theorem of Line Integrals. Using the technique demonstrated in Example 4, you can find the potential function to be $f(x, y) = xy^3 + x + y + K$, and, by the Fundamental Theorem,

$$W = \int_{C_1} \mathbf{F} \cdot d\mathbf{r} = f(2, 0) - f(0, 0) = 2.$$

c. Knowing that $\mathbf{F}$ is conservative, you have a third option. Because the value of the line integral is independent of path, you can replace the semicircular path with a *simpler path*. Choose the straight-line path C_2 from $(0, 0)$ to $(2, 0)$. Let $\mathbf{r}(t) = t\mathbf{i}$ for $0 \le t \le 2$, so that

$$d\mathbf{r} = \mathbf{i}\, dt \quad \text{and} \quad \mathbf{F}(x, y) = \mathbf{i} + \mathbf{j}.$$

Then, the integral is

$$\int_{C_1} \mathbf{F} \cdot d\mathbf{r} = \int_{C_2} \mathbf{F} \cdot d\mathbf{r} = \int_0^2 1\, dt = t\Big]_0^2 = 2.$$

Of the three options, obviously the third one is the easiest.

C_1: $\mathbf{r}(t) = (1 - \cos t)\mathbf{i} + \sin t\mathbf{j}$

C_2: $\mathbf{r}(t) = t\mathbf{i}$

Figure 15.24

Conservation of Energy

In 1840, the English physicist Michael Faraday wrote, "Nowhere is there a pure creation or production of power without a corresponding exhaustion of something to supply it." This statement represents the first formulation of one of the most important laws of physics—the **Law of Conservation of Energy.** In modern terminology, the law is stated as follows: *In a conservative force field, the sum of the potential and kinetic energies of an object remains constant from point to point.*

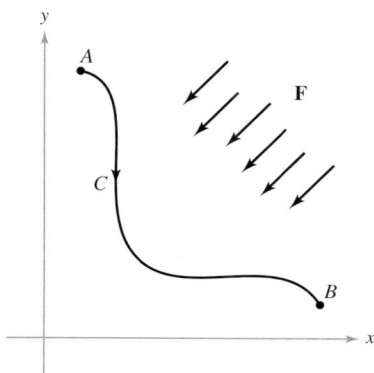

You can use the Fundamental Theorem of Line Integrals to derive this law. From physics, the **kinetic energy** of a particle of mass m and speed v is

$$k = \frac{1}{2}mv^2. \qquad \text{Kinetic energy}$$

The **potential energy** p of a particle at point (x, y, z) in a conservative vector field $\mathbf{F}$ is defined as $p(x, y, z) = -f(x, y, z)$, where f is the potential function for $\mathbf{F}$. Consequently, the work done by $\mathbf{F}$ along a smooth curve C from A to B is

$$W = \int_C \mathbf{F} \cdot d\mathbf{r} = f(x, y, z)\Big]_A^B = -p(x, y, z)\Big]_A^B = p(A) - p(B)$$

as shown in Figure 15.25. In other words, work W is equal to the difference in the potential energies of A and B. Now, suppose that $\mathbf{r}(t)$ is the position vector for a particle moving along C from $A = \mathbf{r}(a)$ to $B = \mathbf{r}(b)$. At any time t, the particle's velocity, acceleration, and speed are $\mathbf{v}(t) = \mathbf{r}'(t)$, $\mathbf{a}(t) = \mathbf{r}''(t)$, and $v(t) = \|\mathbf{v}(t)\|$, respectively. So, by Newton's Second Law of Motion, $\mathbf{F} = m\mathbf{a}(t) = m(\mathbf{v}'(t))$, and the work done by $\mathbf{F}$ is

$$\begin{aligned}
W &= \int_C \mathbf{F} \cdot d\mathbf{r} \\
&= \int_a^b \mathbf{F} \cdot \mathbf{r}'(t)\, dt \\
&= \int_a^b \mathbf{F} \cdot \mathbf{v}(t)\, dt \\
&= \int_a^b [m\mathbf{v}'(t)] \cdot \mathbf{v}(t)\, dt \\
&= \int_a^b m[\mathbf{v}'(t) \cdot \mathbf{v}(t)]\, dt \\
&= \frac{m}{2} \int_a^b \frac{d}{dt}[\mathbf{v}(t) \cdot \mathbf{v}(t)]\, dt \\
&= \frac{m}{2} \int_a^b \frac{d}{dt}[\|\mathbf{v}(t)\|^2]\, dt \\
&= \frac{m}{2}\Big[\|\mathbf{v}(t)\|^2\Big]_a^b \\
&= \frac{m}{2}\Big[[v(t)]^2\Big]_a^b \\
&= \frac{1}{2}m[v(b)]^2 - \frac{1}{2}m[v(a)]^2 \\
&= k(B) - k(A).
\end{aligned}$$

Equating these two results for W produces

$$p(A) - p(B) = k(B) - k(A)$$
$$p(A) + k(A) = p(B) + k(B)$$

which implies that the sum of the potential and kinetic energies remains constant from point to point.

MICHAEL FARADAY (1791–1867)

Several philosophers of science have considered Faraday's Law of Conservation of Energy to be the greatest generalization ever conceived by humankind. Many physicists have contributed to our knowledge of this law. Two early and influential ones were James Prescott Joule (1818–1889) and Hermann Ludwig Helmholtz (1821–1894).

The work done by $\mathbf{F}$ along C is
$$W = \int_C \mathbf{F} \cdot d\mathbf{r} = p(A) - p(B).$$
Figure 15.25

15.3 Exercises

Evaluating a Line Integral for Different Parametrizations In Exercises 1–4, show that the value of $\int_C \mathbf{F} \cdot d\mathbf{r}$ is the same for each parametric representation of C.

1. $\mathbf{F}(x, y) = x^2\mathbf{i} + xy\mathbf{j}$

  (a) $\mathbf{r}_1(t) = t\mathbf{i} + t^2\mathbf{j}, \quad 0 \le t \le 1$

  (b) $\mathbf{r}_2(\theta) = \sin\theta\mathbf{i} + \sin^2\theta\mathbf{j}, \quad 0 \le \theta \le \dfrac{\pi}{2}$

2. $\mathbf{F}(x, y) = (x^2 + y^2)\mathbf{i} - x\mathbf{j}$

  (a) $\mathbf{r}_1(t) = t\mathbf{i} + \sqrt{t}\mathbf{j}, \quad 0 \le t \le 4$

  (b) $\mathbf{r}_2(w) = w^2\mathbf{i} + w\mathbf{j}, \quad 0 \le w \le 2$

3. $\mathbf{F}(x, y) = y\mathbf{i} - x\mathbf{j}$

  (a) $\mathbf{r}_1(\theta) = \sec\theta\mathbf{i} + \tan\theta\mathbf{j}, \quad 0 \le \theta \le \dfrac{\pi}{3}$

  (b) $\mathbf{r}_2(t) = \sqrt{t + 1}\mathbf{i} + \sqrt{t}\mathbf{j}, \quad 0 \le t \le 3$

4. $\mathbf{F}(x, y) = y\mathbf{i} + x^2\mathbf{j}$

  (a) $\mathbf{r}_1(t) = (2 + t)\mathbf{i} + (3 - t)\mathbf{j}, \quad 0 \le t \le 3$

  (b) $\mathbf{r}_2(w) = (2 + \ln w)\mathbf{i} + (3 - \ln w)\mathbf{j}, \quad 1 \le w \le e^3$

Testing for Conservative Vector Fields In Exercises 5–10, determine whether the vector field is conservative.

5. $\mathbf{F}(x, y) = e^x(\sin y\mathbf{i} + \cos y\mathbf{j})$ **6.** $\mathbf{F}(x, y) = 15x^2y^2\mathbf{i} + 10x^3y\mathbf{j}$

7. $\mathbf{F}(x, y) = \dfrac{1}{y^2}(y\mathbf{i} + x\mathbf{j})$

8. $\mathbf{F}(x, y, z) = y\ln z\mathbf{i} - x\ln z\mathbf{j} + \dfrac{xy}{z}\mathbf{k}$

9. $\mathbf{F}(x, y, z) = y^2z\mathbf{i} + 2xyz\mathbf{j} + xy^2\mathbf{k}$

10. $\mathbf{F}(x, y, z) = \sin yz\mathbf{i} + xz\cos yz\mathbf{j} + xy\sin yz\mathbf{k}$

Evaluating a Line Integral of a Vector Field In Exercises 11–24, find the value of the line integral

$$\int_C \mathbf{F} \cdot d\mathbf{r}.$$

(*Hint:* If F is conservative, the integration may be easier on an alternative path.)

11. $\mathbf{F}(x, y) = 2xy\mathbf{i} + x^2\mathbf{j}$

  (a) $\mathbf{r}_1(t) = t\mathbf{i} + t^2\mathbf{j}, \quad 0 \le t \le 1$

  (b) $\mathbf{r}_2(t) = t\mathbf{i} + t^3\mathbf{j}, \quad 0 \le t \le 1$

12. $\mathbf{F}(x, y) = ye^{xy}\mathbf{i} + xe^{xy}\mathbf{j}$

  (a) $\mathbf{r}_1(t) = t\mathbf{i} - (t - 3)\mathbf{j}, \quad 0 \le t \le 3$

  (b) The closed path consisting of line segments from $(0, 3)$ to $(0, 0)$, from $(0, 0)$ to $(3, 0)$, and then from $(3, 0)$ to $(0, 3)$

13. $\mathbf{F}(x, y) = y\mathbf{i} - x\mathbf{j}$

  (a) $\mathbf{r}_1(t) = t\mathbf{i} + t\mathbf{j}, \quad 0 \le t \le 1$

  (b) $\mathbf{r}_2(t) = t\mathbf{i} + t^2\mathbf{j}, \quad 0 \le t \le 1$

  (c) $\mathbf{r}_3(t) = t\mathbf{i} + t^3\mathbf{j}, \quad 0 \le t \le 1$

14. $\mathbf{F}(x, y) = xy^2\mathbf{i} + 2x^2y\mathbf{j}$

  (a) $\mathbf{r}_1(t) = t\mathbf{i} + \dfrac{1}{t}\mathbf{j}, \quad 1 \le t \le 3$

  (b) $\mathbf{r}_2(t) = (t + 1)\mathbf{i} - \tfrac{1}{3}(t - 3)\mathbf{j}, \quad 0 \le t \le 2$

15. $\displaystyle\int_C y^2\,dx + 2xy\,dy$

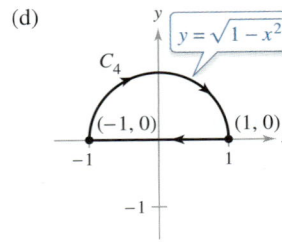

16. $\displaystyle\int_C (2x - 3y + 1)\,dx - (3x + y - 5)\,dy$

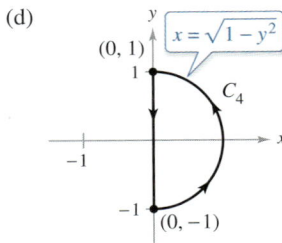

17. $\displaystyle\int_C 2xy\,dx + (x^2 + y^2)\,dy$

  (a) C: ellipse $\dfrac{x^2}{25} + \dfrac{y^2}{16} = 1$ from $(5, 0)$ to $(0, 4)$

  (b) C: parabola $y = 4 - x^2$ from $(2, 0)$ to $(0, 4)$

18. $\int_C (x^2 + y^2)\, dx + 2xy\, dy$

   (a) $\mathbf{r}_1(t) = t^3\mathbf{i} + t^2\mathbf{j}, \quad 0 \le t \le 2$

   (b) $\mathbf{r}_2(t) = 2\cos t\,\mathbf{i} + 2\sin t\,\mathbf{j}, \quad 0 \le t \le \dfrac{\pi}{2}$

19. $\mathbf{F}(x, y, z) = yz\mathbf{i} + xz\mathbf{j} + xy\mathbf{k}$

   (a) $\mathbf{r}_1(t) = t\mathbf{i} + 2\mathbf{j} + t\mathbf{k}, \quad 0 \le t \le 4$

   (b) $\mathbf{r}_2(t) = t^2\mathbf{i} + t\mathbf{j} + t^2\mathbf{k}, \quad 0 \le t \le 2$

20. $\mathbf{F}(x, y, z) = \mathbf{i} + z\mathbf{j} + y\mathbf{k}$

   (a) $\mathbf{r}_1(t) = \cos t\,\mathbf{i} + \sin t\,\mathbf{j} + t^2\mathbf{k}, \quad 0 \le t \le \pi$

   (b) $\mathbf{r}_2(t) = (1 - 2t)\mathbf{i} + \pi^2 t\mathbf{k}, \quad 0 \le t \le 1$

21. $\mathbf{F}(x, y, z) = (2y + x)\mathbf{i} + (x^2 - z)\mathbf{j} + (2y - 4z)\mathbf{k}$

   (a) $\mathbf{r}_1(t) = t\mathbf{i} + t^2\mathbf{j} + \mathbf{k}, \quad 0 \le t \le 1$

   (b) $\mathbf{r}_2(t) = t\mathbf{i} + t\mathbf{j} + (2t - 1)^2\mathbf{k}, \quad 0 \le t \le 1$

22. $\mathbf{F}(x, y, z) = -y\mathbf{i} + x\mathbf{j} + 3xz^2\mathbf{k}$

   (a) $\mathbf{r}_1(t) = \cos t\,\mathbf{i} + \sin t\,\mathbf{j} + t\mathbf{k}, \quad 0 \le t \le \pi$

   (b) $\mathbf{r}_2(t) = (1 - 2t)\mathbf{i} + \pi t\mathbf{k}, \quad 0 \le t \le 1$

23. $\mathbf{F}(x, y, z) = e^z(y\mathbf{i} + x\mathbf{j} + xy\mathbf{k})$

   (a) $\mathbf{r}_1(t) = 4\cos t\,\mathbf{i} + 4\sin t\,\mathbf{j} + 3\mathbf{k}, \quad 0 \le t \le \pi$

   (b) $\mathbf{r}_2(t) = (4 - 8t)\mathbf{i} + 3\mathbf{k}, \quad 0 \le t \le 1$

24. $\mathbf{F}(x, y, z) = y\sin z\mathbf{i} + x\sin z\mathbf{j} + xy\cos x\mathbf{k}$

   (a) $\mathbf{r}_1(t) = t^2\mathbf{i} + t^2\mathbf{j}, \quad 0 \le t \le 2$

   (b) $\mathbf{r}_2(t) = 4t\mathbf{i} + 4t\mathbf{j}, \quad 0 \le t \le 1$

Using the Fundamental Theorem of Line Integrals In Exercises 25–34, evaluate the line integral using the Fundamental Theorem of Line Integrals. Use a computer algebra system to verify your results.

25. $\int_C (3y\mathbf{i} + 3x\mathbf{j}) \cdot d\mathbf{r}$

   C: smooth curve from $(0, 0)$ to $(3, 8)$

26. $\int_C [2(x + y)\mathbf{i} + 2(x + y)\mathbf{j}] \cdot d\mathbf{r}$

   C: smooth curve from $(-1, 1)$ to $(3, 2)$

27. $\int_C \cos x\sin y\, dx + \sin x\cos y\, dy$

   C: line segment from $(0, -\pi)$ to $\left(\dfrac{3\pi}{2}, \dfrac{\pi}{2}\right)$

28. $\int_C \dfrac{y\, dx - x\, dy}{x^2 + y^2}$

   C: line segment from $(1, 1)$ to $\left(2\sqrt{3}, 2\right)$

29. $\int_C e^x\sin y\, dx + e^x\cos y\, dy$

   C: cycloid $x = \theta - \sin\theta, y = 1 - \cos\theta$ from $(0, 0)$ to $(2\pi, 0)$

30. $\int_C \dfrac{2x}{(x^2 + y^2)^2}\, dx + \dfrac{2y}{(x^2 + y^2)^2}\, dy$

   C: circle $(x - 4)^2 + (y - 5)^2 = 9$ clockwise from $(7, 5)$ to $(1, 5)$

31. $\int_C (z + 2y)\, dx + (2x - z)\, dy + (x - y)\, dz$

   (a) C: line segment from $(0, 0, 0)$ to $(1, 1, 1)$

   (b) C: line segments from $(0, 0, 0)$ to $(0, 0, 1)$ to $(1, 1, 1)$

   (c) C: line segments from $(0, 0, 0)$ to $(1, 0, 0)$ to $(1, 1, 0)$ to $(1, 1, 1)$

32. Repeat Exercise 31 using the integral

$$\int_C zy\, dx + xz\, dy + xy\, dz.$$

33. $\int_C -\sin x\, dx + z\, dy + y\, dz$

   C: smooth curve from $(0, 0, 0)$ to $\left(\dfrac{\pi}{2}, 3, 4\right)$

34. $\int_C 6x\, dx - 4z\, dy - (4y - 20z)\, dz$

   C: smooth curve from $(0, 0, 0)$ to $(3, 4, 0)$

Work In Exercises 35 and 36, find the work done by the force field F in moving an object from P to Q.

35. $\mathbf{F}(x, y) = 9x^2y^2\mathbf{i} + (6x^3y - 1)\mathbf{j}; \quad P(0, 0), Q(5, 9)$

36. $\mathbf{F}(x, y) = \dfrac{2x}{y}\mathbf{i} - \dfrac{x^2}{y^2}\mathbf{j}; \quad P(-1, 1), Q(3, 2)$

37. Work A stone weighing 1 pound is attached to the end of a two-foot string and is whirled horizontally with one end held fixed. It makes 1 revolution per second. Find the work done by the force $\mathbf{F}$ that keeps the stone moving in a circular path. [*Hint:* Use Force = (mass)(centripetal acceleration).]

38. Work Let $\mathbf{F}(x, y, z) = a_1\mathbf{i} + a_2\mathbf{j} + a_3\mathbf{k}$ be a constant force vector field. Show that the work done in moving a particle along any path from P to Q is $W = \mathbf{F} \cdot \overrightarrow{PQ}$.

39. Work

A zip line is installed 50 meters above ground level. It runs to a point on the ground 50 meters away from the base of the installation. Show that the work done by the gravitational force field for a 175-pound person moving the length of the zip line is the same for each path.

   (a) $\mathbf{r}(t) = t\mathbf{i} + (50 - t)\mathbf{j}$

   (b) $\mathbf{r}(t) = t\mathbf{i} + \frac{1}{50}(50 - t)^2\mathbf{j}$

40. Work Can you find a path for the zip line in Exercise 39 such that the work done by the gravitational force field would differ from the amounts of work done for the two paths given? Explain why or why not.

WRITING ABOUT CONCEPTS

41. Fundamental Theorem of Line Integrals State the Fundamental Theorem of Line Integrals.

42. Independence of Path What does it mean that a line integral is independent of path? State the method for determining whether a line integral is independent of path.

43. Think About It Let $\mathbf{F}(x, y) = \dfrac{y}{x^2 + y^2}\mathbf{i} - \dfrac{x}{x^2 + y^2}\mathbf{j}$. Find the value of the line integral $\int_C \mathbf{F} \cdot d\mathbf{r}$.

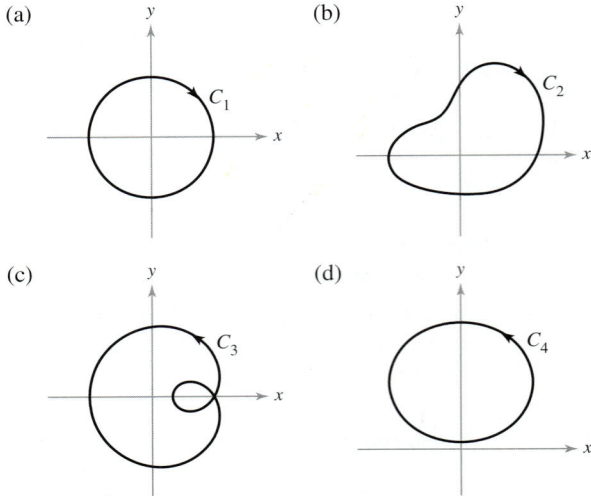

(a)

(b)

(c)

(d)

44. HOW DO YOU SEE IT? Consider the force field shown in the figure. To print an enlarged copy of the graph, go to *MathGraphs.com*.

(a) Give a verbal argument that the force field is not conservative because you can identify two paths that require different amounts of work to move an object from $(-4, 0)$ to $(3, 4)$. Of the two paths, which requires the greater amount of work?

(b) Give a verbal argument that the force field is not conservative because you can find a closed curve C such that $\int_C \mathbf{F} \cdot d\mathbf{r} \neq 0$.

Graphical Reasoning In Exercises 45 and 46, consider the force field shown in the figure. Is the force field conservative? Explain why or why not.

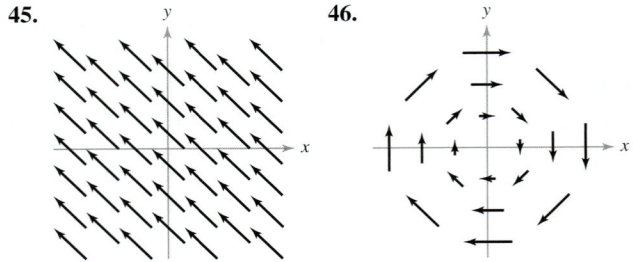

45.

46.

True or False? In Exercises 47–50, determine whether the statement is true or false. If it is false, explain why or give an example that shows it is false.

47. If C_1, C_2, and C_3 have the same initial and terminal points and $\int_{C_1} \mathbf{F} \cdot d\mathbf{r}_1 = \int_{C_2} \mathbf{F} \cdot d\mathbf{r}_2$, then $\int_{C_1} \mathbf{F} \cdot d\mathbf{r}_1 = \int_{C_3} \mathbf{F} \cdot d\mathbf{r}_3$.

48. If $\mathbf{F} = y\mathbf{i} + x\mathbf{j}$ and C is given by $\mathbf{r}(t) = (4 \sin t)\mathbf{i} + (3 \cos t)\mathbf{j}$, for $0 \leq t \leq \pi$, then $\int_C \mathbf{F} \cdot d\mathbf{r} = 0$.

49. If $\mathbf{F}$ is conservative in a region R bounded by a simple closed path and C lies within R, then $\int_C \mathbf{F} \cdot d\mathbf{r}$ is independent of path.

50. If $\mathbf{F} = M\mathbf{i} + N\mathbf{j}$ and $\partial M/\partial x = \partial N/\partial y$, then $\mathbf{F}$ is conservative.

51. Harmonic Function A function f is called *harmonic* when $\dfrac{\partial^2 f}{\partial x^2} + \dfrac{\partial^2 f}{\partial y^2} = 0$. Prove that if f is harmonic, then

$$\int_C \left(\frac{\partial f}{\partial y}\, dx - \frac{\partial f}{\partial x}\, dy \right) = 0$$

where C is a smooth closed curve in the plane.

52. Kinetic and Potential Energy The kinetic energy of an object moving through a conservative force field is decreasing at a rate of 15 units per minute. At what rate is the potential energy changing?

53. Investigation Let $\mathbf{F}(x, y) = \dfrac{y}{x^2 + y^2}\mathbf{i} - \dfrac{x}{x^2 + y^2}\mathbf{j}$.

(a) Show that

$$\frac{\partial N}{\partial x} = \frac{\partial M}{\partial y}$$

where

$$M = \frac{y}{x^2 + y^2} \quad \text{and} \quad N = \frac{-x}{x^2 + y^2}.$$

(b) Let $\mathbf{r}(t) = \cos t\mathbf{i} + \sin t\mathbf{j}$ for $0 \leq t \leq \pi$. Find $\int_C \mathbf{F} \cdot d\mathbf{r}$.

(c) Let $\mathbf{r}(t) = \cos t\mathbf{i} - \sin t\mathbf{j}$ for $0 \leq t \leq \pi$. Find $\int_C \mathbf{F} \cdot d\mathbf{r}$.

(d) Let $\mathbf{r}(t) = \cos t\mathbf{i} + \sin t\mathbf{j}$ for $0 \leq t \leq 2\pi$. Find $\int_C \mathbf{F} \cdot d\mathbf{r}$. Why doesn't this contradict Theorem 15.7?

(e) Show that $\nabla\left(\arctan \dfrac{x}{y} \right) = \mathbf{F}$.

15.4 Green's Theorem

■ Use Green's Theorem to evaluate a line integral.
■ Use alternative forms of Green's Theorem.

Green's Theorem

In this section, you will study **Green's Theorem,** named after the English mathematician George Green (1793–1841). This theorem states that the value of a double integral over a *simply connected* plane region R is determined by the value of a line integral around the boundary of R.

A curve C given by $\mathbf{r}(t) = x(t)\mathbf{i} + y(t)\mathbf{j}$, where $a \le t \le b$, is **simple** when it does not cross itself—that is, $\mathbf{r}(c) \ne \mathbf{r}(d)$ for all c and d in the open interval (a, b). A connected plane region R is **simply connected** when every simple closed curve in R encloses only points that are in R (see Figure 15.26). Informally, a simply connected region cannot consist of separate parts or holes.

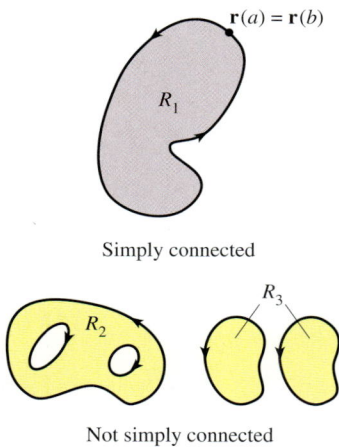

Simply connected

Not simply connected

Figure 15.26

THEOREM 15.8   Green's Theorem

Let R be a simply connected region with a piecewise smooth boundary C, oriented counterclockwise (that is, C is traversed *once* so that the region R always lies to the *left*). If M and N have continuous first partial derivatives in an open region containing R, then

$$\int_C M\,dx + N\,dy = \iint_R \left(\frac{\partial N}{\partial x} - \frac{\partial M}{\partial y} \right) dA.$$

Proof A proof is given only for a region that is both vertically simple and horizontally simple, as shown in Figure 15.27.

$$\int_C M\,dx = \int_{C_1} M\,dx + \int_{C_2} M\,dx$$

$$= \int_a^b M(x, f_1(x))\,dx + \int_b^a M(x, f_2(x))\,dx$$

$$= \int_a^b [M(x, f_1(x)) - M(x, f_2(x))]\,dx$$

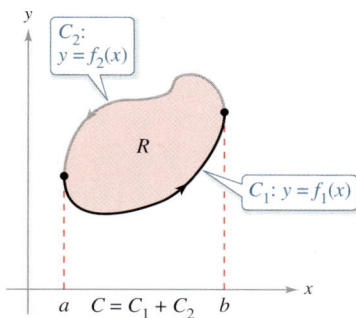

R is vertically simple.

On the other hand,

$$\iint_R \frac{\partial M}{\partial y}\,dA = \int_a^b \int_{f_1(x)}^{f_2(x)} \frac{\partial M}{\partial y}\,dy\,dx$$

$$= \int_a^b M(x, y) \Big]_{f_1(x)}^{f_2(x)}\,dx$$

$$= \int_a^b [M(x, f_2(x)) - M(x, f_1(x))]\,dx.$$

Consequently,

$$\int_C M\,dx = -\iint_R \frac{\partial M}{\partial y}\,dA.$$

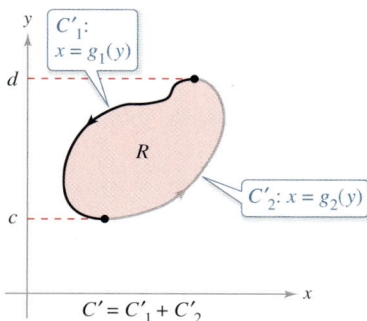

R is horizontally simple.
Figure 15.27

Similarly, you can use $g_1(y)$ and $g_2(y)$ to show that $\int_C N\,dy = \iint_R \partial N/\partial x\,dA$. By adding the integrals $\int_C M\,dx$ and $\int_C N\,dy$, you obtain the conclusion stated in the theorem.

See LarsonCalculus.com for Bruce Edwards's video of this proof.

An integral sign with a circle is sometimes used to indicate a line integral around a simple closed curve, as shown below. To indicate the orientation of the boundary, an arrow can be used. For instance, in the second integral, the arrow indicates that the boundary C is oriented counterclockwise.

1. $\oint_C M\,dx + N\,dy$ **2.** $\oint_C M\,dx + N\,dy$

EXAMPLE 1   **Using Green's Theorem**

Use Green's Theorem to evaluate the line integral

$$\int_C y^3\,dx + (x^3 + 3xy^2)\,dy$$

where C is the path from $(0, 0)$ to $(1, 1)$ along the graph of $y = x^3$ and from $(1, 1)$ to $(0, 0)$ along the graph of $y = x$, as shown in Figure 15.28.

Solution   Because $M = y^3$ and $N = x^3 + 3xy^2$, it follows that

$$\frac{\partial N}{\partial x} = 3x^2 + 3y^2 \quad \text{and} \quad \frac{\partial M}{\partial y} = 3y^2.$$

Applying Green's Theorem, you then have

$$\int_C y^3\,dx + (x^3 + 3xy^2)\,dy = \iint_R \left(\frac{\partial N}{\partial x} - \frac{\partial M}{\partial y}\right) dA$$

$$= \int_0^1 \int_{x^3}^x [(3x^2 + 3y^2) - 3y^2]\,dy\,dx$$

$$= \int_0^1 \int_{x^3}^x 3x^2\,dy\,dx$$

$$= \int_0^1 3x^2 y\Big]_{x^3}^x dx$$

$$= \int_0^1 (3x^3 - 3x^5)\,dx$$

$$= \left[\frac{3x^4}{4} - \frac{x^6}{2}\right]_0^1$$

$$= \frac{1}{4}.$$

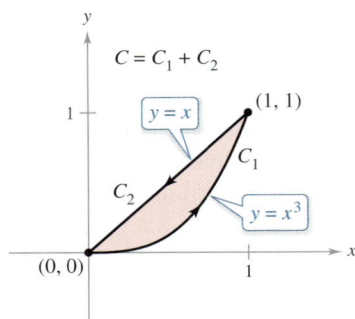

$C = C_1 + C_2$

$y = x$ $(1, 1)$

C_1

C_2

$y = x^3$

$(0, 0)$

C is simple and closed, and the region *R* always lies to the left of *C*.

Figure 15.28

Green's Theorem cannot be applied to every line integral. Among other restrictions stated in Theorem 15.8, the curve C must be simple and closed. When Green's Theorem does apply, however, it can save time. To see this, try using the techniques described in Section 15.2 to evaluate the line integral in Example 1. To do this, you would need to write the line integral as

$$\int_C y^3\,dx + (x^3 + 3xy^2)\,dy$$

$$= \int_{C_1} y^3\,dx + (x^3 + 3xy^2)\,dy + \int_{C_2} y^3\,dx + (x^3 + 3xy^2)\,dy$$

where C_1 is the cubic path given by

$$\mathbf{r}(t) = t\mathbf{i} + t^3\mathbf{j}$$

from $t = 0$ to $t = 1$, and C_2 is the line segment given by

$$\mathbf{r}(t) = (1 - t)\mathbf{i} + (1 - t)\mathbf{j}$$

from $t = 0$ to $t = 1$.

EXAMPLE 2 **Using Green's Theorem to Calculate Work**

While subject to the force

$$\mathbf{F}(x, y) = y^3\mathbf{i} + (x^3 + 3xy^2)\mathbf{j}$$

a particle travels once around the circle of radius 3 shown in Figure 15.29. Use Green's Theorem to find the work done by $\mathbf{F}$.

Solution From Example 1, you know by Green's Theorem that

$$\int_C y^3\,dx + (x^3 + 3xy^2)\,dy = \iint_R 3x^2\,dA.$$

In polar coordinates, using $x = r\cos\theta$ and $dA = r\,dr\,d\theta$, the work done is

$$
\begin{aligned}
W &= \iint_R 3x^2\,dA \\
&= \int_0^{2\pi}\int_0^3 3(r\cos\theta)^2 r\,dr\,d\theta \\
&= 3\int_0^{2\pi}\int_0^3 r^3\cos^2\theta\,dr\,d\theta \\
&= 3\int_0^{2\pi}\frac{r^4}{4}\cos^2\theta\bigg]_0^3 d\theta \\
&= 3\int_0^{2\pi}\frac{81}{4}\cos^2\theta\,d\theta \\
&= \frac{243}{8}\int_0^{2\pi}(1 + \cos 2\theta)\,d\theta \\
&= \frac{243}{8}\left[\theta + \frac{\sin 2\theta}{2}\right]_0^{2\pi} \\
&= \frac{243\pi}{4}.
\end{aligned}
$$

$\blacksquare$

When evaluating line integrals over closed curves, remember that for conservative vector fields (those for which $\partial N/\partial x = \partial M/\partial y$), the value of the line integral is 0. This is easily seen from the statement of Green's Theorem:

$$\int_C M\,dx + N\,dy = \iint_R \left(\frac{\partial N}{\partial x} - \frac{\partial M}{\partial y}\right)dA = 0.$$

EXAMPLE 3 **Green's Theorem and Conservative Vector Fields**

Evaluate the line integral

$$\int_C y^3\,dx + 3xy^2\,dy$$

where C is the path shown in Figure 15.30.

Solution From this line integral, $M = y^3$ and $N = 3xy^2$. So, $\partial N/\partial x = 3y^2$ and $\partial M/\partial y = 3y^2$. This implies that the vector field $\mathbf{F} = M\mathbf{i} + N\mathbf{j}$ is conservative, and because C is closed, you can conclude that

$$\int_C y^3\,dx + 3xy^2\,dy = 0.$$

$\blacksquare$

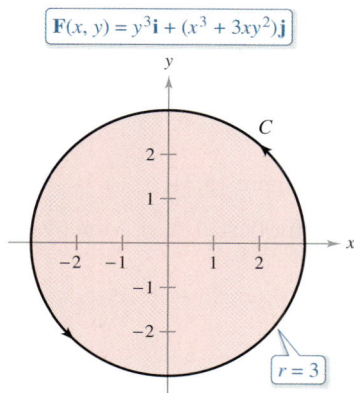

$\boxed{\mathbf{F}(x, y) = y^3\mathbf{i} + (x^3 + 3xy^2)\mathbf{j}}$

C

$r = 3$

Figure 15.29

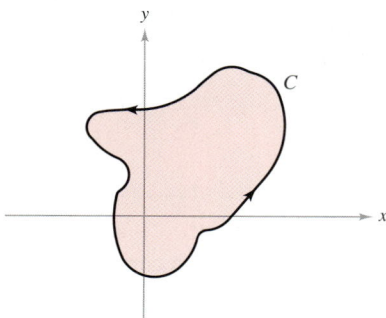

C

C is closed.
Figure 15.30

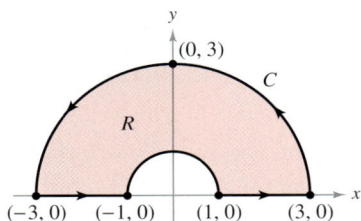

C is piecewise smooth.
Figure 15.31

EXAMPLE 4 **Using Green's Theorem**

⋮ **▷** *See LarsonCalculus.com for an interactive version of this type of example.*

Evaluate

$$\int_C (\arctan x + y^2) \, dx + (e^y - x^2) \, dy$$

where C is the path enclosing the annular region shown in Figure 15.31.

Solution In polar coordinates, R is given by $1 \le r \le 3$ for $0 \le \theta \le \pi$. Moreover,

$$\frac{\partial N}{\partial x} - \frac{\partial M}{\partial y} = -2x - 2y = -2(r \cos \theta + r \sin \theta).$$

So, by Green's Theorem,

$$\int_C (\arctan x + y^2) \, dx + (e^y - x^2) \, dy = \iint_R -2(x + y) \, dA$$

$$= \int_0^\pi \int_1^3 -2r(\cos \theta + \sin \theta) r \, dr \, d\theta$$

$$= \int_0^\pi -2(\cos \theta + \sin \theta)\frac{r^3}{3}\bigg]_1^3 \, d\theta$$

$$= \int_0^\pi \left(-\frac{52}{3}\right)(\cos \theta + \sin \theta) \, d\theta$$

$$= -\frac{52}{3}\bigg[\sin \theta - \cos \theta\bigg]_0^\pi$$

$$= -\frac{104}{3}.$$

In Examples 1, 2, and 4, Green's Theorem was used to evaluate line integrals as double integrals. You can also use the theorem to evaluate double integrals as line integrals. One useful application occurs when $\partial N/\partial x - \partial M/\partial y = 1$.

$$\int_C M \, dx + N \, dy = \iint_R \left(\frac{\partial N}{\partial x} - \frac{\partial M}{\partial y}\right) dA$$

$$= \iint_R 1 \, dA \qquad \frac{\partial N}{\partial x} - \frac{\partial M}{\partial y} = 1$$

$$= \text{area of region } R$$

Among the many choices for M and N satisfying the stated condition, the choice of

$$M = -\frac{y}{2} \quad \text{and} \quad N = \frac{x}{2}$$

produces the following line integral for the area of region R.

THEOREM 15.9 **Line Integral for Area**

If R is a plane region bounded by a piecewise smooth simple closed curve C, oriented counterclockwise, then the area of R is given by

$$A = \frac{1}{2}\int_C x \, dy - y \, dx.$$

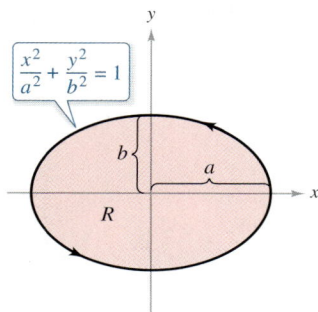

Figure 15.32

EXAMPLE 5   Finding Area by a Line Integral

Use a line integral to find the area of the ellipse $(x^2/a^2) + (y^2/b^2) = 1$.

Solution   Using Figure 15.32, you can induce a counterclockwise orientation to the elliptical path by letting $x = a \cos t$ and $y = b \sin t$, $0 \le t \le 2\pi$. So, the area is

$$A = \frac{1}{2}\int_C x\,dy - y\,dx = \frac{1}{2}\int_0^{2\pi} [(a\cos t)(b\cos t)\,dt - (b\sin t)(-a\sin t)\,dt]$$

$$= \frac{ab}{2}\int_0^{2\pi}(\cos^2 t + \sin^2 t)\,dt$$

$$= \frac{ab}{2}\Big[\,t\,\Big]_0^{2\pi}$$

$$= \pi ab.$$

Green's Theorem can be extended to cover some regions that are not simply connected. This is demonstrated in the next example.

EXAMPLE 6   Green's Theorem Extended to a Region with a Hole

Let R be the region inside the ellipse $(x^2/9) + (y^2/4) = 1$ and outside the circle $x^2 + y^2 = 1$. Evaluate the line integral

$$\int_C 2xy\,dx + (x^2 + 2x)\,dy$$

where $C = C_1 + C_2$ is the boundary of R, as shown in Figure 15.33.

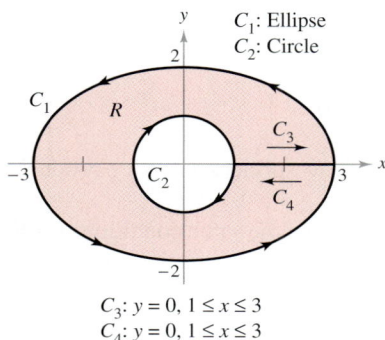

$C_3: y = 0, 1 \le x \le 3$
$C_4: y = 0, 1 \le x \le 3$

Figure 15.33

Solution   To begin, introduce the line segments C_3 and C_4, as shown in Figure 15.33. Note that because the curves C_3 and C_4 have opposite orientations, the line integrals over them cancel. Furthermore, apply Green's Theorem to the region R using the boundary $C_1 + C_4 + C_2 + C_3$ to obtain

$$\int_C 2xy\,dx + (x^2 + 2x)\,dy = \iint_R \left(\frac{\partial N}{\partial x} - \frac{\partial M}{\partial y}\right)dA$$

$$= \iint_R (2x + 2 - 2x)\,dA$$

$$= 2\iint_R dA$$

$$= 2(\text{area of } R)$$

$$= 2(\pi ab - \pi r^2)$$

$$= 2[\pi(3)(2) - \pi(1^2)]$$

$$= 10\pi.$$

In Section 15.1, a necessary and sufficient condition for conservative vector fields was listed. There, only one direction of the proof was shown. You can now outline the other direction, using Green's Theorem. Let $\mathbf{F}(x, y) = M\mathbf{i} + N\mathbf{j}$ be defined on an open disk R. You want to show that if M and N have continuous first partial derivatives and $\partial M/\partial y = \partial N/\partial x$, then $\mathbf{F}$ is conservative. Let C be a closed path forming the boundary of a connected region lying in R. Then, using the fact that $\partial M/\partial y = \partial N/\partial x$, apply Green's Theorem to conclude that

$$\int_C \mathbf{F} \cdot d\mathbf{r} = \int_C M\,dx + N\,dy = \iint_R \left(\frac{\partial N}{\partial x} - \frac{\partial M}{\partial y}\right)dA = 0.$$

This, in turn, is equivalent to showing that $\mathbf{F}$ is conservative (see Theorem 15.7).

Alternative Forms of Green's Theorem

This section concludes with the derivation of two vector forms of Green's Theorem for regions in the plane. The extension of these vector forms to three dimensions is the basis for the discussion in the remaining sections of this chapter. For a vector field $\mathbf{F}$ in the plane, you can write

$$\mathbf{F}(x, y, z) = M\mathbf{i} + N\mathbf{j} + 0\mathbf{k}$$

so that the curl of $\mathbf{F}$, as described in Section 15.1, is given by

$$\text{curl } \mathbf{F} = \nabla \times \mathbf{F} = \begin{vmatrix} \mathbf{i} & \mathbf{j} & \mathbf{k} \\ \dfrac{\partial}{\partial x} & \dfrac{\partial}{\partial y} & \dfrac{\partial}{\partial z} \\ M & N & 0 \end{vmatrix} = -\frac{\partial N}{\partial z}\mathbf{i} + \frac{\partial M}{\partial z}\mathbf{j} + \left(\frac{\partial N}{\partial x} - \frac{\partial M}{\partial y}\right)\mathbf{k}.$$

Consequently,

$$(\text{curl } \mathbf{F}) \cdot \mathbf{k} = \left[-\frac{\partial N}{\partial z}\mathbf{i} + \frac{\partial M}{\partial z}\mathbf{j} + \left(\frac{\partial N}{\partial x} - \frac{\partial M}{\partial y}\right)\mathbf{k}\right] \cdot \mathbf{k} = \frac{\partial N}{\partial x} - \frac{\partial M}{\partial y}.$$

With appropriate conditions on $\mathbf{F}$, C, and R, you can write Green's Theorem in the vector form

$$\int_C \mathbf{F} \cdot d\mathbf{r} = \iint_R \left(\frac{\partial N}{\partial x} - \frac{\partial M}{\partial y}\right) dA$$

$$= \iint_R (\text{curl } \mathbf{F}) \cdot \mathbf{k} \, dA. \qquad \textcolor{red}{\text{First alternative form}}$$

The extension of this vector form of Green's Theorem to surfaces in space produces **Stokes's Theorem,** discussed in Section 15.8.

For the second vector form of Green's Theorem, assume the same conditions for $\mathbf{F}$, C, and R. Using the arc length parameter s for C, you have $\mathbf{r}(s) = x(s)\mathbf{i} + y(s)\mathbf{j}$. So, a unit tangent vector $\mathbf{T}$ to curve C is given by $\mathbf{r}'(s) = \mathbf{T} = x'(s)\mathbf{i} + y'(s)\mathbf{j}$. From Figure 15.34, you can see that the *outward* unit normal vector $\mathbf{N}$ can then be written as

$$\mathbf{N} = y'(s)\mathbf{i} - x'(s)\mathbf{j}.$$

Consequently, for $\mathbf{F}(x, y) = M\mathbf{i} + N\mathbf{j}$, you can apply Green's Theorem to obtain

$$\int_C \mathbf{F} \cdot \mathbf{N} \, ds = \int_a^b (M\mathbf{i} + N\mathbf{j}) \cdot (y'(s)\mathbf{i} - x'(s)\mathbf{j}) \, ds$$

$$= \int_a^b \left(M\frac{dy}{ds} - N\frac{dx}{ds}\right) ds$$

$$= \int_C M \, dy - N \, dx$$

$$= \int_C -N \, dx + M \, dy$$

$$= \iint_R \left(\frac{\partial M}{\partial x} + \frac{\partial N}{\partial y}\right) dA \qquad \textcolor{red}{\text{Green's Theorem}}$$

$$= \iint_R \text{div } \mathbf{F} \, dA.$$

Therefore,

$$\int_C \mathbf{F} \cdot \mathbf{N} \, ds = \iint_R \text{div } \mathbf{F} \, dA. \qquad \textcolor{red}{\text{Second alternative form}}$$

The extension of this form to three dimensions is called the **Divergence Theorem** and will be discussed in Section 15.7. The physical interpretations of divergence and curl will be discussed in Sections 15.7 and 15.8.

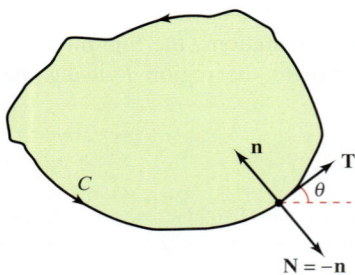

$\mathbf{T} = \cos\theta\mathbf{i} + \sin\theta\mathbf{j}$

$\mathbf{n} = \cos\left(\theta + \dfrac{\pi}{2}\right)\mathbf{i} + \sin\left(\theta + \dfrac{\pi}{2}\right)\mathbf{j}$

$\quad = -\sin\theta\mathbf{i} + \cos\theta\mathbf{j}$

$\mathbf{N} = \sin\theta\mathbf{i} - \cos\theta\mathbf{j}$

Figure 15.34

15.4 Exercises

See CalcChat.com for tutorial help and worked-out solutions to odd-numbered exercises.

Verifying Green's Theorem In Exercises 1–4, verify Green's Theorem by evaluating both integrals

$$\int_C y^2\, dx + x^2\, dy = \int\int_R \left(\frac{\partial N}{\partial x} - \frac{\partial M}{\partial y}\right) dA$$

for the given path.

1. C: boundary of the region lying between the graphs of $y = x$ and $y = x^2$

2. C: boundary of the region lying between the graphs of $y = x$ and $y = \sqrt{x}$

3. C: square with vertices $(0, 0)$, $(1, 0)$, $(1, 1)$, and $(0, 1)$

4. C: rectangle with vertices $(0, 0)$, $(3, 0)$, $(3, 4)$, and $(0, 4)$

Verifying Green's Theorem In Exercises 5 and 6, verify Green's Theorem by using a computer algebra system to evaluate both integrals

$$\int_C xe^y\, dx + e^x\, dy = \int\int_R \left(\frac{\partial N}{\partial x} - \frac{\partial M}{\partial y}\right) dA$$

for the given path.

5. C: circle given by $x^2 + y^2 = 4$

6. C: boundary of the region lying between the graphs of $y = x$ and $y = x^3$ in the first quadrant

Evaluating a Line Integral Using Green's Theorem In Exercises 7–10, use Green's Theorem to evaluate the integral

$$\int_C (y - x)\, dx + (2x - y)\, dy$$

for the given path.

7. C: boundary of the region lying between the graphs of $y = x$ and $y = x^2 - 2x$

8. C: $x = 2\cos\theta$, $y = \sin\theta$

9. C: boundary of the region lying inside the rectangle bounded by $x = -5$, $x = 5$, $y = -3$, and $y = 3$, and outside the square bounded by $x = -1$, $x = 1$, $y = -1$, and $y = 1$

10. C: boundary of the region lying inside the semicircle $y = \sqrt{25 - x^2}$ and outside the semicircle $y = \sqrt{9 - x^2}$

Evaluating a Line Integral Using Green's Theorem In Exercises 11–20, use Green's Theorem to evaluate the line integral.

11. $\displaystyle\int_C 2xy\, dx + (x + y)\, dy$

C: boundary of the region lying between the graphs of $y = 0$ and $y = 1 - x^2$

12. $\displaystyle\int_C y^2\, dx + xy\, dy$

C: boundary of the region lying between the graphs of $y = 0$, $y = \sqrt{x}$, and $x = 9$

13. $\displaystyle\int_C (x^2 - y^2)\, dx + 2xy\, dy$

C: $x^2 + y^2 = 16$

14. $\displaystyle\int_C (x^2 - y^2)\, dx + 2xy\, dy$

C: $r = 1 + \cos\theta$

15. $\displaystyle\int_C e^x\cos 2y\, dx - 2e^x\sin 2y\, dy$

C: $x^2 + y^2 = a^2$

16. $\displaystyle\int_C 2\arctan\frac{y}{x}\, dx + \ln(x^2 + y^2)\, dy$

C: $x = 4 + 2\cos\theta$, $y = 4 + \sin\theta$

17. $\displaystyle\int_C \cos y\, dx + (xy - x\sin y)\, dy$

C: boundary of the region lying between the graphs of $y = x$ and $y = \sqrt{x}$

18. $\displaystyle\int_C (e^{-x^2/2} - y)\, dx + (e^{-y^2/2} + x)\, dy$

C: boundary of the region lying between the graphs of the circle $x = 6\cos\theta$, $y = 6\sin\theta$ and the ellipse $x = 3\cos\theta$, $y = 2\sin\theta$

19. $\displaystyle\int_C (x - 3y)\, dx + (x + y)\, dy$

C: boundary of the region lying between the graphs of $x^2 + y^2 = 1$ and $x^2 + y^2 = 9$

20. $\displaystyle\int_C 3x^2 e^y\, dx + e^y\, dy$

C: boundary of the region lying between the squares with vertices $(1, 1)$, $(-1, 1)$, $(-1, -1)$, and $(1, -1)$, and $(2, 2)$, $(-2, 2)$, $(-2, -2)$, and $(2, -2)$

Work In Exercises 21–24, use Green's Theorem to calculate the work done by the force **F** on a particle that is moving counterclockwise around the closed path C.

21. $\mathbf{F}(x, y) = xy\mathbf{i} + (x + y)\mathbf{j}$

C: $x^2 + y^2 = 1$

22. $\mathbf{F}(x, y) = (e^x - 3y)\mathbf{i} + (e^y + 6x)\mathbf{j}$

C: $r = 2\cos\theta$

23. $\mathbf{F}(x, y) = (x^{3/2} - 3y)\mathbf{i} + (6x + 5\sqrt{y})\mathbf{j}$

C: boundary of the triangle with vertices $(0, 0)$, $(5, 0)$, and $(0, 5)$

24. $\mathbf{F}(x, y) = (3x^2 + y)\mathbf{i} + 4xy^2\mathbf{j}$

C: boundary of the region lying between the graphs of $y = \sqrt{x}$, $y = 0$, and $x = 9$

Area In Exercises 25–28, use a line integral to find the area of the region R.

25. R: region bounded by the graph of $x^2 + y^2 = a^2$

26. R: triangle bounded by the graphs of $x = 0$, $3x - 2y = 0$, and $x + 2y = 8$

27. R: region bounded by the graphs of $y = 5x - 3$ and $y = x^2 + 1$

28. R: region inside the loop of the folium of Descartes bounded by the graph of

$$x = \frac{3t}{t^3 + 1}, \quad y = \frac{3t^2}{t^3 + 1}$$

WRITING ABOUT CONCEPTS

29. Green's Theorem State Green's Theorem.

30. Area Give the line integral for the area of a region R bounded by a piecewise smooth simple curve C.

Using Green's Theorem to Verify a Formula In Exercises 31 and 32, use Green's Theorem to verify the line integral formulas.

31. The centroid of the region having area A bounded by the simple closed path C is

$$\bar{x} = \frac{1}{2A} \int_C x^2 \, dy, \quad \bar{y} = -\frac{1}{2A} \int_C y^2 \, dx.$$

32. The area of a plane region bounded by the simple closed path C given in polar coordinates is

$$A = \frac{1}{2} \int_C r^2 \, d\theta.$$

Centroid In Exercises 33–36, use the results of Exercise 31 to find the centroid of the region.

33. R: region bounded by the graphs of $y = 0$ and $y = 4 - x^2$

34. R: region bounded by the graphs of $y = \sqrt{a^2 - x^2}$ and $y = 0$

35. R: region bounded by the graphs of $y = x^3$ and $y = x$, $0 \le x \le 1$

36. R: triangle with vertices $(-a, 0)$, $(a, 0)$, and (b, c), where $-a \le b \le a$

Area In Exercises 37–40, use the results of Exercise 32 to find the area of the region bounded by the graph of the polar equation.

37. $r = a(1 - \cos \theta)$

38. $r = a \cos 3\theta$

39. $r = 1 + 2 \cos \theta$ (inner loop)

40. $r = \dfrac{3}{2 - \cos \theta}$

41. Maximum Value

(a) Evaluate $\displaystyle\int_{C_1} y^3 \, dx + (27x - x^3) \, dy$,

where C_1 is the unit circle given by $\mathbf{r}(t) = \cos t\mathbf{i} + \sin t\mathbf{j}$, for $0 \le t \le 2\pi$.

(b) Find the maximum value of $\displaystyle\int_C y^3 \, dx + (27x - x^3) \, dy$, where C is any closed curve in the xy-plane, oriented counterclockwise.

42. HOW DO YOU SEE IT? Use Green's Theorem to explain why

$$\int_C f(x) \, dx + g(y) \, dy = 0$$

where f and g are differentiable functions and C is a piecewise smooth simple closed path (see figure).

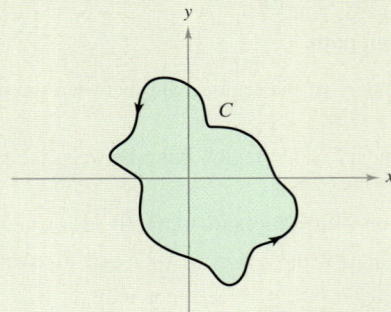

43. Green's Theorem: Region with a Hole Let R be the region inside the circle $x = 5 \cos \theta$, $y = 5 \sin \theta$ and outside the ellipse $x = 2 \cos \theta$, $y = \sin \theta$. Evaluate the line integral

$$\int_C (e^{-x^2/2} - y) \, dx + (e^{-y^2/2} + x) \, dy$$

where $C = C_1 + C_2$ is the boundary of R, as shown in the figure.

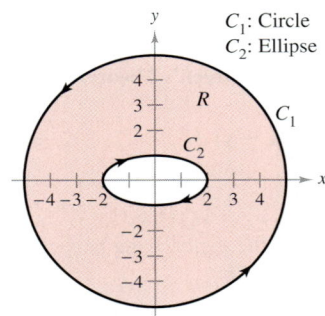

44. Green's Theorem: Region with a Hole Let R be the region inside the ellipse $x = 4 \cos \theta$, $y = 3 \sin \theta$ and outside the circle $x = 2 \cos \theta$, $y = 2 \sin \theta$. Evaluate the line integral

$$\int_C (3x^2y + 1) \, dx + (x^3 + 4x) \, dy$$

where $C = C_1 + C_2$ is the boundary of R, as shown in the figure.

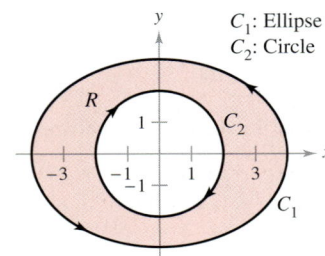

45. Think About It Let

$$I = \int_C \frac{y\,dx - x\,dy}{x^2 + y^2}$$

where C is a circle oriented counterclockwise. Show that $I = 0$ when C does not contain the origin. What is I when C does contain the origin?

46. Think About It For each given path, verify Green's Theorem by showing that

$$\int_C y^2\,dx + x^2\,dy = \int\int_R \left(\frac{\partial N}{\partial x} - \frac{\partial M}{\partial y}\right) dA.$$

For each path, which integral is easier to evaluate? Explain.

(a) C: triangle with vertices $(0, 0)$, $(4, 0)$, and $(4, 4)$

(b) C: circle given by $x^2 + y^2 = 1$

47. Proof

(a) Let C be the line segment joining (x_1, y_1) and (x_2, y_2). Show that $\int_C -y\,dx + x\,dy = x_1y_2 - x_2y_1$.

(b) Let (x_1, y_1), (x_2, y_2), . . . , (x_n, y_n) be the vertices of a polygon. Prove that the area enclosed is

$$\tfrac{1}{2}[(x_1y_2 - x_2y_1) + (x_2y_3 - x_3y_2) + \cdots +$$
$$(x_{n-1}y_n - x_ny_{n-1}) + (x_ny_1 - x_1y_n)].$$

48. Area Use the result of Exercise 47(b) to find the area enclosed by the polygon with the given vertices.

(a) Pentagon: $(0, 0)$, $(2, 0)$, $(3, 2)$, $(1, 4)$, and $(-1, 1)$

(b) Hexagon: $(0, 0)$, $(2, 0)$, $(3, 2)$, $(2, 4)$, $(0, 3)$, and $(-1, 1)$

Proof In Exercises 49 and 50, prove the identity, where R is a simply connected region with boundary C. Assume that the required partial derivatives of the scalar functions f and g are continuous. The expressions $D_{\mathbf{N}}f$ and $D_{\mathbf{N}}g$ are the derivatives in the direction of the outward normal vector $\mathbf{N}$ of C, and are defined by $D_{\mathbf{N}}f = \nabla f \cdot \mathbf{N}$, and $D_{\mathbf{N}}g = \nabla g \cdot \mathbf{N}$.

49. Green's first identity:

$$\int\int_R (f\nabla^2 g + \nabla f \cdot \nabla g)\,dA = \int_C f D_{\mathbf{N}}g\,ds$$

[*Hint:* Use the second alternative form of Green's Theorem and the property $\text{div}(f\,\mathbf{G}) = f\,\text{div}\,\mathbf{G} + \nabla f \cdot \mathbf{G}.$]

50. Green's second identity:

$$\int\int_R (f\nabla^2 g - g\nabla^2 f)\,dA = \int_C (f D_{\mathbf{N}}g - g D_{\mathbf{N}}f)\,ds$$

(*Hint:* Use Green's first identity from Exercise 49 twice.)

51. Proof Let $\mathbf{F} = M\mathbf{i} + N\mathbf{j}$, where M and N have continuous first partial derivatives in a simply connected region R. Prove that if C is simple, smooth, and closed, and $N_x = M_y$, then $\int_C \mathbf{F} \cdot d\mathbf{r} = 0$.

PUTNAM EXAM CHALLENGE

52. Find the least possible area of a convex set in the plane that intersects both branches of the hyperbola $xy = 1$ and both branches of the hyperbola $xy = -1$. (A set S in the plane is called *convex* if for any two points in S the line segment connecting them is contained in S.)

SECTION PROJECT

Hyperbolic and Trigonometric Functions

(a) Sketch the plane curve represented by the vector-valued function $\mathbf{r}(t) = \cosh t\,\mathbf{i} + \sinh t\,\mathbf{j}$ on the interval $0 \le t \le 5$. Show that the rectangular equation corresponding to $\mathbf{r}(t)$ is the hyperbola $x^2 - y^2 = 1$. Verify your sketch by using a graphing utility to graph the hyperbola.

(b) Let $P = (\cosh \phi, \sinh \phi)$ be the point on the hyperbola corresponding to $\mathbf{r}(\phi)$ for $\phi > 0$. Use the formula for area

$$A = \frac{1}{2}\int_C x\,dy - y\,dx$$

to verify that the area of the region shown in the figure is $\frac{1}{2}\phi$.

(c) Show that the area of the indicated region is also given by the integral

$$A = \int_0^{\sinh \phi} \left[\sqrt{1 + y^2} - (\coth \phi)y\right] dy.$$

Confirm your answer in part (b) by numerically approximating this integral for $\phi = 1, 2, 4,$ and 10.

(d) Consider the unit circle given by $x^2 + y^2 = 1$. Let θ be the angle formed by the x-axis and the radius to (x, y). The area of the corresponding sector is $\frac{1}{2}\theta$. That is, the trigonometric functions $f(\theta) = \cos \theta$ and $g(\theta) = \sin \theta$ could have been defined as the coordinates of that point $(\cos \theta, \sin \theta)$ on the unit circle that determines a sector of area $\frac{1}{2}\theta$. Write a short paragraph explaining how you could define the hyperbolic functions in a similar manner, using the "unit hyperbola" $x^2 - y^2 = 1.$

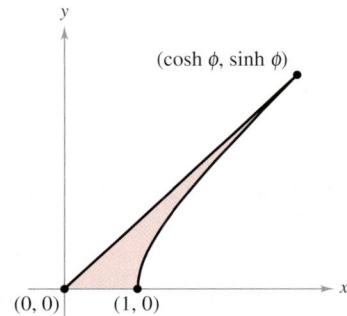

15.5 Parametric Surfaces

- Understand the definition of a parametric surface, and sketch the surface.
- Find a set of parametric equations to represent a surface.
- Find a normal vector and a tangent plane to a parametric surface.
- Find the area of a parametric surface.

Parametric Surfaces

You already know how to represent a curve in the plane or in space by a set of parametric equations—or, equivalently, by a vector-valued function.

$$\mathbf{r}(t) = x(t)\mathbf{i} + y(t)\mathbf{j} \qquad\qquad \text{Plane curve}$$
$$\mathbf{r}(t) = x(t)\mathbf{i} + y(t)\mathbf{j} + z(t)\mathbf{k} \qquad\qquad \text{Space curve}$$

In this section, you will learn how to represent a surface in space by a set of parametric equations—or by a vector-valued function. For curves, note that the vector-valued function **r** is a function of a *single* parameter t. For surfaces, the vector-valued function is a function of *two* parameters u and v.

Definition of Parametric Surface

Let x, y, and z be functions of u and v that are continuous on a domain D in the uv-plane. The set of points (x, y, z) given by

$$\mathbf{r}(u, v) = x(u, v)\mathbf{i} + y(u, v)\mathbf{j} + z(u, v)\mathbf{k} \qquad \text{Parametric surface}$$

is called a **parametric surface.** The equations

$$x = x(u, v), \quad y = y(u, v), \quad \text{and} \quad z = z(u, v) \qquad \text{Parametric equations}$$

are the **parametric equations** for the surface.

If S is a parametric surface given by the vector-valued function **r**, then S is traced out by the position vector $\mathbf{r}(u, v)$ as the point (u, v) moves throughout the domain D, as shown in Figure 15.35.

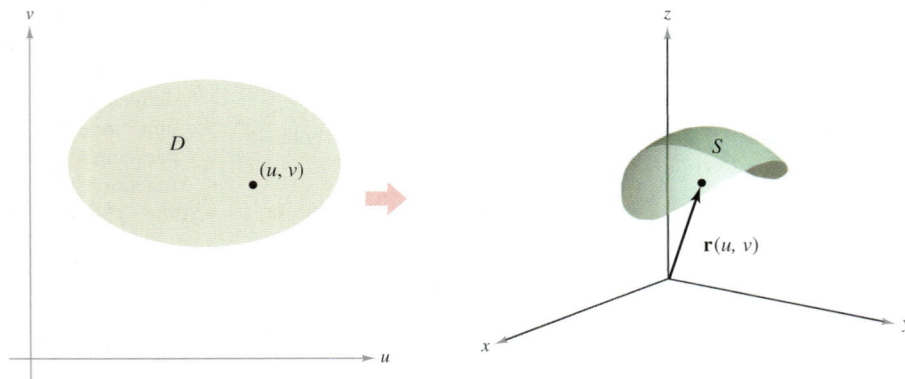

Figure 15.35

▷ **TECHNOLOGY** Some computer algebra systems are capable of graphing surfaces that are represented parametrically. If you have access to such software, use it to graph some of the surfaces in the examples and exercises in this section.

EXAMPLE 1 **Sketching a Parametric Surface**

Identify and sketch the parametric surface S given by

$$\mathbf{r}(u, v) = 3 \cos u\mathbf{i} + 3 \sin u\mathbf{j} + v\mathbf{k}$$

where $0 \le u \le 2\pi$ and $0 \le v \le 4$.

Solution Because $x = 3 \cos u$ and $y = 3 \sin u$, you know that for each point (x, y, z) on the surface, x and y are related by the equation

$$x^2 + y^2 = 3^2.$$

In other words, each cross section of S taken parallel to the xy-plane is a circle of radius 3, centered on the z-axis. Because $z = v$, where

$$0 \le v \le 4$$

you can see that the surface is a right circular cylinder of height 4. The radius of the cylinder is 3, and the z-axis forms the axis of the cylinder, as shown in Figure 15.36.

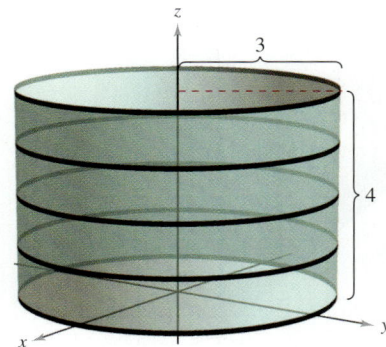
Figure 15.36

As with parametric representations of curves, parametric representations of surfaces are not unique. That is, there are many other sets of parametric equations that could be used to represent the surface shown in Figure 15.36.

EXAMPLE 2 **Sketching a Parametric Surface**

Identify and sketch the parametric surface S given by

$$\mathbf{r}(u, v) = \sin u \cos v\mathbf{i} + \sin u \sin v\mathbf{j} + \cos u\mathbf{k}$$

where $0 \le u \le \pi$ and $0 \le v \le 2\pi$.

Solution To identify the surface, you can try to use trigonometric identities to eliminate the parameters. After some experimentation, you can discover that

$$\begin{aligned} x^2 + y^2 + z^2 &= (\sin u \cos v)^2 + (\sin u \sin v)^2 + (\cos u)^2 \\ &= \sin^2 u \cos^2 v + \sin^2 u \sin^2 v + \cos^2 u \\ &= \sin^2 u(\cos^2 v + \sin^2 v) + \cos^2 u \\ &= \sin^2 u + \cos^2 u \\ &= 1. \end{aligned}$$

So, each point on S lies on the unit sphere, centered at the origin, as shown in Figure 15.37. For fixed $u = d_i$, $\mathbf{r}(u, v)$ traces out latitude circles

$$x^2 + y^2 = \sin^2 d_i, \quad 0 \le d_i \le \pi$$

that are parallel to the xy-plane, and for fixed $v = c_i$, $\mathbf{r}(u, v)$ traces out longitude (or meridian) half-circles.

To convince yourself further that $\mathbf{r}(u, v)$ traces out the entire unit sphere, recall that the parametric equations

$$x = \rho \sin \phi \cos \theta, \quad y = \rho \sin \phi \sin \theta, \quad \text{and} \quad z = \rho \cos \phi$$

where $0 \le \theta \le 2\pi$ and $0 \le \phi \le \pi$, describe the conversion from spherical to rectangular coordinates, as discussed in Section 11.7.

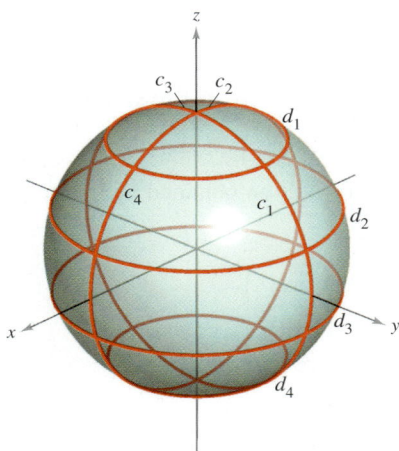
Figure 15.37

Finding Parametric Equations for Surfaces

In Examples 1 and 2, you were asked to identify the surface described by a given set of parametric equations. The reverse problem—that of writing a set of parametric equations for a given surface—is generally more difficult. One type of surface for which this problem is straightforward, however, is a surface that is given by $z = f(x, y)$. You can parametrize such a surface as

$$\mathbf{r}(x, y) = x\mathbf{i} + y\mathbf{j} + f(x, y)\mathbf{k}.$$

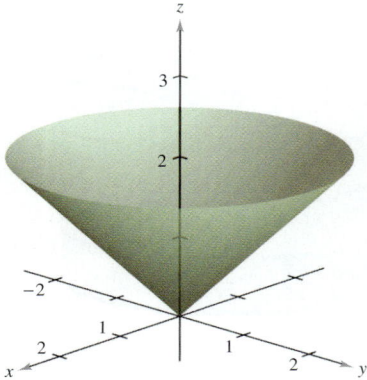

Figure 15.38

EXAMPLE 3   Representing a Surface Parametrically

Write a set of parametric equations for the cone given by

$$z = \sqrt{x^2 + y^2}$$

as shown in Figure 15.38.

Solution   Because this surface is given in the form $z = f(x, y)$, you can let x and y be the parameters. Then the cone is represented by the vector-valued function

$$\mathbf{r}(x, y) = x\mathbf{i} + y\mathbf{j} + \sqrt{x^2 + y^2}\,\mathbf{k}$$

where (x, y) varies over the entire xy-plane.   ◼

A second type of surface that is easily represented parametrically is a surface of revolution. For instance, to represent the surface formed by revolving the graph of

$$y = f(x), \quad a \le x \le b$$

about the x-axis, use

$$x = u, \quad y = f(u) \cos v, \quad \text{and} \quad z = f(u) \sin v$$

where $a \le u \le b$ and $0 \le v \le 2\pi$.

EXAMPLE 4   Representing a Surface of Revolution Parametrically

$\cdots\cdots\triangleright$ *See LarsonCalculus.com for an interactive version of this type of example.*

Write a set of parametric equations for the surface of revolution obtained by revolving

$$f(x) = \frac{1}{x}, \quad 1 \le x \le 10$$

about the x-axis.

Figure 15.39

Solution   Use the parameters u and v as described above to write

$$x = u, \quad y = f(u) \cos v = \frac{1}{u} \cos v, \quad \text{and} \quad z = f(u) \sin v = \frac{1}{u} \sin v$$

where

$$1 \le u \le 10 \quad \text{and} \quad 0 \le v \le 2\pi.$$

The resulting surface is a portion of *Gabriel's Horn*, as shown in Figure 15.39.   ◼

The surface of revolution in Example 4 is formed by revolving the graph of $y = f(x)$ about the x-axis. For other types of surfaces of revolution, a similar parametrization can be used. For instance, to parametrize the surface formed by revolving the graph of $x = f(z)$ about the z-axis, you can use

$$z = u, \quad x = f(u) \cos v, \quad \text{and} \quad y = f(u) \sin v.$$

Normal Vectors and Tangent Planes

Let S be a parametric surface given by

$$\mathbf{r}(u, v) = x(u, v)\mathbf{i} + y(u, v)\mathbf{j} + z(u, v)\mathbf{k}$$

over an open region D such that x, y, and z have continuous partial derivatives on D. The **partial derivatives of r** with respect to u and v are defined as

$$\mathbf{r}_u = \frac{\partial x}{\partial u}(u, v)\mathbf{i} + \frac{\partial y}{\partial u}(u, v)\mathbf{j} + \frac{\partial z}{\partial u}(u, v)\mathbf{k}$$

and

$$\mathbf{r}_v = \frac{\partial x}{\partial v}(u, v)\mathbf{i} + \frac{\partial y}{\partial v}(u, v)\mathbf{j} + \frac{\partial z}{\partial v}(u, v)\mathbf{k}.$$

Each of these partial derivatives is a vector-valued function that can be interpreted geometrically in terms of tangent vectors. For instance, if $v = v_0$ is held constant, then $\mathbf{r}(u, v_0)$ is a vector-valued function of a single parameter and defines a curve C_1 that lies on the surface S. The tangent vector to C_1 at the point

$$(x(u_0, v_0), y(u_0, v_0), z(u_0, v_0))$$

is given by

$$\mathbf{r}_u(u_0, v_0) = \frac{\partial x}{\partial u}(u_0, v_0)\mathbf{i} + \frac{\partial y}{\partial u}(u_0, v_0)\mathbf{j} + \frac{\partial z}{\partial u}(u_0, v_0)\mathbf{k}$$

as shown in Figure 15.40. In a similar way, if $u = u_0$ is held constant, then $\mathbf{r}(u_0, v)$ is a vector-valued function of a single parameter and defines a curve C_2 that lies on the surface S. The tangent vector to C_2 at the point $(x(u_0, v_0), y(u_0, v_0), z(u_0, v_0))$ is given by

$$\mathbf{r}_v(u_0, v_0) = \frac{\partial x}{\partial v}(u_0, v_0)\mathbf{i} + \frac{\partial y}{\partial v}(u_0, v_0)\mathbf{j} + \frac{\partial z}{\partial v}(u_0, v_0)\mathbf{k}.$$

If the normal vector $\mathbf{r}_u \times \mathbf{r}_v$ is not $\mathbf{0}$ for any (u, v) in D, then the surface S is called **smooth** and will have a tangent plane. Informally, a smooth surface is one that has no sharp points or cusps. For instance, spheres, ellipsoids, and paraboloids are smooth, whereas the cone given in Example 3 is not smooth.

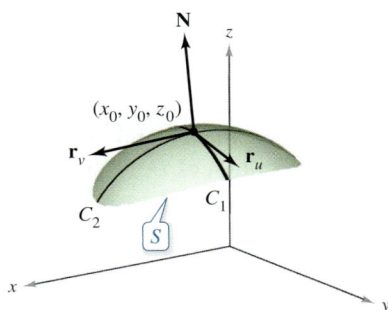

Figure 15.40

Normal Vector to a Smooth Parametric Surface

Let S be a smooth parametric surface

$$\mathbf{r}(u, v) = x(u, v)\mathbf{i} + y(u, v)\mathbf{j} + z(u, v)\mathbf{k}$$

defined over an open region D in the uv-plane. Let (u_0, v_0) be a point in D. A normal vector at the point

$$(x_0, y_0, z_0) = (x(u_0, v_0), y(u_0, v_0), z(u_0, v_0))$$

is given by

$$\mathbf{N} = \mathbf{r}_u(u_0, v_0) \times \mathbf{r}_v(u_0, v_0) = \begin{vmatrix} \mathbf{i} & \mathbf{j} & \mathbf{k} \\ \dfrac{\partial x}{\partial u} & \dfrac{\partial y}{\partial u} & \dfrac{\partial z}{\partial u} \\ \dfrac{\partial x}{\partial v} & \dfrac{\partial y}{\partial v} & \dfrac{\partial z}{\partial v} \end{vmatrix}.$$

Figure 15.40 shows the normal vector $\mathbf{r}_u \times \mathbf{r}_v$. The vector $\mathbf{r}_v \times \mathbf{r}_u$ is also normal to S and points in the opposite direction.

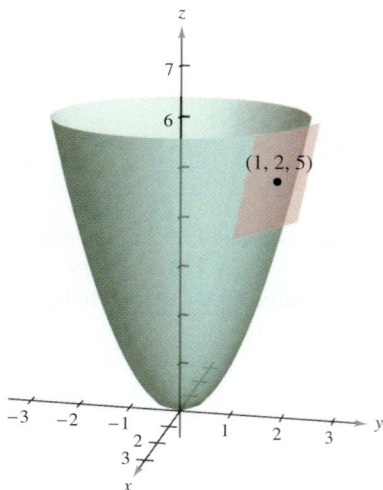

Figure 15.41

| EXAMPLE 5 | **Finding a Tangent Plane to a Parametric Surface** |

Find an equation of the tangent plane to the paraboloid

$$\mathbf{r}(u, v) = u\mathbf{i} + v\mathbf{j} + (u^2 + v^2)\mathbf{k}$$

at the point $(1, 2, 5)$.

Solution   The point in the uv-plane that is mapped to the point $(x, y, z) = (1, 2, 5)$ is $(u, v) = (1, 2)$. The partial derivatives of $\mathbf{r}$ are

$$\mathbf{r}_u = \mathbf{i} + 2u\mathbf{k} \quad \text{and} \quad \mathbf{r}_v = \mathbf{j} + 2v\mathbf{k}.$$

The normal vector is given by

$$\mathbf{r}_u \times \mathbf{r}_v = \begin{vmatrix} \mathbf{i} & \mathbf{j} & \mathbf{k} \\ 1 & 0 & 2u \\ 0 & 1 & 2v \end{vmatrix} = -2u\mathbf{i} - 2v\mathbf{j} + \mathbf{k}$$

which implies that the normal vector at $(1, 2, 5)$ is

$$\mathbf{r}_u \times \mathbf{r}_v = -2\mathbf{i} - 4\mathbf{j} + \mathbf{k}.$$

So, an equation of the tangent plane at $(1, 2, 5)$ is

$$-2(x - 1) - 4(y - 2) + (z - 5) = 0$$
$$-2x - 4y + z = -5.$$

The tangent plane is shown in Figure 15.41.   ◼

Area of a Parametric Surface

To define the area of a parametric surface, you can use a development that is similar to that given in Section 14.5. Begin by constructing an inner partition of D consisting of n rectangles, where the area of the ith rectangle D_i is $\Delta A_i = \Delta u_i \Delta v_i$, as shown in Figure 15.42. In each D_i, let (u_i, v_i) be the point that is closest to the origin. At the point $(x_i, y_i, z_i) = (x(u_i, v_i), y(u_i, v_i), z(u_i, v_i))$ on the surface S, construct a tangent plane T_i. The area of the portion of S that corresponds to D_i, ΔT_i, can be approximated by a parallelogram in the tangent plane. That is, $\Delta T_i \approx \Delta S_i$. So, the surface of S is given by $\Sigma \, \Delta S_i \approx \Sigma \, \Delta T_i$. The area of the parallelogram in the tangent plane is

$$\|\Delta u_i \mathbf{r}_u \times \Delta v_i \mathbf{r}_v\| = \|\mathbf{r}_u \times \mathbf{r}_v\| \, \Delta u_i \, \Delta v_i$$

which leads to the next definition.

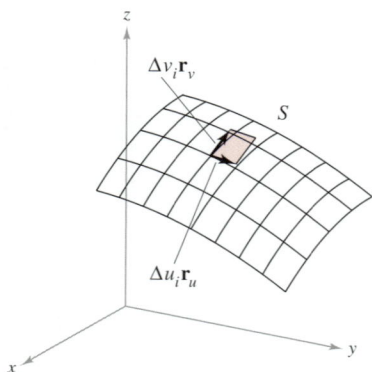

Figure 15.42

Area of a Parametric Surface

Let S be a smooth parametric surface

$$\mathbf{r}(u, v) = x(u, v)\mathbf{i} + y(u, v)\mathbf{j} + z(u, v)\mathbf{k}$$

defined over an open region D in the uv-plane. If each point on the surface S corresponds to exactly one point in the domain D, then the **surface area** of S is given by

$$\text{Surface area} = \iint_S dS = \iint_D \|\mathbf{r}_u \times \mathbf{r}_v\| \, dA$$

where

$$\mathbf{r}_u = \frac{\partial x}{\partial u}\mathbf{i} + \frac{\partial y}{\partial u}\mathbf{j} + \frac{\partial z}{\partial u}\mathbf{k} \quad \text{and} \quad \mathbf{r}_v = \frac{\partial x}{\partial v}\mathbf{i} + \frac{\partial y}{\partial v}\mathbf{j} + \frac{\partial z}{\partial v}\mathbf{k}.$$

For a surface S given by $z = f(x, y)$, this formula for surface area corresponds to that given in Section 14.5. To see this, you can parametrize the surface using the vector-valued function

$$\mathbf{r}(x, y) = x\mathbf{i} + y\mathbf{j} + f(x, y)\mathbf{k}$$

defined over the region R in the xy-plane. Using

$$\mathbf{r}_x = \mathbf{i} + f_x(x, y)\mathbf{k} \quad \text{and} \quad \mathbf{r}_y = \mathbf{j} + f_y(x, y)\mathbf{k}$$

you have

$$\mathbf{r}_x \times \mathbf{r}_y = \begin{vmatrix} \mathbf{i} & \mathbf{j} & \mathbf{k} \\ 1 & 0 & f_x(x, y) \\ 0 & 1 & f_y(x, y) \end{vmatrix} = -f_x(x, y)\mathbf{i} - f_y(x, y)\mathbf{j} + \mathbf{k}$$

and

$$\|\mathbf{r}_x \times \mathbf{r}_y\| = \sqrt{[f_x(x, y)]^2 + [f_y(x, y)]^2 + 1}.$$

This implies that the surface area of S is

$$\text{Surface area} = \iint_R \|\mathbf{r}_x \times \mathbf{r}_y\| \, dA$$
$$= \iint_R \sqrt{1 + [f_x(x, y)]^2 + [f_y(x, y)]^2} \, dA.$$

EXAMPLE 6   Finding Surface Area

REMARK The surface in Example 6 does not quite fulfill the hypothesis that each point on the surface corresponds to exactly one point in D. For this surface, $\mathbf{r}(u, 0) = \mathbf{r}(u, 2\pi)$ for any fixed value of u. However, because the overlap consists of only a semicircle (which has no area), you can still apply the formula for the area of a parametric surface.

Find the surface area of the unit sphere

$$\mathbf{r}(u, v) = \sin u \cos v\mathbf{i} + \sin u \sin v\mathbf{j} + \cos u\mathbf{k}$$

where the domain D is $0 \le u \le \pi$ and $0 \le v \le 2\pi$.

Solution   Begin by calculating $\mathbf{r}_u$ and $\mathbf{r}_v$.

$$\mathbf{r}_u = \cos u \cos v\mathbf{i} + \cos u \sin v\mathbf{j} - \sin u\mathbf{k}$$
$$\mathbf{r}_v = -\sin u \sin v\mathbf{i} + \sin u \cos v\mathbf{j}$$

The cross product of these two vectors is

$$\mathbf{r}_u \times \mathbf{r}_v = \begin{vmatrix} \mathbf{i} & \mathbf{j} & \mathbf{k} \\ \cos u \cos v & \cos u \sin v & -\sin u \\ -\sin u \sin v & \sin u \cos v & 0 \end{vmatrix}$$
$$= \sin^2 u \cos v\mathbf{i} + \sin^2 u \sin v\mathbf{j} + \sin u \cos u\mathbf{k}$$

which implies that

$$\|\mathbf{r}_u \times \mathbf{r}_v\| = \sqrt{(\sin^2 u \cos v)^2 + (\sin^2 u \sin v)^2 + (\sin u \cos u)^2}$$
$$= \sqrt{\sin^4 u + \sin^2 u \cos^2 u}$$
$$= \sqrt{\sin^2 u}$$
$$= \sin u. \qquad \sin u > 0 \text{ for } 0 \le u \le \pi$$

Finally, the surface area of the sphere is

$$A = \iint_D \|\mathbf{r}_u \times \mathbf{r}_v\| \, dA$$
$$= \int_0^{2\pi} \int_0^\pi \sin u \, du \, dv$$
$$= \int_0^{2\pi} 2 \, dv$$
$$= 4\pi.$$

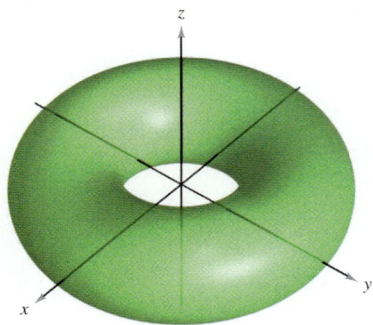

Figure 15.43

EXAMPLE 7   **Finding Surface Area**

Find the surface area of the torus given by

$$\mathbf{r}(u, v) = (2 + \cos u) \cos v\mathbf{i} + (2 + \cos u) \sin v\mathbf{j} + \sin u\mathbf{k}$$

where the domain D is given by $0 \le u \le 2\pi$ and $0 \le v \le 2\pi$. (See Figure 15.43.)

Solution   Begin by calculating $\mathbf{r}_u$ and $\mathbf{r}_v$.

$$\mathbf{r}_u = -\sin u \cos v\mathbf{i} - \sin u \sin v\mathbf{j} + \cos u\mathbf{k}$$
$$\mathbf{r}_v = -(2 + \cos u) \sin v\mathbf{i} + (2 + \cos u) \cos v\mathbf{j}$$

The cross product of these two vectors is

$$\mathbf{r}_u \times \mathbf{r}_v = \begin{vmatrix} \mathbf{i} & \mathbf{j} & \mathbf{k} \\ -\sin u \cos v & -\sin u \sin v & \cos u \\ -(2 + \cos u) \sin v & (2 + \cos u) \cos v & 0 \end{vmatrix}$$

$$= -(2 + \cos u)(\cos v \cos u\mathbf{i} + \sin v \cos u\mathbf{j} + \sin u\mathbf{k})$$

which implies that

$$\|\mathbf{r}_u \times \mathbf{r}_v\| = (2 + \cos u)\sqrt{(\cos v \cos u)^2 + (\sin v \cos u)^2 + \sin^2 u}$$
$$= (2 + \cos u)\sqrt{\cos^2 u(\cos^2 v + \sin^2 v) + \sin^2 u}$$
$$= (2 + \cos u)\sqrt{\cos^2 u + \sin^2 u}$$
$$= 2 + \cos u.$$

Finally, the surface area of the torus is

$$A = \int\!\!\int_D \|\mathbf{r}_u \times \mathbf{r}_v\| \, dA$$
$$= \int_0^{2\pi}\int_0^{2\pi} (2 + \cos u) \, du \, dv$$
$$= \int_0^{2\pi} 4\pi \, dv$$
$$= 8\pi^2.$$

■

Exploration

For the torus in Example 7, describe the function $\mathbf{r}(u, v)$ for fixed u. Then describe the function $\mathbf{r}(u, v)$ for fixed v.

For a surface of revolution, you can show that the formula for surface area given in Section 7.4 is equivalent to the formula given in this section. For instance, suppose f is a nonnegative function such that f' is continuous over the interval $[a, b]$. Let S be the surface of revolution formed by revolving the graph of f, where $a \le x \le b$, about the x-axis. From Section 7.4, you know that the surface area is given by

$$\text{Surface area} = 2\pi\int_a^b f(x)\sqrt{1 + [f'(x)]^2} \, dx.$$

To represent S parametrically, let

$$x = u, \quad y = f(u) \cos v, \quad \text{and} \quad z = f(u) \sin v$$

where $a \le u \le b$ and $0 \le v \le 2\pi$. Then,

$$\mathbf{r}(u, v) = u\mathbf{i} + f(u) \cos v\mathbf{j} + f(u) \sin v\mathbf{k}.$$

Try showing that the formula

$$\text{Surface area} = \int\!\!\int_D \|\mathbf{r}_u \times \mathbf{r}_v\| \, dA$$

is equivalent to the formula given above (see Exercise 58).

15.5 Exercises

See CalcChat.com for tutorial help and worked-out solutions to odd-numbered exercises.

Matching In Exercises 1–6, match the vector-valued function with its graph. [The graphs are labeled (a), (b), (c), (d), (e), and (f).]

(a)

(b)

(c)

(d)

(e)

(f)

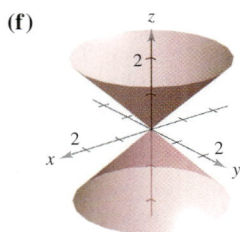

1. $\mathbf{r}(u, v) = u\mathbf{i} + v\mathbf{j} + uv\mathbf{k}$

2. $\mathbf{r}(u, v) = u \cos v\mathbf{i} + u \sin v\mathbf{j} + u\mathbf{k}$

3. $\mathbf{r}(u, v) = u\mathbf{i} + \frac{1}{2}(u + v)\mathbf{j} + v\mathbf{k}$

4. $\mathbf{r}(u, v) = u\mathbf{i} + \frac{1}{4}v^3\mathbf{j} + v\mathbf{k}$

5. $\mathbf{r}(u, v) = 2 \cos v \cos u\mathbf{i} + 2 \cos v \sin u\mathbf{j} + 2 \sin v\mathbf{k}$

6. $\mathbf{r}(u, v) = 4 \cos u\mathbf{i} + 4 \sin u\mathbf{j} + v\mathbf{k}$

Sketching a Parametric Surface In Exercises 7–10, find the rectangular equation for the surface by eliminating the parameters from the vector-valued function. Identify the surface and sketch its graph.

7. $\mathbf{r}(u, v) = u\mathbf{i} + v\mathbf{j} + \dfrac{v}{2}\mathbf{k}$

8. $\mathbf{r}(u, v) = 2u \cos v\mathbf{i} + 2u \sin v\mathbf{j} + \frac{1}{2}u^2\mathbf{k}$

9. $\mathbf{r}(u, v) = 2 \cos u\mathbf{i} + v\mathbf{j} + 2 \sin u\mathbf{k}$

10. $\mathbf{r}(u, v) = 3 \cos v \cos u\mathbf{i} + 3 \cos v \sin u\mathbf{j} + 5 \sin v\mathbf{k}$

Graphing a Parametric Surface In Exercises 11–16, use a computer algebra system to graph the surface represented by the vector-valued function.

11. $\mathbf{r}(u, v) = 2u \cos v\mathbf{i} + 2u \sin v\mathbf{j} + u^4\mathbf{k}$

$0 \le u \le 1, \quad 0 \le v \le 2\pi$

12. $\mathbf{r}(u, v) = 2 \cos v \cos u\mathbf{i} + 4 \cos v \sin u\mathbf{j} + \sin v\mathbf{k}$

$0 \le u \le 2\pi, \quad 0 \le v \le 2\pi$

13. $\mathbf{r}(u, v) = 2 \sinh u \cos v\mathbf{i} + \sinh u \sin v\mathbf{j} + \cosh u\mathbf{k}$

$0 \le u \le 2, \quad 0 \le v \le 2\pi$

14. $\mathbf{r}(u, v) = 2u \cos v\mathbf{i} + 2u \sin v\mathbf{j} + v\mathbf{k}$

$0 \le u \le 1, \quad 0 \le v \le 3\pi$

15. $\mathbf{r}(u, v) = (u - \sin u) \cos v\mathbf{i} + (1 - \cos u) \sin v\mathbf{j} + u\mathbf{k}$

$0 \le u \le \pi, \quad 0 \le v \le 2\pi$

16. $\mathbf{r}(u, v) = \cos^3 u \cos v\mathbf{i} + \sin^3 u \sin v\mathbf{j} + u\mathbf{k}$

$0 \le u \le \dfrac{\pi}{2}, \quad 0 \le v \le 2\pi$

Think About It In Exercises 17–20, determine how the graph of the surface $s(u, v)$ differs from the graph of $\mathbf{r}(u, v) = u \cos v\mathbf{i} + u \sin v\mathbf{j} + u^2\mathbf{k}$ (see figure), where $0 \le u \le 2$ and $0 \le v \le 2\pi$. (It is not necessary to graph s.)

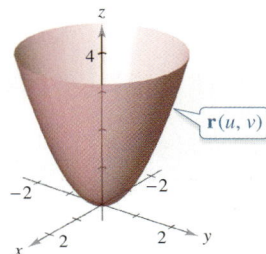

17. $s(u, v) = u \cos v\mathbf{i} + u \sin v\mathbf{j} - u^2\mathbf{k}$

$0 \le u \le 2, \quad 0 \le v \le 2\pi$

18. $s(u, v) = u \cos v\mathbf{i} + u^2\mathbf{j} + u \sin v\mathbf{k}$

$0 \le u \le 2, \quad 0 \le v \le 2\pi$

19. $s(u, v) = u \cos v\mathbf{i} + u \sin v\mathbf{j} + u^2\mathbf{k}$

$0 \le u \le 3, \quad 0 \le v \le 2\pi$

20. $s(u, v) = 4u \cos v\mathbf{i} + 4u \sin v\mathbf{j} + u^2\mathbf{k}$

$0 \le u \le 2, \quad 0 \le v \le 2\pi$

Representing a Surface Parametrically In Exercises 21–30, find a vector-valued function whose graph is the indicated surface.

21. The plane $z = y$

22. The plane $x + y + z = 6$

23. The cone $y = \sqrt{4x^2 + 9z^2}$

24. The cone $x = \sqrt{16y^2 + z^2}$

25. The cylinder $x^2 + y^2 = 25$

26. The cylinder $4x^2 + y^2 = 16$

27. The cylinder $z = x^2$

28. The ellipsoid $\dfrac{x^2}{9} + \dfrac{y^2}{4} + \dfrac{z^2}{1} = 1$

29. The part of the plane $z = 4$ that lies inside the cylinder $x^2 + y^2 = 9$

30. The part of the paraboloid $z = x^2 + y^2$ that lies inside the cylinder $x^2 + y^2 = 9$

Surface of Revolution In Exercises 31–34, write a set of parametric equations for the surface of revolution obtained by revolving the graph of the function about the given axis.

Function	Axis of Revolution
31. $y = \dfrac{x}{2}$, $\quad 0 \le x \le 6$	x-axis
32. $y = \sqrt{x}$, $\quad 0 \le x \le 4$	x-axis
33. $x = \sin z$, $\quad 0 \le z \le \pi$	z-axis
34. $z = y^2 + 1$, $\quad 0 \le y \le 2$	y-axis

Tangent Plane In Exercises 35–38, find an equation of the tangent plane to the surface represented by the vector-valued function at the given point.

35. $\mathbf{r}(u, v) = (u + v)\mathbf{i} + (u - v)\mathbf{j} + v\mathbf{k}$, $(1, -1, 1)$

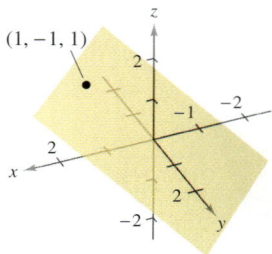

Figure for 35   Figure for 36

36. $\mathbf{r}(u, v) = u\mathbf{i} + v\mathbf{j} + \sqrt{uv}\,\mathbf{k}$, $(1, 1, 1)$

37. $\mathbf{r}(u, v) = 2u \cos v\mathbf{i} + 3u \sin v\mathbf{j} + u^2\mathbf{k}$, $(0, 6, 4)$

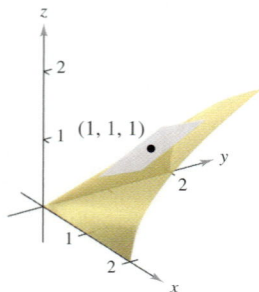

38. $\mathbf{r}(u, v) = 2u \cosh v\mathbf{i} + 2u \sinh v\mathbf{j} + \tfrac{1}{2}u^2\mathbf{k}$, $(-4, 0, 2)$

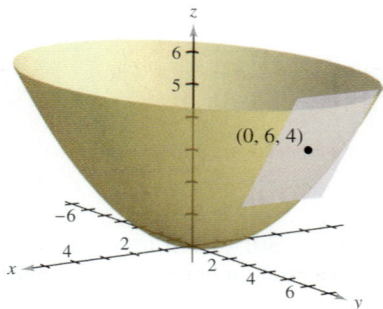

Area In Exercises 39–46, find the area of the surface over the given region. Use a computer algebra system to verify your results.

39. The part of the plane $\mathbf{r}(u, v) = 4u\mathbf{i} - v\mathbf{j} + v\mathbf{k}$, where $0 \le u \le 2$ and $0 \le v \le 1$

40. The part of the paraboloid $\mathbf{r}(u, v) = 2u \cos v\mathbf{i} + 2u \sin v\mathbf{j} + u^2\mathbf{k}$, where $0 \le u \le 2$ and $0 \le v \le 2\pi$

41. The part of the cylinder $\mathbf{r}(u, v) = a \cos u\mathbf{i} + a \sin u\mathbf{j} + v\mathbf{k}$, where $0 \le u \le 2\pi$ and $0 \le v \le b$

42. The sphere $\mathbf{r}(u, v) = a \sin u \cos v\mathbf{i} + a \sin u \sin v\mathbf{j} + a \cos u\mathbf{k}$, where $0 \le u \le \pi$ and $0 \le v \le 2\pi$

43. The part of the cone $\mathbf{r}(u, v) = au \cos v\mathbf{i} + au \sin v\mathbf{j} + u\mathbf{k}$, where $0 \le u \le b$ and $0 \le v \le 2\pi$

44. The torus $\mathbf{r}(u, v) = (a + b \cos v) \cos u\mathbf{i} + (a + b \cos v) \sin u\mathbf{j} + b \sin v\mathbf{k}$, where $a > b$, $0 \le u \le 2\pi$, and $0 \le v \le 2\pi$

45. The surface of revolution $\mathbf{r}(u, v) = \sqrt{u} \cos v\mathbf{i} + \sqrt{u} \sin v\mathbf{j} + u\mathbf{k}$, where $0 \le u \le 4$ and $0 \le v \le 2\pi$

46. The surface of revolution $\mathbf{r}(u, v) = \sin u \cos v\mathbf{i} + u\mathbf{j} + \sin u \sin v\mathbf{k}$, where $0 \le u \le \pi$ and $0 \le v \le 2\pi$

WRITING ABOUT CONCEPTS

47. Parametric Surface Define a parametric surface.

48. Surface Area Give the double integral that yields the surface area of a parametric surface over an open region D.

49. Representing a Cone Parametrically Show that the cone in Example 3 can be represented parametrically by $\mathbf{r}(u, v) = u \cos v\mathbf{i} + u \sin v\mathbf{j} + u\mathbf{k}$, where $0 \le u$ and $0 \le v \le 2\pi$.

50. HOW DO YOU SEE IT? The figures below are graphs of $\mathbf{r}(u, v) = u\mathbf{i} + \sin u \cos v\mathbf{j} + \sin u \sin v\mathbf{k}$, where $0 \le u \le \pi/2$ and $0 \le v \le 2\pi$. Match each of the graphs with the point in space from which the surface is viewed. The points are $(10, 0, 0)$, $(-10, 10, 0)$, $(0, 10, 0)$, and $(10, 10, 10)$.

(a)   (b)

(c)   (d)

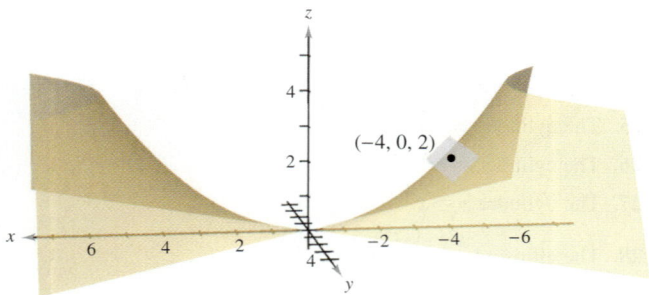

51. Astroidal Sphere   An equation of an **astroidal sphere** in x, y, and z is

$$x^{2/3} + y^{2/3} + z^{2/3} = a^{2/3}.$$

A graph of an astroidal sphere is shown below. Show that this surface can be represented parametrically by

$$\mathbf{r}(u, v) = a \sin^3 u \cos^3 v \mathbf{i} + a \sin^3 u \sin^3 v \mathbf{j} + a \cos^3 u \mathbf{k}$$

where $0 \leq u \leq \pi$ and $0 \leq v \leq 2\pi$.

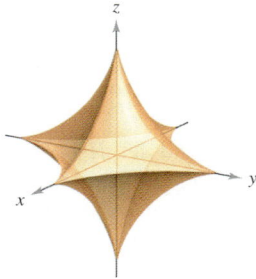

52. Different Views of a Surface   Use a computer algebra system to graph three views of the graph of the vector-valued function

$$\mathbf{r}(u, v) = u \cos v \mathbf{i} + u \sin v \mathbf{j} + v \mathbf{k}, \quad 0 \leq u \leq \pi, \; 0 \leq v \leq \pi$$

from the points $(10, 0, 0)$, $(0, 0, 10)$, and $(10, 10, 10)$.

53. Investigation   Use a computer algebra system to graph the torus

$$\mathbf{r}(u, v) = (a + b \cos v) \cos u \mathbf{i} +$$
$$(a + b \cos v) \sin u \mathbf{j} + b \sin v \mathbf{k}$$

for each set of values of a and b, where $0 \leq u \leq 2\pi$ and $0 \leq v \leq 2\pi$. Use the results to describe the effects of a and b on the shape of the torus.

(a) $a = 4$, $b = 1$ (b) $a = 4$, $b = 2$

(c) $a = 8$, $b = 1$ (d) $a = 8$, $b = 3$

54. Investigation   Consider the function in Exercise 14.

(a) Sketch a graph of the function where u is held constant at $u = 1$. Identify the graph.

(b) Sketch a graph of the function where v is held constant at $v = 2\pi/3$. Identify the graph.

(c) Assume that a surface is represented by the vector-valued function $\mathbf{r} = \mathbf{r}(u, v)$. What generalization can you make about the graph of the function when one of the parameters is held constant?

55. Surface Area   The surface of the dome on a new museum is given by

$$\mathbf{r}(u, v) = 20 \sin u \cos v \mathbf{i} + 20 \sin u \sin v \mathbf{j} + 20 \cos u \mathbf{k}$$

where $0 \leq u \leq \pi/3, 0 \leq v \leq 2\pi$, and $\mathbf{r}$ is in meters. Find the surface area of the dome.

56. Hyperboloid   Find a vector-valued function for the hyperboloid

$$x^2 + y^2 - z^2 = 1$$

and determine the tangent plane at $(1, 0, 0)$.

57. Area   Graph and find the area of one turn of the spiral ramp

$$\mathbf{r}(u, v) = u \cos v \mathbf{i} + u \sin v \mathbf{j} + 2v \mathbf{k}$$

where $0 \leq u \leq 3$ and $0 \leq v \leq 2\pi$.

58. Surface Area   Let f be a nonnegative function such that f' is continuous over the interval $[a, b]$. Let S be the surface of revolution formed by revolving the graph of f, where $a \leq x \leq b$, about the x-axis. Let $x = u$, $y = f(u) \cos v$, and $z = f(u) \sin v$, where $a \leq u \leq b$ and $0 \leq v \leq 2\pi$. Then, S is represented parametrically by $\mathbf{r}(u, v) = u\mathbf{i} + f(u) \cos v \mathbf{j} + f(u) \sin v \mathbf{k}$. Show that the following formulas are equivalent.

$$\text{Surface area} = 2\pi \int_a^b f(x) \sqrt{1 + [f'(x)]^2} \, dx$$

$$\text{Surface area} = \int\int_D \|\mathbf{r}_u \times \mathbf{r}_v\| \, dA$$

59. Open-Ended Project   The parametric equations

$$x = 3 + \sin u[7 - \cos(3u - 2v) - 2\cos(3u + v)]$$
$$y = 3 + \cos u[7 - \cos(3u - 2v) - 2\cos(3u + v)]$$
$$z = \sin(3u - 2v) + 2\sin(3u + v)$$

where $-\pi \leq u \leq \pi$ and $-\pi \leq v \leq \pi$, represent the surface shown below. Try to create your own parametric surface using a computer algebra system.

60. Möbius Strip   The surface shown in the figure is called a **Möbius strip** and can be represented by the parametric equations

$$x = \left(a + u \cos \frac{v}{2}\right) \cos v, \; y = \left(a + u \cos \frac{v}{2}\right) \sin v, \; z = u \sin \frac{v}{2}$$

where $-1 \leq u \leq 1, 0 \leq v \leq 2\pi$, and $a = 3$. Try to graph other Möbius strips for different values of a using a computer algebra system.

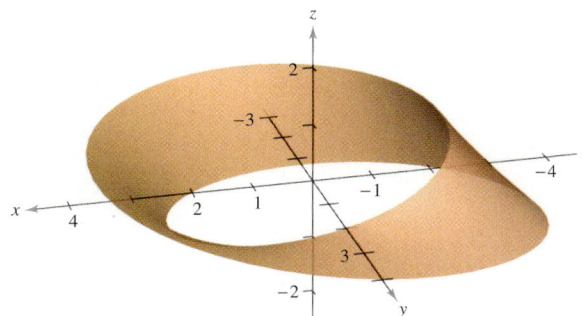

15.6 Surface Integrals

■ Evaluate a surface integral as a double integral.
■ Evaluate a surface integral for a parametric surface.
■ Determine the orientation of a surface.
■ Understand the concept of a flux integral.

Surface Integrals

The remainder of this chapter deals primarily with **surface integrals.** You will first consider surfaces given by $z = g(x, y)$. Later in this section, you will consider more general surfaces given in parametric form.

Let S be a surface given by $z = g(x, y)$ and let R be its projection onto the xy-plane, as shown in Figure 15.44. Let g, g_x, and g_y be continuous at all points in R and let f be a scalar function defined on S. Employing the procedure used to find surface area in Section 14.5, evaluate f at (x_i, y_i, z_i) and form the sum

$$\sum_{i=1}^{n} f(x_i, y_i, z_i) \, \Delta S_i$$

where

$$\Delta S_i \approx \sqrt{1 + [g_x(x_i, y_i)]^2 + [g_y(x_i, y_i)]^2} \, \Delta A_i.$$

Provided the limit of this sum as $\|\Delta\|$ approaches 0 exists, the **surface integral of f over** S is defined as

$$\int_S \int f(x, y, z) \, dS = \lim_{\|\Delta\| \to 0} \sum_{i=1}^{n} f(x_i, y_i, z_i) \, \Delta S_i.$$

This integral can be evaluated by a double integral.

Scalar function f assigns a number to each point of S.
Figure 15.44

THEOREM 15.10   Evaluating a Surface Integral

Let S be a surface with equation $z = g(x, y)$ and let R be its projection onto the xy-plane. If g, g_x, and g_y are continuous on R and f is continuous on S, then the surface integral of f over S is

$$\int_S \int f(x, y, z) \, dS = \int_R \int f(x, y, g(x, y)) \sqrt{1 + [g_x(x, y)]^2 + [g_y(x, y)]^2} \, dA.$$

For surfaces described by functions of x and z (or y and z), you can make the following adjustments to Theorem 15.10. If S is the graph of $y = g(x, z)$ and R is its projection onto the xz-plane, then

$$\int_S \int f(x, y, z) \, dS = \int_R \int f(x, g(x, z), z) \sqrt{1 + [g_x(x, z)]^2 + [g_z(x, z)]^2} \, dA.$$

If S is the graph of $x = g(y, z)$ and R is its projection onto the yz-plane, then

$$\int_S \int f(x, y, z) \, dS = \int_R \int f(g(y, z), y, z) \sqrt{1 + [g_y(y, z)]^2 + [g_z(y, z)]^2} \, dA.$$

If $f(x, y, z) = 1$, the surface integral over S yields the surface area of S. For instance, suppose the surface S is the plane given by $z = x$, where $0 \le x \le 1$ and $0 \le y \le 1$. The surface area of S is $\sqrt{2}$ square units. Try verifying that

$$\int_S \int f(x, y, z) \, dS = \sqrt{2}.$$

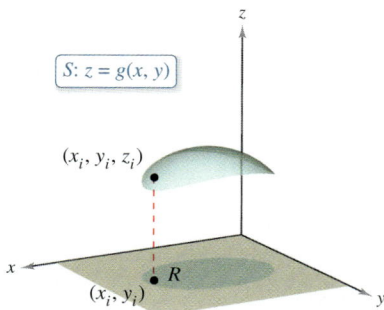

EXAMPLE 1   **Evaluating a Surface Integral**

Evaluate the surface integral

$$\iint_S (y^2 + 2yz)\, dS$$

where S is the first-octant portion of the plane

$$2x + y + 2z = 6.$$

Solution   Begin by writing S as

$$z = \frac{1}{2}(6 - 2x - y)$$

$$g(x, y) = \frac{1}{2}(6 - 2x - y).$$

Using the partial derivatives $g_x(x, y) = -1$ and $g_y(x, y) = -\frac{1}{2}$, you can write

$$\sqrt{1 + [g_x(x, y)]^2 + [g_y(x, y)]^2} = \sqrt{1 + 1 + \frac{1}{4}} = \frac{3}{2}.$$

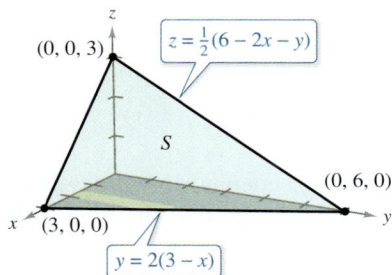

$z = \frac{1}{2}(6 - 2x - y)$

$(0, 0, 3)$

S

$(0, 6, 0)$

$(3, 0, 0)$

$y = 2(3 - x)$

Figure 15.45

Using Figure 15.45 and Theorem 15.10, you obtain

$$\iint_S (y^2 + 2yz)\, dS = \iint_R f(x, y, g(x, y)) \sqrt{1 + [g_x(x, y)]^2 + [g_y(x, y)]^2}\, dA$$

$$= \iint_R \left[y^2 + 2y\left(\frac{1}{2}\right)(6 - 2x - y) \right]\left(\frac{3}{2}\right) dA$$

$$= 3 \int_0^3 \int_0^{2(3-x)} y(3 - x)\, dy\, dx$$

$$= 6 \int_0^3 (3 - x)^3\, dx$$

$$= -\frac{3}{2}(3 - x)^4 \Big]_0^3$$

$$= \frac{243}{2}.$$

An alternative solution to Example 1 would be to project S onto the yz-plane, as shown in Figure 15.46. Then, $x = \frac{1}{2}(6 - y - 2z)$, and

$$\sqrt{1 + [g_y(y, z)]^2 + [g_z(y, z)]^2} = \sqrt{1 + \frac{1}{4} + 1} = \frac{3}{2}.$$

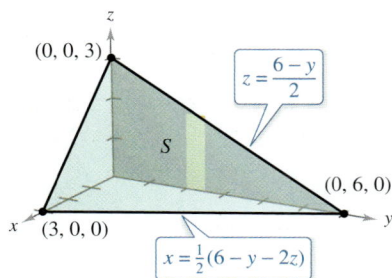

$(0, 0, 3)$

$z = \frac{6 - y}{2}$

S

$(0, 6, 0)$

$(3, 0, 0)$

$x = \frac{1}{2}(6 - y - 2z)$

Figure 15.46

So, the surface integral is

$$\iint_S (y^2 + 2yz)\, dS = \iint_R f(g(y, z), y, z) \sqrt{1 + [g_y(y, z)]^2 + [g_z(y, z)]^2}\, dA$$

$$= \int_0^6 \int_0^{(6-y)/2} (y^2 + 2yz)\left(\frac{3}{2}\right) dz\, dy$$

$$= \frac{3}{8} \int_0^6 (36y - y^3)\, dy$$

$$= \frac{243}{2}.$$

Try reworking Example 1 by projecting S onto the xz-plane.

In Example 1, you could have projected the surface S onto any one of the three coordinate planes. In Example 2, S is a portion of a cylinder centered about the x-axis, and you can project it onto either the xz-plane or the xy-plane.

EXAMPLE 2   Evaluating a Surface Integral

•••▷ *See LarsonCalculus.com for an interactive version of this type of example.*

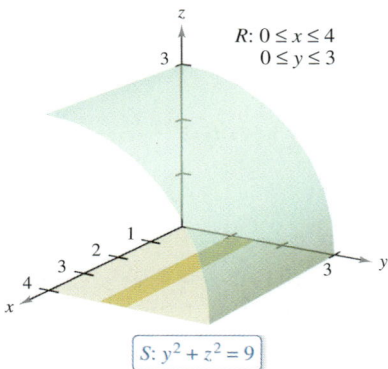

Evaluate the surface integral

$$\iint_S (x + z)\, dS$$

where S is the first-octant portion of the cylinder

$$y^2 + z^2 = 9$$

between $x = 0$ and $x = 4$, as shown in Figure 15.47.

Solution   Project S onto the xy-plane, so that

$$z = g(x, y) = \sqrt{9 - y^2}$$

and obtain

$$\sqrt{1 + [g_x(x, y)]^2 + [g_y(x, y)]^2} = \sqrt{1 + \left(\frac{-y}{\sqrt{9 - y^2}}\right)^2}$$

$$= \frac{3}{\sqrt{9 - y^2}}.$$

Theorem 15.10 does not apply directly, because g_y is not continuous when $y = 3$. However, you can apply Theorem 15.10 for $0 \le b < 3$ and then take the limit as b approaches 3, as follows.

$$\iint_S (x + z)\, dS = \lim_{b \to 3^-} \int_0^b \int_0^4 \left(x + \sqrt{9 - y^2}\right)\frac{3}{\sqrt{9 - y^2}}\, dx\, dy$$

$$= \lim_{b \to 3^-} 3\int_0^b \int_0^4 \left(\frac{x}{\sqrt{9 - y^2}} + 1\right) dx\, dy$$

$$= \lim_{b \to 3^-} 3\int_0^b \frac{x^2}{2\sqrt{9 - y^2}} + x \Big]_0^4 \, dy$$

$$= \lim_{b \to 3^-} 3\int_0^b \left(\frac{8}{\sqrt{9 - y^2}} + 4\right) dy$$

$$= \lim_{b \to 3^-} 3\left[4y + 8\arcsin\frac{y}{3}\right]_0^b$$

$$= \lim_{b \to 3^-} 3\left(4b + 8\arcsin\frac{b}{3}\right)$$

$$= 36 + 24\left(\frac{\pi}{2}\right)$$

$$= 36 + 12\pi$$ ■

$R: 0 \le x \le 4$
$0 \le y \le 3$

$S: y^2 + z^2 = 9$

Figure 15.47

▷ **TECHNOLOGY**   Some computer algebra systems are capable of evaluating improper integrals. If you have access to such computer software, use it to evaluate the improper integral

$$\int_0^3 \int_0^4 (x + \sqrt{9 - y^2})\frac{3}{\sqrt{9 - y^2}}\, dx\, dy.$$

Do you obtain the same result as in Example 2?

You have already seen that when the function f defined on the surface S is simply $f(x, y, z) = 1$, the surface integral yields the *surface area* of S.

$$\text{Area of surface} = \iint_S 1 \, dS$$

On the other hand, when S is a lamina of variable density and $\rho(x, y, z)$ is the density at the point (x, y, z), then the *mass* of the lamina is given by

$$\text{Mass of lamina} = \iint_S \rho(x, y, z) \, dS.$$

EXAMPLE 3   **Finding the Mass of a Surface Lamina**

A cone-shaped surface lamina S is given by

$$z = 4 - 2\sqrt{x^2 + y^2}, \quad 0 \le z \le 4$$

as shown in Figure 15.48. At each point on S, the density is proportional to the distance between the point and the z-axis. Find the mass m of the lamina.

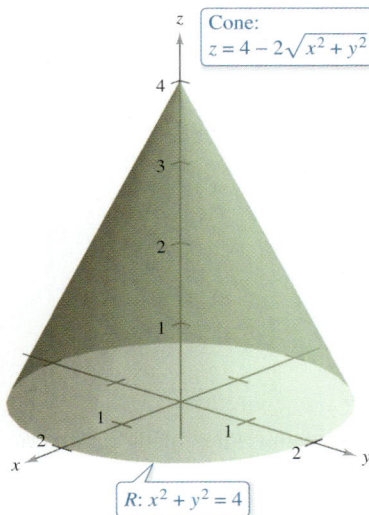

Cone:
$z = 4 - 2\sqrt{x^2 + y^2}$

$R: x^2 + y^2 = 4$

Figure 15.48

Solution   Projecting S onto the xy-plane produces

$$S: z = 4 - 2\sqrt{x^2 + y^2} = g(x, y), \quad 0 \le z \le 4$$
$$R: x^2 + y^2 \le 4$$

with a density of $\rho(x, y, z) = k\sqrt{x^2 + y^2}$. Using a surface integral, you can find the mass to be

$$
\begin{aligned}
m &= \iint_S \rho(x, y, z) \, dS \\
&= \iint_R k\sqrt{x^2 + y^2}\sqrt{1 + [g_x(x, y)]^2 + [g_y(x, y)]^2} \, dA \\
&= k\iint_R \sqrt{x^2 + y^2}\sqrt{1 + \frac{4x^2}{x^2 + y^2} + \frac{4y^2}{x^2 + y^2}} \, dA \\
&= k\iint_R \sqrt{5}\sqrt{x^2 + y^2} \, dA \\
&= k\int_0^{2\pi}\int_0^2 \left(\sqrt{5}r\right)r \, dr \, d\theta \qquad \text{\color{red}Polar coordinates} \\
&= \frac{\sqrt{5}k}{3}\int_0^{2\pi} r^3\bigg]_0^2 \, d\theta \\
&= \frac{8\sqrt{5}k}{3}\int_0^{2\pi} d\theta \\
&= \frac{8\sqrt{5}k}{3}\bigg[\theta\bigg]_0^{2\pi} \\
&= \frac{16\sqrt{5}k\pi}{3}.
\end{aligned}
$$

▷ **TECHNOLOGY**   Use a computer algebra system to confirm the result shown in Example 3. The computer algebra system *Mathematica* evaluated the integral as follows.

$$k\int_{-2}^2\int_{-\sqrt{4-y^2}}^{\sqrt{4-y^2}} \sqrt{5}\sqrt{x^2 + y^2} \, dx \, dy = k\int_0^{2\pi}\int_0^2 \left(\sqrt{5}r\right)r \, dr \, d\theta = \frac{16\sqrt{5}k\pi}{3}$$

Parametric Surfaces and Surface Integrals

For a surface S given by the vector-valued function

$$\mathbf{r}(u, v) = x(u, v)\mathbf{i} + y(u, v)\mathbf{j} + z(u, v)\mathbf{k} \qquad \text{Parametric surface}$$

defined over a region D in the uv-plane, you can show that the surface integral of $f(x, y, z)$ over S is given by

$$\iint_S f(x, y, z) \, dS = \iint_D f(x(u, v), y(u, v), z(u, v))\|\mathbf{r}_u(u, v) \times \mathbf{r}_v(u, v)\| \, dA.$$

Note the similarity to a line integral over a space curve C.

$$\int_C f(x, y, z) \, ds = \int_a^b f(x(t), y(t), z(t))\|\mathbf{r}'(t)\| \, dt \qquad \text{Line integral}$$

Also, notice that ds and dS can be written as

$$ds = \|\mathbf{r}'(t)\| \, dt \quad \text{and} \quad dS = \|\mathbf{r}_u(u, v) \times \mathbf{r}_v(u, v)\| \, dA.$$

EXAMPLE 4 Evaluating a Surface Integral

Example 2 demonstrated an evaluation of the surface integral

$$\iint_S (x + z) \, dS$$

where S is the first-octant portion of the cylinder

$$y^2 + z^2 = 9$$

between $x = 0$ and $x = 4$ (see Figure 15.49). Reevaluate this integral in parametric form.

Solution In parametric form, the surface is given by

$$\mathbf{r}(x, \theta) = x\mathbf{i} + 3 \cos \theta \mathbf{j} + 3 \sin \theta \mathbf{k}$$

where $0 \le x \le 4$ and $0 \le \theta \le \pi/2$. To evaluate the surface integral in parametric form, begin by calculating the following.

$$\mathbf{r}_x = \mathbf{i}$$
$$\mathbf{r}_\theta = -3 \sin \theta \mathbf{j} + 3 \cos \theta \mathbf{k}$$

$$\mathbf{r}_x \times \mathbf{r}_\theta = \begin{vmatrix} \mathbf{i} & \mathbf{j} & \mathbf{k} \\ 1 & 0 & 0 \\ 0 & -3 \sin \theta & 3 \cos \theta \end{vmatrix} = -3 \cos \theta \mathbf{j} - 3 \sin \theta \mathbf{k}$$

$$\|\mathbf{r}_x \times \mathbf{r}_\theta\| = \sqrt{9 \cos^2 \theta + 9 \sin^2 \theta} = 3$$

So, the surface integral can be evaluated as follows.

$$\iint_D (x + 3 \sin \theta)3 \, dA = \int_0^4 \int_0^{\pi/2} (3x + 9 \sin \theta) \, d\theta \, dx$$

$$= \int_0^4 \left[3x\theta - 9 \cos \theta \right]_0^{\pi/2} dx$$

$$= \int_0^4 \left(\frac{3\pi}{2}x + 9 \right) dx$$

$$= \left[\frac{3\pi}{4}x^2 + 9x \right]_0^4$$

$$= 12\pi + 36$$

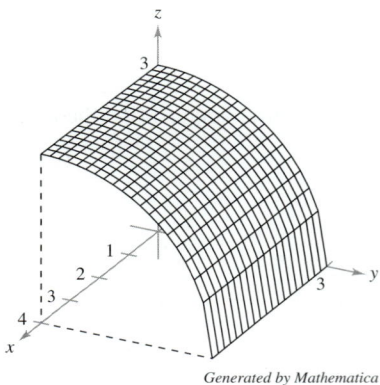

Generated by Mathematica

Figure 15.49

Orientation of a Surface

Unit normal vectors are used to induce an orientation to a surface S in space. A surface is **orientable** when a unit normal vector $\mathbf{N}$ can be defined at every nonboundary point of S in such a way that the normal vectors vary continuously over the surface S. The surface S is called an **oriented surface.**

An orientable surface S has two distinct sides. So, when you orient a surface, you are selecting one of the two possible unit normal vectors. For a closed surface such as a sphere, it is customary to choose the unit normal vector $\mathbf{N}$ to be the one that points outward from the sphere.

Most common surfaces, such as spheres, paraboloids, ellipses, and planes, are orientable. (See Exercise 43 for an example of a surface that is *not* orientable.) Moreover, for an orientable surface, the gradient vector provides a convenient way to find a unit normal vector. That is, for an orientable surface S given by

$$z = g(x, y) \qquad \text{Orientable surface}$$

let

$$G(x, y, z) = z - g(x, y).$$

Then, S can be oriented by either the unit normal vector

$$\mathbf{N} = \frac{\nabla G(x, y, z)}{\|\nabla G(x, y, z)\|}$$

$$= \frac{-g_x(x, y)\mathbf{i} - g_y(x, y)\mathbf{j} + \mathbf{k}}{\sqrt{1 + [g_x(x, y)]^2 + [g_y(x, y)]^2}} \qquad \text{Upward unit normal vector}$$

or the unit normal vector

$$\mathbf{N} = \frac{-\nabla G(x, y, z)}{\|\nabla G(x, y, z)\|}$$

$$= \frac{g_x(x, y)\mathbf{i} + g_y(x, y)\mathbf{j} - \mathbf{k}}{\sqrt{1 + [g_x(x, y)]^2 + [g_y(x, y)]^2}} \qquad \text{Downward unit normal vector}$$

as shown in Figure 15.50. If the smooth orientable surface S is given in parametric form by

$$\mathbf{r}(u, v) = x(u, v)\mathbf{i} + y(u, v)\mathbf{j} + z(u, v)\mathbf{k} \qquad \text{Parametric surface}$$

then the unit normal vectors are given by

$$\mathbf{N} = \frac{\mathbf{r}_u \times \mathbf{r}_v}{\|\mathbf{r}_u \times \mathbf{r}_v\|}$$

and

$$\mathbf{N} = \frac{\mathbf{r}_v \times \mathbf{r}_u}{\|\mathbf{r}_v \times \mathbf{r}_u\|}.$$

For an orientable surface given by

$$y = g(x, z) \quad \text{or} \quad x = g(y, z)$$

you can use the gradient vector

$$\nabla G(x, y, z) = -g_x(x, z)\mathbf{i} + \mathbf{j} - g_z(x, z)\mathbf{k} \qquad G(x, y, z) = y - g(x, z)$$

or

$$\nabla G(x, y, z) = \mathbf{i} - g_y(y, z)\mathbf{j} - g_z(y, z)\mathbf{k} \qquad G(x, y, z) = x - g(y, z)$$

to orient the surface.

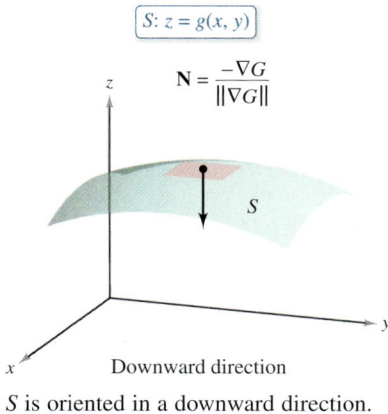

$S: z = g(x, y)$

$$\mathbf{N} = \frac{\nabla G}{\|\nabla G\|}$$

Upward direction

S is oriented in an upward direction.

$S: z = g(x, y)$

$$\mathbf{N} = \frac{-\nabla G}{\|\nabla G\|}$$

Downward direction

S is oriented in a downward direction.

Figure 15.50

Flux Integrals

One of the principal applications involving the vector form of a surface integral relates to the flow of a fluid through a surface. Consider an oriented surface S submerged in a fluid having a continuous velocity field $\mathbf{F}$. Let ΔS be the area of a small patch of the surface S over which $\mathbf{F}$ is nearly constant. Then the amount of fluid crossing this region per unit of time is approximated by the volume of the column of height $\mathbf{F} \cdot \mathbf{N}$, as shown in Figure 15.51. That is,

$$\Delta V = (\text{height})(\text{area of base})$$
$$= (\mathbf{F} \cdot \mathbf{N}) \, \Delta S.$$

Consequently, the volume of fluid crossing the surface S per unit of time (called the **flux of F across** S) is given by the surface integral in the next definition.

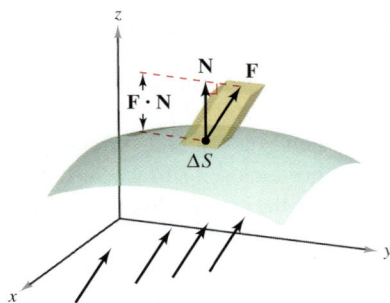

The velocity field $\mathbf{F}$ indicates the direction of the fluid flow.

Figure 15.51

Definition of Flux Integral

Let $\mathbf{F}(x, y, z) = M\mathbf{i} + N\mathbf{j} + P\mathbf{k}$, where M, N, and P have continuous first partial derivatives on the surface S oriented by a unit normal vector $\mathbf{N}$. The **flux integral of F across** S is given by

$$\iint_S \mathbf{F} \cdot \mathbf{N} \, dS.$$

Geometrically, a flux integral is the surface integral over S of the *normal component* of $\mathbf{F}$. If $\rho(x, y, z)$ is the density of the fluid at (x, y, z), then the flux integral

$$\iint_S \rho \mathbf{F} \cdot \mathbf{N} \, dS$$

represents the *mass* of the fluid flowing across S per unit of time.

To evaluate a flux integral for a surface given by $z = g(x, y)$, let

$$G(x, y, z) = z - g(x, y).$$

Then, $\mathbf{N} \, dS$ can be written as follows.

$$\mathbf{N} \, dS = \frac{\nabla G(x, y, z)}{\|\nabla G(x, y, z)\|} \, dS$$

$$= \frac{\nabla G(x, y, z)}{\sqrt{(g_x)^2 + (g_y)^2 + 1}} \sqrt{(g_x)^2 + (g_y)^2 + 1} \, dA$$

$$= \nabla G(x, y, z) \, dA$$

THEOREM 15.11 Evaluating a Flux Integral

Let S be an oriented surface given by $z = g(x, y)$ and let R be its projection onto the xy-plane.

$$\iint_S \mathbf{F} \cdot \mathbf{N} \, dS = \iint_R \mathbf{F} \cdot [-g_x(x, y)\mathbf{i} - g_y(x, y)\mathbf{j} + \mathbf{k}] \, dA \qquad \text{Oriented upward}$$

$$\iint_S \mathbf{F} \cdot \mathbf{N} \, dS = \iint_R \mathbf{F} \cdot [g_x(x, y)\mathbf{i} + g_y(x, y)\mathbf{j} - \mathbf{k}] \, dA \qquad \text{Oriented downward}$$

For the first integral, the surface is oriented upward, and for the second integral, the surface is oriented downward.

EXAMPLE 5   **Using a Flux Integral to Find the Rate of Mass Flow**

Let S be the portion of the paraboloid

$$z = g(x, y) = 4 - x^2 - y^2$$

lying above the xy-plane, oriented by an upward
unit normal vector, as shown in Figure 15.52.
A fluid of constant density ρ is flowing through
the surface S according to the vector field

$$\mathbf{F}(x, y, z) = x\mathbf{i} + y\mathbf{j} + z\mathbf{k}.$$

Find the rate of mass flow through S.

Solution   Begin by computing the partial
derivatives of g.

$$g_x(x, y) = -2x$$

and

$$g_y(x, y) = -2y$$

The rate of mass flow through the surface S is

$$\iint_S \rho \mathbf{F} \cdot \mathbf{N} \, dS = \rho \iint_R \mathbf{F} \cdot [-g_x(x, y)\mathbf{i} - g_y(x, y)\mathbf{j} + \mathbf{k}] \, dA$$

$$= \rho \iint_R [x\mathbf{i} + y\mathbf{j} + (4 - x^2 - y^2)\mathbf{k}] \cdot (2x\mathbf{i} + 2y\mathbf{j} + \mathbf{k}) \, dA$$

$$= \rho \iint_R [2x^2 + 2y^2 + (4 - x^2 - y^2)] \, dA$$

$$= \rho \iint_R (4 + x^2 + y^2) \, dA$$

$$= \rho \int_0^{2\pi} \int_0^2 (4 + r^2) r \, dr \, d\theta \qquad \text{Polar coordinates}$$

$$= \rho \int_0^{2\pi} 12 \, d\theta$$

$$= 24\pi\rho.$$

Figure 15.52

For an oriented surface S given by the vector-valued function

$$\mathbf{r}(u, v) = x(u, v)\mathbf{i} + y(u, v)\mathbf{j} + z(u, v)\mathbf{k} \qquad \text{Parametric surface}$$

defined over a region D in the uv-plane, you can define the flux integral of $\mathbf{F}$ across
S as

$$\iint_S \mathbf{F} \cdot \mathbf{N} \, dS = \iint_D \mathbf{F} \cdot \left(\frac{\mathbf{r}_u \times \mathbf{r}_v}{\|\mathbf{r}_u \times \mathbf{r}_v\|}\right) \|\mathbf{r}_u \times \mathbf{r}_v\| \, dA$$

$$= \iint_D \mathbf{F} \cdot (\mathbf{r}_u \times \mathbf{r}_v) \, dA.$$

Note the similarity of this integral to the line integral

$$\int_C \mathbf{F} \cdot d\mathbf{r} = \int_C \mathbf{F} \cdot \mathbf{T} \, ds.$$

A summary of formulas for line and surface integrals is presented on page 1103.

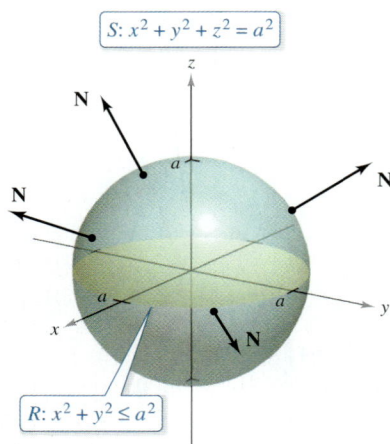

$S: x^2 + y^2 + z^2 = a^2$

$R: x^2 + y^2 \le a^2$

Figure 15.53

EXAMPLE 6 **Finding the Flux of an Inverse Square Field**

Find the flux over the sphere S given by

$$x^2 + y^2 + z^2 = a^2 \qquad \text{Sphere } S$$

where $\mathbf{F}$ is an inverse square field given by

$$\mathbf{F}(x, y, z) = \frac{kq}{\|\mathbf{r}\|^2} \frac{\mathbf{r}}{\|\mathbf{r}\|} = \frac{kq\mathbf{r}}{\|\mathbf{r}\|^3} \qquad \text{Inverse square field } \mathbf{F}$$

and

$$\mathbf{r} = x\mathbf{i} + y\mathbf{j} + z\mathbf{k}.$$

Assume S is oriented outward, as shown in Figure 15.53.

Solution The sphere is given by

$$\begin{aligned}
\mathbf{r}(u, v) &= x(u, v)\mathbf{i} + y(u, v)\mathbf{j} + z(u, v)\mathbf{k} \\
&= a \sin u \cos v\mathbf{i} + a \sin u \sin v\mathbf{j} + a \cos u\mathbf{k}
\end{aligned}$$

where $0 \le u \le \pi$ and $0 \le v \le 2\pi$. The partial derivatives of $\mathbf{r}$ are

$$\mathbf{r}_u(u, v) = a \cos u \cos v\mathbf{i} + a \cos u \sin v\mathbf{j} - a \sin u\mathbf{k}$$

and

$$\mathbf{r}_v(u, v) = -a \sin u \sin v\mathbf{i} + a \sin u \cos v\mathbf{j}$$

which implies that the normal vector $\mathbf{r}_u \times \mathbf{r}_v$ is

$$\begin{aligned}
\mathbf{r}_u \times \mathbf{r}_v &= \begin{vmatrix} \mathbf{i} & \mathbf{j} & \mathbf{k} \\ a \cos u \cos v & a \cos u \sin v & -a \sin u \\ -a \sin u \sin v & a \sin u \cos v & 0 \end{vmatrix} \\
&= a^2(\sin^2 u \cos v\mathbf{i} + \sin^2 u \sin v\mathbf{j} + \sin u \cos u\mathbf{k}).
\end{aligned}$$

Now, using

$$\begin{aligned}
\mathbf{F}(x, y, z) &= \frac{kq\mathbf{r}}{\|\mathbf{r}\|^3} \\
&= kq\frac{x\mathbf{i} + y\mathbf{j} + z\mathbf{k}}{\|x\mathbf{i} + y\mathbf{j} + z\mathbf{k}\|^3} \\
&= \frac{kq}{a^3}(a \sin u \cos v\mathbf{i} + a \sin u \sin v\mathbf{j} + a \cos u\mathbf{k})
\end{aligned}$$

it follows that

$$\begin{aligned}
\mathbf{F} \cdot (\mathbf{r}_u \times \mathbf{r}_v) &= \frac{kq}{a^3}[(a \sin u \cos v\mathbf{i} + a \sin u \sin v\mathbf{j} + a \cos u\mathbf{k}) \cdot \\
&\qquad a^2(\sin^2 u \cos v\mathbf{i} + \sin^2 u \sin v\mathbf{j} + \sin u \cos u\mathbf{k})] \\
&= kq(\sin^3 u \cos^2 v + \sin^3 u \sin^2 v + \sin u \cos^2 u) \\
&= kq \sin u.
\end{aligned}$$

Finally, the flux over the sphere S is given by

$$\begin{aligned}
\iint_S \mathbf{F} \cdot \mathbf{N} \, dS &= \iint_D (kq \sin u) \, dA \\
&= \int_0^{2\pi} \int_0^{\pi} kq \sin u \, du \, dv \\
&= 4\pi kq.
\end{aligned}$$

The result in Example 6 shows that the flux across a sphere S in an inverse square field is independent of the radius of S. In particular, if $\mathbf{E}$ is an electric field, then the result in Example 6, along with Coulomb's Law, yields one of the basic laws of electrostatics, known as **Gauss's Law:**

$$\iint_S \mathbf{E} \cdot \mathbf{N} \, dS = 4\pi k q \qquad \text{Gauss's Law}$$

where q is a point charge located at the center of the sphere and k is the Coulomb constant. Gauss's Law is valid for more general closed surfaces that enclose the origin, and relates the flux out of the surface to the total charge q inside the surface.

Surface integrals are also used in the study of **heat flow.** Heat flows from areas of higher temperature to areas of lower temperature in the direction of greatest change. As a result, measuring **heat flux** involves the gradient of the temperature. The flux depends on the area of the surface. It is the normal direction to the surface that is important, because heat that flows in directions tangential to the surface will produce no heat loss. So, assume that the heat flux across a portion of the surface of area ΔS is given by $\Delta H \approx -k\nabla T \cdot \mathbf{N} \, dS$, where T is the temperature, $\mathbf{N}$ is the unit normal vector to the surface in the direction of the heat flow, and k is the thermal diffusivity of the material. The heat flux across the surface is given by

$$H = \iint_S -k\nabla T \cdot \mathbf{N} \, dS.$$

This section concludes with a summary of different forms of line integrals and surface integrals.

SUMMARY OF LINE AND SURFACE INTEGRALS

Line Integrals

$$ds = \|\mathbf{r}'(t)\| \, dt$$
$$= \sqrt{[x'(t)]^2 + [y'(t)]^2 + [z'(t)]^2} \, dt$$

$$\int_C f(x, y, z) \, ds = \int_a^b f(x(t), y(t), z(t)) \, ds \qquad \text{Scalar form}$$

$$\int_C \mathbf{F} \cdot d\mathbf{r} = \int_C \mathbf{F} \cdot \mathbf{T} \, ds$$
$$= \int_a^b \mathbf{F}(x(t), y(t), z(t)) \cdot \mathbf{r}'(t) \, dt \qquad \text{Vector form}$$

Surface Integrals $[z = g(x, y)]$

$$dS = \sqrt{1 + [g_x(x, y)]^2 + [g_y(x, y)]^2} \, dA$$

$$\iint_S f(x, y, z) \, dS = \iint_R f(x, y, g(x, y))\sqrt{1 + [g_x(x, y)]^2 + [g_y(x, y)]^2} \, dA \qquad \text{Scalar form}$$

$$\iint_S \mathbf{F} \cdot \mathbf{N} \, dS = \iint_R \mathbf{F} \cdot [-g_x(x, y)\mathbf{i} - g_y(x, y)\mathbf{j} + \mathbf{k}] \, dA \qquad \text{Vector form (upward normal)}$$

Surface Integrals (parametric form)

$$dS = \|\mathbf{r}_u(u, v) \times \mathbf{r}_v(u, v)\| \, dA$$

$$\iint_S f(x, y, z) \, dS = \iint_D f(x(u, v), y(u, v), z(u, v)) \, dS \qquad \text{Scalar form}$$

$$\iint_S \mathbf{F} \cdot \mathbf{N} \, dS = \iint_D \mathbf{F} \cdot (\mathbf{r}_u \times \mathbf{r}_v) \, dA \qquad \text{Vector form}$$

15.6 Exercises

See **CalcChat.com** for tutorial help and worked-out solutions to odd-numbered exercises.

Evaluating a Surface Integral In Exercises 1–4, evaluate

$$\iint_S (x - 2y + z) \, dS.$$

1. $S: z = 4 - x, \quad 0 \le x \le 4, \quad 0 \le y \le 3$
2. $S: z = 15 - 2x + 3y, \quad 0 \le x \le 2, \quad 0 \le y \le 4$
3. $S: z = 2, \quad x^2 + y^2 \le 1$
4. $S: z = \frac{2}{3}x^{3/2}, \quad 0 \le x \le 1, \quad 0 \le y \le x$

Evaluating a Surface Integral In Exercises 5 and 6, evaluate

$$\iint_S xy \, dS.$$

5. $S: z = 3 - x - y$, first octant
6. $S: z = h, \quad 0 \le x \le 2, \quad 0 \le y \le \sqrt{4 - x^2}$

Evaluating a Surface Integral In Exercises 7 and 8, use a computer algebra system to evaluate

$$\iint_S xy \, dS.$$

7. $S: z = 9 - x^2, \quad 0 \le x \le 2, \quad 0 \le y \le x$
8. $S: z = \frac{1}{2}xy, \quad 0 \le x \le 4, \quad 0 \le y \le 4$

Evaluating a Surface Integral In Exercises 9 and 10, use a computer algebra system to evaluate

$$\iint_S (x^2 - 2xy) \, dS.$$

9. $S: z = 10 - x^2 - y^2, \quad 0 \le x \le 2, \quad 0 \le y \le 2$
10. $S: z = \cos x, \quad 0 \le x \le \dfrac{\pi}{2}, \quad 0 \le y \le \dfrac{1}{2}x$

Mass In Exercises 11 and 12, find the mass of the surface lamina S of density ρ.

11. $S: 2x + 3y + 6z = 12$, first octant, $\rho(x, y, z) = x^2 + y^2$
12. $S: z = \sqrt{a^2 - x^2 - y^2}, \quad \rho(x, y, z) = kz$

Evaluating a Surface Integral In Exercises 13–16, evaluate

$$\iint_S f(x, y) \, dS.$$

13. $f(x, y) = y + 5$
    $S: \mathbf{r}(u, v) = u\mathbf{i} + v\mathbf{j} + 2v\mathbf{k}, \quad 0 \le u \le 1, \quad 0 \le v \le 2$
14. $f(x, y) = xy$
    $S: \mathbf{r}(u, v) = 2 \cos u \mathbf{i} + 2 \sin u \mathbf{j} + v\mathbf{k}$
    $0 \le u \le \dfrac{\pi}{2}, \quad 0 \le v \le 1$

15. $f(x, y) = x + y$
    $S: \mathbf{r}(u, v) = 2 \cos u \mathbf{i} + 2 \sin u \mathbf{j} + v\mathbf{k}$
    $0 \le u \le \dfrac{\pi}{2}, \quad 0 \le v \le 1$
16. $f(x, y) = x + y$
    $S: \mathbf{r}(u, v) = 4u \cos v \mathbf{i} + 4u \sin v \mathbf{j} + 3u\mathbf{k}$
    $0 \le u \le 4, \quad 0 \le v \le \pi$

Evaluating a Surface Integral In Exercises 17–22, evaluate

$$\iint_S f(x, y, z) \, dS.$$

17. $f(x, y, z) = x^2 + y^2 + z^2$
    $S: z = x + y, \quad x^2 + y^2 \le 1$
18. $f(x, y, z) = \dfrac{xy}{z}$
    $S: z = x^2 + y^2, \quad 4 \le x^2 + y^2 \le 16$
19. $f(x, y, z) = \sqrt{x^2 + y^2 + z^2}$
    $S: z = \sqrt{x^2 + y^2}, \quad x^2 + y^2 \le 4$
20. $f(x, y, z) = \sqrt{x^2 + y^2 + z^2}$
    $S: z = \sqrt{x^2 + y^2}, \quad (x - 1)^2 + y^2 \le 1$
21. $f(x, y, z) = x^2 + y^2 + z^2$
    $S: x^2 + y^2 = 9, \quad 0 \le x \le 3, \quad 0 \le y \le 3, \quad 0 \le z \le 9$
22. $f(x, y, z) = x^2 + y^2 + z^2$
    $S: x^2 + y^2 = 9, \quad 0 \le x \le 3, \quad 0 \le z \le x$

Evaluating a Flux Integral In Exercises 23–28, find the flux of F through S,

$$\iint_S \mathbf{F} \cdot \mathbf{N} \, dS$$

where N is the upward unit normal vector to S.

23. $\mathbf{F}(x, y, z) = 3z\mathbf{i} - 4\mathbf{j} + y\mathbf{k}$
    $S: z = 1 - x - y$, first octant
24. $\mathbf{F}(x, y, z) = x\mathbf{i} + y\mathbf{j}$
    $S: z = 6 - 3x - 2y$, first octant
25. $\mathbf{F}(x, y, z) = x\mathbf{i} + y\mathbf{j} + z\mathbf{k}$
    $S: z = 1 - x^2 - y^2, \quad z \ge 0$
26. $\mathbf{F}(x, y, z) = x\mathbf{i} + y\mathbf{j} + z\mathbf{k}$
    $S: x^2 + y^2 + z^2 = 36$, first octant
27. $\mathbf{F}(x, y, z) = 4\mathbf{i} - 3\mathbf{j} + 5\mathbf{k}$
    $S: z = x^2 + y^2, \quad x^2 + y^2 \le 4$
28. $\mathbf{F}(x, y, z) = x\mathbf{i} + y\mathbf{j} - 2z\mathbf{k}$
    $S: z = \sqrt{a^2 - x^2 - y^2}$

Evaluating a Flux Integral   In Exercises 29 and 30, find the flux of F over the closed surface. (Let N be the outward unit normal vector of the surface.)

29. $\mathbf{F}(x, y, z) = (x + y)\mathbf{i} + y\mathbf{j} + z\mathbf{k}$

$S: z = 16 - x^2 - y^2, \quad z = 0$

30. $\mathbf{F}(x, y, z) = 4xy\mathbf{i} + z^2\mathbf{j} + yz\mathbf{k}$

$S:$ unit cube bounded by $x = 0,\ x = 1,\ y = 0,\ y = 1,\ z = 0,$ $z = 1$

31. Electrical Charge   Let $\mathbf{E} = yz\mathbf{i} + xz\mathbf{j} + xy\mathbf{k}$ be an electrostatic field. Use Gauss's Law to find the total charge enclosed by the closed surface consisting of the hemisphere $z = \sqrt{1 - x^2 - y^2}$ and its circular base in the xy-plane.

32. Electrical Charge   Let $\mathbf{E} = x\mathbf{i} + y\mathbf{j} + 2z\mathbf{k}$ be an electrostatic field. Use Gauss's Law to find the total charge enclosed by the closed surface consisting of the hemisphere $z = \sqrt{1 - x^2 - y^2}$ and its circular base in the xy-plane.

Moment of Inertia   In Exercises 33 and 34, use the following formulas for the moments of inertia about the coordinate axes of a surface lamina of density ρ.

$$I_x = \iint_S (y^2 + z^2)\rho(x, y, z)\, dS$$

$$I_y = \iint_S (x^2 + z^2)\rho(x, y, z)\, dS$$

$$I_z = \iint_S (x^2 + y^2)\rho(x, y, z)\, dS$$

33. Verify that the moment of inertia of a conical shell of uniform density about its axis is $\frac{1}{2}ma^2$, where m is the mass and a is the radius and height.

34. Verify that the moment of inertia of a spherical shell of uniform density about its diameter is $\frac{2}{3}ma^2$, where m is the mass and a is the radius.

Moment of Inertia   In Exercises 35 and 36, find I_z for the given lamina with uniform density of 1. Use a computer algebra system to verify your results.

35. $x^2 + y^2 = a^2, \quad 0 \le z \le h$ **36.** $z = x^2 + y^2, \quad 0 \le z \le h$

Flow Rate   In Exercises 37 and 38, use a computer algebra system to find the rate of mass flow of a fluid of density ρ through the surface S oriented upward when the velocity field is given by $\mathbf{F}(x, y, z) = 0.5z\mathbf{k}$.

37. $S: z = 16 - x^2 - y^2, \quad z \ge 0$

38. $S: z = \sqrt{16 - x^2 - y^2}$

WRITING ABOUT CONCEPTS

39. Surface Integral   Define a surface integral of the scalar function f over a surface $z = g(x, y)$. Explain how to evaluate the surface integral.

40. Orientable Surface   Describe an orientable surface.

41. Flux Integral   Define a flux integral and explain how it is evaluated.

---

42.   HOW DO YOU SEE IT?   Is the surface shown in the figure orientable? Explain why or why not.

Double twist

43. Investigation

(a) Use a computer algebra system to graph the vector-valued function

$$\mathbf{r}(u, v) = (4 - v \sin u)\cos(2u)\mathbf{i} + (4 - v \sin u)\sin(2u)\mathbf{j} + v \cos u\,\mathbf{k}, \quad 0 \le u \le \pi, \quad -1 \le v \le 1.$$

This surface is called a Möbius strip.

(b) Is the surface orientable? Explain why or why not.

(c) Use a computer algebra system to graph the space curve represented by $\mathbf{r}(u, 0)$. Identify the curve.

(d) Construct a Möbius strip by cutting a strip of paper, making a single twist, and pasting the ends together.

(e) Cut the Möbius strip along the space curve graphed in part (c), and describe the result.

SECTION PROJECT

Hyperboloid of One Sheet

Consider the parametric surface given by the function

$$\mathbf{r}(u, v) = a \cosh u \cos v\,\mathbf{i} + a \cosh u \sin v\,\mathbf{j} + b \sinh u\,\mathbf{k}.$$

(a) Use a graphing utility to graph $\mathbf{r}$ for various values of the constants a and b. Describe the effect of the constants on the shape of the surface.

(b) Show that the surface is a hyperboloid of one sheet given by

$$\frac{x^2}{a^2} + \frac{y^2}{a^2} - \frac{z^2}{b^2} = 1.$$

(c) For fixed values $u = u_0$, describe the curves given by

$$\mathbf{r}(u_0, v) = a \cosh u_0 \cos v\,\mathbf{i} + a \cosh u_0 \sin v\,\mathbf{j} + b \sinh u_0\mathbf{k}.$$

(d) For fixed values $v = v_0$, describe the curves given by

$$\mathbf{r}(u, v_0) = a \cosh u \cos v_0\mathbf{i} + a \cosh u \sin v_0\mathbf{j} + b \sinh u\mathbf{k}.$$

(e) Find a normal vector to the surface at $(u, v) = (0, 0)$.

15.7 Divergence Theorem

■ Understand and use the Divergence Theorem.
■ Use the Divergence Theorem to calculate flux.

Divergence Theorem

Recall from Section 15.4 that an alternative form of Green's Theorem is

$$\int_C \mathbf{F} \cdot \mathbf{N} \, ds = \int\int_R \left(\frac{\partial M}{\partial x} + \frac{\partial N}{\partial y} \right) dA$$

$$= \int\int_R \text{div } \mathbf{F} \, dA.$$

In an analogous way, the **Divergence Theorem** gives the relationship between a triple integral over a solid region Q and a surface integral over the surface of Q. In the statement of the theorem, the surface S is **closed** in the sense that it forms the complete boundary of the solid Q. Regions bounded by spheres, ellipsoids, cubes, tetrahedrons, or combinations of these surfaces are typical examples of closed surfaces. Let Q be a solid region on which a triple integral can be evaluated, and let S be a closed surface that is oriented by *outward* unit normal vectors, as shown in Figure 15.54. With these restrictions on S and Q, the Divergence Theorem can be stated as shown below the figure.

**CARL FRIEDRICH GAUSS
(1777–1855)**

The *Divergence Theorem* is also called *Gauss's Theorem*, after the famous German mathematician Carl Friedrich Gauss. Gauss is recognized, with Newton and Archimedes, as one of the three greatest mathematicians in history. One of his many contributions to mathematics was made at the age of 22, when, as part of his doctoral dissertation, he proved the *Fundamental Theorem of Algebra*. *See LarsonCalculus.com to read more of this biography.*

S_1: Oriented by
upward unit normal vector

S_2: Oriented by
downward unit normal vector

Figure 15.54

THEOREM 15.12   The Divergence Theorem

Let Q be a solid region bounded by a closed surface S oriented by a unit normal vector directed outward from Q. If $\mathbf{F}$ is a vector field whose component functions have continuous first partial derivatives in Q, then

$$\int\int_S \mathbf{F} \cdot \mathbf{N} \, dS = \int\int\int_Q \text{div } \mathbf{F} \, dV.$$

REMARK   As noted at the left above, the Divergence Theorem is sometimes called Gauss's Theorem. It is also sometimes called Ostrogradsky's Theorem, after the Russian mathematician Michel Ostrogradsky (1801–1861).

•••••••••••••••••••▷
•
•
•
••**REMARK**   This proof is
restricted to a *simple* solid
region. The general proof is
best left to a course in
advanced calculus.

Proof   For $\mathbf{F}(x, y, z) = M\mathbf{i} + N\mathbf{j} + P\mathbf{k}$, the theorem takes the form

$$\iint_S \mathbf{F} \cdot \mathbf{N} \, dS = \iint_S (M\mathbf{i} \cdot \mathbf{N} + N\mathbf{j} \cdot \mathbf{N} + P\mathbf{k} \cdot \mathbf{N}) \, dS$$

$$= \iiint_Q \left(\frac{\partial M}{\partial x} + \frac{\partial N}{\partial y} + \frac{\partial P}{\partial z} \right) dV.$$

You can prove this by verifying that the following three equations are valid.

$$\iint_S M\mathbf{i} \cdot \mathbf{N} \, dS = \iiint_Q \frac{\partial M}{\partial x} \, dV$$

$$\iint_S N\mathbf{j} \cdot \mathbf{N} \, dS = \iiint_Q \frac{\partial N}{\partial y} \, dV$$

$$\iint_S P\mathbf{k} \cdot \mathbf{N} \, dS = \iiint_Q \frac{\partial P}{\partial z} \, dV$$

Because the verifications of the three equations are similar, only the third is discussed. Restrict the proof to a **simple solid** region with upper surface

$$z = g_2(x, y) \qquad \text{\textcolor{red}{Upper surface}}$$

and lower surface

$$z = g_1(x, y) \qquad \text{\textcolor{red}{Lower surface}}$$

whose projections onto the xy-plane coincide and form region R. If Q has a lateral surface like S_3 in Figure 15.55, then a normal vector is horizontal, which implies that $P\mathbf{k} \cdot \mathbf{N} = 0$. Consequently, you have

$$\iint_S P\mathbf{k} \cdot \mathbf{N} \, dS = \iint_{S_1} P\mathbf{k} \cdot \mathbf{N} \, dS + \iint_{S_2} P\mathbf{k} \cdot \mathbf{N} \, dS + 0.$$

On the upper surface S_2, the outward normal vector is upward, whereas on the lower surface S_1, the outward normal vector is downward. So, by Theorem 15.11, you have

$$\iint_{S_1} P\mathbf{k} \cdot \mathbf{N} \, dS = \iint_R P(x, y, g_1(x, y))\mathbf{k} \cdot \left(\frac{\partial g_1}{\partial x}\mathbf{i} + \frac{\partial g_1}{\partial y}\mathbf{j} - \mathbf{k} \right) dA$$

$$= -\iint_R P(x, y, g_1(x, y)) \, dA$$

and

$$\iint_{S_2} P\mathbf{k} \cdot \mathbf{N} \, dS = \iint_R P(x, y, g_2(x, y))\mathbf{k} \cdot \left(-\frac{\partial g_2}{\partial x}\mathbf{i} - \frac{\partial g_2}{\partial y}\mathbf{j} + \mathbf{k} \right) dA$$

$$= \iint_R P(x, y, g_2(x, y)) \, dA.$$

Adding these results, you obtain

$$\iint_S P\mathbf{k} \cdot \mathbf{N} \, dS = \iint_R [P(x, y, g_2(x, y)) - P(x, y, g_1(x, y))] \, dA$$

$$= \iint_R \left[\int_{g_1(x, y)}^{g_2(x, y)} \frac{\partial P}{\partial z} \, dz \right] dA$$

$$= \iiint_Q \frac{\partial P}{\partial z} \, dV.$$

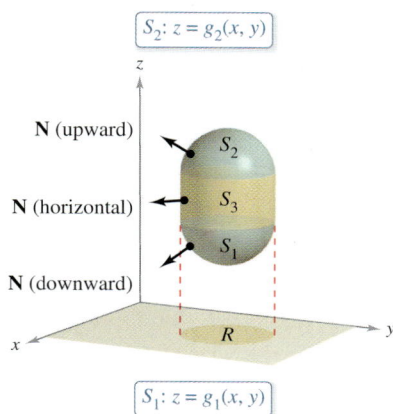

$S_2: z = g_2(x, y)$

$\mathbf{N}$ (upward)

$\mathbf{N}$ (horizontal)

$\mathbf{N}$ (downward)

S_2

S_3

S_1

R

$S_1: z = g_1(x, y)$

Figure 15.55

See LarsonCalculus.com for Bruce Edwards's video of this proof.

Let Q be the solid region bounded by the coordinate planes and the plane

$$2x + 2y + z = 6$$

and let $\mathbf{F} = x\mathbf{i} + y^2\mathbf{j} + z\mathbf{k}$. Find

$$\iint_S \mathbf{F} \cdot \mathbf{N}\, dS$$

where S is the surface of Q.

Solution   From Figure 15.56, you can see that Q is bounded by four subsurfaces. So, you would need four *surface integrals* to evaluate

$$\iint_S \mathbf{F} \cdot \mathbf{N}\, dS.$$

However, by the Divergence Theorem, you need only one triple integral. Because

$$\text{div } \mathbf{F} = \frac{\partial M}{\partial x} + \frac{\partial N}{\partial y} + \frac{\partial P}{\partial z}$$

$$= 1 + 2y + 1$$

$$= 2 + 2y$$

you have

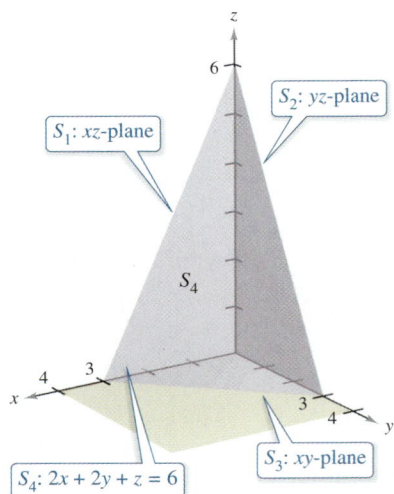

S_1: xz-plane

S_2: yz-plane

S_4

S_3: xy-plane

S_4: $2x + 2y + z = 6$

Figure 15.56

$$\iint_S \mathbf{F} \cdot \mathbf{N}\, dS = \iiint_Q \text{div } \mathbf{F}\, dV$$

$$= \int_0^3 \int_0^{3-y} \int_0^{6-2x-2y} (2 + 2y)\, dz\, dx\, dy$$

$$= \int_0^3 \int_0^{3-y} (2z + 2yz)\Big]_0^{6-2x-2y} dx\, dy$$

$$= \int_0^3 \int_0^{3-y} (12 - 4x + 8y - 4xy - 4y^2)\, dx\, dy$$

$$= \int_0^3 \left[12x - 2x^2 + 8xy - 2x^2y - 4xy^2 \right]_0^{3-y} dy$$

$$= \int_0^3 (18 + 6y - 10y^2 + 2y^3)\, dy$$

$$= \left[18y + 3y^2 - \frac{10y^3}{3} + \frac{y^4}{2} \right]_0^3$$

$$= \frac{63}{2}.$$

▷ **TECHNOLOGY**   If you have access to a computer algebra system that can evaluate triple-iterated integrals, use it to verify the result in Example 1. When you are using such a utility, note that the first step is to convert the triple integral to an iterated integral—this step may be done by hand. To give yourself some practice with this important step, find the limits of integration for the following iterated integrals. Then use a computer to verify that the value is the same as that obtained in Example 1.

$$\int_?^? \int_?^? \int_?^? (2 + 2y)\, dy\, dz\, dx, \qquad \int_?^? \int_?^? \int_?^? (2 + 2y)\, dx\, dy\, dz$$

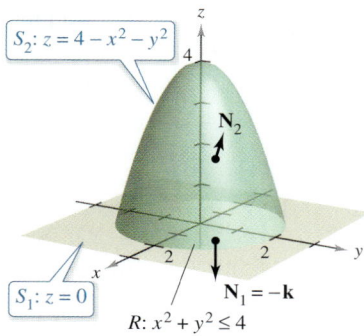

$S_2: z = 4 - x^2 - y^2$

N_2

2   2

$S_1: z = 0$ $N_1 = -k$

$R: x^2 + y^2 \leq 4$

Figure 15.57

EXAMPLE 2   **Verifying the Divergence Theorem**

Let Q be the solid region between the paraboloid

$$z = 4 - x^2 - y^2$$

and the xy-plane. Verify the Divergence Theorem for

$$\mathbf{F}(x, y, z) = 2z\mathbf{i} + x\mathbf{j} + y^2\mathbf{k}.$$

Solution   From Figure 15.57, you can see that the outward normal vector for the surface S_1 is $\mathbf{N}_1 = -\mathbf{k}$, whereas the outward normal vector for the surface S_2 is

$$\mathbf{N}_2 = \frac{2x\mathbf{i} + 2y\mathbf{j} + \mathbf{k}}{\sqrt{4x^2 + 4y^2 + 1}}.$$

So, by Theorem 15.11, you have

$$\iint_S \mathbf{F} \cdot \mathbf{N} \, dS$$

$$= \iint_{S_1} \mathbf{F} \cdot \mathbf{N}_1 \, dS + \iint_{S_2} \mathbf{F} \cdot \mathbf{N}_2 \, dS$$

$$= \iint_{S_1} \mathbf{F} \cdot (-\mathbf{k}) \, dS + \iint_{S_2} \mathbf{F} \cdot \frac{(2x\mathbf{i} + 2y\mathbf{j} + \mathbf{k})}{\sqrt{4x^2 + 4y^2 + 1}} \, dS$$

$$= \iint_R -y^2 \, dA + \iint_R (4xz + 2xy + y^2) \, dA$$

$$= -\int_{-2}^{2} \int_{-\sqrt{4-y^2}}^{\sqrt{4-y^2}} y^2 \, dx \, dy + \int_{-2}^{2} \int_{-\sqrt{4-y^2}}^{\sqrt{4-y^2}} (4xz + 2xy + y^2) \, dx \, dy$$

$$= \int_{-2}^{2} \int_{-\sqrt{4-y^2}}^{\sqrt{4-y^2}} (4xz + 2xy) \, dx \, dy$$

$$= \int_{-2}^{2} \int_{-\sqrt{4-y^2}}^{\sqrt{4-y^2}} [4x(4 - x^2 - y^2) + 2xy] \, dx \, dy$$

$$= \int_{-2}^{2} \int_{-\sqrt{4-y^2}}^{\sqrt{4-y^2}} (16x - 4x^3 - 4xy^2 + 2xy) \, dx \, dy$$

$$= \int_{-2}^{2} \left[8x^2 - x^4 - 2x^2y^2 + x^2y \right]_{-\sqrt{4-y^2}}^{\sqrt{4-y^2}} dy$$

$$= \int_{-2}^{2} 0 \, dy$$

$$= 0.$$

On the other hand, because

$$\text{div } \mathbf{F} = \frac{\partial}{\partial x}[2z] + \frac{\partial}{\partial y}[x] + \frac{\partial}{\partial z}[y^2] = 0 + 0 + 0 = 0$$

you can apply the Divergence Theorem to obtain the equivalent result

$$\iint_S \mathbf{F} \cdot \mathbf{N} \, dS = \iiint_Q \text{div } \mathbf{F} \, dV$$

$$= \iiint_Q 0 \, dV$$

$$= 0.$$

Plane:
x + z = 6

Cylinder:
x² + y² = 4

Figure 15.58

EXAMPLE 3   **Using the Divergence Theorem**

Let Q be the solid bounded by the cylinder $x^2 + y^2 = 4$, the plane $x + z = 6$, and the xy-plane, as shown in Figure 15.58. Find

$$\iint_S \mathbf{F} \cdot \mathbf{N} \, dS$$

where S is the surface of Q and

$$\mathbf{F}(x, y, z) = (x^2 + \sin z)\mathbf{i} + (xy + \cos z)\mathbf{j} + e^y\mathbf{k}.$$

Solution   Direct evaluation of this surface integral would be difficult. However, by the Divergence Theorem, you can evaluate the integral as follows.

$$\iint_S \mathbf{F} \cdot \mathbf{N} \, dS = \iiint_Q \text{div } \mathbf{F} \, dV$$

$$= \iiint_Q (2x + x + 0) \, dV$$

$$= \iiint_Q 3x \, dV$$

$$= \int_0^{2\pi} \int_0^2 \int_0^{6 - r\cos\theta} (3r\cos\theta)r \, dz \, dr \, d\theta$$

$$= \int_0^{2\pi} \int_0^2 (18r^2\cos\theta - 3r^3\cos^2\theta) \, dr \, d\theta$$

$$= \int_0^{2\pi} (48\cos\theta - 12\cos^2\theta) \, d\theta$$

$$= \left[48\sin\theta - 6\left(\theta + \frac{1}{2}\sin 2\theta\right) \right]_0^{2\pi}$$

$$= -12\pi$$

Notice that cylindrical coordinates with

$$x = r\cos\theta \quad \text{and} \quad dV = r \, dz \, dr \, d\theta$$

were used to evaluate the triple integral.

Even though the Divergence Theorem was stated for a simple solid region Q bounded by a closed surface, the theorem is also valid for regions that are the finite unions of simple solid regions. For example, let Q be the solid bounded by the closed surfaces S_1 and S_2, as shown in Figure 15.59. To apply the Divergence Theorem to this solid, let $S = S_1 \cup S_2$. The normal vector $\mathbf{N}$ to S is given by $-\mathbf{N}_1$ on S_1 and by $\mathbf{N}_2$ on S_2. So, you can write

$$\iiint_Q \text{div } \mathbf{F} \, dV = \iint_S \mathbf{F} \cdot \mathbf{N} \, dS$$

$$= \iint_{S_1} \mathbf{F} \cdot (-\mathbf{N}_1) \, dS + \iint_{S_2} \mathbf{F} \cdot \mathbf{N}_2 \, dS$$

$$= -\iint_{S_1} \mathbf{F} \cdot \mathbf{N}_1 \, dS + \iint_{S_2} \mathbf{F} \cdot \mathbf{N}_2 \, dS.$$

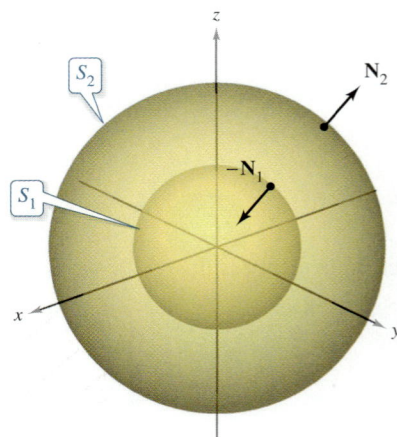

Figure 15.59

Flux and the Divergence Theorem

To help understand the Divergence Theorem, consider the two sides of the equation

$$\iint_S \mathbf{F} \cdot \mathbf{N} \, dS = \iiint_Q \operatorname{div} \mathbf{F} \, dV.$$

You know from Section 15.6 that the flux integral on the left determines the total fluid flow across the surface S per unit of time. This can be approximated by summing the fluid flow across small patches of the surface. The triple integral on the right measures this same fluid flow across S, but from a very different perspective—namely, by calculating the flow of fluid into (or out of) small *cubes* of volume ΔV_i. The flux of the ith cube is approximately $\operatorname{div} \mathbf{F}(x_i, y_i, z_i) \, \Delta V_i$ for some point (x_i, y_i, z_i) in the ith cube. Note that for a cube in the interior of Q, the gain (or loss) of fluid through any one of its six sides is offset by a corresponding loss (or gain) through one of the sides of an adjacent cube. After summing over all the cubes in Q, the only fluid flow that is not canceled by adjoining cubes is that on the outside edges of the cubes on the boundary. So, the sum

$$\sum_{i=1}^{n} \operatorname{div} \mathbf{F}(x_i, y_i, z_i) \, \Delta V_i$$

approximates the total flux into (or out of) Q, and therefore through the surface S.

To see what is meant by the divergence of $\mathbf{F}$ at a point, consider ΔV_α to be the volume of a small sphere S_α of radius α and center (x_0, y_0, z_0) contained in region Q, as shown in Figure 15.60. Applying the Divergence Theorem to S_α produces

$$\text{Flux of } \mathbf{F} \text{ across } S_\alpha = \iiint_{Q_\alpha} \operatorname{div} \mathbf{F} \, dV \approx \operatorname{div} \mathbf{F}(x_0, y_0, z_0) \, \Delta V_\alpha$$

where Q_α is the interior of S_α. Consequently, you have

$$\operatorname{div} \mathbf{F}(x_0, y_0, z_0) \approx \frac{\text{flux of } \mathbf{F} \text{ across } S_\alpha}{\Delta V_\alpha}$$

and, by taking the limit as $\alpha \to 0$, you obtain the divergence of $\mathbf{F}$ at the point (x_0, y_0, z_0).

$$\operatorname{div} \mathbf{F}(x_0, y_0, z_0) = \lim_{\alpha \to 0} \frac{\text{flux of } \mathbf{F} \text{ across } S_\alpha}{\Delta V_\alpha} = \text{flux per unit volume at } (x_0, y_0, z_0)$$

The point (x_0, y_0, z_0) in a vector field is classified as a source, a sink, or incompressible, as shown in the list below.

1. **Source,** for div $\mathbf{F} > 0$ See Figure 15.61(a).
2. **Sink,** for div $\mathbf{F} < 0$ See Figure 15.61(b).
3. **Incompressible,** for div $\mathbf{F} = 0$ See Figure 15.61(c).

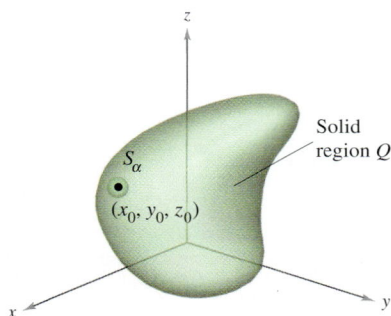

Solid region Q

S_α

(x_0, y_0, z_0)

Figure 15.60

REMARK In hydrodynamics, a *source* is a point at which additional fluid is considered as being introduced to the region occupied by the fluid. A *sink* is a point at which fluid is considered as being removed.

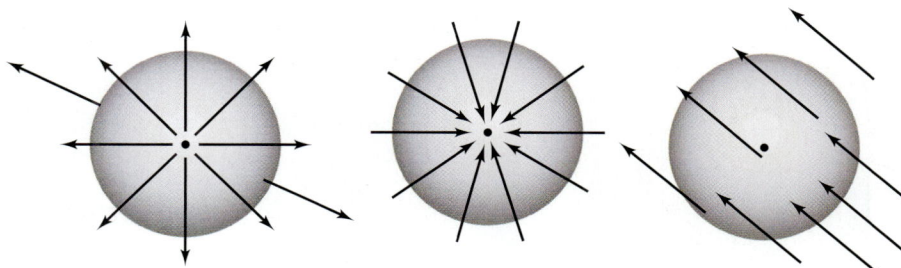

(a) Source: div $\mathbf{F} > 0$

(b) Sink: div $\mathbf{F} < 0$

(c) Incompressible: div $\mathbf{F} = 0$

Figure 15.61

EXAMPLE 4   Calculating Flux by the Divergence Theorem

• • • • ▷ *See LarsonCalculus.com for an interactive version of this type of example.*

Let Q be the region bounded by the sphere $x^2 + y^2 + z^2 = 4$. Find the outward flux of the vector field $\mathbf{F}(x, y, z) = 2x^3\mathbf{i} + 2y^3\mathbf{j} + 2z^3\mathbf{k}$ through the sphere.

Solution   By the Divergence Theorem, you have

$$
\begin{aligned}
\text{Flux across } S &= \iint_S \mathbf{F} \cdot \mathbf{N} \, dS \\
&= \iiint_Q \text{div } \mathbf{F} \, dV \\
&= \iiint_Q 6(x^2 + y^2 + z^2) \, dV \\
&= 6 \int_0^2 \int_0^\pi \int_0^{2\pi} \rho^4 \sin \phi \, d\theta \, d\phi \, d\rho \qquad \text{Spherical coordinates} \\
&= 6 \int_0^2 \int_0^\pi 2\pi \rho^4 \sin \phi \, d\phi \, d\rho \\
&= 12\pi \int_0^2 2\rho^4 \, d\rho \\
&= 24\pi \left(\frac{32}{5} \right) \\
&= \frac{768\pi}{5}.
\end{aligned}
$$

15.7 Exercises

See CalcChat.com for tutorial help and worked-out solutions to odd-numbered exercises.

Verifying the Divergence Theorem In Exercises 1–6, verify the Divergence Theorem by evaluating

$$\iint_S \mathbf{F} \cdot \mathbf{N} \, dS$$

as a surface integral and as a triple integral.

1. $\mathbf{F}(x, y, z) = 2x\mathbf{i} - 2y\mathbf{j} + z^2\mathbf{k}$

   S: cube bounded by the planes $x = 0$, $x = a$, $y = 0$, $y = a$, $z = 0$, $z = a$

2. $\mathbf{F}(x, y, z) = 2x\mathbf{i} - 2y\mathbf{j} + z^2\mathbf{k}$

   S: cylinder $x^2 + y^2 = 4$, $0 \le z \le h$

Figure for 1

Figure for 2

3. $\mathbf{F}(x, y, z) = (2x - y)\mathbf{i} - (2y - z)\mathbf{j} + z\mathbf{k}$

   S: surface bounded by the plane $2x + 4y + 2z = 12$ and the coordinate planes

4. $\mathbf{F}(x, y, z) = xy\mathbf{i} + z\mathbf{j} + (x + y)\mathbf{k}$

   S: surface bounded by the planes $y = 4$ and $z = 4 - x$ and the coordinate planes

Figure for 3

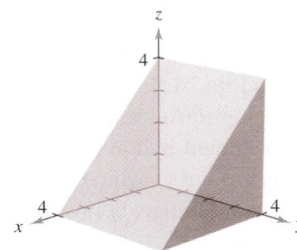

Figure for 4

5. $\mathbf{F}(x, y, z) = xz\mathbf{i} + zy\mathbf{j} + 2z^2\mathbf{k}$

   S: surface bounded by $z = 1 - x^2 - y^2$ and $z = 0$

6. $\mathbf{F}(x, y, z) = xy^2\mathbf{i} + yx^2\mathbf{j} + e\mathbf{k}$

   S: surface bounded by $z = \sqrt{x^2 + y^2}$ and $z = 4$

Using the Divergence Theorem In Exercises 7–16, use the Divergence Theorem to evaluate

$$\iint_S \mathbf{F} \cdot \mathbf{N}\, dS$$

and find the outward flux of F through the surface of the solid bounded by the graphs of the equations. Use a computer algebra system to verify your results.

7. $\mathbf{F}(x, y, z) = x^2\mathbf{i} + y^2\mathbf{j} + z^2\mathbf{k}$

   $S:\ x = 0, x = a, y = 0, y = a, z = 0, z = a$

8. $\mathbf{F}(x, y, z) = x^2 z^2\mathbf{i} - 2y\mathbf{j} + 3xyz\mathbf{k}$

   $S:\ x = 0, x = a, y = 0, y = a, z = 0, z = a$

9. $\mathbf{F}(x, y, z) = x^2\mathbf{i} - 2xy\mathbf{j} + xyz^2\mathbf{k}$

   $S:\ z = \sqrt{a^2 - x^2 - y^2}, z = 0$

10. $\mathbf{F}(x, y, z) = xy\mathbf{i} + yz\mathbf{j} - yz\mathbf{k}$

    $S:\ z = \sqrt{a^2 - x^2 - y^2}, z = 0$

11. $\mathbf{F}(x, y, z) = x\mathbf{i} + y\mathbf{j} + z\mathbf{k}$

    $S:\ x^2 + y^2 + z^2 = 9$

12. $\mathbf{F}(x, y, z) = xyz\mathbf{j}$

    $S:\ x^2 + y^2 = 4, z = 0, z = 5$

13. $\mathbf{F}(x, y, z) = x\mathbf{i} + y^2\mathbf{j} - z\mathbf{k}$

    $S:\ x^2 + y^2 = 25, z = 0, z = 7$

14. $\mathbf{F}(x, y, z) = (xy^2 + \cos z)\mathbf{i} + (x^2 y + \sin z)\mathbf{j} + e^z\mathbf{k}$

    $S:\ z = \frac{1}{2}\sqrt{x^2 + y^2}, z = 8$

15. $\mathbf{F}(x, y, z) = xe^z\mathbf{i} + ye^z\mathbf{j} + e^z\mathbf{k}$

    $S:\ z = 4 - y, z = 0, x = 0, x = 6, y = 0$

16. $\mathbf{F}(x, y, z) = xy\mathbf{i} + 4y\mathbf{j} + xz\mathbf{k}$

    $S:\ x^2 + y^2 + z^2 = 16$

Using the Divergence Theorem In Exercises 17 and 18, evaluate

$$\iint_S \text{curl } \mathbf{F} \cdot \mathbf{N}\, dS$$

where S is the closed surface of the solid bounded by the graphs of $x = 4$ and $z = 9 - y^2$, and the coordinate planes.

17. $\mathbf{F}(x, y, z) = (4xy + z^2)\mathbf{i} + (2x^2 + 6yz)\mathbf{j} + 2xz\mathbf{k}$

18. $\mathbf{F}(x, y, z) = xy\cos z\mathbf{i} + yz\sin x\mathbf{j} + xyz\mathbf{k}$

WRITING ABOUT CONCEPTS

19. **Divergence Theorem** State the Divergence Theorem.

20. **Classifying a Point in a Vector Field** How do you determine whether a point (x_0, y_0, z_0) in a vector field is a source, a sink, or incompressible?

---

21. **Closed Surface** Verify that

$$\iint_S \text{curl } \mathbf{F} \cdot \mathbf{N}\, dS = 0$$

for any closed surface S.

22. **HOW DO YOU SEE IT?** The graph of a vector field F is shown. Does the graph suggest that the divergence of F at P is positive, negative, or zero?

23. **Volume**

    (a) Use the Divergence Theorem to verify that the volume of the solid bounded by a surface S is

    $$\iint_S x\, dy\, dz = \iint_S y\, dz\, dx = \iint_S z\, dx\, dy.$$

    (b) Verify the result of part (a) for the cube bounded by $x = 0$, $x = a$, $y = 0$, $y = a$, $z = 0$, and $z = a$.

24. **Constant Vector Field** For the constant vector field $\mathbf{F}(x, y, z) = a_1\mathbf{i} + a_2\mathbf{j} + a_3\mathbf{k}$, verify the following integral for any closed surface S.

    $$\iint_S \mathbf{F} \cdot \mathbf{N}\, dS = 0$$

25. **Volume** For the vector field $\mathbf{F}(x, y, z) = x\mathbf{i} + y\mathbf{j} + z\mathbf{k}$, verify the following integral, where V is the volume of the solid bounded by the closed surface S.

    $$\iint_S \mathbf{F} \cdot \mathbf{N}\, dS = 3V$$

26. **Verifying an Identity** For the vector field $\mathbf{F}(x, y, z) = x\mathbf{i} + y\mathbf{j} + z\mathbf{k}$, verify that

    $$\frac{1}{\|\mathbf{F}\|} \iint_S \mathbf{F} \cdot \mathbf{N}\, dS = \frac{3}{\|\mathbf{F}\|} \iiint_Q dV.$$

Proof In Exercises 27 and 28, prove the identity, assuming that Q, S, and N meet the conditions of the Divergence Theorem and that the required partial derivatives of the scalar functions f and g are continuous. The expressions $D_N f$ and $D_N g$ are the derivatives in the direction of the vector N and are defined by

$$D_N f = \nabla f \cdot \mathbf{N}, \quad D_N g = \nabla g \cdot \mathbf{N}.$$

27. $\displaystyle\iiint_Q (f\nabla^2 g + \nabla f \cdot \nabla g)\, dV = \iint_S f D_N g\, dS$

    [*Hint:* Use div$(f\mathbf{G}) = f\,$div $\mathbf{G} + \nabla f \cdot \mathbf{G}$.]

28. $\displaystyle\iiint_Q (f\nabla^2 g - g\nabla^2 f)\, dV = \iint_S (f D_N g - g D_N f)\, dS$

    (*Hint:* Use Exercise 27 twice.)

15.8 Stokes's Theorem

■ Understand and use Stokes's Theorem.
■ Use curl to analyze the motion of a rotating liquid.

Stokes's Theorem

A second higher-dimension analog of Green's Theorem is called **Stokes's Theorem,** after the English mathematical physicist George Gabriel Stokes. Stokes was part of a group of English mathematical physicists referred to as the Cambridge School, which included William Thomson (Lord Kelvin) and James Clerk Maxwell. In addition to making contributions to physics, Stokes worked with infinite series and differential equations, as well as with the integration results presented in this section.

Stokes's Theorem gives the relationship between a surface integral over an oriented surface S and a line integral along a closed space curve C forming the boundary of S, as shown in Figure 15.62. The positive direction along C is counterclockwise relative to the normal vector $\mathbf{N}$. That is, if you imagine grasping the normal vector $\mathbf{N}$ with your right hand, with your thumb pointing in the direction of $\mathbf{N}$, then your fingers will point in the positive direction C, as shown in Figure 15.63.

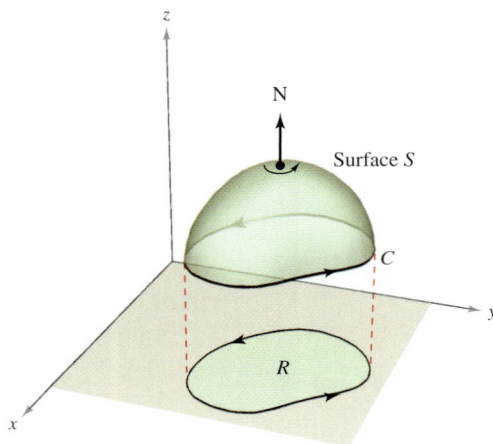

**GEORGE GABRIEL STOKES
(1819–1903)**

Stokes became a Lucasian professor of mathematics at Cambridge in 1849. Five years later, he published the theorem that bears his name as a prize examination question there.
See LarsonCalculus.com to read more of this biography.

Figure 15.62

Direction along C is counterclockwise relative to $\mathbf{N}$.
Figure 15.63

THEOREM 15.13   Stokes's Theorem

Let S be an oriented surface with unit normal vector $\mathbf{N}$, bounded by a piecewise smooth simple closed curve C with a positive orientation. If $\mathbf{F}$ is a vector field whose component functions have continuous first partial derivatives on an open region containing S and C, then

$$\int_C \mathbf{F} \cdot d\mathbf{r} = \int\int_S (\text{curl } \mathbf{F}) \cdot \mathbf{N} \, dS.$$

In Theorem 15.13, note that the line integral may be written in the differential form $\int_C M \, dx + N \, dy + P \, dz$ or in the vector form $\int_C \mathbf{F} \cdot \mathbf{T} \, ds$.

Bettmann/Corbis

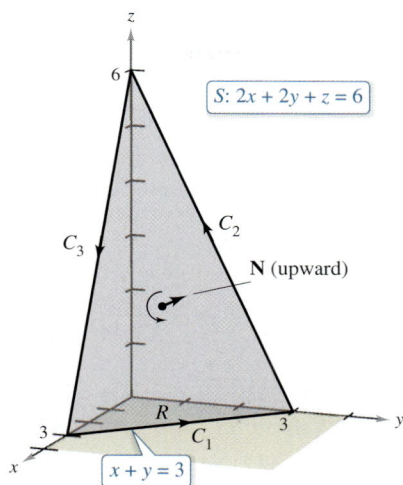

S: $2x + 2y + z = 6$

$x + y = 3$

Figure 15.64

EXAMPLE 1 **Using Stokes's Theorem**

Let C be the oriented triangle lying in the plane

$$2x + 2y + z = 6$$

as shown in Figure 15.64. Evaluate

$$\int_C \mathbf{F} \cdot d\mathbf{r}$$

where $\mathbf{F}(x, y, z) = -y^2\mathbf{i} + z\mathbf{j} + x\mathbf{k}$.

Solution Using Stokes's Theorem, begin by finding the curl of $\mathbf{F}$.

$$\text{curl } \mathbf{F} = \begin{vmatrix} \mathbf{i} & \mathbf{j} & \mathbf{k} \\ \dfrac{\partial}{\partial x} & \dfrac{\partial}{\partial y} & \dfrac{\partial}{\partial z} \\ -y^2 & z & x \end{vmatrix} = -\mathbf{i} - \mathbf{j} + 2y\mathbf{k}$$

Considering

$$z = g(x, y) = 6 - 2x - 2y$$

you can use Theorem 15.11 for an upward normal vector to obtain

$$\int_C \mathbf{F} \cdot d\mathbf{r} = \int_S \int (\text{curl } \mathbf{F}) \cdot \mathbf{N} \, dS$$

$$= \int_R \int (-\mathbf{i} - \mathbf{j} + 2y\mathbf{k}) \cdot [-g_x(x, y)\mathbf{i} - g_y(x, y)\mathbf{j} + \mathbf{k}] \, dA$$

$$= \int_R \int (-\mathbf{i} - \mathbf{j} + 2y\mathbf{k}) \cdot (2\mathbf{i} + 2\mathbf{j} + \mathbf{k}) \, dA$$

$$= \int_0^3 \int_0^{3-y} (2y - 4) \, dx \, dy$$

$$= \int_0^3 (-2y^2 + 10y - 12) \, dy$$

$$= \left[-\frac{2y^3}{3} + 5y^2 - 12y \right]_0^3$$

$$= -9.$$

Try evaluating the line integral in Example 1 directly, *without* using Stokes's Theorem. One way to do this would be to consider C as the union of C_1, C_2, and C_3, as follows.

$$C_1: \mathbf{r}_1(t) = (3 - t)\mathbf{i} + t\mathbf{j}, \quad 0 \le t \le 3$$
$$C_2: \mathbf{r}_2(t) = (6 - t)\mathbf{j} + (2t - 6)\mathbf{k}, \quad 3 \le t \le 6$$
$$C_3: \mathbf{r}_3(t) = (t - 6)\mathbf{i} + (18 - 2t)\mathbf{k}, \quad 6 \le t \le 9$$

The value of the line integral is

$$\int_C \mathbf{F} \cdot d\mathbf{r} = \int_{C_1} \mathbf{F} \cdot \mathbf{r}_1'(t) \, dt + \int_{C_2} \mathbf{F} \cdot \mathbf{r}_2'(t) \, dt + \int_{C_3} \mathbf{F} \cdot \mathbf{r}_3'(t) \, dt$$

$$= \int_0^3 t^2 \, dt + \int_3^6 (-2t + 6) \, dt + \int_6^9 (-2t + 12) \, dt$$

$$= 9 - 9 - 9$$

$$= -9.$$

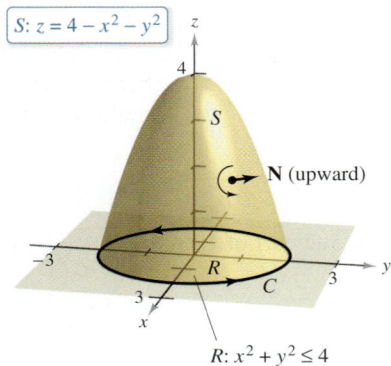

$S: z = 4 - x^2 - y^2$

$R: x^2 + y^2 \le 4$

Figure 15.65

EXAMPLE 2 Verifying Stokes's Theorem

⋯▷ *See LarsonCalculus.com for an interactive version of this type of example.*

Let S be the portion of the paraboloid

$$z = 4 - x^2 - y^2$$

lying above the xy-plane, oriented upward (see Figure 15.65). Let C be its boundary curve in the xy-plane, oriented counterclockwise. Verify Stokes's Theorem for

$$\mathbf{F}(x, y, z) = 2z\mathbf{i} + x\mathbf{j} + y^2\mathbf{k}$$

by evaluating the surface integral and the equivalent line integral.

Solution As a *surface integral*, you have $z = g(x, y) = 4 - x^2 - y^2$, $g_x = -2x$, $g_y = -2y$, and

$$\text{curl } \mathbf{F} = \begin{vmatrix} \mathbf{i} & \mathbf{j} & \mathbf{k} \\ \dfrac{\partial}{\partial x} & \dfrac{\partial}{\partial y} & \dfrac{\partial}{\partial z} \\ 2z & x & y^2 \end{vmatrix} = 2y\mathbf{i} + 2\mathbf{j} + \mathbf{k}.$$

By Theorem 15.11, you obtain

$$\iint_S (\text{curl } \mathbf{F}) \cdot \mathbf{N} \, dS = \iint_R (2y\mathbf{i} + 2\mathbf{j} + \mathbf{k}) \cdot (2x\mathbf{i} + 2y\mathbf{j} + \mathbf{k}) \, dA$$

$$= \int_{-2}^{2} \int_{-\sqrt{4-x^2}}^{\sqrt{4-x^2}} (4xy + 4y + 1) \, dy \, dx$$

$$= \int_{-2}^{2} \left[2xy^2 + 2y^2 + y \right]_{-\sqrt{4-x^2}}^{\sqrt{4-x^2}} dx$$

$$= \int_{-2}^{2} 2\sqrt{4 - x^2} \, dx$$

$$= \text{Area of circle of radius 2}$$

$$= 4\pi.$$

As a *line integral*, you can parametrize C as

$$\mathbf{r}(t) = 2\cos t\mathbf{i} + 2\sin t\mathbf{j} + 0\mathbf{k}, \quad 0 \le t \le 2\pi.$$

For $\mathbf{F}(x, y, z) = 2z\mathbf{i} + x\mathbf{j} + y^2\mathbf{k}$, you obtain

$$\int_C \mathbf{F} \cdot d\mathbf{r} = \int_C M \, dx + N \, dy + P \, dz$$

$$= \int_C 2z \, dx + x \, dy + y^2 \, dz$$

$$= \int_0^{2\pi} [0 + 2\cos t(2\cos t) + 0] \, dt$$

$$= \int_0^{2\pi} 4\cos^2 t \, dt$$

$$= 2\int_0^{2\pi} (1 + \cos 2t) \, dt$$

$$= 2\left[t + \frac{1}{2}\sin 2t \right]_0^{2\pi}$$

$$= 4\pi.$$

Physical Interpretation of Curl

Stokes's Theorem provides insight into a physical interpretation of curl. In a vector field $\mathbf{F}$, let S_α be a *small* circular disk of radius α, centered at (x, y, z) and with boundary C_α, as shown in Figure 15.66. At each point on the circle C_α, $\mathbf{F}$ has a normal component $\mathbf{F} \cdot \mathbf{N}$ and a tangential component $\mathbf{F} \cdot \mathbf{T}$. The more closely $\mathbf{F}$ and $\mathbf{T}$ are aligned, the greater the value of $\mathbf{F} \cdot \mathbf{T}$. So, a fluid tends to move along the circle rather than across it. Consequently, you say that the line integral around C_α measures the **circulation of F around C_α.** That is,

$$\int_{C_\alpha} \mathbf{F} \cdot \mathbf{T}\, ds = \text{circulation of } \mathbf{F} \text{ around } C_\alpha.$$

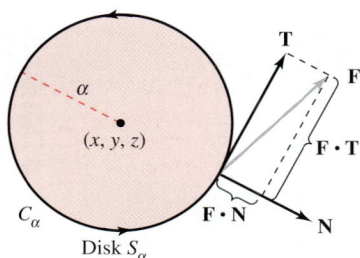

Figure 15.66

Now consider a small disk S_α to be centered at some point (x, y, z) on the surface S, as shown in Figure 15.67. On such a small disk, curl $\mathbf{F}$ is nearly constant, because it varies little from its value at (x, y, z). Moreover, curl $\mathbf{F} \cdot \mathbf{N}$ is also nearly constant on S_α because all unit normals to S_α are about the same. Consequently, Stokes's Theorem yields

$$\int_{C_\alpha} \mathbf{F} \cdot \mathbf{T}\, ds = \int\int_{S_\alpha} (\text{curl } \mathbf{F}) \cdot \mathbf{N}\, dS$$

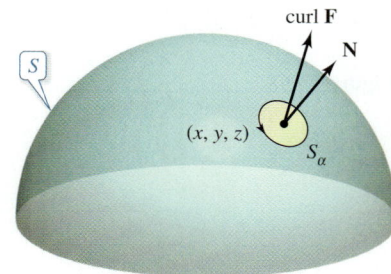

Figure 15.67

$$\approx (\text{curl } \mathbf{F}) \cdot \mathbf{N} \int\int_{S_\alpha} dS$$

$$\approx (\text{curl } \mathbf{F}) \cdot \mathbf{N}(\pi\alpha^2).$$

So,

$$(\text{curl } \mathbf{F}) \cdot \mathbf{N} \approx \frac{\int_{C_\alpha} \mathbf{F} \cdot \mathbf{T}\, ds}{\pi\alpha^2}$$

$$= \frac{\text{circulation of } \mathbf{F} \text{ around } C_\alpha}{\text{area of disk } S_\alpha}$$

$$= \text{rate of circulation.}$$

Assuming conditions are such that the approximation improves for smaller and smaller disks ($\alpha \to 0$), it follows that

$$(\text{curl } \mathbf{F}) \cdot \mathbf{N} = \lim_{\alpha \to 0} \frac{1}{\pi\alpha^2} \int_{C_\alpha} \mathbf{F} \cdot \mathbf{T}\, ds$$

which is referred to as the **rotation of F about N.** That is,

$$\text{curl } \mathbf{F}(x, y, z) \cdot \mathbf{N} = \text{rotation of } \mathbf{F} \text{ about } \mathbf{N} \text{ at } (x, y, z).$$

In this case, the rotation of $\mathbf{F}$ is maximum when curl $\mathbf{F}$ and $\mathbf{N}$ have the same direction. Normally, this tendency to rotate will vary from point to point on the surface S, and Stokes's Theorem

$$\underbrace{\int\int_S (\text{curl } \mathbf{F}) \cdot \mathbf{N}\, dS}_{\text{Surface integral}} = \underbrace{\int_C \mathbf{F} \cdot d\mathbf{r}}_{\text{Line integral}}$$

says that the collective measure of this *rotational* tendency taken over the entire surface S (surface integral) is equal to the tendency of a fluid to *circulate* around the boundary C (line integral).

Figure 15.68

| EXAMPLE 3 | **An Application of Curl** |

A liquid is swirling around in a cylindrical container of radius 2, so that its motion is described by the velocity field

$$\mathbf{F}(x, y, z) = -y\sqrt{x^2 + y^2}\,\mathbf{i} + x\sqrt{x^2 + y^2}\,\mathbf{j}$$

as shown in Figure 15.68. Find

$$\int_S \int (\text{curl } \mathbf{F}) \cdot \mathbf{N}\, dS$$

where S is the upper surface of the cylindrical container.

Solution   The curl of $\mathbf{F}$ is given by

$$\text{curl } \mathbf{F} = \begin{vmatrix} \mathbf{i} & \mathbf{j} & \mathbf{k} \\ \dfrac{\partial}{\partial x} & \dfrac{\partial}{\partial y} & \dfrac{\partial}{\partial z} \\ -y\sqrt{x^2+y^2} & x\sqrt{x^2+y^2} & 0 \end{vmatrix} = 3\sqrt{x^2 + y^2}\,\mathbf{k}.$$

Letting $\mathbf{N} = \mathbf{k}$, you have

$$\int_S \int (\text{curl } \mathbf{F}) \cdot \mathbf{N}\, dS = \int_R \int 3\sqrt{x^2 + y^2}\, dA$$

$$= \int_0^{2\pi} \int_0^2 (3r) r\, dr\, d\theta$$

$$= \int_0^{2\pi} r^3 \Big]_0^2 d\theta$$

$$= \int_0^{2\pi} 8\, d\theta$$

$$= 16\pi.$$

If curl $\mathbf{F} = \mathbf{0}$ throughout region Q, then the rotation of $\mathbf{F}$ about each unit normal $\mathbf{N}$ is 0. That is, $\mathbf{F}$ is irrotational. From Section 15.1, you know that this is a characteristic of conservative vector fields.

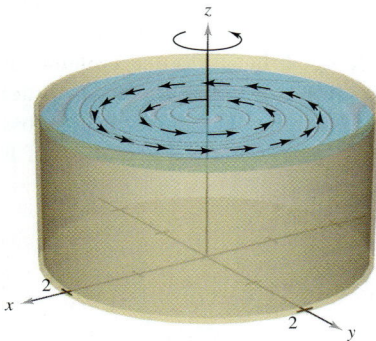

SUMMARY OF INTEGRATION FORMULAS

Fundamental Theorem of Calculus

$$\int_a^b F'(x)\, dx = F(b) - F(a)$$

Fundamental Theorem of Line Integrals

$$\int_C \mathbf{F} \cdot d\mathbf{r} = \int_C \nabla f \cdot d\mathbf{r} = f(x(b), y(b)) - f(x(a), y(a))$$

Green's Theorem

$$\int_C M\, dx + N\, dy = \int_R \int \left(\frac{\partial N}{\partial x} - \frac{\partial M}{\partial y} \right) dA = \int_C \mathbf{F} \cdot \mathbf{T}\, ds = \int_C \mathbf{F} \cdot d\mathbf{r} = \int_R \int (\text{curl } \mathbf{F}) \cdot \mathbf{k}\, dA$$

$$\int_C \mathbf{F} \cdot \mathbf{N}\, ds = \int_R \int \text{div } \mathbf{F}\, dA$$

Divergence Theorem

$$\int_S \int \mathbf{F} \cdot \mathbf{N}\, dS = \int\int\int_Q \text{div } \mathbf{F}\, dV$$

Stokes's Theorem

$$\int_C \mathbf{F} \cdot d\mathbf{r} = \int_S \int (\text{curl } \mathbf{F}) \cdot \mathbf{N}\, dS$$

15.8 Exercises

See CalcChat.com for tutorial help and worked-out solutions to odd-numbered exercises.

Finding the Curl of a Vector Field In Exercises 1–4, find the curl of the vector field F.

1. $\mathbf{F}(x, y, z) = (2y - z)\mathbf{i} + e^z\mathbf{j} + xyz\mathbf{k}$

2. $\mathbf{F}(x, y, z) = x \sin y\mathbf{i} - y \cos x\mathbf{j} + yz^2\mathbf{k}$

3. $\mathbf{F}(x, y, z) = e^{x^2+y^2}\mathbf{i} + e^{y^2+z^2}\mathbf{j} + xyz\mathbf{k}$

4. $\mathbf{F}(x, y, z) = \arcsin y\mathbf{i} + \sqrt{1 - x^2}\mathbf{j} + y^2\mathbf{k}$

Verifying Stokes's Theorem In Exercises 5–8, verify Stokes's Theorem by evaluating $\int_C \mathbf{F} \cdot \mathbf{T}\, ds = \int_C \mathbf{F} \cdot d\mathbf{r}$ as a line integral and as a double integral.

5. $\mathbf{F}(x, y, z) = (-y + z)\mathbf{i} + (x - z)\mathbf{j} + (x - y)\mathbf{k}$

   $S: z = 9 - x^2 - y^2, \quad z \geq 0$

6. $\mathbf{F}(x, y, z) = (-y + z)\mathbf{i} + (x - z)\mathbf{j} + (x - y)\mathbf{k}$

   $S: z = \sqrt{1 - x^2 - y^2}$

7. $\mathbf{F}(x, y, z) = xyz\mathbf{i} + y\mathbf{j} + z\mathbf{k}$

   $S: 6x + 6y + z = 12$, first octant

8. $\mathbf{F}(x, y, z) = z^2\mathbf{i} + x^2\mathbf{j} + y^2\mathbf{k}$

   $S: z = y^2, \quad 0 \leq x \leq a, \quad 0 \leq y \leq a$

Using Stokes's Theorem In Exercises 9–18, use Stokes's Theorem to evaluate $\int_C \mathbf{F} \cdot d\mathbf{r}$. In each case, C is oriented counterclockwise as viewed from above.

9. $\mathbf{F}(x, y, z) = 2y\mathbf{i} + 3z\mathbf{j} + x\mathbf{k}$

   C: triangle with vertices $(2, 0, 0)$, $(0, 2, 0)$, and $(0, 0, 2)$

10. $\mathbf{F}(x, y, z) = \arctan \dfrac{x}{y}\mathbf{i} + \ln\sqrt{x^2 + y^2}\mathbf{j} + \mathbf{k}$

    C: triangle with vertices $(0, 0, 0)$, $(1, 1, 1)$, and $(0, 0, 2)$

11. $\mathbf{F}(x, y, z) = z^2\mathbf{i} + 2x\mathbf{j} + y^2\mathbf{k}$

    $S: z = 1 - x^2 - y^2, \quad z \geq 0$

12. $\mathbf{F}(x, y, z) = 4xz\mathbf{i} + y\mathbf{j} + 4xy\mathbf{k}$

    $S: z = 9 - x^2 - y^2, \quad z \geq 0$

13. $\mathbf{F}(x, y, z) = z^2\mathbf{i} + y\mathbf{j} + z\mathbf{k}$

    $S: z = \sqrt{4 - x^2 - y^2}$

14. $\mathbf{F}(x, y, z) = x^2\mathbf{i} + z^2\mathbf{j} - xyz\mathbf{k}$

    $S: z = \sqrt{4 - x^2 - y^2}$

15. $\mathbf{F}(x, y, z) = -\ln\sqrt{x^2 + y^2}\mathbf{i} + \arctan \dfrac{x}{y}\mathbf{j} + \mathbf{k}$

    $S: z = 9 - 2x - 3y$ over $r = 2 \sin 2\theta$ in the first octant

16. $\mathbf{F}(x, y, z) = yz\mathbf{i} + (2 - 3y)\mathbf{j} + (x^2 + y^2)\mathbf{k}, \quad x^2 + y^2 \leq 16$

    S: the first-octant portion of $x^2 + z^2 = 16$ over $x^2 + y^2 = 16$

17. $\mathbf{F}(x, y, z) = xyz\mathbf{i} + y\mathbf{j} + z\mathbf{k}$

    $S: z = x^2, \quad 0 \leq x \leq a, \quad 0 \leq y \leq a$

    $\mathbf{N}$ is the downward unit normal to the surface.

18. $\mathbf{F}(x, y, z) = xyz\mathbf{i} + y\mathbf{j} + z\mathbf{k}, \quad x^2 + y^2 \leq a^2$

    S: the first-octant portion of $z = x^2$ over $x^2 + y^2 = a^2$

Motion of a Liquid In Exercises 19 and 20, the motion of a liquid in a cylindrical container of radius 1 is described by the velocity field $\mathbf{F}(x, y, z)$. Find $\int_S \int (\text{curl } \mathbf{F}) \cdot \mathbf{N}\, dS$, where S is the upper surface of the cylindrical container.

19. $\mathbf{F}(x, y, z) = \mathbf{i} + \mathbf{j} - 2\mathbf{k}$

20. $\mathbf{F}(x, y, z) = -z\mathbf{i} + y\mathbf{k}$

WRITING ABOUT CONCEPTS

21. **Stokes's Theorem** State Stokes's Theorem.

22. **Curl** Give a physical interpretation of curl.

23. **Proof** Let $\mathbf{C}$ be a constant vector. Let S be an oriented surface with a unit normal vector $\mathbf{N}$, bounded by a smooth curve C. Prove that

$$\int_S \int \mathbf{C} \cdot \mathbf{N}\, dS = \frac{1}{2}\int_C (\mathbf{C} \times \mathbf{r}) \cdot d\mathbf{r}.$$

24. **HOW DO YOU SEE IT?** Let S_1 be the portion of the paraboloid lying above the xy-plane, and let S_2 be the hemisphere, as shown in the figures. Both surfaces are oriented upward.

For a vector field $\mathbf{F}(x, y, z)$ with continuous partial derivatives, does

$$\int_{S_1} \int (\text{curl } \mathbf{F}) \cdot \mathbf{N}\, dS_1 = \int_{S_2} \int (\text{curl } \mathbf{F}) \cdot \mathbf{N}\, dS_2?$$

Explain your reasoning.

PUTNAM EXAM CHALLENGE

25. Let $\mathbf{G}(x, y) = \left(\dfrac{-y}{x^2 + 4y^2}, \dfrac{x}{x^2 + 4y^2}, 0\right)$.

    Prove or disprove that there is a vector-valued function $\mathbf{F}(x, y, z) = (M(x, y, z), N(x, y, z), P(x, y, z))$ with the following properties:

    (i) M, N, P have continuous partial derivatives for all $(x, y, z) \neq (0, 0, 0)$;

    (ii) Curl $\mathbf{F} = 0$ for all $(x, y, z) \neq (0, 0, 0)$;

    (iii) $\mathbf{F}(x, y, 0) = \mathbf{G}(x, y)$.

    This problem was composed by the Committee on the Putnam Prize Competition.
    © The Mathematical Association of America. All rights reserved.

Review Exercises

See **CalcChat.com** for tutorial help and worked-out solutions to odd-numbered exercises.

Sketching a Vector Field In Exercises 1 and 2, find $\|\mathbf{F}\|$ and sketch several representative vectors in the vector field. Use a computer algebra system to verify your results.

1. $\mathbf{F}(x, y, z) = x\mathbf{i} + \mathbf{j} + 2\mathbf{k}$ **2.** $\mathbf{F}(x, y) = \mathbf{i} - 2y\mathbf{j}$

Finding a Conservative Vector Field In Exercises 3 and 4, find the conservative vector field for the potential function by finding its gradient.

3. $f(x, y, z) = 2x^2 + xy + z^2$ **4.** $f(x, y, z) = x^2 e^{yz}$

Finding a Potential Function In Exercises 5–12, determine whether the vector field is conservative. If it is, find a potential function for the vector field.

5. $\mathbf{F}(x, y) = -\dfrac{y}{x^2}\mathbf{i} + \dfrac{1}{x}\mathbf{j}$ **6.** $\mathbf{F}(x, y) = \dfrac{1}{y}\mathbf{i} - \dfrac{y}{x^2}\mathbf{j}$

7. $\mathbf{F}(x, y) = (xy^2 - x^2)\mathbf{i} + (x^2y + y^2)\mathbf{j}$

8. $\mathbf{F}(x, y) = (-2y^3 \sin 2x)\mathbf{i} + 3y^2(1 + \cos 2x)\mathbf{j}$

9. $\mathbf{F}(x, y, z) = 4xy^2\mathbf{i} + 2x^2\mathbf{j} + 2z\mathbf{k}$

10. $\mathbf{F}(x, y, z) = (4xy + z^2)\mathbf{i} + (2x^2 + 6yz)\mathbf{j} + 2xz\mathbf{k}$

11. $\mathbf{F}(x, y, z) = \dfrac{yz\mathbf{i} - xz\mathbf{j} - xy\mathbf{k}}{y^2z^2}$

12. $\mathbf{F}(x, y, z) = \sin z(y\mathbf{i} + x\mathbf{j} + \mathbf{k})$

Divergence and Curl In Exercises 13–20, find (a) the divergence of the vector field $\mathbf{F}$ and (b) the curl of the vector field $\mathbf{F}$.

13. $\mathbf{F}(x, y, z) = x^2\mathbf{i} + xy^2\mathbf{j} + x^2z\mathbf{k}$

14. $\mathbf{F}(x, y, z) = y^2\mathbf{j} - z^2\mathbf{k}$

15. $\mathbf{F}(x, y, z) = (\cos y + y \cos x)\mathbf{i} + (\sin x - x \sin y)\mathbf{j} + xyz\mathbf{k}$

16. $\mathbf{F}(x, y, z) = (3x - y)\mathbf{i} + (y - 2z)\mathbf{j} + (z - 3x)\mathbf{k}$

17. $\mathbf{F}(x, y, z) = \arcsin x\mathbf{i} + xy^2\mathbf{j} + yz^2\mathbf{k}$

18. $\mathbf{F}(x, y, z) = (x^2 - y)\mathbf{i} - (x + \sin^2 y)\mathbf{j}$

19. $\mathbf{F}(x, y, z) = \ln(x^2 + y^2)\mathbf{i} + \ln(x^2 + y^2)\mathbf{j} + z\mathbf{k}$

20. $\mathbf{F}(x, y, z) = \dfrac{z}{x}\mathbf{i} + \dfrac{z}{y}\mathbf{j} + z^2\mathbf{k}$

Evaluating a Line Integral In Exercises 21–26, evaluate the line integral along the given path(s).

21. $\displaystyle\int_C (x^2 + y^2)\, ds$

(a) C: line segment from $(0, 0)$ to $(3, 4)$

(b) C: $x^2 + y^2 = 1$, one revolution counterclockwise, starting at $(1, 0)$

22. $\displaystyle\int_C xy\, ds$

(a) C: line segment from $(0, 0)$ to $(5, 4)$

(b) C: counterclockwise around the triangle with vertices $(0, 0)$, $(4, 0)$, and $(0, 2)$

23. $\displaystyle\int_C (x^2 + y^2)\, ds$

C: $\mathbf{r}(t) = (1 - \sin t)\mathbf{i} + (1 - \cos t)\mathbf{j}$, $0 \le t \le 2\pi$

24. $\displaystyle\int_C (x^2 + y^2)\, ds$

C: $\mathbf{r}(t) = (\cos t + t \sin t)\mathbf{i} + (\sin t - t \cos t)\mathbf{j}$, $0 \le t \le 2\pi$

25. $\displaystyle\int_C (2x - y)\, dx + (x + 2y)\, dy$

(a) C: line segment from $(0, 0)$ to $(3, -3)$

(b) C: one revolution counterclockwise around the circle $x = 3 \cos t$, $y = 3 \sin t$

26. $\displaystyle\int_C (2x - y)\, dx + (x + 3y)\, dy$

C: $\mathbf{r}(t) = (\cos t + t \sin t)\mathbf{i} + (\sin t - t \sin t)\mathbf{j}$, $0 \le t \le \pi/2$

Evaluating a Line Integral In Exercises 27 and 28, use a computer algebra system to evaluate the line integral over the given path.

27. $\displaystyle\int_C (2x + y)\, ds$

$\mathbf{r}(t) = a\cos^3 t\mathbf{i} + a\sin^3 t\mathbf{j}$,

$0 \le t \le \pi/2$

28. $\displaystyle\int_C (x^2 + y^2 + z^2)\, ds$

$\mathbf{r}(t) = t\mathbf{i} + t^2\mathbf{j} + t^{3/2}\mathbf{k}$,

$0 \le t \le 4$

Lateral Surface Area In Exercises 29 and 30, find the lateral surface area over the curve C in the xy-plane and under the surface $z = f(x, y)$.

29. $f(x, y) = 3 + \sin(x + y)$; C: $y = 2x$ from $(0, 0)$ to $(2, 4)$

30. $f(x, y) = 12 - x - y$; C: $y = x^2$ from $(0, 0)$ to $(2, 4)$

Evaluating a Line Integral of a Vector Field In Exercises 31–36, evaluate $\displaystyle\int_C \mathbf{F} \cdot d\mathbf{r}$.

31. $\mathbf{F}(x, y) = xy\mathbf{i} + 2xy\mathbf{j}$

C: $\mathbf{r}(t) = t^2\mathbf{i} + t^2\mathbf{j}$, $0 \le t \le 1$

32. $\mathbf{F}(x, y) = (x - y)\mathbf{i} + (x + y)\mathbf{j}$

C: $\mathbf{r}(t) = 4\cos t\mathbf{i} + 3\sin t\mathbf{j}$, $0 \le t \le 2\pi$

33. $\mathbf{F}(x, y, z) = x\mathbf{i} + y\mathbf{j} + z\mathbf{k}$

C: $\mathbf{r}(t) = 2\cos t\mathbf{i} + 2\sin t\mathbf{j} + t\mathbf{k}$, $0 \le t \le 2\pi$

34. $\mathbf{F}(x, y, z) = (2y - z)\mathbf{i} + (z - x)\mathbf{j} + (x - y)\mathbf{k}$

C: curve of intersection of $x^2 + z^2 = 4$ and $y^2 + z^2 = 4$ from $(2, 2, 0)$ to $(0, 0, 2)$

35. $\mathbf{F}(x, y, z) = (y + z)\mathbf{i} + (x + z)\mathbf{j} + (x + y)\mathbf{k}$

C: curve of intersection of $z = x^2 + y^2$ and $y = x$ from $(0, 0, 0)$ to $(2, 2, 8)$

36. $\mathbf{F}(x, y, z) = (x^2 - z)\mathbf{i} + (y^2 + z)\mathbf{j} + x\mathbf{k}$

C: curve of intersection of $z = x^2$ and $x^2 + y^2 = 4$ from $(0, -2, 0)$ to $(0, 2, 0)$

Evaluating a Line Integral In Exercises 37 and 38, use a computer algebra system to evaluate the line integral.

37. $\displaystyle\int_C xy\,dx + (x^2 + y^2)\,dy$

C: $y = x^2$ from $(0, 0)$ to $(2, 4)$ and $y = 2x$ from $(2, 4)$ to $(0, 0)$

38. $\displaystyle\int_C \mathbf{F}\cdot d\mathbf{r}$

$\mathbf{F}(x, y) = (2x - y)\mathbf{i} + (2y - x)\mathbf{j}$

C: $\mathbf{r}(t) = (2\cos t + 2t\sin t)\mathbf{i} + (2\sin t - 2t\cos t)\mathbf{j}$, $\quad 0 \le t \le \pi$

39. Work Find the work done by the force field $\mathbf{F} = x\mathbf{i} - \sqrt{y}\,\mathbf{j}$ along the path $y = x^{3/2}$ from $(0, 0)$ to $(4, 8)$.

40. Work A 20-ton aircraft climbs 2000 feet while making a $90°$ turn in a circular arc of radius 10 miles. Find the work done by the engines.

Using the Fundamental Theorem of Line Integrals In Exercises 41 and 42, evaluate the integral using the Fundamental Theorem of Line Integrals.

41. $\displaystyle\int_C 2xyz\,dx + x^2z\,dy + x^2y\,dz$

C: smooth curve from $(0, 0, 0)$ to $(1, 3, 2)$

42. $\displaystyle\int_C y\,dx + x\,dy + \frac{1}{z}\,dz$

C: smooth curve from $(0, 0, 1)$ to $(4, 4, 4)$

43. Evaluating a Line Integral Evaluate the line integral

$\displaystyle\int_C y^2\,dx + 2xy\,dy.$

(a) C: $\mathbf{r}(t) = (1 + 3t)\mathbf{i} + (1 + t)\mathbf{j}, \quad 0 \le t \le 1$

(b) C: $\mathbf{r}(t) = t\mathbf{i} + \sqrt{t}\,\mathbf{j}, \quad 1 \le t \le 4$

(c) Use the Fundamental Theorem of Line Integrals, where C is a smooth curve from $(1, 1)$ to $(4, 2)$.

44. Area and Centroid Consider the region bounded by the x-axis and one arch of the cycloid with parametric equations $x = a(\theta - \sin\theta)$ and $y = a(1 - \cos\theta)$. Use line integrals to find (a) the area of the region and (b) the centroid of the region.

Evaluating a Line Integral In Exercises 45–50, use Green's Theorem to evaluate the line integral.

45. $\displaystyle\int_C y\,dx + 2x\,dy$

C: boundary of the square with vertices $(0, 0)$, $(0, 1)$, $(1, 0)$, and $(1, 1)$

46. $\displaystyle\int_C xy\,dx + (x^2 + y^2)\,dy$

C: boundary of the square with vertices $(0, 0)$, $(0, 2)$, $(2, 0)$, and $(2, 2)$

47. $\displaystyle\int_C xy^2\,dx + x^2y\,dy$

C: $x = 4\cos t, \quad y = 4\sin t$

48. $\displaystyle\int_C (x^2 - y^2)\,dx + 2xy\,dy$

C: $x^2 + y^2 = a^2$

49. $\displaystyle\int_C xy\,dx + x^2\,dy$

C: boundary of the region between the graphs of $y = x^2$ and $y = 1$

50. $\displaystyle\int_C y^2\,dx + x^{4/3}\,dy$

C: $x^{2/3} + y^{2/3} = 1$

Graphing a Parametric Surface In Exercises 51 and 52, use a computer algebra system to graph the surface represented by the vector-valued function.

51. $\mathbf{r}(u, v) = \sec u\cos v\,\mathbf{i} + (1 + 2\tan u)\sin v\,\mathbf{j} + 2u\mathbf{k}$

$0 \le u \le \dfrac{\pi}{3}, \quad 0 \le v \le 2\pi$

52. $\mathbf{r}(u, v) = e^{-u/4}\cos v\,\mathbf{i} + e^{-u/4}\sin v\,\mathbf{j} + \dfrac{u}{6}\mathbf{k}$

$0 \le u \le 4, \quad 0 \le v \le 2\pi$

53. Investigation Consider the surface represented by the vector-valued function

$\mathbf{r}(u, v) = 3\cos v\cos u\,\mathbf{i} + 3\cos v\sin u\,\mathbf{j} + \sin v\,\mathbf{k}.$

Use a computer algebra system to do the following.

(a) Graph the surface for $0 \le u \le 2\pi$ and $-\dfrac{\pi}{2} \le v \le \dfrac{\pi}{2}$.

(b) Graph the surface for $0 \le u \le 2\pi$ and $\dfrac{\pi}{4} \le v \le \dfrac{\pi}{2}$.

(c) Graph the surface for $0 \le u \le \dfrac{\pi}{4}$ and $0 \le v \le \dfrac{\pi}{2}$.

(d) Graph and identify the space curve for $0 \le u \le 2\pi$ and $v = \dfrac{\pi}{4}$.

(e) Approximate the area of the surface graphed in part (b).

(f) Approximate the area of the surface graphed in part (c).

54. Evaluating a Surface Integral Evaluate the surface integral $\iint_S z\,dS$ over the surface S:

$\mathbf{r}(u, v) = (u + v)\mathbf{i} + (u - v)\mathbf{j} + \sin v\,\mathbf{k}$

where $0 \le u \le 2$ and $0 \le v \le \pi$.

55. Approximating a Surface Integral Use a computer algebra system to graph the surface S and approximate the surface integral

$\iint_S (x + y)\,dS$

where S is the surface

S: $\mathbf{r}(u, v) = u\cos v\,\mathbf{i} + u\sin v\,\mathbf{j} + (u - 1)(2 - u)\mathbf{k}$
over $0 \le u \le 2$ and $0 \le v \le 2\pi$.

56. Mass A cone-shaped surface lamina S is given by

$$z = a\left(a - \sqrt{x^2 + y^2}\right), \quad 0 \le z \le a^2.$$

At each point on S, the density is proportional to the distance between the point and the z-axis.

(a) Sketch the cone-shaped surface.

(b) Find the mass m of the lamina.

Verifying the Divergence Theorem **In Exercises 57 and 58, verify the Divergence Theorem by evaluating**

$$\iint_S \mathbf{F} \cdot \mathbf{N}\, dS$$

as a surface integral and as a triple integral.

57. $\mathbf{F}(x, y, z) = x^2\mathbf{i} + xy\mathbf{j} + z\mathbf{k}$

Q: solid region bounded by the coordinate planes and the plane $2x + 3y + 4z = 12$

58. $\mathbf{F}(x, y, z) = x\mathbf{i} + y\mathbf{j} + z\mathbf{k}$

Q: solid region bounded by the coordinate planes and the plane $2x + 3y + 4z = 12$

Verifying Stokes's Theorem **In Exercises 59 and 60, verify Stokes's Theorem by evaluating**

$$\int_C \mathbf{F} \cdot d\mathbf{r}$$

as a line integral and as a double integral.

59. $\mathbf{F}(x, y, z) = (\cos y + y \cos x)\mathbf{i} + (\sin x - x \sin y)\mathbf{j} + xyz\mathbf{k}$

S: portion of $z = y^2$ over the square in the xy-plane with vertices $(0, 0)$, $(a, 0)$, (a, a), and $(0, a)$

$\mathbf{N}$ is the upward unit normal vector to the surface.

60. $\mathbf{F}(x, y, z) = (x - z)\mathbf{i} + (y - z)\mathbf{j} + x^2\mathbf{k}$

S: first-octant portion of the plane $3x + y + 2z = 12$

61. Proof Prove that it is not possible for a vector field with twice-differentiable components to have a curl of $x\mathbf{i} + y\mathbf{j} + z\mathbf{k}$.

SECTION PROJECT

The Planimeter

You have learned many calculus techniques for finding the area of a planar region. Engineers use a mechanical device called a *planimeter* for measuring planar areas, which is based on the area formula given in Theorem 15.9 (page 1078). As you can see in the figure, the planimeter is fixed at point O (but free to pivot) and has a hinge at A. The end of the tracer arm $\overline{AB}$ moves counterclockwise around the region R. A small wheel at B is perpendicular to $\overline{AB}$ and is marked with a scale to measure how much it rolls as B traces out the boundary of region R. In this project, you will show that the area of R is given by the length L of the tracer arm $\overline{AB}$ multiplied by the distance D that the wheel rolls.

Assume that point B traces out the boundary of R for $a \le t \le b$. Point A will move back and forth along a circular arc around the origin O. Let $\theta(t)$ denote the angle in the figure and let $(x(t), y(t))$ denote the coordinates of A.

(a) Show that the vector $\overrightarrow{OB}$ is given by the vector-valued function

$$\mathbf{r}(t) = [x(t) + L \cos \theta(t)]\mathbf{i} + [y(t) + L \sin \theta(t)]\mathbf{j}.$$

(b) Show that the following two integrals are equal to zero.

$$I_1 = \int_a^b \frac{1}{2}L^2 \frac{d\theta}{dt}\, dt \qquad I_2 = \int_a^b \frac{1}{2}\left(x\frac{dy}{dt} - y\frac{dx}{dt}\right) dt$$

(c) Use the integral $\displaystyle\int_a^b [x(t)\sin\theta(t) - y(t)\cos\theta(t)]'\, dt$ to show that the following two integrals are equal.

$$I_3 = \int_a^b \frac{1}{2}L\left(y\sin\theta\frac{d\theta}{dt} + x\cos\theta\frac{d\theta}{dt}\right) dt$$

$$I_4 = \int_a^b \frac{1}{2}L\left(-\sin\theta\frac{dx}{dt} + \cos\theta\frac{dy}{dt}\right) dt$$

(d) Let $\mathbf{N} = -\sin\theta\mathbf{i} + \cos\theta\mathbf{j}$. Explain why the distance D that the wheel rolls is given by

$$D = \int_C \mathbf{N} \cdot \mathbf{T}\, ds.$$

(e) Show that the area of region R is given by

$$I_1 + I_2 + I_3 + I_4 = DL.$$

■ FOR FURTHER INFORMATION For more information about Green's Theorem and planimeters, see the article "As the Planimeter's Wheel Turns: Planimeter Proofs for Calculus Class" by Tanya Leise in *The College Mathematics Journal*. To view this article, go to *MathArticles.com*.

P.S. Problem Solving

See **CalcChat.com** for tutorial help and worked-out solutions to odd-numbered exercises.

1. Heat Flux Consider a single heat source located at the origin with temperature

$$T(x, y, z) = \frac{25}{\sqrt{x^2 + y^2 + z^2}}.$$

(a) Calculate the heat flux across the surface

$$S = \left\{ (x, y, z): z = \sqrt{1 - x^2}, -\frac{1}{2} \le x \le \frac{1}{2}, 0 \le y \le 1 \right\}$$

as shown in the figure.

(b) Repeat the calculation in part (a) using the parametrization

$$x = \cos u, \quad y = v, \quad z = \sin u$$

where

$$\frac{\pi}{3} \le u \le \frac{2\pi}{3} \quad \text{and} \quad 0 \le v \le 1.$$

2. Heat Flux Consider a single heat source located at the origin with temperature

$$T(x, y, z) = \frac{25}{\sqrt{x^2 + y^2 + z^2}}.$$

(a) Calculate the heat flux across the surface

$$S = \left\{ (x, y, z): z = \sqrt{1 - x^2 - y^2}, x^2 + y^2 \le 1 \right\}$$

as shown in the figure.

(b) Repeat the calculation in part (a) using the parametrization

$$x = \sin u \cos v, \quad y = \sin u \sin v, \quad z = \cos u$$

where

$$0 \le u \le \frac{\pi}{2} \quad \text{and} \quad 0 \le v \le 2\pi.$$

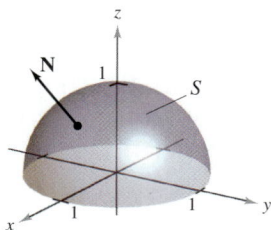

3. Moments of Inertia Consider a wire of density $\rho(x, y, z)$ given by the space curve

$$C: \mathbf{r}(t) = x(t)\mathbf{i} + y(t)\mathbf{j} + z(t)\mathbf{k}, \quad a \le t \le b.$$

The **moments of inertia** about the x-, y-, and z-axes are given by

$$I_x = \int_C (y^2 + z^2)\rho(x, y, z)\, ds$$

$$I_y = \int_C (x^2 + z^2)\rho(x, y, z)\, ds$$

$$I_z = \int_C (x^2 + y^2)\rho(x, y, z)\, ds.$$

Find the moments of inertia for a wire of uniform density $\rho = 1$ in the shape of the helix

$$\mathbf{r}(t) = 3\cos t\,\mathbf{i} + 3\sin t\,\mathbf{j} + 2t\mathbf{k}, \quad 0 \le t \le 2\pi \text{ (see figure).}$$

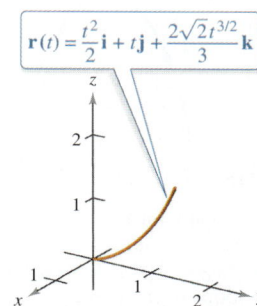

Figure for 3   Figure for 4

4. Moments of Inertia Find the moments of inertia for a wire of density $\rho = \dfrac{1}{1 + t}$ given by the curve

$$C: \mathbf{r}(t) = \frac{t^2}{2}\mathbf{i} + t\mathbf{j} + \frac{2\sqrt{2}\,t^{3/2}}{3}\mathbf{k}, \quad 0 \le t \le 1 \text{ (see figure).}$$

5. Laplace's Equation Let $\mathbf{F}(x, y, z) = x\mathbf{i} + y\mathbf{j} + z\mathbf{k}$, and let $f(x, y, z) = \|\mathbf{F}(x, y, z)\|$.

(a) Show that $\nabla(\ln f) = \dfrac{\mathbf{F}}{f^2}$.

(b) Show that $\nabla\left(\dfrac{1}{f}\right) = -\dfrac{\mathbf{F}}{f^3}$.

(c) Show that $\nabla f^n = nf^{n-2}\mathbf{F}$.

(d) The **Laplacian** is the differential operator

$$\nabla^2 = \nabla \cdot \nabla = \frac{\partial^2}{\partial x^2} + \frac{\partial^2}{\partial y^2} + \frac{\partial^2}{\partial z^2}$$

and **Laplace's equation** is

$$\nabla^2 w = \frac{\partial^2 w}{\partial x^2} + \frac{\partial^2 w}{\partial y^2} + \frac{\partial^2 w}{\partial z^2} = 0.$$

Any function that satisfies this equation is called **harmonic.** Show that the function $w = 1/f$ is harmonic.

6. Green's Theorem Consider the line integral

$$\int_C y^n \, dx + x^n \, dy$$

where C is the boundary of the region lying between the graphs of $y = \sqrt{a^2 - x^2}$ $(a > 0)$ and $y = 0$.

(a) Use a computer algebra system to verify Green's Theorem for n, an odd integer from 1 through 7.

(b) Use a computer algebra system to verify Green's Theorem for n, an even integer from 2 through 8.

(c) For n an odd integer, make a conjecture about the value of the integral.

7. Area Use a line integral to find the area bounded by one arch of the cycloid $x(\theta) = a(\theta - \sin\theta)$, $y(\theta) = a(1 - \cos\theta)$, $0 \le \theta \le 2\pi$, as shown in the figure.

Figure for 7

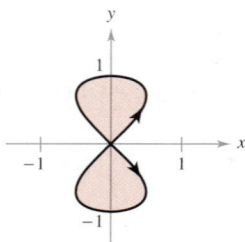

Figure for 8

8. Area Use a line integral to find the area bounded by the two loops of the eight curve

$$x(t) = \frac{1}{2}\sin 2t, \quad y(t) = \sin t, \quad 0 \le t \le 2\pi$$

as shown in the figure.

9. Work The force field $\mathbf{F}(x, y) = (x + y)\mathbf{i} + (x^2 + 1)\mathbf{j}$ acts on an object moving from the point $(0, 0)$ to the point $(0, 1)$, as shown in the figure.

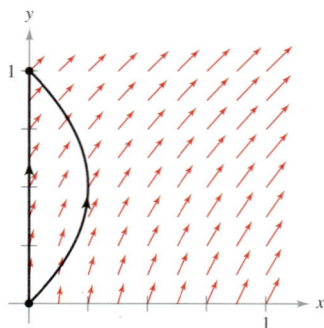

(a) Find the work done when the object moves along the path $x = 0, 0 \le y \le 1$.

(b) Find the work done when the object moves along the path $x = y - y^2, 0 \le y \le 1$.

(c) The object moves along the path $x = c(y - y^2), 0 \le y \le 1$, $c > 0$. Find the value of the constant c that minimizes the work.

10. Work The force field $\mathbf{F}(x, y) = (3x^2y^2)\mathbf{i} + (2x^3y)\mathbf{j}$ is shown in the figure below. Three particles move from the point $(1, 1)$ to the point $(2, 4)$ along different paths. Explain why the work done is the same for each particle, and find the value of the work.

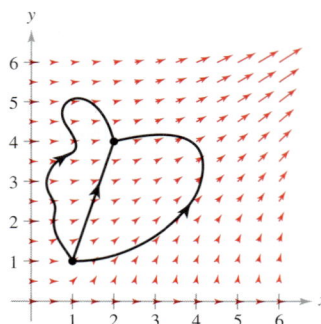

11. Proof Let S be a smooth oriented surface with normal vector $\mathbf{N}$, bounded by a smooth simple closed curve C. Let $\mathbf{v}$ be a constant vector, and prove that

$$\iint_S (2\mathbf{v} \cdot \mathbf{N}) \, dS = \int_C (\mathbf{v} \times \mathbf{r}) \cdot d\mathbf{r}.$$

12. Area and Work How does the area of the ellipse $\dfrac{x^2}{a^2} + \dfrac{y^2}{b^2} = 1$ compare with the magnitude of the work done by the force field

$$\mathbf{F}(x, y) = -\frac{1}{2}y\mathbf{i} + \frac{1}{2}x\mathbf{j}$$

on a particle that moves once around the ellipse (see figure)?

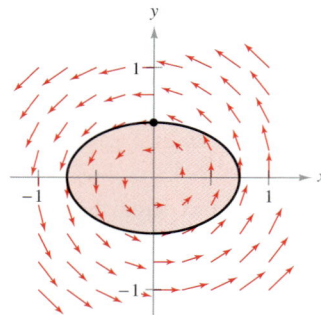

13. Verifying Identities

(a) Let f and g be scalar functions with continuous partial derivatives, and let C and S satisfy the conditions of Stokes's Theorem. Verify each identity.

(i) $\displaystyle\int_C (f\nabla g) \cdot d\mathbf{r} = \iint_S (\nabla f \times \nabla g) \cdot \mathbf{N} \, dS$

(ii) $\displaystyle\int_C (f\nabla f) \cdot d\mathbf{r} = 0$ (iii) $\displaystyle\int_C (f\nabla g + g\nabla f) \cdot d\mathbf{r} = 0$

(b) Demonstrate the results of part (a) for the functions $f(x, y, z) = xyz$ and $g(x, y, z) = z$. Let S be the hemisphere $z = \sqrt{4 - x^2 - y^2}$.

Appendices

A Proofs of Selected Theorems

For this edition, we have made Appendix A, Proofs of Selected Theorems, available in video format at *LarsonCalculus.com*. When you navigate to that website, you will find a link to Bruce Edwards explaining each proof in the text, including those in this appendix. We hope these videos enhance your study of calculus. The text version of this appendix is available at *CengageBrain.com*.

Proofs of Selected Theorems sample at *LarsonCalculus.com*

2.2 Basic Differentiation Rules and Rates of Change **107**

The Power Rule

Before proving the next rule, it is important to review the procedure for expanding a binomial.

$(x + \Delta x)^2 = x^2 + 2x\Delta x + (\Delta x)^2$

$(x + \Delta x)^3 = x^3 + 3x^2\Delta x + 3x(\Delta x)^2 + (\Delta x)^3$

$(x + \Delta x)^4 = x^4 + 4x^3\Delta x + 6x^2(\Delta x)^2 + 4x(\Delta x)^3 + (\Delta x)^4$

$(x + \Delta x)^5 = x^5 + 5x^4\Delta x + 10x^3(\Delta x)^2 + 10x^2(\Delta x)^3 + 5x(\Delta x)^4 + (\Delta x)^5$

The general binomial expansion for a positive

$(x + \Delta x)^n = x^n + nx^{n-1}(\Delta x) + \dfrac{n(n-1)}{2}$

This binomial expansion is used in proving a s

THEOREM 2.3 The Power Rule

If n is a rational number, then the function f

$$\dfrac{d}{dx}[x^n] = nx^{n-1}.$$

For f to be differentiable at $x = 0$, n must be on an interval containing 0.

REMARK From Example 7 in Section 2.1, you know that the function $f(x) = x^{1/3}$ is defined at $x = 0$, but is not differentiable at $x = 0$. This is because $x^{-2/3}$ is not defined on an interval containing 0.

Proof If n is a positive integer greater than 1

$$\dfrac{d}{dx}[x^n] = \lim_{\Delta x \to 0} \dfrac{(x + \Delta x)^n - x^n}{\Delta x}$$

$$= \lim_{\Delta x \to 0} \dfrac{x^n + nx^{n-1}(\Delta x) + \dfrac{n(n-1)}{2}}{}$$

$$= \lim_{\Delta x \to 0} \left[nx^{n-1} + \dfrac{n(n-1)x^{n-2}}{2} \right]$$

$$= nx^{n-1} + 0 + \cdots + 0$$

$$= nx^{n-1}.$$

This proves the case for which n is a positive integer the case for $n = 1$. Example 7 in Section 2.3 proves the case for which negative integer. (In Exercise 71 in Section 2.5, you are asked to prove the for which n is rational. (In Section 5.5, the Power Rule will be extended to cover irrational values of n.)
See LarsonCalculus.com for Bruce Edwards's video of this proof.

When using the Power Rule, the case for which $n = 1$ is best thought of as a separate differentiation rule. That is,

$$\dfrac{d}{dx}[x] = 1.$$ Power Rule when $n = 1$

This rule is consistent with the fact that the slope of the line $y = x$ is 1, as shown in Figure 2.15.

The slope of the line $y = x$ is 1.
Figure 2.15

LARSON CALCULUS

Home About Contact Purchase

Calculus 10e   Easy Access Study Guide   SEARCH

Chapter 2 ▼   Section 2 ▼   Proof: The P ▼   GO

Worked-out Solutions

Calculus Videos

Watch instructional videos by Dana Mosely and proof videos by co-author Bruce Edwards.

Interactive Examples

Rotatable Graphs

Printable Graphs

Data Downloads

Math Articles

Biographies

◄ Back   Chapter 2 > Section 2 > Proof: The Power Rule   Next ►

Credits Privacy Terms of Use
Copyright © Larson Texts Inc. All Rights Reserved

Bruce Edwards's Proof of the Power Rule at *LarsonCalculus.com*

Forms Involving u^n

1. $\displaystyle\int u^n \, du = \frac{u^{n+1}}{n+1} + C, \ n \neq -1$

2. $\displaystyle\int \frac{1}{u} \, du = \ln|u| + C$

Forms Involving $a + bu$

3. $\displaystyle\int \frac{u}{a+bu} \, du = \frac{1}{b^2}\big(bu - a\ln|a+bu|\big) + C$

4. $\displaystyle\int \frac{u}{(a+bu)^2} \, du = \frac{1}{b^2}\left(\frac{a}{a+bu} + \ln|a+bu|\right) + C$

5. $\displaystyle\int \frac{u}{(a+bu)^n} \, du = \frac{1}{b^2}\left[\frac{-1}{(n-2)(a+bu)^{n-2}} + \frac{a}{(n-1)(a+bu)^{n-1}}\right] + C, \quad n \neq 1, 2$

6. $\displaystyle\int \frac{u^2}{a+bu} \, du = \frac{1}{b^3}\left[-\frac{bu}{2}(2a - bu) + a^2\ln|a+bu|\right] + C$

7. $\displaystyle\int \frac{u^2}{(a+bu)^2} \, du = \frac{1}{b^3}\left(bu - \frac{a^2}{a+bu} - 2a\ln|a+bu|\right) + C$

8. $\displaystyle\int \frac{u^2}{(a+bu)^3} \, du = \frac{1}{b^3}\left[\frac{2a}{a+bu} - \frac{a^2}{2(a+bu)^2} + \ln|a+bu|\right] + C$

9. $\displaystyle\int \frac{u^2}{(a+bu)^n} \, du = \frac{1}{b^3}\left[\frac{-1}{(n-3)(a+bu)^{n-3}} + \frac{2a}{(n-2)(a+bu)^{n-2}} - \frac{a^2}{(n-1)(a+bu)^{n-1}}\right] + C, \quad n \neq 1, 2, 3$

10. $\displaystyle\int \frac{1}{u(a+bu)} \, du = \frac{1}{a}\ln\left|\frac{u}{a+bu}\right| + C$

11. $\displaystyle\int \frac{1}{u(a+bu)^2} \, du = \frac{1}{a}\left(\frac{1}{a+bu} + \frac{1}{a}\ln\left|\frac{u}{a+bu}\right|\right) + C$

12. $\displaystyle\int \frac{1}{u^2(a+bu)} \, du = -\frac{1}{a}\left(\frac{1}{u} + \frac{b}{a}\ln\left|\frac{u}{a+bu}\right|\right) + C$

13. $\displaystyle\int \frac{1}{u^2(a+bu)^2} \, du = -\frac{1}{a^2}\left[\frac{a+2bu}{u(a+bu)} + \frac{2b}{a}\ln\left|\frac{u}{a+bu}\right|\right] + C$

Forms Involving $a + bu + cu^2, \ b^2 \neq 4ac$

14. $\displaystyle\int \frac{1}{a+bu+cu^2} \, du = \begin{cases} \dfrac{2}{\sqrt{4ac - b^2}} \arctan \dfrac{2cu + b}{\sqrt{4ac - b^2}} + C, & b^2 < 4ac \\[2.5ex] \dfrac{1}{\sqrt{b^2 - 4ac}} \ln\left|\dfrac{2cu + b - \sqrt{b^2 - 4ac}}{2cu + b + \sqrt{b^2 - 4ac}}\right| + C, & b^2 > 4ac \end{cases}$

15. $\displaystyle\int \frac{u}{a+bu+cu^2} \, du = \frac{1}{2c}\left(\ln|a+bu+cu^2| - b\int \frac{1}{a+bu+cu^2} \, du\right)$

Forms Involving $\sqrt{a + bu}$

16. $\displaystyle\int u^n\sqrt{a+bu} \, du = \frac{2}{b(2n+3)}\left[u^n(a+bu)^{3/2} - na\int u^{n-1}\sqrt{a+bu} \, du\right]$

17. $\displaystyle\int \frac{1}{u\sqrt{a+bu}} \, du = \begin{cases} \dfrac{1}{\sqrt{a}}\ln\left|\dfrac{\sqrt{a+bu} - \sqrt{a}}{\sqrt{a+bu} + \sqrt{a}}\right| + C, & a > 0 \\[2.5ex] \dfrac{2}{\sqrt{-a}}\arctan\sqrt{\dfrac{a+bu}{-a}} + C, & a < 0 \end{cases}$

18. $\displaystyle\int \frac{1}{u^n\sqrt{a+bu}} \, du = \frac{-1}{a(n-1)}\left[\frac{\sqrt{a+bu}}{u^{n-1}} + \frac{(2n-3)b}{2}\int \frac{1}{u^{n-1}\sqrt{a+bu}} \, du\right], \ n \neq 1$

19. $\displaystyle\int \frac{\sqrt{a + bu}}{u}\,du = 2\sqrt{a + bu} + a\int \frac{1}{u\sqrt{a + bu}}\,du$

20. $\displaystyle\int \frac{\sqrt{a + bu}}{u^n}\,du = \frac{-1}{a(n - 1)}\left[\frac{(a + bu)^{3/2}}{u^{n-1}} + \frac{(2n - 5)b}{2}\int \frac{\sqrt{a + bu}}{u^{n-1}}\,du\right],\ n \ne 1$

21. $\displaystyle\int \frac{u}{\sqrt{a + bu}}\,du = \frac{-2(2a - bu)}{3b^2}\sqrt{a + bu} + C$

22. $\displaystyle\int \frac{u^n}{\sqrt{a + bu}}\,du = \frac{2}{(2n + 1)b}\left(u^n\sqrt{a + bu} - na\int \frac{u^{n-1}}{\sqrt{a + bu}}\,du\right)$

Forms Involving $a^2 \pm u^2$, $a > 0$

23. $\displaystyle\int \frac{1}{a^2 + u^2}\,du = \frac{1}{a}\arctan\frac{u}{a} + C$

24. $\displaystyle\int \frac{1}{u^2 - a^2}\,du = -\int \frac{1}{a^2 - u^2}\,du = \frac{1}{2a}\ln\left|\frac{u - a}{u + a}\right| + C$

25. $\displaystyle\int \frac{1}{(a^2 \pm u^2)^n}\,du = \frac{1}{2a^2(n - 1)}\left[\frac{u}{(a^2 \pm u^2)^{n-1}} + (2n - 3)\int \frac{1}{(a^2 \pm u^2)^{n-1}}\,du\right],\ n \ne 1$

Forms Involving $\sqrt{u^2 \pm a^2}$, $a > 0$

26. $\displaystyle\int \sqrt{u^2 \pm a^2}\,du = \frac{1}{2}\left(u\sqrt{u^2 \pm a^2} \pm a^2\ln\left|u + \sqrt{u^2 \pm a^2}\right|\right) + C$

27. $\displaystyle\int u^2\sqrt{u^2 \pm a^2}\,du = \frac{1}{8}\left[u(2u^2 \pm a^2)\sqrt{u^2 \pm a^2} - a^4\ln\left|u + \sqrt{u^2 \pm a^2}\right|\right] + C$

28. $\displaystyle\int \frac{\sqrt{u^2 + a^2}}{u}\,du = \sqrt{u^2 + a^2} - a\ln\left|\frac{a + \sqrt{u^2 + a^2}}{u}\right| + C$

29. $\displaystyle\int \frac{\sqrt{u^2 - a^2}}{u}\,du = \sqrt{u^2 - a^2} - a\,\text{arcsec}\frac{|u|}{a} + C$

30. $\displaystyle\int \frac{\sqrt{u^2 \pm a^2}}{u^2}\,du = \frac{-\sqrt{u^2 \pm a^2}}{u} + \ln\left|u + \sqrt{u^2 \pm a^2}\right| + C$

31. $\displaystyle\int \frac{1}{\sqrt{u^2 \pm a^2}}\,du = \ln\left|u + \sqrt{u^2 \pm a^2}\right| + C$

32. $\displaystyle\int \frac{1}{u\sqrt{u^2 + a^2}}\,du = \frac{-1}{a}\ln\left|\frac{a + \sqrt{u^2 + a^2}}{u}\right| + C$
   33. $\displaystyle\int \frac{1}{u\sqrt{u^2 - a^2}}\,du = \frac{1}{a}\,\text{arcsec}\frac{|u|}{a} + C$

34. $\displaystyle\int \frac{u^2}{\sqrt{u^2 \pm a^2}}\,du = \frac{1}{2}\left(u\sqrt{u^2 \pm a^2} \mp a^2\ln\left|u + \sqrt{u^2 \pm a^2}\right|\right) + C$

35. $\displaystyle\int \frac{1}{u^2\sqrt{u^2 \pm a^2}}\,du = \mp\frac{\sqrt{u^2 \pm a^2}}{a^2 u} + C$
   36. $\displaystyle\int \frac{1}{(u^2 \pm a^2)^{3/2}}\,du = \frac{\pm u}{a^2\sqrt{u^2 \pm a^2}} + C$

Forms Involving $\sqrt{a^2 - u^2}$, $a > 0$

37. $\displaystyle\int \sqrt{a^2 - u^2}\,du = \frac{1}{2}\left(u\sqrt{a^2 - u^2} + a^2\arcsin\frac{u}{a}\right) + C$

38. $\displaystyle\int u^2\sqrt{a^2 - u^2}\,du = \frac{1}{8}\left[u(2u^2 - a^2)\sqrt{a^2 - u^2} + a^4\arcsin\frac{u}{a}\right] + C$

39. $\displaystyle\int \frac{\sqrt{a^2 - u^2}}{u}\, du = \sqrt{a^2 - u^2} - a \ln\left|\frac{a + \sqrt{a^2 - u^2}}{u}\right| + C$ **40.** $\displaystyle\int \frac{\sqrt{a^2 - u^2}}{u^2}\, du = \frac{-\sqrt{a^2 - u^2}}{u} - \arcsin \frac{u}{a} + C$

41. $\displaystyle\int \frac{1}{\sqrt{a^2 - u^2}}\, du = \arcsin \frac{u}{a} + C$ **42.** $\displaystyle\int \frac{1}{u\sqrt{a^2 - u^2}}\, du = \frac{-1}{a} \ln\left|\frac{a + \sqrt{a^2 - u^2}}{u}\right| + C$

43. $\displaystyle\int \frac{u^2}{\sqrt{a^2 - u^2}}\, du = \frac{1}{2}\left(-u\sqrt{a^2 - u^2} + a^2 \arcsin \frac{u}{a}\right) + C$ **44.** $\displaystyle\int \frac{1}{u^2\sqrt{a^2 - u^2}}\, du = \frac{-\sqrt{a^2 - u^2}}{a^2 u} + C$

45. $\displaystyle\int \frac{1}{(a^2 - u^2)^{3/2}}\, du = \frac{u}{a^2\sqrt{a^2 - u^2}} + C$

Forms Involving sin u or cos u

46. $\displaystyle\int \sin u\, du = -\cos u + C$ **47.** $\displaystyle\int \cos u\, du = \sin u + C$

48. $\displaystyle\int \sin^2 u\, du = \frac{1}{2}(u - \sin u \cos u) + C$ **49.** $\displaystyle\int \cos^2 u\, du = \frac{1}{2}(u + \sin u \cos u) + C$

50. $\displaystyle\int \sin^n u\, du = -\frac{\sin^{n-1} u \cos u}{n} + \frac{n-1}{n}\int \sin^{n-2} u\, du$ **51.** $\displaystyle\int \cos^n u\, du = \frac{\cos^{n-1} u \sin u}{n} + \frac{n-1}{n}\int \cos^{n-2} u\, du$

52. $\displaystyle\int u \sin u\, du = \sin u - u \cos u + C$ **53.** $\displaystyle\int u \cos u\, du = \cos u + u \sin u + C$

54. $\displaystyle\int u^n \sin u\, du = -u^n \cos u + n \int u^{n-1} \cos u\, du$ **55.** $\displaystyle\int u^n \cos u\, du = u^n \sin u - n \int u^{n-1} \sin u\, du$

56. $\displaystyle\int \frac{1}{1 \pm \sin u}\, du = \tan u \mp \sec u + C$ **57.** $\displaystyle\int \frac{1}{1 \pm \cos u}\, du = -\cot u \pm \csc u + C$

58. $\displaystyle\int \frac{1}{\sin u \cos u}\, du = \ln|\tan u| + C$

Forms Involving tan u, cot u, sec u, or csc u

59. $\displaystyle\int \tan u\, du = -\ln|\cos u| + C$ **60.** $\displaystyle\int \cot u\, du = \ln|\sin u| + C$

61. $\displaystyle\int \sec u\, du = \ln|\sec u + \tan u| + C$

62. $\displaystyle\int \csc u\, du = \ln|\csc u - \cot u| + C$ or $\displaystyle\int \csc u\, du = -\ln|\csc u + \cot u| + C$

63. $\displaystyle\int \tan^2 u\, du = -u + \tan u + C$ **64.** $\displaystyle\int \cot^2 u\, du = -u - \cot u + C$

65. $\displaystyle\int \sec^2 u\, du = \tan u + C$ **66.** $\displaystyle\int \csc^2 u\, du = -\cot u + C$

67. $\displaystyle\int \tan^n u\, du = \frac{\tan^{n-1} u}{n-1} - \int \tan^{n-2} u\, du, \; n \neq 1$ **68.** $\displaystyle\int \cot^n u\, du = -\frac{\cot^{n-1} u}{n-1} - \int (\cot^{n-2} u)\, du, \; n \neq 1$

69. $\displaystyle\int \sec^n u\, du = \frac{\sec^{n-2} u \tan u}{n-1} + \frac{n-2}{n-1}\int \sec^{n-2} u\, du, \; n \neq 1$

70. $\displaystyle\int \csc^n u\, du = -\frac{\csc^{n-2} u \cot u}{n-1} + \frac{n-2}{n-1}\int \csc^{n-2} u\, du, \; n \neq 1$

71. $\displaystyle\int \frac{1}{1 \pm \tan u}\, du = \frac{1}{2}\left(u \pm \ln|\cos u \pm \sin u|\right) + C$

72. $\displaystyle\int \frac{1}{1 \pm \cot u}\, du = \frac{1}{2}\left(u \mp \ln|\sin u \pm \cos u|\right) + C$

73. $\displaystyle\int \frac{1}{1 \pm \sec u}\, du = u + \cot u \mp \csc u + C$

74. $\displaystyle\int \frac{1}{1 \pm \csc u}\, du = u - \tan u \pm \sec u + C$

Forms Involving Inverse Trigonometric Functions

75. $\displaystyle\int \arcsin u\, du = u \arcsin u + \sqrt{1 - u^2} + C$

76. $\displaystyle\int \arccos u\, du = u \arccos u - \sqrt{1 - u^2} + C$

77. $\displaystyle\int \arctan u\, du = u \arctan u - \ln\sqrt{1 + u^2} + C$

78. $\displaystyle\int \operatorname{arccot} u\, du = u \operatorname{arccot} u + \ln\sqrt{1 + u^2} + C$

79. $\displaystyle\int \operatorname{arcsec} u\, du = u \operatorname{arcsec} u - \ln\left|u + \sqrt{u^2 - 1}\right| + C$

80. $\displaystyle\int \operatorname{arccsc} u\, du = u \operatorname{arccsc} u + \ln\left|u + \sqrt{u^2 - 1}\right| + C$

Forms Involving e^u

81. $\displaystyle\int e^u\, du = e^u + C$

82. $\displaystyle\int u e^u\, du = (u - 1)e^u + C$

83. $\displaystyle\int u^n e^u\, du = u^n e^u - n\int u^{n-1} e^u\, du$

84. $\displaystyle\int \frac{1}{1 + e^u}\, du = u - \ln(1 + e^u) + C$

85. $\displaystyle\int e^{au} \sin bu\, du = \frac{e^{au}}{a^2 + b^2}(a \sin bu - b \cos bu) + C$

86. $\displaystyle\int e^{au} \cos bu\, du = \frac{e^{au}}{a^2 + b^2}(a \cos bu + b \sin bu) + C$

Forms Involving $\ln u$

87. $\displaystyle\int \ln u\, du = u(-1 + \ln u) + C$

88. $\displaystyle\int u \ln u\, du = \frac{u^2}{4}(-1 + 2 \ln u) + C$

89. $\displaystyle\int u^n \ln u\, du = \frac{u^{n+1}}{(n+1)^2}[-1 + (n+1) \ln u] + C,\ n \neq -1$

90. $\displaystyle\int (\ln u)^2\, du = u\left[2 - 2 \ln u + (\ln u)^2\right] + C$

91. $\displaystyle\int (\ln u)^n\, du = u(\ln u)^n - n\int (\ln u)^{n-1}\, du$

Forms Involving Hyperbolic Functions

92. $\displaystyle\int \cosh u\, du = \sinh u + C$

93. $\displaystyle\int \sinh u\, du = \cosh u + C$

94. $\displaystyle\int \operatorname{sech}^2 u\, du = \tanh u + C$

95. $\displaystyle\int \operatorname{csch}^2 u\, du = -\coth u + C$

96. $\displaystyle\int \operatorname{sech} u \tanh u\, du = -\operatorname{sech} u + C$

97. $\displaystyle\int \operatorname{csch} u \coth u\, du = -\operatorname{csch} u + C$

Forms Involving Inverse Hyperbolic Functions (in logarithmic form)

98. $\displaystyle\int \frac{du}{\sqrt{u^2 \pm a^2}} = \ln\left(u + \sqrt{u^2 \pm a^2}\right) + C$

99. $\displaystyle\int \frac{du}{a^2 - u^2} = \frac{1}{2a}\ln\left|\frac{a + u}{a - u}\right| + C$

100. $\displaystyle\int \frac{du}{u\sqrt{a^2 \pm u^2}} = -\frac{1}{a}\ln\frac{a + \sqrt{a^2 \pm u^2}}{|u|} + C$

Answers to Odd-Numbered Exercises

Chapter P

Section P.1   *(page 8)*

1. b   **2.** d   **3.** a   **4.** c

5.   **7.**

9.   **11.**

13.   **15.**

(a) $y \approx 1.73$ (b) $x = -4$

17. $(0, -5), \left(\frac{5}{2}, 0\right)$ **19.** $(0, -2), (-2, 0), (1, 0)$

21. $(0, 0), (4, 0), (-4, 0)$ **23.** $(0, 2), (4, 0)$ **25.** $(0, 0)$

27. Symmetric with respect to the y-axis

29. Symmetric with respect to the x-axis

31. Symmetric with respect to the origin   **33.** No symmetry

35. Symmetric with respect to the origin

37. Symmetric with respect to the y-axis

39.   **41.**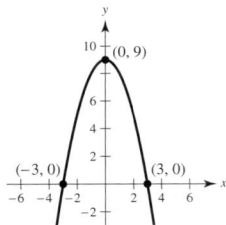

Symmetry: none   Symmetry: y-axis

43.   **45.**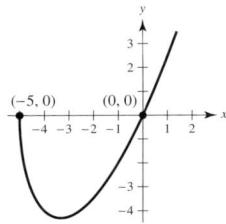

Symmetry: none   Symmetry: none

47.   **49.**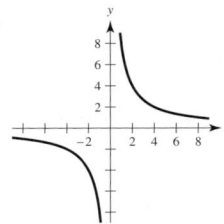

Symmetry: origin   Symmetry: origin

51.   **53.**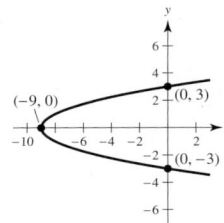

Symmetry: y-axis   Symmetry: x-axis

55. 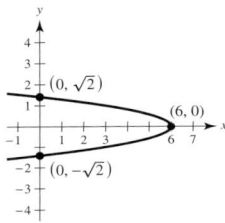   **57.** $(3, 5)$

Symmetry: x-axis

59. $(-1, 5), (2, 2)$ **61.** $(-1, -2), (2, 1)$

63. $(-1, -5), (0, -1), (2, 1)$ **65.** $(-2, 2), \left(-3, \sqrt{3}\right)$

67. (a) $y = 0.005t^2 + 0.27t + 2.7$

(b)

The model is a good fit for the data.

(c) \$21.5 trillion

69. 4480 units

71. (a) $k = 4$ (b) $k = -\frac{1}{8}$

(c) All real numbers k (d) $k = 1$

73. Answers will vary. Sample answer: $y = (x + 4)(x - 3)(x - 8)$

75. (a) Proof   (b) Proof

77. False. $(4, -5)$ is not a point on the graph of $x = y^2 - 29$.

79. True

Section P.2   *(page 16)*

1. $m = 2$ **3.** $m = -1$

5.

$m = 3$

7.

m is undefined.

9.

$m = 2$

11.

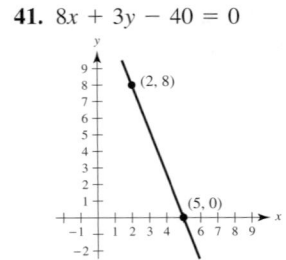

13. Answers will vary. Sample answers: $(0, 2), (1, 2), (5, 2)$

15. Answers will vary. Sample answers: $(0, 10), (2, 4), (3, 1)$

17. $3x - 4y + 12 = 0$ **19.** $2x - 3y = 0$

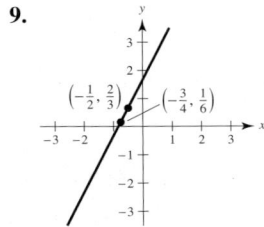

21. $3x - y - 11 = 0$ **23.** (a) $\frac{1}{3}$ (b) $10\sqrt{10}$ ft

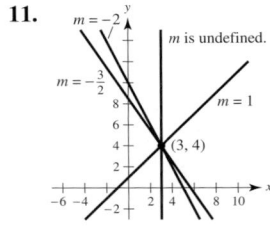

25. $m = 4, (0, -3)$ **27.** $m = -\frac{1}{5}, (0, 4)$

29. m is undefined, no y-intercept

31.

33.

35.

37.

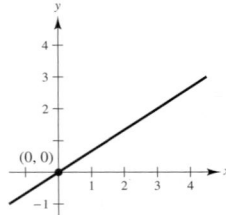

39. $2x - y = 0$

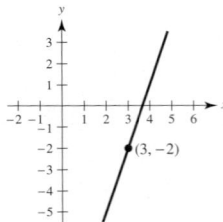

41. $8x + 3y - 40 = 0$

43. $x - 6 = 0$

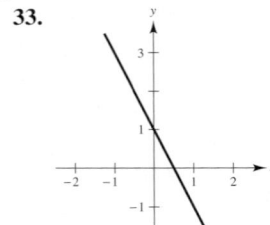

45. $22x - 4y + 3 = 0$

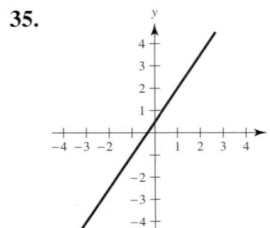

47. $x - 3 = 0$ **49.** $3x + 2y - 6 = 0$ **51.** $x + y - 3 = 0$

53. $x + 2y - 5 = 0$ **55.** (a) $x + 7 = 0$ (b) $y + 2 = 0$

57. (a) $x - y + 3 = 0$ (b) $x + y - 7 = 0$

59. (a) $2x - y - 3 = 0$ (b) $x + 2y - 4 = 0$

61. (a) $40x - 24y - 9 = 0$ (b) $24x + 40y - 53 = 0$

63. $V = 250t + 1350$ **65.** $V = -1600t + 20{,}400$

67. Not collinear, because $m_1 \neq m_2$

69. $\left(0, \dfrac{-a^2 + b^2 + c^2}{2c}\right)$ **71.** $\left(b, \dfrac{a^2 - b^2}{c}\right)$

73. (a) The line is parallel to the x-axis when $a = 0$ and $b \neq 0$.

(b) The line is parallel to the y-axis when $b = 0$ and $a \neq 0$.

(c) Answers will vary. Sample answer: $a = -5$ and $b = 8$

(d) Answers will vary. Sample answer: $a = 5$ and $b = 2$

(e) $a = \frac{5}{2}$ and $b = 3$

75. $5F - 9C - 160 = 0$; $72°F \approx 22.2°C$

77. (a) Current job: $W = 2000 + 0.07s$

Job offer: $W = 2300 + 0.05s$

(b)

You will make more money at the job offer until you sell $15,000. When your sales exceed $15,000, your current job will pay you more.

(c) No, because you will make more money at your current job.

79. (a) $x = (1530 - p)/15$

(b)

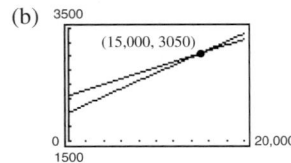

45 units

(c) 49 units

81. $12y + 5x - 169 = 0$ **83.** $(5\sqrt{2})/2$ **85.** $2\sqrt{2}$

87–91. Proofs   **93.** True   **95.** True

Section P.3   *(page 27)*

1. (a) -4 (b) -25 (c) $7b - 4$ (d) $7x - 11$
3. (a) 5 (b) 0 (c) 1 (d) $4 + 2t - t^2$
5. (a) 1 (b) 0 (c) $-\frac{1}{2}$ (d) 1
7. $3x^2 + 3x\,\Delta x + (\Delta x)^2,\ \Delta x \neq 0$
9. $\left(\sqrt{x-1} - x + 1\right)/[(x-2)(x-1)]$
11. Domain: $(-\infty, \infty)$; Range: $[0, \infty)$
13. Domain: $(-\infty, \infty)$; Range: $(-\infty, \infty)$
15. Domain: $[0, \infty)$; Range: $[0, \infty)$
17. Domain: $[-4, 4]$; Range: $[0, 4]$
19. Domain: All real numbers t such that $t \neq 4n + 2$, where n is an integer; Range: $(-\infty, -1] \cup [1, \infty)$
21. Domain: $(-\infty, 0) \cup (0, \infty)$; Range: $(-\infty, 0) \cup (0, \infty)$
23. Domain: $[0, 1]$
25. Domain: All real numbers x such that $x \neq 2n\pi$, where n is an integer
27. Domain: $(-\infty, -3) \cup (-3, \infty)$
29. (a) -1 (b) 2 (c) 6 (d) $2t^2 + 4$
    Domain: $(-\infty, \infty)$; Range: $(-\infty, 1) \cup [2, \infty)$
31. (a) 4 (b) 0 (c) -2 (d) $-b^2$
    Domain: $(-\infty, \infty)$; Range: $(-\infty, 0] \cup [1, \infty)$

33.
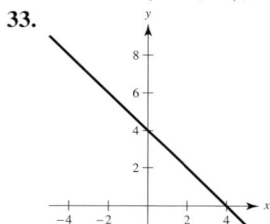
Domain: $(-\infty, \infty)$
Range: $(-\infty, \infty)$

35.
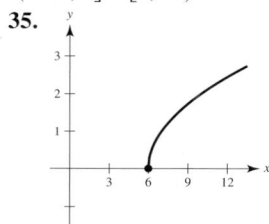
Domain: $[6, \infty)$
Range: $[0, \infty)$

37.
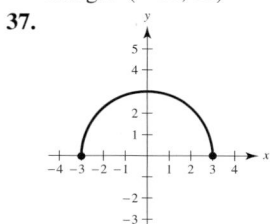
Domain: $[-3, 3]$
Range: $[0, 3]$

39.
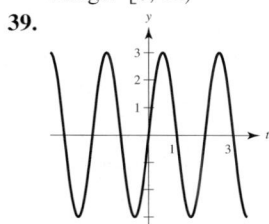
Domain: $(-\infty, \infty)$
Range: $[-3, 3]$

41. The student travels $\frac{1}{2}$ mile/minute during the first 4 minutes, is stationary for the next 2 minutes, and travels 1 mile/minute during the final 4 minutes.
43. y is not a function of x.   45. y is a function of x.
47. y is not a function of x.   49. y is not a function of x.
51. Horizontal shift to the right two units
    $y = \sqrt{x - 2}$
53. Horizontal shift to the right two units and vertical shift down one unit
    $y = (x - 2)^2 - 1$
55. d 56. b 57. c 58. a 59. e 60. g

61. (a)

(b)

(c)

(d)

(e)

(f)

(g)

(h)
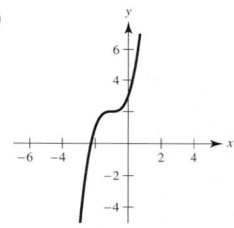

63. (a) $3x$ (b) $3x - 8$ (c) $12x - 16$ (d) $\frac{3}{4}x - 1$
65. (a) 0 (b) 0 (c) -1 (d) $\sqrt{15}$
    (e) $\sqrt{x^2 - 1}$ (f) $x - 1\ (x \geq 0)$
67. $(f \circ g)(x) = x$; Domain: $[0, \infty)$
    $(g \circ f)(x) = |x|$; Domain: $(-\infty, \infty)$
    No, their domains are different.
69. $(f \circ g)(x) = 3/(x^2 - 1)$;
    Domain: $(-\infty, -1) \cup (-1, 1) \cup (1, \infty)$
    $(g \circ f)(x) = (9/x^2) - 1$; Domain: $(-\infty, 0) \cup (0, \infty)$
    No
71. (a) 4 (b) -2
    (c) Undefined. The graph of g does not exist at $x = -5$.
    (d) 3 (e) 2
    (f) Undefined. The graph of f does not exist at $x = -4$.
73. Answers will vary.
    Sample answer: $f(x) = \sqrt{x}$; $g(x) = x - 2$; $h(x) = 2x$
75. (a) $\left(\frac{3}{2}, 4\right)$ (b) $\left(\frac{3}{2}, -4\right)$
77. f is even. g is neither even nor odd. h is odd.
79. Even; zeros: $x = -2, 0, 2$
81. Odd; zeros: $x = 0, \dfrac{\pi}{2} + n\pi$, where n is an integer
83. $f(x) = -5x - 6,\ -2 \leq x \leq 0$ 85. $y = -\sqrt{-x}$

87. Answers will vary.
Sample answer:

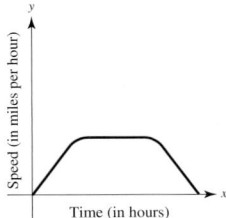

89. Answers will vary.
Sample answer:

91. $c = 25$

93. (a) $T(4) = 16°C$, $T(15) \approx 23°C$
(b) The changes in temperature occur 1 hour later.
(c) The temperatures are 1° lower.

95. (a)

(b) $A(25) \approx 443$ acres/farm

97. $f(x) = |x| + |x - 2| = \begin{cases} 2x - 2, & x \geq 2 \\ 2, & 0 < x < 2 \\ -2x + 2, & x \leq 0 \end{cases}$

99–101. Proofs   **103.** $L = \sqrt{x^2 + \left(\dfrac{2x}{x-3}\right)^2}$

105. False. For example, if $f(x) = x^2$, then $f(-1) = f(1)$.
107. True
109. False. $f(x) = 0$ is symmetric with respect to the x-axis.
111. Putnam Problem A1, 1988

Section P.4   *(page 34)*

1. (a) and (b)

(c) $790

3. (a) $d = 0.066F$
(b)

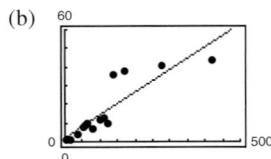

The model fits well.
(c) 3.63 cm

5. (a) $y = 0.122x + 2.07$, $r \approx 0.87$

(b)

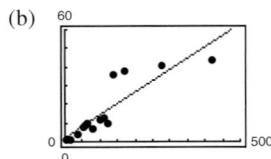

(c) Greater per capita energy consumption by a country tends to correspond to greater per capita gross national product of the country. The three countries that differ most from the linear model are Canada, Italy, and Japan.
(d) $y = 0.142x - 1.66$, $r \approx 0.97$

7. (a) $S = 180.89x^2 - 205.79x + 272$
(b)

(c) When $x = 2$, $S \approx 583.98$ pounds.
(d) About 4 times greater
(e) About 4.37 times greater; No; Answers will vary.

9. (a) $y = -1.806x^3 + 14.58x^2 + 16.4x + 10$
(b)

(c) 214 hp

11. (a) $y_1 = -0.0172t^3 + 0.305t^2 - 0.87t + 7.3$
$y_2 = -0.038t^2 + 0.45t + 3.5$
$y_3 = 0.0063t^3 - 0.072t^2 + 0.02t + 1.8$
(b)

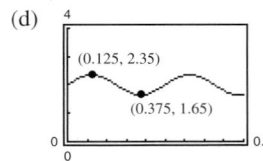

About 15.31 cents/mi

13. (a) Yes. At time t, there is one and only one displacement y.
(b) Amplitude: 0.35; Period: 0.5
(c) $y = 0.35 \sin(4\pi t) + 2$
(d)

(0.125, 2.35)
(0.375, 1.65)

The model appears to fit the data well.

15. Answers will vary.   **17.** Putnam Problem A2, 2004

Review Exercises for Chapter P   *(page 37)*

1. $\left(\frac{8}{5}, 0\right)$, $(0, -8)$ **3.** $(3, 0)$, $\left(0, \frac{3}{4}\right)$ **5.** Not symmetric
7. Symmetric with respect to the x-axis, the y-axis, and the origin

9.

Symmetry: none

11.

Symmetry: origin

13.

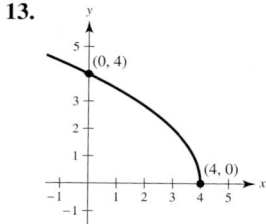

Symmetry: none

15. $(-2, 3)$ **17.** $(-2, 3), (3, 8)$

19.

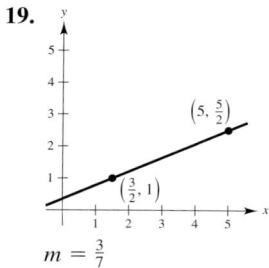

$m = \frac{3}{7}$

21. $7x - 4y - 41 = 0$

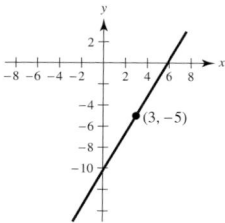

23. $2x + 3y + 6 = 0$

25.

27.

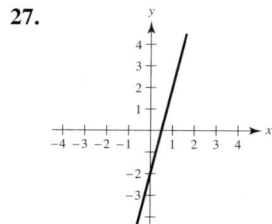

29. $x - 4y = 0$

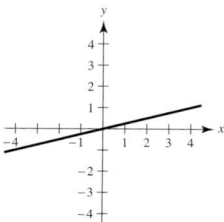

31. (a) $7x - 16y + 101 = 0$
(b) $5x - 3y + 30 = 0$
(c) $4x - 3y + 27 = 0$
(d) $x + 3 = 0$

33. $V = 12,500 - 850t$; $9950

35. (a) 4   (b) 29   (c) -11 (d) $5t + 9$

37. $8x + 4\,\Delta x, \ \Delta x \neq 0$

39. Domain: $(-\infty, \infty)$; Range: $[3, \infty)$

41. Domain: $(-\infty, \infty)$; Range: $(-\infty, 0]$

43.

Not a function

45.

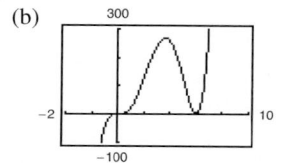

Function

47. $f(x) = x^3 - 3x^2$

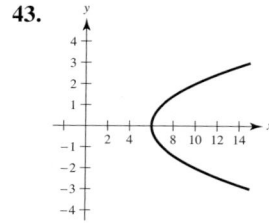

(a) $g(x) = -x^3 + 3x^2 + 1$
(b) $g(x) = (x - 2)^3 - 3(x - 2)^2 + 1$

49. (a)

(b)

(c)

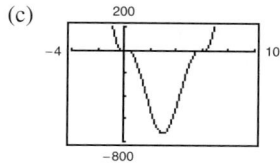

51. (a) $y = -1.204x + 64.2667$

(b)

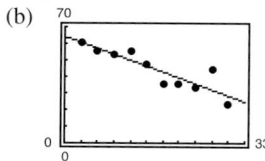

(c) The data point $(27, 44)$ is probably an error. Without this point, the new model is $y = -1.4344x + 66.4387$.

53. (a) Yes. For each time t, there corresponds one and only one displacement y.
(b) Amplitude: 0.25; Period: 1.1   (c) $y \approx \frac{1}{4}\cos(5.7t)$
(d)

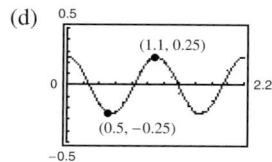

The model appears to fit the data.

P.S. Problem Solving   *(page 39)*

1. (a) Center: $(3, 4)$; Radius: 5
(b) $y = -\frac{3}{4}x$ (c) $y = \frac{3}{4}x - \frac{9}{2}$ (d) $\left(3, -\frac{9}{4}\right)$

3.

(a)

(b)

(c)

(d)

(e)

(f)

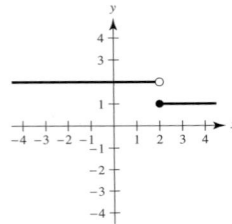

5. (a) $A(x) = x[(100 - x)/2]$; Domain: $(0, 100)$

(b)

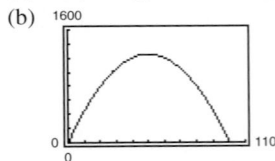

Dimensions 50 m × 25 m
yield maximum area of
1250 m².

(c) 50 m × 25 m; Area = 1250 m²

7. $T(x) = \left[2\sqrt{4 + x^2} + \sqrt{(3 - x)^2 + 1}\right]/4$

9. (a) 5, less (b) 3, greater (c) 4.1, less
(d) $4 + h$ (e) 4; Answers will vary.

11. (a) Domain: $(-\infty, 1) \cup (1, \infty)$; Range: $(-\infty, 0) \cup (0, \infty)$

(b) $f(f(x)) = \dfrac{x - 1}{x}$

Domain: $(-\infty, 0) \cup (0, 1) \cup (1, \infty)$

(c) $f(f(f(x))) = x$

Domain: $(-\infty, 0) \cup (0, 1) \cup (1, \infty)$

(d)

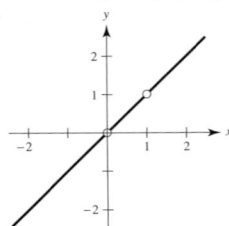

The graph is not a line
because there are holes at
$x = 0$ and $x = 1$.

13. (a) $x \approx 1.2426, -7.2426$

(b) $(x + 3)^2 + y^2 = 18$

15. Proof

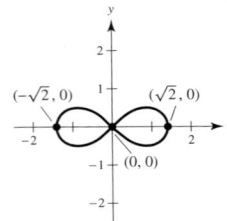

Chapter 1

Section 1.1   *(page 47)*

1. Precalculus: 300 ft
3. Calculus: Slope of the tangent line at $x = 2$ is 0.16.
5. (a) Precalculus: 10 square units
(b) Calculus: 5 square units
7. (a)

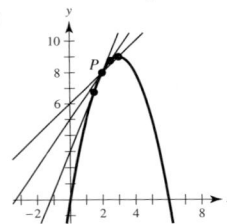

(b) $1; \frac{3}{2}; \frac{5}{2}$
(c) 2. Use points closer to P.

9. Area ≈ 10.417; Area ≈ 9.145; Use more rectangles.

Section 1.2   *(page 55)*

1.

x	3.9	3.99	3.999	4
$f(x)$	0.2041	0.2004	0.2000	?

x	4.001	4.01	4.1
$f(x)$	0.2000	0.1996	0.1961

$\displaystyle \lim_{x \to 4} \frac{x - 4}{x^2 - 3x - 4} \approx 0.2000$ $\left(\text{Actual limit is } \dfrac{1}{5}.\right)$

3.

x	-0.1	-0.01	-0.001	0
$f(x)$	0.5132	0.5013	0.5001	?

x	0.001	0.01	0.1
$f(x)$	0.4999	0.4988	0.4881

$\displaystyle \lim_{x \to 0} \frac{\sqrt{x + 1} - 1}{x} \approx 0.5000$ $\left(\text{Actual limit is } \dfrac{1}{2}.\right)$

5.

x	-0.1	-0.01	-0.001	0
$f(x)$	0.9983	0.99998	1.0000	?

x	0.001	0.01	0.1
$f(x)$	1.0000	0.99998	0.9983

$\lim\limits_{x \to 0} \dfrac{\sin x}{x} \approx 1.0000$ (Actual limit is 1.)

7.

x	0.9	0.99	0.999	1
$f(x)$	0.2564	0.2506	0.2501	?

x	1.001	1.01	1.1
$f(x)$	0.2499	0.2494	0.2439

$\lim\limits_{x \to 1} \dfrac{x-2}{x^2+x-6} \approx 0.2500 \left(\text{Actual limit is } \dfrac{1}{4}.\right)$

9.

x	0.9	0.99	0.999	1
$f(x)$	0.7340	0.6733	0.6673	?

x	1.001	1.01	1.1
$f(x)$	0.6660	0.6600	0.6015

$\lim\limits_{x \to 1} \dfrac{x^4-1}{x^6-1} \approx 0.6666 \left(\text{Actual limit is } \dfrac{2}{3}.\right)$

11.

x	-6.1	-6.01	-6.001	-6
$f(x)$	-0.1248	-0.1250	-0.1250	?

x	-5.999	-5.99	-5.9
$f(x)$	-0.1250	-0.1250	-0.1252

$\lim\limits_{x \to -6} \dfrac{\sqrt{10-x}-4}{x+6} \approx -0.1250 \left(\text{Actual limit is } -\dfrac{1}{8}.\right)$

13.

x	-0.1	-0.01	-0.001	0
$f(x)$	1.9867	1.9999	2.0000	?

x	0.001	0.01	0.1
$f(x)$	2.0000	1.9999	1.9867

$\lim\limits_{x \to 0} \dfrac{\sin 2x}{x} \approx 2.0000$ (Actual limit is 2.)

15. 1   **17.** 2

19. Limit does not exist. The function approaches 1 from the right side of 2, but it approaches -1 from the left side of 2.

21. Limit does not exist. The function oscillates between 1 and -1 as x approaches 0.

23. (a) 2
(b) Limit does not exist. The function approaches 1 from the right side of 1, but it approaches 3.5 from the left side of 1.
(c) Value does not exist. The function is undefined at $x = 4$.
(d) 2

25.

27.

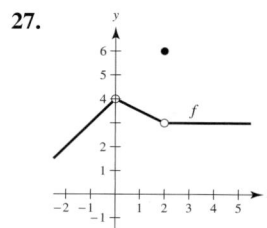

$\lim\limits_{x \to c} f(x)$ exists for all points on the graph except where $c = 4$.

29. $\delta = 0.4$ **31.** $\delta = \frac{1}{11} \approx 0.091$
33. $L = 8$. Let $\delta = 0.01/3 \approx 0.0033$.
35. $L = 1$. Let $\delta = 0.01/5 = 0.002$.   **37.** 6   **39.** -3
41. 3   **43.** 0   **45.** 10   **47.** 2   **49.** 4

51.

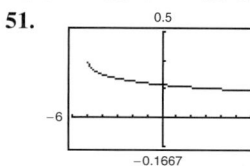

$\lim\limits_{x \to 4} f(x) = \frac{1}{6}$
Domain: $[-5, 4) \cup (4, \infty)$
The graph has a hole at $x = 4$.

53.

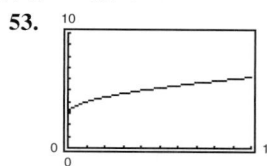

$\lim\limits_{x \to 9} f(x) = 6$
Domain: $[0, 9) \cup (9, \infty)$
The graph has a hole at $x = 9$.

55. (a)

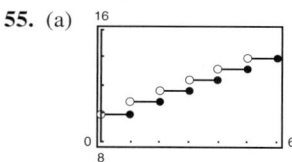

(b)

t	3	3.3	3.4	3.5
C	11.57	12.36	12.36	12.36

t	3.6	3.7	4
C	12.36	12.36	12.36

$\lim\limits_{t \to 3.5} C(t) = 12.36$

(c)

t	2	2.5	2.9	3
C	10.78	11.57	11.57	11.57

t	3.1	3.5	4
C	12.36	12.36	12.36

The limit does not exist because the limits from the right and left are not equal.

57. Answers will vary. Sample answer: As x approaches 8 from either side, $f(x)$ becomes arbitrarily close to 25.

59. (i) The values of f approach different numbers as x approaches c from different sides of c.

(ii) The values of f increase or decrease without bound as x approaches c.

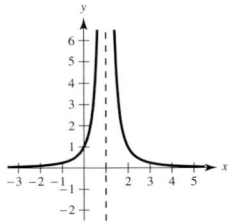

(iii) The values of f oscillate between two fixed numbers as x approaches c.

61. (a) $r = \dfrac{3}{\pi} \approx 0.9549$ cm

(b) $\dfrac{5.5}{2\pi} \le r \le \dfrac{6.5}{2\pi}$, or approximately $0.8754 < r < 1.0345$

(c) $\lim\limits_{r \to 3/\pi} 2\pi r = 6$; $\varepsilon = 0.5$; $\delta \approx 0.0796$

63.

x	-0.001	-0.0001	-0.00001
$f(x)$	2.7196	2.7184	2.7183

x	0.00001	0.0001	0.001
$f(x)$	2.7183	2.7181	2.7169

$\lim\limits_{x \to 0} f(x) \approx 2.7183$

65.

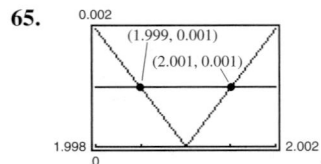

$\delta = 0.001$, $(1.999, 2.001)$

67. False. The existence or nonexistence of $f(x)$ at $x = c$ has no bearing on the existence of the limit of $f(x)$ as $x \to c$.

69. False. See Exercise 17.

71. Yes. As x approaches 0.25 from either side, $\sqrt{x}$ becomes arbitrarily close to 0.5.

73. $\lim\limits_{x \to 0} \dfrac{\sin nx}{x} = n$ **75–77.** Proofs

79. Putnam Problem B1, 1986

Section 1.3   *(page 67)*

1.

3.

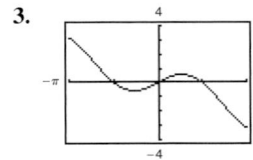

(a) 0   **(b)** -5

(a) 0   **(b)** About 0.52 or $\pi/6$

5. 8   **7.** -1 **9.** 0   **11.** 7   **13.** 2   **15.** 1

17. $1/2$ **19.** $1/5$ **21.** 7   **23. (a)** 4   **(b)** 64   **(c)** 64

25. (a) 3   **(b)** 2   **(c)** 2   **27.** 1   **29.** $1/2$ **31.** 1

33. $1/2$ **35.** -1 **37. (a)** 10   **(b)** 5   **(c)** 6   **(d)** $3/2$

39. (a) 64   **(b)** 2   **(c)** 12   **(d)** 8

41. $f(x) = \dfrac{x^2 + 3x}{x}$ and $g(x) = x + 3$ agree except at $x = 0$.

$\lim\limits_{x \to 0} f(x) = \lim\limits_{x \to 0} g(x) = 3$

43. $f(x) = \dfrac{x^2 - 1}{x + 1}$ and $g(x) = x - 1$ agree except at $x = -1$.

$\lim\limits_{x \to -1} f(x) = \lim\limits_{x \to -1} g(x) = -2$

45. $f(x) = \dfrac{x^3 - 8}{x - 2}$ and $g(x) = x^2 + 2x + 4$ agree except at $x = 2$.

$\lim\limits_{x \to 2} f(x) = \lim\limits_{x \to 2} g(x) = 12$

47. -1 **49.** $1/8$ **51.** $5/6$ **53.** $1/6$ **55.** $\sqrt{5}/10$

57. $-1/9$ **59.** 2   **61.** $2x - 2$ **63.** $1/5$ **65.** 0

67. 0   **69.** 0   **71.** 1   **73.** $3/2$

75.

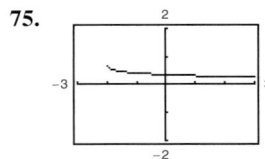

The graph has a hole at $x = 0$.

Answers will vary. Sample answer:

x	-0.1	-0.01	-0.001	0.001	0.01	0.1
$f(x)$	0.358	0.354	0.354	0.354	0.353	0.349

$\lim\limits_{x \to 0} \dfrac{\sqrt{x + 2} - \sqrt{2}}{x} \approx 0.354$; Actual limit is $\dfrac{1}{2\sqrt{2}} = \dfrac{\sqrt{2}}{4}$.

77.

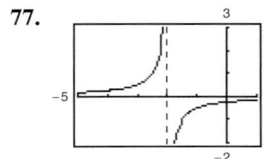

The graph has a hole at $x = 0$.

Answers will vary. Sample answer:

x	-0.1	-0.01	-0.001
$f(x)$	-0.263	-0.251	-0.250

x	0.001	0.01	0.1
$f(x)$	-0.250	-0.249	-0.238

$\lim\limits_{x \to 0} \dfrac{[1/(2 + x)] - (1/2)}{x} \approx -0.250$; Actual limit is $-\dfrac{1}{4}$.

79.

The graph has a hole at $t = 0$.

Answers will vary. Sample answer:

t	-0.1	-0.01	0	0.01	0.1
$f(t)$	2.96	2.9996	?	2.9996	2.96

$\lim\limits_{t \to 0} \dfrac{\sin 3t}{t} \approx 3.0000$; Actual limit is 3.

81.

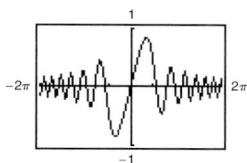

The graph has a hole at $x = 0$.

Answers will vary. Sample answer:

x	-0.1	-0.01	-0.001	0	0.001	0.01	0.1
$f(x)$	-0.1	-0.01	-0.001	?	0.001	0.01	0.1

$\lim\limits_{x \to 0} \dfrac{\sin x^2}{x} = 0$; Actual limit is 0.

83. 3   **85.** $2x - 4$ **87.** $-1/(x + 3)^2$ **89.** 4

91.

93.

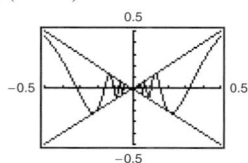

The graph has a hole at $x = 0$.

95. (a) f and g agree at all but one point if c is a real number such that $f(x) = g(x)$ for all $x \neq c$.

(b) Sample answer: $f(x) = \dfrac{x^2 - 1}{x - 1}$ and $g(x) = x + 1$ agree at all points except $x = 1$.

97. If a function f is squeezed between two functions h and g, $h(x) \leq f(x) \leq g(x)$, and h and g have the same limit L as $x \to c$, then $\lim\limits_{x \to c} f(x)$ exists and equals L.

99.

The magnitudes of $f(x)$ and $g(x)$ are approximately equal when x is close to 0. Therefore, their ratio is approximately 1.

101. -64 ft/sec (speed = 64 ft/sec)   **103.** -29.4 m/sec

105. Let $f(x) = 1/x$ and $g(x) = -1/x$.

$\lim\limits_{x \to 0} f(x)$ and $\lim\limits_{x \to 0} g(x)$ do not exist. However,

$\lim\limits_{x \to 0} \left[f(x) + g(x) \right] = \lim\limits_{x \to 0} \left[\dfrac{1}{x} + \left(-\dfrac{1}{x} \right) \right] = \lim\limits_{x \to 0} 0 = 0$

and therefore does exist.

107–111. Proofs

113. Let $f(x) = \begin{cases} 4, & x \geq 0 \\ -4, & x < 0 \end{cases}$.

$\lim\limits_{x \to 0} |f(x)| = \lim\limits_{x \to 0} 4 = 4$

$\lim\limits_{x \to 0} f(x)$ does not exist because for $x < 0$, $f(x) = -4$ and for $x \geq 0$, $f(x) = 4$.

115. False. The limit does not exist because the function approaches 1 from the right side of 0 and approaches -1 from the left side of 0.

117. True.

119. False. The limit does not exist because $f(x)$ approaches 3 from the left side of 2 and approaches 0 from the right side of 2.

121. Proof

123. (a) All $x \neq 0$, $\dfrac{\pi}{2} + n\pi$

(b)

The domain is not obvious. The hole at $x = 0$ is not apparent from the graph.

(c) $\dfrac{1}{2}$ (d) $\dfrac{1}{2}$

Section 1.4 *(page 79)*

1. (a) 3   (b) 3   (c) 3; $f(x)$ is continuous on $(-\infty, \infty)$.

3. (a) 0   (b) 0   (c) 0; Discontinuity at $x = 3$

5. (a) -3 (b) 3   (c) Limit does not exist.
Discontinuity at $x = 2$

7. $\frac{1}{16}$ **9.** $\frac{1}{10}$

11. Limit does not exist. The function decreases without bound as x approaches -3 from the left.

13. -1 **15.** $-1/x^2$ **17.** 5/2   **19.** 2

21. Limit does not exist. The function decreases without bound as x approaches π from the left and increases without bound as x approaches π from the right.

23. 8

25. Limit does not exist. The function approaches 5 from the left side of 3 but approaches 6 from the right side of 3.

27. Discontinuities at $x = -2$ and $x = 2$

29. Discontinuities at every integer

31. Continuous on $[-7, 7]$ **33.** Continuous on $[-1, 4]$

35. Nonremovable discontinuity at $x = 0$

37. Continuous for all real x

39. Nonremovable discontinuities at $x = -2$ and $x = 2$

41. Continuous for all real x

43. Nonremovable discontinuity at $x = 1$
Removable discontinuity at $x = 0$

45. Continuous for all real x

47. Removable discontinuity at $x = -2$
Nonremovable discontinuity at $x = 5$

49. Nonremovable discontinuity at $x = -7$

51. Continuous for all real x

53. Nonremovable discontinuity at $x = 2$

55. Continuous for all real x

57. Nonremovable discontinuities at integer multiples of $\pi/2$

59. Nonremovable discontinuities at each integer

61. $a = 7$ **63.** $a = 2$ **65.** $a = -1$, $b = 1$

67. Continuous for all real x

69. Nonremovable discontinuities at $x = 1$ and $x = -1$

71. Continuous on the open intervals
$\ldots, (-3\pi, -\pi), (-\pi, \pi), (\pi, 3\pi), \ldots$

73.

75.

Nonremovable discontinuity   Nonremovable discontinuity

at each integer   at $x = 4$

77. Continuous on $(-\infty, \infty)$ **79.** Continuous on $[0, \infty)$

81. Continuous on the open intervals $\ldots, (-6, -2), (-2, 2),$
$(2, 6), \ldots$

83. Continuous on $(-\infty, \infty)$

85.

The graph has a hole at $x = 0$. The graph appears to be continuous, but the function is not continuous on $[-4, 4]$. It is not obvious from the graph that the function has a discontinuity at $x = 0$.

87. Because $f(x)$ is continuous on the interval $[1, 2]$ and $f(1) = 37/12$ and $f(2) = -8/3$, by the Intermediate Value Theorem there exists a real number c in $[1, 2]$ such that $f(c) = 0$.

89. Because $f(x)$ is continuous on the interval $[0, \pi]$ and $f(0) = -3$ and $f(\pi) \approx 8.87$, by the Intermediate Value Theorem there exists a real number c in $[0, \pi]$ such that $f(c) = 0$.

91. 0.68, 0.6823   **93.** 0.56, 0.5636

95. $f(3) = 11$ **97.** $f(2) = 4$

99. (a) The limit does not exist at $x = c$.

(b) The function is not defined at $x = c$.

(c) The limit exists, but it is not equal to the value of the function at $x = c$.

(d) The limit does not exist at $x = c$.

101. If f and g are continuous for all real x, then so is $f + g$ (Theorem 1.11, part 2). However, f/g might not be continuous if $g(x) = 0$. For example, let $f(x) = x$ and $g(x) = x^2 - 1$. Then f and g are continuous for all real x, but f/g is not continuous at $x = \pm 1$.

103. True

105. False. A rational function can be written as $P(x)/Q(x)$, where P and Q are polynomials of degree m and n, respectively. It can have, at most, n discontinuities.

107. The functions differ by 1 for non-integer values of x.

109. $C = \begin{cases} 0.40, & 0 < t \le 10 \\ 0.40 + 0.05 [\![t - 9]\!], & t > 10, t \text{ is not an integer} \\ 0.40 + 0.05(t - 10), & t > 10, t \text{ is an integer} \end{cases}$

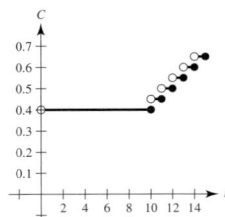

There is a nonremovable discontinuity at each integer greater than or equal to 10.

111–113. Proofs   **115.** Answers will vary.

117. (a)

(b) There appears to be a limiting speed, and a possible cause is air resistance.

119. $c = \left(-1 \pm \sqrt{5}\right)/2$

121. Domain: $[-c^2, 0) \cup (0, \infty)$; Let $f(0) = 1/(2c)$

123. $h(x)$ has a nonremovable discontinuity at every integer except 0.

125. Putnam Problem B2, 1988

Section 1.5 *(page 88)*

1. $\lim\limits_{x \to -2^+} 2 \left| \dfrac{x}{x^2 - 4} \right| = \infty$, $\lim\limits_{x \to -2^-} 2 \left| \dfrac{x}{x^2 - 4} \right| = \infty$

3. $\lim\limits_{x \to -2^+} \tan(\pi x/4) = -\infty$, $\lim\limits_{x \to -2^-} \tan(\pi x/4) = \infty$

5. $\lim\limits_{x \to 4^+} \dfrac{1}{x - 4} = \infty$, $\lim\limits_{x \to 4^-} \dfrac{1}{x - 4} = -\infty$

7. $\lim\limits_{x \to 4^+} \dfrac{1}{(x - 4)^2} = \infty$, $\lim\limits_{x \to 4^-} \dfrac{1}{(x - 4)^2} = \infty$

9.

x	-3.5	-3.1	-3.01	-3.001	-3
$f(x)$	0.31	1.64	16.6	167	?

x	-2.999	-2.99	-2.9	-2.5
$f(x)$	-167	-16.7	-1.69	-0.36

$\lim\limits_{x \to -3^+} f(x) = -\infty$; $\lim\limits_{x \to -3^-} f(x) = \infty$

11.

x	-3.5	-3.1	-3.01	-3.001	-3
$f(x)$	3.8	16	151	1501	?

x	-2.999	-2.99	-2.9	-2.5
$f(x)$	-1499	-149	-14	-2.3

$$\lim_{x\to-3^+} f(x) = -\infty; \quad \lim_{x\to-3^-} f(x) = \infty$$

13. $x = 0$ **15.** $x = \pm 2$ **17.** No vertical asymptote
19. $x = -2, x = 1$ **21.** $x = 0, x = 3$
23. No vertical asymptote   **25.** $x = n$, n is an integer.
27. $t = n\pi$, n is a nonzero integer.
29. Removable discontinuity at $x = -1$
31. Vertical asymptote at $x = -1$
33. ∞ **35.** ∞ **37.** $-\frac{1}{5}$ **39.** $-\infty$ **41.** $-\infty$
43. ∞ **45.** 0   **47.** ∞
49.

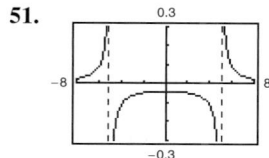

51.

$$\lim_{x\to 1^+} f(x) = \infty \qquad\qquad \lim_{x\to 5^-} f(x) = -\infty$$

53. Answers will vary.

55. Answers will vary. Sample answer: $f(x) = \dfrac{x-3}{x^2 - 4x - 12}$

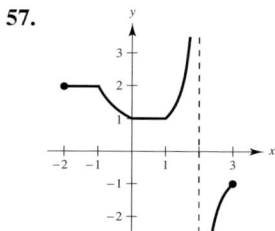

57.

59. (a)

x	1	0.5	0.2	0.1
$f(x)$	0.1585	0.0411	0.0067	0.0017

x	0.01	0.001	0.0001
$f(x)$	≈ 0	≈ 0	≈ 0

$$\lim_{x\to 0^+} \frac{x - \sin x}{x} = 0$$

(b)

x	1	0.5	0.2	0.1
$f(x)$	0.1585	0.0823	0.0333	0.0167

x	0.01	0.001	0.0001
$f(x)$	0.0017	≈ 0	≈ 0

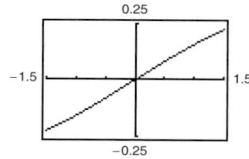

$$\lim_{x\to 0^+} \frac{x - \sin x}{x^2} = 0$$

(c)

x	1	0.5	0.2	0.1
$f(x)$	0.1585	0.1646	0.1663	0.1666

x	0.01	0.001	0.0001
$f(x)$	0.1667	0.1667	0.1667

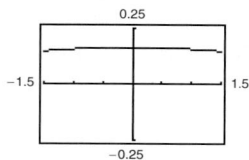

$$\lim_{x\to 0^+} \frac{x - \sin x}{x^3} = 0.1667 \ (1/6)$$

(d)

x	1	0.5	0.2	0.1
$f(x)$	0.1585	0.3292	0.8317	1.6658

x	0.01	0.001	0.0001
$f(x)$	16.67	166.7	1667.0

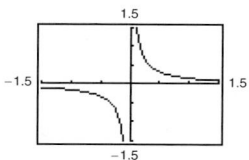

$$\lim_{x\to 0^+} \frac{x - \sin x}{x^4} = \infty$$

For $n > 3$, $\displaystyle\lim_{x\to 0^+} \frac{x - \sin x}{x^n} = \infty$.

61. (a) $\dfrac{7}{12}$ ft/sec   (b) $\dfrac{3}{2}$ ft/sec

(c) $\displaystyle\lim_{x\to 25^-} \frac{2x}{\sqrt{625 - x^2}} = \infty$

63. (a) $A = 50 \tan \theta - 50\theta$; Domain: $(0, \pi/2)$

(b)

θ	0.3	0.6	0.9	1.2	1.5
$f(\theta)$	0.47	4.21	18.0	68.6	630.1

(c) $\lim\limits_{\theta \to \pi/2} A = \infty$

65. False; let $f(x) = (x^2 - 1)/(x - 1)$

67. False; let $f(x) = \tan x$

69. Let $f(x) = \dfrac{1}{x^2}$ and $g(x) = \dfrac{1}{x^4}$, and let $c = 0$. $\lim\limits_{x \to 0} \dfrac{1}{x^2} = \infty$ and

$\lim\limits_{x \to 0} \dfrac{1}{x^4} = \infty$, but $\lim\limits_{x \to 0} \left(\dfrac{1}{x^2} - \dfrac{1}{x^4} \right) = \lim\limits_{x \to 0} \left(\dfrac{x^2 - 1}{x^4} \right) = -\infty \neq 0$.

71. Given $\lim\limits_{x \to c} f(x) = \infty$, let $g(x) = 1$. Then $\lim\limits_{x \to c} \dfrac{g(x)}{f(x)} = 0$ by Theorem 1.15.

73. Answers will vary.

Review Exercises for Chapter 1 (page 91)

1. Calculus   Estimate: 8.3

3.

x	2.9	2.99	2.999	3
$f(x)$	-0.9091	-0.9901	-0.9990	?

x	3.001	3.01	3.1
$f(x)$	-1.0010	-1.0101	-1.1111

$\lim\limits_{x \to 0} \dfrac{x - 3}{x^2 - 7x + 12} \approx -1.0000$

5. (a) 4   (b) 5   **7.** 5; Proof   **9.** -3; Proof   **11.** 36

13. $\sqrt{6} \approx 2.45$ **15.** 16   **17.** $\frac{4}{3}$ **19.** $-\frac{1}{4}$ **21.** $\frac{1}{2}$

23. -1 **25.** 0   **27.** $\sqrt{3}/2$ **29.** -3 **31.** -5

33.

The graph has a hole at $x = 0$.

x	-0.1	-0.01	-0.001	0
$f(x)$	0.3352	0.3335	0.3334	?

x	0.001	0.01	0.1
$f(x)$	0.3333	0.3331	0.3315

$\lim\limits_{x \to 0} \dfrac{\sqrt{2x + 9} - 3}{x} \approx 0.3333$; Actual limit is $\frac{1}{3}$.

35.

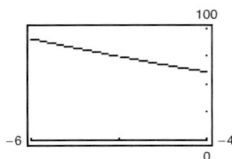

The graph has a hole at $x = -5$.

x	-5.1	-5.01	-5.001	-5
$f(x)$	76.51	75.15	75.02	?

x	-4.999	-4.99	-4.9
$f(x)$	74.99	74.85	73.51

$\lim\limits_{x \to -5} \dfrac{x^3 + 125}{x + 5} \approx 75.00$; Actual limit is 75.

37. -39.2 m/sec   **39.** $\frac{1}{6}$ **41.** $\frac{1}{4}$ **43.** 0

45. Limit does not exist. The limit as t approaches 1 from the left is 2, whereas the limit as t approaches 1 from the right is 1.

47. 3   **49.** Continuous for all real x

51. Nonremovable discontinuity at $x = 5$

53. Nonremovable discontinuities at $x = -1$ and $x = 1$
Removable discontinuity at $x = 0$

55. $c = -\frac{1}{2}$ **57.** Continuous for all real x

59. Continuous on $[4, \infty)$

61. Removable discontinuity at $x = 1$
Continuous on $(-\infty, 1) \cup (1, \infty)$

63. Proof   **65.** (a) -4 (b) 4   (c) Limit does not exist.

67. $x = 0$ **69.** $x = \pm 3$ **71.** $x = \pm 8$ **73.** $-\infty$

75. $\frac{1}{3}$ **77.** $-\infty$ **79.** $\frac{4}{5}$ **81.** ∞

83. (a) \$14,117.65   (b) \$80,000.00   (c) \$720,000.00
(d) ∞

P.S. Problem Solving (page 93)

1. (a) Perimeter $\triangle PAO = 1 + \sqrt{(x^2 - 1)^2 + x^2} + \sqrt{x^4 + x^2}$
Perimeter $\triangle PBO = 1 + \sqrt{x^4 + (x - 1)^2} + \sqrt{x^4 + x^2}$

(b)

x	4	2	1
Perimeter $\triangle PAO$	33.0166	9.0777	3.4142
Perimeter $\triangle PBO$	33.7712	9.5952	3.4142
$r(x)$	0.9777	0.9461	1.0000

x	0.1	0.01
Perimeter $\triangle PAO$	2.0955	2.0100
Perimeter $\triangle PBO$	2.0006	2.0000
$r(x)$	1.0475	1.0050

1

3. (a) Area (hexagon) $= (3\sqrt{3})/2 \approx 2.5981$
Area (circle) $= \pi \approx 3.1416$
Area (circle) $-$ Area (hexagon) ≈ 0.5435

(b) $A_n = (n/2) \sin(2\pi/n)$

(c)

n	6	12	24	48	96
A_n	2.5981	3.0000	3.1058	3.1326	3.1394

3.1416 or π

5. (a) $m = -\frac{12}{5}$ (b) $y = \frac{5}{12}x - \frac{169}{12}$

(c) $m_x = \dfrac{-\sqrt{169 - x^2} + 12}{x - 5}$

(d) $\frac{5}{12}$; It is the same as the slope of the tangent line found in (b).

7. (a) Domain: $[-27, 1) \cup (1, \infty)$

(b)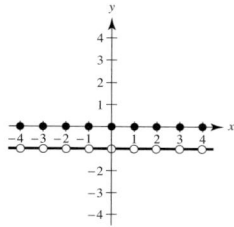

(c) $\frac{1}{14}$ (d) $\frac{1}{12}$

The graph has a hole at $x = 1$.

9. (a) g_1, g_4 (b) g_1 (c) g_1, g_3, g_4

11.

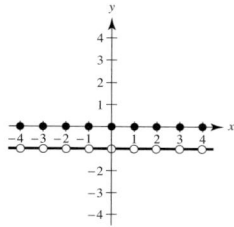

The graph jumps at every integer.

(a) $f(1) = 0,\ \ f(0) = 0,\ \ f\left(\frac{1}{2}\right) = -1,\ \ f(-2.7) = -1$

(b) $\lim\limits_{x \to 1^-} f(x) = -1,\ \ \lim\limits_{x \to 1^+} f(x) = -1,\ \ \lim\limits_{x \to 1/2} f(x) = -1$

(c) There is a discontinuity at each integer.

13. (a)

(b) (i) $\lim\limits_{x \to a^+} P_{a,b}(x) = 1$

(ii) $\lim\limits_{x \to a^-} P_{a,b}(x) = 0$

(iii) $\lim\limits_{x \to b^+} P_{a,b}(x) = 0$

(iv) $\lim\limits_{x \to b^-} P_{a,b}(x) = 1$

(c) Continuous for all positive real numbers except a and b

(d) The area under the graph of U and above the x-axis is 1.

Chapter 2

Section 2.1 *(page 103)*

1. $m_1 = 0, m_2 = 5/2$

3. (a)–(c)

$y = \dfrac{f(4) - f(1)}{4 - 1}(x - 1) + f(1) = x + 1$

5. $m = -5$

7. $m = 4$

9. $m = 3$ **11.** $f'(x) = 0$ **13.** $f'(x) = -10$

15. $h'(s) = \frac{2}{3}$ **17.** $f'(x) = 2x + 1$ **19.** $f'(x) = 3x^2 - 12$

21. $f'(x) = \dfrac{-1}{(x - 1)^2}$ **23.** $f'(x) = \dfrac{1}{2\sqrt{x + 4}}$

25. (a) Tangent line:
$y = -2x + 2$

(b)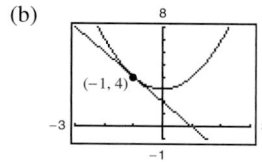

27. (a) Tangent line:
$y = 12x - 16$

(b)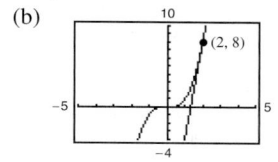

29. (a) Tangent line:
$y = \frac{1}{2}x + \frac{1}{2}$

(b)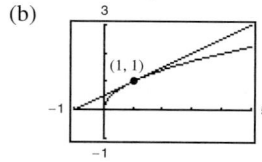

31. (a) Tangent line:
$y = \frac{3}{4}x - 2$

(b)

33. $y = 2x - 1$ **35.** $y = 3x - 2; y = 3x + 2$

37. $y = -\frac{1}{2}x + \frac{3}{2}$

39.

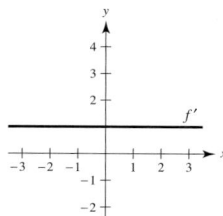

The slope of the graph of f is 1 for all x-values.

41.

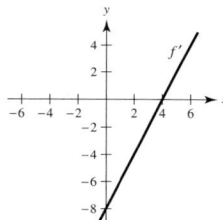

The slope of the graph of f is negative for $x < 4$, positive for $x > 4$, and 0 at $x = 4$.

43.

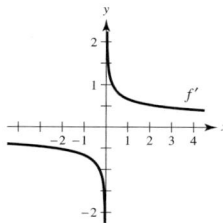

The slope of the graph of f is negative for $x < 0$ and positive for $x > 0$. The slope is undefined at $x = 0$.

45. Answers will vary.
Sample answer: $y = -x$

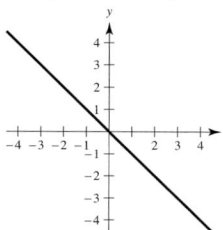

47. $g(4) = 5; g'(4) = -\frac{5}{3}$

49. $f(x) = 5 - 3x$
$c = 1$

51. $f(x) = -x^2$
$c = 6$

53. $f(x) = -3x + 2$

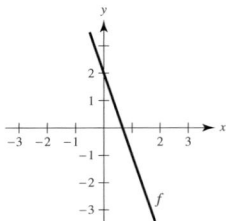

55. $y = 2x + 1; y = -2x + 9$

57. (a)

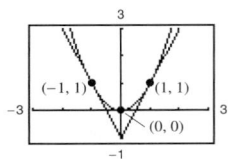

For this function, the slopes of the tangent lines are always distinct for different values of x.

(b)

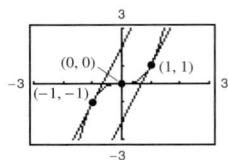

For this function, the slopes of the tangent lines are sometimes the same.

59. (a)

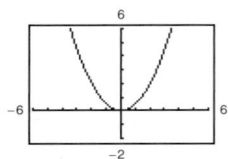

$f'(0) = 0, f'\left(\frac{1}{2}\right) = \frac{1}{2}, f'(1) = 1, f'(2) = 2$

(b) $f'\left(-\frac{1}{2}\right) = -\frac{1}{2}, f'(-1) = -1, f'(-2) = -2$

(c)

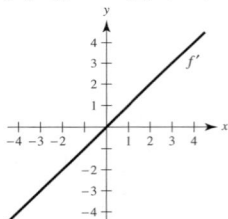

(d) $f'(x) = x$

61.

$g(x) \approx f'(x)$

63. $f(2) = 4; f(2.1) = 3.99; f'(2) \approx -0.1$ **65.** 6   **67.** 4

69. $g(x)$ is not differentiable at $x = 0$.

71. $f(x)$ is not differentiable at $x = 6$.

73. $h(x)$ is not differentiable at $x = -7$.

75. $(-\infty, 3) \cup (3, \infty)$ **77.** $(-\infty, -4) \cup (-4, \infty)$

79. $(1, \infty)$

81.

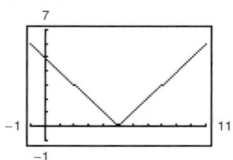

$(-\infty, 5) \cup (5, \infty)$

83.

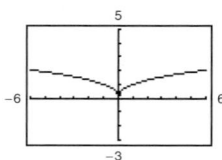

$(-\infty, 0) \cup (0, \infty)$

85. The derivative from the left is -1 and the derivative from the right is 1, so f is not differentiable at $x = 1$.

87. The derivatives from both the right and the left are 0, so $f'(1) = 0$.

89. f is differentiable at $x = 2$.

91. (a) $d = \left(3|m + 1|\right)/\sqrt{m^2 + 1}$

(b)

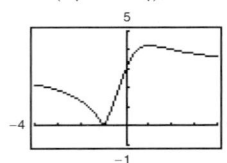

Not differentiable at $m = -1$

93. False. The slope is $\lim\limits_{\Delta x \to 0} \dfrac{f(2 + \Delta x) - f(2)}{\Delta x}$.

95. False. For example, $f(x) = |x|$. The derivative from the left and the derivative from the right both exist but are not equal.

97. Proof

Section 2.2   (page 114)

1. (a) $\frac{1}{2}$ (b) 3   **3.** 0   **5.** $7x^6$ **7.** $-5/x^6$

9. $1/(5x^{4/5})$ **11.** 1   **13.** $-4t + 3$ **15.** $2x + 12x^2$

17. $3t^2 + 10t - 3$ **19.** $\dfrac{\pi}{2}\cos\theta + \sin\theta$ **21.** $2x + \dfrac{1}{2}\sin x$

23. $-\dfrac{1}{x^2} - 3\cos x$

	Function	Rewrite	Derivative	Simplify
25.	$y = \dfrac{5}{2x^2}$	$y = \dfrac{5}{2}x^{-2}$	$y' = -5x^{-3}$	$y' = -\dfrac{5}{x^3}$
27.	$y = \dfrac{6}{(5x)^3}$	$y = \dfrac{6}{125}x^{-3}$	$y' = -\dfrac{18}{125}x^{-4}$	$y' = -\dfrac{18}{125x^4}$
29.	$y = \dfrac{\sqrt{x}}{x}$	$y = x^{-1/2}$	$y' = -\dfrac{1}{2}x^{-3/2}$	$y' = -\dfrac{1}{2x^{3/2}}$

31. -2 **33.** 0   **35.** 8   **37.** 3   **39.** $2x + 6/x^3$

41. $2t + 12/t^4$ **43.** $8x + 3$ **45.** $(x^3 - 8)/x^3$

47. $3x^2 + 1$ **49.** $\dfrac{1}{2\sqrt{x}} - \dfrac{2}{x^{2/3}}$ **51.** $\dfrac{3}{\sqrt{x}} - 5\sin x$

53. (a) $2x + y - 2 = 0$

(b)

55. (a) $3x + 2y - 7 = 0$

(b)

57. $(-1, 2), (0, 3), (1, 2)$ **59.** No horizontal tangents

61. (π, π) **63.** $k = -8$ **65.** $k = 3$ **67.** $k = 4/27$

69.

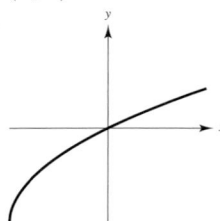

71. $g'(x) = f'(x)$ **73.** $g'(x) = -5f'(x)$

75.

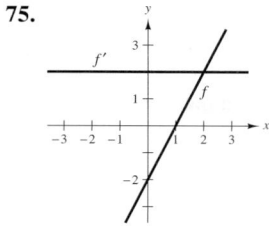

The rate of change of f is constant, and therefore f' is a constant function.

77. $y = 2x - 1$ $y = 4x - 4$

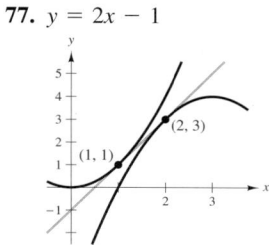

79. $f'(x) = 3 + \cos x \neq 0$ for all x.   **81.** $x - 4y + 4 = 0$

83.

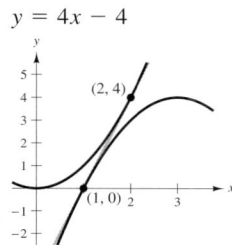

$f'(1)$ appears to be close to -1.
$f'(1) = -1$

85. (a)

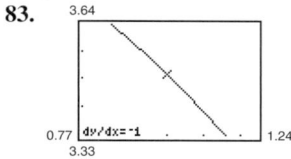

$(3.9, 7.7019)$,
$S(x) = 2.981x - 3.924$

(b) $T(x) = 3(x - 4) + 8 = 3x - 4$

The slope (and equation) of the secant line approaches that of the tangent line at $(4, 8)$ as you choose points closer and closer to $(4, 8)$.

(c)

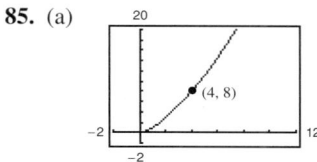

The approximation becomes less accurate.

(d)

Δx	-3	-2	-1	-0.5	-0.1	0
$f(4 + \Delta x)$	1	2.828	5.196	6.548	7.702	8
$T(4 + \Delta x)$	-1	2	5	6.5	7.7	8

Δx	0.1	0.5	1	2	3
$f(4 + \Delta x)$	8.302	9.546	11.180	14.697	18.520
$T(4 + \Delta x)$	8.3	9.5	11	14	17

87. False. Let $f(x) = x$ and $g(x) = x + 1$.
89. False. $dy/dx = 0$ **91.** True
93. Average rate: 4   **95.** Average rate: $\frac{1}{2}$
Instantaneous rates: Instantaneous rates:
$f'(1) = 4; f'(2) = 4$ $f'(1) = 1; f'(2) = \frac{1}{4}$

97. (a) $s(t) = -16t^2 + 1362; v(t) = -32t$ (b) -48 ft/sec
(c) $s'(1) = -32$ ft/sec; $s'(2) = -64$ ft/sec
(d) $t = \dfrac{\sqrt{1362}}{4} \approx 9.226$ sec   (e) -295.242 ft/sec

99. $v(5) = 71$ m/sec; $v(10) = 22$ m/sec
101.

103.

105. $V'(6) = 108$ cm^3/cm
107. (a) $R(v) = 0.417v - 0.02$
(b) $B(v) = 0.0056v^2 + 0.001v + 0.04$
(c) $T(v) = 0.0056v^2 + 0.418v + 0.02$
(d)

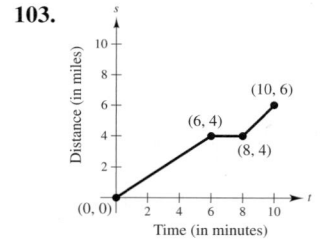

(e) $T'(v) = 0.0112v + 0.418$
$T'(40) = 0.866$
$T'(80) = 1.314$
$T'(100) = 1.538$

(f) Stopping distance increases at an increasing rate.
109. Proof   **111.** $y = 2x^2 - 3x + 1$
113. $9x + y = 0, 9x + 4y + 27 = 0$ **115.** $a = \frac{1}{3}, b = -\frac{4}{3}$
117. $f_1(x) = |\sin x|$ is differentiable for all $x \neq n\pi$, n an integer.
$f_2(x) = \sin|x|$ is differentiable for all $x \neq 0$.
119. Putnam Problem A2, 2010

Section 2.3   *(page 125)*

1. $2(2x^3 - 6x^2 + 3x - 6)$ **3.** $(1 - 5t^2)/(2\sqrt{t})$
5. $x^2(3 \cos x - x \sin x)$ **7.** $(1 - x^2)/(x^2 + 1)^2$
9. $(1 - 5x^3)/[2\sqrt{x}(x^3 + 1)^2]$ **11.** $(x \cos x - 2 \sin x)/x^3$
13. $f'(x) = (x^3 + 4x)(6x + 2) + (3x^2 + 2x - 5)(3x^2 + 4)$
$= 15x^4 + 8x^3 + 21x^2 + 16x - 20$
$f'(0) = -20$

15. $f'(x) = \dfrac{x^2 - 6x + 4}{(x - 3)^2}$ **17.** $f'(x) = \cos x - x \sin x$

$f'(1) = -\dfrac{1}{4}$ $f'\left(\dfrac{\pi}{4}\right) = \dfrac{\sqrt{2}}{8}(4 - \pi)$

	Function	Rewrite	Differentiate	Simplify
19.	$y = \dfrac{x^2 + 3x}{7}$	$y = \dfrac{1}{7}x^2 + \dfrac{3}{7}x$	$y' = \dfrac{2}{7}x + \dfrac{3}{7}$	$y' = \dfrac{2x + 3}{7}$
21.	$y = \dfrac{6}{7x^2}$	$y = \dfrac{6}{7}x^{-2}$	$y' = -\dfrac{12}{7}x^{-3}$	$y' = -\dfrac{12}{7x^3}$
23.	$y = \dfrac{4x^{3/2}}{x}$	$y = 4x^{1/2},$ $x > 0$	$y' = 2x^{-1/2}$	$y' = \dfrac{2}{\sqrt{x}},$ $x > 0$

25. $\dfrac{3}{(x + 1)^2}$, $x \neq 1$ **27.** $(x^2 + 6x - 3)/(x + 3)^2$
29. $(3x + 1)/(2x^{3/2})$ **31.** $6s^2(s^3 - 2)$
33. $-(2x^2 - 2x + 3)/[x^2(x - 3)^2]$
35. $10x^4 - 8x^3 - 21x^2 - 10x - 30$
37. $-\dfrac{4xc^2}{(x^2 - c^2)^2}$ **39.** $t(t \cos t + 2 \sin t)$

41. $-(t \sin t + \cos t)/t^2$ **43.** $-1 + \sec^2 x = \tan^2 x$

45. $\dfrac{1}{4t^{3/4}} - 6 \csc t \cot t$ **47.** $\dfrac{3}{2} \sec x(\tan x - \sec x)$

49. $\cos x \cot^2 x$ **51.** $x(x \sec^2 x + 2 \tan x)$

53. $4x \cos x + (2 - x^2) \sin x$

55. $\dfrac{2x^2 + 8x - 1}{(x + 2)^2}$ **57.** $\dfrac{1 - \sin \theta + \theta \cos \theta}{(1 - \sin \theta)^2}$

59. $y' = \dfrac{-2 \csc x \cot x}{(1 - \csc x)^2}$, $\quad -4\sqrt{3}$

61. $h'(t) = \sec t(t \tan t - 1)/t^2, \quad 1/\pi^2$

63. (a) $y = -3x - 1$ **65.** (a) $y = 4x + 25$

(b)   (b)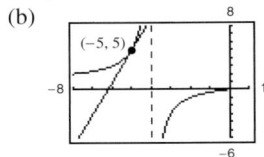

67. (a) $4x - 2y - \pi + 2 = 0$ **69.** $2y + x - 4 = 0$

(b)

71. $25y - 12x + 16 = 0$ **73.** $(1, 1)$ **75.** $(0, 0), (2, 4)$

77. Tangent lines: $2y + x = 7; 2y + x = -1$

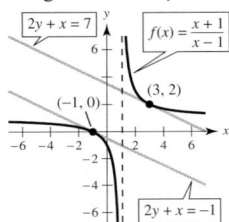

79. $f(x) + 2 = g(x)$ **81.** (a) $p'(1) = 1$ (b) $q'(4) = -1/3$

83. $(18t + 5)/(2\sqrt{t})$ cm²/sec

85. (a) $-\$38.13$ thousand/100 components

(b) $-\$10.37$ thousand/100 components

(c) $-\$3.80$ thousand/100 components

The cost decreases with increasing order size.

87. Proof

89. (a) $h(t) = 112.4t + 1332$

$p(t) = 2.9t + 282$

(b)

(c) $A = \dfrac{112.4t + 1332}{2.9t + 282}$

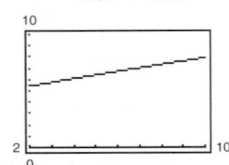

A represents the average health care expenditures per person (in thousands of dollars).

(d) $A'(t) = \dfrac{27{,}834}{8.41t^2 + 1635.6t + 79{,}524}$

$A'(t)$ represents the rate of change of the average health care expenditures per person for the given year t.

91. $12x^2 + 12x - 6$ **93.** $3/\sqrt{x}$ **95.** $2/(x - 1)^3$

97. $2 \cos x - x \sin x$ **99.** $2x$ **101.** $1/\sqrt{x}$ **103.** 0

105. -10

107. Answers will vary.   **109.**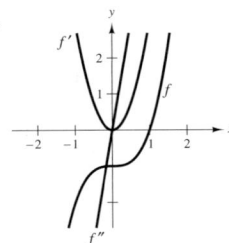

Sample answer:

$f(x) = (x - 2)^2$

111.   **113.**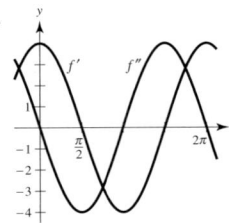

115. $v(3) = 27$ m/sec

$a(3) = -6$ m/sec²

The speed of the object is decreasing.

117.

t	0	1	2	3	4
$s(t)$	0	57.75	99	123.75	132
$v(t)$	66	49.5	33	16.5	0
$a(t)$	-16.5	-16.5	-16.5	-16.5	-16.5

The average velocity on $[0, 1]$ is 57.75, on $[1, 2]$ is 41.25, on $[2, 3]$ is 24.75, and on $[3, 4]$ is 8.25.

119. $f^{(n)}(x) = n(n - 1)(n - 2) \cdots (2)(1) = n!$

121. (a) $f''(x) = g(x)h''(x) + 2g'(x)h'(x) + g''(x)h(x)$

$f'''(x) = g(x)h'''(x) + 3g'(x)h''(x) + 3g''(x)h'(x) + g'''(x)h(x)$

$f^{(4)}(x) = g(x)h^{(4)}(x) + 4g'(x)h'''(x) + 6g''(x)h''(x) + 4g'''(x)h'(x) + g^{(4)}(x)h(x)$

(b) $f^{(n)}(x) = g(x)h^{(n)}(x) + \dfrac{n!}{1!(n - 1)!} g'(x)h^{(n - 1)}(x) +$

$\dfrac{n!}{2!(n - 2)!} g''(x)h^{(n - 2)}(x) + \cdots +$

$\dfrac{n!}{(n - 1)!1!} g^{(n - 1)}(x)h'(x) + g^{(n)}(x)h(x)$

123. $n = 1: f'(x) = x \cos x + \sin x$

$n = 2: f'(x) = x^2 \cos x + 2x \sin x$

$n = 3: f'(x) = x^3 \cos x + 3x^2 \sin x$

$n = 4: f'(x) = x^4 \cos x + 4x^3 \sin x$

General rule: $f'(x) = x^n \cos x + nx^{(n - 1)} \sin x$

125. $y' = -1/x^2$, $y'' = 2/x^3$,
$\quad x^3y'' + 2x^2y' = x^3(2/x^3) + 2x^2(-1/x^2)$
$\qquad\qquad = 2 - 2 = 0$

127. $y' = 2\cos x$, $y'' = -2\sin x$,
$\quad y'' + y = -2\sin x + 2\sin x + 3 = 3$

129. False. $dy/dx = f(x)g'(x) + g(x)f'(x)$ **131.** True

133. True   **135.** $f'(x) = 2|x|$; $f''(0)$ does not exist.

137. Proof

Section 2.4 (page 136)

$y = f(g(x))$	$u = g(x)$	$y = f(u)$
1. $y = (5x - 8)^4$	$u = 5x - 8$	$y = u^4$
3. $y = \sqrt{x^3 - 7}$	$u = x^3 - 7$	$y = \sqrt{u}$
5. $y = \csc^3 x$	$u = \csc x$	$y = u^3$

7. $12(4x - 1)^2$ **9.** $-108(4 - 9x)^3$ **11.** $-1/(2\sqrt{5 - t})$

13. $4x/\sqrt[3]{(6x^2 + 1)^2}$ **15.** $-x/\sqrt[4]{(9 - x^2)^3}$

17. $-1/(x - 2)^2$ **19.** $-2/(t - 3)^3$

21. $-3/[2\sqrt{(3x + 5)^3}]$ **23.** $2x(x - 2)^3(3x - 2)$

25. $\dfrac{1 - 2x^2}{\sqrt{1 - x^2}}$ **27.** $\dfrac{1}{\sqrt{(x^2 + 1)^3}}$

29. $\dfrac{-2(x + 5)(x^2 + 10x - 2)}{(x^2 + 2)^3}$ **31.** $\dfrac{-9(1 - 2v)^2}{(v + 1)^4}$

33. $20x(x^2 + 3)^9 + 2(x^2 + 3)^5 + 20x^2(x^2 + 3)^4 + 2x$

35. $(1 - 3x^2 - 4x^{3/2})/[2\sqrt{x}(x^2 + 1)^2]$

The zero of y' corresponds to the point on the graph of the function where the tangent line is horizontal.

37. $-\dfrac{\sqrt{\dfrac{x + 1}{x}}}{2x(x + 1)}$

y' has no zeros.

39. $-[\pi x \sin(\pi x) + \cos(\pi x) + 1]/x^2$

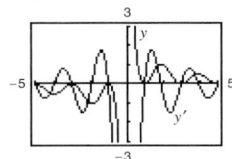

The zeros of y' correspond to the points on the graph of the function where the tangent lines are horizontal.

41. (a) 1   (b) 2; The slope of sin ax at the origin is a.

43. $-4\sin 4x$ **45.** $15\sec^2 3x$ **47.** $2\pi^2 x \cos(\pi x)^2$

49. $2\cos 4x$ **51.** $(-1 - \cos^2 x)/\sin^3 x$

53. $8\sec^2 x \tan x$ **55.** $10\tan 5\theta \sec^2 5\theta$

57. $\sin 2\theta \cos 2\theta = \frac{1}{2}\sin 4\theta$ **59.** $\dfrac{6\pi \sin(\pi t - 1)}{\cos^3(\pi t - 1)}$

61. $\dfrac{1}{2\sqrt{x}} + 2x\cos(2x)^2$ **63.** $2\sec^2 2x \cos(\tan 2x)$

65. $y' = \dfrac{x + 4}{\sqrt{x^2 + 8x}}, \frac{5}{3}$ **67.** $f'(x) = \dfrac{-15x^2}{(x^3 - 2)^2}, -\frac{3}{5}$

69. $f'(t) = \dfrac{-5}{(t - 1)^2}, -5$ **71.** $y' = -12\sec^3 4x \tan 4x, 0$

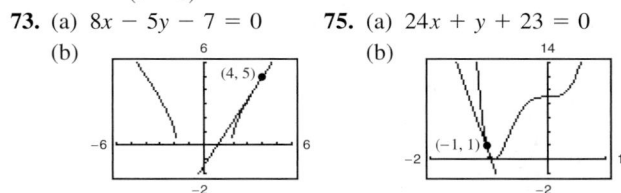

73. (a) $8x - 5y - 7 = 0$ **75.** (a) $24x + y + 23 = 0$

(b)   (b)

77. (a) $2x - y - 2\pi = 0$ **79.** (a) $4x - y + (1 - \pi) = 0$

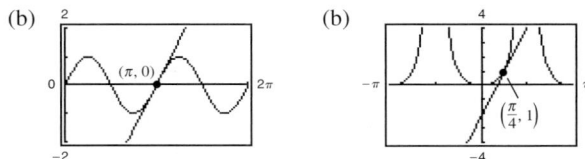

(b)   (b)

81. $3x + 4y - 25 = 0$

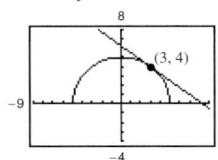

83. $\left(\dfrac{\pi}{6}, \dfrac{3\sqrt{3}}{2}\right), \left(\dfrac{5\pi}{6}, -\dfrac{3\sqrt{3}}{2}\right), \left(\dfrac{3\pi}{2}, 0\right)$ **85.** $2940(2 - 7x)^2$

87. $\dfrac{2}{(x - 6)^3}$ **89.** $2(\cos x^2 - 2x^2 \sin x^2)$

91. $h''(x) = 18x + 6, 24$

93. $f''(x) = -4x^2 \cos(x^2) - 2\sin(x^2), 0$

95.   **97.**

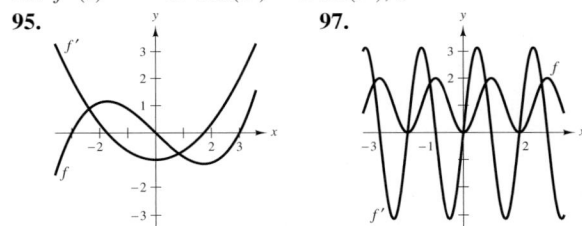

The zeros of f' correspond to the points where the graph of f has horizontal tangents.   The zeros of f' correspond to the points where the graph of f has horizontal tangents.

99. The rate of change of g is three times as fast as the rate of change of f.

101. (a) $g'(x) = f'(x)$ (b) $h'(x) = 2f'(x)$
(c) $r'(x) = -3f'(-3x)$ (d) $s'(x) = f'(x + 2)$

x	-2	-1	0	1	2	3
$f'(x)$	4	$\frac{2}{3}$	$-\frac{1}{3}$	-1	-2	-4
$g'(x)$	4	$\frac{2}{3}$	$-\frac{1}{3}$	-1	-2	-4
$h'(x)$	8	$\frac{4}{3}$	$-\frac{2}{3}$	-2	-4	-8
$r'(x)$		12	1			
$s'(x)$	$-\frac{1}{3}$	-1	-2	-4		

103. (a) $\frac{1}{2}$
(b) $s'(5)$ does not exist because g is not differentiable at 6.

105. (a) 1.461   (b) -1.016 **107.** 0.2 rad, 1.45 rad/sec

109. (a)

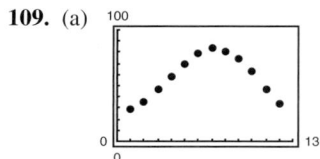

$$T(t) = 56.1 + 27.6 \sin(0.48t - 1.86)$$

(b)

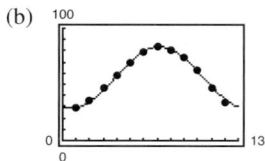

The model is a good fit.

(c) $T'(t) \approx 13.25 \cos(0.48t - 1.86)$

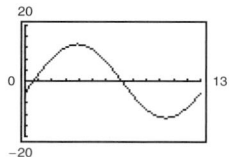

(d) The temperature changes most rapidly around spring (March–May) and fall (Oct.–Nov.)

The temperature changes most slowly around winter (Dec.–Feb.) and summer (Jun.–Aug.)

Yes. Explanations will vary.

111. (a) 0 bacteria per day   (b) 177.8 bacteria per day
(c) 44.4 bacteria per day   (d) 10.8 bacteria per day
(e) 3.3 bacteria per day
(f) The rate of change of the population is decreasing as time passes.

113. (a) $f'(x) = \beta \cos \beta x$
$f''(x) = -\beta^2 \sin \beta x$
$f'''(x) = -\beta^3 \cos \beta x$
$f^{(4)}(x) = \beta^4 \sin \beta x$
(b) $f''(x) + \beta^2 f(x) = -\beta^2 \sin \beta x + \beta^2(\sin \beta x) = 0$
(c) $f^{(2k)}(x) = (-1)^k \beta^{2k} \sin \beta x$
$f^{(2k-1)}(x) = (-1)^{k+1} \beta^{2k-1} \cos \beta x$

115. (a) $r'(1) = 0$ (b) $s'(4) = \frac{5}{8}$

117. (a) and (b) Proofs

119. $g'(x) = 3\left(\dfrac{3x-5}{|3x-5|}\right), \quad x \neq \dfrac{5}{3}$

121. $h'(x) = -|x|\sin x + \dfrac{x}{|x|}\cos x, \quad x \neq 0$

123. (a) $P_1(x) = 2(x - \pi/4) + 1$
$P_2(x) = 2(x - \pi/4)^2 + 2(x - \pi/4) + 1$
(b)   (c) P_2

(d) The accuracy worsens as you move away from $x = \pi/4$.

125. False. If $y = (1 - x)^{1/2}$, then $y' = \frac{1}{2}(1 - x)^{-1/2}(-1)$.

127. True   **129.** Putnam Problem A1, 1967

Section 2.5   (page 145)

1. $-x/y$ **3.** $-\sqrt{y/x}$ **5.** $(y - 3x^2)/(2y - x)$

7. $(1 - 3x^2y^3)/(3x^3y^2 - 1)$
9. $(6xy - 3x^2 - 2y^2)/(4xy - 3x^2)$ **11.** $\cos/[4 \sin(2y)]$
13. $(\cos x - \tan y - 1)/(x \sec^2 y)$
15. $[y \cos(xy)]/[1 - x \cos(xy)]$
17. (a) $y_1 = \sqrt{64 - x^2}; y_2 = -\sqrt{64 - x^2}$
(b)

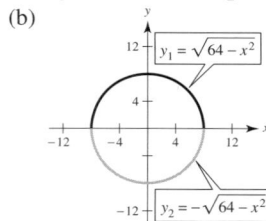

(c) $y' = \mp \dfrac{x}{\sqrt{64 - x^2}} = -\dfrac{x}{y}$ (d) $y' = -\dfrac{x}{y}$

19. (a) $y_1 = \dfrac{\sqrt{x^2 + 16}}{4}; y_2 = \dfrac{-\sqrt{x^2 + 16}}{4}$
(b)

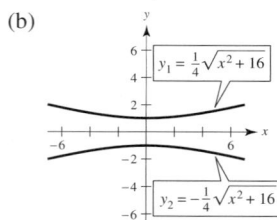

(c) $y' = \dfrac{\pm x}{4\sqrt{x^2 + 16}} = \dfrac{x}{16y}$ (d) $y' = \dfrac{x}{16y}$

21. $-\dfrac{y}{x}, -\dfrac{1}{6}$ **23.** $\dfrac{98x}{y(x^2 + 49)^2}$, Undefined

25. $-\dfrac{y(y + 2x)}{x(x + 2y)}, -1$ **27.** $-\sin^2(x + y)$ or $-\dfrac{x^2}{x^2 + 1}, 0$

29. $-\dfrac{1}{2}$ **31.** 0   **33.** $y = -x + 7$ **35.** $y = -x + 2$

37. $y = \sqrt{3}x/6 + 8\sqrt{3}/3$ **39.** $y = -\frac{2}{11}x + \frac{30}{11}$

41. (a) $y = -2x + 4$ (b) Answers will vary.

43. $\cos^2 y, -\dfrac{\pi}{2} < y < \dfrac{\pi}{2}, \dfrac{1}{1 + x^2}$ **45.** $-4/y^3$

47. $-36/y^3$ **49.** $(3x)/(4y)$

51. $2x + 3y - 30 = 0$

53. At $(4, 3)$:
Tangent line: $4x + 3y - 25 = 0$
Normal line: $3x - 4y = 0$

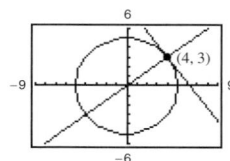

At $(-3, 4)$:
Tangent line: $3x - 4y + 25 = 0$
Normal line: $4x + 3y = 0$

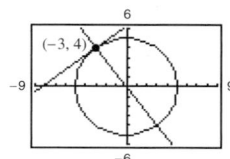

55. $x^2 + y^2 = r^2 \Rightarrow y' = -x/y \Rightarrow y/x =$ slope of normal line. Then for (x_0, y_0) on the circle, $x_0 \neq 0$, an equation of the normal line is $y = (y_0/x_0)x$, which passes through the origin. If $x_0 = 0$, the normal line is vertical and passes through the origin.

57. Horizontal tangents: $(-4, 0), (-4, 10)$
Vertical tangents: $(0, 5), (-8, 5)$

59.

61.

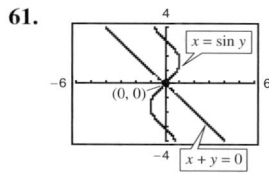

At $(1, 2)$:
Slope of ellipse: -1
Slope of parabola: 1
At $(1, -2)$:
Slope of ellipse: 1
Slope of parabola: -1

At $(0, 0)$:
Slope of line: -1
Slope of sine curve: 1

63. Derivatives: $\dfrac{dy}{dx} = -\dfrac{y}{x}, \dfrac{dy}{dx} = \dfrac{x}{y}$

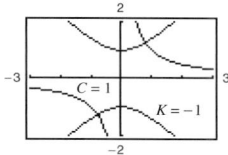

65. Answers will vary. In the explicit form of a function, the variable is explicitly written as a function of x. In an implicit equation, the function is only implied by an equation. An example of an implicit function is $x^2 + xy = 5$. In explicit form, it would be $y = (5 - x^2)/x$.

67.

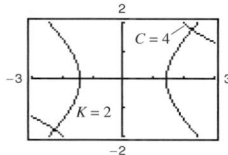

Use starting point B.

69. (a)

(b)

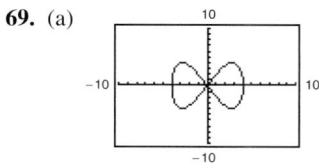

$y_1 = \frac{1}{3}\left[(\sqrt{7} + 7)x + (8\sqrt{7} + 23)\right]$

$y_2 = -\frac{1}{3}\left[(-\sqrt{7} + 7)x - (23 - 8\sqrt{7})\right]$

$y_3 = -\frac{1}{3}\left[(\sqrt{7} - 7)x - (23 - 8\sqrt{7})\right]$

$y_4 = -\frac{1}{3}\left[(\sqrt{7} + 7)x - (8\sqrt{7} + 23)\right]$

(c) $\left(\dfrac{8\sqrt{7}}{7}, 5\right)$

71. Proof   **73.** $y = -\dfrac{\sqrt{3}}{2}x + 2\sqrt{3}, y = \dfrac{\sqrt{3}}{2}x - 2\sqrt{3}$

75. (a) $y = 2x - 6$
(b)

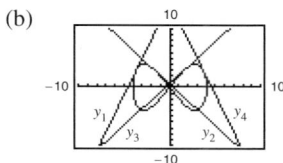

(c) $\left(\frac{28}{17}, -\frac{46}{17}\right)$

Section 2.6   *(page 153)*

1. (a) $\frac{3}{4}$ (b) 20 **3.** (a) $-\frac{5}{8}$ (b) $\frac{3}{2}$
5. (a) -8 cm/sec   (b) 0 cm/sec   (c) 8 cm/sec
7. (a) 12 ft/sec   (b) 6 ft/sec   (c) 3 ft/sec
9. In a linear function, if x changes at a constant rate, so does y. However, unless $a = 1$, y does not change at the same rate as x.
11. (a) 64π cm²/min   (b) 256π cm²/min
13. (a) 972π in.³/min; $15{,}552\pi$ in.³/min
(b) If dr/dt is constant, dV/dt is proportional to r^2.
15. (a) 72 cm³/sec   (b) 1800 cm³/sec
17. $8/(405\pi)$ ft/min   **19.** (a) 12.5% (b) $\frac{1}{144}$ m/min
21. (a) $-\frac{7}{12}$ ft/sec; $-\frac{3}{2}$ ft/sec; $-\frac{48}{7}$ ft/sec
(b) $\frac{527}{24}$ ft²/sec   (c) $\frac{1}{12}$ rad/sec
23. Rate of vertical change: $\frac{1}{5}$ m/sec
Rate of horizontal change: $-\sqrt{3}/15$ m/sec
25. (a) -750 mi/h   (b) 30 min
27. $-50/\sqrt{85} \approx -5.42$ ft/sec
29. (a) $\frac{25}{3}$ ft/sec   (b) $\frac{10}{3}$ ft/sec
31. (a) 12 sec   (b) $\frac{1}{2}\sqrt{3}$ m   (c) $\sqrt{5}\pi/120$ m/sec
33. Evaporation rate proportional to $S \Rightarrow \dfrac{dV}{dt} = k(4\pi r^2)$

$V = \left(\dfrac{4}{3}\right)\pi r^3 \Rightarrow \dfrac{dV}{dt} = 4\pi r^2 \dfrac{dr}{dt}$. So $k = \dfrac{dr}{dt}$.

35. 0.6 ohm/sec   **37.** $\dfrac{dv}{dt} = \dfrac{16r}{v}\sec^2\theta\dfrac{d\theta}{dt}, \dfrac{d\theta}{dt} = \dfrac{v}{16r}\cos^2\theta\dfrac{dv}{dt}$

39. $\dfrac{2\sqrt{21}}{525} \approx 0.017$ rad/sec

41. (a) $\dfrac{200\pi}{3}$ ft/sec   (b) 200π ft/sec
(c) About 427.43π ft/sec
43. About 84.9797 mi/h
45. (a) $\dfrac{dy}{dt} = 3\dfrac{dx}{dt}$ means that y changes three times as fast as x changes.
(b) y changes slowly when $x \approx 0$ or $x \approx L$. y changes more rapidly when x is near the middle of the interval.
47. -18.432 ft/sec²   **49.** About -97.96 m/sec

Review Exercises for Chapter 2   *(page 157)*

1. $f'(x) = 0$ **3.** $f'(x) = 2x - 4$ **5.** 5
7. f is differentiable at all $x \neq 3$.   **9.** 0 **11.** $3x^2 - 22x$
13. $\dfrac{3}{\sqrt{x}} + \dfrac{1}{\sqrt[3]{x^2}}$ **15.** $-\dfrac{4}{3t^3}$ **17.** $4 - 5\cos\theta$
19. $-3\sin\theta - (\cos\theta)/4$ **21.** -1 **23.** 0
25. (a) 50 vibrations/sec/lb   (b) 33.33 vibrations/sec/lb

27. (a) $s(t) = -16t^2 - 30t + 600$
  $v(t) = -32t - 30$
  (b) -94 ft/sec
  (c) $v'(1) = -62$ ft/sec; $v'(3) = -126$ ft/sec
  (d) About 5.258 sec   (e) About -198.256 ft/sec

29. $4(5x^3 - 15x^2 - 11x - 8)$ **31.** $\sqrt{x} \cos x + \sin x/(2\sqrt{x})$

33. $\dfrac{-(x^2 + 1)}{(x^2 - 1)^2}$ **35.** $\dfrac{4x^3 \cos x + x^4 \sin x}{\cos^2 x}$

37. $3x^2 \sec x \tan x + 6x \sec x$ **39.** $-x \sin x$

41. $y = 4x + 10$ **43.** $y = -8x + 1$ **45.** $-48t$

47. $\frac{225}{4}\sqrt{x}$ **49.** $6 \sec^2 \theta \tan \theta$

51. $v(3) = 11$ m/sec; $a(3) = -6$ m/sec^2 **53.** $28(7x + 3)^3$

55. $-\dfrac{2x}{(x^2 + 4)^2}$ **57.** $-45 \sin(9x + 1)$

59. $\frac{1}{2}(1 - \cos 2x) = \sin^2 x$ **61.** $(36x + 1)(6x + 1)^4$

63. $\dfrac{3}{(x^2 + 1)^{3/2}}$ **65.** $\dfrac{-3x^2}{2\sqrt{1 - x^3}}; -2$ **67.** $-\dfrac{8x}{(x^2 + 1)^2}; 2$

69. $-\csc 2x \cot 2x; 0$ **71.** $384(8x + 5)$ **73.** $2 \csc^2 x \cot x$

75. (a) $-18.667°/h$ (b) $-7.284°/h$
  (c) $-3.240°/h$ (d) $-0.747°/h$

77. $-\dfrac{x}{y}$ **79.** $\dfrac{y(y^2 - 3x^2)}{x(x^2 - 3y^2)}$ **81.** $\dfrac{y \sin x + \sin y}{\cos x - x \cos y}$

83. Tangent line: $3x + y - 10 = 0$
  Normal line: $x - 3y = 0$

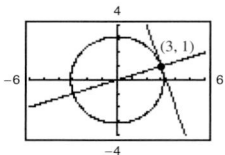

85. (a) $2\sqrt{2}$ units/sec   (b) 4 units/sec   (c) 8 units/sec
87. 450π km/h

P.S. Problem Solving   (page 159)

1. (a) $r = \frac{1}{2}; x^2 + (y - \frac{1}{2})^2 = \frac{1}{4}$
  (b) Center: $(0, \frac{5}{4}); x^2 + (y - \frac{5}{4})^2 = 1$
3. $p(x) = 2x^3 + 4x^2 - 5$
5. (a) $y = 4x - 4$ (b) $y = -\frac{1}{4}x + \frac{9}{2}; (-\frac{9}{4}, \frac{81}{16})$
  (c) Tangent line: $y = 0$ (d) Proof
  Normal line: $x = 0$

7. (a) Graph $\begin{cases} y_1 = \dfrac{1}{a}\sqrt{x^2(a^2 - x^2)} \\ y_2 = -\dfrac{1}{a}\sqrt{x^2(a^2 - x^2)} \end{cases}$ as separate equations.

  (b) Answers will vary. Sample answer:

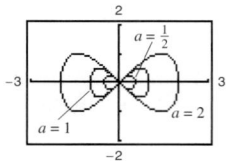

  The intercepts will always be $(0, 0)$, $(a, 0)$, and $(-a, 0)$, and the maximum and minimum y-values appear to be $\pm \frac{1}{2}a$.

  (c) $\left(\dfrac{a\sqrt{2}}{2}, \dfrac{a}{2}\right)$, $\left(\dfrac{a\sqrt{2}}{2}, -\dfrac{a}{2}\right)$, $\left(-\dfrac{a\sqrt{2}}{2}, \dfrac{a}{2}\right)$, $\left(-\dfrac{a\sqrt{2}}{2}, -\dfrac{a}{2}\right)$

9. (a) When the man is 90 ft from the light, the tip of his shadow is $112\frac{1}{2}$ ft from the light. The tip of the child's shadow is $111\frac{1}{9}$ ft from the light, so the man's shadow extends $1\frac{7}{18}$ ft beyond the child's shadow.
  (b) When the man is 60 ft from the light, the tip of his shadow is 75 ft from the light. The tip of the child's shadow is $77\frac{7}{9}$ ft from the light, so the child's shadow extends $2\frac{7}{9}$ ft beyond the man's shadow.
  (c) $d = 80$ ft
  (d) Let x be the distance of the man from the light, and let s be the distance from the light to the tip of the shadow.
  If $0 < x < 80$, then $ds/dt = -50/9$.
  If $x > 80$, then $ds/dt = -25/4$.
  There is a discontinuity at $x = 80$.

11. (a) $v(t) = -\frac{27}{5}t + 27$ ft/sec   (b) 5 sec; 73.5 ft
  $a(t) = -\frac{27}{5}$ ft/sec^2
  (c) The acceleration due to gravity on Earth is greater in magnitude than that on the moon.

13. Proof. The graph of L is a line passing through the origin $(0, 0)$.
15. (a) j would be the rate of change of acceleration.
  (b) $j = 0$. Acceleration is constant, so there is no change in acceleration.
  (c) a: position function, d: velocity function, b: acceleration function, c: jerk function

Chapter 3
Section 3.1   (page 167)

1. $f'(0) = 0$ **3.** $f'(2) = 0$ **5.** $f'(-2)$ is undefined.
7. 2, absolute maximum (and relative maximum)
9. 1, absolute maximum (and relative maximum);
  2, absolute minimum (and relative minimum);
  3, absolute maximum (and relative maximum)
11. $x = 0, x = 2$ **13.** $t = 8/3$ **15.** $x = \pi/3, \pi, 5\pi/3$
17. Minimum: $(2, 1)$ **19.** Minimum: $(2, -8)$
  Maximum: $(-1, 4)$ Maximum: $(6, 24)$
21. Minimum: $\left(-1, -\frac{5}{2}\right)$ **23.** Minimum: $(0, 0)$
  Maximum: $(2, 2)$ Maximum: $(-1, 5)$
25. Minimum: $(0, 0)$ **27.** Minimum: $(1, -1)$
  Maxima: $\left(-1, \frac{1}{4}\right)$ and $\left(1, \frac{1}{4}\right)$ Maximum: $\left(0, -\frac{1}{2}\right)$
29. Minimum: $(-1, -1)$
  Maximum: $(3, 3)$
31. Minimum value is -2 for $-2 \le x < -1$.
  Maximum: $(2, 2)$
33. Minimum: $(3\pi/2, -1)$ **35.** Minimum: $(\pi, -3)$
  Maximum: $(5\pi/6, 1/2)$ Maxima: $(0, 3)$ and $(2\pi, 3)$
37. (a) Minimum: $(0, -3)$;   **39.** (a) Minimum: $(1, -1)$;
  Maximum: $(2, 1)$ Maximum: $(-1, 3)$
  (b) Minimum: $(0, -3)$ (b) Maximum: $(3, 3)$
  (c) Maximum: $(2, 1)$ (c) Minimum: $(1, -1)$
  (d) No extrema   (d) Minimum: $(1, -1)$

41.

  Minimum: $(4, 1)$

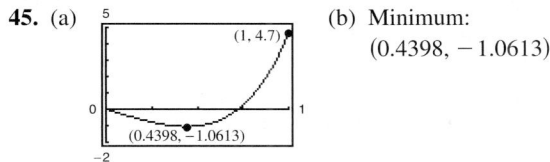

43.

Minima: $\left(\dfrac{-\sqrt{3}+1}{2}, \dfrac{3}{4}\right)$ and $\left(\dfrac{\sqrt{3}+1}{2}, \dfrac{3}{4}\right)$

Maximum: $(3, 31)$

45. (a)

(b) Minimum:
$(0.4398, -1.0613)$

47. Maximum: $\left|f''\left(\sqrt[3]{-10+\sqrt{108}}\right)\right| = f''\left(\sqrt{3}-1\right) \approx 1.47$

49. Maximum: $\left|f^{(4)}(0)\right| = \dfrac{56}{81}$

51. Answers will vary. Sample answer: Let $f(x) = 1/x$. f is continuous on $(0, 1)$ but does not have a maximum or minimum.

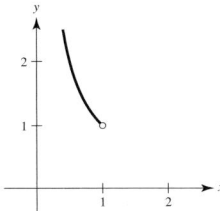

53. Answers will vary. Sample answer:

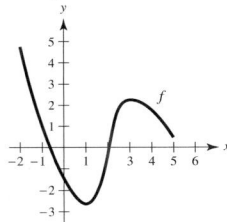

55. (a) Yes   (b) No   **57.** (a) No   (b) Yes
59. Maximum: $P(12) = 72$; No. P is decreasing for $I > 12$.
61. $\theta = \operatorname{arcsec}\sqrt{3} \approx 0.9553$ rad
63. True   **65.** True   **67.** Proof
69. Putnam Problem B3, 2004

Section 3.2   *(page 174)*

1. $f(-1) = f(1) = 1$; f is not continuous on $[-1, 1]$.
3. $f(0) = f(2) = 0$; f is not differentiable on $(0, 2)$.
5. $(2, 0), (-1, 0); f'\left(\frac{1}{2}\right) = 0$ **7.** $(0, 0), (-4, 0); f'\left(-\frac{8}{3}\right) = 0$
9. $f'\left(\frac{3}{2}\right) = 0$ **11.** $f'\left(\dfrac{6-\sqrt{3}}{3}\right) = 0; f'\left(\dfrac{6+\sqrt{3}}{3}\right) = 0$
13. Not differentiable at $x = 0$ **15.** $f'\left(-2+\sqrt{5}\right) = 0$
17. $f'\left(\dfrac{\pi}{2}\right) = 0; f'\left(\dfrac{3\pi}{2}\right) = 0$ **19.** $f'\left(\dfrac{\pi}{6}\right) = 0$
21. Not continuous on $[0, \pi]$
23.

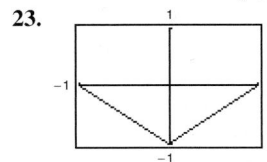

25.

Rolle's Theorem does not apply.

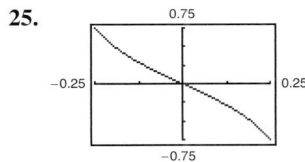

Rolle's Theorem does not apply.

27. (a) $f(1) = f(2) = 38$
(b) Velocity $= 0$ for some t in $(1, 2)$; $t = \frac{3}{2}$ sec

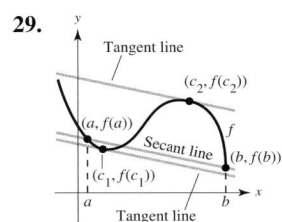

29.

31. The function is not continuous on $[0, 6]$.
33. The function is not continuous on $[0, 6]$.
35. (a) Secant line: $x + y - 3 = 0$ (b) $c = \frac{1}{2}$
(c) Tangent line: $4x + 4y - 21 = 0$
(d)

37. $f'(-1/2) = -1$ **39.** $f'\left(1/\sqrt{3}\right) = 3, f'\left(-1/\sqrt{3}\right) = 3$
41. $f'\left(\frac{8}{27}\right) = 1$ **43.** f is not differentiable at $x = -\frac{1}{2}$.
45. $f'(\pi/2) = 0$
47. (a)–(c)

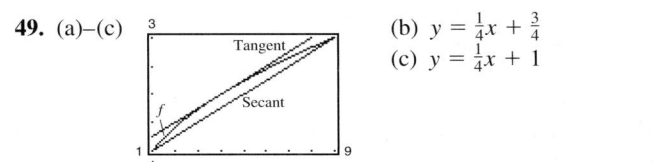

(b) $y = \frac{2}{3}(x-1)$
(c) $y = \frac{1}{3}\left(2x + 5 - 2\sqrt{6}\right)$

49. (a)–(c)

(b) $y = \frac{1}{4}x + \frac{3}{4}$
(c) $y = \frac{1}{4}x + 1$

51. (a) -14.7 m/sec   (b) 1.5 sec
53. No. Let $f(x) = x^2$ on $[-1, 2]$.
55. No. $f(x)$ is not continuous on $[0, 1]$. So it does not satisfy the hypothesis of Rolle's Theorem.
57. By the Mean Value Theorem, there is a time when the speed of the plane must equal the average speed of 454.5 miles/hour. The speed was 400 miles/hour when the plane was accelerating to 454.5 miles/hour and decelerating from 454.5 miles/hour.
59. Proof
61. (a)

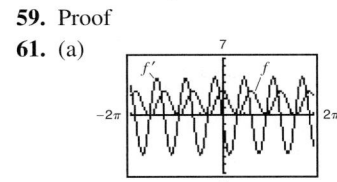

(b) Yes; yes
(c) Because $f(-1) = f(1) = 0$, Rolle's Theorem applies on $[-1, 1]$. Because $f(1) = 0$ and $f(2) = 3$, Rolle's Theorem does not apply on $[1, 2]$.
(d) $\lim\limits_{x \to 3^-} f'(x) = 0$; $\lim\limits_{x \to 3^+} f'(x) = 0$

63.

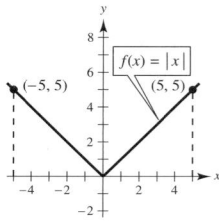

65–67. Proofs   **69.** $f(x) = 5$ **71.** $f(x) = x^2 - 1$
73. False. f is not continuous on $[-1, 1]$.   **75.** True
77–85. Proofs

Section 3.3   *(page 183)*

1. (a) $(0, 6)$ (b) $(6, 8)$
3. Increasing on $(3, \infty)$; Decreasing on $(-\infty, 3)$
5. Increasing on $(-\infty, -2)$ and $(2, \infty)$; Decreasing on $(-2, 2)$
7. Increasing on $(-\infty, -1)$; Decreasing on $(-1, \infty)$
9. Increasing on $(1, \infty)$; Decreasing on $(-\infty, 1)$
11. Increasing on $\left(-2\sqrt{2}, 2\sqrt{2}\right)$;
    Decreasing on $\left(-4, -2\sqrt{2}\right)$ and $\left(2\sqrt{2}, 4\right)$
13. Increasing on $(0, \pi/2)$ and $(3\pi/2, 2\pi)$;
    Decreasing on $(\pi/2, 3\pi/2)$
15. Increasing on $(0, 7\pi/6)$ and $(11\pi/6, 2\pi)$;
    Decreasing on $(7\pi/6, 11\pi/6)$
17. (a) Critical number: $x = 2$
    (b) Increasing on $(2, \infty)$; Decreasing on $(-\infty, 2)$
    (c) Relative minimum: $(2, -4)$
19. (a) Critical number: $x = 1$
    (b) Increasing on $(-\infty, 1)$; Decreasing on $(1, \infty)$
    (c) Relative maximum: $(1, 5)$
21. (a) Critical numbers: $x = -2, 1$
    (b) Increasing on $(-\infty, -2)$ and $(1, \infty)$;
        Decreasing on $(-2, 1)$
    (c) Relative maximum: $(-2, 20)$;
        Relative minimum: $(1, -7)$
23. (a) Critical numbers: $x = -\frac{5}{3}, 1$
    (b) Increasing on $\left(-\infty, -\frac{5}{3}\right), (1, \infty)$;
        Decreasing on $\left(-\frac{5}{3}, 1\right)$
    (c) Relative maximum: $\left(-\frac{5}{3}, \frac{256}{27}\right)$;
        Relative minimum: $(1, 0)$
25. (a) Critical numbers: $x = \pm 1$
    (b) Increasing on $(-\infty, -1)$ and $(1, \infty)$;
        Decreasing on $(-1, 1)$
    (c) Relative maximum: $\left(-1, \frac{4}{5}\right)$; Relative minimum: $\left(1, -\frac{4}{5}\right)$
27. (a) Critical number: $x = 0$
    (b) Increasing on $(-\infty, \infty)$
    (c) No relative extrema
29. (a) Critical number: $x = -2$
    (b) Increasing on $(-2, \infty)$; Decreasing on $(-\infty, -2)$
    (c) Relative minimum: $(-2, 0)$
31. (a) Critical number: $x = 5$
    (b) Increasing on $(-\infty, 5)$; Decreasing on $(5, \infty)$
    (c) Relative maximum: $(5, 5)$

33. (a) Critical numbers: $x = \pm\sqrt{2}/2$; Discontinuity: $x = 0$
    (b) Increasing on $(-\infty, -\sqrt{2}/2)$ and $(\sqrt{2}/2, \infty)$;
        Decreasing on $\left(-\sqrt{2}/2, 0\right)$ and $\left(0, \sqrt{2}/2\right)$
    (c) Relative maximum: $\left(-\sqrt{2}/2, -2\sqrt{2}\right)$;
        Relative minimum: $\left(\sqrt{2}/2, 2\sqrt{2}\right)$
35. (a) Critical number: $x = 0$; Discontinuities: $x = \pm 3$
    (b) Increasing on $(-\infty, -3)$ and $(-3, 0)$;
        Decreasing on $(0, 3)$ and $(3, \infty)$
    (c) Relative maximum: $(0, 0)$
37. (a) Critical number: $x = 0$
    (b) Increasing on $(-\infty, 0)$; Decreasing on $(0, \infty)$
    (c) Relative maximum: $(0, 4)$
39. (a) Critical number: $x = 1$
    (b) Increasing on $(-\infty, 1)$; Decreasing on $(1, \infty)$
    (c) Relative maximum: $(1, 4)$
41. (a) Critical numbers: $x = \pi/6, 5\pi/6$;
        Increasing on $(0, \pi/6), (5\pi/6, 2\pi)$;
        Decreasing on $(\pi/6, 5\pi/6)$
    (b) Relative maximum: $(\pi/6, (\pi + 6\sqrt{3})/12)$;
        Relative minimum: $\left(5\pi/6, (5\pi - 6\sqrt{3})/12\right)$
43. (a) Critical numbers: $x = \pi/4, 5\pi/4$;
        Increasing on $(0, \pi/4), (5\pi/4, 2\pi)$;
        Decreasing on $(\pi/4, 5\pi/4)$
    (b) Relative maximum: $(\pi/4, \sqrt{2})$;
        Relative minimum: $(5\pi/4, -\sqrt{2})$
45. (a) Critical numbers:
        $x = \pi/4, \pi/2, 3\pi/4, \pi, 5\pi/4, 3\pi/2, 7\pi/4$;
        Increasing on $(\pi/4, \pi/2), (3\pi/4, \pi), (5\pi/4, 3\pi/2)$,
            $(7\pi/4, 2\pi)$;
        Decreasing on $(0, \pi/4), (\pi/2, 3\pi/4), (\pi, 5\pi/4)$,
            $(3\pi/2, 7\pi/4)$;
    (b) Relative maxima: $(\pi/2, 1), (\pi, 1), (3\pi/2, 1)$;
        Relative minima: $(\pi/4, 0), (3\pi/4, 0)$,
                $(5\pi/4, 0), (7\pi/4, 0)$
47. (a) Critical numbers: $\pi/2, 7\pi/6, 3\pi/2, 11\pi/6$;
        Increasing on $(0, \pi/2), (7\pi/6, 3\pi/2), (11\pi/6, 2\pi)$;
        Decreasing on $(\pi/2, 7\pi/6), (3\pi/2, 11\pi/6)$
    (b) Relative maxima: $(\pi/2, 2), (3\pi/2, 0)$;
        Relative minima: $(7\pi/6, -1/4), (11\pi/6, -1/4)$
49. (a) $f'(x) = 2(9 - 2x^2)/\sqrt{9 - x^2}$
    (b)   (c) Critical numbers:
            $x = \pm 3\sqrt{2}/2$

    (d) $f' > 0$ on $\left(-3\sqrt{2}/2, 3\sqrt{2}/2\right)$;
        $f' < 0$ on $(-3, -3\sqrt{2}/2), (3\sqrt{2}/2, 3)$
        f is increasing when f' is positive and decreasing when f' is
        negative.

51. (a) $f'(t) = t(t \cos t + 2 \sin t)$

(b)

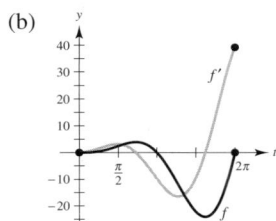

(c) Critical numbers:
$t = 2.2889, 5.0870$

(d) $f' > 0$ on $(0, 2.2889)$, $(5.0870, 2\pi)$;
$f' < 0$ on $(2.2889, 5.0870)$
f is increasing when f' is positive and decreasing when f' is negative.

53. (a) $f'(x) = -\cos(x/3)$

(b)

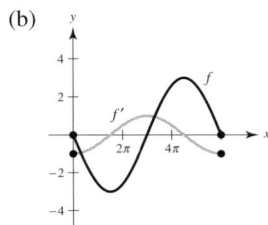

(c) Critical numbers:
$x = 3\pi/2, 9\pi/2$

(d) $f' > 0$ on $\left(\frac{3\pi}{2}, \frac{9\pi}{2}\right)$;
$f' < 0$ on $\left(0, \frac{3\pi}{2}\right)$, $\left(\frac{9\pi}{2}, 6\pi\right)$
f is increasing when f' is positive and decreasing when f' is negative.

55. $f(x)$ is symmetric with respect to the origin.
Zeros: $(0, 0), (\pm\sqrt{3}, 0)$

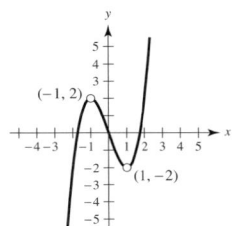

$g(x)$ is continuous on $(-\infty, \infty)$, and $f(x)$ has holes at $x = 1$ and $x = -1$.

57.

59.

61.

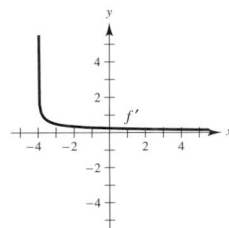

63. $g'(0) < 0$ **65.** $g'(-6) < 0$ **67.** $g'(0) > 0$

69. Answers will vary. Sample answer:

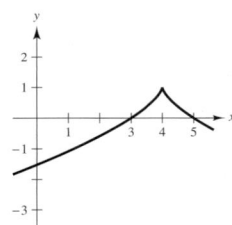

71. $(5, f(5))$ is a relative minimum.

73. (a)

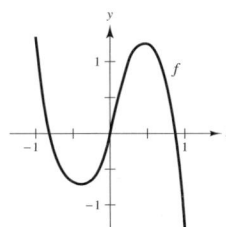

(b) Critical numbers: $x \approx -0.40$ and $x \approx 0.48$
(c) Relative maximum: $(0.48, 1.25)$;
Relative minimum: $(-0.40, 0.75)$

75. (a) $s'(t) = 9.8(\sin\theta)t$; speed $= |9.8(\sin\theta)t|$

(b)

θ	0	$\pi/4$	$\pi/3$	$\pi/2$	$2\pi/3$	$3\pi/4$	π
$s'(t)$	0	$4.9\sqrt{2}t$	$4.9\sqrt{3}t$	$9.8t$	$4.9\sqrt{3}t$	$4.9\sqrt{2}t$	0

The speed is maximum at $\theta = \pi/2$.

77. (a)

t	0	0.5	1	1.5	2	2.5	3
$C(t)$	0	0.055	0.107	0.148	0.171	0.176	0.167

$t = 2.5$ h

(b)

$t \approx 2.38$ h   (c) $t \approx 2.38$ h

79. $r = 2R/3$

81. (a) $v(t) = 6 - 2t$ (b) $[0, 3)$ (c) $(3, \infty)$ (d) $t = 3$

83. (a) $v(t) = 3t^2 - 10t + 4$

(b) $\left[0, (5 - \sqrt{13})/3\right)$ and $\left((5 + \sqrt{13})/3, \infty\right)$

(c) $\left(\frac{5 - \sqrt{13}}{3}, \frac{5 + \sqrt{13}}{3}\right)$ (d) $t = \frac{5 \pm \sqrt{13}}{3}$

85. Answers will vary.

87. (a) Minimum degree: 3

(b) $a_3(0)^3 + a_2(0)^2 + a_1(0) + a_0 = 0$
$a_3(2)^3 + a_2(2)^2 + a_1(2) + a_0 = 2$
$3a_3(0)^2 + 2a_2(0) + a_1 = 0$
$3a_3(2)^2 + 2a_2(2) + a_1 = 0$

(c) $f(x) = -\frac{1}{2}x^3 + \frac{3}{2}x^2$

89. (a) Minimum degree: 4

(b) $a_4(0)^4 + a_3(0)^3 + a_2(0)^2 + a_1(0) + a_0 = 0$
$a_4(2)^4 + a_3(2)^3 + a_2(2)^2 + a_1(2) + a_0 = 4$
$a_4(4)^4 + a_3(4)^3 + a_2(4)^2 + a_1(4) + a_0 = 0$
$4a_4(0)^3 + 3a_3(0)^2 + 2a_2(0) + a_1 = 0$
$4a_4(2)^3 + 3a_3(2)^2 + 2a_2(2) + a_1 = 0$
$4a_4(4)^3 + 3a_3(4)^2 + 2a_2(4) + a_1 = 0$

(c) $f(x) = \frac{1}{4}x^4 - 2x^3 + 4x^2$

91. True   **93.** False. Let $f(x) = x^3$.

95. False. Let $f(x) = x^3$. There is a critical number at $x = 0$, but not a relative extremum.

97–99. Proofs   **101.** Putnam Problem A3, 2003

Section 3.4   *(page 192)*

1. $f' > 0, f'' < 0$ **3.** Concave upward: $(-\infty, \infty)$

5. Concave upward: $(-\infty, 2)$; Concave downward: $(2, \infty)$

7. Concave upward: $(-\infty, -2), (2, \infty)$;
Concave downward: $(-2, 2)$

9. Concave upward: $(-\infty, -1), (1, \infty)$;
Concave downward: $(-1, 1)$

11. Concave upward: $(-2, 2)$;
Concave downward: $(-\infty, -2), (2, \infty)$

13. Concave upward: $(-\pi/2, 0)$; Concave downward: $(0, \pi/2)$

15. Point of inflection: $(2, 8)$; Concave downward: $(-\infty, 2)$;
Concave upward: $(2, \infty)$

17. Points of inflection: $(-2, -8), (0, 0)$;
Concave upward: $(-\infty, -2), (0, \infty)$;
Concave downward: $(-2, 0)$

19. Points of inflection: $(2, -16), (4, 0)$;
Concave upward: $(-\infty, 2), (4, \infty)$;
Concave downward: $(2, 4)$

21. Concave upward: $(-3, \infty)$

23. Points of inflection: $\left(-\sqrt{3}/3, 3\right), \left(\sqrt{3}/3, 3\right)$;
Concave upward: $\left(-\infty, -\sqrt{3}/3\right), \left(\sqrt{3}/3, \infty\right)$;
Concave downward: $\left(-\sqrt{3}/3, \sqrt{3}/3\right)$

25. Point of inflection: $(2\pi, 0)$;
Concave upward: $(2\pi, 4\pi)$; Concave downward: $(0, 2\pi)$

27. Concave upward: $(0, \pi), (2\pi, 3\pi)$;
Concave downward: $(\pi, 2\pi), (3\pi, 4\pi)$

29. Points of inflection: $(\pi, 0), (1.823, 1.452), (4.46, -1.452)$
Concave upward: $(1.823, \pi), (4.46, 2\pi)$
Concave downward: $(0, 1.823), (\pi, 4.46)$

31. Relative maximum: $(3, 9)$

33. Relative maximum: $(0, 3)$; Relative minimum: $(2, -1)$

35. Relative minimum: $(3, -25)$

37. Relative minimum: $(0, -3)$

39. Relative maximum: $(-2, -4)$; Relative minimum: $(2, 4)$

41. No relative extrema, because f is nonincreasing.

43. (a) $f'(x) = 0.2x(x - 3)^2(5x - 6)$;
$f''(x) = 0.4(x - 3)(10x^2 - 24x + 9)$
(b) Relative maximum: $(0, 0)$;
Relative minimum: $(1.2, -1.6796)$;
Points of inflection: $(0.4652, -0.7048)$,
$(1.9348, -0.9048), (3, 0)$

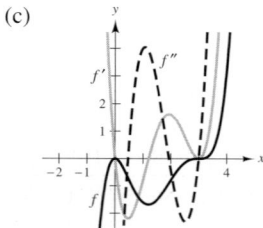

(c)

f is increasing when f' is positive and decreasing when f' is negative. f is concave upward when f'' is positive and concave downward when f'' is negative.

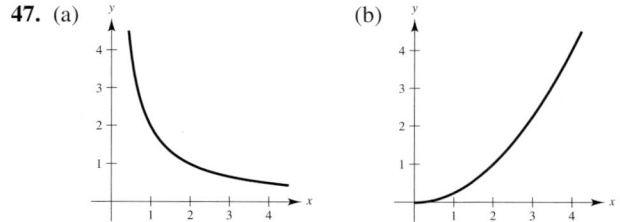

45. (a) $f'(x) = \cos x - \cos 3x + \cos 5x$;
$f''(x) = -\sin x + 3 \sin 3x - 5 \sin 5x$
(b) Relative maximum: $(\pi/2, 1.53333)$;
Points of inflection: $(\pi/6, 0.2667), (1.1731, 0.9637)$,
$(1.9685, 0.9637), (5\pi/6, 0.2667)$

(c)

f is increasing when f' is positive and decreasing when f' is negative. f is concave upward when f'' is positive and concave downward when f'' is negative.

47. (a)

(b)

49. Answers will vary. Sample answer: $f(x) = x^4$; $f''(0) = 0$, but $(0, 0)$ is not a point of inflection.

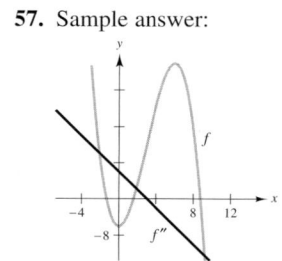

51.

53.

55.

57. Sample answer:

59. (a) $f(x) = (x - 2)^n$ has a point of inflection at $(2, 0)$ if n is odd and $n \geq 3$.

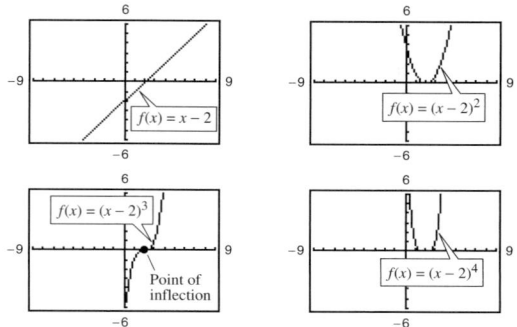

(b) Proof

61. $f(x) = \frac{1}{2}x^3 - 6x^2 + \frac{45}{2}x - 24$

63. (a) $f(x) = \frac{1}{32}x^3 + \frac{3}{16}x^2$ (b) Two miles from touchdown

65. $x = 100$ units

67. (a)

t	0.5	1	1.5	2	2.5	3
S	151.5	555.6	1097.6	1666.7	2193.0	2647.1

$1.5 < t < 2$

(b)

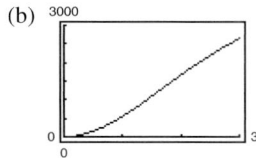

(c) About 1.633 yr

$t \approx 1.5$

69. $P_1(x) = 2\sqrt{2}$

$P_2(x) = 2\sqrt{2} - \sqrt{2}(x - \pi/4)^2$

The values of f, P_1, and P_2 and their first derivatives are equal when $x = \pi/4$. The approximations worsen as you move away from $x = \pi/4$.

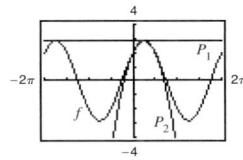

71. $P_1(x) = 1 - x/2$

$P_2(x) = 1 - x/2 - x^2/8$

The values of f, P_1, and P_2 and their first derivatives are equal when $x = 0$. The approximations worsen as you move away from $x = 0$.

73.

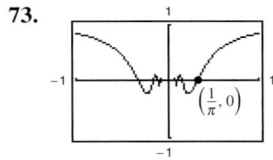

75. True

77. False. f is concave upward at $x = c$ if $f''(c) > 0$.

79. Proof

Section 3.5   *(page 202)*

1. f   **2.** c   **3.** d   **4.** a   **5.** b   **6.** e

7.

x	10^0	10^1	10^2	10^3
$f(x)$	7	2.2632	2.0251	2.0025

x	10^4	10^5	10^6
$f(x)$	2.0003	2.0000	2.0000

$\lim_{x \to \infty} \dfrac{4x + 3}{2x - 1} = 2$

9.

x	10^0	10^1	10^2	10^3
$f(x)$	-2	-2.9814	-2.9998	-3.0000

x	10^4	10^5	10^6
$f(x)$	-3.0000	-3.0000	-3.0000

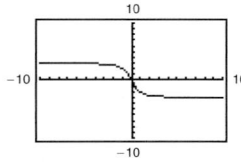

$\lim_{x \to \infty} \dfrac{-6x}{\sqrt{4x^2 + 5}} = -3$

11.

x	10^0	10^1	10^2	10^3
$f(x)$	4.5000	4.9901	4.9999	5.0000

x	10^4	10^5	10^6
$f(x)$	5.0000	5.0000	5.0000

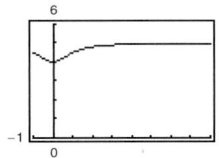

$\lim_{x \to \infty} \left(5 - \dfrac{1}{x^2 + 1} \right) = 5$

13. (a) ∞ (b) 5   (c) 0   **15.** (a) 0   (b) 1   (c) ∞

17. (a) 0   (b) $-\frac{2}{3}$ (c) $-\infty$ **19.** 4   **21.** $\frac{2}{3}$ **23.** 0

25. $-\infty$ **27.** -1 **29.** -2 **31.** $\frac{1}{2}$ **33.** ∞

35. 0   **37.** 0

39.

41.

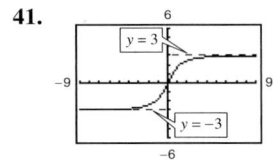

43. 1   **45.** 0   **47.** $\frac{1}{6}$

49.

x	10^0	10^1	10^2	10^3	10^4	10^5	10^6
$f(x)$	1.000	0.513	0.501	0.500	0.500	0.500	0.500

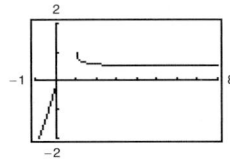

$\lim_{x \to \infty} \left[x - \sqrt{x(x - 1)} \right] = \frac{1}{2}$

51.

x	10^0	10^1	10^2	10^3	10^4	10^5	10^6
$f(x)$	0.479	0.500	0.500	0.500	0.500	0.500	0.500

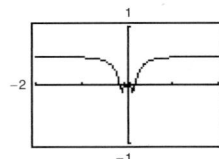

The graph has a hole at $x = 0$.

$\lim_{x \to \infty} x \sin \dfrac{1}{2x} = \dfrac{1}{2}$

53. As x becomes large, $f(x)$ approaches 4.

55. Answers will vary. Sample answer: Let

$$f(x) = \frac{-6}{0.1(x-2)^2 + 1} + 6.$$

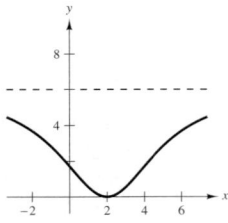

57. (a) 5   (b) -5

59.

61.

63.

65.

67.

69.

71.

73.

75.

77.

79.

81.

83. (a)

(c)

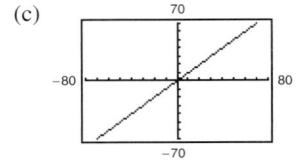

(b) Proof   Slant asymptote: $y = x$

85. 100%   **87.** $\lim\limits_{t \to \infty} N(t) = +\infty$; $\lim\limits_{t \to \infty} E(t) = c$

89. (a)

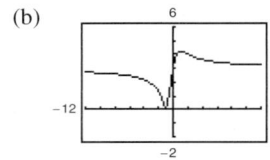

(b) Yes. $\lim\limits_{t \to \infty} S = \dfrac{100}{1} = 100$

91. (a) $\lim\limits_{x \to \infty} f(x) = 2$

(b) $x_1 = \sqrt{\dfrac{4 - 2\varepsilon}{\varepsilon}}$, $x_2 = -\sqrt{\dfrac{4 - 2\varepsilon}{\varepsilon}}$

(c) $\sqrt{\dfrac{4 - 2\varepsilon}{\varepsilon}}$ (d) $-\sqrt{\dfrac{4 - 2\varepsilon}{\varepsilon}}$

93. (a) Answers will vary. $M = \dfrac{5\sqrt{33}}{11}$ **95–97.** Proofs

(b) Answers will vary. $M = \dfrac{29\sqrt{177}}{59}$

99. (a) $d(m) = \dfrac{|3m + 3|}{\sqrt{m^2 + 1}}$

(b)

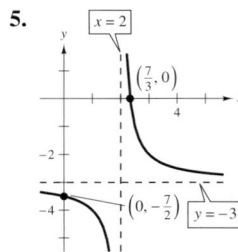

(c) $\lim\limits_{m \to \infty} d(m) = 3$;

$\lim\limits_{m \to -\infty} d(m) = 3$;

As m approaches $\pm\infty$, the distance approaches 3

101. Proof

103. False. Let $f(x) = \dfrac{2x}{\sqrt{x^2 + 2}}$. $f'(x) > 0$ for all real numbers.

Section 3.6   (page 212)

1. d   **2.** c   **3.** a   **4.** b

5.

7.

9.

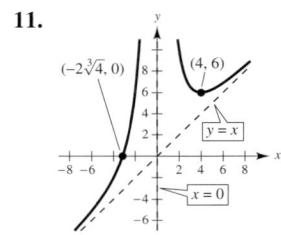

11.

13.

$x = 4$

$y = x - 2$

$(6, 6)$

$(0, -3)$

$(2, -2)$

15.

$\left(\dfrac{8}{3}, \dfrac{16\sqrt{3}}{9}\right)$

$(0, 0)$ $(4, 0)$

17.

$(1, 1)$ $\left(\dfrac{27}{8}, 0\right)$

$(0, 0)$

19.

$(0, 2)$

$(1, 0)$

21.

$\left(-\dfrac{4}{3}, 0\right)$ $(0, 0)$

$\left(-\dfrac{2}{3}, -\dfrac{16}{27}\right)$

$(-1, -1)$

23.

$(-\sqrt[4]{5}, 0)$

$(-1, 4)$

$(0, 0)$ $(\sqrt[4]{5}, 0)$

$(1, -4)$

25.

Minimum: $(-1.10, -9.05)$;
Maximum: $(1.10, 9.05)$;
Points of inflection:
$(-1.84, -7.86)$, $(1.84, 7.86)$;
Vertical asymptote: $x = 0$;
Horizontal asymptote: $y = 0$

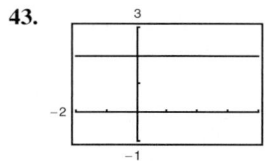

27.

Point of inflection: $(0, 0)$;
Horizontal asymptotes: $y = \pm 2$

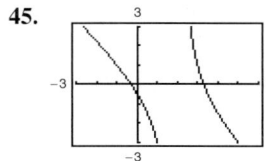

29.

Relative minimum:
$\left(\dfrac{\pi}{3}, \dfrac{2\pi}{3} - 2\sqrt{3}\right)$;
Relative maximum:
$\left(\dfrac{5\pi}{3}, \dfrac{10\pi}{3} + 2\sqrt{3}\right)$;
Points of inflection: $(0, 0)$,
$(\pi, 2\pi)$, $(2\pi, 4\pi)$

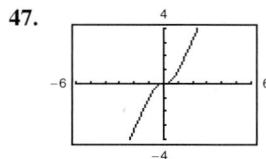

31.

Relative minimum: $\left(\pi, -\dfrac{5}{4}\right)$;
Points of inflection:
$\left(\dfrac{2\pi}{3}, -\dfrac{3}{8}\right)$, $\left(\dfrac{4\pi}{3}, -\dfrac{3}{8}\right)$

33.

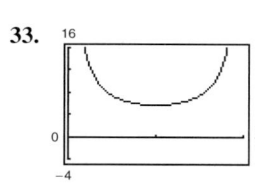

Relative minimum: $\left(\dfrac{\pi}{4}, 4\sqrt{2}\right)$;
Vertical asymptotes: $x = 0, \dfrac{\pi}{2}$

35. f is decreasing on $(2, 8)$, and therefore $f(3) > f(5)$.

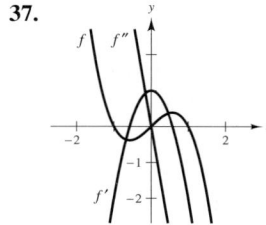

37.

The zeros of f' correspond to the points where the graph of f has horizontal tangents. The zero of f'' corresponds to the point where the graph of f' has a horizontal tangent.

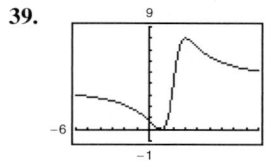

39.

The graph crosses the horizontal asymptote $y = 4$.
The graph of a function f does not cross its vertical asymptote $x = c$ because $f(c)$ does not exist.

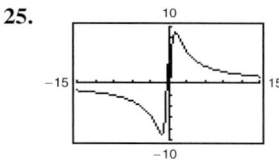

41.

The graph has a hole at $x = 0$.
The graph crosses the horizontal asymptote $y = 0$.
The graph of a function f does not cross its vertical asymptote $x = c$ because $f(c)$ does not exist.

43.

The graph has a hole at $x = 3$.
The rational function is not reduced to lowest terms.

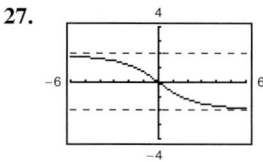

45.

The graph appears to approach the line $y = -x + 1$, which is the slant asymptote.

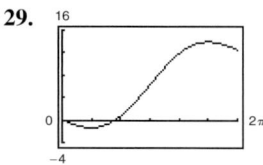

47.

The graph appears to approach the line $y = 2x$, which is the slant asymptote.

49.

f

f''

51.

53. (a)

The graph has holes at $x = 0$ and at $x = 4$.

Visually approximated critical numbers: $\frac{1}{2}, 1, \frac{3}{2}, 2, \frac{5}{2}, 3, \frac{7}{2}$

(b) $f'(x) = \dfrac{-x \cos^2(\pi x)}{(x^2 + 1)^{3/2}} - \dfrac{2\pi \sin(\pi x) \cos(\pi x)}{\sqrt{x^2 + 1}}$;

Approximate critical numbers: $\frac{1}{2}, 0.97, \frac{3}{2}, 1.98, \frac{5}{2}, 2.98, \frac{7}{2}$;
The critical numbers where maxima occur appear to be integers in part (a), but by approximating them using f', you can see that they are not integers.

55. Answers will vary. Sample answer: $y = 1/(x - 3)$

57. Answers will vary.
Sample answer: $y = (3x^2 - 7x - 5)/(x - 3)$

59. (a) x_0, x_2, x_4 **(b)** x_2, x_3 **(c)** x_1 **(d)** x_1 **(e)** x_2, x_3

61. (a)–(h) Proofs

63. Answers will vary. Sample answer: The graph has a vertical asymptote at $x = b$. If a and b are both positive or both negative, then the graph of f approaches ∞ as x approaches b, and the graph has a minimum at $x = -b$. If a and b have opposite signs, then the graph of f approaches $-\infty$ as x approaches b, and the graph has a maximum at $x = -b$.

65. $y = 4x, \ y = -4x$ **67.** Putnam Problem 13(i), 1939

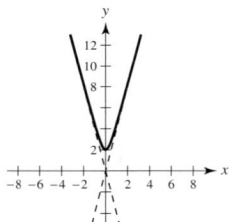

Section 3.7   *(page 220)*

1. (a) and (b)

First Number, x	Second Number	Product, P
10	$110 - 10$	$10(110 - 10) = 1000$
20	$110 - 20$	$20(110 - 20) = 1800$
30	$110 - 30$	$30(110 - 30) = 2400$
40	$110 - 40$	$40(110 - 40) = 2800$
50	$110 - 50$	$50(110 - 50) = 3000$
60	$110 - 60$	$60(110 - 60) = 3000$
70	$110 - 70$	$70(110 - 70) = 2800$
80	$110 - 80$	$80(110 - 80) = 2400$
90	$110 - 90$	$90(110 - 90) = 1800$
100	$110 - 100$	$100(110 - 100) = 1000$

The maximum is attained near $x = 50$ and 60.

(c) $P = x(110 - x)$

(d)

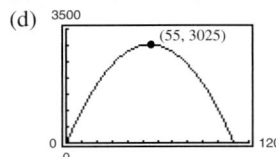

(e) 55 and 55

3. $S/2$ and $S/2$ **5.** 21 and 7   **7.** 54 and 27

9. $l = w = 20$ m   **11.** $l = w = 4\sqrt{2}$ ft   **13.** $(1, 1)$

15. $\left(\frac{7}{2}, \sqrt{\frac{7}{2}}\right)$

17. Dimensions of page: $(2 + \sqrt{30})$ in. $\times (2 + \sqrt{30})$ in.

19. 700×350 m

21. Rectangular portion: $16/(\pi + 4) \times 32/(\pi + 4)$ ft

23. (a) $L = \sqrt{x^2 + 4 + \dfrac{8}{x - 1} + \dfrac{4}{(x - 1)^2}}, \quad x > 1$

(b)

Minimum when $x \approx 2.587$

(c) $(0, 0), (2, 0), (0, 4)$

25. Width: $5\sqrt{2}/2$; Length: $5\sqrt{2}$

27. (a)

(b)

Length, x	Width, y	Area, xy
10	$2/\pi (100 - 10)$	$(10)(2/\pi)(100 - 10) \approx 573$
20	$2/\pi (100 - 20)$	$(20)(2/\pi)(100 - 20) \approx 1019$
30	$2/\pi (100 - 30)$	$(30)(2/\pi)(100 - 30) \approx 1337$
40	$2/\pi (100 - 40)$	$(40)(2/\pi)(100 - 40) \approx 1528$
50	$2/\pi (100 - 50)$	$(50)(2/\pi)(100 - 50) \approx 1592$
60	$2/\pi (100 - 60)$	$(60)(2/\pi)(100 - 60) \approx 1528$

The maximum area of the rectangle is approximately 1592 m².

(c) $A = 2/\pi(100x - x^2), \quad 0 < x < 100$

(d) $\dfrac{dA}{dx} = \dfrac{2}{\pi}(100 - 2x)$
$= 0$ when $x = 50$;
The maximum value is approximately 1592 when $x = 50$.

(e)

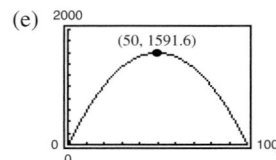

29. $18 \times 18 \times 36$ in.

31. No. The volume changes because the shape of the container changes when it is squeezed.

33. $r = \sqrt[3]{21/(2\pi)} \approx 1.50$ ($h = 0$, so the solid is a sphere.)

35. Side of square: $\dfrac{10\sqrt{3}}{9 + 4\sqrt{3}}$; Side of triangle: $\dfrac{30}{9 + 4\sqrt{3}}$

37. $w = \left(20\sqrt{3}\right)/3$ in., $h = \left(20\sqrt{6}\right)/3$ in.

39.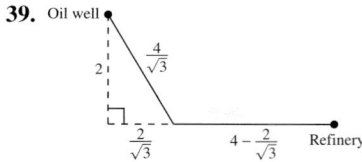

The path of the pipe should go underwater from the oil well to the coast following the hypotenuse of a right triangle with leg lengths of 2 miles and $2/\sqrt{3}$ miles for a distance of $4/\sqrt{3}$ miles. Then the pipe should go down the coast to the refinery for a distance of $\left(4 - 2/\sqrt{3}\right)$ miles.

41. One mile from the nearest point on the coast

43.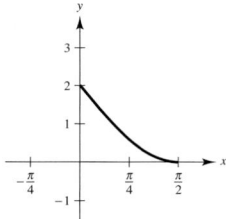

(a) Origin to y-intercept: 2;
Origin to x-intercept: $\pi/2$
(b) $d = \sqrt{x^2 + (2 - 2\sin x)^2}$

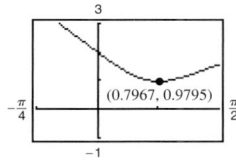

(c) Minimum distance is 0.9795 when $x \approx 0.7967$.

45. About 1.153 radians or 66°   **47.** 8%
49. $y = \frac{64}{141}x$; $S \approx 6.1$ mi   **51.** $y = \frac{3}{10}x$; $S_3 \approx 4.50$ mi
53. Putnam Problem A1, 1986

Section 3.8   (page 229)

In the answers for Exercises 1 and 3, the values in the tables have been rounded for convenience. Because a calculator and a computer program calculates internally using more digits than they display, you may produce slightly different values from those shown in the tables.

1.

n	x_n	$f(x_n)$	$f'(x_n)$	$\dfrac{f(x_n)}{f'(x_n)}$	$x_n - \dfrac{f(x_n)}{f'(x_n)}$
1	2.2000	−0.1600	4.4000	−0.0364	2.2364
2	2.2364	0.0015	4.4728	0.0003	2.2361

3.

n	x_n	$f(x_n)$	$f'(x_n)$	$\dfrac{f(x_n)}{f'(x_n)}$	$x_n - \dfrac{f(x_n)}{f'(x_n)}$
1	1.6	−0.0292	−0.9996	0.0292	1.5708
2	1.5708	0	−1	0	1.5708

5. −1.587   **7.** 0.682   **9.** 1.250, 5.000
11. 0.900, 1.100, 1.900   **13.** 1.935   **15.** 0.569
17. 4.493   **19.** (a) Proof   (b) $\sqrt{5} \approx 2.236$; $\sqrt{7} \approx 2.646$
21. $f'(x_1) = 0$ **23.** 0.74   **25.** Proof
27. (a)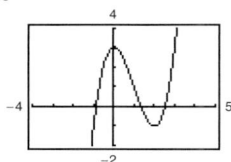

(b) 1.347   (c) 2.532

(d) 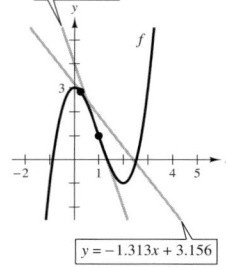 x-intercept of $y = -3x + 4$ is $\frac{4}{3}$.
x-intercept of
$y = -1.313x + 3.156$
is approximately 2.404.

(e) If the initial estimate $x = x_1$ is not sufficiently close to the desired zero of a function, then the x-intercept of the corresponding tangent line to the function may approximate a second zero of the function.

29. Answers will vary. Sample answer:
If f is a function continuous on $[a, b]$ and differentiable on (a, b), where $c \in [a, b]$ and $f(c) = 0$, then Newton's Method uses tangent lines to approximate c. First, estimate an initial x_1 close to c. (See graph.) Then determine x_2 using $x_2 = x_1 - f(x_1)/f'(x_1)$. Calculate a third estimate x_3 using $x_3 = x_2 - f(x_2)/f'(x_2)$. Continue this process until $|x_n - x_{n+1}|$ is within the desired accuracy, and let x_{n+1} be the final approximation of c.

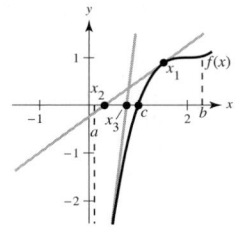

31. (1.939, 0.240)   **33.** $x \approx 1.563$ mi
35. False; let $f(x) = \dfrac{x^2 - 1}{x - 1}$.   **37.** True   **39.** 0.217

Section 3.9   (page 236)

1. $T(x) = 4x - 4$

x	1.9	1.99	2	2.01	2.1
$f(x)$	3.610	3.960	4	4.040	4.410
$T(x)$	3.600	3.960	4	4.040	4.400

3. $T(x) = 80x - 128$

x	1.9	1.99	2	2.01	2.1
$f(x)$	24.761	31.208	32	32.808	40.841
$T(x)$	24.000	31.200	32	32.800	40.000

5. $T(x) = (\cos 2)(x - 2) + \sin 2$

x	1.9	1.99	2	2.01	2.1
$f(x)$	0.946	0.913	0.909	0.905	0.863
$T(x)$	0.951	0.913	0.909	0.905	0.868

7. $\Delta y = 0.331$; $dy = 0.3$ **9.** $\Delta y = -0.039$; $dy = -0.040$
11. $6x\, dx$ **13.** $(x \sec^2 x + \tan x)\, dx$
15. $-\dfrac{13}{(2x - 1)^2}\, dx$ **17.** $\dfrac{-x}{\sqrt{9 - x^2}}\, dx$ **19.** $(3 - \sin 2x)\, dx$

21. (a) 0.9　(b) 1.04　　**23.** (a) 8.035　　(b) 7.95

25. (a) $\pm\dfrac{5}{8}$ in.^{2}　　(b) 0.625%

27. (a) ±10.75 cm^2　　(b) about 1.19%

29. (a) ±20.25 in.^{3}　　(b) ±5.4 in.^{2}　　(c) 0.6%; 0.4%

31. 27.5 mi; About 7.3%　　**33.** (a) $\frac{1}{4}$%　　(b) 216 sec $= 3.6$ min

35. 6407 ft

37. $f(x) = \sqrt{x}, dy = \dfrac{1}{2\sqrt{x}}\,dx$

$f(99.4) \approx \sqrt{100} + \dfrac{1}{2\sqrt{100}}(-0.6) = 9.97$

Calculator: 9.97

39. $f(x) = \sqrt[4]{x}, dy = \dfrac{1}{4x^{3/4}}\,dx$

$f(624) \approx \sqrt[4]{625} + \dfrac{1}{4(625)^{3/4}}(-1) = 4.998$

Calculator: 4.998

41. $y - f(0) = f'(0)(x - 0)$

$y - 2 = \frac{1}{4}x$

$y = 2 + x/4$

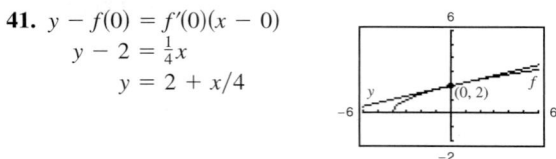

43. The value of dy becomes closer to the value of Δy as Δx decreases.

45. $f(x) = \sqrt{x}$; $dy = \dfrac{1}{2\sqrt{x}}\,dx$

$f(4.02) \approx \sqrt{4} + \dfrac{1}{2\sqrt{4}}(0.02) = 2 + \frac{1}{4}(0.02)$

47. True　　**49.** True

Review Exercises for Chapter 3　*(page 238)*

1. Maximum: $(0, 0)$;　　**3.** Maximum: $(4, 0)$;

Minimum: $\left(-\frac{5}{2}, -\frac{25}{4}\right)$　　　Minimum: $(0, -2)$

5. Maximum: $\left(3, \frac{2}{3}\right)$;　　**7.** Maximum: $(2\pi, 17.57)$;

Minimum: $\left(-3, -\frac{2}{3}\right)$　　　Minimum: $(2.73, 0.88)$

9. $f(0) \neq f(4)$　　**11.** Not continuous on $[-2, 2]$

13. $f'\left(\dfrac{2744}{729}\right) = \dfrac{3}{7}$　　**15.** f is not differentiable at $x = 5$.

17. $f'(0) = 1$

19. No; The function has a discontinuity at $x = 0$, which is in the interval $[-2, 1]$.

21. Increasing on $\left(-\frac{3}{2}, \infty\right)$; Decreasing on $\left(-\infty, -\frac{3}{2}\right)$

23. Increasing on $(-\infty, 1)$, $\left(\frac{7}{3}, \infty\right)$; Decreasing on $\left(1, \frac{7}{3}\right)$

25. Increasing on $(1, \infty)$; Decreasing on $(0, 1)$

27. (a) Critical number: $x = 3$

(b) Increasing on $(3, \infty)$; Decreasing on $(-\infty, 3)$

(c) Relative minimum: $(3, -4)$

(d)

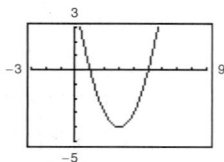

29. (a) Critical number: $t = 2$

(b) Increasing on $(2, \infty)$; Decreasing on $(-\infty, 2)$

(c) Relative minimum: $(2, -12)$

(d)

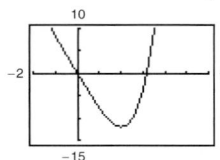

31. (a) Critical number: $x = -8$; Discontinuity: $x = 0$

(b) Increasing on $(-8, 0)$;

Decreasing on $(-\infty, -8)$ and $(0, \infty)$

(c) Relative minimum: $\left(-8, -\frac{1}{16}\right)$

(d)

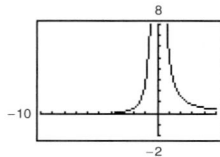

33. (a) Critical numbers: $x = \dfrac{3\pi}{4}, \dfrac{7\pi}{4}$

(b) Increasing on $\left(\dfrac{3\pi}{4}, \dfrac{7\pi}{4}\right)$;

Decreasing on $\left(0, \dfrac{3\pi}{4}\right)$ and $\left(\dfrac{7\pi}{4}, 2\pi\right)$

(c) Relative minimum: $\left(\dfrac{3\pi}{4}, -\sqrt{2}\right)$;

Relative maximum: $\left(\dfrac{7\pi}{4}, \sqrt{2}\right)$

(d)

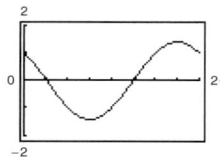

35. $(3, -54)$; Concave upward: $(3, \infty)$;

Concave downward: $(-\infty, 3)$

37. No point of inflection; Concave upward: $(-5, \infty)$

39. $(\pi/2, \pi/2)$, $(3\pi/2, 3\pi/2)$; Concave upward: $(\pi/2, 3\pi/2)$;

Concave downward: $(0, \pi/2)$, $(3\pi/2, 2\pi)$

41. Relative minimum: $(-9, 0)$

43. Relative maxima: $\left(\sqrt{2}/2, 1/2\right)$, $\left(-\sqrt{2}/2, 1/2\right)$;

Relative minimum: $(0, 0)$

45. Relative maximum: $(-3, -12)$; Relative minimum: $(3, 12)$

47.　　　　　　　　　　　**49.** Increasing and concave down

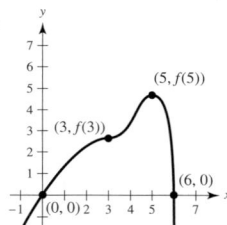

51. (a) $D = 0.00188t^4 - 0.1273t^2 + 2.672t^2 - 7.81t + 77.1$

(b)

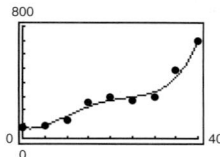

(c) Maximum in 2010; Minimum in 1970　　(d) 2010

53. 8   **55.** $\frac{2}{3}$ **57.** $-\infty$ **59.** 0   **61.** 6

63.

65.

67.

69.

71.

73.

75.

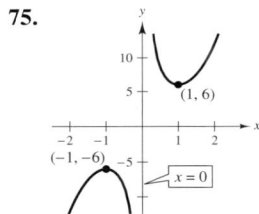

77. $x = 50$ ft and $y = \frac{200}{3}$ ft   **79.** $(0, 0), (5, 0), (0, 10)$

81. 14.05 ft   **83.** $32\pi r^3/81$ **85.** $-1.532, -0.347, 1.879$

87. $-2.182, -0.795$ **89.** -0.755

91. $\Delta y = 0.03005; dy = 0.03$

93. $dy = (1 - \cos x + x \sin x)\,dx$ **95.** (a) $\pm 8.1\pi$ cm^3

(b) $\pm 1.8\pi$ cm^2 (c) About 0.83%; About 0.56%

P.S. Problem Solving   *(page 241)*

1. Choices of a may vary.

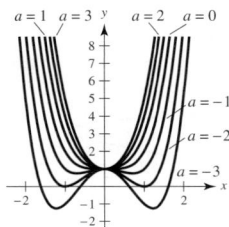

(a) One relative minimum at $(0, 1)$ for $a \geq 0$

(b) One relative maximum at $(0, 1)$ for $a < 0$

(c) Two relative minima for $a < 0$ when $x = \pm\sqrt{-a/2}$

(d) If $a < 0$, then there are three critical points; if $a \geq 0$, then there is only one critical point.

3. All c, where c is a real number   **5.** Proof

7. The bug should head towards the midpoint of the opposite side. Without calculus, imagine opening up the cube. The shortest distance is the line PQ, passing through the midpoint as shown.

9. $a = 6, b = 1, c = 2$ **11.** Proof

13. Greatest slope: $\left(-\frac{\sqrt{3}}{3}, \frac{3}{4}\right)$; Least slope: $\left(\frac{\sqrt{3}}{3}, \frac{3}{4}\right)$

15. Proof   **17.** Proof; Point of inflection: $(1, 0)$

19. (a) $P(x) = x - x^2$

(b)

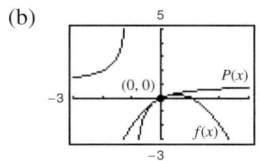

Chapter 4

Section 4.1   *(page 251)*

1. Proof   **3.** $y = 3t^3 + C$ **5.** $y = \frac{2}{5}x^{5/2} + C$

Original Integral	Rewrite	Integrate	Simplify
7. $\displaystyle\int \sqrt[3]{x}\,dx$	$\displaystyle\int x^{1/3}\,dx$	$\dfrac{x^{4/3}}{4/3} + C$	$\dfrac{3}{4}x^{4/3} + C$
9. $\displaystyle\int \dfrac{1}{x\sqrt{x}}\,dx$	$\displaystyle\int x^{-3/2}\,dx$	$\dfrac{x^{-1/2}}{-1/2} + C$	$-\dfrac{2}{\sqrt{x}} + C$

11. $\frac{1}{2}x^2 + 7x + C$ **13.** $\frac{1}{6}x^6 + x + C$

15. $\frac{2}{5}x^{5/2} + x^2 + x + C$ **17.** $\frac{3}{5}x^{5/3} + C$

19. $-1/(4x^4) + C$ **21.** $\frac{2}{3}x^{3/2} + 12x^{1/2} + C$

23. $x^3 + \frac{1}{2}x^2 - 2x + C$ **25.** $5 \sin x - 4 \cos x + C$

27. $t + \csc t + C$ **29.** $\tan \theta + \cos \theta + C$ **31.** $\tan y + C$

33. Answers will vary. Sample answer:

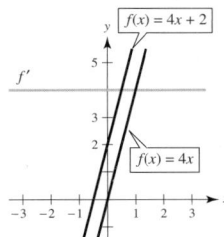

35. $f(x) = 3x^2 + 8$ **37.** $h(t) = 2t^4 + 5t - 11$

39. $f(x) = x^2 + x + 4$ **41.** $f(x) = -4\sqrt{x} + 3x$

43. (a) Answers will vary. Sample answer: (b) $y = \dfrac{x^3}{3} - x + \dfrac{7}{3}$

45. (a)

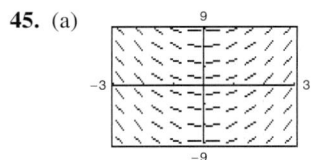

(b) $y = x^2 - 6$

(c)

47. When you evaluate the integral $\int f(x)\,dx$, you are finding a function $F(x)$ that is an antiderivative of $f(x)$. So, there is no difference.

49.

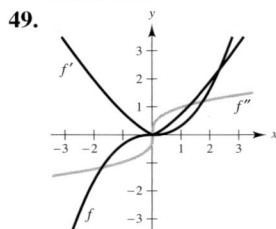

51. (a) $h(t) = \frac{3}{4}t^2 + 5t + 12$ (b) 69 cm   **53.** 62.25 ft

55. (a) $t \approx 2.562$ sec   (b) $v(t) \approx -65.970$ ft/sec

57. $v_0 \approx 62.3$ m/sec   **59.** 320 m; -32 m/sec

61. (a) $v(t) = 3t^2 - 12t + 9$; $a(t) = 6t - 12$
    (b) $(0, 1), (3, 5)$ (c) -3

63. $a(t) = -1/(2t^{3/2})$; $x(t) = 2\sqrt{t} + 2$

65. (a) 1.18 m/sec^2 (b) 190 m

67. (a) 300 ft   (b) 60 ft/sec ≈ 41 mi/h

69. False. f has an infinite number of antiderivatives, each differing by a constant.

71. True   **73.** True   **75.** $f(x) = \dfrac{x^3}{3} - 4x + \dfrac{16}{3}$

77. Proof

Section 4.2   *(page 263)*

1. 75   **3.** $\dfrac{158}{85}$ **5.** $4c$ **7.** $\displaystyle\sum_{i=1}^{11} \dfrac{1}{5i}$

9. $\displaystyle\sum_{j=1}^{6}\left[7\left(\dfrac{j}{6}\right) + 5\right]$ **11.** $\dfrac{2}{n}\displaystyle\sum_{i=1}^{n}\left[\left(\dfrac{2i}{n}\right)^3 - \left(\dfrac{2i}{n}\right)\right]$ **13.** 84

15. 1200   **17.** 2470   **19.** 12,040

21. $(n + 2)/n$

    $n = 10$: $S = 1.2$
    $n = 100$: $S = 1.02$
    $n = 1000$: $S = 1.002$
    $n = 10{,}000$: $S = 1.0002$

23. $[2(n + 1)(n - 1)]/n^2$

    $n = 10$: $S = 1.98$
    $n = 100$: $S = 1.9998$
    $n = 1000$: $S = 1.999998$
    $n = 10{,}000$: $S = 1.99999998$

25. $13 < (\text{Area of region}) < 15$

27. $55 < (\text{Area of region}) < 74.5$

29. $0.7908 < (\text{Area of region}) < 1.1835$

31. The area of the shaded region falls between 12.5 square units and 16.5 square units.

33. $A \approx S \approx 0.768$ **35.** $A \approx S \approx 0.746$
    $A \approx s \approx 0.518$ $A \approx s \approx 0.646$

37. $\displaystyle\lim_{n\to\infty}\left[\dfrac{12(n + 1)}{n}\right] = 12$

39. $\displaystyle\lim_{n\to\infty}\dfrac{1}{6}\left(\dfrac{2n^3 - 3n^2 + n}{n^3}\right) = \dfrac{1}{3}$

41. $\displaystyle\lim_{n\to\infty}\left[(3n + 1)/n\right] = 3$

43. (a)   (b) $\Delta x = (2 - 0)/n = 2/n$

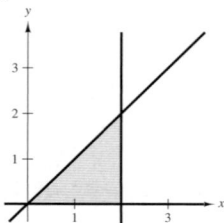

(c) $s(n) = \displaystyle\sum_{i=1}^{n} f(x_{i-1})\,\Delta x = \sum_{i=1}^{n}[(i - 1)(2/n)](2/n)$

(d) $S(n) = \displaystyle\sum_{i=1}^{n} f(x_i)\,\Delta x = \sum_{i=1}^{n}[i(2/n)](2/n)$

(e)

n	5	10	50	100
$s(n)$	1.6	1.8	1.96	1.98
$S(n)$	2.4	2.2	2.04	2.02

(f) $\displaystyle\lim_{n\to\infty}\sum_{i=1}^{n}[(i - 1)(2/n)](2/n) = 2$;

$\displaystyle\lim_{n\to\infty}\sum_{i=1}^{n}[i(2/n)](2/n) = 2$

45. $A = 3$ **47.** $A = \dfrac{7}{3}$

    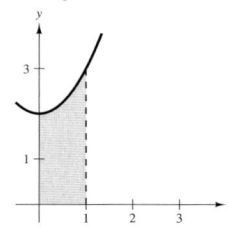

49. $A = 54$ **51.** $A = 34$

    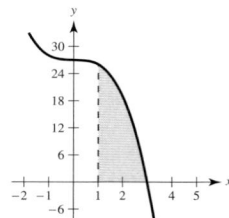

53. $A = \dfrac{2}{3}$ **55.** $A = 8$

    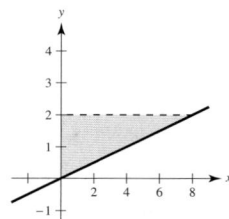

57. $A = \dfrac{125}{3}$ **59.** $A = \dfrac{44}{3}$

    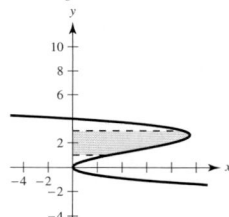

61. $\dfrac{69}{8}$ **63.** 0.345 **65.** b

67. You can use the line $y = x$ bounded by $x = a$ and $x = b$. The sum of the areas of the inscribed rectangles in the figure below is the lower sum.

The sum of the areas of the circumscribed rectangles in the figure below is the upper sum.

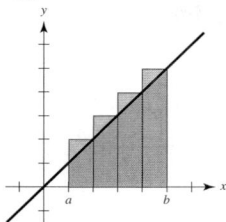

The rectangles in the first graph do not contain all of the area of the region, and the rectangles in the second graph cover more than the area of the region. The exact value of the area lies between these two sums.

69. (a)

(b)

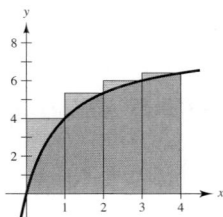

$s(4) = \frac{46}{3}$

$S(4) = \frac{326}{15}$

(c)

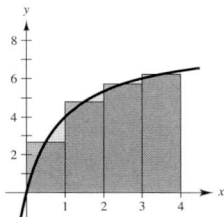

(d) Proof

$M(4) = \frac{6112}{315}$

(e)

n	4	8	20	100	200
$s(n)$	15.333	17.368	18.459	18.995	19.060
$S(n)$	21.733	20.568	19.739	19.251	19.188
$M(n)$	19.403	19.201	19.137	19.125	19.125

(f) Because f is an increasing function, $s(n)$ is always increasing and $S(n)$ is always decreasing.

71. True

73. Suppose there are n rows and $n + 1$ columns. The stars on the left total $1 + 2 + \cdots + n$, as do the stars on the right. There are $n(n + 1)$ stars in total. So, $2[1 + 2 + \cdots + n] = n(n + 1)$ and $1 + 2 + \cdots + n = [n(n + 1)]/2$.

75. For n odd, $\left(\frac{n + 1}{2}\right)^2$ blocks;

For n even, $\frac{n^2 + 2n}{4}$ blocks

77. Putnam Problem B1, 1989

Section 4.3 *(page 273)*

1. $2\sqrt{3} \approx 3.464$ **3.** 32   **5.** 0   **7.** $\frac{10}{3}$

9. $\int_{-1}^{5} (3x + 10)\, dx$ **11.** $\int_{0}^{3} \sqrt{x^2 + 4}\, dx$ **13.** $\int_{0}^{4} 5\, dx$

15. $\int_{-4}^{4} (4 - |x|)\, dx$ **17.** $\int_{-5}^{5} (25 - x^2)\, dx$

19. $\int_{0}^{\pi/2} \cos x\, dx$ **21.** $\int_{0}^{2} y^3\, dy$

23.

$A = 12$

25.

$A = 8$

27.

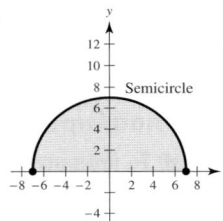

$A = 14$

29.

$A = 1$

31.

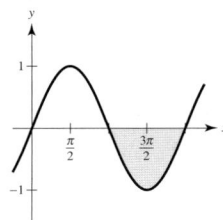

$A = 49\pi/2$

33. -6 **35.** 48   **37.** -12

39. 16   **41.** (a) 13   (b) -10 (c) 0   (d) 30

43. (a) 8   (b) -12 (c) -4 (d) 30   **45.** $-48, 88$

47. (a) $-\pi$ (b) 4   (c) $-(1 + 2\pi)$ (d) $3 - 2\pi$
(e) $5 + 2\pi$ (f) $23 - 2\pi$

49. (a) 14   (b) 4   (c) 8   (d) 0   **51.** 40   **53.** a   **55.** d

57. No. There is a discontinuity at $x = 4$.

59. $a = -2$, $b = 5$

61. Answers will vary. Sample answer: $a = \pi$, $b = 2\pi$

$$\int_{\pi}^{2\pi} \sin x\, dx < 0$$

63. True   **65.** True   **67.** False. $\int_{0}^{2} (-x)\, dx = -2$

69. 272   **71.** Proof

73. No. No matter how small the subintervals, the number of both rational and irrational numbers within each subinterval is infinite, and $f(c_i) = 0$ or $f(c_i) = 1$.

75. $a = -1$ and $b = 1$ maximize the integral.   **77.** $\frac{1}{3}$

Section 4.4   (page 288)

1.

Positive

3.

Zero

5. 12   **7.** -2 **9.** $-\frac{10}{3}$ **11.** $\frac{1}{3}$ **13.** $\frac{1}{2}$ **15.** $\frac{2}{3}$

17. -4 **19.** $-\frac{1}{18}$ **21.** $-\frac{27}{20}$ **23.** $\frac{25}{2}$ **25.** $\frac{64}{3}$

27. $\pi + 2$ **29.** $\pi/4$ **31.** $2\sqrt{3}/3$ **33.** 0   **35.** $\frac{1}{6}$

37. 1   **39.** $\frac{52}{3}$ **41.** 20   **43.** $\frac{32}{3}$

45. $3\sqrt[3]{2}/2 \approx 1.8899$ **47.** $2\sqrt{3} \approx 3.4641$

49. $\pm \arccos \sqrt{\pi}/2 \approx \pm 0.4817$

51. Average value $= 6$ **53.** Average value $= \frac{1}{4}$
$x = \pm\sqrt{3} \approx \pm 1.7321$ $x = \sqrt[3]{2}/2 \approx 0.6300$

55. Average value $= 2/\pi$ **57.** About 540 ft
$x \approx 0.690, x \approx 2.451$

59. (a) 8   (b) $\frac{4}{3}$ (c) $\int_1^7 f(x)\,dx = 20$; Average value $= \frac{10}{3}$

61. (a) $F(x) = 500 \sec^2 x$ (b) $1500\sqrt{3}/\pi \approx 827$ N

63. About 0.5318 L

65. (a) $v = -0.00086t^3 + 0.0782t^2 - 0.208t + 0.10$
(b)   (c) 2475.6 m

67. $F(x) = 2x^2 - 7x$ **69.** $F(x) = -20/x + 20$
$F(2) = -6$ $F(2) = 10$
$F(5) = 15$ $F(5) = 16$
$F(8) = 72$ $F(8) = \frac{35}{2}$

71. $F(x) = \sin x - \sin 1$
$F(2) = \sin 2 - \sin 1 \approx 0.0678$
$F(5) = \sin 5 - \sin 1 \approx -1.8004$
$F(8) = \sin 8 - \sin 1 \approx 0.1479$

73. (a) $g(0) = 0, g(2) \approx 7, g(4) \approx 9, g(6) \approx 8, g(8) \approx 5$
(b) Increasing: $(0, 4)$; Decreasing: $(4, 8)$
(c) A maximum occurs at $x = 4$.
(d)

75. $\frac{1}{2}x^2 + 2x$ **77.** $\frac{3}{4}x^{4/3} - 12$ **79.** $\tan x - 1$

81. $x^2 - 2x$ **83.** $\sqrt{x^4 + 1}$ **85.** $x \cos x$ **87.** 8

89. $\cos x \sqrt{\sin x}$ **91.** $3x^2 \sin x^6$

93.

An extremum of g occurs
at $x = 2$.

95. (a) $\frac{3}{2}$ ft to the right
(b) $\frac{113}{10}$ ft

97. (a) 0 ft   (b) $\frac{63}{2}$ ft   **99.** (a) 2 ft to the right   (b) 2 ft

101. 28 units   **103.** 8190 L

105. $f(x) = x^{-2}$ has a nonremovable discontinuity at $x = 0$.

107. $f(x) = \sec^2 x$ has a nonremovable discontinuity at $x = \pi/2$.

109. $2/\pi \approx 63.7\%$ **111.** True

113. $f'(x) = \dfrac{1}{(1/x)^2 + 1}\left(-\dfrac{1}{x^2}\right) + \dfrac{1}{x^2 + 1} = 0$
Because $f'(x) = 0, f(x)$ is constant.

115. (a) 0   (b) 0   (c) $xf(x) + \int_0^x f(t)\,dt$ (d) 0

Section 4.5   (page 301)

$\displaystyle\int f(g(x))g'(x)\,dx$	$u = g(x)$	$du = g'(x)\,dx$
1. $\displaystyle\int (8x^2 + 1)^2(16x)\,dx$	$8x^2 + 1$	$16x\,dx$
3. $\displaystyle\int \tan^2 x \sec^2 x\,dx$	$\tan x$	$\sec^2 x\,dx$

5. $\frac{1}{5}(1 + 6x)^5 + C$ **7.** $\frac{2}{3}(25 - x^2)^{3/2} + C$

9. $\frac{1}{12}(x^4 + 3)^3 + C$ **11.** $\frac{1}{15}(x^3 - 1)^5 + C$

13. $\frac{1}{3}(t^2 + 2)^{3/2} + C$ **15.** $-\frac{15}{8}(1 - x^2)^{4/3} + C$

17. $1/[4(1 - x^2)^2] + C$ **19.** $-1/[3(1 + x^3)] + C$

21. $-\sqrt{1 - x^2} + C$ **23.** $-\frac{1}{4}(1 + 1/t)^4 + C$

25. $\sqrt{2x} + C$ **27.** $2x^2 - 4\sqrt{16 - x^2} + C$

29. $-1/[2(x^2 + 2x - 3)] + C$

31. (a) Answers will vary.   (b) $y = -\frac{1}{3}(4 - x^2)^{3/2} + 2$
Sample answer:

33. $-\cos(\pi x) + C$

35. $\displaystyle\int \cos 8x\,dx = \dfrac{1}{8}\int (\cos 8x)(8)\,dx = \dfrac{1}{8}\sin 8x + C$

37. $-\sin(1/\theta) + C$

39. $\frac{1}{4}\sin^2 2x + C$ or $-\frac{1}{4}\cos^2 2x + C_1$ or $-\frac{1}{8}\cos 4x + C_2$

41. $\frac{1}{2}\tan^2 x + C$ or $\frac{1}{2}\sec^2 x + C_1$ **43.** $f(x) = 2\cos(x/2) + 4$

45. $f(x) = \frac{1}{12}(4x^2 - 10)^3 - 8$

47. $\frac{2}{5}(x + 6)^{5/2} - 4(x + 6)^{3/2} + C = \frac{2}{5}(x + 6)^{3/2}(x - 4) + C$

49. $-\left[\frac{2}{3}(1 - x)^{3/2} - \frac{4}{5}(1 - x)^{5/2} + \frac{2}{7}(1 - x)^{7/2}\right] + C =$
$-\frac{2}{105}(1 - x)^{3/2}(15x^2 + 12x + 8) + C$

51. $\frac{1}{8}\left[\frac{2}{5}(2x-1)^{5/2} + \frac{4}{3}(2x-1)^{3/2} - 6(2x-1)^{1/2}\right] + C =$
$\left(\sqrt{2x-1}/15\right)\left(3x^2 + 2x - 13\right) + C$

53. $-x - 1 - 2\sqrt{x+1} + C$ or $-\left(x + 2\sqrt{x+1}\right) + C_1$

55. 0   **57.** $12 - \frac{8}{9}\sqrt{2}$ **59.** 2   **61.** $\frac{1}{2}$

63. $f(x) = (2x^3 + 1)^3 + 3$ **65.** 1209/28   **67.** $2\left(\sqrt{3} - 1\right)$

69. $\frac{272}{15}$ **71.** $\frac{2}{3}$ **73.** (a) $\frac{64}{3}$ (b) $\frac{128}{3}$ (c) $-\frac{64}{3}$ (d) 64

75. $2\displaystyle\int_0^3 (4x^2 - 6)\, dx = 36$

77. If $u = 5 - x^2$, then $du = -2x\, dx$ and
$\int x(5 - x^2)^3\, dx = -\frac{1}{2}\int (5 - x^2)^3(-2x)\, dx = -\frac{1}{2}\int u^3\, du.$

79. (a) $\displaystyle\int x^2 \sqrt{x^3 + 1}\, dx$ (b) $\displaystyle\int \tan(3x) \sec^2(3x)\, dx$

81. \$340,000

83. (a) 102.532 thousand units   (b) 102.352 thousand units
(c) 74.5 thousand units

85. (a) $P_{0.50,\,0.75} \approx 35.3\%$ (b) $b \approx 58.6\%$

87. (a)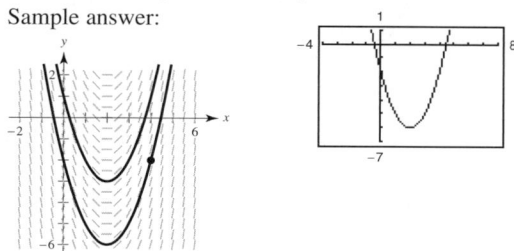
(b) g is nonnegative, because the graph of f is positive at the beginning and generally has more positive sections than negative ones.

(c) The points on g that correspond to the extrema of f are points of inflection of g.

(d) No, some zeros of f, such as $x = \pi/2$, do not correspond to extrema of g. The graph of g continues to increase after $x = \pi/2$, because f remains above the x-axis.

(e)
The graph of h is that of g shifted 2 units downward.

89. (a) and (b) Proofs

91. False. $\displaystyle\int (2x + 1)^2\, dx = \frac{1}{6}(2x + 1)^3 + C$ **93.** True

95. True   **97–99.** Proofs   **101.** Putnam Problem A1, 1958

Section 4.6   *(page 310)*

	Trapezoidal	Simpson's	Exact
1.	2.7500	2.6667	2.6667
3.	4.2500	4.0000	4.0000
5.	20.2222	20.0000	20.0000
7.	12.6640	12.6667	12.6667
9.	0.3352	0.3334	0.3333
	Trapezoidal	**Simpson's**	**Graphing Utility**
11.	3.2833	3.2396	3.2413
13.	0.3415	0.3720	0.3927
15.	0.5495	0.5483	0.5493
17.	−0.0975	−0.0977	−0.0977
19.	0.1940	0.1860	0.1858

21. Trapezoidal: Linear (1st-degree) polynomials
Simpson's: Quadratic (2nd-degree) polynomials

23. (a) 1.500   (b) 0.000   **25.** (a) $\frac{1}{4}$ (b) $\frac{1}{12}$

27. (a) $n = 366$ (b) $n = 26$ **29.** (a) $n = 77$ (b) $n = 8$

31. (a) $n = 130$ (b) $n = 12$ **33.** (a) $n = 643$ (b) $n = 48$

35. (a) 24.5   (b) 25.67   **37.** 0.701   **39.** 89,250 m²

41. 10,233.58 ft-lb   **43.** 3.1416   **45.** 2.477   **47.** Proof

Review Exercises for Chapter 4   *(page 312)*

1. $\dfrac{x^2}{2} - 6x + C$ **3.** $\dfrac{4}{3}x^3 + \dfrac{1}{2}x^2 + 3x + C$

5. $x^2/2 - 4/x^2 + C$ **7.** $x^2 + 9\cos x + C$

9. $y = 1 - 3x^2$ **11.** $f(x) = 4x^3 - 5x - 3$

13. (a) Answers will vary.   (b) $y = x^2 - 4x - 2$
Sample answer:

15. (a) 3 sec; 144 ft   (b) $\frac{3}{2}$ sec   (c) 108 ft

17. 240 ft/sec   **19.** 60   **21.** $\displaystyle\sum_{n=1}^{10} \dfrac{1}{3n}$ **23.** 192

25. 420   **27.** 3310

29. 9.038 < (Area of region) < 13.038

31. $A = 15$ **33.** $A = 12$

35. $\frac{27}{2}$ **37.** $\displaystyle\int_{-4}^0 (2x + 8)\, dx$

39.

$A = \dfrac{25}{2}$

41. (a) 17   (b) 7   (c) 9   (d) 84

43. 56   **45.** 0   **47.** $\frac{422}{5}$ **49.** $\left(\sqrt{2} + 2\right)/2$

51. $-\cos 2 + 1 \approx 1.416$ **53.** 30   **55.** $\frac{1}{4}$

57. Average value $= \frac{2}{5}, x = \frac{25}{4}$

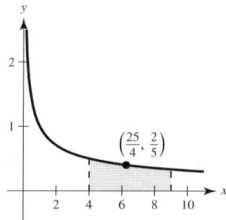

59. $x^2 \sqrt{1 + x^3}$ **61.** $x^2 + 3x + 2$ **63.** $\frac{2}{3}\sqrt{x^3 + 3} + C$

65. $-\frac{1}{30}(1 - 3x^2)^5 + C = \frac{1}{30}(3x^2 - 1)^5 + C$

67. $\frac{1}{4}\sin^4 x + C$　　**69.** $-2\sqrt{1 - \sin\theta} + C$

71. $\dfrac{1}{3\pi}(1 + \sec\pi x)^3 + C$

73. (a) Answers will vary.　　(b) $y = -\frac{1}{3}(9 - x^2)^{3/2} + 5$
Sample answer:

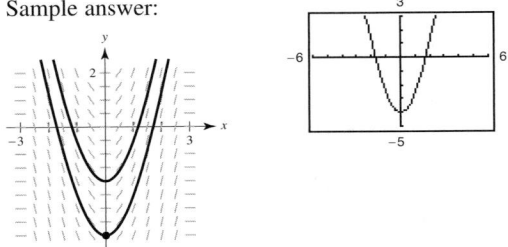

75. $\frac{455}{2}$　**77.** 2　**79.** $28\pi/15$　**81.** 2　**83.** $\frac{468}{7}$

85. (a) $\frac{64}{5}$　(b) $\frac{32}{5}$　(c) $\frac{96}{5}$　(d) -32

87. Trapezoidal Rule: 0.285　　**89.** Trapezoidal Rule: 0.637
Simpson's Rule: 0.284　　　　　Simpson's Rule: 0.685
Graphing Utility: 0.284　　　　Graphing Utility: 0.704

P.S. Problem Solving　(page 315)

1. (a) $L(1) = 0$　(b) $L'(x) = 1/x$, $L'(1) = 1$
(c) $x \approx 2.718$　(d) Proof

3. (a) $\displaystyle\lim_{n\to\infty}\left[\frac{32}{n^5}\sum_{i=1}^{n}i^4 - \frac{64}{n^4}\sum_{i=1}^{n}i^3 + \frac{32}{n^3}\sum_{i=1}^{n}i^2\right]$
(b) $(16n^4 - 16)/(15n^4)$　(c) $16/15$

5. (a)

(b)

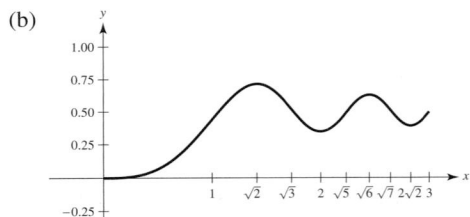

(c) Relative maxima at $x = \sqrt{2}$, $\sqrt{6}$
Relative minima at $x = 2, 2\sqrt{2}$
(d) Points of inflection at $x = 1, \sqrt{3}, \sqrt{5}, \sqrt{7}$

7. (a)

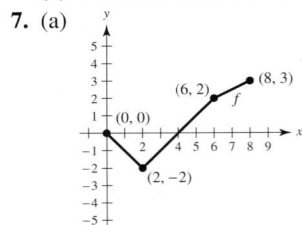

(b)

x	0	1	2	3	4	5	6	7	8
$F(x)$	0	$-\frac{1}{2}$	-2	$-\frac{7}{2}$	-4	$-\frac{7}{2}$	-2	$\frac{1}{4}$	3

(c) $x = 4, 8$　(d) $x = 2$

9. Proof　**11.** $\frac{2}{3}$　**13.** $1 \le \displaystyle\int_0^1\sqrt{1 + x^4}\,dx \le \sqrt{2}$

15. (a)

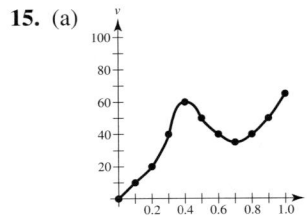

(b) $(0, 0.4)$ and $(0.7, 1.0)$　(c) 150 mi/h^2
(d) Total distance traveled in miles; 38.5 mi
(e) Sample answer: 100 mi/h^2

17. (a)–(c) Proofs

19. (a) $R(n), I, T(n), L(n)$
(b) $S(4) = \frac{1}{3}[f(0) + 4f(1) + 2f(2) + 4f(3) + f(4)] \approx 5.42$

Chapter 5
Section 5.1　(page 325)

1. (a) 3.8067　(b) $\ln 45 = \displaystyle\int_1^{45}\frac{1}{t}\,dt \approx 3.8067$

3. (a) -0.2231　(b) $\ln 0.8 = \displaystyle\int_1^{0.8}\frac{1}{t}\,dt \approx -0.2231$

5. b　**6.** d　**7.** a　**8.** c

9.

11.

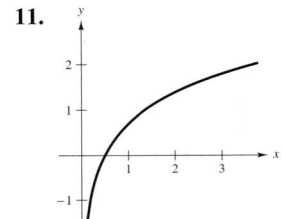

Domain: $x > 0$　　　　Domain: $x > 0$

13.

15.

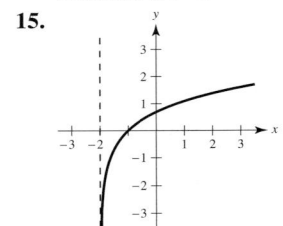

Domain: $x > 3$　　　　Domain: $x > -2$

17. (a) 1.7917　(b) -0.4055　(c) 4.3944　(d) 0.5493

19. $\ln x - \ln 4$　　**21.** $\ln x + \ln y - \ln z$

23. $\ln x + \frac{1}{2}\ln(x^2 + 5)$　　**25.** $\frac{1}{2}[\ln(x - 1) - \ln x]$

27. $\ln z + 2\ln(z - 1)$　　**29.** $\ln\dfrac{x - 2}{x + 2}$

31. $\ln\sqrt[3]{\dfrac{x(x + 3)^2}{x^2 - 1}}$　　**33.** $\ln\dfrac{9}{\sqrt{x^2 + 1}}$

35. (a)

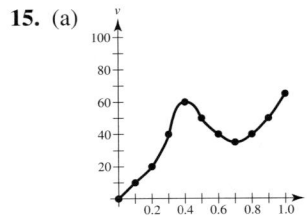

(b) $f(x) = \ln\dfrac{x^2}{4} = \ln x^2 - \ln 4$
$= 2\ln x - \ln 4$
$= g(x)$

37. $-\infty$　　**39.** $\ln 4 \approx 1.3863$　　**41.** $1/x$　　**43.** $2/x$

45. $4(\ln x)^3/x$ **47.** $2/(t + 1)$ **49.** $\dfrac{2x^2 - 1}{x(x^2 - 1)}$

51. $\dfrac{1 - x^2}{x(x^2 + 1)}$ **53.** $\dfrac{1 - 2\ln t}{t^3}$ **55.** $\dfrac{2}{x\ln x^2} = \dfrac{1}{x\ln x}$

57. $\dfrac{1}{1 - x^2}$ **59.** $\dfrac{-4}{x(x^2 + 4)}$ **61.** $\cot x$

63. $-\tan x + \dfrac{\sin x}{\cos x - 1}$

65. (a) $y = 4x - 4$ **67.** (a) $5x - y - 2 = 0$
(b)   (b)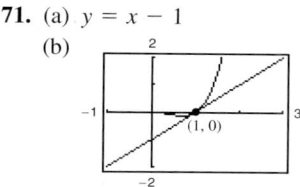

69. (a) $y = \frac{1}{3}x - \frac{1}{12}\pi + \frac{1}{2}\ln\left(\frac{3}{2}\right)$
(b)

71. (a) $y = x - 1$
(b)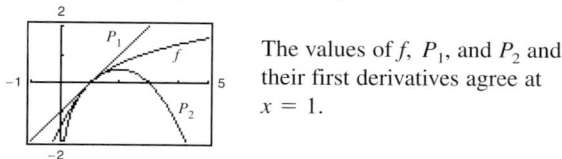

73. $\dfrac{2xy}{3 - 2y^2}$ **75.** $\dfrac{y(1 - 6x^2)}{1 + y}$

77. $xy'' + y' = x(-2/x^2) + (2/x) = 0$
79. Relative minimum: $\left(1, \frac{1}{2}\right)$
81. Relative minimum: $(e^{-1}, -e^{-1})$
83. Relative minimum: (e, e); Point of inflection: $(e^2, e^2/2)$
85. $P_1(x) = x - 1$; $P_2(x) = x - 1 - \frac{1}{2}(x - 1)^2$

The values of f, P_1, and P_2 and their first derivatives agree at $x = 1$.

87. $x \approx 0.567$ **89.** $(2x^2 + 1)/\sqrt{x^2 + 1}$
91. $\dfrac{3x^3 + 15x^2 - 8x}{2(x + 1)^3\sqrt{3x - 2}}$ **93.** $\dfrac{(2x^2 + 2x - 1)\sqrt{x - 1}}{(x + 1)^{3/2}}$
95. The domain of the natural logarithmic function is $(0, \infty)$, and the range is $(-\infty, \infty)$. The function is continuous, increasing, and one-to-one, and its graph is concave downward. In addition, if a and b are positive numbers and n is rational, then $\ln(1) = 0$, $\ln(a \cdot b) = \ln a + \ln b$, $\ln(a^n) = n\ln a$, and $\ln(a/b) = \ln a - \ln b$.
97. (a) Yes. If the graph of g is increasing, then $g'(x) > 0$. Because $f(x) > 0$, you know that $f'(x) = g'(x)f(x)$ and thus $f'(x) > 0$. Therefore, the graph of f is increasing.
(b) No. Let $f(x) = x^2 + 1$ (positive and concave up), and let $g(x) = \ln(x^2 + 1)$ (not concave up).
99. False. $\ln x + \ln 25 = \ln 25x$

101. False. π is a constant, so $\dfrac{d}{dx}[\ln \pi] = 0$.

103. (a) 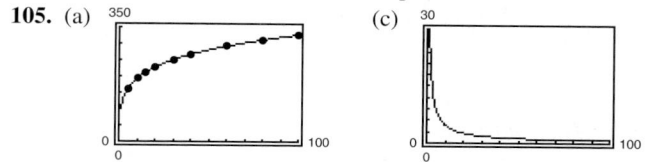   (b) 30 yr; $503,434.80
(c) 20 yr; $386,685.60

(d) When $x = 1398.43$, $dt/dx \approx -0.0805$. When $x = 1611.19$, $dt/dx \approx -0.0287$.
(e) Two benefits of a higher monthly payment are a shorter term and a lower total amount paid.

105. (a)   (c)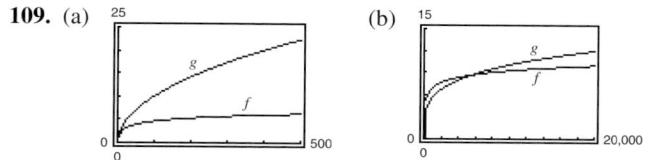

(b) $T'(10) \approx 4.75°/\text{lb/in.}^2$ $\displaystyle\lim_{p \to \infty} T'(p) = 0$
$T'(70) \approx 0.97°/\text{lb/in.}^2$ Answers will vary.

107. (a) 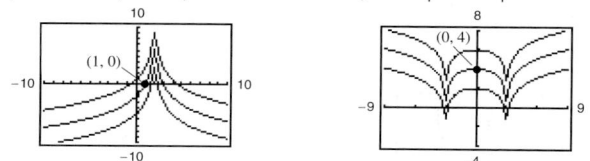   (b) When $x = 5$,
$dy/dx = -\sqrt{3}$.
When $x = 9$,
$dy/dx = -\sqrt{19}/9$.
(c) $\displaystyle\lim_{x \to 10^-} \dfrac{dy}{dx} = 0$

109. (a)   (b)

For $x > 4$, $g'(x) > f'(x)$.   For $x > 256$, $g'(x) > f'(x)$.
g is increasing at a faster g is increasing at a faster
rate than f for large values   rate than f for large values
of x.   of x.
$f(x) = \ln x$ increases very slowly for large values of x.

Section 5.2   *(page 334)*

1. $5\ln|x| + C$ **3.** $\ln|x + 1| + C$ **5.** $\frac{1}{2}\ln|2x + 5| + C$
7. $\frac{1}{2}\ln|x^2 - 3| + C$ **9.** $\ln|x^4 + 3x| + C$
11. $x^2/2 - \ln(x^4) + C$ **13.** $\frac{1}{3}\ln|x^3 + 3x^2 + 9x| + C$
15. $\frac{1}{2}x^2 - 4x + 6\ln|x + 1| + C$ **17.** $\frac{1}{3}x^3 + 5\ln|x - 3| + C$
19. $\frac{1}{3}x^3 - 2x + \ln\sqrt{x^2 + 2} + C$ **21.** $\frac{1}{3}(\ln x)^3 + C$
23. $-\frac{2}{3}\ln|1 - 3\sqrt{x}| + C$
25. $2\ln|x - 1| - 2/(x - 1) + C$
27. $\sqrt{2x} - \ln|1 + \sqrt{2x}| + C$
29. $x + 6\sqrt{x} + 18\ln|\sqrt{x} - 3| + C$ **31.** $3\ln\left|\sin\dfrac{\theta}{3}\right| + C$
33. $-\frac{1}{2}\ln|\csc 2x + \cot 2x| + C$ **35.** $\frac{1}{3}\sin 3\theta - \theta + C$
37. $\ln|1 + \sin t| + C$ **39.** $\ln|\sec x - 1| + C$
41. $y = -3\ln|2 - x| + C$ **43.** $y = \ln|x^2 - 9| + C$

45. $f(x) = -2\ln x + 3x - 2$

47. (a)

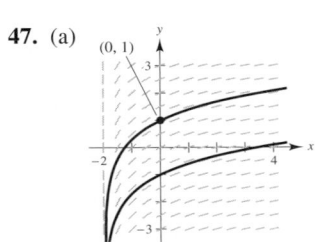

(b) $y = \ln\left(\dfrac{x+2}{2}\right) + 1$

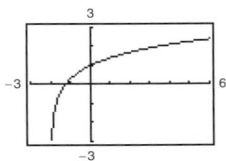

49. $\frac{5}{3}\ln 13 \approx 4.275$ **51.** $\frac{7}{3}$ **53.** $-\ln 3 \approx -1.099$

55. $\ln\left|\dfrac{2 - \sin 2}{1 - \sin 1}\right| \approx 1.929$ **57.** $2\left[\sqrt{x} - \ln\left(1 + \sqrt{x}\right)\right] + C$

59. $\ln\left(\dfrac{\sqrt{x} - 1}{\sqrt{x} + 1}\right) + 2\sqrt{x} + C$ **61.** $\ln\left(\sqrt{2} + 1\right) - \dfrac{\sqrt{2}}{2} \approx 0.174$

63. $1/x$ **65.** $1/x$ **67.** $6 \ln 3$ **69.** $\frac{1}{2}\ln 2$

71. $\frac{15}{2} + 8 \ln 2 \approx 13.045$ **73.** $(12/\pi)\ln\left(2 + \sqrt{3}\right) \approx 5.03$

75. Trapezoidal Rule: 20.2   **77.** Trapezoidal Rule: 5.3368
Simpson's Rule: 19.4667   Simpson's Rule: 5.3632

79. Power Rule   **81.** Log Rule   **83.** d   **85.** $x = 2$

87. Proof

89. $-\ln|\cos x| + C = \ln|1/\cos x| + C = \ln|\sec x| + C$

91. $\ln|\sec x + \tan x| + C = \ln\left|\dfrac{\sec^2 x - \tan^2 x}{\sec x - \tan x}\right| + C$
$\qquad\qquad\qquad\qquad = -\ln|\sec x - \tan x| + C$

93. 1   **95.** $1/(e - 1) \approx 0.582$

97. $P(t) = 1000(12 \ln|1 + 0.25t| + 1); P(3) \approx 7715$

99. About 4.15 min

101.

(a) $A = \frac{1}{2}\ln 2 - \frac{1}{4}$
(b) $0 < m < 1$
(c) $A = \frac{1}{2}(m - \ln m - 1)$

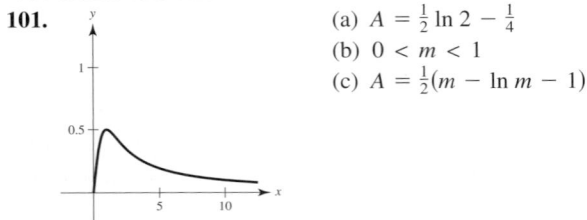

103. False. $\frac{1}{2}\ln x = \ln x^{1/2}$ **105.** True   **107.** Proof

Section 5.3 *(page 343)*

1. (a) $f(g(x)) = 5[(x - 1)/5] + 1 = x;$
$\qquad g(f(x)) = [(5x + 1) - 1]/5 = x$
(b)

3. (a) $f(g(x)) = \left(\sqrt[3]{x}\right)^3 = x; g(f(x)) = \sqrt[3]{x^3} = x$
(b)

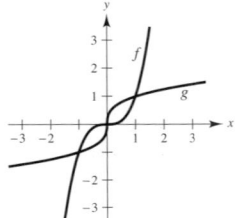

5. (a) $f(g(x)) = \sqrt{x^2 + 4} - 4 = x;$
$\qquad g(f(x)) = \left(\sqrt{x - 4}\right)^2 + 4 = x$
(b)

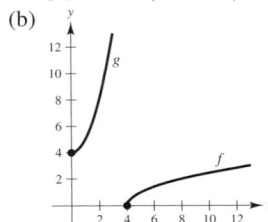

7. (a) $f(g(x)) = \dfrac{1}{1/x} = x; g(f(x)) = \dfrac{1}{1/x} = x$
(b)

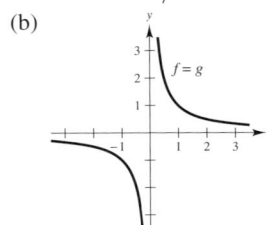

9. c   **10.** b   **11.** a   **12.** d

13.

One-to-one, inverse exists.

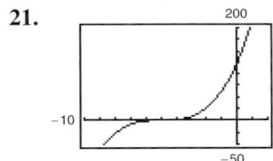

15.

Not one-to-one,
inverse does not exist.

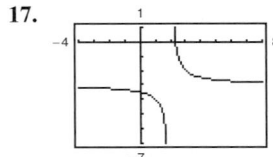

17.

One-to-one, inverse exists.

19.

One-to-one, inverse exists.

21.

One-to-one, inverse exists.

23. Inverse exists.   **25.** Inverse does not exist.

27. Inverse exists.   **29.** $f'(x) = 2(x - 4) > 0$ on $(4, \infty)$

31. $f'(x) = -8/x^3 < 0$ on $(0, \infty)$

33. $f'(x) = -\sin x < 0$ on $(0, \pi)$

35. (a) $f^{-1}(x) = (x + 3)/2$
(b)

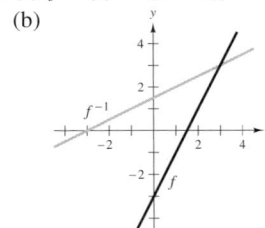

(c) f and f^{-1} are symmetric about $y = x$.
(d) Domain of f and f^{-1}: all real numbers
Range of f and f^{-1}: all real numbers

37. (a) $f^{-1}(x) = x^{1/5}$
(b)

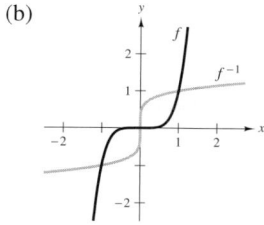

(c) f and f^{-1} are symmetric about $y = x$.
(d) Domain of f and f^{-1}: all real numbers
Range of f and f^{-1}: all real numbers

39. (a) $f^{-1}(x) = x^2, \quad x \geq 0$
(b)

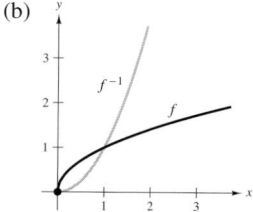

(c) f and f^{-1} are symmetric about $y = x$.
(d) Domain of f and f^{-1}: $x \geq 0$
Range of f and f^{-1}: $y \geq 0$

41. (a) $f^{-1}(x) = \sqrt{4 - x^2}, \quad 0 \leq x \leq 2$
(b)

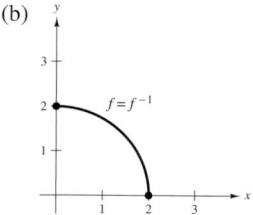

(c) f and f^{-1} are symmetric about $y = x$.
(d) Domain of f and f^{-1}: $0 \leq x \leq 2$
Range of f and f^{-1}: $0 \leq y \leq 2$

43. (a) $f^{-1}(x) = x^3 + 1$
(b)

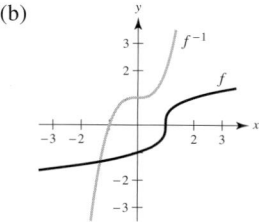

(c) f and f^{-1} are symmetric about $y = x$.
(d) Domain of f and f^{-1}: all real numbers
Range of f and f^{-1}: all real numbers

45. (a) $f^{-1}(x) = \sqrt{7}x/\sqrt{1 - x^2}, \quad -1 < x < 1$
(b)

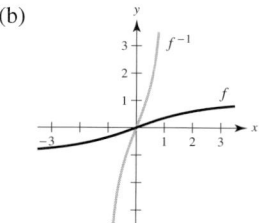

(c) f and f^{-1} are symmetric about $y = x$.
(d) Domain of f: all real numbers
Domain of f^{-1}: $-1 < x < 1$
Range of f: $-1 < y < 1$
Range of f^{-1}: all real numbers

47.

x	0	1	2	4
$f(x)$	1	2	3	4

x	1	2	3	4
$f^{-1}(x)$	0	1	2	4

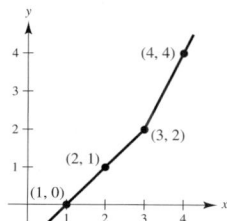

49. (a) Proof
(b) $y = \frac{20}{7}(80 - x)$
x: total cost
y: number of pounds of the less expensive commodity
(c) $[62.5, 80]$ (d) 20 lb
51. One-to-one
$f^{-1}(x) = x^2 + 2, \quad x \geq 0$
53. One-to-one
$f^{-1}(x) = 2 - x, \quad x \geq 0$
55. Sample answer: $f^{-1}(x) = \sqrt{x} + 3, \quad x \geq 0$
57. Sample answer: $f^{-1}(x) = x - 3, \quad x \geq 0$
59. Inverse exists. Volume is an increasing function, and therefore is one-to-one. The inverse function gives the time t corresponding to the volume V.
61. Inverse does not exist.   **63.** $-1/6$ **65.** $1/17$
67. $2\sqrt{3}/3$ **69.** -2
71. (a) Domain of f: $(-\infty, \infty)$ (b) Range of f: $(-\infty, \infty)$
Domain of f^{-1}: $(-\infty, \infty)$ Range of f^{-1}: $(-\infty, \infty)$
(c)

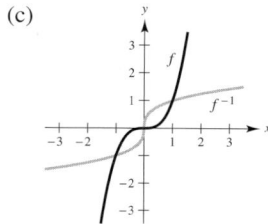

(d) $f'\left(\frac{1}{2}\right) = \frac{3}{4}, \left(f^{-1}\right)'\left(\frac{1}{8}\right) = \frac{4}{3}$

73. (a) Domain of f: $[4, \infty)$ (b) Range of f: $[0, \infty)$
Domain of f^{-1}: $[0. \infty)$ Range of f^{-1}: $[4, \infty)$
(c)

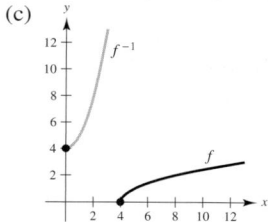

(d) $f'(5) = \frac{1}{2}, (f^{-1})'(1) = 2$
75. 32   **77.** 600   **79.** $(g^{-1} \circ f^{-1})(x) = (x + 1)/2$
81. $(f \circ g)^{-1}(x) = (x + 1)/2$
83. Let $y = f(x)$ be one-to-one. Solve for x as a function of y. Interchange x and y to get $y = f^{-1}(x)$. Let the domain of f^{-1} be the range of f. Verify that $f(f^{-1}(x)) = x$ and $f^{-1}(f(x)) = x$.
Sample answer: $f(x) = x^3; y = x^3; x = \sqrt[3]{y}; y = \sqrt[3]{x};$
$f^{-1}(x) = \sqrt[3]{x}$
85. Many x-values yield the same y-value. For example, $f(\pi) = 0 = f(0)$. The graph is not continuous at $[(2n - 1)\pi]/2$, where n is an integer.
87. $\frac{1}{4}$ **89.** False. Let $f(x) = x^2$.   **91.** True
93. (a)

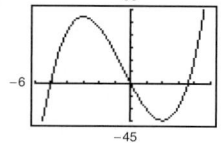

(b) $c = 2$

f does not pass the horizontal line test.
95–97. Proofs   **99.** Proof; concave upward
101. Proof; $\sqrt{5}/5$

103. (a) Proof   (b) $f^{-1}(x) = \dfrac{b - dx}{cx - a}$

(c) $a = -d$, or $b = c = 0$, $a = d$

Section 5.4   *(page 352)*

1. $x = 4$ **3.** $x \approx 2.485$ **5.** $x = 0$ **7.** $x \approx 0.511$

9. $x \approx 8.862$ **11.** $x \approx 7.389$ **13.** $x \approx 10.389$

15. $x \approx 5.389$

17.

19.

21.

23. (a)

Translation two units
to the right

(b)

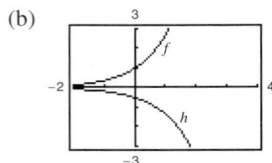

Reflection in the x-axis
and a vertical shrink

(c)

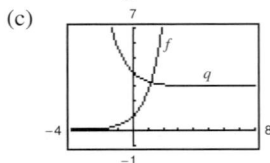

Reflection in the y-axis
and a translation three
units upward

25. c   **26.** d   **27.** a   **28.** b

29.

31.

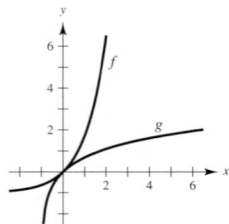

33. $2e^{2x}$ **35.** $e^{\sqrt{x}}/(2\sqrt{x})$ **37.** e^{x-4} **39.** $e^x\left(\dfrac{1}{x} + \ln x\right)$

41. $e^x(x^3 + 3x^2)$ **43.** $3(e^{-t} + e^t)^2(e^t - e^{-t})$

45. $2e^{2x}/(1 + e^{2x})$ **47.** $-2(e^x - e^{-x})/(e^x + e^{-x})^2$

49. $-2e^x/(e^x - 1)^2$ **51.** $2e^x \cos x$ **53.** $\cos(x)/x$

55. $y = 3x + 1$ **57.** $y = -x + 2$ **59.** $y = (1/e)x - 1/e$

61. $y = ex$ **63.** $\dfrac{10 - e^y}{xe^y + 3}$ **65.** $y = (-e - 1)x + 1$

67. $3(6x + 5)e^{-3x}$

69. $y'' - y = 0$

$4e^{-x} - 4e^{-x} = 0$

71. Relative minimum: $(0, 1)$

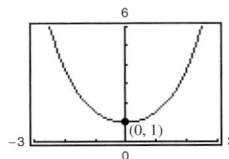

73. Relative maximum:

$\left(2, 1/\sqrt{2\pi}\right)$

Points of inflection:

$\left(1, \dfrac{e^{-0.5}}{\sqrt{2\pi}}\right), \left(3, \dfrac{e^{-0.5}}{\sqrt{2\pi}}\right)$

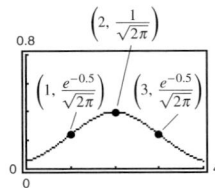

75. Relative minimum: $(0, 0)$

Relative maximum: $(2, 4e^{-2})$

Points of inflection:

$\left(2 \pm \sqrt{2}, (6 \pm 4\sqrt{2})e^{-(2 \pm \sqrt{2})}\right)$

77. Relative maximum:

$(-1, 1 + e)$

Point of inflection: $(0, 3)$

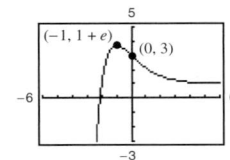

79. $A = \sqrt{2}e^{-1/2}$

81. $\left(\dfrac{1}{2}, e\right)$

83. (a)

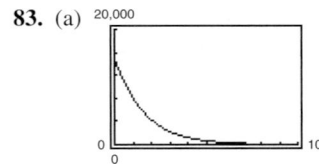

(b) When $t = 1$,

$\dfrac{dV}{dt} \approx -5028.84$.

When $t = 5$,

$\dfrac{dV}{dt} \approx -406.89$.

(c)

85. (a)

$\ln P = -0.1499h + 9.3018$

(c)

(b) $P = 10,957.7e^{-0.1499h}$

(d) $h = 5$: -776

$h = 18$: -111

87. $P_1 = 1 + x; P_2 = 1 + x + \frac{1}{2}x^2$

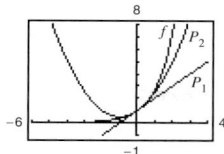

The values of f, P_1, and P_2 and their first derivatives agree at $x = 0$.

89. $12! = 479,001,600$
Stirling's Formula: $12! \approx 475,687,487$

91. $e^{5x} + C$ **93.** $\frac{1}{2}e^{2x-1} + C$ **95.** $\frac{1}{3}e^{x^3} + C$

97. $2e^{\sqrt{x}} + C$

99. $x - \ln(e^x + 1) + C_1$ or $-\ln(1 + e^{-x}) + C_2$

101. $-\frac{2}{3}(1 - e^x)^{3/2} + C$ **103.** $\ln|e^x - e^{-x}| + C$

105. $-\frac{5}{2}e^{-2x} + e^{-x} + C$ **107.** $\ln|\cos e^{-x}| + C$

109. $(e^2 - 1)/(2e^2)$ **111.** $(e - 1)/(2e)$

113. $(e/3)(e^2 - 1)$ **115.** $\ln\left(\dfrac{1 + e^6}{2}\right)$

117. $(1/\pi)[e^{\sin(\pi^2/2)} - 1]$

119. (a)

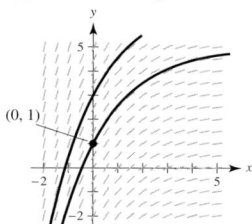

(b) $y = -4e^{-x/2} + 5$

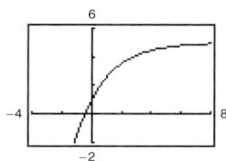

121. $[1/(2a)]e^{ax^2} + C$ **123.** $f(x) = \frac{1}{2}(e^x + e^{-x})$

125. $e^5 - 1 \approx 147.413$ **127.** $2(1 - e^{-3/2}) \approx 1.554$

      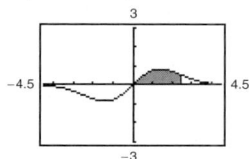

129. Midpoint Rule: 92.190; Trapezoidal Rule: 93.837;
Simpson's Rule: 92.7385

131. The probability that a given battery will last between 48 months and 60 months is approximately 47.72%.

133. $a = \ln 3$

135. $f(x) = e^x$
The domain of $f(x)$ is $(-\infty, \infty)$, and the range of $f(x)$ is $(0, \infty)$. $f(x)$ is continuous, increasing, one-to-one, and concave upward on its entire domain.
$\lim\limits_{x \to -\infty} e^x = 0$ and $\lim\limits_{x \to \infty} e^x = \infty$

137. (a) Log Rule   (b) Substitution

139. $\displaystyle\int_0^x e^t \, dt \geq \int_0^x 1 \, dt; e^x - 1 \geq x; e^x \geq x + 1$ for $x \geq 0$

141. (a) $t = \dfrac{1}{2k}\ln\dfrac{B}{A}$

(b) $x''(t) = k^2(Ae^{kt} + Be^{-kt})$, k^2 is the constant of proportionality.

143. Proof

Section 5.5   *(page 362)*

1. -3 **3.** 0 **5.** (a) $\log_2 8 = 3$ (b) $\log_3(1/3) = -1$

7. (a) $10^{-2} = 0.01$ (b) $\left(\frac{1}{2}\right)^{-3} = 8$

9.

11.

13.

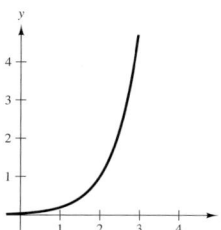

15. d   **16.** c   **17.** b   **18.** a

19. (a) $x = 3$ (b) $x = -1$ **21.** (a) $x = \frac{1}{3}$ (b) $x = \frac{1}{16}$

23. (a) $x = -1, 2$ (b) $x = \frac{1}{3}$ **25.** 1.965   **27.** -6.288

29. 12.253   **31.** 33.000   **33.** ± 11.845

35.

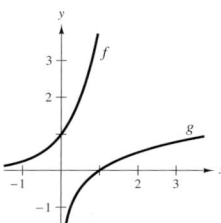

37. $(\ln 4)4^x$ **39.** $(-4 \ln 5)5^{-4x}$ **41.** $9^x(x \ln 9 + 1)$

43. $t2^t(t \ln 2 + 2)$ **45.** $-2^{-\theta}[(\ln 2) \cos \pi\theta + \pi \sin \pi\theta]$

47. $5/[(\ln 4)(5x + 1)]$ **49.** $2/[(\ln 5)(t - 4)]$

51. $x/[(\ln 5)(x^2 - 1)]$ **53.** $(x - 2)/[(\ln 2)x(x - 1)]$

55. $(3x - 2)/[(2x \ln 3)(x - 1)]$ **57.** $5(1 - \ln t)/(t^2 \ln 2)$

59. $y = -2x \ln 2 - 2 \ln 2 + 2$

61. $y = [1/(27 \ln 3)]x + 3 - 1/\ln 3$ **63.** $2(1 - \ln x)x^{(2/x) - 2}$

65. $(x - 2)^{x+1}[(x + 1)/(x - 2) + \ln(x - 2)]$

67. $y = x$ **69.** $y = \dfrac{\cos e}{e}x - \cos e + 1$

71. $3^x/\ln 3 + C$ **73.** $\dfrac{1}{3}x^3 - \dfrac{2^{-x}}{\ln 2} + C$

75. $[-1/(2 \ln 5)](5^{-x^2}) + C$ **77.** $\ln(3^{2x} + 1)/(2 \ln 3) + C$

79. $7/(2 \ln 2)$ **81.** $4/\ln 5 - 2/\ln 3$ **83.** $26/\ln 3$

85. (a) $x > 0$ (b) 10^x (c) $3 \leq f(x) \leq 4$
(d) $0 < x < 1$ (e) 10   (f) 100^n

87. (a) \$40.64   (b) $C'(1) \approx 0.051P, C'(8) \approx 0.072P$
(c) $\ln 1.05$

89.

n	1	2	4	12
A	\$1410.60	\$1414.78	\$1416.91	\$1418.34

n	365	Continuous
A	\$1419.04	\$1419.07

91.

n	1	2	4	12
A	$4321.94	$4399.79	$4440.21	$4467.74

n	365	Continuous
A	$4481.23	$4481.69

93.

t	1	10	20	30
P	$95,122.94	$60,653.07	$36,787.94	$22,313.02

t	40	50
P	$13,533.53	$8208.50

95.

t	1	10	20	30
P	$95,132.82	$60,716.10	$36,864.45	$22,382.66

t	40	50
P	$13,589.88	$8251.24

97. c

99. (a) 6.7 million ft³/acre

(b) $t = 20$: $\dfrac{dV}{dt} = 0.073$; $t = 60$: $\dfrac{dV}{dt} = 0.040$

101. (a)   (b) 10,000 fish

(c) 1 month: About 114 fish/mo

10 months: About 403 fish/mo

(d) About 15 mo

103. (a) $y_1 = -40x + 743$, $y_2 = 968 - 265.5 \ln x$,

$y_3 = 836.817(0.9169)^x$, $y_4 = 1344.8884x^{-0.5689}$

(b)

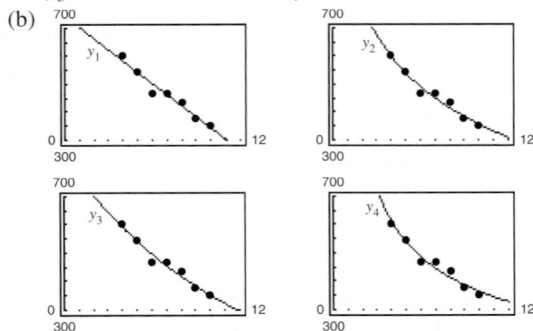

(c) The number of pancreas transplants is decreasing by about 40 transplants each year.

(d) $y_1{}'(8) = -40.04$, $y_2{}'(8) = -33.18$, $y_3{}'(8) = -36.27$, $y_4{}'(8) = -29.30$; y_1 is decreasing at the greatest rate.

105. $y = 1200(0.6^t)$ **107.** e **109.** e^2

111. False. e is an irrational number.   **113.** True   **115.** True

117. (a) $(2^3)^2 = 2^6 = 64$

$2^{(3^2)} = 2^9 = 512$

(b) No. $f(x) = (x^x)^x = x^{(x^2)}$ and $g(x) = x^{(x^x)}$

(c) $f'(x) = x^{x^2}(x + 2x \ln x)$

$g'(x) = x^{x^x + x - 1}[x(\ln x)^2 + x \ln x + 1]$

119. Proof

121. (a) $\dfrac{dy}{dx} = \dfrac{y^2 - yx \ln y}{x^2 - xy \ln x}$

(b) (i) 1 when $c \neq 0$, $c \neq e$ (ii) -3.1774

(iii) -0.3147

(c) (e, e)

123. Putnam Problem B3, 1951

Section 5.6   (page 372)

1. $\left(-\sqrt{2}/2, 3\pi/4\right)$, $(1/2, \pi/3)$, $\left(\sqrt{3}/2, \pi/6\right)$ **3.** $\pi/6$

5. $\pi/3$ **7.** $\pi/6$ **9.** $-\pi/4$ **11.** 2.50

13. $\arccos(1/1.269) \approx 0.66$ **15.** x **17.** $\sqrt{1 - x^2}/x$

19. $1/x$ **21.** (a) 3/5   (b) 5/3

23. (a) $-\sqrt{3}$ (b) $-\frac{13}{5}$ **25.** $\sqrt{1 - 4x^2}$

27. $\sqrt{x^2 - 1}/|x|$ **29.** $\sqrt{x^2 - 9}/3$ **31.** $\sqrt{x^2 + 2}/x$

33. $x = \frac{1}{3}\left[\sin\left(\frac{1}{2}\right) + \pi\right] \approx 1.207$ **35.** $x = \frac{1}{3}$

37. (a) and (b) Proofs   **39.** $2/\sqrt{2x - x^2}$

41. $-3/\sqrt{4 - x^2}$ **43.** $e^x/(1 + e^{2x})$

45. $\left(3x - \sqrt{1 - 9x^2} \arcsin 3x\right)/\left(x^2\sqrt{1 - 9x^2}\right)$

47. $-t/\sqrt{1 - t^2}$ **49.** $2 \arccos x$ **51.** $1/(1 - x^4)$

53. $\arcsin x$ **55.** $x^2/\sqrt{16 - x^2}$ **57.** $2/(1 + x^2)^2$

59. $y = \frac{1}{3}\left(4\sqrt{3}x - 2\sqrt{3} + \pi\right)$ **61.** $y = \frac{1}{4}x + (\pi - 2)/4$

63. $y = (2\pi - 4)x + 4$

65. $P_1(x) = x$; $P_2(x) = x$

67. $P_1(x) = \dfrac{\pi}{6} + \dfrac{2\sqrt{3}}{3}\left(x - \dfrac{1}{2}\right)$

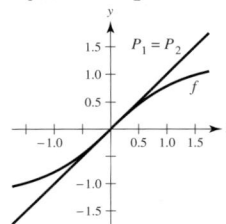

$P_2(x) = \dfrac{\pi}{6} + \dfrac{2\sqrt{3}}{3}\left(x - \dfrac{1}{2}\right) + \dfrac{2\sqrt{3}}{9}\left(x - \dfrac{1}{2}\right)^2$

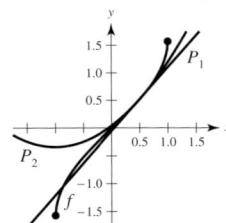

69. Relative maximum: $(1.272, -0.606)$

Relative minimum: $(-1.272, 3.747)$

71. Relative maximum: $(2, 2.214)$

73.

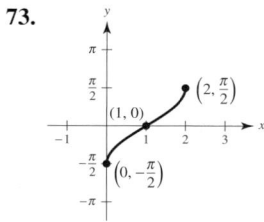

Maximum: $\left(2, \dfrac{\pi}{2}\right)$

Minimum: $\left(0, -\dfrac{\pi}{2}\right)$

Point of inflection: $(1, 0)$

75.

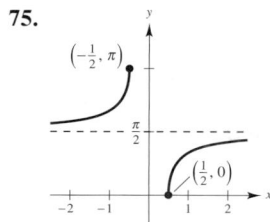

Maximum: $\left(-\dfrac{1}{2}, \pi\right)$

Minimum: $\left(\dfrac{1}{2}, 0\right)$

Asymptote: $y = \dfrac{\pi}{2}$

77. $y = -2\pi x/(\pi + 8) + 1 - \pi^2/(2\pi + 16)$
79. $y = -x + \sqrt{2}$
81. If the domains were not restricted, then the trigonometric functions would have no inverses, because they would not be one-to-one.
83. (a) $\arcsin(\arcsin 0.5) \approx 0.551$
   $\arcsin(\arcsin 1)$ does not exist.
   (b) $\sin(-1) \le x \le \sin(1)$
85. False. The range of arccos is $[0, \pi]$.
87. True   **89.** True
91. (a) $\theta = \text{arccot}(x/5)$
   (b) $x = 10$: 16 rad/h; $x = 3$: 58.824 rad/h
93. (a) $h(t) = -16t^2 + 256$; $t = 4$ sec
   (b) $t = 1$: -0.0520 rad/sec; $t = 2$: -0.1116 rad/sec
95. $50\sqrt{2} \approx 70.71$ ft   **97.** (a) and (b) Proofs
99. (a)

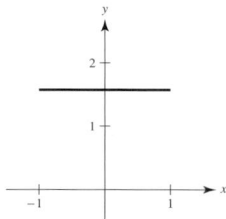

(b) The graph is a horizontal line at $\dfrac{\pi}{2}$.
(c) Proof

101. $c = 2$ **103.** Proof

Section 5.7   *(page 380)*

1. $\arcsin \dfrac{x}{3} + C$ **3.** $\text{arcsec}|2x| + C$
5. $\arcsin(x + 1) + C$ **7.** $\dfrac{1}{2}\arcsin t^2 + C$
9. $\dfrac{1}{10}\arctan \dfrac{t^2}{5} + C$ **11.** $\dfrac{1}{4}\arctan(e^{2x}/2) + C$
13. $\arcsin\left(\dfrac{\tan x}{5}\right) + C$ **15.** $2\arcsin\sqrt{x} + C$
17. $\dfrac{1}{2}\ln(x^2 + 1) - 3\arctan x + C$
19. $8\arcsin[(x - 3)/3] - \sqrt{6x - x^2} + C$ **21.** $\pi/6$
23. $\pi/6$ **25.** $\dfrac{1}{5}\arctan\dfrac{3}{5} \approx 0.108$
27. $\arctan 5 - \pi/4 \approx 0.588$ **29.** $\pi/4$ **31.** $\dfrac{1}{32}\pi^2 \approx 0.308$
33. $\pi/2$ **35.** $\ln|x^2 + 6x + 13| - 3\arctan[(x + 3)/2] + C$
37. $\arcsin[(x + 2)/2] + C$ **39.** $4 - 2\sqrt{3} + \dfrac{1}{6}\pi \approx 1.059$
41. $\dfrac{1}{2}\arctan(x^2 + 1) + C$

43. $2\sqrt{e^t - 3} - 2\sqrt{3}\arctan(\sqrt{e^t - 3}/\sqrt{3}) + C$ **45.** $\pi/6$
47. a and b   **49.** a, b, and c
51. No. This integral does not correspond to any of the basic integration rules.
53. $y = \arcsin(x/2) + \pi$
55. (a)

(b) $y = \dfrac{2}{3}\arctan\dfrac{x}{3} + 2$

57.

59.

61. $\pi/3$ **63.** $\pi/8$ **65.** $3\pi/2$
67. (a)

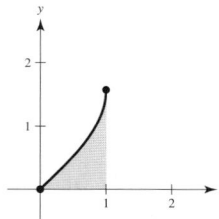

(b) 0.5708
(c) $(\pi - 2)/2$

69. (a) $F(x)$ represents the average value of $f(x)$ over the interval $[x, x + 2]$. Maximum at $x = -1$
   (b) Maximum at $x = -1$
71. False. $\displaystyle\int \dfrac{dx}{3x\sqrt{9x^2 - 16}} = \dfrac{1}{12}\text{arcsec}\dfrac{|3x|}{4} + C$
73. True   **75–77.** Proofs
79. (a) $\displaystyle\int_0^1 \dfrac{1}{1 + x^2}\,dx$ (b) About 0.7847
   (c) Because $\displaystyle\int_0^1 \dfrac{1}{1 + x^2}\,dx = \dfrac{\pi}{4}$, you can use the Trapezoidal Rule to approximate $\dfrac{\pi}{4}$. Multiplying the result by 4 gives an estimation of π.

Section 5.8   *(page 390)*

1. (a) 10.018   (b) -0.964 **3.** (a) $\dfrac{4}{3}$ (b) $\dfrac{13}{12}$
5. (a) 1.317   (b) 0.962   **7–13.** Proofs
15. $\cosh x = \sqrt{13}/2$; $\tanh x = 3\sqrt{13}/13$; $\text{csch } x = 2/3$;
   $\text{sech } x = 2\sqrt{13}/13$; $\coth x = \sqrt{13}/3$
17. ∞ **19.** 0   **21.** 1   **23.** $3\cosh 3x$
25. $-10x[\text{sech}(5x^2)\tanh(5x^2)]$ **27.** $\coth x$ **29.** $\sinh^2 x$
31. $\text{sech } t$ **33.** $y = -2x + 2$ **35.** $y = 1 - 2x$
37. Relative maxima: $(\pm\pi, \cosh \pi)$; Relative minimum: $(0, -1)$
39. Relative maximum: $(1.20, 0.66)$;
   Relative minimum: $(-1.20, -0.66)$

41. (a)

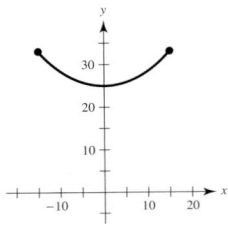

(b) 33.146 units; 25 units

(c) $m = \sinh(1) \approx 1.175$

43. $\frac{1}{2}\sinh 2x + C$ **45.** $-\frac{1}{2}\cosh(1 - 2x) + C$

47. $\frac{1}{3}\cosh^3(x - 1) + C$ **49.** $\ln|\sinh x| + C$

51. $-\coth(x^2/2) + C$ **53.** $\operatorname{csch}(1/x) + C$ **55.** $\ln(5/4)$

57. $\frac{1}{5}\ln 3$ **59.** $\pi/4$ **61.** Answers will vary.

63. $\cosh x,\ \operatorname{sech} x$ **65.** $3/\sqrt{9x^2 - 1}$ **67.** $\dfrac{1}{2\sqrt{x}(1 - x)}$

69. $|\sec x|$ **71.** $\dfrac{-2\operatorname{csch}^{-1} x}{|x|\sqrt{1 + x^2}}$ **73.** $2\sinh^{-1}(2x)$

75. $\dfrac{\sqrt{3}}{18}\ln\left|\dfrac{1 + \sqrt{3}x}{1 - \sqrt{3}x}\right| + C$ **77.** $\ln(\sqrt{e^{2x} + 1} - 1) - x + C$

79. $2\sinh^{-1}\sqrt{x} + C = 2\ln(\sqrt{x} + \sqrt{1 + x}) + C$

81. $\dfrac{1}{4}\ln\left|\dfrac{x - 4}{x}\right| + C$ **83.** $\ln\left(\dfrac{3 + \sqrt{5}}{2}\right)$ **85.** $\dfrac{\ln 7}{12}$

87. $\dfrac{1}{4}\arcsin\left(\dfrac{4x - 1}{9}\right) + C$

89. $-\dfrac{x^2}{2} - 4x - \dfrac{10}{3}\ln\left|\dfrac{x - 5}{x + 1}\right| + C$

91. $8\arctan(e^2) - 2\pi \approx 5.207$ **93.** $\frac{5}{2}\ln(\sqrt{17} + 4) \approx 5.237$

95. $\frac{52}{31}$ kg   **97.** (a) $-\sqrt{a^2 - x^2}/x$ (b) Proof

99–107. Proofs   **109.** Putnam Problem 8, 1939

Review Exercises for Chapter 5   *(page 393)*

1.

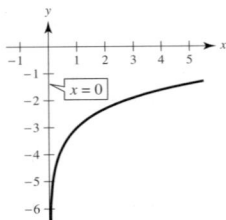

Domain: $x > 0$

3. $\frac{1}{5}[\ln(2x + 1) + \ln(2x - 1) - \ln(4x^2 + 1)]$

5. $\ln(3\sqrt[3]{4 - x^2}/x)$ **7.** $1/(2x)$ **9.** $(1 + 2\ln x)/(2\sqrt{\ln x})$

11. $-\dfrac{8x}{x^4 - 16}$ **13.** $y = -x + 1$ **15.** $\frac{1}{7}\ln|7x - 2| + C$

17. $-\ln|1 + \cos x| + C$ **19.** $3 + \ln 2$ **21.** $\ln(2 + \sqrt{3})$

23. (a) $f^{-1}(x) = 2x + 6$

(b)

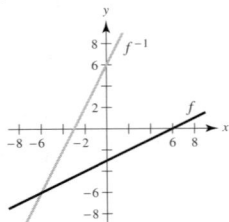

(c) Proof

(d) Domain of f and f^{-1}: all real numbers
Range of f and f^{-1}: all real numbers

25. (a) $f^{-1}(x) = x^2 - 1,\quad x \geq 0$

(b)

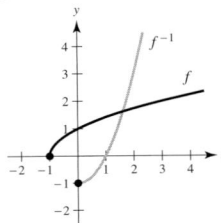

(c) Proof

(d) Domain of f: $x \geq -1$; Domain of f^{-1}: $x \geq 0$
Range of f: $y \geq 0$; Range of f^{-1}: $y \geq -1$

27. (a) $f^{-1}(x) = x^3 - 1$

(b)

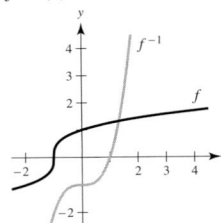

(c) Proof

(d) Domain of f and f^{-1}: all real numbers
Range of f and f^{-1}: all real numbers

29. $1/[3(\sqrt[3]{-3})^2] \approx 0.160$ **31.** $3/4$ **33.** $x \approx 1.134$

35. $e^4 - 1 \approx 53.598$ **37.** $te^t(t + 2)$

39. $(e^{2x} - e^{-2x})/\sqrt{e^{2x} + e^{-2x}}$ **41.** $x(2 - x)/e^x$

43. $y = 6x + 1$ **45.** $-y/[x(2y + \ln x)]$ **47.** $-\frac{1}{2}e^{1 - x^2} + C$

49. $(e^{4x} - 3e^{2x} - 3)/(3e^x) + C$ **51.** $(1 - e^{-3})/6 \approx 0.158$

53. $\ln(e^2 + e + 1) \approx 2.408$ **55.** About 1.729

57.

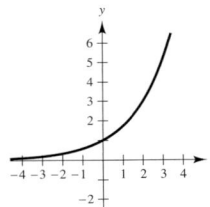

59. $3^{x-1}\ln 3$ **61.** $x^{2x+1}(2\ln x + 2 + 1/x)$

63. $-1/[\ln 3(2 - 2x)]$ **65.** $5^{(x+1)^2}/(2\ln 5) + C$

67. (a) Domain: $0 \leq h < 18{,}000$

(b)

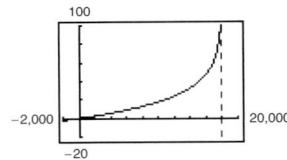

(c) $t = 0$

Vertical asymptote: $h = 18{,}000$

69. (a) $1/2$ (b) $\sqrt{3}/2$ **71.** $(1 - x^2)^{-3/2}$

73. $\dfrac{x}{|x|\sqrt{x^2 - 1}} + \operatorname{arcsec} x$ **75.** $(\arcsin x)^2$

77. $\frac{1}{2}\arctan(e^{2x}) + C$ **79.** $\frac{1}{2}\arcsin x^2 + C$

81. $\frac{1}{4}[\arctan(x/2)]^2 + C$ **83.** $\frac{2}{3}\pi + \sqrt{3} - 2 \approx 1.826$

85. $y' = -4\operatorname{sech}(4x - 1)\tanh(4x - 1)$

87. $y' = -16x\operatorname{csch}^2(8x^2)$ **89.** $y' = \dfrac{4}{\sqrt{16x^2 + 1}}$

91. $\frac{1}{3}\tanh x^3 + C$ **93.** $\ln|\tanh x| + C$

95. $\dfrac{1}{12}\ln\left|\dfrac{3 + 2x}{3 - 2x}\right| + C$

P.S. Problem Solving   *(page 395)*

1. $a = 1, b = \frac{1}{2}, c = -\frac{1}{2}$

$$f(x) = \frac{1 + x/2}{1 - x/2}$$

3. (a)   (b) 1   (c) Proof

5.

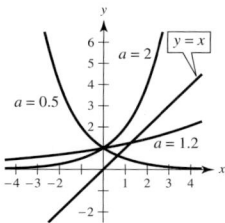

$y = 0.5^x$ and $y = 1.2^x$
intersect the line $y = x$;
$0 < a < e^{1/e}$

7. $e - 1$

9. (a) Area of region $A = \left(\sqrt{3} - \sqrt{2}\right)/2 \approx 0.1589$
    Area of region $B = \pi/12 \approx 0.2618$
    (b) $\frac{1}{24}\left[3\pi\sqrt{2} - 12\left(\sqrt{3} - \sqrt{2}\right) - 2\pi\right] \approx 0.1346$
    (c) 1.2958   (d) 0.6818

11. Proof   **13.** $2 \ln \frac{3}{2} \approx 0.8109$

15. (a) (i)

(ii)

(iii)

(b) Pattern: $y_n = 1 + \dfrac{x}{1!} + \dfrac{x^2}{2!} + \cdots + \dfrac{x^n}{n!} + \cdots$

$y_4 = 1 + \dfrac{x}{1!} + \dfrac{x^2}{2!} + \dfrac{x^3}{3!} + \dfrac{x^4}{4!}$

(c) The pattern implies that $e^x = 1 + \dfrac{x}{1!} + \dfrac{x^2}{2!} + \dfrac{x^3}{3!} + \cdots$.

Chapter 6

Section 6.1   *(page 403)*

1–11. Proofs   **13.** Not a solution   **15.** Solution
17. Solution   **19.** Solution   **21.** Not a solution
23. Solution   **25.** Not a solution   **27.** Not a solution
29. $y = 3e^{-x/2}$ **31.** $4y^2 = x^3$

33.

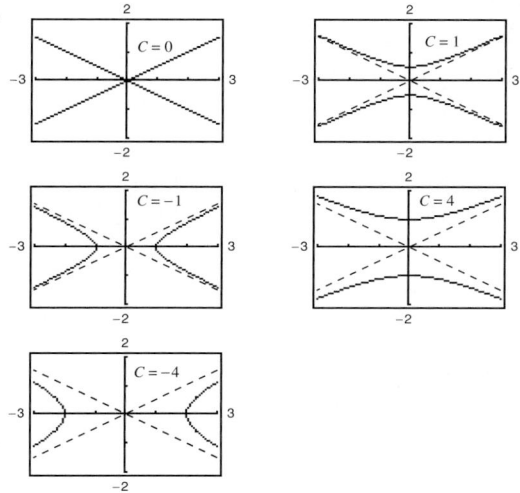

35. $y = 3e^{-2x}$ **37.** $y = 2 \sin 3x - \frac{1}{3}\cos 3x$
39. $y = -2x + \frac{1}{2}x^3$ **41.** $2x^3 + C$
43. $y = \frac{1}{2}\ln(1 + x^2) + C$ **45.** $y = x - \ln x^2 + C$
47. $y = -\frac{1}{2}\cos 2x + C$
49. $y = \frac{2}{5}(x - 6)^{5/2} + 4(x - 6)^{3/2} + C$ **51.** $y = \frac{1}{2}e^{x^2} + C$

53.

x	-4	-2	0	2	4	8
y	2	0	4	4	6	8
dy/dx	-4	Undef.	0	1	$\frac{4}{3}$	2

55.

x	-4	-2	0	2	4	8
y	2	0	4	4	6	8
dy/dx	$-2\sqrt{2}$	-2	0	0	$-2\sqrt{2}$	-8

57. b   **58.** c   **59.** d   **60.** a
61. (a) and (b)   **63.** (a) and (b)

    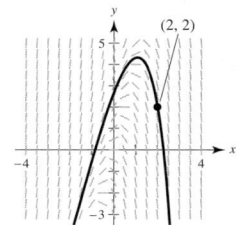

(c) As $x \to \infty$, $y \to -\infty$;   (c) As $x \to \infty$, $y \to -\infty$;
    as $x \to -\infty$, $y \to -\infty$ as $x \to -\infty$, $y \to -\infty$

65. (a)

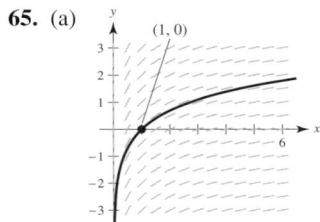

As $x \to \infty$, $y \to \infty$

(b)

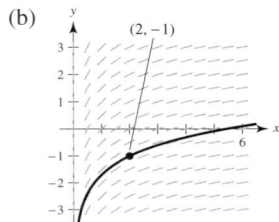

As $x \to \infty$, $y \to \infty$

67. (a) and (b)

69. (a) and (b)

71. (a) and (b)

73.

n	0	1	2	3	4	5	6
x_n	0	0.1	0.2	0.3	0.4	0.5	0.6
y_n	2	2.2	2.43	2.693	2.992	3.332	3.715

n	7	8	9	10
x_n	0.7	0.8	0.9	1.0
y_n	4.146	4.631	5.174	5.781

75.

n	0	1	2	3	4	5	6
x_n	0	0.05	0.1	0.15	0.2	0.25	0.3
y_n	3	2.7	2.438	2.209	2.010	1.839	1.693

n	7	8	9	10
x_n	0.35	0.4	0.45	0.5
y_n	1.569	1.464	1.378	1.308

77.

n	0	1	2	3	4	5	6
x_n	0	0.1	0.2	0.3	0.4	0.5	0.6
y_n	1	1.1	1.212	1.339	1.488	1.670	1.900

n	7	8	9	10
x_n	0.7	0.8	0.9	1.0
y_n	2.213	2.684	3.540	5.958

79.

x	0	0.2	0.4	0.6	0.8	1
$y(x)$ (exact)	3.0000	3.6642	4.4755	5.4664	6.6766	8.1548
$y(x)$ ($h = 0.2$)	3.0000	3.6000	4.3200	5.1840	6.2208	7.4650
$y(x)$ ($h = 0.1$)	3.0000	3.6300	4.3923	5.3147	6.4308	7.7812

81.

x	0	0.2	0.4	0.6	0.8	1
$y(x)$ (exact)	0.0000	0.2200	0.4801	0.7807	1.1231	1.5097
$y(x)$ ($h = 0.2$)	0.0000	0.2000	0.4360	0.7074	1.0140	1.3561
$y(x)$ ($h = 0.1$)	0.0000	0.2095	0.4568	0.7418	1.0649	1.4273

83. (a) $y(1) = 112.7141°$; $y(2) = 96.3770°$; $y(3) = 86.5954°$

(b) $y(1) = 113.2441°$; $y(2) = 97.0158°$; $y(3) = 87.1729°$

(c) Euler's Method: $y(1) = 112.9828°$; $y(2) = 96.6998°$;
$y(3) = 86.8863°$

Exact solution: $y(1) = 113.2441°$; $y(2) = 97.0158°$;
$y(3) = 87.1729°$

The approximations are better using $h = 0.05$.

85. The general solution is a family of curves that satisfies the differential equation. A particular solution is one member of the family that satisfies given conditions.

87. Begin with a point (x_0, y_0) that satisfies the initial condition $y(x_0) = y_0$. Then, using a small step size h, calculate the point $(x_1, y_1) = (x_0 + h, y_0 + hF(x_0, y_0))$. Continue generating the sequence of points $(x_n + h, y_n + hF(x_n, y_n))$ or (x_{n+1}, y_{n+1}).

89. False. $y = x^3$ is a solution of $xy' - 3y = 0$, but $y = x^3 + 1$ is not a solution.

91. True

93. (a)

x	0	0.2	0.4	0.6	0.8	1
y	4	2.6813	1.7973	1.2048	0.8076	0.5413
y_1	4	2.56	1.6384	1.0486	0.6711	0.4295
y_2	4	2.4	1.44	0.864	0.5184	0.3110
e_1	0	0.1213	0.1589	0.1562	0.1365	0.1118
e_2	0	0.2813	0.3573	0.3408	0.2892	0.2303
r		0.4312	0.4447	0.4583	0.4720	0.4855

(b) If h is halved, then the error is approximately halved because r is approximately 0.5.

(c) The error will again be halved.

95. (a)

(b) $\lim\limits_{t\to\infty} I(t) = 2$

97. $\omega = \pm 4$ **99.** Putnam Problem 3, Morning Session, 1954

Section 6.2   *(page 412)*

1. $y = \frac{1}{2}x^2 + 3x + C$ **3.** $y = Ce^x - 3$
5. $y^2 - 5x^2 = C$ **7.** $y = Ce^{(2x^{3/2})/3}$ **9.** $y = C(1 + x^2)$
11. $dQ/dt = k/t^2$
   $Q = -k/t + C$
13. (a)

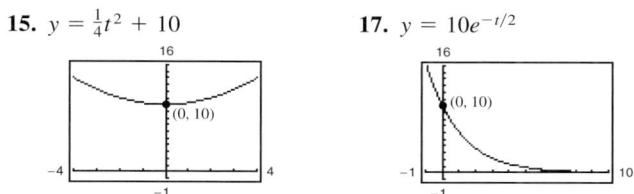

(b) $y = 6 - 6e^{-x^2/2}$

15. $y = \frac{1}{4}t^2 + 10$

17. $y = 10e^{-t/2}$

19. $\frac{8192}{4}$ **21.** $y = (1/2)e^{[(\ln 10)/5]t} \approx (1/2)e^{0.4605t}$
23. $y = 5(5/2)^{1/4}e^{[\ln(2/5)/4]t} \approx 6.2872e^{-0.2291t}$
25. C is the initial value of y, and k is the proportionality constant.
27. Quadrants I and III; dy/dx is positive when both x and y are positive (Quadrant I) or when both x and y are negative (Quadrant III).
29. Amount after 1000 yr: 12.96 g;
   Amount after 10,000 yr: 0.26 g
31. Initial quantity: 7.63 g;
   Amount after 1000 yr: 4.95 g
33. Amount after 1000 yr: 4.43 g;
   Amount after 10,000 yr: 1.49 g
35. Initial quantity: 2.16 g;
   Amount after 10,000 yr: 1.62 g
37. 95.76%
39. Time to double: 11.55 yr; Amount after 10 yr: $7288.48
41. Annual rate: 8.94%; Amount after 10 yr: $1833.67
43. Annual rate: 9.50%; Time to double: 7.30 yr
45. $224,174.18   **47.** $61,377.75
49. (a) 10.24 yr   (b) 9.93 yr   (c) 9.90 yr   (d) 9.90 yr
51. (a) $P = 2.21e^{-0.006t}$ (b) 2.08 million
   (c) Because $k < 0$, the population is decreasing.
53. (a) $P = 33.38e^{0.036t}$ (b) 47.84 million
   (c) Because $k > 0$, the population is increasing.
55. (a) $N = 100.1596(1.2455)^t$ (b) 6.3 h
57. (a) $N \approx 30(1 - e^{-0.0502t})$ (b) 36 days

59. (a) Because the population increases by a constant each month, the rate of change from month to month will always be the same. So, the slope is constant, and the model is linear.
   (b) Although the percentage increase is constant each month, the rate of growth is not constant. The rate of change of y is $dy/dt = ry$, which is an exponential model.
61. (a) $P_1 = 106e^{0.01487t} \approx 106(1.01499)^t$
   (b) $P_2 = 107.2727(1.01215)^t$
   (c)

   (d) 2029

63. (a) 20 dB   (b) 70 dB   (c) 95 dB   (d) 120 dB
65. 379.2°F
67. False. The rate of growth dy/dx is proportional to y.
69. False. The prices are rising at a rate of 6.2% per year.

Section 6.3   *(page 421)*

1. $y^2 - x^2 = C$ **3.** $15y^2 + 2x^3 = C$ **5.** $r = Ce^{0.75s}$
7. $y = C(x + 2)^3$ **9.** $y^2 = C - 8\cos x$
11. $y = -\frac{1}{4}\sqrt{1 - 4x^2} + C$ **13.** $y = Ce^{(\ln x)^2/2}$
15. $y^2 = 4e^x + 5$ **17.** $y = e^{-(x^2+2x)/2}$
19. $y^2 = 4x^2 + 3$ **21.** $u = e^{(1 - \cos v^2)/2}$ **23.** $P = P_0 e^{kt}$
25. $4y^2 - x^2 = 16$ **27.** $y = \frac{1}{3}\sqrt{x}$ **29.** $f(x) = Ce^{-x/2}$
31.

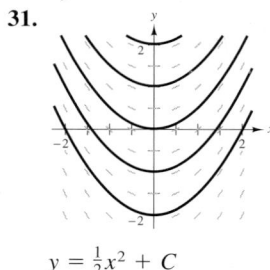

$y = \frac{1}{2}x^2 + C$

33. (a) $dy/dx = k(y - 4)$ (b) a   (c) Proof
34. (a) $dy/dx = k(x - 4)$ (b) b   (c) Proof
35. (a) $dy/dx = ky(y - 4)$ (b) c   (c) Proof
36. (a) $dy/dx = ky^2$ (b) d   (c) Proof
37. 97.9% of the original amount
39. (a) $w = 1200 - 1140e^{-kt}$
   (b) $w = 1200 - 1140e^{-0.8t}$ $w = 1200 - 1140e^{-0.9t}$

   $w = 1200 - 1140e^{-t}$

   (c) 1.31 yr; 1.16 yr; 1.05 yr   (d) 1200 lb

41. Circles: $x^2 + y^2 = C$
Lines: $y = Kx$
Graphs will vary.

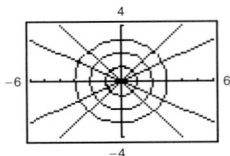

43. Parabolas: $x^2 = Cy$
Ellipses: $x^2 + 2y^2 = K$
Graphs will vary.

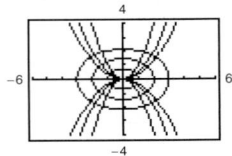

45. Curves: $y^2 = Cx^3$
Ellipses: $2x^2 + 3y^2 = K$
Graphs will vary.

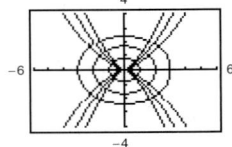

47. d   **48.** a   **49.** b   **50.** c
51. (a) 0.75   (b) 2100   (c) 70   (d) 4.49 yr
(e) $dP/dt = 0.75P(1 - P/2100)$
53. (a) 3   (b) 100
(c)   (d) 50

55. $y = 36/(1 + 8e^{-t})$ **57.** $y = 120/(1 + 14e^{-0.8t})$
59. (a) $P = \dfrac{200}{1 + 7e^{-0.2640t}}$ (b) 70 panthers   (c) 7.37 yr
(d) $dP/dt = 0.2640P(1 - P/200)$; 65.6   (e) 100 yr
61. Answers will vary.   **63.** Proof
65. (a) $v = 20(1 - e^{-1.386t})$
(b) $s \approx 20t + 14.43(e^{-1.386t} - 1)$
67. Homogeneous of degree 3   **69.** Homogeneous of degree 3
71. Not homogeneous   **73.** Homogeneous of degree 0.
75. $|x| = C(x - y)^2$ **77.** $|y^2 + 2xy - x^2| = C$
79. $y = Ce^{-x^2/(2y^2)}$
81. False. $y' = x/y$ is separable, but $y = 0$ is not a solution.
83. True

Section 6.4 *(page 428)*

1. Linear; can be written in the form $dy/dx + P(x)y = Q(x)$
3. Not linear; cannot be written in the form $dy/dx + P(x)y = Q(x)$
5. $y = 2x^2 + x + C/x$ **7.** $y = -16 + Ce^x$
9. $y = -1 + Ce^{\sin x}$ **11.** $y = (x^3 - 3x + C)/[3(x - 1)]$
13. $y = e^{x^3}(x + C)$
15. (a) Answers will vary.   (b) $y = \frac{1}{2}(e^x + e^{-x})$
(c)

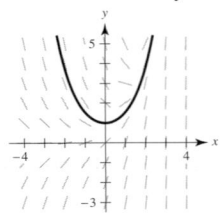

17. $y = 1 + 4/e^{\tan x}$ **19.** $y = \sin x + (x + 1)\cos x$
21. $xy = 4$ **23.** $y = -2 + x\ln|x| + 12x$
25. $P = -N/k + (N/k + P_0)e^{kt}$

27. (a) \$4,212,796.94   (b) \$31,424,909.75
29. (a) $\dfrac{dN}{dt} = k(75 - N)$ (b) $N = 75 + Ce^{-kt}$
(c) $N = 75 - 55.9296e^{-0.0168t}$
31. $v(t) = -159.47(1 - e^{-0.2007t})$; -159.47 ft/sec
33. $I = \dfrac{E_0}{R} + Ce^{-Rt/L}$ **35.** Proof
37. (a) $Q = 25e^{-t/20}$ (b) $-20\ln\left(\frac{3}{5}\right) \approx 10.2$ min   (c) 0
39. Answer (a)   **41.** $\dfrac{dy}{dx} + P(x)y = Q(x)$; $u(x) = e^{\int P(x)\,dx}$
43. c   **44.** d   **45.** a   **46.** b
47. (a)   (c)

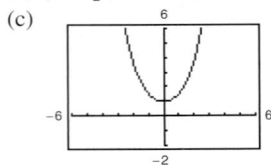

(b) $(-2, 4)$: $y = \frac{1}{2}x(x^2 - 8)$
$(2, 8)$: $y = \frac{1}{2}x(x^2 + 4)$
49. (a)

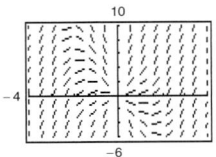

(b) $(1, 1)$: $y = (2\cos 1 + \sin 1)\csc x - 2\cot x$
$(3, -1)$: $y = (2\cos 3 - \sin 3)\csc x - 2\cot x$
(c)

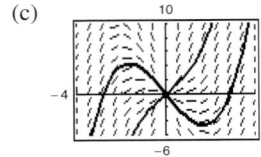

51. $2e^x + e^{-2y} = C$ **53.** $y = Ce^{-\sin x} + 1$
55. $y = [e^x(x - 1) + C]/x^2$ **57.** $y = \frac{12}{5}x^2 + C/x^3$
59. $1/y^2 = Ce^{2x^3} + \frac{1}{3}$ **61.** $y = 1/(Cx - x^2)$
63. $1/y^2 = 2x + Cx^2$ **65.** $y^{2/3} = 2e^x + Ce^{2x/3}$
67. False. $y' + xy = x^2$ is linear.

Review Exercises for Chapter 6 *(page 431)*

1. Yes   **3.** $y = \frac{4}{3}x^3 + 7x + C$ **5.** $y = \frac{1}{2}\sin 2x + C$
7. $y = -e^{2-x} + C$
9.

x	-4	-2	0	2	4	8
y	2	0	4	4	6	8
dy/dx	-10	-4	-4	0	2	8

11. (a) and (b)

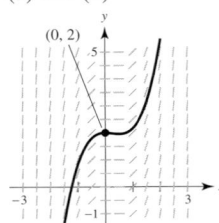

13.

n	0	1	2	3	4	5	6
x_n	0	0.05	0.1	0.15	0.2	0.25	0.3
y_n	4	3.8	3.6125	3.4369	3.2726	3.1190	2.9756

n	7	8	9	10
x_n	0.35	0.4	0.45	0.5
y_n	2.8418	2.7172	2.6038	2.4986

15. $y = -\frac{5}{3}x^3 + x^2 + C$

17. $y = -3 - 1/(x + C)$ **19.** $y = Ce^x/(2 + x)^2$

21. $\dfrac{dy}{dt} = \dfrac{k}{t^3}; \quad y = -\dfrac{k}{2t^2} + C$ **23.** $y \approx \frac{3}{4}e^{0.379t}$

25. $y = \dfrac{9}{20} e^{(1/2)\ln(10/3)t}$ **27.** About 7.79 in.

29. About 37.5 yr

31. (a) $S \approx 30e^{-1.7918/t}$ (b) 20,965 units

(c)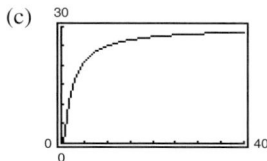

33. $y^2 = 5x^2 + C$ **35.** $y = Ce^{8x^2}$

37. $y^4 = 6x^2 - 8$ **39.** $y^4 = 2x^4 + 1$

41.

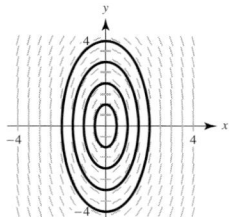

Graphs will vary.
$4x^2 + y^2 = C$

43. (a) 0.55   (b) 5250   (c) 150   (d) 6.41 yr

(e) $\dfrac{dP}{dt} = 0.55P\left(1 - \dfrac{P}{5250}\right)$

45. $y = \dfrac{80}{1 + 9e^{-t}}$

47. (a) $P(t) = \dfrac{20{,}400}{1 + 16e^{-0.553t}}$ (b) 17,118 trout   (c) 4.94 yr

49. $y = -10 + Ce^x$ **51.** $y = e^{x/4}\left(\frac{1}{4}x + C\right)$

53. $y = (x + C)/(x - 2)$ **55.** $y = \frac{1}{10}e^{5x} + \frac{29}{10}e^{-5x}$

P.S. Problem Solving   *(page 433)*

1. (a) $y = 1/(1 - 0.01t)^{100}; \quad T = 100$

(b) $y = 1/\left[\left(\dfrac{1}{y_0}\right)^\varepsilon - k\varepsilon t\right]^{1/\varepsilon}; \quad$ Explanations will vary.

3. (a) $y = Le^{-Ce^{-kt}}$

(b)

(c) As $t \to \infty$, $y \to L$, the carrying capacity.

(d) $y_0 = 500 = 5000e^{-C} \Rightarrow e^C = 10 \Rightarrow C = \ln 10$

The graph is concave upward on $(0, 41.7)$ and downward on $(41.7, \infty)$.

5. 1481.45 sec $\approx$ 24 min, 41 sec

7. 2575.95 sec $\approx$ 42 min, 56 sec

9. (a) $s = 184.21 - Ce^{-0.019t}$

(b)   (c) As $t \to \infty$, $Ce^{-0.019t} \to 0$, and $s \to 184.21$.

11. (a) $C = 0.6e^{-0.25t}$ (b) $C = 0.6e^{-0.75t}$

    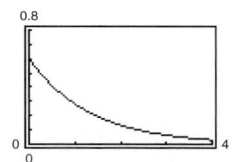

Chapter 7

Section 7.1   *(page 442)*

1. $-\displaystyle\int_0^6 (x^2 - 6x)\, dx$ **3.** $\displaystyle\int_0^3 (-2x^2 + 6x)\, dx$

5. $-6\displaystyle\int_0^1 (x^3 - x)\, dx$

7.   **9.**

11.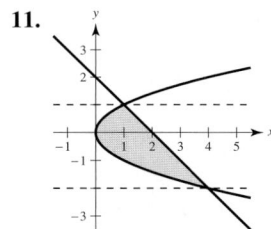

13. d

15. (a) $\frac{125}{6}$ (b) $\frac{125}{6}$

(c) Integrating with respect to y; Answers will vary.

17.

$\frac{32}{3}$

19.

$\frac{9}{2}$

21.

1

23.

$\frac{4}{3}$

25.

$\frac{9}{2}$

27.

6

29.

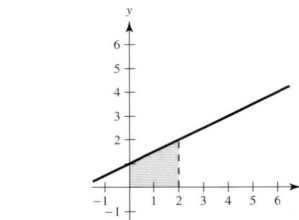

$10 \ln 5 \approx 16.094$

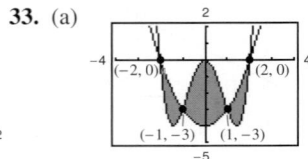

31. (a)

(b) $\frac{37}{12}$

33. (a)

(b) 8

35. (a)

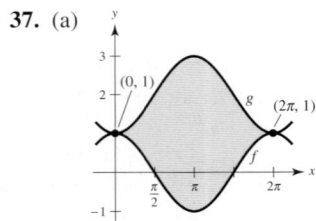

(b) $\pi/2 - 1/3 \approx 1.237$

37. (a)

$4\pi \approx 12.566$

39.

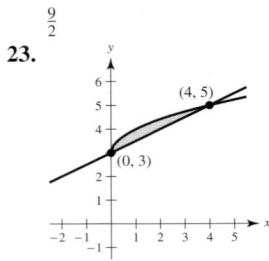

$2(1 - \ln 2) \approx 0.614$

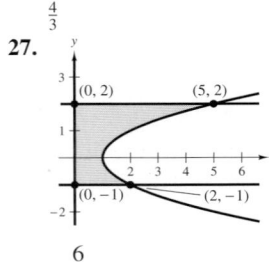

41.

$(1/2)(1 - 1/e) \approx 0.316$

43. (a)

(b) 4

45. (a)

(b) About 1.323

47. (a)

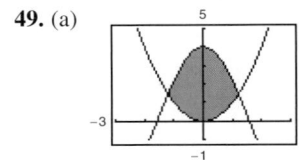

(b) The function is difficult to integrate.

(c) About 4.7721

49. (a)

(b) The intersections are difficult to find.

(c) About 6.3043

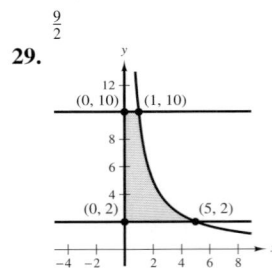

51. $F(x) = \frac{1}{4}x^2 + x$

(a) $F(0) = 0$

(b) $F(2) = 3$

(c) $F(6) = 15$

53. $F(\alpha) = (2/\pi)[\sin(\pi\alpha/2) + 1]$

(a) $F(-1) = 0$

(b) $F(0) = 2/\pi \approx 0.6366$

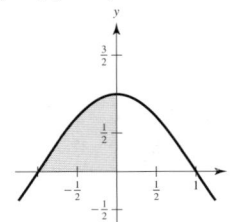

(c) $F(1/2) = (\sqrt{2} + 2)/\pi \approx 1.0868$

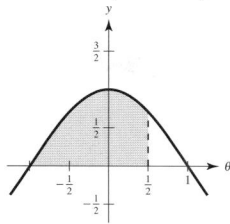

55. 14   **57.** 16

59. Answers will vary. Sample answers:
(a) About 966 ft^2 (b) About 1004 ft^2

61. $\displaystyle\int_{-2}^{1} [x^3 - (3x - 2)]\, dx = \frac{27}{4}$

63. $\displaystyle\int_{0}^{1} \left[\frac{1}{x^2 + 1} - \left(-\frac{1}{2}x + 1 \right) \right] dx \approx 0.0354$

65. Answers will vary.
Example: $x^4 - 2x^2 + 1 \le 1 - x^2$ on $[-1, 1]$
$$\int_{-1}^{1} [(1 - x^2) - (x^4 - 2x^2 + 1)]\, dx = \frac{4}{15}$$

67. (a) The integral $\int_0^5 [v_1(t) - v_2(t)]\, dt = 10$ means that the first car traveled 10 more meters than the second car between 0 and 5 seconds.
The integral $\int_0^{10} [v_1(t) - v_2(t)]\, dt = 30$ means that the first car traveled 30 more meters than the second car between 0 and 10 seconds.
The integral $\int_{20}^{30} [v_1(t) - v_2(t)]\, dt = -5$ means that the second car traveled 5 more meters than the first car between 20 and 30 seconds.
(b) No. You do not know when both cars started or the initial distance between the cars.
(c) The car with velocity v_1 is ahead by 30 meters.
(d) Car 1 is ahead by 8 meters.

69. $b = 9\left(1 - 1/\sqrt[3]{4}\right) \approx 3.330$ **71.** $a = 4 - 2\sqrt{2} \approx 1.172$

73. Answers will vary. Sample answer: $\frac{1}{6}$

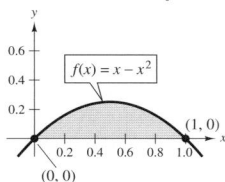

75. R_1; \$11.375 billion

77. (a) $y = 0.0124x^2 - 0.385x + 7.85$
(b)

(c)

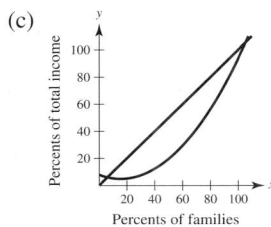

(d) About 2006.7

79. (a) About 6.031 m^2 (b) About 12.062 m^3 (c) 60,310 lb

81. $\sqrt{3}/2 + 7\pi/24 + 1 \approx 2.7823$ **83.** True

Answers to Odd-Numbered Exercises   A57

85. False. Let $f(x) = x$ and $g(x) = 2x - x^2$. f and g intersect at $(1, 1)$, the midpoint of $[0, 2]$, but
$$\int_a^b [f(x) - g(x)]\, dx = \int_0^2 [x - (2x - x^2)]\, dx = \frac{2}{3} \ne 0.$$

87. Putnam Problem A1, 1993

Section 7.2   *(page 453)*

1. $\pi \displaystyle\int_0^1 (-x + 1)^2\, dx = \frac{\pi}{3}$ **3.** $\pi \displaystyle\int_1^4 (\sqrt{x})^2\, dx = \frac{15\pi}{2}$

5. $\pi \displaystyle\int_0^1 [(x^2)^2 - (x^5)^2]\, dx = \frac{6\pi}{55}$ **7.** $\pi \displaystyle\int_0^4 (\sqrt{y})^2\, dy = 8\pi$

9. $\pi \displaystyle\int_0^1 (y^{3/2})^2\, dy = \frac{\pi}{4}$

11. (a) $9\pi/2$ (b) $(36\pi\sqrt{3})/5$ (c) $(24\pi\sqrt{3})/5$
(d) $(84\pi\sqrt{3})/5$

13. (a) $32\pi/3$ (b) $64\pi/3$ **15.** 18π

17. $\pi\left(48 \ln 2 - \frac{27}{4}\right) \approx 83.318$ **19.** $124\pi/3$

21. $832\pi/15$ **23.** $\pi \ln 5$ **25.** $2\pi/3$

27. $(\pi/2)(1 - 1/e^2) \approx 1.358$ **29.** $277\pi/3$ **31.** 8π

33. $\pi^2/2 \approx 4.935$ **35.** $(\pi/2)(e^2 - 1) \approx 10.036$

37. 1.969   **39.** 15.4115   **41.** $\pi/3$ **43.** $2\pi/15$

45. $\pi/2$ **47.** $\pi/6$

49. A sine curve on $[0, \pi/2]$ revolved about the x-axis

51. The parabola $y = 4x - x^2$ is a horizontal translation of the parabola $y = 4 - x^2$. Therefore, their volumes are equal.

53. (a) This statement is true. Explanations will vary.
(b) This statement is false. Explanations will vary.

55. $2\sqrt{2}$ **57.** $V = \frac{4}{3}\pi(R^2 - r^2)^{3/2}$ **59.** Proof

61. $\pi r^2 h[1 - (h/H) + h^2/(3H^2)]$

63.

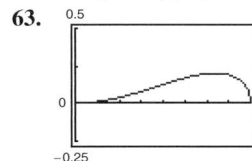

$\pi/30$

65. (a) 60π (b) 50π

67. (a) $V = \pi\left(4b^2 - \frac{64}{3}b + \frac{512}{15}\right)$

(b)

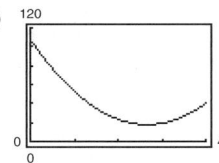

(c) $b = \frac{8}{3} \approx 2.67$

$b \approx 2.67$

69. (a) ii; right circular cylinder of radius r and height h
(b) iv; ellipsoid whose underlying ellipse has the equation $(x/b)^2 + (y/a)^2 = 1$
(c) iii, sphere of radius r
(d) i; right circular cone of radius r and height h
(e) v; torus of cross-sectional radius r and other radius R

71. (a) $\frac{81}{10}$ (b) $\frac{9}{2}$ **73.** $\frac{16}{3}r^3$

75. (a) $\frac{2}{3}r^3$ (b) $\frac{2}{3}r^3 \tan\theta$; As $\theta \to 90°$, $V \to \infty$.

Section 7.3   (page 462)

1. $2\pi \int_0^2 x^2 \, dx = \dfrac{16\pi}{3}$ **3.** $2\pi \int_0^4 x\sqrt{x} \, dx = \dfrac{128\pi}{5}$

5. $2\pi \int_0^4 \dfrac{1}{4}x^3 \, dx = 32\pi$ **7.** $2\pi \int_0^2 x(4x - 2x^2) \, dx = \dfrac{16\pi}{3}$

9. $2\pi \int_0^2 x(x^2 - 4x + 4) \, dx = \dfrac{8\pi}{3}$

11. $2\pi \int_2^4 x\sqrt{x - 2} \, dx = \dfrac{128\pi}{15}\sqrt{2}$

13. $2\pi \int_0^1 x\left(\dfrac{1}{\sqrt{2\pi}}e^{-x^2/2}\right) dx = \sqrt{2\pi}\left(1 - \dfrac{1}{\sqrt{e}}\right) \approx 0.986$

15. $2\pi \int_0^2 y(2 - y) \, dy = \dfrac{8\pi}{3}$

17. $2\pi \left[\displaystyle\int_0^{1/2} y \, dy + \int_{1/2}^1 y\left(\dfrac{1}{y} - 1\right) dy\right] = \dfrac{\pi}{2}$

19. $2\pi \int_0^8 y^{4/3} \, dy = \dfrac{768\pi}{7}$

21. $2\pi \int_0^2 y(4 - 2y) \, dy = 16\pi/3$ **23.** 8π **25.** 16π

27. Shell method; it is much easier to put x in terms of y rather than vice versa.

29. (a) $128\pi/7$ (b) $64\pi/5$ (c) $96\pi/5$

31. (a) $\pi a^3/15$ (b) $\pi a^3/15$ (c) $4\pi a^3/15$

33. (a)   (b) 1.506

y = (1 − x^{4/3})^{3/4}

35. (a)   (b) 187.25

y = ∛(x − 2)²(x − 6)²

37. (a) The rectangles would be vertical.
(b) The rectangles would be horizontal.

39. Both integrals yield the volume of the solid generated by revolving the region bounded by the graphs of $y = \sqrt{x - 1}$, $y = 0$, and $x = 5$ about the x-axis.

41. a, c, b

43. (a) Region bounded by $y = x^2$, $y = 0$, $x = 0$, $x = 2$
(b) Revolved about the y-axis

45. (a) Region bounded by $x = \sqrt{6 - y}$, $y = 0$, $x = 0$
(b) Revolved about $y = -2$

47. Diameter $= 2\sqrt{4 - 2\sqrt{3}} \approx 1.464$ **49.** $4\pi^2$

51. (a) Proof   (b) (i) $V = 2\pi$ (ii) $V = 6\pi^2$ **53.** Proof

55. (a) $R_1(n) = n/(n + 1)$ (b) $\displaystyle\lim_{n \to \infty} R_1(n) = 1$
(c) $V = \pi a b^{n+2}[n/(n + 2)]$; $R_2(n) = n/(n + 2)$
(d) $\displaystyle\lim_{n \to \infty} R_2(n) = 1$
(e) As $n \to \infty$, the graph approaches the line $x = b$.

57. (a) and (b) About 121,475 ft³   **59.** $c = 2$

61. (a) $64\pi/3$ (b) $2048\pi/35$ (c) $8192\pi/105$

Section 7.4   (page 473)

1. (a) and (b) 17   **3.** $\dfrac{5}{3}$ **5.** $\dfrac{2}{3}(2\sqrt{2} - 1) \approx 1.219$

7. $5\sqrt{5} - 2\sqrt{2} \approx 8.352$ **9.** 309.3195

11. $\ln\left[(\sqrt{2} + 1)/(\sqrt{2} - 1)\right] \approx 1.763$

13. $\dfrac{1}{2}(e^2 - 1/e^2) \approx 3.627$ **15.** $\dfrac{76}{3}$

17. (a)   **19.** (a)

(b) $\displaystyle\int_0^2 \sqrt{1 + 4x^2} \, dx$ (b) $\displaystyle\int_1^3 \sqrt{1 + \dfrac{1}{x^4}} \, dx$

(c) About 4.647   (c) About 2.147

21. (a)   **23.** (a)

(b) $\displaystyle\int_0^\pi \sqrt{1 + \cos^2 x} \, dx$ (b) $\displaystyle\int_0^2 \sqrt{1 + e^{-2y}} \, dy$

(c) About 3.820 $= \displaystyle\int_{e^{-2}}^1 \sqrt{1 + \dfrac{1}{x^2}} \, dx$

(c) About 2.221

25. (a)   (b) $\displaystyle\int_0^1 \sqrt{1 + \left(\dfrac{2}{1 + x^2}\right)^2} \, dx$

(c) About 1.871

27. b

29. (a) 64.125   (b) 64.525   (c) 64.666   (d) 64.672

31. $20[\sinh 1 - \sinh(-1)] \approx 47.0$ m   **33.** About 1480

35. $3 \arcsin \dfrac{2}{3} \approx 2.1892$

37. $2\pi \displaystyle\int_0^3 \dfrac{1}{3}x^3\sqrt{1 + x^4} \, dx = \dfrac{\pi}{9}(82\sqrt{82} - 1) \approx 258.85$

39. $2\pi \displaystyle\int_1^2 \left(\dfrac{x^3}{6} + \dfrac{1}{2x}\right)\left(\dfrac{x^2}{2} + \dfrac{1}{2x^2}\right) dx = \dfrac{47\pi}{16} \approx 9.23$

41. $2\pi \displaystyle\int_{-1}^1 2 \, dx = 8\pi \approx 25.13$

43. $2\pi \displaystyle\int_1^8 x\sqrt{1 + \dfrac{1}{9x^{4/3}}} \, dx = \dfrac{\pi}{27}(145\sqrt{145} - 10\sqrt{10}) \approx 199.48$

45. $2\pi \displaystyle\int_0^2 x\sqrt{1 + \dfrac{x^2}{4}} \, dx = \dfrac{\pi}{3}(16\sqrt{2} - 8) \approx 15.318$

47. 14.424

49. A rectifiable curve is a curve with a finite arc length.

51. The integral formula for the area of a surface of revolution is derived from the formula for the lateral surface area of the frustum of a right circular cone. The formula is $S = 2\pi r L$, where $r = \frac{1}{2}(r_1 + r_2)$, which is the average radius of the frustum, and L is the length of a line segment on the frustum. The representative element is $2\pi f(d_i)\sqrt{1 + (\Delta y_i/\Delta x_i)^2}\,\Delta x_i$.

53. (a)

(b) y_1, y_2, y_3, y_4

(c) $s_1 \approx 5.657$; $s_2 \approx 5.759$; $s_3 \approx 5.916$; $s_4 \approx 6.063$

55. 20π **57.** $6\pi(3 - \sqrt{5}) \approx 14.40$

59. (a) Answers will vary. Sample answer: 5207.62 in.^{3}

(b) Answers will vary. Sample answer: 1168.64 in.^{2}

(c) $r = 0.0040y^3 - 0.142y^2 + 1.23y + 7.9$

(d) 5279.64 in.^{3}; 1179.5 in.^{2}

61. (a) $\pi(1 - 1/b)$ (b) $2\pi \int_1^b \sqrt{x^4 + 1}/x^3\,dx$

(c) $\displaystyle\lim_{b\to\infty} V = \lim_{b\to\infty} \pi(1 - 1/b) = \pi$

(d) Because $\dfrac{\sqrt{x^4 + 1}}{x^3} > \dfrac{\sqrt{x^4}}{x^3} = \dfrac{1}{x} > 0$ on $[1, b]$,

you have $\displaystyle\int_1^b \frac{\sqrt{x^4 + 1}}{x^3}\,dx > \int_1^b \frac{1}{x}\,dx = \Big[\ln x\Big]_1^b = \ln b$

and $\displaystyle\lim_{b\to\infty} \ln b \to \infty$. So, $\displaystyle\lim_{b\to\infty} 2\pi \int_1^b \frac{\sqrt{x^4 + 1}}{x^3}\,dx = \infty$.

63. Fleeing object: $\frac{2}{3}$ unit

Pursuer: $\dfrac{1}{2}\displaystyle\int_0^1 \frac{x + 1}{\sqrt{x}}\,dx = \frac{4}{3} = 2\Big(\frac{2}{3}\Big)$

65. $384\pi/5$ **67–69.** Proofs

Section 7.5 *(page 483)*

1. 48,000 ft-lb   **3.** 896 N-m   **5.** 40.833 in.-lb $\approx$ 3.403 ft-lb

7. 160 in.-lb $\approx$ 13.3 ft-lb   **9.** 37.125 ft-lb

11. (a) 487.805 mile-tons $\approx 5.151 \times 10^9$ ft-lb

(b) 1395.349 mile-tons $\approx 1.473 \times 10^{10}$ ft-lb

13. (a) 2.93×10^4 mile-tons $\approx 3.10 \times 10^{11}$ ft-lb

(b) 3.38×10^4 mile-tons $\approx 3.57 \times 10^{11}$ ft-lb

15. (a) 2496 ft-lb   (b) 9984 ft-lb   **17.** $470{,}400\pi$ N-m

19. 2995.2π ft-lb   **21.** $20{,}217.6\pi$ ft-lb   **23.** 2457π ft-lb

25. 600 ft-lb   **27.** 450 ft-lb   **29.** 168.75 ft-lb

31. If an object is moved a distance D in the direction of an applied constant force F, then the work W done by the force is defined as $W = FD$.

33. The situation in part (a) requires more work. There is no work required for part (b) because the distance is 0.

35. (a) 54 ft-lb   (b) 160 ft-lb   (c) 9 ft-lb   (d) 18 ft-lb

37. $2000 \ln(3/2) \approx 810.93$ ft-lb   **39.** 3249.4 ft-lb

41. 10,330.3 ft-lb

43. (a) $16{,}000\pi$ ft-lb   (b) 24,888.889 ft-lb

(c) $F(x) = -16{,}261.36x^4 + 85{,}295.45x^3 - 157{,}738.64x^2 + 104{,}386.36x - 32.4675$

(d) 0.524 ft   (e) 25,180.5 ft-lb

Section 7.6 *(page 494)*

1. $\bar{x} = -\frac{4}{3}$ **3.** $\bar{x} = 4$ **5.** (a) $\bar{x} = 8$ (b) $\bar{x} = -\frac{3}{4}$

7. $x = 6$ ft   **9.** $(\bar{x}, \bar{y}) = \big(\frac{10}{9}, -\frac{1}{9}\big)$ **11.** $(\bar{x}, \bar{y}) = \big(2, \frac{48}{25}\big)$

13. $M_x = \rho/3, M_y = 4\rho/3,\quad (\bar{x}, \bar{y}) = (4/3, 1/3)$

15. $M_x = 4\rho, M_y = 64\rho/5,\quad (\bar{x}, \bar{y}) = (12/5, 3/4)$

17. $M_x = \rho/35, M_y = \rho/20,\quad (\bar{x}, \bar{y}) = (3/5, 12/35)$

19. $M_x = 99\rho/5, M_y = 27\rho/4,\quad (\bar{x}, \bar{y}) = (3/2, 22/5)$

21. $M_x = 192\rho/7, M_y = 96\rho,\quad (\bar{x}, \bar{y}) = (5, 10/7)$

23. $M_x = 0, M_y = 256\rho/15,\quad (\bar{x}, \bar{y}) = (8/5, 0)$

25. $M_x = 27\rho/4, M_y = -27\rho/10,\quad (\bar{x}, \bar{y}) = (-3/5, 3/2)$

27.

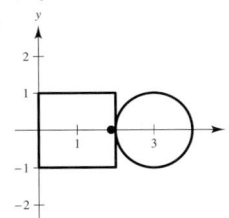

$(\bar{x}, \bar{y}) = (3.0, 126.0)$

29.

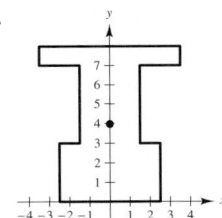

$(\bar{x}, \bar{y}) = (0, 16.2)$

31.

$(\bar{x}, \bar{y}) = \Big(\dfrac{4 + 3\pi}{4 + \pi}, 0\Big)$

33.

$(\bar{x}, \bar{y}) = \Big(0, \dfrac{135}{34}\Big)$

35. $(\bar{x}, \bar{y}) = \Big(\dfrac{2 + 3\pi}{2 + \pi}, 0\Big)$ **37.** $160\pi^2 \approx 1579.14$

39. $128\pi/3 \approx 134.04$

41. The center of mass $(\bar{x}, \bar{y})$ is $\bar{x} = M_y/m$ and $\bar{y} = M_x/m$, where:

1. $m = m_1 + m_2 + \cdots + m_n$ is the total mass of the system.

2. $M_y = m_1 x_1 + m_2 x_2 + \cdots + m_n x_n$ is the moment about the y-axis.

3. $M_x = m_1 y_1 + m_2 y_2 + \cdots + m_n y_n$ is the moment about the x-axis.

43. See Theorem 7.1 on page 493.   **45.** $(\bar{x}, \bar{y}) = \Big(\dfrac{b}{3}, \dfrac{c}{3}\Big)$

47. $(\bar{x}, \bar{y}) = \Big(\dfrac{(a + 2b)c}{3(a + b)}, \dfrac{a^2 + ab + b^2}{3(a + b)}\Big)$

49. $(\bar{x}, \bar{y}) = (0, 4b/(3\pi))$

51. (a)

(b) $\bar{x} = 0$ by symmetry

(c) $M_y = \displaystyle\int_{-\sqrt{b}}^{\sqrt{b}} x(b - x^2)\,dx = 0$ because $x(b - x^2)$ is an odd function.

(d) $\bar{y} > b/2$ because the area is greater for $y > b/2$.

(e) $\bar{y} = (3/5)b$

53. (a) $(\bar{x}, \bar{y}) = (0, 12.98)$

(b) $y = (-1.02 \times 10^{-5})x^4 - 0.0019x^2 + 29.28$

(c) $(\bar{x}, \bar{y}) = (0, 12.85)$

55. $(\bar{x}, \bar{y}) = (0, 2r/\pi)$

57. $(\bar{x}, \bar{y}) = \left(\dfrac{n+1}{n+2}, \dfrac{n+1}{4n+2}\right)$; As $n \to \infty$, the region shrinks toward the line segments $y = 0$ for $0 \le x \le 1$ and $x = 1$ for $0 \le y \le 1$; $(\bar{x}, \bar{y}) \to \left(1, \dfrac{1}{4}\right)$.

Section 7.7   (page 501)

1. 1497.6 lb   **3.** 4992 lb   **5.** 748.8 lb   **7.** 1123.2 lb

9. 748.8 lb   **11.** 1064.96 lb   **13.** 117,600 N

15. 2,381,400 N   **17.** 2814 lb   **19.** 6753.6 lb

21. 94.5 lb   **23–25.** Proofs   **27.** 960 lb

29. Answers will vary. Sample answer (using Simpson's Rule): 3010.8 lb

31. $3\sqrt{2}/2 \approx 2.12$ ft; The pressure increases with increasing depth.

33. Because you are measuring total force against a region between two depths

Review Exercises for Chapter 7   (page 503)

1.

64/3

3.

$\pi/2$

5.

$\dfrac{1}{2}$

7.

$e^2 + 1$

9.

$2\sqrt{2}$

11.

$\dfrac{512}{3}$

13.

$\dfrac{1}{6}$

15. (a) 9920 ft^2 (b) $10{,}413\frac{1}{3}$ ft^2

17. (a) 9π (b) 18π (c) 9π (d) 36π **19.** $\pi^2/4$

21. $2\pi \ln 2.5 \approx 5.757$ **23.** 1.958 ft

25. $\frac{8}{15}\left(1 + 6\sqrt{3}\right) \approx 6.076$ **27.** 4018.2 ft   **29.** 15π

31. 62.5 in.-lb $\approx$ 5.208 ft-lb

33. $122{,}980\pi$ ft-lb $\approx$ 193.2 foot-tons   **35.** 200 ft-lb

37. $a = 15/4$ **39.** 3.6   **41.** $(\bar{x}, \bar{y}) = \left(1, \frac{17}{5}\right)$

43. $(\bar{x}, \bar{y}) = \left(\dfrac{2(9\pi + 49)}{3(\pi + 9)}, 0\right)$ **45.** 3072 lb

47. Wall at shallow end: 15,600 lb
Wall at deep end: 62,400 lb
Side wall: 72,800 lb

P.S. Problem Solving   (page 505)

1. 3   **3.** $y = 0.2063x$

5.

$\dfrac{5\sqrt{2}\pi}{3}$

7. $V = 2\pi\left[d + \frac{1}{2}\sqrt{w^2 + l^2}\right]lw$

9. $f(x) = 2e^{x/2} - 2$ **11.** 89.3%

13.

(a) $(\bar{x}, \bar{y}) = \left(\dfrac{63}{43}, 0\right)$

(b) $(\bar{x}, \bar{y}) = \left(\dfrac{3b(b + 1)}{2(b^2 + b + 1)}, 0\right)$

(c) $\left(\dfrac{3}{2}, 0\right)$

15. Consumer surplus: 1600; Producer surplus: 400

17. Wall at shallow end: 9984 lb
Wall at deep end: 39,936 lb
Side wall: $19{,}968 + 26{,}624 = 46{,}592$ lb

Chapter 8

Section 8.1 *(page 512)*

1. b   **3.** c

5. $\int u^n\, du$ **7.** $\int \dfrac{du}{u}$ **9.** $\int \dfrac{du}{\sqrt{a^2 - u^2}}$

$u = 5x - 3, n = 4$ $u = 1 - 2\sqrt{x}$ $u = t, a = 1$

11. $\int \sin u\, du$ **13.** $\int e^u\, du$ **15.** $2(x - 5)^7 + C$

$u = t^2$ $u = \sin x$

17. $-7/[6(z - 10)^6] + C$ **19.** $\frac{1}{2}v^2 - 1/[6(3v - 1)^2] + C$

21. $-\frac{1}{3}\ln|-t^3 + 9t + 1| + C$

23. $\frac{1}{2}x^2 + x + \ln|x - 1| + C$ **25.** $\ln(1 + e^x) + C$

27. $\dfrac{x}{15}(48x^4 + 200x^2 + 375) + C$ **29.** $\sin(2\pi x^2)/(4\pi) + C$

31. $-2\sqrt{\cos x} + C$ **33.** $2\ln(1 + e^x) + C$

35. $(\ln x)^2 + C$ **37.** $-\ln|\csc\alpha + \cot\alpha| + \ln|\sin\alpha| + C$

39. $-\frac{1}{4}\arcsin(4t + 1) + C$ **41.** $\frac{1}{2}\ln|\cos(2/t)| + C$

43. $6\arcsin[(x - 5)/5] + C$ **45.** $\frac{1}{4}\arctan[(2x + 1)/8] + C$

47. (a)   (b) $\frac{1}{2}\arcsin t^2 - \frac{1}{2}$

49. $y = 4e^{0.8x}$

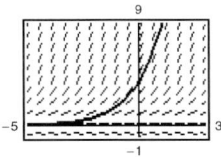

51. $y = \frac{1}{2}e^{2x} + 10e^x + 25x + C$ **53.** $r = 10\arcsin e^t + C$

55. $y = \frac{1}{2}\arctan(\tan x/2) + C$ **57.** $\frac{1}{2}$

59. $\frac{1}{2}(1 - e^{-1}) \approx 0.316$ **61.** 8   **63.** $\pi/18$

65. $18\sqrt{6}/5 \approx 8.82$ **67.** $\frac{4}{3} \approx 1.333$

69. $\frac{1}{3}\arctan\left[\frac{1}{3}(x + 2)\right] + C$ **71.** $\tan\theta - \sec\theta + C$

Graphs will vary.   Graphs will vary.

Example: Example:

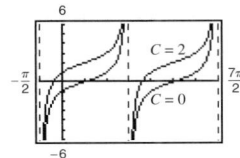

One graph is a vertical   One graph is a vertical
translation of the other.   translation of the other.

73. Power Rule: $\int u^n\, du = \dfrac{u^{n+1}}{n + 1} + C$; $u = x^2 + 1, n = 3$

75. Log Rule: $\int \dfrac{du}{u} = \ln|u| + C$; $u = x^2 + 1$

77. $a = \sqrt{2}, b = \dfrac{\pi}{4}$; $-\dfrac{1}{\sqrt{2}}\ln\left|\csc\left(x + \dfrac{\pi}{4}\right) + \cot\left(x + \dfrac{\pi}{4}\right)\right| + C$

79. $a = \frac{1}{2}$

81. (a) They are equivalent because
$$e^{x + C_1} = e^x \cdot e^{C_1} = Ce^x, C = e^{C_1}.$$
(b) They differ by a constant.
$$\sec^2 x + C_1 = (\tan^2 x + 1) + C_1 = \tan^2 x + C$$

83. a

85. (a)   (b)

(c)

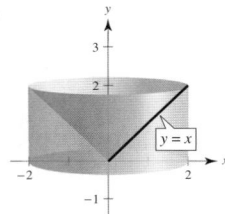

87. (a) $\pi(1 - e^{-1}) \approx 1.986$

(b) $b = \sqrt{\ln\left(\dfrac{3\pi}{3\pi - 4}\right)} \approx 0.743$

89. $\ln(\sqrt{2} + 1) \approx 0.8814$

91. $(8\pi/3)(10\sqrt{10} - 1) \approx 256.545$ **93.** $\frac{1}{3}\arctan 3 \approx 0.416$

95. About 1.0320

97. (a) $\frac{1}{3}\sin x(\cos^2 x + 2)$

(b) $\frac{1}{15}\sin x(3\cos^4 x + 4\cos^2 x + 8)$

(c) $\frac{1}{35}\sin x(5\cos^6 x + 6\cos^4 x + 8\cos^2 x + 16)$

(d) $\int \cos^{15} x\, dx = \int (1 - \sin^2 x)^7 \cos x\, dx$

You would expand $(1 - \sin^2 x)^7$.

99. Proof

Section 8.2 *(page 521)*

1. $u = x, dv = e^{2x}\, dx$ **3.** $u = (\ln x)^2, dv = dx$

5. $u = x, dv = \sec^2 x\, dx$ **7.** $\frac{1}{16}x^4(4\ln x - 1) + C$

9. $\dfrac{1}{9}\sin 3x - \dfrac{1}{3}x\cos 3x + C$ **11.** $-\dfrac{1}{16e^{4x}}(4x + 1) + C$

13. $e^x(x^3 - 3x^2 + 6x - 6) + C$

15. $\frac{1}{4}[2(t^2 - 1)\ln|t + 1| - t^2 + 2t] + C$ **17.** $\frac{1}{3}(\ln x)^3 + C$

19. $e^{2x}/[4(2x + 1)] + C$ **21.** $\frac{2}{15}(x - 5)^{3/2}(3x + 10) + C$

23. $x\sin x + \cos x + C$

25. $(6x - x^3)\cos x + (3x^2 - 6)\sin x + C$

27. $x\arctan x - \frac{1}{2}\ln(1 + x^2) + C$

29. $-\frac{3}{34}e^{-3x}\sin 5x - \frac{5}{34}e^{-3x}\cos 5x + C$ **31.** $x\ln x - x + C$

33. $y = \dfrac{2}{5}t^2\sqrt{3 + 5t} - \dfrac{8t}{75}(3 + 5t)^{3/2} + \dfrac{16}{1875}(3 + 5t)^{5/2} + C$

$= \dfrac{2}{625}\sqrt{3 + 5t}(25t^2 - 20t + 24) + C$

35. (a)

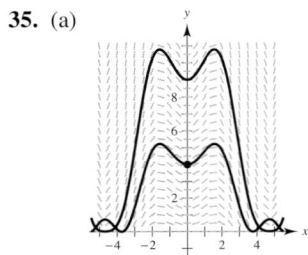

(b) $2\sqrt{y} - \cos x - x \sin x = 3$

37.

39. $2e^{3/2} + 4 \approx 12.963$

41. $\dfrac{\pi}{8} - \dfrac{1}{4} \approx 0.143$

43. $(\pi - 3\sqrt{3} + 6)/6 \approx 0.658$

45. $\frac{1}{2}[e(\sin 1 - \cos 1) + 1] \approx 0.909$

47. $8 \operatorname{arcsec} 4 + \sqrt{3}/2 - \sqrt{15}/2 - 2\pi/3 \approx 7.380$

49. $(e^{2x}/4)(2x^2 - 2x + 1) + C$

51. $(3x^2 - 6) \sin x - (x^3 - 6x) \cos x + C$

53. $x \tan x + \ln|\cos x| + C$

55. $2(\sin \sqrt{x} - \sqrt{x} \cos \sqrt{x}) + C$

57. $\frac{1}{2}(x^4 e^{x^2} - 2x^2 e^{x^2} + 2e^{x^2}) + C$

59. (a) Product Rule

(b) Answers will vary. Sample answer: You want dv to be the most complicated portion of the integrand.

61. (a) No, substitution   (b) Yes, $u = \ln x$, $dv = x \, dx$

(c) Yes, $u = x^2$, $dv = e^{-3x} \, dx$ (d) No, substitution

(e) Yes, $u = x$ and $dv = \dfrac{1}{\sqrt{x+1}} \, dx$ (f) No, substitution

63. $\frac{1}{3}\sqrt{4 + x^2}(x^2 - 8) + C$

65. $n = 0$: $x(\ln x - 1) + C$

$n = 1$: $\frac{1}{4}x^2(2 \ln x - 1) + C$

$n = 2$: $\frac{1}{9}x^3(3 \ln x - 1) + C$

$n = 3$: $\frac{1}{16}x^4(4 \ln x - 1) + C$

$n = 4$: $\frac{1}{25}x^5(5 \ln x - 1) + C$

$\displaystyle \int x^n \ln x \, dx = \dfrac{x^{n+1}}{(n+1)^2}[(n+1) \ln x - 1] + C$

67–71. Proofs   **73.** $-x^2 \cos x + 2x \sin x + 2 \cos x + C$

75. $\frac{1}{36}x^6(6 \ln x - 1) + C$

77. $\dfrac{e^{-3x}(-3 \sin 4x - 4 \cos 4x)}{25} + C$

79.

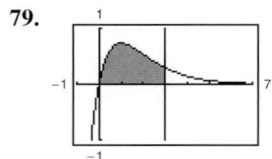

$2 - \dfrac{8}{e^3} \approx 1.602$

81.

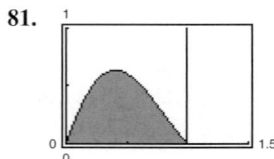

$\dfrac{\pi}{1 + \pi^2}\left(\dfrac{1}{e} + 1\right) \approx 0.395$

83. (a) 1   (b) $\pi(e - 2) \approx 2.257$ (c) $\frac{1}{2}\pi(e^2 + 1) \approx 13.177$

(d) $\left(\dfrac{e^2 + 1}{4}, \dfrac{e - 2}{2}\right) \approx (2.097, 0.359)$

85. In Example 6, we showed that the centroid of an equivalent region was $(1, \pi/8)$. By symmetry, the centroid of this region is $(\pi/8, 1)$.

87. $[7/(10\pi)](1 - e^{-4\pi}) \approx 0.223$ **89.** \$931,265

91. Proof   **93.** $b_n = [8h/(n\pi)^2] \sin(n\pi/2)$

95. (a) $y = \frac{1}{4}(3 \sin 2x - 6x \cos 2x)$

(b)

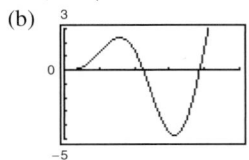

(c) You obtain the following points.

n	x_n	y_n
0	0	0
1	0.05	0
2	0.10	7.4875×10^{-4}
3	0.15	0.0037
4	0.20	0.0104
⋮	⋮	⋮
80	4.00	1.3181

(d) You obtain the following points.

n	x_n	y_n
0	5	0
1	0.1	0
2	0.2	0.0060
3	0.3	0.0293
4	0.4	0.0801
⋮	⋮	⋮
40	4.0	1.0210

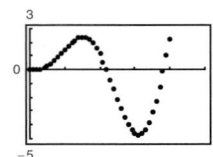

97. The graph of $y = x \sin x$ is below the graph of $y = x$ on $[0, \pi/2]$.

99. For any integrable function, $\int f(x) \, dx = C + \int f(x) \, dx$, but this cannot be used to imply that $C = 0$.

Section 8.3   (page 530)

1. $-\frac{1}{6}\cos^6 x + C$ **3.** $\frac{1}{16}\sin^8 2x + C$

5. $-\frac{1}{3}\cos^3 x + \frac{1}{5}\cos^5 x + C$

7. $-\frac{1}{3}(\cos 2\theta)^{3/2} + \frac{1}{7}(\cos 2\theta)^{7/2} + C$

9. $\frac{1}{12}(6x + \sin 6x) + C$

11. $\frac{1}{8}(2x^2 - 2x \sin 2x - \cos 2x) + C$ **13.** $\frac{16}{35}$

15. $63\pi/512$ **17.** $5\pi/32$ **19.** $\frac{1}{4}\ln|\sec 4x + \tan 4x| + C$

21. $(\sec \pi x \tan \pi x + \ln|\sec \pi x + \tan \pi x|)/(2\pi) + C$

23. $\frac{1}{2}\tan^4(x/2) - \tan^2(x/2) - 2 \ln|\cos(x/2)| + C$

25. $\dfrac{1}{2}\left[\dfrac{\sec^5 2t}{5} - \dfrac{\sec^3 2t}{3}\right] + C$ **27.** $\dfrac{1}{24}\sec^6 4x + C$

29. $\frac{1}{7}\sec^7 x - \frac{1}{5}\sec^5 x + C$

31. $\ln|\sec x + \tan x| - \sin x + C$

33. $(12\pi\theta - 8 \sin 2\pi\theta + \sin 4\pi\theta)/(32\pi) + C$

35. $y = \frac{1}{9}\sec^3 3x - \frac{1}{3}\sec 3x + C$

37. (a)

(b) $y = \frac{1}{2}x - \frac{1}{4}\sin 2x$

39.

41. $\frac{1}{16}(2 \sin 4x + \sin 8x) + C$

43. $\frac{1}{12}(3 \cos 2x - \cos 6x) + C$ **45.** $\frac{1}{8}(2 \sin 2\theta - \sin 4\theta) + C$

47. $\frac{1}{4}(\ln|\csc^2 2x| - \cot^2 2x) + C$

49. $-\frac{1}{3}\cot 3x - \frac{1}{9}\cot^3 3x + C$

51. $\ln|\csc t - \cot t| + \cos t + C$

53. $\ln|\csc x - \cot x| + \cos x + C$ **55.** $t - 2\tan t + C$

57. π **59.** $3(1 - \ln 2)$ **61.** $\ln 2$ **63.** 4

65. (a) Save one sine factor and convert the remaining factors to cosines. Then expand and integrate.

(b) Save one cosine factor and convert the remaining factors to sines. Then expand and integrate.

(c) Make repeated use of the power reducing formulas to convert the integrand to odd powers of the cosine. Then proceed as in part (b).

67. (a) $\frac{1}{2}\sin^2 x + C$ (b) $-\frac{1}{2}\cos^2 x + C$

(c) $\frac{1}{2}\sin^2 x + C$ (d) $-\frac{1}{4}\cos 2x + C$

The answers are all the same; they are just written in different forms. Using trigonometric identities, you can rewrite each answer in the same form.

69. (a) $\frac{1}{18}\tan^6 3x + \frac{1}{12}\tan^4 3x + C_1$, $\frac{1}{18}\sec^6 3x - \frac{1}{12}\sec^4 3x + C_2$

(b)

(c) Proof

71. $\frac{1}{3}$ **73.** 1 **75.** $2\pi(1 - \pi/4) \approx 1.348$

77. (a) $\pi^2/2$ (b) $(\bar{x}, \bar{y}) = (\pi/2, \pi/8)$ **79–81.** Proofs

83. $-\frac{1}{15}\cos x(3\sin^4 x + 4\sin^2 x + 8) + C$

85. $\dfrac{5}{6\pi}\tan\dfrac{2\pi x}{5}\left(\sec^2\dfrac{2\pi x}{5} + 2\right) + C$

87. (a) $H(t) \approx 57.72 - 23.36\cos(\pi t/6) - 2.75\sin(\pi t/6)$

(b) $L(t) \approx 42.04 - 20.91\cos(\pi t/6) - 4.33\sin(\pi t/6)$

(c)

The maximum difference is at $t \approx 4.9$, or late spring.

89. Proof

Section 8.4   (page 539)

1. $x = 3\tan\theta$ **3.** $x = 5\sin\theta$ **5.** $x/(16\sqrt{16 - x^2}) + C$

7. $4\ln\left|(4 - \sqrt{16 - x^2})/x\right| + \sqrt{16 - x^2} + C$

9. $\ln\left|x + \sqrt{x^2 - 25}\right| + C$

11. $\frac{1}{15}(x^2 - 25)^{3/2}(3x^2 + 50) + C$

13. $\frac{1}{3}(1 + x^2)^{3/2} + C$ **15.** $\frac{1}{2}[\arctan x + x/(1 + x^2)] + C$

17. $\frac{1}{2}x\sqrt{9 + 16x^2} + \frac{9}{8}\ln\left|4x + \sqrt{9 + 16x^2}\right| + C$

19. $\frac{25}{4}\arcsin(2x/5) + \frac{1}{2}x\sqrt{25 - 4x^2} + C$

21. $\arcsin(x/4) + C$ **23.** $4\arcsin(x/2) + x\sqrt{4 - x^2} + C$

25. $-\dfrac{(1 - x^2)^{3/2}}{3x^3} + C$ **27.** $-\dfrac{1}{3}\ln\left|\dfrac{\sqrt{4x^2 + 9} + 3}{2x}\right| + C$

29. $3/\sqrt{x^2 + 3} + C$

31. $\frac{1}{2}\left(\arcsin e^x + e^x\sqrt{1 - e^{2x}}\right) + C$

33. $\frac{1}{4}\left[x/(x^2 + 2) + \left(1/\sqrt{2}\right)\arctan\left(x/\sqrt{2}\right)\right] + C$

35. $x\,\mathrm{arcsec}\,2x - \frac{1}{2}\ln\left|2x + \sqrt{4x^2 - 1}\right| + C$

37. $\arcsin[(x - 2)/2] + C$

39. $\sqrt{x^2 + 6x + 12} - 3\ln\left|\sqrt{x^2 + 6x + 12} + (x + 3)\right| + C$

41. (a) and (b) $\sqrt{3} - \pi/3 \approx 0.685$

43. (a) and (b) $9\left(2 - \sqrt{2}\right) \approx 5.272$

45. (a) and (b) $-(9/2)\ln(2\sqrt{7}/3 - 4\sqrt{3}/3 - \sqrt{21}/3 + 8/3)$
$+ 9\sqrt{3} - 2\sqrt{7} \approx 12.644$

47. (a) Let $u = a\sin\theta$, $\sqrt{a^2 - u^2} = a\cos\theta$, where $-\pi/2 \le \theta \le \pi/2$.

(b) Let $u = a\tan\theta$, $\sqrt{a^2 + u^2} = a\sec\theta$, where $-\pi/2 < \theta < \pi/2$.

(c) Let $u = a\sec\theta$, $\sqrt{u^2 - a^2} = \tan\theta$ if $u > a$ and $\sqrt{u^2 - a^2} = -\tan\theta$ if $u < -a$, where $0 \le \theta < \pi/2$ or $\pi/2 < \theta \le \pi$.

49. (a) $\frac{1}{2}\ln(x^2 + 9) + C$; The answers are equivalent.

(b) $x - 3\arctan(x/3) + C$; The answers are equivalent.

51. True

53. False. $\displaystyle\int_0^{\sqrt{3}} \dfrac{dx}{(1 + x^2)^{3/2}} = \int_0^{\pi/3} \cos\theta\,d\theta$

55. πab

57. (a) $5\sqrt{2}$ (b) $25(1 - \pi/4)$ (c) $r^2(1 - \pi/4)$

59. $6\pi^2$ **61.** $\ln\left[\dfrac{5\left(\sqrt{2} + 1\right)}{\sqrt{26} + 1}\right] + \sqrt{26} - \sqrt{2} \approx 4.367$

63. Length of one arch of sine curve: $y = \sin x$, $y' = \cos x$

$$L_1 = \int_0^\pi \sqrt{1 + \cos^2 x}\,dx$$

Length of one arch of cosine curve: $y = \cos x$, $y' = -\sin x$

$$L_2 = \int_{-\pi/2}^{\pi/2} \sqrt{1 + \sin^2 x}\,dx$$

$$= \int_{-\pi/2}^{\pi/2} \sqrt{1 + \cos^2(x - \pi/2)}\,dx, \quad u = x - \pi/2, du = dx$$

$$= \int_{-\pi}^0 \sqrt{1 + \cos^2 u}\,du = \int_0^\pi \sqrt{1 + \cos^2 u}\,du = L_1$$

65. $(0, 0.422)$

67. $(\pi/32)\left[102\sqrt{2} - \ln(3 + 2\sqrt{2})\right] \approx 13.989$

69. (a) 187.2π lb   (b) $62.4\pi d$ lb   **71.** Proof

73. $12 + 9\pi/2 - 25\arcsin(3/5) \approx 10.050$

75. Putnam Problem A5, 2005

Section 8.5   (page 549)

1. $\dfrac{A}{x} + \dfrac{B}{x - 8}$ **3.** $\dfrac{A}{x} + \dfrac{Bx + C}{x^2 + 10}$

5. $\frac{1}{6}\ln|(x - 3)/(x + 3)| + C$ **7.** $\ln|(x - 1)/(x + 4)| + C$

9. $5\ln|x-2| - \ln|x+2| - 3\ln|x| + C$

11. $x^2 + \frac{3}{2}\ln|x-4| - \frac{1}{2}\ln|x+2| + C$

13. $1/x + \ln|x^4 + x^3| + C$

15. $2\ln|x-2| - \ln|x| - 3/(x-2) + C$

17. $\ln|(x^2+1)/x| + C$

19. $\frac{1}{6}[\ln|(x-2)/(x+2)| + \sqrt{2}\arctan(x/\sqrt{2})] + C$

21. $\ln|x+1| + \sqrt{2}\arctan[(x-1)/\sqrt{2}] + C$

23. $\ln 3$ **25.** $\frac{1}{2}\ln(8/5) - \pi/4 + \arctan 2 \approx 0.557$

27. $\ln|1 + \sec x| + C$ **29.** $\ln\left|\dfrac{\tan x + 2}{\tan x + 3}\right| + C$

31. $\frac{1}{5}\ln\left|\dfrac{e^x - 1}{e^x + 4}\right| + C$ **33.** $2\sqrt{x} + 2\ln\left|\dfrac{\sqrt{x}-2}{\sqrt{x}+2}\right| + C$

35–37. Proofs   **39.** First divide x^3 by $(x-5)$.

41. (a) Substitution: $u = x^2 + 2x - 8$ (b) Partial fractions
(c) Trigonometric substitution (tan) or inverse tangent rule

43. $12\ln\left(\frac{9}{8}\right) \approx 1.4134$ **45.** 4.90 or \$490,000

47. $V = 2\pi\left(\arctan 3 - \frac{3}{10}\right) \approx 5.963$; $(\bar{x}, \bar{y}) \approx (1.521, 0.412)$

49. $x = n[e^{(n+1)kt} - 1]/[n + e^{(n+1)kt}]$ **51.** $\pi/8$

Section 8.6   (page 555)

1. $-\frac{1}{2}x(10 - x) + 25\ln|5 + x| + C$ **3.** $-\sqrt{1 - x^2}/x + C$

5. $\frac{1}{24}(3x + \sin 3x \cos 3x + 2\cos^3 3x \sin 3x) + C$

7. $-2(\cot\sqrt{x} + \csc\sqrt{x}) + C$ **9.** $x - \frac{1}{2}\ln(1 + e^{2x}) + C$

11. $\frac{1}{16}x^8(8\ln x - 1) + C$

13. (a) and (b) $\frac{1}{27}e^{3x}(9x^2 - 6x + 2) + C$

15. (a) and (b) $\ln|(x+1)/x| - 1/x + C$

17. $\frac{1}{2}[(x^2 + 1)\text{arccsc}(x^2 + 1) + \ln(x^2 + 1 + \sqrt{x^4 + 2x^2})] + C$

19. $\sqrt{x^2 - 4}/(4x) + C$

21. $\frac{4}{25}[\ln|2 - 5x| + 2/(2 - 5x)] + C$

23. $e^x \arccos(e^x) - \sqrt{1 - e^{2x}} + C$

25. $\frac{1}{2}(x^2 + \cot x^2 + \csc x^2) + C$

27. $(\sqrt{2}/2)\arctan[(1 + \sin\theta)/\sqrt{2}] + C$

29. $-\sqrt{2 + 9x^2}/(2x) + C$

31. $\frac{1}{4}(2\ln|x| - 3\ln|3 + 2\ln|x||) + C$

33. $(3x - 10)/[2(x^2 - 6x + 10)] + \frac{3}{2}\arctan(x - 3) + C$

35. $\frac{1}{2}\ln|x^2 - 3 + \sqrt{x^4 - 6x^2 + 5}| + C$

37. $2/(1 + e^x) - 1/[2(1 + e^x)^2] + \ln(1 + e^x) + C$

39. $\frac{1}{2}(e - 1) \approx 0.8591$ **41.** $\frac{32}{5}\ln 2 - \frac{31}{25} \approx 3.1961$

43. $\pi/2$ **45.** $\pi^3/8 - 3\pi + 6 \approx 0.4510$ **47–51.** Proofs

53. $\dfrac{1}{\sqrt{5}}\ln\left|\dfrac{2\tan(\theta/2) - 3 - \sqrt{5}}{2\tan(\theta/2) - 3 + \sqrt{5}}\right| + C$ **55.** $\ln 2$

57. $\frac{1}{2}\ln(3 - 2\cos\theta) + C$ **59.** $-2\cos\sqrt{\theta} + C$ **61.** $4\sqrt{3}$

63. (a) $\displaystyle\int x\ln x\, dx = \frac{1}{2}x^2\ln x - \frac{1}{4}x^2 + C$

$\displaystyle\int x^2\ln x\, dx = \frac{1}{3}x^3\ln x - \frac{1}{9}x^3 + C$

$\displaystyle\int x^3\ln x\, dx = \frac{1}{4}x^4\ln x - \frac{1}{16}x^4 + C$

65. (a) Arctangent Formula, Formula 23,

$\displaystyle\int \frac{1}{u^2 + 1}\, du, \; u = e^x$

(b) Log Rule: $\displaystyle\int \frac{1}{u}\, du, \; u = e^x + 1$

(c) Substitution: $u = x^2, \; du = 2x\, dx$
Then Formula 81.

(d) Integration by parts   (e) Cannot be integrated
(f) Formula 16 with $u = e^{2x}$

67. False. Substitutions may first have to be made to rewrite the integral in a form that appears in the table.

69. 1919.145 ft-lb   **71.** $32\pi^2$ **73.** About 401.4

Section 8.7   (page 564)

1.

x	-0.1	-0.01	-0.001	0.001	0.01	0.1
$f(x)$	1.3177	1.3332	1.3333	1.3333	1.3332	1.3177

$\frac{4}{3}$

3.

x	1	10	10^2	10^3	10^4	10^5
$f(x)$	0.9900	90,483.7	3.7×10^9	4.5×10^{10}	0	0

0

5. $\frac{3}{8}$ **7.** $\frac{1}{8}$ **9.** $\frac{5}{3}$ **11.** 4   **13.** 0   **15.** ∞ **17.** $\frac{11}{4}$

19. $\frac{3}{5}$ **21.** 1   **23.** $\frac{5}{4}$ **25.** ∞ **27.** 0   **29.** 1

31. 0   **33.** 0   **35.** ∞ **37.** $\frac{5}{9}$ **39.** 1   **41.** ∞

43. (a) Not indeterminate   **45.** (a) $0 \cdot \infty$
(b) ∞ (b) 1
(c)   (c)

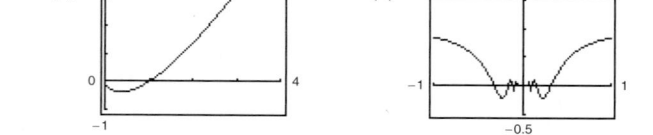

47. (a) Not indeterminate   **49.** (a) ∞^0
(b) 0   (b) 1
(c)   (c)

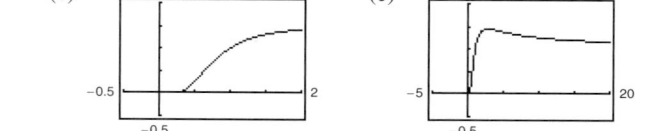

51. (a) 1^∞ (b) e **53.** (a) 0^0 (b) 3
(c)   (c)

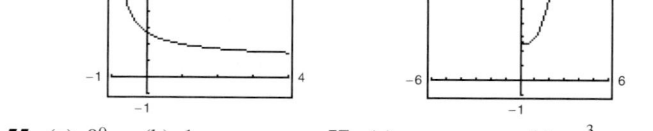

55. (a) 0^0 (b) 1   **57.** (a) $\infty - \infty$ (b) $-\frac{3}{2}$
(c)   (c)

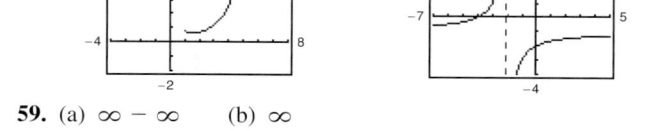

59. (a) $\infty - \infty$ (b) ∞
(c)

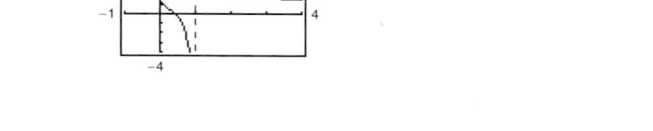

61. $\dfrac{0}{0}, \dfrac{\infty}{\infty}, 0 \cdot \infty, 1^{\infty}, 0^0, \infty - \infty$

63. Answers will vary. Examples:
  (a) $f(x) = x^2 - 25, g(x) = x - 5$
  (b) $f(x) = (x - 5)^2, g(x) = x^2 - 25$
  (c) $f(x) = x^2 - 25, g(x) = (x - 5)^3$

65. (a) Yes: $\dfrac{0}{0}$ (b) No: $\dfrac{0}{-1}$ (c) Yes: $\dfrac{\infty}{\infty}$ (d) Yes: $\dfrac{0}{0}$

  (e) No: $\dfrac{-1}{0}$ (f) Yes: $\dfrac{0}{0}$

67.

x	10	10^2	10^4	10^6	10^8	10^{10}
$\dfrac{(\ln x)^4}{x}$	2.811	4.498	0.720	0.036	0.001	0.000

69. 0   **71.** 0   **73.** 0

75. Horizontal asymptote: **77.** Horizontal asymptote:
  $y = 1$ $y = 0$
  Relative maximum: $(e, e^{1/e})$ Relative maximum: $(1, 2/e)$

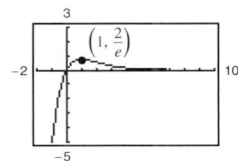

79. Limit is not of the form $0/0$ or ∞/∞.
81. Limit is not of the form $0/0$ or ∞/∞.
83. (a) $\displaystyle \lim_{x \to \infty} \dfrac{x}{\sqrt{x^2 + 1}} = \lim_{x \to \infty} \dfrac{\sqrt{x^2 + 1}}{x} = \lim_{x \to \infty} \dfrac{x}{\sqrt{x^2 + 1}}$

  Applying L'Hôpital's Rule twice results in the original limit, so L'Hôpital's Rule fails.
  (b) 1
  (c)

85.

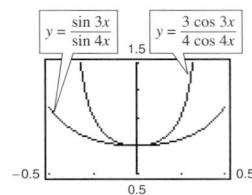

  As $x \to 0$, the graphs get closer together (they both approach 0.75). By L'Hôpital's Rule,
  $$\lim_{x \to 0} \dfrac{\sin 3x}{\sin 4x} = \lim_{x \to 0} \dfrac{3 \cos 3x}{4 \cos 4x} = \dfrac{3}{4}.$$

87. $v = 32t + v_0$ **89.** Proof   **91.** $c = \dfrac{2}{3}$ **93.** $c = \pi/4$
95. False. L'Hôpital's Rule does not apply because $\lim_{x \to 0}(x^2 + x + 1) \neq 0$.
97. True   **99.** $\dfrac{3}{4}$ **101.** $\dfrac{4}{3}$ **103.** $a = 1, b = \pm 2$
105. Proof   **107.** (a) $0 \cdot \infty$ (b) 0   **109.** Proof
111. (a)–(c) 2
113. (a)

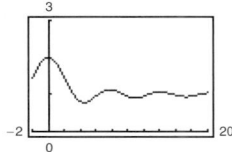

  (b) $\lim_{x \to \infty} h(x) = 1$
  (c) No

115. Putnam Problem A1, 1956

Section 8.8   *(page 575)*

1. Improper; $0 \le \dfrac{3}{5} \le 1$
3. Not improper; continuous on $[0, 1]$
5. Not improper; continuous on $[0, 2]$
7. Improper; infinite limits of integration
9. Infinite discontinuity at $x = 0$; 4
11. Infinite discontinuity at $x = 1$; diverges
13. Infinite discontinuity at $x = 0$; diverges
15. Infinite limit of integration; converges to 1   **17.** $\dfrac{1}{2}$
19. Diverges   **21.** Diverges   **23.** 2   **25.** $1/[2(\ln 4)^2]$
27. π **29.** $\pi/4$ **31.** Diverges   **33.** Diverges
35. 0   **37.** $-\dfrac{1}{4}$ **39.** Diverges   **41.** $\pi/3$ **43.** $\ln 3$
45. $\pi/6$ **47.** $2\pi\sqrt{6}/3$ **49.** $p > 1$ **51.** Proof
53. Diverges   **55.** Converges   **57.** Converges
59. Diverges   **61.** Converges
63. An integral with infinite integration limits, an integral with an infinite discontinuity at or between the integration limits
65. The improper integral diverges.   **67.** e **69.** π
71. (a) 1   (b) $\pi/2$ (c) 2π
73.

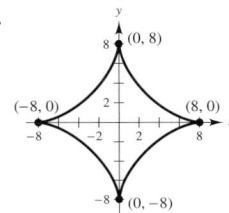

  Perimeter = 48

75. $8\pi^2$ **77.** (a) $W = 20{,}000$ mile-tons   (b) 4000 mi
79. (a) Proof   (b) $P = 43.53\%$ (c) $E(x) = 7$
81. (a) \$757,992.41   (b) \$837,995.15   (c) \$1,066,666.67
83. $P = \left[2\pi NI\left(\sqrt{r^2 + c^2} - c\right)\right]/\left(kr\sqrt{r^2 + c^2}\right)$
85. False. Let $f(x) = 1/(x + 1)$.   **87.** True
89. (a) and (b) Proofs

  (c) The definition of the improper integral $\displaystyle \int_{-\infty}^{\infty}$ is not $\displaystyle \lim_{a \to \infty} \int_{-a}^{a}$
  but rather that if you rewrite the integral that diverges, you can find that the integral converges.

91. (a) $\displaystyle \int_{1}^{\infty} \dfrac{1}{x^n} \, dx$ will converge if $n > 1$ and diverge if $n \le 1$.
  (b)   (c) Converges

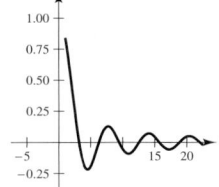

93. (a)   (b) About 0.2525
                                  (c) 0.2525; same by symmetry

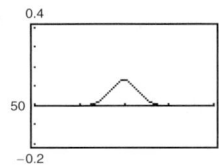

95. $1/s, s > 0$ **97.** $2/s^3, s > 0$ **99.** $s/(s^2 + a^2), s > 0$
101. $s/(s^2 - a^2), s > |a|$
103. (a) $\Gamma(1) = 1, \Gamma(2) = 1, \Gamma(3) = 2$ (b) Proof
  (c) $\Gamma(n) = (n - 1)!$

105. $c = 1$; $\ln(2)$

107. $8\pi[(\ln 2)^2/3 - (\ln 4)/9 + 2/27] \approx 2.01545$

109. $\displaystyle\int_0^1 2\sin(u^2)\,du$; 0.6278   **111.** Proof

Review Exercises for Chapter 8   *(page 579)*

1. $\frac{1}{3}(x^2 - 36)^{3/2} + C$ **3.** $\frac{1}{2}\ln|x^2 - 49| + C$

5. $\ln(2) + \frac{1}{2} \approx 1.1931$ **7.** $100\arcsin(x/10) + C$

9. $\frac{1}{9}e^{3x}(3x - 1) + C$ **11.** $\frac{1}{13}e^{2x}(2\sin 3x - 3\cos 3x) + C$

13. $-\frac{1}{2}x^2\cos 2x + \frac{1}{2}x\sin 2x + \frac{1}{4}\cos 2x + C$

15. $\frac{1}{16}\left[(8x^2 - 1)\arcsin 2x + 2x\sqrt{1 - 4x^2}\right] + C$

17. $\sin(\pi x - 1)[\cos^2(\pi x - 1) + 2]/(3\pi) + C$

19. $\frac{2}{3}[\tan^3(x/2) + 3\tan(x/2)] + C$ **21.** $\tan\theta + \sec\theta + C$

23. $3\pi/16 + \frac{1}{2} \approx 1.0890$ **25.** $3\sqrt{4 - x^2}/x + C$

27. $\frac{1}{3}(x^2 + 4)^{1/2}(x^2 - 8) + C$ **29.** $256 - 62\sqrt{17} \approx 0.3675$

31. (a), (b), and (c) $\frac{1}{3}\sqrt{4 + x^2}(x^2 - 8) + C$

33. $6\ln|x + 3| - 5\ln|x - 4| + C$

35. $\frac{1}{4}[6\ln|x - 1| - \ln(x^2 + 1) + 6\arctan x] + C$

37. $x - \frac{64}{11}\ln|x + 8| + \frac{9}{11}\ln|x - 3| + C$

39. $\frac{1}{25}[4/(4 + 5x) + \ln|4 + 5x|] + C$ **41.** $1 - \sqrt{2}/2$

43. $\frac{1}{2}\ln|x^2 + 4x + 8| - \arctan[(x + 2)/2] + C$

45. $\ln|\tan\pi x|/\pi + C$ **47.** Proof

49. $\frac{1}{8}(\sin 2\theta - 2\theta\cos 2\theta) + C$

51. $\frac{4}{3}[x^{3/4} - 3x^{1/4} + 3\arctan(x^{1/4})] + C$

53. $2\sqrt{1 - \cos x} + C$ **55.** $\sin x\ln(\sin x) - \sin x + C$

57. $\frac{5}{2}\ln|(x - 5)/(x + 5)| + C$

59. $y = x\ln|x^2 + x| - 2x + \ln|x + 1| + C$ **61.** $\frac{1}{5}$

63. $\frac{1}{2}(\ln 4)^2 \approx 0.961$ **65.** π **67.** $\frac{128}{15}$

69. $(\bar{x}, \bar{y}) = (0, 4/(3\pi))$ **71.** 3.82   **73.** 0   **75.** ∞

77. 1   **79.** $1000e^{0.09} \approx 1094.17$ **81.** Converges; $\frac{32}{3}$

83. Diverges   **85.** Converges; 1   **87.** Converges; $\pi/4$

89. (a) \$6,321,205.59   (b) \$10,000,000

91. (a) 0.4581   (b) 0.0135

P.S. Problem Solving   *(page 581)*

1. (a) $\frac{4}{3}, \frac{16}{15}$ (b) Proof   **3.** $\ln 3$ **5.** Proof

7. (a)

Area ≈ 0.2986

(b) $\ln 3 - \frac{4}{5}$ (c) $\ln 3 - \frac{4}{5}$

9. $\ln 3 - \frac{1}{2} \approx 0.5986$

11. (a) ∞ (b) 0   (c) $-\frac{2}{3}$

The form $0 \cdot \infty$ is indeterminant.

13. About 0.8670   **15.** $\dfrac{1/12}{x} + \dfrac{1/42}{x - 3} + \dfrac{1/10}{x - 1} + \dfrac{111/140}{x + 4}$

17–19. Proofs   **21.** About 0.0158

Chapter 9

Section 9.1   *(page 592)*

1. 3, 9, 27, 81, 243   **3.** 1, 0, −1, 0, 1   **5.** 2, −1, $\frac{2}{3}$, $-\frac{1}{2}$, $\frac{2}{5}$

7. 3, 4, 6, 10, 18   **9.** c   **10.** a   **11.** d   **12.** b

13. 14, 17; add 3 to preceding term.

15. 80, 160; multiply preceding term by 2.   **17.** $n + 1$

19. $1/[(2n + 1)(2n)]$ **21.** 5   **23.** 2

25.

27.

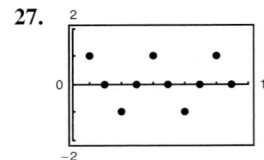

Converges to 4   Diverges

29. Converges to 0   **31.** Diverges   **33.** Converges to 5

35. Converges to 0   **37.** Diverges   **39.** Converges to 0

41. Converges to 1   **43.** Converges to 0

45. Answers will vary. Sample answer: $6n - 4$

47. Answers will vary. Sample answer: $n^2 - 3$

49. Answers will vary. Sample answer: $(n + 1)/(n + 2)$

51. Answers will vary. Sample answer: $(n + 1)/n$

53. Monotonic, bounded   **55.** Not monotonic, bounded

57. Monotonic, bounded   **59.** Not monotonic, bounded

61. (a) $\left|7 + \frac{1}{n}\right| \geq 7 \implies$ bounded

$a_n > a_{n+1} \implies$ monotonic

So, $\{a_n\}$ converges.

(b)

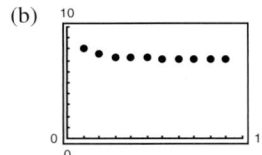

Limit $= 7$

63. (a) $\left|\frac{1}{3}\left(1 - \frac{1}{3^n}\right)\right| < \frac{1}{3} \implies$ bounded

$a_n < a_{n+1} \implies$ monotonic

So, $\{a_n\}$ converges.

(b)

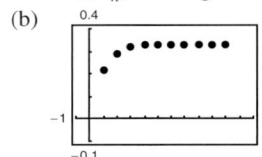

Limit $= \frac{1}{3}$

65. $\{a_n\}$ has a limit because it is bounded and monotonic; because $2 \leq a_n \leq 4$, $2 \leq L \leq 4$.

67. (a) No. $\displaystyle\lim_{n\to\infty} A_n$ does not exist.

(b)

n	1	2	3	4
A_n	\$10,045.83	\$10,091.88	\$10,138.13	\$10,184.60

n	5	6	7
A_n	\$10,231.28	\$10,278.17	\$10,325.28

n	8	9	10
A_n	\$10,372.60	\$10,420.14	\$10,467.90

69. No. A sequence is said to converge when its terms approach a real number.

71. (a) $10 - \frac{1}{n}$

(b) Impossible. The sequence converges by Theorem 9.5.

(c) $a_n = \dfrac{3n}{4n + 1}$

(d) Impossible. An unbounded sequence diverges.

73. (a) $\$4{,}500{,}000{,}000(0.8)^n$

(b)

Year	1	2
Budget	$\$3{,}600{,}000{,}000$	$\$2{,}880{,}000{,}000$

Year	3	4
Budget	$\$2{,}304{,}000{,}000$	$\$1{,}843{,}200{,}000$

(c) Converges to 0

75. 1, 1.4142, 1.4422, 1.4142, 1.3797, 1.3480; Converges to 1

77. Proof   **79.** True   **81.** True

83. (a) 1, 1, 2, 3, 5, 8, 13, 21, 34, 55, 89, 144

(b) 1, 2, 1.5, 1.6667, 1.6, 1.6250, 1.6154, 1.6190, 1.6176, 1.6182   (c) Proof

(d) $\rho = \left(1 + \sqrt{5}\right)/2 \approx 1.6180$

85. (a) 1.4142, 1.8478, 1.9616, 1.9904, 1.9976

(b) $a_n = \sqrt{2 + a_{n-1}}$ (c) $\lim\limits_{n \to \infty} a_n = 2$

87. (a) Proof

(b)

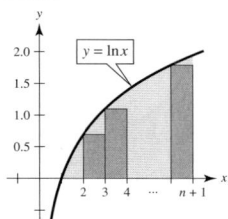

(c) Proof   (d) Proof

(e) $\dfrac{\sqrt[20]{20!}}{20} \approx 0.4152;$

$\dfrac{\sqrt[50]{50!}}{50} \approx 0.3897;$

$\dfrac{\sqrt[100]{100!}}{100} \approx 0.3799$

89–91. Proofs   **93.** Putnam Problem A1, 1990

Section 9.2   *(page 601)*

1. 1, 1.25, 1.361, 1.424, 1.464

3. 3, -1.5, 5.25, -4.875, 10.3125

5. 3, 4.5, 5.25, 5.625, 5.8125   **7.** Geometric series: $r = \frac{7}{6} > 1$

9. $\lim\limits_{n \to \infty} a_n = 1 \neq 0$ **11.** $\lim\limits_{n \to \infty} a_n = 1 \neq 0$

13. $\lim\limits_{n \to \infty} a_n = \frac{1}{2} \neq 0$ **15.** Geometric series: $r = \frac{5}{6} < 1$

17. Geometric series: $r = 0.9 < 1$

19. Telescoping series: $a_n = 1/n - 1/(n+1)$; Converges to 1.

21. (a) $\frac{11}{3}$

(b)

n	5	10	20	50	100
S_n	2.7976	3.1643	3.3936	3.5513	3.6078

(c)

(d) The terms of the series decrease in magnitude relatively slowly, and the sequence of partial sums approaches the sum of the series relatively slowly.

23. (a) 20

(b)

n	5	10	20	50	100
S_n	8.1902	13.0264	17.5685	19.8969	19.9995

(c)

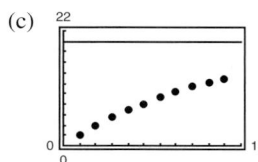

(d) The terms of the series decrease in magnitude relatively slowly, and the sequence of partial sums approaches the sum of the series relatively slowly.

25. 15   **27.** 3   **29.** 32   **31.** $\dfrac{1}{2}$ **33.** $\dfrac{\sin(1)}{1 - \sin(1)}$

35. (a) $\displaystyle\sum_{n=0}^{\infty} \dfrac{4}{10}(0.1)^n$ **37.** (a) $\displaystyle\sum_{n=0}^{\infty} \dfrac{81}{100}(0.01)^n$

(b) $\dfrac{4}{9}$ (b) $\dfrac{9}{11}$

39. (a) $\displaystyle\sum_{n=0}^{\infty} \dfrac{3}{40}(0.01)^n$ (b) $\dfrac{5}{66}$ **41.** Diverges   **43.** Diverges

45. Converges   **47.** Diverges   **49.** Diverges

51. Diverges   **53.** Diverges   **55.** See definitions on page 595.

57. The series given by

$$\sum_{n=0}^{\infty} ar^n = a + ar + ar^2 + \cdots + ar^n + \cdots, a \neq 0$$

is a geometric series with ratio r. When $0 < |r| < 1$, the series

converges to the sum $\displaystyle\sum_{n=0}^{\infty} ar^n = \dfrac{a}{1 - r}$.

59. The series in (a) and (b) are the same. The series in (c) is different unless $a_1 = a_2 = \cdots = a$ is constant.

61. $|x| < \dfrac{1}{3}; \dfrac{3x}{1 - 3x}$ **63.** $0 < x < 2; (x - 1)/(2 - x)$

65. $-1 < x < 1; 1/(1 + x)$

67. (a) x (b) $f(x) = 1/(1 - x)$, $|x| < 1$

(c)

Answers will vary.

69. The required terms for the two series are $n = 100$ and $n = 5$, respectively. The second series converges at a higher rate.

71. $160{,}000(1 - 0.95^n)$ units

73. $\displaystyle\sum_{i=0}^{\infty} 200(0.75)^i$; Sum $= \$800$ million   **75.** 152.42 feet

77. $\dfrac{1}{8}; \displaystyle\sum_{n=0}^{\infty} \dfrac{1}{2}\left(\dfrac{1}{2}\right)^n = \dfrac{1/2}{1 - 1/2} = 1$

79. (a) $-1 + \displaystyle\sum_{n=0}^{\infty}\left(\dfrac{1}{2}\right)^n = -1 + \dfrac{a}{1 - r} = -1 + \dfrac{1}{1 - 1/2} = 1$

(b) No   (c) 2

81. (a) 126 in.^{2} (b) 128 in.^{2}

83. The $\$2{,}000{,}000$ sweepstakes has a present value of $\$1{,}146{,}992.12$. After accruing interest over the 20-year period, it attains its full value.

85. (a) $\$5{,}368{,}709.11$ (b) $\$10{,}737{,}418.23$ (c) $\$21{,}474{,}836.47$

87. (a) $\$14{,}773.59$ (b) $\$14{,}779.65$

89. (a) $\$91{,}373.09$ (b) $\$91{,}503.32$

91. False. $\lim\limits_{n \to \infty} \dfrac{1}{n} = 0$, but $\displaystyle\sum_{n=1}^{\infty} \dfrac{1}{n}$ diverges.

93. False. $\displaystyle\sum_{n=1}^{\infty} ar^n = \left(\dfrac{a}{1 - r}\right) - a$; The formula requires that the geometric series begins with $n = 0$.

95. True   **97.** Answers will vary. Example: $\displaystyle\sum_{n=0}^{\infty} 1, \displaystyle\sum_{n=0}^{\infty} (-1)$

99–101. Proofs   **103.** 2

Section 9.3   *(page 609)*

1. Diverges   **3.** Converges   **5.** Converges
7. Converges   **9.** Diverges   **11.** Diverges
13. Converges   **15.** Converges   **17.** Converges
19. Diverges   **21.** Converges   **23.** Diverges
25. $f(x)$ is not positive for $x \geq 1$.
27. $f(x)$ is not always decreasing.   **29.** Converges
31. Diverges   **33.** Diverges   **35.** Converges
37. Converges
39. (a)

n	5	10	20	50	100
S_n	3.7488	3.75	3.75	3.75	3.75

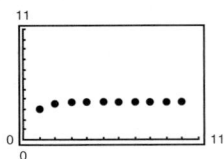

The partial sums approach the sum 3.75 very quickly.

(b)

n	5	10	20	50	100
S_n	1.4636	1.5498	1.5962	1.6251	1.635

The partial sums approach the sum $\pi^2/6 \approx 1.6449$ more slowly than the series in part (a).

41. See Theorem 9.10 on page 605. Answers will vary. For example, convergence or divergence can be determined for the series $\sum_{n=1}^{\infty} \dfrac{1}{n^2 + 1}$.

43. No. Because $\sum_{n=1}^{\infty} \dfrac{1}{n}$ diverges, $\sum_{n=10,000}^{\infty} \dfrac{1}{n}$ also diverges. The convergence or divergence of a series is not determined by the first finite number of terms of the series.

45. (a)

The area under the rectangles is greater than the area under the curve. Because $\int_{1}^{\infty} \dfrac{1}{\sqrt{x}}\, dx = \left[2\sqrt{x} \right]_{1}^{\infty} = \infty$ diverges, $\sum_{n=1}^{\infty} \dfrac{1}{\sqrt{n}}$ diverges.

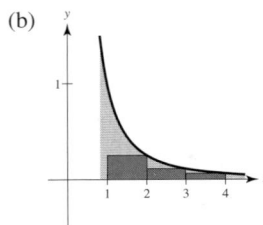

(b)

The area under the rectangles is less than the area under the curve. Because $\int_{1}^{\infty} \dfrac{1}{x^2}\, dx = \left[-\dfrac{1}{x} \right]_{1}^{\infty} = 1$ converges, $\sum_{n=2}^{\infty} \dfrac{1}{n^2}$ converges $\left(\text{and so does } \sum_{n=1}^{\infty} \dfrac{1}{n^2}\right)$.

47. $p > 1$ **49.** $p > 1$ **51.** $p > 3$ **53.** Proof
55. $S_5 = 1.4636$ **57.** $S_{10} \approx 0.9818$ **59.** $S_4 \approx 0.4049$
 $R_5 = 0.20$ $R_{10} \approx 0.0997$ $R_4 \approx 5.6 \times 10^{-8}$
61. $N \geq 7$ **63.** $N \geq 16$
65. (a) $\sum_{n=2}^{\infty} \dfrac{1}{n^{1.1}}$ converges by the p-Series Test because $1.1 > 1$.

 $\sum_{n=2}^{\infty} \dfrac{1}{n \ln n}$ diverges by the Integral Test because $\int_{2}^{\infty} \dfrac{1}{x \ln x}\, dx$ diverges.

(b) $\sum_{n=2}^{\infty} \dfrac{1}{n^{1.1}} = 0.4665 + 0.2987 + 0.2176 + 0.1703 + 0.1393 + \cdots$

 $\sum_{n=2}^{\infty} \dfrac{1}{n \ln n} = 0.7213 + 0.3034 + 0.1803 + 0.1243 + 0.0930 + \cdots$

(c) $n \geq 3.431 \times 10^{15}$

67. (a) Let $f(x) = 1/x$. f is positive, continuous, and decreasing on $[1, \infty)$.

 $S_n - 1 \leq \int_{1}^{n} \dfrac{1}{x}\, dx = \ln n$

 $S_n \geq \int_{1}^{n+1} \dfrac{1}{x}\, dx = \ln(n + 1)$

 So, $\ln(n + 1) \leq S_n \leq 1 + \ln n$.

(b) $\ln(n + 1) - \ln n \leq S_n - \ln n \leq 1$
 Also, $\ln(n + 1) - \ln n > 0$ for $n \geq 1$. So,
 $0 \leq S_n - \ln n \leq 1$, and the sequence $\{a_n\}$ is bounded.

(c) $a_n - a_{n+1} = [S_n - \ln n] - [S_{n+1} - \ln(n + 1)]$

 $= \int_{n}^{n+1} \dfrac{1}{x}\, dx - \dfrac{1}{n + 1} \geq 0$

 So, $a_n \geq a_{n+1}$.

(d) Because the sequence is bounded and monotonic, it converges to a limit, γ.

(e) 0.5822

69. (a) Diverges   (b) Diverges

(c) $\sum_{n=2}^{\infty} x^{\ln n}$ converges for $x < 1/e$.

71. Diverges   **73.** Converges   **75.** Converges
77. Diverges   **79.** Diverges   **81.** Converges

Section 9.4   *(page 616)*

1. (a)

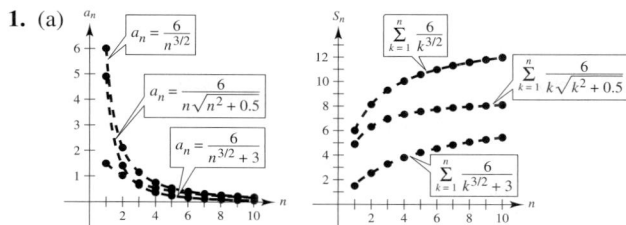

(b) $\displaystyle\sum_{n=1}^{\infty} \frac{6}{n^{3/2}}$; Converges

(c) The magnitudes of the terms are less than the magnitudes of the terms of the *p*-series. Therefore, the series converges.

(d) The smaller the magnitudes of the terms, the smaller the magnitudes of the terms of the sequence of partial sums.

3. Diverges   5. Diverges   7. Diverges   9. Converges

11. Converges   13. Diverges   15. Diverges

17. Converges   19. Converges   21. Diverges

23. Diverges; *p*-Series Test

25. Converges; Direct Comparison Test with $\displaystyle\sum_{n=1}^{\infty} \left(\frac{1}{5}\right)^n$

27. Diverges; *n*th-Term Test   29. Converges; Integral Test

31. $\displaystyle\lim_{n\to\infty} \frac{a_n}{1/n} = \lim_{n\to\infty} na_n$; $\displaystyle\lim_{n\to\infty} na_n \neq 0$, but is finite.

The series diverges by the Limit Comparison Test.

33. Diverges   35. Converges

37. $\displaystyle\lim_{n\to\infty} n\left(\frac{n^3}{5n^4 + 3}\right) = \frac{1}{5} \neq 0$; So, $\displaystyle\sum_{n=1}^{\infty} \frac{n^3}{5n^4 + 3}$ diverges.

39. Diverges   41. Converges

43. Convergence or divergence is dependent on the form of the general term for the series and not necessarily on the magnitudes of the terms.

45. See Theorem 9.13 on page 614. Answers will vary. For example,

$\displaystyle\sum_{n=2}^{\infty} \frac{1}{\sqrt{n} - 1}$ diverges because $\displaystyle\lim_{n\to\infty} \frac{1/\sqrt{n} - 1}{1/\sqrt{n}} = 1$ and

$\displaystyle\sum_{n=2}^{\infty} \frac{1}{\sqrt{n}}$ diverges (*p*-series).

47. (a) Proof

(b)

n	5	10	20	50	100
S_n	1.1839	1.2087	1.2212	1.2287	1.2312

(c) 0.1226   (d) 0.0277

49. False. Let $a_n = 1/n^3$ and $b_n = 1/n^2$.   51. True

53. True   55. Proof   57. $\displaystyle\sum_{n=1}^{\infty} \frac{1}{n^2}, \sum_{n=1}^{\infty} \frac{1}{n^3}$ 59–65. Proofs

67. Putnam Problem B4, 1988

Section 9.5   *(page 625)*

1. (a)

n	1	2	3	4	5
S_n	1.0000	0.6667	0.8667	0.7238	0.8349

n	6	7	8	9	10
S_n	0.7440	0.8209	0.7543	0.8131	0.7605

(b)

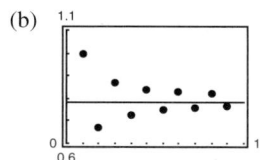

(c) The points alternate sides of the horizontal line $y = \pi/4$ that represents the sum of the series. The distances between the successive points and the line decrease.

(d) The distance in part (c) is always less than the magnitude of the next term of the series.

3. (a)

n	1	2	3	4	5
S_n	1.0000	0.7500	0.8611	0.7986	0.8386

n	6	7	8	9	10
S_n	0.8108	0.8312	0.8156	0.8280	0.8180

(b)

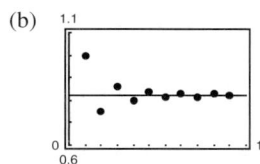

(c) The points alternate sides of the horizontal line $y = \pi^2/12$ that represents the sum of the series. The distances between the successive points and the line decrease.

(d) The distance in part (c) is always less than the magnitude of the next term of the series.

5. Converges   7. Converges   9. Diverges   11. Diverges

13. Converges   15. Diverges   17. Diverges

19. Converges   21. Converges   23. Converges

25. Converges   27. $1.8264 \leq S \leq 1.8403$

29. $1.7938 \leq S \leq 1.8054$ 31. 10   33. 7

35. 7 terms (Note that the sum begins with $n = 0$.)

37. Converges absolutely   39. Converges absolutely

41. Converges conditionally   43. Diverges

45. Converges conditionally   47. Converges absolutely

49. Converges absolutely   51. Converges conditionally

53. Converges absolutely

55. An alternating series is a series whose terms alternate in sign.

57. $|S - S_N| = |R_N| \leq a_{N+1}$

59. (a) False. For example, let $a_n = \dfrac{(-1)^n}{n}$.

Then $\displaystyle\sum a_n = \sum \frac{(-1)^n}{n}$ converges

and $\displaystyle\sum (-a_n) = \sum \frac{(-1)^{n+1}}{n}$ converges.

But, $\displaystyle\sum |a_n| = \sum \frac{1}{n}$ diverges.

(b) True. For if $\sum |a_n|$ converged, then so would $\sum a_n$ by Theorem 9.16.

61. True   63. $p > 0$

65. Proof; The converse is false. For example: Let $a_n = 1/n$.

67. $\displaystyle\sum_{n=1}^{\infty} \frac{1}{n^2}$ converges, hence so does $\displaystyle\sum_{n=1}^{\infty} \frac{1}{n^4}$.

69. (a) No. $a_{n+1} \leq a_n$ is not satisfied for all n. For example, $\frac{1}{9} < \frac{1}{8}$.

(b) Yes. 0.5

71. Converges; *p*-Series Test   **73.** Diverges; *n*th-Term Test
75. Converges; Geometric Series Test
77. Converges; Integral Test
79. Converges; Alternating Series Test
81. The first term of the series is 0, not 1. You cannot regroup series terms arbitrarily.

Section 9.6 *(page 633)*

1–3. Proofs   **5.** d   **6.** c   **7.** f   **8.** b   **9.** a
10. e
11. (a) Proof
(b)

n	5	10	15	20	25
S_n	13.7813	24.2363	25.8468	25.9897	25.9994

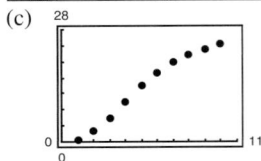

(c)   (d) 26

(e) The more rapidly the terms of the series approach 0, the more rapidly the sequence of partial sums approaches the sum of the series.

13. Converges   **15.** Diverges   **17.** Diverges
19. Converges   **21.** Converges   **23.** Converges
25. Diverges   **27.** Converges   **29.** Converges
31. Diverges   **33.** Converges   **35.** Converges
37. Converges   **39.** Diverges   **41.** Converges
43. Diverges   **45.** Converges   **47.** Converges
49. Converges   **51.** Converges; Alternating Series Test
53. Converges; *p*-Series Test   **55.** Diverges; *n*th-Term Test
57. Diverges; Geometric Series Test
59. Converges; Limit Comparison Test with $b_n = 1/2^n$
61. Converges; Direct Comparison Test with $b_n = 1/3^n$
63. Diverges; Ratio Test   **65.** Converges; Ratio Test
67. Converges; Ratio Test   **69.** a and c   **71.** a and b
73. $\displaystyle\sum_{n=0}^{\infty} \frac{n+1}{7^{n+1}}$ **75.** (a) 9   (b) -0.7769

77. Diverges; $\displaystyle\lim_{n\to\infty}\left|\frac{a_{n+1}}{a_n}\right| > 1$

79. Converges; $\displaystyle\lim_{n\to\infty}\left|\frac{a_{n+1}}{a_n}\right| < 1$ **81.** Diverges; $\lim a_n \neq 0$

83. Converges   **85.** Converges   **87.** $(-3, 3)$
89. $(-2, 0]$ **91.** $x = 0$
93. See Theorem 9.17 on page 627.
95. No; the series $\displaystyle\sum_{n=1}^{\infty} \frac{1}{n + 10,000}$ diverges.
97. Absolutely; by Theorem 9.17   **99–105.** Proofs
107. (a) Diverges   (b) Converges   (c) Converges
(d) Converges for all integers $x \geq 2$
109. Putnam Problem 7, morning session, 1951

Section 9.7 *(page 658)*

1. d   **2.** c   **3.** a   **4.** b

5. $P_1 = \frac{1}{16}x + \frac{1}{4}$ **7.** $P_1 = \sqrt{2}x + \sqrt{2}(4 - \pi)/4$

P_1 is the first-degree Taylor polynomial for f at 4. P_1 is the first-degree Taylor polynomial for f at $\pi/4$.

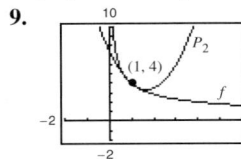

9.

x	0	0.8	0.9	1	1.1
$f(x)$	Error	4.4721	4.2164	4.0000	3.8139
$P_2(x)$	7.5000	4.4600	4.2150	4.0000	3.8150

x	1.2	2
$f(x)$	3.6515	2.8284
$P_2(x)$	3.6600	3.5000

11. (a)   (b) $f^{(2)}(0) = -1$ $P_2^{(2)}(0) = -1$
$f^{(4)}(0) = 1$ $P_4^{(4)}(0) = 1$
$f^{(6)}(0) = -1$ $P_6^{(6)}(0) = -1$
(c) $f^{(n)}(0) = P_n^{(n)}(0)$

13. $1 + 4x + 8x^2 + \frac{32}{3}x^3 + \frac{32}{3}x^4$
15. $1 - \frac{1}{2}x + \frac{1}{8}x^2 - \frac{1}{48}x^3 + \frac{1}{384}x^4$ **17.** $x - \frac{1}{6}x^3 + \frac{1}{120}x^5$
19. $x + x^2 + \frac{1}{2}x^3 + \frac{1}{6}x^4$ **21.** $1 - x + x^2 - x^3 + x^4 - x^5$
23. $1 + \frac{1}{2}x^2$ **25.** $2 - 2(x - 1) + 2(x - 1)^2 - 2(x - 1)^3$
27. $2 + \frac{1}{4}(x - 4) - \frac{1}{64}(x - 4)^2 + \frac{1}{512}(x - 4)^3$
29. $\ln 2 + \frac{1}{2}(x - 2) - \frac{1}{8}(x - 2)^2 + \frac{1}{24}(x - 2)^3 - \frac{1}{64}(x - 2)^4$

31. (a) $P_3(x) = \pi x + \frac{\pi^3}{3}x^3$

(b) $Q_3(x) = 1 + 2\pi\left(x - \frac{1}{4}\right) + 2\pi^2\left(x - \frac{1}{4}\right)^2 + \frac{8\pi^3}{3}\left(x - \frac{1}{4}\right)^3$

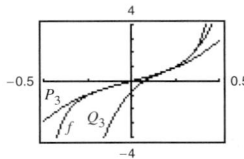

33. (a)

x	0	0.25	0.50	0.75	1.00
$\sin x$	0	0.2474	0.4794	0.6816	0.8415
$P_1(x)$	0	0.25	0.50	0.75	1.00
$P_3(x)$	0	0.2474	0.4792	0.6797	0.8333
$P_5(x)$	0	0.2474	0.4794	0.6817	0.8417

(b)

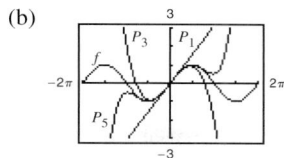

(c) As the distance increases, the polynomial approximation becomes less accurate.

35. (a) $P_3(x) = x + \frac{1}{6}x^3$

(b)

x	-0.75	-0.50	-0.25	0	0.25
$f(x)$	-0.848	-0.524	-0.253	0	0.253
$P_3(x)$	-0.820	-0.521	-0.253	0	0.253

x	0.50	0.75
$f(x)$	0.524	0.848
$P_3(x)$	0.521	0.820

(c)

37.

39.

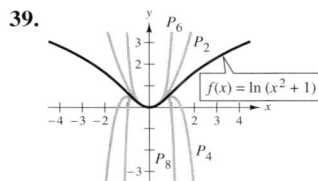

41. 2.7083   43. 0.7419   45. $R_4 \leq 2.03 \times 10^{-5}$; 0.000001
47. $R_3 \leq 7.82 \times 10^{-3}$; 0.00085   49. 3   51. 5
53. $n = 9$; $\ln(1.5) \approx 0.4055$ 55. $-0.3936 < x < 0$
57. $-0.9467 < x < 0.9467$
59. The graphs of the approximating polynomial P and the elementary function f both pass through the point $(c, f(c))$, and the slope of the graph of P is the same as the slope of the graph of f at the point $(c, f(c))$. If P is of degree n, then the first n derivatives of f and P agree at c. This allows the graph of P to resemble the graph of f near the point $(c, f(c))$.
61. See "Definitions of nth Taylor Polynomial and nth Maclaurin Polynomial" on page 638.
63. As the degree of the polynomial increases, the graph of the Taylor polynomial becomes a better and better approximation of the function within the interval of convergence. Therefore, the accuracy is increased.
65. (a) $f(x) \approx P_4(x) = 1 + x + (1/2)x^2 + (1/6)x^3 + (1/24)x^4$
    $g(x) \approx Q_5(x) = x + x^2 + (1/2)x^3 + (1/6)x^4 + (1/24)x^5$
    $Q_5(x) = xP_4(x)$
    (b) $g(x) \approx P_6(x) = x^2 - x^4/3! + x^6/5!$
    (c) $g(x) \approx P_4(x) = 1 - x^2/3! + x^4/5!$
67. (a) $Q_2(x) = -1 + (\pi^2/32)(x + 2)^2$
    (b) $R_2(x) = -1 + (\pi^2/32)(x - 6)^2$

(c) No. Horizontal translations of the result in part (a) are possible only at $x = -2 + 8n$ (where n is an integer) because the period of f is 8.
69. Proof
71. As you move away from $x = c$, the Taylor polynomial becomes less and less accurate.

Section 9.8   (page 654)

1. 0   3. 2   5. $R = 1$ 7. $R = \frac{1}{4}$ 9. $R = \infty$
11. $(-4, 4)$ 13. $(-1, 1]$ 15. $(-\infty, \infty)$ 17. $x = 0$
19. $(-6, 6)$ 21. $(-5, 13]$ 23. $(0, 2]$ 25. $(0, 6)$
27. $\left(-\frac{1}{2}, \frac{1}{2}\right)$ 29. $(-\infty, \infty)$ 31. $(-1, 1)$ 33. $x = 3$
35. $R = c$ 37. $(-k, k)$ 39. $(-1, 1)$
41. $\displaystyle\sum_{n=1}^{\infty} \frac{x^{n-1}}{(n-1)!}$ 43. $\displaystyle\sum_{n=1}^{\infty} \frac{x^{2n-1}}{(2n-1)!}$
45. (a) $(-3, 3)$ (b) $(-3, 3)$ (c) $(-3, 3)$ (d) $[-3, 3)$
47. (a) $(0, 2]$ (b) $(0, 2)$ (c) $(0, 2)$ (d) $[0, 2]$
49. A series of the form
$$\sum_{n=0}^{\infty} a_n(x - c)^n = a_0 + a_1(x - c) + a_2(x - c)^2 + \cdots$$
$$+ a_n(x - c)^n + \cdots$$
is called a power series centered at c, where c is a constant.
51. The interval of convergence of a power series is the set of all values of x for which the power series converges.
53. You differentiate and integrate the power series term by term. The radius of convergence remains the same. However, the interval of convergence might change.
55. Many answers possible.
    (a) $\displaystyle\sum_{n=1}^{\infty} \left(\frac{x}{2}\right)^n$ Geometric: $\left|\frac{x}{2}\right| < 1 \implies |x| < 2$
    (b) $\displaystyle\sum_{n=1}^{\infty} \frac{(-1)^n x^n}{n}$ converges for $-1 < x \leq 1$
    (c) $\displaystyle\sum_{n=1}^{\infty} (2x + 1)^n$ Geometric:
    $|2x + 1| < 1 \implies -1 < x < 0$
    (d) $\displaystyle\sum_{n=1}^{\infty} \frac{(x - 2)^n}{n4^n}$ converges for $-2 \leq x < 6$
57. (a) For $f(x)$: $(-\infty, \infty)$; For $g(x)$: $(-\infty, \infty)$
    (b) Proof   (c) Proof   (d) $f(x) = \sin x$; $g(x) = \cos x$
59–63. Proofs
65. (a) Proof   (b) Proof
    (c)

    (d) 0.92

67. (a) $\frac{8}{3}$

    (b) $\frac{8}{13}$

(c) The alternating series converges more rapidly. The partial sums of the series of positive terms approach the sum from below. The partial sums of the alternating series alternate sides of the horizontal line representing the sum.

(d)

M	10	100	1000	10,000
N	5	14	24	35

69.

71.

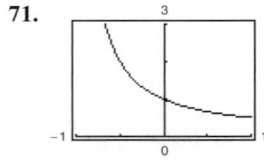

$f(x) = \cos x$ $f(x) = 1/(1 + x)$

73. False. Let $a_n = (-1)^n/(n2^n)$.   75. True   77. Proof
79. (a) $(-1, 1)$ (b) $f(x) = (c_0 + c_1 x + c_2 x^2)/(1 - x^3)$
81. Proof

Section 9.9   (page 662)

1. $\sum\limits_{n=0}^{\infty} \dfrac{x^n}{4^{n+1}}$ 3. $\sum\limits_{n=0}^{\infty} \dfrac{4}{3}\left(\dfrac{-x}{3}\right)^n$

5. $\sum\limits_{n=0}^{\infty} \dfrac{(x-1)^n}{2^{n+1}}$ 7. $\sum\limits_{n=0}^{\infty} (3x)^n$ 9. $-\dfrac{5}{9}\sum\limits_{n=0}^{\infty}\left[\dfrac{2}{9}(x+3)\right]^n$

$(-1, 3)$ $\left(-\dfrac{1}{3}, \dfrac{1}{3}\right)$ $\left(-\dfrac{15}{2}, \dfrac{3}{2}\right)$

11. $\sum\limits_{n=0}^{\infty} \dfrac{(-1)^n 3^{n+1} x^n}{4^{n+1}}$ 13. $\sum\limits_{n=0}^{\infty}\left[\dfrac{1}{(-3)^n} - 1\right]x^n$

$\left(-\dfrac{4}{3}, \dfrac{4}{3}\right)$ $(-1, 1)$

15. $\sum\limits_{n=0}^{\infty} x^n[1 + (-1)^n] = 2\sum\limits_{n=0}^{\infty} x^{2n}$ 17. $2\sum\limits_{n=0}^{\infty} x^{2n}$

$(-1, 1)$ $(-1, 1)$

19. $\sum\limits_{n=1}^{\infty} n(-1)^n x^{n-1}$ 21. $\sum\limits_{n=0}^{\infty} \dfrac{(-1)^n x^{n+1}}{n+1}$

$(-1, 1)$ $(-1, 1]$

23. $\sum\limits_{n=0}^{\infty} (-1)^n x^{2n}$ 25. $\sum\limits_{n=0}^{\infty} (-1)^n (2x)^{2n}$

$(-1, 1)$ $\left(-\dfrac{1}{2}, \dfrac{1}{2}\right)$

27.

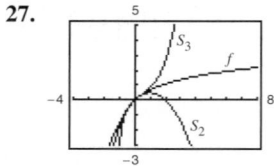

x	0.0	0.2	0.4	0.6	0.8	1.0
S_2	0.000	0.180	0.320	0.420	0.480	0.500
$\ln(x+1)$	0.000	0.182	0.336	0.470	0.588	0.693
S_3	0.000	0.183	0.341	0.492	0.651	0.833

29. (a)

(b) $\ln x, 0 < x \le 2, R = 1$
(c) -0.6931
(d) $\ln(0.5)$; The error is approximately 0.

31. 0.245   33. 0.125   35. $\sum\limits_{n=1}^{\infty} nx^{n-1}, -1 < x < 1$

37. $\sum\limits_{n=0}^{\infty} (2n+1)x^n, -1 < x < 1$

39. $E(n) = 2$. Because the probability of obtaining a head on a single toss is $\frac{1}{2}$, it is expected that, on average, a head will be obtained in two tosses.

41. Because $\dfrac{1}{1+x} = \dfrac{1}{1-(-x)}$, substitute $(-x)$ into the geometric series.

43. Because $\dfrac{5}{1+x} = 5\left(\dfrac{1}{1-(-x)}\right)$, substitute $(-x)$ into the geometric series and then multiply the series by 5.

45. Proof   47. (a) Proof   (b) 3.14
49. $\ln\frac{3}{2} \approx 0.4055$; See Exercise 21.
51. $\ln\frac{7}{5} \approx 0.3365$; See Exercise 49.
53. $\arctan\frac{1}{2} \approx 0.4636$; See Exercise 52.
55. The series in Exercise 52 converges to its sum at a lower rate because its terms approach 0 at a much lower rate.
57. The series converges on the interval $(-5, 3)$ and perhaps also at one or both endpoints.
59. $\sqrt{3}\pi/6$ 61. $S_1 = 0.3183098862, 1/\pi \approx 0.3183098862$

Section 9.10   (page 673)

1. $\sum\limits_{n=0}^{\infty} \dfrac{(2x)^n}{n!}$ 3. $\dfrac{\sqrt{2}}{2}\sum\limits_{n=0}^{\infty} \dfrac{(-1)^{n(n+1)/2}}{n!}\left(x - \dfrac{\pi}{4}\right)^n$

5. $\sum\limits_{n=0}^{\infty} (-1)^n(x-1)^n$ 7. $\sum\limits_{n=0}^{\infty} \dfrac{(-1)^n(x-1)^{n+1}}{n+1}$

9. $\sum\limits_{n=0}^{\infty} \dfrac{(-1)^n(3x)^{2n+1}}{(2n+1)!}$ 11. $1 + x^2/2! + 5x^4/4! + \cdots$

13–15. Proofs   17. $\sum\limits_{n=0}^{\infty} (-1)^n(n+1)x^n$

19. $1 + \sum\limits_{n=1}^{\infty} \dfrac{1 \cdot 3 \cdot 5 \cdots (2n-1)x^n}{2^n n!}$

21. $\dfrac{1}{2}\left[1 + \sum\limits_{n=1}^{\infty} \dfrac{(-1)^n 1 \cdot 3 \cdot 5 \cdots (2n-1)x^{2n}}{2^{3n} n!}\right]$

23. $1 + \dfrac{x}{2} + \sum\limits_{n=2}^{\infty} \dfrac{(-1)^{n+1} 1 \cdot 3 \cdot 5 \cdots (2n-3)x^n}{2^n n!}$

25. $1 + \dfrac{x^2}{2} + \sum\limits_{n=2}^{\infty} \dfrac{(-1)^{n+1} 1 \cdot 3 \cdot 5 \cdots (2n-3)x^{2n}}{2^n n!}$

27. $\sum\limits_{n=0}^{\infty} \dfrac{x^{2n}}{2^n n!}$ 29. $\sum\limits_{n=1}^{\infty} \dfrac{(-1)^{n-1} x^n}{n}$ 31. $\sum\limits_{n=0}^{\infty} \dfrac{(-1)^n(3x)^{2n+1}}{(2n+1)!}$

33. $\sum\limits_{n=0}^{\infty} \dfrac{(-1)^n 4^{2n} x^{2n}}{(2n)!}$ 35. $\sum\limits_{n=0}^{\infty} \dfrac{(-1)^n x^{3n}}{(2n)!}$

37. $\sum\limits_{n=0}^{\infty} \dfrac{x^{2n+1}}{(2n+1)!}$ 39. $\dfrac{1}{2}\left[1 + \sum\limits_{n=0}^{\infty} \dfrac{(-1)^n(2x)^{2n}}{(2n)!}\right]$

41. $\sum\limits_{n=0}^{\infty} \dfrac{(-1)^n x^{2n+2}}{(2n+1)!}$ 43. $\begin{cases} \sum\limits_{n=0}^{\infty} \dfrac{(-1)^n x^{2n}}{(2n+1)!}, & x \ne 0 \\ 1, & x = 0 \end{cases}$

45. Proof
47. $P_5(x) = x + x^2 + \frac{1}{3}x^3 - \frac{1}{30}x^5$

49. $P_5(x) = x - \frac{1}{2}x^2 - \frac{1}{6}x^3 + \frac{3}{40}x^5$

51. $P_4(x) = x - x^2 + \frac{5}{6}x^3 - \frac{5}{6}x^4$

53. $\displaystyle\sum_{n=0}^{\infty} \frac{(-1)^{(n+1)}x^{2n+3}}{(2n+3)(n+1)!}$ **55.** 0.6931   **57.** 7.3891

59. 0   **61.** 1   **63.** 0.8075   **65.** 0.9461   **67.** 0.4872

69. 0.2010   **71.** 0.7040   **73.** 0.3412

75. $P_5(x) = x - 2x^3 + \frac{2}{3}x^5$

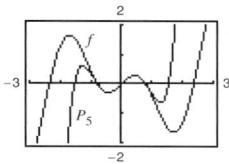 $\left[-\frac{3}{4}, \frac{3}{4}\right]$

77. $P_5(x) = (x-1) - \frac{1}{24}(x-1)^3 + \frac{1}{24}(x-1)^4 - \frac{71}{1920}(x-1)^5$

    $\left[\frac{1}{4}, 2\right]$

79. See "Guidelines for Finding a Taylor Series" on page 668.

81. (a) Replace x with $(-x)$.   (b) Replace x with $3x$.
(c) Multiply series by x.

83. Proof

85. (a)

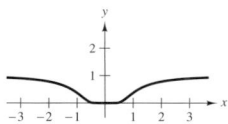

(b) Proof
(c) $\displaystyle\sum_{n=0}^{\infty} 0x^n = 0 \neq f(x)$

87. Proof   **89.** 10   **91.** -0.0390625 **93.** $\displaystyle\sum_{n=0}^{\infty} \binom{k}{n} x^n$

95. Proof

Review Exercises for Chapter 9   *(page 676)*

1. 5, 25, 125, 625, 3125   **3.** $-\frac{1}{4}, \frac{1}{16}, -\frac{1}{64}, \frac{1}{256}, -\frac{1}{1024}$ **5.** a

6. c   **7.** d   **8.** b

9.   Converges to 5

11. Converges to 5   **13.** Diverges   **15.** Converges to 0

17. Converges to 0   **19.** $a_n = 5n - 2$ **21.** $a_n = \dfrac{1}{(n!+1)}$

23. (a)

n	1	2	3	4
A_n	\$8100.00	\$8201.25	\$8303.77	\$8407.56

n	5	6	7	8
A_n	\$8512.66	\$8619.07	\$8726.80	\$8835.89

(b) \$13,148.96

25. 3, 4.5, 5.5, 6.25, 6.85

27. (a)

n	5	10	15	20	25
S_n	13.2	113.3	873.8	6648.5	50,500.3

(b)

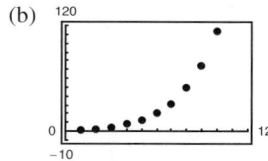

29. (a)

n	5	10	15	20	25
S_n	0.4597	0.4597	0.4597	0.4597	0.4597

(b)

31. $\frac{5}{3}$ **33.** 5.5   **35.** (a) $\displaystyle\sum_{n=0}^{\infty}(0.09)(0.01)^n$ (b) $\frac{1}{11}$

37. Diverges   **39.** Diverges   **41.** $45\frac{1}{3}$ m   **43.** Diverges

45. Converges   **47.** Diverges   **49.** Diverges

51. Converges   **53.** Diverges   **55.** Converges

57. Converges   **59.** Diverges   **61.** Diverges

63. Converges   **65.** Diverges

67. (a) Proof

(b)

n	5	10	15	20	25
S_n	2.8752	3.6366	3.7377	3.7488	3.7499

(c)   (d) 3.75

69. $P_3(x) = 1 - 2x + 2x^2 - \frac{4}{3}x^3$

71. $P_3(x) = 1 - 3x + \frac{9}{2}x^2 - \frac{9}{2}x^3$ **73.** 3 terms

75. $(-10, 10)$ **77.** $[1, 3]$ **79.** Converges only at $x = 2$

81. (a) $(-5, 5)$ (b) $(-5, 5)$ (c) $(-5, 5)$ (d) $[-5, 5)$

83. Proof   **85.** $\displaystyle\sum_{n=0}^{\infty} \frac{2}{3}\left(\frac{x}{3}\right)^n$ **87.** $\displaystyle\sum_{n=0}^{\infty} 2\left(\frac{x-1}{3}\right)^n; (-2, 4)$

89. $\ln\frac{5}{4} \approx 0.2231$ **91.** $e^{1/2} \approx 1.6487$

93. $\cos\frac{2}{3} \approx 0.7859$ **95.** $\dfrac{\sqrt{2}}{2} \displaystyle\sum_{n=0}^{\infty} \frac{(-1)^{n(n+1)/2}}{n!}\left(x - \frac{3\pi}{4}\right)^n$

97. $\displaystyle\sum_{n=0}^{\infty} \frac{(x \ln 3)^n}{n!}$ **99.** $-\displaystyle\sum_{n=0}^{\infty}(x+1)^n$

101. $1 + x/5 - 2x^2/25 + 6x^3/125 - 21x^4/625 + \cdots$

103. (a)–(c) $1 + 2x + 2x^2 + \dfrac{4}{3}x^3$ **105.** $\displaystyle\sum_{n=0}^{\infty} \dfrac{(6x)^n}{n!}$

107. $\displaystyle\sum_{n=0}^{\infty} \dfrac{(-1)^n (2x)^{2n+1}}{(2n+1)!}$ **109.** 0

P.S. Problem Solving   *(page 679)*

1. (a) 1   (b) Answers will vary. Example: $0, \frac{1}{3}, \frac{2}{3}$ (c) 0
3. Proof   **5.** (a) Proof   (b) Yes   (c) Any distance
7. (a) $\displaystyle\sum_{n=0}^{\infty} \dfrac{x^{n+2}}{(n+2)n!}; \dfrac{1}{2}$ (b) $\displaystyle\sum_{n=0}^{\infty} \dfrac{(n+1)x^n}{n!}; 5.4366$
9. For $a = b$, the series converges conditionally. For no values of a and b does the series converge absolutely.
11. Proof   **13.** (a) Proof   (b) Proof
15. (a) The height is infinite.   (b) The surface area is infinite.
(c) Proof

Chapter 10

Section 10.1   *(page 692)*

1. a   **2.** e   **3.** c   **4.** b   **5.** f   **6.** d
7. Vertex: $(0, 0)$ **9.** Vertex: $(-5, 3)$
Focus: $(-2, 0)$ Focus: $\left(-\frac{21}{4}, 3\right)$
Directrix: $x = 2$ Directrix: $x = -\frac{19}{4}$

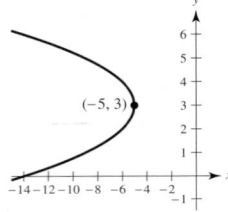

11. Vertex: $(-1, 2)$ **13.** Vertex: $(-2, 2)$
Focus: $(0, 2)$ Focus: $(-2, 1)$
Directrix: $x = -2$ Directrix: $y = 3$

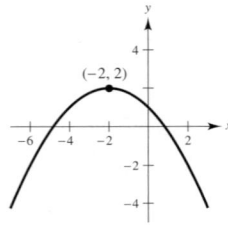

15. $y^2 - 8y + 8x - 24 = 0$ **17.** $x^2 - 32y + 160 = 0$
19. $x^2 + y - 4 = 0$ **21.** $5x^2 - 14x - 3y + 9 = 0$
23. Center: $(0, 0)$ **25.** Center: $(3, 1)$
Foci: $\left(0, \pm\sqrt{15}\right)$ Foci: $(3, 4), (3, -2)$
Vertices: $(0, \pm 4)$ Vertices: $(3, 6), (3, -4)$
$e = \sqrt{15}/4$ $e = \frac{3}{5}$

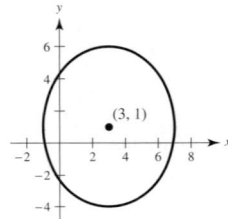

27. Center: $(-2, 3)$
Foci: $\left(-2, 3 \pm \sqrt{5}\right)$
Vertices: $(-2, 6), (-2, 0)$
$e = \sqrt{5}/3$

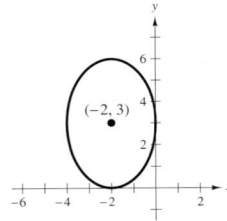

29. $x^2/36 + y^2/11 = 1$
31. $(x - 3)^2/9 + (y - 5)^2/16 = 1$
33. $x^2/16 + 7y^2/16 = 1$
35. Center: $(0, 0)$ **37.** Center: $(2, -3)$
Vertices: $(\pm 5, 0)$ Foci: $\left(2 \pm \sqrt{10}, -3\right)$
Foci: $(\pm\sqrt{41}, 0)$ Vertices: $(1, -3), (3, -3)$
Asymptotes: $y = \pm\dfrac{b}{a}x$
$\qquad\qquad = \pm\dfrac{4}{5}x$

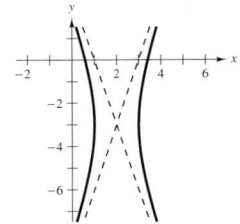

39. Degenerate hyperbola
Graph is two lines: $y = -3 \pm \frac{1}{3}(x + 1)$, intersecting at $(-1, -3)$.

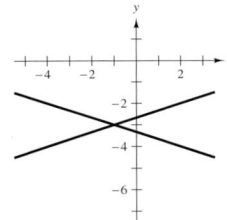

41. $x^2/1 - y^2/25 = 1$ **43.** $y^2/9 - (x - 2)^2/(9/4) = 1$
45. $y^2/4 - x^2/12 = 1$ **47.** $(x - 3)^2/9 - (y - 2)^2/4 = 1$
49. (a) $\left(6, \sqrt{3}\right): 2x - 3\sqrt{3}y - 3 = 0$
$\qquad\left(6, -\sqrt{3}\right): 2x + 3\sqrt{3}y - 3 = 0$
(b) $\left(6, \sqrt{3}\right): 9x + 2\sqrt{3}y - 60 = 0$
$\qquad\left(6, -\sqrt{3}\right): 9x - 2\sqrt{3}y - 60 = 0$
51. Ellipse   **53.** Parabola   **55.** Circle   **57.** Hyperbola
59. (a) A parabola is the set of all points (x, y) that are equidistant from a fixed line and a fixed point not on the line.
(b) For directrix $y = k - p$: $(x - h)^2 = 4p(y - k)$
For directrix $x = h - p$: $(y - k)^2 = 4p(x - h)$
(c) If P is a point on a parabola, then the tangent line to the parabola at P makes equal angles with the line passing through P and the focus, and with the line passing through P parallel to the axis of the parabola.

61. (a) A hyperbola is the set of all points (x, y) for which the absolute value of the difference between the distances from two distinct fixed points is constant.

(b) Transverse axis is horizontal: $\dfrac{(x - h)^2}{a^2} - \dfrac{(y - k)^2}{b^2} = 1$

Transverse axis is vertical: $\dfrac{(y - k)^2}{a^2} - \dfrac{(x - h)^2}{b^2} = 1$

(c) Transverse axis is horizontal:
$y = k + (b/a)(x - h)$ and $y = k - (b/a)(x - h)$
Transverse axis is vertical:
$y = k + (a/b)(x - h)$ and $y = k - (a/b)(x - h)$

63. (a) Ellipse   (b) Hyperbola   (c) Circle
(d) Sample answer: Eliminate the y^2-term.

65. $\frac{9}{4}$ m   **67.** (a) Proof   (b) Point of intersection: $(3, -3)$

69.

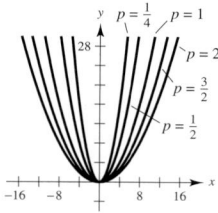

As p increases, the graph of $x^2 = 4py$ gets wider.

71. $[16(4 + 3\sqrt{3} - 2\pi]/3 \approx 15.536$ ft^2

73. Minimum distance: 147,099,713.4 km
Maximum distance: 152,096,286.6 km

75. About 0.9372   **77.** $e \approx 0.9671$

79. (a) Area $= 2\pi$ (b) Volume $= 8\pi/3$
Surface area $= [2\pi(9 + 4\sqrt{3}\pi)]/9 \approx 21.48$
(c) Volume $= 16\pi/3$
Surface area $= \dfrac{4\pi[6 + \sqrt{3}\ln(2 + \sqrt{3})]}{3} \approx 34.69$

81. 37.96   **83.** 40   **85.** $(x - 6)^2/9 - (y - 2)^2/7 = 1$

87. $x \approx 110.3$ mi   **89.** Proof

91. False. See the definition of a parabola   **93.** True   **95.** True

97. Putnam Problem B4, 1976

Section 10.2   (page 703)

1.

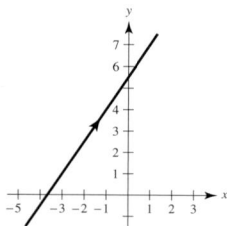

$3x - 2y + 11 = 0$

3.

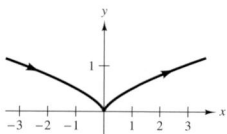

$y = (x - 1)^2$

5.

$y = \frac{1}{2}x^{2/3}$

7.

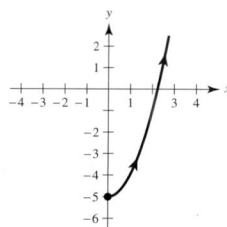

$y = x^2 - 5, \quad x \geq 0$

9.

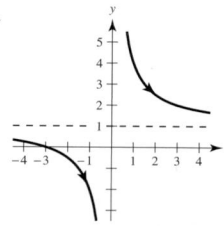

$y = (x + 3)/x$

11.

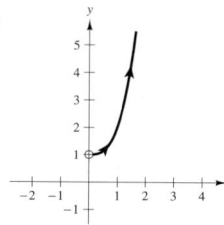

$y = |x - 4|/2$

13.

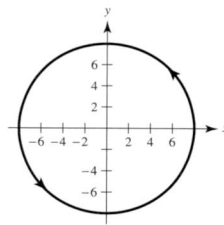

$y = x^3 + 1, \quad x > 0$

15.

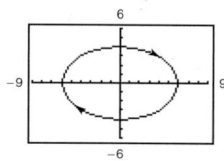

$y = 1/x, \quad |x| \geq 1$

17.

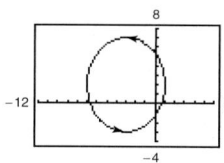

$x^2 + y^2 = 64$

19.

$\dfrac{x^2}{36} + \dfrac{y^2}{16} = 1$

21.

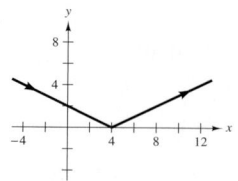

$\dfrac{(x - 4)^2}{4} + \dfrac{(y + 1)^2}{1} = 1$

23.

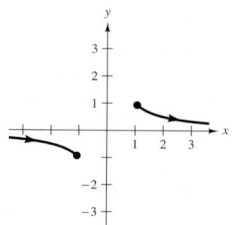

$\dfrac{(x + 3)^2}{16} + \dfrac{(y - 2)^2}{25} = 1$

25.

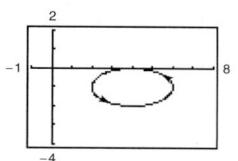

$\dfrac{x^2}{16} - \dfrac{y^2}{9} = 1$

27.

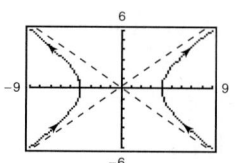

$y = \ln x$

29.

$y = \dfrac{1}{x^3}, \quad x > 0$

31. Each curve represents a portion of the line $y = 2x + 1$.

Domain	Orientation	Smooth
(a) $-\infty < x < \infty$	Up	Yes
(b) $-1 \le x \le 1$	Oscillates	No, $\dfrac{dx}{d\theta} = \dfrac{dy}{d\theta} = 0$ when $\theta = 0, \pm\pi, \pm 2\pi, \dots$
(c) $0 < x < \infty$	Down	Yes
(d) $0 < x < \infty$	Up	Yes

33. (a) and (b) represent the parabola $y = 2(1 - x^2)$ for $-1 \le x \le 1$. The curve is smooth. The orientation is from right to left in part (a) and in part (b).

35. (a)

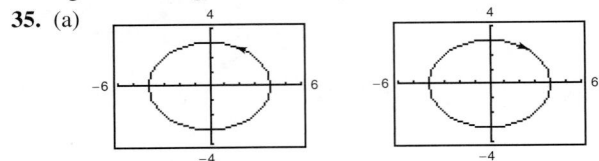

(b) The orientation is reversed.

(c) The orientation is reversed.

(d) Answers will vary. For example,

$x = 2 \sec t$ $x = 2 \sec(-t)$

$y = 5 \sin t$ $y = 5 \sin(-t)$

have the same graphs, but their orientations are reversed.

37. $y - y_1 = \dfrac{y_2 - y_1}{x_2 - x_1}(x - x_1)$ **39.** $\dfrac{(x - h)^2}{a^2} + \dfrac{(y - k)^2}{b^2} = 1$

41. $x = 4t$
$y = -7t$
(Solution is not unique.)

43. $x = 3 + 2\cos\theta$
$y = 1 + 2\sin\theta$
(Solution is not unique.)

45. $x = 10\cos\theta$
$y = 6\sin\theta$
(Solution is not unique.)

47. $x = 4\sec\theta$
$y = 3\tan\theta$
(Solution is not unique.)

49. $x = t$
$y = 6t - 5$;
$x = t + 1$
$y = 6t + 1$
(Solution is not unique.)

51. $x = t$
$y = t^3$;
$x = \tan t$
$y = \tan^3 t$
(Solution is not unique.)

53. $x = t + 3, \ y = 2t + 1$ **55.** $x = t, \ y = t^2$

57.

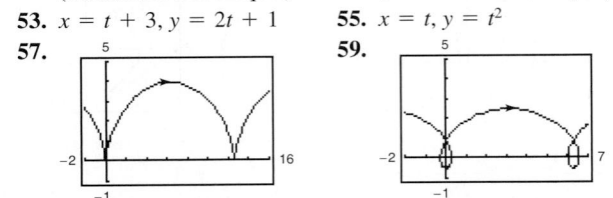

Not smooth at $\theta = 2n\pi$

59.

Smooth everywhere

61.

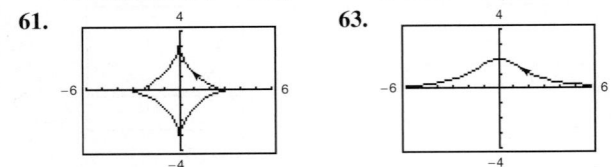

Not smooth at $\theta = \frac{1}{2}n\pi$

63.

Smooth everywhere

65. A plane curve C is a set of parametric equations, $x = f(t)$ and $y = g(t)$, and the graph of the parametric equations.

67. A curve C represented by $x = f(t)$ and $y = g(t)$ on an interval I is called smooth when f' and g' are continuous on I and not simultaneously 0, except possibly at the endpoints of I.

69. d; $(4, 0)$ is on the graph.   **71.** b; $(1, 0)$ is on the graph.

73. $x = a\theta - b\sin\theta; \ y = a - b\cos\theta$

75. False. The graph of the parametric equations is the portion of the line $y = x$ when $x \ge 0$.

77. True

79. (a) $x = \left(\frac{440}{3}\cos\theta\right)t; \ y = 3 + \left(\frac{440}{3}\sin\theta\right)t - 16t^2$

(b)

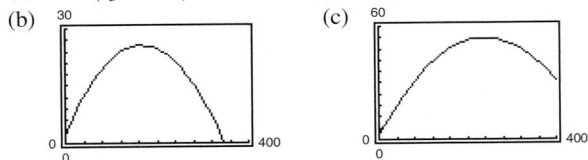

Not a home run

(c)

Home run

(d) $19.4°$

Section 10.3   *(page 711)*

1. $-3/t$ **3.** -1

5. $\dfrac{dy}{dx} = \dfrac{3}{4}, \dfrac{d^2y}{dx^2} = 0$; Neither concave upward nor concave downward

7. $dy/dx = 2t + 3, \ d^2y/dx^2 = 2$
At $t = -1, dy/dx = 1, \ d^2y/dx^2 = 2$; Concave upward

9. $dy/dx = -\cot\theta, \ d^2y/dx^2 = -(\csc\theta)^3/4$
At $\theta = \pi/4, dy/dx = -1, \ d^2y/dx^2 = -\sqrt{2}/2$;
Concave downward

11. $dy/dx = 2\csc\theta, \ d^2y/dx^2 = -2\cot^3\theta$
At $\theta = \pi/6, dy/dx = 4, \ d^2y/dx^2 = -6\sqrt{3}$;
Concave downward

13. $dy/dx = -\tan\theta, \ d^2y/dx^2 = \sec^4\theta\csc\theta/3$
At $\theta = \pi/4, dy/dx = -1, \ d^2y/dx^2 = 4\sqrt{2}/3$;
Concave upward

15. $\left(-2/\sqrt{3}, 3/2\right)$: $3\sqrt{3}x - 8y + 18 = 0$
$(0, 2)$: $y - 2 = 0$
$\left(2\sqrt{3}, 1/2\right)$: $\sqrt{3}x + 8y - 10 = 0$

17. $(0, 0)$: $2y - x = 0$
$(-3, -1)$: $y + 1 = 0$
$(-3, 3)$: $2x - y + 9 = 0$

19. (a) and (d)

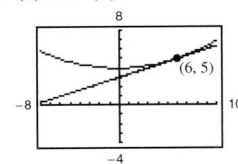

(b) At $t = 1, \ dx/dt = 6,$
$dy/dt = 2,$ and $dy/dx = 1/3.$
(c) $y = \frac{1}{3}x + 3$

21. (a) and (d)

(b) At $t = -1, \ dx/dt = -3,$
$dy/dt = 0,$ and $dy/dx = 0.$
(c) $y = 2$

23. $y = \pm\frac{3}{4}x$ **25.** $y = 3x - 5$ and $y = 1$

27. Horizontal: $(1, 0), (-1, \pi), (1, -2\pi)$
Vertical: $(\pi/2, 1), (-3\pi/2, -1), (5\pi/2, 1)$

29. Horizontal: $(4, 0)$
Vertical: None

31. Horizontal: $(5, -2), (3, 2)$
Vertical: None

33. Horizontal: $(0, 3), (0, -3)$
    Vertical: $(3, 0), (-3, 0)$

35. Horizontal: $(5, -1), (5, -3)$ **37.** Horizontal: None
    Vertical: $(8, -2), (2, -2)$ Vertical: $(1, 0), (-1, 0)$

39. Concave downward: $-\infty < t < 0$
    Concave upward: $0 < t < \infty$

41. Concave upward: $t > 0$

43. Concave downward: $0 < t < \pi/2$
    Concave upward: $\pi/2 < t < \pi$

45. $4\sqrt{13} \approx 14.422$ **47.** $\sqrt{2}(1 - e^{-\pi/2}) \approx 1.12$

49. $\frac{1}{12}[\ln(\sqrt{37} + 6) + 6\sqrt{37}] \approx 3.249$ **51.** $6a$ **53.** $8a$

55. (a) 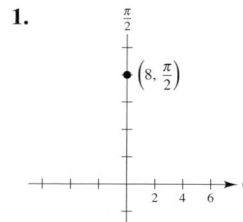   (b) 219.2 ft
                          (c) 230.8 ft

57. (a) 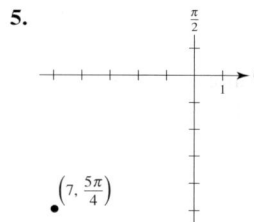   (b) $(0, 0), (4\sqrt[3]{2}/3, 4\sqrt[3]{4}/3)$
                          (c) About 6.557

59. (a)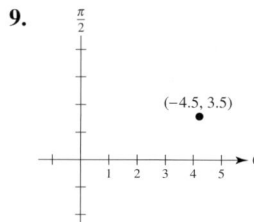

    (b) The average speed of the particle on the second path is twice the average speed of the particle on the first path.

    (c) 4π

61. $S = 2\pi \displaystyle\int_0^4 \sqrt{10}(t + 2)\,dt = 32\pi\sqrt{10} \approx 317.907$

63. $S = 2\pi \displaystyle\int_0^{\pi/2} \left(\sin\theta\cos\theta\sqrt{4\cos^2\theta + 1}\right) d\theta$

    $= \dfrac{(5\sqrt{5} - 1)\pi}{6}$

    ≈ 5.330

65. (a) $27\pi\sqrt{13}$ (b) $18\pi\sqrt{13}$ **67.** 50π **69.** $12\pi a^2/5$

71. See Theorem 10.7, Parametric Form of the Derivative, on page 706.

73. 6

75. (a) $S = 2\pi \displaystyle\int_a^b g(t) \sqrt{\left(\dfrac{dx}{dt}\right)^2 + \left(\dfrac{dy}{dt}\right)^2}\,dt$

    (b) $S = 2\pi \displaystyle\int_a^b f(t) \sqrt{\left(\dfrac{dx}{dt}\right)^2 + \left(\dfrac{dy}{dt}\right)^2}\,dt$

77. Proof   **79.** $3\pi/2$ **81.** d   **82.** b   **83.** f   **84.** c

85. a   **86.** e   **87.** $\left(\frac{3}{4}, \frac{8}{5}\right)$ **89.** 288π

91. (a) $dy/dx = \sin\theta/(1 - \cos\theta)$; $d^2y/dx^2 = -1/[a(\cos\theta - 1)^2]$

    (b) $y = (2 + \sqrt{3})[x - a(\pi/6 - \frac{1}{2})] + a(1 - \sqrt{3}/2)$

    (c) $(a(2n + 1)\pi, 2a)$

    (d) Concave downward on $(0, 2\pi), (2\pi, 4\pi)$, etc.

    (e) $s = 8a$

93. Proof

95. (a)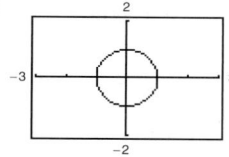

    (b) Circle of radius 1 and center at $(0, 0)$ except the point $(-1, 0)$

    (c) As t increases from -20 to 0, the speed increases, and as t increases from 0 to 20, the speed decreases.

97. False. $\dfrac{d^2y}{dx^2} = \dfrac{\dfrac{d}{dt}\left[\dfrac{g'(t)}{f'(t)}\right]}{f'(t)} = \dfrac{f'(t)g''(t) - g'(t)f''(t)}{[f'(t)]^3}.$

Section 10.4 *(page 722)*

1.

    $(0, 8)$

3.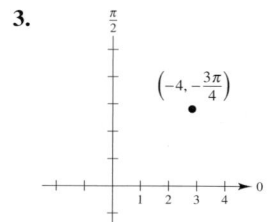

    $(2\sqrt{2}, 2\sqrt{2}) \approx (2.828, 2.828)$

5.

    $(-4.95, -4.95)$

7.

    $(-1.004, 0.996)$

9.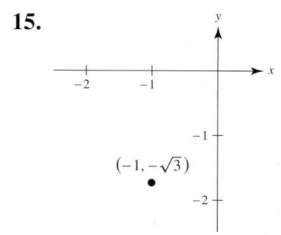

    $(4.214, 1.579)$

11.

    $(2\sqrt{2}, \pi/4), (-2\sqrt{2}, 5\pi/4)$

13.

    $(5, 2.214), (-5, 5.356)$

15.

    $(2, 4\pi/3), (-2, \pi/3)$

17.

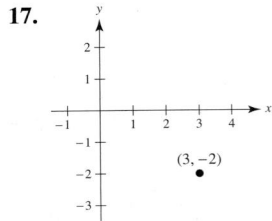

(3.606, −0.588)
(−3.606, 2.554)

19.

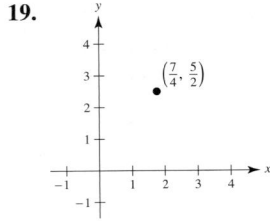

$\left(\frac{7}{4}, \frac{5}{2}\right)$

(3.052, 0.960)
(−3.052, 4.102)

21. (a)

(4, 3.5)

(b)

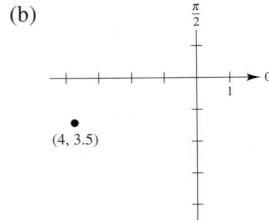

(4, 3.5)

23. $r = 3$

25. $r = a$

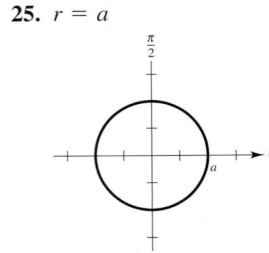

27. $r = 8 \csc \theta$

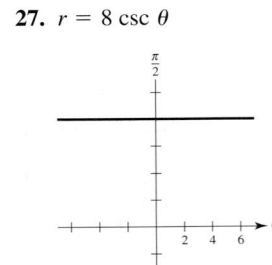

29. $r = \dfrac{-2}{3 \cos \theta - \sin \theta}$

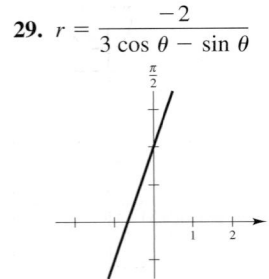

31. $r = 9 \csc^2 \theta \cos \theta$

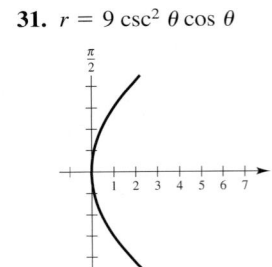

33. $x^2 + y^2 = 16$

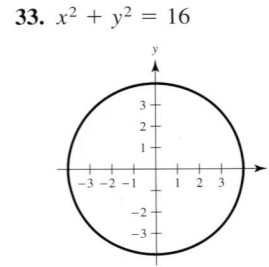

35. $x^2 + y^2 - 3y = 0$

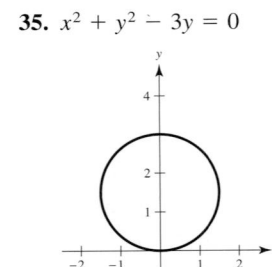

37. $\sqrt{x^2 + y^2} = \arctan(y/x)$

39. $x - 3 = 0$

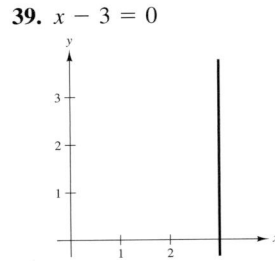

41. $x^2 - y = 0$

43.

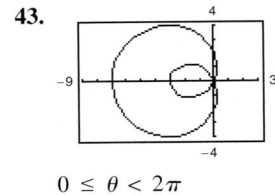

$0 \le \theta < 2\pi$

45.

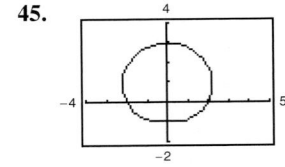

$0 \le \theta < 2\pi$

47.

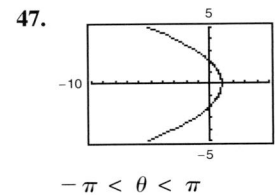

$-\pi < \theta < \pi$

49.

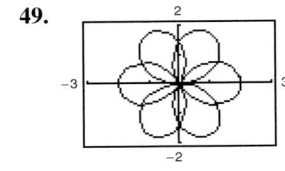

$0 \le \theta < 4\pi$

51.

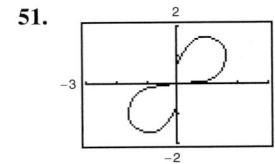

$0 \le \theta < \pi/2$

53. $(x - h)^2 + (y - k)^2 = h^2 + k^2$
Radius: $\sqrt{h^2 + k^2}$
Center: (h, k)

55. $\sqrt{17}$ **57.** About 5.6

59. $\dfrac{dy}{dx} = \dfrac{2 \cos \theta (3 \sin \theta + 1)}{6 \cos^2 \theta - 2 \sin \theta - 3}$
$(5, \pi/2)$: $dy/dx = 0$
$(2, \pi)$: $dy/dx = -2/3$
$(-1, 3\pi/2)$: $dy/dx = 0$

61. (a) and (b)   (c) $dy/dx = -1$

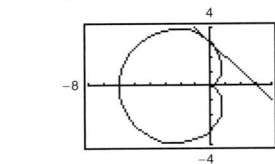

63. (a) and (b)   (c) $dy/dx = -\sqrt{3}$

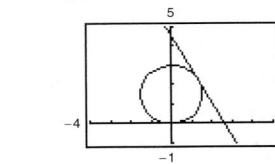

65. Horizontal: $(2, 3\pi/2), \left(\frac{1}{2}, \pi/6\right), \left(\frac{1}{2}, 5\pi/6\right)$
Vertical: $\left(\frac{3}{2}, 7\pi/6\right), \left(\frac{3}{2}, 11\pi/6\right)$

67. $(5, \pi/2), (1, 3\pi/2)$

69.

$\theta = 0$

73.

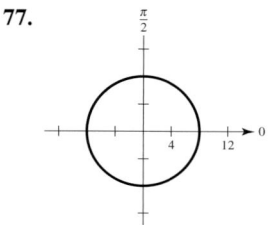

$\theta = \pi/6, \pi/2, 5\pi/6$

77.

81.

85.

89.

71.

$\theta = \pi/2$

75.

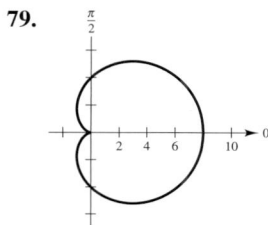

$\theta = 0, \pi/2$

79.

83.

87.

91.

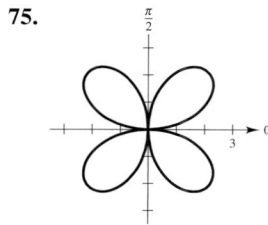

93. The rectangular coordinate system is a collection of points of the form (x, y), where x is the directed distance from the y-axis to the point and y is the directed distance from the x-axis to the point. Every point has a unique representation.

The polar coordinate system is a collection of points of the form (r, θ), where r is the directed distance from the origin O to a point P and θ is the directed angle, measured counterclockwise, from the polar axis to the segment $\overline{OP}$. Polar coordinates do not have unique representations.

95. Slope of tangent line to graph of $r = f(\theta)$ at (r, θ) is
$$\frac{dy}{dx} = \frac{f(\theta)\cos\theta + f'(\theta)\sin\theta}{-f(\theta)\sin\theta + f'(\theta)\cos\theta}.$$
If $f(\alpha) = 0$ and $f'(\alpha) \neq 0$, then $\theta = \alpha$ is tangent at the pole.

97. (a)   (b)

(c)

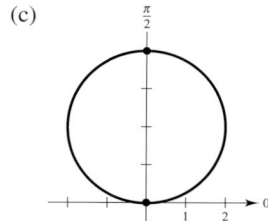

99. Proof

101. (a) $r = 2 - \sin(\theta - \pi/4)$ (b) $r = 2 + \cos\theta$
$$= 2 - \frac{\sqrt{2}(\sin\theta - \cos\theta)}{2}$$

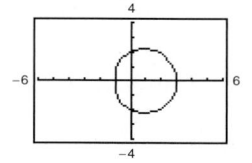

(c) $r = 2 + \sin\theta$ (d) $r = 2 - \cos\theta$

103. (a)   (b)

105.

$\psi = \pi/2$

107.

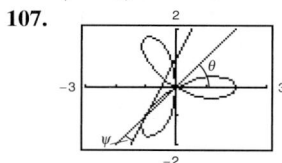

$\psi = \arctan\frac{1}{3} \approx 18.4°$

109.

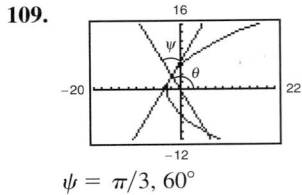

$\psi = \pi/3, 60°$

111. True　　**113.** True

Section 10.5　*(page 731)*

1. $8\displaystyle\int_0^{\pi/2} \sin^2 \theta \, d\theta$　**3.** $\dfrac{1}{2}\displaystyle\int_{\pi/2}^{3\pi/2} (3 - 2\sin\theta)^2 \, d\theta$　**5.** 9π

7. $\pi/3$　**9.** $\pi/8$　**11.** $3\pi/2$　**13.** 27π　**15.** 4

17.

$\left(2\pi - 3\sqrt{3}\right)/2$

19.

$\left(2\pi - 3\sqrt{3}\right)/2$

21.

$\pi + 3\sqrt{3}$

23.

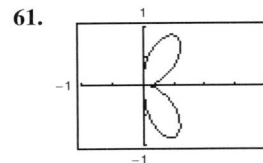

$9\pi + 27\sqrt{3}$

25. $(1, \pi/2), (1, 3\pi/2), (0, 0)$

27. $\left(\dfrac{2 - \sqrt{2}}{2}, \dfrac{3\pi}{4}\right), \left(\dfrac{2 + \sqrt{2}}{2}, \dfrac{7\pi}{4}\right), (0, 0)$

29. $\left(\dfrac{3}{2}, \dfrac{\pi}{6}\right), \left(\dfrac{3}{2}, \dfrac{5\pi}{6}\right), (0, 0)$　**31.** $(2, 4), (-2, -4)$

33.

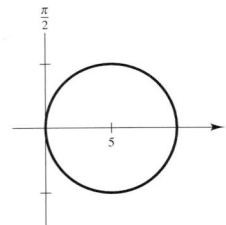

$(0, 0), (0.935, 0.363),$
$(0.535, -1.006)$
The graphs reach the pole
at different times (θ-values).

35.

$\dfrac{4}{3}\left(4\pi - 3\sqrt{3}\right)$

37.

$11\pi - 24$

39.

$\dfrac{2}{3}\left(4\pi - 3\sqrt{3}\right)$

41.

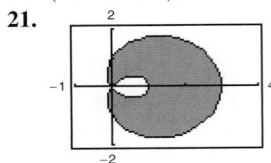

$\pi/3 + \sqrt{3}/2$

43. $5\pi a^2/4$　　**45.** $(a^2/2)(\pi - 2)$

47. (a) $(x^2 + y^2)^{3/2} = ax^2$

(b)

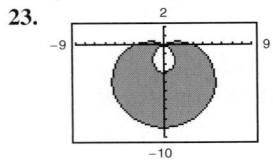

(c) $15\pi/2$

49. The area enclosed by the function is $\pi a^2/4$ if n is odd and is $\pi a^2/2$ if n is even.

51. 16π　**53.** 4π　**55.** 8

57.

About 4.16

59.

About 0.71

61.

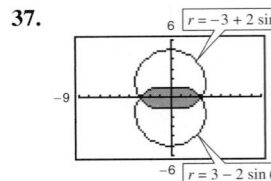

About 4.39

63. 36π　**65.** $\dfrac{2\pi\sqrt{1 + a^2}}{1 + 4a^2}(e^{\pi a} - 2a)$　**67.** 21.87

69. You will only find simultaneous points of intersection. There may be intersection points that do not occur with the same coordinates in the two graphs.

71. (a) Circle of radius 5
Area $= 25\pi$

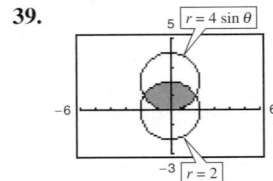

(b) Circle of radius 5/2
Area $= \dfrac{25}{4}\pi$

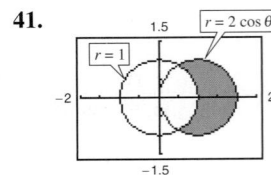

73. $40\pi^2$

75. (a) 16π

(b)

θ	0.2	0.4	0.6	0.8	1.0	1.2	1.4
A	6.32	12.14	17.06	20.80	23.27	24.60	25.08

(c) and (d) For $\frac{1}{4}$ of area ($4\pi \approx 12.57$): 0.42

For $\frac{1}{2}$ of area ($8\pi \approx 25.13$): 1.57($\pi/2$)

For $\frac{3}{4}$ of area ($12\pi \approx 37.70$): 2.73

(e) No. The results do not depend on the radius. Answers will vary.

77. Circle

79. (a)

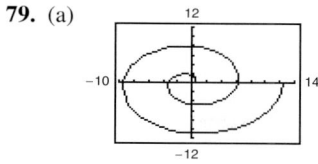

The graph becomes larger and more spread out. The graph is reflected over the *y*-axis.

(b) $(an\pi, n\pi)$, where $n = 1, 2, 3, \ldots$

(c) About 21.26   (d) $4/3\pi^3$

81. $r = \sqrt{2}\cos\theta$

83. False. The graphs of $f(\theta) = 1$ and $g(\theta) = -1$ coincide.

85. Proof

Section 10.6   *(page 739)*

1.

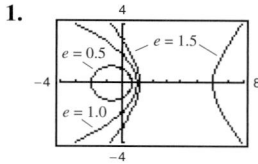

(a) Parabola
(b) Ellipse
(c) Hyperbola

3.

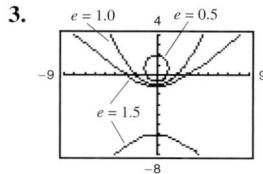

(a) Parabola
(b) Ellipse
(c) Hyperbola

5. (a)

Ellipse
As $e \to 1^-$, the ellipse becomes more elliptical, and as $e \to 0^+$, it becomes more circular.

(b)

Parabola

(c)

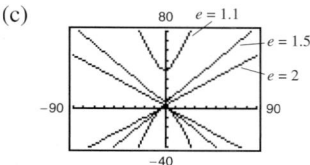

Hyperbola
As $e \to 1^+$, the hyperbola opens more slowly, and as $e \to \infty$, it opens more rapidly.

7. c   **8.** f   **9.** a   **10.** e   **11.** b   **12.** d

13. $e = 1$
Distance $= 1$

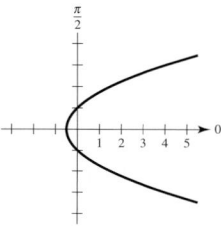

Parabola

15. $e = 3$
Distance $= \frac{1}{2}$

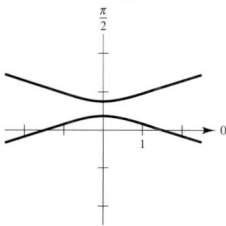

Hyperbola

17. $e = 2$
Distance $= \frac{5}{2}$

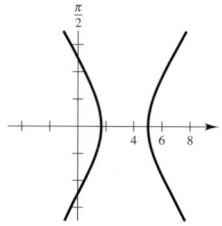

Hyperbola

19. $e = \frac{1}{2}$
Distance $= 6$

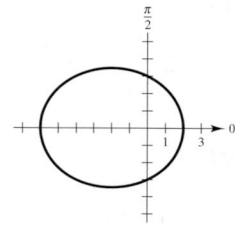

Ellipse

21. $e = \frac{1}{2}$
Distance $= 50$

Ellipse

23.

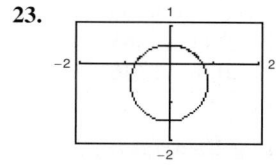

Ellipse
$e = \frac{1}{2}$

25.

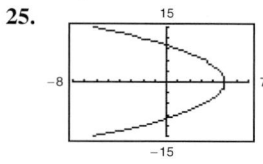

Parabola
$e = 1$

27.

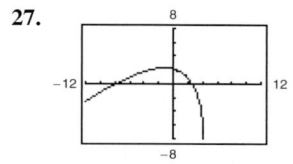

Rotated $\pi/3$ radian counterclockwise.

29.

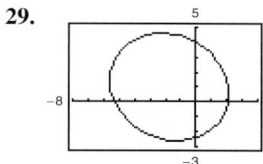

Rotated $\pi/6$ radian clockwise.

31. $r = \dfrac{8}{8 + 5\cos\left(\theta + \dfrac{\pi}{6}\right)}$

33. $r = 3/(1 - \cos\theta)$ **35.** $r = 1/(2 + \sin\theta)$

37. $r = 2/(1 + 2\cos\theta)$ **39.** $r = 2/(1 - \sin\theta)$

41. $r = 16/(5 + 3\cos\theta)$ **43.** $r = 9/(4 - 5\sin\theta)$

45. $r = 4/(2 + \cos\theta)$

47. If $0 < e < 1$, the conic is an ellipse.
If $e = 1$, the conic is a parabola.
If $e > 1$, the conic is a hyperbola.

49. If the foci are fixed and $e \to 0$, then $d \to \infty$. To see this, compare the ellipses

$$r = \frac{1/2}{1 + (1/2)\cos\theta}, e = \frac{1}{2}, d = 1 \text{ and}$$

$$r = \frac{5/16}{1 + (1/4)\cos\theta}, e = \frac{1}{4}, d = \frac{5}{4}.$$

51. Proof

53. $r^2 = \dfrac{9}{1 - (16/25)\cos^2\theta}$ **55.** $r^2 = \dfrac{-16}{1 - (25/9)\cos^2\theta}$

57. About 10.88   **59.** 3.37

61. $\dfrac{7979.21}{1 - 0.9372\cos\theta}$; 11,015 mi

63. $r = \dfrac{149,558,278.0560}{1 - 0.0167\cos\theta}$

Perihelion: 147,101,680 km

Aphelion: 152,098,320 km

65. $r = \dfrac{4,497,667,328}{1 - 0.0086\cos\theta}$

Perihelion: 4,459,317,200 km

Aphelion: 4,536,682,800 km

67. Answers will vary. Sample answers:

(a) 3.591×10^{18} km²; 9.322 yr

(b) $\alpha \approx 0.361 + \pi$; Larger angle with the smaller ray to generate an equal area

(c) Part (a): 1.583×10^9 km; 1.698×10^8 km/yr

Part (b): 1.610×10^9 km; 1.727×10^8 km/yr

69. Proof

Review Exercises for Chapter 10   *(page 742)*

1. e   **2.** c   **3.** b   **4.** d   **5.** a   **6.** f

7. Circle

Center: $\left(\frac{1}{2}, -\frac{3}{4}\right)$

Radius: 1

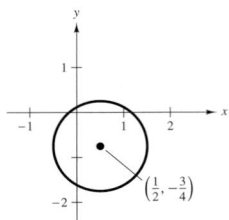

9. Hyperbola

Center: $(-4, 3)$

Vertices: $(-4 \pm \sqrt{2}, 3)$

Foci: $(-4 \pm \sqrt{5}, 3)$

$e = \sqrt{\dfrac{5}{2}}$

Asymptotes:

$y = 3 + \dfrac{\sqrt{3}}{\sqrt{2}}(x + 4)$;

$y = 3 - \dfrac{\sqrt{3}}{\sqrt{2}}(x + 4)$

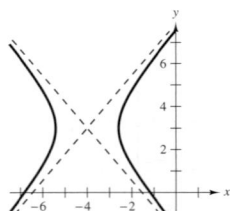

11. Ellipse

Center: $(2, -3)$

Vertices: $(2, -3 \pm \sqrt{2}/2)$

$e = \sqrt{\dfrac{1}{3}}$

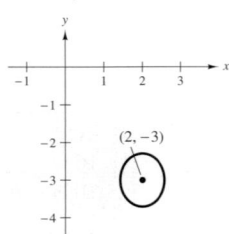

13. Parabola

Vertex: $(3, -1)$

Focus: $(3, 1)$

Directrix: $y = -3$

$e = 1$

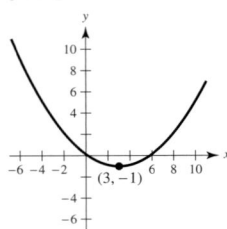

15. $y^2 - 4y - 12x + 4 = 0$

17. $\dfrac{x^2}{49} + \dfrac{y^2}{24} = 1$

19. $\dfrac{(x-3)^2}{5} + \dfrac{(y-4)^2}{9} = 1$

21. $\dfrac{y^2}{64} - \dfrac{x^2}{16} = 1$

23. $\dfrac{x^2}{49} - \dfrac{(y+1)^2}{32} = 1$

25. (a) $(0, 50)$ (b) About 38,294.49

27.

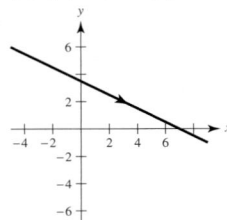

$x + 2y - 7 = 0$

29.

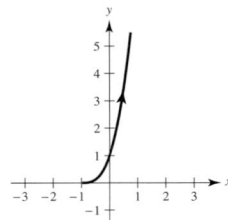

$y = (x + 1)^3, x > -1$

31.

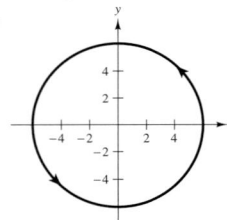

$x^2 + y^2 = 36$

33.

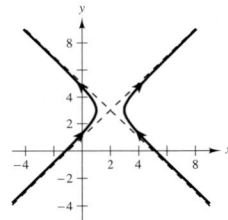

$(x - 2)^2 - (y - 3)^2 = 1$

35. $x = t, y = 4t + 3; x = t + 1, y = 4t + 7$

(Solution is not unique.)

37.

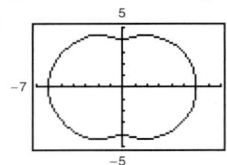

39. $\dfrac{dy}{dx} = -\dfrac{4}{5}, \dfrac{d^2y}{dx^2} = 0$

At $t = 3$, $\dfrac{dy}{dx} = -\dfrac{4}{5}, \dfrac{d^2y}{dx^2} = 0$; Neither concave upward or concave downward

41. $\dfrac{dy}{dx} = -2t^2, \dfrac{d^2y}{dx^2} = 4t^3$

At $t = -1, \dfrac{dy}{dx} = -2, \dfrac{d^2y}{dx^2} = -4$; Concave downward

43. $\dfrac{dy}{dx} = -4\cot\theta, \dfrac{d^2y}{dx^2} = -4\csc^3\theta$

At $\theta = \dfrac{\pi}{6}, \dfrac{dy}{dx} = -4\sqrt{3}, \dfrac{d^2y}{dx^2} = -32$; Concave downward

45. $\dfrac{dy}{dx} = -4\tan\theta,\ \dfrac{d^2y}{dx^2} = \dfrac{4}{3}\sec^4\theta\csc\theta$

    At $\theta = \dfrac{\pi}{3},\ \dfrac{dy}{dx} = -4\sqrt{3},\ \dfrac{d^2y}{dx^2} = \dfrac{128\sqrt{3}}{9}$; Concave upward

47. (a) and (d)

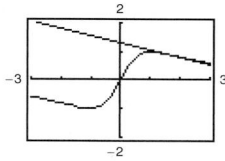

    (b) $dx/d\theta = -4,\ dy/d\theta = 1,\ dy/dx = -\dfrac{1}{4}$

    (c) $y = -\dfrac{1}{4}x + \dfrac{3\sqrt{3}}{4}$

49. Horizontal: $(5, 0)$ **51.** Horizontal: $(2, 2), (2, 0)$
     Vertical: None   Vertical: $(4, 1), (0, 1)$

53. $\dfrac{1}{54}(145^{3/2} - 1) \approx 32.315$

55. (a) $s = 12\pi\sqrt{10} \approx 119.215$ **57.** $A = 3\pi$
     (b) $s = 4\pi\sqrt{10} \approx 39.738$

59.

61.

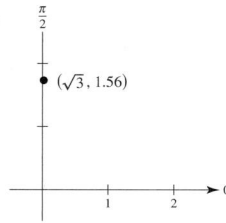
    $(\sqrt{3}, 1.56)$

Rectangular: $(0, -5)$ Rectangular: $(0.0187, 1.7320)$

63.

65.

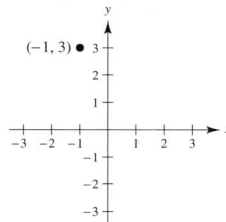
   $(-1, 3)$

$\left(4\sqrt{2}, \dfrac{7\pi}{4}\right), \left(-4\sqrt{2}, \dfrac{3\pi}{4}\right)$ $\left(\sqrt{10}, 1.89\right), \left(-\sqrt{10}, 5.03\right)$

67. $r = 5$

69. $r = 9\csc\theta$

71. $r = 4\tan\theta\sec\theta$

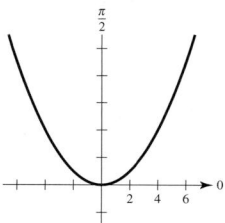

73. $x^2 + y^2 - 3x = 0$

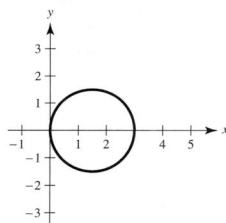

75. $x^2 + (y - 3)^2 = 9$

77. $y = -\dfrac{1}{2}x^2$

79.

81.

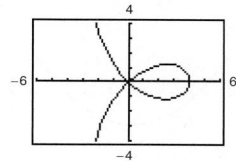

83. Horizontal: $\left(\dfrac{3}{2}, \dfrac{2\pi}{3}\right), \left(\dfrac{3}{2}, \dfrac{4\pi}{3}\right)$

    Vertical: $\left(\dfrac{1}{2}, \dfrac{\pi}{3}\right), (2, \pi), \left(\dfrac{1}{2}, \dfrac{5\pi}{3}\right)$

85.

87. Circle

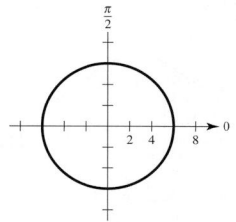

$\theta = 0, \dfrac{\pi}{3}, \dfrac{2\pi}{3}$

89. Line

91. Rose curve

93. Limaçon

95. Rose curve

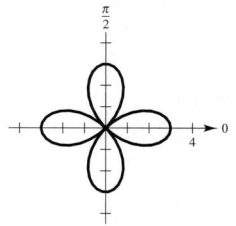

97. $\dfrac{9\pi}{20}$ **99.** $\dfrac{9\pi}{2}$ **101.** 4

103.

$9\pi - \dfrac{27\sqrt{3}}{2}$

105.

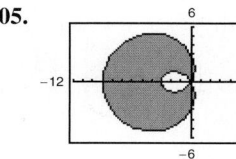

$9\pi + 27\sqrt{3}$

107. $\left(1 + \dfrac{\sqrt{2}}{2}, \dfrac{3\pi}{4}\right), \left(1 - \dfrac{\sqrt{2}}{2}, \dfrac{7\pi}{4}\right), (0, 0)$ **109.** $\dfrac{5\pi}{2}$

111. $S = 2\pi \displaystyle\int_0^{\pi/2} (1 + 4\cos\theta)\sin\theta\sqrt{17 + 8\cos\theta}\,d\theta$

$= 34\pi\sqrt{17}/5 \approx 88.08$

113. Parabola
$e = 1$; Distance $= 6$;

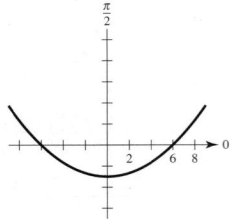

115. Ellipse
$e = \frac{2}{3}$; Distance $= 3$;

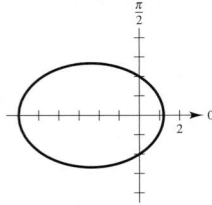

117. Hyperbola
$e = \frac{3}{2}$; Distance $= \frac{4}{3}$;

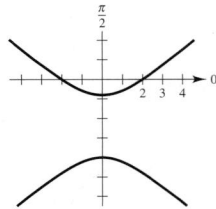

119. $r = \dfrac{4}{1 + \cos\theta}$ **121.** $r = \dfrac{9}{1 + 3\sin\theta}$

123. $r = \dfrac{5}{3 - 2\cos\theta}$

P.S. Problem Solving   *(page 745)*

1. (a)

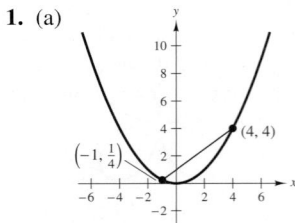

(b) and (c) Proofs

5. (a) $y^2 = x^2[(1 - x)/(1 + x)]$
(b) $r = \cos 2\theta \cdot \sec\theta$
(c)

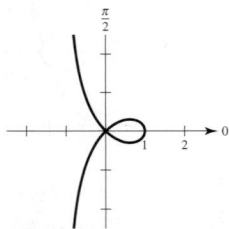

(d) $y = x, y = -x$
(e) $\left(\dfrac{\sqrt{5} - 1}{2}, \pm\dfrac{\sqrt{5} - 1}{2}\sqrt{-2 + \sqrt{5}}\right)$

3. Proof

7. (a)

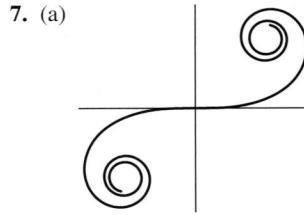

Generated by Mathematica

(b) Proof
(c) $a, 2\pi$

9. $A = \frac{1}{2}ab$ **11.** $r^2 = 2\cos 2\theta$

13. $r = \dfrac{d}{\sqrt{2}}e^{((\pi/4) - \theta)}, \theta \geq \dfrac{\pi}{4}$

15. (a) $r = 2a\tan\theta\sin\theta$
(b) $x = 2at^2/(1 + t^2)$
$y = 2at^3/(1 + t^2)$
(c) $y^2 = x^3/(2a - x)$

17.

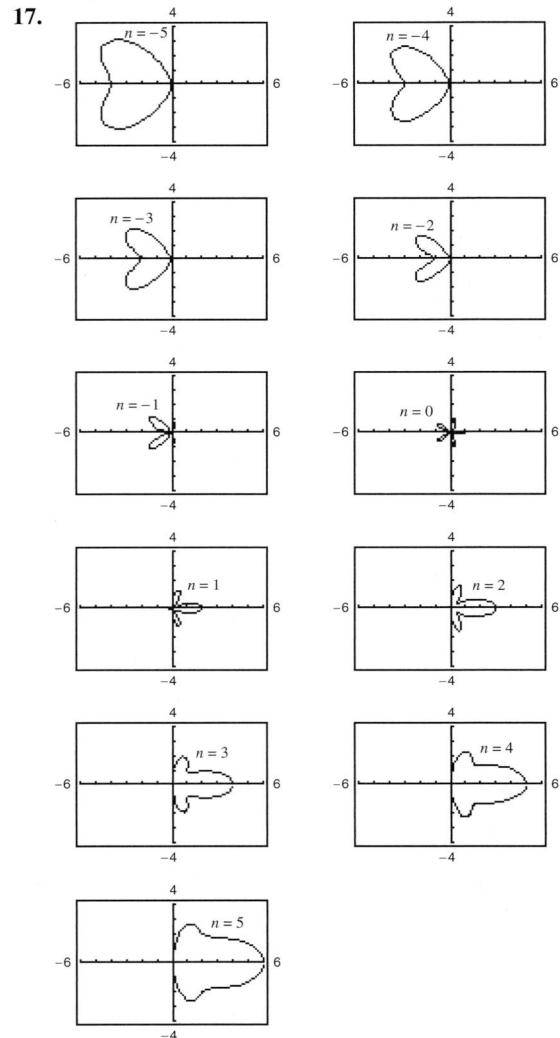

$n = 1, 2, 3, 4, 5$ produce "bells"; $n = -1, -2, -3, -4, -5$
produce "hearts."

Chapter 11

Section 11.1 (page 755)

1. (a) $\langle 4, 2 \rangle$
(b)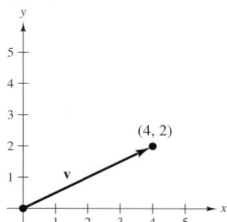

3. (a) $\langle -6, 0 \rangle$
(b)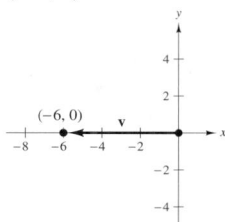

5. $\mathbf{u} = \mathbf{v} = \langle 2, 4 \rangle$ **7.** $\mathbf{u} = \mathbf{v} = \langle 6, -5 \rangle$

9. (a) and (d)
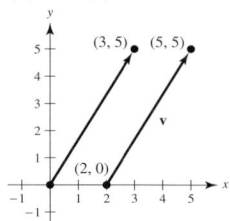
(b) $\langle 3, 5 \rangle$
(c) $\mathbf{v} = 3\mathbf{i} + 5\mathbf{j}$

11. (a) and (d)
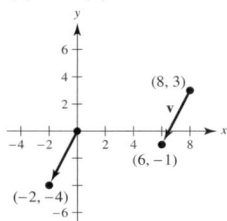
(b) $\langle -2, -4 \rangle$
(c) $\mathbf{v} = -2\mathbf{i} - 4\mathbf{j}$

13. (a) and (d)
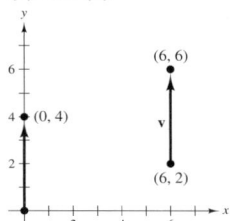
(b) $\langle 0, 4 \rangle$ (c) $\mathbf{v} = 4\mathbf{j}$

15. (a) and (d)
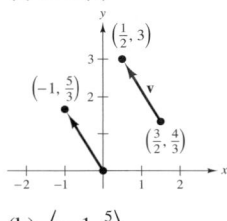
(b) $\left\langle -1, \frac{5}{3} \right\rangle$
(c) $\mathbf{v} = -\mathbf{i} + \frac{5}{3}\mathbf{j}$

17. (a) $\langle 6, 10 \rangle$
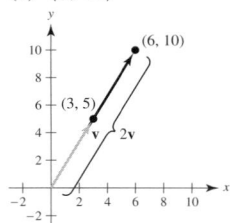
(c) $\left\langle \frac{21}{2}, \frac{35}{2} \right\rangle$
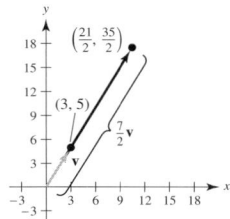

(b) $\langle -9, -15 \rangle$
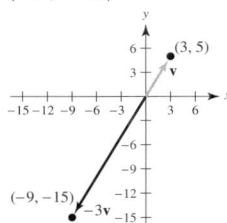
(d) $\left\langle 2, \frac{10}{3} \right\rangle$
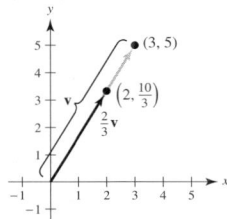

19. (a) $\left\langle \frac{8}{3}, 6 \right\rangle$ (b) $\langle 6, -15 \rangle$ (c) $\langle -2, -14 \rangle$ (d) $\langle 18, -7 \rangle$

21.

23.

25.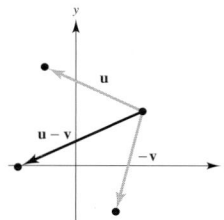

27. $(3, 5)$ **29.** 7
31. 5   **33.** $\sqrt{61}$
35. $\left\langle \sqrt{17}/17, 4\sqrt{17}/17 \right\rangle$
37. $\left\langle 3\sqrt{34}/34, 5\sqrt{34}/34 \right\rangle$
39. (a) $\sqrt{2}$ (b) $\sqrt{5}$ (c) 1
(d) 1   (e) 1   (f) 1

41. (a) $\sqrt{5}/2$ (b) $\sqrt{13}$ (c) $\sqrt{85}/2$ (d) 1   (e) 1   (f) 1

43.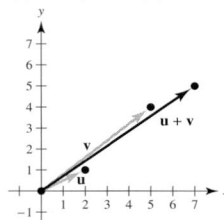

$\|\mathbf{u}\| + \|\mathbf{v}\| = \sqrt{5} + \sqrt{41}$ and $\|\mathbf{u} + \mathbf{v}\| = \sqrt{74}$
$\sqrt{74} \le \sqrt{5} + \sqrt{41}$

45. $\langle 0, 6 \rangle$ **47.** $\left\langle -\sqrt{5}, 2\sqrt{5} \right\rangle$ **49.** $\langle 3, 0 \rangle$

51. $\left\langle -\sqrt{3}, 1 \right\rangle$ **53.** $\left\langle \dfrac{2 + 3\sqrt{2}}{2}, \dfrac{3\sqrt{2}}{2} \right\rangle$

55. $\langle 2\cos 4 + \cos 2, 2\sin 4 + \sin 2 \rangle$

57. Answers will vary. Example: A scalar is a single real number, such as 2. A vector is a line segment having both direction and magnitude. The vector $\left\langle \sqrt{3}, 1 \right\rangle$, given in component form, has a direction of $\pi/6$ and a magnitude of 2.

59. $(-4, -1), (6, 5), (10, 3)$ **61.** $a = 1, b = 1$
63. $a = 1, b = 2$ **65.** $a = \frac{2}{3}, b = \frac{1}{3}$
67. (a) $\pm \left(1/\sqrt{37} \right)\langle 1, 6 \rangle$ **69.** (a) $\pm \left(1/\sqrt{10} \right)\langle 1, 3 \rangle$
(b) $\pm \left(1/\sqrt{37} \right)\langle 6, -1 \rangle$ (b) $\pm \left(1/\sqrt{10} \right)\langle 3, -1 \rangle$

    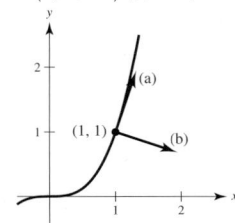

71. (a) $\pm\frac{1}{5}\langle -4, 3 \rangle$
(b) $\pm\frac{1}{5}\langle 3, 4 \rangle$
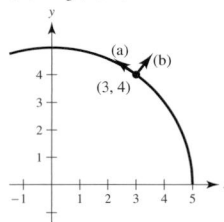

73. $\left\langle -\sqrt{2}/2, \sqrt{2}/2 \right\rangle$
75. $10.7°, 584.6$ lb
77. $71.3°, 228.5$ lb
79. (a) $\theta = 0°$ (b) $\theta = 180°$
(c) No, the resultant can only be less than or equal to the sum.
81. Horizontal: 1193.43 ft/sec
Vertical: 125.43 ft/sec
83. $38.3°$ north of west
882.9 km/h

85. True   **87.** True   **89.** False. $\|a\mathbf{i} + b\mathbf{j}\| = \sqrt{2}|a|$
91–93. Proofs   **95.** $x^2 + y^2 = 25$

Section 11.2   (page 763)

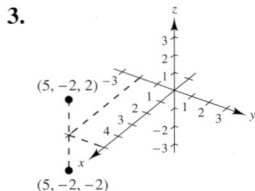

1.

3.

5. $(-3, 4, 5)$ **7.** $(12, 0, 0)$ **9.** 0
11. Six units above the xy-plane
13. Three units behind the yz-plane
15. To the left of the xz-plane
17. Within three units of the xz-plane
19. Three units below the xy-plane, and below either quadrant I or quadrant III
21. Above the xy-plane and above quadrants II or IV, *or* below the xy-plane and below quadrants I or III
23. $\sqrt{69}$ **25.** $\sqrt{61}$ **27.** $7, 7\sqrt{5}, 14$; Right triangle
29. $\sqrt{41}, \sqrt{41}, \sqrt{14}$; Isosceles triangle
31. $(0, 0, 9), (2, 6, 12), (6, 4, -3)$ **33.** $(2, 6, 3)$
35. $\left(\frac{3}{2}, -3, 5\right)$ **37.** $(x - 0)^2 + (y - 2)^2 + (z - 5)^2 = 4$
39. $(x - 1)^2 + (y - 3)^2 + (z - 0)^2 = 10$
41. $(x - 1)^2 + (y + 3)^2 + (z + 4)^2 = 25$
Center: $(1, -3, -4)$ Radius: 5
43. $\left(x - \frac{1}{3}\right)^2 + (y + 1)^2 + z^2 = 1$
Center: $\left(\frac{1}{3}, -1, 0\right)$ Radius: 1
45. (a) $\langle -2, 2, 2 \rangle$ **47.** (a) $\langle -3, 0, 3 \rangle$
(b) $\mathbf{v} = -2\mathbf{i} + 2\mathbf{j} + 2\mathbf{k}$ (b) $\mathbf{v} = -3\mathbf{i} + 3\mathbf{k}$
(c)

(c)

49. $\mathbf{v} = \langle 1, -1, 6 \rangle$ **51.** (a) and (d)
$\|\mathbf{v}\| = \sqrt{38}$
$\mathbf{u} = \dfrac{1}{\sqrt{38}}\langle 1, -1, 6 \rangle$

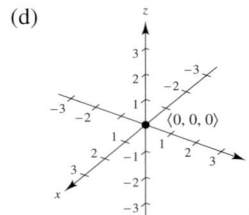

(b) $\langle 4, 1, 1 \rangle$
(c) $\mathbf{v} = 4\mathbf{i} + \mathbf{j} + \mathbf{k}$
53. $(3, 1, 8)$
55. (a)

(b)

(c)

(d)

57. $\langle 7, 0, -4 \rangle$ **59.** $\left\langle \frac{7}{2}, 3, \frac{5}{2} \right\rangle$ **61.** a and b
63. a   **65.** Collinear   **67.** Not collinear
69. $\overrightarrow{AB} = \langle 1, 2, 3 \rangle$, $\overrightarrow{CD} = \langle 1, 2, 3 \rangle$, $\overrightarrow{BD} = \langle -2, 1, 1 \rangle$, $\overrightarrow{AC} = \langle -2, 1, 1 \rangle$; Because $\overrightarrow{AB} = \overrightarrow{CD}$ and $\overrightarrow{BD} = \overrightarrow{AC}$, the given points form the vertices of a parallelogram.
71. 0   **73.** $\sqrt{34}$ **75.** $\sqrt{14}$
77. (a) $\frac{1}{3}\langle 2, -1, 2 \rangle$ (b) $-\frac{1}{3}\langle 2, -1, 2 \rangle$
79. (a) $\dfrac{2\sqrt{2}}{5}\mathbf{i} - \dfrac{\sqrt{2}}{2}\mathbf{j} + \dfrac{3\sqrt{2}}{10}\mathbf{k}$ (b) $-\dfrac{2\sqrt{2}}{5}\mathbf{i} + \dfrac{\sqrt{2}}{2}\mathbf{j} - \dfrac{3\sqrt{2}}{10}\mathbf{k}$
81. The terminal points of the   **83.** $\left\langle 0, 10/\sqrt{2}, 10/\sqrt{2} \right\rangle$
vectors $t\mathbf{u}, \mathbf{u} + t\mathbf{v}$, and   **85.** $\left\langle 1, -1, \frac{1}{2} \right\rangle$
$s\mathbf{u} + t\mathbf{v}$ are collinear.

87.

$\langle 0, \sqrt{3}, \pm1 \rangle$ **89.** $(2, -1, 2)$

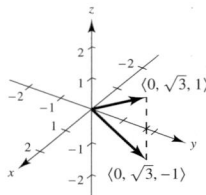

91. (a)

(b) $a = 0, a + b = 0, b = 0$
(c) $a = 1, a + b = 2, b = 1$
(d) Not possible

93. x_0 is directed distance to yz-plane.
y_0 is directed distance to xz-plane.
z_0 is directed distance to xy-plane.
95. $(x - x_0)^2 + (y - y_0)^2 + (z - z_0)^2 = r^2$ **97.** 0
99. $(\sqrt{3}/3)\langle 1, 1, 1 \rangle$
101. (a) $T = 8L/\sqrt{L^2 - 18^2}, L > 18$
(b)

L	20	25	30	35	40	45	50
T	18.4	11.5	10	9.3	9.0	8.7	8.6

(c)

(d) Proof   (e) 30 in.

103. Tension in cable AB: 202.919 N
 Tension in cable AC: 157.909 N
 Tension in cable AD: 226.521 N
105. $\left(x - \frac{4}{3}\right)^2 + (y - 3)^2 + \left(z + \frac{1}{3}\right)^2 = \frac{44}{9}$

 Sphere; center: $\left(\frac{4}{3}, 3, -\frac{1}{3}\right)$, radius: $\frac{2\sqrt{11}}{3}$

Section 11.3 *(page 773)*

1. (a) 17 (b) 25 (c) 25 (d) $\langle -17, 85 \rangle$ (e) 34
3. (a) -26 (b) 52 (c) 52 (d) $\langle 78, -52 \rangle$ (e) -52
5. (a) 2 (b) 29 (c) 29 (d) $\langle 0, 12, 10 \rangle$ (e) 4
7. (a) 1 (b) 6 (c) 6 (d) $\mathbf{i} - \mathbf{k}$ (e) 2
9. (a) $\pi/2$ (b) $90°$ **11.** (a) 1.7127 (b) $98.1°$
13. (a) 1.0799 (b) $61.9°$ **15.** (a) 2.0306 (b) $116.3°$
17. 20 **19.** Orthogonal **21.** Neither **23.** Orthogonal
25. Right triangle; answers will vary.
27. Acute triangle; answers will vary.
29. $\cos \alpha = \frac{1}{3}, \alpha \approx 70.5°$ **31.** $\cos \alpha = \frac{3}{\sqrt{17}}, \alpha \approx 43.3°$

 $\cos \beta = \frac{2}{3}, \beta \approx 48.2°$ $\cos \beta = \frac{2}{\sqrt{17}}, \beta \approx 61.0°$

 $\cos \gamma = \frac{2}{3}, \gamma \approx 48.2°$ $\cos \gamma = -\frac{2}{\sqrt{17}}, \gamma \approx 119.0°$

33. $\cos \alpha = 0, \alpha \approx 90°$
 $\cos \beta = 3/\sqrt{13}, \beta \approx 33.7°$
 $\cos \gamma = -2/\sqrt{13}, \gamma \approx 123.7°$
35. (a) $\langle 2, 8 \rangle$ (b) $\langle 4, -1 \rangle$ **37.** (a) $\left\langle \frac{5}{2}, \frac{1}{2} \right\rangle$ (b) $\left\langle -\frac{1}{2}, \frac{5}{2} \right\rangle$
39. (a) $\langle -2, 2, 2 \rangle$ (b) $\langle 2, 1, 1 \rangle$
41. (a) $\left\langle 0, \frac{33}{25}, \frac{44}{25} \right\rangle$ (b) $\left\langle 2, -\frac{8}{25}, \frac{6}{25} \right\rangle$
43. See "Definition of Dot Product" page 766.
45. (a) and (b) are defined. (c) and (d) are not defined because it is not possible to find the dot product of a scalar and a vector or to add a scalar to a vector.
47. See Figure 11.29 on page 770.
49. Yes.

51. $17,490.25; Total revenue
$$\left\| \frac{\mathbf{u} \cdot \mathbf{v}}{\|\mathbf{v}\|^2} \mathbf{v} \right\| = \left\| \frac{\mathbf{v} \cdot \mathbf{u}}{\|\mathbf{u}\|^2} \mathbf{u} \right\|$$
$$|\mathbf{u} \cdot \mathbf{v}| \frac{\|\mathbf{v}\|}{\|\mathbf{v}\|^2} = |\mathbf{v} \cdot \mathbf{u}| \frac{\|\mathbf{u}\|}{\|\mathbf{u}\|^2}$$
$$\frac{1}{\|\mathbf{v}\|} = \frac{1}{\|\mathbf{u}\|}$$
$$\|\mathbf{u}\| = \|\mathbf{v}\|$$

53. Answers will vary. Example: $\langle 12, 2 \rangle$ and $\langle -12, -2 \rangle$
55. Answers will vary. Example: $\langle 2, 0, 3 \rangle$ and $\langle -2, 0, -3 \rangle$
57. $\arccos(1/\sqrt{3}) \approx 54.7°$
59. (a) 8335.1 lb (b) 47,270.8 lb
61. 425 ft-lb **63.** 2900.2 km-N
65. False. For example, $\langle 1, 1 \rangle \cdot \langle 2, 3 \rangle = 5$ and $\langle 1, 1 \rangle \cdot \langle 1, 4 \rangle = 5$, but $\langle 2, 3 \rangle \neq \langle 1, 4 \rangle$.
67. (a) $(0, 0), (1, 1)$
 (b) To $y = x^2$ at $(1, 1)$: $\left\langle \pm \sqrt{5}/5, \pm 2\sqrt{5}/5 \right\rangle$
 To $y = x^{1/3}$ at $(1, 1)$: $\left\langle \pm 3\sqrt{10}/10, \pm \sqrt{10}/10 \right\rangle$
 To $y = x^2$ at $(0, 0)$: $\langle \pm 1, 0 \rangle$
 To $y = x^{1/3}$ at $(0, 0)$: $\langle 0, \pm 1 \rangle$
 (c) At $(1, 1)$: $\theta = 45°$
 At $(0, 0)$: $\theta = 90°$

69. (a) $(-1, 0), (1, 0)$
 (b) To $y = 1 - x^2$ at $(1, 0)$: $\left\langle \pm \sqrt{5}/5, \mp 2\sqrt{5}/5 \right\rangle$
 To $y = x^2 - 1$ at $(1, 0)$: $\left\langle \pm \sqrt{5}/5, \pm 2\sqrt{5}/5 \right\rangle$
 To $y = 1 - x^2$ at $(-1, 0)$: $\left\langle \pm \sqrt{5}/5, \pm 2\sqrt{5}/5 \right\rangle$
 To $y = x^2 - 1$ at $(-1, 0)$: $\left\langle \pm \sqrt{5}/5, \mp 2\sqrt{5}/5 \right\rangle$
 (c) At $(1, 0)$: $\theta = 53.13°$
 At $(-1, 0)$: $\theta = 53.13°$
71. Proof
73. (a) (b) $k\sqrt{2}$ (c) $60°$ (d) $109.5°$

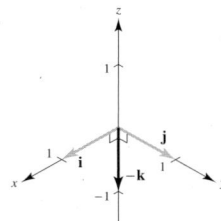

75–77. Proofs

Section 11.4 *(page 781)*

1. $-\mathbf{k}$ **3.** $\mathbf{i}$

5. $-\mathbf{j}$

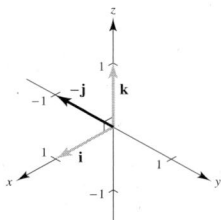

7. (a) $20\mathbf{i} + 10\mathbf{j} - 16\mathbf{k}$ (b) $-20\mathbf{i} - 10\mathbf{j} + 16\mathbf{k}$ (c) $\mathbf{0}$
9. (a) $17\mathbf{i} - 33\mathbf{j} - 10\mathbf{k}$ (b) $-17\mathbf{i} + 33\mathbf{j} + 10\mathbf{k}$ (c) $\mathbf{0}$
11. $\langle 0, 0, 54 \rangle$ **13.** $\langle -1, -1, -1 \rangle$ **15.** $\langle -2, 3, -1 \rangle$
17. $\left\langle -\frac{7}{9\sqrt{3}}, -\frac{5}{9\sqrt{3}}, \frac{13}{9\sqrt{3}} \right\rangle$ or $\left\langle \frac{7}{9\sqrt{3}}, \frac{5}{9\sqrt{3}}, -\frac{13}{9\sqrt{3}} \right\rangle$
19. $\left\langle \frac{3}{\sqrt{59}}, \frac{7}{\sqrt{59}}, \frac{1}{\sqrt{59}} \right\rangle$ or $\left\langle -\frac{3}{\sqrt{59}}, -\frac{7}{\sqrt{59}}, -\frac{1}{\sqrt{59}} \right\rangle$
21. 1 **23.** $6\sqrt{5}$ **25.** $9\sqrt{5}$ **27.** $\frac{11}{2}$
29. $10 \cos 40° \approx 7.66$ ft-lb
31. (a) $\mathbf{F} = -180(\cos \theta \mathbf{j} + \sin\theta \mathbf{k})$
 (b) $\|\overrightarrow{AB} \times \mathbf{F}\| = |225 \sin \theta + 180 \cos \theta|$
 (c) $\|\overrightarrow{AB} \times \mathbf{F}\| = 225(1/2) + 180(\sqrt{3}/2) \approx 268.38$
 (d) $\theta = 141.34°$
 $\overrightarrow{AB}$ and $\mathbf{F}$ are perpendicular.
 (e)

From part (d), the zero is $\theta \approx 141.34°$, when the vectors are parallel.

33. 1 **35.** 6 **37.** 2 **39.** 75
41. (a) = (b) = (c) = (h) and (e) = (f) = (g)
43. See "Definition of Cross Product of Two Vectors in Space" on page 775.

45. The magnitude of the cross product will increase by a factor of 4.

47. False. The cross product of two vectors is not defined in a two-dimensional coordinate system.

49. False. Let $\mathbf{u} = \langle 1, 0, 0 \rangle$, $\mathbf{v} = \langle 1, 0, 0 \rangle$, and $\mathbf{w} = \langle -1, 0, 0 \rangle$. Then $\mathbf{u} \times \mathbf{v} = \mathbf{u} \times \mathbf{w} = \mathbf{0}$, but $\mathbf{v} \neq \mathbf{w}$.

51–59. Proofs

Section 11.5   *(page 790)*

1. (a) Yes   (b) No

Parametric Equations (a)	Symmetric Equations (b)	Direction Numbers
3. $x = 3t$ $y = t$ $z = 5t$	$\dfrac{x}{3} = y = \dfrac{z}{5}$	$3, 1, 5$
5. $x = -2 + 2t$ $y = 4t$ $z = 3 - 2t$	$\dfrac{x+2}{2} = \dfrac{y}{4} = \dfrac{z-3}{-2}$	$2, 4, -2$
7. $x = 1 + 3t$ $y = -2t$ $z = 1 + t$	$\dfrac{x-1}{3} = \dfrac{y}{-2} = \dfrac{z-1}{1}$	$3, -2, 1$
9. $x = 5 + 17t$ $y = -3 - 11t$ $z = -2 - 9t$	$\dfrac{x-5}{17} = \dfrac{y+3}{-11} = \dfrac{z+2}{-9}$	$17, -11, -9$
11. $x = 7 - 10t$ $y = -2 + 2t$ $z = 6$	Not possible	$-10, 2, 0$

13. $x = 2$ $y = 3$ $z = 4 + t$

15. $x = 2 + 3t$ $y = 3 + 2t$ $z = 4 - t$

17. $x = 5 + 2t$ $y = -3 - t$ $z = -4 + 3t$

19. $x = 2 - t$ $y = 1 + t$ $z = 2 + t$

21. $P(3, -1, -2)$; $\mathbf{v} = \langle -1, 2, 0 \rangle$

23. $P(7, -6, -2)$; $\mathbf{v} = \langle 4, 2, 1 \rangle$

25. $L_1 = L_2$ and is parallel to L_3.   **27.** L_1 and L_3 are identical.

29. $(2, 3, 1)$; $\cos \theta = 7\sqrt{17}/51$ **31.** Not intersecting

33. (a) Yes   (b) Yes   **35.** $y - 3 = 0$

37. $2x + 3y - z = 10$ **39.** $2x - y - 2z + 6 = 0$

41. $3x - 19y - 2z = 0$ **43.** $4x - 3y + 4z = 10$

45. $z = 3$ **47.** $x + y + z = 5$ **49.** $7x + y - 11z = 5$

51. $y - z = -1$ **53.** $x - z = 0$

55. $9x - 3y + 2z - 21 = 0$ **57.** Orthogonal

59. Neither; $83.5°$ **61.** Parallel

63.

65.

67.

69.

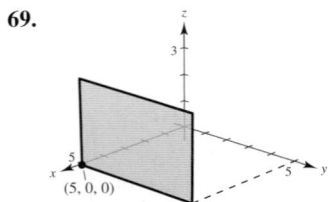

71. P_1 and P_2 are parallel.   **73.** $P_1 = P_4$ and is parallel to P_2.

75. (a) $\theta \approx 65.91°$

(b) $x = 2$
$y = 1 + t$
$z = 1 + 2t$

77. $(2, -3, 2)$; The line does not lie in the plane.

79. Not intersecting   **81.** $6\sqrt{14}/7$ **83.** $11\sqrt{6}/6$

85. $2\sqrt{26}/13$ **87.** $27\sqrt{94}/188$ **89.** $\sqrt{2533}/17$

91. $7\sqrt{3}/3$ **93.** $\sqrt{66}/3$

95. Parametric equations: $x = x_1 + at$, $y = y_1 + bt$, and $z = z_1 + ct$

Symmetric equations: $\dfrac{x - x_1}{a} = \dfrac{y - y_1}{b} = \dfrac{z - z_1}{c}$

You need a vector $\mathbf{v} = \langle a, b, c \rangle$ parallel to the line and a point $P(x_1, y_1, z_1)$ on the line.

97. Simultaneously solve the two linear equations representing the planes and substitute the values back into one of the original equations. Then choose a value for t and form the corresponding parametric equations for the line of intersection.

99. Yes. If $\mathbf{v}_1$ and $\mathbf{v}_2$ are the direction vectors for the lines L_1 and L_2, then $\mathbf{v} = \mathbf{v}_1 \times \mathbf{v}_2$ is perpendicular to both L_1 and L_2.

101. (a)

Year	2005	2006	2007	2008	2009	2010
z (approx.)	16.39	17.98	19.78	20.87	19.94	21.04

The approximations are close to the actual values.

(b) An increase

103. (a) $\sqrt{70}$ in.

(b)

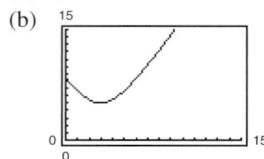

(c) The distance is never zero.

(d) 5 in.

105. $\left(\frac{77}{13}, \frac{48}{13}, -\frac{23}{13} \right)$ **107.** $\left(-\frac{1}{2}, -\frac{9}{4}, \frac{1}{4} \right)$ **109.** True   **111.** True

113. False. Plane $7x + y - 11z = 5$ and plane $5x + 2y - 4z = 1$ are both perpendicular to plane $2x - 3y + z = 3$ but are not parallel.

Section 11.6   *(page 802)*

1. c   **2.** e   **3.** f   **4.** b   **5.** d   **6.** a

7. Plane

9. Right circular cylinder

11. Elliptic cylinder

13. Ellipsoid

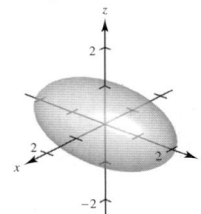

15. Hyperboloid of one sheet   **17.** Hyperboloid of two sheets

19. Elliptic paraboloid   **21.** Hyperbolic paraboloid

23. Elliptic cone

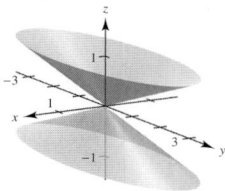

25. Let C be a curve in a plane and let L be a line not in a parallel plane. The set of all lines parallel to L and intersecting C is called a cylinder. C is called the generating curve of the cylinder, and the parallel lines are called rulings.

27. See pages 796 and 797.

29. xy-plane: ellipse; three-space: hyperboloid of one sheet

31. $x^2 + z^2 = 4y$ **33.** $4x^2 + 4y^2 = z^2$

35. $y^2 + z^2 = 4/x^2$ **37.** $y = \sqrt{2z}$ (or $x = \sqrt{2z}$)   **39.** $128\pi/3$

41. (a) Major axis: $4\sqrt{2}$ (b) Major axis: $8\sqrt{2}$
         Minor axis: 4   Minor axis: 8
         Foci: $(0, \pm2, 2)$ Foci: $(0, \pm4, 8)$

43. $x^2 + z^2 = 8y$; Elliptic paraboloid

45. $x^2/3963^2 + y^2/3963^2 + z^2/3950^2 = 1$

47. $x = at, y = -bt, z = 0$;
         $x = at, y = bt + ab^2, z = 2abt + a^2b^2$

49. True   **51.** False. A trace of an ellipsoid can be a single point.

53. The Klein bottle does not have both an "inside" and an "outside." It is formed by inserting the small open end through the side of the bottle and making it contiguous with the top of the bottle.

Section 11.7   *(page 809)*

1. $(-7, 0, 5)$ **3.** $(3\sqrt{2}/2, 3\sqrt{2}/2, 1)$ **5.** $(-2\sqrt{3}, -2, 3)$

7. $(5, \pi/2, 1)$ **9.** $(2\sqrt{2}, -\pi/4, -4)$ **11.** $(2, \pi/3, 4)$

13. $z = 4$ **15.** $r^2 + z^2 = 17$ **17.** $r = \sec\theta\tan\theta$

19. $r^2 \sin^2\theta = 10 - z^2$

21. $x^2 + y^2 = 9$ **23.** $x - \sqrt{3}y = 0$

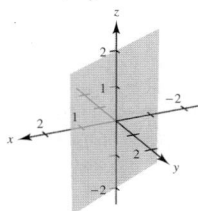

25. $x^2 + y^2 + z^2 = 5$ **27.** $x^2 + y^2 - 2y = 0$

29. $(4, 0, \pi/2)$ **31.** $(4\sqrt{2}, 2\pi/3, \pi/4)$

33. $(4, \pi/6, \pi/6)$ **35.** $(\sqrt{6}, \sqrt{2}, 2\sqrt{2})$ **37.** $(0, 0, 12)$

39. $(\frac{5}{2}, \frac{5}{2}, -5\sqrt{2}/2)$ **41.** $\rho = 2 \csc\phi\csc\theta$ **43.** $\rho = 7$

45. $\rho = 4 \csc\phi$ **47.** $\tan^2\phi = 2$

49. $x^2 + y^2 + z^2 = 25$ **51.** $3x^2 + 3y^2 - z^2 = 0$

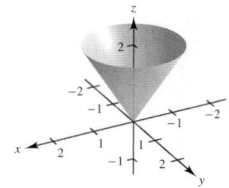

53. $x^2 + y^2 + (z - 2)^2 = 4$ **55.** $x^2 + y^2 = 1$

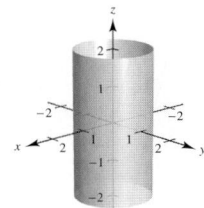

57. d   **58.** e   **59.** c   **60.** a   **61.** f   **62.** b

63. $(4, \pi/4, \pi/2)$ **65.** $(4\sqrt{2}, \pi/2, \pi/4)$

67. $(2\sqrt{13}, -\pi/6, \arccos[3/\sqrt{13}])$

69. $(13, \pi, \arccos[5/13])$ **71.** $(10, \pi/6, 0)$ **73.** $(36, \pi, 0)$

75. $(3\sqrt{3}, -\pi/6, 3)$ **77.** $(4, 7\pi/6, 4\sqrt{3})$

79. Rectangular to cylindrical:
         $r^2 = x^2 + y^2, \tan\theta = y/x, z = z$
         Cylindrical to rectangular:
         $x = r\cos\theta, y = r\sin\theta, z = z$

81. Rectangular to spherical:
         $\rho^2 = x^2 + y^2 + z^2, \tan\theta = y/x, \phi = \arccos(z/\sqrt{x^2 + y^2 + z^2})$
         Spherical to rectangular:
         $x = \rho\sin\phi\cos\theta, y = \rho\sin\phi\sin\theta, z = \rho\cos\phi$

83. (a) $r^2 + z^2 = 25$ (b) $\rho = 5$

85. (a) $r^2 + (z - 1)^2 = 1$ (b) $\rho = 2\cos\phi$

87. (a) $r = 4\sin\theta$ (b) $\rho = 4\sin\theta/\sin\phi = 4\sin\theta\csc\phi$

89. (a) $r^2 = 9/(\cos^2\theta - \sin^2\theta)$
         (b) $\rho^2 = 9\csc^2\phi/(\cos^2\theta - \sin^2\theta)$

91.

93.

95.

97.

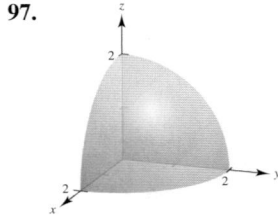

99. Rectangular: $0 \le x \le 10; 0 \le y \le 10; 0 \le z \le 10$
101. Spherical: $4 \le \rho \le 6$
103. Cylindrical: $r^2 + z^2 \le 9, r \le 3 \cos \theta, 0 \le \theta \le \pi$
105. False. $r = z$ represents a cone.
107. False. See page 805.   **109.** Ellipse

Review Exercises for Chapter 11   (page 811)

1. (a) $\mathbf{u} = \langle 3, -1 \rangle, \mathbf{v} = \langle 4, 2 \rangle$ (b) $\mathbf{u} = 3\mathbf{i} - \mathbf{j}, \mathbf{v} = 4\mathbf{i} + 2\mathbf{j}$
   (c) $\|\mathbf{u}\| = \sqrt{10}, \|\mathbf{v}\| = 2\sqrt{5}$ (d) $10\mathbf{i}$
3. $\mathbf{v} = \langle 4, 4\sqrt{3} \rangle$ **5.** $(-5, 4, 0)$ **7.** $\sqrt{22}$
9. $(x - 3)^2 + (y + 2)^2 + (z - 6)^2 = \frac{225}{4}$
11. $(x - 2)^2 + (y - 3)^2 + z^2 = 9$; Center: $(2, 3, 0)$; Radius: 3
13. (a) and (d)

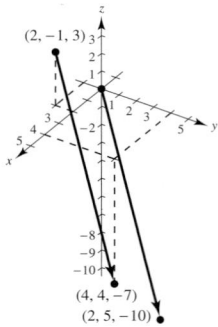

(b) $\mathbf{u} = \langle 2, 5, -10 \rangle$
(c) $\mathbf{u} = 2\mathbf{i} + 5\mathbf{j} - 10\mathbf{k}$

15. Collinear
17. $(1/\sqrt{38})\langle 2, 3, 5 \rangle$
19. (a) $\mathbf{u} = \langle -1, 4, 0 \rangle$
         $\mathbf{v} = \langle -3, 0, 6 \rangle$
     (b) 3   (c) 45
21. (a) $\dfrac{\pi}{12}$ (b) $15°$
23. (a) π (b) $180°$
25. Orthogonal
27. $\langle 2, 10 \rangle$
29. $\langle 1, 0, 1 \rangle$

31. Answers will vary. Example: $\langle -6, 5, 0 \rangle, \langle 6, -5, 0 \rangle$
33. (a) $-9\mathbf{i} + 26\mathbf{j} - 7\mathbf{k}$ (b) $9\mathbf{i} - 26\mathbf{j} + 7\mathbf{k}$ (c) $\mathbf{0}$
35. (a) $-8\mathbf{i} - 10\mathbf{j} + 6\mathbf{k}$ (b) $8\mathbf{i} + 10\mathbf{j} - 6\mathbf{k}$ (c) $\mathbf{0}$
37. $\left\langle \dfrac{8}{\sqrt{377}}, \dfrac{12}{\sqrt{377}}, \dfrac{13}{\sqrt{377}} \right\rangle$ or $\left\langle -\dfrac{8}{\sqrt{377}}, -\dfrac{12}{\sqrt{377}}, -\dfrac{13}{\sqrt{377}} \right\rangle$
39. $100 \sec 20° \approx 106.4$ lb
41. (a) $x = 3 + 6t, y = 11t, z = 2 + 4t$
     (b) $(x - 3)/6 = y/11 = (z - 2)/4$
43. $x = 1, y = 2 + t, z = 3$ **45.** $x = t, y = -1 + t, z = 1$
47. $27x + 4y + 32z + 33 = 0$ **49.** $x + 2y = 1$ **51.** $\frac{8}{7}$
53. $\sqrt{35}/7$
55. Plane

57. Plane

59. Ellipsoid   **61.** Hyperboloid of two sheets

63. Cylinder

65. $x^2 + z^2 = 2y$

67. (a) $(4, 3\pi/4, 2)$ (b) $\left(2\sqrt{5}, 3\pi/4, \arccos\left[\sqrt{5}/5\right]\right)$
69. $\left(50\sqrt{5}, -\pi/6, \arccos\left[1/\sqrt{5}\right]\right)$
71. $\left(25\sqrt{2}/2, -\pi/4, -25\sqrt{2}/2\right)$
73. (a) $r^2 \cos 2\theta = 2z$ (b) $\rho = 2 \sec 2\theta \cos \phi \csc^2 \phi$
75. $\left(x - \frac{5}{2}\right)^2 + y^2 = \frac{25}{4}$ **77.** $x = y$

P.S. Problem Solving   (page 813)

1–3. Proofs   **5.** (a) $3\sqrt{2}/2 \approx 2.12$ (b) $\sqrt{5} \approx 2.24$
7. (a) $\pi/2$ (b) $\frac{1}{2}(\pi abk)k$
   (c) $V = \frac{1}{2}(\pi ab)k^2$
       $V = \frac{1}{2}$(area of base)height
9. Proof
11. (a)

(b)

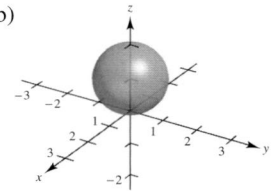

13. (a) Tension: $2\sqrt{3}/3 \approx 1.1547$ lb
       Magnitude of $\mathbf{u}$: $\sqrt{3}/3 \approx 0.5774$ lb
   (b) $T = \sec \theta$; $\|\mathbf{u}\| = \tan \theta$; Domain: $0° \le \theta \le 90°$
   (c)

θ	$0°$	$10°$	$20°$	$30°$
T	1	1.0154	1.0642	1.1547
$\|\mathbf{u}\|$	0	0.1763	0.3640	0.5774

θ	$40°$	$50°$	$60°$
T	1.3054	1.5557	2
$\|\mathbf{u}\|$	0.8391	1.1918	1.7321

   (d)

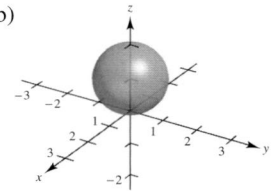

   (e) Both are increasing functions

   (f) $\lim\limits_{\theta \to \pi/2^-} T = \infty$ and $\lim\limits_{\theta \to \pi/2^-} \|\mathbf{u}\| = \infty$
       Yes. As θ increases, both T and $\|\mathbf{u}\|$ increase.
15. $\langle 0, 0, \cos \alpha \sin \beta - \cos \beta \sin \alpha \rangle$; Proof
17. $D = \dfrac{|\overrightarrow{PQ} \cdot \mathbf{n}|}{\|\mathbf{n}\|} = \dfrac{|\mathbf{w} \cdot (\mathbf{u} \times \mathbf{v})|}{\|\mathbf{u} \times \mathbf{v}\|} = \dfrac{|(\mathbf{u} \times \mathbf{v}) \cdot \mathbf{w}|}{\|\mathbf{u} \times \mathbf{v}\|} = \dfrac{|\mathbf{u} \cdot (\mathbf{v} \times \mathbf{w})|}{\|\mathbf{u} \times \mathbf{v}\|}$
19. Proof

Chapter 12

Section 12.1   *(page 821)*

1. $(-\infty, -1) \cup (-1, \infty)$ 3. $(0, \infty)$

5. $[0, \infty)$ 7. $(-\infty, \infty)$

9. (a) $\frac{1}{2}\mathbf{i}$ (b) $\mathbf{j}$ (c) $\frac{1}{2}(s+1)^2\mathbf{i} - s\mathbf{j}$
   (d) $\frac{1}{2}\Delta t(\Delta t + 4)\mathbf{i} - \Delta t\mathbf{j}$

11. (a) $\ln 2\mathbf{i} + \frac{1}{2}\mathbf{j} + 6\mathbf{k}$ (b) Not possible

    (c) $\ln(t-4)\mathbf{i} + \frac{1}{t-4}\mathbf{j} + 3(t-4)\mathbf{k}$

    (d) $\ln(1 + \Delta t)\mathbf{i} - \frac{\Delta t}{1 + \Delta t}\mathbf{j} + 3\Delta t\mathbf{k}$

13. $\mathbf{r}(t) = 3t\mathbf{i} + t\mathbf{j} + 2t\mathbf{k}, \quad 0 \le t \le 1$
    $x = 3t, \ y = t, \ z = 2t, \quad 0 \le t \le 1$

15. $\mathbf{r}(t) = (-2 + t)\mathbf{i} + (5 - t)\mathbf{j} + (-3 + 12t)\mathbf{k}, \quad 0 \le t \le 1$
    $x = -2 + t, \ y = 5 - t, \ z = -3 + 12t, \quad 0 \le t \le 1$

17. $t^2(5t - 1)$; No, the dot product is a scalar.

19. b   20. c   21. d   22. a

23.

25.

27.

29.

31.

33.

35.

37.

39.

Parabola

41.

Helix

43.
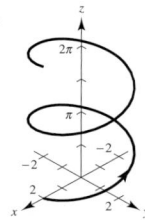

(a) The helix is translated two units back on the x-axis.
(b) The height of the helix increases at a greater rate.
(c) The orientation of the graph is reversed.
(d) The axis of the helix is the x-axis.
(e) The radius of the helix is increased from 2 to 6.

45–51. Answers will vary.

53.
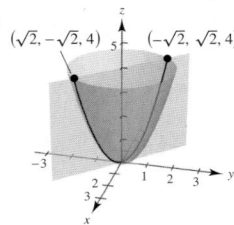

$\mathbf{r}(t) = t\mathbf{i} - t\mathbf{j} + 2t^2\mathbf{k}$

55.
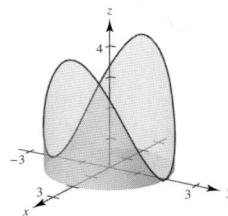

$\mathbf{r}(t) = 2\sin t\mathbf{i} + 2\cos t\mathbf{j} + 4\sin^2 t\mathbf{k}$

57.
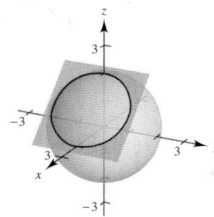

$\mathbf{r}(t) = (1 + \sin t)\mathbf{i} + \sqrt{2}\cos t\mathbf{j} + (1 - \sin t)\mathbf{k}$ and
$\mathbf{r}(t) = (1 + \sin t)\mathbf{i} - \sqrt{2}\cos t\mathbf{j} + (1 - \sin t)\mathbf{k}$

59.
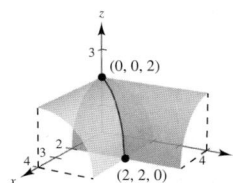

$\mathbf{r}(t) = t\mathbf{i} + t\mathbf{j} + \sqrt{4 - t^2}\mathbf{k}$

61. Let $x = t$, $y = 2t \cos t$, and $z = 2t \sin t$. Then

$$y^2 + z^2 = (2t \cos t)^2 + (2t \sin t)^2$$
$$= 4t^2 \cos^2 t + 4t^2 \sin^2 t$$
$$= 4t^2(\cos^2 t + \sin^2 t)$$
$$= 4t^2.$$

Because $x = t$, $y^2 + z^2 = 4x^2$.

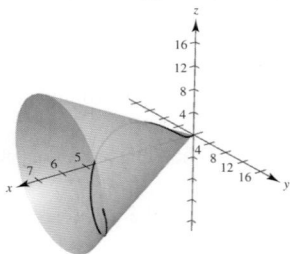

63. $\pi \mathbf{i} - \mathbf{j}$ **65.** 0   **67.** $\mathbf{i} + \mathbf{j} + \mathbf{k}$
69. $(-\infty, 0), (0, \infty)$ **71.** $[-1, 1]$
73. $(-\pi/2 + n\pi, \pi/2 + n\pi)$, n is an integer.
75. $\mathbf{s}(t) = t^2\mathbf{i} + (t - 3)\mathbf{j} + (t + 3)\mathbf{k}$
77. $\mathbf{s}(t) = (t^2 - 2)\mathbf{i} + (t - 3)\mathbf{j} + t\mathbf{k}$
79. A vector-valued function $\mathbf{r}$ is continuous at $t = a$ if the limit of $\mathbf{r}(t)$ exists as $t \to a$ and $\lim_{t \to a} \mathbf{r}(t) = \mathbf{r}(u)$. The function

$$\mathbf{r}(t) = \begin{cases} \mathbf{i} + \mathbf{j}, & t \geq 2 \\ -\mathbf{i} + \mathbf{j}, & t < 2 \end{cases} \text{ is not continuous at } t = 0.$$

81. Answers will vary. Sample answer:

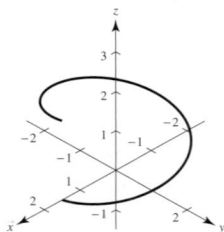

$$\mathbf{r}(t) = 1.5 \cos t\mathbf{i} + 1.5 \sin t\mathbf{j} + \frac{1}{\pi}t\mathbf{k}, \quad 0 \leq t \leq 2\pi$$

83–85. Proofs   **87.** Yes; Yes   **89.** Not necessarily
91. True   **93.** True

Section 12.2   *(page 830)*

1. $\mathbf{r}'(t) = 2t\mathbf{i} + \mathbf{j}$
$\mathbf{r}(2) = 4\mathbf{i} + 2\mathbf{j}$
$\mathbf{r}'(2) = 4\mathbf{i} + \mathbf{j}$

3. $\mathbf{r}'(t) = -\sin t\mathbf{i} + \cos t\mathbf{j}$
$\mathbf{r}(\pi/2) = \mathbf{j}$
$\mathbf{r}'(\pi/2) = -\mathbf{i}$

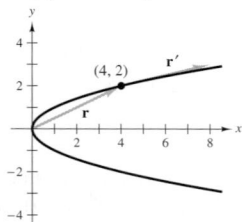

$\mathbf{r}'(t_0)$ is tangent to the curve at t_0.

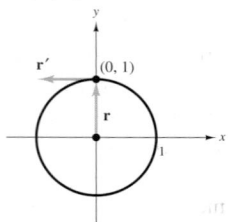

$\mathbf{r}'(t_0)$ is tangent to the curve at t_0.

5. $\mathbf{r}'(t) = \langle e^t, 2e^{2t} \rangle$
$\mathbf{r}(0) = \mathbf{i} + \mathbf{j}$
$\mathbf{r}'(0) = \mathbf{i} + 2\mathbf{j}$

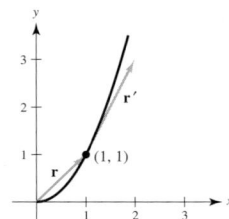

$\mathbf{r}'(t_0)$ is tangent to the curve at t_0.

7. $\mathbf{r}'(t) = -2 \sin t\mathbf{i} + 2 \cos t\mathbf{j} + \mathbf{k}$

$$\mathbf{r}\left(\frac{3\pi}{2}\right) = -2\mathbf{j} + \left(\frac{3\pi}{2}\right)\mathbf{k}$$

$$\mathbf{r}'\left(\frac{3\pi}{2}\right) = 2\mathbf{i} + \mathbf{k}$$

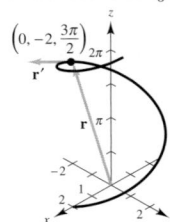

9. $3t^2\mathbf{i} - 3\mathbf{j}$ **11.** $-2 \sin t\mathbf{i} + 5 \cos t\mathbf{j}$
13. $6\mathbf{i} - 14t\mathbf{j} + 3t^2\mathbf{k}$ **15.** $-3a \sin t \cos^2 t\mathbf{i} + 3a \sin^2 t \cos t\mathbf{j}$
17. $-e^{-t}\mathbf{i} + (5te^t + 5e^t)\mathbf{k}$
19. $\langle \sin t + t \cos t, \cos t - t \sin t, 1 \rangle$
21. (a) $3t^2\mathbf{i} + t\mathbf{j}$ (b) $6t\mathbf{i} + \mathbf{j}$ (c) $18t^3 + t$
23. (a) $-4 \sin t\mathbf{i} + 4 \cos t\mathbf{j}$ (b) $-4 \cos t\mathbf{i} - 4 \sin t\mathbf{j}$ (c) 0
25. (a) $t\mathbf{i} - \mathbf{j} + \frac{1}{2}t^2\mathbf{k}$ (b) $\mathbf{i} + t\mathbf{k}$ (c) $t^3/2 + t$
    (d) $-t\mathbf{i} - \frac{1}{2}t^2\mathbf{j} + \mathbf{k}$
27. (a) $\langle t \cos t, t \sin t, 1 \rangle$
    (b) $\langle \cos t - t \sin t, \sin t + t \cos t, 0 \rangle$ (c) t
    (d) $\langle -\sin t - t \cos t, \cos t - t \sin t, t^2 \rangle$
29. $(-\infty, 0), (0, \infty)$ **31.** $(n\pi/2, (n + 1)\pi/2)$
33. $(-\infty, \infty)$ **35.** $(-\infty, 0), (0, \infty)$
37. $(-\pi/2 + n\pi, \pi/2 + n\pi)$, n is an integer.
39. (a) $\mathbf{i} + 3\mathbf{j} + 2t\mathbf{k}$ (b) $-\mathbf{i} + (9 - 2t)\mathbf{j} + (6t - 3t^2)\mathbf{k}$
    (c) $40t\mathbf{i} + 15t^2\mathbf{j} + 20t^3\mathbf{k}$ (d) $8t + 9t^2 + 5t^4$
    (e) $8t^3\mathbf{i} + (12t^2 - 4t^3)\mathbf{j} + (3t^2 - 24t)\mathbf{k}$
    (f) $2\mathbf{i} + 6\mathbf{j} + 8t\mathbf{k}$
41. (a) $7t^6$ (b) $12t^5\mathbf{i} - 5t^4\mathbf{j}$ **43.** $t^2\mathbf{i} + t\mathbf{j} + t\mathbf{k} + \mathbf{C}$
45. $\ln|t|\mathbf{i} + t\mathbf{j} - \frac{2}{5}t^{5/2}\mathbf{k} + \mathbf{C}$
47. $(t^2 - t)\mathbf{i} + t^4\mathbf{j} + 2t^{3/2}\mathbf{k} + \mathbf{C}$ **49.** $\tan t\mathbf{i} + \arctan t\mathbf{j} + \mathbf{C}$
51. $4\mathbf{i} + \frac{1}{2}\mathbf{j} - \mathbf{k}$ **53.** $a\mathbf{i} + a\mathbf{j} + (\pi/2)\mathbf{k}$
55. $2\mathbf{i} + (e^2 - 1)\mathbf{j} - (e^2 + 1)\mathbf{k}$
57. $2e^{2t}\mathbf{i} + 3(e^t - 1)\mathbf{j}$ **59.** $600\sqrt{3}t\mathbf{i} + (-16t^2 + 600t)\mathbf{j}$
61. $((2 - e^{-t^2})/2)\mathbf{i} + (e^{-t} - 2)\mathbf{j} + (t + 1)\mathbf{k}$
63. See "Definition of the Derivative of a Vector-Valued Function" and Figure 12.8 on page 824.
65. The three components of $\mathbf{u}$ are increasing functions of t at $t = t_0$.
67–73. Proofs
75. (a)

    The curve is a cycloid.

    (b) The maximum of $\|\mathbf{r}'\|$ is 2; the minimum of $\|\mathbf{r}'\|$ is 0. The maximum and the minimum of $\|\mathbf{r}''\|$ are 1.
77. Proof   **79.** True
81. False. Let $\mathbf{r}(t) = \cos t\mathbf{i} + \sin t\mathbf{j} + \mathbf{k}$, then $d/dt[\|\mathbf{r}(t)\|] = 0$, but $\|\mathbf{r}'(t)\| = 1$.

Section 12.3 *(page 838)*

1. (a) $\mathbf{v}(t) = 3\mathbf{i} + \mathbf{j}$
$\|\mathbf{v}(t)\| = \sqrt{10}$
$\mathbf{a}(t) = \mathbf{0}$
(b) $\mathbf{v}(1) = 3\mathbf{i} + \mathbf{j}$
$\mathbf{a}(1) = \mathbf{0}$
(c)

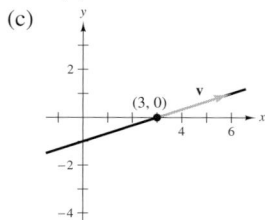

3. (a) $\mathbf{v}(t) = 2t\mathbf{i} + \mathbf{j}$
$\|\mathbf{v}(t)\| = \sqrt{4t^2 + 1}$
$\mathbf{a}(t) = 2\mathbf{i}$
(b) $\mathbf{v}(2) = 4\mathbf{i} + \mathbf{j}$
$\mathbf{a}(2) = 2\mathbf{i}$
(c)

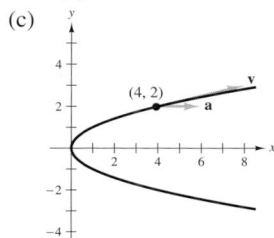

5. (a) $\mathbf{v}(t) = -2\sin t\mathbf{i} + 2\cos t\mathbf{j}$
$\|\mathbf{v}(t)\| = 2$
$\mathbf{a}(t) = -2\cos t\mathbf{i} - 2\sin t\mathbf{j}$
(b) $\mathbf{v}(\pi/4) = -\sqrt{2}\mathbf{i} + \sqrt{2}\mathbf{j}$
$\mathbf{a}(\pi/4) = -\sqrt{2}\mathbf{i} - \sqrt{2}\mathbf{j}$
(c)

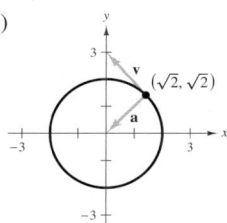

7. (a) $\mathbf{v}(t) = \langle 1 - \cos t, \sin t \rangle$
$\|\mathbf{v}(t)\| = \sqrt{2 - 2\cos t}$
$\mathbf{a}(t) = \langle \sin t, \cos t \rangle$
(b) $\mathbf{v}(\pi) = \langle 2, 0 \rangle$
$\mathbf{a}(\pi) = \langle 0, -1 \rangle$
(c)

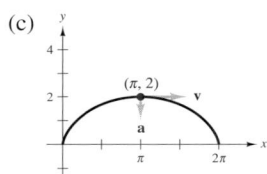

9. (a) $\mathbf{v}(t) = \mathbf{i} + 5\mathbf{j} + 3\mathbf{k}$
$\|\mathbf{v}(t)\| = \sqrt{35}$
$\mathbf{a}(t) = \mathbf{0}$
(b) $\mathbf{v}(1) = \mathbf{i} + 5\mathbf{j} + 3\mathbf{k}$
$\mathbf{a}(1) = \mathbf{0}$

11. (a) $\mathbf{v}(t) = \mathbf{i} + 2t\mathbf{j} + t\mathbf{k}$
$\|\mathbf{v}(t)\| = \sqrt{1 + 5t^2}$
$\mathbf{a}(t) = 2\mathbf{j} + \mathbf{k}$
(b) $\mathbf{v}(4) = \mathbf{i} + 8\mathbf{j} + 4\mathbf{k}$
$\mathbf{a}(4) = 2\mathbf{j} + \mathbf{k}$

13. (a) $\mathbf{v}(t) = \mathbf{i} + \mathbf{j} - \left(t/\sqrt{9 - t^2}\right)\mathbf{k}$
$\|\mathbf{v}(t)\| = \sqrt{(18 - t^2)/(9 - t^2)}$
$\mathbf{a}(t) = \left(-9/(9 - t^2)^{3/2}\right)\mathbf{k}$
(b) $\mathbf{v}(0) = \mathbf{i} + \mathbf{j}$
$\mathbf{a}(0) = -\tfrac{1}{3}\mathbf{k}$

15. (a) $\mathbf{v}(t) = 4\mathbf{i} - 3\sin t\mathbf{j} + 3\cos t\mathbf{k}$
$\|\mathbf{v}(t)\| = 5$
$\mathbf{a}(t) = -3\cos t\mathbf{j} - 3\sin t\mathbf{k}$
(b) $\mathbf{v}(\pi) = \langle 4, 0, -3 \rangle$
$\mathbf{a}(\pi) = \langle 0, 3, 0 \rangle$

17. (a) $\mathbf{v}(t) = (e^t\cos t - e^t\sin t)\mathbf{i} + (e^t\sin t + e^t\cos t)\mathbf{j} + e^t\mathbf{k}$
$\|\mathbf{v}(t)\| = e^t\sqrt{3}$
$\mathbf{a}(t) = -2e^t\sin t\mathbf{i} + 2e^t\cos t\mathbf{j} + e^t\mathbf{k}$
(b) $\mathbf{v}(0) = \langle 1, 1, 1 \rangle$
$\mathbf{a}(0) = \langle 0, 2, 1 \rangle$

19. $\mathbf{v}(t) = t(\mathbf{i} + \mathbf{j} + \mathbf{k})$
$\mathbf{r}(t) = (t^2/2)(\mathbf{i} + \mathbf{j} + \mathbf{k})$
$\mathbf{r}(2) = 2(\mathbf{i} + \mathbf{j} + \mathbf{k})$

21. $\mathbf{v}(t) = \left(t^2/2 + \tfrac{9}{2}\right)\mathbf{j} + \left(t^2/2 - \tfrac{1}{2}\right)\mathbf{k}$
$\mathbf{r}(t) = \left(t^3/6 + \tfrac{9}{2}t - \tfrac{14}{3}\right)\mathbf{j} + \left(t^3/6 - \tfrac{1}{2}t + \tfrac{1}{3}\right)\mathbf{k}$
$\mathbf{r}(2) = \tfrac{17}{3}\mathbf{j} + \tfrac{2}{3}\mathbf{k}$

23. $\mathbf{v}(t) = -\sin t\mathbf{i} + \cos t\mathbf{j} + \mathbf{k}$
$\mathbf{r}(t) = \cos t\mathbf{i} + \sin t\mathbf{j} + t\mathbf{k}$
$\mathbf{r}(2) = (\cos 2)\mathbf{i} + (\sin 2)\mathbf{j} + 2\mathbf{k}$

25. Maximum height: 45.5 ft; The ball will clear the fence.

27. $v_0 = 40\sqrt{6}$ ft/sec; 78 ft   **29.** Proof

31. (a) $\mathbf{r}(t) = \left(\tfrac{440}{3}\cos\theta_0\right)t\mathbf{i} + \left[3 + \left(\tfrac{440}{3}\sin\theta_0\right)t - 16t^2\right]\mathbf{j}$
(b)

The minimum angle appears to be $\theta_0 = 20°$.
(c) $\theta_0 \approx 19.38°$

33. (a) $v_0 = 28.78$ ft/sec; $\theta = 58.28°$ (b) $v_0 \approx 32$ ft/sec

35. 1.91°

37. (a)

(b)

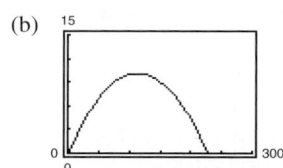

Maximum height: 2.1 ft
Range: 46.6 ft

Maximum height: 10.0 ft
Range: 227.8 ft

(c)

(d)

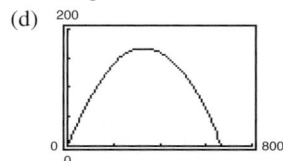

Maximum height: 34.0 ft
Range: 136.1 ft

Maximum height: 166.5 ft
Range: 666.1 ft

(e)

(f)

Maximum height: 51.0 ft
Range: 117.9 ft

Maximum height: 249.8 ft
Range: 576.9 ft

39. Maximum height: 129.1 m; Range: 886.3 m

41. Proof

43. $\mathbf{v}(t) = b\omega[(1 - \cos\omega t)\mathbf{i} + \sin\omega t\mathbf{j}]$
$\mathbf{a}(t) = b\omega^2(\sin\omega t\mathbf{i} + \cos\omega t\mathbf{j})$
(a) $\|\mathbf{v}(t)\| = 0$ when $\omega t = 0, 2\pi, 4\pi, \ldots$
(b) $\|\mathbf{v}(t)\|$ is maximum when $\omega t = \pi, 3\pi, \ldots$

45. $\mathbf{v}(t) = -b\omega\sin\omega t\mathbf{i} + b\omega\cos\omega t\mathbf{j}$
$\mathbf{v}(t) \cdot \mathbf{r}(t) = 0$

47. $\mathbf{a}(t) = -b\omega^2(\cos\omega t\mathbf{i} + \sin\omega t\mathbf{j}) = -\omega^2\mathbf{r}(t)$; $\mathbf{a}(t)$ is a negative multiple of a unit vector from $(0, 0)$ to $(\cos\omega t, \sin\omega t)$, so $\mathbf{a}(t)$ is directed toward the origin.

49. $8\sqrt{10}$ ft/sec

51. The velocity of an object involves both magnitude and direction of motion, whereas speed involves only magnitude.

53. Proof

55. (a) $\mathbf{v}(t) = -6 \sin t\mathbf{i} + 3 \cos t\mathbf{j}$
$\|\mathbf{v}(t)\| = 3\sqrt{3 \sin^2 t + 1}$
$\mathbf{a}(t) = -6 \cos t\mathbf{i} - 3 \sin t\mathbf{j}$

(b)

t	0	$\pi/4$	$\pi/2$	$2\pi/3$	π
Speed	3	$3\sqrt{10}/2$	6	$3\sqrt{13}/2$	3

(c)

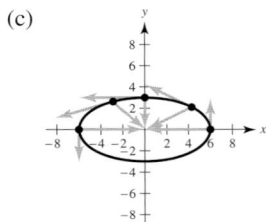

(d) The speed is increasing when the angle between $\mathbf{v}$ and $\mathbf{a}$ is in the interval $[0, \pi/2)$, and decreasing when the angle is in the interval $(\pi/2, \pi]$.

57. Proof
59. False. Acceleration is the derivative of the velocity.
61. True

Section 12.4   (page 848)

1. $\mathbf{T}(1) = (\sqrt{2}/2)(\mathbf{i} + \mathbf{j})$ **3.** $\mathbf{T}(\pi/4) = (\sqrt{2}/2)(-\mathbf{i} + \mathbf{j})$
5. $\mathbf{T}(e) = (3e\mathbf{i} - \mathbf{j})/\sqrt{9e^2 + 1} \approx 0.9926\mathbf{i} - 0.1217\mathbf{j}$
7. $\mathbf{T}(0) = (\sqrt{2}/2)(\mathbf{i} + \mathbf{k})$ **9.** $\mathbf{T}(0) = (\sqrt{10}/10)(3\mathbf{j} + \mathbf{k})$
$x = t$ $x = 3$
$y = 0$ $y = 3t$
$z = t$ $z = t$
11. $\mathbf{T}(\pi/4) = \frac{1}{2}\langle -\sqrt{2}, \sqrt{2}, 0 \rangle$
$x = \sqrt{2} - \sqrt{2}t$
$y = \sqrt{2} + \sqrt{2}t$
$z = 4$
13. $\mathbf{N}(2) = (\sqrt{5}/5)(-2\mathbf{i} + \mathbf{j})$
15. $\mathbf{N}(2) = (-\sqrt{5}/5)(2\mathbf{i} - \mathbf{j})$
17. $\mathbf{N}(1) = (-\sqrt{14}/14)(\mathbf{i} - 2\mathbf{j} + 3\mathbf{k})$
19. $\mathbf{N}(3\pi/4) = (\sqrt{2}/2)(\mathbf{i} - \mathbf{j})$
21. $\mathbf{T}(1) = (\sqrt{2}/2)(\mathbf{i} - \mathbf{j})$ **23.** $\mathbf{T}(1) = (-\sqrt{5}/5)(\mathbf{i} - 2\mathbf{j})$
$\mathbf{N}(1) = (\sqrt{2}/2)(\mathbf{i} + \mathbf{j})$ $\mathbf{N}(1) = (-\sqrt{5}/5)(2\mathbf{i} + \mathbf{j})$
$a_T = -\sqrt{2}$ $a_T = 14\sqrt{5}/5$
$a_N = \sqrt{2}$ $a_N = 8\sqrt{5}/5$
25. $\mathbf{T}(0) = (\sqrt{5}/5)(\mathbf{i} - 2\mathbf{j})$ **27.** $\mathbf{T}(\pi/2) = (\sqrt{2}/2)(-\mathbf{i} + \mathbf{j})$
$\mathbf{N}(0) = (\sqrt{5}/5)(2\mathbf{i} + \mathbf{j})$ $\mathbf{N}(\pi/2) = (-\sqrt{2}/2)(\mathbf{i} + \mathbf{j})$
$a_T = -7\sqrt{5}/5$ $a_T = \sqrt{2}e^{\pi/2}$
$a_N = 6\sqrt{5}/5$ $a_N = \sqrt{2}e^{\pi/2}$
29. $\mathbf{T}(t) = -\sin(\omega t)\mathbf{i} + \cos(\omega t)\mathbf{j}$
$\mathbf{N}(t) = -\cos(\omega t)\mathbf{i} - \sin(\omega t)\mathbf{j}$
$a_T = 0$
$a_N = a\omega^2$
31. $\|\mathbf{v}(t)\| = a\omega$; The speed is constant because $a_T = 0$.
33. $\mathbf{r}(2) = 2\mathbf{i} + \frac{1}{2}\mathbf{j}$
$\mathbf{T}(2) = (\sqrt{17}/17)(4\mathbf{i} - \mathbf{j})$
$\mathbf{N}(2) = (\sqrt{17}/17)(\mathbf{i} + 4\mathbf{j})$

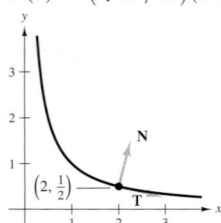

35. $\mathbf{r}(2) = 5\mathbf{i} - 4\mathbf{j}$
$\mathbf{T}(2) = \dfrac{\mathbf{i} - 2\mathbf{j}}{\sqrt{5}}$
$\mathbf{N}(2) = \dfrac{-2\mathbf{i} - \mathbf{j}}{\sqrt{5}}$, perpendicular to $\mathbf{T}(2)$

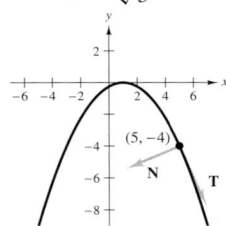

37. $\mathbf{T}(1) = (\sqrt{14}/14)(\mathbf{i} + 2\mathbf{j} - 3\mathbf{k})$
$\mathbf{N}(1)$ is undefined.
a_T is undefined.
a_N is undefined.
39. $\mathbf{T}(1) = (\sqrt{6}/6)(\mathbf{i} + 2\mathbf{j} + \mathbf{k})$
$\mathbf{N}(1) = (\sqrt{30}/30)(-5\mathbf{i} + 2\mathbf{j} + \mathbf{k})$
$a_T = 5\sqrt{6}/6$
$a_N = \sqrt{30}/6$
41. $\mathbf{T}(0) = (\sqrt{3}/3)(\mathbf{i} + \mathbf{j} + \mathbf{k})$
$\mathbf{N}(0) = (\sqrt{2}/2)(\mathbf{i} - \mathbf{j})$
$a_T = \sqrt{3}$
$a_N = \sqrt{2}$
43. Let C be a smooth curve represented by $\mathbf{r}$ on an open interval I. The unit tangent vector $\mathbf{T}(t)$ at t is defined as
$$\mathbf{T}(t) = \frac{\mathbf{r}'(t)}{\|\mathbf{r}'(t)\|}, \quad \mathbf{r}'(t) \neq 0.$$
The principal unit normal vector $\mathbf{N}(t)$ at t is defined as
$$\mathbf{N}(t) = \frac{\mathbf{T}'(t)}{\|\mathbf{T}'(t)\|}, \quad \mathbf{T}'(t) \neq 0.$$
The tangential and normal components of acceleration are defined as $\mathbf{a}(t) = a_T\mathbf{T}(t) + a_N\mathbf{N}(t)$.
45. (a) The particle's motion is in a straight line.
(b) The particle's speed is constant.
47. $\mathbf{v}(t) = \mathbf{r}'(t) = 3\mathbf{i} + 4\mathbf{j}$
$\|\mathbf{v}(t)\| = \sqrt{9 + 16} = 5$
$\mathbf{a}(t) = \mathbf{v}'(t) = 0$
$\mathbf{T}(t) = \dfrac{\mathbf{v}(t)}{\|\mathbf{v}(t)\|} = \dfrac{3}{5}\mathbf{i} + \dfrac{4}{5}\mathbf{j}$
$\mathbf{T}'(t) = 0 \Rightarrow \mathbf{N}(t)$ does not exist.
The path is a line. The speed is constant (5).
49. (a) $t = \frac{1}{2}$: $a_T = \sqrt{2}\pi^2/2, a_N = \sqrt{2}\pi^2/2$
$t = 1$: $a_T = 0, a_N = \pi^2$
$t = \frac{3}{2}$: $a_T = -\sqrt{2}\pi^2/2, a_N = \sqrt{2}\pi^2/2$
(b) $t = \frac{1}{2}$: Increasing because $a_T > 0$.
$t = 1$: Maximum because $a_T = 0$.
$t = \frac{3}{2}$: Decreasing because $a_T < 0$.
51. $\mathbf{T}(\pi/2) = (\sqrt{17}/17)(-4\mathbf{i} + \mathbf{k})$
$\mathbf{N}(\pi/2) = -\mathbf{j}$
$\mathbf{B}(\pi/2) = (\sqrt{17}/17)(\mathbf{i} + 4\mathbf{k})$
53. $\mathbf{T}(\pi/4) = (\sqrt{2}/2)(\mathbf{j} - \mathbf{k})$
$\mathbf{N}(\pi/4) = -(\sqrt{2}/2)(\mathbf{j} + \mathbf{k})$
$\mathbf{B}(\pi/4) = -\mathbf{i}$

55. $T(\pi/3) = \left(\sqrt{5}/5\right)\!\left(i - \sqrt{3}j + k\right)$
$N(\pi/3) = -\frac{1}{2}\!\left(\sqrt{3}i + j\right)$
$B(\pi/3) = \left(\sqrt{5}/10\right)\!\left(i - \sqrt{3}j - 4k\right)$

57. $N(t) = \dfrac{1}{\sqrt{16t^2 + 9}}(-4ti + 3j)$

59. $N(t) = \dfrac{1}{\sqrt{5t^2 + 25}}(-ti - 2tj + 5k)$

61. $a_T = \dfrac{-32(v_0 \sin\theta - 32t)}{\sqrt{v_0^2 \cos^2\theta + (v_0 \sin\theta - 32t)^2}}$

$a_N = \dfrac{32 v_0 \cos\theta}{\sqrt{v_0^2 \cos^2\theta + (v_0 \sin\theta - 32t)^2}}$

At maximum height, $a_T = 0$ and $a_N = 32$.

63. (a) $r(t) = 60\sqrt{3}\,ti + (5 + 60t - 16t^2)j$

(b)

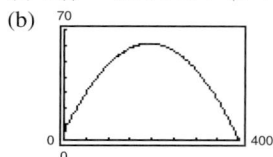

Maximum height ≈ 61.245 ft
Range ≈ 398.186 ft

(c) $v(t) = 60\sqrt{3}\,i + (60 - 32t)j$
$\|v(t)\| = 8\sqrt{16t^2 - 60t + 225}$
$a(t) = -32j$

(d)

t	0.5	1.0	1.5
Speed	112.85	107.63	104.61

t	2.0	2.5	3.0
Speed	104	105.83	109.98

(e)

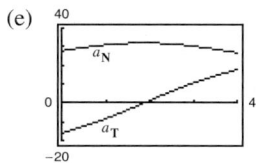

The speed is decreasing when a_T and a_N have opposite signs.

65. (a) $4\sqrt{625\pi^2 + 1} \approx 314$ mi/h
(b) $a_T = 0$, $a_N = 1000\pi^2$
$a_T = 0$ because the speed is constant.

67. (a) The centripetal component is quadrupled.
(b) The centripetal component is halved.

69. 4.74 mi/sec   **71.** 4.67 mi/sec

73. False; centripetal acceleration may occur with constant speed.

75. (a) Proof   (b) Proof   **77–79.** Proofs

Section 12.5   *(page 860)*

1.

$3\sqrt{10}$

3.

$\left(13\sqrt{13} - 8\right)/27$

5.

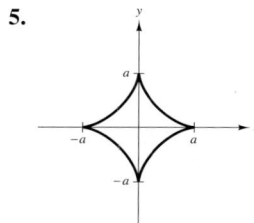

$6a$

7. (a) $r(t) = \left(50t\sqrt{2}\right)i + \left(3 + 50t\sqrt{2} - 16t^2\right)j$
(b) $\frac{649}{8} \approx 81$ ft   (c) 315.5 ft   (d) 362.9 ft

9.

$\sqrt{26}$

11.

$3\sqrt{17}\,\pi/2$

13.

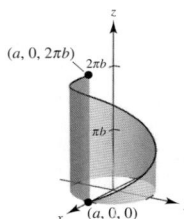

$2\pi\sqrt{a^2 + b^2}$

15. (a) $2\sqrt{21} \approx 9.165$ (b) 9.529
(c) Increase the number of line segments.   (d) 9.571

17. (a) $s = \sqrt{5}\,t$ (b) $r(s) = 2\cos\dfrac{s}{\sqrt{5}}i + 2\sin\dfrac{s}{\sqrt{5}}j + \dfrac{s}{\sqrt{5}}k$
(c) $s = \sqrt{5}$: $(1.081, 1.683, 1.000)$
$s = 4$: $(-0.433, 1.953, 1.789)$
(d) Proof

19. 0   **21.** $\frac{2}{5}$ **23.** 0   **25.** $\sqrt{2}/2$ **27.** 1   **29.** $\frac{1}{4}$

31. $1/a$ **33.** $\sqrt{5}/(1 + 5t^2)^{3/2}$ **35.** $\frac{3}{25}$ **37.** $\frac{12}{125}$

39. $7\sqrt{26}/676$ **41.** $K = 0$, $1/K$ is undefined.

43. $K = 4/17^{3/2}$, $1/K = 17^{3/2}/4$ **45.** $K = 4$, $1/K = 1/4$

47. $K = 12/145^{3/2}$, $1/K = 145^{3/2}/12$ **49.** (a) $(1, 3)$ (b) 0

51. (a) $K \to \infty$ as $x \to 0$ (No maximum)   (b) 0

53. (a) $\left(1/\sqrt{2}, -\ln 2/2\right)$ (b) 0

55. $(0, 1)$ **57.** $(\pi/2 + K\pi, 0)$

59. $s = \displaystyle\int_a^b \sqrt{[x'(t)]^2 + [y'(t)]^2 + [z'(t)]^2}\, dt = \int_a^b \|r'(t)\|\, dt$

61. The curve is a line.

63. (a) $K = \dfrac{2|6x^2 - 1|}{(16x^6 - 16x^4 + 4x^2 + 1)^{3/2}}$

(b) $x = 0$: $x^2 + \left(y + \dfrac{1}{2}\right)^2 = \dfrac{1}{4}$

$x = 1$: $x^2 + \left(y - \dfrac{1}{2}\right)^2 = \dfrac{5}{4}$

(c)

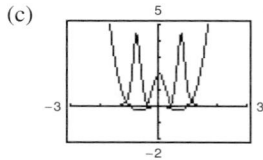

The curvature tends to be greatest near the extrema of the function and decreases as $x \to \pm\infty$. However, f and K do not have the same critical numbers.

Critical numbers of f: $x = 0, \pm\sqrt{2}/2 \approx \pm 0.7071$

Critical numbers of K: $x = 0, \pm 0.7647, \pm 0.4082$

65. Proof   **67.** (a) 12.25 units   (b) $\frac{1}{2}$ **69–71.** Proofs

73. (a) 0   (b) 0   **75.** $\frac{1}{4}$ **77.** Proof

79. $K = [1/(4a)]|\csc(\theta/2)|$ **81.** 3327.5 lb

Minimum: $K = 1/(4a)$

There is no maximum.

83. Proof   **85.** False. See Exploration on page 851.

87. True   **89–95.** Proofs

Review Exercises for Chapter 12   (page 863)

1. (a) All reals except $(\pi/2) + n\pi$, n is an integer.

(b) Continuous except at $t = (\pi/2) + n\pi$, n is an integer.

3. (a) $(0, \infty)$ (b) Continuous for all $t > 0$

5. (a) $\mathbf{i} - \sqrt{2}\mathbf{k}$ (b) $-3\mathbf{i} + 4\mathbf{j}$

(c) $(2c - 1)\mathbf{i} + (c - 1)^2\mathbf{j} - \sqrt{c + 1}\mathbf{k}$

(d) $2\Delta t\mathbf{i} + \Delta t(\Delta t + 2)\mathbf{j} - (\sqrt{\Delta t + 3} - \sqrt{3})\mathbf{k}$

7. $\mathbf{r}(t) = (3 - t)\mathbf{i} - 2t\mathbf{j} + (5 - 2t)\mathbf{k}$, $0 \le t \le 1$

$x = 3 - t, y = -2t, z = 5 - 2t$, $0 \le t \le 1$

9.

11.

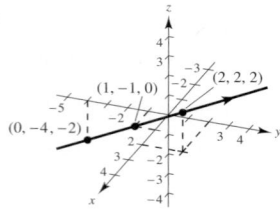

13. $\mathbf{r}(t) = t\mathbf{i} + \left(-\frac{3}{4}t + 3\right)\mathbf{j}$ **15.**

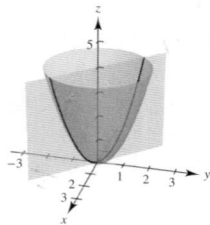

$x = t, y = -t, z = 2t^2$

17. $4\mathbf{i} + \mathbf{k}$

19. (a) $(2t + 4)\mathbf{i} - 6t\mathbf{j}$ **21.** (a) $6t^2\mathbf{i} + 4\mathbf{j} - 2t\mathbf{k}$

(b) $2\mathbf{i} - 6\mathbf{j}$ (b) $12t\mathbf{i} - 2\mathbf{k}$

(c) $40t + 8$ (c) $72t^3 + 4t$

(d) $-8\mathbf{i} - 12t^2\mathbf{j} - 48t\mathbf{k}$

23. (a) $3\mathbf{i} + \mathbf{j}$ (b) $-5\mathbf{i} + (2t - 2)\mathbf{j} + 2t^2\mathbf{k}$

(c) $18t\mathbf{i} + (6t - 3)\mathbf{j}$ (d) $4t + 3t^2$

(e) $\left(\frac{8}{3}t^3 - 2t^2\right)\mathbf{i} - 8t^3\mathbf{j} + (9t^2 - 2t + 1)\mathbf{k}$

(f) $2\mathbf{i} + 8t\mathbf{j} + 16t^2\mathbf{k}$

25. $t\mathbf{i} + 3t\mathbf{j} + 2t^2\mathbf{k} + \mathbf{C}$ **27.** $2t^{3/2}\mathbf{i} + 2\ln|t|\mathbf{j} + t\mathbf{k} + \mathbf{C}$

29. $\frac{32}{3}\mathbf{j}$ **31.** $2(e - 1)\mathbf{i} - 8\mathbf{j} - 2\mathbf{k}$

33. $\mathbf{r}(t) = (t^2 + 1)\mathbf{i} + (e^t + 2)\mathbf{j} - (e^{-t} + 4)\mathbf{k}$

35. (a) $\mathbf{v}(t) = 4\mathbf{i} + 3t^2\mathbf{j} - \mathbf{k}$

$\|\mathbf{v}(t)\| = \sqrt{17 + 9t^4}$

$\mathbf{a}(t) = 6t\mathbf{j}$

(b) $\mathbf{v}(1) = 4\mathbf{i} + 3\mathbf{j} - \mathbf{k}$

$\mathbf{a}(1) = 6\mathbf{j}$

37. (a) $\mathbf{v}(t) = \langle -3\cos^2 t \sin t, 3\sin^2 t \cos t, 3\rangle$

$\|\mathbf{v}(t)\| = 3\sqrt{\sin^2 t \cos^2 t + 1}$

$\mathbf{a}(t) = \langle 3\cos t(2\sin^2 t - \cos^2 t), 3\sin t(2\cos^2 t - \sin^2 t), 0\rangle$

(b) $\mathbf{v}(\pi) = \langle 0, 0, 3\rangle$

$\mathbf{a}(\pi) = \langle 3, 0, 0\rangle$

39. About 191.0 ft   **41.** About 38.1 m/sec

43. $\mathbf{T}(1) = \dfrac{\sqrt{10}}{10}\mathbf{i} + \dfrac{3\sqrt{10}}{10}\mathbf{j}$

45. $\mathbf{T}\left(\dfrac{\pi}{3}\right) = -\dfrac{\sqrt{15}}{5}\mathbf{i} + \dfrac{\sqrt{5}}{5}\mathbf{j} + \dfrac{\sqrt{5}}{5}\mathbf{k}$;

$x = -\sqrt{3}t + 1, y = t + \sqrt{3}, z = t + \dfrac{\pi}{3}$

47. $\mathbf{N}(1) = -\dfrac{3\sqrt{10}}{10}\mathbf{i} + \dfrac{\sqrt{10}}{10}\mathbf{j}$ **49.** $\mathbf{N}\left(\dfrac{\pi}{4}\right) = -\mathbf{j}$

51. $\mathbf{T}(3) = -\dfrac{\sqrt{13}}{65}\mathbf{i} - \dfrac{18\sqrt{13}}{65}\mathbf{j}$

$\mathbf{N}(3) = \dfrac{18\sqrt{13}}{65}\mathbf{i} - \dfrac{\sqrt{13}}{65}\mathbf{j}$

$a_\mathbf{T} = -\dfrac{2\sqrt{13}}{585}$

$a_\mathbf{N} = \dfrac{4\sqrt{13}}{65}$

53.

$5\sqrt{13}$

55.

60

57.

$3\sqrt{29}$

59.

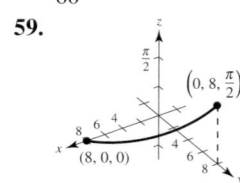

$\sqrt{65}\pi/2$

61. 0   **63.** $(2\sqrt{5})/(4 + 5t^2)^{3/2}$ **65.** $\sqrt{2}/3$

67. $K = \sqrt{17}/289; r = 17\sqrt{17}$ **69.** $K = \sqrt{2}/4; r = 2\sqrt{2}$

71. 2016.7 lb

P.S. Problem Solving   (page 865)

1. (a) a (b) πa (c) $K = \pi a$

3. Initial speed: 447.21 ft/sec; $\theta \approx 63.43°$

5–7. Proofs

9. Unit tangent: $\left\langle -\frac{4}{5}, 0, \frac{3}{5} \right\rangle$
Unit normal: $\langle 0, -1, 0 \rangle$
Binormal: $\left\langle \frac{3}{5}, 0, \frac{4}{5} \right\rangle$

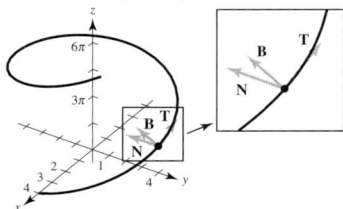

11. (a) Proof   (b) Proof

13. (a)   (b) 6.766

(c) $K = [\pi(\pi^2 t^2 + 2)]/(\pi^2 t^2 + 1)^{3/2}$
$K(0) = 2\pi$
$K(1) = [\pi(\pi^2 + 2)]/(\pi^2 + 1)^{3/2} \approx 1.04$
$K(2) \approx 0.51$

(d)   (e) $\lim\limits_{t \to \infty} K = 0$

(f) As $t \to \infty$, the graph spirals outward and the curvature decreases.

Chapter 13

Section 13.1   (page 876)

1. Not a function because for some values of x and y (for example $x = y = 0$), there are two z-values.

3. z is a function of x and y.   **5.** z is not a function of x and y.

7. (a) 6   (b) -4 (c) 150   (d) $5y$ (e) $2x$ (f) $5t$

9. (a) 5   (b) $3e^2$ (c) $2/e$ (d) $5e^y$ (e) xe^2 (f) te^t

11. (a) $\frac{2}{3}$ (b) 0   (c) $-\frac{3}{2}$ (d) $-\frac{10}{3}$

13. (a) $\sqrt{2}$ (b) $3 \sin 1$ (c) $-3\sqrt{3}/2$ (d) 4

15. (a) -4 (b) -6 (c) $-\frac{25}{4}$ (d) $\frac{9}{4}$

17. (a) $2, \Delta x \neq 0$ (b) $2y + \Delta y, \Delta y \neq 0$

19. Domain: $\{(x, y): x$ is any real number, y is any real number$\}$
Range: $z \geq 0$

21. Domain: $\{(x, y): y \geq 0\}$
Range: all real numbers

23. Domain: $\{(x, y): x \neq 0, y \neq 0\}$
Range: all real numbers

25. Domain: $\{(x, y): x^2 + y^2 \leq 4\}$
Range: $0 \leq z \leq 2$

27. Domain: $\{(x, y): -1 \leq x + y \leq 1\}$
Range: $0 \leq z \leq \pi$

29. Domain: $\{(x, y): y < -x + 4\}$
Range: all real numbers

31. (a) $(20, 0, 0)$ (b) $(-15, 10, 20)$
(c) $(20, 15, 25)$ (d) $(20, 20, 0)$

33.

35.

37.

39.

41.

43.

45. c   **46.** d   **47.** b   **48.** a

49. Lines: $x + y = c$

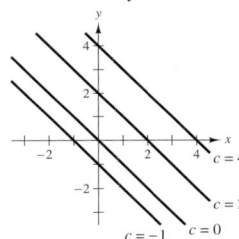

51. Ellipses: $x^2 + 4y^2 = c$
(except $x^2 + 4y^2 = 0$ is the point $(0, 0)$.)

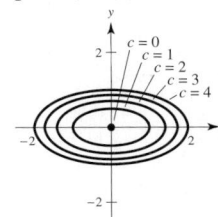

53. Hyperbolas: $xy = c$

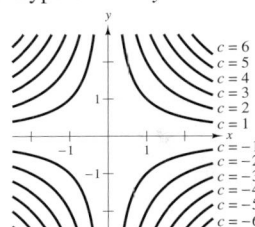

55. Circles passing through $(0, 0)$
Centered at $(1/(2c), 0)$

57.

59.

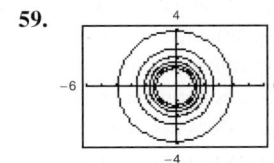

61. The graph of a function of two variables is the set of all points (x, y, z) for which $z = f(x, y)$ and (x, y) is in the domain of f. The graph can be interpreted as a surface in space. Level curves are the scalar fields $f(x, y) = c$, where c is a constant.

63. $f(x, y) = x/y$; the level curves are the lines $y = (1/c)x$.

65. The surface may be shaped like a saddle. For example, let $f(x, y) = xy$. The graph is not unique; any vertical translation will produce the same level curves.

67.

	Inflation Rate		
Tax Rate	0	0.03	0.05
0	$1790.85	$1332.56	$1099.43
0.28	$1526.43	$1135.80	$937.09
0.35	$1466.07	$1090.90	$900.04

69.

71.

73.

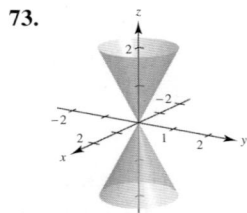

75. (a) 243 board-ft   (b) 507 board-ft

77.

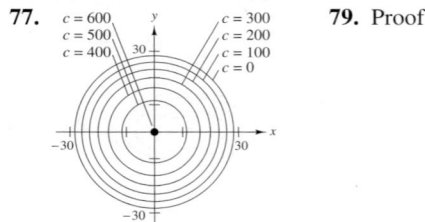

79. Proof

81. (a) $k = \frac{520}{3}$
(b) $P = 520T/(3V)$
   The level curves are lines.

83. (a) C (b) A (c) B

85. $C = 1.20xy + 1.50(xz + yz)$

87. False. Let $f(x, y) = 4$.

89. True   **91.** Putnam Problem A1, 2008

Section 13.2   (page 887)

1–3. Proofs   **5.** 1   **7.** 12   **9.** 9, continuous

11. e^2, continuous   **13.** 0, continuous for $y \neq 0$

15. $\frac{1}{2}$, continuous except at $(0, 0)$ **17.** 0, continuous

19. 0, continuous for $xy \neq 1$, $|xy| \leq 1$

21. $2\sqrt{2}$, continuous for $x + y + z \geq 0$ **23.** 0

25. Limit does not exist.   **27.** Limit does not exist.

29. Limit does not exist.   **31.** 0

33. Limit does not exist.   **35.** Continuous, 1

37.

(x, y)	$(1, 0)$	$(0.5, 0)$	$(0.1, 0)$	$(0.01, 0)$	$(0.001, 0)$
$f(x, y)$	0	0	0	0	0

$y = 0$: 0

(x, y)	$(1, 1)$	$(0.5, 0.5)$	$(0.1, 0.1)$
$f(x, y)$	$\frac{1}{2}$	$\frac{1}{2}$	$\frac{1}{2}$

(x, y)	$(0.01, 0.01)$	$(0.001, 0.001)$
$f(x, y)$	$\frac{1}{2}$	$\frac{1}{2}$

$y = x$: $\frac{1}{2}$
Limit does not exist.
Continuous except at $(0, 0)$

39.

(x, y)	$(1, 0)$	$(0.5, 0)$	$(0.1, 0)$	$(0.01, 0)$	$(0.001, 0)$
$f(x, y)$	0	0	0	0	0

$y = 0$; 0

(x, y)	$(1, 1)$	$(0.5, 0.5)$	$(0.1, 0.1)$
$f(x, y)$	$\frac{1}{2}$	1	5

(x, y)	$(0.01, 0.01)$	$(0.001, 0.001)$
$f(x, y)$	50	500

$y = x$; ∞
The limit does not exist.
Continuous except at $(0, 0)$

41. f is continuous. g is continuous except at $(0, 0)$. g has a removable discontinuity at $(0, 0)$.

43. 0   **45.** 0   **47.** 1   **49.** 1   **51.** 0

53. Continuous except at $(0, 0, 0)$ **55.** Continuous

57. Continuous   **59.** Continuous

61. Continuous for $y \neq 2x/3$ **63.** (a) $2x$ (b) -4

65. (a) $1/y$ (b) $-x/y^2$ **67.** (a) $3 + y$ (b) $x - 2$

69. True

71. False. Let $f(x, y) = \begin{cases} \ln(x^2 + y^2), & x \neq 0, y \neq 0 \\ 0, & x = 0, y = 0 \end{cases}$.

73. (a) $(1 + a^2)/a$, $a \neq 0$ (b) Limit does not exist.
(c) No, the limit does not exist. Different paths result in different limits.

75. 0   **77.** $\pi/2$ **79.** Proof

81. See "Definition of the Limit of a Function of Two Variables" on page 881; show that the value of $\displaystyle\lim_{(x, y)\to(x_0, y_0)} f(x, y)$ is not the same for two different paths to (x_0, y_0).

83. (a) No. The existence of $f(2, 3)$ has no bearing on the existence of the limit as $(x, y) \to (2, 3)$.
(b) No. $f(2, 3)$ can equal any number, or not even be defined.

Section 13.3   (page 896)

1. No. Because you are finding the partial derivative with respect to x, you consider y to be constant. So, the denominator is considered a constant and does not contain any variables.

3. No. Because you are finding the partial derivative with respect to y, you consider x to be constant. So, the denominator is considered a constant and does not contain any variables.

5. Yes. Because you are finding the partial derivative with respect to x, you consider y to be constant. So, both the numerator and denominator contain variables.

7. $f_x(x, y) = 2$
$f_y(x, y) = -5$

9. $f_x(x, y) = 2xy^3$
$f_y(x, y) = 3x^2y^2$

11. $\partial z/\partial x = \sqrt{y}$
$\partial z/\partial y = x/(2\sqrt{y})$

13. $\partial z/\partial x = 2x - 4y$
$\partial z/\partial y = -4x + 6y$

15. $\partial z/\partial x = ye^{xy}$
$\partial z/\partial y = xe^{xy}$

17. $\partial z/\partial x = 2xe^{2y}$
$\partial z/\partial y = 2x^2e^{2y}$

19. $\partial z/\partial x = 1/x$
$\partial z/\partial y = -1/y$

21. $\partial z/\partial x = 2x/(x^2 + y^2)$
$\partial z/\partial y = 2y/(x^2 + y^2)$

23. $\partial z/\partial x = (x^3 - 3y^3)/(x^2y)$
$\partial z/\partial y = (-x^3 + 12y^3)/(2xy^2)$

25. $h_x(x, y) = -2xe^{-(x^2+y^2)}$
$h_y(x, y) = -2ye^{-(x^2+y^2)}$

27. $f_x(x, y) = x/\sqrt{x^2 + y^2}$
$f_y(x, y) = y/\sqrt{x^2 + y^2}$

29. $\partial z/\partial x = -y \sin xy$
$\partial z/\partial y = -x \sin xy$

31. $\partial z/\partial x = 2 \sec^2(2x - y)$
$\partial z/\partial y = -\sec^2(2x - y)$

33. $\partial z/\partial x = ye^y \cos xy$
$\partial z/\partial y = e^y(x \cos xy + \sin xy)$

35. $\partial z/\partial x = 2 \cosh(2x + 3y)$
$\partial z/\partial y = 3 \cosh(2x + 3y)$

37. $f_x(x, y) = 1 - x^2$
$f_y(x, y) = y^2 - 1$

39. $f_x(x, y) = 3$
$f_y(x, y) = 2$

41. $f_x(x, y) = 1/(2\sqrt{x + y})$
$f_y(x, y) = 1/(2\sqrt{x + y})$

43. $f_x = -1$
$f_y = 0$

45. $f_x = -1$
$f_y = \frac{1}{2}$

47. $f_x = \frac{1}{4}$
$f_y = \frac{1}{4}$

49. $f_x = -\frac{1}{4}$
$f_y = \frac{1}{4}$

51. $g_x(1, 1) = -2$
$g_y(1, 1) = -2$

53. $H_x(x, y, z) = \cos(x + 2y + 3z)$
$H_y(x, y, z) = 2 \cos(x + 2y + 3z)$
$H_z(x, y, z) = 3 \cos(x + 2y + 3z)$

55. $\dfrac{\partial w}{\partial x} = \dfrac{x}{\sqrt{x^2 + y^2 + z^2}}$
$\dfrac{\partial w}{\partial y} = \dfrac{y}{\sqrt{x^2 + y^2 + z^2}}$
$\dfrac{\partial w}{\partial z} = \dfrac{z}{\sqrt{x^2 + y^2 + z^2}}$

57. $F_x(x, y, z) = \dfrac{x}{x^2 + y^2 + z^2}$
$F_y(x, y, z) = \dfrac{y}{x^2 + y^2 + z^2}$
$F_z(x, y, z) = \dfrac{z}{x^2 + y^2 + z^2}$

59. $f_x = 3; f_y = 1; f_z = 2$

61. $f_x = 1; f_y = 1; f_z = 1$

63. $f_x = 0; f_y = 0; f_z = 1$

65. $x = 2, y = -2$

67. $x = -6, y = 4$

69. $x = 1, y = 1$

71. $x = 0, y = 0$

73. $\dfrac{\partial^2 z}{\partial x^2} = 0$
$\dfrac{\partial^2 z}{\partial y^2} = 6x$
$\dfrac{\partial^2 z}{\partial y \partial x} = \dfrac{\partial^2 z}{\partial x \partial y} = 6y$

75. $\dfrac{\partial^2 z}{\partial x^2} = 2$
$\dfrac{\partial^2 z}{\partial y^2} = 6$
$\dfrac{\partial^2 z}{\partial y \partial x} = \dfrac{\partial^2 z}{\partial x \partial y} = -2$

77. $\dfrac{\partial^2 z}{\partial x^2} = \dfrac{y^2}{(x^2 + y^2)^{3/2}}$
$\dfrac{\partial^2 z}{\partial y^2} = \dfrac{x^2}{(x^2 + y^2)^{3/2}}$
$\dfrac{\partial^2 z}{\partial y \partial x} = \dfrac{\partial^2 z}{\partial x \partial y} = \dfrac{-xy}{(x^2 + y^2)^{3/2}}$

79. $\dfrac{\partial^2 z}{\partial x^2} = e^x \tan y$
$\dfrac{\partial^2 z}{\partial y^2} = 2e^x \sec^2 y \tan y$
$\dfrac{\partial^2 z}{\partial y \partial x} = \dfrac{\partial^2 z}{\partial x \partial y} = e^x \sec^2 y$

81. $\dfrac{\partial^2 z}{\partial x^2} = -y^2 \cos xy$
$\dfrac{\partial^2 z}{\partial y^2} = -x^2 \cos xy$
$\dfrac{\partial^2 z}{\partial y \partial x} = \dfrac{\partial^2 z}{\partial x \partial y} = -xy \cos xy - \sin xy$

83. $\partial z/\partial x = \sec y$
$\partial z/\partial y = x \sec y \tan y$
$\partial^2 z/\partial x^2 = 0$
$\partial^2 z/\partial y^2 = x \sec y(\sec^2 y + \tan^2 y)$
$\partial^2 z/\partial y \partial x = \partial^2 z/\partial x \partial y = \sec y \tan y$
No values of x and y exist such that $f_x(x, y) = f_y(x, y) = 0$.

85. $\partial z/\partial x = (y^2 - x^2)/[x(x^2 + y^2)]$
$\partial z/\partial y = -2y/(x^2 + y^2)$
$\partial^2 z/\partial x^2 = (x^4 - 4x^2y^2 - y^4)/[x^2(x^2 + y^2)^2]$
$\partial^2 z/\partial y^2 = 2(y^2 - x^2)/(x^2 + y^2)^2$
$\partial^2 z/\partial y \partial x = \partial^2 z/\partial x \partial y = 4xy/(x^2 + y^2)^2$
No values of x and y exist such that $f_x(x, y) = f_y(x, y) = 0$.

87. $f_{xyy}(x, y, z) = f_{yxy}(x, y, z) = f_{yyx}(x, y, z) = 0$

89. $f_{xyy}(x, y, z) = f_{yxy}(x, y, z) = f_{yyx}(x, y, z) = z^2e^{-x} \sin yz$

91. $\partial^2 z/\partial x^2 + \partial^2 z/\partial y^2 = 0 + 0 = 0$

93. $\partial^2 z/\partial x^2 + \partial^2 z/\partial y^2 = e^x \sin y - e^x \sin y = 0$

95. $\partial^2 z/\partial t^2 = -c^2 \sin(x - ct) = c^2(\partial^2 z/\partial x^2)$

97. $\partial^2 z/\partial t^2 = -c^2/(x + ct)^2 = c^2(\partial^2 z/\partial x^2)$

99. $\partial z/\partial t = -e^{-t} \cos x/c = c^2(\partial^2 z/\partial x^2)$

101. Yes, $f(x, y) = \cos(3x - 2y)$.

103. If $z = f(x, y)$, then to find f_x, you consider y constant and differentiate with respect to x. Similarly, to find f_y, you consider x constant and differentiate with respect to y.

105.

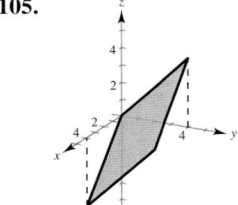

107. The mixed partial derivatives are equal. See Theorem 13.3.

109. (a) 72   (b) 72

111. $IQ_M = \dfrac{100}{C}, IQ_M(12, 10) = 10$

IQ increases at a rate of 10 points per year of mental age when the mental age is 12 and the chronological age is 10.

$IQ_C = -\dfrac{100M}{C^2}, IQ_C(12, 10) = -12$

IQ decreases at a rate of 12 points per year of chronological age when the mental age is 12 and the chronological age is 10.

113. An increase in either the charge for food and housing or the tuition will cause a decrease in the number of applicants.

115. $\partial T/\partial x = -2.4°/\text{m}, \partial T/\partial y = -9°/\text{m}$

117. $T = PV/(nR) \implies \partial T/\partial P = v/(nR)$
$P = nRT/V \implies \partial P/\partial V = -nRT/V^2$
$V = nRT/P \implies \partial V/\partial T = nR/P$
$\partial T/\partial P \cdot \partial P/\partial V \cdot \partial V/\partial T =$
$-nRT/(VP) = -nRT/(nRT) = -1$

119. (a) $\dfrac{\partial z}{\partial x} = 0.461; \dfrac{\partial z}{\partial y} = 0.301$

(b) As the expenditures on amusement parks and campgrounds (x) increase, the expenditures on spectator sports (z) increase. As the expenditures on live entertainment (y) increase, the expenditures on spectator sports (z) also increase.

121. False. Let $z = x + y + 1$.   **123.** True

125. (a) $f_x(x, y) = \dfrac{y(x^4 + 4x^2y^2 - y^4)}{(x^2 + y^2)^2}$

$f_y(x, y) = \dfrac{x(x^4 - 4x^2y^2 - y^4)}{(x^2 + y^2)^2}$

(b) $f_x(0, 0) = 0,\ f_y(0, 0) = 0$

(c) $f_{xy}(0, 0) = -1,\ f_{yx}(0, 0) = 1$

(d) f_{xy} or f_{yx} or both are not continuous at $(0, 0)$.

127. Proof

Section 13.4   *(page 905)*

1. $dz = 4xy^3\, dx + 6x^2y^2\, dy$

3. $dz = 2(x\, dx + y\, dy)/(x^2 + y^2)^2$

5. $dz = (\cos y + y \sin x)\, dx - (x \sin y + \cos x)\, dy$

7. $dz = (e^x \sin y)\, dx + (e^x \cos y)\, dy$

9. $dw = 2z^3 y \cos x\, dx + 2z^3 \sin x\, dy + 6z^2 y \sin x\, dz$

11. (a) $f(2, 1) = 1, f(2.1, 1.05) = 1.05, \Delta z = 0.05$

(b) $dz = 0.05$

13. (a) $f(2, 1) = 11, f(2.1, 1.05) = 10.4875, \Delta z = -0.5125$

(b) $dz = -0.5$

15. (a) $f(2, 1) = e^2 \approx 7.3891, f(2.1, 1.05) = 1.05e^{2.1} \approx 8.5745,$

$\Delta z \approx 1.1854$

(b) $dz \approx 1.1084$

17. 0.44   **19.** 0.094

21. In general, the accuracy worsens as Δx and Δy increase.

23. If $z = f(x, y)$, then $\Delta z \approx dz$ is the propagated error and

$\dfrac{\Delta z}{z} \approx \dfrac{dz}{z}$ is the relative error.

25. $dA = h\, dl + l\, dh$

$\Delta A - dA = dl\, dh$

27.

Δr	Δh	dV	ΔV	$\Delta V - dV$
0.1	0.1	8.3776	8.5462	0.1686
0.1	-0.1	5.0265	5.0255	-0.0010
0.001	0.002	0.1005	0.1006	0.0001
-0.0001	0.0002	-0.0034	-0.0034	0.0000

29. ± 3.92 cubic inches; 0.82%

31. $dC = \pm 2.4418; dC/C = 19\%$ **33.** 10%

35. (a) $V = 18 \sin \theta$ ft^3; $\theta = \pi/2$ (b) 1.047 ft^3

37. $L \approx 8.096 \times 10^{-4} \pm 6.6 \times 10^{-6}$ microhenrys

39. Answers will vary.

Example:

$\varepsilon_1 = \Delta x$

$\varepsilon_2 = 0$

41. Answers will vary.

Example:

$\varepsilon_1 = y\, \Delta x$

$\varepsilon_2 = 2x\, \Delta x + (\Delta x)^2$

43. Proof

Section 13.5   *(page 913)*

1. $26t; 52$ **3.** $e^t(\sin t + \cos t); 1$ **5.** (a) and (b) $-e^{-t}$

7. (a) and (b) $2e^{2t}$ **9.** (a) and (b) $3(2t^2 - 1)$

11. $-11\sqrt{29}/29 \approx -2.04$

13. $\partial w/\partial s = 4s, 4$

$\partial w/\partial t = 4t, 0$

15. $\partial w/\partial s = 5 \cos(5s - t), 0$

$\partial w/\partial t = -\cos(5s - t), 0$

17. (a) and (b)

$\dfrac{\partial w}{\partial s} = t^2(3s^2 - t^2)$

$\dfrac{\partial w}{\partial t} = 2st(s^2 - 2t^2)$

19. (a) and (b)

$\dfrac{\partial w}{\partial s} = te^{s^2 - t^2}(2s^2 + 1)$

$\dfrac{\partial w}{\partial t} = se^{s^2 - t^2}(1 - 2t^2)$

21. $\dfrac{y - 2x + 1}{2y - x + 1}$

23. $-\dfrac{x^2 + y^2 + x}{x^2 + y^2 + y}$

25. $\dfrac{\partial z}{\partial x} = \dfrac{-x}{z}$

$\dfrac{\partial z}{\partial y} = \dfrac{-y}{z}$

27. $\dfrac{\partial z}{\partial x} = -\dfrac{x}{y + z}$

$\dfrac{\partial z}{\partial y} = -\dfrac{z}{y + z}$

29. $\dfrac{\partial z}{\partial x} = \dfrac{-\sec^2(x + y)}{\sec^2(y + z)}$

$\dfrac{\partial z}{\partial y} = -1 - \dfrac{\sec^2(x + y)}{\sec^2(y + z)}$

31. $\dfrac{\partial z}{\partial x} = -\dfrac{(ze^{xz} + y)}{xe^{xz}}$

$\dfrac{\partial z}{\partial y} = -e^{-xz}$

33. $\dfrac{\partial w}{\partial x} = -\dfrac{y + w}{x - z}$

$\dfrac{\partial w}{\partial y} = -\dfrac{x + z}{x - z}$

$\dfrac{\partial w}{\partial z} = \dfrac{w - y}{x - z}$

35. $\dfrac{\partial w}{\partial x} = \dfrac{y \sin xy}{z}$

$\dfrac{\partial w}{\partial y} = \dfrac{x \sin xy - z \cos yz}{z}$

$\dfrac{\partial w}{\partial z} = -\dfrac{y \cos yz + w}{z}$

37. (a) $f(tx, ty) = \dfrac{(tx)(ty)}{\sqrt{(tx)^2 + (ty)^2}}$

$= t\left(\dfrac{xy}{\sqrt{x^2 + y^2}}\right) = tf(x, y); \ n = 1$

(b) $xf_x(x, y) + yf_y(x, y) = \dfrac{xy}{\sqrt{x^2 + y^2}} = 1f(x, y)$

39. (a) $f(tx, ty) = e^{tx/ty} = e^{x/y} = f(x, y); \ n = 0$

(b) $xf_x(x, y) + yf_y(x, y) = \dfrac{xe^{x/y}}{y} - \dfrac{xe^{x/y}}{y} = 0$

41. 47   **43.** $dw/dt = (\partial w/\partial x \cdot dx/dt) + (\partial w/\partial y \cdot dy/dt)$

45. $\dfrac{dy}{dx} = -\dfrac{f_x(x, y)}{f_y(x, y)}$

$\dfrac{\partial z}{\partial x} = -\dfrac{f_x(x, y, z)}{f_z(x, y, z)}$

$\dfrac{\partial z}{\partial y} = -\dfrac{f_y(x, y, z)}{f_z(x, y, z)}$

47. 4608π in.^{3}/min; 624π in.^{2}/min   **49.** $28m$ cm^2/sec

51–55. Proofs

Section 13.6   *(page 924)*

1. $-\sqrt{2}$ **3.** $\dfrac{2 + \sqrt{3}}{2}$ **5.** 1   **7.** $-\dfrac{7}{25}$ **9.** 6

11. $2\sqrt{5}/5$ **13.** $3\mathbf{i} + 10\mathbf{j}$ **15.** $4\mathbf{i} - \mathbf{j}$
17. $6\mathbf{i} - 10\mathbf{j} - 8\mathbf{k}$ **19.** -1 **21.** $2\sqrt{3}/3$ **23.** $3\sqrt{2}$
25. $-8/\sqrt{5}$ **27.** $2[(x + y)\mathbf{i} + x\mathbf{j}]; 2\sqrt{2}$
29. $\tan y\mathbf{i} + x \sec^2 y\mathbf{j}; \sqrt{17}$ **31.** $e^{-x}(-y\mathbf{i} + \mathbf{j}); \sqrt{26}$
33. $\dfrac{x\mathbf{i} + y\mathbf{j} + z\mathbf{k}}{\sqrt{x^2 + y^2 + z^2}}; 1$ **35.** $yz(yz\mathbf{i} + 2xz\mathbf{j} + 2xy\mathbf{k}); \sqrt{33}$

37.

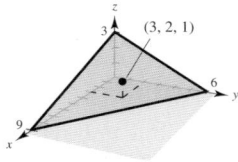

39. (a) $-5\sqrt{2}/12$
(b) $3/5$
(c) $-1/5$
(d) $-11\sqrt{10}/60$

41. $\sqrt{13}/6$
43. (a) Answers will vary. Example: $-4\mathbf{i} + \mathbf{j}$
(b) $-\frac{2}{5}\mathbf{i} + \frac{1}{10}\mathbf{j}$ (c) $\frac{2}{5}\mathbf{i} - \frac{1}{10}\mathbf{j}$
The direction opposite that of the gradient
45. (a)

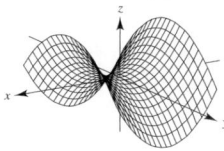

(b) $D_{\mathbf{u}}f(4, -3) = 8 \cos \theta + 6 \sin \theta$

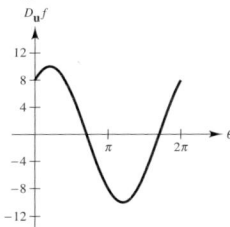

Generated by Mathematica

(c) $\theta \approx 2.21, \theta \approx 5.36$
Directions in which there is no change in f
(d) $\theta \approx 0.64, \theta \approx 3.79$
Directions of greatest rate of change in f
(e) 10; Magnitude of the greatest rate of change
(f)

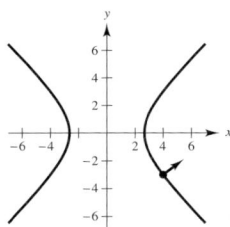

Generated by Mathematica

Orthogonal to the level curve
47. $-2\mathbf{i} - 3\mathbf{j}$ **49.** $3\mathbf{i} - \mathbf{j}$
51. (a) $16\mathbf{i} - \mathbf{j}$ (b) $(\sqrt{257}/257)(16\mathbf{i} - \mathbf{j})$
(c) $y = 16x - 22$
(d)

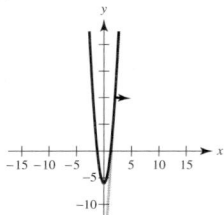

53. (a) $6\mathbf{i} - 4\mathbf{j}$ (b) $(\sqrt{13}/13)(3\mathbf{i} - 2\mathbf{j})$ (c) $y = \frac{3}{2}x - \frac{1}{2}$
(d)

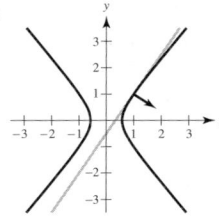

55. The directional derivative of $z = f(x, y)$ in the direction of $\mathbf{u} = \cos \theta\mathbf{i} + \sin \theta\mathbf{j}$ is
$$D_{\mathbf{u}}f(x, y) = \lim_{t \to 0} \frac{f(x + t \cos \theta, y + t \sin \theta) - f(x, y)}{t}$$
if the limit exists.
57. See the definition on page 918. See the properties on page 919.
59. The gradient vector is normal to the level curves.
61. $5\nabla h = -(5\mathbf{i} + 12\mathbf{j})$ **63.** $\frac{1}{625}(7\mathbf{i} - 24\mathbf{j})$
65. $6\mathbf{i} - 10\mathbf{j}; 11.66°$ per centimeter   **67.** $y^2 = 10x$
69. True   **71.** True   **73.** $f(x, y, z) = e^x \cos y + \frac{1}{2}z^2 + C$
75. (a) Proof   (b) Proof
(c)

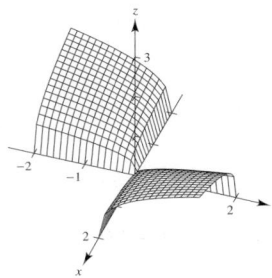

Section 13.7   (page 933)

1. The level surface can be written as $3x - 5y + 3z = 15$, which is an equation of a plane in space.
3. The level surface can be written as $4x^2 + 9y^2 - 4z^2 = 0$, which is an elliptic cone that lies on the z-axis.
5. $\frac{1}{13}(3\mathbf{i} + 4\mathbf{j} + 12\mathbf{k})$ **7.** $\frac{1}{13}(4\mathbf{i} + 3\mathbf{j} + 12\mathbf{k})$
9. $4x + 2y - z = 2$ **11.** $3x + 4y - 5z = 0$
13. $2x - 2y - z = 2$ **15.** $3x + 4y - 25z = 25(1 - \ln 5)$
17. $x - 4y + 2z = 18$ **19.** $6x - 3y - 2z = 11$
21. $x + y + z = 9$ **23.** $2x + 4y + z = 14$
$x - 3 = y - 3 = z - 3$ $\dfrac{x - 1}{2} = \dfrac{y - 2}{4} = \dfrac{z - 4}{1}$
25. $6x - 4y - z = 5$ **27.** $10x + 5y + 2z = 30$
$\dfrac{x - 3}{6} = \dfrac{y - 2}{-4} = \dfrac{z - 5}{-1}$ $\dfrac{x - 1}{10} = \dfrac{y - 2}{5} = \dfrac{z - 5}{2}$
29. $x - y + 2z = \pi/2$
$\dfrac{(x - 1)}{1} = \dfrac{(y - 1)}{-1} = \dfrac{z - (\pi/4)}{2}$
31. (a) $\dfrac{x - 1}{1} = \dfrac{y - 1}{-1} = \dfrac{z - 1}{1}$ (b) $\dfrac{1}{2}$, not orthogonal
33. (a) $\dfrac{x - 3}{4} = \dfrac{y - 3}{4} = \dfrac{z - 4}{-3}$ (b) $\dfrac{16}{25}$, not orthogonal
35. (a) $\dfrac{x - 3}{1} = \dfrac{y - 1}{5} = \dfrac{z - 2}{-4}$ (b) 0, orthogonal
37. $86.0°$ **39.** $77.4°$ **41.** $(0, 3, 12)$ **43.** $(2, 2, -4)$
45. $(0, 0, 0)$ **47.** Proof   **49.** (a) Proof   (b) Proof

51. $(-2, 1, -1)$ or $(2, -1, 1)$

53. $F_x(x_0, y_0, z_0)(x - x_0) + F_y(x_0, y_0, z_0)(y - y_0)$
$$+ F_z(x_0, y_0, z_0)(z - z_0) = 0$$

55. Answers will vary.

57. (a) Line: $x = 1, y = 1, z = 1 - t$
   Plane: $z = 1$
   (b) Line: $x = -1, y = 2 + \frac{6}{25}t, z = -\frac{4}{5} - t$
   Plane: $6y - 25z - 32 = 0$
   (c)

59. (a) $x = 1 + t$
   $y = 2 - 2t$
   $z = 4$
   $\theta \approx 48.2°$
   (b)

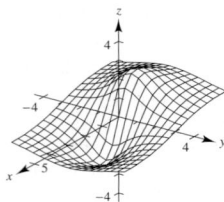

61. $F(x, y, z) = \dfrac{x^2}{a^2} + \dfrac{y^2}{b^2} + \dfrac{z^2}{c^2} - 1$

$F_x(x, y, z) = \dfrac{2x}{a^2}$

$F_y(x, y, z) = \dfrac{2y}{b^2}$

$F_z(x, y, z) = \dfrac{2z}{c^2}$

Plane: $\dfrac{2x_0}{a^2}(x - x_0) + \dfrac{2y_0}{b^2}(y - y_0) + \dfrac{2z_0}{c^2}(z - z_0) = 0$

$\dfrac{x_0 x}{a^2} + \dfrac{y_0 y}{b^2} + \dfrac{z_0 z}{c^2} = 1$

63. $F(x, y, z) = a^2x^2 + b^2y^2 - z^2$
$F_x(x, y, z) = 2a^2x$
$F_y(x, y, z) = 2b^2y$
$F_z(x, y, z) = -2z$
Plane: $2a^2x_0(x - x_0) + 2b^2y_0(y - y_0) - 2z_0(z - z_0) = 0$
$a^2x_0 x + b^2y_0 y - z_0 z = 0$
Therefore, the plane passes through the origin.

65. (a) $P_1(x, y) = 1 + x - y$
   (b) $P_2(x, y) = 1 + x - y + \frac{1}{2}x^2 - xy + \frac{1}{2}y^2$
   (c) If $x = 0$, $P_2(0, y) = 1 - y + \frac{1}{2}y^2$.
   This is the second-degree Taylor polynomial for e^{-y}.
   If $y = 0$, $P_2(x, 0) = 1 + x + \frac{1}{2}x^2$.
   This is the second-degree Taylor polynomial for e^x.
   (d)

x	y	$f(x, y)$	$P_1(x, y)$	$P_2(x, y)$
0	0	1	1	1
0	0.1	0.9048	0.9000	0.9050
0.2	0.1	1.1052	1.1000	1.1050
0.2	0.5	0.7408	0.7000	0.7450
1	0.5	1.6487	1.5000	1.6250

(e)

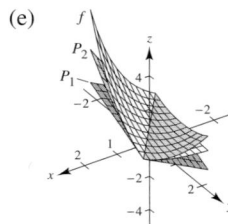

67. Proof

Section 13.8   *(page 942)*

1. Relative minimum:
   $(1, 3, 0)$

3. Relative minimum:
   $(0, 0, 1)$

5. Relative minimum:
   $(-1, 3, -4)$

7. Relative maximum:
   $(40, 40, 3200)$

9. Saddle point:
   $(0, 0, 0)$

11. Relative maximum:
   $\left(\frac{1}{2}, -1, \frac{31}{4}\right)$

13. Relative minimum:
   $(3, -4, -5)$

15. Relative minimum:
   $(0, 0, 0)$

17. Saddle point:
   $(1, -1, -1)$

19. There are no critical numbers.

21.

23.

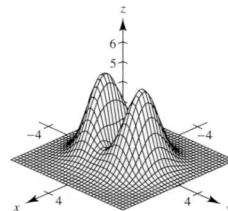

Relative maximum: $(-1, 0, 2)$
Relative minimum: $(1, 0, -2)$

Relative minimum: $(0, 0, 0)$
Relative maxima: $(0, \pm 1, 4)$
Saddle points: $(\pm 1, 0, 1)$

25. z is never negative. Minimum: $z = 0$ when $x = y \neq 0$.

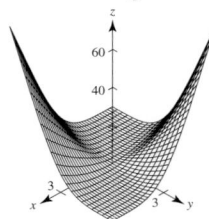

27. Insufficient information

29. Saddle point

31. $-4 < f_{xy}(3, 7) < 4$

33. (a) $(0, 0)$
   (b) Saddle point: $(0, 0, 0)$
   (c) $(0, 0)$
   (d)

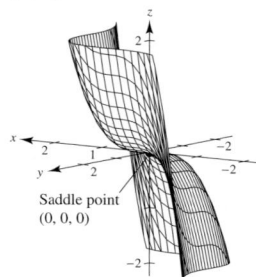

Saddle point
$(0, 0, 0)$

35. (a) $(1, a), (b, -4)$
(b) Absolute minima: $(1, a, 0), (b, -4, 0)$
(c) $(1, a), (b, -4)$
(d)

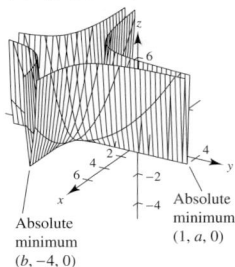

Absolute minimum $(b, -4, 0)$

Absolute minimum $(1, a, 0)$

37. (a) $(0, 0)$
(b) Absolute minimum: $(0, 0, 0)$
(c) $(0, 0)$
(d)

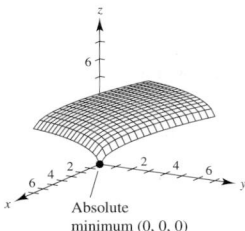

Absolute minimum $(0, 0, 0)$

39. Relative minimum: $(0, 3, -1)$

41. Absolute maximum: **43.** Absolute maximum:
$(4, 0, 21)$ $(0, 1, 10)$
Absolute minimum: Absolute minimum:
$(4, 2, -11)$ $(1, 2, 5)$

45. Absolute maxima: **47.** Absolute maxima:
$(\pm 2, 4, 28)$ $(-2, -1, 9), (2, 1, 9)$
Absolute minimum: Absolute minima:
$(0, 1, -2)$ $(x, -x, 0), |x| \le 1$

49. (a) See the definition on page 936.
(b) See the definition on page 936.
(c) See the definition on page 937.
(d) See the definition on page 939.

51. Answers will vary.
Sample answer:

No extrema

53. (a) $f_x = 2x = 0, f_y = -2y = 0 \Rightarrow (0, 0)$ is a critical point.
$g_x = 2x = 0, g_y = 2y = 0 \Rightarrow (0, 0)$ is a critical point.
(b) $d = 2(-2) - 0 < 0 \Rightarrow (0, 0)$ is a saddle point.
$d = 2(2) - 0 > 0 \Rightarrow (0, 0)$ is a relative minimum.

55. False. Let $f(x, y) = 1 - |x| - |y|$ at the point $(0, 0, 1)$.

57. False. Let $f(x, y) = x^2 y^2$ (see Example 4 on page 940).

Section 13.9   *(page 949)*

1. $\sqrt{3}$ **3.** $\sqrt{7}$ **5.** $x = y = z = 3$ **7.** 10, 10, 10

9. 9 ft $\times$ 9 ft $\times$ 8.25 ft; $26.73

11. Let x, y, and z be the length, width, and height, respectively, and let V_0 be the given volume. Then $V_0 = xyz$ and $z = V_0/xy$. The surface area is
$S = 2xy + 2yz + 2xz = 2(xy + V_0/x + V_0/y)$.
$\left. \begin{aligned} S_x &= 2(y - V_0/x^2) = 0 \\ S_y &= 2(x - V_0/y^2) = 0 \end{aligned} \right\} \begin{aligned} x^2 y - V_0 &= 0 \\ xy^2 - V_0 &= 0 \end{aligned}$
So, $x = \sqrt[3]{V_0}$, $y = \sqrt[3]{V_0}$, and $z = \sqrt[3]{V_0}$.

13. $x_1 = 3; x_2 = 6$ **15.** Proof

17. $x = \sqrt{2}/2 \approx 0.707$ km
$y = (3\sqrt{2} + 2\sqrt{3})/6 \approx 1.284$ km

19. Write the equation to be maximized or minimized as a function of two variables. Take the partial derivatives and set them equal to zero or undefined to obtain the critical points. Use the Second Partials Test to test for relative extrema using the critical points. Check the boundary points.

21. (a) $y = \frac{3}{4}x + \frac{4}{3}$ (b) $\frac{1}{6}$ **23.** (a) $y = -2x + 4$ (b) 2

25. $y = \frac{37}{43}x + \frac{7}{43}$ **27.** $y = -\frac{175}{148}x + \frac{945}{148}$

    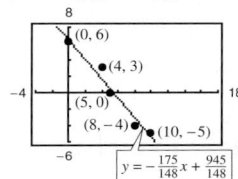

29. (a) $y = 1.6x + 84$ (b) 1.6

31. $a\sum_{i=1}^{n} x_i^4 + b\sum_{i=1}^{n} x_i^3 + c\sum_{i=1}^{n} x_i^2 = \sum_{i=1}^{n} x_i^2 y_i$
$a\sum_{i=1}^{n} x_i^3 + b\sum_{i=1}^{n} x_i^2 + c\sum_{i=1}^{n} x_i = \sum_{i=1}^{n} x_i y_i$
$a\sum_{i=1}^{n} x_i^2 + b\sum_{i=1}^{n} x_i + cn = \sum_{i=1}^{n} y_i$

33. $y = \frac{3}{7}x^2 + \frac{6}{5}x + \frac{26}{35}$ **35.** $y = x^2 - x$

    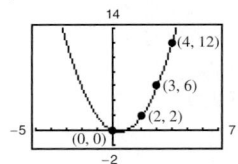

37. (a) $y = -0.22x^2 + 9.66x - 1.79$
(b)

39. (a) $\ln P = -0.1499h + 9.3018$ (b) $P = 10{,}957.7e^{-0.1499h}$
(c)   (d) Proof

41. Proof

Section 13.10   (page 958)

1. $f(5, 5) = 25$ **3.** $f(1, 2) = 5$ **5.** $f(25, 50) = 2600$
7. $f(1, 1) = 2$ **9.** $f(3, 3, 3) = 27$ **11.** $f\left(\frac{1}{3}, \frac{1}{3}, \frac{1}{3}\right) = \frac{1}{3}$
13. Maxima: $f\left(\sqrt{2}/2, \sqrt{2}/2\right) = 5/2$
$\qquad\qquad f\left(-\sqrt{2}/2, -\sqrt{2}/2\right) = 5/2$
$\quad$ Minima: $f\left(-\sqrt{2}/2, \sqrt{2}/2\right) = -1/2$
$\qquad\qquad f\left(\sqrt{2}/2, -\sqrt{2}/2\right) = -1/2$
15. $f(8, 16, 8) = 1024$ **17.** $\sqrt{2}/2$ **19.** $3\sqrt{2}$
21. $\sqrt{11}/2$ **23.** 2   **25.** $\sqrt{3}$ **27.** $(-4, 0, 4)$
29. Optimization problems that have restrictions or constraints on
$\quad$ the values that can be used to produce the optimal solutions are
$\quad$ called constrained optimization problems.
31. $\sqrt{3}$ **33.** $x = y = z = 3$
35. 9 ft × 9 ft × 8.25 ft; \$26.73   **37.** Proof
39. $2\sqrt{3}a/3 \times 2\sqrt{3}b/3 \times 2\sqrt{3}c/3$
41. $\sqrt[3]{360} \times \sqrt[3]{360} \times \frac{4}{3}\sqrt[3]{360}$ ft

43. $r = \sqrt[3]{\dfrac{v_0}{2\pi}}$ and $h = 2\sqrt[3]{\dfrac{v_0}{2\pi}}$ **45.** Proof

47. $P(15,625/18,3125) \approx 226,869$
49. $x \approx 191.3$
$\quad y \approx 688.7$
$\quad$ Cost $\approx$ \$55,095.60
51. Putnam Problem 2, morning session, 1938

Review Exercises for Chapter 13   (page 960)

1. (a) 9   (b) 3   (c) 0   (d) $6x^2$
3. Domain: $\{(x, y): x \ge 0 \text{ and } y \ne 0\}$
$\quad$ Range: all real numbers
5. Lines: $y = 2x - 3 + c$

7. (a)

$\quad$ (b) g is a vertical translation of f
$\qquad$ two units upward.
$\quad$ (c) g is a horizontal translation of f
$\qquad$ two units to the right.

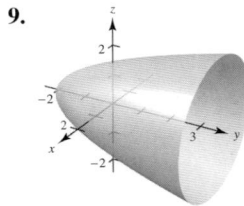

$\quad$ (d)

9.

11. Limit: $\frac{1}{2}$ **13.** Limit: 0
$\quad$ Continuous except at $(0, 0)$ $\quad$ Continuous
15. $f_x(x, y) = 15x^2$ **17.** $f_x(x, y) = e^x \cos y$
$\quad f_y(x, y) = 7$ $\qquad f_y(x, y) = -e^x \sin y$
19. $f_x(x, y) = 4y^3 e^{4x}$
$\quad f_y(x, y) = 3y^2 e^{4x}$
21. $f_x(x, y, z) = 2z^2 + 6yz - 5y^3$
$\quad f_y(x, y, z) = 6xz - 15xy^2$
$\quad f_z(x, y, z) = 4xz + 6xy$
23. $f_{xx}(x, y) = 6$
$\quad f_{yy}(x, y) = 12y$
$\quad f_{xy}(x, y) = f_{yx}(x, y) = -1$
25. $h_{xx}(x, y) = -y \cos x$
$\quad h_{yy}(x, y) = -x \sin y$
$\quad h_{xy}(x, y) = h_{yx}(x, y) = \cos y - \sin x$
27. Slope in x-direction: 0
$\quad$ Slope in y-direction: 4
29. $(xy \cos xy + \sin xy)\,dx + (x^2 \cos xy)\,dy$
31. $dw = (3y^2 - 6x^2yz^2)\,dx + (6xy - 2x^3z^2)\,dy + (-4x^3yz)\,dz$
33. (a) $f(2, 1) = 10$ $\qquad$ (b) $dz = 0.5$
$\qquad f(2.1, 1.05) = 10.5$
$\qquad \Delta z = 0.5$
35. $\pm\pi$ cubic inches; 15%
37. $dw/dt = (8t - 1)/(4t^2 - t + 4)$
39. $\partial w/\partial r = (4r^2t - 4rt^2 - t^3)/(2r - t)^2$
$\quad \partial w/\partial t = (4r^2t - rt^2 + 4r^3)/(2r - t)^2$
41. $\partial z/\partial x = (-2x - y)/(y + 2z)$
$\quad \partial z/\partial y = (-x - 2y - z)/(y + 2z)$
43. -50 **45.** $\frac{2}{3}$ **47.** $\langle 4, 4\rangle, 4\sqrt{2}$ **49.** $\left\langle -\frac{1}{2}, 0\right\rangle, \frac{1}{2}$
51. (a) $54\mathbf{i} - 16\mathbf{j}$ (b) $\dfrac{27}{\sqrt{793}}\mathbf{i} - \dfrac{8}{\sqrt{793}}\mathbf{j}$ (c) $y = \dfrac{27}{8}x - \dfrac{65}{8}$
$\quad$ (d)

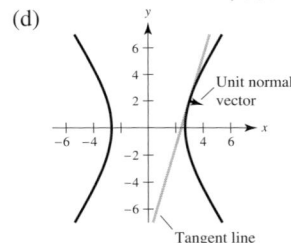

53. $2x + 6y - z = 8$
55. $z = 4$
57. Tangent plane: $4x + 4y - z = 8$
$\quad$ Normal line: $x = 2 + 4t, y = 1 + 4t, z = 4 - t$
59. $\theta \approx 36.7°$ **61.** Relative maximum: $(4, -1, 9)$
63. Relative minimum: $\left(-4, \frac{4}{3}, -2\right)$
65. Relative minimum: $(1, 1, 3)$

67. $\sqrt{3}$ **69.** $x_1 = 2, x_2 = 4$ **71.** $y = \frac{161}{226}x + \frac{456}{113}$

73. (a) $y = 0.138x + 22.1$ (b) 46.25 bushels per acre

75. $f(4, 4) = 32$ **77.** $f(15, 7) = 352$ **79.** $f(3, 6) = 36$

81. $x = \sqrt{2}/2 \approx 0.707$ km; $y = \sqrt{3}/3 \approx 0.577$ km;

$z = (60 - 3\sqrt{2} - 2\sqrt{3})6 \approx 8.716$ km

P.S. Problem Solving   *(page 963)*

1. (a) 12 square units   (b) Proof   (c) Proof

3. (a) $y_0 z_0(x - x_0) + x_0 z_0(y - y_0) + x_0 y_0(z - z_0) = 0$

(b) $x_0 y_0 z_0 = 1 \implies z_0 = 1/x_0 y_0$

Then the tangent plane is

$$y_0\left(\frac{1}{x_0 y_0}\right)(x - x_0) + x_0\left(\frac{1}{x_0 y_0}\right)(y - y_0) + x_0 y_0\left(z - \frac{1}{x_0 y_0}\right) = 0.$$

Intercepts: $(3x_0, 0, 0), (0, 3y_0, 0), \left(0, 0, \dfrac{3}{x_0 y_0}\right)$

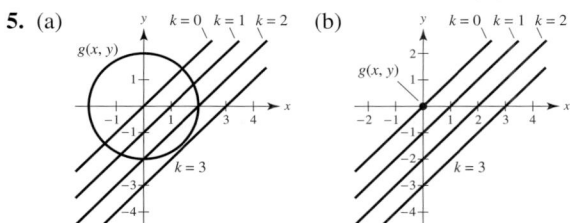

5. (a)

$k = 0 \quad k = 1 \quad k = 2$

$k = 3$

Maximum value: $2\sqrt{2}$

(b)

$k = 0 \quad k = 1 \quad k = 2$

$k = 3$

Maximum and minimum value: 0

The method of Lagrange multipliers does not work because $\nabla g(x_0, y_0) = \mathbf{0}$.

7. $2\sqrt[3]{150} \times 2\sqrt[3]{150} \times 5\sqrt[3]{150}/3$

9. (a) $x\dfrac{\partial f}{\partial x} + y\dfrac{\partial f}{\partial y} = xCy^{1-a}ax^{a-1} + yCx^a(1-a)y^{1-a-1}$

$= ax^aCy^{1-a} + (1-a)x^aC(y^{1-a})$

$= Cx^a y^{1-a}[a + (1-a)]$

$= Cx^a y^{1-a}$

$= f(x, y)$

(b) $f(tx, ty) = C(tx)^a(ty)^{1-a}$

$= Ctx^a y^{1-a}$

$= tCx^a y^{1-a}$

$= tf(x, y)$

11. (a) $x = 32\sqrt{2}t$

$y = 32\sqrt{2}t - 16t^2$

(b) $\alpha = \arctan\left(\dfrac{y}{x + 50}\right) = \arctan\left(\dfrac{32\sqrt{2}t - 16t^2}{32\sqrt{2}t + 50}\right)$

(c) $\dfrac{d\alpha}{dt} = \dfrac{-16(8\sqrt{2}t^2 + 25t - 25\sqrt{2})}{64t^4 - 256\sqrt{2}t^3 + 1024t^2 + 800\sqrt{2}t + 625}$

(d)

No; The rate of change of α is greatest when the projectile is closest to the camera.

(e) α is maximum when $t = 0.98$ second.

No; the projectile is at its maximum height when $t = \sqrt{2} \approx 1.41$ seconds.

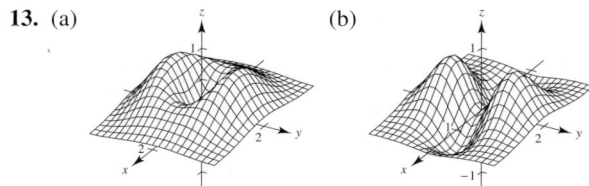

13. (a)

(b)

Minimum: $(0, 0, 0)$

Maxima: $(0, \pm 1, 2e^{-1})$

Saddle points: $(\pm 1, 0, e^{-1})$

Minima: $(\pm 1, 0, -e^{-1})$

Maxima: $(0, \pm 1, 2e^{-1})$

Saddle point: $(0, 0, 0)$

(c) $\alpha > 0$

Minimum: $(0, 0, 0)$

Maxima: $(0, \pm 1, \beta e^{-1})$

Saddle points:

$(\pm 1, 0, \alpha e^{-1})$

$\alpha < 0$

Minima: $(\pm 1, 0, \alpha e^{-1})$

Maxima: $(0, \pm 1, \beta e^{-1})$

Saddle point: $(0, 0, 0)$

15. (a)

(b)

(c) Height

(d) $dl = 0.01, dh = 0$: $dA = 0.01$

$dl = 0, dh = 0.01$: $dA = 0.06$

17–21. Proofs

Chapter 14

Section 14.1   *(page 972)*

1. $2x^2$ **3.** $y \ln(2y)$ **5.** $(4x^2 - x^4)/2$

7. $(y/2)[(\ln y)^2 - y^2]$ **9.** $x^2(1 - e^{-x^2} - x^2 e^{-x^2})$ **11.** 3

13. $\frac{8}{3}$ **15.** $\frac{1}{2}$ **17.** 2   **19.** $\frac{1}{3}$ **21.** 1629   **23.** $\frac{2}{3}$

25. 4   **27.** $\pi/2$ **29.** $(\pi^2/32) + (1/8)$ **31.** $\frac{1}{2}$

33. Diverges   **35.** 24   **37.** $\frac{16}{3}$ **39.** $\frac{8}{3}$ **41.** 5   **43.** $\frac{9}{2}$

45.

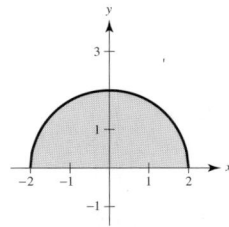

$\int_0^4 \int_x^4 f(x, y)\, dy\, dx$

47.

$\int_0^2 \int_{-\sqrt{4-y^2}}^{\sqrt{4-y^2}} f(x, y)\, dx\, dy$

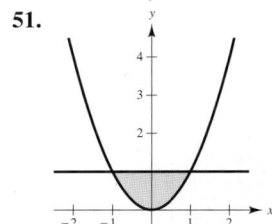

49.

$\int_0^{\ln 10} \int_{e^x}^{10} f(x, y)\, dy\, dx$

51.

$\int_0^1 \int_{-\sqrt{y}}^{\sqrt{y}} f(x, y)\, dx\, dy$

53.

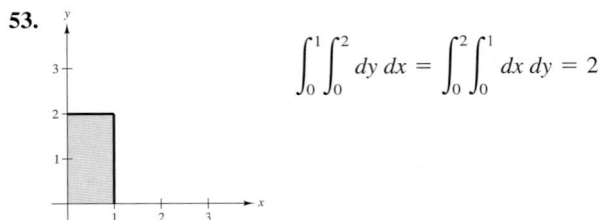

$$\int_0^1 \int_0^2 dy\, dx = \int_0^2 \int_0^1 dx\, dy = 2$$

55.

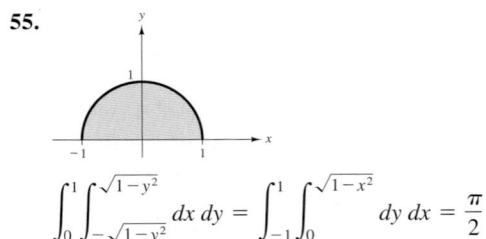

$$\int_0^1 \int_{-\sqrt{1-y^2}}^{\sqrt{1-y^2}} dx\, dy = \int_{-1}^1 \int_0^{\sqrt{1-x^2}} dy\, dx = \frac{\pi}{2}$$

57.

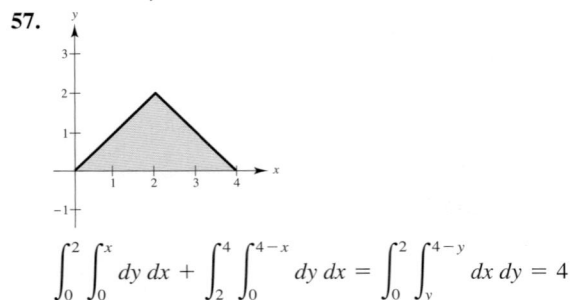

$$\int_0^2 \int_0^x dy\, dx + \int_2^4 \int_0^{4-x} dy\, dx = \int_0^2 \int_y^{4-y} dx\, dy = 4$$

59.

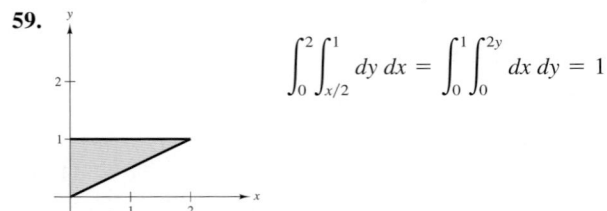

$$\int_0^2 \int_{x/2}^1 dy\, dx = \int_0^1 \int_0^{2y} dx\, dy = 1$$

61.

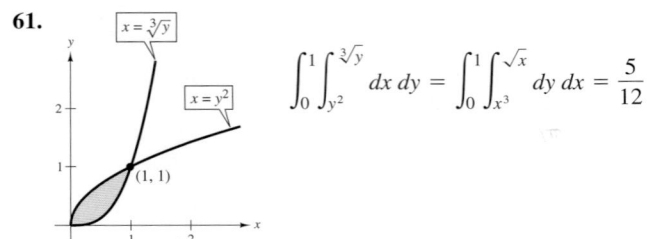

$$\int_0^1 \int_{y^2}^{\sqrt[3]{y}} dx\, dy = \int_0^1 \int_{x^3}^{\sqrt{x}} dy\, dx = \frac{5}{12}$$

63. The first integral arises using vertical representative rectangles. The second two integrals arise using horizontal representative rectangles.

Value of the integrals: $15{,}625\pi/24$

65.

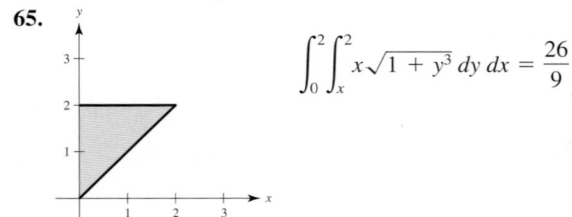

$$\int_0^2 \int_x^2 x\sqrt{1+y^3}\, dy\, dx = \frac{26}{9}$$

67.

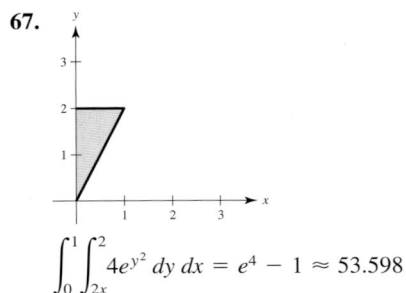

$$\int_0^1 \int_{2x}^2 4e^{y^2}\, dy\, dx = e^4 - 1 \approx 53.598$$

69.

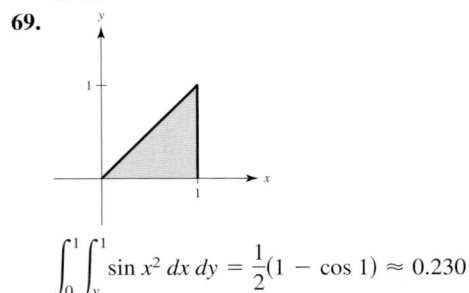

$$\int_0^1 \int_y^1 \sin x^2\, dx\, dy = \frac{1}{2}(1 - \cos 1) \approx 0.230$$

71. $\frac{1664}{105}$ **73.** $(\ln 5)^2$ **75.** 20.5648   **77.** $15\pi/2$

79. (a)

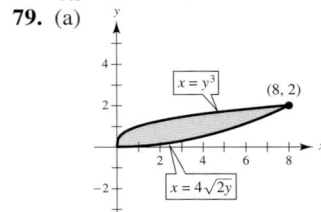

(b) $\int_0^8 \int_{x^2/32}^{\sqrt[3]{x}} (x^2 y - xy^2)\, dy\, dx$ (c) 67,520/693

81. An iterated integral is an integral of a function of several variables. Integrate with respect to one variable while holding the other variables constant.

83. If all four limits of integration are constant, the region of integration is rectangular.

85. True

Section 14.2   *(page 983)*

1. 24 (approximation is exact)

3. Approximation: 52; Exact: $\frac{160}{3}$

5.

8

7.

36

9.

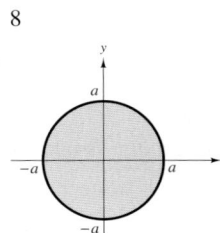

0

11. $\int_0^3 \int_0^5 xy \, dy \, dx = \dfrac{225}{4}$

$\int_0^5 \int_0^3 xy \, dx \, dy = \dfrac{225}{4}$

13. $\int_1^2 \int_x^{2x} \dfrac{y}{x^2 + y^2} \, dy \, dx = \dfrac{1}{2} \ln \dfrac{5}{2}$

$\int_1^2 \int_1^y \dfrac{y}{x^2 + y^2} \, dx \, dy + \int_2^4 \int_{y/2}^2 \dfrac{y}{x^2 + y^2} \, dx \, dy = \dfrac{1}{2} \ln \dfrac{5}{2}$

15. $\int_0^1 \int_{4-x}^{4-x^2} -2y \, dy \, dx = -\dfrac{6}{5}$

$\int_3^4 \int_{4-y}^{\sqrt{4-y}} -2y \, dx \, dy = -\dfrac{6}{5}$

17. $\int_0^3 \int_{4y/3}^{\sqrt{25-y^2}} x \, dx \, dy = 25$

$\int_0^4 \int_0^{3x/4} x \, dy \, dx + \int_4^5 \int_0^{\sqrt{25-x^2}} x \, dy \, dx = 25$

19. 4   **21.** 4   **23.** $\frac{3}{8}$ **25.** 1

27. $\int_0^1 \int_0^x xy \, dy \, dx = \dfrac{1}{8}$

29. $2 \int_0^1 \int_0^x \sqrt{1-x^2} \, dy \, dx = \dfrac{2}{3}$

31. $\int_0^2 \int_0^{\sqrt{4-x^2}} (x+y) \, dy \, dx = \dfrac{16}{3}$

33. $2 \int_0^2 \int_0^{\sqrt{1-(x-1)^2}} (2x - x^2 - y^2) \, dy \, dx$

35. $4 \int_0^2 \int_0^{\sqrt{4-x^2}} (x^2 + y^2) \, dy \, dx$

37. $\int_0^2 \int_{-\sqrt{2-2(y-1)^2}}^{\sqrt{2-2(y-1)^2}} (4y - x^2 - 2y^2) \, dx \, dy$

39. $81\pi/2$ **41.** 1.2315   **43.** Proof

45.

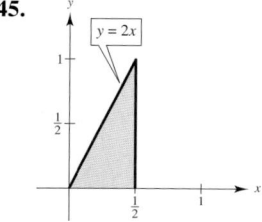

$\int_0^1 \int_{y/2}^{1/2} e^{-x^2} \, dx \, dy = 1 - e^{-1/4} \approx 0.221$

47.

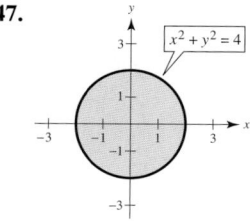

$\int_{-2}^2 \int_{-\sqrt{4-x^2}}^{\sqrt{4-x^2}} \sqrt{4-y^2} \, dy \, dx = \dfrac{64}{3}$

49.

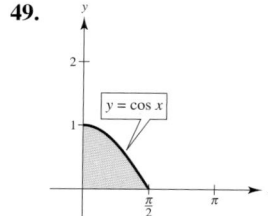

$\int_0^1 \int_0^{\text{arccos } y} \sin x \sqrt{1 + \sin^2 x} \, dx \, dy = \dfrac{1}{3}(2\sqrt{2} - 1)$

51. 2   **53.** $\frac{8}{3}$ **55.** $(e-1)^2$ **57.** 25,645.24

59. See "Definition of Double Integral" on page 976. The double integral of a function $f(x, y) \geq 0$ over the region of integration yields the volume of that region.

61. No; 6π is the greatest possible value.   **63.** Proof; $\frac{1}{5}$

65. Proof; $\frac{7}{27}$ **67.** 400; 272

69. False. $V = 8 \int_0^1 \int_0^{\sqrt{1-y^2}} \sqrt{1 - x^2 - y^2} \, dx \, dy$

71. $R: x^2 + y^2 \leq 9$ **73.** $\frac{1}{2}(1 - e)$

75. Putnam Problem A2, 1989

Section 14.3   *(page 991)*

1. Rectangular   **3.** Polar

5. The region R is a half-circle of radius 8. It can be described in polar coordinates as

$R = \{(r, \theta): 0 \leq r \leq 8, 0 \leq \theta \leq \pi\}$.

7. $R = \{(r, \theta): 4 \leq r \leq 8, 0 \leq \theta \leq \pi/2\}$

9. $\pi/4$ **11.** 0

     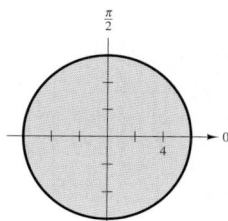

13. $5\sqrt{5}\pi/6$ **15.** $(9/8) + (3\pi^2/32)$

     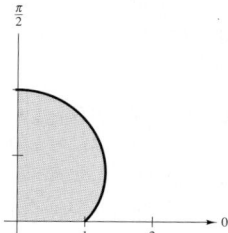

17. $a^3/3$ **19.** 4π **21.** $243\pi/10$ **23.** $\frac{2}{3}$

25. $(\pi/2) \sin 1$ **27.** $\int_0^{\pi/4} \int_0^{2\sqrt{2}} r^2 \, dr \, d\theta = \dfrac{4\sqrt{2}\pi}{3}$

29. $\int_0^{\pi/2} \int_0^2 r^2(\cos \theta + \sin \theta) \, dr \, d\theta = \dfrac{16}{3}$

31. $\int_0^{\pi/4} \int_1^2 r\theta \, dr \, d\theta = \dfrac{3\pi^2}{64}$ **33.** $\dfrac{1}{8}$ **35.** $\dfrac{250\pi}{3}$

37. $\frac{64}{9}(3\pi - 4)$ **39.** $2\sqrt{4 - 2\sqrt[3]{2}}$ **41.** 9π

43. $3\pi/2$ **45.** π

47.

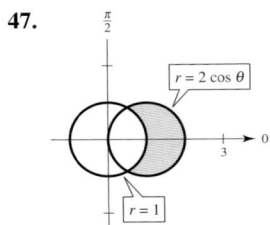

$r = 2\cos\theta$

$r = 1$

$\dfrac{\pi}{3} + \dfrac{\sqrt{3}}{2}$

49.

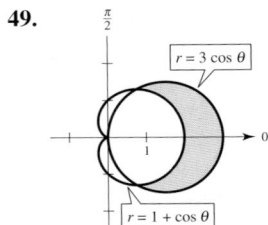

$r = 3\cos\theta$

$r = 1 + \cos\theta$

π

51.

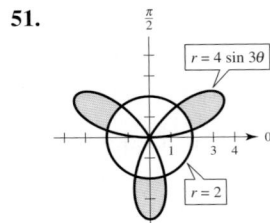

$r = 4\sin 3\theta$ $\dfrac{4\pi}{3} + 2\sqrt{3}$

$r = 2$

53. Let R be a region bounded by the graphs of $r = g_1(\theta)$ and $r = g_2(\theta)$ and the lines $\theta = a$ and $\theta = b$. When using polar coordinates to evaluate a double integral over R, R can be partitioned into small polar sectors.

55. r-simple regions have fixed bounds for θ and variable bounds for r.

θ-simple regions have variable bounds for θ and fixed bounds for r.

57. 486,788   **59.** 1.2858   **61.** 56.051

63. False. Let $f(r, \theta) = r - 1$ and let R be a sector where $0 \le r \le 6$ and $0 \le \theta \le \pi$.

65. (a) 2π (b) $\sqrt{2\pi}$

67. (a) $\displaystyle\int_{2}^{4}\int_{y/\sqrt{3}}^{y} f\, dx\, dy$

(b) $\displaystyle\int_{2/\sqrt{3}}^{2}\int_{2}^{\sqrt{3}x} f\, dy\, dx + \int_{2}^{4/\sqrt{3}}\int_{x}^{\sqrt{3}x} f\, dy\, dx + \int_{4/\sqrt{3}}^{4}\int_{x}^{4} f\, dy\, dx$

(c) $\displaystyle\int_{\pi/4}^{\pi/3}\int_{2\csc\theta}^{4\csc\theta} fr\, dr\, d\theta$

69. $\dfrac{4}{\pi}$

Section 14.4   (page 1000)

1. $m = 4$ **3.** $m = \dfrac{1}{8}$

5. (a) $m = ka^2, (a/2, a/2)$ (b) $m = ka^3/2, (a/2, 2a/3)$

(c) $m = ka^3/2, (2a/3, a/2)$

7. (a) $m = ka^2/2, (a/3, 2a/3)$ (b) $m = ka^3/3, (3a/8, 3a/4)$

(c) $m = ka^3/6, (a/2, 3a/4)$

9. (a) $\left(\dfrac{a}{2} + 5, \dfrac{a}{2}\right)$ (b) $\left(\dfrac{a}{2} + 5, \dfrac{2a}{3}\right)$

(c) $\left(\dfrac{2(a^2 + 15a + 75)}{3(a + 10)}, \dfrac{a}{2}\right)$

11. $m = k/4, (2/3, 8/15)$ **13.** $m = 30k, (14/5, 4/5)$

15. $m = k(e - 1), \left(\dfrac{1}{e - 1}, \dfrac{e + 1}{4}\right)$

17. $m = \dfrac{256k}{15}, \left(0, \dfrac{16}{7}\right)$ **19.** $m = \dfrac{2kL}{\pi}, \left(\dfrac{L}{2}, \dfrac{\pi}{8}\right)$

21. $m = \dfrac{k\pi a^2}{8}, \left(\dfrac{4\sqrt{2}a}{3\pi}, \dfrac{4a(2 - \sqrt{2})}{3\pi}\right)$

23. $m = \dfrac{k}{8}(1 - 5e^{-4}), \left(\dfrac{e^4 - 13}{e^4 - 5}, \dfrac{8}{27}\left[\dfrac{e^6 - 7}{e^6 - 5e^2}\right]\right)$

25. $m = k\pi/3, \left(81\sqrt{3}/(40\pi), 0\right)$

27. $\bar{x} = \sqrt{3}b/3$ **29.** $\bar{x} = a/2$ **31.** $\bar{x} = a/2$

$\bar{y} = \sqrt{3}h/3$ $\bar{y} = a/2$ $\bar{y} = a/2$

33. $I_x = 32k/3$ **35.** $I_x = 16k$

$I_y = 16k/3$ $I_y = 512k/5$

$I_0 = 16k$ $I_0 = 592k/5$

$\bar{\bar{x}} = 2\sqrt{3}/3$ $\bar{\bar{x}} = 4\sqrt{15}/5$

$\bar{\bar{y}} = 2\sqrt{6}/3$ $\bar{\bar{y}} = \sqrt{6}/2$

37. $\displaystyle 2k\int_{-b}^{b}\int_{0}^{\sqrt{b^2 - x^2}} (x - a)^2\, dy\, dx = \dfrac{k\pi b^2}{4}(b^2 + 4a^2)$

39. $\displaystyle \int_{-a}^{a}\int_{0}^{\sqrt{a^2 - x^2}} ky(y - a)^2\, dy\, dx = ka^5\left(\dfrac{56 - 15\pi}{60}\right)$

41. $\dfrac{L}{3}$ **43.** $\dfrac{L}{2}$ **45.** See definitions on page 996.

47. Answers will vary.   **49.** Proof

Section 14.5   (page 1007)

1. 24   **3.** 12π **5.** $\frac{1}{2}\left[4\sqrt{17} + \ln(4 + \sqrt{17})\right]$

7. $\frac{4}{27}(31\sqrt{31} - 8)$ **9.** $\sqrt{2} - 1$ **11.** $\sqrt{2}\pi$

13. $2\pi a(a - \sqrt{a^2 - b^2})$ **15.** $48\sqrt{14}$ **17.** 20π

19. $\displaystyle\int_{0}^{1}\int_{0}^{x}\sqrt{5 + 4x^2}\, dy\, dx = \dfrac{27 - 5\sqrt{5}}{12} \approx 1.3183$

21. $\displaystyle\int_{-3}^{3}\int_{-\sqrt{9 - x^2}}^{\sqrt{9 - x^2}}\sqrt{1 + 4x^2 + 4y^2}\, dy\, dx$

$= \dfrac{\pi}{6}(37\sqrt{37} - 1) \approx 117.3187$

23. $\displaystyle\int_{0}^{1}\int_{0}^{1}\sqrt{1 + 4x^2 + 4y^2}\, dy\, dx \approx 1.8616$

25. $\displaystyle\int_{0}^{4}\int_{0}^{10}\sqrt{1 + e^{2xy}(x^2 + y^2)}\, dy\, dx$

27. $\displaystyle\int_{-2}^{2}\int_{-\sqrt{4 - x^2}}^{\sqrt{4 - x^2}}\sqrt{1 + e^{-2x}}\, dy\, dx$

29. If f and its first partial derivatives are continuous on the closed region R in the xy-plane, then the area of the surface S given by $z = f(x, y)$ over R is

$$\iint_{R}\sqrt{1 + [f_x(x, y)]^2 + [f_y(x, y)]^2}\, dA.$$

31. No. The size and shape of the graph stay the same; just the position is changed. So, the surface area does not increase.

33. (a) $812\pi\sqrt{609}$ cm^3 (b) $100\pi\sqrt{609}$ cm^2 **35.** 16

Section 14.6   (page 1017)

1. 18   **3.** $\frac{1}{10}$ **5.** $(15/2)(1 - 1/e)$ **7.** $-\frac{40}{3}$ **9.** $\frac{324}{5}$

11. $\displaystyle V = \int_{0}^{5}\int_{0}^{5-x}\int_{0}^{5-x-y} dz\, dy\, dx$

13. $\displaystyle V = \int_{-\sqrt{6}}^{\sqrt{6}}\int_{-\sqrt{6 - y^2}}^{\sqrt{6 - y^2}}\int_{0}^{6 - x^2 - y^2} dz\, dx\, dy$

15. $\displaystyle V = \int_{-4}^{4}\int_{-\sqrt{16 - x^2}}^{\sqrt{16 - x^2}}\int_{(x^2 + y^2)/2}^{\sqrt{80 - x^2 - y^2}} dz\, dy\, dx$

17. $\frac{256}{15}$ **19.** $4\pi a^3/3$ **21.** $\frac{256}{15}$ **23.** 10

25.

$$\int_0^1 \int_0^1 \int_{-1}^{-\sqrt{z}} dy\, dz\, dx$$

27.

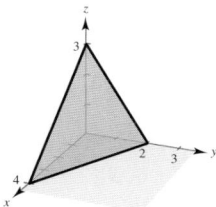

$$\int_0^3 \int_0^{(12-4z)/3} \int_0^{(12-4z-3x)/6} dy\, dx\, dz$$

29.

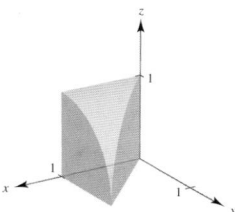

$$\int_0^1 \int_0^x \int_0^{\sqrt{1-y^2}} dz\, dy\, dx$$

31. $\displaystyle\int_0^1 \int_0^x \int_0^3 xyz\, dz\, dy\, dx, \quad \int_0^1 \int_y^1 \int_0^3 xyz\, dz\, dx\, dy,$

$\displaystyle\int_0^1 \int_0^3 \int_0^x xyz\, dy\, dz\, dx, \quad \int_0^3 \int_0^1 \int_0^x xyz\, dy\, dx\, dz,$

$\displaystyle\int_0^3 \int_0^1 \int_y^1 xyz\, dx\, dy\, dz, \quad \int_0^3 \int_0^1 \int_y^1 xyz\, dx\, dz\, dy$

33. $\displaystyle\int_{-3}^3 \int_{-\sqrt{9-x^2}}^{\sqrt{9-x^2}} \int_0^4 xyz\, dz\, dy\, dx, \quad \int_{-3}^3 \int_{-\sqrt{9-y^2}}^{\sqrt{9-y^2}} \int_0^4 xyz\, dz\, dx\, dy,$

$\displaystyle\int_{-3}^3 \int_0^4 \int_{-\sqrt{9-x^2}}^{\sqrt{9-x^2}} xyz\, dy\, dz\, dx, \quad \int_0^4 \int_{-3}^3 \int_{-\sqrt{9-x^2}}^{\sqrt{9-x^2}} xyz\, dy\, dx\, dz,$

$\displaystyle\int_0^4 \int_{-3}^3 \int_{-\sqrt{9-y^2}}^{\sqrt{9-y^2}} xyz\, dx\, dy\, dz, \quad \int_{-3}^3 \int_0^4 \int_{-\sqrt{9-y^2}}^{\sqrt{9-y^2}} xyz\, dx\, dz\, dy$

35. $\displaystyle\int_0^1 \int_0^{1-z} \int_0^{1-y^2} dx\, dy\, dz, \quad \int_0^1 \int_0^{1-y} \int_0^{1-y^2} dx\, dz\, dy,$

$\displaystyle\int_0^1 \int_0^{2z-z^2} \int_0^{1-z} 1\, dy\, dx\, dz + \int_0^1 \int_{2z-z^2}^1 \int_0^{\sqrt{1-x}} 1\, dy\, dx\, dz,$

$\displaystyle\int_0^1 \int_{1-\sqrt{1-x}}^1 \int_0^{1-z} 1\, dy\, dz\, dx + \int_0^1 \int_0^{1-\sqrt{1-x}} \int_0^{\sqrt{1-x}} 1\, dy\, dz\, dx,$

$\displaystyle\int_0^1 \int_0^{\sqrt{1-x}} \int_0^{1-y} dz\, dy\, dx$

37. $m = 8k, \bar{x} = \frac{3}{2}$ **39.** $m = 128k/3, \bar{z} = 1$

41. $m = k \displaystyle\int_0^b \int_0^b \int_0^b xy\, dz\, dy\, dx$

$M_{yz} = k \displaystyle\int_0^b \int_0^b \int_0^b x^2 y\, dz\, dy\, dx$

$M_{xz} = k \displaystyle\int_0^b \int_0^b \int_0^b xy^2\, dz\, dy\, dx$

$M_{xy} = k \displaystyle\int_0^b \int_0^b \int_0^b xyz\, dz\, dy\, dx$

43. $\bar{x}$ will be greater than 2, and $\bar{y}$ and $\bar{z}$ will be unchanged.
45. $\bar{x}$ and $\bar{z}$ will be unchanged, and $\bar{y}$ will be greater than 0.
47. $(0, 0, 3h/4)$ **49.** $\left(0, 0, \frac{3}{2}\right)$ **51.** $\left(5, 6, \frac{5}{4}\right)$
53. (a) $I_x = 2ka^5/3$ **55.** (a) $I_x = 256k$
    $I_y = 2ka^5/3$ $I_y = 512k/3$
    $I_z = 2ka^5/3$ $I_z = 256k$
  (b) $I_x = ka^8/8$ (b) $I_x = 2048k/3$
    $I_y = ka^8/8$ $I_y = 1024k/3$
    $I_z = ka^8/8$ $I_z = 2048k/3$
57. Proof
59. $\displaystyle\int_{-1}^1 \int_{-1}^1 \int_0^{1-x} (x^2 + y^2)\sqrt{x^2 + y^2 + z^2}\, dz\, dy\, dx$

61. (a) $m = \displaystyle\int_{-2}^2 \int_{-\sqrt{4-x^2}}^{\sqrt{4-x^2}} \int_0^{4-x^2-y^2} kz\, dz\, dy\, dx$

  (b) $\bar{x} = \bar{y} = 0$, by symmetry.

$\bar{z} = \dfrac{1}{m} \displaystyle\int_{-2}^2 \int_{-\sqrt{4-x^2}}^{\sqrt{4-x^2}} \int_0^{4-x^2-y^2} kz^2\, dz\, dy\, dx$

  (c) $I_z = \displaystyle\int_{-2}^2 \int_{-\sqrt{4-x^2}}^{\sqrt{4-x^2}} \int_0^{4-x^2-y^2} kz(x^2 + y^2)\, dz\, dy\, dx$

63. $\frac{13}{3}$ **65.** $\frac{3}{2}$
67. See "Definition of Triple Integral" on page 1009 and Theorem 14.4, "Evaluation by Iterated Integrals," on page 1010.
69. (a)   **71.** $Q: 3z^2 + y^2 + 2x^2 \le 1; 4\sqrt{6}\pi/45 \approx 0.684$
73. Putnam Problem B1, 1965

Section 14.7 *(page 1025)*

1. 27   **3.** $\frac{52}{45}$ **5.** $\pi/8$ **7.** $\pi(e^4 + 3)$

9.

$(1 - e^{-9})\pi/4$

11.

$64\sqrt{3}\pi/3$

13. Cylindrical: $\displaystyle\int_0^{2\pi} \int_0^2 \int_{r^2}^4 r^2 \cos\theta\, dz\, dr\, d\theta = 0$

Spherical: $\displaystyle\int_0^{2\pi} \int_0^{\arctan(1/2)} \int_0^{4\sec\phi} \rho^3 \sin^2\phi \cos\theta\, d\rho\, d\phi\, d\theta$

$+ \displaystyle\int_0^{2\pi} \int_{\arctan(1/2)}^{\pi/2} \int_0^{\cot\phi\csc\phi} \rho^3 \sin^2\phi \cos\phi\, d\rho\, d\phi\, d\theta = 0$

15. Cylindrical: $\displaystyle\int_0^{2\pi} \int_0^a \int_a^{a+\sqrt{a^2-r^2}} r^2 \cos\theta\, dz\, dr\, d\theta = 0$

Spherical: $\displaystyle\int_0^{\pi/4} \int_0^{2\pi} \int_{a\sec\phi}^{2a\cos\phi} \rho^3 \sin^2\phi \cos\theta\, d\rho\, d\theta\, d\phi = 0$

17. $(2a^3/9)(3\pi - 4)$ **19.** $\pi/16$ **21.** $(2a^3/9)(3\pi - 4)$
23. $48k\pi$ **25.** $\pi r_0^2 h/3$ **27.** $(0, 0, h/5)$
29. $I_z = 4k \displaystyle\int_0^{\pi/2} \int_0^{r_0} \int_0^{h(r_0-r)/r_0} r^3\, dz\, dr\, d\theta = 3mr_0^2/10$
31. Proof   **33.** $9\pi\sqrt{2}$ **35.** $16\pi^2$ **37.** $k\pi a^4$
39. $(0, 0, 3r/8)$ **41.** $k\pi/192$

43. Rectangular to cylindrical: Cylindrical to rectangular:

$r^2 = x^2 + y^2$ $x = r\cos\theta$

$\tan\theta = y/x$ $y = r\sin\theta$

$z = z$ $z = z$

45. $\displaystyle\int_{\theta_1}^{\theta_2}\int_{g_1(\theta)}^{g_2(\theta)}\int_{h_1(r\cos\theta,\,r\sin\theta)}^{h_2(r\cos\theta,\,r\sin\theta)} f(r\cos\theta,\,r\sin\theta,\,z)r\,dz\,dr\,d\theta$

47. (a) r constant: right circular cylinder about z-axis

      θ constant: plane parallel to z-axis

      z constant: plane parallel to xy-plane

   (b) ρ constant: sphere

      θ constant: plane parallel to z-axis

      ϕ constant: cone

49. Putnam Problem A1, 2006

Section 14.8   *(page 1032)*

1. $-\frac{1}{2}$ **3.** $1 + 2v$ **5.** 1   **7.** $-e^{2u}$

9.

11.

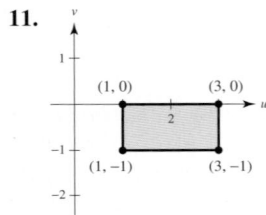

13. $\displaystyle\iint_R 3xy\,dA = \int_{-2/3}^{2/3}\int_{1-x}^{(1/2)x+2} 3xy\,dy\,dx$

$\displaystyle + \int_{2/3}^{4/3}\int_{(1/2)x}^{(1/2)x+2} 3xy\,dy\,dx + \int_{4/3}^{8/3}\int_{(1/2)x}^{4-x} 3xy\,dy\,dx = \frac{164}{9}$

15. $\frac{8}{3}$ **17.** 36   **19.** $(e^{-1/2} - e^{-2})\ln 8 \approx 0.9798$ **21.** 96

23. $12(e^4 - 1)$ **25.** $\frac{100}{9}$ **27.** $\frac{2}{5}a^{5/2}$ **29.** One

31. (a)

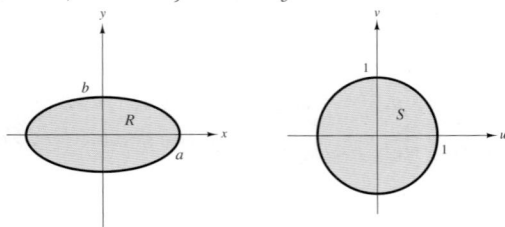

   (b) ab (c) πab

33. See "Definition of the Jacobian" on page 1027.   **35.** u^2v

37. $-uv$ **39.** $-\rho^2\sin\phi$ **41.** Putnam Problem A2, 1994

Review Exercises for Chapter 14   *(page 1034)*

1. $4x^5$ **3.** $\frac{29}{6}$ **5.** 36   **7.** $\frac{3}{2}$ **9.** 16

11.

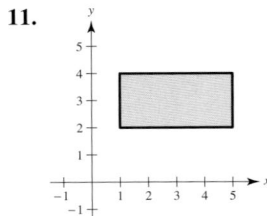

$\displaystyle\int_2^4\int_1^5 dx\,dy = \int_1^5\int_2^4 dy\,dx = 8$

13.

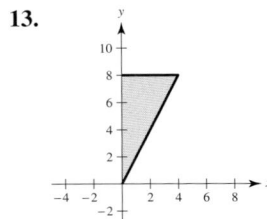

$\displaystyle\int_0^4\int_{2x}^8 dy\,dx = \int_0^8\int_0^{y/2} dx\,dy = 16$

15. $\displaystyle\int_0^2\int_0^4 4xy\,dy\,dx = \int_0^4\int_0^2 4xy\,dx\,dy = 64$ **17.** 21

19. $\frac{40}{3}$ **21.** $\frac{40}{3}$ **23.** 13.67°C

25. $(h^3/6)\left[\ln(\sqrt{2} + 1) + \sqrt{2}\right]$ **27.** $\frac{81}{5}$ **29.** $9\pi/2$

31.

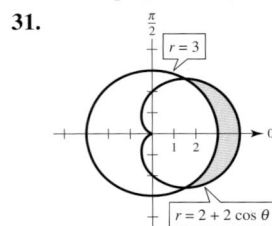

$\dfrac{9\sqrt{3}}{2} - \pi$

33. (a) $r = 3\sqrt{\cos 2\theta}$

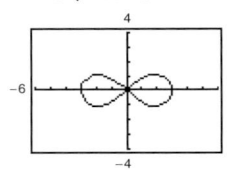

   (b) 9   (c) $3(3\pi - 16\sqrt{2} + 20) \approx 20.392$

35. $m = \dfrac{32k}{5},\ \left(\dfrac{5}{3}, \dfrac{5}{2}\right)$ **37.** $m = \dfrac{k}{4},\ \left(\dfrac{32}{45}, \dfrac{64}{55}\right)$

39. $I_x = ka^2b^3/6$

     $I_y = ka^4b/4$

     $I_0 = (2ka^2b^3 + 3ka^4b)/12$

     $\bar{\bar{x}} = a/\sqrt{2}$

     $\bar{\bar{y}} = b/\sqrt{3}$

41. $\dfrac{(101\sqrt{101} - 1)\pi}{6}$ **43.** $\dfrac{1}{6}(37\sqrt{37} - 1)$

45. (a) 30,415.74 ft³   (b) 2081.53 ft²   **47.** 56

49. $\dfrac{abc}{3}(a^2 + b^2 + c^2)$ **51.** $\dfrac{8\pi}{5}$ **53.** 36

55.

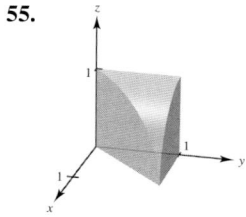

$$\int_0^1 \int_x^1 \int_0^{\sqrt{1-x^2}} dz\, dy\, dx$$

57. $m = \dfrac{500k}{3}, \bar{x} = \dfrac{5}{2}$ **59.** $12\sqrt{3}$ **61.** $\dfrac{2\pi^2}{3}$

63. $\pi\left[3\sqrt{13} + 4\ln\left(\dfrac{3 + \sqrt{13}}{2}\right)\right] \approx 48.995$ **65.** 16π

67. -9 **69.** $\sin^2\theta - \cos^2\theta$

71. $5\ln 5 - 3\ln 3 - 2 \approx 2.751$ **73.** 81

P.S. Problem Solving *(page 1037)*

1. $8(2 - \sqrt{2})$ **3.** $\dfrac{1}{3}$ **5.** (a)–(g) Proofs

7. The results are not the same. Fubini's Theorem is not valid because f is not continuous on the region $0 \le x \le 1$, $0 \le y \le 1$.

9. $\sqrt{\pi}/4$ **11.** If $a, k > 0$, then $1 = ka^2$ or $a = 1/\sqrt{k}$.

13. Answers will vary.

15. The greater the angle between the given plane and the xy-plane, the greater the surface area. So $z_2 < z_1 < z_4 < z_3$.

17.

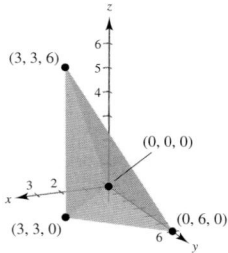

$$\int_0^3 \int_0^{2x} \int_x^{6-x} dy\, dz\, dx = 18$$

Chapter 15

Section 15.1 *(page 1049)*

1. d   **2.** c   **3.** a   **4.** b

5. $\sqrt{2}$ **7.** $3|y|$

9. $\sqrt{3}$ **11.**

13.

15. $2x\mathbf{i} + 4y\mathbf{j}$ **17.** $(10x + 3y)\mathbf{i} + (3x + 2y)\mathbf{j}$

19. $6yz\mathbf{i} + 6xz\mathbf{j} + 6xy\mathbf{k}$ **21.** $2xye^{x^2}\mathbf{i} + e^{x^2}\mathbf{j} + \mathbf{k}$

23. $[xy/(x + y) + y\ln(x + y)]\mathbf{i} + [xy/(x + y) + x\ln(x + y)]\mathbf{j}$

25. Conservative   **27.** Conservative   **29.** Conservative

31. Not conservative   **33.** Conservative: $f(x, y) = xy + K$

35. Conservative: $f(x, y) = x^2y + K$

37. Not conservative   **39.** Not conservative

41. Conservative: $f(x, y) = e^x \cos y + K$

43. $4\mathbf{i} - \mathbf{j} - 3\mathbf{k}$ **45.** $-2\mathbf{k}$ **47.** $2x/(x^2 + y^2)\mathbf{k}$

49. $\cos(y - z)\mathbf{i} + \cos(z - x)\mathbf{j} + \cos(x - y)\mathbf{k}$

51. Conservative: $f(x, y, z) = \frac{1}{2}(x^2y^2z^2) + K$

53. Not conservative   **55.** Conservative: $f(x, y, z) = xz/y + K$

57. $2x + 4y$ **59.** $\cos x - \sin y + 2z$ **61.** 4   **63.** 0

65. See "Definition of Vector Field" on page 1040. Some physical examples of vector fields include velocity fields, gravitational fields, and electric force fields.

67. See "Definition of Curl of a Vector Field" on page 1046.

69. $9x\mathbf{j} - 2y\mathbf{k}$ **71.** $z\mathbf{j} + y\mathbf{k}$ **73.** $3z + 2x$ **75.** 0

77. (a)–(h) Proofs   **79.** True

81. False. Curl f is meaningful only for vector fields, when direction is involved.

83. $M = 3mxy(x^2 + y^2)^{-5/2}$
$\partial M/\partial y = 3mx(x^2 - 4y^2)/(x^2 + y^2)^{7/2}$
$N = m(2y^2 - x^2)(x^2 + y^2)^{-5/2}$
$\partial N/\partial x = 3mx(x^2 - 4y^2)/(x^2 + y^2)^{7/2}$
Therefore, $\partial N/\partial x = \partial M/\partial y$ and $\mathbf{F}$ is conservative.

Section 15.2 *(page 1061)*

1. $\mathbf{r}(t) = \begin{cases} t\mathbf{i} + t\mathbf{j}, & 0 \le t \le 1 \\ (2 - t)\mathbf{i} + \sqrt{2 - t}\,\mathbf{j}, & 1 \le t \le 2 \end{cases}$

3. $\mathbf{r}(t) = \begin{cases} t\mathbf{i}, & 0 \le t \le 3 \\ 3\mathbf{i} + (t - 3)\mathbf{j}, & 3 \le t \le 6 \\ (9 - t)\mathbf{i} + 3\mathbf{j}, & 6 \le t \le 9 \\ (12 - t)\mathbf{j}, & 9 \le t \le 12 \end{cases}$

5. $\mathbf{r}(t) = 3\cos t\mathbf{i} + 3\sin t\mathbf{j}, \quad 0 \le t \le 2\pi$

7. 20   **9.** $5\pi/2$

11. (a) C: $\mathbf{r}(t) = t\mathbf{i} + t\mathbf{j}, 0 \le t \le 1$ (b) $2\sqrt{2}/3$

13. (a) C: $\mathbf{r}(t) = \cos t\,\mathbf{i} + \sin t\,\mathbf{j}$, $0 \le t \le \pi/2$ (b) $\pi/2$

15. (a) C: $\mathbf{r}(t) = t\,\mathbf{i}$, $0 \le t \le 1$ (b) $1/2$

17. (a) C: $\mathbf{r}(t) = \begin{cases} t\,\mathbf{i}, & 0 \le t \le 1 \\ (2-t)\mathbf{i} + (t-1)\mathbf{j}, & 1 \le t \le 2 \\ (3-t)\mathbf{j}, & 2 \le t \le 3 \end{cases}$

(b) $\frac{19}{6}(1 + \sqrt{2})$

19. (a) C: $\mathbf{r}(t) = \begin{cases} t\,\mathbf{i}, & 0 \le t \le 1 \\ \mathbf{i} + t\mathbf{k}, & 0 \le t \le 1 \\ \mathbf{i} + t\mathbf{j} + \mathbf{k}, & 0 \le t \le 1 \end{cases}$ (b) $\frac{23}{6}$

21. $8\sqrt{5}\pi(1 + 4\pi^2/3) \approx 795.7$ **23.** $2\pi + 2$

25. $(k/12)(41\sqrt{41} - 27)$ **27.** 1   **29.** $\frac{1}{2}$ **31.** $\frac{9}{4}$

33. About 249.49   **35.** 66   **37.** 0   **39.** $-10\pi^2$

41. Positive   **43.** Zero

45. (a) $\frac{236}{3}$; Orientation is from left to right, so the value is positive.

(b) $-\frac{236}{3}$; Orientation is from right to left, so the value is negative.

47. $\mathbf{F}(t) = -2t\mathbf{i} - t\mathbf{j}$
$\mathbf{r}\,'(t) = \mathbf{i} - 2\mathbf{j}$
$\mathbf{F}(t) \cdot \mathbf{r}\,'(t) = -2t + 2t = 0$
$\displaystyle\int_C \mathbf{F} \cdot d\mathbf{r} = 0$

49. $\mathbf{F}(t) = (t^3 - 2t^2)\mathbf{i} + (t - t^2/2)\mathbf{j}$
$\mathbf{r}\,'(t) = \mathbf{i} + 2t\mathbf{j}$
$\mathbf{F}(t) \cdot \mathbf{r}\,'(t) = t^3 - 2t^2 + 2t^2 - t^3 = 0$
$\displaystyle\int_C \mathbf{F} \cdot d\mathbf{r} = 0$

51. 1010   **53.** $\frac{190}{3}$ **55.** 25   **57.** $\frac{63}{2}$ **59.** $-\frac{11}{6}$

61. $\frac{316}{3}$ **63.** $5h$ **65.** $\frac{1}{2}$ **67.** $(h/4)[2\sqrt{5} + \ln(2 + \sqrt{5})]$

69. $\frac{1}{120}(25\sqrt{5} - 11)$

71. (a) $12\pi \approx 37.70$ cm^2

(b) $12\pi/5 \approx 7.54$ cm^3

(c)

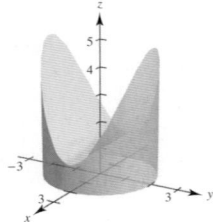

73. $I_x = I_y = a^3\pi$

75. (a)

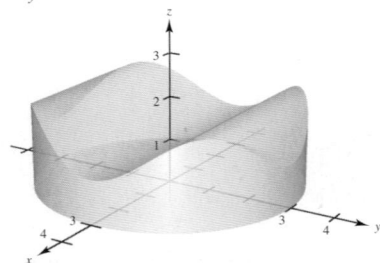

(b) 9π cm$^2 \approx 28.274$ cm^2

(c) Volume $= 2\displaystyle\int_0^3 2\sqrt{9 - y^2}\left[1 + 4\frac{y^2}{9}\left(1 - \frac{y^2}{9}\right)\right] dy$

$= 27\pi/2 \approx 42.412$ cm^3

77. 1750 ft-lb

79. See "Definition of Line Integral" on page 1052 and Theorem 15.4, "Evaluation of a Line Integral as a Definite Integral," on page 1053.

81. z_3, z_1, z_2, z_4; The greater the height of the surface over the curve $y = \sqrt{x}$, the greater the lateral surface area.

83. False. $\displaystyle\int_C xy\,ds = \sqrt{2}\int_0^1 t^2\,dt$

85. False. The orientations are different.   **87.** -12

Section 15.3   (page 1072)

1. (a) $\displaystyle\int_0^1 (t^2 + 2t^4)\,dt = \frac{11}{15}$

(b) $\displaystyle\int_0^{\pi/2} (\sin^2\theta\cos\theta + 2\sin^4\theta\cos\theta)\,d\theta = \frac{11}{15}$

3. (a) $\displaystyle\int_0^{\pi/3} (\sec\theta\tan^2\theta - \sec^3\theta)\,d\theta \approx -1.317$

(b) $\displaystyle\int_0^3 \left[\frac{\sqrt{t}}{2\sqrt{t+1}} - \frac{\sqrt{t+1}}{2\sqrt{t}}\right] dt \approx -1.317$

5. Conservative   **7.** Not conservative

9. Conservative   **11.** (a) 1   (b) 1

13. (a) 0   (b) $-\frac{1}{3}$ (c) $-\frac{1}{2}$

15. (a) 64   (b) 0   (c) 0   (d) 0   **17.** (a) $\frac{64}{3}$ (b) $\frac{64}{3}$

19. (a) 32   (b) 32   **21.** (a) $\frac{2}{3}$ (b) $\frac{17}{6}$ **23.** (a) 0   (b) 0

25. 72   **27.** -1 **29.** 0   **31.** (a) 2   (b) 2   (c) 2

33. 11   **35.** 30,366   **37.** 0

39. (a) $d\mathbf{r} = (\mathbf{i} - \mathbf{j})\,dt \Longrightarrow \displaystyle\int_0^{50} 175\,dt = 8750$ ft-lb

(b) $d\mathbf{r} = \left(\mathbf{i} - \frac{1}{25}(50 - t)\mathbf{j}\right)dt \Longrightarrow 7\displaystyle\int_0^{50} (50 - t)\,dt$

$= 8750$ ft-lb

41. See Theorem 15.5, "Fundamental Theorem of Line Integrals," on page 1066.

43. (a) 2π (b) 2π (c) -2π (d) 0

45. Yes, because the work required to get from point to point is independent of the path taken.

47. False. It would be true if $\mathbf{F}$ were conservative.

49. True   **51.** Proof

53. (a) Proof   (b) $-\pi$ (c) π

(d) -2π; does not contradict Theorem 15.7 because $\mathbf{F}$ is not continuous at $(0, 0)$ in R enclosed by C.

(e) $\nabla\left(\arctan\dfrac{x}{y}\right) = \dfrac{1/y}{1 + (x/y)^2}\mathbf{i} + \dfrac{-x/y^2}{1 + (x/y)^2}\mathbf{j}$

Section 15.4   (page 1081)

1. $\frac{1}{30}$ **3.** 0   **5.** About 19.99   **7.** $\frac{9}{2}$ **9.** 56

11. $\frac{4}{3}$ **13.** 0   **15.** 0   **17.** $\frac{1}{12}$ **19.** 32π

21. π **23.** $\frac{225}{2}$ **25.** πa^2 **27.** $\frac{9}{2}$

29. See Theorem 15.8 on page 1075.   **31.** Proof

33. $\left(0, \frac{8}{5}\right)$ **35.** $\left(\frac{8}{15}, \frac{8}{21}\right)$ **37.** $3\pi a^2/2$

39. $\pi - 3\sqrt{3}/2$ **41.** (a) $51\pi/2$ (b) $243\pi/2$

43. 46π

45. $\int_C \mathbf{F} \cdot d\mathbf{r} = \int_C M\,dx + N\,dy = \int\int_R \left(\dfrac{\partial N}{\partial x} - \dfrac{\partial M}{\partial y}\right) dA = 0;$
$I = -2\pi$ when C is a circle that contains the origin.

47–51. Proofs

Section 15.5 *(page 1091)*

1. e   **2.** f   **3.** b   **4.** a   **5.** d   **6.** c
7. $y - 2z = 0$ **9.** $x^2 + z^2 = 4$
Plane   Cylinder

11.   **13.**

15.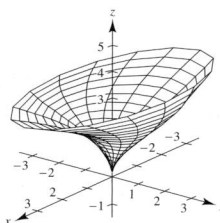

17. The paraboloid is reflected (inverted) through the xy-plane.
19. The height of the paraboloid is increased from 4 to 9.
21. $\mathbf{r}(u, v) = u\mathbf{i} + v\mathbf{j} + v\mathbf{k}$
23. $\mathbf{r}(u, v) = \frac{1}{2}u \cos v\mathbf{i} + u\mathbf{j} + \frac{1}{3}u \sin v\mathbf{k},\ u \geq 0, 0 \leq v \leq 2\pi$ or
$\mathbf{r}(x, y) = x\mathbf{i} + \sqrt{4x^2 + 9y^2}\mathbf{j} + z\mathbf{k}$
25. $\mathbf{r}(u, v) = 5 \cos u\mathbf{i} + 5 \sin u\mathbf{j} + v\mathbf{k}$
27. $\mathbf{r}(u, v) = u\mathbf{i} + v\mathbf{j} + u^2\mathbf{k}$
29. $\mathbf{r}(u, v) = v \cos u\mathbf{i} + v \sin u\mathbf{j} + 4\mathbf{k},\quad 0 \leq v \leq 3$
31. $x = u, y = \dfrac{u}{2} \cos v, z = \dfrac{u}{2} \sin v,\quad 0 \leq u \leq 6, 0 \leq v \leq 2\pi$
33. $x = \sin u \cos v, y = \sin u \sin v, z = u$
$0 \leq u \leq \pi, 0 \leq v \leq 2\pi$
35. $x - y - 2z = 0$ **37.** $4y - 3z = 12$ **39.** $8\sqrt{2}$
41. $2\pi ab$ **43.** $\pi ab^2 \sqrt{a^2 + 1}$
45. $(\pi/6)\left(17\sqrt{17} - 1\right) \approx 36.177$
47. See "Definition of Parametric Surface" on page 1084.
49–51. Proofs
53. (a)   (b)

(c)

(d)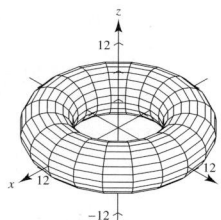

The radius of the generating circle that is revolved about the z-axis is b, and its center is a units from the axis of revolution.
55. 400π m^2
57.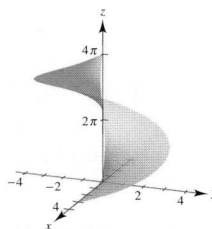

$2\pi\left[\frac{3}{2}\sqrt{13} + 2 \ln\left(3 + \sqrt{13}\right) - 2 \ln 2\right]$
59. Answers will vary. Sample answer: Let
$x = (2 - u)(5 + \cos v) \cos 3\pi u$
$y = (2 - u)(5 + \cos v) \sin 3\pi u$
$z = 5u + (2 - u) \sin v$
where $-\pi \leq u \leq \pi$ and $-\pi \leq v \leq \pi$.

Section 15.6 *(page 1104)*

1. $12\sqrt{2}$ **3.** 2π **5.** $27\sqrt{3}/8$
7. $\left(391\sqrt{17} + 1\right)/240$ **9.** About -11.47
11. $\frac{364}{3}$ **13.** $12\sqrt{5}$ **15.** 8   **17.** $\sqrt{3}\pi$
19. $32\pi/3$ **21.** 486π **23.** $-\frac{4}{3}$ **25.** $3\pi/2$
27. 20π **29.** 384π **31.** 0   **33.** Proof
35. $2\pi a^3 h$ **37.** $64\pi\rho$
39. See Theorem 15.10, "Evaluating a Surface Integral," on page 1094.
41. See "Definition of Flux Integral" on page 1100; see Theorem 15.11, "Evaluating a Flux Integral," on page 1100.
43. (a)

(b) No. If a normal vector at a point P on the surface is moved around the Möbius strip once, it will point in the opposite direction.

(c)

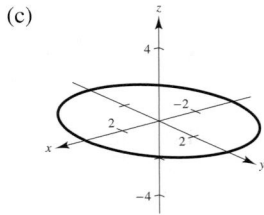

Circle

(d) Construction

(e) A strip with a double twist that is twice as long as the Möbius strip.

Section 15.7   *(page 1112)*

1. a^4 **3.** 18   **5.** π **7.** $3a^4$ **9.** 0
11. 108π **13.** 0   **15.** $18(e^4 - 5)$ **17.** 0
19. See Theorem 15.12, "The Divergence Theorem," on page 1106.
21–27. Proofs

Section 15.8   *(page 1119)*

1. $(xz - e^z)\mathbf{i} - (yz + 1)\mathbf{j} - 2\mathbf{k}$
3. $z(x - 2e^{y^2+z^2})\mathbf{i} - yz\mathbf{j} - 2ye^{x^2+y^2}\mathbf{k}$ **5.** 18π **7.** 0
9. -12 **11.** 2π **13.** 0   **15.** $\frac{8}{3}$ **17.** $a^5/4$ **19.** 0
21. See Theorem 15.13, "Stokes's Theorem," on page 1114.
23. Proof   **25.** Putnam Problem A5, 1987

Review Exercises for Chapter 15   *(page 1120)*

1. $\sqrt{x^2 + 5}$

3. $(4x + y)\mathbf{i} + x\mathbf{j} + 2z\mathbf{k}$
5. Conservative: $f(x, y) = y/x + K$
7. Conservative: $f(x, y) = \frac{1}{2}x^2y^2 - \frac{1}{3}x^3 + \frac{1}{3}y^3 + K$
9. Not conservative
11. Conservative: $f(x, y, z) = x/(yz) + K$
13. (a) div $\mathbf{F} = 2x + 2xy + x^2$ (b) curl $\mathbf{F} = -2xz\mathbf{j} + y^2\mathbf{k}$
15. (a) div $\mathbf{F} = -y \sin x - x \cos y + xy$
        (b) curl $\mathbf{F} = xz\mathbf{i} - yz\mathbf{j}$
17. (a) div $\mathbf{F} = \dfrac{1}{\sqrt{1 - x^2}} + 2xy + 2yz$
        (b) curl $\mathbf{F} = z^2\mathbf{i} + y^2\mathbf{k}$
19. (a) div $\mathbf{F} = \dfrac{2x + 2y}{x^2 + y^2} + 1$ (b) curl $\mathbf{F} = \dfrac{2x - 2y}{x^2 + y^2}\mathbf{k}$
21. (a) $\frac{125}{3}$ (b) 2π **23.** 6π **25.** (a) 18   (b) 18π
27. $9a^2/5$ **29.** $\left(\sqrt{5}/3\right)(19 - \cos 6) \approx 13.446$
31. 1   **33.** $2\pi^2$ **35.** 36   **37.** $\frac{4}{3}$
39. $\frac{8}{3}\left(3 - 4\sqrt{2}\right) \approx -7.085$ **41.** 6
43. (a) 15   (b) 15   (c) 15
45. 1   **47.** 0   **49.** 0

51.

53. (a)

(b)

(c)

(d)

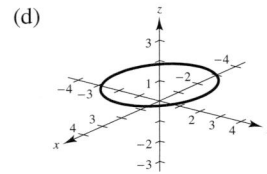

Circle

(e) About 14.436
(f) About 4.269

55.

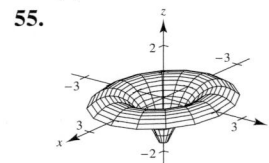

0

57. 66   **59.** $2a^6/5$ **61.** Proof

P.S. Problem Solving   *(page 1123)*

1. (a) $\left(25\sqrt{2}/6\right)k\pi$ (b) $\left(25\sqrt{2}/6\right)k\pi$
3. $I_x = \left(\sqrt{13}\pi/3\right)(27 + 32\pi^2)$;
      $I_y = \left(\sqrt{13}\pi/3\right)(27 + 32\pi^2)$;
      $I_z = 18\sqrt{13}\pi$
5. (a)–(d) Proofs   **7.** $3a^2\pi$
9. (a) 1   (b) $\frac{13}{15}$ (c) $\frac{5}{2}$
11. Proof   **13.** (a)–(b) Proofs

Index

ALGEBRA

Factors and Zeros of Polynomials

Let $p(x) = a_n x^n + a_{n-1} x^{n-1} + \cdots + a_1 x + a_0$ be a polynomial. If $p(a) = 0$, then a is a *zero* of the polynomial and a solution of the equation $p(x) = 0$. Furthermore, $(x - a)$ is a *factor* of the polynomial.

Fundamental Theorem of Algebra

An *n*th degree polynomial has n (not necessarily distinct) zeros. Although all of these zeros may be imaginary, a real polynomial of odd degree must have at least one real zero.

Quadratic Formula

If $p(x) = ax^2 + bx + c$, and $0 \leq b^2 - 4ac$, then the real zeros of p are $x = \left(-b \pm \sqrt{b^2 - 4ac}\right)/2a$.

Special Factors

$$x^2 - a^2 = (x - a)(x + a) \qquad\qquad x^3 - a^3 = (x - a)(x^2 + ax + a^2)$$

$$x^3 + a^3 = (x + a)(x^2 - ax + a^2) \qquad\qquad x^4 - a^4 = (x^2 - a^2)(x^2 + a^2)$$

Binomial Theorem

$$(x + y)^2 = x^2 + 2xy + y^2 \qquad\qquad (x - y)^2 = x^2 - 2xy + y^2$$

$$(x + y)^3 = x^3 + 3x^2 y + 3xy^2 + y^3 \qquad\qquad (x - y)^3 = x^3 - 3x^2 y + 3xy^2 - y^3$$

$$(x + y)^4 = x^4 + 4x^3 y + 6x^2 y^2 + 4xy^3 + y^4 \qquad\qquad (x - y)^4 = x^4 - 4x^3 y + 6x^2 y^2 - 4xy^3 + y^4$$

$$(x + y)^n = x^n + nx^{n-1}y + \frac{n(n-1)}{2!}x^{n-2}y^2 + \cdots + nxy^{n-1} + y^n$$

$$(x - y)^n = x^n - nx^{n-1}y + \frac{n(n-1)}{2!}x^{n-2}y^2 - \cdots \pm nxy^{n-1} \mp y^n$$

Rational Zero Theorem

If $p(x) = a_n x^n + a_{n-1} x^{n-1} + \cdots + a_1 x + a_0$ has integer coefficients, then every *rational zero* of p is of the form $x = r/s$, where r is a factor of a_0 and s is a factor of a_n.

Factoring by Grouping

$$acx^3 + adx^2 + bcx + bd = ax^2(cx + d) + b(cx + d) = (ax^2 + b)(cx + d)$$

Arithmetic Operations

$$ab + ac = a(b + c) \qquad \frac{a}{b} + \frac{c}{d} = \frac{ad + bc}{bd} \qquad \frac{a + b}{c} = \frac{a}{c} + \frac{b}{c}$$

$$\frac{\left(\dfrac{a}{b}\right)}{\left(\dfrac{c}{d}\right)} = \left(\frac{a}{b}\right)\left(\frac{d}{c}\right) = \frac{ad}{bc} \qquad \frac{\left(\dfrac{a}{b}\right)}{c} = \frac{a}{bc} \qquad \frac{a}{\left(\dfrac{b}{c}\right)} = \frac{ac}{b}$$

$$a\left(\frac{b}{c}\right) = \frac{ab}{c} \qquad \frac{a - b}{c - d} = \frac{b - a}{d - c} \qquad \frac{ab + ac}{a} = b + c$$

Exponents and Radicals

$$a^0 = 1, \quad a \neq 0 \qquad (ab)^x = a^x b^x \qquad a^x a^y = a^{x+y} \qquad \sqrt{a} = a^{1/2} \qquad \frac{a^x}{a^y} = a^{x-y} \qquad \sqrt[n]{a} = a^{1/n}$$

$$\left(\frac{a}{b}\right)^x = \frac{a^x}{b^x} \qquad \sqrt[n]{a^m} = a^{m/n} \qquad a^{-x} = \frac{1}{a^x} \qquad \sqrt[n]{ab} = \sqrt[n]{a}\,\sqrt[n]{b} \qquad (a^x)^y = a^{xy} \qquad \sqrt[n]{\frac{a}{b}} = \frac{\sqrt[n]{a}}{\sqrt[n]{b}}$$

FORMULAS FROM GEOMETRY

Triangle

$h = a \sin \theta$

$\text{Area} = \dfrac{1}{2}bh$

(Law of Cosines)

$c^2 = a^2 + b^2 - 2ab \cos \theta$

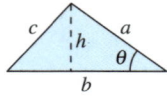

Right Triangle

(Pythagorean Theorem)

$c^2 = a^2 + b^2$

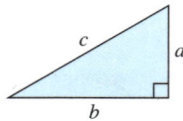

Equilateral Triangle

$h = \dfrac{\sqrt{3}s}{2}$

$\text{Area} = \dfrac{\sqrt{3}s^2}{4}$

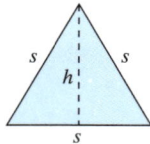

Parallelogram

$\text{Area} = bh$

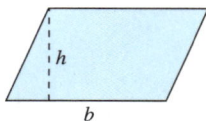

Trapezoid

$\text{Area} = \dfrac{h}{2}(a + b)$

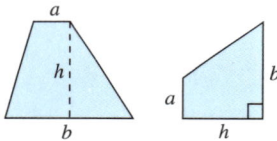

Circle

$\text{Area} = \pi r^2$

$\text{Circumference} = 2\pi r$

Sector of Circle

(θ in radians)

$\text{Area} = \dfrac{\theta r^2}{2}$

$s = r\theta$

Circular Ring

(p = average radius,
w = width of ring)

$\text{Area} = \pi(R^2 - r^2)$

$\qquad = 2\pi pw$

Sector of Circular Ring

(p = average radius,
w = width of ring,
θ in radians)

$\text{Area} = \theta pw$

Ellipse

$\text{Area} = \pi ab$

$\text{Circumference} \approx 2\pi \sqrt{\dfrac{a^2 + b^2}{2}}$

Cone

(A = area of base)

$\text{Volume} = \dfrac{Ah}{3}$

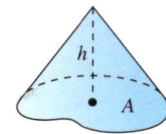

Right Circular Cone

$\text{Volume} = \dfrac{\pi r^2 h}{3}$

$\text{Lateral Surface Area} = \pi r \sqrt{r^2 + h^2}$

Frustum of Right Circular Cone

$\text{Volume} = \dfrac{\pi(r^2 + rR + R^2)h}{3}$

$\text{Lateral Surface Area} = \pi s(R + r)$

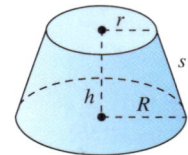

Right Circular Cylinder

$\text{Volume} = \pi r^2 h$

$\text{Lateral Surface Area} = 2\pi rh$

Sphere

$\text{Volume} = \dfrac{4}{3}\pi r^3$

$\text{Surface Area} = 4\pi r^2$

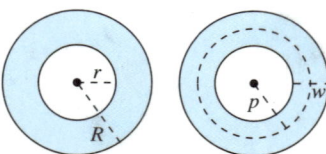

Wedge

(A = area of upper face,
B = area of base)

$A = B \sec \theta$

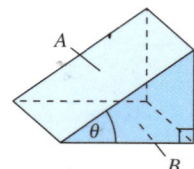